Essential Oils

CONTACT ALLERGY AND CHEMICAL COMPOSITION

Essential Oils
CONTACT ALLERGY AND CHEMICAL COMPOSITION

Anton C. de Groot • Erich Schmidt

CRC Press
Taylor & Francis Group
Boca Raton London New York

CRC Press is an imprint of the
Taylor & Francis Group, an **informa** business

First published in paperback 2024

First published 2016 by CRC Press
2385 NW Executive Center Drive, Suite 320, Boca Raton FL 33431

and by CRC Press
4 Park Square, Milton Park, Abingdon, Oxon, OX14 4RN

CRC Press is an imprint of Taylor & Francis Group, LLC

© 2016, 2024 Taylor & Francis Group, LLC

ISBN: 978-1-4822-4640-7 (hbk)
ISBN: 978-1-03-291734-4 (pbk)
ISBN: 978-0-203-75660-7 (ebk)
ISBN: 978-1-00-320435-0 (eBook+)

**Visit the Taylor & Francis Web site at
http://www.taylorandfrancis.com**

**and the CRC Press Web site at
http://www.crcpress.com**

Contents

Preface

Essential oils are complex substances, which are usually obtained by steam-distillation of plant material or – in the case of citrus fruits – by cold-pressing. As a group, they have widespread applications in foods, drinks, perfumes, cosmetics, and other products. Most have long been used for therapeutic purposes; currently, aromatherapy with essential oils enjoys great popularity. Under normal conditions of established use, most oils appear to have a good safety profile, but toxicity may sometimes be observed. Allergic reactions have long been known to occur, e.g., from cassia oil, cinnamon bark oil, and costus root oil. Dermatologists, notably those who are interested in dermato-allergy, have been confronted in increasing frequency with allergic reactions to essential oils in the past 25 years. Possible explanations include Increase in use of essential oils for skin application, availability of commercial preparations for patch testing, more liberal patch testing of essential oils and such test materials in patients with suspected contact dermatitis ('seek and you shall find') and current awareness among dermatologists of possible allergic reactions to essential oils.

When contact allergy to one or more essential oils has been detected by patch testing, its relevance is sometimes obvious (aromatherapists with hand dermatitis, patients who have applied tea tree oil for therapeutic purposes), but far more often is difficult to assess: the dermatologist and the patient cannot find a source of contact with the oil. One of the explanations for this may be that the oils or their ingredients can be present in a large array of applications and that they are 'hidden' in products where no details on composition are available, e.g., in perfumes or in household products. Another observation is that, in patients with established allergic contact dermatitis from essential oils, virtually never attempts are made to identify the allergenic component(s). Doing this is important, however, as the exact allergens must be known, in order to be able to adequately instruct the allergic patient, which chemicals and products should be avoided to prevent recurrences of allergic contact dermatitis. In some cases, co-reactivity (other positive patch test reactions) to fragrance materials, which are present in high concentrations in the reacting oils, hint at the offending chemical, e.g., in the case of positive reactions to geraniol and rose or geranium oil, to eugenol and clove or bay oil, to citral and citronella oil, or to cinnamaldehyde and cassia or cinnamon oil. However, with the exception of tea tree oil, the allergenic ingredients in essential oils are largely unknown. Besides that, it may be assumed that the composition of essential oils is virtually unknown to most dermatologists, as such information is never published in dermatological journals.

The goal of this book, initiated by the senior author Anton de Groot, dermatologist, primarily was to improve the dermatologists' level of knowledge and quality of patient care, by fully reviewing the literature on contact allergy to essential oils and their chemical composition. He soon found that for the section of Chemical composition, additional expertise was needed, which was found in the person of the second author, Erich Schmidt.

In our literature search, we have found 79 essential oils which have caused contact allergy or allergic contact dermatitis. For some oils, only one or two such reports are available; for others, however, there is a considerable amount of literature, e.g., tea tree oil, ylang-ylang oil, lavender oil, rose oil, turpentine oil and sandalwood oil.

Next to fully reviewing contact allergy to essential oils, we also present the (possible) chemical composition of these essential oils. Such data are essential to dermatologists for adequately diagnosing and counseling their patients with (suspected) contact allergy to and allergic contact dermatitis from essential oils. Where limited information, e.g., on the main components and their concentrations, would be sufficient for most dermatologists, we decided to present detailed and extensive compositional data to fit the need for information of other professionals, such as chemists working in the perfume industry.

There are 93 chapters (91 oils and 2 jasmine absolutes) on chemical composition, which means that the number of chapters exceeds that of the oils which have caused contact allergy. This is due to the fact that some oils may be obtained from different species (e.g., cedarwood oil Atlas, cedarwood oil China, cedarwood oil Himalaya, cedarwood oil Texas, cedarwood oil Virginia), from different parts of the same plant (e.g., cinnamon bark oil and cinnamon leaf oil; clove bud oil, clove stem oil, and clove leaf oil) or from various cultivars producing different oils (lavandin abrial oil, lavandin grosso oil). In dermatological literature, however, such data are virtually never provided ('cedarwood oil', 'clove oil', 'lavandin oil') and therefore we chose to present the data from all possible source species and plant parts.

We have used two sources of information on the chemical composition of essential oils. The first is literature data of analytical studies. The search for relevant scientific articles has resulted in a very heterogeneous set of data for individual oils. For some, we found only a few analyses; others have been investigated in 60 to over 100 published reports. Only few studies report the results of analyzing a large number of commercial essential oils (notably with citrus oils); in the majority, only one, two or a few samples of oil hydrodistilled in the laboratory with Clevenger-type apparatus have been analyzed. The number or articles reviewed exceeds 2,500.

The second source of information on the composition of essential oils is the very large and unique set of analytical data (nearly 6,400 samples investigated), generated by GC-MS equipment by the second author between 1998 and 2014, while he was working in a German company, which designs and produces natural compositions of fine essential oils and perfume oils. The samples of essential oils came from all over the world. The number of analyses per oil discussed in this book varies from 0 (neem oil, cassia leaf oil) to 422 (sweet orange oil). All analytical results (chemicals identified, range of concentrations) are presented in this book. They have not previously been published and represent by far the largest set of essential oils analyses ever reported in scientific literature.

The extensive and in-depth information provided in this book, most of which cannot be found *as such* anywhere else in literature, should be of interest not only to dermatologists and their patients allergic to essential oils or individual fragrances, but also to academic scientists working in the field of essential oils and fragrances, cosmetic chemists, analytical chemists, perfumers, aromatherapists, other professionals working with (products containing) essential oils, legislators, relevant non-governmental organizations, and to individuals involved in the producing, selling and acquisition of essential oils.

Anton C. de Groot
Erich Schmidt

Acknowledgments

The authors wish to thank Suzanne E. de Groot LLM, for her help with data implementation in Chapter 6.

We are grateful to Kurt Kitzing GmbH, Hinterm Alten Schloss 21, 86757 Wallerstein, Germany, for permission to publish the data of the analyses performed by Erich Schmidt during his employment with the company (www.kurtkitzing.de).

The authors thank the American Chemical Society, 1155 Sixteenth Street, NW, Washington, DC 20036, USA, for allowing us to publish the CAS (Chemical Abstract Service) numbers of the chemicals mentioned in this book; CAS Registry Number is a Registered Trademark of the American Chemical Society (www.cas.org; www.acs.org).

We appreciate that the International Organization for Standardization (ISO), Chemin de Blandonnet 8, CP 401, 1214 Vernier, Geneva, Switzerland, has agreed that we publish quantitative ISO norms for essential oils for a reasonable fee (www.iso.org).

ABOUT THE AUTHORS

Anton C. de Groot (1951) received his medical and specialist training at the University of Groningen, The Netherlands. In 1980, he started his career as dermatologist in private practice in 's-Hertogenbosch. At that time, he had already become interested in contact allergy and side effects of drugs by writing the chapter 'Drugs used on the skin' with his mentor prof. Johan Nater, for the famous series 'Meyler's Side Effects of Drugs'. Soon, the subject of the chapter in new Editions and the yearly 'Side Effects of Drugs Annuals' would be expanded to include oral drugs used in dermatology and cosmetics (1980-2000). Contact allergy to cosmetics would become de Groot's main area of interest and expertise and in 1988, he received his PhD degree on his Thesis entitled 'Adverse Reactions to Cosmetics', supervised by prof. Nater. Frustrated by the lack of easily accessible information on the ingredients of cosmetic products, and convinced that compulsory ingredient labelling of cosmetics (which at that time was already implemented in the USA) would benefit both consumers and allergic patients and would lead to only slight and temporary disadvantages to the cosmetics industry, De Groot approached the newly founded European Society of Contact Dermatitis and became Chairman of the Working Party Community Affairs. The European Commission and its committees, elected legislators, national trade, health departments and the cosmetics industries were extensively lobbied. This resulted in new legislation by the Commission of the European Communities in 1991, making ingredient labelling mandatory for all cosmetic products sold or disposed of in EC Member States by December 31, 1997.

Anton has been the chairman of the 'Contact Dermatitis Group' of the Dutch Society for Dermatology and Venereology from 1984 to 1998. In 1990, he was one the founders of the Dutch Journal of Dermatology and Venereology and was Editor of this scientific journal for 20 years, of which he served 10 years as Editor-in-chief. De Groot has authored 12 books, mostly dermatology books for medical students, general practitioners, 'skin therapists' (huidtherapeuten, a paramedical profession largely restricted to The Netherlands) and podotherapists; in addition, he wrote a booklet 'Living with Eczema' for patients with atopic eczema and their parents. His two international books ('Unwanted Effects of Cosmetics and Drugs used in Dermatology' and 'Patch Testing') have both had three editions, the most recent one for Patch Testing in 2008 (www.patchtesting.info). In addition, Anton has written more than 60 book chapters and over 300 articles in Dutch and international journals. He served as board member of several journals and is currently member of the Editorial Advisory Board of the Journal 'Dermatitis'. His most recent project is a new book 'Cosmetic Allergy – Cosmetic Allergens'. He has retired from dermatology practice, but regularly teaches general dermatology to junior medicals doctors at the University of Groningen.

Erich Schmidt (1950) completed his study of business economics at the Landshut University of Applied Sciences, Bavaria, Germany. In 1977 he started his career in a middle sized German company, which designs and produces perfume compositions on conventional or natural base for all kinds of applications and also trades essential oils. Since first working with essential oils, his interest in analyzing their composition increased. After training as perfumer and having become manager responsible for purchase and quality control, he established an analytical laboratory and by autodidactic training, results soon were equal to those found in science. With spreading use of aromatherapy in Germany, he provided this field with information about risks, safety evaluation and genuineness of essential oils. Seminars as well as many lectures in symposiums on the subject of aromatherapy have been helpful to teach aromatherapists and nursing staff of hospitals the proper handling and application of the oils. In 1989 he was appointed member of the German delegation and in 1995 became the German responsible representative of ISO (International Organization for Standardization) Technical Committee 54, Essential oils. Several ISO standards have been established under his responsibility. As manager of the company, he was member of the board of the German organization of the fragrance industry (DVRH; Deutscher Verband der Riechstoff-Hersteller; www.riechstoffverband.de) from 2002 to 2012. His special expertise was the field of natural substances and their composition, which led to a closer cooperation between the DVRH and the IVDK (Informationsverbund Dermatologischer Kliniken; www.ivdk.org). His interest then concentrated on the safety of natural chemical substances, both including essential oils and extracts. A number of articles on essential oils were published in the journals Forum Essenzia and Euro Cosmetics. From 2004, he decided to focus more on scientific work in cooperation with the division of Clinical Pharmacy and Diagnostics of the University of Vienna, Austria, in the fields of clinical pharmacy, biological activity of essential oils and single odorant compounds and analyses of pharmaceutically used lipophilic plant compounds. Since 2005, he authored or co-authored more than 90 publications in journals such as Natural Product Communications, Journal of Essential Oil Research, Journal of Essential Oil Bearing Plants, Perfumer & Flavorist, Bioscience, Biotechnology, and Biochemistry, Ernährung, Österreichische Zeitschrift für Wissenschaft,Technik, Recht und Wirtschaft, The International Journal of Essential Oil Therapeutics, Journal of Agricultural and Food Chemistry, Recent Research Developments in Agronomy & Horticulture, Scientia Pharmaceutica, Planta Medica, Recent Progress in Medicinal Plants and Contact Dermatitis. He also contributed a chapter to the 'Handbook of Essential Oils'.

Chapter 1   INTRODUCTION

1.1   WHY A BOOK ON CONTACT ALLERGY TO AND CHEMICAL COMPOSITION OF ESSENTIAL OILS?

Essential oils are complex substances that are usually obtained by steam-distillation or hydrodistillation of plant material. Citrus oils, originating from the peels of *Citrus* species such as grapefruit, lemon, mandarin, orange, or tangerine, are produced by a process called cold-pressing (Chapter 2). Various plant parts can be used for essential oil production, e.g., fruits (oils of angelica fruit, aniseed, carrot seed), roots (oil of angelica root, lovage, valerian, vetiver), flowering aerial tops (oils of sweet basil, clary sage, lavandin, lavender, sage, thyme), leaves (oils of bay, cassia leaf, cinnamon leaf, citronella, clove leaf, laurel leaf), flowers (oils of cananga, chamomile, neroli, rose, ylang-ylang), peels (citrus oils: grapefruit, lemon, mandarin, orange, tangerine), wood (oils of cedarwood, guaiacwood, rosewood, sandalwood), or exudate or oleoresin (oils of elemi, galbanum, olibanum, turpentine). Sometimes more than one part of the plant is used for obtaining the oil, e.g., for dwarf pine oil (needles, terminal branchlets), rosemary oil (flowering tops and leaves) and petitgrain bigarade oil (leaves, twigs, little green fruits). Some plants can be utilized for producing different essential oils, for example, cinnamon bark and cinnamon leaf oil, angelica fruit and angelica root oil, and oils of clove from bud, stem or leaves of the clove, *Syzygium aromaticum* (L.). Three entirely different oils can be obtained from *Citrus aurantium* L., the bitter orange: neroli oil from the flowers, bitter orange oil from the fruit peels and petitgrain bigarade oil from the leaves, twigs and little green fruit.

Essential oils are produced in many parts and countries of the world and are very important trade commodities. As a group, they have widespread applications in foods, soft drinks, alcoholic beverages, perfumes, cosmetics, toothpastes and dentifrices. Many are added to topical pharmaceutical preparations; some are used in products such as incense or cigarettes. Most have long been used for therapeutic purposes to treat a wide range of diseases and ailments. Some are employed in food preservation, in farm animal health, and in agriculture, where they may be applied as pesticides (4). A rapidly growing application of essential oils is in aromatherapy (1–3). Aromatherapy has been defined as the therapeutic application of essential oils and aromatic plant extracts in a holistic context, to maintain or improve physical, emotional and mental well-being (1) or briefly the therapeutic use of essential oils (2). The oils are usually applied to the skin, but can also be given orally, by inhalation, by distribution through the air or by other means.

Under normal conditions of established use, most oils appear to have a good safety profile. However, toxicity may sometimes occur. The majority of adverse events are mild, but serious toxic reactions from some essential oils have been observed, including abortions or abnormalities in pregnancy, neurotoxicity manifesting as seizures or retardation of infant development, bronchial hyper-reactivity and hepatotoxicity; accidental ingestion by young children has occasionally proved fatal (4). Allergic reactions to essential oils may also occur, e.g., from cassia oil, cinnamon bark oil, and costus root oil, which are considered too hazardous for use in aromatherapy by some authors (3).

The interest of the senior author (Anton de Groot, dermatologist) in allergic contact dermatitis, especially allergic reactions to cosmetic products, is the basis of this book. Dermatologists, notably those who are interested in dermato-allergy, have been confronted in increasing frequency with allergic reactions to essential oils in the past 25 years. This development may be explained by the following.

Increasing use of essential oils on the skin

The use of essential oils applied to the skin, either in undiluted form (rare, usually self-medication) or in products containing considerable quantities of essential oils, appears to be increasing. Oils may be applied by the user himself or through massages performed by professionals such as aromatherapists, masseurs, or physiotherapists.

Although in lower concentrations, essential oils may be present in a wide range of consumer products with which we have regular contact, including perfumes and (other) cosmetics.

Availability of commercial test preparations

In the last two decades, a large number of essential oils have become available as commercial preparations for patch testing (Chapter 3). Patch testing is the diagnostic procedure performed in patients with suspected allergic contact dermatitis, and the availability of essential oil test materials has greatly facilitated the detection of contact allergy to these oils.

Increase in patch testing of essential oils by dermatologists in selected patients or in routine testing

Many dermatologists test a series of essential oil test substances, often combined with single fragrance chemicals, in a 'fragrance series'. This series is usually patch tested in the case of suspected contact allergy to fragrances or in patients who previously had positive patch tests to other fragrances or fragrance markers such as the fragrance mix (I, II or both) or *Myroxylon pereirae* resin (balsam of Peru). Moreover, in the USA, five essential oils (tea tree oil 5%, lavender oil 2%, peppermint oil 2%, jasmine absolute 2% and ylang-ylang oil 2%) are currently present in the screening series of the North American Contact Dermatitis Group (NACDG), which means that they are routinely tested in all patients suspected of contact dermatitis (5). This inevitably results in more cases of contact allergy to essential oils being detected.

Current awareness of possible allergic reactions to essential oils

Many articles on contact allergy to essential oils have recently been published in various dermatological journals, and cases have been presented during congresses.

Two societies of dermatologists and other professionals interested in contact dermatitis (European Society of Contact Dermatitis, American Contact Dermatitis Society) are very active and have their own scientific journals, *Contact Dermatitis* and *Dermatitis*. The large number of publications on tea tree oil allergy certainly boosted awareness and interest in essential oils allergy.

While preparing a manuscript 'Cosmetic allergy – Cosmetic allergens', the senior author made an observation on relevance data in studies on essential oil allergy. The finding of a positive patch test reaction to an essential oil should—as with any test substance—always be followed by determination of its relevance: can the dermatitis of the patient be (partly) explained by contact with the oil or products containing it? In many studies providing results of patch testing with essential oils, no relevance data were provided. In the screening studies of the NACDG, where relevance can be assessed as definite, probable or possible, relevance scores for definite + probable are often <40%, with 'definite relevance' usually being lower than 15% and sometimes even zero. This does not mean, of course, that the patients do not have contact allergy to the oil (or at least the essential oil test substance), but that the dermatologist and the patient cannot find a source of contact with the oil. One of the explanations for this may be that the oils or their ingredients can be present in a large array of applications and that they are hidden in products where no details on composition are available, e.g., in perfumes or in household products. On the other hand, in some cases, contact with the incriminated oil(s) may be easily established, for example, in aromatherapists with hand dermatitis, in patients with dermatitis at the site of therapeutic application of tea tree oil or other oils or in patients developing skin reactions at the site of application of topical pharmaceutical preparations containing essential oils.

Another observation was that, in patients with established allergic contact dermatitis from essential oils, virtually never were attempts made to identify the allergenic component(s). This is important, however, as the exact allergens must be known in order to be able to adequately instruct the allergic about patient which chemicals and products should be avoided to prevent recurrences of allergic contact dermatitis. In some cases, co-reactivity (other positive patch test reactions) to fragrance materials, which are present in high concentrations in the reacting oils, could hint at the offending chemical, e.g., in the case of positive reactions to geraniol and rose or geranium oil, to eugenol and clove or bay oil, to citral and citronella oil, to cinnamaldehyde and cassia or cinnamon oil, to menthol and peppermint oil or to carvone and spearmint oil. Only a few investigators have done analyses of oils that caused allergic contact dermatitis (6-10). In fact, with the exception of tea tree oil which has been extensively investigated by Dr. Björn M. Hausen and his co-workers (11-13) and other German and Austrian investigators (14), the allergens in essential oils are largely unknown. Moreover, we think that the composition of essential oils is virtually unknown to most dermatologists, as such information

is not published in dermatological journals, but in journals such as the *Journal of Essential Oil Research, Journal of Essential Oil Bearing Plants, Journal of Agriculture and Food Chemistry, Flavour and Fragrance Journal, Industrial Crops and Products, Natural Products Research* and many others, which never come to the attention of the dermatological community. Also, no major review articles on contact allergy to essential oils have been published in scientific journals.

Aiming at improving the level of knowledge and quality of patient care, the senior author (Anton de Groot) then considered fully reviewing the literature on contact allergy to essential oils and their chemical composition. An example oil chapter was written and sent to approximately 70 dermatologists all over the world with expertise in contact dermatitis, to verify whether this was considered useful and to ask them for focus points. Virtually all responded positively and enthusiastically and the decision to write the book was made. It was soon realized that for the section on chemical composition, additional other expertise was needed and was found in the person of the second author. Erich Schmidt has worked for decades as a perfumer, analyst, and essential oils specialist in a German fragrance company, has long been a member of an ISO technical committee, and has cooperated with German dermatologists studying contact allergy to essential oils (15).

1.2   CONTACT ALLERGY TO ESSENTIAL OILS: A SURVEY OF OILS AND PLANTS

In our literature search (for specifics see Chapter 5.0 Introduction), we have found 79 essential oils which have caused contact allergy (positive patch test reactions) or allergic contact dermatitis. For some oils, only one or two such reports are available. For others, however, there is a considerable amount of literature on contact allergy and allergic contact dermatitis, e.g., in the case of tea tree oil, ylang-ylang oil, lavender oil, peppermint oil, jasmine absolute, geranium oil, rose oil, turpentine oil and sandalwood oil. Due to the selection criterion (contact allergy reported), the group of 79 oils is very heterogeneous. Some are high volume essential oils, such as orange oil (worldwide production 60,000 tons per year), corn mint oil (35,000 tons) and lemon oil (nearly 9,000 tons); others are produced and commercialized in very small quantities. The estimated commercial quantity of zdravetz oil, for example, which is produced in Bulgaria only, is about 20 kilograms. Costus root oil is apparently still used in aromatherapy (2), but has been forbidden in fragrances and cosmetics in the EU for decades because of its sensitizing properties. Some oils are prohibited by IFRA (International Fragrance Association), including costus root oil, verbena oil, boldo oil, massoia bark oil, rue oil and santolina oil. Others are prohibited or restricted by governmental regulations and laws, as in the case of sassafras oil, rosewood oil, guaiacwood oil and East Indian sandalwood oil. In the case of the latter three, production and export are prohibited as they are endangered species (16).

Table 1.1 provides a list of 91 essential oils and two absolutes (jasmine absolute, which was included because

Table 1.1 List of essential oils

Chapter	Common name	Botanical source	Part(s) of plant used	ISO[a]
5.1	Angelica fruit oil	*Angelica archangelica* L.	Fruit	
5.2	Angelica root oil	*Angelica archangelica* L.	Rhizome and root	
5.3	Aniseed oil	*Pimpinella anisum* L.	Fruit	3475
5.4	Basil oil, sweet	*Ocimum basilicum* L.	Flowering aerial top	11043
5.5	Bay oil	*Pimenta racemosa* (Mill.) J.W. Moore	Leaf	3045
5.6	Bergamot oil	*Citrus bergamia* (Risso et Poit.)	Pericarp (peel)	3520
5.7	Black cumin oil	*Nigella sativa* L.	Seed	
5.8	Black pepper oil	*Piper nigrum* L.	Fruit	3061
5.9	Cajeput oil	*Melaleuca cajuputi* Powell	Leaf, terminal branchlet	
5.10	Calamus oil	*Acorus calamus* L.	Rhizome	
5.11	Cananga oil	*Cananga odorata* (Lam.) Hook f. et Thomson, forma *macrophylla*	Flower	3523
5.12	Carrot seed oil	*Daucus carota* L.	Fruit	
5.13	Cassia bark oil	*Cinnamomum cassia* (Nees & T. Nees) J. Presl	Bark	3216
5.14	Cassia leaf oil	*Cinnamomum cassia* (Nees & T. Nees) J. Presl	Leaf	
5.15	Cedarwood oil, Atlas	*Cedrus atlantica* (Endl.) G. Manetti ex Carrière	Wood	
5.16	Cedarwood oil, China	*Cupressus funebris* (Endl.)	Wood	9843
5.17	Cedarwood oil, Texas	*Juniperus ashei* J. Buchholz	Wood	4725
5.18	Cedarwood oil, Virginia	*Juniperus virginiana* L.	Wood	4724
5.19	Chamomile oil, German	*Chamomilla recutita* (L.) Rauschert	Flowering tops	19332
5.20	Chamomile oil, Roman	*Chamaemelum nobile* (L.)	Flowering tops	
5.21	Cinnamon bark oil, Sri Lanka	*Cinnamomum zeylanicum* Blume	Twig and bark of stem	
5.22	Cinnamon leaf oil, Sri Lanka	*Cinnamomum zeylanicum* Blume	Leaf	3524
5.23	Citronella oil, Java	*Cymbopogon winterianus* Jowitt.	Aerial part (leaves)	3848
5.24	Citronella oil, Sri Lanka	*Cymbopogon nardus* (L.) Rendle	Aerial part (leaves)	3849
5.25	Clary sage oil	*Salvia sclarea* L.	Flowering top (and leaf)	
5.26	Clove bud oil	*Syzygium aromaticum* (L.) Merr. & L.M. Perry	Bud	3142
5.27	Clove leaf oil	*Syzygium aromaticum* (L.) Merr. & L.M. Perry	Leaf	3141
5.28	Clove stem oil	*Syzygium aromaticum* (L.) Merr. & L.M. Perry	Stem	3143
5.29	Coriander fruit oil	*Coriandrum sativum* L.	Fruit	3516
5.30	Costus root oil	*Saussurea costus* (Falc.) Lipsch.	Root	
5.31	Cypress oil	*Cupressus sempervirens* L.	Twig with leaves	
5.32	Dwarf pine oil	*Pinus mugo* Turra	Leaf (needle), terminal branchlets	21093
5.33	Elemi oil	*Canarium luzonicum* (Blume) A. Gray	Wood exudate	10624
5.34	Eucalyptus citriodora oil	*Eucalyptus citriodora* Hook.	Leaf, terminal branch	3044
5.35	Eucalyptus globulus oil	*Eucalyptus globulus* Labill.	Leaf, terminal branch	770
5.36	Galbanum resin oil	*Ferula gummosa* Boiss.	Root exudate	40716
5.37	Geranium oil	*Pelargonium* x spp.	Herbaceous part	4731
5.38	Ginger oil	*Zingiber officinale* Roscoe.	Rhizome	16928
5.39	Grapefruit oil	*Citrus paradisi* Macfad.	Pericarp (peel)	3053
5.40	Guaiacwood oil	*Bulnesia sarmientoi* Lorentz ex Griseb.	Wood	
5.41	Hyssop oil	*Hyssopus officinalis* L. ssp. *officinalis*	Flowering top and leaf	9841
5.42	Jasminum grandiflorum absolute	*Jasminum grandiflorum* L.	Flower	
5.43	Jasminum sambac absolute	*Jasminum sambac* (L.) Aiton.	Flower	
5.44	Juniper berry oil	*Juniperus communis* L.	Fruit, terminal branchlets	8897
5.45	Laurel leaf oil	*Laurus nobilis* L.	Leaf	
5.46	Lavandin abrial oil	*Lavandula angustifolia* Mill. x *Lavandula latifolia* Medik. 'Abrial'	Flowering top	3054
5.47	Lavandin grosso oil	*Lavandula angustifolia* Mill. x *Lavandula latifolia* Medik. 'Grosso'	Flowering top	8902
5.48	Lavandin oil	*Lavandula angustifolia* Mill. x *Lavandula latifolia* Medik.	Flowering top	
5.49	Lavender oil	*Lavandula angustifolia* Mill.	Flowering top	3515
5.50	Lemon oil	*Citrus limon* (L.) Burm. f.	Pericarp (peel)	855
5.51	Lemongrass oil, East Indian	*Cymbopogon flexuosus* (Nees ex Steudel) J.F. Watson	Aerial part (leaves)	4718
5.52	Lemongrass oil, West Indian	*Cymbopogon citratus* (DC) Stapf.	Whole aerial part (leaves)	3217

Table 1.1 List of essential oils (*continued*)

Chapter	Common name	Botanical source	Part(s) of plant used	ISO[a]
5.53	Litsea cubeba oil	*Litsea cubeba* (Lour) Pers.	Fruit	3214
5.54	Lovage oil	*Levisticum officinale* W.D.J. Koch	Root	11019
5.55	Mandarin oil	*Citrus reticulata* Blanco	Pericarp (peel)	3528
5.56	Marjoram oil (sweet)	*Origanum majorana* L.	Flowering top	
5.57	Melissa oil (lemon balm oil)	*Melissa officinalis* L.	Aerial parts	
5.58	Myrrh oil	*Commiphora myrrha* (Nees) Engl. and related *Commiphora* species	Wood exudate	
5.59	Neem oil	*Azadirachta indica* A. Juss.	Seed	
5.60	Neroli oil	*Citrus aurantium* L.	Flower	3517
5.61	Niaouli oil	*Melaleuca quinquenervia* (Cav.) S.T. Blake	Leaves, terminal branchlets	
5.62	Nutmeg oil	*Myristica fragrans* Houtt.	Seed	3215
5.63	Olibanum (frankincense) oil	*Boswellia sacra* Flueck.	Wood exudate	
5.64	Orange oil, bitter	*Citrus aurantium* L.	Pericarp (peel)	9844
5.65	Orange oil, sweet	*Citrus sinensis* (L.) Osbeck	Pericarp (peel)	3140
5.66	Palmarosa oil	*Cymbopogon martini* (Roxb.) Will. Watson	Aerial part (leaves)	4727
5.67	Patchouli oil	*Pogostemon cablin* (Blanco) Benth.	Leaf	3757
5.68	Peppermint oil	*Mentha x piperita* L.	Aerial parts, leaf	856
5.69	Petitgrain bigarade oil	*Citrus aurantium* L.	Leaf, twig, little green fruit	8901
5.70	Pine needle oil (Scots pine oil)	*Pinus sylvestris* L.	Needle, twig	
5.71	Ravensara oil	*Ravensara aromatica* Sonn.	Twig with leaves	
5.72	Rosemary oil	*Rosmarinus officinalis* L.	Flowering top, leaf	1342
5.73	Rose oil	*Rosa x damascena* Mill.	Flower	9842
5.74	Rosewood oil	*Aniba rosaeodora* Ducke, *Aniba parviflora* (Meisn.) Mez.	Wood	3761
5.75	Sage oil, Dalmatian	*Salvia officinalis* L.	Flowering top	9909
5.76	Sage oil, Spanish	*Salvia lavandulifolia* Vahl	Flowering top	3526
5.77	Sandalwood oil	*Santalum album* L.	Wood	3518
5.78	Silver fir oil	*Abies alba* Mill.	Needles	
5.79	Spearmint oil	*Mentha spicata* L.	Flowering aerial part, leaf	3033
5.80	Spike lavender oil	*Lavandula latifolia* Medik.	Flowering top	4719
5.81	Star anise oil	*Illicium verum* Hook. f.	Fruit	11016
5.82	Tangerine oil	*Citrus tangerina* Hort. ex Tan.	Peel	
5.83	Tea tree oil	*Melaleuca alternifolia* (Maiden et Betche) Cheel; *Melaleuca linariifolia* Smith; *Melaleuca dissitiflora* F. Muell.	Leaf, terminal branchlet	4730
5.84	Thuja oil	*Thuja occidentalis* L.	Twig with leaves	
5.85	Thyme oil	*Thymus vulgaris* L.	Flowering top	
5.86	Thyme oil, Spanish	*Thymus zygis* L.	Flowering top	14715
5.87	Turpentine oil, Iberian type	*Pinus pinaster* Aiton	Oleoresin	11020
	Turpentine oil, Chinese type	*Pinus massoniana* Lamb.	Oleoresin	21389
5.88	Valerian oil	*Valeriana officinalis* L.	Rhizome, root	
5.89	Vetiver oil	*Chrysopogon zizanioides* (L.) Roberty	Root	4716
5.90	Ylang-ylang oil	*Cananga odorata* (Lam.) Hook f. et Thomson, *forma genuina*	Flower	3063
5.91	Zdravetz oil	*Geranium macrorrhizum* L.	Aerial part	
5.92	Cardamom oil	*Elettaria cardamomum* (*L.*) Maton	Fruit	4733
5.93	Cedarwood oil Himalaya	*Cedrus deodara* (Roxb. ex D.Don) G.Don	Wood	

[a] number of ISO standard; International Organization for Standardization, Geneva, Switzerland, www.iso.org

there is quite extensive literature on contact allergy to it) with their common name, botanical source, part of the plant used for obtaining the oil and – if available – numbers of ISO standards (International Organization for Standardization, Geneva, Switzerland, www.iso. org). The number in the left column corresponds to their number in Chapter 5, where all oils are discussed, both contact allergy and chemical composition. The number of chapters exceeds the number of essential oils for which contact allergy has been described. This is due to the fact that some oils may be obtained from different species (e.g., cedarwood oil Atlas, cedarwood oil China, cedarwood oil Himalaya, cedarwood oil Texas, cedarwood oil Virginia; the same goes for lemongrass oil, thyme oil, sage oil), from different parts of the same plant (e.g., cinnamon bark oil and cinnamon leaf oil; clove bud oil, clove stem oil, and clove leaf oil) or from various cultivars producing different oils (lavandin abrial oil, lavandin grosso oil). In dermatological literature, however, such data are virtually never provided ('cedarwood oil', 'clove oil', 'lavandin oil', 'thyme oil') and therefore we chose to present the data from all possible source species and plant parts. Full literature data on contact allergy to / allergic contact dermatitis from individual essential oils are presented in Chapters 5.1-5.93. General aspects (e.g., frequency of contact allergy, relevance of observed positive patch test reactions, clinical picture of allergic contact dermatitis, products responsible for allergic reactions, co-reactivity to other oils and products, the allergens in essential oils, and diagnostic procedures) are discussed in Chapter 3.

1.3   CHEMICAL COMPOSITION OF ESSENTIAL OILS: DATA PROVIDED

The primary goal of this part of the book originally was to aid dermatologists in adequately diagnosing and counseling their patients with (suspected) contact allergy to and allergic contact dermatitis from essential oils. To achieve this, knowledge of the (possible) composition of essential oils used by allergic patients is needed. In the diagnostic phase, compositional information enables or facilitates the identification of the causative allergenic ingredients. In the second phase, counseling allergic patients how to proceed, information on the constituents of essential oils is equally important, as contact with these chemicals has to be avoided to prevent recurrences of allergic contact dermatitis. The advice often given to patients reacting to one or more oils, to simply avoid contact with all essential oils, fragrances and fragranced products, is less than optimal patient care. Every patient who, for whatever reason, wants to continue having contact with (certain) essential oils, needs to be informed which oils are safe to use and which should be avoided. This is certainly critical for patients who work professionally with essential oils, such as masseurs, physiotherapists and aromatherapists, so they can select safe oils and can continue their work activities. Therefore, this book provides not only extensive literature data on reported analyses of essential oils (Chapters 5.1-5.93), but also an alphabetical list of all chemicals identified in this group of essential oils (approximately 4,350) specifying in which they may be present (Chapter 6). Thus, patients who have allergies to specific components (e.g., limonene, linalool, geraniol, citral et cetera) can easily find in which oils these chemicals are present and in which concentrations. Oils not containing the allergens should be safe, unless cross-reactivity can occur, which should be considered by the dermatologist. However, not all oils (possibly) containing the allergen should necessarily be avoided, depending on the concentration range; such data are available in this book.

We have used two sources of information on the chemical composition of essential oils. The first is literature data of analytical studies of essential oils. The search for relevant scientific articles has resulted in a very heterogeneous set of data for individual oils. For some oils, we found only a few analyses; others have been investigated in 60 to over 100 published reports. Only a few studies report the results of analyzing a large number of commercial essential oils (notably with citrus oils); in the majority, only one, two or a few samples of oil hydrodistilled in the laboratory with Clevenger-type apparatus have been analyzed. Some were prepared from plants that were cultivated, but for others, plants growing in the wild or material (plants, fruits, roots) purchased 'at a local market' were used as biomass for oil production. Data collection, selection and interpretation are specified in Chapter 5.0 Introduction. The results of all articles reviewed (over 2,500) are shown in Tables 5.1-5.93.

The second source of information on the composition of essential oils is the very large and unique set of analytical data, generated by GC-MS equipment by the second author between 1998 and 2014, while he was working in a German company which designs and produces natural compositions of fine essential oils and perfume oils.

Although his analytical data reach back to the 1980s, it was decided to include only data from 1998 and later, as the quality of analytical equipment, mass spectra databases (including his own database of >2,500 chemicals) and identification software (17-19) at that time had reached a level which could ensure correct analyses of the oil samples under investigation. With every new database or database version, older data were corrected, if necessary. The results of nearly 6,400 analyses of essential oils and jasmine absolute samples are presented in this book. Oils containing synthetic compounds, indicating clear and intentional adulteration (in which case they are no longer essential oils but perfume mixtures), were excluded. Oils in which chemicals were identified in abnormal concentrations as compared to standards or our own experience were included. Such deviant concentrations can have various causes, including the use of the wrong botanical species, the mixing of biomasses, and intentional adulteration by adding other natural components. Some oils commercialized under the same name can be produced from different species, e.g., eucalyptus oil from *Eucalyptus globulus* and from *Cinnamomum longipaniculatum*, and tea tree oil from various *Melaleuca* species. These were all included in the dataset.

The samples of essential oils (and jasmine absolutes) that have been analyzed came from all over the world and were sent by producers, their agents, wholesalers or other persons or organizations. In most cases, analyses were used to decide whether or not the oil will be purchased. Sometimes analyses were made of 8-10 samples of a particular essential oil from various origins for purchase decisions. Samples were also sent by companies using oils, when they either possess no analytical equipment or do not have enough experience in essential oils analysis. Numerous analyses were commissioned by the German authorities for aiding foreign countries in opening the German market for their agricultural products. Samples were also sent by supervising authorities and organizations to confirm their naturalness and exclude adulterations. Finally, data were generated in so-called round robin tests. In this procedure, various independent laboratories receive an identical group of oil samples, which are analyzed with standardized test methods. The results are then collected and combined to establish the quantity profile for each component of the oil. This is the procedure performed by ISO (International Organization for Standardization; www.iso.org), as well as by other organizations dealing with essential oils, including the European Federation of Essential Oils (EFEO, http://efeo-org.org/) and the International Fragrance Association (IFRA, www.ifraorg.org). Most samples investigated were conventionally produced market oils, but a considerable number were obtained from plants bred in bio culture.

All essential oil samples have been analyzed by the same methods and routine. Gas chromatography–mass spectrometry (GC-MS) was the analytical technique employed using highly sophisticated mass spectra libraries (17-19) and automatic evaluation. This allowed achieving adequate analytical results quantified down to 0.01% on the basis of mass spectra, retention indices and retention times. If necessary, chemicals of special interest (for whatever reason) could be detected down to 1 ppm, although working near the baseline does increase the risk of wrong identifications. All concentrations above 0.09% were given one decimal.

The number of analyses per oil discussed in this book varies from 0 (neem oil, cassia leaf oil) to 422 (sweet orange oil); numbers for each oil are shown in Table 5.0.2 (Chapter 5.0 Introduction). All analytical results (chemicals identified, range of concentrations) can be found in columns A of the main tables of Chapters 5.1-5.93. These results have not previously been published and represent by far the most extensive set of essential oils analyses ever reported in scientific literature.

The extensive and in-depth information provided in this book, most of which cannot be found *as such* anywhere else in literature, should be of interest to dermatologists, academic scientists working in the field of essential oils and fragrances, cosmetic chemists, analytical chemists, perfumers, aromatherapists, other professionals working with (products containing) essential oils, legislators, relevant non-governmental organizations, and patients who are contact allergic to fragrances or essential oils.

LITERATURE

1 Rhind JP. Essential oils. A handbook for aromatherapy practice, 2nd Edition. London: Singing Dragon, 2012

2 Lawless J. The encyclopedia of essential oils, 2nd Edition. London: Harper Thorsons, 2014

3 Davis P. Aromatherapy. An A-Z, 3rd Edition. London: Vermilion, 2005

4 Tisserand R, Young R. Essential oil safety, 2nd Edition. London: Churchill Livingstone Elsevier, 2014

5 Cheng J, Zug KA. Fragrance allergic contact dermatitis. Dermatitis 2014;25:232-245

6 Dharmagunawardena B, Takwale A, Sanders KJ, Cannan S, Roger A, Ilchyshyn A. Gas chromatography: an investigative tool in multiple allergies to essential oils. Contact Dermatitis 2002;47:288-292

7 Zacher KD, Ippen H. Kontaktekzem durch Bergamottöl. Derm Beruf Umwelt 1984;32:95-97

8 Hayakawa R, Matsunaga K, Arima Y. Depigmented contact dermatitis due to incense. Contact Dermatitis 1987;16:272-274

9 De Groot AC, Weijland JW. Systemic contact dermatitis from tea tree oil. Contact Dermatitis 1992;27:279-280

10 Vilaplana J, Romaguera C, Grimalt F. Contact dermatitis from geraniol in Bulgarian rose oil. Contact Dermatitis 1991;24:301

11 Knight TE, Hausen BM. Melaleuca oil (tea-tree oil) dermatitis. J Am Acad Dermatol 1994;30:423-427

12 Hausen BM, Reichling J, Harkenthal M. Degradation products of monoterpenes are the sensitizing agents in tea tree oil. Am J Cont Derm 1999;10:68-77

13 Hausen BM. Evaluation of the main contact allergens in oxidized tea tree oil. Dermatitis 2004;15:213-214

14 Pirker C, Hausen BM, Uter W, Hillen U, Brasch J, Bayerl C, et al. Sensitization to tea tree oil in Germany and Austria. A multicenter study of the German Contact Dermatitis group. J Dtsch Dermatol Ges 2003;1:629-634

15 Uter W, Schmidt E, Geier J, Lessmann H, Schnuch A, Frosch P. Contact allergy to essential oils: current patch test results (2000–2008) from the Information Network of Departments of Dermatology (IVDK). Contact Dermatitis 2010;63:277-283

16 CITES, Convention on International Trade in Endangered Species of Wild Fauna and Flora. www.cites.org

17 NIST/EPA/NIH Mass Spectral Library 2008. Wiley (www.wiley.com), former editions and later upgrades

18 Adams RP. Identification of essential oil components by Gas Chromatography/Mass Spectrometry. Carol Stream, IL, USA: Allured Publishing Corporation, 2001 and later editions

19 Massfinder 4. Hamburg, Germany: Dr. Hochmuth Scientific Consulting, 2014 and previous versions; http://massfinder.com/wiki/MassFinder_4

Chapter 2 ESSENTIAL OILS: GENERAL ASPECTS

2.1 INTRODUCTION

In everyday life man has frequent contact with essential oils or products containing their source materials. It starts in the morning with toothpaste, deodorant, perfume or aftershave, and cosmetics. At breakfast we peel oranges, drink fruit juice, and eat marmalade, and at dinner there are vegetables, spices and foods flavored with essential oils. Environmental contact with essential oil source materials occurs when handling flowers, plants, conifers, wood, cut grass and hay.

It is estimated that essential oils can be obtained from around 30,000 plant species; about 600 of these are or have been used in the past. Currently, the number of essential oils produced is limited to approximately 150, of which 70-80 are high-volume products. In 2009, the world production of essential oils was over 120,000 metric tons (1). Essential oils have many applications in foods, beverages, perfumes and cosmetics. They are also widely employed as pharmacological agents in various forms of medicine, e.g., traditional medicine, folk medicine, Ayurveda and more recently aromatherapy. The largest quantity of essential oils is probably used for flavoring purposes, followed by their use in the fragrance and cosmetics industries.

Many books on essential oils have been published, some focusing on chemistry, others on their use in aromatherapy (e.g., 25-27) or the oils' safety (30); however, the chemical composition of essential oils has – with the exception of *Citrus* essential oils (24) – not recently been presented in extenso. In this book, all possible chemical profiles of 91 essential oils and 2 absolutes (jasmine) are presented. Both commercial oils (from a large dataset of over 6,300 analyses performed by one of us [Erich Schmidt]) and laboratory-produced essential oils (which can differ considerably in their qualitative and quantitative composition) are discussed.

2.2 WHAT ARE ESSENTIAL OILS?

An essential oil is defined by the International Organization for Standardization (ISO) as a product obtained from a natural raw material of plant origin, by steam distillation (which includes hydrodistillation), by mechanical processes from the pericarp (peel) of citrus fruits, or by dry distillation, after separation of the aqueous phase—if—any by physical processes (11). Essential oils may undergo physical treatments, if they do not result in any significant change in their composition, such as filtration, decantation or centrifugation. For essential oils obtained by steam-distillation, which is the usual production method for most commercial oils, there is a separate definition: essential oil which is obtained by steam distillation with addition of water to the still (hydrodistillation) or without addition of water to the still (directly by steam) (11). Citrus oils (bergamot, grapefruit, lemon, mandarin, tangerine, sweet orange, bitter orange) are so-called cold-pressed essential oils; they are obtained by mechanical processes from the peels of the fruit of a citrus. Essential oils should come from one clearly defined plant source and species; breeding, cultivation and harvest must be optimized, production has to be done according to GMP (Good Manufacturing Practices) (28), and they have to fulfill the requirements as laid down in relevant ISO standards.

Modification of essential oils after production

Sometimes, essential oils undergo one or more physical processes (post-treatment) to alter their chemical composition; these are called post-treatment essential oils. One such treatment is fractional distillation (in this situation called rectification), resulting in rectified oils. The objective of post-treatments can be to eliminate certain individual chemicals (e.g., phototoxic furocoumarins [psoralens] from bergamot oil, methyl eugenol from rose oils), lower the concentration of certain fractions, such as the terpenes or sesquiterpenes by rectification (terpeneless essential oil, terpeneless and sesquiterpeneless essential oil), to concentrate one or more chemicals (concentrated or folded essential oils) or change the color of the essential oil. Reducing the amount of the monoterpenes limonene, γ-terpinene, α- and β-pinene and terpinolene, for example, results in a – desired – higher concentration of 1,8 cineole in eucalyptus oil. Especially citrus oils undergo treatments of concentration by rectification. The main objective here is to *reduce* the concentration of limonene because of its bitterness and limited solubility in flavor preparations, and to *increase* the amounts of C_7-C_{12} fatty aldehydes, citral and further minor components including α-terpineol, citronellol, nerol, linalool, linalyl acetate, geraniol, geranyl acetate and neryl acetate.

2.3 PRODUCTS OBTAINED FROM PLANTS WHICH ARE *NOT* ESSENTIAL OILS

In older literature, but still today, both in the lay press and in scientific literature, the term essential oil is often incorrectly used for a variety of products obtained from plant material by methods other than distillation or cold-pressing, e.g., by solvent extraction, supercritical fluid extraction (which is a well-accepted method to produce highly sophisticated solvent-free products for the flavor industry) or simultaneous distillation and extraction. Reports of essential oils that were extracted are by definition wrong. Examples of other products obtained from plant material which are *not* essential oils include the following.

Absolutes are products obtained from concretes, resinoids, supercritical fluid extracts or pomades by extraction with ethanol. These products contain not only fragrance compounds, but also fatty acids, their methyl esters and paraffins (up to 50% by weight). As these fatty materials do not contribute to the fragrance but do cause solubility problems (perfumes in winter may become cloudy when cooling off), they have to be removed, which is done by solubilizing in a large excess of ethyl alcohol. The solution is then cooled to below 5°C and filtered to eliminate the precipitated waxes. After removing the ethyl alcohol by distillation and thereafter concentration of the solution by vacuum distillation, a product remains an absolute. This product contains all the fragrance compounds, but still also some fatty materials. These absolutes are the products used in the fragrance industries, of which jasmine absolutes and violet leaf absolutes are prominent examples.

Aromatic waters (synonym: hydrolates) are aqueous distillates which remain after steam-distillation of plant material and removal (separation) of the essential oil. These products (e.g., rose water) are often added to cosmetic products. They differ from essential oils in both composition and concentration of chemicals.

Concretes are products obtained from fresh plant material by extraction with one or more solvents, which are subsequently (partly or totally) removed. Conventional steam-distillation is generally considered unsuitable to process some flowers and other plant material from, for example, jasmine, cassie (*Vachellia farnesiana* (L.) Wight & Arn., fragrant acacia), narcissus, *Osmanthus* species, tuberose and violet leaves, since it induces thermal degradation of many compounds contained in the plant. Therefore, solvents, usually hexane or supercritical fluids such as CO_2, are used to extract the fragrant chemicals. After the solvent has evaporated, a viscose product remains called a concrete.

Extracts are products obtained from plant biomass by treating them with one or more solvents.

Pomades are fragrant products obtained with a very old method using fat. In cold enfleurage (enfleurage à froid) flowers are pressed into a layer of animal fat for diffusion of the odoriferous compounds of the flower in the fat. Flowers can also be immersed in warm melted fat (enfleurage à chaude). Nowadays, pomades are hardly produced anymore because of very high cost.

Resinoids are products obtained by extracting dry plant material, usually gum resins like myrrh or olibanum, with one or more solvents.

These products are not discussed in this book, with the exception of jasmine absolutes. They are of great interest to dermatologists, as there is considerable literature on contact allergic reactions to jasmine absolute.

2.4 HISTORY

The process of distillation may date back to around 5,000 years ago, but its origin is unknown. Some point at China as the cradle of hydrodistillation, while others think it was the Indus culture; some authors credit the Arabs as being the inventors of the technique of distillation (6). There are indications that the art of distillation was already a serious science in ancient Mesopotamia (4). However, no essential oils were produced, but distillation was rather used for alchemy or medicinal purposes, and an important function appeared to be concentration and purification of alcohol. The ancient Egyptians expanded and improved Mesopotamian techniques, largely because of their uses in embalming (4). Transfer of the practice of distillation to the West was brought on by the Crusades. Crusaders returning home brought with them distilled alcohol for extracting plant material, fragrant raw materials and refined glass for the production of distillation equipment. The first documented distilled plant material was rose water (21). Hieronymus Brunschwyk in the 12th

century, in his treatise The true art to distil, listed some 25 essential oils produced at that time. However, these essential oils were probably fragrant alcohols or aromatic waters rather than essential oils in the current sense. This age of alchemy brought progress in distilling alcohol and alcohol together with plant material (5). The thirteenth century Arabian writer Ad-Dimaschki also provided a description of the distillation process, and wrote about the production of distilled rose water as well as of a cooling system in the distillation plant.

In the 16th century, the availability of printed books facilitated 'scientists' seeking guidance on the distillation of essential oils. At that time, discrimination between volatile oils and odorous fatty oils could be made. In the 19th century, distilling plants was further developed and the procedure was now based on higher knowledge of chemistry and analysis of single components of the distilled products, which by now were real essential oils. The oils were obtained from cultivations of rose and jasmine, but also from medicinal plants, herbs, spices and citrus fruits, growing in the countries around the Mediterranean Sea, such as France, Spain, Italy, Morocco, Tunisia and Egypt. A major development was the invention of industrial scale vessels in France around 1930. These were walled in and fixed, and were heated with wood or the dried, exhausted biomass previously used for essential oil production. However, a source of fresh water had to be there for the cooling unit of the distillation plant. This stimulated the development of the first mobile distillation units. Working near the fields of cultivation or greater natural resources of wild growing plants, long transports to a distillation plant were not necessary and water from brooks or rivers was everywhere in the Provence. This area would soon become the center of improvement of technology for larger production units, which at that time had become necessary to satisfy consumer demand for lavender and lavandin oil. Not only lavender and lavandin were cultivated in the Provence, but also fennel, basil, thyme, sage, clary sage and tarragon. Ships brought spices and exotic plants such as patchouli to France, which were distilled there and prepared for use. The technical improvements spread all over the northern hemisphere to Bulgaria, Turkey, Italy, Spain, Portugal, as well as to northern Africa. A very important chapter in the history of distillation of plant material was the invention of the alambic à bainmarie, technically speaking a double-walled distillation plant. Steam was not only passed through the biomass, but was also used to heat the wall of the still. This new method improved the speed of the distillation as well as the quality of the top notes of the essential oils thus produced. More recently, the distillation process has become computer controlled with many adjustable parameters, leading to the production of high-quality essential oils.

Citrus oils were already produced by manual expression in the 16th and 17th centuries. Real progress was made in the 20th century with the development of machines which would fulfill the requirements for the production of large quantities of citrus essential oils, such as the apparatus for the Pellatrice and Sfumatrice methods of citrus oil production. Before that, peels were manually pressed with wooden or iron tools; juice and oil were captured and separated by centrifugation. At the end of the 19th century and the beginning of the 20th, hundreds of women and men would sit in great halls in Sicily, Italy, working on endless quantities of fruits.

More information on the history of essential oils can be found in ref. 6.

2.5   WHAT ARE ESSENTIAL OILS USED FOR?

Essential oils are an important trade commodity. The 10 essential oils with the highest estimated trade volumes in 2007 were sweet orange oil (51,000 tons), cornmint oil (32,000 tons), lemon oil (9,200 tons), eucalyptus oil (4,000 tons), peppermint oil (3,300 tons), citronella oil (1,800 tons), clove leaf oil (1,800 tons), Chinese sassafras oil (1,800 tons), lime oil distilled (1,800 tons) and lavandin oil (1,200 tons) (1). This is more than 105,000 metric tons (a metric ton is 1,000 kilograms). All other essential oils are traded in smaller quantities, yet they are important, being frequently used in flavors, cosmetics and perfumery. The prices for essential oils depend mainly on the complexity of cultivated plant species, the mode of harvesting (which is sometimes done by hand, e.g., flowers from rose, bitter orange and jasmine) and the yield of oil. The yield varies from 0.01% in the case of orris oil (prepared from the rhizome of *Iris pallida*, which is the most expensive essential oil with a price of around 150,000 USD/kg) up to 40% for wild thyme. Examples of prices for some of the top 10 products are 4.50 USD/kg for sweet orange oil, 50 USD/kg for lemon oil and around 100 USD/kg for peppermint oil. The total value of the essential oils trade amounts to many billions of USD (2).

Essential oils have a wide field of applications; examples are shown in Table 2.1. The largest buyer of essential oils is the flavor industry, making flavors which are widely used by the food, fragrance and cosmetics industries. In the food industry, essential oils and chemicals derived from them are added to a vast array of products including non-alcoholic (fruit juice preparations, flavored waters, many other soft drinks) and alcoholic beverages, and foods (e.g., milk drinks, yoghurt, candies, chocolate, quick meals, baking mixtures, meats, sausages, teas and even spices and herbs) for flavoring purposes. Essential oils may also be added to foods (e.g., to meat) for (longer) preservation and are widely employed by the tobacco industry. Essential oils are also used as masking agents for topical pharmaceutical preparations. Another major application of essential oils is in perfumes (perfume, eau de parfum, eau de toilette, eau de cologne), cosmetics, household products such as detergents, softeners, cleaning products and other consumer commodities including room scents,

Table 2.2 Concentrations of fragrance chemicals used in finished cosmetic products (12,31)

Product	Concentration range
Perfumes	
Eau de cologne	3-7%
Eau de parfum	12-15%
Eau de toilette	7-12%
Extrait	up to 30%
Perfume	18-25%
Splash cologne	1-3%
Leave-on cosmetics	
Body creams	0.1-0.3%
Creams	0.05-0.4%
Deodorant cream	0.5-0.8%
Deodorant spray	0.1-0.5%
Deodorant stick	0.8-1.0%
Fragrancing cream	0.5-1.5%
Hair deodorant	0.02-0.05%
Hair styling aid	0.05-0.2%
Lipstick	0.005-0.03%
Lotions	0.4-1.2%
Rinse-off products	
Bar soap	1-7%
Bath salt	0.2-0.4%
Bubble bath	0.8-1.5%
Conditioner (hair)	0.1-0.3%
Face cleansers	0.02-0.06%
Liquid soap	0.2- 0.5%
Shampoo	0.1-0.3%
Shaving creams and gels	0.2-0.4%
Shower gel and body washes	0.5- 1.0%

candles, incense and bouquets (a bouquet is a potpourri existing of dried flowers, barks and leaves, perfumed by adding perfume mixtures supplied separately).

Essential oils and fragrance chemicals are often incorporated to give the desired scent to cosmetics products. However, they are also employed as masking fragrance. Certain ingredients of cosmetics, such as fatty acids, lipid balancing bases, aqueous plant extracts, fatty oils and surfactants, have an inherent smell. To mask this, an efficient and suitable perfume composition is added to these products. All cosmetics can be presumed to contain fragrances, unless they are specifically labeled as 'fragrance-free', 'contains no perfume', 'non-scented' or have similar indications. However, even then, they may sometimes contain (masking) fragrance materials (29). Approximate concentrations of fragrances in perfumes and cosmetics are shown in Table 2.2. In general,

Table 2.1 Examples of essential oil applications

Flavor industry: flavors for the food industry
Food industry: soft drinks, foods, alcoholic beverages, spices, herbs, tea, food preservation
Fragrance industry: perfumes, fragrances for other products
Cosmetics industry
Household products: detergents, softeners, room scents, candles, incense
Tobacco industry: cigarettes
Medicinal use: folk medicine, traditional medicine, phytotherapy, balneotherapy, aromatherapy
Pharmaceutical industry: masking agent
Animal food

fragrances contain between 5 and 10% of essential oils. Most commonly used are orange, bergamot, lemon, lavender, patchouli and cedarwood oils. Spice oils are employed in small concentrations and others in traces. Expensive oils are only used in fine fragrances.

Essential oils are also widely employed to promote health and combat diseases, e.g., in forms of traditional and folk medicine, phytotherapy (the use of extracts of natural origin as medicines or health-promoting agents) and aromatherapy. Aromatherapy, which appears to be gaining in popularity, has been defined by aromatherapy and essential oils book authors as 'The therapeutic application of essential oils and aromatic plant extracts in a holistic context, to maintain or improve physical, emotional and mental well-being' (25), 'the use of essential oils, applied topically, orally, by inhalation or by other means, to promote health, hygiene and psychological wellbeing' (30) or - briefly - 'the therapeutic use of essential oils' (26).

Essential oils are also used in animal feeding (32). They may be employed as appetizer and for preservation purposes, as certain essential oils that contain high concentrations of carvacrol such as origanum and thyme oil have been shown to have antimicrobial effects (10). Herbs and essential oils may, according to some authors, even be able to partly substitute for antibiotic growth promoters (33).

2.6   PRODUCTION OF ESSENTIAL OILS

Biomass for production of most essential oils can grow in various countries and on more than one continent. Successful cultivation depends inter alia on the climate, soil, stress by insects, and water. For most plants, a warm climate with sufficient rain or water supply benefits high-quality crops. A balanced content of mineral elements (fertilizing) benefits the growth of the plants and the content of essential oil. Stress by insects will lead to higher oil yield, as the oil is used as a repellent by the plant. Some plants can grow only in higher areas, e.g., lavender in France (above 800 meters) and certain herbs in the Himalayas. For sufficient crops of herbs and spices, ample exposure to sunlight is necessary. Several Asian countries, e.g., India, benefit from their sunny climate and plants can be harvested two or sometimes three times per year. More information on these aspects can be found in ref. 6.

To produce high-quality commercial essential oils, the biomass should be perfect and the production method optimal. Most essential oils are produced by steam-distillation, but citrus oils are obtained by expression methods. For many crops, it is very important that distillation takes place very soon after harvesting. Melissa, for example, will start withering shortly after being cut; for these plants, field distillation, where harvested plants are immediately placed in the distillation vessel and distilled in situ, can best be used. The same applies to the production of rose essential oils. The rose petals have to be picked very early in the morning, when the essential oil yield is at its highest level; distillation takes place until noon. Another important feature in essential oils production is adequate pre-distillation biomass

preparation. Drying will increase the yield of oil, as a larger volume of biomass can be processed; it also saves energy. Some plants have to be dried to result in a perfect quality essential oil, e.g., in the cases of cinnamon, orris and vetiver. Fruits and seeds need to be ground to guarantee complete exhaustion of the oil present in the plants during distillation, for example, pepper, pimento, coriander, anise and fennel. Chipping is necessary when distilling wood and heartwood (e.g., cedarwood, cypress, rosewood). In the case of sandalwood oil, the wood even needs to be pulverized. Most herbs are cut into smaller pieces before distillation. This is usually already done automatically on the field by the harvesting equipment; the biomass is then loaded into containers, which can be connected to a steam source and cooling unit in the distillation plant. Shortly before the harvest, a few kilograms of biomass are steam-distilled and analyzed by GC/MS to determine at what timeframe the harvest results in optimal composition of the oil. More information on these aspects can be found in ref. 6.

Distillation methods

There are two methods of distillation: steam-distillation and water-steam distillation. In the steam-distillation method, there is a direct supply of steam to the biomass. The distillation apparatus consists of a still pot (kettle) containing the biomass, a cooling system (condenser), an oil separator and a high-capacity steam generator. The still pot looks like a cylindrical vertical storage tank and has steam pipes located at the bottom. The biomass is placed on multiple perforated plates resembling sieves located over each other; this separates the plant material, prevents compaction and allows the steam unimpeded access to the biomass. The walls of the oil-bearing plant cells are destroyed by pressure and heat, and volatile chemicals are set free. The oil-laden steam leaves the kettle through an outlet in the still pot lid and enters the alambic. This is the head of the distillation plant and here the vaporous mix of water and oil starts to condense. The steam is then passed through the cooling system, which is either a plate heat exchanger or a surface heat exchanger, such as a cold-water condenser. The condensate is subsequently separated into essential oil and distillation water in an oil separator, e.g., a Florentine flask. The distillation water is sometimes redistilled and the essential oil dried and stored (6).

The water-steam distillation method is used mostly for rose oil production. Here, water is present in the double-walled vessel, and the water contains the biomass. The water plus biomass is heated by steam, which is led through the walls of the still. When the water starts to boil, steam is liberated containing the volatiles. The rest of the procedure is identical to steam-distillation, described above.

Currently, distillation is a computer-controlled process with programmed parameters, which include steam quantity, temperature, speed of heating and the pressure of the steam flowing through the vessel. These parameters can vary during the distillation process. Usually, steam-distillation is performed under slightly higher than atmospheric pressure. Some oils like cistus oil and some conifer oils are distilled with superheated steam, at a temperature higher than its vaporization point (around 105°C) to increase the yield and enrich the oil with a higher quantity of high boiling components. The size of the vessel is also an important factor. Most vessels have a capacity of 4 tons or higher. For the production of some oils, such as pepper and orris oil, however, smaller vessels are needed, as the biomass has to be stirred during the distillation process to prevent accretion to the wall, which results in burning. For the production of most oils, steel or iron vessels are used. However, sometimes glass is preferred to avoid any reaction with iron particles. More information on these aspects can be found in ref. 6.

Production of citrus essential oils

A very important category of essential oils is the citrus oils, obtained from citrus fruits of the Rutaceae family. The main fruits are sweet orange, lemon, bergamot, lime, bitter orange, mandarin and grapefruit. Their oil cells are located in the epicarps (outermost layer of the peel, the pericarp), and their smell can be released by scratching or cutting. All citrus oils are prepared by expression of the pericarps with one exception: lime oil can be produced by both expression and distillation, which results in two oils differing in composition, smell and flavor. Distillation for other citrus oils is impossible, as heating will destroy important but unstable molecules, such as fatty aldehydes. These chemicals are also unstable when in contact with acids (from the fruit juice) and therefore the oil from the peels should be removed in a way that such contact is avoided.

Three important methods are used: the Sfumatrice method, the Pellatrice method and the FMC method (Food Machinery Corporation), also called the brown method. In the Sfumatrice method, the fruits are cut in half and the endocarp (the flesh) is removed. The peel then is squeezed in the Sfumatrice machines and the mixture of oil, water and juice is decanted and the oil phase later separated by centrifugation. The Pellatrice method uses spiked Archimedes screws to bruise the oil cells in the epicarp and initiate the flow of oil. The peel oil is washed out by a permanent water spray and then is filtered and centrifuged. The FMC process is designed for the processing of large quantities of fruits. It is a combination of juice pressing and peel pressing. Through a hole in the bottom of the fruit, cut by a knife, a tube is placed within the fruit. By pressing each single fruit, juice runs through the tube and is led to a separate system. At the same time, the epicarp is also pressed and the oil will run outside the fruit in a separate tube, thus avoiding contact between the (acid) juice and the oil. The oil from the oil–water emulsion pressed from the peels is separated by centrifugation. More information on these procedures can be found in ref. 6.

The last process to produce essential oils is the so-called dry distillation, which applies to distillation of wood, barks, roots or gums, without addition of water or steam (11). Without the supply of oxygen, this process is in fact a form of carbonizing. The chemicals that emerge from the biomass are mainly phenolic compounds that have a leathery odor. An example of a dry-distilled essential oil is birch tar oil, which was formerly used in perfumery. Because of toxicity issues, however, it is no longer used in this application. More information can be found in ref. 6.

2.7 CHEMISTRY OF ESSENTIAL OILS

Essential oils are multi-component mixtures; in individual oils, up to 400 chemicals or even more may be identified when sophisticated analytical equipment and the proper detection methods are used. In mass spectra databases, over 5,500 entries for chemicals can be found, of which more than 3,000 may be present in essential oils. The largest group of chemicals found in essential oils consists of terpenes. Terpenes are hydrocarbon chemicals produced from five-carbon isoprene units (C_5H_8). From these units, numerous molecules can be constructed in biosynthesis, both linear-chained chemicals and molecules with one or more ring structures. There are several classes of terpenes based on their number of isoprene units; in essential oils, the most important ones are the monoterpenes and the sesquiterpenes. Monoterpenes possess 10 carbon atoms and are built from two isoprene units ($C_{10}H_{16}$) which can form various carbon skeletons. There are three forms of monoterpenes: linear (acyclic), monocyclic and bicyclic. Sesquiterpenes possess 15 carbon atoms; they are built from three isoprene units ($C_{15}H_{24}$) and may occur in acyclic, monocyclic, bicyclic and tricyclic forms. These are the terpenes with higher boiling temperatures. Mono- and sesquiterpenes are the main components in essential oils, sesquiterpenes comprising about 25% of the terpene fraction. Volatility and the typical odor of essential oils are created through both groups. Chemical modification of terpenes and sesquiterpenes, e.g., by rearranging carbon skeletons or by oxidation, produces compounds generally termed terpenoids. The oxidation products are the most important, creating subgroups like alcohols, aldehydes, phenols, ethers and ketones. Infrequently, during biosynthesis, functional groups like sulfur or nitrogen are bond to or integrated in the carbon skeleton.

The most important chemical classes and some examples of compounds belonging to the various groups found in essential oils are shown in Table 2.3. Also shown is the number of oils/absolutes in which they were present in a group of 91 oils and two jasmine absolutes analyzed by one of us (ES) between 1998 and 2014 (total of nearly 6,400 samples) and the oils/absolutes containing the chemicals in the highest concentrations.

2.8 FACTORS INFLUENCING THE COMPOSITION OF ESSENTIAL OILS

The composition of essential oils can vary considerably from country to country, from producer to producer, and even from year to year for the same producer and crop. Factors that may influence the oils' chemical composition are shown in Table 2.4. Important parameters

Table 2.3 Chemical classes, examples of compounds, and data on their presence in essential oils/absolutes (34-37)

Chemical class and important chemicals	Nr. oils[a]	Oils / absolutes with highest concentrations
Acyclic terpenes		
Myrcene	73	Rosemary (46.0%), bay (32.0%), dwarf pine (30.2%), juniper berry (21.6%)
(*E*)-Ocimene	51	Melissa (9.3%), neroli (9.3%)
(*Z*)-Ocimene	1	Bay (0.1%)
Monocyclic monoterpenes		
p-Cymene	66	Black cumin (57.5%), thyme (25.7%), Spanish thyme (25.4%)
α-Fenchene	10	Valerian (4.0%), thuja (1.4%), spike lavender (1.0%)
Limonene	84	Sweet orange (95.7%), bitter orange (95.5%), grapefruit (95.5%)
β-Phellandrene	47	Elemi (23.9%), angelica root (22.0%), black pepper (17.4%)
α-Terpinene	50	Tea tree (11.7%), marjoram (10.3%), ravensara (8.2%)
γ-Terpinene	66	Marjoram (50.2%), tea tree (23.0%), thyme (21.2%)
Terpinolene	65	Tea tree (45.7%), dwarf pine (8.3%), marjoram (6.8%)
Bicyclic monoterpenes		
Camphene	59	Pine needle (27.4%), valerian (24.1%), silver fir (17.3%)
δ3-Carene	33	Dwarf pine (34.4%), cypress (25.9%), black pepper (17.6%)
α-Pinene	81	Cypress (68.0%), juniper berry (66.6%), olibanum (46.5%)
β-Pinene	73	Galbanum resin (81.2%), silver fir (31.7%), dwarf pine (29.0%)
Sabinene	61	Nutmeg (36.7%), ravensara (25.5%), juniper berry (17.7%)
α-Thujene	48	Olibanum (ex *Boswellia serrata*) (65.3%), thyme (25.9 %), olibanum (ex *Boswellia sacra*) (25.8%)
Tricyclic monoterpenes		
Tricyclene	22	Pine needle (2.7%), silver fir (2.6%), juniper berry (2.0%)
Acyclic sesquiterpenes		
α-Farnesene	8	Melissa (9.1%), ginger (5.7%), citronella Sri Lanka (4.6%)
(*E,E*)-α-Farnesene	15	Ylang-ylang third (28.9%), ylang-ylang second (19.7%), ylang-ylang extra (10.3%), cananga (7.1%)
(*E*)-β-Farnesene	23	Chamomile German (35.4%), lavender (4.5%), spike lavender (4.3%)
Monocyclic sesquiterpenes		
α-Bisabolene	4	Carrot seed (2.9%), chamomile German (2.7%), spike lavender (1.5%)
β-Bisabolene	22	Carrot seed (7.8%), ginger (6.3%), black pepper (5.9%)
Curcumene	12	Ginger (5.8%), costus root (2.3%), valerian (1.8%)
β-Elemene	41	Myrrh (8.7%), costus root (7.7%), basil (5.5%), melissa (4.7%)
α-Humulene	60	Sage Dalmatian (12.4%), cananga (9.4%), myrrh (8.8%)
α-Zingiberene	4	Ginger (17.0%), black pepper (1.2%), aniseed (0.5%)
Bicyclic sesquiterpenes		
α-Cadinene	2	Myrrh (2.2%), cananga (0.4%)
γ-Cadinene	25	Pine needle (5.3%), hyssop (2.9%), vetiver (2.8%)
δ-Cadinene	55	Cananga (6.1%), ylang-ylang third (5.6%), ylang-ylang second (4.6%), calamus (4.4%)
β-Caryophyllene	85	Cananga (38.0%), black pepper (32.4%), melissa (29.4%), ylang-ylang second (19.5%)
β-Selinene	11	Costus root (4.2%), black pepper (3.2%), carrot seed (3.0%)
Valencene	2	Vetiver (1.9%), sweet orange (0.1%)
Tricyclic sesquiterpenes		
Aromadendrene	10	Tea tree (2.0%), eucalyptus (1.8%), thyme (0.8%)
β-Bourbonene	18	Hyssop (2.4%), myrrh (2.1%), spearmint (1.7%)
α-Cedrene	8	Cedarwood China (38.4%), cedarwood Virginia (27.2%), cedarwood Texas (21.3%)
β-Cedrene	6	Cedarwood China (8.2%), cedarwood Virginia (7.7%), cedarwood Texas (5.2%)
α-Gurjunene	11	Myrrh (3.8%), niaouli (2.8%), tea tree (2.8%)
β-Gurjunene	4	Valerian (0.9%), calamus (0.8%), hyssop (0.3%)
α-Patchoulene	1	Patchouli (8.0%)
β-Patchoulene	2	Patchouli (4.1%), guaiacwood (0.5%)
α-Santalene	8	Sandalwood New Caledonia (6.5%), sandalwood Australia (4.9%), sandalwood (1.9%)
Viridiflorene	4	Tea tree (2.1%), niaouli (2.1%), sage Dalmatian (0.2%)

Table 2.3 Chemical classes, examples of compounds, and data on their presence in essential oils/absolutes (34-37) (*continued*)

Chemical class and important chemicals	Nr. oils[a]	Oils / absolutes with highest concentrations
Acyclic monoterpene alcohols		
Citronellol	15	Geranium (49.4%), rose (44.8%), citronella Java (13.5%)
Geraniol	40	Palmarosa (86.9%), citronella Sri Lanka (48.7%), geranium (31.8%)
Lavandulol	6	Lavender (2.4%), lavandin grosso (1.2%), lavandin abrial (0.2%)
Linalool	74	Rosewood (88.4%), coriander fruit (81.2%), thyme (68.5%)
Nerol	25	Melissa (16.2%), rose (11.0%), ravensara (1.7%)
Acyclic monoterpene esters		
Citronellyl acetate	12	Thyme (17.5%), citronella Java (4.5%), melissa (3.9%)
Geranyl acetate	40	Thyme (21.8%), carrot seed (12.9%), palmarosa (12.5%)
Linalyl acetate	22	Clary sage (64.2%), petitgrain bigarade (54.0%), lavender (43.3%)
Neryl acetate	20	Melissa (2.9%), petitgrain bigarade (2.8%), neroli (2.0%)
Terpinyl acetate	17	Laurel leaf (12.0%), Spanish sage (9.0%), cajeput (3.5%)
Acyclic monoterpene aldehydes		
Citronellal	15	Eucalyptus citriodora (84.4%), citronella Java (49.6%), citronella Sri Lanka (12.0%)
Geranial	22	Lemongrass West India (46.3%), lemongrass East India (46.3%), litsea cubeba (42.3%)
Neral	19	Lemongrass East India (35.5%), litsea cubeba (34.5%), lemongrass West India (33.3%)
Monocyclic monoterpene alcohols		
(Z)-Carveol	3	Spearmint (2.0%), olibanum (0.2%), elemi (0.05%)
Menthol	5	Peppermint (47.9%), geranium (2.3%), spearmint (2.2%)
Pulegol	4	Peppermint (1.1%), citronella Java (0.9%), melissa (0.6%)
Terpineol	71	Ravensara (14.7%), cajeput (11.6%), niaouli (9.2%)
Monocyclic monoterpene phenols		
Carvacrol	4	Thyme (77.8%), black cumin (5.8%), olibanum (0.1%)
Eugenol	23	Clove bud (90.4%), clove stem (90.1%), clove leaf (89.1%)
Thymol	5	Thyme Spanish (56.2%), thyme (47.8%), black cumin (2.2%)
Monocyclic monoterpene ketones		
Carvone	10	Spearmint (82.3%), peppermint (0.9%), olibanum (0.6%)
Menthone	5	Peppermint (38.5%), geranium (2.3%), melissa (2.1%)
Piperitone	4	Peppermint (5.4%), spearmint (0.6%), basil (0.06%)
Pulegone	1	Peppermint (5.4%)
Bicyclic monoterpene alcohols		
Borneol	39	Citronella Sri Lanka (7.6%), sage Spanish (7.0%), rosemary (6.3%)
Myrtenol	7	Galbanum resin (2.2%), hyssop (2.1%), olibanum (1.3%)
Pinocarveol	5	Hyssop (19.2%), olibanum (3.2%), laurel leaf (0.1%)
Verbenol (*cis-* and *trans-*)	13	Litsea cubeba (1.8%), angelica root (1.2%), melissa (0.9%)
Bicyclic monoterpene aldehydes		
Myrtenal	8	Galbanum resin (1.2%), chamomile Roman (0.8%), hyssop (0.7%)
Bicyclic terpene ketones		
Camphor (α-, *d-* and *dl-*)	31	Sage Spanish (36.0%), spike lavender (35.1%), rosemary (24.2%)
Fenchone	3	Thuja (16.7%), basil (0.7%), dwarf pine (0.06%)
Monocyclic sesquiterpene alcohols		
Bisabolol (α-, β-, γ-)	15	Chamomile German (38.3%), sandalwood Australia (12.8%), zdravetz (2.8%)
Elemol	19	Elemi (22.1%), citronella Java (7.0%), hyssop (1.9%)
Lanceol	3	Sandalwood oils (3.3%-15.2%)
Bicyclic sesquiterpene alcohols		
Cadinol (α-, δ- and τ-)	27	Basil (7.1%), vetiver (3.5%), ylang-ylang third (2.6%)
Chrysanthemol	1	Chamomile German (0.1%)
Eudesmol (α-, β-, γ-, 10-epi-γ-)	22	Geranium (5.6%), zdravetz (5.3%), guaiacwood (4.7%)
Muurolol (α- and τ-)	11	Ylang-ylang second (2.0%), cananga (1.8%), melissa (1.2%)

Table 2.3 Chemical classes, examples of compounds, and data on their presence in essential oils/absolutes (34-37) (*continued*)

Chemical class and important chemicals	Nr. oils[a]	Oils / absolutes with highest concentrations
Bicyclic sesquiterpene aldehydes		
Isovalencenal	–	
Zizanal	–	
Bicyclic sesquiterpene ketones		
Germacrone (*E,E*-)	1	Zdravetz (66.6%)
Khushimone	1	Vetiver (3.3%)
Nootkatone	4	Grapefruit (0.4%), vetiver (0.3%), bitter orange (0.05%)
Valeranone	1	Valerian (4.0%)
α-Vetivone	1	Vetiver (6.1%)
β-Vetivone	1	Vetiver (5.7%)
Tricyclic sesquiterpene alcohols		
Patchoulol	1	Patchouli (33.9%)
α-Santalol	3	Sandalwood oils (42.9%-53.3%)
β-Santalol	3	Sandalwood oils (17.7%-23.6%)
Viridiflorol	5	Niaouli (22.6%), Dalmatian sage (2.7%), tea tree (0.8%)
Aromatic alcohols		
Benzyl alcohol	8	Jasmine sambac (6.3%), jasmine (3.7%), ylang-ylang extra (0.3%)
Cinnamyl alcohol	3	Cassia (0.4%), jasmine sambac (0.2%), cinnamon leaf (0.08%)
2-Phenethyl alcohol	4	Rose (2.7%), Jasmine sambac (1.2%), cassia (0.7%)
Aromatic esters		
(*Z*)-Methylcinnamate	1	Basil (23.6%)
Methyl salicylate	8	Jasmine (2.8%), Jasmine sambac (1.6%), clove bud, clove leaf and clove stem (each 0.3%)
Aromatic aldehydes		
Benzaldehyde	7	Cassia (2.3%), niaouli (1.8%), cinnamon bark (1.0%)
Cinnamic aldehyde ((*E*)-, (*Z*)-)	5	Cassia (83.1%), cinnamon bark (72.7%), cinnamon leaf (2.8%)
Cuminaldehyde	–	
Vanillin	–	
Phenolic ethers		
(*E*)-Anethole	8	Aniseed (98.6%), star anise (90.1%), black cumin (0.9%)
(*Z*)-Anethole	2	Star anise (0.4%), aniseed (0.3%)
Methyl chavicol (estragole)	14	Basil (87.0%), ravensara (19.9%), star anise (5.9%)
Safrole	4	Ravensara (2.0%), nutmeg (1.8%), cinnamon leaf (1.3%)
Oxides		
Bisabolol oxide A	1	Chamomile German (46.0%)
Bisabolol oxide B	1	Chamomile German (12.1%)
1,8-Cineole	44	Eucalyptus (88.7%), cajeput (70.2%), ravensara (68.0%), niaouli (61.2%)
Linalool oxide (*cis*-, *trans*-, furanoid, pyranoid)	37	Rosewood (2.1%), coriander fruit (1.4%), geranium (1.0%)
Menthofuran	1	Peppermint (7.0%)
Acids		
Geranic acid	1	Litsea cubeba (0.1%)
Isovaleric acid	1	Valerian (2.4%)
Zizanoic (khusenic) acid	1	Vetiver (4.3%)

[a] number of essential oils / jasmine absolutes in which the chemicals have been identified, in a group of 91 essential oils and two jasmine absolutes investigated between 1998 and 2014 (nearly 6,400 samples investigated)

Table 2.4 Factors that may influence the chemical composition of essential oils[a]

Plant parameters
Species
Cultivar or variety
Chemotype
Age of plant
Cultivated or wild plants

Environmental parameters
Origin of plant
Climate: temperature, rain, sun, altitude
Soil conditions
Fertilization
Use of herbicides, pesticides
Water supply or water stress
Insect stress
Monocultures or interbreeding with other species
Presence of weeds (especially bio culture)

Harvest and post-harvest / pre-distillation parameters
Plant part or plant parts used
Season of harvest
Year of harvest
Number of harvests per year
Pretreatment of biomass (fresh, dried, particle size, mode of drying [sun, shade, oven], drying time)
Storage conditions of source plant material and storage time
Weeds unintentionally harvested with essential oil source species (amount, species)

Production parameters
Mode of production: steam-distillation, hydrodistillation, steam/hydrodistillation, cold-pressing (citrus oils)
Distillation parameters: temperature, flow, pressure, distillation time, fractional distillation
Commercial oil or laboratory-produced in Clevenger-type apparatus
Rectification (see Modification of essential oils after production)
Adulteration with natural materials or synthetic chemical compounds
Inadequate removal of capture solvents (laboratory-prepared oils)

Other parameters
Storage condition of essential oils
Storage time of essential oils
Age of essential oil
Aging: exposure to oxygen and ultraviolet light
Analytical equipment used, accuracy of analysis, availability of mass spectra libraries, automated software

[a] examples. The table is not meant to be exhaustive.

---

are those related to the plant (species, cultivar, variety), to the environment (climatic and soil conditions, water, fertilization), to harvest and post-harvest conditions (part of the plant harvested, season, mode and duration of storage and pre-distillation treatments such as drying), to the mode of production (differences between laboratory-produced oils and large-scale commercial steam-distilled essential oils) and other factors, of which the analytical procedures used to identify the composition qualitatively and quantitatively are very important.

Various plant species are known to have several so-called chemotypes. A clear and widely accepted definition for this well-known phenomenon is lacking. In practice, it means that, within a population of one plant species with the same morphological features, groups exist with different composition of their secondary plant products. This is probably regulated by one or only a few genes (13). The oils produced from the various genotypes are qualitatively very similar, i.e., they contain the same

spectrum of chemicals, but there are major differences in the quantities of one or several of these. Thus, a specific chemical may be absent or present in trace quantities in one chemotype and be the dominant component in concentrations of >50% in another. In 25 commercial thyme oils from *Thymus vulgaris*, for example, the following ranges in concentrations were observed for the major components: carvacrol trace-77.8%, linalool 0.03-68.5%, thymol 0.2-47.8%, 1,8-cineole 0.2-36.5%, *cis*-sabinene hydrate 0.07-32.7% and geraniol 0-26.0% (Chapter 5.85). The chemotypes are usually assigned on the basis of single dominant chemicals (the compound with the highest concentrations) or a few major components in essential oils, although this method may well be criticized (14,16). Corroboration of chemotypes by genetic coding has not been done. In fact, the term 'chemotype' is applied rather loosely by various authors, and some suggested they had found a new chemotype on the basis of one laboratory-hydrodistilled oil containing an unusual dominant chemical. Plant species that produce oils with several chemotypes include thyme (both from *Thymus vulgaris* [3,15, Chapter 5.85] and from *Thymus zygis* [Chapter 5.86]), *Melaleuca alternifolia* (tea tree oil, Chapter 5.83), rosemary (Chapter 5.72), and ravensara (Chapter 5.71).

The mode of production also heavily influences the composition of essential oils. Thus, laboratory-prepared oils, which are produced with a Clevenger-type apparatus, in which the biomass is cooked in boiling water, show major differences from the same oils produced in large industrial plants with steam-distillation. Long distillation times may alter the quality of odor and the whole composition. Plant parts used in scientific studies are often carefully selected and totally clean of any byproducts. Thus, when only the flowers from lavender are hydrodistilled in a Clevenger apparatus, it will often contain no coumarin. In industry, several centimeters of stalks remain with the flowers after harvesting and these contain coumarin. Other byproducts may also influence the composition. In conventional culture, herbicides are used to combat weed. In the bio culture, however, such products are forbidden and, consequently, weeds can grow excessively. As a result, organically produced plant material can include up to 30% extraneous herbs and weed, which can change the composition of the oil. Some weeds contain acids, which can react with free alcohols generated by the basic biomass. In the case of lavender, for example, this may result in higher values of lavandulyl acetate and linalyl acetate. The use of higher pressure in commercial steam-distillation of biomass such as wood, root and plants containing resins results in higher values of sesquiterpenes. The composition of the oil, or actually its *reported* composition, also depends on the analytical procedures performed. For the industry, determining concentrations down to 0.01% for individual chemicals is usually sufficient. In scientific publications, it is often tried to identify chemicals in lower concentrations also.

2.9 ANALYSIS OF ESSENTIAL OILS

The analysis of essential oils has made tremendous progress in the last 35 years. In the early years of analytical chemistry, analytical assays were simple, with chemical reaction methods to detect single components. Isolation of the compound by separation through distillation was helpful for its recognition. Reaction experiments gradually improved and other tests including gravimetric analysis, photometry and titration were soon standard. With the invention gas-chromatography, a huge step was taken forward in identifying chemicals not only qualitatively, but also their quantities. Today this method is highly sophisticated and is performed in every good laboratory. Currently, in essential oil chemistry, several tests are used to determine the chemical composition of essential oils and to verify their naturalness.

Organoleptic methods are still very important. The human nose is able to detect some molecules in traces and even a few parts per million (ppm) in air, whereas analytical equipment has to be very finely adjusted to get proper results. Vanillin and maltol are examples of

such chemicals as are sulfuric compounds like *p*-menthanone-3-thiol, which gives black currant its typical odor. Essential oils are always checked olfactorily, even when analytical results confirm their identity and naturalness, as odor is still their most important quality! Appearance and color are also important factors for commercial use. Citrus fruits can vary widely in color depending on the time of harvesting. Clementine oil produced from fruits at the beginning of the ripening stage is pale yellowish or orange and gradually changes to orange or orangered when riper fruits are used. A constant color is a precondition (consumer demand) for use of essential oils in cosmetics as well as for food flavors.

Physicochemical test methods are widely used in the chemical industry. Currently, these tests are all automated; connectivity with databases ensures the standard of quality. To obtain full essential oils quality control, the following parameters are established and tests performed: relative density, refractive index, solubility tests and determination of evaporation residues (to rule out adulteration with fatty or high boiling substances). Also helpful are older methods such as phenol content, carbonyl value, ester and saponification value and the content of free and total alcohol.

Essential oils are often analyzed with chromatographic methods, e.g., thin layer chromatography (TLC), a fast and inexpensive method for identifying substances and testing the purity of compounds. The principle of chromatography is based on the distribution of the constituents to be separated between two immiscible phases, the stationary phase with a large surface area and a mobile phase that ascends through the stationary phase by capillary forces. A glass plate covered with a thin layer of adsorbent (usually silica gel or alumina) is used as the stationary phase. One drop of essential oil is applied near its edge. The plate is then positioned in a developing chamber with one end immersed in the developing solvent, which is the mobile phase. When the mobile phase with the oil constituents has reached about two-thirds of the plate length, the plate is removed, dried, and the separated components are located. In some cases the spots are directly visible; others must be visualized (colored) by the application of a solution of iodine or sulfuric acid, or with the aid of ultraviolet light. Identification of chemicals is based on the calculation of the so-called ratio of fronts (Rf) value for each spot (38).

Gas chromatography (GC) is also based on the principle of separation of chemicals using a stationary and a mobile phase. A carrier gas is used to transport small amounts of essential oil within the gas phase through a column at a constant speed. In general, essential oil GC analyses are carried out on 25-50 meter columns, with 0.20-0.32 mm internal diameters, and 0.25 µm stationary liquid phase film thicknesses. By constant or rising temperature in the oven, in which the column is placed, and permanent absorption and desorption of the oil components between liquid and mobile (gas) phase, separation of the various components of the oil is enabled. After reproducible times (retention time, RT) the various components reach the outlet of the column and are registered by the flame ionization detector

(FID). Components are burned by a small hydrogen flame using synthetic air (to avoid misinterpretation by polluted environmental air), whereby a signal is recorded that generates a chromatogram, which is a signal versus time graphic (19). The chromatogram contains information for both qualitative and quantitative analysis of the essential oil (39). The peak area and height are a function of the amount of chemical present. The identity of the compounds is determined by the retention time. Unfortunately, several chemicals can have the same retention time, which necessitates increasing separation performance. In addition, confirmation of the identification based on the retention time by other tests is necessary.

After mass spectrometry (MS) was invented, gas chromatography coupled with MS (GC-MS) soon became the standard procedure for analyzing essential oils. In mass spectrometry, molecules are bombarded with ions under vacuum and high-temperature conditions. This generates charged molecules or molecule fragments, of which mass-to-charge ratios are determined, supplying a unique fingerprint ('mass spectrum') for the chemical. Mass spectra are stored in databases, of which several are commercially available (40-42). These libraries can also be self-established by numerous runs with molecules of highest purity. In GC-MS, spectra of chemicals are automatically compared by computer with their retention time or index, which results in a correct allocation and identification of the chemicals

A revolution was the development of two-dimensional GC with time of flight MS (GCxGC-TOF-MS). Very often, in spite of using high-quality separation equipment, different molecules elute (leave the column) in very small time frames and therefore cannot be reliably separated (co-elution). In this new method, two columns are used, which are connected. The system for comprehensive two-dimensional GC takes every peak (component) and slices small cuts, before transferring them to the second column, which works 50 times faster than the first column. A fast detector collects the data at a rate of 100 Hz (Hertz, cycles per second) or more. This results in a plot picture showing a three-dimensional display or bird's eye view of the substances identified. An ordinary mass spectra unit is equipped with an acquisition rate of only 1 or 1.5 spectra per second, so mass spectra needed to be established much faster. A so-called time of flight (TOF) detector is able to handle narrow peaks from GCxGC up to 500 spectra per second. This increases the number of detected components tremendously.

The naturalness of an essential oil can also be investigated by ^{13}C NMR (nuclear magnetic resonance). This method is based on the detection of the magnetic properties of ^{13}C nuclei. Separation of the oil is unnecessary. It is an expensive method but provides very accurate results (19).

For most of the applications described in this section, standards on the technique of testing have been created by ISO (International Organization for Standardization, www.iso.org) (20). More information on the analysis of essential oils can be found in ref. 17.

2.10 QUALITY, PURITY AND ADULTERATION OF ESSENTIAL OILS

Essential oils sold on the internet often bear labels as 'pure', 'natural', or '100% natural'. Such labelling does not guarantee that the oils are of good quality. In fact, such labelling should not be necessary as, according to ISO standards (11), essential oils must be natural and pure. Indeed, good-quality oils on the market conform to the standards laid down by ISO for individual essential oils. However, it cannot be denied that lesser-quality products are commercially available. Poor or lesser quality of essential oils can have various causes, including adulteration, contamination, inadequate oil production and aging (Table 2.5).

Table 2.5 Causes of lesser quality of essential oils

Adulteration
Addition of foreign substances

- Fat oils, ethanol, water (formerly)
- Other essential oils (turpentine oil to pine needle of fir oil; cananga oil to ylang-ylang oil)
- Fractions of essential oils (including foreruns) and defined natural isolates
- Steam-distilled citrus oil added to expressed (genuine) citrus oils
- Synthetic chemicals known to occur in nature: linalool, menthol, limonene

Prohibited non-physical methods for rectifying oils
Intentional mixing of biomass before distillation (cananga blossoms with ylang-ylang, formerly lavandin flowers with lavender)

Contamination
Chemicals from packing material
Chemical residues from inadequately cleansed distillation stills
Chemical residues from inadequately cleansed drums and other containers

Inadequate production
Use of wrong species
Use of more than one species or subspecies
Use of several chemotypes of one species

Aging
Formation of allergenic oxidation products (tea tree oil, citrus oils, lavender oil, turpentine oil)

Adulteration

Adulteration may be described as making essential oils impure or inferior by adding foreign substances. The methods and materials for adulteration are manifold. Often terms as cutting, stretching or bouqetting are used as euphemisms. Adulteration can have many reasons, but the main reason is to make more profit. Demands of clients of the essential oil industry are another reason. For some parties, lower prices are more important than genuineness and quality of the oils. Sometimes, adulteration is performed to ensure sufficient supply of oils to clients, e.g., in the case of a bad harvest, rapidly rising demand from the market or speculation with oil, and to avoid large variations in market prices.

Materials used for adulteration

Formerly, *fat oils*, *ethanol* and even *water* were used for adulteration. The presence of ethanol can be an adulteration, but it may also be a normal ingredient of certain essential oils. Rose oils, for example, may have up to 7% of ethanol according to its ISO standard (23). In other oils, ethanol concentrations of <0.1% are accepted. While water used to be an adulterant, it is currently considered doubtful whether its presence indicates adulteration. Many essential oils are able to bind a certain amount of water. In the production of citrus oils, large quantities of water are used. Although the oil is separated by centrifugation, some water will remain in the oils. Conifer oils such as pine and fir oil can also contain some water. The water-binding capacity of oils increases with higher temperatures and, thus, especially oils produced in the summer may contain appreciable amounts of water. To check this, a larger sample is placed in a measuring glass cylinder in the refrigerator for some days, after which the water is visualized at the bottom and its content can be quantified.

Adulteration by adding other essential oils is common. Turpentine oil, for example, is used to stretch pine needle or fir needle oil. The main components of all three oils are α- and β-pinene, and as turpentine oil is far cheaper than the others, this is a perfect blending material. Cananga oil may be added to the – far more expensive – ylang-ylang oil.

Fractions of essential oils and *defined natural isolates* are also useful for adulteration, notably for cutting other, more expensive ones. Many essential oils undergo fractional distillation for flavor purposes and these fractions are used to 'lift up' the concentration of chemicals in other essential oils to the desired or maximum allowed concentration. Although they are from natural sources, the mingling with oils from other plant species prohibits the oil from being termed essential oil. Limonene is such a component, which is available in high purity at a moderate price. Another is citral, which is distilled from litsea cubeba oil and can be added to lemon oil. However, it has a different ratio of geranial to neral and is, compared to synthetic citral, much more expensive. Geraniol can be distilled in high purity from palmarosa oil or citronella Sri Lanka oil and may be added to the more expensive geranium or rose oils. Eugenol can easily be distilled from clove leaf oil, the cheapest clove oil, and is used to increase its concentration in clove bud or pimento berry

oil. Linalool from ho leaf oil (ex *Cinnamomum camphora*) is sometimes added to lavender, clary sage or rosewood oil. So-called foreruns, multicomponent mixtures (monoterpenes like tricyclene, limonene, pinenes, camphene, etc.) can also be used to adulterate other oils.

The peels of citrus fruits still contain some essential oil after cold-pressing. Therefore, these peels are often hydrodistilled and the *hydrodistilled citrus oils* thus produced (which are – with the exception of lime oil – not essential oils) are often added to the genuine, cold-pressed citrus essential oils.

Finally, adulteration may be performed by adding chemically produced *synthetic compounds*, which are known to be present in nature. They are often termed 'nature identical' compounds, but this term may not be used anymore in Europe in the context of flavors and fragrances. They are not recovered from plants, but from other sources, usually from petroleum fractions. However, in the production process, usually by-products are generated, which can easily be detected by GC-MS. The finding of dehydro- and dihydrolinalool and -linalyl acetate, for example, indicates adulteration with synthetic linalool/linalyl acetate. In addition, synthetic linalool and linalyl acetate are racemic mixtures ((S)- and (R)-), whereas in nature only the 3R-(-)- enantiomers are found.

Sometimes, oils from the same species, but coming from different geographic areas, are mixed. This is *not* considered adulteration. Lemon oils originating from Italy, Spain, USA and Argentina can be mixed and are still lemon essential oils. However, the origin of the oil is sometimes specifically mentioned, as it has increased value, exclusivity or its name is protected, e.g., in the case of Italian lemon oil. However, when Italian lemon oil is mixed with lemon oil from Argentina, it cannot be sold anymore as lemon oil Italian.

Sometimes essential oils have to be post-treated (rectified) to diminish the concentration of one or more chemicals or remove it/them completely, when they are considered dangerous (phototoxicity, carcinogenic properties). For some chemicals, allowed concentrations are regulated by IFRA (International Fragrance Association) or by governmental laws, e.g., European Union legislation. In bergamot oil, for example, the phototoxic psoralens are reduced or eliminated, by using either physical or chemical methods. The chemical methods are not allowed according ISO rules and to avoid the loss of naturalness, users did not agree to such a treatment.

A last form of what might be considered adulteration is intentional mixing of biomass from two species before distillation. Cananga blossoms, for example, may be added to ylang-ylang flowers, as they have a higher content of oil. In former times, lavandin blossoms were added to lavender blossoms to increase the (desired) linalyl content.

All forms of adulteration can be detected by analytical techniques (17,18).

Contamination

Contamination is the presence of low concentrations of foreign chemicals in essential oils, which have entered the product unintentionally. Phthalates such as ethyl phthalate may be contaminants from their presence as softeners in plastic packing material or containers. Storage of essential oil in drums that previously contained other chemicals and that have not been cleaned properly will lead to quality loss. Improper cleaning of the distillation unit may also cause problems. A well-known example is melissa oil produced in a plant that had previously been used for thyme oil distillation. The melissa oil was contaminated with thymol and could not be commercialized anymore.

Inadequate production

Most essential oils must be produced from one species. When cultivated plants are used, this demand can easily be fulfilled. However, sometimes biomass is collected from plants growing in the wild. This bears the risk, especially in the case of producers that have insufficient knowledge and experience, that the oil will be prepared from not only the intended species, but also from other, similar-looking, plant (sub)species. In addition, several chemotypes of identical-looking plants may be mixed, which means that the concentration of the major ingredients will most likely not conform to the ISO standard. Examples are oils produced from *Thymus* and *Cinnamomum* species.

Especially in developing countries, production plants are not always equipped to control parameters that guarantee optimal distillation. Perfect biomass which is distilled incorrectly, e.g., too much steam, too high temperatures, or a badly functioning cooling unit, will inevitably lead to lower-quality oils.

Aging

When essential oils are stored under the wrong conditions (too warm, exposure to light and oxygen), a chemical process occurs called aging. As a result of esterification, reduction and oxidization of chemicals, the composition of the oil changes. In aged lemon oil, for example, the concentrations of limonene, γ-terpinene and citral are reduced and the amounts of *p*-cymene, *cis*-8,9-limonene oxide and 2,3-epoxygeraniol increase (22). In fact, the concentration of *p*-cymene in lemon oil can be considered a marker for the aging process. The main problem with this form of aging is that peroxides and hydroperoxides are formed, which may be strong allergens. It is well known, for example, that important essential oil ingredients such as linalool, limonene, geraniol and citronellol are weak contact allergens, but that their auto-oxidation products have far stronger sensitizing capacities (7,8,9,43). Such oxidation products are well-known causes of contact allergy to tea tree oil (Chapter 5.83) and turpentine oil (Chapter 5.87).

LITERATURE

1 Lawrence BM. A preliminary report on the world production of some selected essential oils and countries. Perfum Flavor 2009;34(1):38-44

2 Lawrence BM. A planning scheme to evaluate new aromatic plants for the flavor and fragrance industries. In: J Janick and JE Simon, Eds. New crops. New York: Wiley, 1993:620-627

3 Schmidt E, Wanner J, Höferl M, Jirovetz L, Buchbauer G, Gochec V, et al. Chemical composition, olfactory analysis and antibacterial activity of *Thymus vulgaris* L. chemo-types geraniol, thujanol-4/terpinen-4-ol, thymol and linalool cultivated in Southern France. Nat Prod Commun 2012;7:1095-1098

4 Levey M. Evidence of ancient distillation in Mesopotamia. Centaurus 1959;4(1):23-33

5 Gildemeister E, Hoffmann F. Die Ätherische Öle. Leipzig, Germany: Verlag Schimmel & Co, 1931

6 Schmidt E. Production of essential oils. In: HC Baser and G Buchbauer, Eds. Handbook of essential oils: Science, technology and applications. Boca Raton, London, New York: CRC Press, Taylor & Francis Group, 2010:83-119

7 Rudbäck J, Hagvall L, Börje A, Nilsson U, Karlberg A-T. Characterization of skin sensitizers from autoxidized citronellol – impact of the terpene structure on the autoxidation process. Contact Dermatitis 2014;70:329-339

8 Hagvall L, Backtorp C, Svensson S, Nyman G, Börje A, Karlberg AT. Fragrance compound geraniol forms contact allergens on air exposure. Identification and quantification of oxidation products and effect on skin sensitization. Chem Res Toxicol 2007;20: 807-814

9 Nilsson U, Bergh M, Shao LP, Karlberg A-T. Analysis of contact allergenic compounds in oxidized d-limonene. Chromatographia 1996;42:199-205

10 Franz C, Baser KHC, Windisch W. Essential oils and aromatic plants in animal feeding – a European perspective. A review. Flavour Fragr J 2010;25:327-340

11 ISO 9235:2013 Aromatic natural raw materials – Vocabulary. International Organization for Standardization, Geneva, Switzerland. www.iso.org

12 Müller MP, Lamparsky D. Perfumes: Art, science, and technology. London, New York: Elsevier Science Publishers Ltd, 1991:348-362

13 Franz C. Genetics. In: RKM Hay and PG Waterman, Eds. Volatile oil crops: Their biology, biochemistry and production. Harlow, UK: Longman Scientific & Technical, 1993:63-96

14 Erdtmann H. Some aspects of chemotaxonomy. In: T Swain, Ed. Chemical plant taxonomy, 2nd Ed. New York: Academic Press, 1963:89-125

15 Kaloustian J. Southern French thyme oils: chromatographic study of chemo-types. J Sci Food Agric 2005;85:2437-2444

16 Desjardins AE. Natural product chemistry meets genetics: When is a genotype a chemotype? J Agric Food Chem 2008;56:7587-7592

17 Schmidt E, Wanner J. Adulteration of essential oils. In: HC Baser and G Buchbauer, Eds. Handbook of essential oils: Science, technology and applications, 2nd Edition. Boca Raton, London: CRC Press, Taylor & Francis Group, 2015

18 Busch KW, Busch MA. Chiral analysis. Amsterdam: Elsevier, 2006:47-48

19 Kubezka KH, Formácek V. Essential oil analysis by capillary gas chromatography and carbon-13 NMR spectroscopy. Chichester, UK: John Wiley & Sons, 2002

20 http://www.iso.org/iso/home/search.htm?qt=Esse ntial+oils&published=on&active

21 Gerbers K. Buch über die Chemie des Parfüms und seine Destillationen. Ein Beitrag zur Geschichte der Arabischen Parfümchemie und Drogenkunde aus dem 9. Jahrhundert P.C. Leipzig, Germany: Kommissionsverlag F.A. Brockhaus, 1948. Translated from KITAB KIMIYA' AL-'ITR WAT-TAS'IDAT

22 Dugo G, Cotroneo A, Bonaccorsi I, Trozzi A. Composition of the volatile fraction of citrus peel oils. In: G Dugo and L Mondello, Eds. Citrus oils. Composition, advanced analytical techniques, contaminants, and biological activity. Boca Raton, Fl., USA: CRC Press, Taylor & Francis Group, 2011:147

23 ISO/DIS 9842:2003 Oil of rose (*Rosa x damascena* Miller), International Organization for Standardization, Geneva, Switzerland, www.iso.org

24 Dugo G, Mondello L, Eds. Citrus Oils. Composition, advanced analytical techniques, contaminants, and biological activity. Boca Raton, Fl., USA: CRC Press, Taylor & Francis Group, 2011

25 Rhind JP. Essential oils. A handbook for aromatherapy practice, 2nd Edition. London: Singing Dragon, 2012

26 Lawless J. The encyclopedia of essential oils, 2nd Edition. London: Harper Thorsons, 2014

27 Davis P. Aromatherapy. An A-Z, 3rd Edition. London: Vermilion, 2005

28 ISO 22716:2007 Cosmetics − Good Manufacturing Practices (GMP) − Guidelines on Good Manufacturing Practices. International Organization for Standardization, Geneva, Switzerland, www.iso.org

29 Nardelli A, Thijs L, Janssen K, Goossens A. Rosa centifolia in a 'non-scented' moisturizing body lotion as a cause of allergic contact dermatitis. Contact Dermatitis 2009;61:306-309

30 Tisserand R, Young R. Essential oil safety, 2nd Edition. Edinburgh, UK: Churchill Livingstone Elsevier, 2014

31 Schmidt E. Unpublished observations

32 Becker PM, Galetti S. Food and feed components for gut health-promoting adhesion of E. coli and *Salmonella enterica*. J Sci Food Agric 2008;88:2026-2035

33 Ehrlinger M. Phytogene Zusatzstoffe in der Tierernährung. Inaugural-Dissertation, Tierärztliche Fakultät der Ludwig-Maximilians-Universität, München, Germany, 2007:235-247

34 Breitmaier E. Terpenes: Flavors, fragrances, pharmaca, pheromones. Weinheim, Germany: Wiley-VCH Verlag GmbH & Co. KGaA, 2008:10-85

35 Surburg H, Panten J. Common fragrance and flavor materials, 5th Edition. Weinheim, Germany: Wiley-VCH Verlag GmbH & Co. KGaA, 2006:8-162

36 Terpenoids. Cyperlipid Center. http://www.cyperlipid.org/simple/simp0004.htm

37 Rowe DJ. Chemistry and technology of flavors and fragrances. Oxford, UK: Blackwell Publishing Ltd, 2005:56-131

38 Sherma J. Thin-layer chromatography in food and agricultural analysis. J. Chromatogr A 2000;880:129-147

39 d'Acampora Zellner B, Dugo P, Dugo G, Mondello L. Analysis of essential oils. In: HC Baser and G Buchbauer, Eds. Handbook of essential oils: Science, technology and applications. Boca Raton, London, New York: CRC Press, Taylor & Francis Group, 2010:151-184

40 NIST/EPA/NIH Mass Spectral Library. Wiley. www.wiley.com

41 Adams RP. Identification of essential oil components by gas chromatography/mass spectrometry, 4th Edition. Carol Stream, IL, USA: Allured Publishing Corporation, 2007

42 Massfinder 4. Hamburg, Germany: Dr. Hochmuth Scientific Consulting, 2014. http://massfinder.com/wiki/MassFinder_4

43 Karlberg A-T, Börje A, Johansen JD, Lidén C, Rastogi S, Roberts D, et al. Activation of non-sensitizing or low-sensitizing fragrance substances into potent sensitizers − prehaptens and prohaptens. Contact Dermatitis 2013;69:323-334

Chapter 3 CONTACT ALLERGY TO ESSENTIAL OILS: GENERAL ASPECTS

In this chapter, general aspects of contact allergy to essential oils are discussed. Literature data on specific aspects of contact allergy to and allergic contact dermatitis from individual oils are presented in Chapters 5.1-5.93.

3.1 ESSENTIAL OILS WHICH HAVE CAUSED CONTACT ALLERGY

In our literature search, we have found 79 essential oils (essential oils as per ISO definitions) which have caused contact allergy (positive patch test reactions) or allergic contact dermatitis. They are shown in Table 3.1; the numbers in the first column refer to chapter numbers of the individual oils. The number of Chapters (n=93) exceeds the number of oils which have caused contact allergy, which is explained as follows. Some essential oils can be obtained from more than one part of a source plant: cassia oil from the bark or from the leaves of *Cinnamomum cassia*; the same goes for cinnamon oil. Clove oil is usually obtained from the leaves of the plant *Syzygium aromaticum* L., but can also have buds or even the stem as source biomass. In other cases, essential oils with the same commonly used name can be obtained from different plant *species*. Thus, cedarwood oil may be produced from five plant species resulting in cedarwood oil Atlas, cedarwood oil China, cedarwood oil Himalaya, cedarwood oil Texas and cedarwood oil Virginia. Eucalyptus oil can be obtained from *Corymbia citriodora* or from *Eucalyptus globulus*. The same situation applies to lemongrass oil (*Cymbopogon flexuosus* yields East Indian lemongrass oil, whereas West Indian lemongrass oil is produced from *Cymbopogon citratus*), sage oil (Dalmatian sage oil from *Salvia officinalis*, Spanish sage oil from *Salvia lavandulifolia*) and thyme oil can be produced from *Thymus vulgaris* (thyme oil) or from *Thymus zygis* (Spanish thyme oil). Jasmine absolutes can be produced from *Jasminum grandiflorum* or from *Jasminum sambac*. Lavandin oils, finally, are all obtained by steam-distilling the flowering tops of the lavandin (bastard lavender), *Lavandula angustifolia* Mill. x *Lavandula latifolia* Medik., but from different cultivars such as 'Abrial', 'Grosso' and 'Super'. In literature on contact allergy to essential oils, the exact nature of the oil (plant species, plant part used) is rarely mentioned and oils are termed cassia oil, cedarwood oil, cinnamon oil, clove oil, eucalyptus oil etc. However, the composition of these oils, which are presented under the same name, can differ considerably and depends on the source species and the plant parts used. For some, there are specific ISO standards, e.g., lavandin 'Abrial', lavandin 'Grosso', clove bud, clove leaf, clove stem, both eucalyptus oils, two jasmine absolutes, both lemongrass oils, both sage oils and an ISO file for Spanish thyme oil. Therefore, we have decided to include full chapters on the chemical composition of these oils, although they have not been reported to cause contact allergy *under their specific names*.

Of the 79 oils that have caused allergic contact dermatitis or contact allergy, 45 have been tested in groups of consecutive patients suspected of contact dermatitis. Testing in groups of *selected* patients has been performed with 53 essential oils, usually in patients suspected of fragrance allergy, of cosmetic dermatitis, or patients who previously had a positive reaction to the fragrance mix or one of the other indicators of fragrance allergy such as *Myroxylon pereirae* resin (balsam of Peru). Case reports of allergic contact dermatitis were found for 67 oils. For 37 essential oils, data on positive patch test reactions that had no relevance, or the relevance of which was uncertain or not mentioned, are provided; these are usually reactions to oils tested in a fragrance series. General aspects of routine testing data and case reports are discussed below. Specific data on all aspects of contact allergy / allergic contact dermatitis can be found in the individual oil files in Chapters 5.1-5.93.

3.2 FREQUENCY OF CONTACT ALLERGY TO ESSENTIALS OILS

For the frequency of contact allergy to essential oils, prevalence rates of positive patch test reactions, tested in consecutive patients suspected of contact dermatitis, can be used as an indication. Relevant data are summarized in Table 3.2. Forty-five essential oils have been tested in consecutive patients; the number of studies per oil ranges from 1 to 19. For 21 oils of this group (47%), information is limited to one study; in the case of 15 of these 21 oils, this was a study from Poland published in 1976, in which a very small group of 200 dermatitis patients was tested with 35 essential oils (2).

For 17 oils, the highest prevalence rate observed was 1.0% or lower. Eleven oils have had maximum rates of positive patch test reactions between 1% and 1.5% and in only nine have rates over 2% been observed. These are, in order of descending maximum values, laurel oil 6.9% (observed between 1953 and 1962 in a period with many reactions in Germany, France and Switzerland from its presence in ointments, which most likely was *vegetable laurel berries* oil; data cited in ref. 3), turpentine oil 4.2% (study from Spain from before 1986; in Germany in 1998 a 4.4% prevalence was observed, thereafter declining to 1.5-2%), orange oil 3.2% (study from 1967 to 1970), tea tree oil 2.7%, citronella oil 2.5% (very small 1976 study), ylang-ylang oil 2.6%, sandalwood oil 2.4%, clove oil 2.1% (observed in a weak study with a high rate of macular erythema and weak reactions) and costus root oil 2.1% (small study from 1982, historical allergen). These are the highest prevalence rates observed for these oils; in other studies (far) lower percentages have been found. Thus, it appears that contact allergy to essential oils as a group is not frequent (although in the general German adult population, 2.5% of 1141 test subjects reacted to oil of turpentine, 4.3% of the women and 0.7% of the men) (133). Currently the highest prevalences of positive

Table 3.1 Essential oils which have cause contact allergy/allergic contact dermatitis

Chapter	Name of essential oil	Number of publications found in literature				
		Routine testing[a]	Selective testing[b]	Case reports[c]	Positive patch tests[d]	Probable allergen(s) identified[e]
5.1	Angelica oil (unspecified)	1	1	1		
5.1	Angelica fruit oil			1		α-pinene
5.2	Angelica root oil		1	1		
5.3	Aniseed oil		1	4	1	anethole
5.4	Basil oil, sweet			2		linalool
5.5	Bay oil		1			eugenol
5.6	Bergamot oil	5	2	4	1	limonene, linalool, β-pinene
5.7	Black cumin oil			6		
5.8	Black pepper oil			1		caryophyllene, α-pinene
5.9	Cajeput oil			1		caryophyllene, α-pinene
5.10	Calamus oil		2			
5.11	Cananga oil	3	12	2	8	
5.92	Cardamom oil			1		
5.12	Carrot seed oil			1		
5.13	Cassia oil (unspecified)	2	2	5	1	cinnamic aldehyde
5.13	Cassia bark oil					
5.14	Cassia leaf oil					
5.15	Cedarwood oil (unspecified)	3	3	3		
5.15	Cedarwood oil, Atlas	2				
5.16	Cedarwood oil, China	2				
5.93	Cedarwood oil, Himalaya	1	1			
5.17	Cedarwood oil, Texas					
5.18	Cedarwood oil, Virginia		1			
5.19	Chamomile oil (unspecified)	2	2	2		α-pinene
5.19	Chamomile oil, German		3			
5.20	Chamomile oil, Roman		1	2		
5.21	Cinnamon oil (unspecified)		1	13	3	cinnamic aldehyde, eugenol
5.21	Cinnamon bark oil, Sri Lanka					
5.22	Cinnamon leaf oil, Sri Lanka	1	1		1	eugenol
5.23	Citronella oil (unspecified)[f]	1	2	4	1	citronellal, citronellol, geraniol, geranyl acetate, limonene
5.23	Citronella oil, Java				1	limonene
5.24	Citronella oil, Sri Lanka			1		limonene
5.25	Clary sage oil	1	2	4	1	geraniol, linalool
5.26	Clove oil (unspecified)[f]	3	8	8	3	eugenol
5.26	Clove bud oil		2			
5.27	Clove leaf oil					
5.28	Clove stem oil					
5.29	Coriander fruit oil	1	2	1	1	linalool, α-pinene
5.30	Costus root oil	1		2	2	
5.31	Cypress oil			3		α-pinene
5.32	Dwarf pine oil	2	1	1	2	
5.33	Elemi oil			1		
5.34	Eucalyptus oil (unspecified)	2	7	7	1	α-pinene
5.34	Eucalyptus citriodora oil					
5.35	Eucalyptus globulus oil			1		
5.36	Galbanum resin oil			1		α-pinene
5.37	Geranium oil	1	12	14	3	caryophyllene, citral (neral + geranial), citronellol, geraniol, linalool
5.38	Ginger oil			1		
5.39	Grapefruit oil			1		
5.40	Guaiacwood oil		2			
5.41	Hyssop oil			1		
5.42	Jasmine absolute (unspecified) + Jasminum grandiflorum abs.	12	13	5	8	eugenol, linalool
5.43	Jasminum sambac absolute					

Table 3.1 Essential oils which have cause contact allergy/allergic contact dermatitis (*continued*)

Chapter	Name of essential oil	Routine testing[a]	Selective testing[b]	Case reports[c]	Positive patch tests[d]	Probable allergen(s) identified[e]
5.44	Juniper berry oil[f]	1	2	1		α-pinene
5.45	Laurel leaf oil[f, g]	1	5	7	4	
5.46	Lavandin abrial oil					
5.47	Lavandin grosso oil					
5.48	Lavandin oil (unspecified)	1	2	1		
5.49	Lavender oil	3	5	19	8	caryophyllene, linalool
5.50	Lemon oil	2	7	9	4	limonene, β-pinene
5.51	Lemongrass oil (unspecified) + Lemongrass oil, East Indian	4	5	8	5	citral (geranial + neral), geraniol
5.52	Lemongrass oil, West Indian					
5.53	Litsea cubeba oil	1	2			
5.54	Lovage oil			2		
5.55	Mandarin oil			1		
5.56	Marjoram oil (sweet)			2		caryophyllene, linalool, α-pinene
5.57	Melissa oil (lemon balm oil)			2		caryophyllene, geraniol
5.58	Myrrh oil			2	1	
5.59	Neem oil			3		
5.60	Neroli oil	1	4	9	4	geraniol
5.61	Niaouli oil[f]			3		α-pinene
5.62	Nutmeg oil				1	
5.63	Olibanum (frankincense) oil			2	1	caryophyllene, α-pinene
5.64	Orange oil (unspecified)[f]	2	3	4	1	
5.64	Orange oil, bitter	1	2			
5.65	Orange oil, sweet[f]	1	2	2	1	limonene
5.66	Palmarosa oil				2	
5.67	Patchouli oil	3	5		3	
5.68	Peppermint oil[f]	10	11	18	9	caryophyllene, limonene, menthol, α-pinene, piperitone, pulegone
5.69	Petitgrain bigarade oil	1	2	2		geraniol, linalool
5.70	Pine needle oil	2	3	4	2	α-pinene
5.71	Ravensara oil			2		linalool, α-pinene
5.72	Rosemary oil		1	2	1	linalool, α-pinene
5.73	Rose oil	1	7	8	8	citronellol, geraniol, linalool
5.74	Rosewood oil			2		linalool
5.75	Sage oil (unspecified)			1		
5.75	Sage oil, Dalmatian					
5.76	Sage oil, Spanish					
5.77	Sandalwood oil[f]	8	17	4	3	santalol
5.78	Silver fir oil	1	2			
5.79	Spearmint oil	3	8	11		carvone
5.80	Spike lavender oil	1	2			
5.81	Star anise oil	1				anethole, methyl chavicol, limonene
5.82	Tangerine oil			1		
5.83	Tea tree oil	19	5	30	NR	aromadendrene, ascaridole, *d*-carvone, *l*-carvone, 1,8-cineole, *p*-cymene, *d*-limonene, myrcene, α-phellandrene, α-terpinene, terpinen-4-ol, terpino-lene, 1,2,4-trihydroxymenthane, sabinene and viridiflorene
5.84	Thuja oil			1		
5.85	Thyme oil (unspecified)[f]		3			
5.85	Thyme oil				1	linalool, α-pinene
5.86	Thyme oil, Spanish					
5.87	Turpentine oil	13	8	12	2	dipentene, α-phellandrene, α-pinene, β-pinene, α-terpineol
5.88	Valerian oil				1	

Table 3.1 Essential oils which have cause contact allergy/allergic contact dermatitis (*continued*)

Chapter	Name of essential oil	Routine testing[a]	Selective testing[b]	Case reports[c]	Positive patch tests[d]	Probable allergen(s) identified[e]
5.89	Vetiver oil	1	2	2		
5.90	Ylang-ylang oil	14	15	10	11	caryophyllene, linalool
5.91	Zdravetz oil	1	2			
5.92	see Cardamom oil					
5.93	see Cedarwood oil, Himalaya					

[a] testing in consecutive patients suspected of contact dermatitis

[b] testing in groups of selected patients, e.g., patients suspected of fragrance allergy, patients with previous positive patch tests to the fragrance mix or *Myroxylon pereirae* resin (balsam of Peru)

[c] cases of allergic contact dermatitis with extensive or limited clinical description or positive patch tests indicated by the authors as relevant for the patient's dermatitis

[d] positive patch tests, the relevance of which was unknown, uncertain or not stated

[e] patch test reactivity of one or more chemicals which have been demonstrated in the essential oil used by the patient by chemical analysis, or known to be present in such oils in high concentrations (e.g., geraniol in geranium and rose oils, citronellal and citronellol in citronella oils, cinnamaldehyde in cinnamon oils), or which have been demonstrated in commercial samples of the essential oil in question in concentrations >3% (analytical data from Erich Schmidt, shown in the oil chapters 5.1-5.93, usually in column A of Table 2)

[f] additional references are given for these oils; their data are not included in this table

[g] laurel oil in older publications was not the essential oil from laurel leaves, but mostly fatty oil expressed from laurel berries

NR not recorded: data on positive patch test reactions with no, unknown or uncertain relevance have not been collected because of the abundance of other literature on tea tree oil

Table 3.2 Prevalence rates of sensitization to essential oils in routine testing and relevance data

Name of essential oil	Nr. of studies	Nr. of patients (total)	Time frame of studies/publications	Prevalence range	Relevance data (% relevant)
Angelica oil (unspecified)	1	200	1976	1.0%	NS
Bergamot oil	5	2,490	1977-2007	0.2-1.5%	NS; 100%[a]
Cananga oil	3	1,108	1976-2007	0.5-1.2%	NS; 100%[a]
Cassia oil (unspecified)	2	950	1976-1985	0.7-1.0%	NS
Cedarwood oil (unspecified)	3	2,124	1976-2000	0.6-1.5%	NS
Cedarwood oil, Himalaya	1	200	1976	1.5%	NS
Chamomile oil (unspecified)	2	490	1967-1976	0.3-0.5%	NS
Cinnamon leaf oil	1	1,382	1975	1.1%	NS
Citronella oil (unspecified)	1	200	1976	2.5%	NS
Clary sage oil	1	200	1976	0.5%	NS
Clove oil (unspecified)	3	906	1967-2007	0.3-2.1%	NS; 100%[a]
Coriander fruit oil	1	200	1976	1.0%	NS
Costus root oil	1	282	1982	2.1%	NS
Dwarf pine oil	2	1,606	1998-2000	0.6-0.7%	NS
Eucalyptus oil (unspecified)	2	879	1976-2007	0.6-1.5%	NS; 100%[a]
Geranium oil	1	486	2000-2007	1.2%	100%[a]
Jasmine absolute (unspecified)[b] + Jasminum grandiflorum abs.	12	32,708	1983-2012	0.3-1.5%	NS; 9-37%[c]; 92%[a]
Juniper berry oil	1	200	1976	0.5%	NS
Laurel leaf oil	1	>1,000	1953-1962	3.1-6.9%[d]	NS
Lavandin oil (unspecified)	1	200	1976	0.5%	NS
Lavender oil[b]	3	8,953	2002-2012	0.1-1.2%	NS; 30-69%[c]
Lemon oil	2	545	1976-2003	0.5-0.9%	NS; 100%[a]

Table 3.2 Prevalence rates of sensitization to essential oils in routine testing and relevance data (*continued*)

Name of essential oil	Nr. of studies	Nr. of patients (total)	Time frame of studies/publications	Prevalence range	Relevance data (% relevant)
Lemongrass oil (unspecified) + Lemongrass oil, East Indian	4	4,909	1998-2008	0.6-1.6%	NS; 100%[a]
Litsea cubeba oil	1	200	1976	1.5%	NS
Neroli oil	1	324	2000-2007	0.3%	100%[a]
Orange oil (unspecified)	2	1,268	1967-2007	0.6-3.2%	NS[e]
Orange oil, bitter	1	200	1976	1.5%	NS
Orange oil, sweet	1	200	1976	0.5%	NS
Patchouli oil	3	4,052	1998-2008	0.6-0.9%	NS
Peppermint oil[b]	10	14,326	1970-2012	0.3-1.8%	NS; 36-39%[c]; 100%[a]
Petitgrain bigarade oil	1	200	1976	0.5%	NS
Pine needle oil	2	3,700	1973-1976	0.4-2.0%	NS
Rose oil	1	679	2000-2007	1.6%	100%[a]
Sandalwood oil	8	10,921	1983-2008	0.1-2.4%	NS; 82%[a]
Silver fir oil	1	200	1976	1.0%	NS
Spearmint oil	3	2,106	1998-2007	0.8-1.6%	NS; 100%[a]
Spike lavender oil	1	200	1976	0.5%	NS
Tea tree oil[b]	19	44,964	1997-2013	0.1-2.7%	NS; average 40%
Turpentine oil	13	253,065	1972-2008	1.2-4.2%	NS
Vetiver oil	1	200	1976	0.5%	NS
Ylang-ylang oil[b]	14	37,356	1976-2012	0.7-2.6%	NS; 0-27%[c]; 100%[a]
Zdravetz oil	1	200	1976	0.5%	NS

[a] rather weak study; relevance data included 'questionable' and 'past' relevance (132)

[b] currently present in the North American Contact Dermatitis Group (NACDG) screening series

[c] definite + probable relevance in the NACDG studies

[d] very old data from Germany, when laurel oil caused many reactions from its presence in ointments; however, it was most likely *vegetable* oil from laurel *berries* (data cited in ref. 3)

[e] in an Italian study with a 3.2% prevalence rate of sensitization, many patients had contact with oranges

NS: not specified, not stated or unknown

patch test reactions are observed – depending on the country to tea tree oil, ylang-ylang oil and sandalwood oil (which for sandalwood oil is rather curious, because its price is so high that it is virtually only used in fine fragrances), but these cannot – on the basis of current data – be regarded as frequent and important sensitizers. In the past, however, ylang-ylang oil, jasmine absolute, cananga oil and some other essential oils have caused many cases of allergic reactions to cosmetics in Japan (see below in the section 'Pigmented cosmetic dermatitis').

3.3 CLINICAL RELEVANCE OF POSITIVE PATCH TEST REACTIONS TO ESSENTIAL OILS

The finding of a positive patch test reaction to an essential oil should – as with any test substance – be followed by determination of its relevance: can the dermatitis of the patient be (partly) explained by contact with the oil or products containing it? As all dermatologists know, this is the most difficult part of the diagnostic procedure. This unfortunately is evident from relevance data of the studies shown in Table 3.2. In >80% of all studies, no information on relevance was provided. In nearly 2/3

of all oils, no information on relevance is available at all. In one study, the observed patch test reactions to the tested essential oils were considered to be relevant in 82-100%. Unfortunately, this was a rather weak study, where questionable and past relevance were included (132). The NACDG provides relevance data for the essential oils which are included in the NACDG screening series. Current relevance is defined as definite (use test with the suspected item was positive, or a patch test to the object or product was positive), probable (the antigen could be verified as present in known skin contactants and clinical presentation was consistent), or possible (patient was exposed to circumstances in which skin contact with materials known to contain the allergen was likely to occur (4). Percentages for definite + probable relevance were as follows: lavender oil 30-69%, tea tree oil 20-56%, peppermint oil 36-39%, jasmine absolute 14-37% and ylang-ylang oil 0-27%. Rates for definite relevance were usually <10-15% and sometimes zero. It may thus be concluded that reliable data on the relevance of positive patch test reactions to essential oils as reported in literature are largely lacking or inadequate. Insufficient knowledge of the chemical composition of essential oils and difficulties in ascertaining whether oils or their ingredients are present in materials with which the patients have contact may partly be responsible. It should be realized that previously acquired contact allergy to a fragrance chemical per se may result in a (non-relevant) positive patch test to an essential oil containing it, e.g., when tested in the fragrance series. For example, a patient sensitized to geraniol in a cream showed positive reactions to Bulgarian rose oil and geranium oil Bourbon, to which the patient had apparently not been exposed, but both of which contain high concentrations of geraniol, likely explaining the positive patch test reactions (131).

Of 14 patients with occupational contact dermatitis from d-limonene and patch tested with tea tree oil 5% in petrolatum, 5 (36%) had a positive (n=4) or doubtful positive (n=1) reaction to tea tree oil. This indicates that previously acquired contact allergy to limonene may result in a non-relevant positive patch test to tea tree oil (134). In fact, we think it is likely that many positive patch test reactions to essential oils in routine testing, for which no relevance can be ascertained, may well be explained this way and that sensitization is not the result of contact with the essential oil itself.

3.4   REPORTS OF ALLERGIC CONTACT DERMATITIS FROM ESSENTIAL OILS

Case reports of allergic contact dermatitis have been described for 67 essential oils; the numbers per oil range from 1 (21 oils) to >10: spearmint oil 11, turpentine oil 12, cinnamon oil 13, geranium oil 14, peppermint oil 18, lavender oil 19, tea tree oil 30. These include both cases with proper clinical data and some cases where positive patch tests were considered by the authors to be relevant (testing in groups of patients excluded). It should, however, be realized that some patients with allergic contact dermatitis have reacted to a large number of essential oils with which they had contact and that these particular case reports are shown in all of these oils. Of two aromatherapists with occupational allergic contact dermatitis, for example, one reacted to 32 and the other to 21 essential oils (5).

3.5   CLINICAL PICTURE OF ALLERGIC CONTACT DERMATITIS FROM ESSENTIAL OILS

There are no descriptions of the clinical picture of allergic contact dermatitis from essential oils in larger groups of patients. As the products in which the oils may be present vary considerably (pure oils, perfumes, toothpastes, other cosmetics, topical pharmaceutical preparations, other products: see the section 'Products responsible for allergic contact dermatitis to essential oils' below), as do the sites of application and the goals of applying the oils to the skin, the presentation of patients with allergic contact dermatitis may take many forms. These are sometimes easy, in other cases difficult to recognize as essential oil related.

There have been many case reports of patients who developed occupational allergic contact dermatitis from oils used for massaging clients, including aromatherapists, masseuses, and physiotherapists. The patients are usually women and present with dermatitis of the hands and sometimes of the forearms; in a number of them, dermatitis spreads to other parts of the body and may become generalized (96) (see the section 'Occupational allergic contact dermatitis' below). In clients receiving the massages, generalized allergic contact dermatitis may develop (81).

Undiluted oils are also frequently applied for therapeutic purposes, notably tea tree oil. Dermatitis will appear at the site of application and may stay limited to the primary site, but spreading of dermatitis is not infrequent and even generalization occurs occasionally (13,80). Unilateral dermatitis of the cheeks may be caused by lavender drops applied on the pillow for their presumed hypnotic effects (19).

Many topical pharmaceutical products contain essential oils, usually as fragrance rather than as an active ingredient. Allergic contact dermatitis from such products will develop at the site of application. In a large group of 127 patients with iatrogenic allergic contact dermatitis from essential oils (and other fragrances) in topical pharmaceutical drugs, women were more affected than men, and legs, hands, and face were the most commonly affected body sites (59). The use of vaginal suppositories has led to erythematous vulvitis, leucorrhoea and patches of dermatitis on the buttocks (66).

Allergic reactions of the oral mucosa and lips

Essential oils are frequently added to toothpastes, especially spearmint, peppermint and cinnamon oils. Contact allergy may lead to symptoms of the oral mucosa, the lips and the perioral skin. Possible oral manifestations include burning/ sore mouth (40,53), stomatitis (40,43,50,51,58), swelling of the tongue, lips and gingival mucosa (40) and ulceration of the mouth (40,44). There are some indications that oral lichen planus may be worsened by contact allergy to essential oils (48,74,75). Oral symptoms can also be caused by contact allergy to other products for oral use such as antiseptic spray, mouthwash (47,52) and chewing gum (52). The lips often show cheilitis (allergic contact dermatitis of the lips) (40,42,43,49,50,51), which may be erosive (54); fissuring can sometime be observed (53) as is angular cheilitis (44). The allergic reaction may also result in dermatitis of the surrounding skin (40,44,53). Rarely, dermatitis of the fingers or the palm has been described from toothpaste running along the toothbrush on the hand (43,46); cheilitis or stomatitis may even be absent then (46). Of course, cheilitis can also be caused by products applied to the lips, such as lip balms (34). Recurrence of cinnamon oil-induced cheilitis by eating cinnamon has been observed (82).

Pigmented cosmetic dermatitis

In Japan, in the 1960s and 1970s, many female Japanese patients developed facial pigmentation following dermatitis of the face (6). This so-called pigmented cosmetic dermatitis was shown to be caused by contact allergy to components of cosmetic products, notably essential oils (ylang-ylang, cananga, lavender, sandalwood), jasmine absolute, other fragrance materials, antimicrobials, preservatives and coloring materials (6,9). The number of patients decreased strongly after

1978, when major cosmetic companies began to eliminate strong contact sensitizers from their products, including dihydro-isoeugenol from ylang-ylang oil (7,8,9). Pigmented contact dermatitis from essential oil allergy is still seen occasionally (39,80).

Other forms of dermatitis

Airborne allergic contact dermatitis has been observed from inhaling the vapors of a hot aqueous solution of tea tree oil (83) and the spreading of essential oils in the air through aromatherapy lamps (84). Systemic administration of tea tree oil has worsened dermatitis from topical application (systemic contact dermatitis) (125).

Sometimes allergic contact dermatitis from essential oils may present itself as erythema multiforme-like reactions (14,76). In two cases, topical application and ingestion of black cumin oil resulted in generalized erythema multiforme and bullous erythema multiforme / toxic epidermal necrolysis (77,78). In one patient, contact allergy to tea tree oil applied undiluted to a piercing wound may have precipitated linear IgA disease (79).

3.6 PRODUCTS RESPONSIBLE FOR ALLERGIC CONTACT DERMATITIS FROM ESSENTIAL OILS

Various categories of products responsible for allergic contact dermatitis from essential oils can be distinguished:

Pure oils

Case reports of allergic contact dermatitis from contact with pure oils have been reported repeatedly, e.g., from patchouli oil (10), black cumin oil (11,13), citronella oil (12), laurel oil (14,16), neem oil (17), lovage oil (18), and – through indirect contact by application on a pillow – lavender oil (19). These reactions can be expected especially with oils that have alleged therapeutic actions when applied to the skin. This is certainly the case with

tea tree oils, which are used topically for therapeutic purposes on a variety of skin conditions including acne, eczema, sunburn, wounds (of any cause), warts, herpes and fungal infections. At least 85 patients with allergic contact dermatitis from tea tree oil have been described in case reports. Of the cases where the products responsible for the allergic reactions were specified, most (63%) related to pure tea tree oil (for references see Chapter 5.83). A separate category of patients who become sensitized to pure oils or diluted oils in high concentrations is that of people who have frequent contact with such products at work, e.g., aromatherapists and masseurs. Many such cases of occupational contact dermatitis to essential oils have been described (see the section 'Occupational contact dermatitis' below).

Cosmetic products

Essential oils are widely used in cosmetic products and the development of allergic cosmetic dermatitis from them has been reported in a number of publications. However, related to their widespread use, the role of cosmetics in essential oil contact allergy is modest. Examples are shown in Table 3.3.

Toothpastes and other oral preparations

Although toothpastes are diluted under normal use circumstances and the contact time with the oral mucosa, lips, and perioral skin is short, essential oils in these products have been reported to cause contact allergic reactions repeatedly: cinnamon oil (40,42,43,44,45,46), spearmint oil (51,53,54,56,57), peppermint oil (49,50,51), cassia oil (40,42,43), aniseed oil (58) and laurel oil (35). Oil of cassia and oil of cloves have caused contact allergy in a dental tablet (41), peppermint oil in antiseptic spray (47), in mint-flavored mouthwashes and foods (48) and spearmint oil in mouth rinses and chewing gum (52).

Table 3.3 Reported cases of contact allergy to essential oils in cosmetic products

Essential oil	Cosmetic products and number of patients[b]	References
Angelica root oil	Eye cream	29
Black cumin oil	Skin care product	30
Citronella oil	Deodorant	31
Eucalyptus oil	Bath/shower product, skin care product	21
Geranium oil	Aftershave, hair dye, face cream, nail polish, lip balm	26,34
Jasmine absolute	Face cream	26
Laurel oil	Face mask	35
Lavender oil	Bath/shower product (2), skin care product, deodorant, massage product, shampoo, perfume, gel	21,36,37,38
Lemon oil	Face powder	39
Neroli oil	Skin care product, facial moisturizer	21,28
Niaouli oil	Skin care product	21
Orange oil	Skin care product	21
Peppermint oil	Depilatory product, lip balm	23,27
Rose oil	Perfume (2), face cream, skin care product (3), hand soap ('fragrance-free'), body lotion[a]	21,25,26,34, 119
Tea tree oil	Shaving gel, shampoo (2), skin care product (6), soap and cream, cream, shaving oil	20,21,22,23,
Ylang-ylang oil	Eye cream	29

[a] *Rosa centifolia* oil; [b] the number of patients is one for each product category per oil unless indicated otherwise

Pharmaceutical products

In the period 1978-2008, allergic contact dermatitis from essential oils present in topical pharmaceutical products was diagnosed in Leuven, Belgium as follows: lavender oil (n=24), neroli oil (n=14), geranium oil (n=12), eucalyptus oil (n=10), rose oil (n=5), pine needle oil (n=4), laurel oil (n=2) and tea tree oil (n=2) (59). There may be overlap with the two cases of contact allergy to pine needle oil, five to neroli oil, and two to eucalyptus oil documented in another Belgian study (60). Other case reports of essential oil allergy from its presence in topical pharmaceutical preparations include: tea tree oil in wart paint (61); peppermint oil in a transdermal therapeutic system (62); two cases of neroli oil in a topical NSAID preparation (63, may also overlap with refs. 59 and 60); two cases of Roman chamomile oil in Kamillosan® (64) and one in a homeopathic preparation (65); cinnamon oil in a vaginal suppository (66); dwarf pine oil in a topical NSAID preparation (67); eucalyptus oil in Vicks Vaporub (68) and in an anti-inflammatory cream (69); geranium oil in a topical preparation to promote wound healing and prevent scar formation (70) and two cases of lavender oil allergy from their presence in antihistamine creams (71,72).

Other products

Other products in which essential oils have caused allergic contact dermatitis have included immersion oil for dermatoscopy (the essential oil was a contaminant) (85), cedarwood oil used as the vehicle for applying a black henna temporary tattoo (86), cinnamon oil added to a mud bath in a spa (87), aromatherapy lamps (whereby the water vapour produced distributes the essential oils as an aerosol in the air (84) and wax polish (88).

3.7   OCCUPATIONAL ALLERGIC CONTACT DERMATITIS FROM ESSENTIAL OILS

Massage professionals

Occupational allergic contact dermatitis from essential oils has been reported repeatedly in professionals who massage their clients. Frequent contact, contact with highly concentrated essential oil products and the use of multiple oils, often containing the same allergens, all contribute to the risk of sensitization. The potential relationship of hand dermatitis with the use of aromatherapy products among massage therapists has been investigated (89). The 12-month rate of hand dermatitis was 15% by self-reported criteria and 23% by symptom-based criteria. A significant independent risk factor was the use of products such as massage oils, creams or lotions containing essential oils (89). A study from Germany showed that massage therapists and physiotherapists had the highest occupational risk of fragrance contact allergy (90). Patients are usually female and present with hand dermatitis, which may spread to the forearms and other parts of the body. When patch-tested, they usually react to a (large) number of essential oils used at work, oils tested in a fragrance series and individual fragrances therein; in 70% of the cases, there is co-reactivity to the fragrance mix I and often also to the fragrance mix II and *Myroxylon pereirae*

resin (balsam of Peru) (see the section 'Co-reactivity to other test substances in patients reacting to essential oils' below). Occupational allergic contact dermatitis in professionals performing massages has been observed in aromatherapists (5,94,95,96,97,100), beauticians/beauty therapists (20,91,92,99), complementary therapists (20,28), masseurs (73,93), a naturopathic therapist (73), physiotherapists (73,98) and a reflexologist (73).

Workers in the fragrance and cosmetic industries

Lovage oil sensitized one worker in the fragrance industry (104). Six bottle fillers in a perfume factory became sensitized to the perfumes they worked with, various fragrance materials and essential oils (101). A patient working in a cosmetic factory had occupational dermatitis from contact allergy to ylang-ylang oil in a fragrance mixture he was handling daily (105). A woman packing cosmetics developed occupational allergic contact dermatitis from ylang-ylang oil (106). Bergamot oil sensitized one worker in the fragrance industry (111).

Clary sage oil caused dermatitis in an unknown number of perfumery workers (114).

Workers in the food industries

Occupational hand dermatitis from lemon oil was diagnosed in a cook/barman/fruit grower (102). One case of occupational contact allergy to spearmint oil in a chewing gum finisher has been reported (107). Occupational allergic contact dermatitis from eucalyptus oil (ex *E. globulus*) was observed in one patient with hand dermatitis working in the food industry; an unknown number of food handlers were sensitized to peppermint oil (108). Two patients working in the food industry developed occupational allergic contact dermatitis from aniseed oil and its main ingredient anethole (109). One patient had cheilitis from contact allergy to the vapour of cinnamon oil to which she was occupationally exposed while making bubble gum (112). Hand dermatitis in a baker was ascribed to cinnamon oil (113).

Other occupations

A porter became sensitized to oil of lemon, which he used occupationally (12). Eight of 30 men working on a boat developed dermatitis which was caused by accidental contact with lemongrass oil which had been spilled (103).

A hairdresser who had occupational contact dermatitis of the hands developed contact allergy to lavender oil which was present in a shampoo used at work (36). Another hairdresser developed occupational contact allergy to lavender oil from its presence in an eau de cologne (37). A pedicurist developed dermatitis of the left side of the face from tea tree oil she used to stop small point bleedings in her patients (92). One patient working as painter and car mechanic developed occupational contact dermatitis from pine oil in a wax polish (88). One case of occupational allergic contact dermatitis in a porcelain painter due to anise, turpentine and lavender oils, which were mixed with pigments for painting, was reported (110). Oil from juniper berries was an occupational allergen for the skin and respiratory tract (115, specifics unknown).

3.8   CO-REACTIVITY TO OTHER TEST SUBSTANCES IN PATIENTS REACTING TO ESSENTIAL OILS

In most publications, patients reacting positively to an essential oil co-reacted – if tested – to other essential oils, the fragrance mix I, the fragrance mix II, *Myroxylon pereirae* resin (balsam of Peru), individual fragrance chemicals, or a combination of these. There are (at least) three possible explanations. In many such cases, co-reactivity probably results from pseudo-cross-reactivity, meaning that the test substances share common components responsible for the positive patch test reactions. In a number of cases, real cross-reactivity may play a role. Third, as many patients (e.g., in the case of occupational contact allergy) are exposed to a large number of essential oils and possibly individual fragrances, concomitant sensitization to various products and chemicals can be anticipated.

Co-reactivity to other essential oils

Many patients in case reports react to a great number of essential oils, which is certainly true for aromatherapists and other professionals using oils and essential oil-containing products for massaging their clients (28,96,97,100).

Of two aromatherapists, for example, one reacted to 32 and the other to 21 essential oils (5). Another aromatherapist was patch test positive to 17 of 20 oils used at her work (95). In a group of 19 patients allergic to fragrances and tested with a series of essential oils, 6 reacted to one oil only, 3 to two oils, 3 to three, 3 to four, 1 to five, 1 reacted to seven oils and 1 even had positive patch test reactions to nine oils (116).

Co-reactivity between cananga oil and ylang-ylang oil

Of eight patients with dermatitis from fragrances and reacting to ylang-ylang oil, seven (88%) also reacted to cananga oil (116). Conversely, of seven patients with dermatitis from fragrances and reacting to cananga oil, all co-reacted to ylang-ylang oil (116). Simultaneous reactivity has also been observed in many case reports. Both oils are obtained from the tree *Cananga odorata*, ylang-ylang oil from its *forma genuina* and cananga oil from its *forma macryphylla* and they have similarities in their compositions.

Co-reactions to the fragrance mixes

Co-reactivity to the fragrance mix I (FM I) (and to a lesser degree to the fragrance mix II (FM II), which was introduced more recently) in patients with contact allergy to or allergic contact dermatitis from essential oils is very frequent, both in case reports (e.g., 5,28,73,95,96,97,100,119) and in investigations in groups of patients. Of 637 patients reacting to at least one essential oil seen by the IVDK (Germany, Austria, Switzerland), 55% co-reacted to FM I, 35% to FM II and 63% to one of these two (117). Of 15 patients reacting to ylang-ylang oil in The Netherlands, 15 (83%) co-reacted

to FM I (120). Co-reactivity to FM for individual essential oils as observed in a European multi-center study was as follows: ylang-ylang oil 36/42 (86%), lemongrass oil 18/25 (72%), jasmine absolute 11/20 (55%) and patchouli oil 9/13 (69%) (121). Co-reactivity to FM I in individuals with positive patch tests to turpentine oil (which is nearly exclusively tested by the IVDK) was 46% (122).

Co-reactions to *Myroxylon pereirae* resin and colophony

Of 431 patients reacting to turpentine oil, *Myroxylon pereirae* resin co-reacted in 29% (versus 7% in turpentine-negative patients) and colophony in 23% (versus 3% in turpentine-negative patients), which makes these associations statistically significant (122). Positive reactions to colophony (which is the material that remains after hydrodistillation of gum resin of *Pinus* species to produce turpentine oils) are associated especially with essential oils of woody origin such as dwarf pine needle oil, sandalwood oil and cedarwood oil (123). A statistically significant association between positive patch test reactions to essential oils and the Compositae-mix has also been reported (123).

Co-reactivity to important components of essential oils (individual fragrance chemicals)

This is discussed below in the section 'The allergens in essential oils'.

3.9   ANALYTICAL INVESTIGATION OF THE COMPONENTS OF ESSENTIAL OILS WHICH HAVE CAUSED ALLERGIC CONTACT DERMATITIS

In few published investigations, analyses of essential oils used by patients allergic to them have been performed. This is especially useful when patients also react to individual fragrance chemicals; the analysis can show the presence (or absence) of such chemicals in the oils. Quantitative analysis can determine the concentration of the chemicals and thus give an indication of whether the positive patch test reaction to the oils can be caused by these components.

In the most extensive analytical investigation performed thus far, one sample of 27 essential oils each and two samples of 11 other essential oils each, used by one or two aromatherapists, who reacted to many of these oils, were investigated by GC-MS (5). One of the patients was also allergic to geraniol, linalool, linalyl acetate and α-pinene, the other to geraniol, α-pinene and caryophyllene. α-Pinene was demonstrated in 42 oil samples, of which 13 had concentrations of 1-10% (v/v) and 5 of >10%. Linalool was found in 22 samples, of which 9 had concentrations of 1-10% (v/v) and 11 of >10%. Geraniol was present in 16 oil samples, of which 7 had concentrations of 1-10% (v/v) and 8 of >10%. Linalool was found in 22 samples, of which 9 had concentrations of 1-10% (v/v) and 11 of >10%. Caryophyllene was identified in 37 samples, of which 20 had concentrations of 1-10% (v/v) and 9 of >10% (5).

Bergamot oil sensitized one female worker in the fragrance industry, who also reacted to α-pinene and β-pinene. Commercial bergamot oils were investigated for the presence of these chemicals and they were identified in maximum concentrations of 1.7% (α-pinene) and 9.3% (β-pinene) (111).

One patient suspected to be allergic to incense had positive patch tests to two brands of incense, sandalwood oil, musk ambrette and santalol; gas chromatography of pentane:ether extracts of the incense showed 9% and 34% musk ambrette and 8% santalol in both incenses; the sandalwood oil extract had 73% santalol, which is the dominant component of sandalwood oil (124).

One patient developed contact allergy to a perfume; she reacted to the fragrance mix, Bulgarian rose oil and geraniol; chromatographic analysis of the perfume showed it to contain 33% citronellol and 20% geraniol (126).

One patient became sensitized from the topical application of pure tea tree oil for therapeutic purposes (125). Gas chromatography/mass spectrometry identified a number of components, including 1,8-cineole. The patient was tested with 1,8 cineole and retested with tea tree oil and reacted to both test substances. The concentration of this chemical was not determined (125). For analytical investigations of tea tree oil samples see Chapter 5.83.

3.10   THE ALLERGENS IN ESSENTIAL OILS

With the exception of tea tree oil and turpentine oil, large-scale patch testing with components of essential oils in allergic patients in order to determine the major sensitizers has not been performed. Tea tree oil, stored in open bottles or in a bottle opened several times, suffers an aging process resulting in photo-oxidation leading to degradation products (peroxides, epoxides and endoperoxides), which are strong sensitizers. The most important sensitizers in tea tree oil appear to be terpinolene, ascaridole, α-terpinene and its oxidation products, 1,2,4-trihydroxymenthane, α-phellandrene, d-limonene and myrcene. Other chemicals which may be responsible for tea tree oil allergy, albeit less frequently, include aromadendrene, d-carvone, l-carvone, terpinen-4-ol, viridiflorene, sabinene, p-cymene and 1,8-cineole. Most of these have been found in low concentrations or not at all in commercial tea tree oils, which can be explained by the fact that these were fresh oil samples (see Chapter 5.83). In turpentine oil, the allergens are δ3-carene hydroperoxides (in oils from Scandinavia [probably hardly used anymore] and Indonesia), α-pinene, oxidized limonene, β-pinene, α-terpineol, and rarely α-phellandrene (see Chapter 5.87). In Poland, 30% of patients allergic to turpentine oil also reacted to pine needle oil; α-pinene is the major ingredient (and important allergen) of turpentine oil and also an important component of pine needle oil, with concentrations in commercial oils ranging from 13 to 20% (133).

For some other oils, their major allergens can tentatively be identified on the basis of their general composition and co-reactions to important ingredients in patch testing as documented in literature. For example, 50% of 94 patients allergic to clove (bud) oil co-reacted to eugenol, which is the major ingredient (>82%) of this essential oil (117). Of 13 patients known to be allergic to *Myroxylon pereirae*, propolis and clove oil, 6 (46%) reacted to eugenol, which is a component of all three test substances (128). Eugenol is also an important component of bay oil (40-55%). Of seven patients reacting to bay oil and tested with eugenol, five (71%) co-reacted to eugenol (116).

In the case of lemongrass oil, of 67 patients reacting to this oil, 34 (51%) also reacted to citral; citral consists of geranial + neral, both of which are the dominant ingredients of lemongrass oils (117). Commercial geranium oils contain 6-32% geraniol. Co-reactivity to geraniol in patients reacting to geranium oil has been observed in at least 50% of those tested with geraniol (129,131,132). Geraniol sensitivity is also observed in patients reacting to rose oils (131,132); commercial rose oils contain 5-24% geraniol. In cinnamon oil, cinnamaldehyde is the major component (43-73%) and this chemical has co-reacted repeatedly in patients allergic to cinnamon oils (Chapter 5.21). Menthol (23-48% in commercial oils) is probably the most important sensitizer in peppermint oils (Chapter 5.68). Spearmint oils contain 60-80% carvone. In about 50% of the patients allergic to spearmint oil in who carvone is tested, co-reactivity occurs (Chapter 5.79). In citrus peel oils (bergamot, grapefruit, lemon, orange, mandarin, tangerine), limonene is by far the most important constituent, but has thus far not been identified as the main allergen. The reason may be that patients allergic to citrus peel oils have not yet been tested with oxidized limonene, which is the allergenic form of limonene (127). Probably for the same reason, linalool has as yet not been found as a major allergenic ingredient of lavender and ylang-ylang oils, in which high concentrations of linalool and linalyl acetate are present.

Possible or probable allergens in essential oils are shown in Table 3.1 (right column); for specific data see Chapters 5.1-5.93.

3.11   PATCH TESTING WITH ESSENTIAL OILS AND THEIR INGREDIENTS

Essential oils should be patch tested in any patient suspected of contact allergy to these oils on the basis of the history and the clinical picture. Suspicion is high in cases of hand dermatitis in aromatherapists, masseuses etc., in patients working in the fragrance industry, the food industry and in individuals who have applied essential oils to their skin, e.g., for therapeutic purposes (notably tea tree oil). In addition, in any patient with symptoms of the oral mucosa, lips (cheilitis) and the perioral skin, contact allergy to essential oils in toothpastes or other oral preparations should be considered. However, the range of products in which essential oils can be present is broad and localization and aspect of essential oil-induced allergic contact dermatitis are often not specific.

Several essential oils are available as commercial patch test substances (Table 3.4) In the USA four essential oils and one absolute (tea tree oil 5%, lavender oil 2%, peppermint oil 2%, jasmine absolute 2% and ylang-ylang

Table 3.4 Essential oils, fragrance components and other chemicals identified in essential oils which are commercially available for patch testing

Patch test allergen	Suppliers			
	Trolab	Chemo	Brial	Allergeaze
ESSENTIAL OILS				
Bergamot oil			2%	2%
Cananga oil		2%[f]		
Cedarwood oil	10%		10%[k]	10%[k]
Cinnamon oil			0.5%[l]	0.5%[l]
Clove oil	2%		2%[m]	2%[m]
Eucalyptus oil	2%		2%[n]	
Geranium oil, Bourbon		2%[g]		
Jasmine absolute	5%	2%[e]	2%[o]	2%[b,o]
Laurel leaf oil	2%		2%	2%
Lavender oil		2%[h]	2%	2%
Lemongrass oil	2%		2%[p]	2%[p]
Lemon oil	2%		2%	2%
Neroli oil	2%		5%	2 and 5%
Orange oil	2%		2%	2%
Patchouli oil	10%			10%
Peppermint oil	2%	2%	2%	2%
Rosemary oil			0.5%	0.5%
Rose oil		2%, *extract*	0.5%[q]	0.5%[q]
Sandalwood oil	10%	2%[c,d]		10%[r]
Tea tree oil		5% ox.		5% ox.
Turpentine oil	10%	0.4% ox.[i]	10%[r]	10%[s]
Ylang-ylang oil	10%	2%[j]		2%
FRAGRANCES / OTHER CHEMICALS IDENTIFIED IN ESSENTIAL OILS				
Fragrances				
Amyl cinnamal (α-amylcinnamic aldehyde)	1%	2%	1%	1%
α-Amylcinnamic alcohol	1%	5%	1%	1%
Anethole		5%		
Anisyl alcohol	1%	10% Soft.	1%	1%
Benzaldehyde	5%		5%	5%
Benzoic acid		5%	5%	1 and 5%
Benzyl alcohol	1%	10% Soft.	1 and 5%	1 and 5%
Benzyl benzoate	1%	10%	1%	1%
Benzyl cinnamate	5%	10%	5%	5%
Benzyl salicylate	1%	10%	1%	1%
Camphor			1%	1%
Carvone		5%		5%
Cinnamal (cinnamic aldehyde)	1%	1%	1%	1%
Cinnamyl alcohol	1%	2%	1%	1%
Citral	2%	2%	2%	2%
Citronellal			2%	2%
Citronellol	1%	1%	1%	1%
Coumarin	5%	5%	5%	5%
Dipentene (*dl*-limonene)	2%			
Eugenol	1%	2%	1%	1%
Farnesol	5%	5%	5%	5%
Geraniol	1%	2%	1%	1%
Hexyl cinnamal (α-hexylcinnamic aldehyde)	10%	10%	10%	10%
Hydroxycitronellal[t]	1%	2%	1%	1%
Isoeugenol	1%	2%	1%	1%
α-Isomethyl ionone (γ-methylionone)[a,t]	1%	10%		0.1%
d-Limonene		10%	2%	2 and 3%
Limonene hydroperoxides		0.3%		0.3%
Linalool	10%	10%		
Linalool hydroperoxides		1%		1%
Menthol	1%	2%	1%	1%
Methyl anthranilate		5%		
Methyl salicylate			2%	2%
α-Pinene			15%	15%
Salicylaldehyde	2%		2%	2%

Table 3.4 Essential oils, fragrance components and other chemicals identified in essential oils which are commercially available for patch testing (*continued*)

Patch test allergen	Suppliers			
	Trolab	**Chemo**	**Brial**	**Allergeaze**
Thymol			1%	1%
Vanillin	10%	10%	10%	10%
Chemicals identified in essential oils, so far not found in nature				
Butylphenyl methylpropional (Lilial®, *p-tert*-butyl-α-methyl-hydrocinnamic aldehyde)	10%	10%		
Dibutyl phthalate	5%	5%	5%	5%
Di-(2-ethylhexyl) phthalate (dioctyl phthalate)	5%	5%	5%	5%
Diethyl phthalate			5%	5%
Dimethyl phthalate	5%		5%	5%
Hydroxycitronellal	1%	2%	1%	1%
Hydroxyisohexyl 3-cyclohexene carboxaldehyde (Lyral®)	5%	5%	5%	5%
α-Isomethyl ionone (γ-methylionone)[a]	1%	10%		0.1%
Musk ambrette			5%	5%
Indicators of fragrance / essential oils allergy				
Colophonium (rosin)	20%	20%	20%	20%
Compositae mix		2.5 and 5%	6%	6%
Fragrance mix I (cinnamyl alcohol, cinnamal, hydroxycitronellal, amyl cinnamal, geraniol, eugenol, isoeugenol, *Evernia prunastri*)	8%	8%	8%	8%
Fragrance mix II (citral, citronellol, coumarin, farnesol, hexyl cinnamal, hydroxyisohexyl 3-cyclohexene carboxaldehyde [Lyral®])	14%	14%	14%	14%
Myroxylon pereirae (Balsam of Peru)	25%	25%	25%	25%
Perfume mix (cinnamyl alcohol, cinnamal, hydroxycitronellal, eugenol, Isoeugenol, geraniol)		6%		
Sesquiterpene lactone mix (alantolactone, costunolide, dehydrocostus lactone)	0.1%	0.1%	0.1%	0.1%

Trolab: www.smartpracticeurope.com.de; Chemotechnique: www.chemotechnique.se; Brial: www.brial.com; Allergeaze: www.allergeaze.com

All test substances are in petrolatum unless otherwise indicated

ox.: oxidized; Soft.: Softisan

[a] the chemical which was found in essential oils is 'methylionone'

[b] 'jasmine oil officinale (grandiflorum)'

[c] produced from *Santalum album*: main ingredients are β-santalol 21% and α-santalol 50% (information from Chemotechnique)

[d] in some patients, 2% may be too low to detect sensitization (121)

[e] jasmine absolute *ex J. grandiflorum*; main ingredients are benzyl benzoate 10-20%, phytol 10-20%, isophytol 5-10%, linalool 5-10%, eugenol 5-10%, benzyl alcohol 1-5% and benzyl salicylate 0.1-1% (information from Chemotechnique)

[f] main ingredients: β-caryophyllene 30-40%, geranyl acetate 5-10%, benzyl benzoate 1-5%, linalool 1-5%, methyl benzoate 1-5%, benzyl salicylate 1-5%, farnesol 1-5%, *p*-cresyl methyl ether 1-5%, geraniol 1-5% (information from Chemotechnique)

[g] main ingredients: DL-citronellol 30-40%, geraniol 10-20%, linalool 10-20%, menthone 5-10% and citral 1-5% (information from Chemotechnique)

[h] main ingredients are linalool 20-30%, linalyl acetate 20-30%, coumarin 10-20%, terpinen-4-ol 1-5%, geraniol 0.1-1% and caryophyllene oxide 0.1-1% (information from Chemotechnique); the test substance is prepared from lavender *absolute*

[i] main ingredients are δ3-carene 40-65% and α-pinene <20% (information from Chemotechnique); obtained from *Pinus roxburghii* Sarg.

[j] main ingredients are 'other hydrocarbons' 30-40%, linalool 10-20%, *p*-cresyl methyl ether 10-20%, β-caryophyllene 5-10%, geranyl acetate 5-10%, methyl benzoate 5-10%, benzyl benzoate 5-10%, benzyl salicylate 1-5%, farnesol 1-5%, geraniol 1-5% and isoeugenol 1-5% (information from Chemotechnique)

[k] prepared from Virginian cedarwood oil (information from Brial/Allergeaze)

[l] cinnamon bark oil from *Cinnamomum zeylanicum* (information from Brial/Allergeaze)

[m] clove bud oil (information from Brial/Allergeaze)

[n] eucalyptus oil ex *E. globulus* (information from Brial)

[o] prepared from *Jasminum sambac* (information from Brial/Allergeaze)

[p] West Indian lemongrass oil produced from *Cymbopogon citratus* (information from Brial/Allergeaze)

[q] *synthetic* rose oil (information from Brial/Allergeaze)

[r] prepared from *Pinus pinaster* (information from Brial/Allergeaze)

[s] prepared from *Santalum album* (information from Brial/Allergeaze)

[t] has not been found in nature so far

Table 3.5 Suggested patch test concentrations for chemicals for which there are no commercial test preparations

Chemical	CAS number	Patch test concentration(s) and vehicles
Acetyl cedrene	32388-55-9	1 and 5% pet.; 10% DIPP
Ambrettolide	7779-50-2	5% pet.
Aromadendrene	489-39-4	1% alc.; 5% DEP
Ascaridole	512-85-6	2% pet.; 5% DEP or aq.
Azulene	275-51-4	1% pet.
Benzyl acetate	140-11-4	1 and 5% pet.
Benzyl propionate	122-63-4	3% pet.
Bisabolol	515-69-5	5% pet.
δ3-Carene	13466-78-9	5% pet.; 15% o.o.
Carvacrol	499-75-2	5% pet.
β-Caryophyllene	87-44-5	5% pet.; 3% pet., ox.
Caryophyllene oxide	1139-30-6	3.9% pet.
1,8-Cineole	470-82-6	5% pet., DEP or alc.
Cinnamic acid	621-82-9	5% pet.
Cinnamyl cinnamate	122-69-0	5% pet.
Coniferyl alcohol	458-35-5	1% pet.
Coniferyl benzoate	4159-29-9	1% pet.
Cuminaldehyde	122-03-2	10% pet.
p-Cymene	99-87-6	1% alc.; 5% DEP
β-Damascone	23726-91-2	0.2% pet.
Dihydrocoumarin	119-84-6	5% pet.
Falcarinol	21852-80-2	0.03-2% pet.
Geranial	141-27-5	1.5-3.5% pet.
Geranyl acetate	105-87-3	3% pet.
Guaiazulene	489-84-9	1% pet.
Heptanal	111-71-7	3% pet.
Hexadecanolide	109-29-5	5% pet.
(Z)-3-Hexenyl salicylate	65405-77-8	3% pet.
Hydroxycitronellol	107-74-4	7% pet.
α-Ionone	127-41-3	5-10% pet.
β-Ionone	79-77-6	5-10% pet.
Isoeugenyl acetate	93-29-8	1.2% alc.
Isopulegol	89-79-2	5% pet.
Isosafrole	120-58-1	5% pet.
Isothymol	4427-56-9	2% pet.
Limonene oxide	1195-92-2	1% pet.
Linalyl acetate	115-95-7	5-10% pet.
o-Methoxycinnamaldehyde (2)	1504-74-1	4% pet.
7-Methoxycoumarin	531-59-9	1% pet.
8-Methoxypsoralen	298-81-7	0.15% pet.
Methyl p-anisate	121-98-2	4% pet.
Methyl cinnamate	103-26-4	10% pet.
Methyl dihydrojas-monate	24851-98-7	1 and 5% pet.
Methyl eugenol	93-15-2	5% pet.
Methyl isoeugenol	93-16-3	5% pet.
Myrcene	123-35-3	1 and 5% o.o.; 5% pet.; 3% pet., ox.
Neral	106-26-3	1.5-3.5% pet.
Nerol	106-25-2	5% pet.
Nerolidol	7212-44-4	1% pet.
Nonanal	124-19-6	1% pet.
Nonanol	28473-21-4	2% pet.
Nopyl acetate	128-51-8	10% pet.
Octanol	29063-28-3	2% pet.
α-Phellandrene	99-83-2	1% alc.; 5% DEP or aq.
2-Phenethyl alcohol	60-12-8	1-10% pet.
Phenylacetaldehyde	122-78-1	2% pet.
β-Pinene	127-91-3	1% pet.; 15% o.o.
Piperitone	89-81-6	10% pet.
Piperonal	120-57-0	1 and 5% pet.
Propylidene phthalide	17369-59-4	2% pet.
Pulegone	89-82-7	10% pet.
Sabinene	3387-41-5	5% DEP
Safrole	94-59-7	1% pet.
Santalol		2-10% pet.

Table 3.5 Suggested patch test concentrations for chemicals for which there are no commercial test preparations (*continued*)

Chemical	CAS number	Patch test concentration(s) and vehicles
α-Terpinene	99-86-5	1% alc.; 5% DEP or aq.
Terpinen-4-ol	562-74-3	1 and 5% alc. or DEP; 10% alc., DEP or aq.
α-Terpineol	98-55-5	1 and 5% pet.; 1 and 10% alc.
Terpinolene	586-62-9	1% alc.; 5% pet.; 10% DEP or aq.
Terpinyl acetate	8007-35-0	5% pet.
1,2,4-Trihydroxymenthane		5% DEP or pet.
Viridiflorene	21747-46-6	5% DEP

alc.: alcohol; aq.: aqua, water; DEP: diethyl phthalate; DIPP: diisopropylene glycol; o.o.: olive oil; ox.: oxidized; pet.: petrolatum

oil 2%) are currently present in the screening series of the North American Contact Dermatitis Group (NACDG) and tested in all consecutive patients suspected of contact dermatitis (1). Many other investigators test a series of essential oils and other fragrance materials (available commercial test substances are shown in Table 3.4) only when fragrance contact allergy is suspected; another indication for further testing is when a patient has previously shown positive patch test reactions to one or more of the fragrance indicator allergens, such as the fragrance mix(es) and *Myroxylon pereirae* resin, to other fragrance materials, or to the patient's personal fragrances or fragranced cosmetics. For screening purposes, these series of commercial allergens are very useful. However, when patients have a history of contact with one or more oils, it is preferable (if not imperative) to test these products themselves, because of the strong variability that may occur in the composition of essential oils. In addition, in aged oils that have been exposed to light, oxygen and temperature changes, new allergenic chemicals may have formed (which is well known in, for example, tea tree oil and lavender oil) which are not present in adequately stored (cool, dark, unexposed to air) commercial test substances.

For most essential oils, testing at 2-5% in petrolatum (or both, unless many oils are tested, risk of great number of reactions with false-positives due to the excited skin syndrome) will likely be adequate. Lower concentrations may be appropriate for costus root oil (0.1%, very unlikely to encounter, historical allergen), black cumin oil (0.5%), star anise oil (0.5%, it may be preferable to test the main ingredient anethole 5%), and cassia and cinnamon (bark) oils (1%, because of the very high concentrations of cinnamic aldehyde). The various cedarwood oils may be tested as follows: cedarwood oil Atlas 8%, cedarwood oil China 10%, cedarwood oil Virginia 8%; for silver fir oil, 10% may be appropriate. When obtaining positive patch tests to one or more essential oils, a search for the allergenic ingredients should preferably be initiated. Likely candidates are discussed above in the section 'The allergens in essential oils' and in Table 3.1, right column. It is advisable to consult the individual oil chapters (Chapters 5.1-5.93) to identify important components. A list of all chemicals which can be present in essential oils and which have caused contact allergy (usually not from

their presence in essential oils) is provided in Chapter 4. The chemicals from this list which are commercially available are shown in Table 3.4. Suggested patch test concentrations for chemicals in this list which cannot be purchased as commercial test preparations are shown in Table 3.5.

LITERATURE

1 Cheng J, Zug KA. Fragrance allergic contact dermatitis. Dermatitis 2014;25:232-245

2 Rudzki E, Grzywa Z, Bruo WS. Sensitivity to 35 essential oils. Contact Dermatitis 1976;2:196-200

3 Foussereau J, Benezra C, Ourisson G. Contact dermatitis from laurel. I. Clinical aspects. Transactions of the St John's Hospital Dermatological Society 1967;53:141-146

4 Warshaw EM, Maibach HI, Taylor JS, Sasseville D, DeKoven JG, Zirwas MJ, et al. North American Contact Dermatitis Group patch test results: 2011-2012. Dermatitis 2015;26:49-59

5 Dharmagunawardena B, Takwale A, Sanders KJ, Cannan S, Roger A, Ilchyshyn A. Gas chromatography: an investigative tool in multiple allergies to essential oils. Contact Dermatitis 2002;47:288-292

6 Nakayama H, Harada R, Toda M. Pigmented cosmetic dermatitis. Int J Dermatol 1976;15:673-675

7 Sugawara M, Nakayama H, Watanabe S. Contact hypersensitivity to ylang-ylang oil. Contact Dermatitis 1990;23:248-249

8 Toyoda T, Watanabe S, Kawasaki M, et al. Dihydro-isoeugenol found in ylang-ylang oil. Skin Res 1989;31 (suppl.7):35-43 (in Japanese)

9 Nakayama H, Matsuo S, Hayakawa K, Takhashi K, Shigematsu T, Ota S. Pigmented cosmetic dermatitis. Int J Dermatol 1984;23:299-305

10 Hausen BM, Kunze B. Kontaktallergie auf Patchouli-Öl. Akt Dermatol 1991;17:199-202

11 Steinmann A, Schätzle M, Agathos M, Breit R. Allergic contact dermatitis from black cumin (*Nigella sativa*) oil after topical use. Contact Dermatitis 1997;36:268-269

12 Keil H. Contact dermatitis due to oil of citronellal. J Invest Dermatol 1947;8:327-334

13 Lleonart R, Andrés B, Molinero J, Corominas M. Systemic allergic contact dermatitis due to black cumin oil. Contact Dermatitis 2014;70(Suppl. 1):45

14 Athanasiadis GI, Pfab F, Klein A, Braun-Falco M, Ring J, Ollert M. Erythema multiforme due to contact with laurel oil. Contact Dermatitis 2007;57:116-118

15 Hausen BM. Lorbeer-Allergie Ursache, Wirkung und Folgen der äusserlichen Anwendung eines sogenannten Naturheilmittels. Dtsch Med Wschr 1985;110: 634-638

16 Özden MG, Öztaş P, Öztaş MO, Önder M. Allergic contact dermatitis from Laurus nobilis (laurel) oil. Contact Dermatitis 2001;45:178

17 Reutemann P, Ehrlich A. Neem oil: an herbal therapy for alopecia areata causes dermatitis. Dermatitis 2008;19:E12-E15

18 Lapeere H, Boone B, Verhaeghe E, Ongenae K, Lambert J. Contact dermatitis caused by lovage (Levisticum officinalis) essential oil. Contact Dermatitis 2013;69:181-182

19 Coulson IH, Ali Khan AS. Facial 'pillow' dermatitis due to lavender oil allergy. Contact Dermatitis 1999;41:111

20 Coutts I, Shaw S, Orton D. Patch testing with pure tea tree oil - 12 months experience. Br J Dermatol 2002;147(Suppl. 62):70

21 Nardelli A, Drieghe J, Claes L, Boey L, Goossens A. Fragrance allergens in 'specific' cosmetic products. Contact Dermatitis 2011;64:212-219

22 Christoffers WA, Blömeke B, Coenraads P-J, Schuttelaar M-LA. Co-sensitization to ascaridole and tea tree oil. Contact Dermatitis 2013;69:187-189

23 Travassos AR, Claes L, Boey L, Drieghe J, Goossens A. Non-fragrance allergens in specific cosmetic products. Contact Dermatitis 2011;65:276-285

24 Williams JD, Nixon RL, Lee A. Recurrent allergic contact dermatitis due to allergen transfer by sunglasses. Contact Dermatitis 2007;57:120-121

25 Vilaplana J, Romaguera C, Grimalt F. Contact dermatitis from geraniol in Bulgarian rose oil. Contact Dermatitis 1991;24:301

26 Penchalaiah K, Handa S, Lakshmi SB, Sharma VK, Kumar B. Sensitizers commonly causing allergic contact dermatitis from cosmetics. Contact Dermatitis 2000;43:311-313

27 Tran A, Pratt M, DeKoven J. Acute allergic contact dermatitis of the lips from peppermint oil in a lip balm. Dermatitis 2010;21:111-115

28 Newsham J, Rai S, Williams JDL. Two cases of allergic contact dermatitis to neroli oil. Br J Dermatol 2011;165(Suppl.1):76

29 Larsen WG. Cosmetic dermatitis due to a perfume. Contact Dermatitis 1975;1:142-145

30 Zedlitz S, Kaufmann R, Boehncke W. Allergic contact dermatitis from black cumin (Nigella sativa) oil-containing ointment. Contact Dermatitis 2002;46:188

31 Davies MG, Hodgson GA, Evans E. Contact dermatitis from an ostomy deodorant. Contact Dermatitis 1978;4:11-13

32 Starke JC. Photoallergy to sandalwood oil. Arch Dermatol 1967;96:62-63

33 Scheinman PL. Is it really fragrance free? Am J Cont Derm 1997;8:239-242

34 Chang Y-C, Maibach HI. Pseudo flautist's lip: allergic contact cheilitis from geraniol. Contact Dermatitis 1997;37:39

35 Spier HW, Sixt I. Lorbeer als Träger eines wenig beachteten kontaktekzemotogenen Allergens. Derm Wochenschr 1953;128:805-810

36 Brandao FM. Occupational allergy to lavender oil. Contact Dermatitis 1986;15:249-250

37 Ménard E. Les dermatoses professionelles. Concours Médicale 1961;83:4308-4311

38 Varma S, Blackford S, Statham BN, Blackwell A. Combined contact allergy to tea tree oil and lavender oil complicating chronic vulvovaginitis. Contact Dermatitis 2000;42:309-310

39 Serrano G, Pujol C, Cuadra J, Gallo S, Aliaga A. Riehl's melanosis: Pigmented contact dermatitis caused by fragrances. J Am Acad Dermatol 1989;21:1057-1060

40 Magnusson B, Wilkinson DS. Cinnamic aldehyde in toothpaste. 1. Clinical aspects and patch tests. Contact Dermatitis 1975;1:70-76

41 Silvers SH. Stomatitis and dermatitis venenata with purpura, resulting from oil of cloves and oil of cassia. Dental Items of Interest 1939;61:649-651. Data cited in ref. 40

42 Laubach JL, Malkenson FD, Ringrose EJ. Cheilitis caused by cinnamon (cassia) oil in toothpaste. JAMA 1953;152:404-405. Data cited in ref. 40

43 Drake TE, Maibach HI. Allergic contact dermatitis and stomatitis caused by a cinnamic aldehyde-flavored toothpaste. Arch Dermatol 1976;112:202-203

44 Millard LG. Acute contact sensitivity to a new toothpaste. J Dentistry 1973;1:168-170

45 Millard LG. Contact sensitivity to toothpaste. Brit Med J 1973;1:676

46 Cummer CL. Dermatitis due to oil of cinnamon. Arch Dermatol Syphilol 1940;42:674-675

47 Dooms-Goossens A, Degreef H, Holvoet C, Maertens M. Turpentine-induced hypersensitivity to peppermint oil. Contact Dermatitis 1977;3:304-308

48 Fleming CJ, Forsyth A. D5 patch test reactions to menthol and peppermint. Contact Dermatitis 1998;38:337

49 Freeman S, Stephens R. Cheilitis: Analysis of 75 cases referred to a contact dermatitis clinic. Am J Cont Derm 1999;10:198-200

50 Downs AMR, Lear JT, Sansom JE. Contact sensitivity in patients with oral symptoms. Contact Dermatitis 1998; 39:258-259

51 Hjorth N, Jervoe P. Allergisk Kontaktstomatitis og Kontaktdermatitis fremkaldt of smagsstoffer i tandpasta. Tandlaegebladet 1967;71:937-942. Data cited in ref. 40

52 Clayton R, Orton D. Contact allergy to spearmint oil in a patient with oral lichen planus. Contact Dermatitis 2004;51:314-315

53 Skrebova N, Brocks K, Karlsmark T. Allergic contact cheilitis from spearmint oil. Contact Dermatitis 1998;39:35-36

54 Worm M, Jeep S, Sterry W, Zuberbier T. Perioral contact dermatitis caused by L-carvone in toothpaste. Contact Dermatitis 1998;38:338

55 Sainio E-L, Kanerva L. Contact allergens in toothpastes and a review of their hypersensitivity. Contact Dermatitis 1995;33:100-105

56 Grattan CEH, Peachy RD. Contact sensitization to toothpaste flavouring. J Royal Coll Gen Pract 1985;35:498. data cited in ref. 55

57 Baer ON. Toothpaste allergies. J Clin Pediatr Dent 1992;16:230-231. Data cited in ref. 55

58 Loveman AB. Stomatitis venenata. Report of a case of sensitivity of the mucous membranes and the skin to oil of anise. Arch Derm Syph 1938;37:70-81

59 Nardelli A, D'Hooge E, Drieghe J, Dooms M, Goossens A. Allergic contact dermatitis from fragrance components in specific topical pharmaceutical products in Belgium. Contact Dermatitis 2009;60:303-313

60 Devleeschouwer V, Roelandts R, Garmyn M, Goossens A. Allergic and photoallergic contact dermatitis from ketoprofen: results of (photo) patch testing and follow-up of 42 patients. Contact Dermatitis 2008;58:159-166

61 Bhushan M, Beck MH. Allergic contact dermatitis from tea tree oil in a wart paint. Contact Dermatitis 1997;36:117-118

62 Foti C, Conserva A, Antelmi A, Lospalluti L, Angelini G. Contact dermatitis from peppermint and menthol in a local action transcutaneous patch. Contact Dermatitis 2003;49:312-313

63 Matthieu L, Meuleman L, van Hecke E, Blondeel A, Dezfoulian B, Constandt L, Goossens A. Contact and photocontact allergy to ketoprofen. The Belgian experience. Contact Dermatitis 2004;50:238-241

64 McGeorge BCL, Steele MC. Allergic contact dermatitis of the nipple from Roman chamomile ointment. Contact Dermatitis 1991;24:139-140

65 Giordano-Labadie F, Schwarze HP, Bazex J. Allergic contact dermatitis from chamomile used in phytotherapy. Contact Dermatitis 2000;42:247

66 Lauriola MM, De Bitonto A, Sena P. Allergic contact dermatitis due to cinnamon oil in galenic vaginal suppositories. Acta Derm Venereol 2010;90:187-188

67 Knöll R, Ulrich R, Spallek W. Allergic contact eczema to etofenamate and dwarf pine oil. Sportverletz Sportschaden 1990;4(2):96-98 (article in German)

68 Noiles K, Pratt M. Contact dermatitis to Vicks VapoRub. Dermatitis 2010;21:167-169

69 Vilaplana J, Romaguera C. Allergic contact dermatitis due to eucalyptol in an anti-inflammatory cream. Contact Dermatitis 2000;43:118

70 Eun HC, Lee AY. Contact dermatitis due to Madecassol. Contact Dermatitis 1985;13:310-313

71 Zina G, Bony G. Phenergan cream (role of base constituents). Contact Dermatitis Newsletter 1969;6:117

72 Le Coulant P, Texier L, Malleville J, Doussy NN. Sensibilization à une crème antihistaminique. Bull Soc Franç Derm Syph 1964;71:234-237

73 Trattner A, David M, Lazarov A. Occupational contact dermatitis due to essential oils. Contact Dermatitis 2008;58:282-284

74 Gunatheesan S, Tam MM, Tate B, Tversky J, Nixon R. Retrospective study of oral lichen planus and allergy to spearmint oil. Australas J Dermatol 2012;53:224-228

75 Cahill J, Gunatheesan S, Tam M, Tate B, Nixon R. Oral lichen planus and allergy to spearmint oil. Contact Dermatitis 2012;66 (Suppl. 2):38 (FC1.03)

76 Khanna M, Qasem K, Sasseville D. Allergic contact dermatitis to tea tree oil with erythema multiforme-like id reaction. Dermatitis 2000;11:238-242

77 Gelot P, Bara-Passot C, Gimenez-Arnau E, Beneton N, Maillard H, Celerier P. Bullous drug eruption with Nigella sativa oil. Ann Dermatol Venereol 2012;139:287-291

78 Nosbaum A, Ben Said B, Halpern S-J, Nicolas J-F, Bérard F. Systemic allergic contact dermatitis to black cumin essential oil expressing as generalized erythema multiforme. Eur J Dermatol 2011;21:447-448

79 Perrett CM, Evans AV, Russell-Jones R. Tea tree oil dermatitis associated with linear IgA disease. Clin Exp Dermatol 2003;28:167-170

80 Gad El-Rab MO, Al-Sheikh OA. Is the European standard series suitable for patch testing in Riyadh, Saudi Arabia? Contact Dermatitis 1995;33:310-314

81 Adişen E, Önder M. Allergic contact dermatitis from Laurus nobilis oil induced by massage. Contact Dermatitis 2007;56:360-361

82 Leifer W. Contact dermatitis due to cinnamon. Recurrence of dermatitis following oral administration of cinnamon oil. Arch Dermatol Syphylol 1951;64:52-55

83 de Groot AC. Airborne allergic contact dermatitis from tea tree oil. Contact Dermatitis 1996;35:304-305

84 Schaller M, Korting HC. Allergic airborne contact dermatitis from essential oils used in aromatherapy. Clin Exp Dermatol 1995;20:143-145

85 Franz H, Frank R, Rytter M, Haustein UF. Allergic contact dermatitis due to cedarwood oil after dermatoscopy. Contact Dermatitis 1998;38:182-183

86 Temesvári E, Podányi B, Pónyai G, Németh I. Fragrance sensitization caused by temporary henna tattoo. Contact Dermatitis 2002;47:240

87 Garcia-Abujeta JL, de Larramendi CH, Pomares Berna J, Munoz Palomino E. Mud bath dermatitis due to cinnamon oil. Contact Dermatitis 2005;52:234

88 Martins C, Gonçalo M, Gonçalo S. Allergic contact dermatitis from dipentene in wax polish. Contact Dermatitis 1995;33:126-127

89 Crawford GH, Katz KA, Ellis E, James WD. Use of aromatherapy products and increased risk of hand dermatitis in massage therapists. Arch Dermatol 2004;140:991-996

90 Uter W, Schnuch A, Geier J, Pfahlberg A, Gefeller O. IVDK study group; Information Network of Departments of Dermatology. Association between

occupation and contact allergy to the fragrance mix: a multifactorial analysis of national surveillance data. Occup Environ Med 2001;58:392-398

91 Romaguera C, Vilaplana J. Occupational contact dermatitis from ylang-ylang oil. Contact Dermatitis 2000;43:251

92 Van der Valk PG, de Groot AC, Bruynzeel DP, Coenraads PJ, Weijland JW. Allergic contact eczema due to tea tree oil. Ned Tijdschr Geneeskd 1994;138:823-825

93 Jung P, Sesztak-Greinecker G, Wantke F, Götz M, Jarisch R, Hemmer W. Mechanical irritation triggering allergic contact dermatitis from essential oils in a masseur. Contact Dermatitis 2006;54:297-299

94 Keane FM, Smith HR, White IR, Rycroft RJG. Occupational allergic contact dermatitis in two aromatherapists. Contact Dermatitis 2000;43:49-51

95 Selvaag E, Holm J, Thune P. Allergic contact dermatitis in an aromatherapist with multiple sensitizations to essential oils. Contact Dermatitis 1995;33:354-355

96 Boonchai W, Lamtharachai P, Sunthonpalin P. Occupational allergic contact dermatitis from essential oils in aromatherapists. Contact Dermatitis 2007;56:181-182

97 Cockayne SE, Gawkrodger DJ. Occupational contact dermatitis in an aromatherapist. Contact Dermatitis 1997;37:306-307

98 Sánchez-Pérez J, García-Díez A. Occupational allergic contact dermatitis from eugenol, oil of cinnamon and oil of cloves in a physiotherapist. Contact Dermatitis 1999;41:346-347

99 De Mozzi P, Johnston GA. An outbreak of allergic contact dermatitis caused by citral in beauticians working in a health spa. Contact Dermatitis 2014;70:377-379

100 Bleasel N, Tate B, Rademaker M. Allergic contact dermatitis following exposure to essential oils. Australas J Dermatol 2002;43:211-213

101 Schubert HJ. Skin diseases in workers at a perfume factory. Contact Dermatitis 2006;55:81-83

102 Audicana M, Bernaola G. Occupational contact dermatitis from citrus fruits: Lemon essential oils. Contact Dermatitis 1994;31:183-185

103 Mendelsohn HV. Dermatitis from lemon grass oil (*Cymbopogon citratus* or *Andropogon citratus*). Arch Dermatol 1944;50:34-35

104 Calnan CD. Lovage sensitivity. Contact Dermatitis Newsletter 1969;5:99

105 Rudzki E, Rebandel P, Grzywa Z. Occupational dermatitis from cosmetic creams. Contact Dermatitis 1993;29:210

106 Kanerva L, Estlander T, Jolanki R. Occupational allergic contact dermatitis caused by ylang-ylang oil. Contact Dermatitis 1995;33:198-199

107 Morris GE. Dermatoses among food handlers. Ind Med Surg 1954;23:343

108 Peltonen L, Wickstrom G, Vaahtoranta M. Occupational dermatoses in the food industry. Dermatosen 1985;33:166-169

109 Garcia-Bravo B, Pérez Bernal A, Garcia-Hernandez MJ, Camacho F. Occupational contact dermatitis from anethole in food handlers. Contact Dermatitis 1997;37:38

110 Vente C, Fuchs T. Contact dermatitis due to oil of turpentine in a porcelain painter. Contact Dermatitis 1997;37:187

111 Zacher KD, Ippen H. Kontaktekzem durch Bergamottöl. Derm Beruf Umwelt 1984;32:95-97

112 Miller J. Cheilitis from sensitivity to oil of cinnamon present in bubble gum. JAMA 1941;116:131-132

113 Nethercott JR, Holness DL. Occupational dermatitis in food handlers and bakers. J Am Acad Dermatol 1989;21:485-490

114 Gutman SG, Somov BA. Allergic reactions caused by components of perfumery preparations. Vestnik Dermatologii i Venerologii 1968;42:62 (in Russian)

115 Rothe A, Heine A, Rebohle E. Oil from juniper berries as an occupational allergen for the skin and respiratory tract. Berufsdermatosen 1973;21:11-16 (in German)

116 Meynadier JM, Meynadier J, Peyron JL, Peyron L. Formes cliniques des manifestations cutanées d'allergie aux parfums. Ann Dermatol Venereol 1986;113:31-39

117 Uter W, Schmidt E, Geier J, Lessmann H, Schnuch A, Frosch P. Contact allergy to essential oils: current patch test results (2000–2008) from the Information Network of Departments of Dermatology (IVDK). Contact Dermatitis 2010;63:277-283

118 Mitchell JC. Contact hypersensitivity to some perfume materials. Contact Dermatitis 1975;1:197-199

119 Nardelli A, Thijs L, Janssen K, Goossens A. *Rosa centifolia* in a 'non-scented' moisturizing body lotion as a cause of allergic contact dermatitis. Contact Dermatitis 2009;61:306-309

120 de Groot AC, Coenraads PJ, Bruynzeel DP, Jagtman BA, van Ginkel CJW, Noz K, van der Valk PGM et al. Routine patch testing with fragrance chemicals in The Netherlands. Contact Dermatitis 2000;42:184-185

121 Frosch PJ, Johansen JD, Menné T, Pirker C, Rastogi SC, Andersen KE et al. Further important sensitizers in patients sensitive to fragrances. II. Reactivity to essential oils. Contact Dermatitis 2002;47:279-287

122 Treudler R, Richter G, Geier J, Schnuch A, Orfanos CE, Tebbe B. Increase in sensitization to oil of turpentine: recent data from a multicenter study on 45,005 patients from the German-Austrian Information Network of Departments of Dermatology (IVDK). Contact Dermatitis 2000;42:68-73

123 Paulsen E, Andersen KE. Colophonium and Compositae mix as markers of fragrance allergy: Cross-reactivity between fragrance terpenes, colophonium and Compositae plant extracts. Contact Dermatitis 2005;53:285-291

124 Hayakawa R, Matsunaga K, Arima Y. Depigmented contact dermatitis due to incense. Contact Dermatitis 1987;16:272-274

125 de Groot AC, Weijland JW. Systemic contact dermatitis from tea tree oil. Contact Dermatitis 1992;27:279-280

126 Vilaplana J, Romaguera C, Grimalt F. Contact dermatitis from geraniol in Bulgarian rose oil. Contact Dermatitis 1991;24:301

127 Karlberg A-T, Börje A, Johansen JD, Lidén C, Rastogi S, Roberts D, et al. Activation of non-sensitizing or low-sensitizing fragrance substances into potent sensitizers – prehaptens and prohaptens. Contact Dermatitis 2013;69:323-334

128 Rudzki E, Grzywa Z. Dermatitis from propolis. Contact Dermatitis 1983;9:40-45

129 Nakayama H, Hanaoka H, Ohshiro A. Allergen Controlled System (ACS). Tokyo, Japan: Kanehara Shuppan, 1974:42. Data cited in ref. 118

130 Pesonen M, Suomela S, Kuuliala O, Henriks-Eckerman M-L, Aalto-Korte K. Occupational contact dermatitis caused by D-limonene. Contact Dermatitis 2014;71:273-279

131 Juarez A, Goiriz R, Sanchez-Perez J, Garcia-Diez A. Disseminated allergic contact dermatitis after exposure to a topical medication containing geraniol. Dermatitis 2008;19:163

132 Wetter DA, Yiannias JA, Prakash AV, Davis MD, Farmer SA, el-Azhary RA, et al. Results of patch testing to personal care product allergens in a standard series and a supplemental cosmetic series: an analysis of 945 patients from the Mayo Clinic Contact Dermatitis Group, 2000-2007. J Am Acad Dermatol 2010;63:789-798

133 Schäfer T, Böhler E, Ruhdorfer S, Weigl L, Wessner D, Filipiak B, et al. Epidemiology of contact allergy in adults. Allergy 2001;56:1192-1196

134 Pesonen M, Suomela S, Kuuliala O, Henriks-Eckerman M-L, Aalto-Korte K. Occupational contact dermatitis caused by D-limonene. Contact Dermatitis 2014;71:273-279

Chapter 4 CHEMICALS IDENTIFIED IN ESSENTIAL OILS WHICH HAVE CAUSED CONTACT ALLERGY

In this chapter, a list of over 110 chemicals identified in essential oils which have reportedly caused contact allergy or allergic contact dermatitis is presented (Table 4.1). Some are well-known allergens; others have caused allergic reactions infrequently, rarely, or even only once. The list includes some compounds that are known allergens in essential oils per se, for example, terpinolene, ascaridole, α-terpinene, 1,2,4-trihydroxymenthane, α-phellandrene, limonene and myrcene in tea tree oil; α-pinene, δ3-carene hydroperoxides and limonene in turpentine oil; geraniol in geranium and rose oils; cinnamic aldehyde in cassia and cinnamon oils; menthol in peppermint oil; carvone in spearmint oil and eugenol in clove oils. The majority of chemicals, however, have caused contact allergy only by their presence in products other than essential oils. When treating patients with allergic reactions to essential oils, the (possible) composition of the oils should be checked first in the individual oil files (Chapters 5.1-5.93). Following that, the list presented here may give some direction to facilitate the search for the responsible allergen(s) in these patients, especially when lack of concomitant reactions to oil ingredients and the nature of the oil composition fail to give proper guidance.

A number of chemicals reported to be present in essential oils may not have been identified correctly; some have apparently not been found in nature so far and are known to be synthetic compounds. Examples include the esters of phthalic acid (dibutyl phthalate, bis-(2-ethylhexyl)phthalate (dioctyl phthalate), diethyl phthalate, dimethyl phthalate), butylphenyl methylpropional (Lilial®) and hydroxyisohexyl 3-cyclohexene carboxaldehyde, trade name Lyral. Nevertheless, as it is impossible to definitely *exclude* their presence in nature, we have opted to include these chemicals in the list. In addition, these chemicals may have entered the oils as a result of contamination or adulteration (and have been identified correctly).

In Table 4.1 the following data of the chemicals are presented: name of the compound, CAS number, availability as commercial test preparation (for suppliers and test concentrations see Chapter 3, Table 4), important synonyms, sometimes some additional information and one or more relevant literature references. It falls outside the scope of this book to provide extensive additional data on these chemicals such as chemical/IUPAC names, other synonyms, EINECS numbers, chemical formulas, chemical structures, chemical classes, toxicology reviews, more extensive literature references et cetera; this will be the subject of a planned future publication (A.C. de Groot. *Cosmetic Allergy – Cosmetic Allergens*).

LITERATURE

1 Handley J, Burrows D. Allergic contact dermatitis from the synthetic fragrances Lyral and acetyl cedrene in separate underarm deodorant preparations. Contact Dermatitis 1994;31:288-290

2 Frosch PJ, Pilz B, Andersen KE, Burrows D, Camarasa JG, Dooms-Goosens A, et al. Patch testing with fragrances: results of a multicenter study of the European Environmental and Contact Dermatitis Research Group with 48 frequently used constituents of perfumes. Contact Dermatitis 1995;33:333-342

3 Larsen W, Nakayama H, Fischer T, Elsner P, Frosch P, Burrows D, et al. Fragrance contact dermatitis: a worldwide multicenter investigation (Part II). Contact Dermatitis 2001;44:344-346

4 Mann J, McFadden JP, White JML, White IR, Banerjee P. Baseline series fragrance markers fail to predict contact allergy. Contact Dermatitis 2014;70:276-281

5 Uter W, Geier J, Frosch P, Schnuch A. Contact allergy to fragrances: current patch test results (2005–2008) from the Information Network of Departments of Dermatology. Contact Dermatitis 2010;63:254-261

6 Andersen KE. Contact allergy to toothpaste flavors. Contact Dermatitis 1978;4:195-198

7 Rudzki E, Grzywa Z. Sensitizing and irritating properties of star anise oil. Contact Dermatitis 1976;2:305-306

8 Larsen WG, Nakayama H, Lindberg M, Fischer T, Elsner P, Burrows D, et al. Fragrance contact dermatitis: a worldwide multicenter investigation (Part I). Am J Contact Dermat 1996;7:77-83

9 Hausen BM. Evaluation of the main contact allergens in oxidized tea tree oil. Dermatitis 2004;15:213-214

10 Knight TE, Hausen BM. Melaleuca oil (tea-tree oil) dermatitis. J Am Acad Dermatol 1994;30:423-427

11 Christoffers WA, Blömeke B, Coenraads P-J, Schuttelaar M-LA. The optimal patch test concentration for ascaridole as a sensitizing component of tea tree oil. Contact Dermatitis 2014;71:129-137

12 Pirker C, Hausen BM, Uter W, Hillen U, Brasch J, Bayerl C, et al. Sensitization to tea tree oil in Germany and Austria. A multicenter study of the German Contact Dermatitis group. J Dtsch Dermatol Ges 2003;1:629-634

13 Francalanci S, Sertoli A, Giorgini S, Pigatto P, Santucci B, Valsecchi R. Multicentre study of allergic contact cheilitis from toothpastes. Contact Dermatitis 2000; 43:216-222

14 Balato N, Leimbo G, Nappa D, Ayala F. Allergic cheilitis to azulene. Contact Dermatitis 1985;13:39-40

15 Hausen BM. Contact allergy to balsam of Peru. II. Patch test results in 102 patients with selected balsam of Peru constituents. Am J Contact Derm 2001;12:93-102

Table 4.1 Chemicals identified in essential oils which have caused contact allergy/allergic contact dermatitis

Chemical	CAS	CTP	Synonyms and additional information	Refs.
Acetyl cedrene	32388-55-9		Trade name: Vertofix; not found in nature so far	1,2
Ambrettolide	7779-50-2		Synonym: ω-6-hexadecenlactone	3
(E)-Amylcinnamyl alcohol		+	The commercial test preparation is α-amylcinnamic alcohol	4,5
Anethole	104-46-1	+		6,7
Anisyl alcohol	1331-81-3	+		4,5,8
Aromadendrene	489-39-4			9,10
Ascaridole	512-85-6			9,11,12
Azulene	275-51-4			13,14
Benzaldehyde	100-52-7	+		5
Benzoic acid	65-85-0	+		15,16
Benzyl acetate	140-11-4			17
Benzyl alcohol	100-51-6	+		4,5,15,18
Benzyl benzoate	120-51-4	+		4,5,15
Benzyl cinnamate	103-41-3	+		4,5,15
Benzyl propionate	122-63-4			96
Benzyl salicylate	118-58-1	+		4,8,15
Bisabolol	515-69-5			19,20
Bis-(2-ethylhexyl) phthalate	117-81-7	+	Not found in nature so far	104
Butylphenyl methylpropional	80-54-6	+	Trade name: Lilial; not found in nature so far	4
Camphor	76-22-2	+		21,22,23
δ3-Carene	13466-78-9		The allergen is oxidized δ3-carene	24,25,26
Carvacrol	499-75-2			27
Carvone	99-49-0	+		6,9,28,29
β-Caryophyllene	87-44-5		Auto-oxidation increases the sensitizing potency (30,33)	31,32,34
Caryophyllene oxide	1139-30-6			31
1,8-Cineole	470-82-6			35
Cinnamic acid	621-82-9			15
Cinnamic aldehyde	104-55-2	+	Synonyms: cinnamaldehyde, cinnamal; present in the fragrance mix I	4,5,36,37
Cinnamyl alcohol	104-54-1	+	Present in the fragrance mix I	5,15,36
Cinnamyl benzoate	5320-75-2			38
Cinnamyl cinnamate	122-69-0			15
Citral	5392-40-5	+	Combination of neral and geranial; present in the fragrance mix II	39,40,41 42,43
Citronellal	106-23-0	+		44
Citronellol	106-22-9	+	Present in the fragrance mix II; auto-oxidation leads to more sensitizing hydroperoxides (45)	4,36,42
Coniferyl alcohol	458-35-5			15
Coniferyl benzoate	4159-29-9			15
Costunolide	553-21-9		Present in the sesquiterpene lactone mix	46
Coumarin	91-64-5	+	Present in the fragrance mix II	4,36,42
Cuminaldehyde	122-03-2			47
p-Cymene	99-87-6			10
β-Damascone			Contact allergy was to a mixture of α- and β-damascone	32
Dehydrocostus lactone	477-43-0	+	Present in the sesquiterpene lactone mix	48
Dibutyl phthalate	84-74-2	+	Not found in nature so far	101
Diethyl phthalate	84-66-2	+	Not found in nature so far	102,103
Dihydrocoumarin	119-84-6			49
Dimethyl phthalate	131-11-3	+	Not found in nature so far	103
Eugenol	97-53-0	+	Present in the fragrance mix I	4,5,15,36
Falcarinol	21852-80-2			50,51
Farnesol	4602-84-0	+	Present in the fragrance mix II	36,37,42
Geranial	141-27-5		Present in citral (with neral)	40,41
Geraniol	106-24-1	+	Present in the fragrance mix I; auto-oxidation increases the sensitizing potency (30,52) and leads to the formation of neral and geranial (40)	5,36,37 40,41
Geranyl acetate	105-87-3			44
Guaiazulene	489-84-9			14,53
Heptanal	111-71-7			54
Hexadecanolide	109-29-5			3,55
(Z)-3-Hexenyl salicylate	65405-77-8			56

Table 4.1 Chemicals identified in essential oils which have caused contact allergy/allergic contact dermatitis (*continued*)

Chemical	CAS	CTP	Synonyms and additional information	Refs.
α-Hexylcinnamaldehyde	101-86-0	+	Not found in nature so far; present in the fragrance mix II	36,37,42
Hydroxycitronellal	107-75-5	+	Not found in nature so far; present in the fragrance mix I	5,36,37
Hydroxycitronellol	107-74-4		Not found in nature so far	57
Hydroxyisohexyl 3-cyclohexene carboxaldehyde		+	Trade name: Lyral®; not found in nature so far; present in the fragrance mix II	4,36,97,98, 99
α-Ionone	127-41-3		Patients were tested with mixed ionone isomers (47,58)	47,58
β-Ionone	79-77-6		Patients were tested with mixed ionone isomers (47,58)	47,58
Isoeugenol	97-54-1	+	Present in the fragrance mix I	5,36,37
Isoeugenyl acetate	93-29-8			59
Isopulegol	89-79-2			17
Isosafrole	120-58-1			96
Isothymol	4427-56-9			60
Limonene	138-86-3	+	Auto-oxidation increases the sensitizing potency of D-limonene (30) by forming limonene hydroperoxides (63)	37,61,62,64,65
Limonene oxide	1195-92-2			66
Linalool	78-70-6	+	Auto-oxidation increases the sensitizing potency (30) by forming hydroperoxides	4,67,68
Linalyl acetate	115-95-7		Auto-oxidation increases the sensitizing potency (30)	69,70,87
Menthol	89-78-1	+		71,72,73
o-Methoxycinnamaldehyde (2-)	1504-74-1			38
7-Methoxycoumarin	531-59-9		Synonym: herniarin	74
8-Methoxypsoralen	298-81-7			75
Methyl p-anisate	121-98-2			76
Methyl anthranilate	134-20-3	+		77,78
Methyl cinnamate	103-26-4			15,79
Methyl dihydrojasmonate	24851-98-7		Synonym: hedione	32
Methyl eugenol	93-15-2			57
Methylionone	1335-46-2	+	The commercial test preparation is γ-methylionone (α-isomethyl ionone); the patient reacted to 'methylionone γ', containing 65-75% α-methylionone (58)	58
Methyl isoeugenol	93-16-3			57
Methyl salicylate	119-36-8	+		80,81
Musk ambrette	123-69-3	+	Not found in nature so far	82,83,84
Myrcene	123-35-3			9,12,31
Neral	106-26-3		Present in citral (with geranial)	40,41
Nerol	106-25-2			57
Nerolidol	7212-44-4			15
Nonanal	124-19-6			54
Nonanol	28473-21-4			54
Nopyl acetate	128-51-8			47
Octanol	29063-28-3			54
α-Phellandrene	99-83-2			9,12,85
2-Phenethyl alcohol	60-12-8			17,47
Phenylacetaldehyde	122-78-1			76,86
α-Pinene	80-56-8	+	Oxidation increases the sensitizing potential	26,85,87
β-Pinene	127-91-3			26,44,82
Piperitone	89-81-6			100
Piperonal	120-57-0		Synonym: heliotropine	17,32
Propylidene phthalide	17369-59-4			76
Pulegone	89-82-7			100
Retinyl (vitamin A) acetate	127-47-9		Not found in nature so far	105
Sabinene	3387-41-5			9
Safrole	94-59-7			7
Salicylaldehyde	90-02-8	+		5,88
Santalol				89-92
α-Terpinene	99-86-5		Auto-oxidation increases the sensitizing potency (30,95)	9,12
Terpinen-4-ol	562-74-3			9,93
α-Terpineol	98-55-5			82,87,94

Table 4.1 Chemicals identified in essential oils which have caused contact allergy/allergic contact dermatitis (*continued*)

Chemical	CAS	CTP	Synonyms and additional information	Refs.
Terpinolene	586-62-9			9,12,25
Terpinyl acetate	8007-35-0			17
Thymol	89-83-8	+		23,36,95
1,2,4-Trihydroxymenthane				9,12
Vanillin	121-33-5	+		5
Viridiflorene	21747-46-6		Synonym: ledene	9,12

CTP: Commercial test preparation available (Chapter 3, Table 4)

16 Jacob SE, Stechschulte S. Eyelid dermatitis associated with balsam of Peru constituents: benzoic acid and benzyl alcohol. Contact Dermatitis 2008;58:111-112

17 Larsen WG. Perfume dermatitis. A study of 20 patients. Arch Dermatol 1977;113:623-626

18 Curry EJ, Warshaw EM. Benzyl alcohol allergy: importance of patch testing with personal products. Dermatitis 2005;16:203-208

19 Jacob SE, Matiz C, Herro EM. Compositae-associated allergic contact dermatitis from bisabolol. Dermatitis 2011;22:102-105

20 Russell K, Jacob SE. Bisabolol. Dermatitis 2010;21:57-58

21 Stevenson OE, Finch TM. Allergic contact dermatitis from rectified camphor oil in Earex ear drops. Contact Dermatitis 2003;49:51

22 Noiles K, Pratt M. Contact dermatitis to Vicks VapoRub. Dermatitis 2010;21:167-169

23 Nardelli A, D'Hooghe E, Drieghe J, Dooms M, Goossens A. Allergic contact dermatitis from fragrance components in specific topical pharmaceutical products in Belgium. Contact Dermatitis 2009;60:303-313

24 Hellerstrom S, Thyresson N, Blohm SG, Widmark G. On the nature of the eczematogenic component of oxidized delta 3-carene. J Invest Dermatol 1955;24:217-224

25 Castelain PY, Camoin JP, Jouglard J. Contact dermatitis to terpene derivatives in a machine cleaner. Contact Dermatitis 1980;6:358-360

26 Romaguera C, Alomar A, Conde-Salazar L, Camarasa JMG, Grimalt F, Martin Pascual A, et al. Turpentine sensitization. Contact Dermatitis 1986;14:197

27 Andersen A. Final report on the safety assessment of sodium *p*-chloro-*m*-cresol, *p*-chloro-*m*-cresol, chlorothymol, mixed cresols, *m*-cresol, *o*-cresol, *p*-cresol, isopropyl cresols, thymol, *o*-cymen-5-ol, and carvacrol. Int J Toxicol 2006;25 (Suppl. 1):29-127

28 Quertermous J, Fowler JF Jr. Allergic contact dermatitis from carvone in hair conditioners. Dermatitis 2010;21:116-117

29 Paulsen E, Andersen KE, Carsen L, Egsgaard H. Carvone: an overlooked contact allergen cross-reacting with sesquiterpene lactones? Contact Dermatitis 1993;29:138-143

30 Karlberg A-T, Börje A, Johansen JD, Lidén C, Rastogi S, Roberts D, et al. Activation of non-sensitizing or low-sensitizing fragrance substances into potent sensitizers – prehaptens and prohaptens. Contact Dermatitis 2013;69:323-334

31 Matura M, Sköld M, Börje A, Andersen KE, Bruze M, Frosch P, et al. Selected oxidized fragrance terpenes are common contact allergens. Contact Dermatitis 2005;52:320-328

32 Frosch PJ, Johansen JD, Menné T, Pirker C, Rastogi SC, Andersen KE, et al. Further important sensitizers in patients sensitive to fragrances. II. Reactivity to essential oils. Contact Dermatitis 2002;47:279-287

33 Sköld M, Karlberg AT, Matura M, Börje A. The fragrance chemical betacaryophyllene—air oxidation and skin sensitization. Food Chem Toxicol 2006;44:538-545

34 Paulsen E, Andersen KE. Colophonium and Compositae mix as markers of fragrance allergy: Cross-reactivity between fragrance terpenes, colophonium and Compositae plant extracts. Contact Dermatitis 2005;53:285-291

35 de Groot AC, Weyland JW. Systemic contact dermatitis from tea tree oil. Contact Dermatitis 1992;27:279-280

36 Nardelli A, Carbonez A, Drieghe J, Goossens A. Results of patch testing with fragrance mix 1, fragrance mix 2, and their ingredients, and *Myroxylon pereirae* and colophonium, over a 21-year period. Contact Dermatitis 2013;68:307-313

37 Heisterberg MV, Menné T, Johansen JD. Contact allergy to the 26 specific fragrance ingredients to be declared on cosmetic products in accordance with the EU cosmetics directive. Contact Dermatitis 2011;65:266-275

38 Malten KE. Four bakers showing positive patch-tests to a number of fragrance materials, which can also be used as flavors. Acta Dermatovenereologica 1979;59 (Suppl.85):117-121

39 De Mozzi P, Johnston GA. An outbreak of allergic contact dermatitis caused by citral in beauticians working in a health spa. Contact Dermatitis 2014;70:377-379

40 Hagvall L, Christensson JB. Cross-reactivity between citral and geraniol – can it be attributed to oxidized geraniol? Contact Dermatitis 2014;71:280-288

41 Hagvall L, Karlberg A-T, Brared Christensson J. Contact allergy to air-exposed geraniol: clinical observations and report of 14 cases. Contact Dermatitis 2012;67:20-27

42 Krautheim A, Uter W, Frosch P, Schnuch A, Geier J. Patch testing with fragrance mix II: results of the IVDK 2005–2008. Contact Dermatitis 2010; 63:262-269

43 Heydorn S, Menné T, Andersen KE, Bruze M, Svedman C, White IR, Basketter DA. Citral, a fragrance allergen and irritant. Contact Dermatitis 2003;49:32-36

44 Keil H. Contact dermatitis due to oil of citronellal. J Invest Dermatol 1947;8:327-334

45 Rudbäck J, Hagvall L, Börje A, Nilsson U, Karlberg A-T. Characterization of skin sensitizers from autoxidized citronellol – impact of the terpene structure on the autoxidation process. Contact Dermatitis 2014;70:329-339

46 Le Coz C-J, Lepoittevin J-P. Occupational erythema-multiforme-like dermatitis from sensitization to costus resinoid, followed by flare-up and systemic contact dermatitis from β-cyclocostunolide in a chemistry student. Contact Dermatitis 2001;44:310-311

47 de Groot AC, Liem DH, Nater JP, van Ketel WG. Patch tests with fragrance materials and preservatives. Contact Dermatitis 1985;12: 87-92

48 Paulsen E, Sogaard J, Andersen KE. Occupational dermatitis in Danish gardeners and greenhouse workers (III). Compositae-related skin symptoms. Contact Dermatitis 1998;38:140-146

49 Wilkinson JD, Andersen KE, Camarasa JG, Ducombs G, Frosch P, Lahti A, et al. Preliminary results of effectiveness of two forms of fragrance mix as screening agents for fragrance sensitivity. In: Frosch PJ, Dooms-Goossens A, Lachapelle J-M, Rycroft RJG, ScheperRJ, Eds. Current topics in contact dermatitis. Heidelberg: Springer Verlag, 1989:127-131

50 Machado S, Silva E, Massa A. Occupational allergic contact dermatitis from falcarinol. Contact Dermatitis 2002;47:113-114

51 Hansen L, Hammershøy O, Boll PM. Allergic contact dermatitis from falcarinol isolated from *Schefflera arboricola*. Contact Dermatitis 1986;14:91-93

52 Hagvall L, Backtorp C, Svensson S, Nyman G, Börje A, Karlberg AT. Fragrance compound geraniol forms contact allergens on air exposure. Identification and quantification of oxidation products and effect on skin sensitization. Chem Res Toxicol 2007;20: 807-814

53 Angelini G, Vena GA. Allergic contact cheilitis to guaiazulene. Contact Dermatitis 1984;10:311

54 Larsen WG. Cosmetic dermatitis due to a perfume. Contact Dermatitis 1975;1:142-145

55 An S, Lee AY, Lee CH, Kim D-W, Hahm JH, Kim K-J, et al. Fragrance contact dermatitis in Korea: a joint study. Contact Dermatitis 2005;53:320-323

56 Shaw DW. Allergic contact dermatitis from octisalate and *cis*-3-hexenyl salicylate. Dermatitis 2006;17:152-155

57 Larsen W, Nakayama H, Fischer T, Elsner P, Frosch P, Burrows D, et al. Fragrance contact dermatitis – a worldwide multicenter investigation (Part III). Contact Dermatitis 2002;46:141-144

58 Bernaola G, Escayol P, Fernandez E, Fernandez de Corrés L. Contact dermatitis from methylionone fragrance. Contact Dermatitis 1989;20:71-72

59 White IR, Johansen JD, Arnau EG, Lepoittevin JP, Rastogi S, Bruze M, et al. Isoeugenol is an important contact allergen: can it be safely replaced with isoeugenyl acetate? Contact Dermatitis 1999; 41:272-275

60 Meynadier JM, Meynadier J, Peyron JL, Peyron L. Formes cliniques des manifestations cutanées d'allergie aux parfums. Ann Dermatol Venereol 1986;113:31-39

61 Bråred Christensson J, Andersen KE, Bruze M, Johansen JD, Garcia-Bravo B, et al. An international multicentre study on the allergenic activity of air-oxidized R-limonene. Contact Dermatitis 2013;68:214-223

62 Christensson JB, Andersen KE, Bruze M, Johansen JD, Garcia-Bravo B, Gimenez Arnau A, et al. Positive patch test reactions to oxidized limonene: exposure and relevance. Contact Dermatitis 2014;71:264-272

63 Nilsson U, Bergh M, Shao LP, Karlberg A-T. Analysis of contact allergenic compounds in oxidized *d*-limonene. Chromatographia 1996;42:199-205

64 Matura M, Goossens A, Bordalo O, Garcia-Bravo B, Magnusson K, Wrangsjö K, Karlberg A-T. Oxidized citrus oil (R-limonene): a frequent sensitizer in Europe. J Am Acad Dermatol 2002;47:709-714

65 Pesonen M, Suomela S, Kuuliala O, Henriks-Eckerman M-L, Aalto-Korte K. Occupational contact dermatitis caused by D-limonene. Contact Dermatitis 2014;71:273-279

66 Matura M, Goossens A, Bordalo O, Garcia-Bravo B, Magnusson K, Wrangsjö K, et al. Patch testing with oxidized R-(+)-limonene and its hydroperoxide fraction. Contact Dermatitis 2003;49:15-21

67 Brared Christensson J, Andersen KE, Bruze M, Johansen JD, Garcia-Bravo B, Gimenez Arnau A, et al. Air-oxidized linalool – a frequent cause of fragrance contact allergy. Contact Dermatitis 2012;67:247-259

68 Björkman YA, Hagvall L, Siwmark C, Niklasson B, Karlberg A-T, Bråred Christensson J. Air-oxidized linalool elicits eczema in allergic patients – a repeated open application test study. Contact Dermatitis 2014;70:129-138

69 Sköld M, Hagvall L, Karlberg A-T. Autoxidation of linalyl acetate, the main component of lavender oil, creates potent contact allergens. Contact Dermatitis 2008;58:9-14

70 de Groot AC, Bruynzeel DP, Bos JD, van der Meeren HLM, van Joost T, Jagtman BA, Weyland JW. The allergens In cosmetics. Arch Dermatol 1988;124:1525-1529

71 Morton CA, Garioch J, Todd P, Lamey PJ, Forsyth A. Contact sensitivity to menthol and peppermint in patients with intra-oral symptoms. Contact Dermatitis 1995;32:281-284

72 Lewis FM, Shah M, Gawkrodger DJ. Contact sensitivity to food additives can cause oral and perioral symptoms. Contact Dermatitis 1995;33:429-430

73 Nakagawa S, Tagami H, Aiba S. Erythema multiforme-like generalized contact dermatitis to l-menthol contained in anti-inflammatory medical compresses as an ingredient. Contact Dermatitis 2009;61:178-179

74 Paulsen E, Otkjær A, Andersen KE. The coumarin herniarin as a sensitizer in German chamomile [Chamomilla recutita (L.) Rauschert, Compositae]. Contact Dermatitis 2010;62:338-342

75 Korffmacher H, Hartwig R, Matthes U, Dirschka T, Albassamt A, Weindorf N, Altmeyer P. Contact allergy to 8-methoxypsoralen. Contact Dermatitis 1994;30:283-285

76 Malten KE, van Ketel WG, Nater JP, Liem DH. Reactions in selected patients to 22 fragrance materials. Contact Dermatitis 1984;11:1-10

77 Wenk KS, Ehrlich AE. Fragrance series testing in eyelid dermatitis. Dermatitis 2012;23:22-26

78 Greenspoon J, Ahluwalia R, Juma N, Rosen CF. Allergic and photoallergic contact dermatitis: A 10-year experience. Dermatitis 2013;24:29-32

79 Mitchell JC, Calnan CD, Clendenning WE, Cronin E, Hjorth N, Magnusson B, et al. Patch testing with some components of balsam of Peru. Contact Dermatitis 1976;2: 57-58

80 Oiso N, Fukai K, Ishii M. Allergic contact dermatitis due to methyl salicylate in a compress. Contact Dermatitis 2004;51:34-35

81 de Groot AC, Coenraads PJ, Bruynzeel DP, Jagtman BA, van Ginkel CJW, Noz K, van der Valk PGM, et al. Routine patch testing with fragrance chemicals in The Netherlands. Contact Dermatitis 2000;42:184-185

82 Santucci B, Cristaudo A, Cannistraci C, Picardo M. Contact dermatitis to fragrances. Contact Dermatitis 1987;16:93-95

83 Hayakawa R, Matsunaga K, Arima Y. Depigmented contact dermatitis due to incense. Contact Dermatitis 1987;16:272-274

84 Hayakawa R, Matsunaga K, Arima Y. Airborne pigmented contact dermatitis due to musk ambrette in incense. Contact Dermatitis 1987;16:96-98

85 Dooms-Goossens A, Degreef H, Holvoet C, Maertens M. Turpentine-induced hypersensitivity to peppermint oil. Contact Dermatitis 1977;3:304-308

86 Sanchez-Politta S, Campanelli A, Pashe-Koo F, Saurat JH, Piletta P. Allergic contact dermatitis to phenylacetaldehyde: a forgotten allergen? Contact Dermatitis 2007;56:171-172

87 Dharmagunawardena B, Takwale A, Sanders KJ, Cannan S, Roger A, Ilchyshyn A. Gas chromatography: an investigative tool in multiple allergies to essential oils. Contact Dermatitis 2002;47:288-292

88 Aalto-Korte K, Välimaa J, Henriks-Eckerman M-L, Jolanki R. Allergic contact dermatitis from salicyl alcohol and salicylaldehyde in aspen bark (Populus tremula). Contact Dermatitis 2005;52:93-95

89 Goossens A, Merckx L. Allergic contact dermatitis from farnesol in a deodorant. Contact Dermatitis 1997;37:179-180

90 Hayakawa R, Matsunaga K, Arima Y. Depigmented contact dermatitis due to incense. Contact Dermatitis 1987;16:272-274

91 Utsumi M, Sugai T, Shoji A, Watanabe K, Asoh S, Hashimoto Y. Incidence of positive reactions to sandalwood oil and its related fragrance materials in patch tests and a case of contact allergy to natural and synthetic sandalwood oil in a museum worker. Skin Res 1992;34 (Suppl.14):209-213

92 Sugai T. Group study IV – farnesol and lily aldehyde. Environ Dermatol 1994;1:213-214

93 Lippert U, Walter A, Hausen BM, Fuchs Th. Increasing incidence of contact dermatitis to tea tree oil. J Allergy Clin Immunol 2000;105;S43 (abstract 127)

94 Cachao P, Menezes Brandao F, Carmo M, Frazao S, Silva M. Allergy to oil of turpentine in Portugal. Contact Dermatitis 1986;14:205-208

95 Lorenzi S, Placucci F, Vincenzi C, Bardazzi F, Tosti A. Allergic contact dermatitis due to thymol. Contact Dermatitis 1995;33:439-440

96 Nakayama H. Fragrance hypersensitivity and its control. In: Frosch PJ, Dooms-Goossens A, Lachapelle J-M, Rycroft RJG, Scheper RJ, Eds. Current topics in contact dermatitis. Heidelberg: Springer Verlag, 1989:83-91

97 Isaksson M, Inerot A, Lidén C, Lindberg M, Matura M, Möller H, et al. Multicentre patch testing with fragrance mix II and hydroxyisohexyl 3-cyclohexene carboxaldehyde by the Swedish Contact Dermatitis Research Group. Contact Dermatitis 2014;70:187-189

98 Uter W, Geier J, Schnuch A, Gefeller O. Risk factors associated with sensitization to hydroxyisohexyl 3-cyclohexene carboxaldehyde. Contact Dermatitis 2013;69:72-77

99 Heisterberg MV, Laurberg G, Veien NK, Menné T, Avnstorp C, Kaaber K, et al. Prevalence of allergic contact dermatitis caused by hydroxyisohexyl 3-cyclohexene carboxaldehyde has not changed in Denmark. Contact Dermatitis 2012;67:49-51

100 Saito F, Oka K. Allergic contact dermatitis due to peppermint oil. Skin Res 1990;32:161-167

101 Chowdhury MMU, Statham BN. Allergic contact dermatitis from dibutyl phthalate and benzalkonium chloride in Timodine® cream. Contact Dermatitis 2002;46:57

102 Oliwiecki S, Beck MH, Chalmers RJ. Contact dermatitis from spectacle frames and hearing aid containing diethyl phthalate. Contact Dermatitis 1991;25:264-265

103 Capon F, Camble MP, Bernardeau K, Kalis B. Occupational contact dermatitis caused by computer mice. Contact Dermatitis 1996;35:57-58

104 Walker SL, Smith HR, Rycroft RJG, Broome C. Occupational contact dermatitis from headphones containing diethylhexyl phthalate. Contact Dermatitis 2000;42:164

105 Heidenheim M, Jemec GBE. Occupational allergic contact dermatitis from vitamin A acetate. Contact Dermatitis 1995;33:439

Chapter 5   CHEMICAL COMPOSITION OF AND CONTACT ALLERGY TO ESSENTIAL OILS

Chapter 5.0 INTRODUCTION TO THE OIL CHAPTERS

In this chapter, literature data on contact allergy to and chemical composition of essential oils are presented. There are 91 subchapters of individual essential oils and 2 of an absolute (jasmine absolute); these were selected on the sole criterion that they have caused contact allergy. For some oils, only one or two such reports are available. For others, however, there is a considerable amount of literature on contact allergy and allergic contact dermatitis, e.g., in the case of tea tree oil, ylang-ylang oil, lavender oil, peppermint oil, jasmine absolute, geranium oil, rose oil, turpentine oil and sandalwood oil. The relevant literature on contact allergy to/allergic contact dermatitis from any essential oil is discussed in its individual oil chapter. General aspects (e.g., frequency of contact allergy to essential oils, relevance of observed positive patch test reactions, clinical picture of allergic contact dermatitis, products responsible for allergic reactions, co-reactivity to other oils and products, the allergens in essential oils, and diagnostic procedures) are discussed in Chapter 3. A complete alphabetical list of the oils with their common names, botanical names of the source plant species, part of the plant used and ISO standards is given in Chapter 1.

All subchapters, which have the commonly used oil name as title, have a standardized format and present the following sections.

DEFINITION
Here, a definition of the oil is given, including its common name, ISO name (see below), the source species' botanical name and the plant part(s) used for obtaining the essential oil.

INCI NOMENCLATURE
This section provides the definition or description of the oils as provided in CosIng, the European Commission database with information on cosmetic substances and ingredients (http://ec.europa.eu/consumers/cosmetics/cosing/), the names used in the INCI (International Nomenclature Cosmetic Ingredient) systems of the European Union (CosIng) and of the USA (http://online.personalcarecouncil.org/jsp/Home.jsp; paid subscription only), their CAS (Chemical Abstract Service) number(s) (SciFinder, paid subscription only [www.cas.org]; CAS Registry Number is a Registered Trademark of the American Chemical Society; permission has been granted to use the CAS registry numbers) and their EINECS number (European Inventory of Existing Chemical Substances; http://echa.europa.eu/information-on-chemicals/ec-inventory). It should be appreciated that the CosIng database contains a considerable number of mistakes, e.g., the wrong botanical names or CAS numbers.

ISO (INTERNATIONAL ORGANIZATION FOR STANDARDIZATION) STANDARD
This section provides the number of the essential oil's ISO standard (if available), the ISO name of the oil, botanical origin, plant part used and the minimum and maximum acceptable concentrations of the main oil ingredients. We have found some mistakes in these standards, notably the wrong botanical names or aberrant plant parts used. The copyright holder (ISO, Geneva, Switzerland; www.iso.org) has given us permission to reproduce their data.

THE PLANT, THE OIL AND THEIR USES
This section provides general information on the oil's source plant, its origin, cultivation, what the plant is used for and the applications of the essential oil. Many of the plants can be eaten or used as a condiment or spice. Quite often, they are used as medicinal plants, usually in traditional or folk medicine, for a variety of diseases and ailments. A range of pharmacological activities is often attributed to plants or plant parts. The essential oils usually have similar applications, and many are used in foods and drinks, perfumes, cosmetics, topical pharmaceuticals or in other applications such as household products, incense and cigarettes. Virtually all essential oils are used in aromatherapy, which has been defined as 'The therapeutic application of essential oils and aromatic plant extracts in a holistic context, to maintain or improve physical, emotional and mental well-being' (1) or 'the therapeutic use of essential oils' (2). We often give some details of the claimed pharmacological properties and reported therapeutic indications. However, it should be realized that proof of health promoting properties in humans is, according to current standards, virtually always lacking. We wish to emphasize that, by mentioning these data, we do *not* support the various claims made, indications given and uses as described.

CHEMICAL COMPOSITION
This is the main part of the oil chapters, where literature data on the chemical composition of essential oils (both laboratory prepared and commercial) are presented.

Data selection
We have included articles only when oils were analyzed which conform to our definition of 'essential oils' (see Chapter 2): commercial essential oils, laboratory-produced oils obtained by hydrodistillation, steam-distillation, or microwave-assisted hydrodistillation and – in the case of *Citrus* oils – by cold pressing. Oils obtained by simultaneous distillation and extraction, by extraction with solvents or by supercritical fluid extraction (which are often incorrectly termed 'essential oils') were excluded, with the exception of jasmine absolutes. We can, however, not be absolutely certain that some data cited in other publications were not from essential oils sensu stricto.

Data were collected and processed by one of us (ACdG). Most were acquired online through the Library of the University of Groningen, the Netherlands: PubMed (US National Library of Medicine National Institutes of Health), the databases of ScienceDirect (Elsevier),

EBSCOhost, ACS Publications (American Chemical Society), Taylor and Francis Online (including the very important journals *Journal of Essential Oil Research* and *Journal of Essential Oil Bearing Plants*), Wiley Online Library, Springer Link, and the online available issues of *Perfumer and Flavorist* (back to November 2008). Other important sources were the three most recent books of Brian M. Lawrence's famous articles 'Progress in essential oils' as reprinted from *Perfumer and Flavorist* (3-5). They contain a wealth of information, also from journals that are difficult to access (his older books are out of print and are very hard to acquire). The literature lists of relevant articles were searched for other useful publications and sometimes obtained from or ordered by the University library, if not available online, including quite a few 'Progress in Essential Oil' articles published in earlier issues of *Perfumer and Flavorist*. Finally the internet was searched for relevant data (most important search terms: 'name of essential oil' chemical composition), which often yielded additional articles, e.g., from open access journals not cited in PubMed. The depth of the search depended on the amount of available data: we tried to obtain all we could find for oils where few analytical data were available, but have not searched exhaustively in the case of oils where much published analytical data are available. Nevertheless, for most oils, we believe we present the great majority (>80-90%) of (more recent) data published in English language scientific literature.

We have found that many analytical studies present data of sub-optimal or even poor quality. In many articles, experimental data were insufficiently described. Quite often, retention times and indices were not provided. Sometimes, analytical methods were inadequate, e.g., the identification of chemicals was based only on retention times. Usually, only one library of mass spectra (mostly NIST) was employed; this increases the risk of misidentification, which can be avoided (or decreased) by also consulting other libraries such as MassFinder, which contains a larger number of spectra for chemicals found in essential oils. In a number of cases, wrong identifications were made on the basis of elution order. In only a few cases, triplicate testing for reliable quantifications had been performed. Rarely, explanations were given for the presence of chemicals which should not be part of essential oils, for example, solvents used for collection of the oil after distillation in a Clevenger apparatus, or synthetic chemicals.

In some studies, names for chemicals were used that we could not find in any chemical database or with the aid of any search engine, and some were probably or certainly wrong. We have chosen *not* to exclude these chemicals from presentation in the tables. The same goes for compounds, which are synthetic and therefore should not be present in essential oils. Obviously, their 'presence' may have been the result of a misidentification by the investigator. However, they can also have entered the oil as the result of adulteration (up to 2004 large amounts of phthalates were sometimes added), contamination (e.g., phthalates from plastic containers, phthalates from disinfectants containing alcohol denatured with diethyl phthalate) or be impurities from the production method, and have been correctly identified. Finally, the possibility that a chemical which is known as a synthetic compound will be found in nature one day, e.g., in an essential oil, although very unlikely, cannot be excluded. Therefore, we have included these chemicals in the data presentation. In Chapter 6 (but mostly not in the individual oil chapters), they are indicated as 'Chemicals which have not been found in nature up to now' and are shown here in Table 5.0.1.

Presentation of analytical data

The heart of the section 'Chemical composition' is a table (usually Table 2) presenting an alphabetical list of chemicals which have been identified in published analytical studies in the essential oil under discussion. For each chemical, the concentrations in which it has been found is given (relative percentage of peak areas in GC/MS; one decimal for all concentrations >0.09%) and in which publications. The table has several columns. Column A always presents unpublished analytical data from one of

---

Table 5.0.1 Chemicals identified in essential oils which have not been found in nature up to now[a]

Benzodiazepine	Lilial®
5-Benzodiazepine	Lyral®
Bis(2-ethylhexyl)phthalate	Methyl ethyl phthalate
Butylcyclohexyl phthalate	α-Methylionone
Cyclamen aldehyde	6-Methyl-γ-ionone
Dehydrolinalyl acetate	6-Methyl-γ-(*E*)-ionone
Dibutyl phthalate	Musk ambrette
Dichloroacetic acid, 1-adamantylmethyl ester	(*E*)-9-Octadecenoic acid, trimethylsilyl ester
Diethyl phthalate	7,di-*n*-Octyl phthalate
Dihydrolinalyl acetate	Phthalate
Dihydromyrcenyl acetate	Phthalic acid
Dimethyl-*o*-phthalate	Retinal (= Vitamin A aldehyde)
Ftalate	9-*cis*-Retinal
Heptafluorobutanoic acid, 2-(1-adamantyl) ethyl ester	Retinyl acetate
α-Hexylcinnamaldehyde	β-Terpinene
Hydroxycitronellal	Testosteron (not in flora)
Hydroxycitronellol	Tetrahydrolinalyl acetate
Ibuprofen	Vitamin A aldehyde (=Retinal)

[a] we do not pretend this list to be exhaustive

us (Erich Schmidt; see 'Analytical data of commercial oils' below). Then there is a (variable) number of other columns (e.g., B, C, D, E), in which per column all analytical data from one particular study are presented. These are usually the studies with the largest number of ingredients found and/or with the largest number of oil samples analyzed; both commercial oils (which are a small minority) and laboratory-produced oil samples (usually oils hydrodistilled with a Clevenger-type apparatus) qualify.

For a few oils, there is so little published analytical material (that we could find and obtain) that we could show *all* data in this table: cedarwood oil Texas, clove stem oil, elemi oil, guaiacwood oil and neem oil. For the others, showing full analytical data is impossible: for rosemary oils, for example, we show data from over 100 publications.

To be able to show the most important data material (chemicals, concentrations), a wider right column has been created (usually E, F or G), in which data of all publications not shown in the other columns are summarized. There, concentrations for each chemical have a letter (a-z) or a letter plus a number (e.g., y1, z5) attached in superscript. In the legend below the table, these symbols refer to the source study. In most cases, information is also provided on the number of oils investigated, the production mode (commercial, laboratory produced, hydrodistilled or steam-distilled), the plant (cultivated, wild, cultivar, season of harvesting etc.) and the country where the oil(s) was/were produced.

Most data on chemicals found in literature could all be included in this column; in this case the row of the column is not completely filled. For many others, however, there are so much data that they cannot all be shown in the right column. There, we show the *highest* concentrations found in any analytical study. In these cases, the row in the right column is full (sometimes two rows for chemicals with very high concentrations). Thus, it should be realized that such data are not representative of the general composition, as we have selected only the highest concentrations.

Nomenclature
In various studies, chemicals were quite often indicated with different names (common name, chemical name, IUPAC name). In such cases, we have chosen a 'preferred' name to show in the tables. Chapter 6 presents an alphabetical list of all chemicals found in the essential oils with their preferred names; synonyms are included and refer to the preferred name.

Further information provided in the section 'Chemical composition'
This section standard also provides the following information:

- A physical and olfactory description of the oil, its yield (in percentage of biomass) and the main producing countries
- The total number of chemicals identified in the essential oils from various origins and the percentage of these chemicals which have been found in a single study only
- The most important chemicals found in the essential oil (from all literature data) with their maximum concentrations; their details are also shown in a separate table. We realize that these data are heavily influenced by the very heterogeneous results of analyses performed in scientific studies; therefore, we have *also* summarized the ten most important chemicals found in the commercial essential oils analyzed by the second author with minimum and maximum concentrations found
- Chemicals which are known ingredients of the essential oil discussed which were found in unusually high amounts (concentrations are mentioned)
- Chemicals which are unusual or rare constituents of the oil, or have not previously been identified in the essential oil, and which were found in a high concentration in one or two studies
- The ten most important chemicals found in the commercial essential oils analyzed by Erich Schmidt with concentration ranges
- Published information on (possible) chemotypes of the essential oils/source plants

Analytical data of commercial essential oils
In this book, analytical data from one of us (Erich Schmidt) are presented for the first time. Between 1998 and 2014, he analyzed 6,400 samples of the essential oils discussed here. The number of analyses per oil varied from 0 (neem oil, cassia leaf oil) to 422 (sweet orange oil); numbers for each oil are shown in Table 5.0.2. All analytical results (chemicals identified, range of concentrations) can be found in columns A of the main tables of Chapters 5.1-5.93. These results have not previously been published and represent by far the most extensive set of essential oils analyses ever reported in scientific literature. More detailed information can be found in Chapter 1.

Overview of some results presented in Chapters 5.1-5.93 and interpretation of data

Number of chemicals in individual oils
The (approximate) number of chemicals identified in specific essential oils varies considerably (Table 5.0.2). On the low end of the spectrum (turpentine oils and cardamom oil excluded, as we did not perform a literature search on them) are guaiacwood oil (25 chemicals identified), clove stem oil (n=35), cedarwood oil Texas (n=45), elemi oil (n=45), cedarwood oil Virginia (n=55), cedarwood oil Himalaya (n=60), rosewood oil (n=60) and neem oil (n=65). For most of these, this can be explained by the small number of published analytical reports (n=2-5). In the majority, the number of identified components ranges from 100 to 250. Thirteen oils had 250-300 chemical identifications, and ten ranged from 300 to 400. Five oils scored 400-500 identified chemicals: basil oil, laurel leaf oil, lavender oil, rose oil and vetiver oil. At the high end of the spectrum, finally, in geranium oil some 500 chemicals have been identified and 505 in rosemary essential oils.

Chemicals found in one study only
Many chemicals found in an essential oil were identified in one study only; percentages range from 22 to 79. Low percentages (i.e., congruent results of various analyses) are rare: cananga oil 22%, elemi oil 30%, ravensara oil 39%. In the case of cananga oil, the explanation is

Table 5.0.2 Number of commercial oils investigated (A), number of components identified in individual oils (B) and percentage of components found in one study only (C)

Essential oil	A	B	C	Essential oil		A	B	C
Angelica fruit oil	11	120	57%	Lavender oil		374	450	64%
Angelica root oil	31	165	40%	Lemon oil		178	245	NK[b]
Aniseed oil	81	120	47%	Lemongrass oil, East Indian		44	175	52%
Basil oil, sweet	47	435	53%	Lemongrass oil, West Indian		32	245	57%
Bay oil	33	110	51%	Litsea cubeba oil		101	170	61%
Bergamot oil	103	250	49%	Lovage oil		17	95	52%
Black cumin oil	13	340	62%	Mandarin oil		98	255	NK[b]
Black pepper oil	46	305	58%	Marjoram oil (sweet)		49	240	55%
Cajeput oil	51	130	51%	Melissa oil (lemon balm oil)		53	310	49%
Calamus oil	14	275	47%	Myrrh oil		46	110	69%
Cananga oil	25	100	22%	Neem oil		0	65	NK[c]
Cardamom oil	101	28[a]	NK	Neroli oil		79	190	50%
Carrot seed oil	41	315	60%	Niaouli oil		39	150	46%
Cassia bark oil	38	245	62%	Nutmeg oil		51	120	NK[b]
Cassia leaf oil	0	115	69%	Olibanum (frankincense) oil		28	245	58%
Cedarwood oil, Atlas	24	135	53%	Orange oil, bitter		72	215	NK[b]
Cedarwood oil, China	21	135	58%	Orange oil, sweet		422	335	NK[b]
Cedarwood oil, Himalaya	8	60	60%	Palmarosa oil		34	155	49%
Cedarwood oil, Texas	119	45	43%	Patchouli oil		52	210	66%
Cedarwood oil, Virginia	36	55	54%	Peppermint oil		157	335	59%
Chamomile oil, German	85	280	56%	Petitgrain bigarade oil		47	157	45%
Chamomile oil, Roman	22	165	58%	Pine needle oil		112	255	56%
Cinnamon bark oil, Sri Lanka	43	160	46%	Ravensara oil		41	95	39%
Cinnamon leaf oil, Sri Lanka	30	160	52%	Rosemary oil		108	505	53%
Citronella oil, Java	132	165	53%	Rose oil		51	440	56%
Citronella oil, Sri Lanka	29	145	52%	Rosewood oil		36	60	42%
Clary sage oil	34	295	59%	Sage oil, Dalmatian		55	310	47%
Clove bud oil	31	200	67%	Sage oil, Spanish		42	120	41%
Clove leaf oil	201	110	65%	Sandalwood oil, East India		39	125	69%
Clove stem oil	29	35	73%	Sandalwood oil, Australia		23	90	70%
Coriander fruit oil	38	200	48%	Sandalwood, New Caledonia		39	90	73%
Costus root oil	6	135	59%	Silver fir oil		16	110	42%
Cypress oil	33	205	57%	Spearmint oil		71	250	45%
Dwarf pine oil	283	170	57%	Spike lavender oil		24	395	63%
Elemi oil	39	45	30%	Star anise oil		41	160	54%
Eucalyptus citriodora oil	57	220	60%	Tangerine oil		28	125	46%
Eucalyptus globulus oil	185	250	58%	Tea tree oil		97	220	55%
Galbanum resin oil	21	225	75%	Thuja oil		44	120	42%
Geranium oil	97	500	60%	Thyme oil		25	325	50%
Ginger oil	41	295	51%	Thyme oil, Spanish		38	170	47%
Grapefruit oil	122	210	NK[b]	Turpentine oil, Iberian type		47	31[a]	NK
Guaiacwood oil	27	25	52%	Turpentine oil, Chinese type		42	37[a]	NK
Hyssop oil	26	285	52%	Valerian oil		11	330	50%
Jasminum grandiflorum abs.	41	220	68%	Vetiver oil		51	445	55%
Jasminum sambac absolute	11	270	61%	Ylang-ylang oil	extra	51	190	46%
Juniper berry oil	395	295	45%		first	12	145	55%
Laurel leaf oil	23	425	54%		second	42	120	52%
Lavandin abrial oil	110	88	45%		third	22	170	58%
Lavandin grosso oil	148	100	44%	Zdravetz oil		8	275	79%
Lavandin oil	26	180	61%					

A number of commercial oils analyzed by Erich Schmidt in the period 1998-2014
B approximate number of chemicals identified in essential oils of various origins
C approximate percentage of chemicals found in one study only
NK not known

[a] low number, because no literature review has been performed
[b] the percentage of chemicals found in one study only cannot be calculated as a result of the data collection method
[c] only one major analytical investigation published

simple: there were two (different?) studies from the same authors with largely the same results, which means that most chemicals were already found in two studies. The same goes for ravensara oil: two reports from the same authors and also from the same source material.

In >80% of the oils, between 40 and 65% of the chemicals are identified only in one study. In 11 oils, this applied to over 2/3 of all chemicals. In some of these cases, a low number of analytical studies may be the explanation (clove stem oil 3 studies, zdravetz oil 4, sandalwood oil New Caledonia 5, cassia leaf oil 7). However, clove bud oil has 40 analytical publications and patchouli oil more than 50, but still 67% resp. 66% of the chemicals in these oils were found in one study only! In the majority of cases, the singles are found in lab-hydrodistilled oils from wild plants or sometimes from material purchased at a local market. Possible explanations include genetic diversity of the source plant material from adaptation to local circumstances, the use of different plant species for the same oil (e.g., geranium oil, rose oil, myrrh oil, rosewood oil, tea tree oil, turpentine oil), the use of different cultivars, botanical misinterpretation or mistakes made in analysis. In some studies, only certain fractions of the oil were investigated, which yields chemicals which have such low concentrations that they go unnoticed when the entire oil is analyzed (and thus presented in one study only). Factors affecting the composition of essential oils are discussed in Chapter 2.

Atypical compositions

Atypical compositions have been published for most oils. Some are extremely atypical, which might indicate a botanical misidentification, but genetic variations can often not be excluded. Also, many studies report well-known constituents in atypically high (or low) concentrations and some present chemicals which are rare or even previously absent in the essential oils, in (very) high concentrations. These findings can either be correct or wrong, which is usually difficult to assess on the basis of the data presented. Factors affecting the composition of essential oils are discussed in Chapter 2.

Contact allergy/allergic contact dermatitis

This section provides a detailed review of the literature on contact allergy to the essential oil, starting with the subsection General (a surveying summary), followed by (if applicable) Testing in groups of patients, Case reports, Positive patch test reactions and Allergens in essential oils. General aspects of contact allergy to essential oils are presented in Chapter 3.

Literature

In this section, full bibliographical data of all reviewed articles are given. In some references of BM Lawrence's articles 'Progress in Essential Oils' in *Perfumer and Flavorist*, the number of the last page is missing ('last page unknown'). We *did* read those articles, though not in the journal, but in the Essential Oil books (3-5). The articles have been reprinted there, with only their first page number given; we have not been able to find a full bibliographic data source for these articles.

LITERATURE

1 Rhind JP. Essential oils. A handbook for aromatherapy practice, 2nd Edition. London: Singing Dragon, 2012
2 Lawless J. The encyclopedia of essential oils, 2nd Edition. London: Harper Thorsons, 2014
3 Lawrence BM. Essential oils 2001-2004. Carol Stream, USA: Allured Publishing Corporation, 2006
4 Lawrence BM. Essential oils 2005-2007. Carol Stream, USA: Allured Publishing Corporation, 2008
5 Lawrence BM. Essential oils 2008-2011. Carol Stream, USA: Allured Publishing Corporation, 2012

Chapter 5.1   ANGELICA FRUIT OIL

There are two forms of Angelica essential oils, fruit and root oil. In this chapter, the fruit oil is discussed. Angelica root oil is presented in Chapter 5.2.

DEFINITION

Angelica fruit oil is the essential oil obtained from the fruit of the angelica (wild parsnip), *Angelica archangelica* L. (synonyms: *Angelica officinalis* Moench, *Archangelica officinalis* (Moench) Hoffm.).

INCI NOMENCLATURE

Description/definition: Angelica archangelica seed oil is the volatile oil obtained from the seeds of the holy ghost, *Angelica archangelica* L., Umbelliferae (Apiaceae)

INCI name EU & USA: Angelica archangelica seed oil

CAS registry number(s): 8015-64-3; 84775-41-7
EINECS number(s): 283-871-1

The fruits and seeds are often considered as synonymous, but the seeds are contained within the fruits. The seed oil mentioned in INCI is an essential oil, not a fixed oil, and probably synonymous with the fruit oil.

ISO (INTERNATIONAL ORGANIZATION FOR STANDARDIZATION) STANDARD

There is currently no ISO standard for Angelica fruit oil.

THE PLANT, THE OIL, AND THEIR USES

For general information on *Angelica archangelica* L. see under Angelica root oil. Commercial angelica root and fruit (=seed) oils are mostly obtained from cultivated *Angelica archangelica* L. ssp. *angelica* var. *sativa* (8).

CHEMICAL COMPOSITION

Angelica fruit oil is a clear, mobile, pale yellow to amber liquid which has a fresh, pungent, terpenic odor. The yield of essential oil from the fruits of *Angelica archangelica* L. generally varies from 0.8 to 1.2%. The main producing countries of this oil are France, the Netherlands, Hungary and USA.

Literature data (up to November 1, 2014) on the chemical composition of angelica fruit oils and unpublished analytical data from one of us (E.S.) are shown in Table 5.1.1 in alphabetical order. In angelica fruit oils from various origins, over 120 chemicals have been identified. About 57% of these were found in a single reviewed publication only. The major compounds found in angelica fruit oils from different sources are shown in Table 5.1.2. They include (highest concentrations in any study given): β-phellandrene (76.0%), α-pinene (41.4%), β-pinene (7.8%), α-phellandrene (7.4%), myrcene (6.3%) and limonene (2.9%). Well-known ingredients of angelica fruit oils that were present in high concentrations in one study only were sabinene (20.4%) and bicyclogermacrene (10.1%).

Commercial oils

The ten chemicals that had the highest maximum concentrations in 11 commercial angelica fruit essential oil samples (concentration ranges provided) are the following: β-phellandrene (52.1-76.0%), α-pinene (2.3-13.3%), myrcene (0.6-5.0%), α-phellandrene (0.7-3.7%), limonene (2.3-2.9%), (*E*)-β-ocimene (0.5-1.8%), bicyclogermacrene (0.4-1.5%), camphene (0.1-1.2%), cryptone (0.01-1.2%) and α-humulene (0.5-1.1%) (Erich Schmidt, unpublished analytical data).

CONTACT ALLERGY/ALLERGIC CONTACT DERMATITIS

Angelica oil (unspecified)

General

Contact allergy to/allergic contact dermatitis from angelica oil (plant part unspecified) has been reported in two publications only.

Testing in groups of patients

Two hundred dermatitis patients from Poland were tested with angelica oil 2% in petrolatum and two (1%) reacted; relevance data were not provided (12). In a group of 51 patients allergic to *Myroxylon pereirae* resin (balsam of Peru) and/or turpentine and/or wood tar and/or colophony and tested with angelica oil 2% in petrolatum, three (5.9%) had a positive patch test; relevance data were not provided (12).

Case report

A female aromatherapist developed occupational contact dermatitis of the hands from contact allergy to angelica oil; she also reacted to ylang-ylang oil and geraniol, which were also used at her work (11).

Angelica fruit oil

General

Contact allergy to/possible allergic contact dermatitis from angelica fruit oil has been reported in one publication only. A false-positive patch test reaction due to the excited skin syndrome cannot be excluded. In the same case report, α-pinene may have been an allergen in angelica fruit.

Case reports

One case of non-occupational contact dermatitis in an aromatherapist with allergies to multiple essential oils used at work, including angelica seed oil. The patient also reacted to geraniol, linalool, linalyl acetate, α-pinene, the fragrance mix and various other fragrance materials. α-Pinene, linalool, and geraniol were demonstrated by GC-MS in angelica seed oil (13); α-pinene may be an important constituent in angelica fruit oils and has been found in concentrations of up to 13.3% in commercial angelica fruit oils (Table 5.1.1, column A).

Table 5.1.1 Constituents identified in angelica fruit oils

Constituent	CAS	Percentage and range						
		A	B (2)	C (1)	D (7)	E (3)	F (4)	G
(E)-Anethole	4180-23-8					0.7		
Angelic acid	565-63-9	0.01-0.1						
Artemisia ketone	546-49-6	0.07-0.08						
Benzyl isovalerate	103-38-8		0-0.1			0.1		
Bergaptene	484-20-8		tr-0.4					
Bicycloelemene	32531-56-9							0.8-3.2ᶜ
Bicyclogermacrene	24703-35-3	0.4-1.5			0.4	1.5	2.3	3.0-10.1ᶜ
β-Bisabolene	495-61-4		0.6-1.6		0.3ᵃ			
α-Bisabolol	515-69-5		tr-0.5					
epi-α-Bisabolol	78148-59-1		0-0.2					
(6R,7R)-Bisabolone	72441-71-5		tr-0.2					
Borneol	507-70-0			0.1				
Bornyl acetate	76-49-3	0.0-0.1	0.1-0.2	0.06		<0.5		
β-Bourbonene	5208-59-3		0.1-0.4	tr	0.1	<0.5	0.9	
n-Butyl angelate	7785-64-0	0.0-0.07			0.06			
α-Cadinene	24406-05-1		0-0.1				2.1	
γ-Cadinene	39029-41-9		tr-0.2			0.5		
δ-Cadinene	483-76-1	0.08-0.6	tr-1.3	tr	0.2	0.6		
α-Cadinol	481-34-5		0-0.1				0.5	
Camphene	79-92-5	0.1-1.2	0.1-0.5	0.3	0.4	0.1	0.4	
Camphor	76-22-2		tr					
δ3-Carene	13466-78-9	0.06-0.1	tr-0.5	0.2	0.07	<0.5	7.5	
β-Caryophyllene	87-44-5	0.05-0.08	0.1-0.4	0.08	0.07	<0.5		
Caryophyllene oxide	1139-30-6		tr-0.2	tr				
α-Copaene	3856-25-5	0.2-0.8	tr-0.3	0.2	0.8			3.3ᵇ; 0.3-0.6ᶜ
β-Copaene	18252-44-3			0.06			0.3	
Cryptone	500-02-7	0.01-1.2	0.2-0.4	0.1		1.1		2.6ᵇ
β-Cubebene	13744-15-5				0.03			
10-epi-Cubebol	176589-53-0		0-0.1					
Cuminaldehyde	122-03-2						0.1	
ar-Curcumene	644-30-4		0-0.7		0.1			0-0.9ᶜ
γ-Curcumene	28976-68-3		0-3.2					
p-Cymene	99-87-6	0.06-1.0	tr-0.1	0.6	0.5	0.8		
p-Cymen-7-ol	536-60-7		0-tr					
p-Cymen-8-ol	1197-01-9		0-tr					
β-Elemene	33880-83-0		0.1-0.3		0.1	3.4		
γ-Elemene	29873-99-2		0.6-1.4		0.08			0.2-2.4ᶜ
δ-Elemene	20307-84-0		tr-0.2					
cis-β-Elemenone	32663-57-3		0-0.1					
Elemol (α-)	639-99-6		0.2-1.3					
(E)-β-Farnesene	18794-84-8		0.2-0.4	tr				
Farnesol	4602-84-0					<0.5		
(E,Z)-Farnesol	3879-60-5		0.1					
(E,E)-Farnesyl acetate	4128-17-0		0-tr					
Germacrene B	15423-57-1		0.1	0.3	0.7		0.6	0.1-1.3ᶜ
Germacrene D	23986-74-5	0.2-0.7	0.2-3.0		0.7	0.3	2.6	
Heptadecanolide	5637-97-8			tr				
Heptanal	111-71-7					<0.5		
(Z)-2-Hexenyl iso-valerate			0-0.1					
Hexyl isovalerate	10032-13-0		tr-0.1			<0.5		
α-Humulene	6753-98-6	0.5-1.1	1.0-3.4	0.6	1.1	0.8	1.5	
Humulene epoxide II	19888-34-7		0-0.5					
Isoamyl benzyl ether	122-73-6		0-0.6					
Isoamyl isovalerate	659-70-1	0.08-0.3	0.1-0.9			0.3		
Isoamyl 2-methylbuty-rate	27625-35-0	0.08-0.1	tr-0.5					
Isobergaptene	482-48-4		tr-0.2					
Isobutyl isovalerate	589-59-3	0.09-0.7						
Isopropyl angelate	61692-76-0	0.0-0.5						
Isopropyl isovalerate	32665-23-9	0.0-0.6						
Isopropyl 2-methyl- butyrate	66576-71-4	0.02-0.1						
Isopropyl tiglate	1733-25-1	0.0-0.2						
Isoterpinolene	586-63-0			tr			0.3	

Table 5.1.1 Constituents identified in angelica fruit oils (*continued*)

Constituent	CAS	Percentage and range						
		A	B (2)	C (1)	D (7)	E (3)	F (4)	G
Isovaleraldehyde	590-86-3					<0.5		
Limonene	138-86-3	2.3-2.9	2.7	2.3		1.8		0.5-1.7[c]
Linalool	78-70-6			0.3		<0.5		
Linalyl acetate	115-95-7			0.2				
Longicyclene	1137-12-8		0-tr			0.6		
Longifolene (junipene)	475-20-7		0-0.9					
Longipinanol	66141-14-8		0.1-1.6					
α-Longipinene	5989-08-2						0.7	
p-Mentha-1,5-dien-8-ol	1686-20-0		0-0.1					
p-Menth-1-en-7-al	21391-98-0		0-0.1					
p-Menth-2-en-1-ol	619-62-5					<0.5		
cis-p-Menth-2-en-1-ol	29803-82-5		0-0.1					
Menthol	89-78-1		0-0.1					
3-Methyl-3-butenyl isovalerate	54410-94-5		0-0.4					
2-Methylbutyl 2-me-thylbutyrate	2445-78-5					0.07		
2-Methylbutyl valerate	55590-83-5				0.09			
Methylcarvacrol	6379-73-3					<0.5		
Methyl hexadecanoate	112-30-0		tr-0.2					
Methyl octadecanoate	112-61-8		tr-0.2					
Methyl thymol	1076-56-9					<0.5		
trans-Muurola-3,5-diene	262352-88-5		0-0.2					
α-Muurolene	10208-80-7		0.1-1.6	0.07	0.3[a]			
γ-Muurolene	30021-74-0		0.2-2.1	0.5				
α-Muurolol	104245-48-9		tr-1.2					
Myrcene	123-35-3	0.6-5.0	2.0-6.3	2.9	2.9	3.2	3.6	0-2.1[c]
(*E*)-Nerolidol	40716-66-3		0-0.1					
(*E*)-β-Ocimene	3779-61-1	0.5-1.8	tr-0.6	0.3	0.5	1.8		
(*Z*)-β-Ocimene	3338-55-4	0.2-0.6	0-0.1	0.2	0.3	0.6	0.6	
3-Octanone	106-68-3						1.5	
Osthole	484-12-8		tr-0.2					
15-Pentadecanolide	106-02-5	0.07-0.3	0.4-1.1	0.2	0.2	0.6		
α-Phellandrene	99-83-2	0.7-3.7	2.6-7.4	3.7	2.7	1.9		2.3[b]; 0-3.4[c]
β-Phellandrene	555-10-2	52.1-76.0	33.6-63.4	74.7	72.1	65.8	49.3	59.4[b]; 0-55.2[c]
2-Phenylethyl 2-methylbutyrate	24817-51-4		0-0.4					
α-Pinene	80-56-8	2.3-13.3	4.2-12.8	6.6	8.8	6.6	15.1	2.9[b]; 14.4-41.4[c]
β-Pinene	127-91-3	0.3-0.8	0.6-3.7	0.6	0.7	0.6	7.8	0.7-1.7[c]
trans-Pinocarvyl acetate	1686-15-3		0-0.7					
Pseudolimonene	499-97-8	0.1-0.6			0.6		2.6	
Sabinene	3387-41-5	0.2-0.7	1.3-20.4	0.4	0.7	0.2		0.3-2.4[c]
cis-Sabinene hydrate	15537-55-0		0-0.3					
trans-Sabinene hydrate	17699-16-0		0-0.2					
trans-Sabinyl acetate	139757-62-3		tr-0.2					
Selina-3,7(11)-diene	6813-21-4		0-0.3					
(*E*)-Sesquilavandulol	104121-84-8		0-0.5					
β-Sesquiphellandrene	20307-83-9		0-0.7		0.2			0-0.7[c]
Spathulenol	6750-60-3		tr-0.3			2.5	0.5	0.3-1.6[c]
α-Terpinene	99-86-5			tr				
γ-Terpinene	99-85-4	0.01-0.02	tr-0.5	tr		<0.5	0.5	
Terpinen-4-ol	562-74-3		0.2-0.6	0.2			0.7	
α-Terpineol	98-55-5		tr					
Terpinolene	586-62-9	0.06-0.09	0.2-0.4	0.09	0.09	<0.5		
Tetradecanal	124-25-4		0-0.3					
α-Thujene	2867-05-2		tr-0.1	0.2		<0.5		
13-Tridecanolide	1725-04-8	0.0-0.3	0.6-1.3	0.3	0.2			
cis-Verbenol	1845-30-3		0-0.2					
trans-Verbenol	1820-09-3		0.2-0.3					
α-Ylangene	14912-44-8		tr-0.2		0.1			
β-Ylangene	20479-06-5		0-0.3					
α-Zingiberene	495-60-3		0-2.4		0.6			0.4-3.6[c]

Table 5.1.1 Constituents identified in angelica fruit oils (*continued*)

A eleven essential oil samples from France, Hungary and Germany, analyzed between 1997 and 2008; lowest and highest concentrations given (E. Schmidt, unpublished data)
B five lab-hydrodistilled seed oils from wild growing *Angelica archangelica* in three habitats in Lithuania; lowest and highest concentrations given (ref. 2); parts of these data had been previously published in ref. 5.
C one laboratory steam-distilled oil from dried angelica seeds (ref. 1)
D one oil of angelica seed (ref. 7)
E two lab-hydrodistilled fruit oils from wild growing and transplanted French *Angelica archangelica* ssp. *archangelica* var. *sativa*; highest concentrations given (ref. 3)
F one lab-distilled oil from wild growing angelica from Siberia (not absolutely certain that it is the correct species) (ref. 4)
G data from other studies (indicated with superscript letters); highest concentrations found in any study reviewed here are given; when two or more oils were investigated, only the highest concentrations are mentioned, unless indicated otherwise

[a] β-bisabolene and α-muurolene combined; [b] five main constituents of the seed oils from Central Station of Seed Production in Bydgoszcz, Poland (ref. 9); [c] three laboratory steam-distilled oils from *A. archangelica* fruits from Iceland; lowest and highest concentrations given (ref. 10)

tr: trace (in column B: <0.1; in column C: <0.06)

Table 5.1.2 Major constituents of angelica fruit oils

Constituent	CAS	Percentage and range						
		A	B (2)	C (1)	D (7)	E (3)	F (4)	G
β-Phellandrene	555-10-2	52.1-76.0	33.6-63.4	74.7	72.1	65.8	49.3	59.4[b]; 0-55.2[c]
α-Pinene	80-56-8	2.3-13.3	4.2-12.8	6.6	8.8	6.6	15.1	2.9[b]; 14.4-41.4[c]
β-Pinene	127-91-3	0.3-0.8	0.6-3.7	0.6	0.7	0.6	7.8	0.7-1.7[c]
α-Phellandrene	99-83-2	0.7-3.7	2.6-7.4	3.7	2.7	1.9		2.3[b]; 0-3.4[c]
Myrcene	123-35-3	0.6-5.0	2.0-6.3	2.9	2.9	3.2	3.6	0-2.1[c]
Limonene	138-86-3	2.3-2.9	2.7	2.3	1.8			0.5-1.7[c]

LEGEND: SEE UNDER TABLE 5.1.1

LITERATURE

1 Lopes D, Strobl H, Kolodziejczyk P. 14-Meth ylpen-tadecano-15-lactone (Muscolide): A new macrocyclic lactone from the oil of *Angelica archangelica* L. Chem Biodivers 2004;1:1880-1887
2 Nivinskiene O, Butkiene R, Mockute D. The seed (fruit) essential oils of *Angelica archangelica* L. growing wild in Lithuania. J Essent Oil Res 2007;19:477-481
3 Bernard C. Essential oils of three *Angelica* L. species growing in France. Part II: Fruit oils. J Essent Oil Res 2001;13:260-263
4 Shchipitsyna OS, Efremov AA. Composition of ethereal oil isolated from various vegetative parts of *Angelica* from the Siberian region. Russ J Bioorg Chem 2011;37:888-892
5 Nivinskienė O, Butkienė R, Mockutė D. Chemical composition of seed (fruit) essential oils of *Angelica archangelica* L. growing wild in Lithuania. Chemia 2005;16:51-54
6 Lawrence BM. Essential oils 2005-2007. Carol Stream, USA: Allured Publishing Corporation, 2008:64-65
7 Kubeczka K-H, Formacek V. Essential oils analysis by capillary gas chromatography and carbon-13 NMR Spectroscopy, 2nd edition. New York, USA: John Wiley and Sons, 2002:3-9; data cited in ref. 6
8 Lawrence BM. Progress in essential oils. Angelica root and seed oils. Perfum Flavor 1999;24:47-49
9 Wolski T, Najda A, Ludwichuk A. The content and composition of essential oils and fatty acids obtained from the fruits of angelica (*Angelica officinalis* Hoffm.). Herba Polonica 2003;49:151-156
10 Sigurdsson S, Ögmundsdottir HM, Gudbjarnason S. The cytotoxic effect of two chemotypes of essential oils from the fruits of *Angelica archangelica* L. Anticancer Res 2005;25:1877-1880
11 Keane FM, Smith HR, White IR, Rycroft RJG. Occupational allergic contact dermatitis in two aromatherapists. Contact Dermatitis 2000;43:49-51
12 Rudzki E, Grzywa Z, Bruo WS. Sensitivity to 35 essential oils. Contact Dermatitis 1976;2:196-200
13 Dharmagunawardena B, Takwale A, Sanders KJ, Cannan S, Roger A, Ilchyshyn A. Gas chromatography: an investigative tool in multiple allergies to essential oils. Contact Dermatitis 2002;47:288-292

Chapter 5.2 ANGELICA ROOT OIL

There are two forms of Angelica essential oils, fruit and root oil. In this chapter, the root oil is discussed. Angelica fruit oil is presented in Chapter 5.1.

DEFINITION

Angelica root oil is the essential oil obtained from the roots and rhizomes of the angelica (wild parsnip), *Angelica archangelica* L. (synonyms: *Angelica officinalis* Moench, *Archangelica officinalis* (Moench) Hoffm., *Angelica sativa* Mill.).

INCI NOMENCLATURE

Description/definition: Angelica archangelica root oil is the volatile oil obtained from the roots of the plant holy ghost, *Angelica archangelica* L., Umbelliferae (Apiaceae)

INCI name EU & USA: Angelica archangelica root oil

CAS registry number(s): 8015-64-3; 84775-41-7

EINECS number(s): 283-871-1

ISO (INTERNATIONAL ORGANIZATION FOR STANDARDIZATION) STANDARD

There is currently no ISO standard for angelica root oil.

THE PLANT, THE OIL, AND THEIR USES

Angelica archangelica L. is an aromatic, perennial herb that grows up to 2 meters tall. The plant is native to the temperate regions of Asia (Caucasus, Siberia), northern, middle and east Europe, and the Himalayas, and has become widely naturalized in northern temperate regions. The plant is cultivated in Italy, Germany, Finland, Hungary, and several other countries including Korea, India and North America (GRIN Taxonomy for Plants; www.kew.org). Commercial angelica root and fruit oils are mostly obtained from cultivated *Angelica archangelica* L. ssp. *angelica* var. *sativa* (12).

Angelica is used on a large scale in the grocery trade. The roots and fruits of the angelica and their essential oils are popular flavoring agents for confectionary, liqueurs and other beverages such as Bénédictine, Chartreuse, gin and absinthe. *A. archangelica* is believed to possess 'angelic' healing power and has many medicinal applications. This plant has been used in traditional and folk medicine as a remedy for nervous headaches, anxiety, fever, skin rashes, wounds, rheumatism, and toothaches (1,4). *A. archangelica* is employed as an antiseptic, expectorant, emmenagogue (induces or hastens menstruation), and a diuretic. In the Chinese system of medicine, the plant is commonly used for cerebral diseases (1). A leaf extract of the angelica may have a beneficial effect on nocturia in individuals with decreased nocturnal bladder capacity (2).

Essential oils of the fruits (often incorrectly referred to as 'seeds', as the seeds are contained within the fruits) and roots of *Angelica archangelica* are used for healing purposes, as a spice and as a fragrance component in perfumery (3) and cosmetics. These oils are believed to exhibit antispasmodic, stimulant, carminative, diuretic, nervine, tonic and some other activities (6). Angelica root oils are also part of aromatherapy practices (9). The biological activities and medicinal uses of *Angelica archangelica* L. are reviewed in references 1 and 4.

CHEMICAL COMPOSITION

Angelica root oil is, when freshly distilled, a colorless, mobile liquid; with storage, its color changes from yellowish to brownish. The odor is aromatic, spicy with a peppery touch and a typical earthy-rooty character. The yield of essential oil from the roots of *Angelica archangelica* L. generally varies from 0.2 to 0.6% for fresh roots and 0.8 to 1.4% for dried roots. The main producing countries of this oil are Hungary, France, Poland, Lithuania, Germany, Netherlands and India.

Literature data (up to November 1, 2014) on the chemical composition of angelica root oils and unpublished analytical data from one of us (E.S.) are shown in Table 5.2.1 in alphabetical order. In angelica root oils from various origins, over 165 chemicals have been identified. About 40% of these were found in a single reviewed publication only. The major compounds found in angelica root oils from different sources are shown in Table 5.2.2. They include (highest concentrations in any study given) α-pinene (32.2%), β-phellandrene (30.5%), α-phellandrene (22.0%), δ3-carene (17.5%), limonene (16.4%), myrcene (13.1%), sabinene (11.3%) and *p*-cymene (10.6%). Well-known ingredients of angelica root oils that were present in concentrations of 4% or higher in one or two studies were osthole (5.3 and 8.8%), cyclopentadecanolide (14.9%), germacrene D (9.1%), 13-tridecanolide (6.1%), terpinen-4-ol (5.1%), and α-copaen-11-ol (4.5%). Uncommon or rare constituents found in high concentrations in single studies include (Z)-β-farnesene (7.1%).

Commercial oils

The ten chemicals that had the highest maximum concentrations in 31 commercial angelica root essential oil samples (concentration ranges provided) are the following: α-pinene (11.4-27.0%), α-phellandrene (4.5-22.0%), β-phellandrene (9.4-18.7%), δ3-carene (6.4-17.5%), limonene (6.4-15%), sabinene (0.6-11.3%), *p*-cymene (1.3-8.4%), (E)-ocimene (0-5.9%), myrcene (1.4-5.4%) and (Z)-β-ocimene (0.7-4.6%) (Erich Schmidt, unpublished analytical data).

Chemotypes

The composition of Indian angelica oil is quite different from oils produced elsewhere. Its main compounds are α-pinene (>80%) and myrcene (1.2–11.4%). β-Phellandrene (0.2%) and α-phellandrene (<0.1 %) are present in very low concentrations, contrasting to the high concentrations found in oils from other origins (Table 5.2.1); δ3-carene and the ocimenes may be absent (E.S., unpublished analytical data from two Indian samples of angelica oils).

Table 5.2.1 Constituents identified in angelica root oils

Constituent	CAS	Percentage and range					
		A	B (9)	C (11)	D (5)	E (3)	F
(E)-Anethole	4180-23-8						+[f]
Angelicin	523-50-2						+[a]
Aromadendrene	489-39-4			0-0.7			
γ-Atlantone	532-66-1				0.03		
cis-α-Bergamotene	18252-46-5						0.4[b]
Bergapten	484-20-8						+[f]
Bicyclogermacrene	24703-35-3		0.1-0.5	0.1-0.6			2.0[b]
β-Bisabolene	495-61-4	0.1-1.7	0.3-2.3	0.3-1.3		0.2	+[f]
Bisabolone			tr-0.4	0-0.5			
Borneol	507-70-0				0.2	tr	0.9[c]
Bornyl acetate	76-49-3	0.3-3.2	1.8-4.3	2.4-4.2	1.8	0.1	+[f]; 0.4[e]; 0.8[c]
Bornyl isovalerate	76-50-6						+[f]
β-Bourbonene	5208-59-3						0.9[b]
Cadina-1,4-diene	29837-12-5		tr-0.5	tr-0.5			
α-Cadinene	24406-05-1						0.8[b]
γ-Cadinene	39029-41-9		tr-0.6	0-0.4			+[f]; 0.5[b]
δ-Cadinene	483-76-1	0.06-0.9	0.2-0.9	0.2-0.9	0.07	0.3	+[f]; 0.1[e]; 0.2[g]; 0.8[c]; 1.4[b]
α-Cadinol	481-34-5				0.06		0.9[b]
δ-Cadinol	19435-97-3				0.04		
Camphene	79-92-5	0.7-1.8	0.7-1.5	0.2-1.5	1.9	0.4	+[f]; 0.9[b,e]; 1.5[g]
δ2-Carene	554-61-0	0.08-0.4			0.2	0.1	0.1[e]; 0.2[g]
δ3-Carene	13466-78-9	6.4-17.5	10.6-16.9	3.4-16.0	16.3	5.7	+[f]; 0.3[b]; 13.7[c]; 13.8[e]; 16.5[g]
(E)-Carveol	1197-07-5						+[f]
Carvone	99-49-0						+[f]
Carvyl acetate	97-42-7			0-0.1	0.2		
cis-Carvyl acetate	1205-42-1		0-0.3	0-0.2			
trans-Carvyl acetate	1134-95-8		tr-0.1				
β-Caryophyllene	87-44-5	0.06-0.4	tr-1.2	tr-0.3		0.2	0.2[g]
γ-Caryophyllene	118-65-0		tr-0.4	0.1-0.7			+[f]
Caryophyllene oxide	1139-30-6		0.1-0.3	0.2-0.4	0.04		
α-Cedrene	469-61-4						0.7[b]
β-Cedrene	546-28-1		0.1-0.7	0.1-0.4			0.7[b]
cis-Chrysanthenyl acetate	67999-48-8		tr-0.4	0.1-0.6			+[f]
trans-Chrysanthenyl acetate	50764-55-1		tr-0.5	0.1-1.1			+[f]
α-Copaene	3856-25-5	0.1-1.7	0.1-0.7	0.1-0.6	0.6	0.9	+[f]; 0.4[e]; 0.9[c,g]
β-Copaene	18252-44-3				0.2	0.2	0.2[g]; 0.3[b]; 0.6[c]
Copaene alcohol					0.04		
Copaen-4α-ol (β-)	124753-76-0			tr-0.3			
α-Copaen-8-ol (cis-)	58569-25-8		0.1-0.7	0.1-2.5	0.2		
α-Copaen-11-ol	41370-56-3		0.3-1.8	0.3-4.5	tr		+[f]; 1.2[c]
Cryptone	500-02-7				0.2		1.1[e]
α-Cubebene	17699-14-8						+[f]
β-Cubebene	13744-15-5	0.08-0.3	0.1-0.5	0-0.4			
Cuparene	16982-00-6		0.1-0.4				
α-Cuparene				0.1-0.4	0.03		
Curcumene (ar-; α-)	644-30-4						+[f]
Cyclopentadecanolide	106-02-5	0.06-3.6	1.9-3.6	2.0-14.9	0.9	0.5	+[f]; +[d]
m-Cymene	535-77-3						0.1[e];
o-Cymene	527-84-4	0.03-0.1					
p-Cymene	99-87-6	1.3-8.4	1.5-10.6	0.6-3.9	6.4	5.0	+[f]; 2.2[g]; 6.2[c]; 8.8[e]
p-Cymenene	1195-32-0	0.1-0.3	0-1.1	0-4.6			0.1[e]
m-Cymen-8-ol	5208-37-7	0.06-0.3			0.2		+[f]; 0.2[g];
p-Cymen-7-ol	536-60-7		0.1-1.0	0.1-0.7			
p-Cymen-8-ol	1197-01-9		tr-1.0	0.1-1.1	0.7		0.2[e]
β-Elemene	33880-83-0	0.1-1.6	0.1-0.6	0.1-0.6			+[f]; 0.1[e]; 0.2[g]; 3.4[b]
γ-Elemene	29873-99-2		0.4-1.7	0.6-2.2	0.06		
cis-γ-Elemene							0.9[b]
δ-Elemene	20307-84-0		tr-0.2	tr-0.2	0.4		
Elemol	639-99-6		0.2-1.6	0.3-2.9	0.05		
4,8-Epoxyterpinolene	4584-23-0						0.1[e];
α-Eudesmol	473-16-5		0.1-1.0	0.1-1.1			
β-Eudesmol	473-15-4		0.4-2.2	0.4-1.8	0.5		+[f]

Table 5.2.1 Constituents identified in angelica root oils (*continued*)

Constituent	CAS	Percentage and range					
		A	B (9)	C (11)	D (5)	E (3)	F
γ-Eudesmol	1209-71-8		tr-0.2	0.1-1.2			
7-epi-γ-Eudesmol	117066-77-0		0.1-0.8	0.1-0.7			
(*E,E*)-α-Farnesene	502-61-4						2.6[b]
β-Farnesene	502-60-3			0.1-0.7			
(*E*)-β-Farnesene	18794-84-8		0.1-0.8				+[f]
(*Z*)-β-Farnesene	28973-97-9						<0.1[e]; 7.1[b]
α-Fenchene	471-84-1	0.01-0.05			0.05		
Germacrene B	15423-57-1		tr-0.3	0.1-0.6	0.4		0.1[g];
Germacrene D	23986-74-5	0.2-1.4	0.2-0.9	0.2-1.0	0.6	9.1[b];	+[f]; 0.1[e]; 0.3[g]
lactone							
Heptadecanolide	5637-97-8		0.2-0.5	0.2-2.9	0.4	0.06	+[f]; +[d]
Heptanal	111-71-7	0.03-0.06				tr	
Hexanal	66-25-1	0.04-0.08				tr	
Hexadecanolide	109-29-5		0.1	0.1-0.5			+[d]
β-Himachalene	1461-03-6						+[f]; 0.2[g]
α-Humulene	6753-98-6	0.4-1.7	0.9-1.7	0.4-1.4	0.5	1.1	+[f]; 0.4[e]; 0.6[c,g]
β-Humulene	116-04-1						0.3[b]
Humulene epoxide II	19888-34-7		0-0.3			0.1	
α-Humulene oxide	96638-51-6	0.09-0.3		0.2-0.5	0.4		
Humulenol (II)	19888-00-7				0.09		
Imperatorin	482-44-0						+[a]
Isoamyl benzoate	94-46-2		0.1-0.4				
Isoamyl benzyl ether	122-73-6			0.1-0.3			
Isobornyl acetate	125-12-2						1.2[g]
Isobutylbenzene	538-93-2				0.5		
Isoterpinolene	586-63-0					0.3	0.3[e]
Limonene	138-86-3	6.4-15.0	0-9.2	0-8.8	6.6	5.9	+[f]; 8.4[e]; 10.0[c]; 16.4[g]
cis-Limonene epoxide	13837-75-7						0.1[e]
Limonene-1,2-oxide	1195-92-2						+[f]
Linalool	78-70-6				0.1		+[f]; 0.1[e]
Longicyclene	1137-12-8				0.07		
Longipinanol	66141-14-8			0.4-4.2			
p-Mentha-1(7),5-di-en-2-ol	30681-15-3						0.3[e]
p-Mentha-2,8-diene	499-99-0				1.6		
trans-m-Mentha-2,8-diene			0-1.5				
trans-p-Mentha-2,8-diene	5113-87-1	0.2-0.8		0-1.5			
p-Mentha-1,5-dien-8-ol	1686-20-0		tr-0.5	0.1-2.6			
cis-p-Menth-4-ene-1,2-diol							0.2[e]
cis-p-Menth-2-en-1-ol	29803-82-5	0.06-0.1	0-0.3	0-0.5	0.5	0.1	+[f]; 0.1[e]; 0.3[g]
trans-p-Menth-2-en-1-ol	29803-81-4	0.1-0.6	0-0.4	0-0.4	0.09	0.07	+[f]; 0.1[e]; 0.2[g];
Menthol	89-78-1		tr-0.2	0.1-0.3			
Menthone	89-80-5				0.08		
3-Methylcyclopenta-decanone	541-91-3						0.9[c]
13-lactone							
12-Methyl-13-trideca-nolide	57092-32-7		0.1-0.5	0.1-0.9	tr	0.06	+[f]; +[d]
α-Muurolene	10208-80-7		0.2-0.9	0.2-1.1	0.1	0.4	+[f]; 0.1[e]; 0.3[g]; 1.5[c]
γ-Muurolene	30021-74-0		0.1-0.5	0.2-0.4		0.6	
α-Muurolol	104245-48-9		tr-0.6	0.1-0.4	0.08		
τ-Muurolol	19912-62-0				0.03		
Myrcene	123-35-3	1.4-5.4	1.4-3.4	0.4-3.0	5.3	2.8	+[f]; 4.6[b]; 4.8[e]; 5.5[g]; 13.1[c]
Myrtenal	564-94-3		tr-0.6	0.1-0.5			
Myrtenol	515-00-4		tr-0.5	tr-0.4			
Myrtenyl acetate	1079-01-2		0-0.2	0-0.4			
(*E*)-Ocimene	27400-72-2						+[f]
(*Z*)-Ocimene	27400-71-1						+[f]
(*E*)-β-Ocimene	3779-61-1	0-5.9	0.1-0.7	0.1-0.5	tr	2.3	1.7[e]; 3.1[b]; 3.4[c]; 5.1[g];
(*Z*)-β-Ocimene	3338-55-4	0.7-4.6				0.05	0.9	0.8[e]; 1.3[b]; 1.8[g]
n-Octanal	124-13-0	0.04-1.0					
Osthole	484-12-8		1.2-5.3	1.5-8.8			+[f]; 1.0[c]
14-Pentadecanolide							0.4[e]
2-Pentylfuran	3777-69-3						+[f]
Perillyl alcohol	536-59-4				0.1		

Table 5.2.1 Constituents identified in angelica root oils (*continued*)

Constituent	CAS	Percentage and range					
		A	B (9)	C (11)	D (5)	E (3)	F
Phellandral	21391-98-0						+[f]
α-Phellandrene	99-83-2	4.5-22.0	2.1-9.1	0.7-9.1	2.0	19.1	+[f]; 1.7[e]; 1.9[b]; 4.3[c]; 8.7[g]
β-Phellandrene	555-10-2	9.4-18.7	0.1-18.5	tr-15.4	1.3	26.6	+[f]; 10.1[e]; 14.5[c]; 30.5[b]
α-Pinene	80-56-8	11.4-27.0	11.2-20.8	3.8-19.4	32.2	15.7	+[f]; 21.0[c]; 21.3[g]; 23.6[b]; 24.5[e]
β-Pinene	127-91-3	0.2-3.2	0.2-3.2	0.2-3.1	1.5	1.1	+[f]; 1.3[g]; 1.4[b,e]
α-Pinene oxide	1686-14-2						0.1[e]
cis-Piperitol	16721-38-3		0-1.0	0-0.1		tr	+[f]; 0.1[e]; 0.2[g]
trans-Piperitol	16721-39-4				0.3	tr	+[f]; 0.2[g]
Piperitone	89-81-6						+[f]
Psoralen	66-97-7						+[f]; 0.3-1.3[a]
Rosefuran	15186-51-3						0.1[e]
Sabina ketone	513-20-2						+[f]
Sabinene	3387-41-5	0.6-11.3	1.1-7.5	2.1-7.5	0.5	0.7	+[f]; 5.1[g]; 6.3[e]
trans-Sabinene hy-drate	17699-16-0		0.03-0.4			0.06	
Sabinol							+[f]
Sabinyl acetate	53833-85-5			0.1-0.5	0.2		+[f]; 0.2[e]
trans-Sabinyl acetate	139757-62-3		0-0.4				
β-Selinene	17066-67-0		tr-0.5	tr			
β-Sesquiphellandrene	20307-83-9						+[f]
Spathulenol	6750-60-3		0.1-0.4	0.2-0.6	0.1		0.1[c]
Sylvestrene	1461-27-4						0.2[e]
α-Terpinene	99-86-5	0.08-0.9	0.1-2.7	0.1-2.0	0.3	0.1	+[f]; 0.7[g]
γ-Terpinene	99-85-4	0.4-3.0	0.5-3.0	0.4-2.4	3.1	0.3	+[f]; 0.4[e]; 1.3[g]
Terpinen-4-ol	562-74-3	0.07-1.5	0.3-1.5	0.9-5.1	0.08	0.09	+[f]; 0.5[g]; 1.0[e]
α-Terpineol	98-55-5		tr-0.4	0.1-0.7	0.4	tr	+[f]
Terpinolene	586-62-9	0.3-2.3	0.9-2.3	0.5-1.9	0.9	0.5	+[f]; 0.3[e]; 1.4[g]
(14-)Tetradecanolide	3537-83-5						+[d]
Thuja-2,4(10)-diene	36262-09-6					0.1	0.3[g]; 0.5[e];
α-Thujene	2867-05-2	0.04-0.8	tr-0.3	t-0.2	0.1	0.4	+[f]; 0.8[g]; 1.3[e];
Toluene	108-88-3	0.04-0.1					
Tricyclene	508-32-7	0.01-0.04			0.04		
13-Tridecanolide	1725-04-8	0.08-0.8	1.4-2.9	1.5-6.1	0.6	0.7	+[f]; +[d]; 0.5[c]
15-Tridecanolide							+[a]
(3E,5Z)-1,3,5-Undeca-triene	19883-27-3						0.1[e]
5-Undecen-3-yne	74744-31-3						+[f]
cis-Verbenol	1845-30-3	0.07-0.5	0.1-0.5	0.2-0.8		tr	0.2[g]
trans-Verbenol	1820-09-3	0.1-1.2	0.2-0.6	0.4-1.1		0.2	
Verbenyl acetate	33522-69-9				0.4		
Zingiberene	495-60-3		0.1-0.8	0.1-0.5			+[f]

A thirty-one angelica root essential oil samples from France, Hungary, Poland, Germany and Netherlands analyzed between 1997 and 2013; lowest and highest concentrations given (E. Schmidt, unpublished data)

B thirteen lab-distilled essential oils from the roots of wild Lithuanian angelica from four habitats collected between 1996 and 2002; lowest and highest concentrations given (ref. 9)

C six lab-distilled oils from wild angelica collected in two places in Lithuania, 2 were stored for 2.5 months; lowest and highest concentrations given (ref. 11)

D one lab steam-distilled angelica root oil from angelica harvested in the Auvergne, France (ref. 5)

E one batch of oil obtained by steam-distillation of fresh roots in a commercial Canadian distillation unit (ref. 3)

F data from other studies (indicated with superscript letters); highest concentrations found in any study reviewed here given; when two or more oils were investigated, only the highest concentrations are mentioned, unless indicated otherwise

[a] data cited in ref. 11; [b] one lab steam-distilled oil from wild growing Siberian *Angelica*; the composition of this oil was different from usual angelica root oils, so possibly the plant examined may have been another species from the genus *Angelica* than *Angelica archangelica* (ref. 10); [c] four lab-hydrodistilled oils obtained from the roots of young cultivated *Angelica archangelica* plants in Brazil with distillation times varying from 2 to 24 hours (ref. 7); [d] one commercial angelica root oil from France (ref. 8); [e] one commercial *Angelica archangelica* root oil from China (ref. 13); [f] one oil from dried angelica roots purchased in Germany obtained by steam-distillation followed by liquid-liquid extraction with a mixture of dichloromethane and pentane 1:9; as the amounts were expressed in mg/kg, the presence of the chemicals in the oil is indicated with +[f] (ref. 14); [g] one lab-hydrodistilled oil from the roots of Italian cultivated *A. archangelica* (ref. 16)

tr: trace (columns B and C: <0.1; column D: <0.03; column E: <0.06); + present in the oil investigated, but quantity not stated or in a manner which cannot be compared to the other data

Table 5.2.2 Major constituents of angelica root oils

Constituent	CAS	Percentage and range					
		A	B (9)	C (11)	D (5)	E (3)	F
α-Pinene	80-56-8	11.4-27.0	11.2-20.8	3.8-19.4	32.2	15.7	21.0[c]; 21.3[g]; 23.6[b]; 24.5[e]
β-Phellandrene	555-10-2	9.4-18.7	0.1-18.5	tr-15.4	1.3	26.6	10.1[e]; 14.5[c]; 30.5[b]
α-Phellandrene	99-83-2	4.5-22.0	2.1-9.1	0.7-9.1	2.0	19.1	1.7[e]; 1.9[b]; 4.3[c]; 8.7[g]
δ3-Carene	13466-78-9	6.4-17.5	10.6-16.9	3.4-16.0	16.3	5.7	0.3[b]; 13.7[c]; 13.8[e]; 16.5[g]
Limonene	138-86-3	6.4-15.0	0-9.2	0-8.8	6.6	5.9	8.4[e]; 10.0[c]; 16.4[g]
Myrcene	123-35-3	1.4-5.4	1.4-3.4	0.4-3.0	5.3	2.8	4.6[b]; 4.8[e]; 5.5[g]; 13.1[c]
Sabinene	3387-41-5	0.6-11.3	1.1-7.5	2.1-7.5	0.5	0.7	5.1[g]; 6.3[e]
p-Cymene	99-87-6	1.3-8.4	1.5-10.6	0.6-3.9	6.4	5.0	2.2[g]; 6.2[c]; 8.8[e]
(E)- β-Ocimene	3779-61-1	0-5.9	0.1-0.7	0.1-0.5	tr	2.3	1.7[e]; 3.1[b]; 3.4[c]; 5.1[g];
(Z)-β-Ocimene	3338-55-4	0.7-4.6			0.05	0.9	0.8[e]; 1.3[b]; 1.8[g]
Bornyl acetate	76-49-3	0.3-3.2	1.8-4.3	2.4-4.2	1.8	0.1	0.4[e]; 0.8[c]

LEGEND: SEE UNDER TABLE 5.2.1

CONTACT ALLERGY/ALLERGIC CONTACT DERMATITIS

General
Contact allergy to/allergic contact dermatitis from angelica root oil has been reported in two publications only. Data on angelica oil (plant part used not specified) are presented in Chapter 5.1 Angelica fruit oil.

Testing in groups of patients
A group of 86 patients from Poland previously reacting to the fragrance mix was tested with angelica root oil and two (2.3%) had a positive patch test reaction; relevance data were not provided (16).

Case reports
A patient developed allergic contact dermatitis from the perfume in an eye cream; she was patch tested with all 94 components of the perfume and reacted to angelica root oil (test concentration unknown) and eleven of the other chemicals in the perfume (17).

LITERATURE

1 Bhat ZA, Kumar D, Shah MY. *Angelica archangelica* Linn. is an angel on earth for the treatment of diseases: a review. Int J Nutr Pharm Neurol Dis 2011;1:35-49

2 Sigurdsson S, Geirsson G, Gudmundsdottir H, Egilsdottir PB, Gudbjarnason S. A parallel, randomized, double-blind, placebo-controlled study to investigate the effect of SagaPro on nocturia in men. Scand J Urol 2013;47:26-32

3 Lopes D, Strobl H, Kolodziejczyk P. 14-Methylpentadecano-15-lactone (Muscolide): A new macrocyclic lactone from the oil of *Angelica archangelica* L. Chem Biodivers 2004;1:1880-1887

4 Sarker SD, Nahar L. Natural medicine: the genus *Angelica*. Curr Med Chem 2004;11:1479-500

5 Chalchat J-C, Garry R-P. Essential oil of angelica roots (*Angelica archangelica* L.): Optimization of distillation, location in plant and chemical composition. J Essent Oil Res 1997;9:311-319

6 Nivinskiene O, Butkiene R, Mockute D. The seed (fruit) essential oils of *Angelica archangelica* L. growing wild in Lithuania. J Essent Oil Res 2007;19:477-481

7 Paroul N, Rota L, Frizzo C, Atti dos Santos AC, Moyna P, Gower AE, Atti Serafini L, Cassel E. Chemical composition of the volatiles of *Angelica* root obtained by hydrodistillation and supercritical CO_2 extraction. J Essent Oil Res 2002;14:282-285

8 Schultz K, Kraft P. Characterization of the macrolide fraction of *Angelica* root oil and enantiomeric composition of 12-methyl-13-tridecanolide. J Essent Oil Res 1997;9:509-514

9 Nivinskienė O, Butkienė R, Mockutė D. The chemical composition of the essential oil of *Angelica archangelica* L. roots growing wild in Lithuania. J Essent Oil Res 2005;17:373-377

10 Shchipitsyna OS, Efremov AA. Composition of ethereal oil isolated from various vegetative parts of *Angelica* from the Siberian region. Russ J Bioorg Chem 2011;37:888-892

11 Nivinskienė O, Butkienė R, Mockutė D. Changes in the chemical composition of essential oil of *Angelica archangelica* L. roots during storage. Chemija (Vilnius) 2003;14:52-56

12 Lawrence BM. Progress in essential oils. Angelica root and seed oils. Perfum Flavor 1999;24:47-49

13 Wedge DE, Klun JA, Tabanca N, Demirci B, Ozek T, Can Baser KH, et al. Bioactivity-guided fractionation and GC/MS fingerprinting of *Angelica sinensis* and *Angelica archangelica* root components for antifungal and mosquito deterrent activity. J Agric Food Chem 2009;57:464-470

14 Nykänen I, Nykänen L, Alkio M. Composition of angelica root oils obtained by supercritical CO_2 extraction and steam distillation. J Essent Oil Res 1991;3:229-236

15 Fraternale D, Flamini G, Ricci D. Essential oil composition and antimicrobial activity of *Angelica archangelica* L. (Apiaceae) roots. J Med Food 2014;17:1043–1047

16 Rudzki E, Grzywa Z. Allergy to perfume mixture. Contact Dermatitis 1986;15:115-116

17 Larsen WG. Cosmetic dermatitis due to a perfume. Contact Dermatitis 1975;1:142-145

Chapter 5.3 ANISEED OIL

DEFINITION

Aniseed oil (essential oil of aniseed) is the essential oil obtained from the fruit of the anise, *Pimpinella anisum* L.

INCI NOMENCLATURE

Description/definition: Pimpinella anisum fruit oil is the essential oil obtained from the dried ripe fruits of the anise, *Pimpinella anisum* L., Umbelliferae (Apiaceae)

INCI name EU: Pimpinella anisum fruit oil

INCI name USA: Pimpinella anisum (anise) fruit oil

CAS registry number(s): 8007-70-3; 84775-42-8

EINECS number(s): 283-872-7

ISO (INTERNATIONAL ORGANIZATION FOR STANDARDIZATION) STANDARD

ISO number: 3475

ISO name: Essential oil of aniseed

Botanical origin: *Pimpinella anisum* L.

Parts of plant used: Fruit

ISO values: ISO values (minimum and maximum concentrations) are shown in Table 5.3.1.

Aniseed oil should not be confused with star anise oil, obtained from *Illicium verum* (Chapter 5.81).

THE PLANT, THE OIL, AND THEIR USES

Pimpinella anisum L. is a herbaceous annual plant growing to one meter tall or higher. It probably originates from the eastern Mediterranean region and south-west Asia and is widely cultivated in the Mediterranean rim, Russia, South Africa, and Brazil (1,3, GRIN Taxonomy for Plants). Western cuisines have long used anise to flavor some dishes, drinks, and candies. Anise is an essential ingredient in certain spirits such as arak, absinthe, anisette, pastis, ouzo, sambuca, and raki (4). The fruits (commercially called seeds [aniseeds]) are said to be antiseptic, antispasmodic, carminative, digestive, expectorant, stimulant, muscle relaxant, analgesic, anticonvulsant and stomachic (11). They are used in herbal medicine for the treatment of asthma, coughs and pulmonary afflictions as well as digestive disorders such as wind, bloating, colic, nausea and indigestion. It was also reported that extracts from anise fruits have therapeutic effects on several conditions, such as gynecological and neurological disorders. It has mild estrogenic effects, which explains the use of this plant in folk medicine for increasing milk secretion, dysmenorrhea and menopausal hot flushes in women. In diabetic patients, aniseeds may show hypoglycemic and hypolipidemic effect (1,6,9,11). Aniseed is also used as a breath freshener.

The essential oil obtained from the fruits is used in perfumery, toothpastes, medicinally, as a food flavoring and as an insecticide against head-lice and mites (3). It is also utilized in aromatherapy (37), though some warn about its toxicity (38). The pharmacological activities of *Pimpinella anisum* have been reviewed (6,13). The

European Medicines Agency recently concluded that 'Medicinal use of aniseed and anise oil is not supported by clinical evidence' (32).

CHEMICAL COMPOSITION

Aniseed oil is a clear, more or less mobile liquid or solid crystalline mass, which has a spicy and sweet odor from the high amounts of anethole. The yield of essential oil from the fruits of *Pimpinella anisum* L. generally varies from 1.5 to 6.0%. The main producing countries of this oil are China, Spain, Egypt, France and Italy. Literature data (up to November 2, 2014) on the chemical composition of aniseed oils and unpublished analytical data from one of us (E.S.) are shown in Table 5.3.2 in alphabetical order. In aniseed oils from various origins, over 120 chemicals have been identified. About 47% of these were found in a single reviewed publication only.

The major compounds found in aniseed oils from different sources are shown in Table 5.3.3. Aniseed oil is always dominated by (*E*)-anethole with concentrations up to 96.3%. Other important components (maximum concentrations shown) may be methyl chavicol (20.2%; an extremely high 85.3% is probably erroneous), γ-himachalene (15.2%) and *p*-anisaldehyde (5.4%). Well-known ingredients of aniseed oils that were present in concentrations of 4% or higher in one or two studies were *trans*-pseudoisoeugenyl 2-methylbutyrate (6.4% and 12.7%), limonene (9.8%) and (*Z*)-anethole (7.4%). Uncommon or rare constituents of aniseed oil found in high concentrations in single studies include α-longipinene (10.1%), fenchone (5.0 and 6.2%), 3,4-dimethoxystyrene (5.2%) and cyclosativene (5.2%).

Commercial oils

The ten chemicals that had the highest maximum concentrations in 81 commercial aniseed essential oil samples (concentration ranges provided) are the following: (*E*)-anethole (91.0-98.6%), limonene (0.01-3.5%), methyl chavicol (0.01-3.2%), γ-himachalene (1.4-2.9%), *p*-anisaldehyde (0.1-1.4%), *p*-methoxyphenylacetone (0-1.4%), *trans*-pseudoisoeugenyl 2-methylbutyrate (0.4-1.3%), *p*-anisic acid (0-0.9%), α-zingiberene (0.2-0.5%) and (*Z*)-anethole (0.03-0.3%) (Erich Schmidt, unpublished analytical data).

As (*E*)-anethole dominates aniseed oil, all oxidation products of anethole (anisyl alcohol, anisaldehyde, anisic acid, anethole epoxide, *p*-methoxyphenylacetone, *p*-methoxypropiophenone, anisyl ketone) may be present in aniseed oils. The presence of these compounds can be an indication of the deterioration of the oil (3).

CONTACT ALLERGY/ALLERGIC CONTACT DERMATITIS

General

Contact allergy to/allergic contact dermatitis from aniseed oil has been reported in a few studies, including some patients with occupational allergic contact dermatitis. In a case report, anethole may have been the allergen in aniseed oil.

Table 5.3.1 ISO values (%) for aniseed oil[a]

Compound	CAS	Minimum	Maximum
(E)-Anethole	4180-23-8	87.0	94.0
γ-Himachalene	53111-25-4	1.0	5.0
Methyl chavicol	140-67-0	0.5	3.0
trans-Pseudoisoeugenyl 2-methylbutyrate	58989-20-1	0.3	2.0
p-Anisaldehyde	123-11-5	0.1	1.4
(Z)-Anethole	25679-28-1	0.1	0.4

[a] ISO 3475 Essential oil of aniseed ©ISO 2002; Geneva, Switzerland, www.iso.org

Testing in groups of patients

In a group of 21 patients with dermatitis caused by fragrances and tested with a series of essential oils, one (5%) reacted to 'anise oil' 2%; relevance data were not provided (44).

Case reports and positive patch tests

One patient had psoriasis-like dermatitis from contact allergy to aniseed oil (39). Two patients working in the food industry developed occupational allergic contact dermatitis from aniseed oil (tested 5% olive oil) and its main ingredient anethole (41). One case of dermatitis and stomatitis was caused by contact allergy to aniseed

Table 5.3.2 Constituents identified in aniseed oils

Constituent	CAS	A	B (14)	C (3)	D (9)	E (16)	F
Acetanisole	5451-83-2					0.2	0.9[m]
Alismol	87827-55-2						0.6[b]
α-Amorphane			0-0.2		0.2		0.2[s,w]
(E)-Anethole	4180-23-8	91.0-98.6	81.2-91.5	76.9-93.7	90.2	89.5	87.6[z]; 90.1[y]; 93.9[h]; 94.2[g]; 94.4[v]; 95.3[q]; 96.2[u]; 96.3[j]
(Z)-Anethole	25679-28-1	0.03-0.3	0.1-0.2	0-2.0	0.2	0.1	0.2[w]; 0.5[j]; 0.6[h]; 2.1[b]; 2.3[m]; 7.4[o]
p-Anisaldehyde	123-11-5	0.1-1.4		0-5.4		0.5	0.6[r]; 0.7[f]; 0.9[m]; 1.9[e]; 2.1[a]; 2.9[b]
p-Anisaldehyde di-methyl acetal	2186-92-7						0.8[j] (artifact)
Anisic acid	1335-08-6					0.1	
p-Anisic acid	100-09-4	0.0-0.9					
p-Anisyl acetone	104-20-1						0.2[b]; 0.3[l]; 1.1[a]
Anisyl alcohol	1331-81-3					0.09	
Aromadendrene	489-39-4						0.1[c]
1H-Benzocycloheptene							0.2[z]
cis-α-Bergamotene	18252-46-5						0.1[a]
trans-α-Bergamotene	13474-59-4	0.03-0.2					tr[q]; 0.6[a]
Biisocrotyl	764-13-6						0.09[f]
(Z)-α-Bisabolene	29837-07-8						0.2[c]; 1.8[c]
β-Bisabolene	495-61-4	0.2-0.3	0.1-0.9	0-0.6	0.5		0.1[a]; 0.2[g]; 0.3[h]; 0.4[w]; 0.6[k]; 0.7[b]
β-Bourbonene	5208-59-3			0-0.9			
Butanoic acid, 2 methyl-, 4-methoxy-2-(3-methyloxiranyl) phenyl ester	97180-28-4						0.3[l]
γ-Cadinene	39029-41-9			0-0.3			
δ-Cadinene	483-76-1			0-0.3			tr[q]
ω-Cadinene	17627-21-3						0.07[a]
α-Cadinol	481-34-5						0.08[w]; 0.1[s]; 0.2[a]
α-Calacorene	21391-99-1						tr[q]
Camphene	79-92-5					<0.01	tr[x]; 0.1[m]
Camphor	76-22-2						0.2[o,s,y]
δ3-Carene	13466-78-9					0.01	tr[q]; <0.05[m]; 0.1[a]; 0.3[t]
Carvone	99-49-0						0.7[c]
Carvotanacetone	499-71-8						2.5[z]
β-Caryophyllene	87-44-5	0.0-0.2					<0.01[q]; 0.2[f]; 0.3[b]; 1.3[a]
1,8-Cineole	470-82-6					<0.01	0.02[r]; 0.1[m,p,y]; 0.2[a]
α-Copaene	3856-25-5						<0.01[q]
α-Cuparene							0.1[h,i]
Curcumene	644-30-4	0.1-0.2		0-0.4			<0.01[q]; 0.2[c]; 1.4[b]
Cyclosativene	22469-52-9		0-0.1		0.1		0.1[w]; 5.2[o]
p-Cymene	99-87-6	0.01-0.02				0.1	0.01[q]; <0.1[b]; 0.1[m,p]; 0.2[a]; 0.5[t]
5,6-Diethenyl-1-methylcyclohexene							0.06[f]
Di-α-furylmethane	1197-40-6						0.2[f]

Table 5.3.2 Constituents identified in aniseed oils (*continued*)

Constituent	CAS	Percentage and range					
		A	B (14)	C (3)	D (9)	E (16)	F
Dihydrocarvone	5948-04-9						0.3^z
cis-Dihydrocarvone	3792-53-8						0.1^c
Dill apiole	484-31-1						1.1^z;
3,4-Dimethoxystyrene	6380-23-0						5.2^f
1,2-Dimethylindan	17057-82-8						0.08^f
Elemene	11029-06-4						0.2^f
β-Elemene	33880-83-0		0-0.1		0.1		0.02^q; 0.04^a; 0.06^r; 0.1^w; 0.2^s
δ-Elemene	20307-84-0	0.05-0.2	0-0.5		0.5		0.06^q; 0.1^c; 0.3^k; 0.5^{s,w}; 1.3^b
trans-Epoxypseudo-isoeugenyl 2-methyl-butyrate	125028-84-4			0-2.3			0.1^q; 0.3^r; 2.1^b
α-Ethyl-*p*-anisyl alcohol							0.3^f
Ethyl hexadecanoate	628-97-7						0.04^f
Ethyl oleate	111-62-6						0.9^f
γ-Eudesmol (selinenol)	1209-71-8						0.08^a
Eugenyl acetate	93-28-7						3.9^k
α-Farnesene	502-61-4			0-0.4			
(*E,E*)-α-Farnesene	502-61-4						0.2^a
β-Farnesene	502-60-3			0-0.7			
(*E*)-β-Farnesene	18794-84-8	0.02-0.06					
(*E*)-β-Farnesol	106-28-5						<0.05^m
Fenchone	1195-79-5						5.0^{p,y}; 6.2^x;
α-Fenchyl acetate	111821-74-0						0.1^{o,y}
Foeniculin (*E*-)	78259-41-3						3.3^a
Geijerene	6902-73-4						0.02^q; 0.04^r; 0.4^b
Geraniol	106-24-1						0.06^a; 0.4^b
Germacrene B	15423-57-1						<0.01^q
Germacrene D	23986-74-5			0-1.1			0.8^d
α-Gurjunene	489-40-7						4.0^f
Hexadecanoic acid	57-10-3						0.3^g
α-Himachalene	3853-83-6	0.1-0.3	0.1-0.8	0-0.4	0.9		0.2^h; 0.3^k; 0.5^f; 0.7^w; 1.0^s; 1.7^b
β-Himachalene	1461-03-6	0.05-0.2	0.1-0.5		0.5		0.2^r; 0.3^k; 0.4^w; 0.5^s; 0.8^z; 1.8^b
γ-Himachalene	53111-25-4	1.4-2.9	1.9-7.9	0.4-8.2	8.3		1.8^z; 2.3^h; 3.5^k; 7^w; 12.3^s;15.2^b
α-Humulene	6753-98-6						0.2^a; 1.1^t
3-Hydroxycarbofuran	16655-82-6						0.8^f
Isoeugenol	97-54-1					0.2	
o-Isoeugenol	1076-55-7						1.9^l
Isogeijerene	5975-49-5			0-0.9			<0.01^q; <0.1^b
Isoledene	95910-36-4						3.3^o
Limonene	138-86-3	0.01-3.5					0.3^b; 0.8^c; 1.3^t; 1.4^z; 1.5^a; 9.8^x
Linalool	78-70-6	0.01-0.1	0-0.6		0.2	0.8	0.2^m; 0.3^{f,l}; 0.6^b; 0.9^s; 2.3^a
Linalyl acetate	115-95-7					<0.01	
Linoleic acid	60-33-3						tr^x
α-Longipinene	5989-08-2						0.01^q; 0.08^f; 10.1^o
Menthol	89-78-1						0.1^j
p-Methoxyphenyl-acetone	122-84-9	0.0-1.4					0.1^r
Methyl chavicol	140-67-0	0.01-3.2	0.3-1.2	0.5-2.3	0.8	0.8	3.6^f; 7.3^a; 8.8^t; 20.2^b; 85.3^x
Methyl eugenol	93-15-2		0-0.1			0.6	0.09^z; 0.1^i
1-Methylguanine	938-85-2						0.1^f
(*E*)-Methylisoeugenol	6379-72-2		0.1-0.3		0.2		0.1^s; 0.2^w
Methyl 1-phenylallyl ether	22665-13-0						1.7^f
β-Monopalmitin	23470-00-0						0.2^f
α-Muurolene	10208-80-7		0-0.1		0.2		0.2^{s,w}
γ-Muurolene	30021-74-0						0.5^b; 1.1^d
Myrcene	123-35-3	0.0-0.07				<0.01	tr^y; <0.05^{m,p}; 0.06^a; 0.3^t
(*E*)-Nerolidol	40716-66-3						0.1^a
Nonanal	124-19-6			0-1.4			
(*E*)-β-Ocimene	3779-61-1					<0.01	<0.05^m
(*Z*)-β-Ocimene	3338-55-4					<0.01	tr^y; <0.05^{m,p}
Oleic acid	112-80-1						0.5^f

Table 5.3.2 Constituents identified in aniseed oils (*continued*)

Constituent	CAS	Percentage and range					
		A	B (14)	C (3)	D (9)	E (16)	F
Osmorhizole	3698-23-5						0.8[b]
α-Phellandrene	99-83-2					0.02	tr[y]; <0.05[p]; 0.1[m]; 0.4[a]; 0.6[t]
β-Phellandrene	555-10-2					<0.01	tr[q]; 0.1[m]; 0.2[a]
α-Pinene	80-56-8	0.01-0.1				<0.01	0.01[q]; 0.1[p,s]; 0.2[m]; 0.5[a]; 1.1[t]
β-Pinene	127-91-3	0.01-0.02				<0.01	tr[q]; 0.04[a]; <0.05[m]
Pregeijerene	20082-17-1			0-0.5			0.02[q,r]; 0.1[b]
cis-Pseudoisoeugenyl 2-methylbutyrate							2.3[e]
trans-Pseudoisoeuge-nyl 2-methylbutyrate	58989-20-1	0.4-1.3		0.4-6.4			0.4[e]; 0.7[g,q]; 1.6[r]; 12.7[b]
Sabinene	3387-41-5					<0.01	tr[q,y]; <0.05[m,p]
Safrole	94-59-7						0.6[m,n]; 2.5[o]
β-Sesquiphellandrene	20307-83-9	0.05-0.2	0-0.2	0-0.4	0.1		<0.01[q]; 0.09[s]; 0.1[c,w,z]; 0.3[k]
Spathulenol	6750-60-3						0.08[s]; 0.1[w]
(-)-Spathulenol (β-)	77171-55-2		0-0.2		0.1		<0.1[b]; 0.2[f]
α-Terpinene	99-86-5					<0.01	<0.05[m]; 0.1[a]; 0.2[f]
γ-Terpinene	99-85-4					0.09	0.01[q]; 0.08[a]; 0.1[b,s]; 0.6[t]
Terpinen-4-ol	562-74-3					0.5	<0.05[m]; <0.1[b]; 0.3[a]
α-Terpineol	98-55-5					1.0	0.1[m]; 0.2[a]
Terpinolene	586-62-9					0.04	<0.05[m]; 0.08[a]
1,2,4-Trimethylene-cyclohexane	14296-81-2						0.1[f]
Valencene	4630-07-3						0.2[g]
Viridiflorene (ledene)	21747-46-6						0.1[a]
α-Ylangene	14912-44-8						0.01[q]; 0.2[f]; 0.3[b]
α-Zingiberene	495-60-3	0.2-0.5	0.2-1.3	0-1.1	1.0		0.4[c,s]; 0.8[w]; 1.2[d]; 2.1[b]; 2.9[k]

A eighty-one aniseed essential oil samples from China, Italy, Spain and France analyzed between 1998 and 2013; lowest and highest concentrations given (E. Schmidt, unpublished data)

B fifteen lab-hydrodistilled oils from the aniseeds of 15 accessions of *P. anisum* cultivated in Germany; lowest and highest concentrations given (ref. 14)

C fourteen lab-hydrodistilled aniseed oils from *Pimpinella anisum* plants from 11 European countries (ref. 3)

D six oils from fruits of *P. anisum* of three cultivars grown on two localities in Germany; highest concentrations given (ref. 9)

E twenty-nine oils from seed sources from various parts of Turkey; average concentrations for all oils given (ref. 16)

F data from other studies (indicated with superscript letters); highest concentrations found in any study reviewed here given; when two or more oils were investigated, only the highest concentrations are mentioned, unless indicated otherwise

[a] one lab-distilled oil from fruit of *P. anisum* cultivated in Poland (ref. 10); [b] four lab-distilled (two classical, two microwave-assisted) oils from Turkish *P. anisum* fruit, whole and ground (ref. 15); [c] one lab-hydrodistilled aniseed oil from fruits of plants cultivated in Iran (ref. 26); [d] one lab-hydrodistilled oil from the fruits of plants cultivated in Iran from Hungarian seeds (ref. 2); [e] one steam-distilled aniseed oil from Turkey (ref. 5); [f] one lab-hydrodistilled oil from Turkish aniseeds (ref. 7); [g] one lab-hydrodistilled oil from aniseeds of wild growing *P. anisum* in Turkey (ref. 8); [h] one lab-distilled oil from aniseeds purchased in Egypt (ref. 12); [i] one lab-hydrodistilled oil from Turkish aniseeds (ref. 11); [j] one lab-hydrodistilled oil from Argentinian *P. anisum* fruits (ref. 25); [k] one lab-hydrodistilled oil from fruits in the ripening phase from plants grown in Iran of Hungarian seeds (ref. 24); [l] one lab-hydrodistilled oil of seeds of *P. anisum* of Algerian origin (ref. 1); [m] one aniseed oil (ref. 17); [n] incorrect identification (ref. 20); [o] one commercial aniseed oil; on the basis of the high concentration of (Z)-anethole (7.4%), the authors suggested the oil had been obtained from the whole plant rather than from the fruits (ref. 18); BM Lawrence (ref. 20) suggested that safrole, α-longipinene, cyclosativene and isoledene had been incorrectly identified; [p] one lab hydrodistilled oil from fruits purchased at a local Indian market (ref. 19); [q] one lab-prepared aniseed oil (ref. 22); [r] one lab-distilled oil and a Spanish commercial oil (ref. 23); [s] one lab-hydrodistilled oil from dried fruits of *P. anisum* cultivated in the Czech republic (ref. 27); [t] one lab-hydrodistilled aniseed oil from South Korea (ref. 28); [u] one oil sample from the USA (ref. 29); [v] one lab-distilled aniseed oil from Greece (ref. 35); [w] three lab-prepared oils from two *P. anisum* cultivars grown in Germany (ref. 30); [x] two lab-distilled oils from aniseeds collected in the wild in Morocco and Yemen; both had an extremely atypical composition with methyl chavicol (77-85%) as dominant ingredient; a mistake (botanical misidentification?) seems very likely (ref. 31); [y] one lab-hydrodistilled oil from aniseeds purchased at a local market in India (ref. 33); [z] one lab-distilled aniseed oil from Iran (ref. 36)

tr: trace

Table 5.3.3 Major constituents of aniseed oils

Constituent	CAS	Percentage and range					
		A	B (14)	C (3)	D (9)	E (16)	F
(E)-Anethole	4180-23-8	91.0-98.6	81.2-91.5	76.9-93.7	90.2	89.5	94.4[v]; 95.3[q]; 96.2[u]; 96.3[j]
Methyl chavicol	140-67-0	0.01-3.2	0.3-1.2	0.5-2.3	0.8	0.8	3.6[f]; 7.3[a]; 8.8[t]; 20.2[b]; 85.3[x]
γ-Himachalene	53111-25-4	1.4-2.9	1.9-7.9	0.4-8.2	8.3		1.8[z]; 2.3[h]; 3.5[k]; 7[w]; 12.3[s];15.2[b]
p-Anisaldehyde	123-11-5	0.1-1.4		0-5.4		0.5	0.6[r]; 0.7[f]; 0.9[m]; 1.9[e]; 2.1[a]; 2.9[b]

LEGEND: SEE UNDER TABLE 5.3.2

oil in toothpaste (43). One case of occupational allergic contact dermatitis in a porcelain painter due to oil of anise, oil of turpentine and lavender oil that were mixed with pigments for painting (42). One positive patch test to aniseed oil in a patient working in a cosmetic factory, who had occupational dermatitis from a fragrance mixture he was handling daily (40).

LITERATURE

1 Saibi S, Belhadj M, El-Hadi B. Essential oil composition of *Pimpinella anisum* from Algeria. Anal Chem Lett 2012;2:401-404

2 Yamini Y, Bahramifar N, Sefidkon F, Saharkhiz MJ, Salamifar E. Extraction of essential oil from *Pimpinella anisum* using supercritical carbon dioxide and comparison with hydrodistillation. Nat Prod Res 2008;22:212-218

3 Orav A, Raal A, Arak E. Essential oil composition of *Pimpinella anisum* L. fruits from various European countries. Nat Prod Res 2008;22:227-232

4 Ertan Anli R, Bayram M. Traditional aniseed-flavored spirit drinks. Food Rev Internat 2010;26:246-269

5 Rodrigues VM, Rosa PTV, Marques MOM, Petenate AJ, Meireles MAA. Supercritical extraction of essential oil from aniseed (*Pimpinella anisum* L) using CO$_2$: solubility, kinetics, and composition data. J Agric Food Chem 2003;51:1518-1523

6 Shojaii A, Abdollahi Fard M. Review of pharmacological properties and chemical constituents of *Pimpinella anisum*. International Scholarly Research Network. ISRN Pharmaceutics, Volume 2012, Article ID 510795, 8 pages. doi:10.5402/2012/510795

7 Topal U, Sasaki M, Goto M, Otles S. Chemical compositions and antioxidant properties of essential oils from nine species of Turkish plants obtained by supercritical carbon dioxide extraction and steam distillation. Int J Food Sci Nutr 2008;59:619-634

8 Tabanca N, Demirci B, Ozek T, Kirimer N, Baser KHC, Bedir E, et al. Gas chromatographic–mass spectrometric analysis of essential oils from *Pimpinella* species gathered from Central and Northern Turkey. J Chromatogr A 2006;1117:194-205

9 Ullah H, Honermeier B. Fruit yield, essential oil concentration and composition of three anise cultivars (*Pimpinella anisum* L.) in relation to sowing date, sowing rate and locations. Ind Crops Prod 2013;42:489-499

10 Skalicka-Wozniak K, Walasek M, Ludwiczuk A, Głowniak G. Isolation of terpenoids from *Pimpinella anisum* essential oil by high-performance counter-current chromatography. J Sep Sci 2013;36:2611-2614

11 Özcan MM, Chalchat JC. Chemical composition and antifungal effect of anise (*Pimpinella anisum* L.) fruit oil at ripening stage. Ann Microbiol 2006;56:353-358

12 Nenaah GE, Ibrahim SIA. Chemical composition and the insecticidal activity of certain plants applied as powders and essential oils against two stored-products *Coleopteran* beetles. J Pest Sci 2011;84:393-402

13 Silano V, Delbò M. Assessment Report on *Pimpinella Anisum* L. European Medicines Agency, Evaluation of Medicines for Human Use. Doc. Ref: EMEA/HMPC/137421/2006.

14 Ullah H, Mahmood A, Ijaz M, Tadesse B, Honermeier B. Evaluation of anise (*Pimpinella anisum* L.) accessions with regard to morphological characteristics, fruit yield, oil contents and composition. J Med Plants Res 2013;7:2177-2186

15 Kürkçüoğlu M, Koşar M, Başer KHC. Comparison of microwave-assisted hydrodistillation and hydrodistillation methods for *Pimpinella anisum* L. Available at: http://bildiri.anadolu.edu.tr/papers/bildirimakale/399_417u73.pdf

16 Arslan N, Gurbuz BB, Sarihan EO, Bayrak A, Gumuscu A. Variation in the essential oil content and composition of Turkish anise (*Pimpinella anisum* L.) populations. Turk J Agric Forest 2004;28:173-177. Data cited in ref. 20

17 Ranade G. Profile. Aniseed *Pimpinella anisum* Linn. PAFAI 2007;9:81. Data cited in ref. 20

18 Dawidar AM, Mogib MA, El-Ghorab AH, Mahfouz M, Elsaid FG, Hussein KH. Chemical composition and effect of photo-oxygenation on biological activities of Egyptian commercial anise and fennel essential oils. J Essent Oil Bear Plants 2008;11:124-136

19 Singh G, Kapoor IPS, Singh P, Heluani CS, Catalan CAN. Chemical composition and antioxidant potential of essential oil and oleoresins from anise seeds (*Pimpinella anisum* L.). Internat J Essent Oil Ther 2008;2:122-130. Data cited in ref. 20

20 Lawrence BM. Progress in essential oils. Anise oil. Perfum Flavor 2011;36(October):75-76

21 Lawrence BM. Essential oils 2001-2004. Carol Stream, USA: Allured Publishing Corporation, 2006:207-212

22 Kubeczka K-H. The essential oil composition of *Pimpinella* species. In: KHC Baser and N Kirimer, Eds. Progress in essential oil research. Eskisehir, Turkey: Anadolou University Press, 1998:35-56. Data cited in ref. 21

23 Kubeczka K-H, Formacek V. Essential oils analysis by capillary gas chromatography and carbon-13 NMR Spectroscopy, 2nd edition. New York, USA: John Wiley and Sons, 2002:11-20. Data cited in ref. 21

24 Omidbaigi R, Hadjiakhoondi A, Saharkhiz M. Changes in content and chemical composition of *Pimpinella anisum* oil at various harvest time. J Essent Oil Bear Plants 2003;6:46-50

25 Gende LB, Maggi MD, Fritz R, Eguaras MJ. Antimicrobial activity of *Pimpinella anisum* and *Foeniculum vulgare* essential oils against *Paenibaccilus larvae*. J Essent Oil Res 2009;21:91-93

26 Sharifi R, Kiani H, Farzaneh M, Ahmadzadeh M. Chemical composition of essential oils of Iranian *Pimpinella anisum* L. and *Foeniculum vulgare* Miller and their antifungal activity against postharvest pathogens. J Essent Oil Bear Plants 2008;11:514-522

27 Pavela R. Insecticidal properties of *Pimpinella anisum* essential oils against the *Culex quinquefasciatus* and the non-target organism *Daphnia magna*. J Asia-Pacific Entomol 2014;17:287-293

28 Lee H-S. *p*-Anisaldehyde: Acaricidal component of *Pimpinella anisum* seed oil against the house dust mites *Dermatophagoides farina* and *Dermatophagoides pteronyssinus*. Planta Medica 2004;70:279-281

29 Zheljazkov VD, Astatkie T, O'Brocki B, Jeliazkova E. Essential oil composition and yield of anise from different distillation times. Hort Science 2013;48:1393-1396

30 Ullah H. Fruit yield and quality of anise (*Pimpinella anisum* L.) in relation to agronomic and environmental factors. PhD Thesis, Justus Liebig University Giessen, Germany. Giessen, Germany: VVB Laufersweiler Verlag, 2012

31 Al Maofari A, El Hajjaji S, Debbab A, Zaydoun S, Ouaki B, Charof R, et al. Chemical composition and antibacterial properties of essential oils of *Pimpinella anisum* L. growing in Morocco and Yemen. St Cerc St CICBIA 2013;14:11-16

32 European Medicines Agency, Committee on Herbal Medicinal Products (HMPC). Assessment report on *Pimpinella anisum* L., fructus and *Pimpinella anisum* L., aetheroleum. EMEA/HMPC/321181/2012, November 2013.

33 Singh G, Kapoor IPS, Singh P, de Heluani CS, Catalan CAN. Chemical composition and antioxidant potential of essential oil and oleoresins from anise seeds (*Pimpinella anisum* L.). Int J Essent Oil Ther 2008;2:122-130

34 Naher S, Ghosh A, Aziz S. Comparative studies on physicochemical properties and GC-MS analysis of essential oil of the two varieties of the aniseed (*Pimpinella anisum* Linn.) in Bangladesh. Int J Pharm Phytopharmacol Res 2012;2:92-95 (data not shown)

35 Kimbaris AC, Koliopoulos G, Michaelakis A, Konstantopoulou MA. Bioactivity of *Dianthus caryophyllus*, *Lepidium sativum*, *Pimpinella anisum*, and *Illicium verum* essential oils and their major components against the West Nile vector *Culex pipiens*. Parasitol Res 2012;111:2403-2410

36 Jamshidzadeh A, Hamedi A, Altalqi A, Najibi A. Comparative evaluation of analgesic activities of aniseed essential and fixed oils. Int J Pharm Res Scholars (IJPRS) 2014;3:227-235

37 Rhind JP. Essential oils. A handbook for aromatherapy practice, 2nd Edition. London: Singing Dragon, 2012

38 Davis P. Aromatherapy. An A-Z, 3rd Edition. London: Vermilion, 2005

39 Assalve D, Caraffini S, Lisi P. Psoriasis-like allergic contact dermatitis from aniseed oil. Annali Italiani di Dermatologia Clinica e Sperimentale 1987;41:411-414

40 Rudzki E, Rebandel P, Grzywa Z. Occupational dermatitis from cosmetic creams. Contact Dermatitis 1993;29:210

41 Garcia-Bravo B, Pérez Bernal A, Garcia-Hernandez MJ, Camacho F. Occupational contact dermatitis from anethole in food handlers. Contact Dermatitis 1997;37:38

42 Vente C, Fuchs T. Contact dermatitis due to oil of turpentine in a porcelain painter. Contact Dermatitis 1997;37:187

43 Loveman AB. Stomatitis venenata. Report of a case of sensitivity of the mucous membranes and the skin to oil of anise. Arch Derm Syph 1938;37:70-81

44 Meynadier JM, Meynadier J, Peyron JL, Peyron L. Formes cliniques des manifestations cutanées d'allergie aux parfums. Ann Dermatol Venereol 1986;113:31-39

Chapter 5.4   BASIL OIL, SWEET

DEFINITION

Sweet basil oil (essential oil of basil) is the essential oil obtained from the flowering aerial tops of the (sweet) basil, *Ocimum basilicum* L.

INCI NOMENCLATURE

Description/definition: Ocimum basilicum herb oil is an essential oil obtained from the herbs of the sweet basil, *Ocimum basilicum* L., Labiatae

INCI name EU: Ocimum basilicum herb oil (perfuming name, not an INCI name proper)

INCI name USA: Not in the Personal Care Products Council Ingredient Database

CAS registry number(s): 84775-71-3; 8015-73-4

EINECS number(s): 283-900-8

Description/definition: Ocimum basilicum flower/leaf extract is an extract of the flowers and leaves of the basil, *Ocimum basilicum* L., Labiatae

INCI name EU: Ocimum basilicum flower/leaf extract

INCI name USA: Ocimum basilicum (basil) flower/leaf extract

CAS registry number(s): 84775-71-3; 8015-73-4

EINECS number(s): 283-900-8

ISO (INTERNATIONAL ORGANIZATION FOR STANDARDIZATION) STANDARD

ISO number: 11043

ISO name: Essential oil of basil, methyl chavicol type

Botanical origin: *Ocimum basilicum* L.

Parts of plant used: Flowering aerial top

ISO values: ISO values (minimum and maximum concentrations) are shown in Table 5.4.1.

AFNOR (Association Française de Normalisation) values for the linalool chemotype of basil oil are shown in Table 5.4.2).

THE PLANT, THE OIL, AND THEIR USES

Ocimum basilicum L. is an aromatic, erect, almost glabrous annual herb, which grows to between 0.3 and 0.5 meters. The plant is native to India, Iran and tropical Asia, and now grows wild in tropical and sub-tropical regions (33). It is cultivated for commercial use in many countries around the world, including France, Hungary, Greece, Italy, Egypt, Morocco, Indonesia and several states in the USA (1). Basil leaves are widely used to flavor soups, meat pies, fish dishes, certain cheeses, tomato salads et cetera. It is an important seasoning in, for example, tomato paste. The leaves can be eaten as a salad. Basil is also used in perfumery, soap-making, and to flavor liqueurs (1).

The herb is widely used in systems of traditional medicine, including Ayurveda and traditional Chinese medicine.

Reported medicinal uses include the treatment of hyperlipidemia, headache, cough, cold, bronchitis, rhinitis, stomach ache, nausea, diarrhea, dysentery, constipation, flatulence, worms, kidney complaints, infections, various types of fever, rheumatism, muscle aches, gout, mental fatigue and to sooth the nerves (1,31, www.kew.org). It is also used to eliminate toxins and as a first aid treatment for wasp stings and snake bites (31). Ointments made of basil leaves are utilized for the treatment of warts, insect bites and acne (1).

Ocimum basilicum L. is the major essential oil crop around the world (3). The essential oil, which may be obtained from the leaves, the aerial parts or the flowering tops, is used to flavor various food products such as confectionery, baked goods, ice creams, puddings, liquors and non-alcoholic beverages. It may also be used as a flavoring for certain dental and oral hygiene products (1,31). In addition, *Ocimum basilicum* essential oils, notably those from the flowering tops, are widely used in the pharmaceutical, cosmetic, aromatherapy and perfumery industries (2,31,33). Basil oils are reported to possess a range of biological activities such as immunostimulant, hypnotic, local anesthetic, anticonvulsant, galactogogue, stomachic, antiviral, sedative, antitussive, diuretic, carminative, spasmodic, insect repellent, nematicidal, antibacterial, antifungal and antioxidant (3,21,31,33).

CHEMICAL COMPOSITION

Basil oil is a pale yellow to ambery yellow clear mobile liquid, which has a spicy, slightly anisic to spicy woody or to spicy cinnamic odor. The yield of essential oil from the flowering aerial tops of *Ocimum basilicum* L. generally varies from 0.01 to 0.3%. The main producing countries of this oil are India, Vietnam, France, Hungary, Egypt, Nepal, Indonesia, Morocco and USA.

Literature data (up to August 5, 2014) on the chemical composition of basil oils and unpublished analytical data from one of us (E.S.) are shown in Table 5.4.3 in alphabetical order. In basil oils from various origins, over 435 chemicals have been identified. About 53% of these were found in a single reviewed publication only.

The major compounds found in basil oils from different sources are shown in Table 5.4.4. They include (highest concentrations in any study given) linalool (98.9%), methyl chavicol (94.6%), methyl eugenol (91.1%), (*E*)-methyl cinnamate (82.4%), 1,8-cineole (54.3%), geranial (39.4%), neral (29.4%), eugenol (29.0%) and epi-α-cadinol (27.5%). Well-known ingredients of basil oils that were present in high concentrations (>15%) in one or two studies were (*E*)-anethole (74.6%), borneol (31.0% and 42.1%), (*Z*)-methyl cinnamate (38.2%), menthone (33.1%), β-bisabolene (25.6%), α-cadinol (21.1%) and α-farnesene (16.4%). Uncommon or rare constituents of basil oil found in high concentrations (>11%) in single studies include linalyl acetate (55.2%), elemene (39.1%), α-bergamotene (23.1%), (*E*)-myroxide (19.6%), α-linalool (16.0%) (adulteration), naphthalene (13.7%), α-cadinene (11.9%) and humulene (ep)oxide II (11.0%).

Table 5.4.1 ISO values (%) for basil oil, methyl chavicol type[a]

Compound	CAS	Minimum	Maximum
Methyl chavicol	140-67-0	75.0	87.0
1,8-Cineole	470-82-6	1.0	3.5
Linalool	78-70-6	0.5	3.0
(E)-β-Ocimene	3779-61-1	0.9	2.8
Methyl eugenol	93-15-2	0.3	2.5
Camphor	76-22-2	0.15	0.8
Terpinen-4-ol	562-74-3	0.2	0.6

[a] ISO 11043 Essential oil of basil, methyl chavicol type ©ISO 1998; Geneva, Switzerland, www.iso.org

Table 5.4.2 AFNOR values (%) for basil oil, linalool chemotype[a]

Compound	CAS	Minimum	Maximum
Linalool	78-70-6 4	5.0	62.0
Methyl chavicol	140-67-0	tr	30.0
Eugenol	97-53-0	2.0	15.0
1,8-Cineole	470-82-6	2.0	8.0
Terpinen-4-ol	562-74-3	tr	4.0
(E)-β-Ocimene	3779-61-1	0.2	2.0
Camphor	76-22-2	0.2	1.5

[a] AFNOR NF T 75-244 Huile essentielle de basilic, type linalol © AFNOR 1998; 11, rue de Francis de Pressensé, 93571 La Plaine Saint-Denis Cedex, France; www.afnor.org

Commercial oils

The ten chemicals that had the highest maximum concentrations in 47 commercial sweet basil essential oil samples (concentration ranges provided) are the following: methyl chavicol (0.2-87.0%), linalool (0.6-55.8%), methyl eugenol (0-24.7%), (Z)-methyl cinnamate (0-23.6%), trans-α-bergamotene (0.01-19.8%), eugenol (0.03-15.3%), 1,8-cineole (0.03-13.7%), τ-cadinol (0-7.1%), β-elemene (0-5.5%) and α-guaiene (0-4.3%) (Erich Schmidt, unpublished analytical data).

Chemotypes

There are several types of basil oil in international commerce, each derived principally from different cultivars or chemotypes of sweet basil. The oils of commerce are known as European (French or sweet basil), Egyptian, Reunion or Comoro, and tropical basil oils (2,3,21). The high-quality European type of basil oil from Italy, France, USA and South Africa characteristically contains linalool, methyl chavicol (estragole) and smaller quantities of 1,8-cineole, α-pinene, β-pinene, and myrcene. Egyptian basil oil of commerce is similar to the European oil, except that the concentration of linalool is significantly lower while the concentration of methyl chavicol is significantly higher. In contrast, Reunion or Comoro basil oils from Reunion, Comoros, Seychelles, Thailand, Madagascar and Vietnam contain little if any linalool and have a harsher, spicy aroma due to the very high concentration of methyl chavicol (2,3,5). The tropical oils from Java, India, Pakistan and Guatemala are rich in methyl cinnamate (3,5,21). There is also a eugenol-rich type of basil oil from North Africa and Russia, Eastern Europe, and parts of Asia (3,14,21).

Table 5.4.3 Constituents identified in basil oils

Constituent	CAS	Percentage and range						
		A	B (64)	C (2)	D (11)	E (3)	F (30)	G
Acetic acid	64-19-7							+^{z35}
α-Acoradiene	24048-44-0							tr^f; 0.05^{z15}
β-Acoradiene	28477-64-7			tr				tr^{f,l}; 0.07^{z15}; 0.2^{z4}
10-epi-β-Acoradiene	847374-86-1			0.1				
7-epi-Amiteol	147383-87-7					0.2		
α-Amorphene	20085-19-2		2.8					1.1^z; 1.6^t; 3.1^j; 4.1^{z2}; 4.8^{z32}
δ-Amorphene	189165-79-5							0.1^f; 0.3^{l,p}
(E)-Anethole	4180-23-8		0.5	tr				0.1^{z5}; 0.2^{z12}; 0.6^{z14}; 1.5^v; 74.6^e
(Z)-Anethole	25679-28-1			0.1				
p-Anisaldehyde	123-11-5		0.5	tr	0.1			0.07^{z19}; 0.2^o
Aromadendrene	489-39-4		4.3		0.7		0.2	tr^{f,l}; 0.09^{z7}; 0.2^{z2}; 0.3^v; 1.3^{z24}
allo-Aromadendrene	25246-27-9		0.2		0.1			0.2^t; 0.8^r
Azulene	275-51-4							0.7^{z9}
Benzaldehyde	100-52-7		0.3					<0.01^{z15}; 0.07^{z3}
Benzoic acid	65-85-0							+^{z35}
Benzyl alcohol	100-51-6							+^{z35}
Benzyl benzoate	120-51-4							+^{z35}
α-Bergamotene	17699-05-7		23.1					3.7^k; 4.1^{z10}; 5.7^{z23}; 8.0^t; 9.2^s
cis-α-Bergamotene	18252-46-5			tr				0.5^r; 1.1^{z5}; 5.8^{z24}; 10.0^n
trans-α-Bergamotene	13474-59-4	0.01-19.8		3.1	5.9		3.2	1.1^{z1}; 2.4^p; 3.4^c; 3.9^{z2}; 6.9^r; 7.5^f; 7.6^e; 14.9^j; 15.8^{z4}; 17.5^v
Bicycloelemene	32531-56-9							0.01^{z15}
Bicyclogermacrene	24703-35-3	0-1.2	0.9	1.0	0.9	2.0		1.7^b; 2.0^u; 3.1^{z24}; 4.0^v; 7.8^{z4}

Table 5.4.3 Constituents identified in basil oils (*continued*)

Constituent	CAS	A	B (64)	C (2)	D (11)	E (3)	F (30)	G
epi-Bicyclophellandrene								7.4[z28]
epi-Bicyclosesquiphellandrene	54274-73-6						0.4	0.4[k]; 0.6[z9]
α-Bisabolene	17627-44-0		2.2			3.4	1.0	1.1[t]; 3.6[z8]
(E)-α-Bisabolene	25532-79-0	0,09-2.0		0.6				0.5[z1]; 0.9[b]; 9.5[n]
(Z)-α-Bisabolene	29837-07-8							0.1[o]; 0.7[v]; 0.8[p]; 10.1[z31]
β-Bisabolene	495-61-4		0.4	tr	25.6	0.3		0.1[k]; 0.3[z8]; 0.5[t]; 0.6[n]; 2.5[m]
γ-Bisabolene	495-62-5		0.2					
(E)-γ-Bisabolene	53585-13-0				3.9			
(Z)-γ-Bisabolene	13062-00-5							1.1[z28]
(Z)-δ-Bisabolene								1.1[y1]
α-Bisabolol	515-69-5			0.1				tr[f,l]; 0.3[z2]; 0.4[s]; 0.9[z4]
epi-α-Bisabolol	78148-59-1			0.1	0.3			
β-Bisabolol	15352-77-9				0.5	0.3		0.2[z4]
Borneol	507-70-0	0.03-0.2	0.9	0.4	1.4	0.9		0.7[f]; 1.9[x]; 2.1[k]; 3.7[h]; 60.1[z9]
Bornyl acetate	76-49-3	tr-4.2	4.9		1.2			1.1[p]; 1.2[z18]; 1.3[r]; 1.8[z3]; 2.0[d]
β-Bourbonene	5208-59-3			0.1	0.1	0.2		<0.1[t]; 0.2[n,w]; 0.4[z2]; 0.6[v]
Bulnesene	164108-17-2							1.5[z18,z20]
α-Bulnesene	3691-11-0		1.9	2.0			1.1	0.7[j]; 0.8[f]; 0.9[l]; 1.9[p,z4]; 2.3[z11]
Butanal	123-72-8							+[z35]
n-Butyric acid	107-92-6							+[z35]
Cadina-1,4-diene	29837-12-5		0.9					0.03[z7]; 0.2[t]; 1.0[k]
cis-Cadina-1(6),4-diene	1187195-00-1			0.5				
Cadina-3,5-diene	267665-20-3							0.1[z15]; 0.9[z4]
α-Cadinene	24406-05-1			tr	11.9			tr[l]; 0.3[v]; 0.4[c,r]; 1.9[o]
β-Cadinene	523-47-7		7.2					
γ-Cadinene	39029-41-9		3.1	2.5	0.7	5.0	2.5	3.5[z4]; 4.1[e]; 4.4[n]; 5.4[s]; 6.3[z2]
δ-Cadinene	483-76-1	0.01-0.9	0.3	0.1		8.7	0.7	2.1[z4]; 3.6[y]; 6.7[z14]; 7.7[z2]; 8.1[z10]
d-Cadinene	880143-55-5		2.2					
Cadinol	11070-72-7							6.9[h]
α-Cadinol	481-34-5		5.9	0.6	3.4	2.9		2.6[z4]; 2.9[z]; 6.5[p]; 6.8[t]; 21.1[z2]
epi-α-Cadinol	5937-11-1	0-7.1		9.3	1.8		8.3	5.7[l]; 6.2[z11]; 6.5[z24]; 6.7[f]; 7.1[z14] 7.6[c]; 12.4[s]; 17.3[n]; 27.5[z4]
γ-Cadinol	50895-55-1							3.7[e]
δ-Cadinol	19435-97-3							0.6[z11]
α-Calacorene	21391-99-1			tr				+[z33]
β-Calacorene	50277-34-4							+[z33]
cis-Calamene	72937-55-4							0.2[z]; 0.3[r]; 0.4[z24]; 1.1[z2]
trans-Calamene								0.5[f]
Calamenene	483-77-2					0.4		0.02[z3]; 0.2[z14,z20]; 0.8[c]; 1.0[s]
cis-Calamenene	72937-55-4		0.2					
trans-Calamenene	73209-42-4			0.5				0.3[z18]; 0.4[l]; 0.5[p,v]
Calarene	17334-55-3		1.8					<0.1[z4]
Camphene	79-92-5	tr-1.2		tr	0.4		0.3	0.2[g]; 0.4[h]; 0.5[w]; 0.8[c]; 4.7[z17]
Camphenol	3570-04-5						0.3	
E-Camphenone								0.2[y]
Camphor	76-22-2	0-2.1	3.9	0.7	9.1	3.2	1.4	3.4[d]; 4.3[m]; 4.5[h]; 31.0[x]; 42.1[z17]
δ2-Carene (=δ4-)	554-61-0				0.1			
δ3-Carene	13466-78-9							0.04[z12]; 0.08[n]; 0.7[z2]
Carvacrol	499-75-2			0.7	0.1	tr		0.7[z14]; 0.9[z9]; 1.8[n]; 2.7[e]
(E)-Carveol	1197-07-5							0.4[w]
Carvone	99-49-0				tr		0.5	0.2[w]; 0.4[z2]; 0.5[z5]; 1.6[z14]
Caryophyllene	87-44-5		2.0					
β-Caryophyllene	87-44-5	0.02-1.4	1.0	0.5	5.2	4.2	43.0	2.5[w]; 3.2[d]; 4.5[y]; 6.2[z31]; 7.1[x]; 7.5[v]; 8.0[j]; 10.7[z31]; 14.9[n]
Caryophyllene oxide	1139-30-6	0-0.2	0.5	0.1	1.1	2.0	13.9	1.1[b]; 2.5[z2]; 4.0[z9]; 8.0[n]; 11.4[w]
Caryophyllenol	38284-26-3							0.3[y]
Cedrene	11028-42-5							tr[f]

Table 5.4.3 Constituents identified in basil oils (*continued*)

Constituent	CAS	Percentage and range						
		A	B (64)	C (2)	D (11)	E (3)	F (30)	G
α-Cedrene	469-61-4							0.3[z2]
β-Cedrene	546-28-1			tr				tr[j]; 2.9[z2]
Chavicol	501-92-8		1.1					0.02[z7]; 0.1[z]; 0.3[t]; 0.8[o]; 1.0[z2]
trans-Chrysanthemal	20104-05-6							2.5[b]
trans-Chrysanthenol	38043-83-3							1.4[z1]
Chrysanthenyl acetate								0.3[z6]
1,4-Cineole	470-67-7						0.2	
1,8-Cineole	470-82-6	0.03-13.7	12.1	8.1	17.5	14.5	14.6	9.8[d]; 10.1[v]; 12.9[c]; 13.7[z13]; 14.8[m]; 15.3[h]; 16.5[r]; 54.3[w]
Cinnamyl acetate	103-54-8							1.2[z3]
(*E*)-Cinnamyl acetate	21040-45-9							0.09[z20]
Citral	5392-40-5					65.6		0.6[g]
Citronellal	106-23-0							1.6[o]
Citronellol (β-, DL-)	106-22-9				0.4	0.2		0.2[g]
Citronellyl acetate	150-84-5				0.6			
α-Copaene	3856-25-5	0-0.6	0.7	0.1	1.9	0.6	0.4	0.6[z2]; 0.9[n]; 1.4[z26]; 1.5[v]; 7.5[w]
β-Copaene	18252-44-3			tr	0.5			0.09[z15]
Coumarin	91-64-5							+[z35]
α-Cubebene	17699-14-8		0.1	tr	0.4	1.6	0.1	0.1[t,z]; 0.3[z14]; 0.4[o]; 0.8[w]; 6.2[z15]
β-Cubebene	13744-15-5		1.1	tr	0.1	0.5		0.6[x]; 0.9[n]; 1.1[t]; 1.8[w]; 2.3[m]
10-epi-Cubebol	176589-53-0							tr[l]
Cubenol	21284-22-0							0.7[j,z5]; 0.8[r,z11,z24]
1-epi-Cubenol	81939-29-9						1.9	tr[l]
1,10-di-epi-Cubenol	73365-77-2			1.3				0.3[z19]; 0.7[l]; 0.8[f,p,z18]; 4.2[z4]
Cuminaldehyde	122-03-2							+[z35]
β-Curcumene	28976-67-2			tr				
Cyclobazzanene	88661-61-4							0.03[z12]
1,6-Cyclodecadiene	1124-79-4							0.6[z9]
Cyclohexane	110-82-7							2.1[z9]
Cyclohexanol	108-93-0							0.4[k]
Cyclohexene	110-83-8							0.5[k]
p-Cymene	99-87-6		0.7	0.1	0.2		0.4	0.2[n]; 0.6[v]; 0.9[e]; 1.1[j]; 1.2[w]; 1.3[x]
p-Cymenene	1195-32-0	0-0.3						
p-Cymen-4-ol			0.1					
p-Cymen-8-ol	1197-01-9			tr				0.2[o]
β-Damascenone (*E*)-	23726-93-4		0.1	tr				0.2[z14]
β-Damascone	23726-91-2							+[z35]
2,4-Decadienal	2363-88-4							+[z35]
Decanal	112-31-2							0.7[r]
n-Decane	124-18-5			tr				tr[z34]
n-Decanoic acid	334-48-5							+[z35]
Decanol	36729-58-5							+[z35]
2,3-Dehydro-1,8-cineole	92760-25-3			tr				
Dibutyl octanedioate	16090-77-0							<0.01[z7]
Dihydroanethole	104-45-0			tr				0.2[y1]
Dihydroedulan II	41678-32-4							0.4[y]
Dill ether	74410-10-9					0.2		
2,6-Dimethyl-1-hep- tanol	2768-12-9							0.3[y]; 1.0[n]
(*E*)-2,6-Dimethyl-3,7-octadien-2,6-diol	13741-21-4			0.1				
3,7-Dimethyl-1,5-octadien-3,7-diol	13741-21-4							0.3[z15]
3,7-Dimethyl-1,7-octadiene-3,6-diol								0.03[z15]
Docosane	629-97-0							<0.1[y1]
1-Docosene	1599-67-3							tr[z28,y1]
δ-Dodecalactone	713-95-1							+[z35]
Dodecanol	112-53-8							+[z35]
3-Dodecanone	1534-27-6							0.2[y]
Eicosane	112-95-8							<0.1[y1]; 0.2[y];
(*E*)-1-Eicosene	3452-07-1							0.2[y]

Table 5.4.3 Constituents identified in basil oils (*continued*)

Constituent	CAS	A	B (64)	C (2)	D (11)	E (3)	F (30)	G
Elemene	11029-06-4							39.1[h]
β-Elemene	33880-83-0	0-5.5	2.5	2.0	0.8	1.6		2.7[f]; 3.2[p]; 3.5[v]; 3.6[b]; 8.2[z14]
γ-Elemene	29873-99-2		0.2					0.6[z8]; 1.1[z9]; 1.9[z14]; 10.9[m]
δ-Elemene	20307-84-0							tr[c,l]
Elemol	639-99-6						0.5	0.1[o]; 0.3[z2]; 2.1[z16]; 2.9[z6]
Elixene	3242-08-8							0.05[z7]; 0.3[z11]
Epizonarene	41702-63-0							0.5[k]
(E)-Epoxyocimene	255832-06-5		0.3					
(Z)-Epoxyocimene								0.4[z24]
Eremophilene	10219-75-7							+[z33]
2-Ethylfuran	3208-16-0							+[z35]: 0.02[z14]
Ethyl isovalerate	108-64-5				0.7			
Ethyl-2-methyl butyrate	7452-79-1							+[z35]
Eudesma-4(15),7-dien-1β-ol	119120-23-9			0.1				
α-Eudesmol	473-16-5							0.1[z16]; 0.2[z6]; 4.7[z1]; 7.1[b]
β-Eudesmol	473-15-4			0.3	2.5	0.7		0.3[s]; 0.4[z5]; 0.6[q]; 1.1[z2]; 5.7[r]
γ-Eudesmol	1209-71-8				1.9			0.2[z6]; 0.4[z16]
10-epi-γ-Eudesmol	15051-81-7							0.1[t]
Eugenol	97-53-0	0.03-15.3	20.1	2.4	5.9	21.1	4.1	8.7[h]; 11.1[z4]; 12.3[z18]; 13.5[z21]; 14.2[i]; 24.8[j]; 27.6[p]; 29.0[d]
Eugenyl acetate	93-28-7							+[z35]; 0.3[z20]
α-Farnesene	502-61-4		16.4					0.9[c]; 2.0[m]; 5.5[j]
(E,E)-α-Farnesene	502-61-4				0.3			
(Z,E)-α-Farnesene	26560-14-5			tr				
β-Farnesene	502-60-3		1.8					0.3[t]; 1.4[z9]
(E)-β-Farnesene	18794-84-8	0.05-4.1		0.1	1.9	1.1	0.7	0.4[k]; 0.7[b]; 1.1[z]; 3.9[z2]; 4.1[v]
(Z)-β-Farnesene	28973-97-9			tr	0.1			0.2[v]; 1.0[r]; 1.3[f]; 1.6[n]; 1.7[j]; 2.9[p]
Farnesol	4602-84-0							0.2[b,k]
(E,E)-Farnesol	106-28-5							0.1[v]
Farnesyl acetate	29548-30-9							0.1[z4]
endo-Fenchol (α-)	512-13-0		0.6	tr				tr[l,z33]
exo-Fenchol (β-)	22627-95-8			tr				
Fenchone	1195-79-5	0-0.7	2.1	0.1		0.5	0.9	1.9[j]; 2.6[n]; 3.1[z9]; 7.0[v]; 10.1[z12]
Fenchyl acetate	13851-11-1	0-0.6	3.0			0.1	0.3	tr[l]; 0.6[c]; 1.8[f]
α-Fenchyl acetate (endo-)	111821-74-0							0.04[z15]; 0.4[j]; 0.6[z12,z13]
β-Fenchyl acetate (exo-)	76109-40-5							1.2[z12]
Fenchyl alcohol	1632-73-1	0-0.3	1.7					0.3[c,z8]
Furfural	98-01-1			tr				
Geranial	141-27-5	0-0.6	0.7	0.2	36.9			1.6[j]; 24.6[n]; 25.7[z31]; 39.4[o]
Geraniol	106-24-1		0.9	5.1	6.8	3.9		0.1[b,g]; 0.3[q]; 1.5[j]; 1.7[z10]; 2.1[n] 4.3[p]; 5.2[o]; 12.5[v]; 13.4[f]; 16.5[l]
(E)-Geraniol	106-24-1						1.1	
(Z)-Geraniol	106-25-2						0.2	1.3[s]
Geranoic acid	459-80-3							+[z35]
Geranyl acetate	105-87-3		0.4	0.1	0.2			0.5[r]; 0.7[f,n]; 1.4[z6]; 1.7[l]; 4.0[z16]
Geranyl acetone	3796-70-1			tr				
(Z)-Geranylacetone	3879-26-3							0.1[z14]
Geranyl formate	105-86-2				0.2			0.5[e]
Germacrene	28028-64-0							5.7[z10]
Germacrene A	28387-44-2							0.6[e]; 1.2[z14]; 1.4[z26]; 2.3[k]; 3.2[h]
Germacrene B	15423-57-1						1,4	0.3[p]; 0.9[b]; 1.0[k]; 1.3[c]
Germacrene D	23986-74-5	0.01-3.6	1.7	3.0	2.6	4.3		4.1[p]; 4.4[h]; 5.7[z24]; 6.0[v]; 8.7[n]
Globulol	489-41-8							tr[l]; 0.2[z17]
α-Guaiene	3691-12-1	0-4.3		0.6			0.7	1.0[p]; 1.1[k]; 1.4[f]; 2.1[z1]; 2.6[b]; 2.9[n]
β-Guaiene	88-84-6		2.9			1.7		
trans-β-Guaiene	192053-49-9							2.5[v]
γ-Guaiene	145267-53-4							0.8[z13]; 2.1[z24]; 4.3[z14]
Guaiol	489-86-1				0.1			
Guaiyl acetate	134-28-1							3.0[k]

Table 5.4.3 Constituents identified in basil oils (*continued*)

Constituent	CAS	A	B (64)	C (2)	D (11)	E (3)	F (30)	G
					Percentage and range			
α-Gurjunene	489-40-7			tr	0.1			tr[z16]; 0.2[z5]; 0.3[o]
β-Gurjunene	73464-47-8							<0.1[t]; 0.1[o]; 0.6[n]; 1.4[z17]
γ-Gurjunene	22567-17-5							+[z35]
Heneicosane	629-94-7							<0.1[y1]
Hentriacontane	630-04-6							tr[y1]
Heptadecane	629-78-7			tr				
(*E,E*)-2,4-Heptadienal	4313-03-5			tr				
Heptanal	111-71-7							+[z35]
Heptanoic acid	111-14-8							+[z35]
n-Hexadecane	544-76-3			tr				
Hexadecanoic acid	57-10-3		0.6			0.7		
1-Hexadecene	629-73-2			tr				tr[z28]
Hexahydrofarnesyl acetone	502-69-2			0.1				0.4[z14]
Hexanal	66-25-1							0.03[z14]
Hexanoic acid	142-62-1							+[z35]
1-Hexanol (*n*-)	111-27-3			tr	tr			<0.01[z7]
(*E*)-3-Hexenol	928-97-2							0.4[o]
(*Z*)-3-Hexen-1-ol	928-96-1	0-0.04		0.1	0.1			0.05[z3]; 0.06[z7]
(*E*)-3-Hexenyl acetate	3681-82-1			tr				0.2[o]
(*Z*)-3-Hexenyl acetate	3681-71-8	0-0.06			0.2			<0.01[z15]; 0.01[z14]; 0.1[z17]
(*Z*)-3-Hexenyl angelate	84060-80-0			tr				
(*Z*)-3-Hexenyl benzoate	25152-85-6							+[z35]
(*Z*)-2-Hexenyl butyrate								0.07[z14]
(*Z*)-3-Hexenyl (*Z*)-3-hexenoate	61444-38-0			tr	0.2			
(*Z*)-3-Hexenyloxy-acetaldehyde	68133-72-2			tr				
α-Himachalene	3853-83-6							0.2[z2]
Himachalol	1891-45-8							tr[f]
(*E*)-Hotrienol	53834-70-1			0.3				
α-Humulene	6753-98-6	0.04-1.2	2.6	0.7	2.7	1.6		2.9[z4]; 3.0[v]; 3.2[z26]; 6.3[w]; 13.6[n]
β-Humulene	116-04-1							1.8[z20]
α-Humulene (ep)oxide	96638-51-6					0.5		0.9[z2]; 2.0[b]
Humulene (ep)oxide I	19888-33-6			0.2				0.1[o]
Humulene (ep)oxide II	19888-34-7				0.2			0.08[z19]; 0.1[z4]; 0.7[z1]; 11.0[w]
exo-2-Hydroxycine-ole acetate	72257-53-5							0.05[z15]
Intermedeol	6168-59-8							tr[l]; 0.5[z4]
(*E*)-β-Ionone	79-77-6					0.2		0.1[z]; 0.2[z14]; 1.0[z2]
Isoamyl acetate	123-92-2				tr			
Isoamyl alcohol	123-51-3							+[z35]
Isoamyl isovalerate	659-70-1							0.07[z12]
Isoborneol	124-76-5							0.3[z17]
Isobornyl acetate	125-12-2			1.3	1.5		0.6	
(*Z*)-Isocitral	72203-97-5							0.9[o]
Isoeugenol	97-54-1		10.6					
trans-Isolimonene	6876-12-6							2.0[y]
Isolongifolene	1135-66-6							0.4[k]
Isomenthone	491-07-6	0-1.9		tr				tr[z34]; 0.3[z27]
Isoneomenthol								0.4[z2]; 7.5[z]
Isopulegol acetate	57576-09-7							0.3[z]
Isospathulenol	88395-46-4		0.9					
Isoterpinolene	586-63-0					1.5		
Isovaleric acid	503-74-2							+[z35]
(*E*)-Jasmone	6261-18-3							+[z35]
(*Z*)-Jasmone	488-10-8							0.2[z16]
Lauric acid	143-07-7							+[z35]
Lavandulyl acetate	25905-14-0							0.3[z6]; 0.8[z16]
Ledol	577-27-5							+[z35]; 0.2[y]
Lilac aldehyde A	53447-46-4			0.1				

Table 5.4.3 Constituents identified in basil oils (*continued*)

Constituent	CAS	Percentage and range						
		A	B (64)	C (2)	D (11)	E (3)	F (30)	G
Limonenal	6784-13-0						0.4	
Limonene	138-86-3	0.03-1.1	0.6	0.1	0.8	0.4	3.6	2.3[w]; 4.7[x]; 6.2[e]; 7.6[z17]; 10.4[z21]
cis-Limonene oxide	13837-75-7							0.2[w]
α-Linaloool	598-07-2			16.0[a]				
Linalool	78-70-6	0.6-55.8	66.4	79.5	74.3	60.2	33.0	67.9[i]; 73.2[d]; 74.0[l]; 74.5[m]; 75.9[f]; 77.5[p]; 83.6[z23]; 98.9[x]
Linalool oxide	1365-19-1							1.1[s]
cis-Linalool oxide	11063-77-7		1.7				0.3	tr[l]; 0.08[z3]; 0.3[z19]; 0.4[b,f]
cis-Linalool oxide, furanoid	11063-77-7			0.7	tr			0.1[z1]
cis-Linalool oxide, pyranoid	14009-71-3			tr				
trans-Linalool oxide	11063-78-8		1.6					0.08[z3]; 0.1[y]; 0.2[b,u]; 0.3[f,t]
trans-Linalool oxide, furanoid	34995-77-2			0.7	0.2			
Linalyl acetate	115-95-7				0.5	0.6		0.2[z14]; 0.5[s]; 14.0[z16]; 55.2[z6]
Linalyl formate	115-99-1							0.9[k]
Linalyl propionate	144-39-8					2.4		1.7[h]
Longipinanol	66141-14-8							tr[l]
γ-Maaliene	20071-49-2							0.2[z17]
Maaliol	527-90-2			0.3				
p-Mentha-2,4(8)-diene	586-63-0							0.2[z6]
p-Menthene	5502-88-5							0.3[z27]
p-Menth-2-en-1-ol	619-62-5							<0.01[z15]; 0.5[c]
cis-p-Menth-2-en-1-ol	29803-82-5							0.2[t]; 0.6[o]
trans-p-Menth-2-en-1-ol	29803-81-4							0.5[o]
Menthofuran	494-90-6							0.1[y]
Menthol	89-78-1	0-0.7		0.1				0.6[n]; 6.1[z]
Menthone	89-80-5	0-0.3		0.1				0.3[z34]; 0.6[y]; 1.2[z2]; 33.1[z]
Menthyl acetate	16409-45-3							tr[z34]; 5.6[z]
p-Methoxyaceto-phenone	100-06-1			tr				+[z35]
p-Methoxybenzoic acid	100-09-4							+[z35]
3-Methoxycinnamal-dehyde	56578-36-0							0.2[z1]; 0.3[b]
o-Methoxycinnamal-dehyde (2-)	1504-74-1							0.3[n]
(E)-p-Methoxycinna-maldehyde	24680-50-0			0.1				
2-Methoxycinnamic alcohol		0-0.5						
2-Methoxy-3-methyl-pyrazine	2847-30-5							+[z35]
3-Methylbutanal	590-86-3							0.05[z14]
2-Methylbutanoic acid	116-53-0				0.1			+[z35]
Methylcarvacrol	6379-73-3			tr				
Methyl chavicol	140-67-0	0.2-87.0	87.8	22.8	88.2	76.3	12.3	71.5[d]; 85.5[e]; 87.2[g]; 87.8[m]; 89.9[j]; 90.0[o]; 94.3[q]; 94.6[t]
Methyl cinnamate	103-26-4					63.1		11.2[g]; 21.6[z20]; 59.3[z25]
(E)-Methyl cinnamate	1754-62-7		7.4		38.7		74.5	16.7[z14]; 31.9[z23]; 41.9[z31]; 82.4[m]
(Z)-Methyl cinnamate	19713-73-6	0-23.6	38.2		6.1		9.8	0.1[t]; 0.4[p]; 0.6[b]; 4.7[z3]; 5.9[m]
2-Methyldecane	6975-98-0			tr				
Methyl eugenol	93-15-2	0-24.7	38.8	0.2	14.9	34.2	26.0	0.8[p]; 2.0[n]; 2.3[f]; 2.5[m]; 5.6[r] 40.5[j]; 74.5[e]; 78.0[z15]; 91.1[d]
Methylfuran	27137-41-3					0.5		+[z35]
Methylheptenone	409-02-9	0-0.1						
6-Methyl-3-heptenone						1.3		
6-Methyl-5-hepten-2-one	110-93-0			tr	0.5			tr[o]; 0.3[b,z1]; 1.6[v]

Table 5.4.3 Constituents identified in basil oils (*continued*)

Constituent	CAS	Percentage and range						
		A	B (64)	C (2)	D (11)	E (3)	F (30)	G
Methyl isoeugenol	93-16-3							2.2[z29]
(*E*)-Methylisoeugenol	6379-72-2							0.5[z9]
Methyl isovalerate	556-24-1							+[z35]
Methyl jasmine						0.4		
Methyl jasmonate	1211-29-6							+[z35]
Methyl epi-jasmonate	42536-97-0							+[z35]
Methyl 2-methylbu-tyrate	868-57-5							+[z35]
Methyl salicylate	119-36-8							+[z35]
Methyl thymol	1076-56-8			tr				
Mintsulfide	72445-42-2							0.01[z27]
cis-Muurola-3,5-diene	157374-44-2			0.2				tr[f]; 0.4[p]; 0.5[l]
cis-Muurola-4(14),5-diene	157477-72-0							0.1[z15]; 0.6[f,l]; 0.7[p]; 1.3[z4]
trans-Muurola-4(14), 5-diene	262352-87-4			tr				
α-Muurolene	10208-80-7	0-1.1		tr				tr[q]; 0.1[o,z2]; 0.3[z7]; 0.9[e]; 1.1[z13]
γ-Muurolene	30021-74-0			tr	0.9			0.2[z11]; 0.4[c]; 0.6[u]; 0.9[o,s]; 2.2[r]
ε-Muurolene	30021-46-6							0.7[z15]
τ-Muurolene	152287-43-9							0.2[b]
cis-14-nor-Muurol-5-en-4-one				tr				
α-Muurolol	104245-48-9			0.1	0.1			
τ-Muurolol (epi-α-)	19912-62-0							0.2[z1]
Myrcene	123-35-3	0.02-2.9	1.0	0.5	1.1	0.9	1.4	2.0[g]; 2.9[r]; 3.3[z6]; 3.6[z9]; 5.6[z16]
Myristic acid	544-63-8							+[z35]
(*E*)-Myroxide	28977-57-3							<0.01[z15]; 19.6[w]
(*Z*)-Myroxide	33281-83-3						0.3	tr[f,l]; 0.1[z19]
Myrtenal	564-94-3							<0.1[x]; 0.1[z17]
Myrtenol	515-00-4							<0.1[x]; 3.3[z17]
Myrtenyl acetate	1079-01-2							0.2[z16,z17]
Naphthalene	91-20-3			tr				0.7[z9]; 13.7[k]
Neointermedeol	5945-72-2							tr[l]
Neomenthol	3623-51-6			0.1				
Neophytadiene	504-96-1							0.2[k]
Neral	106-26-3	0.08-0.5		0.1	27.6			10.0[v]; 20.8[z31]; 22.7[n]; 29.4[o]
Neranoic acid								+[z35]
Neric acid	37349-29-4							+[z35]
Nerol	106-25-2		0.4	0.1	9.7	1.8		0.9[z16]; 2.7[n]; 11.2[o]; 12.6[z24]
Nerolidol	7212-44-4		0.8			0.7	0.6	0.4[z2]; 0.5[j,z14]
(*E*)-Nerolidol	40716-66-3			0.2	0.4			0.4[z4,z5]; 1.6[z1]; 2.4[b]
(*Z*)-Nerolidol	3790-78-1							0.7[z18,z20]
Neryl acetate	141-12-8		0.2	tr	0.7	0.3		0.3[z14]; 0.7[o]; 0.9[z6]; 2.4[z16]
Nonacosane	630-03-5							tr[z28,y1]
Nonadecane	629-92-5							<0.1[y1]
Nonadecene	27400-77-7							0.1[y]
β-Ocimene	13877-91-3		0.3					3.5[z11]; 10.5[z9]
(*E*)-β-Ocimene	3779-61-1	0.02-2.7	0.5	1.4	3.5			2.3[z26]; 2.8[p]; 3.0[j]; 3.4[q]; 3.9[n]
(*Z*)-β-Ocimene	3338-55-4	tr-1.8		tr	0.8	2.0		1.8[z26]; 2.0[z6]; 2.1[b]; 2.2[z7]; 3.2[e]
(*E*)-Ocimene	27400-72-2						2.4	1.0[c]
(*Z*)-Ocimene	27400-71-1						0.5	tr[c]
allo-Ocimene	673-84-7							0.07[z18]; 0.1[y]; 2.4[z16]
neo-allo-Ocimene	7216-56-0		0.2					<0.01[z15]; 0.2[t]; 0.3[z16]; 3.0[z6]
Octadecane	593-45-3							0.3[y1]
1-Octadecene	112-88-9			tr				tr[z28]
2,4-Octadienal	30361-28-5							+[z35]
Octanal	124-13-0							+[z35]
n-Octane	111-65-9`			tr				
Octanoic acid	124-07-2							+[z35]
1-Octanol (*n*-)	111-87-5			tr		0.2		0.04[z14]; 0.1[z5]
3-Octanol	589-98-0			0.1	tr			tr[w]; 0.03[z7]; 0.1[z]; 0.7[z2]
3-Octanone	106-68-3				tr			

Table 5.4.3 Constituents identified in basil oils (*continued*)

Constituent	CAS	A	B (64)	C (2)	D (11)	E (3)	F (30)	G
Octan-3-yl acetate	103-09-3							0.2^{z16}; 0.3^{z6}
trans-Oct-2-en-1-al	2548-87-0							$+^{z35}$
1-Octen-3-ol	3391-86-4	0-0.06	0.2	0.4	0.2			0.3^{z5}; 1.0^w; 1.2^n; 1.4^{z17}
3-Octenol	18185-81-4							0.1^y
7-Octen-4-ol	53907-72-5							0.2^{y1}
Octenyl acetate	37366-04-4							0.03^{z14}
1-Octen-3-yl acetate	2442-10-6			0.1	0.3			0.07^{z18}; 0.1^{z16}; 0.2^{z6}
Octyl acetate	112-14-1	0-0.2	0.4	0.1			0.2	0.02^{z15}; 0.05^{z3}; 0.2^{z14}; 0.4^k
Palustrol	5986-49-2							0.3^w
Pentadecane	629-62-9			tr				
2-Pentylfuran	3777-69-3			tr				
Perillaldehyde	2111-75-3							$+^{z35}$
Phellandrene	1329-99-3					0.2		
α-Phellandrene	99-83-2				0.2			0.2^{z2}; 0.4^{z17}; 0.6^n; 2.2^x; 4.4^{z12}
α-Phellandrene-8-ol	1686-20-0						0.5	
β-Phellandrene	555-10-2							0.06^{z7}; 0.1^{z5}; 0.6^{z17}; 1.2^{z22}
Phenethyl alcohol	60-12-8							$+^{z35}$
Phenylacetaldehyde	122-78-1			tr				
Phenylethyl acetate	93-92-5							$+^{z35}$
Phytol	7541-49-3							0.1^z
trans-Pinane (*E*-)	10281-53-5							0.2^y
α-Pinene	80-56-8	0.02-2.7	0.3	0.1	0.5		0.9	0.9^h; 1.2^v; 1.7^m; 4.4^x; 5.4^{z17}
β-Pinene	127-91-3	0.04-2.1	0.7	0.4	1.4	1.2	1.6	1.3^j; 1.9^m; 2.0^c; 2.3^h; 8.2^w
trans-Pinocamphone	547-60-4			tr				
Pinocarvone	30460-92-5							tr^l; $<0.01^{z15}$; 0.9^r
cis-Piperitol	16721-38-3							0.7^{z30}
trans-Piperitol	16721-39-4							0.1^{z30}
Piperitone	89-81-6	0-0.06						0.3^z
Plinol	72402-00-7			tr				
Plinol C	4028-60-8			tr				
Propanoic acid	79-09-4							$+^{z35}$
Pulegone	89-82-7							$0.1^{y,z2}$; 0.3^n; 3.7^z
Quinoline	91-22-5							$+^{z35}$
Rosefuran epoxide	92356-06-4							6.0^w
cis-Rose oxide	3033-23-6							$0.2^{z,z22}$
Sabinene	3387-41-5	0.01-1.2	0.7	0.2	0.2	0.4	0.7	$0.4^{c,j}$; 0.5^v; 0.6^e; 1.0^x; 1.8^r
Sabinene hydrate	546-79-2							0.07^n
cis-Sabinene hydrate	15537-55-0			0.1	1.2		0.8	0.1^o; $0.3^{l,z17}$; 0.6^p; 0.9^{z26}; 1.9^w
trans-Sabinene hydrate	17699-16-0		0.5		tr	0.3		tr^w; $0.2^{r,z17}$; 0.9^{z26}
Salicylic acid	69-72-7							$+^{z35}$
Salvial-4(14)-en-1-one	73809-82-2			tr				
α-Santalene	512-61-8						0.3	0.4^{z20}
β-Santalene	511-59-1						0.8	
cis-β-Santalene								0.1^{z20}
α-Selinene	473-13-2			tr	1.5		0.7	0.8^o; 1.0^{z10}; 2.1^r; 3.6^w; 4.3^{z17}
β-Selinene	17066-67-0			0.2	4.4	2.8	1.1	1.7^w; 2.1^{z20}; 3.6^{z10}; 5.6^{z17}
Selin-11-en-4-ol	16641-47-7			1.3				
β-Sesquiphellandrene	20307-83-9				0.3	0.4	0.1	0.2^p; 2.6^{z4}
Sesquisabinene B	1367879-38-6			tr				
7-epi-Sesquithujene	159407-35-9			tr				
β-Sinensal (*cis*-)	17909-87-4							1.0^b
Spathulenol	6750-60-3		1.6	0.6	0.7	0.4	1.2	0.6^y; 1.6^m; 1.7^{z14}; 2.9^{z2}; 4.1^n
α-Terpinene	99-86-5			tr	0.1	0.2	0.3	tr^f; 0.05^c; 0.09^n; $0.2^{g,t}$; 0.3^j
γ-Terpinene	99-85-4		0.4	tr	1.9		0.9	0.4^p; 0.5^x; 0.9^j; 1.5^{z2}; 1.9^k
Terpinen-4-ol	562-74-3	tr-1.4	8.0	0.3	4.8	0.8	6.4	3.1^c; 3.5^{z21}; 5.3^j; 6.6^o; 9.0^x
1-Terpineol	586-82-3							$<0.01^{z3}$
α-Terpineol	98-55-5	0.04-1.7	2.1	1.0	1.0		2.6	1.4^{z13}; 2.4^j; 2.5^o; 5.1^{z16}; 6.6^w
γ-Terpineol	586-81-2							2.6^m

Table 5.4.3 Constituents identified in basil oils (*continued*)

Constituent	CAS	Percentage and range						
		A	B (64)	C (2)	D (11)	E (3)	F (30)	G
δ-Terpineol	7299-42-5	0-0.2		tr	0.3			tr[f,l]; 0.1[z4]; 0.2[z19]; 0.5[r]
Terpinolene	586-62-9	0-0.2	2.0	0.2	0.5	0.2		0.06[r]; 0.4[p]; 0.5[j]; 1.1[z17]; 1.5[c,e]
Terpinyl acetate	8007-35-0							0.9[k]
α-Terpinyl acetate	80-26-2			tr			0.8	tr[z16]; 0.1[z4]; 0.7[n]
Tetradecane	629-59-4			tr				
Tetradecene	26952-13-6			tr				
Tetramethylpyrazine	1124-11-4							+[z35]
β-Thujaplicine	499-44-5							0.3[y1,y2]
α-Thujene	2867-05-2			tr	tr			0.2[j,n]; 0.3[w,z21]; 0.4[z26]; 0.5[r]
α-Thujone	546-80-5		0.1					0.4[t]; 0.5[y]
Thymol	89-83-8		0.3	tr				+[z35]; 0.1[z14]; 0.3[n]; 0.5[e]
Thymyl acetate	528-79-0							0.4[z2]
Torreyol	19435-97-3							<0.1[z20]
n-Triacontane	638-68-6							tr[z28]
Tricosane	638-67-5							<0.1[y1]
Tricyclene	508-32-7							0.1[z17]
Tritriacontane	630-05-7							tr[z28,y1]
Undecane	1120-21-4		tr					
(9Z)-Undecen-1-al					0.1			
Valencene	4630-07-3							0.5[k]
Valeric acid	109-52-4							+[z35]
Vanillin	121-33-5							+[z35]
Verbenol	473-67-6							0.5[r]
trans-Verbenol	1820-09-3							0.8[v]
Viridiflorene (ledene)	21747-46-6							0.1[p]; 0.4[f]
Viridiflorol	552-02-3		0.5					0.07[n]; 0.1[y]; 0.4[z6]; 0.5[z16]; 1.8[s]
α-Ylangene	14912-44-8							0.02[z15]; 0.2[k]
α-Zingiberene	495-60-3		2.7			3.3		0.4[e]

A forty-seven basil essential oil samples from Egypt, India, France, Hungary, Morocco, Nepal and Vietnam, analyzed between 1998 and 2013; lowest and highest concentrations given (E. Schmidt, unpublished data)

B twenty-seven lab-hydrodistilled oils from the flowers and leaves of 27 cultivars of *O. basilicum* L. grown in Croatia; highest concentrations given (ref. 64); all chemicals were absent in one or more oil samples and therefore their lower concentration was zero; 1,8-cineole was found in all but one oil and the range in the other 26 was 0.4-12.1; linalool was present in all but two oils and its range in the other 25 oils was 1.7-66.4 (ref. 64)

C fifteen lab-hydrodistilled oils from the aerial parts of *O. basilicum* in the flowering stage, nine from the wild in Serbia and six obtained from two local pharmacies, plus three commercial oils from Serbia adulterated with α-linalool; mean values were presented for the wild group, both pharmacies groups separately and the commercial oils; highest mean values given (ref. 2); all oils were of the linalool chemotype

D fifteen lab-hydrodistilled oils from the aerial parts in bloom of basil plants growing wild in the foot- and mid-hills of northern India; highest concentrations given (ref. 11); most chemicals were absent in one or more oils and therefore their lower concentration was zero; the following components were found in all oils (range in brackets): myrcene (<0.05-0.7), 1,8-cineole (<0.05-17.5), linalool (0.1-74.3) and β-caryophyllene (tr-6.2) (ref. 11)

E eighteen lab-hydrodistilled oils from the aerial parts with flowers of *O. basilicum* cultivated in Turkey from eighteen Turkish landraces; highest concentrations given (ref. 3); most chemicals were absent in one or more oil samples and therefore their lower concentration was zero; the following components were found in all (but one, in the case of 1,8-cineole) oils (range in brackets): 1,8-cineole (<0.1-14.5), linalool 1.7-60.2, zingiberene (0.9-3.3), β-selinene (<0.1-2.8), germacrene D (0.7-4.3), γ-cadinene (0.4-5.0) and hexadecanoic acid (tr-0.7) (ref. 3)

F twelve lab-hydrodistilled oils from the whole flowering plant of 12 basil cultivars grown in Colombia; highest concentrations given (ref. 30); most chemicals were absent in one or more oils and therefore their lower concentration was zero; the following components were found in all oil samples (range in brackets): 1,8-cineole (0.3-14.6), linalool (0.9-33.0), methyl chavicol (0.5-12.3), (Z)-methylcinnamate (0.4-9.7) and (E)-methyl cinnamate (8.3-74.5) (ref. 30)

G data from other studies (indicated with superscript letters); highest concentrations found in any study reviewed here given; when two or more oils were investigated, only the highest concentrations are mentioned, unless indicated otherwise

Table 5.4.3 Constituents identified in basil oils (*continued*)

[a] adulteration of commercial oils with α-linaloool; [b] twelve lab-hydrodistilled oils from the aerial parts in the flowering stage of 12 accessions of *O. basilicum* var. *purpurascens* cultivated in Iran (ref. 52); [c] ten lab-hydrodistilled oils from the whole plant at the beginning of flowering of ten basil cultivars grown in Italy (ref. 7); [d] thirty-eight lab-hydrodistilled oils from 38 *O. basilicum* accessions of various genotypes in full bloom, cultivated in Mississippi, USA (ref. 5); only 10 components were investigated; [e] five lab-hydrodistilled oils from the leaves and inflorescences of wild growing basil collected at five localities in Togo (ref. 32); [f] eight essential oils from the whole over-ground part above the lignified parts of the sprout of four basil cultivars from Poland in full bloom with two regimes of nitrogen application to the leaves (ref. 19); [g] five commercial oil samples and six lab-hydrodistilled oils from three basil cultivars in mid-flowering cultivated in Australia, three from frozen fresh and three from dried plant material (ref. 58); [h] fifteen lab-hydrodistilled oils from three *O. basilicum* cultivars harvested monthly in a period 5 months in Morocco, partly in full bloom (ref. 44); [i] twelve lab-hydrodistilled oils from flowering basil plants collected in the wild and in gardens at several places in Benin; only eight components were investigated (ref. 36); [j] twenty-four steam-distilled oils from the aerial parts of *O. basilicum* harvested at three locations in Mali and prepared from both fresh and dried plant material (ref. 14); [k] eight lab-distilled oils from Turkey prepared from the aerial parts of basil grown under four different nitrogen levels (ref. 24); [l] sixteen lab-hydrodistilled oils from two cultivars from Turkey grown under 8 different fertilization regimes (ref. 23); [m] thirteen lab-hydrodistilled oils from the aerial parts in full bloom of 13 basil cultivars grown in the USA (ref. 21); [n] fourteen lab-hydrodistilled oils from the aerial parts of two Iranian landraces, from fresh material and biomass dried in six different manners (ref. 63); [o] four lab-hydrodistilled oils from four basil cultivars in full bloom grown in India (ref. 62); [p] four lab-hydrodistilled oils from the above-ground parts over ligneous shoot fragments of four cultivars of *O. basilicum* at the beginning of flowering, cultivated in Poland (ref. 45); [q] four lab-hydrodistilled oils from the above ground parts in full bloom of two cultivars harvested in two seasons in India (ref. 4); [r] four lab-hydrodistilled oils from the aerial parts of *O. basilicum* at the beginning of the flowering cultivated in Iran under different irrigation regimes (ref. 50); [s] four lab-hydrodistilled oils from the aerial parts in full bloom of *O. basilicum* harvested in Pakistan in 4 seasons (ref. 56); [t] three lab-distilled oils from one cultivar of *O. basilicum* aerial parts in the flowering phase and from the varieties *purpurascens* and *diffiforme* (which are considered synonymous with *O. basilicum* L. by the Plant List) from Croatia (13,37); [u] two lab-hydrodistilled oils from the aerial parts in the blooming stage of two basil cultivars grown in northern India (ref. 17); [v] two lab-hydrodistilled oils from the aerial parts of flowering wild basil in Iran (ref. 16); [w] two lab-hydrodistilled oils from the leaves and flowering tops of basils collected in the wild in Tanzania, both with very atypical compositions (ref. 57);

[x] two samples of lab-hydrodistilled oil from the flowering tops of basil growing wild in Kenya (ref. 9); [y] one lab-hydrodistilled oil from the aerial parts of flowering *O. basilicum* cultivated in Serbia and Montenegro (ref. 60); [y1] data from various studies cited in ref. 69; because of the abundance of literature on basil oils, only chemicals which have *not* been found in any other study are mentioned here (ref. 69); [y2] incorrect identification (ref. 69); [z] one lab-hydrodistilled oil from the flowering above ground parts of endemic purple green leaved basil cultivated in north-west Iran (ref. 31); [z1] one lab-hydrodistilled oil from *O. basilicum* var. *purpurascens* aerial parts in the flowering stage cultivated in Iran (ref. 12); [z2] four lab-hydrodistilled oils from the aerial flowering parts of basil harvested in the wild in Iran and dried before distillation in four different manners (ref. 33); data which are partly also published in ref. 31 (under [z]), are not presented here again; [z3] one laboratory steam-distilled oil from basil in full bloom cultivated in Israel (ref. 23); [z4] one laboratory-prepared oil from the aerial parts in full bloom of basil cultivated in Italy (ref. 8); [z5] one lab-hydrodistilled oil from a basil cultivated in Mongolia and in full bloom (ref. 20); [z6] one lab-hydrodistilled oil from the aerial parts at the flowering stage of basil collected in the wild in Turkey (ref. 48); [z7] one lab-hydrodistilled basil oil from the flowering aerial parts of plants grown near a university campus in Taiwan (ref. 6); [z8] one steam-distilled oil from the aerial parts in bloom from basil cultivated in Cuba (ref. 28); [z9] one extremely atypical basil oil sample obtained by distillation from the aerial parts in the flowering stage of *O. basilicum* cultivated in Iran (ref. 43); [z10] one lab-hydrodistilled oil from the leaves and flowering tops of basil cultivated in Serbia (ref. 55); [z11] one lab-hydrodistilled oil from the flowering aerial parts of basil cultivated in Romania (ref. 61); [z12] one lab-hydrodistilled basil flower oil from Turkey (ref. 10); [z13] one lab-hydrodistilled oil from the aerial parts in full bloom of a basil cultivar from Iran (ref. 29); [z14] one lab-hydrodistilled and one steam-distilled oil sample from the aerial parts (flowering?) of basil from Turkey (ref. 39); [z15] one lab-hydrodistilled oil from the over-ground parts of basil (flowering?) cultivated in Turkey (ref. 18); [z16] one lab-hydrodistilled oil from basil collected in the wild in Algeria (ref. 46); [z17] one lab-hydrodistilled oil from basil growing wild in north-east India with a camphoreous odor (ref. 15); [z18] one commercial basil oil sample produced in Egypt (ref. 59); [z19] one lab-hydrodistilled oil from the aerial parts of basil purchased at a local market in Turkey (ref. 51); [z20] one commercial oil of unknown origin (ref. 38); [z21] one lab-hydrodistilled oil from a whole (flowering?) basil plant from Cameroon (ref. 22); [z22] one lab-hydrodistilled oil from the aerial (flowering?) parts of basil growing in the wild in southern Italy (ref. 49); [z23] nine lab-hydrodistilled oils from the flowering aerial parts of 9 *O. basilicum* cultivars from Brazil (ref. 34); only 4 components were investigated; [z24] several lab-hydrodistilled oils from the flowering aerial parts of basil cultivated in Iran; mean concentrations given (ref. 47); [z25] one lab-hydrodistilled oil from the inflorescences of basil from India; only linalool and methyl cinnamate were investigated (ref. 42); [z26] five lab-hydrodistilled oils from the aerial parts in the flowering stage of basils collected in vegetable gardens at five locations in Togo (ref. 53); [z27] data from various studies cited in ref. 65; [z28] data cited in ref. 27; [z29] data taken from ref. 25; [z30] data cited in ref. 28; [z31] lab-hydrodistilled oils from the above ground parts of flowering basil of 17 cultivars grown in Poland (ref. 35); [z32] one lab steam-distilled oil from the semi-wilted flowering tops of basil produced in Nepal (ref. 66); [z33] one steam-distilled oil from Madagascar; the sesquiterpene fraction was investigated; chemicals are indicated with +[z33], as their concentration (in the sesquiterpene fraction) cannot be compared with the other data (ref. 67); [z34] one commercial oil sample purchased from a German manufacturer (ref. 68); [z35] detailed examination of a lab-hydrodistilled methyl chavicol-rich oil from basil stem and leaves from the Philippines by solvent extraction, distillation and column chromatography (ref. 54)

tr: trace (in columns C, D and G[f]: <0.05; in column G[o]: <0.1); + present in the oil investigated, but quantity not stated

Table 5.4.4 Major constituents of basil oils

Constituent	CAS	Percentage and range							
	A	B (64)	C (2)	D (11)	E (3)	F (30)	G		
Linalool	78-70-6	0.6-55.8	66.4	79.5	74.3	60.2		33.0	75.9[f]; 77.5[p]; 83.6[z23]; 98.9[x]
Methyl chavicol	140-67-0	0.2-87.0	87.8	22.8	88.2	76.3		12.3	89.9[i]; 90.0[o]; 94.3[q]; 94.6[t]
Methyl eugenol	93-15-2	0-24.7	38.8	0.2	14.9	34.2		26.0	40.5[j]; 74.5[e]; 78.0[z15]; 91.1[d]
(E)-Methyl cinnamate	1754-62-7		7.4			38.7		74.5	16.7[z14]; 31.9[z3]; 41.9[z31]; 82.4[m]
1,8-Cineole	470-82-6	0.03-13.7	12.1	8.1	17.5	14.5		14.6	14.8[m]; 15.3[h]; 16.5[r]; 54.3[w]
Geranial	141-27-5	0-0.6	0.7	0.2		36.9			1.6[j]; 24.6[n]; 25.7[z31]; 39.4[o]
Neral	106-26-3	0.08-0.5			0.1		27.6		10.0[v]; 20.8[z31]; 22.7[n]; 29.4[o]
Eugenol	97-53-0	0.03-15.3	20.1		2.4	5.9	21.1	4.1	14.2[i]; 24.8[j]; 27.6[p]; 29.0[d]
epi-α-Cadinol	5937-11-1	0-7.1			9.3	1.8		8.3	7.6[c]; 12.4[s]; 17.3[n]; 27.5[z4]
α-trans-Bergamotene	13474-59-4	0.01-19.8			3.1	5.9		3.2	7.5[f]; 7.6[e]; 14.9[i]; 15.8[z4]; 17.5[v]
Geraniol	106-24-1			0.9	5.1	6.8	3.9		4.3[p]; 5.2[o]; 12.5[v]; 13.4[f]; 16.5[l]
β-Caryophyllene	87-44-5	0.02-1.4	1.0	0.5		5.2	4.2	43.0	7.5[v]; 8.0[i]; 10.7[z31]; 14.9[n]
Caryophyllene oxide	1139-30-6	0-0.2	0.5	0.1	1.1		2.0	13.9	1.1[b]; 2.5[z2]; 4.0[z9]; 8.0[n]; 11.4[w]
Nerol	106-25-2			0.4	0.1	9.7	1.8		0.9[z16]; 2.7[n]; 11.2[o]; 12.6[z24]

LEGEND: SEE UNDER TABLE 5.4.3

Many chemotypes of essential oil of basil have been proposed, usually based on the amount of the prevailing ingredients, mostly linalool, methyl chavicol, eugenol, methyl eugenol and methyl cinnamate (3,4,5,7,11,12,14,64). Some are based on one (dominant) ingredient only (e.g., linalool type, methyl eugenol type, methyl chavicol type), others on combinations (e.g., methyl cinnamate/linalool type, methyl eugenol/linalool chemotype, linalool/eugenol type). In some studies, other chemicals have been found to be the major ingredient, including (E)-anethole (32), bergamotene (5), trans-α-bergamotene (16), β-bisabolene (11), borneol (43), τ-cadinol (8), camphor (9,15), β-caryophyllene (30), 1,8-cineole (57), citral (3,11), elemene (44), geranial (11,35,62), linalyl acetate (48), menthone (31,33), and (E)-myroxide (57). The main chemotypes have been reviewed in refs. 11 and 14. The linalool present in basil oils is always β-linalool. The presence of α-linalool may indicate that the oil in question is either forged or is synthetic basil oil (2).

Because of the abundance of literature on basil oils, we have focused on oils that were prepared from the flowering tops (few studies) or from the aerial parts of basil in full bloom (most studies), as these likely are the qualities commonly used in the pharmaceutical, cosmetic, aromatherapy and perfumery industries. Studies specifically mentioning basil *leaves* only as source material for essential oils (of which there are dozens) are not discussed (with the exception of ref. 54, as many components not found in any other study were mentioned there). However, some authors have shown that the chemical profiles of oils obtained from the leaves and those from the inflorescences (of the same plants) are largely comparable (4,6,9) and obviously, the aerial parts consist for a large part of leaves. All cultivars have been included in the search as were the *Ocimum basilicum* varieties *diffiforme* and *purpurascens*, which are synonyms of *O. basilicum* L. according to the Plant List. From review articles on basil oils (27,28,65,69), only data of chemicals *not* mentioned in other sources, reviewed here, are cited.

CONTACT ALLERGY/ALLERGIC CONTACT DERMATITIS

General
Contact allergy to/allergic contact dermatitis from basil oil has been reported in two publications only, both from occupational exposure in aromatherapists. In neither can a false-positive reaction due to the excited skin syndrome be excluded. In one case report, linalool may have been an allergen in basil oil (71).

Case reports
An aromatherapist had chronic hand dermatitis and was patch test positive to 17 of 20 oils used at her work (tested 1% and 5% in petrolatum), including basil oil (70). Two other aromatherapists had contact dermatitis (occupational in one) with allergies to multiple essential oils used at their work, including basil oil. Both patients also reacted to geraniol, α-pinene, the fragrance mix and various other fragrance materials. In addition, one proved to be allergic to linalool and linalyl acetate, the other to caryophyllene; α-pinene, linalool, geraniol and caryophyllene were demonstrated by GC-MS in basil oil (71). Linalool may be an important component of basil oil and has been found in concentrations of up to 55.8% in commercial basil oils (Table 5.4.3, column A).

LITERATURE

1 Pushpangadan P, George V. Basil. In: Peter KV, Ed. Handbook of herbs and spices, 2nd Ed., Vol. 1. Oxford-Cambridge-Philadelphia-New Delhi: Woodhead Publishing Ltd, 2012: Chapter 4, 55-72

2 Radulovíc, Blagojevíc PD, Miltojevíc AB. α-Linalool – a marker compound of forged/synthetic sweet basil (*Ocimum basilicum* L.) essential oils. J Sci Food Agric 2013; 93:3292-3303

3 Telci I, Bayram E, Yılmaz G, Avci B. Variability in essential oil composition of Turkish basils (*Ocimum basilicum* L.). Biochem System Ecol 2006;34:489-497

4   Verma RS, Padalia RC, Chauhana A. Variation in the volatile terpenoids of two industrially important basil (*Ocimum basilicum* L.) cultivars during plant ontogeny in two different cropping seasons from India. J Sci Food Agric 2012;92:626-631

5   Zheljazkov VD, Callahan A, Cantrell CL. Yield and oil composition of 38 basil (*Ocimum basilicum L.*) accessions grown in Mississippi. J Agr Food Chem 2008;56:241-245

6   Sheen L-Y, Tsai Ou Y-H, Tsai S-J. Flavor characteristic compounds found in the essential oil of *Ocimum basilicum* L. with sensory evaluation and statistical analysis. J Agric Food Chem 1991;39:939-943

7   Marotti M, Piccaglia R, Giovanelli E. Differences in essential oil composition of basil (*Ocimum basilicum* L.) Italian cultivars related to morphological characteristics. J Agric Food Chem 1996;44:3926-3929

8   Occhipinti A, Capuzzo A, Bossi S, Milanesi C, Maffei ME. Comparative analysis of supercritical CO_2 extracts and essential oils from an *Ocimum basilicum* chemotype particularly rich in T-cadinol. J Essent Oil Res 2013;25:272-277

9   Dambolena JS, Zunino MP, López AG, Rubinstein HR, Zygadlo JA, Mwangi JW, et al. Essential oils composition of *Ocimum basilicum* L. and *Ocimum gratissimum* L. from Kenya and their inhibitory effects on growth and fumonisin production by *Fusarium verticillioides*. Innov Food Sci Emerg Technol 2010;11:410-414

10 Chalchat J-C, Özcan MM. Comparative essential oil composition of flowers, leaves and stems of basil (*Ocimum basilicum* L.) used as herb. Food Chem 2008;110:501-503

11 Verma RS, Padalia RC, Chauhana A, Thul ST. Exploring compositional diversity in the essential oils of 34 *Ocimum* taxa from Indian flora. Ind Crops Prod 2013;45:7-19

12 Dolatabad SS, Moghaddam M, Chalajour H. Essential oil composition of four *Ocimum* species and varieties growing in Iran. J Essent Oil Res 2014;26:315-321

13 Carović-Stanko K, Orlić S, Politeo O, Strikić F, Kolak I, Milos M, et al. Composition and antibacterial activities of essential oils of seven *Ocimum* taxa. Food Chem 2010;119:196-201. Data also presented in ref. 37

14 Chalchat J-C, Garry R-P, Sidibé L, Harama M. Aromatic plants of Mali (I): Chemical composition of essential oils of *Ocimum basilicum* L. J Essent Oil Res1999;11:375-380

15 Purkayastha J, Nath SC. Composition of the camphor-rich essential oil of *Ocimum basilicum* L. native to northeast India. J Essent Oil Res 2006;18:332-334

16 Yavari M, Mirdamadi S, Masoudi S, Tabatabaei-Anaraki M, Larijani K, Rustaiyan A. Composition and antibacterial activity of the essential oil of a green type and a purple type of *Ocimum basilicum* L. from Iran. J Essent Oil Res 2011;23:1-4

17 Padalia RC, Verma RS. Comparative volatile oil composition of four *Ocimum* species from northern India. Nat Product Res 2011;25:569-575

18 Özcan M, Chalchat J-C. Essential oil composition of *Ocimum basilicum* L. and *Ocimum minimum* L. in Turkey. Czech J Food Sci 2002;20:223-228

19 Nurzyńska-Wierdak R. Sweet basil essential oil composition: relationship between cultivar, foliar feeding with nitrogen and oil content. J Essent Oil Res 2012;24:217-227

20 Shatar S, Altantsetseg S. Essential oil composition of some plants cultivated in Mongolian climate. J Essent Oil Res 2000;12:745-750

21 Vieira RF, Simon JE. Chemical characterization of basil (*Ocimum* spp.) based on volatile oils. Flavour Fragr J 2006;21:214-221

22 Amvam Zollo PH, Biyiti L, Tchoumbougnang F, Menut C, Lamaty G, Bouchet Ph. Aromatic plants of tropical central Africa. Part XXXII. Chemical composition and antifungal activity of thirteen essential oils from aromatic plants of Cameroon. Flavour Fragr J 1998;13:107-114

23 Nurzyńska-Wierdak R, Borowski B, Dzida K, Zawiślak G, Kowalski R. Essential oil composition of sweet basil cultivars as affected by nitrogen and potassium fertilization. Turk J Agric Forest 2013;37:427-436

24 Daneshian A, Gurbuz B, Cosge B, Ipek A. Chemical components of essential oils from basil (*Ocimum basilicum* L.) grown at different nitrogen levels. Int J Nat Engin Sci 2009;3:9-13

25 Bahl JR, Garg SN, Bansal RP, Naqvi AA, Singh V, Kumar S. Yield and quality of shoot essential oil from the vegetative, flowering and fruiting stage crops of *Ocimum basilicum* cv. Kusumohak. J Med Arom Plan Sci 2000;22:743-746. Data cited in ref. 27

26 Kamada T, Casali VWD, Barbosa LCA, Fortes ICP, Fingal FL. Phenotype plasticity of the essential oil in basil accesses (*Ocimum basilicum* L). Rev Bras Plant Med 1999;1:12-22. Data cited in in ref. 27

27 Lawrence BM. Progress in essential oils. Basil oil. In: Lawrence BM. Essential oils 2001-2004. Carol Stream, USA: Allured Publishing Corporation, 2006:293-301 (reprinted from Perfumer and Flavorist 2004;29(6):80-90)

28 Lawrence BM. Progress in essential oils. Basil oil. In: Lawrence BM. Essential oils 2005-2007. Carol Stream, USA: Allured Publishing Corporation, 2008:11-16 (reprinted from Perfumer and Flavorist 2005;30(2):72-79)

29 Omidbaigi R, Mirzai M, Moghadam MS. Chemical investigation of a new cultivar of purple basil (*Ocimum basilicum* cv. Opal) from Iran. J Essent Oil Bear Plants 2007;10:209-214

30 Vina A, Murillo E. Essential oil composition from twelve varieties of basil (*Ocimum spp*) grown in Colombia. J Braz Chem Soc 2003;14:744-749

31 Hassanpouraghdam MB, Hassani A, Shalamzari MS. Menthone and estragole rich essential oil of cultivated *Ocimum basilicum* L. from Northwest Iran. Chemija 2010;21:59-62. Data also published in ref. 33

32 Koba K, Poutouli PW, Christine R, Jean-Pierre C, Komla S. Chemical composition and antimicrobial properties of different basil essential oils chemotypes from Togo. Bangladesh J Pharmacol 2009;4:1-8

33 Hassanpouraghdam MB, Hassani A, Vojodi L, Farsad-Akhtar N. Drying method affects essential oil content and composition of basil (*Ocimum basilicum* L.) J Essent Oil Bear Plants 2010;13:759-766. Data also partly published in ref. 31

34 Serafinia LA, Fernandes Pauletti G, Duarte Rota L, Atti dos Santos AC, Agostini F, Zattera F, Moyna P. Evaluation of the essential oils from nine basil (*Ocimum basilicum* L.) cultivars planted in southern Brazil. J Essent Oil Bear Plants 2009;12:471-475

35 Nurzyńska-Wierdak R. Morphological and chemical variability of *Ocimum basilicum* L. (Lamiaceae). Mod Phytomorphol 2013;3:115-118

36 Yayi E, Moudachirou M, Chalchat JC. Chemotyping of three *Ocimum* species from Benin: *O. basilicum*, *O. canum* and *O. gratissimum*. J Essent Oil Res 2001;13:13-17

37 Carović-Stanko K, Fruk G, Satovic Z, Ivić D, Politeo O, Sever Z, et al. Effects of *Ocimum* spp. essential oil on *Monilinia laxa in vitro*. J Essent Oil Res 2013;25:143-148. Data also presented in ref. 13

38 Tognolini M, Barocelli E, Ballabeni V, Bruni B, Bianchi A, Chiavarini M, Impicciatore M. Comparative screening of plant essential oils: Phenylpropanoid moiety as basic core for antiplatelet activity. Life Sciences 2006;78:1419-1432

39 Özek T, Beis SH, Demirçakmak B, Baser KHC. Composition of the essential oil of *Ocimum basilicum* L. cultivated in Turkey. J Essent Oil Res 1995;7:203-205

40 Pino JA, Roncal E, Rosado A, Goire I. The essential oil of *Ocimum basilicum* L. from Cuba. J Essent Oil Res 1994;6:89-90

41 Fleisher Z, Fleisher A. Volatiles of *Ocimum basilicum* traditionally grown in Israel. J Essent Oil Res 1992;4:97-99

42 Singh RS, Bordoloi DN. Changes in the linalool and methyl cinnamate amounts in a methyl cinnamate-rich clone of *Ocimum basilicum* at different growth stages. J Essent Oil Res 1991;3:475-476

43 Farhang V, Amini J, Ebadollahi A, Sadeghi GR. *Ocimum basilicum* L. essential oil cultivated in Iran: chemical composition and antifungal activity against three *Phytophthora* species. Arch Phytopathol Plant Protect 2014;47:1696-1703

44 Belkamel A, Bammi J, Janneot V, Belkamel A, Dehbi Y, Douira A. Évaluation de la biomasse et analyse des huiles essentielles de trois variétés de basilic (*Ocimum basilicum* L.) cultivées au Maroc. Acta Botanica Gallica: Botany Letters 2008;155:467-476

45 Nurzyñska-Wierdak R. Morphological variability and essential oil composition of four *Ocimum basilicum* L. cultivars. J Essent Oil Bear Plants 2014;17:112-119

46 Brada M, Hadj Khelifa L, Achour D, Wathelet JP, Lognay G. Essential oil composition of *Ocimum basilicum* L. and *Ocimum gratissimum* L. from Algeria. J Essent Oil Bear Plants 2011;14:810-814

47 Omidbaigi R, Mirzaei M, Moghadam MS. Difference of growth traits, essential oil content and compositions between diploid and induced tetraploid plants of basil (*Ocimum basilicum* L.). J Essent Oil Bear Plants 2010;13:579-587

48 Ozcan M, Chalchat J-C. Essential oil composition of a new chemotype of basil (*Ocimum basilicum* L.) cultivating in Turkey. J Essent Oil Bear Plants 2004;7:155-159

49 Senatore F, De Fusco R, Grassia A, Moro CO, Rigano ED, Napolitano F. Chemical composition and antibacterial activity of essential oils from five culinary herbs of the Lamiaceae family growing in Campania, Southern Italy. J Essent Oil Bear Plants 2003;6:166-173

50 Omidbaigi R, Hassani A, Sefidkon F. Essential oil content and composition of sweet basil (*Ocimum basilicum*) at different irrigation regimes. J Essent Oil Bear Plants 2003;6:104-108

51 Figueredo G, Ünver A, Chalchat JC, Arslan D, Öcan MM. A research on the composition of essential oil isolated from some aromatic plants by microwave and hydrodistillation. J Food Biochem 2012;36:334-343

52 Pirmoradi MR, Moghaddam M, Farhadi N. Chemotaxonomic analysis of the aroma compounds in essential oils of two different *Ocimum basilicum* L. varieties from Iran. Chem Biodivers 2013;10:1361-1371

53 Sanda K, Koba K, Nambo P, Gaset A. Chemical investigation of *Ocimum* species growing in Togo. Flavour Fragr J 1998;13:226-232

54 Hasegawa Y, Tajima K, Toi N, Sugimura Y. Characteristic components found in the essential oil of *Ocimum basilicum* L. Flavour Fragr J 1997;12:195-200

55 Filip S, Vidovic S, Adamovic D, Zekovic Z. Fractionation of non-polar compounds of basil (*Ocimum basilicum* L.) by supercritical fluid extraction (SFE). J Supercrit Fluids 2014;86:85-90

56 Hussain AI, Anwar F, Sherazi STH, Przybylski R. Chemical composition, antioxidant and antimicrobial activities of basil (*Ocimum basilicum*) essential oils depends on seasonal variations. Food Chem 2008;108:986-995

57 Runyoro D, Ngassapa O, Vagionas K, Aligiannis N, Graikou K, Chinou I. Chemical composition and antimicrobial activity of the essential oils of four *Ocimum* species growing in Tanzania. Food Chem 2010;119:311-316

58 Lachowicz KJ, Jones GP, Briggs DR, Bienvenu FE, Palmer MV, Ting SST, et al. Characteristics of essential oil from basil (*Ocimum basilicum* L.) grown in Australia. J Agric Food Chem 1996;44:877-881

59 Sacchetti G, Medici A, Maietti S, Radice M, Muzzoli M, Manfredini S, et al. Composition and functional properties of the essential oil of Amazonian basil, *Ocimum micranthum* Willd., Labiatae in comparison with commercial essential oils. J Agric Food Chem 2004;52:3486-3491

60 Bozin B, Mimica-Dukic N, Simin N, Anackov G. Characterization of the volatile composition of essential oils of some Lamiaceae spices and the antimicrobial and antioxidant activities of the entire oils. J Agric Food Chem 2006;54:1822-1828

61 Stefan M, Zamfirache MM, Padurariu C, Trută E, Gostin I. The composition and antibacterial activity of essential oils in three *Ocimum* species growing in Romania. Cent Eur J Biol 2013;8:600-608

62 Padalia RC, Verma RS, Chauhan A, Chanotiya CS. Changes in aroma profiles of 11 Indian *Ocimum* taxa during plant ontogeny. Acta Physiol Plant 2013;35:2567-2587

63 Pirbalouti AG, Mahdad E, Craker L. Effects of drying methods on qualitative and quantitative properties of essential oil of two basil landraces. Food Chem 2013;141:2440-2449

64 Liber Z, Carovic-Stanko K, Politeo O, Strikic F, Kolak I, Milos M, Satovic Z. Chemical characterization and genetic relationships among *Ocimum basilicum* L. cultivars. Chem Biodivers 2011;8:1978-1989

65 Lawrence BM. Progress in essential oils. Perfum Flavor 1995;20(July/August):29-41

66 Yonzon M, Lee DJ, Yokochi T, Kawano Y, Nakahara T. Antimicrobial activities of essential oils of Nepal. J Essent Oil Res 2005;17:107-111

67 Gaydou EM, Faure R, Bianchini J-P, Lamaty G, Rakotonirainy O, Randriamiharisoa R. Sesquiterpene composition of basil oil. Assignment of the ^{1}H and ^{13}C NMR spectra of β-elemene with two-dimensional NMR. J Agric Food Chem 1989;37:1032-1037

68 Baratta MT, Dorman HJD, Deans SG, Figueiredo AC, Barroso JAG, Ruberto G. Antimicrobial and antioxidant properties of some commercial essential oils. Flavour Fragr J 1998;13:235-244

69 Lawrence BM. Progress in essential oils. Perfum Flavor 1998;23(Nov/Dec):35-50

70 Selvaag E, Holm J, Thune P. Allergic contact dermatitis in an aromatherapist with multiple sensitizations to essential oils. Contact Dermatitis 1995;33:354-355

71 Dharmagunawardena B, Takwale A, Sanders KJ, Cannan S, Roger A, Ilchyshyn A. Gas chromatography: an investigative tool in multiple allergies to essential oils. Contact Dermatitis 2002;47:288-292

Chapter 5.5 BAY OIL

DEFINITION

Bay oil (essential oil of bay) is the essential oil obtained from the leaves of the bay rum tree (West Indian bay), *Pimenta racemosa* (Mill.) J.W. Moore.

INCI NOMENCLATURE

Description/definition: Pimenta acris leaf oil is the volatile oil distilled from the leaves of the bay, *Pimenta acris*, Myrtaceae

INCI name EU: Pimenta acris leaf oil

INCI name USA: Pimenta acris (bay) leaf oil

Other names: Bay leaf oil

CAS registry number(s): 8006-78-8; 85085-61-6; 91721-75-4

EINECS number(s): 294-376-5

Pimenta racemosa var. *racemosa* and *Pimenta acris* (Sw.) Kostel are synonyms for *Pimenta racemosa* (Mill.) J.W. Moore (GRIN taxonomy for plants, The Plant List, 2).

ISO (INTERNATIONAL ORGANIZATION FOR STANDARDIZATION) STANDARD

ISO number: 3045

ISO name: Essential oil of bay

Botanical origin: *Pimenta racemosa* (Mill.) J.W. Moore

Parts of plant used: Leaf

ISO values: ISO values (minimum and maximum concentrations) are shown in Table 5.5.1.

Bay oil should not be confused with *sweet* bay oil, which is obtained from the leaves of *Laurus nobilis* L. (see Laurel leaf oil, Chapter 5.45).

THE PLANT, THE OIL, AND THEIR USES

Pimenta racemosa (Mill.) J.W. Moore, commonly known as bay, bayrum tree or West Indian bay, is a shrub or small slender tree 7.5-15 meters tall with very aromatic leaves. It is native to the Caribbean (West Indies) and northern South America, and is cultivated widely in tropical countries including Indonesia, West Indies, Venezuela, Mexico, Puerto Rico, Guyana, Jamaica, Tanzania (Zanzibar and Pemba) and Cameroon (2). It is used for producing bay rum, which is not for drinking, but is a fragrance for external use (GRIN Taxonomy for Plants; http://plants.jstor.org/). In the Caribbean, bay leaves are used for cooking rice dishes, soups and stews and for making tea. Other applications include its uses as air freshener and insect repellent. Bay leaves are also used in Caribbean folk medicine, e.g., to lower blood pressure, or for the treatment of digestive problems or headache (http://Latinfood.about.com). The dried fruits of the *P. racemosa* tree are used around the world as spice.

Essential oils may be obtained from the leaves and the fruits. They are used in perfumes, aftershaves, lotions enhancing hair growth and strength or acting against hair loss, and for commercial food flavoring. In addition, the essential bay leaf oil is important in Caribbean folk medicine and used for the treatment of rheumatism, toothache and other ailments because of its apparent anti-inflammatory and analgesic properties (2,3). It is also employed in aromatherapy practices (22).

CHEMICAL COMPOSITION

Bay oil is a clear mobile liquid of brownish to dark brown color, which has a spicy odor, reminding of cloves. The yield of essential oil from the leaves of *Pimenta racemosa* (Mill.) J.W. Moore generally varies from 0.8 to 1.4% of fresh leaves and from 2.5 to 3.5% when obtained from dried leaves. The main producing countries of this oil are West India and some countries in South America including Venezuela and Guyana.

Literature data (up to October 29, 2014) on the chemical composition of bay oils and unpublished analytical data from one of us (E.S.) are shown in Table 5.5.2 in alphabetical order. In geranium oils from various origins, over 110 chemicals have been identified. About 51% of these were found in a single reviewed publication only.

The major compounds found in bay oils from different sources are shown in Table 5.5.3; they include (highest concentrations in any study given) eugenol (92.9%), geranial (53.2%), methyl eugenol (48.1%), methyl chavicol (32.8%), neral (32.6%), myrcene (30.9%), 1,8-cineole (20.4%), and chavicol (17.1%). Well-known ingredients of bay oils that were present in high concentrations in one study were β-pinene (22.9%), terpinen-4-ol (20.7%), α-terpineol (10.0%), *p*-cymene (8.0%) and geraniol (7.5%).

Commercial oils

The ten chemicals that had the highest maximum concentrations in 33 commercial bay essential oil samples (concentration ranges provided) are the following: eugenol (41.4-54.0%), myrcene (20.5-32.0%), chavicol (6.6-10.8%), limonene (2.4-3.2%), linalool (0.5-2.7%), terpinolene (0.1-2.1%), β-caryophyllene (0.4-1.9%), β-pinene (0.05-1.8%), dimyrcene (0.2-1.5%) and (*E*)-ocimene (0.7-1.3%) (Erich Schmidt, unpublished analytical data).

Chemotypes

There appear to be (at least) three chemotypes of bay oil. In Guadeloupe, three varieties of *Pimenta racemosa* (Mill.) J.W. Moore may be found (16). They are identical in their external morphology, but can easily be distinguished from the smell of their leaves. One of them, known as the common bay tree, is particularly abundant and is characterized by the smell of clove from its leaves and fruits. The other two varieties are rather rare and have smells of lemon and anise. The common clove type proved to contain eugenol (56%) and chavicol (17%) as main ingredients (the eugenol-chavicol chemotype). This is the commercial type essential oil of bay. The lemon type was dominated by geranial (40%) and neral (32%) (geranial-neral type), whereas the anise type of the bay tree had methyl eugenol (48%) and methyl chavicol (33%) as main ingredients (methyl eugenol-methyl chavicol type) (16). The anise and lemon type had previously also been demonstrated in bay oils from the Dominican Republic (15,17) and in even far older studies (cited in

Table 5.5.1 ISO values (%) for bay oil[a]

Compound	CAS	Minimum	Maximum
Eugenol	97-53-0	42	56
Myrcene	123-35-3	20	30
Chavicol	501-92-8	8.0	13.0
Limonene	138-86-3	1	4
Linalool	78-70-6	1	3
Methyl eugenol	93-15-2	0.1	2

[a] ISO Essential oil of bay ©ISO 2004; Geneva, Switzerland, www.iso.org

ref. 15). Early reports of high concentrations of citral (up to 65%) (cited in ref. 15) and α-humulene, β-caryophyllene and limonene (cited in refs. 9,10) have remained unsubstantiated. In one report from Cuba, terpinen-4-ol (21%) was the ingredient with the highest concentration (14, data also reported in refs. 7, 12 and 21). By far, most oils of which the analyses have been reported and which are discussed here had eugenol as main ingredient accompanied by high concentrations of chavicol, myrcene or both (Table 5.5.2).

Table 5.5.2 Constituents identified in bay oils

Constituent	CAS	A	B (3)	C (2)	D (9)	E (16)	F
							Percentage and range
Acetone	67-64-1						tr[f]
Acetyl chavicol	61499-22-7						0.3[j]
α-Amorphene	20085-19-2				0.4		
Anethole	104-46-1					0.2	0.1[h]
Aromadendrene	489-39-4				0.09		
allo-Aromadendrene	25246-27-9				0.3		
Bergamotene							0.2[d]
Borneol	507-70-0						0.8[c]
α-Cadinene	24406-05-1			0.08			0.5[c]
β-Cadinene	523-47-7				tr		
γ-Cadinene	39029-41-9	0.03-0.2		0.2	0.2		
δ-Cadinene	483-76-1	0.4-0.7	0-0.2	0.6	1.0	0.8	0.01[f]; 0.5[d]; 1.5[b]
α-Cadinol	481-34-5				0.3	0.2	0.02[f]
γ-Cadinol	50895-55-1				tr		
τ-Cadinol	5937-11-1		0-0.2				0.2[c]
Calamenene	483-77-2					0.2	
cis-Calamenene	72937-55-4				0.01		
Camphene	79-92-5				0.1		
Camphor	76-22-2				tr	1.5	
δ3-Carene	13466-78-9						0.2[c]
trans-2-Caren-2-ol							0.1[f]
Carvacrol	499-75-2						0.01[f]
Carveol	99-48-9						0.1[f]
β-Caryophyllene	87-44-5	0.4-1.9	0-0.3	0.5	7.2	0.6	0.02[f]; 0.2[j]; 0.4[a]; 0.7[b]; 4.0[d]; 4.9[i]
Caryophyllene oxide	1139-30-6	0.02-1.0				0.1	0.03[f]
β-Chamigrene	18431-82-8				0.2		
Chavicol	501-92-8	6.6-10.8	7.1-9.3	9.3	15.5	17.1	7.7[b]; 8.9[e]; 10.1[h]; 10.3[j]; 10.4[d]; 10.5[a]
Chrysanthenone	473-06-3					1.6	
1,8-Cineole	470-82-6	0.1-0.6	2.1-3.2	0.7	1.4	1.2	1.1[a]; 1.3[c]; 3.2[j]; 4.7[j]; 9.7[e]; 20.4[h]
Cinnamaldehyde	104-55-2				0.1		
Citronellal	106-23-0						0.04[f]
Citronellyl acetate	150-84-5						0.08[f]
α-Copaene	3856-25-5	0.2-0.6		0.3	0.4	0.4	0.3[a]; 1.3[d]
α-Cubebene	17699-14-8			tr			
Cyclocitral	52844-21-0						0.2[f]
α-Cyclocitral	432-24-6					0.2	
p-Cymene	99-87-6	0.4-0.7	0.4-0.9	0.5	1.0	0.5	0.2[f]; 0.4[j]; 0.5[c]; 0.8[a]; 1.7[e]; 8.0[h]
p-Cymenene	1195-32-0				0.3		0.2[h]
p-Cymen-8-ol	1197-01-9				tr	0.1	0.6[c]
Decanal	112-31-2	0.02-0.1					
(Z)-4-Decenal	21662-09-9						<0.06[f]
2,3-Dehydro-1,8-cineole	92760-25-3					0.1	0.3[f]
Dimyrcene	532-87-6	0.2-1.5					
β-Elemene	33880-83-0				0.2		
Eugenol	97-53-0	41.4-54.0	45.2-52.7	45.6	68.9	56.1	46.3[a]; 52.7[c]; 54.5[j]; 60[d]; 64[i]; 92.9[b]
Eugenyl acetate	93-28-7				0.5		0.2[c]
α-Farnesene	502-61-4	0.1-0.8			0.6		

Table 5.5.2 Constituents identified in bay oils (*continued*)

Constituent	CAS	Percentage and range					
		A	B (3)	C (2)	D (9)	E (16)	F
(*E,E*)-α-Farnesene	502-61-4		0-0.3	0.5		0.2	0.3[l]
Geranial	141-27-5	0.06-0.3		0.06	0.2	40.3	0.2[h]; 41.3[f]; 53.2[g]
Geranic acid	459-80-3					1.0	0.07[f]
Geraniol	106-24-1	0.05-0.1		0.08	0.1	3.5	0.2[a]; 2.8[g]; 7.5[f]
Geranyl acetate	105-87-3						0.4[f]
Geranyl formate	105-86-2						0.01[f]
Germacrene D	23986-74-5					0.3	0.05[f]; 0.5[b]
Globulol	489-41-8				0.4		
α-Gurjunene	489-40-7			0.01	0.1		
Hexanol	111-27-3						0.05[f]
(*E*)-2-Hexenal	6728-26-3						0.02[f]
(*Z*)-3-Hexenol	928-96-1						0.2[f]
γ-Homogeraniol							0.3[f]
α-Humulene	6753-98-6	0.1-0.3	0-0.1	0.4	1.3	0.2	0.01[f]; 0.3[b]
Ibuprofen	15687-27-1						0.2[j,k]
Isoeugenol	97-54-1			0.7			0.3[c]
Limonene	138-86-3	2.4-3.2	3.0-4.0	2.9	3.9	5.3	1.9[g]; 2[i]; 3.1[a,c]; 3.4[l]; 3.8[b]; 6.0[e]
Linalool	78-70-6	0.5-2.7	0.1-2.1	2.3	3.6	6.0	1.6[i]; 1.9[c]; 2.2[d]; 2.6[a]; 3.0[g]; 3.2[e]; 3.4[f]
cis-Linalool oxide	11063-77-7	0.01-0.03		0.09			
cis-Linalool oxide, furanoid	5989-33-3						0.06[f]
trans-Linalool oxide	11063-78-8					0.1	
trans-Linalool oxide, furanoid	34995-77-2						0.07[f]
Menthadienol							2.1[f]
6-Methoxyeugenol	6627-88-9						0.3[h]
2-Methyl-3-buten-2-ol	115-18-4						tr[f]
Methyl chavicol	140-67-0	0.01-0.4		0.3	0.05	32.8	0.1[a]; 0.8[h]; 1.3[d]; 31.6[g]
Methyl cinnamate	103-26-4						0.1[h]
Methyl eugenol	93-15-2	0.04-1.3		1.0	11.9	48.1	0.2[e]; 0.3[h]; 0.5[a]; 5.4[d]; 43.1[g]
6-Methyl-5-hepten-2-one	110-93-0					0.3	0.2[g]; 0.9[a]; 1.1[f]
α-Muurolene	10208-80-7				0.5		
γ-Muurolene	30021-74-0						0.02[f]
γ-Muurolol	138068-73-2				tr		
Myrcene	123-35-3	20.5-32.0	21.9-30.9	25.0	16.2	12.8	9.6[b]; 14.6[i]; 21.3[e]; 25.9[a]; 26.6[c]; 30.9[l]
Neral	106-26-3	0.02-0.2		0.09		31.7	0.2[h]; 31.7[f]; 32.6[g]
Neric acid	37349-29-4					0.2	0.1[f]
Nerol	106-25-2			0.4			
(*E*)-Nerolidol	40716-66-3			0.5			
Neryl acetate	141-12-8						0.09[f]
(*E*)-2-Nonenal	18829-56-6						<0.06[f]
Ocimene	13877-91-3						2.3[e]
(*E*)-Ocimene	27400-72-2	0.7-1.3					
(*E*)-β-Ocimene	3779-61-1		0-0.3	0.9	1.4	0.4	0.2[c,f]; 0.3[g,l]
(*Z*)-Ocimene	27400-71-1	0.05-0.1					0.4[j]
(*Z*)-β-Ocimene	3338-55-4			0.1	0.1	0.1	0.04[f]; 0.2[b]; 0.3[c]
n-Octanol (1-)	111-87-5	0.01-0.2		0.2			0.3[h]
3-Octanol	589-98-0	0.3-0.7		0.7	0.8	3.0	0.1[g]; 0.3[e]; 0.4[b]
3-Octanone	106-68-3	0.1-1.1		0.8	1.1		0.4[d]; 0.7[b]; 0.8[e]
1-Octen-3-ol	3391-86-4	0.7-1.1	1.3-2.4	0.9	1.7		0.2[f]; 1.3[g]; 1.5[h]; 1.6[c,e]; 1.9[l]
α-Phellandrene	99-83-2	0.3-0.6		0.4	0.5	0.3	0.1[a,f,g]; 0.7[b,c]; 0.8[j]; 1.6[h]
β-Phellandrene	555-10-2	0.04-1.0		0.4			0.1[g]; 1.1[e]
α-Pinene	80-56-8	0.3-0.4	0.3-0.6	0.5	0.5	0.5	0.1[b,g]; 0.2[f]; 0.4[a]; 0.5[j,l]; 1.9[e]; 2.0[h]
β-Pinene	127-91-3	0.05-1.8	0-0.1	0.07	0.09		0.3[h]; 0.4[e]; 22.9[j]
Sabinene	3387-41-5				tr		0.3[h]; 1.4[j]
α-Selinene	473-13-2				0.6		
β-Selinene	17066-67-0				0.3		
Spathulenol	6750-60-3				0.1		
α-Terpinene	99-86-5	0.09-0.2	0.1-0.6	0.2	0.2		0.02[f]; 0.1[b,h]; 0.3[j]; 0.6[l]
γ-Terpinene	99-85-4	0.07-0.2	0-0.2	0.2	0.2	0.1	0.04[f]; 0.1[b]; 0.3[j]; 4.6[h]
Terpinen-4-ol	562-74-3	0.2-0.6	0.7-0.9	0.5	0.9	1.2	0.5[j]; 0.6[d]; 0.7[a]; 0.9[e]; 1.9[l]; 20.7[h]
α-Terpineol	98-55-5	0.06-0.7	0.7-0.9	0.07	0.5	0.5	0.3[a,c]; 0.4[j]; 0.9[l]; 2.3[e]; 10.0[h]

Table 5.5.2 Constituents identified in bay oils (*continued*)

Constituent	CAS	Percentage and range					
		A	B (3)	C (2)	D (9)	E (16)	F
Terpinolene	586-62-9	0.1-2.1	0-0.3	0.4	0.2	0.1	0.03[f]; 0.1[g]; 3.1[h]
α-Thujene	2867-05-2	0.01-0.3		0.08	tr		0.02[f]; 0.1[j]; 0.2[c]; 1.1[h]; 1.9[e]
α-Thujone	546-80-5					0.1	0.1[g]
Thymol	89-83-8						0.3[j]
Torreyol	19435-97-3		0-0.3				
Verbenol	473-67-6					1.0	1.1[f]
Viridiflorene (ledene)	21747-46-6						0.01[f]

A thirty-three essential oil samples from West India analyzed between 2003 and 2013; lowest and highest concentrations given (E. Schmidt, unpublished data)

B six lab-prepared oil samples from leaves of *P. racemosa* growing wild in Benin, collected in two places in the years 2003, 2004 and 2005; lowest and highest concentrations given (ref. 3)

C one commercial oil from Jamaican bay rum tree prepared in Germany (ref. 2)

D two commercial bay leaf oils from unknown origin; highest concentrations given (ref. 9)

E one lab-hydrodistilled oil from leaves of 'clove-type' *P. racemosa* from Guadeloupe (eugenol-chavicol chemotype), one from the lemon variety (geranial-neral chemotype) and one from leaves of the anise variety (methyleugenol-methylchavicol type) (ref. 16)

F data from other studies (indicated with superscript letters); highest concentrations found in any study reviewed here given; when two or more oils were investigated, only the highest concentrations are mentioned, unless indicated otherwise

[a] one commercial oil sample purchased in South Korea (ref. 1); [b] two lab-prepared oil samples from leaves harvested in spring and autumn in North India (ref. 4); [c] one lab-hydrodistilled oil collected from leaves of a bay rum tree growing wild in Benin (ref. 10); [d] one commercial oil of unknown origin (ref. 11); [e] one bay oil sample, origin unknown (ref. 6); [f] one hydrodistilled oil from leaves harvested in Guadeloupe of the geranial-neral chemotype (ref. 8); [g] one steam-distilled oil from leaves from the lemon variety and one from the anise variety of *P. racemosa* harvested in the Dominican Republic (ref. 15); [h] one lab-hydrodistilled oil from leaves of *P. racemosa* growing wild in western Cuba (ref. 14; data apparently also presented in refs. 7, 12 and 21); [i] one lab-hydrodistilled leaf oil from plant material purchased from a Czech company (ref. 18); [j] one lab-hydrodistilled *P. racemosa* leaf oil from plant material harvested in the wild in Benin (ref. 19); [k] misidentification; [l] one lab-hydrodistilled oil from the leaves of *P. racemosa* grown at a farm in Benin (ref. 20);

tr: trace (in columns D and F[f]: <0.01);

CONTACT ALLERGY/ALLERGIC CONTACT DERMATITIS

General

Contact allergy to bay oil has been reported in one publication, but no cases of allergic contact dermatitis from the oil have been identified. In a group of fragrance sensitive patients reacting to bay oil, eugenol may have been an important allergen.

Testing in groups of patients

In a group of 21 patients with dermatitis caused by fragrances and tested with a series of essential oils, nine (43%) reacted to oil of bay 1.5% in petrolatum; relevance data were not provided (23). Bay oil consists of 40-55% of eugenol, and as eugenol is an important cause of fragrance sensitivity, the high percentage of positive reactions to bay oil (43%) in this fragrance-sensitive population may be ascribed to eugenol. Indeed, of the seven patients reacting

Table 5.5.3 Major constituents of bay oils

Constituent	CAS	Percentage and range					
		A	B (3)	C (2)	D (9)	E (16)	F
Eugenol	97-53-0	41.4-54.0	45.2-52.7	45.6	68.9	56.1	46.3[a]; 52.7[c]; 54.5[j]; 60[d]; 64[i]; 92.9[b]
Geranial	141-27-5	0.06-0.3		0.06	0.2	40.3	0.2[h]; 41.3[f]; 53.2[g]
Methyl eugenol	93-15-2	0.04-1.3		1.0	11.9	48.1	0.2[e]; 0.3[h]; 0.5[a]; 5.4[d]; 43.1[g]
Methyl chavicol	140-67-0	0.01-0.4		0.3	0.05	32.8	0.1[a]; 0.8[h]; 1.3[d]; 31.6[g]
Neral	106-26-3	0.02-0.2		0.09		31.7	0.2[h]; 31.7[f]; 32.6[g]
Myrcene	123-35-3	20.5-32.0	21.9-30.9	25.0	16.2	12.8	9.6[b]; 14.6[i]; 21.3[e]; 25.9[a]; 26.6[c]; 30.9[l]
1,8-Cineole	470-82-6	0.1-0.6	2.1-3.2	0.7	1.4	1.2	1.1[a]; 1.3[c]; 3.2[l]; 4.7[j]; 9.7[e]; 20.4[h]
Chavicol	501-92-8	6.6-10.8	7.1-9.3	9.3	15.5	17.1	7.7[b]; 8.9[e]; 10.1[h]; 10.3[j]; 10.4[d]; 10.5[a]
β-Caryophyllene	87-44-5	0.4-1.9	0-0.3	0.5	7.2	0.6	0.02[f]; 0.2[j]; 0.4[a]; 0.7[b]; 4.0[d]; 4.9[i]
Limonene	138-86-3	2.4-3.2	3.0-4.0	2.9	3.9	5.3	1.9[g]; 2[i]; 3.1[a,c]; 3.4[l]; 3.8[b]; 6.0[e]
Linalool	78-70-6	0.5-2.7	0.1-2.1	2.3	3.6	6.0	1.6[i]; 1.9[c]; 2.2[d]; 2.6[a]; 3.0[g]; 3.2[e]; 3.4[f]

LEGEND: SEE UNDER TABLE 5.5.2

to bay oil and tested with eugenol, five (71%) co-reacted to eugenol (23).

LITERATURE

1 Kim J, Lee Y-S, Lee S-G, Shin S-C, Park I-K. Fumigant antifungal activity of plant essential oils and components from West Indian bay (*Pimenta racemosa*) and thyme (*Thymus vulgaris*) oils against two phytopathogenic fungi. Flavour Fragr J 2008;23:272-277

2 Jirovetz L, Buchbauer G, Stoilova I, Krastanov A, Stoyanova A, Schmidt E. Spice plants: Chemical composition and antioxidant properties of *Pimenta* Lindl. essential oils, part 2: *Pimenta racemosa* (Mill.) J.W. Moore leaf oil from Jamaica. Ernährung/Nutrition 2007;31: 293-300

3 Alitonou GA, Noudogbessi J-P, Sessou P, Tonouhewa A, Avlessi F, Menut C, et al. Chemical composition and biological activities of essential oils of *Pimenta racemosa* (Mill.) J.W. Moore from Benin. Int J Biosc 2012;2:1-12

4 Pragadheesh VS, Yadav A, Singh SC, Gupta N, Chanotiya CS. Leaf essential oil of cultivated *Pimenta racemosa* (Mill.) J.W. Moore from North India: Distribution of phenylpropanoids and chiral terpenoids. Med Aromat Plants 2013;2:118. Available at http://dx.doi.org/10.4172/2167-0412.1000118

5 Lawrence BM. Progress in essential oils. Perfum Flavor 2000;25 (July/Aug):55-70

6 Buttery RG, Black DR, Guadangi DG, Ling LC, Connolly G, Teranishi R. California bay oil. 1. Constituents, odor properties. J Agric Food Chem 1974;22:773-777 (cited in ref. 5)

7 Bello A, Urquiola A, Garcia JJ, Rosado A, Pino JA. Essential oil from leaves of *Pimenta racemosa* (Mill.) J.W. Moore (Myrtaceae) from Western Cuba. Ing Cienc Quim 1998;18:21-23 (cited in ref. 5)

8 Chapron N, Parfait A, Bourgeois P. Industrial importance of essential oils of 'Bois d'Inde' - *Pimenta racemosa* var. *racemosa* (M. Miller) J.W. Moore. Rivista Ital. EPPOS (Numero Speciale) 1998;467-474 (cited in ref. 5)

9 Tucker AM, Maciarello MJ, Adams RP, Landrum LR, Zanoni TA. Volatile leaf oils of Caribbean Myrtaceae. I. Three varieties of *Pimenta racemosa* (Milller) J. Moore of the Dominican Republic and the commercial bay oil. J Essent Oil Res 1991;3:323-329

10 Ayedoun AM, Adeoti BS, Setondji J, Menut C, Lamaty G, Bessière J-M. Aromatic plants from tropical West Africa. IV. Chemical composition of leaf oil of *Pimenta racemosa* (Miller) J.W. Moore var. *racemosa* from Benin. J Essent Oil Res 1996;8:207-209

11 Delespaul Q, de Billerbeck VG, Roques CG, Michel G, Marquier-Vinuales C, Bessière J-M. The antifungal activity of essential oils as determined by different screening methods. J Essent Oil Res 2000;12:256-266

12 Bello A, Pino J, Marbot R, Urquiola A, Aguero J. Components volátiles des plantas de la famila Myrtaceae de la región occidental de Cuba. Revista CENIC Cienc Quim 2001;32:143-147 (cited in ref. 13)

13 Lawrence BM. Progress in essential oils. Bay oil. Perfum Flav 2007;32 (July):46-52

14 Bello A, Rodriguez ML, Castiñeira N, Urquiola A, Rosado A, Pino JA. Chemical compositon of the leaf oil of *Pimenta racemosa* (Mill.) J. Moore from Western Cuba. J Essent Oil Res 1995;7:423-424

15 McHale D, Laurie WA, Woof MA. Composition of West Indian bay oils. Food Chem 1977;2:19-25

16 Abaul J, Bourgeois P, Bessiere JM. Chemical composition of the essential oils of chemotypes of *Pimenta racemosa* var. *racemosa* (P. Miller) J. W. Moore (Bois d'Inde) of Guadeloupe (F.W.I.). Flavour Fragr J 1995;10:319–321

17 Ames GR, Barrow M, Borton C, Casey, T, Matthews WS, Nabney J. Bay oil distillation in Dominica. Tropical Science 1971;13:13

18 Zabka M, Pavela R, Prokinova E. Antifungal activity and chemical composition of twenty essential oils against significant indoor and outdoor toxigenic and aeroallergenic fungi. Chemosphere 2014;112:443-448

19 Tchobo FP, Alitonou GA, Soumanou MM, Barea B, Bayrasy C, Laguerre M, et al. Chemical composition and ability of essential oils from six aromatic plants to counteract lipid oxidation in emulsions. J Am Oil Chem Soc 2014;91:471-479

20 Keke M, Yehouenou B, de Souza C, Sohounhloue D. Evaluation of hygienic and nutritional quality of peulh cheese treated by *Sorghum vulgaris* (L) and *Pimenta racemosa* (Miller) extracts. Scientific Study and Research 2009;10:29-46

21 Leyva M, Tacoronte JE, Marquetti Mdel C. Chemical composition and lethal effect of essential oil from *Pimenta racemosa* (Myrtales: Myrtaceae) on *Blatella germanica* (Dictyoptera: Blattellidae) (article in Spanish). Rev Cubana Med Trop 2007;59:154-158

22 Rhind JP. Essential oils. A handbook for aromatherapy practice, 2nd Edition. London: Singing Dragon, 2012

23 Meynadier JM, Meynadier J, Peyron JL, Peyron L. Formes cliniques des manifestations cutanées d'allergie aux parfums. Ann Dermatol Venereol 1986;113:31-39.

Chapter 5.6   BERGAMOT OIL

DEFINITION
Bergamot oil (essential oil of bergamot) is the essential oil obtained from the pericarp (peel) of the bergamot orange, *Citrus bergamia* (Risso et Poit.)

INCI NOMENCLATURE

Bergamot fruit oil

Description/definition: Citrus aurantium bergamia fruit oil is the psoralen-free volatile oil obtained from the fruit of *Citrus aurantium* L. var. *bergamia*, Rutaceae

INCI name EU: Citrus aurantium bergamia fruit oil

INCI name USA: Citrus aurantium bergamia (bergamot) fruit oil

Other names: bergamot oil bergaptene free; bergamot oil rectified

CAS registry number(s): 8007-75-8; 68648-33-9

EINECS number(s): 616-915-9; 614-687-5 (not EINECS numbers, but numbers in the EC format)

Bergamot peel oil

Description/definition: Citrus aurantium bergamia peel oil is the volatile oil obtained from the peel of *Citrus aurantium bergamia*, Rutaceae

INCI name EU: Citrus aurantium bergamia peel oil

INCI name USA: Citrus aurantium bergamia (bergamot) peel oil

CAS registry number(s): 89957-91-5

EINECS number(s): 289-612-9

BERGAMOT PEEL OIL EXPRESSED

Description/definition: Citrus bergamia peel oil expressed is the essential oil expressed from the epicarps of the bergamot, *Citrus bergamia risso*, Rutaceae

INCI name EU: Citrus aurantium bergamia peel oil expressed (perfuming name, not an INCI name proper)

INCI name USA: not in the Personal Care Products Council Ingredient Database

Other names: Bergamot orange oil

CAS registry number(s): 89957-91-5; 85049-52-1

EINECS number(s): 289-612-9

ISO (INTERNATIONAL ORGANIZATION FOR STANDARDIZATION) STANDARD

ISO number: 3520

ISO name: Essential oil of bergamot, Italian type

Botanical origin: *Citrus bergamia* (Risso et Poit.) (synonym: *Citrus aurantium* L. subsp. *bergamia* (Wight et Arnott) Engler)

Parts of plant used: Pericarp (peel)

ISO values: ISO values (minimum and maximum concentrations) are shown in Table 5.6.1.

By ISO definition, all citrus essential oils except lime oil are produced by expression; oils obtained from citrus fruits by distillation may not be called essential oils according to ISO criteria (except lime oil). In industry, however, sometimes residues from expression of the citrus peels undergo steam distillation, to obtain the remaining oil; these volatile oils may then be added to the expressed oil. As some of the compounds undergo changes forced by high temperature during distillation, this addition (which is an adulteration) changes the composition of the essential oil. Because of this, and also because it cannot be excluded that oils entirely produced by hydrodistillation reach the market, both the ingredients found in 'genuine' bergamot essential oils obtained by expression and those that may be present in hydrodistilled oils are presented here.

THE PLANT, THE OIL, AND THEIR USES
Bergamot is the common name of the fruit and plant of *Citrus bergamia* Risso et Poiteau. It is a small tree that blossoms during the winter, producing a fragrant pale yellow, spherical fruit 7.5-10 cm in diameter, which has a bitter taste but a pleasant odor (31). Its origin is uncertain, but probably lies in the Mediterranean region. *Citrus bergamia* is commercially grown mainly in the southern Italian region of Calabria in the area of Reggio Calabria, limited to a narrow strip of the coast, along the Ionian and the Thyrrenian seas, of about 150 km. Here, more than 90% of the world production is realized (23,25). Bergamot is also cultivated and industrially processed in Ivory Coast, Brazil and China. Small cultivations are registered in Turkey, Argentina, Uruguay, Cameroon, and in some countries in North Africa (27,31). The fruit is almost exclusively grown for the production of essential peel oil, not for juice consumption. Recently, however, bergamot has been reconsidered, along with other *Citrus* fruits, as an ingredient of traditional Mediterranean cooking (27).

Bergamot essential oils are obtained by cold extraction, generally using 'pelatrice'-type machines. From the residues of the cold extraction, oils of lower quality can be recovered with distillation techniques and chemical modification such as 'torchiati', 'ricicli', 'pulizia dischi', 'distilled' and 'bergapten-free' bergamot oils (see below, under Phototoxicity of bergamot oil) (27, 35).

Among the *Citrus* peel oils, because of its unique fragrance and freshness, bergamot essential oil is the most valuable and is therefore mainly employed in the perfumery (e.g., in the original Eau de Cologne) and cosmetic industries, where it may be used in skin care creams and lotions for its cooling and refreshing nature (15,25,38). The oil also serves for the flavoring of sweets, tobacco, liqueur (curaçao), tea, e.g., Earl Grey tea (37), baked goods, desserts, chewing gum and soft drinks (23,27). Because of its antiseptic and antibacterial properties, bergamot oil is inserted in the pharmacopoeia of several countries and is used in the pharmaceutical industry (23,25) and in sanitary preparations (15). Both in folk medicine and in aromatherapy (where it is a popular oil [15,32]), bergamot oil is perceived as analgesic,

Table 5.6.1 ISO values (%) for bergamot oil[a]

Compound	CAS	Minimum	Maximum
Limonene	138-86-3	30.0	45.0
Linalyl acetate	115-95-7	22.0	36.0
Linalool	78-70-6	3.0	15.0
γ-Terpinene	99-85-4	6.0	10.0
β-Pinene	127-91-3	5.5	9.5
β-Bisabolene	495-61-4	0.3	0.55
Geranial	141-27-5	0.25	0.5

[a] ISO 3520 Essential oil of bergamot, Italian type ©ISO 1998; Geneva, Switzerland, www.iso.org

antidepressant, antimicrobial, carminative, digestive, sedative and febrifuge (26).

Phototoxicity of bergamot oil

Already nearly a century ago, it was discovered that bergamot oil is phototoxic, causing skin reactions (burns) and secondary hyperpigmentation when the skin to which products (notably fragrances) containing bergamot oils had been applied were exposed to sunlight (26). These reactions were caused by furocoumarins (psoralens) present in the non-volatile fraction of the oils. Bergapten (5-methoxypsoralen), which is also photomutagenic and photocarcinogenic (32), is considered the main culprit. To avoid such problems, and also to adhere to legislation (in Europe, 5-MOP-containing cosmetics have been banned or restricted to certain concentrations) and regulations from the fragrance industry, furocoumarin-free extracts were developed for preparing commercial products instead of the original oils (25). Such 'furocoumarin-free' or 'bergaptenfree' oils or 'bergamot FCF' (furocoumarin-free) (32) are

obtained mainly by alkaline treatment or modern distillation techniques (27,30,43) and maintain the composition of the volatile fraction and a bouquet very similar to cold-pressed oils (27) (although some state that such procedures result in a considerable impoverishment of the composition of the oil thus obtained [15]).

For aromatherapy oils, however, there are currently no official limits to 5-MOP concentrations and no strict legal requirement for placing warning labels on these products in many countries, and consequently the use of such oils still results in occasional cases of phototoxic dermatitis from exposure to sunlight or artificial ultraviolet radiation (32).

CHEMICAL COMPOSITION

Bergamot oil is a greenish to yellow, clear mobile liquid which has a fresh, citrusy, soft green and fruity odor with floral accent. The yield of essential oil from the fruits and peels of Citrus bergamia (Risso et Poit.) generally varies from 0.4 to 0.6%. The main producing countries of this oil are Italy, Argentina and Ivory Coast.

Cold-pressed Citrus essential oils consist of a volatile fraction, which represents 85-99% of the oil, and a non-volatile fraction 1-15%, which mainly contains oxygen heterocyclic compounds, especially coumarins, psoralens and polymethoxyflavones (28) (Table 5.6.4). These components play an important role in the characterization of cold-pressed Citrus oils, since the composition of this fraction is characteristic of each oil. Moreover, many of the pharmacological and toxicological activities possessed by Citrus oils have been demonstrated to be related to these components (28).

Literature data (up to June 18, 2014) on the chemical composition of bergamot oils and unpublished analytical data from one of us (E.S.) are shown in Table 5.6.2

Table 5.6.2 Constituents identified in bergamot oils

Constituent	CAS	Percentage and range					
		A	B (1)	C (1)	D (1-18)	E (1,19-24)	F
Acetic acid	64-19-7		tr				
γ-Acoradiene	28400-12-6		tr				
Aromadendrene	489-39-4		0.4				
Ascaridole	512-85-6				0-0.01		tr-0.01[c]
Bergamotene					0-0.3	0-0.3	
α-Bergamotene	17699-05-7		0.9	0.2	0-0.3	0-0.3	0.6[m]; 0.9[i]
cis-α-Bergamotene	18252-46-5	0-0.01	0.05		0-0.3	0-0.4	tr[e]; 0.02[c]; 0.3[d]
trans-α-Bergamotene	13474-59-4	0.02-1.2	0.9		0-0.9	0-0.4	0.2-0.4[c]; 0.3[d,k,l]; 0.4[b]; 0.9[e]
trans-β-Bergamotene	15438-94-5				0-0.6		0.01-0.02[c]
Bergapten	484-20-8	0-0.2			tr		0.1[k]; 0.2[e]
Bicyclogermacrene	24703-35-3		0.08		0-0.03	0-0.03	0.01-0.07[c]; 0.01[h]
α-Bisabolene	17627-44-0		tr				0.1[i]
(E)-α-Bisabolene	25532-79-0				0-0.01		0.01[c]
(Z)-α-Bisabolene	29837-07-8		tr		0-0.03	0-tr	0.03-0.04[c]
β-Bisabolene	495-61-4	0.01-0.5	1.3	1.0	0-0.7	0-0.9	0.3-0.5[c]; 0.4[d]; 0.8[m]; 1.2[e]; 1.5[i]
(E)-γ-Bisabolene	53585-13-0				0-0.02		tr-0.03[c]
(Z)-γ-Bisabolene	13062-00-5		0.01		0-0.01	0-tr	tr-0.01[c]

Table 5.6.2 Constituents identified in bergamot oils (*continued*)

Constituent	CAS	Percentage and range					
		A	B (1)	C (1)	D (1-18)	E (1,19-24)	F
α-Bisabolol	515-69-5		0.1		0-0.03	0-0.03	tr-0.03[c]; 0.03[b]; 0.1[e,i]
epi-α-Bisabolol	78148-59-1		tr				
β-Bisabolol	15352-77-9		tr				
epi-β-Bisabolol	235421-59-7				0-0.01		
Borneol	507-70-0		tr				
Bornyl acetate	76-49-3		0.04		0-0.1	0-0.1	tr[e]; 0.01-0.02[c]; 0.02[b]
Butyl acetate	123-86-4			tr			
δ-Cadinene	483-76-1		tr	0.3	0-0.03		tr-0.01[c]
α-Cadinol	481-34-5		tr				
τ-Cadinol	5937-11-1		tr				
Camphene	79-92-5	0.02-1.1	0.1	0.2	0-0.05	0-0.04	tr[e]; 0.02-0.03[c]; 0.04[b]
Campherenol	18530-03-5		0.03		0-0.02	0-0.02	tr-0.01[c]; 0.2[b]
Camphor	76-22-2		0.01		0-0.01	0-0.1	tr-0.01[c]
δ2-Carene (= δ4-)	554-61-0		0.1				
δ3-Carene	13466-78-9		0.9	2.0	0-0.02	0-0.01	tr[c]; 0.01[b]; <0.05[i]; 0.9[e]
Carvacrol	499-75-2		tr				
Carveol	99-48-9						
(*E*)-Carveol	1197-07-5		tr			0-0.01	
(*Z*)-Carveol	1197-06-4		tr				
Carvone	99-49-0		0.1		0-0.1		tr-0.01[c]
Carvyl acetate	97-42-7		tr				
β-Caryophyllene	87-44-5	0.05-0.9	0.7	0.5	0-0.6	0-0.8	0.3-0.4[c]; 0.3[d,k,l]; 0.5[b]; 0.6[i]; 0.7[e]
Caryophyllene alcohol	56747-96-7		tr				
Caryophyllene oxide	1139-30-6		tr		0-tr	0-tr	0.2[m]
Caryophyllenol I	32214-88-3		tr				
Caryophyllenol II	32214-89-4		tr				
Cedrol	77-53-2				0-tr		
1,4-Cineole	470-67-7			tr			
1,8-Cineole	470-82-6		0.02	0.3	0-0.04	0-tr	+[k]
Citronellal	106-23-0	0-0.1	0.06	0.07	0-0.03	0-0.04	tr[e]; 0.01-0.02[c]; 0.02[b];
Citronellol	106-22-9		0.3	0.06	0-0.02	0-0.01	tr-0.02[c]
Citronellyl acetate	150-84-5	·	0.1	0.2	0-0.1	0-0.05	0.01-0.3[c]; 0.03[b]; 0.1[e]
Citronellyl formate	105-85-1			0.02			
Cuminyl alcohol	536-60-7		tr				
Curcumene (ar-; α-)	644-30-4		tr				
γ-Curcumene	28976-68-3				0-tr		tr-0.01[c]
m-Cymene	535-77-3		0.3			tr	0.3[e]
p-Cymene	99-87-6	0.1-0.7	3.6	1.6	0-1.4	0-0.2	0.07-0.6[c]; 0.4[b]; 0.5[n]; 0.6[k]; 2.0[d]
p-Cymenene	1195-32-0		tr				
p-Cymen-8-ol	1197-01-9		tr			0-tr	
Decanal	112-31-2	0.02-0.3	0.3	0.1	0-0.08	0-0.3	0.03-0.07[c]; 0.07[b]; 0.1[e,m]
Decanol	36729-58-5		tr		0-tr		
(*E*)-2-Decenal	3913-81-3		0.01		0-tr	0-tr	
Decyl acetate	112-17-4		0.1	0.1	0-0.04	0-0.06	0.02-0.05[c]; 0.1[e]
Dihydrocitronellol	106-21-8		0.05				
2,3-Dihydro epoxy-geranyl acetate							tr[f]
2,3-Dihydro epoxy-neryl acetate							tr[f]
Dihydrolinalool	18479-51-1			0.07			
Dimethoxycoumarin							0.1[e]
5,7-Dimethoxycou-marin	487-06-9	tr-0.1					

Table 5.6.2 Constituents identified in bergamot oils (*continued*)

Constituent	CAS	Percentage and range					
		A	B (1)	C (1)	D (1-18)	E (1,19-24)	F
2,6-Dimethyl-6-acetoxyocta-1,7-dien-3-one			tr				
2,6-Dimethyl-6-acetoxyocta-7-en-3-one			tr				
3,7-Dimethyl-3-acetoxyocta-1,5-dien-7-ol							+ (ref. 33)
3,7-Dimethyl-3-acetoxyocta-1,7-dien-6-ol	41610-78-0		tr				
2,6-Dimethyl-6-acetoxyocta-1-en-7-one			tr				
3,7-Dimethyl-3-hydroxy-1,6-octadienyl formate						0-0.1	
2,3-Dimethyl-3-(4-methyl-3-pentenyl)-2-norbornanol	98205-40-4						0.01[d]
4,8-Dimethyl-1,3(E),7-nonatriene	19945-61-0				0-0.01		tr-0.01
4,8-Dimethyl-1,3(Z),7-nonatriene	21214-62-0				0-tr		
3,7-Dimethyl-1,5-octadiene-3,7-diol	13741-21-4		tr				
3,7-Dimethyl-1,7-octadiene-3,6-diol	51276-33-6		tr				
2,7-Dimethyl-2,6-octadien-1-ol	22410-74-8					0-tr	
cis-2,6-Dimethyl-1,5,7-octatrien-3-ol			tr				
trans-2,6-Dimethyl-1,5,7-octatrien-3-ol			tr				
1,3-Divinylbenzene	108-57-6					0-0.2	
Dodecanal	112-54-9			0.2	0-0.03	0-0.05	tr[c,e]
Dodecane	112-40-3		0.01	0.03	0-tr	0-tr	
Dodecanol	112-53-8		0.01		tr	0-tr	
Dodecenal	82107-89-9	tr-0.09					
β-Elemene	33880-83-0				0-0.04	0-tr	tr-0.03[c]
γ-Elemene	29873-99-2		tr				
δ-Elemene	20307-84-0		0.06		0-0.04	0-0.1	tr-0.02[c]; 0.05[b]
β-Eudesmol	473-15-4		tr				
(E,E)-α-Farnesene	502-61-4		tr		0-0.01	0-0.08	tr-0.02[c]
(E)-β-Farnesene	18794-84-8		tr	0.6	0-0.07	0-0.1	0.05-0.07[c]; 0.1[m]
(Z)-β-Farnesene	28973-97-9	0.01-0.4	0.09		0-0.5	0-0.2	0.01-0.09[c]; 0.1[e]
Farnesol	4602-84-0		tr				
(E,E)-α-Farnesol			0.01				
endo-Fenchol (α-)	512-13-0			0.01			
Fenchyl alcohol	1632-73-1				0-0.01		tr[c]
Geranial	141-27-5	0.2-0.6	1.3	0.5	0-0.7	tr-0.8	0.2-0.4[c]; 0.3[d,l]; 0.7[b]; 0.8[i]
Geraniol	106-24-1	tr-0.1	0.07	5.7	0-0.1	0-1.7	
Geranyl acetate	105-87-3	0.04-0.6	1.6	1.9	0-2.4	0.02-2.7	0.03-0.6[c]; 0.4[b,d,l]; 0.7[m]; 1.6[e]
2,3-Geranyl acetate oxide	76638-49-8		tr				
Geranyl formate	105-86-2			0.02			
Geranyl propanoate	105-90-8		tr				
Germacrene A	28387-44-2						0.01[h]
Germacrene B	15423-57-1		0.04		tr-0.01	0-0.01	0.01[h]; 0.02[b]
Germacrene C	34323-15-4						0.02[h]

Table 5.6.2 Constituents identified in bergamot oils (*continued*)

Constituent	CAS	Percentage and range					
		A	B (1)	C (1)	D (1-18)	E (1,19-24)	F
Germacrene D	23986-74-5		0.1		0-0.1	0-0.2	tr[b]; 0.03-0.06[c]; 0.06[h]; 0.1[e]
γ-Heptalactone	105-21-5			0.04			
Heptyl acetate	112-06-1		0.02		0-0.01	0-tr	tr-0.01[c]
Hexanal	66-25-1		0.02				
Hexanol	111-27-3		tr	0.01			
(E)-2-Hexenal	6728-26-3		tr				
(E)-2-Hexen-1-ol	928-95-0		tr				
(Z)-3-Hexenol	928-96-1		0.01	tr			
(Z)-3-Hexenyl acetate	3681-71-8		tr				
Hexyl acetate	142-92-7		0.2	0.2	0-0.02	0-0.1	tr-0.03[c]; 0.2[e]
Hotrienol	20053-88-7		tr				
Hotrienyl acetate	150447-00-0		tr				
α-Humulene	6753-98-6		0.1	tr	0-0.02	0-0.2	0.02-0.03[c]; 0.04[b,i]; 0.1[e,m]
Humulene oxide I	19888-33-6		tr				
Humulene oxide II	19888-34-7		tr				
Hydroxylinalool	256418-61-8		0.1		0-+	0-0.1	0.1[e]
Indole	120-72-9		tr		0-tr	0-tr	
Isobornyl acetate	125-12-2				0-0.01		
Isocaryophyllene oxide			tr				
(Z)-Isocitral	72203-97-5						tr[c]
Isomenthone	491-07-6		0.01				
Isomenthyl acetate	20777-45-1		0.1				
Isopulegol	89-79-2		0.01	0.01	0-0.1	0-tr	tr[b,c]
Lavandulol	498-16-8			0.02			
Lilial	80-54-6		tr				
Limonene	138-86-3	31.4-45.1	54.9	45.1	23.5-50.5	17.0-60.0	28.7-45.8[c]; 42.4[b]; 43.7[i]; 49.1[d]
cis-Limonene oxide	13837-75-7		0.02	0.07	0-0.02	0-0.02	tr[b]; tr-0.02[c]
trans-Limonene oxide	4959-35-7		0.01		0-0.02	0-0.02	tr[b]; tr-0.02[c]
Limonen-4-ol	3419-02-1		tr				
Limonen-10-ol	3269-90-7						tr[f]
Linalool	78-70-6	4.7-16.7	24.2	17.9	5.6-36.1	1.2-29.0	4.3-26.0[c]; 20.2[i]; 26.0[b]; 36.1[d]
cis-Linalool oxide	11063-77-7	0-0.05			0-tr	0-tr	
cis-Linalool oxide, furanoid	5989-33-3		tr				0.1[m]
cis-Linalool oxide, pyranoid	14009-71-3		tr				
trans-Linalool oxide	11063-78-8	0-0.04			0-0.1	0-tr	
trans-Linalool oxide, furanoid	34995-77-2		tr				
trans-Linalool oxide, pyranoid	39028-58-5		tr				
Linalyl acetate	115-95-7	20.3-35.6	41.4	31.7	11.8-35.5	17.3-42.1	24.2-34.7[c]; 37.8[b]; 38.7[e]; 40.5[i]
Linalyl acetate oxide	477705-86-5		tr				
Linalyl propionate	144-39-8	0.01-0.06	0.07		0-0.1	0-0.2	0.02-0.06[c]; 0.06[b]; 0.2[i]
p-Mentha-1,8-dien-9-ol	1946-01-6		tr				
cis-p-Mentha-2,8-dien-1-ol	3886-78-0	tr					
trans-p-Mentha-2,8-dien-1-ol	4017-77-0	tr					
p-Mentha-1,3-dien-7-ol, acetate	81893-40-5	tr					
p-Mentha-1,7-dien-4-yl acetate		tr					
p-Mentha-1,7(10)-dien-2-yl acetate		tr					
p-Mentha-1,8(1)-dien-9-yl acetate	15111-97-4	tr					

Table 5.6.2 Constituents identified in bergamot oils (*continued*)

Constituent	CAS	Percentage and range					
		A	B (1)	C (1)	D (1-18)	E (1,19-24)	F
p-Mentha-1,3,8-triene	18368-95-1	0.2					
cis-p-Menth-2-en-1-ol	29803-82-5	tr					
trans-p-Menth-2-en-1-ol	29803-81-4	tr					
p-Menth-1-en-4,5-oxide		tr					
p-Menth-4-en-1,2-oxide		tr					
p-Menth-1-en-9-yl acetate	28839-13-6		tr			0-0.2	
Menthol	89-78-1	0.2					
Menthone (*p*-)	89-80-5	0.06					
Methylacetophenone	26444-19-9						tr[f]
Methyl geranate	2349-14-6		0.02	0.05	0-0.05	0-0.01	tr-0.01[c]; 0.03[b]
6-Methyl-3-heptanol	18720-66-6		tr				
6-Methyl-5-hepten-2-one	110-93-0		0.01	0.02	0-0.05	0-tr	0.01-0.08; 0.02[b]
Methyl *N*-methyl anthranilate	85-91-6		tr				
α-Muurolene	10208-80-7		tr				
γ-Muurolene	30021-74-0		0.07				
δ-Muurolene	120021-96-7				0-+	0-tr	
Myrcene	123-35-3	0.7-1.8	2.3	2.0	0.7-5.1	0.4-2.3	0.5-1.2[c]; 1.0[d,k]; 1.1[i]; 1.2[b]; 2.0[e]
Myrtenal	564-94-3						0.2[f]
Myrtenol	515-00-4						0.3[f]
Neomenthol	3623-51-6		0.02				
Neomenthyl acetate	2230-87-7		tr				
Neral	106-26-3	0.05-0.3	0.7	0.4	0-0.3	0-0.4	0.2-0.3[c]; 0.2[d,k]; 0.4[b,e]; 0.5[i]
Nerol	106-25-2	0.02-0.2	0.3	0.6	0-0.2	0-0.8	tr-0.1[c]; 0.01[i]; 0.1[d]; 0.2[b]
Nerolidol	7212-44-4				0-0.02	0-0.01	
(*E*)-Nerolidol	40716-66-3		0.1	tr	0-0.02	0-0.2	0.02-0.06[c]; 0.07[b]; 0.1[e]
(*Z*)-Nerolidol	3790-78-1			tr	0-+		
Neryl acetate	141-12-8	0.1-0.7	1.6	1.3	0-1.2	0-2.1	0.3-0.4[c]; 0.5[k]; 0.6[b]; 0.8[m]; 1.6[e]
2,3-Neryl acetate oxide			tr				
Neryl propionate	105-91-9		tr				
Nonanal	124-19-6	0.01-0.06	0.09	0.04	0-0.03	0-0.05	tr[e]; 0.01-0.09[c]; 0.06[b]
Nonanol	28473-21-4		tr	0.02	tr		
Nonyl acetate	143-13-5		0.05	0.05	0-0.1	0-0.06	tr[e]; 0.01-0.03[c]; 0.02[b]
Nootkatone	4674-50-4	0.01-0.03	0.5		0-0.1	0-0.2	0.01-0.1[c]; 0.05[b]; 0.5[e]
Norbornanol	86368-39-0		0.02		0-0.01	0-0.02	0.01-0.02[c]; 0.04[b]
β-Ocimene	13877-91-3				0-0.5	0-0.5	
(*Z*)-Ocimene	27400-71-1						0.3[i]
(*E*)-β-Ocimene	3779-61-1	0.03-0.6	1.1	0.7	0-2.7	0-0.4	0.1[k,m]; 0.2-0.3[c]; 0.2[d,i]; 0.3[b]
(*Z*)-β-Ocimene	3338-55-4	0.02-0.3	0.4		0-1.5	0-0.06	tr[e]; 0.02-0.08[c]; 0.09[b]
allo-Ocimene	673-84-7		tr				
cis-2,3-Ocimene oxide			tr				
trans-2,3-Ocimene oxide			tr				
Octadienyl formate			0.1				0.1[e]
Octanal	124-13-0	0.01-0.09	0.08	0.1	0-0.09	0-0.09	0.03-0.06[c]
Octanoic acid	124-07-2		tr		0-tr		
Octanol	111-87-5		0.03		0-tr	0-tr	tr[b]
1-Octen-3-ol	3391-86-4		0.08				
Octyl acetate	112-14-1	0.01-0.3	0.2	0.2	0-0.2	0-0.2	0.06-0.1[c]; 0.1[b]; 0.2[e]
Pentadecane (*n*-)	629-62-9			0.2			
Pentanol	30899-19-5		tr				
2-Pentanol	6032-29-7		tr				
Perillaldehyde	2111-75-3	0-0.05	0.1		0-0.01		tr-0.01[c]
Perillene	539-52-6						tr[f]

Table 5.6.2 Constituents identified in bergamot oils (*continued*)

Constituent	CAS	Percentage and range					
		A	B (1)	C (1)	D (1-18)	E (1,19-24)	F
Perillyl acetate	15111-96-3		tr		0-tr		
Perillyl alcohol	536-59-4		tr			0-0.01	
α-Phellandrene	99-83-2	0.01-0.07	0.2	0.04	0-0.1	0-0.1	0.01[i]; 0.02-0.03[c]
β-Phellandrene	555-10-2	0-0.6	0.2		0-0.2	0-0.4	0.1[k]
Phenethyl alcohol	60-12-8		0.5				0.5[e]
α-Pinene	80-56-8	0.6-1.7	1.9	1.3	0.5-1.6	0.2-1.5	0.9-1.3[c]; 0.9[n]; 1.1[j,k]; 1.2[b]; 1.4[l]
β-Pinene	127-91-3	4.8-9.3	10.6	5.8	3.0-9.6	2.7-10.1	4.9-7.1[c]; 6.5[k]; 7.2[l]; 7.3[b]; 9.5[d]
trans-Pinocarveol	1674-08-4		tr				
trans-Pinocarvyl acetate	1686-15-3		tr				
Piperitone	89-81-6		tr				
Pulegone	89-82-7			tr			
Sabinene	3387-41-5	0.03-1.3	1.7	1.0	tr-1.3	0.3-1.5	0.9-1.2[c]; 0.9[n]; 1.0[d]; 1.1[k]; 1.2[b,l]
cis-Sabinene hydrate	15537-55-0	0.01-0.06	0.07	0.04	0-0.8	0-0.07	0.1[i]
trans-Sabinene hydrate	17699-16-0		tr		0-0.1	0-0.09	0.05[b]
cis-Sabinene hydrate acetate	77318-48-0		0.07		0-0.1	0-0.1	0.09[b]
trans-Sabinene hydrate acetate	77318-47-9		0.1		0.04-0.06	0-0.09	
α-Santalene	512-61-8		tr				
β-Santalene	511-59-1		0.02		0-0.03	0-0.1	tr-0.01[c]
epi-β-Santalene	25532-78-9		tr				
(Z)-β-Santalol	42495-69-2		0.01				
α-Selinene	473-13-2		tr				
β-Selinene	17066-67-0		0.04				
β-Sesquiphellandrene	20307-83-9		tr		0-0.01	0-0.07	tr-0.03[c]; 0.02[b]
cis-Sesquisabinene hydrate	58319-05-4				0-tr		tr-0.01[c]
(E)-Sesquisabinene hydrate	145512-84-1				0-tr		tr-0.01[c]
α-Sinensal	17909-77-2					0-tr	
Sitosterol	12002-39-0						+[g]
Solanone	1937-54-8					0-0.1	
(E)-Solanone	54868-48-3				0-0.1		
Spathulenol	6750-60-3		tr		0-0.01	0-0.01	tr-0.01[c]
α-Terpinene	99-86-5	0.06-0.2	0.3		0-0.3	0.1-0.7	0.05-0.2[c]; 0.1[m]; 0.2[d,e,i]; 0.3[b]
γ-Terpinene	99-85-4	5.1-8.5	12.6	6.0	2.5-10.2	1.0-9.2	5.9-7.8[c]; 7.7[b]; 8.1[d]; 8.7[l]; 10.3[i]
Terpinen-4-ol	562-74-3	tr-0.07	0.3	0.2	0-0.2	tr-0.6	0.02-0.03[c]; 0.09[i]; 0.1[b]; 0.2[d,m]
Terpinen-4-yl acetate	4821-04-9		tr				
α-Terpineol	98-55-5	0.04-0.5	0.8	3.1	0-0.4	0.02-3.0	0.06-0.1[c]; 0.1[e,l]; 0.3[i]; 0.4[d]; 0.6[b]
Terpinolene	586-62-9	0.1-0.4	0.7	0.4	tr-0.6	0-0.4	0.2-0.3[c]; 0.3[d]; 0.4[e]; 0.5[b]; 0.6[i]
α-Terpinyl acetate	80-26-2	0.01-0.3	0.5	0.3	0-0.3	0-0.3	tr[e]; 0.07-0.2[c]; 0.2[b,d]; 0.4[i]
α-Terpinyl isobutyrate	7774-65-4					0-0.2	
Tetradecanal	124-25-4		0.01		0-0.01	0-0.06	0.02[b]
α-Thujene	2867-05-2	0.08-0.4	0.5	0.2	0-0.4	0-0.4	0.2-0.3[c]; 0.2[e,k]; 0.3[b,d,j,l]; 0.4[l,m]
Thymol	89-83-8		tr				
Tricyclene	508-32-7		0.01		0-0.03	0-0.01	tr-0.01[c]; tr[b]
Tridecanal	10486-19-8					0-0.06	0.2[b]
Tridecane	629-50-5		0.05				
3-(3,4,5-Trimethoxy-phenyl)-propenyl acetate			tr				
Undecanal	112-44-7		0.02	0.2	0-0.05	0-0.04	tr-0.04[c]; 0.01[b]
Undecyl acetate	1731-81-3		tr				
Valencene	4630-07-3						0.6[i]
Verbenol	473-67-6					0-+	0-0.01

Table 5.6.2 Constituents identified in bergamot oils (*continued*)

A one hundred and three bergamot essential oil samples from Italy, Argentina and Ivory Coast, analyzed between 1998 and 2013; lowest and highest concentrations given (E. Schmidt, unpublished data)
B review of studies published between 1979 and 1999 on the composition of industrial and laboratory prepared cold-pressed bergamot oils; data cited in ref. 1; highest concentrations given
C review of studies published between 1979 and 1999 on the composition of laboratory-distilled bergamot oils; data cited in ref. 1; highest concentrations given
D review of studies published between 1998 and 2009 on the composition of industrial cold-pressed bergamot oils (refs. 2-18), among which are a number of bergapten-free oils (2,11,15,16,18); data cited in ref. 1; lowest and highest concentrations given
E review of studies published between 1998 and 2009 on the composition of laboratory cold-pressed (refs. 19-23) and distilled (21,24) bergamot oils; data cited in ref. 1; lowest and highest concentrations given
F data from other studies (indicated with superscript letters); highest concentrations found in any study reviewed here given; when two or more oils were investigated, only the highest concentrations are mentioned, unless indicated otherwise

[a] incorrect identity; [b] ten cold-pressed bergamot essential oils (three lab-prepared, seven industrial oils) from three cultivars of *C. bergamia* and from various parts of South Italy (ref. 25); [c] forty-two samples of industrially cold-pressed Calabrian (South Italy) bergamot essential oils produced in three seasons (2008-2011); lowest and highest concentrations given (ref. 27); [d] twenty-two Italian genuine cold-pressed oils and two oils from Ivory Coast; highest concentrations given, but only if at least 0.1% or not mentioned in (several) other studies (ref. 29); [e] one laboratory cold-pressed oil from south Turkey (ref 31); [f] six commercial oils from Italy, three freshly prepared and three aged for 2 years; this article was discussed in ref. 1 (column B), but some compounds found and not mentioned by the authors of ref. 1 are presented here (ref. 33); [g] one commercial bergamot oil (ref. 34); [h] one commercial Italian bergamot oil; only the germacrenes were investigated (ref. 36); [i] two laboratory cold-pressed bergamot oils from fruits cultivated on the island of Kefalonia, Greece (ref. 39); [j] one Italian bergamot oil sample (ref. 40); [k] one bergamot oil from Italy (ref. 46); [l] several samples of Calabrian (South Italy) bergamot oils (ref. 47); [m] one lab-hydrodistilled oil from Italian bergamot fruit peels (ref. 48); [n] one commercial oil from Brazil (ref. 50);

tr: trace (in column F[b] and F[c]: <0.01; in column F[e]: <0.1); + present in the oil investigated, but quantity not stated

in alphabetical order. In bergamot oils from various origins, over 250 chemicals have been identified. About 49% of these were found in a single reviewed publication only. The major compounds found in bergamot oils from different sources are shown in Table 5.6.3. They include (highest concentrations in any study given) limonene (60.0%), linalyl acetate (42.1%), linalool (36.1%), γ-terpinene (12.6%), β-pinene (10.6%), myrcene (5.1%), *p*-cymene (3.6%) and geranyl acetate (2.7%).

Commercial oils

The ten chemicals that had the highest maximum concentrations in 103 commercial bergamot essential oil samples (concentration ranges provided) are the following: limonene (31.4-45.1%), linalyl acetate (20.3-35.6%), linalool (4.7-16.7%), β-pinene (4.8-9.3%), γ-terpinene (5.1-8.5%), myrcene (0.7-1.8%), α-pinene (0.6-1.7%), sabinene (0.03-1.3%), α-*trans*-bergamotene (0.02-1.2%) and camphene (0.02-1.1%) (Erich Schmidt, unpublished analytical data).

It should be realized that values in hydrodistilled oils may (but not necessarily do) vary considerably from genuine cold-pressed bergamot oils because of the many hydrolytic reactions that take place during oil isolation. In addition, the hydrodistilled oils lack the non-volatile oxygenated heterocyclic compounds which may be present in the cold-pressed oils (Table 5.6.4).

The results of some recent studies on oils other than industrial cold-pressed bergamot oils are not discussed in Table 5.6.2: bergapten-free oils (25,27); oils recovered by traditional or modern distillation procedures from the residue immediately after cold extraction (27,29,39);

and laboratory hand-pressed oils (27). In general, the composition of furocoumarin/bergapten-free bergamot oils, obtained by alkaline treatment or distillation, is similar to that of cold-pressed oils (27). The recovered distilled oils, however, contain less linalyl acetate and more linalool than cold-pressed oils, which is caused by the hydrolytic degradation of the ester during distillation (1,29,39). Terpinen-4-ol and α-terpineol are present in higher amounts compared with the cold-pressed oils (1,27). Ranges for the chemicals in cold-pressed oils, bergapten-free bergamot oils and distilled bergamot oils based on literature review are provided in ref. 29. The history, global distribution, and nutritional importance of *Citrus* fruits have been reviewed in ref. 49.

CONTACT ALLERGY/ALLERGIC CONTACT DERMATITIS

General

Contact allergy to/allergic contact dermatitis from bergamot oil has been reported repeatedly. In groups of consecutive patients suspected of contact dermatitis, prevalence rates of up to 1.5% positive patch test reactions have been observed, but reliable data on relevance are lacking. Most cases of allergic contact dermatitis to bergamot oil have been the result of occupational exposure in masseuses/aromatherapists and the fragrance industry. Linalool and β-pinene may have been allergens in bergamot oils in two case reports (52,60); co-reactions to limonene, the main component of bergamot oil, has been observed once (51). Phototoxicity due to the psoralens in bergamot oils used to be frequent, but has become rare nowadays.

Table 5.6.3 Major constituents of bergamot oils

Constituent	CAS	Percentage and range					
		A	B (1)	C (1)	D (1-18)	E (1,19-24)	F
Limonene	138-86-3	31.4-45.1	54.9	45.1	23.5-50.5	17.0-60.0	28.7-45.8[c]; 42.4[b]; 43.7[l]; 49.1[d]
Linalyl acetate	115-95-7	20.3-35.6	41.4	31.7	11.8-35.5	17.3-42.1	24.2-34.7[c]; 37.8[b]; 38.7[e]; 40.5[i]
Linalool	78-70-6	4.7-16.7	24.2	17.9	5.6-36.1	1.2-29.0	4.3-26.0[c]; 20.2[i]; 26.0[b]; 36.1[d]
γ-Terpinene	99-85-4	5.1-8.5	12.6	6.0	2.5-10.2	1.0-9.2	5.9-7.8[c]; 7.7[b]; 8.1[d]; 8.7[l]; 10.3[i]
β-Pinene	127-91-3	4.8-9.3	10.6	5.8	3.0-9.6	2.7-10.1	4.9-7.1[c]; 6.5[k]; 7.2[l]; 7.3[b]; 9.5[d]
Myrcene	123-35-3	0.7-1.8	2.3	2.0	0.7-5.1	0.4-2.3	0.5-1.2[c]; 1.0[d,k]; 1.1[l]; 1.2[b]; 2.0[e]
p-Cymene	99-87-6	0.1-0.7	3.6	1.6	0-1.4	0-0.2	0.07-0.6[c]; 0.4[b]; 0.5[n]; 0.6[k]; 2.0[d]
Geranyl acetate	105-87-3	0.04-0.6	1.6	1.9	0-2.4	0.02-2.7	0.03-0.6[c]; 0.4[b,d,l]; 0.7[m]; 1.6[e]

LEGEND: SEE UNDER TABLE 5.6.2

Testing in groups of patients

The results of patch tests with bergamot oil in routine testing (consecutive patients suspected of contact dermatitis) and in groups of selected patients are shown in Table 5.6.5. In routine testing, rates of positive reactions ranged from 0.2% to 1.5%, whereas between 1.5% and 2.0% of patients in selected groups had positive patch tests.

Case reports

Two aromatherapists had contact dermatitis (occupational in one) with allergies to multiple essential oils used at their work, including bergamot oil. Both patients also reacted to geraniol, α-pinene, the fragrance mix and various other fragrance materials. In addition, one proved to be allergic to linalool and linalyl acetate, the other to caryophyllene; α-pinene, linalool, geraniol and caryophyllene were demonstrated by GC-MS in bergamot oil (52). Linalool may be an important component of bergamot oil and has been found in concentrations of up to 16.7% in commercial bergamot oils (Table 5.6.2, column A). One masseuse had occupational contact dermatitis from contact allergy to bergamot oil; she also reacted to other essential oils and fragrance materials (53). Bergamot oil sensitized one worker in the fragrance industry; she also reacted to α-pinene and β-pinene; commercial bergamot oils were investigated for the presence of these chemicals and they were identified in maximum concentrations of 1.7% (α-pinene) and 9.3% (β-pinene) (60). In a group of

Table 5.6.4 Heterocyclic oxygenated compounds present in cold-pressed bergamot oil (27,30,34,38,41,43,44)[a]

Name	Synonym(s)	CAS
Hydroxylated polymethoxyflavones		
5-Hydroxy-3',4',7,8-tetramethoxyflavone		
Polymethoxyflavones		
3',4',5,6,7,8-Hexamethoxyflavone	Nobiletin	478-01-3
3',4',5,6,7-Pentamethoxyflavone	Sinensetin	2306-27-6
4',5,6,7,8-Pentamethoxyflavone	Tangeretin	481-53-8
4',5,6,7-Tetramethoxyflavone	Tetra-O-methylscutellarein	1168-42-9
Coumarins and psoralens		
Byakangelicol		26091-79-2
5,7-Dimethoxycoumarin	Citropten; Limetin	487-06-9
5,8-Dimethoxy-6,7-furanocoumarin[b]	Isopimpinellin	482-27-9
5-(6',7'-Epoxy)geranyloxypsoralen	Epoxybergamottin	206978-14-5
5-Geranyloxy-7-methoxycoumarin		7380-39-4
5-Geranyloxy-8-methoxypsoralen[b]		69239-53-8
5-Geranoxypsoralen	Bergamottin; Bergaptin	7380-40-7
5-Hydroxy-6,7-furanocoumarin	Bergaptol	486-60-2
5-Isopentenyloxy-7-methoxycoumarin		35590-41-1
5-Isopentenyloxy-8-methoxypsoralen	Cnidilin; Isophellopterin	14348-22-2
7-Methoxycoumarin	Herniarin	531-59-9
5-Methoxy-8-(2,3-dihydroxy-3-methylbutoxy)psoralen	Bjakangelicin; Byakangelicin	482-25-7
5-Methoxy-7-hydroxycoumarin		3067-10-5
7-Methoxy-6-(3-methyl-2-butenyl)coumarin	Suberosin	581-31-7
5-Methoxypsoralen	Bergapten	484-20-8
Oxypeucedanin		26091-73-6

[a] the 1999-2013 literature on the non-volatiles in bergamot oil has been reviewed in ref. 45

[b] according to the authors of ref. 41, the presence of these compounds indicates that the commercial oils in which they were demonstrated were adulterated with lime oil

Table 5.6.5 Results of testing groups of patients with bergamot oil

Years and Country	Test conc. & vehicle	Number of patients tested | positive (%)			Selection of patients (S); Relevance (R); Comments (C)	Ref.
Routine testing						
2000-2007 USA	2% pet.	500	6	(1.2%)	R: 100%; C: see Comments below; tested was 'bergamot – natural'	55
2000-2007 USA	2% pet.	500	4	(0.8%)	R: 100%; C: weak study: a. high rate of macular erythema and weak reactions, b. relevance figures include 'questionable' and 'past' relevance; tested was 'bergamot – synthetic'	55
1983-1984 Italy	2% pet.	1,200	2	(0.2%)	R: not stated	61
<1976 Poland	2% pet.	200	3	(1.5%)	R: not stated	54
1967-1970	10% pet.	590	3	(0.5%)	R: unknown	62
Testing in selected groups of patients						
<1994 Japan	?	?	?	(1.5%)	S: unknown; R: unknown	59
<1976 Poland	2% pet.	51	1	(2.0%)	S: patients allergic to *Myroxylon pereirae* (balsam of Peru) and/ or turpentine and/or wood tar and/or colophony; R: not stated	54

pet.: petrolatum

70 patients with proven allergic cosmetic dermatitis, bergamot oil was the allergen in one (63).

Positive patch tests (relevance unknown, uncertain or not stated)

One positive patch test to bergamot oil and its main ingredient limonene in a confectioner with occupational contact dermatitis from cardamom (which also contains limonene) (51).

Photosensitivity

Formerly, the most frequent side effect of bergamot oil was phototoxicity due to its content of furocoumarines, notably 5-methoxypsoralen (bergapten). Currently, psoralens are (largely) removed from the crude oil before being used in fragrances and other cosmetics, including bronzing and sun-protecting products (which should not contain bergapten in quantities >1 ppm in the European Union). There are no regulations, however, for its content in essential oils used in aromatherapy. Indeed, several cases of phototoxicity from such use have been reported (57,58). One patient developed phototoxicity from bath oil, which was shown to contain 5-methoxypsoralen (57). Two patients developed bullous phototoxic dermatitis from aromatherapy with bergamot oil, of which one from its use in a sauna; one of the preparations proved to contain 2400 ppm (0.24%) bergapten (58). Two cases of photo*allergy* from bergamot oil were reported from Italy (56).

LITERATURE

1 Dugo G, Cotroneo A, Bonaccorsi I, Trozzi A. Composition of the volatile fraction of citrus peel oils. In: G Dugo and L Mondello, Eds. Citrus oils: composition, advanced analytical techniques, contaminants, and biological activity. Boca Raton, USA: CRC Press, Taylor & Francis Group, 2011:90-115

2 Ferrini AM, Mannoni V, Hodzic S, Salvatore G, Aureli P. Activité antimicrobienne de l'huile de bergamotte par rapport à la composition chimique et l'origine. Riv Ital EPPOS 1998;(Numero speciale):139-153. Data cited in ref. 1

3 Oberhofer B, Nikiforov A, Buchbauer G, Jirovetz L, Bicchi C. Investigation of the alteration of the composition of various essential oils used in aroma lamp applications. Flavour Fragr J 1999;14:293-299

4 Sawamura M, Poiana M, Kawamura A, et al. Volatile components of peel oils of Italian and Japanese lemon and bergamot. Ital J Food Sci 1999;11:121-130. Data cited in ref. 1

5 Gionfriddo F, Mangiola C, Siano F, Castaldo D. Le caratteristiche dell'essenza di bergamotto prodotta nella campagna 1998/99. Essenz Deriv Agrum 2000;70:133-145. Data cited in ref. 1

6 Russo MT, Antonelli A, Carnacini A. Experiences with solid CO_2 concentration of bergamot cold pressed essential oil (*Citrus bergamia* Risso). J Essent Oil Res 2001;13:247-249

7 Kubeczka K-H, Formacek V. Essential oils analysis by capillary gas chromatography and carbon-13 NMR Spectroscopy, 2nd edition. New York, USA: John Wiley and Sons, 2002:21-26. Data cited in ref. 1

8 Mondello L, Casilli A, Tranchida PQ, Cicero L, Dugo P, Dugo G. Comparison of fast and conventional GC analysis for citrus essential oils. J Agric Food Chem 2003;51:5602-5606

9 Mondello L, Casilli A, Tranchida PQ, Costa R, Dugo P, Dugo G. Fast GC for the analysis of citrus oils. J Chromatogr Sci 2004;42:410-416

10 Tranchida PQ, Lo Presti M, Costa R, Dugo P, Dugo G, Mondello L. High-throughput analysis of bergamot essential oil by fast solid-phase microextraction-capillary gas-chromatography flame ionization detection. J Chromatogr A 2006;1103:162-165

11 Poiana M, Mincione A, Gionfriddo F, Castaldo D. Supercritical carbon dioxide separation of bergamot essential oil by a countercurrent process. Flavour Fragr J 2003;18:429-435

12 Franceschi E, Grings MB, Frizzo CD, Oliveira JV, Dariva C. Phase behavior of lemon and bergamot peel oils in supercritical CO_2. Fluid Phase Equilibr 2004;226:1-8

13 Sawamura M, Onishi Y, Ikemoto J, Minh Tu NT, Lan Phi NT. Characteristic odour components of bergamot (*Citrus bergamia* Risso) essential oil. Flavour Fragr J 2006;21:609-615

14 Onishi Y, Minh Tu NT, Ikemoto J, Ukeda H, Sawamura M. Studies on characteristic odor components of bergamot essential oil. 47th TEAC, Tokyo, 2003. Data cited in ref. 1

15 Belsito EL, Carbone C, Di Gioia ML, Leggio A, Liguori A, Perri F, et al. Comparison of the volatile constituents in cold pressed bergamot oil and a volatile oil isolated by vacuum distillation. J Agric Food Chem 2007;55:7847-7851

16 Costa R, Dugo P, Navarra M, Dugo G, Mondello L. Study on the chemical composition variability of some processed bergamot (*Citrus bergamia*) essential oils. Flavour Fragr J 2010;25:4-12

17 Sciarrone, 2009, personal communication to the authors of ref. 1. Data cited in ref. 1

18 Verzera A, Trozzi A, Stagno d'Alcontres I, Mondello L, Dugo G, Sebastiani E. The composition of the volatile fraction of Calabrian bergamot essential oil. Riv Ital EPPOS 1998;25:17-38. Data cited in ref. 1

19 Sawamura M. Volatile components of essential oils of the Citrus genus. Recent Res Dev Agric Food Chem 2000;4:131-164. Data cited in ref. 1

20 Verzera A, La Rosa G, Zappalà M, Cotroneo A. Essential oil composition of different cultivars of bergamot grown in Sicily. Ital J Food Sci 2000;12:493-501. Data cited in ref. 1

21 Kirbaslar FG, Kirbaslar SI, Dramur U. The composition of Turkish bergamot oils produced by cold pressing and steam distillation. J Essent Oil Res 2001;13:411-415

22 Gionfriddo F, Catalfamo M, Siano F, Mangiola C, Cautela D, Castaldo D. Determinazione delle caratteristiche analitiche e della composizione enantiomerica di oli essenziali agrumari ai fini dell'accertamento della purezza e della qualità. Nota I – Essenze di arancia amara, arancia dolce e bergamotto. Essenz Deriv Agrum 2003;73:29-39. Data cited in ref. 1

23 Verzera A, Trozzi A, Gazea F, Cicciarello G, Cotroneo A. Effect of rootstock on the composition of bergamot (*Citrus bergamia* Risso et Poiteau) essential oil. J Agric Food Chem 2003;51:206-210

24 Kiwanuka P, Mottram DS, Baigrie BD. The effects of processing on the constituents and enantiomeric composition of bergamot essential oil. In: V Lanzotti and O Taglialatela-Scafati, Eds. Flavour and fragrance chemistry. Dordrecht, The Netherlands: Kluwer Academic Publisher, 2000:67-75. Data cited in ref. 1

25 Russo M, Serra D, Suraci F, Postorino S. Effectiveness of electronic nose systems to detect bergamot (*Citrus bergamia* Risso et Poiteau) essential oil quality and genuineness. J Essent Oil Res 2012;24:137-151

26 Forlot P, Pevet P. Bergamot (*Citrus bergamia* Risso et Poiteau) essential oil: Biological properties, cosmetic and medical use. A review. J Essent Oil Res 2012;24:195-201

27 Dugo G, Bonaccorsi I, Sciarrone D, Schipilliti L, Russo M, Cotroneo A, et al. Characterization of cold-pressed and processed bergamot oils by using GC-FID, GC-MS, GC-C-IRMS, enantio-GC, MDGC, HPLC and HPLC-MS-IT-TOF. J Essent Oil Res 2012;24:93-117

28 Russo M, Torre G, Carnovale C, Bonaccorsi I, Mondello L, Dugo P. A new HPLC method developed for the analysis of oxygen heterocyclic compounds in Citrus essential oils. J Essent Oil Res 2012;24:119-129

29 Schipilliti L, Dugo G, Santi L, Dugo P, Mondello L. Authentication of bergamot essential oil by gas chromatography-combustion-isotope ratio mass spectrometer (GC-C-IRMS). J Essent Oil Res 2011;23:60-71

30 Dugo P, Russo M. The oxygen heterocyclic components of Citrus essential oils. In: G Dugo and L Mondello, Eds. Citrus oils: composition, advanced analytical techniques, contaminants, and biological activity. Boca Raton, USA: CRC Press, Taylor & Francis Group, 2011:405-444

31 Kirbaslar SI, Kirbaslar FG, Dramur U. Volatile constituents of Turkish bergamot oil. J Essent Oil Res 2000;12:216-220

32 Kaddu S, Kerl H, Wolf P. Accidental bullous phototoxic reactions to bergamot aromatherapy oil. J Am Acad Dermatol 2001;45:458-461

33 Mazza G. Etude sur la composition aromatique de l'huile essentielle de bergamote (*Citrus aurantium* subsp. bergamia Risso et Poiteau Engler) par chromatographie gazeuse et spectrometrie de masse. J Chromatogr 1986;362:87-99

34 Ehret C, Maupetit P. Two sinapyl alcohol derivatives from bergamot essential oil. Phytochem 1982;21:2984-2985

35 Mondello L, Verzera A, Previti P, Crispo F, Dugo G. Multidimensional capillary GC-GC for the analysis of complex samples. 5. Enantiomeric distribution of monoterpene hydrocarbons, monoterpene alcohols, and linalyl acetate of bergamot (*Citrus bergamia* Risso et Poiteau) oils. J Agric Food Chem 1998;46:4275-4282

36 Feger W, Brandauer H, Ziegler H. Germacrenes in citrus peel oils. J Essent Oil Res 2001;13:274-277

37 Orth A-M, Yu L, Engel K-H. Assessment of dietary exposure to flavouring substances via consumption of flavoured teas. Part 1: occurrence and contents of monoterpenes in Earl Grey teas marketed in the European Union. Food Addit Contam Part A 2013;30:1701-1714

38 Donato P, Bonaccorsi I, Russo M, Dugo P. Determination of new bioflavonoids in bergamot (*Citrus bergamia*) peel oil by liquid chromatography coupled to tandem ion trap–time-of-flight mass spectrometry. Flavour Fragr J 2014;29:131-136

39 Melliou E, Michaelakis A, Koliopoulos G, Skaltsounis A-L, Magiatis P. High quality bergamot oil from Greece: Chemical analysis using chiral gas chromatography and larvicidal activity against the West Nile virus vector. Molecules 2009;14:839-849

40 Zani F, Massimo G, Benvenuti S, Biandini A, Albasini A, Melegari M, et al. Studies on the genotoxic properties of essential oils with *Bacillus subtilis* recassay and *Salmonella*/microsome reversion assay. Planta Med 1991;57:237-241

41 Dugo G, Bartle KD, Bonaccorsi I, Catalfamo M, Cotroneo A, Dugo P, et al. Advanced analytical techniques for the analysis of Citrus essential oils. Part 3. Oxygen heterocyclic compounds: HPLC, HPLC/MS, OPLC, SFC, Fast HPLC analysis. Essenz Deriv Agrum 1999;69:251-283. Data cited in ref. 42

42 Lawrence BM. Progress in essential oils. Perfum Flavor 2002;27(6):46-? (last page unknown)

43 Gionfriddo F, Postorino E, Calabro G. Elimination of furocoumarines in bergamot peel oil. Perfum Flavor 2004;29(July/August):48-52

44 Frérot E, Decorzant E. Quantification of total furocoumarines in citrus oils by HPLC coupled with UV, fluorescence and mass detection. J Agric Food Chem 2004;52:6879-6886

45 Lawrence BM. Progress in essential oils. Perfum Flavor 2013;38(September):42-54

46 Williams DG. The chemistry of essential oils, 2nd Ed. Port Washington, NY, USA: Micelle Press, 2008:168-169

47 Mangiola C, Postorino E, Gionfriddo F, Catalfamo M, Manganaro R. Evaluation of the genuineness of cold-pressed bergamot oil. Perfum Flavor 2009;34(10):26-32

48 Menichini F, Tundis R, Loizzo MR, Bonesi M, Provenzano E, de Cindio B, et al. In vitro photoinduced cytotoxic activity of *Citrus bergamia* and *C. medica* L. cv. Diamante peel essential oils and identified active coumarins. Pharm Biol 2010;48:1059-1065

49 Liu YQ, Heying E, Tanumihardjo SA. History, global distribution, and nutritional importance of Citrus fruits. Compreh Rev Food Sci Food Saf 2012;11:530–545

50 Murbach Teles Andrade BF, Nunes Barbosa L, da Silva Probst I, Fernandes Júnior A. Antimicrobial activity of essential oils. J Essent Oil Res 2014;26:34-40

51 Mobacken H, Fregert S. Allergic contact dermatitis from cardamom. Contact Dermatitis 1975;1:175-176

52 Dharmagunawardena B, Takwale A, Sanders KJ, Cannan S, Roger A, Ilchyshyn A. Gas chromatography: an investigative tool in multiple allergies to essential oils. Contact Dermatitis 2002;47:288-292

53 Trattner A, David M, Lazarov A. Occupational contact dermatitis due to essential oils. Contact Dermatitis 2008;58:282-284

54 Rudzki E, Grzywa Z, Bruo WS. Sensitivity to 35 essential oils. Contact Dermatitis 1976;2:196-200

55 Wetter DA, Yiannias JA, Prakash AV, Davis MD, Farmer SA, el-Azhary RA, et al. Results of patch testing to personal care product allergens in a standard series and a supplemental cosmetic series: an analysis of 945 patients from the Mayo Clinic Contact Dermatitis Group, 2000-2007. J Am Acad Dermatol 2010;63:789-798

56 Pigatto PD, Legori A, Bigardi AS, Guarrera M, Tosti A, Santucci B, et al. Gruppo Italiano recerca dermatiti da contatto ed ambientali Italian multicenter study of allergic contact photodermatitis: epidemiological aspects. Am J Contact Dermatitis 1996;7:158-163

57 Clark SM, Wilkinson SM. Phototoxic contact dermatitis from 5-methoxypsoralen in aromatherapy oil. Contact Dermatitis 1998;38:289-290

58 Kaddu S, Kerl H, Wolf P. Accidental bullous phototoxic reactions to bergamot aromatherapy oil. J Am Acad Dermatol 2001;45:458-461

59 Sugai T. Group study IV – farnesol and lily aldehyde. Environ Dermatol 1994;1:213-214

60 Zacher KD, Ippen H. Kontaktekzem durch Bergamottöl. Derm Beruf Umwelt 1984;32:95-97

61 Santucci B, Cristaudo A, Cannistraci C, Picardo M. Contact dermatitis to fragrances. Contact Dermatitis 1987;16:93-95

62 Meneghini CL, Rantuccio F, Lomuto M. Additives, vehicles and active drugs of topical medicaments as causes of delayed-type allergic dermatitis. Dermatologica 1971;143:137-147

63 Schorr WF. Cosmetic allergy: Diagnosis, incidence, and management. Cutis 1974;14:844-850

Chapter 5.7   BLACK CUMIN OIL

DEFINITION

Black cumin oil is the essential oil obtained from the seeds of the black caraway (black cumin), *Nigella sativa* L.

INCI NOMENCLATURE

Description/definition: Nigella sativa seed extract is an extract of the seeds of the black caraway, *Nigella sativa* L., Ranunculaceae
INCI name EU &USA: Nigella sativa seed extract

CAS registry number(s): 90064-32-7
EINECS number(s): 290-094-1

ISO (INTERNATIONAL ORGANIZATION FOR STANDARDIZATION) STANDARD

There is currently no ISO standard for black cumin oil.

Black cumin essential oil should not be confused with Nigella sativa *seed* oil (INCI name EU and USA), which is the *fixed* (vegetable) oil expressed from the seeds of *Nigella sativa* L. (CAS 90064-32-7, EINECS 290-094-1). The seeds of the black caraway contain approximately 35% by weight of a crude oil, which is composed of 98.2–99.9% of this fixed oil (8) and which is rich in linoleic acid, oleic acid, palmitic acid and stearic acid (20,23) and is used in cosmetics as emollient and skin conditioner.

The nomenclature of the plant is also confusing. In English, *Nigella sativa* is variously called fennel flower, nutmeg flower, and Roman coriander. Other names used, sometimes misleadingly, are onion seed and black sesame, both of which are similar-looking, but unrelated plants. The names black seed and black caraway may also refer to *Bunium persicum*. The seeds are frequently referred to as black cumin, but original black cumin is obtained from *Bunium bulbocastanum* (synonym: *Carum bulbocastanum* L.), a member of the Apiaceae (Umbelliferae) family (Wikipedia, www.theplantlist.org).

THE PLANT, THE OIL, AND THEIR USES

Nigella sativa L. (black caraway, black cumin) is an annual flowering plant, which grows to 20–30 cm tall. The plant is native to Turkey and Iraq and is sometimes naturalized from the Mediterranean region to central Asia; it is widely cultivated in different parts of the world, mainly in countries bordering the Mediterranean Sea (10); Egypt is one of the main producers of *N. sativa* seeds (8, GRIN Taxonomy for Plants). The seeds of the black cumin are used as a spice and a condiment for flavouring a variety of foods such as bread, bakery products and cheese. They have long been used also in folk medicine as a traditional remedy for a wide range of illnesses, including bronchial asthma, cough, bronchitis, influenza, fever, eczema (4), diabetes, headache, dizziness, dysentery, infections, obesity, back pain, hypertension, inflammation, and gastrointestinal problems (5). It has also been reported that *N. sativa* seeds prevent dental plaques and caries (11) and may be useful as a carminative, diuretic, lactagogue and vermifuge (10).

The essential black cumin oil is used as condiment, carminative, analgesic and food preservative (11) and has been recommended as a remedy for cough, asthma, flu, vertigo and headache (9). *Nigella sativa* oil is reported to possess antioxidant, anti-inflammatory, antihistaminic, antibacterial, antifungal, immune-stimulant, hypoglycemic, antihypertensive, antiviral and antimalarial properties while also being beneficial for the respiratory system (5,9,15). Many of these activities have been attributed to thymoquinone, thymol, carvacrol, carvone and *p*-cymene contained within the seeds. It is apparently not used in aromatherapy (35). The possible health effects of *Nigella sativa* (products) have been reviewed (3,14,22). Pharmacological properties were summarized in ref. 26.

CHEMICAL COMPOSITION

Black cumin oil is a yellow to brown, clear mobile liquid, which has a spicy, herbaceous and phenolic odor. The yield of essential oil from the seeds of *Nigella sativa* L. generally varies from 0.5 to 1.4%. The main producing countries of this oil are India, Turkey, Egypt and Morocco.

Literature data (up to November 6, 2014) on the chemical composition of black cumin oils and unpublished analytical data from one of us (E.S.) are shown in Table 5.7.1 in alphabetical order. In black cumin oils from various origins, over 340 chemicals have been identified. About 62% of these were found in a single reviewed publication only.

The major compounds found in black cumin oils from different sources are shown in Table 5.7.2. They include (highest concentrations in any study given) thymoquinone (63.3%), *p*-cymene (62.3%), γ-terpinene (45.7%), α-thujene (17.5%), α-pinene (13.8%), longifolene (10.2%), β-pinene (6.2%) and terpinen-4-ol (4.3%). The amounts of *p*-cymene and thymoquinone heavily depend on the extraction method (8,10). Well-known ingredients of *Nigella sativa* essential oils that were present in high concentrations in one or two studies were (*E*)-anethole (27.1%), thymol (12.1% and 26.8%), carvacrol (12.9%), cuminaldehyde (12.7%), thymohydroquinone (12.2%), limonene (10.6%) and terpinolene (9.1%). Uncommon or rare constituents of black cumin oil found in high concentrations in single studies include linoleic acid (29.5%), 2-methyl-3-phenylpropanal (13.2%) (possible misidentification), hexadecanoic acid (15.6%), 2,4(10)-thujadiene (14.0%) (possible misidentification), (*Z*)-nerolidol (12.3%), dihydrocarveol (10.4%), ethyl linoleate (9.4%), thujyl alcohol (7.4%), *trans*-*p*-menth-2-en-1-ol (6.1%), cuminyl acetate (5.9%) (possible misidentification) and neothujol (5.7%).

Commercial oils

The ten chemicals that had the highest maximum concentrations in 13 commercial black cumin essential oil samples (concentration ranges provided) are the following: *p*-cymene (19.9-57.5%), thymoquinone (0.6-24.5%), α-thujene (8.9-17.0%), *trans*-*p*-menth-2-en-1-ol (0-6.1%), carvacrol (tr-5.8%), β-pinene (0.7-4.4%), α-pinene (0.7-4.1%), limonene (0.08-3.7%), longifolene (0.4-0.6%) and γ-terpinene (0.1-2.4%) (Erich Schmidt, unpublished analytical data).

Table 5.7.1 Constituents of black cumin oil

Constituent	CAS	Percentage and range					
		A	B (1)	C (11)	D (15)	E (19)	F
4-Acetyl-1-methyl-cyclohexane							tr[t,u]
β-Acoradienol	149496-35-5			tr			
α-Amorphene	20085-19-2						0.6[d]
Anethemol							0.2[t,u]
(E)-Anethole	4180-23-8	0.02-0.9	0-0.2	tr	0.7		0.1[f,k,o]; 0.6[i]; 27.1[c]
p-Anisaldehyde	123-11-5						<0.1[k]; 1.7[c]
Anymol							tr[t,u]
Apiole	523-80-8					0.1	tr[u]; 1.0[c]
Artemisia ketone	546-49-6						0.1[k]
Benzaldehyde	100-52-7				tr		
cis-α-Bergamotene	18252-46-5			tr			
trans-α-Bergamotene	13474-59-4						tr[u]
α-Bisabolene	17627-44-0						tr[d]
β-Bisabolene	495-61-4		0-0.1		0.1	tr	<0.05[q]; 0.03[b]; 0.07[i]; 0.2[u];
(Z)-α-Bisabolene	29837-07-8						tr[u]
(Z)-γ-Bisabolene	13062-00-5						tr[u]; 0.3[o]
α-Bisabolol	515-69-5				0.2		
β-Bisabolol	15352-77-9				0.3		0.7[r]
Borneol	507-70-0			tr		tr	tr[u]; <0.1[k]
Bornyl acetate	76-49-3	0.1-0.3	0-0.5		0.2	0.1	tr[d]; 0.1[o]; 0.2[p,q]; 0.4[i,j]; 0.7[u]; 1.3[l]
β-Bourbonene	5208-59-3						tr[u]
Butyl octadecanoate	123-95-5						tr[r]
Cadalene	483-78-3			tr			
α-Cadinene	24406-05-1			tr			
γ-Cadinene	39029-41-9			0.3		0.1	<0.05[q]
δ-Cadinene	483-76-1		0-0.2	0.4	0.1		tr[r,u]; 0.2[b]
τ-Cadinol	5937-11-1						0.1[r];
α-Calacorene	21391-99-1			0.2			
Camphene	79-92-5	tr-0.3	tr-0.1	0.2		tr	0.06[i]; 0.1[e,k]; 0.2[b]; 0.3[g,u]
Camphor	76-22-2		0-0.2	tr	0.8	0.1	tr[u]; <0.05[q]; 0.1[i,k]; 0.2[b,l]
2-Caren-10-al	124752-20-1						0.8[t,u]
δ3-Carene	13466-78-9						tr[r]; 0.1[u]
Carvacrol	499-75-2	tr-5.8	0-1.9	0.6	12.9	3.0	1.5[f]; 2.1[i]; 2.3[b]; 2.5[d,n]; 3.7[c]; 4.2[j]; 4.3[g]
Carvacryl acetate	6380-28-5		0-0.3				
L-Carvenol							0.8[d]
Carvenone	499-74-1		0-1.9				
cis-Carveol	1197-06-4				0.4		0.1[r]
(E)-Carveol	1197-07-5				0.7		
Carvone	99-49-0		tr-0.1		4.4	0.2	<0.05[p]; 0.1[b]; 0.2[i]; 0.4[j]; 1.1[q]; 1.8[r]; 2.0[c]
Carvotanacetone	499-71-8						<0.1[k]
cis-Carvyl acetate	1205-42-1				0.3		
β-Caryophyllene	87-44-5	0-0.1	tr-0.2	0.3	0.1		0.07[i]; 0.1[b,r]; 0.4[h,o,q]; 0.5[u]
γ-Caryophyllene (Z-)	118-65-0			tr	0.1		<0.1[o]
9-epi-(E)-Caryophyllene	68832-35-9				0.1		
Caryophyllene acetate	32214-91-8						0.4[r]
Caryophyllene oxide	1139-30-6					tr	0.1[u]; 0.7[h]
Caryophyllenyl alcohol	913176-41-7			tr			
Chrysanthenyl acetate							
cis-Chrysanthenyl acetate	67999-48-8					0.1	
1,8-Cineole	470-82-6			tr			tr[j]; 0.1[k,n,p]; 0.4[o]
(E)-Cinnamaldehyde	14371-10-9				0.3		
Citronellyl butyrate	141-16-2		0-0.1		0.1		0.1[b]
α-Copaene	3856-25-5		0-0.2	tr		tr	tr[u]; 0.1[b]
Coumarin	91-64-5				0.1		
α-Cubebene	17699-14-8			tr			3.0[j]
β-Cubebene	13744-15-5						tr[r]
o-Cumenol	88-69-7						tr[u]
Cuminaldehyde	122-03-2		0-0.1		2.9		<0.05[p]; 0.9[r]; 12.7[s]
Cuminyl acetate	59230-57-8						5.9[t,u]

Table 5.7.1 Constituents of black cumin oil (*continued*)

Constituent	CAS	Percentage and range					
		A	B (1)	C (11)	D (15)	E (19)	F
Cuminyl alcohol	536-60-7						0.3[u]; 6.4[s]
Curcumene	644-30-4				0.1		0.03[b]; 0.1[o,u]
Cyclosativene	22469-52-9					tr	
o-Cymene	527-84-4						0.1[s]; 3.2[m]
p-Cymene	99-87-6	19.9-57.5	37-73	49.5	8.9	60.2	47.4[g]; 54.1[h]; 56.2[q]; 56.4[e]; 57.6[b]; 62.3[l]
p-Cymenene	1195-32-0		0-0.2				<0.05[q]; 0.1[b]; 0.2[o,r]
p-Cymen-7-ol	536-60-7		tr		2.7		
p-Cymen-8-ol	1197-01-9		0-6.2	tr	2.7		<0.05[p]; 0.09[i]; 0.1[k,u]; 0.2[n,r]; 0.4[c]
p-Cymen-9-ol	4371-50-0						1.1[o]
p-Cymen-7-ol acetate	59230-57-8		0-0.1				
(E,E)-2,4-Decadienal	25152-84-5				0.1		
Decanal	112-31-2			tr			
n-Decane	124-18-5			tr			0.4[c]
n-Decanoic acid	334-48-5				0.1		
Dehydro-1,8-cineole	92760-25-3						tr[u]
γ-Dehydro-arhimachalene	51766-65-5		0-0.1				
Dehydrosabinaketone	147043-52-5						4.5[o]
1,10-Diepicubenol	73365-77-2		tr				
Dihydrocarveol	38049-26-2						10.4[h]
trans-Dihydrocarvone	5948-04-9		0-1.0		2.7	0.7	0.3[c]
2,3-Dihydrofarnesol	51411-24-6		tr				
Dihydrotagetone	1879-00-1			tr			
Dill apiole	484-31-1				tr		0.2[r,u]
Dill ether	74410-10-9						tr[t,u]
2-(3,3-Dimethylcyclohexadiene) ethanol							tr[t,u]
1-(1,4-Dimethyl-3-cyclohexen-1-yl)- ethanone	43219-68-7						tr[t,u]; 0.5[i]; 3.9[r]
Dimethyl ionene				0.4			
Dodecanal	112-54-9		0-0.1		tr		tr[r]
Dodecanoic acid	143-07-7				0.4		
α-Duprezianene	79801-29-9			tr			
β-Elemene	33880-83-0						5.5[m]
γ-Elemene	29873-99-2		tr		0.1		
Elemicin	487-11-6						0.1[u]
Epizonarene	41702-63-0			tr			0.1[i]
4,5-Epoxy-1-isopropyl-4-methyl-1- cyclohexene							0.9[i]
Erucic acid	1072-39-5						tr[r]
Ethyl decanoate	110-38-3						<0.1[k]
1-Ethyl-2,3-dimethyl benzene	933-98-2						0.2[c]
Ethyl dodecanoate	106-33-2						<0.1[k]
Ethyl heptanoate	106-30-9						1.2[d]
Ethyl hexadecanoate	628-97-7						tr[d]; <0.1[k]; 0.3[l]; 2.8[k]
Ethyl hexanoate	123-66-0						1.0[d]
Ethyl linoleate	544-35-4						9.4[k]
Ethyl linolenate	1141-91-9						0.1[k]
Ethyl nonanoate	123-29-5						<0.1[k]
Ethyl octadecanoate	111-61-5						0.2[k]; 0.8[d]
Ethyl octanoate	106-32-1						tr[d]; 0.1[k]
Ethyl oleate	111-62-6						0.6[d]; 2.7[k]
Ethyl pentadecanoate	41114-00-5						<0.1[k]
Ethyl tetradecanoate	124-06-1						0.2[k]; 2.1[d]
α-Eudesmol	473-16-5						0.4[p]
β-Eudesmol	473-15-4						0.5[p]
10-epi-γ-Eudesmol	15051-81-7						0.3[p];
Eugenol	97-53-0				0.5		0.5[o]
Eugenyl acetate	93-28-7						0.1[o]
Farnesal D (cis,trans-)						0.2	
β-Farnesene	502-60-3				tr	tr	

Table 5.7.1 Constituents of black cumin oil (*continued*)

Constituent	CAS	Percentage and range					
		A	B (1)	C (11)	D (15)	E (19)	F
(*Z*)-β-Farnesene	28973-97-9			tr			
(*E,Z*)-Farnesol	3879-60-5						0.1[r];
(*Z,Z*)-Farnesol	16106-95-9			tr			
(*E,E*)-Farnesyl acetate	4128-17-0				0.2		0.7[r]
(*Z,E*)-Farnesyl acetate	40266-29-3				0.2		
α-Fenchene	471-84-1						0.1[o]
β-Fenchol	22627-95-8						0.1[o]
Fenchone	1195-79-5						1.1[c]
α-Fenchyl alcohol	512-13-0						0.1[u]
Geranial	141-27-5		0-0.2	0.3			tr[u]
Geraniol	106-24-1						tr[u]
Geranyl acetate	105-87-3						0.2[t,u]
(*E*)-Geranylacetone	3796-70-1			tr	0.1		
Germacrene	28028-64-0						1.1[d]
Germacrene B	15423-57-1		tr				
Germacrene D	23986-74-5						tr[u]
Heptacosane	593-49-7						tr[r]
Heptadecane	629-78-7						0.2[r];
Heptanal	111-71-7						tr[d]
2-Heptenal	2463-63-0						0.1[k]
Hepten-2-one	30640-40-5			tr			
(*E*)-2-Heptenyl acetate	16939-73-4			tr			
Hexacosane	630-01-3						tr[r]
n-Hexadecane	544-76-3						0.2[c]
Hexadecanoic acid	57-10-3				11.5		<0.1[o]; 0.02[b]; 0.2[k]; 15.6[r]
Hexanal	66-25-1			tr			0.1[k]
(*Z*)-3-Hexenyl acetate	3681-71-8			tr			
α-Himachalene	3853-83-6		tr				0.03[b]
γ-Himachalene	53111-25-4			tr			tr[r]
Himachalol	1891-45-8		0-0.2				
α-Humulene	6753-98-6		0-0.1		0.1		0.1[b]
14-Hydroxy-9-epi-(*E*)-caryophyllene	244226-09-3			tr			
3α-Hydroxymanool							0.2[r];
4-Hydroxy-4-methyl-acetophenone							0.1[t,u]
4-Hydroxy-4-methyl-2-cyclohexenone	60565-80-2						<0.1[k]
α-Ionone	127-41-3				tr		
Isobornyl acetate	125-12-2		0-0.3	tr			0.3[b,h]
Isocaryophyllene	118-65-0						
Isocumene	74296-31-4						0.2[r];
Isodihydrocarveol	18675-35-9		0-0.7		0.1		
(*E*)-Isoeugenol	5932-68-3		0-0.1				
(*Z*)-Isoeugenyl acetate					0.2		
Isolongifolene	1135-66-6		0-0.4		0.1		<0.1[o]; 0.2[b]
allo-Isolongifolene	87064-18-4					tr	
Isolongifolol	1139-17-9		0-0.2		0.5		
Isopentyl butyrate	106-27-4			tr			
2-Isopropyl-5-methyl-(2*Z*)-hexanal	66656-67-5						0.1[o]
Isopropyl myristate	110-27-0				0.4		
Isothujol	513-23-5						1.5[h]
Karahanaenone	19822-67-4	0.2-0.7	0-0.7		2.0		tr[r]; 0.5[b]
Ledol	577-27-5						tr[u]
Limonene	138-86-3	0.08-3.7	0-0.2	2.9		1.3	2.3[j]; 2.6[d]; 2.9[j]; 3.2[f]; 3.7[e]; 4.3[c]; 10.6[s]
cis-Limonene oxide	13837-75-7		tr		0.1		tr[u]; 0.2[b]; 0.8[h]
trans-Limonene oxide	4959-35-7		0-0.3				tr[u]; 0.3[h]
Linalool	78-70-6	0-0.7	tr	0.3	1.2		<0.05[q]; 0.1[i,k,o]; 0.2[r,u]; 0.8[l]
Linalyl acetate	115-95-7						<0.05[q]; 0.1[o]
Linoleic acid	60-33-3				7.9		0.2[l]; 29.5[r]

Table 5.7.1 Constituents of black cumin oil (*continued*)

Constituent	CAS	Percentage and range					
		A	B (1)	C (11)	D (15)	E (19)	F
Longicyclene	1137-12-8		0-0.1			0.4	0.06[i]; 0.08[b]
Longifolene (junipene)	475-20-7	0.4-3.6	0.4-3.1	0.3	1.8	tr	2.6[q]; 3.1[m]; 5.7[c]; 6.3[i]; 6.7[b]; 7.0[h]; 10.2[j]
α-Longipinene	5989-08-2	0-0.1	0.2-1.1	tr	0.7	0.1	0.5[e,p]; 0.7[q]; 0.8[o]; 1.0[d]; 1.5[i]; 1.6[h]; 1.9[b]
β-Longipinene	41432-70-6		0-0.1				<0.1[o]; 1.2[l]
p-Mentha-1,5-dien-8-ol	1686-20-0			tr			
cis-p-Mentha-2,8-dien-1-ol	3886-78-0						0.1[o]
p-Mentha-1,8-dien-4-ol	3419-02-1						<0.1[k]
trans-p-Mentha-2,8-dien-1-ol	4017-77-0						0.5[e]
p-Mentha-1,3,8-triene	18368-95-1		tr-0.4		0.1		tr[r]; 0.2[b]
p-Mentha-1,5,8-triene	21195-59-5						tr[t,u]
trans-Pinocarveol	1674-08-4						
trans-p-Menth-2-en-1-ol	29803-81-4	0-6.1					6.1[e]
p-Menth-2-en-7-ol							tr[u]
cis-4-Methoxythujane	1100111-04-3					?	
trans-4-Methoxythujane	1100111-06-5					4.0	
p-Methyl acetophenone	122-00-9			tr			
3-Methylcatechol	1189946-33-5						0.2[o]
Methyl chavicol	140-67-0		tr		0.3		0.6[r]; 0.9[d]; 1.9[c]
Methyl cumyl ether	935-67-1						tr[t,u]
Methyl decanoate	110-42-9						0.2[r];
(1-Methylethylidene)-cyclohexane	5749-72-4						tr[t,u]
Methyl eugenol	93-15-2			0.1	1.0		0.8[r]
Methyl geranate	2349-14-6				0.1		0.2[r]
Methyl hexadecanoate	112-30-0						0.1[k,r]
6-Methyl-α-ionone	79-69-6		0-0.1		0.2		
6-Methyl-α-(E)-ionone							0.1[o]
6-Methyl-γ-(E)-ionone	79-68-5				0.4		0.2[b]
2-Methylisovaleraldehyde							0.1[r]
Methyl linoleate	112-63-0		0-0.4		0.5		0.2[k]; 0.4[r]
3-Methyl nonane	5911-04-6						0.6[c]
Methyl octadecanoate	112-61-8						<0.1[k]
Methyl oleate	112-62-9						0.1[k]
2-Methyl-3-phenyl-propanal	5445-77-2						13.2[t,u]
1-Methyl-3-propyl benzene	1074-43-7						0.5[i]; 0.7[c]
Methyl thymol	1076-56-8						tr[u]; <0.1[o]
Methyl tetradecanoate	124-10-7						tr[l]
cis-Muurola-3,5-diene	157374-44-2			tr			
cis-Muurola-4(14),5-diene	157477-72-0		tr				
γ-Muurolene	30021-74-0		0-0.2	0.2			0.2[b]
Myrcene	123-35-3	0.03-0.3	tr	0.2		0.4	0.07[b,i]; 0.1[k,f,k]; 0.3[m]; 0.6[c]; 0.9[d]; 1.0[s]; 1.2[u]
Myristicin	607-91-0					tr	1.0[u]; 1.4[c]
(Z)-Myroxide	33281-83-3		tr				
Myrtenal	564-94-3						tr[u]
Myrtenol	515-00-4				3.1		tr[u]; 2.4[m]
Myrtenyl acetate	1079-01-2			0.5			
Naphthalene	91-20-3						0.4[r]
2(1H)-Naphthalenone	136156-72-4						2.6[c]
5-Neocedranol	13567-44-7				0.2		
α-Neoclovene	4545-68-0			tr			
Neothujol	35732-36-6						5.7[f]
Neral	106-26-3			tr			
Nerol	106-25-2						1.3[c]
(E)-Nerolidol	40716-66-3			0.2			
(Z)-Nerolidol	3790-78-1						12.3[r];
Neryl acetate	141-12-8				tr		
Nonacosane	630-03-5						tr[r]
Nonadecane	629-92-5						0.5[r]
γ-Nonalactone	104-61-0			0.2			
Nonanal	124-19-6			tr			

Table 5.7.1 Constituents of black cumin oil (*continued*)

Constituent	CAS	Percentage and range					
		A	B (1)	C (11)	D (15)	E (19)	F
n-Nonane	111-84-2			1.3			0.2[r]
2-(*Z*)-Nonen-1-al				tr			
(*E*)-β-Ocimene	3779-61-1			tr		tr	0.3[u]; 2.0[o]
(*Z*)-β-Ocimene	3338-55-4			tr		tr	1.1[u]
allo-Ocimene	673-84-7						tr[u]
Octadecane	593-45-3						1.0[r];
Octadecanoic acid	57-11-4				0.2		
Octacosane	630-02-4						tr[r]
(*E*)-9-Octadecenoic acid	112-79-8				1.8		
n-Octanal	124-13-0			tr			
Octanoic acid	124-07-2						0.2[r]
2-Octanol	123-96-6			tr			
n-Octanol	111-87-5			tr			
1,3,6-Octatriene	929-20-4						tr[u]
Oleic acid	112-80-1				6.3		0.04[b]; 0.1[l]
Oxidohimachalene	64825-84-9			0.4			
β-Panasinsene	56684-97-0			tr			
α-Patchoulene	560-32-7		tr				
β-Patchoulene	514-51-2		0-0.1	tr			
γ-Patchoulene	508-55-4			tr			
Patchoulenol	17806-54-1			tr			
Pentacosane	629-99-2						tr[r]
Pentadecanoic acid	1002-84-2				1.6		
2-Pentadecanone	2345-28-0		0-0.1			tr	
Pentylbenzene	538-68-1						0.1[r]
Perillyl alcohol	536-59-4						tr[u]
Phellandral	21391-98-0						tr[u]
α-Phellandrene	99-83-2	tr-0.06	tr	0.5	0.2	0.2	tr[u]; <0.05[q]; 0.1[o,p]; 0.6[c]
β-Phellandrene	555-10-2	0-0.2					<0.1[k]
Phenylacetaldehyde	122-78-1						<0.1[o]; 0.1[r]
1-Phenyl-1-butanol	614-14-2						0.9[t,u]
2-Phenylpropanal	93-53-8			tr			
Pimaradiene	1686-61-9		0-0.1		0.5		<0.01[b]; 0.2[i]
trans-Pinane	10281-53-5						<0.1[o]
α-Pinene	80-56-8	0.7-4.1	0-3.2	5.4		2.0	3.7[l]; 3.8[q]; 4.0[u]; 4.1[e]; 7.6[o]; 9.3[k]; 13.8[d]
β-Pinene	127-91-3	0.7-4.4	0-3.8	4.3		2.1	3.0[j]; 3.1[h]; 3.3[o]; 3.7[s]; 3.8[l]; 4.2[q]; 4.4[e]; 6.2[u]
α-Pinene oxide	1686-14-2						tr[u]
β-Pinene oxide	6931-54-0		tr		1.9		0.4[o]
cis-Pinocarveol	6712-79-4				0.1		
trans-Pinocarveol	1674-08-4						tr[u]
cis-Piperitol	16721-38-3				0.2		
trans-Piperitol	16721-39-4				0.1		
Pulegone	89-82-7						tr[e]
Rimuene	1686-67-5						0.2[r]
Sabinene	3387-41-5	0.1-2.2	0-7.5	1.4	0.4	0.8	1.4[c]; 1.5[l]; 1.6[j]; 1.7[d]; 1.9[e]; 2.0[q]; 2.5[o]; 4.3[b]
Sabinene hydrate	546-79-2						0.6[l]
cis-Sabinene hydrate	15537-55-0	0-1.2	0-0.1		0.5	tr	tr[u]; 0.02[b]; <0.1[k]; 0.2[i]; 1.0[e]
trans-Sabinene hydrate	17699-16-0		0-2.2		1.0	0.5	tr[r]; 0.09[b]; 0.1[k,p]; 0.2[i]; 0.8[o]; 1.1[m]
cis-Sabinene hydrate acetate	77318-48-0		0-0.1				<0.1[o]
trans-Sabinene hydrate acetate	77318-47-9		0-0.2		0.2		0.1[o,p]
cis-Sabinene hydrate methyl ether							+[a]
trans-Sabinene hydrate methyl ether							+[a]
α-Selinene	473-13-2						2.2[m]
β-Selinene	17066-67-0					0.1	0.2[u]; 0.4[m]
7-epi-α-Selinene	123123-37-5						0.3[m]
Selin-7(11)-en-4-yl acetate			tr				
Silphiperfola-4,7(14)-diene	210637-49-3			tr			

Table 5.7.1 Constituents of black cumin oil (*continued*)

Constituent	CAS	Percentage and range					
		A	B (1)	C (11)	D (15)	E (19)	F
Sinularene							tr[u]
Spathulenol	6750-60-3						tr[u]
Squalene	111-02-4						tr[r]
Styrallyl alcohol	13323-81-4						3.5[s,t]
α-Terpinen-7-al	1197-15-5						1.4[r]
γ-Terpinen-7-al	22580-90-1			tr			
α-Terpinene	99-86-5	tr-0.8	0-0.8	2.5	0.2	tr	0.3[u]; 0.5[i,j]; 0.6[e,m]; 0.7[b]; 1.0[n]; 1.1[l]; 4.2[r]
γ-Terpinene	99-85-4	0.1-2.4	0.4-3.0	3.7	0.6	12.9	1.4[d]; 2.0[b]; 2.4[m]; 3.5[n]; 11.8[f]; 24.4[u]; 45.7[s]
γ-Terpinene 1,2-epoxide	17023-74-4						+[a]
Terpinen-4-ol	562-74-3	0.2-0.8	0.3-0.7	0.4	8.9	0.9	0.8[o]; 0.9[q]; 1.0[p]; 1.8[h]; 2.1[l,n]; 2.4[i]; 4.3[d]
1-Terpineol	586-82-3						tr[u]; 0.2[i]
α-Terpineol	98-55-5			0.2	0.8	tr	tr[u]; <0.05[p]; 0.08[i]; 0.8[f]
(Z)-β-Terpineol	7299-40-3		tr				
Terpinolene	586-62-9	tr-0.8	0-0.9	1.2	0.8	0.6	tr[r]; 0.06[i,o]; 0.1[k,o]; 0.2[b]; 0.9[u]; 9.1[d]
Terpinolene 1,2-epoxide	6784-10-7						tr[u]; +[a]
Tetracosane	646-31-1						tr[r]
Tetradecanal	124-25-4				0.1		0.08
n-Tetradecane	629-59-4						0.2[c]
Tetradecanoic acid	544-63-8				0.7		tr[l]; 1.4[r]
(E)-α-Tetradecenal							0.6[r]
(E)-7-Tetradecenol	37011-95-3						0.6[r]
2,3,4,5-Tetramethyl-2-cyclopenten-1-ol							0.1[t,u]
2,4(10)-Thujadiene	36262-09-6			tr	0.2		0.06[i]; 0.2[r]; 14.0[t,u]
β-Thujaplicine	499-44-5		tr				
γ-Thujaplicine	672-76-4		tr				
Thuj-3-en-10-al	57129-54-1		0-0.1				
α-Thujene	2867-05-2	8.9-17.0	6.5-16.5	18.9	0.7	7.2	10.1[j]; 10.4[h]; 17.0[l]; 17.3[b,e]; 17.5[q]
α-Thujone	546-80-5		0-0.1				0.1[t,u]
β-Thujone	471-15-8					tr	<0.1[k]
Thujopsene	470-40-6						0.1[b]
cis-Thujopsene	32435-95-3		0-0.1				
Thujyl alcohol	513-23-5						7.4[m]
Thymohydroquinone	2217-60-9	0.2-0.6	tr-0.7		12.2	tr	0.4[n]; 0.6[k,o]; 0.8[b]; 3.4[p]
Thymol	89-83-8	tr-2.2			1.5	tr	<0.1[k]; 0.1[i]; 0.2[b,e]; 1.7[d]; 1.8[l]; 12.1[r]; 26.8[m]
Thymoquinone	490-91-5	0.6-24.5	6.2-18.4	0.8	21.8	tr	24.5[k]; 37.6[p]; 48.2[e]; 54.8[j]; 56.1[o]; 63.3[b,f]
o-Tolualdehyde	529-20-4						<0.1[o]
n-Triacontane	638-68-6						tr[r];
Tricosane	638-67-5						0.3[r]
Tricyclene	508-32-7					0.8[f]	tr[u]
Tridecanal	10486-19-8		0-0.2				
Tridecanoic acid	638-53-9				0.1		
2-Tridecanone	593-08-8		0-0.1		0.2	0.1	tr[r]; <0.1[k]; 0.1[b,i]
2,7,7-Trimethylbicy-clo[2.2.1] hept-2-ene	514-14-7						tr[t,u]
1-(1,2,3-Trimethylcy-clopent-2-enyl)-me-thanone							tr[t,u]
Umbellulone	24545-81-1				0.3		
Undecane (*n*-)	1120-21-4						0.9[r];
2-Undecanone	112-12-9		0-0.2		0.1		0.09[j]; 0.1[k]; 0.2[b]
Uvidine							1.3[c]
trans-Verbenol	1820-09-3	0-0.3				0.3	tr[u]; 0.3[e]
cis-Verbenyl acetate	29135-27-1			tr			
Widdrol	6892-80-4			tr			

Table 5.7.1 Constituents of black cumin oil (*continued*)

Constituent	CAS	Percentage and range					
		A	B (1)	C (11)	D (15)	E (19)	F
α-Ylangene	14912-44-8		tr	tr			
β-Ylangene	20479-06-5			tr			
α-Zingiberene	495-60-3				0.2		tr[u];

A thirteen black cumin essential oil samples from India, Morocco and Turkey, analyzed between 1998 and 2005; lowest and highest concentrations given (E. Schmidt, unpublished data)
B five samples of *Nigella sativa* essential oil from Algeria, Syria, Jordan, Iran and Ethiopia obtained with the technique of microwave-assisted hydrodistillation; lowest and highest concentrations given (ref. 1)
C one lab-hydrodistilled oil from seeds of *N. sativa* collected in Tunisia (ref. 11)
D two lab-hydrodistilled oils from two locations in Algeria; highest concentration given (ref. 15)
E one lab hydrodistilled *Nigella sativa* essential oil from black cumin seeds obtained from a local market in Poland (ref.19)
F data from other studies (indicated with superscript letters); highest concentrations found in any study reviewed here given; when two or more oils were investigated, only the highest concentrations are mentioned, unless indicated otherwise

[a] one lab-hydrodistilled oil from the seeds of *N. sativa* cultivated in Tunisia (ref. 5); [b] ten oils obtained from Algerian seeds by microwave hydrodistillation with distillation times varying from 0.5 to 10 minutes (ref. 7); [c] one lab-hydrodistilled oil from seeds purchased at a local market in India (ref. 13); [d] data taken from ref. 12; [e] one lab-hydrodistilled and one steam-distilled oil from the seeds of *N. sativa* grown in Bulgaria (ref 17); [f] three lab-hydrodistilled oils from seeds of *N. sativa* cultivated at two locations in Egypt (ref. 8); [g] one lab-hydrodistilled oil from seeds purchased in Morocco (ref. 9); [h] one lab-hydrodistilled oil from *N. sativa* seeds bought at a local market in India (ref. 21); [i] one lab-hydrodistilled oil from seeds purchased at a local market in India (ref. 6); [j] four steam-distilled oils of *N. sativa* seeds cultivated in Iran under different irrigation treatments (ref. 16); [k] one lab-distilled oil (ref. 24); [l] two oils produced from seeds purchased in France (ref. 18); [m] one sample of *Nigella sativa* essential oil (ref. 25); [n] one lab-hydrodistilled oil from seeds of *N. sativa* cultivated in Tunisia (ref. 10); [o] four oils from *Nigella sativa* seeds cultivated in Tunisia and obtained with different steam-distillation techniques (ref. 27); [p] one lab-hydrodistilled oil from black cumin seeds obtained from a local Indian market (ref. 28); [q] one hydrodistilled and one steam-distilled oil from black cumin seeds obtained from a local market in Istanbul, Turkey (ref. 29); [r] oil from *N. sativa* seeds from Algeria, obtained by hydrodistillation with a very atypical composition with low *p*-cymene and thymoquinone and very high linoleic acid content (ref. 31); [s] one lab-hydrodistilled oil from black cumin seed from Iran (ref. 32); [t] (probably) incorrect identification (ref. 30); [u] one lab-hydrodistilled oil from black cumin seeds cultivated in Iran (ref. 33)

tr: traces (in column B <0.03%; in columns C, F[r] and F[u]: <0.05%); + present in black cumin oil, but amount not quantified

---

Ref. 34 (cited in ref. 30) is not discussed, as the sum of the ingredients was > 150% (34).

CONTACT ALLERGY/ALLERGIC CONTACT DERMATITIS

General
Eight cases of allergic contact dermatitis from black cumin oil have been reported. The dermatitis tends to be extensive and concomitant ingestion of the oil may lead to serious erythema multiforme-like eruptions (36,37). Patch test reactions are usually strong.

Case reports
One patient developed allergic contact dermatitis from the application of undiluted black cumin oil on the neck for a sore throat; positive patch test reactions were obtained with the pure oil and with 1% and 0.5% black cumin oil (38). A woman developed allergic contact dermatitis from a skin care product containing black cumin

Table 5.7.2 Major constituents of black cumin oil

Constituent	CAS	Percentage and range					
		A	B (1)	C (11)	D (15)	E (19)	F
Thymoquinone	490-91-5	0.6-24.5	6.2-18.4	0.8	21.8	tr	24.5[k]; 37.6[p]; 48.2[e]; 54.8[j]; 56.1[o]; 63.3[b,f]
p-Cymene	99-87-6	19.9-57.5	37-73	49.5	8.9	60.2	47.4[g]; 54.1[h]; 56.2[q]; 56.4[e]; 57.6[b]; 62.3[l]
γ-Terpinene	99-85-4	0.1-2.4	0.4-3.0	3.7	0.6	12.9	1.4[d]; 2.0[b]; 2.4[m]; 3.5[n]; 11.8[f]; 24.4[u]; 45.7[s]
α-Thujene	2867-05-2	8.9-17.0	6.5-16.5	18.9	0.7	7.2	10.1[j]; 10.4[h]; 17.0[l]; 17.3[b,e]; 17.5[q]
α-Pinene	80-56-8	0.7-4.1	0-3.2	5.4		2.0	3.7[l]; 3.8[q]; 4.0[u]; 4.1[e]; 7.6[o]; 9.3[k]; 13.8[d]
Longifolene (junipene)	475-20-7	0.4-3.6	0.4-3. 1	0.3	1.8	tr	2.6[q]; 3.1[m]; 5.7[c]; 6.3[l]; 6.7[b]; 7.0[h]; 10.2[j]
β-Pinene	127-91-3	0.7-4.4	0-3.8	4.3		2.1	3.0[j]; 3.1[h]; 3.3[o]; 3.7[s]; 3.8[l]; 4.2[q]; 4.4[e]; 6.2[u]
Terpinen-4-ol	562-74-3	0.2-0.8	0.3-0.7	0.4	8.9	0.9	0.8[o]; 0.9[q]; 1.0[p]; 1.8[h]; 2.1[l,n]; 2.4[i]; 4.3[d]

LEGEND: SEE UNDER TABLE 5.7.1

oil; upon patch testing she reacted to the cosmetic and to black cumin. There was a negative reaction, however, to black cumin oil itself, but this can easily be explained by the fact that a *cold-pressed* oil, not the essential oil, was used for testing (39). A female patient developed generalized dermatitis from contact allergy to black cumin oil applied to the face (50). One patient had pigmented dermatitis of the face and a second had generalized dermatitis from contact allergy to *Nigella sativa* black seed oil, which was confirmed by an open test (41). One case of a bullous erythema multiforme/toxic epidermal necrolysis-like eruption was reported in a patient who had ingested *Nigella sativa* oil and had applied it to the skin; there was a strongly positive patch test reaction to the oil (36). A patient had allergic contact dermatitis expressing as generalized erythema multiforme with both ingestion and topical application of the oil; a patch test and ROAT with the pure oil were strongly positive (37). One unpublished case of Stevens-Johnson syndrome from black cumin oil (contact allergy?) was mentioned as a personal communication (37).

LITERATURE

1 Benkaci-Ali F, Baaliouamer A, Wathelet J-P, Marlier M. Etude comparative de la composition chimique de la *Nigella sativa* Linn. de quelques régions du monde, extraites par micro-ondes. Rivista Ital EPPOS 2006;41:23-32. Data cited in ref. 2

2 Lawrence BM. Progress in essential oils. Perfum Flavor 2008;33(January):38-? (last page unknown)

3 Butt MS. *Nigella sativa*: Reduces the risk of various maladies. Crit Rev Food Sci Nutrit 2010;50:654-665

4 Yousefi M, Barikbin B, Kamalinejad M, Abolhasani E, Ebadi A, Younespour S, et al. Comparison of therapeutic effect of topical *Nigella* with betamethasone and eucerin in hand eczema. JEADV 2013;27:1498-1504

5 Bourgou S, Pichette A, Lavoie S, Marzouk B, Legault J. Terpenoids isolated from Tunisian *Nigella sativa* L. essential oil with antioxidant activity and the ability to inhibit nitric oxide production. Flavour Fragr J 2012;27:69-74

6 Singh G, Marimuthu P, de Heluani CS, Catalan C. Chemical constituents and antimicrobial and antioxidant potentials of essential oil and acetone extract of *Nigella sativa* seeds. J Sci Food Agric 2005;85:2297-2306

7 Benkaci-Ali F, Baaliouamer A, Meklati BY. Kinetic study of microwave extraction of essential oil of *Nigella sativa* L. seeds. Chromatographia 2006;64:227-231

8 Edris A. Evaluation of the volatile oils from different local cultivars of *N. sativa* L. grown in Egypt with emphasis on the effect of extraction method on thymoquinone. J Essent Oil Bear Plants 2010;13:154-164

9 Rchid H, Nmila R, Bessière JM, Sauvaire Y, Chokaïri M. Volatile components of *Nigella damascena* L. and *Nigella sativa* L. seeds. J Essent Oil Res 2004;16:585-587

10 Bourgou S, Pichette A, Marzouk B, Legault J. Bioactivities of black cumin essential oil and its main terpenes from Tunisia. South Afr J Bot 2010;76:210-216

11 Harzalla HJ, Kouidhi B, Flamini G, Bakhrouf A, Mahjoub T. Chemical composition, antimicrobial potential against cariogenic bacteria and cytotoxic activity of Tunisian *Nigella sativa* essential oil and thymoquinone. Food Chem 2011;129:1469-1474

12 Toma C-C, Simu GM, Hanganu D, Olah N, Georgiana FM, Hammami VC, Hammami M. Chemical composition of the Tunesian *Nigella sativa*. Note I. Profile on essential oil. Farmacia 2010;58:458-464

13 Gerige SJ, Gerige MKY, Rao M. GC-MS analysis of *Nigella sativa* seeds and antimicrobial activity of its volatile oil. Braz Arch Biol Technol 2009;52:1189-1192

14 Ahmad A, Husain A, Mujeeb M, Khan SA, Najmi AK, Siddique NA, et al. A review on therapeutic potential of *Nigella sativa*: A miracle herb. Asian Pac J Trop Biomed 2013;3:337-352

15 Benkaci-Ali F, Baaliouamer A, Meklati BY, Chemat F. Chemical composition of seed essential oils from Algerian *Nigella sativa* extracted by microwave and hydrodistillation. Flavour Fragr J 2007;22:148-153

16 Mozaffari F-S, Ghorbanli M, Babai A, Farzami Sepehr M. The effect of water stress on the seed oil of *Nigella sativa* L. J Essent Oil Res 2000;12:36-38

17 Stoyanova A, Georgiev E, Wajs A, Kalemba D. A comparative investigation on the composition of the volatiles of *Nigella sativa* L. from Bulgaria. J Essent Oil Bear Plants 2003;6:207-209

18 Bourrel C, Vilarem G, Perineau F. Etude des composes aromatiques des graines de Nigelle (*Nigella sativa* L.); Evaluation des proprietées antibacteriennes et antifongiques. Rivista Ital EPPOS 1993;10:21-27 Data cited in ref. 2

19 Wajs A, Bonikowski R, Kalemba D. Composition of essential oil from seeds of *Nigella sativa* L. cultivated in Poland. Flavour Fragr J 2008;23:126-132

20 Benkaci-Ali F, Baaliouamer A, Wathelet JP, Marlier M. Chemical composition and physicochemical characteristics of fixed oils from Algerian *Nigella Sativa* seeds. Chem Nat Comp 2012;47:925-931

21 Sunita M, Meenakshi S. Chemical composition and antidermatophytic activity of *Nigella sativa* essential oil. Afr J Pharm Pharmacol 2013;7:1286-1292

22 El-Din Hussein El-Tahir K, Bakeet DM. The black seed *Nigella sativa* Linnaeus - A mine for multi cures: a plea for urgent clinical evaluation of its volatile oil. JTU Med Sc 2006;1:1-19; available at http://jtaibahumedsc.net/issues/volum1_issue1/1.pdf

23 Nickavar B, Mojab F, Javidnia K, Roodgar Amoli MA. Chemical composition of the fixed and volatile oils of *Nigella sativa* L. from Iran. Z Naturforsch c 2003;58:629-631

24 Aboutabl EA, El-Azzouny AA, Hammerschmidt F-J. Aroma volatiles of *Nigella sativa* L. seeds. In: E-J Brunke, Ed. Progress in essential oil research. New York: Walter de Gruyter, 1986:48-55. Data cited in ref. 2

25 D'Antuono LF, Moretti A, Lovato FSA. Seed yield, yield components, oil content and essential oil content and composition of *Nigella sativa* L. and *Nigella damascena* L. Indust Crops Prod 2002;15:59-69

26 Ali BH, Blunden G. Pharmacological and toxicological properties of *Nigella sativa*. Phytother Res 2003;17:299-305

27 Benkaci-Ali F, Akloul R, Boukenouche A, De Pauw E. Chemical composition of the essential oil of *Nigella sativa* seeds extracted by microwave steam distillation. J Essent Oil Bear Plants 2013;16:781-794

28 Singh S, Das SS, Singh G, Schuff C, de Lampasona MP, Catalán CAN. Composition, *in vitro* antioxidant and antimicrobial activities of essential oil and oleoresins obtained from black cumin seeds (*Nigella sativa* L.). BioMed Res Int Vol. 2014, Article ID 918209, 10 pages. http://dx.doi.org/10.1155/2014/918209

29 Kokoska L, Havlik J, Valterova I, Soviva H, Sajfrtova M, Jankovska I. Comparison of chemical composition and antibacterial activity of *Nigella sativa* seed essential oils obtained by different extraction methods. J Food Prot 2008;71:2475-2480

30 Lawrence BM. Progress in essential oils. Perfum Flavor 2014;39(Feb):50-58

31 Benkaci-Ali F, Baaliouamer A, Meklati BY. Etude comparative de la composition chimique de la *Nigella sativa* Linn. de la région de media extraite par hydrodistillation et par micro-ondes. Rivista Ital EPPOS 2005;40:15-24. Data cited in ref. 30

32 Pourmortazavi SM, Hajimirsadeghi SS. Supercritical fluid extraction in plant essential and volatile oil analysis. J Chromatogr A 2007;1163:2-24

33 Jalai-Heravi M, Zekavat B, Sereshi H. Use of gas chromatography-mass spectrometry combined with resolution methods to characterize the essential oil components of Iranian cumin and caraway. J Chromatogr A 2007;1143:215-226

34 Adamu HM, Ekanem EO, Bulama S. Identification of essential oil components from *Nigella sativa* seed by gas chromatography-mass spectrometry. Pak J Nutrition 2010;9:966-967. Data cited in ref. 30

35 Lawless J. The encyclopedia of essential oils, 2nd Edition. London: Harper Thorsons, 2014

36 Gelot P, Bara-Passot C, Gimenez-Arnau E, Beneton N, Maillard H, Celerier P. Bullous drug eruption with *Nigella sativa* oil. Ann Dermatol Venereol 2012;139:287-291

37 Nosbaum A, Ben Said B, Halpern S-J, Nicolas J-F, Bérard F. Systemic allergic contact dermatitis to black cumin essential oil expressing as generalized erythema multiforme. Eur J Dermatol 2011;21:447-448

38 Steinmann A, Schätzle M, Agathos M, Breit R. Allergic contact dermatitis from black cumin (*Nigella sativa*) oil after topical use. Contact Dermatitis 1997;36:268-269

39 Zedlitz S, Kaufmann R, Boehncke W. Allergic contact dermatitis from black cumin (*Nigella sativa*) oil-containing ointment. Contact Dermatitis 2002;46:188

40 Lleonart R, Andrés B, Molinero J, Corominas M. Systemic allergic contact dermatitis due to black cumin oil. Contact Dermatitis 2014;70(Suppl. 1):45

41 Gad El-Rab MO, Al-Sheikh OA. Is the European standard series suitable for patch testing in Riyadh, Saudi Arabia? Contact Dermatitis 1995;33:310-314

Chapter 5.8 BLACK PEPPER OIL

DEFINITION

Black pepper oil (essential oil of black pepper) is the essential oil obtained from the fruit of the (black) pepper, *Piper nigrum* L.

INCI NOMENCLATURE

Description/definition: Piper nigrum fruit oil is the volatile oil distilled from the dried ripe fruit of black pepper, *Piper nigrum* L., Piperaceae

INCI name EU: Piper nigrum fruit oil

INCI name USA: Piper nigrum (pepper) fruit oil

CAS registry number(s): 8006-82-4; 84929-41-9

EINECS number(s): 284-524-7

ISO (INTERNATIONAL ORGANIZATION FOR STANDARDIZATION) STANDARD

ISO number: 3061

ISO name: Essential oil of black pepper

Botanical origin: *Piper nigrum* L.

Parts of plant used: Fruit

ISO (International Organization for Standardization) values are shown in Table 5.8.1.

Black pepper essential oil should not be confused with Piper nigrum (pepper) seed oil (also CAS 84929-41-9, EINECS 284-524-7), which is a fatty, expressed (vegetable, fixed) oil from black pepper seeds.

THE PLANT, THE OIL, AND THEIR USES

Black pepper (*Piper nigrum* L.) is a perennial woody evergreen plant that can grow to a height of 50-60 cm, which is native to India (15). The tropical forest of the Malabar region of southern India is considered to be the center of its origin; it is now widely cultivated in the tropics. The most important exporters of black pepper are India, Indonesia, Brazil, Malaysia, Sri Lanka and Vietnam. 'Black pepper' is obtained by sundrying green berries harvested before full maturity is reached, whereas fully ripe dried fruits devoid of pericarp form the commercial 'white pepper'. Pepper is the most widely used spice throughout the world, appreciated for both its aroma and its pungency. The 'king of spices' is not only used for culinary, but also for medicinal purposes (1,2,10). Medicinally black pepper may be used for digestive disorders like large intestine toxins, different gastric problems, diarrhea, and indigestion and also against respiratory disorders including cold, fever, and asthma (15,24). In addition, black peppers have applications as preservatives and as biocontrol agents.

The essential oil of black pepper is obtained by hydrodistillation of the (powdered) black pepper fruits. This produces a colorless volatile oil with characteristic odor and sharp taste. Pepper essential oil has important applications in the food and pharmacological industries, perfumery, cosmetics, home remedies (10,15,19).

Externally, the oil is used as a rubefacient and anti-rheumatic, and as gargling agent for sore throat (28). It is also employed in aromatherapy practice (40).

The biological activities (24,28) and health claims (2) of black pepper have been reviewed.

CHEMICAL COMPOSITION

Black pepper oil is a colorless or light yellow to blue, mobile liquid which has a terpenic, aromatic, somewhat spicy and herbaceous odor. The yield of essential oil from the fruit corn of *Piper nigrum* L. generally varies from 1.0 to 2.6%. The main producing countries of this oil are India, Madagascar, Vietnam, Sri Lanka and Indonesia.

Literature data (up to November 6, 2014) on the chemical composition of black pepper oils and unpublished analytical data from one of us (ES) are shown in Table 5.8.2 in alphabetical order. In black pepper oils from various origins, over 305 chemicals have been identified. About 58% of these were found in a single publication reviewed only. The major compounds found in black pepper oils are shown in Table 5.8.3. They include (highest concentrations in any study given) β-caryophyllene (65.7%), sabinene (41.5%), δ3-carene (33.2%), α-pinene (29.5%), caryophyllene oxide (29.3%), β-pinene (29.3%), limonene (26.8%), myrcene (23.3%), α-phellandrene (17.4%) and p-cymene (13.0%). Well-known ingredients of black pepper oils that were present in high concentrations in one or two studies were α-terpinene (7.3% and 13.3%), germacrene D (11.0%), terpinen-4-ol (10.1%), terpinolene (8.5%) and camphene (8.0%). Uncommon or rare constituents of black pepper oil found in high concentrations in single studies include limonen-6-ol, pivalate (25.4%), (*E,E*)-α-farnesene (16.3%), germacrene (12.8%), eudesmol (9.6%), (*E*)-3(10)-caren-4-ol (8.4%) and eugenol (5.6%).

Commercial oils

The ten chemicals that had the highest maximum concentrations in 46 commercial black pepper essential oil samples (concentration ranges provided) are the following: β-caryophyllene (0.9-32.4%), α-pinene (5.1-29.5%), limonene (10.2-24.7%), β-pinene (6.7-20.3%), δ3-carene (4.3-17.6%), α-phellandrene (0.8-17.4%), sabinene (0.1-15.4%), camphene (0.1-8.0%), myrcene (1.5-6.3%) and p-cymene (0.2-6.0%) (Erich Schmidt, unpublished analytical data).

Chemotypes

According to the different amounts of the main essential oil components caryophyllene, sabinene, δ3-carene, and limonene, three pepper chemotypes have been suggested: a sabinene/caryophyllene type (amounts of sabinene 9.3–20.9%, of β-caryophyllene 17.6–34.6%), a caryophyllene type (amounts of caryophyllene 39.8–65.7%) and a δ3-carene/caryophyllene/limonene type (amounts of δ3-carene 13.9–32.7%, of β-caryophyllene 13.9–46.6% and of limonene 10.9–19.5%) (3).

Table 5.8.1 ISO values (%) for black pepper oil[a,b]

Compound	CAS	Minimum	Maximum
β-Caryophyllene	87-44-5	10.0	40.0
α-Pinene	80-56-8	2.5	26.0
Limonene	138-86-3	7.0	25.0
δ3-Carene	13466-78-9	3.0	20.0
Sabinene	3387-41-5	nd	17.0
β-Pinene	127-91-3	nd	15.0
Germacrene D	23986-74-5	nd	6.5
β-Selinene	17066-67-0	nd	6.0
α-Selinene	473-13-2	nd	5.0
α-Copaene	3856-25-5	nd	4.5
δ-Elemene	20307-84-0	0.5	4.5
Caryophyllene oxide	1139-30-6	nd	1.0

[a] ISO 3061 Essential oil of black pepper ©ISO 2008; Geneva, Switzerland, www.iso.org

[b] there are separate ISO norms for black pepper oils from India, Sri Lanka, Indonesia and Madagascar; the lowest and highest values for any of these four are given

nd = not detectable

CONTACT ALLERGY/ALLERGIC CONTACT DERMATITIS

General

Contact allergy to and possible allergic contact dermatitis from black pepper oil have been reported in one publication only. A false-positive patch test reaction due to the excited skin syndrome cannot be excluded. In the same case report, α-pinene and caryophyllene may have been allergens in black pepper oil (41).

Case reports

Two aromatherapists had contact dermatitis (occupational in one) with allergies to multiple essential oils used at their work, including black pepper oil. Both patients also reacted to geraniol, α-pinene, the fragrance mix and various other fragrance materials. In addition, one proved to be allergic to linalool and linalyl acetate, the other to caryophyllene; α-pinene and caryophyllene were demonstrated by GC-MS in black pepper oil (41). Both chemicals are the main components of black pepper oils with concentrations

Table 5.8.2 Constituents identified in black pepper oils

Constituent	CAS	Percentage and range					
		A	B (11)	C (5-9)	D (1)	E (18)	F
Acetaldehyde	75-07-0		0.01				
Adamantane	281-23-2						2.8[z5]
α-Amorphene	20085-19-2			0-3.2	0.1	1.3-1.5	1.7[k]
(E)-Anethole	4180-23-8						0.4[v]
Aromadendrene	489-39-4						+[z1]
Azulene	275-51-4						0.6[z5]
Benzaldehyde	100-52-7						<0.1[z6]
Benzene, 1-methyl-2-(1-methylethenyl)-	7399-49-7						1.6[w]
Benzene, 1-methyl-4-(1-methylethenyl)-	1195-32-0						0.3[w]
5-Benzodiazepine							0.2[z5]
Benzoic acid	65-85-0						+[z2]
Benzyl alcohol	100-51-6						0.2[z5]
Benzyl benzoate	120-51-4		tr				
cis-α-Bergamotene	18252-46-5	0.05-0.9	0.1				+[z2]
trans-α-Bergamotene	13474-59-4	0.2-1.2	0-1.8		0-0.3		+[z2]
Bicyclo[5.3.0]decane, 2-methylene-5-(1-methylvinyl)-8-methyl-							0.8[v]
Bicyclogermacrene	24703-35-3						+[z6]; 0.4[h]
α-Bisabolene	17627-44-0	0.2-1.0				0.5-4.3[b]	2.5[k]
β-Bisabolene	495-61-4	0.6-5.9	0.8	0-6.0	3.9	0.5-4.3[b]	0.5[j]; 0.6[z7]; 1.1[g]; 1.8[f,q]; 7.2[i]; 7.7[h]
α-Bisabolol	515-69-5	tr			0.2		
β-Bisabolol	15352-77-9	1.3				0.09-0.2	
Borneol	507-70-0	0.2	0-0.2			tr	
Bornyl acetate	76-49-3						0.9[v]
Bornylene	464-17-5						<0.05[j]
Bufa-20,22-dienolide, 14-hydroxy-3-oxo-, (5α-)-	39845-12-0						0.05[v]
α-Bulnesene	3691-11-0					0.09-1.9	+[z6]; tr[z7]
n-Butyric acid	107-92-6		tr				+[z2]
Cadalene	483-78-3						+[z6];
Cadina-1,4-diene	29837-12-5	0.2					+[z6]; 0.07[h]; 0.3[w]
Cadina-4(15),6-diene		0.2					+[z]
Cadina-5,10(15)-dien-4-ol							0.2[z6]

Table 5.8.2 Constituents identified in black pepper oils (*continued*)

Constituent	CAS	Percentage and range					
		A	B (11)	C (5-9)	D (1)	E (18)	F
Cadinene	29350-73-0						0.6[j]
β-Cadinene	523-47-7						0.6[q]; 0.7[x]; 1.1[f]
γ-Cadinene	39029-41-9	0.2-0.8					2.8[k]
δ-Cadinene	483-76-1	0.2-1.8	0.1	0-1.8	1.3	0-0.1	0.2[f,m,p]; 0.3[j]; 0.7[g]; 2.4[h]; 3.9[o]
α-Cadinol	481-34-5			tr-2.7			
δ-Cadinol	19435-97-3	0.05-0.9			1.4		
τ-Cadinol	5937-11-1			0-3.3			0.3[g]
α-Calacorene	21391-99-1						0.2[m]
Calamenene	483-77-2		0.5				+[z6]
l-Calamenene	483-77-2						0.3[f]
cis-Calamenene	72937-55-4				0.2		
trans-Calamenene	73209-42-4						0.4[m]
Camphene	79-92-5	0.1-8.0	tr	0-0.6		0.1-0.2	0.1[z7]; 0.2[k,z4]; 0.3[f]; 0.4[y]; 0.5[j]; 0.8[u]
Camphor	76-22-2						<0.05[j]; <0.1[z6]; 0.08[h,w]
Carene	74806-04-5						4.2[z5]
δ2-Carene (= δ4)	554-61-0						+[z6]; 0.1[j]
δ3-Carene	13466-78-9	4.3-17.6	2.8	0-23.4	4.4	0-2.8	22.0[z4]; 22.2[w]; 27.9[j]; 32.7[s]; 33.2[j]
(*E*)-3(10)-Caren-4-ol	22626-38-6						8.4[v]
2-Caren-3-one							<0.1[z6]
3-Caren-2-one							<0.1[z6]
Carvacrol	499-75-2						<0.1[z6]
Carveol	99-48-9		0.3				
(*E*)-Carveol	1197-07-5			0-0.1		0-0.02	0.3[m]
(*Z*)-Carveol	1197-06-4			0-0.3		0.01-0.03[c]	0.2[m]
Carvone	99-49-0		0.2			0.01-0.03[c]	0.2[h,z6]; 0.3[f]; 0.4[m]
L-Carvone	6485-40-1						1.7[v]
Carvone oxide	33204-74-9			0-0.4		0-0.01	0.1[k]
Carvotanacetone	499-71-8						+[z2]
β-Caryophyllene	87-44-5	0.9-32.4	7.3	6.4-47.5	29.9	21.2-27.7	29.2[r]; 39.7[o]; 39.8[v]; 41.5[p]; 65.7[s]
Caryophyllene alcohol	913176-41-7		1.2			0.02-0.07	+[z2]
Caryophyllen-5-ol II							<0.1[z6]
Caryophyllene oxide	1139-30-6	0.2-1.2	0.9	0.2-6.0	3.9	0.3-0.9	4.1[o]; 6.2[t]; 8.0[z5]; 9.9[j]; 14.2[s]; 29.3[m]
Caryophyllenol	38284-26-3			0-1.0			0.1[k]; 0.5[m]
Cedrene	11028-42-5						0.5[v]
α-Cedrene	469-61-4						0.3[g]
Cedrenol	28231-03-0						0.6[q]
Cedrol	77-53-2			0-4.1		0-0.07	0.1[k]; 3.2[t]
α-Chamigrene	19912-83-5						+[z1]
1-Chloroeicosane	42217-02-7						0.03
Cholest-22-ene-21-ol, 3,5-dehydro-6-methoxy-, pivalate							0.09[v]
1,8-Cineole	470-82-6						+[z,z2]; 0.2[m,z6]; 1.0[z4]
Cinnamic acid	621-82-9						+[z2]
Citronellal	106-23-0			0-0.8		0.01-0.03	1.0[k]
Citronellol	106-22-9		0.1				
Clovene	469-92-1					0.07-0.1	
α-Copaene	3856-25-5	0.2-4.4	1.4	0-0.2	2.8	0.4-0.8	1.9[j]; 2.3[y]; 3.1[m]; 3.7[f]; 3.8[o]; 4.5[l]; 6.3[h]
β-Copaene	18252-44-3				0.1		tr[k]
m-Cresol	108-39-4						<0.1[z6]
o-Cresol	95-48-7						tr[m]
p-Cresol	106-44-5						<0.1[z6]
Cryptone	500-02-7				tr		+[z2]; tr[m]
Cubebene	11012-64-9						1.9[q]
α-Cubebene	17699-14-8		0.1	tr-6.8	0.1	0.2-3.3[d]	tr[z7]; 0.09[f,p]; 0.2[y]; 0.3[h]; 1.9[k]; 3.5[t]

Table 5.8.2 Constituents identified in black pepper oils *(continued)*

Constituent	CAS	Percentage and range					
		A	B (11)	C (5-9)	D (1)	E (18)	F
β-Cubebene	13744-15-5		tr		0.3		0.1^j; 0.2^f; 0.3^i; 0.5^h; 0.7^w
Cubebol	23445-02-5				0.8		
α-Cubenene	205537-26-4	0.03-0.4					0.4^{z5};
Cubenol	21284-22-0						0.1^f; 0.2^v; 2.5^v
Cuminaldehyde	122-03-2		0.01				
Cuminyl alcohol	536-60-7		0.07				
Cuparene	16982-00-6			0-4.6		0.04-1.4	
ar-Curcumene	644-30-4	0.07-0.4				0.04-0.3	+^{z6}; tr^{z7}; 0.2^m; 0.3^k; 1.5^v
β-Curcumene	28976-67-2						0.1^g
1,6-Cyclodecadiene	1124-79-4						0.4^{z5}
Cyclohexene	110-83-8						1.1^{z5}
Cyclosativene	22469-52-9						0.08^h
m-Cymene	535-77-3						+^{z6}; 0.1^j; 0.4^{m,n}; 1.1^x
o-Cymene	527-84-4	tr-0.1					+^{z}
p-Cymene	99-87-6	0.2-6.0	2.1	0-6.2	1.2	0-9.7	0.8^m; 0.9^{f,p}; 1.0^u; 1.4^{z4}; 8.2^t; 13.0^k
p-Cymene 8-methyl ether							+^{z2}
m-Cymenene	1124-20-5						+^{z6}; 0.1^m
p-Cymenene	1195-32-0						+^{z6}; tr^h; 0.1^m
m-Cymen-8-ol	5208-37-7	tr-0.09					tr^m
p-Cymen-8-ol	1197-01-9	0.1					+^{z2}; <0.1^u; 1.4^m
Decanol	36729-58-5		1.0				
Dihydrocarveol	38049-26-2			0-0.5		0.01-0.04	+^{z2}
1,2-Dihydropyridine,1-(1-oxobutyl)-	849947-72-4						0.07^v
1,1-Dimethyl-2-(3-methyl-1,3-butadien-yl) cyclopropane	68998-21-0						+^{z1}
1,5-Dimethyl-6-meth-ylenespiro[2.4]heptane	62238-24-8						+^{z1}
7-Dimethyl-1-(1-me-thylethyl)							1.2^{z5}
(E,E)-2,6-Dimethyl-3,5,7-octatriene-2-ol							0.8^v
Docosane	629-97-0				0.1		
Dodecanal	112-54-9		0.2				
Dodecanol	112-53-8		2.2				
Eicosane	112-95-8				0.3		+^{z}
(E)-3-Eicosene	74685-33-9						+^{z}
Elemene	11029-06-4						3.5^{z5}
α-Elemene	5951-67-7						2.6^{z5}
β-Elemene	33880-83-0	0.3-2.2	2.6	0-2.4	0.9	0.05-0.09	0.1^k; 0.3^m; 0.4^h; 0.8^g; 0.9^{f,i}; 1.6^{o,w}
(E)-β-Elemene							+^{z6}
(Z)-β-Elemene							+^{z6}
γ-Elemene	29873-99-2		0.06				+^{z6}; 0.04^h; 0.2^f; 1.5^v
δ-Elemene	20307-84-0	0.2-2.4	tr		0.5		1.8^j; 1.9^g; 2.0^f; 2.5^o; 2.7^x; 4.0^p
Elemicin	487-11-6		tr				
Elemol	639-99-6	0.04-1.5	0.2	0-10.5	0.2	0.06-0.1	0.3^g; 2.7^q
10-Epizonarene	41702-63-0						+^{z1}
4,5-Epoxycarene							<0.1^{z6}
Eremophilene	10219-75-7						+^{z1}; 0.4^x
1-Ethenyloxyhexade-cane	822-28-6						+^{z1}
Eucarvone	503-93-5						0.4^{z5}
Eudesmol	51317-08-9						9.6^r
α-Eudesmol	473-16-5						0.1^{k,q}; 1.2^g
β-Eudesmol	473-15-4		1.4				0.3^q
γ-Eudesmol	1209-71-8						0.1^q
Eugenol	97-53-0		0.07				0.6^{z4}; 4.0^w; 5.6^y
(E,E)-α-Farnesene	502-61-4		tr			0-0.7	+^{z6}; 16.3^t
β-Farnesene	502-60-3						0.2^f; 0.6^w; 0.9^i; 1.3^v
(E)-β-Farnesene	18794-84-8	0.09-0.9	0.4	0-2.2	0.4	0.03-0.2	+^{z6}; tr^{z7}; 0.3^h
(E,E)-α-Farnesol			0.4				

Table 5.8.2 Constituents identified in black pepper oils (*continued*)

Constituent	CAS	Percentage and range					
		A	B (11)	C (5-9)	D (1)	E (18)	F
(*Z,E*)-α-Farnesol			1.1				
α-Fenchene	471-84-1	0-0.05					+[z6]; +[z]
Fenchone	1195-79-5						<0.1[z6]
Furanodiene	19912-61-9						2.1[g]
Geraniol	106-24-1						0.5[y]
Geranyl acetate	105-87-3					0.01-0.1	0.2[k]
Germacrene	28028-64-0						0.5[z5]; 12.8[r]
Germacrene A	28387-44-2						+[z6]
Germacrene B	15423-57-1				0.2		+[z6]; 0.3[h]; 1.4[g]
Germacrene C	34323-15-4						+[z6]
Germacrene D	23986-74-5	0.2-1.6	11.0		0.3	0.03-0.3	+[z6]; 0.1[k]; 0.2[f]; 0.4[h]; 0.9[z7]; 1.5[w]
α-Guaiene	3691-12-1	0.05-0.2	0.3	0-3.0		0-0.1	+[z6]; +[z]; tr[h,z7]
β-Guaiene	88-84-6		0.2				
trans-β-Guaiene	192053-49-9			0-4.1			
γ-Guaiene	145267-53-4						+[z6]
α-Gurjunene	489-40-7	0.04-0.3	0.3				+[z6]; +[z]; 0.1[h]; 0.2[g]; 0.4[f]
β-Gurjunene	73464-47-8						+[z]; 0.1[h]
γ-Gurjunene	22567-17-5						+[z6]; +[z]; 0.1[j]; 1.5[f]
Heneicosane	629-94-7				0.2		
Heptadecane	629-78-7				0.3		
3,5-Heptadienal, 2-ethylidene-6-methyl-	99172-18-6						0.5[v]
3-Heptanol	589-82-2		tr				
Hexadecane	544-76-3				0.2		
Hexadecanol	51260-59-4		0.1				
Hexahydronaphthalene	41375-99-9						+[z1]
Hexanoic acid	142-62-1						+[z2]
(*E*)-2-Hexenol	928-95-0		tr				
α-Himachalene	3853-83-6						1.3[v]
α-Humulene	6753-98-6	0.4-2.1	0.6	0-2.7	1.9	0.1-0.3	1.2[f]; 1.4[h]; 1.8[x]; 2.0[p]; 2.5[z5]; 2.8[o]
α-Humulene epoxide	96638-51-6	tr-0.07					
Humulene epoxide I	19888-33-6						tr[k]
Humulene epoxide II	19888-34-7						<0.1[z6]; 1.4[m]
8-Hydroxycarvone	7712-46-1						0.2[m]
6-Hydroxypiperitol							0.6[m]
Isoaromadendrene epoxide	499134-59-7						0.08[v]
Isoborneol	124-76-5		tr				tr[m]
Isocaryophyllene	118-65-0						+[z6]; +[z2]; 0.1[v]
Isocaryophyllene oxide							<0.1[z6]; 1.6[m]
Isoledene	95910-36-4						0.3[v]
Isospathulenol	88395-46-4						3.1[m]
Isoterpinolene	586-63-0	0.07-0.7					+[z1]
Isothujol	513-23-5						0.2[f]
Ledene oxide II							2.4[v]
Levomenol	23089-26-1						+[z1]
Limonene	138-86-3	10.2-24.7	10.3	8.3-23.8	13.2	16.7-22.7	23.8[z4]; 24.1[i]; 26.2[l]; 26.5[t]; 26.8[r]
Limonene oxide	1195-92-2						3.0[k]
Limonen-6-ol, pivalate							25.4[v]
Linalool	78-70-6	0.1-0.7	2.5	0-1.3	0.5	0.2-0.5	0.3[h,i]; 0.4[j]; 0.5[k]; 0.6[f,m]; 0.8[z6]; 1.1[g]
cis-Linalool oxide, furanoid	5989-33-3						3.0[k]
trans-Linalool oxide	11063-78-8			tr-1.2			
trans-Linalool oxide, furanoid	34995-77-2					0-0.2	tr[m]; 1.0[k]
α-*p*-Menthadiene							+[z6]
m-Mentha-3(8),6-diene	25946-29-6						+[z3]
α,*p*-Menthadienol							<0.1[z6]
p-Mentha-1(7),2-di-en-8-ol	65293-09-6						+[z]
m-Mentha-1,3-dien-8-ol							0.2[z6]

Table 5.8.2 Constituents identified in black pepper oils (*continued*)

Constituent	CAS	Percentage and range					
		A	B (11)	C (5-9)	D (1)	E (18)	F
p-Mentha-1,5-dien-8-ol	1686-20-0						<0.1[z6]
cis-*p*-Mentha-2,8-dien-1-ol	3886-78-0					0.02-0.05[b]	<0.1[z6]
trans-*p*-Mentha-2,8-dien-1-ol	4017-77-0						<0.1[z6]; 0.7[v]
trans-*p*-Menthan-8-ol							<0.1[z6]
α,*p*-Menthatriene							+[z6]
p-Mentha-1,3,8-triene	18368-95-1						+[z6]
cis-*p*-Menth-2-en-1-ol	29803-82-5			0-0.3		0.02-0.05[b]	0.05[v]; 0.1[h]
trans-*p*-Menth-2-en-1-ol	29803-81-4			0-0.2		0.01	0.5[k]; 2.8[v]
cis-p-Menth-8-en- 2-ol				0-0.2			
trans-p-Menth-8-en-2-ol				0-0.1			
9-Methoxycalamenene							0.05[v]
p-Methylacetophenone	122-00-9						<0.1[z6]
Methyl citronellate	2270-60-2						<0.1[z6]
Methyl eugenol	93-15-2	0.9					+[z2]
Methyl piperinate							<0.1[z6]
Mintsulfide	72445-42-2						0.002[z7]
α-Muurolene	10208-80-7				1.6		0.7[h]
γ-Muurolene	30021-74-0			0-5.1		0.2-0.9	+[z6]; 0.1[j]; 0.5[g]; 0.6[f]
α-Muurolol	104245-48-9						+[z]; 0.1[h]
τ-Muurolol (epi-α-)	19912-62-0				0.2		0.2[g]
Myrcene	123-35-3	1.5-6.3	1.4	0-18.6	1.0	2.2-8.4	2.8[w]; 2.9[z4]; 6.3[q]; 9.1[o]; 15.2[t]; 23.3[r]
Myristicin	607-91-0						+[z2]; 0.1[j]
Myrtenal	564-94-3						0.3[m]
Myrtenol	515-00-4			0-0.2		0.04-0.2	0.5[m]
Naphthalene	91-20-3						4.2[z5]
Neral	106-26-3						0.3[v]
Nerolidol	7212-44-4						+[z2]; 4.5[r]
(*E*)-Nerolidol	40716-66-3		1.6	0-1.3	0.4	0.03-0.1	tr[k]; 0.1[f]; 0.3[g]
(*Z*)-Nerolidol	3790-78-1					0.05-0.2	tr[k]
Nerolidyl acetate	56001-43-5						0.2[v]
Neryl acetate	141-12-8			0-0.3		0.05-0.2	
Nonadecane	629-92-5				0.4		
Nonanal	124-19-6		tr				
Nonane	111-84-2						tr[k]
n-Nonanoic acid	12-05-0		0.06				
(*E*)-2-Nonenal	18829-56-6		1.1				4.0[k]
Nopinone	24903-95-5						+[z2]
β-Ocimene	13877-91-3			0-12.0			0.3[q]; 0.5[f]
(*E*)-β-Ocimene	3779-61-1	0.1-1.0	1.3		tr	0.2-2.8	+[z6]; 0.06[h]; 0.1[z7]; 0.5[g]
(*Z*)-Ocimene	27400-71-1						<0.05[j]
(*Z*)-β-Ocimene	3338-55-4	0.02-0.07	3.2			0-0.4[a]	+[z6]; 0.1[z7]
allo-Ocimene	673-84-7						+[z1]
(*Z*)-Ocimenol	39900-51-1						0.9[u]
Octadecane	593-45-3				0.4		
16-Octadecenal	56554-87-1						+[z]
(*E*)-9-Octadecenoic acid, trimethylsilyl ester	96851-47-7						0.1[v,z8]
Octahydronaphthalene	31244-58-3						+[z1]
Octanoic acid	57-11-4		0.2				
7-Octen-4-ol	53907-72-5						0.2[v]
12-Oxabicyclo[9.1.0]-dodeca							3.4[z5];
Patchoulene	1405-16-9						0.1[j]
α-Patchoulene	560-32-7						+[z2]
α-Phellandrene	99-83-2	0.8-17.4	8.6		0.6	0-2.3	2.7[v]; 3.3[k]; 3.7[x,z4]; 5.7[w]; 5.9[s]; 17.4[q]
α-Phellandrene epoxide	288393-04-4						+[z]
β-Phellandrene	555-10-2	0.2-3.6	0.4			0-0.4[a]	0.02[h]; 0.3[j]; 2.9[g]
Phellandrenol							<0.1[z6]
β-Phellandren-6-ol							0.1[z6]
2-Phenethyl benzoate	94-47-3		tr				

Table 5.8.2 Constituents identified in black pepper oils (*continued*)

Constituent	CAS	Percentage and range					
		A	B (11)	C (5-9)	D (1)	E (18)	F
Pinadiene							+^{z1}
Pinan-3-one							<0.1^{z6}
α-Pinene	80-56-8	5.1-29.5	6.4	1.7-14.6	4.5	5.1-6.2	12.9^{z4}; 8.6^l; 8.8^q; 9.1^i; 20.9^t; 25.4^u
β-Pinene	127-91-3	6.7-20.3	10.0	0-29.3	7.9	6.4-11.1	14.0^j; 14.4^l; 15.2^k; 15.7^u; 24.4^t
α-Pinene epoxide	1686-14-2						<0.1^{z6};
Pinocarveol	5947-36-4	0.02-0.08					
trans-Pinocarveol	1674-08-4						<0.1^{z6}; 0.7^v
Pinocarvone	30460-92-5						<0.1^{z6}; 0.1^m
Piperidine	110-89-4						+^{z2}
cis-Piperitol	16721-38-3						tr^h; <0.1^{z6}
Piperitone	89-81-6		tr	0-0.2		tr-0.04	<0.1^{z6}; 0.1^{k,m}; 0.3^w
Piperitone 1-oxide	5286-38-4						0.2^m
Piperonal	120-57-0						+^{z2}; 0.1^{z6}; 0.2^w; 1.0^p
Piperonic acid	136-72-1						+^{z2}
Sabinene	3387-41-5	0.1-15.4	2.5	0-27.3	5.9	1.9-17.2	16.5^g; 19.0^{z7}; 20.9^s; 24.5^l; 41.5^r
cis-Sabinene hydrate	15537-55-0	0.03-0.2			0.2		<0.1^{z6}; 0.1^g; 0.3^f; 0.4^h
trans-Sabinene hydrate	17699-16-0	0.03-0.1	tr		tr	0-0.3	<0.1^{z6}; 0.2^f; 0.3^{g,h}; 2.0^k
trans-Sabinol	471-16-9						0.5^m
Safrole	94-59-7		0.5				+^{z2}; 0.2^{z4}
α-Santalene	512-61-8	0.03-0.3					+^{z2}
Sarisan	18607-93-7		0.4				
α-Selinene	473-13-2	0.2-3.6		0-4.2	1.1	0.07-0.5	0.3^j; 0.5^h; 0.6^{g,z7}; 0.8^f; 1.1^o; 1.3^o
β-Selinene	17066-67-0	0.2-3.2		0-0.9	1.5	0.6-1.4	0.05^p; 0.3^j; 0.4^m; 0.7^{f,g}; 0.8^h; 1.1^{o,z4}
γ-Selinene	515-17-3						+^{z1}
Selin-11-en-4-ol	16641-47-7						<0.1^{z6}
β-Sesquiphellandrene	20307-83-9						+^{z6}
Spathulenol	6750-60-3	tr-0.1	1.1				0.2^f; 0.6^j; 1.5^{z5}
α-Terpinene	99-86-5	0.03-2.5	tr		tr	0-1.1	0.2^g; 0.3^h; 0.4^v; 1.0^{t,z4}; 7.3^m; 13.3^{z5}
γ-Terpinene	99-85-4	0.05-4.4	tr	0-0.5	0.1	0-0.5	0.2^{j,k,w}; 0.3^f; 0.4^{g,i}; 0.5^h; 0.7^p; 4.7^t
Terpinen-4-ol	562-74-3	0.02-3.2	1.6	0-0.9	0.7	0.2-0.5	0.3^{z6}; 0.5^{z4}; 0.6^f; 1.7^g; 2.0^h; 10.1^v
1-Terpinen-5-ol	55708-42-4						<0.1^{z6}; +^{z2}
1-Terpineol	586-82-3						0.05^h
α-Terpineol	98-55-5	0.04-0.5	1.6	0-0.3	0.2	0.07-0.2	0.2^{h,z6}; 0.3^w; 0.4^m; 0.5^k; 0.7^f
β-Terpineol	138-87-4					0-0.03	0.5^k; 3.8^v
δ-Terpineol	7299-42-5						0.5^g
Terpinolene	586-62-9	0.2-1.7	1.3	0-0.3	0.1	0.1-0.2	0.2^{f,h,q}; 0.4^g; 0.9^j; 1.1^w; 1.2^{z4}; 8.5^v
α-Terpinyl acetate	80-26-2			tr-1.7		0.9-1.3	0.2^q
Tetradecane	629-59-4				tr		
α-Thujene	2867-05-2	0.06-2.3	tr	0-3.6	0.8	0.7-1.6	0.1^j; 0.2^i; 0.4^v; 0.8^{z7}; 1.4^{f,g}; 2.0^l; 3.0^t
β-Thujene	28634-89-1						+^{z1}
Tricyclene	508-32-7						+^{z6}; 1.7^h
1,4,4-Trimethylcyclo-heptadienone							<0.1^{z6}
4,8,8-Trimethylspiro-[2.6] nona-4,6-diene	81532-24-3						+^{z1}
ar-Turmerone	532-65-0						0.7^m
2-Undecanone	112-12-9						<0.1^{z6}
Valencene	4630-07-3						+^{z}
Verbenol	473-67-6						0.5^v
cis-Verbenol	1845-30-3						0.2^m
Verbenone	80-57-9						2.0^m

Table 5.8.2 Constituents identified in black pepper oils (*continued*)

Constituent	CAS	Percentage and range					
		A	B (11)	C (5-9)	D (1)	E (18)	F
α-Ylangene	14912-44-8				0.1		0.1[m]
α-Zingiberene	495-60-3	0.07-1.2	tr				tr[k]; 0.2[g]

A forty-six black pepper essential oil samples from India, Vietnam, Sri Lanka, Madagascar and Indonesia analyzed between 1998 and 2013 (E. Schmidt, unpublished data)
B one lab-hydrodistilled oil from Cameroon (ref. 11)
C essential oils from 18 cultivars from India and 1 from Indonesia produced in various seasons; lowest and highest concentrations given for any cultivar in any season (refs. 5,6,7,8,9)
D one lab-hydrodistilled oil sample from India (ref. 1)
E four lab-hydrodistilled oils from 4 Indian cultivars; lowest and highest concentrations given (ref. 18)
F data from other studies (indicated with superscript letters); highest concentrations found in any study reviewed here given; when two or more oils were investigated, only the highest concentrations are mentioned, unless indicated otherwise

[a] β-phellandrene and (Z)-β-ocimene combined; [b] *cis-p*-menth-2-en-1-ol and *cis-p*-mentha-2,8-dien-1-ol combined; [c] (Z)-carveol and carvone combined; [d] α-cubebene and δ-elemene combined; [e] α-bisabolene and β-bisabolene combined; [f] one lab-hydrodistilled oil from Malaysian black pepper (ref. 15); [g] one lab-hydrodistilled oil from black pepper from St. Tomé e Príncipe (ref. 25); [h] one lab-hydrodistilled oil from black peppers purchased at a local Indian market (ref. 10); [i] one lab-hydrodistilled oil from black peppers purchased at a local Egyptian market (ref. 19); [j] one hydrodistilled and one microwave-assisted hydrodistilled oil from Chinese black peppers (ref. 20); [k] one lab-produced oil from India (ref. 14); [l] three oils of ground Indian black pepper of various particle sizes (ref. 16); [m] one lab-prepared oil from black pepper grown in Cuba (ref. 12); [n] incorrect identification based on GC elution order according to ref. 17; [o] one lab-hydrodistilled oil from Malaysian black pepper (ref. 21); [p] one lab-hydrodistilled and one steam-distilled (superheated steam) oil from France (ref. 22); [q] one lab-hydrodistilled oil from India (ref. 26); [r] twenty-six lab-hydrodistilled oils from 26 cultivars grown in India; only the main components were investigated (ref. 23); [s] (probably) fifty-one lab-hydrodistilled oils from black peppers of unknown origin, purchased from a German spice company; only the main components were investigated (ref. 3); [t] three lab-hydrodistilled oils and one commercial black pepper oil from India (ref. 27); [u] one oil locally produced in Madagascar (ref. 29); [v] three lab-hydrodistilled oils from three Chinese provinces (ref. 30); [w] one microwave-assisted hydrodistilled oil from black peppers purchased at a local Shanghai market (ref. 31); [x] one oil from China (ref. 32); [y] one lab-hydrodistilled oil; origin of peppers unknown (ref. 34); [z] several black and white pepper oils from Brazil, Indonesia, Malaysia and Vietnam; concentrations not specified (ref. 33); [z1] several black and white pepper oils from Brazil, Indonesia, Malaysia and Vietnam; concentrations not specified (ref. 33); incorrect identification according to ref. 13; [z2] data from literature from before 1985, cited in ref. 4; [z3] data taken from ref. 35; [z4] one commercial oil from Brazil (ref. 36); [z5] one commercial black pepper oil from India (ref. 39); [z6] one lab-hydrodistilled oil from Muntok black pepper; percentages in the monoterpenes fraction and the sesquiterpene hydrocarbon fraction were expressed as quantity *in that particular* fraction and are therefore indicated with +[z6], as the percentages cannot be compared with the other data; the percentages of the oxygenated monoterpenoids, oxygenated sesquiterpenes and miscellaneous compounds were valid, though (ref. 38); [z7] various publications from 1981 to 1990, cited in ref. 37; [z8] originates from the gas chromatography column

tr: trace (in columns B and E: <0.01; in column D: <0.05; in columns C, F[k] and F[m]: <0.1); + present in the oil investigated, but quantity not stated or presented in a manner which cannot be compared with the other data

Table 5.8.3 Major constituents of black pepper oils

Constituent	CAS	Percentage and range					
		A	B (11)	C (5-9)	D (1)	E (18)	F
β-Caryophyllene	87-44-5	0.9-32.4	7.3	6.4-47.5	29.9	21.2-27.7	29.2[r]; 39.7[o]; 39.8[v]; 41.5[p]; 65.7[s]
Sabinene	3387-41-5	0.1-15.4	2.5	0-27.3	5.9	1.9-17.2	16.5[g]; 19.0[z7]; 20.9[s]; 24.5[l]; 41.5[r]
δ3-Carene	13466-78-9	4.3-17.6	2.8	0-23.4	4.4	0-2.8	22.0[z4]; 22.2[w]; 27.9[j]; 32.7[s]; 33.2[j]
α-Pinene	80-56-8	5.1-29.5	6.4	1.7-14.6	4.5	5.1-6.2	12.9[o]; 8.6[l]; 8.8[q]; 9.1[i]; 20.9[t]; 25.4[u]
Caryophyllene oxide	1139-30-6	0.2-1.2	0.9	0.2-6.0	3.9	0.3-0.9	4.1[o]; 6.2[t]; 8.0[z5]; 9.9[i]; 14.2[s]; 29.3[m]
β-Pinene	127-91-3	6.7-20.3	10.0	0-29.3	7.9	6.4-11.1	14.0[j]; 14.4[l]; 15.2[k]; 15.7[u]; 24.4[t]
Limonene	138-86-3	10.2-24.7	10.3	8.3-23.8	13.2	16.7-22.7	23.8[z4]; 24.1[i]; 26.2[l]; 26.5[t]; 26.8[r]
Myrcene	123-35-3	1.5-6.3	1.4	0-18.6	1.0	2.2-8.4	2.8[w]; 2.9[z4]; 6.3[q]; 9.1[o]; 15.2[t]; 23.3[r]
α-Phellandrene	99-83-2	0.8-17.4	8.6		0.6	0-2.3	2.7[y]; 3.3[k]; 3.7[x,z4]; 5.7[w]; 5.9[s]; 17.4[q]
p-Cymene	99-87-6	0.2-6.0	2.1	0-6.2	1.2	0-9.7	0.8[m]; 0.9[f,p]; 1.0[u]; 1.4[z4]; 8.2[t]; 13.0[k]
β-Bisabolene	495-61-4	0.6-5.9	0.8	0-6.0	3.9	0.5-4.3[b]	0.5[i]; 0.6[z7]; 1.1[g]; 1.8[f,q]; 7.2[l]; 7.7[h]
α-Cubebene	17699-14-8		0.1	tr-6.8	0.1	0.2-3.3[d]	tr[z7]; 0.09[f,p]; 0.2[y]; 0.3[h]; 1.9[k]; 3.5[t]
α-Copaene	3856-25-5	0.2-4.4	1.4	0-0.2	2.8	0.4-0.8	1.9[j]; 2.3[y]; 3.1[m]; 3.7[f]; 3.8[o]; 4.5[l]; 6.3[h]

LEGEND: SEE UNDER TABLE 5.8.2

found in commercial oils in maximum concentrations of 32% for caryophyllene and 29% for α-pinene (Table 5.8.2, column A).

REFERENCES

1 Kapoor IPS, Singh B, Singh G, De Heluani CS, De Lampasona MO, Catalan CAN. Chemistry and in vitro antioxidant activity of volatile oil and oleoresins of black pepper (Piper nigrum). J Agric Food Chem 2009;57:5358-5364

2 Butt MS, Pasha I, Tauseef Sultan M, Randhawa MA, Saeed F, Ahmed W. Black pepper and health claims: a comprehensive treatise. Crit Rev Food Sci Nutrit 2013;53:875-886

3 Schulz H, Baranska M, Quilitzsch R, Schütze, Lösing G. Characterization of peppercorn, pepper oil, and pepper oleoresin by vibrational spectroscopy methods. J Agric Food Chem 2005;53:3358-3363

4 Buckle KA, Rathnawathie M, Brophy JJ. Compositional differences of black, green and white pepper (Piper nigrum L.) oil from three cultivars. J Food Technol 1985;20:599-613

5 Nirmala Menon A, Padmakumari KP, Jayalekshmy A, Gopalakrishnan M, Narayanan CS. Essential oil composition of four popular Indian cultivars of black pepper (Piper nigrum L.). J Essent Oil Res 2000;12:431-434

6 Nirmala Menon A, Padmakumari KP, Jayalekshmy AJ. Essential oil composition of four major cultivars of black pepper (Piper nigrum L.). J Essent Oil Res 2002;14:84-86

7 Nirmala Menon A, Padmakumari KP, Jayalekshmy AJ. Essential oil composition of four major cultivars of black pepper (Piper nigrum L.). III. J Essent Oil Res 2003;15:155-157

8 Nirmala Menon A, Padmakumari KP. Essential oil composition of four major cultivars of black pepper (Piper nigrum L.). IV. J Essent Oil Res 2005;17:206-208

9 Nirmala Menon A, Padmakumari KP. Studies on essential oil composition of cultivars of black pepper (Piper nigrum L.). V. J Essent Oil Res 2005;17:153-155

10 Singh G, Marimuthu P, Catalan C, deLampasona MP. Chemical, antioxidant and antifungal activities of volatile oil of black pepper and its acetone extract. J Sci Food Agric 2004;84:1878-1884

11 Jirovetz L, Buchbauer G, Ngassoum MB, Geissler M. Aroma compound analysis of Piper nigrum and Piper guineense essential oils from Cameroon using solid-phase microextraction-gas chromatography, solid-phase microextraction-gas chromatography-mass spectrometry and olfactometry. J Chromatogr A 2002;976:265-275

12 Pino JA, Ramada RM, Fuentes V. Essential oil composition of black pepper from Cuba. Revisita Cienc Quim 2001;32:59-60. Data cited in ref. 17

13 Lawrence BM. Essential oils 2001-2004. Carol Stream, USA: Allured Publishing Corporation, 2006:97-103 (reprinted from Perfum Flavor 2002;27(3):48-? [last page unknown])

14 Nirmala Menon A. The aromatic compounds of pepper. J Med Arom Plant Sci 2000;22:185-200

15 Bagheri H, Bin Abdul Manap MY, Solati Z. Antioxidant activity of Piper nigrum L. essential oil extracted by supercritical CO_2 extraction and hydrodistillation. Talanta 2014;121:220-228

16 Mohan Rao LJ. Quality of essential oils and processed materials of selected spices and herbs. J Med Arom Plant Sci 2000;22:808-816

17 Lawrence BM. Progress in essential oils. Perfum Flavor 2010;35(4):48-57

18 Gopalakrishnan M, Menon N, Padmakumari KP, Jayalekshmy A, Narayanan CS. GC analysis and odor profiles of four new Indian genotypes of Piper nigrum L. J Essent Oil Res 1993;5:247-253

19 Abd El Mageed MA, Mansour AF, El Massry KF, Ramadan MM, Shaheen MS. The effect of microwaves on essential oils of white and black pepper (Piper nigrum L.) and their antioxidant activities. J Essent Oil Bear Plants 2011;14:214-223

20 Wang Y, Jiang Z-T, Li R. Composition comparison of essential oils extracted by hydrodistillation and microwave-assisted hydrodistillation from black pepper (Piper nigrum L.) grown in China. J Essent Oil Bear Plants 2009;12:374-380

21 Tewtrakul S, Hase K, Kadota S, Namba T, Komatsu K, Tanaka K. Fruit oil composition of Piper chaba Hunt., P. longum L. and P. nigrum L. J Essent Oil Res 2000;12:603-608

22 Rouatbi M, Duquenoy A, Giampaoli P. Extraction of the essential oil of thyme and black pepper by superheated steam. J Food Engin 2007;78:708-714

23 Zachariah TJ, Safeer AL, Jayarajan K, Leela NK, Vipin TM, Saji KV. Correlation of metabolites in the leaf and berries of selected black pepper varieties. Scientia Horticulturae 2010;123:418-422

24 Ahmad N, Fazal H, Haider Abbasi B, Farooq S, Ali M, Ali Khan M. Biological role of Piper nigrum L. (Black pepper): A review. Asian Pac J Trop Biomed 2012;2(Suppl.):S1945-S1953

25 Martins AP, Salgueiro L, Vila R, Tomi F, Canigueral S, Casanova J, et al. Essential oils from four Piper species. Phytochem 1998;49:2019-2023

26 McCarron M, Mills A, Whittaker D, Kurian T, Verghese J. Comparison between green and black pepper oils from Piper nigrum L. berries of Indian and Sri Lankan origin. Flavour Fragr J 1995;10:47-50

27 Mamatha BS, Prakash M, Nagarajan S, Bhat KK. Evaluation of the flavor quality of pepper (Piper nigrum L.) cultivars by GCc–MS, electronic nose and sensory analysis techniques. J Sens Stud 2008;23:498-513

28 Meghwal M, Goswami TK. Piper nigrum and piperine: An update. Phytother Res 2013;27:1121-1130

29 Möllenbeck S, König T, Schreier P, Schwab W, Rajaonarivony J, Ranarivedo L. Chemical composition and analyses of enantiomers of essential oils from Madagascar. Flavour Fragr J 1997;12:63-69

30 Renjie L, Shidi S, Yongjun M. Analysis of volatile oil composition of the peppers from different production areas. Med Chem Res 2010;19:157-165

31 Liu L, Song G, Hu Y. GC–MS analysis of the essential oils of *Piper nigrum* L. and *Piper longum* L. Chromatographia 2007;66:785-790

32 Zhu L-F, Li Y-H, Li B-L, Lu B-Y, Xia N-H. Aromatic plants and essential constituents. South China Institute of Botany, Chinese Academy of Sciences, Hai Feng Publishing Company, 1993:41-42. Distributed by Peace Book Company Ltd, Hong Kong. Data cited in ref. 13

33 Korány K, Amtmann M. Gas chromatography/mass spectrometry measurements in the investigation of pepper aroma structures. Rapid Comm Mass Spectr 1997;11:686-690

34 Dorman HJ, Surai P, Deans SG. *In vitro* antioxidant activity of a number of plant essential oils and phytocontituents. J Essent Oil Res 2000;12:241-248

35 Nussbaumer C, Cadalbert R, Kraft P. Identification of *m*-Mentha-3(8),6-diene (isosylveterpinolene) in black pepper oil. Helv Chim Acta 1999;82:53-58

36 Murbach Teles Andrade BF, Nunes Barbosa L, da Silva Probst I, Fernandes Júnior A. Antimicrobial activity of essential oils. J Essent Oil Res 2014;26:34-40

37 Lawrence BM. Progress in essential oils. Perfum Flavor 1995;20(March/April):49-59

38 Kollmansberger H, Nitz S, Drawert F. Über die Aromastoffzusammensetzung von Hochdruckextrakten I. Pfeffer (*Piper nigrum* var. Muntok). Zeit Lebensmitt Untersuch Forsch 1992;194:545-551. Data cited in ref. 37

39 Jeena K, Liju VB, Umadevi NP, Kuttan R. Antioxidant, antiinflammatory and antinociceptive properties of black pepper essential oil (*Piper nigrum* Linn). J Essent Oil Bear Plants 2014;17:1-12

40 Lawless J. The encyclopedia of essential oils, 2nd Edition. London: Harper Thorsons, 2014

41 Dharmagunawardena B, Takwale A, Sanders KJ, Cannan S, Roger A, Ilchyshyn A. Gas chromatography: an investigative tool in multiple allergies to essential oils. Contact Dermatitis 2002;47:288-292

Chapter 5.9   CAJEPUT OIL

DEFINITION

Cajeput oil is the essential oil obtained from the leaves and the terminal branchlets (twigs) of the cajaput tree, *Melaleuca cajuputi* Powell (synonyms: *Melaleuca minor* Smith., *Melaleuca leucodendra* auct. nonn.). The terms cajaput, cajuput and cajeput are synonymously used for both the tree and the oil.

INCI NOMENCLATURE

Description/definition: Melaleuca leucadendron cajaputi oil is the volatile oil obtained from *Melaleuca leucadendron* L. var. *cajaputi*, Myrtaceae

INCI name EU & USA: Melaleuca leucadendron cajaputi oil

Other name(s): cajuput oil

CAS registry number(s): 8008-98-8; 85480-37-1

EINECS number(s): 287-316-4

Description/definition: Melaleuca leucadendron cajuputi leaf oil is the essential oil distilled from the leaves of the cajeput, *Melaleuca leucadendron* L. var. *cajuputi*, Myrtaceae

INCI name EU: Melaleuca leucadendron cajuputi leaf oil (perfuming name, not an INCI name proper)

INCI name USA: Not in the Personal Care Products Council Ingredient Database

CAS registry number(s): 8008-98-8; 85480-37-1

EINECS number(s): 287-316-4

The source of cajeput oil is *Melaleuca cajuputi* Powell, not *Melaleuca leucadendra* (sometimes termed *Melaleuca leucadendron*) (4). This earlier nomenclature confusion stems from the fact that the trees of both *Melaleuca* species are known in Indonesia and Malaysia under a same name, gelam (4,15). However, the composition of the oil of *Melaleuca leucadendra* differs considerably from that of cajeput oil (15,16).

ISO (INTERNATIONAL ORGANIZATION FOR STANDARDIZATION) STANDARD

There is currently no ISO standard for cajeput oil.

THE PLANT, THE OIL, AND THEIR USES

Melaleuca cajuputi Powell (swamp tea tree, paperbark tea tree) is an evergreen shrub or tree which may measure up to 25 meters in height. It has a single flexible trunk with a white spongy bark that flakes off easily. The tree is native to tropical Asia and northern Australia and is cultivated in Indonesia (especially Sulawezi, the former Celebes), Vietnam and Malaysia (GRIN taxonomy for plants). The name 'cajeput' is derived from its Indonesian name, 'kayu putih' or 'white wood'. The essential oil made from the leaves and twigs of the tree is extremely pungent and has the odor of a mixture of turpentine and camphor. It is frequently employed externally as a counterirritant and as an ingredient in some liniments for sore muscles such as Tiger Balm and the Indonesian traditional medicine Minyak Telon. Cajeput oil has also been used in Vietnam, Indonesia, China and elsewhere for the treatment of purulent skin lesions, as inhalant in the treatment of nasal catarrh, and as painkiller for headache, toothache, rheumatism, and convulsions in the form of applied plaster. In addition, it is employed as an ingredient in expectorants, and the oil, which is also used in aromatherapy, has antiseptic and antifungal properties. Other applications include its use as a treatment against skin mites, as insect repellent and the oil may have anti-termite activity and shows some promise in dengue vector control (1,12,13,15, Wikipedia).

CHEMICAL COMPOSITION

Cajeput oil is a colorless to yellowish easily mobile clear liquid which has a fresh, camphoraceus, minty and eucalyptus-like odor. The yield of essential oil from the leaves and terminal branches of *Melaleuca cajuputi* Powell generally varies from 0.8 to 1.7%. The main producing countries of this oil are Indonesia (notably Java), the Philippines and Vietnam.

Literature data (up to November 7, 2014) on the chemical composition of cajeput oils and unpublished analytical data from one of us (E.S.) are shown in Table 5.9.1 in alphabetical order. In cajeput oils from various origins, over 130 chemicals have been identified. About 51% of these were found in a single reviewed publication only.

The major compounds found in cajeput oils from different sources are shown in Table 5.9.2. They include (highest concentrations in any study given) 1,8-cineole (70.8%), α-terpineol (22.6%), terpinolene (21.3%), β-caryophyllene (20.1%), γ-terpinene (18.0%), *p*-cymene (15.0%), α-humulene (11.9%), limonene (10.2%), α-pinene (7.6%) and terpinen-4-ol (6.2%). Well-known ingredients of cajeput oils that were present in high concentrations in one or two studies were β-elemene (14.0%), (*E*)-β-ocimene (13.2%), α-terpinene (11.1%), globulol (7.6%) and alloaromadendrene (5.5%). Uncommon or rare constituents of cajeput oil found in high concentrations in single studies include geranyl acetate (27.4%), viridiflorol (13.4%), 3,5-dimethyl-4,6-di-*o*-methylchloroacetophenone (10%), (*E*)-β-farnesene (7.7%), cyclohexanecarboxaldehyde (7.5%), 1-methyl-2-(1-methylethenyl)benzene (6.7%), *p*-mentha-1,3,8-triene (6.3%) and farnesyl acetone (5.8%).

Commercial oils

The ten chemicals that had the highest maximum concentrations in 51 commercial cajeput essential oil samples (concentration ranges provided) are the following: 1,8-cineole (46.0-70.2%), α-terpineol (2.5-11.6%), limonene (1.8-10.2%), *p*-cymene (0.1-9.5%),

Table 5.9.1 Constituents identified in cajeput oils

Constituent	CAS	Percentage and range				
		A	B (15)	C (2)	D (3)	E
Aromadendrene	489-39-4					0.2[f]
Aromadendrene homologue			0-1.5[a]			
allo-Aromadendrene	25246-27-9	0.09-1.0	0-4.4			0.3-5.5[n]
allo-Aromadendrene homologue			0-1.1			
Benzaldehyde	100-52-7	0.04-0.2			0.1	
Benzyl benzoate	120-51-4	0.2-1.2				
Benzyl salicylate	118-58-1	0.5-3.4				
α-Bisabolol	515-69-5					0-1.5[m]
α-Bulnesene	3691-11-0		0-2.7			
Bulnesol	22451-73-6				0.2	0-0.7[m]
p-tert-Butylbenzoic acid	98-73-7			tr		0-0.2[m]
Cadina-1,4-diene	29837-12-5	tr-0.4	0.4			0-0.2[m]
γ-Cadinene	39029-41-9			tr		
δ-Cadinene	483-76-1	0.05-1.3	0-0.1	tr	0.05	0-0.9[m]; 0.3-1.5[n]
α-Cadinol	481-34-5			tr		1.1[o];
epi-α-Cadinol	5937-11-1			tr		0.6[h]; 0.9[o];
δ-Cadinol	19435-97-3		0-0.1	tr		0-0.04[n]; 0-1.9[m]
Cajolone						2.8[k]
α-Calacorene	21391-99-1					
Calamenene	483-77-2			tr		
Camphene	79-92-5	0.01-0.5	0-0.05	tr	0.04	0-0.03[n]; 0.2[d]
Camphor	76-22-2	0.01-0.4				
δ2-Carene (δ4-)	554-61-0				4.7	
δ3-Carene	13466-78-9	0.01-0.2	tr		3.7	
β-Caryophyllene	87-44-5	0.08-5.2	1.9-9.1	2.5	1.7	0-3.6[b,m]; 3.9[f]; 4.1[o]; 5.4[e]; 5.9[k]; 11.0-20.1[n]
Caryophyllene oxide	1139-30-6	0.02-0.9	0-0.8	0.3	0.2	0-0.4[n]; 2.3[o]
1,4-Cineole	470-67-7	0.2				
1,8-Cineole	470-82-6	46.0-70.2	31.8-69.4	41.1	49.7	0-2.9[m]; 1.1[o]; 61.2[d]; 62[e]; 70.8[g]
Citral	5392-40-5					0.5[o]
Citronellol	106-22-9	0.01-0.3				
α-Copaene	3856-25-5	0.05-0.2	0-1.2[a]	tr	0.2	0-0.03[a,n]; 0-1.0[m]
Cyclohexanecarboxal-dehyde	2043-61-0					0-7.5[m]
3-Cyclohexene-1-carboxaldehyde	100-50-5					0-1.9[m]; 0.6[o];
3-Cyclohexen-1-ol, 4-methyl-3-(1-methyl-ethyl)-	654053-64-2					1.4-2.7[n]
o-Cymene	527-84-4	0.02-0.08	0-0.3			0-0.9[n]
p-Cymene	99-87-6	0.1-9.5		6.8	5.3	0-1.4[n]; 0.9[j]; 2.2[f]; 2.4-15.0[m]; 4.2[j]; 7.5[k]
p-Cymenene	1195-32-0			0.2		
p-Cymen-8-ol	1197-01-9					0.5-3.0[m]
1-Decanol	112-30-1					0-1.9[m]
1,3-Dimethoxybutane	10143-66-5					0-0.8[m]
3,4-Dimethyl-3-cyclo-hexene-1-carboxal-dehyde	18022-66-7					0.9[o]
3,5-Dimethyl-4,6-di-O-methylchloroaceto-phenone	21722-31-6					10[l]
1-Dodecanol	112-53-8					0-1.6[m]
1-Dodecene	112-41-4					0-0.7[m]
β-Elemene	33880-83-0		0.03-0.2	0.3	0.1	1.6-14.0[m]; 1.8[o]
γ-Elemene	29873-99-2			tr		
δ-Elemene	20307-84-0			tr		
Elemol (α-)	639-99-6			tr		
Eremophilene	10219-75-7					0-1.9[n]
Eudesma-4(14),11-diene	17066-67-0		0-0.3			0.2-1.6[n]
α-Eudesmol	473-16-5			0.7	1.6[c]	
β-Eudesmol	473-15-4	0.06-1.1		0.7	0.8	
γ-Eudesmol	1209-71-8			0.6		
epi-γ-Eudesmol (7-)	117066-77-0				0.7	
Eugenol	97-53-0				0.05	
Eugenyl acetate	93-28-7					0-0.4[m]
α-Farnesene	502-61-4			0.3		
(E)-β-Farnesene	18794-84-8	tr-0.3				6.5-7.7[m]

Table 5.9.1 Constituents identified in cajeput oils (*continued*)

Constituent	CAS	Percentage and range				
		A	B (15)	C (2)	D (3)	E
Farnesol	4602-84-0					0-1.9[m]
Farnesyl acetone	1117-52-8					0-5.8[m]
Fenchyl alcohol	1632-73-1	0.08-0.4				
Geraniol	106-24-1	tr-0.4		0.4	0.2	
Geranyl acetate	105-87-3				0.06	
Geranyl acetone	3796-70-1					9.4-27.4[m]
Germacrene B	15423-57-1				0.1	0-1.3[m]
Germacrene D	23986-74-5	0.4-0.4				0-2.1[m]
Globulol	489-41-8		0-10.4		0.1	0.8[o]; 7.6[h]
α-Guaiene	3691-12-1	0.04-0.3	0.03-1.3		1.4	0-4.8[n]; 0.5[m]
β-Guaiene	88-84-6			0.3		
Guaiol	489-86-1	0.1-1.2		1.2	1.2	0-2.5[m]; 1.6[i]; 3.8[j]
α-Gurjunene	489-40-7	0.7-1.0				
γ-Gurjunene	22567-17-5	tr-0.5	0.5			
Hexadecanoic acid	57-10-3					0-1.1[m]
α-Humulene	6753-98-6	0.05-2.3	0.9-5.0	1.6	1.3	0.5[g]; 1.8[f]; 2.1[o]; 2.8-4.5[m]; 5.3-11.9[n]
α-Humulene epoxide	96638-51-6				0.2	
Isocaryophyllene	118-65-0		0-4.9	2.3		0-3.4[n]
8-Isopropenyl-1,5-di-methyl-1,5-cyclode-cadiene						1.1[o]
Kauran-18-al						0-2.1[m]
Ledol	577-27-5		0-0.7			0-3.7[n]; 0.8[h]
Limonene	138-86-3	1.8-10.2	1.4-2.6	4.1	5.1	0.03-0.2[n]; 0.8[o]; 3.8[g]; 4.3[i]; 7.0[f]
dl-Limonene	138-86-3					1.0-2.9[m]
Limonene dioxide	96-08-2					0-0.4[m]
Linalool	78-70-6	0.3-3.7		3.6	2.5	0-1.5[m]; 0.9[o]; 2.8[j]; 2.9[i]; 3.7[e]
Linalyl acetate	115-95-7	0.1-0.2				
Linalyl propionate	144-39-8		0-8.8			
Longifolene	475-20-7					0.4[d]
α-Maaliene	489-28-1			tr		
p-Mentha-2,4(8)-diene	586-63-0					0.9[g]
p-Mentha-1,3,8-triene	18368-95-1				6.3	
p-Menth-8-en-2-ol, acetate	20777-49-5					0-3.9[m]
Methyl eugenol	93-15-2	0.04-0.06				
1-Methyl-2-(1-methyl-ethenyl) benzene						6.7[o]
2-Methyltridecane	1560-96-9					0-0.7[m]
Myrcene	123-35-3	0.2-2.8	0.6-1.6	0.9	1.6	0.3[d]; 0.4-2.7[n]; 0.7-1.9[m]; 2.0[g]
Naphthalene,decahy-dro-4a-methyl-1-methylene-7-(2-methylethenyl)-						0-8.1[n]
(E)-β-Ocimene	3779-61-1	tr-0.06	0-5.8[a]	tr		0-13.2[a,n]; 1.3-4.4[m]
(Z)-β-Ocimene	3338-55-4	tr-0.08		tr		
1,2,3,4,4a,5,6,8a-Octa-hydronaphthalene	4276-46-4				1.3	
Patchoulane	25491-20-7			0.2		
α-Phellandrene	99-83-2	0.03-0.7		0.5	1.0	0.4[d]; 0.6[o]; 0.7-5.0[m]
β-Phellandrene	555-10-2			0.2		0-0.4[m]
Phenanthrene,1-methyl-7-(1-methylethyl)-	483-65-8					1.3[o]
α-Pinene	80-56-8	0.6-7.6	0-1.8	3.2	2.4	3,8[o]; 5.3[e]; 5.4[d]; 5.5[f]
β-Pinene	127-91-3	0.7-4.4	0.7-3.6	0.8	1.7	0.5[o]; 1.5[g]; 2.2-5.6[n]; 3.2[f]; 4.4[e]
cis-Rose oxide	3033-23-6	0.01-0.4				
Sabinene	3387-41-5	0.01-4.5		tr	0.03	
trans-Sabinol	471-16-9					
Selin-6-en-4-ol						2.4[o]
α-Selinene	473-13-2	0.1-1.6		1.5	1.1	1.3[f]
β-Selinene	17066-67-0	0.1-1.8		1.5	1.2	1.6[f]
β-Sesquiphellandrene	20307-83-9			tr		
(-)-Spathulenol (β-)	77171-55-2		0-0.2			0.6-4.6[m]; 4.1[o]
α-Terpinene	99-86-5	0.2-1.9	0-0.2	0.6	0.5	0.4[f]; 0.5-3.0[m]; 0.8[o]; 1.4[n]; 1.7[d]; 11.1[e]

Table 5.9.1 Constituents identified in cajeput oils (*continued*)

Constituent	CAS	A	B (15)	C (2)	D (3)	E
γ-Terpinene	99-85-4	0.7-5.4	0-1.5	4.6	2.6	2.0-15.1[m]; 2.9[e]; 5.1[o]; 7.3[i]; 8.9-10.0[n]; 18.0[k]
Terpinen-4-ol	562-74-3	0.2-1.4	0-0.4	1.5	1.3	0.7[f]; 1.3[e]; 1.4[i]; 1.6-6.2[m]; 6.2[o]
1-Terpineol	586-82-3	0.6-0.7				
α-Terpineol	98-55-5	2.5-11.6	0-16.6	8.7		2.6[o]; 0.4-27.9[n]; 0.7-2.4[m]; 10.6[d]; 15.3[j]; 22.6[h]
β-Terpineol	138-87-4	0.1-0.3				
γ-*Terpineol*	586-81-2	1.2-2.7				3.2[d]
δ-Terpineol	7299-42-5	0.06-0.3			0.2	
Terpinolene (α-)	586-62-9	0.4-5.3	0-1.0	11.0	3.3	1.1[f]; 1.9[d]; 6.9[o]; 2.7-18.3[m]; 3.9-4.1[n]; 21.3[k]
Terpinyl acetate	8007-35-0	0.05-3.5				
2,6,10,14-Tetramethylheptadecane	18344-37-1					0-1.1[m]
3-Thujanol	513-23-5					0-0.6[m]
α-Thujene	2867-05-2	0.07-1.3			1.1	0.7[f]; 1.6[o]
Thujopsene	470-40-6		tr			
2,5,6-Trimethyl-1,3,6-heptatriene	42123-66-0				3.8	
Viridiflorene	21747-46-6		0-4.9[a]			0.05-1.2[a,n]
Viridiflorol	552-02-3					13.4[h]
α-Ylangene	14912-44-8	tr-0.3		0.3		

A fifty-one essential oil samples from Indonesia (notably Java) and Vietnam analyzed between 1998 and 2013; lowest and highest concentrations given (E. Schmidt, unpublished data)
B four cajeput oils from leaves of *M. cajuputi* growing wild in various places in Indonesia (ref. 15)
C one commercial oil from Vietnam (ref. 2)
D one commercial oil from Vietnam prepared from leaves and twigs of *M. cajuputi* investigated by two laboratories; highest concentrations given (ref. 3)
E data from other studies (indicated with superscript letters); highest concentrations found in any study reviewed here given; when two or more oils were investigated, only the highest concentrations are mentioned, unless indicated otherwise

[a] incorrect identity based on GC elution order (ref. 4); [b] includes γ-caryophyllene; [c] α- and β-eudesmol combined; [d] one sample of commercial cajeput oil (ref. 6); [e] one commercial cajeput oil purchased in Hungary (ref. 7); [f] one commercial cajeput oil sample (ref. 8); [g] one commercial sample of Vietnamese cajeput oil (ref. 10); [h] one lab-distilled oil from *M. cajuputi* ssp. *cajuputi* from Brazil (ref. 11); [i] cited as 'main components of Vietnamese cajuput oil' (ref. 13); [j] two lab-hydrodistilled oils from Vietnam (ref. 2); [k] one Malaysian leaf oil with low 1,8-cineole content (ref. 14); [l] one sample of Malaysian cajeput oil (ref. 9); [m] six lab-hydrodistilled oils from leaves of *M. cajuputi* growing wild in various parts of Thailand; lowest and highest concentrations given; low cineole content (ref. 17); [n] two cajeput leaf oils from Indonesia; chemotype with no 1,8-cineole; both concentrations given (ref. 15); [o] one steam-distilled cajuput oil sample from Thailand with low 1,8-cineole content (ref. 18)

tr: trace (in columns C and H: <0.1)

Table 5.9.2 Major constituents of cajeput oils

Constituent	CAS	A	B (15)	C (2)	D (3)	E
1,8-Cineole	470-82-6	46.0-70.2	31.8-69.4	41.1	49.7	0-2.9[m]; 1.1[o]; 61.2[d]; 62[e]; 70.8[g]
α-Terpineol	98-55-5	2.5-11.6	0-16.6	8.7		2.6[o]; 0.4-27.9[n]; 0.7-2.4[m]; 10.6[d]; 15.3[j]; 22.6[h]
Terpinolene (α-)	586-62-9	0.4-5.3	0-1.0	11.0	3.3	1.1[f]; 1.9[d]; 6.9[o]; 2.7-18.3[m]; 3.9-4.1[n]; 21.3[k]
β-Caryophyllene	87-44-5	0.08-5.2	1.9-9.1	2.5	1.7	0-3.6[b,m]; 3.9[f]; 4.1[o]; 5.4[e]; 5.9[k]; 11.0-20.1[n]
γ-Terpinene	99-85-4	0.7-5.4	0-1.5	4.6	2.6	2.0-15.1[m]; 2.9[e]; 5.1[o]; 7.3[i]; 8.9-10.0[n]; 18.0[k]
p-Cymene	99-87-6	0.1-9.5		6.8	5.3	0-1.4[n]; 0.9[j]; 2.2[f]; 2.4-15.0[m]; 4.2[l]; 7.5[k]
α-Humulene	6753-98-6	0.05-2.3	0.9-5.0	1.6	1.3	0.5[g]; 1.8[f]; 2.1[o]; 2.8-4.5[m]; 5.3-11.9[n]
Limonene	138-86-3	1.8-10.2	1.4-2.6	4.1	5.1	0.03-0.2[n]; 0.8[o]; 3.8[g]; 4.3[i]; 7.0[f]
α-Pinene	80-56-8	0.6-7.6	0-1.8	3.2	2.4	3,8[o]; 5.3[e]; 5.4[d]; 5.5[f]
Terpinen-4-ol	562-74-3	0.2-1.4	0-0.4	1.5	1.3	0.7[f]; 1.3[e]; 1.4[i]; 1.6-6.2[m]; 6.2[o]

LEGEND: SEE UNDER TABLE 5.9.1

α-pinene (0.6-7.6%), γ-terpinene (0.7-5.4%), terpinolene (0.4-5.3%), β-caryophyllene (0.08-5.2%), sabinene (0.01-4.5%) and β-pinene (0.7-4.4%) (Erich Schmidt, unpublished analytical data).

Chemotypes

The commercial value of cajeput oil depends on its 1,8-cineole (eucalyptol, cajeputol) content (1). In Indonesia, cajeput essential oil with 1,8-cineole content higher than 55% is graded as 'prime quality', and that with less than 55% is graded as 'standard' (15). Thus, commercial samples of cajeput oil of Indonesian origin had concentrations of 1,8-cineole ranging from 55.6 to 69.1% (5). However, there are also oils with very low 1,8-cineole concentrations (14,17,18) or even with no 1,8-cineole at all, for example, the leaf oils from some trees growing wild in Thailand (17) and in Indonesia (15). In such oils, the chemicals with high concentrations were found to be α-terpineol, terpinolene, γ-terpinene, β-caryophyllene, geranyl acetone, β-elemene, (E)-β-ocimene and α-humulene, but there was no consistent pattern in these oils (14,15,17).

CONTACT ALLERGY / ALLERGIC CONTACT DERMATITIS

General

Allergic contact dermatitis from cajeput oil has been observed in two publications only. In both reports, false-positive patch test reactions due to the excited skin syndrome cannot be excluded. In one case report, α-pinene and caryophyllene may have been allergens in cajeput oil (20).

Case reports

An aromatherapist had chronic hand dermatitis and was patch test positive to 17 of 20 oils used at her work (tested 1% and 5% in petrolatum), including cajeput oil; cajeput was given as a synonym for tea tree oil, which is incorrect (19). Two other aromatherapists had contact dermatitis (occupational in one) with allergies to multiple essential oils used at their work including cajeput oil. Both patients also reacted to geraniol, α-pinene, the fragrance mix and various other fragrance materials. In addition, one proved to be allergic to linalool and linalyl acetate, the other to caryophyllene; α-pinene and caryophyllene were demonstrated by GC-MS in cajeput oil (20). Both chemicals have been found in commercial cajeput oils in maximum concentrations of 5.2% (caryophyllene) and 7.6% (α-pinene) (Table 5.9.1, column A).

LITERATURE

1 Barbosa LCA, Silva CJ, Teixeira RR, Alves Meira MAS, Pinheiro AL. Chemistry and biological activities of essential oils from *Melaleuca* L. species. Agriculturae Conspectus Scientificus 2013;78:11-23 Available at: http://www.agr.unizg.hr/smotra/pdf_78/acs78_02.pdf

2 Motl O, Hodačová J, Ubik K. Composition of Vietnamese cajuput essential oil. Flavour Fragr J 1990;5:39-43

3 Minh CP, Dürbeck K. Cajeput aus Bio-Anbau in Vietnam. F·O·R·U·M 2006;29:33-37

4 Lawrence BM. Progress in essential oils. Perfum Flav 2012;37(April):56-57

5 Lawrence BM. Unpublished data, 1992. Data cited in ref. 4

6 Reichling J, Harkenthal M, Saller R. Wirkung ausgewählter ätherischer Öle. Erfahrungsheilkunde 1999;6:357-366. Data cited in ref. 4

7 Hethelyi E, Takacs G, Palfine Ledniczky M, Domokos J. Gas chromatographic investigation of the biologically active components of *Melaleuca* species and of natural cosmetic components containing tea tree oil. Olaj Szappan Kozmet 2000;49:25-37. Data cited in ref. 4

8 Christoph F, Stahl-Biskup E, Kaulfers P-M. Death kinetics of *Staphylococcus aureus* exposed to commercial tea tree oils. J Essent Oil Res 2001;13:98-102

9 Lowry JB. New constituents of biogenetic, pharmacological and historical interest from *Melaleuca cajuputi* oil. Nature 1973;241:61-62. Data cited in ref. 13

10 Milchard MJ, Clery R, DaCosta N, Flowerdew M, Gates L, Moss N, et al. Application of gas-liquid chromatography to the analysis of essential oils. Fingerprints of 12 essential oils. Perfum Flavor 2004;29:28-36 (issue number unknown)

11 Silva CJ, Barbosa LCA, Maltha CRA, Pinheiro AL, Ismail FD. Comparative study of the essential oils of seven *Melaleuca* (*Myrtaceae*) species grown in Brazil. Flavour Fragr J 2007;22:474-478

12 Abu Bakar A, Sulaiman S, Omar B, Mat Ali R. Evaluation of *Melaleuca cajuputi* (Family: Myrtaceae) essential oil in aerosol spray cans against dengue vectors in low cost housing flats. J Arthropod-Borne Dis 2012;6:28-35

13 Cuong ND, Xuyen TT, Motl O, Stransky K, Presslova J, Jedlikova Z, et al. Antibacterial properties of Vietnamese cajuput oil. J Essent Oil Res 1994;6:63-67

14 Roszaini K, Nor Azah MA, Mailina J, Zaini S, Faridz ZM. Toxicity and antitermite activity of the essential oils from *Cinnamomum camphora*, *Cymbopogon nardus*, *Melaleuca cajuputi* and *Dipterocarpus* sp. against Coptotermes *curvignathus*. Wood Sci Technol 2013;47:1273-1284

15 Sakasegawa M, Hori K, Yatagai M. Composition and antitermite activities of essential oils from *Melaleuca* species. J Wood Sci 2003;49:181-187

16 Brophy JJ, Lassak EV. *Melaleuca leucadendra* L. leaf oil: Two phenylpropanoid chemotypes. Flavour Fragr J 1988;3:43-46

17 Kim JH, Liu KH, Yoon Y, Sornnuwat Y, Kitirattrakarn Y, Anantachoke C. Essential leaf oils from *Melaleuca cajuputi*. In: UR Palaniswamy, LE Craker, ZE Gardner, Eds. Proc WOCMAP III, Vol.6: Traditional Medicine

& Nutraceuticals. Leuven, Belgium: International Society for Horticultural Science, 2005:65-72. Available at: http://wwwlib.teiep.gr/images/stories/acta/Acta%20680/680_8.pdf

18 Tawatsin A, Asavadachanukorn P, Thavara U, Wongsinkongman P, Bansidhi J, Boonruad T, et al. Repellency of essential oils extracted from plants in Thailand against four mosquito vectors (Diptera: Culicidae) and oviposition deterrent effects against Aedes aegypti (Diptera: Culicidae). Southeast Asian J Trop Med Public Health 2006;37:915-931

19 Selvaag E, Holm J, Thune P. Allergic contact dermatitis in an aromatherapist with multiple sensitizations to essential oils. Contact Dermatitis 1995;33:354-355

20 Dharmagunawardena B, Takwale A, Sanders KJ, Cannan S, Roger A, Ilchyshyn A. Gas chromatography: an investigative tool in multiple allergies to essential oils. Contact Dermatitis 2002;47:288-292

Chapter 5.10 CALAMUS OIL

DEFINITION

Calamus oil is the essential oil obtained from the rhizomes of the calamus (flagroot, myrtle flag, sweet flag), *Acorus calamus* L.

INCI NOMENCLATURE

Description/definition: Acorus calamus root oil is an essential oil obtained from the rhizomes of the calamus ('sweet flag'), *Acorus calamus* L., Araceae

INCI name EU: Acorus calamus root oil (perfuming name, not an INCI name proper)

INCI name USA: Not in the Personal Care Products Council Ingredient Database

Other names: Sweet flag oil

CAS registry number(s): 8015-79-0; 84775-39-3

EINECS number(s): 283-869-0

ISO (INTERNATIONAL ORGANIZATION FOR STANDARDIZATION) STANDARD

There is currently no ISO standard for calamus oil.

THE PLANT, THE OIL, AND THEIR USES

Acorus calamus L. or 'sweet flag' (recently removed from the family Araceae [34]) is a reed-like semiaquatic perennial plant, which is native to India. It is found growing wild in abundance there, ascending to 2200 meters in the Himalayas. It also grows in the temperate zones of Europe, East Asia and North America. Calamus inhabits perpetually wet areas such as the banks of streams and rivers and around ponds, lakes, and swamps. The stout aromatic roots of *A. calamus* spread horizontally (rhizomes) and can grow to almost 0.5-1.25 m in length (1,2,3). It is cultivated in South Africa (GRIN Taxonomy for Plants).

Acorus calamus includes four cytotypes distinguished by chromosome number: diploid (2x = 24), triploid (3x = 36), tetraploid (4x = 48) and hexaploid (6x = 72). The distinction is important, as the ploidy determines the amount of the toxic chemical β-asarone in the rhizomes and the rhizome oil (see below). The rhizome in various forms, such as powder, decoction, aqueous and solvent extract, and its volatile oil are reported to possess insecticidal, antibacterial, fungicidal, larvicidal, and antitermite properties, and repel larvae and insects (3). Indeed, in India, *A. calamus* and its essential oil serve mainly as an insecticide and insect repellent (4). The rhizomes and their oils are in addition used in Ayurvedic, Unani and folk medicines. In the Ayurvedic system of medicine, the rhizomes are considered to possess antispasmodic, antidiarrhoeic, carminative, anthelminthic, antidepressant and anxiolytic properties. The rhizomes and their oils are used to treat insomnia, melancholia, neurosis, epilepsy, hysteria, loss of memory, remittent fever, rheumatism, toothache, and respiratory ailments (3,5). Calamus oils have also been used as a folk remedy for the treatment of arthritis, neuralgia, diarrhoea, dyspepsia, hair loss and other disorders (6).

The essential oil from the rhizomes may sometimes be found in beer and aromatic cordial and liqueur preparations (3,5) and the rhizomes of the European *A. calamus* and calamus essential oil are used in the flavoring industry. The uses in Europe and India are different because the composition of the essential oils is different. The phenylpropanoid, β-asarone, is dominant in the tetraploid Indian type calamus oil (up to 96%), while it is found in lower percentages in the European oils and in those from temperate regions in India, such as Kashmir. *In vitro* studies have shown the rhizome and its oil to be toxic and induce malignant tumors, which is attributed to β-asarone (2,4,7). Therefore, calamus products for human use should contain no or negligible amount of β-asarone (7). In several countries, including the USA, *A. calamus* and its oil have been prohibited as a food additive, and in the EU it is not used anymore in perfumery. Also, in recent years many herbal shops have stopped recommending or dispensing it (2). It is also considered too toxic for aromatherapy (45). However, calamus products are available for recreational (hallucinogenic) use on the internet and acute intoxications from abuse of such substances, mainly characterized by prolonged vomiting, have been reported and are not rare (38). Several review articles on pharmacological activities, medicinal applications and biological properties have been published (7,25,31,39,41).

CHEMICAL COMPOSITION

Calamus oil is a yellowish to yellowish brown, clear liquid which has an aromatic, herbaceous, spicy and creamy odor. The yield of essential oil from the rhizomes of *Acorus calamus* L. generally varies from 0.5 to 0.8% from fresh roots and from 1.4 to 8.0% from dried roots and is also dependent on the origin of the plant material. The main producing countries of this oil are India, Nepal, Japan and Russia.

Literature data (up to December 11, 2014) on the chemical composition of calamus oils and unpublished analytical data from one of us (E.S.) are shown in Table 5.10.2 in alphabetical order. In calamus oils from various origins, over 275 chemicals have been identified. About 47% of these were found in a single reviewed publication only.

The major compounds found in calamus oils from different sources are shown in Table 5.10.3. They include (highest concentrations in any study given) β-asarone (96.3%), (Z)-methylisoeugenol (87.3%), α-asarone (58.0%), preisocalamenediol (34.9%), acorenone (28.8%), shyobunone (22.1%), isoshyobunone (13.1%) and methyl eugenol (8.6%). Well-known ingredients of calamus oils that were present in high concentrations in one or two studies were elemicin (72.7%), (E)-methylisoeugenol (7.9% and 14.1%), limonene (13.1% and 13.4%), isocalamenediol (12.8%), α-pinene (12.4%), linalool (12%), α-terpineol (11.8%), α-calacorene (8.5%), 1,8-cineole (7.0%) and calamenediol (6.5%). Uncommon or rare constituents of calamus oil found in high concentrations in single studies include methyl isoeugenol (41.5% [methyl isoeugenol is the main component of *Acorus gramineus*]), isoeugenol (25.0%), cyclohexanone (21.3%),

Table 5.10.1 Ploidy of *Acorus calamus*: geographic locations and content of β-asarone (1,2,3,5,30)

Ploidy	Geographic locations	Content of β-asarone
Diploid	North America, parts of Asia	no β-asarone present
Triploid	Europe, North America, Himalayan region of India	3-21%
Tetraploid	India, Indonesia, Taiwan	73-96%
	Japan, East Siberia	10-40%
	Thailand, Singapore, Vietnam	higher than from Japan and East Siberia
Hexaploid	Kashmir, India	5%

Table 5.10.2 Constituents identified in calamus rhizome oils

Constituent	CAS	Percentage and range					
		A	B (8)	C (32)	D (8)	E (14)	F
Acarone							0.1[g]
Acetic acid	64-19-7		tr		tr		
α-Acoradiene	24048-44-0				0.5	0.4	+[b]; 0.2[v]; 0.3[g]
β-Acoradiene	28477-64-7	0-0.4			0.5	0.4	+[b]; 0.6[f,v]; 0.7[g]
γ-Acoradiene	28400-12-6				0.7		
Acora-3(10),14-diene	28908-21-6					0.4	
Acora-7(11),9-dien-2-one						0.6	
Acorafuran							+[p]
Acoragermacrone	50281-45-3				tr		
Acorenone	5956-05-8				8.1	14.2	0.2[k]; 10.5[g,h]; 14.4[o]; 16.1[v]; 20.9[b]; 21.6[f]; 28.8[x]
4-epi-Acorenone (B)	56363-00-9					0.7	
Acorone	10121-28-5				0.7	3.0	0.1[k]; 0.5[v]; 0.7[f]; 2.0[b]
epi-Acorone	185303-18-8					1.0	
Amyl isovalerate	25415-62-7						4.9[l]
trans-Anethole	4180-23-8				0.6		
4,5-di-epi-Aristolo-chene	54868-40-5				tr		
5-epi-Aristolochene	115888-31-8				0.5		
Aristolene	6831-16-9						2.5[i]
Aristolen-1α-ol	34143-95-8						+[b]
Aristolone	160568-09-2	0.08-0.7					+[b]; tr[m,v]; 1.6[f]
Aromadendrene	489-39-4						1.1[f]
allo-Aromadendrene	25246-27-9						tr[e]; 0.3[i]; 0.6[f]
Asaronaldehyde	4460-86-0		0.4		tr		8.1[t]
α-Asarone ((*E*)-; *trans*-)	2883-98-9	0.4-0.9	6.8	0.1-4.7	0.5		0.5[e]; 1.6[u]; 2.6[d]; 4.9[z]; 5.6[m]; 9.7[c]; 13.8[r]; 15.0[s]; 16.8[j]; 50.1[k]; 58.0[t]
β-Asarone ((*Z*)-; *cis*-)	5273-86-9	1.6-79.6	77.7	69.8-95.6	5.2		74.8[m]; 76.5[v]; 81.3[e]; 83.2[c]; 91.2[j]; 91.6[z]; 91.7[u]; 92.1[r]; 92.4[d,z3]; 96.3[s]
γ-Asarone (euasarone)	5353-15-1		tr	tr-1.6	0.3		0.3[g]; 1.6[i]
Benzaldehyde	100-52-7		tr				
Benzenaminium	17032-11-0						4.9[m]
Benzyl benzoate	120-51-4						0.2[c]
α-Bergamotene	17699-05-7						1.0[o]
trans-α-Bergamotene	13474-59-4				tr	0.8	1.2[f,v]
(*E*)-α-Bergamotol							0.7[l]
Bicyclogermacrene	24703-35-3					0.2	0.04[e]
epi-Bicyclosesquiphel-landrene	54274-73-6						1.0[t]
α-Bisabolene	17627-44-0		0.07				
(*Z*)-α-Bisabolene	29837-07-8						0.1[c]
β-Bisabolene	495-61-4		0.1				1.2[c]; 4.1[m]
α-Bisabolol	515-69-5			0-1.6			0.1[k]; 0.6[c]
Borneol	507-70-0	0.07-0.2					<0.05[v]; 0.1[b]; 0.6[q]; 1.1[w]
Bornyl acetate	76-49-3	0.1-1.1		0-0.2	tr	0.5	+[b]; 0.1[k,q,v]; 0.2[f]; 1.0[v]; 4.6[o]
α-Bulnesene	3691-11-0						0.3[k]; 5.5[w]
Cadala-1,4,9-triene	71609-04-6		tr		0.2		
Cadalene	483-78-3						+[b]; 1.0[f]
α-Cadinene	24406-05-1			0-0.3			tr[e]; 0.1[c]; 1.4[k]

Table 5.10.2 Constituents identified in calamus rhizome oils (*continued*)

Constituent	CAS	Percentage and range					
		A	B (8)	C (32)	D (8)	E (14)	F
β-Cadinene	523-47-7						tr^e
γ-Cadinene	39029-41-9						0.01^e
δ-Cadinene	483-76-1	0.4-4.4	0.2	0-0.8	1.7	0.5	0.3^y; 0.5^v; 1.0^z2; 1.1^m; 1.3^o; 1.7^f
α-Cadinol	481-34-5					0.8	<0.05^c; 0.2^g,y; 0.3^k; 0.7^o; 1.0^z2; 1.3^b
10-α-Cadinol			0.1		0.5		
10-epi-α-Cadinol	5937-11-1		0.05		0.3		
τ-Cadinol	58580-31-7		0.04		0.2		
α-Calacorene	21391-99-1	1.7-8.5	0.9	0-0.2	3.6		+^b; 0.08^e; 0.1^c; 0.2^y; 1.1^z; 2.0^f
β-Calacorene	50277-34-4			0-0.4			+^b; tr^e; 0.1^y; 0.5^f,k
Calacorene hydrate					2.1		
trans-Calamene	40772-39-2	2.5-5.4					tr^f
Calamenediol	30167-28-3		0.4		5.2	0.5	1.0^z2; 1.4^o; 6.5^v
Calamenene	483-77-2		tr				1.5^t; 5.1^z2
cis-Calamenene	72937-55-4	0.8-2.5		0-0.3			+^b; 0.8^l
trans-Calamenene	73209-42-4			tr-1.0			
Calamenoic acid							0.5^t
Calamenone							0.5^t
Calamol	66219-01-0						2.2^t; 7.8^z2
Calamusenone	71305-96-9				3.2		0.5^o
Calarene	17334-55-3				0.4		1.5^m; 4.1^i
Camphene	79-92-5	tr-7.3	tr	0-0.4	1.5	0.2	0.5^o; 0.6^k; 0.8^g; 1.0^z; 1.2^w; 1.6^f; 7.4^x
Camphene hydrate	465-31-6				tr		<0.05^y
6-Campholenal							0.3^z
Camphor	76-22-2	0.02-5.9		0-0.7	3.2	0.1	0.8^t; 1.2^f; 1.5^o; 2.4^k; 5.1^b; 5.9^g; 8.1^x
Capronic acid	142-62-1		tr				
Caprylic acid	124-07-2		tr				
δ2-Carene (δ4-)	554-61-0						0.03^e
δ3-Carene	13466-78-9						+^z2
Carvacrol	499-75-2				1.4		0.3^i
Carvone	99-49-0				tr		5.6^l
β-Caryophyllene	87-44-5	0.2-3.2	0.01	tr-0.8	tr		0.08^e; 0.2^c; 0.6^g; 1.0^t; 1.6^w; 4.4^z
Caryophyllene oxide	1139-30-6	0-1.4	0.02		0.2		0.1^c; 0.7^g; 1.3^i; 1.4^b,f; 1.7^z
α-Cedrene	469-61-4				tr	0.5	3.1^k
β-Cedrene	546-28-1					1.0	tr^g; 1.1^b; 1.3^f; 1.5^k; 1.6^v
Cedrol	77-53-2				0.2		+^b; 0.7^v
1,8-Cineole	470-82-6	0.1-1.8	tr	0-0.1	tr		0.2^k; 0.3^i; 0.5^q; 0.6^g; 1.1^o; 7.0^t
α-Copaene	3856-25-5	0-1.1	tr	0-0.2	tr		tr^m; 0.01^e; 0.2^f; 0.3^o; 1.1^k
β-Copaene	18252-44-3						2.5^l
Cryptoacorone	5989-62-8					7.5	2.0^v
α-Cubebene	17699-14-8						tr^f; 1.3^l
β-Cubebene	13744-15-5			0-0.5			
Cuparene	16982-00-6					0.2	0.7^k; 1.1^m
Curcumene (ar-; α-)	644-30-4		0.4		0.5	0.6	+^b; 0.1^c; 0.4^g; 0.6^f; 0.8^v
β-Curcumene	28976-67-2		0.1				
γ-Curcumene	28976-68-3					0.2	1.2^f
2,5-Cyclohexadiene							1.1^m
Cyclohexanol	108-93-0						2.3^m
Cyclohexanone	108-94-1						21.3^m
o-Cymene	527-84-4						tr^f
p-Cymene	99-87-6	0.1-2.3	tr	tr-1.9	1.4		tr^e; 0.1^g,z; 0.2^i; 0.6^w; 1.1^t
p-Cymen-8-ol	1197-01-9				tr		
Decanal	112-31-2			0-0.2		0.1	<0.05^c; 0.04^e
Decanoic acid	334-48-5						1.0^t
Decanol	36729-58-5						0.2^q
(E)-4-Decenal	65405-70-1						0.03^e
(Z)-4-Decenal	21662-09-9						0.09^e
Decyl acetate	112-17-4					0.1	
Dehydroaromadendrene	698388-95-3						1.6^z
Dehydroxyisocalamenediol		0.6-2.9					3.5^f

Table 5.10.2 Constituents identified in calamus rhizome oils (*continued*)

Constituent	CAS	A	B (8)	C (32)	D (8)	E (14)	F
1,2-Dicyclopropylcyclobutane	61141-62-6						0.02[i]
4,4-Diethyl-2,5-octadiyne	61227-87-0						0.9[i]
2,3-Dimethoxytoluene	4463-33-6					0.2	0.1[v]
β-Elemene	33880-83-0	0.1-1.3		0-1.5	tr	0.3	0.07[e]; 0.1[v]; 0.2[b]; 0.3[v]; 0.4[f]; 0.5[o]
γ-Elemene	29873-99-2		0.03		tr		
δ-Elemene	20307-84-0		tr				+[b]; 0.3[c]; 0.5[o]
β-Elemenol	65018-04-4		tr				
Elemicin	487-11-6	0-0.3	0.6	tr-0.8			<0.05[c]; 0.1[v]; 0.3[d,e]; 3.7[r]; 72.7[s]
Elemol	639-99-6		0.06		0.3		0.1[v]; 0.6[g]; 2.0[q]
Eudesma-3,11-dien-2-one	86917-79-5					0.1	
α-Eudesmol	473-16-5						2.8[q]
γ-Eudesmol	1209-71-8				tr		0.1[v]; 0.7[l]; 0.9[q]
Eudesmyl acetate	51317-10-3						1.5[q]
Eugenol	97-53-0		0.1				0.2[t]; 5.7[i]
(E)-β-Farnesene	18794-84-8					1.1	1.0[o]
α-Farnesol	58181-75-2	0-1.1					
Fenchone	1195-79-5				tr		
α-Funebrene	50894-66-1						0.2[f,k]; 0.4[v]
2-epi-α-Funebrene	854154-70-4					0.2	
β-Funebrene	79120-98-2					1.6	1.7[b]; 3.8[f]; 3.9[v]
Furfural	98-01-1		1.0		tr		
Furyl methyl ketone			0.1				
Geraniol	106-24-1						0.2[z]; 0.3[e]
Geranyl acetate	105-87-3						tr[z]; 0.1[b]
Germacrene A	28387-44-2						0.02[e]; 0.9[i]
Germacrene B	15423-57-1						<0.05[c]; 0.04[e]; 0.3[k]
Germacrene D	23986-74-5		0.07	0-1.4		0.4	2.1[k]
Germacrene D-4-ol	198991-79-6					0.1	0.7[b]; 1.2[m]; 1.7[z]
Globulol	489-41-8						3.3[l]
Guaiazulene	489-84-9		tr		tr		
Guaiazulene isomer			0.2				
cis-β-Guaiene	372162-07-7						1.5[f]
trans-β-Guaiene	372162-07-7						0.4[f]
α-Gurjunene	489-40-7				tr		tr[e]; 0.7[k]; 0.8[g]; 1.3[m]
β-Gurjunene	73464-47-8	0-0.8	0.2	0-0.6	6.7		0.1[v]; 0.2[e]; 0.4[c]; 1.3[b]; 2.5[g]; 2.7[f]
γ-Gurjunene	22567-17-5				0.5		
n-Heptadecane	629-78-7						0.3[c,t]
Hexanal	66-25-1						0.05[e]
Hexanol	111-27-3						0.01[e]; 0.8[i]
Hexenal	1335-39-3						1.4[g]
α-Himachalene	3853-83-6						2.6[f]
Hinesene	123484-18-4					0.2	
Hotrienol	20053-88-7				tr		
α-Humulene	6753-98-6	0-0.9	0.04		0.5		0.02[e]; 0.2[k]; 0.5[l]; 0.8[t]
Humulene epoxide I	19888-33-6				tr		
Humulene (ep)oxide II	19888-34-7			tr-3.5	0.5		+[b]; 0.9[g]
Humulenol II	19888-00-7				0.3		
6α-Hydroxygermacra-1(10),4-diene	20674-02-6					0.2	
2-Hydroxy-3-methylvalerianic acid methyl ester			tr		tr		
Isoacarone						0.5	0.1[g]; 1.4[f]
Isoacorone	6168-64-5						0.8[v]; 5.7[o]
Isoamyl alcohol	123-51-3						+[z2];
Isocalamenediol	25330-21-6	0-4.3	0.1		0.2	3.1	+[z1]; 2.0[z2]; 12.8[b]
Isocomenene							0.4[g]
(E)-Isoelemicin	5273-85-8	tr-1.5				tr	
(Z)-Isoelemicin	5273-84-7		1.3		tr		<0.1[b]; 1.0[q]; 1.1[c]; 1.3[d]; 1.7[e,m]
Isoeugenol	97-54-1						25.0[z2]
(E)-Isoeugenol	5932-68-3					0.1	
(Z)-Isoeugenol	5912-86-7					0.2	

Table 5.10.2 Constituents identified in calamus rhizome oils (*continued*)

Constituent	CAS	Percentage and range					
		A	B (8)	C (32)	D (8)	E (14)	F
(*E*)-Isoeugenyl acetate	5912-87-8			0-3.7			
Isogermacrene A	783322-20-3					0.2	
Isomenthyl acetate	20777-45-1				tr		
4-Isopropyl-6-methyl-1,2,3,4-tetrahydro-naphthalen-1-one	57494-10-7				tr		
Isoshyobunone	21698-46-4	0.2-7.2	0.5		6.3	1.3	9.4[z2]; 9.9[v]; 13.0[g]; 13.1[o]
Isospathulenol	88395-46-4				tr		
Isovelleral	37841-91-1						0.7[i]
Kessane	3321-66-2						0.06[e]; 0.1[v]; 0.2[d]
Khusinol	24268-34-6						tr[m]
Khusiol	66512-56-9					0.1	
Limonene	138-86-3	0.07-3.9	tr		0.2	tr	0.1[b]; 0.2[m]; 0.6[k]; 1.1[o]; 13.1[l]; 13.4[w]
Linalool	78-70-6	0.2-1.3	0.1	0-1.2	1.1		0.9[d]; 1.0[b]; 1.1[k]; 1.3[g]; 2.0[t]; 2.5[l]; 12[z2]
cis-Linalool oxide	11063-77-7				tr		+[b]; <0.01
trans-Linalool oxide	11063-78-8				tr		<0.01[g]
Linalyl acetate	115-95-7			0-0.7	tr		0.2[c]
Linalyl propionate				0-0.2			0.1[c]
Longifolene	475-20-7						tr[m]; 1.2[f]; 8.2[w]
p-Mentha-2,4(8)-diene	586-63-0						+[b]
cis-p-Mentha-2,8-dien-1-ol	3886-78-0						<0.05[y]
cis-p-Menth-2-en-1-ol	29803-82-5						0.01[e]
trans-p-Menth-2-en-1-ol	29803-81-4						0.01[e]
Menthol	89-78-1				tr		
Menthone	89-80-5				tr		
Menthyl acetate	16409-45-3				tr		
1*H*-3a,7-Methanoazulene	173-07-9						6.9[m]
2-Methylbutyl acetate	624-41-9						1.5[l]
Methyl chavicol	140-67-0	0-0.4					0.9[k]
Methyl eugenol	93-15-2	0.03-6.5	0.02			tr	0.03[e]; 0.1[v]; 0.2[m]; 1.6[s]; 2.1[w]; 8.6[k]
5-Methyl-2-furaldehyde	620-02-0		0.2				
3-Methylfurfural	33342-48-2	0-0.1					
Methyl isoeugenol	93-16-3						0.6[q]; 1.6[z3]; 1.9[t]; 41.5[m]
(*E*)-Methylisoeugenol	6379-72-2	0.2-0.9	0.2	0.1-2.9	0.3		0.4[d]; 0.8[o]; 0.9[i]; 1.6[s]; 7.9[z2]; 14.1[k]
(*Z*)-Methylisoeugenol	6380-24-1	0.3-13.7	1.3	tr-4.4	0.6	0.6	0.6[c]; 0.8[b]; 1.1[e]; 2.1[d]; 2.6[m]; 4.2[y,z]; 11.3[r]; 48.9[z2]; 87.3[s]
4-Methylisopropenylbenzene	1195-32-0		tr		tr		
Methyl 2-methylbutyrate	868-57-5						0.4[i]
α-Muurolene	10208-80-7					0.2	0.02[e]; 0.4[c]
γ-Muurolene	30021-74-0						tr[e]; 0.1[v]; 0.9[l]; 3.2[z]
Muurol-5-en-4α-ol	157374-45-3						0.4[g]
τ-Muurolol	19912-62-0					0.1	+[b]
Myrcene	123-35-3	0.05-0.2	tr	0.0.1	tr	tr	0.01[e]; 0.3[w]; 0.7[o]
Myrtenol	515-00-4			0-0.1			tr[z]
Naphthalene	91-20-3						0.5[t]
α-Neocallitropsene	729602-94-2					0.1	
(*E*)-Nerolidol	40716-66-3					1.5	+[b]; 0.8[f]; 2.0[v]
Nonanal	124-19-6						tr[e]
n-Nonane	111-84-2						<0.05[c]
Nonyl acetate	143-13-5					tr	
β-Ocimene	13877-91-3						2.5[t]
(*E*)-Ocimene	27400-72-2		tr		tr		0.08[e]
(*Z*)-Ocimene	3338-55-4		0.1		0.3		0.1[c]; 0.4[e]; 2.0[d]
(*E*)- β-Ocimene	27400-72-2	0.08-0.6		0-0.2			0.6[y]; 0.8[o]
(*Z*)-β-Ocimene	3338-55-4	0.2-2.1		0-tr			tr[m]; 0.3[u]; 0.6[o]
allo-Ocimene	673-84-7		tr		tr		<0.05[y]
(*E,E*)-allo-Ocimene	3016-19-1						tr[e]
(*E,Z*)-allo-Ocimene	7216-56-0						tr[e]
n-Octadecane	593-45-3						0.1[c]
Octanal	124-13-0						0.02[e]

Table 5.10.2 Constituents identified in calamus rhizome oils (*continued*)

Constituent	CAS	Percentage and range					
		A	B (8)	C (32)	D (8)	E (14)	F
n-Octane	111-65-9						<0.05[c]
3-Octanol	589-98-0						0.2[q]
1-Octen-3-ol	3391-86-4				tr		
7-Octen-4-ol	53907-72-5						1.1[q];
Octenyl acetate	2442-10-6						0.1[q];
Octyl acetate	112-14-1					0.1	2.3[l];
Oplopanone	1911-78-0				tr		
6,11-Oxido-acor-4-ene							1.8[v]; 2.0[b]
Oxidohimachalene	64825-84-9						+[b]
Patchoulene	1405-16-9						0.3[t]
2-Pentylfuran	3777-69-3	0-0.02	tr		tr		
Perillene	539-52-6						0.9[l]
α-Phellandrene	99-83-2						tr[e]; 0.08[l]; 0.2[o]
β-Phellandrene	555-10-2			0-tr			tr[e,z]
α-Pinene	80-56-8	0.02-12.4	tr	0-0.7	0.3	0.1	0.2[k,l]; 0.3[f,g]; 0.4[t]; 0.7[o]; 0.8[v]; 3.0[w]
β-Pinene	127-91-3	0.01-0.9	tr	0-0.3	0.2	0.2	0.01[e]; <0.1[d]; 0.3[g,l]; 0.7[t]; 0.9[o]; 1.3[v]
trans-Pinocarvyl acetate	1686-15-3				0.2		
Piperitone	89-81-6				tr		
Preisocalamenediol	25645-19-6	0.9			1.0	18.0	10.0[z2]; 12.1[o]; 17.3[g]; 18.8[v]; 34.9[r,s]
Prezizaene	31145-21-8					1.7	1.5[v]
Prezizaene 2							2.2[b]
Prezizaene isomer						0.8	+[b]
Sabinene	3387-41-5	0.1-0.7			tr		tr[e]; 0.1[f]; 0.2[g]; 0.7[t]; 1.5[w]
cis-Sabinene hydrate	15537-55-0						<0.05[v]; 0.1[g]; 0.7[i]
trans-Sabinene hydrate	17699-16-0						<0.05[v]
α-Selinene	473-13-2			0.1-7.2	3.8		+[b]; 0.7[o]; 1.5[g]; 3.9[f]
β-Selinene	17066-67-0				0.6		+[b]; 0.02[e]
δ-Selinene	473-14-3		tr		1.9		
Selin-11-4α-ol	16641-47-7						2.8[f]
(Z)-Sesquilavandulol	121521-16-2						14.7[f,z4]
β-Sesquiphellandrene	20307-83-9		0.1		0.2	2.1	3.0[z2]
trans-β-Sesquiphellandrol	56144-27-5		0.5				
cis-Sesquisabinene hydrate	58319-05-4						0.6[f]
Sesquithuriferol	117468-55-0					0.3	
Shyobunone	21698-44-2		0.4		2.6	13.3	0.5[g,o]; 1.0[e]; 4.4[v]; 7.8[b,f]; 14.3[s]; 22.1[r]
Shyobunone epimer			0.1		0.6		2.9[f]
2,6-di-epi-Shyobunone			0.2		2.3		2.6[g]
epi-Shyobunone	39020-72-9						3.4[s]; 12.1[r]
6-epi-Shyobunone	65794-23-2		0.1			3.1	2.1[v]; 3.3[e]; 3.7[o]
Spathulenol	6750-60-3		tr		0.7	0.7	0.3[v]; 1.1[m]; 1.4[b,g]; 1.7[z]; 1.9[f]
[3]Staffane-3,3-dicarboxylic acid							0.8[i]
Sylvestrene	1461-27-4						0.2[f]
α-Terpinene	99-86-5				tr		0.4[o]
γ-Terpinene	99-85-4	0.06-0.2	tr		0.2		tr[e,m]
Terpinen-4-ol	562-74-3	0.2-1.2		0-0.1	tr		+[b]; 0.1[m,v]; 0.2[g]; 1.1[o]; 1.2[k]; 1.7[t]
α-Terpineol	98-55-5	0.2-0.7		0-tr	tr		0.2[g]; 0.4[c]; 0.7[k]; 0.9[l]; 2.0[t]; 11.8[w]
(Z)-β-Terpineol	7299-40-3						23.4[l]
γ-Terpineol	586-81-2						3.9[w]
Terpinolene	586-62-9	0.06-0.2	tr		tr		<0.05[v]
Tetradecanal	124-25-4						2.6[f]
1,2,3,6-Tetrahydropyridine	694-05-3						0.04[i]
α-Thujene	2867-05-2	0-0.09			tr		0.8[l]; 0.9[i]
Torreyol	19435-97-3				tr		0.7[l]
Tricyclene	508-32-7	0.03-0.2					tr[f]; 0.01[e]
2,4,5-Trimethoxy-phenylacetone	16603-18-2		0.3		tr		
1-(2,4,5-Trimethoxy-phenyl)-1-methoxy-propan-2-ol	98205-47		tr				

Table 5.10.2 Constituents identified in calamus rhizome oils (*continued*)

Constituent	CAS	A	B (8)	C (32)	D (8)	E (14)	F
2,4,5-Trimethoxypro-piophenone			0.03		tr		
10-Undecenal	112-45-8						0.03[e]
Valencene	4630-07-3						0.06[e]; 0.5[g]
Viridiflorene	21747-46-6		0.04		tr		0.3[g]
Viridiflorol	552-02-3		0.01		tr		
Vulgarone A	62065-10-5					0.3	
Zingiberene	495-60-3					0.2	+[b]

A fourteen calamus essential oil samples from India, Nepal and Japan, analyzed between 1998 and 2006; lowest and highest concentrations given (E. Schmidt, unpublished data)
B one high β-asarone commercial oil sample from India (tetraploid) (ref. 8)
C twenty-one lab-hydrodistilled oils from calamus rhizomes collected at different locations in Uttarakhand, India; lowest and highest concentrations given (ref. 32)
D one low β-asarone oil sample from Europe (triploid) (ref. 8)
E one industrially steam-distilled oil from rhizomes of *A. calamus* cultivated in Canada of the diploid variety (without β-asarone) (ref. 14)
F data from other studies (indicated with superscript letters); highest concentrations found in any study reviewed here given; when two or more oils were investigated, only the highest concentrations are mentioned, unless indicated otherwise

[b] one lab-hydrodistilled rhizome oil from calamus harvested in Lithuania (ref. 2); [c] one lab-hydrodistilled oil from India (ref. 3); [d] three lab-hydrodistilled calamus rhizome oils from India (ref. 9); [e] one oil produced in Bangladesh (ref. 13); [f] one lab-hydrodistilled and one steam-distilled oil sample from the rhizomes of triploid *A. calamus* from North India (ref. 4); [g] one lab-hydrodistilled calamus oil of the European type from Turkey (ref. 1); [h] in the form of 1,4-trans-1,7-*trans*-acorenone; [i] one lab-hydrodistilled oil from the rhizomes of calamus grown in a botanical garden in India (ref. 33); this report contains some incomplete chemical names, which were not taken into account; [j] twenty-seven lab-hydrodistilled oils from rhizomes collected in various parts of India; only the asarones were investigated; all were of the tetraploid variety with concentrations of β-asarone ranging from 66 to 91.2% (ref. 34); [k] one lab-hydrodistilled oil from dried rhizomes purchased in a herb market in China; very atypical composition with a percentage of 50.1 for α-asarone (ref. 35); [l] one lab-hydrodistilled oil from India (ref. 36); [m] one lab-hydrodistilled oil from India (ref. 21); [n] one lab-hydrodistilled oil from Korea that is so atypical that a botanical misidentification or other error seems likely (ref. 6); [o] one lab-hydrodistilled calamus rhizome oil from Mongolia (ref. 24); [p] one industrially steam-distilled oil sample of the rhizomes of *A. calamus* collected in Kazakhstan (ref. 15); [q] one oil sample from China (ref. 26); [r] twenty oil samples from calamus rhizomes collected in Japan (ref. 11); [s] nineteen oils obtained from rhizomes cultivated in Japan, China and various other Asian countries (ref. 12); [t] one oil from India; the high content of α-asarone is atypical (ref. 27); [u] one calamus root oil sample from India (ref. 23); [v] one oil from rhizomes harvested in Canada, from a diploid variety without β-asarone (ref. 28); [w] one commercial calamus oil purchased in Germany with an atypical composition (ref. 29); [x] one calamus oil sample of Polish origin (ref. 16); [y] one oil from tetraploid calamus roots from India (ref. 19); [z] six oils from *A. calamus* rhizomes growing wild in India and collected at six different sites (ref. 20); [z1] one steam-distilled oil sample from Japan (ref. 40); [z2] data from various studies, cited in ref. 1; [z3] one lab-hydrodistilled calamus rhizome oil from south India (ref. 42); [z4] misinterpretation, must be γ-asarone

tr traces (in columns B and F[e]: <0.01; in columns C and D: <0.1); + present in the oil investigated, but quantity not stated

Table 5.10.3 Major constituents of calamus rhizome oils

Constituent	CAS	A	B (8)	C (32)	D (8)	E (14)	F
β-Asarone ((Z)-; *cis*-)	5273-86-9	1.6-79.6	77.7	69.8-95.6	5.2		91.6[z]; 91.7[u]; 92.1[r]; 92.4[d,z3]; 96.3[s]
(Z)-Methylisoeugenol	6380-24-1	0.3-13.7	1.3	tr-4.4	0.6	0.6	2.6[m]; 4.2[v,z]; 11.3[r]; 48.9[z2]; 87.3[s]
α-Asarone (*trans*-)	2883-98-9	0.4-0.9	6.8	0.1-4.7	0.5		13.8[r]; 15.0[s]; 16.8[j]; 50.1[k]; 58.0[t]
Preisocalamenediol	25645-19-6		0.9		1.0	18.0	10.0[z2]; 12.1[o]; 17.3[g]; 18.8[v]; 34.9[r,s]
Acorenone	5956-05-8				8.1	14.2	14.4[o]; 16.1[v]; 20.9[b]; 21.6[f]; 28.8[x]
Shyobunone	21698-44-2		0.4		2.6	13.3	0.5[g,o]; 1.0[e]; 4.4[v]; 7.8[b,f]; 14.3[s]; 22.1[r]
Isoshyobunone	21698-46-4	0.2-7.2	0.5		6.3	1.3	9.4[z2]; 9.9[v]; 13.0[g]; 13.1[o]
Methyl eugenol	93-15-2	0.03-6.5	0.02			tr	0.03[e]; 0.1[v]; 0.2[m]; 1.6[s]; 2.1[w]; 8.6[k]
Camphor	76-22-2	0.02-5.9		0-0.7	3.2	0.1	0.8[t]; 1.2[f]; 1.5[o]; 2.4[k]; 5.1[b]; 5.9[g]; 8.1[x]
Camphene	79-92-5	tr-7.3	tr	0-0.4	1.5	0.2	0.5[o]; 0.6[k]; 0.8[g]; 1.0[z]; 1.2[w]; 1.6[f]; 7.4[x]
β-Caryophyllene	87-44-5	0.2-3.2	0.01	tr-0.8	tr		0.08[e]; 0.2[c]; 0.6[g]; 1.0[t]; 1.6[w]; 4.4[z]

LEGEND: SEE UNDER TABLE 5.10.2

(Z)-sesquilavandulol (14.7%; misinterpretation, must be γ-asarone), epishyobunone (12.1%), calamol (7.8%), longifolene (8.2%) and asaronaldehyde (8.1%).

Commercial oils

The ten chemicals that had the highest maximum concentrations in 14 commercial calamus essential oil samples (concentration ranges provided) are the following: α-asarone [(Z)-] (1.8-79.6%), β-asarone (1.6-13.9%), (Z)-methyl isoeugenol (0.3-13.7%), α-pinene (0.02-12.4%), α-calacorene (1.7-8.5%), camphene (tr-7.3%), isoshyobunone (0.2-7.2%), methyl eugenol (0.03-6.5%), camphor (0.02-5.9%) and trans-calamene (2.5-5.4%) (Erich Schmidt, unpublished analytical data).

Chemotypes and ploidy

The chemical composition of individual oils heavily depends on ploidy of the plant source material (2,7). The most characteristic component of calamus oil is β-asarone. The diploid *Acorus calamus* seems to be rather uniform and is characterized by the rhizome oil without β-asarone. The triploid form also shows genetic uniformity with regard to factors governing essential oil production in all European and North American triploid accessions, which are characterized by the presence of 3-21% β-asarone in the rhizome oil. However, the tetraploid *A. calamus* shows significant variations. Indian, Indonesian and Taiwan *A. calamus* oils contain up to 96% of β-asarone, while the tetraploids of Japan and far-east Russia (East Siberia) are characterized by the presence of 10-40% of β-asarone. On the other hand, tetraploid *A. calamus* from Thailand, Singapore and Vietnam differs from the Eastern Asiatic type by the larger amount of β-asarone in the rhizome oils (Table 5.10.1) (1,2,3,5,30).

It should be realized that these differences are not absolute. Thus, in Japan, three chemotypes have been found in which β-asarone content varied from 64.7 to 92.1% (chemotype 1), to 23.5 to 48.7% (chemotype 2) and 2.6 to 4.1% (chemotype 3) (11). Also, in specimens from various parts of China, the concentrations of β-asarone ranged from 10.2 to 53.7% (12). Finally, in the northern parts of India, triploid calamus has been found with concentrations of β-asarone up to 89% (37).

Thus, the chemical profile is strongly dependent on the ploidy of the *A. calamus* plants from which the oils were obtained. The Indian-type calamus oils with high β-asarone content (tetraploid) contain up to 95% β-asarone plus α-asarone; only a few other constituents are found to be present in concentrations >1%. The situation in the oils with lower β-asarone content (or the chemical being absent) is entirely different. Here large quantities may be found of preisocalamenediol (up to 35%), acorenone (up to 29%), shyobunone (up to 22%), isoshyobunone (up to 13%), isocalamenediol (up to 13%) and epi-shyobunone (up to 12%).

The analysis in ref. 17 is not included in Table 5.10.2, as the oil, contrary to what has been suggested (18), was probably not root oil but obtained from the whole plant (17). Additional recent analyses not presented in Table 5.10.2 can be found in refs. 43 and 44.

CONTACT ALLERGY/ALLERGIC CONTACT DERMATITIS

General

Contact allergy to calamus oil has been reported in two publications, but no cases of allergic contact dermatitis from the oil have been identified.

Testing in groups of patients

A group of 86 patients from Poland previously reacting to the fragrance mix was tested with calamus oil and seven (8.1%) had a positive patch test reaction to calamus oil 2% in petrolatum; relevance data were not provided (46). In a group of 21 patients with dermatitis caused by fragrances and tested with a series of essential oils, one (5%) reacted to calamus oil; relevance data were not provided (47).

LITERATURE

1　Özcan M, Akgül A, Chalchat JC. Volatile constituents of the essential oil of *Acorus calamus* L. grown in Konya Province (Turkey). J Essent Oil Res 2002;14:366-368

2　Venskutonis PR, Dagilyte A. Composition of essential oil of sweet flag (*Acorus calamus* L.) leaves at different growing phases. J Essent Oil Res 2003;15:313-318

3　Raina VK, Srivastava SK, Syamasunder KV. Essential oil composition of *Acorus calamus* L. from the lower region of the Himalayas. Flavour Fragr J 2003;18:18-20

4　Marongiu B, Piras A, Porcedda S, Scorciapino A. Chemical composition of the essential oil and supercritical CO_2 extract of *Commiphora myrrha* (Nees) Engl. and of *Acorus calamus* L. J Agric Food Chem 2005;53:7939-7943

5　Ahlawat A, Katoch M, Ram G, Ahuja A. Genetic diversity in *Acorus calamus* L. as revealed by RAPD markers and its relationship with β-asarone content and ploidy level. Scientiae Horticulturae 2010;124:294-297

6　Kim W-J, Hwang K-H, Park D-G, Kim T-J, Kim D-W, Choi D-K, et al. Major constituents and antimicrobial activity of Korean herb *Acorus calamus*. Nat Prod Res 2011;25:1278-1281

7　Singh C, Jamwal U, Sing P. *Acorus calamus* (sweet flag): an overview of oil composition, biological activity and usage. J Medic Arom Plant Sci 2001;23:687-708

8　Mazza G. Gas chromatographic and mass spectrometric studies of the constituents of the rhizome of *Calamus*. I. The volatile constituents of the essential oil. J Chromatogr 1985;328:179-194

9　Bisht D, Pal A, Chanotiya CS, Mishra D, Pandey KN. Terpenoid composition and antifungal activity of three commercially important essential oils against *Aspergillus flavus* and *Aspergillus niger*. Nat Prod Res 2011;25:1993-1998

10　Lawrence BM. Essential oils 2001-2004. Carol Stream, USA: Allured Publishing Corporation, 2006:118-120 (reprinted from Perfumer and Flavorist, September/October 2002;27(5):74-?)

11 Sugimoto N, Mikage M, Ohtsubo H, Kiuchi F, Tsuda Y. Pharmacognostical investigations of Acori rhizomes. (1). Histological and chemical studies of rhizomes of *A. calamus* and *A. graminous* distributed in Japan. Natural Medicines 1997;51:259-264. Data cited in ref. 10

12 Sugimoto H, Ohtsubo M, Mikage M, Kiuchi F, Liu H-M, Tsuda Y. Pharmacognostical investigation of Acori rhizomes in Asian markets. Natural Medicines 1997;51:316-324. Data cited in ref. 10

13 Bonaccorsi I, Cotroneo J, Chowdhury JU, Yusuf M. Studies on essential oil bearing plants of Bangladesh, Part VII. Composition of the rhizomes of *Acorus calamus* L. (sweet flag). Essenz Deriv Agrum 1997;67:394-402. Data cited in ref. 10

14 Garneau F-X, Collin GH, Gagnon H, Bélanger A, Lavoie S, Savard N, et al. Aromas from Quebec. I. Composition of the essential oil of the rhizomes of *Acorus calamus* L. J Essent Oil Res 2008;20:250-254

15 Tkachev AV, Gur'ev AM, Yusubov MS. Acorafuran, a new sesquiterpenoid from *Acorus calamus* essential oil. Chem Nat Comp 2006;42:696-698

16 Gora J, Majda T, Lis A, Tichek A, Kurowska A. Chemical composition of some Polish commercial essential oils. Rivista Ital EPPOS 1997;(Numero Speciale):761-766. Data cited in ref. 22

17 Wilczewska AZ, Ulman M, Chilmończyk Z, Maj J, Koprowicz T, Tomczyk M, Tomczykowa M. Comparison of volatile constituents of *Acorus calamus* and *Asarum europaeum* obtained by different techniques. J Essent Oil Res 2008;20:390-395

18 Lawrence BM. Progress in essential oils. Perfum Flavor 2014;39(February):50-58

19 Rana VS, Verdeguer M, Blazquez MA. Chemical composition of *Acorus calamus* L. leaves and rhizomes from Manipur. Indian Perfum 2008;52:39-43. Data cited in ref. 18

20 Kumar R, Prakash O, Pant AK, Hore SK, Chanotiya CS, Mathela CS. Compositional variations and anthelmintic activity of essential oils from rhizomes of different wild populations of *Acorus calamus* L. and its major component β-asarone. Nat Prod Commun 2009;4:275-278

21 Joshi N, Prakash O, Pant AK. Essential oil composition and *in vitro* antibacterial activity of rhizomes, essential oil and β-asarone from *Acorus calamus* L. collected from lower Himalayan region of Uttarakhand. J Essent Oil Bear Plants 2012;15:33-37

22 Lawrence BM. Progress in essential oils. Perfum Flavor 2010;35(September):42-52

23 Srivastava VK, Singh BM, Negi KS, Pant KC, Suneja P. Gas chromatographic examination of some aromatic plants of Uttar Pradesh hills. Indian Perfum 1997;41(4):129-139. Data cited in ref. 10

24 Todorova MN, Ognyanov IV, Shatar S. Chemical composition of essential oil from Mongolian *Acorus calamus* L. rhizomes. J Essent Oil Res 1995;7:191-193

25 Sharma V, Singh I, Chaudhary P. *Acorus calamus* (The Healing Plant): a review on its medicinal potential, micropropagation and conservation. Nat Prod Res 2014;28:1454-1466

26 Zhu L-F, Li Y-H, Li B-L, Lu B-Y, Zhang W-L. Aromatic plants and essential constituents, Supplement 1. South Institute of Botany, Chinese Academy of Sciences. Hong Kong: Hai Feng Publ Company, Peace Book Company, 1995. Data cited in ref. 10

27 Chowdhury AR, Gupta RC, Sharma MI. Essential oil from the rhizomes of *Acorus calamus* Linn. raised on alkaline soil. Indian Perfum 1997;41(4):154-156. Data cited in ref. 10.

28 Belanger A, Dextraze L, Goudmand H, Garneau F-X, Collin G. Composition de l'huiles essentielle d'*Acorus calamus* du Quebec. Rivista Ital EPPOS 1997;(Numero Speciale):529-534. Data cited in ref. 10

29 Reichling J, Harkenthal M, Saller R. Wirkung ausgewählter ätherischer Öle. Erfahrungsheilkunde 1999;6:357-366. Data cited in ref. 10

30 Dušek K, Galambosi B, Hethelyi EB, Korany K, Karlová K. Morphological and chemical variations of sweet flag (*Acorus calamus* L.) in the Czech and Finnish gene bank collection. Hort Sci (Prague) 2007;34:17-25

31 Rajput SB, Tonge MB, Karuppayil SM. An overview on traditional uses and pharmacological profile of *Acorus calamus* Linn. (sweet flag) and other *Acorus* species. Phytomed 2014;21:268-276

32 Padalia RC, Chauhan A, Verma RS, Bisht M, Thul S, Sundaresan V. Variability in rhizome volatile constituents of *Acorus calamus* L. from Western Himalaya. J Essent Oil Bear Plants 2014;17:32-41

33 Shukla R, Singh P, Prakash B, Dubey NK. Efficacy of *Acorus calamus* L. essential oil as a safe plant-based antioxidant, aflatoxin B1 suppressor and broad spectrum antimicrobial against food-infesting fungi. Int J Food Sci Technol 2013;48:128-135

34 Rana TS, Mahar KS, Pandey MM, Srivastava SK, Rawat AKS. Molecular and chemical profiling of 'sweet flag' (*Acorus calamus* L.) germplasm from India. Physiol Mol Biol Plants 2013;19:231-237

35 Liu XC, Zhou LG, Liu ZL, Du SS. Identification of insecticidal constituents of the essential oil of *Acorus calamus* rhizomes against *Liposcelis bostrychophila* Badonnel. Molecules 2013;18:5684-5696

36 Senthilkumar A, Venkatesalu V. Larvicidal potential of *Acorus calamus* L. essential oil against filarial vector mosquito *Culex quinquefasciatus* (Diptera: Culicidae). Asian Pac J Trop Dis 2012;324-326

37 Ogra RK, Mohanpuria P, Sharma UK, Sharma M, Sinha AK, Ahuja PS. Indian calamus (*Acorus calamus* L.): not a tetraploid. Curr Sci 2009;97:1644-1647

38 Björnstad K, Helander A, Hultén P, Beck O. Bioanalytical investigation of asarone in connection with *Acorus calamus* oil intoxications. J Anal Toxicol 2009;33:604-609

39 Mukherjee PK, Kumar V, Mal M, Houghton PJ. *Acorus calamus*: Scientific validation of Ayurvedic tradition from natural resources. Pharm Biol 2007;45:651-666

40 Koyama H, Hirata Y. Sesquiterpenes from *Acorus calamus* L. Tetrahedron 1971;27:5419-5431

41 Devis A, Bawankar R, Babu S. Current status on biological activities of *Acorus calamus*—a review. Int J Pharm Pharm Sci 2014;6(10):66-71

42 Sujina I, Prabhu V, Hemlal H, Ravi S. Essential oil composition, isolation of β-asarone and its antibacterial and MRSA activity from the rhizome of *Acorus calamus*. J Pharm Res 2012;5:3437-3440

43 Satyal P, Paudel P, Poudel A, Dosoky NS, Moriarity DM, Vogler B, Setzer WN. Chemical compositions, phytotoxicity, and biological activities of *Acorus calamus* essential oils from Nepal. Nat Prod Comm 2013;8:1179-1181

44 Lohani H, Andola HC, Chauhan N, Bhandari U. Variations of essential oil composition of *Acorus calamus* from Uttarakhand Himalaya. J Pharm Res 2012;5:1246-1247

45 Davis P. Aromatherapy. An A-Z, 3rd Edition. London: Vermilion, 2005

46 Rudzki E, Grzywa Z. Allergy to perfume mixture. Contact Dermatitis 1986;15:115-116

47 Meynadier JM, Meynadier J, Peyron JL, Peyron L. Formes cliniques des manifestations cutanées d'allergie aux parfums. Ann Dermatol Venereol 1986;113:31-39

Chapter 5.11   CANANGA OIL

DEFINITION

Cananga oil (essential oil of cananga) is the essential oil obtained from the flowers of the perfume tree (Macassar oil tree), *Cananga odorata* (Lam.) Hook. f. et Thomson, forma *macrophylla*.

INCI NOMENCLATURE

Description/definition: Cananga odorata macrophylla flower extract is an extract obtained from the flowers of the Canadian (?) ylang-ylang, *Cananga odorata* var. *macrophylla*, Anonaceae

INCI name EU: Cananga odorata macrophylla flower extract (perfuming name, not an INCI name proper)

INCI name USA: not in the Personal Care Products Council Ingredient Database

CAS registry number(s): 93686-30-7; 68606-83-7

EINECS number(s): 297-681-1

INCI also recognizes Cananga odorata flower oil, the oil obtained from the flower, *Cananga odorata*, Anonaceae (CAS 83863-30-3; 8006-81-3; 68606-83-7; EINECS 281-092-1). However, it is stated to be defined in ISO 3063, which is the ISO standard for ylang-ylang oil.

ISO (INTERNATIONAL ORGANIZATION FOR STANDARDIZATION) STANDARD

ISO number: 3523

ISO name: Essential oil of cananga

Botanical origin: *Cananga odorata* (Lam.) Hook f. et Thomson, forma *macrophylla*

Parts of plant used: flowers

ISO values: ISO values (minimum and maximum concentrations) are shown in Table 5.11.1.

Cananga oil should not be confused with ylang-ylang oil, *Cananga odorata* (Lam.) Hook f. et Thomson, forma **genuina.** These oils were originally thought to be identical and are still often used – incorrectly – as synonyms. See also Chapter 5.90 on ylang-ylang oil.

THE PLANT, THE OIL, AND THEIR USES

For a general introduction to the *Cananga odorata* tree see under ylang-ylang oil. Ylang-ylang oil, obtained from *Cananga odorata* (Lam.) Hook f. et Thomson, forma *genuina*, is of higher quality than and is generally preferred by the fragrance industry over cananga oil, which is obtained from the *macrophylla* variety of the cananga tree. In general, however, their applications are similar. Cananga oil is produced commercially mainly in Indonesia and to a much lesser extent in Vietnam (6). It is used as a fragrance in perfumes and cosmetics, as flavors in foods and beverages such as ice cream, candy, puddings, baked goods and chewing gum, in pharmaceuticals and in aromatherapy (1,4,5,7,9). In aromatherapy, the essential oil is considered useful for the treatment of depression, breathing problems, high blood pressure, anxiety, and as an aphrodisiac. In Java culture, the flower itself is used against malaria and to treat asthma (5).

CHEMICAL COMPOSITION

Cananga oil is a clear mobile liquid with a faint yellow to darker yellow color, which has a floral and woody, slightly aromatic odor. The yield of essential oil from the flowers of *Cananga odorata* (Lam.) Hook f. et Thomson (forma *macrophylla*) generally varies from 0.5 to 1.0%. The main producing countries of this oil are Indonesia (especially Java) and Madagascar.

Literature data (up to November 7, 2014) (only seven publications) on the chemical composition of cananga oils and unpublished analytical data from one of us (E.S.) are shown in Table 5.11.2 in alphabetical order. In cananga oils from various origins, over 100 chemicals have been identified. About 22% of these were found in a single reviewed publication only. The major compounds found in cananga oils from different sources are shown in Table 5.11.3. They include (highest concentrations in any study given) β-caryophyllene (39.0%), germacrene D (17.2%), α-humulene (11.6%), benzyl benzoate (7.9%), linalool (6.4%) and δ-cadinene (6.1%). Uncommon or rare constituents of cananga oil found in high concentrations in single studies include caryophyllene oxide (15.8% and 13.8%), β-selinene (32.4%), α-cyperone (27.4%), cyperene (15.1%), α-bergamotene (11.3%), isocaryophyllene (10.5%), γ-cadinene (7.6%) and (*E,E*)-α-farnesene (7.1%).

Commercial oils

The ten chemicals that had the highest maximum concentrations in 25 commercial cananga essential oil samples (concentration ranges provided) are the following: β-caryophyllene (32.9-38.0%), germacrene D (4.5-9.5%), α-humulene (8.3-9.4%), (*E,E*)-α-farnesene (2.3-7.1%), δ-cadinene (4.5-6.1%), benzyl benzoate (2.6-5.6%), linalool (1.2-2.8%), γ-muurolene (0.6-2.2%), *p*-cresyl methyl ether (1.2-2.1%) and α-cadinol (0.2-2.1%) (Erich Schmidt, unpublished analytical data).

Commercial cananga oils are obtained by hydro- or steam-distillation of fresh flowers. The composition of cananga oils produced by extraction by various solvents is presented in refs. 4,5,7 and 9. Refs. 11 and 12 are not discussed, as it is uncertain whether the oils analyzed were obtained from the *macrophylla* form of *C. odorata*.

CONTACT ALLERGY/ALLERGIC CONTACT DERMATITIS

General

Contact allergy to/allergic contact dermatitis from cananga oil has been reported in over 25 publications. In groups of consecutive patients suspected of contact dermatitis, prevalence rates of up to 1.2% positive

Table 5.11.1 ISO values (%) for cananga oil[a]

Compound	CAS	Minimum	Maximum
β-Caryophyllene	87-44-5	30.0	40.0
α-Humulene	6753-98-6	7.0	11.0
Germacrene D	23986-74-5	5.0	9.0
δ-Cadinene	483-76-1	4.0	7.0
(E,E)-α-Farnesene	502-61-4	3.0	7.0
Benzyl benzoate	120-51-4	3.0	5.0
Geranyl acetate	105-87-3	1.0	3.0
Linalool	78-70-6	1.0	3.0
α-Cadinol	481-34-5	1.0	2.5
(E,E)-Farnesol	106-28-5	1.0	2.0
p-Cresyl methyl ether	104-93-8	0.5	2.0
Geraniol	106-24-1	0.5	1.5
Benzyl salicylate	118-58-1	0.2	1.0

[a] ISO 3523 Essential Oil of Cananga ©ISO 2002; Geneva, Switzerland, www.iso.org

patch test reactions have been observed, but reliable data on relevance are lacking. Case reports of allergic contact dermatitis to cananga oil are all from occupational exposure in massage therapists/aromatherapists. Co-reactions to the related ylang-ylang oil (also prepared from the *Cananga odorata* tree, but from the *forma genuina*) are frequent. Cananga oil used to be a frequent cause of pigmented cosmetic dermatitis in Japan.

Testing in groups of patients

The results of patch tests with cananga oil in routine testing (consecutive patients suspected of contact dermatitis) and in groups of selected patients are shown in Table 5.11.4. In routine testing, rates of positive reactions ranged from 0.5% to 1.2%, whereas between 0.8% and 33% of patients in selected groups had positive patch tests. The very high positivity rate of 33% was found in a small group of 21 patients previously shown to have allergic contact dermatitis from fragrances (13).

Table 5.11.2 Constituents identified in cananga oils

Constituent	CAS	Percentage and range							
		A	B (9)	C (1)	D (8)	E (5)	F (2)	G (10)	H (12)
Anisole	100-66-3					1.4			
Aromadendrene oxide 1			0.8	0.7					
Aromadendrene oxide 2	85710-39-0		2.9	2.7					
allo-Aromadendrene	25246-27-9		0.3	0.3					
Benzaldehyde	100-52-7		0.1	0.1					
Benzyl acetate	140-11-4	0.03-0.4	0.07	0.1					
Benzyl alcohol	100-51-6		0.07	0.1					
Benzyl benzoate	120-51-4	2.6-5.6	7.1	7.9	2.0	3.0	2.9	7.2	
Benzyl salicylate	118-58-1	0.1-0.5	1.3	1.5	0.07		0.1		
α-Bergamotene	17699-05-7					11.3			
Bicyclogermacrene	24703-35-3	0.09-0.7			0.5				
Bicyclosesquiphellandrene	54324-03-7				1.1				
β-Bisabolene	495-61-4		0.07	0.1					
β-Bisabolol	15352-77-9		2.3	2.4					
Cadina-1,4-diene	29837-12-5		0.3	0.4	0.3				
Cadina-3,5-diene	267665-20-3	0.05-0.2			0.4				
α-Cadinene	24406-05-1	0.2-0.4			0.3				
β-Cadinene	523-47-7		1.4	1.5					
γ-Cadinene	39029-41-9		0.3	0.3			7.6		
δ-Cadinene	483-76-1	4.5-6.1	0.9[b]	0.9[b]	6.0	5.4	5.4		
α-Cadinol	481-34-5	0.2-2.1	0.7	0.7	1.1	2.1			
epi-α-Cadinol	5937-11-1	0.3-1.3	1.0	1.0	0.5				
δ-Cadinol	19435-97-3	0.1-0.3							
Calamenene	483-77-2				0.1				
cis-Calamenene	72937-55-4		0.9[b]	0.9[b]					
trans-Calamenene	73209-42-4		0.1	0.2					
Camphene	79-92-5								3.3
β-Caryophyllene	87-44-5	32.9-38.0	18.2	18.2	38.2	39.0	37.0	36.4	
Caryophyllene oxide	1139-30-6	0.1-0.7		15.8	13.8	0.1	1.2		
8-Cedren-13-ol	18319-35-2		1.7	1.8					
Cedrol	77-53-2		0.5	0.5					
1,8-Cineole	470-82-6								2.2
α-Copaene	3856-25-5	1.2-1.7	1.2	1.3	1.8	2.3			5.8
p-Cresyl methyl ether	104-93-8	1.2-2.1			2.6		1.1		
α-Cubebene	17699-14-8		0.3	0.2	0.3				
β-Cubebene	13744-15-5		0.2	0.2					
Cubebol	23445-02-5				0.4				
Cubenol	21284-22-0		0.4	0.4					
1-epi-Cubenol	81939-29-9				0.2				
Cyperene	2387-78-2								15.1

Table 5.11.2 Constituents identified in cananga oils (*continued*)

Constituent	CAS	Percentage and range							
		A	B (9)	C (1)	D (8)	E (5)	F (2)	G (10)	H (12)
α-Cyperone	473-08-5								27.4
Dihydroeugenol acetate	33943-26-9		0.6	0.6					
α-Elemene	5951-67-7				0.5[c]				
β-Elemene	33880-83-0	0.6-0.8	0.3	0.3					
γ-Elemene	29873-99-2		0.3	0.4					
Elemol (α-)	639-99-6		0.1	0.1	0.4				
α-Eudesmol (α-Selinenol)	473-16-5		1.7	1.6					
β-Eudesmol (β-selinenol)	473-15-4		1.1	1.1					
7-epi-α-Eudesmol	123123-38-6		0.8	0.8					
Eugenol	97-53-0	0.2-0.9	0.3	0.3	0.5				
(E,E)-α-Farnesene	502-61-4	2.3-7.1			3.8				
(Z,Z)-α-Farnesene	28973-99-1				4.4				
(E)-β-Farnesene	18794-84-8		0.8	0.7					
(Z)-β-Farnesene	28973-97-9		0.4	0.3					
Farnesol	4602-84-0					0.7			
(E,E)-Farnesol	106-28-5	0.5-1.3	0.09	0.1	0.8				
(Z,E)-Farnesol	3790-71-4						1.1		
Farnesyl acetate	29548-30-9					0.5			
(E,E)-Farnesyl acetate	4128-17-0	0.02-0.05	0.01	0.8	0.08				
Geranial	141-27-5	0.04-0.1	0.2	0.1					
Geraniol	106-24-1	0.5-1.0	1.7	1.6	1.5	0.5	0.6		
Geranyl acetate	105-87-3	1.0-2.0	1.3	0.2	1.5	0.5	1.8		
Geranyl benzoate	94-48-4		0.6	0.8					
Germacrene D	23986-74-5	4.5-9.5	1.9	2.0	8.3	10.9		17.2	
Globulol	489-41-8		0.9	0.8		0.9			
β-Guaiene	88-84-6				0.9				
cis-β-Guaiene	372162-07-7		0.4	0.4					
trans-β-Guaiene	192053-49-9		3.9	4.3					
(Z)-3-Hexenyl acetate	3681-71-8	0.01-0.02							
(Z)-3-Hexenyl benzoate	25152-85-6		0.4	0.4					
n-Hexyl acetate	142-92-7	0.01-0.04							
α-Humulene	6753-98-6	8.3-9.4	6.7	6.5	9.2	11.6		9.6	
α-Humulene epoxide	96638-51-6					0.2			
Humulene epoxide II	19888-34-7		4.3	4.0					
Isocaryophyllene	118-65-0							10.5	
Limonene	138-86-3								6.8
Linalool	78-70-6	1.2-2.8	2.5	2.2	5.6	6.4	1.7	6.0	
cis-Linalool oxide, furanoid	11063-77-7		0.1	0.1					
trans-Linalool oxide, furanoid	34995-77-2		0.05	0.1					
α-Longipinene	5989-08-2					0.2			
Methyl benzoate	93-58-3	0.01-0.5							
Methyl chavicol	140-67-0	0.06-0.09			0.5[c]				
Methyl eugenol	93-15-2	0.09-0.2	0.6	0.5	0.07				
6-Methyl-5-hepten-2-one	110-93-0		0.2	0.2					
α-Muurolene	10208-80-7	0.4-1.3	0.5	0.6	1.5				
γ-Muurolene	30021-74-0	0.6-2.2	0.5	0.5	2.7				
α-Muurolol	104245-48-9		0.8	0.9	0.4				
τ-Muurolol (epi-α-muurolol)	19912-62-0	1.0-1.8			0.6				
Myrcene	123-35-3	0.2-0.5			0.4				
Naphthalene	91-20-3					0.1			
Neral	106-26-3		0.09						
Nerolidol	7212-44-4				0.06				
(E)-Nerolidol	40716-66-3		0.08	0.1			1.0		
(Z)-Nerolidol	3790-78-1		1.2	1.1					
α-Phellandrene	99-83-2	0.01-0.09							
α-Pinene	80-56-8	0.05-0.2			0.1				2.1
β-Pinene	127-91-3	0.01-0.04							4.4
α-Pinene oxide	1686-14-2		0.2	0.2					
Sabinene	3387-41-5				0.2				
(Z)-α-Santalol	115-71-9		1.2	1.3					
β-Selinene	17066-67-0								32.4
α-Terpinene	99-86-5				0.1				

Table 5.11.2 Constituents identified in cananga oils (*continued*)

Constituent	CAS	Percentage and range							
		A	B (9)	C (1)	D (8)	E (5)	F (2)	G (10)	H (12)
γ-Terpinene	99-85-4	0.1-0.2			0.2				
α-Terpineol	98-55-5	0.03-0.2							
Torreyol	19435-97-3					1.8			
α-Ylangene	14912-44-8	0.1-0.2			0.2				
β-Ylangene	20479-06-5	0.08-0.2			0.3				

A twenty-five essential oil samples from Indonesia (mainly Java) and Madagascar analyzed between 1999 and 2013; lowest and highest concentrations given (E. Schmidt, unpublished data)

B one steam-distilled oil from (dried) flowers picked in East Java (Indonesia) (ref. 9); the results are virtually identical to those presented in ref. 1 by the same authors (column C)

C one lab-steam distilled oil from Indonesia (ref. 1); the results are virtually identical to those presented in ref. 9 by the same authors (column B)

D one oil of cananga of unknown origin and mode of production (ref. 8)

E one cananga oil produced by combined hydro- and steam distillation from Indonesia; although the term cananga oil was used throughout the article (not ylang-ylang), it was not specifically stated that the flowers were from the cananga *macrophylla* variety (ref. 5)

F one oil of cananga of unknown origin and mode of production (ref. 2)

G one lab-hydrodistilled 'cananga odorata' oil from Indonesia; although the term ylang-ylang oil was not used in the article (but neither was cananga oil), it was not specifically stated that the flowers were from the *macrophylla* variety of cananga odorata (ref. 10)

H one oil obtained from flowers harvested from the cananga tree in Port Harcourt, Nigeria with an extremely atypical composition (12)

[a] incorrect identification; [b] *cis*-calamenene and δ-cadinene combined; [c] α-elemene and methyl chavicol combined

tr: trace

Case reports and positive patch tests

An aromatherapist/massage therapist developed occupational contact dermatitis from contact allergy to multiple essential oils; she reacted to both cananga oil and ylang-ylang oil in the fragrance series, which reactions were considered to be relevant (24). A similar case was seen in another aromatherapist (35).

Positive patch tests (relevance unknown, uncertain or not stated)

Two massage therapists/aromatherapists with occupational contact dermatitis from (multiple) essential oils had positive patch tests to cananga oil; it was uncertain whether this oil had been used by these patients (23). Of seven patients allergic to the fragrance farnesol, four

(57%) co-reacted to cananga oil (and various other fragrances) (38). One positive patch test reaction occurred in a patient primarily sensitized to compound tincture of benzoin (25). Among 819 patients suspected of contact dermatitis, two had positive patch tests to cananga oil (15). A naturopathic therapist with occupational contact dermatitis to multiple essential oils also reacted to cananga oil in the fragrance series (33). Positive patch tests to cananga oil occurred in two aromatherapists with occupational allergic contact dermatitis from multiple essential oils who did not use cananga oils at work (32). One positive patch test reaction to cananga oil occurred in a patient with allergic contact dermatitis from lovage oil and jasmine absolute (20). Two patients

Table 5.11.3 Major constituents of cananga oils

Constituent	CAS	Percentage and range							
		A	B (9)	C (1)	D (8)	E (5)	F (2)	G (10)	H (12)
β-Caryophyllene	87-44-5	32.9-38.0	18.2	18.2	38.2	39.0	37.0	36.4	
Germacrene D	23986-74-5	4.5-9.5	1.9	2.0	8.3	10.9		17.2	
α-Humulene	6753-98-6	8.3-9.4	6.7	6.5	9.2	11.6		9.6	
Benzyl benzoate	120-51-4	2.6-5.6	7.1	7.9	2.0	3.0	2.9	7.2	
Linalool	78-70-6	1.2-2.8	2.5	2.2	5.6	6.4	1.7	6.0	
δ-Cadinene	483-76-1	4.5-6.1	0.9[b]	0.9[b]	6.0	5.4	5.4		

LEGEND: SEE UNDER TABLE 5.11.2

Table 5.11.4 Results of testing groups of patients with cananga oil

Years and Country	Test conc. & vehicle	Number of patients tested positive (%)			Selection of patients (S); Relevance (R); Comments (C)	Ref.
Routine testing						
2000-2007 USA	2% pet.	486	5	(1.0%)	R: 100%; C: weak study: a. high rate of macular erythema and weak reactions, b. relevance figures include 'questionable' and 'past' relevance	14
2002-2003 Korea	2% pet.	422	5	(1.2%)	R: not stated	26
<1976 Poland	2% pet.	200	1	(0.5%)	R: not stated	34
Testing in groups of selected patients						
2001-2010 Australia	2% pet.	823	39	(4.7%)	S: not specified; R: 31%	36
2004-2008 Spain	2% pet.	86	3	(3.5%)	S: patients previously reacting to the fragrance mix I or *Myroxylon pereirae* (n=54) or suspected of fragrance contact allergy (n=32); R: not stated	21
<2004 Israel	2% pet.	91	1	(1.1%)	S: patients who had shown a doubtful or positive reaction to the fragrance mix I and/or *Myroxylon pereirae* resin and/or one or two commercial fine fragrances; R: not stated	27
1989-1999 Portugal	2% pet.	67	7	(10.4%)	S: patients who had a positive patch test to the fragrance mix; R: not stated	28
1990-1998 Japan	5% pet.	1,483	16	(1.1%)	S: patients suspected of cosmetic contact dermatitis, virtually all were women; range of annual frequency of sensitization: 0-1.9%; R: not stated	22
1996-1997 UK	2% pet.	10	2	(20%)	S: patients suspected of cosmetic dermatitis and reacting to the fragrance mix; R: not stated	16
<1986 Poland	2% pet.	86	10	(11.6%)	S: patients previously reacting to the fragrance mix; R: not stated	19
<1986 France	2.5% pet.	21	7	(33%)	S: patients with dermatitis caused by fragrances; R: not stated	13
<1983 Poland	2% pet.	16	3	(19%)	S: patients known to be allergic to propolis and *Myroxylon pereirae*; R: not stated	37
1971-1980 Japan	5% pet.	477	4	(0.8%)	S: patients with dermatoses other than pigmented cosmetic dermatitis and volunteers; R: not stated	17
<1976 Poland	2% pet.	51	1	(2.0%)	S: patients allergic to *Myroxylon pereirae* resin (balsam of Peru) and/or turpentine and/or wood tar and/or colophony	34
<1974 Japan	?	183	26	(14.2%)	S: patients suspected of cosmetic dermatitis; R: unknown; in many, there was co-reactivity with benzyl salicylate, which may be present in commercial cananga oils in a concentration of up to 0.5% (Table 5.11.2, column A)	29

pet.: petrolatum

had positive patch tests to cananga oil; they both also reacted to ylang-ylang oil (31).

Pigmented cosmetic dermatitis

In Japan, in the 1960s and 1970s, many female patients developed pigmentation of the face after having facial dermatitis (18). This so-called pigmented cosmetic dermatitis was shown to be caused by contact allergy to components of cosmetic products, notably essential oils, other fragrance materials, antimicrobials, preservatives and coloring materials (17,18). In a group of 620 Japanese patients with this condition investigated between 1970 and 1980, 7-11 % had positive patch test reactions to cananga oil 15% in petrolatum in various time periods (17). The number of patients with pigmented cosmetic dermatitis decreased strongly after 1978, when major cosmetic companies began to eliminate strong contact sensitizers from their products (17).

Co-reactivity to ylang-ylang oil

Of eight patients with dermatitis from fragrances and reacting to ylang-ylang oil, seven (88%) also reacted to cananga oil (13). Of seven patients with dermatitis from fragrances and reacting to cananga oil, all also reacted to ylang-ylang oil (13). In various other reports (e.g., 15,23,24,31,35) patients had positive patch tests to both ylang-ylang oil and cananga oil, which may indicate cross-sensitivity or can be explained by *pseudo*-cross-sensitivity due to the presence of the same component(s).

LITERATURE

1 Kristiawan M, Sobolik V, Allaf K. Isolation of Indonesian cananga oil by instantaneous controlled pressure drop. J Essent Oil Res 2008;20:135-146

2 Buccellato F. Ylang survey. Perfum Flavor 1982;7(4): 9-13. Data cited in ref 3

3 Ekundayo O. A review of the volatiles of the Annonaceae. J Essent Oil Res 1989;1:223-245

4 Kristiawan M, Sobolik V, Al-Haddad M, Allaf K. Effect of pressure-drop rate on the isolation of cananga oil using instantaneous controlled pressure-drop process. Chemical Engineering and Processing 2008;47:66-75

5 Megawati, Saputra SWD. A combination of water-steam distillation and solvent extraction of *Cananga odorata* essential oil. IOSR J Engin 2012;2:5-12

6 Lawrence BM. Progress in essential oils. Perfum Flavor 2004;29(6):80-90

7 Kristiawan M, Sobolik V, Allaf K. Isolation of Indonesian cananga oil using multi-cycle pressure drop process. J Chromatogr A 2008;1192:306-318

8 Kubeczka K-H, Formacek V. Essential oils analysis by capillary gas chromatography and carbon-13 NMR spectroscopy, 2nd Ed. New York, USA: J Wiley and Sons, 2002:27-35. Data cited in ref. 6

9 Kristiawan M, Sobolik V, Allaf K. Yield and composition of Indonesian cananga oil obtained by steam distillation and organic solvent extraction. Int J Food Engin 2012;8(3):article 28 (19 pages). DOI: 10.1515/1556-3758.1412

10 Pujiarti R, Ohtani Y, Widowati TB, Wahyudi, Kasmudjo, Herath NK, Wang CN. Effect of *Melaleuca leucadendron*, *Cananga odorata* and *Pogostemon cablin* oil odors on human physiological responses. Wood Res J 2012;3:100-105

11 Rolli E, Marieschi M, Maietti S, Sacchetti S, Bruni R. Comparative phytotoxicity of 25 essential oils on pre- and post-emergence development of *Solanum lycopersicum* L.: A multivariate approach. Ind Crops Products 2014;60:280-290

12 Obuzor GU, Nwiyoronu KJ. Chemical composition and potentials of the essential oils from flower of *Cananga odorata* from Port Harcourt, Nigeria. Int J Acad Res 2011;3(4, Part I):80-83

13 Meynadier JM, Meynadier J, Peyron JL, Peyron L. Formes cliniques des manifestations cutanées d'allergie aux parfums. Ann Dermatol Venereol 1986;113:31-39

14 Wetter DA, Yiannias JA, Prakash AV, Davis MD, Farmer SA, el-Azhary RA, et al. Results of patch testing to personal care product allergens in a standard series and a supplemental cosmetic series: an analysis of 945 patients from the Mayo Clinic Contact Dermatitis Group, 2000-2007. J Am Acad Dermatol 2010;63:789-798

15 Kohl L, Blondeel A, Song M. Allergic contact dermatitis from cosmetics: retrospective analysis of 819 patch-tested patients. Dermatology 2002;204:334-337

16 Thomson KF, Wilkinson SM. Allergic contact dermatitis to plant extracts in patients with cosmetic dermatitis. Br J Dermatol 2000;142:84-88

17 Nakayama H, Matsuo S, Hayakawa K, Takhashi K, Shigematsu T, Ota S. Pigmented cosmetic dermatitis. Int J Dermatol 1984;23:299-305

18 Nakayama H, Harada R, Toda M. Pigmented cosmetic dermatitis. Int J Dermatol 1976;15:673-675

19 Rudzki E, Grzywa Z. Allergy to perfume mixture. Contact Dermatitis 1986;15:115-116

20 Lapeere H, Boone B, Verhaeghe E, Ongenae K, Lambert J. Contact dermatitis caused by lovage (*Levisticum officinalis*) essential oil. Contact Dermatitis 2013;69:181-182

21 Cuesta L, Silvestre JF, Toledo F, Lucas A, Pérez-Crespo M, Ballester I. Fragrance contact allergy: a 4-year retrospective study. Contact Dermatitis 2010; 63:77-84

22 Sugiura M, Hayakawa R, Kato Y, Sigiura K, Hashimoto R. Results of patch testing with lavender oil in Japan. Contact Dermatitis 2000;43:157-160

23 Bleasel N, Tate B, Rademaker M. Allergic contact dermatitis following exposure to essential oils. Australas J Dermatol 2002;43:211-213

24 Boonchai W, Lamtharachai P, Sunthonpalin P. Occupational allergic contact dermatitis from essential oils in aromatherapists. Contact Dermatitis 2007;56:181-182

25 Sasseville D, Saber M, Lessard L. Allergic contact dermatitis from tincture of benzoin with multiple concomitant reactions. Contact Dermatitis 2009;61:358-360

26 An S, Lee AY, Lee CH, Kim D-W, Hahm JH, Kim K-J, et al. Fragrance contact dermatitis in Korea: a joint study. Contact Dermatitis 2005;53:320-323

27 Trattner A, David M. Patch testing with fine fragrances: comparison with fragrance mix, balsam of Peru and a fragrance series. Contact Dermatitis 2004;49:287-289

28 Manuel Brites M, Goncalo M, Figueiredo A. Contact allergy to fragrance mix—a 10-year study. Contact Dermatitis 2000;43:181-182

29 Nakayama H, Hanaoka H, Ohshiro A. Allergen Controlled System (ACS). Tokyo, Japan: Kanehara Shuppan, 1974:42. Data cited in ref. 18

30 Mitchell JC. Contact hypersensitivity to some perfume materials. Contact Dermatitis 1975;1:197-199

31 Srivastava PK, Bajaj AK. Ylang-ylang oil not an uncommon sensitizer in India. Indian J Dermatol 2014;59:200-201

32 Dharmagunawardena B, Takwale A, Sanders KJ, Cannan S, Roger A, Ilchyshyn A. Gas chromatography: an investigative tool in multiple allergies to essential oils. Contact Dermatitis 2002;47:288-292

33 Trattner A, David M, Lazarov A. Occupational contact dermatitis due to essential oils. Contact Dermatitis 2008;58:282-284

34 Rudzki E, Grzywa Z, Bruo WS. Sensitivity to 35 essential oils. Contact Dermatitis 1976;2:196-200

35 Cockayne SE, Gawkrodger DJ. Occupational contact dermatitis in an aromatherapist. Contact Dermatitis 1997;37:306-307

36 Toholka R, Wang Y-S, Tate B, Tam M, Cahill J, Palmer A, Nixon R. The first Australian Baseline Series: Recommendations for patch testing in suspected contact dermatitis. Australas J Dermatol 2014, Sept. 7. doi: 10.1111/ajd.12186

37 Rudzki E, Grzywa Z. Dermatitis from propolis. Contact Dermatitis 1983;9:40-45

38 Goossens A, Merckx L. Allergic contact dermatitis from farnesol in a deodorant. Contact Dermatitis 1997;37:179-180

Chapter 5.12 CARROT SEED OIL

DEFINITION
Carrot seed oil is the essential oil obtained from the fruits of the carrot, *Daucus carota* L.

INCI NOMENCLATURE
Description/definition: Daucus carota fruit oil is an essential oil obtained from the seeds of the carrot, *Daucus carota* L., Umbelliferae
INCI name EU: Daucus carota fruit oil (perfuming name, not an INCI name proper)
INCI name USA: Not in the Personal Care Council Ingredient Database
CAS registry number (s): 84929-61-3
EINECS number(s): 284-545-1

ISO (INTERNATIONAL ORGANIZATION FOR STANDARDIZATION) STANDARD
There is currently no ISO standard for carrot seed oil.

It should be noted that the Cosing and Personal Care Council Ingredient Database entry: *Daucus carota sativa* seed oil (CAS registry numbers 8015-88-1 and 84929-61-3; EINECS number 284-545-1) is not an essential (volatile) oil but a fixed (vegetable) oil. This oil contains fatty acids such as petroselinic, linoleic, palmitic, stearic, arachidic, palmitoleic, vaccenic and oleic acid (10).

THE PLANT, THE OIL, AND THEIR USES
Daucus carota L. is a biennial flowering herb, which grows to a height of 20-60 cm. It is native to Europe, northern Africa, western Asia, and tropical Asia (Pakistan) and is widely naturalized elsewhere (GRIN taxonomy for Plants). The wild carrot *Daucus carota* L. subsp. *carota* (L.) (also known as Queen Anne's lace and sometimes indicated as 'bird's nest' or 'devil's plague') is the precursor of the cultivated carrot, *Daucus carota* ssp. *sativus* (Hoffm.) Arcang (4,22). The carrot is commercially produced almost all over the world for its nutritive roots (vegetable), which are an important source of carotenoids in daily diet (15). The fruits of *Daucus carota* L. have been used traditionally in the treatment of ancylostomiasis, edema, chronic kidney disease and bladder afflictions. A wide range of pharmacological activities, viz. antibacterial, antifungal, anthelmintic, hepatoprotective and cytotoxic are reported for *D. carota* (15).

Carrot seed oil is obtained from the fruits of *Daucus carota* L., not from the seeds per se. The terms seeds and fruits are often used as synonyms, but the seeds are contained within the fruits (ripe umbels). Most of commercial carrot oil is obtained from the root vegetable carrot of mixed cultivars. However, all *D. carota* subspecies possess essential oils and some of these subspecies, probably mostly the wild carrot *Daucus carota* ssp. *carota*, are also a source of commercial oil production.

Carrot seed oil is widely used as an aromatic and fragrance component in the formulation of alcoholic liquors, food products, perfumes, cosmetics and soaps (1,16). It has been used in folk medicine and aromatherapy for its claimed anthelmintic, antiseptic, diuretic, hepatocellular regenerator, general tonic and stimulant, smooth muscle relaxant, cholesterol regulator and cicatrizant properties (3,16).

CHEMICAL COMPOSITION
In this section, we discuss the chemical composition of essential oils obtained from *Daucus carota* L., *Daucus carota* ssp. *sativus*, and *Daucus carota* ssp. *carota*. It should be mentioned that The Plants List considers *Daucus carota* var. *carota* to be a synonym of *Daucus carota* L. and that the University of Melbourne plant database considers *Daucus carota* subsp. *carota* to be the correct name for *Daucus carota* L., so these two names may well represent a single species. Furthermore, only data from essential oils obtained from ripe fruits (umbels) or seeds (which may in a number of cases have been used as synonyms for the ripe fruits) are presented; reports on the oils acquired from flowering umbels (8,15,16,19,24), pre-flowering stages, aerial parts (18), leaves (22), herbs (8) and stems (22) are *not* discussed here.

Carrot seed oil is a clear mobile, colorless to yellowish liquid which has a terpeny, slightly fatty herbaceous and aromatic odor. The yield of essential oil from the fruits of *Daucus carota* L. varies from 0.4 to 1.5%. The main producing countries of this oil are Hungary, France, and Egypt.

Literature data (up to November 7, 2014) on the chemical composition of carrot seed oils and unpublished analytical data from one of us (E.S.) are shown in Table 5.12.1 in alphabetical order. In carrot seed oils from various origins, over 315 chemicals have been identified. About 60% of these were found in a single reviewed publication only.

The major compounds found in carrot seed oils from different sources are shown in Table 5.12.2. They include (highest concentrations in any study given) β-bisabolene (80.5%), carotol (73.1%), geranyl acetate (65.0%), sabinene (46.6%), (E)-methylisoeugenol (34.0%), α-pinene (31.7%), elemicin (31.5%), β-pinene (13.1%), β-caryophyllene (12.6%) and p-cymene (11.0%). Well-known ingredients of carrot seed oils that were present in high concentrations in one study were myrcene (12.8%) and camphene (10.7%). Uncommon or rare constituents of carrot seed oil found in high concentrations (>6%) in single studies include 3-octen-5-yne, 2,7-dimethyl-, (Z)- (15.7%), *trans*-verbenol (13.4%), 11αH-himachal-4-en-1-β-ol (12.7%), α-cadinol (10.2%), α-asarone (-E) (8.8%), β-bisabolol (8.4%) and benzophenone (7.8%).

Commercial oils
The ten chemicals that had the highest maximum concentrations in 41 commercial carrot seed essential oil samples (concentration ranges provided) are the following: carotol (10.2-36.8%), geranyl acetate (0.9-13.9%), β-caryophyllene (1.9-12.6%), α-pinene (1.9-12.5%), sabinene (0.2-12.2%), β-bisabolene (1.6-7.8%), *p*-cymene

Table 5.12.1 Constituents identified in carrot seed oils

Constituent	CAS	Percentage and range				
		A	B (1)	C (21)	D (27)	E
Acetylcedrene	32388-55-9					0.08[n]
α-Acoradiene	24048-44-0		0-0.3			
β-Acoradiene	28477-64-7		0-0.2			
Acora-4,9-diene	38229-83-3					0.5[d]
Acora-4,10-diene	255062-41-0					0.5[d]
α-Amorphene	20085-19-2			0-3.8		0.1[s]
(E)-Anethole	4180-23-8					0.09[h]; 2.3[n]
Apiole	523-80-8				0-1.2	
Aristolene	6831-16-9					<0.01[k]
Aristolone	160568-09-2				0-0.5	
Aromadendrene	489-39-4				0-0.3	1.9[e]
Aromadendrene oxide						0.4[k]
Artemisia alcohol	29887-38-5					0-tr
α-Asarone (E-)	2883-98-9			0.1-5.4		0.5[r]; 1.5[m]; 1.6[q]; 8.8[h]
Benzaldehyde	100-52-7				tr	
Benzophenone	119-61-9				0-7.8	
Bergamal	106-72-9				0-tr	
α-Bergamotene	17699-05-7			0.2-1.9		0.09[h]; 2.5[t]
cis-α-Bergamotene	18252-46-5	0.3-1.2	tr-0.2		0-0.5	0.1[g]; 0.2[s]; 0.3[n,q]; 0.8[b]; 1.1[d]; 5.5[h]
trans-α-Bergamotene	13474-59-4	0.5-2.0	tr-0.2		tr-0.8	0.1[j]; 1.3[b]; 1.9[d]; 2.4[g]; 2.8[n]; 3.8[q]
Bicyclogermacrene	24703-35-3		0-0.2		0-0.2	0.1[c]; 0.2[j]; 0.3[o]; 0.4[b]; 1.9[g]; 2.4[q]
Bisabolene	495-62-5					0.5[h]
α-Bisabolene	17627-44-0					0.3[s]
(E)-α-Bisabolene	25532-79-0	0.7-2.9			0-tr	1.1[l]; 1.8[q]; 2.8[r]
(Z)-α-Bisabolene	29837-07-8			0-0.3	0-tr	
β-Bisabolene	495-61-4	1.6-7.8	0.3-0.8	1.7-80.5	0.1-1.3	1.9[g]; 4.1[i]; 5.5[m]; 22.6; 51.0[r]; 80.5[h]
(Z)-γ-Bisabolene	13062-00-5		0.1-0.4		0-tr	0.05[g]
Bisabolol	515-69-5				0-0.1	
α-Bisabolol	515-69-5				0-0.4	0.07[h]; 0.08[f]; 0.7[b]
epi-α-Bisabolol	78148-59-1				0-0.2	
β-Bisabolol	15352-77-9				0-0.6	8.4[o]
(E)-Bisabol-11-ol						0.3[j]
Born-5-en-2-ol						1.1[k]
Borneol	507-70-0		tr-0.6		0-0.6	0.4[c]
Bornyl acetate	76-49-3		0.3-1.1	<0.01-2.9		0.2[f]; 0.3[n]; 0.4[d,m]; 0.5[c,q]; 0.6[b,k]
Bulnesol	22451-73-6				0-0.3	
3-n-Butylphthalide	6066-49-5					0.4[n]
Cadina-1,4-diene	29837-12-5			0-0.6		
trans-Cadina-1(6),4-diene	931410-54-7				0-0.2	
α-Cadinene	24406-05-1			0.3-3.7		1.2[a,b]
γ-Cadinene	39029-41-9					0.05[s]; 0.1[c]
δ-Cadinene	483-76-1		0.1-0.4	0-1.2		0.1[m]; 0.3[a,e]
χ-Cadinene	855779-65-6					0.3[e]
cis-Cadinene ether					0-0.1	
trans-Cadinene ether					0-0.1	
α-Cadinol	481-34-5			0.03-10.2	0-0.2	0.1[s]; 0.6[o]; 1.8[q]
τ-Cadinol	5937-11-1				0-tr	
Calamene	1406-50-4					0.01[s]
Calarene	17334-55-3					3.2[e]
Camphene	79-92-5	0.4-1.0	0.2-2.4		0.3-2.8	0.7[d]; 0.9[e,f]; 1.3[b,j]; 1.8[c]; 2.2[i]; 10.7[t]
Camphenol (6-)	3570-04-5				0-3.2	

Table 5.12.1 Constituents identified in carrot seed oils (*continued*)

Constituent	CAS	Percentage and range				
		A	B (1)	C (21)	D (27)	E
6-Camphenone	53803-33-1				0-0.6	
6-Campholenal						0.1[j]
α-Campholenal	4501-58-0		0.2-2.5		0.3-1.8	0.09[s]; 0.2[c]
δ2-Carene (=δ4-)	554-61-0					0.1[s]; 0.6[k]
δ3-Carene	13466-78-9					0.8[a,e]
(*E*)-3-Caren-2-ol	139563-36-3					0.5[k]
3(10)-Caren-2-ol	93905-77-2					0.3[k]
Carotol	465-28-1	10.2-36.8		0.04-27.8	4.7-49.8	48.0[q]; 48.9[b]; 52.4[o]; 66.8[g]; 73.1[d]
Carvacrol	499-75-2				0-tr	0.06[s]
Carvenone	499-74-1				0-tr	
Carveol	99-48-9					0.09[s]
(*E*)-Carveol	1197-07-5		0.1-0.5		tr-2.0	0.1[d]; 0.3[k]
(*Z*)-Carveol	1197-06-4				0-tr	
Carvomenthone	499-70-7					0.08[n]
Carvone	99-49-0		0-0.3		tr-1.1	0.03[g]
cis-Carvone oxide					0-tr	
trans-Carvone oxide	18383-49-8				0-tr	
Caryophylla-4(14), 8(15)-dien-5-ol	644981-74-8				0-tr	
β-Caryophyllene	87-44-5	1.9-12.6	0.2-2.0	0.1-2.7	0.2-1.6	4.6[t]; 5.0[k]; 5.3[b]; 5.6[d]; 10.3[i]; 10.7[e]
9-epi-(*E*)-Caryophyllene	68832-35-9				0-tr	
Caryophyllene oxide	1139-30-6	0.1-3.1	0.4-4.7		0.8-4.4	0.5[o]; 2.8[d]; 3.1[b]; 4.2[s]; 4.3[e]; 4.4[k]
Caryophyllenyl alcohol	913176-41-7				0-0.1	
β-Cedrene	546-28-1	0.5-0.6				0.05[s]
β-Cedren-9-one					0-0.3	
Cedrene oxide	29597-36-2					1.1[n]
α-Cedrol	77-53-2				0-0.2	
α-Chamigrene	19912-83-5					0.09[s]
Chavicol	501-92-8				0-0.2	
cis-Chrysanthenol	55722-60-6		0.1-0.8			
cis-Chrysanthenyl acetate	67999-48-8		tr-0.2			
trans-Chrysanthenyl acetate	50764-55-1		tr-0.2			
1,8-Cineole	470-82-6				0-tr	
(*E*)-Cinnamaldehyde	14371-10-9				0-tr	
Copaene						0.4[n]
α-Copaene	3856-25-5		tr-0.2		0-tr	0.2[q]
15-Copaenol	115728-41-1					0.3[g]
Cryptone	500-02-7				0-tr	
α-Cubebene	17699-14-8					0.5[e]; 1.6[o]
β-Cubebene	13744-15-5			0.02-0.5	0-tr	0.2[q,s]; 0.5[e]
Cubenol	21284-22-0					0.1[s]
epi-Cubenol	19912-67-5				0-0.1	
Cuminaldehyde	122-03-2		0.1-0.5		0-0.8	0.09[s]
Cuminyl alcohol	536-60-7					0.9[k]
Curcumene	644-30-4	0.1-0.2				0.06[s]; 0.2[g]; 0.4[d]; 0.8[k]
γ-Curcumene	28976-68-3		0-0.4			
Cyclosativene	22469-52-9				0-tr	
o-Cymene	527-84-4					1.3[e]; 1.4[k]
p-Cymene	99-87-6	0.4-6.0	0.3-1.9	0.06-1.6	0.3-7.2	0.9[d]; 1.0[n,t]; 1.9[s]; 5.6[b]; 6.1[i]; 11.0[p]
p-Cymenene	1195-32-0				0-tr	
p-Cymen-7-ol	536-60-7		tr-0.7		0-tr	
p-Cymen-8-ol	1197-01-9		tr-0.4		0-0.5	0.02[s]; 0.07[g]
p-Cymen-9-ol	4371-50-0				0-tr	

Table 5.12.1 Constituents identified in carrot seed oils (*continued*)

Constituent	CAS	Percentage and range				
		A	B (1)	C (21)	D (27)	E
α-Cyperone	473-08-5					0.2[s]
Dauca-5,8-diene	142928-08-3					0.3[d]
Daucene	16661-00-0	1.1-1.7	1.6-5.9		0-0.6	0.2[j]; 0.5[n]; 1.6[s]; 4.0[b]; 8.7[g]
trans-Dauc-8-en-4β-ol	255062-40-9					4.1[d]
Daucol	887-08-1	0.1-1.0			0.3-5.0	1.7[d]; 2.0[e]; 2.9[b]; 3.9[o]; 5.3[p]; 7.4[s]
γ-Decalactone	706-14-9				0-0.2	
Dehydrolinalool	29171-20-8				0-1.3	
Dehydrosabinaketone	147043-52-5					0-0.2
Dihydrocarvyl acetate	57287-13-5				0-tr	
(E)-β-10,11-Dihydro-10,11-epoxyfarnesene	255062-42-1					0.2[d]
Dimethyl ionone	68555-94-2				0-0.1	
β-Elemene	33880-83-0	0.1-0.7			0-tr	0.07[s]; 0.8[q]
γ-Elemene	29873-99-2			0.2-1.5	0-tr	0.01[s]
Elemicin	487-11-6			0.06-15.9		4.1[c]; 5.2[r]; 10.9[m]; 16.3[j]; 31.5[q]
Elemol	639-99-6				0-0.2	
Elemol acetate	60031-93-8				0-0.2	
Epoxy allo-aromaden-drene	85760-81-2				0-0.2	
Eudesm-7(11)-en-4α-ol	473-04-1					0.3[m]; 0.6[r]; 8.2[q]
α-Eudesmol	473-16-5					0.04[s]
β-Eudesmol	473-15-4				0-tr	0.6[s]
γ-Eudesmol	1209-71-8					0.1[o]
Eugenol	97-53-0		0-0.1			
α-Farnesene	502-61-4					0.5[h]; 3.0[n]; 3.4[e]
(E,E)-α-Farnesene	502-61-4					0.7[o]
(Z,Z)-α-Farnesene	28973-99-1					5.9[g]
β-Farnesene	502-60-3			0-0.4		1.2[n]; 4.0[e]
(E)-β-Farnesene	18794-84-8	0.4-2.8	0.1-0.4		0.2-2.5	0.5[k]; 0.6[f]; 0.7[r]; 1.8[s]; 2.5[d]; 2.9[q]; 4.5[t]
(Z)-β-Farnesene	28973-97-9				0-tr	0.1[j]; 0.2[f]; 0.6[k,s]; 2.0[b]; 4.5[q]
(Z,E)-Farnesol	3790-71-4					0.2[o]
Farnesyl acetate	29548-30-9					0.2[s]
(E,E)-Farnesyl acetone	1117-52-8					1.3[o]
Farnesyl alcohol	4602-84-0					0.8[h]
α-Fenchyl alcohol	512-13-0				0-tr	
β-Funebrene	79120-98-2					0.08[s]
Geranial	141-27-5		0-0.1		0-0.4	0.3[r]
Geraniol	106-24-1	0.5-2.3			0-2.7	0.4[b]; 0.7[s]; 0.8[k]; 1.2[r]; 2.2[d]; 2.3[n]
Geranyl acetate	105-87-3	0.9-13.9	0.1-1.0	0-48.8	0-23.0	10.4[n]; 13.8[b]; 24.6[j]; 28.1[c]; 65.0[r]
Germacradien-5-ol				<0.01-1.1		
Germacrene	28028-64-0					0.3[n]
Germacrene B	15423-57-1				0-tr	0.3[o,r]; 0.6[q]; 1[t]
Germacrene D	23986-74-5	0.1-0.3	tr-0.8	0.4-6.4	0-0.1	0-tr[b]; 0.1[f,j]; 0.2[o]; 0.3[c]; 0.7[m]; 2.3[g]
Germacrene D-4-ol	198991-79-6					0.1[c]
α-Gurjunene	489-40-7			0-0.4		0.3[r]
β-Gurjunene	73464-47-8					3.2[a,e]; 5.8[o]
Heptanal	111-71-7				0-tr	
2-Heptanone	110-43-0				0-tr	
Hexanal	66-25-1				0-1.0	
Hexanol (1-; *n*-)	111-27-3				0-tr	
(E)-3-Hexenyl acetate	3681-82-1				0-tr	
α-Himachalene	3853-83-6			0-0.2		0.4[r]; 0.6[e]; 0.7[s]
β-Himachalene	1461-03-6			0-1.7	0-0.2	tr[m]; 1.3[r]

Table 5.12.1 Constituents identified in carrot seed oils (*continued*)

Constituent	CAS	Percentage and range				
		A	B (1)	C (21)	D (27)	E
11αH-Himachal-4-en-1-β-ol						3.2[m]; 9.0[r]; 12.7[q]
α-Humulene	6753-98-6	0.9-1.8	0.1-0.3	0.6-4.2	0-0.1	0.2[j]; 0.3[m]; 0.4[i,k]; 0.5[h]; 0.9[d]; 1.3[q]
Humulene epoxide I	19888-33-6	0.1-0.2				<0.01[k]
Humulene epoxide II	19888-34-7		0.1-0.4		tr-0.5	0.6[s]
14-Hydroxy-9-epi-(E)-caryophyllene	244226-09-3				0-0.4	
Isocaryophyllene	118-65-0					0.5[i]
Isodaucene	142878-08-8					0.8[d]
Isobornyl acetate	125-12-2				0.4-1.6	
Isoitalicene	94482-89-0					0-tr
Isoledene	95910-36-4					1.1[k]
Isolimonene	499-99-0					3.2[a,e]
Isolongifolene	1135-66-6				0-0.4	
Isopinocamphone	15358-88-0					0.05[s]
Isoterpinolene	586-63-0				0-0.2	
Italicene	94535-52-1					0-tr
Italicene epoxide	104188-24-1				0-0.1	
Juniper camphor	473-04-1				0-0.6	0.4[s]; 0.7[k]
Lanceol	10067-29-5					0.4[h]
Ledene oxide	882187-44-2					0.2[s]
Levomenol	23089-26-1					0.3[e]
Limonene	138-86-3	0.5-1.9	2.3-4.2	0-<0.01	1.0-6.0	1.2[r]; 1.7[k]; 2.0[e]; 2.1[m]; 2.6[c]; 3.6[f]; 8.4[j]
cis-Limonene oxide	13837-75-7				0-0.3	
Linalool	78-70-6	0.2-1.3	1.4-4.0		1.0-6.7	0.4[s]; 0.5[e]; 0.8[b]; 1.2[c]; 1.4[n]; 1.6[f]; 3.4[i]
cis-Linalool oxide	11063-77-7				0-0.3	0.01[s]
trans-Linalool oxide	11063-78-8				0-0.8	0.02[s]
Linalyl acetate	115-95-7				0-0.3	
Longicamphenylone	38647-26-6				0-tr	
Longifolenaldehyde	66537-42-6					0.6[s]; 0.7[k]
Longifolene (junipene)	475-20-7				0-tr	3.3[n]
α-Longipinene	5989-08-2			0-4.4		0.8[a,e]; 1.3[q]; 3.1[r]
p-Mentha-1,5-dien-8-ol	1686-20-0		0.2-1.8		0-1.1	0.09[s]
p-Mentha-1,4-dien-7-ol	22539-72-6					0.05[s]
cis-p-Mentha-1(7),8-dien-1-ol					0-tr	
trans-p-Mentha-1(7),8-dien-1-ol					0-tr	
p-Mentha-1(7),8(10)-dien-9-ol	29548-13-8					0.3[h]
cis-p-Mentha-2,8-dien-1-ol	3886-78-0				0-tr	
trans-p-Mentha-2,8-dien-1-ol	4017-77-0				0-0.3	
cis-p-Menth-2-en-1-ol	29803-82-5		0.1-0.5		0-0.3	tr[j]
Methyl eugenol	93-15-2		0-0.3		0-tr	0.7[q]
6-Methyl-5-hepten-2-ol	1569-60-4				0-tr	
6-Methyl-5-hepten-2-one	110-93-0				0-0.1	0.1[r]
(E)-Methylisoeugenol	6379-72-2		0-0.3	0-6.7	0-0.4	0.07[h]; 0.1[d]; 10.0[r]; 21.8[j]; 34.0[m]
3-Methylnonane	5911-04-6					0.08[s]
cis-Muurola-4(14),5-diene	157477-72-0				0-tr	
Muurola-4,10(14)-dien-1β-ol						0.7[s]

Table 5.12.1 Constituents identified in carrot seed oils (*continued*)

Constituent	CAS	Percentage and range				
		A	B (1)	C (21)	D (27)	E
γ-Muurolene	30021-74-0			0-0.6	0-tr	0.4[s]; 1.2[q]; 1.3[k]
Myrcene	123-35-3	0.9-1.9	tr-0.2	0.05-1.0	0.5-10.5	1.3[t]; 1.5[e]; 1.6[n]; 2.0[i]; 3.0[i]; 3.9[c]; 12.8[j]
Myrcene epoxide	29414-55-9				0-0.2	
Myristicin	607-91-0					0-1.3
trans-Myrtanol	15358-91-5				0-tr	
Myrtenal	564-94-3		0.1-1.5		0.2-3.7	0.3[s]
Myrtenol	515-00-4		0.3-1.6		0-0.8	0.1[d]; 1.2[k]
Myrtenyl acetate	1079-01-2		tr-0.2		0-tr	
Neocallitropsene	729602-94-2					0.5[r]
Neral	106-26-3				0-tr	0.5[n]
Nerol	106-25-2	0.5-1.2			0-0.3	0.1[r]; 0.4[i]; 1.7[n]
(*E*)-Nerolidol	40716-66-3				0-0.2	0.5[b]
Neryl acetate	141-12-8				0-tr	0.02[s]; 0.1[r]; 0.5[i]; 1.2[e]
Nonadecane	629-92-5					4.6[o]
Nonanal	124-19-6				0-tr	0.05[g]
(*E*)-2-Nonenal	18829-56-6				0-tr	
Nopinone	24903-95-5					0.1[s]
(*E*)-α-Ocimene	6874-10-8					0.4[n]
(*E*)-β-Ocimene	3779-61-1	0.05-0.4	tr-0.8		0-tr	0.1[j,m]; 0.2[r]; 0.4[f]
(*Z*)-β-Ocimene	3338-55-4				0-0.2	0.2[j]; 0.3[c,m]; 1.0[f]
Octacosane	630-02-4					1.0[o]
Octadecane	593-45-3					0.1[o]
n-Octanal	124-13-0				0-tr	
3-Octanone	106-68-3				0-tr	
3-Octen-5-yne, 2,7-dimethyl-, (*Z*)-	28935-76-4					15.7[t]
α-Patchoulene	560-32-7				0-tr	
β-Patchoulene	514-51-2				0-tr	
γ-Patchoulene	508-55-4				0-tr	
Pentylbenzene	538-68-1					0.4[n]
Pentadecane	629-62-9				0-0.5	
α-Phellandrene	99-83-2				0-tr	0.1[j]; 0.2[b]
β-Phellandrene	555-10-2	0.04-0.2	tr-0.2		0-tr	0.2[j]; 0.5[m]
Phenylacetaldehyde	122-78-1					0-tr
(*E*)-Phytol	150-86-7					1.3[o]
α-Pinene	80-56-8	1.9-12.5	16.0-24.5	0-2.4	2.5-21.7	18.0[b]; 21.1[f]; 24.5[i]; 30.4[c]; 31.7[j]
β-Pinene	127-91-3	0.4-4.4	1.0-3.0	0.4-7.1	0.3-3.8	1.6[i]; 1.9[e]; 4.1[f]; 4.4[n]; 4.5[b]; 13.1[c]
α-Pinene oxide	1686-14-2				0-3.0	
β-Pinene oxide	6931-54-0				0-0.7	
Pinen-4-ol						0.4[a,e]
cis-Pinocamphone					0-tr	
trans-Pinocamphone	547-60-4					0-0.4
Pinocarveol	5947-36-4					1.3[n]; 1.4[k]
trans-Pinocarveol	1674-08-4		0.1-1.9		0.4-6.1	0.02[s]; 0.04[g]; 0.06[f]; 0.3[d]; 1.2[b]
Pinocarvone	30460-92-5		0.1-1.7		0.3-2.8	0.3[s]; 0.5[k]; 1.0[b]
trans-Pinocarvyl acetate	1686-15-3				0-tr	
Piperitenone	491-09-8				0-tr	
trans-Piperitol	16721-39-4				0-tr	
Piperitone	89-81-6				0-tr	
Pulegone	89-82-7				0-tr	
Retinal	116-31-4					0.7[e]
Sabina ketone	513-20-2		0.1-1.0		0-1.7	1.3[k]
Sabinene	3387-41-5	0.2-12.2	28.2-37.5	0.4-42.0	0.6-28.8	12.7[p]; 14.5[q]; 18.7[k]; 45.3[f]; 46.6[c]
cis-Sabinene hydrate	15537-55-0	0.05-0.5	0.1-1.1		tr-1.6	0.1[s]; 1.8[c]

Table 5.12.1 Constituents identified in carrot seed oils (*continued*)

Constituent	CAS	Percentage and range				
		A	B (1)	C (21)	D (27)	E
trans-Sabinene hydrate	17699-16-0	0.05-0.1	0.2-1.0			0.2[s]; 0.3[j]; 1.2[c]
Sabinene ketone						0.3[s]
cis-Sabinol	3310-02-9				0-tr	0.1[s]; 0.5[q]
trans-Sabinol	471-16-9		0.1-1.6		0-1.6	
trans-Sabinyl acetate	139757-62-3				0-tr	
α-Santalene	512-61-8				0-tr	1.9[n]
β-Santalene	511-59-1					0.8[n]
epi-β-Santalene	25532-78-9				0-0.4	0.2[g]
(Z)-α-Santalol	115-71-9				0-0.2	
Sativene	6813-05-4					0.02[s]
Sedanenolide	63038-10-8					0.1[n]
Sedanolide	6415-59-4					0.08[n]
α-Selinene	473-13-2	0.3-2.4		0-4.7	0-0.2	0.4[d]; 0.7[b]; 0.9[g]; 1.1[k]; 2.4[n]; 7.4[q]
β-Selinene	17066-67-0	0.4-3.0	0-0.6	0.04-9.3	0-0.9	0.3[r]; 0.6[f]; 1.1[d]; 2.2[g]; 3.5[b]; 4.2[k]
δ-Selinene	473-14-3					1.0[q]
Selin-7(11)-en-4α-ol	16641-47-7			0.06-1.7		
α-Sesquiphellandrene	495-60-3					2.1[e]
β-Sesquiphellandrene	20307-83-9	0.4-0.6	0.1-0.3	0-2.9	0-0.2	0.1[h]; 0.2[s]; 0.3[b,d]; 0.4[r]; 0.5[g]; 0.9[k]
7-epi-Sesquithujene	159407-35-9				0-tr	
Sesquithuriferol	117468-55-0				0-tr	
(E)-Sesquisabinene hydrate	145512-84-1				0-tr	
(Z)-Sesquisabinene hydrate	58319-05-4		0.1-0.4		0-0.3	
Shyobunone	21698-44-2					tr[m]
6-epi-Shyobunone	65794-23-2					tr[m]
Silphiperfol-5-en-3-ol A					0-0.4	
α-Sinensal	17909-77-2					0.4[a,o]
Spathulenol	6750-60-3		tr-0.5		0-0.3	0.1[c]; 0.5[h]; 0.8[o]; 1.1[s]
(Z)-Tagetenone	33746-71-3				0-tr	
α-Terpinen-7-al	1197-15-5		tr-0.4		0-0.3	
α-Terpinene	99-86-5	0.1-1.0	1.5-2.4		0-0.3	tr[j]; 0.2[m]; 1.1[c]; 1.4[e,f]; 1.5[b]
β-Terpinene	99-84-3					5.1[i]
γ-Terpinene	99-85-4	0.3-2.6	tr-0.8		tr-1.0	0.6[q]; 1.4[e]; 2.3[i]; 2.4[f]; 3.2[b]; 4.1[c]
Terpinen-4-ol	562-74-3	0.4-2.3	4.6-7.5		0.2-1.4	0.7[m]; 1.0[s]; 1.2[q]; 2.2[c]; 3.0[b]; 4.5[f]
γ-Terpinen-7-ol						0.4[k]
α-Terpineol	98-55-5	0.2-2.3	0.4-1.3		tr-0.2	0.1[j,m,r]; 0.3[f]; 0.4[q]; 0.7[c]; 1.0[b,n]
δ-Terpineol	7299-42-5				0-1.2	
Terpinolene	586-62-9	0.07-0.9	1.0-1.8		0-0.5	0.2[b]; 0.4[k]; 0.6[f,q]; 0.7[e]; 0.8[c]; 0.9[n]
Terpinyl acetate	8007-35-0	0.2-0.5				
α-Terpinyl acetate	80-26-2		0.1-0.7		tr-0.6	0.4[k]; 0.5[c]
γ-Terpinyl acetate	10235-63-9					0.06[s]
Tetracosane	646-31-1					5.8[o]
Thuja-2,4(10)-diene	36262-09-6				tr-1.0	0.5[k]
α-Thujenal	57129-54-1				0-0.9	
α-Thujene	2867-05-2	0.08-1.9			tr-0.7	0.1[j]; 0.2[o]; 0.3[d,k]; 0.7[f]; 0.8[c]; 1.9[b,e]
Thujol	35732-37-7					0.2[s]
α-Thujone (*cis*-)	546-80-5					0.03[s]
β-Thujone (*trans*-)	471-15-8				0-0.2	
trans-Thuj-3-en-10-al	57129-54-1		tr-0.5			
Thymol	89-83-8				0-0.1	

Table 5.12.1 Constituents identified in carrot seed oils (*Continued*)

Constituent	CAS	A	B (1)	C (21)	D (27)	E
Tricosane	638-67-5					1.3[o]
Tricyclene	508-32-7				0-0.1	
Umbellulone	24545-81-1					0-tr
Undecanal	112-44-7				0-tr	
Verbenene	4080-46-0					0.1[s]
Verbenol	473-67-6	0.4-0.8				0.5[n]
cis-Verbenol	1845-30-3				0-7.7	0.4[s]; 2.8[k]; 3.2[b]
trans-Verbenol	1820-09-3		0.1-2.8		tr-13.4	0.08[g]; 0.1[f,r]; 0.2[c]; 0.4[d]; 1.7[s]
Verbenone	80-57-9	0.1-0.3	0.6-1.4		tr-3.9	0.03[g]; 0.1[r]; 0.7[s]; 1.1[k]
cis-Verbenyl acetate	29135-27-1				0-tr	
β-Vetivene	27840-40-0		0-1.6			
β-Vetivenene	27840-40-0					0.5[j]
α-Ylangene	14912-44-8					0.1[r]

A forty-one carrot seed essential oil samples from Hungary, France, Egypt and Germany obtained from Daucus carota ssp. sativus, analyzed between 2001 and 2013; lowest and highest concentrations given (E. Schmidt, unpublished data)
B seventeen lab-hydrodistilled oils prepared from fruits of Daucus carota ssp. carota growing wild near and far from roads at five locations in Lithuania, harvested between 1995 and 2000; lowest and highest concentrations given (ref. 1)
C ten lab-hydrodistilled oils from mature seeds of ten natural Tunisian populations of Daucus carota over four different bioclimatic zones; lowest and highest concentrations given (ref 21)
D nine lab-hydrodistilled oils from commercial D. carota ssp. sativus fruits from Italy (ref. 27)
E data from other studies (indicated with superscript letters); highest concentrations found in any study reviewed here given; when two or more oils were investigated, only the highest concentrations are mentioned, unless indicated otherwise

[a] incorrect identification (ref. 2); [b] one commercial oil and two oils prepared from the fruits of two cultivated breeds of *Daucus carota* ssp. *sativus* (ref. 3); [c] three lab-hydrodistilled oils from *D. carota* ssp. *carota* fruits found wild in Vienna (ref. 4); [d] three commercial *D. carota* seed oils, origin unknown (ref. 7); [e] two commercial carrot seed oils from Poland investigated by the same group of researchers; possibly the same oil was investigated twice because of nearly identical chemical compositions (refs. 5 and 9); [f] one lab-hydrodistilled oil from Polish *D. carota* ssp. *carota* (ref. 8); [g] one lab-hydrodistilled oil from the seeds (probably not the fruits) of carrots bought at a local market in Turkey (ref. 10); [h] one lab-steam-distilled oil from commercially available carrot seeds (ref. 11); [i] one sample of carrot seed oil from cultivated carrots (ref. 12); [j] one lab-hydrodistilled oil from ripe fruits of *Daucus carota* ssp. *sativus* cultivated in India (ref. 15); [k] one lab-hydrodistilled oil from the fruits of *Daucus carota* L. 'Chanteney' from Serbia (ref. 17); [l] one commercial carrot seed oil from Corsica (ref. 20); [m] one oil from ripe umbels of wild Corsican *D. carota*, vapor-distilled in an industrial apparatus (ref. 22); [n] one carrot seed oil of unknown origin (ref. 23); [o] one oil produced from North Indian carrot seed (ref. 6); [p] one carrot seed oil produced commercially in Poland (ref. 14); [q] two lab-hydrodistilled oils from *D. carota* ssp. *carota* with ripe umbels and mature seeds growing wild at two different bioclimatic sites in Tunisia (ref. 25); [r] two lab-hydrodistilled oils from wild *D. carota* subspecies *carota* fruits growing wild in Portugal and Italy (ref. 16); [s] one lab-hydrodistilled oil from Polish *D. carota* L. seeds (ref. 26); one lab-hydrodistilled carrot seed oil sample from Iran (ref. 28)

tr: trace (in column B: <0.1)

(0.4-6.0%), β-pinene (0.4-4.4%), caryophyllene oxide (0.1-3.1%) and β-selinene (0.4-3.0%) (Erich Schmidt, unpublished analytical data).

Chemotypes

Four compositions of carrot seed oil, which include oils of all the subspecies of *Daucus carota* L., are usually distinguished (1,7,22). Three of these are characterized by the occurrence of a main component, viz. sabinene, geranyl acetate and carotol, and the fourth contains the three components with an approximately equal ratio (1,7). However, oils with β-bisabolene as the dominant ingredient with concentrations up to 55-80% have been reported repeatedly (11,16,21). In addition, some chemical compositions for carrot seed oil, characterized by γ-bisabolene (87.0%, China), geraniol esters (81%), α-pinene (55.0%), geraniol (24-50%), (*E*)-asarone (40.3%, Japan), or β-caryophyllene (29.0%) as the major constituents, have

apparently also been reported (1,7,27). Another proposed classification system based on chemotypes and dominant ingredients is as follows: chemotype I (sabinene), chemotype II (β-bisabolene), chemotype III (β-bisabolene/geranyl acetate or geranyl acetate/β-bisabolene), chemotype IV (carotol) (21).

CONTACT ALLERGY/ALLERGIC CONTACT DERMATITIS

General

Only one case of allergic contact dermatitis from carrot seed oil has been reported.

Case reports

A female 'complementary therapist' developed occupational contact dermatitis from a multitude of essential oils used at work, including carrot seed oil (29).

Table 5.12.2 Major constituents of carrot seed oils

Constituent	CAS	Percentage and range				
		A	B (1)	C (21)	D (27)	E
β-Bisabolene	495-61-4	1.6-7.8	0.3-0.8	1.7-80.5	0.1-1.3	1.9[g]; 4.1[i]; 5.5[m]; 22.6; 51.0[r]; 80.5[h]
Carotol	465-28-1	10.2-36.8		0.04-27.8	4.7-49.8	48.0[q]; 48.9[b]; 52.4[o]; 66.8[g]; 73.1[d]
Geranyl acetate	105-87-3	0.9-13.9	0.1-1.0	0-48.8	0-23.0	10.4[n]; 13.8[b]; 24.6[i]; 28.1[c]; 65.0[r]
Sabinene	3387-41-5	0.2-12.2	28.2-37.5	0.4-42.0	0.6-28.8	12.7[p]; 14.5[q]; 18.7[k]; 45.3[f]; 46.6[c]
(E)-Methylisoeugenol	6379-72-2		0-0.3	0-6.7		0.07[h]; 0.1[d]; 10.0[r]; 21.8[l]; 34.0[m]
α-Pinene	80-45-8	1.9-12.5	16.0-24.5	0-2.4	2.5-21.7	18.0[b]; 21.1[f]; 24.5[i]; 30.4[c]; 31.7[j]
Elemicin	487-11-6			0.06-15.9		4.1[c]; 5.2[r]; 10.9[m]; 16.3[l]; 31.5[q]
β-Pinene	127-91-3	0.4-4.4	1.0-3.0	0.4-7.1	0.3-3.8	1.6[j]; 1.9[e]; 4.1[f]; 4.4[n]; 4.5[b]; 13.1[c]
β-Caryophyllene	87-44-5	1.9-12.6	0.2-2.0	0.1-2.7	0.2-1.6	4.6[t]; 5.0[k]; 5.3[b]; 5.6[d]; 10.3[i]; 10.7[e]
p-Cymene	99-87-6	0.4-6.0	0.3-1.9	0.06-1.6	0.3-7.2	0.9[d]; 1.0[n,t]; 1.9[s]; 5.6[b]; 6.1[i]; 11.0[p]
β-Selinene	17066-67-0	0.4-3.0	0-0.6	0.04-9.3	0-0.9	0.3[r]; 0.6[f]; 1.1[d]; 2.2[g]; 3.5[b]; 4.2[k]
Daucene	16661-00-0	1.1-1.7	1.6-5.9		0-0.6	0.2[j]; 0.5[n]; 1.6[s]; 4.0[b]; 8.7[g]
Limonene	138-86-3	0.5-1.9	2.3-4.2	0-<0.01	1.0-6.0	1.2[r]; 1.7[k]; 2.0[e]; 2.1[m]; 2.6[c]; 3.6[f]; 8.4[j]

LEGEND: SEE UNDER TABLE 5.12.1

LITERATURE

1 Mockute D, Nivinskiene O. The sabinene chemotype of essential oil of seeds of *Daucus carota* L. ssp. *carota* growing wild in Lithuania. J Essent Oil Res 2004;16:277-281

2 Lawrence BM. Progress in essential oils. Perfum Flavor 2006;31(7):39-? (last page unknown)

3 Staniszewska M, Kula J, Wieczorkiewicz M, Kusewicz D. Essential oils of wild and cultivated carrots—the chemical composition and antimicrobial activity. J Essent Oil Res 2005;17:579-583

4 Chizzola R. Composition of the essential oil from *Daucus carota* ssp. *carota* growing wild in Vienna. J Essent Oil Bear Plants 2010;13:12-19

5 Jasicka-Misiak I, Lipok J, Nowakowska EM, Wieczorek PP, Młynarz P, Kafarski P. Antifungal activity of the carrot seed oil and its major sesquiterpenes compounds. Z Naturforsch C 2004;59:791-796

6 Raina VK, Kumar A, Naqvi AA, Tandon S, Aggarwal KK, Kahol AP. Composition of North Indian carrot seed oil. Indian Perfumer 2004;48:425-428. Data cited in ref. 2

7 Mazzoni V, Tomi F, Casanova J. A daucane-type sesquiterpene from *Daucus carota* seed oil. Flavour Fragr J 1999;14:268-272

8 Góra J, Lis A, Kula J, Staniszewska M, Wołoszynet A. Chemical composition variability of essential oils in the ontogenesis of some plants. Flavour Fragr J 2002;17:445-451

9 Jasicka-Misiak I, Lipok J, Kafarski P. Phytotoxic activity of the carrot (*Daucus carota* L) seed oil and its major components. J Essent Oil Bear Plants 2002;5:132-143

10 Özcan MM, Chalchat JC. Chemical composition of carrot seeds (*Daucus carota* L.) cultivated in Turkey: characterization of the seed oil and essential oil. Grasas y Aceites 2007;58:359-365

11 Imamu X, Yili A, Aisa HA, Maksimov VV, Veshkurova ON, Salikhov S. Chemical composition and antimicrobial activity of essential oil from *Daucus carota sativa* seeds. Chem Nat Comp 2007;43:495-496

12 Kilbarda V, Nanusevic N, Dogovic N, Ivanić R, Savin K. Content of the essential oil of carrot and its antibacterial activity. Pharmazie 1996;51:777-778

13 Lawrence BM. Progress in essential oils. Perfum Flavor 2003;28(5):70-? (last page unknown)

14 Góra J, Majda T, Lis A, Tichek A, Kurowska A. Chemical composition of some Polish commercial essential oils. Revista Ital EPPOS 1997;(Numero Speciale):761-766. Data cited in ref. 13

15 Verma RS, Padalia RC, Chauhan A. Chemical composition variability of essential oil during ontogenesis of *Daucus carota* L. subsp. *sativus* (Hoffm.) Arcang. Ind Crops Prod 2014;52:809-814

16 Maxia A, Marongiu B, Piras A, Porcedda S, Tuveri E, Gonçalves MJ, et al. Chemical characterization and biological activity of essential oils from *Daucus carota* L. subsp. *carota* growing wild on the Mediterranean coast and on the Atlantic coast. Fitoterapia 2009;80:57-61

17 Glisić SB, Misić DR, Stamenić MD, Zizović IT, Asanin RM, Skala DU. Supercritical carbon dioxide extraction of carrot fruit essential oil: Chemical composition and antimicrobial activity. Food Chem 2007;105:346-352

18 Rossi PG, Bao L, Luciani A, Panighi J, Desjobert J-M, Costa J, et al. (E)-Methylisoeugenol and elemicin: antibacterial components of *Daucus carota* L. essential oil against *Campylobacter jejuni*. J Agric Food Chem 2007;55:7332-7336

19 Staniszewska M, Kula J. Composition of the essential oil from wild carrot umbels (*Daucus carota* L. ssp. *carota*) growing in Poland. J Essent Oil Res 2001;13:439-441

20 Rossi P-G, Berti L, Panighi J, Luciani A, Maury J, Muselli A, et al. Antibacterial action of essential oils from Corsica. J Essent Oil Res 2007;19:176-182

21 Rokbeni N, M'rabeta Y, Dziri S, Chaabane H, Jemli M, Fernandez X, et al. Variation of the chemical composition and antimicrobial activity of the essential oils of natural populations of Tunisian *Daucus carota* L. (Apiaceae). Chem Biodivers 2013;10:2278-2290

22 Gonny M, Bradesi P, Casanova J. Identification of the components of the essential oil from wild Corsican *Daucus carota* L. using 13C-NMR spectroscopy. Flavour Fragr J 2004;19:424-433

23 Cu J-Q, Perineau F, Delmas M, Gaset A. Comparison of chemical composition of carrot seed essential oil extracted by different solvents. Flavour Fragrance J 1989;4:225-231

24 Kula J, Izydorczyk K, Czajkowska A, Bonikowski R. Chemical composition of carrot umbel oils from *Daucus carota* L. ssp. *sativus* cultivated in Poland. Flavour Fragr J 2006;21: 667-669

25 Marzouki H, Khaldi A, Falconieri D, Piras A, Marongiu B, Molicotti P, Zanetti S. Essential oils of *Daucus carota* subsp. *carota* of Tunisia obtained by supercritical carbon dioxide extraction. Nat Prod Comm 2010;5:1955-1958

26 Smigielski KB, Majewska M, Kunicka-Styczynska A, Gruska R, Stanczyk L. The effect of commercial enzyme preparation-assisted maceration on the yield, quality, and bioactivity of essential oil from waste carrot seeds (*Daucus carota* L.). Grasas Aceites 2014;65(4):e047 (11 pages). doi: http://dx.doi.org/10.3989/gya.0467141

27 Flamini G, Cosimi E, Cioni PL, Molfetta I, Braca A. Essential-oil composition of *Daucus carota* ssp. *major* (Pastinocello carrot) and nine different commercial varieties of *Daucus carota* ssp. *sativus* fruits. Chem Biodivers 2014;11:1022-1033

28 Mahboubi M, Kazempour N, Mahboubi A. The efficacy of essential oils as natural preservatives in vegetable oil. J Diet Suppl 2014;11:334-346

29 Newsham J, Rai S, Williams JDL. Two cases of allergic contact dermatitis to neroli oil. Br J Dermatol 2011;165(Suppl. 1):76

Chapters 5.13 and 5.14 CASSIA OIL

Cassia oil (essential oil of cassia) is the essential oil obtained from the leaves, twigs and terminal branchlets of the Chinese cinnamon, *Cinnamomum cassia* (Nees & T. Nees) J. Presl (synonyms: *Cinnamomum aromaticum* Nees; *Cinnamomum cassia* Nees ex Blume, *Cinnamomum cassia* auct). Enquiries by one of us (E.S.) with commercial parties – those providing cassia oils to the fragrance industries — generally confirm that the plant parts used for producing cassia oils are the leaves, twigs and terminal branchlets. In spite of this, here we discuss two separate oils with difference source material, cassia *bark* oil (Chapter 5.13) and cassia *leaf* oil (Chapter 5.14), for the following reasons:

1. many authors of studies analyzing cassia oils specifically mention bark as the source of the cassia oils, and in a number of studies only the leaves were investigated; in some reports, the source material is not specified and only the term 'cassia oil' was used. However: 'leaves, twigs and terminal branchlets' are never mentioned as the plant parts from which the cassia oils were obtained
2. on-line commercial parties offer both cassia bark oils and cassia leaf oils

ISO DATA FOR CASSIA OIL

ISO (INTERNATIONAL ORGANIZATION FOR STANDARDIZATION) STANDARD
ISO number: 3216
ISO name: Essential oil of cassia
Botanical origin[a]**:** *Cinnamomum cassia* (Nees & T. Nees) J. Presl, syn. *Cinnamomum cassia* auct.
Parts of plant used: Leaf, twig, and terminal branchlet

Table 5.13.1 ISO values (%) for cassia oil[a]

Compound	CAS	Minimum	Maximum
(*E*)-Cinnamaldehyde	14371-10-9	70.0	88.0
(*E*)-2-Methoxycinnamaldehyde	60125-24-8	3.0	15.0
(*E*)-Cinnamyl acetate	21040-45-9	0.0	6.0
Coumarin	91-64-5	1.5	4.0
Benzaldehyde	100-52-7	0.5	2.0
o-Methoxycinnamyl acetate (2-)	110823-66-0	0.0	2.0
Salicylaldehyde	90-02-8	0.2	1.0
Cinnamyl alcohol	104-54-1	0.0	1.0
(*Z*)-Cinnamaldehyde	57194-69-1	0.0	0.7
Phenylacetaldehyde	122-78-1	0.0	0.7
Eugenol	97-53-0	0.0	0.5
2-Phenethyl alcohol	60-12-8	0.0	0.5
Styrene	100-42-5	0.0	0.15
Acetophenone	98-86-2	0.0	0.1

[a] ISO 3216 Essential oil of cassia ©ISO 1997; Geneva, Switzerland, www.iso.org

ISO values: ISO values (minimum and maximum concentrations) are shown in Table 5.13.1.

[a] In the ISO standard, *Cinnamomum tsumu* Helms is mentioned as the botanical origin of the essential oil of cassia, but this name is incorrect.

Chapter 5.13 CASSIA BARK OIL

DEFINITION

Cassia bark oil (essential oil of cassia bark) is the essential oil obtained from the bark of the Chinese cinnamon, *Cinnamomum cassia* (Nees & T. Nees) J.Presl (synonyms: *Cinnamomum aromaticum* Nees; *Cinnamomum cassia* Nees ex Blume, *Cinnamomum cassia* auct).

INCI NOMENCLATURE
Description/definition: Cinnamomum cassia oil is the volatile oil obtained from the whole plant of the Chinese cinnamon, *Cinnamomum cassia* (L.), Lauraceae
INCI name EU & USA: Cinnamomum cassia oil
Other names: Chinese cinnamon oil
CAS registry number(s): 84961-46-6; 8007-80-5 (Personal Care Products Council Ingredient Database)
EINECS number(s): 284-635-0

In non-botanical literature, cassia oil is often termed 'cinnamon oil'. In fact, CosIng mentions 'cinnamon oil chinense' as a synonym for cassia leaf oil and the RIFM uses the term 'cinnamon bark oil'. In the Personal Care Products Council Ingredient Database, cinnamomum cassia oil is also termed 'Chinese cinnamon oil'.

However, cassia oil should not be confused with cinnamon bark oil and cinnamon leaf oil Sri Lanka type, which are obtained from the 'true cinnamon', *Cinnamomum zeylanicum* Blume (see Chapters 5.21 and 5.22). The chemical profile of the oils and also their uses (including medicinal) have many similarities, though.

THE PLANT, THE OIL, AND THEIR USES
Cinnamomum cassia (Nees & T. Nees) J. Presl, also called Chinese cinnamon, is a medium-sized (10-15 meter tall) evergreen tree, with greyish bark and hard elongated leaves that are 10-15 cm long, belonging to the Lauraceae. It is native to China and is widely distributed in China, Vietnam, Sri Lanka, Madagascar, Seychelles and India (4). The tree is cultivated in China, Laos, Thailand, Vietnam, Indonesia and Malaysia (GRIN Taxonomy for Plants). It is one of the most important economic plant resources in tropical and subtropical areas.

The dried bark of the tree is the source of the spice cassia (32). Cassia bark is widely used in pharmaceutical preparations, seasonings, cosmetics, foods, drinks, commodity essences, and chemical industries (32). It is considered to have medicinal properties, such as antioxidant, anti-allergic, antimicrobial, anti-tumorigenic, carminative, anti-inflammatory, and anti-diabetic (3,4,32).

It is commonly used as traditional Chinese medicine for treating chronic bronchitis, gastritis, impotence, dyspnea, blood circulation disturbances, inflammatory diseases, rheumatism and neurodynia (4,7,32). In fact, it is one of the most important Chinese medicinal materials as recorded in the Pharmacopoeia of the People's Republic of China (7,32).

Essential oils are important products from *C. cassia*, and they may be obtained from barks, twigs, leaves, calyces and seeds, the most important one apparently being cassia bark oil (32). The bark essential oil is used as a food and drink flavoring agent, in the cosmetics industry and for medicinal purposes, including the treatment of diarrhea and other problems of the digestive system (4,5). *C. cassia* leaves also contain large amounts of oils (cassia leaf oil), and are used similarly to cassia bark oil in flavoring, medicine and especially cola-type drinks (4).

Cassia oils are not used in aromatherapy because of the risk of dermal sensitization (35,36). Possible health effects of 'cinnamon' have been reviewed (1). 'Cinnamon' has been considered promising for its anti-diabetic effect, but in a Cochrane review it was concluded that there is insufficient evidence to support the use of cinnamon for type 1 or type 2 diabetes mellitus (22). A comprehensive review of all aspects of *Cinnamomum cassia* is provided in ref. 19. It should be realized that 'cinnamon' may be obtained either from *Cinnamomum cassia* (Nees & T. Nees) J. Presl or from *Cinnamomum zeylanicum* Blume (true cinnamon).

CHEMICAL COMPOSITION

Cassia oil is a mobile liquid with yellowish to reddish brown color, which has a spicy sweet odor, reminding of cinnamon bark. The yield of essential oil from the bark of *Cinnamomum cassia* (Nees & T. Nees) J. Presl varies from 1.5 to 4.0%, depending on whether fresh or dried bark is used as source material. The main producing countries of this oil are Indonesia, China, Vietnam, Laos, Thailand, Malaysia and India.

Literature data (up to November 10, 2014) on the chemical composition of cassia oils (including some samples of cassia oils where bark was not specifically mentioned as the source material [9,10,11,14,15,23,24,27,28,30] and unpublished analytical data from one of us (E.S.) are shown in Table 5.13.2 in alphabetical order. The latter

Table 5.13.2 Constituents identified in cassia bark oils and oils where the plant part used is unknown

Constituent	CAS	Percentage and range					
		A	B (2)	C (7)	D (8)	E (13)	F
Acetic acid	64-19-7						<0.1[l]
Acetophenone	98-86-2	0.03-0.09	tr-1.0		0.1	0-0.6	0.08[d]; <0.1[l]; 0.1[p]; 0.2[s]
Acetyl eugenol	93-28-7				0.2		
Acoradiene	24048-44-0						0.1[p]
Adamantane	281-23-2						0.2[c]
α-Amorphene	20085-19-2			0.08-0.3			
γ-Amorphene	6980-46-7						2.9[p]
(E)-Anethole	4180-23-8						1.4[p]
p-Anisaldehyde	123-11-5				0.3	tr	<0.05[a,j]; <0.1[l]
Anisole	100-66-3						0.2[p]
Aromadendrene	489-39-4				0.5		<0.1[l]
Azulene	275-51-4				0.6		
Benzaldehyde	100-52-7	0.9-2.3	0.2-0.6	0.1-2.1	0.5	0.5-1.1	0.8[q]; 1.3[m]; 1.4[r]; 1.5[j]; 4.9[l]
Benzofuran	271-89-6						0.3[c]
Benzoic acid	65-85-0					0.07-0.1	+[b]; <0.1[l]
Benzyl alcohol	100-51-6						0.02[c]; <0.1[l]; 0.2[o]
Benzyl benzoate	120-51-4	0.01-0.3			0.3	tr-0.4	0.1[l]; 0.6[j]; 3.2[n]; 10.2[o]
Benzyl formate	104-57-4						0.08[c]
Benzylidenemalon-aldehyde	82700-43-4						0.1[c]; 1.1[m]
α-Bergamotene	17699-05-7						0.09[p]
trans-α-Bergamotene	13474-59-4			0-0.7			1.5[g]
(E)-α-Bisabolene	25532-79-0			0-2.6			
(Z)-α-Bisabolene	29837-07-8			0-1.0			
β-Bisabolene	495-61-4			0.2-1.8		tr-0.2	0.07[s]; 0.2[p]
α-Bisabolol	515-69-5		0-1.0	tr-1.3	0.8		0.5[p]
β-Bisabolol	15352-77-9					tr-0.4	0.1[j]
2-Bornanol, 2-methyl-	91278-70-5				0.6		
Borneol	507-70-0	0.05-0.1	0-0.2	tr-0.1		0.06-1.3	0.1[d,e]; 0.2[p]; 0.4[j,s]
Bornyl acetate	76-49-3						<0.1[l]

Table 5.13.2 Constituents identified in cassia bark oils and oils where the plant part used is unknown *(continued)*

Constituent	CAS	Percentage and range					
		A	B (2)	C (7)	D (8)	E (13)	F
Butyl 2-methylbuty-rate	15706-73-7						<0.1[l]
Cadalene	483-78-3						0.6[g]; 0.7[p]; 1.8[m]
β-Cadinene	523-47-7					tr-0.1	
γ-Cadinene	39029-41-9			0.2-0.5			0.07[c]; 0.5[j]; 1.9[g]; 3.4[s]
δ-Cadinene	483-76-1			0.4-3.8		tr-0.1	0.7[q]; 2.0[m]; 2.6[g]; 4.1[p]
Cadinene-5,8-diene							0.2[s]
Cadinol				0.08-0.4			
α-Cadinol	481-34-5		0-0.5		2.7		<0.1[l]; 0.4[p]; 0.6[q]; 1.9[s]
epi-α-Cadinol	5937-11-1			tr-0.4			<0.1[l]; 0.2[p]
α-Calacorene	21391-99-1		0.6-1.3		0.5		0.5[p]
cis-Calamene	72937-55-4			0.1-0.7			4.5[m]
Calamenene	483-77-2						<0.1[l]; 1.2[p]
Camphene	79-92-5	0.05-0.8	0.2-0.6			0.05-0.1	0.03[d]; <0.1[l]; 0.2[n]; 0.5[s]
Camphor	76-22-2		0.4-1.0			0-0.08	0.3[s]
δ3-Carene	13466-78-9					tr-0.07[a]	<0.1[l]
Carvacrol	499-75-2						<0.1[l]
Carvone	99-49-0					0-0.3	
Carvotanacetone	499-71-8						<0.1[l]
β-Caryophyllene	87-44-5	0.02-0.1	tr-0.4	0-1.3	2.2	tr-0.3	0.4[h]; 1.2[g]; 2.9[k]; 3.5[n]; 5.6[c]
Caryophyllene oxide	1139-30-6	0.02-0.2		tr-0.3	1.3	0-0.1	0.06[p]; <0.1[l]; 0.7[r]; 1.5[k]
Caryophyllenyl alcohol	913176-41-7			tr-0.5			0.09[p]
Cedrene	11028-42-5		0.1-0.9				
β-Cedrene	546-28-1				1.6		
di-epi-α-Cedrene epoxide							0.09[p]
Chavicol	501-92-8				0.3		+[b]; <0.1[l]; 0.3[l]
2-Chlorocyclohexanol	1561-86-0						1.4[r]
1,8-Cineole	470-82-6	0.02-0.3			0.9	0.06-1.1	0.08[s]; 0.3[n]; 0.7[l]
(E)-Cinnamaldehyde	14371-10-9	75.4-83.1	66.3-77.2	33.9-76.4	53.9	80.4-88.5	60.0[r]; 64.6[k]; 68.5[s]; 78.6[d]; 82.4[p]; 87.0[l]; 87.2[q]; 93.5[j]
(Z)-Cinnamaldehyde	57194-69-1	0.04-0.6	0.7-2.1	0.5-2.2			0.1[l]; 0.6[d]; 0.9[p]; 1.1[j]; 2.2[s]
Cinnamic acid	621-82-9			0-7.5			
(E)-Cinnamic acid	140-10-3	0.2-0.7				0.1-3.0	+[b]; <0.1[l]; 0.2[c]
(Z)-Cinnamic acid	102-94-3						+[b]
Cinnamyl acetate	103-54-8		0.1-0.9	0.1-49.6			+[l]; 0.1[l]; 0.3[p]; 0.5[s]; 1.8[h]
(E)-Cinnamyl acetate	21040-45-9	1.1-4.7				0.6-5.1	3.6[l]; 4.0[o]; 6.0[f]; 7.6[j]; 11.7[r]
(E)-Cinnamyl alcohol	4407-36-7	0.1-0.4		0-0.4	0.8	0.05-0.1	+[b]; 0.2[l]; 0.9[f]; 1.0[c]; 1.4[q]
α-Copaene	3856-25-5	0.09-0.2	2.6-3.9	1.1-14.3		0.2-0.7	0.7[q]; 0.9[j]; 1.2[k]; 4.7[s]; 10[m]
Coumarin	91-64-5	0.6-2.6				tr-0.5	+[l]; 1.3[h]; 1.9[d]; 8.7[l]
o-Cresol	95-48-7						+[b]
α-Cubebene	17699-14-8				2.0		0.1[c]; 1.1[j]; 8.7[p]
β-Cubebene	13744-15-5						0.07[c]
Cubenol	21284-22-0			tr-0.7	1.1		0.4[p]
Cumene	98-82-8			0.3-2.0			
Cuminaldehyde	122-03-2						0.4[l]
Cuminyl alcohol	536-60-7				0.1		
Cuparene	16982-00-6				0.2		
Curcumene (ar-; α-)	644-30-4		0.3-0.5		0.5		<0.1[l]; 0.3[p]
(+)-Cycloisosativene				tr-0.3			
Cyclosativene	22469-52-9						0.5[p]
p-Cymene	99-87-6	0.04-0.3				0.04-0.2	<0.1[l]; 0.8[n]; 1.0[c]; 3.2[k]
p-Cymen-8-ol	1197-01-9						<0.1[l]
Decanal	112-31-2		0-0.6				
Decanoic acid	334-48-5					0-tr	+[b]
Dimethoxy allylphenol							<0.1[l]
Dimethoxycinnamal-dehyde							<0.1[l]

Table 5.13.2 Constituents identified in cassia bark oils and oils where the plant part used is unknown *(continued)*

Constituent	CAS	Percentage and range					
		A	B (2)	C (7)	D (8)	E (13)	F
1,2-Dimethoxy-4-(3- methoxy-1-prope-nyl) benzene	58045-87-7						2.1[h]
3,4-Dimethoxy-phenethyl alcohol-	7417-21-2						0.8[h]
cis-1,4-Dimethylada mantane							0.3[s]
1,3-Dimethylbenzene	108-38-3		tr-2.0				
α,4-Dimethylbenz-enemethanol							0.07[s]
α,p-Dimethylstyrene	1195-32-0						<0.1[l]
2,5-Dimethylundecane	17301-22-3						0.3[h]
2,6-Dimethylundecane	17301-23-4		0-0.2				
Dodecane	112-40-3						0.2[s]
Dodecanoic acid	143-07-7					0-tr	
Eicosane	112-95-8				0.6		
α-Elemene	5951-67-7						<0.1[l]
β-Elemene	33880-83-0					tr-0.06	<0.1[l]; 0.1[p]
γ-Elemene	29873-99-2					0-0.4	0.09[s]
Epiglobulol	88728-58-9						0.05[p]
Ethyl cinnamate	103-36-6		tr-0.5				<0.1[l]; 0.3[c]; 0.5[l]
Ethyl (E)-cinnamate	4192-77-2					tr-0.1	
4-Ethylguaiacol	2785-89-9						+[b]; 0.5[l]
Ethyl-p-methoxycin-namate	1929-30-2						0.06[p]
1-Ethyl-2-methyl-benzene	611-14-3			0.4-2.5			
1-Ethyl-4-methyl-benzene	622-96-8			0-0.9			
3-Ethyl-3-methyl-decane	17312-66-2		0-0.3				
2-Ethyl-5-propyl-phenol	72386-20-0						0.2[h]
Eugenol	97-53-0	0.04-0.4	tr-0.2		5.4	0.03-1.1	0.1[l]; 1.0[c]; 1.6[k]; 7.7[o]; 72.1[n]
Eugenyl acetate	93-28-7						<0.1[l]; 0.2[l]; 0.5[o]; 3.8[n]
α-Farnesene	502-61-4		0-0.8				0.05[s]
β-Farnesene	502-60-3				0.1		
(E)-β-Farnesene	18794-84-8			tr-0.6			<0.1[l]
Farnesol	4602-84-0						<0.1[l]
(Z,E)-Farnesol	3790-71-4				0.8		
Fenchone	1195-79-5						<0.1[l]
Fenchyl alcohol	1632-73-1						0.1[s]
9H-Fluoren-9-ol	1689-64-1						0.06[c]
Furfural	98-01-1						0.07[s]
Geranial	141-27-5						<0.1[l]
Geraniol	106-24-1					0.08-0.3	<0.1[l]
Geranyl acetate	105-87-3		0-0.4				
Germacrene	28028-64-0				0.6		
Germacrene D	23986-74-5						0.3[h]
Globulol ((-)-)	489-41-8			tr-0.4			0.07[p]
Guaiacol	90-05-1					0-0.08	+[b]; <0.1[l]
Guaiacyl cinnamate						tr	
α-Guaiene	3691-12-1		4.3-7.6				
β-Guaiene	88-84-6						0.1[p]
Heptanoic acid	111-14-8						+[b]
Hexadecanal	629-80-1						0.05[p]
Hexadecanoic acid	57-10-3				1.2		
Hexadecanol	51260-59-4				0.2		
Hexadecenal	27104-14-9				0.4		
(Z)-7-Hexadecenal	56797-40-1						0.1[p]
Hexanal	66-25-1						0.2[s]

Table 5.13.2 Constituents identified in cassia bark oils and oils where the plant part used is unknown *(continued)*

Constituent	CAS	Percentage and range					
		A	B (2)	C (7)	D (8)	E (13)	F
Hexanoic acid	142-62-1						+[b]
Hexanol	111-27-3						<0.1[l]
3-Hexen-1-ol	544-12-7						<0.1[l]
α-Himachalene (α-*cis-*)	3853-83-6						<0.1[l]
Humulane-1,6-dien-3-ol	915392-38-0						0.07[p]
α-Humulene	6753-98-6				0.4	0-0.2	<0.1[l]; 0.2[p]; 0.3[s]; 0.6[n]
Hydrocinnamaldehyde	104-53-0	0.2-0.7	0.1-0.7	0.2-0.6		0-0.2	0.7[j]; 1.7[q]; 2.0[l]; 3.7[s]
Hydrocinnamic acid	501-52-0					0-0.2	+[b]; <0.1[l]
Hydrocinnamyl alcohol	122-97-4	0.3-0.5					0.09[c]
2-Hydroxyaceto-phenone (α-) (ω-)	582-24-1						+[b]
2-Hydroxycinnamaldehyde	3541-42-2						+[i]
4-Hydroxy-2-phenethyl alcohol	501-94-0					0-0.1	
Isoamyl benzoate	94-46-2						<0.1[l]
Isoamyl isovalerate	659-70-1						<0.1[l]
Isoaromadendrene epoxide	499134-59-7						0.08[p]
Isoborneol	124-76-5					0-0.3	
Isocaryophyllene	118-65-0						<0.1[l]
(Z)-Isoeugenol	5912-86-7	0.0-0.2				0.1-0.7	<0.1[l]
Isoledene	95910-36-4		0.1-1.0		1.7		0.2[s]
Isosativene	24959-83-9				0.2		0.1[p]
Ledol	577-27-5				1.9		
Limonene	138-86-3	0.03-0.1				0.1-0.3	<0.1[l]; 0.2[j,n]; 0.7[s]; 1.0[k]
Linalool	78-70-6		0.7-0.9		0.1	0.08-0.2	0.1[l]; 1.2[n]; 4.9[k]; 9.2[o]
cis-Linalool oxide, furanoid	5989-33-3						<0.1[l]
trans-Linalool oxide, furanoid	34995-77-2						<0.1[l]
Linalyl acetate	115-95-7						<0.1[l]
Longifolene-(V4)							0.2[p]
Menthene	29350-67-2						<0.1[l]
Menthol	89-78-1						0.1[c]
2-Methoxybenzaldehyde	135-02-4	0.2-0.7			0.2	0-0.1	+[b]; 0.4[d]; 0.9[c]; 1.0[r]
o-Methoxycinnamaldehyde (2-)	1504-74-1		1.0-2.6	0.09-6.7			0.5[j]; 0.7[l]; 1.0[q]; 2.7[p]; 5.1[h] 0.4[s]; 20.1[r]
(E)-2-Methoxycinnamaldehyde (*o*-)	60125-24-8	6.8-11.1				tr-2.5	2.7[l]; 3.6[g]; 7.4[c]; 9.4[f]; 12.3[d]
(Z)-2-Methoxycinnamaldehyde	76760-43-5						0.2[d]
(E)-2-Methoxycinnamic acid	1011-54-7						+[b]
(Z)-2-Methoxycinnamic acid	14737-91-8						+[b]; 43.1[h]
o-Methoxycinnamyl acetate (2-)	110823-66-0	0.2-0.7					1.6[r]
(E)-2-methoxycinnamyl acetate	38822-47-8						+[b]; 0.06[d]
2-Methoxycinnamic alcohol							<0.1[l]
2-Methoxydihydrocinnamic acid	6342-77-4						+[b]
Methyl benzoate	93-58-3						+[b]
2-Methylbenzofuran	4265-25-2						0.1[c,d,e]
2-Methylbutyric acid	116-53-0						+[b]
3-Methylbutyric acid	503-74-2						+[b]
Methyl cinnamate	103-26-4				0.5		<0.1[l]
Methyl eugenol	93-15-2				0.1	tr-0.05	<0.1[l]

Table 5.13.2 Constituents identified in cassia bark oils and oils where the plant part used is unknown *(continued)*

Constituent	CAS	Percentage and range					
		A	B (2)	C (7)	D (8)	E (13)	F
Methyl salicylate	119-36-8						6.2[c]
α-Muurolene	10208-80-7		0.5-1.8	0.5-2.5		0-0.2	<0.1[l]; 0.3[s]; 0.5[p]; 2.3[m]
γ-Muurolene	30021-74-0		0.6-1.5			tr-0.5	<0.1[l]; 1.0[j,p]
α-Muurolol	104245-48-9						0.4[p]
τ-Muurolol	19912-62-0			0.2-0.8			0.2[j]; 0.7[p]
Myrcene	123-35-3					tr-0.1	<0.1[l]
Myrcenol	543-39-5				0.1		
Naphthalene, 1,2,3, 4,4a,7-hexahydro- 1,6-dimethyl-4-(1- methylethyl)-	16728-99-7		0-1.2				
Nerol	106-25-2						<0.1[l]
Nerolidol	7212-44-4				0.6		0.1[d]
(E)-Nerolidol	40716-66-3			tr-0.1			
(Z)-Nerolidol	3790-78-1						0.3[c]
Nonanal	124-19-6						0.1[l]
Nonanoic acid	12-05-0					0-tr	+[b]
9,12-Octadecadienoic acid	2197-37-7				0.4		
(E)-β-Ocimene	3779-61-1						<0.1[l]
(Z)-β-Ocimene	3338-55-4						<0.1[l]
Octanoic acid	124-07-2					0-tr	+[b]
8-Oxo-neoisolongi- folene							0.08[p]
Palustrol	5986-49-2						<0.1[l]
Patchoulene	1405-16-9					0-0.04[a]	
α-Patchoulene	560-32-7						0.4[p]
Patchouli alcohol	5986-55-0				0.2		
Pentadecanoic acid	1002-84-2				0.4		
2,2,4,6,6-Pentame- thylheptane	13475-82-6						0.2[h]
α-Phellandrene	99-83-2					tr-0.1	0.06[s]; <0.1[l]; 0.5[n]
β-Phellandrene	555-10-2						<0.1[l]; 2.2[k]
2-Phenethyl alcohol	60-12-8	0.2-0.7			0.1	tr-0.2	+[b]; <0.1[l]; 0.1[c]; 0.3[h]; 0.4[d]
β-Phenethyl cinna- mate	103-53-7						0.2[h]; 0.9[r]
Phenol	108-95-2						+[b]
Phenylacetaldehyde	122-78-1	0.0-0.09				tr-0.3	0.3[c]
2-Phenylalcohol	60-12-8						2.5[l]
Phenylethyl acetate	93-92-5	0.1-0.5					+[b]; <0.05[j]; 0.2[c]; 0.7[r]; 2.3[d]
2-Phenylethyl ben- zoate	94-47-3	0.02-0.06					0.1[l]
2-Phenylethyl formate	104-62-1						<0.1[l]
2-Phenylpropanal acid	93-53-8						0.05[c]; 0.1[l]
3-Phenylpropanol	122-97-4						0.5[d]
3-Phenylpropyl ace- tate	122-72-5					0-05-0.2	+[b]; <0.1[l]
Phthalic acid	88-99-3				0.1		
Phytol	7541-49-3				0.2		
α-Pinene	80-56-8	0.06-0.2				0.1-0.3	0.3[s]; 0.6[n]; 1.2[m]; 1.5[k]
β-Pinene	127-91-3	0.03-0.1	tr-0.3			0.1-0.2	<0.1[l]; 0.2[n]
Propylbenzene	103-65-1			0.5-2.8			
Sabinene	3387-41-5						<0.1[l]
Safrole	94-59-7					tr-0.2	<0.1[l]
Salicylaldehyde	90-02-8	0.04-0.3				0.04-0.9	+[b]; <0.1[l]; 1.0[c]; 0.2[d]; 0.4[j]
Salicylic acid	69-72-7					0.1-0.2	
Sativene	6813-05-4			tr-0.4			0.3[p]; 1.8[g]
β-Selinene	17066-67-0						<0.1[l]
Seychellene	20085-93-2						0.09[p]
Spathulenol	6750-60-3	0.09-0.1					0.5[p]
(-)-Spathulenol (β-)	77171-55-2			tr-0.4			<0.1[l]

Table 5.13.2 Constituents identified in cassia bark oils and oils where the plant part used is unknown *(continued)*

Constituent	CAS	Percentage and range					
		A	B (2)	C (7)	D (8)	E (13)	F
Styrene	100-42-5	0.09-0.2`	0.1-0.2				0.07[c]; 0.09[d]; <0.1[l]; 0.1[s]
α-Terpinene	99-86-5						<0.1[l]
γ-Terpinene	99-85-4						0.1-0.3
Terpinen-4-ol	562-74-3						0.09[c]; <0.1[l]; 1.0[k]; 0.3[s]
α-Terpineol	98-55-5		tr-0.2		1.0	0.07-2.1	0.06[p]; <0.1[l]; 0.2[j]; 1.0[k]
cis-α-Terpineol			0-0.2				
Terpinolene	586-62-9					0-0.04[a]	<0.1[l]
Tetradecanal	124-25-4			tr-1.0			
Tetradecanoic acid	544-63-8				0.5		
Tetradecenal	54264-02-7				0.08		
p-Tolylacetaldehyde	104-09-6						<0.1[l]
Tricosene	56924-46-0				0.5		
2,5,9-Trimethyldecane	62108-22-9						0.5[h]
2,6,10-Trimethyl-dodecane	3891-98-3		0-0.3				
2,2,4-Trimethyl-1,3-pentanediol	144-19-4		tr-0.2				
Undecanoic acid	112-37-8					0-0.1	
Vanillin	121-33-5					tr-0.1[a]	<0.1[l]; 0.3[c]
2-Vinylbenzaldehyde	28272-96-0						<0.1[l]
4-Vinylbenzaldehyde	1791-26-0						<0.1[l]
2-Vinylphenol	695-84-1						+[b]
Viridiflorene (ledene)	21747-46-6			tr-0.8			
Viridiflorol	552-02-3				2.1		
α-Ylangene	14912-44-8			tr-0.2			2.5[g]

A thirty-eight cassia essential oil samples from China, Vietnam, Indonesia and Thailand analyzed between 1998 and 2013; lowest and highest concentrations given (E. Schmidt, unpublished data); it is likely that most of these oils have been prepared from leaves, twigs and terminal branchlets
B seventeen lab-hydrodistilled oils from the bark of *C. cassia* from 7 parts of China; in the article, mean values were given for triplicate analysis results for each sample; lowest and highest (mean) concentrations given (ref. 2)
C eleven lab-distilled oils from Chinese branch bark (n=3) and stem bark (n=8) at different stages of growth; lowest and highest concentrations given (ref. 7)
D one lab-hydrodistilled oil from the bark of *Cinnamomum cassia* purchased in Bangladesh; it should be noted that the source tree was identified by its bark! (ref. 8)
E unknown number of cassia bark oils from China; lowest and highest concentrations given (13)
F data from other studies (indicated with superscript letters); highest concentrations found in any study reviewed here given; when two or more oils were investigated, only the highest concentrations are mentioned, unless indicated otherwise

[a] incorrect identification (ref. 12); [b] one commercial oil of cassia from China; only qualitative data provided (ref. 10); [c] an oil of *C. cassia* produced in China (ref. 14); [d] one 'authentic' sample of cassia oil, probably from China (ref. 11); [e] borneol and 2-methylbenzofuran combined; [f] one sample of Chinese cassia oil (ref. 15); [g] one laboratory steam-distilled cassia bark oil from China (ref. 6); [h] one commercial steam-distilled cassia bark oil from Taiwan (ref. 3); [i] one bark oil from Korea (ref. 9); [j] data taken from various publications on cassia oils reviewed in ref. 16; [k] one commercial oil sample obtained in Corsica, France (ref. 27); it was suggested (ref. 16) that, as the oil contains eugenol, which is not a normal constituent of cassia oil, the oil was not 100% natural; [l] two cassia bark oils from China and Australia (ref. 21); [m] one lab-hydrodistilled oil from Chinese cassia bark (ref. 5); [n] one commercial 'cinnamon oil' ex *Cinnamomum cassia* from Brazil; this oil is extremely atypical, as is contains >70% eugenol and no or very little cinnamaldehyde and can therefore not be a genuine cassia oil (ref. 23); [o] one oil *ex Cinnamomum cassia* obtained from a research society in India (ref. 24,25); this oil may have been adulterated with eugenol and eugenyl- and benzyl-derivatives (ref. 16); [p] thirty lab-hydrodistilled oils from 30 cassia bark samples and 19 from twig samples collected in China; the average concentrations were presented for the bark and the twig group; highest average concentration given (ref. 32); [q] one commercial *C. cassia* oil from China (ref. 30); [r] one commercial oil sample *ex C. cassia* (ref. 28); [s] one lab-hydrodistilled oil from Chinese *C. cassia* bark (ref. 31)

tr: trace (in columns B and C: <0.1%; in column E: <0.01); + present in the oil investigated, but quantity not stated

Table 5.13.3 Major constituents of cassia bark oils

Constituent	CAS	Percentage and range					
		A	B (2)	C (7)	D (8)	E (13)	F
(E)-Cinnamaldehyde	14371-10-9	75.4-83.1	66.3-77.2	33.9-76.4	53.9	80.4-88.5	82.4[p]; 87.0[l]; 87.2[q]; 93.5[j]
(E)-2-Methoxycinna-maldehyde (o-)	60125-24-8	6.8-11.1				tr-2.5	2.7[l]; 3.6[g]; 7.4[c]; 9.4[f]; 12.3[d]
(E)-Cinnamyl acetate	21040-45-9	1.1-4.7				0.6-5.1	3.6[l]; 4.0[o]; 6.0[f]; 7.6[j]; 11.7[r]
Benzaldehyde	100-52-7	0.9-2.3	0.2-0.6	0.1-2.1	0.5	0.5-1.1	0.8[q]; 1.3[m]; 1.4[r]; 1.5[j]; 4.9[l]

LEGEND: SEE UNDER TABLE 5.13.2

data (column A) are not specifically from cassia bark, but are from commercial oils which are generally obtained from leaves, twigs and terminal branchlets of the cassia.

In cassia oils from various origins, over 245 chemicals have been identified. About 62% of these were found in a single reviewed publication only. The major compounds found in cassia oils from different sources are shown in Table 5.13.3. They include (highest concentrations in any study given) (E)-cinnamaldehyde (93.5%), 2-methoxy-cinnamaldehyde (including the (E)-isomer) (20.1%) and (E)-cinnamyl acetate (11.7%). Well-known ingredients of cassia oils that were present in high concentrations (>6%) in one or two studies were eugenol (72.1%), α-copaene (14.3% and 10%), benzyl benzoate (10.2%), linalool (9.2%) and coumarin (8.7%). Uncommon or rare constituents of cassia oil found in high concentrations (>6%) in single studies include α-cubebene (8.7%), α-guaiene (7.6%), cinnamic acid (7.5%) and methyl salicylate (6.2%).

Commercial oils

The ten chemicals that had the highest maximum concentrations in 38 commercial cassia essential oil samples, probably prepared from leaves, twigs and terminal branchlets (concentration ranges provided), are the following: (E)-cinnamaldehyde (75.4-83.1%), (E)-o-methoxycinnamaldehyde (6.8-11.1%), (E)-cinnamyl acetate (1.1-4.7%), coumarin (0.6-2.6%), benz-aldehyde (0.9-2.3%), limonene (0.03-1.1%), camphene (0.05-0.8%), (E)-cinnamic acid (0.2-0.7%), 2-methoxy-benzaldehyde (0.2-0.7%) and 2-phenethyl alcohol (0.2-0.7%) (Erich Schmidt, unpublished analytical data).

Not discussed are the results of analysis of 13 bark oils from China, Vietnam and Korea in ref. 26, as the amounts of the major ingredients were expressed in a way that cannot be compared to the other data (26).

CONTACT ALLERGY/ALLERGIC CONTACT DERMATITIS

Cassia oil (unspecified)

General
Contact allergy to/allergic contact dermatitis from cassia oil has been reported in a few publications. In groups

of consecutive patients suspected of contact dermatitis, prevalence rates of up to 1% positive patch test reactions have been observed, but the relevance is unknown. Most case reports were patients with oral symptoms, cheilitis and/or perioral dermatitis from contact allergy to cassia oil in toothpastes. Although unproven, the most important sensitizer is likely to be (E)-cinnamaldehyde, the major component in both the cassia bark oil (75-85%) and cassia leaf oil (60-70%). There are no reports on contact allergy to cassia oil specified to be either cassia bark oil or cassia leaf oil. In literature, no good discrimination is made between cassia oil and cinnamon oil (from Cinnamomum zeylanicum), and sometimes they are used as synonyms.

Testing in groups of patients

Routine testing
Two hundred dermatitis patients from Poland were patch tested with cassia oil 2% in petrolatum and two (1%) reacted (37). Of 750 consecutive dermatitis patients tested with 2% cassia oil, 5 (0.7%) had a positive patch test (42). In neither report were relevance data provided.

Testing in groups of selected patients
Fifty-one patients allergic to Myroxylon pereirae resin (balsam of Peru) and/or turpentine and/or wood tar and/or colophony were tested with cassia oil 2% in petrola-tum and ten (19.6%) had a positive patch test; relevance data were not provided (37). A group of 86 patients from Poland previously reacting to the fragrance mix was tested with cassia oil and 24 (27.9%) had a positive patch test reaction; relevance data were not provided (38). In this group of 86, there were 16 reactions to cinnamic aldehyde, which is by far the most important constituent of cassia oil (38).

Case reports
One patient was described with stomatitis from contact allergy to cassia oil and cinnamon oil in toothpaste, who was negative, however, to their main component cinnamic aldehyde (39). One patient had burning mouth and soreness and swelling of the tongue from contact

allergy to her toothpaste; the patient also reacted to cassia oil and its main ingredient cinnamic aldehyde, but it was not stated whether these compounds were present in the toothpaste (39). One patient had mucosal inflammation and purpuric perioral macules associated with the use of a prophylactic dental tablet and was found to be contact allergic to its components oil of cassia and oil of cloves (40). One case of cheilitis from allergy to cinnamon (cassia) oil in toothpaste (41). One patient developed contact stomatitis and dermatitis from contact allergy to cinnamaldehyde and cinnamon (cassia) oil in toothpaste (43).

Positive patch tests (relevance unknown, uncertain or not stated)

In 16 patients known to be allergic to propolis and *Myroxylon pereirae*, three (19%) had positive patch tests to cassia oil 2% in petrolatum (44).

Cassia bark oil

No reports on contact allergy to cassia bark oil, specifically mentioned to be obtained from the bark of *Cinnamomum cassia*, have been found.

LITERATURE

References are shown in Chapter 5.14 Cassia leaf oil

Chapter 5.14 CASSIA LEAF OIL

DEFINITION
Cassia leaf oil is the essential oil obtained from the leaves and twigs of the Chinese cinnamon, *Cinnamomum cassia* (Nees & T. Nees) J. Presl (synonyms: *Cinnamomum aromaticum* Nees, *Cinnamomum cassia* Nees ex Blume, *Cinnamomum cassia* auct).

INCI NOMENCLATURE
Description/definition: Cinnamomum cassia leaf oil is the volatile oil obtained by steam distillation from the leaves and twigs of the Chinese cinnamon, *Cinnamomum cassia* (L.), Lauraceae
INCI name EU & USA: Cinnamomum cassia leaf oil
Other names: Cassia oil; cinnamon oil chinense (CosIng); cinnamon bark oil (RIFM, Research Institute for Fragrance Materials)
CAS registry number (s): 8007-80-5; 84961-46-6
EINECS number(s): 284-635-0

ISO (INTERNATIONAL ORGANIZATION FOR STANDARDIZATION) STANDARD
There is currently no ISO standard for cassia leaf oil.

THE PLANT, THE OIL, AND THEIR USES
This section is discussed under cassia (bark) oil (Chapter 5.13).

CHEMICAL COMPOSITION
Cassia leaf oil is a yellow to brown liquid with the characteristic odor and taste of cassia cinnamon. On aging or exposure to air it darkens and thickens.

Literature data (up to November 10, 2014) on the chemical composition of cassia leaf oils are shown in Table 5.14.4 in alphabetical order. Unpublished analytical data from one of us (E.S.) on oils obtained from leaves, twigs and terminal branchlets are shown in Table 5.13.2, column A. In cassia leaf oils from various origins (we could find only 7 publications), over 115 chemicals have been identified. About 69% of these were found in a single reviewed publication only. The major compounds found in cassia leaf oils from different sources are shown in Table 5.14.5. They include (highest concentrations in any study given) (*E*)-cinnamaldehyde (78.4%), 2-methoxycinnamaldehyde (including its (*E*)-isomer (25.4%) and cinnamyl acetate including its (*E*)-isomer (12.5%). Ingredients of cassia leaf oils that were present in high concentrations (>6%) in one or two studies were 3-methoxy-1,2-propanediol (29.3%) and coumarin (6.4% and 15.3%),

Table 5.14.4 Constituents identified in cassia leaf oils

Constituent	CAS	Percentage and range					
		A (4)	B (13)	C (17)	D (20)	E (19)	F
Acetaldehyde	75-07-0			0.5			
Acetic acid	64-19-7			0.1			
Acetophenone	98-86-2	2.1-3.0	tr-0.1			0.1	
p-Anisaldehyde	123-11-5		0.6-1.0				
Benzaldehyde	100-52-7	0-2.0	1.4-1.5	0.1	1.2	1.1	
Benzenemethanol	100-51-6			0.07			
Benzoic acid	65-85-0		0.07-0.1				
Benzyl alcohol	100-51-6		tr-0.05				
Benzyl benzoate	120-51-4		0.07-0.2		0.1		
Benzylidenemalonaldehyde	82700-43-4			0.3			
β-Bisabolene	495-61-4		tr-0.06				
Bisabolol (α-)	515-69-5	0-0.6					
β-Bisabolol	15352-77-9		tr				
Borneol	507-70-0	0-1.2	0.2-0.4				
4-Butylbenzyl alcohol	60834-63-1			0.1			
δ-Cadinene	483-76-1		tr				
α-Cadinol	481-34-5	0-0.9					
α-Calacorene	21391-99-1	0-0.9					
Camphene	79-92-5	0.5-4.8	0.04-0.05		tr		
Camphor	76-22-2	2.4-4.0	0.07-0.2				
δ3-Carene	13466-78-9		0.03-0.05				
Carvone	99-49-0		0.6				
β-Caryophyllene	87-44-5	0-0.9	0.2				0.2[c]
Caryophyllene oxide	1139-30-6		0.2				
Cedrene	11028-42-5	2.8-5.9					
Chavicol	501-92-8				tr		
1,8-Cineole	470-82-6		0.05-0.08		tr		
(*E*)-Cinnamaldehyde	14371-10-9	57.9-66.5	64.1-68.3	30.4	77.2	74.1	77.9[b]; 78.4[c]
(*Z*)-Cinnamaldehyde	57194-69-1	0-0.7					
(*E*)-Cinnamic acid	140-10-3		0.8-2.5				
Cinnamyl acetate	103-54-8	0.4-0.8			3.6	6.6	
(*E*)-Cinnamyl acetate	21040-45-9		4.5-12.5				9.2[c]

167

Table 5.14.4 Constituents identified in cassia leaf oils (*continued*)

Constituent	CAS	Percentage and range					
		A (4)	B (13)	C (17)	D (20)	E (19)	F
Cinnamyl alcohol	104-54-1			0.7	tr	0.2	
(*E*)-Cinnamyl alcohol	4407-36-7		0.2				
α-Copaene	3856-25-5	0-0.9					
Coumarin	91-64-5		0.03-0.08	6.4	15.3	1.2	
Cuminaldehyde	122-03-2				tr		
Curcumene	644-30-4	0-0.7					
p-Cymene	99-87-6		0.1-0.2				
Decanoic acid	334-48-5		tr				
1,1-Diethoxyethane	105-57-7			0.2			
2,6-Dimethylundecane	17301-23-4	0-0.7					
Dodecane	112-40-3			0.1			
Dodecanoic acid	143-07-7		tr-0.04				
γ-Elemene	29873-99-2		0-tr				
δ-Elemene	20307-84-0	1.0-2.2					
Ethyl acetate	141-78-6			0.8			
Ethyl alcohol	64-17-5			0.1			
Ethylbenzene	100-41-4	0-5.4					
Ethyl (*E*)-cinnamate	4192-77-2		0.1-0.3		0.4		
Ethyl formate	109-94-4			0.2			
4-Ethylguaiacol	2785-89-9				0.8		
3-Ethyl-3-methyldecane	17312-66-2	1.2-1.7					
Eugenol	97-53-0	0-0.5	0.04-0.06		tr		4.5[c]
Eugenyl acetate	93-28-7				0.1		
Farnesol	4602-84-0				tr		
Geranial	141-27-5				tr		
Geraniol	106-24-1		tr				
Geranyl acetate	105-87-3	0-1.2					
Glycerin				3.0			
Guaiacol	90-05-1		tr		tr		
Guaiacyl cinnamate			tr[a]				
α-Guaiene	3691-12-1	0-0.8					
Hexanol	111-27-3				tr		
α-Humulene	6753-98-6		tr-0.03		tr		
Hydrocinnamaldehyde	104-53-0	0-0.8	0.7-0.8[a]		tr	0.2	
Hydrocinnamic acid	501-52-0		0.2-0.5				
Isoamyl benzoate	94-46-2	2.2-3.6					
Isoborneol	124-76-5		0-0.2				
Isoeugenol	97-54-1				tr		
(*Z*)-Isoeugenol	5912-86-7		0.1-0.3				
Isoledene	95910-36-4	0-0.9					
Isopropyl acetate	108-21-4			0.1			
Limonene	138-86-3		0.1-0.2				
Linalool	78-70-6	0-1.3	0.1-0.2		tr		
2-Methoxybenzaldehyde	135-02-4		0-0.1	0.7		0.6	
o-Methoxycinnamaldehyde (2-)	1504-74-1	<0.1-0.9		25.4		10.5	
(*E*)-2-Methoxycinnamaldehyde (*o*-)	60125-24-8		8.4-10.5				10.5[b]
3-Methoxy-1,2-propanediol	623-39-2			29.3			
1-Methoxy-2-propanol	107-98-2			0.07			
Methyl alaninate	10065-72-2		tr-0.05[a]				
Methyl benzoate	93-58-3				tr		
2-Methylbenzofuran	4265-25-2					0.2	
α-Muurolene	10208-80-7	0-0.6					
γ-Muurolene	30021-74-0		tr				
α-Muurolol	104245-48-9		0-0.08				
Myrcene	123-35-3		0.02-0.03				
Nerolidol	7212-44-4					0.2	
2-Nitroethanol	625-48-9			0.1			

Table 5.14.4 Constituents identified in cassia leaf oils (*continued*)

Constituent	CAS	Percentage and range					
		A (4)	B (13)	C (17)	D (20)	E (19)	F
Nonanal	124-19-6				0.1		
Nonanoic acid	12-05-0		tr-0.1				
Octanoic acid	124-07-2		tr				
Patchoulene	1405-16-9		0.06-0.07[a]				
α-Phellandrene	99-83-2		0.01-0.06				
2-Phenethyl alcohol	60-12-8		0.1-0.3	1.3	0.2	0.6	
Phenylethyl acetate	93-92-5		tr-1.6		0.2	0.7	
3-Phenylpropyl acetate	122-72-5	0-0.5	0.2-0.4				
α-Pinene	80-56-8		0.05-0.4		tr	0.1	
β-Pinene	127-91-3	0-1.2	0.04-0.2	tr			
Salicylaldehyde	90-02-8		0.05-0.4			0.2	
Salicylic acid	69-72-7		tr-0.1				
Styrene	100-42-5	1.1-3.9					
Terpinen-4-ol	562-74-3				0.1		
1-Terpineol	586-82-3						0.2[c]
α-Terpineol	98-55-5	0-0.8	tr-0.1		0.1		4.2[c]
(*E*)-β-Terpineol	7299-41-4						0.4[c]
(*Z*)-β-Terpineol	7299-40-3						0.8[c]
γ-Terpineol	586-81-2						2.2[c]
Terpinolene	586-62-9		tr[a]		tr		
α-Thujene	2867-05-2	0.8-2.9					
2,2,4-Trimethyl-1,3-pentane-diol	144-19-4	0-1.1					
Undecanoic acid	112-37-8		0-0.05				
Vanillin	121-33-5		tr[a]				
2-Vinylphenol	695-84-1				tr		

A five lab-distilled oils from leaves of *C. cassia* of various ages obtained from branches between 1 and 4 years old; data were means of three determinations of each sample; lowest and highest concentrations given (ref. 4)
B unknown number of cassia leaf oils from China; lowest and highest concentrations given (ref. 13)
C one lab-hydrodistilled leaf oil from China (ref. 17)
D one leaf oil from Australia (ref. 20)
E one leaf oil from China (cited in ref. 19, page 171)
F data from other studies (indicated with superscript letters); highest concentrations found in any study reviewed here given; when two or more oils were investigated, only the highest concentrations are mentioned, unless indicated otherwise

[a] incorrect identification (ref. 12); [b] one commercial oil from leaves and branches of *C. cassia* (ref. 29); [c] one lab-hydrodistilled leaf oil from Thailand (ref. 33)

tr: trace (in column C: <0.01; in column D: <0.1)

CONTACT ALLERGY/ALLERGIC CONTACT DERMATITIS

No reports on contact allergy to cassia leaf oil, specifically mentioned to be obtained from the leaves of *Cinnamomum cassia*, have been found. Literature on contact allergy to/ allergic contact dermatitis from 'cassia oil' (botanical source not specified) is discussed in Chapter 5.13 Cassia bark oil.

Table 5.15.5 Major constituents of cassia oils

Constituent	CAS	Percentage and range					
		A (4)	B (13)	C (17)	D (20)	E (19)	F
(*E*)-Cinnamaldehyde	14371-10-9	57.9-66.5	64.1-68.3	30.4	77.2	74.1	77.9[b]; 78.4[c]
2-Methoxycinnamaldehyde	1504-74-1	<0.1-0.9		25.4		10.5	
(*E*)-Cinnamyl acetate	21040-45-9		4.5-12.5				9.2[c]
(*E*)-2-Methoxycinnamaldehyde	60125-24-8		8.4-10.5				10.5[b]
Cinnamyl acetate	103-54-8	0.4-0.8			3.6	6.6	

LEGEND: SEE UNDER TABLE 5.14.4

LITERATURE

1 Gruenwald J, Freder J, Armbruester N. Cinnamon and health. Crit Rev Food Sc Nutrit 2010;50:822-834

2 Li Y-Q, Kong D-X, Wu H. Analysis and evaluation of essential oil components of cinnamon barks using GC–MS and FTIR spectroscopy. Ind Crops Prod 2013;41:269-278

3 Chang C-T, Chang W-L, Hsu J-C, Shih Y, Chou S-T. Chemical composition and tyrosinase inhibitory activity of Cinnamomum cassia essential oil. Botanical Studies 2013;54:10 (7 pages). http://www.as-botanicalstudies.com/content/54/1/10. Data also presented in ref. 34

4 Li Y-Q, Kong D-X, Huang R-S, Liang H-L, Xua C-G, Wu H. Variations in essential oil yields and compositions of Cinnamomum cassia leaves at different developmental stages. Ind Crops Prod 2013;47:92-101

5 Dong Y, Lu N, Cole RB. Analysis of the volatile organic compounds in Cinnamomum cassia bark by direct sample introduction thermal desorption gas chromatography–mass spectrometry. J Essent Oil Res 2013;25:458-463

6 Jiang Z, Jiang H, Xie P. Antifungal activities against Sclerotinia sclerotiorum by Cinnamomum cassia oil and its main components. J Essent Oil Res 2013;25:444-451

7 Geng SL, Cui ZX, Huang XC, Chen YF, Xu D, Xiong P. Variations in essential oil yield and composition during Cinnamomum cassia bark growth. Ind Crops Prod 2011;33:248-252

8 Islam R, Islam Khan R, Al-Reza SM, Jeong YT, Song CH, Halequzzamanc M. Chemical composition and insecticidal properties of Cinnamomum aromaticum (Nees) essential oil against the stored product beetle Callosobruchus maculatus (F.). J Sci Food Agric 2009;89:1241-1246

9 Choi J, Lee K-T, Ka H, Jung W-T, Jung H-J, Park H-J. Constituents of the essential oil of the Cinnamomum cassia stem bark and the biological properties. Arch Pharm Res 2001;24:418-423

10 ter Heide R. Qualitative analysis of the essential oil of cassia (Cinnamomum cassia Blume). J Agr Food Chem 1972;20:747-751

11 Zhu M-S, Liu S, Lao R-J, Bu Y-L. GC/MS detection of synthetic cinnamic aldehyde added to cassia oil. Acta Pharm Sinica 1996;31:461-465. Data cited in ref. 12.

12 Lawrence BM. Essential oils 2001-2004. Carol Stream, USA: Allured Publishing Corporation, 2006:60-62

13 Li Z-Q, Luo L, Huang R, Xia Y-Q. Chemical studies of cinnamon, true plants from Yunnan province. Yunnan Daxue, Xuebao Ziran Kexueban 1998;20 (Suppl.):377-379. Data cited in ref. 12

14 Li LL, Yuan WJ. Cinnamon oil analysis by GC and GC/MS. Fujian Fenxi Ceshi 1999;8:1121-1125. Data cited in ref. 12.

15 Ehlers D, Hilmerands S, Bartholomae S. Hochdruck-flüssig-chromatographische Untersuchung von Zimt-CO$_2$-Hochdruckextrakten im Vergleich mit Zimtölen. Z Lebensm Unter Forsch 1995;200:282-288

16 Lawrence BM. Progress in essential oils. Perfum Flavor 2012;37(April):56-62

17 Wang R, Wang R-J, Yang B. Extraction of essential oils from five cinnamon leaves and identification of their volatile compound compositions. Innov Food Sci Emerg Technol 2009;10:289-292

18 Shen Q, Chen FL, Luo JB. Comparison studies on chemical constituents of essential oil from Ramulus Cinnamomi and Cortex Cinnamomi by GC–MS. J Chin Med Mater 2002;25:257–258 (In Chinese with English abstract). Cited in ref. 32

19 Cinnamon and Cassia. PN Ravindran, K Nirmal Babu and M Shylaja, Eds. Boca Raton – London – New York – Washington: CRC Press, 2004

20 Senanayake UM. The nature, description and biosynthesis of volatiles in Cinnamomum spp. Ph.D. Thesis, University of New South Wales, Australia, 1977 (data cited in ref. 19, page 171).

21 Vernin C, Vernin G, Metzger J, Puigol I. La canelle, première partie. Analyse CPG/Sm Banuqe SPECMA d'huile essentielle de canelle de Ceylan et de Chine. Parfumes, Cosmetiques, Aromes 1990;93:85-90. Data cited in ref. 19.

22 Leach MJ, Kumar S. Cinnamon for diabetes mellitus. Cochrane Database of Systematic Reviews 2012, Issue 9. Art. No. CD007170. DOI: 10.1002/14651858.CD007170.pub2.

23 Murbach Teles Andrade BF, Nunes Barbosa L, da Silva Probst I, Fernandes Júnior A. Antimicrobial activity of essential oils. J Essent Oil Res 2014;26:34-40

24 Pawar VC, Thaker VS. In vitro efficacy of 75 essential oils against Aspergillus niger. Mycoses 2006;49:316-323. Data also presented in ref. 25

25 Pawar VC, Thaker VS. Evaluation of the anti-Fusarium oxysporum f. sp. cicer and anti-Alternaria porri effects of some essential oils. World J Microbiol Biotechnol 2007;23:1099-1106. Data also presented in ref. 24

26 He Z-D, Qiao C-F, Han Q-B, Cheng C-L, Xu H-X, Jiang R-W, et al. Authentication and quantitative analysis on the chemical profile of cassia bark (Cortex cinnamomi) by high-pressure liquid chromatography. J Agric Food Chem 2005;53:2424-2428

27 Rossi P-G, Berti L, Panighi J, Luciani A, Maury J, Muselli A, et al. Antibacterial action of essential oils from Corsica. J Essent Oil Res 2007;19:176-182

28 Sheng L, Zhu MJ. Inhibitory effect of Cinnamomum cassia oil on non-O157 Shiga toxin-producing Escherichia coli. Food Control 2014;46:374-381

29 Mith H, Duré R, Delcenserie V, Zhiri A, Daube G, Clinquart A. Antimicrobial activities of commercial essential oils and their components against foodborne pathogens and food spoilage bacteria. Food Sci Nutrit 2014;2:403-416

30 Kocevski D, Du M, Kan J, Jing C, Lacanin I, Pavlovic H. Antifungal effect of Allium tuberosum, Cinnamomum cassia, and Pogostemon cablin essential oils and their components against population of Aspergillus species. J Food Sci 2013;78:M731-M737

31 Huang DF, Xu J-G, Liu J-X, Zhang H, Hu QP. Chemical constituents, antibacterial activity and mechanism of action of the essential oil from *Cinnamomum cassia* bark against four food-related bacteria. Microbiol 2014;83:357-365

32 Deng X, Liao Q, Xu X, Yao M, Zhou Y, Lin M, Zhang P, Xie Z. Analysis of essential oils from cassia bark and cassia twig samples by GC-MS combined with multivariate data analysis. Food Anal Methods 2014;7:1840-1847

33 Pannee C, Chandhanee I, Wacharee L. Antiinflammatory effects of essential oil from the leaves of *Cinnamomum cassia* and cinnamaldehyde on lipopolysaccharide-stimulated J774A.1 cells. J Adv Pharm Technol Res 2014;5(4):164-170

34 Chou S-T, Chang W-L, Chang C-T, Hsu S-L, Lin Y-C, Shih Y. Cinnamomum cassia essential oil inhibits α-MSH-induced melanin production and oxidative stress in murine B16 melanoma cells. Int J Mol Sci 2013;14:19186-19201. Data also presented in ref. 3

35 Rhind JP. Essential oils. A handbook for aromatherapy practice, 2nd Edition. London: Singing Dragon, 2012

36 Lawless J. The encyclopedia of essential oils, 2nd Edition. London: Harper Thorsons, 2014

37 Rudzki E, Grzywa Z, Bruo WS. Sensitivity to 35 essential oils. Contact Dermatitis 1976;2:196-200

38 Rudzki E, Grzywa Z. Allergy to perfume mixture. Contact Dermatitis 1986;15:115-116

39 Magnusson B, Wilkinson DS. Cinnamic aldehyde in toothpaste. 1. Clinical aspects and patch tests. Contact Dermatitis 1975;1:70-76

40 Silvers SH. Stomatitis and dermatitis venenata with purpura, resulting from oil of cloves and oil of cassia. Dental Items of Interest 1939;61:649-651. Data cited in ref. 39

41 Laubach JL, Malkenson FD, Ringrose EJ. Cheilitis caused by cinnamon (cassia) oil in toothpaste. JAMA 1953;152:404-405. Data cited in ref. 39

42 Rudzki E, Grzywa Z. The value of a mixture of cassia and citronella oils for detection of hypersensitivity to essential oils. Dermatosen 1985;33:59-62

43 Drake TE, Maibach HI. Allergic contact dermatitis and stomatitis caused by a cinnamic aldehyde-flavored toothpaste. Arch Dermatol 1976;112:202-203

44 Rudzki E, Grzywa Z. Dermatitis from propolis. Contact Dermatitis 1983;9:40-45

Chapter 5.15 CEDARWOOD OIL, ATLAS

There are five major cedarwood essential oils: cedarwood oil Atlas, cedarwood oil Himalaya, cedarwood oil Texas, cedarwood oil Virginia and cedarwood oil China. These are obtained from different botanical species, and as a consequence, their chemical compositions differ both qualitatively and quantitatively. Unfortunately, in non-botanical literature, usually the term 'cedarwood oil' is used, lacking information on the botanical origin.

DEFINITION
Cedarwood oil, Atlas, is the essential oil obtained from the wood of the Atlantic cedar (Atlas cedar), *Cedrus atlantica* (Endl.) G. Manetti ex Carrière.

INCI NOMENCLATURE
Description/definition: Cedrus atlantica wood oil is an essential oil obtained from the wood of the tree, *Cedrus atlantica*, Pinaceae
INCI name EU: Cedrus atlantica wood oil (perfuming name, not an INCI name proper)
INCI name USA: Not in the Personal Care Products Council Ingredient Database
CAS registry number (s): 92201-55-3
EINECS number(s): 295-985-9

ISO (INTERNATIONAL ORGANIZATION FOR STANDARDIZATION) STANDARD
Currently there is no ISO standard for cedarwood oil, Atlas.

THE PLANT, THE OIL, AND THEIR USES
Cedrus atlantica is an evergreen coniferous tree which can grow up to 40 meters high and 2 meters in diameter. It is native to the Atlas Mountains in Morocco and Algeria (GRIN Taxonomy for Plants, 9). Atlas cedar is the principal species in Moroccan forests used for production of timber. It is also cultivated in southern France for this purpose.

The machining waste, sawdust (estimated to be 8-30%) is the ground material for the production of Atlas cedarwood essential oil, which is used in the perfumery and flavor industry and for its medicinal properties, especially for their content of himachalenes (1). *C. atlantica* wood oil is said to possess antiseptic, diuretic, astringent, fungicidal, sedative, antimicrobial and stimulant properties (9). It is also used in aromatherapy (16,17).

CHEMICAL COMPOSITION
Cedarwood oil, Atlas is a pale to dark yellow, clear mobile liquid which has a balsamic, soft woody odor reminding of sandalwood. The yield of essential oil from the wood of *Cedrus atlantica* (Endl.) G. Manetti ex Carrière generally varies from 0.2 to 0.5%. The main producing countries of this oil are Morocco and Algeria.

Table 5.15.1 Constituents identified in cedarwood oils Atlas

Constituent	CAS	Percentage and range					
		A	B (4)	C (1)	D (7)	E (3)	F
Abietadiene	36312-33-1				1.5		
Abietatriene	19407-28-4				1.3		
Acetone	67-64-1						+[e]
4-Acetyl-1-methyl-cyclohexane				0-0.6			
Acorenone	5956-05-8			0-0.3		0.1	
allo-Aromadendrene	25246-27-9		0.6-0.7				
Aromadendrene	489-39-4				3.3		
(E)-α-Atlantone	26294-59-7	0.03-2.7	1.6-3.1	5.2-29.5	1.6	11.2	0.3-2.0[i]; 10.8[f]; 28.8[h]; 30.8[g]
(Z)-α-Atlantone	56192-70-2	0.05-1.2	0.4-0.6	1.0-5.9	0.4	2.5	0-0.2[i]; 0.2[f]; 0.6[m]; 5.2[g,h]
β-Atlantone	38331-79-2		0.2-0.7				
γ-Atlantone	532-66-1						1.6[f]
(E)-γ-Atlantone	108549-47-9	0.4-1.8	1.4-2.8[d]	1.2-4.2	0.1	1.9	2.3[m]
(Z)-γ-Atlantone	108549-48-0	0.4-1.9	0.8-1.4				
Benzyl benzoate	120-51-4			0-0.5		0.3	
(Z)-trans-α-Bergamotol	88034-74-6			0-0.2		1.8	0.1[h]; 1.1[g]
Cadalene	483-78-3		tr	0-1.4	0.9	1.2	
δ-Cadinene	483-76-1	0.7-2.2	1.8-2.1	0.5-2.6	0.8	2.0	1.3[h]; 1.6[g]
Calacorene	38599-17-6		0.6-0.8				
α-Calacorene	21391-99-1	0.4-0.7		0.5-1.6	0.1	1.1	1.0[h]; 1.6[g]
β-Calacorene	50277-34-4					0.3	
Calamenene	483-77-2		0.1-0.6				
cis-Calamenene	72937-55-4				0.1		
Calamenen-1-ol			0.3				

Table 5.15.1 Constituents identified in cedarwood oils Atlas *(continued)*

Constituent	CAS	Percentage and range					
		A	B (4)	C (1)	D (7)	E (3)	F
Camphene	79-92-5		tr-0.1		0.4		0.1-0.5[i]
α-Campholenal	4501-58-0				0.2		
Carotol	465-28-1			0-0.8			1.2[g]
(E)-Carveol	1197-07-5				tr		
Carvone	99-49-0				0.1		
Caryophyllene			0.3-0.6				
β-Caryophyllene	87-44-5				6.0		0.4-0.6[i]
α-Caryophyllene alcohol	4586-22-5						0.4[e]
Caryophyllene oxide	1139-30-6				tr		0.2-0.8[i]
Cedranone	13567-40-3			0.7-2.3		1.0	1.6[g,h]
α-Cedrene	469-61-4	0.5-2.3	0.1-0.2				15.7[k]; 19.8[j]
Cedrol (α-)	77-53-2					0.6	3.8[h]; 22.1[j,n]; 22.4[k,n]
epi-Cedrol	19903-73-2						3.3[g]
Cedroxyde	13786-79-3						1.8[g]; 2.0[h]
1,4-Cineole	470-67-7		tr				
1,8-Cineole	470-82-6	0.02-0.2			0.3		
α-Copaene	3856-25-5		tr-0.1		0.3		
Cubenol	21284-22-0		0.5-0.8				2.5[g]
1-epi-Cubenol	81939-29-9			1.1-2.5		2.2	2.7[h]
Cuparene	16982-00-6			0-0.6		0.3	
Curcumene			tr-0.2				
ar-Curcumene (α-)	644-30-4		0.6-1.0				
γ-Curcumene	28976-68-3		0.4-0.5	0.7-1.5	0.4	1.3	0.8[h]
o-Cymene	527-84-4				0.1		
p-Cymene	99-87-6		0-0.1		0.1		
Decane	124-18-5		tr				
Dehydroabietal	13601-88-2				0.8		
Dehydroaroma-dendrene	698388-95-3		tr-0.1				
Dehydro-β-atlantone							+[e]
Dehydrohimachalene							1.1[f]
α-Dehydro-ar-hima-chalene	78204-62-3			0.4-1.2	0.2	0.8	
γ-Dehydro-ar-hima-chalene	51766-65-5	0.2-3.0		0-0.9	0.6	0.08	
8,9-Dehydroiso-longifolene			0.6-0.9	0-0.5			
8,9-Dehydroneoiso-longifolene						0.5	
Deodarone	41943-81-1	1.1-2.3	1.4-2.8[d]	1.2-7.7		4.2	4.4[g,h]
10,11-Dihydroatlan-tone							+[e]
Dihydrocalamenenol			tr-0.2				
4,4-Dimethyl-3-(3-methyl-2-buten-1-yliden)-2-methylide-nebicyclo[4.1.0]heptane			0.1-1.4				
Dodecane	112-40-3		tr				
Elemicin	487-11-6			0-0.7		0.4	
6,7-Epoxy-β-hima-chalene							+[e]
epi-Epoxy-β-himacha-lene							2.0[e]
(E)-β-Farnesene	18794-84-8				1.4		0.8-3.1[i]
(Z)-β-Farnesene	28973-97-9			0-0.2		0.2	
Fenchone	1195-79-5				0.4		
Furfuryl octanoate	39252-03-4						1.9[g]
α-Gurjunene	489-40-7	0.5-0.7	1.4-3.9[c]				
Himachala-2,4-diene				0.3-0.7		0.3	0.3-1.6[i]
Himachalene			0.1-0.3				

Table 5.15.1 Constituents identified in cedarwood oils Atlas

Constituent	CAS	Percentage and range					
		A	B (4)	C (1)	D (7)	E (3)	F
α-Himachalene	3853-83-6	15.5-19.3	14.4-16.9	7.4-16.4	2.1	10.9	0.9-6.0[i]; 3.1[f]; 5.7[h]; 8.0[m]; 15.8[g]
α-ar-Himachalene		0.8-1.0	0.9-1.5				
β-Himachalene	1461-03-6	40.0-52.0	41.3-46.0	23.4-40.4	3.7	33.8	1.7-7.6[i]; 7[m]; 14.6[h]; 21.2[f]; 39.7[g]
γ-Himachalene	53111-25-4	9.5-12.0	10.1-11.2	5.1-9.7	2.3	6.9	0.9-5.5[i]; 9.6[g]; 10.8[f]; 15.8[m]
γ-ar-Himachalene			0.7-1.2				
α-Himachalene epoxide							0.9[m]
β-Himachalene oxide	31560-66-4	0.4-0.9	0.3-0.6	0-1.6		0.2	8.7[f]
Himachalol	1891-45-8	0.4-3.1	0.2-0.3	1.7-3.7	28.1	1.9	6.5[g]; 7.1[h]; 8.3-48.8[i]; 46.3[m]
allo-Himachalol	19435-77-9	0.5-0.7					0.9-3.6[i]; 4.7[m]
Hinesol acetate	88494-77-3						1.3[g,h]
α-Humulene	6753-98-6	0.02-0.6			1.3		0.1-0.3[i]
14-Hydroxy-δ-cadinene	153408-92-5				1.0		
14-Hydroxy-α-muurolene	105661-29-8			0.2-1.0		0.1	
α-Ionone	127-41-3						0.1[e]
(E)-β-Ionone	79-77-6			0-0.4		0.3	
Isocedranol	13567-45-8			1.2-3.1		2.8	3.5[h]
5-Isocedrol							1.5[g,h]
Isoledene	95910-36-4			0.1-0.4			
Isolongifolene	1135-66-6						0.3[m]
3-Isothujopsanone	25966-81-8			0-0.6		0.5	
Khusimol	16223-63-5			0-0.6		0.6	1.1[g,h]
Limonene	138-86-3	0-0.05	0.1-0.2		0.5		0.3-2.0[i]; 6.0[l]
cis-Linalool oxide	11063-77-7				0.1		
Longiborneol	465-24-7			0-1.0		0.7	0.1-0.6[i]; 1.2[h]; 1.4[m]
Longifolene	475-20-7			0-0.5	0.3	0.4	0.4[i]
α-Longipinene	5989-08-2	0-0.08					0.1-0.3[i]
Manool	596-85-0						1.2[m]
Manoyl oxide	596-84-9				0.8		
1-Methyl-4-acetyl-cyclohex-1-ene							+[e]
4-(4-Methylcyclohex-3-enyl)pent-3-en-2-one	94390-70-2						0.8[e]
Methyl dehydro-abietate	1235-74-1				0.9		
6-Methyl-γ-ionone	79-68-5						0.7[g]
α-Muurolene	10208-80-7						4.6[j]
Myrcene	123-35-3				0.1		0.3-0.6[i]; 4.8[l]
Myrtenal	564-94-3				0.1		
Myrtenol	515-00-4				0.8		
(E)-Nerolidol	40716-66-3						0-0.1[i]
Oxidohimachalene	64825-84-9			0.4-1.0	1.2	0.8	1.9[h]
Pentadecane	629-62-9		tr				
β-Phellandrene	555-10-2						0-0.2[i]
Pimaradiene	1686-61-9				0.1		
α-Pinene	80-56-8	0.01-0.3	0.1-0.3		5.6		6.0-55.0[i]; 63.1[l]
β-Pinene	127-91-3		tr-0.2		0.7		1.3-3.6[i]; 4.5[l]
Pinocamphone (trans-)	547-60-4				0.1		
trans-Pinocarveol	1674-08-4				tr		
Podocephalol	66656-01-7		0.1-0.5[a]				
trans-Rose oxide	5258-11-7						0.7[g,h]
Sabinene	3387-41-5				tr		
β-Sesquiphellandrene	20307-83-9						1.1[g,h]
β-Spathulene	53526-64-0		tr-0.1				
Spathulenol	6750-60-3				0.7		
α-Terpineol	98-55-5	0-6.0			0.8		0.3-1.2[i]
(Z)-β-Terpineol	7299-40-3				1.7		
Tetradecane	629-59-4		tr	0-0.8		0.7	

Table 5.15.1 Constituents identified in cedarwood oils Atlas

Constituent	CAS	Percentage and range					
		A	B (4)	C (1)	D (7)	E (3)	F
Thuj-3-en-10-al	57129-54-1				tr		
α-Thujene	2867-05-2				0.3		
Thujopsadiene	24048-40-6			0-0.4		0.2	
Thujopsene	470-40-6						27.8[j]
Undecane	1120-21-4		tr-0.1				
Veratrole	91-16-7		1.4-3.9[c]				
Verbenene	4080-46-0				0.1		
Verbenone	80-57-9				0.1		
β-Vetivenene	27840-40-0			0-1.4	0.2	0.9	
Widdrol	6892-80-4						3.8[j]
α-Ylangene	14912-44-8				0.2		

A twenty-four cedarwood Atlas essential oil samples from Morocco and Algeria analyzed between 1998 and 2013; lowest and highest concentrations given (E. Schmidt, unpublished data)
B five samples of industrially produced steam-distilled oils from sawdust from Moroccan *C. atlantica* specimens; lowest and highest concentrations given (ref. 4)
C seven oils produced from trees harvested at different sites in Morocco; lowest and highest concentrations given (ref. 1)
D one lab-hydrodistilled wood oil from Tunisian *Cedrus atlantica* (ref. 7)
E one laboratory steam-distilled oil sample from Morocco (ref. 3)
F data from other studies (indicated with superscript letters); highest concentrations found in any study reviewed here given; when two or more oils were investigated, only the highest concentrations are mentioned, unless indicated otherwise

[a] incorrect identity based on GC elution order (ref. 2); [b] correct isomer not identified; [c] veratrole and α-gurjunene combined; [d] (*E*)-γ-atlantone and deodarone combined; [e] older literature (<1976) cited in ref. 2; [f] major components of some Algerian Atlas cedarwood oils (ref. 5); [g] one laboratory steam-distilled and one lab-hydrodistilled cedarwood oil Atlas from Morocco; (ref. 6); [h] one lab-hydrodistilled oil from Moroccan sawdust (ref. 8); [i] 48 lab-hydrodistilled wood oils from Corsican wild growing *C. atlantica*; there were two chemotypes, one dominated by α-pinene (21 samples), the other by himachalol (27 samples); average concentrations of both groups given (ref. 9); [j] one commercial *Cedrus atlantica* oil purchased from a Brazilian company with a very atypical composition with high concentrations of widdrene, α-cedrol and α-cedrene; very unlikely to be *Cedrus atlantica* oil, possibly mixture of China and Virginia cedarwood oils (ref. 13); [k] one commercial steam-distilled oil (probably adulterated) of *Cedrus atlantica* wood prepared in Korea with an atypical composition and α-cedrol and α-cedrene as main components (ref. 14); [l] one commercial oil sample from Corsica, France, of the α-pinene chemotype (ref. 12); [m] two lab-hydrodistilled oils from Lebanese *Cedrus atlantica* heartwood (ref. 15); [n] cedrol in this high concentration must be adulteration

tr: trace (column B: <0.05%; column D: <0.1%); + present in the oil investigated, but quantity not stated

Literature data (up to October 25, 2014) on the chemical composition of cedarwood oil Atlas and unpublished analytical data from one of us (E.S.) are shown in Table 5.15.1 in alphabetical order. In cedarwood oils Atlas from various origins, over 135 chemicals have been identified. About 53% of these were found in a single reviewed publication only. The major compounds found in cedarwood oils Atlas from different sources are shown in Table 5.15.2. They include (highest concentrations in any study given) β-himachalene (52.0%), himachalol (48.8%), (*E*)-α-atlantone (30.8%), α-himachalene (19.3%), γ-himachalene (15.8%), deodarone (7.7%) and (*Z*)-α-atlantone (5.9%).

Well-known ingredients of cedarwood oils Atlas that were present in high concentrations in one or two studies were α-pinene (63.1%, chemotype found only in Corsica, France), α-cedrene (19.8%) and β-himachalene oxide (8.7%). Uncommon or rare constituents of cedarwood oils Atlas found in high concentrations in single studies include thujopsene (27.8%) and cedrol (22.1% and 22.4%).

Commercial oils

The ten chemicals that had the highest maximum concentrations in 24 commercial cedarwood Atlas essential oil samples (concentration ranges provided) are the following: β-himachalene (40.0-52.0%), α-himachalene (α-*cis*-) (15.5-19.3%), γ-himachalene (9.5-12.0%), α-terpineol (0-6.0%), himachalol (0.4-3.1%), γ-dehydro-ar-himachalene (0.2-3.0%), (*E*)-α-atlantone (0.03-2.7%), deodarone (1.1-2.3%), α-cedrene (0.5-2.3%) and δ-cadinene (0.7-2.2%) (Erich Schmidt, unpublished analytical data).

Chemotypes

In most cedarwood Atlas oils, the himachalenes, (*E*)-α-atlantone and to a lesser degree himachalol are the main ingredients. In wild growing *C. atlantica* populations on the island of Corsica (France), however, there are two distinct chemotypes, one dominated by himachalol (average concentration in 27 samples 48.8±11.7%), the other one by α-pinene (average concentration in 21 samples

Table 5.15.2 Major constituents of cedarwood oils Atlas

Constituent	CAS	Percentage and range					
		A	B (4)	C (1)	D (7)	E (3)	F
β-Himachalene	1461-03-6	40.0-52.0	41.3-46.0	23.4-40.4	3.7	33.8	1.7-7.6[i]; 7[m]; 14.6[h]; 21.2[f]; 39.7[g]
Himachalol	1891-45-8	0.4-3.1	0.2-0.3	1.7-3.7	28.1	1.9	6.5[g]; 7.1[h]; 8.3-48.8[i]; 46.3[m]
(E)-α-Atlantone	26294-59-7	0.03-2.7	1.6-3.1	5.2-29.5	1.6	11.2	0.3-2.0[i]; 10.8[f]; 28.8[h]; 30.8[g]
α-Himachalene	3853-83-6	15.5-19.3	14.4-16.9	7.4-16.4	2.1	10.9	0.9-6.0[i]; 3.1[f]; 5.7[h]; 8.0[m]; 15.8[g]
γ-Himachalene	53111-25-4	9.5-12.0	10.1-11.2	5.1-9.7	2.3	6.9	0.9-5.5[i]; 9.6[g]; 10.8[f]; 15.8[m]
Deodarone	41943-81-1	1.1-2.3	1.4-2.8[d]	1.2-7.7	4.2		4.4[g,h]
(Z)-α-Atlantone	56192-70-2	0.05-1.2	0.4-0.6	1.0-5.9	0.4	2.5	0-0.2[i]; 0.2[f]; 0.6[m]; 5.2[g,h]

LEGEND: SEE UNDER TABLE 5.15.1

55.0±15.3%) (9). The high concentration of α-pinene is highly unusual for cedarwood Atlas *wood* oil, but it is the dominant chemical in most *needle* oil compositions (18-63%) (10,11). A commercial oil from Corsica also had a very high (63%) α-pinene content (ref. 12).

CONTACT ALLERGY/ALLERGIC CONTACT DERMATITIS

Cedarwood oil, unspecified or partly specified

General
Contact allergy to/allergic contact dermatitis from cedarwood oil has been reported in several publications. In groups of consecutive patients suspected of contact dermatitis, prevalence rates of up to 1.5% positive patch test reactions have been observed, but relevance data are lacking. In most publications, the botanical origin of the cedarwood oil was not mentioned. In two publications (23,25, both parts of the same study) the test substance was prepared from 50% Moroccan cedarwood oil (presumably cedarwood oil, Atlas, *ex Cedrus atlantica*) and 50% Chinese cedarwood oil (presumably *ex Cupressus funebris*). One large study from Germany, Austria and Switzerland used Virginian cedarwood oil (*ex Juniperus virginiana*) for patch testing. Reports of contact allergy to cedarwood specified to be obtained from *Juniperus ashei* J. Buchholz (cedarwood oil, Texas) have not been found.

Testing in groups of patients
The results of patch tests with cedarwood oil in routine testing (consecutive patients suspected of contact dermatitis) and in groups of selected patients are shown in Table 5.15.3. In routine testing, rates of positive reactions ranged from 0.6% to 1.5%, whereas between 0.7% and 5.9% of patients in selected groups had positive patch tests.

Table 5.15.3 Results of testing groups of patients with cedarwood oil

Years and Country	Test conc. & vehicle	Number of patients		Selection of patients (S); Relevance (R); Comments (C)	Ref.
		tested	positive (%)		
Routine testing					
1999-2000 Denmark	10% pet.	318	3 (0.9%)	R: not specified; C: this study was part of the international study mentioned below (23); C: the test substance was prepared from 50% Moroccan and 50% Chinese cedarwood oil	25
1998-2000 six European countries	10% pet.	1,606	10 (0.6%)	R: not specified for individual oils/chemicals; C: the test substance was prepared from 50% Moroccan and 50% Chinese cedarwood oil	23
<1976 Poland	2% pet.	200	3 (1.5%)	R: not stated	20
Testing in groups of selected patients					
2000-2008 IVDK	10% pet.	6,223	52 (0.8%)	S: patients with dermatitis suspected of causal exposure to fragrances; R: not stated; C: the cedarwood oil was prepared from *Juniperus virginiana* (cedarwood oil Virginia)	22
1997-2000 Austria	2% pet.	747	5 (0.7%)	S: patients suspected of fragrance allergy; R: not stated	18
<1976 Poland	2% pet.	51	3 (5.9%)	S: patients allergic to *Myroxylon pereirae* resin and/or turpentine and/or wood tar and/or colophony; R: not stated	20

IVDK Information Network of Departments of Dermatology, Germany, Switzerland, Austria (www.ivdk.org). pet.: petrolatum

Case reports

An aromatherapist had occupational contact dermatitis with allergies to multiple essential oils used at work, including cedarwood oil. The patient also reacted to geraniol, α-pinene, caryophyllene, the fragrance mix and various other fragrance materials; caryophyllene was demonstrated by GC-MS in cedarwood oil (19), but in none of the cedarwood oils is caryophyllene an important component. One case of allergic contact dermatitis was caused by cedarwood oil present as a contaminant in immersion oil for dermatoscopy (21). One patient developed allergic contact dermatitis from cedarwood oil used as the vehicle for applying a black henna temporary tattoo (24).

Cedarwood oil, Atlas

General

Possible contact allergy to cedarwood oil, Atlas, has been reported in two publications, where patients were tested with a mixture of cedarwood oil, Morocco (presumably a synonym of Atlas) and Chinese cedarwood oil. This is discussed in Table 5.15.3.

LITERATURE

1   Aberchane M, Fechtal M, Chaouch A. Analysis of Moroccan Atlas cedarwood oil (*Cedrus atlantica* Manetti). J Essent Oil Res 2004;16:542-547

2   Lawrence BM. Atlas cedarwood oil. Perfum Flav 2009;34 (November):57-59

3   Aberchane M, Satrani B, Fechtal M, Chaouch A. Effet de l'infection du bois de cèdre de l'Atlas par *Trametes pini* et *Ungulina officinalis* sur la composition chimique et l'activité antibactérienne et antifongique des huiles essentielles. Acta Bot Gall 2003;150:223-229

4   Chalchat J-C, Garry R-Ph, Michet A, Benjilali B. Essential oil components in sawdust of *Cedrus atlantica* from Morocco. J Essent Oil Res 1994;6:323-325

5   Dahoun A, Derriche R, Belabbes R. Influence du mode d' extraction et de la composition de l'huile essentielle et de la concrète du bois de cèdre de l'Atlas Algérien. Rivista Ital EPPOS 1993;10:29-32. Data cited in ref. 2

6   Aberchane M, Fechtal M, Chaouch A, Bouayoune T. Influence de la durée et de la technique d'extraction sur le rendement et la qualité des huiles essentielles du Cèdre de l'Atlas (*Cedrus atlantica* Manetti). Ann Rech For Maroc 2001;34:110-118. Data cited in ref. 2.

7   Boudarene L, Rahim L, Baaliouamer A, Meklati BV. Analysis of Algerian essential oils from twigs, needles and wood of *Cedrus atlantica* G. Manetti by GC/MS. J Essent Oil Res 2004;16:531-534

8   Satrani B, Aberchane M, Farah A, Chaouch A, Talbi M. Composition chimique et activité antimicrobienne des huiles essentielles extraites par hydrodistillation fractionnée du bois de *Cedrus atlantica* Manetti. Acta Bot Gall 2006;153:97-104

9   Paoli M, Nam A-M, Castola V, Casanova J, Bighelli A. Chemical variability of the wood essential oil of *Cedrus atlantica* Manetti from Corsica. Chem Biodivers 2011;8:344-351

10 Lawrence BM. Progress in essential oils. Atlas cedar needle oil. Perfum Flav 2011;36 (December):58-59

11 Derwich E, Benziane Z, Boukir A. Chemical composition and in vitro antibacterial activity of the essential oil of *Cedrus atlantica*. Int J Agric Biol 2010;12:381-385

12 Rossi PG, Berti L, Panighi J, Luciani A, Maury J, Muselli A, et al. Antibacterial action of essential oils from Corsica. J Essent Oil Res 2007;19:176-182

13 Murbach Teles Andrade BF, Nunes Barbosa L, da Silva Probst I, Fernandes Júnior A. Antimicrobial activity of essential oils. J Essent Oil Res 2014;26:34-40

14 Shin S. Anti-*Aspergillus* activities of plant essential oils and their combination effects with ketoconazole or amphotericin B. Arch Pharm Res 2003;26:389-393

15 Saab AM, Harb FY, Koenig WA. Essential oils components in heartwood of *Cedrus libani* and *Cedrus atlantica* from Lebanon. Minerva Biotecnologica 2005;17:159-161

16 Rhind JP. Essential oils. A handbook for aromatherapy practice, 2nd Edition. London: Singing Dragon, 2012

17 Lawless J. The encyclopedia of essential oils, 2nd Edition. London: Harper Thorsons, 2014

18 Wöhrl S, Hemmer W, Focke M, Götz M, Jarisch R. The significance of fragrance mix, balsam of Peru, colophonium and propolis as screening tools in the detection of fragrance allergy. Br J Dermatol 2001;145:268-273

19 Dharmagunawardena B, Takwale A, Sanders KJ, Cannan S, Roger A, Ilchyshyn A. Gas chromatography: an investigative tool in multiple allergies to essential oils. Contact Dermatitis 2002;47:288-292

20 Rudzki E, Grzywa Z, Bruo WS. Sensitivity to 35 essential oils. Contact Dermatitis 1976;2:196-200

21 Franz H, Frank R, Rytter M, Haustein UF. Allergic contact dermatitis due to cedarwood oil after dermatoscopy. Contact Dermatitis 1998;38:182-183

22 Uter W, Schmidt E, Geier J, Lessmann H, Schnuch A, Frosch P. Contact allergy to essential oils: current patch test results (2000–2008) from the Information Network of Departments of Dermatology (IVDK). Contact Dermatitis 2010;63:277-283

23 Frosch PJ, Johansen JD, Menné T, Pirker C, Rastogi SC, Andersen KE, et al. Further important sensitizers in patients sensitive to fragrances. II. Reactivity to essential oils. Contact Dermatitis 2002;47:279-287

24 Temesvári E, Podányi B, Pónyai G, Németh I. Fragrance sensitization caused by temporary henna tattoo. Contact Dermatitis 2002;47:240

25 Paulsen E, Andersen KE. Colophonium and Compositae mix as markers of fragrance allergy: Cross-reactivity between fragrance terpenes, colophonium and Compositae plant extracts. Contact Dermatitis 2005;53:285-291

Chapter 5.16 CEDARWOOD OIL, CHINA

There are five major cedarwood essential oils: cedarwood oil Atlas, cedarwood oil Himalaya, cedarwood oil Texas, cedarwood oil Virginia and cedarwood oil China. These are obtained from different botanical species, and as a consequence, their chemical compositions differ both qualitatively and quantitatively. Unfortunately, in non-botanical literature, usually the term 'cedarwood oil' is used, lacking information on the botanical origin.

DEFINITION

Cedarwood oil, China, (essential oil of cedarwood, Chinese type), is the essential oil obtained from the wood of the Chinese weeping cypress (mourning cypress), *Cupressus funebris* (Endl.).

INCI NOMENCLATURE

Description/definition: Cupressus funebris wood oil is an essential oil obtained from the twigs of the cypress, *Cupressus funebris*, Cupressaceae
INCI name EU: Cupressus funebris wood oil
INCI name USA: Not in the Personal Care Products Council Ingredients Database
Other names: Chinese cedarwood oil
CAS registry number (s): 85085-29-6
EINECS number(s): 285-360-9

ISO (INTERNATIONAL ORGANIZATION FOR STANDARDIZATION) STANDARD

ISO number: 9843
ISO name: Essential oil of cedarwood, Chinese type
Botanical origin: *Chamaecyparis funebris* (Endl.) Franco (synonym: *Cupressus funebris* Endl.)
Parts of plant used: Wood
ISO values: ISO values (minimum and maximum concentrations) are shown in Table 5.16.1.

Table 5.16.1 ISO values (%) for cedarwood oil, Chinese type[a]

Compound	CAS	Minimum	Maximum
Thujopsadiene	24048-40-6	18.0	39.0
α-Cedrene + β-funebrene	469-61-4	13.0	29.0
	79120-98-2		
Cedrol	77-53-2	10.0	20.0
β-Cedrene	546-28-1	4.0	11.0
Cuparene	16982-00-6	1.0	3.0
Widdrol	6892-80-4	0.5	3.0

[a] ISO 9843 Essential oil of cedarwood, Chinese type ©ISO 2002; Geneva, Switzerland, www.iso.org

The name *Chamaecyparis funebris* (Endl.) Franco, used by ISO, is a synonym for the accepted name *Cupressus funebris* Endl. (2, GRIN Taxonomy for Plants, The Plant List).

THE PLANT, THE OIL, AND THEIR USES

Cupressus funebris Endl. (synonyms: *Chamaecyparis funebris* (Endl.) Franco; Chinese weeping cypress) is an evergreen coniferous tree up to 35 meters tall and with a trunk up to 2 meters in diameter, which is native to China and Vietnam and is cultivated widely in south China (the Gymnosperm Database, http://www.conifers.org). The general source of Chinese cedarwood oil is from stumps after logging. Workers dig the stumps, cut off the cortex (or sapwood, that contains little or no oil), and distill the heartwood. However, there are limited natural stands of mature *C. funebris*, and it has been suggested that some commercial oils may be produced from a mixture of *C. funebris* and *Platycladus orientalis*, and possibly *C. funebris* and other Cupressaceae species (2).

Chinese cedarwood oil and wood are used to prepare incense in China. In medicine, specially thickened varieties of cedarwood oils are used as immersion oil and for clarification in microscopy, because they have the same refraction index as glass (5). It is not used in aromatherapy (10).

CHEMICAL COMPOSITION

Cedarwood oil China is an almost colorless to light yellow, clear mobile liquid which has a smoky, crude woody odor. The yield of essential oil from the wood of *Cupressus funebris* Endl. generally varies from 1.7 to 3.4%. The main producing country of this oil is China.

Literature data (up to November 11, 2014) on the chemical composition of cedarwood oils China and unpublished analytical data from one of us (E.S.) are shown in Table 5.16.2 in alphabetical order. In cedarwood oils China from various origins (we could find only eight publications), over 135 chemicals have been identified. About 58% of these were found in a single reviewed publication only. The major compounds found in cedarwood oils China from different sources are shown in Table 5.16.3. They include (highest concentrations in any study given) α-cedrene (44.2%), thujopsene (40.9%), cedrol (26.1%), cuparene (15.1%), β-cedrene (11.5%) and β-himachalene (6.4%). A well-known ingredient of cedarwood oils China present in high concentrations in one study was widdrol (9.5%).

Commercial oils

The ten chemicals that had the highest maximum concentrations in 21 commercial cedarwood China essential oil samples (concentration ranges provided) are the following: thujopsene (16.5-40.9%), α-cedrene (10.3-38.4%), cedrol (9.5-14.5%), β-cedrene (3.4-8.2%), β-funebrene (1.8-3.4%), cuparene (1.9-2.8%), methylcarvacrol (0.1-2.1%), β-himachalene (1.2-2.1%), γ-acoradiene (0.8-1.9%) and β-chamigrene (0.6-1.8%) (Erich Schmidt, unpublished analytical data).

Table 5.16.2 Constituents identified in cedarwood oils China

Constituent	CAS	Percentage and range					
		A	B (2)	C (1)	D (3)	E (7)	F
Abietadiene	36312-33-1		0-0.2			0.3	
Abietatriene	19407-28-4		0-0.1				
α-Acoradiene	24048-44-0	0.1-0.6	0-0.7	0.4		0.6	
β-Acoradiene	28477-64-7	0.5-0.7	0-0.9	1.2		0.3	
10-epi-β-Acoradiene, isomer			0-0.4				
γ-Acoradiene	28400-12-6	0.8-1.9	0.4-4.8	3.3		2.6	
α-Acorenol	28296-85-7	0.2-0.7	0-0.6		0.1		
β-Acorenol	28400-11-5	0.09-0.5	0-1.0				
β-Alaskene	28908-21-6	0.2-0.3	0.1-1.7	1.0		0.1	
allo-Aromadendrene	25246-27-9			0.4	0.2		
α-Barbatene	53060-59-6		0-0.1				
β-Barbatene	39863-73-5		0-0.4				
trans-α-Bergamotene	13474-59-4			0.1			
Bicyclogermacrene	24703-35-3		0-1.5				
β-Biotol	19902-26-2		0-2.0		0.4		
β-Biotone	19902-29-5		0-0.6				
α-Bisabolene	17627-44-0		0.3-0.6				
β-Bisabolene	495-61-4		0-0.3			0.4	
γ-Bisabolene	495-62-5			1.2			
(E)-γ-Bisabolene	53585-13-0		0-0.8				
α-Bisabolol	515-69-5	0.2-0.5		0.5	0.1	0.8	tr[f]
epi-α-Bisabolol	78148-59-1		0-0.9				
epi-β-Bisabolol	235421-59-7		0-0.3				
Borneol	507-70-0				tr		
Bornyl acetate	76-49-3				tr		
Cadalene	483-78-3		0-0.2	0.4			
α-Cadinene	24406-05-1		0-1.2	0.5			
γ-Cadinene	39029-41-9		0-tr	0.9	3.9[b]		
δ-Cadinene	483-76-1		0-2.6	2.7	1.0		
α-Cadinol	481-34-5		0-0.3		0.2		
epi-α-Cadinol	5937-11-1		0-0.2				
α-Calacorene	21391-99-1		0-3.1	1.5	0.2		
β-Calacorene	50277-34-4			0.1			
Calamenene	483-77-2			1.0	0.9		
trans-Calamenene	73209-42-4		0-3.9				
Camphor	76-22-2		0-tr		tr		
δ3-Carene	13466-78-9		0-tr		tr		
Carvacrol	499-75-2		0-tr				
Caryolan-8-ol	178737-45-6		0-1.0				
β-Caryophyllene	87-44-5	0-1.1	0-tr	1.0	0.9		2.8[g]
Caryophyllene oxide	1139-30-6		0-0.1		tr		
α-Cedrenal	28387-62-4			0.2			
1,7-di-epi-α-Cedrenal			0-0.2				
α-Cedrene	469-61-4	10.3-38.4	3.7-44.2	16.9	24.1	26.4	8.2[g]; 27.1[d]; 33.8[f]; 39.1[c]
β-Cedrene	546-28-1	3.4-8.2	3.1-11.5	5.7	7.4	9.2	2.5[g]; 8.1[f]; 8.8[d]; 10.5[c]
Cedr-8(15)-en-9-α-ol	13567-41-4		0-0.3				
2-epi-α-Cedren-3-one	288249-25-2		0-0.9				
Cedrol	77-53-2	9.5-14.5	1.7-23.4	7.6	11.3	9.6	8.5[f]; 12.0[d]; 18.6[c]; 26.1[g]
allo-Cedrol	50657-30-2		0.2-2.1				
epi-Cedrol	19903-73-2		0-0.5				
Cedryl acetate	77-54-3		0-0.2			0.1	
Cembrene	1898-13-1					0.1	
α-Chamigrene	19912-83-5	0.04-0.2	tr-1.4	0.6		1.4	1.6[f]
β-Chamigrene	18431-82-8	0.6-1.8	0.6-2.5	0.9		2.2	
β-Chamigrene, isomer			0-0.6				
α-Chamipinene	847374-85-0		0-0.9				
1,8-Cineole	470-82-6			0.1[a]			
α-Copaene	3856-25-5		0-tr				

Table 5.16.2 Constituents identified in cedarwood oils China *(continued)*

Constituent	CAS	Percentage and range					
		A	B (2)	C (1)	D (3)	E (7)	F
Cubenol	21284-22-0			0.3			
1-epi-Cubenol	81939-29-9			0.3	0.2		
Cuparene	16982-00-6	1.9-2.8	1.8-10.2	5.4	1.0	3.4	2.2[f]; 3.3[d]; 3.7[c]; 15.1[g]
α-Cuprenene	29621-78-1			1.1			
β-Cuprenene	119683-81-7			2.0			
γ-Cuprenene	4895-23-2		0.3-2.8		0.3		
δ-Cuprenene	66389-22-8		0-0.5				
4-Cuprenen-1-ol			0-0.1				
Curcumene (ar-; α-)	644-30-4	0.2-0.3	0.4-1.8	2.5	1.3	0.4	0.3[f]; 0.4[d]; 3.2[g]
γ-Curcumene	28976-68-3					0.2	
Cyclosativene	22469-52-9			0.1			
p-Cymene	99-87-6		0-tr				
Daucene	16661-00-0				0.1		
cis-Dihydromayurone	7129-16-0		0-0.1				
α-Duprezianene	79801-29-9		0.1-1.5				
β-Elemene	33880-83-0	0.1-0.5	0-1.6	0.5	1.6		
Eremophila ketone			0-0.5				
α-Eudesmol	473-16-5		0-0.3	0.1			
β-Eudesmol	473-15-4		0-0.3				
γ-Eudesmol	1209-71-8			0.8			
(E)-β-Farnesene	18794-84-8		0-0.5			0.1	
(Z)-β-Farnesene	28973-97-9					0.1	
Fenchol	1632-73-1				tr		
α-Funebrene	50894-66-1	0.1-0.3	0-0.3	0.4	1.1		2.4[c]
2-epi-α-Funebrene	854154-70-4		0.1-1.8				
β-Funebrene	79120-98-2	1.8-3.4		1.4			3.0[c]
Gleenol	72203-99-7		0.2				
α-Gurjunene	489-40-7			0.1			
γ-Gurjunene	22567-17-5		0-0.3				
α-Himachalene	3853-83-6		0-0.4			0.2	0.5[d]; 0.7[f]
β-Himachalene	1461-03-6	1.2-2.1	1.2-6.4			1.4	0.5[d]; 0.7[f]; 2.7[g]
α-Humulene	6753-98-6			0.2			
Isoabienol	10207-79-1				tr		
Isobazzanene	88661-59-0		0-0.4				
Isoledene	95910-36-4		0-0.1				
Isolongifolene	1135-66-6		0-0.3				
3-Isothujopsanone	25966-81-8		0-0.1				
Italicene	94535-52-1				0.5		
Junenol	472-07-1		0-1.0	0.1			
Junicedranol	168180-13-0		0-0.1				
Junicedranone			0-0.3				
Junipercedrol	175448-28-9				0.4		
Khusian-2-ol (khusiol)	66512-56-9				0.3		
Limonene	138-86-3		0-0.5	0.1[a]			
Longicamphenylone	38647-26-6		0-tr				
Longifolene	475-20-7		0-0.8				
Manoyl oxide	596-84-9				0.1		
Mayurone	4677-90-1		0-0.2				
Methylcarvacrol	6379-73-3	0.1-2.1	0-2.3		0.1	0.7	0.6[d]
α-Muurolene	10208-80-7		0-tr	2.0	3.9[b]		
γ-Muurolene	30021-74-0		0-0.6	1.4	0.7		
τ-Muurolol (epi-α-)	19912-62-0		0-0.3	0.3	0.2		
Myltayl-4(12)-ene	79562-97-3		0-0.2				
Naphthalene	91-20-3		0-tr				
α-Neocallitropsene	729602-94-2		0-0.6				
(E)-Nerolidol	40716-66-3		0-0.1	0.1			
allo-Ocimene	673-84-7				tr		
Palustrol	5986-49-2				0.1		
Pentadecane (n-)	629-62-9			<0.1			
α-Pinene	80-56-8		0-tr		tr		
Prezizaan-15-al	87059-20-9				0.2		
Prezizaene	31145-21-8		0-0.7		0.6		

Table 5.16.2 Constituents identified in cedarwood oils China *(continued)*

Constituent	CAS	Percentage and range					
		A	B (2)	C (1)	D (3)	E (7)	F
Pseudowiddrene	32540-28-6		0-1.7				
Selina-4,11-diene	17627-30-4			0.9			
α-Selinene	473-13-2			1.7	1.1	3.1	3.0[d]
β-Selinene	17066-67-0		0.2-1.1	1.5	0.6	0.2	
β-Sesquiphellandrene	20307-83-9	0-0.1					
Sesquithujene	58319-06-5	0-0.1					
Sesquithuriferol	117468-55-0				0.8		
Terpinen-4-ol	562-74-3		0-0.1		tr		
α-Terpineol	98-55-5		0-0.1		0.4		
Terpinolene (α-)	586-62-9		0-0.5				
α-Terpinyl acetate	80-26-2		0-tr		tr		
Thujopsadiene	24048-40-6	0.3-0.6	0-1.2				
3-Thujopsanone	25966-79-4		0-0.5				
cis-Thujopsenal	470-41-7		0-0.2				
Thujopsene	470-40-6	16.5-40.9	1.9-37.4[e]		25.9	29.9	7.8[c]; 18.3[g]; 30[d]; 36.0[f]
Thymol	89-83-8		0-0.1				
1,4,6-Trimethyl naph-thalene	2131-42-2		0-0.8				
Widdrol	6892-80-4	0-1.5	0-0.8	0.8		9.5	1.1[f]; 4.9[d]
β-Ylangene	20479-06-5				1.7		
Zonarene	41929-05-9			0.5			

A twenty-one cedarwood China essential oil samples from China analyzed between 1998 and 2013; lowest and highest concentrations given (E. Schmidt, unpublished data)
B nine commercial Chinese cedarwood oils; lowest and highest concentrations given (ref. 2)
C one commercial *C. funebris* oil (ref. 1)
D one commercial Chinese cedarwood essential oil (ref. 3)
E one sample of Chinese cedarwood oil (ref. 7)
F data from other studies (indicated with superscript letters); highest concentrations found in any study reviewed here given; when two or more oils were investigated, only the highest concentrations are mentioned, unless indicated otherwise

[a] limonene and 1,8-cineole combined; [b] γ-cadinene and α-muurolene combined; [c] one lab-distilled oil from Vietnamese *C. funebris* (ref. 4); [d] data taken from ref. 5; [e] cis-thujopsene, CAS 32435-95-3; [f] one commercial Chinese cedarwood oil (ref. 8); [g] one steam-distilled oil from Chinese *C. funebris* chips (ref. 9)

tr: trace (in column B: <0.1%; in column D: <0.05%)

CONTACT ALLERGY/ALLERGIC CONTACT DERMATITIS

Possible contact allergy to cedarwood oil, China, has been reported in two publications, where patients were tested with a mixture of cedarwood oil, Morocco (Atlas) and Chinese cedarwood oil. This is discussed in Table 5.15.3 of Chapter 5.15 Cedarwood oil, Atlas.

Table 5.16.3 Major constituents of cedarwood oils China

Constituent	CAS	Percentage and range					
		A	B (2)	C (1)	D (3)	E (7)	F
α-Cedrene	469-61-4	10.3-38.4	3.7-44.2	16.9	24.1	26.4	8.2[g]; 27.1[d]; 33.8[f]; 39.1[c]
Thujopsene	470-40-6	16.5-40.9	1.9-37.4[e]		25.9	29.9	7.8[c]; 18.3[g]; 30[d]; 36.0[f]
Cedrol	77-53-2	9.5-14.5	1.7-23.4	7.6	11.3	9.6	8.5[f]; 12.0[d]; 18.6[c]; 26.1[g]
Cuparene	16982-00-6	1.9-2.8	1.8-10.2	5.4	1.0	3.4	2.2[f]; 3.3[d]; 3.7[c]; 15.1[g]
β-Cedrene	546-28-1	3.4-8.2	3.1-11.5	5.7	7.4	9.2	2.5[g]; 8.1[f]; 8.8[d]; 10.5[c]
β-Himachalene	1461-03-6	1.2-2.1	1.2-6.4			1.4	0.5[d]; 0.7[f]; 2.7[g]

LEGEND: SEE UNDER TABLE 5.16.2

LITERATURE

1 Carroll JF, Tabanca N, Kramer M, Elejalde NM, Wedge DE, Bernier UR, et al. Essential oils of *Cupressus funebris*, *Juniperus communis*, and *J. chinensis* (Cupressaceae) as repellents against ticks (Acari: Ixodidae) and mosquitoes (Diptera: Culicidae) and as toxicants against mosquitoes. J Vector Ecol 2011;36:258-268

2 Adams RP, Li S. The botanical source of Chinese cedarwood oil: *Cupressus funebris* or Cupressaceae species? J Essent Oil Res 2008;20:235-242

3 Duquesnoy E, Dinh NH, Castola V, Casanova J. Composition of a pyrolytic oil from *Cupressus funebris* Endl. of Vietnamese origin. Flavour Fragr J 2006;21:453-457

4 Hoi TM, Moi LD, Muselli A, Bighelli A, Casanova J. Analyse de l'huile essentielle de *Cupressus funebris* du Vietnam par RMN du carbone-13. Rivista Ital EPPOS (special issue) 1996;633-637. Data cited in refs. 2 and 6. In ref. 6, the bibliography is slightly different: Rivista Ital EPPOS (special issue) 1997;632-637.

5 Burfield T. Odour profiling (of essential oils) and subjectivity. Presentation given at the RQA's 12th Annual Conference, Saturday, 9th March 2002, Regent's College Conference Centre, London, UK. Data cited in: Cedarwood Oils, Part 1. Copyright Tony Burfield, September 2002. Available at: http://www.users.globalnet.co.uk/~nodice/new/magazine/cedar/cedar.htm

6 Lawrence BM. Progress in essential oils. Perfum Flavor 1998;23 (September/October):67-72

7 Adams RP. Cedarwood oil – analyses and properties. In: HF Linskens and JF Jackson, Eds. Modern methods of plant analysis new series, volume 12. New York: Springer Verlag, 1991:159-173. Data available at http://www.juniperus.org/uploads/2/2/6/3/22639912/99-1991sprverlag159-173.pdf

8 Shu C-K, Lawrence BM. Reasons for the variation in composition of some commercial oils. In: SJ Risch and C-T Ho, Eds. Spices, flavor chemistry and antioxidant properties. American Chemical Society Symposium Series 660. Washington, DC, USA: American Chemical Society, 1997:138-159. Data cited in ref. 6

9 Hou Y-S, Jiang Y-Y, Ming Y, Yu J, Zhang L. Chemical composition, antibacterial and antioxidant activities of essential oil from *Cupressus funebris* chips. Journal of Sichuan Agricultural University, China 2013-03 (in Chinese). Available at: http://en.cnki.com.cn/Article_en/CJFDTOTAL-SCND201303014.htm

10 Lawless J. The encyclopedia of essential oils, 2nd Edition. London: Harper Thorsons, 2014

Chapter 5.17 CEDARWOOD OIL, TEXAS

There are five major cedarwood essential oils: cedarwood oil Atlas, cedarwood oil Himalaya, cedarwood oil Texas, cedarwood oil Virginia and cedarwood oil China. These are obtained from different botanical species, and as a consequence, their chemical compositions differ both qualitatively and quantitatively. Unfortunately, in non-botanical literature, usually the term 'cedarwood oil' is used, lacking information on the botanical origin.

DEFINITION
Cedarwood oil, Texas (essential oil of cedarwood oil, Texas type) is the essential oil obtained from the wood of the Mexican juniper (Mexican cedar, mountain cedar), *Juniperus ashei* J. Buchholz (synonym: *Juniperus mexicana* auct.).

INCI NOMENCLATURE
Description/definition: Juniperus mexicana oil is the volatile oil obtained from *Juniperus mexicana*, Cupressaceae
INCI name EU & USA: Juniperus mexicana oil
CAS registry number(s): 68990-83-0; 91722-61-1
EINECS number(s): 294-461-7

Description/definition: Juniperus mexicana wood oil is an essential oil obtained from the wood of the Juniper, *Juniperus mexicana*, Cupressaceae
INCI name EU: Juniperus mexicana wood oil (perfuming name, not an INCI name proper)
INCI name USA: Not in the Personal Care Products Council Ingredient database
CAS registry number(s): 91722-61-1

ISO (INTERNATIONAL ORGANIZATION FOR STANDARDIZATION) STANDARD
ISO number: 4725
ISO name: Essential oil of cedarwood, Texas type
Botanical origin: *Juniperus ashei* J. Buchholz (synonym: *Juniperus Mexicana* auct.) Parts of plant used: Wood
ISO values: ISO values (minimum and maximum concentrations) are shown in Table 5.17.1.

THE PLANT, THE OIL, AND THEIR USES
Juniperus ashei J. Buchholz is a dioecious (biparental reproductive) large evergreen shrub or small tree, 6-15 meters tall, usually single-stemmed for the basal 1-3 meters, and up to 50 cm in diameter. The tree is native to the USA and northern Mexico. Its wood is steam-distilled to produce Texas cedarwood oil, a pleasant fragrance used (sometimes after rectification) in a range of fragrance applications such as soap perfumes, candles, cosmetics, household sprays, and floor polishes. The oil is especially employed to give dry nuances in citrus and woody compositions (2). It is also used in aromatherapy and as an insecticide. Another very important application is as feedstock for the manufacture of chemical derivatives, such as cedrol, cedryl methyl ether, acetyl

Table 5.17.1 ISO values (%) for cedarwood oil Texas[a]

Compound	CAS	Minimum	Maximum
Cedrol	77-53-2	20.0	—
Widdrol	6892-80-4	2.5	—
Thujopsene	470-40-6	25.0	35.0
Cuparene	16982-00-6	1.5	32.0
α-Cedrene	469-61-4	15.0	25.0
β-Cedrene	546-28-1	3.0	8.0

[a] ISO 4725 Essential oil of cedarwood, Texas type ©ISO 2004; Geneva, Switzerland, www.iso.org

cedrene, and cedryl acetate. In medicine, small quantities of specially thickened varieties of cedarwood oils are employed as immersion oil and for clarification in microscopy, because they share the same refraction index as glass (www.conifers.org; GRIN Taxonomy for Plants, Wikipedia, Cedarwood Oil – National Toxicology Program, http://ntp.niehs.nih.gov/ntp/htdocs/Chem_Background/ExSumPdf/cedarwood_oil_508.pdf).

CHEMICAL COMPOSITION
Cedarwood oil Texas is a brown to reddish viscous liquid, which has a woody, dry and soft odor with a slightly smoky top note. The yield of essential oil from the wood of *Juniperus ashei* J. Buchholz generally varies from 1.7 to 4.0%. This oil is produced only in the state of Texas, USA.

Literature data (up to October 28, 2014) on the chemical composition of cedarwood oils Texas and unpublished analytical data from one of us (E.S.) are shown in Table 5.17.2 in alphabetical order. In cedarwood oils Texas from various origins (we could find only seven publications), over 45 chemicals have been identified. About 43% of these were found in a single reviewed publication only. The major compounds found in cedarwood oils Texas from different sources are shown in Table 5.17.3. They include (highest concentrations in any study given) thujopsene, including its *cis*-isomer (60.4%), α-cedrene (41.2%), cedrol (33.7%), β-cedrene (12.2%), cuparene (3.6%) and widdrol (3.2%).

Commercial oils
The ten chemicals that had the highest maximum concentrations in 119 commercial cedarwood Texas essential oil samples (concentration ranges provided) are the following: thujopsene (31.6-49.2%), cedrol (14.3-33.7%), α-cedrene (1.8-21.3%), β-cedrene (1.5-5.2%), widdrol (0.4-3.2%), cuparene (1.3-2.9%), β-caryophyllene (0.08-1.9%), β-himachalene (0.5-1.9%), α-acorenol (0.7-1.8%) and α-acoradiene (0.5-1.4%) (Erich Schmidt, unpublished analytical data).

CONTACT ALLERGY/ALLERGIC CONTACT DERMATITIS
No reports on contact allergy to cedarwood oil, Texas specifically mentioned to be obtained from the wood of *Juniperus ashei* J. Buchholz have been found. Literature

Table 5.17.2 Constituents identified in cedarwood oils Texas

Constituent	CAS	Percentage and range							
		A	B (1)	C (2)	D (3)	E (4)	F (6)	G (7)	H (8)
α-Acoradiene	24048-44-0	0.5-1.4		0.3			0.7		
β-Acoradiene	28477-64-7		0.7	0.6			0.6		
γ-Acoradiene	28400-12-6		0.6				0.7		
α-Acorenol	28296-85-7	0.7-1.8	1.2	1.6					
β-Acorenol	28400-11-5	0.4-1.3	0.6	0.7					
β-Alaskene	28908-21-6		0.2				0.2		
α-Bisabolol	515-69-5						0.4	tr	
epi-α-Bisabolol	78148-59-1		0.2						
Camphor	76-22-2	0-0.5					0.2		
Carvacrol	499-75-2		<0.1						
β-Caryophyllene	87-44-5	0.08-1.9							
α-Cedrene	469-61-4	1.8-21.3	21.8	8.6	20.5	1.8	30.7	22.6	41.2
β-Cedrene	546-28-1	1.5-5.2	5.9	3.4	6.0	1.6	5.5	5.5	12.2
8-Cedren-13-ol	18319-35-2	0.2-0.8						0.9	
Cedrol	77-53-2	14.3-33.7	30.5	28.6	26.6[a]	19.0	19.1	12.2	4.9
allo-Cedrol	50657-30-2		0.3						
epi-Cedrol	19903-73-2		0.3				0.4		
α-Chamigrene	19912-83-5	0.8-1.4	1.2	1.6			1.2	1.5	
β-Chamigrene	18431-82-8		1.2	0.1			1.1		
Cubenol	21284-22-0						0.2		
Cuparenal	16982-01-7		0.2						
Cuparene	16982-00-6	1.3-2.9	3.6	1.8		2.8	1.7	1.9	
ar-Curcumene	644-30-4		0.4				0.1	tr	
γ-Curcumene	28976-68-3						0.1		
γ-Dehydro-ar-hima-chalene	51766-65-5		0.1						
α-Duprezianene	79801-29-9	0.05-0.3	0.3						
β-Duprezianene	178443-10-2	0.2-0.5							
2-epi-α-Funebrene	854154-70-4		<0.1						
β-Funebrene	79120-98-2	0-0.3							
α-Himachalene	3853-83-6	0.4-0.8	0.6	0.2			0.5	0.8	
β-Himachalene	1461-03-6	0.5-1.9	1.8	3.5	3.9		1.4	1.1	
γ-Himachalene	53111-25-4			1.2			0.1		
Intermedeol	6168-59-8		0.2						
Isoitalicene	94482-89-0		<0.1						
Limonene	138-86-3			0.3					
Mayurone	4677-90-1		1.1						
(E)-Nerolidol	40716-66-3			0.3					
α-Pinene	80-56-8	0.01-0.5		0.5					0.8
Pseudowiddrene	32540-28-6		0.6						
α-Selinene	473-13-2				2.1	1.5			
7-epi-Sesquithujene	159407-35-9		<0.1						
Terpinen-4-ol	562-74-3		<0.1						
Thujopsadiene	24048-40-6		1.0				0.1		
trans-3-Thujopsanone	25966-79-4						0.8		
Thujopsene	470-40-6	31.6-49.2		38.4	29.4	60.4	25.0	46.8	24.3
cis-Thujopsene	32435-95-3		24.6						
cis-Thujopsenic acid			0.1						
Valencene	4630-07-3						0.1		
Widdrol	6892-80-4	0.4-3.2				1.1	1.6	1.1	1.7

A one hundred and nineteen cedarwood Texas essential oil samples from Texas, USA, analyzed between 1998 and 2013; lowest and highest concentrations given (E. Schmidt, unpublished data)
B one commercial cedarwood Texas oil (ref. 1)
C one commercial oil from a German supplier (ref. 2)
D one commercial Juniperus ashei oil sample (ref. 3)
E one laboratory steam-distilled Texas cedarwood oil (ref. 4)
F one Texas cedarwood oil sample (ref. 6)
G one commercial cedarwood Texas oil (ref. 7)
H one cedarwood oil Texas oil sample; only six ingredients were investigated (ref. 8)

[a] combined with widdrol, a very minor component; tr: trace

Table 5.17.3 Major constituents of cedarwood oils Texas

Constituent	CAS	Percentage and range							
		A	B (1)	C (2)	D (3)	E (4)	F (6)	G (7)	H
Thujopsene	470-40-6	31.6-49.2		38.4	29.4	60.4	25.0	46.8	24.3[b]
α-Cedrene	469-61-4	1.8-21.3	21.8	8.6	20.5	1.8	30.7	22.6	41.2[b]
Cedrol	77-53-2	14.3-33.7	30.5	28.6	26.6[a]	19.0	19.1	12.2	4.9[b]
cis-Thujopsene	32435-95-3		24.6						
β-Cedrene	546-28-1	1.5-5.2	5.9	3.4	6.0	1.6	5.5	5.5	12.2[b]
Cuparene	16982-00-6	1.3-2.9	3.6	1.8		2.8	1.7	1.9	
Widdrol	6892-80-4	0.4-3.2				1.1	1.6	1.1	1.7[b]

LEGEND: SEE UNDER TABLE 5.17.2

on contact allergy to/allergic contact dermatitis from 'cedarwood oil' (botanical source not specified) is discussed in Chapter 5.15 Cedarwood oil, Atlas.

LITERATURE

1 Adams RP, Li S. The botanical source of Chinese cedarwood oil: *Cupressus funebris* or Cupressaceae species? J Essent Oil Res 2008;20:235-242

2 Wanner J, Schmidt E, Bail S, Jirovetz L, Buchbauer G, Gochev V, et al. Chemical composition and antibacterial activity of selected essential oils and some of their main compounds. Nat Prod Commun 2010;5:1359-1364

3 Payne KW, Wittwer R, Anderson S, Eisenbraun EJ. Use of a modified abderhalden apparatus for comparing laboratory and industrial methods of isolating eastern red cedar essential oil. Forest Product J 1999;49:90-92

4 Adams RP. Investigation of *Juniperus* species of the United States for new sources of cedarwood oil. Economic Botany 1987;41:48-54

5 Lawrence BM. Progress in essential oils. Perfum Flavor 1998;23 (Sept/Oct):67-72

6 Adams RP. Cedarwood oil – analyses and properties. In: HF Linskens and JF Jackson, Eds. Modern methods of plant analysis new series, volume 12. New York: Springer Verlag, 1991:159-173. Data available at http://www.juniperus.org/uploads/2/2/6/3/22639912/99-1991sprverlag159-173.pdf

7 Shu C-K, Lawrence BM. Reasons for the variation in composition of some commercial oils. In: SJ Risch and C-T Ho, Eds. Spices, flavor chemistry and antioxidant properties. American Chemical Society Symposium Series 660. Washington, DC: Am Chem Soc, 1997:138-159. Data cited in ref. 5

8 Coleman WM, Lawrence BM. A comparison of selected analytical approaches to the analysis of an essential oil. Flav Fragr J 1997;12:1-8

Chapter 5.18 CEDARWOOD OIL, VIRGINIA

There are five major cedarwood essential oils: cedarwood oil Atlas, cedarwood oil Himalaya, cedarwood oil Texas, cedarwood oil Virginia and cedarwood oil China. These are obtained from different botanical species, and as a consequence, their chemical compositions differ both qualitatively and quantitatively. Unfortunately, in non-botanical literature, usually the term 'cedarwood oil' is used, lacking information on the botanical origin.

DEFINITION
Cedarwood oil, Virginia, (essential oil of cedarwood, Virginian type; Virginian cedarwood oil; red cedarwood oil) is the essential oil obtained from the wood of the (eastern) red cedar (coastal cedar), *Juniperus virginiana* L.

INCI NOMENCLATURE
Description/definition: Juniperus virginiana wood oil is an essential oil obtained from the wood and twigs of the red cedar, *Juniperus virginiana* L., Cupressaceae
INCI name EU: Juniperus virginiana wood oil (perfuming name, not an INCI name proper)
INCI name USA: Juniperus virginiana oil (extracted from the *whole* plant, *Juniperus virginiana*)
CAS registry number (s): 8000-27-9; 85085-41-2
EINECS number(s): 285-370-3

ISO (INTERNATIONAL ORGANIZATION FOR STANDARDIZATION) STANDARD
ISO number: 4724
ISO name: Essential oil of cedarwood, Virginian type
Botanical origin: *Juniperus virginiana* L.
Parts of plant used: Wood
ISO values: ISO values (minimum and maximum concentrations) are shown in Table 5.18.1.

THE PLANT, THE OIL, AND THEIR USES
Juniperus virginiana L., commonly called eastern red cedar, is a single-stemmed coniferous evergreen tree which can grow to 30 meters tall. The tree is native to the USA and Canada, naturalized elsewhere and is also cultivated. Virginia cedarwood oils, which are obtained by steam-distillation of sawdust, waste shavings, old stumps, and chipped logs of eastern red cedar, have many commercial uses. They are used in polishes to restore the smell of cedar to furniture and as fragrance in cosmetic formulations, including shampoos for humans and animals, aftershave lotions, soap bars, and perfumes. They are also found in insect repellents, massage oils, cleaning products, room deodorants, disinfectants, liniments, incense oils and numerous other industrial products and may be used for medicinal purposes and in aromatherapy (13). In medicine, small quantities of specially thickened varieties of cedarwood

Table 5.18.1 ISO values (%) for cedarwood oil Virginia[a]

Compound	CAS	Minimum	Maximum
α-Cedrene + β-funebrene	469-61-4 79120-98-2	22.0	35.0
Cedrol	77-53-2	16.0	25.0
Thujopsene	470-40-6	10.0	25.0
β-Cedrene + β-caryophyllene	546-28-1 87-44-5	4.0	8.0
Cuparene	16982-00-6	1.5	7.0
Widdrol	6892-80-4	2.0	5.0

[a] ISO 4724 Essential oil of cedarwood, Virginian type ©ISO 2004; Geneva, Switzerland, www.iso.org

oils are employed as immersion oil and for clarification in microscopy, because they have the same refraction index as glass (1; GRIN taxonomy for Plants; www.wildflower.org; Cedarwood Oil – National Toxicology Program, http://ntp.niehs.nih.gov/ntp/htdocs/Chem_Background/ExSumPdf/cedarwood_oil_508.pdf).

CHEMICAL COMPOSITION
Cedarwood oil, Virginia is a slightly viscous liquid, sometimes containing crystals from cedrol, colorless to pale yellow, which has a mild woody, warm and dry odor. The yield of essential oil from the wood of *Juniperus virginiana* L. generally varies from 1.5 to 3.0%. The only producing country of this oil is the USA.

Literature data (up to October 29, 2014) on the chemical composition of cedarwood oils Virginia and unpublished analytical data from one of us (E.S.) are shown in Table 5.18.2 in alphabetical order. In cedarwood oils Virginia from various origins, over 55 chemicals have been identified. About 54% of these were found in a single reviewed publication only. The major compounds found in cedarwood oils Virginia from different sources are shown in Table 5.18.2. They include (highest concentrations in any study given) cedrol (68.2%, includes widdrol as minor component), α-cedrene (38.0%), thujopsene (30.0%), β-cedrene (9.2%), cuparene (6.3%), β-funebrene (4.8%) and widdrol (4.0%). Cedrene (not further specified) was identified in a single study only, in a concentration of 80%.

Commercial oils
The ten chemicals that had the highest maximum concentrations in 36 commercial cedarwood Virginia essential oil samples (concentration ranges provided) are the following: thujopsene (22.2-33.2%), α-cedrene (12.8-27.2%), cedrol (15.8-25.9%), β-cedrene (3.6-7.7%), cuparene (1.1-6.3%), β-funebrene (1.6-4.8%), widdrol (1.0-4.0%), α-acoradiene (0.5-2.3%), β-caryophyllene (0.5-1.9%) and β-chamigrene (1.1-1.8%) (Erich Schmidt, unpublished analytical data).

A review of various aspects of Virginian cedarwood oils is provided in ref. 1.

Table 5.18.1 Constituents identified in cedarwood oils Virginia

Constituent	CAS	Percentage and range				
		A	B (2)	C (3)	D (10)	E
α-Acoradiene	24048-44-0	0.5-2.3	<0.1		0.2	
β-Acoradiene	28477-64-7		0.4		0.3	
10-epi-β-Acoradiene	847374-86-1		0.3			
γ-Acoradiene	28400-12-6		0.7		0.9	
α-Acorenol	28296-85-7	0.5-1.2	0.8			
β-Acorenol	28400-11-5	0.4-0.9	0.4			
β-Alaskene	28908-21-6		0.2		0.1	
α-Bisabolol	515-69-5	0.2-0.9			0.6	tr[k]
epi-α-Bisabolol	78148-59-1		0.3			
Camphor	76-22-2	0.1-0.2				
β-Caryophyllene	87-44-5	0.5-1.9		1.5[d]		+[j]
Cedrene	11028-42-5			80[i]		
α-Cedrene	469-61-4	12.8-27.2	27.7	2.9-8.1	21.1	3.2[h]; 13.3[f]; 30.1[e]; 35.0[c,g]; 38.0[k]
β-Cedrene	546-28-1	3.6-7.7	6.3	0.5-0.8	8.2	3.3[f]; 7.7[h]; 7.8[e]; 9.2[k]
Cedrenol	28231-03-0					'small amounts'[i]
Cedrol	77-53-2	15.8-25.9	26.5	29.3-49.9[a]	22.2	3-14[i]; 4.0[c,g]; 12.3[k]; 15.8[h]; 24.3[a,e]; 68.2[a,f]
allo-Cedrol	50657-30-2		0.2			
epi-Cedrol	19903-73-2		0.3		0.2	
α-Chamigrene	19912-83-5	0-0.02	1.0		1.6	
β-Chamigrene	18431-82-8	1.1-1.8	1.3		1.8	1.4[k]
α-Chamipinene	847374-85-0		0.1			
Cubenol	21284-22-0				0.1	
Cuparenal	16982-01-7		<0.1			
Cuparene	16982-00-6	1.1-6.3	1.7		1.6	+[j]; 0.9[k]; 2.0[c,g]; 2.6[f]; 6.3[h]
γ-Cuprenene	4895-23-2		1.2			
δ-Cuprenene	66389-22-8		0.4			
ar-Curcumene	644-30-4		0.2		0.1	
γ-Curcumene	28976-68-3				0.1	
α-Duprezianene	79801-29-9		0.6			
β-Duprezianene	178443-10-2		0.5			
β-Elemene	33880-83-0					+[j]
(E)-β-Farnesene	18794-84-8		0.1			
(Z)-β-Farnesene	28973-97-9				0.1	
α-Funebrene	50894-66-1		0.3			
2-epi-α-Funebrene	854154-70-4		0.5			
β-Funebrene	79120-98-2	1.6-4.8	<0.1			
α-Himachalene	3853-83-6		0.1		0.2	
β-Himachalene	1461-03-6		0.6		2.1	2.1[k]; 3.5[e]
α-Humulane						+[j]
β-Humulane						+[j]
α-Humulene	6753-98-6			0.3[d]		
Isobazzanene	88661-59-0		0.3			
Isolongifolene	1135-66-6		0.1			
Junicedranone			0.3			
Limonene	138-86-3	0.01-0.08				
Mayurone	4677-90-1		0.7			
Prezizaene	31145-21-8		0.2			
Pseudowiddrene	32540-28-6		0.3			
α-Selinene	473-13-2				3.0	2.1[e]
β-Selinene	17066-67-0		<0.1		0.1	
7-epi-Sesquithujene	159407-35-9		0.3			
Thujopsadiene	24048-40-6		0.4			
cis-Thujopsenal	470-41-7		0.7			
Thujopsene	22.2-33.2			2.8-6.0	21.3	17.7[e]; 22.7[f]; 23.4[k]; 27.6[h]; 30.0[c,g]
cis-Thujopsene	32435-95-3		17.9			
Valencene	4630-07-3				0.1	+[j]
Widdrol	6892-80-4	1.0-4.0	<0.1	[b]	2.3	1[h]; 1.9[k]; 2.0[c,g]

Table 5.18.1 Constituents identified in cedarwood oils Virginia *(continued)*

A thirty-six cedarwood Virginia essential oil samples from the Virginia, USA, analyzed between 1998 and 2013; lowest and highest concentrations given (E. Schmidt, unpublished data)
B one commercial oil sample (ref. 2)
C 15 lab-hydrodistilled oil samples from five trees with ages ranging from 26 to 63 years and taken from lower, center and top parts of the heartwood of the eastern red cedar trees; lowest and highest concentrations given (ref. 3)
D one sample of Virginia cedarwood oil (ref. 10)
E data from other studies (indicated with superscript letters); highest concentrations found in any study reviewed here given; when two or more oils were investigated, only the highest concentrations are mentioned, unless indicated otherwise

[a] cedrol + widdrol; widdrol is a very minor component; [b] see cedrol: widdrol is a very minor component of cedrol + widdrol; [c] one commercial Virginian cedarwood oil (ref. 12); [e] one commercial Virginia cedarwood oil (ref. 4); [f] one steam-distilled laboratory oil and one steam-distilled commercial oil (ref. 4); [g] data from ref. 5 (and ref. 12, identical data); [h] data from ref. 6 (cited in ref. 1); [i] data from ref. 7 (cited in ref. 1); [j] data from ref. 8 (cited in ref. 1); [k] one commercial cedarwood oil Virginia (ref. 11)

tr: trace; + present in the oil investigated, but quantity not stated

Table 5.18.3 Major constituents of cedarwood oils Virginia

Constituent	CAS	Percentage and range				
		A	**B (2)**	**C (3)**	**D (10)**	**E**
Cedrol	77-53-2	15.8-25.9	26.5	29.3-49.9[a]	22.2	3-14[i]; 4.0[c,g]; 12.3[k]; 15.8[h]; 24.3[a,e]; 68.2[a,f]
α-Cedrene	469-61-4	12.8-27.2	27.7	2.9-8.1	21.1	3.2[h]; 13.3[f]; 30.1[e]; 35.0[c,g]; 38.0[k]
Thujopsene		22.2-33.2		2.8-6.0	21.3	17.7[e]; 22.7[f]; 23.4[k]; 27.6[h]; 30.0[c,g]
β-Cedrene	546-28-1	3.6-7.7	6.3	0.5-0.8	8.2	3.3[f]; 7.7[h]; 7.8[e]; 9.2[k]
Cuparene	16982-00-6	1.1-6.3	1.7		1.6	+[j]; 0.9[k]; 2.0[c,g]; 2.6[f]; 6.3[h]
β-Funebrene	79120-98-2	1.6-4.8	<0.1			
Widdrol	6892-80-4	1.0-4.0	<0.1	[b]	2.3	1[h]; 1.9[k]; 2.0[c,g]

LEGEND: SEE UNDER TABLE 5.18.2

CONTACT ALLERGY/ALLERGIC CONTACT DERMATITIS

Contact allergy to cedarwood oil, Virginia, has been reported in one publication and is discussed in Table 5.18.3 of Chapter 5.15 Cedarwood oil, Atlas.

LITERATURE

1 Semen E, Hiziroglu S. Production, yield and derivatives of volatile oils from eastern red cedar (*Juniperus Virginiana* L.). Am J Environ Sci 2005;1:133-138

2 Adams RP, Li S. The botanical source of Chinese cedarwood oil: *Cupressus funebris* or Cupressaceae species? J Essent Oil Res 2008;20:235-242

3 Dunford NT, Hiziroglu S, Holcomb R. Effect of age on the distribution of oil in eastern red cedar tree segments. Bioresour Technol 2007;98:2636-2640

4 Payne KW, Wittwer R, Anderson S, Eisenbraun EJ. Use of a modified abderhalden apparatus for comparing laboratory and industrial methods of isolating eastern red cedar essential oil. Forest Product J 1999;49:90-92

5 Adams RP. Investigation of arid land junipers as sources of cedarwood oil, biocides and fuel. NTIS Report Number: PB88-102074, 1985. Data cited in: Cosmetic Ingredient Review Expert Panel. Final report on the safety assessment of *Juniperus communis* extract, *Juniperus oxycedrus* extract, *Juniperus oxycedrus* tar, *Juniperus phoenicea* extract, and *Juniperus virginiana* extract. Int J Toxicol 2001;20 (Suppl. 2):41–56

6 Adams RP. Yields and seasonal variation of photochemicals from *Juniperus* species of the United States. Biomass 1987;12:129-139. Data cited in ref. 1

7 Flake RH, Von Rudloff E, Turner BL. Confirmation of a clinal pattern of chemical differentiation in *Juniperus vir- giniana* from terpenoid data obtained in successive years. Rec Adv Phytochem 1973;6:215-228. Data cited in ref. 1

8 Walker GT. Cedarwood oil. Amer Perfum Essent Oil Rev 1968;5:347-352. Data cited in ref. 1

9 Lawrence BM. Progress in essential oils. Perfum Flavor 1998;23(Sept/Oct):67-72

10 Adams RP. Cedarwood oil – analyses and properties. In: HF Linskens and JF Jackson, Eds. Modern methods of plant analysis new series, volume 12. New York: Springer Verlag, 1991:159-173. Data available at http://www.juniperus.org/uploads/2/2/6/3/22639912/99-1991sprverlag159-173.pdf

11 Shu C-K, Lawrence BM. Reasons for the variation in composition of some commercial oils. In: SJ Risch and C-T Ho, Eds. Spices, flavor chemistry and antioxidant properties. American Chemical Society Symposium Series 660. Washington, DC: Am Chem Soc, 1997:138-159. Data cited in ref. 9

12 Adams RP. Investigation of *Juniperus* species of the United States for new sources of cedarwood oil. Economic Botany 1987;41:48-54

13 Lawless J. The encyclopedia of essential oils, 2nd Edition. London: Harper Thorsons, 2014

Chapter 5.19 CHAMOMILE OIL, GERMAN

Chamomile oil (may also be spelled camomile sometimes) may refer either to German chamomile oil obtained from *Chamomilla recutita* L. Rauschert or to chamomile oil, Roman (synonym: English) obtained from *Chamaemelum nobile* (L.) All., better known as *Anthemis nobilis* L. In non-botanical literature, often the nature of the oil and the botanical source are not specified. This chapter discusses German chamomile oil (also termed blue chamomile oil); Roman chamomile oil is presented in Chapter 5.20.

DEFINITION

German chamomile oil (essential oil of German chamomile, essential oil of blue chamomile) is the essential oil obtained from the flowering tops of the (blue, common, Hungarian, German) chamomile, *Chamomilla recutita* (L.) Rauschert (synonym: *Matricaria chamomilla* auct., *Matricaria recutita*).

INCI NOMENCLATURE

Description/definition: Chamomilla recutita flower oil is the volatile oil distilled from the dried flower heads of the *Matricaria recutita*, syn. *Chamomilla recutita* (L.), Compositae
INCI name EU: Chamomilla recutita flower oil
INCI name USA: Chamomilla recutita (matricaria) flower oil
Other names: Chamomile oil, Hungarian; matricaria oil; camomile oil German, Hungarian
CAS registry number(s): 8002-66-2; 8053-34-7
EINECS number(s): Not available

Description/definition: Matricaria recutita flower oil is the essential oil distilled from the flowers of the plant, *Matricaria recutita*, Compositae (syn: *Chamomilla recutita*)
INCI name EU: Matricaria recutita flower oil (perfuming name, not an INCI name proper)
INCI name USA: Not in the Personal Care Products Council Ingredient Database
CAS registry number(s): 84082-60-0; 8053-34-7
EINECS number(s): 282-006-5

ISO (INTERNATIONAL ORGANIZATION FOR STANDARDIZATION) STANDARD

ISO number: 19332
ISO name: Essential oil of German chamomile, essential oil of blue chamomile
Botanical origin: *Chamomilla recutita* (L.) Rauschert (synonym: *Matricaria chamomilla* auct.)
Parts of plant used: flowering tops
ISO values: ISO values (minimum and maximum concentrations) are shown in Table 5.19.1.

THE PLANT, THE OIL, AND THEIR USES

Chamomilla recutita (L.) Rauschert (syn. *Matricaria chamomilla* L., *Matricaria recutita* L.), popularly known as German chamomile, is a perennial plant belonging to the Asteraceae family. Native of Europe and adjoining Asian countries, the plant is cultivated all over the

Table 5.19.1 ISO values (%) for German chamomile oil (essential oil of blue chamomile)[a]

Compound	CAS	Minimum	Maximum
(*E*)-β-Farnesene	18794-84-8	15.0	51.0
α-Bisabolol oxide A	22567-36-8	2.0	50.0
α-Bisabolol	515-69-5	1.0	40.0
Chamazulene	529-05-5	2.0	22.0
Bisabolol oxide B	55399-12-7	2.0	21.0
α-Bisabolone oxide A	22567-38-0	1.0	6.5

[a] ISO 19332 Essential oil of German/blue chamomile ©ISO 2007; Geneva, Switzerland, www.iso.org

world for the flowers and the flower oil, particularly in European countries such as Hungary, France, Germany, Czech Republic, Slovakia and the former Yugoslavia, and in Egypt and Argentina (3,12,16).

German chamomile is an important medicinal and aromatic plant of both traditional and modern systems of medicine (12). Chamomile flowers are still an official drug in the pharmacopoeia of some 20 countries. Both the flowers and the essential oil have been reported to possess anti-inflammatory, spasmolytic, antiseptic, carminative, sedative and ulcer-protecting properties (12). Especially Europeans use chamomile in a wide variety of products. Compresses and gargles are used externally for the treatment of inflammations and irritations of the skin and mucosa, including the mouth and gums, the respiratory tract and hemorrhoids. In Bulgarian folk medicine, chamomile flower extract is used against insomnia, hysteria, gastritis, headache, stomach pain, for wound epithelialization, gas relief, and as antispasmodic and anti-sweating agent (24). German chamomile is very commonly utilized to make herbal teas, which are said to relieve spasms and inflammatory conditions of the gastrointestinal tract, ulcers, and menstrual disorders. The tea is also renowned as a gentle sleep aid, particularly for children (2,8,25).

The essential oil of *Chamomilla recutita*, which is obtained by steam distillation of the flowering tops, has a blue color, hence the common term 'blue chamomile oil'. The color is caused by the compound chamazulene, which is actually an artefact component formed during the steam distillation of the oil from matricin (proazulene) (2,28). The oil has wide application in medicine, cosmetics, aromatherapy and foodstuffs and in the flavoring of alcoholic and non-alcoholic beverages. The possible health effects of 'chamomile' and chamomile tea have been reviewed (1,15,36).

CHEMICAL COMPOSITION

German chamomile oil is a greenish blue to dark blue clear and slightly viscous liquid which has a herbaceous, fresh and slightly fruity odor. The yield of essential oil from the flowers of *Chamomilla recutita* (L.) Rauschert generally varies from 0.15 to 0.5% depending on whether fresh or dried flowers are used. The main producing countries of this oil are Hungary, France, Germany,

Serbia, Bosnia-Herzegovina, Czech Republic, Slowakia, Egypt and Argentina.

Literature data (up to November 15, 2014) on the chemical composition of German chamomile oils and unpublished analytical data from one of us (E.S.) are shown in Table 5.19.2 in alphabetical order. In German chamomile oils from various origins, over 280 chemicals have been identified. About 56% of these were found in a single reviewed publication only. The major compounds found in German chamomile oils from different sources are shown in Table 5.19.3. They include (highest concentrations in any study given) α-bisabolol (71.9%), α-bisabolol oxide A (63.0%), β-farnesene (52.3%), (E)-β-farnesene (43.5%), α-bisabolol oxide B (35.6%), α-bisabolone oxide A (31.5%), chamazulene (31.4%), cis-spiroether (26.1%), spiroethers (18.3%), trans-spiroether(16.8%) and spathulenol (11.3%). Well-known ingredients of German chamomile oils that were present in high concentrations (>7%) in one or two studies were β-bisabolene (21.9%), 1,8-cineole (15.2%), β-caryophyllene (7.2% and 13.0%), artemisia ketone (10.6%), β-pinene (10.1%), (E,E)-α-farnesene (8.3%), α-pinene (8.1%) and (E)-nerolidol (7.4%). Uncommon or rare constituents of German chamomile oils found in high concentrations (>7%) in single studies include (E,E)-farnesol (15.6%), butyl phthalate (15.1%, erroneous), 8-isobutyryloxy isobornyl isobutyrate (14.0%), camphor (12.9%), tetracosane (12.5%), α-farnesene (9.7%), (E)-γ-bisabolene (8.5%), (Z)-β-farnesene (7.1%) and prochamazulene (7.1%).

Commercial oils

The ten chemicals that had the highest maximum concentrations in 85 commercial German chamomile essential oil samples (concentration ranges provided) are the following: α-bisabolol oxide A (1.0-46.0%), α-bisabolol (1.5-38.3%), (E)-β-farnesene (18.4-35.4%), chamazulene (0.6-21.6%), trans-spiroether (0.2-16.8%), bisabolol oxide B (3.1-12.1%), cis-spiroether (2.5-10.5%), α-bisabolone oxide A (0.1-8.5%), (E,E)-α-farnesene (0.1-5.8%) and spathulenol (0.5-5.8%) (Erich Schmidt, unpublished analytical data).

Chemotypes

A large variation in the composition of German chamomile oils has been observed. Based on the concentrations

Table 5.19.2 Constituents identified in German chamomile oils

Constituent	CAS	Percentages and range					
		A	B (2)	C (51)	D (7)	E (52)	F
Ambrettolide	7779-50-2						1.4[h]
α-Amorphene	20085-19-2						0.1[c]
Anethole	104-46-1						6.2[t]
(E)-Anethole	4180-23-8						+[z12]; 0.1[c]; 1.3[h]
Aromadendrene	489-39-4						0.07[g]; 0.1[c]
allo-Aromadendrene	25246-27-9	0.0-0.2	0-0.6	0.1			0.8[h]
allo-Aromadendrene epoxide	85760-81-2			tr			
Artemisia alcohol	29887-38-5	0.1-0.5	0-0.5		<0.1	0.1	0.06[g]; 0.2[h]; 0.3[y4]; 0.6[b,d]; 1.5[f]; 1.6[c]; 1.9[e]
Artemisia ketone	546-49-6	0.2-0.7	tr-1.2	0.8	<0.1	0.2	1.4[y4]; 2.1[c]; 3.0[x]; 3.1[d]; 3.4[f]; 10.6[e]
Artemisyl acetate	3465-88-1	0.03-0.1					0.1[y4]; 1.2[c]
β-Atlantol	420109-31-5						0.2[c]
Azulene	275-51-4						1.1[x]
Benzaldehyde	100-52-7				<0.1		0.2[c]; 0.3[y4]
Benzyl alcohol	100-51-6				<0.1		
cis-α-Bergamotene	18252-46-5						0.2[c]
trans-α-Bergamotene	13474-59-4	0.0-0.3					
(Z)-α-Bergamotol	88034-74-6					0.1	
Bicyclogermacrene	24703-35-3	0.3-1.8		tr	0.1	0.6	0.9[h]; 2.0[g]; 2.1[z11]; 4.1[y]; 4.4[c]; 5.2[j]; 6.6[o]
1,1'-Biphenyl, 3,3'-di-methyl-	612-75-9					0.2	
(E)-α-Bisabolene	25532-79-0	0.0-2.7					
(Z)-α-Bisabolene	29837-07-8						0.1[h]
β-Bisabolene	495-61-4	0.0-1.5	0-0.4		19.6	0.01	0.1[j]; 0.2[y2]; 3.9[o]; 5.2[k]; 21.9[h]
(E)-γ-Bisabolene	53585-13-0	0.0-0.3				0.02	8.5[j]
(Z)-γ-Bisabolene	13062-00-5	0.0-0.5			0.5		
α-Bisabolol	515-69-5	1.5-38.3	0.1-44.2	5.6	1.7		50.6[z1]; 56.9[y2]; 57.91[z8]; 61.5[z3]; 71.9[z2]
epi-α-Bisabolol	78148-59-1	0.0-0.5					0.4[y3]
β-Bisabolol	15352-77-9		0-1.1				0.1[y2]; 1.1[n]
α-Bisabolol oxide A	22567-36-8	1.0-46.0	3.1-56.0	39.4	43.8	35.4	46.6[i]; 56.5[y1]; 57.1[z2]; 59.0[z3]; 63.0[y5]
α-Bisabolol oxide B	55399-12-7	3.1-12.1	3.9-27.2	9.9	3.8	35.6	23.3[r]; 23.5[z5]; 28.0[x]; 30.6[w]; 30.9[b]; 33.0[i]
α-Bisabolone oxide A	58985-73-2	0.1-8.5	0.5-24.8	13.9	13.6	9.1	6.1[i]; 7.4[y5]; 9.2[q]; 11.2[c]; 17.4[e]; 31.5[h]
Borneol	507-70-0	0.02-0.05	0.2		0.7	0.05	0.2[y4]; 0.3[w]; 0.4[b]; 0.8[f]; 1.1[m]; 1.2[e]
l-Borneol	464-45-9						0.3[h]
Bornyl acetate	76-49-3		0-1.3				

Table 5.19.2 Constituents identified in German chamomile oils *(Continued)*

Constituent	CAS	Percentages and range					
		A	B (2)	C (51)	D (7)	E (52)	F
β-Bourbonene	5208-59-3				0.2		<0.05[w]; 0.1[e]; 0.3[b]; 0.4[f]
Butanoic acid, 2-methyl-, ethyl ester	7452-79-1					0.07	
Butyl 2-methylbutyrate	15706-73-7					0.01	0.1[c]
Butyl 3-methylbutyrate	109-19-3					0.09	
Butyl phthalate	84-74-2						15.1[a,z10]
(E)-Cadina-1(6),4-diene	931410-54-7				<0.1		0.1[c]
α-Cadinene	24406-05-1				0.2		3.8[p]
γ-Cadinene	39029-41-9	0.0-0.2		0.1	0.1		0.1[b,c,e]; 0.2[d]; 0.8[z5]; 1.5[n]; 2.3[z10]
δ-Cadinene	483-76-1	0.04-0.4	0-0.5	0.1	0.1	0.02	0.2[e,f,g]; 0.4[b,z11]; 0.6[c]; 1.5[n]
γ-Cadinol	50895-55-1			0.2			
δ-Cadinol	19435-97-3					0.01	
τ-Cadinol	58580-31-7	0.2-0.8				0.06	0.5[b]; 0.8[d,l]; 1.9[e]; 2.4[f]
α-Calacorene	21391-99-1				<0.1		
trans-Calamenene	73209-42-4				<0.1		
trans-Calamenen-10-ol	828923-23-5						0.9[y5]
Calarene	17334-55-3					0.06	0.3[m]
Camphene	79-92-5	0.0-1.4					0.1[b,j]; 0.2[e]; 2.0[n]
Camphor	76-22-2				<0.1	0.01	0.2[c]; 0.4[y4]; 0.7[e,f]; 1.0[m]; 1.1[b]; 4.4[n]; 12.9[z4]
δ3-Carene	13466-78-9						2.9[x]
Carvacrol	499-75-2						0.2[m]
(E)-Carveol	1197-07-5				0.1		
Carvone	99-49-0		0-0.9				0.3[c,y4]
β-Caryophyllene	87-44-5	0.07-0.4	0-0.6	0.1	<0.1	0.06	0.4[j,y4]; 0.5[w]; 0.9[p]; 1.6[c]; 7.2[z4]; 13.0[x]
Caryophyllene oxide	1139-30-6	0.0-0.1		0.1		0.05	0.2[e,f]; 0.3[c,d]; 0.7[y4]; 0.9[w]; 1.9[z4]; 2.8[x]
α-Cedrene epoxide	29597-36-2						0.2[c]
Cedrol	77-53-2		0.1-1.1				
12-epi-Cedrol							0.8[m]
Cedryl acetate	77-54-3						0.9[m]
Chamazulene	529-05-5	0.6-21.6	0.8-15.3	4.7	2.4	19.3	16.6[y4]; 20.6[z1]; 21.6[v]; 24.5[p,z7]; 31.4[m]
(E)-Chrysanthemol	5617-92-5	0.0-0.1					0.3[c]
cis-Chrysanthenol	55722-60-6						0.3[c]
1,8-Cineole	470-82-6	0.0-1.0	tr-0.5	0.2	<0.1	0.09	0.1[d,g,y4]; 0.2[r,w]; 0.4[j]; 1.2[c]; 15.2[m]
α-Copaene	3856-25-5	0.0-0.1		tr		<0.01	0.04[g]; 0.06[z11]; 0.1[c]; 0.2[e]; 1.5[y4]
β-Costol	515-20-8					0.1	
α-Cubebene	17699-14-8						2.7[y2];
Cubebol	23445-02-5						0.2[m]
Cubenol	21284-22-0			0.1			
ar-Curcumene (α-)	644-30-4	0.0-0.8			<0.1		0.4[h]
p-Cymene	99-87-6	0.1-0.2	0-0.3	0.2	<0.1	0.04	<0.1[o]; 0.1[b,c,d,g,k]; 0.2[f,j,y4]; 0.3[e,h]; 1.9[n]; 2.0[r]
(E)-β-Damascenone	23726-93-4						0.2[m]
(E,E)-2,4-Decadienal	25152-84-5						0.2[c]
Decanal	112-31-2						0.1[c]
Decane	124-18-5						0.1[c]
Decanoic acid	334-48-5	0.1-0.6	tr-5.1	0.2			0.1[c]; 0.4[y4]; 0.5[z11]; 0.8[d]; 1.0[l]; 5.6[i]
Decanol	112-30-1		0-0.4				
1-Decene	872-05-9						0.1[c]
Dehydronerolidol	2387-68-0						0.09[g]
Dehydrosabina ketone	147043-52-5						0.5[c]
Dehydrosesquicineole	211237-38-6	0.2-0.6					
Dendrolasin	23262-34-2						0.2[g]
(E)-Dendrolasin	23262-34-2	0.1-0.5		tr	0.5		
Di(2-ethylhexyl) phthalate	117-81-7					0.3	+[z]
2,5-Dihydro-2,5-dimethylfuran					<0.1		
Dihydrolinalyl acetate	61476-73-1						3.4[a,z10]
Dihydronerolidol	20685-70-5			0.2			

Table 5.19.2 Constituents identified in German chamomile oils *(Continued)*

Constituent	CAS	Percentages and range					
		A	B (2)	C (51)	D (7)	E (52)	F
2,5-Dimethoxy-p-cymene	14753-08-3	0.0-0.3					
2,6-Dimethyl-5-heptenal	106-72-9				<0.1		
(E)-4,8-Dimethyl-3,7-nonadien-2-one	27539-94-2	0.0-1.2					
(Z)-4,8-Dimethyl-3,7-nonadien-2-one	27576-61-7	0.0-0.4					
4,8-Dimethyl-3,8-nonadien-2-one							0.04[g]
4,8-Dimethyl-7-nonen-2-one	3664-64-0	0.0-0.1					
Dodecane	112-40-3						0.2[c]
1-Dodecanol	112-53-8	0-0.8					
n-Eicosane	112-95-8			0.1			6.1[h]
β-Elemene	33880-83-0	0.09-0.2			<0.1	0.03	0.07[g]; 0.1[d]; 0.2[z11]; 0.3[c]; 0.8[y4]; 2.3[n]
δ-Elemene	20307-84-0	0.0-0.2			<0.1	0.01	2.1[c]
Eremophilene	10219-75-7						0.5[h]
Ethyl decanoate	110-38-3				<0.1		0.8[i]
Ethyl hexadecanoate	628-97-7						<0.1[i]; 0.4[i]
Ethyl hexanoate	123-66-0				<0.1		
Ethyl isovalerate	108-64-5				<0.1		
Ethyl-2-methylbutyrate	7452-79-1	0.1-0.3			<0.1		
α-Eudesmol	473-16-5			0.2			0.5[c]; 0.8[m]
β-Eudesmol	473-15-4						0.4[d]; 0.6[b]; 1.3[e]; 4.5[i]
γ-Eudesmol	1209-71-8			0.3			
γ-Eudesmol acetate	67996-61-6						1.0[m]
Eugenol	97-53-0						0.1[m]; 0.4[c]
α-Farnesene	502-61-4						1.0[i]; 1.8[x]; 2.4[p]; 9.7[y]
(E,E)-α-Farnesene	502-61-4	0.1-5.8	0-1.3		3.1	0.1	0.4[y4]; 0.6[d]; 0.7[f]; 0.8[b]; 3.0[c]; 4.8[z11]; 8.3[g]
(E,Z)-α-Farnesene	28973-98-0	0.0-0.9					0.5[h]
(Z,E)-α-Farnesene	26560-14-5	0.0-0.6					0.7[z11]; 0.8[g]
β-Farnesene	502-60-3						2.7[y3]; 4.9[y1]; 52.3[z10]
(E)-β-Farnesene	18794-84-8	18.4-35.4	2.3-6.6	2.3		4.0	13.5[y4]; 21.9[i]; 36.9[y]; 37.2[o]; 42.6[g]; 43.5[j,k]
(Z)-β-Farnesene	28973-97-9	0.1-0.8			<0.1		4.6[y5]; 7.1[y2]
α-Farnesol	58181-75-2		0-0.5				
(E,E)-Farnesol	106-28-5	0.0-5.6					0.1[m]; 15.6[y2]
(E,Z)-Farnesol	3879-60-5	0.0-4.0				0.06	
(Z,E)-Farnesol	3790-71-4						
Furfural	98-01-1				<0.1		
Furfuryl methyl sulfide	40228-18-0						0.1[c]
Geraniol	106-24-1				<0.1		0.1[b,d,f]; 0.3[y4]
Geranyl acetate	105-87-3						<0.05[e]; 0.1[d]
Geranyl acetone	3796-70-1						0.2[m]
Geranyl isovalerate	109-20-6			0.3			
Geranyl tiglate	7785-33-3			0.5			
Germacrene A	28387-44-2					0.03	+[z]
Germacrene B	15423-57-1	0-1.7					0.1[m]; 2.3[n]
Germacrene D	23986-74-5	1.1-3.2	0-1.9	0.2		0.8	2.4[k,o]; 2.9[f,g j]; 4.6[c]; 5.1[y,z11]; 5.5[z4]; 5.8[p]
Globulol	489-41-8						0.2[g]
Guaiazulene	489-84-9	0.0-0.2					4.2[y2]
β-Guaiene	88-84-6					0.03	
trans-β-Guaiene	192053-49-9						0.3[c]
α-Gurjunene	489-40-7						0.04[g]
Heptacosane	593-49-7					0.2	1.2[c]
2-Heptanone	110-43-0				<0.1		
(Z)-2-Heptenal							0.1[c]
Hexacosane	630-01-3					0.02	0.2[c]
Hexadecanoic acid	57-10-3	0.0-0.3		tr			0.4[c]; 1.6[i]; 1.9[i]; 4.9[i]; 6.0[y4]
Hexahydrofarnesyl acetone	502-69-2			0.1		0.07	0.1[y4]

Table 5.19.2 Constituents identified in German chamomile oils *(Continued)*

Constituent	CAS	Percentages and range					
		A	B (2)	C (51)	D (7)	E (52)	F
Hexanal	66-25-1				<0.1		
Hexanedioic acid, di-octyl ester	123-79-5					0.08	
Hexatriacontane	630-06-8						2.2[i]
(E)-2-Hexenal	6728-26-3				<0.1		
(E)-2-Hexenol	928-95-0						1.0[e]
(Z)-3-Hexenol	928-96-1	0.0-0.1			0.1		0.1[e]; 0.3[b]
(Z)-3-Hexenyl acetate	3681-71-8						0.1[d]; 0.2[b]; 0.6[f]; 0.7[e]
(Z)-2-Hexenyl isovalerate							0.1[c]
(Z)-3-Hexenyl isovalerate	35154-45-1				tr		
Hexyl butanoate	2639-63-6	0.0-0.1					
Hexyl 2-methylbutyrate	10032-15-2				0.2		
α-Humulene	6753-98-6				0.01		0.1[w]; 0.2[m]
Humulene epoxide II	19888-34-7						0.6[w]
14-Hydroxy-9-epi-(E)-caryophyllene	244226-09-3						0.4[c]
(E)-β-Ionone	79-77-6				0.1		
Isoamyl acetate	123-92-2				0.05		+[z]
Isoborneol	124-76-5						0.03[g]; 0.2[b]; 0.5[f]
Isobornyl formate	1200-67-5						0.2[c]
8-Isobutyryloxy isobornyl isobutyrate					14.0		
Isocomene	65372-78-3						0.2[z11]
α-Isocomene	65372-78-3	0.0-0.4					0.3[g,h]
Isofaurinone	87038-80-0		0.2				
Isomenthone	491-07-6						0.5[h]
Isophytol	505-32-8						0.2[e]
Isopropyl-2-methyl-butyrate	66576-71-4	0.0-0.1					
Lavandulol	498-16-8					0.02	
Lavandulyl acetate	25905-14-0	0.0-0.1					
Ledol	577-27-5				<0.1		
Lepidozene	133005-43-3					0.07	
Limonene	138-86-3	0.1-0.3	0-0.4	tr	0.1	1.2	0.1[b,c,g,h]; 0.2[e]; 0.3[d,j]; 0.8[r]
Linalool	78-70-6	0.0-2.2	0-0.3		0.1		0.1[d,f]; 0.2[b,e]; 0.3[c,w]; 0.9[y4]; 2.0[m]
cis-Linalool oxide	11063-77-7				<0.1		
trans-Linalool oxide	11063-78-8				<0.1		0.5[m]
Linoleic acid	60-33-3					0.1	+[z]
cis-Linoleic acid				0.1			
epi-Longipinanol							0.1[c]
α-Longipinene	5989-08-2						0.1[c]
β-Maaliene	489-29-2						0.07[g]
Matricin	29041-35-8						<0.5[q]
cis-p-Menth-2-en-1-ol	29803-82-5		0-1.7				
Menthone	89-80-5						+[z12]
L-Menthone	14073-97-3						0.4[h]
Menthyl acetate	16409-45-3						0.2[a,y2];
2-Methylbutyl 2-methylbutyrate	2445-78-5			0.2			
Methyl decanoate	110-42-9				<0.1		0.2[c]
Methylguaiacol	91-16-7				<0.1		
5-Methyl-3-heptanone	541-85-5						0.1[c]
6-Methyl-5-hepten-2-ol	1569-60-4				<0.1		
6-Methyl-5-hepten-2-one	110-93-0	0.0-0.1		0.1	0.1		0.2[d]; 0.3[c]; 0.5[y4]; 0.7[f]; 1.2[b]; 2.4[n]; 5.4[e]
5-Methyl-2-hexanal					<0.1		
Methyl linoleate	112-63-0						0.3[c]
2-Methyloctane	3221-61-2					0.07	

Table 5.19.2 Constituents identified in German chamomile oils *(Continued)*

Constituent	CAS	Percentages and range					
		A	B (2)	C (51)	D (7)	E (52)	F
3-Methyloctane	2216-33-3					0.08	+[z]
Methyl salicylate	119-36-8						0.1[c]
α-Muurolene	10208-80-7		0-0.5	0.2	1.1		0.06[h]; 0.2[c,m]; 0.9[y4]; 2.9[p]
γ-Muurolene	30021-74-0`	0-0.8					0.1[f]; 0.6[x]
δ-Muurolene	120021-96-7						1.0[d]
α-Muurolol	104245-48-9				0.3		
τ-Muurolol	19912-62-0						0.7[b]
Myrcene	123-35-3	0.0-0.1		0.1	<0.1	<0.01	0.1[d,e,f]; 0.2[b]; 0.3[r]; 1.0[m]; 1.3[n]
Neoiso(iso)pulegyl acetate	57576-10-0						0.2[c]
Neral	106-26-3					0.04	
Nerol	106-25-2						0.1[b,y4]
(E)-Nerolidol	40716-66-3	0.2-1.4	tr-7.4	0.3	0.2	0.4	0.1[e]; 0.2[d,g]; 0.4[b,y2]; 0.5[f,y4]
Nerolidyl acetate	56001-43-5					0.05	
Nonacosane	630-03-5					0.07	+[z]
Nonadecane	629-92-5			0.2			
Nonanal	124-19-6	0.0-0.1		0.2	<0.1		0.2[c]; 3.0[n]
Nonane	111-84-2					0.04	
3-Nonen-2-one	14309-57-0				<0.1		
trans-Ocimene	27400-72-2	0.2-0.7					
(E)-β-Ocimene	3779-61-1	0.2-0.6		0.2		0.1	0.1[c,g]; 0.2[b]; 0.4[d]; 0.5[f]; 0.7[e]; 1.9[j]
(Z)-β-Ocimene	3338-55-4	0.1-0.2				0.02	0.1[d,f]; 0.2[b,e]; 0.7[g]; 1.7[j]
Octacosane	630-02-4						0.2[c]
n-Octadecane	593-45-3			0.2			0.9[h]
(E,E)-3,5-Octadien-2-one	30086-02-3				<0.1		
n-Octanal	124-13-0	0.0-0.1		0.1	<0.1		0.1[c]
Octane	111-65-9					0.2	
2-Octanol	123-96-6				<0.1		
3-Octanol	589-98-0			0.1			
1-Octen-3-ol	3391-86-4			<0.1			
3-Octen-2-one	1669-44-9			<0.1			
γ-Palmitolactone	730-46-1		0.1				
α-Patchoulene	560-32-7						0.1[e]; 0.2[f]; 0.3[d]; 0.6[b]
Pentacosane	629-99-2		0.7			0.8	1.0[l]; 2.1[c]
Pentadecane	629-62-9	0.0-0.1					
2-Pentylfuran	3777-69-3	0.0-0.1					0.05[g]
α-Phellandrene	99-83-2						<0.05[e]; 0.1[y2]; 1.0[m]
2-Phenethyl alcohol	60-12-8				0.2		
(E)-Phytol	150-86-7						0.2[g]
Phytol acetate	5016-85-3						0.2[c]
Phytone	16825-16-4	0.0-1.2					
α-Pinene	80-56-8	0.0-1.2	0-0.1	tr	<0.1	0.01	0.04[h]; 0.1[b]; 0.3[y4]; 0.5[e]; 1.9[x]; 4.7[n]; 8.1[m]
β-Pinene	127-91-3	0.0-2.2		0.2	<0.1		<0.1[o]; 0.1[y2]; 0.3[k]; 3.7[n]; 10.1[m]
trans-Pinocarveol	1674-08-4	0.0-0.1					
Pinocarvone	30460-92-5				<0.1		1.0[o]; 2.9[k]
Piperitone	89-81-6						1.1[m]
Prochamazulene	489-87-2						7.1[z4]
Pulegone	89-82-7						0.2[b]; 0.4[d]; 0.5[e,f]
Pyrethrin	88108-26-3						0.07[h]
Sabinene	3387-41-5				<0.1	0.09	<0.1[o]; 0.04[g]; 0.1[c,y4]; 0.2[k]; 0.3[w]; 2.0[m]
trans-Sabinene hydrate	17699-16-0						0.1[c]; 0.3[j]
Safrole	94-59-7				<0.1		
Salicylic acid	69-72-7						3.0[n]
Salvial-4(14)-en-1-one	73809-82-2						0.6[c]
(Z)-α-Santalol	115-71-9						0.1[c]
(Z)-β-Santalol	42495-69-2				1.0		
Sativene	6813-05-4						0.04[g]
β-Selinene	17066-67-0					0.05	0.2[g]; 0.6[h,y4]; 0.7[c]
β-Sesquiphellandrene	20307-83-9	0.0-0.5					
Sesquirosefuran	39007-93-7						0.2[g]

Table 5.19.2 Constituents identified in German chamomile oils *(Continued)*

Constituent	CAS	Percentages and range					
		A	B (2)	C (51)	D (7)	E (52)	F
(*E*)-Sesquisabinene hydrate	145512-84-1						1.5[c]
Spathulenol	6750-60-3	0.5-5.8	1.7-4.8	2.4	3.4	0.7	2.6[q]; 3.0[i]; 3.7[i]; 4.0[w]; 6.8[z6]; 10.8[h]; 11.3[c]
cis-Spiroether	16863-61-9	2.5-10.5	8.8-26.1	11.9			7.6[b]; 7.7[f]; 8.6[i]; 10.4[z11]; 13.2[y4]; 13.4[c]
trans-Spiroether	50257-98-2	0.2-16.8	0-1.2	0.4			0.3[c]; 0.4[b]; 0.6[f]; 0.7[i,z11]; 1.0[g]; 1.2[y4]; 6.0[p]
Spiroethers							3.0[z2]; 4.7[z5]; 8.0[u]; 11.5[z3]; 16.7[z6]; 18.3[t]
α-Terpinene	99-86-5			tr		0.01	<0.1[o]; 0.1[j,y2]; 0.5[r]; 0.9[m]
γ-Terpinene	99-85-4	0.1-0.2	0-0.1	0.2	<0.1	0.1	0.1[b,j,k]; 0.3[d]; 0.5[f]; 0.6[o]; 1.6[e]; 2.4[n]
Terpinen-4-ol	562-74-3	0.0-0.1		0.1	<0.1	0.02	0.1[b,d,e,f]; 0.2[c]; 1.7[r]; 4.1[m]
1-Terpineol	586-82-3				<0.1		
α-Terpineol	98-55-5	0.0-1.8	0-0.7	0.1	0.1	0.05	0.1[y4]; 0.2[e]; 0.3[d,f]; 0.4[c]; 6.0[z4]
γ-Terpineol	586-81-2	0.0-0.3					0.1[b]
Terpinolene	586-62-9						0.06[z11]
Tetracosane	646-31-1					0.05	0.4[c]; 12.5[h]
n-Tetradecane	629-59-4	0.0-0.1					
α-Thujene	2867-05-2						0.1[k]; 0.6[m]
α-Thujone	546-80-5				<0.1		0.1[h]; 0.2[c]
cis-Thujone	471-15-8						<0.05[w]
trans-Thujone	33766-30-2						<0.05[w]
Torilenol	84071-85-2						1.9[m]
Tricosane	638-67-5			0.1		0.07	0.9[c]
Tricyclene	508-32-7						0.1[e]; 0.2[j]
Tridecanal	10486-19-8			tr			
2,2,6-Trimethylcyclo-hexanone	2408-37-9				<0.1		
Undecanoic acid	112-37-8			0.2			
Valencene	4630-07-3					0.03	
Viridiflorol	552-02-3			tr			0.2[y2]
p-Xylene	106-42-3						0.1[c]
α-Ylangene	14912-44-8				0.1		0.2[y4]
Yomogi alcohol	26127-98-0				<0.1	0.03	1.3[c]
Yomogi alcohol ace-tate	40018-26-6	0.0-0.1					

A eighty-five German chamomile essential oil samples from Hungary, Serbia, France, Egypt, India, South Africa and Germany analyzed between 1993 and 2013; lowest and highest concentrations given (E. Schmidt, unpublished data)

B thirteen lab-hydrodistilled oils from flowers bought in retail pharmacies or cultivated in eleven European countries; minimum and maximum concentrations given (ref. 2)

C one lab-hydrodistilled oil from plants cultivated in Estonia (ref. 51)

D one sample of chamomile oil obtained by lab-hydrodistillation from wild plants collected in Esphahan, Iran (ref. 7)

E four lab-hydrodistilled oils from plants cultivated in Iran under different irrigation regimes; highest concentrations given (ref. 52)

F data from other studies (indicated with superscript letters); highest concentrations found in any study reviewed here given; when two or more oils were investigated, only the highest concentrations are mentioned, unless indicated otherwise

[a] incorrect identification/the compound does not exist naturally (ref. 48,54); [b] two lab-hydrodistilled oils from two Chamomilla recutita cultivars from India (ref. 10); [c] two oils from tubular and ligulate florets from Italian wild Chamomilla recutita (ref.4); [d] one sample of lab-hydrodistilled chamomile essential oil from the lower region of the Himalayas, India (ref. 8); [e] one lab-hydrodistilled oil from flowers (whole capitula) of chamomile plants cultivated in India; in the same article, also the results of oils obtained from disc florets, ray florets, leaves, stems and roots of Chamomilla recutita were presented (ref. 9); [f] three lab-hydrodistilled oils from three German chamomile accessions produced in India (ref. 12); [g] one commercial chamomile oil from Nepal (ref. 30); [h] ten lab-hydrodistilled oils from flowers bought in Romanian pharmacies, nine samples originating from Romania and one from Germany of which six in the form of tea bags (ref. 11); the presence of L-menthone, isomenthone, (E)-anethole, pyrethrin and ambrettolide in some oils indicated, according to the authors, 'inadequate processing or contamination'; [i] five samples of chamomile oil produced from plants grown in Brazil plus one Egyptian oil (ref. 39); [j] one lab-hydrodistilled oil produced from the flowering aerial parts of chamomile collected in Serbia (ref. 42); [k] samples of oil from wild chamomile and *C.* recutita cultivated in Iran under different fertilization schemes (ref. 43); [l] one commercial chamomile oil from Egypt (ref. 31); [m] one lab-hydrodistilled oil from flowers and head branches of chamomile growing wild in Iran (ref. 17); [n] one lab-hydrodistilled oil from flower heads of chamomile growing wild in Greece (ref. 21); also, the results of analysis of an oil produced from disk florets were presented in the article; [o] two lab-hydrodistilled oils obtained from plants cultivated in Iran in two climatic zones out of seeds of wild growing chamomile (ref. 19); [p] four oils from flowers of two wild growing and two cultivated chamomile populations in Hungary (ref. 28); [q] one steam-distilled oil from Italy (ref. 27); [r] one lab-hydrodistilled oil from Egypt (ref. 26); [s] one commercial oil purchased from a German producer (ref. 5); [t] ten laboratory steam-distilled oils from dried Egyptian flowers with distillation times varying between half an hour to 12 hours (ref. 24); [u] one lab-distilled flower oil from Croatia; the authors also produced and investigated leaf oil, petal oil and yellow floret oil (ref. 45); [v] one oil from fresh and one from shade-dried flowers (ref. 46);

Table 5.19.2 Constituents identified in German chamomile oils *(continued)*

[w] one lab-hydrodistilled oil from Polish dried chamomile flowers (ref. 37); the authors also investigated an oil prepared from granulated chamomile material; [x] five lab-hydrodistilled oils from wild chamomile flowers, both fresh and dried for varying times (ref. 35); [y] one commercial oil from Nepal (ref. 34); [y1] one lab-hydrodistilled oil from chamomile flowers grown in Brazil (ref. 55); [y2] one lab-hydrodistilled oil from the flowers of *C. recutita* collected in Iran (ref. 56); [y3] one lab-hydrodistilled oil from a wild Hungarian population (ref. 57); [y4] ten lab-hydrodistilled oils from 10 farms cultivating *C. recutita* in India (ref. 59); [y5] oils produced from chamomile flowers cultivated in Iran with nitrogen application and plant density as parameters (ref. 60); [z] various lab-hydrodistilled oils from Iran obtained from plants cultivated at different irrigation regimes and plant densities; indicated with +[z], as the concentrations were not mentioned (ref. 6); [z1] fifteen lab-hydrodistilled oils from cultivated plants from seeds of wild chamomile collected in various parts of Italy; only α-bisabolol and chamazulene were analyzed (ref. 33); [z2] eight lab-hydrodistilled oils from two diploid and two tetraploid chamomile cultivars from Italy with two harvest times (ref. 32); [z3] nine lab-hydrodistilled oils from wild chamomile populations collected at seven sites in Iran (ref. 18); [z4] one lab-hydrodistilled oil from the aerial parts of wild Moroccan chamomile (ref. 20); [z5] one lab-hydrodistilled oil from flowers collected at an abandoned chamomile plantation in Brazil (ref. 22); [z6] sixty-eight lab-hydrodistilled oils from seventeen chamomile cultivars from the former Yugoslavia with different sowing and harvesting times; only chamazulene, (E)-β-farnesene, spathulenol and the spiroethers were analyzed (ref. 23); [z7] oils of 32 accessions of chamomile grown in Hungary; only farnesene, bisabolol oxide A, bisabolol oxide B, α-bisabolol and chamazulene were investigated (ref. 40); [z8] oils from eight chamomile populations over two seasons growing wild in Hungary; only only farnesene, bisabolol oxide A, α-bisabolol and chamazulene were investigated (ref. 41); [z9] nine oils from chamomile selections from different regions of Iran; only the major constituents were investigated (ref. 44); [z10] one steam-distilled oil from dried flowers grown commercially in Brazil (ref. 49); [z11] two oils of Egyptian and Spanish origin (ref. 50); [z12] four essential oils of ligulate and tubular flowers of diploid and tetraploid chamomile plants; components indicated with +[z12], as the amounts cannot be compared with the other data (ref. 13)

tr: trace (in column C: <0.05); + present in the oil investigated, but quantity not stated or expressed in a manner which cannot be compared to the other data

of the major constituents, six chemotypes were proposed in 1987, which classification is frequently cited (53): 1 Type A: bisabolol oxide A as main component; 2 Type B: bisabolol oxide B as main component; 3 Type C: α-bisabolol as main component; 4 Type D: comparable amounts of α-bisabolol and bisabolol oxide A and B; 5: α-bisabolone oxide A as main component; 6: green essential oil with low amount of matricine.

Another classification was proposed more recently (3): A: α-bisabolol oxide A; B: α-bisabolol oxide B; C: α-bisabolol and α-bisabolol oxide A; D: α-bisabolol and α-bisabolol oxide B; E: α-bisabolone oxide; F: α-bisabolol, chamazulene and *trans*-spiroether. However, neither of these chemotype classifications mentions (E)-β-farnesene, which has been found in very high concentrations (up to 43.5%) and sometimes as the dominant ingredient in several publications (19,30,34,42,43,

unpublished data from E.S) and which has the highest allowed concentrations in ISO norms (15-51%).

The data of ref. 29 are not presented in Table 5.19.2, as they contain many mistakes.

CONTACT ALLERGY/ALLERGIC CONTACT DERMATITIS

Chamomile oil (unspecified)

General

Contact allergy to/allergic contact dermatitis from chamomile oil (botanical source not specified) has been reported in a few publications only. In groups of consecutive patients suspected of contact dermatitis (and in one study [67] also other forms of dermatitis), prevalence rates of up to 0.5% positive patch test reactions have been

Table 5.19.3 Major constituents of German chamomile oils

Constituent	CAS	Percentages and range					
		A	B (2)	C (51)	D (7)	E (52)	F
α-Bisabolol	515-69-5	1.5-38.3	0.1-44.2	5.6	1.7		50.6[z1]; 56.9[y2]; 57.91[z8]; 61.5[z3]; 71.9[z2]
α-Bisabolol oxide A	22567-36-8	1.0-46.0	3.1-56.0	39.4	43.8	35.4	46.6[i]; 56.5[y1]; 57.1[z2]; 59.0[z3]; 63.0[y5]
β-Farnesene	502-60-3						2.7[y3]; 4.9[y1]; 52.3[z10]
(E)-β-Farnesene	18794-84-8	18.4-35.4	2.3-6.6	2.3		4.0	13.5[y4]; 21.9[i]; 36.9[y]; 37.2[o]; 42.6[g]; 43.5[j,k]
α-Bisabolol oxide B	55399-12-7	3.1-12.1	3.9-27.2	9.9	3.8	35.6	23.3[r]; 23.5[z5]; 28.0[x]; 30.6[w]; 30.9[b]; 33.0[i]
α-Bisabolone oxide A	58985-73-2	0.1-8.5	0.5-24.8	13.9	13.6	9.1	6.1[j]; 7.4[y5]; 9.2[q]; 11.2[c]; 17.4[e]; 31.5[h]
Chamazulene	529-05-5	0.6-21.6	0.8-15.3	4.7	2.4	19.3	16.6[y4]; 20.6[z1]; 21.6[v]; 24.5[p,z7]; 31.4[m]
cis-Spiroether	16863-61-9	2.5-10.5	8.8-26.1	11.9			7.6[b]; 7.7[f]; 8.6[i]; 10.4[z11]; 13.2[y4]; 13.4[c]
Spiroethers							3.0[z2]; 4.7[z5]; 8.0[u]; 11.5[z3]; 16.7[z6]; 18.3[t]
trans-Spiroether	50257-98-2	0.2-16.8	0-1.2	0.4			0.3[c]; 0.4[b]; 0.6[f]; 0.7[i,z11]; 1.0[g]; 1.2[y4]; 6.0[p]
Spathulenol	6750-60-3	0.5-5.8	1.7-4.8	2.4	3.4	0.7	2.6[q]; 3.0[i]; 3.7[i]; 4.0[w]; 6.8[z6]; 10.8[h]; 11.3[c]
Bicyclogermacrene	24703-35-3	0.3-1.8		tr	0.1	0.6	0.9[h]; 2.0[g]; 2.1[z11]; 4.1[v]; 4.4[c]; 5.2[i]; 6.6[o]

LEGEND: see under Table 5.19.2

observed, but relevance data are lacking. Both cases of (possible) allergic contact dermatitis were from occupational exposure in aromatherapists. In one case report, α-pinene may have been an allergen in chamomile oil.

TESTING IN GROUPS OF PATIENTS

Routine testing

Two hundred dermatitis patients from Poland were patch tested with 'camomile oil' 2% in petrolatum and one (0.5%) reacted (63). Between 1967 and 1970, 290 patients with various forms of dermatitis were tested in south Italy and 1 (0.3%) had a positive patch test (67). In neither of these studies was the relevance mentioned,

Testing in groups of selected patients

Fifty-one patients allergic to *Myroxylon pereirae* resin (balsam of Peru) and/or turpentine and/or wood tar and/or colophony were tested with 'camomile oil' 2% in petrolatum and one (2.0%) had a positive patch test; relevance data were not provided (63). One relevant positive patch test reaction to chamomile oil 1% in petrolatum was found among 122 patients (0.8%) who reported adverse cutaneous reactions to products (notably cosmetics) containing botanical ingredients in a questionnaire and who were subsequently tested with a 'botanical series' (64).

Case reports

An aromatherapist had chronic hand dermatitis and was patch test positive to 17 of 20 oils used at her work (tested 1% and 5% in petrolatum), including chamomile oil (61). Another aromatherapist had non-occupational contact dermatitis with allergies to multiple essential oils used at work, including chamomile oil. The patient also reacted to geraniol, linalool, linalyl acetate, α-pinene, the fragrance mix and various other fragrance materials; α-pinene was demonstrated by GC-MS in the chamomile oil (62). In commercial *Roman* chamomile oil, α-pinene has been found in a maximum concentration of 11.5% (Table 5.20.2, column A in Chapter 5.20, Roman chamomile oil).

Chamomile oil, German

General

Contact allergy to German chamomile oil has been reported in a few publications, but relevance data were lacking and no cases of allergic contact dermatitis have been identified.

Testing in groups of patients

A group of 51 patients from Poland allergic to *Myroxylon pereirae* resin (balsam of Peru) and/or wood tar and/or colophony was tested with German chamomile oil 2% in petrolatum and one (2.0%) had a positive patch test; relevance data were not provided (63). Another group of 86 patients from Poland previously reacting to the fragrance mix was tested with chamomile oil and three (3.4%) had a positive patch test reaction; relevance data were not provided (65). Nine patients from Denmark who had previously tested

positive to a short ether extract of German chamomile were tested with blue chamomile oil 1% and 4% in petrolatum and one (11%) had a positive reaction; the relevance of this was not mentioned (66).

LITERATURE

1 Srivastava JK, Shankar E, Gupta S. Chamomile: a herbal medicine of the past with bright future. Mol Med Report 2010;3:895-901

2 Orav A, Raal A, Arak E. Content and composition of the essential oil of *Chamomilla recutita* (L.) Rauschert from some European countries. Nat Prod Res 2010;24:48-55

3 Schmidt E, Höferl M, Wanner J, Jirovetz L, Buchbauer G, Gochev V, Geissler M. Chemical composition and antibacterial activity of five chemotypes of *Chamomilla recutita* (L.) Rauschert cultivated in Europe, India and South Africa. Data presented at the 43rd International Symposium on Essential Oils, Lisboa, Portugal, September 5-8, 2012

4 Tirillini B, Pagiotti R, Menghini L, Pintore G. Essential oil composition of ligulate and tubular flowers and receptacles from wild *Chamomilla recutita* (L.) Rausch grown in Italy. J Essent Oil Res 2006;18:42-45

5 Stević T, Berić T, Savikin K, Soković M, Godevac D, Dimkić I, et al. Antifungal activity of selected essential oils against fungi isolated from medicinal plant. Ind Crops Prod 2014;55:116-122

6 Pirzad A, Reza Shakiba M, Zehtab-Salmasi S, Mohammadi A, Darvishzadeh R, Hassani A. Effects of irrigation regime and plant density on essential oil composition of German chamomile (*Matricaria chamomilla*). J Herbs Spices Med Plants 2011;17:107-118

7 Pino JA, Bayat F, Marbot R, Aguero J. Essential oil of chamomile *Chamomilla recutita* (L.) Rausch. from Iran. J Essent Oil Res 2002;14:407-408

8 Sashidhara KV, Verma RS, Ram P. Essential oil composition of *Matricaria recutita* L. from the lower region of the Himalayas. Flavour Fragr J 2006;21:274-276

9 Das M, Ram G, Singh A, Mallavarapu GR, Ramesh S, Ram M, Kumar S. Volatile constituents of different plant parts of *Chamomilla recutita* L. Rausch. grown in the Indo-Gangetic plains. Flavour Fragr J 2002;17:9-12

10 Das M, Mallavarapu GR, Gupta SK, Kumar S. Prospects of cultivation of chamomile *Chamomilla recutita* and production of chamomile oil in India. J Med Arom Plant Sci 2000;23:747-750. Data also shown in ref. 47

11 Cioanca O, Aprotosoaie AC, Spac A, Hancianu M, Stanescu UH. Contribution to the study of the pharmaceutical quality of some chamomile commercial samples. Note I. The analysis of the volatile oil. Farmacia (Bucharest, Romania) 2010;58:308-314

12 Das M, Kumar S, Mallavarapu GR, Ramesh S. Composition of the essential oils of the flowers of three accessions of *Chamomilla recutita* (L.) Rausch. J Essent Oil Res 1999;11:615-618

13 Pekic B, Zekovic Z, Petrovic L, Adamovic D. Essential oil of chamomile ligulate and tubular flowers. J Essent Oil Res 1999;11:16-18

14 Lawrence BM. Progress in essential oils. Perfum Flavor 2008;33(7):44-? (last page unknown)

15 Singh O, Khanam Z, Misra N, Srivastava MK. Chamomile (Matricaria chamomilla L.): an overview. Pharmacog Rev 2011;5:82-95

16 Ehlert D, Adamek R, Giebel A, Horn H-J. Influence of comb parameters on picking properties for chamomile flowers (Matricaria recutita). Ind Crop Prod 2011;33:242-247

17 Farhoudi R. Chemical constituents and antioxidant properties of Matricaria recutita and Chamaemelum nobile essential oil growing wild in the south west of Iran. J Essent Oil Bear Plants 2013;16:531-537

18 Šalamon I, Ghanavati M, Abrahimpour F. Potential of medicinal plant production in Iran and variability of chamomile (Matricaria recutita L.) essential oil quality. J Essent Oil Bear Plants 2010;13:638-643

19 Karami A, Khush-Khui M, Saharkhiz MJ, Sefidkon F. Essential oil content and compositions of German chamomile (Chamomilla recutita L. Rauschert) cultivated in temperate and subtropical zones of Iran. J Essent Oil Bear Plants 2009;12:703-707

20 Chebli B, Hmamouchi M, Achouri M, Idrissi Hassani LM. Composition and in vitro fungitoxic activity of 19 essential oils against two post-harvest pathogens. J Essent Oil Res 2004;16:507-511

21 Papazoglou V, Anastassaki T, Demetzos C, Loukis A. Composition of the essential oils of wild Chamomilla recutita (L.) Rausch. grown in Greece. J Essent Oil Res 1998;10:635-636

22 Matos FJA, Machado MIL, Alencar JW, Craveiro AA. Constituents of Brazilian chamomile oil. J Essent Oil Res 1993;5:337-339

23 Gasic O, Lukic V, Adamovic D. The influence of sowing and harvest time on the essential oils of Chamomilla recutita (L.) Rausch. J Essent Oil Res 1991;3:295-302

24 Gawde A, Cantrell CL, Zheljazkov VD, Astatkie T, Schlegeld V. Steam distillation extraction kinetics regression models to predict essential oil yield, composition, and bioactivity of chamomile oil. Ind Crops Prod 2014;58:61-67

25 Raal A, Orav A, Püssa T, Valner C, Malmiste B, Arak E. Content of essential oil, terpenoids and polyphenols in commercial chamomile (Chamomilla recutita L. Rauschert) teas from different countries. Food Chem 2012;131:632-638

26 Hamdy Roby MH, Sarhana MA, Abdel-Hamed Selima K, Khalela KI. Antioxidant and antimicrobial activities of essential oil and extracts of fennel (Foeniculum vulgare L.) and chamomile (Matricaria chamomilla L.). Ind Crops Prod 2013;44:437-445

27 Scalia S, Giuffreda L, Pallado P. Analytical and preparative supercritical fluid extraction of Chamomile flowers and its comparison with conventional methods. J Pharm Biomed Anal 1999;21:549-558

28 Szöke E, Mádaya E, Tyihák E, Kuzovkina IN, Lemberkovics E. New terpenoids in cultivated and wild chamomile (in vivo and in vitro). J Chromatogr B 2004;800:231-238

29 Jamalian A, Shams-Ghahfarokhi M, Jaimand K, Pashootan N, Amani A, Razzaghi-Abyaneh M. Chemical composition and antifungal activity of Matricaria recutita flower essential oil against medically important dermatophytes and soil-borne pathogens. J Mycol Méd 2012;22:308-315

30 Heuskin S, Godin B, Leroy P, Capella Q, Wathelet J-P, Verheggen F, et al. Fast gas chromatography characterisation of purified semiochemicals from essential oils of Matricaria chamomilla L. (Asteraceae) and Nepeta cataria L. (Lamiaceae). J Chromatogr A 2009;1216:2768-2775

31 Mitoshi M, Kuriyama I, Nakayama H, Miyazato H, Sugimoto K, Kobayashi Y, et al. Effects of essential oils from herbal plants and citrus fruits on DNA polymerase inhibitory, cancer cell growth inhibitory, antiallergic, and antioxidant activities. J Agric Food Chem 2012;60:11343–11350

32 D' Andrea L. Variation of morphology, yield and essential oil components in common chamomile (Chamomilla recutita (L.) Rauschert) cultivars grown in Southern Italy. J Herbs Spices Med Plants 2002;9:359-365

33 Taviani P, Rosellini D, Veronesi F. Variation for agronomic and essential oil traits among wild populations of Chamomilla recutita (L.) Rauschert from Central Italy. J Herbs Spices Med Plants 2002;9:353-358

34 Yonzon M, Lee DJ, Yokochi T, Kawano Y, Nakahara T. Antimicrobial activities of essential oils of Nepal. J Essent Oil Res 2005;17:107-111

35 Borsato AV, Doni-Filho L, Rakocevic M, Coccio LC, Paglia EC. Chamomile essential oils extracted from flower heads and recovered water during drying process. J Food Process Preserv 2009;33:500-512

36 McKay DL, Blumberg JB. A review of the bioactivity and potential health benefits of chamomile tea (Matricaria recutita L.). Phytother Res 2006;20:519-530

37 Kowalski R, Wawrzykowski J. Essential oils analysis in dried materials and granulates obtained from Thymus vulgaris L., Salvia officinalis L., Mentha piperita L. and Chamomilla recutita L. Flavour Fragr J 2009;24:31-35

38 Lawrence BM. Progress in essential oils. Perfum Flavor 2012;37(May):54-60

39 Presibella MM, DeBaggi Villas-Boas L, da Silva Belletti KM, de Moraes Santos CA, Weffort-Santos AM. Comparison of chemical constituents of Chamomilla recutita (L.) Rauschert essential oils and its antichemotactic activity. Braz Arch Biol Technol 2006;49:717-724. Data cited in ref. 38

40 Gosztola B, Nemeth E, Sarozi Sz, Szabo H, Kozak A. Comparative evaluation of chamomile (Matricaria recutita L.) populations from different origins. Intern J Hort Sci 2006;12:91-95. Data cited in ref. 38

41 Gosztola B, Nemeth E, Kozak A, Sarozi Sz, Szabo K. Comparative evaluation of Hungarian chamomile (*Matricaria recutita* L.) populations. In: I Salamon, Ed. Proceedings of the First International Symposium on Chamomile Research, Development and Production. Acta Hort 2007;749:157-162. Data cited in ref. 38

42 Soković M, Marin PD, Brkić D, van Griensven LJLD. Chemical composition and antibacterial activity of essential oils of ten aromatic plants against human pathogenic bacteria. Food 2007;1:220-226

43 Karami A, Khush-Khui M, Sefidkon F. Effects of nitrogen and potassium on yield, essential oil content and composition of cultivated and wild populations of *Chamomilla recutita* L. Rauschert. J Med Arom Plant Sci 2008;30:113-116. Data cited in ref. 38

44 Salamon I, Ghanavati M, Abrahimpour F. Potential of medicinal plant production in Iran and variability of chamomile (*Matricaria recutita* L.) essential oil quality. J Essent Oil Bear Plants 2010;13:638-643

45 Grgesina D, Mandic ML, Karuza L, Klapec T, Bockinac D. Chemical composition of different parts of *Matricaria chamomilla*. Prehrambeno-Technol Biotechnol Rev 1995;33:111-113. Data cited in ref. 14

46 Mishra DK, Naik SN, Srivastava VK, Prasad R. Effect of drying *Matricaria chamomilla* flowers on chemical composition of essential oil. J Med Arom Plant Sci 1999;21:1020-1025. Data cited in ref. 14

47 Kumar S, Das M, Singh A, Ram G, Mallavarapu GR, Ramesh S. Composition of the essential oils of the flowers, shoots and roots of two cultivars of *Chamomilla recutita*. J Med Arom Plant Sci 2001;23:617-623. Some data also shown in ref. 10

48 Lawrence BM. Progress in essential oils. Perfum Flavor 2005;30(8):56-? (last page unknown)

49 Povh NP, Garcia CA, Marquez MOM, Meireles MAA. Extraction of essential oil and oleoresin from chamomile (*Chamomilla recutita* (L) Rauschert) by steam distillation and extraction with organic solvents: a process design approach. Rev Bras Pl Med 2001;4:1-8. Data cited in ref. 48

50 Kubeczka K-H, Formacek V. Essential oils analysis by capillary gas chromatography and carbon-13 NMR spectroscopy, 2nd edition. New York, USA: John Wiley and Sons, 2002:49-60. Data cited in ref. 48

51 Raal A, Kaur H, Orav A, Arak E, Kailas T, Müürisepp M. Content and composition of essential oils in some Asteraceae species. Proceedings of the Estonian Academy of Sciences 2011;60:55-63

52 Pirzad A, Alyari H, Shakiba M, Zehtab-Salmasi S, Mohammadi A. Essential oil content and composition of German chamomile (*Matricaria chamomilla* L.) at different irrigation regimes. Agronomy J 2006;5:451-455

53 Schilcher H. Die Kamille. In: Handbuch für Ärtze, Apotheker und andere Naturwissenschaftler. Wissenschaftliche Verlagsgesellschaft mbH: Stuttgart, Germany, 1987:152

54 Lawrence BM. Progress in essential oils. Perfum Flavor 2014;39(November):40-50

55 Maia NB, Bovi OA, Perecin MB, Marques MOM, Granja NP, Le Roy Trujillo A. New crops with potential to produce essential oil with high linalool content helping preserve rosewood—An endangered Amazon species. Acta Hort 2004;629:39-43. Data cited in ref 54

56 Tolouee M, Alinezhad S, Saberi R, Eslamifar A, Zad SJ, Jaimand K, et al. Effect of *Matricaria chamomilla* L. flower essential oil on the growth and ultrastructure of *Aspergillus niger* van Tieghm. Int J Food Microbiol 2010;139:127-133

57 Gosztola B, Sarosi S, Németh E. Variability of the essential oil content and composition of chamomile (*Matricaria recutita* L.) affected by weather conditions. Nat Prod Commun 2010;5:465-470

58 Lal RK, Chandra R, Misra HO, Shankar H, Lal C, Singh AK, et al. Registration of a new released high yielding variety 'CIMAP Sammohak' of chamomile (*Chamomilla recutita* L. Rauschert). J Med Arom Plant Sci 2010;32:450-482. Data cited in ref. 54

59 Lohani H, Chauhan NK, Mohan M, Kumar K. Evaluation of essential oil of *Matricaria chamomilla* cultivated in farmers field of Uttarakhand. J Med Arom Plant Sci 2011;33:411-414. Data cited in ref. 54

60 Rahmati M, Azizi M, Khayyat MH, Nemati H, Asili J. Yield and oil constituents of chamomile (*Matricaria chamomilla* L.) flower depending on nitrogen application, plant density and climatic conditions. J Essent Oil Bear Plants 2011;14:731-741

61 Selvaag E, Holm J, Thune P. Allergic contact dermatitis in an aromatherapist with multiple sensitizations to essential oils. Contact Dermatitis 1995;33:354-355

62 Dharmagunawardena B, Takwale A, Sanders KJ, Cannan S, Roger A, Ilchyshyn A. Gas chromatography: an investigative tool in multiple allergies to essential oils. Contact Dermatitis 2002;47:288-292

63 Rudzki E, Grzywa Z, Bruo WS. Sensitivity to 35 essential oils. Contact Dermatitis 1976;2:196-200

64 Corazza M, Borghi A, Gallo R, Schena D, Pigatto P, Lauriola MM, et al. Topical botanically derived products: use, skin reactions, and usefulness of patch tests. A multicentre Italian study. Contact Dermatitis 2014;70:90-97

65 Rudzki E, Grzywa Z. Allergy to perfume mixture. Contact Dermatitis 1986;15:115-116

66 Paulsen E, Christensen LP, Andersen KE. Cosmetics and herbal remedies with Compositae plant extracts – are they tolerated by Compositae-allergic patients? Contact Dermatitis 2008;58:15-23

67 Meneghini CL, Rantuccio F, Lomuto M. Additives, vehicles and active drugs of topical medicaments as causes of delayed-type allergic dermatitis. Dermatologica 1971;143:137-147

Chapter 5.20 CHAMOMILE OIL, ROMAN

Chamomile (may also be spelled camomile sometimes) oil may refer either to German chamomile oil obtained from *Chamomilla recutita* L. or to Roman (synonym: English) chamomile oil obtained from *Chamaemelum nobile* (L.) All., better known as *Anthemis nobilis* L. In non-botanical literature, often the nature of the oil and the botanical source are not specified. This chapter discusses Roman chamomile oil; the German chamomile oil (also termed blue chamomile oil) is presented in Chapter 5.19.

DEFINITION

Chamomile oil Roman (English) is the essential oil obtained from the flowering tops of the (common, corn, garden, English, Roman) chamomile, *Chamaemelum nobile* (L.) All. (synonyms: *Anthemis nobilis* L., *Ortmenis nobilis* (L.) J. Gay ex Coss. et Germ.).

INCI NOMENCLATURE

Description/definition: Anthemis nobilis flower oil is the volatile oil distilled from the dried flower heads of the Roman chamomile, *Anthemis nobilis* L., Asteraceae
INCI name EU & USA: Anthemis nobilis flower oil
Other names: Roman chamomile oil; English chamomile oil; camomile oil Roman
CAS registry number(s): 8015-92-7; 84649-86-5
EINECS number(s): 283-467-5

ISO (INTERNATIONAL ORGANIZATION FOR STANDARDIZATION) STANDARD

There is currently no ISO standard for Roman chamomile oil. AFNOR (Association Française de normalisation) values are shown in Table 5.20.1.

Table 5.20.1 AFNOR values (%) for Roman chamomile oil[a]

Compound	CAS	Minimum	Maximum
Isobutyl angelate	7779-81-9	30.0	45.0
Isoamyl angelate	10482-55-0	12.0	22.0
Methallyl angelate	61692-78-2	6.0	10.0
Isobutyl isobutyrate	97-85-5	2.0	9.0
Methylpentyl angelate	53082-58-9	3.0	7.0
trans-Pinocarveol	1674-08-4	2.0	7.0
Pinocarvone	30460-92-5	1.3	6.0
Isoamyl isobutyrate	2050-01-3	2.5	5.0
Isobutyl methacrylate	97-86-9	0.5	3.0
3-Methylbutyl methacrylate	7336-27-8	0.5	1.5

[a] AFNOR NF T 75-253 Huile essentielle de camomille romaine, ©AFNOR 2008; 11, rue de Francis de Pressensé, 93571 La Plaine Saint-Denis Cedex, France

THE PLANT, THE OIL, AND THEIR USES

Chamaemelum nobile (L.) All. is a low perennial herb with daisy-like white flowers, native to the Portuguese Azores, northern Africa (Algeria, Morocco), and Europe (Ireland, United Kingdom, France, Portugal, Spain). It is naturalized in Australia, New Zealand, other parts of Europe and the USA and also widely cultivated (GRIN Taxonomy for Plants; 12). One variety with full flowers and containing a large amount of essential oil is preferred for cultivation: *C. nobile* var. *flora plena* (synonym: var. *ligulosa*) (2).

Traditionally, chamomile is considered to be an antiseptic, disinfectant, bactericidal, fungicidal and vermifuge agent. It has been used for centuries as anti-inflammatory, antioxidant, mild astringent, mild sedative, antispasmodic, antibacterial and healing medicine. Oral dosage forms (decoctions and infusions) are used for the symptomatic treatment of gastrointestinal disorders and of the painful component of functional digestive symptoms. External applications of extracts and lotions are recommended as repellent, emollient, in the treatment of skin disorders and for eye irritation or discomfort of various etiologies. Furthermore, it is used as an analgesic agent in diseases of the oral cavity, oropharynx or both and as a mouthwash for oral hygiene (12). Many different preparations of chamomile have been developed, the most popular of which is in the form of herbal tea.

The possible health effects of chamomile (not properly differentiated between the German and Roman varieties) have been reviewed (1). The European Medicines Agency in 2011 concluded that 'The provided clinical and non-clinical data do not fulfil the requirements of a well-established medicinal use with recognised efficacy and an acceptable level of safety of Roman chamomile products' (19).

Essential oils of chamomile are used extensively in the cosmetics and perfume industries (2). It is also popular in aromatherapy, whose practitioners believe it to be a calming agent to end stress and aid in sleep (9).

CHEMICAL COMPOSITION

Roman chamomile oil is a clear mobile liquid which may be colorless or have a yellowish color with a tinge of blue and which has a fruity, etheric and herbal green odor. The yield of essential oil from the flowers of *Chamaemelum nobile* (L.) All. generally varies from 0.3 to 1.0%, depending on whether the whole plant or the flowers alone are used for oil production. The main producing countries of this oil are France, Tunisia, Morocco, Italy, Spain and Egypt.

Literature data (up to November 19, 2014) on the chemical composition of Roman chamomile oils and unpublished analytical data from one of us (E.S.) are shown in Table 5.20.2 in alphabetical order. In Roman chamomile oils from various origins, over 165 chemicals have been identified. About 58% of these were found in a single reviewed publication only. The major compounds found in Roman chamomile oils from different sources are shown in Table 5.20.3. They include (highest concentrations in any study given) isobutyl angelate (45.9%), isoamyl angelate (33.9%), 2-methylbutyl angelate (22.9%), 3-methylpentyl angelate (19.1%), methallyl

Table 5.20.2 Constituents identified in Roman chamomile oils

Constituent	CAS	Percentage and range			
		A	B (4)	C (5)	D
n-Amyl angelate	7785-63-9	0-0.7			0.07[t]
n-Amyl methacrylate	2849-98-1	0-1.5			2.1[p]
Amyl propionate	624-54-5				0.5-10[r]
Angelyl acetate	41414-68-0				3.7[e,o]
Angelyl angelate			0.9-1.4		1.4[j]
Benzaldehyde	100-52-7				0.02[t]
Benzyl angelate	37526-87-7				+[i]
Benzyl isobutyrate	103-28-6				0.1[t]
β-Bisabolene	495-61-4				+[k]
Borneol	507-70-0	0.1-0.3		0.3	0.2[t]; 0.3[p]; 0.4[j]; <0.5[r]
β-Bourbonene	5208-59-3				0.07[k]
α-Bulnesene	3691-11-0				<0.5[r]
Butanoic acid, 3-hydroxy-2-me-thylene-, 2-methylpropyl ester	80758-68-5				1.0[j]
2-Butenyl angelate					7.3[j]
n-Butyl angelate	7785-64-0	0.07-0.7	0.7-0.8	0.5	0.4[t]; 0.5-10.0[r]; 0.6[p]; 0.9[m]; 36.8[s]
Butyl 2-butenoate	7299-91-4				1.3[g]
n-Butyl butyrate	109-21-7		0.8-0.9		1.2[j]
Butyl methacrylate	97-88-1				0.3[p]
2,4-di-t-Butyl-6-methylphenol	616-55-7				+[i]
n-Butyl tiglate	7785-66-2			0.1	
γ-Cadinene	39029-41-9				+[k]
δ-Cadinene	483-76-1				<0.5[r]
Camphene	79-92-5	0.4-1.4	0.7-0.8	0.4	<0.05[b]; 0.4[e,m,q,t]; <0.5[r]; 1.1[p]
α-Campholenal	4501-58-0				0.1[p]
Camphor	76-22-2				0.02[t]; <0.05[p]
β-Caryophyllene	87-44-5				0.5-10[r]
Caryophyllene oxide	1139-30-6				0.3[j]
1,8-Cineole	470-82-6		0.1		0.1[t]; <0.5[r]
α-Copaene	3856-25-5				<0.5[r]
β-Copaene	18252-44-3				<0.5[r]
Cubebene	11012-64-9				+[k]
1,10-di-epi-Cubenol	73365-77-2				0.1[p]
Curcumene (ar-; α-)	644-30-4				tr[t]
Cyclodecane	293-96-9				2.6[s]
p-Cymene	99-87-6	0.05-0.2			0.2[p]; 0.5[q]
p-Cymenene	1195-32-0				<0.5[r]
p-Cymen-8-ol	1197-01-9				0.05[t]
β-Damascenone (E)-	23726-93-4				+[i]
Decanoic acid	334-48-5				0.01[t]
3-Decen-5-one	32064-73-6				8.9[s]
(E)-1-(2,6-Dimethylphenyl)-2-buten-1-one	80445-59-6				+[a]
Ethyl angelate					0.08[t]
Ethyl isobutyrate	97-62-1				tr[t]
Ethyl-2-methylbutyrate	7452-79-1	0.09-0.3			
β-Farnesene	502-60-3				0.1[t]
(E,E)-α-Farnesene	502-61-4	0-0.2			
Geranyl isobutyrate	2345-26-8				+[i]
Germacrene D	23986-74-5	0.9-1.3			+[k]; 0.3[p]
Heptanol	53535-33-4				1.0[b]; 1.2[c]
(Z)-3-Hexenyl angelate	84060-80-0		0.2		0.2[p]
(Z)-3-Hexenyl tiglate	67883-79-8	0.0-0.4			+[i]
Hexyl acetate	142-92-7				0.5-10[r]
Hexyl angelate	65652-33-7		0.9-1.4		
Hexyl butyrate (butanoate)	2639-63-6				3.9[m]
Hexyl methacrylate	142-09-6	0.2-2.1			
Hexyl-2-methylbutyrate	10032-15-2	0.0-0.6			0.4[p]
Hexyl propanoate	2445-76-3	0.01-0.3			

Table 5.20.2 Constituents identified in Roman chamomile oils *(continued)*

Constituent	CAS	Percentage and range			
		A	B (4)	C (5)	D
Hexyl tiglate	16930-96-4				+[l]
α-Humulene	6753-98-6				0.01[k]
2-Hydroxy-2-methyl-3-butenyl angelate			0.3-0.4	0.2	0.1[p]; 1.0[j]
Isoamyl acetate	123-92-2	0.02-0.1	0.1	0.1	+[l]; 0.4[j]; 0.9[t]
Isoamyl angelate	10482-55-0	2.2-22.3	4.6-5.3	5.5	17.9[m]; 20.5[l]; 22.0[h]; 22.1[e]; 25[r]; 33.9[f]
Isoamyl butyrate	106-27-4	0.8-3.2	0.4		0.5-10[r]; 2.0[e]; 2.6[m]; 2.7[q]; 7.6[f]
Isoamyl 3-hydroxy-2-methyl-enebutyrate					0.5[t]
Isoamyl isobutyrate	2050-01-3	0.6-2.4		0.8	0.6[j]; 3.1[p,t]; 4.6[f]; 5.2[b]; 6.3[c]
Isoamyl 2-methylbutyrate	27625-35-0	0.3-0.8			0.1[p]; 2.8[m]; 4.1[b]; 4.5[c]
Isoamyl propionate	105-68-0	0.0-0.07			0.06[t]
Isoamyl tiglate	41519-18-0				21.4[s]
Isobutyl acetate	110-19-0	0.03-0.2		3.6	0.02[t]
Isobutyl angelate	7779-81-9	11.2-34.4	36.3-38.5	21.6	32.1[d]; 35.8[q]; 36.0[m]; 37.4[l]; 45.9[f]
Isobutyl butyrate	539-90-2				0.04[t]; 6.2[l]; 9.0[h]
Isobutyl crotonate	589-66-2	0.0-1.8			
Isobutyl 3-hydroxy-2-me-thylenebutyrate			0.3		0.08[t]
Isobutyl isobutyrate	97-85-8	1.5-7.1	3.9-5.3		2.4[p]; 3.7[m]; 4.3[s]; 4.4[j]; 5.0[h]; 5.3[d]; 6.6[g]
Isobutyl isovalerate	589-59-3	0-0.2	0.1-0.2		0.5-10[r]
Isobutyl methacrylate	97-86-9	0.9-2.8	1.4-2.1	1.9	0.3[t]; 2.2[p]; 2.3[j]; 3.0[h]; 5.8[f]
Isobutyl-2-methylbutyrate	2445-67-2	0.6-1.7		0.4	0.5[e]; 0.7[m,p]; 2.5[g]; 10.0-25.0[r]
Isobutyl propionate	540-42-1				+[l]
Isobutyl tiglate	61692-84-0	0.09-0.5	0.3-0.4		1.7[j]
Isobutyric acid	79-31-2				0.03[t]
4-Isopropenylbenzaldehyde	10133-50-3				+[a]
Isopropyl angelate	61692-76-0			1.0	
Isopropyl 2-methylbutyrate	66576-71-4	0.4-1.2			
Lepalone	80445-58-5				+[a]
Limonene	138-86-3	0.05-0.1	0.1		0.1[g]; 0.2[m]; 0.7[q]; 1.8[j]
Methallyl angelate	61692-78-2	6.8-8.0		9.1	10.0[h]; 10.8[g]; 12.4[p]; 13.1[t]
Methallyl methacrylate	816-74-0				1.7[p]; 2.3[e]
3-Methylamyl angelate	53082-58-9				+[l,n]; 0.5-10[r]
3-Methylamyl isobutyrate	84254-84-2				+[l,n]
3-Methylamyl methacrylate					+[l,n]
3-Methylamyl valerate	113615-01-3				+[n]
2-Methyl-2-buten-1-ol	4675-87-0				8.4[g]
2-Methyl-2-butenyl acetate	33425-30-8			1.6	3.1[g]
3-Methyl-2-butenyl acetate	1191-16-8	0.3-2.3			
2-Methyl-2-butenyl angelate					+[l]; 1.9[p]
(E)-2-Methyl-2-butenyl angelate				2.3	
3-Methyl-2-butenyl angelate	83783-82-8		7.9-8.4[a]		
(E)-2-Methyl-2-butenyl isobutyrate	95654-17-4			5.1	0.2[t]
(E)-2-Methyl-2-butenyl metha-crylate	88142-95-4			0.3	+[n]
2-Methyl-2-butenyl-2-methyl-butanoate	95654-18-5				+[l]
2-Methylbutyl acetate	624-41-9	0.1-1.2		0.2	1.2[d]
2-Methylbutyl angelate	61692-77-1	2.0-9.7	18.2-20.3	14.4	7.0[h]; 13.0[b]; 16.2[d]; 17.2[c]; 17.4[j]; 22.9[q]
2-Methylbutyl isobutyrate	2445-69-4	0.5-4.7	3.1-3.5	5.2	<0.05[b]; 0.5[p]; 1.1[c]; 4.3[j,s]; 4.6[g]
2-Methylbutyl methacrylate	60608-94-8		0.7-0.8	1.2	1.0[e]; 1.5[h]; 3.7[f]
3-Methylbutyl methacrylate	7336-27-8		0.2	0.5	1.3[p]; 1.5[t]
2-Methylbutyl 2-methylbu-tyrate	2445-78-5	0.9-1.2	0.5	0.4	0.4[q]; 0.6[p]; 0.9[j]; 1.2[d]; 2.7[m]; 10.0-25.0[r]
2-Methylbutyl tiglate	2445-78-5				0.2[t]
Methyl chavicol	140-67-0				5.0[j]

Table 5.20.2 Constituents identified in Roman chamomile oils *(continued)*

Constituent	CAS	Percentage and range			
		A	B (4)	C (5)	D
2-Methylenepropane-1,3-diyl 1-angelate-3-isobutyrate					+[i]
4-Methylhexanol	818-49-5				1.7[j]
Methyl isobutyrate	547-63-7				82.4[l] (obvious mistake)
Methyl 2-methylbutyrate	868-57-5	0.0-0.1			1.9[d]
3-Methyl-1-pentanol	589-35-5				0.2[t]
3-Methylpentyl acetate	35897-13-3				0.9[t]
2-Methylpentyl angelate					22.7[t]
3-Methylpentyl angelate	53082-58-9	0.4-12.8		8.4	12.2[p]; 19.1[f]
3-Methylpentyl 3-hydroxy-2-methylenebutyrate					0.7[t]
3-Methylpentyl isobutyrate	84254-84-2	0.6-1.4		4.7	1.4[p]; 12.5[t]
3-Methylpentyl isovalerate	35852-41-6				0.04[t]
3-Methylpentyl methacrylate		0.0-1.8		1.1	
2-Methylpentyl propionate					0.1[t]
3-Methylpentyl valerate				0.3	
2-Methylpropane-1,3-diyl 1-angelate-3-isobutyrate					+[i]
2-Methyl-2-propenyl isobutyrate				0.2	0.9[t]; 1.4[p]
2-Methyl-2-propenyl methacrylate	816-74-0			0.7	0.8[t]; 2.3[e]
2-Methyl-2-propenyl tiglate	7493-71-2				0.8[t]
Methyl tiglate	6622-76-0	0.01-0.03			
Myrcene	123-35-3				0.1[p]; <0.5[r]; 1.9[l]
Myrtenal	564-94-3	0.5-0.8		0.5	0.5[p]; 0.5-10.0[r]; 0.7[t]
Myrtenol	515-00-4				+[n]; 0.4[p,t]; <0.5[r]
Myrtenyl acetate	1079-01-2			1.6	
Myrtenyl angelate					+[i]
Myrtenyl isobutyrate	29021-37-2				+[i]; 0.05[t]
Myrtenyl isovalerate	33900-84-4				+[i]
Myrtenyl 2-methylbutyrate	138530-44-6				+[i]
Nonanoic acid	12-05-0				0.04[t]
(Z)-Ocimene	27400-71-1				0.5[q]
1-Octen-3-ol	3391-86-4				0.05[t]
Pentadecanal	2765-11-9				+[i]
Pentyl 2-methylbutanoate	68039-26-9	0-0.08			
Pentyl 2-methylcrotonate	7785-65-1	0.2-2.1			
2-Phenethyl propionate	122-70-3				+[i]
2-Phenylethyl isobutyrate	103-48-0				+[i]
3-Phenylpropyl isobutyrate	103-58-2				+[i]
α-Pinene	80-56-8	2.2-11.5	1.2-1.6	1.3	3.5[s]; 5.9[g,l]; 7.1[p]; 7.3[f]; 8.3[e]; 10.0[r]
β-Pinene	127-91-3	0.3-1.1	0.1-0.2		0.5[j]; 0.6[e]; 0.7[p]; 1.1[g]; 1.6[b]; 1.8[c]; 10.0[r]
Pinocamphone	547-60-4				10.0[r]
trans-Pinocamphone	547-60-4	0.3-0.5			1.0[j]
Pinocarveol	5947-36-4				<0.5[r]
trans-Pinocarveol	1674-08-4	1.7-5.2	3.1-4.5	0.8	1.3[b]; 3.0[g]; 3.1[p]; 4.4[t]; 4.5[j]; 5.0[h]; 9.3[f]
Pinocarvone	30460-92-5	2.3-5.6		0.8	2.4[t]; 2.8[l]; 4.0[h]; 4.4[p]; 5.1[s]; 10-25.0[r]
Prenyl acetate	191-16-8				0.3[p]; 1.4[d]
Prenyl isobutyrate	76649-23-5	0.2-1.6			1.6[p]
Propyl angelate	53082-57-8	0.6-1.7	1.1-1.3		0.5-10[r]; 0.8[c]; 1.0[t]; 1.1[m]; 1.3[e]; 1.8[j]
Propyl isobutyrate	644-49-5	0-0.1			
Propyl methacrylate	2210-28-8	0-0.08	0.1-0.2		
Propyl tiglate	61692-83-9				1.6[s]; 12.0[b]; 13.1[c]
Sabinene	3387-41-5	0.09-0.2			0.1[p]; 0.5-10.0[r]; 1.2[q]
β-Selinene	17066-67-0				+[k]; 0.2[k]
γ-Terpinene	99-85-4				+[n]; <0.5[r]
Terpinen-4-ol	562-74-3			0.3	
α-Terpineol	98-55-5				0.03[t]

Table 5.20.2 Constituents identified in Roman chamomile oils *(continued)*

Constituent	CAS	Percentage and range			
		A	B (4)	C (5)	D
α-Thujenal	57129-54-1				0.9[j]
α-Thujene	2867-05-2				1.2[q]
Tiglyl acetate	19248-94-3				3.7[e,o]
Tridecanal	10486-19-8				+[i]
Valeranone	55528-90-0				+[n]

A twenty-two Roman chamomile essential oil samples from Egypt, Tunisia, Italy, France and Germany analyzed between 2006 and 2014; lowest and highest concentrations given (E. Schmidt, unpublished data)
B two commercial samples of Roman chamomile oil produced from Italian flowers (ref. 4)
C one lab-hydrodistilled oil from the Slovak Republic prepared from wild growing *C. nobile* in full bloom (ref. 5)
D data from other studies (indicated with superscript letters); highest concentrations found in any study reviewed here given; when two or more oils were investigated, only the highest concentrations are mentioned, unless indicated otherwise

[a] one commercial oil of unknown origin (ref. 16); [b] one lab-hydrodistilled oil from *Chamaemelum nobile* var. *flora plena* flowers culti-vated in Iran (ref. 3); [c] three lab-hydrodistilled oils from *C. nobile* var. *flora plena* flowers cultivated in Iran and dried in three different ways: shade drying, sun drying and oven drying (ref. 2); [d] one commercial Roman chamomile oil from the Provence, France (ref. 6); [e] one oil of Lithuanian origin (ref. 8); [f] four lab-hydrodistilled oils, three from cultivars and one from plants growing in the wild (ref. 7); [g] one sample of chamomile flower oil of unknown origin (ref. 31); [h] unknown number of lab-hydrodistilled oils from wild growing flowers in Morocco (ref. 14); [i] investigation of the fraction containing 2-phenylethyl isobutyrate of a commercial Roman chamomile oil of unknown origin (ref. 10); [j] one commercial oil from the Provence, France (ref. 11); [k] one commercial oil of unknown origin (ref. 15); [l] three steam-distilled oils from three *C. nobile* batches from Slovak Republic (ref. 20); [m] one oil of unknown origin (ref. 24); [n] one lab-hydrodistilled oil from flowers collected in a university garden in Italy (ref. 13); [o] either angelyl acetate or tiglyl acetate; [p] one industrially steam-distilled oil from chamomile flower heads cultured in Poland (ref. 25); [q] one commercial oil sample from France (ref. 27); [r] unknown number of chamomile oils *ex C. nobile* L. from Italy; highest concentration or range given; it should be realized that this study dates back to 1976 and may be inaccurate according to current standards; for many chemicals, a range of (exactly) 0.5-10% was mentioned (ref. 28); [s] >140 samples of Roman chamomile oil of unknown origin; average concentrations (probably the highest for a particular subgroup) given (ref. 29); [t] one lab-hydrodistilled oil from the flowers of *C. nobile* (ref. 30); + present in the oil investigated, but quantity not stated

angelate (13.1%), α-pinene (11.5%), *trans*-pinocarveol (9.3%) and isobutyl isobutyrate (7.1%). Well-known ingredients of Roman chamomile oils that were pres-ent in high concentrations (>10%) in one or two studies were *n*-butyl angelate (36.8%), isobutyl-2-methylbutyr-ate (25.0%), 2-methylbutyl 2-methylbutyrate (25.0%), pinocarvone (25.0%), 3-methylpentyl isobutyrate (12.5%), isoamyl butyrate (10%) and β-pinene (10.0%). Uncommon or rare constituents of Roman chamomile oils found in high concentrations (>10%) in single stud-ies include methyl isobutyrate (82.4%), 2-methylpentyl angelate (22.7%), isoamyl tiglate (21.4%), propyl tiglate (13.1%), amyl propionate (10%), hexyl acetate (10%), isobutyl isovalerate (10%), 3-methylamyl angelate (10%) and pinocamphone (10.0%).

Commercial oils

The ten chemicals that had the highest maximum con-centrations in 22 commercial Roman chamomile essen-tial oil samples (concentration ranges provided) are the following: isobutyl angelate (11.2-34.4%), isoamyl angel-ate (2.2-22.3%), 3-methylpentyl angelate (0.4-12.8%), α-pinene (2.2-11.5%), 2-methylbutyl angelate (2.0-9.7%), methallyl angelate (6.8-8.0%), isobutyl isobutyrate (1.5-7.1%), pinocarvone (2.3-5.6%), *trans*-pinocarveol

Table 5.20.3 Major constituents of Roman chamomile oils

Constituent	CAS	Percentage and range			
		A	B (4)	C (5)	D
Isobutyl angelate	7779-81-9	11.2-34.4	36.3-38.5	21.6	32.1[d]; 35.8[q]; 36.0[m]; 37.4[l]; 45.9[f]
Isoamyl angelate	10482-55-0	2.2-22.3	4.6-5.3	5.5	17.9[m]; 20.5[l]; 22.0[h]; 22.1[e]; 25[r]; 33.9[f]
2-Methylbutyl angelate	61692-77-1	2.0-9.7	18.2-20.3	14.4	7.0[h]; 13.0[b]; 16.2[d]; 17.2[c]; 17.4[j]; 22.9[q]
3-Methylpentyl angelate	53082-58-9	0.4-12.8		8.4	12.2[p]; 19.1[f]
Methallyl angelate	61692-78-2	6.8-8.0		9.1	10.0[h]; 10.8[g]; 12.4[p]; 13.1[t]
α-Pinene	80-56-8	2.2-11.5	1.2-1.6	1.3	3.5[s]; 5.9[g,l]; 7.1[p]; 7.3[f]; 8.3[e]; 10.0[r]
trans-Pinocarveol	1674-08-4	1.7-5.2	3.1-4.5	0.8	1.3[b]; 3.0[g]; 3.1[p]; 4.4[t]; 4.5[j]; 5.0[h]; 9.3[f]
Isobutyl isobutyrate	97-85-8	1.5-7.1	3.9-5.3		2.4[p]; 3.7[m]; 4.3[s]; 4.4[j]; 5.0[h]; 5.3[d]; 6.6[g]

LEGEND: SEE UNDER TABLE 5.20.2

(1.7-5.2%) and 2-methylbutyl isobutyrate (0.5-4.7%) (Erich Schmidt, unpublished analytical data).

The analyses of oils of *C. nobile* in refs. 9 and 18 are not discussed here, as there was probably a botanical misidentification; the source plants were most likely *Chamomilla recutita* L. Rauschert, the *German* chamomile.

In neither study, any angelate esters, characteristic for Roman chamomile, were found, but in ref. 9 chamazulene was identified in a 27.8% concentration and in ref. 18 >50% α-bisabolol oxide A was found, chemicals which are both characteristic for German chamomile but not the Roman variety. In ref. 26, the composition of a *Chamaemelum nobile* oil sample is supplied, but it contains chamazulene (12.6%) and prochamazulene (7.1%) and is virtually identical to the also presented composition of a German chamomile oil, which must be erroneous (26).

Older literature from before 1992 has been reviewed in refs. 22 and 23.

CONTACT ALLERGY/ALLERGIC CONTACT DERMATITIS

General
Contact allergy to/allergic contact dermatitis from Roman chamomile oil has been reported in a few studies only. The literature on 'chamomile oil' (botanical origin not specified) is discussed in Chapter 5.19 Chamomile oil, German.

Testing in groups of patients
In a group of 271 patient with dermatitis, who used products containing *Anthemis nobilis* and were tested with *Anthemis nobilis* 1% in petrolatum, there were eight positive patch test reactions (3.0%); these were 'at least possibly relevant'; not certain it was essential oil (34).

Case reports
A patient developed generalized dermatitis from *Anthemis nobilis* oil in a homeopathic preparation; she also applied chamomile compresses and drank chamomile tea (32). Two cases of allergic contact dermatitis were caused by Roman chamomile oil in the topical pharmaceutical preparation Kamillosan® (33).

LITERATURE

1 Srivastava JK, Shankar E, Gupta S. Chamomile: a herbal medicine of the past with bright future. Mol Med Report 2010;3:895-901

2 Omidbaigi R, Sefidkon F, Kazemi F. Influence of drying methods on the essential oil content and composition of Roman chamomile. Flavour Fragr J 2004;19:196-198

3 Omidbaigi R, Sefidkon F, Kazemi F. Roman chamomile oil: comparison between hydrodistillation and supercritical fluid extraction. J Essent Oil Bear Plants 2003;6:191-194

4 Antonelli A, Fabbri C. Study on Roman chamomile (*Chamaemelum nobile* L.) oil. J Essent Oil Res 1998;10:571-574

5 Farkas P, Holla M, Vaverkova S, et al. Composition of the essential oil from the flowerheads of *Chamaemelum nobile* (L) All. (Asteraceae) cultivated in Slovak Republic. J Essent Oil Res 2003;15:83-85

6 Bail S, Buchbauer G, Jirovetz L, et al. Antimicrobial activities of Roman chamomile oil from France and its main compounds. J Essent Oil Res 2009;21:283-286

7 Hethelyi E, Palinkas G, Palinkas J. Roman chamomile (*Anthemis nobilis*) oil analysis by GC and GC/MS. Olaj Szappan Kozmetika 1999;48:116-123. Data cited in ref. 17

8 Povilaityte V, Venskutonis PR, Jukneviciene G. Aroma and antioxidant properties of Roman chamomile (*Anthemis nobilis* L.). In: AM Spannier et al, Eds. Food flavors and chemistry – Advances of the new millennium. London: Royal Society of Chemistry, 2001:567-577. Data cited in ref. 17

9 Farhoudi R. Chemical constituents and antioxidant properties of *Matricaria recutita* and *Chamaemelum nobile* essential oil growing wild in the South West of Iran. J Essent Oil Bear Plants 2013;16:531-537

10 Bassols F, Thomas AF. The occurrence of 3-phenylpropyl isobutyrate in Roman camomile oil. J Essent Oil Res 1991;3:309-312

11 Tognolini M, Barocelli E, Ballabeni V, Bruni R, Bianchi A, Chiavarini M, et al. Comparative screening of plant essential oils: Phenylpropanoid moiety as basic core for antiplatelet activity. Life Sciences 2006;78:1419-1432

12 Guimarães R, Barros L, Dueñas M, Calhelha RC, Carvalho AM, Santos-Buelga C, et al. Nutrients, phytochemicals and bioactivity of wild Roman chamomile: A comparison between the herb and its preparations. Food Chem 2013;136:718-725

13 Bicchi C, Frattini C, Raverdino V. Considerations and remarks on the analysis of *Anthemis nobilis* L. essential oil by capillary gas chromatography and "hyphenated" techniques. J Chromatogr 1987;411:237-249

14 Lahlou M, Berrada R, Agoumi A, Hmamouchi M. The potential effectiveness of essential oils in the control of human head lice in Morocco. Int J Aromather 2001;10:108-123

15 Klimes I, Lamparsky D, Scholz E. Vorkommen neuer bifunktioneller Ester im Römisch-Kamillenöl (*Anthemis nobilis* L.). Helv Chim Acta 1981;64:2338-2349

16 Thomas AF, Egger JC. Novel ketones from Roman camomile oil. Helv Chim Acta 1981;64:2393-2396

17 Lawrence BM. Progress in essential oils. Perfum Flavor 2003;28(5):70-? (last page unknown)

18 Dezfooli NA, Hasanzadeh N, Rezaee MB, Ghasemi A. Antibacterial activity and chemical compositions of *Chamaemelum nobile* essential oil/extracts against *Pseudomonas tolaasii*, the causative agent of mushroom brown blotch. Ann Biol Res 2012;3:2602-2608

19 European Medicines Agency. Assessment report on *Chamaemelum nobile* (L.) All., flos. EMA/HMPC/560906/2010, Committee on Herbal Medicinal Products (HMPC), 27 January 2011

20 Vaverková S, Mikulásová M, Mistríková, Habán M. Content and composition of the essential oil from flowerheads of *Chamaemelum nobile* (L.) All. Acta Fytotechnica et Zootechnica 2011; special number, 53-55. Available at: http://spu.fem.uniag.sk/acta/download.php?id=976

21 Lawrence BM. Progress in essential oils. Perfum Flavor 1998;23(6):35-50

22 Lawrence BM. Progress in essential oils. Perfum Flavor 1992;17(5):131-146

23 Lawrence BM. Progress in essential oils. Perfum Flavor 1990;15(4):63-71

24 Chialva F, Gabri G, Liddle PAP, Ulian F. Qualitative evaluation of aromatic herbs by direct headspace GC analysis. Applications of the method and comparison with the traditional analysis of essential oils. J High Res Chrom & Chromatograph Commun 1982;5:182-188

25 Wozniak M, Maciag A, Kalemba D. Essential oil and hydrolate from *Chamaemelum nobile* (L.) All. flowers. Data presented at the 41st International Symposium on Essential Oils, Wrozlaw, Poland, September 5-8, 2010

26 Chebli B, Hmamouchi M, Achouri M, Idrissi Hassani LM. Composition and *in vitro* fungitoxic activity of 19 essential oils against two post-harvest pathogens. J Essent Oil Res 2004;16:507-511

27 Seo S-M, Kim J, Kang J, Koh S-H, Ahn Y-J, Kang K-S, Park IK. Fumigant toxicity and acetylcholinesterase inhibitory activity of 4 Asteraceae plant essential oils and their constituents against Japanese termite (*Reticulitermes speratus* Kolbe). Pest Biochem Physiol 2014;113:55-61

28 Nano GM, Sacco T, Frattini C. Ricerche botanische e chimiche su *Anthemis nobilis* L. ed alcune sue cultivar. Essenz Deriv Agrum 1976;46:171-175. Data cited in ref. 21

29 Damiani P, Menghini A, Tirillini B, Daddy P. A study on the essential oil of *Chamaemelum nobile* (L) Allioni. Fitoterapia 1983;54:213-222. Data cited in ref. 21

30 Brunke EJ, Hammerschmidt FJ, Schmaus G. Flower scent of some traditional medicinal plants. In: R Teranishi, RG Buttery and H Sugisawa, Eds. Bioactive volatile compounds from plants. ACS Symposium Series no. 525. Washington, DC, USA: American Chemical Society, 1993:282-296. Data cited in ref. 21

31 Fauconnier M-L, Jazrri M, Holmes J, Shimomura K, Marlier M. *Anthemis nobilis* L. (Roman chamomile): *In vitro* culture, micropropagation and the production of essential oils. In: YPS Bajaj, Ed. Medicinal and aromatic plants, IX Edition. Berlin, Germany: Springer Verlag, 1996:16-36. Data cited in ref. 21

32 Giordano-Labadie F, Schwarze HP, Bazex J. Allergic contact dermatitis from chamomile used in phytotherapy. Contact Dermatitis 2000;42:247

33 McGeorge BCL, Steele MC. Allergic contact dermatitis of the nipple from Roman chamomile ointment. Contact Dermatitis 1991;24:139-140

34 Guin JD. Use of consumer product ingredients for patch testing. Dermatitis 2005;16:71-77

Chapter 5.21 CINNAMON BARK OIL, SRI LANKA

There are two Sri Lanka types of cinnamon oils, bark oil and leaf oil. In this chapter, the bark oil is discussed. Cinnamon leaf oil is presented in Chapter 5.22.

DEFINITION

Cinnamon bark oil, Sri Lanka type (essential oil of cinnamon bark, Sri Lanka type), is the essential oil obtained from the twigs and bark of stem of the cinnamon, *Cinnamomum zeylanicum* Blume (synonym: *Cinnamomum verum* J. Presl).

INCI NOMENCLATURE

Description/definition: Cinnamomum zeylanicum bark oil is the volatile oil expressed from the bark of the Ceylon cinnamon, *Cinnamomum zeylanicum*, Lauraceae
INCI name EU & USA: Cinnamomum zeylanicum bark oil
Other names: Cinnamon oil Ceylon
CAS registry number(s): 8015-91-6; 84649-98-9
EINECS number(s): 283-479-0

ISO (INTERNATIONAL ORGANIZATION FOR STANDARDIZATION) STANDARD

There is currently no ISO standard for cinnamon bark oil.

In most publications, the name *Cinnamomum zeylanicum* is used. However, both GRIN Taxonomy for Plants and The Plant List give *Cinnamomum verum* J. Presl as preferred name (and *C. zeylanicum* as synonym). Cinnamon oils should not be confused with cassia oils, which are obtained from the Chinese cinnamon, *Cinnamomum cassia* (Nees & T. Nees) J. Presl (see Chapters 5.13 and 5.14).

THE PLANT, THE OIL, AND THEIR USES

Cinnamomum zeylanicum Blume is an evergreen small tree, up to 10 meters tall with a black-brown bark, and an inner bark with cinnamon flavor. The tree is native to Sri Lanka. It is naturalized on islands in the western Indian Ocean, the Pacific and the Caribbean (West Indies). The cinnamon is cultivated in Africa (Madagascar, Seychelles), Asia (China, Taiwan, India, Sri Lanka, Vietnam, Indonesia, Malaysia), Mexico, the Caribbean (Jamaica, Martinique), French Guiana and Brazil (GRIN Taxonomy for Plants; Mansfeld's World Database of Agriculture and Horticultural Crops, http://mansfeld.ipk-gatersleben.de).

The dried bark is the source of the spice cinnamon, which is widely used in seasonings, cosmetics, foods, drinks, commodity essences and the chemical industries (1). Cinnamon is also a major medicinal substance in China and many other countries and is present in a multitude of pharmaceutical preparations (26). It has been found to possess anti-inflammatory, antioxidant, antibacterial, antifungal, insecticidal, acaricidal, anti-mutagenic and anti-cancer properties (1,29). *C. zeylanicum* is used as a remedy for a variety of illnesses including diarrhea, chills, influenza, parasitic worms and diabetes (26). In Ayurvedic medicine, cinnamon bark has been used as an antiemetic, antidiarrheal, anti-flatulent, and general stimulant (27,29). Possible health effects of cinnamon have been reviewed in refs. 3 and 4. Cinnamon has been considered promising for its anti-diabetic effect, but in a Cochrane review it was concluded that there is insufficient evidence to support the use of cinnamon for type 1 or type 2 diabetes mellitus (6).

The essential oil, extracted from the bark by distillation (cinnamon bark oil), serves as flavor for bakery and confectionary products, meat, liqueur, chewing-gum, ice-cream, beverages, soup and pharmaceutical products. Cinnamon bark oil is not used in aromatherapy because of the risk of skin sensitization (61,62). It is liable to adulteration with the leaf oil. The leaf oil is used as an antibacterial and anti-fungal agent, and a uterine stimulant. It is also important in the cosmetic and perfumery industries (26) and is used in aromatherapy (62).

It should be realized that 'cinnamon' may be obtained either from *Cinnamomum zeylanicum* Blume (true cinnamon) or from *Cinnamomum cassia* (Nees & T. Nees) J. Presl. (cassia, often referred to as [Chinese] cinnamon) and other *Cinnamomum* species. A comprehensive review of all aspects of *Cinnamomum zeylanicum* is provided in ref. 5. Other useful reviews include ref. 27, 28, and 29.

CHEMICAL COMPOSITION

Cinnamon bark oil is an amber to brown, clear mobile liquid which has a strong sweet spicy, characteristic odor. The yield of essential oil from twigs and bark of stem of *Cinnamomum zeylanicum* Blume generally varies from 0.5 to 1.3%, depending on the dryness of the bark. The main producing countries of this oil are Sri Lanka, Indonesia, Seychelles, Vietnam and Madagascar.

Literature data (up to November 17, 2014) on the chemical composition of cinnamon bark oils and unpublished analytical data from one of us (E.S.) are shown in Table 5.21.1 in alphabetical order. In cinnamon bark oils from various origins, over 160 chemicals have been identified. About 46% of these were found in a single reviewed publication only. The major compounds found in cinnamon bark oils from different sources are shown in Table 5.21.2. They include (highest concentrations in any study given) (*E*)-cinnamaldehyde (97.7%), eugenol (74.3%), β-caryophyllene (48.3%), β-phellandrene (8.4%), linalool (13.8%) and (*E*)-cinnamyl acetate (13.6%). Well-known ingredients of cinnamon bark oils that were present in high concentrations (>7%) in one or two studies were benzyl benzoate (3 studies: 9.3%, 15.1% and 84.7%), α-terpineol (15.0%), limonene (8.3% and 13.2%), benzaldehyde (9.9% and 12.2%), *p*-cymene (9.9%), camphene (9.6%), cinnamyl acetate (8.7%) and α-copaene (7.2%). Uncommon or rare constituents of cinnamon bark oils found in high concentrations (>7%) in single studies include camphor (15.2%), cinnamyl cinnamate (14.0%), isoquinoline (8.9%) and methyl isoeugenol (7.8%).

Commercial oils

The ten chemicals that had the highest maximum concentrations in 43 commercial cinnamon bark

Table 5.21.1 Constituents identified in cinnamon bark oils

Constituent	CAS	Percentage and range				
		A	B (18)	C (11)	D (14)	E
Acetophenone	98-86-2					<0.1[m]
Acetyl eugenol	93-28-7				0.2	
α-Amorphene	20085-19-2					0.5[j]
(E)-Anethole	4180-23-8					1.7[r]
Aromadendrene	489-39-4					0.1[y5]
Benzaldehyde	100-52-7	0.1-1.0		0.3	0.3	0.6[t]; 0.7[x4]; 0.9[x3]; 1.3[y]; 1.7[c]; 2.3[p]; 9.9[n]; 12.2[h]
Benzoic acid	65-85-0					0.8[h]
Benzyl alcohol	100-51-6				tr	1.0[c]
Benzyl benzoate	120-51-4	0.2-2.5	1.5-2.2	1.6	0.7	0.7[x]; 1.4[g]; 2.0[t]; 4.0[r]; 9.3[q]; 15.1[e]; 84.7[k]
Benzyl cinnamate	103-41-3					0.6[h]; 1.4[x4]
Benzyl formate	104-57-4					0.5[x4]
trans-α-Bergamotene	13474-59-4					0.2[y5]
β-Bergamotene	6895-56-3					0.1[f]
β-Bisabolene	495-61-4					0.2[f]
α-Bisabolol	515-69-5					0.2[m]; 0.5[x1]
Borneol	507-70-0	0.06-0.3	0.09-0.4		0.02	0.1[m]; 0.7[k]
Bornyl acetate	76-49-3		0-tr		0.1	0.5[x7]
Cadalene	483-78-3					0.6[x1]
γ-Cadinene	39029-41-9		0-tr			3.1[y]
δ-Cadinene	483-76-1	0.03-0.6	tr			0.4[x1]; 0.6[f]; 0.7[y5]; 0.9[j]; 4.7[y]; 6.3[x6]
α-Cadinol	481-34-5		0.1-0.3			1.0[y]
epi-α-Cadinol	5937-11-1		0.06-0.2			0.2[x1]
α-Calacorene	21391-99-1					0.3[x1]
cis-Calamene	72937-55-4					0.4[f]
Calamenene	483-77-2					0.8[x1]
cis-Calamenene	72937-55-4		0-tr			
Camphene	79-92-5	0.06-0.7	0-0.7	0.4	0.2	0.2[i,m]; 0.3[x2]; 0.4[x]; 0.6[k]; 0.8[t]; 0.9[j]; 1.0[r]; 9.6[x9]
Camphor	76-22-2	0-0.1		15.2	tr	tr[j]; 0.1[x7]; <0.2[g]; 0.5[m]
δ3-Carene	13466-78-9	0.01-0.1	0-0.1	0.1	0.03	0.09[x2]; 0.1[v]
Carvacrol	499-75-2					1.6[g]
β-Caryophyllene	87-44-5	2.7-7.5	1.5-5.6	2.4	3.3	5.8[d]; 6.4[j]; 6.9[y5]; 7.5[y1]; 10.4[e]; 13.6[r]; 48.3[x5]
γ-Caryophyllene	118-65-0					tr[r]
Caryophyllene oxide	1139-30-6	0.2-0.7	0.4-0.5		tr	tr[t]; 0.2[f,s]; 0.3[g]; 1.8[r]
Caryophyllenyl alcohol	913176-41-7					0.9[x1]; 1.1[r]
α-Cedrene	469-61-4					0.2[c]; 0.4[m]
1,8-Cineole	470-82-6	0.05-0.2	2.2-2.8		2.0	0.2[g]; 0.4[x2]; 1.6[n]; 2.1[e]; 2.3[i]; 3.5[l]; 3.7[o]; 5.2[y1]
(E)-Cinnamaldehyde	14371-10-9	43.0-72.7	59.3-60.5	52.2	75	76.2[x3]; 77.3[x4]; 79.1[x8]; 79.3[x7]; 79.4[d]; 79.6[x6]; 79.8[x1]; 81.5[y3]; 81.9[x9]; 90.1[t]; 90.2[y4]; 97.7[j]
(Z)-Cinnamaldehyde	57194-69-1	0.3-1.0	0.06-0.07	0.5		<0.2[g]; 0.2[c]; 0.4[u]; 0.7[x6]; 0.8[x4]; 2.3[m]
Cinnamaldehyde pro-pyleneglycol acetal	4353-01-9					5.6[b,c]
(E)-Cinnamic acid	140-10-3					<0.1[s]; 1.2[n]
Cinnamyl acetate	103-54-8				5.0	0.2[m]; 0.5[m]; 2.0[y4]; 2.6[o]; 3.5[x3]; 8.7[y1]
(E)-Cinnamyl acetate	21040-45-9	0.9-7.8	tr-13.6			2.8[g]; 3.8[x2]; 4.5[q]; 4.6[x]; 4.7[v]; 6.8[s]; 7.4[n]; 12.9[d]
(Z)-Cinnamyl acetate				4.1		
Cinnamyl alcohol	104-54-1	0.03-0.06			0.3	0.5[x7]
(E)-Cinnamyl alcohol	4407-36-7		0.2			<0.1[e]; 0.2[c]; 0.3[x]; 1.0[u]; 2.0[s]
Cinnamyl cinnamate	122-69-0					14.0[x4]
Citronellal	106-23-0					0.9[l]
α-Copaene	3856-25-5	0.1-1.3	0.1-0.2			0.1[c]; 0.3[x]; 0.7[g,m]; 0.8[j]; 1.2[f]; 4.8[y5]; 7.2[x6]
Coumarin	91-64-5		0-0.4		0.7	
Cubenene	29837-12-5					1.5[x6]
1-epi-Cubenol	81939-29-9		tr-0.05			0.5[x1]
Cuminaldehyde	122-03-2		0-tr		0.04	0.2[u,x1]
Curcumene	644-30-4					0.2[m,x1]
o-Cymene	527-84-4					5.4[x8]
p-Cymene	99-87-6	2.2-2.8	0.3-4.7	2.3	1.1	1.6[l]; 1.7[i]; 1.9[o]; 2.5[x2]; 3.9[t]; 4.0[y5]; 5.4[r]; 9.9[v]
Decanal	112-31-2					0.2[m]
Dihydrolinalool	18479-51-1					tr[u]
Dill apiole	484-31-1					0.4[a,f]
1,3-Dimethylbenzene	108-38-3					0.2[m]
β-Elemene	33880-83-0		0-tr			2.7[y]

Table 5.21.1 Constituents identified in cinnamon bark oils *(continued)*

Constituent	CAS	Percentage and range				
		A	B (18)	C (11)	D (14)	E
Elemol (α-)	639-99-6		0-tr			
Ethyl benzoate	93-89-0					0.08[k]
Ethyl cinnamate	103-36-6					0.2[m]
Ethyl (E)-cinnamate	4192-77-2					1.8[e]; 2.6[y]
Ethyl-p-methoxy-cinnamate	1929-30-2					0.7[b,c]
3-Ethyl-3-methyl-decane	17312-66-2					0.3[m]
Eugenol	97-53-0	0.2-16.4	1.4-1.7	2.9	2.2	5.3[v]; 6.1[x2]; 7.0[y2]; 8.1[x5]; 9.0[q]; 10[e]; 11.8[p]; 11.9[x7]; 14.9[t]; 16.7[y3]; 33.9[s]; 46.5[g]; 74.3[l]
Eugenyl acetate	93-28-7	0.03-0.6	tr			0.2[i,k]; 0.4[c,e]; 0.8[q]; 0.9[u]; 2.2[g]
α-Farnesene	502-61-4					0.3[m]
Farnesol	4602-84-0				0.03	0.1[x7]; 0.2[f]
Fenchone	1195-79-5				tr	1.4[p]
Furfural	98-01-1				0.03	
Geranial	141-27-5				tr	
Geraniol	106-24-1				0.1	0.2[x7]; 2.0[y]
Geranyl acetate	105-87-3				tr	0.1[m]; 5.8[w]
Germacrene D	23986-74-5					0.3[f]; 0.9[x6]
α-Guaiene	3691-12-1					1.5[m]
Hexadecanoic acid	57-10-3					3.3[r]
Hexanol	111-27-3				tr	
α-Humulene	6753-98-6	0.3-1.2	0.3-1.0	0.5	0.6	tr[t]; 0.1[c,y3]; 0.6[x]; 0.9[g]; 1.0[x2]; 1.5[y5]; 4.0[r]
Humulene oxide I	19888-33-6		0.08-0.09			
Humulene oxide II	19888-34-7		tr-0.07			0.5[r]
Hydrocinnamaldehyde	104-53-0	0.1-0.5	0.1-0.3	0.1	0.4	0.3[c,u]; 0.5[m]; 0.6[x]; 0.7[x1]
Hydrocinnamyl acetate	122-72-5	0.03-0.1				
Hydrocinnamyl alcohol	122-97-4	0.05-0.4				
4-Hydroxycinnamaldehyde	2538-87-6					1.8[x7]
Isoeugenol	97-54-1				0.02	4.1[e]
(E)-Isoeugenol	5932-68-3					1.1[x9]
Isoledene	95910-36-4					0.3[m]
Isoquinoline	119-65-3					8.9[x8]
Limonene	138-86-3	0.4-2.0	0.5-0.6	1.0	0.5	0.9[l]; 1.6[t]; 2.4[y4]; 3.2[k]; 4.1[x]; 4.4[n]; 8.3[c]; 13.2[u]
Linalool	78-70-6	2.1-7.2	4.4-13.8	3.2	2.4	3.5[o]; 4.6[i]; 4.8[y5]; 6.3[e]; 6.9[t]; 7.0[x2]; 9.4[r]; 10.3[q]
cis-Linalool oxide, furanoid	5989-33-3		tr-0.4			
trans-Linalool oxide, furanoid	34995-77-2		0.05-0.4		0.1	
Linalyl acetate	115-95-7				tr	0.6[h]; 2.7[u]; 3.1[v]
Linalyl isobutyrate	78-35-3					0.1[x7]
2-Methoxybenzaldehyde	135-02-4					0.1[x]
(E)-2-Methoxycinnamaldehyde (o-)	60125-24-8			0.5		0.3[x2]; 0.7[s]
o-Methoxycinnamaldehyde (2-)	1504-74-1					1.1[m]; 12[x3]
p-Methoxycinnamaldehyde	1963-36-6					1.9[y]; 2.3[f]
(E)-p-Methoxycinnamaldehyde	24680-50-0	0.1-0.4				
2-Methoxyphenol	90-05-1					0.9[x4]
Methyl chavicol	140-67-0		1.3		tr	0.3[k]
Methyl cinnamate	103-26-4				tr	
(E)-Methyl cinnamate	1754-62-7					0.6[y5]
(Z)-Methyl cinnamate	19713-73-6			0.7		
3-Methylcoumarin	2445-82-1					1.9[y]
Methyl eugenol	93-15-2			0.3	tr	0.06[k]; 0.1[x7]; 3.4[e]; 6.7[y]
6-Methyl-5-hepten-2-one	110-93-0		0-tr			

Table 5.21.1 Constituents identified in cinnamon bark oils *(continued)*

Constituent	CAS	Percentage and range				
		A	B (18)	C (11)	D (14)	E
Methyl isoeugenol	93-16-3					7.8^e
(E)-Methylisoeugenol	6379-72-2					0.1^{b,c}
α-Muurolene	10208-80-7					0.1^m; 0.4^f; 1.0^{x1}; 3.1^{x6}
γ-Muurolene	30021-74-0					tr^{x1}; 0.3^{y5}
α-Muurolol	104245-48-9					0.3^f; 1.9^{x1}
τ-Muurolol	19912-62-0					1.5^{x1}
Myrcene	123-35-3	0.01-0.4	0.09	0.1	0.1	tr^r; 0.04^v; 0.4^c; 1.4^l; 1.7^x
1-Naphthalenol, 2-methyl						0.8^{x4}
Neral	106-26-3		0-0.1			
Nerol	106-25-2		tr-0.1		tr	
(E)-Nerolidol	40716-66-3		0.06-0.1			
(E)-Ocimene	27400-72-2				tr	
(Z)-Ocimene	27400-71-1				0.03	
(E)-β-Ocimene	3779-61-1	0.02-0.09	tr-0.1	tr		0.1^x
(Z)-β-Ocimene	3338-55-4	0.01-3.6	tr-0.07	tr		
Pentadecanol	629-76-5					0.1^{x1}
α-Phellandrene	99-83-2	0.5-3.2	0.2-0.4	0.5	0.6	0.2^w; 0.4^c; 0.8^{x2}; 1.0^x; 1.1^h; 2.8^{y5}; 3.6^r; 5.2^d
β-Phellandrene	555-10-2	1.1-4.6		3.3		0.1^x; 0.6^p; 1.5^{x2}; 8.0^{y5}; 11.3^r; 18.4^d
2-Phenethyl alcohol	60-12-8				0.4	0.8^{x3}
2-Phenethyl benzoate	94-47-3		tr			
Phenol	108-95-2				tr	
Phenylacetaldehyde	122-78-1				tr	0.9^{x4}
Phenylethyl acetate	93-92-5				0.1	<0.5^{x3}
2-Phenylethyl benzoate	94-47-3				tr	
3-Phenylpropyl acetate	122-72-5		tr-0.2		0.1	
α-Pinene	80-56-8	0.2-1.8	0.3-1.9	0.7	0.6	1.1^k; 1.5^o; 1.6^n; 1.8^c; 2.0^{y5}; 2.2^r; 3.2^l; 5.7^d
β-Pinene	127-91-3	0.07-0.8	0-0.2	0.3	0.1	0.1^l; 0.3^{x2}; 0.4^m; 0.6^t; 1.0^l; 1.5^k; 2.5^{x9}
Piperitone	89-81-6				0.1	
Sabinene	3387-41-5	0.03-0.6	0-0.08	tr	0.02	tr^{y5}; 0.1^x; 0.6^r
Safrole	94-59-7	0.01-0.8	tr		tr	tr^t; 0.04^l; 0.4^c; 0.6^l
Salicylaldehyde	90-02-8					<0.5^{x3}; 0.5^l
α-Selinene	473-13-2				tr	
β-Selinene	17066-67-0		tr-0.1			
Spathulenol	6750-60-3		tr-0.1			
Styrene	100-42-5	0-0.2		`		0.1^m; 0.6^{y5}
α-Terpinene	99-86-5	0.3-2.6	0.2-0.6	0.4	0.4	0.08^v; 0.3^{x2}; 0.7^u; 0.9^x; 1.4^t; 2.2^r
γ-Terpinene	99-85-4	0.05-0.2	0.1	0.1	0.03	0.09^l; 0.1^x; 0.2^{y5}; 0.3^l
Terpinen-4-ol	562-74-3	0.04-0.1	0.7-1.6	1.2	0.4	0.1^{x1}; 0.3^g; 0.4^{x2}; 0.8^{i,p}; 0.9^{y5}; 1.2^{t,y}; 2.2^r
α-Terpineol	98-55-5	0.3-1.4		1.6	0.7	0.7^g; 0.8^l; 0.9^l; 1.6^{y5}; 1.7^p; 1.8^t; 2.3^r; 15.0^w
Terpinolene	586-62-9	0.06-0.3	tr-0.1	0.1	0.1	<0.01^t; 0.06^l; 0.1^{x2}; 0.4^{y5}
α-Terpinyl acetate	80-26-2					2.5^{x1}
Tetradecanal	124-25-4					0.8^r
Tetrahydrolinalyl acetate	20780-48-7					1.2^y
α-Thujene	2867-05-2	0.03-1.0	0.05-0.07	0.1		0.04^l; 0.1^{x2}; 0.3^r; 0.6^w; 1.1^{x7}
Thymol	89-83-8					0.3^c
2,2,4-Trimethyl-1,3-pentanediol	144-19-4					<0.1^m
ar-Turmerone	532-65-0					1.0^y
Vanillin	121-33-5				tr	
2-Vinylphenol	695-84-1				0.03	
α-Ylangene	14912-44-8				0.3	<0.01^t

Table 5.21.1 Constituents identified in cinnamon bark oils *(continued)*

A forty-three cinnamon bark oil essential oil samples from Sri Lanka, Madagascar, Seychelles and Indonesia analyzed between 1998 and 2013; lowest and highest concentrations given (E. Schmidt, unpublished data)
B two oils obtained from cinnamon barks produced in two different locations in India; lowest and highest concentrations given (ref. 18)
C one lab-produced oil from bark of wild growing *C. zeylanicum* in Madagascar (ref. 11)
D one cinnamon bark oil, uncertain whether it was commercial oil purchased in the UK or a steam-distilled oil from cinnamon bark of unknown origin (ref. 14)
E data from other studies (indicated with superscript letters); highest concentrations found in any study reviewed here given; when two or more oils were investigated, only the highest concentrations are mentioned, unless indicated otherwise

[a] incorrect identification according to ref. 17; [b] adulterant according to ref. 17; [c] one commercial oil obtained in Serbia (ref. 10); [d] one oil of cinnamon bark produced in Madagascar and two commercial oils (ref. 19); [e] one oil produced from the bark of *C. zeylanicum* from India (ref. 20); [f] one lab-distilled cinnamon bark oil (ref. 21); [g] one lab-hydrodistilled oil from cinnamon bark powder purchased at a local market in Italy (ref. 8); [h] one commercial oil from Korea (ref. 9); [i] one lab-hydrodistilled cinnamon bark oil (ref. 13); [j] one lab-hydrodistilled oil from cinnamon bark purchased at a local market in India (ref. 16); [k] one lab-hydrodistilled bark oil from India (ref. 12); [l] two bark oils, which may have been leaf oils (ref. 15); [m] one lab-hydrodistilled oil from a Chinese study (ref. 1); [n] one lab-hydro-distilled oil produced in Turkey (ref. 2); [o] one sample of lab-hydrodistilled bark oil from Sri Lanka (refs. 7,43); [p] one sample of cinnamon bark oil (ref. 22); [q] one sample of Indian cinnamon oil (ref. 23); [r] one hydrodistilled oil from Sri Lanka (ref. 24); [s] unknown number of cinnamon bark oils (ref. 31); [t] unknown number of bark oils from *C. zeylanicum* cultivated in Brazil (ref. 32); [u] one commercial cinnamon oil obtained from a German company (ref. 33); because of the occurrence of dihydrolinalool and a high concentration (13.2%) of limonene, this oil was suggested to have been adulterated (ref. 25); [v] one cinnamon bark oil of unknown origin (ref. 35); [w] one lab-produced oil from cinnamon bark from Morocco (ref. 36); [x] one oil sample from Sri Lankan *C. zeylanicum* (BM Lawrence, unpublished data, cited in ref. 25); [x1] one lab-hydrodistilled oil from cinnamon bark purchased at a local market in Pakistan (ref. 37); [x2] one cinnamon bark oil of Sri Lankan origin (ref. 38); [x3] one commercial cinnamon bark oil sample from an Italian company (ref. 39); [x4] one commercial bark oil from China (ref. 40); [x5] one commercial *Cinnamomum verum* oil with an extremely atypical composition with 48.3% β-caryophyllene as the dominant ingredient (ref. 41); [x6] one lab-hydrodistilled bark oil from India (ref. 42); [x7] one commercial oil sample from the bark of *C. zeylanicum* (ref. 44); [x8] one commercial cinnamon bark oil from India (ref. 45); [x9] one commercial oil sample purchased from a German company (ref. 46); [y] one sample of cinnamon bark oil purchased from a Medicinal Plant Research Center in Iran (ref. 47); [y1] one commercial bark oil sample from Sri Lanka (ref. 48); [y2] one commercial oil sample from a Belgian supplier (ref. 49); [y3] one commercial oil purchased in Brazil (ref. 50); [y4] one lab-hydrodistilled oil from cinnamon bark obtained at a local herb shop in Thailand (ref. 51); [y5] one lab-hydrodistilled cinnamon bark oil from Malaysia (ref. 52)

tr: trace (in columns B and C: <0.05; in column D: <0.02)

essential oil samples (concentration ranges provided) are the following: (*E*)-cinnamaldehyde (43.0-72.7%), eugenol (0.2-16.4%), (*E*)-cinnamyl acetate (0.9-7.8%), β-caryophyllene (2.7-7.5%), linalool (2.1-7.2%), β-phellandrene (1.1-4.6%), (*Z*)-β-ocimene (0.01-3.6%), α-phellandrene (0.5-3.2%), *p*-cymene (2.2-2.8%) and α-terpinene (0.3-2.6%) (Erich Schmidt, unpublished analytical data).

Chemotypes

The major component of cinnamon bark oil is virtually always (*E*)-cinnamaldehyde, with percentages of the total oil composition of up to 97.7% (Table 5.21.1). In one study of two samples of cinnamon bark oils from Madagascar, eugenol was the dominant ingredient (74.3% and 65.2%) (15). However, the profiles found corresponded very well to the leaf oils and a mistake in the study design cannot be excluded and may well be likely. Eugenol was also the dominant chemical in a lab-hydrodistilled oil obtained from dried cinnamon bark purchased at a local market in Italy with a concentration of 46.5%, versus 32.7% for (*E*)-cinnamaldehyde (8); adulteration with leaf powder might be a possible explanation. Yet, concentrations of eugenol of >10% and in one case up to 34% have been reported in other oils (Table 5.21.1). In one analysis, a hydrodistilled oil from cinnamon bark harvested in northeast India proved to contain 84.7% benzyl benzoate; the leaf oil of the source tree also contained high amounts of benzyl benzoate (12). In another investigation, β-caryophyllene was found to be the dominant ingredient with a 48.3% concentration (41).

Table 5.21.2 Major constituents of cinnamon bark oils

Constituent	CAS	Percentage and range				
		A	B (18)	C (11)	D (14)	E
(*E*)-Cinnamaldehyde	14371-10-9	43.0-72.7	59.3-60.5	52.2	75	79.8[x1]; 81.5[y3]; 81.9[x9]; 90.1[f]; 90.2[y4]; 97.7[j]
Eugenol	97-53-0	0.2-16.4	1.4-1.7	2.9	2.2	11.9[x7]; 14.9[t]; 16.7[y3]; 33.9[s]; 46.5[g]; 74.3[l]
β-Caryophyllene	87-44-5	2.7-7.5	1.5-5.6	2.4	3.3	5.8[d]; 6.4[i]; 6.9[y5]; 7.5[y1]; 10.4[e]; 13.6[r]; 48.3[x5]
β-Phellandrene	555-10-2	1.1-4.6		3.3		0.1[x]; 0.6[p]; 1.5[x2]; 8.0[y5]; 11.3[r]; 18.4[d]
Linalool	78-70-6	2.1-7.2	4.4-13.8	3.2	2.4	3.5[o]; 4.6[i]; 4.8[y5]; 6.3[e]; 6.9[t]; 7.0[x2]; 9.4[r]; 10.3[q]
(*E*)-Cinnamyl acetate	21040-45-9	0.9-7.8	tr-13.6			2.8[g]; 3.8[x2]; 4.5[q]; 4.6[x]; 4.7[v]; 6.8[s]; 7.4[n]; 12.9[d]

LEGEND: SEE UNDER TABLE 5.21.1

Additional analyses of cinnamon bark oils not discussed in Table 5.21.1 are provided in refs. 53-58 and 60; these can all be accessed on-line.

CONTACT ALLERGY/ALLERGIC CONTACT DERMATITIS

Cinnamon oil, unspecified

General
Contact allergy to/allergic contact dermatitis from cinnamon has been reported in over 15 publications. Routine testing in unselected dermatitis patients has not been performed. Many case reports of allergy to cinnamon oil relate to its presence in toothpastes, and there were also some patients with occupational allergic contact dermatitis. We assume (but cannot be certain) that most reports of contact allergy to 'cinnamon oil' concern cinnamon *bark* oil, dominated by (E)-cinnamaldehyde (versus the *leaf* oil, which has eugenol as the major component). Indeed, several authors have found concomitant reactions to cinnamon oil and cinnamaldehyde (70,71,76). However, there must be other allergenic ingredients in cinnamon oil (e.g., eugenol [78], which can be present in concentrations of up to 16% in commercial cinnamon bark oils [Table 5.21.1, column A]), as cinnamaldehyde is sometimes tested negative (72,77). In literature, no good distinction is made between cinnamon oil and cassia oil; sometimes they are used as synonyms.

Testing in groups of patients
In a group of 21 patients with dermatitis caused by fragrances and tested with a series of essential oils, nine (43%) reacted to oil of cinnamon, Ceylon (=Sri Lanka). In four cinnamon oil-allergic patients who were also tested with eugenol, three (75%) had a concomitant positive patch test reaction; cinnamaldehyde was not tested. Because of the high percentage of positive reactions to eugenol and it is unknown whether cinnamaldehyde would be positive, it is possible that the oil used for testing was cinnamon *leaf* oil (78).

Case reports
A physiotherapist developed occupational contact dermatitis from a massage cream and its ingredients cinnamon oil and clove oil; the patient was also allergic to eugenol, which may well have been the causative allergen, as it is the major ingredient of clove oil and can be present in commercial cinnamon bark oils in a concentration of up to 16% (Table 5.21.1, column A) (79). One patient had cheilitis from contact allergy to the vapor of cinnamon oil to which she was occupationally exposed while making bubble gum (73). One patient with dermatitis and stomatitis attributed to cinnamon oil and clove oil (63, cited in ref. 79). A woman developed erythematous vulvitis, thick leucorrhoea and patches of dermatitis on the buttocks after the use of a galenic vaginal suppository containing 3% cinnamon oil; there were positive patch test reactions to the fragrance mix, the suppository, its ingredient cinnamon oil (3% and 1% positive,

0.5% negative) and cinnamyl alcohol (not an important component of cinnamon oil). The reaction to cinnamal, its major ingredient was, however, negative (77). A baker with hand dermatitis had a positive patch test to cinnamon oil 5% in petrolatum, judged to be relevant (69). One case of generalized allergic contact dermatitis from cinnamon oil added to a mud bath in a spa; the patient reacted to cinnamon 'essence', eugenol, cinnamic alcohol and cinnamic aldehyde (70). One patient with cheilitis from contact allergy to cinnamon oil developed a recurrence of dermatitis after eating cinnamon (66).

Toothpastes
One patient with stomatitis from contact allergy to cinnamon oil and cassia oil in toothpaste, who was negative, however, to their main component cinnamic aldehyde (72). One case of cheilitis from allergy to cinnamon (cassia) oil in toothpaste (67). Three patients became sensitized to cinnamon oil in their toothpaste; one had perioral dermatitis, the second unilateral angular cheilitis, while case 3 had marked ulceration of the mucous membranes without external involvement of the lips (74,75). One patient with allergic contact stomatitis and dermatitis from cinnamic aldehyde and cinnamon (cassia) oil in toothpaste (76). One patient had dermatitis of some fingers from cinnamon oil in toothpaste, which ran along the tooth-brush on his hand; there was no cheilitis or stomatitis (65).

Positive patch tests (relevance unknown, uncertain or not stated)
One positive patch test to cinnamon oil in a patient working in a cosmetic factory who had occupational dermatitis from 2-bromo-2-nitropropane-1,3-diol (64). Two patients with contact dermatitis reacting to both cinnamon oil and its main ingredient, cinnamic aldehyde (71). One positive patch test reaction to cinnamon oil, possibly from its presence in a topical pharmaceutical preparation (68).

Cinnamon bark oil, Sri Lanka type
No reports on contact allergy to cinnamon bark oil, specifically mentioned to be obtained from the *bark* of *Cinnamomum zeylanicum*, have been found. Literature on contact allergy to/allergic contact dermatitis from 'cinnamon oil' (plant part used not specified) is discussed above.

LITERATURE

1 Li Y-Q, Kong DX, Wua H. Analysis and evaluation of essential oil components of cinnamon barks using GC–MS and FTIR spectroscopy. Ind Crops Prod 2013;41:269-278

2 Unlu M, Ergene E, Unlu GV, Zeytinoglu HS, Vural N. Composition, antimicrobial activity and in vitro cytotoxicity of essential oil from *Cinnamomum zeylanicum* Blume (Lauraceae). Food Chem Toxicol 2010;48:3274-3280

3 Gruenwald J, Freder J, Armbruester N. Cinnamon and health. Crit Rev Food Sc Nutrit 2010;50:822-834

4   Ranasinghe P, Pigera S, Premakumara GAS, Galappaththy P, Constantine GR, Katulanda P. Medicinal properties of 'true' cinnamon (*Cinnamomum zeylanicum*): a systematic review. BMC Compl Alter Med 2013;13:275

5   Cinnamon and Cassia. PN Ravindran, K Nirmal Babu and M Shylaja, eds. Boca Raton – London – New York – Washington, DC: CRC Press, 2004

6   Leach MJ, Kumar S. Cinnamon for diabetes mellitus. Cochrane Database of Systematic Reviews 2012, Issue 9. Art. No. CD007170. DOI: 10.1002/14651858. CD007170.pub2.

7   Samarasekera R, Kalhari KS, Weerasinghe IS. Insecticidal activity of essential oils of Ceylon *Cinnamomum* and *Cymbopogon* species against *Musca domestica*. J Essent Oil Res 2006;18:352-354. Data also presented in ref. 43

8   Chericoni S, Prieto JM, Iacopini P, Cioni P, Morelli I. In vitro activity of the essential oil of *Cinnamomum zeylanicum* and eugenol in peroxynitrite induced oxidative process. J Agric Food Chem 2005;53:4762-4765

9   Yang Y-C, Lee H-S, Lee SH, Clark JM, Ahn Y-J. Ovicidal and adulticidal activities of *Cinnamomum zeylanicum* bark essential oil compounds and related compounds against *Pediculus humanus capitis* (Anoplura: Pediculicidae). Int J Parasitol 2005;35:1595–1600

10 Simić A, Soković MD, Ristić M, Grujić-Jovanović S, Vukojević J, Marin PD. The chemical composition of some Lauraceae essential oils and their antifungal activities. Phytother Res 2004;18:713-717

11 Chalchat J-C, Valade I. Chemical composition of leaf oils of *Cinnamomum* from Madagascar: *C. zeylanicum* Blume, *C. camphora* L., *C. fragrans* Baillon and *C. angustifolium*. J Essent Oil Res 2000;12:537-540. Data also presented in ref. 34

12 Nath SC, Pathak MG, Baruah A. Benzyl benzoate, the major component of the leaf and stem bark oil of *Cinnamomum zeylanicum* Blume. J Essent Oil Res 1996;8:327-328

13 Tateo F, Chizzini F. The composition and quality of supercritical CO_2 extracted cinnamon. J Essent Oil Res 1989;1:165-168

14 Senanayake UM, Lee TH, Wills RBH. Volatile constituents of cinnamon (*Cinnamomum zeylanicum*) oils. J Agric Food Chem 1978;26:822-824

15 De Medici D, Pieretti S, Salvatore G, Nicoletti M, Rasoanaivo P. Chemical analysis of oils of Malagasy medicinal plants by gas chromatography and NMR Spectroscopy. Flavour Fragr J 1992;7:275-281

16 Singh G, Maurya S, deLampasona MP, Catalan CAN. A comparison of chemical, antioxidant and antimicrobial studies of cinnamon leaf and bark volatile oils, oleoresins and their constituents. Food Chem Toxicol 2007;45:1650-1661

17 Lawrence BM. Progress in essential oils. Perfum Flavor 2013;38(December):44-48

18 Mallavarapu GR, Rajeswara Rao BR. Chemical constituents and uses of *Cinnamomum zeylanicum* Blume. In: L Jirovetz, NX Dung and VK Varshney, Eds. Aromatic plants from Asia: their chemistry and application in food therapy. Dehradun, India: HK Bhalla and Sons, 2007:49-75. Data cited in refs. 17 and 30

19 Koroch A, Ranarivelo L, Behra O, Juliani HR, Simon JE. Quality attributes of ginger and cinnamon essential oils from Madagascar. In: J Janick and A Whipkey, Eds. Issues in new crops and new uses. Alexandria, VA, USA: ASHS Press, 2007:338-341. Data cited in ref. 17

20 Baruah A, Nath SC, Hazarika AK. Investigation of the essential oils of *Cinnamomum verum* Presl. grown at lower Brahmaputra Valley of Assam. Indian Perfum 2010;54:21-23. Data cited in ref. 17

21 Fahti A, Hassani A, Ghosta Y, Abddlahi A, Meshkatalsadat MH. The potential of thyme, clove, cinnamon and ajowan essential oils in inhibiting the growth of *Botrytis cinerea* and *Monilinia fructicola*. J Essent Oil Bear Plants 2012;15:38-47

22 Ferhout A, Bohatier J, Guillot J, Chalchat J-C. Antifungal activity of selected essential oils, cinnamaldehyde and carvacrol against *Malassezia furfur* and *Candida albicans*. J Essent Oil Res 1999;11:119-129

23 Pawar VC, Thaker VS. Evaluation of the anti-*Fusarium oxysporum* f. sp. *cicer* and anti-*Alternaria porri* effects of some essential oils. World J Microbiol Biotechnol 2007;23:1099-1106. Data also presented in ref. 59

24 Marongiu B, Piras A, Porcedda S, Tuveri E, Sanjsut E, Meli M, et al. Supercritical CO_2 extract of *Cinnamomum zeylanicum*: Chemical characterization and anti-tyrosinase activity. J Agric Food Chem 2007;55:10022–10027

25 Lawrence BM. Essential oils 2001-2004. Carol Stream, USA: Allured Publishing Corporation, 2006: 71-73, 305-306

26 Tira-Picos V, Gbolade AA, Nogueria JMF, Ajibesin KK. Analysis of leaf essential oil constituents of *Cinnamomum zeylanicum* grown in Nigeria. J Essent Oil Bear Plants 2009;12:76-80

27 Jayaprakasha GK, Mohan Rao LJ. Chemistry, biogenesis, and biological activities of *Cinnamomum zeylanicum*. Crit Rev Food Sc Nutrit 2011;51:547-562

28 Barceloux DG. Cinnamon (Cinnamomum species). DM-Dis Mon 2009;55:327-335

29 Manosi D, Suvra M, Budhimanta N, Jayram H. Ethnobotany, phytochemical and pharmacological aspects of *Cinnamomum zeylanicum* Blume. Int Res J Pharm 2013;4:58-63

30 Lawrence BM. Progress in essential oils. Perfum Flavor 2009;34(6):54-58

31 Ehlers D, Hilmerand S, Bartholomae S. Hochdruckflüssig-chromatographische Untersuchung von Zimt-CO_2-Hochdruckextrakten in Vergleich mit Zimtölen. Z Lebensm Unter Forsch 1995;200:282-288

32 Koketsu M, Goncalves SL, de Oliveira Godoy RL, Lopes D, Morsbach N. Oleos essenciais de cascas e folhas de canels (*Cinnamomum verum* Presl.) cultivada no Parana. Cienc Technol Aliment 1997;17:281-285. Data cited in ref. 25

33   Baratta MT, Dorman HJD, Deans SG, Figueiredo AC, Barroso JG, Ruberto G. Antimicrobial and antioxidant properties of some commercial essential oils. Flavour Fragr J 1998;13:235-244

34   Chalchat JC, Valade I. Plantes à huiles essentielles d'origine Malagache: composition chimique des huiles essentielles de *Cinnamomum zeylanicum, angustifolium, camphorata* et *fragrans*. Rivista Ital EPPOS 1998;Numero Speciale:729-740. Data also presented in ref. 11 and cited in ref. 25

35   Reichling J, Harkenthal M, Saller R. Wirkung ausgewählter ätherischer Öle. Erfahrungsheilkunde 1999;6:357-366. Data cited in ref. 25

36   Lahlou M, Berrada R, Agoumi A, Hmamouchi M. The potential effectiveness of essential oils in the control of human head lice in Morocco. Internat J Aromather 2001;10:108-123

37   ur-Rahman A, Choudhary MI, Farooq A, Ahmed A, Iqbal MZ, Demirci B, et al. Antifungal activities and essential oil constituents of some spices from Pakistan. Third Electron Conf Synth Org Chem (ECSOC-3) 1999, Sept 1-13:1-10. Data cited in ref. 25

38   Kubeczka K-H, Formacek V. Essential oils analysis by capillary gas chromatography and carbon-13 NMR Spectroscopy, 2nd edition. New York, USA: John Wiley and Sons, 2002:61-66. Data cited in ref. 25

39   Mazzarrino G, Paparella A, Chaves-Lopez C, Faberi A, Sergi M, Sigismondi C, et al. *Salmonella enterica* and *Listeria monocytogenes* inactivation dynamics after treatment with selected essential oils. Food Control 2015;50:794-803

40   Lu F, Ding Y-C, YE X-Q, Ding Y-T. Antibacterial effect of cinnamon oil combined with thyme or clove oil. Agricult Sci China 2011;10:1482-1487

41   Pandey A, Chattopadhyay P, Banerjee S, Pakshirajan K, Singh L. Antitermitic activity of plant essential oils and their major constituents against termite *Odontotermes assamensis* Holmgren (Isoptera: Termitidae) of North East India. Int Biodeter Biodegrad 2014;75:63-67

42   Eikani MH, Golmohammad F, Sadr ZB, Amoli HS, Mirza M. Optimization of superheated water extraction of essential oils from cinnamon bark using response surface methodology. J Essent Oil Bear Plants 2013;16:740-748

43   Samarasekera R, Kalhari KS, Weerasinghe IS. Mosquitocidal activity of leaf and bark essential oils of Ceylon *Cinnamomum zeylanicum*. J Essent Oil Res 2005;17:301-303. Data also presented in ref. 7

44   Moretti MDL, Bazzoni E, Passino GS, Prota R. Antifeedant effects of some essential oils on *Ceratitis capitata* Wied. (Diptera, Tephritidae). J Essent Oil Res 1998;10:405-412

45   Khan MSA, Ahmad I. *In vitro* inhibition of growth and virulence factors production in azole-resistant strains of non-albicans *Candida* by *Cinnamomum verum, Cymbopogon citratus, Cymbopogon martini* and *Syzygium aromaticum* essential oils. J Biol Active Prod Nature 2013;3:139-153

46   Akhtar Y , Le Page E, Stevens A, Bradbury R, da Camara CAG, Isman BM. Effect of chemical complexity of essential oils on feeding deterrence in larvae of the cabbage looper. Physiol Entomol 2012;37:81-91

47   Moarefiani M, Barzegari M, Sattari M. *Cinnamomum zeylanicum* essential oil as a natural antioxidant and antibactrial in cooked sausage. J Food Biochem 2013;37:62-69

48   Ranasinghe L, Jayawardena B, Abeywickrama K. Fungicidal activity of essential oils of *Cinnamomum zeylanicum* (L.) and *Syzygium aromaticum* (L.) Merr et L.M. Perry against crown rot and anthracnose pathogens isolated from banana. Lett Appl Microbiol 2002;35:208-211

49   Mayaud L, Carricajo A, Zhiri A, Aubert G. Comparison of bacteriostatic and bactericidal activity of 13 essential oils against strains with varying sensitivity to antibiotics. Lett Appl Microbiol 2008;47:167-173

50   Azeredo CMO, Santos TG, Lameiro de Noronha Sales Maia BH, Soares MJ. *In vitro* biological evaluation of eight different essential oils against *Trypanosoma cruzi*, with emphasis on *Cinnamomum verum* essential oil. BMC Compl Alter Med 2014;14:309

51   Rattanachaikunsopon P, Phumkhachorn P. Potential of cinnamon (*Cinnamomum verum*) oil to control *Streptococcus iniae* infection in tilapia (*Oreochromis niloticus*). Fish Sci 2010;76:287-293

52   Bin Jantan I, Moharam BAK, Santhanam J, Jamal JA. Correlation between chemical composition and antifungal activity of the essential oils of eight *Cinnamomum* species. Pharmac Biol 2008;46:406-412

53   Abbasipour H, Mahmoudvand M, Rastegar F, Hosseinpour MH. Chemical composition, fumigant toxicity and oviposition deterrence of *Cinnamomum zeylanicum* essential oil to three stored-product insects. Archives Des Sciences 2012;65:533-542

54   Paranayama PA, Wimalasena S, Jayatilake GS, Jayawardena AL, Senanayake UM, Mubarak AM. A comparison of essential oil constituents of bark, leaf, root and fruit of cinnamon (*Cinnamomum zeylanicum* Blum) grown in Sri Lanka. J Natn Sci Foundation Sri Lanka 2001;29:147-153

55   Hadri Z, Allem R, Perry MG. Effect of essential oil of *Cinnamomum zeylanicum* on some pathogenic bacteria. Afric J Microbiol Res 2014;8:1026-1031

56   Pooja A, Arun N, Maninder K. GC-MS profile of volatile oils of *Cinnamomum zeylanicum* Blume and *Ocimum kilimandscharicum* Baker ex Gurke. Int J Pharm Sci Rev Res 2013;19:124-126

57   Hanafi EM, Maghraby NA, Ramadan MM, Mahmoud MA, El-Allawy HM. Aromatherapy of *Cinnamomum zeylanicum* bark oil for treatment of scabies in rabbits with emphasis on the productive performance. Amer-Eurasian J Agric Environ Sci 2010;7:719-727

58   Husain SS, Ali M. Analysis of volatile oil of the stem bark of *Cinnamomum zeylanicum* and its antimicrobial activity. Int J Res Pharm Sc 2013;3(4):40-49

59 Pawar VC, Thaker VS. *In vitro* efficacy of 75 essential oils against *Aspergillus niger*. Mycoses 2006;49:316-323. Data also presented in ref. 23

60 Queiroga Sarmento Guerra F, Moura Mendes J, Araújo de Oliveira W, Martins da Costa JG, Douglas Melo Coutinho H, Oliveira Lima E. Chemical composition and antimicrobial activity of Cinnamomum zeylanicum Blume essential oil on multi-drug resistant *Acinetobacter* spp. strains. Biofar 2012;8:62-70

61 Rhind JP. Essential oils. A handbook for aromatherapy practice, 2nd Edition. London: Singing Dragon, 2012

62 Lawless J. The encyclopedia of essential oils, 2nd Edition. London: Harper Thorsons, 2014

63 Silvers SH. Stomatitis and dermatitis venenata with purpura, resulting from oil of cloves and oil of cassia. Dental Items of Interest 1939;61:649-651

64 Rudzki E, Rebandel P, Grzywa Z. Occupational dermatitis from cosmetic creams. Contact Dermatitis 1993;29:210

65 Cummer CL. Dermatitis due to oil of cinnamon. Arch Dermatol Syphilol 1940;42:674-675

66 Leifer W. Contact dermatitis due to cinnamon. Recurrence of dermatitis following oral administration of cinnamon oil. Arch Dermatol Syphylol 1951;64:52-55. Data cited in ref. 72

67 Laubach JL, Malkinson FD, Ringrose EJ. Cheilitis caused by cinnamon (cassia) oil in toothpaste. JAMA 1953;152:404-405. Data cited in ref. 72

68 Nardelli A, D'Hooge E, Drieghe J, Dooms M, Goossens A. Allergic contact dermatitis from fragrance components in specific topical pharmaceutical products in Belgium. Contact Dermatitis 2009;60:303-313

69 Nethercott JR, Holness DL. Occupational dermatitis in food handlers and bakers. J Am Acad Dermatol 1989;21:485-490

70 Garcia-Abujeta JL, de Larramendi CH, Pomares Berna J, Munoz Palomino E. Mud bath dermatitis due to cinnamon oil. Contact Dermatitis 2005;52:234

71 Kirton V, Wilkinson DS. Sensitivity to cinnamic aldehyde in a toothpaste. 2. Further studies. Contact Dermatitis 1975;1:77-80

72 Magnusson B, Wilkinson DS. Cinnamic aldehyde in toothpaste. 1. Clinical aspects and patch tests. Contact Dermatitis 1975;1:70-76

73 Miller J. Cheilitis from sensitivity to oil of cinnamon present in bubble gum. JAMA 1941;116:131-132. Data cited in ref. 72

74 Millard LG. Acute contact sensitivity to a new toothpaste. J Dentistry 1973;1:168-170

75 Millard LG. Contact sensitivity to toothpaste. Brit Med J 1973;1:676

76 Drake TE, Maibach HI. Allergic contact dermatitis and stomatitis caused by a cinnamic aldehyde-flavored toothpaste. Arch Dermatol 1976;112:202-203

77 Lauriola MM, De Bitonto A, Sena P. Allergic contact dermatitis due to cinnamon oil in galenic vaginal suppositories. Acta Derm Venereol 2010;90:187-188

78 Meynadier JM, Meynadier J, Peyron JL, Peyron L. Formes cliniques des manifestations cutanées d'allergie aux parfums. Ann Dermatol Venereol 1986;113:31-39

79 Sánchez-Pérez J, García-Díez A. Occupational allergic contact dermatitis from eugenol, oil of cinnamon and oil of cloves in a physiotherapist. Contact Dermatitis 1999;41:346-347

Chapter 5.22 CINNAMON LEAF OIL, SRI LANKA

There are two Sri Lanka types of cinnamon oils, bark oil and leaf oil. In this chapter, the leaf oil is discussed. Cinnamon bark oil is presented in Chapter 5.21.

DEFINITION

Cinnamon leaf oil, Sri Lanka type (essential oil of cinnamon leaf, Sri Lanka type), is the essential oil obtained from the leaves of the cinnamon, *Cinnamomum zeylanicum* Blume (synonym: *Cinnamomum verum* J. Presl.).

INCI NOMENCLATURE

Description/definition: Cinnamomum zeylanicum leaf oil is the volatile oil obtained from the leaves of the Ceylon cinnamon, *Cinnamomum zeylanicum*, Lauraceae
INCI name EU & USA: Cinnamomum zeylanicum leaf oil
Other names: Ceylon cinnamon leaf oil
CAS registry number(s): 8015-91-6; 84649-98-9
EINECS number(s): 283-479-0

ISO (INTERNATIONAL ORGANIZATION FOR STANDARDIZATION) STANDARD

ISO number: 3524
ISO name: Essential oil of cinnamon leaf, Sri Lanka type
Botanical origin: *Cinnamomum zeylanicum* Blume (synonym: *Cinnamomum verum* J. Presl.)
Parts of plant used: Leaf
ISO values: ISO values (minimum and maximum concentrations) are shown in Table 5.22.1.

Cinnamon oils should not be confused with cassia oils, which are obtained from the Chinese cinnamon, *Cinnamomum cassia* (Nees & T. Nees) J. Presl. (see Chapters 5.13 and 5.14).

THE PLANT, THE OIL, AND THEIR USES

This section is discussed under Cinnamon *bark* oil.

Table 5.22.1 ISO values (%) for cinnamon leaf oil, Sri Lanka type[a]

Compound	CAS	Minimum	Maximum
Eugenol	97-53-0	70.0	83.0
Benzyl benzoate	120-51-4	2.0	4.0
Eugenyl acetate	93-28-7	1.3	3.0
(*E*)-Cinnamyl acetate	21040-45-9	1.1	1.8
(*E*)-Cinnamaldehyde	14371-10-9	0.8	1.5

[a] ISO 3524 Essential oil of cinnamon leaf, Sri Lanka type ©ISO 2003; Geneva, Switzerland, www.iso.org

CHEMICAL COMPOSITION

Cinnamon leaf oil is a light to dark amber clear and mobile liquid, which has a strong spicy odor, reminding of fresh crushed cloves with tender cinnamic notes. The yield of essential oil from the leaves of *Cinnamomum zeylanicum* Blume generally varies from 1.2 to 1.8%. The main producing countries of this oil are Sri Lanka, Madagascar, Vietnam and the Seychelles.

Literature data (up to November 17, 2014) on the chemical composition of cinnamon leaf oils and unpublished analytical data from one of us (E.S.) are shown in Table 5.22.2 in alphabetical order. In cinnamon leaf oils from various origins, over 160 chemicals have been identified. About 52% of these were found in a single reviewed publication only. The major compounds found in cinnamon leaf oils from different sources are shown in Table 5.22.3. They include (highest concentrations in any study given) eugenol (93.5%), linalool (10.8%, one report with extremely high concentration), benzyl benzoate (8.3%, 2 reports with extremely high concentrations), and β-caryophyllene (6.9%, one report with very high concentration). Well-known ingredients of cinnamon leaf oils that were present in high concentrations (>10%) in one or two studies were linalool (85.7%), benzyl benzoate (65.4% and 74.8%), (*E*)-cinnamaldehyde (41.3% and 58.5%), β-caryophyllene (53.2%), safrole (39.5%), 1,8-cineole (22.6%), terpinen-4-ol (11.6%), benzaldehyde (11.2%), α-phellandrene (10.0%) and eugenyl acetate (8.0% and 10.0%). Uncommon or rare constituents of cinnamon leaf oils found in high concentrations in single studies include *o*-cymene (19.4%) and cinnamic acid (16.6%).

Commercial oils

The ten chemicals that had the highest maximum concentrations in 30 commercial cinnamon leaf essential oil samples (concentration ranges provided) are the following: eugenol (72.2-81.6%), β-caryophyllene (1.3-4.6%), eugenyl acetate (1.2-4.1%), benzyl benzoate (0.6-3.6%), (*E*)-cinnamaldehyde (0.8-2.8%), linalool (1.8-2.8%), (*E*)-cinnamyl acetate (1.0-2.0%), α-phellandrene (0.5-1.5%), α-pinene (0.4-1.4%) and safrole (0.8-1.3%) (Erich Schmidt, unpublished analytical data).

Chemotypes

In nearly all analytical studies, eugenol is the major ingredient of cinnamon leaf oils with percentages up to 93.5% of the total oil composition. However, in a number of investigations, benzyl benzoate was the dominant chemical, with concentrations up to 74.8% (11,16,17,18). (*E*)-Cinnamaldehyde, the predominant chemical in almost all cinnamon *bark* oils, was found to be abundant in an older study (25, cited in ref. 26), was the major ingredient (41.3%) of a commercial leaf oil from Madagascar in one study (10) and was found in a concentration of 58.5% in a leaf oil from India in another (twice published) study (29,44). However, these may well have been bark oils. Linalool was predominant (85.7%) in an analysis of a cinnamon leaf oil sample from South India (8). In another study from India, the major ingredient of an oil obtained from 'the aerial parts'

(which probably are leaves with twigs) of a cultivated *C. zeylanicum*, proved to be (*E*)-cinnamyl acetate (58.5%) (27). β-Caryophyllene (53.2%) was found to be the dominant ingredient in a commercial leaf oil sample from a USA company (ref. 41).

Table 5.22.2 Constituents identified in cinnamon leaf oils

Constituent	CAS	Percentage and range				
		A	B (15)	C (9)	D (12)	E
Acetophenone	98-86-2					1.3[u]
Acetyleugenol	93-28-7				2.5	1.7[x]
α-Amorphene	20085-19-2					<0.01[j]
Aromadendrene	489-39-4					0.1[f]; 1.1[j]
allo-Aromadendrene	25246-27-9					0.1[d]
Benzaldehyde	100-52-7	0.09-0.2			0.7	0.1[b,c,v,y3]; 0.5[m]; 11.2[u]
Benzene, (3-ethoxy-1,5-hexadien-1-yl)-	67323-95-9					1.1[i]
Benzoic acid	65-85-0					3.6[u]
Benzyl acetate	140-11-4					0.1[b]
Benzyl alcohol	100-51-6	0-0.2			tr	0.2[b,i]
Benzyl benzoate	120-51-4	0.6-3.6	0.1-8.3		3.5	3.9[f]; 4.0[v]; 5.5[m]; 5.6[x1]; 6.8[k]; 65.4[p]; 74.8[o]
Bicyclogermacrene	24703-35-3					0.2[d]; 0.9[o]; 3.6[j]
Borneol	507-70-0		tr-0.1	0.02	tr	0.05[y1]; 0.1[q]; 0.3[m,o]; 0.9[i]; 1.5[p]
Bornyl acetate	76-49-3				tr	
γ-Cadinene	39029-41-9		0.05-0.08		0.01	<0.1[n]; 0.3[m]; <0.5[o]
δ-Cadinene	483-76-1			0.06		tr[y3]; 0.1[o]; 0.2[c]; 0.4[j]; 0.6[w]; 3.6[f]
α-Cadinol	481-34-5			0.06		<0.1[n]; 0.1[o]
epi-α-Cadinol	5937-11-1			0.03		
(-)-Calamenene						0.3[w]
Camphene	79-92-5	0.1-0.5	0.2	0.05	0.4	0.3[b,c,y8]; 0.4[r]; 0.5[y5]; 0.6[m]; 1.2[p]; 1.3[o]
Camphor	76-22-2		tr-0.1	0.1	tr	<0.1[r]; 0.1[o]
δ3-Carene	13466-78-9	0.2-0.6	tr-0.05	0.06	0.1	0.09[c]; 0.2[m]; 0.6[b]
β-Caryophyllene	87-44-5	1.3-4.6	0.5-2.3	0.9	5.8	4.5[c]; 5.0[y1]; 5.4[y5]; 5.5[f]; 5.7[d]; 6.7[y7]; 6.9[p]; 53.2[w]
(Z)-β-Caryophyllene						5.0[w]
Caryophylleneoxide	1139-30-6	0.3-0.6	0.1	0.2	0.5	0.2[y8]; 0.4[o]; 0.5[b,r]; 0.6[f]; 0.7[c]; 0.9[y5]; 1.2[w]; 1.4[t]
α-Cedrene	469-61-4					0.3[s]
Cedrol	77-53-2					<0.1[n]
1,8-Cineole	470-82-6	0.1-0.6	0.2-0.4		0.6	0.04[p]; 0.1[q]; 0.4[t]; 0.5[v]; 0.6[b]; 0.7[j]; 1.1[m]; 22.6[y2]
Cinnamaldehyde	104-55-2					0.6[p]; 1.2[h,y4]; 1.4[o]; 2.7[x,y]; 3.8[l]; 4.9[m]
(E)-Cinnamaldehyde	14371-10-9	0.8-2.8	0.6-1.5		2.0	1.4[k]; 1.7[q]; 2.2[t]; 2.7[v]; 4.9[n]; 16.3[j]; 41.3[r]; 58.5[u]
(Z)-Cinnamaldehyde	57194-69-1		tr	0.05		<0.1[r]
Cinnamicacid	621-82-9					16.6[u]
Cinnamyl acetate	103-54-8				1.7	0.09[w]; 1.1[y8]; 1.7[f]; 1.8[b]; 2.5[y5]; 2.9[l]; 3.2[r]; 4.1[o]
(E)-Cinnamyl acetate	21040-45-9	1.0-2.0	0.7-2.5			0.1[n]; 0.3[m]; 0.7[y7]; 0.9[q]; 2.7[m]
(Z)-Cinnamyl acetate				2.6		0.1[k]; 1.0[v]
Cinnamyl alcohol	104-54-1	0-0.08			0.4	0.07[i]; 0.09[v]; 0.3[c]; 0.9[r]
(E)-Cinnamyl alcohol	4407-36-7		0-tr			0.1[q]; 0.2[k,m]
Citronellal	106-23-0		tr-0.4	0.07		
β-Citronellene	2436-90-0					0.1[e]
α-Copaene	3856-25-5					0.1[s]; 0.3[o]; 0.4[d]; 0.5[w]; 0.7; 0.8[l]; 1.2[f,y7]
Coumarin	91-64-5				tr	0.05[i]
α-Cubebene	17699-14-8	0.6-0.9				<0.01[j]; 0.9[b]
β-Cubebene	13744-15-5					<0.1[n]; <0.5[o]
Cuminaldehyde	122-03-2		tr		0.1	0.8[y8]
o-Cymene	527-84-4					19.4[y2]
p-Cymene	99-87-6	0.2-0.8	0.2-0.3	0.5	1.2	0.6[p]; 0.8[y1]; 0.9[v]; 1.0[y8]; 1.2[y5]; 2.0[m]; 2.3[m]; 3.5[r]
p-Cymenene	1195-32-0					<0.1[n]

Table 5.22.2 Constituents identified in cinnamon leaf oils (*continued*)

Constituent	CAS	Percentage and range				
		A	B (15)	C (9)	D (12)	E
Diethyl phthalate	84-66-2					1.4[a,u]
Dihydrocarveol	38049-26-2					<0.1[q]
Dipropyl sulfone	598-03-8					0.01[i]
β-Elemene	33880-83-0		0.3			0.1[o]
γ-Elemene	29873-99-2					0.08[c]; 0.2[m]
Elemol (α-)	639-99-6					<0.1[n]
Ethyl benzoate	93-89-0					0.2[p]
Ethyl cinnamate	103-36-6				0.02	
Ethyl(E)-cinnamate	4192-77-2					0.3[t]
Eugenol	97-53-0	72.2-81.6	81.4-84.5	76.6	70.1	76.1[y7]; 89.1[s]; 90.2[y3]; 91[k]; 92.7[t]; 93.1[e]; 93.5[f]
Eugenyl acetate	93-28-7	1.2-4.1	0.1-2.9	2.7		0.1[q]; 0.6[v]; 0.7[p,t]; 1.1[y5]; 2.8[y4]; 3.7[f]; 8.0[k]; 10.0[l] 1.1[b]
(E,E)-α-Farnesene	502-61-4	0-1.1				
(Z,E)-α-Farnesene	26560-14-5		0-tr			
β-Farnesene	502-60-3					0.3[e]
(E)-β-Farnesene	18794-84-8			0.2		
Farnesol	4602-84-0				0.1	<0.1[n]
(E,E)-Farnesol	106-28-5		tr			
Fenchone	1195-79-5				tr	
Furfural	98-01-1				0.02	
Geranial	141-27-5				0.02	
Geraniol	106-24-1				0.04	<0.1[n]; 0.2[k]; 0.3[y8]
Geranyl acetate	105-87-3					0.2[f]
Geranyl benzoate	94-48-4			0.05		
Germacrene D	23986-74-5					0.08[c]; 0.6[j]; 1.1[o]
2-Heptanone	110-43-0			0.02		
Hexanal	66-25-1					0.3[d]
Hexanal	66-25-1					0.3[d]
Hexanol (1-; *n*-)	111-27-3				tr	0.1[q,y8]
2-Hexanol	626-93-7			0.01		
(Z)-3-Hexenal	6789-80-6					0.3[d]
1-Hexen-3-ol	4798-44-1					0.1[q]
(E)-2-Hexen-1-ol	928-95-0					0.1[q]
(Z)-3-Hexenol	928-96-1	0-0.1				0.1[b,q]
3-Hexen-1-ol	544-12-7					<0.1[n]
(E)-2-Hexenyl acetate	2497-18-9					0.1[q]
(Z)-3-Hexenyl acetate	3681-71-8	0-0.08				0.1[b,q]
(Z)-3-Hexenyl benzoate	25152-85-6	0-0.06				0.1[b]
α-Hexylcinnamaldehyde	101-86-0			0.3		
α-Humulene	6753-98-6	0.4-0.8	0.1-0.5		0.9	0.6[v]; 0.7[c]; 0.9[c,o]; 1.0[f,y5]; 1.1[y7]; 3.5[x]; 6.1[w]
α-Humulene epoxide	96638-51-6	0.02-0.04				
Humulene epoxide II	19888-34-7					0.2[y8]
Humulene oxide I	19888-33-6			0.05		
Hydrocinnamaldehyde	104-53-0	0-0.1			0.2	0.04[i]; <0.1[r]; 0.1[v]
Hydrocinnamyl acetate	122-72-5					0.4[l]
Hydrocinnamyl alcohol	122-97-4					0.07[i]
Isoeugenol	97-54-1	0-0.1			0.1	0.07[v]; 0.1[b]
(Z)-Isoeugenol	5912-86-7					0.05[i]
Isoeugenol acetate	93-29-8					0.5[e]
Limonene	138-86-3	0.2-0.5			0.4	0.2[c,q]; 0.3[v]; 0.4[y1,y7]; 0.5[b,m,s]; 1.3[o]; 3.4[p]
Linalool	78-70-6	1.8-2.8	1.6-3.7	8.5	3.4	3.1[f]; 3.3[y5]; 3.7[y7]; 3.9[r]; 4.3[s]; 5.8[c]; 10.8[p]; 85.7[q]
Linalool oxide	1365-19-1					0.1[e,n]
cis-Linalooloxide, furanoid	5989-33-3		tr	0.04		0.2[q]; 0.4[m]
trans-Linalool oxide, furanoid	34995-77-2		tr	0.1	0.1	0.05[d]; 0.1[m,q]
Linalyl acetate	115-95-7		0-0.5		tr	<0.1[n]; 0.1[q]; 0.8[k]
Linalyl propionate	144-39-8					<0.1[n]
p-Mentha-1(7),8-diene	499-97-8					<0.01[j]
3-Methylbenzothio-phene	1455-18-1					0.01[i]
Methyl chavicol	140-67-0		tr-0.2	0.06	tr	0.7[p]
Methyl cinnamate	103-26-4				0.03	0.09[v]
(E)-Methylcinnamate	1754-62-7		tr			0.2[t]

Table 5.22.2 Constituents identified in cinnamon leaf oils (*continued*)

Constituent	CAS	Percentage and range				
		A	B (15)	C (9)	D (12)	E
(Z)-Methylcinnamate	19713-73-6		tr	0.2		
Methyl dodecanoate	111-82-0					0.8[e]
Methyl eugenol	93-15-2				0.01	0.1[y3]; 0.2[p]
Methyl isoeugenol	93-16-3				tr	
(E)-Methylisoeugenol	6379-72-2		tr			
(Z)-Methyl isoeugenol	6380-24-1		0-tr			
Mintsulfide	72445-42-2					<0.5[o]
α-Muurolene	10208-80-7					0.1[f]
α-Muurolol	104245-48-9					<0.1[n]
Myrcene	123-35-3	0.05-0.8	0.09-0.1	0.09	0.1	0.1[y3]; 0.6[m]; 0.8[v]; 1.1[k]; 1.2[h,y4]; 1.3[o]
Myrtenal	564-94-3					<0.1[n]
Neohexane	75-83-2					0.01[i]
Nerol	106-25-2		tr-0.3	0.02	tr	
Nerolidol	7212-44-4					0.09[d]
(E)-Nerolidol	40716-66-3	0-0.06		0.02		0.1[b]
Neryl acetate	141-12-8					<0.1[n]
Nonanal	124-19-6				tr	0.3[q]
(E)-β-Ocimene	3779-61-1	0.04-0.4	0-0.05	0.04	0.02	<0.1[q]; 0.1[b]; 0.6[m]
(Z)-β-Ocimene	3338-55-4		0-tr		0.02	
(Z)-Ocimenol	39900-51-1					1.8[r]
n-Octanal	124-13-0		0-tr			
Perillaldehyde	2111-75-3					<0.1[n]
α-Phellandrene	99-83-2	0.5-1.5	0.5-1.0	1.2	0.1	0.7[v]; 0.9[b]; 1.2[o]; 1.3[y5]; 1.9[j]; 2.5[m]; 3.0[r]; 10.0[y2]
β-Phellandrene	555-10-2	0.2-1.0	tr			<0.1[q]; 0.2[b,w]; 0.4[c]; 0.5[d,y3]; 1.6[y5]; 7.4[r]
Phenethyl acetate	103-45-7				0.03	
2-Phenethyl alcohol	60-12-8		0-tr	0.02	tr	0.1[b,i]
2-Phenethyl anthranilate	133-18-6			0.02		
2-Phenethyl decanoate	61810-55-7			0.03		
2-Phenethyl propionate	122-70-3		0-tr			
Phenol	108-95-2				0.02	
Phenylacetaldehyde	122-78-1				tr	2.4[u]
2-Phenylethyl benzoate	94-47-3		tr-0.08		tr	
3-Phenylpropyl acetate	122-72-5				1.0	
α-Pinene	80-56-8	0.4-1.4	0.4-0.5	0.5	1.0	1.0[m]; 1.1[k]; 1.2[b]; 1.3[y5]; 1.8[x]; 2.0[p]; 2.6[o]; 5.2[y2]
β-Pinene	127-91-3	0.1-0.4	0.2	0.08	0.3	0.1[y3]; 0.3[b,c]; 0.5[r,y5]; 0.7[m]; 1.4[o]; 1.8[p]
Piperitone	89-81-6		0-0.03	3.3	0.04	0.3[k]
4-(2-Propenyl)phenol	501-92-8					0.3[i]
Sabinene	3387-41-5		tr	0.02	tr	0.04[d]
Sabinol						0.2[v]
Safrole	94-59-7	0.8-1.3	tr-0.2	0.06	2.3	0.08[v]; 0.3[w]; 1.2[y4]; 1.3[b]; 1.4[y8]; 2.0[f,y7]; 39.5[m]
Safrone						1.8[y5]
α-Selinene	473-13-2			0.1		
β-Selinene	17066-67-0		tr		tr	0.4[y3]
Spathulenol	6750-60-3	0-0.07		0.1		0.1[d]; 0.2[c,q]; 0.5[j]
Styrene	100-42-5	0.01-0.1				0.1[b]
α-Terpinene	99-86-5	0.09-0.3	tr-0.03	0.5	0.1	0.05[y8]; 0.1[b]; 0.4[c]; 0.7[m]; 1.1[v]; 1.8[r]
γ-Terpinene	99-85-4	0-0.07	tr-0.05	0.03	0.1	0.08[c]; 0.1[b]; 0.4[m]; <0.5[o]
Terpinen-4-ol	562-74-3	0.01-0.1	0.04-0.05	0.09	0.1	0.1[v]; 0.2[k]; 0.3[q]; <0.5[o]; 1.2[m]; 4.3[m]; 11.6[r]
α-Terpineol	98-55-5	0.2-0.5	0.1		0.4	tr[y3]; 0.3[b,k,v]; 0.4[o,y1]; 0.9[m]; 1.1[q]
Terpinolene	586-62-9	0.08-0.6	0.05-0.1	0.04	0.1	tr[y3]; 0.06[y8]; 0.1[o,q]; 0.2[b,m]; 0.3[m]; 0.6[v]
Thiazole	288-47-1					0.4[e]
α-Thujene	2867-05-2	0.03-0.2	0.04-0.06	0.1		0.07[y8]; 0.1[j,y3]; 0.2[b]; 0.3[m]; 1[o]
α-Thujone	546-80-5					<0.1[n]
Valencene	4630-07-3					0.2[d]
Vanillin	121-33-5					0.1[i]
p-Vinylbenzohydrazide						0.1[i]
2-Vinylphenol	695-84-1				tr	
Viridiflorol	552-02-3					0.1[o]; 0.7[m]
α-Ylangene	14912-44-8				1.0	<0.01[m]; 0.1[v]

Table 5.22.2 Constituents identified in cinnamon leaf oils (*continued*)

A thirty cinnamon leaf essential oil samples from Sri Lanka, Madagascar and Vietnam, analyzed between 1998 and 2013; lowest and highest concentrations given (E. Schmidt, unpublished data)
B two lab-distilled leaf oils from experimental farms in India; both concentrations given (ref. 15)
C one lab-hydrodistilled oil from cinnamon leaves from a plantation in India (9)
D one oil, uncertain whether it was a commercial oil sample purchased in London or a lab-hydrodistilled leaf oil (ref. 12)
E data from other studies (indicated with superscript letters); highest concentrations found in any study reviewed here given; when two or more oils were investigated, only the highest concentrations are mentioned, unless indicated otherwise

[a] incorrect identification/does not exist in nature (ref. 20); [b] one commercial oil purchased in Germany (ref. 7); [c] one lab-hydrodistilled cinnamon leaf oil from India (ref. 2); [d] one steam-distilled oil from the leaves from the Fiji Islands (ref. 22); [e] one lab-hydrodistilled oil from *C. zeylanicum* from Malaysia (ref. 1); [f] one commercial cinnamon leaf oil (ref. 3); [g] one lab-hydrodistilled essential oil obtained from leaves collected in an experimental forest in Taiwan (ref. 4); [h] one lab-hydrodistilled oil from leaves of wild growing *C. zeylanicum* in Sri Lanka (ref. 6); [i] one lab-hydrodistilled oil from leaves obtained in a botanical garden in China (ref. 13); [j] one lab-hydrodistilled oil from cinnamon leaves purchased in India (ref. 14); [k] oils from different sections of cinnamon leaves from India (ref. 19); [l] one cinnamon oil from Bangladesh (ref. 21); [m] data from various studies cited in ref. 23; [n] one oil sample produced from the dried leaves of *C. zeylanicum* growing in Cameroon (ref. 24); [o] one oil made from leaves collected in a medicinal plant garden in Nigeria with benzyl benzoate as major ingredient (ref. 5); [p] one lab-hydrodistilled oil from India with benzyl benzoate as dominant ingredient (ref. 11); [q] one lab-hydrodistilled oil from cinnamon leaves from South India with linalool as dominant ingredient (ref. 8); [r] one commercial leaf oil from Madagascar with (*E*)-cinnamaldehyde as major ingredient; this may also have been a bark oil or a strongly adulterated leaf oil (ref. 10); [s] one lab-hydrodistilled cinnamon oil from leaves collected in a botanical garden in Cameroon (ref. 35); [t] one lab-hydrodistilled leaf oil from India (ref. 28); [u] one cinnamon leaf oil obtained from an Indian research society; the composition was very atypical with high concentrations of (*E*)-cinnamaldehyde (58.5%), cinnamic acid (16.6%) and benzaldehyde (11.2%) and was probably a bark oil (refs. 29,44); [v] one lab-hydrodistilled *C. zeylanicum* leaf oil from Sri Lanka (ref. 42); [w] one leaf oil sample purchased from a USA company with a very atypical composition dominated by β-caryophyllene (53.2%) (ref. 41); [x] one commercial cinnamon leaf oil from Sri Lanka (ref. 40); [y] one commercial leaf oil *ex Cinnamomum zeylanicum* from Sri Lanka (ref. 38); [y1] one commercial *C. verum* leaf oil purchased from a Belgian company (ref. 39); [y2] one lab-hydrodistilled cinnamon oil from leaves collected in the wild in Brazil with an extremely atypical composition with high concentrations of 1,8-cineole (22.6%) and *o*-cymene (19.4%); probably botanical misinterpretation (ref. 37); [y3] one lab-hydrodistilled cinnamon leaf oil from Malaysia (ref. 36); [y4] one lab-hydrodistilled leaf oil from Sri Lanka (ref. 34); [y5] one commercial cinnamon leaf oil *ex Cinnamomum zeylanicum* from Brazil (ref. 32); [y6] one cinnamon leaf oil obtained from a German company (ref. 33); [y7] one commercial cinnamon leaf oil purchased from an Italian company (ref. 31); [y8] one commercial oil purchased in Italy (ref. 30);

tr: trace (in column B: <0.02; in column D: <0.01)

Table 5.22.3 Major constituents of cinnamon leaf oils

Constituent	CAS	Percentage and range				
		A	B (15)	C (9)	D (12)	E
Eugenol	97-53-0	72.2-81.6	81.4-84.5	76.6	70.1	76.1[y7]; 89.1[s]; 90.2[y3]; 91[k]; 92.7[t]; 93.1[e]; 93.5[f]
Linalool	78-70-6	1.8-2.8	1.6-3.7	8.5	3.4	3.1[f]; 3.3[y5]; 3.7[y7]; 3.9[r]; 4.3[s]; 5.8[c]; 10.8[p]; 85.7[q]
Benzyl benzoate	120-51-4	0.6-3.6	0.1-8.3		3.5	3.9[f]; 4.0[v]; 5.5[m]; 5.6[x1]; 6.8[k]; 65.4[p]; 74.8[o]
β-Caryophyllene	87-44-5	1.3-4.6	0.5-2.3	0.9	5.8	4.5[c]; 5.0[y1]; 5.4[y5]; 5.5[f]; 5.7[d]; 6.7[y7]; 6.9[p]; 53.2[w]

LEGEND: SEE UNDER TABLE 5.22.2

CONTACT ALLERGY/ALLERGIC CONTACT DERMATITIS

General

Contact allergy to cinnamon leaf oil has been rarely reported and there are no cases of proven allergic contact dermatitis/stomatitis. The literature on 'cinnamon oil' (not specified as bark or leaf oil) is discussed in Chapter 5.21 Cinnamon bark oil.

Testing in groups of patients

Routine testing

The members of the International Contact Dermatitis Research Group before 1975 tested 1382 dermatitis patients with cinnamon leaf oil 1% in petrolatum and saw 15 (1.1%) positive reactions; three (20%) were also positive to eugenol, the major component of cinnamon leaf oil; relevance is unknown (data cited in ref. 45).

Testing in groups of selected patients

In a group of 21 patients with dermatitis caused by fragrances and tested with a series of essential oils, nine (43%) reacted to oil of cinnamon, Ceylon (=Sri Lanka) (not specified whether it was bark or of leaf oil). In four cinnamon oil-allergic patients who were also tested with eugenol, three (75%) had a concomitant positive patch test reaction. Cinnamaldehyde was not tested. Because of the high percentage of positive reactions to eugenol and as it was unknown whether there were positive reactions to cinnamaldehyde, it is possible that the oil used for testing was cinnamon *leaf* oil (46).

Positive patch tests (relevance unknown, uncertain or not stated)

One positive patch test reaction to cinnamon leaf oil in a patient with an allergic reaction to toothpaste; the presence of the oil was not ascertained (45).

LITERATURE

1 Subki SYM, Jamal JA, Husain K, Manshoor N. Characterisation of leaf essential oils of three *Cinnamomum* species from Malaysia by gas chromatography and multivariate data analysis. Pharmacogn J 2013;5:22-29

2 Sunil Kumar KN, Sangeetha B, Rajalekshmi M, Ravishankar B, Muralidhar R, Yashovarma B. Chemoprofile of *tvakpatra*; leaves of *Cinnamomum verum* J.S. Presl. Pharmacogn J 2012;4:26-30

3 Nogueira Trajano V, de Oliveira Lima E, Santos de Souza F. Antifungal activity of the essential oil of *Cinnamomum zeylanicum* Blume and eugenol on *Aspergillus flavus*. J Essent Oil Bear Plants 2012;15:785-793

4 Cheng S-S, Chung M-J, Chen Y-J, Chang S-T. Antipathogenic activities and chemical composition of *Cinnamomum osmophloeum* and *Cinnamomum zeylanicum* leaf essential oils. J Wood Chem Technol 2011;31:73-87

5 Tira-Picos V, Gbolade AA, Nogueria JMF, Ajibesin KK. Analysis of leaf essential oil constituents of *Cinnamomum zeylanicum* grown in Nigeria. J Essent Oil Bear Plants 2009;12:76-80

6 Samarasekera R, Kalhari KS, Weerasinghe IS. Insecticidal activity of essential oils of Ceylon *Cinnamomum* and *Cymbopogon* species against *Musca domestica*. J Essent Oil Res 2006;18:352-354

7 Schmidt E, Jirovetz L, Buchbauer G, Eller GA, Stoilova I, Krastanov A, et al. Composition and antioxidant activities of the essential oil of Cinnamon (*Cinnamomum zeylanicum* Blume) leaves from Sri Lanka. J Essent Oil Bear Plants 2006;9:170-182

8 Jirovetz L, Buchbauer G, Ruzicka J, Shafi MP, Rosamma MK. Analysis of *Cinnamomum zeylanicum* Blume leaf oil from South India. J Essent Oil Res 2001;13:442-443

9 Raina VK, Srivastava SK, Aggarwal KK, Ramesh S, Kumar S. Essential oil composition of *Cinnamomum verum* Blume leaves from Little Andaman, India. Flavour Fragr J 2001;16:374-376

10 Möllenbeck S, König T, Schreier P, Schwab W, Rajaonarivony J, Ranarivelo L. Chemical composition and analyses of enantiomers of essential oils from Madagascar. Flavour Fragr J 1997;12:63-69

11 Nath SC, Pathak MG, Baruah A. Benzyl benzoate, the major component of the leaf and stem bark oil of *Cinnamomum zeylanicum* Blume. J Essent Oil Res 1996;8:327-328

12 Senanayake UM, Lee TH, Wills RBH. Volatile constituents of cinnamon *(Cinnamomum zeylanicum)* oils. J Agric Food Chem 1978;26:822-824

13 Wang R, Wang R, Yang B. Extraction of essential oils from five cinnamon leaves and identification of their volatile compound compositions. Inn Food Sci Emerg Technol 2009;10:289-292

14 Singh G, Maurya S, deLampasona MP, Catalan CAN. A comparison of chemical, antioxidant and antimicrobial studies of cinnamon leaf and bark volatile oils, oleoresins and their constituents. Food Chem Toxicol 2007;45:1650-1661

15 Mallavarapu GR, Ramesh S, Chandrasekhara RS, Rajeswara-Rao BR, Kaul PN, Bhattacharya AK. Investigation of the essential oil of cinnamon leaf grown at Bangalore and Hyderabad. Flavour Fragr J 1995;10:239-242

16 Ramachandran RY, Pal SC, Dutta PK. Major constituents of essential oils of *Cinnamomum zeylanicum*. Indian Perfum 1988;32:86-89. Data cited in ref. 11

17 Lyenger MA, Ghosh IK, Nay SGK. Evalution of locally growing *Cinnamomum* species. Indian Drugs 1994;31:87-89. Data cited in ref. 11

18 Cheng BQ, Yu XJ. Introduced Ceylon cinnamon cultivars in Xishangbanna and chemical components of their essential oils. Chem Ind Forest Prod 1991;11:325-332. Data cited in ref. 11

19 Rajeswara Rao BR, Rajput DK, Sastry KP, Kothari SK, Bhattacharya AK. Leaf essential oil profiles of *Cinnamomum zeylanicum* Blume. Indian Perfum 2006;50:44-46. Data cited in ref. 20

20 Lawrence BM. Progress in essential oils. Perfum Flavor 2009;34(6):54-58

21 Begum J, Yusuf M, Chowdhury JU, Hossain MM, Hossain ME, Ahmed S, Anwar MN. Composition and antifungal activity of essential oil of leaves of *Cinnamomum verum* Presl grown in Bangladesh. Indian Perfum 2007;51(2):69-71, also published Indian Perfum 2007;51(4):15-18. Data cited in ref. 20. Multiple incorrect data were found and are not mentioned (20)

22 Patel K, Ali S, Sotheeswaran S, Dufour J-P. Composition of the leaf essential oil of *Cinnamomum verum* (Lauraceae) from Fiji Islands. J Essent Oil Bear Plants 2007;10:374-377

23 Lawrence BM. Essential oils 2001-2004. Carol Stream, USA: Allured Publishing Corporation, 2006:105-107, 312-313

24 Jirovetz L, Buchbauer G, Ngassoum MB, Essia-Ngang JJ, Tatsadjieu LN, Adjoudji O. Chemical composition and antibacterial activities of the essential oils of *Plectranthus glandulosus* and *Cinnamomum zeylanicum* from Cameroon. Sci Pharm 2002;70:93-99

25 Variyar PS, Bandyopadhyay C. On some chemical aspects of *Cinnamomum zeylanicum*. PAFAI J 1989:Oct-Dec:35-37

26 Chalchat J-C, Valade I. Chemical composition of leaf oils of *Cinnamomum* from Madagascar: *C. zeylanicum* Blume, *C. camphora* L., *C. fragrans* Baillon and *C. angustifolium*. J Essent Oil Res 2000;12:537-540

27 Agarwal R, Pant AK, Prakash O. Chemical composition and biological activities of essential oils of Cinnamomum tamala, Cinnamomum zeylenicum and Cinnamomum camphora growing in

Uttarakhand. In: MM Srivastava, LD Khemani and S Srivastava, Eds. Chemistry of Phytopotentials: Health, Energy and Environmental Perspectives. Berlin, Heidelberg: Springer-Verlag, 2012: chapter 18, 87-92

28 Baruah A, Nath SC, Hazarika AK. Investigation of the essential oils of Cinnamomum verum Presl. grown at lower Brahmaputra valley of Assam. Indian Perfum 2010;54(3):21-23. Data cited in ref. 43

29 Pawar VC, Thaker VS. *In vitro* efficacy of 75 essential oils against *Aspergillus niger*. Mycoses 2006;49:316-323. Data also presented in ref. 48

30 Fratini F, Casella S, Leonardi M, Pisseri F, Ebani VV, Pistelli L, Pistelli L. Antibacterial activity of essential oils, their blends and mixtures of their main constituents against some strains supporting livestock mastitis. Fitoterapia 2014;96:1-7

31 Fichi G, Flamini G, Zaralli LJ, Perruccia S. Efficacy of an essential oil of *Cinnamomum zeylanicum* against *Psoroptes cuniculi*. Phytomedicine 2007;14:227-231

32 Dias de Castro R, Oliveira Lima E. Anti-*Candida* activity and chemical composition of *Cinnamomum zeylanicum* Blume essential oil. Braz Arch Biol Technol 2013;56:749-755

33 Schmidt E, Jirovetz L, Wlcek K, Buchbauer G, Gochev V, Girova T, et al. Antifungal activity of eugenol and various eugenol-containing essential oils against 38 clinical isolates of *Candida albicans*. J Essent Oil Bear Plants 2007;10:421-429

34 Samarasekera R, Kalhari KS, Weerasinghe IS. Mosquitocidal activity of leaf and bark essential oils of Ceylon *Cinnamomum zeylanicum*. J Essent Oil Res 2005;17:301-303

35 Jazet-Dongmo PM, Tatsadjieu LT, Tchoumougnang F, Sameza ML, Ndongson B, Amvam Zollo PH, Menut C. Chemical composition, antiradical and antifungal activities of essential oil of the leaves of *Cinnamomum zeylanicum* Blume from Cameroon. Nat Prod Commun 2007;2:1-4

36 Bin Jantan I, Moharam BAK, Santhanam J, Jamal JA. Correlation between chemical composition and antifungal activity of the essential oils of eight *Cinnamomum* species. Pharmac Biol 2008;46:406-412

37 Mello da Silveira S, Cunha Júnior A, Scheuermann GN, Luiz Secchi F, Werneck Vieira CR. Chemical composition and antimicrobial activity of essential oils from selected herbs cultivated in the South of Brazil against food spoilage and foodborne pathogens. Ciência Rural 2012;42:1300-1306

38 Ranasinghe L, Jayawardena B, Abeywickrama K. Fungicidal activity of essential oils of *Cinnamomum zeylanicum* (L.) and *Syzygium aromaticum* (L.) Merr et L.M.Perry against crown rot and anthracnose pathogens isolated from banana. Lett Appl Microbiol 2002;35:208-211

39 Mayaud L, Carricajo A, Zhiri A, Aubert G. Comparison of bacteriostatic and bactericidal activity of 13 essential oils against strains with varying sensitivity to antibiotics. Lett Appl Microbiol 2008;47:167-173

40 Paranagama P, Abeysekera T, Nugaliyadde L, Abeywickrama K. Effect of the essential oils of *Cymbopogon citratus*, *C. nardus* and *Cinnamomum zeylanicum* on pest incidence and grain quality of rough rice (paddy) stored in an enclosed seed box. Food Agricult Environm 2003;1:134-136

41 Wei A, Shibamoto T. Antioxidant/lipoxygenase inhibitory activities and chemical compositions of selected essential oils. J Agric Food Chem 2010;58:7218-7225

42 Paranagama PA, Wimalasena S, Jayatilake GS, Jayawardena AL, Senanayake UM, Mubarak AM. A comparison of essential oil constituents of bark, leaf, root and fruit of cinnamon (Cinnamomum zeylanicum Blum) grown in Sri Lanka. J Natn Sci Foundation Sri Lanka 2001;29:147-153

43 Lawrence BM. Progress in essential oils. Perfum Flavor 2014;39(October):46-52

44 Pawar VC, Thaker VS. Evaluation of the anti-*Fusarium oxysporum* f. sp. *cicer* and anti-*Alternaria porri* effects of some essential oils. World J Microbiol Biotechnol 2007;23:1099-1106. Data also presented in ref. 29

45 Magnusson B, Wilkinson DS. Cinnamic aldehyde in toothpaste. 1. Clinical aspects and patch tests. Contact Dermatitis 1975;1:70-76

46 Meynadier JM, Meynadier J, Peyron JL, Peyron L. Formes cliniques des manifestations cutanées d'allergie aux parfums. Ann Dermatol Venereol 1986;113:31-39.

Chapter 5.23 CITRONELLA OIL, JAVA

There are two major citronella oils, citronella oil, Java type obtained from *Cymbopogon winterianus* Jowitt and citronella oil, Sri Lanka type, which is obtained from *Cymbopogon nardus* (L.). In this chapter, citronella oil Java is discussed. The Sri Lanka type oil is presented in Chapter 5.24.

DEFINITION

Citronella oil Java is the essential oil obtained from the aerial parts (leaves) of the Java (Burma) citronella, *Cymbopogon winterianus* Jowitt.

INCI NOMENCLATURE

Description/definition: Cymbopogon winterianus herb oil is an essential oil obtained from the herbs of the plant, *Cymbopogon winterianus*, Gramineae
INCI name EU: Cymbopogon winterianus herb oil (perfuming name, not an INCI name proper)
INCI name USA: Not in the Personal Care Products Council Ingredient Database
CAS registry number(s): 8000-29-1; 91771-61-8
EINECS number(s): 294-954-7

ISO (INTERNATIONAL ORGANIZATION FOR STANDARDIZATION) STANDARD

ISO number: 3848
ISO name: Essential oil of citronella, Java type
Botanical origin: *Cymbopogon winterianus* Jowitt
Parts of plant used: Aerial parts (leaves)
ISO values: ISO values (minimum and maximum concentrations) are shown in Table 5.23.1.

Table 5.23.1 ISO values (%) for citronella oil Java[a]

Compound	CAS	Minimum	Maximum
Citronellal	106-23-0	31.0	40.0
Geraniol	106-24-1	20.0	25.0
Citronellol	106-22-9	8.5	14.0
Geranyl acetate	105-87-3	2.5	5.5
Limonene	138-86-3	2.0	5.0
Citronellyl acetate	150-84-5	2.0	4.0
Elemol	639-99-6	1.3	4.0
Germacrene D	23986-74-5	1.5	3.0
δ-Cadinene	483-76-1	1.5	2.5
β-Elemene	33880-83-0	0.7	2.5
Isopulegol	89-79-2	0.5	1.7
Linalool	78-70-6	0.5	1.5
Eugenol	97-53-0	0.5	1.0
Geranial	141-27-5	0.3	1.0

[a] ISO 3848 Essential oil of citronella, Java type ©ISO 2001; Geneva, Switzerland, www.iso.org

THE PLANT, THE OIL, AND THEIR USES

Cymbopogon winterianus Jowitt (Java citronella, also known as Java lemongrass) is an aromatic perennial herb (grass) with fibrous roots, which seems to have arisen as a distinct form of the Sri Lanka citronella, *Cymbopogon nardus* (33). It is over two meters tall with smooth and shiny leaves that are up to one meter long; the plant has a fresh, green, lemony odor. The Java citronella is cultivated in Africa (Ghana), Asia (China, India, Vietnam, Indonesia, Malaysia), South America (Guatemala, Honduras, Brazil), and elsewhere; the wild distribution is unknown (GRIN Taxonomy for Plants, 3,33). The leaves can be dried and powdered, or used fresh as an herb in Asian, African and Latin American cuisines (3). In folk medicine of the Brazilian Northeast, the *Cymbopogon winterianus* is used for the treatment of epilepsy and anxiety (36).

'Citronella oil' is the essential oil obtained by steam distillation from the leaves of different species of *Cymbopogon*, of which *Cymbopogon winterianus* Jowitt and *Cymbopogon nardus* L. (Rendle) (produces citronella oil, Sri Lanka type) are the most important. The oil is used on a large scale as a source of perfumery chemicals such as citronellal, citronellol and geraniol. These aroma chemicals, isolated from the essential oil, are further employed for producing high value semi-synthetic aroma chemicals including hydroxycitronellal, synthetic menthol and esters of geraniol and citronellol such as geranyl acetate and citronellyl acetate (1,3,5,7). The chemicals find extensive use in soap, candles and incense, perfumery, and in the cosmetic, flavoring and pharmaceutical industries (1,3,7). Citronella oil is also a renowned plant-based insect repellent (1,4,5). Traditional medicinal uses of citronella oil include treatment of fever, intestinal parasites, and digestive and menstrual problems. In Chinese medicine, it is used for rheumatic pain. In addition, the essential oil is claimed to exhibit a variety of biological properties, such as analgesic, antifungal, anticonvulsant and anxiolytic (1,5). Citronella oils are used extensively in aromatherapy, where they are considered to have warming and activating qualities on both mental and physical levels (1).

CHEMICAL COMPOSITION

Citronella oil Java is a pale yellow to yellowish, clear mobile liquid which has a fresh floral and sweet rosy lemon-like odor. The yield of essential oil from the aerial parts (leaves) of *Cymbopogon winterianus* Jowitt. generally varies from 0.5 to 1.2%. The main producing countries of this oil are China, India, Vietnam and Java.

Literature data (up to November 13, 2014) on the chemical composition of citronella oils Java and unpublished analytical data from one of us (E.S.) are shown in Table 5.23.2 in alphabetical order. In citronella oils Java from various origins, over 165 chemicals have been identified. About 53% of these were found in a single reviewed publication only. The major compounds found in citronella oils Java from different sources are shown in Table 5.23.3. They include (highest concentrations in any study given) citronellal (61%), geraniol (52.3%), citronellol (19.2%), elemol (17.2%), geranyl acetate (16.0%), geranial (12.9%), neral (8.9%), citronellyl acetate (6.8%) and limonene (6.1%). Well-known ingredients of citronella oils Java that were present in high concentrations (>7%) in one study were linalool (27.4%), β-elemene (13.0%), eugenol (10.3%) and α-cadinol (8.0%).

Commercial oils

The ten chemicals that had the highest maximum concentrations in 138 commercial citronella oil Java essential oil samples (concentration ranges provided) are the following: citronellal (31.5-49.6%), geraniol (17.3-25.2%), citronellol (8.7-13.5%), geranyl acetate (1.9-8.7%), elemol (α-) (1.3-7.0%), limonene (2.5-5.9%), citronellyl acetate (1.6-4.5%), β-elemene (0.7-3.2%), germacrene D (0.4-2.9%) and δ-cadinene (1.4-2.8%) (Erich Schmidt, unpublished analytical data).

Table 5.23.2 Constituents identified in citronella oils, Java

Constituent	CAS	Percentage and range					
		A	B (28)	C (29)	D (27)	E (3)	F
γ-Amorphene	6980-46-7						0.7[h]
(E)-Anethole	4180-23-8		0.7				
Apiole	523-80-8					0.3-1.2	
Aromadendrene	489-39-4				0.1		0.1[h]
Bergamal	106-72-9	0.01-0.05	0.05			0-0.1	0.1[h,x]
trans-α-Bergamotene	13474-59-4						0.08[f]; 0.1[n]
Bicyclogermacrene	24703-35-3						1.8[t]
β-Bisabolene	495-61-4						0.3[f,g]
β-Bisabolol	15352-77-9						0.03[h]
Borneol	507-70-0			tr-1.0			0.1[f]; 0.2[z2,y2]; 2.6[o]
Bornyl acetate	76-49-3						0.8[o]
endo-1-Bourbonalol							1.0[d]
β-Bourbonene	5208-59-3	tr-0.1		0-0.1			0.05[f]; 0.1[z2]; 0.2[m]; 0.4[j,o]; 2.9[h]
Bulnesol	22451-73-6		0.2			0-0.09	
Cadina-1,4-diene	29837-12-5						0.05[j]
trans-Cadina-1,4-diene	20085-13-6		0.05				
cis-Cadina-1(6),4-diene	1187195-00-1				0.1		
(E)-Cadina-1(6),4-diene	931410-54-7		0.05				
α-Cadinene	24406-05-1		0.08		0.1		0.1[z5]
β-Cadinene	523-47-7					2.5-5.7	
γ-Cadinene	39029-41-9	0.3-2.5	0.4		0.4		1.0[e,z6]; 1.1[f]; 1.3[j,k]; 1.7[z2]; 2.2[y2]
δ-Cadinene	483-76-1	1.4-2.8		0-tr	2.5		2.3[d,j]; 3.0[t]; 3.2[j]; 3.4[o]; 4.2[z4]
α-Cadinol	481-34-5	0.08-0.6	1.6	0-0.2	2.7	2.0-3.8	1.4[o]; 2.0[n]; 2.3[j]; 4.8[b]; 5.1[z4]; 8.0[z]
epi-α-Cadinol(τ-)	5937-11-1	0.2-0.4		0.2-1.9	0.5		0.4[z5]; 0.6[j]; 0.8[j]; 1.6[o]; 2.0[n]
δ-Cadinol	19435-97-3						0.05[j]
α-Calacorene	21391-99-1						0.02[j]; 6.0[o]
β-Calacorene	50277-34-4						1.5[o]
Calamenene	483-77-2						1.3[j,k]
cis-Calamenene	72937-55-4						4.3[o]
(+)-Calarene	17334-55-3					0-0.08	
Camphor	76-22-2			0-0.3			1.3[o]
β-Caryophyllene	87-44-5	0.02-0.5	0.1	tr-0.4		0.2-0.2	0.4[w]; 0.6[x]; 1.0[j]; 1.3[y2]; 1.4[t]; 2.9[z3]
Caryophyllene oxide	1139-30-6			tr-2.8			1.0[g]; 1.2[f]; 1.5[o]; 2.2[y2]; 3.5[n];
Cedrol	77-53-2						0.2[a,h]
β-Chamigrene	18431-82-8						1.2[o]
1,8-Cineole	470-82-6					0-0.2	0.2[t]; 1.3[o]
Citronellal	106-23-0	31.5-49.6	35.5	11.8-35.4	26.5	3.1-23.0	41.6[t]; 42.5[y2]; 42.8[g]; 45.0[s]; 45.1[z8]; 48.5[h,z4]; 50.9[f]; 61[q,v]
Citronellol	106-22-9	8.7-13.5	10.9	7.4-11.0	7.3	10.3-13.8	14.4[s]; 15.3[z8]; 15.7[j]; 16.7[t]; 17.4[u]; 18.2[p]; 18.4[b]; 19.2[r]
Citronellyl acetate	150-84-5	1.6-4.5	2.5	0.1-4.4	2.5	0.4-1.2	5.3[d,y]; 5.9[y2]; 6.2[y1]; 6.5[z8]; 6.8[s]
Citronellyl butyrate	141-16-2			tr-0.5			+[z9]; <0.01[c]
Citronellyl formate	105-85-1						0.1[c]; 0.2[y2]
Citronellyl propionate	141-14-0						0.1[c]
α-Copaene	3856-25-5		0.02				
β-Copaene	18252-44-3		0.03				
α-Cubebene	17699-14-8						tr[o]
β-Cubebene	13744-15-5					1.6-4.5	0.3[o]

Table 5.23.2 Constituents identified in citronella oils, Java (*continued*)

Constituent	CAS	Percentage and range					
		A	B (28)	C (29)	D (27)	E (3)	F
epi-Cubebol	38230-60-3		0.1				
Cubenol	21284-22-0						1.0[z]
1-epi-Cubenol	81939-29-9		0.05		0.1		1.9[z]
1,10-di-epi-Cubenol	73365-77-2						2.0[z]
o-Cymene	527-84-4		0.02				0.1[o]
p-Cymene	99-87-6			0-0.1			0.01[z5]; 0.04[f,j]; 0.1[g,n,t]; 0.3[z2]
Dauca-5,8-diene	142928-08-3		0.03				
Decanal	112-31-2		0.08				0.1[f,m,z2]; 0.4[x]; 0.5[w]
Decenal	25447-70-5						0.1[a,h]
cis-Dihydrocarvone	3792-53-8						1.3[l]
Elemene	11029-06-4						0.6[e]
β-Elemene	33880-83-0	0.7-3.2	1.7	0-0.3	4.4	3.0-3.5	2.3[z6]; 3.4[l]; 3.9[j,o]; 6.4[d]; 13.0[z4]
γ-Elemene	29873-99-2						0.3[a,f]
δ-Elemene	20307-84-0						0.1[h,o]
Elemol(α-)	639-99-6	1.3-7.0	3.7	0.8-3.8	14.5	10.0-17.1	6.2[l]; 6.7[d,j]; 6.8[z5]; 7.2[b]; 9.6[y2]; 12.3[l]; 15.3[n]; 15.4[z4]; 17.2[s]
Ethanol	64-17-5	0.03-0.3					
Eudesmol	51317-08-9					1.9-3.1	
α-Eudesmol	473-16-5				0.2		0.9[h]; 1.6[m]; 2.3[o]; 2.6[j]
5-epi-7-epi-α-Eudesmol	446050-56-2		0.03				
β-Eudesmol	473-15-4	0.1-0.5	0.3	0-0.3	0.2		0.4[h]; 0.8[b,j]; 1.1[l]; 1.2[o]
γ-Eudesmol	1209-71-8	0.2-0.3	0.6		0.8	0.7-1.3	1.0[d]; 1.1[b]; 1.2[z]; 1.4[o]; 2.2[z4]
10-epi-γ-Eudesmol	15051-81-7		0.07	tr-2.0	0.1		0.2[m,n]
Eugenol	97-53-0	0.2-1.5	0.8			0.4-1.0	1.2[t]; 1.3[z7]; 1.5[y1]; 5.3[e]; 10.3[d]
(E)-β-Farnesene	18794-84-8			0-0.1			
Farnesol	4602-84-0					0-0.8	0.2[m]; 0.6[d]; 1.1[n]
(E,E)-Farnesol	106-28-5						0.1[h]; 0.3[y1]; 0.6[j]; 3.0[d]
(E,Z)-Farnesol	3879-60-5				0.2		
(Z,E)-Farnesol	3790-71-4						0.07[j]
α-Fenchene	471-84-1						0.3[o]
Geranial	141-27-5	0.4-1.1	0.5	0.9-12.9	0.7	0.4-0.5	1.1[y1]; 1.5[z]; 5.8[w]; 8.1[x]; 11.8[y2]
Geraniol	106-24-1	17.3-25.2	21.8	22.4-50.1	16.2	19.9-31.7	28.1[z5]; 28.8[u]; 32.4[y1]; 32.8[w]; 40.1[x]; 45.7[y2]; 50.1[s]; 52.3[z3]
Geranyl acetate	105-87-3	1.9-8.7	3.2	tr-4.6	3.4	0.3-1.1	6.6[z4]; 8.3[r,z8]; 8.6[y]; 9.3[y2]; 16.0[y1]
Geranyl butyrate	106-29-6						<0.01[c]
Geranyl formate	105-86-2						+[z9]; 1.5[i]; 1.8[y2]; 5.2[z3]
Germacrene	28028-64-0						2.6[d]
Germacrene A	28387-44-2		0.4				0.8[z]
Germacrene B	15423-57-1						6.8[z]
Germacrene D	23986-74-5	0.4-2.9	1.9	0-0.8	1.1		1.5[h,l]; 1.7[z5]; 2.4[t]; 2.6[m]; 3.5[z2]
Germacrene D-4-ol	198991-79-6	0.4-0.7	0.5	0-2.6			0.3[z5]; 1.3[b]; 1.7[m]
β-Germacrenol							1.9[y1]
α-Guaiene	3691-12-1						0.04[f]
β-Guaiene	88-84-6					0-0.1	
α-Gurjunene	489-40-7						1.2[o]
β-Gurjunene	73464-47-8				0.2		0.02[h]
α-Humulene	6753-98-6	0.06-0.2	0.1	0-0.5	0.1	0.2-0.3	0.07[f]; 0.1[m,n]; 0.2[g,j,z2]; 0.3[o]
Hydroxycitronellal	107-75-5						<0.01[c]
8-Hydroxyneomenthol	3564-95-2		0.2				
Isocaryophyllene oxide							1.5[t]
(E)-Isocitral	72203-98-6				0.2		
Isoisopulegol	18674-65-2	0.3-2.7		0-0.3		0.2-0.3	0.07[j]; 0.2[x]; 0.3[w]; 0.5[z5,z7]
Isomenthone	491-07-6		0.03				0.04[a,h]
Isopulegol	89-79-2		1.2				1.0[b]; 1.1[w,z4]; 1.4[z5]; 1.6[e]; 3.7[l]
Juniperc amphor	473-04-1						1.3[j]
Ledene alcohol	1197210-11-9					0-0.5	
Ledene oxide	882187-44-2					0-0.1	
Limonene	138-86-3	2.5-5.9	3.9	tr-2.7	2.2		2.3[e,j,o]; 3.1[y1]; 3.4[d,h]; 3.6[r]; 3.9[l]; 4.2[z2]; 4.4[z4]; 5.5[g]; 5.7[t,y2]; 6.1[f]
cis-Limonene oxide	13837-75-7						tr[o]

Table 5.23.2 Constituents identified in citronella oils, Java (*continued*)

Constituent	CAS	A	B (28)	C (29)	D (27)	E (3)	F
Linalool	78-70-6	0.5-1.4	1.2	tr-0.4	0.7	0.5-0.9	1.1[g]; 1.3[d,j]; 1.4[w]; 1.5[p,z2]; 27.4[o]
cis-Linalool oxide, furanoid	5989-33-3						0.1[o]
Linalyl acetate	115-95-7			tr-0.6			0.3[f,i; ,y2]; 1.0[z3]
Longifolene (junipene)	475-20-7					0-0.3	3.5[o]
p-Mentha-3,8-diene	586-67-4						0.05[j]
p-Menth-3-en-8-ol	18479-65-7		0.06				
Methylchavicol	140-67-0		0.04				
Methyleugenol	93-15-2	0.01-2.5					+[z9]; 0.1[z7]
6-Methyl-5-hepten-2-one	110-93-0	0.04-0.2	0.06	0-0.1			0.09[f]; 0.1[d,z5]; 0.2[g,w,z2]; 0.4[y2]
Methylisoeugenol	93-16-3						+[z9]
trans-Muurola-3,5-diene	262352-88-5		0.03				
cis-Muurola-4(14),5-diene	157477-72-0				0.1		
trans-Muurola-4(14),5-diene	262352-87-4		0.05		0.1		
α-Muurolene	10208-80-7	tr-0.2	0.5		0.4	0.9-1.5	0.4[m]; 0.5[n,z5]; 0.8[l]; 1.3[j]
γ-Muurolene	30021-74-0	0.01-2.3	0.1	0-0.8	0.1	0.3-0.4	0.2[z5]; 0.4[h,l]; 1.8[b]
τ-Muurolene	152287-43-9						0.9[n]
α-Muurolol	104245-48-9		0.2				0.05[z5]; 0.4[f]; 1.6[z4]; 1.8[o]; 2.0[z]
τ-Muurolol (epi-α-)	19912-62-0	0.2-0.5	0.9		0.7	1.6-2.8	0.2[z5]; 1.0[m,y1]; 2.6[b]
Myrcene	123-35-3	0.02-0.2	0.07		3.3		0.2[h,n]; 0.4[o,t]; 0.5[f]; 0.7[j]; 1.3[z2]
Neoisoisopulegol	21290-09-5		0.08				
Neoisopulegol	29141-10-4	tr-0.1					0.1[z5]
Neral	106-26-3	0.2-0.6	0.3	tr-8.9	0.5	0-0.4	0.6[h]; 0.9[y1]; 4.5[w]; 6.0[x]; 8.1[y2]
Nerol	106-25-2	0.2-0.3		tr-0.3	0.4		0.2[t,z5]; 0.3[i,m]; 0.5[c,z7,y2]
Neryl acetate	141-12-8		0.03		0.1		<0.1[z2]; 0.1[c]; 0.8[q,v]; 3.2[l]
Neryl propionate	105-91-9				0.1		
Nonanal	124-19-6						0.2[f]
(E)-β-Ocimene	3779-61-1			0-0.1	0.7		0.09[x]; 0.1[f,j,n]; 0.2[w]; 0.4[z2]
(Z)-β-Ocimene	3338-55-4		0.01	0-tr			0.08[j]; 0.1[o]; 0.2[f,x]; 0.3[z2]; 0.4[w]
allo-Ocimene	673-84-7				0.2		
3-Octanone	106-68-3						0.1[t]
Perillaldehyde	2111-75-3						0.02[h]
α-Phellandrene	99-83-2		0.02				0.04[f]; 0.1[z2]
β-Phellandrene	555-10-2	0.04-0.1					0.04[z5]; 0.1[m]; 1.1[o]
Phytol	7541-49-3					0-0.09	
Phytone	16825-16-4					0-0.2	
α-Pinene	80-56-8	0.04-0.1	0.01				0.03[f,z5]
β-Pinene	127-91-3			0-1.8			
Pulegol	529-02-2	0.3-0.9					
Pulegone	89-82-7						0.03[j]
cis-Rose oxide	3033-23-6		0.03	tr-0.1		0.2-0.8	0.08[z5]
trans-Rose oxide	5258-11-7		0.01				
Sabinene	3387-41-5						0.08[z5]; 0.3[o]
trans-Sabinene hydrate	17699-16-0			0-tr			
Sclareol	515-03-7					0.9-2.5	
Selinene	27104-12-7						1.4[f]
α-Selinene	473-13-2						0.5[j,o]
β-Selinene	17066-67-0		0.06				0.04[f]
Spathulenol	6750-60-3			0.1-1.5		0-0.3	2.1[o]
γ-Terpinene	99-85-4		0.02				
δ-Terpinene	586-62-9					0-0.06	
Terpinen-4-ol	562-74-3		0.06	0-0.1			<0.01[m]; 0.05[g,j]; 0.07[f]; 0.1[z2]
Terpinen-4-yl acetate	4821-04-9						0.1[f,g]
α-Terpineol	98-55-5	0.05-2.7	0.06	0-0.6		0.1-0.2	0.09[g]; 0.1[z2]; 0.7[y2]; 1.3[o]; 2.9[t]
Terpinolene(α-)	586-62-9	0.01-0.8	0.07	0-tr			0.04[j]; 0.08[z5]; 0.09[f,g]; 0.1[m,o,z2]
α-Thujene	2867-05-2						0.1[o]
α-Thujone	546-80-5						tr[o]
β-Thujone(*cis*-)	471-15-8						tr[o]
Thymol	89-83-8		0.03				
Torreyol	19435-97-3						1.7[d]

Table 5.23.2 Constituents identified in citronella oils, Java (*continued*)

Constituent	CAS	A	B (28)	C (29)	D (27)	E (3)	F
Tricyclene	508-32-7						0.1[o]
Valencene	4630-07-3					0-0.04	
Viridiflorene(ledene)	21747-46-6				0.1		
Viridiflorol	552-02-3						0.08[f];0.8[n]
α-Ylangene	14912-44-8						0.4[o]
β-Ylangene	20479-06-5				0.3		
Zonarene	41929-05-9				0.1		

A one hundred and thirty-two citronella Java essential oil samples from China, India and Vietnam, analyzed between 1998 and 2013; lowest and highest concentrations given (E. Schmidt, unpublished data)
B one lab-hydrodistilled oil from Brazil (ref. 28)
C six lab-hydrodistilled citronella oils from six Indian *C. winterianus* cultivars; lowest and highest concentrations given (ref. 29)
D one lab-hydrodistilled oil from Brazil (ref. 27)
E four citronella oils from plant material originating from China and Brazil, of which two were lab-hydrodistilled and two microwave-assisted hydrodistilled; lowest and highest concentrations given (ref. 3)
F data from other studies (indicated with superscript letters); highest concentrations found in any study reviewed here given; when two or more oils were investigated, only the highest concentrations are mentioned, unless indicated otherwise

[a] incorrect identity based on GC elution order/incorrect identification (ref. 21); [b] one lab-hydrodistilled oil from *C. winterianus* leaves cultivated in Brazil (ref. 26); [c] one commercial oil from China (ref. 12); [d] one lab-hydrodistilled oil from Brazil (ref. 31); [e] one sample of Java citronella oil (ref. 22); [f] one citronella oil *ex C. winterianus* produced in India (ref. 23); [g] one oil produced from Java citronella grown above 900 meter in India (ref. 24); [h] one oil from plants grown in an experimental garden in coastal Orissa, India (ref. 25); [i] one lab-hydrodistilled oil from *C. winterianus* cultivated in India (ref. 40); [j] one steam-distilled oil from Cuba (ref. 9); [k] calamenene and γ-cadinene combined; [l] one commercial oil sample from citronella ex *Cymbopogon winterianus* obtained from a German company (ref. 30); [m] one lab-hydrodistilled oil from the aerial parts of Java citronella grown in Brazil (ref. 33); [n] one lab-hydrodistilled oil from cultivar Java-2 from India (ref. 32); [o] one lab-hydrodistilled oil from leaves of *Cymbopogon winterianus* growing wild in Tanzania; this oil has a very atypical composition with very high linalool content and no citronellal (ref. 5); [p] one citronella oil prepared in a field distillation unit on a research farm in India (ref. 2); [q] one lab-hydrodistilled citronella oil from the leaves of *C. winterianus* collected in a Brazilian medicinal and aromatic plant garden (ref. 4); [r] one lab-hydrodistilled oil from citronella cultivated at a regional research lab in India (ref. 6); [s] thirty lab-hydrodistilled oils from *C. winterianus* accessions from various parts of India; only the main chemicals were investigated (ref. 7); [t] three lab-hydrodistilled oils from citronella harvested in Zimbabwe in three consecutive years (ref. 8); [u] ten lab-hydrodistilled oils from two harvests per year in five successive years in south India; only the data for citronellal, citronellol and geraniol were given (ref. 10); [v] one lab-hydrodistilled oil from Brazil (ref. 35); [w] one lab-hydrodistilled oil sample from Brazil (ref. 34); [x] one lab-hydrodistilled oil from Brazil (ref. 36); [y] one sample of lab-hydrodistilled oil from India (ref. 38); [y1] thirteen lab-hydrodistilled oils from the fresh herbage of *C. winterianus* cultivated in India and harvested at 13 different dates (ref. 41); [y2] lab-hydrodistilled oils from 5 cultivars of *C. winterianus* cultivated in India and harvested at three different times in one year (ref. 42); [z] one lab-distilled oil sample prepared from the aerial parts of *C. winterianus* growing wild in Brazil (ref. 39); [z2] one lab-hydrodistilled oil from Java citronella cultivated in India (ref. 37); [z3] one Java citronella oil produced in India (ref. 17); [z4] various fractions obtained every 15 minutes by hydrodistillation of citronella grass (ref. 19); as these are fractions, the quantities cannot be compared to data of entire oils; [z5] one sample of citronella oil ex *Cymbopogon winterianus* (ref. 18); [z6] one oil from India (ref. 20); [z7] one oil with the 'typical composition' of Java citronella oil (ref. 14); [z8] five oils from five *C. winterianus* cultivars grown in India (ref. 15); [z9] older literature cited in ref. 11

tr: trace (in column C: <0.1); + present in the oil investigated, but quantity not stated

Table 5.23.3 Major constituents of citronella oils Java

Constituent	CAS	A	B (28)	C (29)	D (27)	E (3)	F
Citronellal	106-23-0	31.5-49.6	35.5	11.8-35.4	26.5	3.1-23.0	45.1[z8]; 48.5[h,z4]; 50.9[f]; 61[q,v]
Geraniol	106-24-1	17.3-25.2	21.8	22.4-50.1	16.2	19.9-31.7	40.1[x]; 45.7[y2]; 50.1[s]; 52.3[z3]
Citronellol	106-22-9	8.7-13.5	10.9	7.4-11.0	7.3	10.3-13.8	17.4[u]; 18.2[p]; 18.4[b]; 19.2[r]
Elemol (α-)	639-99-6	1.3-7.0	3.7	0.8-3.8	14.5	10.0-17.1	12.3[l]; 15.3[n]; 15.4[z4]; 17.2[s]
Geranyl acetate	105-87-3	1.9-8.7	3.2	tr-4.6	3.4	0.3-1.1	6.6[z4]; 8.3[r,z8]; 8.6[y]; 9.3[y2]; 16.0[y1]
Geranial	141-27-5	0.4-1.1	0.5	0.9-12.9	0.7	0.4-0.5	1.1[y1]; 1.5[z]; 5.8[w]; 8.1[x]; 11.8[y2]
Neral	106-26-3	0.2-0.6	0.3	tr-8.9	0.5	0-0.4	0.6[h]; 0.9[y1]; 4.5[w]; 6.0[x]; 8.1[y2]
Citronellyl acetate	150-84-5	1.6-4.5	2.5	0.1-4.4	2.5	0.4-1.2	5.3[d,v]; 5.9[y2]; 6.2[y1]; 6.5[z8]; 6.8[s]
Limonene	138-86-3	2.5-5.9	3.9	tr-2.7	2.2		4.2[z2]; 4.4[z4]; 5.5[g]; 5.7[t,y2]; 6.1[f]

LEGEND: SEE UNDER TABLE 5.23.2

CONTACT ALLERGY/ALLERGIC CONTACT DERMATITIS

Citronella oil (unspecified)

General

Contact allergy to/allergic contact dermatitis from citronella oil has been reported infrequently. In one group of consecutive patients suspected of contact dermatitis, the prevalence rate of positive patch test reactions was 2.5%, but relevance data are lacking. In all but one publication, the botanical source of the oil (Java, Sri Lanka = Ceylon) was not mentioned. In case reports, geraniol, citronellal, citronellol, geranyl acetate and limonene may have been allergens in citronella oils (43,49).

Testing in groups of patients

Routine testing

Two hundred dermatitis patients from Poland were patch tested with citronella oil 2% in petrolatum and five (2.5%) reacted; relevance data were not provided (44).

Testing in groups of selected patients

In a group of 51 patients allergic to *Myroxylon pereirae* resin (balsam of Peru) and/or turpentine and/or wood tar and/or colophony and tested with citronella oil 2% in petrolatum, two (3.9%) had a positive patch test; relevance data were not provided (44). A group of 86 patients from Poland previously reacting to the fragrance mix was tested with citronella oil and one (1.1%) had a positive patch test reaction; relevance data were not provided (45).

Case reports

An aromatherapist had occupational contact dermatitis with allergies to multiple essential oils used at work, including citronella oil. The patient also reacted to geraniol, α-pinene, caryophyllene, the fragrance mix and various other fragrance materials; geraniol and caryophyllene were demonstrated by GC-MS in the citronella oil. Geraniol may be present in commercial citronella oils in concentrations up to 49% (Sri Lanka) (Chapter 5.24) and 25% (Java) (Table 5.23.2, column A) (43). A woman changed the urinary bag from her child daily and thereby used an ostomy deodorant; she developed dermatitis on the hands and the face; on patch testing she reacted to the deodorant and subsequently to one ingredient, which consisted mainly of citronella oil and citronellal (but according to the manufacturer also pine oil derivatives) (47). One patient allergic to a topical pharmaceutical product intended to treat osteomuscular pain reacted to the active ingredients and to 'citronella', one of the product's base ingredients (48). Two patients had used oil of citronella for protection against mosquitos and developed dermatitis. They reacted to oil of citronella (pure and 50% in mineral oil, concentrations are slightly irritant), citronellal, citronellol, hydroxycitronellal, citral and geranyl acetate; citronellal and citronellol are both important components of citronella oils and geranyl acetate may be present in concentrations up to 11% in commercial citronella oil Sri lanka (Chapter 5.24) (49).

Positive patch tests (relevance unknown, uncertain or not stated)

One positive patch test to citronella oil in a patient working in a cosmetic factory who had occupational dermatitis from 2-bromo-2-nitropropane-1,3-diol (46).

Additional information on allergy to citronella oil may be found in ref. 50 (article not read).

Citronella oil, Java

A patient occupationally sensitized to lemon oil and limonene, also had positive patch test reactions to oil of citronella, Java, oil of citronella, Ceylon (= Sri Lanka) (both probably undiluted, may be irritant), turpentine and its component β-pinene (49). Limonene is the dominant component of lemon oil and may be present in citronella oils in concentrations up to 6% (Table 5.23.2, column A) (Java) and 10.4% (Chapter 5.24) (Sri Lanka).

LITERATURE

1 Katiyar R, Gupta S, Yadav KR. *Cymbopogon winterianus*: An important species for essential Java citronella oil and medicinal value. National Conference on Forest Biodiversity: Earth's Living Treasure, 22 May, 2011. Available at: http://www.upsbdb.org/pdf/Souvenir2011/16.pdf

2 Rajeswara Rao BR, Rajput DK, Patel RP. Storage of essential oils: Influence of presence of water for short periods on the composition of major constituents of the essential oils of four economically important aromatic crops. J Essent Oil Bear Plants 2011;14:673-678

3 Yang X, Jiang Z-T, Wang Y, Li R. Composition comparison of volatile oils of *Cymbopogon winterianus* Jowitt obtained by different extraction methods. J Essent Oil Bear Plants 2010;13:721-726

4 Gonçalves TB, de Sousa EO, Rodrigues FFG, da Costa JGM. Chemical composition and antibacterial evaluation of the essential oil from *Cymbopogon winterianus* Jowitt (Gramineae). J Essent Oil Bear Plants 2010;13:426-431

5 Malele RS, Mwangi JW, Thoithi GN, Kibwage IO, López ML, Zunino MP, et al. Essential oil of *Cymbopogon winterianus* Jowitt from Tanzania: Composition and antimicrobial activity. J Essent Oil Bear Plants 2007;10:83-87

6 Saikia RC, Sarma A, Sarma TC, Baruah PK. Comparative study of essential oils from leaf and inflorescence of Java citronella (*Cymbopogon winterianus* Jowitt). J Essent Oil Bear Plants 2006;9:85-87

7 Lal RK, Sharma JR, Misra HO, Sharma S, Naqvi AA. Genetic variability and relationship in quantitative and qualitative traits of Java citronella (*Cymbopogon winterianus* Jowitt). J Essent Oil Res 2001;13:158-162

8 Chagonda LS, Makanda C, Chalchat J-C. Essential oils of cultivated *Cymbopogon winterianus* (Jowitt) and of *C. citratus* (DC) (Stapf) from Zimbabwe. J Essent Oil Res 2000;12:478-480

9 Pino JA, Rosado A, Correa MT. Chemical composition of the essential oil of *Cymbopogon winterianus* Jowitt from Cuba. J Essent Oil Res 1996;8:693-694

10 Prakasa Rao ECS, Singh M. Long-term studies on yield and quality of Java citronella (*Cymbopogon winterianus* Jowitt) in relation to nitrogen application. J Essent Oil Res 1991;3:419-424

11 Lawrence BM. Progress in essential oils. Perfum Flavor 1977;2(2):4-5

12 Jirovetz L, Eller G, Buchbauer G, Schmidt E, Denkova Z, Stoyanova AS, et al. Chemical composition, antimicrobial activities and some odor descriptions of some essential oils with characteristic floral-rosy scent and of their principal aroma compounds. Recent Res Devel Agron Hort 2006;2:1-12

13 Lawrence BM. Progress in essential oils. Perfum Flavor 2011;35(4):52-59

14 Ranade GS. Profile essential oils: Citronella oil. FAFAI 2008;(Oct/Dec):69. Data cited in ref. 13

15 Lal RK, Khanuja SPS, Chandra R, Misra HO, Singh A, Bansal K, et al. Registration of a high quality oil yielding new variety with enhanced field establishment capacity: CIM Jeeva of Java citronella (*Cymbopogon winterianus*). J Med Arom Plant Sci 2008;30:225-226

16 Lawrence BM. Progress in essential oils. Perfum Flavor 2006;31(9):40-50

17 Aggarwal KK, Ahmad A, Kumar TRS, Jain N, Gupta VK, Kumar S, et al. Antimicrobial activity spectra of *Pelargonium graveolens* and *Cymbopogon winterianus* Jowitt oil constituents and acyl derivatives. J Med Arom Plant Sci 2000;22:544-548

18 Kubeczka K-H, Formacek V. Essential oils analysis by capillary gas chromatography and carbon-13 NMR spectroscopy, 2nd Ed. New York, USA: John Wiley and Sons, 2002:67-80. Data cited in ref. 16

19 Rout PK, Jena KS, Rao YR. Fractional distillation of citronella (*Cymbopogon winterianus* Jowitt). FAFAI 2001;3/4:57-59, 61-62. Data cited in ref. 16

20 Narayanan P. Comparison of quality of essential oils from South India. FAFAI 2003;5(1):47-56. Data cited in ref. 16

21 Lawrence BM. Progress in essential oils. Perfum Flavor 2003;28(1):60-? (last page unknown)

22 Buchbauer G, Jirovetz L, Nikiforow A. Use of GC-FID, GC-FTIR-MS and olfactory characterization in the analyses of essential oils and plant extracts. In: H-F Linskens, JF Jackson, Eds. Modern methods of plant analysis. Volume 19. Plant volatile analysis. Berlin: Springer-Verlag, 1997:97-117. Data cited in ref. 21

23 Kaul PN, Bhattacharya AK, Singh K, Rajeswara Rao BR. Chemical composition of the essential oil of Java citronella (*Cymbopogon winterianus* Jowitt) grown in Andhra Pradesh. PAFAI 1997;19(3):29-33. Data cited in ref. 21

24 Rajeswara Rao BR, Kaul PN, Bhattacharya AK. Java citronella (*Cymbopogon winterianus* Jowitt) cultivation in a tribal area of Andhra Pradesh. J Essent Oil Bear Plants 1998;1:114-118

25 Rao YR, Rout PK, Jena KS, Sahu SB. Composition of essential oil of citronella (*Cymbopogon winterianus* Jowitt) grown in coastal Orissa. FAFAI 2000;2(4):29-31. Data cited in ref. 21

26 Duarte MCT, Figueira GM, Sartoratto A, Rehder VLG, Delarmelina C. Anti-Candida activity of Brazilian medicinal plants. J Ethnopharmacol 2005;97:305-311

27 da F. Rodrigues KA, Dias CN, do Amaral FMM, Moraes DFC, Mouchrek Filho VE, Andrade EHA, et al. Molluscicidal and larvicidal activities and essential oil composition of *Cymbopogon winterianus*. Pharm Biol 2013;51:1293-1297

28 Gusmão NMS, de Oliveira JV, do A.F. Navarro DM, Dutra KA, da Silva WA, Wanderley MJA. Contact and fumigant toxicity and repellency of *Eucalyptus citriodora* Hook., *Eucalyptus staigeriana* F., *Cymbopogon winterianus* Jowitt and *Foeniculum vulgare* Mill. essential oils in the management of *Callosobruchus maculatus* (FABR.) (Coleoptera: Chrysomelidae, Bruchinae). J Stored Prod Res 2013;54:41-47

29 Padalia RC, Verma RS, Chanotiya CS, Yadav A. Chemical fingerprinting of the fragrant volatiles of nineteen Indian cultivars of *Cymbopogon Spreng*. (Poaceae). Rec Nat Prod 2011;5:290-299

30 Simic A, Rancic A, Sokovic MD, Ristic M, Grujic-Jovanovic S, Vukojevic J, Marin PD. Essential oil composition of *Cymbopogon winterianus* and *Carum carvi* and their antimicrobial activities. Pharm Biol 2008;46:437-441

31 Araújo de Oliveira W, de Oliveira Pereira F, Gomes de Luna GCD, Oliveira Lima I, Wanderley PA, Baltazar de Lima R, et al. Antifungal activity of *Cymbopogon winterianus* Jowitt ex Bor against *Candida albicans*. Brazil J Microbiol 2011;42:433-441

32 Rajeswara Rao BR, Bhattacharya AK, Mallavarapu GR, Ramesh S. Yellowing and crinkling disease and its impact on the yield and composition of the essential oil of citronella (*Cymbopogon winterianus* Jowitt). Flavour Fragr J 2004;19:344-350

33 Lorenzo D, Dellacassa E, Atti-Serani L, Santos AC, Frizzo C, Paroul N, et al. Composition and stereoanalysis of *Cymbopogon winterianus* Jowitt oil from southern Brazil. Flavour Fragr J 2000;15:177-181

34 Leite BLS, Souza TT, Antoniolli AR, Guimarães AG, Siqueira RS, Quintans JSS, et al. Volatile constituents and behavioral change induced by *Cymbopogon winterianus* leaf essential oil in rodents. Afr J Biotechnol 2011;10:8312-8319

35 Ribeiro Silva M, Matos Ximenes R, Galberto Martins da Costa J, Leal LKAM, de Lopes AA, Socorro de Barros Viana G. Comparative anticonvulsant activities of the essential oils (EOs) from *Cymbopogon winterianus* Jowitt and *Cymbopogon citratus* (DC) Stapf. in mice. Naunyn-Schmied Arch Pharmacol 2010;381:415-426

36 Quintans-Junior LJ, Souza TT, Leite BS, Lessa NMN, Bonjardim LR, Santos MRV, et al. Phythochemical screening and anticonvulsant activity of *Cymbopogon winterianus* Jowitt (Poaceae) leaf essential oil in rodents. Phytomed 2008;15:619-624

37 Rajeswara Rao BR, Kaul PN, Bhattacharya AK, Mallavarapu GR, Ramesh S. Yield and chemical composition of the essential oils of three *Cymbopogon* species suffering from iron chlorosis. Flavour Fragr J 1996;11:289-293

38 Chutia M, Mahanta JJ, Saikia RC, Baruah AKS, Sarma TC. Influence of leaf blight disease on yield of oil and its constituents of Java citronella and *in-vitro* control of the pathogen using essential oils. World J Agric Sci 2006;2:319-321

39 Cassel E, Vargas RMF. Experiments and modeling of the *Cymbopogon winterianus* essential oil extraction by steam distillation. J Mex Chem Soc 2006;50:126-129

40 Naqvi, AA, Mandal S, Chattopadhyay A, Prasad A. Salt effect on the quality and recovery of essential oil of citronella (*Cymbopogon winterianus* Jowitt). Flavour Fragr J 2002;17:109-110

41 Kakaraparthi PS, Srinivas KVNS, Kumar JK, Kumar AN, Rajput DK, Sarma UM. Variation in the essential oil content and composition of Citronella (*Cymbopogon winterianus* Jowitt.) in relation to time of harvest and weather conditions. Ind Crops Prod 2014;61:240-248

42 Verma RS, Ur-Rahman L, Verma RK, Chauhan A, Singh A, Kukreja AK, Khanuja SPS. Qualitative performance of Java citronella (*Cymbopogon winterianus* Jowitt) cultivars in Kumaon Himalaya. J Med Arom Plant Sci 2009;31:321-325

43 Dharmagunawardena B, Takwale A, Sanders KJ, Cannan S, Roger A, Ilchyshyn A. Gas chromatography: an investigative tool in multiple allergies to essential oils. Contact Dermatitis 2002;47:288-292

44 Rudzki E, Grzywa Z, Bruo WS. Sensitivity to 35 essential oils. Contact Dermatitis 1976;2:196-200

45 Rudzki E, Grzywa Z. Allergy to perfume mixture. Contact Dermatitis 1986;15:115-116

46 Rudzki E, Rebandel P, Grzywa Z. Occupational dermatitis from cosmetic creams. Contact Dermatitis 1993;29:210

47 Davies MG, Hodgson GA, Evans E. Contact dermatitis from an ostomy deodorant. Contact Dermatitis 1978;4:11-13

48 Bilbao I, Aguirre A, Zabala R, Gonzalez R, Raton J, Diaz Perez JL. Allergic contact dermatitis from butoxyethyl nicotinic acid and *Centella asiatica* extract. Contact Dermatitis 1995;33:435-436

49 Keil H. Contact dermatitis due to oil of citronellal. J Invest Dermatol 1947;8:327-334

50 Rudzki E, Grzywa Z. The value of a mixture of cassia and citronella oils for detection of hypersensitivity to essential oils. Dermatosen 1985;33:59-62

Chapter 5.24 CITRONELLA OIL, SRI LANKA

There are two citronella oils, citronella oil Java obtained from *Cymbopogon winterianus* Jowitt and citronella oil, Sri Lanka type, which is obtained from *Cymbopogon nardus* (L.) Rendle. In this chapter, citronella oil Sri Lanka (formerly Ceylon citronella) is discussed. The Java type oil is presented in Chapter 5.23.

DEFINITION

Citronella oil, Sri Lanka type, is the essential oil obtained from the aerial parts (leaves) of the Ceylon citronella (citronella grass), *Cymbopogon nardus* (L.) Rendle.

INCI NOMENCLATURE

Description/definition: Cymbopogon nardus oil is the essential oil obtained by direct steam-distillation of the dried fresh grass citronella, *Cymbopogon nardus* (L.), Gramineae

INCI name EU: Cymbopogon nardus oil

INCI name USA: Cymbopogon nardus (citronella) oil

CAS registry number(s): 8000-29-1; 89998-15-2

EINECS number(s): 289-753-6

ISO (INTERNATIONAL ORGANIZATION FOR STANDARDIZATION) STANDARD

ISO number: 3849

ISO name: Essential oil of citronella, Sri Lanka type

Botanical origin: *Cymbopogon nardus* (L.) W. Watson var. *lenabatu* Stapf.

Parts of plant used: Aerial parts (leaves)

ISO values: ISO values (minimum and maximum concentrations) are shown in Table 5.24.1.

It should be noted that the name used by ISO for Sri Lanka lemongrass, *Cymbopogon nardus* (L.) W. Watson var. *lenabatu* Stapf., is not an acknowledged botanical name. We prefer to use the name *Cymbopogon nardus*

(L.) Rendle, which will indeed be the botanical name in the next ISO 3849 update.

THE PLANT, THE OIL, AND THEIR USES

Cymbopogon nardus (L.) Rendle (syn. *Andropogon nardus* L.), commonly known as citronella grass or Ceylon (Sri Lanka) citronella, is a perennial from the Poaceae (Gramineae) grass family, which has stems of 75-300 cm and aromatic leaves 20-60 cm long. It is native to tropical Africa, India and Sri Lanka and is cultivated in China, Sri Lanka, Myanmar and Indonesia (Grin Taxonomy for Plants). *C. nardus* is found in two cultivated forms: (a) Lena Batu and (b) Maha Pengiri or old citronella grass, also known as Winter's grass. The latter has been raised to the status of a separate species, *C. winterianus* Jowitt, and is cultivated in many parts of the world for its valuable citronella oil called citronella oil Java (10). The uses of these two citronella plants (*C. winterianus* Jowitt and *C. nardus* (L.) Rendle) and their essential oils are fairly similar and are discussed in the chapter Citronella oil, Java.

CHEMICAL COMPOSITION

Citronella oil Sri Lanka is a pale yellowish to pale brownish, clear mobile liquid which has a floral, rosy warm and slightly woody odor. The yield of essential oil from the aerial parts (leaves) of *Cymbopogon nardus* L. (Rendle) generally varies from 0.3 to 0.5%. The main producing countries of this oil are Sri Lanka, Indonesia (Java), China and Vietnam.

Literature data (up to November 13, 2014) on the chemical composition of citronella oils Sri Lanka and unpublished analytical data from one of us (E.S.) are shown in Table 5.24.2 in alphabetical order. In citronella oils Sri Lanka from various origins, over 145 chemicals have been identified. About 52% of these were found in a single reviewed publication only. The major compounds found in citronella oils Sri Lanka from different sources are shown in Table 5.24.3. They include (highest concentrations in any study given) geraniol (48.7%), citronellal (47.0%), methyl isoeugenol (26.4%), geranial (22.7%), citronellol (15.3%), borneol (11.8%), geranyl acetate (11.6%), limonene (11%), camphene (10.9%), (*E*)-methylisoeugenol (10.7%), β-caryophyllene (6.5%) and δ-cadinene (4.2%). Well-known ingredients of citronella oils Sri Lanka that were present in high concentrations in one or two studies were neral (14.2% and 28.6%), (*Z*)-methylisoeugenol (9.9%), elemol (7.0% and 8.2%), eugenol (8.1%) and geranyl butyrate (7.7%). Uncommon or rare constituents of citronella oils Sri Lanka found in high concentrations (>7%) in single studies include nerol (24.3%; misinterpretation: is in fact geraniol), α-bisabolol (14.8%), γ-*terpineol* (9.2%) and elemicin (7.3%).

Commercial oils

The ten chemicals that had the highest maximum concentrations in 29 commercial citronella oil Sri Lanka essential oil samples (concentration ranges provided) are the following: geraniol (18.0-48.7%), citronellal

Table 5.24.1 ISO values (%) for citronella oil, Sri Lanka type[a]

Compound	CAS	Minimum	Maximum
Geraniol	106-24-1	15.0	23.0
Limonene	138-86-3	7.0	11.5
Methyl isoeugenol	93-16-3	7.0	11.0
Camphene	79-92-5	7.0	10.0
Citronellol	106-22-9	3.0	8.5
Borneol	507-70-0	4.0	7.0
Citronellal	106-23-0	3.0	6.0

[a] ISO 3849 Essential oil of citronella, Sri Lanka type ©ISO 2002; Geneva, Switzerland, www.iso.org

(1.0-12%), geranial (0.5-11.3%), camphene (7.0-10.9%), (*E*)-methylisoeugenol (6.7-10.7%), limonene (3.0-10.4%), (*Z*)-methylisoeugenol (0.4-9.9%), citronellol (1.7-9.6%), borneol (5.0-7.6%) and methyl eugenol (0.4-7.2%) (Erich Schmidt, unpublished analytical data).

The data in ref. 7 are not discussed here, as the identity of the plant from which the oil was obtained

was in error (ref. 6). Ref. 12 is not discussed, as the results are most likely incorrect (limonene 22%, isopulegol 71%, squalene 2%). Ref. 13 is not discussed, as the sum of the ingredients appears to be >150% and ref. 24 contains a large number of nomenclature mistakes and is therefore exempt from discussion.

Table 5.24.2 Constituents identified in citronella oils, Sri Lanka

Constituent	CAS	Percentage and range					
		A	B (9)	C (11)	D (18)	E (17)	F
α-Amorphene	20085-19-2			1.8			0.2[q]
Arachidic acid	506-30-9						0.1[a,b]
Behenic acid	112-85-6						0.1[a,b]
Bergamal	106-72-9			0.2			0.02[q]; 0.3[e]
cis-α-Bergamotene	18252-46-5						0.2[q]
trans-α-Bergamotene	13474-59-4	tr-0.9					0.9[s]
Bicyclogermacrene	24703-35-3						0.4[c]
α-Bisabolol	515-69-5						14.8[i]
Bis(2-ethylhexyl) phthalate	117-81-7						2.0[a,b]
2-Bornene (α-)	464-17-5						0.05[q]
Borneol	507-70-0	5.0-7.6	5.6				2.5[b]; 4.7[k]; 5.5[q]; 5.9[s]; 7[r]; 11.8[i]
Bornyl acetate	76-49-3	0.4-0.6					0.5[s]; 1.1[q]
β-Bourbonene	5208-59-3	0.1-1.2	0.3				0.02[q]; 0.1[m]
Cadinene	29350-73-0				3.1		1.1[d]
α-Cadinene	24406-05-1					0.3	0.01[q]
γ-Cadinene	39029-41-9						0.2[a,m]; 0.4[q]; 0.3[e]
δ-Cadinene	483-76-1	0.5-4.1	0.5	1.1		4.2	0.4[o]; 0.8[q]; 1.1[h]; 1.5[e]; 1.6[m]; 1.8[g]; 2.3[n]
α-Cadinol	481-34-5		0.4	2.0	1.1	2.8	0.6[e,q]; 0.9[n]; 1.8[o]
epi-α-Cadinol	5937-11-1		0.3	1.1		1.4	
Camphene	79-92-5	7.0-10.9	9.8				0.1[m]; 4.5[i]; 8.4[k]; 8.8[s]; 8.9[q]; 10[r]
Camphenilone	13211-15-9						0.09[q]
α-Campholenal	4501-58-0						0.02[q]
δ2-Carene (= δ4-)	554-61-0						0.2[q]
δ3-Carene	13466-78-9	0.01-1.5	0.1				
β-Caryophyllene	87-44-5	0.4-3.2	1.7	0.1			0.8[f]; 1.4[q]; 1.9[k]; 2.0[i]; 2.2[b]; 2.5[d]; 6.5[g]
Caryophyllene oxide	1139-30-6	0.2-0.4					
1,8-Cineole	470-82-6						0.2[q]; 0.5[s]; 2.1[i]
Citral	5392-40-5					0.7	
Citronellal	106-23-0	1.0-12.0	4.1	35.9	27.6	26.2	36.5[h]; 37.5[d]; 38.5[m]; 41.3[c]; 42.0[e]; 47.0[p]
Citronellic acid	502-47-6					0.8	
Citronellol	106-22-9	1.7-9.6	3.7	11.6	13.6	13.0	10.6[o]; 13.0[p]; 13.1[h]; 14.2[m]; 14.5[e]; 15.3[n]
Citronellyl acetate	150-84-5	0.6-1.9			4.6		2.0[p]; 2.2[h]; 2.6[e]; 3.3[i]; 3.9[m]; 4.4[c,n]; 5.7[j]
Citronellyl formate	105-85-1						0.2[e]
Citronellyl propionate	141-14-0	0.9-3.9					
α-Cubebene	17699-14-8						0.02[q]
β-Cubebene	13744-15-5	1.1-3.8		0.1	3.2		0.02[q]; 1.2[a,m]
Cubenene [s]	29837-12-5						0.01[q]
Cubenol	21284-22-0			1.8			
1-epi-Cubenol	81939-29-9						0.01[q]
1,6-Cyclodecadiene, 1-methyl-5-methyl-ene-8-(1-methylethyl)-	37839-63-7						2.3[g]
Cyclohexane, 1-ethe-nyl-1-methyl-2,4-*bis*-(1-methylethenyl)	110823-68-2						3.3[g]
Cyclohexane, 1-ethe-nyl-1-methyl-2,4-*bis*-(1-methylethenyl)-, [1S-(1α,2α,4α)]							1.3[g]

Table 5.24.2 Constituents identified in citronella oils, Sri Lanka (*continued*)

Constituent	CAS	Percentage and range					
		A	B (9)	C (11)	D (18)	E (17)	F
Cyclohexanemetha-nol, 4-ethenyl-a,a-4-trimethyl-3-(1-methyl-ethenyl)-,[1R-(1a,3a,4a)]-							1.3[g]
p-Cymene	99-87-6	0.01-0.2	0.2				0.2[q]
Decanal	112-31-2	0.1-0.2					0.1[q]
2,6-Dimethyl-2,6-octadiene	2792-39-4						1.6[g]
(*Z*)-2,6-Dimethyl-2,6-octadiene	2492-22-0					4.3	6.9[g]
Elemene	11029-06-4						1.5[d]; 3.9[n]
β-Elemene	33880-83-0	0.5-1.3	0.9	1.9	2.9	2.8	0.2[b]; 0.3[q]; 0.5[o]; 1.0[m,s]; 1.9[e]
Elemicin	487-11-6						0.02[q]; 7.3[o]
Elemol (α-)	639-99-6	0.1-1.7	1.0	9.0	3.9	5.1	3.7[o]; 4.0[j]; 4.8[c]; 5.1[n]; 5.5[i]; 7.0[p]; 8.2[h]
bis-(2-Ethylhexyl)-1,2-benzene dicarboxylic acid ester	132969-07-4						2.0[b]
β-Eudesmol	473-15-4				0.3		0.1[q]
γ-Eudesmol	1209-71-8				0.4	0.7	0.06[q]
10-epi-γ-Eudesmol	15051-81-7	0.2-0.6					
τ-Eudesmol				0.6			
Eugenol	97-53-0	0.05-0.3			1.3	1.5	0.5[b]; 0.9[m]; 1.0[e,o]; 1.5[g]; 8.1[d]
Farnesene	502-61-4						1.5[c]
α-Farnesene	502-61-4	0.6-4.6					0.4[q]; 4.2[s]
(*Z,E*)-α-Farnesene	26560-14-5						2.9[q]
(*Z*)-β-Farnesene	28973-97-9						0.04[q]
Farnesol	4602-84-0						0.02[q]
(*E,E*)-Farnesol	106-28-5						0.9[c]
(*Z,E*)-Farnesol	3790-71-4			0.2			
Geranial	141-27-5	0.5-11.3	0.5	0.6	0.5		1.2[d]; 1.4[o]; 2.0[e]; 2.1[c]; 9.8[l]; 11.0[g]; 22.7[f]
Geraniol	106-24-1	18.0-48.7	18.3	24.3[t]	22.5	19.8	27.9[q]; 29.4[d]; 30.7[n]; 35.7[f,i]; 36.5[l]; 41.0[p]
Geranyl acetate	105-87-3	0.5-5.2	2.9	1.3	5.2	3.6	4.6[o]; 4.8[l]; 5.3[n]; 5.7[j]; 8.9[k]; 9.7[f]; 11.6[q]
Geranyl butyrate	106-29-6	0.5-1.5					0.8[q]; 1.1[s]; 7.7[i]
Geranyl formate	105-86-2						0.05[q]; 0.5[e]; 4.4[k]
Geranyl hexanoate	10032-02-7						0.1[q]
Geranyl isobutyrate	2345-26-8						6.9[g]
Geranyl propanoate	105-90-8						1.1[k]
Germacradienol							0.3[e]
Germacrene D	23986-74-5	0.2-1.5	1.2	1.5		1.0	0.03[q]; 0.5[c,o]; 0.7[h]; 0.8[n]; 1.6[s]
Germacrene D-4-ol	198991-79-6						0.3[n]; 1.3[c]; 1.5[b]; 1.6[h]
Guaiene	88-84-6						1.9[a,m]
γ-Gurjunene	22567-17-5			0.2			
Hedycaryol	21657-90-9						1.3[g]
Hexadecanol	51260-59-4						0.5[b]
Hexahydrofarnesyl acetone	502-69-2						0.01[q]
α-Humulene	6753-98-6	0.2-0.4		0.1			0.2[q]; 0.3[b]
Isoborneol	124-76-5						0.07[q]; 1.1[b]
Isoelemicin	487-12-7						0.05[q]
Isopulegol	89-79-2	0.06-0.5	0.04	0.2	1.7	0.6	0.07[q]; 0.1[d]; 0.9[m]; 1.4[h]; 2.7[o]
Isopulegol II	1370348-44-9					1.4	
Lavandulol	498-16-8						0.7[b]
Ledol	577-27-5			0.6			
Limonene	138-86-3	3.0-10.4	9.5	2.2	3.6	3.0	3.2[o]; 4.6[l]; 4.8[i]; 7.8[k]; 9.3[s]; 9.7[q]; 11[r]
Linalool	78-70-6	0.5-3.3	0.6	0.4	1.0	0.7	0.7[b,e]; 0.9[d]; 1.1[h]; 1.3[f]; 1.4[o]; 1.7[l]; 2.5[i]
cis-Linalool oxide	11063-77-7						0.2[g]
cis-p-Menth-2-en-1-ol	29803-82-5						0.06[q]
trans-p-Menth-2-en-1-ol	29803-81-4						0.06[q]

Table 5.24.2 Constituents identified in citronella oils, Sri Lanka (*continued*)

Constituent	CAS	Percentage and range					
		A	B (9)	C (11)	D (18)	E (17)	F
2-Methoxy-3-(2-propenyl)phenol	1941-12-4						4.5[g]
exo-Methyl-camphenilol							0.3[q];
Methyl eugenol	93-15-2	0.4-7.2	1.0				0.2[q]; 1.1[s]; 5.7[i]
Methylheptenone	409-02-9						0.1[l]
6-Methyl-5-hepten-2-one	110-93-0	0.07-1.2	0.07				
Methyl isoeugenol	93-16-3						5.2[k]; 11[r]; 26.4[i]
(*E*)-Methylisoeugenol	6379-72-2	6.7-10.7	6.9				0.9[q]; 8.8[s]
(*Z*)-Methylisoeugenol	6380-24-1	0.4-9.9					0.03[q]
Muurolene	69671-15-4					1.3	
α-Muurolene	10208-80-7	0.5-0.9		0.2		1.2	0.2[e]; 0.4[m]; 0.8[n]
γ-Muurolene	30021-74-0	0.1-0.3			0.7	0.6	0.4[m]
τ-Muurolene	152287-43-9			0.4			
α-Muurolol	104245-48-9		0.04				
τ-Muurolol (epi-α-)	19912-62-0	0.05-0.2	0.2		0.8		0.3[e]
Myrcene	123-35-3	0.1-1.5	0.9	0.2			0.1[d]; 0.4[i]; 0.5[o]; 0.9[s]; 1.2[k,q]; 2.9[b]
Myrtenol	515-00-4						0.09[q]
Myrtenyl acetate	1079-01-2						0.02[q]
Naphthalene, 1,2,3,4, 4a,5,6,8a-octahydro-7-methyl-4-methylene-1-(1-methylethyl)-, (1α,4aα,8aα)-							0.6[g]
Neoisopulegol	29141-10-4		0.3				1.2[s]
Neophytadiene	504-96-1						0.02[q]
Neral	106-26-3	0.2-6.9	0.2	0.4	0.3		0.4[q]; 0.7[d]; 2.5[j]; 6.6[l]; 14.2[f]; 28.6[o]
Nerol	106-25-2	0.3-1.0	0.5	24.3[t]			0.2[d]; 1.3[i]; 1.5[b]
Neryl acetate	141-12-8						0.03[q]
(*E*)-Nerolidol	40716-66-3						4.8[b]
β-Ocimene	13877-91-3						1.0[i]; 1.1[l]
(*E*)-β-Ocimene	3779-61-1	0.3-1.8	1.0				1.1[s]; 1.8[q]
(*Z*)-β-Ocimene	3338-55-4	0.2-2.2	2.1				1.8[s]; 2.8[k]
(*E,E*)-allo-Ocimene	3016-19-1						0.1[q]
Octanol	29063-28-3						0.05[q]
α-Phellandrene	99-83-2	0.1-0.8	0.1				0.3[q]
β-Phellandrene	555-10-2	0.03-0.8	0.5				0.7[q]
α-Pinene	80-56-8	1.9-3.1	2.7				0.5[b]; 2.3[s]; 2.5[k,q]; 4.8[m]
β-Pinene	127-91-3	0.01-1.2	0.08				0.3[q]; 0.4[m]; 1.5[b]
cis-Piperitol	16721-38-3						0.02[q]
Phthalate							1.7[a,b]
Pulegol	529-02-2						
cis-Rose oxide	3033-23-6		0.04				
Sabinene	3387-41-5	0.06-0.3	0.2				0.3[q]; 1.1[o]
cis-Sabinene hydrate	15537-55-0						3.8[b]
α-Selinene	473-13-2						1.0[b]
β-Selinene	17066-67-0						0.8[b]
β-Sesquiphellandrene	20307-83-9						0.01[q]
Seychellene	20085-93-2			0.2			
Spathulenol	6750-60-3						0.01[q]; 0.2[b]
α-Terpinene	99-86-5	0.05-0.2					0.2[q]; 1.0[i]
γ-Terpinene	99-85-4	0.03-0.1	0.03				0.2[q]
Terpinen-4-ol	562-74-3	0.4-0.9					0.5[q]; 1.3[s]
α-Terpineol	98-55-5	0.3-2.2	2.7				0.5[b]; 1.1[q]; 1.3[s]; 1.5[k]
γ-Terpineol	586-81-2						9.2[b]
Terpinolene	586-62-9	0.7-1.0	0.6				0.1[q]; 0.8[s]
α-Terpinyl acetate	80-26-2						0.04[q]
Tricyclene	508-32-7	1.1-1.7	1.5				1.3[q,s]

Table 5.24.2 Constituents identified in citronella oils, Sri Lanka (*continued*)

Constituent	CAS	Percentage and range					
		A	B (9)	C (11)	D (18)	E (17)	F
3,3,5-Trimethyl-1,5-heptadiene	74630-29-8						0.7[a,b]
Valencene	4630-07-3						0.7[b]

A twenty-nine citronella Sri Lanka essential oil samples from Sri Lanka, Nepal and Vietnam, analyzed between 1998 and 2014; lowest and highest concentrations given (E. Schmidt, unpublished data)
B one Sri Lankan citronella essential oil sample (ref. 9)
C one lab-hydrodistilled oil from the leaves of *C. nardus* grown in a botanical garden in Benin (ref. 11); very atypical, as geraniol appears to be absent, but this can be explained by a misinterpretation: nerol should be geraniol
D one commercial citronella oil ex *C. nardus* purchased in South Africa (ref. 18)
E one commercial Sri Lankan citronella oil from China (ref. 17)
F data from other studies (indicated with superscript letters); highest concentrations found in any study reviewed here given; when two or more oils were investigated, only the highest concentrations are mentioned, unless indicated otherwise

[a] incorrect identity based on GC elution order/incorrect identification/does not exist naturally (refs. 2,6,25); [b] one lab-hydrodistilled oil from India (ref. 10); [c] one lab-hydrodistilled oil from Benin (ref. 21); [d] one lab-hydrodistilled oil from Congo (ref. 21); [e] one commercial *C. nardus* essential oil purchased from a French company (ref. 22); [f] one steam-distilled oil from Thailand (ref. 23); [g] one steam-distilled oil from Malaysia (ref. 20); [h] one lab-hydrodistilled oil from Brazil (ref. 19); [i] eight steam-distilled oils from four selections of *C. nardus* cultivated in Sri Lanka and grown at two temperature regimes (ref. 16); [j] one lab-hydrodistilled oil from Malaysia (ref. 15); [k] one lab-hydrodistilled oil from *C. nardus* collected in southern Sri Lanka (ref. 1); [l] two lab-hydrodistilled oils from India obtained from *C. nardus* var. *nardus* and from *C. nardus* var. *Java II*; highest concentrations given (ref. 14); [m] one oil sample from *Cymbopogon nardus* (L.) Rendle produced in China (ref. 3); [n] one commercial Sri Lankan citronella oil (ref. 4); [o] one sample of citronella oil ex *Cymbopogon nardus* produced from plants grown in Brazil (ref. 5); [p] unknown number of oils from Togo (ref. 8); [q] one sample of Ceylon citronella oil (ref. 26); [r] unknown number of Ceylon citronella oils (ref. 27); [s] one commercial citronella oil *ex C. nardus* produced in Sri Lanka (ref. 28); [t] indicated as nerol, but must be geraniol

tr: trace; + present in the oil investigated, but quantity not stated

Table 5.24.3 Major constituents of citronella oils, Sri Lanka

Constituent	CAS	Percentage and range					
		A	B (9)	C (11)	D (18)	E (17)	F
Geraniol	106-24-1	18.0-48.7	18.3	24.3[t]	22.5	19.8	27.9[q]; 29.4[d]; 30.7[n]; 35.7[f,i]; 36.5[l]; 41.0[p]
Citronellal	106-23-0	1.0-12.0	4.1	35.9	27.6	26.2	36.5[h]; 37.5[d]; 38.5[m]; 41.3[c]; 42.0[e]; 47.0[p]
Methyl isoeugenol	93-16-3						5.2[k]; 11[r]; 26.4[i]
Geranial	141-27-5	0.5-11.3	0.5	0.6	0.5		1.2[d]; 1.4[o]; 2.0[e]; 2.1[c]; 9.8[l]; 11.0[g]; 22.7[f]
Citronellol	106-22-9	1.7-9.6	3.7	11.6	13.6	13.0	10.6[o]; 13.0[p]; 13.1[h]; 14.2[m]; 14.5[e]; 15.3[n]
Borneol	507-70-0	5.0-7.6	5.6				2.5[b]; 4.7[k]; 5.5[s]; 5.9[s]; 7[r]; 11.8[i]
Geranyl acetate	105-87-3	0.5-5.2	2.9	1.3	5.2	3.6	4.6[o]; 4.8[l]; 5.3[n]; 5.7[i]; 8.9[k]; 9.7[f]; 11.6[q]
Limonene	138-86-3	3.0-10.4	9.5	2.2	3.6	3.0	3.2[o]; 4.6[l]; 4.8[i]; 7.8[k]; 9.3[s]; 9.7[q]; 11[r]
Camphene	79-92-5	7.0-10.9	9.8				0.1[m]; 4.5[l]; 8.4[k]; 8.8[s]; 8.9[q]; 10[r]
(*E*)-Methylisoeugenol	6379-72-2	6.7-10.7	6.9				0.9[q]; 8.8[s]
β-Caryophyllene	87-44-5	0.4-3.2	1.7	0.1			0.8[f]; 1.4[q]; 1.9[k]; 2.0[i]; 2.2[b]; 2.5[d]; 6.5[g]
δ-Cadinene	483-76-1	0.5-4.1	0.5	1.1		4.2	0.4[o]; 0.8[q]; 1.1[h]; 1.5[e]; 1.6[m]; 1.8[g]; 2.3[n]

LEGEND: SEE UNDER TABLE 5.24.2

CONTACT ALLERGY/ALLERGIC CONTACT DERMATITIS

General

There is only one documented case of contact allergy to citronella oil, Sri Lanka.

Case report

A patient occupationally sensitized to lemon oil and limonene also had positive patch test reactions to oil of citronella, Ceylon (= Sri Lanka), oil of citronella, Java (both probably undiluted, may be irritant), turpentine and its component β-pinene (29). Limonene is the dominant component of lemon oil and may be present in citronella oils in concentrations up to 6% (Java) (Chapter 5.23) and 10.4% (Sri Lanka) (Table 5.24.2, column A).

LITERATURE

1 Samarasekera R, Kalhari KS, Weerasinghe IS. Insecticidal activity of essential oils of Ceylon *Cinnamomum* and *Cymbopogon* species against *Musca domestica*. J Essent Oil Res 2006;18:352-354

2 Lawrence BM. Progress in essential oils. Perfum Flavor 2003;28(1):60-? (last page unknown)

3 Zhu L-F, Li Y-H, Li B-L, Lu B-Y, Xia N-H. Aromatic plants and essential constituents. South China Institute of Botany, Chinese Academy of Sciences, Hai Feng Publish Co, 1993:112. Distributed by Peace Book Company Ltd, Hong Kong. Data cited in ref. 2

4 Delespaul Q, de Billerbeck VG, Roques CG, Michel G, Marquier-Viñuales C, Bessière J-M. The antifungal activity of essential oils as determined by different screening methods. J Essent Oil Res 2000;12:256-266

5 Lemos TLG, Monte FJQ, Matos FJA, Alencar JW, Craveiro AA, Barbosa RCSB, et al. Chemical composition and antibacterial activity of essential oils from Brazilian plants. Fitoterapia 1992;63:266-268. Data cited in ref. 6

6 Lawrence BM. Progress in essential oils. Perfum Flavor 2006;31(9):40-50

7 Hifnawy MS, Rashwan OA, Rabeh MA. Comparative chemical and biological investigations of certain essential oils belonging to the families Asteraceae, Lamiaceae and Graminae. Bull Fac Pharm Cairo Univ 2001;39(2):35-53 (not discussed because of mistaken identity of the plant: ref. 6)

8 Koumaglo HK. Bilan et perspectives de la production des huiles essentielles au Togo. In: G Collin, F-X Garneau, Eds. 5e Colloque, Produits naturelles d'origine végétale. Quebec, Canada: Université Chicoutimi, 2002:173-175. Data cited in ref. 6

9 Kubeczka K-H, Formacek V. Essential oils analysis by capillary gas chromatography and carbon-13 NMR spectroscopy, 2nd Ed. New York, USA: John Wiley and Sons, 2002:67-80. Data cited in ref. 6

10 Mahalwal VS, Ali M. Volatile constituents of *Cymbopogon nardus* (Linn.) Rendle. Flavour Fragr J 2003;18:73-76

11 Kpoviessi S, Bero J, Agbani P, Gbaguidi B, Kpadonou-Kpoviessi B, Sinsin B, et al. Chemical composition, cytotoxicity and *in vitro* antitrypanosomal and antiplasmodial activity of the essential oils of four *Cymbopogon* species from Benin. J Ethnopharmacol 2014;151:652-659

12 Kanko C, El-Hadj Sawaliho B, Kone S, Koukoua G, N'Guessan YT. Étude des propriétés physico-chimiques des huiles essentielles de *Lippia multiflora*, *Cymbopogon citratus*, *Cymbopogon nardus*, *Cymbopogon giganteus*. CR Chimie 2004;7:1039-1042 (Not discussed, as the results must be incorrect)

13 Tyagi BK, Shahi AK, Kaul BL. Evaluation of repellent activities of *Cymbopogon* essential oils against mosquito vectors of malaria, filariasis and dengue fever in India. Phytomed 1998;5:324-329. (Not discussed, as the sum of the ingredients appears to be >150%)

14 Khanuja SPS, Shasany AK, Pawar A, Lal RK, Darokar MP, Naqvi AA, et al. Essential oil constituents and RAPD markers to establish species relationship in *Cymbopogon* Spreng. (Poaceae). Biochem Syst Ecol 2005;33:171-186

15 Roszaini K, Nor Azah MA, Mailina J, Zaini S, Faridz ZM. Toxicity and antitermite activity of the essential oils from *Cinnamomum camphora*, *Cymbopogon nardus*, *Melaleuca cajuputi* and *Dipterocarpus* sp. against *Coptotermes curvignathus*. Wood Sci Technol 2013;47:1273-1284

16 Herath HMW, Iruthayathas EE, Ormrod DP. Temperature effcts on essential oil composition of *Citronella* selections. Economic Botany 1979;33:425-430

17 Chen Q, Xu S, Wu T, Guo J, Sha S, Zheng X, Yu T. Effect of citronella essential oil on the inhibition of postharvest *Alternaria alternata* in cherry tomato. J Sci Food Agric 2014; 94: 2441-2447

18 Sellamuthu PS, Sivakumar D, Soundy P. Antifungal activity and chemical composition of thyme, peppermint and citronella oils in vapor phase against avocado and peach postharvest pathogens. J Food Saf 2013;33:86-93

19 Wagner de S. Aguiar R, Ootani MA, Donizeti Ascencio S, Ferreira TPS, dos Santos MM, dos Santos GR. Fumigant antifungal activity of *Corymbia citriodora* and *Cymbopogon nardus* essential oils and citronellal against three fungal species. Sci World J 2014, Article ID 492138, 8 pages. Available at: http://dx.doi.org/10.1155/2014/492138

20 Wei LS, Wee W. Chemical composition and antimicrobial activity of *Cymbopogon nardus* citronella essential oil against systemic bacteria of aquatic animals. Iran J Microbiol 2013;5:147-152

21 Abena AA, Gbenou JD, Yayi E, Moudachirou M, Ongoka RP, Ouamba JM, Silou T. Comparative chemical and analgesic properties of essential oils of *Cymbopogon nardus* (L) Rendle of Benin and Congo. Afr J Trad CAM 2007;4:267-272

22 Billerbeck VG, Roques CG, Bessière JM, Fonvieille JL, Dargent R. Effects of *Cymbopogon nardus* (L.) W. Watson essential oil on the growth and morphogenesis of *Aspergillus niger*. Canadian J Microb 2001;47:9-17

23 Nakahara K, Alzoreky NS, Yoshihashi T, Nguyen HTT, Trakoontivakorn G. Chemical composition and antifungal activity of essential oil from *Cymbopogon nardus* (Citronella grass). Japan Agric Res Quar 2003;37:249-252

24 Setiawati W, Murtiningsih R, Hasyim A. Laboratory and field evaluation of essential oils from *Cymbopogon nardus* as oviposition deterrent and ovicidal activities against *Helicoverpa armigera* Hubner on chili pepper. Indon J Agric Sci 2011;12:9-16. (Not discussed because of large number of nomenclature mistakes)

25 Lawrence BM. Progress in essential oils. Perfum Flavor 2011;36(November):52-58

26 Anonymous. GC-MS analysis of citronella oil (Ceylon). Indian Perfum 2005;162. Data cited in ref. 25

27 Ranade GS. Profile: essential oils. Citronella Ceylon. FAFAI 2008;Oct/Nov:70. Data cited in ref. 25

28 Milchard M, Clery R, Esdale R, Gates L, Judge F, Moss N, Moyler DA, et al. Application of gas-liquid chromatography to the analysis of essential oils. GLC fingerprint chromatograms of six essential oils. Perfum Flavor 2010;35(6):34-40

29 Keil H. Contact dermatitis due to oil of citronellal. J Invest Dermatol 1947;8:327-334

Chapter 5.25 CLARY SAGE OIL

DEFINITION
Clary sage oil is the essential oil obtained from the flowering tops of the clary (clary sage), *Salvia sclarea* L. (synonym: *Salvia sclarea* var. *turkestaniana* Mottet).

INCI NOMENCLATURE
Description/definition: Salvia sclarea flower oil is an essential oil obtained from the flowers and foliage of the clary sage, *Salvia sclarea* L., Lamiaceae
INCI name EU: Salvia sclarea flower oil (perfuming name, not an INCI name proper)
INCI name USA: Not in the Personal Care Products Council Ingredient Database
CAS registry number(s): 84775-83-7
EINECS number(s): 283-911-8

Description/definition: Salvia sclarea oil is a volatile oil obtained from the clary sage, *Salvia sclarea* L., Lamiaceae
INCI name EU: Salvia sclarea oil
INCI name USA: Salvia sclarea (clary) oil
Other name(s): Clary oil
CAS registry number(s): 8016-63-5; 84775-83-7
EINECS number(s): 283-911-8

ISO (INTERNATIONAL ORGANIZATION FOR STANDARDIZATION) STANDARD
Currently there is no ISO standard for clary sage oil. AFNOR (Association Française de Normalisation) values are shown in Table 5.25.1.

Quality clary sage oil is preferably made solely from the flowering tops of *Salvia sclarea* L. but many commercially available oils are produced from the flowering tops *and* the leaves (aerial parts, above-ground plant). Oils purely made from leaves are not discussed here.

THE PLANT, THE OIL, AND THEIR USES
Salvia sclarea L., commonly known as clary sage or clary, is a biennial or short-lived herbaceous perennial, which reaches 0.9 to 1.2 meters in height and is one of the most valuable aromatic plants of the temperate region. It is native to southern Europe, central Asia and to Africa up to the Atlantic Ocean and is cultivated all over the world, commonly in the former USSR, Bulgaria, France, Morocco, USA, England, central Europe and West China (17).

The whole plant, and notably the inflorescences, possesses a very strong aromatic scent. Fresh leaves of the clary sage are sometimes used in soups, and in the Middle Ages the seeds were used for clearing vision, from which it received its popular name, 'clary', or 'clear eye' (10). In Greece, *S. sclarea* is locally used for coughs, colds, and blood cleaning, on wounds and sore eyes, and as a diuretic (16). In traditional herbal medicine, the plant is employed as an antispasmodic, carminative and estrogenic agent (36).

Sclareol, the main component of the clary sage *concrete*, is the most common starting material for the synthesis of Ambrox®, which is a valuable and sustainable synthetic ambergris-like material (10,17,19,23). Ambergris is a waxy substance secreted by sperm whales. It has historically been appreciated for its musky and sweet earthy odor and has been used for many years as a fixative in high-end perfumes (27).

The essential oil of the clary sage serves as flavoring agent in food, liqueurs, non-alcoholic beverages and tobacco, and is used in perfumery, in the cosmetic industries and for pharmaceutical and medicinal purposes (1,6,10,17,19,23). It is considered to have analgesic, antifungal, antimicrobial, antioxidant, anti-inflammatory, antitumor, anti-tuberculosis and larvicidal properties (1,10,36) and has been used for treating inflammatory diseases in the oral cavity (6). In aromatherapy and (folk) medicine, clary sage oil is employed to treat stress, asthma, digestive and menstrual problems and as an antidepressant, antiseptic, antispasmodic, carminative, and aphrodisiac (5,29,36).

CHEMICAL COMPOSITION
Clary sage oil is a mobile, clear and colorless to light yellowish or light brown liquid which has a fresh herbaceous, soft floral to woody ambery changing odor. The yield of essential oil from the flowering tops of *Salvia sclarea* L. generally varies from 0.05 to 0.15% for fresh herbs and from 0.1 to 0.12% for dried herbs. The main producing countries of this oil are Bulgaria, France, Hungary, Russia, Spain, Morocco, India, USA and Egypt.

Table 5.25.1 AFNOR values (%) for clary sage oil[a]

Compound	CAS	Classic method[b]		Method *Broyée en vert*[c]	
		Minimum	Maximum	Minimum	Maximum
Linalyl acetate	115-95-7	62	78	56	70.5
Linalool	78-70-6	6.5	13.5	13	24
Germacrene D	23986-74-5	1.5	12	1.2	7.5
Sclareol	515-03-7	0.4	2.6	0.4	2.6
α-Terpineol	98-55-5	traces	1.2	1	5

[a]AFNOR NF T 75-255 Huile essentielle de sauge sclarée (*Salvia sclarea* Linnaeus) dite 'traditionelle' et huile essentielle de sauge sclarée (*Salvia sclarea* Linnaeus) broyée en vert, © AFNOR 1992; 11, rue de Francis de Pressensé, 93571 La Plaine Saint-Denis Cedex, France
[b] the plants are cut and dried in the fields; distillation takes place later (sometimes after months)
[c] the plants are cut and transported immediately in containers, in which hydrodistillation takes place

Literature data (up to November 13, 2014) on the chemical composition of clary sage oils and unpublished analytical data from one of us (E.S.) are shown in Table 5.25.2 in alphabetical order. In clary sage oils from various origins, over 295 chemicals have been identified. About 59% of these were found in a single reviewed publication only. The major compounds found in clary sage oils from different sources are shown in Table 5.25.3. They include (highest concentrations in any study given) linalyl acetate (81.1%), linalool (42.3%), germacrene D (24.7%), α-terpineol (17.2%, one report with a very high concentration), sclareol (15.7%), geranyl acetate (12.1%, one report with a very high concentration), geraniol (9.6%, one report with a very high concentration), and nerol (7.4%). Well-known ingredients of clary sage oils that were present in high concentrations in one or two studies were α-terpineol (47.4%), geranyl acetate (36.8%), geraniol (24.5%), caryophyllene oxide (8.5% and 24.1%), β-caryophyllene (three studies: 9.0%, 16.2%, 21.3%), spathulenol (11.4%), bicyclogermacrene (9.6%), myrcene (7.3% and 8.4%) and α-copaene (7.8%). Uncommon or rare constituents of clary sage oils found in high concentrations (>7%) in single studies include diethyl phthalate (34.9%, error, not demonstrated to occur in nature), α-terpinyl acetate (22.1%), geranial (19.4%), neral (11.3%), 1H-naphtho[2.1.6]pyran (8.6%), hexatriacontane (7.4%) and β-humulene (7.1%).

Commercial oils

The ten chemicals that had the highest maximum concentrations in 34 commercial clary sage essential oil samples (concentration ranges provided) are the following: linalyl acetate (31.2-64.2%), linalool (18.6-31.0%), α-copaene (0.2-7.8%), germacrene D (0.4-5.4%), α-terpineol (0.6-5.1%), (Z)-β-ocimene (0.2-4.8%), bicyclogermacrene (0.2-4.8%), geranyl acetate (0.5-4%), β-caryophyllene (0.6-4.0%) and geraniol (0.2-3.6%) (Erich Schmidt, unpublished analytical data).

The composition of clary sage *leaf* oils considerably differs from that of the oils obtained from the flowering tops. Usually, the main components of leaf oils are germacrene D (with concentrations up to 69%) and β-caryophyllene (3,6,21); also, the concentration of caryophyllene oxide is usually far higher than in the flower oil (6,21). High levels of germacrene D in clary sage flower oils may thus indicate that a certain amount of leaves was mixed with the flowering tops to produce the oil (41).

Table 5.25.2 Constituents identified in clary sage oils

Constituent	CAS	Percentage and range					
		A	B (30)	C (13)	D (9)	E (16)	F
Acetic acid	64-19-7			tr			
Acetoxylinalool	1301265-88-2						0.04[i]
3(1-Adamantyl)sydnone							0.3[y3]
α-Amorphene	20085-19-2						0.7[y3]
δ-Amorphene	189165-79-5						0.3[p]
Amyl isobutyrate	2445-72-9				tr		
(E)-Anethole	4180-23-8						2.0[z5]; 0-30mg/ml[z17]
Aromadendrene	489-39-4		0.1			tr	0.03[i]; 0.1[z2]; 0.3[r]
allo-Aromadendrene	25246-27-9						0.08[r]
Benzaldehyde	100-52-7		tr[c]			tr	
Benzeneacetaldehyde	122-78-1					tr	tr[j]
Benzyl benzoate	120-51-4					tr	
Benzyl tiglate	37526-88-8		tr				
trans- α-Bergamotene	13474-59-4					tr	
Bicycloelemene	32531-56-9						0.2[k]; 1.1[y3]
Bicyclogermacrene	24703-35-3	0.2-4.8			0.5	tr	1.4[i]; 2.0[q]; 2.3[j]; 3.4[y3]; 3.8[y4]; 9.6[y2]
epi-Bicyclosesquiphel-landrene	54274-73-6						0.4[z2]
β-Bisabolene	495-61-4						0.6[r]
α-Bisabolol	515-69-5						0.1[k]
Borneol	507-70-0	0.04-0.06			0.1		tr[j]; 0.03[z5]; 0.1[k,x]; 1.2[z1]
Bornyl acetate	76-49-3	tr-0.07	tr		1.1		tr[j]; 0.02[y2]; 0.1[z1]; 0.8[t]
1-Bromoadamantane	768-90-1						0.4[y3]
β-Bourbonene	5208-59-3	0.1-0.5	0.2	tr	0.3		tr-0.2[g]; 0.2[m]; 0.3[p]; 0.9[t]; 2.5[q]; 4.8[y4]
Cadalene	483-78-3						0.2[y3]
α-Cadinene	24406-05-1				0.1		1.9[z15]
β-Cadinene	523-47-7						0.4[m]
γ-Cadinene	39029-41-9				0.2		0.1[z5]; 1.5[y2]
δ-Cadinene	483-76-1	0.1-0.5	tr	0-tr		0.5	0.4[n]; 0.5[j]; 0.6[z5,]; 0.7[q]; 1.1[z18]; 1.6[y3]
α-Cadinol	481-34-5				1.1		
τ-Cadinol	5937-11-1						0.2[k]
α-Calacorene	21391-99-1					tr	0.01[z5]; 0.2[k]

Table 5.25.2 Constituents identified in clary sage oils (*continued*)

Constituent	CAS	A	B (30)	C (13)	D (9)	E (16)	F
Calacorene	38599-17-6						0.3[y3]
β-Calacorene	50277-34-4				tr		0.1[k]
(1*S*)-*cis*-Calamenene	483-77-2				0.8		
trans-Calamenene	73209-42-4						tr[k]
Calarene	17334-55-3						tr[k]; 1.0[z2]; 1.3[y3]
Camphene	79-92-5	0.02-0.2	tr		tr	tr	tr[j]; 0.06[z7]; 0.1[k,q,r]; 0.9[z1]
Camphor	76-22-2	tr-0.04					tr[k]; <0.1[y]; 0.3[t]; 0.7[z7]; 1.0[z]
δ3-Carene	13466-78-9					0.4	0.3[y3]
Carvacrol	499-75-2		tr				tr[z1]; 0.07[z5]; 1.2[w]; 1.3[x]; 4.1[h]
Carvone	99-49-0						0.1[x]
(2*R*,5*E*)-Caryophyll-5-en-12-al							+[z19]; 0.3[x]
(2*S*,5*E*)-Caryophyll-5-en-12-al							+[z19]; 0.2[x]
β-Caryophyllene	87-44-5		1.2	0.6-1.3	4.8	2.3	5.1[y1]; 7.6[y3]; 9.0[z1]; 16.2[y2]; 21.3[y4]
Caryophyllene oxide	1139-30-6	0.1-0.7	1.5	0.5-0.8	1.0	2.3	2.2[k]; 3.8[z1]; 5.3[z18]; 8.5[y4]; 24.1[y1]
α-Cedrene	469-61-4	0.6-4.0			0.1		0.1[k]
Cedrol	77-53-2						0.2[k]
12-epi-Cedrol							0.1[k]
Cembrene A	31570-39-5						0.4[y3]
1,8-Cineole	470-82-6	tr-0.08	tr	0.1	0.8[f]	tr	0.1[x,z8]; 0.3[z1,z4]; 0.4[z]; 0.9[w]; 2.3[t]
Citral	5392-40-5			0.1-0.2			
Citronellol	106-22-9						0.4[z5]
α-Copaene	3856-25-5	0.2-7.8	0.5	0.1-0.2		1.7	1.3[k]; 1.5[j]; 2.1[z5]; 2.6[q]; 3.6[y4]; 4.2[y3]
β-Copaene	18252-44-3	tr-0.06					0.06[i]; 0.1[x]; 0.6[p]; 3.8[y2]
α-Copaen-8-ol (*cis*-)	58569-25-8						0.7[y3]
β-Copaen-4α-ol	124753-76-0					tr	
(+)-β-Costol							0.2[y3]
α-Cubebene	17699-14-8				0.4	0.3	0.05[i]; 0.1[k]; 0.2[y2]; 0.6[y3]; 1.6[y4];
β-Cubebene	13744-15-5		0.1	tr-0.1		0.5	tr-0.2[g]; 0.5[j]; 0.6[x]; 0.7[t]; 1.4[y4]; 4.5[z2]
Cubebol	23445-02-5						0.1[k]
epi-Cubebol	38230-60-3					tr	
Cubenol	21284-22-0				0.2		
8-Cumenol			0.2				
Cuparene	16982-00-6						0.2[a,z5]
α-Cuprenene	29621-78-1						0.1[x]
β-Cyclocitral	432-25-7						tr[k]
Cyercene 1	136669-16-4						0.4[y3]
p-Cymene	99-87-6	0.02-0.06		0-0.1		tr	tr[s]; 0.03[i]; 0.1[h,k]; 0.2[t,y]; 0.5[x]; 1.5[l]
β-Damascenone (*E*-)	23726-93-4		0.1		0.3		tr[k]
cis-Dehydroxylinalool oxide	73413-94-2						0.05[i]
trans-Dehydroxylinalool oxide							0.04[i]
Deutenyl curcumene			1.3				
3,6-Diacetoxy-2,6-di-methyl-1,7-octadiene							+[z19]
(2*R*,5*R*,6*R*)-2,12:5,6-Diepoxycaryophyllane	60444-80-6 (?)						+[z19]
Diethyl phthalate	84-66-2			tr-34.9			
cis-Dihydroagarofuran	150652-94-1						0.02[i]; 0.1[x]; 0.4[j]
Dihydro-α-agarofuran	20053-66-1						+[z19]
Dihydro-β-agarofuran	5956-09-2			tr-0.1			
Dihydrocarveol	38049-26-2				0.1		
cis-Dihydrocarvone	3792-53-8						0.01[a,z5]
Dihydro-8-cumenol			tr				
Dimethylbenzene butanal							0.9[n]
Dimethyl *o*-phthalate	131-11-3			tr-0.2			
(*E*)-2,6-Dimethyl-10-(*p*-tolyl)-undeca-2,6-diene	55968-43-9						0.3[k]

Table 5.25.2 Constituents identified in clary sage oils (*continued*)

Constituent	CAS	Percentage and range					
		A	B (30)	C (13)	D (9)	E (16)	F
Docosane	629-97-0						0.7[m]; 3.2[l]
1-Docosene	1599-67-3						0.7[j]
Dodecahydro-3α,6,6, 9α-tetramethyl(2,1β)-furan			0.1				
δ-Dodecalactone	713-95-1						0.1[x]
(E)-2-Dodecenal	20407-84-5						0.1[x]
α-Elemene	5951-67-7						1.8[y2]
β-Elemene	33880-83-0			tr-0.1	0.4	tr	tr-0.2[g]; 0.2[n]; 0.3[k]; 0.4[t,u]; 0.5[j]; 0.8[y4]
δ-Elemene	20307-84-0				tr		0.09[r]; 0.2[x]; 0.3[j,p]; 0.4[y4]
Elemol	639-99-6		0.2				0.2[j]
Epoxy allo-aromaden-drene	85760-81-2						2.1[z1]
(2R,5E)-2,12-Epoxy-caryophyll-5-ene							+[z19]
8,13-Epoxy-15,16-dinorlabd-12-ene	14752-13-7		0.2			3.0	
1,5-Epoxysalvial-4(14)-ene	88395-47-5					0.4	+[z19]; 0.1[n]; 0.6[y3]
Estrone	53-16-7						0.7[y3]
Ethyl benzoate	93-89-0			tr-0.1			
Eudesma-4(15),7-dien-1β-ol	119120-23-9						0.1[n]; 0.4[k]
Eudesmol	51317-08-9		tr				
α-Eudesmol	473-16-5		0.2	tr-0.3		0.8	0.1[a,z5]; 0.3[k,x]; 0.4[n]; 0.5[j]; 0.6[y4]
7-epi-α-Eudesmol	123123-38-6					tr	
β-Eudesmol	473-15-4		1.5	tr-0.3	0.7	1.3	0.8[j]; 0.9[y4]; 1.0[z2]; 1.1[y2]; 2.0[o]; 3.6[m]
γ-Eudesmol	1209-71-8						0.1[a,z5]; 1.2[y4]
10-epi-γ-Eudesmol	15051-81-7		tr				
Eugenol	97-53-0						0.9[z1]
α-Farnesene	502-61-4	0.1-0.8		tr-0.1			
(E,E)-α-Farnesene	502-61-4				0.3	tr	0-0.6[g]; 0.1[i,j]; 0.2[s,z5]; 0.3[x]
(Z,Z)-α-Farnesene	28973-99-1						0.2[p]
β-Farnesene	502-60-3		0.1[d]				
(E)-β-Farnesene	18794-84-8						tr[k]; 0.01[i]; 0.3[z2]
(Z)-β-Farnesene	28973-97-9				0.1		tr[k]
(E,E)-Farnesol	106-28-5						0.1[k]
(Z,Z)-Farnesol	16106-95-9						0.2[k]; 0.4[y3]
Farnesyl acetate	29548-30-9					0.5	
(5E,9E)-Farnesyl acetone	1117-52-8						0.1[x]
(5E,9Z)-Farnesyl acetone							0.3[x]
endo-Fenchol	512-13-0						0.2[k]
Fenchone	1195-79-5						0.1[k]
Formic acid	64-18-6		tr[c]				
Geranial	141-27-5		2.2		0.9	1.0	19.4[z14]
Geraniol	106-24-1	0.2-3.6	6.5	1.4-2.5	1.1	4.2	1.2[z18]; 4.6[z]; 5.3[v]; 6.1[z12]; 6.8[z1]; 7.7[k] 9.6[z15]; 24.5[z14]
Geranyl acetate	105-87-3	0.5-4.0	7.5	1.7-2.8	1.6	12.1	2.2-3.3[g]; 5.4[h]; 5.5[l]; 5.8[k]; 6.6[j12]; 8.4[w]; 36.8[z14]
Geranyl acetone	3796-70-1						0.1[k]
Geranyl formate	105-86-2	0.05-0.8	1.3	3.2-5.1[e]	0.3	0.5	tr-0.1[g]; 0.08[s]; 0.1[i,j]; 0.2[z2]
(E,E)-Geranyllinalool	1113-21-9						0.1[x]
(E,Z)-Geranyllinalool							0.2[x]
(Z,E)-Geranyllinalool							0.1[x]
Geranyl propionate	105-90-8						tr[s]
Germacrene	28028-64-0			0.3-1.3			0.2[a,z5]
Germacrene A	28387-44-2					tr	0.1[x]; 0.2[q] 5.8[y4]
Germacrene D	23986-74-5	0.4-5.4			7.6	2.6	8.7[m]; 9.8[z1]; 10.6[r]; 11.0[z11]; 11.4[x]; 12.7[y3]; 13.3[j]; 17.7[y4]; 19.8[q]; 24.7[y2]
Germacrene D-4-ol	198991-79-6					tr	
α-Guaiene	3691-12-1						0.2[a,z5]

Table 5.25.2 Constituents identified in clary sage oils (*continued*)

Constituent	CAS	Percentage and range					
		A	B (30)	C (13)	D (9)	E (16)	F
cis-β-Guaiene	372162-07-7						2[y4]
α-Gurjunene	489-40-7						0.3[z5]
β-Gurjunene	73464-47-8				0.1	tr	0.08[r]
γ-Gurjunene	22567-17-5						tr[r]; 0.4[z]; 2.5[y4]
Hedycaryol	21657-90-9						0.1[z2]
Hexadecanoic acid	57-10-3						1.4[y3]
Hexanol	111-27-3	0.05-0.1	0.1				+[z6]; 0.1[i]
Hexatriacontane	630-06-8						7.4[l]
2-Hexenal	505-57-7		0.2				
(*E*)-2-Hexenal	6728-26-3						0.04[i]
2-Hexenol			0.1				
(*E*)-2-Hexenol	928-95-0						0.2[i]
3-Hexenol			0.2				
(*Z*)-3-Hexenol	928-96-1	0.09-0.3				tr	0-tr[g]; 0.1[z3]; 0.3[i]
3-Hexen-1-ol	544-12-7			0-0.1			
Hexenyl acetate	28933-77-9		tr				
(*Z*)-3-Hexenyl acetate	3681-71-8	0.01-0.04					0.02[i]
(*Z*)-3-Hexenyl benzoate	25152-85-6					tr	
(*Z*)-3-Hexenyl butyrate	16491-36-4						0.03[i]
Humulane-1,6-dien-3-ol	915392-38-0						0.6[m]
α-Humulene	6753-98-6	0.02-0.2	0.1	0-0.1		tr	0.2[i,j]; 0.3[y4]; 0.4[q]; 0.9[y3]; 1.3[y2]; 1.4[z1]
β-Humulene	116-04-1						7.1[z15]
Humulene epoxide II	19888-34-7						0.1[k]
Humulene epoxide III	21624-36-2						0.2[k]
14-Hydroxy-9-epi-(*E*)-caryophyllene	244226-09-3						0.3[y3]
Hydroxycitronellal	107-75-5		tr				
endo-8-Hydroxycyclo-isolongifolene			0.1				
3-Hydroxydodecanoic acid methyl ester	76835-66-0				0.3		
Hydroxylinalool	256418-61-8						1.2
β-Ionone	79-77-6						0.2[k]
Isoabienol	10207-79-1						tr[k]
Isoamyl alcohol	123-51-3						0.01[i]
Isoaromadendrene epoxide	499134-59-7						0.5[y3]
Isobornyl acetate	125-12-2						0.06[z5]
Isobutyric acid	79-31-2						3.3[y3]
Isocaryophyllene	118-65-0						+[z19]
Isocaryophyllene oxide							+[z19]
Isogermacrene D	317819-80-0						0.1[k]
Isospathulenol	88395-46-4						+[z19]; 0.1[n]
Isoterpinolene	586-63-0						0.4[t]
Labda-7,14,-dien-13-ol	40185-30-6						0.1[k]
Lanceol acetate	199273-99-9						0.1[a,z5]
Lavandulol	498-16-8						0.1[d]
Limonene	138-86-3	0.1-0.6	0.2	0.1-0.3	0.8[f]	0.6	tr-0.3[g]; 1.3[s]; 1.6[t]; 1.7[z1]; 2.1[z2]; 3.1[y]
Linalool	78-70-6	18.6-31.0	17.2	17.0-28.8	33.0	30.4	14.5-29.8[g]; 30.0[w]; 31.9[m]; 36.0[h]; 36.5[v]; 39.4[z12]; 40.2[y]; 42.2[z15]; 42.3[k]
cis-Linalool oxide	11063-77-7	0-0.3		tr-0.2		tr	tr[k]; 0.03[i]; <0.1[z14]; 0.8[y]
cis-Linalool oxide, furanoid	5989-33-3		0.2		tr		0.4[z7]
trans-Linalool oxide	11063-78-8	tr-0.5		0.1		tr	tr[k]; 0.03[i]
trans-Linalool oxide, furanoid	34995-77-2		1.1		tr		0.4[z7]
Linalyl acetate	115-95-7	31.2-64.2	14.3	29.5-51.6	16.9	31.1	45.7-60.8[g]; 60.0[s]; 60.9[z3]; 66.7[z8]; 67.4[z16]; 67.5[p]; 69.9[z7]; 71.6[u]; 81.1[z9]
Linalyl formate	115-99-1	tr-0.3	tr	0.2-0.3			0.1[t,w]; 0.3[i,j,p]
Longiborneol	465-24-7						0.1[a,z5]
Longifolene	475-20-7						0.06[z5]

Table 5.25.2 Constituents identified in clary sage oils (*continued*)

Constituent	CAS	Percentage and range					
		A	B (30)	C (13)	D (9)	E (16)	F
trans-Longipinocarveol	889109-69-7						0.6[y3]
Longipinocarvone	65556-52-7						0.3[y3]
Manool	596-85-0		2.5		0.6	1.2	0.07[z5]; 0.2[x]; 0.3[k]; 1.0[j]; 1.2[o]; 1.8[m]
13-epi-Manool	138-62-6						0.4[k]; 1.3[m]; 1.8[o]; 2.3[n]; 5.7[l]
Manoyl oxide	596-84-9		0.1		tr	1.1	0.6[z2,z3,k]; 0.8[m]; 2.4[n]; 2.8[l]
13-epi-Manoyl oxide	1227-93-6		0.2		1.1	0.6	0.2[k]; 0.3[m]; 1.1[n]
p-Menth-1-en-9-al	29548-14-9						tr[k]
o-Menth-8-ene	15193-25-6						0.6[y3]
cis-p-Menth-2-en-1-ol	29803-82-5						tr[j]
Menthol	89-78-1			tr-0.1			
Menthone	89-80-5						0.1[x]
1,4-Methanoazulen-7-one							0.5[y3]
1-Methoxyhexane-3-thiol							+[z6]
Methylcarvacrol	6379-73-3		tr				
Methyl ethyl phthalate	34006-77-4			tr-0.3			
Methyl eugenol	93-15-2						tr[x]
Methylheptenone	409-02-9		tr				
2-Methyl-2-vinyl-4-iso-propenyltetrahydro-furan			tr				
Mintoxide							tr[k]
Mintsulfide	72445-42-2						+[z19]
cis-Muurola-4(14),5-diene	157477-72-0				tr		
Muurolene	69671-15-4		tr				
α-Muurolene	10208-80-7				0.4	tr	0.1[r]; 3.3[w]
γ-Muurolene	30021-74-0						0.1[r]; 0.9[z5]
Muurolol	119757-72-1		tr				
τ-Muurolol (epi-α-)	19912-62-0						1.5[y3]
Myrcene	123-35-3	0.2-1.5	0.3	0.2-0.6	1.6	1.8	2.6[p]; 3.0[w]; 3.3[z5]; 3.4[k]; 7.3[y]; 8.4[z2,z18]
Myrtenol	515-00-4						tr[a,z5,k]
1H-Naphtho[2.1.6]-pyran							8.6[y1]
Neopentylidene-cyclohexane	39546-80-0						0.3[y3]
Neral	106-26-3		1.5				1.4[z15]; 1.9[z1]; 2.1[y]; 4.5[h]; 11.3[z14]
Nerol	106-25-2	0.06-1.3	5.5	0.6-1.0	0.7	1.2	1.6[z]; 2.0[v]; 2.1[z12]; 2.5[k]; 5.1[nz15]; 7.4[z14]
Nerolidol			tr				
(*Z*)-Nerolidol	3790-78-1				tr		
Nerol oxide	1786-08-9		tr	tr			0.03[i]
Neryl acetate	141-12-8	0.3-1.4	5.2	1.0-1.6	1.2	7.8	3.0[k]; 3.2[h]; 3.6[z18]; 3.7[o]; 4.1[z2]; 4.7[m]
Neryl formate	2142-94-1						tr[j]; 0.06[i]; 0.1[k]
(*2E,6Z*)-Nonadienal	557-48-2						tr[j]
Nonanal	124-19-6		tr				tr[j]
12-Nor-caryophyll-5-en-2-one							0.3[y3]
(-)-(*E*)-12-Norcaryo-phyll-5-en-2-one							+[z19]
12-Norcyercene-B							0.5[y3]
(*E*)-Ocimene	27400-72-2						1.2[p]; 1.3[q]; 1.5[r]
(*Z*)-Ocimene	27400-71-1						0.3[y3]; 0.6[q]; 0.9[r]
(*Z*)-α-Ocimene	6874-44-8						0.5[m]; 1.0[o]
β-Ocimene	13877-91-3						0.7[z1]
(*E*)-β-Ocimene	3779-61-1	0.3-1.3	0.2	tr-0.4	0.9	1.3	1.8[o]; 3.0[z5]; 3.2[w]; 4.4[z2]; 4.8[y]; 5.7[z18]
(*Z*)-β-Ocimene	3338-55-4	0.2-4.8	0.1	0.1-0.2	0.6	0.8	1.0[k]; 1.7[z5]; 1.8[w]; 1.9[z2]; 2.0[y]; 2.6[z18]
allo-Ocimene	673-84-7						tr[j]; 0.4[p]; 0.8[z2]
3-Octanone	106-68-3						0.02[i]
1-Octen-3-ol	3391-86-4	0.03-0.08	0.1[b]	0-tr			tr[g]; 0.07[i]
1-Octen-3-yl acetate	2442-10-6						tr[j]
β-Oplopenone	28305-60-4					tr	
Perillene	539-52-6		0.1				

Table 5.25.2 Constituents identified in clary sage oils (*continued*)

Constituent	CAS	A	B (30)	C (13)	D (9)	E (16)	F
α-Phellandrene	99-83-2						0.1[k,x,z5]
β-Phellandrene	555-10-2	0.08-0.2					0.8[f]
1-Phenyl-2,4-pentadiyne							1.2[x]
Phytol	7541-49-3						0.2-0.3[g]; 1.1[m]
α-Pinene	80-56-8	0.01-0.3	0.1		0.3	tr	0.04[s]; 0.1[p,q,r]; 0.3[u,z7]; 0.8[z1]; 4.6[t]
β-Pinene	127-91-3	0.02-0.3	0.1	0-tr	0.4	tr	0.1[p,x]; 0.2[j,k,q,r]; 0.3[z7]; 0.6[z1]; 0.9[t]
Pinocamphone	547-60-4						0.1[k]
Piperitone	89-81-6						tr[k]
Pulegone	89-82-7						0.4[x]
α-Pyronene	514-94-3						0.09[a,z5]
Sabinene	3387-41-5	0.01-0.2			tr		tr[j,r]; 0.01[i]; 0.08[y4]; 0.2[q]
trans-Sabinene hydrate	17699-16-0						tr[k]
Sabinene hydrate acetate	87553-42-2						2.2[a,z5]
Salviadienol	1064085-05-7						0.2[k]
Salvial-4(14)-en-1-one	73809-82-2					0.4	+[z19]; 0.1[n]; 0.2[q]; 0.4[k]; 0.6[y3]
(Z)-α-Santalol	115-71-9						0.1[a,z5]
Santolina alcohol	35671-15-9						0.1[x]
Sclareol	515-03-7	0.08-1.0	5.2		0.5	5.6	1.3-2.3[z15]; 1.8[z15]; 2.7[q]; 5.3[z3]; 5.9[z13]; 9.3[j]; 11.0[z1]; 11.5[y1]; 15.4[m]; 15.7[n]
Sclareol oxide	5153-92-4			tr-0.2			0.5[j]; 0.6[z3]; 0.8[m]; 1.4[o]; 2.5[n]; 4[y3]
β-Selinene	17066-67-0						0.1[k]; 0.2[j,x]
γ-*cis*-Sesquicycloge-raniol							0.4[y3]
Spathulenol	6750-60-3	0.2-1.0	0.2	0.1-0.3	0.1	0.7	1.9[y2]; 2.0[z18]; 5.0[y4]; 6.9[y3]; 11.4[y1]
α-Terpinene	99-86-5			0-0.1		tr	0-tr[g]; tr[k]; 0.01[i]; 0.1[x,z5]; 0.3[q]; 6.6[z10]
γ-Terpinene	99-85-4				tr	tr	0-0.1[g]; tr[j]; 0.09[z5]; 0.2[k]; 0.3[x]
Terpinen-4-ol	562-74-3	0.02-0.1	0.2		tr		0-tr[g]; 0.04[i]; 0.06[r]; 0.1[x]; 0.2[k]; 0.9[z1]
Terpinen-4-yl acetate	4821-04-9						0.4[a,z5]
α-Terpineol	98-55-5	0.6-5.1	15.1	3.2-5.1[e]	5.6	7.6	1.8-5.3[g]; 11.2[z]; 12.0[h]; 14.3[k]; 17.2[z12]; 47.4[z20]
β-Terpineol	138-87-4						1.2[t]
trans-β-Terpineol	7299-41-4						1.3[z15]
(Z)-β-Terpineol	7299-40-3				tr		
γ-Terpineol	586-81-2	0.05-0.2					0.05[i]; 0.9[t]
δ-Terpineol	7299-42-5	0-0.2					
Terpinolene	586-62-9	0.09-0.2	tr	0-tr	0.4	0.6	0.2[p]; 0.3[z5]; 0.4[j,m]; 0.6[k]; 0.9[z2]; 1.0[z18]
Terpinyl acetate	8007-35-0	0.02-0.2					tr[s]; 0.03[i]; <0.1[l]
α-Terpinyl acetate	80-26-2						<0.1[y]; 0.2[j,t]; 0.4[z7]; 2.0[a,z5]; 22.1[z20]
Tetracosane	646-31-1						4.7
α-Thujene	2867-05-2						0.1[k]
α-Thujone	546-80-5						0.4[x]; 0.5[z]
β-Thujone	471-15-8						0.1[z]; 0.5[x]
Thymol	89-83-8		tr				0.4[z5]; 1.5[x]; 1.6[h]
Torilenol	84071-85-2						0.1[k]
Torreyol	19435-97-3				0.4		
Tricosane	638-67-5						4.0[l]
Tridecane	629-50-5						0.7[z1]
Valencene	4630-07-3				2.8	0.5	0-0.3[g]; 0.07[i]; 0.4[n]; 0.8[z5]; 1.0[m]
Valeranone	55528-90-0						0.8[r]
Valerianol (kusunol)	20489-45-6		0.1				
β-Vetivenene	27840-40-0						0.7[y3]
Viridiflorene (ledene)	21747-46-6			tr			1.6[r]
Vulgarol B							0.3[y3]
Ylangene							1.2[q]
α-Ylangene	14912-44-8	0-0.3	0.1[b]			tr	0.5[i]
β-Ylangene	20479-06-5						1.8[w]
Zingiberene	495-60-3						0.3[z2]

Table 5.25.2 Constituents identified in clary sage oils (*continued*)

A thirty-four clary sage essential oil samples from France, Bulgaria, Hungary, Egypt and Russia, analyzed between 1998 and 2013 (E. Schmidt, unpublished data)
B one lab-hydrodistilled oil from flowers and leaves of *S. sclarea* collected in Greece in the wild (ref. 30)
C four commercial clary sage oils from China; lowest and highest concentrations given (ref. 13)
D one lab-hydrodistilled oil from *S. sclarea* in the flowering stage collected in the wild in Spain (ref. 9)
E two lab-hydrodistilled oils from the flowering aerial parts of sage growing wild in Greece at two locations; highest concentrations given (ref. 16)
F data from other studies (indicated with superscript letters); highest concentrations found in any study reviewed here given; when two or more oils were investigated, only the highest concentrations are mentioned, unless indicated otherwise

[a] incorrect identity based on GC elution order (ref. 44); [b] α-ylangene and 1-octen-3-ol combined; [c] formic acid and benzaldehyde combined; [d] β-farnesene and lavandulol combined; [e] α-terpineol and geranyl formate combined; [f] β-phellandrene, 1,8-cineole and limonene combined; [g] three lab-hydrodistilled oils from flowering tops of cultivated *S. sclarea* from different bioclimatic zones in India; lowest and highest concentrations given (ref. 7); [h] unknown number of flower oils produced by lab-hydrodistillation of the inflorescences from wild and cultivated clary sage plants harvested in a four year period in Uzbekistan (ref. 3); [i] one commercial oil sample purchased in Bulgaria, analyzed on two different columns (ref. 37); [j] two lab-hydrodistilled oils from the aerial parts of clary sage plants in full bloom, one obtained from a wild population in Iran and the other from plants cultivated from the seeds of this wild population (ref. 38); [k] two lab-hydrodistilled oils from the aerial parts of in vivo and in vitro cultivated *S. sclarea* in Poland; highest concentrations given; the oils were obtained by hydrodistillation for 5 hours, which explains why the concentration of linalyl acetate was very low (1.1%) and that of linalool very high (42%) (ref. 12); [l] one laboratory steam-distilled oil from flowering tops from Hungary (ref. 2); [m] four lab-hydrodistilled oils from clary sage plants cultivated in India under different shade levels (ref. 1); [n] one lab-hydrodistilled oil from flowering *S. sclarea* cultivated in the Slovak republic (ref. 6); [o] six lab-hydrodistilled oils from flowering plants cultivated in the northwestern Himalayas, India, and transplanted at different times (ref. 10); [p] one steam-distilled commercial clary sage oil sample from south France (ref. 11); [q] four laboratory steam-distilled oils from the aerial parts of clary sage plants grown in Uruguay from European seeds, harvested in two years during full flowering and at the beginning of seed ripening (ref. 18); [r] one lab-hydrodistilled oil from the inflorescences of *S. sclarea* cultivated from the seeds of a wild biotype found in Sicily, Italy (ref. 21); [s] one commercial oil from Russia, obtained from the 'leaves of the flowering top' (ref. 28, data also reported in ref. 5); [t] one (commercial?) oil obtained from a medicinal plant research institute in Belgrade, Serbia (ref. 29); [u] one commercial clary sage oil (ref. 32); [v] unknown number of clary sage oils from plants on different spacing and fertilizer level schemes in India (ref. 33); [w] one lab-hydrodistilled oil from flowering *S. sclarea* found growing wild in Iran (ref. 34); [x] one lab-hydrodistilled oil from the aerial parts of flowering *S. sclarea* growing wild in Tajikistan (ref. 35); [y] one commercial clary sage oil and one lab-hydrodistilled oil from flowering spikes of *S. sclarea* from India (ref. 36); [y1] one lab-hydrodistilled oil from clary sage aerial parts collected in Turkey with an extremely atypical composition (ref. 53; only summary accessed); [y2] one lab-hydrodistilled oil from the flowering aerial parts of *S. sclarea* growing wild in Turkey (ref. 52); [y3] one lab-hydrodistilled oil from the aerial parts of flowering clary sage collected in their natural habitat in Iran (ref. 51); [y4] four lab-hydrodistilled oils from the aerial parts of flowering *S. sclarea* accessions collected in the wild in Iran at 4 different locations (ref. 50); [z] one clary sage flower oil from either Spain or Brazil (ref. 39); [z1] one lab-hydrodistilled oil from a population of clary sage growing wild in Iran (ref. 40); [z2] one lab-hydrodistilled oil from the flowering tops of clary sage growing wild in Croatia (ref. 42); [z3] one clary sage oil from India (ref. 43); [z4] one clary sage oil of unknown origin (ref. 45); [z5] one lab-prepared clary sage oil and one commercial oil from Russia (ref. 46); the author of ref. 44 found many mistakes in this study; [z6] one oil sample produced in Switzerland (ref. 47); [z7] one clary sage oil sample, probably commercial (ref. 48); [z8] one commercial oil sample purchased from a Brazilian company (ref. 4); [z9] one lab-hydrodistilled oil from the flowering aerial parts of wild clary sage collected in South France (ref. 8); [z10] one commercial clary sage oil from Hungary (ref. 14); [z11] one commercial oil obtained from a German company (ref. 15); [z12] six lab-hydrodistilled oils from the flowers of *S. sclarea* plants cultivated in India under different growth regulator schemes (ref. 17); [z13] three lab-hydrodistilled oils from inflorescences and leaves of clary sage grown in Austria; only the mean values were shown (ref. 19); [z14] two oils obtained by hydrodistillation of *S. sclarea* cultivated in Israel from seeds of wild populations in Upper Galilee and harvested early and late in the flowering stage; very atypical composition with low linalool and linalyl acetate levels and with geranyl acetate, geraniol, and geranial as the dominant ingredients (ref. 22); [z15] two lab-hydrodistilled clary sage flower oils from plants cultivated in India and harvested in two years (ref. 23); [z16] one laboratory steam-distilled oil from the inflorescences of *S. sclarea* cultivated in Serbia (ref. 24); [z17] ten commercial clary sage oils investigated for the presence of (*E*)-anethole, which was found in 5/10 oils in concentrations ranging from 0.16 to 30 mg/ml oil (ref. 25); [z18] one lab-hydrodistilled oil from the flowering tops of *S. sclarea* growing wild in Central Italy (ref. 26); [z19] one commercial oil, probably from France (ref. 20); [z20] one lab-hydrodistilled oil from wild clary sage harvested in Sicily, Italy; plant parts used unknown) (ref. 49)

tr: trace (in columns B, F[g] and F[j]: <0.1; in columns C, F[k] and F[s]: <0.01; in columns D and E: <0.05); + present in the oil investigated, but quantity not stated

Table 5.25.3 Major constituents of clary sage (flower) oils

Constituent	CAS	Percentages and range					
		A	B (30)	C (13)	D (9)	E (16)	F
Linalyl acetate	115-95-7	31.2-64.2	14.3	29.5-51.6	16.9	31.1	67.4[z16]; 67.5[p]; 69.9[z7]; 71.6[u]; 81.1[z9]
α-Terpineol	98-55-5	0.6-5.1	15.1	3.2-5.1[e]	5.6	7.6	11.2[z]; 12.0[h]; 14.3[k]; 17.2[z12]; 47.4[z20]
Linalool	78-70-6	18.6-31.0	17.2	17.0-28.8	33.0	30.4	36.5[v]; 39.4[z12]; 40.2[y]; 42.2[z15]; 42.3[k]
Geranyl acetate	105-87-3	0.5-4.0	7.5	1.7-2.8	1.6	12.1	5.4[h]; 5.5[l]; 5.8[k]; 6.6[j12]; 8.4[w]; 36.8[z14]
Germacrene D	23986-74-5	0.4-5.4			7.6	2.6	12.7[y3]; 13.3[j]; 17.7[y4]; 19.8[q]; 24.7[y2]
Geraniol	106-24-1	0.2-3.6	6.5	1.4-2.5	1.1	4.2	6.1[z12]; 6.8[z1]; 7.7[k]; 9.6[z15]; 24.5[z14]
β-Caryophyllene	87-44-5		1.2	0.6-1.3	4.8	2.3	5.1[y1]; 7.6[y3]; 9.0[z1]; 16.2[y2]; 21.3[y4]
Sclareol	515-03-7	0.08-1.0	5.2		0.5	5.6	9.3[j]; 11.0[z1]; 11.5[y1]; 15.4[m]; 15.7[n]
Nerol	106-25-2	0.06-1.3	5.5	0.6-1.0	0.7	1.2	1.6[z]; 2.0[v]; 2.1[z12]; 2.5[k]; 5.1[nz15]; 7.4[z14]

LEGEND: SEE UNDER TABLE 5.25.2

CONTACT ALLERGY/ALLERGIC CONTACT DERMATITIS

General
Contact allergy to/allergic contact dermatitis from clary sage oil has been reported in a few publications only. In a group of consecutive patients suspected of contact dermatitis, the prevalence rate of positive patch test reactions was 0.5%, but relevance data were lacking. All four case reports concerned occupational allergic contact dermatitis in aroma-/massage-/complementary therapists and perfume factory workers. In one case report, linalool and geraniol may have been allergens in clary sage oil (54).

Testing in groups of patients

Routine testing
Two hundred dermatitis patients from Poland were tested with clary sage 2% in petrolatum and one (0.5%) reacted; relevance data were not provided (55).

Testing in groups of selected patients
In a group of 51 patients allergic to *Myroxylon pereirae* resin (balsam of Peru) and/or turpentine and/or wood tar and/or colophony and tested with clary sage oil 2% in petrolatum, four (4.9%) had a positive patch test; relevance data were not provided (55). A group of 86 patients from Poland previously reacting to the fragrance mix was tested with clary sage oil and four (4.6%) had a positive patch test reaction; relevance data were not provided (58).

Case reports
Two aromatherapists had contact dermatitis (one occupational) with allergies to multiple essential oils used at their work, including clary sage oil. Both patients also reacted to geraniol, α-pinene, the fragrance mix and various other fragrance materials. In addition, one proved to be allergic to linalool and linalyl acetate, the other to caryophyllene; α-pinene, linalool, geraniol and caryophyllene were demonstrated by GC-MS in clary sage oils (54). Linalool and to a lesser degree geraniol may be important components of clary sage oil, with maximum concentrations of 31.0% and 3.6% resp. found in clary sage essential oils (Table 5.25.2, column A). Occupational contact dermatitis developed in a massage therapist with allergies to various essential oils, including clary sage oil (56). A 'complementary therapist' developed occupational contact dermatitis from a multitude of essential oils used at work, including clary sage oil (57). Clary sage oil caused dermatitis in an unknown number of perfumery workers (60, article not read).

Positive patch tests (relevance unknown, uncertain or not stated)
Three patients in a group of 16 (19%) known to be allergic to propolis and *Myroxylon pereirae* had positive patch test reactions to clary sage oil; relevance data were not provided (59).

LITERATURE
1 Kumar R, Sharma S, Pathania V. Effect of shading and plant density on growth, yield and oil composition of clary sage (*Salvia sclarea* L.) in north western Himalaya. J Essent Oil Res 2013;25:23-32

2 Ronyai E, Simandi B, Lemberkovics E, Veress T, Patiaka D. Comparison of the volatile composition of clary sage oil obtained by hydrodistillation and supercritical fluid extraction. J Essent Oil Res 1999;11:69-71

3 Dzumayev KhK, Tsibulskaya IA, Zenkevich IG, Tkachenko KG, Satzyperova IF. Essential oils of *Salvia sclarea* L. produced from plants grown in southern Uzbekistan. J Essent Oil Res 1995;7:597-604

4 Murbach Teles Andrade BF, Nunes Barbosa L, da Silva Probst L, Fernandes Júnior A. Antimicrobial activity of essential oils. J Essent Oil Res 2014;26:34-40

5 Jirovetz L, Wlcek K, Buchbauer G, Gochev V, Girova T, Stoyanova A, Schmidt E, Geissler M. Antifungal activities of essential oils of *Salvia lavandulifolia*, *Salvia officinalis* and *Salvia sclarea* against various pathogenic *Candida* species. J Essent Oil Bear Plants 2007;10:430-439. Data also presented in ref. 28

6 Farkas P, Hollá M, Tekel J, Mellen S, Vaverková S. Composition of the essential oils from the flowers and leaves of *Salvia sclarea* L. (Lamiaceae) cultivated in Slovak Republic. J Essent Oil Res 2005;17:141-144

7 Yadav A, Chanotiya CS, Singh AK. Terpenoid compositions and enantio-differentiation of linalool and sclareol in *Salvia sclarea* L. from three different climatic regions in India. J Essent Oil Res 2010;22:589-592

8 Foray L, Bertrand C, Pinguet F, Soulier M, Astre C, Marion C, et al. In vitro cytotoxic activity of three essential oils from *Salvia* species. J Essent Oil Res 1999;11:522-526

9 Torres ME, Velasco-Negueruela A, Pérez-Alonso MJ, Gil Pinilla M. Volatile constituents of two *Salvia* species grown wild in Spain. J Essent Oil Res 1997;9:27-33

10 Kumar R, Sharma S, Singh B. Evaluation of transplanting time effect on characteristic growth, essential oil and its composition in clary sage (*Salvia sclarea* L.) in North-Western Himalayas. J Essent Oil Bear Plants 2011;14:260-265

11 Tognolini M, Barocelli E, Ballabeni V, Bruni R, Bianchi A, Chiavarini M, Impicciatore M. Comparative screening of plant essential oils: Phenylpropanoid moiety as basic core for antiplatelet activity. Life Sciences 2006;78:1419-1432

12 Kuźma L, Kalemba D, Różalski M, Różalska B, Więckowska-Szakiel M, Krajewska U, Wysokińska H. Chemical composition and biological activities of essential oil from *Salvia sclarea* plants regenerated *in vitro*. Molecules 2009;14:1438-1447

13 Cai J, Lin P, Zhu X, Su Q. Comparative analysis of clary sage (*S. sclarea* L.) oil volatiles by GC–FTIR and GC–MS. Food Chem 2006;99:401-407

14 Tserennadmid R, Takó M, Galgóczy L, Papp T, Pesti M, Vágvölgyi C, et al. Anti yeast activities of some essential oils in growth medium, fruit juices and milk. Int J Food Microbiol 2011;44:480-486

15 Schwiertz A, Duttke C, Hild J, Muller HJ. *In vitro* activity of essential oils on microorganisms isolated from vaginal infections. Int J Aromather 2006;16:169-174

16 Pitarokili D, Couladis M, Petsikos-Panayotarou N, Tzakou O. Composition and antifungal activity on soil-borne pathogens of the essential oil of *Salvia sclarea* from Greece. J Agric Food Chem 2002;50:6688-6691

17 Singh V, Sood R, Ramesh K, Singh B. Effects of growth regulator application on growth, flower, oil yield, and quality of clary sage (*Salvia sclarea* L.). J Herbs Spices Med Plants 2008;14:29-36

18 Lorenzo D, Paz D, Davies P, Villamil J, Vila R, Cañigueral S, Dellacassa E. Characterization and enantiomeric distribution of some terpenes in the essential oil of a Uruguayan biotype of *Salvia sclarea* L. Flavour Fragr J 2004;19:303-307

19 Schmiderer C, Grassi P, Novak J, Weber M, Franz C. Diversity of essential oil glands of clary sage (*Salvia sclarea* L., Lamiaceae). Plant Biol 2008;10:433-440

20 Maurer B, Hauser A. New sesquiterpenoids from clary sage oil (*Salvia sclarea* L.). Helvetica Chimica Acta 1983;66:2223-2235

21 Carrubba A, la Torre R, Piccaglia R, Marotti M. Characterization of an Italian biotype of clary sage (*Salvia sclarea* L.) grown in a semi-arid Mediterranean environment. Flavour Fragr J 2002;17:191-194

22 Elnir O, Ravid U, Putievsky E, Dudai N, Ladizinsky G. The chemical composition of two clary sage chemotypes and their hybrids. Flavour Fragr J 1991;6:153-155

23 Lattoo SK, Dhar RS, Dhar AK, Sharma PR, Agarwal SG. Dynamics of essential oil biosynthesis in relation to inflorescence and glandular ontogeny in *Salvia sclarea*. Flavour Fragr J 2006;21:817-821

24 Pesic PZ, Bankovic VM. Investigation on the essential oil of cultivated *Salvia sclarea* L. Flavour Fragr J 2003;18:228-230

25 Fiori J, Hudaib M, Valgimigli L, Gabbanini S, Cavrini V. Determination of *trans*-anethole in *Salvia sclarea* essential oil by liquid chromatography and GC-MS. J Sep Sci 2002;25:703-709

26 Fraternale D, Giamperi L, Bucchini A, Ricci D, Epifano F, Genovese S, Curini M. Composition and antifungal activity of essential oil of *Salvia sclarea* from Italy. Chem Nat Comp 2005;41:604-606

27 Caniard A, Zerbe P, Legrand S, Cohade A, Valot N, Magnard J-L, Bohlmann J, Legendre L. Discovery and functional characterization of two diterpene synthases for sclareol biosynthesis in *Salvia sclarea* (L.) and their relevance for perfume manufacture. BMC Plant Biol 2012;12:119 (13 pages)

28 Jirovetz L, Buchbauer G, Denkova Z, Slavchev A, Stoyanova A, Schmidt E. Chemical composition, antimicrobial activities and odor descriptions of various *Salvia* sp. and *Thuja* sp. essential oils. Ernährung/Nutrition 2006;30:152-159. Data also presented in ref. 5

29 Dzamic A, Sokovic M, Ristic M, Grujic-Jovanovic S, Vukojevic J, Marin PD. Chemical composition and antifungal activity of *Salvia sclarea* (Lamiaceae) essential oil. Arch Biol Sci Belgrade 2008;60:233-237

30 Souleles C, Argyriadou N. Constituents of the essential oil of *Salvia sclarea* growing wild in Greece. Int J Pharmacogn 1997;35:218-220

31 Lawrence BM. Progress in essential oils. Perfum Flavor 2012;31(May):54-60

32 Williams DG. The chemistry of essential oils, 2nd Ed. Port Washington, NY, USA: Micelle Press, 2008:174-175. Data cited in ref. 31

33 Singh V, Sood R, Ramesh K, Singh B. Effect of plant spacing and organic manures on growth, yield and quality of clary sage (*Salvia sclarea* L.). Indian Perfum 2008;52:54-56. Data cited in ref. 31

34 Saharkhiz MJ, Ghani A, Hassanzadeh-Khayyat M. Changes in essential oil content and composition of clary sage (*Salvia sclarea*) aerial parts during different phenological stages. Med Arom Plant Sci Biotech 2009;3(special issue):90-93. Data cited in ref. 31

35 Sharopov FS, Setzer WN. The essential oil of *Salvia sclarea* L. from Tajikistan. Rec Nat Prod 2012;6:75-79

36 Verma RS. Chemical investigation of decanted and hydrophilic fractions of *Salvia sclarea* essential oil. Asian J Tradit Med 2010;5:102-108

37 Hristova Y, Gochev V, Wanner J, Jirovetz L, Schmidt E, Girova T, Kuzmanov A. Chemical composition and antifungal activity of essential oil of *Salvia sclarea* L. from Bulgaria against clinical isolates of *Candida* species. J BioSci Biotech 2013;2:39-44

38 Nasermoadeli S, Rowshan V. Comparison of *Salvia sclarea* L. essential oil components in wild and field population. Intl J Agri Crop Sci 2013;5:828-831

39 Pierozan MK, Pauletti GF, Rorota L, Atti dos Santos AC, Lerin LA, Di Luccio M, et al. Chemical characterization and antimicrobial activity of essential oils of *Salvia* L. species. Ciênc Tecnol Aliment Campinas 2009;29:764-770

40 Paknejadi M, Foroohi F, Yousefzadi M. Antimicrobial activities of the essential oils of five *Salvia* species from Tehran province, Iran. J Paramed Sci (JPS) 2012;3:12-18

41 Lawrence BM. Progress in essential oils. Perfum Flavor 2008;33(1):27-? (last page unknown)

42 Mastelic J, Jerkovic I. Application of co-distillation with superheated pentane vapour to the isolation of unstable essential oils. Flavour Fragr J 2003;18:521-526

43 Yaseen M, Sattar A, Naqvi AA, Khanuja SPS. Clary sage (*Salvia sclarea*): high value essential oil crop in north Indian plains. Indian Perfum 2006;50:35-38. Data cited in ref. 41

44 Lawrence BM. Progress in essential oils. Perfum Flavor 2004;29(1):44-? (last page unknown)

45 Zani F, Masssimo G, Benvenuti S, Bianchi A, Albasini A, Melegari M, et al. Studies on the genotoxic properties of essential oils with *Bacillus subtilis* rec-assay and *Salmonella*/Microsome reversion assay. Planta Med 1991;57:237-241. Data cited in ref. 44

46 Hudaib M, Bellardi MG, Rubies-Autonell C, Fiori J, Cavrini V. Chromatographic (GC-MS, HPLC) and virological evaluation of *Salvia sclarea* infected by BBWV-1. Il Farmaco 2001;56:219-227. Data cited in ref. 44

47 van der Waal M, Niclass Y, Snowden RL, Bernadinelli G, Escher S. 1-Methoxyhexane-3-thiol, a powerful odorant of clary sage (*Salvia sclarea*). Helv Chim Acta 2002;85:1246-1260. Data cited in ref. 44

48 Kubeczka K-H, Formacek V. Essential oils analysis by capillary gas chromatography and carbon-13 NMR Spectroscopy, 2nd edition. New York, USA: John Wiley and Sons, 2002:315-319. Data cited in ref 44

49 Peana AT, Moretti MDL, Juliano C. Chemical composition and antimicrobial action of the essential oils from *Salvia desoleana* and *S. sclarea*. Planta Med 1999;65:752-754

50 Rajabi Z, Ebrahimi M, Farajpour M, Mirza M, Ramshini H. Compositions and yield variation of essential oils among and within nine *Salvia* species from various areas of Iran. Ind Crops Prod 2014;61:233-239

51 Salimpour S, Mazooji A, Darzikolaei SA. Chemotaxonomy of six *Salvia* species using essential oil composition markers. J Med Plants Res 2011;5:1795-1805

52 Öğütçü A, Sökmen A, Sökmen M, Polissiou M, Serkedjieva J, Daferera D, et al. Bioactivities of the various extracts and essential oils of *Salvia limbata* C.A.Mey. and *Salvia sclarea* L. Turk J Biol 2008;32:181-192

53 Yuce E, Yildirim N, Yildirim NC, Paksoy MY, Bagci E. Essential oil composition, antioxidant and antifungal activities of *Salvia sclarea* L. from Munzur Valley in Tunceli, Turkey. Cell Mol Biol (Noisy-le-grand) 2014;60:1-5

54 Dharmagunawardena B, Takwale A, Sanders KJ, Cannan S, Roger A, Ilchyshyn A. Gas chromatography: an investigative tool in multiple allergies to essential oils. Contact Dermatitis 2002;47:288-292

55 Rudzki E, Grzywa Z, Bruo WS. Sensitivity to 35 essential oils. Contact Dermatitis 1976;2:196-200

56 Cockayne SE, Gawkrodger DJ. Occupational contact dermatitis in an aromatherapist. Contact Dermatitis 1997;37:306-307

57 Newsham J, Rai S, Williams JDL. Two cases of allergic contact dermatitis to neroli oil. Br J Dermatol 2011;165(Suppl.1):76

58 Rudzki E, Grzywa Z. Allergy to perfume mixture. Contact Dermatitis 1986;15:115-116

59 Rudzki E, Grzywa Z. Dermatitis from propolis. Contact Dermatitis 1983;9:40-45

60 Gutman SG, Somov BA. Allergic reactions caused by components of perfumery preparations. Vestnik Dermatologii i Venerologii 1968;42:62 (in Russian)

Chapter 5.26 CLOVE BUD OIL

There are three types of clove essential oils: those made from the buds (clove bud oil), leaves (clove leaf oil) and from the stems (clove stem oil) of *Syzygium aromaticum* L. In this chapter, the bud oil is discussed. The clove leaf oil and clove stem oil are presented in Chapters 5.27 and 5.28.

DEFINITION
Clove bud oil (essential oil of clove buds) is the essential oil obtained from the buds of the clove (clove tree) *Syzygium aromaticum* (L.) Merr. & L.M. Perry.

INCI NOMENCLATURE
Description/definition: Eugenia caryophyllus bud oil is an essential oil steam-distilled from the dried flower buds of the clove, *Syzygium aromaticum*, syn. *Eugenia caryophyllus*, Myrtaceae
INCI name EU: Eugenia caryophyllus bud oil
INCI name USA: Eugenia caryophyllus (clove) bud oil
CAS registry number(s): 8000-34-8; 84961-50-2
EINECS number(s): 284-638-7

ISO (INTERNATIONAL ORGANIZATION FOR STANDARDIZATION) STANDARD
ISO number: 3142
ISO name: Essential oil of clove buds
Botanical origin: *Syzygium aromaticum* (L.) Merr. & L.M. Perry (synonym: *Eugenia caryophyllus* (Spreng.) Bullock & S.G. Harrison)
Parts of plant used: Bud
ISO values: ISO values (minimum and maximum concentrations) are shown in Table 5.26.1.

Table 5.26.1 ISO values (%) for clove bud oil[a]

Compound	CAS	Minimum	Maximum
Eugenol	97-53-0	75.0	87.0
Eugenyl acetate	93-28-7	8.0	15.0
β-Caryophyllene	87-44-5	2.0	7.0

[a] ISO 3142 Essential oil of clove buds ©ISO 1997; Geneva, Switzerland, www.iso.org

THE PLANT, THE OIL, AND THEIR USES
The clove tree *Syzygium aromaticum* (L.) Merril and Perry (synonyms: *Eugenia caryophyllus* (Spreng.) Bullock and S.G. Harrison; *Caryophyllus aromaticus* L.; *Eugenia aromatica* (L.) Baill.; *Eugenia caryophyllata* Thunb.), is a perennial tropical plant which grows to a height ranging from 10 to 20 meters. The clove tree is native to the Maluku Islands in Indonesia, also known as the Moluccas or the Spice Islands. Two major products are available and marketed from clove tree: the cloves, which are the unopened dried fully grown buds, and the essential oil extracted either from bud, leaf or stem (16). Today the most important producers of cloves are Tanzania (Zanzibar and Pemba islands), Madagascar, Indonesia, Sri Lanka and Brazil (3,4,16,17,22); the clove trees also grow in India, Kenya, Malaysia (Isle of Penang), Mauritius, Reunion, Mexico, Seychelles, Haiti, and the south of China (5,9,13).

Cloves, the aromatic dried flower buds of the clove tree, are used as a spice in virtually all of the world's cuisine, either whole or in ground form. In Indonesia, cloves are smoked in cigarettes known as 'kreteks', which typically contain 60% tobacco and 40% ground cloves. Also, cloves are an important incense material in Chinese and Japanese cultures. The clove is known to be a traditional medicinal plant used as an expectorant, anti-emetic, stimulant, antiflatulent and for treatment of dyspepsia (4).

Clove essential oils are widely used and well known for their medicinal properties. Traditional uses of clove oil include use in dental care as an antiseptic and analgesic, where the undiluted oil may be rubbed on the gums to treat toothache. It is active against oral bacteria associated with dental caries and periodontal disease and effective against a large number of other bacteria. Antifungal, anti-carcinogenic, anti-allergic, anti-mutagenic, anaesthetic (8), antiphlogistic, anti-vomiting, antispasmodic, carminative, antioxidant, anthelminthic, and insect repellent (especially against mosquitoes) activities from clove oils have all been reported (2,3,5,6,7,17). It has shown efficacy in a cream in the treatment of chronic anal fissure (23). Most of these biological activities are ascribed to its main ingredient eugenol, which is also used as a starting material for the production of vanilin (1,2,3,22). An overview of potential health benefits of clove is provided in ref. 30.

Clove essential oils are also used as flavoring agents and antimicrobial materials in food and alcoholic and soft drinks (17) and in cosmetic products. Owing to their various biological activities, clove oil finds extensive use in dental formulations, toothpaste, breath freshener, mouth washes, soaps, and insect repellents (2,3). In addition, clove *bud* oils are used in aromatherapy (22), but the leaf and stem oils are not utilized because of the risk of sensitization from the very high concentrations of eugenol (55,56,57).

CHEMICAL COMPOSITION
Clove bud oil is a yellowish to brown, clear, mobile, sometimes slightly viscous liquid which has an intense spicy and aromatic odor, reminding of fresh crushed cloves. The yield of essential oil from the buds of *Syzygium aromaticum* (L.) Merr. & L.M. Perry generally varies from 15 to 20%. The main producing countries of this oil are Madagascar, Indonesia, Sri Lanka, Malaysia and India.

Literature data (up to November 13, 2014) on the chemical composition of clove bud oils (clove oils which were not specifically mentioned to be obtained from buds, leaves or stems were considered to be bud oils and are included here) and unpublished analytical data from one of us (E.S.) are shown in Table 5.26.2 in alphabetical order. In clove bud oils from various origins, over 200 chemicals have been identified. About 67% of these were found in a single reviewed publication only. The major compounds found in clove bud oils from different sources are shown in Table 5.26.3. They include (highest concentrations in any study given) eugenol (90.4%), β-caryophyllene (43.0%), eugenyl acetate (29.2%) and α-humulene (4.6%). Uncommon or rare constituents of clove bud oils found in high concentrations (>7%) in single studies include myrtenone (49.1%), benzyl alcohol (34.1%), benzyl salicylate (14.8%), 1-ethyl-3-nitrobenzene (11.1%), isoeugenol (10.4%), 3-(1-methylethyl)benzoic acid (9.0%) and δ-cadinene (7.0%).

Commercial oils

The ten chemicals that had the highest maximum concentrations in 31 commercial clove bud essential oil samples (concentration ranges provided) are the following: eugenol (82.0-90.4%), β-caryophyllene (2.6-12.0%), eugenyl acetate (0.3-6.4%), α-humulene (0.5-2.1%), caryophyllene oxide (0.06-0.4%), caryophyllene alcohol (0.01-0.3%), methyl salicylate (0.02-0.3%), chavicol (0.04-0.2%), methyl eugenol (0.04-0.2%) and furfural (0.01-0.1%) (Erich Schmidt, unpublished analytical data).

Comparative chemical composition of clove essential oils

Eugenol is the main ingredient in all three (bud, leaf, stem) clove oils. Although there is a wide variation in concentrations, generally increasing percentages of eugenol were observed from bud (72-82%) to leaf (75-84%) and to stem oils (87-97%) in a large number of essential oil samples from Indonesia, Madagascar and Zanzibar (16). In the bud essential oil this compound was followed by eugenyl acetate (8.6-21.3%). In leaf and stem oils, however, this compound was either absent or present in low concentrations (leaf oil maximum 1.5%, in stem oil maximum 2.5%) (16). In the leaf essential oil, the second and third main compounds were β-caryophyllene (mean 14%) and α-humulene (mean 1.6%), less represented in the bud essential oil (mean 4.4% and 0.5%, respectively) and in the stem essential oil (mean 5.2% and 0.7%, respectively). Lawrence (31) also indicates that eugenol concentrations are usually higher in leaf oil (77.0-91.0%) and in stem oil (76.4-90.0%) than in clove bud oil (69.8-84.4%). Conversely, in the case of eugenyl acetate, concentrations are generally (far) higher in the bud oils (4.5-22.1%) than in stem (1.5-8.0%) and leaf oils (0.8-1.6%) (31).

Table 5.26.2 Constituents identified in clove bud oils

Constituent	CAS	Percentage and range						
		A	B (16)	C (14)	D (12)	E (3)	F (11)	G
Acetophenone	98-86-2							0.008[w]; 0.2[w]
2-Acetoxydodecane							0.01	
1-Acetoxy-2-propanol	627-69-0							+[a]
9β-Acetoxy-3,5α,8-trimethyltricyclo[6.3.1.0(1,5)]dodec-3-ene					0.06			
α-Amorphene	20085-19-2			0.05				tr[h]; 0.1[w]
(E)-Anethole	4180-23-8					tr		tr[h]
Aromadendrene	489-39-4							0.4[w]
allo-Aromadendrene	25246-27-9			0.3		0.3		0.1[z2]; 1.3[z6]
allo-Aromadendrene epoxide	85760-81-2							0.3[o]
α-Asarone	2883-98-9							1.5[k]
Benzaldehyde	100-52-7				tr			+[a]; 0.07[k]; 0.2[w]
Benzoic acid	65-85-0				tr			
Benzyl acetate	140-11-4	tr-0.01			0.07			tr[h,z3]; 0.01[t]; 0.09[w]
Benzyl alcohol	100-51-6							+[a]; tr[h]; 0.1[w]; 0.6[k]; 34.1[q]
Benzyl benzoate	120-51-4			0.1				tr[z2,z4]
Benzyl salicylate	118-58-1							14.8[z1]
Benzyl tiglate	37526-88-8							tr[n]
(E)-α-Bergamotene	13474-59-4					1.3		0.2[o]
epi-Bicyclosesquiphellandrene	54274-73-6							0.2[z2]
cis-α-Bisabolene epoxide	121467-35-4							0.05[v]
Bornyl acetate	76-49-3							0.1[z2]
2-Butenal	4170-30-3							+[a]
n-Butyl benzoate	136-60-7					1.3		
2-Butyl-1-methylpyrrolidone								0.1[g]
Cadinene	29350-73-0							1.5[c]
α-Cadinene	24406-05-1							0.3[w]
β-Cadinene	523-47-7							0.3[w]; 0.6[z7]; 1.6[y2,z2]
γ-Cadinene	39029-41-9					0.8		tr[h,m]; 0.3[d]; 0.5[k]

Table 5.26.2 Constituents identified in clove bud oils (*continued*)

Constituent	CAS	Percentage and range						
		A	B (16)	C (14)	D (12)	E (3)	F (11)	G
δ-Cadinene	483-76-1			0.2		0.2	0.3	0.2[v]; 0.3[w]; 0.4[k]; 0.7[x]; 7.0[z5]
α-Cadinol	481-34-5					0.1		tr[z2]
epi-α-Cadinol	5937-11-1					0.1		
t-Cadinol	58580-31-7					0.1		
Calacorene	38599-17-6							0.1[n]
Calamenene	483-77-2					0.1		tr[m,z4]; 0.1[n]; 0.2[z7]; 0.3[x]
Camphene	79-92-5				tr			0.7[v]
Camphor	76-22-2							tr[z2]
l-Camphor	464-48-2							0.2[x]
Carvacrol	499-75-2						2.4	
Carvone	99-49-0					0.2		tr[h]
Caryophylla-2(12),6(13)-dien-5-ol								0.02[b]
Caryophylla-4(12),8(13)-dien-5β-ol	19431-80-2							0.2[z2]
β-Caryophyllene	87-44-5	2.6-12.0	2.8-8.6	18.9	13.0	19.5	43.0	12.4[s]; 13.7[z7]; 14.4[z2]; 16.8[l] 19.0[u]; 25.0[j]; 31.5[w]; 36.9[c]
γ-Caryophyllene	118-65-0							0.5[z6]; 6.0[z5]
Caryophyllene oxide	1139-30-6	0.06-0.4	0.06-0.4	1.5		0.4	0.2	1.1[x]; 1.3[m]; 1.7[l]; 1.9[e]; 2.4[z5]
Caryophyllenyl alcohol	913176-41-7	0.01-0.3						0.2[o]
Cedran-8β-ol								tr[z2]
Chavicol	501-92-8	0.04-0.2	0-0.2	0.3		0.1	0.5	0.2[z4,b]; 0.3[l]; 0.4[h,x]; 1.1[w]
1,8-Cineole	470-82-6			0.06				0.06[w]; 0.1[k]
Cinnamaldehyde	104-55-2							tr[n]
β-Citronellene	2436-90-0							0.02[t]
Citronellol	106-22-9							0.7[z7]
α-Copaene	3856-25-5		0-0.3	0.4		0.1	0.8	0.2[i]; 0.3[o]; 0.5[g]; 1.0[z2]; 1.9[c]
β-Copaen-4α-ol	124753-76-0							tr[z3]
α-Cubebene	17699-14-8					tr	0.8	0.1[i]; 0.4[p]; 0.6[w]; 0.7[k]; 0.9[c]
Cubenene	29837-12-5							0.2[z7]
Cubenol	21284-22-0					0.1		0.5[c]
p-Cresol	106-44-5							tr[z2]
1,3-Cyclohexadiene, 1,3,3,5-tetramethyl-	4724-89-4							0.7[p]
Cyclosativene	22469-52-9			0.2				
m-Cymene	535-77-3							0.2[p]
p-Cymene	99-87-6				tr			<0.01[b]; 0.09[k]; 0.5[v]
2,4-Diacetylphloroglucinol	2161-86-6							0.6[x]
2,5-Dimethylanisole	1706-11-2							+[a]
2-(1,1-Dimethylethyl)-2,5-cyclohexadiene-1,4-dione								+[a]
1,2-Dimethyl-3,5-bis(1-methylethenyl)-cyclo-hexane	62337-99-9			0.2				
Dodecane	112-40-3							1.3[f]
β-Elemene	33880-83-0							0.02[w]
δ-Elemene	20307-84-0				tr			
Elixene	3242-08-8			3.9				
Epiglobulol	88728-58-9							tr[z2]
Eremophilene	10219-75-7							0.1[w]
2-Ethenyl-1,1-dimethyl-3-methylcyclohexane							0.04	
Ethyl benzoate	93-89-0						0.04	+[a]; tr[h,z2]; 0.02[w]
Ethyl cinnamate	103-36-6							tr[n]
Ethyl hexanoate	123-66-0				tr		0.02	0.2[w]; 0.7[n]
1-Ethyl-3-nitrobenzene	7369-50-8			11.1				
Ethyl octanoate	106-32-1							tr[n]
Eugenol	97-53-0	82.0-90.4	72.1-82.4	49.7	69.8	82.6	46.5	83.6[s]; 84.0[r]; 84.1[z]; 85.1[i]; 86.7[z6]; 87[b]; 88.6[n]; 89.5[z3]
m-Eugenol	501-19-9							5.0[q]
Eugenyl acetate	93-28-7	0.3-6.4	8.6-21.3		16.1	6.0	2.6	9.5[t]; 10.8[r]; 19.7[z2]; 20.9[h]; 21[h]; 24.6[k]; 27.8[l]; 29.2[x]

Table 5.26.2 Constituents identified in clove bud oils (*continued*)

Constituent	CAS	Percentage and range						
		A	B (16)	C (14)	D (12)	E (3)	F (11)	G
α-Farnesene	502-61-4			1.1			0.7	0.2[z7]; 0.3[z2]; 0.4[c]; 0.5[w]
(*E,E*)-α-Farnesene	502-61-4	0.01-0.04						
(*Z,E*)-α-Farnesene	26560-14-5							0.2[l]
(*E,E*)-Farnesol	106-28-5							1.6[k]
(*Z,E*)-Farnesol	3790-71-4							0.9[k]
Formic acid	64-18-6							6.6[e]
Furfural	98-01-1	0.01-0.1			tr			0.03[w]
Geraniol	106-24-1							1.3[z5]
Germacrene D	23986-74-5			0.08	0.1		0.4	0.2[z2]
Guaiazulene	489-84-9							0.3[p]
α-Guaiene	3691-12-1			0.02				
Heneicosane (*n-*)	629-94-7							0.5[x]
n-Heptadecane	629-78-7				0.1			
(*E,E*)-2,4-Heptadienal	4313-03-5							+[a]
2-Heptanol	543-49-7				tr		0.02	+[a]; tr[n]; 0.03[k]; 0.1[w]
2-Heptanone	110-43-0				tr			0.003[w]; 0.04[b]; 0.8[w]; 0.9[n]
Heptyl acetate	112-06-1							tr[z3]
2-Heptyl acetate	5921-82-4			0.05	tr		0.2	0.01[t]; 0.04[b]; 0.08[w]
Hexadecanoic acid	57-10-3							4.3[e]
1,2,3,4,4a,7-Hexahydro-1,6-dimethyl-4(1-methyl-ethyl)naphthalene	16728-99-7			0.2				
Hexanol	111-27-3				tr			
2-Hexanone	591-78-6							0.2[k]
(*E*)-2-Hexenal	6728-26-3				tr			
(*Z*)-3-Hexenol	928-96-1				tr			
α-Humulene	6753-98-6	0.5-2.1	0.3-1.0		0.6	1.9	4.6	2.4[l]; 4.3[w]; 3.1[s]; 4.1[z5]; 4.4[c]
α-Humulene epoxide	96638-51-6	tr-0.03						tr[n]; 0.1[o]; 0.2[w]
Humulene epoxide I	19888-33-6					0.1		
Humulene epoxide II	19888-34-7				tr	0.1		tr[m]
Humulenol	19888-00-7							0.3[n]
4-Hydroxy-*cis*-caryophyl-lene								0.2[o]
Isochavicol								0.1[z3]
Isoeugenol	97-54-1		0-0.2			0.8		0.9[p]; 1.0[k]; 10.4[z]
(*E*)-Isoeugenol	5932-68-3							+[a]; 0.1[f]
Isoledene	95910-36-4							0.6[c]
Isophthaldehyde	626-19-7							+[a]
Ledol	577-27-5							tr[z2]
Limonene	138-86-3	0.01-0.06		0.05	tr			tr[z2]; 0.06[t]; 0.5[v]
cis-Limonene oxide	13837-75-7							0.2[x]
Linalool	78-70-6	0.01-0.2			tr	0.1		tr[h]; 0.01[b]; 0.1[f,v]
trans-Linalool oxide	11063-78-8				tr			
Linalyl acetate	115-95-7					0.1		
Megastigma-4,6(*E*),8(*Z*)-triene	71186-24-8			0.05				
2-Methoxyfuran	25414-22-6							0.4[p]
2-Methoxy-3-(2-propenyl)-phenol	1941-12-4							<0.05[d]
m-Methylacetophenone	585-74-0					0.1		
Methyl benzoate	93-58-3					tr	0.04	+[a]; tr[h]; 0.2[w]
3-Methyl-3-butenol	763-32-6				tr			
3-Methyl-4-butenyl acetate					tr			
2-Methyl butyrate					tr			
Methyl chavicol	140-67-0							tr[n]
Methylcyclopentane	96-37-7							4.0[g]
4-Methylene-1-methyl-2-(2-methyl-1-propen-1-yl)-1-vinylcycloheptane	826337-63-7			0.08				
3-(1-Methylethyl)benzoic acid	5651-47-8			9.0				

Table 5.26.2 Constituents identified in clove bud oils (*continued*)

Constituent	CAS	A	B (16)	C (14)	D (12)	E (3)	F (11)	G
Methyl eugenol	93-15-2	0.04-0.2	0-0.08			tr		tr[h]; 0.1[i,o]; 1.0[m]
5-Methyl-2-furaldehyde	620-02-0	0.01-0.05						0.07[w]
p-Methylguaiacol	93-51-6				tr			
6-Methyl-5-hepten-2-one	110-93-0				tr			0.2[w]
Methyl hexanoate	106-70-7				tr			
2-Methyl-5-(1-methylethe-nyl)cyclohexyl acetate								+[a]
Methyl octanoate	111-11-5							tr[n]
2-Methylpentanal	123-15-9							+[a]
Methyl salicylate	119-36-8	0.02-0.3	0-0.3	0.3	tr	0.3	0.2	0.07[b]; 0.1[d]; 0.2[h]; 0.3[i]
α-Muurolene	10208-80-7							tr[n]; 0.1[z2]
γ-Muurolene	30021-74-0							0.1[z2]; 0.2[c]
t-Muurolol	19912-62-0					tr		
Myrcene	123-35-3				tr			1.8[v]
Myrtenone								49.1[e]
Naphthalene, 1,2,4a,5,8,8a-hexahydro-4,7-dime-thyl-1-(1-methylethyl)-	16509-53-8							0.3[p]
Nerol	106-25-2					0.1		
Nerolidol	7212-44-4							0.05[w]
(*E*)-Nerolidol	40716-66-3					0.4		
Nonacosane	630-03-5							1.4[x]
2-Nonanol	628-99-9							tr[h]; 0.05[k]
2-Nonanone	821-55-6				tr		0.1	tr[h,z3]; 0.02[b]; 0.3[w]
β-Ocimene	13877-91-3							0.5[w]
(*E*)-β-Ocimene	3779-61-1				tr			0.1[w]; 0.2[v]; 0.3[b]
(*Z*)-β-Ocimene	3338-55-4				tr		0.05	
Octadecanoic acid	57-11-4							0.4[v]
n-Octane	111-65-9					0.1		
2-Octanone	111-13-7				tr			
Oleic acid	112-80-1							0.6[v]
Perillene	539-52-6				tr			
α-Phellandrene	99-83-2							0.5[v]
β-Phellandrene	555-10-2							0.5[v]
Phenylacetaldehyde	122-78-1							0.003[w]
2-Phenylethyl acetate	103-45-7							tr[n]
Phenylmethyl acetate	140-11-4						0.04	+[a]
α-Pinene	80-56-8					0.1		0.004[w]; 0.01[b]; 0.02[k]; 0.5[v]
β-Pinene	127-91-3							0.1[v]; 0.2[k]
2-Propanone,1-(acetyloxy)-	592-20-1							+[a]
2-Propanone, 2-methyl-hydrazone	5771-02-8							5.6[g]
Pulegone	89-82-7							4.1[e]
Propylene glycol	57-55-6							6.0[z1]
2*H*-Pyran-2-one, tetrahy-dro-6,6-dimethyl-								0.4[g]
3(2*H*)-Pyridazinone	504-30-3							+[a]
3-Pyrrolidinol	40499-83-0							+[a]
trans-Sabinene hydrate	17699-16-0							0.1[v];
Salicylic acid	69-72-7							+[a]
α-Selinene	473-13-2					0.3		
β-Selinene	17066-67-0							0.9[w]
(*E*)-Sesquilavandulol	104121-84-8							0.2[o]
Squalene	111-02-4			0.7				
α-Terpinene	99-86-5							1.7[v]
γ-Terpinene	99-85-4							0.9[p]
Terpinen-4-ol	562-74-3				tr			0.02[k]; 0.9[v]
α-Terpineol	98-55-5				tr			0.06[v]; 1.0[k]
β-Terpineol	138-87-4							0.07[k]
Terpinolene	586-62-9							0.6[v]
α-Terpinyl acetate	80-26-2							tr[n]

Table 5.26.2 Constituents identified in clove bud oils (*continued*)

Constituent	CAS	Percentage and range						
		A	B (16)	C (14)	D (12)	E (3)	F (11)	G
Tetracyclo[6.3.2.0(2,5).0(1,8)]tridecan-9-ol, 4,4-dimethyl-	100469-05-4(?)			0.7				
Tetrahydro-3-methylfuran								2.5[g]
1,5,5,8-Tetramethyl-12-oxabicyclo[9.1.0]dodeca-3,7-diene	90820-79-4			0.1				
α-Thujene	2867-05-2							0.3[v];
Thymol	89-83-8						0.4	0.3[l]; 0.4[v]; 2.6[y]
Tricyclo[3.2.2.0]nonane-2-carboxylic acid								0.07[v]
(*E,Z,Z*)-2,4,7-Tridecatrienal	13552-96-0							0.7[p]
2',3',4'-Trimethoxyaceto-phenone	13909-73-4			0.1				0.5[x]; 1.4[y2,z2]
2',4',6'-Trimethoxyaceto-phenone	832-58-6							tr[z2]
2-Undecanone	112-12-9						0.03	tr[h]; 0.2[w]
Vanillin	121-33-5					tr		0.9[q]; 1.2[k]
Viridiflorol	552-02-3							0.2[o]; 0.4[x]
α-Ylangene	14912-44-8					tr		0.5[y]

A thirty-one clove bud essential oil samples from Madagascar, Indonesia, Sri Lanka and Malaysia analyzed between 1998 and 2013; lowest and highest concentrations given (E. Schmidt, unpublished data)
B forty-five commercial samples of clove bud oils from Indonesia, Madagascar and Zanzibar; lowest and highest concentrations given (ref. 16)
C one lab-hydrodistilled clove bud essential oil sample from Bangladesh (ref. 14)
D one lab-hydrodistilled clove bud essential oil from Cuba (ref. 12)
E one lab-hydrodistilled oil of clove buds purchased at a local Indian market and one commercial oil from Madagascar; highest concentrations given (ref. 3)
F two lab-hydrodistilled oils, one from Indian and the other from Indonesian cloves; highest concentrations given (ref. 11)
G data from other studies (indicated with superscript letters); highest concentrations found in any study reviewed here given; when two or more oils were investigated, only the highest concentrations are mentioned, unless indicated otherwise

[a] one lab-steam distilled oil from cloves purchased at a local USA market; only presence indicated here (with +), as the amounts were given in mg/g weight of dried clove buds and they are therefore not comparable to the other data (ref. 20); [b] one industrially steam-distilled oil from Turkish cloves (ref. 10); [c] one lab-hydrodistilled and one lab-steam distilled oil from cloves purchased at a Chinese factory (ref. 5); [d] one laboratory steam-distilled oil from Egyptian clove buds (ref. 8); [e] one lab-hydrodistilled oil from clove buds obtained at a local Indian market with a very atypical composition (ref. 9); [f] one lab-hydrodistilled oil (in 1% NaCl solution) from China (ref. 13); [g] one lab-prepared oil from Malaysia (ref. 15); [h] one commercial clove bud oil purchased in Korea (ref. 18); [i] data from various analytical studies on clove bud oil reviewed in ref. 21; [j] data taken from ref. 19; [k] three lab-hydrodistilled oils from clove buds harvested in local gardens in India at various stages of maturity (ref. 24); [l] one lab-hydrodistilled oil from Iranian clove buds (ref. 25); [m] one clove bud essential oil of unknown origin (ref. 26); [n] one clove oil (most likely clove bud oil) from unknown origin (ref. 1); [o] one clove oil of unknown origin, most likely clove bud oil (ref. 17); [p] one commercial clove oil from China (ref. 28); [q] one commercial clove oil from India (ref. 29, also presented in ref. 43); this oil is adulterated (ref. 31); [r] one commercial clove bud oil from Madagascar (ref. 32); [s] one commercial oil from a Brazilian aromatherapy oils supplier (ref. 33); [t] one lab-hydrodistilled clove oil from buds purchased in a Brazilian spice store (ref. 34); [u] one laboratory steam-distilled oil from cloves purchased at a local market in Poland (ref. 35); [v] one lab-hydrodistilled clove oil from buds purchased at a local market in Nigeria (ref. 36); [w] three lab-hydrodistilled oils from clove buds purchased in local supermarkets in Saudi Arabia, United Arab Emirate and Jordan (ref. 37); [x] one laboratory steam-distilled clove oil from buds purchased at a local market in India (ref. 38); [y] one commercial clove oil from Indonesia (ref. 39); [y2] incorrect identification (ref. 31); [z] one lab-hydrodistilled clove bud oil from Egypt (ref. 40); [z1] one commercial clove oil sample from a Turkish producer (ref. 41); [z2] one lab-hydrodistilled bud oil from China (ref. 42); [z3] one lab-hydrodistilled oil from clove buds from India (ref. 44); [z4] one commercial clove bud oil sample (ref. 45); [z5] one commercial oil from India (ref. 46); [z6] one laboratory steam-distilled oil sample (ref. 47); [z7] one lab-hydrodistilled clove bud oil from Thailand (ref. 48); [z8] doubtful whether this occurs in plants

tr: traces (in columns E and G[z2]: <0.05; in column G[z3]: <0.1); + present in the oil investigated, but quantity not stated or indicated in a manner that they cannot be compared with the other data

Table 5.26.3 Major constituents of clove bud oils

Constituent	CAS	Percentage and range						
		A	B (16)	C (14)	D (12)	E (3)	F (11)	G
Eugenol	97-53-0	82.0-90.4	72.1-82.4	49.7	69.8	82.6	46.5	86.7[z6]; 87[b]; 88.6[n]; 89.5[z3]
β-Caryophyllene	87-44-5	2.6-12.0	2.8-8.6	18.9	13.0	19.5	43.0	19.0[u]; 25.0[j]; 31.5[w]; 36.9[c]
Eugenyl acetate	93-28-7	0.3-6.4	8.6-21.3		16.1	6.0	2.6	21[h]; 24.6[k]; 27.8[l]; 29.2[x]
α-Humulene	6753-98-6	0.5-2.1	0.3-1.0		0.6	1.9	4.6	2.4[l]; 4.3[w]; 3.1[s]; 4.1[z5]; 4.4[c]

LEGEND: SEE UNDER TABLE 5.26.2

Additional analyses of clove bud essential oils not presented in Table 5.26.2 can be found in refs. 49-54; these can all be accessed on-line.

CONTACT ALLERGY/ALLERGIC CONTACT DERMATITIS

Clove oil (unspecified)

General
Contact allergy to/allergic contact dermatitis from clove oil has been reported in over 25 publications. In groups of consecutive patients suspected of contact dermatitis, prevalence rates of up to 2.1% positive patch test reactions have been observed, but adequate relevance data are lacking. Case reports have included patients with cosmetic allergy, stomatitis and two cases of occupational contact dermatitis in a physiotherapist and a masseuse. In several studies, there were concomitant reactions to eugenol, the main component of clove oil.

Testing in groups of patients
The results of patch tests with clove oil in routine testing (consecutive patients suspected of contact dermatitis) and in groups of selected patients are shown in Table 5.26.4. In routine testing, rates of positive reactions ranged from 0.5% to 2.1%, whereas between 1.6% and

Table 5.26.4 Results of testing groups of patients with clove oil

Years and Country	Test conc. & vehicle	Number of patients tested	positive (%)	Selection of patients (S); Relevance (R); Comments (C)	Ref.
Routine testing					
2000-2007 USA	2% pet.	326	7 (2.1%)	R: 100%; C: weak study: a. high rate of macular erythema and weak reactions, b. relevance figures include 'questionable' and 'past' relevance	59
<1976 Poland	2% pet.	200	1 (0.5%)	R: not stated	64
1967-1970 Italy	1% pet.	380	1 (0.3%)	R: not stated; C: patients who had forms of dermatitis other than contact dermatitis were also tested	82
Testing in groups of selected patients					
2000-2008 IVDK	2% pet.	6,893	105 (1.5%)	S: patients with dermatitis suspected of causal exposure to fragrances; R: not stated; C: 50% of the patients allergic to clove oil who were also tested with eugenol co-reacted to this major (>82%) component of the oil; the oil tested was a clove bud oil	71
<2002 Japan, USA and 5 European countries	10% pet.	218	42 (19.3%)	S: patients previously shown to be allergic to fragrances; R: not stated; C: tested was clove bud oil	72
1997-2000 Austria	2% pet.	747	12 (1.6%)	S: patients suspected of fragrance allergy; R: not stated	60
1989-1999 Portugal	2% pet.	67	9 (13.4%)	S: patients who had a positive patch test to the fragrance mix; R: not stated	73
1994-1995 UK	2% pet.	40	2 (5%)	S: patients previously reacting to the fragrance mix; R: not stated	74
<1986 Poland	2% pet.	86	12 (14.0%)	S: patients previously reacting to the fragrance mix; R: not stated; in the group of 86, 25% reacted to eugenol, which constitutes >80% of the oil	66
<1983 Poland	2% pet.	16	13 (81%)	S: patients known to be allergic to propolis and *Myroxylon pereirae*; R: not stated; C: 6 patients also reacted to eugenol, which is the main ingredient of clove oil and is present in *Myroxylon pereirae* and may also be present in propolis (80)	79
<1976 Poland	2% pet.	51	1 (2.0%)	S: patients allergic to *Myroxylon pereirae* resin (balsam of Peru) and/or turpentine and/or wood tar and/or colophony; R: not stated	64

IVDK Information Network of Departments of Dermatology, Germany, Switzerland, Austria (www.ivdk.org); pet.: petrolatum

81% of patients in selected groups had positive patch tests. The very high positivity rate of 81% was seen in a small group of 16 patients previously reacting to *Myroxylon pereirae* and propolis (79).

Case reports

Clove oil was responsible for 1 out of 399 cases of cosmetic allergy where the causal allergen was identified in a study of the NACDG, USA, 1977-1983 (58). In a group of 70 patients with proven allergic cosmetic dermatitis, clove oil was the allergen in one (83). A physiotherapist had occupational contact dermatitis from a massage cream and its ingredients cinnamon oil and clove oil; the patient was also allergic to eugenol, the main component of clove oil (61). One patient with dermatitis and stomatitis attributed to clove oil and cinnamon oil (62, cited in ref. 61). A masseuse/reflexologist developed occupational contact dermatitis from clove oil and two other essential oils (63).

In a group of 63 patients with facial dermatitis, one had a relevant positive patch test reaction to clove oil (75).

One patient with mucosal inflammation and purpuric perioral macules associated with the use of a prophylactic dental tablet from contact allergy to its components oil of cloves and oil of cassia was reported (77). One patient suffering from oral lichenoid mucositis had relevant positive patch test reactions to clove oil, lemon oil and metals (81).

Positive patch tests (relevance unknown, uncertain or not stated)

One patient with burning mouth syndrome reacting to clove oil, eugenol, isoeugenol, *Myroxylon pereirae* resin and the fragrance mix (69). One positive patch test to clove oil in 20 perfume factory workers without dermatitis (65). Two positive patch test reactions in patients allergic to ascaridole, an allergen in tea tree oil (negative to tea tree oil itself) (70).

Additional information in older literature on contact allergy to clove oil can be found in refs. 67, 68 and 78 (articles not read).

Clove bud oil

In only two articles, it was specified that the clove oil used for testing was obtained from clove buds (clove bud oil) (71,72). Their data are discussed in Table 5.26.4.

LITERATURE

1 Chaieb K, Hajlaoui H, Zmantar T, Kahla-Nakbi AB, Rouabhia M, Mahdouani K, et al. The chemical composition and biological activity of clove essential oil, *Eugenia caryophyllata* (*Syzygium aromaticum* L. Myrtaceae): A short review. Phytother Res 2007;21:501-506

2 Jirovetz L, Buchbauer G, Stoilova I, Stoyanova A, Krastanov A, Schmidt E. Chemical composition and antioxidant properties of clove leaf essential oil. J Agric Food Chem 2006;54:6303-6307

3 Srivastava AK, Srivastava SK, Syamsundar KV. Bud and leaf essential oil composition of *Syzygium aromaticum* from India and Madagascar. Flavour Fragr J 2005;20: 51-53

4 Raina VK, Srivastava SK, Aggarwal KK, Syamasundar KV, Kumar S. Essential oil composition of *Syzygium aromaticum* leaf from Little Andaman, India. Flavour Fragr J 2001;16:334-336

5 Wenqiang G, Shufen L, Ruixiang Y, Shaokun T, Can Q. Comparison of essential oils of clove buds extracted with supercritical carbon dioxide and other three traditional extraction methods. Food Chem 2007;101:1558-1564

6 Shapiro R. Prevention of vector transmitted diseases with clove oil insect repellent. J Ped Nurs 2012;27: 346-349

7 Alqareer A, Alyahya A, Andersson L. The effect of clove and benzocaine versus placebo as topical anesthetics. J Dent 2006;34:747-750

8 El-Mesallamy AMD, El-Gerby M, Abd El Azim MHM, Awad A. Antioxidant, antimicrobial activities and volatile constituents of clove flower buds oil. J Essent Oil Bear Plants 2012;15:900-907

9 Kapoor R, Ali M, Alkhtar MS, Kaskoos RA, Siddiqui AW, Mir SR. Composition of volatile oil of *Syzygium aromaticum* buds. J Essent Oil Bear Plants 2005;8:196-199

10 Alma MH, Ertas M, Nitz S, Koomannsberger H. Chemical composition and content of essential oil from the bud of cultivated Turkish clove (*Syzygium aromaticum* L.). BioResources 2007;2:265-269

11 Hossain MA, Al-Hashmi RA, Weli AM, Al-Riyami Q, Al-Sabahib JN. Constituents of the essential oil from different brands of *Syzigium caryophyllatum* L by gas chromatography-mass spectrometry. As Pac J Trop Biomed 2012;2:S1446-S1449

12 Pino JA, Marbot R, Agero J, Fuentes V. Essential oil from buds and leaves of clove (*Syzygium aromaticum* (L.) Merr. et Perry) grown in Cuba. J Essent Oil Res 2001;13:278-279

13 Huang W-W, Feng Y-C, Huang Y, Li H-L. Chemical composition, antioxidant and the possible use as skincare ingredient of clove oil (*Syzygium aromaticum* (L.) Merr. & Perry) and citronella oil (*Cymbopogon goeringii*) from China. J Essent Oil Res 2013;25:315-323

14 Bhuiyan NI, Begum J, Nandi NC, Akter F. Constituents of the essential oil from leaves and buds of clove (*Syzigium caryophyllatum* (L.) Alston). Afr J Plant Sc 2010;4:451-454

15 Seongwei L, Musa N, Wee W, Nadirah M. Chemical composition and antimicrobial activity of the essential oil of *Syzygium aromaticum* flower bud (clove) against fish systemic bacteria isolated from aquaculture sites. Front Agric China 2009;3:332-336

16 Razafimamonjison G, Jahiel M, Duclos T, Ramanoelina P, Fawbush F, Danthu P. Bud, leaf and stem essential oil composition of clove (*Syzygium aromaticum* L.) from Indonesia, Madagascar and Zanzibar. Int J Basic Appl Sci 2014;3(3):224-233

17 Dzamić A, Soković M, Ristić MS, Grijić-Jovanović S, Vukojević J, Marin PD. Chemical composition and antifungal activity of *Illicium verum* and *Eugenia caryophyllata* essential oils. Chem Nat Comp 2009;45:259-261

18 Yang J-C, Lee S-H, Lee WJ, Choi D-H, Ahn YJ. Ovicidal and adulticidal effects of *Eugenia caryophyllata* bud and leaf oil compounds on *Pediculus capitis*. J Agric Food Chem 2003;51:4884-4888

19 Zachariah TJ, Krishnamoorthy B, Rema J, Mathew PA. Oil constituents in bud and pedicel of clove (*Syzygium aromaticum*). Indian Perfum 2005;49:313-316

20 Lee K-G, Shibamoto T. Antioxidant property of aroma extract isolated from clove buds [*Syzygium aromaticum* (L.) Merr. et Perry]. Food Chem 2001;74:443-448

21 Lawrence BM. Essential oils 2001-2004. Carol Stream, USA: Allured Publishing Corporation, 2006:195-196

22 Marzouk TMF, El-Nemer AMR, Baraka HN. The effect of aromatherapy abdominal massage on alleviating menstrual pain in nursing students: A prospective randomized cross-over study. Evid Based Complement Alternat Med 2013; article number 742421, 6 pages

23 Elwakeel HA, Moneim HA, Farid M, Gohar AA. Clove oil cream: a new effective treatment for chronic anal fissure. Colorect Dis 2006;9:549-552

24 Golapakrishnan M, Narayanan CS, Mathew AG. Chemical composition of Indian clove bud, stem and leaf oils. Indian Perfum 1988;32:229-235

25 Fathi Z, Hassani A, Ghosta Y, Abdollahi A, Meshkatalsadat MH. The potential of thyme, clove, cinnamon and ajowan essential oils in inhibiting the growth of *Botrytis cinerea* and *Monilinia fructicola*. J Essent Oil Bear Plants 2012;15:38-47

26 Vernin G, Vernin E, Metzger J, Pujol L, Parkanyi C. GC/MS analysis of clove essential oils. In: G Charalambous, Ed. Spices, herbs and edible fungi. Amsterdam: Elsevier Science Publishers, 1994:483-500. Data cited in ref. 27

27 Lawrence BM. Progress in essential oils. Perfum Flavor 1995;20(Sept/Oct):102-103

28 Xing YG, Xu QL, Li XH, Che ZM, Yun J. Antifungal activities of clove oil against *Rhizopus nigricans*, *Aspergillus flavus* and *Penicillium citrunum* in vitro and in wounded fruit test. J Food Saf 2012;32:84-93

29 Pawar VC, Thaker VS. In vitro efficacy of 75 essential oils against *Aspergillus niger*. Mycoses 2006;49:316-323. Data also presented in ref. 43

30 Singletary K. Clove: Overview of potential health benefits. Nutrition Today 2014;49(4):207-224

31 Lawrence BM. Progress in essential oils. Perfum Flavor 2013;38(November):46-52

32 Azizkhani M, Elizaquível P, Sánchez G, Selma MV, Aznar R. Comparative efficacy of *Zataria multiflora* Boiss., *Origanum compactum* and *Eugenia caryophyllus* essential oils against *E. coli* O157:H7, feline calicivirus and endogenous microbiota in commercial baby-leaf salads. Int J Food Microbiol 2013;166:249-255

33 Murbach Teles Andrade BF, Nunes Barbosa L, da Silva Probst I, Fernandes Júnior A. Antimicrobial activity of essential oils. J Essent Oil Res 2014;26:34-40

34 Cruz GS, Wanderley-Teixeira V, Oliveira JV, Correia AA, Breda MO, Alves TJS, et al. Bioactivity of *Piper hispidinervum* (Piperales: Piperaceae) and *Syzygium aromaticum* (Myrtales: Myrtaceae) oils, with or without formulated Bta on the biology and immunology of *Spodoptera frugiperda* (Lepidoptera: Noctuidae). J Econ Entomol 2014;107:144-153

35 Dawidowicz AL, Olszowy M. Does antioxidant properties of the main component of essential oil reflect its antioxidant properties? The comparison of antioxidant properties of essential oils and their main components, Nat Prod Res 2014;28:1952-1963

36 Fayemiwo KA, Adeleke MA, Okoro OP, Awojide SH, Awoniyi IO. Larvicidal efficacies and chemical composition of essential oils of *Pinus sylvestris* and *Syzygium aromaticum* against mosquitoes. Asian Pac J Trop Biomed 2014;4:30-34

37 Hossain MA, AL Harbi SA, Weli AM, Al-Riyami Q, Al-Sabahi JN. Comparison of chemical constituents and antimicrobial activities of three essential oils from three different brands' clove samples collected from Gulf Region. Asian Pac J Trop Dis 2014;4:262-268

38 Islamuddin M, Sahal D, Afrin F. Apoptosis-like death in *Leishmania donovani* promastigotes induced by eugenol-rich oil of *Syzygium aromaticum*. J Med Microbiol 2014;63:74-85

39 Rolli E, Marieschi M, Maietti S, Sacchetti G, Bruni R. Comparative phytotoxicity of 25 essential oils on pre- and post-emergence development of *Solanum lycopersicum* L.: A multivariate approach. Ind Crops Prod 2014;60:280-290

40 Tarek N, Hassan HM, AbdelGhani SMM, Radwan IA, Hammouda O, El-Gendy AO. Comparative chemical and antimicrobial study of nine essential oils obtained from medicinal plants growing in Egypt. Beni-Suef University Journal of Basic and Applied Sciences 2014;3:149-156

41 Zengin H, Baysal AH. Antioxidant and antimicrobial activities of thyme and clove essential oils and application in minced beef. J Food Process Preserv 2014;11 pages. doi:10.1111/jfpp.12344

42 Zhao C-X, Liang Y-Z, Li X-N. Chemical composition and antimicrobial activity of the essential oils from clove. Nat Prod Res Dev (China) 2006;18:381-385. Data cited in ref. 31

43 Pawar VG, Thaker VS. Evaluation of the anti-*Fusarium oxysporum* f. sp. *cicer* and anti-*Alternaria porri* effects of some essential oils. World J Microbiol Biotechnol 2007;23:1099-1106. Data also presented in ref. 28. Data cited in ref. 31.

44 Pal M, Kumar A, Srivastava M, Banerji R. Phytochemical investigation of *Syzygium aromaticum*. J Med Arom Plant Sci 2008;30:304-305

45 Williams DG. The chemistry of essential oils, 2nd Ed. Port Washington, NY, USA: Micelle Press 2008:176-177. Data cited in ref. 31

46 Khan MSA, Zahin M, Hasan S, Hussain FM, Ahmad I. Inhibition of quorum sensing regulated bacterial functions by plant essential oils with special reference to clove oil. Lett Microbiol 2009;49:354-360. Data cited in ref. 31

47 Maggi MD, Ruffinengo SR, Gende LB, Sarlo EG, Eguaras MJ, Bailac PN, et al. Laboratory evaluations of *Syzygium aromaticum* (L.) Merr. et Perry essential oil against *Varroa destructor*. J Essent Oil Res 2010;22:119-122

48 Sulthanont N, Choochote W, Tuetun B, Junkum A, Jitpakdi A, Chaithong U, et al. Chemical composition and lavicidal activity of edible plant-derived essential oils against the pyrethroid-susceptible and resistant-strains of *Aedes aegypti* (Diptera:Culicidae). J Vector Ecol 2010;35:106-115

49 Zeng L, Lao CZ, Cen YJ, Liang GW. Study on the insecticidal activity compounds of the essential oil from *Syzygium aromaticum* against stored grain insect pests. 10th International Working Conference on Stored Product Protection. Julius-Kühn-Archiv 2010; 425:766-771

50 Politeo O, Jukić M, Milos M. Chemical composition and antioxidant activity of essential oils of twelve spice plants. Croatica Chemica Acta 2006;79:545-552

51 Dègnon GR, Faton AN, Adjou ES, Noudogbessi J-P, Dahouenon-Ahoussi E, Soumanou MM, Sohounhloue DCK. Antifungal potential of clove essential oil (*Syzigium aromaticum* L.) in the post-smoking preservation of mackerel (*Trachurus trachurus*) in Benin. Int Res J Biological Sci 2013;2(10):36-42

52 Pinto E, Vale-Silva L, Cavaleiro C, Salgueiro L. Antifungal activity of the clove essential oil from *Syzygium aromaticum* on *Candida*, *Aspergillus* and dermatophyte species. J Med Microbiol 2009;58:1454-1462

53 Sessou P, Farougou S, Yèhouenou B, Agniwo B, Alitonou G, Azokpota P, et al. Biological control of spoilage and pathogens moulds in culture medium and Beninese traditional cheese wagashi by *Syzygium aromaticum* essential oil. Afr J Microbiol Res 2013;7:2454-2463

54 Abed Karkosh AS. Study of *in vitro* antibacterial activity of the essential oils of cloves (*Syzygium aromaticum*) and the effect of temperature on antibacterial activity. Euphrates J Agricult Sci 2012;4:15-19

55 Rhind JP. Essential oils. A handbook for aromatherapy practice, 2nd Edition. London: Singing Dragon, 2012

56 Lawless J. The encyclopedia of essential oils, 2nd Edition. London: Harper Thorsons, 2014

57 Davis P. Aromatherapy. An A-Z, 3rd Edition. London: Vermilion, 2005

58 Adams RM, Maibach HI. A five-year study of cosmetic reactions. J Am Acad Dermatol 1985;13:1062-1069

59 Wetter DA, Yiannias JA, Prakash AV, Davis MD, Farmer SA, el-Azhary RA, et al. Results of patch testing to personal care product allergens in a standard series and a supplemental cosmetic series: an analysis of 945 patients from the Mayo Clinic Contact Dermatitis Group, 2000-2007. J Am Acad Dermatol 2010;63:789-798

60 Wöhrl S, Hemmer W, Focke M, Götz M, Jarisch R. The significance of fragrance mix, balsam of Peru, colophonium and propolis as screening tools in the detection of fragrance allergy. Br J Dermatol 2001; 145:268-273

61 Sánchez-Pérez J, García-Díez A. Occupational allergic contact dermatitis from eugenol, oil of cinnamon and oil of cloves in a physiotherapist. Contact Dermatitis 1999;41:346-347

62 Silvers SH. Stomatitis and dermatitis venenata with purpura, resulting from oil of cloves and oil of cassia. Dental Items of Interest 1939;61:649-651

63 Trattner A, David M, Lazarov A. Occupational contact dermatitis due to essential oils. Contact Dermatitis 2008;58:282-284

64 Rudzki E, Grzywa Z, Bruo WS. Sensitivity to 35 essential oils. Contact Dermatitis 1976;2:196-200

65 Schubert HJ. Skin diseases in workers at a perfume factory. Contact Dermatitis 2006;55:81-83

66 Rudzki E, Grzywa Z. Allergy to perfume mixture. Contact Dermatitis 1986;15:115-116

67 Sternberg L. Contact Dermatitis. Cases caused by oil of cloves and by oil of camomile tea (*Anthemis Cotula*). J Allergy 1937;8:185-186

68 Gaul LE. Dermatitis of the hands from oil of cloves. Skin (Los Angeles) 1963;2:314

69 Steele JC, Bruce AJ, Davis MDP, Torgerson RR, Drage LA, Rogers RS III. Clinically relevant patch test results in patients with burning mouth syndrome. Dermatitis 2012;23:61-70

70 Christoffers WA, Blömeke B, Coenraads P-J, Schuttelaar M-LA. The optimal patch test concentration for ascaridole as a sensitizing component of tea tree oil. Contact Dermatitis 2014;71:129-137

71 Uter W, Schmidt E, Geier J, Lessmann H, Schnuch A, Frosch P. Contact allergy to essential oils: current patch test results (2000–2008) from the Information Network of Departments of Dermatology (IVDK). Contact Dermatitis 2010;63:277-283

72 Larsen W, Nakayama H, Fischer T, Elsner P, Frosch P, Burrows D, et al. Fragrance contact dermatitis – a worldwide multicenter investigation (Part III). Contact Dermatitis 2002;46:141-144

73 Manuel Brites M, Goncalo M, Figueiredo A. Contact allergy to fragrance mix — a 10-year study. Contact Dermatitis 2000;43:181-182

74 Katsarma G, Gawkrodger DJ. Suspected fragrance allergy requires extended patch testing to individual fragrance allergens. Contact Dermatitis 1999;41:193-197

75 Sha M, Lewis FM, Gawkrodger DJ. Facial dermatitis and eyelid dermatitis: a comparison of patch test results and final diagnoses. Contact Dermatitis 1996;34:140-141

76 Magnusson B, Wilkinson DS. Cinnamic aldehyde in toothpaste. 1. Clinical aspects and patch tests. Contact Dermatitis 1975;1:70-76

77 Silvers SH. Stomatitis and dermatitis venenata with purpura, resulting from oil of cloves and oil of cassia. Dental Items of Interest 1939;61:649-651. Data cited in ref. 76

78 Calnan CD. Oil of cloves, laurel, lavender, peppermint. Contact Dermatitis Newsletter 1970;7:148

79 Rudzki E, Grzywa Z. Dermatitis from propolis. Contact Dermatitis 1983;9:40-45

80 de Groot AC. Propolis: A review of properties, applications, chemical composition, contact allergy, and other adverse effects. Dermatitis 2013;24:263-282

81 Yiannias JA, el-Azhary RA, Hand JH, Pakzad SY, Rogers RS III. Relevant contact sensitivities in patients with the diagnosis of oral lichen planus. J Am Acad Dermatol 2000;42:177-182

82 Meneghini CL, Rantuccio F, Lomuto M. Additives, vehicles and active drugs of topical medicaments as causes of delayed-type allergic dermatitis. Dermatologica 1971;143:137-147

83 Schorr WF. Cosmetic allergy: Diagnosis, incidence, and management. Cutis 1974;14:844-850

Chapter 5.27   CLOVE LEAF OIL

There are three types of clove essential oils: those made from the buds (clove bud oil), leaves (clove leaf oil) and from the stems (clove stem oil) of *Syzygium aromaticum* L. In this chapter, clove leaf oil is discussed. The bud and stem oils are presented in Chapters 5.26 and 5.28.

DEFINITION

Clove leaf oil (essential oil of clove leaves) is the essential oil obtained from the leaves of the clove (clove tree), *Syzygium aromaticum* (L.) Merr. & L.M. Perry.

INCI NOMENCLATURE

Description/definition: Eugenia caroyphyllus leaf oil is an essential oil steam-distilled from the leaves of the clove, *Eugenia caryophyllus*, Myrtaceae
INCI name EU: Eugenia caryophyllus leaf oil
INCI name USA: Eugenia caryophyllus (clove) leaf oil
CAS registry number(s): 8000-34-8; 84961-50-2; 8015-97-2 (old CAS number, mentioned in CosIng)
EINECS number(s): 284-638-7

ISO (INTERNATIONAL ORGANIZATION FOR STANDARDIZATION) S.TANDARD

ISO number: 3141
ISO name: Essential oil of clove leaves
Botanical origin: *Syzygium aromaticum* (L.) Merr. & L.M. Perry (synonym: *Eugenia caryophyllus* (Spreng.) Bullock & S.G. Harrison)
Parts of plant used: Leaf
ISO values: ISO values (minimum and maximum concentrations) are shown in Table 5.27.1.

Table 5.27.1 ISO values (%) for clove leaf oil[a]

Compound	CAS	Minimum	Maximum
Eugenol	97-53-0	80.0	92.0
β-Caryophyllene	87-44-5	4.0	17.0
Eugenyl acetate	93-28-7	0.2	1.0

[a] ISO Essential oil of clove leaf ©ISO 1997; Geneva, Switzerland, www.iso.org

THE PLANT, THE OIL, AND THEIR USES

For a general introduction to clove and clove oils see Chapter 5.26 Clove bud oil.

CHEMICAL COMPOSITION

Clove leaf oil is a yellow to slightly brown mobile to slight viscous liquid which has a spicy, aromatic and sweet odor. The yield of essential oil from the leaves of *Syzygium aromaticum* (L.) Merr. & L.M. Perry generally varies from 1 to 4%. The main producing countries of this oil are Madagascar, Indonesia, Sri Lanka and India.

Literature data (up to November 13, 2014) on the chemical composition of clove leaf oils and unpublished analytical data from one of us (E.S.) are shown in Table 5.27.2 in alphabetical order. In clove leaf oils from various origins, over 110 chemicals have been identified. About 65% of these were found in a single reviewed publication only. The major compounds found in clove leaf oils from different sources are shown in Table 5.27.3. They include (highest concentrations in any study given) eugenol (94.4%), β-caryophyllene (20.5%), eugenyl acetate (5.9%) and α-humulene (3.4%). Uncommon or rare constituents of clove leaf oils found in high concentrations (>4%) in single studies include 1,8-cineole (4.6% and 5.8%).

Commercial oils

The ten chemicals that had the highest maximum concentrations in 201 commercial clove leaf essential oil samples (concentration ranges provided) are the following: eugenol (75.7-89.1%), β-caryophyllene (2.6-19.5%), eugenyl acetate (0.08-5.9%), α-humulene (0.5-3.0%), (*E,E*)-farnesyl acetate (0.01-0.7%), benzyl benzoate (0.03-0.6%), caryophyllene oxide (0.2-0.6%), α-copaene (0.01-0.6%), δ-cadinene (0.1-0.6%) and caryophyllene alcohol (0.02-0.4%) (Erich Schmidt, unpublished analytical data).

CONTACT ALLERGY/ALLERGIC CONTACT DERMATITIS

Articles on contact allergy to clove oil, specified as being clove *leaf* oil, have not been found. The literature on contact allergy to/allergic contact dermatitis from 'clove oil' '(plant part unspecified, most likely the buds) is discussed in Chapter 5.26 Clove bud oil.

Table 5.27.2 Constituents identified in clove leaf oils

Constituent	CAS	Percentages and range						
		A	B (6)	C (5)	D (4)	E (2)	F (1)	G
2-Acetylfuran	1192-62-7				tr			
17-(Acetyloxy)-kauran-18-al	1421058-94-7	0.1						
allo-Aromadendrene	25246-27-9			0.05				+^{e}(tr)
allo-Aromadendrene oxide	85760-81-2			0.1				
α-Asarone	2883-98-9							0.5^c
Benzaldehyde	100-52-7				tr			0.07^c
Benzoic acid	65-85-0				tr			
Benzyl alcohol	100-51-6							1.4^c
Benzyl benzoate	120-51-4	0.03-0.6						
Benzyl n-octanoate	10276-85-4						0.1	
(E)-α-Bergamotene	13474-59-4					0.4		
β-Bisabolene	495-61-4			0.06				
β-Bourbonene	5208-59-3							+^{e}(tr)
Cadalene	483-78-3							+^{e}(0.03); 0.2^a
γ-Cadinene	39029-41-9	0-0.4				0.3		0.2^a; 0.4^{d,f}; 0.6^c
δ-Cadinene	483-76-1	0.1-0.6		0.2				<0.01^c; 0.3^h; 0.4^{d,f}
α-Cadinol	481-34-5			2.4		0.1		+^{e}(0.03)
δ-Cadinol	19435-97-3							+^{e}(0.05)
τ-Cadinol	5937-11-1					0.2		0.07^a
α-Calacorene	21391-99-1							+^{e}(0.08)
Calamenene	483-77-2							0.2^h; 0.3^d
Calamenenol	52658-10-3							+^{e}(0.2)
Camphene	79-92-5				tr	0.1		
d-Camphor	464-49-3	tr-0.06						
β-Caryophyllene	87-44-5	2.6-19.5	11.7-19.5	3.9	20.5	13.0	17.4	6.4^c; 14.0^d; 15.6^b; 17.2^h
Caryophyllene alcohol	56747-96-7	0.02-0.4					0.1	
Caryophyllene oxide	1139-30-6	0.2-0.6	0.05-0.6	0.8	tr	0.5	0.4	0.5^h; 0.6^b; 0.7^a; 0.9^d
Cedr-9-ene				0.2				
Chavicol	501-92-8	tr-0.2	0-0.1	0.08		0.1	0.1	0.1^b
1,8-Cineole	470-82-6			5.8			0.1	0.06^c; 4.6^i
α-Clovene							tr	
α-Copaene	3856-25-5	0.01-0.6	0-0.2	0.2				0.04^a; 0.6^{d,h}
α-Cubebene	17699-14-8			0.02				0.08^c; 0.2^{d,h}
Cubenol	21284-22-0			0.2				+^{e}(tr)
m-Cymene	535-77-3			0.2				
p-Cymene	99-87-6				tr			0.05^c
p-Cymenene	1195-32-0				tr			
γ-Elemene	29873-99-2			0.2				
Ethyl cinnamate	103-36-6							0.5^i
Eugenol	97-53-0	75.7-89.1	75.0-83.6	74.3	78.1	82.0	76.8	79.5^b; 81.0^c; 84.9^i; 94.4^a
Eugenyl acetate	93-28-7	0.08-5.9	0-1.5			0.4	1.2	tr^d; <0.01^c; 0.4^i;
(E,E)-α-Farnesene	502-61-4						0.1	0.06^a
(E)-β-Farnesene	18794-84-8				tr			
(E,E)-Farnesol	106-28-5							0.6^c
(Z,E)-Farnesol	3790-71-4							0.6^c
(E,E)-Farnesyl acetate	4128-17-0	0.01-0.7						
Furfural	98-01-1	0.01-0.4			tr			
Germacrene A	28387-44-2				tr			
Germacrene D	23986-74-5			0.4				
Globulol	489-41-8			0.4				
o-Guaiacol	90-05-1				tr			
α-Guaiene	3691-12-1			0.06				
β-Guaiene	88-84-6			0.09				
n-Heptadecane	629-78-7					0.2		
Hexadecyl acetate	629-70-9							0.09^a
Hexanol	111-27-3				tr			
2-Hexanone	591-78-6							0.03^c
(E)-2-Hexenal	6728-26-3				tr			
(Z)-3-Hexenol	928-96-1	tr-0.02			tr			
Humuladienone	24405-90-1							+^{e}(0.5)
Humulane-1,6-dien-3-ol	915392-38-0			0.5				
α-Humulene	6753-98-6	0.5-3.0	1.4-2.2	1.5	1.1	1.5	2.1	0.4^a; 1.4^c; 1.8^d; 2.1^h; 3.4^b

Table 5.27.2 Constituents identified in clove leaf oils (*continued*)

Constituent	CAS	Percentages and range						
		A	B (6)	C (5)	D (4)	E (2)	F (1)	G
α-Humulene epoxide	96638-51-6	0.04-0.1					0.1	
Humulene epoxide II	19888-34-7				tr	0.1		0.07[a]; 0.1[d]
Isoeugenol	97-54-1	0.08-0.2				0.1		<0.01[c]; 1.5[i]
(E)-Isoeugenol	5932-68-3		0-0.2				0.1	
(Z)-Isoeugenol	5912-86-7						tr	
(E)-Isoeugenyl acetate	5912-87-8						tr	
Isopropyl myristate	110-27-0					0.1		
Juniper camphor	473-04-1			0.2				
Ledol	577-27-5			0.2				+[e](tr)
Limonene	138-86-3			2.1	tr		0.1	
cis-Limonene oxide	13837-75-7						tr	
trans-Limonene oxide	4959-35-7						tr	
Linalool	78-70-6	0.01-0.3		0.1	tr			0.08[a]; 0.4[i];
cis-Linalool oxide	11063-77-7				tr			
trans-Linalool oxide	11063-78-8				tr	0.1		
Linalyl acetate	115-95-7	0.04-0.3						
Methyl chavicol	140-67-0						0.2	
Methyl eugenol	93-15-2	0.01-0.2	0-0.2				tr	0.1[d]
(E)-Methylisoeugenol	6379-72-2						tr	
(Z)-Methylisoeugenol	6380-24-1						tr	
Methyl salicylate	119-36-8	0.01-0.2		0.2		0.1	0.1	
α-Muurolene	10208-80-7							+[e](0.02)
γ-Muurolene	30021-74-0							+[e,g](0.1)
Myristic acid	544-63-8					0.1		
Nerol	106-25-2							0.8[a]
(E)-Nerolidol	40716-66-3	0.01-0.1				0.2		0.03[a]
Nerolidyl acetate	56001-43-5			0.06				
2-Nonanol	628-99-9							0.03[c]
2-Nonanone	821-55-6				tr			
(E)-β-Ocimene	3779-61-1							0.03[a]
n-Octanol	111-87-5				tr			
2-Octyl acetate	2051-50-5				tr			
Oleic acid	112-80-1					0.1		
Palustrol	5986-49-2							+[e](tr)
Perillene	539-52-6				tr			
α-Phellandrene	99-83-2			0.09				
β-Phellandrene	555-10-2			0.1				
Phenylacetaldehyde	122-78-1				tr			
Phytol					tr			
α-Pinene	80-56-8			0.3				0.03[c]
β-Pinene	127-91-3			0.5				0.03[c]
α-Terpinene	99-86-5			0.3				
γ-Terpinene	99-85-4			0.2				
Terpinen-4-ol	562-74-3			0.5				0.03[a]
α-Terpineol	98-55-5							0.4[c]
β-Terpineol	138-87-4							0.1[c]
α-Terpinyl acetate	80-26-2			0.6				
α-Ylangene	14912-44-8							+[e](1.5)
Zingiberene (α-)	495-60-3							+[e](0.1); +[g]

A two hundred and one clove leaf essential oil samples from Madagascar, Indonesia, Sri Lanka and India analyzed between 1998 and 2013; lowest and highest concentrations given (E. Schmidt, unpublished data)

B thirty-two samples of commercial clove leaf oils from Indonesia, Madagascar and Zanzibar; lowest and highest concentrations given (ref. 6)

C one lab-hydrodistilled clove leaf essential oil sample from Bangladesh (ref. 5)

D one lab-hydrodistilled leaf oil from Cuba (ref. 4)

E one commercial clove leaf essential oil sample from Madagascar, purchased in India (ref. 2)

F one sample of rectified clove leaf essential oil prepared in Germany, origin unknown (ref. 1)

G data from other studies (indicated with superscript letters); highest concentrations found in any study reviewed here given; when two or more oils were investigated, only the highest concentrations are mentioned, unless indicated otherwise

Table 5.27.2 Constituents identified in clove leaf oils (*continued*)

[a] one lab-hydrodistilled oil from Indian *S. aromaticum* leaves (ref. 3); [b] one commercial clove leaf oil purchased in Korea (ref. 7); [c] one lab-hydrodistilled oil from clove leaves harvested in local gardens in India (ref. 8); [d] one sample of Indonesian clove leaf oil; only the data of the major compounds were given (ref. 9); [e] one sample of Indonesian clove leaf oil; the chemicals found in the oil and presented here are indicated with +, as the figures were concentrations *in the sesquiterpenes fractions* of the oil (indicated in brackets), and therefore cannot be compared with the other data in the table (ref. 9); [f] γ-cadinene and δ-cadinene combined; [g] γ-muurolene and zingiberene combined; [h] one sample of commercial clove leaf oil (ref. 11); [i] one lab-hydrodistilled clove leaf oil from India (ref. 13)

tr: trace (in column D: <0.1); + present in the oil investigated, but quantity not stated or indicated in a manner that they cannot be compared with the other data

Table 5.27.3 Major constituents of clove leaf oils

Constituent	CAS	Percentages and range						
		A	B (6)	C (5)	D (4)	E (2)	F (1)	G
Eugenol	97-53-0	75.7-89.1	75.0-83.6	74.3	78.1	82.0	76.8	79.5[b]; 81.0[c]; 84.9[i]; 94.4[a]
β-Caryophyllene	87-44-5	2.6-19.5	11.7-19.5	3.9	20.5	13.0	17.4	6.4[c]; 14.0[d]; 15.6[b]; 17.2[h]
Eugenyl acetate	93-28-7	0.08-5.9	0-1.5			0.4	1.2	tr[d]; <0.01[c]; 0.4[i];
α-Humulene	6753-98-6	0.5-3.0	1.4-2.2	1.5	1.1	1.5	2.1	0.4[a]; 1.4[c]; 1.8[d]; 2.1[h]; 3.4[b]

LEGEND: SEE UNDER TABLE 5.27.2

LITERATURE

1 Jirovetz L, Buchbauer G, Stoilova I, Stoyanova A, Krastanov A, Schmidt E. Chemical composition and antioxidant properties of clove leaf essential oil. J Agric Food Chem 2006;54:6303-6307

2 Srivastava AK, Srivastava SK, Syamsundar KV. Bud and leaf essential oil composition of *Syzygium aromaticum* from India and Madagascar. Flavour Fragr J 2005;20: 51-53

3 Raina VK, Srivastava SK, Aggarwal KK, Syamasundar KV, Kumar S. Essential oil composition of *Syzygium aromaticum* leaf from · Little Andaman, India. Flavour Fragr J 2001;16:334-336

4 Pino JA, Marbot R, Agero J, Fuentes V. Essential oil from buds and leaves of clove (*Syzygium aromaticum* (L.) Merr. et Perry) grown in Cuba. J Essent Oil Res 2001;13:278-279

5 Bhuiyan NI, Begum J, Nandi NC, Akter F. Constituents of the essential oil from leaves and buds of clove (*Syzigium caryophyllatum* (L.) Alston). Afr J Plant Sc 2010;4:451-454

6 Razafimamonjison G, Jahiel M, Duclos T, Ramanoelina P, Fawbush F, Danthu P. Bud, leaf and stem essential oil composition of clove (*Syzygium aromaticum* L.) from Indonesia, Madagascar and Zanzibar. Int J Basic Appl Sci 2014;3(3):224-233.

7 Yang J-C, Lee S-H, Lee WJ, Choi DH, Ahn YJ. Ovicidal and adulticidal effects of *Eugenia caryophyllata* bud and leaf oil compounds on *Pediculus capitis*. J Agric Food Chem 2003;51:4884-4888

8 Golapakrishnan M, Narayanan CS, Mathew AG. Chemical composition of Indian clove bud, stem and leaf oils. Indian Perfum 1988;32:229-235

9 Vernin G, Vernin E, Metzger J, Pujol L, Parkanyi C. GC/MS analysis of clove essential oils. In: G Charalambous, Ed. Spices, herbs and edible fungi. Amsterdam: Elsevier Science Publishers, 1994:483-500. Data cited in ref. 10

10 Lawrence BM. Progress in essential oils. Perfum Flavor 1995 (Sept/Oct);20:102-103

11 Wei A, Shibamoto T. Antioxidant/lipoxygenase inhibitory activities and chemical compositions of selected essential oils. J Agric Food Chem 2010;58:7218-7225

12 Lawrence BM. Progress in essential oils. Perfum Flavor 2013;38(November):46-52

13 Baruah A, Nath SC, Hazarika AK. Leaf essential oil of *Syzygium aromaticum* (L.) Merrill and Perry grown at Diphu area of Assam, Northeast India. PAFAI 2010; 12(2):89-93. Data cited in ref. 12

Chapter 5.28 CLOVE STEM OIL

There are three types of clove essential oils: those made from the buds (clove bud oil), leaves (clove leaf oil) and from the stems (clove stem oil) of *Syzygium aromaticum* L. In this chapter, clove stem oil is discussed. The bud and leaf oils are presented in Chapters 5.26 and 5.27.

DEFINITION

Clove stem oil (essential oil of clove stem) is the essential oil obtained from the stem of the clove (clove tree), *Syzygium aromaticum* (L.) Merr. & L.M. Perry.

INCI NOMENCLATURE

Description/definition: Eugenia caroyphyllus stem oil is an essential oil steam-distilled from the stems of the clove, *Eugenia caryophyllus*, Myrtaceae
INCI name EU: Eugenia caryophyllus stem oil
INCI name USA: Eugenia caryophyllus (clove) stem oil
CAS registry number(s): 8000-34-8; 84961-50-2
EINECS number(s): 284-638-7

ISO (INTERNATIONAL ORGANIZATION FOR STANDARDIZATION) STANDARD

ISO number: 3143
ISO name: Essential oil of clove stem
Botanical origin: *Syzygium aromaticum* (L.) Merr. & L.M. Perry (synonym: *Eugenia caryophyllus* (Spreng.) Bullock & S.G. Harrison)
Parts of plant used: Stem
ISO values: ISO values (minimum and maximum concentrations) are shown in Table 5.28.1.

Table 5.28.1 ISO values (%) for clove stem oil[a]

Compound	CAS	Minimum	Maximum
Eugenol	97-53-0	83.0	92.0
Eugenyl acetate	93-28-7	4.0	12.0
β-Caryophyllene	87-44-5	0.5	4.0

[a] ISO 3143 Essential oil of clove stem ©ISO 1997; Geneva, Switzerland, www.iso.org

THE PLANT, THE OIL, AND THEIR USES

For a general introduction to clove and clove oils see under Clove bud oil.

CHEMICAL COMPOSITION

Clove stem oil is a yellowish to light brown, clear mobile, sometimes slightly viscous liquid which has a spicy, aromatic odor, reminding of fresh crushed cloves. The yield of essential oil from the stem of *Syzygium aromaticum* (L.) Merr. & L.M. Perry generally varies from 5 to 10%. The main producing countries of this oil are Madagascar, Indonesia, Sri Lanka, Malaysia and India.

Literature data (up to November 13, 2014) on the chemical composition of clove stem oils (only two publications found) and unpublished analytical data from one of us (E.S.) are shown in Table 5.28.2 in alphabetical order. In clove stem oils from various origins over 35 chemicals have been identified. About 73% of these were found in a single reviewed publication only; this high percentage may partly be explained by the low number of analytical studies performed. The major compounds found in clove stem oils from different sources are shown in Table 5.28.3. They include (highest concentrations in any study given) eugenol (97%), β-caryophyllene (9.5%), eugenyl acetate (5.9%) and α-humulene (1.5%).

Commercial oils

The ten chemicals that had the highest maximum concentrations in 29 commercial clove stem essential oil samples (concentration ranges provided) are the following: eugenol (86.5-90.1%), β-caryophyllene (2.4-9.5%), eugenyl acetate (0.2-5.9%), α-humulene (0.5-1.4%), caryophyllene oxide (0.2-0.6%), methyl salicylate (0.03-0.3%), furfural (0.01-0.2%), methyl eugenol (0.04-0.1%), caryophyllene alcohol (0.03-0.1%) and (E,E)-α-farnesene (0.01-0.1%) (Erich Schmidt, unpublished analytical data).

CONTACT ALLERGY/ALLERGIC CONTACT DERMATITIS

Articles on contact allergy to clove oil specified to be clove *stem* oil have not been found. The literature on contact allergy to/allergic contact dermatitis from 'clove oil' '(plant part unspecified, most likely the buds) is discussed in Chapter 5.26 Clove bud oil.

Table 5.28.2 Constituents identified in clove stem oils

Constituent	CAS	Percentage and range		
		A	B (1)	C (2)
α-Asarone (E-)	2883-98-9			0.5
Benzaldehyde	100-52-7			0.03
Benzyl acetate	140-11-4	tr-0.01		
Benzyl alcohol	100-51-6			1.5
γ-Cadinene	39029-41-9			0.4
δ-Cadinene	483-76-1			0.6
β-Caryophyllene	87-44-5	2.4-9.5	1.7-9.7	6.6
Caryophyllene alcohol	56747-96-7	0.03-0.1		
Caryophyllene oxide	1139-30-6	0.2-0.6	0.1-0.7	
Chavicol	501-92-8	0.05-0.3	0-0.2	
1,8-Cineole	470-82-6			0.04
α-Copaene	3856-25-5		0-0.3	
α-Cubebene	17699-14-8			0.5
p-Cymene	99-87-6			0.05
Eugenol	97-53-0	86.5-90.1	87.5-97	80.2
Eugenyl acetate	93-28-7	0.2-5.9	0.1-2.5	0.4
(E,E)-α-Farnesene	502-61-4	0.01-0.1		
(E,E)-Farnesol	106-28-5			0.7
(Z,E)-Farnesol	3790-71-4			0.8
Furfural	98-01-1	0.01-0.2		
2-Heptanol	543-49-7			0.01
2-Hexanone	591-78-6			0.05
α-Humulene	6753-98-6	0.5-1.4	0.2-1.3	1.5
α-Humulene epoxide	96638-51-6	tr		
Isoeugenol	97-54-1		0-0.8	0.4
Limonene	138-86-3	0.01-0.05		0.02[a]
Linalool	78-70-6	0.01-0.04		
Methyl eugenol	93-15-2	0.04-0.1	0-0.2	
5-Methyl-2-furaldehyde	620-02-0	tr		
Methyl salicylate	119-36-8	0.03-0.3	0-0.6	
2-Nonanol	628-99-9			0.06
α-Pinene	80-56-8			0.03
β-Pinene	127-91-3			0.06
α-Terpinene	99-86-5			0.02[a]
α-Terpineol	98-55-5			0.8
(E)-β-Terpineol				0.1
Vanillin	121-33-5			0.7

A twenty-nine clove stem essential oil samples from Madagascar, Indonesia, Sri Lanka and Malaysia analyzed between 1998 and 2013; lowest and highest concentrations given (E. Schmidt, unpublished data)

B forty-four samples of commercial clove stem oils from Indonesia, Madagascar and Zanzibar; lowest, highest concentrations given (ref. 1)

C one lab-hydrodistilled oil from fresh clove stems harvested in local gardens in India (ref. 2)

[a] α-terpinene and limonene combined; tr: trace

Table 5.28.3 Major constituents of clove stem oils

Constituent	CAS	Percentage and range		
		A	B (1)	C (2)
Eugenol	97-53-0	86.5-90.1	87.5-97	80.2
β-Caryophyllene	87-44-5	2.4-9.5	1.7-9.7	6.6
Eugenyl acetate	93-28-7	0.2-5.9	0.1-2.5	0.4
α-Humulene	6753-98-6	0.5-1.4	0.2-1.3	1.5

LEGEND: SEE UNDER TABLE 5.28.2

LITERATURE

1 Razafimamonjison G, Jahiel M, Duclos T, Ramanoelina P, Fawbush F, Danthu P. Bud, leaf and stem essential oil composition of clove (*Syzygium aromaticum* L.) from Indonesia, Madagascar and Zanzibar. Int J Basic Appl Sci 2014;3(3):224-233.

2 Golapakrishnan M, Narayanan CS, Mathew AG. Chemical composition of Indian clove bud, stem and leaf oils. Indian Perfum 1988;32:229-35

Chapter 5.29 CORIANDER FRUIT OIL

Most coriander oils are prepared from the ripe dried fruits of the coriander, *Coriandrum sativum* L. Essential oils can also be obtained from the leaves (coriander leaf oil, also known as cilantro oil). However, their smell is very unpleasant, and the leaf oils are mainly added to food and sometimes to natural cosmetics in extremely low doses to give a special effect; they are also employed in the pharmaceutical industry to produce natural (*E*)-2-decenal. Cilantro oils are probably hardly ever used in fine fragrances, cosmetics or aromatherapy and are therefore not discussed here.

DEFINITION

Coriander fruit oil (essential oil of coriander fruits) is the essential oil obtained from the fruit of the coriander, *Coriandrum sativum* L.

INCI NOMENCLATURE

Description/definition: Coriandrum sativum fruit oil is the volatile oil obtained from the dried fruit of the coriander, *Coriandrum sativum* L., Apiaceae (Umbelliferae)
INCI name EU: Coriandrum sativum fruit oil
INCI name USA: Coriandrum sativum (coriander) fruit oil
CAS registry number(s): 8008-52-4; 84775-50-8
EINECS number(s): 283-880-0

ISO (INTERNATIONAL ORGANIZATION FOR STANDARD-IZATION) STANDARD

ISO number: 3516
ISO name: Essential oil of coriander fruits
Botanical origin: *Coriandrum sativum* L.
Parts of plant used: Fruit
ISO values: ISO values (minimum and maximum concentrations) are shown in Table 5.29.1.

The fruit is often (also) termed seed, but they are not synonymous: the seeds are contained within the ripe fruits. Thus, the fruit essential oil should not be confused with Coriandrum sativum *seed* oil (INCI name, CAS 84775-50-8, EINECS 283-880-0), which is the *fixed* (vegetable) oil of

Table 5.29.1 ISO values (%) for coriander fruit oil[a]

Compound	CAS	Minimum	Maximum
Linalool	78-70-6	65.0	78.0
α-Pinene	80-56-8	3.0	7.0
γ-Terpinene	99-85-4	2.0	7.0
Limonene	138-86-3	2.0	5.0
Geranyl acetate	105-87-3	1.0	3.5
Geraniol	106-24-1	0.5	3.0
Myrcene	123-35-3	0.5	1.5
α-Terpineol	98-55-5	0.5	1.5

[a] ISO 3516 Essential oil of Coriander fruits ©ISO 1997; Geneva, Switzerland, www.iso.org

the coriander seed and is used as an emollient. The dried fruits contain approximately 20% fatty material and the fatty acids in coriander seeds are mostly characterized by petroselinic acid (66-75%), followed by linoleic acid (14-20%), oleic acid (5-8%), palmitic acid (4-5.5%), stearic acid (1%), palmitoleic acid (0.3%), arachidic acid (0.1%), α-linolenic acid (0.2%), myristic acid (0.1-0.2%) and gadoleic acid (0.1-0.2%) (37; see also ref. 45).

In literature on contact allergy to coriander oil, both the terms coriander oil (1) and coriander seed oil (2) are used; also in the latter publication, however, the essential oil (fruit oil) is meant, not the fixed oil, which is very unlikely to cause allergic reactions.

THE PLANT, THE OIL, AND THEIR USES

Coriander (synonyms: Chinese parsley, cilantro) is a glabrous, aromatic, herbaceous annual plant with a height of 30–90 cm. It is native to Mediterranean Europe and Western Asia and naturalized worldwide. It is now extensively cultivated as an important vegetable (leaves), spice and medicinal plant (fruits). The main exporters of coriander fruits are Russia, Egypt, India, Bulgaria, Morocco, Canada, China, Romania, Poland, the USA and Italy (3). The coriander is commonly used as a condiment or spice in the Mediterranean area, as a flavoring agent in food (confectionary, bread, curry powder [which contains 25-40% coriander fruits]), in perfumes, cosmetics and pharmaceutical preparations. In the Caucasus, Iran, Iraq and most parts of Latin America, the leaves and young plants are consumed in soups, salads and dressings (Mansfeld's World Database of Agriculture and Horticultural Crops, http://mansfeld.ipk-gatersleben.de). Coriander is a frequent ingredient in the preparation of Ayurvedic medicines and is a traditional home therapy for a variety of ailments (3). *C. sativum* fruits have been used as an aromatic carminative, stomachic, antispasmodic, galactagogue, laxative, antibilious, refrigerant and aphrodisiac (3,4,12). It is claimed to have sedative, anxiolytic and blood pressure lowering activities (5) and has been used for rheumatic complaints, diabetes, gastro-intestinal complaints, worms, insomnia, convulsions, anxiety and loss of appetite (6,12).

Coriander fruit essential oil, obtained by steam distillation of the ripe fruits containing seeds, is included among the 20 major essential oils in the world market. The oil is extensively used as a flavoring agent in all types of food products, including alcoholic beverages, candy, pickles and meat sauce, in tobacco, in seasonings and in perfumes. Many medicinal properties have been attributed to coriander essential oil, including antibacterial, antioxidant, antidiabetic, anticancer and anti-mutagenic activities (7,8,30). It may have some effectiveness in the treatment of fungal infections (9). The oil is also used in aromatherapy (10). Reviews of coriander, its products, their properties and their uses are provided in refs. 3,11,42, and 47.

CHEMICAL COMPOSITION

Coriander fruit oil is a colorless to pale yellow, clear mobile liquid which has a floral, fresh and herbaceous odor, reminding of linalool. The yield of essential oil from the fruits of *Coriandrum sativum* L. generally varies from 0.3 to 1.9%. The main producing countries of this oil are Russia, Egypt, India, Morocco, USA, Poland and Hungary.

Literature data (up to November 18, 2014) on the chemical composition of coriander fruit oils and unpublished analytical data from one of us (E.S.) are shown in Table 5.29.2 in alphabetical order. In coriander fruit oils from various origins, over 200 chemicals have been identified. About 48% of these were found in a single reviewed publication only. The major compounds found in coriander fruit oils from different sources are shown in Table 5.29.3. They include (highest concentrations in any study given) linalool (90.6%), α-pinene (31.5%), *p*-cymene (26.6%), geranyl acetate (24.5%), limonene (19.8%), γ-terpinene (18.2%), geraniol (17.9%), camphor (8.9%) and camphene (6.1%). Well-known ingredients of coriander fruit oils that were present in high concentrations (>7%) in one or two studies were neryl acetate

(8.7% and 12.2%), tetradecanoic acid (10.5%) and nerol (7.9%). Uncommon or rare constituents of coriander fruit oils found in high concentrations (>7%) in single studies include δ3-carene (60.5%; misidentification, is linalool), hexadecanoic acid (15.9%), (*E*)-2-dodecanal (two studies: 10.6% and 8.1%), α-thujene (7.9%) and linalyl acetate (7.1%).

Commercial oils

The ten chemicals that had the highest maximum concentrations in 38 commercial coriander essential oil samples (concentration ranges provided) are the following: linalool (55.0-81.2%), α-pinene (0.9-8.7%), camphor (2.2-8.0%), *p*-cymene (0.2-7.8%), geraniol (0.2-7.0%), γ-terpinene (0.4-6.6%), geranyl acetate (1.4-5.0%), camphene (0.2-4.8%), limonene (1.0-3.6%) and hexadecanoic acid (0-2.8%) (Erich Schmidt, unpublished analytical data).

Table 5.29.2 Constituents identified in coriander fruit oils

Constituent	CAS	Percentage and range				
		A	B (53)	C (12)	D (52)	E
Anethole	104-46-1					0.05[z2]; 0.1[p]; 0.2[x]; 0.5[v6]; 0.7[q]; 2.0[h]
(*E*)-Anethole	4180-23-8					0.3[z19]; 1.8[z18]; 3.5[z27]; 4.0[z22]; 19.8[z4]
(Z)-Anethole	25679-28-1					<0.1[y]
Benzyl benzoate	120-51-4					4.5[c]
Bicyclogermacrene	24703-35-3					0.04[z28]; 0.1[d]
β-Bisabolene	495-61-4					0.09[z]
β-Bisabolol	15352-77-9					0.9[w]
Borneol (endo-)	507-70-0	0.2-1.6	0.04-0.2		tr-0.2	0.7[u]; 1.3[x]; 1.0[z6]; 1.2[z9,z18]; 1.6[z28]; 2.4[y3]
Bornyl acetate	76-49-3					0.1[r]; 2.5[z22]
γ-Cadinene	39029-41-9					0.1[z22]
δ-Cadinene	483-76-1					<0.05[z]
α-Cadinol	481-34-5					0.1[w]
epi-α-Cadinol	5937-11-1					tr[d]
Camphene	79-92-5	0.2-4.8	0.02-0.8	0.1	tr-0.9	0.2-1.2[j]; 1.6[d]; 1.8z[8]; 2.0[y4]; 2.3[f]; 6.1[l]
α-Campholenal	4501-58-0					+[z25]; <0.05[z]
γ-Campholenic acid	67246-55-3					+[z25]
Camphor	76-22-2	2.2-8.0	0.07-5.7	0.3	0.5-5.2	1.2-5.3[e,k]; 3.2-5.4[j]; 5.5[n]; 5.6[g,q]; 5.7[d]; 5.8[z1]; 5.9[z18]; 6.4[p,z19]; 6.5[y2]; 6.6[l]; 8.9[z28]
δ2-Carene (δ4-carene)	554-61-0	tr-0.9				0.2[w]
δ3-Carene	13466-78-9					0.1[x]; 0.3[z6]; 0.4[o]; 0.5[w]; 1.1[q]; 60.5[y2,a1]
Carvacrol	499-75-2					0.06[z1]; 0.08[u]; 0.1[q,w]; 0.3[x]; 0.5[o,v]; 0.8[c]
Carveol	99-48-9					0.1[s]
(*E*)-Carveol	1197-07-5					0.06[z3]
Carvone	99-49-0	tr-0.4				0.7[o]; 1.0[q]; 1.1[x]; 1.2[s]; 1.6[e]; 1.7[y1]; 2.5[h]
β-Caryophyllene	87-44-5	0.08-1.3		0.1	tr-0.5	0.6[d]; 0.7[o,z6]; 0.9[y1]; 1.5[g]; 1.7[c]; 2.1[q]; 3.3[x]
Caryophyllene oxide	1139-30-6					tr[d,q]; <0.05[z19]; 0.05[z]; 0.1[s]
trans-Chrysanthenyl acetate	50764-55-1					tr[d]
1,8-Cineole	470-82-6	0.04-1.3	0-1.0		0-0.2	0.1-0.2[j]; 0.2[o,s,v]; 0.3[z2]; 0.4[z]; 1.0[x]; 3.3[w]
Citral	5392-40-5		0.05-0.2			0.2[z22]
Citronellal	106-23-0			0.4		tr[d]; 0.09[z]; 0.1[s,z1,z5]; 0.2[r,z11]; 0.4[c,x]; 0.5[r]
Citronellol	106-22-9	0.01-1.1	0.6-3.6			0.1[z]; 0.2[s,z1]; 0.3[p,q]; 0.5[d]; 0.7[h]; 1.1[o]; 1.7[i]
Citronellyl acetate	150-84-5			0.1		tr[d]; 0.2[z12]; 0.3[z]; 0.6[c]
Cuminaldehyde	122-03-2					0.5[i]; 0.6[z]; 1.1[z11]
Curcumene	644-30-4					0.1[z]
Cyclogeranyl acetate						0.06[z3]
p-Cymene	99-87-6	0.2-7.8	0-1.5		tr-2.7	0.1-8.1[e]; 0.3-1.0[j]; 0.8-3.6[c]; 3.5[s]; 4.4[y4]; 4.6[f]; 4.9[q]; 5.8[h]; 6.2[k]; 7.0[n]; 12.4[z22]; 26.6[y3]
p-Cymenol	25497-27-2					0.1[z3]

Table 5.29.2 Constituents identified in coriander fruit oils (*continued*)

Constituent	CAS	Percentage and range				
		A	B (53)	C (12)	D (52)	E
p-Cymen-8-ol	20834059-7					tr[v]; 0.1[q]; 0.2[w]; 0.3[u,x]; 0.4[z2,z6]; 0.6[p]; 1.9[o]
Decanal	112-31-2	0-0.1	0-0.1	1.6	tr-1.2	0.6[j,n]; 0.9[l]; 1.0[c]; 1.4[w]; 2.5[z9]; 4.7[z8]
(*E*)-2-Decanal				0.4		0.2[z14]; 0.6[c]
Decanoic acid	334-48-5					0.2[z4]; 1.2[s]
n-Decanol (1-)	112-30-1					0.1[r,z3]; 0.3[c]; 0.6[s]; 0.8[z8]; 1.0[l,z27]
3-Decanol	1565-81-7					0.1[z3]
Decenal	25447-70-5					0.2[r]; 0.3[s]
2-Decenal	3913-71-1		0.06-0.4			0.6[z8]
(*E*)-2-Decenal	3913-81-3	0-0.05				0.1[z19]; 0.3[m,z27]; 1.3[z9]
Decene	25339-53-1					0.3[r]
(*E*)-2-Decenol	18409-18-2					0.1[z27]; 0.2[s]
Dehydrolinalool	29171-20-8					0.4[w]
Dihydrocarveol	38049-26-2					0.1[s]; 0.3[y1]
cis-Dihydrocarvone	3792-53-8					<0.1[p]; 0.1[q,s]; 0.4[x]; 0.5[z6]; 0.6[z2]; 2.4[o,v]
Dihydrocarvone (*trans*-)	5948-04-9					0.5[h]
Dill apiole	484-31-1	0-0.08				0.5[z28]; 2.7[s,t]
Dimethoxycitral						0.2[r]
n-Docosane	629-97-0			1.6		
Dodecanal	112-54-9			0.4	0-0.1	0.05[z1]; 0.1[z14,z27]; 0.2[z28]; 0.3[r]; 1.0[c]; 2.4[z9]
(*E*)-2-Dodecanal				10.6		0.6[z14]; 8.1[c]
(*Z*)-2-Dodecanal						0.1[z14]
Dodecane	112-40-3					0.4[c]; 0.5[z15]
Dodecanoic acid	143-07-7					0.08[r]; 0.8[s]
Dodecenal	82107-89-9	tr-0.2				2.9[r]
2-Dodecenal	4826-62-4					0.8[s]
(*E*)-2-Dodecenal	20407-84-5					0.2[d]; 0.5[l]; 0.8[z8]; 1.0[z9]
3-Dodecenal	68083-57-8					
n-Eicosane	112-95-8			1.5		
β-Elemene	33880-83-0					
δ-Elemene	20307-84-0					0.01[o,v]; 0.4[z2]; 0.5[z6]; 0.6[x]
Elemol	639-99-6					0.09[z]
Ethyl tetradecanoate	124-06-1					4.5[r]
β-Eudesmol	473-15-4					<0.1[w]
Eugenol	97-53-0					0.1[p]; 0.2[z19]; 0.4[z2]; 0.6[z6]; 0.8[x]; 2.1[h,o]; 2.6[q]
Eugenyl acetate	93-28-7					tr[q]; 0.07[v]; 0.4[z2,z6]; 0.5[x]; 2.6[o]
(*E*)-β-Farnesene	18794-84-8					tr-0.1
β-Fenchyl alcohol	22627-95-8					0.1[z4]
Geranial	141-27-5					0.2[d,z3]; 0.5[z6]; 0.7[z2]; 1.0[x]; 1.3[e]; 6.3[w]
Geraniol	106-24-1	0.2-7.0	0.1-7.4		1.1-5.7	1.7[r,z16]; 2.0[y]; 2.0-3.2[j]; 2.8[q]; 3.3[d]; 3.4[y4]; 3.5[z28]; 3.6[e,o]; 3.9[j,m]; 8.2[l]; 17.9[z9]
Geranyl acetate	105-87-3	1.4-5.0	1.0-8.6		0.2-4.3	0.2-5.4[e]; 3.6[n]; 4.4[f]; 4.6[z18]; 5.7[z12]; 6.2[d]; 8.5[r]; 10.2[l]; 10.6[z22]; 13.8[o]; 15.1[x]; 24.5[i]
Geranyl formate	105-86-2					tr[q]; <0.05[z3]
Germacrene B	15423-57-1					tr[d]
Germacrene D	23986-74-5					0.05[v]; <0.1[w]; 0.2[z2]; 0.4[h,z6]; 0.9[o]; 1.0[x]
β-Gurjunene	73464-47-8					<0.1[w]
n-Heneicosane	629-94-7			0.2		
n-Heptadecane	629-78-7			0.4		tr[d]
Heptanal	111-71-7		0.01-0.2			tr[v]; 0.2[o,z6,z27]; 0.3[z2]; 0.5[u]; 0.6[p]
Heptanol	53535-33-4					0.8[x]
Hexadecanal	629-80-1					0.9[i]
n-Hexadecane	544-76-3			2.4		3.8[c]
Hexadecanoic acid	57-10-3	0-2.8				1.8[i]; 4[r]; 5.8[z28]; 15.9[s]
Hexahydrofarnesyl acetate	99624-94-9					0.2[d]
Hexahydrofarnesyl acetone	502-69-2			0.3		
Hexanal	66-25-1	0-0.02				
Hexanol	111-27-3					0.01[u]
(*Z*)-3-Hexanol						0.08[u]
(*E*)-2-Hexenal	6728-26-3					0.2[z13]
(*E*)-3-Hexenyl butyrate	53398-84-8					0.4[z3]

Table 5.29.2 Constituents identified in coriander fruit oils (*continued*)

Constituent	CAS	Percentage and range				
		A	B (53)	C (12)	D (52)	E
(*Z*)-3-Hexenyl butyrate	16491-36-4					0.01[v]; 0.1[q]; 0.2[z2]; 0.4[o]; 0.7[z6]
α-Humulene	6753-98-6	0.09-0.4	0-0.2		tr-0.3	0.4[z6]; 0.7[z2]; 0.8[z15]; 1.2[o]; 1.3[e]; 5.3[w]
β-Humulene	116-04-1					0.1[w]
Humulene epoxide II	19888-34-7					tr[d]
8-Hydroxylinalool	64142-78-5					0.1[z3]
(*Z*)-Isoapiole						2.7[s,t]
Isoborneol	124-76-5					0.1[i]
Isobornyl acetate	125-12-2					0.1[z3]
Isovaleraldehyde	590-86-3					<0.05[z]
Lavandulol	498-16-8					0.4[z10]
Limonene	138-86-3	1.0-3.6	0.04-1.6	0.3	0.1-2.4	0.1-3.2[e]; 0.9-5.1[j]; 2.6[o]; 4.0[y4]; 4.9[g]; 6.0[z28] 6.3[d]; 6.9[h]; 7.3[l]; 9.4[k]; 15.7[o]; 19.8[y3]
Linalool	78-70-6	55.0-81.2	49.3-83.2	89.5	62.8-90.0	40.9-79.9[c]; 58.0-80.3[e]; 68.9-83.7[j]; 83.2[z12]; 83.6[z6]; 86.1[u]; 87.5[o,v]; 90.6[z17]
Linalool oxide	1365-19-1		0.2-2.8			3.3[f]
cis-Linalool oxide	11063-77-7					0.1[z2]; 0.2[w]; 0.5[z]; 0.6[q,z12]; 0.9[z18]; 2.6[i]
cis-Linalool oxide, furanoid	5989-33-3	0.05-1.1			tr-0.3	tr-0.4[j]; 0.1[s,u]; 0.3[g,v,z6]; 0.4[z23]; 2.5[o]; 2.8[z3]
cis-Linalool oxide, pyranoid	14009-71-3	0.05-0.4				
trans-Linalool oxide	11063-78-8					0.1[z2,z7]; 0.2[z1]; 0.7[w]; 0.8[q,x]; 0.9[z18]; 2.2[z12]
trans-Linalool oxide, furanoid	34995-77-2	0.09-1.4			tr-0.3	tr-0.3[j]; 0.1[z23]; 0.2[o,u]; 0.4[g,z6]; 0.6[p]; 3.1[z3]
trans-Linalool oxide, pyranoid	39028-58-5	0.08-0.5				0.3[z3]
Linalyl acetate	115-95-7					0.3[z27]; 1.1[e]; 1.3[w]; 1.8[d]; 2.7[f]; 3.1[n]; 7.1[p]
Linalyl propionate	144-39-8					0.2[z7]
p-Mentha-1,4-dien-7-ol	22539-72-6					6.5[z10]
Menthol	89-78-1					0.01[u]; 0.05[v]; 0.1[p,q]; 0.3[h]; 0.2[o]; 0.5[w,x,z6]
Menthyl acetate	16409-45-3					4.5[z22]
Methyl chavicol	140-66-9					0.2[d]; 0.4[z4]
Methyl citronellate	2270-60-2					0.1[c]
Methyl eugenol	93-15-2					0.2[w]
6-Methyl-5-hepten-2-ol	1569-60-4					tr[y1]; <0.05[z3]
6-Methyl-5-hepten--2-one	409-02-9					<0.05[z,z3]; 0.1[q]
2-Methyl-5-phenol						0.1[s]
2-Methyl-3-phenyl-propanal	5445-77-2					1.5[z10]
Myrcene	123-35-3	0.2-2.1	0-0.2	0.2	0.1-1.0	0.4-1.0[j]; 1.4[g]; 1.8[y4]; 2.0[d]; 2.5[f]; 3.4[l]; 5.5[w]
Myrcene epoxide	29414-55-9					0.3[y1]
Myristicin	607-91-0					1.0[s]
Myrtenal	564-94-3	0-0.07	0.06-0.7			0.2[z28]
Myrtenol	19894-97-4	0.03-0.2				tr[d]; 0.2[w]
Myrtenyl acetate	1079-01-2			0.1		0.1[d]; 0.2[r]; 1.0[y1]
Neral	106-26-3		0-0.2			0.1[d,o,q,v]; 0.2[p]; 0.3[x]; 0.4[i]; 0.8[z6]; 4.5[h]
Nerol	106-25-2	0.03-1.0	0-1.5	0.7	tr-0.2	0.4[g]; 0.5[w]; 0.8[z6]; 1.0[o]; 1.8[c]; 2.0[x]; 7.9[y3]
Nerolidol	7212-44-4	0.2-0.5				0.2[z4]
(*E*)-Nerolidol	40716-66-3				tr-0.4	tr[d]; 2.3[g]
Neryl acetate	141-12-8	0.1-0.3	0.01-0.09	9.1		2.3-14.7[c]; 3.5[z17]; 4.4[d]; 8.7[z22]; 12.2[z9]
n-Nonadecane	629-92-5			0.9		
Nonanal	124-19-6	0.05-0.09	0.03-0.4		0-0.1	0-0.1[j]; 0.1[s]; 0.2[z1]
n-Nonane	111-84-2			0.1		
n-Nonanoic acid	12-05-0					0.2[s]
β-Ocimene	13877-91-3					0.09[y1]
(*E*)-β-Ocimene	3779-61-1	0.02			0-tr	0.04[z1]; 0.06[z]; 0.1[z7]; 0.2[g]; 1.7[z5,z24]
(*Z*)-β-Ocimene	3338-55-4	0.08				0.02[z2]; 0.08[z7]; 0.2[z11,z13]; 0.4[o,x,z6]
n-Octadecane	593-45-3			0.8		
Octadecanoic acid	57-11-4					0.2[z28]; 1.1[s]
Octanal (*n*-)	124-13-0	0-0.02	0-0.3			0.01[u]; 0.2[z27]; 2.2[z15]

Table 5.29.2 Constituents identified in coriander fruit oils (*continued*)

Constituent	CAS	A	B (53)	C (12)	D (52)	E
Octanoic acid	124-07-2	tr-0.3				0.1[s]
n-Octanol (1-)	111-87-5		0.08-0.4			<0.05[z5]; 0.1[z]; 0.2[z1]; 0.4[s]; 1.0[z12]; 2.0[i]
1-Octen-3-ol	3391-86-4		0.03-0.5			
Osthole	484-12-8					1.3[r]
Pentacosane	629-99-2					0.1[s]
Pentadecane (*n*-)	629-62-9			0.4		
Pentadecanoic acid	1002-84-2					1.1[s]
Phellandrene	1329-99-3		0.09-0.7			4.0[n]
α-Phellandrene	99-83-2				0-0.3	tr[y1]; 0.07[z]; 0.3[w]; 2.2[z16,y2]
β-Phellandrene	555-10-2	0.04-0.5			0-0.2[b]	tr[y1]; 0.04[z28]; 0.1[z1]; 0.2[y,z21]; 1.7[w]
2-Phenylethyl acetate	103-45-7					0.1[z23]
α-Pinene	80-56-8	0.9-8.7	2.0-9.1	7.8	0.1-6.7	1.2-6.5[j]; 1.2-6.6[c]; 7.3[n]; 8.5[z21]; 9.0[g]; 9.3[d]; 9.4[k]; 9.6[f]; 10.9[e]; 15.5[y4]; 23.2[i]; 31.5[l]
β-Pinene	127-91-3	0.1-0.8	0.3-1.1	0.6	tr-0.6	0.1-0.6[j]; 1.0[q]; 1.1[i]; 1.2[c,f,l]; 2.5[d]; 4.5[z26]
cis-Pinocarveol	6712-79-4					0.7[y1];
trans-Pinocarveol	547-61-5					0.08[z]
Pinocarvone	30460-92-5					tr[d]
Pinocarvyl acetate	1078-95-1					<0.1[y]
cis-Pinocarvyl acetate	73366-18-4					0.1[z3]
Sabina ketone	513-20-2					0.3[w]
Sabinene	3387-41-5	0.05-0.8	0.04-0.3	0.3	tr-0.5	0.1-0.3[j]; 0.5[g]; 0.6[x]; 0.7[h]; 0.8[c,f]; 0.9[e]; 1.2[l]
cis-Sabinene hydrate	15537-55-0					0.2[w]; 0.5[y1]
trans-Sabinene hydrate	17699-16-0					<0.01[u]; 0.04[z1]; <0.05[z5,z24]
cis-Sabinene hydrate acetate	77318-48-0					0.3[y1]
trans-Sabinyl acetate	139757-62-3					0.06[z1]
β-Sesquiphellandrene	20307-83-9					0.2[z]
Spathulenol	6750-60-3					tr[d]
α-Terpinen-7-al	1197-15-5					<0.1[z11]
γ-Terpinen-7-al	22580-90-1					0.2[z11]
α-Terpinene	99-86-5				tr-0.1	0.09[u]; 0.1[g,p]; 0.3[c-o]; 0.4[x]; 0.5[m,z24]; 0.7[h,z7]
γ-Terpinene	99-85-4	0.4-6.6	0.2-3.8	13.2	0.1-10.7	0.3-11.2[e]; 2.2-5.1[j]; 8.8[z13]; 8.9[p]; 9.2[k]; 9.3[g]; 10.5[z23]; 11.9[f,y4]; 12.1[d]; 15.4[l]; 18.2[y2]
δ-Terpinene	586-62-9					4.6[q]
Terpinen-4-ol	562-74-3	0.1-1.6		0.1	tr-0.4	0.1-0.3[j]; 0.1-1.2[e]; 0.7[g]; 0.8[z6]; 1.4[z28]
α-Terpineol	98-55-5	0.3-2.3			tr-3.7	0.2-0.8[e]; 1.5[z28]; 2.3[z28]; 2.6[g]; 3.2[z26]
β-Terpineol	138-87-4		0.2-5.9			0.5[z18]
Terpinolene	586-62-9	0.09-1.8	0.3-5.9		tr-0.6	0.3-0.6[j]; 1.0[z24]; 1.3[d,m]; 1.7[z23]; 2.9[o]; 5.4[z28]
α-Terpinyl acetate	80-26-2		0-0.2			
n-Tetracosane	646-31-1			0.6		
Tetradecanal	124-25-4					tr[d]; 0.1[z1]; 0.2[z28]; 0.6[c]
n-Tetradecane	629-59-4					1.7[c]
Tetradecanoic acid	544-63-8	tr-0.4				0.6[z28]; 1.0[z10]; 1.4[i]; 2.9[r]; 5.9[z15]; 10.5[s]
Tetradecanol	27196-00-5					0.3[r]
2-Tetradecenal	64461-99-0					0.4[d]
Tetradecene	26952-13-6					tr[d]
Tetrahydroionol	4361-23-3					0.1[z14]; 0.3[c]
α-Thujene	2867-05-2		0-0.2			tr[y1]; 0.3[c,z8]; 0.5[p]; 0.6[z6]; 0.7[o]; 1.4[z7]; 7.9[z11]
Thymol	89-83-8					0.05[u]; 0.8[c]; 0.9[z6]; 1.4[z2]; 1.9[v]; 2.3[o]; 3.4[z1]
Tricyclene	508-32-7					tr[d,y1]; 0.04[u]; <0.05[z3]; <0.1[y]; 0.1[p]
Tridecanal	10486-19-8					tr[d]; 1.0[i]
Tridecanoic acid	638-53-9					0.3[s]
α-Turmerone	82508-15-4					0.3[z]
ar-Turmerone	532-65-0					0.6[z]
β-Turmerone	82508-14-3					0.6[z]
Undecanal	112-44-7			0.2	0-0.2	tr[d]; 0.08[r]; 0.1[z,z28]; 0.5[c,j,z14]; 0.7[l]
Undecane	1120-21-4					0.1[c]
Undecanoic acid	112-37-8					0.3[i]; 0.4[s]
Verbenene	4080-46-0					<0.05[z]
Verbenone	80-57-9					tr[d]; 0.1[z]
p-Vinylguaiacol	7786-61-0					0.1[s]

Table 5.29.2 Constituents identified in coriander fruit oils (*continued*)

Constituent	CAS	Percentage and range				
		A	B (53)	C (12)	D (52)	E
Viridiflorol	552-02-3					0.1[d]
α-Ylangene	14912-44-8					<0.05[z]
Zingiberene	495-60-3					0.1[z]

A thirty-eight coriander fruit essential oil samples from Russia, Hungary, Egypt, India, Poland and the USA analyzed between 1998 and 2013; lowest and highest concentrations given (E. Schmidt, unpublished data)
B sixty lab-hydrodistilled fruit oils from 60 populations of *C. sativum* cultivated in the USA from seeds originating from 29 countries all over the world; lowest and highest concentrations given (ref. 53)
C twenty-six lab-hydrodistilled fruit oil samples from 26 Iranian landraces cultivated in a common environment in Iran; highest concentrations given (ref. 12); the lowest concentration was 0% (absent in one or more oils) for nearly all chemicals with the exception of linalool (69.7%), neryl acetate (1.3%) and γ-terpinene (0.1%)
D one hundred lab-hydrodistilled oils from seeds of plants cultivated in the USA from 100 different seed sources, mainly from European Botanical Garden collections; lowest and highest concentrations given (ref. 52)
E data from other studies (indicated with superscript letters); highest concentrations found in any study reviewed here given; when two or more oils were investigated, only the highest concentrations are mentioned, unless indicated otherwise

[a] incorrect identification; [a1] misidentification, is linalool; [b] β-phellandrene and 1,8-cineole combined; [c] nineteen lab-hydrodistilled oils from 19 *C. sativum* accessions from various parts of Iran; lowest and highest concentrations given, unless the lowest concentration was zero (chemical absent in one or more oils), in which case only the highest concentration is given (ref. 4); [d] six lab-hydrodistilled fruit oils from *C. sativum* cv. Sandra (ref. 5); [e] seventeen lab-hydrodistilled coriander seed oils, 12 from seeds purchased in various European countries and Turkey and five from cultivation in Estonia; lowest and highest concentrations given, unless the lowest concentration was zero (chemical absent in one or more oils), in which case only the highest concentration is given (ref.1 4); [f] eighteen industrially produced coriander seed oils from Georgia and Russia (ref. 16); [g] sixteen commercial coriander fruits oils from unknown origin (ref. 52); [h] nine lab-hydrodistilled oils from corianders cultivated in Egypt under different fertilization treatments (ref. 17); [i] eight samples of lab-hydrodistilled coriander seed oil from plants from various regions in India (ref. 19); [j] six lab-hydrodistilled oils from the fruits of *C. sativum* cultivated in various parts of Argentina, five commercial oils from Argentina and three from Russia; lowest and highest concentrations given (ref. 29); [k] seven lab-hydrodistilled oils and eleven commercial oil samples available in Germany (ref. 61); [l] various fruit oils from *C. sativum* var. *microcarpum* L. under different regimes of crushing the fruits and hydrodistillation times (ref. 43); [m] four lab-hydrodistilled fruit oils from a European and an Argentinian landrace cultivated under two levels of nitrogen fertilization and weediness (ref. 24); [n] two lab-hydrodistilled oils of coriander of two cultivars grown in Canada and two commercial oils from Hungary and India (ref. 6); [o] two fruit oils from Tunisia, one obtained by hydrodistillation, the other with steam-distillation (ref. 22); [p] two lab-hydrodistilled oil samples, one from Tunisian and one from Canadian coriander fruit (ref. 28); [q] one commercial and one steam-distilled fruit oil from Romania (ref. 33); [r] two fruit oil samples from India, one obtained by hydrodistillation, the second with microwave-assisted distillation (ref. 48); [s] four lab-hydrodistilled oils from mature fruits of two coriander varieties grown in Turkey at two locations (ref. 64); [t] dill apiole and (Z)-isoapiole combined; [u] one lab-hydrodistilled coriander fruit oil from Tunisia (ref. 7); [v] one lab-hydrodistilled oil from ripe coriander fruits cultivated in Tunisia (ref. 8); [w] one sample of lab-hydrodistilled seed oil from India (ref. 13); [x] one lab-hydrodistilled oil from coriander fruits growing wild in Iran (ref. 21); [y] one commercial oil sample from Sicily, Italy (ref. 26); [y1] one lab-hydrodistilled oil from coriander fruits purchased from a commercial Egyptian company (ref. 68); [y2] one commercial coriander seed oil from Austria with an extremely atypical composition: no linalool, but δ3-carene (60.5%) as dominant ingredient (ref. 40); [y3] five lab-hydrodistilled coriander fruit oils from (not entirely ripe) fruits cultivated in Egypt under different water treatments (ref. 69); [y4] nine lab-hydrodistilled coriander fruit oils from the USA with distillation times varying from 1.25 minutes to 4 hours; the high value for α-pinene (15.5%) was seen with the shortest distillation time (ref. 70); [z] one lab-hydrodistilled coriander oil obtained from seeds purchased in India (ref. 31); [z1] one lab-hydrodistilled oil from seeds purchased at a local market in Ethiopia (ref. 35); [z2] one lab-hydrodistilled oil from fully ripe fruits of *C. sativum* cultivated in Tunisia (ref. 38); [z3] one commercial fruit oil of unknown origin (ref. 39); [z4] one lab-hydrodistilled oil from fruits obtained in Lebanon (ref. 41); [z5] one lab-hydrodistilled seed oil from Italy (ref. 44); [z6] one lab-hydrodistilled oil from mature fruits from Tunisia (ref. 47); [z7] one lab-hydrodistilled coriander fruit oil from Egypt (ref. 18); [z8] one lab-hydrodistilled seed oil from Pakistan (ref. 10); [z9] one lab-hydrodistilled fruit oil from Brazil (ref. 20); [z10] one lab-hydrodistilled fruit oil from Algeria (ref. 23); [z11] one lab-hydrodistilled coriander seed oil from Iran, possibly from wild plants (ref. 25); [z12] one lab-hydrodistilled oil from fruits obtained at a local market in India (ref. 27); [z13] one lab-hydrodistilled oil from coriander purchased at a local market in Egypt (ref. 32); [z14] one bulked oil from lab-hydrodistilled oils from an Iranian landrace (ref. 34); [z15] one lab-hydrodistilled coriander oil from Algeria (ref. 36); [z16] one lab-hydrodistilled oil from Polish coriander fruits (ref. 49); [z17] two lab-hydrodistilled oils from Turkish *C. sativum* var. *microcarpum* and var. *macrocarpum* (ref. 50); [z18] one lab-hydrodistilled coriander fruit oil from Serbia (ref. 62); [z19] one lab-hydrodistilled fruit oil from Italy (ref. 63); [z20] one commercial oil purchased in Canada (ref. 30); [z21] one commercial coriander seed oil sample (ref. 55); [z22] one lab-hydrodistilled oil of coriander seed from Romania (ref. 56); [z23] one sample of coriander seed oil (ref. 58); [z24] oils of coriander seed produced by hydrodistillation of seed ground to particle sizes of 0.4-0.8 mm (ref. 59); [z25] data taken from ref. 65; [z26] one sample of coriander seed oil produced in the USA (ref. 54); [z27] one commercial coriander seed oil sample from the USSR (ref. 67); [z28] data from various older studies cited in ref. 66

tr: trace (in column E[d]: <0.1; in column E[v] and E[y1]: <0.01); + present in the oil investigated, but quantity not stated

Table 5.29.3 Major constituents of coriander fruit oils

Constituent	CAS	Percentage and range				
		A	B (53)	C (12)	D (52)	E
Linalool	78-70-6	55.0-81.2	49.3-83.2	89.5	62.8-90.0	83.2[z12]; 83.6[z6]; 86.1[u]; 87.5[o,v]; 90.6[z17]
α-Pinene	80-56-8	0.9-8.7	2.0-9.1	7.8	0.1-6.7	9.3[d]; 9.4[k]; 9.6[f]; 10.9[e]; 15.5[y4]; 23.2[i]; 31.5[l]
p-Cymene	99-87-6	0.2-7.8	0-1.5		tr-2.7	4.6[f]; 4.9[q]; 5.8[h]; 6.2[k]; 7.0[n]; 12.4[z22]; 26.6[y3]
Geranyl acetate	105-87-3	1.4-5.0	1.0-8.6		0.2-4.3	8.5[r]; 10.2[l]; 10.6[z22]; 13.8[o]; 15.1[x]; 24.5[i]
Limonene	138-86-3	1.0-3.6	0.04-1.6	0.3	0.1-2.4	6.0[z28]; 6.3[d]; 6.9[h]; 7.3[l]; 9.4[k]; 15.7[o]; 19.8[y3]
γ-Terpinene	99-85-4	0.4-6.6	0.2-3.8	13.2	0.1-10.7	9.3[g]; 10.5[z23]; 11.9[f,y4]; 12.1[d]; 15.4[l]; 18.2[y2]
Geraniol	106-24-1	0.2-7.0	0.1-7.4		1.1-5.7	3.4[y4]; 3.5[z28]; 3.6[e,o]; 3.9[j,m]; 8.2[l]; 17.9[z9]
Camphor	76-22-2	2.2-8.0	0.07-5.7	0.3	0.5-5.2	5.8[z1]; 5.9[z18]; 6.4[p,z19]; 6.5[y2]; 6.6[l]; 8.9[z28]
Camphene	79-92-5	0.2-4.8	0.02-0.8	0.1	tr-0.9	0.2-1.2[j]; 1.6[d]; 1.8z[8]; 2.0[y4]; 2.3[f]; 6.1[l]

LEGEND: SEE UNDER TABLE 5.29.2

The results of ref. 15 are not discussed here, as a two-year-old plant was apparently used (*C. sativum* is annual).

CONTACT ALLERGY/ALLERGIC CONTACT DERMATITIS

General
Contact allergy to/allergic contact dermatitis from coriander oil has been reported in a few studies only. In a group of consecutive patients suspected of contact dermatitis, the prevalence rate of positive patch test reactions was 1.0%, but relevance data were lacking. In one case report of occupational allergic contact dermatitis in an aromatherapist, the patient co-reacted to linalool and α-pinene, the two major ingredients of coriander oil.

Testing in groups of patients

Routine testing
Two hundred dermatitis patients from Poland were tested with coriander oil 2% in petrolatum and two (1%) reacted; relevance data were not provided (71).

Testing in groups of selected patients
In a group of 51 patients allergic to *Myroxylon pereirae* resin (balsam of Peru) and/or turpentine and/or wood tar and/or colophony and tested with coriander oil 2% in petrolatum, there were four (7.8%) positive patch test reactions (71). A group of 86 patients from Poland previously reacting to the fragrance mix was tested with coriander oil and three (3.4%) had a positive patch test reaction (73). In neither of these studies were relevance data provided.

Case reports
An aromatherapist had occupational contact dermatitis with allergies to multiple essential oils used at work, including coriander seed oil. The patient also reacted to geraniol, α-pinene, caryophyllene, the fragrance mix and various other fragrance materials; α-pinene and linalool were demonstrated by GC-MS in the coriander seed oil and are in fact the major ingredients of commercial coriander oils (Table 5.29.2, column A) (72).

Positive patch tests (relevance unknown, uncertain or not stated)
In a group of 16 patients known to be allergic to propolis and *Myroxylon pereirae* there were four (25%) reactions to coriander oil; no relevance data were provided (74).

LITERATURE
1 Rudzki E, Grzywa Z, Bruo WS. Sensitivity to 35 essential oils. Contact Dermatitis 1976;2:196-200
2 Dharmagunawardena B, Takwale A, Sanders KJ, Cannan S, Rodger A, Ilchyshyn A. Gas chromatography: an investigative tool in multiple allergies to essential oils. Contact Dermatitis 2002;47:288-292
3 Sharma MM, Sharma RK. Coriander. In: Peter KV, Ed. Handbook of herbs and spices, 2nd Ed., Vol. 1. Oxford-Cambridge-Philadelhpia-New Delhi: Woodhead Publishing Ltd, 2012: Chapter 12, 216-249
4 Ebrahimi SN, Hadian J, Ranjbar H. Essential oil compositions of different accessions of *Coriandrum sativum* L. from Iran. Nat Prod Res 2010;24:1287-1294
5 Tsagkti A, Hancianu M, Aprotosoaie C, Cioanca O, Tzakou O. Volatile constituents of Romanian coriander fruit. Rec Nat Prod 2012;6:156-160
6 Zheljazkov VD, Pickett KM, Caldwell CD, Pincock JA, Roberts JC, Mapplebeck L. Cultivar and sowing date effects on seed yield and oil composition of coriander in Atlantic Canada. Ind Crops Prod 2008;28:88-94
7 Sriti J, Talou T, Wannes WA, Cerny M, Marzouk B. Essential oil, fatty acid and sterol composition of Tunisian coriander fruit different parts. J Sci Food Agric 2009;89:1659-1664
8 Msaada K, Hosni K, Ben Taarid M, Chahed T, Kchouk ME, Marzouk B. Changes on essential oil composition of coriander (*Coriandrum sativum* L.) fruits during three stages of maturity. Food Chem 2007;102:1131-1134
9 Beikert FC, Anastasiadou Z, Fritzen B, Frank U, Augustin M. Topical treatment of tinea pedis using 6% coriander oil in unguentum leniens: a randomized, controlled, comparative study. Dermatology 2013;226:47-51
10 Shahwar MK, El-Ghorab AH, Anjum FM, Butt MS, Hussain S, Nadeem M. Characterization of coriander (*Coriandrum sativum* L.) seeds and leaves: volatile and non volatile extracts. Int J Food Prop 2012;15:736-747

11 Shavandi MA, Haddadian Z, Shah Ismail MH. *Eryngium foetidum* L., *Coriandrum sativum* and *Persicaria odorata* L.: a review. J Asian Sci Res 2012;2:410-426

12 Hadian J, Ebrahimi SN, Akramian M, Mumivand H. Variability in the essential oil content and composition of Iranian landraces of coriander (*Coriandrum sativum* L.), cultivated in a common environment. J Essent Oil Bear Plants 2012;15:89-96

13 Padalia RC, Karki N, Sah AN, Verma RS. Volatile constituents of leaf and seed essential oil of *Coriandrum sativum* L. J Essent Oil Bear Plants 2011;14:610-616

14 Orav A, Arak E, Raal A. Essential oil composition of *Coriandrum sativum* L. fruits from different countries. J Essent Oil Bear Plants 2011;14:118-123

15 Bhuiyan NI, Begum J, Sultana M. Chemical composition of leaf and seed essential oil of *Coriandrum sativum* L. from Bangladesh. Bangladesh J Pharmac 2009;4:150-153

16 Misharina TA. Influence of the duration and conditions of storage on the composition of the essential oil from coriander seeds. Appl Biochem Microbiol 2001;37:622-628

17 Said-Al Ahl HAH, Khalid KA. Response of *Coriandrum sativum* L. essential oil to organic fertilizers. J Essent Oil Bear Plants 2010;13:37-44

18 Romeilah RM, Fayed SA, Mahmoud GI. Chemical compositions, antiviral and antioxidant activities of seven essential oils. J Appl Sci Res 2010;6:50-62

19 Ravi R, Prakash M, Bhatt KK. Aroma characterization of coriander (*Coriandrum sativum* L.). Eur Food Res Technol 2007;225:367-374

20 Soares BV, Morais SM, Oliveira dos Santos Fontenelle R, Queiroz VA, Vila-Nova NS, Pereira CMC, et al. Antifungal activity, toxicity and chemical composition of the essential oil of *Coriandrum sativum* L. fruits. Molecules 2012;17:8439-8448

21 Khani A, Rahdari T. Chemical composition and insecticidal activity of essential oil from *Coriandrum sativum* seeds against *Tribolium confusum* and *Callosobruchus maculatus*. ISRN Pharmaceutics 2012, Article ID 263517, 5 pages. doi:10.5402/2012/263517

22 Msaada K, Ben Taârit M, Hosni K, Salem N, Tammar S, Bettaieb I, et al. Comparison of different extraction methods for the determination of essential oils and related compounds from coriander (*Coriandrum sativum* L.). Acta Chim Slov 2012;59:803-813

23 Zoubiri S, Baaliouamer A. Essential oil composition of *Coriandrum sativum* seed cultivated in Algeria as food grains protectant. Food Chem 2010;122:1226-1228

24 Gil A, De La Fuente EB, Lenardis AE, López Pereira M, Suárez SA, Bandoni A, et al. Coriander essential oil composition from two genotypes grown in different environmental conditions. J Agric Food Chem 2002;50:2870-2877

25 Eikani MH, Golmohammad F, Rowshanzamir S. Subcritical water extraction of essential oils from coriander seeds (*Coriandrum sativum* L.). J Food Eng 2007;80:735-740

26 Baratta MT, Dorman HJD, Deans SG, Biondi DM, Ruberto G. Chemical composition, antimicrobial and antioxidative activity of laurel, sage, rosemary, oregano and coriander essential oils. J Essent Oil Res 1998;10:618-627

27 Bankar R, Kumar A, Puri S, Sharma A, Khan IA. Chemical composition and antimicrobial activity of essential oil from seed of *Coriandrum sativum* L. Anal Chem Lett 2011;1:189-193

28 Sriti J, Wannes WA, Talou T, Vilarem G, Marzouk B. Chemical composition and antioxidant activities of Tunisian and Canadian coriander (*Coriandrum sativum* L.) fruit. J Essent Oil Res 2011;23:7-15

29 Bandoni AL, Mizrahi I, Juarez MA. Composition and quality of the essential oil of coriander (*Coriandrum sativum* L.) from Argentina. J Essent Oil Res 1998;10:581-584

30 Delaquis PJ, Stanich K, Girard B, Mazza G. Antimicrobial activity of individual and mixed fractions of dill, cilantro, coriander and eucalyptus essential oils. Int J Food Microbiol 2002;74:101-109

31 Singh G, Maurya S, de Lampasona MP, Catalan CAN. Studies on essential oils, Part 41. Chemical composition, antifungal, antioxidant and sprout suppressant activities of coriander (*Coriandrum sativum*) essential oil and its oleoresin. Flavour Fragr J 2006;21:472-479

32 Abd El Mageed MA, Mansour AF, El Massry KF, Ramadan MM, Shaheen MS, Shaaban H. Effect of microwaves on essential oils of coriander and cumin seeds and on their antioxidant and antimicrobial activities. J Essent Oil Bear Plants 2012;15:614-627

33 Anitescu G, Doneanu C, Radulescu V. Isolation of coriander oil: comparison between steam distillation and supercritical CO_2 extraction. Flavour Fragr J 1997;12:173-176

34 Nadjafi F, Mahdavi Damghani AM, Ebrahimi SN. Effect of irrigation regimes on yield, yield components, content and composition of the essential oil of four Iranian land races of coriander (*Coriandrum sativum*). J Essent Oil Bear Plants 2009;12:300-309

35 Mikre W, Rohloff J, Hymete A. Volatile constituents and antioxidant activity of essential oils obtained from important aromatic plants of Ethiopia. J Essent Oil Bear Plants 2007;10:465-474

36 Benyoussef E-H, Beddek N, Zouaghi N, Belabbes R, Bessière JM. Isolation of coriander oils by different processes. J Essent Oil Bear Plants 2004;7:129-135

37 Sriti J, Msaada K, Talou T, Faye M, Kartika IA, Marzouk B. Extraction of coriander oil by twin-screw extruder: Screw configuration and operating conditions effect. Ind Crops Prod 2012;40:355-360

38 Msaada K, Ben Taarit M, Hosni K, Hammami M, Marzouk B. Regional and maturational effects on essential oils yields and composition of coriander (*Coriandrum sativum* L.) fruits. Scientia Horticulturae 2009;122:116-124

39 Silva F, Ferreira S, Duarte A, Mendonca DI, Domingues FC. Antifungal activity of *Coriandrum sativum* essential oil, its mode of action against *Candida* species and potential synergism with amphotericin B. Phytomed 2011;9:42-47

40 Teixeira B, Marques A, Ramos C, Neng NR, Nogueira JMF, Saraiva JA, Nunes ML. Chemical composition and antibacterial and antioxidant properties of commercial essential oils. Ind Crops Prod 2013;43:587-595

41 Knio KM, Usta J, Dagher S, Zournajian H, Kreydiyyeh S. Larvicidal activity of essential oils extracted from commonly used herbs in Lebanon against the seaside mosquito, *Ochlerotatus caspius*. Bioresource Technol 2008;99:763-768

42 Mahendra P, Bisht S. *Coriandrum sativum*: A daily use spice with great medicinal effect. Pharmacogn J 2011;3:84-88

43 Smallfield BM, van Klink JW, Perry NB, Dodds KG. Coriander spice oil: effects of fruit crushing and distillation time on yield and composition. J Agric Food Chem 2001;49:118-123

44 Grosso C, Ceilho JA, Urieta JS, Palavra AMF, Barroso JG. Herbicidal activity of volatiles from coriander, winter savory, cotton lavender, and thyme isolated by hydrodistillation and supercritical fluid extraction. J Agric Food Chem 2010;58:11007-11013

45 Sriti J, Talou T, Msaada K, Marzouk B. Comparative analysis of fatty acid, sterol and tocol composition of Tunisian and Canadian coriander (*Coriandrum sativum* L.) fruit. Anal Chem Lett 2011;1:375-383

46 Sahib NG, Anwar F, Gilani A-H, Hamid AA, Saari N, Alkharfy KM. Coriander (*Coriandrum sativum* L.): A potential source of high-value components for functional foods and nutraceuticals — A Review. Phytother Res 2013;27:1439-1456

47 Msaada K, Hosni K, Ben Taarit M, Ouchikh O, Marzouk B. Variations in essential oil composition during maturation of coriander (*Coriandrum sativum* L.) fruits. J Food Biochem 2009;33:603-612

48 Salehi Sourmaghi MH, Kiaee G, Golfakhrabadi F, Jamalifar H, Khanavi M. Comparison of essential oil composition and antimicrobial activity of *Coriandrum sativum* L. extracted by hydrodistillation and microwave-assisted hydrodistillation. J Food Sci Technol 2014, DOI 10.1007/s13197-014-1286-x

49 Benelli G, Flamini G, Fiore G, Cioni PL, Conti B. Larvicidal and repellent activity of the essential oil of *Coriandrum sativum* L. (Apiaceae) fruits against the filariasis vector *Aedes albopictus* Skuse (Diptera: Culicidae). Parasitol Res 2013;112:1155-1161

50 Duman AD, Telci I, Dayisoylu KS, Digrak M, Demirtas I, Alma MH. Evaluation of bioactivity of linalool-rich essential oils from *Ocimum basilicum* and *Coriandrum sativum* varieties. Nat Prod Commun 2010;5:969-974

51 Lawrence BM. Progress in essential oils. Perfum Flavor 2013;38 (February):50-53

52 Lawrence BM. Unpublished data (1980). Data cited in ref. 51

53 Lopez PA, Widrlechner MP, Simon PW, Rai S, Boylston TD, Isbell TA, et al. Assessing phenotypic, biochemical, and molecular diversity in coriander (*Coriandrum sativum* L.) germplasm. Genet Resour Crop Evol 2008;55:247-275

54 Pavela R. Sublethal effects of some essential oils on the cotton leaf worm *Spodoptera littosalis* (Boisduval). J Essent Oil Bear Plants 2012;15:144-156

55 Giamperi L, Fraternale D, Ricci D. The *in vitro* action of essential oils on different organisms. J Essent Oil Res 2002;14:312-318

56 Zorca M, Gäinar I, Bala D. Supercritical CO_2 extraction of essential oil from coriander fruits. An Univ Bucharest Chimie 2006;2:79-83. Data cited in ref. 57

57 Lawrence BM. Progress in essential oils. Perfum Flavor 2011;36(1):48-52

58 Williams DG. The chemistry of essential oils, 2nd Ed. Port Washington, NY, USA: Micelle Press, 2008:178. Data cited in ref. 57

59 Grosso C, Ferraro V, Figueiredo AC, Barroso JG, Coelho JA, Palavra AM. Supercritical carbon dioxide extraction of volatile oil from Italian coriander seeds. Food Chem 2008;111:197-203

60 Lawrence BM. Progress in essential oils. Perfum Flavor 2008;33(1):38-? (last page unknown)

61 Braun M, Frantz G. Chirale Säulen decken Verfalschungen auf. Pharm Ztg 2001;146:2493-2499. Data cited in ref. 60

62 Samojlik I, Lakic N, Mimica-Dukic N, Dakovic-Svajcer K, Bozin B. Antioxidant and hepatoprotective potential of essential oils of coriander (*Coriandrum sativum* L.) and caraway (*Carum carvi* L.) (Apiaceae). J Agric Food Chem 2010;58:8848-8853

63 Lo Cantore P, Iacobellis NS, De Marco A, Capasso F, Senatore F. Antibacterial activity of *Coriandrum sativum* L. and *Foeniculum vulgare* Miller var. *vulgare* (Miller) essential oils. J Agric Food Chem 2004;52:7862-7866

64 Telci I, Toncer OG, Sahbaz N. Yield, essential oil content and composition of *Coriandrum sativum* varieties (var. *vulgare* Alef and var. *microcarpum* DC.) grown in two different locations. J Essent Oil Res 2006;18:189-193

65 de Rijke D, ter Heide R, Boelens H. New compounds with small rings in essential oils. Paper no. 179, 8th International Congress of Essential Oils, Cannes, France, 1980. Data cited in: Lawrence BM. Progress in essential oils. Perfum Flavor 1984;9(April/May):23-24

66 Lawrence BM. Progress in essential oils. Perfum Flavor 1997;22(1):43-56

67 Lawrence BM. Unpublished data, 1979. Data cited in ref. 51

68 Hanafi RS, Sobeh M, Ashour ML, El-Readi MZ, Desoukey SY, Niess R, et al. Chemical composition and biological activity of essential oils of cumin and coriander fruits from Egypt. Nat Prod J 2014;4:63-69

69 Hassan FAS, Ali EF. Impact of different water regimes based on class-A pan on growth, yield and oil content of *Coriandrum sativum* L. plant. J Saudi Soc Agric Sc 2014;13:155–161

70 Zheljazkov VD, Astatkie T, Schlegel V. Hydrodistillation extraction time effect on essential oil yield, composition, and bioactivity of coriander oil. J Oleo Sci 2014;63:857-865

71 Rudzki E, Grzywa Z, Bruo WS. Sensitivity to 35 essential oils. Contact Dermatitis 1976;2:196-200

72 Dharmagunawardena B, Takwale A, Sanders KJ, Cannan S, Roger A, Ilchyshyn A. Gas chromatography: an investigative tool in multiple allergies to essential oils. Contact Dermatitis 2002;47:288-292

73 Rudzki E, Grzywa Z. Allergy to perfume mixture. Contact Dermatitis 1986;15:115-116

74 Rudzki E, Grzywa Z. Dermatitis from propolis. Contact Dermatitis 1983;9:40-45

Chapter 5.30   COSTUS ROOT OIL

DEFINITION
Costus root oil is the essential oil obtained from the roots of the costus, *Saussurea costus* (Falc.) Lipsch. (synonyms: *Aplotaxis lappa* Decne.; *Aucklandia costus* Falck.; *Saussurea lappa* (Decne.) C.B. Clarke).

INCI NOMENCLATURE
Description/definition: Costus root oil is the volatile oil obtained from the root of the herbaceous plant costus, *Saussurea lappa* Clarke, Asteraceae
INCI name EU: Costus root oil (perfuming name, not an INCI name proper)
INCI name USA: Not in the Personal Care Products Council Ingredient Database
CAS registry number(s): 8023-88-9; 90106-55-1
EINECS number(s): 290-278-1

ISO (INTERNATIONAL ORGANIZATION FOR STANDARDIZATION) STANDARD
There is currently no ISO standard for costus root oil.

THE PLANT, THE OIL, AND THEIR USES
Saussurea costus (Falc.) Lipsch. is a perennial herb with a stout simple stem 1-2 meters high. The costus is endemic In India (where it is called kuth in Hindi) in the sub-alpine regions of Jammu and Kashmir, Himachal Pradesh and Uttaranchal, in altitudes of 2500-4000 m. Costus can also be found in Pakistan. The major producers of cultivated *Saussurea costus* are India, China and Vietnam. The costus is a well-known medicinal plant used in the indigenous systems of medicine in India, Korea, Tibet, China, and Japan. In the Indian Ayurveda, Siddha, and Unani systems, it is used either as a single drug or in combination with other drugs. Its roots are used mainly as an antispasmodic in asthma, cough and also in the treatment of cholera, chronic skin diseases and rheumatism. Other indications for different preparations include cold, gastric ulcers, quartan malaria, leprosy, persistent hiccups, hair loss, stomach ache, toothache, typhoid fever, gout, erysipelas and many other conditions including cancer (1,2,12). Costus roots and its products have been reported to possess anti-inflammatory, antitumor, hepatoprotective, anti-ulcer, cholagogue, immune-modulatory, hypolipidemic, hypoglycemic, antimicrobial, anti-parasitic, insect repellent and central nervous system depressant properties (1,2,12). Reviews of botanical, chemical, pharmacological and therapeutic aspects are provided in refs. 2, 10 and 15.

To produce costus root oils, prior to distillation, the roots are not only comminuted, but also macerated in water. The distillation is then a combination of water- and steam-distillation to produce the yellow to yellow brownish viscous essential oil. It may apparently be used as a flavoring component in foods and beverages. In cosmetics, costus root oil used to serve as a fixative

and fragrance. Forty years ago it was demonstrated that the oil (through its major ingredients costunolide and dehydrocostus lactone) is a very powerful sensitizer (3). Dermal application may also cause (severe) skin irritation (3). Because of these dangers, IFRA (International Fragrances Association) banned the use of costus root oil in cosmetics and fragrances in 1974. In the EU, the use of costus root essential oil as a fragrance ingredient is prohibited. Quite curiously, according to some, the oil can be used in aromatherapy (18); others, however, consider its use (quite rightfully) contra-indicated (19).

CHEMICAL COMPOSITION
Costus root oil is a yellow to brown, viscous liquid which has a woody, erogenic animal, balsamic and slightly sweet odor. The yield of essential oil from the roots of *Saussurea costus* (Falc.) Lipsch. generally varies from 0.3 to 1.2%. The main producing countries of this oil are India, Pakistan and Nepal.

Whereas the composition of costus roots per se and their solvent extracts has been studied extensively, little recent information on the ingredients of costus root oil can be found in literature. In fact, no mention of it is made in any of the three most recent reviews on essential oils by BM Lawrence (5,6,7). This may indicate that its use is very limited, which may conveniently be explained by the well-known sensitizing capacity of the oil when applied to the skin, the ban of its use by IFRA and to the prohibition of its use in cosmetics in the EU (3).

Literature data (up to November 19, 2014) on the chemical composition of costus root oils and unpublished analytical data from one of us (ES) are shown in Table 5.30.1 in alphabetical order. In costus root oils from various origins, over 135 chemicals have been identified. About 59% of these were found in a single reviewed publication only. The major compounds found in costus root oils from different sources are shown in Table 5.30.2. They include (highest concentrations in any study given) dehydrocostus lactone (46.8%), β-costol (13.6%), costunolide (9.3%), α-costol (7.4%), α-selinene (5.0%), curcumene (4.9%), β-elemene (4.8%), β-selinene (4.5%) and β-caryophyllene (3.7%). Uncommon or rare constituents of costus root oils found in high concentrations (>7%) in single studies include *n*-heptadeca-1,8-11-14-tetraene (35.4%), *Z,Z,Z,*-hexadeca-7,10,13-trienal (two studies: 25.5% and 23.3%), 1,3-cyclooctadiene (19.2%), δ-elemene (12.7%), cedrenol (9.9%), hexadecatrienal (9.2%), dihydrodehydrocostus lactone (7.9%) and β-elemol (7.7%).

Commercial oils
The ten chemicals that had the highest maximum concentrations in 6 commercial costus root essential oil samples (concentration ranges provided) are the following: *n*-heptadeca-1,8,11,14-tetraene (26.1-35.4%), dehydrocostunolide (1.8-20.5%), dihydrodehydrocostus lactone (2.5-7.9%), β-elemenol (2.6-7.7%), γ-costol (1.5-7.6%), β-elemene (0.1-4.8%), β-costol (3.5-4.7%), *trans*-caryophyllene oxide (0.5-4.3%), β-selinene (1.8-4.2%) and β-caryophyllene (1.1-3.7%) (Erich Schmidt, unpublished analytical data).

Table 5.30.1 Constituents identified in costus root oils

Constituent	CAS	Percentage and range					
		A	B (1)	C (11)	D (12)	E (16)	F
Acetic acid	64-19-7						+[c]
α-Amorphenic acid	69793-64-2						+[c]
Anethole	104-46-1			tr	0-0.09		
allo-Aromadendrene	25246-27-9				0.2		
2,12-Bergamotadien-14-al							0.2[e]
trans-α-Bergamotene	13474-59-4	0.4-1.3	0.4				0.2[e]
(Z)-trans-α-Bergamotol	88034-74-6					0.2	1.1[b]
(Z)-α-Bisabolene	29837-07-8				0.2-0.4		
Borneol	507-70-0					0.1	
Butylidenephthalide (3-)	551-08-6						0.2[b]
δ-Cadinene	483-76-1				0.2-0.3		
Camphene	79-92-5			tr	tr	tr	
Camphor	76-22-2			0.3	0-tr		
Carvacrol	499-75-2					0.1	
Carvotanacetone	499-71-8				0.04-0.06		
β-Caryophyllene	87-44-5	1.1-3.7	1.2	2.3	0-0.08	tr	0.9[e]; 1.7[b]
Caryophyllene oxide	1139-30-6	0.5-4.3	0.8	2.0	0.5-0.8	1.7	0.8[e]
Cedrenol	28231-03-0						9.9[b]
8-Cedren-13-ol	18319-35-2		5.1				
Chrysanthenone	473-06-3			0.8	0-tr		
1,8-Cineole	470-82-6		1.4	1.0	0-tr		
(E)-Cinnamaldehyde	14371-10-9		0.5				
Citronellal	106-23-0			0.5	0-0.02		
Citronellyl propionate	141-14-0				0-0.03		
α-Copaene	3856-25-5		0.6		tr-0.05		
α-Costal	4586-01-0						0.1[e]
β-Costal							0.1[e]
γ-Costal							0.1[e]
α-Costol	65018-15-7	2.1-2.9				7.4	3.8[e]; 4.0[d]
β-Costol	515-20-8	3.5-4.7	0.2			0.3	6.0[e]; 13.6[d]
γ-Costol				1.8	2.2-2.5		3.3[e]
Costunolide	553-21-9	1.5-7.6	9.3	2.6	0.1-0.2		4.3-8.3[a]
Cryptone	500-02-7			tr	0-tr		
Curcumene (ar-; α-)	644-30-4	0.9-2.3	4.3		2.9-4.9	0.1	0.7[e]; 2.8[b]
γ-Curcumene	28976-68-3		0.1				
1,3-Cyclooctadiene	1700-10-3						19.2[b]
Cyercene 4	136669-19-7				0-0.5		
p-Cymene	99-87-6	0.08-0.4		1.4	0-0.03	0.3	2.0[d]
Cyperene	2387-78-2						0.2[b]
α-Cyperene	17627-30-4	0.3-0.6					0.2[e]
Dehydrocostus lactone	477-43-0	1.8-20.5	46.8	16.7	10.3-11.0	28.9	13.7[b]; 16.3-25.4[a]
Dehydro-β-ionone					0-0.2		
Dehydrosaussurea lactone	28290-35-9			3.2	0-1.1		
Dihydrodehydrocostus lactone	4955-03-7	2.5-7.9					
Dihydro-α-ionone	31499-72-6	0.4-1.0	1.2			1.6	
Elema-1,3,11(13)-trien-12-al							0.2[e]
α-Elemene	5951-67-7						4.7[b]
β-Elemene	33880-83-0	0.1-4.8	0.3	2.1	4.1-4.7	1.2	1.2[e]; 3.0[d]
γ-Elemene	29873-99-2						2.1[d]
δ-Elemene	20307-84-0						12.7[d]
β-Elemenol	65018-04-4						4.9[e]
Elemicin	487-11-6				0-0.08		
α-Elemol	639-99-6	0.4-1.3	0.4	5.8	1.8-3.0	0.6	1.0[e]; 3.2[d]
β-Elemol	32142-08-8	2.6-7.7					
Ethyl linoleate	544-35-4				0.02-0.03		
4-Ethyloctanoic acid	16493-80-4						+[c]
β-Eudesmene	17066-67-0						1.8[b]
Eudesmol	51317-08-9						1.5[b]
α-Eudesmol	473-16-5				0-0.3		
β-Eudesmol	473-15-4		1.3	0.5	0-4.6		
γ-Eudesmol (selinenol)	1209-71-8		0.5			tr	

Table 5.30.1 Constituents identified in costus root oils (*continued*)

Constituent	CAS	Percentage and range					
		A	B (1)	C (11)	D (12)	E (16)	F
(*E,Z*)-α-Farnesene	28973-98-0				0-3.3		
α-Fenchene	471-84-1				0-3.5		
Furfural	98-01-1				0-tr		
Geranyl acetone	3796-70-1	0.8-1.7	0.8	3.8	1.2-1.4	2.3	
Globulol	489-41-8					tr	
β-Guaiene	88-84-6				1.3-2.7		
γ-Gurjunene	22567-17-5		0.2				
Heneicosane	629-94-7				0-0.02		
n-Heptadeca-1,8-11-14-tetraene	71046-96-3	26.1-35.4					
Heptanoic acid	111-14-8						+[c]
Hexadecatrienal						9.2	
Z,Z,Z-Hexadeca-7,10,13-trienal	56797-43-4			25.5	21.2-23.3		
Hexanal	66-25-1			tr	0-0.07		
Hexanoic acid	142-62-1						+[c]
β-Himachalene	1461-03-6				0.7-0.9		
α-Humulene	6753-98-6	0.2-0.6			0.4-0.7		
α-Ionene	475-03-6		2.5	3.9	2.4-2.9		
(*E*)-α-Ionol	25312-34-9					0.1	
α-Ionone	127-41-3	1.1-2.6				1.6	3.3[d]
β-Ionone	79-77-6	0.1-1.3		2.3	1.1-2.0	0.4	
(*E*)-β-Ionone	79-77-6		1.3				
Isocritonilide	62458-57-5				0-1.7		
3-Isopropylpentanoic acid	60308-89-6						+[c]
Italicene ether	104188-25-2					0.5	
Lepidozenal						0.1	
Limonene	138-86-3	0.1-0.2	0.3	0.1	0-tr	0.1	
Linalool	78-70-6		0.3	0.6	tr-0.05	0.1	
Linoleic acid	60-33-3				0-tr		
α-Longifolene (junipene)	475-20-7						1.0[b]
cis-p-Menth-2-en-1-ol	29803-82-5					tr	
Menthone	89-80-5				0-tr		
4-Methylbenzaldehyde (*p*-)	104-87-0					tr	
3-Methylbutyric acid	503-74-2						+[c]
Methyl chavicol	140-67-0				0-0.9		
Methyl linoleate	112-63-0				0.04-0.1		
7-Methyl-3,4-octadiene	37050-05-8						1.5[b]
Methyl salicylate	119-36-8					0.1	
Myrcene	123-35-3			tr	tr		
5-epi-Neointermedeol	136734-27-5				0-1.0		
Nerolidol	7212-44-4			0.3	0-0.3		
(*E*)-Nerolidol	40716-66-3		0.3			tr	
Octanoic acid	124-07-2						+[c]
7-Octenoic acid	18719-24-9						+[c]
(*E*)-β-Ocimene	3779-61-1			2.3	0-tr		
Paeonol	552-41-0		1.0			1.4	
1-Pentadecene	13360-61-7	0.7-1.5					
Phellandrene	1329-99-3				0-0.02		
α-Phellandrene	99-83-2	0.02-0.7	0.2	0.6	0.2		
β-Phellandrene	555-10-2		0.2	tr	0.03-0.1	tr	
Phenylacetaldehyde	122-78-1					tr	
α-Pinene	80-56-8	0.04-0.3	2.5	0.06	tr-0.07	0.1	
β-Pinene	127-91-3	0.1-0.9	1.5	0.2	0-0.2	0.5	1.6[d]
Pogostol	21698-41-9					2.3	
Pregnane					tr-0.03		
14β-Pregnane					tr		
Sabinene	3387-41-5	0.08-0.2		0.06	0-0.2	0.1	
Safrole	94-59-7		0.2				
(*Z*)-α-Santalol	115-71-9		0.3				
α-Selinene	473-13-2	0.6-2.2	2.8	1.2	1.7-2.7	1.5	0.1[e]; 5.0[d]
β-Selinene	17066-67-0	1.8-4.2	0.4	0.6	0.6-2.7		1.4[e]; 4.5[d]

Table 5.30.1 Constituents identified in costus root oils (*continued*)

Constituent	CAS	Percentage and range					
		A	B (1)	C (11)	D (12)	E (16)	F
γ-Selinene	515-17-3						1.6[b]
Spathulenol	6750-60-3		2.3				
α-Terpinene	99-86-5	0.01-0.1		0.6	0-0.04	0.1	
γ-Terpinene	99-85-4	0.03-0.2	0.1		tr-0.1	0.2	
δ-Terpinene	586-62-9					0.1	
Terpinen-4-ol	562-74-3	0.1-0.2	1.8	1.6	0.1-0.6	0.7	3.4[d]
α-Terpineol	98-55-5		0.4		0.1	0.1	
δ-Terpineol	7299-42-5		0.1				
α-Terpinolene	586-62-9	0.02-0.07			tr-0.04		
α-Thujene	2867-05-2	0.01-0.1		0.2	0-0.02	tr	
Thymol	89-83-8			0.1	0-0.2		
2,4,6-Trimethylbenzal-dehyde	487-68-3						1.2[b]
Valerenal	4176-16-3				0.4		
Valerenol	101628-22-2			4.2	4.2-5.3		
Viridiflorol	552-02-3		1.1				
Vulgarol B				3.1	0-5.1		

A six costus root essential oil samples from India and Nepal analyzed in 2013 and 2014; lowest and highest concentrations given (E. Schmidt, unpublished data)
B one lab-hydrodistilled sample of costus root essential oil from plant material purchased in China (1)
C one lab-hydrodistilled oil from costus roots cultivated in India (ref. 11)
D two lab-hydrodistilled oils, one from Korean and one from Chinese costus roots; both concentrations given (ref. 12)
E one lab-hydrodistilled oil from Chinese costus roots (ref. 16)
F data from other studies (indicated with superscript letters); highest concentrations found in any study reviewed here given; when two or more oils were investigated, only the highest concentrations are mentioned, unless indicated otherwise

[a] unknown number of costus root essential oils from seven producing areas (provinces) in China; only the data of the main constituents dehydrocostus lactone and costunolide are known (article in Chinese) (ref. 9); [b] one lab-hydrodistilled oil from *Radix saussureae* (costus root) purchased from a commercial herb factory in China (ref. 8); [c] one costus root oil of unknown origin; the acidic fraction was investigated; no quantitative data available (ref. 4); [d] one lab-hydrodistilled oil of Indian cultivated costus roots (ref. 14); [e] one commercial oil purchased from a French manufacturer (ref. 17)

tr traces (in column C: <0.02; in column D: <0.01; in column E: <0.1); + present in the oil, but quantity not given

Table 5.30.2 Major constituents of costus root oils

Constituent	CAS	Percentage and range					
		A	B (1)	C (11)	D (12)	E (16)	F
Dehydrocostus lactone	477-43-0	1.8-20.5	46.8	16.7	10.3-11.0	28.9	13.7[b]; 16.3-25.4[a]
β-Costol	515-20-8	3.5-4.7	0.2			0.3	6.0[e]; 13.6[d]
Costunolide	553-21-9	1.5-7.6	9.3	2.6	0.1-0.2		4.3-8.3[a]
α-Costol	65018-15-7	2.1-2.9				7.4	3.8[e]; 4.0[d]
α-Selinene	473-13-2	0.6-2.2	2.8	1.2	1.7-2.7	1.5	0.1[e]; 5.0[d]
Curcumene (ar-; α-)	644-30-4	0.9-2.3	4.3		2.9-4.9	0.1	0.7[e]; 2.8[b]
β-Elemene	33880-83-0	0.1-4.8	0.3	2.1	4.1-4.7	1.2	1.2[e]; 3.0[d]
β-Selinene	17066-67-0	1.8-4.2	0.4	0.6	0.6-2.7		1.4[e]; 4.5[d]
β-Caryophyllene	87-44-5	1.1-3.7	1.2	2.3	0-0.08	tr	0.9[e]; 1.7[b]

LEGEND: SEE UNDER TABLE 5.30.1

CONTACT ALLERGY/ALLERGIC CONTACT DERMATITIS

General

Contact allergy to/allergic contact dermatitis from costus root oil has been reported in a few older studies only. Costus root oil is an irritant and has a strong sensitizing capacity due to the presence of the sesquiterpene lactones costunolide and dehydrocostus lactone. Therefore, IFRA (International Fragrances Association) banned the use of costus root oil in cosmetics and fragrances in 1974 (27). Costus oil pure causes irritation and patch test sensitization; 1% in petrolatum was considered suitable for patch testing (25).

Testing in groups of patients

The members of the North American Contact Dermatitis Group (NACDG) saw 6 positive patch test reactions to 'essence costus notre distillate' in a group of 282 (2.1%) patch tested dermatitis patients. Despite the high rate of sensitization, patch testing with it was abandoned (26).

Case reports

Costus root oil in cosmetics was reported to be the cause of contact dermatitis of the face in some Japanese women (22,24). 'Costus oil' was responsible for 1 out of 399 cases of cosmetic allergy where the causal allergen was identified in a study of the NACDG, USA, 1977-1983 (20).

Positive patch tests (relevance unknown, uncertain or not stated)

One positive patch test reaction to 'costus oil' was observed in a group of 460 patients with positive patch tests related to cosmetics; relevance data were not provided (21). Five patients reacted to costus root oil 1% in petrolatum, but had never had contact with the oil. However, they were all allergic to *Chrysanthemum* x *morifolium* (Compositae) or to the liverwort *Frullania* (Jubuilaceae). These plants are known to yield sesquiterpene lactones, and the five patients also showed positive patch test reactions to alantolactone and to some related lactones, which may indicate cross-sensitization to costus root oil (25).

LITERATURE

1 Liu ZL, He Q, Chu SS, Wang CF, Du SS, Deng ZW. Essential oil composition and larvicidal activity of *Saussurea lappa* roots against the mosquito *Aedes albopictus* (Diptera: Culicidae). Parasitol Res 2012;110:2125-2130

2 Pandey MM, Rastogi S, Rawat AKS. *Saussurea costus*: botanical, chemical and pharmacological review of an ayurvedic medicinal plant. J Ethnopharmacol 2007;110:379-390

3 RIFM. Fragrance raw materials monographs. Costus root, essential oil, absolute and concrete. Food Cosmet Toxicol 1974;12:867-868

4 Rijke D de, Traas PC, ter Heide R, Boelens H, Takken HJ. Acidic components in essential oils of costus root, patchouli and olibanum. Phytochem 1978;17:1664-1666

5 Lawrence BM. Progress in essential oils 2008-2011. Carol Stream, USA: Allured Publishing Corporation, 2012

6 Lawrence BM. Progress in essential oils 2005-2007. Carol Stream, USA: Allured Publishing Corporation, 2008

7 Lawrence BM. Progress in essential oils 2001-2004. Carol Stream, USA: Allured Publishing Corporation, 2006

8 Su S-I, Duan J-A, Zhao X-H, Hua Y-Q, Hou P-F, Shang E-X, et al. Bioactive components from oils of Siwu decoction and Xiangfu Siwu decoction by gas chromatography: mass spectrometry and principle components analysis. Mode Tradit Chin Med Mat Med 2008;10:50-57

9 Chen FL, Tan XM, Tang QF, Xing XF. GC-MS analysis of volatile oil from *Aucklandia lappa* from different producing areas. China Pharmacy 2011;22:2187-2189 (in Chinese). Data cited in ref. 1

10 Zahara K, Tabassum S, Sabir S, Arshad M, Qureshi R, Amjad MS, Chaudhari SK. A review of therapeutic potential of *Saussurea lappa* — An endangered plant from Himalaya. Asian Pac J Trop Med 2014; 7(Suppl.1):S60-S69

11 Gwari G, Bhandari U, Andola HC, Lohani H, Chauhan N. Volatile constituents of *Saussurea costus* roots cultivated in Uttarakhand Himalayas, India. Pharmacognosy Res 2013;5(3):179-182

12 Chang K-M, Kim G-H. Comparison of volatile aroma components from *Saussurea lappa* C.B. Clarke root oils. J Food Sci Nutr 2008;13:128-133. Data partly also presented in ref. 13

13 Chang KM, Kim GH. Volatile flavor compounds of *Saussurea lappa* C.B. Clarke root oils by hydro distillation-GC and GC/MS. Food Quality Culture 2007;1:13-17. Data also presented in ref. 12

14 Negi JS, Bisht VK, Bhandari AK, Bhatt VP, Sati MK, Mohanty JP, et al. Antidiarrheal activity of methanol extract and major essential oil contents of *Saussurea lappa* Clarke. Afr J Pharm Pharmacol 2013;7:474-477

15 Wei H, Yan LH, Feng WH, Ma GX, Peng Y, Wang ZM, Xiao PG. Research progress on active ingredients and pharmacologic properties of *Saussurea lappa*. Curr Opin Complement Alternat Med 2014;1:1-7

16 Zhao MT. Caracterisations chimiques et biologiques d'extraits de plantes aromatiques et médicinales oubliées ou sousutilisées de Midi-Pyrenées (France) et Chongqing (Chine). Doctoral Thesis, Université de Toulose, France, 2014:70-71. Available at: http://ethesis.inp-toulouse.fr/archive/00002794/01/zhao.pdf

17 Maurer B, Grieder A. Sesquiterpenoids from costus root oil (*Saussurea lappa* Clarke). Helv Chim Acta 1977;60:2177-2190

18 Lawless J. The encyclopedia of essential oils, 2nd Edition. London: Harper Thorsons, 2014

19 Davis P. Aromatherapy. An A-Z, 3rd Edition. London: Vermilion, 2005

20 Adams RM, Maibach HI. A five-year study of cosmetic reactions. J Am Acad Dermatol 1985;13:1062-1069

21 Romaguera C, Camarasa JMG, Alomar A, Grimalt F. Patch tests with allergens related to cosmetics. Contact Dermatitis 1983;9:167-168

22 Nakayama H. Fragrance hypersensitivity and its control. In: Frosch PJ, Dooms-Goossens A, Lachapelle J-M, Rycroft RJG, Scheper RJ, Eds. Current topics in contact dermatitis. Heidelberg: Springer Verlag, 1989:83-91

23 Mitchell JC, Epstein WL. Contact hypersensitivity to a perfume material, Costus absolute. Arch Dermatol 1974;110:871

24 Nakayama H. Cosmetic dermatitis in Japan. Read before the First Annual Clinical Chemical Correlation Seminar, Pacific Dermatological Association, Vancouver, British Columbia, Canada, 1973. Data cited in ref. 4

25 Mitchell JC. Contact sensitivity to costus root oil, an ingredient of some perfumes. Arch Dermatol 1974;109:572

26 Mitchell JC, Adams RM, Glendenning WE, Fisher A, Kanof N, Larsen W, et al. Results of standard patch tests with substances abandoned. Contact Dermatitis 1982;8:336-337

27 RIFM. Fragrance raw materials monographs. Costus root, essential oil, absolute and concrete. Food Cosmet Toxicol 1974;12:867-868

Chapter 5.31   CYPRESS OIL

DEFINITION

Cypress oil is the essential oil obtained from the twigs with leaves of the Italian (Mediterranean) cypress, *Cupressus sempervirens* L.

INCI NOMENCLATURE

Description/definition: Cupressus sempervirens leaf oil is the volatile oil obtained from leaves of the cypress, *Cupressus sempervirens* L., Pinaceae

INCI name EU & USA: Cupressus sempervirens leaf oil

CAS registry number(s): 84696-07-1

EINECS number(s): 283-626-9

Description/definition: Cupressus sempervirens oil is the volatile oil obtained from the whole plant, the cypress, *Cupressus sempervirens* L., Pinaceae

INCI name EU & USA: Cupressus sempervirens oil

CAS registry number(s): 8013-86-3; 84696-07-1

EINECS number(s): 283-626-9

ISO (INTERNATIONAL ORGANIZATION FOR STANDARDIZATION) STANDARD

Currently, there is no ISO standard for cypress oil.

AFNOR (Association Française de Normalisation) values are shown in Table 5.31.1.

THE PLANT, THE OIL, AND THEIR USES

The Italian (common) cypress (*Cupressus sempervirens* L.) is native to northern Africa (Libya), Asia (Cyprus, Iran, Israel, Jordan, Lebanon, Syria, Turkey), Italy and Greece; it is also cultivated (GRIN Taxonomy for Plants). This tree is mainly used as an ornamental tree due to its conical crown shape, but it can also be used for timber, as a privacy screen, and protection against wind as well. Moreover, cypress has proved to be very suitable as a pioneer species for reforestation as it can tolerate poor, barren, and superficial soils.

Phytopreparations obtained from the core and young branches of the cypress are reported to have antiseptic, adstringent, anti-inflammatory, antispasmodic, deodorant, and diuretic effects, to promote circulation to the kidneys and bladder area and to improve bladder tone (1). Cypress products are used in traditional medicine for the treatment of coughs, colds, flu, parasitic infections, rheumatism, haemorrhoids, and as a strong hair tonic (2,10). Commercial cypress essential oils, which are obtained from the young twigs with leaves of the cypress, are mainly used in perfumes, but can be used as flavors as well; another application is in aromatherapy (34).

CHEMICAL COMPOSITION

Cypress oil is a very pale to yellowish orange, clear, easily mobile liquid which has a terpeny, fresh and coniferous, slightly herbal odor. The yield of essential oil from the twigs with leaves of *Cupressus sempervirens* L. generally varies from 0.2 to 1.2%, depending on the freshness of the material and the harvest season. The main producing countries of this oil are France, Spain, Italy and Algeria.

Literature data (up to November 21, 2014) on the chemical composition of cypress oils and unpublished analytical data from one of us (E.S.) are shown in Table 5.31.2 in alphabetical order. In cypress oils from various origins, over 205 chemicals have been identified. About 57% of these were found in a single reviewed publication only.

The major compounds found in cypress oils from different sources are shown in Table 5.31.3. They include (highest concentrations in any study given) α-pinene (79.2%), sabinene (39.6%), δ3-carene (32.6%), limonene (32%), cedrol (23.7%), 1,8-cineole (23.2%), α-terpinyl acetate (12%), camphene (9.9%), terpinolene (9.4%) and β-pinene (7.1%).

Well-known ingredients of cypress oils that were present in high concentrations (>7%) in one or two studies were α-terpinene (23.6%), terpinen-4-ol (11.4% and 21.1%), myrcene (15.7%), germacrene D (12.1% and 13.0%) and γ-terpinene (8.0%). Uncommon or rare constituents of cypress oils found in high concentrations (>7%) in single studies include D-limonene (84.6%), (+)-2-carene (40.6%), allo-ocimene (24%), β-cyclocitral (8.9%), isopinocamphone (8.3%), dehydroabietane (8.0%) and phenanthrene (7.8%).

Commercial oils

The ten chemicals that had the highest maximum concentrations in 33 commercial cypress essential oil samples (concentration ranges provided) are the following: α-pinene (43.2-68.0%), δ3-carene (7.2-25.9%), limonene (0.1-10.8%), cedrol (0.4-4.2%), terpinolene (0.8-3.5%), terpinyl acetate (1.2-3.2%), myrcene (1.0-2.9%), sabinene (0.5-2.2%), β-caryophyllene (0.2-2.1%) and germacrene D (0.02-2%) (Erich Schmidt, unpublished analytical data).

Chemotypes

In nearly all analyses of cypress oils, α-pinene is the dominant ingredient, often followed by δ3-carene. In some publications, other chemicals were found to have the highest concentrations including sabinene (39.6%) (16), D-limonene (84.6%) (31) and (+)-2-carene (40.6%) (31).

CONTACT ALLERGY/ALLERGIC CONTACT DERMATITIS

General

Reports on routine testing with cypress oil have not been found. Allergic contact dermatitis to cypress oil has been reported in three studies only, describing occupational allergic contact dermatitis from cypress oils in two aromatherapists and a naturopathic therapist. One of these patients was allergic to α-pinene, which is the dominant component of cypress oils and may have been the allergen in that case.

Case reports

A naturopathic therapist had occupational contact dermatitis from cypress oil; she also reacted to other

Table 5.31.1 AFNOR values (%) for cypress oil[a]

Compound	CAS	Minimum	Maximum
α-Pinene	80-56-8	40.0	65.0
δ3-Carene	13466-78-9	12.0	25.0
Cedrol	77-53-2	0.8	7.0
Limonene	138-86-3	1.8	5.0
Terpinyl acetate	8007-35-0	1.0	4.0
Myrcene	123-35-3	1.0	3.5
Germacrene D	23986-74-5	0.5	3.0
β-Pinene	127-91-3	0.5	3.0
Terpinen-4-ol	562-74-3	0.2	2.0

[a] AFNOR NF T 75-254 Huile essentialle de cyprès; © AFNOR 1992, La Plaine Saint-Denis Cedex, France, www.afnor.org

Table 5.31.2 Constituents identified in cypress oils

Constituent	CAS	A	B (17)	C (1)	D (6)	E (8)	F (2)	G
Abienol	1616-86-0				tr			
Abietadiene	36312-33-1		tr	0.4-0.5				0.2[i]
Abietatriene	19407-28-4		1.0	0.4-0.8	tr			0.3[s]; 1.7[i]
Acetophenone	98-86-2							0.6[z2]
α-Acorenol	28296-85-7							3.5[s]
β-Acorenol	28400-11-5							0.5[s]
Aromadendrene	489-39-4		0.2					0.1[u]; 0.6[i]; 4.5[z2]
allo-Aromadendrene	25246-27-9		0.5				0.5	
trans-α-Bergamotene	13474-59-4							5.2[z2]
Bicyclogermacrene	24703-35-3							tr[u]
Bicyclosesquiphellan-drene	54324-03-7							2.3[f]
epi-Bicyclosesqui-phellandrene	54274-73-6							0.8[j]
β-Bisabolene	495-61-4		tr	0.5-0.4				
Borneol	507-70-0	0.3-0.4	0.1	0.2-0.0	tr		tr	tr[p,u]; 1.3[j]
Bornyl acetate	76-49-3	0.06-0.7	0.4			2.0	0.3	0.6[k]; 0.7[i]; 0.9[s]; 1.0[b]; 1.8[j]; 2.2[r]
Bornylene	464-17-5							tr[p]
β-Bourbonene	5208-59-3		0.1					0.1[u]
t-Butylbenzene	98-06-6							tr[a,p]
Cadalene	483-78-3							0.6[s]
Cadina-1,4-diene	29837-12-5		0.1					tr[p]
α-Cadinene	24406-05-1		0.3					
γ-Cadinene	39029-41-9		0.1		tr	tr	0.1	1.0[j]; 1.3[h]; 1.7[k]
δ-Cadinene	483-76-1	0.1-1.2	0.1	0.2-0.6	tr	1.2	0.6	0.5[w]; 0.6[d]; 0.8[b]; 1.6[i]; 1.9[f]; 2.2[h]
Cadinenol	17910-08-6							tr[p]
α-Cadinol	481-34-5				tr	0.8		tr[p]; 0.1[u]; 0.4[o]; 1.2[b]
δ-Cadinol	19435-97-3							tr[p]
τ-Cadinol	58580-31-7		0.1	0.5-1.3				
α-Calacorene	21391-99-1		0.1	0.2-0.1		tr		tr[p,u]
β-Calacorene	50277-34-4		0.3	0.6-1.0				tr[u]
Calamenene	483-77-2							3.5[s]
cis-Calamenene	72937-55-4		0.9	0.2-0.0		tr		0.3[i]; 1.7[f]
trans-Calamenen-10-ol	828923-23-5							0.8[s]
Camphene	79-92-5	0.4-1.7	tr		tr		0.5	0.3[q]; 0.5[e]; 1.4[h]; 2.4[j]; 7.3[b]; 9.9[r]
6-Camphenone	53803-33-1							3.7[z2]
α-Campholenal	4501-58-0		0.1	0.2-0.9	tr			
Camphor	76-22-2	0.02-0.5	0.1	0.1-0.1		tr		tr[p]; 0.1[w]; 0.4[d]; 0.5[e]; 0.8[j]; 1.2[s]
(+)-2-Carene	4497-92-1							40.6[z2]
δ2-Carene (= δ4-)	554-61-0							0.5[s]

Table 5.31.2 Constituents identified in cypress oils (*continued*)

Constituent	CAS	Percentage and range						
		A	B (17)	C (1)	D (6)	E (8)	F (2)	G
δ3-Carene	13466-78-9	7.2-25.9	22.9	7.2-13.2	10.8	10.6	0.2	22.1^w; 24^r; 24.4^u; 24.5^x; 32.6^k
cis-Caren-3-one								tr^p
Carvacrol	499-75-2		0.1				tr	0.1^w; 0.4^d; 0.8^j; 1.8^u
Carveol	99-48-9							0.6^x
Carvone	99-49-0							1.5^x
β-Caryophyllene	87-44-5	0.2-2.1	tr		tr		0.3	tr^u; 0.1^w; 0.4^p; 0.5^d; 1.2^z1; 4.2^l
γ-Caryophyllene	118-65-0			2.2-1.1				
Caryophyllene oxide	1139-30-6		0.1	0.3-1.1		tr	0.1	tr^p,u
α-Cedrene	469-61-4	0.04-0.6	0.1	0.6-1.3		tr		0.2^p; 0.6^b,m; 1.3^s
β-Cedrene	546-28-1	0.02-0.5				tr		0.1^b
Cedrenol	28231-03-0					tr		tr^p
Cedrol (α-)	77-53-2	0.4-4.2	5.2	19.3-7.7	4.9	tr	8.3	6.7^o; 9.5^j; 12.9^k; 21.3^b; 23.7^h
α-Chamigrene	19912-83-5							0.3^s
β-Chamigrene	18431-82-8							0.4^z2
cis-Chrysanthenyl acetate	67999-48-8							tr^u
1,8-Cineole	470-82-6	0.3-0.6		1.0-0.0		tr		0.3^q; 1.2^d ; 23.2^z2
Citronellol	106-22-9							0.06^p; 0.2^e
α-Copaene	3856-25-5	0.04-0.2	0.4		tr	tr		tr^m; 0.06^p; 0.4^i
α-Copaen-11-ol	41370-56-3			0.3-0.1				
β-Copaene	18252-44-3							0.3^u
α-Cubebene	17699-14-8		0.1			tr		tr^m,u; 0.2^i,p
β-Cubebene	13744-15-5		tr					tr^p; 0.3^e; 0.4^i; 0.7^a,o
Cubebol	23445-02-5			0.1-0.3				
Cubenol	21284-22-0							0.07^p
epi-Cubenol	19912-67-5							tr^u
Cuparene	16982-00-6					tr		
β-Cyclocitral	432-25-7							8.9^z2
Cyclofenchene	488-97-1							tr^p
Cycloisolongifolene	28380-07-6							1.0^b
o-Cymene	527-84-4							1.2^d
p-Cymene	99-87-6	0.2-0.5	0.5	0.2-1.1	tr		0.2	0.3^m; 0.4^u; 0.6^x; 0.7^q; 2.8^f; 3.1^e
p-Cymenene	1195-32-0				tr			tr^p; 0.03^o; 0.7^u
p-Cymen-8-ol	1197-01-9	0.02-0.2	0.1		tr	tr		tr^p; 0.6^s; 0.8^j
Cyperene	2387-78-2							0.1^e
(E,E)-2,4-Decadienal	25152-84-5							tr^u
Dehydroabietane	19407-28-4					4.2		0.3^p; 8.0^h
cis-Dihydrocarvone	3792-53-8		0.1					
trans-Dihydrocarvone	5948-04-9		0.1					
cis-2,7-Dimethyl oct-5-yn-3-ene								4.4^h
2,5-Dimethylstyrene	2039-89-6		tr					
Dodecane	112-40-3					tr		
β-Elemene	33880-83-0		0.1					0.05^e; 0.2^i; 0.9^u;
γ-Elemene	29873-99-2						tr	
Elemol	639-99-6		0.1	0.1-0.0				
Epizonarene	41702-63-0			0.2-0.6				0.2^o; 0.4^s
trans-4,5-Epoxy-(E)-2-decenal	134454-31-2							5.4^n
Eucumene								1.4^f
4(15),6-Eudesmadiene								0.9^h
α-Eudesmol	473-16-5						0.3	
β-Eudesmol	473-15-4						0.2	
γ-Eudesmol	1209-71-8							0.2^u
8-epi-γ-Eudesmol								0.1^u
Eugenol	97-53-0							0.7^o
(E,E)-α-Farnesene	502-61-4							1.0^u
α-Fenchene	471-84-1	0-0.8	0.5	0.6-0.7	tr	0.5		tr^m,u; 0.1^o; 0.5^j; 0.8^p,q; 1.3^d

Table 5.31.2 Constituents identified in cypress oils (*continued*)

Constituent	CAS	A	B (17)	C (1)	D (6)	E (8)	F (2)	G
Fenchone	1195-79-5							tr[p]
Fenchyl acetate	111821-74-0		tr					
α-Fenchyl alcohol	1632-73-1							0.2[j]
Ferruginol	514-62-5		tr					0.2[b]; 0.6[i]
α-Funebrene	50894-66-1							1.0[b]
β-Funebrene	79120-98-2							0.6[s]
Geranyl acetate	105-87-3							0.4[q]
Germacrene B	15423-57-1		0.2	1.5-1.2				0.3[i]
Germacrene D	23986-74-5	0.02-2.0	2.5	12.1-3.9	2.4	2.7	2.3	1.4[d]; 1.6[w]; 1.8[i]; 2.8[b]; 4[z1]; 13.0[l]
Globulol	489-41-8							tr[u]
α-Gurjunene	489-40-7		tr					0.3[i]
β-Gurjunene	73464-47-8		0.1					
Hexanoic acid	142-62-1							0.5[u]
(Z)-3-Hexen-1-ol	928-96-1							tr[p]
(Z)-3-Hexenyl acetate	3681-71-8							tr[u]
α-Himachalene	3853-83-6		tr					
β-Himachalene	1461-03-6							3.5[z2]
α-Humulene	6753-98-6	0.03-1.1	0.4	2.1-1.9	tr	tr	0.3	0.5s; 0.8b; 1.2u; 1.3z1; 2.8k; 4.2l
γ-Humulene	26259-79-0							0.1[u]
Humulene epoxide II	19888-34-7							tr[u]
Isoborneol	124-76-5		0.1					
Isobornyl acetate	125-12-2			0.3-0.7	tr	tr		tr[p]; 1.9[h]
β-Isocomene								tr[u]
Isopimara-9(11),15-diene	39702-28-8		tr					2.0[i]; 3.9[k]
Isopinocamphone	15358-88-0							8.3[z2]
p-Isopropyl anisole	4132-48-3					tr		tr[p]
Karahanaenone	19822-67-4	0.08-0.4						0.4[m]
Kaurane	1573-40-6		0.2					
(E,Z)-11,13-Labdadien-8-ol			0.2					0.1[i]
13(16),14-Labdadien-8-ol		0.4						1.3[i]; 2.4[b]
Limonene	138-86-3	0.1-10.8	5.1	2.2-1.9	3.2	4.5	4.6	5.8[b]; 6.2[j]; 6.4[s]; 7.0[k]; 7.3[e]; 32[y]
β-Limonene	5989-54-8							4.4[d]
D-Limonene	5989-27-5							84.6[z2]
Linalool	78-70-6	0.2-0.8	0.3	0.1-0.0	tr	tr	tr	0.3[p]; 0.7[j]; 0.8[m]; 0.9[s]; 2.6[i]
Linalyl acetate	115-95-7		tr				tr	0.2[p]
Longifolene	475-20-7		tr	0.6-0.6				0.6[i]
β-Longipinene	41432-70-6							0.2[s]
Manool	596-85-0		tr					tr[p]
Manoyl oxide	596-84-9		3.3	0.9-1.7		6.7		0.2[p,u]; 2.8[k]; 3.9[i]
13-epi-Manoyl oxide	1227-93-6		tr					1.3[i]
p-Mentha-1,5-dien-8-ol	1686-20-0		0.1					
p-Menthan-8-yl acetate	20777-41-7							3.6[c,j]
cis-p-Menth-2-en-1-ol	29803-82-5		0.1					
trans-p-Menth-2-en-1-ol	29803-81-4		0.1		tr			
p-Menth-1-en-3-yl acetate								tr[p]
7-Methoxybenzofuran	7168-85-6							0.2[o]
Methyl carvacrol	6379-73-3	0.01-0.2	3.3					0.5[o]; 0.8[i]; 0.9[b]
Methyl eugenol	93-15-2							tr[p]
Methyl thymol	1076-56-8	0.07-0.4	tr		tr		0.2	0.6[k]
Muurola-3,5-diene	157374-44-2							0.3[s]
Muurola-4,10(14)-dien-8β-ol								1.8[z2]
α-Muurolene	10208-80-7		tr	0.5-0.5	tr	tr	0.2	0.1[p]; 0.2[i]; 0.4[j]
γ-Muurolene	30021-74-0		0.3				0.2	0.1[u]; 0.3[i,j,p]; 3.2[s]
Muurol-5-en-4-one								3.1[s]
α-Muurolol	104245-48-9				tr			
epi-α-Muurolol	19912-62-0				tr			
τ-Muurolol	19912-62-0		tr	0.6-0.1				
Myrcene	123-35-3	1.0-2.9	2.5	1.0-1.9	4.6	1.4	3.9	4[w]; 5.0[l]; 5.1[t]; 5.5[z]; 6.6[z2]; 15.7[n]
Myrtenal	564-94-3		0.1	0.1-0.0				tr[p]

Table 5.31.2 Constituents identified in cypress oils (*continued*)

Constituent	CAS	Percentage and range						
		A	B (17)	C (1)	D (6)	E (8)	F (2)	G
Myrtenol	515-00-4		0.1	0.2-0.1		tr		tr[p]; 0.1[j]
Myrtenyl acetate	1079-01-2							tr[u]
(E)-Nerolidol	40716-66-3							tr[u]
Nezukol	14699-32-2			0.3-0.6				
2-Nonanone	821-55-6							tr[u]
(E)-β-Ocimene	3779-61-1	0.03-0.08	0.1		tr	tr		tr[p]; 0.4[u]
(Z)-β-Ocimene	3338-55-4	0.01-0.2	0.1					tr[p]
allo-Ocimene	673-84-7							24[h]
1-Octen-3-ol	3391-86-4							1.7[u]
2-Pentyl-2-cyclopen-ten-1-one	25564-22-1							1.6[h]
Perillene	539-52-6							tr[p]
α-Phellandrene	99-83-2		0.1	1.4-0.0		tr		0.2[u]; 1.1[e]
β-Phellandrene	555-10-2	0.2-0.5		0.1-0.0	tr	0.4		tr[m]; 1.5[f]
Phenanthrene	85-01-8							7.8[b]
Phenanthrenol	30774-95-9							1.8[b]
Phenol	108-95-2							0.5[b]
(E)-Phytol	150-86-7							1[u]
Pimaradiene	1686-61-9				2.9			
α-Pinene	80-56-8	43.2-68.0	20.0	27.5-35.8	54.1	44.9	60.5	58.5[z2]; 61.3[m]; 64[u]; 64.6[i]; 79.2[k]
β-Pinene	127-91-3	0.9-1.7	1.1	0.8-2.5	2.5	1.6	2.9	2.2[b,i]; 2.5[w]; 2.6[i]; 2.8[u]; 6.0[r]; 7.1[j]
Pinocarveol	5947-36-4						tr	
trans-Pinocarveol	1674-08-4							tr[p]
Pinocarvone	30460-92-5				tr			
cis-Piperitol	16721-38-3					tr		tr[p]
Piperonal	120-57-0							5.3[n]
Pulegone	89-82-7							0.2[u]
Sabinene	3387-41-5	0.5-2.2	0.3	0.2-1.3	2.4	0.6	1.3	14.8[f]; 24.4[t]; 30.6[z2]; 30.8[r]; 39.6[e]
Sandaracopimaradiene	1686-56-2		0.3			1.2		0.5[i]; 0.9[k]; 1.0[p]
β-Santalene	511-59-1							2.5[z2]
Santene	529-16-8							0.02[x]
β-Selinene	17066-67-0		0.7	0.6-1.8				0.4[i]
Sempervirol	1857-11-0			0.1-0.4				
Spathulenol	6750-60-3							1.3[s]; 2.1[u]
Terpinene	8013-00-1							5.7[f]
α-Terpinene	99-86-5	0.1-1.1	0.3		tr		0.2	0.7[d]; 3.3[x]; 3.9[e]; 4.2[f]; 23.6[z2]
γ-Terpinene	99-85-4	0.3-1.6	0.6		1.0	0.5	0.5	0.8[q]; 0.9[m]; 1.0[i]; 1.2[b]; 6.1[e]; 8.0[t]
Terpinen-4-ol	562-74-3	0.4-1.9	1.5	1.8-1.5	1.5	1.9		1.7[s]; 2.3[d]; 7[h]; 7.3[e]; 11.4[f]; 21.1[t]
Terpinen-4-yl acetate	4821-04-9		1.7					0.9[i]
α-Terpineol	98-55-5	0.2-0.5	0.6	1.1-0.0	tr	tr	tr	0.5[k]; 0.7[j]; 0.8[s]; 2.2[r]; 2.7[f]; 3.1[b]
δ-Terpineol	7299-42-5		tr	0.1-1.7				
Terpinolene	586-62-9	0.8-3.5	9.4	1.3-1.9	4.1	2.7	2.0	4.6[j]; 4.7[d]; 5.9[i]; 6.6[z1]; 6.9[v]; 9.2[b]
α-Terpinyl acetate	80-26-2	1.2-3.2	7.5	0.2-0.5	5.5	12.0		5.9[l]; 6.5[p]; 6.6[z1]; 7.1[j]; 9.1[r], 12[z2]
α-Terpinyl formate	2153-26-6							1.0[p]
α-Thujene	2867-05-2	0.4-1.3	0.1	0.1-0.0	tr		0.4	1[q]; 1.1[z2]; 1.3[j]; 1.4[w]; 1.7[d]; 2.2[e]
Thujopsene	470-40-6							0.7[s]
Thujyl alcohol	513-23-5							0.8[r]
Thymol	89-83-8						tr	
Totarol	511-15-9		0.3					4.8[i]
Tricyclene	508-32-7	0.05-0.2	0.1	0.1-0.1		tr	0.5	0.1[p]; 0.2[e,l,s]; 0.3[j,w]; 0.4[d]
Tridecane	629-50-5					tr		
Tridecanol	26248-42-0							tr[u]
Umbellulone	24545-81-1					tr		0.2[p]; 0.7[e]; 4.0[s]
Undecane	1120-21-4					tr		
1-Undecanol	112-42-5					tr		
2-Undecanol	1653-30-1				tr			1.3[u]
Verbenene	4080-46-0							

Table 5.31.2 Constituents identified in cypress oils (*continued*)

Constituent	CAS	Percentage and range						
		A	B (17)	C (1)	D (6)	E (8)	F (2)	G
Viridiflorol	552-02-3							0.1[u]
α-Ylangene	14912-44-8		0.4					tr[p]
Zingiberene	495-60-3							0.1[a,o]

A thirty-three cypress essential oil samples from France, Spain and Algeria analyzed between 1998 and 2013; lowest and highest concentrations given (E. Schmidt, unpublished data)
B one sample of oil from branches of Tunisian cypress (ref. 17)
C one lab-prepared oil from leaves and one from branches from Tunisian cypress; in front of the hyphen: leaf oil, after the hyphen: oil from branches (ref. 1)
D one lab-distilled cypress leaf oil sample from Greece (ref. 6)
E two lab-distilled oils from cypress twigs and leaves from Algeria harvested in two different years; highest concentrations given (ref. 8)
F one lab-distilled leaf oil from Algerian *Cupressus sempervirens* (ref. 2)
G data from other studies (indicated with superscript letters); highest concentrations found in any study reviewed here given; when two or more oils were investigated, only the highest concentrations are mentioned, unless indicated otherwise

[a] incorrect identification (ref. 22); [b] one lab-distilled twigs and needles oil from Egypt (ref. 4); [c] probably misidentification of α-terpinyl acetate (ref. 19); [d] one lab-distilled oil from the aerial parts of a cypress from Tunisia (ref. 3); [e] one commercial sample of *Cupressus sempervirens* essential oil purchased in Ecuador (ref. 16); [f] one lab-distilled oil from the leaves of *C. sempervirens* from Cameroon (ref. 7); [h] unspecified cypress leaf oil with an atypical chemical profile (ref. 10); [i] ten Tunisian branch cypress oil specimens produced over ten different hydrodistillation times varying from 15 minutes to 4 hours (ref. 11); [j] one commercially available cypress oil purchased in Italy (ref. 13); [k] twelve lab-distilled oils from Croatian cypress leaves harvested each month during one year (ref. 9); [l] one lab-prepared oil of *Cupressus sempervirens* leaves grown in Iran (ref. 12); [m] one French and one Spanish commercial cypressoil (ref. 14); [n] one lab-distilled leaf oil from Tunisian cypress; only the main components were mentioned (ref. 5); [o] one cypress oil produced in Croatia (ref. 15); [p] one lab-prepared and one industrially hydrodistilled Algerian cypress oil (ref. 18); [q] one commercial cypress oil *ex Cupressus sempervirens* purchased from a Brazilian company (ref. 21); [r] one cypress oil prepared from twigs and leaves of Egyptian *C. sempervirens* (ref. 23); [s] one commercial cypress oil from Provence, France (ref. 20); [t] one lab-hydrodistilled cypress oil from Morocco (ref. 24); [u] three lab-hydrodistilled leaf oils from Tunisia from 2 collection sites in the vicinity of a cement factory and one site unexposed to cement dust (ref. 25); [v] one lab-hydrodistilled cypress oil from Egypt (ref. 26); [w] one lab-hydrodistilled oil from Saudi Arabia (ref. 28); [x] one lab-hydrodistilled oil from leaves and twigs of *C. sempervirens* from Pakistan (ref. 29); [y] one commercial oil provided by a Moroccan company (ref. 30); [z] one lab-hydrodistilled cypress leaf oil from Egypt (ref. 33); [z1] one steam-distilled oil from an Iranian *C. sempervirens* cultivar (ref. 32); [z2] five lab-hydrodistilled oils from cypress needles collected in 4 European countries (Spain, Greece, Poland, Finland) (ref. 31)

tr: traces (in columns D, F and G[m]: <0.1%; in columns E, G[p] and G[u]: <0.05%)

essential oils and fragrance materials (37). An aromatherapist had chronic hand dermatitis and was patch test positive to 17 of 20 oils used at her work (tested 1% and 5% in petrolatum), including cypress oil (35). Another aromatherapist had non-occupational contact dermatitis with allergies to multiple essential oils used at work, including cypress oil. The patient also reacted to geraniol, linalool, linalyl acetate, α-pinene, the fragrance mix and various other fragrance materials; α-pinene, linalool, terpineol and caryophyllene were demonstrated by GC-MS in the cypress oil (36). α-Pinene is indeed the dominant ingredient in commercial cypress essential oils with concentrations ranging from 43% to 68% (Table 5.31.2, column A).

Table 5.31.3 Major constituents of cypress oils

Constituent	CAS	Percentage and range						
		A	B (17)	C (1)	D (6)	E (8)	F (2)	G
α-Pinene	80-56-8	43.2-68.0	20.0	27.5-35.8	54.1	44.9	60.5	58.5[z2]; 61.3[m]; 64[u]; 64.6[i]; 79.2[k]
Sabinene	3387-41-5	0.5-2.2	0.3	0.2-1.3	2.4	0.6	1.3	14.8[f]; 24.4[t]; 30.6[z2]; 30.8[r]; 39.6[e]
δ3-Carene	13466-78-9	7.2-25.9	22.9	7.2-13.2	10.8	10.6	0.2	22.1[w]; 24[r]; 24.4[u]; 24.5[x]; 32.6[k]
Limonene	138-86-3	0.1-10.8	5.1	2.2-1.9	3.2	4.5	4.6	5.8[b]; 6.2[j]; 6.4[s]; 7.0[k]; 7.3[e]; 32[y]
Cedrol (α-)	77-53-2	0.4-4.2	5.2	19.3-7.7	4.9	tr	8.3	6.7[o]; 9.5[i]; 12.9[k]; 21.3[b]; 23.7[h]
1,8-Cineole	470-82-6	0.3-0.6		1.0-0.0		tr		0.3[q]; 1.2[d]; 23.2[z2]
α-Terpinyl acetate	80-26-2	1.2-3.2	7.5	0.2-0.5	5.5	12.0		5.9[i]; 6.5[p]; 6.6[z1]; 7.1[i]; 9.1[r]; 12[z2]
Camphene	79-92-5	0.4-1.7	tr		tr		0.5	0.3[q]; 0.5[e]; 1.4[h]; 2.4[j]; 7.3[b]; 9.9[r]
Terpinolene	586-62-9	0.8-3.5	9.4	1.3-1.9	4.1	2.7	2.0	4.6[j]; 4.7[d]; 5.9[i]; 6.6[z1]; 6.9[v]; 9.2[b]
β-Pinene	127-91-3	0.9-1.7	1.1	0.8-2.5	2.5	1.6	2.9	2.2[b,i]; 2.5[w]; 2.6[i]; 2.8[u]; 6.0[r]; 7.1[j]

LEGEND: SEE UNDER TABLE 5.31.2

LITERATURE

1 Ismail A, Lamia H, Mohsen H, Samia G, Bassem J. Chemical composition, bio-herbicidal and antifungal activities of essential oils isolated from Tunisian common cypress (*Cupressus sempervirens* L.). J Med Plants Res 2013;7:1070-1080

2 Mazari K, Bendimerad N, Bekhechi C, Fernandez X. Chemical composition and antimicrobial activity of essential oils isolated from Algerian *Juniperus phoenicea* L. and *Cupressus sempervirens* L. J Med Plants Res 2010;4:959-964

3 Boukhris M, Regane G, Yangui T, Sayadi S, Bouaziz M. Chemical composition and biological potential of essential oil from Tunisian *Cupressus sempervirens* L.J Arid Land Stud 2012;22:329-332

4 Elansary HO, Salem MZM, Ashmawy NA, Yacout MM. Chemical composition, antibacterial and antioxidant activities of leaves essential oils from *Syzygium cumini* L., *Cupressus sempervirens* L. and *Lantana camara* L. from Egypt. J Agricult Sci 2012;4:144-152

5 Herzi N, Camy S, Bouajila J, Romdhane M, Condoret JS. Extraction of essential oil from *Cupressus sempervirens*: comparison of global yields, chemical composition and antioxidant activity obtained by hydrodistillation and supercritical extraction. Nat Prod Res 2013;27:1795-1799

6 Giatropoulos A, Pitarokili D, Papaioannou F, Papachristos DP, Kiliopoulos G, Emmanouel N, et al. Essential oil composition, adult repellency and larvicidal activity of eight Cupressaceae species from Greece against *Aedes albopictus* (Diptera: Culicidae). Parasitol Res 2013;112:1113–1123

7 Tapondjou AL, Adler C, Fontem DA, Bouda H, Reichmuth C. Bioactivities of cymol and essential oils of *Cupressus sempervirens* and *Eucalyptus saligna* against *Sitophilus zeamais* Motschulsky and *Tribolium confusum* du Val. J Stored Prod Res 2005;41:91-102

8 Chanegriha N, Baâliouamer A, Meklati BY, Chretien JR, Keraviset G. GC and GC/MS leaf oil analysis of four Algerian cypress species. J Essent Oil Res 1997;9:555-559

9 Milos M, Radonic A, Mastelic J. Seasonal variation in essential oil compositions of *Cupressus sempervirens* L. J Essent Oil Res 2002;14:222-223

10 Ibrahim NA, El-Seedi HR, Mohammed MMD. Constituents and biological activity of the chloroform extract and essential oil of *Cupressus sempervirens*. Chem Nat Comp 2009;45:309-313

11 Cheraif I, Ben Jannet H, Hammami M, Mighri Z. Hydrodistillation kinetic investigation of essential oil from the Tunisian *Cupressus sempervirens* L. J Essent Oil Bear Plants 2005;8:165-172

12 Emami SA, Assili J, Rahimizadeh M, Fazly-Bazzaz BS, Hassanzadeh-Khayyat M. Chemical and antimicrobial studies of *Cupressus sempervirens* L. and *C. horizontalis* Mill. essential oils. Iran J Pharm Sci 2006;2:103-108. Data cited in ref. 19, also available at: http://webcache.googleusercontent.com/search?q=cache:cVwG775BcaAJ:ijps. sums.ac.ir/files/PDFfiles/7th.pdf1609568303.pdf+&cd=8&hl=nl&ct=clnk&gl=nl

13 Romeo FV, De Luca S, Piscopo A, Poiana M. Antimicrobial effect of some essential oils. J Essent Oil Res 2008;20:373-379

14 Williams DG. The chemistry of essential oils, 2nd ed. Port Washington, NY: Micelle Press, 2008:179-181. Data cited in ref. 19

15 Milos M, Radonic A. Essential oil and glycosidically bound volatile compounds from Croatian *Cupressus sempervirens* L. Acta Pharm 1996;46:309-314

16 Sacchetti G, Maietti S, Muzzoli M, Scaglanti M, Manfredini S, Radice M, Bruni R. Comparative evaluation of 11 essential oils of different origin as functional antioxidants, antiradicals and antimicrobials in foods. Food Chem 2005;91:621-632

17 Chéraif I, Ben Jannet H, Hammami M, Mighri Z. Contribution à l'étude de la composition chimique de l'huile essentielle des rameaux de *Cupressus sempervirens* L. poussant en Tunisie. J Soc Chim Tun 2005;7:75-82

18 Chanegriha N, Baâliouamer A, Meklati BY, Favre-Bonvin J, Alamercey S. Chemical composition of Algerian cypress essential oil. J Essent Oil Res 1993;5:671-674

19 Lawrence BM. Progress in essential oils. Perfum Flavor 2009;34(December):53-56

20 Tognolini M, Barocelli E, Ballabeni V, Bruni R, Bianchi A, Chiavarini M, Impicciatore M. Comparative screening of plant essential oils: Phenylpropanoid moiety as basic core for antiplatelet activity. Life Sciences 2006;78:1419-1432. Data also presented in ref. 27

21 Murbach Teles Andrade BF, Nunes Barbosa L, da Silva Probst I, Fernandes Júnior A. Antimicrobial activity of essential oils. J Essent Oil Res 2014;26:34-40

22 Lawrence BM. Progress in essential oils. Perfum Flavor 1995;20(Jul/Aug):34-35

23 Afifi MS, El-Sharkawy SH, Maatoog GT, El-Sohly M, Rosazza JPN. Essential oils of *Thuja occidentalis, Thuja orientalis, Cupressus sempervirens* and *Juniperus phoenicea* Mans. J Pharm Sci 1992;8:37-46. Data cited in ref. 22

24 Chebli B, Hmamouchi M, Achouri M, Idrissi Hassani LM. Composition and *in vitro* fungitoxic activity of 19 essential oils against two post-harvest pathogens. J Essent Oil Res 2004;16:507-511

25 Hosni K, Hassen I, M'Rabet Y, Casabianca H. Biochemical response of *Cupressus sempervirens* to cement dust: Yields and chemical composition of its essential oil. Arab J Chem 2014, doi: http://dx.doi.org/10.1016/j.arabjc.2014.10.042

26 Badawy MEI, Abdelgaleil SAM. Composition and antimicrobial activity of essential oils isolated from Egyptian plants against plant pathogenic bacteria and fungi. Ind Crops Prod 2014;52:776-782. Data also presented in ref. 33

27 Rolli E, Marieschi M, Maietti S, Sacchetti M, Bruni R. Comparative phytotoxicity of 25 essential oils on pre-and post-emergence development of *Solanum lycopersicum* L.: A multivariate approach. Ind Crops Prod 2014;60:280-290. Data also presented in ref. 20

28 Selim SA, Adam ME, Hassan SM, Albalawi AR. Chemical composition, antimicrobial and antibiofilm activity of the essential oil and methanol extract of the Mediterranean cypress (*Cupressus sempervirens* L.). Compl Altern Med 2014;14:179 (8 pages)

29 Mahmood Z, Ahmed I, Saeed MUQ, Sheikh MA. Investigation of physico-chemical composition and antimicrobial activity of essential oil extracted from lignin-containing *Cupressus sempervirens*. BioResources 2013;8:1625-1633

30 Aazza S, LyoussiB, Miguel MG. Antioxidant and anti-acetylcholinesterase activities of some commercial essential oils and their major compounds. Molecules 2011;16:7672-7690

31 Dawidowicz AL, Czapczynska NB, Wianowska D. Relevance of the sea sand disruption method (ssdm) for the biometrical differentiation of the essential-oil composition from conifers. Chem Biodivers 2013;10:241-250

32 Emami SA, Khayyat MH, Rahimizadeh M, Fazly-Bazzaz BS, Assili J. Chemical constituents of *Cupressus sempervirens* L. cv. Cereiformis Rehd. essential oils. Iran J Pharm Sci 2004:1:39-42

33 Saad MMG, Abdelgaleil SAM. Allelopathic potential of essential oils isolated from aromatic plants on *Silybum marianum* L. Glob Adv Res J Agricult Sci 2014;3(9):289-297. Data partly also presented in ref. 26

34 Rhind JP. Essential oils. A handbook for aromatherapy practice, 2nd Edition. London: Singing Dragon, 2012

35 Selvaag E, Holm J, Thune P. Allergic contact dermatitis in an aromatherapist with multiple sensitizations to essential oils. Contact Dermatitis 1995;33:354-355

36 Dharmagunawardena B, Takwale A, Sanders KJ, Cannan S, Roger A, Ilchyshyn A. Gas chromatography: an investigative tool in multiple allergies to essential oils. Contact Dermatitis 2002;47:288-292

37 Trattner A, David M, Lazarov A. Occupational contact dermatitis due to essential oils. Contact Dermatitis 2008;58:282-284

Chapter 5.32 DWARF PINE OIL

DEFINITION

Dwarf pine oil (essential oil of dwarf pine) is the essential oil obtained from needles with twigs (leaves and terminal branchlets) of the (dwarf) mountain pine, *Pinus mugo* Turra.

INCI NOMENCLATURE

Description/definition: Pinus mugo leaf oil is an essential oil obtained from the leaves of the pine, *Pinus mugo*, Pinaceae

INCI name EU & USA: Pinus mugo leaf oil

CAS registry number(s): 8000-26-8; 90082-72-7

EINECS number(s): 290-163-6

Description/definition: Pinus pumilio oil is the volatile oil distilled from the needles and branches of the mountain pine, *Pinus pumilio*, Pinaceae

INCI name EU & USA: Pinus pumilio branch/leaf oil

CAS registry number(s): 97676-05-6 (indicated by CosIng, but wrongly, refers to *Pinus pumila*, the Siberian dwarf pine); 90082-73-8 (*Pinus mugo pumilio*)

EINECS number(s): 307-681-6 (indicated by CosIng, but wrongly, refers to *Pinus pumila*, the Siberian dwarf pine); 290-164-1 (*Pinus mugo pumilio*)

ISO (INTERNATIONAL ORGANIZATION FOR STANDARDIZATION) STANDARD:

ISO number: 21093
ISO name: Essential oil of dwarf pine
Botanical origin: *Pinus mugo* Turra, syn. *Pinus montana* Mill.
Parts of plant used: leaf (needle), terminal branchlets
ISO values : ISO values (minimum and maximum concentrations) are shown in Table 5.32.1.

THE PLANT, THE OIL, AND THEIR USES

The dwarf mountain pine (synonyms: mountain pine, mugo pine) *Pinus mugo* Turra (synonyms: *Pinus montana* Mill., *Pinus montana* var. *pumilio* (Haenke) Willk., *Pinus mugo* var. *pumilio*, *Pinus pumilio* Haenke, and *Pinus mugo* Turra var. *pumilio* (Haenke) Zenari) is an evergreen shrub (rarely a tree) 1-5 meters tall, with one or more curved trunks and long branches. It is found in the Pyrenees, Alps, Erzgebirge, Carpathians, northern Apennines and the Balkan Peninsula mountains, mostly at altitudes between 1,000 and 2,200 meters. Dwarf pine oil (synonym: mountain pine oil) is obtained from the needles and twigs of the dwarf mountain pine. It has been used to treat acute and chronic respiratory diseases by steam or cold inhalation. It is also added to bath oils and may be used as air freshener (11). In Bulgaria, the leaves are used for making herbal tea. A recent trend is the use of the mugo pine in cooking. Buds and young cones are harvested from the wild in the spring and left to dry in the sun over the summer and into the fall. The cones and buds gradually drip syrup, which is then boiled down to a concentrate and combined with sugar to make pine syrup (GRIN Taxonomy for Plants; www.conifers.org; Wikipedia). Dwarf pine oil is considered a dermal irritant and sensitizer (from its high δ3-carene content, air-oxidized chemicals of which are important allergens in some turpentine oils) and is therefore not used in aromatherapy (22,23).

CHEMICAL COMPOSITION

Dwarf pine oil is a clear and transparent, mobile liquid which has a discreet green, slightly fatty odor with terpenicbackground (turpentine-like smell) (10,11). The yield of essential oil from the twigs with needles of *Pinus mugo* Turra generally varies from 0.12 to 0.8%. The main producing countries of this oil are Italy (South Tirol), Austria, Germany and Slovenia.

Literature data (up to November 23, 2014) on the chemical composition of dwarf pine oils and unpublished analytical data from one of us (E.S.) are shown in Table 5.32.2 in alphabetical order. In dwarf pine oils from various origins, over 170 chemicals have been identified. About 57% of these were found in a single reviewed publication only. The major compounds found in dwarf pine oils from different sources are shown in Table 5.32.3. They include (highest concentrations in any study given) α-pinene (52.8%), limonene (37.2%), δ3-carene (35.2%), myrcene (30.2%), terpinolene (29.2%), β-pinene (29.0%), bornyl acetate (27.0%), β-phellandrene (20.6%) and camphene (11.7%). Well-known ingredients of dwarf pine oils that were present in high concentrations (>7%) in one or two studies were α- + β-phellandrene (75.0%), α-pinene + β-pinene (74.1%), germacrene D (12.1%), γ-cadinene (9.0%) and α-terpineol (7.3%). Uncommon or rare constituents of dwarf pine oils found in high concentrations (>7%) in single studies include α-thujene (27.6%, but is combined with α-pinene) and β-fenchene (11.6%).

Commercial oils

The ten chemicals that had the highest maximum concentrations in 283 commercial dwarf pine essential oil samples (concentration ranges provided) are the following: α-pinene (0.3-38.8%), δ3-carene (0.6-34.4%), myrcene (1.6-30.2%), β-pinene (0.01-29.0%), limonene (0.6-24.2%), β-phellandrene (0.8-20.6%), terpinolene (0.4-8.3%), β-caryophyllene (0.5-5.7%), *p*-cymene (0.2-4.3%) and α-terpinene (0.07-4.2%) (Erich Schmidt, unpublished analytical data).

CONTACT ALLERGY/ALLERGIC CONTACT DERMATITIS

General

Contact allergy to/allergic contact dermatitis from dwarf pine oil has been reported in a few publications only. In groups of consecutive patients suspected of contact dermatitis, prevalence rates of up to 0.7% positive patch test reactions have been observed, but relevance data are lacking.

Table 5.32.1 ISO values (%) for dwarf pine oil[a]

Compound	CAS	Minimum	Maximum
α-Pinene	80-56-8	10.0	30.0
δ3-Carene	13466-78-9	5.0	25.0
β-Phellandrene	555-10-2	8.0	17.0
Limonene	138-86-3	8.0	14.0
β-Pinene	127-91-3	3.0	14.0
Myrcene	123-35-3	3.0	11.0
Terpinolene	586-62-9	1.0	8.0
Bornyl acetate	76-49-3	1.0	5.0
β-Caryophyllene	87-44-5	0.5	5.0
p-Cymene	99-87-6	0.01	2.5

[a] ISO 21093 Essential oil of dwarf pine ©ISO 2002; Geneva, Switzerland, www.iso.org

Testing in groups of patients

Between 1998 and 2000, dwarf pine oil 2% in petrolatum was tested in 1606 consecutive patient suspected of contact dermatitis in six European countries and there were 12 (0.7%) positive patch test reactions (27). In the same period and as part of this international study, 318 patients were tested in Denmark and two (0.6%) reacted (29). The relevance was not specified in either study. In a group of 21 patients with dermatitis caused by fragrances and tested with a series of essential oils, four (19%) reacted to dwarf pine oil; relevance data were not provided (28).

Case reports and positive patch test reactions

Allergic contact dermatitis (probably one case) to a pharmaceutical cream containing the NSAID etofenamate was

Table 5.32.2 Constituents identified in dwarf pine oils

Constituent	CAS	Percentage and range					
		A	B (9)	C (3)	D (11,12)	E (11,12)	F
Abienol	1616-86-0						1.6[m]
Abieta-8,13(15)-dien-18-al							0.7[m]
Abietadiene	36312-33-1						0.5[m]; 0.8[s]
Abieta-8(14),13(15)-diene							<0.05[s]; 0.6[m]
8,13-Abietadien-18-ol							0.3[m]
Abietal	6704-50-3						0.8[m]
Abietatriene	19407-28-4						0.7[m]
δ-Amorphene	189165-79-5						<0.05[s]
(E)-Anethole	4180-23-8						0.04[o]
Aromadendrene	489-39-4						<0.05[s]; 0.1[m]
Benzaldehyde	100-52-7						0.04[o]
Benzyl benzoate	120-51-4						0.3[m]; 1.5[h]
Bicycloelemene	32531-56-9						0.4[s]
Bicyclogermacrene	24703-35-3			1.9			0.3[m]; 4.2[s]; 6.8[g]
α-Bisabolol	515-69-5	0.0-0.1					
Borneol	507-70-0	0.05-0.1		0.6			tr[s]; <0.05[l]; 0.2[o]; 1.5[m]
Bornyl acetate	76-49-3	0.2-8.2	0.5-6.6	8.2	1.5-8.0	1.4-7.9	4.5[l]; 6.8[k]; 7.6[s]; 9.2[d]; 11.5[m] 22.0[n]; 27.0[e]
Bornyl isovalerate	76-50-6						0.08[o]; 0.3[m]
β-Bourbonene	5208-59-3	0.04-0.06					0.2[m,s]
trans-Cadina-1,4-diene	20085-13-6						0.2[s]
cis-Cadina-1(6),4-diene	1187195-00-1						0.3[s]
Cadinene	29350-73-0						4.0[l];
α-Cadinene	24406-05-1						0.1[m]; 0.2[s]
γ-Cadinene	39029-41-9	0.05-2.2		0.7	0.2-1.3	<0.1-0.4	0.3[m]; 0.6[p]; 0.8[l]; 1.3[s]; 9.0[h]
δ-Cadinene	483-76-1	0.04-2.1		2.9	0.6-4.3	<0.1-1.0	1.1[l]; 1.2[m]; 2.2[s]; 2.9[p]; 6.6[g]
α-Cadinol	481-34-5	0.08-0.3		1.1			0.4[o]; 1.9[s]; 4.1[m]; 4.4[h]; 5.0[g]
δ-Cadinol	19435-97-3	0.07-0.09					
τ-Cadinol	58580-31-7			0.4			tr[q]; 0.2[o]; 0.6[s]; 0.7[l]; 1.3[m]
Camphene	79-92-5	0.2-10.2	1.3-7.8	8.1	1.1-4.4	1.3-6.7	3.8[l]; 4.8[e]; 6.3[s]; 8.0[n]; 11.7[p]
Camphene hydrate	465-31-6						0.3[m]
Camphor	76-22-2						0.2[m]
δ3-Carene	13466-78-9	0.6-34.4	1.0-34.4	16.2	7.6-33.2	0.5-30.1	12.0[s]; 17.7[k]; 23.9[i]; 26.6[j] 28.3[l]; 35.2[q]
Car-3-en-2-one							0.1[m]
Carvacrol	499-75-2						1.6[h]
(E)-Carveol	1197-07-5						0.2[m]

Table 5.32.2 Constituents identified in dwarf pine oils (*continued*)

Constituent	CAS	Percentage and range					
		A	B (9)	C (3)	D (11,12)	E (11,12)	F
(Z)-Carveol	1197-06-4						0.3[m]
Carvone	99-49-0						tr[m]; 0.02[o]
Caryophylla-4(12),8(13)-dien-5-ol							0.1[m]
β-Caryophyllene	87-44-5	0.5-5.7	0-3.0	2.8	1.6-9.0	<0.1-4.2	5.4[j]; 5.9[m]; 6.0[l]; 6.4[g]; 6.8[s]
Caryophyllene oxide	1139-30-6	0.01-0.3			<0.1-0.5	0-0.8	0.2[s]; 1.6[i]; 2.0[m]; 3.7[l]; 6.1[h]
α-Cedrene	469-61-4	0.2-0.3					
Chamazulene	529-05-5						3.1[p]
cis-Chrysanthenol	55722-60-6						1.0[h]
1,8-Cineole	470-82-6						tr[p];
Citronellol	106-22-9						0.2[m,o]
Citronellyl acetate	150-84-5	0.07-0.09					0.06[o]; 0.1[m]
Citronellyl isobutyrate	97-89-2						0.02[o]
Citronellyl isovalerate	68922-10-1						0.03[o]
Citronellyl valerate	7540-53-6						0.02[o]
Coahuilensol methyl ether							0.1[m]
α-Copaene	3856-25-5	0.05-0.1		0.06	0-0.2	<0.1-0.2	0.1[m]; 0.2[l,s]
β-Copaene	18252-44-3						0.2[s]
Cryptone	500-02-7						0.4[m,o]; 1.3[i]
α-Cubebene	17699-14-8	0.09-0.1		0.2			tr[s]
β-Cubebene	13744-15-5						2.3[a,l]
Cubebol	23445-02-5						0.3[s]
1,10-di-epi-Cubenol	73365-77-2						<0.05[s]; 0.3[m]
1-epi-Cubenol	81939-29-9						0.1[s]
Cuminaldehyde	122-03-2						0.08[o]; 0.6[i]
Cuminyl alcohol	536-60-7						0.3[m]
p-Cymene	99-87-6	0.2-4.3	0.2-2.1	0.2	0-0.2	0.5-6.9	tr[m]; 0.5[l]; 0.6[e]; 0.7[q]; 1.1[i]
p-Cymenene	1195-32-0	0.02-0.2					
m-Cymen-8-ol	5208-37-7						0.5[m];
p-Cymen-8-ol	1197-01-9	0.03-0.4					0.1[o]; 0.9[i]; 1.6[m]
2-Decanone	693-54-9						0.02[o]
2-Decenal	3913-71-1						0.04[o]
Dehydroabietal	13601-88-2						4.3[m]
Dehydroabietol							0.4[m]
β-Elemene	33880-83-0	0.03-0.2		0.3	<0.1-0.9	0-0.2	<0.05[l]; 0.3[m]; 2.8[s]
Elemol (α-)	639-99-6	0.02-0.03					
Epicubenol	19912-67-5						1.8[h]
Ethyl cinnamate	103-36-6						0.03[o]
(E)-β-Farnesene	18794-84-8	0.0-0.09					0.1[m,s]
α-Fenchene	471-84-1						6.4[h]
β-Fenchene	497-32-5						11.6[e,f]
Fenchone	1195-79-5	0.02-0.06					
Fenchyl alcohol (fenchol)	1632-73-1	0.0-0.05					
α-Fenchyl alcohol	512-13-0						0.07[o]; 0.1[m]
Furopelargone A	1143-45-9						0.2[m]
Geranyl acetate	105-87-3	0.06-0.07					
Germacrene D	23986-74-5	0.04-1.2		3.9	0.6-6.0	0-0.6	0.6[m]; 0.7[i]; 11.8[g]; 12.1[s]
Germacrene D-4-ol	198991-79-6			0.4			6.1[s]
Guaia-6,9-diene	37839-64-8						1.4[m]
Heptadecane	629-78-7						0.2[m]
Hexadecane	544-76-3						tr[m]
Hexanal	66-25-1	0.02-0.03					
Hexanol (1-)	111-27-3	0.01-0.02					
(Z)-3-Hexen-1-ol	928-96-1	0.02-0.03					
α-Humulene	6753-98-6	0.03-0.9		0.4	0.3-1.3	<0.1-0.9	0.4[i]; 0.5[q]; 0.9[l]; 1.2[m]; 1.6[p]
α-Humulene epoxide	96638-51-6	0.01-0.04					0.03[o]
Humulene epoxide II	19888-34-7						0.4[m]
14-Hydroxy-epi-(E)-caryophyllene							0.6[m]
Isoborneol	124-76-5						0.1[m];
Isobornyl acetate	125-12-2						2.0[h]

Table 5.32.2 Constituents identified in dwarf pine oils (*continued*)

Constituent	CAS	Percentage and range					
		A	B (9)	C (3)	D (11,12)	E (11,12)	F
Isobornyl isovalerate	7779-73-9						0.2[m]; 1.0[s]
Isodihydrocarvyl acetate	220329-20-4						0.1[m]
Isopimara-9(11),15-diene	39702-28-8						0.1[m]
Isoterpinolene	586-63-0	0.04-0.05					
Limonene	138-86-3	0.6-24.2	2.2-37.2	0.9	0.7-6.8	6.1-37.1	3.0[s]; 5.6[q]; 7.8[k]; 8.2[e]; 8.5[j]
Linalool	78-70-6	0.02-0.4			0-0.2	0-0.2	0.3[i,m]
Linalyl acetate	115-95-7						0.2[s]
Longifolene	475-20-7	0.1-0.2					tr[p]; 0.2[m]; 1.5[l]
Manoyl oxide	596-84-9						1.0[m]; 2.1[s]
13-epi-Manoyl oxide	1227-93-6						2.9[m]
2-keto-Manoyl oxide							2.0[s]
p-Mentha-1,5-dien-8-ol	1686-20-0						0.5[m]
cis-p-Menth-2-en-1-ol	29803-82-5						0.05[o]; 0.3[m]
trans-p-Menth-2-en-1-ol	29803-81-4						0.08[o]; 0.5[m]
Methyl dehydroabietate	1235-74-1						0.2[m]
Methyl levopimarate							<0.05[s]
Methyl thymol	1076-56-8	0.03-0.6			<0.1-0.7	0-0.4	<0.05[s]; 0.2[i,o]; 0.7[m]; 0.8[l]
α-Multijugenol	34298-31-2						0.2[o]
trans-Muurola-4(14),5-diene	262352-87-4						0.1[s]
α-Muurolene	10208-80-7	0.4-0.6			0.5-3.7	0-0.8	0.3[m]; 0.6[s]; 0.9[l]; 2.4[h]
γ-Muurolene	30021-74-0						tr[p,s]; 0.1[m]; 3.5[h]
α-Muurolol	104245-48-9			0.1			0.08[o]; 0.2[h]; 0.4[s]; 0.5[m]
τ-Muurolol	19912-62-0			0.5			0.2[o]; 0.9[s]; 1.1[i]; 1.3[m]; 4.6[h]
Myrcene	123-35-3	1.6-30.2	0.3-10.1	3.9	1.7-21.5	0.5-10.2	3.3[j]; 5.0[p,q]; 6.9[s]; 7.0[k]; 9.7[n]
Myrtenal	564-94-3						0.05[o]
Myrtenol	515-00-4						0.2[m]
(E)-Nerolidol	40716-66-3						0.1[m]
Nonadecane	629-92-5						0.1[m]
β-Ocimene	13877-91-3						0.3[l]
(E)-β-Ocimene	3779-61-1	0.08-0.8		1.6			0.1[i]; 0.3[q]; 1.0[s]
(Z)-β-Ocimene	3338-55-4	0.03-0.2		0.07			<0.05[s]
n-Octadecane	593-45-3						0.2[m]
n-Octanol	111-87-5	0.02-0.02					
Oplopanone	1911-78-0						0.1[m]
β-Oplopenone	28305-60-4						0.8[h]
Oxomanoyl oxide							0.6[m]
Phellandral	21391-98-0						0.1[o]; 0.2[m]
α-Phellandrene	99-83-2	0.01-1.1			0-1.1	0.1-0.9	tr[m]; <0.05[l]; 0.2[i]; 0.5[q]
β-Phellandrene	555-10-2	0.8-20.6	0.1-16.1	1.5	2.4-18.0	0.5-16.0	1.2[m]; 2.6[s]; 7.2[i]; 8.3[j]; 9.9[k]; 15.0[q]; 15.2[l]
α- + β-Phellandrene							1.1-75.0[n]
Pimaradiene	1686-61-9						0.1[m]; 0.2[s]
α-Pinene	80-56-8	0.3-52.8	0.5-46.9	27.6[c]	7.9-22.8	4.1-31.5	4.5[l]; 13.1[q]; 14.1[s]; 20.6[k]; 32.0[p]; 33.3[h]; 37.6[e]
β-Pinene	127-91-3	0.01-29.0	2.4-27.5	3.4	1.5-19.0	1.3-20.7	2.1[s]; 2.5[l]; 6.9[k]; 7.6[q]; 7.8[l]; 15.8[j]; 28.4[p]
α-Pinene + β-Pinene							11.6-74.1[n]
α-Pinene (ep)oxide	1686-14-2	0.01-0.07					1.5[h]
Pinocamphone	547-60-4						0.1[o]
trans-Pinocamphone	547-60-4						tr[m];
Pinocarveol	5947-36-4						0.2[o]
trans-Pinocarveol	1674-08-4						tr[m]
cis-Piperitol	16721-38-3						0.2[m]
trans-Piperitol	16721-39-4						0.4[m]
Piperitenone	491-09-8						0.05[o]; 1.2[i]
Piperitone	89-81-6						1.0[m]
Pregeijerene B							0.6[m]
Sabinene	3387-41-5	0.2-2.6		1.0	0.5-1.2	<0.1-1.7	0.6[q]; 1.3[i]; 4.8[s]

Table 5.32.2 Constituents identified in dwarf pine oils (*continued*)

Constituent	CAS	Percentage and range					
		A	B (9)	C (3)	D (11,12)	E (11,12)	F
Sandaracopimaradiene	1686-56-2						0.3[m]
Sandaracopimarinal	3855-14-9						0.3[m]
Santene	529-16-8	0.01-0.8				0-0.2	
α-Selinene	473-13-2						1.3[l]
β-Selinene	17066-67-0	0.1-0.2					tr[m]; 0.2[s]; 2.1[h]
Sesquisabinene	58319-04-3						0.3[s]
Spathulenol	6750-60-3	0.0-1.1			<0.1-3.8	0-<0.1	0.8[m]; 1.2[i]; 1.5[s]; 3.5[h]
Sylvestrene	1461-27-4						3.0[s]
α-Terpinene	99-86-5	0.07-4.2		0.3	0.2-0.7	0.2-4.5	tr[m]; <0.05[l]; 0.5[s]; 0.6[q]; 6.4[p]
γ-Terpinene	99-85-4	0.08-1.8		0.4	0.3-0.9	0.1-5.0	0.1[m]; 0.3[l]; 0.4[s]; 0.6[q]
Terpinen-4-ol	562-74-3	0.04-2.6			0.4-1.3	0.2-1.4	0.2[s]; 0.8[o]; 0.9[l]; 1.2[i]; 3.6[m]
α-Terpineol	98-55-5	0.03-0.7		0.4	0.3-1.5	0.3-3.6	0.5[i]; 0.9[o]; 1.7[p]; 7.3[m]
β-Terpineol	138-87-4						tr[p]
Terpinolene	586-62-9	0.4-8.3	0-14.1	3.2	0.1-5.6	1.0-29.2	3.0[g]; 3.1[l]; 3.6[q]; 3.8[s]; 5.8[j]
Terpinyl acetate	8007-35-0	0.05-1.0	0-0.5		0.4-2.1	0-0.4	1.6[l]; 2.5[i]; 3.6[h]
α-Terpinyl acetate	80-26-2						0.6[o]; 1.3[s]; 2.4[m]
δ-Terpinyl acetate							tr[m]
α-Terpinyl isobutyrate	7774-65-4						+[r]
α-Thujene	2867-05-2	0.03-2.8		27.6[c]			0.4[l]; 0.8[l]; 1.2[s]
α-Thujone	546-80-5						1.2[h]
Thujopsene	470-40-6	0.1-1.1					
Tricyclene	508-32-7	0.07-0.6		1.7	0.3-2.3	0.2-1.5	<0.05[l]; 0.4[i,q]; 1.2[h]; 1.7[s]
2-Undecanone	112-12-9						0.08[o]; 0.2[m]; 0.8[h]
1-Undecen-10-one							0.05[o]
cis-Verbenol	1845-30-3	0.02-0.04					

A 283 dwarf pine essential oil samples from Italy (South Tirol), Austria, Germany, Slovenia and New Zealand analyzed between 1990 and 2013; lowest and highest concentrations given (E. Schmidt, unpublished data)

B twenty-two commercial oils purchased in Germany; lowest and highest concentrations given (ref. 9)

C one sample of dwarf pine oil from Austria (ref. 3)

D twenty-one lab-distilled oils from *Pinus mugo* twigs and needles collected in various parts and at various altitudes ranging from 350 to 2200 meter in Austria; lowest and highest concentrations given (refs. 11,12)

E twenty-one commercial oils obtained from Austrian and German producers; lowest and highest concentrations given (refs. 11,12)

F data from other studies (indicated with superscript letters); highest concentrations found in any study reviewed here given; when two or more oils were investigated, only the highest concentrations are mentioned, unless indicated otherwise

[a] incorrect identity based on GC elution order (ref. 6); [b] correct isomer not identified; [c] α-pinene and α-thujene combined; [d] one lab-distilled needle oil from P. *mugo* var. *pumilio* (ref. 7, data cited in ref. 6); [e] one lab-distilled needle oil from Mongolian P. *mugo* var. *pumilio* (ref. 8, data cited in ref. 6); [f] identification requires corroboration (ref. 6); [g] unknown number of lab-distilled oils from needles (without twigs) from *Pinus mugo* from Macedonia; the article appears to be available as abstract only (ref. 10); [h] one lab-distilled oil from the Royal Botanic Garden of Edinburgh, UK (ref. 1); [i] one lab-distilled oil from twigs and needles of Serbian *Pinus mugo* (ref. 2); [j] three lab-distilled oils from Germany (ref. 9); [k] one oil produced from the needles and twigs of P. *mugo* var. *mugo* (syn. P. *mughus* Scop.) in Norway; this is a synonym of P. *mugo* Turra (ref. 13); [l] one lab-produced oil from *Pinus mugo* from the Botanic Garden in Tübingen, Germany (ref. 4); [m] one lab-hydrodistilled oil from leaves of P. *mugo* Turra subsp. *mugo* growing in the Apennines, Italy; this oil is atypical, as it contains low concentrations of monoterpenes such as α- and β-pinenes, δ-3-carene and β-phellandrene.; however, it contains fair amounts of abietane derivatives which are rare in species of the genus *Pinus* (ref. 20); [n] unknown number of commercial dwarf pine oils of unknown origin (ref. 15); [o] investigation of the oxygenated and sesquiterpenoid fraction of a sample of dwarf pine oil (ref. 16); [p] one *Pinus mugo* essential oil sample of unknown origin (ref. 17); [q] one commercial oil sample from P. *mugo* ssp. *pumilio* (ref. 18); [r] one *Pinus mugo* leaf oil sample (ref. 19); [s] two lab-hydrodistilled oils from needles of P. *mugo* and P. *mugo* var. *pumilio* harvested in the botanical garden of Cambridge University (ref. 21)

tr traces (in column F[m]: <0.1)

Table 5.32.3 Major constituents of dwarf pine oils

Constituent	CAS	Percentage and range					
		A	B (9)	C (3)	D (11,12)	E (11,12)	F
α-Pinene	80-56-8	0.3-52.8	0.5-46.9	27.6[c]	7.9-22.8	4.1-31.5	20.6[k]; 32.0[p]; 33.3[h]; 37.6[e]
Limonene	138-86-3	0.6-24.2	2.2-37.2	0.9	0.7-6.8	6.1-37.1	3.0[s]; 5.6[q]; 7.8[k]; 8.2[e]; 8.5[j]
δ3-Carene	13466-78-9	0.6-34.4	1.0-34.4	16.2	7.6-33.2	0.5-30.1	23.9[i]; 26.6[i]; 28.3[l]; 35.2[q]
Myrcene	123-35-3	1.6-30.2	0.3-10.1	3.9	1.7-21.5	0.5-10.2	3.3[j]; 5.0[p,q]; 6.9[s]; 7.0[k]; 9.7[n]
Terpinolene	586-62-9	0.4-8.3	0-14.1	3.2	0.1-5.6	1.0-29.2	3.0[g]; 3.1[l]; 3.6[q]; 3.8[s]; 5.8[j]
β-Pinene	127-91-3	0.01-29.0	2.4-27.5	3.4	1.5-19.0	1.3-20.7	7.6[q]; 7.8[l]; 15.8[j]; 28.4[p]
Bornyl acetate	76-49-3	0.2-8.2	0.5-6.6	8.2	1.5-8.0	1.4-7.9	9.2[d]; 11.5[m]; 22.0[n]; 27.0[e]
β-Phellandrene	555-10-2	0.8-20.6	0.1-16.1	1.5	2.4-18.0	0.5-16.0	1.2[m]; 2.6[s]; 7.2[l]; 8.3[j]; 9.9[k]; 15.0[q]; 15.2[l]
Camphene	79-92-5	0.2-10.2	1.3-7.8	8.1	1.1-4.4	1.3-6.7	3.8[j]; 4.8[e]; 6.3[s]; 8.0[n]; 11.7[p]
β-Caryophyllene	87-44-5	0.5-5.7	0-3.0	2.8	1.6-9.0	<0.1-4.2	5.4[j]; 5.9[m]; 6.0[l]; 6.4[g]; 6.8[s]
p-Cymene	99-87-6	0.2-4.3	0.2-2.1	0.2	0-0.2	0.5-6.9	tr[m]; 0.5[l]; 0.6[e]; 0.7[q]; 1.1[i]

LEGEND: SEE UNDER TABLE 5.32.2

caused by its base ingredient dwarf pine oil (26). One positive patch test to dwarf pine oil in a patient with contact allergy to *Melaleuca alternifolia* (tea tree) oil (24). One positive patch test reactions among 20 workers in a perfume factory without dermatitis (25).

LITERATURE

1　Tsitsimpikou C, Petrakis PV, Ortiz A, Harvalad C, Roussis V. Volatile needle terpenoids of six *Pinus* species. J Essent Oil Res 2001;13:174-178

2　Stevanovic T, Garneau F-X, Jean FI, Gagnon H, Vilotic D, Petrovic S, et al. The essential oil composition of *Pinus mugo* Turra from Serbia. Flavour Fragr J 2005;20:96-97

3　Kubeczka K-H, Formacek V. Essential oils analysis by capillary gas chromatography and carbon-13 NMR Spectroscopy, 2nd Ed. New York: John Wiley and Sons, 2002:99-105. Data cited in ref. 5

4　Reichling J, Harkenthal M. Mountain pine oil. Pharmaceutical quality of *Pini pumilionis aetheroleum*. Dtsch Apoth Ztg 1998;138:3503-3510. Data cited in ref. 6

5　Lawrence BM. Essential oils 2005-2007. Carol Stream, USA: Allured Publishing Corporation, 2008:48

6　Lawrence BM. Progress in essential oils. Perfum Flavor 2010;35(July):50-57

7　Kolesnikova RD. Essential oils of some conifers. Rastit Resur 1985;21:130-140. Data cited in ref. 6

8　Satar S. Analyse der ätherischen Öle aus den Nadeln einiger Koniferen aus der Mongolischen Volksrepublik. Pharmazie 1986;41:155-156. Data cited in ref. 6

9　Braun M. Verbesserung der Arzneibuchvorschriften und ihre Angleichung an das Europäische Arzneibuch am Beispiel von Ätherischen Ölen. Thesis, University of Regensburg, Germany, 2002. Available at: http://epub.uni-regensburg.de/9910/1/Hauptdoku.PDF

10　Karapandzova M, Stefkov G, Trajkovska Dokik E, Kadifkova Panovska T, Kaftandzieva A, Kulevanova S. Chemical characterization and antimicrobial activity of the needle essential oil of *Pinus mugo* (Pinaceae) from Macedonian flora. Planta Med 2011;77-PL59; DOI: 10.1055/s-0031-1282708 (available as abstract)

11　Kartnig T, Fischer U, Bucar F. Vergleichende gaschromatographische Untersuchungen an Latschenkieferölen. Sci Pharm 1996; 64:487-496

12　Kartnig T, Fischer U, Bucar F. Vergleichende gaschromatographische Untersuchungen an Latschenkieferölen, 2. Mitteilung. Sci Pharm 1997;65:289-297

13　Rohloff J, Langleite BO. Monoterpene patterns of industrially produced needle tree oils. In: L Jirovetz and G Buchbauer, Eds. Processing, analysis and application of essential oils. Dehradun, India: Harkrishan Bhalla & Sons, 2005:155-168

14　Lawrence BM. Progress in essential oils. Perfum Flavor 1999;24(May/June):47-64

15　Glasl H, Wagner H. Gaschromatographische Untersuchung von Arzneibuchdrogen (1) 7. Mitteilung. GC Untersuchung von Pinaceen-Ölen des Handels und Versuche zu ihrer Standardisierung. Deut Apoth Ztg 1980;120:64-67. Data cited in ref. 14

16　Bambagiotti M, Coran SA, Giannellini V, Vincieri FF, Moneti G. Investigation of *Pinus mugo* essential oil oxygenated fraction by combined use of gas chromatography and dry column chromatography. Planta Med 1981;43:39-45. Data cited in ref. 14

17　Latish VG, Kolesnikova RD, Derjuzhkin RI. Essential oils of some *Pinus* species. In: 9th International congress of essential oils, Essential oils technical paper book 2, Singapore, 1983:83-89. Data cited in ref. 14

18　Kubeczka KH, Schultze W. Biology and chemistry of conifer oils. Flavour Fragr J 1987;2:137-148

19　Maurer B, Hanser A. Identification de certains produits aromatisants dans les arômes. Parfum Cosmet Arômes 1988;no.2:69-72. Data cited in ref. 14

20　Venditti A, Serrilli AM, Vittori S, Papa F, Maggi F, Di Cecco M, et al. Secondary metabolites from *Pinus mugo* Turra subsp. *mugo* growing in the Majella National Park (Central Apennines, Italy). Chem Biodivers 2013;10:2091-2100

21 Ioannou E, Koutsaviti A, Tzakou O, Roussis V. The genus *Pinus*: a comparative study on the needle essential oil composition of 46 pine species. Phytochem Rev 2014;13:741-768

22 Lawless J. The encyclopedia of essential oils, 2nd Edition. London: Harper Thorsons, 2014

23 Davis P. Aromatherapy. An A-Z, 3rd Edition. London: Vermilion, 2005

24 Hausen BM, Reichling J, Harkenthal M. Degradation products of monoterpenes are the sensitizing agents in tea tree oil. Am J Contact Dermat 1999;10:68-77

25 Schubert HJ. Skin diseases in workers at a perfume factory. Contact Dermatitis 2006;55:81-83

26 Knöll R, Ulrich R, Spallek W. Allergic contact eczema to etofenamate and dwarf pine oil. Sportverletz Sportschaden 1990;4(2):96-98 (article in German)

27 Frosch PJ, Johansen JD, Menné T, Pirker C, Rastogi SC, Andersen KE, et al. Further important sensitizers in patients sensitive to fragrances. II. Reactivity to essential oils. Contact Dermatitis 2002;47:279-287

28 Meynadier JM, Meynadier J, Peyron JL, Peyron L. Formes cliniques des manifestations cutanées d'allergie aux parfums. Ann Dermatol Venereol 1986;113:31-39

29 Paulsen E, Andersen KE. Colophonium and Compositae mix as markers of fragrance allergy: Cross-reactivity between fragrance terpenes, colophonium and Compositae plant extracts. Contact Dermatitis 2005;53:285-291

Chapter 5.33   ELEMI OIL

DEFINITION
Elemi oil (essential oil of elemi) is the essential oil obtained from the wood exudate of the Manila elemi tree (elemi canary tree), *Canarium luzonicum* (Blume) A. Gray (synonym: *Canarium luzonicum* Miq.).

INCI NOMENCLATURE
Description/definition: Canarium luzonicum gum oil is an essential oil obtained from the resin exudate of the elemi, *Canarium luzonicum* (syn. *C. commune*[a]), Burseraceae

INCI name EU: Canarium luzonicum gum oil (perfuming name, not an INCI name proper)

INCI name USA: not in the Personal Care Products Council Ingredient Database

Other name(s): elemi resin oil (CosIng)

CAS registry number(s): none mentioned in CosIng; 8031-63-8

EINECS number(s): none mentioned in CosIng

[a]*Canarium commune* is *not* a synonym for *Canarium luzonicum*.

ISO (INTERNATIONAL ORGANIZATION FOR STANDARDIZATION) STANDARD
ISO number: 10624

ISO name: Essential oil of elemi

Botanical origin: *Canarium luzonicum* Miq.

Parts of plant used: wood exudate

ISO values: ISO values (minimum and maximum concentrations) are shown in Table 5.33.1.

We prefer the name *Canarium luzonicum* (Blume) A. Gray over *Canarium luzonicum* Miq. (accepted name according to GRIN Taxonomy for Plants and The Plant List)

THE PLANT, THE OIL, AND THEIR USES
Canarium luzonicum (Blume) A. Gray (also indicated as *Canarium luzonicum* Miq., *Canarium luzonicum* (Miq.) A. Gray; synonym: *Pimela luzonica* Blume), commonly known as Manila elemi, is an evergreen tree native to the Philippines. An oleoresin is harvested from it, a yellow substance of honey-like consistency. Elemi resin is chiefly used commercially in varnishes and lacquers, and certain printing inks. It is also used as a herbal medicine to treat bronchitis, catarrh, extreme coughing, aged skin, scars, wounds and stress. It should be stressed that the name elemi is a term applied to a variety of resinous products obtained from different countries and having different botanical origins (1). The species concerned all belong to the family Burseraceae. However, the greater part of the world's supply comes from the Philippine Islands (Manila elemi) and is obtained from the trunk of *Canarium luzonicum*.

Steam distillation of the elemi resin yields the essential oil of elemi. A variety of foodstuffs are flavored with elemi oil and in Europe it is used in spices and seasonings. In the US, elemi oil is also used in fragrances (3). It is also utilized in aromatherapy (5).

CHEMICAL COMPOSITION
Elemi oil is a colorless to pale yellow clear mobile liquid which has a terpenic, woody somewhat spicy and balsamic odor. The yield of essential oil from the wood exudate of *Canarium luzonicum* (Blume) A. Gray generally varies from 20.0 to 30.0%. The main producing country of this oil is the Philippines.

Literature data (up to November 15, 2014) on the chemical composition of elemi oils (only two publications and one reference to studies performed before 1986 found) and unpublished analytical data from one of us (E.S.) are shown in Table 5.33.2 in alphabetical order. In elemi oils from various origins, over 45 chemicals have been identified. About 30% of these were found in a single reviewed publication only. The major compounds found in elemi oils from different sources are shown in Table 5.33.3. They include (highest concentrations in any study given) limonene (80%), α-phellandrene (23.9%), elemol (22.1%), elemicin (8.7%), *p*-cymene (7.1%), sabinene (6.6%), β-phellandrene (3.9%), α-terpineol (3.1%) and terpinolene (3.1%).

Commercial oils
The ten chemicals that had the highest maximum concentrations in 39 commercial elemi essential oil samples (concentration ranges provided) are the following: limonene (33.0-58.4%), α-phellandrene (10.3-23.9%), elemol (6.1-22.1%), elemicin (0.5-8.7%), *p*-cymene (0.9-7.1%), sabinene (2.5-6.6%), β-phellandrene (1.7-3.9%), terpinolene (0.7-3.1%), α-terpineol (0.4-3.1%) and α-eudesmol (0.3-1.6%) (Erich Schmidt, unpublished analytical data).

CONTACT ALLERGY/ALLERGIC CONTACT DERMATITIS

General
Contact allergy to and possible allergic contact dermatitis from elemi oil have been reported in one publication only. A false-positive patch test reaction due to the excited skin syndrome cannot be excluded.

Table 5.33.1 ISO values (%) for elemi oil[a]

Compound	CAS	Minimum	Maximum
Limonene	138-86-3	40.0	72.0
Elemol (α-)	639-99-6	1.0	25.0
α-Phellandrene	99-83-2 1	0.0	24.0
Elemicin	487-11-6	0.5	8.0
Sabinene	3387-41-5	3.0	8.0
α-Terpineol	98-55-5	0.4	3.0

[a] ISO 10624 Essential oil of elemi ©ISO 1998; Geneva, Switzerland, www.iso.org

Case reports

An aromatherapist had occupational contact dermatitis with allergies to multiple essential oils used at work, including elemi oil. The patient also reacted to geraniol, α-pinene, caryophyllene, the fragrance mix and various other fragrance materials. α-Pinene and caryophyllene were demonstrated by GC-MS in the elemi oil (6), but are not important components of commercial elemi oils (Table 5.33.2, column A).

Table 5.33.2 Constituents identified in elemi oils

Constituent	CAS	Percentages and range			
		A	B (1)	C (2)	D
Bulnesol	22451-73-6		0.02-0.04		
δ-Cadinene	483-76-1		0.02-0.04		
Camphene	79-92-5		0.03-0.02		
Camphor	76-22-2	0.03-0.3	0.04-0.04		
(E)-Carveol	1197-07-5	0.06-0.08	0.03-0.08		
(Z)-Carveol	1197-06-4	0.04-0.05	0.02-0.06		
Carvone	99-49-0	tr-0.08	0.08-0.2		
β-Caryophyllene	87-44-5	0.2-0.4			0.1
1,8-Cineole	470-82-6				2.5
α-Copaene	3856-25-5	0.1-0.2	0.06-0.09		
p-Cymene	99-87-6	0.9-7.1	1.1-1.8		1.4
p-Cymenene	1195-32-0	0-0.1			
p-Cymen-8-ol	1197-01-9	0.07-0.1	0.2-0.4		
β-Elemene	33880-83-0		0.08-0.2	0.2-0.3	
Elemicin	487-11-6	0.5-8.7	2.4-4.7	1.8	
Elemol (α-)	639-99-6	6.1-22.1	6.3-13.7	2.8	15-17.3
Eudesmol			0.1-0.2		
α-Eudesmol (α-Selinenol)	473-16-5	0.3-1.6			
β-Eudesmol	473-15-4	0.3-0.6	0.2-0.4		
10-epi-γ-Eudesmol	15051-81-7	0.4-0.7			
Germacrene D	23986-74-5		0.1-0.1		
Guaiol	489-86-1	0.1-0.3	0.1-0.2		
α-Gurjunene	489-40-7	0-0.3			
α-Humulene	6753-98-6	0.06-0.1	0.07-0.2	<0.01	
Limonene	138-86-3	33.0-58.4	56.0-54.6	65.0	23-80
Limonene oxide	1195-92-2	0-0.1			
Linalool	78-70-6		0.02-0.04		
Methyl eugenol	93-15-2	0.3-0.6	0.2-0.3	0.2	
Myrcene	123-35-3	0.6-1.0	0.05-0.03	0.6	
(E)-β-Ocimene	3779-61-1	0.3-0.5	0.03-0.03	0.3	
(Z)-β-Ocimene	3338-55-4	0.3-0.7	0.04-0.02	0.5	
α-Phellandrene	99-83-2	10.3-23.9	17.6-8.8	11.7	
α-Phellandrene-δ-dimer			0.05-0.1		
β-Phellandrene	555-10-2	1.7-3.9	2.3-1.8	1.6	
α-Pinene	80-56-8	0.3-0.7	0.5-0.3	0.4	
β-Pinene	127-91-3	0.2-0.4	0.1-0.08	0.3	
Piperitone	89-81-6	0-0.05	0.06-0.08		
Sabinene	3387-41-5	2.5-6.6	5.7-3.4	5.9	
cis-Sabinene hydrate	15537-55-0	0.05-0.6	0.1-0.1		
trans-Sabinene hydrate	17699-16-0	0.01-0.3	0.5-0.4	0.2	
cis-Sabinol	3310-02-9		0.1-0.2		
α-Terpinene	99-86-5	0.4-1.1	0.2-0.08		
γ-Terpinene	99-85-4	0.3-0.8	0.3-0.2	0.3	

Table 5.33.2 Constituents identified in elemi oils (*continued*)

Constituent	CAS	Percentages and range			
		A	B (1)	C (2)	D
Terpinen-4-ol	562-74-3	0.5-1.2	0.5-0.5	0.5	
α-Terpineol	98-55-5	0.4-3.1	1.1-2.7		
Terpinolene	586-62-9	0.7-3.1	2.8-2.5	0.3	
α-Thujene	2867-05-2	0.1-0.3		0.1	

A thirty-nine elemi essential oil samples from Philippines analyzed between 1998 and 2013; lowest and highest concentrations given (E. Schmidt, unpublished data)
B two lab produced elemi oils from the Philippines; fresh and aged sample examined; before the hyphen: fresh oil, after the hyphen: oil kept in a colorless glass bottle in the refrigerator for one year (ref. 1)
C one sample of elemi oil of unknown origin (ref. 2)
D data cited in in ref. 1 (literature from before 1986)

tr: trace

Table 5.33.3 Major constituents of elemi oils

Constituent	CAS	Percentages and range			
		A	B (1)	C (2)	D
Limonene	138-86-3	33.0-58.4	56.0-54.6	65.0	23-80
α-Phellandrene	99-83-2	10.3-23.9	17.6-8.8	11.7	
Elemol	639-99-6	6.1-22.1	6.3-13.7	2.8	15-17.3
Elemicin	487-11-6	0.5-8.7	2.4-4.7	1.8	
p-Cymene	99-87-6	0.9-7.1	1.1-1.8	1.4	
Sabinene	3387-41-5	2.5-6.6	5.7-3.4	5.9	
β-Phellandrene	555-10-2	1.7-3.9	2.3-1.8	1.6	
α-Terpineol	98-55-5	0.4-3.1	1.1-2.7		
Terpinolene	586-62-9	0.7-3.1	2.8-2.5	0.3	

LEGEND: SEE UNDER TABLE 5.33.2

LITERATURE

1 Villanueva MA, Torres RC, Can Başer KH, Öztek T, Kürkçüoğlu M. The composition of Manila elemi oil. Flavour Fragr J 1993;8:35-37
2 Moyler DA, Clery RA. The aromatic resins: their chemistry and uses. In: KAD Swift, ed. Flavors and Fragrances. Cambridge, UK: Royal Society of Chemistry, 1997:46-115. Data cited in ref. 4
3 Mogana R, Wiart C. *Canarium* L. A phytochemical and pharmacological review. J Pharm Res 2011;4:2482-2489.
4 Lawrence BM. Essential oils 2001-2004. Carol Stream, USA: Allured Publishing Corporation, 2006:215-216
5 Lawless J. The encyclopedia of essential oils, 2nd Edition. London: Harper Thorsons, 2014
6 Dharmagunawardena B, Takwale A, Sanders KJ, Cannan S, Roger A, Ilchyshyn A. Gas chromatography: an investigative tool in multiple allergies to essential oils. Contact Dermatitis 2002;47:288-292

Chapter 5.34   EUCALYPTUS CITRIODORA OIL

There are two major eucalyptus oils: eucalyptus citriodora oil obtained from *Corymbia citriodora* and eucalyptus globulus oil, obtained from *Eucalyptus globulus* Labill. In this chapter, Eucalyptus citriodora oil is discussed. Eucalyptus globulus oil is presented in Chapter 5.35.

DEFINITION

Eucalyptus citriodora oil (essential oil of eucalyptus citriodora) is the essential oil obtained from the leaves and terminal branches of the citron-scent gum (lemon-scent gum), *Corymbia citriodora* (Hook.) K.D. Hill & L.A.S. Johnson (synonym: *Eucalyptus citriodora* Hook.).

INCI NOMENCLATURE

Description/definition: Eucalyptus citriodora oil is the volatile oil obtained from *Eucalyptus citriodora*, Myrtaceae

INCI name EU & USA: Eucalyptus citriodora oil

CAS registry number(s): 8000-48-4; 85203-56-1

EINECS number(s): 286-249-8

ISO (INTERNATIONAL ORGANIZATION FOR STANDARDIZATION) STANDARD

ISO number: 3044

ISO name: Essential oil of eucalyptus citriodora

Botanical origin: *Corymbia citriodora* (Hook.) K.D. Hill & L.A.S. Johnson (synonym: *Eucalyptus citriodora* Hook.)

Parts of plant used: Leaf, terminal branch

ISO values: ISO values (minimum and maximum concentrations) are shown in Table 5.34.1.

Both in literature and the essential oil trade, virtually always the name *Eucalyptus citriodora* Hook. is used.

Other species of the genus *Eucalyptus* from which essential oils are mentioned in CosIng include *Eucalyptus globulus* (see Chapter 5.35), *Eucalyptus dives* (CAS 90028-48-1, EINECS 289-839-3), *Eucalyptus radiata* (CAS 92201-64-4, EINECS 295-995-3) and *Eucalyptus Smithii* (CAS 91771-68-5, EINECS 294-962-0).

THE PLANT, THE OIL, AND THEIR USES

Corymbia citriodora (Hook.) K.D. Hill & L.A.S. Johnson, also known as broad leaved peppermint tree, citron-scent gum, lemon-scent gum or spotted gum, is a tall tree, growing to 40-60 meters in height (17). This tree is native to Australia and is cultivated mainly in Brazil, south China, India, Sri Lanka, Zaire, Kenya, South Africa, Fiji and in other (sub)tropical countries (GRIN Taxonomy for Plants). It is a graceful tree having strong lemon-scented leaves, used mainly as timber plant (9). The leaves and terminal branches serve for the production of the essential eucalyptus citriodora oil, which has wide applications in perfumery, pharmaceutical, chemical and other industries (17). It contains high (60-90%) concentrations of citronellal and is used in perfumes, soaps, hair oils and other cosmetics, disinfectants and room fresheners (9,15,18). The chemical industries use the oil as source material to produce citronellol, hydroxycitronellol and menthol (9,25,37). A rectified form of this oil, in which the major component citronellal has been turned into *cis*- and *trans*-isomers of *p*-menthane-3,8-diol, is used in insect repellents (19).

The oil also has pharmacological applications in anti-inflammatory and antipyretic remedies and against the symptoms of respiratory infections, such as colds, flu and sinus congestion (1,17,18). In addition, the oil is applied externally to cuts and skin infections and can be administered internally for a wide range of complaints (17). In traditional medicine, it is used as an antispasmodic and to relieve joint pains (37). The essential oil from the leaves is purported to have a number of biological properties, including antibacterial, antifungal, antispasmodic, nematicidal, allelopathic (suppresses growth of certain other plant species), insecticidal and antioxidant (1,17,18).

Eucalyptus citriodora oil has applications in aromatherapy (65,66).

CHEMICAL COMPOSITION

Eucalyptus citriodora oil is a mobile, colorless to yellow or yellowish green liquid which has a fresh, slightly citrusy odor reminding of citronellal. The yield of essential oil from the leaves and terminal branches of *Corymbia citriodora* (Hook.) K.D. Hill & L.A.S. Johnson generally varies from 0.5 to 2.0%. The main producing countries of this oil are China, India, Sri Lanka, Indonesia (Java), Madagascar, Guinea and Haiti.

Literature data (up to November 24, 2014) on the chemical composition of eucalyptus citriodora oils and unpublished analytical data from one of us (E.S.) are shown in Table 5.34.2 in alphabetical order. In eucalyptus citriodora oils from various origins, over 220 chemicals have been identified. About 60% of these were found in a single reviewed publication only. The major compounds found in eucalyptus citriodora oils from different sources are shown in Table 5.34.3. They include (highest concentrations in any study given) citronellal (90.1%), citronellyl acetate (25.1%), isopulegol (22.9%), citronellol (21.9%), neoisopulegol (9.4%) and β-caryophyllene (5.4%). Well-known ingredients of eucalyptus citriodora oils that were present in high concentrations (>7%) in one or two studies were 1,8-cineole (4 studies: 44.1%, 45.2%, 54.0%, 55.4%, see under chemotypes), β-pinene (25.7%), α-pinene (18.3% and 23.7%; the highest concentration was found in a 1,8-cineole dominated oil), sabinene (19.6%), isoisopulegol (8.5% and 12.7%), α-thujene (11.9%) and α-terpinene (11.2%). Uncommon or rare constituents of eucalyptus citriodora oils found in high concentrations (>7%) in single studies include 6-octenal (two studies: 55.3% and 77.1%; misidentification, must be citronellal), 3-hexen-1-ol (31.3%), camphene (two studies, but from the same botanical source material: 29.7% and 30.0%), (*Z*)-geraniol (19.7%), 3,6-dimethyl-5-hepten-1-ol (13.1%), ledol (two studies, but from the

Table 5.34.1 ISO values (%) for eucalyptus citriodora oil[a]

Compound	CAS	Minimum	Maximum
Citronellal	106-23-0	75.0	100
Neoisopulegol + isoisopulegol	29141-10-4 18674-65-2	0	10.0

[a] ISO 3044 Essential oil of eucalyptus citriodora ©ISO 1997; Geneva, Switzerland, www.iso.org

same botanical source material: 9.1% and 9.2%), methyl eugenol (8.2%), α-citronellol (7.3%) and trans-isopulegol (7.0%).

Commercial oils

The ten chemicals that had the highest maximum concentrations in 57 commercial eucalyptus citriodora essential oil samples (concentration ranges provided) are the following: citronellal (68.6-84.4%), neoisopulegol

Table 5.34.2 Constituents identified in eucalyptus citriodora oils

Constituent	CAS	Percentage and range				
		A	B (3,49)	C (5)	D (37)	E
Acetaldehyde	75-07-0			tr		
γ-Amorphene	6980-46-7					0.02^{z10}
Aromadendrene	489-39-4				0-0.3	0.03^{z14}; 0.04^n; $0.1^{f,r}$; 0.3^y
allo-Aromadendrene	25246-27-9					0.1^f
β-Asarone	5273-86-9					0.02^i
β-Atlantol	420109-31-5					tr^e
Benzyl isobutyrate	103-28-6		tr			
Bergamal	106-72-9	0.06-0.1	0.2-0.8	tr		$0.2^{e,p,w}$; 0.3^i
Bicyclogermacrene	24703-35-3		0-0.2	0.1		$0.1^{q,w}$; 0.2^n; 1.0^j
Bicyclo[7.2.0]undec-4-ene	6671-82-5					1.3^{z26}
β-Bisabolene	495-61-4					0.2^{z14}; 1.3^{z24}
Borneol	507-70-0	0.01-0.1		0.1	0.2-0.6	tr^e; 0.1^j; 0.8^r; 1.3^{z19}
Cadina-1,4-diene	29837-12-5					3.0^{z10}
trans-Cadina-1,4-diene	20085-13-6					0.1^e
α-Cadinene	24406-05-1					0.1^{z10}; 0.5^{z7}; 1.0^{z17}
γ-Cadinene	39029-41-9		0-tr		0-tr	$tr^{r,z10}$; 0.08^{z14}
δ-Cadinene	483-76-1			tr	0-0.1	tr^x; 0.1^r; 0.7^{z10}
α-Cadinol	481-34-5					0.6^{z10}
δ-Cadinol	19435-97-3					0.3^{z10}
τ-Cadinol (10-epi-α-)	58580-31-7		0-0.4			
α-Calacorene	21391-99-1					0.06^{z10}
cis-Calamene	72937-55-4					0.1^e
Calamenene	483-77-2					0.4^{z10}
(1S)-cis-Calamenene	483-77-2					$<1\%^{z25}$
trans-Calamenen-10-ol	828923-23-5					0.5^e
Camphene	79-92-5	0.01-0.1		0.1		0.1^{z8}; 0.3^l; 29.7^{z7}; 30.0^{z17}
α-Campholenal	4501-58-0					0.1^e
γ-Carene						0.2^z
δ3-Carene	13466-78-9					0.06^z; 0.7^{z3}
Carvacrol	499-75-2					4.5^m
Carvenone	499-74-1					0.06^{z10}
Carvomenthene	5502-88-5			0.1		
β-Caryophyllene	87-44-5	0.5-2.6	0.2-0.7	1.2^d	0-0.2	1.9^q; $2.6^{l,r}$; 2.7^{z21}; 2.9^n; 3.5^j; 3.9^{z3}; 5.4^{z14}
Caryophyllene oxide	1139-30-6		0-0.5	0.1	tr-0.1	0.06^{z14}; $0.1^{e,l,j}$; 0.2^{z10}; $0.3^{n,t}$; 0.4^s; 6.9^{z3}
β-Cedrene	546-28-1			1.2^d		
1,4-Cineole	470-67-7			tr		
1,8-Cineole	470-82-6	0.1-2.0	0-0.7	0.6	0.1-0.8	1.5^w; 1.8^n; 2.0^k; 2.9^p; 3.0^s; $3.4^{t,z23}$; 3.5^{z3}; 4.9^j; 44.1^{z7}; 45.2^{z17}; 54.0^{z19}; 55.4^{z10}
Citral	5392-40-5					0.07^n
Citronellal	106-23-0	68.6-84.4	57.1-75-4	63.6	69.6-87.4	78.5^{z9}; 79.2^{z13}; 81.2^z; 82.7^{z6}; $83.5^{v,z30}$; 84.9^m; 86.0^e; 89.6^p; 89.9^h; 90.1^x
α-Citronellal	141-26-4			0.1		
β-Citronellene	2436-90-0					4.8^y
Citronellic acid	502-47-6			0.7		0.7^{z21}; 0.8^q; 1.1^{z5}; 1.9^t; 2.4^{z20}; 4.7^k
Citronellol (β-, DL-)	106-22-9	3.9-8.0	8.0-14.1		5.1-9.9	11.9^g; 12.6^n; 12.8^h; 13.0^t; 13.1^{z11}; 17.0^{z3} 17.3^{z14}; 17.9^{z8}; 20.0^w; 20.4^j; 21.9^y
α-Citronellol				7.3		

Table 5.34.2 Constituents identified in eucalyptus citriodora oils (*continued*)

Constituent	CAS	A	B (3,49)	C (5)	D (37)	E
Citronellyl acetate	150-84-5	0.3-3.7		2.6	0.4-1.2	2.2[q,s]; 2.3[l]; 2.4[n]; 2.5[m]; 3.3[p]; 4.2[h]; 9.0[w]; 10.3[j]; 11.4[k]; 12.7[u]; 13.7[z24]; 14.2[z3]; 25.1[z5]
Citronellyl formate	105-85-1		0-0.2	tr		tr[e]; 0.2[t]
α-Cubebene	17699-14-8					0.04[z10]
β-Cubebene	13744-15-5					0.3[z10]
Cubenol	21284-22-0					0.08[z10]
epi-Cubenol	19912-67-5					0.1[z10]
Cuminaldehyde	122-03-2					4.2[m]
Cycloenane						0.5[z]
Cyclohexane, 1-methyl-4-(1-methyl-ethenyl)-, *cis*-	6252-33-1			0.1		
Cyclohexane, 1-methyl-4-(1-methyl-ethenyl)-, *trans*-	6252-33-1			0.1		
Cyclohexanol	108-93-0					1.1[z26]; 3.4[z25]; 6.7[z22]
Cyclohexene, 3-methyl-6-(1-isopropyl)-	5256-65-5			0.3		
Cyclosativene	22469-52-9			tr		
p-Cymene	99-87-6	0.02-0.5	tr-0.1	2.5	0.1-0.3	0.3[f]; 0.5[r,u,z4]; 0.6[n]; 1.0[t]; 1.4[z10]; 3.0[z19]
p-Cymenene	1195-32-0					0.07[z10]
p-Cymen-8-ol	1197-01-9					0.4[z10]; 0.7[z7,z17]
Decane	124-18-5					0.2[z12]
Dihydrocarvyl acetate	57287-13-5					4.9[z24]
Dihydrocitronellol	106-21-8					1.5[a,z3]
Dihydrofarnesal	32480-08-3					0.06[n]
1,5-Dimethylcyclopentene	16491-15-9			tr[b]		
2,6-Dimethyl-5-heptenal	see Bergamal					
3,6-Dimethyl-5-hepten-1-ol	51673-46-2					13.1[z24]
(Z)-2,6-Dimethyl-2,6-octadiene	2492-22-0					2.9[z22]
3,7-Dimethyl-2,6-octadien-1-ol (citrol)	624-15-7					0.5[z24]
1,6,10-Dodecatrien-3-ol, 3,7,11-trime-thyl-, S-(Z)-	142-50-7					4.4[z24]
Eicosane (n-)	112-95-8					0.3[z7]; 0.8[z1]
n-Eicosan-5-ol						0.5[z1]
n-Eicosan-6-ol						0.3[z1]
n-Eicos-14-en-2-ol						0.3[z1]
β-Elemene	33880-83-0			0.1		0.04[i,n]; 0.1[e]
γ-Elemene	29873-99-2					0.2[z9]; 0.3[v]
Elemol	639-99-6		0-3.1			0.04[z14]
Ethanol	64-17-5			0.1		
α-Eudesmol	473-16-5		0-0.1			tr[e]; 0.5[y]
β-Eudesmol	473-15-4					0.05[n]; 0.1[e]; 0.8[y]
γ-Eudesmol	1209-71-8		0-2.2			0.1[e]; 1.1[y]
Eugenol	97-53-0		0-2.6	0.1	0-0.3	0.1[e,r,z12]; 0.2[y]; 0.3[u]; 0.4[z30]; 3.5[l]
Farnesal	19317-11-4					+[z28]
Farnesol	4602-84-0			tr		0.2[z3]
(E,E)-Farnesol	106-28-5					0.9[z10]
α-Fenchene	471-84-1			tr		
Geranial	141-27-5		0-0.1	tr	0-0.1	tr[e,r]; 0.03[o]; 0.05[i]; 0.8[z10]
Geraniol	106-24-1	0.02-0.1	0-0.3	0.2	0.1-0.9	0.02[i]; 0.1[o,q,r]; 0.2[e,n]; 0.4[g,s,y,z29]; 2.0[z14]
(Z)-Geraniol (β-Nerol)	106-25-2					19.7[z24]
Geranyl acetate	105-87-3			<0.2	0-0.2	<0.1[m]; 0.1[r]; 0.2[e]; 0.3[v,z9]; 0.4[z14]; 2.1[z3]
Geranylgeraniol	24034-73-9					+[z28]
Germacrene B	15423-57-1					0.04[n]
Germacrene D	23986-74-5			0.1		0.04[i]; 0.1[q]; 0.2[n]
Germacrene D-4-ol	198991-79-6					0.2[z10]
Globulol	489-41-8		0-0.1			0.04[n]; <0.09[q]; 0.3[z]
α-Guaiene	3691-12-1					<1[z25]
Guaiol	489-86-1					0.1[a,z3]
α-Gurjunene	489-40-7					0.06[z10]
γ-Gurjunene	22567-17-5					0.3[f]
Hedycaryol	21657-90-9					0.09[p]
Heneicosane	629-94-7					0.2[z7]; 0.3[z1]

Table 5.34.2 Constituents identified in eucalyptus citriodora oils (*continued*)

Constituent	CAS	Percentage and range				
		A	B (3,49)	C (5)	D (37)	E
n-Heneicos-3-ene						0.3[z1]
n-Heneicos-4-ene						0.4[z1]
n-Heneicos-8-ene						0.3[z1]
n-Heneicos-10-ene						0.2[z1]
(*E,E*)-2,4-Heptadiene	2384-94-3			tr[b]		
(*E*)-2-Hexenal	6728-26-3					<0.09[q]
3-Hexenoic acid, butyl ester, (*Z*)-	69668-84-4					1.5[z24]
3-Hexen-1-ol	544-12-7					31.3[z24]
(*Z*)-3-Hexen-1-ol	928-96-1					tr[q]
α-Himachalene (α-*cis*-)	3853-83-6					+[z28]
α-Humulene	6753-98-6		tr-0.2	0.1		0.04[z10]; 0.05[i]; 0.1[e,r,t]; 0.2[j,n]; 0.3[z14]
Hydroxycitronellol	107-74-4					1.4[a,z3]
4-Hydroxy-α,*p*-dimethylcyclohexaneme-thanol				0.4		
(*Z*)-(2-Hydroxy-2-isopropyl-5-methyl)-cyclohexanol				0.9		
Isoamyl isobutyrate	2050-01-3			tr		
Isobutyl isobutyrate	97-85-8	0.02-0.1		0.1		0.1[z29]; 0.2[z10]
Isocaryophyllene oxide				tr		0.1[z14]
Isoisopulegol	18674-65-2	0.9-3.7	0.6-4.1	0.7		0.1[z9]; 0.6[n]; 0.7[j]; 3.0[w]; 4.1[i]; 8.5[z15]; 12.7[y]
Isolimonene	499-99-0			0.1		0.5[y]
Isopulegol	89-79-2		2.5-5.1	4.5	0.9-3.1	7.3[z18]; 7.4[z2]; 8.2[n]; 8.9[g]; 9.6[q]; 10.3[z11]; 10.4[z4]; 14.6[t]; 15.0[h]; 15.5[z23]; 22.9[z21]
cis-Isopulegol						4.1[z20]
trans-Isopulegol						7.0[z20]
Isopulegol acetate	57576-09-7			tr		0.04[n]; <0.1[z3]
Isospathulenol	88395-46-4					0.1[z3]
(*Z*)-Jasmone	488-10-8			0.2		0.07[i]; 0.1[e]; 0.2[w]; 0.3[n]; 0.4[t]
Juniper camphor	473-04-1					0.1[z]
Ledol	577-27-5					9.1[z17]; 9.2[z7]
Limonene	138-86-3	0.1-0.5	tr-0.4	0.4	0.1-0.2	0.5[j]; 0.8[z4]; 1.1[t]; 1.3[s]; 2.3[z19]; 2.4[z8]; 5.7[z10]
Linalool	78-70-6	0.1-0.5	0.5-0.6	0.2	2.1-6.4	1.3[z8,z16]; 1.4[s]; 1.9[z13]; 2.2[t]; 2.8[h]; 4.2[z4]
Linalool oxide	1365-19-1			0.2		2.4[z3]
cis-Linalool oxide	11063-77-7					1.3[l]
cis-Linalool oxide, furanoid	11063-77-7			0.2		0.7[z14]
trans-Linalool oxide	11063-78-8					0.2[s];
trans-Linalool oxide, furanoid	34995-77-2			0.7[c]		0.09[z14]; 0.1[r]
Linalyl acetate	115-95-7					0.3[y]; 1.6[z30]; 4.4[z7]; 4.5[z17]; 5.6[u]
Longicyclene	1137-12-8			0.2		
Longifolene (junipene)	475-20-7	0.08-0.6		0.7		1.3[z29]
α-Longipinene	5989-08-2			0.1		
p-Mentha-3,8-diene	586-67-4					0.05[i]; 0.1[q,x]; 0.2[s]; 0.3[n]; 0.6[t]
cis-p-Mentha-1(7),8-dien-2-ol	22626-43-3					0.03[z10]
p-Menthane-3,8-diol	42822-86-6					0.08[v]; 2.9[n]; 4.7[t]
α-Menthene						4.1[z3]
p-Menth-1-en-9-ol	18479-68-0					<0.1[z3]
cis-p-Menth-2-en-8-ol						<0.01[z13]; <0.05[h]
trans-p-Menth-2-en-8-ol						0.2[z13]; 0.3[h]
Menthol	89-78-1				0.1-0.2	0.3[r]; 6.1[z27]
1,4-Methanoazulene	249-73-0					1.5[z22]
Methyl chavicol	140-67-0					1.4[u]
Methyl citronellate	2270-60-2					0.1[p]; 2.0[t]
Methylcyclopentane	96-37-7			tr		
3-Methylcyclopentene	1120-62-3			tr		
6-Methyl-2,4-di-*tert*butylphenol						+[z28]
Methyl eugenol	93-15-2		0-2.1			0.6[j]; 2.2[v]; 2.3[z29]; 8.2[f]
Methylionone	1335-46-2					0.1[n]
cis-1-Methyl-3-iso-propylcyclopentane				0.1		
trans-1-Methyl-3-iso-propylcyclopentane				0.1		

Table 5.34.2 Constituents identified in eucalyptus citriodora oils (*continued*)

Constituent	CAS	Percentage and range				
		A	B (3,49)	C (5)	D (37)	E
1-Methyl-4-isopro-pylidene cyclohexane	1124-27-2			0.2		
α-Muurolene	10208-80-7					1.2[z10]
γ-Muurolene	30021-74-0					0.1[z10]
α-Muurolol	104245-48-9					0.07[p]
τ-Muurolol(epi-α-)	19912-62-0					0.3[z10]
Myrcene	123-35-3	0.1-0.2	tr-0.1	0.1	tr-0.2	0.2[f,h,n]; 0.3[z14]; 0.4[t]; 0.6[j]; 0.7[z8]
cis-Myrtanol	15358-92-6					0.1[e]
Myrtenol	515-00-4					0.1[z7,z17]
2(3*H*)-Naphthalenone,4,4a,5,6,7,8-hexahydro-4a-methyl-	826-56-2					+[z28]
Neoisoisopulegol	21290-09-5	0.2-0.5	0.2-0.4			tr[v]; 0.2[z9]; 0.3[e,z29]; 0.4[i,q]
Neoisopulegol	29141-10-4	3.7-9.4	0.7-1.3			1.9[v,z9]; 2.3[o]; 4.8[q]; 5.5[z29]; 7.3[i]; 7.8[w]
Neoisopulegyl acetate	57576-10-0					0.3[w]
Neral	106-26-3		0-0.9			0.03[i]; 0.1[r]
Nerol	106-25-2				0.1-0.2	0.06[i,o]; <0.1[z30]; 0.1[q]
(*E*)-Nerolidol	40716-66-3					0.1[p]
Neryl acetate	141-12-8			tr		
1,7-Nonadiene	13150-91-9					0.4[z26]
β-Ocimene	13877-91-3		0-tr			
(*E*)-Ocimene	27400-72-2					0.1[m]
(*E*)-β-Ocimene	3779-61-1	0.04-0.1		0.9	0.1-0.8	0.02[z10]; 0.05[n]; 0.07[p]; <0.09[q]; 0.2[r,s,z14]
(*Z*)-β-Ocimene	3338-55-4			0.1	tr-0.2	0.08[n]; <0.09[q]; 0.09[o]; 0.5[z14]
Octadecane-3,12-diol						0.3[z1]
2,6-Octadiene	4974-27-0					2.6[z26]
1,6-Octadien-3-ol	51361-43-4					3.2[z22]
6-Octenal	63826-25-5					55.3[z22,z31]; 77.1[z26,z31]
1-Octen-3-ol	3391-86-4					0.2[z10]
6-Octen-1-ol	63768-12-7					8.3[z22]; 14.1[z26]
Palustrol	5986-49-2					0.2[z]
β-Patchoulene	514-51-2					0.09[z]
α-Phellandrene	99-83-2				tr-0.1	<0.05[r]; 0.06[z14]; 0.1[e,q,z29]; 0.3[n]
β-Phellandrene	555-10-2	0.07-0.8				0.2[z29]; 0.4[z3]
2-Phenylethyl acetate	103-45-7			0.7[c]		0.1[v,z9]
2-Phenylethyl isobutyrate	103-48-0			tr		
α-Pinene	80-56-8	0.2-3.5	0.1-0.7	2.2	tr-0.3	1.3[f]; 1.4[u]; 1.5[k]; 2.1[o]; 2.2[z8]; 2.3[j]; 3.3[z22]; 8.2[z7]; 8.4[z17]; 18.3[z1]; 23.7[z19]
α-Pinene oxide	1686-14-2			tr		
β-Pinene	127-91-3	0.4-1.3	0.3-0.5	0.8		1.2[q]; 1.7[j,k,l]; 2.2[t]; 2.8[z23]; 8.2[z7]; 25.7[z1]
trans-Pinocarveol	1674-08-4					2.3[z19]
Pregn-5-en-20-one, 3,7-dihydroxy-, 3-acetate	1863-39-4					0.3[z24]
Pulegol	529-02-2					2.5[z13]; 4.7[h]
(*E*)-Pulegol	22472-79-3					0.07[y]
cis-Rose oxide	3033-23-6		tr-0.2	0.1		tr[q]; 0.05[i]; 0.07[n]; 0.1[w]; 0.2[t]; 0.3[p]; 0.9[f]
trans-Rose oxide	5258-11-7	0.02-0.08	0-0.1			0.05[z10]; 0.08[i]; 0.1[p]; 0.3[f]
Sabinene	3387-41-5	0.02-0.2		0.1	0.1-0.4	0.1[q,t]; 0.2[n,s]; 0.4[r]; 0.6[x]; 2.3[z10]; 19.6[z1]
cis-Sabinene hydrate	15537-55-0					0.1[z7]
α-Santalene	512-61-8			tr		
Sativene	6813-05-4			tr		
Spathulenol	6750-60-3		0-0.2	tr		0.01[i]; 0.06[z10]; 0.1[e,j,w]; 1.1[z3]
Terpenediol	80-53-5			tr		
α-Terpinene	99-86-5	0.01-0.04	0-0.2	0.2	0-0.4	0.04[n]; 0.08[o]; 0.1[f,q,r]; 0.2[z14]; 11.2[z8]
γ-Terpinene	99-85-4	0.08-0.4	tr-0.5			0.1[w]; 0.2[f,o,q]; 0.3[l,n]; 0.4[u]; 0.6[k]; 1.2[j]; 1.5[t]
Terpinen-4-ol	562-74-3			0.3		0.1[i,j,r]; 0.2[o]; 0.3[e,s]; 0.5[z10]; 0.6[z14]; 1.0[t]
α-Terpineol	98-55-5	0.06-0.2	tr-0.5	0.2	0-0.3	0.4[j,u]; 0.5[f]; 0.7[l]; 0.9[k]; 1.0[t]; 1.6[z8]; 3.0[z19]
δ-Terpineol	7299-42-5					0.2[i]
Terpinolene	586-62-9	0.07-0.2	tr-0.1	0.3		0.02[i]; 0.07[p]; 0.1[m,x]; 0.2[o,q]; 0.3[j,n,s]; 1.6[t]
Terpinyl acetate	8007-35-0					0.1[m]
α-Terpinyl acetate	80-26-2					0.5[z10]; 1.5[l]; 1.8[k]

Table 5.34.2 Constituents identified in eucalyptus citriodora oils (*continued*)

Constituent	CAS	Percentage and range				
		A	B (3,49)	C (5)	D (37)	E
1,5,5,8-Tetramethyl-12-oxabicyclo[9.1.0] dodeca-3,7-diene	90820-79-4					+[z28]
Thujene	58037-87-9					0.2[s]
α-Thujene	2867-05-2	0.01-0.03				
Torreyol	19435-97-3	0.09[z10]		tr	tr-0.2	0.05[z14]; 0.2[r]; 0.3[m,z3]; 0.5[x]; 1.1[z7]; 11.9[z1]
Tricyclene	508-32-7					<0.01[i]
Verbenone	80-57-9					0.3[t]
trans-Verbenyl acetate	1203-21-0					tr[e]
Viridiflorene (ledene)	21747-46-6					1.2[e]
Viridiflorol	552-02-3					0.2[z]; 0.3[n]
α-Ylangene	14912-44-8					tr[e]

A fifty-seven *Corymbia citriodora* essential oil samples from China, India, Indonesia (Java), Madagascar and Ivory Coast, analyzed between 1999 and 2013; lowest and highest concentrations given (E. Schmidt, unpublished data)
B six lab-hydrodistilled oils from the leaves of *C. citriodora* cultivated in Congo and harvested in two seasons and in two separate years; lowest and highest concentrations given (ref. 3, data also published in ref. 49)
C one oil produced at Reunion Island and purchased at a local market (ref. 5)
D twelve lab-hydrodistilled oils from leaves of *C. citriodora* grown in an experimental field in India and harvested each month for a year; lowest and highest concentrations given (ref. 37)
E data from other studies (indicated with superscript letters); highest concentrations found in any study reviewed here given;when two or more oils were investigated, only the highest concentrations are mentioned, unless indicated otherwise

[a]incorrect identity based on GC elution order/(possibly) incorrect identification (ref. 50); [b]1,5-dimethylcyclopentene *or* (*E,E*)-2,4-heptadiene; [c]2-phenylethyl acetate and linalool oxide II combined; [d]β-caryophyllene combined with β-cedrene (?); [e]one lab-hydrodistilled oil prepared from fresh leaves harvested at a research center in India (ref. 1); [f]one steam-distilled leaf oil from Mali (ref. 2); [g]one lab-hydrodistilled leaf oil from an arboretum in Ethiopia (ref. 7); [h]thirty-five oil samples from leaves harvested in 7 months from two localities in Benin and obtained from either fresh or dry leaf material and with different distillation times (ref. 8); [i]one lab-hydrodistilled leaf oil from Bangladesh (ref. 9); [j]one lab-hydrodistilled leaf oil from Burundi (ref. 12); [k]one lab-hydrodistilled leaf and twig oil from an arboretum in Morocco (ref. 13); [l]one lab-hydrodistilled oil from the Democratic Republic of Congo (ref. 14, data also presented in ref. 35); [m]two lab-hydrodistilled leaf oils, one from Ethiopia, one from Kenya (ref. 16); [n]one lab-hydrodistilled leaf oil from India (ref. 18); [o]one commercial *C. citriodora* oil from China (ref. 19); [p]one lab-hydrodistilled oil from Brazil (ref. 20); [q]two lab-prepared oils, one obtained by hydrodistillation and the other by steam distillation of leaves collected at an experimental field in India (ref. 24); [r]one industrially steam-distilled leaf oil from an experimental farm in India (ref. 25); [s]one commercial oil purchased from a USA manufacturer (ref. 26); [t]one microwave-assisted lab-hydrodistilled leaf oil from *C. citriodora* from an experimental station in Colombia (ref. 27); [u]two commercial steam-distilled oil samples from Madagascar from two harvest years (ref. 29); [v]one lab-hydrodistilled leaf oil from trees of a university botanical garden in Benin (ref. 30); [w]one lab-hydrodistilled leaf oil from *C. citriodora* growing wild in Benin (ref. 32); [x]one commercial leaf oil purchased in Germany (ref. 34); [y]two lab-hydrodistilled oils from adult and juvenile leaves of *C. citriodora* collected in India (ref. 40); [z]various essential oils obtained from *C. citriodora* leaves growing wild in Crete (Greece) with different distillation parameters and integral versus comminuted leaves (ref. 42); [z1]one lab-hydrodistilled oil from India with an extremely atypical composition; the authors suggested that, with 15 components, they had analyzed 100% of the oil (ref. 43); probably botanical misinterpretation; [z2]one commercial oil obtained from an Australian company (ref. 45); [z3]four oils from leaves harvested in four seasons in Egypt (ref. 51); [z4]two oils from the Ivory Coast prepared from fresh and from dried leaves; highest concentrations given (ref. 52); the data from the fresh leaves are also presented in ref. 61; [z5]one lab-hydrodistilled leaf oil from Morocco (ref. 53); [z6]two commercial oils, one from China and one from India (ref. 54); [z7]one lab-hydrodistilled leaf oil from India with 1,8-cineole (44.1%) as dominant ingredient and otherwise also extremely atypical composition; probably a botanical misinterpretation rather than specific chemotype (ref. 63); [z8]one lab-hydrodistilled oil from the leaves of a tree on a university campus in Hawaii (ref. 4); [z9]two steam-distilled oil samples from Mali (ref. 6); [z10]one leaf oil from an arboretum in Rwanda with 1,8-cineole (55.4%) as dominant ingredient; unknown whether this represents chemotype 4 or a botanical misinterpretation; as the oil contains many other ingredients not characteristic of *C. citriodora* oil, the latter possibility should certainly be considered (ref. 10); see the section Chemotypes above Table 5.34.2; [z11]one lab-hydrodistilled leaf oil from Benin (ref. 11); [z12]one oil of Ethiopian origin (ref. 57); [z13]two oils produced in Benin; highest concentrations given (ref. 58); [z14]one fresh and one aged *C. citriodora* oil from South India; highest concentrations given (ref. 59); in the aged oil, the concentration of citronellal had decreased considerably (from 79.8 to 50.4%), whereas the amounts of isopulegol + borneol, citronellol and β-caryophyllene increased; [z15]one sample of steam-distilled eucalyptus citriodora oil (ref. 60); [z16]six field-distilled oils from an Indian research farm with various amounts of water added to the oil and various storage times (1-15 days) before analyzing; only citronellal, isopulegol, citronellol and linalool were investigated (ref. 15); [z17]six lab-hydrodistilled leaf oils from India, of which five had been subjected to various physical treatments; only the data of the control (untreated) oil are presented here (ref. 17); same source and same composition with high 1,8-cineole, camphene, α-pinene, ledol and linaly acetate content as in ref. 63; see the section on Chemotypes just above Table 5.34.2;

Table 5.34.2 Constituents identified in eucalyptus citriodora oils (*continued*)

[z18] one steam-distilled commercial oil from Brazil (ref. 21); [z19] one lab-hydrodistilled leaf oil from an arboretum in Tunisia; atypical composition with high 1.8-cineole content (ref. 22); see the section on Chemotypes above Table 5.34.2; [z20] one commercial oil purchased from a South African company (ref. 23); [z21] one steam-distilled oil from leaves and twigs of *C. citriodora* from Uruguay (ref. 28); [z22] one commercial oil sample from China with an extremely atypical composition and 6-octenal as major ingredient; see the section on Chemotypes above Table 5.34.2 (ref. 31); [z23] one lab-hydrodistilled leaf oil from Brazil (ref. 36); [z24] one lab-hydrodistilled oil from powdered leaves from Egypt with an extremely atypical composition with 3-hexen-1-ol as dominant (31.3%) ingredient (ref. 38); [z25] one lab-hydrodistilled leaf oil from a Brazilian tree farm (ref. 39); [z26] one lab-hydrodistilled leaf oil from a tree growing wild in Nigeria; extremely atypical composition with 6-octenal (77.1%) as dominant ingredient; see the section Chemotypes above Table 5.34.2 (ref. 41); [z27] one commercial oil purchased in Brazil (ref. 62); [z28] one commercial eucalyptus citriodora oil sample from a Swiss company (ref. 64); [z29] one commercial oil from China (ref. 44); [z30] unknown number of lab-hydrodistilled oils from mature leaves from India (ref. 48); [z31] misidentification, must be citronellal

tr: trace (in columns B, C and D: <0.1; in column E[e]: <0.05); + present in the oil investigated, but quantity not stated

(3.7-9.4%), citronellol (3.9-8.0%), citronellyl acetate (0.3-3.7%), isoisopulegol (0.9-3.7%), α-pinene (0.2-3.5%), β-caryophyllene (0.5-2.6%), 1,8-cineole (0.1-2.0%), β-pinene (0.4-1.3%) and β-phellandrene (0.07-0.8%) (Erich Schmidt, unpublished analytical data).

Chemotypes

By far, most essential oils of *C. citriodora* contain citronellal as major constituent in concentrations of 65% or higher. However, some authors have suggested the existence of four chemotypic forms (3,6):

Chemotype 1: citronellal >65%
Chemotype 2: citronellal 20-50%
Chemotype 3: citronellol and its acetate ester in the majority, citronellal 1-14%
Chemotype 4: mainly monoterpenic hydrocarbons and 1,8-cineole, low levels of citronellal (< 10%).

Up to now, there have been four reports of *C. citriodora* oils with 1,8-cineole as main ingredient (10,17,22,63). Two of these oils were prepared from the same source, the garden of the Jamia Hamdard University, Delhi, India (17,63). These oils both also had high concentrations of camphene (30%), α-pinene (8%), ledol (9%) and linalyl acetate (4.5%), which was not the case in the other two reports, which were quite different, also from each other. Because of this, and as there are many other *Eucalyptus* species with high 1,8-cineole content, we and others

(46,50,56) feel that, although it cannot be excluded that these reports may represent a separate chemotype, it is equally possible – if not more likely—that there was a botanical misidentification in these studies and that chemotype 4 in fact does not exist.

In addition, in two reports, 6-octenal (55% and 77%), 6-octen-1-ol (8% and 14%) and cyclohexanol (3.4% and 6.6%) were the major ingredients, citronellal being absent (refs. 31 and 41). In one of these studies a commercial leaf and wood oil from China had been investigated (31); the other was a lab-hydrodistilled leaf oil from *C. citriodora* growing wild in Nigeria (41). Whether this is a separate chemotype or the result of botanical misinterpretation is unclear at the moment.

The composition of the oil in ref. 33 is not discussed because of obscurities in the data presentation. For the composition of the leaf oils from very young pot-grown *C. citriodora* plants see ref. 47.

CONTACT ALLERGY/ALLERGIC CONTACT DERMATITIS

Eucalyptus citriodora oil

No reports on contact allergy to eucalyptus oil, specifically mentioned to be obtained from *Eucalyptus citriodora*, have been found. Literature on contact allergy to/ allergic Contact dermatitis from 'eucalyptus oil' (botanical source not specified) is discussed in Chapter 5.35.

Table 5.34.3 Major constituents of eucalyptus citriodora oils

Constituent	CAS	Percentage and range				
		A	B (3,49)	C (5)	D (37)	E
Citronellal	106-23-0	68.6-84.4	57.1-75-4	63.6	69.6-87.4	83.5[v]; 84.9[m]; 86.0[e]; 89.6[p]; 89.9[h]; 90.1[x]
Citronellyl acetate	150-84-5	0.3-3.7		2.6	0.4-1.2	10.3[j]; 11.4[k]; 12.7[u]; 13.7[z24]; 14.2[z3]; 25.1[z5]
Isopulegol	89-79-2		2.5-5.1	4.5	0.9-3.1	10.4[z4]; 14.6[t]; 15.0[h]; 15.5[z23]; 22.9[z21]
Citronellol	106-22-9	3.9-8.0	8.0-14.1		5.1-9.9	17[z3]; 17.3[z14]; 17.9[z8]; 20.0[w]; 20.4[j]; 21.9[v]
Neoisopulegol	29141-10-4	3.7-9.4	0.7-1.3			1.9[v,z9]; 2.3[o]; 4.8[q]; 5.5[z29]; 7.3[i]; 7.8[w]
β-Caryophyllene	87-44-5	0.5-2.6	0.2-0.7	1.2[d]	0-0.2	1.9[q]; 2.6[l,r]; 2.7[z21]; 2.9[n]; 3.5[j]; 3.9[z3]; 5.4[z14]

LEGEND: SEE UNDER TABLE 5.34.2

LITERATURE

1 Verma RS, Padalia RC, Pandey V, Chauhan A. Volatile oil composition of vegetative and reproductive parts of lemon-scented gum (*Eucalyptus citriodora* Hook.). J Essent Oil Res 2013;25:452-457

2 Traoré N, Sidibé L, Figuérédo G, Chalchat J-C. Chemical composition of five essential oils of *Eucalyptus* species from Mali: *E. houseana* F.V. Fitzg. ex Maiden, *E. citriodora* Hook., *E. raveretiana* F. Muell., *E. robusta* Smith and *E. urophylla* S.T. Blake.J Essent Oil Res 2010;22:510-513

3 Silou T, Loumouamou AN, Loukakou E, Chalchat J-C, Figuérédo G. Intra and interspecific variations of yield and chemical composition of essential oils from five *Eucalyptus* species growing in the Congo-Brazzaville. *Corymbia* Subgenus. J Essent Oil Res 2009;21:203-211. Data also published in ref. 49

4 Chen J, Lichwa J, Ray C. Essential oils of selected Hawaiian plants and associated litters. J Essent Oil Res 2007;19:276-278

5 Vernin GA, Parkanyi C, Cozzolino F, Fellous R. GC/MS analysis of the volatile constituents of *Corymbia citriodora* Hook. from Réunion Island. J Essent Oil Res 2004;16:560-565

6 Chalchat J-C, Garry R-P, Sidibé L, Harama M. Aromatic plants of Mali (V): Chemical composition of essential oils of four *Eucalyptus* species implanted in Mali: *Eucalyptus camaldulensis, E. citriodora, E. torelliana* and *E. tereticornis*. J Essent Oil Res 2000;12:695-701

7 Dagne E, Bisrat D, Alemayehu M, Worku T. Essential oils of twelve *Eucalyptus* species from Ethiopia. J Essent Oil Res 2000;12:467-470

8 Moudachirou M, Gbénou JD, Chalchat JC, Chabard JL, Lartigue C. Chemical composition of essential oils of *Eucalyptus* from Bénin: *Eucalyptus citriodora* and *E. camaldulensis*. Influence of location, harvest time, storage of plants and time of steam distillation. J Essent Oil Res 1999;11:109-118

9 Mondello L, Verzera A, Bonaccorsi Y, Chowdhury JU, Yusef M, Begum J. Studies in the essential oil bearing plants of Bangladesh. Part V. Composition of the leaf oils of *Eucalyptus citriodora* Hook and *E. alba* Reinw. ex Blume. J Essent Oil Res 1998;10:185-188

10 Chalchat J-C, Muhayimana A, Habimana JB, Chabard JL. Aromatic plants of Rwanda. II. Chemical composition of essential oils of ten *Eucalyptus* species growing in Ruhande Arboretum, Butare, Rwanda. J Essent Oil Res 1997;9:159-165

11 Sohounhloue DK, Dangou J, Gnomhossou B, Garneau F-X, Gagnon H, Jean F-I. Leaf oils of three *Eucalyptus* species from Benin: *E. torelliana* F. Muell., *E. citriodora* Hook, and *E. tereticornis* Smith. J Essent Oil Res 1996;8:111-113

12 Dethier M, Nduwimana A, Cordier Y, Menut C, Lamaty G. Aromatic plants of tropical Central Africa. XVI. Studies on essential oils of five *Eucalyptus* species grown in Burundi. J Essent Oil Res 1994;6:469-473

13 Zrira SS, Benjilali BB, Fechtal MM, Richard HH. Essential oils of twenty-seven *Eucalyptus* species grown in Morocco. J Essent Oil Res 1992;4:259-264

14 Cimanga OK, Apers S, De Bruyne T, Van Miert S, Hermans N , Totté J, et al. Chemical composition and antifungal activity of essential oils of some aromatic medicinal plants growing in the Democratic Republic of Congo. J Essent Oil Res 2002;14:382-387. Data also presented in ref. 35

15 Rajeswara Rao BR, Rajput DK, Patel RP. Storage of essential oils: influence of presence of water for short periods on the composition of major constituents of the essential oils of four economically important aromatic crops. J Essent Oil Bear Plants 2011;14:673-678

16 Manguro LOA, Opiyo SA, Asefa A, Dagne E, Muchori PW. Chemical constituents of essential oils from three *Eucalyptus* species acclimatized in Ethiopia and Kenya. J Essent Oil Bear Plants 2010;13:561-567

17 Sultana S, Ali M, Ansari SH, Bagri P. The effect of physical factors on chemical composition of the essential oil of *Eucalyptus citriodora* Hook. leaves. J Essent Oil Bear Plants 2008;11:69-74

18 Singh HP, Kaur S, Negi K, Kumari S, Saini V, Batish DR, Kohli RK. Assessment of in vitro antioxidant activity of essential oil of *Eucalyptus citriodora* (lemon-scented Eucalypt; Myrtaceae) and its major constituents. LWT - Food Sci Technol 2012;48:237-241

19 Drapeau J, Rossano M, Touraud D, Obermayr U, Geier M, Rose A, Kunz W. Green synthesis of para-menthane-3,8-diol from *Eucalyptus citriodora*: Application for repellent products. C. R. Chimie 2011;14:629-635

20 Gusmão NMS, de Oliveira JV, do A.F. Navarro DM, Dutra KA, da Silva WA, Wanderley MJA. Contact and fumigant toxicity and repellency of *Eucalyptus citriodora* Hook., *Eucalyptus staigeriana* F., *Cymbopogon winterianus* Jowitt and *Foeniculum vulgare* Mill. essential oils in the management of *Callosobruchus maculatus* (FABR.) (Coleoptera: Chrysomelidae, Bruchinae). J Stored Prod Res 2013;54:41-47

21 Maciel MV, Morais SM, Bevilaqua CML, Silva RA, Barros RS, Sousa RN, Sousa LC, et al. Chemical composition of *Eucalyptus* spp. essential oils and their insecticidal effects on *Lutzomyia longipalpis*. Vet Parasitol 2010;167:1–7

22 Elaissi A, Hadj Salah K, Mabrouk S, Larbi KM, Chemli R, Harzallah-Skhiri F. Antibacterial activity and chemical composition of 20 *Eucalyptus* species essential oils. Food Chem 2011;129:1427-1434

23 Combrinck S, Regnier T, Kamatou GPP. *In vitro* activity of eighteen essential oils and some major components against common postharvest fungal pathogens of fruit. Ind Crops Prod 2011;33:344-349

24 Mann TS, Kiran Babu GD, Guleria S, Singh B. Variation in the volatile oil composition of *Eucalyptus citriodora* produced by hydrodistillation and supercritical fluid extraction techniques. Nat Prod Res 2013;27:675-679

25 Rajeswara Rao BR, Kaul PN, Syamasundar KV, Ramesh S. Comparative composition of decanted and recovered essential oils of *Eucalyptus citriodora* Hook. Flavour Fragr J 2003;18:133-135

26 Han J, Kim S-I, Choi B-R, Lee S-G, Ahn Y-J. Fumigant toxicity of lemon eucalyptus oil constituents to acaricide-susceptible and acaricide-resistant *Tetranychus urticae*. Pest Manag Sci 2011;67:1583-1588

27 Olivero-Verbel J, Nerio LS, Stashenko E. Bioactivity against *Tribolium castaneum* Herbst (Coleoptera: Tenebrionidae) of *Cymbopogon citratus* and *Eucalyptus citriodora* essential oils grown in Colombia. Pest Manag Sci 2010;66:664-668

28 Dellacassa E, Menéndez P, Moyna P, Soler E. Chemical composition of *Eucalyptus* essential oils grown in Uruguay. Flavour Fragr J 1990;5:91-95

29 de Medici D, Pieretti S, Salvatore G. Chemical analysis of essential oils of Malagasy medicinal plants by GC and NMR spectra. Flavour Fragr J 1992;7:275-281

30 Gbenou JD, Ahounou JF, Akakpo HB, Laleye A, Yayi E, Gbaguidi F, et al. Phytochemical composition of *Cymbopogon citratus* and *Eucalyptus citriodora* essential oils and their anti-inflammatory and analgesic properties on Wistar rats. Mol Biol Rep 2013;40:1127-1134

31 George DR, Masic D, Sparagano OAE, Guy JH. Variation in chemical composition and acaricidal activity against *Dermanyssus gallinae* of four eucalyptus essential oils. Exp Appl Acarol 2009;48:43-50

32 Bossou AD, Mangelinckx S, Yedomonhan H, Boko PM, Akogbeto MC, De Kimpe N, et al. Chemical composition and insecticidal activity of plant essential oils from Benin against *Anopheles gambiae* (Giles). Parasites & Vectors 2013;6:337 (17 pages)

33 Smith Vera S, Zambrano DF, Méndez-Sanchez SC, Rodríguez-Sanabria F, Stashenko EE, Duque Luna JE. Essential oils with insecticidal activity against larvae of *Aedes aegypti* (Diptera: Culicidae). Parasitol Res 2014;113:2647-2654

34 Mulyaningsih S, Sporer F, Reichling J, Wink M. Antibacterial activity of essential oils from *Eucalyptus* and of selected components against multidrug-resistant bacterial pathogens. Pharm Biology 2011;49:893-899

35 Cimanga K, Kambu K, Tona L, Apers S, De Bruyne T, Hermans N, et al. Correlation between chemical composition and antibacterial activity of essential oils of some aromatic medicinal plants growing in the Democratic Republic of Congo. J Ethnopharmacol 2002;79:213-220. Data also presented in ref. 14

36 Wagner de S. Aguiar R, Ootani MA, Donizeti Ascencio S, Ferreira TPS, dos Santos MM, dos Santos GR. Fumigant antifungal activity of *Corymbia citriodora* and *Cymbopogon nardus* essential oils and citronellal against three fungal species. Scientific World Journal Volume 2014, Article ID 492138, 8 pages

37 Manika N, Mishra P, Kumar N, Chanotiya CS, Bagchi GD. Effect of season on yield and composition of the essential oil of *Eucalyptus citriodora* Hook. leaf grown in sub-tropical conditions of North India. J Med Plant Res 2012;6:2875-2879

38 Abd El Mageed AA, Osman AK, Tawfik AQ, Mohammed HA. Chemical composition of essential oils of four *Eucalyptus* species (Myrtaceae) from Egypt. Res J Phytochem 2011;5:115-122

39 Zini CA, Zanin KD, Christensen E, Caramao EB, Pawliszyn J. Solid-phase micro extraction of volatile compounds from the chopped leaves of three species of *Eucalyptus*. J Agric Food Chem 2003;51:2679-2686

40 Batish DR, Singh HP, Setia N, Kaur S, Kohli RK. Chemical composition and inhibitory activity of essential oil from decaying leaves of *Eucalyptus citriodora*. Z Naturforsch c 2006;61:52-56

41 Habila N, Agbaji AS, Ladan Z, Bello IA, Haruna E, Dakare MA, et al. Evaluation of in vitro activity of essential oils against *Trypanosoma brucei brucei* and *Trypanosoma evansi*. J Parasitol Res 2010, Article ID 534601, 5 pages

42 Mejdoub R, Katsiotis ST. Factors influencing the yield and the quality of the obtaining essential oil from the leaves of *Eucalyptus citriodora* growing in Crete. Sci Pharm 1998;66:93-105

43 Mittal A, Ali M. Volatile composition of the leaves of *Eucalyptus citriodora* Hook. Int J Res Ayurveda Pharm 2011;2:509-511

44 Jirovetz L, Bail S, Buchbauer G, Stoilova I, Krastanov A, Stoyanova A, et al. Chemical composition, olfactory evaluation and antioxidant effects of the leaf essential oil of *Corymbia citriodora* (Hook.) from China. Nat Prod Commun 2007;2:599-606

45 Lee Y-S, Kim J, Shin S-C, Lee S-G, Park IK. Antifungal activity of Myrtaceae essential oils and their components against three pathogenic fungi. Flavour Fragr J 2008;23:23-28

46 Lawrence BM. Progress in essential oils. Perfum Flavor 2011;36(8):56-59

47 Vaknin Y, Dudai N, Murkhovsky L, Gelfandbein L, Fischer R, Degari A. Effects of pot size on leaf production and essential oil content and composition of *Eucalyptus citriodora* Hook. (Lemon-scented gum). J Herbs Spices Med Plants 2009;15:164-176

48 Verma RS, Verma RK, Chauhan A, Yadav AK. Changes in the essential oil content and composition of *Eucalyptus citriodora* Hook. during leaf ontogeny and leaf storage. Indian Perfum 2009;53:22-25. Data cited in ref. 46

49 Loumouamou AN, Silou Th, Mapola G, Chalchat JC, Figuérédo G. Yield and composition of essential oils from *Eucalyptus citriodora* x *Eucalyptus torelliana*, a hybrid species growing in Congo-Brazzaville. J Essent Oil Res 2009;21:295-299. Data also published in ref. 3

50 Lawrence BM. Progress in essential oils. Perfum Flavor 2013;38(May):44-54

51 El-Zalabani SM, Koheil MM, Meselhy KM, El-Gizawy HA, Sleem AA. Effect of seasonal variation on composition and bioactivities of the essential oil of *Eucalyptus citriodora* Hook. grown in Egypt. Egypt J Biomed Sci 2007;24:260-276. Data cited in ref. 50

52 Tonzibo ZF, N'Guessan YT, Chalcat J-C. Effect of drying on leaf oil production from *Eucalyptus citriodora* from the Ivory Coast. J Essent Oil Bear Plants 1998;1:56-65. Data partly also presented in ref. 61

53 Lahlou M, Berrada R, Agoumi A, Hmamouchi M. The potential effectiveness of essential oils in the control of head lice in Morocoo. Int J Aromather 2001;10:108-123

54 Narayanan P. Comparison of quality of essential oils from South India. FAFAI 2003;5:47-56. Data cited in ref 55

55 Lawrence BM. Progress in essential oils. Perfum Flavor 2006;31(5):60-? (last page unknown)

56 Lawrence BM. Progress in essential oils. Perfum Flavor 2001;26(4):68-? (last page unknown)

57 Asefa A, Dagne E. Essential oils of three *Eucalyptus* species acclimatized in Ethiopia. Bull Chem Soc Ethiopia 1997;11:47-50. Data cited in ref. 56

58 Chalcat JC, Moudachirou M, Gbenou JD, Chabard JL, Lartigue C. Essential oils of *Eucalyptus citriodora* and *Eucalyptus camaldulensis* from Benin: chemical composition, optimization. Rivista Ital EPPOS 1997;(Numero Speciale):642-649. Data cited in ref. 56

59 Kaul PN, Bhattacharya AK, Rajeswara Rao BR, Mallavarapu GR, Ramesh S. Ätherisches Öl aus Eucalyptus citriodora. Zusammensetzung des in Indien erzeugten Produkts. Parfum Kosmet 1997;78(10):38-40. Data cited in ref. 56

60 Betts TJ. Solid phase microextraction of volatile constituents from individual fresh *Eucalyptus* leaves of three species. Planta Med 2000;66:193-195. Data cited in ref. 56

61 Bedi Sahouo G, Tonzibo GZ, Boti B, Chopard C, Mahy JP, N'Guessan YT. Anti-inflammatory and analgesic activities: chemical constituents of essential oils of *Ocimum gratissimum*, *Eucalyptus citriodora* and *Cymbopogon giganteus* inhibited lipoxygenase I-1 and cyclooxygenase of PGHS. Bull Chem Soc Ethiop 2003;17:191-197. Data also presented in ref. 52

62 Ribeiro JC, Ribeiro WLC, Camurça-Vasconcelos ALF, Macedo ITF, Santos JML, et al. Efficacy of free and nanoencapsulated *Eucalyptus citriodora* essential oils on sheep gastrointestinal nematodes and toxicity for mice. Vet Parasitol 2014;204:243-248

63 Husain SS, Ali M. Volatile oil constituents of the leaves of *Eucalyptus citriodora* and influence on clinically isolated pathogenic microorganisms. J Scientif Innov Res 2013;2: 852-858

64 Ramos Alvarenga RF, Wan B, Inui T, Franzblau SG, Pauli GF, Jaki BU. Airborne antituberculosis activity of *Eucalyptus citriodora* essential oil. J Nat Prod 2014;77:603–610

65 Rhind JP. Essential oils. A handbook for aromatherapy practice, 2nd Edition. London: Singing Dragon, 2012

66 Lawless J. The encyclopedia of essential oils, 2nd Edition. London: Harper Thorsons, 2014

Chapter 5.35   EUCALYPTUS GLOBULUS OIL

There are two major eucalyptus oils: Eucalyptus citriodora oil obtained from *Corymbia citriodora* and eucalyptus globulus oil, obtained from *Eucalyptus globulus* Labill. In this chapter, eucalyptus globulus oil is discussed. Eucalyptus citriodora essential oil is presented in Chapter 5.34.

DEFINITION

Eucalyptus globulus oil (essential oil of eucalyptus globulus) is the essential oil obtained from the leaves and terminal branches of the (southern, Victorian) blue gum, *Eucalyptus globulus* Labill.

INCI NOMENCLATURE

Description/definition: Eucalyptus globulus leaf/twig oil is the volatile oil obtained from the leaves and twigs of *Eucalyptus globulus*, Myrtaceae

INCI name EU & USA: Eucalyptus globulus leaf/twig oil

CAS registry number(s): 8000-48-4; 84625-32-1

EINECS number(s): 283-406-2

Description/definition: Eucalyptus globulus leaf oil is the volatile oil obtained from the fresh leaves of the eucalyptus, *Eucalyptus globulus* and other species of eucalyptus, *Myrtaceae*

INCI name EU & USA: Eucalyptus globulus leaf oil

CAS registry number(s): 8000-48-4; 84625-32-1

EINECS number(s): 283-406-2

ISO (INTERNATIONAL ORGANIZATION FOR STANDARDIZATION) STANDARD

ISO number: 770
ISO name: Essential oil of eucalyptus globulus
Botanical origin: *Eucalyptus globulus* Labill.
Parts of plant used: Leaf, terminal branch
ISO values: ISO values (minimum and maximum concentrations) are shown in Table 5.35.1.

Other species of the genus *Eucalyptus* from which essential oils are mentioned in CosIng include *Eucalyptus citriodora* (see Chapter 5.34), *Eucalyptus dives* (CAS 90028-48-1, EINECS 289-839-3), *Eucalyptus radiata* (CAS 92201-64-4, EINECS 295-995-3) and *Eucalyptus Smithii* (CAS 91771-68-5, EINECS 294-962-0). It may be assumed that some commercial eucalyptus oils are obtained from *Eucalyptus* species other than *E. globulus* (46,47).

THE PLANT, THE OIL, AND THEIR USES

Eucalyptus globulus (Tasmanian blue gum) is a tall, evergreen tree native to Tasmania and Victoria (Australia). It has been widely planted in temperate South America, China and sub-Saharan Africa. Vast plantations have been established in southern Australia, Spain, Chile and elsewhere, notably for paper pulp and timber production (20, GRIN Taxonomy for plants; www.kew.org).

The leaves (with twigs) of *Eucalyptus globulus* (and to a lesser degree other *Eucalyptus* species) are the principal sources of eucalyptus oil. These essential oils, which have a high 1,8-cineole (eucalyptol) content, are widely used in pharmaceutical, confectionery and cosmetic industries (2,26). Essential oils from *Eucalyptus* exhibit antibacterial, antifungal, analgesic, anti-inflammatory and insecticide properties (26). The oil is medicinally used as a decongestant for treating catarrh, bronchitis, sore throat and influenza, but may also be employed in the treatment of pulmonary tuberculosis, diabetes and asthma. Other applications include in liniments for bruises, sprains and muscular pains (5,15,26,44). Eucalyptus oils are also used as a general disinfectant, cleaner and deodorizer about the house and are utilized in aromatherapy (53,54).

CHEMICAL COMPOSITION

Eucalyptus globulus oil is a mobile, colorless to pale yellow colored liquid which has a fresh, aromatic, and minty odor. The yield of essential oil from the leaves and terminal branches of *Eucalyptus globulus* Labill. generally varies from 0.8 to 2.4%. The main producing countries of this oil are China, Australia, Portugal, Spain, South Africa, India and Brazil. The oils from China are mostly obtained from *Cinnamomum longipaniculatum* (Gamble) N. Chao ex H.W.Li. This means that they are strictly speaking not eucalyptus oils according to ISO criteria, but their composition generally conforms to the standard.

Literature data (up to November 24, 2014) on the chemical composition of eucalyptus globulus oils and unpublished analytical data from one of us (E.S.) are shown in Table 5.35.2 in alphabetical order. In eucalyptus globulus oils from various origins, over 250 chemicals have been identified. About 58% of these were found in a single reviewed publication only. The major compounds found in eucalyptus globulus oils from different sources are shown in Table 5.35.3. They include (highest concentrations in any study given) 1,8-cineole (90.0%), α-pinene (32.7%), *p*-cymene (9.1%, in one report a concentration >25%), limonene (18.8%), aromadendrene (13.5%), globulol (9.8%), *trans*-pinocarveol (6.9%) and α-terpineol (6.5%). Well-known ingredients of eucalyptus globulus oils that were present in high concentrations (>7%) in one or two studies were *p*-cymene (27.2%), camphene (23.1%), β-pinene (10.9% and 14.1%), terpinen-4-ol (10.1%) and γ-terpinene (8.8%). Uncommon or rare constituents of eucalyptus globulus oils found in high concentrations (>7%) in single studies include eugenol (22.6%), cryptone (17.8%), *m*-cymene (10.2%, misidentification, must be *p*-cymene), phellandral (9.4%), geranyl acetate (7.3%) and β-ionone epoxide (7.0%, misidentified).

Table 5.35.1 ISO values (%) for eucalyptus globulus oil[a]

Compound	CAS	Minimum	Maximum
1,8-Cineole	470-82-6	60.0	80.0
α-Pinene	80-56-8	1.0	22.0
Limonene	138-86-3	1.0	15.0
Aromadendrene	489-39-4	traces	10.0
p-Cymene	99-87-6	1.0	6.0
trans-Pinocarveol	1674-08-4	traces	6.0
α-Phellandrene	99-83-2	0.1	1.5
Globulol	489-41-8	traces	1.5

[a] ISO 770 Essential oil of eucalyptus globulus ©ISO 2002; Geneva, Switzerland, www.iso.org

Commercial oils

The ten chemicals that had the highest maximum concentrations in 185 commercial eucalyptus globulus essential oil samples (concentration ranges provided) are the following: 1,8-cineole (61.6-88.7%), limonene (4.5-12.9%), α-pinene (0.3-8.2%), β-pinene (0.3-5.8%), γ-terpinene (0.2-4.9%), terpinolene (0.02-3.6%), p-cymene (1.1-3.1%), α-terpineol (0.02-1.9%), aromadendrene (0.01-1.8%) and trans-pinocarveol (0.01-1.7%) (Erich Schmidt, unpublished analytical data).

Literature data on the chemical composition of eucalyptus globulus oils from before 1990 can be found in refs. 1, 24 and 45.

Table 5.35.2 Constituents identified in eucalyptus globulus oils

Constituent	CAS	Percentage and range				
		A	B (27)	C (10)	D (15)	E
1-Acetyl-3-isopropyl-cyclopent-5-ene						0.1[t]
Agarospirol	1460-73-7		0.2			
(E)-Anethole	4180-23-8					<0.1[s]
Anhydrolinalool oxide	84616-87-5					0.02[t]
p-Anisaldehyde	123-11-5					<0.1[s]
Aristolene	6831-16-9					0.7[y2]
Aromadendrene	489-39-4	0.01-1.8	3.4	0-10.0	0.9-13.5	2.1[z7]; 2.2[t,z2]; 2.6[w]; 3.0[z8]; 3.6[i]; 7.1p; 10[y2]
allo-Aromadendrene	25246-27-9	0.08-0.3	0.8	0-2.4		0.4[m]; 0.6[h,z7]; 0.7[u]; 0.8[i]; 2.2[y2]; 2.5[c]; 2.7[z2]
Benzene	71-43-2					3.4[z5]
trans-α-Bergamotene	13474-59-4					0.4[y4]; 1.1[s]
Bicyclo[2.2.1]heptan-2-one, 5,5-dimethyl-3-methylene-	499126-76-0					0.1[c]
α-Bisabolol	515-69-5					3.8[p]
Borneol	507-70-0	0.02-0.06	0.2		0-0.3	0.09[d,w]; 0.1[s,z1]; 0.2[i]; 0.3[e,h,v]; 1.8[b]; 3.5[y5]
Bornyl acetate	76-49-3					+[r]
β-Bourbonene	5208-59-3			0-0.05		
Bulnesene	164108-17-2			0-0.3		
α-Bulnesene	3691-11-0		0.1			0.02[t]; 0.05[u]
α-Cadinene	24406-05-1		tr	0-0.6		
β-Cadinene	523-47-7					+[r]
γ-Cadinene	39029-41-9		tr	0-0.2	0.4-3.1	0.1[i]
δ-Cadinene	483-76-1		0.1	0-0.09	tr-0.2	
α-Cadinol	481-34-5		tr	0-0.1		0.6[c]
δ-Cadinol	19435-97-3		0.3			
α-Calacorene	21391-99-1		tr			
Calarene	17334-55-3					0.06[j]
Camphene	79-92-5	0.01-0.1	0.1	0.3-1.2	tr-0.2	0.09[o]; 0.1[a,m]; 0.2[s]; 0.5[v,x]; 0.6[y5]; 23.1[f]
Camphene hydrate	465-31-6					0.2[y5]
α-Campholenal	4501-58-0		0.1			0.06[t]; 0.09[u]; 0.1[s]; 0.2[y2,y3]; 0.6[y5]
Camphor	76-22-2	0.01-0.3	tr	0-tr		0.3[w]; 0.4[y5]; 1.7[s]
δ3-Carene	13466-78-9					0.2[y5]
4-Carene	29050-33-7					6.9[y2]
trans-Caren-2-ol						0.4[y5]
Carvacrol	499-75-2		0.1		0-0.2	tr[v]; 0.1[a]; 0.2[y5]
Carveol	99-48-9					0.4[y2]
(E)-Carveol	1197-07-5		0.3	0-0.3		0.05[w]; 0.08[v]; 0.09[c]; 0.1[s]; 1.2[b]
(Z)-Carveol	1197-06-4		0.1			0.1[v]; 0.4[s]; 0.5[b]
Carvone	99-49-0		0.2			0.05[v]; 0.08[t]; 0.1[s]; 0.2[u,y2]; 0.4[z2]; 0.5[y3]
Carvotanacetone	499-71-8		tr			
cis-Carvyl acetate	1205-42-1					0.03[c]
β-Caryophyllene	87-44-5	0.4-0.6	0.1	0-0.08		0.2[b]; 0.6[z2]; 0.8[y2]; 0.9[z8]; 1.5[y]; 1.8[z7]; 2.5[n]
Caryophyllene oxide	1139-30-6		0.1	0-0.5	0.7-2.5	0.3[i]; 1.1[z2]

Table 5.35.2 Constituents identified in eucalyptus globulus oils (*continued*)

Constituent	CAS	Percentage and range				
		A	B (27)	C (10)	D (15)	E
Chrysanthenone	473-06-3					1.7[y3]
1,4-Cineole	470-67-7					4.1[e]
1,8-Cineole	470-82-6	61.6-88.7	53.8	4.1-50.3	63.1-85.8	83.9[z]; 84.9[y4]; 86.5[d]; 86.9[g]; 89.0[z5]; 90.0[z3]
Citral	5392-40-5					1.2[e]
Citronellal	106-23-0					tr[d]
Citronellol	106-22-9				tr-0.2	0.1[f]; 0.2[d]
α-Copaene	3856-25-5		tr	tr-0.06	0-0.2	0.03[w]; 0.2[i]
Cryptone	500-02-7		0.1	0-17.8		1.3[f]
α-Cubebene	17699-14-8			tr-0.2		0.03[j]; 0.1[s]
β-Cubebene	13744-15-5					0.1[s]
Cubenol	21284-22-0					0.08[c]
Cuminaldehyde	122-03-2			0-5.1		2.8[n]
1,4-Cyclohexadiene	628-41-1					1.9[z5]
Cyclolongifolene	164108-26-3			0-0.04		
m-Cymene	535-77-3					4.0[y2,z9]; 10.2[y5,z9]
o-Cymene	527-84-4					0.2[s]; 2.9[z]; 3.5[z5]; 3.7[e]
p-Cymene	99-87-6	1.1-3.1	0.7	tr-27.2		2.3[z7]; 2.7[p]; 5.1[a]; 5.3[y4]; 7.0[z2]; 8.2[j,y]; 9.1[z1]
m-Cymenene	1124-20-5					1.5[z3]
p-Cymenene	1195-32-0		0.1	0-0.8		0.02[u]; 0.05[a,j]; 0.07[w]; 0.3[h]; 0.5[s]
p-Cymen-7-ol	536-60-7			0-0.4		
p-Cymen-8-ol	1197-01-9		0.1	0-1.8		tr[v]; 0.04[t]; 0.2[u,y2]; 0.3[e]; 1.1[b]
Cyperene	2387-78-2					0.3[s]
Dehydroaromadendrene	698388-95-3					0.09[c]
3,7-Dimethyl-2,6-octadien-1-ol (citrol)	624-15-7					0.1[c]
2,7-Dimethyl-3,5-oc-tadione						0.2[y5]
2,6-Dimethyl-1,5,7-octatrien-3-ol	29414-56-0					+[r]; 0.06[c]
Dodecane	112-40-3					0.07[t]
Durene	95-93-2					1.0[y5]
Eicosane	112-95-8					0.03[t]
α-Elemene	5951-67-7			0-tr		
β-Elemene	33880-83-0			0-tr		
δ-Elemene	20307-84-0					0.05
Epiglobulol	88728-58-9		0.9	0-1.5		tr[d]; 0.2[t]; 0.3[q]; 0.4[c,m,w]; 0.8[u]
Epoxymenthyl acetate	29815-69-8					0.08[t]
Eremophilene	10219-75-7			0-1.4		0.1[t]
Eudesma-4(14),11-diene (β-eudesmene)	17066-67-0					0.05[c]
α-Eudesmol	473-16-5		0.4			0.02[u]; 0.09[v]; 0.2[k]; 0.3[z8]; 0.4[b,c]; 1.7[i]
β-Eudesmol	473-15-4		0.1	0-0.4	0.1-0.7	0.09[w]; 0.1[u]; 0.3[n]; 0.4[b,k]; 0.8[z8]; 1.3[i]; 5.5[p]
γ-Eudesmol	1209-71-8		tr	0-0.5	0.9-3.4	0.06[u]; 0.08[v]; 0.1[k]; 0.2[c]; 0.3[b,z8]; 0.6[i]; 3.7[p]
Eugenol	97-53-0				0.2-1.5	22.6[y]
Farnesyl acetate	29548-30-9		tr			
α-Fenchene	471-84-1		tr			0.01[j]; 0.02[w]; 0.04[t]
Fenchone	1195-79-5					tr[u]
Fenchyl alcohol	1632-73-1	0.01-0.1	0.1			tr[e,g]; 0.05[c]; <0.1[s]; 0.1[t,z1]; 1.0[e]; 1.1[y5]; 1.5[b]
Geranial	141-27-5					0.1[k]
Geraniol	106-24-1		0.1		0.1-0.5	tr[g,v]; 0.1[z1]; 0.2[f]; 0.3[k]; 0.4[q]
(*E*)-Geraniol	106-24-1					0.2[e]
Geranyl acetate	105-87-3					0.2[e,w]; 0.7[c]; 7.3[p]
Geranyl propanoate	105-90-8					+[r]
Globulol	489-41-8	0.07-0.9	4.5	0.3-8.0	0.9-3.4	3.4[x]; 3.5[y3]; 4.6[z8]; 5.3[u]; 5.7[z7]; 7.3[f]; 9.8[p]
α-Guaiene	3691-12-1					+[r]; 0.07[c]; 4.8[y3]
β-Guaiene	88-84-6					0.1[w]
Guaiol	489-86-1					0.2[v]; 1.0[b]; 4.6[p]
α-Gurjunene	489-40-7	0.01-0.04	tr	0-1.1		0.02[j]; 0.04[t]; 0.1[c,d]; 0.3[s,w]; 0.6[i]; 1.3[y2]
β-Gurjunene	73464-47-8		0.2	0-0.3		0.07[c]; 0.1[t]; 0.8[s]
Heptadecane	629-78-7					+[r]; 0.06[t]
1-Heptanol	111-70-6					0.1[y4]
Hexadecane	544-76-3					0.1[t]
(*E*)-2-Hexenal	6728-26-3					0.1[j]; 0.2[y4]
1-Hexen-3-ol	4798-44-1			0-0.1		

Table 5.35.2 Constituents identified in eucalyptus globulus oils (*continued*)

Constituent	CAS	Percentage and range				
		A	B (27)	C (10)	D (15)	E
(*Z*)-3-Hexen-1-ol	928-96-1					<0.1[k]; 0.2[j]
Hexenyl acetate	28933-77-9					<0.1[k]
α-Humulene	6753-98-6		0.1	0-0.1		0.05[t]; 0.1[z1]; 0.5[y2]
exo-2-Hydroxycineole	66965-45-5					0.04[c]
4-Hydroxy-4-methyl-2-pentanone	123-42-2					4.3[y1]
β-Ionone epoxide	23267-57-4					7.0[y5,z9]
Isoamyl alcohol	123-51-3					0.2[t]
Isoamyl isobutyrate	2050-01-3					0.07[w]
Isoamyl isovalerate	659-70-1		tr	0-0.09		tr[u]; 0.02[t]; 0.2[s]
Isoamyl propionate	105-68-0					0.2[y5]
Isobicyclogermacral			tr			
Isobornyl acetate	125-12-2					0.3[s]
Isobornyl formate	1200-67-5					0.1[c]
Isobornyl propionate	2756-56-1					1.1[y5]
Isofenchone	6541-58-8					0.4[x]
Isoledene	95910-36-4		0.1			0.06[c]
Isolongifolane						0.2[y5]
Isolongifolen-8-ol						0.2[y5]
Isopentenyl isovalerate	231623-80-6					0.2[o]
Isopinocamphone	15358-88-0					0.01[t]
2-Isopropyl-5-methyl-9-methylene-bicyclo [4.4.0]dec-1-ene	150320-52-8					+[r]
Isopulegol	89-79-2		tr			tr[d]; 2.2[b]
Isopulegol acetate	57576-09-7					0.04[c]
Isospathulenol	88395-46-4		1.9	0-0.8		
Isovaleraldehyde	590-86-3			0-0.1		0.1[u,w]; 0.6[k]; 0.7[p]
Jacksone	1228757-70-7		0.1			
Jasmone	488-10-8					3.6[y5]
Ledol	577-27-5		0.2			+[r]; 0.06[c]; 0.4[b]; 0.1[u,v]; 0.2[h]; 4.9[p]
Limonen-10-al	57074-31-4					+[r]
Limonene	138-86-3	4.5-12.9	2.4	0.2-4.2		6.7[a]; 7.6[w]; 7.8[l]; 8.2[z]; 10.1[e]; 13.2[p]; 18.8[z1]
cis-Limonene oxide	13837-75-7					0.2[y3]
α-Linalool	598-07-2	0.01-0.02				0.2[y5]
Linalool (β-)	78-70-6	0.02-0.3	tr	tr-0.8	tr-0.5	0.09[a,w]; 0.1[g,n,v]; 0.2[c,h,i]; 0.3[f]; 0.5[b];
Linalool oxide	1365-19-1					0.3[z1]
cis-Linalool oxide, furanoid	5989-33-3			0-tr	0.1-0.6	
trans-Linalool oxide, furanoid	34995-77-2			0-tr		
Linalyl acetate	115-95-7					2.4[e]
Longifolene (junipene)	475-20-7					0.2[s]
p-Mentha-2,4(8)-diene	586-63-0					1.6[p]
p-Mentha-1,5-dien-7-ol	19876-45-0			0-0.4		
cis-p-Mentha-1(7), 5-dien-2-ol	30681-15-3			tr-0.7		
trans-p-Mentha-1(7), 5-dien-2-ol	30681-15-3			0-0.6		0.2[w]
p-Mentha-1(7),8-dien-2-ol (isocarveol)	35907-10-9			0-0.2		+[r]; 0.08[c]; 0.3[t]
cis-p-Mentha-1(7),8-dien-2-ol	22626-43-3		0.7			0.1[w]; 0.2[u]
trans-p-Mentha-1(7), 8-dien-2-ol	21391-84-4		0.7			0.4[u]
p-Mentha-1,8-dien-4-ol	3419-02-1		0.1			
cis-p-Mentha-1,8-dien-6-ol	1197-06-4		tr			0.03[u]
trans-p-Mentha-1,8-dien-6-ol	1197-07-5					0.1[u]
trans-p-Mentha-2,8-dienol	4017-77-0					0.2[y5]
p-Menthatriene	116868-92-9			0-0.03		
p-Mentha-1,3,8-triene	18368-95-1					3.8[s]
cis-p-Menth-2-en-1-ol	29803-82-5			0-1.2		tr[z8]
trans-p-Menth-2-en-1-ol	29803-81-4			0-1.4		
Menthol	89-78-1					0.1[z1]; 0.3[e]
Menthone	89-80-5					0.3[e]
p-Methylacetophe-none	122-00-9		0.1			
Methyl chavicol	140-67-0					0.1[s]
Methyldihydrojasmonate	24851-98-7					1.6[y5]
Methyl eugenol	93-15-2			0.2-0.5		
6-Methyl-5-hepten-2-one	110-93-0		tr			
2-Methylhexadecan-1-ol	68526-87-4					0.5[y5]

Table 5.35.2 Constituents identified in eucalyptus globulus oils (*continued*)

Constituent	CAS	A	B (27)	C (10)	D (15)	E
o-Methyl-α-methyl-styrene	7399-49-7					0.4[e]
Methyl 15-oxoeicosanoate						0.3[y5]
α-Muurolene	10208-80-7			0-0.3		
α-Muurolol	104245-48-9				0-0.3	
τ-Muurolol(epi-α-)	19912-62-0			0-0.2		
Myrcene	123-35-3	0.2-1.3	0.1	0-0.2	tr-0.2	0.7[g,k]; 0.9[d]; 1.1[y]; 1.2[p]; 1.5[z1]; 1.6[l]; 1.9[e]
Myrtenal	564-94-3		0.1			0.05[v]; 0.07[t]; 0.5[b]
Myrtenol	515-00-4		0.1			tr[v]; 0.04[t]
Myrtenyl propionate						+[r]
5-epi-Neointermedeol	136734-27-5			0-0.5		
Neral	106-26-3					tr[k]; 1.1[o]; 2.7[z2]
Nerol	106-25-2					0.2[v]
(*E*)-Nerolidol	40716-66-3		0.2			0.3[v]
(*Z*)-Nerolidol	3790-78-1					1.0[b]
Neryl acetate	141-12-8					0.2[e]; 0.4[o]; 3.0[p]
Nonadecane	629-92-5					+[r]; 0.05[t]
4,6-Nonanedione	14090-88-1					0.3[y5]
β-Ocimene	13877-91-3					0.2[e]; 1.0[z1]
o-Ocimene						1.3[y3]
(*Z*)-Ocimene	27400-71-1					0.1[o]
(*E*)-β-Ocimene	3779-61-1	0.03-0.3	0.1	0-tr	tr-0.2	<0.1[k,s]; 0.02[y4]; 0.05[u]; 0.07[a]; 0.1[g,l,w,z8]
(*Z*)-β-Ocimene	3338-55-4	0.02-0.3		0-0.06		<0.1[k,s]; 0.03[c]; 0.09[w]; 0.1[g]; 0.3[j8]; 0.4[z7]
allo-Ocimene	673-84-7	0.01-0.2				tr[g]
Octadecane	593-45-3					+[r]; 0.05[t]
β-Oplopenone	28305-60-4		tr			
Palustrol	5986-49-2		0.1	0-0.2		0.09[u]
di-epi-Palustrol						+[r]
β-Panasinsene	56684-97-0					0.03[c]
Pentacosane	629-99-2		tr			
Pentadecane	629-62-9					0.05[t]
Perillyl alcohol	536-59-4		tr			
Phellandral	21391-98-0			0-9.4		
α-Phellandrene	99-83-2	0.2-1.6				0.5[a]; 0.6[g]; 0.7[l]; 0.9[e]; 1.3[z1]; 1.4[d]; 6.2[o]
β-Phellandrene	555-10-2			0-1.0		tr[z8]; 0.04[a]; 0.8[l]
L-Phellandrene	6153-17-9					0.5[z3]
α-Phellandrene epoxide	288393-04-4			0-1.5		+[r]
2-Phenethyl alcohol	60-12-8					+[r]
Phenethyl isovalerate	140-26-1					0.09[c]; 0.1[t]
2-Phenethyl propionate	122-70-3					0.05[u]
2-Phenylethyl isovalerate	140-26-1					2.5[p]
β-Phenylpropanoate			0.2			
α-Pinene	80-56-8	0.3-8.2	12.1	0.05-17.9	1.2-3.1	20.0[o]; 21.2[z7]; 23.9[z3]; 29.2[z2]; 26.7[p]; 32.7[y]
β-Pinene	127-91-3	0.3-5.8	0.3	0.05-0.4		1.4[y]; 1.6[l]; 2.7[f]; 3.4[p]; 3.7[e]; 10.9[z2]; 14.1[y5]
α-Pinene oxide	1686-14-2					0.4[e]
Pinocarveol	5947-36-4					1.0[y5]
trans-Pinocarveol	1674-08-4	0.01-1.7	3.7	0-6.9		1.0[w]; 1.6[h,t]; 2.5[z8]; 2.7[z7]; 3.3[m]; 3.8[y2]; 4.3[b]
Pinocarvone	30460-92-5		1.7	0-1.1		0.7[m]; 0.8[y3]; 1.0[t]; 1.2[z8]; 1.9[z7]; 2.2[z2]; 3.0[y2]
cis-Piperitol	16721-38-3		0.1	0-0.2		0.1[z8]
trans-Piperitol	16721-39-4		0.1			<0.5[z7]; 0.2[z8];
Piperitone	89-81-6					tr[d]; 0.05[v]; 0.1[h]; 0.2[b,z8]; <0.5[z7]
Pulegone	89-82-7					0.2[y2]
Sabina ketone	513-20-2			0-1.0		
Sabinene	3387-41-5	0.02-0.07		0.03-0.2		tr[g]; 0.01[u]; 0.02[j]; <0.05[m]; 1.9[z2]; 0.3[c]
cis-Sabinol	3310-02-9					+[r]; 0.4[y3]
α-Selinene	473-13-2		0.5	0-tr	tr-0.4	0.2[i,v]
β-Selinene	17066-67-0		0.1	0-0.2		0.3[y2]
δ-Selinene	473-14-3			0-tr		
Spathulene	116845-09-1			0-0.9		
Spathulenol	6750-60-3		0.2	0.1-17.0	0.2-0.7	0.1[c]; 0.3[y5]; 0.4[z8]; 0.7[z7]; 1.5[v]; 1.7[p]; 6.9[b]
(*Z*)-Tagetenone	33746-71-3					0.1[s]
α-Terpinene	99-86-5	0.05-0.1	tr	tr-1.2		tr[z8]; 0.02[j]; <0.1[k]; 0.1[g]; 0.2[a,s]; 0.6[y3]; 0.9[o]
γ-Terpinene	99-85-4	0.2-4.9	tr	tr-1.5	0.2-1.2	1.3[z2]; 2.5[z2]; 2.6[d,n]; 2.9[e]; 3.2[j]; 3.9[a,g]; 8.8[z1]

Table 5.35.2 Constituents identified in eucalyptus globulus oils (*continued*)

Constituent	CAS	Percentage and range				
		A	B (27)	C (10)	D (15)	E
Terpinen-4-ol	562-74-3	0.01-0.8	0.2	0-6.0	1.4-2.6	0.6[b]; 0.7[d,l]; 0.8[i]; 1.1[y5]; 1.4[z1]; 1.5[j,n]; 10.1[p]
1-Terpineol	586-82-3		0.1			
α-Terpineol	98-55-5	0.02-1.9	3.3	0-2.3	0.2-1.4	2.9[b]; 3.2[z7]; 3.6[z1,z2]; 3.8[n]; 4.7[e]; 5.8[i]; 6.5[y5]
β-Terpineol	138-87-4					0.8[e]
δ-Terpineol	7299-42-5	0.01-0.1				0.07[u]
Terpinolene (α-)	586-62-9	0.02-3.6	tr	tr-0.3		0.1[i]; 0.2[a,c,g]; 0.4[p]; 0.5[y2]; 0.6[o]; 0.8[d]; 6.0[e]
α-Terpinyl acetate	80-26-2	0.01-0.8				2.1[x]; 2.7[h]; 2.8[b]; 3.1[c,j]; 3.7[o]; 4.7[p]; 5.3[l]
β-Terpinyl acetate	10198-23-9					0.2[e]
Tetradecane	629-59-4					+[r]; 0.08[t]
1,2,3,3-Tetramethylcyclopenten-4-one						0.7[y5]
6,10,11,11-Tetramethyltricyclo[5.3.0.1 (2,3)]undec-1(7)-ene						+[r]
6,10,11,11-Tetramethyltricyclo[5.3.0.1 (2,3)]undec-7-ene						+[r]
Thuja-2,4(10)-diene	36262-09-6		tr			
α-Thujene	2867-05-2	0.04-0.1		0-1.1	0.1-0.4	tr[e,g]; 0.03[j]; 0.05[o]; 0.6[l]
α-Thujone	546-80-5					0.1[s]
β-Thujone (*cis*-)	471-15-8				tr-0.2	2.2[y3]
Thujopsene	470-40-6					0.3[s]
Thymol	89-83-8		0.1	0-3.5	0.1-0.4	0.04[a]; 0.2[y5]
p-Tolylmethylcarbinol	536-50-5					0.3[e]
Torquatone	3567-96-2		0.1			
Tridecane	629-50-5					+[r]; 0.07[t]
1,7,7-Trimethylbicyclo [2.2.1]hept-5-en-2-ol	91055-72-0					0.03[c]
Umbellulone	24545-81-1					0.6[s]
Undecane	1120-21-4					0.1[t]
Verbenene	4080-46-0		0.1			0.03[w]
Verbenol	473-67-6					0.04[t]
cis-Verbenol	1845-30-3					2.7[y3]
Verbenone	80-57-9		tr			0.04[t]; 0.1[e,f]; 0.9[y3]
β-Vetivene	27840-40-0			0-0.5		
Viridiflorene (ledene)	21747-46-6			0-1.1		tr[d]; 0.2[c]; 0.3[y5]; 1.1[y2]; 5.5[p]
Viridiflorol	552-02-3		0.8	0-1.6		tr[d]; 0.5[k]; 0.6[u]; 0.7[z7,z8]
α-Ylangene	14912-44-8					0.1[s]

A 185 essential oil samples from China, Portugal and Australia analyzed between 1998 and 2014; lowest and highest concentrations given (E. Schmidt, unpublished data)

B one lab-hydrodistilled oil from *E. globulus* leaves collected in a Tunisian arboretum (ref. 27)

C two steam-distilled leaf oils from Spain and two from Montenegro; lowest and highest concentrations given (ref. 10)

D five lab-hydrodistilled leaf oils from *E. globulus* cultivated in India, leaves collected in various seasons, and five oils from leaves after coppicing the trees; lowest and highest concentrations given (ref. 15)

E data from other studies (indicated with superscript letters); highest concentrations found in any study reviewed here given; when two or more oils were investigated, only the highest concentrations are mentioned, unless indicated otherwise

[a] one lab-hydrodistilled oil from leaves of wild growing *E. globulus* trees in Morocco (ref. 23); [b] twelve steam-distilled leaf oils from Morocco, leaves harvested in 12 months (ref. 9); [c] one lab-hydrodistilled oil from leaves from the province of Yunnan, China (ref. 44); [d] one commercial eucalyptus leaf oil sample purchased from a German manufacturer (ref. 26); [e] one commercial oil from India (ref. 16); [f] one lab-hydrodistilled leaf oil from the Democratic Republic of Congo (ref. 14); [g] one commercial oil sample from China (ref. 13); [h] one lab-distilled oil from leaves and twigs collected in an arboretum in Morocco (ref. 12); [i] one lab-hydrodistilled oil from Burundian *E. globulus* leaves (ref. 11); [j] one steam-distilled oil from leaves collected in an arboretum in Rwanda (ref. 8); [k] one lab-hydrodistilled leaf oil from *E. globulus* cultivated in Zambia (ref. 7); [l] one lab-hydrodistilled oil from leaves collected in an Ethiopian arboretum (ref. 6); [m] one steam-distilled leaf oil from *E. globulus* growing wild in Algeria (ref. 4); [n] two lab-hydrodistilled oils from leaves from Ethiopia and Kenya (ref. 2);

Table 5.35.2 Constituents identified in eucalyptus globulus oils (*continued*)

[o] one commercial oil from Ecuador (ref. 3); [p] fifteen lab-hydrodistilled leaf oils produced in Portugal from leaves collected in two regions and three seasons (different leaf ages); with the exception of α-pinene (1.5%) and 1,8-cineole (47.7%) all lowest concentrations were zero (ref. 20); [q] one hydrodistilled leaf oil sample (ref. 41); [r] one oil from China obtained from the fractional distillation of *E. globulus* leaf oil; as the quantities cannot be compared with the other data, they are indicated with +[r] (ref. 42); [s] one lab-hydrodistilled oil from Italy (ref. 43); [t] data cited in ref. 33; uncertain whether an oil was investigated or whether the previous literature was reviewed (ref. 34); [u] one oil of Australian origin (ref. 35); [v] one lab-prepared oil from Morocco (ref. 36); [w] one oil of Spanish and one of Chinese origin (ref. 38); [x] one sample of Spanish eucalyptus oil (ref. 31); [y] two oil samples from Madagascar (ref. 32); [y1] one commercial oil *ex Eucalyptus globulus* purchased in Brazil (ref. 48); [y2] one lab-hydrodistilled oil from leaves collected in an arboretum in Tunisia (ref. 49); [y3] one lab-hydrodistilled oil from leaves of *E. globulus* from Brazil (ref. 50); [y4] one commercial oil sample obtained from an Italian company (ref. 51); [y5] one lab-hydrodistilled oil sample from leaves of *E. globulus* harvested in Cameroon (ref. 52); [z] one commercial oil purchased in Brazil (ref. 18); [z1] one commercial oil purchased in India (ref. 17); [z2] four lab-hydrodistilled eucalyptus leaf oils from different geoclimatic regions of India (ref. 5); [z3] one commercial eucalyptus oil purchased in Australia (ref. 22); [z4] one steam-distilled oil from Uruguayan eucalyptus twigs and leaves (ref. 25); [z5] one commercial oil obtained from a UK company (ref. 28); [z6] one commercial oil probably purchased in Poland (ref. 29); [z7] ten leaf oils from trees cultivated in Tasmania and ten from plants cultivated in Victoria (Australia); the results were shown as means for both groups; highest mean concentrations given (ref. 19); [z8] six steam-distilled eucalyptus oils from adult leaves and six from juvenile leaves collected in natural forests in Tasmania; the results were shown as means for both groups; highest mean concentrations given (ref. 21); [z9] misidentification (in the case of *m*-cymene: must be *p*-cymene)

tr: trace (in columns B and D: <0.1; in columns E[g] and E[v]: <0.05); + present in the oil investigated, but quantity not stated; +[r] one oil from China obtained from the fractional distillation of *E. globulus* leaf oil; as the quantities cannot be compared with the other data, they are indicated with +[r] (ref. 42)

CONTACT ALLERGY/ALLERGIC CONTACT DERMATITIS

Eucalyptus oil (unspecified)

General

Contact allergy to/allergic contact dermatitis from eucalyptus oil has been reported in seventeen publications. In all but one (55; *Eucalyptus globulus* oil) the botanical origin of the eucalyptus oil was not specified. In groups of consecutive patients suspected of contact dermatitis, prevalence rates of up to 1.5% positive patch test reactions have been observed, but reliable relevance data are lacking. Some cases of allergic contact dermatitis were in patients sensitized to eucalyptus oil in topical pharmaceutical products; in others, cosmetics were the culprit. There were also two aromatherapists with occupational allergic contact dermatitis, in one of whom α-pinene may have been an allergen in eucalyptus oil.

Testing in groups of patients

The results of patch tests with eucalyptus oil in routine testing (consecutive patients suspected of contact dermatitis) and in groups of selected patients are shown in Table 5.35.4. In routine testing, rates of positive reactions ranged from 0.6% to 1.5%, whereas between 0.2% and 5.9% of patients in selected groups had positive patch tests.

Case reports

An aromatherapist had chronic hand dermatitis and was patch test positive to 17 of 20 oils used at her work (tested 1% and 5% in petrolatum), including eucalyptus oil (62). Another aromatherapist had non-occupational contact dermatitis with allergies to multiple essential oils used at work, including eucalyptus oil. The patient also reacted to geraniol, linalool, linalyl acetate, α-pinene, the fragrance mix and various other fragrance materials; α-pinene was demonstrated by GC-MS in the eucalyptus oil (60). α-Pinene may be an important

Table 5.35.3 Major constituents of eucalyptus globulus oils

Constituent	CAS	Percentage and range				
		A	B (27)	C (10)	D (15)	E
1,8-Cineole	470-82-6	61.6-88.7	53.8	4.1-50.3	63.1-85.8	83.9[z]; 84.9[y4]; 86.5[d]; 86.9[g]; 89.0[z5]; 90.0[z3]
α-Pinene	80-56-8	0.3-8.2	12.1	0.05-17.9	1.2-3.1	20.0[o]; 21.2[z7]; 23.9[z3]; 29.2[z2]; 26.7[p]; 32.7[y]
p-Cymene	99-87-6	1.1-3.1	0.7	tr-27.2		2.3[z7]; 2.7[p]; 5.1[a]; 5.3[y4]; 7.0[z2]; 8.2[j,v]; 9.1[z1]
Limonene	138-86-3	4.5-12.9	2.4	0.2-4.2		6.7[a]; 7.6[w]; 7.8[i]; 8.2[z]; 10.1[e]; 13.2[p]; 18.8[z1]
Aromadendrene	489-39-4	0.01-1.8	3.4	0-10.0	0.9-13.5	2.1[z7]; 2.2[t,z2]; 2.6[w]; 3.0[z8]; 3.6[i]; 7.1p; 10[y2]
Globulol	489-41-8	0.07-0.9	4.5	0.3-8.0	0.9-3.4	3.4[x]; 3.5[y3];4.6[z8]; 5.3[u]; 5.7[z7]; 7.3[f]; 9.8[p]
trans-Pinocarveol	1674-08-4	0.01-1.7	3.7	0-6.9		1.0[w]; 1.6[h,t]; 2.5[z8]; 2.7[z7]; 3.3[m]; 3.8[y2]; 4.3[b]
α-Terpineol	98-55-5	0.02-1.9	3.3	0-2.3	0.2-1.4	2.9[b]; 3.2[z7]; 3.6[z1,z2]; 3.8[n]; 4.7[e]; 5.8[i]; 6.5[y5]

LEGEND: SEE UNDER TABLE 5.35.2

Table 5.35.4 Results of testing groups of patients with eucalyptus oil

Years and Country	Test conc. & vehicle	Number of patients tested/positive (%)			Selection of patients (S); Relevance (R); Comments (C)	Ref.
Routine testing						
2000-2007 USA	2% pet.	679	4	(0.6%)	R: 100%; C: weak study: a. high rate of macular erythema and weak reactions, b. relevance figures include 'questionable' and 'past' relevance	56
<1976 Poland	2% pet.	200	3	(1.5%)	R: not stated	61
Testing in groups of selected patients						
2000-2008 IVDK	2% pet.	6,680	16	(0.2%)	S: patients with dermatitis suspected of causal exposure to fragrances; R: not stated	67
<2005 USA	2% pet.	96	3	(3.1%)	S: patients using consumer products containing eucalyptus oil; R: 'at least possibly relevant'	65
<2002 Japan, USA and 5 European countries	10% pet.	218	4	(1.8%)	S: patients previously shown to be allergic to fragrances; R: not stated	68
1997-2000 Austria	2% pet.	747	4	(0.5%)	S: patients suspected of fragrance allergy; R: not stated	57
1994-1995 UK	2% pet.	40	1	(2.5%)	S: patients previously reacting to the fragrance mix; R: not stated; C: there was also one positive reaction in a patient negative to the fragrance mix	70
<1986 Poland	2% pet.	86	1	(1.2%)	S: patients previously reacting to the fragrance mix; R: not stated	63
<1976 Poland	2% pet.	51	3	(5.9%)	S: patients allergic to *Myroxylon pereirae* resin (balsam of Peru) and/or turpentine and/or wood tar and/or colophony	61

IVDK Information Network of Departments of Dermatology, Germany, Switzerland, Austria (www.ivdk.org); pet.: petrolatum

---

component of *E. globulus* oil (maximum concentration of 8.2% found in commercial oils) and to a lesser degree of *E. citriodora* oil (maximum concentration found 3.5%) (columns A of Tables 5.35.2 and 5.34.2). Two patients had positive patch test reactions to eucalyptus oil, which was present in cosmetics that had given a positive patch test or had been positive in a usage test, seen in a 9-year period in one clinic in Belgium (66). In a group of 70 patients with proven allergic cosmetic dermatitis, eucalyptus oil was the allergen in one (71). One patient had allergic contact dermatitis from eucalyptus oil in Vicks Vaporub (64). One patient with allergic contact dermatitis from eucalyptus oil present in an anti-inflammatory cream (69). Two cases of contact sensitization to eucalyptus oil in topical pharmaceutical products have been reported (58).

Positive patch tests (relevance unknown, uncertain or not stated)

One positive patch test to eucalyptus oil in a case of airborne allergic contact dermatitis from aromatherapy caused by other essential oils (59).

Eucalyptus globulus oil

Occupational allergic contact dermatitis from eucalyptus oil (ex *E. globulus*) was observed in one patient with hand dermatitis working in the food industry (55).

LITERATURE

1 Lawrence BM. Progress in essential oils. Eucalyptus oil (cineole type). Perfum Flav 1990;15(6):58-61
2 Manguro LOA, Opiyo SA, Asefa A, Dagne E, Muchori PW. Chemical constituents of essential oils from three *Eucalyptus* species acclimatized in Ethiopia and Kenya. J Essent Oil Bear Plants 2010;13:561-567
3 Sacchetti G, Maietti S, Muzzoli M, Scaglianti M, Manfredini S, Radice M, et al. Comparative evaluation of 11 essential oils of different origin as functional antioxidants, antiradicals and antimicrobials in foods. Food Chem 2005;91:621-632
4 Benayache S, Benayache F, Benyahia S, Chalchat J-C, Garry R-P. Leaf oils of some *Eucalyptus* species growing in Algeria. J Essent Oil Res 2001;13:210-213
5 Mandal S, Dwivedi PD, Singh A, Naqvi AA, Bagchi GD. Capillary gas chromatographic analysis of *Eucalyptus globulus* from different geoclimatic zones in India. J Essent Oil Res 2001;13:196-197
6 Dagne E, Bisrat D, Alemayehu M, Worku T. Essential oils of twelve *Eucalyptus* species from Ethiopia. J Essent Oil Res 2000;12:467-470
7 Chisowa EH. Chemical composition of essential oils of three *Eucalyptus* species grown in Zambia. J Essent Oil Res 1997;9:653-655
8 Chalchat J-C, Muhayimana A, Habimana JB, Chabard JL. Aromatic plants of Rwanda. II. Chemical composition of essential oils of ten *Eucalyptus* species growing in Ruhande Arboretum, Butare, Rwanda. J Essent Oil Res 1997;9:159-165

9   Zrira SS, Benjilali BB. Seasonal changes in the volatile oil and cineole contents of five *Eucalyptus* species growing in Morocco. J Essent Oil Res 1996;8:19-24

10 Chalchat J-C, Chabard JL, Gorunovic MS, Djermanovic V, Bulatovic V. Chemical composition of *Eucalyptus globulus* oils from the Montenegro coast and east coast of Spain. J Essent Oil Res 1995;7:147-152

11 Dethier M, Nduwimana A, Cordier Y, Menut C, Lamaty G. Aromatic plants of tropical Central Africa. XVI. Studies on essential oils of five *Eucalyptus* species grown in Burundi. J Essent Oil Res 1994;6:469-473

12 Zrira SS, Benjilali BB, Fechtal MM, Richard HH. Essential oils of twenty-seven *Eucalyptus* species grown in Morocco. J Essent Oil Res 1992;4:259-264

13 Cornwell CP, Lassak EV. Camphor in commercial and potentially commercial oils of *Eucalyptus* species and other Myrtaceae. J Essent Oil Res 2010;22:59-65

14 Cimanga K, Apers S, de Bruyne T, Van Miert S, Hermans N, Totté J, et al. Chemical composition and antifungal activity of essential oils of some aromatic medicinal plants growing in the Democratic Republic of Congo. J Essent Oil Res 2002;14:382-387. Data also published in ref. 37

15 Manika N, Chanotiya CS, Negi MPS, Bagchi GD. Copious shoots as a potential source for the production of essential oil in *Eucalyptus globulus*. Ind Crops Prod 2013;46:80-84

16 Kumar P, Mishra S, Malik A, Satya S. Compositional analysis and insecticidal activity of *Eucalyptus globulus* (family: Myrtaceae) essential oil against housefly (*Musca domestica*). Acta Tropica 2012;122:212-218

17 Tyagi AK, Malik A. Antimicrobial potential and chemical composition of *Eucalyptus globulus* oil in liquid and vapour phase against food spoilage microorganisms. Food Chem 2011;126:228-235

18 Maciel MV, Morais SM, Bevilaqua CML, Silva RA, Barros RS, Sousa RN, et al. Chemical composition of *Eucalyptus* spp. essential oils and their insecticidal effects on *Lutzomyia longipalpis*. Vet Parasitol 2010;167:1-7

19 Li J, Madden JL. Analysis of leaf oils from a *Eucalyptus* species trial. Biochem System Ecol 1995;23:167-177

20 Silvestre AJD, Cavaleiro JAS, Delmond B, Filliatre C, Bourgeois G. Analysis of the variation of the essential oil composition of *Eucalyptus globulus* Labill. from Portugal using multivariate statistical analysis. Ind Crops Prod 1997;6:27-33

21 Li H, Madden JL, Potts BM. Variation in volatile leaf oils of the Tasmanian *Eucalyptus* Species. II. Subgenus *Symphyomyrtus*. Biochem System Ecol 1996;24:547-569

22 Yang Y-C, Choi H-Y, Choi W-S, Clark JM, Ahn Y-J. Ovicidal and adulticidal activity of *Eucalyptus globulus* leaf oil terpenoids against *Pediculus humanus capitis* (Anoplura: Pediculidae). J Agric Food Chem 2004;52:2507-2511

23 Ait-Ouazzou A, Lorán S, Bakkali M, Laglaoui A, Rota C, Herrera A, et al. Chemical composition and antimicrobial activity of essential oils of *Thymus algeriensis*, *Eucalyptus globulus* and *Rosmarinus officinalis* from Morocco. J Sci Food Agric 2011;91:2643-2651

24 Boelens MH. Essential oils and aroma chemicals from *Eucalyptus globulus* L. Perfum Flav 1985;9(6):1-14

25 Dellacassa E, Menéndez P, Moyna P, Soler E. Chemical composition of *Eucalyptus* essential oils grown in Uruguay. Flavour Fragr J 1990;5:91-95

26 Mulyaningsih S, Sporer F, Reichling J, Wink M. Antibacterial activity of essential oils from *Eucalyptus* and of selected components against multidrug-resistant bacterial pathogens. Pharm Biol 2011;49:893-899

27 Elaissi A, Medini H, Khouja ML, Simmonds M, Lynen F, Farhat F, et al. Variation in volatile leaf oils of five *Eucalyptus* species harvested from Jbel Abderrahman Arboreta (Tunisia). Chem Biodivers 2011;8:352-361

28 George DR, Masic D, Sparagano OAE, Guy JH. Variation in chemical composition and acaricidal activity against *Dermanyssus gallinae* of four eucalyptus essential oils. Exp Appl Acarol 2009;48:43-50

29 Golebiowski M, Paszkiewicz M, Halinski L, Malinski E, Stepnowski P.. Chemical composition of commercially available essential oils from eucalyptus, pine, ylang, and juniper. Chem Nat Comp 2009;45:278-279

30 Lawrence BM. Progress in essential oils. Perfum Flavor 1993;18(3):61-72

31 Renedo J, Otero JA, Mira JR. Huile essentielle d' *Eucalyptus globulus* L. de Cantabrie (Espagne). Variation au cours de la distillation. Plant Medicin Phytother 1990;24:31-35. Data cited in ref. 30

32 De Medici D, Pieretti S, Salvatore G, Nicoletti M, Rasoanaivo P. Chemical analysis of essential oils of Malagasy medicinal plants by gas chromatography and NMR spectroscopy. Flavour Fragr J 1992;7:275-281

33 Lawrence BM. Progress in essential oils. Perfum Flavor 1997;22(1):43-56

34 Tewari R, Akhila A. Essential oil from *Eucalyptus globulus* Labill.: A review. Curr Res Med Aroma Plants 1985;7:94-102. Data cited in ref. 33

35 Boland DJ, Brophy JJ, House APN. Eucalyptus leaf oils. Use, chemistry, distillation and marketing. Melbourne, Australia: Inkata Press, 1991. Data cited in ref. 33

36 Farah A, Fechtal M, Chaouch A. Effet de l'hybridation interspécifique sur la teneur et la composition chimique des huiles essentielles d'eucalyptus cultivées au Maroc. Biotechnol Agron Soc Environ 2002;6:163-169. Data cited in ref. 39

37 Cimanga K, Kambu K, Tona L, Apers S, De Bruyne T, Hermans N, et al. Correlation between chemical composition and antibacterial activity of essential oils of some aromatic medicinal plants growing in the Democratic Republic of Congo. J Ethnopharmacol 2002;79:213-220

38 Kubeczka K-H, Formacek V. Essential oil analysis by capillary gas chromatography and carbon-13 NMR spectroscopy, 2nd Edition. New York: J Wiley and Sons, 2002:107-110. Data cited in ref. 39

39 Lawrence BM. Progress in essential oils. Perfum Flavor 2006;31(9): 40-? (last page unknown)

40 Lawrence BM. Progress in essential oils. Perfum Flavor 2001;26(2):22-? (last page unknown)

41 Muselli A, Bighelli A, Corticchiato M, Acquarone L, Casanova J. Composition chimique d'huiles essentielles d'Eucalyptus globulus hydrodistillées et hydro-diffusées. Rivista Ital EPPOS (Numero Speciale) 1997:638-643. Data cited in ref. 40

42 Zhao Z-D, Sun Z, Liang Z-Q, Wang Y. Gas chromatography of residue from fractional distillation of Eucalyptus globulus leaf oil. Chem Forest Prod 1997;17:37-40. Data cited in ref. 40

43 Della Porta G, Porcedda S, Marongiu B, Reverchon E. Isolation of eucalyptus oil by supercritical fluid extraction. Flavour Fragr J 1999;14:214-218

44 Song A, Wang Y, Liu Y. Study on the chemical constituents of the essential oil of the leaves of Eucalyptus globulus Labill from China. Asian J Trad Med 2009;4:134-140

45 Lawrence BM. Progress in essential oils. Eucalyptus oil (cineole-rich). Perfum Flav 1986;11(6):39-40

46 Coppen JJW, Hone GA. Eucalyptus oils: a review of production and markets. Natural Resources Institute (NRI), University of Greenwich, Bulletin 56, 1992. Available at: http://gala.gre.ac.uk/11084

47 Li L. The outlook for Chinese essential oils. Paper presented at the IFEAT International Conference in Shanghai, 18-22 Oct. 2009. Essential China: A Major Consuming Market and Sourcing Partner in a Turbulent World. Pages 55-67 in the printed Conference Proceedings. Available at: http://www.ifeat.org/wp-content/uploads/2012/10/Le+Li+-+China+Essential+Oils.pdf

48 Murbach Teles Andrade BF, Nunes Barbosa L, da Silva Probst I, Fernandes Júnior A. Antimicrobial activity of essential oils. J Essent Oil Res 2014;26:34-40

49 Ben Slimane B, Ezzine O, Dhahri S, Ben Jamaa ML. Essential oils from two Eucalyptus from Tunisia and their insecticidal action on Orgyia trigotephras (Lepidotera, Lymantriidae). Biological Research 2014;47:29 http://www.biolres.com/content/47/1/29

50 Goldbeck JC, do Nascimento JE, Jacob RG, Fiorentini AM, Padilha da Silva W. Bioactivity of essential oils from Eucalyptus globulus and Eucalyptus urograndis against planktonic cells and biofilms of Streptococcus mutans. Ind Crops Prod 2014;60:304-309

51 Fratini F, Casella S, Leonardi M, Pisseri F, Ebani VV, Pistelli L, Pistelli L. Antibacterial activity of essential oils, their blends and mixtures of their main constituents against some strains supporting livestock mastitis. Fitoterapia 2014;96:1-7

52 Lambert Sameza M, Bedine Boat MA, Nguemezi ST, Nguemnang Mabou LC, Jazet Dongmo PM, Fekam Boyom F, Menut C. Potential use of Eucalyptus globulus essential oil against Phytophthora colocasiae the causal agent of taro leaf blight. Eur J Plant Pathol 2014;140:243-250

53 Rhind JP. Essential oils. A handbook for aromatherapy practice, 2nd Edition. London: Singing Dragon, 2012

54 Lawless J. The encyclopedia of essential oils, 2nd Edition. London: Harper Thorsons, 2014

55 Peltonen L, Wickstrom G, Vaahtoranta M. Occupational dermatoses in the food industry. Dermatosen 1985;33:166-169

56 Wetter DA, Yiannias JA, Prakash AV, Davis MD, Farmer SA, el-Azhary RA, et al. Results of patch testing to personal care product allergens in a standard series and a supplemental cosmetic series: an analysis of 945 patients from the Mayo Clinic Contact Dermatitis Group, 2000-2007. J Am Acad Dermatol 2010;63:789-798

57 Wöhrl S, Hemmer W, Focke M, Götz M, Jarisch R. The significance of fragrance mix, balsam of Peru, colophonium and propolis as screening tools in the detection of fragrance allergy. Br J Dermatol 2001;145:268-273

58 Devleeschouwer V, Roelandts R, Garmyn M, Goossens A. Allergic and photoallergic contact dermatitis from ketoprofen: results of (photo) patch testing and follow-up of 42 patients. Contact Dermatitis 2008;58:159-166

59 Schaller M, Korting HC. Allergic airborne contact dermatitis from essential oils used in aromatherapy. Clin Exp Dermatol 1995;20:143-145

60 Dharmagunawardena B, Takwale A, Sanders KJ, Cannan S, Roger A, Ilchyshyn A. Gas chromatography: an investigative tool in multiple allergies to essential oils. Contact Dermatitis 2002;47:288-292

61 Rudzki E, Grzywa Z, Bruo WS. Sensitivity to 35 essential oils. Contact Dermatitis 1976;2:196-200

62 Selvaag E, Holm J, Thune P. Allergic contact dermatitis in an aromatherapist with multiple sensitizations to essential oils. Contact Dermatitis 1995;33:354-355

63 Rudzki E, Grzywa Z. Allergy to perfume mixture. Contact Dermatitis 1986;15:115-116

64 Noiles K, Pratt M. Contact dermatitis to Vicks VapoRub. Dermatitis 2010;21:167-169

65 Guin JD. Use of consumer product ingredients for patch testing. Dermatitis 2005;16:71-77

66 Nardelli A, Drieghe J, Claes L, Boey L, Goossens A. Fragrance allergens in 'specific' cosmetic products. Contact Dermatitis 2011;64:212-219

67 Uter W, Schmidt E, Geier J, Lessmann H, Schnuch A, Frosch P. Contact allergy to essential oils: current patch test results (2000–2008) from the Information Network of Departments of Dermatology (IVDK). Contact Dermatitis 2010;63:277-283

68 Larsen W, Nakayama H, Fischer T, Elsner P, Frosch P, Burrows D, et al. Fragrance contact dermatitis – a worldwide multicenter investigation (Part III). Contact Dermatitis 2002;46:141-144

69 Vilaplana J, Romaguera C. Allergic contact dermatitis due to eucalyptol in an anti-inflammatory cream. Contact Dermatitis 2000;43:118

70 Katsarma G, Gawkrodger DJ. Suspected fragrance allergy requires extended patch testing to individual fragrance allergens. Contact Dermatitis 1999;41:193-197

71 Schorr WF. Cosmetic allergy: Diagnosis, incidence, and management. Cutis 1974;14:844-850

Chapter 5.36 GALBANUM RESIN OIL

DEFINITION

Galbanum resin oil is the essential oil obtained from the resin of the galbanum, *Ferula gummosa* Boiss. (synonym: *Ferula galbaniflua* Boiss. & Buhse).

INCI NOMENCLATURE

Description/definition: Ferula galbaniflua resin oil is the essential oil obtained from the resin of the galbanum, *Ferula galbaniflua*, Apiaceae

INCI name EU: Ferula galbaniflua resin oil

INCI name USA: Ferula galbaniflua (galbanum) resin oil

CAS registry number(s): 8023-91-4; 93165-40-3

EINECS number(s): 296-925-4

ISO (INTERNATIONAL ORGANIZATION FOR STANDARDIZATION) STANDARD

ISO number: 14716

ISO name: Essential oil of galbanum

Botanical origin: *Ferula galbaniflua* Boiss. & Buhse, syn. *Ferula gummosa* Boiss.

Parts of plant used: root exudate

ISO values: ISO values (minimum and maximum concentrations) are shown in Table 5.36.1.

THE PLANT, THE OIL, AND THEIR USES

Ferula gummosa Boiss. is a perennial, herbaceous, very resinous plant of the Apiaceae family, indigenous to Iran and possibly Turkmenistan, generally growing up to 1 meter tall and wide. It can be found in the wild mainly in Iran, Turkey, Afghanistan and neighboring countries, and can become 3 meters tall when growing naturally (5,8). The major producer of *Ferula gummosa* is Iran. The resin of this plant, called 'galbanum', has a strong scent and is used in the preparation of various types of incense. It is collected either by exposing the upper part of the root and cutting it into strips or by making incisions in the trunk (GRIN Taxonomy for Plants). The galbanum from *F. gummosa* is said to have many bioactive properties including antimicrobial, anti-inflammatory, anticonvulsant, carminative, expectorant, anti-catarrh, anti-rheumatic, anti-nociceptive, anti-hysteric, laxative, aphrodisiac, antiseptic, and analgesic, and is therefore widely used for numerous afflictions in folk medicine (5). In addition, galbanum is also used in the manufacture of glues, textiles and cosmetics, and due to its transparency and high-power bond, it is employed to glue gems to jewelry (5,6). The resin can also be utilized in food flavoring, where it contributes to the savory notes of curries and sauces (8). The essential oil of galbanum (galbanum essential oil), obtained from the resin by steam distillation, is widely used in aromatherapy (13,14) and in perfumery.

CHEMICAL COMPOSITION

Galbanum resin oil is a colorless to light yellow, clear, easily mobile liquid which has a terpenic green, aromatic and woody note with sometimes sulfurous odor. The yield of essential oil from the resin of *Ferula gummosa* Boiss. generally varies from 9.0 to 20.0%. The main producing country of this oil is Iran.

Literature data (up to November 24, 2014) on the chemical composition of galbanum resin oils and unpublished analytical data from one of us (E.S.) are shown in Table 5.36.2 in alphabetical order. In galbanum resin oils from various origins, over 225 chemicals have been identified. About 75% of these were found in a single reviewed publication only. Most of these 'single' chemicals were identified in one of two studies (5,6), in which the sesquiterpene composition and the monoterpene composition of a lab-hydrodistilled oil from galbanum, harvested from 50 *F. gummosa* plants in north Iran, were investigated; the chemical composition of this oil was entirely different from all other studies (Table 5.36.2, column B).

The major compounds found in galbanum resin oils from different sources are shown in Table 5.36.3. They include (highest concentrations in any study given) β-pinene (81.2%), α-pinene (18.3%), limonene (14.0%), δ3-carene (13.1%), myrcene (10.0%) and bulnesol (7.1%). Well-known ingredients of galbanum resin oils that were present in high concentrations (>7%) in one or two studies were terpinolene (10.0%), linalool (9.0%) and *p*-cymene (8.5%).

Commercial oils

The ten chemicals that had the highest maximum concentrations in 21 commercial galbanum resin essential oil samples (concentration ranges provided) are the following: β-pinene (51.0-81.2%), α-pinene (2.3-10.1%), δ3-carene (1.6-9.5%), myrcene (0.1-6.6%), α-phellandrene (0.2-3.6%), bulnesol (0.4-3.4%), *trans*-pinocarveol (0.1-3.3%), limonene (0.4-3.2%), β-phellandrene (0.3-2.8%) and myrtenol (0.06-2.2%) (Erich Schmidt, unpublished analytical data).

CONTACT ALLERGY/ALLERGIC CONTACT DERMATITIS

General

Contact allergy to and possible allergic contact dermatitis from galbanum resin oil has been reported in one publication only. A false-positive patch test reaction due to the excited skin syndrome cannot be excluded, but α-pinene may have been an allergen in galbanum resin oil in this case.

Table 5.36.1 ISO values (%) for galbanum resin oil (essential oil of galbanum)[a]

Compound	CAS	Minimum	Maximum	Ratio
β-Pinene	127-91-3	40.0	70.0	
α-Pinene	80-56-8	5.0	21.0	
δ3-Carene	13466-78-9	2.0	16.0	
Myrcene	123-35-3	2.5	3.5	
Sabinene	3387-41-5	0.3	3.0	
(3E,5Z)-1,3,5-Undecatriene	19883-27-3	0.4	2.0	
Ratio of (3E,5Z)-1,3,5-Undecatriene to (E,E)-1,3,5-Undecatriene	19883-27-3 19883-29-5			2.0-5.5

[a] ISO 14716 Essential oil of galbanum ©ISO 1998; Geneva, Switzerland, www.iso.org

Case reports

An aromatherapist had non-occupational allergic contact dermatitis with allergies to multiple essential oils used at work, including 'galbanum oil'. The patient also reacted to geraniol, linalool, linalyl acetate, α-pinene, the fragrance mix and various other fragrance materials; α-pinene and linalool were demonstrated by GC-MS in the galbanum oil (15). α-Pinene is an important component of galbanum resin oil and has been found in a maximum concentration of 10.1% in commercial galbanum resin oils (Table 5.36.2, column A).

Table 5.36.2 Constituents identified in galbanum resin oils

Constituent	CAS	Percentage and range					
		A	B (5,6)	C (4)	D (9)	E (12)	F
α-Acorenol	28296-85-7		0.1				
α-Amorphene	20085-19-2				0.6	1.8	1.1[f]
S-sec-Amyl 3-methyl-2-butenthioate							+[c]
S-sec-Amyl thiotiglate							+[c]
p-Anethole	104-46-1		0.01				
allo-Aromadendrene	25246-27-9					0.8	0.5[f]
iso-Aromadendrene	499134-59-7		0.5				
Artemisia ketone	546-49-6		tr				
Bicycloelemene	32531-56-9		0.2				
Bicyclogermacrene	24703-35-3		0.4				
epi-Bicyclosesquiphellandrene	54274-73-6					4.4	
α-Bisabolene	17627-44-0		0.4				
β-Bisabolene	495-61-4		0.2				0.8
γ-Bisabolene	495-62-5		0.4				
α-Bisabolol	515-69-5		3.6				
Bisabolol oxide A	22567-36-8		0.4				
α-Bisabolol oxide B	26184-88-3		0.2				
α-Bisabolone oxide	22567-38-0		0.6				
Borneol (= endo-)	507-70-0	0.04-0.1	0.2				
Bornyl acetate	76-49-3	0.07-0.9					
β-Bourbonene	5208-59-3		0.4		0.4		
Bulnesol	22451-73-6	0.4-3.4	7.1	0.2		1.2	
S-sec-Butyl benzothioate							+[c]
Butyl isovalerate	109-19-3						3.0[e]
S-sec-Butyl 3-methylbutanthioate							+[c]
S-sec-Butyl 3-methyl-2-but-2-enethioate	34322-09-3	0.3-1.9					+[c]
S-sec-Butyl 2-methyl-2-hexenethioate							+[c]
S-sec-Butyl 2-methyl-2-pentenethioate							+[c]
S-sec-Butyl thiotiglate							+[c]
Cadina-1,4-diene	29837-12-5		0.6			0.8	
β-Cadinene	523-47-7		0.6				
γ-Cadinene	39029-41-9	0.3-0.6	1.0	0.3		4.1	
δ-Cadinene	483-76-1	0.3-1.5	1.6	0.4		2.0	1.5[f]
14-nor-Cadin-5-en 4-one isomer A						1.4	
α-Cadinol	481-34-5		0.5				
epi-α-Cadinol	5937-11-1	0.2-0.7	2.4				
α-Calacorene	21391-99-1		0.5				
Calamenene	483-77-2		0.8				
cis-Calamenene	72937-55-4					1.0	
Camphene	79-92-5	0.2-1.0	0.1	0.3	0.4		0.2[d]
Camphene hydrate	465-31-6		0.1				

Table 5.36.2 Constituents identified in galbanum resin oils (*continued*)

Constituent	CAS	Percentage and range					
		A	B (5,6)	C (4)	D (9)	E (12)	F
6-Camphenone	53803-33-1		0.01				
α-Campholenal	4501-58-0		0.2				1.0[e]
Camphor	76-22-2		0.01				
2-Caren-10-al	124752-20-1		tr				
3-Caren-10-al	14595-13-2		0.01				
δ3-Carene	13466-78-9	1.6-9.5	0.1	12.1	0.3	0.6	6.7[a]; 7.5[f]; 9.0[e]; 13.1[d]
2-Caren-4-ol	6617-35-2		0.01				
3-Caren-2-ol	93905-79-4		tr				
Carotol	465-28-1		0.2				
Carveol	99-48-9		0.3				
Carvone	99-49-0		0.2				
Carvone oxide	33204-74-9		0.01				
Carvotanacetone	499-71-8		0.02				
β-Caryophyllene	87-44-5	0.1-0.4	0.2	0.1			0.1[f]
Caryophyllene oxide	1139-30-6		0.1				
α-Cedrene	469-61-4		0.5				
β-Cedrene	546-28-1		0.6				
Cedrenol	28231-03-0		0.7				
Cedrol	77-53-2		0.09				
epi-Cedrol	19903-73-2		1.0				
β-Chamigrene	18431-82-8		0.6				
1,4-Cineole	470-67-7	0.05-0.2					
1,8-Cineole	470-82-6		0.06				
Citronellal	106-23-0		0.07				
Citronellol	106-22-9		0.2				
Clovene	469-92-1		0.1				
α-Copaene	3856-25-5	0.2-0.3	0.9		0.5	0.7	
α-Cubebene	17699-14-8		0.01				
β-Cubebene	13744-15-5		0.3			4.9	
Cubebol	23445-02-5		0.2				
Cubenol	21284-22-0		0.1				
1-epi-Cubenol	81939-29-9		0.2				
Cuminaldehyde	122-03-2		0.1				
Cuparene	16982-00-6		0.4				
Cyclosativene	22469-52-9					0.6	
m-Cymene	535-77-3		0.1				
o-Cymene	527-84-4		0.02				
p-Cymene	99-87-6	0.2-0.9	tr	0.1	8.5	4.1	
p-Cymen-8-ol	1197-01-9		0.2				
Dendralasine	23262-34-2		0.7				
Dihydroanethole	104-45-0		tr				
Dihydrocarvone	5948-04-9		0.01				
Dihydrocurcumene	1461-02-5		0.9				
3,3-Dimethylbutyric acid	1070-83-3						1.5[e]
β-Elemene	33880-83-0	0.1-0.2	0.1		0.01	0.6	
γ-Elemene	29873-99-2		0.1	2.4			
δ-Elemene	20307-84-0		0.1				
Elemol	639-99-6	0.07-0.4	2.4				
2,3-Epoxycarane	62413-92-7		tr				
2,3-Epoxygeranial			0.01				
Eremophilene	10219-75-7					0.7	
Eucarvone	503-93-5		0.05				
α-Eudesmol	473-16-5		4.3				
β-Eudesmol	473-15-4		1.3	0.1			
γ-Eudesmol	1209-71-8		0.3				
10-epi-γ-Eudesmol	15051-81-7		1.0				
(E,E)-Farnesal	502-67-0		0.08				
β-Farnesene	502-60-3		0.5				
(E,E)-Farnesol	106-28-5		0.1				
(Z,Z)-Farnesol	16106-95-9		0.2				
Fenchone	1195-79-5		tr				

Table 5.36.2 Constituents identified in galbanum resin oils (*continued*)

Constituent	CAS	Percentage and range					
		A	B (5,6)	C (4)	D (9)	E (12)	F
α-Fenchyl acetate	111821-74-0	0.08-0.9		0.1			
Fenchyl alcohol	1632-73-1	0.02-0.06	0.08				
Geraniol	106-24-1		0.08				
Germacrene A	28387-44-2		0.3				
Germacrene B	15423-57-1		0.01	0.6			
Germacrene D	23986-74-5	0.3-1.0	3.5	0.7	0.4		
Guaia-3,7-diene	6754-04-7		1.0				
α-Guaiene	3691-12-1		0.2				
Guaiol	489-86-1	0.6-1.9	3.4	0.5		0.9	
β-Gurjunene	73464-47-8		0.09				
Hedycaryol	21657-90-9		2.6				
Hexyl isovalerate	10032-13-0						2.0[e]
β-Himachalene	1461-03-6		0.2				
Hotrienol	20053-88-7		0.01				
α-Humulene	6753-98-6		0.9	0.3			
Ipsdienol	35628-00-3		0.1				
Isoborneol	124-76-5		0.02				
2-Isobutyl-3-methoxy-pyrazine	24683-00-9	0.03-0.07					
Isoledene	95910-36-4		0.02			0.6	1.6[f]
trans-Isolongifolanone	14727-47-0		0.4				
Isolongifolene	1135-66-6		0.5				
Isopinocamphone	15358-88-0		0.05				
S-Isopropyl benzothioate							+[c]
2-Isopropyl-3-methoxypyrazine	25773-40-4						+[b]
S-Isopropyl 3-methyl-2-butenthioate							+[c]
S-Isopropyl thiotiglate							+[c]
Isopulegol	89-79-2		0.01				
Isoterpinolene	586-63-0	tr-0.1					
Jaeschkeanadiol	41690-67-9		0.2				
Ledene oxide	882187-44-2		0.2				
Ledol	577-27-5		0.1				
Liguloxide	21764-22-7		0.09				
Limonene	138-86-3	0.4-3.2	0.1	4.0	2.2	0.8	1.2[f]; 1.8[d]; 14.0[e]
Limonene dioxide	96-08-2		tr				
Limonene oxide	1195-92-2		tr				
Linalool	78-70-6	0.04-0.6	0.1	0.5			9.0[e]
Linalool oxide	1365-19-1		0.02				
Longifolene (junipene)	475-20-7		0.03				
α-Longipinene	5989-08-2		0.1		0.5		
p-Mentha-3,8-diene	586-67-4	0.04-0.06					
p-Mentha-1,5-dien-7-ol	19876-45-0		tr				
p-Mentha-1,5-dien-8-ol	1686-20-0		0.4				
p-Mentha-1(7),8-dien-2-ol (isocarveol)	35907-10-9		0.08				
Mentha-2,8-dien-1-ol	58940-40-2		0.08				
p-Mentha-1,8-dien-3-one	529-01-1		0.01				
p-Menth-1-en-9-al	29548-14-9		0.01				
Menthene	29350-67-2		0.02				
p-Menth-2-ene-1,8-diol	57030-53-2		0.2				
p-Menth-2-en-1-ol	619-62-5		0.1				
Methylcarvacrol	6379-73-3	0.07-1.8					
Methyl chavicol	140-67-0		0.03				
α-Muurolene	10208-80-7		0.1				
τ-Muurolol	19912-62-0		0.2				
Myrcene	123-35-3	0.1-6.6	0.2	4.6	0.3	0.6	2.9[d]; 10.0[e]
Myrcenol	543-39-5		0.05				
Myrtanal	4764-14-1		0.1				
Myrtanol	514-99-8		0.01				
Myrtenal	564-94-3	0.08-0.2	0.2		1.3		
Myrtenol	515-00-4	0.06-2.2	3.8	0.2			

Table 5.36.2 Constituents identified in galbanum resin oils (*continued*)

Constituent	CAS	A	B (5,6)	C (4)	D (9)	E (12)	F
Myrtenyl acetate	1079-01-2	0.08-0.5					
Neoclovene	4545-68-0		0.6				
Neral	106-26-3		0.1				
Nerol	106-25-2		tr				
Nerolidol	7212-44-4		2.0				
Nopinone	24903-95-5				0.3		
β-Ocimene	13877-91-3		0.2				
(*E*)-β-Ocimene	3779-61-1	0.08-1.1		0.2			1.3[d]
(*Z*)-β-Ocimene	3338-55-4	0.2-1.2		1.2			2.3[d]
allo-Ocimene	673-84-7			0.03			
Perillaldehyde	2111-75-3		0.3				
Perillene	539-52-6		0.09				
Perillyl alcohol	536-59-4		1.2				
Phellandral	21391-98-0		0.03				5.0[e]
α-Phellandrene	99-83-2	0.2-3.6		0.6			
β-Phellandrene	555-10-2	0.3-2.8	0.04			0.3	0.3[a]
2-Pinanol	473-54-1		0.3				
α-Pinene	80-56-8	2.3-10.1	0.1	5.7	14.3	5.4	9.5[f]; 9.6[d]; 13.0[e]; 18.3[a]
α-Pinene (ep)oxide	1686-14-2		0.03				
β-Pinene	127-91-3	51.0-81.2	0.04	58.8	14.1	43.1	2.0[e]; 50.1[a]; 53.7[d]; 62.7[f]
Pinocamphone	547-60-4		0.04				
Pinocarveol	5947-36-4		1.0				
trans-Pinocarveol	1674-08-4	0.1-3.3					
Pinocarvone	30460-92-5		0.3				
Pinocarvyl acetate	1078-95-1	0.06-0.5					
Piperitol	491-04-3		0.09				
Piperitone	89-81-6		0.01				
Rose oxide	16409-43-1		0.01				
Sabinene	3387-41-5	0.3-1.4	0.04		40.1		1.4[d]; 3[f]; 3.1[a]
Sabinene hydrate	546-79-2		0.08				
cis-Sabinene hydrate	15537-55-0	0.04-0.2				1.2	
trans-Sabinene hydrate	17699-16-0					1.2	
Sabinol	471-16-9		0.05				
α-Santalol	115-71-9		1.7				
Santolina alcohol	35671-15-9		tr				
Sativene	6813-05-4		0.03				
α-Selinene	473-13-2		0.4				1[f]
β-Selinene	17066-67-0		0.6				
δ-Selinene	473-14-3					1.7	
α-Selin-11-en-4-ol	16641-47-7		0.1				
β-Sinensal	17909-87-4		0.07				
Spathulenol	6750-60-3		0.8			2.7	
α-Terpinene	99-86-5		0.05		0.6		1.2[f]
γ-Terpinene	99-85-4	0.04-0.4	0.09		1.1		0.53[f]; 6.0[e]
Terpinen-4-ol	562-74-3	0.1-0.6	0.4		1.5	4.1	0.7[f]
α-Terpineol	98-55-5	0.06-0.3	0.3				
β-Terpineol	138-87-4		0.09				
Terpinolene	586-62-9	0.2-0.5	0.2	0.5	0.4		0.5[d]; 1.1[f]; 10.0[e]
Terpinyl acetate	8007-35-0	0.3-0.7					
α-Terpinyl acetate	80-26-2			0.5			
α-Thujenal	57129-54-1		0.01				
α-Thujene	2867-05-2	0.5-1.3	0.2		8.1	2.5	1.3[f]; 3.3[a]
Thujenol			0.01				
Thujen-2-one (3-)	546-78-1		tr				
α-Thujone	546-80-5				0.8		
β-Thujone (*cis*-)	471-15-8		0.01				
Thujopsene	470-40-6		0.2				
Tricyclene	508-32-7						2.3[d]
(*E,E*)-1,3,5-Undecatriene	19883-29-5	0.3-1.0					0.2[d]
(*3E,5Z*)-1,3,5-Undecatriene	19883-27-3	0.3-1.2		1.8			+[b]; 0.5[d]
(*6Z,8E*)-Undeca-6,8,10-trien-3-one	1123751-39-2						+[b]

Table 5.36.2 Constituents identified in galbanum resin oils (*continued*)

Constituent	CAS	Percentage and range					
		A	B (5,6)	C (4)	D (9)	E (12)	F
(6Z,8E)-Undeca-6,8,10-trien-4-one							+[b]
Verbenene	4080-46-0		0.07				
trans-Verbenol	1820-09-3			0.2	1.1		
Verbenone	80-57-9		0.2				
Viridiflorol	552-02-3					0.6	
Widdrol	6892-80-4		0.6				
β-Ylangene	20479-06-5		0.6				

A twenty-one galbanum essential oil samples from Iran, analyzed between 1998 and 2013; lowest and highest concentrations given (E. Schmidt, unpublished data)

B one lab-hydrodistilled oil from galbanum, harvested from 50 *F. gummosa* plants in north Iran; in ref. 5, the sesquiterpene composition was investigated and reported and in ref. 6 the monoterpene composition; the chemical composition of this oil was entirely different from all other studies, also those from Iran (refs. 5,6)

C one lab-hydrodistilled oil from galbanum resin harvested in Iran (ref. 4)

D one lab-hydrodistilled oil from galbanum purchased at an Iranian market (ref. 9)

E one commercial galbanum resin oil from Iran (ref. 12)

F data from other studies (indicated with superscript letters); highest concentrations found in any study reviewed here given; when two or more oils were investigated, only the highest concentrations are mentioned, unless indicated otherwise

[a] one lab-hydrodistilled oil from Iran (ref. 10); [b] one commercial galbanum resin oil purchased in Tokyo (ref. 8); [c] one galbanum oil sample of unknown origin (ref. 2); [d] one galbanum oil sample of unknown origin (ref. 3); [e] one lab-hydrodistilled oil from Iranian galbanum (ref. 7); [f] one lab-hydrodistilled oil from galbanum resin from Iran (ref. 11)

tr: trace (in column B: <0.01); + present in the oil investigated, but quantity not stated

Table 5.36.3 Major constituents of galbanum resin oils

Constituent	CAS	Percentage and range					
		A	B (5,6)	C (4)	D (9)	E (12)	F
β-Pinene	127-91-3	51.0-81.2	0.04	58.8	14.1	43.1	2.0[e]; 50.1[a]; 53.7[d]; 62.7[f]
α-Pinene	80-56-8	2.3-10.1	0.1	5.7	14.3	5.4	9.5[f]; 9.6[d]; 13.0[e]; 18.3[a]
Limonene	138-86-3	0.4-3.2	0.1	4.0	2.2	0.8	1.2[f]; 1.8[d]; 14.0[e]
δ3-Carene	13466-78-9	1.6-9.5	0.1	12.1	0.3	0.6	6.7[a]; 7.5[f]; 9.0[e]; 13.1[d]
Myrcene	123-35-3	0.1-6.6	0.2	4.6	0.3	0.6	2.9[d]; 10.0[e]
Bulnesol	22451-73-6	0.4-3.4	7.1	0.2		1.2	

LEGEND: SEE UNDER TABLE 5.36.2

LITERATURE

1 Lawrence BM. Progress in essential oils. Perfum Flavor 2007;32(September):44-? (last page number unknown)

2 Fellous R, George G, Rochard S, Schippa C. Detection and authentication of sulphur compounds in natural raw materials. In: H Woidich, G Buchbauer, Eds. Proceedings 12th International Congress of Flavours, Fragrances and Essential Oils, Vienna, October 4-8, 1992. Vienna, Austria: Fachzeitschriftenverlag, 1993:38-52. Data cited in ref. 1

3 Moyler DA, Clery RA. The aromatic resins: their chemistry and uses. In: KAD Swift, Ed. Flavours and Fragrances. Cambridge: Royal Society of Chemistry, 1997:46-115. Data cited in ref. 1

4 Ghannadi A, Amree S. Volatile oil constituents of *Ferula gummosa* Boiss. from Kashan, Iran. J Essent Oil Res 2002;14:420-421

5 Jalali HT, Petronilho S, Villaverde JJ, Coimbra MA, Domingues MRM, Ebrahimian ZJ et al. Assessment of the sesquiterpenic profile of *Ferula gummosa* oleo-gum-resin (galbanum) from Iran. Contributes to its valuation as a potential source of sesquiterpenic compounds. Ind Crops Prod 2013;44:185-191

6 Jalali HT, Petronilho S, Villaverde JJ, Coimbra MA, Domingues MRM, Ebrahimian ZJ, et al. Deeper insight into the monoterpenic composition of *Ferula gummosa* oleo-gum-resin from Iran. Ind Crops Prod 2012;36:500-507

7 Sadraei H, Asghari GR, Hajhashemi V, Kolagar A, Ebrahimi M. Spasmolytic activity of essential oil and various extracts of *Ferula gummosa* Boiss. on ileum contractions. Phytomed 2001;8:370-376

8 Miyazawa N, Nakanishi A, Tomita N, Ohkubo Y, Maeda T, Fujita A. Novel key aroma components of galbanum oil. J Agric Food Chem 2009;57:1433-1439

9 Abedi D, Jalali M, Asghari G, Sadeghi N. Composition and antimicrobial activity of oleogumresin of *Ferula gumosa* Bioss. essential oil using Alamar Blue™. Res Pharm Sciences 2008;3:41-45. Available at: http://webcache.googleusercontent.com/search?q=cache:teJSHIcp_yYJ:rps.mui.ac.ir/index.php/jrps/article/viewFile/40/42+&cd=3&hl=nl&ct=clnk&gl=nl

10 Mortazaienezhad F, Sadeghian MM. Investigation of compounds from Galbanum (*Ferula gummosa*) Bioss. Asian J Plant Sci 2006;5:905–906

11 Mahboubi M, Kazempour N, Mahboubi A. The efficacy of essential oils as natural preservatives in vegetable oil. J Diet Suppl, Early Online: 2014:1-13. DOI: 10.3109/19390211.2014.887603

12 Mahboubi M, Kazempour N, Mahboubi M. Antimicrobial activity of rosemary, fennel and galbanum essential oils against clinical isolates of *Staphylococcus aureus*. Biharean Biologist 2011;5:4-7

13 Rhind JP. Essential oils. A handbook for aromatherapy practice, 2nd Edition. London: Singing Dragon, 2012

14 Lawless J. The encyclopedia of essential oils, 2nd Edition. London: Harper Thorsons, 2014

15 Dharmagunawardena B, Takwale A, Sanders KJ, Cannan S, Roger A, Ilchyshyn A. Gas chromatography: an investigative tool in multiple allergies to essential oils. Contact Dermatitis 2002;47:288-292

Chapter 5.37   GERANIUM OIL

DEFINITION

Geranium oil (essential oil of geranium) is the essential oil obtained from the herbaceous part of *Pelargonium* x ssp.

INCI NOMENCLATURE

Description/definition: Pelargonium graveolens oil is the volatile oil obtained from the whole plant of the Bourbon geranium, *Pelargonium graveolens* (L.), Geraniaceae

INCI name EU & USA: Pelargonium graveolens oil

CAS registry number(s): 8000-46-2; 90082-51-2

EINECS number(s): 290-140-0

ISO (INTERNATIONAL ORGANIZATION FOR STANDARDIZATION) STANDARD

ISO number: 4731

ISO name: Essential oil of geranium

Botanical origin: Perlargonium x ssp.

Parts of plant used: Herbaceous part (leaf, stem, stalk, petiole, flower)

ISO values: ISO values (minimum and maximum concentrations) are shown in Table 5.37.1.

ISO gives as botanical source for essential oils of geranium *Pelargonium* x *ssp*. (*ssp*. = subspecies). We presume that *spp*. = species is meant (although subspecies also may qualify, of course). Other species of the genus *Pelargonium*, from which geranium oils are mentioned in CosIng, include *Pelargonium roseum* (CAS 90082-55-6, EINECS 290-144-2) and *Pelargonium crispum* (CAS 90082-46-5; EINECS 290-134-8). Geranium oils may also be prepared from species of the genus *Geranium*, including *Geranium maculatum* (CAS 84650-10-2, EINECS 283-491-6) and *Geranium macrorrhizum* (Zdravetz oil, Chapter 5.91).

THE PLANT, THE OIL, AND THEIR USES

Pelargonium graveolens L' Hér. is an erect, multi-branched shrub which can reach a height of up to 1.3 meters and a spread of 1 meter. The hairy stems are herbaceous when young, becoming woody with age. The deeply incised leaves are velvety and soft to the touch owing to the presence of numerous glandular hairs. The leaves scent strongly like roses (9). This species and other *Pelargonium* species are native to south tropical Africa (Mozambique, Zimbabwe) and South Africa (GRIN Taxonomy for plants). Plants cultivated under this name differ from wild plants and may be of hybrid origin. In fact, for the production of geranium oils not one plant species is employed, but several *Pelargonium* species and especially their hybrids are used, which are often called rose geraniums or rose-scented geraniums (geranium is in fact a misnomer, which has been passed on from the original Linnaean nomenclature to the introduction of the five

genera in the family Geraniaceae by L'Héritier in 1789, which resulted in the South African *Geranium* species being renamed *Pelargonium*) (22). The most important of these *Pelargonium* species include *P. graveolens* L'Hér., *P. capitatum* (L.) L'Hér., *P. radens* H.E. Moore and *P. odoratissimum* (L.) (apple geranium, sweat-scent pelargonium, 10,35) (24). Currently, for the production of geranium oils, most often hybrids from these or other *Pelargonium* species as parents are used. Many cultivars of geranium exist which have been generally erroneously identified as *Pelargonium graveolens* in the past due to similarities in their morphological characteristics. The most popular cultivars of geranium are the hybrids of *P. capitatum* and *P. radens* (e.g., cultivar 'Rosé' from Reunion Island) or *P. graveolens* and *P. radens* (21,22,25,48).

The scented *Pelargonium* hybrid cultivars used in the production of essential oil are large bush-like plants with small pink flowers and pointed scented leaves (9). They are widely cultivated in Madagascar, China, Algeria, Tunisia, Morocco, South Africa and India and various other countries (25). It should be mentioned that these *Pelargonium* species are distinctly different from the horticultural geraniums, which are basically ornamental and have no commercial usage in perfumery or other industries (9).

Essential oil of geranium is obtained by steam- or hydrodistillation of the leaves plus stems (25,37) of *Pelargonium* cultivars (sometimes during flowering, as the minty fragrance turns to a smell resembling roses (9)), or (according to ISO criteria) from the herbaceous part (leaf, stem, stalk, petiole, flower). It is one of the most valuable natural materials for the perfumery and cosmetic industries (9,12,46,50). It is also used in flavoring tobacco products and pharmaceutical preparations (6,9). The essential oil is sparingly used in food and drinks, not only as flavor but also as a food preservative. It presumably has antibacterial, antifungal and insecticidal properties (6).

Geranium oil is popular in aromatherapy, where it is used for menopausal problems, skin disorders, nervous tension and anxiety (1,6,9). The French medicinal community is said to treat diabetes, diarrhea, gallbladder problems, gastric ulcers, jaundice, liver problems, sterility and urinary stones with this oil (9). In Chinese medicine, it is considered to open up the liver charka and promote the expulsion of toxins that prohibit the achievement of balance within the body (9). Geranium essential oils are highly valued in perfumery for their floral notes; the main constituents are citronellol and geraniol, which have a sweet rose-like (citronellol) and flowery rose-like odor (geraniol) (9). In fact, because of this, geranium oils may sometimes be used as a substitute for the more expensive rose essential oil (48). The oils are sometimes refined by means of a fractionation process to extract fragrant isolates like rhodinol (mixture of citronellol, geraniol and other alcohols such as linalool) (33), or to obtain a high-grade essential oil with the removal of light fractions of terpenes, known as terpeneless oils (1,9,12).

Table 5.37.1 ISO values (%) for geranium oil[a]

Compound	CAS	Minimum	Maximum
Citronellol	106-22-9	18.0	43.0
Geraniol	106-24-1	5.0	20.0
Citronellyl formate	105-85-1	4.0	12.0
Linalool	78-70-6	2.0	11.0
Isomenthone	491-07-6	4.0	10.0
Guaia-6,9-diene	37839-64-8	0.0	9.0
Geranyl formate	105-86-2	1.0	8.0
10-epi-γ-Eudesmol	15051-81-7	0.0	6.2
cis-Rose oxide	3033-23-6	0.3	3.5
Menthone	89-80-5	0.0	2.5
Geranyl tiglate	7785-33-3	0.7	2.0
Citronellyl butyrate	141-16-2	0.4	2.0
trans-Rose oxide	5258-11-7	0.1	1.5
2-Phenylethyl tiglate	55719-85-2	0.4	1.2
α-Terpineol	98-55-5	0.1	1.2

[a] ISO 4731 Essential oil of geranium ©ISO 2012; Geneva, Switzerland, www.iso.org

CHEMICAL COMPOSITION

Geranium oil is an amber to greenish yellow, clear mobile liquid which has an odor with floral rosy notes and green and minty accents. The yield of essential oil from the leaves or herbaceous parts of *Pelargonium* x *ssp.* generally varies from 0.1 to 0.35%. The main producing countries of this oil are Madagascar, China, Algeria, Tunisia, Morocco, South Africa and India.

Articles that mention *Pelargonium graveolens*, *Pelargonium* species indicated as 'rose-scented geranium' or 'rose geranium', *Pelargonium* cultivars such 'Bourbon' or 'Rosé' (known to be rose-scented geraniums) (48), or *Pelargonium capitatum* × *P. radens* as botanical source for the geranium essential oils investigated are included in this review. *Pelargonium odoratissimum* apparently has a quite different composition with very little or no citronellol and geraniol, but high concentrations of isomenthone, methyl eugenol and limonene (10,35,56) and sometimes with concentrations of methyleugenol of >90% (66). *Pelargonium roseum* (which may have a composition similar to rose-scented geraniums [68], but may also contain high concentrations of citronellyl formate and β-caryophyllene [8]) is not discussed, as the status of this species is unresolved according to The Plants List. Analytical data from the 'parent species' *P. radens* and *P. capitatum* are found infrequently in recent studies and are therefore not included (18,22,43,70). Isomenthone-dominated *Pelargonium* oils, not qualifying as 'rose-scented geraniums', are not presented in Table 5.37.2 but shortly cited under Chemotypes.

Literature data (up to December 20, 2014) on the chemical composition of geranium oils and unpublished analytical data from one of us (E.S.) are shown in Table 5.37.2 in alphabetical order. In geranium oils from various origins, over 500 chemicals have been identified. About 60% of these were found in a single reviewed publication only. The major compounds found in geranium oils from different sources are shown in Table 5.37.3. They include (highest concentrations in any study given) citronellol (63.4%), geraniol (45.1%), citronellyl formate 22.4%), 10-epi-γ-eudesmol (22.0%), linalool (18.1%), geranial (17.9%), isomenthone (17.8%), geranyl formate (11.0%) and guaia-6,9-diene (8.4%). Well-known ingredients of geranium oils that were present in high concentrations (>9%) in one or two studies were menthone (47.2%), limonene (36.1%), *p*-cymene (35.8%), citronellal (32.9%), 2-phenethyl alcohol (15.3%), caryophyllene oxide (14.7%), *cis*-rose oxide (13.2%), α-phellandrene (13.0%), geranyl butyrate (9.8%), α-pinene (9.6%) and geranyl acetate (9.1%). Uncommon or rare constituents of geranium oils found in high concentrations (>9%) in single studies include δ-selinene (14.2%) and β-ocimene (10.4%).

Commercial oils

The ten chemicals that had the highest maximum concentrations in 97 commercial geranium essential oil samples (concentration ranges provided) are the following: citronellol (20.1-49.4%), geraniol (5.6-31.8%), citronellyl formate (2.7-15.0%), linalool (1.4-11.1%), geranial (0.1-10.3%), isomenthone (0.1-8.9%), guaia-6,9-diene (0-8.4%), geranyl formate (0.5-7.2%), geranyl propionate (0.4-5.9%) and 10-epi-γ-eudesmol (0-5.6%) (Erich Schmidt, unpublished analytical data).

Chemotypes

As geranium oils may be commercially produced from various *Pelargonium* species and a very large number of hybrids and cultivars, their composition can vary considerably. The main constituents of geranium oils are citronellol (sweet rose-like odor) and geraniol (flowery rose-like odor), which occur in different proportions according to the origin of the oil. The most important cultivars of geranium are the Reunion Island or 'Bourbon' type (named after the former name of the island and considered to give the best quality oil [12,25,31,52]), the Egyptian or North African type, and the Chinese type (48,50). They can be distinguished by the odor of their leaves and chemical profiles of their volatile oils. The Bourbon type oil contains high concentrations of citronellol and geraniol (in an approximate ratio of 1:1 to 1.2:1), citronellyl formate, guaia-6,9-diene and isomenthone. The Egyptian type oil contains citronellol and geraniol (1:1) and citronellyl formate, isomenthone and 10-epi-γ-eudesmol as major constituents. The difference between the Bourbon and Egyptian type oils is that the former contains good amounts of the sesquiterpene guaia-6,9-diene and is practically devoid of 10-epi-γ-eudesmol, whereas the Egyptian type contains fairly good amounts of 10 epi-γ-eudesmol and low amounts of guaia-6,9-diene (48,49,50). The Chinese type oil contains high amounts of citronellol, a low concentration of geraniol (ratio of 3-4:1), high citronellyl formate and guaia-6,9-diene and absence of 10-epi-γ-eudesmol (12,48,49,50). In India,

Table 5.37.2 Constituents identified in geranium oils

Constituent	CAS	Percentage and range				
		A	B (49)	C (69)	D (80,81)	E
Acetaldehyde	75-07-0					+[z10]
Acetic acid	64-19-7					+[z9]
Acetone	67-64-1					+[z10]
α-Agarofuran	5956-12-7		0-0.3		0.4-0	+[z2]; 0.2[h,p]; 1.3[w9]; 1.6[f]
Agarospirol	1460-73-7		0-0.1		0.3-0	0.2[d]; 0.5[f]
α-Amorphene	20085-19-2					tr[h]; 0.1[e]; 0.3[f]
δ-Amorphene	189165-79-5					tr[h]
Anhydrolinalool oxide	84616-87-5					<0.1[z3]; 0.2[z3]
Aristolene	6831-16-9					4.2[y4]
Aromadendrene	489-39-4		0-0.2	0.3	0.4-0.5	0.1[e]; 0.2[d]; 0.3[r]; 0.4[z6]; 0.7[f,w9]; 1.0[j]
Aromadendrene VI						0.03[e]
allo-Aromadendrene	25246-27-9	0.1-0.9	0-0.2			0.1[i,z]; 0.2[c,r]; 0.3[z7]; 0.5[w]; 0.7[d]
allo-Aromadendrene oxide	85760-81-2			0.2		0.4[w5]
Azulene	275-51-4					0.4[f]
Benzaldehyde	100-52-7					<0.1[w2]
Benzoic acid	65-85-0					+[v1,z9]
Benzyl benzoate	120-51-4		0-0.2			
Benzyl tiglate	37526-88-8					+[z8]
trans-α-Bergamotene	13474-59-4					+[z2]; 0.1[j]
α-trans-β-Bergamotene			0-tr			
(E)-α-Bergamotol			0-0.1			
Bicycloelemene	32531-56-9				0.1-0	
Bicyclogermacrene	24703-35-3		0-0.2		0.2-0	tr[g]; 0.02[e]; 0.2[h]; 0.3[z3]; 0.7[c]
β-Bisabolene	495-61-4					+[z8]
Bois de rose oxide	7392-19-0					+[z8]; 0.03[j]
Borneol	507-70-0					+[z8]; 0.1[s]; 0.2[d]
endo-1-Bourbonalol						0.1[z3]
α-Bourbonene	5208-58-2		0-0.1			0.04[e]; 0.5[z3]
β-Bourbonene	5208-59-3	0.04-1.7	0.2-1.1	1.4	1.6-1.0	1.1[c]; 1.2[y9]; 1.3[d]; 1.4[j]; 1.5[u]; 1.8[f]; 3.1[w9]
1,5-di-epi-Bourbonene						0.2[c]
α-Bulnesene	3691-11-0					2.4[s]
Butane	106-97-8					+[z8]
t-Butanol	75-65-0					+[z8]
2-Butanone	78-93-3					+[z10]
(Z)-3-(1-Butenyl)-pyridine						+[v1]
Butyl formate	592-84-7					+[z10]
2-Butylfuran	4466-24-4					<0.1[z3]
Butyric acid	107-92-6					+[v1]; 1.6[p]
Cadalene	483-78-3					1.0[s]
trans-Cadina-1,4-diene	20085-13-6					tr[g]
trans-Cadina-1(6),4-diene	931410-54-7					tr[g]; 1.4[h]
7aH,10bH-Cadina-1(6),4-diene						0.2[c]
β-Cadinene	523-47-7			0.3		0.5[z4]
γ-Cadinene	39029-41-9		0-0.2	0.3	0.2-0.2	0.1[e]; 0.2[j]; 0.3[d]; 0.5[x]; 0.6[c]; 2.4[w9]
δ-Cadinene	483-76-1	0.3-1.4	0-0.6	1.3	1.3-0.8	0.9[c]; 1.0[d]; 1.2[j]; 1.3[w5]; 1.4[o]; 2.6[f]; 3.2[z]
τ-Cadinene	152287-05-3					0.4[w5]
α-Cadinol	481-34-5		0-1.7		0-0.1	0.1[e,g]; 0.2[s]; 0.3[h]
epi-α-Cadinol	5937-11-1		0-0.7			0.1[z7]; 0.4[p,y9]; 1.1[h]
δ-Cadinol	19435-97-3					0.8[y4]
α-Calacorene	21391-99-1		0-0.1		0.09-0.08	0.3[s]; 1.6[y4]
β-Calacorene	50277-34-4					0.1[s]
Calamenene	483-77-2	0.1-0.9		0.6		0.2[w2]; 0.3[d,r]; 0.5[w]; 0.7[w5]; 0.8[z3]; 1.5[z3]

Table 5.37.2 Constituents identified in geranium oils (*continued*)

Constituent	CAS	Percentage and range				
		A	B (49)	C (69)	D (80,81)	E
cis-Calamenene	72937-55-4					0.1[i]; 0.2[s]; 0.4[e]
trans-Calamenene	73209-42-4				0-0.2	0.3[c]
cis-Calemen-10-ol					0.04-0[a]	
Camphene	79-92-5				0.04-0	tr[h]; 0.01[e]; 0.4[d]
α-Campholenic acid	28973-89-9					+[v1]
Camphor	76-22-2					0.4[d]
trans-Carane	554-59-6					0.05[e]
δ2-Carene (= δ4-)	554-61-0					tr[c]
δ3-Carene	13466-78-9					0.07[p]
Carvacrol	499-75-2					0.9[z1]; 1.5[d]
Carvone	99-49-0					0.3[z1]
Caryophylla-4(12), 8(13)-dien-5β-ol	19431-80-2					0.1[e]
β-Caryophyllene	87-44-5	0.5-3.2	0-1.1	1.5	1.2-1.0	0.5[i]; 0.6[g]; 0.9[n]; 1.0[w9]; 1.1[w7]; 1.2[h]; 1.3[d]; 1.5[c]; 1.6[f]; 1.8[o]; 2.5[e]; 3.2[s]; 3.4[y5]; 7.2[v]
γ-Caryophyllene	118-65-0				0.05-0.01	0.02[e]; 0.6[w]
9-epi-(*E*)-Caryophyllene	68832-35-9					0.1[g]
Caryophyllene oxide	1139-30-6	0.1-1.1	0-0.9	0.2	0.2-0	0.6[o]; 0.9[e]; 1.0[y9]; 1.2[y5]; 1.4[w]; 2.0[g]; 14.7[s]
α-Cedrene	469-61-4					<0.01[z7]; 0.03[e]
β-Cedrene	546-28-1		0-0.2			3.5[z4]
α-Cedrol	77-53-2		0-0.5			
(*E*)-Chrysanthemic acid	4638-92-0					+[v1]
(*Z*)-Chrysanthemic acid						+[v1]
1,8-Cineole	470-82-6				0.08-0.07	0.1[o,z7]; 0.2[z3]; 0.7[n]
Citronellal	106-23-0	0.06-1.9	0-0.4		0.2-0.2	0.09[e]; 0.2[h]; 0.3[g]; 0.4[f]; 32.9[z6]
Citronellic acid	502-47-6		0-0.7			0.2[i,z]
Citronellol	106-22-9	20.1-49.4	8.5-50.9	32.1	28.3-36.8	38.9[t]; 39.7[y9]; 40.2[w7]; 42.1[g]; 42.8[y3]; 43.8[x1]; 45.8[i]; 47.3[y8]; 47.5[e]; 58.2[z4]; 63.4[n]
Citronellyl acetate	150-84-5	0.09-1.1	0-0.4	0.4	0.6-0.5	0.9[x]; 1.0[i]; 1.6[n]; 2.1[x1]; 2.4[z4]; 2.5[o]; 3.7[k]
Citronellyl butyrate	141-16-2	0.1-1.8	0-0.4	1.5	0.4-1.1	1.0[w5]; 1.1[x]; 1.2[w7]; 1.3[m]; 1.5[w]; 1.7[w1]
Citronellyl diethylamine						+[v1]
Citronellyl formate	105-85-1	2.7-15.0	1.0-13.3	10.0	5.8-10.0	11.8[f]; 12.1[y5]; 12.7[x2]; 13.2[w9]; 13.9[x4]; 14.0[x3]; 14.2[i]; 16.9[w4]; 20.6[g]; 22.4[w6]
Citronellyl heptanoate			0-0.2			+[z2]; 0.1[i,z]
Citronellyl hexanoate	10580-25-3					0.1[i]
Citronellyl isobutyrate	97-89-2					0.2[m]; 0.3[w5]; 0.7[z4]; 2.8[w4]
Citronellyl isoheptanoate						0.2[i,z]
Citronellyl isohexanoate	71662-18-5		0-0.5			0.2[i,z]
Citronellyl isooctanoate						0.1[i]
Citronellyl isovalerate	68922-10-1		0-tr			
Citronellyl nonanoate						+[z8]
Citronellyl octanoate	72934-05-5		0-0.4		0.03-0	+[z2]; 0.2[i]
Citronellyl propionate	141-14-0	0.1-1.2	0-0.9	0.8	0.6-1.0	0.3[e]; 0.4[w5]; 0.7[x4]; 0.9[m]; 1.0[w]; 1.2[w1]; 1.3[c]
Citronellyl tiglate	24717-85-9	0.02-1.2		0.8	0.2-0.9	0.8[w7]; 0.9[q]; 1.9[k]; 2.0[i]; 2.1[z4]; 2.4[i,z]; 4.7[w2]
(*E*)-Citronellyl tiglate	24717-85-9		0.2-1.1			0.5[h]
(*Z*)-Citronellyl tiglate	84254-89-7					0.3[j]; 0.7[g]
Citronellyl valerate	7540-53-6		0-0.2		0-0.09	0.1[x4]; 0.2[d,e]; 0.3[i,z]; 0.4[g]; 1.4[w]
α-Copaene	3856-25-5	0.08-0.6	0-0.5		1.2-0.6	0.1[w8]; 0.2[k]; 0.3[i]; 0.5[c]; 0.9[d]; 1.0[f]; 1.9[g]
β-Copaene	18252-44-3		0-0.1		0-0.2	0.2[c]; 0.6[u]
α-Costol	65018-15-7					0.2[w2]
p-Cresyl methyl ether	104-93-8					+[v1]
Cryptone	500-02-7					1.7[z1]
α-Cubebene	17699-14-8	0.07-0.8	0-0.2	0.6		0.07[j]; 0.1[p]; 0.2[c,r]; 0.4[w1]; 0.7[w5]; 1.2[f]

Table 5.37.2 Constituents identified in geranium oils (*continued*)

Constituent	CAS	A	B (49)	C (69)	D (80,81)	E
β-Cubebene	13744-15-5			0.1		0.2[w9]; 0.4[f]
Cubebol	23445-02-5			0.1		0.6[s]
epi-Cubebol	38230-60-3					2.3[y9]
Cubenene	29837-12-5				0.08-0.1	0.06[e]; 0.1[c]; 0.3[p]
Cubenol	21284-22-0		0-0.3	0.3		tr[w2]; 0.1[g]; 0.2[j]; 0.3[z4]; 0.6[s]
1-epi-Cubenol	81939-29-9		0-0.2		0.08-0.1	0.1[g,j]
1,10-di-epi-Cubenol	73365-77-2				0.1-0	0.2[h,j]
Cuminaldehyde	122-03-2					1.2[z1]
Curcumene (*ar*-; α-)	644-30-4					0-0.2
α-Cyclogeraniol	6627-74-3					tr[c]
Cyclopentadecane	295-48-7					0.04[e]
m-Cymene	535-77-3					tr[h]
p-Cymene	99-87-6	0.02-1.8			0.2-0.1	0.06[e]; 0.1[c]; 0.2[l]; 0.4[d]; 0.6[n]; 0.8[x]; 35.8[z1]
p-Cymen-8-ol	1197-01-9		0-0.5			
α-Cyperene	17627-30-4					+[z8]
Decanal	112-31-2					+[z8]
Decanoic acid	334-48-5					2.4[i,z]; 3.4[h]
2-Decenoic acid	3913-85-7					+[v1]
Diethyl phthalate	84-66-2					0.7[b,y2]
Dihydroagarofuran					0.1-0	0.1[c]
Dihydrocampholenic acid						+[v1]
1,7-Dihydrofuropelargone						+[z8]
7,8-Dihydrofuropelargone						+[z8]
3,9-Dimethyldecanoic acid						+[v1]
2,6-Dimethyl-5-hepten-1-ol	36806-46-9		0-0.2			
3,5-Dimethylhexanoic acid	60308-87-4					+[v1]
2,5-Dimethyl-2-hexenoic acid						+[v1]
Dimethyl-3-hexenoic acid						+[v1]
2,6-Dimethyl-2,6-octadiene	2792-39-4			0.2		
(*Z*)-2,6-Dimethyl-2,6-octadiene	2492-22-0					0.2[e,p]
3,7-Dimethyloctanoic acid	5698-27-1					+[v1]
Dimethyl octanol	106-21-8		0-0.2			3.3[y1]
(*E*)-3,7-Dimethyl-5-octene-1,7-diol			0-1.4			
3,7-Dimethyl-7-octene-1,6-diol	22460-95-3		0-0.5			
Dimethyloctenoic acid						+[v1]
3,7-Dimethyl-6-oxooctanoic acid	38975-38-1					+[v1,z9]
2,2-Dimethylpropanol	75-84-3					+[z8]
Dimethyl sulfide	75-18-3					+[z10]
Dimethyl sulfone	67-71-0				0-0.03	+[v1]
Dimethyl sulfoxide	67-68-5					+[v1]
Dimethyltetradecanoic acid						+[v1]
2,5-Dimethyl-2-vinylfuran						<0.1[z3]
Diphenyl ether	101-84-8					1.8[y1]
Docosane	629-97-0					+[z8]
6,9,12,15-Docosatetranoic acid						+[v1]
Dodecane	112-40-3				0-0.1	tr[h]
Dodecanoic acid	143-07-7					+[v1,z9]
Eicosane	112-95-8					+[z8]
β-Elemene	33880-83-0	0.09-1.1	0-0.4	0.2	0-0.4	0.3[j]; 0.4[g]; 0.6[h]; 0.7[s]; 1.2[d]
γ-Elemene	29873-99-2			0.4		0.1[z7]; 0.2[h]
δ-Elemene	20307-84-0				0-0.05	

Table 5.37.2 Constituents identified in geranium oils (*continued*)

Constituent	CAS	A	B (49)	C (69)	D (80,81)	E
trans-β-Elemenone	20303-60-0					0.8[h]
Elemol	639-99-6					+[z8]; 0.2[h]; 0.4[f]
Elixene	3242-08-8					0.5[w5]
Epizonalene						6[y8]
Epizonarene	41702-63-0		0-0.1			0.08[e]; 0.4[w9]; 0.6[f]
Eremophilene	10219-75-7					0.3[f]
1,3a-Ethano(1*H*)inden-4-ol, octahydro-2,2,4,7a-tetramethyl-	117591-80-7					0.01[e]
Ethyl alcohol	64-17-5					+[z10]; 0.01[e]
2-Ethylbutyric acid	88-09-5					+[z9]
Ethyl formate	109-94-4					+[z10]
Ethyl geranate	32659-21-5					0.2[h]
Ethyl hexadecanoate	628-97-7					0.8[s]
α-Eudesmol	473-16-5					0.1[i,z]; 0.2[y9]
5-epi-7-epi-α-Eudesmol	446050-56-2		0-0.1			
β-Eudesmol	473-15-4		0-0.5	0.3	0.6-0	0.1[w1]; 0.5[w9]; 0.6[h,y9]; 1.3[z]; 2.0[i]
γ-Eudesmol	1209-71-8		0-0.2		0.09-0	0.1[c]; 0.3[w,y9]; 0.4[h]
10-epi-γ-Eudesmol	15051-81-7	0-5.6	0-4.8	2.1	4.5-0.2	6.3[w5]; 6.4[k]; 6.5[z3]; 6.7[n]; 6.9[x1]; 7.4[z6]; 7.7[f]; 7.9[w9]; 8.2[o]; 8.7[z4]; 9.4[w7]; 10.0[x8]; 22.0[x5]
α-Farnesene	502-61-4				0-0.02	
(*E,E*)-α-Farnesene	502-61-4		0-0.5			
(*E*)-β-Farnesene	18794-84-8					0.2[z7]; 0.4[d]
(*Z*)-β-Farnesene	28973-97-9					0.5[y9]; 3.0[s]
Farnesol	4602-84-0					+[z8]
(*E,E*)-Farnesol	106-28-5					0.2[s]
(*Z,E*)-Farnesol	3790-71-4					0.4[s]; 1.0[h]
(*E,E*)-Farnesyl acetate	4128-17-0		0-tr			
(*Z,E*)-Farnesyl acetate	40266-29-3					0.4[s]
Formic acid	64-18-6					+[z9]
2-epi-β-Funebrene						tr[h]
Furan	110-00-9					+[z10]
Furfural	98-01-1					+[z10]
2-Furfuryl acid						+[v1]
Furopelargone A	1143-45-9		0-0.1			0.3[z3]; 0.5[j]
Furopelargone B			0-1.1			0.1[z4,z7]; 0.4[z4]; 0.8[i,z]; 0.9[w8]; 1.1[x4]
Furopelargonic acetate						+[z8]
Geranial	141-27-5	0.1-10.3	0.2-17.9	0.4		1.3[w8]; 1.6[y6]; 2.5[n]; 2.6[w]; 2.7[x]; 3.0[i]; 4.3[z4]
Geranic acid	459-80-3		0-5.6		0.1-0	<0.1[z4]
(*E*)-Geranic acid	4698-08-2					+[z9]
(*Z*)-Geranic acid						+[z9]
Geraniol	106-24-1	5.6-31.8	0.1-16.0	8.2	15.3-7.9	30.2[x6]; 30.3[x8]; 31.9[i]; 34.6[n]; 36.6[y6]; 38.4[x]; 42.1[z4]; 42.3[y4]; 43.6[k]; 45.1[z2]
Geranyl acetate	105-87-3	0.04-0.8	0-0.4	0.3		1.2[n]; 1.3[z3]; 1.5[y5]; 1.9[w7]; 2.3[f]; 7.3[y1]; 9.1[z5]
(*E*)-Geranylacetone	3796-70-1		0-tr			
Geranyl butyrate	106-29-6	0.2-1.6	0-1.3	1.1	1.1-1.0	1.2[o]; 1.3[w5]; 1.4[c]; 1.5[m]; 1.6[d]; 3.2[i,z]; 9.8[v]
Geranyl formate	105-86-2	0.5-7.2	0-4.9	2.3	3.0-1.4	5.6[j]; 5.7[x4]; 5.8[m]; 5.9[y9]; 6.2[n]; 6.3[u]; 6.4[z4]; 6.7[x2]; 7.1[f]; 7.5[x6]; 8.8[w4]; 9.3[w6]; 11.0[z3]
Geranyl heptanoate	73019-15-5					0.1[q,z7]; 0.4[z4]; 0.6[i,m]
Geranyl hexanoate	10032-02-7	0.04-0.7	0-1.9			0.07[e]; 0.1[z7]; 0.2[g,z4]; 0.4[i,z]
Geranyl isobutyrate	2345-26-8		0-0.3		0.08-0.2	0.1[w1,z7]; 0.3[f,g]; 0.6[z4]; 0.7[z4]; 1.2[y9]
Geranyl isoheptanoate						0.1[i]
Geranyl isohexanoate						0.2[z3]; 0.8[i]
Geranyl isooctanoate						0.2[i]

Table 5.37.2 Constituents identified in geranium oils (*continued*)

Constituent	CAS	Percentage and range				
		A	B (49)	C (69)	D (80,81)	E
Geranyl isovalerate	109-20-6				0-0.2	0.1[c,e]; 0.3[j]; 0.4[z4]; 0.5[w]; 0.6[z4]; 0.7[z7]
Geranyl 2-methylbutyrate	68705-63-5					+[z8]
Geranyl 3-methylpentanoate						+[z8]
Geranyl 3-methylvalerate						+[z8]
Geranyl 4-methylvalerate						+[z8]
Geranyl nonanoate	68039-29-2					<0.1[z3]
Geranyl octanoate	51532-26-4					+[z2]; <0.1[z3,z7]; 0.1[z4]; 0.3[i,z]
Geranyl propionate	105-90-8	0.4-5.9	0-1.0	0.3	0.6-0.5	0.7[w]; 0.9[j]; 1.0[c]; 1.2[x4]; 1.6[h]; 2.3[j]; 5.3[z3]
Geranyl tiglate	7785-33-3	0.3-1.8	0-0.4	1.6	0.9-1.4	2.8[j]; 3.0[n]; 3.3[x8]; 3.5[y4]; 3.9[h]; 4.8[w7]; 7.0[x3]
Geranyl valerate	10402-47-8		0-tr		0-0.2	0.1[i,z]; 0.2[d,z4]; 0.4[h]
Germacrene A	28387-44-2					0.1[h]
Germacrene B	15423-57-1					+[z2]; 0.4[h]
Germacrene D	23986-74-5	0.3-2.5	0-1.1	2.2	0.9-1.1	2.2[d]; 2.3[w5]; 2.4[w7]; 2.5[x8]; 3.1[x3]; 4.3[w9]
Germacrone	6902-91-6					6.1[h,v2]
Gleenol	72203-99-7					0.1[s]
Globulol	489-41-8		0-0.1	0.3		0.2[h]
Guaiacol	90-05-1					+[v1]
4a*H*,10a*H*-Guaia-1(5), 6-diene						0.1[c]
4b*H*,10a*H*-Guaia-1(5), 6-diene						0.5[c]
Guaia-3,7-diene	6754-04-7				1.0-5.7	0.02[e]
Guaia-6,9-diene	37839-64-8	0-8.4	0-0.5	4.1		1.4[j]; 2.2[k]; 2.7[l]; 2.9[y6]; 4.6[h]; 5.4[w]; 6.0[y9]; 6.2[x3]; 6.7[x4]; 6.8[w6]; 7.1[j]; 7.2[x8]; 8.3[w4]
Guaiene	88-84-6					0.4[w5]
α-Guaiene	3691-12-1	0.1-0.9		0.4		+[z8]; 0.1[w]; 0.5[h]; 0.6[j]; 7.2[s]
β-Guaiene	88-84-6			0.2		
cis-β-Guaiene	372162-07-7					0.2[s]
Guaiol	489-86-1			0.1		+[z8]; 0.3[z3]
α-Gurjunene	489-40-7				0.2-0.05	0.07[e]; 0.1[c]; 0.5[w5]; 5.0[z4]
β-Gurjunene	73464-47-8					0.1[z3]
γ-Gurjunene	22567-17-5					0.4[j]; 0.9[h]
Heneicosane	629-94-7					+[z8]
Heptanoic acid	111-14-8					+[v1,z9]
2-Heptanol	543-49-7					<0.1[z7]
2-Heptanone	110-43-0					<0.1[z7]
Heptenoic acid						+[v1]
Hexadecanoic acid	57-10-3					+[v1]; 0.4[f]
Hexahydrofarnesyl acetone	502-69-2		0-0.1		0.01-0.01	<0.1[z3]
Hexane	92112-69-1					+[z8]; 0.1[z3]
Hexanoic acid	142-62-1					+[v1,z9]
Hexanol	111-27-3					0.03[y9]
2-Hexanone	591-78-6					+[z10]
(*E*)-2-Hexenal	6728-26-3					0.1[z3]; 0.2[z4]
(*Z*)-3-Hexenal	6789-80-6					0.3[q]
(*E*)-2-Hexenoic acid	13419-69-7					+[v1]
(*E*)-3-Hexenoic acid	1577-18-0					+[v1]
(*E*)-2-Hexen-1-ol	928-95-0				0-0.04	+[z8]; 0.1[y9]
(*E*)-3-Hexenol	928-97-2					0.01[y9]
3-Hexen-1-ol	544-12-7					+[z8]; <0.1[w2]
(*Z*)-3-Hexen-1-ol	928-96-1	0.01-0.6	0-tr		0.07-0.1	0.1[z4]; 0.2[y9]; 0.4[w7]; 0.8[z4]; 1.0[j]; 1.6[z4]
(*Z*)-3-Hexenyl acetate	3681-71-8					+[z8]
(*Z*)-3-Hexenyl hexanoate	31501-11-8					<0.1[z3]
(*Z*)-3-Hexenyl isovalerate	35154-45-1		0-0.1			

Table 5.37.2 Constituents identified in geranium oils (*continued*)

Constituent	CAS	Percentage and range				
		A	B (49)	C (69)	D (80,81)	E
(Z)-3-Hexenyl tiglate	67883-79-8		0-0.2			
Hexyl formate	629-33-4					+[z10]
Hinesol	23811-08-7					+[z2]; 0.4[h]
α-Humulene	6753-98-6	0.1-1.5	0-0.3	0.4	0.3-0.3	0.6[w1]; 0.7[s]; 0.8[y5]; 0.9[e]; 1.4[l]; 1.5[f]; 6.1[w2]
Humulene epoxide II	19888-34-7		0-0.2			0.09[j]; 1.8[s]
14-Hydroxy-9-epi-(E)-caryophyllene	244226-09-3					0.5[s]
Hydroxycitronellal	107-75-5					0.6[b,m]
4α-Hydroxydihydroagarofuran			0-0.3			
7-Hydroxydihydrocitronellol						+[z8]
7-Hydroxy-6,7-dihydrogeraniol						+[z8]
Isoamyl alcohol	123-51-3					+[z10]; 0.01[e]
Isoamyl formate	110-45-2					+[z10]
Isoborneol	124-76-5					+[z8]
Isobutanal	78-84-2					+[z10]
Isobutyl alcohol	78-83-1					+[z10]
Isobutyl phenylacetate	102-13-6					tr[h]
Isobutyric acid	79-31-2					+[v1,z9]
Isodecanoic acid	26403-17-8					+[v1]
Isogeranic acid						+[v1]
Isogeraniol	5944-20-7					0.06[e]
Isogermacrone epoxide					0-0.6[a]	
Isoheptanoic acid	1330-19-4					+[v1]
Isohexadecanoic acid	32844-67-0					+[v1]
Isohexanoic acid	25103-52-0					+[v1,z9]
Isoisopulegol	18674-65-2					+[z2]
Isoledene	95910-36-4		0.6			0.03[e]; 0.5[w9]; 0.6[w5]; 2.1[f]
Isolongifolene	1135-66-6					0.5[w9]
Isomenthol	3623-52-7	005-0.4				0.1[c]
Isomenthone	491-07-6	0.1-8.9	0-7.9	5.6	6.3-6.4	8.4[x3]; 8.6[x1]; 8.7[w7]; 8.9[z4]; 9.0[y3]; 9.4[z5]; 9.9[g]; 11.1[l]; 12.0[x5]; 13.0[z4]; 17.6[y9]; 17.8[n]
Isooctane	592-27-8					+[z8]
Isooctanoic acid	25103-52-0					+[v1]
Isoprene	78-79-5					+[z10]
(E)-7-Isopropenyl-4-methyl-10-methylene -4-cyclodecen-1-one						0.1[e]
Isopropyl formate	625-55-8					+[z10]
2-Isopropyl-5-methyl-9-methylenebicyclo [4.4.0]dec-1-ene	150320-52-8					1.1[y2]
2-Isopropyl-4-methylpyridine	4855-56-5					+[v1]
Isopulegol	89-79-2	0.03-1.4	0.1-0.7		0.2-0.1	0.05[e]; 0.1[c]; 0.3[y1]
Isoterpinolene	586-63-0				0.03-0.07	
Isoundecanoic acid	2724-56-3					0.6[i]
Isovaleric acid	503-74-2					+[v1,z9]
Junenol	472-07-1		0-0.3			
Juniper camphor	473-04-1					+[z8]
Lavandulol	498-16-8					0.1[w8]
Lavandulyl acetate	25905-14-0					0.5[y2]; 1.0[f]
Ledol	577-27-5				0-0.04	0.04[e]; 0.2[p]
Leoidosene						0.02[e]
Limetol					0.08-0.07	
Limonene	138-86-3	0.06-0.6	0-0.6	0.2	0.3-0.3	0.3[w4]; 0.5[k]; 0.6[o]; 0.8[y4]; 2.5[d]; 7.3[v]; 36.1[y7]

Table 5.37.2 Constituents identified in geranium oils (*continued*)

Constituent	CAS	A	B (49)	C (69)	D (80,81)	E
Linalool	78-70-6	1.4-11.1	0.1-10.9	3.5	7.0-3.6	11.3[z4]; 12.2[z4]; 12.9[x6]; 13.3[x6]; 15.4[y9]; 15.9[w7]; 16.0[w8]; 16.4[y4]; 17.5[k]; 18.1[l]
cis-Linalool oxide	11063-77-7	0.02-0.3				0.09[e]; 0.2[x]; 0.3[w5]
cis-Linalool oxide, furanoid	5989-33-3		0-0.3	0.2	0.4-0.2	0.1[h]; 0.2[i]; 0.3[c]; 0.4[k]; 0.5[q]; 0.6[l]; 1.0[n]
trans-Linalool oxide	11063-78-8	0.03-1.0				0.05[e]; 0.1[h,x]; 0.3[x4]; 2.3[y7]
trans-Linalool oxide, furanoid	34995-77-2		0-0.4		0.2-0.2	0.07[u]; 0.1[c]; 0.2[i]; 0.4[w7]; 0.5[q]; 1.6[n]; 1.8[o]
Linalyl acetate	115-95-7					1.0[y1]
Linoleic acid	60-33-3					+[v1]
Linolenic acid	463-40-1					+[v1]
Longicyclene	1137-12-8					0.09-0
Longifolene (junipene)	475-20-7					0.1[s]; 0.6[z4]
β-Maaliene	489-29-2					+[z8]; 10.0[y2]
trans-p-Mentha-2,8-dien-1-ol	4017-77-0		0-0.2			
p-Menth-1-en-9-al	29548-14-9					+[z8]
cis-p-Menth-2-en-1-ol	29803-82-5					0.3[z1]
Menthol	89-78-1	0.04-2.3	0-tr	0.2	0.3-0	0.08[j]; 0.1[f,z3]; 0.3[h]; 2.2[z5]
Menthomenthol	98167-53-4					0.2[p]
Menthone	89-80-5	0.08-2.3	0-1.1	1.8	2.4-2.1	4.3[j]; 4.5[v]; 4.9[y9]; 5.2[h]; 6.0[w2]; 8.1[y4]; 47.2[x5]
Methacrylic acid	79-71-4					+[v1]
Methanol	67-56-1					+[z10]
2-Methoxyacetophenone	579-74-8					+[v1]
2-Methyl-3-amino-1-pentene						+[v1]
2-Methylbutanal	96-17-3					+[z10]
3-Methylbutanal	590-86-3					+[z10]
3-Methyl-2-butanal						+[z8]
2-Methyl-1-butanol	137-32-6					+[z10]
3-Methyl-2-butanone	563-80-4					+[z8,z10]
(*E*)-2-Methyl-2-butenal	497-03-0					+[z10]
3-Methyl-2-butenal	107-86-8					+[z10]
2-Methyl-3-buten-2-ol	115-18-4		0-0.2			+[z8]
3-Methyl-2-buten-1-ol	556-82-1				0.03-0	
3-Methyl-2-butenyl 2-methylbutanoate						0.3[h]
2-Methylbutyl formate	35073-27-9					+[z8,z10]
Methylbutyrate	623-42-7					+[z8]
2-Methylbutyric acid	116-53-0					+[v1,z9]
Methyl chavicol	140-67-0					0.1[c]
Methyl citronellate	2270-60-2	0-tr				1.0[h]
3-Methylcyclohexanone	591-24-2					+[z8]
2-Methylcyclopentanone	1120-72-5					+[z10]
3-Methylcyclopentanone	1757-42-2					+[z10]
Methyl eugenol	93-15-2					0-tr
Methyl formate	107-31-3					+[z10]
Methyl geranate	2349-14-6	0.09-2.5			0-0.1	tr[g]
6-Methyl-3,5-heptadien-2-one	1604-28-0		0-tr			0.05[e]
6-Methylheptanoic acid	929-10-2					+[z9]
5-Methyl-2-heptanone	18217-12-4				0.09-0.06	
Methylheptenone	409-02-9					+[z8]
6-Methyl-5-hepten-2-ol	1569-60-4		0-tr			
6-Methyl-5-hepten-2-one	110-93-0	0-0.2		0-0.9		+[z2]; tr[g,w2]; 0.03[e]
2-Methylhexenoic acid						+[v1]
2-Methyl-3-hexenoic acid	62243-57-6					+[v1]
Methyl mercaptan	74-93-1					+[z2]

Table 5.37.2 Constituents identified in geranium oils (*continued*)

Constituent	CAS	Percentage and range				
		A	B (49)	C (69)	D (80,81)	E
Methyl-3-methylcyclopentenyl-ketone						+[z8]
8-Methylnonanoic acid	5963-14-4					+[z9]
3-Methyloctanoic acid	6061-10-5					+[v1,z9]
Methyloctenoic acid						+[v1]
2-Methyl-3-octenoic acid						+[v1]
2-Methylpentanoic acid	97-61-0					+[z9]
3-Methyl-1-pentanol	589-35-3					+[z8]; 0.04[e]
3-Methyl-3-pentanol	77-74-7	0.02-1.4				
2-Methyl-3-pentanone	565-69-5					+[z10]
4-Methyl-2-pentanone	108-10-1					+[z10]
4-Methyl-3-penten-2-one	141-79-7					+[z10]
3-Methylpentyl formate						+[z8]
2-Methylpropyl formate	542-55-2					+[z10]
3-Methylvaleric acid	105-43-1					+[z9]
Mintsulfide	72445-42-2					1.8[f]
cis-Muurola-4(14),5-diene	157477-72-0					0.09[j]
α-Muurolene	10208-80-7		0-0.2	0.3	0.1-0	0.03[e]; 0.1[w]; 0.2[c,f]; 0.4[w9]; 0.5[w5]; 0.8[z4]
γ-Muurolene	30021-74-0		0-0.1	0.1	0.4-0	0.1[c,z]; 0.2[d,y9]; 0.8[h]
α-Muurolol	104245-48-9					0.7[z]; 1.0[h]
τ-Muurolol (epi-α-)	19912-62-0		0-0.6			0.4[s]
Myrcene	123-35-3	0.02-0.4	0-0.1	0.2	0.1-0.1	0.2[f]; 0.4[o]; 0.5[k]; 0.6[n]; 0.7[d]; 1.0[y7]; 2.3[l]
Myrtanol	514-99-8			0.2		
cis-Myrtanol	15358-92-6		0-0.1			
trans-Myrtanol	15358-91-5		0-0.5			
Myrtenol	515-00-4				0.1-0	
Neoisoisopulegol	21290-09-5					0.07[e]
Neoisomenthol	491-02-1	0-0.2	0-0.2		0-0.3	+[z2]; 0.2[e]; 0.6[i,z]
Neoisopulegol	29141-10-4		0-0.3			
Neral	106-26-3	0.2-4.7	0.1-7.2	0.5	0.6-0.3	0.6[d]; 0.7[j]; 0.8[w4]; 1.9[n]; 3.4[o]; 3.9[x1]; 4.1[g]
Neric acid	37349-29-4					+[v1]
Nerol	106-25-2	0.08-1.3	0-1.3			1.1[x1]; 1.2[q]; 1.6[n]; 1.7[w8]; 2.1[f]; 2.2[y1]; 8.7[c]
Nerolic acid	4613-38-1		0-1.0			
(*E*)-Nerolidol	40716-66-3		0-23.8			0.1[h,z7]; 0.4[i,z]
Nerolidol oxide-I			0-1.6			
Nerolidol oxide-II			0-1.8			
(*E*)-Nerolidyl acetate	85611-33-2					0.4[s]
(*Z*)-Nerolidyl acetate	91050-14-5					0.9[s]
Neryl acetate	141-12-8			0.3		0.4[w8]; 0.5[w1]; 0.6[p]; 1.0[z]; 1.1[f]; 1.3[y1]; 4.0[z5]
Neryl butyrate	999-40-6					0.1[m]
Neryl formate	2142-94-1		0-0.3		0.06-0.09	tr[h]; 0.03[u]; 0.1[c]; 0.2[k,q]; 0.3[w7]; 0.4[j]
Neryl isobutyrate	2345-24-6					0.5[w8]
Neryl isovalerate	3915-83-1					0.05[j]
Neryl propionate	105-91-9					0.2[m]; 0.6[x4]; 1.5[w8]; 2.2[y4]
Neryl tiglate			0-2.6			
Neryl valerate	10522-33-5		0-tr			
Nonadecane	629-92-5					+[z8]
Nonadecene	27400-77-7					+[z8]
Nonanal	124-19-6					+[z8]
Nonanoic acid	12-05-0					+[v1,z9]
Norbourbonone	13844-03-6		0-0.1			+[z8]
β-Ocimene	13877-91-3					0.3[f]; 10.4[y7]
(*E*)-β-Ocimene	3779-61-1	0.04-0.4	0-0.1		0.1-0.1	0.01[e]; 0.1[c]; 0.2[k]; 0.3[d]; 0.9[l]; 1.1[n]; 1.6[o]

Table 5.37.2 Constituents identified in geranium oils (*continued*)

Constituent	CAS	A	B (49)	C (69)	D (80,81)	E
(Z)-β-Ocimene	3338-55-4	0.03-0.2	0-0.3		0.1-0.2	tr[c,w2]; 0.1[w4]; 0.2[k]; 0.3[d]; 0.4[w8]; 0.6[l]
neo-allo-Ocimene	7216-56-0					0.2[f]
Octadecane	593-45-3					+[z8]
9-Octadecenoic acid	2027-47-6					+[v1]
2,4-Octadienoic acid	83615-26-3					+[v1]
Octane	111-65-9					<0.1[z3]
Octanoic acid	124-07-2					+[v1,z9]
Octanol	29063-28-3				0.06-0	+[z8]
2-Octanol	123-96-6					+[z8]
3-Octanone	106-68-3					0.1[d]
1-Octene	111-66-0					0.4[z4]
1-Octen-3-ol	3391-86-4					+[z8]
6-Oxo-6,7-dihydrocitronellic acid						+[z8]
α-Patchoulene	560-32-7					0.6[s]
Pentacosane	629-99-2					+[z8]
Pentadecane (n-)	629-62-9					tr[h]
Pentadecanoic acid	1002-84-2					+[v1]
Pentadecanol	629-76-5					0.2[h]
1,3-Pentadiene	504-60-9					+[z10]
Pentanol	30899-19-5					+[z8]
2-Pentanone	107-87-9					+[z10]
1-Penten-3-ol	616-25-1					+[z8]
2-Pentylfuran	3777-69-3					<0.1[z3]
Perillene	539-52-6					+[z8]
Phellandral	21391-98-0					0.9[z1]
α-Phellandrene	99-83-2	0.01-1.6			0.06-0.2	tr[g]; 0.03[e]; 0.09[f]; 0.1[c]; 0.2[l]; 0.6[k]; 13.0[z1]
β-Phellandrene	555-10-2	0.04-0.6	0-0.1			tr[c,w2]; 0.2[f]; 5.2[z1]
2-Phenethyl alcohol	60-12-8		0-tr		0.3-0.09	<0.01[x4]; 0.01[e]; 8.0[y1]; 15.3[z5]
2-Phenethyl propionate	122-70-3	0.02-1.2				<0.01[x4]; 0.2[g,w8]; 0.9[i]
Phenol	108-95-2					+[v1]
Phenylacetaldehyde	122-78-1					tr[h,w2]
Phenylacetic acid	103-82-2					+[z9]
2-Phenylethyl acetate	103-45-7					0.01[x4]; <0.1[w1]; 0.3[y7]; 1.6[y1]
2-Phenylethyl butyrate	103-52-6					0.1[z4]; 0.2[x]; 0.5[g]; 0.8[n]
2-Phenylethyl formate	104-62-1					<0.01[x4]
2-Phenylethyl isobutyrate	103-48-0		0-0.1			0.2[w8]; 0.5[h]
2-Phenylethyl isovalerate	140-26-1					+[z8]
2-Phenylethyl 2-methylbutyrate	24817-51-4					0.3[w8]
2-Phenylethyl tiglate	55719-85-2	0.4-1.3	0-2.5	1.0	1.2-0.6	1.3[i]; 1.4[w4]; 1.6[n]; 1.7[o]; 1.9[x8]; 2.3[f,g]; 3.9[x3]
Photocitral A						0.1[z7]
Photocitral B	6040-45-5		0-0.3			
(E,E)-Photocitral			0-0.3			
(Z,Z)-Photocitral	55253-28-6		0-0.2			
Photonerol						0.02[e]
epi-Photonerol A						+[z8]
α-Pinene	80-56-8	0.02-1.5	0-0.8	0.4	0.9-0.3	0.7[c]; 1.0[g]; 1.2[f]; 1.4[o]; 1.6[x]; 6.0[y7]; 9.6[v]
β-Pinene	127-91-3	0-0.3	0-0.2		0.04-0.04	tr[c]; 0.1[k]; 0.01[e]; 0.2[d]; 0.3[y9]; 0.5[z2]; 3.6[v]
cis-Piperitol	16721-38-3		0-0.4			
Piperitone	89-81-6					+[z2]; 0.1[i,m]; 0.3[w,z4]; 0.4[w7]
2-Propanol	67-63-0					+[z10]
Propionic acid	79-09-4					0.2[p]
Propyl formate	110-74-7					+[z10]

Table 5.37.2 Constituents identified in geranium oils (*continued*)

Constituent	CAS	A	B (49)	C (69)	D (80,81)	E
2-Propyl formate						+[z8]
Rose acetate	90-17-5					3.0[y1]
Rosefuran	15186-51-3					+[z8]
Rosefuran epoxide	92356-06-4					+[z8]
cis-Rose oxide	3033-23-6	0.4-4.0	0.5-1.3	0.6	1.3-1.5	2.1[z3]; 2.3[o]; 2.5[n]; 2.8[y5]; 4.6[z4]; 5.5[s]; 13.2[z4]
trans-Rose oxide	5258-11-7	0.2-1.9	0.2-0.5	1.6	0.6-0.6	1.0[m]; 1.1[y5]; 1.7[e]; 1.8[x6]; 2.0[w9]; 2.2[w]; 4.5[s]
Sabina ketone	513-20-2					tr[h]
Sabinene	3387-41-5					tr[i,o]; 0.1[w1]; 0.2[w7,y9]; 0.7[l]
Salicylic acid	69-72-7					+[v1,z9]
Selina-3,7(11)-diene	6813-21-4					0.1[h]
Selina-4(15),7(11)-diene						0.2[c]
α-Selinene	473-13-2					0.1[i]; 0.2[z4]; 0.3[w]; 2.2[w1]; 6.6[w2]
β-Selinene	17066-67-0			0.3	0.2-0	tr[g]; 0.06[e]; 0.1[r]; 0.2[c]; 0.3[s]; 1.0[y9]
γ-Selinene	515-17-3					0.1[c]
δ-Selinene	473-14-3					8.2[p]; 14.2[f,v2]
Selin-11-en-4α-ol	16641-47-7		0-0.1			+[z8]; 0.4[s]
(*E*)-Sesquilavandulol	104121-84-8					0.4[s]
Soledene						0.01[e]
Spathulenol	6750-60-3	0-0.3	0-0.4	0.3		0.1[z3]; 0.2[e,y9]; 0.3[w5]; 0.7[g]; 2.0[s]
α-Terpinene	99-86-5				0.01-0.05	0.2[i]; 0.6[z1]
γ-Terpinene	99-85-4					0.3[d,y9]; 3.9[z1]
Terpinen-4-ol	562-74-3	0-0.3	0-0.3		0.09-0.06	tr[i]; 0.1[d]; 0.2[w1]; 0.4[s]; 0.5[z1]; 1.7[y7]; 6.0[z4]
α-Terpineol	98-55-5	0.1-1.1	0-1.3	0.3	0.7-0.4	1.4[w1]; 1.7[l]; 1.8[y7]; 2.0[y9]; 2.7[z2]; 3.7[w7]
Terpinolene	586-62-9	0-0.2	0-0.1			tr[c]; 0.06[u]; 0.1[w1,w4]; 0.2[d,k]
α-Terpinyl formate	2153-26-6					<0.1[z3]
Tetracosane	646-31-1					+[z8]
n-Tetradecane	629-59-4					tr[h]
Tetradecanoic acid	544-63-8					+[v1]
5,5,9,10-Tetramethyltricyc-lo[7.3.0.0(1,6)] dodecan-11-one						0.5[j]
Theaspirane A	36431-72-8					+[z8]
Theaspirane B	36431-72-8					+[z8]
Thymol	89-83-8					tr[z1]; 0.1[s]
Tiglic acid	80-59-1					+[v1,z9]
Toluene	108-88-3					+[z10]
Tricosane	638-67-5					+[z8]
Tricyclene	508-32-7				0-0.05	<0.1[z3]
Tridecane	629-50-5				0-0.03	tr[h]
Tridecanoic acid	638-53-9					+[v1]
2,5,5-Trimethyl-2,6-heptadienoic acid						+[v1]
Undecane (*n*-)	1120-21-4					tr[h]
Undecanoic acid	112-37-8					0.6[z]
Valencene	4630-07-3				0-0.2	0.8[f]
Valerianol	20489-45-6		0-0.3			0.8[h]
Valeric acid	109-52-4					+[v1,z9]
Viridiflorene (ledene)	21747-46-6		0-0.2	0.8	0.1-1.1	0.3[w5]; 0.5[w2]; 0.7[p]; 1.1[z4]; 2.3[f]; 2.4[w9]; 3.9[f]
Viridiflorol	552-02-3					tr[g]; 0.08[e]; 0.2[d]
Vitispirane	65416-59-3					+[z8]
α-Ylangene	14912-44-8		0-0.1		0.2-0	0.01[e]; 0.09[k]; 0.1[u,z7]; 0.3[w]; 0.4[w1]
β-Ylangene	20479-06-5		0-0.1			
Zonarene	41929-05-9					tr[g]; 0.2[c]

Table 5.37.2 Constituents identified in geranium oils (*continued*)

A seventy-nine geranium essential oil samples from Egypt, Morocco, Tunisia, Reunion, China and Russia analyzed between 1998 and 2013; lowest and highest concentrations given (E. Schmidt, unpublished data)
B three commercial geranium oils of cultivars 'Bourbon', 'China' and 'Egypt' and two lab-hydrodistilled oils from cultivars 'Frensham' and 'Rober's lemon rose'; lowest and highest concentrations given (ref. 49)
C one commercial geranium oil sample from France (ref. 69)
D two commercial geranium oils, one from Egypt, the other from China; both concentrations given; the figures before the hyphen are from the Egyptian oil, the ones after from the oil from China (refs. 80,81)
E data from other studies (indicated with superscript letters); highest concentrations found in any study reviewed here given; when two or more oils were investigated, only the highest concentrations are mentioned, unless indicated otherwise

[a] incorrect identity based on GC elution order/incorrect identification (82); [b] does not occur in nature; [c] one commercial *Pelargonium graveolens* oil sample from Poland (ref. 63); [d] one steam-distilled oil from *P. graveolens* cultivated in Morocco (ref. 4); [e] one lab-hydrodistilled geranium oil sample from Iran (ref. 79); [f] three oil samples distilled from stems (n=1), leaves (n=1) and flowers (n=1) of *P. graveolens* grown in a Tunisian garden; it should be appreciated that these data cannot sensu stricto be compared to the other data, as geranium oil is usually produced from *all* these plant organs together; there is a very high value for δ-selinene in the stem oil (14.2%), for example, with a zero concentration in flower oil, and the stem oil contained 9% germacrene D versus 1.5% in the flower oil (ref. 27); [g] twenty lab-hydrodistilled oils from Iranian *P. graveolens* cultivated in different growing media (ref. 47); [h] one stem and one geranium leaf oil from *P. graveolens* cultivated in Bosnia; as geranium oils are obtained from stems plus leaves plus flowers, these data cannot sensu stricto be considered representative for geranium oil (ref. 28); [i] one lab-hydrodistilled oil from cultivar 'Kunti' from India (ref. 50); [j] one commercial geranium oil sample from Germany (ref. 64); [k] two lab-hydrodistilled oils from India (ref. 21); [l] three lab-hydrodistilled oils from cultivars 'Algerian', 'Bourbon' and 'Kelkar' grown in India (ref. 2); [m] three commercial oils from Africa, China and Reunion; only the principal aroma compounds (floral-rosy odor) were investigated (ref. 5); [n] fifteen lab-hydro-distilled oils from *Pelargonium graveolens* cultivars 'Bourbon', 'CIM-Pawan' and 'Kelkar' harvested five times during annual growth (ref. 6); [o] five lab-hydrodistilled rose-scented geranium oils with storage time (0-4 days) as variable (ref. 7); [p] one lab-hydrodistilled oil from the aerial parts of *P. graveolens* cultivated in Tunisia (ref. 9); [q] one lab-hydrodistilled oil from *P. graveolens* cultivar 'Bourbon' from India (ref. 11); [r] one lab-hydrodistilled oil from *Pelargonium* sp. dried commercial material purchased in Scotland (ref. 13); [s] one lab-hydrodistilled oil from Cuba (ref. 14); [t] sixteen steam-distilled oils (prepared in a commercial distillation unit in India) from rose-scented geranium cultivar 'Bourbon', of which one was freshly distilled and the others had been stored for 6-24 months in various manners; only the nine major components were investigated (ref. 17); [u] one lab-hydrodistilled oil form Indian cultivar 'Bourbon' (ref. 20); [v] one steam-distilled *P. graveolens* oil, probably from the UK, with unusually high concentrations of the pinenes, β-caryophyllene and geranyl butyrate (ref. 22); [v1] analysis of the acidic, phenolic and basic fractions of a Reunion geranium oil (ref. 77); [v2] oil prepared only from the *stems* of *Pelargonium* sp.; [w] one lab-hydrodistilled oil from the aerial parts of rose-scented geranium in the flowering phase and cultivated in Tunisia (ref. 26); [w1] three oils produced in a field steam-distillation unit from fully grown *Pelargonium* species crop plants of the cultivar 'Bourbon' grown in India (ref. 29); [w2] one geranium leaf oil from north India (ref. 32); [w3] twelve lab-hydro-distilled rose-scented geranium cultivar 'Bourbon' oil samples from the leaves and stems harvested each month during a period of one year in India (ref. 33); [w4] sixteen lab-hydrodistilled oils from geranium leaves and stems cultivated in Rwanda with wilting time (5-46 hours) and time of harvesting (during the day) as variables (ref. 36); [w5] one commercial geranium oil sample from Egypt (ref. 42); [w6] sixteen lab-hydrodistilled oils from 4 harvests of South African rose-scented geranium with amounts of lime added to the soil as variable; only the seven main components were investigated (ref. 45); [w7] thirty-two oils of *Pelargonium sp.* cultivar 'Kelkar' from India obtained by 8 different distillation techniques (ref. 46); [w8] one lab-hydrodistilled geranium leaf oil from cultivar 'Rosé' grown at Reunion (ref. 48); [w9] one lab-hydrodistilled oil from the aerial parts of *P. graveolens* grown in Tunisia (ref. 55); [x] five lab-hydrodistilled oils from fresh herbage of *P. graveolens* cultivated in India and transplanted at different dates (ref. 65); [x1] nine oils from *P. graveolens* cultivated in India and planted at 9 different dates (ref. 74); [x2] one commercial geranium oil from Brazil (ref. 3); [x3] five lab-hydrodis-tilled oils from plants cultivated in Portugal with color of the leaves (green or yellow) and preparation (fresh, dried) as variables (ref. 12); [x4] forty-one lab-hydrodistilled oils from 41 monthly harvested leaf samples of the tops of rose-scented geranium from Australia; the average values were given (ref. 15); [x5] five lab-hydrodistilled geranium herb oils from cultivar 'Bourbon' plants cultivated in India and harvested in May (n=4) and winter (n=1) with unusually high concentrations of menthone, isomenthone and 10-epi-γ-eudesmol (ref. 16); [x6] three lab-hydrodistilled oils from three *P. graveolens* cultivation sites in India (ref. 23); [x7] two oils from Indian cultivar 'Bourbon' prepared in a field distillation unit from plants grown with and without intercropping with *Mentha arvensis* (ref. 30); [x8] six lab-hydrodistilled oils from Indian cultivar 'Bourbon', with or without intercropping with *M. piperita* and with row spacing as variables (ref. 34); [x9] two lab-hydrodistilled leaf oils from Portugal (ref. 37); [y] one lab-hydrodistilled leaf oil from Egyptian *P. graveolens* (ref. 38);

Table 5.37.2 Constituents identified in geranium oils (*continued*)

[y1] one commercial geranium leaf oil from Reunion (ref. 41); [y2] one oil sample produced from Tunisian *P. graveolens* (ref. 44); [y3] one lab-hydrodistilled oil from fresh geranium shoots from India (ref. 58); [y4] one lab-hydrodistilled oil sample from *P. graveolens* cultivated in Brazil (ref. 59); [y5] one lab-hydrodistilled oil from the flowering aerial parts of *P. graveolens* cultivated in Iran (ref. 61); [y6] one lab-hydro-distilled oil from Rwandese geranium (ref. 71); [y7] one very atypical lab-hydrodistilled oil sample from *P. graveolens* grown at a Egyptian university farm with low geraniol and citronellol content and dominated by limonene (36%) and ocimene (10%) (ref. 72); [y8] one com-mercial geranium oil obtained from an Italian company (ref. 73); [y9] data from various studies cited in ref. 82; [z] one lab-hydrodistilled oil of the geranium cultivar 'Kunti' from India (ref. 78); [z1] one lab-hydrodistilled oil from a not further defined cultivar of *Pelargonium* species with a very atypical composition, dominated by *p*-cymene and α-phellandrene (ref. 48); [z2] data from various studies cited in ref. 75; [z3] twenty geranium oils from plants harvested in four seasons in Morocco in four geographic locations (ref. 76); [z4] data from various studies cited in ref. 19; [z5] one commercial oil of unknown origin (ref. 83); [z6] one lab-hydrodistilled oil sample from commercially acquired dried geranium leaves of unknown origin (ref. 84); [z7] two steam-distilled geranium oils produced from the same cultivar 'Bourbon' grown in two regions of India (ref. 85); [z8] chemicals tabulated in ref. 1 as constituents identified in rose-scented geranium oils (without references) (ref. 1); [z9] investigation of the acid fraction of geranium oil 'Bourbon' (ref. 40); [z10] qualitative analysis of the most volatile neutral components of Reunion geranium oil (ref. 39)

tr: trace (in columns B, E[g], E[h] and E[w2]: <0.1; in column E[c]: <0.05); + present in the oil investigated, but quantity not stated or presented in a manner which cannot be compared to the other data

a chemotype of geranium exists, the essential oil of which has very high geraniol content (ca. 40%) and low citronellol content (ca. 8%); at the moment, it is appar-ently not used for commercial oil production (21).

Whereas most rose-scented geranium oils are domi-nated by citronellol and geraniol, in a few studies, other chemicals were found to be the dominant component in oils from *Pelargonium* sp., including *p*-cymene (35.8%, ref. 48), limonene (36%, ref. 72), menthone (47.2%, ref. 16), 10-epi-γ-eudesmol (22%, ref. 16), citronellal (32.9%, ref. 84) and 2-phenethyl alcohol (15.3%, ref. 83). Very high (>60%) isomenthone concentrations have been observed in plants derived from leaf cuttings of *Pelargonium* sp. 'Bourbon' (31,51), in a somaclonal mutant isolated in a geraniol-rich rose-scented geranium accession of *Pelargonium graveolens* (50) and in plants derived from stem cutting of *Pelargonium* species 'Kunti' (78), all from India. Moderately high concentrations of isomenthone (26-35%) were found in the essential oil of a somaclonal variant of rose-scented geranium (62). Isomenthone has

also been found as the major compound in the oils of *P. graveolens* (53), *P. radens* (53), hybrids of *P. graveolens* and *P. radens* (53), the cultivar 'Menthe' (hybrid of *P. cap-itatum* and *P. radens*) (52) and other *Pelargonium* species including *P. australe* (cited in ref. 50) and *P. tomentosum* (54). Isomenthone may also be the dominant ingredient of *P. odoratissimum*, together with methyl eugenol and (sometimes) limonene (10,35,56).

A 2006 South African thesis on the influence of har-vesting frequency and plant shoot age on essential oil yield and composition of rose-scented geranium, not dis-cussed in Table 5.37.2, can be accessed on-line (ref. 86).

CONTACT ALLERGY/ALLERGIC CONTACT DERMATITIS

General

Contact allergy to/allergic contact dermatitis from gera-nium oil has been reported in over 30 publications. In a group of consecutive patients suspected of contact

Table 5.37.3 Major constituents of geranium oils

Constituent	CAS	Percentage and range				
		A	B (49)	C (69)	D (80,81)	E
Citronellol	106-22-9	20.1-49.4	8.5-50.9	32.1	28.3-36.8	43.8[x1]; 45.8[l]; 47.3[y8]; 47.5[e]; 58.2[z4]; 63.4[n]
Geraniol	106-24-1	5.6-31.8	0.1-16.0	8.2	15.3-7.9	38.4[x]; 42.1[z4]; 42.3[y4]; 43.6[k]; 45.1[z2]
Citronellyl formate	105-85-1	2.7-15.0	1.0-13.3	10.0	5.8-10.0	14.0[x3]; 14.2[l]; 16.9[w4]; 20.6[g]; 22.4[w6]
10-epi-γ-Eudesmol	15051-81-7	0-5.6	0-4.8	2.1	4.5-0.2	7.9[w9]; 8.2[o]; 8.7[z4]; 9.4[w7]; 10.0[x8]; 22.0[x5]
Linalool	78-70-6	1.4-11.1	0.1-10.9	3.5	7.0-3.6	15.9[w7]; 16.0[w8]; 16.4[y4]; 17.5[k]; 18.1[l]
Geranial	141-27-5	0.1-10.3	0.2-17.9	0.4		1.3[w8]; 1.6[y6]; 2.5[n]; 2.6[w]; 2.7[x]; 3.0[l]; 4.3[z4]
Isomenthone	491-07-6	0.1-8.9	0-7.9	5.6	6.3-6.4	9.9[g]; 11.1[l]; 12.0[x5]; 13.0[z4]; 17.6[y9]; 17.8[n]
Geranyl formate	105-86-2	0.5-7.2	0-4.9	2.3	3.0-1.4	6.7[x2]; 7.1[f]; 7.5[x6]; 8.8[w4]; 9.3[w6]; 11.0[z3]
Guaia-6,9-diene	37839-64-8	0-8.4	0-0.5	4.1		6.2[x3]; 6.7[x4]; 6.8[w6]; 7.1[j]; 7.2[x8]; 8.3[w4]

LEGEND: SEE UNDER TABLE 5.37.2

dermatitis, a prevalence rate of 1.2% positive patch test reactions has been observed, but reliable relevance data are lacking. There have been many case reports of allergic contact dermatitis from geranium oil, mostly in cosmetics and some in pharmaceutical preparations. In addition, there at least five descriptions of occupational allergic contact dermatitis from geranium oil in aromatherapists, a masseuse and a physiotherapist. Often, there was co-reactivity with geraniol, which is an important component of geranium oil, and with rose oils, which also contain high concentrations of geraniol. Other co-reacting chemicals which may be present in concentrations >3% in commercial geranium oils (Table 5.37.2, column A) include citronellol (the dominant ingredient), citral (neral + geranial), linalool and caryophyllene.

Testing in groups of patients

The results of patch tests with geranium oil in routine testing (consecutive patients suspected of contact dermatitis) and in groups of selected patients are shown in Table 5.37.4. In routine testing, the rate of positive reactions was 1.2% (one study only), whereas between 0.8% and 30% of patients in selected groups had positive patch tests. The very high positivity rate of 30% was found in a very small group of 10 patients strongly suspected of fragrance allergy and reacting to the fragrance mix (90).

Case reports

Occupational exposure

Two aromatherapists had contact dermatitis (one occupational) with allergies to multiple essential oils used at their work, including geranium oil. Both patients also

Table 5.37.4 Results of testing groups of patients with geranium oil

Years and Country	Test conc. & vehicle	Number of patients tested | positive (%)			Selection of patients (S); Relevance (R); Comments (C)	Ref.
Routine testing						
2000-2007 USA	2% pet.	486	6	(1.2%)	R: 100%; C: weak study: a. high rate of macular erythema and weak reactions, b. relevance figures include 'questionable' and 'past' relevance	89
Testing in groups of selected patients						
2006-2010 USA	2% pet.	100	1	(1.0%)	S: patients with eyelid dermatitis; R: not stated	102
2004-2008 Spain	2% pet.	86	8	(9.3%)	S: patients previously reacting to the fragrance mix I or *Myroxylon pereirae* (n=54) or suspected of fragrance contact allergy (n=32); R: not stated; C: almost all patients also reacted to geraniol, one of its major components	104
<2004 Israel	2% pet.	91	1	(1.1%)	S: patients who had shown a doubtful or positive reaction to the fragrance mix I and/or *Myroxylon pereirae* resin and/or one or two commercial fine fragrances; R: not stated	107
2000 USA, Japan and 4 European countries	10% pet.	178	15	(8.4%)	S: patients previously shown to be allergic to fragrances; R: not stated	108
1989-1999 Portugal	2% pet.	67	5	(7.5%)	S: patients who had a positive patch test to the fragrance mix; R: not stated	110
1990-1998 Japan	20% pet.	1,483	31	(2.1%)	S: patients suspected of cosmetic contact dermatitis; virtually all were women; range of annual frequency of sensitization: 0-4.0%; R: not stated	91
1996-1997 UK	2% pet.	10	3	(30%)	S: patients strongly suspected of fragrance allergy; all also reacted to the fragrance mix; R: not stated	90
<1986 Poland	2% pet.	86	2	(2.3%)	S: patients previously reacting to the fragrance mix; R: not stated	100
<1986 France	2% pet.	21	3	(14%)	S: patients with dermatitis caused by fragrances; R: not stated; C: the patients reacted to Egyptian geranium oil; two (67%) co-reacted to geraniol; there was also one patient who reacted to Bourbon geranium oil	118
1971-1980 Japan	20% pet.	477	4	(0.8%)	S: patients with dermatoses other than pigmented cosmetic dermatitis and volunteers; R: not stated	98
<1976 Poland	2% pet.	51	2	(3.9%)	S: patients allergic to *Myroxylon pereirae* resin (balsam of Peru) and/or turpentine and/or wood tar and/or colophony	96
<1974 Japan	?	183	3	(1.6%)	S: patients suspected of cosmetic dermatitis; R: unknown; C: in (probably) all there was co-reactivity with geraniol, which may be present in commercial geranium oils in concentrations of up to 32% (Table 5.37.2, column A)	112

pet.: petrolatum

reacted to geraniol, α-pinene, the fragrance mix and various other fragrance materials. In addition, one proved to be allergic to linalool and linalyl acetate, the other to caryophyllene; α-pinene, linalool, geraniol and caryophyllene were demonstrated by GC-MS in geranium oil (94). Linalool and geraniol may be important components of geranium oil (maximum linalool concentration in commercial geranium essential oils found: 11.1%, maximum geraniol concentration: 31.8%) and caryophyllene to a lesser degree (max. 3.2%, Table 5.37.2, column A). Another aromatherapist had chronic hand dermatitis and was patch test positive to 17 of 20 oils used at her work (tested 1% and 5% in petrolatum), including geranium oil (93). Yet another aromatherapist developed occupational contact dermatitis from contact allergy to multiple essential oils; she reacted to geranium oil, Bourbon in the fragrance series, which reaction was considered to be relevant (97). One masseuse with occupational contact dermatitis to (other) essential oils, who also reacted to geranium oil Bourbon and other fragrance materials (95). One beautician had occupational allergic hand dermatitis from products containing citral and essential oils; she reacted to geranium oil, citral, citronellol, geraniol, and limonene (103); citronellol and geraniol are the main components of geranium oil (Table 5.37.2, column A).

Cosmetics

One case of allergic contact dermatitis from geranium oil, Bourbon, in an aftershave; the patient also had *photo*contact allergy to sandalwood (*Santalum album*) oil (88). One patient with hand dermatitis reacted to geranium oil Bourbon, rose oil, geraniol and several other fragrances and essential oils; she used a 'fragrance-free' hand soap containing rose oil (101). One patient with allergic contact dermatitis from 'Rose Absolute' perfume and a body lotion containing *Rosa centifolia*; she reacted to her own products, geranium oil Bourbon, *Rosa centifolia*, rose oil Bulgarian, several indicators of fragrance allergy, lavender oil and various individual fragrance chemicals including linalool, which may be present in commercial geranium oils up to a concentration of 11.1% (Table 5.37.2, column A), citral (combination of neral [maximum 4.7%] and geranial [maximum concentration 10.3%, Table 5.37.2, column A]) and limonene, which may be present in lower concentrations in geranium oil (105). Three patients developed allergic cosmetic dermatitis from geranium oil in hair dye, face cream and nail polish, one reaction each (109). One patient with allergic contact cheilitis from geranium oil in a lip balm has been described; although the title of the article suggests otherwise, geraniol itself was either negative or not tested (111). One patient developed disseminated allergic contact dermatitis from geraniol and lavender essence in a cream; the patient also reacted to geranium oil Bourbon and Bulgarian rose oil (114).

Pharmaceutical preparations and other exposures

Twelve cases of contact allergy to geranium oil present in topical pharmaceutical preparations were reported from Belgium (87). One of three patients allergic to a topical preparation to promote wound healing and prevent scar formation (active ingredient: *Centella asiatica* extract) reacted to geranium oil present in the ointment (116). One patient had allergic contact cheilitis from lime in gin and tonic; she reacted to lime, geranium oil Bulgarian, rose oil, and their important component geraniol (106).

Positive patch tests (relevance unknown, uncertain or not stated)

Three positive patch tests were seen to geranium oil Bourbon in massage therapists/aromatherapists with occupational contact dermatitis from (multiple) essential oils; it was uncertain whether these oils had been used at work by the patients (92). One patient with erythema on the face who reacted upon patch testing to geranium oil and other essential oils (115). Of seven patients allergic to the fragrance farnesol, 3 (43%) co-reacted to geranium oil (and various other fragrances) (117).

Pigmented cosmetic dermatitis

In Japan, in the 1960s and 1970s, many female patients developed pigmentation following dermatitis of the face (99). This so-called pigmented cosmetic dermatitis was shown to be caused by contact allergy to components of cosmetic products, notably essential oils, other fragrance materials, antimicrobials, preservatives and coloring materials (98,99). In a group of 620 Japanese patients with this condition investigated between 1970 and 1980, 2% had positive patch test reactions to geranium oil 20% in petrolatum (98). The number of patients with pigmented cosmetic dermatitis decreased strongly after 1978, when major cosmetic companies began to eliminate strong contact sensitizers from their products (98).

LITERATURE

1 Rajeswara Rao BR. Chemical composition and uses of Indian rose-scented geranium (*Pelargonium* species) essential oil—a review. J Essent Oil Bear Plants 2009;12:381-394

2 Kaul PN, Rajeswara Rao BR, Bhattacharya AK, Singh CP, Singh K. Volatile constituents of three cultivars of rose-scented geranium (*Pelargonium* sp.) as influenced by method of distillation. PAFAI J 1995;17(4):21-26. Data cited in ref. 1

3 Murbach Teles Andrade BF, Nunes Barbosa L, da Silva Probst I, Fernandes Júnior A. Antimicrobial activity of essential oils. J Essent Oil Res 2014;26:34-40

4 El Asbahani A, Jilale A, Voisin SN, Addi EHA, Casabianca H, El Mousadik A, et al. Chemical composition and antimicrobial activity of nine essential oils obtained by steam distillation of plants from the Souss-Massa Region (Morocco). J Essent Oil Res 2015;27:34-44

5 Jirovetz L, Eller G, Buchbauer G, Schmidt E, Denkova Z, Stoyanova AS, et al. Chemical composition, antimicrobial activities and odor descriptions of some essential oils with characteristic floral-rosy scent and of their principal aroma compounds. Recent Res Devel Agronomy & Horticulture 2006;2:1-12 ISBN: 81-308-0054-3

6 Verma RS, ur Rahman L, Verma RK, Chauhan A, Singh A. Essential oil composition of *Pelargonium graveolens* L'Her ex Ait. cultivars harvested in different seasons. J Essent Oil Res 2013;25:372-379

7 Verma RS, Rahman L, Verma RK, Chauhan A, Singh A. Post harvest storage method for rose-scented geranium (*Pelargonium graveolens* L' Herit. ex Ait.). J Essent Oil Bear Plants 2013;16:693-698

8 Dabiri M, Sefidkon F, Yousefi M, Bashiribod S. Volatile components of *Pelargonium roseum* R. Br. J Essent Oil Bear Plants 2011;14:114-117

9 Mnif W, Dhifi W, Jelali N, Baaziz H, Hadded A, Hamdi N. Characterization of leaves essential oil of *Pelargonium graveolens* originating from Tunisia: Chemical composition, antioxidant and biological activities. J Essent Oil Bear Plants 2011;14:761-769. Data also presented in ref. 60

10 Khalid KA, Cai W, Ahmed AMA. Effect of harvesting treatments and distillation methods on the essential oil of lemon balm and apple geranium plants. J Essent Oil Bear Plants 2009;12:120-130

11 Pant KP, Bisht PS, Nautiyal MC. Oil profile of rose scented geranium var. Bourbon grown in the Garhwal Himalaya (India). J Essent Oil Bear Plants 2005;8:28-31

12 Gomes PB, Mata VG, Rodrigues AE. Characterization of Portuguese-grown geranium oil (*Pelargonium* sp.). J Essent Oil Res 2004;16:490-495

13 Dorman HJD, Deans SG. Chemical composition, antimicrobial and *in vitro* antioxidant properties of *Monarda citriodora* var. *citriodora*, *Myristica fragrans*, *Origanum vulgare* ssp. *hirtum*, *Pelargonium* sp. and *Thymus zygis* oils. J Essent Oil Res 2004;16:145-150

14 Pino JA, Rosado A, Fuentes V. Essential oil of rose-scented Geranium (*Pelargonium* sp.) from Cuba. J Essent Oil Res 2001;13:21-22

15 Doimo L, Fletcher RJ, D'Arcy BR. Esters in Australian geranium Oil (*Pelargonium* hybrid). J Essent Oil Res 1999;11:611-614

16 Kaul PN, Rajeswara Rao BR, Mallavarapu GR. Aberrations in the composition of the herb oil of rose-scented geranium (*Pelargonium* species). J Essent Oil Res 1998;10:439-441

17 Kaul PN, Rajeswara Rao BR, Bhattacharya AK, Mallavarapu GR, Ramesh SI. Changes in chemical composition of rose-scented geranium (*Pelargonium* sp.) oil during storage. J Essent Oil Res 1997;9:115-117

18 Viljoen AM, Van der Walt JJA, Swart JPJ, Demarne F-E. A study of the variation in the essential oil of *Pelargonium capitatum* (L.) L'Herit. (Geraniaceae). Part II. The chemo-types of *P. capitatum*. J Essent Oil Res 1995;6:605-611

19 Lawrence BM. Progress in essential oils. Perfum Flavor 2003;28(1):60-? (last page unknown)

20 Bhattacharya AK, Kaul PN, Rajeswara Rao BR, Ramesh SI, Mallavarapu GR. Composition of the oil of rose-scented geranium (*Pelargonium* sp.) grown under the semiarid tropical climate of South India. J Essent Oil Res 1993;5:229-231

21 Mallavarapu GR, Prakasa Rao EVS, Ramesh S, Narayana MR. Chemical and agronomical investigations of a new chemotype of geranium. J Essent Oil Res 1993;5:433-438

22 Lis-Balchin M. Essential oil profiles and their possible use in hybridization of some common scented geraniums. J Essent Oil Res 1991;3:99-105

23 Rajeswara Rao BR, Sastry KP, Prakasa Rao EVS, Ramesh SI. Variation in yields and quality of geranium (*Pelargonium graveolens* L' Hér. ex Aiton) under varied climatic and fertility conditions. J Essent Oil Res 1990;2:73-79

24 Weiss, EA. Essential oil crops. New York, USA: Centre of Agriculture and Biosciences (CAB) International, 1997

25 Anonymous. Rose geranium production. Pretoria, South Africa: Department of Agriculture, Forestry and Fisheries, 2012. Available at: http://www.nda.agric.za/docs/Brochures/ProGuRosegeranium.pdf

26 Boukhatem MN, Kameli A, Saidi F. Essential oil of Algerian rose-scented geranium (*Pelargonium graveolens*): Chemical composition and antimicrobial activity against food spoilage pathogens. Food Control 2013;34:208-213

27 Boukhris M, Ben Nasri-Ayachi M, Mezghani I, Bouaziz M, Boukhris M, Sayadib S. Laboratory trichomes morphology, structure and essential oils of *Pelargonium graveolens* L'Hér. (Geraniaceae). Ind Crops Prod 2013;50:604-610

28 Cavar S, Maksimović. Antioxidant activity of essential oil and aqueous extract of *Pelargonium graveolens* L'Her. Food Control 2012;23:263-267

29 Rajeswara Rao BR, Kaul PN, Syamasundar KV, Ramesh S. Water soluble fractions of rose-scented geranium (*Pelargonium species*) essential oil. Biores Technol 2002;84:243-246

30 Rajeswara Rao BR. Biomass yield, essential oil yield and essential oil composition of rose-scented geranium (*Pelargonium* species) as influenced by row spacings and intercropping with cornmint (*Mentha arvensis* L.f. piperascens Malinv. Ex Holmes). Ind Crops Prod 2002;16:133-144

31 Kulkarni RN, Baskaran K, Ramesh S, Kumar S. Intra-clonal variation for essential oil content and composition in plants derived from leaf cuttings of rose-scented geranium (*Pelargonium* sp.). Ind Crops Prod 1997;6:107-112

32 Rana VS, Juyal JP, Blazquez MA. Chemical constituents of essential oil of *Pelargonium graveolens* leaves. Int J Aromather 2002;12(4):216-218

33 Rajeswara Rao BR, Kaul PN, Mallavarapu GR, Ramesh S. Effect of seasonal climatic changes on biomass yield and terpenoid composition of rose-scented geranium (*Pelargonium* species). Biochem Syst Ecol 1996;24:627-635

34 Verma RK, Chauhan A, Verma, RS, Rahman L-U, Bisht A. Improving production potential and resources use efficiency of peppermint (*Mentha piperita* L.) intercropped with geranium (*Pelargonium graveolens* L. Herit ex Ait) under different plant density. Ind Crops Products 2013;44:577-582

35 Khalid KA, Teixeira da Silva JA, Cai W. Water deficit and polyethylene glycol 6000 affects morphological and biochemical characters of *Pelargonium odoratissimum* (L.). Scientia Horticulturae 2010;125:159-166

36 Malatova K, Hitimana N, Niyibizi T, Simon JE, Juliani HR. Optimization of harvest regime and post-harvest handling in geranium production to

maximize essential oil yield in Rwanda. Ind Crops Prod 2011;34:1348-1352

37 Gomes PB, Mata VG, Rodrigues AE. Production of rose geranium oil using supercritical fluid extraction. J Supercrit Fluids 2007;41:50-60

38 Badawy MEI, Abdelgaleil SAM. Composition and antimicrobial activity of essential oils isolated from Egyptian plants against plant pathogenic bacteria and fungi. Ind Crops Prod 2014;52:776-782

39 Timmer R, Ter Heide R, De Valois PJ, Wobben HJ. Qualitative analysis of the most volatile neutral components of Reunion geranium oil (Pelargonium roseum). J Agric Food Chem 1971;19:1066-1068

40 Ter Heide R, De Valois PJ, Wobben HJ, Timmer R. Analysis of the acid fraction of Reunion geranium oil (Pelargonium graveolens L'Her. ex Ait). J Agric Food Chem 1975;23:57-60

41 Seo S-M, Kim J, Lee S-G, Shin C-H, Shin S-C, Park I-K. Fumigant antitermitic activity of plant essential oils and components from ajowan (Trachyspermum ammi), allspice (Pimenta dioica), caraway (Carum carvi), dill (Anethum graveolens), geranium (Pelargonium graveolens), and litsea (Litsea cubeba) oils against Japanese termite (Reticulitermes speratus Kolbe). J Agric Food Chem 2009;57: 6596-6602

42 Tabanca N, Wang M, Avonto C, Chittiboyina AG, Parcher JF, Carroll JF, et al. Bioactivity-guided investigation of geranium essential oils as natural tick repellents. J Agric Food Chem 2013;61:4101-4107

43 Demarne F-E, Viljoen AM, Van der Walt JJA. A study of the variation in the essential oil and morphology of Pelargonium capitatum (L.) L'Hérit. (Geraniaceae). Part I. The composition of the oil. J Essent Oil Res 1993;5:493-499

44 Bouzenna H, Krichen L. Pelargonium graveolens L'Her. and Artemisia arborescens L. essential oils: Chemical composition, antifungal activity against Rhizoctonia solani and insecticidal activity against Rhysopertha dominica. Nat Prod Res 2013;27:841-846

45 Araya HT, Soundy P, Steyn JM. Liming improves herbage yield, essential oil yield and nutrient uptake of rose-scented geranium (Pelargonium capitatum × P. radens) on acidic soils. NZ J Crop Horticult Sci 2011;39:175-186

46 Babu KGD, Kaul VK. Variation in essential oil composition of rose-scented geranium (Pelargonium sp.) distilled by different distillation techniques. Flavour Fragr J 2005;20:222-231

47 Rezaei Nejad A, Ismaili A. Changes in growth, essential oil yield and composition of geranium (Pelargonium graveolens L.) as affected by growing media. J Sci Food Agric 2014;94:905-910

48 Gauvin A, Lecomte H, Smadja J. Comparative investigations of the essential oils of two scented geranium (Pelargonium spp.) cultivars grown on Reunion Island. Flavour Fragr J 2004;19:455-460

49 Ali A, Murphy CC, Demirci B, Wedge DE, Sampson BJ, Khan IA, Can Baser KH, Tabanca N. Insecticidal and biting deterrent activity of rose-scented geranium (Pelargonium spp.) essential oils and individual compounds against Stephanitis pyrioides

and Aedes aegypti. Pest Manag Sci 2013;69:1385-1392. Analytical data probably also presented in ref. 67

50 Gupta R, Mallavarapu GR, Banerjee S, Kumar S. Characteristics of an isomenthone-rich somaclonal mutant isolated in a geraniol-rich rose-scented geranium accession of Pelargonium graveolens. Flavour Fragr J 2001;16:319-324

51 Kulkarni RN, Mallavarapu GR, Baskaran K, Ramesh S, Kumar S. Composition of the essential oils of two isomenthone-rich variants of geranium (Pelargonium sp.). Flavour Fragr J 1998;13:389-392

52 Demarne F-E. L'amelioration varietale du `Geranium Rosat (Pelargonium sp.). Contribution systematique, caryologique et biochemique. DSc Thesis, Université de Paris-Sud, Centre Dorsay, 1998: 250

53 Van der Walt JJA, Demarne F. Pelargonium graveolens and Pelargonium radens: A comparison of their morphology and essential Oils. S Afr J Bot 1988;54:617-622

54 Demarne FE, van der Walt JJA. [Title is missing]. S Afr J Plant Soil 1990;7:36-? (last page unknown)

55 Boukhris M, Simmonds MSJ, Sayadi S, Bouaziz M. Chemical composition and biological activities of polar extracts and essential oil of rose-scented geranium, Pelargonium graveolens. Phytother Res 2013;27:1206-1213

56 Lis-Balchin M, Roth G. Composition of the essential oils of Pelargonium odoratissimum, P. exstipulatum, and P. fragrans (Geraniaceae) and their bioactivity. Flavour Fragr J 2000;15:391-394

57 Guerrini A, Rossi D, Paganetto G, Tognolini M, Muzzoli M, Romagnoli C, et al. Chemical characterization (GC/MS and NMR fingerprinting) and bioactivities of South-African Pelargonium capitatum (L.) L'Her. (Geraniaceae) essential oil. Chem Biodivers 2011;8:624-642

58 Prasad A, Kumar S, Pandey A, Chand S. Microbial and chemical sources of phosphorus supply modulate the yield and chemical composition of essential oil of rose-scented geranium (Pelargonium species) in sodic soils. Biol Fertil Soils 2012;48:117-122

59 Baldin ELL, Aguiar GP, Fanela TLM, Soares MCE, Groppo M, Crotti AEM. Bioactivity of Pelargonium graveolens essential oil and related monoterpenoids against sweet potato whitefly, Bemisia tabaci biotype B. J Pest Sci 2014, DOI 10.1007/s10340-014-0580-8

60 Ben Hsouna A, Hamdi N. Phytochemical composition and antimicrobial activities of the essential oils and organic extracts from Pelargonium graveolens growing in Tunisia. Lipids in Health and Disease 2012;11:167 (7 pages). Data also presented in ref. 9

61 Ghannadi A, Bagherinejad MR, Abedi D, Jalali M, Absalan B, Sadeghi N. Antibacterial activity and composition of essential oils from Pelargonium graveolens L'Her and Vitex agnus-castus L. Iran J Microbiol 2012;4(4):171-176

62 Kulkarni SS, Ravindra NS, Srinivas KV, Kulkarni RN. A somaclonal variant of rose-scented geranium (Pelargonium spp.) with moderately high

content of isomenthone in its essential oil. Nat Prod Commun 2012;7:1223-1224

63 Bigos M, Wasiela M, Kalemba D, Sienkiewicz M. Antimicrobial activity of geranium oil against clinical strains of *Staphylococcus aureus*. Molecules 2012;17:10276-10291

64 Džamić AM, Soković MD, Ristić MS, Grujić SM, Mileski KS, Marin PD. Chemical composition, antifungal and antioxidant activity of *Pelargonium graveolens* essential oil. J Appl Pharm Sci 2014;4:1-5

65 Verma RS, Verma RK, Yadav AK, Chauhan A. Changes in the essential oil composition of rose-scented geranium (*Pelargonium graveolens* L'Herit ex Ait.) due to date of transplanting under hill conditions of Uttarakhand. Indian J Nat Prod Res 2010;1:367-370

66 Andrade M, Cardoso MG, Batista LR, Freire JM, Nelson DL. Antimicrobial activity and chemical composition of essential oil of *Pelargonium odoratissimum*. Revista Brasileira de Farmacognosia/ Brazilian Journal of Pharmacognosy 2011;21:47-52

67 Murphy CC, Demirci B, Tabanca N, Ali A, Becnel JJ, Sampson BJ, Wedge DE, et al. Chemical composition of rose-scented *Pelargonium* essential oils and their biting deterrence and insecticidal activity. Planta Med 2012;78:88. DOI: 10.1055/s-0032-1307596. Data probably also presented in ref. 49

68 Gâlea C, Hancu G. Antimicrobial and antifungal activity of *Pelargonium roseum* essential oils. Adv Pharm Bull 2014;4(Supp2), 3 pages. doi: 10.5681/apb.2014.075

69 Wang M, Chittiboyina AG, Avonto C, Parcher JF, Khan IA. Comparison of current chemical and stereochemical tests for the identification and differentiation of *Pelargonium graveolens* L'Hér. (Geraniaceae) essential oils: analytical data for (-)-(1S,4R,5S)-guaia-6,9-diene and (-)-(7R,10S)-10-epi-γ-eudesmol. Rec Nat Prod 2014;8:360-372

70 Djeddi S, Djebile K, Hadjbourega G, Achour Z. Composition and anti-microbiological activity of the essential oil of *Pelargonium capitatum* L. (Geraniaceae) from Algeria. Amer-Euras J Sustain Agric 2009;3:1-5

71 Kabera J, Mugiraneza JP, Chalchat JC, Ugirinshuti V. Chemical composition and antimicrobial effect of the essential oil of *Pelargonium graveolens* (Geranium Rosat) grown in Butare (Rwanda) towards formulation of plant-based antibiotics. J Microbiol Res 2013;3(2):87-91

72 Tawfick MM, Gad AS. *In vitro* antimicrobial activities of some Egyptian plants' essential oils with medicinal applications. Am J Drug Discov Developm 2014;4:32-40

73 Rosato A, Piarulli M, Corbo F, Muraglia M, Carone A, Vitali ME, Vitali C. *In vitro* synergistic action of certain combinations of gentamicin and essential oils. Current Medicinal Chemistry 2010;17:3289-3295

74 Singh M, Singh S, Yaseen M. Standardization of planting time for optimum growth and oil production of geranium (*Pelargonium graveolens* L. Her.) under north Indian plains. J Spices Arom Crops 2008;17:247-250

75 Lawrence BM. Progress in essential oils. Perfum Flavor 2005;30(2):66-? (last page unknown)

76 Vernin G, Chakib S, Zamkotsian R-M, Vernin GMF, Larice J-L, Parkanyi C. Classification of geranium essential oils by chemometrics. Rivista Ital EPPOS 2002;34:319. Data cited in ref 75

77 Vernin GA, Chakib S, Vernin GMF, Zamkotsian RM. Geranium (*Pelargonium sp.*) essential oil from Reunion. minor compounds: acids, phenols, pyridines. J Essent Oil Res 2004;16:26-28

78 Saxena G, Banerjee S, Gupta R, Rahman L, Tyagi BR, Kumar S. et al. Composition of the essential oil of a new isomenthone-rich variant of geranium obtained from geraniol-rich cultivar of *Pelargonium* species. J Essent Oil Res 2004;16:85-88

79 Jalali-Heravi M, Zekavat B, Sereshti H. Characterization of essential oil components of Iranian geranium oil using gas chromatography–mass spectrometry combined with chemometric resolution techniques. J Chromatogr A 2006;1114:154-163

80 Anonymous. GC/MS analysis of geranium oil (Egyptian). Indian Perfum 2004;48:237. Data cited in ref. 82

81 Anonymous. GC/MS analysis of geranium oil (Chinese). Indian Perfum 2004;48:240. Data cited in ref. 82

82 Lawrence BM. Progress in essential oils. Perfum Flavor 2007;32(5): 40-? (last page unknown)

83 Reichling J, Harkenthal M, Saller R. Wirkung ausgewählter ätherischer Öle. Erfahrungsheilkunde 1999;6:357-366. Data cited in ref. 19

84 Dorman HFD, Surai P, Deans SG. *In vitro* antioxidant activity of a number of plant essential oils and phytoconstituents. J Essent Oil Res 2000;12:241-248

85 Jain N, Aggarwal KK, Syamasundar KV, Srivastava SK, Kumar S. Essential oil composition of geranium (*Pelargonium* sp.) from the plains of Northern India. Flavour Fragr J 2001;16:44-46

86 Motsa NM. Essential oil yield and composition of rose-scented geranium (*Pelargonium* sp.) as influenced by harvesting frequency and plant shoot age. Ph.D. Thesis, University of Pretoria, South Africa, 2006

87 Nardelli A, D'Hooge E, Drieghe J, Dooms M, Goossens A. Allergic contact dermatitis from fragrance components in specific topical pharmaceutical products in Belgium. Contact Dermatitis 2009;60:303-313

88 Starke JC. Photoallergy to sandalwood oil. Arch Dermatol 1967;96:62-63

89 Wetter DA, Yiannias JA, Prakash AV, Davis MD, Farmer SA, el-Azhary RA, et al. Results of patch testing to personal care product allergens in a standard series and a supplemental cosmetic series: an analysis of 945 patients from the Mayo Clinic Contact Dermatitis Group, 2000-2007. J Am Acad Dermatol 2010;63:789-798

90 Thomson KF, Wilkinson SM. Allergic contact dermatitis to plant extracts in patients with cosmetic dermatitis. Br J Dermatol 2000;142:84-88

91   Sugiura M, Hayakawa R, Kato Y, Sigiura K, Hashimoto R. Results of patch testing with lavender oil in Japan. Contact Dermatitis 2000;43:157-160

92   Bleasel N, Tate B, Rademaker M. Allergic contact dermatitis following exposure to essential oils. Australas J Dermatol 2002;43:211-213

93   Selvaag E, Holm J, Thune P. Allergic contact dermatitis in an aromatherapist with multiple sensitizations to essential oils. Contact Dermatitis 1995;33:354-355

94   Dharmagunawardena B, Takwale A, Sanders KJ, Cannan S, Roger A, Ilchyshyn A. Gas chromatography: an investigative tool in multiple allergies to essential oils. Contact Dermatitis 2002;47:288-292

95   Trattner A, David M, Lazarov A. Occupational contact dermatitis due to essential oils. Contact Dermatitis 2008;58:282-284

96   Rudzki E, Grzywa Z, Bruo WS. Sensitivity to 35 essential oils. Contact Dermatitis 1976;2:196-200

97   Boonchai W, Lamtharachai P, Sunthonpalin P. Occupational allergic contact dermatitis from essential oils in aromatherapists. Contact Dermatitis 2007;56:181-182

98   Nakayama H, Matsuo S, Hayakawa K, Takhashi K, Shigematsu T, Ota S. Pigmented cosmetic dermatitis. Int J Dermatol 1984;23:299-305

99   Nakayama H, Harada R, Toda M. Pigmented cosmetic dermatitis. Int J Dermatol 1976;15:673-675

100   Rudzki E, Grzywa Z. Allergy to perfume mixture. Contact Dermatitis 1986;15:115-116

101   Scheinman PL. Is it really fragrance free? Am J Cont Derm 1997;8:239-242

102   Wenk KS, Ehrlich AE. Fragrance series testing in eyelid dermatitis. Dermatitis 2012;23:22-26

103   De Mozzi P, Johnston GA. An outbreak of allergic contact dermatitis caused by citral in beauticians working in a health spa. Contact Dermatitis 2014;70:377-379

104   Cuesta L, Silvestre JF, Toledo F, Lucas A, Pérez-Crespo M, Ballester I. Fragrance contact allergy: a 4-year retrospective study. Contact Dermatitis 2010; 63:77-84

105   Nardelli A, Thijs L, Janssen K, Goossens A. *Rosa centifolia* in a 'non-scented' moisturizing body lotion as a cause of allergic contact dermatitis. Contact Dermatitis 2009;61:306-309

106   Thomson MA, Preston PW, Prais L, Foulds IS. Lime dermatitis from gin and tonic with a twist of lime. Contact Dermatitis 2007;56:114-115

107   Trattner A, David M. Patch testing with fine fragrances: comparison with fragrance mix, balsam of Peru and a fragrance series. Contact Dermatitis 2004;49:287-289

108   Larsen W, Nakayama H, Fischer T, Elsner P, Frosch P, Burrows D, et al. Fragrance contact dermatitis: a worldwide multicenter investigation (Part II). Contact Dermatitis 2001;44:344-346

109   Penchalaiah K, Handa S, Lakshmi SB, Sharma VK, Kumar B. Sensitizers commonly causing allergic contact dermatitis from cosmetics. Contact Dermatitis 2000;43:311-313

110   Manuel Brites M, Goncalo M, Figueiredo A. Contact allergy to fragrance mix—a 10-year study. Contact Dermatitis 2000;43:181-182

111   Chang Y-C, Maibach HI. Pseudo flautist's lip: allergic contact cheilitis from geraniol. Contact Dermatitis 1997;37:39

112   Nakayama H, Hanaoka H, Ohshiro A. Allergen Controlled System (ACS). Tokyo, Japan: Kanehara Shuppan, 1974:42. Data cited in ref. 113

113   Mitchell JC. Contact hypersensitivity to some perfume materials. Contact Dermatitis 1975;1:197-199

114   Juarez A, Goiriz R, Sanchez-Perez J, Garcia-Diez A. Disseminated allergic contact dermatitis after exposure to a topical medication containing geraniol. Dermatitis 2008;19:163

115   Srivastava PK, Bajaj AK. Ylang-ylang oil not an uncommon sensitizer in India. Indian J Dermatol 2014;59:200-201

116   Eun HC, Lee AY. Contact dermatitis due to Madecassol. Contact Dermatitis 1985;13:310-313

117   Goossens A, Merckx L. Allergic contact dermatitis from farnesol in a deodorant. Contact Dermatitis 1997;37:179-180

118   Meynadier JM, Meynadier J, Peyron JL, Peyron L. Formes cliniques des manifestations cutanées d'allergie aux parfums. Ann Dermatol Venereol 1986;113:31-39

Chapter 5.38   GINGER OIL

DEFINITION
Ginger oil (essential oil of ginger) is the essential oil obtained from the rhizomes of the ginger, *Zingiber officinale* Roscoe.

INCI NOMENCLATURE
Description/definition: Zingiber officinale root oil is the volatile oil obtained from the dried rhizomes of the ginger, *Zingiber officinale* L., *Zingiberaceae*

INCI name EU: Zingiber officinale root oil

INCI name USA: Zingiber officinale (ginger) root oil

CAS registry number(s): 8007-08-7; 84696-15-1

EINECS number(s): 283-634-2

ISO (INTERNATIONAL ORGANIZATION FOR STANDARDIZATION) STANDARD
ISO number: 16928

ISO name: Essential oil of ginger

Botanical origin: *Zingiber officinale* Roscoe

Parts of plant used: rhizome

ISO values: ISO values (minimum and maximum concentrations) are shown in Table 5.38.1.

THE PLANT, THE OIL, AND THEIR USES
Zingiber officinale is a tropical herbaceous plant, possibly native to India. It is widely grown as a commercial crop in south and south east Asia, tropical Africa (especially Sierra Leone and Nigeria), Latin America, the Caribbean (especially Jamaica) and Australia (www.kew.org). The major producers are India, China, Australia and Nigeria (6,18). *Zingiber officinale* is best known as the source of the pungent, aromatic spice called ginger, produced from the knotted and branched rhizome (underground stem, commonly called the 'root'), which is used to add flavor in cooking (12). Fresh and dried ginger very popular in culinary practice and are used in products such as jams, pickles, condiments, sauces, chutneys, candies, beverages and bakery products, and for flavoring tea (9,10,14,17).

Ginger has many applications in traditional medicine (e.g., in China, where ginger products are also recorded in the official Pharmacopoeia [14]), Ayurveda formulations and Arabic herbal traditions (10). Crude or processed ginger has been reported to exhibit anti-inflammatory, antipyretic, antiemetic (8,19), anti-tumor, analgesic, hepatoprotective, hypotensive, antioxidant, anti-hemorrhagic, anti-rheumatic, antifungal and anti-bacterial activities (6,9,14). The fresh or dried rhizome is used in oral or topical preparations to treat a variety of health disorders including rheumatoid arthritis, atherosclerosis, sprains, muscular aches and pains, nausea and vomiting associated with surgery and chemotherapy (8,19), vertigo, travel sickness, morning sickness, coughs, diarrhea, cramps, indigestion, loss of appetite, migraine headaches, ulcers, depression, impotence, painful menstrual periods, sinusitis, sore throats, fever, and flu (6,9,10,12,13,14).

Ginger essential oil is used to flavor ginger beer and ginger ale, and is commonly used as an ingredient in perfumery, cosmetics and pharmaceuticals, including in topical applications as an analgesic (www.kew.org; 6,12).

The oil is also employed by aromatherapists (56,57).

Ginger and *Zingiber officinale* have been extensively reviewed in refs. 20 (chemical constituents, traditional medical uses, pharmacological activities), 15 (technology, chemistry, and biological activities), 8 and 19 (prevention of [chemotherapy-induced] nausea and vomiting), and 16 (botany and horticulture).

CHEMICAL COMPOSITION
Ginger oil is a pale yellow to amber, clear mobile liquid which has a spicy, fresh citrusy and peppery odor. The yield of essential oil from the rhizomes of *Zingiber officinale* Roscoe generally varies from 2.0 to 4.0%, depending on the dryness of the material and the cultivar used. The main producing countries of this oil are China, India, Vietnam, Indonesia, Sri Lanka, Australia and Ecuador.

Literature data (up to November 25, 2014) on the chemical composition of ginger oils and unpublished analytical data from one of us (E.S.) are shown in Table 5.38.2 in alphabetical order. In ginger oils from various origins, over 295 chemicals have been identified. About 51% of these were found in a single reviewed publication only.

The major compounds found in ginger oils from different sources are shown in Table 5.38.3. They include (highest concentrations in any study given) geranial (44.3%), zingiberene (42.2%), camphene (30.8%), neral (26.5%), curcumene (22.1%), 1,8-cineole (18.0%), β-bisabolene (16.3%), β-phellandrene (16.0%), β-sesquiphellandrene (14.3%) and α-farnesene (11.5%). Well-known ingredients of ginger oils that were present in high concentrations (>7%) in one or two studies were elemol (31.1%), geraniol (7.3% and 14.5%), γ-terpinene (12.3%), α-terpineol (10.9%), (*E,E*)-α-farnesene (7.6% and 10.0%), β-farnesene (8.4% and 9.8%), 2-heptanone (9.7%), γ-muurolene (9.5%), geranyl acetate (8.3%) and γ-elemene (7.5%). Uncommon or rare constituents of ginger oils found in high concentrations (>7%) in single studies include cadinol (13.8%), bisabolene (10.6%), 9,12,15-octadecatrienal (9.1%), pentadecanoic acid (8.0%), and geranyl isobutyrate (7.0%).

Commercial oils
The ten chemicals that had the highest maximum concentrations in 41 commercial ginger essential oil samples (concentration ranges provided) are the following: α-zingiberene (17.0-32.4%), geranial (0.06-12.6%), camphene (0.3-12.1%), neral (0.3-10.2%), 2-heptanone (0.01-9.7%), β-sesquiphellandrene (4.5-9.1%), 1,8-cineole (0.4-8.9%), α-copaene (0.2-6.4%), β-bisabolene (5.1-6.3%) and curcumene (2.5-5.8%) (Erich Schmidt, unpublished analytical data).

Table 5.38.1 ISO values (%) for ginger oil[a]

Compound	CAS	Minimum	Maximum
α-Zingiberene	495-60-3	23.0	45.0
β-Sesquiphellandrene	20307-83-9	8.0	17.0
Curcumene (ar-; α-)	644-30-4	3.0	11.0
β-Bisabolene	495-61-4	2.5	9.0
α-Pinene	80-56-8	0.1	6.0
Geranial	141-27-5	0.1	3.5
Geraniol	106-24-1	0.1	3.5
Camphene	79-92-5	0.2	2.0
Neral	106-26-3	0.0	2.0
β-Elemene	33880-83-0	0.0	1.5

[a] ISO 16928 Essential oil of ginger ©ISO 2014; Geneva, Switzerland, www.iso.org

Additional analyses of ginger oils not discussed in Table 5.38.2 can be found in refs. 46-51 and 55; these can all be accessed on-line (sometimes as abstract only).

CONTACT ALLERGY/ALLERGIC CONTACT DERMATITIS

General

Contact allergy to and possible allergic contact dermatitis from ginger oil has been reported in one publication only. A false-positive patch test reaction due to the excited skin syndrome cannot be excluded.

Case reports

An aromatherapist had non-occupational contact dermatitis with allergies to multiple essential oils used at work, including ginger oil. The patient also reacted to geraniol, linalool, linalyl acetate, α-pinene, the fragrance mix and various other fragrance materials. Linalool and geraniol

Table 5.38.2 Constituents identified in ginger oils

Constituent	CAS	Percentage and range					
		A	B (5)	C (36)	D (13)	E (4)	F
Acorenone B	21653-33-8						0.3[c]
Agarospirol	1460-73-7						0.1[b]
α-Amorphene	20085-19-2						1.1[e]
Amyl acetate	628-63-7		tr				
Anethole	104-46-1						0.1[s]
(E)-Anethole	4180-23-8						3.7[n]
Aromadendrene	489-39-4					tr	0.1[b]; 0.5[z3]; 1.2[y]
allo-Aromadendrene	25246-27-9					1.1	0.1[i,q]; 0.5[r,z3]; 1.1[e]
Azulene	275-51-4						0.2[r,a2]
Benzaldehyde	100-52-7					tr	
Bergamal	106-72-9					tr	
Bergamotene							0.5[v]
α-Bergamotene	17699-05-7		1.9	0.5-3.0	2.3		0.1[r]
cis-α-Bergamotene	18252-46-5	0-0.5					0.1[c]; 0.4[e]
trans-α-Bergamotene	13474-59-4	0.05-0.6				0.1	0.07[b]; 0.4[z3]
(Z)-α-Bergamotol	88034-74-6		0.3	0.3-0.8	0.1		
(Z)-trans-α-Berga-motol	88034-74-6						
α-Bergamotyl acetate			0.2				
epi-Bicyclosesqui-phellandrene (1-)	54274-73-6						0.8[r]
Bisabolene	495-62-5						10.6[s]
(E)-α-Bisabolene	25532-79-0						0.7[h]
β-Bisabolene	495-61-4	5.1-6.3	6.8	1.3-7.2	2.5	11.2	2.8-6.2[f]; 11.5[p]; 11.6[y1]; 12.3[n]; 16.3[y2]
(E)-β-Bisabolene							0.2[q]
(E)-γ-Bisabolene	53585-13-0					0.2	0.2[c]; 0.4[b]
(Z)-γ-Bisabolene	13062-00-5						+[u]; 0.6[z3]
trans-α-Bisabolene (ep)oxide	111536-37-9		0.3	0.1-0.9	0.1		
α-Bisabolol	515-69-5						0.1[q]; 0.7[r,a2]; 1.5[k]
β-Bisabolol	15352-77-9		0.3	0.3-1.1	0.6		
Borneol (endo-)	507-70-0	0.4-1.2	0.5	0.1-0.2	0.7	0.2	3.1[i]; 3.2[q]; 3.9[m]; 4.0[h,j,k]; 5.6[x]; 5.8[w]
Bornyl acetate	76-49-3		0.4	0-0.4	tr		0.1[v]; 0.3[c,h,m]; 0.4[b,i]; 0.6[e,q]
β-Bourbonene	5208-59-3					tr	
γ-Bulgarene	68000-46-4						3.8[z5]
Bulnesol	22451-73-6						0.3[s,a2]

Table 5.38.2 Constituents identified in ginger oils (*continued*)

Constituent	CAS	Percentage and range					
		A	B (5)	C (36)	D (13)	E (4)	F
Butanol	35296-72-1						1.1[o]
But-2-en-4-one							0.1[a2,r]
Cadina-1,4-diene	29837-12-5					2.5	
(E)-Cadina-1(6),4-diene	931410-54-7						0.1[c]
α-Cadinene	24406-05-1						0.2[q]
γ-Cadinene	39029-41-9					1.1	1.5-4.5[f]; 1.8[k]; 3.8[z3]; 4.5[g]
δ-Cadinene	483-76-1	0.04-0.2		0.2-0.8	1.8		0.1[c]; 0.2[b]; 0.5[e,s]; 3.5[z7]
Cadinol	11070-72-7						3.0[v]; 13.8[y]
α-Cadinol	481-34-5						0.5[d]
epi-α-Cadinol	5937-11-1						0.6[j]; 2.2[s]
trans-Cadinol							+[u]
β-Calacorene	50277-34-4					0.1	
cis-Calamenene	72937-55-4					tr	
trans-Calamenene	73209-42-4					tr	0.3[e]
Calarene	17334-55-3						0.1[z]
Camphene	79-92-5	0.3-12.1	4.0	1.9-2.5	1.8	0.6	2.9-8.2[f]; 15.9[o]; 18.2[g]; 22.8[p]; 30.8[z6]
Camphene hydrate	465-31-6						0.08[b]; 0.1[q]; 1.9[o]
Camphor	76-22-2	tr-0.2	0.3	tr-0.5		tr	0.1[i,j]; 0.2[b,q]; 0.3[h]; 0.5[d]; 0.7[e]
δ3-Carene	13466-78-9	0.05-0.3		0.1-0.8		tr	0.02[i]; 0.03[b]; <0.05[c]; <0.1[h]; 0.1[v]
trans-3(10)-Caren-2-ol	6909-15-5						1.0[d]
Carvacrol	499-75-2					tr	
Carveol	99-48-9		0.1	tr-1.0			
(Z)-Carveol	1197-06-4					0.3	
trans-Carvone oxide	18383-49-8		0.6				0.4[s]
β-Caryophyllene	87-44-5	0.05-0.8	1.9			0.1	<0.01[b]; 0.1[h,r]; 0.8[s]; 1.4[z7]; 2.2[l]
Caryophyllene oxide	1139-30-6						0.5[v]; 0.8[d]
Cedrene	11028-42-5						0.1[d]; 0.7[v]
α-Cedrene	469-61-4						1.2[d]
β-Cedrene	546-28-1						0.1[r,a2]
Cedrene epoxide (α-)	29597-36-2						0.7[s,a2]; 0.8[v]
Cedr-8(15)-en-9-α-ol	13567-41-4						+[u]; 0.3[d]
Cedr-8(15)-en-10-ol	138117-22-3						+[u]
Cedr-8-en -13-ol	18319-35-2		0.1	0.1-0.4	0.2		
epi-α-Cedrenol			0.2	0.1-0.3	0.1		
Cedrol	77-53-2						1.3[l]; 1.5[s,a2]; 1.7[v]
1,8-Cineole	470-82-6	0.4-8.9	2.4	0-0.5	1.6	0.4	1.4-5.7[f]; 9.7[h]; 10.4[j]; 10.9[x]; 18.0[i]
Cinnamyl cinnamate	122-69-0			0-0.06			
Citronellal	106-23-0	0.2-0.6	0.1	tr-1.3	0.4	tr	0.4[q]; 0.5[h]; 1.2-4.8[f]; 1.9[v]; 3.9[g]
Citronellol	106-22-9	0.05-0.1		tr-0.1	0.7		0.7[c]; 0.9[z4]; 1.3[k]; 1.4[l]; 2.4[q]; 2.5[m]
Citronellyl acetate	150-84-5					tr	0.1[h]; 0.4[e]; 0.6[i,v]; 0.7[m]
Citronellyl butyrate	141-16-2						1.0[z2]
α-Copaene	3856-25-5	0.2-6.4	1.8	0-0.1		0.3	0.4[i]; 0.5[r]; 0.8[e]; 0.9[n]; 1.5[z7]; 1.6[t]
β-Copaene	18252-44-3						0.4[z3]
cis-α-Copaen-8-ol	58569-25-8						+[u]
Cryptone	500-02-7						0.1[c]; 0.6[e]
Cubebene	11012-64-9						0.4[v]
α-Cubebene	17699-14-8					tr	0.1[d]; 0.8[z3]
β-Cubebene	13744-15-5		2.4				
Cubenol	21284-22-0			0.2	0.4-0.9	0.2	0.4[h]; 0.8[v]; 1.0[d]
Cumene	98-82-8						0.1[v]
Curcumene	644-30-4	2.5-5.8	20.0	4.7-9.9	1.5	22.1	2.4-10.7[f]; 14.6[j]; 15.3[p]; 16.7[z5]; 19[y1]
Curcumenyl acetate	19431-85-7		0.1	0-0.4	tr		

364

Essential Oils: Contact Allergy and Chemical Composition

Table 5.38.2 Constituents identified in ginger oils (*continued*)

Constituent	CAS	Percentage and range					
		A	B (5)	C (36)	D (13)	E (4)	F
Cyclohexane, 2-ethe-nyl-1, 1-dimethyl-3-methylene-	95452-08-7						+[u]; 0.5[d]
Cycloisosativene						0.1	0.4[y]
Cyclosativene	22469-52-9						0.1[b,q]; 0.2[d,r]; 0.5[n]
p-Cymene	99-87-6			0.2-1.3	tr		0.1-2.8[f]; 0.3[z5]; 0.6[z4]; 0.9[z6]; 1.4[l]
p-Cymenene	1195-32-0	0.05-0.3					
Decanal	112-31-2	0.05-0.3				0.2	tr[z7]; 0.06[b]; 0.2[v]
(*E*)-2-Decenal	3913-81-3					tr	
Dehydrosabina ketone	147043-52-5					tr	
Dihydro-*cis*-α-copa-en-8-ol							+[u]
cis-Dimethoxycitral							+[x]
trans-Dimethoxycitral							5.0[x]
2,5-Dimethylbenzoic acid	610-72-0		0.2				
3,7-Dimethyl-1,3,7-octatriene	502-99-8						5.7[o]
3,7-Dimethyl-1-octene	4984-01-4						+[u]
3,4-Dimethylstyrene	27831-13-6						0.7[d]
Dodecanal	112-54-9					tr	1.0[s]
β-Elemene	33880-83-0	0.2-1.1	1.7	0.1-0.3	tr	0.4	0.6[r]; 0.7[j]; 1.0[v]; 1.1[e]; 1.2[z3]; 3.4[n]
γ-Elemene	29873-99-2		0.2	0.1-0.3	1.2	tr	0.07[b]; 0.9[z3]; 7.5[y]
δ-Elemene	20307-84-0	0.02-0.2	1.3	tr-0.8	0.8	tr	0.07[b]; 0.1[i]; 0.3[z3]
Elemol	639-99-6		0.2	0.8-1.2	0.4		0.9[h]; 1.0[e]; 1.1[s]; 1.4[z5]; 2.7[k]; 31.1[y]
β-Elemol	32142-08-8						+[u]
Epiglobulol	88728-58-9						+[u]; 0.4[r]; 1.3[d]; 2.0[v]
Epizonarene	41702-63-0						2.6[n]
4,5-Epoxycarane	27867-36-3		tr				
2,3-Epoxygeranial							0.2[a,e]
Eremophilene	10219-75-7						0.6[r]
Ethane	74-84-0						2.2[y]
2-Ethyl-5-methylfuran	1703-52-2					tr	
Eudesma-3,7(11)-diene	6813-21-4		0.2	tr-1.9	2.5		
Eudesmane	473-11-0						0.03-3.0[f]; 3.0[g]
Eudesm-7(11)-en-4-ol	473-04-1		0.1	0.1-0.7	0.7		
Eudesmol	51317-08-9						0.4[r]; 1.9[v]; 2.2[s]
α-Eudesmol	473-16-5		1.4				0.1[b]; 0.3[q]; 0.7[c]
β-Eudesmol	473-15-4		0.1				0.2[z3]; 0.3[b]; 0.4[q,r]; 1.0[c]; 1.5[e,v]; 2.0[h,i]
γ-Eudesmol	1209-71-8		0.5				0.1[q]
10-epi-γ-Eudesmol	15051-81-7						+[u]; 0.06[b]; 0.4[h]
τ-Eudesmol							0.4[d]
Farnesal	19317-11-4						0.3[d]
(*E,E*)-Farnesal	502-67-0						+[u]; 0.2[c]
(*Z,Z*)-Farnesal	3790-68-9						0.4[y]
α-Farnesene	502-61-4	3.8-5.7	3.5				0.3[l]; 4.3[y1]; 4.7[y]; 6.1[q]; 8.8[o]; 11.5[r]
(*E,E*)-α-Farnesene	502-61-4	0.2-0.4		2.7-7.0	4.4		2.4[k]; 4.3[m]; 4.6[h]; 5.5[b]; 7.6[c]; 10.0[z4]
β-Farnesene	502-60-3		3.3		3.0		0.5[r]; 0.6[h,j]; 1.5[z7]; 8.4[l]; 9.8[v]
(*E*)-β-Farnesene	18794-84-8			0.8-2.2			0.1[b]; 0.2[c]; 0.3[q,y1]; 0.5[i]; 0.7[e]
(*Z*)-β-Farnesene	28973-97-9						0.4[c]
Farnesene epoxide			0.2	0.3-0.8	0.1		
Farnesol	4602-84-0						1.9[v]
β-Farnesol	58181-76-3						5.8[l]
(*E,E*)-Farnesol	106-28-5		0.3	0.6-0.9	0.2		0.3[l]; 1.4[h]
(*E,E*)-α-Farnesol							4.1[k]
(*E,Z*)-Farnesol	3879-60-5		0.2				

Table 5.38.2 Constituents identified in ginger oils (*continued*)

Constituent	CAS	Percentage and range					
		A	B (5)	C (36)	D (13)	E (4)	F
(*Z,E*)-Farnesol	3790-71-4		0.6	1.0-1.8	0.4		0.9[s]
(*Z,Z*)-Farnesol	16106-95-9		0.3	0.5-1.2	0.3		
α-Fenchyl acetate	111821-74-0						0.1[r]
α-Fenchyl alcohol	512-13-0		0.2				
Geranial	141-27-5	0.06-12.6	8.5	5.9-20.1	16.5		5.9-19.0[f]; 17.8[p]; 18.3[j]; 25.9[c]; 44.3[m]
Geranic acid	459-80-3				0-0.1	1.0	0.1[q]
Geraniol	106-24-1	0.1-0.7	1.8	0.6-1.1	4.0		3.1[k]; 3.4[c]; 4.7[q]; 7.0[d]; 7.3[m]; 14.5[x]
Geranyl acetate	105-87-3	0.06-3.4	0.3	0.1	18.8		2.2[k,v]; 3.1[e]; 3.5[m]; 6.3[x]; 8.3[p]
Geranyl isobutyrate	2345-26-8						7.0[o]
Germacrene	28028-64-0						1.0[r]
Germacrene A	28387-44-2					0.2	
Germacrene B	15423-57-1					0.4	0.2[q]; 0.3[c]; 0.4[b]; 2.2[d]
Germacrene D	23986-74-5	0.6-2.3	2.3	0-0.5	2.5	0.9	0.7[m]; 1.0[b]; 1.6[y]; 2.7[l]; 4.2[z7]; 6.4[s]
Globulol	489-41-8						+[u]; 0.2[r]; 0.5[y]
β-Guaiacol			tr	tr-0.6	0.2		
α-Guaiene	3691-12-1			0.1-0.3	0.1		+[u]
Guaiol	489-86-1						0.6 [c,f]
β-Gurjunene	73464-47-8					tr	0.1[b]; 0.5[r]
2-Heptanol	543-49-7	0.07-0.2	0.2	0-tr	0.2	tr	0.08[i]; 0.1[c,v]; 0.3[d]; 0.4[q]; 3.0[j]
2-Heptanone	110-43-0	0.01-9.7	0.1			tr	0.03[b,i]; 0.1[l]
2-Heptyl acetate	5921-82-4						0.1[q]
Hexadecanoic acid	57-10-3						0.1[q]
Hexanal	66-25-1	0.04-3.7	0.1			0.2	0.1[b,z7]; 0.2[i]; 0.4[l]; 4.0[o]
Hexane	110-54-3						+[z2]
Hexanol	111-27-3		0.1				<0.1[z7]
2-Hexanone	591-78-6		tr			tr	
α-Himachalene	3853-83-6						0.2[h]
β-Himachalene	1461-03-6						0.1[d]
α-Humulene	6753-98-6					tr	0.7[s]
Isoacorone	6168-64-5						1.1[a,e]
Isoamyl alcohol	123-51-3						0.9[o]
Isoamyl methyl ketone	110-12-3						0.7 [v,a]
Isoborneol	124-76-5			0-0.1		0.8	0.1[h,q]; 1.0[s]
Isobornyl acetate	125-12-2					tr	
Isobornyl formate	1200-67-5						2.0[t]
Isocaryophyllene	118-65-0						2.2[v]
(*Z*)-Isocitral	72203-97-5	0.04-0.4					
Isogeranial	72203-98-6	0.1-0.6					
p-Isopropyl anisole	4132-48-3						+[z2]
Italicene	94535-52-1					0.2	
(*Z*)-Lanceol	859202-95-2		0.1	0.1-0.6	0.2		0.1[d]
Limonene	138-86-3	0.2-1.8	1.9	0-0.8	3.1	0.5	0.8-2.8[f]; 1.3[h]; 2.1[z5]; 2.8[g]; 3.3[o]; 3.5[s]
cis-Limonene epoxide	13837-75-7`						+[z2]
Limonene oxide	1195-92-2						0.3[y]
Linalool	78-70-6	0.1-1.1	tr	tr-0.5	1.0		1.6[m]; 1.7[l]; 1.8[h,w]; 2.5[j]; 2.6[o]; 4.8[x]
cis-Linalool oxide, furanoid	11063-77-7					tr	0.3[s]
trans-Linalool oxide	11063-78-8			tr-0.1			+[x]
trans-Linalool oxide acetate	56469-40-0						+[x]
Linalyl acetate	115-95-7	0.6-2.2	0.4	0-1.0		0.2	0.2[s]
Longicyclene	1137-12-8					tr	
Longifolenaldehyde	66537-42-6						0.5[r,a2]

Table 5.38.2 Constituents identified in ginger oils (*continued*)

Constituent	CAS	Percentage and range					
		A	B (5)	C (36)	D (13)	E (4)	F
trans-Longipinocar-veol	889109-69-7						0.1[d]
cis-p-Mentha-2,8-dienol	3886-78-0						1.8[d]
p-Menth-2-en-1-ol	619-62-5						2.9[y]
cis-p-Menth-2-en-1-ol	29803-82-5						<0.01[b]
Menthol	89-78-1						tr[27a]
Menthone	89-80-5						0.3[h]
2-Methoxy-1,7,7-tri-methylbicyclo[2.2.1] heptane	5331-32-8						0.1[d]
p-Methylaceto-phenone	122-00-9					tr	
2-Methyl-3-buten-2-ol	115-18-4						0.01[b]
2,3-bis(Methylene) bicyclo[3.2.1] octane	49826-54-2						2.1[o]
Methylheptenone	409-02-9	0.07-0.7					0.4[z2]; 1.2[j]
6-Methyl-5-hepten-2-ol	1569-60-4					tr	
6-Methyl-5-hepten-2-one	110-93-0		0.9	0-0.3	2.5	0.2	0.1[c,v]; 0.2[b]; 0.4[h]; 0.7[e]; 1.1[q]
Methyl nerate	1862-61-9					tr	
trans-Muurola-4(14), 5-diene	262352-87-4						0.9[c]
α-Muurolene	10208-80-7		1.2	tr-1.2	2.1	0.3	2.2[27]
γ-Muurolene	30021-74-0		1.4			0.7	0.5[c,e]; 1.8[z3]; 2.6[j]; 3.4[z7]; 9.5[i]
α-Muurolol	104245-48-9		0.2	0.2-0.9	0.2		
τ-Muurolol	19912-62-0						<0.1[d]
Myrcene	123-35-3	0.6-3.7	2.8	0.1-0.4	0.5	0.1	1.4[h]; 1.7[l]; 2.1[37]; 2.4[i]; 3.3[z6]; 7.7[o]
Myrtenal	564-94-3		0.6				0.4[h]; 2.8[d]
Myrtenol	515-00-4		0.2	0-tr		tr	0.1[h]; 0.2[q]
Neral	106-26-3	0.3-10.2	2.7	2.6-9.4	2.2		1.8-7.5[f]; 10.3[k]; 11.2[p]; 16.3[j]; 26.5[m]
Nerol	106-25-2	tr-0.4	1.3	0.4-1.5		0.1	0.1[j]; 0.2[b,r,z7]; 0.6[z3]
Neric acid	37349-29-4			0-0.5	0.4		
Nerolidol	7212-44-4						0.2[i]; 0.5[y]; 0.6[z6]; 2.1[l]; 4.4[o]; 6.0[v]
(*E*)-Nerolidol	40716-66-3	0.4-0.9	1.4	0.9-1.3	1.2	1.6	0.2-1.3[f]; 1.1[e,m]; 1.2[g,n]; 1.4[h]; 1.5[c]
(*Z*)-Nerolidol	3790-78-1			1.1-1.5	1.5		0.2[z7]; 0.5[z4]
Neryl acetate	141-12-8			0.5-2.9	0.9		1.0[h]; 0.2[i]; 0.9[n]
Nonanal	124-19-6					tr	tr[b]; 0.1[v]
2-Nonanol	628-99-9	0.2-0.7		tr-0.4			<0.1[z6]; 0.2[v]; 0.3[j]
2-Nonanone	821-55-6	0.06-0.2				tr	<0.05[c]; 0.1[h]; 0.4[q]
3-Nonen-2-one	14309-57-0					tr	
Nuciferal	25532-74-5						0.1[q]
(*E*)-Nuciferol	1786-15-8						0.1[d]
Ocimene	13877-91-3						6.4[l]
(*Z*)-Ocimene	27400-71-1						0.4[z3]
(*E*)-β-Ocimene	3779-61-1		1.3			tr	<0.01[b]
(*Z*)-β-Ocimene	3338-55-4					tr	<0.01[b]
9,12-Octadecadienal	26537-70-2						4.9[o]
9,12,15-Octadeca-trienal	26537-71-3						9.1[o]
n-Octanal	124-13-0	0.01-3.7				0.1	tr[c]; 0.1[b,q]; 0.2[d]
n-Octanol (1-)	111-87-5						tr[b]
Octenal	25447-69-2		tr	0.06-0.5			
1-Octen-3-ol	3391-86-4					tr	
α-Patchoulene	560-32-7						0.4[r,a2]
Pentadecanoic acid	1002-84-2						8.0[o]
2-Pentanone, 4-hy-droxy-4-methyl-	123-42-2						+[u]

Table 5.38.2 Constituents identified in ginger oils (*continued*)

Constituent	CAS	A	B (5)	C (36)	D (13)	E (4)	F
Pentenyl curcumene							+[z2]
Perillaldehyde	2111-75-3						0.4[n]
α-Phellandrene	99-83-2	0.04-0.5	1.8	0.2-1.5			0.4[b]; 0.7[d]; 0.8[h,n]; 1.4[z4]; 3.9[o]
β-Phellandrene	555-10-2	0.5-5.7	1.8	0.2-2.5	1.8	0.1	1.9-7.6[f]; 7.7[z3]; 8.0[b]; 8.3[g]; 16.0[d]
α-Pinene	80-56-8	1.7-3.7	0.5	0.8-1.2	2.5	0.2	0.7-6.0[f]; 4.7[i]; 6.0[g]; 7.2[z6]; 7.4[p]
β-Pinene	127-91-3	0.2-1.6	1.6			tr	0.5[h,l,z3]; 0.9-3.4[f]; 1.0[z6]; 3.4[f]; 3.9[d]
cis-Piperitol	16721-38-3						tr[q]
Rosefuran	15186-51-3					0.2	
Sabinene	3387-41-5	0.03-0.4	3.0	1-1.6	2.0	tr	<0.05[c]; 0.1[b,i]; 0.2[d,h,z6]; 1.4[l]; 3.3[r]
Sabinene hydrate	546-79-2						0.1[q]
cis-Sabinene hydrate	15537-55-0						<0.01[b]
trans-Sabinene hydrate	17699-16-0						0.1[h]
Sabinol				tr-0.2			
α-Selinene	473-13-2						1.4[v]; 1.5[e]
7-epi-α-Selinene	123123-37-5						0.3[c]; 1.2[i]
β-Selinene	17066-67-0					1.7	0.1[q]
δ-Selinene	473-14-3						0.6[c]
Selin-11-en-4α-ol	16641-47-7						0.3[r]
β-Sesquiphellandrene	20307-83-9	4.5-9.1	6.6	6.2-10.9	3.4	10.5	4.7-12.4[f]; 11.4[s]; 13.1[y1]; 13.5[p]; 14.3[r]
cis-β-Sesquiphellandrol	56144-26-4						0.6[k]
trans-β-Sesquiphellandrol	56144-27-5						2.0[k]
Sesquisabinene hydrate	139341-65-4			tr-0.9	0.6	·	0.5[h]; 7.8[v]
(*E*)-Sesquisabinene hydrate	145512-84-1						0.4[b]; 0.7[c]; 1.4[e,k]
Sesquithujene	58319-06-5	0.06-0.2					0.5[z3]
7-epi-Sesquithujene	159407-35-9						0.1[q]
Seychellene	20085-93-2						+[u]
β-Sinensal (*cis*-)	17909-87-4						0.7[v]
Spathulenol	6750-60-3						1.2[d]
α-Terpinene	99-86-5	0.03-0.5		0.4	1.6	tr	<0.01[b]; <0.1[h]; 0.4[n]
γ-Terpinene	99-85-4	0.02-0.3	0.8	tr-0.8		tr	<0.01[b]; 0.04[i]; 0.5[n]; 0.6[z4]; 12.3[z6]
Terpinen-4-ol	562-74-3	0.06-0.3	0.2	tr-0.2			0.2[j,q]; 0.3[e]; 0.4[i]; 0.6[h]; 1.9[l]; 3.7[d]
1-Terpineol	586-82-3					tr	
α-Terpineol	98-55-5	0.1-0.6	1.3	0.5	0.8	0.5	1.2[e]; 1.8[i]; 2.2[m]; 2.7[j]; 3.3[k]; 10.9[o]
(*Z*)-β-Terpineol	7299-40-3					tr	
Terpinolene	586-62-9	0.1-0.5	0.2			tr	0.1[c]; 0.2[b,q]; 0.5[i]; 0.7[h]; 1.0[v]
α-Terpinyl acetate	80-26-2		tr				0.1[s]; 0.4[r]
α-Thujene	2867-05-2						<0.01[b]; 0.1[s]; 0.2[h]
Thujol	35732-37-7		0.1				
β-Thujone (*cis*-)	471-15-8					tr	
Thujopsene	470-40-6			0.1-0.3			1.2[s,a2]
Toluene	108-88-3	0.01-0.2	0.1		0.2		
Tricyclene	508-32-7	0.1-0.6	tr	0-0.1		tr	<0.05[c]; 0.1[b,e]; 0.2[h]; 0.3[i,z3]; 0.4[d]
2-Tridecanone	593-08-8						+[z2]
ar-Turmerol	38142-57-3						0.9[e]
β-Turmerol							0.3[h]
Undecanal	112-44-7		0.2	0-0.2	0.5		
Undecane	1120-21-4		0.4				
2-Undecanol	1653-30-1						0.1[q]
2-Undecanone	112-12-9	0.5-0.8	0.1	tr-0.7	0.2	tr	0.6[d]; 0.7[q]; 0.8[j]; 0.9[m]; 1.3[k]; 2.4[g]

Table 5.38.2 Constituents identified in ginger oils (*continued*)

Constituent	CAS	Percentage and range					
		A	B (5)	C (36)	D (13)	E (4)	F
Valencene	4630-07-3						+[u]; 1.4[b]
cis-Verbenol	1845-30-3		tr				
Viridiflorene (ledene)	21747-46-6						1.7[h]
Viridiflorol	552-02-3						+[u]
o-Xylene	95-47-6		tr				<0.1[z7]
α-Ylangene	14912-44-8					tr	
β-Ylangene	20479-06-5						0.7[z3]
Zingiberene	495-60-3	17.0-32.4	29.3	5.8-32.2	19.8	11.7	9.8-29.9[f]; 29.2[g]; 35.6[v]; 39.3[r]; 42.2[p]
Zingiberenol	58334-55-7		0.1				1.0[z6]; 1.3[e]; 1.6[v]; 1.7[c]; 2.7[k]; 3.2[y]
epi-Zingiberenol	72346-47-5						+[x]
Zing(ib)erone	122-48-5		0.6	0.1-0.6	0.3		1.9[k]

A forty-one ginger essential oil samples from China, India, Sri Lanka, Ecuador, Vietnam and Indonesia analyzed between 1998 and 2013; lowest and highest concentrations given (E. Schmidt, unpublished data)
B three lab-hydrodistilled oils obtained from the same sample of ginger from a garden in India: fresh, sundried and oven-dried rhizomes of *Z. officinale*; highest concentrations given (ref. 5)
C three lab-hydrodistilled ginger essential oils from ginger cultivars from sub Himalayan India; lowest and highest concentrations given (ref. 36)
D two lab-hydrodistilled oils from two *Z. officinale* cultivars obtained from the agricultural department of Sikkim, India; highest concentrations given (ref. 13)
E one oil obtained by lab-hydrodistillation of ginger rhizomes from a commercial plantation near Havana, Cuba (ref. 4)
F data from other studies (indicated with superscript letters); highest concentrations found in any study reviewed here given; when two or more oils were investigated, only the highest concentrations are mentioned, unless indicated otherwise

[a] incorrect identification (ref. 25); [a2] incorrect identification (ref. 24); [b] one lab-hydrodistilled oil from fresh ginger rhizomes purchased at a local market in India (ref. 17); [c] one lab-hydrodistilled oil from ginger purchased at a local market in India (ref. 7); [d] one lab-hydro-distilled oil from ginger purchased at a local market in Shanghai, China (ref. 14); [e] two oils obtained by steam- and hydrodistillation from Vietnamese crop ginger rhizome (ref. 35); [f] lab-hydrodistilled oils from 17 *Z. officinale* cultivars from different locations in northeast India; lowest and highest concentrations given (ref. 6); [g] lab-hydrodistilled oils from 10 cultivars from different parts of India (ref. 12); [h] oils produced from ginger rhizomes cultivated in two locations in Sao Tomé (ref. 29); [i] two oils produced from fresh and dried Chinese ginger (ref. 30); [j] one lab-hydrodistilled oil from ginger rhizome from a local market in the Central African Republic (ref. 2); [k] one lab-hydrodistilled oil from ginger rhizomes cultivated in Mauritius (ref. 3); [l] four oils from ginger rhizomes purchased in three different places in India (ref. 1); [m] steam-distilled oils from rhizomes of 17 clones of Australian ginger (ref. 11); [n] two oils prepared with hydrodistillation and microwave-assisted hydrodistillation from dried ginger purchased in China (ref. 22); [o] two oils from fresh and dried ginger purchased at a local market in Pakistan (ref. 10); [p] two lab-distilled and three commercial ginger oils, partly of unknown origin (ref. 31); [q] one lab-hydrodistilled oil from fresh Indian ginger rhizomes (ref. 32); [r] one lab-hydrodistilled fresh ginger oil from India (ref. 33); [s] oils from rhizomes of *Z. officinale* from three locations in India (ref. 34); [t] one lab-hydrodistilled oil from a producer of ginger rhizomes in Brazil (ref. 9); [u] six oils from one sample of ginger purchased at a local market in India that had undergone different drying procedures; amounts were presented in µl per 100 gr dry weight and are not mentioned, as they cannot be compared to the other data; their presence in the oil(s) is indicated with +[u] (ref. 18); [v] one Indian ginger oil (ref. 26); [w] lab-distilled ginger rhizome oils from three Indian ginger cultivars (ref. 27); [x] one lab-distilled oil from Nahan, India (ref. 23); [y] one lab-hydrodistilled oil from China and one from Guinea (ref. 37); [y1] one commercial ginger oil purchased from a Brazilian company (ref. 52); [y2] one commercial ginger oil sample purchased in Poland (ref. 53); [z] oils prepared from rhizomes of *Z. officinale* cultivated in Japan (ref. 21); [z2] various studies reviewed in ref. 38; only the chemicals that have not been found in the other studies reviewed here are mentioned (ref. 38); [z3] one commercial steam-distilled oil from Ecuadorian ginger rhizomes (ref. 39); data are also presented in ref. 40; [z4] one lab-hydrodistilled oil from fresh Brazilian ginger rhizomes (ref. 41); [z5] one lab-hydrodistilled oil from ginger rhizomes growing wild in Burkina Fasso (ref. 42); [z6] one ginger essential oil sample produced commercially in Madagascar (ref. 43); [z7] two lab-hydrodistilled oils from fresh and dried Indian ginger rhizomes (ref. 44); the results of the oil from fresh material were already presented in ref. 5

tr: trace (in columns B, C, D and E: <0.1); + present in the oil investigated, but quantity not stated or expressed in a manner which cannot be compared to the other data

Table 5.38.3 Major constituents of ginger oils

Constituent	CAS	Percentage and range					
		A	B (5)	C (36)	D (13)	E (4)	F
Geranial	141-27-5	0.06-12.6	8.5	5.9-20.1	16.5		5.9-19.0[f]; 17.8[p]; 18.3[j]; 25.9[c]; 44.3[m]
Zingiberene	495-60-3	17.0-32.4	29.3	5.8-32.2	19.8	11.7	9.8-29.9[f]; 29.2[g]; 35.6[v]; 39.3[r]; 42.2[p]
Camphene	79-92-5	0.3-12.1	4.0	1.9-2.5	1.8	0.6	2.9-8.2[f]; 15.9[o]; 18.2[g]; 22.8[p]; 30.8[z6]
Neral	106-26-3	0.3-10.2	2.7	2.6-9.4	2.2		1.8-7.5[f]; 10.3[k]; 11.2[p]; 16.3[j]; 26.5[m]
Curcumene	644-30-4	2.5-5.8	20.0	4.7-9.9	1.5	22.1	2.4-10.7[f]; 14.6[j]; 15.3[p]; 16.7[z5]; 19[y1]
1,8-Cineole	470-82-6	0.4-8.9	2.4	0-0.5	1.6	0.4	1.4-5.7[f]; 9.7[h]; 10.4[j]; 10.9[x]; 18.0[i]
β-Bisabolene	495-61-4	5.1-6.3	6.8	1.3-7.2	2.5	11.2	2.8-6.2[f]; 11.5[p]; 11.6[y1]; 12.3[n]; 16.3[y2]
β-Phellandrene	555-10-2	0.5-5.7	1.8	0.2-2.5	1.8	0.1	1.9-7.6[f]; 7.7[z3]; 8.0[b]; 8.3[g]; 16.0[d]
β-Sesquiphellandrene	20307-83-9	4.5-9.1	6.6	6.2-10.9	3.4	10.5	4.7-12.4[f]; 11.4[s]; 13.1[y1]; 13.5[p]; 14.3[r]
α-Farnesene	502-61-4	3.8-5.7	3.5				0.3[j]; 4.3[y1]; 4.7[v]; 6.1[q]; 8.8[o]; 11.5[r]
Myrcene	123-35-3	0.6-3.7	2.8	0.1-0.4	0.5	0.1	1.4[h]; 1.7[j]; 2.1[37]; 2.4[i]; 3.3[z6]; 7.7[o]
α-Pinene	80-56-8	1.7-3.7	0.5	0.8-1.2	2.5	0.2	0.7-6.0[f]; 4.7[i]; 6.0[g]; 7.2[z6]; 7.4[p]

LEGEND: SEE UNDER TABLE 5.38.2

were demonstrated by GC-MS in the ginger oil (58), but neither of these is an important component of commercial ginger oils (Table 5.38.2, column A).

LITERATURE

1 Raina VK, Kumar A, Aggarwal KK. Essential oil composition of ginger (*Zingiber officinale* Roscoe) rhizomes from different places in India. J Essent Oil Bear Plants 2005;8:187-191

2 Menut C, Lamaty G, Bessière J-M, Koudou J. Aromatic plants of tropical Central Africa. XIII. Rhizomes volatile components of two Zingiberales from the Central African Republic. J Essent Oil Res 1994;6:161-164

3 Gurib-Fakim A, Maudarbaccus N, Leach D, Doimo L, Wohlmuth H. Essential oil composition of Zingiberaceae species from Mauritius. J Essent Oil Res 2002;14:271-273

4 Pino JA, Marbot R, Rosado A, Batista A. Chemical composition of the essential oil of *Zingiber officinale* Roscoe L. from Cuba. J Essent Oil Res 2004;16:186-188

5 Nirmala Menon A, Padmakumari KP, Sankari Kutty B, Sumathikutty MA, Sreekumar MM, Arumugham C. Effects of processing on the flavor compounds of Indian fresh ginger (*Zingiber Officinale* Rosc.). J Essent Oil Res 2007;19:105-109. Data also presented in ref. 44

6 Kiran CR, Kumar Chakka A, Padmakumari Amma KP, Nirmala Menon A, Sree Kumar MM, Venugopalan VV. Essential oil composition of fresh ginger cultivars from North-East India. J Essent Oil Res 2013;25:380-387

7 Singh G, Kapoor IPS, Singh P, de Heluani CS, de Lampasona MP, Catalan CAN. Chemistry, antioxidant and antimicrobial investigations on essential oil and oleoresins of *Zingiber officinale*. Food Chem Toxicol 2008;46:3295-3302

8 Palatty P, Haniadka R, Valder B, Arora R, Baliga MS. Ginger in the prevention of nausea and vomiting: A review. Crit Rev Food Sci Nutrit 2013;53:659-669

9 Mesomo MC, Corazza ML, Ndiaye PM, Dalla Santa OR, Cardozo L, de Paula Scheer A. Supercritical CO2 extracts and essential oil of ginger (*Zingiber officinale* R.): Chemical composition and antibacterial activity. J Supercrit Fluids 2013;80:44-49

10 El-Ghorab AH, Nauman M, Anjum FM, Hussain S, Nadeem M. A comparative study on chemical composition and antioxidant activity of ginger (*Zingiber officinale*) and cumin (*Cuminum cyminum*). J Agric Food Chem 2010;58:8231-8237

11 Wohlmuth H, Smith MK, Brooks LO, Myers SP, Leach DN. Essential oil composition of diploid and tetraploid clones of ginger (*Zingiber officinale* Roscoe) grown in Australia. J Agric Food Chem 2006;54:1414-1419

12 Kiran CR, Chakka AK, Padmakumari Amma KP, Nirmala Menon A, Sree Kumar MM, Venugopalan VV. Influence of cultivar and maturity at harvest on the essential oil composition, oleoresin and [6]-gingerol contents in fresh ginger from Northeast India. J Agric Food Chem 2013;61:4145-4154

13 Sasidharan I, Venugopal VV, Nirmala Menon A. Essential oil composition of two unique ginger (*Zingiber officinale* Roscoe) cultivars from Sikkim. Nat Prod Res 2012;26:1759-1764

14 Huang B, Wang G, Chu Z, Qin L. Effect of oven drying, microwave drying, and silica gel drying methods on the volatile components of ginger (*Zingiber officinale* Roscoe) by HS-SPME-GC-MS. Drying Technol 2012;30:248-255

15 Kubra IR, Mohan Rao LJ. An Impression on current developments in the technology, chemistry, and biological activities of ginger (*Zingiber officinale* Roscoe). Crit Rev Food Sci Nutrit 2012;52:651-688

16 Parthasarathy VA, Srinivasan V, Nair RR, Zachariah TJ, Kumar A, Prasath D. Ginger: botany and horticulture. In: J Janick, Ed. Horticultural Reviews, Volume 39, First Edition. Wiley-Blackwell, 2012:273-388

17 Singh G, Maurya S, Catalan C, de Lampasona MP. Studies on essential oils, Part 42: chemical,

antifungal, antioxidant and sprout suppressant studies on ginger essential oil and its oleoresin. Flavour Fragr J 2005;20:1-6

18 Kubra IR, Rao LJM. Effect of microwave drying on the phytochemical composition of volatiles of ginger. Int J Food Sci Technol 2012;47:53-60

19 Marx WM, Teleni L, McCarthy AL, Vitetta L, McKavanagh D, Thomson D, Isenring E. Ginger (*Zingiber officinale*) and chemotherapy-induced nausea and vomiting: a systematic literature review. Nutrit Rev 2013;71:245-254

20 Ross IA. *Zingiber officinale* Roscoe. In: Medicinal plants of the world, vol. 3: Chemical constituents, traditional and modern medicinal uses. Totowa, NJ, USA: Humana Press Inc, 2005:507-560

21 Sakamura F. Changes in volatile constituents of Zingiber officinale rhizomes during storage and cultivation. Phytochem 1987;26:2207-2212

22 Wang Z, Wang L, Li T, Zhou X, Ding L, Yu Y, et al. Rapid analysis of the essential oils from dried *Illicium verum* Hook. f. and *Zingiber officinale* Rosc. by improved solvent-free microwave extraction with three types of microwave-absorption medium. Anal Bioanal Chem 2006;386:1863-1868

23 Gupta S, Pandotra P, Ram G, Anand R, Gupta AP, Husain K, et al. Composition of a monoterpenoid-rich essential oil from the rhizome of *Zingiber officinale* from north western Himalayas. Nat Prod Commun 2011;6:93-96

24 Lawrence BM. Progress in essential oils. Perfum Flavor 2012;37(Dec):42-52

25 Lawrence BM. Progress in essential oils. Perfum Flavor 2008;33(Jan):38-? (last page unknown)

26 Ranade GS. Essential oil profile. Ginger oil. FAFAI 2002;4:73. Data cited in ref. 25

27 Ahmad A, Naqvi AA, Aggarwal KK. Essential oil composition of different rhizomes of ginger (*Zingiber officinale* Rosc.) by GC-MS. FAFAI 2002;4:39-41. Data cited in ref. 25

28 Lawrence BM. Progress in essential oils. Perfum Flavor 2005;30(May):70-? (last page unknown)

29 Martins AP, Salgueiro L, Gonçalves MJ, Proença da Cunha A, Vila R, Canigueral S, et al. Essential oil composition and antimicrobial activity of three Zingiberaceae from S. Tomé Principe. Planta Med 2001;67:580-584. Data cited in ref. 28

30 Li J-P, Wang Y-S, Ma H, Hao J-D, Yang H. Comparison of chemical components between dry and fresh *Zingiber officinale*. Zhongguo Zhongyao Zazhi 2001;26:748-751. Data cited in ref. 28

31 Koroch A, Ranarivelo L, Behra O, Juliani HR, Simon JE. Quality attributes of ginger and cinnamon essential oils from Madagascar. In: J Janick and A Whipkey, Eds. Issues in new crops and new uses. Alexandria, VA, USA: ASHS Press, 2007:338-341. Data cited in ref. 24

32 Rana VS, Verdeguer M, Blazquez MA. A comparative study on the rhizome essential oils of three *Zingiber* species from Manipur. Indian Perfum 2008;52:17-21. Data cited in ref. 24

33 Omanakutty M, Joy B. Cold grinding for retention of fresh flavor components in ginger oil. Indian Perfum 2009;53:35-37. Data cited in ref. 24

34 Padmakumari KP, Sreekumar MM, Sankarakutty B. Composition of volatile oil of ginger (*Zingiber officinale* Roscoe) varieties from Manipur. Indian Perfum 2009;53:16-20. Data cited in ref. 24

35 Stoyanova A, Konakchiev S, Damyanova S, Stoilova I, Suu PT. Composition and antimicrobial activity of ginger essential oil from Vietnam. J Essent Oil Bear Plants 2006;9:93-98

36 Nampoothiri SV, Venugopalan VV, Joy B, Sreekumar MM, Nirmala Menon A. Comparison of essential oil composition of three ginger cultivars from sub Himalayan region. As Pac J Trop Biomed 2012;2(Suppl.):S1347-S1350

37 Toure A, Zhang X. Gas chromatographic analysis of volatile components of Guinean and Chinese ginger oils (*Zingiber officinale*) extracted by steam distillation. J Agron 2007;6:350-355

38 Lawrence BM. Progress in essential oils. Perfum Flavor 1995;20(March-April):49-59

39 Tognolini M, Barocelli E, Ballabeni V, Bruni R, Bianchi A, Chiavarini M, Impicciatore M. Comparative screening of plant essential oils: Phenylpropanoid moiety as basic core for antiplatelet activity. Life Sciences 2006;78:1419-1432. Data also presented in ref. 40 and ref. 54

40 Sacchetti G, Maietti S, Muzzoli M, Scaglanti M, Manfredini S, Radice M, Bruni R. Comparative evaluation of 11 essential oils of different origin as functional antioxidants, antiradicals and antimicrobials in foods. Food Chem 2005;91:621-632. Data also presented in ref. 39 and ref. 54

41 Garcia Yamamoto-Ribeiro MM, Grespan R, Kohiyama CY, Dias Ferreira F, Aparecida Galerani Mossini S, Leite Silva E, et al. Effect of *Zingiber officinale* essential oil on *Fusarium verticillioides* and fumonisin production. Food Chem 2013;141:3147-3152

42 Bayala B, Bassole IHN, Gnoula C, Nebie R, Yonli A, Morel L, et al. Chemical composition, antioxidant, anti-inflammatory and anti-proliferative activities of essential oils of plants from Burkina Faso. PLoS ONE 2014;9(3): e92122. doi:10.1371/journal.pone.0092122

43 Möllenbeck S, König T, Schreier P, Schwab W, Rajaonarivony J, Ranarivelo L. Chemical composition and analyses of enantiomers of essential oils from Madagascar. Flavour Fragr J 1997;12:63-69

44 Sasidharan I, Nirmala Menon A. Comparative chemical composition and antimicrobial activity fresh & dry ginger oils (*Zingiber officinale* Roscoe). Int J Curr Pharm Res 2010;2(4):40-43. Data partly also presented in ref. 5

45 Tang J, Li X-R, Han J. Analysis of volatile components in rhizome zingibers, *Zingiber officinale* Roscoe and ginger peel by GC-MS and chemometric resolution. Journal of Chinese Medicine Research and Development (JCNRD) 2012;1(2):47-53

46 Nour AH, Ranitha M, Nour AH. Extraction and characterization of essential oil from from ginger (*Zingiber officinale* Roscoe) and lemongrass (*Cymbopogon citratus*) by microwave-assisted hydrodistillation. Int J Chem Environ Engin 2013;4:221-226

47 Dieumou FE, Teguia A, Kuiate JR, Tamokou JD, Fonge NB, Dongmo MC. Effects of ginger (*Zingiber officinale*) and garlic (*Allium sativum*) essential oils on growth performance and gut microbial population of broiler chickens. Livestock Res Rural Developm 2009;21:Article #131. http://www.lrrd.org/lrrd21/8/dieu21131.htm

48 Kizhakkayil J, Sasikumar B. Characterization of ginger (*Zingiber officinale* Rosc.) germplasm based on volatile and non-volatile components. Afr J Biotechnol 2012;11:777-786

49 Barman KL, Kumar Jha D. Comparative chemical constituents and antimicrobial activity of normal and organic ginger oils (*Zingiber officinale* Roscoe). Int J Appl Biol Pharm Technol 2013;4:259-266

50 Emmanuel T, Aristide B, Leopold T, Benoît NM, Joseph MT. Phytochemical screening, chemical composition and antimicrobial activity of *Zingiber officinale* essential oil of Adamaoua region (Cameroon). J Chem Pharm Res 2013;5:296-301

51 Philippe S, Souaïbou F, Jean-Pierre N, Brice F, Paulin A, Issaka Y, Dominique S. Chemical composition and in vitro antifungal activity of *Zingiber officinale* essential oil against foodborne pathogens isolated from a traditional cheese wagashi produced in Benin. Int J Biosci (IJB) 2012;2:20-28

52 Murbach Teles Andrade BF, Nunes Barbosa L, da Silva Probst I, Fernandes Júnior A. Antimicrobial activity of essential oils. J Essent Oil Res 2014;26:34-40

53 Golebiowski M, Ostrowski B, Paszkiewicz M, Czerwicka M, Kumirska J, Halinski L, et al. Chemical composition of commercially available essential oils from blackcurrant, ginger, and peppermint. Chem Nat Comp 2008;44:794-796

54 Rolli E, Marieschi M, Maietti S, Sacchetti M, Bruni R. Comparative phytotoxicity of 25 essential oils on pre- and post-emergence development of *Solanum lycopersicum* L.: A multivariate approach. Ind Crops Prod 2014;60:280-290. Data also presented in ref. 39 and ref. 40

55 Hassanpouraghdam MB, Aazami MA, Shalamzari MS, Baneh HD. Essential oil composition of *Zingiber officinale* Rosc. rhizome from the Iranian herb market. Chemija 2011;22:56-59

56 Rhind JP. Essential oils. A handbook for aromatherapy practice, 2nd Edition. London: Singing Dragon, 2012

57 Lawless J. The encyclopedia of essential oils, 2nd Edition. London: Harper Thorsons, 2014

58 Dharmagunawardena B, Takwale A, Sanders KJ, Cannan S, Roger A, Ilchyshyn A. Gas chromatography: an investigative tool in multiple allergies to essential oils. Contact Dermatitis 2002;47:288-292

Chapter 5.39 GRAPEFRUIT OIL

DEFINITION

Grapefruit oil (essential oil of grapefruit) is the essential oil obtained from the pericarp (peel) of the grapefruit, *Citrus paradisi* Macfad.

INCI NOMENCLATURE

Description/definition: Citrus paradisi peel oil is the volatile oil expressed from the peel of the grapefruit, *Citrus paradisi*, Rutaceae

INCI name EU: Citrus paradisi peel oil

INCI name USA: Citrus paradisi (grapefruit) peel oil

CAS registry number(s): 8016-20-4; 90045-43-5

EINECS number(s): 289-904-6

ISO (INTERNATIONAL ORGANIZATION FOR STANDARDIZATION) STANDARD

ISO number: 3053

ISO name: Essential oil of grapefruit

Botanical origin: *Citrus paradisi* Macfad.

Parts of plant used: Pericarp (peel)

ISO values: ISO values (minimum and maximum concentrations) are shown in Table 5.39.1.

By ISO definition, all citrus essential oils except lime oil are produced by expression; oils obtained from citrus fruits by distillation may not be called essential oils according to ISO criteria (except lime oil). In industry, however, sometimes residues from expression of the citrus peels undergo steam distillation, to obtain the remaining oil; these volatile oils may then be added to the expressed oil. As some of the compounds undergo changes forced by high temperature during distillation, this addition (which is an adulteration) changes the composition of the essential oil. Because of this, and also because it cannot be excluded that oils entirely produced by hydrodistillation reach the market, both the ingredients found in 'genuine' grapefruit essential oils obtained by expression as those that may be present in hydrodistilled oils are presented here.

THE PLANT, THE OIL, AND THEIR USES

Citrus paradisi Macfad. is regarded as the result of a spontaneous crossing of *Citrus maxima* (Burm.) Merrill (pumelo, syn. *Citrus grandis* Osbeck) and *Citrus sinensis* (L.) Osbeck (sweet orange). The hybrid probably originated on the West Indies (Barbados, Jamaica), where it was reported growing around 1750 (25). It is a tree, 5-9 m tall, with spreading branches, and sometimes spines which are usually blunt; it produces fruits 10-15 centimeters in diameter. Commercial grapefruit varieties were mainly developed in Florida, and several types are currently cultivated in many tropical and subtropical regions of warm and humid climates. The main producing countries are the USA (Florida, to a lesser extent the west coast), Israel, Mexico, Cuba, Argentina, Morocco, Brazil, the Philippines, the Caribbean Islands, and South Africa (25, Mansfeld's World Database of Agriculture and Horticultural Crops, http://mansfeld.ipk-gatersleben.de; www.eFloras.org). There are two natural groups of grapefruits depending upon the color of the flesh: the white (common) types, including the Duncan, Marsh seedless (White Marsh), and Triumph, and the pigmented types, including redblush, Thompson, and Foster pink (25).

Grapefruits are know for their pleasant and distinctive flavor, which makes them particularly popular for breakfast fruit, juice, salad fruit, or dessert. The juice is rich in potassium, and the seed extract is antimicrobial, particularly against molds (25). The peel of the fruit is utilized candied or exploited as a source of pectin. The fruits, flowers and leaves are also used medicinally (Mansfeld's World Database of Agriculture and Horticultural Crops; www.eFloras.org).

From the byproduct of the grapefruits, the peels (rinds), essential oils of grapefruit are obtained by cold pressing (19). The peel oil has a strong and desirable aroma, useful in industrial flavoring of foods (chewing gum, sweets, baked goods, ice cream et cetera), beverages, perfumes, and cosmetics (9,25). It is also employed in the pharmaceutical industry and in aromatherapy (36,37).

The pharmacological potentials of *Citrus paradisi* are discussed in ref. 35. The history, global distribution, and nutritional importance of *Citrus* fruits have been reviewed in ref. 30.

CHEMICAL COMPOSITION

Grapefruit oil is a yellow to pinkish orange mobile clear liquid, which has a fresh juicy and bitter-sweet citrus peel-like odor. Depending on the quantity of waxes in the oil, it may sometimes have a tight foggy appearance.

The yield of essential oil from the peel of *Citrus paradisi* Macfad. generally varies from 0.3 to 0.9%. The main producing countries of this oil are USA, Cuba, Brazil, India, Mexico, Spain, South Africa, Argentina and Israel.

Cold-pressed *Citrus* essential oils consist of a volatile fraction, which represents 85-99% of the oil, and a non-volatile fraction 1-15%, which mainly contains oxygen heterocyclic compounds, especially coumarins, psoralens and polymethoxyflavones (Table 5.39.4). These components play an important role in the characterization of cold-pressed *Citrus* oils, since the composition of this fraction is characteristic of each oil. Moreover, many of the pharmacological and toxicological activities possessed by *Citrus* oils have been demonstrated to be related to these components (1).

Literature data (up to November 25, 2014) on the chemical composition of grapefruit oils and unpublished analytical data from one of us (E.S.) are shown in Table 5.39.2 in alphabetical order. In grapefruit oils from various origins, over 210 chemicals have been identified.

Table 5.39.1 ISO values (%) for grapefruit oil[a]

	CAS	Minimum	Maximum
Limonene	138-86-3	92.0	96.0
Myrcene	123-35-3	1.5	2.5
n-Octanal	124-13-0	0.2	0.8
Nootkatone	4674-50-4	0.01	0.8
α-Pinene	80-56-8	0.2	0.6
Decanal	112-31-2	0.1	0.6
β-Caryophyllene	87-44-5	0.2	0.5
Sabinene	3387-41-5	0.1	0.4
β-Pinene	127-91-3	0.05	0.1
Nonanal	124-19-6	0.04	0.1
Neral	106-26-3	0.02	0.04

[a] ISO 3053 Essential oil of grapefruit ©ISO 2004; Geneva, Switzerland, www.iso.org

The major compounds found in grapefruit oils from different sources are shown in Table 5.39.3. They include (highest concentrations in any study given) limonene (95.5%), myrcene (28.7%) and α-pinene (5.5%). Well-known ingredients of grapefruit oils that were present in high concentrations (>7%) in one study were δ-cadinene (21.0%), ocimene (β-) (13.4%), nootkatone (10.9%) and (Z)-carveol (7.9%).

Commercial oils

The ten chemicals that had the highest maximum concentrations in 122 commercial grapefruit essential oil samples (concentration ranges provided) are the following: limonene (81.1-95.5%), myrcene (1.3-3.5%), β-caryophyllene (0.1-1.2%), α-pinene (0.2-1.0%), sabinene (0.2-0.8%), (E)-β-ocimene (tr-0.6%), n-octanal (0.1-0.5%), decanal (0.2-0.4%), germacrene D (0.02-0.4%) and neral (0.01-0.4%) (Erich Schmidt, unpublished analytical data).

Table 5.39.2 Constituents identified in grapefruit oils

Constituent	CAS	Percentage and range					
		A	B (2)	C (2)	D (2)	E (cold pressed)	F (distilled)
4-Alloxyimino-2-carene						<0.1[f]	
Aromadendrene	489-39-4					<0.1[f]	
Benzaldehyde	100-52-7					0.07	
α-trans-Bergamotene	13474-59-4		0.01				
Bergaptene	484-20-8	tr-0.2					
Bicyclogermacrene	24703-35-3	0.01-0.05	0.03		0.1		
Bicyclopentan-2-one	4884-24-6				0.01		
β-Bisabolene	495-61-4		0.4		0.02		
(Z)-γ-Bisabolene	13062-00-5						4.1[l]
β-Bourbonene	5208-59-3						2.1[l]
Cadinene	29350-73-0						0.1[h]; 0.2[k]
γ-Cadinene	39029-41-9		0.01	0.1	tr		
δ-Cadinene	483-76-1	0.01-0.2	0.1		0.4		0.1[l,p]; 0.9[q]; 1.3[m]; 21.0[l]
δ-Cadinol	19435-97-3				0.01		
Calamenene	483-77-2						<0.1[k]
Camphene	79-92-5		0.01	tr	tr		tr[o]
δ2-Carene (= δ4-)	554-61-0						0.2[a,k]
δ3-Carene	13466-78-9		0.03	0.3		0.2[f]	tr[o]; 0.2[q]; 0.5[g]
Carveol	99-48-9	tr-0.06		0.1			<0.1[k]; 0.1[o]
(E)-Carveol	1197-07-5		0.2	0.1	0.01	0.2[d]	<0.05[g]; 0.1[i]; 1.9[m]
(Z)-Carveol	1197-06-4		0.1			0.3[d]	<0.01[i]; 7.9[m]
Carvone	99-49-0	tr-0.05	0.2		0.04		<0.01[i]; <0.1[k]
D-Carvone	2244-16-8					1.2[e]	0.7[m]
(E)-Carvone						0.1[d]	
(Z)-Carvone						0.4[d]	
Carvotanacetone	499-71-8						0.1[i]
Carvyl acetate	97-42-7						tr[o]; 0.7[m]
cis-Carvyl acetate	1205-42-1				0.04		
trans-Carvyl acetate	1134-95-8		tr	tr			
β-Caryophyllene	87-44-5	0.1-1.2	0.5	0.3	1.1	<0.1[f]; 0.1[d]; 0.5[e]	0.3[o]; 0.4[g]; 0.6[i]; 0.7[k]; 1.7[m]; 1.9[q]
Caryophyllene oxide	1139-30-6	tr-0.1	tr			0.1[d]	0.6[m]; 4.3[l]
3-Casen-2-ol							<0.1[a,k]
α-Cedrene	469-61-4					0.1[d]	
Cedrol	77-53-2					0.1[d]	

Table 5.39.2 Constituents identified in grapefruit oils (*continued*)

Constituent	CAS	A	B (2)	C (2)	D (2)	E (cold pressed)	F (distilled)
1,8-Cineole	470-82-6				+		
Cinnamaldehyde	104-55-2					0.1[d]	
Citral	5392-40-5						1.2[m]
Citronellal	106-23-0	0.02-0.2	0.08	0.06	0.2	0.2[d]	0.1[h,p]; 0.2[g]; 0.8[m]
Citronellol `	106-22-9		0.06		0.04	0.5[c]	1.8[m]
Citronellyl acetate	150-84-5		0.01	0.01	0.01		0.04[b]
Copaene							0.8[q]
α-Copaene	3856-25-5	0.05-0.2	0.2	0.09	0.4	tr[d]	0.1[o,p]; 0.2[i]; 0.8[m]; 2.5[l]
β-Copaene	18252-44-3		0.06				0.02[h]; 0.1[b]
α-Cubebene	17699-14-8	tr-0.06	+	0.08	<0.05	0.1[d]	0.1[o]
β-Cubebene	13744-15-5	0.02-0.2	+	0.07	0.5		0.1[p]; 0.2[g]; 0.8[m]
Cuminaldehyde	122-03-2					0.1[d]	
p-Cymene	99-87-6	0-0.01	0.3	0.2	1.4		0.4[h]
p-Cymenene	1195-32-0						0.2[b]
(*E,E*)-2,4-Decadienal	25152-84-5		tr		+		
(*E,Z*)-2,4-Decadienal	25152-83-4		tr				
Decanal	112-31-2	0.2-0.4		0.5	1.4	0.3[c,d]	0.1[i,p]; 0.2[h,k,o]; 1.1[q]; 2.5[m]
Decanol	36729-58-5		tr	0.08			tr[o]
(*E*)-2-Decenal	3913-81-3				+		
(*Z*)-2-Decenal	2497-25-8				+		
Decyl acetate	112-17-4	tr-0.06	0.1	0.01	0.01		tr[o]; 0.2[h]
Diacetone alcohol	123-42-2					0.1[f]	
Dihydrocarveol	38049-26-2					0.1[d]	<0.1[k]; 0.1[i]
5,8-Dimethoxy-6,7-furanocoumarin	482-27-9	0-tr					
2,7-Dimethyl-2,6-octadien-1-ol	22410-74-8				0.01		
Dodecanal	112-54-9	tr-0.02	0.06	0.1		tr[d]	tr[o]; 0.2[b]; 0.3[m]
2-Dodecenal	4826-62-4			0.05	0.03		tr[o]
Dodecyl acetate	112-66-3					0.1[d]	
β-Elemene	33880-83-0		0.09	0.2			
γ-Elemene	29873-99-2				0.05		
δ-Elemene	20307-84-0					tr[d]	4.4[l]
Elemol (α-)	639-99-6	0-0.04	0.04		0.2	0.1[d]	0.1[b]; 0.6[m]
6′,7′-Epoxybergamottin	206978-14-5	tr-0.1					
trans-4,5-Epoxide-(*E*)-2-decenal					+		
trans-4,5-Epoxy-(*E*)-2-nonenal					+		
Ethyl butyrate	105-54-4		tr				
2(or5)-Ethyl-4-hydroxy-5(or2)-methyl-3 (2*H*)-furanone					+		
Eugenol	97-53-0				+		1.3[m]
α-Farnesene	502-61-4				0.05		tr[o]
(*E,E*)-α-Farnesene	502-61-4		0.01	0.02		0.1[d]	
β-Farnesene	502-60-3				tr		
(*E*)-β-Farnesene	18794-84-8			0.06			<0.05[g]
(*Z*)-β-Farnesene	28973-97-9		0.03				
Farnesol	4602-84-0						3.8[m]
(*E,E*)-Farnesol	106-28-5		0.01				
(*E,Z*)-Farnesol	3879-60-5		0.01			tr[d]	
Geranial	141-27-5	0.03-0.2	0.1	0.1	0.4	tr[d]	0.1[o]; 0.2[b]; 0.3[g,i]; 0.5[k]
Geraniol	106-24-1		tr	0.05	0.10		tr[o]; 0.1[g]; 0.4[m]
5-Geranoxypsoralen	7380-40-7	tr-0.1					
Geranyl acetate	105-87-3	0.01-0.2	0.1	0.1	0.2		0.1[o]; 0.2[a,k]; 0.7[m]

Table 5.39.2 Constituents identified in grapefruit oils (*continued*)

Constituent	CAS	Percentage and range					
		A	B (2)	C (2)	D (2)	E (cold pressed)	F (distilled)
Geranyl butyrate	106-29-6						0.02[b]; 0.2[m]
Geranyl propanoate	105-90-8					+[d]	
Germacrene A	28387-44-2				0.03		
Germacrene B	15423-57-1		+				2.9[l]
Germacrene C	34323-15-4				tr		
Germacrene D	23986-74-5	0.02-0.4	0.06	0.3			<0.05[g]; <0.1[k]; 0.1[p]; 0.8[m]
Germacrene D-4-ol	198991-79-6		tr		tr		
Globulol	489-41-8					+[d]	
β-Gurjunene	73464-47-8						0.7[m]
γ-Gurjunene	22567-17-5						0.7[m]
Heptyl acetate	112-06-1					0.5[d]	
Hexadecanal	629-80-1		0.01				
Hexadecanol	51260-59-4				0.01		
Hexanal	66-25-1						0.01[b]
Hexanol	111-27-3				0.02		
(*E*)-2-Hexenal	6728-26-3		tr				
1-Hexen-3-ol	4798-44-1			0.01			tr[o]
(*Z*)-3-Hexen-1-ol	928-96-1		tr		tr		
(*E*)-Hexenyl butyrate			tr				
Hexyl acetate	142-92-7	tr-0.01					
Himachalene							<0.1[k]
α-Humulene	6753-98-6	0.01-0.1	0.04	0.06	0.2		tr[o]; <0.1[k]; 0.1[i]; 0.3[q]; 1.0[l]
4-Hydroxy-2,5-dime-thyl-3(2*H*)-furanone	3658-77-3				+		
β-Ionone	79-77-6				+		
Isophorone	78-59-1						1.3[m]
Isopulegol	89-79-2	0-0.02		0.06			tr[o]
Isothujol	513-23-5					+[d]	
Limonene	138-86-3	81.1-95.5	95.4	93.7	94.4	91.1[d]; 94.2[f]	92.5[h]; 93.0[o]; 93.4[j]; 94.2[p]
Limonene dioxide	96-08-2		0.2				
Limonene oxide	1195-92-2		0.3				0.1[k]; 0.9[m]
cis-Limonene oxide	13837-75-7		0.02		0.1	0.1[f]	0.1[i]
trans-Limonene oxide	4959-35-7		0.02		0.1	0.5[e]	<0.01[i]; 0.6[g]
Limonen-10-ol	3269-90-7					+[d]	
Limonen-10-yl-acetate	15111-97-4						0.02[b]
Linalool	78-70-6	0.03-0.3	0.2	0.5	0.2	0.02[c]; 0.1[d]	0.5[q]; 0.7[g]; 1.1[j]; 1.6[j,m]; 4.6[n]
Linalool oxide	1365-19-1			0.7			0.7[o]; 0.9[j]; 4.2[n]; 6.5[m]
cis-Linalool oxide	11063-77-7			0.8			0.3[q]; 0.8[o]
cis-Linalool oxide, furanoid	5989-33-3						1.0[i]
trans-Linalool oxide	11063-78-8			0.6			0.3[o]
trans-Linalool oxide, furanoid	34995-77-2						0.4[i]
Linalyl acetate	115-95-7		0.06			0.1[d]	
Longifolene (junipene)	475-20-7						0.1[i]
cis-*p*-Mentha-2,8-dien-1-ol	3886-78-0			0.2			
trans-*p*-Mentha-2,8-dien-1-ol	4017-77-0					0.2[d]	0.7[m]
p-Mentha-1,8(9)-dien-10-yl acetate			0.01				
p-Mentha-1,3,8-triene	18368-95-1					1.3[e]	
p-Menth-1-en-9-ol	18479-68-0					0.1[d]	
4-Mercapto-4-me-thyl-2-pentanol	31539-84-1				+		
Methional	3268-49-3				+		
6-Methyl-5-hepten-2-one	110-93-0		0.07				

Table 5.39.2 Constituents identified in grapefruit oils (*continued*)

Constituent	CAS	Percentage and range					
		A	B (2)	C (2)	D (2)	E (cold pressed)	F (distilled)
Methyl *N*-methyl anthranilate	85-91-6		tr			+[d]	
4-Methyl-1-octene	13151-12-7					tr[d]	
Methyl oleate	112-62-9						0.2[q]
Methyl palmitate	112-39-0						0.3[q]
α-Muurolene	10208-80-7		+		0.01		0.03[b]
γ-Muurolene	30021-74-0				0.1		
α-Muurolol	104245-48-9				0.01		
Myrcene	123-35-3	1.3-3.5	2.6	2.1	6.9	6.3[c]	2.1[o]; 3.6[k,m]; 7.3[q]; 13.6[i]; 28.7[l]
Myrcene epoxide	29414-55-9					tr[d]	
Neodihydrocarveol	18675-33-7						<0.01[i]
Neral	106-26-3	0.01-0.4	0.06	0.08	0.2	tr[d]	0.04[h]; <0.1[k]; 0.1[o]; 0.3[g,i]
Nerol	106-25-2		tr	0.07	0.08	tr[d]	<0.1[k]; 0.1[o]; 0.2[g]
Nerolidol	7212-44-4			0.02			0.03[b]; 0.6[m]
(*E*)-Nerolidol	40716-66-3	tr-0.02	tr		0.02		
(*Z*)-Nerolidol	3790-78-1					0.1[d]	0.9[l]
Nerolidyl acetate	56001-43-5					tr[d]	
Neryl acetate	141-12-8	0.01-0.09	0.02	0.01	0.1		tr[o]; 0.1[i]; 0.2[b]; 0.3[h]
Neryl formate	2142-94-1						0.1[h]
(2*E*,4*E*)-Nonadienal	5910-87-2				+		
(*E*,*Z*)-2,6-Nonadienol	28069-72-9					tr[d]	
Nonanal	124-19-6	0.02-0.3	0.07	0.2	0.1	0.1[d]	<0.1[k]; 0.1[g,h,o]
Nonane	111-84-2						0.2[q]
Nonanoic acid	12-05-0					tr[d]	
Nonanol	28473-21-4		0.2	0.02			tr[o]; 0.3[b]
(*Z*)-2-Nonenal	60784-31-8				+		
(*Z*)-4-Nonenal	2277-15-8				+		
Nonyl acetate	143-13-5			0.06	tr		
Nootkatone	4674-50-4	0.04-0.4	0.8	0.7	0.06	0.2[d]; 1.3[e]	0.1[k]; 0.7[o]; 10.9[m]
Ocimene (β-)	13877-91-3			0.4	tr		<0.1[k]; 0.6[b]; 13.4[l]
(*E*)-β-Ocimene	3779-61-1	tr-0.6	0.3	0.4	0.5		0.2[o,p]
(*Z*)-β-Ocimene	3338-55-4	0-tr	0.01	0.02	0.02		tr[o]
Octadecanal	638-66-4				tr		
Octanal	124-13-0	0.1-0.5	0.3	0.8	0.5	0.3[d]	0.2[k,o]; 0.4[h]; 0.5[p]; 1.7[q]; 3.6[j]
Octanol	111-87-5	0.01-0.06	0.08	0.3	0.05		0.1[b,g]; 0.2[h]; 0.3[o]; 0.9[i]
2-Octen-4-ol	4798-61-2				0.01		
Octyl acetate	112-14-1	0-0.03	0.07		0.07		<0.1[k]; 0.1[b]
Octyl formate	112-32-3						1.0[j]
α-Panasinsene	56633-28-4						0.4[m]
Paradisiol	148810-80-4					+ (ref. 27)	
Perillaldehyde	2111-75-3	0.01-0.07	0.01		0.1	tr[d]	0.04[b]
Perillene	539-52-6		tr			tr[d]	
Perillyl acetate	15111-96-3					+[d]	
Perillyl alcohol	536-59-4					+[d]	0.2[k]
α-Phellandrene	99-83-2	0-0.09	0.2	0.2	tr	<0.1[f]	0.1[g,o]; 0.5[h]
β-Phellandrene	555-10-2	0-0.1	1.3	0.2	0.2	0.7[f]	tr[o]; 1.2[q]
α-Pinene	80-56-8	0.2-1.0	0.8	0.6	1.7	0.5[d]; 1.0[f]; 1.3[c]	0.7[g,j]; 1.4[k]; 1.5[m]; 2.1[q]; 3.8[i]; 5.5[l]
β-Pinene	127-91-3	0.02-0.4	0.3	0.09	0.7	tr[d]; 2.2[f]	tr[o]; 0.9[b]; 1.2[g]; 1.3[k]
cis-Piperitol	16721-38-3					tr[d]	
Pulegone	89-82-7				0.2		
Sabina ketone	513-20-2					0.2[d]	
Sabinene	3387-41-5	0.2-0.8	1.1	1.2	1.5	0.4[d]	0.2[i]; 0.5[m]; 0.6[g]; 0.7[j]; 0.8[h]; 1.2[o]
cis-Sabinene hydrate	15537-55-0		0.01	0.01	tr		

Table 5.39.2 Constituents identified in grapefruit oils (*continued*)

Constituent	CAS	A	B (2)	C (2)	D (2)	E (cold pressed)	F (distilled)
trans-Sabinene hydrate	17699-16-0			tr	0.03	tr[d]	
Santolina epoxide	60485-45-2						0.4[m]
Santolinatriene	2153-66-4					0.3[f]	
α-Sinensal	17909-77-2		tr			tr[d]	0.2[m]
β-Sinensal (*cis*-)	17909-87-4		0.01		0.05	tr[d]	0.5[m]
α-Terpinene	99-86-5	0-tr	0.01	0.04	tr	1.3[d]; 2.1[c]	tr[o]; 1.0[g]
γ-Terpinene	99-85-4	tr-0.2	0.3	0.2	0.1	tr[d]; 0.07[c]	tr[o]; 0.07[h]; 0.4[p]; 0.9[m]
α-Terpinenol							2.7[m]
Terpinen-4-ol	562-74-3		0.01	0.2	tr		0.1[o]; 0.2[h]; 0.4[b]; 0.8[i]; 1.7[m]
α-Terpineol	98-55-5	0.01-0.4	0.05	0.2	0.2	0.1[d]	0.1[p]; 0.2[g,h,k,o]; 0.7[j]; 0.8[m]; 2.3[i]
(*Z*)-β-Terpineol	7299-40-3		0.01				0.1[i]
Terpinolene	586-62-9	tr-0.1	0.02	0.3	0.1	tr[d]	0.3[o]; 0.5[g]
α-Terpinyl acetate	80-26-2		0.01	0.04			tr[o]
Tetradecanal	124-25-4		tr			+[d]	
Tetradecane	629-59-4						0.5[q]
Tetradecenal	54264-02-7					+[d]	
α-Thujene	2867-05-2	0-tr	0.01	0.01	tr	tr[d]; 0.2[c,f]	tr[o]; 0.03[b]; 0.1[p]
Thymol	89-83-8				tr		
Tridecane	629-50-5						0.3[q]
Undecanal	112-44-7	0-0.01	0.02	0.01	0.02		tr[o]
Undecanoic acid	112-37-8				+	+[d]	
2-Undecenal	2463-77-6			0.03			0.01[b]
Valencene	4630-07-3		tr		tr		0.06[b]; 3.4[m]
Verbenene	4080-46-0						<0.1[a,k]
p-Vinylguaiacol	7786-61-0				+		0.7[m]
Wine lactone	182699-77-0				+		

A one hundred and twenty-two grapefruit essential oil samples from USA, Brazil, Mexico, Spain, Israel, South Africa and Argentina, analyzed between 1998 and 2013; lowest and highest concentrations given (E. Schmidt, unpublished data)
B review of studies published between 1979 and 1999 on the composition of industrial cold-pressed grapefruit oils; data cited in ref. 2; highest concentrations given
C review of studies published between 1979 and 1999 on the composition of laboratory-prepared grapefruit oils, both cold pressed and distilled; data cited in ref. 2; highest concentrations given
D review of studies published between 1998 and 2009 on the composition of cold-pressed grapefruit oils, both industrially (refs. 3-7) and lab prepared (refs. 8,13); data cited in ref. 2; highest concentrations given; only limonene (lowest concentration: 84.8%) and myrcene (lowest concentration: 1.5%) were present in all samples investigated; consequently, the lower limit for all other chemicals mentioned was zero
E data from other studies (indicated with superscript letters) on cold-pressed oils; highest concentrations found in any study reviewed here given; when two or more oils were investigated, only the highest concentrations are mentioned, unless indicated otherwise
F data from other studies on other (non-cold-pressed) grapefruit oils (indicated with superscript letters); highest concentrations found in any study reviewed here given; when two or more oils were investigated, only the highest concentrations are mentioned, unless indicated otherwise

[a] incorrect identification (ref. 15); [b] data from 25 studies and review articles published between 1976 and 2005, cited in ref. 9, not mentioned already in the review of ref. 2 (provided in columns B-D); both hydrodistilled and cold pressed oils; [c] one laboratory cold-pressed oil from Pakistan (ref. 34); [d] one laboratory cold-pressed oil from redblush grapefruit from Kenya (ref. 25); [e] one commercial cold-press oil from Florida (ref. 23); [f] one commercial oil sample from Florida (ref. 29); [g] one lab-hydrodistilled peel oil from Turkey (ref. 32); [h] two steam-distilled oils from Georgian SSR, grapefruit cultivars Duncan and Marsh (ref. 31); [i] one steam-distilled oil from Cuba (ref. 19); [j] one sample of hydrodistilled oil from red and one from white grapefruits from Venezuela (ref. 14); [k] one commercial oil sample (ref. 16); many mistakes were made in identifying the constituents (ref. 15); [l] one sample of lab-hydrodistilled grapefruit oil from Iran; extremely atypical composition with 6.6% limonene and high concentrations of myrcene (28.7%), δ-cadinene (21.0%), β-ocimene (13.4%) and limonene (6.6%) (ref. 18); [m] three lab-hydrodistilled oils from Pakistan prepared from fresh, air-dried and oven-dried grapefruit peels (ref. 17); [n] one lab-hydrodistilled oil from Egypt (ref. 24); [o] two laboratory steam-distilled oils from Sicily, Italy; one from cultivar Duncan, the other not specified (refs. 10,11); [p] one hydrodistilled grapefruit peel oil sample from Nigeria (ref. 12); [q] one hydrodistilled oil from Nigeria (ref. 33)

tr: trace (in column E[d]: <0.05); + present in the oil investigated, but quantity not stated

Table 5.39.3 Major constituents of grapefruit oils

Constituent	CAS	Percentage and range					
		A	B (2)	C (2)	D (2)	E (cold pressed)	F (distilled)
Limonene	138-86-3	81.1-95.5	95.4	93.7	94.4	91.1[d]; 94.2[f]	92.5[h]; 93.0[o]; 93.4[j]; 94.2[p]
Myrcene	123-35-3	1.3-3.5	2.6	2.1	6.9	6.3[c]	2.1[o]; 3.6[k,m]; 7.3[q]; 13.6[i]; 28.7[l]
α-Pinene	80-56-8	0.2-1.0	0.8	0.6	1.7	0.5[d]; 1.0[f]; 1.3[c]	0.7[g,j]; 1.4[k]; 1.5[m]; 2.1[q]; 3.8[i]; 5.5[l]

LEGEND: SEE UNDER TABLE 5.39.2

It should be realized that values in hydrodistilled citrus oils may (but not necessarily do) vary considerably from genuine cold-pressed citrus oils because of the many hydrolytic reactions that take place during oil isolation. In addition, the hydrodistilled oils lack the non-volatile oxygenated heterocyclic compounds which may be present in the cold-pressed oils (Table 5.39.4). However, generally speaking, there appear to be no major differences between the cold-pressed grapefruit oils and the hydrodistilled oils.

The literature on the chemical composition of grapefruit peel oils has (in addition to ref. 2) also been discussed in ref. 9. This is a review of 25 studies and articles on grapefruit peel oil composition published between 1976 and 2005; there is of course a major overlap with the review in ref. 2 (columns B, C, D and E in Table 5.39.2). Data presented in ref. 9 that are not mentioned in ref. 2 are given in column F[b].

Table 5.39.4 Heterocyclic oxygenated compounds present in cold-pressed grapefruit oils (20,21,22,23,26,28)

Name	Synonym(s)	CAS
Hydroxylated polymethoxyflavones		
Hydroxyhexamethoxyflavone (not further specified)		
Hydroxypentamethoxyflavone (not further specified)		
Polymethoxyflavones		
Heptamethoxyflavone		119279-30-0
3,3′,4′,5,6,7,8-Heptamethoxyflavone		1178-24-1
3′,4′,5,6,7,8-Hexamethoxyflavone	Nobiletin	478-01-3
4′,5,6,7,8-Pentamethoxyflavone	Tangeretin	481-53-8
4′,5,6,7-Tetramethoxyflavone	Tetra-O-methylscutellarein	1168-42-9
Coumarins and psoralens		
Auraptenol		108354-46-7
Coumarin, 7-[(6,7-dihydroxy-3,7-dimethyl-2-octenyl)oxy]-	Marmin	14957-38-1
Coumarin, 8-(2,3-dihydroxy-3-methylbutyl)-7-methoxy-	Meranzin hydrate	5875-49-0
Coumarin, 7-[(6,7-epoxy-3,7-dimethyl-2-octenyl)oxy]-, (E)-(+)-	Epoxyaurapten	21499-17-2
Coumarin, 8-(2,3-epoxy-3-methylbutyl)-7-methoxy-, (-)-	Meranzin	23971-42-8
Coumarin, 8-(2-formyl-2-methylpropyl)-7-methoxy-		5980-07-4
Coumarin, 7-methoxy-8-(3-methyl-2-oxobutyl)-	Isomeranzin	1088-17-1
Coumarin, 7-methoxy-8-[(2,2,5,5-tetramethyl-1,3-dioxolan-4-yl)methyl]-	Pranferin	33573-60-3
Dihydroxybergamottin		145414-76-2
6′,7′,-Dihydroxybergamottin		145414-76-2
6′,7′,-Dihydroxybergamottin decanal acetal		1181223-80-2
6′,7′,-Dihydroxybergamottin octanal acetal		1181223-79-9
5,7-Dimethoxycoumarin	Citropten; Limetin	487-06-9
1,5-[(3,6-Dimethyl-6-formyl-2-heptenyl)oxy]psoralen	Aurantiumal [a]	
5-(6′,7′-Epoxy)geranyloxypsoralen	Epoxybergamottin	206978-14-5
5-(6′,7′-Epoxy)geranyloxypsoralen hydrate	Epoxybergamottin hydrate	
7-Geranyloxycoumarin	Aurapten	495-02-3
5-Geranyloxy-7-methoxycoumarin		7380-39-4
5-Geranoxypsoralen	Bergamottin; Bergaptin	7380-40-7
5-Hydroxy-6,7-furanocoumarin	Bergaptol	486-60-2
6′,7′-Marmin decanal acetal		1181223-78-8

Table 5.39.4 Heterocyclic oxygenated compounds present in cold-pressed grapefruit oils (20,21,22,23,26,28)

Name	Synonym(s)	CAS
5-Methoxy-8-(2,3-dihydroxy-3-methylbutoxy)psoralen	Bjakangelicin; Byakangelicin	482-25-7
7-Methoxy-8-isopentenylcoumarin	Osthole	484-12-8
5-Methoxypsoralen	Bergapten	484-20-8
Oxypeucedanin		26091-73-6
Other chemicals		
Citrusal[a]		43145-56-8

[a] artefact (ref. 20)

CONTACT ALLERGY/ALLERGIC CONTACT DERMATITIS

General

Contact allergy to and possible allergic contact dermatitis from grapefruit oil has been reported in one publication only. A false-positive patch test reaction due to the excited skin syndrome cannot be excluded.

Case reports

Two aromatherapists had contact dermatitis (one occupational) with allergies to multiple essential oils used at their work, including grapefruit oil. Both patients also reacted to geraniol, α-pinene, the fragrance mix and various other fragrance materials. In addition, one proved to be allergic to linalool and linalyl acetate, the other to caryophyllene. α-Pinene and caryophyllene were demonstrated by GC-MS in grapefruit oil (38), but neither of these is an important component of grapefruit oil (Table 5.39.2, column A). Limonene, the dominant chemical in essential oils of *Citrus paradisi*, was not tested.

LITERATURE

1 Russo M, Torre G, Carnovale C, Bonaccorsi I, Mondello L, Dugo P. A new HPLC method developed for the analysis of oxygen heterocyclic compounds in Citrus essential oils. J Essent Oil Res 2012;24:119-129

2 Dugo G, Cotroneo A, Bonaccorsi I, Trozzi A. Composition of the volatile fraction of citrus peel oils. In: G Dugo and L Mondello, eds. Citrus oils: composition, advanced analytical techniques, contaminants, and biological activity. Boca Raton, USA: CRC Press, Taylor & Francis Group, 2011:20-29

3 Oberhofer B, Nikiforov A, Buchbauer G, Jirovetz L, Bicchi C. Investigation of the alteration of the composition of various essential oils used in aroma lamp applications. Flavour Fragr J 1999;14:293-299

4 Pino JA, Sánchez M. Chemical composition of grapefruit oil concentrates. J Essent Oil Res 2000;12:167-169

5 Lin J, Rouseff RL. Characterization of aroma-impact compounds in cold pressed grapefruit oil using time-intensity GC-olfactometry and GC-MS. Flavour Fragr J 2001;16:457-463

6 Feger W, Brandauer H, Ziegler H. Analytical investigation of Sweetie peel oil. J Essent Oil Res 2001;13:309-313

7 Viuda-Martos M, Ruiz-Navajas Y, Fernández-López J, Pérez-Álvarez JA. Chemical composition of mandarin (*C. reticulata* L.), grapefruit (*C. paradisi* L.), lemon (*C. limon* L.) and orange (*C. sinensis* L.) essential oils. J Essent Oil Bear Plants 2009;12:236-243

8 Huang Y, Wu Y. Chemical components of essential oils from peels of 25 *Citrus* species and cultivars. Tianran Chanwu Yanjiu Yu Kaifa 1998;10(4):48-54. Data cited in ref. 2

9 Kirbaşlar I, Boz I, Kirbaşlar G. Composition of Turkish lemon and grapefruit peel oils. J Essent Oil Res 2006;18:525-543

10 Ruberto G, Biondi D, Rapisarda P, Renda A, Starrantino A. Essential oil of Cami, a new citrus hybrid. J Agric Food Chem 1997;45:3206-3210

11 Ruberto G, Starrantino A, Rapisarda P. Essential oils from new citrus fruits. Essenz Deriv Agrum 1999;69:15-26. Data cited in ref 2. Same results previously also reported in ref. 10

12 Karioti A, Skaltsa H, Gbolade A. Constituents of the distilled essential oils of *Citrus reticulata* and *C. paradisi* from Nigeria. J Essent Oil Res 2007;19:520-522

13 Sawamura M. Volatile components of essential oils of the *Citrus* genus. Recent Res Develop Agric Food Chem 2000;4:131-164

14 Gonzalez CN, Sanchez F, Quintero A. Chemotaxonomic value of essential oil compounds in *Citrus* species. Acta Hort 2002;576:49-51. Data cited in ref. 15

15 Lawrence BM. Progress in essential oils. Perfum Flavor 2005;30(2):66-? (last page unknown)

16 Veriotti T, McGuigan M, Sacks R. New technologies for the high-speed characterization and analysis of essential oils. Perfum Flavor 2002;27(5):40-49

17 Kamal GM, Anwar F, Hussain AI, Sarri N, Ashraf MY. Yield and chemical composition of *Citrus* essential oils as affected by drying pretreatment of peels. Int Food Res J 2011;18:1275-1282

18 Esmaeili A, Abednazari S, Abdollahzade YM, Abdollahzadeh NM, Mahjoubian R, Tabatabaei-Anaraki M. Peel volatile compounds of apple (*Malus domestica*) and grapefruit (*Citrus paradisi*). J Essent Oil Bear Plants 2012;15:794-799

19 Pino JA, Acevedo A, Rabelo J, González C, Escandon J. Chemical composition of distilled grapefruit oil. J Essent Oil Res 1999;11:75-76

20 Dugo P, Russo M. The oxygen heterocyclic components of *Citrus* essential oils. In: G Dugo and L Mondello, eds. Citrus oils: composition, advanced

analytical techniques, contaminants, and biological activity. Boca Raton, USA: CRC Press, Taylor & Francis Group, 2011:405-444

21 Chebrolu KK, Jayaprakasha GK, Jifon J, Patil BS. Purification of coumarins, including meranzin and pranferin, from grapefruit by solvent partitioning and a hyphenated chromatography. Separ Purif Technol 2013;116:137-144

22 Buiarelli F, Cartoni GP, Coccioli F, Leone T. Analysis of bitter essential oils from orange and grapefruit by high performance liquid chromatography with microbore columns. J Chromatogr A 1996;730:9-16

23 Teixeira B, Marques A, Ramos C, Neng NR, Nogueira JMF, Saraiva GA, et al. Chemical composition and antibacterial and antioxidant properties of commercial essential oils. Ind Crops Prod 2013;43:587-595

24 Badawy MEI, Abdelgaleil SAM. Composition and antimicrobial activity of essential oils isolated from Egyptian plants against plant pathogenic bacteria and fungi. Ind Crops Prod 2014;52:776-782

25 Njoroge SM, Koaze H, Karanja PN, Sawamura M. Volatile constituents of redblush grapefruit (*Citrus paradisi*) and pummelo (*Citrus grandis*) peel essential oils from Kenya. J Agric Food Chem 2005;53:9790-9794

26 César TB, Manthey JA, Myung K. Minor furanocoumarins and coumarins in grapefruit peel oil as inhibitors of human cytochrome P450 3A4. J Nat Prod 2009;72:1702-1704

27 Sulser H, Scherer JR, Stevens KL. The structure of paradisiol, a new sesquiterpene alcohol from grapefruit oil. J Org Chem 1971;36:2422-2426

28 Uckoo RM, Jayaprakasha GK, Patil BS. Hyphenated flash chromatographic separation and isolation of coumarins and polymethoxyflavones from byproduct of *Citrus* juice processing industry. Separ Sci Technol 2013;48:1467-1472

29 Yang S-A, Jeon S-K, Lee E-J, Shim C-H, Lee I-S. Comparative study of the chemical composition and antioxidant activity of six essential oils and their components. Nat Prod Res 2010;24:140-151

30 Liu YQ, Heying E, Tanumihardjo SA. History, global distribution, and nutritional importance of *Citrus* fruits. Compreh Rev Food Sci Food Saf 2012;11:530–545

31 Kekelidze NA, Dzhanikashvili MI, Fishman GM. Essential oil of the grapefruit *Citrus paradisi* growing in the Georgian SSR. Translated by Plenum Publishing Corporation, 1985, from Khimiya Prirodnykh Soedinenii 1985;nr.1:119-120

32 Uysal B, Sozmen F, Aktas O, Oksal BS, Kose EO. Essential oil composition and antibacterial activity of the grapefruit (*Citrus paradisi* L.) peel essential oils obtained by solvent-free microwave extraction: comparison with hydrodistillation. Int J Food Sci Technol 2011;46:1455-1461

33 Okunowo WO, Oyedeji O, Afolabi LO, Matanmi E. Essential oil of grapefruit (*Citrus paradisi*) peels and its antimicrobial activities. Am J Plant Sci 2013;4:1-9

34 Salim-ur-Rehman MMA, Iqbal Z, Iqbal Sultan FMAJ. Genetic variability to essential oil composition in four *Citrus* fruit species. Pak J Bot 2006;38:319-324

35 Gupta V, Kohli K, Ghaiye P, Bansal P, Lather A. Pharmacological potentials of *Citrus paradisi*—an overview. Int J Photother Res 2011;1:8-17

36 Rhind JP. Essential oils. A handbook for aromatherapy practice, 2nd Edition. London: Singing Dragon, 2012

37 Lawless J. The encyclopedia of essential oils, 2nd Edition. London: Harper Thorsons, 2014

38 Dharmagunawardena B, Takwale A, Sanders KJ, Cannan S, Roger A, Ilchyshyn A. Gas chromatography: an investigative tool in multiple allergies to essential oils. Contact Dermatitis 2002;47:288-292

Chapter 5.40   GUAIACWOOD OIL

DEFINITION
Guaiacwood oil is the essential oil obtained from the wood of the guaiacwood (true guaiac; in Spanish: palo santo), *Bulnesia sarmientoi* Lorentz ex Griseb.

INCI NOMENCLATURE
Description/definition: Bulnesia sarmientoi wood oil is the volatile oil obtained from the wood of *Bulnesia Sarmientoi*, Zygophyllaceae
INCI name EU & USA: Bulnesia sarmientoi wood oil
CAS registry number(s): 8016-23-7
EINECS number(s): Not available

ISO (INTERNATIONAL ORGANIZATION FOR STANDARDIZATION) STANDARD
There is currently no ISO standard for guaiacwood oil.

Guaiacwood oil should not be confused with guaiacum officinale wood oil (CAS 84650-13-5; EINECS 283-494-2), the essential oil obtained from *Guaiacum officinale* L. This too is a member of the family Zygophyllaceae and is *also* known as guaiacwood.

THE PLANT, THE OIL, AND THEIR USES
Bulnesia sarmientoi Lorentz ex Griseb. (often misspelled as *Bulnesia sarmienti*) (synonyms: guaiac wood, Paraguaylignum vitae) is a deciduous tree that grows 6-20 meters tall. It inhabits the Gran Chaco region of Argentina, Bolivia, Paraguay and—marginally—Brazil (2,3). The species, called Palo Santo (holy tree) by the local population throughout the Gran Chaco region, is highly valued as a medicinal plant for the many healing powers attributed to infusions brewed from its bark, crust or leaves. It is locally used as a blood cleanser, inducer of perspiration, diuretic, to heal gastric pain, syphilis, leprosy, gout, rheumatism, rheumatoid arthritis, lumbago, wounds and skin diseases; moreover, it is employed to relieve stress and depression, control blood pressure, and prevent atherosclerosis and colds (4). The plant is said to have analgesic, anti-inflammatory, antioxidant, bactericidal, antitumor and serum lipid lowering properties (2).

The essential guaiacwood oil (also known as 'lignum vitae oil', 'guaiac oil', 'guayacol', 'guajol', or 'guayaco'), which is obtained by steam distillation of a mixture of the heartwood and sawdust from the tree, is used in fragrances. The oil has a soft rose-like odor, and has thus been used as an adulterant for rose oil (3,4). In addition, it has been used to perfume luxury soaps by masking the unpleasant smell of synthetic components and as an excipient in the manufacturing of cosmetics (4). The oil of Palo Santo is also highly valued in aromatherapy to which the following advantages are attributed: mood uplifting, helpful for meditation and rest, improvement of mental clarity, calming, relaxing, stress and tension reduction. It is applied through aroma lamps, light bulb rings, massage and mist spray (4). Paraguay currently supplies most of the international demand of guaiacwood oil (4,5). It should be realized that industry may treat guaiac oil in a broad sense, also grouping in this category other Zygophyllaceae species from the genus *Guaiacum* (e.g., *Guaiacum officinale*) (4,5).

CHEMICAL COMPOSITION
Guaiacwood oil is a yellow greenish to ambery semisolid mass which has a woody, balsamic, mild rosy and medicinal odor. The yield of essential oil from the wood of *Bulnesia sarmientoi* Lorentz ex Griseb. generally varies from 3.0 to 5.5 per cent. The main producing countries of this oil are Paraguay, Brazil and Argentina.

Literature data (up to November 25, 2014) on the chemical composition of guaiacwood oils (only two publications found) and unpublished analytical data from one of us (E.S.) are shown in Table 5.40.1 in alphabetical order. In guaiacwood oils from various origins, over 25 chemicals have been identified (please note that column D in Table 5.40.1 presents a *resin oil, not* an essential oil). About 52 per cent of these were found in a single reviewed publication only. The major compounds found in guaiacwood oils from different sources are shown in Table 5.40.2. They include (highest concentrations in any study given) bulnesol (49.8%), guaiol (38.7%), α-eudesmol (4.7%), α-bulnesene (4.5%), β-eudesmol (3.6%) and 10-epi-γ-eudesmol (3.5%).

Commercial oils
The ten chemicals that had the highest maximum concentrations in 22 commercial guaiacwood essential oil samples (concentration ranges provided) are the following: bulnesol (30.5-49.8%), guaiol (25.6-38.7%), α-eudesmol (2.3-4.7%), α-bulnesene (0.8-4.5%), γ-eudesmol (1.6-3.8%), β-eudesmol (0.2-3.6%), 10-epi-γ-eudesmol (1.2-3.5%), guaioxide (0.3-1.9%), elemol (0.5-1.7%) and a guaiol isomer (0-1.0%) (Erich Schmidt, unpublished analytical data).

Table 5.40.1 Constituents identified in guaiacwood oils

Constituent	CAS	Percentage and range			
		A	B (3)	C (6)	D RESIN OIL (1)
α-Bergamotene	17699-05-7			0.07	
(Z)-β-Bisabolene					6.1
α-Bulnesene	3691-11-0	0.8-4.5	1.1		16.7
Bulnesol	22451-73-6	30.5-49.8	34.7	40.5	15.4
Bulnesol isomer		0-0.8	0.8	0.8	
β-Caryophyllene	87-44-5	0-0.2	0.1	0.2	
9-epi-(E)-Caryophyllene	68832-35-9				1.9
β-Chamigrene	18431-82-8				2.2
Dimethyl dodecene				0.09	
β-Elemene	33880-83-0				0.8
Elemol (α-)	639-99-6	0.5-1.7	0.6	1.2	1.2
Elemol isomer				0.5	
Eudesm-5-en-11-ol	337981-29-0		0.7		
α-Eudesmol	473-16-5	2.3-4.7	3.7		
7-epi-α-Eudesmol	123123-38-6		3.3		
β-Eudesmol	473-15-4	0.2-3.6	1.3	0.2	8.9
γ-Eudesmol	1209-71-8	1.6-3.8	2.6		1.3
10-epi-γ-Eudesmol	15051-81-7	1.2-3.5	1.3	2.2	1.0
β-Farnesene	502-60-3			0.1	
Germacrene B	15423-57-1		0.3		4.7
Guaiazulene	489-84-9				0.9
α-Guaiene	3691-12-1	0.2-0.8	0.2		6.9
β-Guaiene	88-84-6		0.1		
cis-β-Guaiene	372162-07-7				6.7
Guaiol	489-86-1	25.6-38.7	20.4	26.8	10.6
Guaiol isomer		0-1.0	1.2	1.4	
Guaioxide	20149-50-2	0.3-1.9	0.6		
α-Gurjunene	489-40-7				0.7
γ-Gurjunene	22567-17-5				1.2
(-)-Hanamyol	94388-63-3		2.5		
Hanamyol isomer 1			0.6		
Hanamyol isomer 2			0.2		
Longiborneol acetate	36204-27-0				0.8
(Z)-Nerolidol	3790-78-1				1.4
β-Patchoulene	514-51-2	0.2-0.5			0.7
γ-Patchoulene	508-55-4				1.4
α-Selinene	473-13-2	0-0.2			3.4
β-Selinene	17066-67-0	0-0.1			
β-Sesquiphellandrene	20307-83-9				3.8
cis-Sesquisabinene hydrate	58319-05-4				0.7
Seychellene	20085-93-2				0.7

A twenty-two guaiacwood essential oil samples from Paraguay and Brazil, analyzed between 1998 and 2013; lowest and highest concentrations given (E. Schmidt, unpublished data)

B one commercial steam-distilled oil sample from Paraguay (ref. 3)

C one guaiacwood oil sample (ref. 6)

D one sample of oil obtained by lab hydrodistillation of the *resin* of *B. sarmientoi* obtained commercially in Italy (ref. 1); note that this is NOT a guaiacwood essential oil

Table 5.40.2 Major constituents of guaiacwood oils

Constituent	CAS	Percentage and range		
		A	B (3)	C (6)
Bulnesol	22451-73-6	30.5-49.8	34.7	40.5
Guaiol	489-86-1	25.6-38.7	20.4	26.8
α-Eudesmol	473-16-5	2.3-4.7	3.7	
α-Bulnesene	3691-11-0	0.8-4.5	1.1	
β-Eudesmol	473-15-4	0.2-3.6	1.3	0.2
10-epi-γ-Eudesmol	15051-81-7	1.2-3.5	1.3	2.2

LEGEND: SEE UNDER TABLE 5.40.1

CONTACT ALLERGY/ALLERGIC CONTACT DERMATITIS

General

Contact allergy to guaiacwood oil has been reported in two publications, but no cases of allergic contact dermatitis from the oil have been identified.

Testing in groups of patients

Two positive patch test reactions to guaiacwood oil 2% petrolatum were observed in 51 patients (3.9%) allergic to *Myroxylon pereirae* resin (balsam of Peru) and/or turpentine and/or wood tar and/or colophony (8). A group of 86 patients from Poland previously reacting to the fragrance mix was tested with guaiacwood oil and one (1.1%) had a positive patch test reaction (9); in neither of these studies were relevance data provided.

LITERATURE

1 Marongiu B, Piras A, Porcedda S, Tuveri E. Isolation of *Guaiacum bulnesia* volatile oil by supercritical carbon dioxide extraction. J Essent Oil Bear Plants 2007;10:221-228

2 Kamruzzaman SM, Endale M, Oh WJ, Park S-C, Kim K-S, Hong JH, et al. Inhibitory effects of *Bulnesia sarmienti* aqueous extract on agonist-induced platelet activation and thrombus formation involves mitogen-activated protein kinases. J Ethnopharmacol 2010;130:614-620

3 Rodilla JM, Silva LA, Martinez N, Lorenzo D, Davyt D, Castillo L, et al. Advances in the identification and agrochemical importance of sesquiterpenoids from *Bulnesia sarmientoi* essential oil. Ind Crops Prod 2011;33:497-503

4 Waller T, Barros M, Draque J, Micucci P. Conservation of the Palo Santo tree, *Bulnesia sarmientoi* Lorentz ex Griseb, in the South America Chaco Region. Medicinal Plant Conservation 2012;15:4-8. Available at: http://www.academia.edu/2120881/Conservation_of_the_Palo_Santo_tree_Bulnesia_sarmientoi_Lorentz_ex_Griseb_in_the_South_American_Chaco_Region

5 Janzen HKJ. Guaiac wood oil, Paraguay and CITES. Paper presented at the IFEAT International Conference in Marrakech, 26-30 Sept. 2010 'North African and Mediterranean essential oils and aromas: 2010 tales and realities of our industry – a new decade of challenges and opportunities'. Pages 317-324 in the Conference Proceedings. Available at: http://www.ifeat.org/wp-content/uploads/2012/10/Janzen+-+Guaiac+wood.pdf

6 Prudent D, Perineau F, Bravo R, Delmas M. Preparation et characterisation d'extraits volatils de bois de guaiac (*Bulnesia sarmientoi* Lor.). Rivist Ital EPPOS 1991;5:35-43. Data cited in ref. 7

7 Lawrence BM. Progress in essential oils. Perfum Flavor 1992;17(6):51

8 Rudzki E, Grzywa Z, Bruo WS. Sensitivity to 35 essential oils. Contact Dermatitis 1976;2:196-200

9 Rudzki E, Grzywa Z. Allergy to perfume mixture. Contact Dermatitis 1986;15:115-116

Chapter 5.41 HYSSOP OIL

DEFINITION
Hyssop oil (essential oil of hyssop) is the essential oil obtained from the leaves and flowering tops (the aerial parts) of the hyssop, *Hyssopus officinalis* L. ssp. *officinalis*.

INCI NOMENCLATURE
Description/definition: Hyssopus officinalis leaf oil is the volatile oil obtained from the leaves of the hyssop, *Hyssopus officinalis* L., Lamiaceae
INCI name EU & USA: Hyssopus officinalis leaf oil
CAS registry number(s): 8006-83-5; 84603-66-7
EINECS number(s): 283-266-2

ISO (INTERNATIONAL ORGANIZATION FOR STANDARDIZATION) STANDARD
ISO number: 9841
ISO name: Essential oil of hyssop
Botanical origin: *Hyssopus officinalis* L. ssp. *officinalis*
Parts of plant used: Leaf and flowering top
ISO values: ISO values (minimum and maximum concentrations) are shown in Table 5.41.1.

According to The Plant List, *Hyssopus officinalis* L. ssp. *officinalis* is a synonym for *Hyssopus officinalis* L.

THE PLANT, THE OIL, AND THEIR USES
Hyssopus officinalis L. is an aromatic perennial subshrub belonging to the Lamiaceae family. Hyssop is native to Northern Africa (Algeria, Morocco), temperate regions of Asia (Iran, Turkey, Caucasus) and middle, southern and eastern Europe (1,9,12). It is an important medicinal and culinary plant extensively cultivated in Russia, Spain, France, Italy, former Yugoslavia, Bulgaria, Hungary, Ukraine (Crimea), the United States and China

(1,3,9,12,17,21). The dried flowering shoots of *Hyssopus officinalis* L. are used (in tea blends and other forms) in folk medicine for the treatment of chronic coughing, laryngitis, bronchitis, wound infections, digestive disorders and to relieve catarrh (5,11,12). It is said to relax peripheral blood vessels, promote sweating and have expectorant, carminative, anti-inflammatory, anti-catarrhal, and antispasmodic activities (12). Despite its bitter taste, the plant is also used as a spice in households and in the food industry for the flavoring of meats and sauces (5).

The leaves and flowering tops produce a pleasant smelling essential oil responsible for most of the biological activities of the plant. The oil is used for flavoring and food preservation, in liqueurs, cosmetic products, in perfumery, in the pharmaceutical field in several antiseptic preparations and for phytotherapeutic uses in folk medicine and aromatherapy (2,5,9,12,14,16). The oil is claimed to possess antibacterial, antimycotic, antiviral, anti-tuberculosis, antioxidant, anthelmintic and muscle-relaxing (spasmolytic) properties (11,14,21). A review on the biological activities of *Hyssopus officinalis* L. is provided in ref. 47.

CHEMICAL COMPOSITION
Hyssop oil is a pale yellow to brownish yellow clear mobile liquid, which has an herbaceous, fresh minty and camphoraceous odor. The yield of essential oil from the aerial parts of *Hyssopus officinalis* L. ssp. *officinalis* generally varies from 0.07 to 0.3 per cent for fresh herb and from 0.3 to 0.9 per cent for dried herb. The main producing countries of this oil are India, France, Morocco and Hungary.

Literature data (up to November 11, 2014) on the chemical composition of hyssop oils and unpublished analytical data from one of us (E.S.) are shown in Table 5.41.2 in alphabetical order. In hyssop oils from various origins, over 285 chemicals have been identified. About 52 per cent of these were found in a single reviewed publication only.

The major compounds found in hyssop oils from different sources are shown in Table 5.41.3. They include (highest concentrations in any study given) pinocamphone (64.8%), isopinocamphone (59.9%), 1,8-cineole (52.9%, not present in 26 commercial oils investigated by one of us [E.S]), β-pinene (32.3%), pinocarvone (29.2%), elemol (17.2%), β-phellandrene (12.7%) and germacrene D (11.2%). Well-known ingredients of hyssop oils that were present in high concentrations (>7%) in one or two studies were α-pinene (70.9%), methyl eugenol (38.3%), limonene (37.4%), *trans*-pinocarveol (19.2%), sabinene (11.0%), terpinen-4-ol (8.3% and 10.5%), α-phellandrene (9.6%), α-terpinene (9.4%) and α-terpineol (7.4%). Uncommon or rare constituents of hyssop oils found in high concentrations (>7%) in single studies include myrtenyl acetate (74.1%), thymol (19.0%), camphor (three studies: 12.5%, 16.3%, 16.4%), *n*-decane (two studies: 8.7% and 11.8%), hedycaryol (two studies: 9.1% and 8.5%), sabinene hydrate (8.5%) and carvacrol (7.7%).

Table 5.41.1 ISO values (%) for hyssop oil[a]

Compound	CAS	Minimum	Maximum
Isopinocamphone	15358-88-0	25	45
Pinocamphone (*trans-*)	547-60-4	8.0	25
β-Pinene	127-91-3	7.0	20
Germacrene D	23986-74-5	1.2	4.5
Limonene	138-86-3	0.6	4.0
Sabinene	3387-41-5	1.0	3.5
allo-Aromadendrene	25246-27-9	1.0	3.0
β-Caryophyllene	87-44-5	1.0	3.0
Myrtenyl methyl ether	10300-03-5	0.9	3.0
β-Bourbonene	5208-59-3	0.8	2.6
Elemol	639-99-6	0.2	2.5
α-Pinene	80-56-8	0.4	1.5
Spathulenol	6750-60-3	0.1	1.5

[a] ISO 9841 Essential oil of hyssop ©ISO 2013; Geneva, Switzerland, www.iso.org

COMMERCIAL OILS

The ten chemicals that had the highest maximum concentrations in 26 commercial hyssop essential oil samples (concentration ranges provided) are the following: isopinocamphone (0.5-47.3%), *trans*-pinocarvone (5.9-37.9%), pinocarveol (0.1-19.2%), β-pinene (8.2-17.2%), β-phellandrene (0.9-6.0%), germacrene D (0.1-3.6%), limonene (1.0-3.5%), caryophyllene oxide (0.2-3.4%), β-caryophyllene (0.8-3%) and γ-cadinene (tr-2.9%) (Erich Schmidt, unpublished analytical data).

Chemotypes

Most oils from *Hyssopus officinalis* are dominated by isopinocamphone or pinocamphone. Some authors have grouped hyssop essential oils into chemotypes, depending upon the contents of the major components (24,28): chemotype A: isopinocamphone >50%; chemotype B: pinocamphone >50%; chemotype A+B: isopinocamphone and pinocamphone 20-50% each; chemotype E: any other component present in a substantial quantity (>30%). Examples of this latter group could include oils with high content of limonene (37%, ref. 8), 1,8-cineole (53%, ref. 6; cited in ref. 40), methyl eugenol (38%, ref. 8), pinocarvone (29,2%, ref. 21), β-phellandrene (43), myrtenyl acetate (74.1%, ref. 42) and α-pinene (70.9%, ref. 40).

Other authors have suggested the following chemotypes: myrtenyl acetate, methyleugenol and limonene, 1,8-cineole, pinocamphone, isopinocamphone, pinocarvone (from *H. officinalis* ssp. *angustifolius* (Bieb.)

Table 5.41.2 Constituents identified in hyssop oils

Constituent	CAS	Percentage and range						
		A	B (19)	C (10)	D (3)	E (4)	F (27)	G
m-Acetanisole	5451-83-2				0.1			
o-Acetanisole	579-74-8							4.7[x7]
γ-Amorphene	6980-46-7						0-0.5	
(*E*)-Anethole	4180-23-8							0.04[i]; 0.3[u]
Aristolene	6831-16-9		0.7					
Aromadendrene	489-39-4	0-0.3	0.1				0.3-0.6	
allo-Aromadendrene	25246-27-9	0.6-2.6	2.4	0.3-1.4	0.8		0-3.7	0.7[w2]; 0.9[x4]; 1.7[s]; 1.9[x]; 3.2[w]
Benzaldehyde	100-52-7				tr			tr[w1,w4]; 0.1[x2]
Benzeneacetaldehyde	122-78-1		0.2		0.2			0.3[x7]
cis-α-Bergamotene	18252-46-5							1.4[j]
trans-α-Bergamotene	13474-59-4							+[w9]
Bicyclogermacrene	24703-35-3	0.1-2.3		0.3-4.3	1.3		2.1-3.3	0.6[w]; 1.2[d]; 1.6[m]; 3.7[x4]; 5.1[s]
Bicycloheptanol								0.7[a,x2,x3]
Bicycloheptene carboxaldehyde								0.6[a,x3]
epi-Bicyclosesquiphellandrene	54274-73-6		2.6					
α-Bisabolene	17627-44-0							0.04[i]
β-Bisabolene	495-61-4				0.06			+[w9]; 0.4[i]; 0.6[f]
α-Bisabolol	515-69-5							1.4[w3]
Borneol	507-70-0							tr[w1]; 3.0[o]
Bornyl acetate	76-49-3							1.4[x5]
β-Bourbonene	5208-59-3	0.1-2.4	0.8		1.1	0.2-0.5	1.1-1.5	1.2[u]; 1.4[n,x]; 1.6[w3]; 1.7[p]; 4.3[w]
Bulnesol	22451-73-6							0.1[w1]
10β-Cadina-1(6),4-diene								+[w9]
α-Cadinene	24406-05-1							2.9[m]
β-Cadinene	523-47-7							3.8[m]
γ-Cadinene	39029-41-9	tr-2.9	0.7		0.09	0.1-0	0-0.2	+[w9]; tr[w4]; 0.05[d]; 0.1[x4]; 0.2[w6]
trans-γ-Cadinene								tr[w1]
δ-Cadinene	483-76-1	0.1-0.9	0.8		tr		0-0.7	+[w9]; 0.2[i,w1]; 0.5[x5]; 0.8[l]
α-Cadinol	481-34-5			0.1-0.4				tr[i,w5]; 0.3[w6]; 0.8[d,x4]
epi-α-Cadinol	5937-11-1					0.3-0.3	0.1-0.8	
cis-Calamenene	72937-55-4							0.8[m]
Calamenene hydrate								0.1[i]
Camphene	79-92-5	tr-0.2		tr-0.1	0.07	0.1-0.1	0-0.1	0.1[d]; 0.2[e]; 0.3[w1]; 0.5[x]; 0.6[w2]
α-Campholenal	4501-58-0							1.1[w1]

Table 5.41.2 Constituents identified in hyssop oils (*continued*)

Constituent	CAS	Percentage and range						
		A	B (19)	C (10)	D (3)	E (4)	F (27)	G
Camphor	76-22-2	0-0.05						6.8[x5]; 12.5[o]; 16.3[r]; 16.4[w5]
δ3-Carene	13466-78-9							0.9[w3]
Carvacrol	499-75-2		0.2					tr[i]; 0.5[p]; 2.5[w3]; 3.0[f]; 5.1[e]; 7.7[x7]
Carveol	99-48-9							3.4[x2,w3]
(E)-Carveol	1197-07-5							0.2[w1]; 0.3[i]
(Z)-Carveol	1197-06-4							0.1[i]
Carvomenthol	499-69-4							0.1[f]
Carvone	99-49-0	0.06-0.3		0.2-0.4	0.1			tr[w1]; 0.2[x]; 0.3[w3]; 1.6[x7]
Carvotanacetone	499-71-8				0.07			+[w9]; 0.4[a,p,w6]
β-Caryophyllene	87-44-5	0.8-3.0	3.1	0.2-2.7	4.0	2.3-0.4	1.1-4.0	2.8[x4]; 2.9[w5]; 4.2[w]; 5.0[x7]; 5.6[m]
9-epi-(E)-Caryophyllene	68832-35-9						0-0.6	+[w9]
γ-Caryophyllene	118-65-0							+[w9]
Caryophyllene oxide	1139-30-6	0.2-3.4	2.1	tr-0.4	5.2	0.2-0.3	0-1.7	0.3[e]; 0.5[n]; 1.2[w6]; 2.1[w1]; 2.6[j]
α-Cedrene	469-61-4							4.4[x]
β-Cedrene	546-28-1	0.2-1.0						
Cedrol	77-53-2							0.3[x]
α-Chamigrene	19912-83-5					1.9-0		
cis-Chrysanthenyl acetate	67999-48-8							tr[w4]
1,8-Cineole	470-82-6		0.6		5.8			6.1[x]; 8.0[w4]; 10.8[x3]; 11.8[x4]; 12.0[o]; 12.2[x9]; 13.0[x2]; 52.9[q]
α-Copaene	3856-25-5	0.09-0.09			0.07		tr-0.1	+[w2]; 0.03[i]; 0.3[w1,x]
β-Copaene	18252-44-3				0.2			tr[w4]; 0.09[w2]; 1.0[x]
Croweacin	484-34-4							0.2[a,w6]
Cryptone	500-02-7							+[w9]; 0.1[w7]; 0.3[x]; 0.4[w4]; 0.6[w2]
α-Cubebene	17699-14-8							+[w9]; 0.5[x5]; 2.2[a]
β-Cubebene	13744-15-5	0.06-0.2	1.5					+[w2,w9]; 0.05[i]
Cubenol	21284-22-0							0.3[x5]
Cumene	98-82-8							0.1[x1]
Cuminaldehyde	122-03-2		0.1					0.7[w3]; 3.2[x7]
Cuminyl alcohol	536-60-7			tr-0.3	0.2			
Curcumene	644-30-4						0-0.9	tr[i]
Cyclosativene	22469-52-9							+[w9]
p-Cymene	99-87-6	0.01-1.1		tr-0.7	tr			1.2[w6]; 1.6[x1]; 1.8[x4]; 2.8[f]; 3.0[e]
p-Cymen-8-ol	1197-01-9							tr[w1]; 0.03[i]; 0.3[f]
p-Cymen-9-ol	4371-50-0							0.2[a,w6]
Cyperene	2387-78-2	0.2-1.8						
Cypertundone							0-0.3	
(E)-β-Damascenone	23726-93-4				0.1			
n-Decane	124-18-5							8.7[w3]; 11.8[x7]
(E)-2-Decenal	3913-81-3						0-tr	
Dehydroaromadendrene	698388-95-3							0.1[a,x2,x3]
Dehydro-1,8-cineole	92760-25-3				tr			
Dehydrosabinaketone	147043-52-5							tr[w4]
Dihydroactinidiolide	17092-92-1		0.2					
Dihydroverbenone	18358-52-6							0.4[a,u]
1,7-Dimethoxy-p-cymene								0.2[i]
(5E)-2,5-Dimethyl-3-methylene-1,5-hepta-diene	1316759-92-8		1.8					
4,7-Dimethyl-4-octa-nol	19781-13-6							0.2[a,s]

Table 5.41.2 Constituents identified in hyssop oils (*continued*)

Constituent	CAS	Percentage and range						
		A	B (19)	C (10)	D (3)	E (4)	F (27)	G
2,3-Dimethyl-1,3-pentadiene	1113-56-0							0.1[a,p,w6]
Dodecane (*n*-)	112-40-3							5.2[x7]
β-Elemene	33880-83-0							+[w2]; tr[w1]; 0.2[i]; 0.9[h]
γ-Elemene	29873-99-2							0.2[a,l]
δ-Elemene	20307-84-0						2.4-3.1	0.4[x]
Elemol	639-99-6	0.2-1.9	17.2	0.5-5.1	7.4			3.4[d]; 5.6[j]; 7.5[x4]; 8.6[x6]; 9.0[x8]
Epiglobulol	88728-58-9							tr[i]
α-Eudesmol	473-16-5		4.3	tr-0.4	1.4	0.2-0		0.2[l,w1]
β-Eudesmol	473-15-4		4.3	tr-0.5			0.6-1.4	tr[w1]; 0.2[l]; 0.3[x]; 0.9[w6]
γ-Eudesmol	1209-71-8		3.7	tr-0.4	1.3	0.2	0.5-1.3	0.3[w6]
10-epi-γ-Eudesmol	15051-81-7	0.06-0.6					0.2-0.7	tr[w1]; 0.3[d]; 0.5[x4]
Eugenol	97-53-0		0.3		0.1		0-0.1	+[w2]; 0.9[w3]; 1.6[x7]
(*E,E*)-α-Farnesene	502-61-4							+[w9]
(*E*)-β-Farnesene	18794-84-8	0.08-0.08						0.04[w2]
(*Z*)-β-Farnesene	28973-97-9							tr[w4]
(*Z*)-β-Farnesol								0.3[l]
Fenchone	1195-79-5							0.3[x]
β-Fenchyl alcohol	22627-95-8							tr[w1]; 0.5[a,p,w6]
Formic acid	64-18-6							0.1[x3]
Geranial	141-27-5	0.02-0.3						0.1[w2]
Geraniol	106-24-1							0.2[w2]; 3.0[o]
Geranyl acetate	105-87-3	0.03-0.1						tr[w1]
Germacra-4(15),5,10-(14)-trien-α-ol							0.1-0.4	
Germacrene	28028-64-0							3.4[x5]
Germacrene B	15423-57-1					0.7-0		
Germacrene D	23986-74-5	0.8-3.6		0.5-6.2	3.4	2.3-1.1	0-5.5	4.7[x8]; 4.8[w]; 5.1[m]; 9.0[s]; 11.2[x4]
Germacrene D-1,10-epoxide	65882-77-1							tr[i]
Germacrene D-11-ol								5.7[j]; 6.1[k]
α-Guaiene	3691-12-1							1.0[w1]
Guaiol	489-86-1		0.4					1.0[w1]
α-Gurjunene	489-40-7		0.1				0.5-1.1	tr[w1]; 0.1[w2]; 0.4[w3]; 0.6[w]
β-Gurjunene	73464-47-8	0.06-0.3					0-0.8	tr[w1]
γ-Gurjunene	22567-17-5		1.1					
Hedycaryol	21657-90-9		2.0				4.0-9.1	8.5[w5]
Heptacosane	593-49-7							0.1[x3]
Heptanal	111-71-7				tr			
Hexadecanal	629-80-1							tr[w1]
Hexadecanoic acid	57-10-3		0.08					
Hexahydrofarnesyl acetone	502-69-2						0.2-0.6	0.2[w1,w6]; 1.5[x7]
Hexahydronaphthalene	41375-99-9							0.1[a,x2,x3]
Hexanal	66-25-1				0.1			
Hexanol	111-27-3				tr			
(*E*)-2-Hexenal	6728-26-3				0.5			
(*Z*)-3-Hexenol	928-96-1	0-tr						+[w9]
α-Humulene	6753-98-6	0.2-1.2	0.9	tr-0.6	0.8		0-1.1	0.4[h]; 0.5[s]; 0.7[x]; 0.9[j]; 3.2[m]
Humulene epoxide II	19888-34-7							tr[w1]
2-Hydroxy-isopinocamphone								0.07[i]
s-(+)-5-(1-Hydroxy-1-methylethyl)-2-cyclohexan-1-one								0.8[a,p,w6]
2-Hydroxypinocamphone	10136-65-9			tr-0.3				

Table 5.41.2 Constituents identified in hyssop oils (*continued*)

Constituent	CAS	Percentage and range						
		A	B (19)	C (10)	D (3)	E (4)	F (27)	G
trans-2-Hydroxypino-camphone	20536-50-9				tr		0-0.3	tr[w4]; 0.08[w2]
(+)-(1*R*,2*R*)-2-Hydroxy-2,6,6-trimethylnor-pinan-3-one								5.4[p]
Isoamyl alcohol	123-51-3							+[w9]
Isoamyl isovalerate	659-70-1							tr[w1]
Isobornyl acetate	125-12-2							tr[w1]
Isogermacrene D	317819-80-0							0.07[w2]
Isopimara-9(11),15-diene	39702-28-8							0.1[w4]
Isopinocampheol	27779-29-9							3.2[a,v]
Isopinocamphone	15358-88-0	22.3-47.3	45.5	5.8-59.9	48.6	26.9-27.7	11.5-33.6	42.0[w8]; 42.7[x3]; 43.3[n]; 43.8[x2]; 55.1[x6]; 56.7[e]; 57.3[f]; 57.7[w4]
2-Isopropyl-5-methyl-9-methylenebicyclo-[4.4.0]dec-1-ene	150320-52-8		2.7					
8-Isopropyl-5-methyl-2-methylene-1,2,3,4,4a,5,6,7-octahydro-naphthalene			3.5					
Isopropyl myristate	110-27-0							0.1[f]
Isospathulenol	88395-46-4							0.1[i]
Isoterpinolene	586-63-0							0.1[f]
(*E*)-Jasmone	6261-18-3				0.1			
(*Z*)-Jasmone	488-10-8		0.2				0-0.4	
Ledol	577-27-5		0.7				0.5-1.1	
Limonene	138-86-3	1.0-3.5		0.6-1.0	0.7	1.7-4.3[c]	0.5-1.1	2.2[w]; 2.7[w1]; 3.0[x7]; 5.6[l]; 7.2[d]; 9.0[x4]; 11.1[w8]; 12.2[n]; 37.4[i]
Limonen-4-ol	3419-02-1							0.2[w7]
Linalool	78-70-6	0.6-1.5	2.5	0.1-2.1	1.3	0.8-0.7	0.9-2.5	1.4[w2]; 1.7[w7]; 1.8[h]; 5.5[x]; 7.9[m]
cis-Linalool oxide	11063-77-7				tr			0.3[p,w6]
trans-Linalool oxide	11063-78-8						0-tr	0.2[p,w6]
Linalyl acetate	115-95-7	0.06-0.8						0.4[x3]; 0.8[h]; 2.9[m]; 5.9[x]
Linalyl propionate	144-39-8							0.4[x2]
β-Maaliene	489-29-2							+[w9]
p-Mentha-1(7),8-diene	499-97-8				0.1			
p-Mentha-1(7),8-dien-2-ol	35907-10-9							0.03[i]
trans-*p*-Mentha-2,8-dien-1-ol	4017-77-0							0.05[i]
Menthanol								0.9[a,x3]
p-Menthanol								0.9[a,x2]
cis-*p*-Menthen-1-ol	35376-39-7							0.05[w2]
trans-*p*-Menthen-1-ol	586-23-2							+[w2]
cis-*p*-Menth-2-en-1-ol	29803-82-5				tr			0.1[w4]
trans-*p*-Menth-2-en-1-ol	29803-81-4		0.2		tr			0.2[w7]
Menthol	89-78-1			0.1-0.3				
Menthone	89-80-5			tr-0.4				0.08[w2]
Methyl *p*-anisate	121-98-2				0.08			0.3[a,p,w6]
Methyl arachidonate	2566-89-4							0.1[a,x3]
Methyl 9-(2-[(2-butyl-cyclopropyl)methyl]-cyclopropyl)nonanoate			0.1					
Methyl chavicol	140-67-0			0.1-0.2				0.09[w2]; 0.2[e]; 0.4[w5]; 2.9[t]

Table 5.41.2 Constituents identified in hyssop oils (*continued*)

Constituent	CAS	A	B (19)	C (10)	D (3)	E (4)	F (27)	G
7-(1-Methylethylidene) bicyclo[4.1.0]heptane	53282-47-6		1.2					
Methyl eugenol	93-15-2	0.1-1.9	1.0	0.1-0.4	0.4	0.3-0.3		0.2[f,l]; 0.4[d,j,p]; 0.5[u,w]; 38.3[i]
Methyl 2-methylbutyrate	868-57-5	0-tr						
Methyl myrtenate	30649-97-9						tr-0.4	tr[w4]; 0.1[w7]; 2.2[w5]
4-Methyl-3-pentenal	5362-50-5							0.4[a,p,w6]
4-Methyl-3-penten-2-one	141-79-7							0.5[a,p,w6]
Mintsulfide	72445-42-2							0.01[y]
trans-Muurola-3,5-diene	262352-88-5						0-0.3	
γ-Muurolene	30021-74-0		0.8				2.4-6.2	+[w2]; tr[w1]
α-Muurolol	104245-48-9			tr-0.3				tr[w4]
Myrcene	123-35-3	0.7-2.6		0.1-1.8	1.3		1.0-2.0	1.7[e]; 2.2[d,s]; 2.9[g]; 3.0[h,o]; 3.5[m,t]
Myristicin	607-91-0							0.9[x7]
Myrtenal	564-94-3	0.2-0.7		0.1-0.4				0.4[x5]; 0.8[x]; 0.9[w7]; 2.0[p]; 2.3[f]
Myrtenol	515-00-4	0.2-2.1	6.0	0.7-3.3		3.1-2.8	2.4-3.0	1.9[w2]; 2.1[e,x8]; 2.6[d]; 2.8[j]; 3.0[x4]
Myrtenyl acetate	1079-01-2	0.09-0.6					0-3.2	+[w2]; tr[w4]; 0.2[x]; 1.0[f]; 74.1[x5]
Myrtenyl formate								0.3[w7]
Myrtenyl methyl ether	10300-03-5	1.0-2.5		1.0-6.1				2.0[h]; 2.5[w7]; 2.8[w4]; 3.4[w]; 5.1[s]
Naphthalene	91-20-3							1.0[x7]
Neral	106-26-3	0.03-0.2						0.07[i]; 0.2[w2]
Nerol	106-25-2							0.1[w2]; 0.3[x]
(*E*)-Nerolidol	40716-66-3							0.3[w5]; 0.4[x7]
Neryl acetate	141-12-8	0.05-0.1						+[w2]; tr[w1]; 0.2[f]; 0.4[e]
Nerylacetone	3879-26-3						tr	
Nitrobicyclononane								0.1[a,x3]
Nonanal	124-19-6							+[w9]; tr[w1,w4]
Nopinone	24903-95-5							0.08[i]; 0.1[g]
(1*R*)-(+)-Norinone								0.8[f]
(*E*)-Ocimene	27400-72-2							0.4[x]
(*Z*)-Ocimene	27400-71-1							0.2[x]
(*E*)-β-Ocimene	3779-61-1	0.1-1.1		tr-0.6		1.2		0.1[g]; 0.4[d]; 0.7[h]; 1.1[s]; 1.6[w]; 2.8[t]
(*Z*)-β-Ocimene	3338-55-4	0.07-1.0		0-0.2	tr		0-1.2	0.1[l]; 0.2[j]; 0.3[w5]; 0.9[w4]; 1.3[g,w3]
allo-Ocimene	673-84-7						0-tr	
9-Octadecenamide	3322-62-1		1.0					
n-Octanal	124-13-0	0.03-0.1						
3-Octanol	589-98-0							+[w9]
3-Octanone	106-68-3				0.09			+[w9]; 0.06[w2]; 0.2[x]
1-Octen-3-ol	3391-86-4		0.3					0.3[w2]; 0.5[f]
3-Oxo-*p*-menth-1-en-7-al	160152-34-1							0.1[w4]
Patchoulane	25491-20-7		0.4					
γ-Patchoulene	508-55-4						0-0.4	
Pentacosadienoic acid								0.1[a,x2,x3]
Perillaldehyde	2111-75-3							+[w2]; tr[i]
Phellandral	21391-98-0				tr			+[w9]
α-Phellandrene	99-83-2	0.02-0.1			tr	1.7-1.7		0.6[d,e]; 1.1[w2]; 1.6[w5]; 9.6[m]
β-Phellandrene	555-10-2	0.9-6.0		0.6-7.9		0.2-4.3[c]	1.5-3.6	2.4[j]; 2.5[x8]; 3.2[w4]; 3.6[x2]; 3.9[w]; 4.0[w5]; 4.2[l]; 9.5[h]; 12.7[t,x6]
Phenandrene								0.1[a,x2]
Phenanthrene	85-01-8							0.1[a,x3]
Phytol acetate	5016-85-3				0.2			
Phytone	16825-16-4	0.08-0.2						
Pinanediol								0.4[f]
cis-2,3-Pinanediol	18680-27-8							tr[w4]
α-Pinene	80-56-8	0.6-1.1		0.1-0.5	0.3	1.1-0.6	0.4-0.8	1.8[j]; 1.9[w]; 2.3[x]; 7.3[x6]; 70.9[w1]
β-Pinene	127-91-3	8.2-17.2	0.2	1.2-13.2	6.1	21.8-11.9	7.0-11.4	14[h]; 16.3[x6]; 16.8[q]; 17.6[g]; 18.4[l]; 19.4[e]; 20.5[w]; 32.3[x4]
Pinocampheol								0.4[w7]
Pinocamphone	547-60-4	5.9-37.9		1.3-64.9	1.9	26.1-39.2		34.0[m]; 38.4[w7]; 46.5[o]; 49[l]; 50.5[y]; 53.5[d]; 60.2[w5]; 64.8[x4]

Table 5.41.2 Constituents identified in hyssop oils (*continued*)

Constituent	CAS	Percentage and range						
		A	B (19)	C (10)	D (3)	E (4)	F (27)	G
trans-Pinocarveol	1674-08-4	0.1-19.2	0.1		0.3		0.3-1.2	0.2[w2]; 0.3[i,p]; 0.5[e]; 0.6[u]; 0.8[w7]
Pinocarvone	30460-92-5			0.1-16.9			9.0-28.1	4.5[w2]; 5.3[w3]; 6.5[f]; 6.8[e]; 7.7[w4]; 11.0[x2]; 11.7[w7]; 20.3[x9]; 29.2[g]
trans-Pinocarvyl acetate	1686-15-3							tr[w1,w4]
cis-Pinonic acid	61826-55-9							0.6[f]; 1.8[w6]
Piperitenone	491-09-8							1.1[x7]
cis-Piperitol	16721-38-3				tr			
trans-Piperitol	16721-39-4							tr[w4]
Piperitone	89-81-6							0.2[w7]; 1.3[x7]; 1.7[w3]
Rosefuran	15186-51-3							0.2[a,x2]; 0.3[a,x3]
Sabinene	3387-41-5	1.3-2.4		0.6-1.8	0.5	2.1-2.0	0.8-2.5	2.2[x3]; 2.3[x2]; 2.9[w]; 5.2[j]; 11.0[x1]
Sabinene hydrate	546-79-2							5.4[x4]; 8.5[w7]
cis-Sabinene hydrate	15537-55-0	0.03-0.2		0.1-4.0				tr[w4]; 0.1[x2]; 0.4[x]; 2.0[d]; 2.7[e]
trans-Sabinene hydrate	17699-16-0	0.02-0.3				0.08-0.3		+[w2]; tr[w4]; 0.05[j]; 0.2[w5]; 0.9[p]
cis-Sabinol	3310-02-9							1.8[x5]
Sabinyl acetate								0.3[u]; 0.5[l]
Salvial-4(14)-en-1-one	73809-82-2							0.04[i]
α-Selinene	473-13-2		1.8					
β-Selinene	17066-67-0						0-1.4	0.2[w1]
δ-Selinene	473-14-3		0.4					0.4
β-Sesquiphellandrene	20307-83-9				0.4			
Seychellene	20085-93-2					0.8-0.4		
Spathulenol	6750-60-3		2.6	0.7-2.0		0.1-0.3	2.0-3.0	2.0[x]; 2.1[x5]; 2.3[m]; 2.8[j]; 3.0[x7]
Succinic acid	110-15-6							0.2[a,x2,x3]
α-Terpinen-7-al	1197-15-5				tr			
α-Terpinene	99-86-5	0.04-0.3		0-0.1	tr	0.05-0.1		0.4[w2]; 0.5[d]; 0.7[x4]; 1.9[e,m]; 9.4[x6]
γ-Terpinene	99-85-4	0.08-0.9		0-0.1	0.08	0.2-1.1	0-tr	0.3[w5]; 0.4[w]; 1.0[d]; 1.3[x4]; 5.6[e]
Terpinen-4-ol	562-74-3	0.02-0.4		0.1-1.0				2.6[d]; 3.4[w7]; 7.1[f]; 8.3[x4]; 10.5[e]
α-Terpineol	98-55-5	0.2-1.4	1.2			0.3-0.3	0.1-0.4	0.5[j]; 0.8[e]; 1.4[w7]; 5.2[x]; 7.4[x6]
(Z)-β-Terpineol	7299-40-3		0.5					
γ-Terpineol	586-81-2							0.4[a,x]
δ-Terpineol	7299-42-5							0.4[w4]; 0.5[w7]
Terpinolene	586-62-9	0.04-0.09			tr	0.08-0.6	tr-0.2	0.1[w4]; 0.2[s,x]; 0.3[d,w1]; 0.7[e]
α-Terpinyl acetate	80-26-2							0.3[w1]; 1.4[x]
n-Tetradecane	629-59-4							1.6[x7]
Tetrahydropyran	142-68-7							0.3[a,x3]
Thuja-2,4(10)-diene	36262-09-6				0.1			+[w9]; 0.2[w1]
α-Thujene	2867-05-2	0.2-0.4		tr-0.2	0.09	0.3-0.3	tr-0.1	0.2[f]; 0.3[d,h]; 0.5[x4]; 0.6[w1]; 1.0[j]
β-Thujene	28634-89-1		0.9					
α-Thujone	546-80-5	0.03-0.4		tr-0.2	0.07	0.04-0.3		0.07[t]; 0.2[w2]; 0.4[e,x4]
β-Thujone (*cis*-)	471-15-8	0.04-0.6	tr-0.3	tr			0-0.1	tr[w1]; 0.1[f]; 0.2[p,t,w6]; 6.1[x]
(-)-β-Thujone (*trans*-)	33766-30-2						0-0.1	tr[w4]; 0.1[g]; 0.3[d]; 0.9[x]
Thymol	89-83-8							0.2[p]; 0.5[w3]; 0.6[f]; 1.3[e]; 19.0[x7]
Torreyol	19435-97-3							tr[i]
Totarene								tr[w4]
Tricosadienoic acid								0.1[a,x3]
Tricyclene	508-32-7							tr[w4]
endo-Trimethylamine								0.1[a,x2,x3]
2,6,6-Trimethylbicyclo [3.1.0]heptan-3-one			0.3					
4,11,11-Trimethyl-8-methylenebicyclo [7.2.0]undec-4-ene			0.5					
Trimethyl norpinen-3-one								5.4[a,w6]

Table 5.41.2 Constituents identified in hyssop oils (*continued*)

Constituent	CAS	Percentage and range						
		A	B (19)	C (10)	D (3)	E (4)	F (27)	G
ar-Turmerone	532-65-0							0.6[x7]
2-Undecanone	112-12-9						0-0.4	
Valencene	4630-07-3		1.0					
Valerianol (kusunol)	20489-45-6	0.08-0.7						
Verbenene	4080-46-0							+[w2]
cis-Verbenol	1845-30-3							tr[w1]
Verbenone	80-57-9							tr[w1]
Viridiflorene (ledene)	21747-46-6	1.2						1.1[w1]
Viridiflorol	552-02-3	1.7					0.1-1.9	0.2[w6]
β-Ylangene	20479-06-5						0.1-0.5	
Zonarene	41929-05-9							+[w9]

A twenty-six hyssop essential oil samples from France, Morocco and Hungary analyzed between 1998 and 2013; lowest and highest concentrations given (E. Schmidt, unpublished data)

B one lab-hydrodistilled oil and one laboratory steam-distilled oil from the flowering aerial parts of *Hyssopus officinalis* cultivated in Poland; highest concentrations given (ref. 19)

C twelve oils produced from three forms of *H. officinalis* ssp. *officinalis* (f. *cyaneus* Alef., f. *ruber* Mill., f. *albus* Alef.) cultivated in Serbia; lowest and highest concentrations given (ref. 10); the color had no influence on the oil composition

D one lab-distilled oil from the aerial parts of *H. officinalis* growing in the garden of the Faculty of Pharmacy of the Medical University of Lublin, Poland (ref. 3)

E 2x ten samples of oils of hyssop cultivated at the experimental station of the Faculty of Horticultural Science of Corvinus University of Budapest, harvested during the entire vegetation period (April-September) in 2001 and 2003; the average concentrations in both years are given (ref. 4)

F six lab-hydrodistilled oils from flowering *Hyssopus officinalis* cultivated at six locations in Lithuania; lowest and highest concentrations given (ref. 27)

G data from other studies (indicated with superscript letters); highest concentrations found in any study reviewed here given; when two or more oils were investigated, only the highest concentrations are mentioned, unless indicated otherwise

[a] incorrect identity based on GC elution order/(probably) misidentification/does not occur naturally (refs. 24,30); [b] correct isomer not identified; [c] limonene +β-phellandrene; [d] one lab-distilled oil from hyssop plants growing wild in India (ref. 14); [e] three samples of cultivated *H. officinalis* harvested in full bloom in 2003, 2004 and 2005 (ref. 12); [f] one sample of lab-distilled oil from wild growing Turkish hyssop (ref. 17); [g] one lab-distilled oil from *Hyssopus officinalis* L. growing wild in Turkey (ref. 21); [h] one commercial oil from Bulgaria (ref. 20); [i] one highly atypical oil of wild growing Montenegrin *Hyssopus officinalis* with very high contents of methyl eugenol and limonene and very low content of pinocamphone and isopinocamphone (ref. 8); [j] one lab-distilled oil from hyssop cultivated in Serbia (ref. 16); [k] one lab-distilled leaf oil from Serbia (ref. 15); [l] one lab-distilled Indian essential oil of cultivated *Hyssopus officinalis* ssp. *officinalis* (ref. 13); [m] two lab oils from wild Italian hyssop (ref. 9); [n] one lab-distilled oil from Italian wild *H. officinalis* (ref. 7); [o] several Bulgarian oils (ref. 18); [p] one lab-distilled oil from Iran (ref. 11); [q] one lab-distilled oil from wild Spanish hyssop (ref. 6); [r] data taken from ref. 22, no details known; [s] one lab-prepared autumn and one spring oil from France (ref. 23); [t] two samples of essential oils from hyssop grown in Lithuania; data cited in ref. 24; [u] one sample of hyssop oil (ref. 25); [v] one oil sample from Moldova (ref. 26); [w] ten samples of essential oils from irrigated and non-irrigated hyssop plants in Spain, harvested and distilled industrially on five different days (ref. 2); [w1] one lab-hydrodistilled leaf oil from Nigeria, probably from wild *H. officinalis*, with an extremely high concentration of α-pinene (ref. 40); [w2] one lab-hydrodistilled oil from the flowering aerial parts of hyssop cultivated in West-Siberia (ref. 51); [w3] one lab-hydrodistilled hyssop leaf oil from Iran; a flower oil was also investigated (ref. 48); [w4] one lab-hydrodistilled leaf oil from wild Indian hyssop; oils from flowers and stems were also analyzed (ref. 41); [w5] three oils from hyssop leaves in various stages of development cultivated in Germany; oils from flowers, stems and roots of *H. officinalis* were also analyzed (ref. 44); [w6] one lab-hydrodistilled oil from the whole flowering plant growing wild in Iran (31); the report contains many mistakes (30); [w7] one oil produced from *H. officinalis* collected in the Himalaya (India) (ref. 32); [w8] one oil of hyssop from plants harvested in South-Italy (ref. 33); [w9] one sample of hyssop oil (ref. 34); [x] one commercial oil from French *H. officinalis* (ref. 29); [x1] one hydrodistilled oil from Iran (ref. 35); [x2] one lab-hydrodistilled oil from wild hyssop collected in India (ref. 36); this report contains many mistakes (30); [x3] one lab-hydrodistilled oil from India (ref. 37); this reports contains quite a few mistakes (30); [x4] eight lab-hydrodistilled oils from hyssop plants cultivated in India and harvested monthly from March to October (ref. 38); [x5] one lab-hydrodistilled oil from flowering hyssop harvested in the wild in Iran with an extremely high concentration of myrtenyl acetate (ref. 42); [x6] data from various studies cited in ref. 40; [x7] one lab-hydrodistilled oil from flowering wild hyssop harvested in Iran with an extremely atypical composition: 19% thymol and no pinocamphone or isopinocamphone (ref. 46); [x8] one lab-hydrodistilled oil from Polish cultivated hyssop (ref. 49); [x9] data from ref. 45, cited in ref. 41; [y] one French hyssop oil sample (ref. 50)

tr: trace (in columns C, G[h] and G[w1]: <0.1; in column D and G[w4]: <0.05)

Table 5.41.3 Major constituents of hyssop oils

Constituent	CAS	Percentage and range						
		A	B (19)	C (10)	D (3)	E (4)	F (27)	G
Pinocamphone	547-60-4	5.9-37.9		1.3-64.9	1.9	26.1-39.2		50.5[v]; 53.5[d]; 60.2[w5]; 64.8[x4]
Isopinocamphone	15358-88-0	22.3-47.3	45.5	5.8-59.9	48.6	26.9-27.7	11.5-33.6	55.1[x6]; 56.7[e]; 57.3[f]; 57.7[w4]
1,8-Cineole	470-82-6		0.6		5.8			12.0[o]; 12.2[x9]; 13.0[x2]; 52.9[q]
β-Pinene	127-91-3	8.2-17.2	0.2	1.2-13.2	6.1	21.8-11.9	7.0-11.4	18.4[j]; 19.4[e]; 20.5[w]; 32.3[x4]
Pinocarvone	30460-92-5			0.1-16.9			9.0-28.1	11.0[x2]; 11.7[w7]; 20.3[x9]; 29.2[g]
Elemol	639-99-6	0.2-1.9	17.2	0.5-5.1	7.4			3.4[d]; 5.6[j]; 7.5[x4]; 8.6[x6]; 9.0[x8]
β-Phellandrene	555-10-2	0.9-6.0		0.6-7.9		0.2-4.3[c]	1.5-3.6	4.0[w5]; 4.2[j]; 9.5[h]; 12.7[t,x6]
Germacrene D	23986-74-5	0.8-3.6		0.5-6.2	3.4	2.3-1.1	0-5.5	4.7[x8]; 4.8[w]; 5.1[m]; 9.0[s]; 11.2[x4]

LEGEND: SEE UNDER TABLE 5.41.2

Arcangeli, a synonym of *Hyssopus officinalis* L. according to The Plant List) and linalool (from *Hyssopus officinalis* L. var. *decumbens*, also a synonym for *Hyssopus officinalis* L. according to The Plant List) (41). Yet other investigators have proposed a similar chemotyping, but include an α-pinene chemotype (their own finding: 70.9%) and β-phellandrene-rich oils (40).

CONTACT ALLERGY/ALLERGIC CONTACT DERMATITIS

General

Contact allergy to and possible allergic contact dermatitis from hyssop oil has been reported in one publication only. A false-positive patch test reaction due to the excited skin syndrome cannot be excluded.

Case reports

An aromatherapist had non-occupational contact dermatitis with allergies to multiple essential oils used at work, including hyssop oil. The patient also reacted to geraniol, linalool, linalyl acetate, α-pinene, the fragrance mix and various other fragrance materials. α-Pinene and geraniol were demonstrated by GC-MS in hyssop oil (52), but neither of these are important components of commercial hyssop oils (Table 5.41.2, column A).

LITERATURE

1 GRIN Taxonomy for Plants; http://www.ars-grin.gov/

2 Moro A, Zalacain A, Hurtado de Mendoza J, Carmona M. Effects of agronomic practices on volatile composition of *Hyssopus officinalis* L. essential oils. Molecules 2011;16:4131-4139

3 Baj T, Kowalski R, Swiatek L, ModzelewskaM, Wolski T. Chemical composition and antioxidant activity of the essentials oil of hyssop (*Hyssopus officinalis* L. ssp. *officinalis*). Annales Univeritatis Mariae Curie-Sklodowska, Lublin, Polonia 2010;23:N3,7.

4 Veres K. Variability and biologically active components of some Lamiaceae species. Thesis, 2007, University of Szeged, Hungary. Available at: http://doktori.bibl.u-szeged.hu/1081/1/Veres_Katalin_2007.pdf

5 Khazaiea HR, Nadjafib F, Bannayana M. Effect of irrigation frequency and planting density on herbage biomass and oil production of thyme (*Thymus vulgaris*) and hyssop (*Hyssopus officinalis*). Ind Crops Prod 2008;27:315-321

6 García Vallejo MC, Herraiz JG, Perez-Alonso MJ, Velasco-Negueruela A. Volatile oil of *Hyssopus officinalis* L from Spain. J Essent Oil Res 1995;7:567-568

7 Renzini G, Scazzocchio F, Lu M, Mazzanti G, Salvatore G. Antibacterial and cytotoxic activity of *Hyssopus officinalis* L. oils. J Essent Oil Res 1999;11:649-654

8 Gorunovic MS, Bogavac PM, Chalcat JC, Chabard JL. Essential oil of *Hyssopus officinalis* of Montenegro origin. J Essent Oil Res 1995;7:39-43

9 Fraternale D, Ricci D, Epifano F, Curini M. Composition and antifungal activity of two essential oils of hyssop (*Hyssopus officinalis* L.). J Essent Oil Res 2004;16:617-622

10 Chalchat J-C, Adamovic D, Gorunovic MS. Composition of oils of three cultivated forms of *Hyssopus officinalis* endemic in Yugoslavia: f. *albus* Alef., f. *cyaneus* Alef. and f. *ruber* Mill. J Essent Oil Res 2001;13:419-421

11 Mahboubi M, Haghi G, Kazempour N. Antimicrobial activity and chemical composition of *Hyssopus officinalis* L. essential oil. J Biol Act Prod Nat 2011;1:132-137. Data also presented in ref. 31

12 Kizil S, Toncer O, Ipek A, Arslan N, Saglam S, Khawar KM. Blooming stages of Turkish hyssop (*Hyssopus officinalis* L.) affect essential oil composition. Acta Agriculturae Scandinavica, Section B—Soil & Plant Science 2008;58:273-279

13 Garg SN, Naqvi AA, Singh A, Ram G, Kumar S. Composition of essential oil from an annual crop of *Hyssopus officinalis* grown in Indian plains. Flavour Fragr J 1999;14:170-172

14 Khan R, Shawl AS, Tantry MA. Determination and seasonal variation of chemical constituents of essential oil of *Hyssopus officinalis* growing in Kashmir valley as incorporated species of western Himalaya. Chem Nat Comp 2012;48:502-504

15 Cvijovic M, Djukic D, Mandic I, Acamovic-Djokovic G, Pesakovic M. Composition and antimicrobial activity of essential oils of some medicinal and spice plants. Chem Nat Comp 2010;46:481-483

16 Mitić V, Đorđević S. Essential oil composition of *Hyssopus officinalis* L. cultivated in Serbia. Facta Universitatis Series: Physics, Chemistry and Technology 2000;2:105-108. Available at: http://facta.junis.ni.ac.rs/phat/phat2000/phat2000-07.pdf

17 Kizil S, Tolan V, Hasimi N, Kilinc E, Kratas H. Chemical composition, antimicrobial and antioxidant activities of hyssop (*Hyssopus officinalis*) essential oil. Not Bot Hort Agrobot Cluj 2010;38:99-103. Available at: http://www.notulaebotanicae.ro/index.php/nbha/article/view/4788/5077

18 Stoyanova A, Grozeva E. Bulgarian essential oils for the future: A review. Indian Perfum 2006;50:42-45

19 Wesolowska A, Jadczak D, Grzeszczuk M. Essential oil composition of hyssop (*Hyssopus officinalis* L.) cultivated in north western Poland. Herba Polonoca 2010;56:57-65. Available at: http://herbapolonica.pl/magazines-files/3653371-art.6-2010.pdf

20 Tsankova ET, Konaktchiev AN, Genova EM. Chemical composition of the essential oils of two *Hyssopus officinalis* taxa. J Essent Oil Res 1993;5:609-611

21 Figueredo G, Ozcan MM, Chalcat JC, Bagci Y, Chalard P. Chemical composition of essential oil of *Hyssopus officinalis* L. and *Origanum acutidens*. J Essent Oil Bear Plants 2012;15:300-306

22 Salma AS. Chemical and physiological studies on anise hyssop (*Agastache foeniculum* Pursh) and hyssop (*Hyssopus officinalis* L.) plants grown in Egypt as new spices. Bulletin of the National Research Centre 2002;27:25-35

23 Jean FI, Collin GJ, Lord D. Essential oils and microwave extracts of cultivated plants. Perfum Flavor 1992;17(3):35-41

24 Lawrence BM. Progress in essential oils. Perfum Flavor 1999;24(3):47-64

25 Bourrel C, Vilarem G, Michel G, Gaset A. Etude des propriétés bacteriostatiques et fongistatiques en milieu solide de 24 huiles essentielles préablement analysées. Rivista Ital EPPOS 1995;16:3-12. Data cited in ref. 24

26 Bodrug M, Miron M, Marcu N, Dragalin I. Biological characteristics and component structure of the oil obtained from *Hyssopus* introduced into Moldova. Bull Acad Stiinte Republ Moldova, Stiinte Biol Chim 1995;3:11-15.Data cited in ref. 24

27 Bernotienė G, Butkienė R. Essential oils of *Hyssopus officinalis* L. cultivated in East Lithuania. Chemija 2010;21:135-138

28 Veres K, Varga E, Dobos Á, Hajdú Z, Máthé I, Pluhár Z, Bernáth J. Investigation on essential oils of *Hyssopus officinalis* L. populations. In: Ch Franz,

A Máthé , G Buchbauer, Eds. Essential oils: basic and applied research. Carol Spring, USA: Allured Publishing Corporation, 1997:217-220. Data cited in ref. 24

29 Tognolini M, Barocelli E, Ballabeni V, Bruni R, Bianchi A, Chiavarini M, Impicciatore M. Comparative screening of plant essential oils: Phenylpropanoid moiety as basic core for antiplatelet activity. Life Sciences 2006;78:1419-1432

30 Lawrence BM. Progress in essential oils. Perfum Flavor 2014;39(November):40-50

31 Mahboubi M, Kazempour N. *In vitro* antimicrobial activity of some essential oils from Labiatae family. J Essent Oil Bear Plants 2009;12:494-508. Data also presented in ref. 11

32 Maheshwari ML, Chandel KPS, Chien M-J. The composition of essential oils from *Hyssopus officinalis* L. and *Cymbopogon jwarancusa* (Jones) Schult. collected in the cold desert of Himalaya. In: BM Lawrence, BD Mookherjee and BJ Willis, Eds. Flavors and Fragrances: A World Perspective. Amsterdam, The Netherlands: Elsevier Science Publishers, 1988: 171-176. Data cited in ref. 30

33 Gionfriddo F, Mangiola C, Pirrello A, Manganaro R. Studie preliminare sulla composizione degli oli essenziali di alcune piante officinali calbresi. Essenz Deriv Agrum 2001;71:67-70. Data cited in ref. 30

34 Costa R, da Fina MR, Valentino MR, Dugo P, Mondello L. Reliable identification of terpenoids and related compounds using linear retention indices interactively with mass spectrometry search. Nat Prod Comm 2007;2:413-418

35 Kazazi H, Rezaei K, Ghotb-Sharif SJ, Emam-Djomeh Z, Yamini Y. Supercritical fluid extraction of flavors and fragrances from *Hyssopus officinalis* L. cultivated in Iran. Food Chem 2007;105:805-811

36 Sah S, Lohani H, Haider SZ, Chauhan NK. Essential oil composition of *Hyssopus officinalis* L. grown in Malari region of Garhwal Himalaya. J Med Arom Plant Sci 2008;30:291-292. Data cited in ref. 30

37 Pal M, Kumar A, Kumar Soni D, Pathre UV. Essential oil composition of *Hyssopus officinalis* by GC-MS. Indian Perfum 2010;54(3):30-32. Data cited in ref. 30

38 Khan R, Rather MA, Shawl AS, Qurishi MA, Alam MS. Seasonal impact on yield and chemical profile of essential oil of *Hyssopus officinalis* L. cultivated in Kashmir valley, India. Indian Perfum 2010;54(3):46-49. Data cited in ref. 30

39 Lawrence BM. Progress in essential oils. Perfum Flavor 2008;33(December):52-57

40 Ogunwande IA, Flamini G, Alese OO, Cioni PL, Ogundajo AL, Setzer WN. A new chemical form of essential oil of *Hyssopus officinalis* L. (Lamiaceae) from Nigeria. Int J Biol Chem Sci 2011;5:46-55

41 Pandey V, Verma RS, Chauhan A, Tiwari R. Compositional variation in the leaf, flower and stem essential oils of hyssop (*Hyssopus officinalis* L.) from Western-Himalaya. J Herbal Med 2014;4:89-95

42 Fathiazad F, Mazandarani M, Hamedeyazdan S. Phytochemical analysis and antioxidant activity of *Hyssopus officinalis* L. from Iran. Adv Pharm Bull 2011;1(2):63-67

43 Lawrence BM. Chemical components of Labiatae oil and their exploitation. In: Harley RM, Reynolds T, Eds. Advances in Labiatae science. Kew, UK: Royal Botanic Gardens, 1992:399-436. Data cited in ref. 41

44 Schultz G, Stahl-Biskup E. Essential oils and glycosidic-bound volatiles from leaves, stems, flowers and roots of *Hyssopus officinalis* L. (Lamiaceae). Flav Fragr J 1991;6:69-73

45 Shah NC. Chemical constituents of *Hyssop officinalis* L.: 'Zufah Yabis, A Unani drug from U.P. Himalaya, India. Indian Perfum 1991;35:49-52. Data cited in ref. 41

46 Dehghanzadeh N, Ketabchi S, Alizadeh A. Essential oil composition and antibacterial activity of *Hyssopus officinalis* L. grown in Iran. Asian J Exper Biol Sci 2012;3:767-771

47 Fathiazad F, Hamedeyazdan S. A review on *Hyssopus officinalis* L.: Composition and biological activities. Afr J Pharm Pharmacol 2011;5:1959-1966

48 Moghtader M. Comparative evaluation of the essential oil composition from the leaves and flowers of *Hyssopus officinalis* L. J Horticult Forestr 2014;6:1-5

49 Zawiślak G. Morphological characters of *Hyssopus officinalis* Ll. and chemical composition of its essential oil. Mod Phytomorphol 2013;4:93-95

50 Takahashi K, Muraki S, Yoshida T. Synthesis and distribution of (-)-mintsulphide, a novel sulphur-containing sesquiterpene. Agric Biol Chem 1981;45:129-131

51 Myadelets MA, Domrachev MV, Cheremushkina VA. A study of the chemical composition of essential oils of some species from the Lamiaceae L. family cultivated in the western Siberian region. Russ J Bioorg Chem 2013;39:733-738

52 Dharmagunawardena B, Takwale A, Sanders KJ, Cannan S, Roger A, Ilchyshyn A. Gas chromatography: an investigative tool in multiple allergies to essential oils. Contact Dermatitis 2002;47:288-292

Chapters 5.42 and 5.43 JASMINE ABSOLUTE

The volatile chemicals in jasmine flowers are much appreciated in the perfumery industries. Conventional steam-distillation is generally considered unsuitable to process such materials, since it induces thermal degradation of many compounds contained in the flowers. Therefore, solvents, usually hexane or supercritical fluids such as CO_2, are used to extract the fragrant chemicals. After the solvent has evaporated, a viscose product remains called a 'concrete'; these concretes may sometimes be used as-is in soaps (1,2). Concretes contain not only fragrance compounds, but also fatty acids, their methyl esters and paraffins (up to 50% by weight), originating from the cuticular waxes covering the surface of the jasmine flowers (1,2). As these fatty materials do not contribute to the fragrance but do cause solubility problems, they have to be removed, which is done by solubilization of the concrete in a large excess of ethyl alcohol. The solution is then cooled and filtered to eliminate the precipitated waxes. After removing the ethyl alcohol and thereafter concentration of the solution by vacuum distillation, a product remains which is called an 'absolute'. This product contains all the fragrance compounds, but still also some fatty acid methyl esters and paraffins (1). These absolutes are the jasmine products used in the fragrance industries.

Jasmine 'volatile oil' can be obtained by treatment of the concrete with superheated steam. This product, which is *not* an essential oil, contains no paraffins and only small quantities of fatty acid methyl esters. Disadvantages are that fragrance compounds of higher molecular weight are not extracted and this technique produces unwanted thermal degradation of the fragrance (1). Volatile jasmine oils thus produced are not used in the fragrance industry, but they are utilized on a small scale in aromatherapy. It should be realized that the term 'oil' or even 'essential oil' is sometimes used for products that may in fact be absolutes (3,4,5). Nevertheless, in certain areas of India, essential oils are obtained by the classical method of hydrodistillation from *J. sambac* flowers (6).

Jasmine absolutes and oils can be obtained from the flowers of several species of the genus *Jasminum*, but are mainly derived from *Jasminum grandiflorum* L. (Spanish jasmine) and *Jasminum sambac* (L.) Aiton (Arabia, China, sambac jasmine) (2,7,8); hence, only these two (and not *J. officinale*, e.g., ref. 4) are discussed here.

J. grandiflorum flowers are mainly grown for production of jasmine absolute in India, Egypt and Morocco. *J. sambac* flowers are harvested in India mostly for ornamental purposes, garlands and religious offerings. Only some 5% of the annual flower harvest in India is used to produce absolute (8), but the sambac absolute appears to be gaining in popularity in the fragrance industries.

In the following chapters, the compositions of jasmine absolutes from *J. grandiflorum* (Chapter 5.42) and *J. sambac* (Chapter 5.43) are discussed. For the chemical composition of *concretes* see references 1,9,10,11,14,17 and 20. Studies not discussed here, as the source species could not be ascertained, include references 12,13,15,19 (partly discussed), 23 and 24. Studies not discussed as the mode of production of the 'oil' (leading to concrete, absolute, or essential oil?) could not be verified include ref. 21.

REFERENCES

1 Reverchon E, Della Porta G, Gorgoglione D. Supercritical CO2 fractionation of jasmine concrete. J Supercr Fluids 1995;8:60-65

2 Hellivan P-J. Jasmine. Reinventing the 'king of perfumes'. Perfum Flavor 2009;34(October):42-52

3 Verzele M, Maes G, Vuye A, Godefroot M, van Alboom M, Vervisch J, et al. Chromatographic investigation of jasmin absolutes. J Chromatogr A 1981;205:367-386

4 Wei A, Shibamoto T. Antioxidant activities and volatile constituents of various essential oils. J Agric Food Chem 2007;55:1737–1742

5 Srivastava HC, Bhupal Rao JVR, Karmakar PG, Angadi SP, Tumar TV, Venkateshwarlu G. A new variety of *Jasminum grandiflorum* Linn. for high yield of essential oil. PAFAI J 1997(Oct/Nov):16-18

6 Rao YR, Rout PK. Geographical location and harvest time dependent variation in the composition of essential oils of *Jasminum sambac*. (L.) Aiton. J Essent Oil Res 2003;15:398-401

7 Braun NA, Kohlenberg B, Sim S, Meier M, Hammerschmidt FJ. *Jasminum flexile* flower absolute from India – a detailed comparison with three other jasmine absolutes. Nat Prod Commun 2009;4:1239-1250

8 Braun NA, Sim S. *Jasminum sambac* flower absolutes from India and China – geographic variations. Nat Prod Commun 2012;7:645-650

9 Prakash O, Sahoo D, Rout PK. Liquid CO2 extraction of *Jasminum grandiflorum* and comparison with conventional processes. Nat Prod Commun 2012;7:89-92

10 Reverchon E, Della Porta G, Gorgoglione D. Supercritical CO2 fractionation of jasmine concrete. J Supercr Fluids 1995;8:60-65

11 Rout PK, Naik SN, Rao YR. Composition of absolutes of *Jasminum sambac* L. flowers fractionated with liquid CO2 and methanol and comparison with liquid CO2 extract. J Essent Oil Res 2010;22:398-406

12 Ranade GS. Essential oil profile Jasmine absolute. FAFAI 2002;4:63

13 Nidry ESJ, Srivastava HC. A comparative study of the antifungal activities of the absolutes of jasmine and tuberose and their constituents. Indian Perfum 2007;51:53-55

14 He C-M, Liang Z-Y, Liu H. Chemical constituents of jasmine absolute extracted by supercritical carbon dioxide. Nat Prod Res Developm 1998;11:53-57

15 Basset F. Journées de Digne. Le jasmin, la fleur de roi. Parf Cosmet Arômes 1994;119:58-64. Data cited in ref. 16

16 Lawrence BM. Progress in essential oils. Perfum Flavor 2005;30(October):60-? (last page unknown)

17 Musalam Y, Kobayashi A, Yamanishi T. Aroma of Indonesian jasmine tea. In: BM Lawrence, BD Mookherjee, BJ Willis, Eds. Flavors and Fragrances: A World Perspective. Amsterdam, The Netherlands: Elsevier Science Publishers, 1988:659-668. Data cited in ref. 18

18 Lawrence BM. Progress in essential oils. Perfum Flavor 1994;19(2):64-69

19 Anaç O. Gas chromatographic analysis of absolutes and volatile oil isolated from Turkish and foreign jasmine concretes. Flavour Fragr J 1986;1:115-119

20 Rawia AE, Lobna ST, Soad MMI. Physiological properties studies on essential oil of *Jasminum grandiflorum* L. as affected by some vitamins. Ozean J Appl Sci 2010;3(1):87-96

21 Zhang Z-G. Studies of volatile constituents of *Jasminum officinale* var. *grandiflorum* L. In: Proceedings International Conference on Essential Oils, Flavors, Fragrances and Cosmetics, IFEAT, London, October 1988 (book published in 1990). Data and reference cited in ref. 22

22 Lawrence BM. Progress in essential oils. Perfum Flavor 1995;20(July/August):29-41

23 Shaath NA, Azzo NR. Egyptian jasmin. Perfum Flavor 1992;17(5):49-55. Data and reference cited in ref. 22

24 Shaath NA, Azzo NR. Essential oils of Egypt. In: G Charalambous, Ed. Food flavors, ingredients and composition. Amsterdam: Elsevier Science Publishers, 1993:591-603. Data and reference cited in ref. 22

Chapter 5.42   JASMINUM GRANDIFLORUM ABSOLUTE

DEFINITION
Jasminum grandiflorum absolute is the absolute obtained from the flowers of the Spanish jasmine, *Jasminum grandiflorum* L.

INCI NOMENCLATURE
Description/definition: Jasminum grandiflorum flower extract is an extract obtained from the flowers of the Spanish jasmine, *Jasminum grandiflorum* L., Oleaceae
INCI name EU: Jasminum grandiflorum flower extract
INCI name USA: Jasminum grandiflorum (jasmine) flower extract
CAS registry number(s): 84776-64-7
EINECS number(s): 283-993-5

Jasminum officinale f. *grandiflorum* (L.) Kobuski, *Jasminum officinale* var. *grandiflorum* (L.) Stokes and *Jasminum officinale* subsp. *grandiflorum* (L.) E. Laguna are synonyms of *Jasminum grandiflorum* L. (www.the-plantlist.org).

Jasminum grandiflorum L. (Spanish jasmine, royal jasmine, Catalonian jasmine) is a scrambling deciduous shrub growing to 2-4 meters tall, with flowers that have an intensely floral, warm, rich and highly diffusive odor. It is native to Africa (Djibouti, Eritrea, Ethiopia, Somalia, Sudan, Kenya, Uganda, Rwanda) and Asia (Oman, Saudi Arabia, Yemen, India, Pakistan). It is occasionally naturalized in the tropics and is cultivated in Africa (Algeria, Egypt, Morocco), India, Italy and France (GRIN Taxonomy for Plants; http://efloras.org). In India, its leaves and flowers are widely used as an Ayurvedic herbal medicine in the treatment of tooth pain, fixing loose teeth, ulcerative stomatitis, leprosy, skin diseases, otorrhea, ear pain, strangury, dysmenorrhea, ulcers, wounds and corns (4).

By the method of solvent extraction the jasmine flowers are converted into jasmine concrete and by separating the waxes with alcohol, filtration and removal of the ethanol, jasmine absolute is obtained, for which there is a great demand in the perfume industry. Jasmine absolute is also employed in aromatherapy (14,15).

CHEMICAL COMPOSITION
Jasmine absolute from *Jasminum grandiflorum* L. is a mobile to viscous light orange to reddish-brown colored liquid, which has an intensive blooming, flowery and long lasting odor. The yield of absolute from the flowers of this species varies from 0.1 to 0.16 per cent. The main producing countries of *Jasminum grandiflorum* absolute are India, China, Algeria, Turkey and Morocco.

Literature data (up to November 25, 2011) on the chemical composition of absolutes of *Jasminum grandiflorum* L. and unpublished analytical data from one of us (E.S.) are shown in Table 5.42.1 in alphabetical order. In *Jasminum grandiflorum* absolutes from various origins,

over 220 chemicals have been identified. About 68 per cent of these were found in a single reviewed publication only. The major compounds found in absolutes of *Jasminum grandiflorum* L. from different sources are shown in Table 5.42.2. They include (highest concentrations in any study given) phytol (isomers) (52.2%), benzyl acetate (32.9%), benzyl benzoate (25.6%), (*E*)-phytol (21.7%), isophytol (12.7%), geranyllinalool (10.2%), (*Z*)-jasmone (9.5%), linalool (9.3%) and eugenol (6.8%). Well-known ingredients of absolutes of *Jasminum grandiflorum* L. that were present in high concentrations (>7%) in one or two studies were (*Z*)-3-hexenyl benzoate (22.0%), methyl linoleate (21.7%), (*E,E*)-α-farnesene (17.0%), squalene 2,3-oxide (9.2% and 11.7%), (*Z*)-phytol (14.2%) and phytyl acetate (11.0%). Uncommon or rare constituents of absolutes of *Jasminum grandiflorum* L. found in high concentrations (>7%) in single studies include paraffins (high) (27.4%), farnesol (12.7%, includes isophytol), *cis*-phytyl acetate (10.5%), isophytyl acetate (7.3%) and methyl jasmonate (7.2%).

Commercial absolutes
The ten chemicals that had the highest maximum concentrations in 41 commercial *Jasminum grandiflorum* absolute samples (concentration ranges provided) are the following: benzyl acetate (9.5-32.9%), benzyl benzoate (7.5-25.6%), (*E*)-phytol (4.1-16.9%), (*Z*)-phytol (0.2-13.3%), isophytol (0.4-11.3%), phytyl acetate (1.4-11.0%), geranyllinalool (1.1-10.2%), linalool (3.0-9.3%), squalene 2,3-oxide (5.9-9.2%) and methyl benzoate (0.06-8.1%) (Erich Schmidt, unpublished analytical data).

Older literature findings have been summarized in references 9 (up to 1992) and 8 (up to 1980).

Differences between the composition of *J. sambac* absolute and *J. grandiflorum* absolute
The most important cited differences between *J. sambac* absolute and *J. grandiflorum* absolute can be seen in Table 5.42.3. Phytol, phytyl acetate, isophytol and isophytyl acetate, important constituents of the *J. grandiflorum* absolutes, are (virtually) absent in sambac absolutes. The same goes for δ-jasmolactone, (*Z*)-jasmone and methyl jasmonate. Tricosene is absent in grandiflorum absolutes. Chemicals generally found in higher concentrations in sambac absolutes are benzyl alcohol, indole, linalool, methyl anthranilate and 2-phenethyl alcohol. Conversely, benzyl acetate, benzyl benzoate, and eugenol concentrations are generally higher in grandiflorum than in sambac absolutes (Table 5.42.3).

CONTACT ALLERGY/ALLERGIC CONTACT DERMATITIS

Jasminum grandiflorum absolute and Jasmine absolute (unspecified)
General
Contact allergy to/allergic contact dermatitis from jasmine absolute has been reported in 40 publications. Jasmine absolute has been included in the screening

Table 5.42.1 Constituents identified in *Jasminum grandiflorum* L. absolutes

Constituent	CAS	Percentage and range					
		A	B (7)	C (9)	D (12)	E (8)	F
Acetone	67-64-1						+[k]
3-Amino-1-phenylbutane	22374-89-6						+[a,b]
α-Amyrin	638-95-9		0.1				
β-Amyrin	559-70-6		0.1				
(*E*)-Anethole	4180-23-8						0.6[m]
Benzaldehyde	100-52-7		tr				
Benzoic acid	65-85-0						tr[n]
Benzyl acetate	140-11-4	9.5-32.9	15.5	24.2-29.7	11.3	8.5-25.1	7.7[m]; 8.1[g]; 18.8[a,n]; 23.7[i]
Benzyl alcohol	100-51-6	0.5-3.7	0.7	1.1-1.8	0.3	0.9-3.0	1.3[i]; 1.7[n]; 2.5[a]
Benzyl benzoate	120-51-4	7.5-25.6	11.4	17.9-21.2	0.6	6.2-13.1	13.3[n]; 20.7[i]; 22.0[m]; 26.2[g]
Benzyl butyrate	103-37-7		tr				
Benzyl crotonate	65416-24-2		tr				
Benzyl docosanoate	85263-74-7		<0.1				
Benzyl eicosanoate	77509-04-7		0.2				
Benzyl formate	104-57-4		tr				+[k]
Benzyl linoleate	47557-83-5		tr				
Benzyl linolenate	77509-02-5		0.2				
Benzyl palmitate	41755-60-6		0.1				
Benzyl propionate	122-63-4		tr				0.05[n]
Benzyl salicylate	118-58-1	0.05-0.6	0.2		0.1		0.1[i]
Benzyl stearate	5531-65-7		0.1				
Benzyl tiglate	37526-88-8		0.1				
Bergamotene						0.3-1.4	+[k]
α-*trans*-Bergamotene	13474-59-4	0.03-0.07					
2,3-Butanediol	513-85-9						tr[n]
Butyl benzoate	136-60-7		tr				
Butyl methyl ether	628-28-4						+[k]
γ-Cadinene	39029-41-9				0.2		
δ-Cadinene	483-76-1				0.5		
α-Cadinol	481-34-5				0.7		
β-Caryophyllene	87-44-5						0.08[n]
Chavicol	501-92-8	0.02-0.09	<0.1				<0.05[i]
Cinnamyl alcohol	104-54-1		0.1				
(*E*)-Cinnamyl benzoate	50555-04-9		0.1				
Coniferyl alcohol	458-35-5		<0.1				
Coniferyl aldehyde	458-36-6		tr				
Coniferyl benzoate	4159-29-9		tr				
p-Cresol	106-44-5	0.07-4.9	0.5			0.3-1.0	0.4[n]; 0.6[i]; 0.8[g]; 1.3[a]
p-Cresyl acetate	140-39-6		tr				
4-Cyclopropyl-2-methoxy-phenol	83356-69-8						tr[n]
n-Decane	124-18-5				<0.1		
trans-7-Decen-5-olide							tr[n]
3,7-Dimethyl-1,5-octa-diene-3,7-diol	13741-21-4		tr				
Diphenylethane	38888-98-1						tr[n]
1,2-Diphenylethanol	614-29-9						tr[n]
Ethanol	64-17-5	0.6-1.1					+[k]
Ethyl acetate	141-78-6	0.03-0.2					
Ethyl benzoate	93-89-0		tr				
Ethyl benzyl ether	539-30-0						tr[n]
2-(2-Ethylcyclopropyl)-3-methyl-2-cyclopenten-1-one	85135-71-3						tr[n]
Ethyl hexadecanoate	628-97-7		tr				
Ethyl jasmonate	54562-26-4						+[l]
Ethyl linolenate	1141-91-9					0-1.5	
Ethyl oleate	111-62-6					0-2.7	
Eugenol	97-53-0	0.9-6.8	1.3	2.0-4.0		0.8-3.5	1.5[a]; 2.0[n]; 2.5[i]; 2.8[g]
Farnesane	3891-98-3						+[k]
α-Farnesene	502-61-4			0.01-0.06		0.8-3.4	+[k,l]; 3.3[m]
(*E,E*)-α-Farnesene	502-61-4	0.8-3.2	1.1		17.0		1.1[i]
(*Z,E*)-α-Farnesene	26560-14-5		0.1		1.0		

Table 5.42.1 Constituents identified in *Jasminum grandiflorum* L. absolutes (*continued*)

Constituent	CAS	Percentage and range					
		A	B (7)	C (9)	D (12)	E (8)	F
(Z,Z)-α-Farnesene	28973-99-1	0.3-0.8					
β-Farnesene	502-60-3						+[l]
(E)-β-Farnesene	18794-84-8				0.5		0.9[n]
Farnesol	4602-84-0			0.2-0.4			12.7[a,e]
(E,E)-Farnesol	106-28-5				0.8		
(E,Z)-Farnesol	3879-60-5				0.5		
(E,E)-Farnesyl acetate	4128-17-0				0.5		
Geraniol	106-24-1	0.01-0.2	tr	0.03-0.1			tr[g]; 0.06[m]; 1.0[i]
Geranyl acetate	105-87-3		tr		0.1		tr[g]; 0.1[m]
Geranyl acetone	3796-70-1	0.02-0.5	<0.1				0.2[m]
Geranyllinalool	77368-82-2	1.1-10.2	3.6	2.8-4.4		3.1-8.0	0.4[n]; 3.0[i]
Germacrene D	23986-74-5				1.9		
Heptadecanal	629-90-3				<0.1		
Heptanal	111-71-7						+[k]
4-Heptanolide	105-21-5						+[l]
n-Heptene	25339-56-4						+[k]
Hexadecane	544-76-3				0.2		
Hexadecanoic acid	57-10-3	2.5-2.5			1.0		
Hexadecyl acetate	629-70-9					0.7-1.7	
Hexanal	66-25-1						+[k]
(E)-2-Hexanal					<0.1		
Hexanol	111-27-3	0.02-0.02	tr				
2-Hexanol	626-93-7						+[k]
3-Hexanol	623-37-0						+[k]
4-Hexanolide	695-06-7						+[l]
(E)-2-Hexenal	6728-26-3		tr				
(Z)-3-Hexenal	6789-80-6						+[k]
(E)-2-Hexen-1-ol	928-95-0					0-0.3	+[k]
(E)-3-Hexenol	928-97-2						+[k]
(Z)-3-Hexen-1-ol	928-96-1	0.01-0.08	<0.1		0.4	0-0.7	+[k]; <0.05[i]; 0.07[n]
cis-4-Hexenol	928-91-6						+[k]
(E)-2-Hexenyl acetate	2497-18-9		tr				+[k]
3-Hexenyl acetate	1708-82-3						+[l]
(E)-3-Hexenyl acetate	3681-82-1				1.2		
(Z)-3-Hexenyl acetate	3681-71-8	0.01-0.2	tr		6.3		+[k]; <0.05[i]; 0.05[n]
(E)-2-Hexenyl benzoate	76841-70-8		<0.1		0.2		
(E)-3-Hexenyl benzoate	75019-52-2						tr[n]; 0.6[m]
(Z)-3-Hexenyl benzoate	25152-85-6	0.8-2.3	0.9		22.0	1.9-3.9	+[k]; 0.9[i]; 1.2[n]; 4.1[m]
(Z)-3-Hexenylbutyrate	16491-36-4		<0.1				
3-Hexenyl isobutyrate	57859-47-9						+[l]
3-Hexenyl propionate							+[l]
Hexyl benzoate	6789-88-4		<0.1				0.4[m]
6-Hydroxyheptenylcyclo-pentane							+[k]
(E)-8-Hydroxylinalool	75991-61-6		tr				
(E)-3-(4-Hydroxy-3-meth-oxyphenyl)allyl acetate			tr				
3-Hydroxy-2-methyl-4-pyrone	118-71-8		tr				
1H-Indole	120-72-9	0.2-3.7	2.6	1.1-1.9	0.1	0.2-3.7	1.2[n]; 1.8[i]; 4.1[a]; 5.5[g]
2-Indolinone	59-48-3		tr				
Isobutyl alcohol	78-83-1		<0.1				
Isoeugenol	97-54-1	0.05-0.2					0.1[i]
(E)-Isoeugenol	5932-68-3		0.1				
Isophytol	505-32-8	0.4-11.3	8.0	2.0-3.0		1.7-7.5	1.5[g]; 5.6[i]; 8.5[n]; 12.7[a,e]
Isophytyl acetate	58425-36-8			2.5-3.1		4.0-7.3	+[l]
Isosqualenol			0.4				
Jasmine ketolactone	70981-24-7		0.4				
δ-Jasmolactone	25524-95-2	0.5-1.3	0.4	0.7-1.5		1.4-5.5	1.1[i]; 1.9[n]
(E)-Jasmone	6261-18-3	0.03-0.04					tr[n]; <0.05[i]
(Z)-Jasmone	488-10-8	1.4-6.1	1.5	7.9-9.5	<0.1	1.4-5.2	1.9[i]; 2.2[n]; 2.9[g]; 4.7[m]
Jasmonyl acetate	149982-46-7						+[l]
Limonene	138-86-3						+[k]
Linalool	78-70-6	3.0-9.3	4.5	4.4-7.3	8.1	1.0-4.0	3.5[a]; 4.6[n]; 7.9[m]; 8.2[i]

Table 5.42.1 Constituents identified in *Jasminum grandiflorum* L. absolutes (*continued*)

Constituent	CAS	Percentage and range					
		A	B (7)	C (9)	D (12)	E (8)	F
cis-Linalool oxide	11063-77-7				0.1		
cis-Linalool oxide, furanoid	5989-33-3	0.1-0.4	tr				0.4[i]
trans-Linalool oxide, furanoid	34995-77-2		tr	0.1-0.4			0.5[i]
Linalyl acetate	115-95-7						0.4[m]
Linolenic acid	463-40-1		0.3				
6-Methoxyeugenol	6627-88-9		tr				
2-Methoxy-4-methyl-phenol	93-51-6		tr				
Methyl acetate	79-20-9	0.04-0.05					
Methyl N-acetylanthranilate	2719-08-6	0.2-1.0	0.5			0.7-3.5	
Methyl anthranilate	134-20-3		tr	0.4-0.8	3.0		tr[g]; 1.0[i]; 1.8[a,d]
Methyl arachinoate			<0.1				
Methyl benzoate	93-58-3	0.06-8.1	0.1				0.05[n]; 0.1[i]; 0.3[g]
Methyl chavicol	140-67-0						0.6[m]
(Z)-Methyl cinnamate	19713-73-6		tr				
Methyl dehydrojasmonate							+[l]
(11Z,14Z,17Z)-Methyl eicosatrienoate	82729-72-4		0.1				
Methyl-Z-11-eicosenoate	2390-09-2		<0.1				
Methyl elaidinate			0.1				
Methyl heptadecanoate	1731-92-6		tr				
6-Methyl-2-heptanone	928-68-7						<0.05[i]
6-Methyl-5-heptanone	13019-20-0				<0.1		
6-Methyl-5-hepten-2-one	110-93-0	0.01-0.02	tr				+[k]; 0.06[m]
4-Methyl-3-hexanol	615-29-2						+[k]
Methyl (Z)-3-hexenoate	13894-62-7						+[k]
5-Methyl-2-hexenol							+[k]
Methyl jasmonate	1211-29-6			5.1-7.2		0.9-4.0	+[k]; tr[g]; 0.9[n]
(E)-Methyl jasmonate		0.07-0.2					
(Z)-Methyl jasmonate	1211-29-6	0.2-1.3	0.5		0.1		0.6[i]
(Z)-methyl epi-jasmonate			0.4		0.1		
Methyl linoleate	112-63-0	0.3-2.0	0.2		3.0		0.3[m]; 2.8[i]; 3.5[n]; 21.7[g,h]
Methyl linolenate	301-00-8	1.7-5.3	2.3		1.2		2.3[a,f]
Methyl N-methyl anthranilate	85-91-6						0.7[n]
Methyl nicotinate	93-60-7						+[a,b]
2-Methylnonane	871-83-0				0.1		
Methyl octadecanoate	112-61-8	1.2-1.2	0.3		0.9		2.2[g]
Methyl oleate	112-62-9	0.8-2.4	0.9				0.8[i,m]; 0.9[n]
Methyl palmitate	112-39-0	0.5-2.8	1.6		1.6	0.5-0.8	1.4[i]; 1.8[a,d]; 3.3[m]; 5.7[g]
3-Methylpentane	96-14-0						+[k]
2-Methyl-2-pentanol	590-36-3						+[k]
2-Methyl-3-pentanol	565-67-3						+[k]
3-Methyl-3-pentanol	77-74-7						+[k]
4-Methyl-2-pentanol	108-11-2						+[k]
2-Methylquinoline	91-63-4						+[l]
Methyl salicylate	119-36-8	0.01-0.08	0.1				+[a,b,k]; 0.1[i]
5-Methyltricosane	22331-09-5				1.8		
τ-Muurolol	19912-62-0				0.6		
Neophytadiene	504-96-1		<0.1				
Neral	106-26-3						+[k]
Nerol	106-25-2	0.07-0.8					0.8[i]
Nerolidol	7212-44-4	0.06-0.3		1.8-2.5			0.5[m]; 0.8[g]; 1.4[a]
(E)-Nerolidol	40716-66-3		0.1				0.3[i]
(Z)-Nerolidol	3790-78-1				2.0		
Nonadecane	629-92-5		tr		0.1		
γ-Nonalactone	104-61-0						+[l]
Nonanol	28473-21-4						+[k]
Octadecanoic acid	57-11-4				1.8		
4-Octanolide	104-50-7						+[l]
Oleic acid	112-80-1				0.3		
Paraffins (high)							27.4[a]
(2E,6E,10E,14E,18E)-2,6,10,15,18-Pentamethyl-2,6,10,14,18-docosapenten-22-al			0.1				

Table 5.42.1 Constituents identified in *Jasminum grandiflorum* L. absolutes (*continued*)

Constituent	CAS	Percentage and range					
		A	B (7)	C (9)	D (12)	E (8)	F
3-Pentanol	584-02-1						$+^k$
3-Penten-1-ol	39161-19-8						$+^k$
Penten-3-ol	77035-93-9						$+^k$
(Z)-2-Pentenyl benzoate	65466-10-6		tr				
2-Phenethyl alcohol	60-12-8	0.01-0.07					$<0.05^i$
2-Phenethyl benzoate	94-47-3		<0.1				
Phenylacetaldehyde	122-78-1		tr				
Phenylacetonitrile	140-29-4	0.01-0.03	<0.1				
2-Phenylethyl acetate	103-45-7	0.02-0.1	<0.1			0.2-0.5	0.1^i
2-Phenylnitroethane	6125-24-2		tr				$+^l$
1-Phenyl-2-propanol	698-87-3						tr^n
Phytadiene	30917-33-0		0.1				
Phytol (isomers)	7541-49-3			7.8-13.3		13.6-52.2	10.9^i; 14.3^m; 41.4^a
(E)-Phytol	150-86-7	4.1-16.9	10.9				tr^n; $21.7^{g,h}$
(Z)-Phytol	5492-30-8	0.2-13.3					14.2^n
Phytone	16825-16-4	0.4-1.0					
Phytyl acetate	10236-16-5	1.4-11.0				0-7.2	5.3^i
cis-Phytyl acetate	5016-85-3		0.2				$10.5^{n,o}$
trans-Phytyl acetate	10236-16-5		6.0				1.1^n
Phytyl benzoate	827598-68-5		0.7				
Pyridine	110-86-1						$+^k$
Quinoline	91-22-5						$+^l$
Sabinene	3387-41-5						$+^k$
β-Sitosterol	83-46-5		0.1				
Squalene	111-02-4	2.7-5.7	4.6				
Squalene 2,3-oxide	7200-26-2	5.9-9.2	11.7				
α-Terpineol	98-55-5	0.8-0.9					0.09^m; 0.9^i
Terpinolene	586-62-9				0.4		
Tetracosane	646-31-1				1.2		
Tetradecane	629-59-4				0.1		
4,8,12,16-Tetramethyl-γ-heptadecalactone			0.2				
α-Tocopherol	59-02-9		0.1				
Trimethyl-6,10,14-penta-decanone	502-69-2		0.5			1.5-2.2	0.8^n; 3.2^m
Ursolic acid	77-52-1						$+^l$
Vanillin	121-33-5		tr				$2.3^{a,f}$
p-Vinylanisole	637-69-4		tr				
4-Vinyl-4-pentanolide							$+^l$
2-Vinylpyridine	100-69-6						$+^l$
3-Vinylpyridine	1121-55-7						$+^l$

A forty-one absolute samples of *Jasminum grandiflorum* analyzed between 2001 and 2013; lowest and highest concentrations given (E. Schmidt, unpublished data)

B one commercial absolute sample from Indian *J. grandiflorum* (ref. 7)

C nine absolutes prepared from flowers picked in India on various days between June and October; lowest and highest concentrations given (ref. 9)

D one absolute from *J. grandiflorum* from India (ref. 12)

E three commercial absolutes from France, Algeria and Italy; lowest and highest concentrations given (ref. 8)

F data from other studies (indicated with superscript letters); highest concentrations found in any study reviewed here given; when two or more absolutes were investigated, only the highest concentrations are mentioned, unless indicated otherwise

[a] five lab-prepared absolutes from Turkey and five commercial absolutes from France, Italy, India, Morocco and Egypt (ref. 11); [b] combined with three other chemicals, therefore hard to quantify; [c] combined with methyl salicylate, methyl nicotinate and 3-amino-1-phenylbutane; [d] methyl anthranilate and methyl palmitate combined; [e] isophytol and farnesol combined; [f] methyl linoleate and vanillin combined; [g] one absolute produced from commercial south Indian concrete (ref. 10); [h] (E)-phytol and methyl linoleate combined; [i] one commercial jasmine absolute sample from India (refs. 2,3); [k] present in the low boiling fraction of a French absolute, an Italian absolute or both; the concentrations were stated, but as they cannot be compared to the other data, the presence of the chemicals is indicated with $+^k$ (ref. 8); [l] literature data from before 1980 cited in ref. 8; not absolutely sure that the species investigated was *J. grandiflorum* (ref. 8); [m] one jasmine absolute of unknown origin (ref. 6); [n] one commercial Egyptian 'absolute oil' (ref. 13); [o] geranyllinalool and cis-phytyl acetate combined

tr: trace (in columns B and F^n: <0.01); + present in the oil investigated, but quantity not stated or presented in a manner which cannot be compared with the other data

Table 5.42.2 Major constituents of *Jasminum grandiflorum* L. absolutes

Constituent	CAS	Percentage and range					
		A	B (7)	C (9)	D (12)	E (8)	F
Phytol (isomers)	7541-49-3			7.8-13.3		13.6-52.2	10.9[i]; 14.3[m]; 41.4[a]
Benzyl acetate	140-11-4	9.5-32.9	15.5	24.2-29.7	11.3	8.5-25.1	7.7[m]; 8.1[g]; 18.8[a,n]; 23.7[i]
Benzyl benzoate	120-51-4	7.5-25.6	11.4	17.9-21.2	0.6	6.2-13.1	13.3[n]; 20.7[i]; 22.0[m]; 26.2[g]
(E)-Phytol	150-86-7	4.1-16.9	10.9				tr[n]; 21.7[g,h]
Isophytol	505-32-8	0.4-11.3	8.0	2.0-3.0		1.7-7.5	1.5[g]; 5.6[i]; 8.5[n]; 12.7[a,e]
Geranyllinalool	77368-82-2	1.1-10.2	3.6	2.8-4.4		3.1-8.0	0.4[n]; 3.0[i]
(Z)-Jasmone	488-10-8	1.4-6.1	1.5	7.9-9.5	<0.1	1.4-5.2	1.9[i]; 2.2[n]; 2.9[g]; 4.7[m]
Linalool	78-70-6	3.0-9.3	4.5	4.4-7.3	8.1	1.0-4.0	3.5[a]; 4.6[n]; 7.9[m]; 8.2[i]
Eugenol	97-53-0	0.9-6.8	1.3	2.0-4.0		0.8-3.5	1.5[a]; 2.0[n]; 2.5[i]; 2.8[g]
1H-Indole	120-72-9	0.2-3.7	2.6	1.1-1.9	0.1	0.2-3.7	1.2[n]; 1.8[i]; 4.1[a]; 5.5[g]
p-Cresol	106-44-5	0.07-4.9	0.5			0.3-1.0	0.4[n]; 0.6[i]; 0.8[g]; 1.3[a]
Benzyl alcohol	100-51-6	0.5-3.7	0.7	1.1-1.8	0.3	0.9-3.0	1.3[i]; 1.7[n]; 2.5[a]

LEGEND: SEE UNDER TABLE 5.42.1

series of the North American Contact Dermatitis Group (NACDG) since 2001. In all but four publications (29,38,42,54), the botanical origin of the jasmine absolute was not mentioned. In the most recent NACDG publications the test substance is called *Jasminum officinale (J. grandiflorum)* oil, which is sensu stricto not correct, as it is not an oil but an absolute and *Jasminum grandiflorum* is not a synonym of *Jasminum officinale* (see above). We assume the material is *Jasminum grandiflorum* absolute.

In groups of consecutive patients suspected of contact dermatitis, prevalence rates of up to 1.5% positive patch test reactions have been observed. Rarely, reactions are considered to be of definite relevance. In the NACDG studies, 'definite + probable relevance' was reported in only 9% (2001-2002) to 37% (2011-2012) of positive patients. In two case reports, linalool and eugenol may have been allergens in jasmine absolute.

Testing in groups of patients

The results of patch tests with jasmine absolute in routine testing (consecutive patients suspected of contact dermatitis) and in groups of selected patients are shown in Table 5.42.4. In routine testing, rates of positive

Table 5.42.3 Differences between the composition of *J. sambac* absolute and *J. grandiflorum* absolute[a]

Chemical	*J. sambac* absolute	*J. grandiflorum* absolute
Benzyl acetate	4.3-21.2	7.7-29.7
Benzyl alcohol	0.3-14.1	0.3-3.0
Benzyl benzoate	0.4-2.1	0.6-26.2
Eugenol	0-0.3	0-4.0
1(10),5-Germacradien-4-ol	0-8.9	absent
Indole	tr-13.4	0.1-5.5
Isophytol	0-0.4	0-12.7
Isophytyl acetate	absent	0-7.3
δ-Jasmolactone	0-<0.1	0-5.5
(Z)-Jasmone	0-0.5	0-9.5
Linalool	2.6-23.1	1.0-8.2
Methyl anthranilate	1.2-12.0	0-3.0
Methyl jasmonate	0-0.2	0-7.2
2-Phenethyl alcohol	0-2.4	0-<0.05
Phytol	0-0.3	0-41.4
Phytyl acetate	0-2.6	0-10.5
Tricosene	0-8.0	absent

[a] source has been lost in writing

reactions ranged from 0.3% to 1.5%, whereas between 1.0% and 35% of patients in selected groups had positive patch tests. The very high positivity rate of 35% was found in a small group of 20 patients previously shown to be allergic to fragrances (48).

Case reports

An aromatherapist developed occupational contact dermatitis from contact allergy to multiple essential oils; she reacted to jasmine absolute, Egyptian, in the fragrance series, which reaction was considered to be relevant (21).

One case of airborne contact dermatitis from 'jasmine oil' and two essential oils from aromatherapy lamps, whereby the water vapor produced distributes the essential oils as an aerosol in the air, was reported; the patient also reacted to linalool (17); linalool is an important constituent of all three oils. One patient developed allergic contact dermatitis from undiluted lovage oil and a mixture of essential oils containing jasmine absolute; she reacted to her own products, jasmine absolute, lovage oil, cananga oil, eugenol and isoeugenol (34); commercial jasmine absolutes (*grandiflorum*, not *sambac*) may contain up to 6.8% eugenol but no or very

Table 5.42.4 Results of testing groups of patients with jasmine absolute

Years and Country	Test conc. & vehicle	Number of patients			Selection of patients (S); Relevance (R); Comments (C)	Ref.
		tested	positive	(%)		
Routine testing						
2011-12 USA, Canada	2% pet.	4,230	19	(0.4%)	R: definite + probable relevance: 37%; C: origin: *Jasminum officinale (J. grandiflorum)* 'oil'; *Jasminum officinale* and *Jasminum grandiflorum* are *not* synonyms	38
2009-10 USA, Canada	2% pet.	4,302	43	(1.0%)	R: definite + probable relevance 18%; C: origin: *Jasminum officinale (J. grandiflorum)* 'oil'; see comments above (38)	29
2000-2008 IVDK	5% pet.	3,668	56	(1.5%)	R: not stated	36
2000-2007 USA	2% pet.	869	12	(1.4%)	R: 92%; C: weak study: a. high rate of macular erythema and weak reactions, b. relevance figures include 'questionable' and 'past' relevance	26
2005-6 USA, Canada	2% pet.	4,447	49	(1.1%)	R: definite + probable relevance 25%	22
<2006 USA, Canada	2% pet.	1,603	7	(0.4%)	R: definite + probable relevance: 14%	32
2003-4 USA, Canada	2% pet.	5,143	44	(0.9%)	R: not stated	23
2002-2003 Korea	2% pet.	422	5	(1.2%)	R: not stated; C: tested was 'Jasmine officinale oil'	54
2001-2 USA, Canada	2% pet.	4,900	34	(0.7%)	R: definite + probable relevance 9.4%	24
1999-2000 Denmark	5% pet.	318	1	(0.3%)	R: not specified; C: this study was part of the international study mentioned below (ref. 40)	53
1998-2000 six European countries	5% pet.	1,606	20	(1.2%)	R: not specified for individual oils/chemicals	40
1983-1984 Italy	2% pet.	1,200	13	(1.1%)	R: not stated	44
Testing in groups of selected patients						
2006-2010 USA	2% pet.	100	2	(2.0%)	S: patients with eyelid dermatitis; R: not stated	31
2000-2008 IVDK	5% pet.	982	11	(1.1%)	S: patients with dermatitis suspected of causal exposure to fragrances; R: not stated	36
2004-2008 Spain	2% pet.	86	3	(3.5%)	S: patients previously reacting to the fragrance mix I or *Myroxylon pereirae* (n=54) or suspected of fragrance contact allergy (n=32); R: not stated	35
<2005 USA	2% or 10% pet.	51	1	(2.0%)	S: patients using consumer products containing jasmine absolute; R: 'at least possibly relevant'	33
<2004 Israel	2% pet.	91	2	(2.2%)	S: patients who had shown a doubtful or positive reaction to the fragrance mix I and/or *Myroxylon pereirae* resin and/or one or two commercial fine fragrances; R: not stated	39
2000 USA, Japan and 4 European countries	10% pet.	178	30	(16.9%)	S: patients previously shown to be allergic to fragrances; R: not stated	41
1990-1998 Japan	20% pet.	1,483	15	(1.0%)	S: patients suspected of cosmetic contact dermatitis, virtually all were women; C: range of annual frequency of sensitization: 0-4.0%; R: not stated	18

Table 5.42.4 Results of testing groups of patients with jasmine absolute (*continued*)

Years and Country	Test conc. & vehicle	Number of patients			Selection of patients (S); Relevance (R); Comments (C)	Ref.
		tested	positive	(%)		
1996-1997 UK	2% pet.	10	2	(20%)	S: patients suspected of cosmetic dermatitis and reacting to the fragrance mix; R: not stated	16
<1994 Japan	?	?	?	(7.3%)	S: unknown; R: unknown	43
1971-1980 Japan	10% pet.	477	10	(2.1%)	S: patients with dermatoses other than pigmented cosmetic dermatitis and volunteers; R: not stated	27
<1977 USA	10% pet.	185	20	(10.8%)	S: not specified; R: not stated	45
1975 USA	10% pet.	20	7	(35%)	S: patients with fragrance contact allergy; R: not stated	48
<1974 Japan	?	183	44	(24.0%)	S: patients suspected of cosmetic dermatitis; R: unknown; C: tested was 'jasmine oil'; in many, there was co-reactivity with benzyl salicylate, which may be present in commercial jasmine absolutes in low concentrations (Table 5.42.1, column A)	46

IVDK Information Network of Departments of Dermatology, Germany, Switzerland, Austria (www.ivdk.org); pet.: petrolatum

low concentrations of isoeugenol (Table 5.42.1, column A). One patient was shown to have allergic cosmetic dermatitis from '*Jasminum officinale*' in a face cream (42). Jasmine absolute was responsible for 3 out of 399 cases of cosmetic (photo)allergy where the causal allergen was identified in a study of the NACDG, USA, 1977-1983 (50).

Positive patch tests (relevance unknown, uncertain or not stated)

Of 7 patients allergic to the fragrance farnesol, 5 (71%) co-reacted to jasmine absolute (and various other fragrances) (49). In a group of 91 patients with cosmetic allergy, there was one patch test reaction to 'jasmin oil'; the relevance was not ascertained, but felt by the authors to be 'indicative' (51). One positive patch test reaction to jasmine absolute in 53 women with chronic anogenital dermatoses (37). Two positive patch test reactions to 'jasmine oil' were seen in 7 patients with allergic contact dermatitis from compound tincture of benzoin (30). One positive patch test reaction was found to jasmine absolute in a patient with occupational hand dermatitis from working in a perfume factory, and one reaction in 20 workers of the same factory without dermatitis (25). A positive patch test to jasmine absolute Egyptian was observed in an aromatherapist allergic to multiple essential oils, who did not use jasmine absolute at work (19). One naturopathic therapist had occupational contact dermatitis from various essential oils; she also reacted to jasmine absolute in the fragrance series, but probably did not use it in her work (20). One positive patch test to 'jasmine oil', possibly from its presence in a topical pharmaceutical preparation (55).

Pigmented cosmetic dermatitis

In Japan, in the 1960s and 1970s, many female patients developed pigmentation following dermatitis of the face (28). This so-called pigmented cosmetic dermatitis was shown to be caused by contact allergy to components of cosmetic products, notably essential oils, other fragrance materials, antimicrobials, preservatives and

coloring materials (27,28). In a group of 620 Japanese patients with this condition investigated between 1970 and 1980, 8-11% had positive patch test reactions to jasmine absolute 10% in petrolatum (27). The number of patients with pigmented cosmetic dermatitis decreased strongly after 1978, when major cosmetic companies began to eliminate strong contact sensitizers from their products (27).

LITERATURE

1 Lawrence BM. Progress in essential oils. Perfum Flavor 2009;34(11):56-59

2 Jirovetz L, Buchbauer G, Eller GA, Stoilova I, Stoyanova A, Krastanov A, et al. Chemical composition and antioxidant properties of *Jasminum grandiflorum* L. absolute from India. In: L Jirovetz, N–X Dung, VK Varshney, Eds. Aromatic plants from Asia: their chemistry and application in food and therapy. Dehradun, India: Har Kishan Bhalla & Sons, 2007:37-48. Data cited in ref. 1

3 Jirovetz L, Buchbauer G, Schweiger T, Denkova Z, Slavchev A, Stoyanova A, et al. Chemical composition, olfactory evaluation and antimicrobial activities of *Jasminum grandiflorum* L. absolute from India. Nat Prod Commun 2007;2:407-412. The data in this study were also provided in ref. 2.

4 Caturvedi AP, Kumar M, Tripathi YB. Efficacy of *Jasminum grandiflorum* L. leaf extract on dermal wound healing in rats. Int Wound J 2012;10:675-682

5 Lawrence BM. Progress in essential oils. Perfum Flavor 2005;30(October):60-? (last page unknown)

6 Cum G, Spadaro A, Citraro T, Gallo R. Processo di estrazione in fase supercritical de fiori di *Jasminum grandifloram* L. Essenze Deriv Agrum 1998;68:384-400. Data cited in ref. 5

7 Braun NA, Kohlenberg B, Sim S, Meier M, Hammerschmidt FJ. *Jasminum flexile* flower absolute from India – a detailed comparison with three other jasmine absolutes. Nat Prod Commun 2009;4:1239-1250

8 Verzele M, Maes G, Vuye A, Godefroot M, van Alboom M, Vervisch J, et al. Chromatographic investigation of jasmin absolutes. J Chromatogr A 1981;205:367-386

9 Verghese J, Sunny TP. Seasonal studies on the concrete and absolute of Indian *Jasminum grandiflorum* L. flowers. Flavour Fragr J 1992;7:323-327

10 Gopalakrishnan N, Narayanan CS. Carbon dioxide extraction of Indian jasmine concrete. Flavour Fragr J 1991;6: 135-138

11 Anaç O. Gas chromatographic analysis of absolutes and volatile oil isolated from Turkish and foreign jasmine concretes. Flavour Fragr J 1986;1:115-119

12 Prakash O, Sahoo D, Rout PK. Liquid CO2 extraction of *Jasminum grandiflorum* and comparison with conventional processes. Nat Prod Commun 2012;7:89-92

13 Toda H, Mihara S, Umano K, Shibamoto T. Photochemical studies on jasmin oil. J Agric Food Chem 1983;31:554-558

14 Rhind JP. Essential oils. A handbook for aromatherapy practice, 2nd Edition. London: Singing Dragon, 2012

15 Lawless J. The encyclopedia of essential oils, 2nd Edition. London: Harper Thorsons, 2014

16 Thomson KF, Wilkinson SM. Allergic contact dermatitis to plant extracts in patients with cosmetic dermatitis. Br J Dermatol 2000;142:84-88

17 Schaller M, Korting HC. Allergic airborne contact dermatitis from essential oils used in aromatherapy. Clin Exp Dermatol 1995;20:143-145

18 Sugiura M, Hayakawa R, Kato Y, Sugiura K, Hashimoto R. Results of patch testing with lavender oil in Japan. Contact Dermatitis 2000;43:157-160

19 Dharmagunawardena B, Takwale A, Sanders KJ, Cannan S, Roger A, Ilchyshyn A. Gas chromatography: an investigative tool in multiple allergies to essential oils. Contact Dermatitis 2002;47:288-292

20 Trattner A, David M, Lazarov A. Occupational contact dermatitis due to essential oils. Contact Dermatitis 2008;58:282-284

21 Boonchai W, Lamtharachai P, Sunthonpalin P. Occupational allergic contact dermatitis from essential oils in aromatherapists. Contact Dermatitis 2007;56:181-182

22 Zug KA, Warshaw EM, Fowler JF Jr, Maibach HI, Belsito DL, Pratt MD, et al. Patch-test results of the North American Contact Dermatitis Group 2005-2006. Dermatitis 2009;20:149-160

23 Warshaw EM, Belsito DV, DeLeo VA, Fowler JF Jr, Maibach HI, Marks JG, et al. North American Contact Dermatitis Group patch-test results, 2003-2004 study period. Dermatitis 2008;19:129-136

24 Pratt MD, Belsito DV, DeLeo VA, Fowler JF Jr, Fransway AF, Maibach HI, et al. North American Contact Dermatitis Group patch-test results, 2001–2002 study period. Dermatitis 2004;15:176-183

25 Schubert HJ. Skin diseases in workers at a perfume factory. Contact Dermatitis 2006;55:81-83

26 Wetter DA, Yiannias JA, Prakash AV, Davis MD, Farmer SA, el-Azhary RA, et al. Results of patch testing to personal care product allergens in a standard series and a supplemental cosmetic series: an analysis of 945 patients from the Mayo Clinic Contact Dermatitis Group, 2000-2007. J Am Acad Dermatol 2010;63:789-798

27 Nakayama H, Matsuo S, Hayakawa K, Takhashi K, Shigematsu T, Ota S. Pigmented cosmetic dermatitis. Int J Dermatol 1984;23:299-305

28 Nakayama H, Harada R, Toda M. Pigmented cosmetic dermatitis. Int J Dermatol 1976;15:673-675

29 Warshaw EM, Belsito DV, Taylor JS, Sasseville D, DeKoven JG, Zirwas MJ, et al. North American Contact Dermatitis Group patch test results: 2009 to 2010. Dermatitis 2013;24:50-59

30 Fettig J, Taylor J, Sood A. Post-surgical allergic contact dermatitis to compound tincture of benzoin and association with reactions to fragrances and essential oils. Dermatitis 2014;25:211-212

31 Wenk KS, Ehrlich AE. Fragrance series testing in eyelid dermatitis. Dermatitis 2012;23:22-26

32 Belsito DV, Fowler JF Jr, Sasseville D, Marks JG Jr, De Leo VA, Storrs FJ. Delayed-type hypersensitivity to fragrance materials in a select North American population. Dermatitis 2006;17:23-28

33 Guin JD. Use of consumer product ingredients for patch testing. Dermatitis 2005;16:71-77

34 Lapeere H, Boone B, Verhaeghe E, Ongenae K, Lambert J. Contact dermatitis caused by lovage (Levisticum officinalis) essential oil. Contact Dermatitis 2013;69:181-182

35 Cuesta L, Silvestre JF, Toledo F, Lucas A, Pérez-Crespo M, Ballester I. Fragrance contact allergy: a 4-year retrospective study. Contact Dermatitis 2010;63:77-84

36 Uter W, Schmidt E, Geier J, Lessmann H, Schnuch A, Frosch P. Contact allergy to essential oils: current patch test results (2000–2008) from the Information Network of Departments of Dermatology (IVDK). Contact Dermatitis 2010;63:277-283

37 Vermaat H, Smienk F, Rustemeyer Th, Bruynzeel DP, Kirtschig G. Anogenital allergic contact dermatitis, the role of spices and flavour allergy. Contact Dermatitis 2008;59:233-237

38 Warshaw EM, Maibach HI, Taylor JS, Sasseville D, DeKoven JG, Zirwas MJ, et al. North American Contact Dermatitis Group patch test results: 2011-2012. Dermatitis 2015;26:49-59

39 Trattner A, David M. Patch testing with fine fragrances: comparison with fragrance mix, balsam of Peru and a fragrance series. Contact Dermatitis 2004;49:287-289

40 Frosch PJ, Johansen JD, Menné T, Pirker C, Rastogi SC, Andersen KE, et al. Further important sensitizers in patients sensitive to fragrances. II. Reactivity to essential oils. Contact Dermatitis 2002;47:279-287

41 Larsen W, Nakayama H, Fischer T, Elsner P, Frosch P, Burrows D, et al. Fragrance contact dermatitis: a worldwide multicenter investigation (Part II). Contact Dermatitis 2001;44:344-346

42 Penchalaiah K, Handa S, Lakshmi SB, Sharma VK, Kumar B. Sensitizers commonly causing allergic contact dermatitis from cosmetics. Contact Dermatitis 2000;43:311-313

43 Sugai T. Group study IV – farnesol and lily aldehyde. Environ Dermatol 1994;1:213-214

44 Santucci B, Cristaudo A, Cannistraci C, Picardo M. Contact dermatitis to fragrances. Contact Dermatitis 1987;16:93-95

45 Rudner EJ. North American Group results. Contact Dermatitis 1977;3:208-209

46 Nakayama H, Hanaoka H, Ohshiro A. Allergen Controlled System (ACS). Tokyo, Japan: Kanehara Shuppan, 1974:42. Data cited in ref. 47

47 Mitchell JC. Contact hypersensitivity to some perfume materials. Contact Dermatitis 1975;1:197-199

48 Larsen WG. Perfume dermatitis. A study of 20 patients. Arch Dermatol 1977;113:623-626

49 Goossens A, Merckx L. Allergic contact dermatitis from farnesol in a deodorant. Contact Dermatitis 1997;37:179-180

50 Adams RM, Maibach HI. A five-year study of cosmetic reactions. J Am Acad Dermatol 1985;13:1062-1069

51 Ngangu Z, Samsoen M, Foussereau J. Einige Aspekte zur Kosmetika-allergie in Strassburg. Dermatosen 1983;31:126-129

52 Broeckx W, Blondeel A, Dooms-Goossens A, Achten G. Cosmetic intolerance. Contact Dermatitis 1987;16: 189-194

53 Paulsen E, Andersen KE. Colophonium and Compositae mix as markers of fragrance allergy: Cross-reactivity between fragrance terpenes, colophonium and Compositae plant extracts. Contact Dermatitis 2005;53:285-291

54 An S, Lee AY, Lee CH, Kim D-W, Hahm JH, Kim K-J, et al. Fragrance contact dermatitis in Korea: a joint study. Contact Dermatitis 2005;53:320-323

55 Nardelli A, D'Hooge E, Drieghe J, Dooms M, Goossens A. Allergic contact dermatitis from fragrance components in specific topical pharmaceutical products in Belgium. Contact Dermatitis 2009;60:303-313

Chapter 5.43 JASMINUM SAMBAC ABSOLUTE

DEFINITION

Jasminum sambac absolute is the absolute obtained from the flowers of the Arabian/China jasmine, *Jasminum sambac* (L.) Aiton.

INCI NOMENCLATURE

Description/definition: Jasminum sambac flower extract is an extract obtained from the flowers of the China jasmine, *Jasminum sambac*, Oleaceae

INCI name EU: Jasminum sambac flower extract

INCI name USA: Jasminum sambac (jasmine) flower extract

CAS registry number(s): 91770-14-8

GENERAL

Jasminum sambac (L.) Aiton (Arabian jasmine, China jasmine, sambac jasmine) is an evergreen vine or shrub reaching up to 0.5-3 meters tall with white, strongly scented flowers that bloom all through the year. It is native to India and widely cultivated, e.g., in India and China. It can be found throughout the tropics from the Arabian Peninsula to southeast Asia and the Pacific Islands used as an ornamental plant with its attractive and strongly scented flowers, which give a floral bouquet consisting of sweet, fruity, heady and sultry notes (1,8, GRIN Taxonomy for Plants). The flowers open only at night and are picked in the morning when the tiny petals are tightly closed. Harvested buds are taken to auction markets and then stored in a cool place until dark. Between 6.00 pm and 8.00 p.m., when the temperature drops, the petals begin to open, releasing the fullness of the fragrance at midnight (13).

In China, the flowers of *J. sambac* are used as flavor for tea leaves to provide the characteristic jasmine impact (8,17). The root of the sambac is a traditional Chinese medicine with anaesthetic and analgesic effects and used for the treatment of insomnia, headache, decayed teeth, and injuries from falls (8). Jasmine sambac absolutes are highly appreciated in the fragrance industries (7). In aromatherapy, the oil is said to be useful for the relief of depression and for uplifting mood (16).

CHEMICAL COMPOSITION

Jasminum sambac (L.) Aiton. absolute is a mobile to viscous light orange to reddish brown colored liquid which has an intensive flowery and slightly animalist and long-lasting odor. The yield of absolute from the flowers of the sambac jasmine generally varies from 0.12 to 0.19 per cent. The main producing countries of this absolute are India and China.

Literature data (up to November 25, 2014) on the chemical composition of *Jasminum sambac* (L.) Aiton. absolutes and unpublished analytical data from one of us (E.S.) are shown in Table 5.43.1 in alphabetical order. In Chinese jasmine absolutes from various origins, over 270 chemicals have been identified. About 61 per cent of these were found in a single reviewed publication only. The major compounds found in *Jasminum sambac* (L.) Aiton. absolutes from different sources are shown in Table 5.43.2. They include (highest concentrations in any study given) linalool (34.2%), benzyl alcohol (32.9%), (*Z*)-3-hexenyl benzoate (22.4%), benzyl acetate (21.2%), (*E,E*)-α-farnesene (19.7%), 1*H*-indole (14.1%), methyl anthranilate (12.0%), methyl linolenate (8.6%) and tricosene (8.0%). Uncommon or rare constituents of *Jasminum sambac* (L.) Aiton. absolutes found in high concentrations (>7%) in single studies include isocaryophyllene (9.5%) and 1(10),5-germacradien-4-ol (two studies: 8.9% and 7.6%).

Commercial absolutes

The ten chemicals that had the highest maximum concentrations in 11 commercial Jasminum sambac absolute samples (concentration ranges provided) are the following: linalool (11.5-34.2%), benzyl acetate (10.5-21.0%), (*E,E*)-α-farnesene (10.6-17.1%), (*Z*)-3-hexenyl benzoate (3.9-14.9%), 1*H*-indole (0.9-12.2%), methyl linolenate (0.4-8.6%), [1(10)*E*,5*E*]-germacradien-4α-ol (0.9-7.6%), methyl anthranilate (0.5-7.6%), benzyl alcohol (1.2-6.3%) and tricosene (0.2-5.4%) (Erich Schmidt, unpublished analytical data).

An additional analytical study of jasmine sambac absolute not discussed in Table 5.43.1 can be found in ref. 20. The most important differences between *J. sambac* absolute and *J. grandiflorum* absolute are discussed in the chapter on *J. grandiflorum* absolute.

CONTACT ALLERGY/ALLERGIC CONTACT DERMATITIS

No publications on contact allergy to or allergic contact dermatitis from jasmine absolute specified to have been obtained from the Arabian/China jasmine, *Jasminum sambac* (L.) Aiton. have been found. The literature on 'jasmine absolute' (botanical source not mentioned) is discussed in Chapter 5.42 Jasminum grandiflorum absolute.

Table 5.43.1 Constituents identified in jasmine extracts

Constituent	CAS	Percentage and range					
		A	B (1)	C (5)	D (3)	E (12)	F
Acetone	67-64-1						+[j]
(E)-1-Acetoxy-2,6-dime-thyl-2,7-octadien-6-ol				0.1			
Aromadendrene	489-39-4						+[g]
Benzaldehyde	100-52-7	0.02-0.2	tr-0.1			tr	+[g,j]; 0.07[l]
Benzoic acid	65-85-0			0.1			+[g]
Benzyl acetate	140-11-4	10.5-21.0	7.5-13.2	8.0	6.6-15.7	21.2	4.3[l]; 7.5[k]; 9.2[g]; 14.2[f]
Benzyl alcohol	100-51-6	1.2-6.3	0.3-1.4	10.7	3.2-14.1	1.1	6.8[k]; 8.4[f]; 10.2[g]; 32.9[m]
Benzyl benzoate	120-51-4	0.5-1.9	0.4-0.9	0.5	0.5-1.7	1.5	+[g]; 0.6[f]; 2.1[k]
Benzyl butyrate	103-37-7			tr		tr	
Benzyl eicosanoate	77509-04-7			tr			
Benzyl formate	104-57-4	0.04-0.09		0.1		tr	
Benzyl isovalerate	103-38-8					tr	
Benzyl linoleate	47557-83-5			0.1			
Benzyl linolenate	77509-02-5			1.1	0.5-2.3		
Benzyl palmitate	41755-60-6			0.2	<0.1-0.6		
Benzyl phenylacetate	102-16-9			tr			
Benzyl propionate	122-63-4					tr	
Benzyl salicylate	118-58-1		0.1-0.2	0.1	<0.1-0.5		+[j]; 0.1[f]
Benzyl stearate	5531-65-7			<0.1			
Benzyl tiglate	37526-88-8			<0.1			
Bergamotene							0.4[k]
Bicyclogermacrene	24703-35-3			0.2			+[j]; 0.2[f]
β-Bisabolene	495-61-4						0.2[k]
Butyl benzoate	136-60-7			<0.1		tr	
β-Cadinene	523-47-7						0.3[k]
γ-Cadinene	39029-41-9		0.1-0.2	0.4			0.4[k]
δ-Cadinene	483-76-1	0.07-1.0	0.3-0.5	0.2			+[g]; 0.2[f]
α-Cadinol	481-34-5	0.6-1.7	0.5-0.1	<0.1			+[j]; 0.1[f]
epi-α-Cadinol	5937-11-1	0.08-1.4		0.1			+[l]
δ-Cadinol	19435-97-3			<0.1			0.3[k]
β-Caryophyllene	87-44-5			<0.1		tr	+[j]; tr[k]; 0.1[f]
(E)-Cinnamic acid	140-10-3			tr			
Cinnamyl acetate	103-54-8			tr			
Cinnamyl alcohol	104-54-1	0.06-0.2	tr-0.1				+[j]; 0.3[f]
(E)-Cinnamyl alcohol	4407-36-7			0.2		tr	
(Z)-Cinnamyl alcohol	4510-34-3					tr	
(E)-Cinnamyl benzoate	50555-04-9			0.2		2.4	
(Z)-Cinnamyl benzoate						0.2	
Citronellol	106-22-9					tr	+[j]
Coniferyl alcohol	458-35-5			<0.1			
p-Cresol	106-44-5	0.01-0.04				<0.1	+[l]
Cubenene	29837-12-5						+[j]
Cyclohexyl benzoate	2412-73-9						+[j]
1-Cyclopentyl-1-hexade-canone	55255-86-2						+[j]
(Z,Z)-3,6-Decadienol							0.01[l]
n-Decane	124-18-5		0-0.1				
(Z)-3-Decen-1-ol	10340-22-4						tr[l]
Dihydrofarnesol	1335-48-4						+[j]
2,6-Dimethyl-5-heptenal	106-72-9						+[j]; 0.6[k]
(Z)-2,6-Dimethyl-2,7-octa-diene-1,6-diol	103619-06-3			0.1			
3,7-Dimethyl-1,5-octa-diene-3,7-diol	13741-21-4			<0.1			
3,7-Dimethyl-1,6-octa-diene-3,5-diol	75654-19-2			tr			

Table 5.43.1 Constituents identified in jasmine extracts (*continued*)

Constituent	CAS	Percentage and range					
		A	B (1)	C (5)	D (3)	E (12)	F
2,2-Dimethyl-1-phenyl-1-propanone	938-16-9						+[g]
Docosane	629-97-0						0.6[k]
1-Docosene	1599-67-3			0.3			
(Z,Z)-3,6-Dodecadienol							0.03[l]
Dodecane	112-40-3					tr	
(Z,Z,Z)-3,6,9-Dodecatri-enol	81345-02-0						0.1[l]
α-Elemene	5951-67-7						+[g]
β-Elemene	33880-83-0	0.02-0.2	0.1-0.2	tr			+[j]; 0.1[f]; 0.4[k]
γ-Elemene	29873-99-2						+[j]
3,4-Epoxyhexanol	67663-02-9						0.2[l]
3,4-Epoxy-(Z)-5-hexenyl benzoate							tr[l]
3,4-Epoxyhexyl acetate	113816-35-6			0.5			0.2[l]
3,4-Epoxyhexyl benzoate	189155-40-6			0.3			1.8[l]
3,4-Epoxyhexyl butyrate	189155-39-3						0.2[l]
Ethyl anthranilate	87-25-2			tr			
trans-2-Ethyl-3-acetoxy-tetrahydrofuran							tr[l]
Ethyl benzoate	93-89-0		tr-0.1	0.1		tr	+[g]
trans-2-Ethyl-3-benzoyl-oxytetrahydrofuran							tr[l]
Ethyl-5-ethylnicotinate	68686-59-9						+[i]
Ethyl hexadecanoate	628-97-7			<0.1		tr	+[j]
Ethyl linoleate	544-35-4			<0.1			
Ethyl linolenate	1141-91-9			<0.1			
Ethyl 4-methyl-5-ethyl-nicotinate							+[i]
Ethyl nicotinate	614-18-6						+[i]
Ethyl phenylacetate	101-97-3			tr			
3-Ethylpyridine	536-78-7						+[i]
Ethyl salicylate	118-61-6		tr-0.2	<0.1		tr	
Ethyl-5-vinylnicotinate							+[i]
Eugenol	97-53-0	0.02-0.08		<0.1	0-0.3	<0.1	+[i]
(E,E)-α-Farnesene	502-61-4	10.6-17.1	2.2-19.7	13.1	7.3-15.6	12.4[d]	10.0[m]; 13.1[f]; 18.4[i]; 19.3[h]
(Z,E)-α-Farnesene	26560-14-5			0.1	0-0.1		0.1[f]
(Z,Z)-α-Farnesene	28973-99-1	0.1-0.5	0.3-0.6				
(E)-β-Farnesene	18794-84-8		0.3-0.9	<0.1			
Farnesol	4602-84-0						0.2[f]
(E,E)-Farnesol	106-28-5		1.3-1.5	0.1	0-1.2	1.3	
(E,Z)-Farnesol	3879-60-5		0.1-1.9				
Farnesyl acetate	29548-30-9						0.2[f]
(E,E)-Farnesyl acetate	4128-17-0		0.2-0.9	0.1	0-0.1	0.5	
(E,E)-Farnesyl formate	85633-25-6					tr	
Geraniol	106-24-1	0.3-0.7	0.1-0.5	0.1		<0.1	+[g]
Geranyl acetate	105-87-3			0.1-0.2	tr	tr	
Geranyl benzoate	94-48-4					tr	
Geranyl formate	105-86-2					tr	
Geranyllinalool	77368-82-2	0.8-4.4		1.4	1.0-2.0	2.6[e]	+[j]; 1.7[f]
Germacradien-4-ol							2.3[f]
1(10),5-Germacra-dien-4-ol	74841-87-5			5.7	1.6-8.9		
[1(10)E,5E]-Germacra-dien-4α-ol	207221-31-6	0.9-7.6					
Germacrene D	23986-74-5		0.1-0.5	0.6			+[i]; 0.5[f]
Heneicosane	629-94-7			0.1			

Table 5.43.1 Constituents identified in jasmine extracts (*continued*)

Constituent	CAS	Percentage and range					
		A	B (1)	C (5)	D (3)	E (12)	F
1-Heneicosene	1599-68-4			0.4			
Heptadecanal	629-90-3		tr-0.1				
Heptadecane	629-78-7						+[g]
Heptadecanoic acid	506-12-7		0.3-1.0				
Heptyl benzoate	7155-12-6					tr	
n-Hexadecane	544-76-3		0-0.1			tr	
Hexadecanoic acid	57-10-3		1.3-1.8	0.3			
(Z)-3,5-Hexadienol							tr[l]
(Z)-3,5-Hexadienyl (Z)-3,5-Hexadienylacetate							0.1[l]
(Z)-3,5-Hexadienylacetate							tr[l]
(Z)-3,5-Hexadienylbenzo-ate							0.05[l]
(Z)-3,5-Hexadienylbuty-rate	69925-34-4						tr[l]
Hexanoic acid	142-62-1	0.04-0.9					
Hexanol	111-27-3			0.1		tr	
2-Hexanol	626-93-7			tr		tr	
3-Hexanol	623-37-0			tr		tr	
(Z)-3-Hexenal	6789-80-6			0.2			
(E)-3-Hexenoicacid	1577-18-0			<0.1			
(E)-2-Hexen-1-ol	928-95-0						0.07[l]
3-Hexenol	544-12-7						+[j]
(E)-3-Hexenol	928-97-2					tr	
(Z)-3-Hexen-1-ol	928-96-1	0.1-1.8	0.3-0.6	0.9		tr	0.4[l]; 1.2[k]
(E)-2-Hexenyl acetate	2497-18-9			<0.1			
(E)-3-Hexenyl acetate	3681-82-1					tr	
(Z)-3-Hexenyl acetate	3681-71-8	0.8-2.7	4.1-6.4	1.6	<0.1-1.6	0.1	0.05[l]; 0.7[f]; 1.2[k]; 1.3[h]; 1.5[g]
(E)-2-Hexenyl benzoate	76841-70-8			0.2			
(E)-3-Hexenyl benzoate	75019-52-2			0.1		tr	0.1[f]
(Z)-3-Hexenyl benzoate	25152-85-6	3.9-14.9	10.5-20.7	5.3	4.7-16.9	2.5	8.8[h]; 9.4[f]; 15.6[k]; 22.4[g]
(E)-3-Hexenylbutyrate	53398-84-8						+[g]
(Z)-3-Hexenylbutyrate	16491-36-4	0.01-0.1		<0.1		tr	+[g]; 0.2[l]; 0.3[k]
(Z)-3-Hexenyl hexanoate	31501-11-8					tr	
3-Hexenyl hexenoate				0.1			
(Z)-3-Hexenyl hydroxy-butyrate				<0.1			
(E)-3-Hexenyl isovalerate						tr	
(Z)-3-Hexenyl isovalerate	35154-45-1			tr		tr	
(Z)-3-Hexenyl methyl-butyrate	53398-85-9			<0.1			
(E)-2-Hexenyl salicylate	68133-77-7			tr			
(Z)-3-Hexenyl salicylate	65405-77-8			tr		tr	
(E)-3-Hexenyl tiglate						tr	
(Z)-3-Hexenyl tiglate	67883-79-8			tr		tr	
Hexyl acetate	142-92-7					tr	
Hexyl benzoate	6789-88-4	0.8-4.2	0.2-0.3	0.1		tr	+[g]
α-Humulene	6753-98-6			0.1			tr[k]; 0.1[f]
Hydrocinnamyl alcohol	122-97-4					tr	
3-Hydroxy-β-ionone	116296-75-4			tr			
8-Hydroxylinalool	64142-78-5						0.1[f]
(E)-8-Hydroxylinalool	75991-61-6			0.3			
3-Hydroxy-2-methyl-4-pyrone	118-71-8			<0.1			
1*H*-Indole	120-72-9	0.9-12.2	0.1-0.5	1.0	0.2-9.7	tr	2.4[k]; 5.1[h]; 13.4[f]; 14.1[l]
2-Indolinone	59-48-3			tr			
Isocaryophyllene	118-65-0		0-1.1				9.5[k]
Isophytol	505-32-8					0.4	
δ-Jasmolactone	25524-95-2			<0.1	0-<0.1	tr	
(Z)-Jasmone	488-10-8	0.02-0.3	tr-0.1		0-<0.1	0.5	0.3[l]
Limonene	138-86-3					tr	
Linalool	78-70-6	11.5-34.2	10.2-14.1	8.4	2.6-9.2	7.4	8.4[k]; 11.0[h]; 13.9[l]; 23.1[g]

Table 5.43.1 Constituents identified in jasmine extracts (*continued*)

Constituent	CAS	Percentage and range					
		A	B (1)	C (5)	D (3)	E (12)	F
Linalool oxide	1365-19-1						0.1[f]; 0.4[k]
cis-Linalool oxide	11063-77-7						0.2[f]
cis-Linalool oxide, furanoid	5989-33-3	0.1-0.6	0.2-3.2	0.2			0.1[l]
cis-Linalool oxide, pyranoid	14009-71-3			0.1			0.6[l]
trans-Linalool oxide, furanoid	34995-77-2	0.09-0.4	0.3-0.6	<0.1			
Linalyl acetate	115-95-7					tr	
Linoleic acid	60-33-3			0.2			
Linolenic acid	463-40-1			0.7			
Methyl *N*-acetylanthra-nilate	2719-08-6			0.1		2.0[c]	+[i]
Methyl anthranilate	134-20-3	0.5-7.6	2.8-6.5	7.6	5.0-12.0	4.9	1.2[k]; 4.7[f]; 5.5[l]; 8.3[h]
Methyl arachidate	1120-28-1					tr	
Methyl arachinoate				0.1			
Methyl benzoate	93-58-3	0.2-0.7	0.1-0.6	0.6	0-0.6	0.3	0.5[k]; 0.6[h]; 0.9[f]; 2.6[l]
2-Methyl-2-butanol	75-85-4						+[j]
(*E*)-Methyl cinnamate	1754-62-7			<0.1		tr	
(*Z*)-Methyl cinnamate	19713-73-6					tr	
6-Methyldocosane	55124-81-7						+[j]
Methyl docosanoate	929-77-1					tr	
Methyl eicosadienoate						tr	
Methyl eicosatrienoate	82729-72-4					tr	
Methyl eicosenoate	76899-35-9					tr	
Methyl elaidinate				tr			
Methyl 5-ethylnicotinate	68686-58-8						+[i]
4-Methyl-3-ethylpyridine	529-21-5						+[i]
Methyl *N*-formylanthrani-late	41270-80-8			tr			
Methyl heneicosadienoate	122768-03-0					tr	
Methyl heneicosanoate	6064-90-0					tr	
Methyl heneicosenoate	146407-38-7					tr	
Methyl heptadecanoate	1731-92-6					tr	
6-Methyl-5-hepten-2-one	110-93-0	0.3-0.5	tr-0.5			tr	
Methyl hexadecanoate	112-39-0		0.1-0.7	0.4	0.1-0.6	1.0	+[j]; 0.6[f,k]; 2.3[l]
Methyl jasmonate	1211-29-6					tr	+[l]
(*E*)-Methyl jasmonate					0-0.2		
(*Z*)-Methyl jasmonate	1211-29-6		0.1	<0.1			
(*Z*)-Methyl epi-jasmonate			0.1-0.4				
Methyl linoleate	112-63-0			0.1	0.1-0.5	tr	0.4[f]
Methyl linolenate	301-00-8	0.4-8.6	1.4-2.8	1.9	1.4-5.2	4.3	3.9[f]
Methyl *N*-methyl anthra-nilate	85-91-6		tr	0.1			+[l]; 0.08[k]
Methyl 4-methyl-5-ethyl-nicotinate							+[i]
Methyl 16-methylhepta-decanoate	5129-61-3						0.3[k]
Methyl 4-methylnicotinate	33402-75-4						+[i]
Methyl 14-methyl penta-decanoate	5129-60-2						+[j]
Methyl 4-methyl-5-vinyl-nicotinate							+[i]
Methyl nicotinate	93-60-7						+[i]
Methyl nonadecanoate	1731-94-8					tr	
Methyl nonadecenoate	19788-74-0					tr	
2-Methylnonane	871-83-0		tr-0.2				
Methyl octadecanoate	112-61-8	0.1-2.5	1.4-3.8	0.3	0.2-0.7	2.0[c]	0.3[f]
Methyl 6,9,12-octade-catrienoate							+[j]
Methyl octanoate	111-11-5					tr	
Methyl oleate	112-62-9			0.2	<0.1-0.6	2.0[c]	
Methyl pentadecanoate	7132-64-1					tr	
Methyl phenylacetate	101-41-7			<0.1			

Table 5.43.1 Constituents identified in jasmine extracts (*continued*)

Constituent	CAS	Percentage and range					
		A	B (1)	C (5)	D (3)	E (12)	F
Methyl phenyl-3-propionate						tr	
Methyl salicylate	119-36-8	0.08-1.6	0.2-0.9	0.2	0.1-1.3	tr	+[j]; 0.2[f]; 0.6[k]
Methyl tetradecanoate	124-10-7					tr	
5-Methyltricosane	22331-09-5		1.5-2.0				
Methyl-5-vinylnicotinate	38940-67-9						+[i]
4-Methyl-3-vinylpyridine							+[i]
α-Muurolene	10208-80-7						+[g]
γ-Muurolene	30021-74-0			0.1			
τ-Muurolol	19912-62-0	0.05-0.3		0.1			+[l]
Nerol	106-25-2	0.2-0.7				tr	
Nerolidol	7212-44-4						0.5[h]
(E)-Nerolidol	40716-66-3	0.2-1.2	0.5-1.0	0.2	0-0.8		0.5[f]
(Z)-Nerolidol	3790-78-1	0.05-0.3	0.1				
1-Nitro-2-phenylethane	6125-24-2			0.1			
Nonadecane	629-92-5		0-0.8				
(E)-β-Ocimene	3779-61-1			<0.1			
9,12-Octadecadienol	506-43-4		0.6-1.2				
Octadecane	593-45-3						2.3[m]
Octyl benzoate	94-50-8					tr	
Oleic acid	112-80-1		0.5-1.5	tr			
Oplopanolol				0.1			
α-Patchoulene	560-32-7						+[g]
Pentyl acetate	628-63-7					tr	
Pentyl benzoate	2049-96-9					tr	
2-Phenethyl alcohol	60-12-8	0.5-1.2		1.5	0-2.0	2.2	1.7[f]; 2.4[l]
1-Phenethyl benzoate						tr	
2-Phenethyl benzoate (β-)	94-47-3			0.2		1.3	0.3[f]
2-Phenylethyl formate	104-62-1					tr	
2-Phenethyl propionate	122-70-3					tr	
Phenol	108-95-2						+[g]
Phenylacetaldehyde	122-78-1			tr		tr	
(E)-Phenylacetaldehyde oxime				0.2[b]	0-1.7[b]		
(Z)-Phenylacetaldehyde oxime				0.2[b]	0-1.7[b]		
syn-Phenylacetaldoxime							1.8[f]
anti-Phenylacetaldoxime							2.6[f]
Phenylacetic acid	103-82-2			<0.1			
Phenylacetonitrile	140-29-4			1.2	0-1.5	1.2	0.3[f]
2-Phenylethyl acetate (β-)	103-45-7	0.3-0.7		0.3		2.4	0.9[f]; 1.0[h]
2-Phenylethyl salicylate	87-22-9				0-<0.1		
Phytol	7541-49-3	0.08-0.3				0.3	
Phytyl acetate	10236-16-5					2.6[e]	
Propyl benzoate	2315-68-6					tr	
Selina-4(15),7(11)-diene							+[g]
Squalene	111-02-4			1.2	0.5-1.9		
Squalene 2,3-oxide	7200-26-2			0.6	0-0.9		
α-Terpineol	98-55-5	0.06-0.3				tr	
Terpinolene (α-)	586-62-9		tr-0.7				0.2[g]
Tetracosane	646-31-1		tr-2.0				
n-Tetradecane	629-59-4		0-0.8			tr	
Tetramethylhexadecane							+[g]
(2E,6E,10E)-3,7,11,15-Tetramethyl-2,6,10,14-hexadecatetraen-1-yl acetate							
Tetramethylpentadecane	80297-59-2						+[g]

Table 5.43.1 Constituents identified in jasmine extracts (*continued*)

Constituent	CAS	A	B (1)	C (5)	D (3)	E (12)	F
α-Tocopherol	59-02-9			0.3			
Tricosane	638-67-5			<0.1			0.3[k]
Tricosene	56924-46-0	0.2-5.4		0.9-8.0			
11-Tricosene	52078-56-5						7.1[k]
(Z)-11-Tricosene	52078-37-2						3.5[f]
Trimethylcyclohexanol	1321-60-4						+[g]
Undecane	1120-21-4			tr			
3-Vinylpyridine	1121-55-7						+[i]

A eleven absolutes of *Jasminum sambac* (L.) Aiton. from India and China analyzed between 2005 and 2013; lowest and highest concentrations given (E. Schmidt, unpublished data)
B three absolutes from flowers picked in various parts of India; lowest and highest concentrations given; the concrete was obtained with pentane (is usually hexane) and the absolute with methanol (is usually ethanol) (ref. 1)
C one commercial absolute from India (ref. 5)
D seven commercial absolutes from India and three from China; lowest and highest concentrations given (ref. 3)
E one sample of Egyptian *J. sambac* absolute (ref. 12)
F data from other studies (indicated with superscript letters); highest concentrations found in any study reviewed here given; when two or more absolutes were investigated, only the highest concentrations are mentioned, unless indicated otherwise

[a] incorrect identitification; [b] (E)- and (Z)-phenylacetaldehyde oxime combined; [c] methyl stearate and methyl oleate and methyl N-acetylanthranilate combined; [d] correct isomer not identified; [e] phytyl acetate and geranyllinalool combined; [f] one lab-prepared Egyptian sambac absolute (ref. 17); [g] several absolutes from different areas of China (ref. 4); [h] one absolute from India (ref. 6); [i] basic fraction of one Chinese sambac absolute sample; percentages were given, but these cannot be compared with the other data, and are therefore not quantified here but indicated with +[i] (ref. 10); [j] one absolute from China (ref. 11); [k] one absolute from China (ref. 14); [l] one *J. sambac* absolute of Chinese origin (ref. 15); [m] one jasmine absolute obtained from Thai *J. sambac* flowers (ref. 19);

tr: trace (in columns B and E: not specified; in column C: <0.01); + present in the oil investigated, but quantity not stated; +[i] these compounds were identified in the basic fraction of an absolute; percentages were given, but these cannot be compared with the other data, and are therefore not quantified here but indicated with +[i] (ref. 10)

Table 5.43.2 Major constituents of *J. sambac* absolutes

Constituent	CAS	A	B (1)	C (5)	D (3)	E (12)	F
Linalool	78-70-6	11.5-34.2	10.2-14.1	8.4	2.6-9.2	7.4	8.4[k]; 11.0[h]; 13.9[l]; 23.1[g]
Benzyl alcohol	100-51-6	1.2-6.3	0.3-1.4	10.7	3.2-14.1	1.1	6.8[k]; 8.4[f]; 10.2[g]; 32.9[m]
(Z)-3-Hexenyl benzoate	25152-85-6	3.9-14.9	10.5-20.7	5.3	4.7-16.9	2.5	8.8[h]; 9.4[f]; 15.6[k]; 22.4[g]
Benzyl acetate	140-11-4	10.5-21.0	7.5-13.2	8.0	6.6-15.7	21.2	4.3[l]; 7.5[k]; 9.2[g]; 14.2[f]
(E,E)-α-Farnesene	502-61-4	10.6-17.1	2.2-19.7	13.1	7.3-15.6	12.4[d]	10.0[m]; 13.1[f]; 18.4[l]; 19.3[h]
1H-Indole	120-72-9	0.9-12.2	0.1-0.5	1.0	0.2-9.7	tr	2.4[k]; 5.1[h]; 13.4[k]; 14.1[l]
Methyl anthranilate	134-20-3	0.5-7.6	2.8-6.5	7.6	5.0-12.0	4.9	1.2[k]; 4.7[f]; 5.5[l]; 8.3[h]
Methyl linolenate	301-00-8	0.4-8.6	1.4-2.8	1.9	1.4-5.2	4.3	3.9[f]
Tricosene	56924-46-0	0.2-5.4		0.9-8.0			

LEGEND: SEE UNDER TABLE 5.43.1

LITERATURE

1 Rout PK, Naik SN, Rao YR. Composition of absolutes of *Jasminum sambac* L. flowers fractionated with liquid CO2 and methanol and comparison with liquid CO2 extract. J Essent Oil Res 2010;22:398-406

2 Lawrence BM. Progress in essential oils. Perfum Flavor 2006;31(10):44-? (last page unknown)

3 Braun NA, Sim S. *Jasminum sambac* flower absolutes from India and China – geographic variations. Nat Prod Commun 2012;7:645-650

4 Guo Y-J, Dai L, Yang L-P, Ren Q. A study on the chemical compositions of *Jasminum sambac* (L.) Aiton in various farming seasons during the blossom period in Fuzhou region. Fujian Fexxi Ceshi 1998;7:785-792. Data cited in ref. 2

5   Braun NA, Kohlenberg B, Sim S, Meier M, Hammerschmidt FJ. *Jasminum flexile* flower absolute from India – a detailed comparison with three other jasmine absolutes. Nat Prod Commun 2009;4:1239-1250

6   Vaze K. Lesser known essential oils of India and their compositions and uses. FAFAI 2003 (3/4):47-58

7   Reverchon E, Della Porta G, Gorgoglione D. Supercritical CO2 fractionation of jasmine concrete. J Supercr Fluids 1995;8:60-65

8   Zeng L-H, Hu M, Yan Y-M, Lu Q, Cheng Y-X. Compounds from the roots of *Jasminum sambac*.J Asian Nat Prod Res 2012;14:1180-1185

9   Lawrence BM. Progress in essential oils. Perfum Flavor 1994;19(2):64-69

10 Toyoda T, Muraki S, Yoshida T. Pyridine and nicotinate derivatives of jasmin. In: Proceedings of the 7th International Essential Oil Congress, Kyoto, 1979. Tokyo: Japan Flav Frag Mfrs, 1979:473-476. Data cited in refs. 9 and 18

11 Zhu L-F, Lu B-Y, Xu D. A study on the chemical constituents of Jasmine essential oil. In: Y Wang, Ed. Proceedings of the Sino/American Symposium on Chemistry of Native Products and Processes, 1980. Beijing, China: Sci Press Beijing Peoples Republic, 1982:309-311. Data cited in refs. 9 and 1

12 Chaput A, Gardou C, Huddadi D, Jullien R, Moustafa Kamel E, Saint-Jaim Y. Etude sur les composants de l'absolue de "foll" Egyptien. In: Proceedings of the 7th International Essential Oil Congress, Cannes. Grasse: FEDAROM, 1982:210-214. Data cited in refs. 9 and 18

13 Hellivan P-J. Jasmine. Reinventing the 'king of perfumes'. Perfum Flavor 2009;34(October):42-52

14 Wu C-S, Zhao D-X, Sun S-W, Ma Y-P, Wang Q-Q, Lu L-C. The minor chemical components of the absolute oil from the flower of *Jasminum sambac* (L.) Aiton. Zhiwu Xuebao 1987;29:636-642. Data cited in refs. 9 and 1

15 Kaiser R. New volatile constituents of *Jasminum sambac* (L.) Aiton. In: BM Lawrence, BD Mookherjee, BJ Willis, Eds. Flavors and Fragrances: A World Perspective. Amsterdam, The Netherlands: Elsevier Science Publishers, 1988: 669-684. Data cited in ref. 9 (and partly in ref. 1)

16 Hongratanaworakit T. Stimulating effect of aromatherapy massage with jasmine oil. Nat Prod Comm 2010;5:157-162

17 Edris AE, Chizzola R, Franz C. Isolation and characterization of the volatile aroma compounds from the concrete headspace and the absolute of *Jasminum sambac* (L.) Ait. (Oleaceae) flowers grown in Egypt. Eur Food Res Technol 2008;226:621-626

18 Lawrence BM. Progress in essential oils. Perfum Flavor 1995;20(July/August):29-41

19 Paibon W, Yimnoi C-A, Tembab N, Boonlue W, Jampachaisri K, Nuengchamnong N, et al. Comparison and evaluation of volatile oils from three different extraction methods for some Thai fragrant flowers. Int J Cosmet Sci 2011;33:150-156

20 Kanlayavattanakul M, Kitsiripaisarn S, Lourith N. Aroma profiles and preferences of *Jasminum sambac* L. flowers grown in Thailand. J Soc Cosm Chem 2013;64:483-494

Chapter 5.44 JUNIPER BERRY OIL

DEFINITION
Juniper berry oil (essential oil of juniper berry) is the essential oil obtained from the fruit and the terminal branchlets of the (common) juniper, *Juniperus communis* L.

INCI NOMENCLATURE
Description/definition: Juniperus communis fruit oil is the volatile oil obtained from the berries of the juniper, *Juniperus communis* L., Cupressaceae
INCI name EU & USA: Juniperus communis fruit oil
CAS registry number(s): 8002-68-4; 84603-69-0; 73049-62-4
EINECS number(s): 283-268-3

ISO (INTERNATIONAL ORGANIZATION FOR STANDARDIZATION) STANDARD
ISO number: 8897
ISO name: Essential oil of juniper berry
Botanical origin: *Juniperus communis* L.
Parts of plant used: Fruit, terminal branchlet
ISO values: ISO values (minimum and maximum concentrations) are shown in Table 5.44.1.

THE PLANT, THE OIL, AND THEIR USES
The common juniper (*Juniperus communis* L.) is an aromatic and evergreen coniferous shrub or tree. It is widely distributed in Europe, North Africa, North America and north Asia southwards to the Himalayas. It produces blue-black 'berries', which are not true berries in the botanical sense, but are the female seed cones which have a berry-like appearance. Juniper berries are widely used in the food industry as flavorings (in tea, beer, brandy and marinades for meat, poultry and fish) and in the production of juniper-flavored spirits, such as gin. Gin is produced by distillation of neutral alcohol and water in the presence of juniper berries and other botanicals, of which coriander and angelica are the most common (13).

The berries (*Juniperi fructus*) and needles (leaves) of juniper contain an essential oil that has a characteristic aromatic flavor and bitter taste. Juniper berry essential oil is used in the pharmaceutical and food industries (also in gin and liqueurs) and perfumery, as well as in cosmetics; the oils are also applied in aromatherapy practices (52,53). Juniper oil is stated to possess a wide range of pharmacological activities. It is said to have diuretic, antiseptic, carminative, and stomachic properties and has traditionally been used for cystitis, flatulence and colic, and as a steam inhalant in the management of bronchitis. Moreover, it has been applied topically for rheumatic pains in joints or muscles. Juniper berry oils are the subject of Pharmacopoeia monographs in various countries. Besides juniper berries and oil, other preparations of juniper such as infusions, decoctions, tinctures and extracts are used in different fields. The junipers are grown commercially in several countries, including the United States, Canada, northern Italy, and Croatia (9,13,21,22,25,26).

CHEMICAL COMPOSITION
Juniper berry oil is a colorless, sometimes greenish clear mobile liquid, which has a fresh and herbaceous, typical 'gin'- note odor with fruity-woody note. The yield of essential oil from the fruit and terminal branches of *Juniperus communis* L. generally varies from 0.6 to 1.8%. The main producing countries of this oil are Austria, Bosnia-Herzegovina, Slovenia, Serbia, Italy and Hungary.

Literature data (up to November 27, 2014) on the chemical composition of juniper berry oils (from *J. communis* and from *J. communis* var. *communis*, a synonym according to The Plant List) and unpublished analytical data from one of us (E.S.) are shown in Table 5.44.2 in alphabetical order. In juniper berry oils from various origins, over 295 chemicals have been identified. About 45 per cent of these were found in a single reviewed publication only.

The major compounds found in juniper berry oils from different sources are shown in Table 5.44.3. They include (highest concentrations in any study given) α-pinene (84.8%), sabinene (36.8%), myrcene (30.9%), terpinen-4-ol (22.9%), limonene (22.1%), β-pinene (17.8%), *p*-cymene (16.2%), germacrene D (14.3%), β-caryophyllene (10.3%) and δ-cadinene (8.7%). Well-known ingredients of juniper berry oils that were present in high concentrations (>7%) in one or two studies were terpinolene (13.6% and 13.6%), citronellol (12.8%), caryophyllene oxide (10.0%), δ3-carene (9.9%) and α-humulene (7.0% and 7.2%). Uncommon or rare constituents of juniper berry oils found in high concentrations (>7%) in single studies include thujopsene (widdrene) (25.2%), δ-cuparene (15.3%), α-cedrol (12.4%) and widdrol (7.7%).

Commercial oils
The ten chemicals that had the highest maximum concentrations in 395 commercial juniper berry essential oil samples (concentration ranges provided) are the following: α-pinene (18.1-66.6%), myrcene (11.6-21.6%),

Table 5.44.1 ISO values (%) for juniper berry oil [a]

Compound	CAS	Minimum	Maximum
α-Pinene	80-56-8	25.0	45.0
Myrcene	123-35-3	3.0	22.0
Sabinene	3387-41-5	4.0	20.0
β-Pinene	127-91-3	1.0	12.0
Limonene	138-86-3	2.0	8.0
Terpinen-4-ol	562-74-3	1.0	6.0
β-Caryophyllene	87-44-5	1.5	5.0
Germacrene D	23986-74-5	1.0	5.0
α-Humulene	6753-98-6	1.0	4.0
δ-Cadinene	483-76-1	1.0	3.5
Bornyl acetate	76-49-3	0.0	0.6

[a] ISO 8897 Essential oil of juniper berry ©ISO 2010; Geneva, Switzerland, www.iso.org

sabinene (1.0-17.7%), β-pinene (1.8-7.8%), limonene (2.5-7.6%), β-caryophyllene (2.0-5.9%), terpinen-4-ol (1.9-5.4%), γ-terpinene (0.4-4.5%), δ-cadinene (0.3-2.7%) and germacrene D (0.07-2.6%) (Erich Schmidt, unpublished analytical data).

For analytical data on oils produced from *unripe* berries from *Juniperus communis* see refs. 5,6,7,14,23,24,33,41 and 44. Analyses of juniper oils purely obtained from *leaves* (needles) of *J. communis* (or leaves plus twigs, e.g., ref. 46) are not discussed.

Chemotypes

In juniper berry oils, α-pinene is nearly always the dominant chemical. In a few cases, however, especially with *J. communis* berries growing at high altitude, sabinene

was found to have the highest concentration in the oils (30,39,50). The finding of widdrene (synonym: thujopsene) (25.2%), δ-cuparene (15.3%) and widdrol (7.7%) as dominant ingredients in a commercial *Juniperus communis* fruit oil purchased from a USA chemical company is therefore highly atypical (ref. 45). A commercial sample of 'Juniperus communis oil' is investigated in ref. 51, but is not discussed here, as it was not clear that it was obtained from the berries and twigs.

CONTACT ALLERGY/ALLERGIC CONTACT DERMATITIS

General

Contact allergy to/allergic contact dermatitis from juniper berry oil has been reported in four publications only. In a

Table 5.44.2 Constituents identified in juniper berry oils

Constituent	CAS	Percentage and range					
		A	B (5)	C (6)	D (37)	E (11)	F
Abietadiene	36312-33-1		tr	tr-0.2			tr[c,e]; 0.1[d]
4-epi-Abietal	34223-60-4	0-0.1					
Abietatriene	19407-28-4		tr-0.1	tr-0.1			+[a2]; tr[c]; 0.1[e]; 0.3[d]
Amorpha-4,9-dien-2-ol	394251-66-2						+[a2]
α-Amorphene	20085-19-2				0-0.4		0.1[y3]; 0.3[y5]
Aromadendrene	489-39-4				0.03-0.1		
allo-Aromadendrene	25246-27-9						0.3[z]
E-Asarone (α-)	2883-98-9						+[a2]
trans-Ascaridole glycol	6790-83-6						+[a2]
β-Atlantol	420109-31-5		0-0.3				
(*E*)-α-Atlantone	26294-59-7			0-0.3			0.3[c]
Benzaldehyde	100-52-7						0.5[h]
cis-α-Bergamotene	18252-46-5						0.3[l]
Bicyclogermacrene	24703-35-3		0-0.6	0-0.4	0-2.1		0.2[d]; 0.4[c]; 0.5[y1]; 1.3[y7]; 3.2[p]
Bicyclosesquiphellan drene	54324-03-7						0.2[y1]
epi-Bicyclosesquiphellan drene	54274-73-6				0.1-0.7		
β-Bisabolene	495-61-4			0-0.2			0.2[c]; 1.1[z]; 2.0[y2]
α-Bisabolol	515-69-5						1.6[y2]
epi-α-Bisabolol	78148-59-1			0-0.5			0.5[c]; 1.3[y5]
Borneol	507-70-0	0.04-0.1	0.3-1.5	0.1-1.1	tr-0.1		1.1[c]; 1.4[u]; 2.3[y]; 2.5[k]; 3.4[t]; 5.2[v]
Bornyl acetate	76-49-3	0.1-0.3	0.3-1.3	0.7-1.3	0.2-0.6	tr-2.0	0.5[h]; 1.1[k]; 1.2[c]; 1.4[y2]; 2.0[v,e]; 3.2[u]
endo-1-Bourbonalol			0-0.1	0.1-0.2			0.1[d]; 0.2[c,e]
β-Bourbonene	5208-59-3			tr			tr[c,e]; 3.8[z]
2,3-Butanediol	513-85-9						+[a2]
Cadalene	483-78-3				0-0.1		+[a2]
cis-Cadina-1,4-diene	29837-12-5		0.1-0.2	tr-0.2			0.1[c]
trans-Cadina-1,4-diene	20085-13-6				tr-0.3		+[a2]; 0.1[d]; 0.2[e]
(*E*)-Cadina-1(6),4-diene	931410-54-7		0-0.2	tr-0.1			tr[d,e]; 0.1[c]
α-Cadinene	24406-05-1		0.1-0.5	0.1-0.4	0.1-0.7		0.1[y3]; 0.2[e]; 0.3[c]; 0.4[l]; 0.5[d]; 1.1[i]
β-Cadinene	523-47-7				0-5.4		
γ-Cadinene	39029-41-9	0.08-0.7	0.2-1.4	0.5-1.1	0.1-2.4	0-1.4	0.5[y3]; 0.6[p]; 0.7[z]; 1.3[i]; 1.8[d]; 2.0[o,y]
δ-Cadinene	483-76-1	0.3-2.7	0.5-2.9	1.4-2.5	tr-4.2	0-3.6	2.5[c]; 2.7[y1]; 2.9[d]; 4.1[y]; 4.4[m]; 8.7[w]
α-Cadinol	481-34-5		1.8-7.6	2.7-8.6	0-2.7		0.9[y2]; 1.7[m]; 2.7[z]; 5.1[e]; 7.1[c]; 7.6[d]
δ-Cadinol	19435-97-3						1.6[y5]

Table 5.44.2 Constituents identified in juniper berry oils (*continued*)

Constituent	CAS	A	B (5)	C (6)	D (37)	E (11)	F
τ-Cadinol	58580-31-7		0.5-2.5	1.1-3.5	0.1-1.7[b]		0.2[y1]; 0.5[z]; 1.0[x]; 1.2[e]; 2.1[d]; 3.2[c]
1,10-di-epi-Cadinol							0.3[z]
α-Calacorene	21391-99-1		0.1-0.2	tr-0.2			+[a2]; 0.1[c]; 0.2[d]
β-Calacorene	50277-34-4						+[a2]
cis-Calamenene	72937-55-4						0.1[y]
trans-Calamenene	40772-39-2						+[a2]
cis-Calamenen-10-ol							+[a2]
trans-Calamenen-10-ol	828923-23-5						+[a2]
Camphene	79-92-5	0.3-1.0	0.3-0.8	0.2-0.5	0-0.6	0.2-1.6	0.6[h]; 0.7[y5]; 0.8[y1]; 0.9[u]; 1.2[g]; 1.9[o]
Camphene hydrate	465-31-6				0-0.2		+[a2]; tr[e]; 0.1[d]; 0.2[y5]
6-Camphenone	53803-33-1		0-0.5				
α-Campholenal	4501-58-0		0.1-4.1	0-1.5	0-0.3		+[a2]; 0.3[e]; 0.4[c]; 0.7[d]; 1.6[y5]; 1.8[y2]
Camphor	76-22-2		0-1.5	tr-0.9			+[a2]; 0.2[c,d]
m-Camphorene	20016-73-3						0.2[y3]
p-Camphorene	532-87-6						0.1[y3]
γ-3-Carene							0.3[y5]
δ2-Carene	554-61-0						0.1[y2]
δ3-Carene	13466-78-9	0.05-0.9	0-0.1	0-1.2	tr-0.2	0-1.8	0.5[y2]; 0.7[g,h]; 1.0[j,k]; 1.1[l]; 1.2[c]; 9.9[i]
4-Carene	29050-33-7						0.9[z5]
Carvacrol	499-75-2						0.2[k]; 0.5[u]; 0.8[i]
Carveol	99-48-9						0.5[y8]
(E)-Carveol	1197-07-5		0-1.2	0-0.2			0.1[d]; 0.2[c,y3]; 0.8[h]; 1.7[y2]
(Z)-Carveol	1197-06-4		0-0.3				0.4[y2]; 0.5[h]
Carvone	99-49-0	0.0-0.08	0-0.2	0-0.2		0-0.9	0.1[y3]; 0.2[c]; 0.3[h,k]; 0.7[i]; 0.8[y2]
trans-Carvyl acetate	1134-95-8						0.9[z]
β-Caryophyllene	87-44-5	2.0-5.9	0.2-0.6	0.1-0.4	1.8-4.1	0-4.0	3.6[y1]; 5.3[p]; 8.2[y]; 9.2[m]; 9.8[z]; 10.3[y]
Caryophyllene oxide	1139-30-6	0.03-0.4	0-1.6	0.3-1.4	0-0.6	0-2.4	0.2[y]; 0.3[y2]; 0.4[x]; 0.5[c]; 0.9[d]; 10.0[z]
Caryophyllenyl alcohol	56747-96-7						+[a2]
α-Cedrene	469-61-4						+[a2]
β-Cedrene	546-28-1						0.2[y6]
α-Cedrol	77-53-2						0.7[y]; 1.2[k]; 12.4[i]
trans-Chrysanthenyl acetate	50764-55-1						tr[e]; 0.1[d]
1,4-Cineole	470-67-7						+a2; 2.0[y8]; 4.0[q]
1,8-Cineole	470-82-6						0.2[y2]; 0.6[k]; 2.9[h]; 5.3[t]
Citronellal	106-23-0						0.7[y2]
Citronellic acid	502-47-6						0.1[y1]
Citronellol	106-22-9	0.0-0.08	0-1.3	0-0.2	0-0.2	0-0.3	0.2[c]; 0.4[d]; 0.5[e]; 1.3[y2]; 12.8[k]
Citronellyl acetate	150-84-5		0-0.1	0-0.1	tr-0.1		+[a2]; 0.1[c,y3]; 0.3[d]; 0.5[g]
Citronellyl butyrate	141-16-2		tr-0.3	0.1-0.2			0.2[c]; 0.3[d]; 0.5[e]
Citronellyl isobutyrate	97-89-2		0-0.5	0-0.3			0.1[c]; 0.5[d]
α-Copaenal							0.1[y2]
α-Copaene	3856-25-5	0.1-0.4	0-0.1	tr-0.1	0.05-1.5	tr-1.8	0.5[z2]; 0.6[y1]; 0.8[n]; 1.0[g]; 1.4[p,y7]
β-Copaene	18252-44-3				0-0.6		+[a2]
α-Corocalene	20129-39-9						+[a2]
α-Cubebene	17699-14-8	0.1-1.0	0-0.1	0-0.1	0.5-1.3	0-1.2[f]	0.6[y1]; 0.8[g]; 0.9[n]; 1.0[y7]; 1.1[p]; 1.3[s]
β-Cubebene	13744-15-5		0-0.1	tr-0.1			0.1[c]; 0.5[y2]; 1.0[y]; 6.4[b,g]
Cubebol	23445-02-5						0.1[y1,y2]
epi-Cubebol	38230-60-3		0-0.4				0.2[y]; 0.8[d]
10-epi-Cubebol	176589-53-0						0.1[d]
1,10-di-epi-Cubenol	73365-77-2		0-0.3	tr-0.3			+[a2]; 0.1[c]; 0.2[e]; 0.3[d]
1-epi-Cubenol	38230-60-3		02-0.8	0.2-0.8	tr-0.4		+[a2]; 0.2[y1]; 0.3[e]; 0.6[c]

Table 5.44.2 Constituents identified in juniper berry oils (*continued*)

Constituent	CAS	Percentage and range					
		A	B (5)	C (6)	D (37)	E (11)	F
Cuminyl alcohol	536-60-7						0.1[c]
Cuparene	16982-00-6						+[a2]
δ-Cuparene							15.3[z1]
Cyclocolorenone	489-45-2						0.2[e]
Cyclofenchene	488-97-1						<0.1[m]
Cyclopentadecanolide	106-02-5		0.1-0.2	tr-0.3			0.1[e]; 0.3[c]; 0.4[d]
Cyclosativene	22469-52-9						+[y]
o-Cymene	527-84-4						+[a2]
p-Cymene	99-87-6	0.4-2.3	0.3-0.5	tr-0.5	0.09-0.6	0.2-11.9	2.1[z2]; 2.3[p]; 5.0[z]; 10.0[u]; 11.9[v]; 16.2[y8]
p-Cymenene	1195-32-0	0.03-0.2	0.2-0.4	tr-0.2			+[a2]; 0.2[c,d,k]
m-Cymen-8-ol	5208-37-7		0.1-0.7	0-0.1			+[a2]; 0.1[y,c]
p-Cymen-8-ol	1197-01-9			tr-0.8	0-0.1		0.2[e,y3]; 0.8[c]; 1.2[y2,y5]; 2.2[h]; 2.3[d]
Cyperene	2387-78-2						0.1[y6]
Decanal	112-31-2						0.1[k]
Dehydroabietal	13601-88-2						tr[d]
trans-Dihydrocarvone	5948-04-9		0-0.4				0.6[d]
Dihydrosabinene	471-12-5						1.3[y5]
(Z)-2,6-Dimethyl-2,6-octadiene	2492-22-0						+[a2]
β-Elemene	33880-83-0	0.2-0.9	0.1-0.8	0.2-0.6	1.1-3.5	0-1.6	0.7[y3]; 0.8[y]; 1.0[p,y1]; 2.1[g]; 2.6[o]
γ-Elemene	29873-99-2	0.3-0.7	0.1-1.1	tr-0.3	1.0-6.4	0-1.2	0.3[y]; 0.8[l]; 0.9[c]; 1.1[d]; 2.6[o]; 3.3[g]
δ-Elemene	20307-84-0		0-0.1		0.1-0.5		+[a2]; 0.1[d,e]
Elemol	639-99-6		0.2-0.5	0.3-0.8	0.06-0.3		+[a2]; 0.5[d,e]; 0.8[c]; 1.0[y2]
Epizonarene	41702-63-0						0.5[g]
Ethyl butyrate	105-54-4						+[a2]
Ethyl decanoate	110-38-3						+[a2]
Ethyl dodecanoate	106-33-2						+[a2]; 0.1[d]
Ethyl heptanoate	106-30-9						+[a2]
Ethyl hexadecanoate	628-97-7						+[a2]
Ethyl hexanoate	123-66-0						+[a2]
Ethyl levulinate	539-88-8						+[a2]
Ethyl octanoate	106-32-1						+[a2]
Ethyl tetradecanoate	124-06-1						+[a2]
Eudesma-4(15),7-dien-1β-ol	119120-23-9		0.1-0.8	0.3-1.1			0.5[e]; 0.8[d]
Eudesm-7(11)-en-4-ol	473-04-1		0-0.6	0.2-0.8	0-0.07		+[a2]; 0.3[c,e]; 0.5[d]
α-Eudesmol	473-16-5		0-0.1				0.1[y2]
β-Eudesmol	473-15-4		0-0.2				
γ-Eudesmol	1209-71-8		0.3-0.5	tr-0.5	0-0.6		+[a2]; 0.2[e]; 0.3[c]; 0.5[d]
Eugenol	97-53-0						1.3[k]
α-Farnesene	502-61-4						5.0[b,v]
(E,E)-α-Farnesene	502-61-4		0-0.1				
(E)-β-Farnesene	18794-84-8	0.2-0.6	tr-0.4	0-0.2			0.2[j]; 0.3[z2,y3]; 0.5[y1]; 1.0[b,g]; 2.2[m]
(Z)-β-Farnesene	28973-97-9						0.1[d,e]; 0.3[y]
(Z,Z)-Farnesol	16106-95-9						tr[d,e]
α-Fenchene	471-84-1						0.1[z2]; 0.2[y3]; 0.3[i]
α- + β-Fenchol							+[a2]
Fenchone	1195-79-5		0-0.1				+[a2]
α-Fenchyl acetate	111821-74-0						+[a2]; 0.1[z2]
α-Fenchyl alcohol	512-13-0	0.01-0.1	tr-0.7	tr			0.1[c]; 0.2[d]; 0.7[h]
Furfural	98-01-1						+[a2]
Furfuryl alcohol	98-00-0						0.3[k]
Geraniol	106-24-1						0.1[y1]; 0.7[k]; 1.6[u]
Geranyl acetate	105-87-3				0-0.2		+[a2]; tr[y3]; 0.02[k]; 0.1[d]
[1(10) E,5E]-Germa-cradien-4α-ol	207221-31-6						0.3[y1]

Table 5.44.2 Constituents identified in juniper berry oils (*continued*)

Constituent	CAS	Percentage and range					
		A	B (5)	C (6)	D (37)	E (11)	F
Germacrene B	15423-57-1		0.2-0.5	0.1-0.3	0.3-2.2		0.5[y2]; 0.6[d]; 0.7[y]; 1.1[y5]; 1.4[j]; 1.8[y1]
Germacrene D	23986-74-5	0.07-2.6	0.1-3.1	0.2-0.8	2.8-10.7	0-8.0	2.2[y5]; 2.7[y4]; 6.3[y1]; 6.7[s]; 8.1[y6]; 9.5[y]; 10[y9]; 10.4[n,x]; 12.1[m]; 14.3[y]
Germacrene D-4-ol	198991-79-6		0-1.1	0.2-1.1	0-0.6		0.6[d,h]; 0.8[e]; 0.9[y5]; 1.1[c]
Gleenol	72203-99-7						+[a2]
cis-β-Guaiene	372162-07-7						+[y]
α-Gurjunene	489-40-7						0.2[y3]
β-Gurjunene	73464-47-8		0-0.2	tr			tr[d]; 0.8[z]
Hexanal	66-25-1						tr[y3]
n-Hexane	110-54-3						+[a2]
2-Hexenal	505-57-7						0.2[d]
Hexyl isovalerate	10032-13-0						+[a2]
α-Humulene	6753-98-6	0.5-2.2	0.1-1.0	0.1-0.6	2.0-4.6		2.1[x]; 2.7[w,y4]; 2.8[u]; 3.7[g]; 7.0[z]; 7.2[y]
Humulene epoxide II	19888-34-7		0.3-2.1	0.3-2.0			0.1[b,y]; 0.2[d]; 0.4[e]; 0.5[y3]; 0.9[c]
2α-Hydroxyamorpha-4,7(11)-diene							+[a2]
17-Hydroxy-δ-cadinene							0.1[d,e]
14-Hydroxy-9-epi-(*E*)-caryophyllene	244226-09-3		0.3-1.1	0.3-1.8			0.7[e]; 1.1[d]; 1.8[c]
14-Hydroxy-α-humulene	108043-85-2		0-0.4	0-0.3			0.3[d]; 0.5[e]
14-Hydroxy-α-muurolene	105661-29-8		0-0.4	0.1-0.3			0.2[c,e]
(*E*)-α-Ionone	127-41-3			tr			tr[c]
(*E*)-β-Ionone	79-77-6		0-0.5	0-0.3			0.2[c]; 0.3[e]; 0.4[d]
Isoborneol	124-76-5						0.2[y5]
Isobornyl butyrate	58479-55-3		0-0.1				
Isocaryophyllene	118-65-0						0.5[h]
(*E*)-Isoeugenol	5932-68-3						0.4[z]
Isopentyl acetate	123-92-2						+[a2]
Junenol	472-07-1						+[a2]
Limonene	138-86-3	2.5-7.6	1.1-5.3	1.6-2.9	2.9-4.5	2.4-22.1	3.1[y1]; 4.5[y4]; 5.1[y2,y3]; 5.5[z3]; 7.9[g]; 10.6[y6]; 11.1[l]; 13.2[h]; 17.5[t]; 22.1[y]
Linalool	78-70-6	0.0-0.2	0-1.3	0-0.5	tr-0.9	0-1.5	0.3[k]; 0.5[y2]; 0.6[e]; 1.1[m]; 1.2[d]; 1.9[y]
trans-Linalool oxide	11063-78-8		0-0.3				
Longiborneol	465-24-7			0-0.3			
Longifolene	475-20-7	1.2-1.4	0-0.1	0-0.2	tr-0.1		tr[e]; 0.1[g,y3]; 0.2[c,d]; 1.2[h]
Manoyl oxide	596-84-9						tr[d,e]
13-epi-Manoyl oxide	1227-93-6						tr[d]
p-Mentha-2,8-diene	499-99-0						0.3[z2]
p-Mentha-1,5-dien-8-ol	1686-20-0		0.2-4.5	0-1.8	0-0.3		0.9[c,y5]; 1.4[d]
cis-p-Mentha-2,8-dien-1-ol	3886-78-0						0.2[d]
p-Menthane	99-82-1						0.1[q,r]
p-Mentha-1,3,8-triene	18368-95-1						+[y]; 0.4[d]
cis-p-Menth-2-en-1-ol	29803-82-5						0.2[y6]; 0.4[d]; 0.6[y2]
trans-p-Menth-2-en-1-ol	29803-81-4						tr[y3]; 0.2[y6]; 0.6[y2]
Methylbenzene	108-88-3						+[a2]; tr[y3]
3-Methyl-3-butenyl valerate							0.4[y5]
Methylcarvacrol	6379-73-3						+[a2]
Methyl citronellate	2270-60-2		0.1-1.5	0.1-1.5	0-0.1		+[a2]; 0.1[y3]; 0.8[d]; 0.9[e]; 1.5[c]; 3.6[y2]
Methylcyclopentane	96-37-7						+[a2]
Methyl geranate	2349-14-6						tr[y3]
6-Methyl-5-hepten-2-one	110-93-0						+[a2]

Table 5.44.2 Constituents identified in juniper berry oils (*continued*)

Constituent	CAS	A	B (5)	C (6)	D (37)	E (11)	F
Methyl thymol	4630-07-3						tr[d]
Muurola-3,5-diene	157374-44-2				0-0.2		
cis-Muurola-3,5-diene	157374-44-2		0-0.3	tr			tr[c,e]
trans-Muurola-3,5-diene	262352-88-5		0-0.1	tr			tr[c,d]; 0.1[e]
cis-Muurola-4(14),5-diene	157477-72-0		0-0.2	tr-0.1			tr[d,e]; 0.1[c]
trans-Muurola-4(14),5-diene	262352-87-4		0-0.1	0-0.2			+[a2]; 0.1[c,d]; 0.4[e]
α-Muurolene	10208-80-7	0.2-1.7	0.1-1.0	0.3-0.8	tr-2.7	0-2.4	0.2[j,m]; 0.3[g,h]; 0.4[y3]; 0.5[y]; 0.6[y1]; 0.9[j]
γ-Muurolene	30021-74-0		0.1-0.5		0-1.3		0.3[y]; 0.4[d,y3,z]; 0.8[y1]; 1.3[c]
14-oxy-α-Muurolene	69394-04-3						tr[e]
α-Muurolol	104245-48-9		0.5-1.2	0.2-1.0	tr-1.7		+[a2]; 0.1[y1]; 0.2[c]; 0.3[y2]; 1.0[e]; 1.2[d]
τ-Muurolol	19912-62-0		0.7-2.4	1.0-2.5	0.1-1.7[b]		+[a2]; 0.3[y1]; 0.9[e]; 1.8[y2]; 1.9[c]; 2.4[d]
Myrcene	123-35-3	11.6-21.6	3.6-17.4	6.3-19.6	2.9-26.5	0.3-22.5	13.8[e]; 14.2[y4]; 15.2[y1]; 18.7[d]; 19.6[c]; 19.9[k]; 21.2[y5]; 24.2[z]; 29.9[y]; 30.9[l]
Myrcenol	543-39-5						1.6[h]
cis-Myrtanol	15358-92-6			0-0.2			0.2[c]
trans-Myrtanol	15358-91-5		0-0.2	0-0.3			tr[e]; 0.3[c]
trans-Myrtanyl acetate			0-0.1	tr-0.1			+[a2]; tr[d,e]; 0.1[c]
Myrtenal	564-94-3						+[a2]; 0.2[y3]; 0.9[y5]
Myrtenol	515-00-4	0.02-0.08	0.1-1.2	0-0.7	0-0.2		+[a2]; 0.1[y3]; 0.3[e]; 0.6[c]; 0.7[k]; 1.5[y2]
Myrtenyl acetate	1079-01-2			tr	0-0.08		0.1[d]; 5.0[y2]
Neoisothujyl acetate	62181-91-3						0.2[g]
Nerol	106-25-2						0.1[d]; 0.4[k]; 2.2[u]
Neryl acetate	141-12-8		0-0.1				0.3[q]; 1.5[g]
Nonadecane	629-92-5			0-0.4			0.4[c]
Nootkatol	50763-67-2						0.1[d]; 0.4[e]
Nootkatone	4674-50-4						0.1[d,e]
β-Ocimene	13877-91-3						0.3[b,l]
(*E*)-Ocimene	27400-72-2						0.1[z2]
(*E*)-β-Ocimene	3779-61-1	0.02-0.07					+[a2]; 0.1[y3]
(*Z*)-β-Ocimene	3338-55-4				0-1.1		+[a2]
allo-Ocimene	673-84-7						2.6[b,h]
(*E,E*)-allo-Ocimene	3016-19-1						0.7[h]
Octadecane	593-45-3			0-0.4			0.4[c]
Octadecanol	26762-44-7						tr[d,e]
3-Octanol	589-98-0						0.5[k]
Oplopanone	1911-78-0		0-0.2				0.3[e]; 0.5[d]
β-Oplopenone	28305-60-4		0-0.2	0-0.4			0.1[d]; 0.3[e]; 0.4[c]
Perillene	539-52-6		0-0.2				+[a2]; 0.1[y3]
α-Phellandrene	99-83-2	0.03-2.0	0-0.3	tr-0.5	0.06-0.8	0-1.1	0.1[j]; 0.2[d,y]; 0.4[k]; 0.5[c]; 1.3[q]; 2.8[y8]
β-Phellandrene	555-10-2	0.4-0.9	0.2-1.2	0.2-1.8		0.1-1.8	0.5[y3]; 0.9[e]; 1.2[d]; 1.8[c]; 2.1[u]; 3.3[q]
β-Phellandrene-8-ol	65293-09-6		0-1.6				tr[d]
α-Pinene	80-56-8	18.1-66.6	21.0-46.3	27.7-52	15.6-43.2	18-57.8	51.4[y3]; 54.3[t]; 57.1[k]; 59.2[c]; 60.1[u]; 62.2[y]; 70.5[y]; 70.8[y1]; 77.5[z4]; 84.8[y8]
β-Pinene	127-91-3	1.8-7.8	0.4-1.1	0.3-2.1	1.7-5.4	0.7-12.6	2.8[d]; 3.2[y5]; 3.4[k]; 3.5[e]; 5.0[y3]; 5.6[u]; 8.7[h]; 12.6[y]; 13.7[y1]; 15.6[z4]; 17.8[z3]
α-Pinene oxide	1686-14-2	0.01-0.08					+[a2]; 0.1[y3]; 0.4[y]
Pinocamphone	547-60-4		0-0.2				

Table 5.44.2 Constituents identified in juniper berry oils (*continued*)

Constituent	CAS	Percentage and range					
		A	B (5)	C (6)	D (37)	E (11)	F
trans-Pinocarveol	1674-08-4	0.07-0.2	tr-2.1	0-0.6	0-0.4		0.2[y8]; 0.4[e]; 0.5[c]; 0.8[d]; 1.9[h]; 2.4[y5]
Pinocarvone	30460-92-5		0-0.5	0-0.1			0.1[c]
cis-Pinocarvyl acetate			0-tr	0-0.3			tr[c]
trans-Pinocarvyl acetate	1686-15-3		0-0.1				<0.1[m]; 0.3[c]
cis-Piperitol	16721-38-3						+[y]; 0.2[d]
Pseudolimonene	499-97-8	0.3-0.7					+[a2]; 0.7[z2]
cis-Rose oxide	3033-23-6						+[a2]
trans-Rose oxide	5258-11-7						+[a2]
Rosifoliol	63891-61-2						+[a2]
Sabinene	3387-41-5	1.0-17.7	0.5-2.5	0.4-1.1	2.8-11.8	tr-18.7	13[y9]; 13.6[z3]; 14.5[y4]; 16.5[k]; 18.0[p,27]; 17.8[j]; 19.4[g]; 27.6[q]; 36.8[y6]
cis-Sabinene hydrate	15537-55-0						0.1[y3]; 0.2[b,g,y2]; 0.3[y6]; 0.5[d]
trans-Sabinene hydrate	17699-16-0						tr[e]; 0.1[y1]; 0.2[b,g]; 0.3[y2]; 0.5[y6]; 1.3[y]
trans-Sabinol	471-16-9		0-0.5				
Salvial-4(14)-en-1-one	73809-82-2		0.1-0.4	0-0.5			+[a2]; 0.1[e]; 0.4[d]
Sandaracopimarinal	3855-14-9		0-0.1				0.1[d]
epi-α-Santalene							+[y]
Sclarene	511-02-4						+[a2]
Selina-3,7(11)-diene	6813-21-4						+[a2]
α-Selinene	473-13-2				0-1.9		0.2[y3]; 0.6[y1,y6]
β-Selinene	17066-67-0		0-0.3	0-0.2	0.2-0.8		+[a2]; 0.1[c,e]; 0.2[y3]; 0.4[y1,m]
Selin-11-en-4α-ol	16641-47-7			0-0.4	0.07-0.6		0.1[d,e]; 0.4[c]
β-Sesquiphellandrene	20307-83-9		0-0.1	0.1-0.4			0.4[c]
Sibirene	14029-18-6				tr-0.7		
Spathulenol	6750-60-3	0.08-0.3	0.4-1.7	tr-3.2	0-1.2	0-1.7	0.2[y]; 0.3[y1]; 0.9[c]; 1.2[d]; 1.4[y5]; 3.1[e]
Styrene	100-42-5						+[a2]
α-Terpinen-7-al	1197-15-5						0.1[d]
α-Terpinene	99-86-5	0.1-2.0	0-0.5	tr-0.4	0.4-1.2	0-1.9	0.8[y5]; 1.1[d,y6]; 1.3[o]; 1.9[q]; 2.2[p]; 2.9[y]
γ-Terpinene	99-85-4	0.4-4.5	0.1-1.9		0.8-2.3	0-3.2	1.4[j,t]; 1.8[m]; 1.9[d]; 2.2[y7]; 2.4[o]; 4.8[y]
Terpinen-4-ol	562-74-3	1.9-5.4	1.4-9.6	0.4-6.1	tr-6.3	0-10.1	1.5[e]; 2.7[h]; 2.9[i]; 3.6[y6]; 4.5[u]; 4.6[y1]; 6.1[c]; 8.1[w]; 8.2[y4]; 8.8[y2]; 9.6[d]; 22.9[y]
Terpinen-4-yl acetate	4821-04-9						+[a2]
α-Terpineol	98-55-5	0.2-1.1	2.9-6.0	0.5-4.6	0.2-1.6	0-4.9	1.2[e]; 2.2[y]; 3.5[c]; 4.9[y]; 5.0[d]; 6.0[h]
trans-β-Terpineol	7299-41-4						0.4[m]
(*Z*)-β-Terpineol	7299-40-3		0-2.5				
γ-Terpineol	586-81-2						1.8[t]
Terpinolene	586-62-9	0.5-2.1	0.5-1.6	0.1-1.4	0.9-2.9	0-13.6	1.5[d]; 1.9[y6]; 2.5[i]; 3.1[m]; 3.9[u]; 13.6[y]
Terpinyl acetate	8007-35-0						0.1[z2,y3]
α-Terpinyl acetate	80-26-2		0-0.1	0-0.2		0-1.2[f]	0.1[e]; 0.2[c,y2]; 0.4[d]
Tetracyclo[6.3.2.0(2,5).0(1,8)]tridecan-9-ol, 4,4-dimethyl-	100469-05-4(?)						+[a2]
Tetradecanol	27196-00-5		0-1.5				1.5[d]
Thuja-2,4(10)-diene	36262-09-6						+[a2]; 0.1[z2]; 0.2[y3]
α-Thujene	2867-05-2	0.1-1.3			0.7-2.3		0.9[y3]; 1.4[w]; 1.5[g]; 2.9[y6]; 3.1[y]; 3.8[z5]
β-Thujene	28634-89-1						0.1[y8]
α-Thujone	546-80-5						0.1[y6]; 0.3[y2]
cis-Thujone (β-)	471-15-8						+[a2]; 0.3[y2]
Thujopsene	470-40-6	0.02-0.4					0.3[z2]; 25.2[z1]
cis-Thujopsene	32435-95-3						+[a2]

Table 5.44.2 Constituents identified in juniper berry oils (*continued*)

Constituent	CAS	Percentage and range					
		A	B (5)	C (6)	D (37)	E (11)	F
Thymol	89-83-8			0-0.6			tr[e]; 0.1[y2]; 0.3[k]; 0.6[c]; 0.7[l]
Thymyl acetate	528-79-0						tr[e]
Tricyclene	508-32-7	0.05-2.0	0-0.3	0-0.1		tr-2.9	+[a2]; 0.1[l,y1,z2]; 0.2[c]
Tridecane	629-50-5		0-tr	0-0.1			+[a2]; tr[d]
3,3,5-Trimethylcyclo-hexene	503-45-7						+[a2]
2-Undecanone	112-12-9		0-0.2	0-0.1			+[a2]; 0.1[c,d,y3]
(3Z,5E)-1,3,5-Unde-catriene							+[a2]
Verbenene	4080-46-0		0-0.4	0-0.3			+[a2]; 0.1[d]; 0.2[y2]; 0.3[c,e]
cis-Verbenol	1845-30-3		0-0.6				0.5[y3]; 0.9[y5]; 3.3[y2]
trans-Verbenol	1820-09-3		0-1.6	0-0.9			0.2[d]; 0.6[e]; 0.7[y8]; 0.9[c]; 2.2[y2]
Verbenone	80-57-9	0.02-0.2	0.4-3.9	0-1.1	0-0.2	0-1.4	0.2[e]; 0.6[y2]; 1.1[c]; 1.6[d]; 2.1[y5]; 2.5[h]
Verbenyl ethyl ether							+[a2]
Viridiflorene (ledene)	21747-46-6						+[y]
Viridiflorol	552-02-3				0-0.2		
Widdrol	6892-80-4						7.7[z1]
Xylene	1330-20-7		0-1.8	tr-0.8			4.2[c]
o-Xylene	95-47-6						+[a2]
p-Xylene	106-42-3						+[a2]
α-Ylangene	14912-44-8				0.03-0.1		+[a2]; 0.1[y6]
β-Ylangene	20479-06-5						tr[e]

A 395 juniper berry essential oil samples from Austria, Bosnia-Herzegovina, Slovenia, Germany, Croatia, France, Italy, Hungary and Turkey, analyzed between 1990 and 2012; lowest and highest concentrations given (E. Schmidt, unpublished data)
B five lab-hydrodistilled essential oils of wild growing juniper berries in northeastern Lithuania; lowest and highest concentrations given (ref. 5)
C five lab-distilled essential oils of ripe berries of *Juniperus communis* growing wild in the Vilnius district, Lithuania; lowest and highest concentrations given (ref. 6)
D fifteen lab-hydrodistilled oils from juniper berries harvested in various parts of the Republic of Macedonia; lowest and highest concentrations given (ref. 37)
E investigation of several commercial and lab-distilled juniper berry oils; highest and lowest concentrations given (ref. 11)
F data from other studies (indicated with superscript letters); highest concentrations found in any study reviewed here given; when two or more oils were investigated, only the highest concentrations are mentioned, unless indicated otherwise

[a] incorrect identification (refs. 1,2,4,28); [a2] chemicals identified in commercial juniper berry oils in the USA, amounts not quantified (ref. 3); [b] τ-cadinol and τ-muurolol combined; [c] five lab-hydrodistilled oils from juniper berries growing in the wild in Lithuania and harvested at various sites (ref. 44); [d] four lab-hydrodistilled oils from berries collected in the wild in Lithuania (ref. 41); [e] two lab-hydrodistilled oils from berries of *J. officinalis* var. *officinalis* from Lithuania (ref. 7); [f] α-terpinyl acetate and α-cubebene combined; [g] six lab-distilled juniper berry oils from Croatia (ref. 17); [h] one commercial juniper berry oil sample (ref. 19); [i] one oil produced from the powdered dried berries of *J. communis* collected in Iran (ref. 18); [j] one lab-distilled oil from Montenegro (ref. 25); [k] three lab-hydrodistilled oils from juniper berries collected in different parts of Greece (ref. 24); [l] one commercial juniper oil from India (ref. 15); [m] three lab-distilled juniper berry oils from Montenegro (ref. 9); [n] one lab-distilled oil from Greece (ref. 8); [o] one commercial juniper oil purchased in the USA (ref. 20); [p] two lab-hydrodistilled oil samples from Serbia (ref. 10); [q] one juniper berry oil sample (ref. 16); [r] compound does not exist naturally (ref. 2); [s] one lab-hydrodistilled oil from juniper berries from Sardinia (Italy) (ref. 23); [t] one lab-distilled oil from wild growing berries in Serbia (ref. 21); [u] two commercial oils and one lab-hydrodistilled oil from Poland (ref. 26); [v] one Polish juniper berry oil sample (ref. 29); [w] oils produced from juniper berries harvested in former Yugoslavia (ref. 31); [x] one hydrodistilled juniper oil (ref. 32); [y] oils investigated in various older publications (cited in ref. 4); [y1] one commercial and one lab-distilled juniper berry oil (ref. 27); [y2] one lab-hydrodistilled oil from India (ref. 12); [y3] one commercial juniper berry oil from Bulgaria (ref. 34); [y4] three lab-hydrodistilled juniper berry oils from three classes of berries harvested in Serbia (ref. 38); [y5] one laboratory steam-distilled oil from berries harvested in the wild in Serbia (ref. 35); [y6] one lab-hydrodistilled oil from Iranian juniper berries with sabinene as the dominant ingredient (ref. 30); [y7] one lab-hydrodistilled juniper berry oil from Serbia (ref. 40); [y8] one industrially steam-distilled berry oil analyzed with two different GC/MS techniques (ref. 47); [y9] one lab-hydrodistilled juniper oil from *J. communis* berries harvested in the wild in Greece (ref. 49); [z] one lab-distilled juniper berry oil from Lithuania (ref. 14); [z1] one commercial juniper fruit oil from a USA chemical company with an extremely atypical composition and widdrene (= thujopsene) (25.2%) as dominant ingredient (ref. 45); [z2] one commercial oil obtained from a German manufacturer (ref. 33); [z3] one commercial oil sample from Croatia (ref. 43); [z4] one commercial juniper oil sample from Poland (ref. 42); [z5] one lab-hydrodistilled oil from Tunisian juniper berries (ref. 36)

tr: trace (in columns B and C: <0.1; in columns F[d] and F[g]: <0.05; in column D: <0.02); + present in the oil investigated, but quantity not stated; +[a2] chemicals identified in commercial juniper berry oils in the USA, amounts not quantified (ref. 3)

Table 5.44.3 Major constituents of juniper berry oils

Constituent	CAS	Percentage and range					
		A	B (5)	C (6)	D (37)	E (11)	F
α-Pinene	80-56-8	18.1-66.6	21.0-46.3	27.7-52	15.6-43.2	18-57.8	62.2[y]; 70.5[y]; 70.8[y1]; 77.5[z4]; 84.8[y8]
Sabinene	3387-41-5	1.0-17.7	0.5-2.5	0.4-1.1	2.8-11.8	tr-18.7	18.0[p]; 17.8[j]; 19.4[g]; 27.6[q]; 36.8[y6]
Myrcene	123-35-3	11.6-21.6	3.6-17.4	6.3-19.6	2.9-26.5	0.3-22.5	19.9[k]; 21.2[y5]; 24.2[z]; 29.9[y]; 30.9[l]
Terpinen-4-ol	562-74-3	1.9-5.4	1.4-9.6	0.4-6.1	tr-6.3	0-10.1	6.1[c]; 8.1[w]; 8.2[y4]; 8.8[y2]; 9.6[d]; 22.9[y]
Limonene	138-86-3	2.5-7.6	1.1-5.3	1.6-2.9	2.9-4.5	2.4-22.1	10.6[y6]; 11.1[l]; 13.2[h]; 17.5[t]; 22.1[y]
β-Pinene	127-91-3	1.8-7.8	0.4-1.1	0.3-2.1	1.7-5.4	0.7-12.6	8.7[h]; 12.6[y]; 13.7[y1]; 15.6[24]; 17.8[z3]
p-Cymene	99-87-6	0.4-2.3	0.3-0.5	tr-0.5	0.09-0.6	0.2-11.9	2.1[z2]; 2.3[p]; 5.0[z]; 10.0[u]; 11.9[y]; 16.2[y8]
Germacrene D	23986-74-5	0.07-2.6	0.1-3.1	0.2-0.8	2.8-10.7	0-8.0	9.5[y]; 10[y9]; 10.4[n,x]; 12.1[m]; 14.3[y]
β-Caryophyllene	87-44-5	2.0-5.9	0.2-0.6	0.1-0.4	1.8-4.1	0-4.0	3.6[y1]; 5.3[p]; 8.2[y]; 9.2[m]; 9.8[z]; 10.3[y]
δ-Cadinene	483-76-1	0.3-2.7	0.5-2.9	1.4-2.5	tr-4.2	0-3.6	2.5[c]; 2.7[y1]; 2.9[d]; 4.1[y]; 4.4[m]; 8.7[w]
α-Cadinol	481-34-5		1.8-7.6	2.7-8.6	0-2.7		0.9[y2]; 1.7[m]; 2.7[z]; 5.1[e]; 7.1[c]; 7.6[d]

LEGEND: SEE UNDER TABLE 5.44.2

group of consecutive patients suspected of contact dermatitis (only one study), the prevalence rate of positive patch test reactions to juniper berry oil was 0.5%; relevance data were not provided. In a case report, α-pinene may have been an allergen in juniper berry oil.

Testing in groups of patients

Two hundred dermatitis patients from Poland were tested with juniper oil 2% in petrolatum and one (0.5%) reacted. The same authors also patch tested 51 patients allergic to *Myroxylon pereirae* resin (balsam of Peru) and/or turpentine and/or wood tar and/or colophony and two (3.9%) had a positive patch test; relevance data were not provided (55). A group of 86 patients from Poland previously reacting to the fragrance mix was tested with juniper berries oil and six (7.0%) had a positive patch test reaction; relevance data were not provided (56).

Case reports

Two aromatherapists had occupational contact dermatitis with allergies to multiple essential oils used at their work, including juniper oil. Both patients also reacted to geraniol, α-pinene, the fragrance mix and various other fragrance materials. In addition, one proved to be allergic to linalool and linalyl acetate, the other to caryophyllene. α-Pinene was demonstrated by GC-MS in the juniper oils (54) and is the dominant ingredient of the oil (Table 5.44.2, column A).

Additional information on allergy to juniper berry oil can be found in ref. 57 (article not read).

LITERATURE

1 Lawrence BM. Progress in essential oils. Perfum Flav 2012;37(October):44-49
2 Lawrence BM. Progress in essential oils. Perfum Flav 2008;33(8):60-? (last page unknown)
3 Robbat A Jr, Kowalsick A, Howell J. Tracking juniper berry content in oils and distillates by spectral deconvolution of gas chromatography/mass spectrometry data. J Chromatogr A 2011;1218:5531-5541
4 Lawrence BM. Juniper berry oil. In: Essential oils 2001-2004. Carol Stream, USA: Allured Publishing Corporation, 2006:36-37
5 Butkienė R, Nivinskienė O, Mockutė D. Volatile compounds of ripe berries (black) of *Juniperus communis* L. growing wild in North-East Lithuania. J Essent Oil Bear Plants 2005;8:140-147
6 Butkienė R, Nivinskienė O, Mockutė D. Differences in the essential oils of the leaves (needles), unripe and ripe berries of *Juniperus communis* L. growing wild in Vilnius district (Lithuania). J Essent Oil Res 2006;18:489-494
7 Butkienė R, Nivinskienė O, Mockutė D. Variety of the essential oils composition of wood, needles (leaves), unripe and ripe berries of *Juniperus communis* var. *communis* growing wild in Druskininkai district. Chemija 2007;18(3):35-40
8 Chatzopoulou PS, Katsiotis KT. Headspace analysis of the volatile constituents from *Juniperus communis* L. berries (cones) grown wild in Greece. Flavour Fragr J 2006;21:492-496
9 Damjanovic B, Skala D, Baras J, Petrovic-Djakov D. Isolation of essential oil and supercritical carbon dioxide extract of *Juniperus communis* L. fruits from Montenegro. Flavour Fragr J 2006;21:875-880
10 Glisic SB, Milojevic SZ, Dimitrijevic SI, Orlovic AM, Skala DU. Antimicrobial activity of the essential oil and different fractions of *Juniperus communis* L. and a comparison with some commercial antibiotics. J Serb Chem Soc 2007;72:311-320. Available at: http://www.doiserbia.nb.rs/ft.aspx?id=0352-51390704311G
11 Kartnig Th, Fischer U, Bucar F. Vergleichende gaschromatographische Untersuchungen an ätherischen Wacholderölen, Fenchelölen und Rosmarinölen. Sci Pharm 1998;66:237-252. Data cited in ref. 2
12 Lohani H, Haider SZ, Chauhan NK, Mohan M. Essential oil composition of the leaves and berries of *Juniperus communis* and *Juniperus indica* from Uttarakhand Himalaya. J Med Arom Plant Sci 2010;33:199-201

13 Tonutti I, Liddle P. Aromatic plants in alcoholic beverages. A review. Flavour Fragr J 2010;25:341-350

14 Loziene K, Labokas J, Venskutonis PR, Maždžieriene R. Chromatographic evaluation of the composition of essential oil and α-pinene enantiomers in *Juniperus communis* L. berries during ripening. J Essent Oil Res 2010;22:453-458

15 Raina VK, Kumar A, Tandon S, Ahmad J, Kahol AP. Composition of juniper berry oil of commerce. Indian Perfum 2005;49:329-332

16 Ranade GS. Essential oil profile. Juniper berry oil. FAFAI 2002;4:51. Data cited in ref. 1

17 Tasic SR, Menkovic NR, Ristić MS, Kovačević NN, Samardžić ZJ. Comparative studies of juniper berry oil from different regions of central Balkan. Acta Hort 1993;344:574-577. Data cited in ref. 1

18 Rezvani S. Analysis of essential oil of *Juniperus communis* and terpenoids dried fruits from Codestan of Iran. Asian J Chem 2010;22:1-3. Data also published in ref. 48

19 Romeo FV, De Luca S, Piscopo A, Poiana M. Antimicrobial effect of some essential oils. J Essent Oil Res 2008;20: 373-379

20 Wei A, Shibamoto T. Antioxidant activities and volatile constituents of various essential oils. J Agric Food Chem 2007;55:1737-1742

21 Milojević SZ, Stojanović TD, Palić R, Lazić ML, Veljković VE. Kinetics of distillation of essential oil from comminuted ripe juniper (*Juniperus communis* L.) berries. Biochem Eng J 2008;39:547-553

22 Orav A, Koel M, Kailas T, Müürisepp M. Comparative analysis of the composition of essential oils and supercritical carbon dioxide extracts from the berries and needles of Estonian juniper (*Juniperus communis* L.). Procedia Chemistry 2010;2:161-167

23 Angioni A, Barra A, Russo MT, Coroneo V, Dessi S, Cabras P. Chemical composition of the essential oils of *Juniperus* from ripe and unripe berries and leaves and their antimicrobial activity. J Agricult Food Chem 2003;51:3073-3078

24 Koukos PK, Papadopoulou KI. Essential oil of *Juniperus communis* L. grown in Northern Greece: variation of fruit oil yield and composition. J Essent Oil Res 1997;9:35-39

25 Damjanovic BM, Skala D, Petrovic-Djakov D, Baras J. A comparison between the oil, hexane extract and supercritical carbon dioxide extract of *Juniperus communis* L. J Essent Oil Res 2003;15:90-92

26 Filipowicz N, Kaminski M, Kurlenda J, Asztemborska M, Ochocka JR. Antibacterial and antifungal activity of juniper berry oil and its selected components. Phytother Res 2003;17:227-231

27 Kubeczka K-H, Formacek V. Essential oils analysis by capillary gas chromatography and carbon-13 NMR spectroscopy, 2nd Ed. New York: John Wiley and Sons, 2002:145-154. Data cited in ref. 28

28 Lawrence BM. Progress in essential oils 2005-2007. Carol Stream, USA: Allured Publishing Corporation, 2008: 60-63

29 Gosa J, Majda T, Lis A, Tichek A, Kurowska A. Chemical composition of some Polish commercial essential oils. Rivista Ital EPPOS (Numero Speciale) 1997:761-766. Data cited in ref. 28

30 Shamir F, Ahmadi L, Mizra M, Korori SAA. Secretory elements of needles and berries of *Juniperus communis* L. ssp. *communis* and its volatile constituents. Flavour Fragr J 2003;18:425-428

31 Matovic M, Lavadinovic V. Essential oil of the fruit of *Juniperus communis* L. growing in Yugoslavia. J Essent Oil Bear Plants 1999;2:101-106

32 Chatzopoulou P, de Haan A, Katsiotis ST. Investigation on the supercritical CO_2 extraction of the volatile constituents from *Juniperus communis* obtained under different treatments of the berries (cones). Planta Med 2002;68:827-831

33 Wanner J, Schmidt E, Bail S, Jirovetz L, Buchbauer G, Gochevd V, et al. Chemical composition and antibacterial activity of selected essential oils and some of their main compounds. Nat Prod Comm 2010;5:1359-1364

34 Höferl M, Stoilova I, Schmidt E, Wanner J, Jirovetz L, Trifonova D, et al. Chemical composition and antioxidant properties of juniper berry (*Juniperus communis* L.) essential oil. Action of the essential oil on the antioxidant protection of *Saccharomyces cerevisiae* model organism. Antioxidants 2014;3:81-98

35 Haziri A, Faiku F, Mehmeti A, Govori S, Abazi S, Daci M, et al. Antimicrobial properties of the essential oil of *Juniperus communis* (L.) growing wild in east part of Kosovo. Am J Pharmacol Toxicol 2013;8:128-133

36 Foudil-Cherif Y, Yassaa N. Enantiomeric and non-enantiomeric monoterpenes of *Juniperus communis* L. and *Juniperus oxycedrus* needles and berries determined by HS-SPME and enantioselective GC/MS. Food Chem 2012;135:1796-1800

37 Sela F, Karapandzova M, Stefkov G, Kulevanova S. Chemical composition of berry essential oils from *Juniperus communis* L. (Cupressaceae) growing wild in Republic of Macedonia and assessment of the chemical composition in accordance to European Pharmacopoeia. Macedonian Pharmaceutical Bulletin 2011;57:43-51

38 Matović M, Bojović B, Jušković M. Composition of essential oils from three classes of juniper fruit from Serbia. J Med Plants Res 2011;5:6160-6163

39 Teuscher E. In: Medicinal Spices. Stuttgart, Germany: Medpharm Scientific Publishing, 2006:205. Data cited in ref. 41

40 Milojević SŽ, Glišić SB, Skala DU. The batch fractionation of *Juniperus communis* l. essential oil: experimental study, mathematical simulation and process economy. Chem Ind Chem Engineer Quart 2010;16:183-191

41 Butkiene R, Nivinskiene O, Mockute D. Two chemotypes of essential oils produced by the same *Juniperus communis* L. growing wild in Lithuania. Chemija 2009;20(3):195-201

42 Golebiowski M, Paszkiewicz M, Halinski L, Malinski E, Stepnowski P. Chemical composition of commercially available essential oils from eucalyptus, pine, ylang, and juniper. Chem Nat Comp 2009;45:278-279

43 Pepeljnjak S, Kosalec I, Kalodera Z, Blazevic N. Antimicrobial activity of juniper berry essential oil (*Juniperus communis* L., Cupressaceae). Acta Pharm 2005;55:417-422

44 Butkienë R, Nivinskienë O, Mockutë D. Chemical composition of unripe and ripe berry essential oils of *Juniperus communis* L. growing wild in Vilnius district. Chemija 2005;15(4):57-63

45 Shin S. Anti-*Aspergillus* activities of plant essential oils and their combination effects with ketoconazole or amphotericin B. Arch Pharm Res 2003;26:389-393

46 Milhau G, Valentin A, Benoit F, Mallié M, Bastide J-M, Pélissier Y, et al. *In vitro* antimalarial activity of eight essential oils. J Essent Oil Res 1997;9:329-333

47 Şerban ES, Socaci SA, Tofană M, Maier SC, Bojiţă MT. Advantages of "headspace" technique for GC/MS analisys of essential oils. Farmacia 2012;60:249-256

48 Rezvani S, Rezai MA, Mahmoodi N. Analysis and antimicrobial activity of the plant *Juniperus communis*. Rasayan J Chem 2009;2:257-260. Data also published in ref. 18

49 Chatzopoulou PS, Katsiotis ST. Study of the essential oil from *Juniperus communis* "berries" (Cones) growing wild in Greece. Planta Med 1993;59:554-556

50 Lawrence BM. Essential Oils 1995-2000. Carol Stream, IL, USA: Allured Publishing Corp, 2003:53. Data cited in ref. 41

51 Carroll JF, Tabanca N, Kramer M, Elejalde NM, Wedge DE, Bernier UR, et al. Essential oils of *Cupressus funebris*, *Juniperus communis*, and *J. chinensis* (Cupressaceae) as repellents against ticks (Acari: Ixodidae) and mosquitoes (Diptera: Culicidae) and as toxicants against mosquitoes. J Vector Ecol 2011;36:258-268

52 Rhind JP. Essential oils. A handbook for aromatherapy practice, 2nd Edition. London: Singing Dragon, 2012

53 Lawless J. The encyclopedia of essential oils, 2nd Edition. London: Harper Thorsons, 2014

54 Dharmagunawardena B, Takwale A, Sanders KJ, Cannan S, Roger A, Ilchyshyn A. Gas chromatography: an investigative tool in multiple allergies to essential oils. Contact Dermatitis 2002;47:288-292

55 Rudzki E, Grzywa Z, Bruo WS. Sensitivity to 35 essential oils. Contact Dermatitis 1976;2:196-200

56 Rudzki E, Grzywa Z. Allergy to perfume mixture. Contact Dermatitis 1986;15:115-116

57 Rothe A, Heine A, Rebohle E. Oil from juniper berries as an occupational allergen for the skin and respiratory tract. Berufsdermatosen 1973;21:11-16 (in German)

Chapter 5.45   LAUREL LEAF OIL

DEFINITION
Laurel leaf oil is the essential oil obtained from the leaves of the laurel (bay, bay laurel), *Laurus nobilis* L.

INCI NOMENCLATURE
Description/definition: Laurus nobilis leaf oil is the volatile oil obtained from the leaves of the laurel, *Laurus nobilis* L., Lauraceae
INCI name EU & USA: Laurus nobilis leaf oil
Other names: bay oil, sweet; sweet bay oil
CAS registry number(s): 8002-41-3; 84603-73-6; 8007-48-5
EINECS number(s): 283-272-5

ISO (INTERNATIONAL ORGANIZATION FOR STANDARDIZATION) STANDARD
There is currently no ISO standard for laurel leaf oil.

Laurel leaf oil is also termed bay oil, sweet (sweet bay oil), not to be confused with 'bay oil', which is obtained from *Pimenta racemosa* (Mill.) J.W. Moore (see Chapter 5.5 Bay oil).

THE PLANT, THE OIL, AND THEIR USES
Laurus nobilis L. is an evergreen shrub up to 8 meters in height and commonly named bay laurel. This tree belongs to the Lauraceae family and is native to the southern Mediterranean region and widely cultivated in Europe and the USA as an ornamental plant. In North African countries (Tunisia, Algeria, Morocco), bay laurel is a common species. It is grown commercially for its aromatic leaves in Turkey, Algeria, Morocco, Portugal, Spain, Italy, France and Mexico. The (fresh and dried) leaves are commonly used as a spicy, aromatic flavoring for soups, fish, meats, stews, puddings, vinegars, and beverages and also has food preserving properties from its antibacterial effects.

Benzene compounds such as eugenol, methyl eugenol and elemicin are responsible for the spicy aroma of bay leaves (1). The leaves are also widely used in folk medicine to treat asthma, cardiac diseases, gastrointestinal problems, rheumatism, urinary problems (it has diuretic properties), stones and many other diseases. The oil is said to cure rheumatic pain and dermatitis and is used by the cosmetic industry in creams, perfumes and soaps (1). It is also used in aromatherapy practices (70,71).

CHEMICAL COMPOSITION
Laurel leaf oil is a slightly yellowish to greenish, clear mobile liquid, which has a fresh, aromatic odor with connotations of eucalyptus and clove leaves. The yield of essential oil from the leaves of *Laurus nobilis* L. generally varies from 0.8 to 2.8 per cent. The main producing countries of this oil are Morocco, Spain, France and Argentina.

Literature data (up to November 30, 2014) on the chemical composition of laurel leaf oils and unpublished analytical data from one of us (E.S.) are shown in Table 5.45.1 in alphabetical order. In laurel leaf oils from various origins, over 425 chemicals have been identified. About 54 per cent of these were found in a single reviewed publication only. The major compounds found in laurel leaf oils from different sources are shown in Table 5.45.2. They include (highest concentrations in any study given) 1,8-cineole (72.1%), linalool (32.0%), α-terpinyl acetate (25.7%), methyl eugenol (21.4%), α-pinene (15.9%), sabinene (15.0%), α-terpineol (9.3%) and β-pinene (6.5%). Well-known ingredients of laurel leaf oils that were present in high concentrations (>7%) in one or two studies were eugenol (three studies: 16.7%, 18.5% and 44.1%), terpinolene (13.6%), borneol (12.8%), α-terpinene (12.5%), (*E*)-sabinene (11.9%), β-elemene (9.6%), camphene (8.9%) and terpinen-4-ol (6.4%). Uncommon or rare constituents of laurel leaf oils found in high concentrations (>7%) in single studies include camphor (34.4%), cinnamaldehyde (30.3%), carene (15.1%), 1-*p*-menthene-8-ethyl acetate (13.5%), δ2-carene (13.1%), valencene (11.0%), isovaleraldehyde (10.5%), (*Z*)-*p*-menth-2-en-1-ol (10.4%), δ-terpinyl acetate (9.9), α-cadinol (7.3%) and (*E*)-β-ocimene (7.1%).

Commercial oils
The ten chemicals that had the highest maximum concentrations in 23 commercial laurel leaf essential oil samples (concentration ranges provided) are the following: 1,8-cineole (38.4-52.0%), terpinyl acetate (6.3-12.0%), sabinene (7.3-11.8%), linalool (1.1-7.8%), α-pinene (4.4-7.5%), β-pinene (3.2-5.0%), limonene (1.0-4.0%), terpinen-4-ol (1.7-3.4%), methyl eugenol (1.0-3.0%) and *p*-cymene (0.04-2.7%) (Erich Schmidt, unpublished analytical data).

Chemotypes
In virtually all analyses of laurel leaf oils, 1,8-cineole (eucalyptol) is the dominant ingredient with concentrations of up to 55-70% (Table 5.45.1). However, in some cases, e.g., in leaves traded as spice in Spain, 1,8-cineole concentrations were only 3.5-7.7%, the dominant components of the lab-hydrodistilled oils being methyl eugenol (18.8-21.4%) and α-terpinyl acetate (17.3-18.3%) (42). In a report from Tunisia, camphor was the dominant constituent with a 34.4% content, whereas 1,8-cineole was second with 20.2% (43). A commercial laurel leaf oil from China contained 44.1% eugenol, 30.3% cinnamaldehyde and only 0.7% 1,8-cineole, which is extremely atypical (ref. 52).

The results of the analyses in refs. 28, 35 and 41 are not discussed. The data from ref. 53 are not shown in Table 5.45.1, as the concentrations of all the chemicals were lower than the lowest one shown in column F of Table 5.45.1. Additional analyses of laurel leaf oils can be found in refs. 45,50 and 59-69; these are not discussed in Table 5.45.1 but can all be accessed on-line.

CONTACT ALLERGY/ALLERGIC CONTACT DERMATITIS

General
Contact allergy to/allergic contact dermatitis from laurel oil has been reported in over 20 publications.

Laurel oil used to be a frequent contact allergen in the middle of the 20th century. At that time, the oil was used to improve the luster of felt hats, which frequently caused hatband dermatitis in the wearer (90). Laurel-based ointments were also a frequent cause of allergic contact dermatitis in those years, especially in France, Germany and Switzerland (92). Between 1953 and 1962, several thousands of dermatitis patients were tested with 'laurel oil' in Germany and positive reactions were seen in 3.1-6.9% of patients (90). However, it should be appreciated that

Table 5.45.1 Constituents identified in laurel leaf oils

Constituent	CAS	Percentage and range					
		A	B (3)	C (12)	D (18)	E (19)	F
2-Acetoxy-1,8-cineole	438619-71-7						0.3[m]
Adamantane tricyclo [3.3.1.1]							0.3[e]
α-Amorphene	20085-19-2		0-0.2				0.2[y2]
δ-Amorphene	189165-79-5						tr[x3]
(E)-Anethole	4180-23-8			0.1			0.1[x5,x6]
Anhydrooplopanone	108654-35-9			0.2			
Aromadendrene	489-39-4		0-0.2				tr[x3]; 0.2[y2]
allo-Aromadendrene	25246-27-9	0.02-0.1					tr[m,x3]; 0.06[y3]; 0.6[n]; 0.9[y7]
Ascaridole	512-85-6						0.1[y4]
Benzaldehyde	100-52-7			<0.01			tr[u]; <0.1[f,g]
Benzene	71-43-2		0-0.3				
Benzyl alcohol	100-51-6						0.2[e]
α-Bergamotene	17699-05-7		0-0.2				
β-Bergamotene	6895-56-3						0.2[y5]
Bicyclogermacrene	24703-35-3			0.1			tr[m]; 0.2[h]; 1.1[y3]; 1.2[y2]; 3.2[y7]
Bicyclo[4.3.1]dec-1(9)-ene							0.4[y2]
Bicyclo[3.1.1] hept-2-ene-2-carboxa							0.6[e]
α-Bisabolene	17627-44-0			0.2			0.4[v]
(Z)-α-Bisabolene	29837-07-8						0.2[y2]; 0.8[x3]; 1.0[x9]
β-Bisabolene	495-61-4					0.03	0.1[u,v]; 0.2[a]
epi-α-Bisabolol	78148-59-1						0.5[a]
β-Bisabolol	15352-77-9						1.4[v]
α-Bisabolol oxide	22567-36-8						1.1[c]
Borneol	507-70-0	0.04-0.2		0.3	0.1		1.1[y6]; 1.5[u]; 2.4[d]; 6.7[y8]; 12.8[i]
Bornyl acetate	76-49-3	0.1-0.4	0-0.5	0.7	0.1	0.3	1.1[m,v]; 1.2[r]; 1.3[d,v]; 2.3[u]; 5.0[y8]
β-Bourbonene	5208-59-3			0.1		0.05	tr[x3] 0.1[x6]
α-Bulnesene	3691-11-0				0.7		tr[n]; 0.6[y3]; 1.8[y7]
Bulnesol	22451-73-6						0.9[n]
Butylbenzene	104-51-8						5.2[y4]
Cadalene	483-78-3			0.1			
Cadina-1,4-diene	29837-12-5			<0.01			
trans-Cadina-1,4-diene	20085-13-6						0.01[y3]
α-Cadinene	24406-05-1						0.2[y8]
γ-Cadinene	39029-41-9		0-0.07	0.3	0.2	0.1	0.2[o]; 0.3[m,v]; 0.4[n]; 0.7[i]; 1.1[y2]
δ-Cadinene	483-76-1	0.04-0.4	0-0.1	0.4	0.4	0.07	0.7[x6]; 0.8[v]; 0.9[y2]; 1.1[y3]; 1.4[y7]
α-Cadinol	481-34-5		0-0.2	0.1			1.1[d]; 1.2[y2]; 2.0[v]; 4.7[x6]; 7.3[y3]
τ-Cadinol	58580-31-7			<0.01			0.5[x6]

Table 5.45.1 Constituents identified in laurel leaf oils (*continued*)

Constituent	CAS	Percentage and range					
		A	B (3)	C (12)	D (18)	E (19)	F
α-Calacorene	21391-99-1			0.1			tr[n]
β-Calacorene	50277-34-4			0.1		0.05	
Calamenene	483-77-2			0.1			0.2[x6,y8]; 0.5[j]
(E)-Calamenene	73209-42-4					0.02	
Camphenilone	13211-15-9						1.1[x1]
Camphene	79-92-5	0.2-0.5	0-8.9	0.7	0.5	0.07	1.1[m,v]; 1.2[z4]; 1.6[x9]; 1.8[d]; 2.9[u]
α-Campholenic aldehyde	4501-58-0			<0.1		0.06	tr[x3]; 0.2[y8]
Camphor	76-22-2		0-2.7				0.1[i]; 0.2[v]; 0.3[y6]; 34.4[y8]
Carene	74806-04-5						15.1[e]
δ2-Carene	554-61-0		0-5.6				0.4[y]; 0.6[c]; 2.0[d]; 13.1[z1]
δ3-Carene	13466-78-9	0.04-0.5	0.1-0.4	<0.1			0.1[h,v]; 0.2[m,u]; 0.5[a]; 0.7[n]; 0.8[x2]
Carvacrol	499-75-2			0.1	0.2	0.05	0.4[i]; 0.5[x6]; 0.8[q]; 1.8[x8]; 3.2[y2]
(Z)-Carveol	1197-06-4			0.1		0.3	tr[x3]; 0.2[i]; 0.3[x1]; 0.9[y2]
trans-Carveol	1197-07-5			0.2		0.2	tr[x3]; 0.1[x5,z5]; 0.2[b,q]; 0.3[y4]
Carvone	99-49-0	0.02-0.05		0.1			tr[x3]; 0.2[v,x5]; 0.3[b]; 0.4[k]
Carvyl acetate	97-42-7						0.1[x5]
(E)-Carvyl acetate	1134-95-8					0.4	
(Z)-Carvyl acetate	1205-42-1			0.1		0.05	
Caryophylladienol II							0.2[z]
Caryophylla-2(12),6-dien-5β-ol							0.1[y4]
Caryophylla-2(12),6(13)-dien-5α-ol							0.1[y4]; 0.2[z]
Caryophylla-3(15),7(14)-dien-6-ol	257293-89-3			0.6			
Caryophylla-4(12),8(13)- dien-5β-ol	19431-80-2						0.2[b]
Caryophylla-4(14),8(15)-dien-5-ol	644981-74-8						0.2[a]; 1.2[x3]
β-Caryophyllene	87-44-5	0.1-1.2	0-0.7	0.4	1.5	0.04	2.7[x6]; 2.8[z6]; 3.2[z2]; 3.8[x9]; 6.4[y7]
Caryophyllene oxide	1139-30-6	0.06-0.4	0-0.5	0.9	0.3	0.2	2.0[x2]; 2.2[x9]; 4.0[y3]; 4.1[x6]; 4.4[y7]
Caryophyllenol I	32214-88-3						0.3[z]
Caryophyllenol II	32214-89-4						0.1[z]
α-Cedrol	77-53-2						0.3[y8]
(Z)-Chrysanthenyl acetate	67999-48-8					0.07	
1,4-Cineole	470-67-7						0.4[v]
1,8-Cineole (eucalyptol)	470-82-6	38.4-52.0	24.6-38.9	21.8	40.9	58.6	58.1[x2]; 60.6[y6]; 60.7[i]; 61.0[q]; 61.6[x5]; 63.2[i]; 68.5[k]; 72.1[e]
1,8-Cineole-2-yl-acetate				0.4			
Cinnamaldehyde	104-55-2						30.3[z2]
Cinnamyl acetate	103-54-8	0.03-0.2	0-0.06	0.7			0.1[y4]; 1.3[n]
(E)-Cinnamyl acetate	21040-45-9						0.1[y2]
Cinnamyl alcohol	104-54-1						3.7[i]
(Z)-Cinnamyl alcohol	4510-34-3						tr[n]
Citral	5392-40-5						0.4[v]
Citronellal	106-23-0						0.5[y6]
β-Citronellene	2436-90-0						0.2[y]
Citronellol	106-22-9						0.8[y6]

Table 5.45.1 Constituents identified in laurel leaf oils (*continued*)

Constituent	CAS	A	B (3)	C (12)	D (18)	E (19)	F
Citronellyl acetate	150-84-5						0.3[v]
α-Copaene	3856-25-5		<0.1	0.8			tr[n,x3]; 0.02[y3]; 0.1[u,v]; 0.3[m]
β-Copaene	18252-44-3						0.2[i]
α-Copaen-8-ol	58569-25-8				0.2		
Cryptone	500-02-7			<0.01			
α-Cubebene	17699-14-8						0.2[u,v]
β-Cubebene	13744-15-5				0.05		tr[n]; 0.1[m]; 0.6[y7]
Cubebol	23445-02-5						tr[m]
epi-Cubebol	38230-60-3						tr[x3]
10-epi-Cubebol	176589-53-0						tr[x3]
Cubenol	21284-22-0						0.3[y8]
1-epi-Cubenol	81939-29-9					0.02	
Cuminaldehyde	122-03-2			0.1	0.2		tr[x3]; 0.1[x6]; 0.2[y4]; 1.0[x1]
Cuminyl alcohol	536-60-7			0.1			0.1[z5]; 0.2[x5]; 0.3[y4]
Curcumene (*ar-*; α-)	644-30-4						0.1[x6]
Cyclobutane, (1,3-butadienyl)-	80344-48-5						0.3[e]
Cyclodecene	3618-12-0		0-0.08				
1,4-Cyclohexadiene-1-methanol	32937-33-0					0.05	
3-Cyclohexene-1-methanol	1679-51-2						0.6[e]
2-Cyclohexen-1-ol	822-67-3		0-0.08				
2-Cyclohexen-1-one, 2-methyl-5-(1-methylethane)-							0.2[e]
Cyclopentane	287-92-3		0-0.1				
Cyclosativene	22469-52-9						tr[n]
o-Cymene	527-84-4					1.3	0.2[n]; 1.0[v]
p-Cymene	99-87-6	0.04-2.7		1.5	0.5	1.8	2.2[x5]; 2.5[r]; 3.0[y6,z3]; 3.1[y4]; 3.6[x1]
p-Cymenene	1195-32-0			<0.1			<0.1[f,g,o]
p-Cymen-7-ol	536-60-7					0.03	tr[m,x3]; 0.6[c]
p-Cymen-8-ol	1197-01-9	0.04-0.1		0.2		0.1	tr[n]; 0.1[y4]; 0.2[z5]; 0.3[q]
p-Cymen-7-ol acetate	59230-57-8			<0.1			
(E)-3-Decenyl acetate	83446-51-9			0.3			
Dehydroaromadendrene	698388-95-3						0.08[m]; 0.1[y2]
2,3-Dehydro-1,8-cineole	92760-25-3	0.04-0.2				0.2	0.1[m]; 0.2[g]; 0.3[x5]; 0.9[z5]; 1.0[y4]
Dehydrocostunolide							tr[m]; 0.1[o]
Dehydrocostus lactone	477-43-0			0.2			0.2[b]
Dibutyl phthalate	84-74-2						0.3[v,w]
Dihydrocarveol	38049-26-2						0.8[x5]
Dihydrocarveol acetate	20777-49-5					0.06	
Dihydro-α-terpineol	498-81-7						0.4[v]
trans-Dihydro-α-terpineol	5114-00-1						0.7[n]
Dihydro-α-terpinyl acetate (cis-)	80-25-1						0.3[y]
2,5-Dimethoxy-p-cymene	14753-08-3						0.2[y]
1,2-Dimethoxy-4-methyl-benzene	494-99-5					0.1	
2,3-Dimethoxytoluene	4463-33-6			<0.1			
6,6-Dimethyl-2-methyl enebicyclo							1.1[e]
2,6-Dimethyl-1,7-octadiene-3,6-diol	51276-33-6						0.1[m]
2,6-Dimethyl-3,7-octadiene-2,6-diol	13741-21-4						tr[m]
3,7-Dimethyl-2,6-octadien-1-ol acetate	16409-44-2						0.1[y2]
Dimethylstyrene	27576-03-0					0.08	
α-p-Dimethylstyrene	p-Cymenene						
2-Dodecanone	6175-49-1			<0.01			

Table 5.45.1 Constituents identified in laurel leaf oils (*continued*)

Constituent	CAS	Percentage and range					
		A	B (3)	C (12)	D (18)	E (19)	F
α-Elemene	5951-67-7						0.1[y2]
β-Elemene	33880-83-0	0.06-0.8	0.1-0.3	0.6	0.5	0.2	0.7[n]; 0.9[r,v]; 1.1[y3]; 1.8[m]; 9.6[x6]
δ-Elemene	20307-84-0				0.3		0.4[y]
Elemicin	487-11-6	0.03-0.1	0-0.1	0.3	0.9		0.7[a]; 0.9[n]; 1.3[y2]; 1.4[d]; 2.0[y7]
α-Elemol	639-99-6						0.1[v,x3]; 0.4[k,o]; 0.6[u]
Endo-2-norborneol acetate							tr[n]
Endrin	72-20-8						0.2[z1]
Epiglobulol	88728-58-9						0.09[b]
Epizonarene	41702-63-0		0-0.06				
1,8-Epoxy-*p*-menth-2-ene				0.2			0.4[e]
Eremanthin	37936-58-6			0.1			0.1[b,x4]
Eremophilene	10219-75-7		0-0.7				0.2[o]; 0.6[y2]
Ethyl cinnamate	103-36-6		0-0.3				
Ethyl hexanoate	123-66-0						tr[n]
Ethyl isovalerate	108-64-5			<0.01			
Ethyl 2-methylbutyrate	7452-79-1			<0.01			
Ethyl 3-phenylpropionate	2021-28-5						0.1[x6]
α-Eudesmol	473-16-5			0.3			0.5[c]; 1.0[j]; 1.3[a]; 2.7[y3]
7-epi-α-Eudesmol	123123-38-6						0.7[c]
β-Eudesmol	473-15-4			1.1		0.08	1.1[x9]; 1.6[j]; 1.8[x2]; 2.2[n]; 3.7[y3]
γ-Eudesmol	1209-71-8			<0.01	0.2		0.2[a]; 2.0[c]
10-epi-γ-Eudesmol	15051-81-7			<0.1	0.2		0.3[a]
Eudesmyl acetate	51317-10-3						tr[m]
α-Eudesmol acetate							tr[x3]
Eugenol	51317-10-3	0.5-1.6	0-2.2	2.9	1.3	0.2	3.5[y6]; 3.6[d]; 4.5[x7]; 5.2[i]; 5.5[a]; 5.6[z4]; 16.7[u]; 18.5[m,v]; 44.1[z2]
Eugenyl acetate	93-28-7						0.2[i]; 0.5[v]; 0.7[z2]; 1.4[x7,y6]
(*E,E*)-α-Farnesene	502-61-4				0.3		
(*Z,E*)-α-Farnesene	26560-14-5			<0.1			
β-Farnesene	502-60-3						0.1[x6]
Farnesol	4602-84-0						0.2[i]; 0.3[i]
(*Z,Z*)-Farnesol	16106-95-9						0.2[z1]
Farnesyl acetone	1117-52-8			<0.1			
α-Fenchene	471-84-1			<0.1			0.5[y8]
endo-Fenchol (α-)	512-13-0						0.2[v]; 0.4[q]
α-Fenchyl acetate	111821-74-0		0-0.4				
Fenchyl alcohol	1632-73-1						0.1[y8]
Geranial	141-27-5						tr[m,x3]
Geraniol	106-24-1	0.01-0.06					0.1[x6]; 0.2[v,x9]; 0.7[i]
cis-Geraniol			0.1				
Geranyl acetate	105-87-3		0-0.08				tr[n]; 0.1[i]; 0.2[y3]; 0.3[v]; 0.7[x7]
Geranyl acetone	3796-70-1			<0.01			tr[n]
Geranyl butyrate	106-29-6						0.1[y2]
Germacrene	28028-64-0		0-0.1				0.1[x4]
Germacrene A	28387-44-2						1.2[m]
Germacrene B	15423-57-1					0.03	
Germacrene D	23986-74-5		0-0.2	0.1		0.02	0.6[m]; 0.7[x9]; 0.8[x3]; 0.9[y7]; 1.0[d]
Germacrene D-4-ol	198991-79-6				0.6		0.2[m]; 0.4[x3]; 0.8[y7]; 0.9[t]; 1.6[a]
Globulol	489-41-8		0-0.2		0.4		0.1[o]; 0.3[y7]
Guaia-6,9-diene	37839-64-8			<0.1			0.07[y3]
α-Guaiene	3691-12-1	0.02-0.09	0-0.8	0.1	0.3	0.03	0.1[m]; 0.2[y2]; 0.3[v,y3]; 0.4[z1]; 1.0[y7]
cis-β-Guaiene	372162-07-7						1.3[y7]; 1.4[n]

Table 5.45.1 Constituents identified in laurel leaf oils (*continued*)

Constituent	CAS	A	B (3)	C (12)	D (18)	E (19)	F
Guaiol	489-86-1						<0.1[o]
α-Gurjunene	489-40-7	0.03-0.1	0-0.1	<0.1	0.4		0.2[y2]; 0.6[y7]
β-Gurjunene	73464-47-8	0.03-0.1	0-0.2				0.3[y8]
γ-Gurjunene	22567-17-5						0.4[z1]
δ-Gurjunene	1786-19-2		0-0.1				
2,4-Heptadienal	5910-85-0			<0.01			
Heptanene			0-0.03				
Hexadecanoic acid	57-10-3						1.7[x6]
Hexamethylbenzene	87-85-4						0.4[y2]
Hexanal	66-25-1			<0.1			tr[x3]; <0.1[f,g]; 0.2[u,y5]
Hexanol	111-27-3			<0.1			tr[x3]
2-Hexenal	505-57-7						0.2[y5]
(*E*)-3-Hexanol							1.0[y]
(*Z*)-3-Hexanol							3.8[y]
Hexenal	505-57-7						0.2[u]
(*E*)-2-Hexenal	6728-26-3			<0.1			tr[x3]; <0.1[f,g]; 0.1[h]; 0.2[m,i]
(*E*)-2-Hexenol	928-95-0			<0.01			tr[x3]; 0.08[i]
3-Hexenol	544-12-7						tr[u]
(*E*)-3-Hexenol	928-97-2			0.1			
(*Z*)-3-Hexenol	928-96-1						tr[x3]; 0.4[m]
3-Hexenyl butyrate	2142-93-0						0.6[r]
α-Himachalene (α-*cis-*)	3853-83-6						0.1[y8]; 0.2[z1]
Homovanillyl alcohol	2380-78-1						0.2[m]
Hotrienol	20053-88-7			0.1			
α-Humulene	6753-98-6	0.02-0.3	0-0.08	0.1	0.05	0.02	0.4[y3]; 0.5[x6]; 1.1[y7]; 2.2[o]; 4.2[z2]
Humulene (ep)oxide	19888-33-6			0.1			
Humulene epoxide II	19888-34-7						tr[m,x3]; 5.9[c]
Hydrocinnamyl acetate	122-72-5						0.08[y3]; 0.4[n]
2-Hydroxy-1,8-cineole	103665-39-0						0.1[m]; 0.4[y4]
3-Hydroxy-1,8-cineole	118013-29-9						0.1[m]
14-Hydroxy-9-epi-β-caryophyllene	123355-03-3						0.2[f]
Intermedeol	6168-59-8			0.3			0.1[y4]
Ipsdienol	35628-00-3			0.1			
Isoborneol	124-76-5						0.4[y6]
Isobornyl acetate	125-12-2						0.4[a,h]; 1.1[y]; 1.7[x9]
Isobutyl isobutyrate	97-85-8			<0.01			
Isobutyl 2-methylbutyrate	2445-67-2			<0.01			
Isoledene	95910-36-4						0.04[y3]
5-Isocedranol	13567-45-8						0.8[y7]; 1.1[d]
Iso-β-elemene	783322-21-4						0.1[m]
(*E*)-Isoelemicin	5273-85-8			0.3			tr[m]
Isoeugenol	5932-68-3		0-0.03				0.3[x9]
(*E*)-Isoeugenol	5932-68-3			0.2			0.6[m]
(*E*)-Isoeugenyl acetate	5912-87-8						2.6[c]
Isohexane	73513-42-5		0.3-04				
Isopinol	110268-86-5			0.1			
Isopropyl 2-methylbutyrate	66576-71-4			<0.1			
cis-Isosafrole	17627-76-8				0.1		
Isospathulenol	88395-46-4		0-0.1				
Isothujol	513-23-5					0.06	
Iso-3-thujyl acetate						1.1	0.5[y]
Isovaleraldehyde	590-86-3		8.8-10.5				
(*Z*)-Jasmone	488-10-8					0.05	
laevo-Jujenol				0.2			
Lavandulol	498-16-8					0.5	
Ledol	577-27-5		0-0.1	0.3			0.5[y2]
Lepidozene	133005-43-3		0-0.4				
Limonene	138-86-3	1.0-4.0		1.5	2.0		1.9[r]; 2.0[h]; 2.3[z4]; 2.5[m]; 2.6[t]

Table 5.45.1 Constituents identified in laurel leaf oils (*continued*)

Constituent	CAS	A	B (3)	C (12)	D (18)	E (19)	F
cis-Limonene oxide	13837-75-7					0.08	0.1[f,g]
trans-Limonene oxide	4959-35-7					0.06	
Linalool	78-70-6	1.1-7.8	9.5-17.7	2.0	6.1	0.2	8.4[y2]; 10.9[y7]; 15.3[x8]; 18.7[u]; 22.1[z6]; 22.7[v]; 26.9[y5]; 32.0[v]
cis-Linalool oxide	11063-77-7			<0.1			0.06[i]; 0.6[n]
trans-Linalool oxide	11063-78-8			<0.1			tr[n]
Linalyl acetate	115-95-7	0.05-0.4	0-0.7				0.4[a]; 0.6[u]; 0.9[x7]; 1.0[v]; 1.4[r]; 2.7[u]
Longicyclene	1137-12-8					0.2	
Longifolene	475-20-7						0.2[z1,z2]
β-Longipinene	41432-70-6						1.2[d]; 1.9[y7]
Lyral	130066-44-3						0.2[gf,g,x]
β-Maaliene	489-29-2		0-0.04				
p-Mentha-1(7),8-diene	499-97-8						tr[n]
p-Mentha-2,4(8)-diene	586-63-0				0.3		
trans-*m*-Mentha-2,8-diene							0.1[y]
p-Mentha-1,4-dien-7-ol	22539-72-6			0.1			0.2[y4]
p-Mentha-1,5-dien-7-ol	19876-45-0						0.2[z5]; 0.3[y4]
p-Mentha-1,5-dien-8-ol	1686-20-0			0.2			0.4[y7]; 0.6[y]
p-Mentha-1(7),2-dien-8-ol	65293-09-6			0.2			
cis-*p*-Mentha-1(7),8-dien-2-ol	22626-43-3			0.3			0.3[x3,z5]; 0.5[y4]
trans-*p*-Mentha-1(7),8-dien-2-ol	21391-84-4			0.3			0.2[x3]; 0.3[b,y5]; 0.5[y4]
cis-*p*-Mentha-2,8-dien-1-ol	3886-78-0			0.1			tr[x3]; 0.6[y4]
trans-*p*-Mentha-2,8-dien-1-ol	4017-77-0			0.1			0.5[b,z5]; 0.6[k]
1,3,8-*p*-Menthatriene	18368-95-1			<0.1			tr[m]
cis-*p*-Menth-2-ene-1,8-diol	54164-91-9						0.2[y4]
trans-*p*-Menth-2-ene-1,8-diol	54164-90-8						0.2[y4,z5]
p-Menth-2-en-1-ol	619-62-5	0.04-0.2					
(*E*)-*p*-Menth-2-en-1-ol	29803-81-4			0.2			tr[n]; <0.05[m]; 0.1[f,g,h]; 0.2[y4]; 0.7[b]
(*Z*)-*p*-Menth-2-en-1-ol	29803-82-5			0.2		0.2	0.07[y3]; 0.1[h,x5]; 0.2[f,g]; 10.4[x2]
1-*p*-Menthene-8-ethyl acetate							13.5[y9]
Menthofuran	494-90-6						0.2[f,g]
Menthone	89-80-5					0.09	tr[x6]
3-Methoxyacetophenone						0.09	
4-Methoxyacetophenone						0.1	
2-Methylbutyl acetate	624-41-9						0.1[o]
Methylcarvacrol	6379-73-3						0.3[y2]
Methyl chavicol	140-67-0		0-0.1	<0.1			0.09[y3]; 0.1[x6,y2]; 0.3[y6]; 0.6[i]
1,3,5-*tris*(Methylene)-cycloheptane							0.3[e]
Methyl eugenol	93-15-2	1.0-3.0	2.9-12.4	3.5	5.1	0.05	6.5[x2]; 8.3[x9]; 8.9[v]; 9.4[a,n]; 9.6[i]; 11.0[y7]; 11.8[t]; 17.8[d] ; 21.4[y3]
Methyl geranate	2349-14-6						1.0[v]
6-Methyl-5-hepten-2-one	110-93-0			<0.1			tr[n]
2-Methyl-2,3-hexadiene	29212-09-7						0.4[e]
3-Methyl-1,3,5-hexatriene	2196-24-9						0.3[e]
Methyl isoeugenol	93-16-3						0.1[g,x5] ; 0.2[x9]; 0.5[z2]; 1.3[z1]; 1.9[x6]
(*E*)-Methylisoeugenol	6379-72-2			0.4			0.1y[u,v,y4]; 0.3[v]; 0.5[n]; 0.8[z1]
(*Z*)-Methylisoeugenol	6380-24-1		0-0.2				0.3[v]; 1.0[z6]
2-Methyl-6-methylene-3,7-octadiene							0.2[e]
2-Methyl-6-methylene-3,7-octadien-2-ol							0.2[y4]

Table 5.45.1 Constituents identified in laurel leaf oils (*continued*)

Constituent	CAS	Percentage and range					
		A	B (3)	C (12)	D (18)	E (19)	F
1-(3-Methylcyclopent-2-enyl) cyclohexene							0.2^{y2}
2-Methyl-3-phenylpro-panal	5445-77-2						0.4^{y2}
Methyl salicylate	119-36-8			<0.01			
Methyl thymol	1076-56-8						0.1^{y2}
(Z)-Muurola-4,5-diene	157477-72-0					0.03	
α-Muurolene	10208-80-7			<0.01	0.3		$0.1^{o,y2}$; 1.3^{v}
γ-Muurolene	30021-74-0			0.1	0.5		tr^n; 0.1^{x5}; $0.2^{v,y8}$
α-Muurolol	104245-48-9				0.5		
τ-Muurolol (epi-α-)	19912-62-0						0.7^{x6}
Myrcene (β-)	123-35-3	0.7-1.4	0.3-0.9	0.6	1.1		1.5^u; 1.6^{z4}; 1.8^{x5}; 2.3^{z6}; 2.4^v
Myrcenol	543-39-5					0.1	
Myrtenal	564-94-3	0.0-0.06		0.4		1.8	0.4^{x6}; 0.8^{z5}; 0.9^{y4}; 1.0^{x5}; 2.4^b
Myrtenol	515-00-4	0.08-0.1					0.5^{x3}; 0.7^{y4}; $0.8^{b,e}$; 1.1^{x5}; 1.5^{x1}
Myrtenyl acetate	1079-01-2	0.03-0.1					0.07^i
Naphtalene	91-20-3		0-0.3				
2-Naphthalenemethanol			0-0.5				
Neoisoisopulegol acetate	256332-34-0						1.1^o
Neophytadiene	504-96-1						0.3^m
Neral	106-26-3						0.2^{x6}
Nerol	106-25-2	0.02-0.5		0.1			tr^m; 0.1^v; 0.3^a; 0.5^v; 0.7^u; 1.0^{x7}
(E)-Nerolidol	40716-66-3			0.1			
(Z)-Nerolidol	3790-78-1						tr^n
Nerol oxide	1786-08-9			0.1			tr^{x3}
Neryl acetate	141-12-8		0-0.1	<0.1			$tr^{m,n}$; 0.06^{y5}; $0.2^{f,g}$; $0.3^{k,x}$ 0.4^{y8}
Nonadecane	629-92-5						0.05^i
2-Nonanone	821-55-6			<0.01			$tr^{m,n}$
2-Norbornanone	497-38-1		0-1.2				
β-Ocimene	13877-91-3		0-0.3				0.1^{y9}
(E)-β-Ocimene	3779-61-1			<0.1			tr^{x3}; $<0.1^{f,g}$; $0.1^{h,u,v}$; $0.3^{i,j}$; 7.1^{z1}
(Z)-β-Ocimene	3338-55-4						$tr^{n,u}$; $<0.1^{f,g}$; 3.1^{z1}
n-Octadecane	593-45-3						tr^{x3}
1-Octen-3-ol	3391-86-4						0.07^{y5}
Octenyl acetate	37366-04-4						tr^m
β-Oplopenone	28305-60-4						0.3^n
Palustrol	5986-49-2			<0.01			
Patchoulene	1405-16-9						0.1^{z1}
1,3-Pentadiene	504-60-9		0-0.02				
Pentane	109-66-0		0-2.1				
Phellandral	21391-98-0			0.1			tr^{x3}; 0.4^{z5}
Phellandrene	1329-99-3	0.2-0.5					
1-Phellandrene	4221-98-1						0.2^{x2}
α-Phellandrene	99-83-2		0-0.1	0.2	0.3	0.2	0.5^h; 0.6^v; 0.8^q; 0.9^v; 2.1^{x9}; 2.5^i
β-Phellandrene	555-10-2		0-5.7				0.4^i; $0.5^{f,g}$
L-Phellandrene	6153-17-9		0-0.2				
α-Phellandrene-8-ol	1686-20-0					0.2	tr^m
Phenol	108-95-2		0-1.7				
γ-Phenylpropyl acetate	122-72-5		0-0.09	0.1			
Phthalic acid	88-99-3						0.4^{x4}
Phytadiene							tr^m
Phytol	7541-49-3						0.2^{x6}

Table 5.45.1 Constituents identified in laurel leaf oils (*continued*)

Constituent	CAS	Percentage and range					
		A	B (3)	C (12)	D (18)	E (19)	F
α-Pinene	80-56-8	4.4-7.5	2.5-4.6	4.6	6.2	3.4	5.6[x2]; 5.9[x8]; 6.0[z4]; 6.5[x7]; 6.7[f]; 7.2[r]; 7.5 [z7]; 7.7[s]; 7.8[o]; 15.9[u]
β-Pinene	127-91-3	3.2-5.0	1.4-2.0	4.0	4.7	3.3	4.1[x8]; 4.2[x7]; 4.3[z3]; 4.6[z6]; 5.0[s]; 5.1[z4]; 5.2[r]; 5.3[f]; 5.9[o]; 6.5[u]
2-Pinen-10-ol	515-00-4			0-0.1			
Pinocarveol	5947-36-4	0.03-0.1					0.6[o]; 0.8[e]
(E)-Pinocarveol	1674-08-4			0.1		0.9	<0.05[m]; 0.5[z5]; 1.0[y4,x2]; 1.5[x5]
cis-Pinocarveol						0.2	
Pinocarvone	30460-92-5			0.2		0.6	0.1[x4]; 0.3[y8]; 0.4[b,k]; 0.5[x2]; 1.0[y4]
(E)-Pinocarvyl acetate	1686-15-3			0.1		0.2	tr[m]
cis-Pinocarvyl acetate	73366-18-4						1.3[x5]
Pinol	2437-97-0			<0.1			
Piperitol	491-04-3	0.01-0.2					
cis-Piperitol	16721-38-3			<0.1		0.09	0.1[y2,y4]
trans-Piperitol	16721-39-4			0.1		0.2	tr[x3]; 0.1[y2]
Piperitone	89-81-6			<0.01			
Pseudolimonene (psi-)	499-97-8						0.4[x4]; 0.7[x2]
Pulegone	89-82-7		0-0.05				
Sabina acetone							0.2[b]
Sabina ketone	513-20-2			0.1		0.2	tr[m]; 0.3[y4]; 0.4[o,x5]
Sabinene	3387-41-5	7.3-11.8		5.3	10.0	3.3	10.1[x4]; 10.2[x7]; 10.6[h]; 11.0[v]; 11.8[u]; 12.1[i]; 14.1[j]; 15.0[z7]
(E)-Sabinene hydrate	17699-16-0	0.05-0.2	0.1-0.3	<0.1			0.6[c]; 0.7[z]; 0.8[b]; 0.9[v]; 1.3[j]; 11.9[a]
(Z)-Sabinene hydrate	15537-55-0	0.08-0.3	0-0.1	<0.01	0.3		0.3[y7]; 0.4[m]; 0.5[e,x8]; 0.6[a]; 0.7[i,y4]
Sabinene hydrate acetate	87553-42-2						1.3[c]
(E)-Sabinene hydrate acetate	77318-47-9						0.6[e,m]; 1.3[c]
(E)-Sabinol	471-16-9			<0.01			tr[m]; 0.1[b]; 0.3[f,g]; 1.0[k]
cis-Sabinol	3310-02-9						0.3[y]
Sabinyl acetate (cis-)	53833-85-5			0.1			0.4[b]
trans-Sabinyl acetate	139757-62-3	0.02-0.2				0.1	0.2[a,f,g]; 0.3[q]
Santalone	59300-51-5					0.3	
Selina-4(15),5-diene	1107026-89-0			<0.01			
Selina-3,11-dien-6α-ol	75521-07-2						1.0[c]
α-Selinene	473-13-2			0.3			0.06[y3]; 0.1[u,v]
7-epi-α-Selinene	123123-37-5						tr[m]; 0.8[n]
β-Selinene	17066-67-0		0-0.2	0.3			0.1[m,v]; 0.3[n,y2]; 0.4[y]; 0.8[y3]; 2.6[y8]
γ-Selinene	515-17-3			<0.01			
Selin-11-en-4α-ol	16641-47-7						tr[n,x3]
Selin-11-en-5α-ol							0.1[y4]
β-Sesquiphellandrene	20307-83-9						0.4[v]
Seychellene	20085-93-2						0.4[z1]
Spathulenol	6750-60-3	0.05-0.2	0.2-1.7	0.8	0.2	0.07	1.2[n]; 1.5[d]; 1.9[x6,y2]; 3.2[y3]; 4.0[y7]
allo-Spathulenol	99147-40-7			0-0.8			
Spirafoliolide							3.7[m]
Squalene	111-02-4						tr[m]
Terpinene	8013-00-1			0-0.9			

Table 5.45.1 Constituents identified in laurel leaf oils (continued)

Constituent	CAS	Percentage and range					
		A	B (3)	C (12)	D (18)	E (19)	F
α-Terpinene	99-86-5	0.2-0.9	0.1-0.3	1.3	0.3	0.5	0.8[x2]; 1.0[v]; 1.5[y3]; 4.1[i]; 12.5[l]
β-Terpinene	99-84-3						0.06[l]
(E)-β-Terpineol	7299-41-4						1.2[x5]
(Z)-β-Terpineol	7299-40-3						0.9[x5]
γ-Terpinene	99-85-4	0.5-2.2	0.2-0.6	1.8	0.6	0.8	1.1[d]; 1.2[x2]; 1.4[x7]; 1.5[v]; 1.6[v]
Terpinen-4-ol	562-74-3	1.7-3.4	0-1.5	4.3	1.8	4.3	4.4[y9]; 4.6[i]; 5.1[e]; 5.7[x1]; 6.4[y]
Terpinen-4-yl acetate	4821-04-9	0.01-0.7					
1-Terpineol	586-82-3		0-1.5				0.2[y]
α-Terpineol	98-55-5	1.5-2.3	0.9-5.8	3.1	2.3	1.4	3.8[q]; 3.9[n]; 4.6[z]; 5.2[x4]; 6.8[j]; 7.2[y8]; 9.3[s,v]; 31[d,p]
β-Terpineol	138-87-4	0.04-0.4					0.2[v,y5]
γ-Terpineol	586-81-2	0.1-0.2					1.8[y]; 1.9[q]
δ-Terpineol	7299-42-5	0.03-0.2		0.6			0.3[h]; 0.6[z]; 0.8[y4]; 0.9[z5]; 2.6[y2]
Terpinolene (α-)	586-62-9	0.1-0.4	0-0.2	0.7		0.3	0.8[v]; 0.9[x7]; 1[u]; 2.2[v]; 13.6[x1]
α-Terpinyl acetate	80-26-2	6.3-12.0		9.8	9.7	8.8	14.6[y]; 14.7[b]; 15.3[d]; 16.6[y7]; 18.0[v]; 18.2[t]; 18.3[y3]; 25.7[j]
δ-Terpinyl acetate				0.9			0.4[m]; 0.5[x8]; 0.8[y3]; 1.0[x2]; 9.9[k]
Terpinyl propionate	80-27-3						0.1[y8]
Tetradecane	629-59-4						+[v]
Tetradecanoic acid	544-63-8						0.2[x6]
Tetradecene	26952-13-6						tr[m]
(E)-5-Tetradecen-3-yne	74744-48-2						0.3[z1]
Tetrahydrocitronellene							0.1[y]
2,4(10)-Thujadiene	36262-09-6	0.01-0.06		0.1			
4-Thujanyl acetate							0.05[y3]
Thuj-3-en-10-al	57129-54-1					0.1	tr[m,x3]; 0.4[y4]; 0.5[z5]
α-Thujene	2867-05-2	0.2-0.7	0.2-0.4	0.6	0.5	0.1	0.4[y9]; 0.5[f]; 0.6[c]; 0.7[v]; 0.9[r]; 1.0[a]
Thujen-2-ol							0.5[o]
Thuj-3-en-10-ol							0.4[o]
4-Thujen-2α-yl acetate				0.5			0.1[m,y2]; 0.8[x1]
α-Thujone	546-80-5						0.5[v]
β-Thujone	471-15-8			<0.01			0.4[v]
neo-3-Thujyl acetate						0.1	
Thymol	89-83-8			<0.01			0.5[f,g]; 0.6[y8]; 1.5[y2]; 3.1[i]
Toluene	108-88-3						0.1[m]
Torreyol	19435-97-3						0.7[v]
Tricyclene	508-32-7			<0.1			tr[x3,y5]; <0.05[m]; <0.1[f,g]
2-Tridecanone	593-08-8			0.1			
Trimethyl-6,10,14-penta-decanone	502-69-2						tr[m]
Umbellulone	24545-81-1			<0.01			
2-Undecanone	112-12-9		0-0.09	0.2		0.06	tr[m]; 0.3[n]
Valencene	4630-07-3						0.4[a]; 0.6[n]; 11.0[i]
Valerenol	101628-22-2						0.1[y2]
Valerianol	20489-45-6			0.2			
Vanillin	121-33-5						0.1[m]
trans-Verbenol	1820-09-3						tr[x3]

Table 5.45.1 Constituents identified in laurel leaf oils (*continued*)

Constituent	CAS	Percentage and range					
		A	B (3)	C (12)	D (18)	E (19)	F
(Z)-Verbenol	1845-30-3					0.04	0.1[y8]; 0.3[q]
Verbenone	80-57-9			<0.1			0.2[y8]
Vetiverol	68129-81-7		0-0.08				
Viridiflorene	21747-46-6		0-0.7			0.08	0.4[n]; 0.5[t]
Viridiflorol	552-02-3		0-0.1	0.3	0.3		0.2[x3]; 0.3[a,x6]; 0.6[y2]; 1.7[y3]
α-Ylangene	14912-44-8			0.1			0.2[m]; 0.6[x6]; 0.8[y7]; 1.0[q]
Zingiberene	495-60-3						0.4[y2]

A twenty-three samples of laurel leaf essential oils from France and Morocco analyzed between 1998 and 2013; lowest and highest concentrations given (E. Schmidt, unpublished data)

B three lab-hydrodistilled laurel leaf oil samples from Morocco, Tunisia and Algeria; lowest and highest concentrations given (ref. 3)

C one lab-hydrodistilled oil sample from Turkish dried laurel leaves purchased in Germany (ref. 12)

D two lab-prepared oil samples from different locations in Montenegro; highest concentrations given (ref. 18)

E one lab-hydrodistilled laurel leaf oil sample from northern Cyprus (ref. 19)

F data from other studies (indicated with superscript letters); highest concentrations found in any study reviewed here given; when two or more oils were investigated, only the highest concentrations are mentioned, unless indicated otherwise

[a] one oil obtained by traditional hydrodistillation and three with microwave-assisted distillation (ref. 4); [b] one lab-hydrodistilled oil from Turkish laurel leaves (ref. 16); [c] one lab-hydrodistilled oil from Jordanian laurel leaves (ref. 5); [d] ten lab-hydrodistilled oils from leaves of *L. nobilis* growing wild, harvested in various parts of Tunisia (ref. 13); [e] one lab-hydrodistilled oil and three microwave-assisted hydrodistilled oils from Turkish laurel leaves (ref. 15); [f] one commercial laurel leaf oil from Italy (ref. 21); the data are virtually identical to those in ref. 14; [g] one lab-hydrodistilled oil from laurel leaves purchased from an Italian company (ref. 14); the data are virtually identical to those in ref. 21; [h] one lab-hydrodistilled oil from leaves of laurels growing wild in Portugal (ref. 11); [i] one lab-hydrodistilled oil from fresh Tunisian laurel leaves and six oils prepared from the same leaves after drying by various methods and times (ref. 1); [j] three lab-hydrodistilled oils from leaves collected in various parts of Turkey (ref. 9); [k] seven lab-prepared oils from laurel leaves harvested from wild growing laurels in 7 locations in Turkey (ref. 20); [l] one lab-hydrodistilled oil from Turkey (ref. 8); [m] two lab-prepared oils from old and young laurel leaves from Turkey (ref. 6); [n] one lab-hydrodistilled oil sample from leaves collected in Sardinia, Italy (ref. 7); [o] data taken from various studies reviewed in ref. 10; [p] probably a mistake, should read 3.1 (ref. 13); [q] four lab-hydrodistilled oils from Iranian laurel leaves dried under various conditions (ref. 2); [r] one commercial laurel leaf oil from Serbia (ref. 22); [s] five steam distilled oils from laurel leaves collected in a university botanical garden in Portugal in five periods in one year; only the main constituents were analyzed (ref. 23); [t] one lab-hydrodistilled oil from France (ref. 17); [u] data from various studies reviewed in ref. 24; [v] data from various studies reviewed in ref. 25; [w] does not occur naturally (ref. 25); [x] Lyral is the trade name for the fragrance hydroxyisohexyl 3-cyclohexene carboxaldehyde, which is a synthetic chemical; [x1] one lab-hydrodistilled laurel leaf oil sample from wild growing *L. nobilis* in Turkey (ref. 26); [x2] eight lab-prepared oil samples from laurel leaves cultivated in Iran, harvested in 4 different phenological stages (seasonal times) and distilled by two methods: hydro- and steam-distillation (ref. 27); [x3] two oils from laurel leaves cultivated in Iran, one prepared by hydrodistillation, the other by steam-distillation; average values were given (ref. 27); [x4] one lab-hydrodistilled laurel leaf oil from Turkey (ref. 30); [x5] one lab-hydrodistilled oil from leaves purchased at a local market in Turkey (ref. 31); [x6] one steam-distilled laurel oil from leaves cultivated in Mongolia (ref. 32); [x7] five lab-hydrodistilled oils from laurel leaves collected in the wild in five provinces of Turkey (ref. 33); [x8] six industrially steam-distilled oils from green branches and leaves of *L. nobilis* cultivated in Argentina and harvested between October and May (ref. 34); [x9] one lab-hydrodistilled leaf oil from Iran (ref. 36); [y] one lab-hydrodistilled leaf oil from Jordan (ref. 37); [y1] one laurel leaf oil prepared in Turkey (ref. 38); [y2] one lab-hydrodistilled oil sample from laurel leaves collected in Portugal (ref. 45); [y3] two lab-hydrodistilled oils from laurel leaves purchased in Spain to be used as spice and one commercial oil intended for medicinal use (ref. 42); [y4] two oils obtained by hydro- and by steam-distillation of the same batch of laurel leaves purchased from a local Turkish supplier (ref. 56); [y5] one lab-hydrodistilled oil sample prepared from leaves from Sicily, Italy (ref. 55); [y6] three lab-hydrodistilled oils from leaves purchased from a Georgian company; one was prepared from fresh material, the other two from leaves stored for a year, of which one in light and one in the dark (ref 48); [y7] two lab-hydrodistilled leaf oil samples, one from Algeria, the other from Morocco (ref. 47); [y8] one lab-hydrodistilled leaf oil sample from Tunisia (ref. 43); [y9] one laurel leaf oil sample, probably from Italy (ref. 46); [z] one lab-hydrodistilled and one microwave-assisted hydrodistilled oil sample from the same batch of commercial Turkish laurel leaves (ref. 49); [z1] one lab-hydrodistilled oil sample from the leaves of *L. nobilis* from Morocco (ref. 51); [z2] one commercial laurel leaf oil from China with an extremely atypical composition: 44.1% eugenol, 30.3% cinnamaldehyde and only 0.7% 1,8-cineole; probably in error (ref. 52); [z3] one lab-hydrodistilled leaf oil sample from Turkey (ref. 55); [z4] one lab-prepared oil sample from Greece (ref. 40); [z5] one lab-hydrodistilled laurel leaf oil from Turkey (ref. 39); [z6] one lab-hydrodistilled oil from leaves harvested in Argentina (ref. 57)

tr: trace (in column F[m]: <0.01; in column F[n]: <0.2; in column F[x3]: <0.1; in column F[y5]: <0.05)

Table 5.45.2 Major constituents of laurel leaf oils

Constituent	CAS	Percentage and range					
		A	B (3)	C (12)	D (18)	E (19)	F
1,8-Cineole (eucalyptol)	470-82-6	38.4-52.0	24.6-38.9	21.8	40.9	58.6	61.6[x5]; 63.2[j]; 68.5[k]; 72.1[e]
Linalool	78-70-6	1.1-7.8	9.5-17.7	2.0	6.1	0.2	22.1[z6]; 22.7[v]; 26.9[y5]; 32.0[v]
α-Terpinyl acetate	80-26-2	6.3-12.0		9.8	9.7	8.8	18.0[v]; 18.2[t]; 18.3[y3]; 25.7[j]
Methyl eugenol	93-15-2	1.0-3.0	2.9-12.4	3.5	5.1	0.05	11.0[y7]; 11.8[t]; 17.8[d]; 21.4[y3]
α-Pinene	80-56-8	4.4-7.5	2.5-4.6	4.6	6.2	3.4	7.2[r]; 7.5[z7]; 7.7[s]; 7.8[o]; 15.9[u]
Sabinene	3387-41-5	7.3-11.8		5.3	10.0	3.3	11.8[u]; 12.1[j]; 14.1[l]; 15.0[z7]
α-Terpineol	98-55-5	1.5-2.3	0.9-5.8	3.1	2.3	1.4	4.6[z]; 5.2[x4]; 6.8[j]; 7.2[y8]; 9.3[s,v];
β-Pinene	127-91-3	3.2-5.0	1.4-2.0	4.0	4.7	3.3	5.0[s]; 5.1[z4]; 5.2[r]; 5.3[f]; 5.9[o]; 6.5[u]

LEGEND: SEE UNDER TABLE 5.45.1

most cases of early reports of contact allergy to laurel oil (<1960) were due to vegetable (not essential) oil from laurel *berries* (78,86,92). Laurel berry oils from *L. nobilis* are forbidden in cosmetics in the EU.

Testing in groups of patients

The results of patch tests with 'laurel oil' and laurel leaf oil in routine testing (consecutive patients suspected of contact dermatitis) and in groups of selected patients are shown in Table 5.45.3. In routine testing, rates of positive reactions ranged from 3.1% to 6.9% (data from half a century ago, probably laurel berries oil), whereas in more recent studies between 0.6% and 50.5% (but this

was also *vegetable* laurel *berries* oil) patients in selected groups had positive patch tests.

Case reports

A patient developed erythema multiforme from contact allergy to laurel oil used for knee arthropathy; the patch test had a target-like appearance and had the histopathology of erythema multiforme (74). Generalized allergic contact dermatitis was caused by contact allergy to a mixture of laurel oil and olive oil used for massage (76). A case of allergic contact dermatitis from the topical use of laurel oil for rheumatic complaints was described (79).

Table 5.45.3 Results of testing groups of patients with laurel (leaf) oil

Years and Country	Test conc. & vehicle	Number of patients			Selection of patients (S); Relevance (R); Comments (C)	Ref.
		tested	positive	(%)		
Routine testing						
1953-1962 Germany	?	>1,000	see right column		Between 1953 and 1962, several thousand dermatitis patients were with 'laurel oil' in Germany; positive reactions were seen in 3.1-6.9% of patients (data cited in ref. 90); most likely *vegetable* oil from laurel *berries*	90
Testing in groups of selected patients						
2001-2010 Australia	2% pet.	681	31	(4.6%)	S: not specified; R: 32%	91
2000-2008 IVDK	2% pet.	6,297	63	(1.0%)	S: patients with dermatitis suspected of causal exposure to fragrances; R: not stated	81
1997-2000 Austria	2% pet.	747	4	(0.6%)	S: patients suspected of fragrance allergy; R: not stated	73
1994-1995 UK	2% pet.	40	2	(5%)	S: patients previously reacting to the fragrance mix; R: not stated	82
1985-1990 Germany	2% pet.	99	50	(50.5%)	S: patients with a positive reaction to the Compositae mix; R: not stated; the reactions to laurel oil were considered as indicators for Compositae allergy; C: laurel oil contains sesquiterpene lactones such as in Compositae, ergo cross-reactivity or pseudo-cross-reactivity can be expected (78); the test substance was not an essential oil, but *vegetable* laurel *berries* oil	86

IVDK Information Network of Departments of Dermatology, Germany, Switzerland, Austria (www.ivdk.org); pet.: petrolatum

One case of dermatitis from contact allergy to laurel oil in a face mask has been reported (84). Contact allergy to laurel oil in toothpaste was described in the 1950s (84, data cited in ref. 83). Two cases of allergic contact dermatitis from laurel oil in a topical pharmaceutical preparation (72).

Positive patch tests (relevance unknown, uncertain or not stated)

One positive patch test to laurel oil in an aromatherapist with occupational contact dermatitis from (multiple) essential oils; it was uncertain whether this oil had been used by the patient (75). One positive patch test reaction to laurel leaf oil in a patient known to be allergic to oak moss (93). A positive patch test reaction to laurel oil was observed in a patient who had airborne allergic contact dermatitis from tea tree oil and lavender oil (80). One positive patch test to laurel oil in a patient suffering from airborne allergic contact dermatitis from aromatherapy with other essential oils (77).

Additional literature on contact allergy to laurel (leaf) oil may be found in refs. 85,87 and 88 (articles not read).

REFERENCES

1 Sellami IH, Aidi Wannes W, Bettaieb I, Berrima S, Chahed T, Marzouk B, et al. Qualitative and quantitative changes in the essential oil of *Laurus nobilis* L. leaves as affected by different drying methods. Food Chem 2011;126:691-697

2 Hadjibagher Kandi MN, Sefidkon F. The influence of drying methods on essential oil content and composition of *Laurus nobilis* L. J Essent Oil Bear Plants 2011;14:302-308

3 Ben Jemâa JM, Tersima N, Taleb Toudert K, Khouja ML. Insecticidal activities of essential oils from leaves of *Laurus nobilis* L. from Tunisia, Algeria and Morocco, and comparative chemical composition. J Stor Prod Res 2012;48:97-104

4 Flamini G, Tebanoa M, Cionia PL, Ceccarini L, Ricci AS, Longo I. Comparison between the conventional method of extraction of essential oil of *Laurus nobilis* L. and a novel method which uses microwaves applied in situ, without resorting to an oven. J Chromatograph A 2007;1143:36-40

5 Al-Kalaldeh JZ, Abu-Dahab R, Afifi FU. Volatile oil composition and antiproliferative activity of *Laurus nobilis*, *Origanum syriacum*, *Origanum vulgare*, and *Salvia triloba* against human breast adenocarcinoma cells. Nutrit Res 2010;30:271-278

6 Kilic A, Hafizoglu H, Kollmannsberger H, Nitz WS. Volatile constituents and key odorants in leaves, buds, flowers and fruits of *Laurus nobilis* L. J Agric Food Chem 2004;52:1601-1606

7 Caredda A, Marongiu B, Porcedda S, Soro C. Supercritical carbon dioxide extraction and characterization of *Laurus nobilis* essential oil. J Agric Food Chem 2002;50:1492-1496

8 Dadalioglu I, Evrendilek GA. Chemical compositions and antibacterial effects of essential oils of Turkish oregano (*Origanum minutiflorum*), bay laurel (*Laurus nobilis*), Spanish lavender (*Lavandula stoechas* L.), and fennel (*Foeniculum vulgare*) on common foodborne pathogens. J Agric Food Chem 2004;52:8255-8260

9 Sangun MK, Aydin E, Timur M, Karadeniz H, Caliskan M, Ozkan A. Comparison of chemical composition of the essential oil of *Laurus nobilis* L. leaves and fruits from different regions of Hatay, Turkey. J Environ Biol 2007;28:731-733

10 Lawrence BM. Essential oils 2001-2004. Carol Stream, USA: Allured Publishing Corporation, 2006:110-114 (reprinted from Perfum Flavor 2002;27(5):74-78)

11 Macchioni F, Perrucci S, Cioni P, Morelli I, Castilho P, Cecchi F. Composition and acaricidal activity of *Laurus novocanariensis* and *Laurus nobilis* essential oils against *Psoroptes cuniculi*. J Essent Oil Res 2006;18:111-114

12 Braun NA, Meier M, Kohlenberg B, Hammerschmidt F-J. δ-Terpinyl acetate. A new natural component from the essential leaf oil of *Laurus nobilis* L. (Lauraceae). J Essent Oil Res 2001;13:95-97

13 Marzouki H, Piras A, Bel Haj Salah K, Medini H, Pivetta T, Bouzid S, et al. Essential oil composition and variability of *Laurus nobilis* L. growing in Tunisia, comparison and chemometric investigation of different plant organs. Nat Prod Res 2009;23:343-354

14 Baratta MT, Dorman HJD, Deans SG, Biondi DM, Ruberto G. Chemical composition, antimicrobial and antioxidative activity of laurel, sage, rosemary, oregano and coriander essential oils. J Essent Oil Res 1998;10:618-627

15 Günes M, Alma MH. The effects of microwave irradiation power on the chemical composition of essential oil from the leaves of Turkish bay laurel. J Electromagn Wav Applicat 2008;22:2205-2216

16 Chalchat JC, Özcan MM, Figueredo G. The composition of essential oils of different parts of laurel, mountain tea, sage and ajowan. J Food Biochem 2011;35:484-499

17 Fiorini C, Fourasté I, David B, Bessière JM. Composition of the flower, leaf, and stem essential oils from *Laurus nobilis* L. Flavour Fragr J 1997;12:91-93

18 Kovacevic NN, Simic MD, Ristic MS. Essential oil of *Laurus nobilis* from Montenegro. Chem Nat Comp 2007;43:408-411

19 Yalçin H, Akin M, Sanda MA, Cakir A. Gas chromatography/mass spectrometry analysis of *Laurus nobilis* essential oil composition of northern Cyprus. J Med Food 2007;10:715-719

20 Özcan M, Chalchat JC. Effect of different locations on the chemical composition of essential oils of laurel (*Laurus nobilis* L.) leaves growing wild in Turkey. J Med Food 2005;8:408-411

21 Giamperi L, Fraternale D, Ricci D. The *in vitro* action of essential oils on different organisms. J Essent Oil Res 2002;14:312-318

22 Simić A, Soković MD, Ristić M, Grujić-Jovanović S, Vukojević J, Marin PD. The chemical composition of some Lauraceae essential oils and their antifungal activities. Phytother Res 2004;18:713-717

23 Roque OR. Seasonal variation in oil composition of *Laurus nobilis* grown in Portugal. J Essent Oil Res 1989;1:199-200

24 Lawrence BM. Progress in essential oils. Perfum Flavor 1995;20(1):47-53

25 Lawrence BM. Progress in essential oils. Perfum Flavor 1993;18(3):61-72

26 Karabörklü S, Ayvaz A, Yilmaz S, Akbulut M. Chemical composition and fumigant toxicity of some essential oils against *Ephestia kuehniella*. J Econ Entomol 2011;104:1212-1219

27 Amin G, Salehi Sourmaghi MH, Jaafari S, Hadjagaee R, Yazdinezhad A. Influence of phenological stages and method of distillation on Iranian cultivated bay leaves volatile oil. Pak J Biol Sci 2007;10:2895-2899

28 Bayramoglu B, Sahin S, Sumnu G. Extraction of essential oil from laurel leaves by using microwaves. Sep Sci Technol 2009;44:722-733

29 Lawrence BM. Progress in essential oils. Perfum Flavor 2012;37(Feb):54-61

30 Yilmaz ES, Timur M, Aslim B. Antimicrobial, antioxidant activity of the essential oil of bay laurel from Hatay, Turkey. J Essent Oil Bear Plants 2013;16:108-116

31 Süntar I, Küpeli Akkol E, Tosun A, Keleş H. Comparative pharmacological and phytochemical investigation on the wound-healing effects of the frequently used essential oils. J Essent Oil Res 2014;26:41-49

32 Shatar S, Altantsetseg S. Essential oil composition of some plants cultivated in Mongolian climate. J Essent Oil Res 2000;12:745-750

33 Akgül A, Kivanç M, Bayrak A. Chemical composition and antimicrobial effect of Turkish laurel leaf oil. J Essent Oil Res 1989;1:277-280

34 Di Leo Lira P, Retta D, Tkacik E, Ringuelet J, Coussio JD, van Baren C, Bandoni AL. Essential oil and by-products of distillation of bay leaves (*Laurus nobilis* L.) from Argentina. Ind Crops Prod 2009;30:259-264

35 Mello da Silveira S, Bittencourt Luciano F, Fronza N, Cunha Jr. A, Neudí Scheuermann G, Werneck Vieira CR. Chemical composition and antibacterial activity of *Laurus nobilis* essential oil towards food-borne pathogens and its application in fresh Tuscan sausage stored at 7°C. LWT - Food Sci Technol 2014;59:86-93

36 Ebrahimi M, Safaralizade MH, Valizadegan O. Contact toxicity of *Azadirachta indica* (Adr. Juss.), *Eucalyptus camaldulensis* (Dehn.) *and Laurus nobilis* (L.) essential oils on mortality cotton aphids, *Aphis gossypii* Glover (Hem.: Aphididae). Arch Phytopathol Plant Prot 2013;46:2153-2162

37 Abu-Dahab R, Kasabri V, Afifi FU. Evaluation of the volatile oil composition and antiproliferative activity of *Laurus nobilis* L. (Lauraceae) on breast cancer cell line models. Rec Nat Prod 2014;8:136-147

38 Özek T. Distillation parameters for pilot plant production of *Laurus nobilis* essential oil. Rec Nat Prod 2012;6:135-143

39 Tabanca N, Avonto C, Wang M, Parcher JF, Ali A, Demirci B, et al. Comparative investigation of *Umbellularia californica* and *Laurus nobilis* leaf essential oils and identification of constituents active against *Aedes aegypti*. J Agric Food Chem 2013;61:12283-12291

40 Hassiotis CN, Dina EI. The effects of laurel (*Laurus nobilis* L.) on development of two mycorrhizal fungi. Int Biodeter Biodegrad 2011;65:628-634

41 Müller-Riebau FJ, Berger BM, Yegen O, Cakir C. Seasonal variations in the chemical compositions of essential oils of selected aromatic plants growing wild in Turkey. J Agric Food Chem 1997;45:4821-4825

42 Peris I, Blázquez MA. Comparative GC-MS analysis of bay leaf (Laurus nobilis L.) essential oils in commercial samples. Int J Food Prop 2014; DOI: 10.1080/10942912.2014.906451

43 Yvon Y, Raoelison EG, Razafindrazaka R, Randriantsoa A, Romdhane M, Chabir N, et al. Relation between chemical composition or antioxidant activity and antihypertensive activity for six essential oils. J Food Sci 2012;77(8):H184-H191

44 Ramos C, Teixeira B, Batista I, Matos O, Serrano C, Neng NR, et al. Antioxidant and antibacterial activity of essential oil and extracts of bay laurel *Laurus nobilis* Linnaeus (Lauraceae) from Portugal. Nat Prod Res 2012;26:518-529

45 Tayoub G, Odeh A, Ghanem I. Chemical composition and fumigation toxicity of Laurus nobilis L. and Salvia officinalis L. essential oils on larvae of khapra beetle (Trogoderma granarium Everts). Herba Polonica 2012;58(2):26-37

46 Saab AM, Tundis R, Loizzo MR, Lampronti I, Borgatti M, Gambari R, et al. Antioxidant and antiproliferative activity of *Laurus nobilis* L. (Lauraceae) leaves and seeds essential oils against K562 human chronic myelogenous leukaemia cells. Nat Prod Research 2012;26:1741-1745

47 Marzouki H, Khaldi A, Chamli R, Bouzid S, Piras A, Falconieri D, et al. Biological activity evaluation of the oils from *Laurus nobilis* of Tunisia and Algeria extracted by supercritical carbon dioxide. Nat Prod Research 2009;23:230-237

48 Misharina TA, Polshkov AN. Antioxidant properties of essential oils: autoxidation of essential oils from laurel and fennel and of their mixtures with essential oil from coriander. Appl Biochem Microbiol 2005;41:610-618

49 Kosar M, Tunalier Z, Özek T, Kürkcüoglu M, Can Baser KH. A simple method to obtain essential oils from *Salvia triloba* L. and *Laurus nobilis* L. by using microwave-assisted hydrodistillation. Z Naturforsch C 2005;60:501-504

50 Shokoohinia Y, Yegdaneh A, Amin G, Ghannadi A. Seasonal variations of *Laurus nobilis* L. leaves volatile oil components in Isfahan, Iran. Res J Pharmacogn 2014;1(3):1-6

51 Cherrat L, Espina L, Bakkali M, Garcıa-Gonzalo D, Pagan R, Laglaoui A. Chemical composition and antioxidant properties of *Laurus nobilis* L. and *Myrtus communis* L. essential oils from Morocco and evaluation of their antimicrobial activity acting alone or in combined processes for food preservation. J Sci Food Agric 2014;94:1197-1204

52 Xu S, Yan F, Ni Z, Chen Q, Zhang H, Zheng X. *In vitro* and *in vivo* control of *Alternaria alternata* in cherry tomato by essential oil from *Laurus nobilis* of Chinese origin. J Sci Food Agric 2014;94:1403-1408

53 Traboulsi AF, El-Haj S, Tueni M, Taoubi K, Nader NA, Mrad A. Repellency and toxicity of aromatic plant extracts against the mosquito *Culex pipiens molestus* (Diptera: Culicidae). Pest Manag Sci 2005;61:597-604

54 Biondi D, Cianci P, Gerad C, Ruberto G, Piattelli M. Antimicrobial activity and chemical composition of essential oils from Sicilian aromatic plants. Flavour Fragr J 1993;8:331-337

55 Isikber AA, Alma MH, Kanat M, Karci A. Fumigant toxicity of essential oils from *Laurus nobilis* and *Rosmarinus officinalis* against all life stages of *Tribolium confusum*. Phytoparasitica 2006;34:167-177

56 Ozek T, Bozan B, Baser KHC. Supercritical CO_2 extraction of volatile components from leaves of *Laurus nobilis* L. Chem Nat Comp 1998;34:668-671

57 Damiani N, Gende LB, Bailac P, Marcangeli JA, Eguaras MJ. Acaricidal and insecticidal activity of essential oils on *Varroa destructor* (Acari: Varroidae) and *Apis mellifera* (Hymenoptera: Apidae). Parasitol Res 2009;106:145-152

58 Soylu EM, Soylu S, Kurt S. Antimicrobial activities of the essential oils of various plants against tomato late blight disease agent *Phytophthora infestans*. Mycopathologia 2006;161:119-128

59 Basak SS, Candan F. Effect of *Laurus nobilis* L. essential oil and its main components on α-glucosidase and reactive oxygen species scavenging activity. Iran J Pharm Res 2013;12:367-379

60 Mello da Silveira S, Cunha A Júnior, Neudí Scheuermann G, Secchi FL, Werneck Vieiral CR. Chemical composition and antimicrobial activity of essential oils from selected herbs cultivated in the South of Brazil against food spoilage and foodborne pathogens. Ciência Rural 2012;42:1300-1306

61 Verdian-rizi M. Variation in the essential oil composition of *Laurus nobilis* L. of different growth stages cultivated in Iran. J Basic Appl Sci 2009;5:33-36

62 El SN, Karagozlu N, Karakaya S, Sahın S. Antioxidant and antimicrobial activities of essential oils extracted from *Laurus nobilis* L. leaves by using solvent-free microwave and hydrodistillation. Food Nutr Sci 2014;5:97-106

63 Quijano CE, Pino J. Characterization of the leaf essential oil from laurel (*Laurus nobilis* L.) grown in Colombia. Revista CENIC Ciencias Quimicas 2007;38:371-374

64 Ivanović J, Mišić D, Ristić M, Pešić O, Žižović I. Supercritical CO_2 extract and essential oil of bay (*Laurus nobilis* L.) − chemical composition and antibacterial activity. J Serb Chem Soc 2010;75:395-404

65 Marzouka H, Elaissi A, Khaldi A, Bouzid S, Falconieri D, Marongiu B, et al. Seasonal and geographical variation of *Laurus nobilis* L. essential oil from Tunisia. The Open Natural Products Journal 2009;2:86-91

66 Moghtader M, Farahmand A. Evaluation of the antibacterial effects of essential oil from the leaves of *Laurus nobilis* L. in Kerman Province. J Microbiol Antimicrobials 2013;5(2):13-17

67 Derwich E, Benziane Z, Boukir A. Chemical composition and antibacterial activity of leaves essential oil of *Laurus nobilis* from Morocco. Austr J Basic Appl Sci 2009;3:3818-3824

68 Politeo O, Jukic M, Milos M. Chemical composition and antioxidant activity of free volatile aglycones from laurel (*Laurus nobilis* L.) compared to its essential oil. Croatica Chemica Acta 2007;80:121-126

69 Dahak K, Bouamama H, Benkhalti F, Taourirte M. Drying methods and their implication on quality, quantity and antimicrobial activity of the essential oil of *Laurus nobilis* L. from Morocco. OnLine J Biol Sci 2014;14:94-101

70 Rhind JP. Essential oils. A handbook for aromatherapy practice, 2nd Edition. London: Singing Dragon, 2012

71 Lawless J. The encyclopedia of essential oils, 2nd Edition. London: Harper Thorsons, 2014

72 Nardelli A, D'Hooge E, Drieghe J, Dooms M, Goossens A. Allergic contact dermatitis from fragrance components in specific topical pharmaceutical products in Belgium. Contact Dermatitis 2009;60:303-313

73 Wöhrl S, Hemmer W, Focke M, Götz M, Jarisch R. The significance of fragrance mix, balsam of Peru, colophonium and propolis as screening tools in the detection of fragrance allergy. Br J Dermatol 2001;145:268-273

74 Athanasiadis GI, Pfab F, Klein A, Braun-Falco M, Ring J, Ollert M. Erythema multiforme due to contact with laurel oil. Contact Dermatitis 2007;57:116-118

75 Bleasel N, Tate B, Rademaker M. Allergic contact dermatitis following exposure to essential oils. Australas J Dermatol 2002;43:211-213

76 Adişen E, Önder M. Allergic contact dermatitis from *Laurus nobilis* oil induced by massage. Contact Dermatitis 2007;56:360-361

77 Schaller M, Korting HC. Allergic airborne contact dermatitis from essential oils used in aromatherapy. Clin Exp Dermatol 1995;20:143-145

78 Hausen BM. Lorbeer-Allergie Ursache, Wirkung und Folgen der äusserlichen Anwendung eines sogenannten Naturheilmittels. Dtsch Med Wschr 1985;110: 634-638

79 Özden MG, Öztaş P, Öztaş MO, Önder M. Allergic contact dermatitis from *Laurus nobilis* (laurel) oil. Contact Dermatitis 2001;45:178

80 De Groot AC. Airborne allergic contact dermatitis from tea tree oil. Contact Dermatitis 1996;35:304-305

81 Uter W, Schmidt E, Geier J, Lessmann H, Schnuch A, Frosch P. Contact allergy to essential oils: current patch test results (2000–2008) from the Information Network of Departments of Dermatology (IVDK). Contact Dermatitis 2010;63:277-283

82 Katsarma G, Gawkrodger DJ. Suspected fragrance allergy requires extended patch testing to individual fragrance allergens. Contact Dermatitis 1999;41:193-197

83 Sainio E-L, Kanerva L. Contact allergens in toothpastes and a review of their hypersensitivity. Contact Dermatitis 1995;33:100-105

84 Spier HW, Sixt I. Lorbeer als Träger eines wenig beachteten kontaktekzemotogenen Allergens. Derm Wochenschr 1953;128:805-810

85 Calnan CD. Oil of cloves, laurel, lavender, peppermint. Contact Dermatitis Newsletter 1970;7:148

86 Hausen BM. A 6-year experience with Compositae mix. Am J Cont Derm 1996;7:94-99

87 Larrègue M, Rat JP, Gallet P, Bressieux JM, Pousset JL. Contact dermatitis caused by dandelion, laurel oil and frullania by cross-allergy. Ann Dermatol Venereol 1978;105:547-548

88 Foussereau J. L'allergie de contact à l'huile de laurier (note préliminaire). Bull Soc Fr Dermatol Syphiligr 1963;70:698-701

89 Bandmann HJ, Dohn W. Laurel oil as a frequent cause of allergic contact eczema. Munch Med Wochenschr 1960;102:680-682 (in German)

90 Foussereau J, Benezra C, Ourisson G. Contact dermatitis from laurel. I. Clinical aspects. Transactions of the St John's Hospital Dermatological Society 1967;53:141-146. Data cited in ref. 92

91 Toholka R, Wang Y-S, Tate B, Tam M, Cahill J, Palmer A, Nixon R. The first Australian Baseline Series: Recommendations for patch testing in suspected contact dermatitis. Australas J Dermatol 2014, Sept. 7. doi: 10.1111/ajd.12186

92 Tisserand R, Young R. Essential oil safety, 2nd Ed. Edinburgh: Churchill Livingstone, 2014:322

93 Benezra C, Schlewer G, Stampf JL. Lactones allergisantes naturelles et synthetiques. Rev Fr Allergol 1978;18:31-33. Data cited in ref. 92

Chapter 5.46   LAVANDIN ABRIAL OIL

DEFINITION

Lavandin abrial oil (essential oil of lavandin Abrial, French type) is the essential oil obtained from the flowering tops of the lavandin (bastard lavender), *Lavandula angustifolia* Mill. x *Lavandula latifolia* Medik. 'Abrial'.

INCI NOMENCLATURE

Description/definition: Lavandula hybrida abrial herb oil is an essential oil distilled from the flowering herbs of the lavandin, *Lavandula hybrida* var. *abrial*, Labiatae
INCI name EU: Lavandula hybrida abrial herb oil (perfuming name, not an INCI name proper)
INCI name USA: Not in the Personal Care Products Council Ingredient Database
CAS registry number(s): 8022-15-9; 93455-96-0
EINECS number(s): 297-384-7

ISO (INTERNATIONAL ORGANIZATION FOR STANDARDIZATION) STANDARD

ISO number: 3054
ISO name: Essential oil of lavandin Abrial, French type
Botanical origin: *Lavandula angustifolia* Mill. x *Lavandula latifolia* Medik. 'Abrial'
Parts of plant used: Flowering top
ISO values: ISO values (minimum and maximum concentrations) are shown in Table 5.46.1.

THE PLANT, THE OIL, AND THEIR USES

The lavandins (*Lavandula x intermedia* Emeric ex Loisel; *Lavandula hybrida*) are a class of sterile hybrids of the true (syn. English) lavender, *Lavandula angustifolia* P. Mill. and the spike lavender, *Lavandula latifolia* Medikus. They are perennial herbs, which may reach a height of 60 cm. The lavandin is native to the Mediterranean area and is widely cultivated for production of essential oils and as a decorative plant in Spain, France, Italy,

Table 5.46.1 ISO values (%) for lavandin abrial oil [a]

Compound	CAS	Minimum	Maximum
Linalool (β-)	78-70-6	26.0	38.0
Linalyl acetate	115-95-7	20.0	29.0
Camphor (*dl*-)	76-22-2	7.0	11.0
1,8-Cineole	470-82-6	6.0	11.0
(*E*)-β-Ocimene	3779-61-1	3.0	7.0
Borneol	507-70-0	1.5	3.5
(*Z*)-β-Ocimene	3338-55-4	1.5	3.0
Lavandulyl acetate	25905-14-0	1.0	2.0
Limonene	138-86-3	0.5	1.5
Lavandulol	498-16-8	0.4	1.2
Terpinen-4-ol	562-74-3	0.3	1.0

[a] ISO 3054 Essential oil of lavandin abrial French type ©ISO 2001; Geneva, Switzerland, www.iso.org

the Balkan Peninsula, Morocco, South Africa, India, Argentina, Australia and Tasmania (2,7,31). In recent years Bulgaria has become a major producer (30,31). The main lavandin cultivars are 'Grosso', 'Abrial' and 'Super' (2,7). In 2002, 'Grosso' accounted for 84% of lavandin production world-wide, 'Abrial' for 8%, 'Super' for 5% and others for 3% (31).

Essential oils are present in the inflorescences of the lavandin, mostly in the calyces of the florets. They are produced commercially by steam distillation from selected cultivars. Lavandin oils are used as fragrances in perfumes, soaps, detergents, cosmetics and air fresheners, and are also employed by the pharmaceutical industries (1,2,8,30,31). In food manufacturing, lavandin essential oil may be used in flavoring beverages, ice cream, baked goods and chewing gum (8,30,31). The oil is believed to combat halitosis, to be antiseptic, antibacterial, antifungal, antispasmodic, carminative, sedative, cholagogue, diuretic, antidepressant, stimulative, anti-inflammatory, healing, and effective for burns and insect bites (2,8,30). These perceived properties render lavandin essential oils highly appreciated in phytotherapy and aromatherapy (8,30). The main components of the essential oil of lavandin are linalyl acetate and linalool. The industrial uses of linalyl acetate and linalool are different. Linalyl acetate is highly appreciated as a food additive because of its flavor, while linalool is used as a natural insecticide or pesticide as well as in food and fragrance applications (9).

Lavandins are cultivated in far (5:1) larger quantities than lavenders (30), because they produce more oil (31), are easier to harvest and are hardier. Lavender and lavandin oils have the same components but in different proportions. The criteria for determining the quality are the levels of linalyl acetate (high), linalool (high) and camphor (low). The scent of the oil deteriorates as the camphor content increases (28). Lavandin oils have a higher content of camphor and lower concentrations of linalyl acetate and linalool and are therefore considered to be of lesser quality (and cheaper) than lavender oils. As a consequence, lavandin oils are commonly blended, either with lavender oil or other commercial essential oils, to create a pleasing fragrance (2,3). Nowadays, lavandin oil is among the top ten most important essential oils by volume (31). A review of various aspects of lavenders (lavender, lavandin, spike lavender), written from the industry's perspective, can be found (on-line) in a dissertation from the International Centre for Aroma Trades Studies, Plymouth, UK (46).

CHEMICAL COMPOSITION

Lavandin abrial oil is a slightly yellow, clear mobile liquid which has an herbal, blooming and camphoraceous odor. The yield of essential oil from the flowering tops of *Lavandula angustifolia* Mill. x *Lavandula latifolia* Medik. 'Abrial' generally varies from 1.5 to 5.0 per cent, depending on the dryness of the plant material. The main producing country of this oil is France.

Literature data (up to December 1, 2014) on the chemical composition of lavandin abrial oils and unpublished analytical data from one of us (E.S.) are shown in Table 5.46.2 in alphabetical order. In lavandin abrial oils from various origins, over 88 chemicals have been identified. About 45 per cent of these were found in a single reviewed publication only. Probably, lavandin abrial oils may also contain many of the chemicals mentioned in Table 5.48.2 (constituents of other cultivars, or cultivar used to obtain the oil unknown). These are not mentioned here, as, although a number of studies presented in table 5.48.2 probably examined lavandin

abrial oils, they were not specifically identified as such. In addition, many studies on lavandin abrial oils have investigated the differences between the oils from cultivars 'Abrial', 'Grosso' and 'Super', thereby focusing on the main constituents only.

The major compounds found in lavandin abrial oils from different sources are shown in Table 5.46.3. They include (highest concentrations in any study given) linalool (56.2%), linalyl acetate (36.2%), camphor (13.3%), 1,8-cineole (12.0%), β-caryophyllene (6.0%), borneol (4.2%), (E)-β-ocimene (3.8%), (Z)-β-ocimene (3.7%) and lavandulyl acetate (2.9%). A well-known ingredient of

Table 5.46.2 Constituents identified in lavandin abrial oils

Constituent	CAS	Percentage and range				
		A	B (1)	C (27)	D (43)	E
Acetone	67-64-1				0.02	
trans-α-Bergamotene	13474-59-4	0.07-0.2				
β-Bergamotene	6895-56-3				0.4	
Bisabolene	495-62-5				0.05	
β-Bisabolene	495-61-4	0.08-0.2				
α-Bisabolol	515-69-5	0-0.04				
β-Bisabolol	15352-77-9				0.2	
Borneol	507-70-0	2.3-4.2	2.4-3.1	2.5-2.6	3.7	1.9[g]; 2.3[a]; 2.6[h]; 3.1[f,h]
Bornyl acetate	76-49-3	0.01-0.2				
Butyl butanoate	109-21-7				0.03	
Butyl isobutanoate	97-87-0				tr	
Butyl tiglate	7785-66-2					
γ-Cadinene	39029-41-9				0.3	
δ-Cadinene	483-76-1	0.1-0.3				
τ-Cadinol	5937-11-1	0.1-0.2			0.2	
Camphene	79-92-5	0.3-0.5			0.6	0.2-0.9[d]; 0.4[c]; 0.5[h]
Camphenilone	13211-15-9				tr	
Camphor	76-22-2	7.6-9.9	10.2-11.5	9.1-11.4	12.2	5.6-10.6[d]; 8.2[e]; 8.9[i]; 9.0[g]; 9.5[h]; 10.0[c]; 11.5[h]; 11.7[a]; 13.3[h]
δ3-Carene	13466-78-9				0.1	0.02[h]
β-Caryophyllene	87-44-5	1.4-3.7			6.0	1.9[b]; 2.3[a]; 2.4[h]; 2.6[h]; 3.1[i]
Caryophyllene oxide	1139-30-6	0-tr			0.1	
1,8-Cineole	470-82-6	5.4-9.9	9.6-10.4	7.4-8.0	10.3	6.8[f]; 8.2[h]; 8.6[e]; 8.8[b]; 9.2[i]; 10.4[h]; 12.0[h]
Cryptone	500-02-7				0.05	
Coumarin	91-64-5	0.04-0.5				
Cubenol	21284-22-0				tr	
Cuminaldehyde	122-03-2				tr	
p-Cymene	99-87-6	0.09-0.3			0.2	0.5[h]
p-Cymen-8-ol	1197-01-9				0.05	
Epoxylinalyl acetate	41610-76-8				0.09	
β-Farnesene	502-60-3				1.1	
(E)-β-Farnesene	18794-84-8	0.09-1.2				
(Z)-β-Farnesene	28973-97-9					0.8[i]
Fenchone	1195-79-5				0.01	
Geraniol	106-24-1	0.1-0.3			0.1	0.8[h]
Geranyl acetate	105-87-3	0.2-0.3			1.2	0.4[h]; 1.9[h]
Germacrene D	23986-74-5	0.1-0.5			1.2	0.5[h]
Hexanol (1-; n-)	111-27-3	0.04-0.1			0.05	0.3[h]
2-Hexenal	505-57-7				tr	
(Z)-3-Hexenol	928-96-1	0.01-0.06				
n-Hexyl acetate	142-92-7	0.08-0.4			0.2	
Hexyl butanoate	2639-63-6	0.2-0.8			0.3	0.4[h]
Hexyl isobutyrate	2349-07-7	0.09-0.3			0.1	0.3[h]
Hexyl isovalerate	10032-13-0	0-0.03				0.5[h]
Hexyl 2-methylbutyrate	10032-15-2	0.09-0.2			0.1	
Hexyl propanoate	2445-76-3				0.06	
Hexyl tiglate	16930-96-4	0.2-0.9			0.2	

Table 5.46.2 Constituents identified in lavandin abrial oils (*continued*)

Constituent	CAS	Percentage and range				
		A	B (1)	C (27)	D (43)	E
Hotrienol	20053-88-7				tr	
α-Humulene	6753-98-6	0.2-0.4			0.4	
Isofenchone	6541-58-8				0.09	
Lavandulol	498-16-8	0.03-0.2		0.4-0.6	0.6	1.0[f,h]; 1.1[h]
Lavandulyl acetate	25905-14-0	0.4-0.6	1.1	1.3-1.7	2.6	0.7[g]; 1.1[h]; 1.5[i]; 1.8[h]; 2.9[b,f]
Lavandulyl butyrate	59550-35-5				tr	
Lavandulyl isobutyrate	51117-20-5					0.6[h]
Limonene	138-86-3	0.7-1.1			1.5	0.7[h]; 1.0[h]; 2.5[i]; 14.1[a]
Linalool	78-70-6	35.1-56.2	39.3-39.6	31.7-33.9	19.6	26.7-43.4[d]; 31.1[b,i]; 33.5[h]; 35.4[c]; 37.9[f]; 39.6[h]; 41.1[g]; 45.1[h]
cis-Linalool oxide	11063-77-7	0.1-0.3			0.1	0.1[h]
trans-Linalool oxide	11063-78-8	0.01-0.1			0.3	0.2[h]
Linalyl acetate	115-95-7	14.9-26.1	22.1-23.2	24.0-25.8	18.6	20.5-35.5[d]; 22.1[h]; 22.3[f]; 23.0[i]; 24.7[g]; 26.2[a]; 27.1[h]; 35.4[c]; 36.2[h]
Methoxyhexane	4747-07-3	0.02-0.1			0.06	
Myrcene	123-35-3	0.5-0.6			1.2	0.6[h]; 2.4[h]
Nerol	106-25-2				tr	1.2[h]
Neryl acetate	141-12-8	0.09-0.2			0.2	0.3[h]
(*E*)-β-Ocimene	3779-61-1	1.4-3.3			4.2	3.0[h]; 3.8[i]
(*Z*)-β-Ocimene	3338-55-4	1.0-3.7			2.6	0.1[a]; 1.8[i]; 2.3[h]
Octanal	124-13-0				tr	
n-Octanol (1-)	111-87-5	0-tr				
3-Octanol	589-98-0				tr	0.3[h]
3-Octanone	106-68-3	0.09-0.2			0.1	3.0[h]
1-Octen-3-ol	3391-86-4	0.1-0.2			0.3	0.3[h]
1-Octen-3-yl acetate	2442-10-6	0.08-0.3			0.5	0.5[h]
3-Octyl acetate	4864-61-3				tr	
α-Phellandrene	99-83-2	0-0.06			0.07	
α-Pinene	80-56-8	0.3-0.9			0.9	0.2-1.1[d]; 0.5[h]; 0.6[c]; 0.8[h]
β-Pinene	127-91-3	0.5-1.9			0.9	0.5[h]; 1.1[h]
Sabinene	3387-41-5	0.06-0.3			0.4	0.2[h]
Sabinene hydrate	546-79-2				0.2	
cis-Sabinene hydrate	15537-55-0	0.08-0.1				
trans-Sabinene hydrate	17699-16-0	tr-0.06				
α-Santalene	512-61-8	0.1-0.2			0.7	
β-Santalene	511-59-1				tr	
α-Terpinene	99-86-5				0.09	
γ-Terpinene	99-85-4	0.08-0.3			0.3	0.4[h]
Terpinen-4-ol	562-74-3	0.2-0.9	0.3-0.4	0.6-0.8	1.2	0.2[a]; 0.4[h]; 0.6[h]; 0.7[g]; 3.8[f]
α-Terpineol	98-55-5	0.4-0.8		0.5-0.6	1.0	0.03[a]; 0.5[h]; 1.3[f]; 4.6[h]
Terpinolene	586-62-9	0.08-0.3			0.5	0.3[h]
α-Thujene	2867-05-2	0-0.05			0.07	
Tricyclene	508-32-7	0.01-0.07			0.01	
5,5,6-Trimethylbicyclo-[2.2.1]heptan-2-one	3292-05-5				0.1	
(3*E*,5*Z*)-1,3,5-Undecatriene	19883-27-3				tr	

A one hundred and ten lavandin abrial essential oil samples from France investigated between 1998 and 2014; lowest and highest concentrations given (E. Schmidt, unpublished data)

B seventeen steam-distilled oil samples from flowers harvested in New Zealand in 1986 (n=8) and 1987 (n=9); lowest and highest mean concentrations for the two years given (ref. 1)

C unknown number of commercial lavandin abrial oils investigated in Argentina (ref. 27)

D one oil sample from lavandin 'Abrialis' (ref. 43)

E data from other studies (indicated with superscript letters); highest concentrations found in any study reviewed here given; when two or more oils were investigated, only the highest concentrations are mentioned, unless indicated otherwise

[a] one steam-distilled oil from Spain (ref. 33); [b] one oil from lavandin 'Abrial' flowers grown in Norway (ref. 3); [c] ten oil samples from France (ref. 12); [d] two hundred commercial lavandin abrial oils from France; lowest and highest concentrations given (ref. 12); [e] one lab-hydrodistilled oil from Italy (ref. 15); [f] one lavandin abrial oil prepared in Argentina (ref. 27); [g] one commercial oil (ref. 1); [h] data from various older (<1995) studies cited in ref. 42; [i] one commercial oil from France (ref. 23)

tr: trace

Table 5.46.3 Major constituents of lavandin abrial oils

Constituent	CAS	Percentage and range				
		A	B (1)	C (27)	D (43)	E
Linalool	78-70-6	35.1-56.2	39.3-39.6	31.7-33.9	19.6	33.5[h]; 35.4[c]; 37.9[f]; 39.6[h]; 41.1[g]; 45.1[h]
Linalyl acetate	115-95-7	14.9-26.1	22.1-23.2	24.0-25.8	18.6	23.0[i]; 24.7[g]; 26.2[a]; 27.1[h]; 35.4[c]; 36.2[h]
Camphor	76-22-2	7.6-9.9	10.2-11.5	9.1-11.4	12.2	8.9[i]; 9.0[g]; 9.5[h]; 10.0[c]; 11.5[h]; 11.7[a]; 13.3[h]
1,8-Cineole	470-82-6	5.4-9.9	9.6-10.4	7.4-8.0	10.3	6.8[f]; 8.2[h]; 8.6[e]; 8.8[b]; 9.2[i]; 10.4[h]; 12.0[h]
β-Caryophyllene	87-44-5	1.4-3.7			6.0	1.9[b]; 2.3[a]; 2.4[h]; 2.6[h]; 3.1[i]
Borneol	507-70-0	2.3-4.2	2.4-3.1	2.5-2.6	3.7	1.9[g]; 2.3[a]; 2.6[h]; 3.1[f,h]
(E)-β-Ocimene	3779-61-1	1.4-3.3			4.2	3.0[h]; 3.8[i]
(Z)-β-Ocimene	3338-55-4	1.0-3.7			2.6	0.1[a]; 1.8[i]; 2.3[h]
Lavandulyl acetate	25905-14-0	0.4-0.6	1.1	1.3-1.7	2.6	0.7[g]; 1.1[h]; 1.5[i]; 1.8[h]; 2.9[b,f]

LEGEND: SEE UNDER TABLE 5.46.2

lavandin abrial oils that was present in a high concentration (>7%) in one study was limonene (14.1%).

Commercial oils

The ten chemicals that had the highest maximum concentrations in 110 commercial lavandin abrial essential oil samples (concentration ranges provided) are the following: linalool (35.1-56.2%), linalyl acetate (14.9-26.1%), 1,8-cineole (5.4-9.9%), camphor (7.6-9.9%), borneol (2.3-4.2%), (Z)-β-ocimene (1.0-3.7%), β-caryophyllene (1.4-3.7%), (E)-β-ocimene (1.4-3.3%), β-pinene (0.5-1.9%) and (E)-β-farnesene (0.09-1.2%) (Erich Schmidt, unpublished analytical data).

Data on lavandin 'Abrial' (essential) oils in ref. 41 are not presented in Table 5.46.2, as it included those of an oil sample obtained by supercritical fluid extraction.

CONTACT ALLERGY/ALLERGIC CONTACT DERMATITIS

No reports on contact allergy to or allergic contact dermatitis from lavandin oil, specified to have been obtained from *Lavandula angustifolia* Mill. x *Lavandula latifolia* Medik. 'Abrial', were found. The literature on contact allergy to 'lavandin oil' (botanical source not specified) is discussed in Chapter 5.48 Lavandin oil.

LITERATURE

For Literature references see Chapter 5.48 Lavandin oil.

Chapter 5.47   LAVANDIN GROSSO OIL

DEFINITION

Lavandin grosso oil (essential oil of lavandin grosso, French type) is the essential oil obtained from the flowering tops of the lavandin (bastard lavender), *Lavandula angustifolia* Mill. x *Lavandula latifolia* Medik. 'Grosso'.

INCI NOMENCLATURE

Description/definition: Lavandula hybrida grosso herb oil is an essential oil distilled from the flowering herbs of the lavandin, *Lavandula hybrida grosso*, Labiatae
INCI name EU: Lavandula hybrida grosso herb oil (perfuming name, not an INCI name proper)
INCI name USA: Not in the Personal Care Products Council Ingredient Database
CAS registry number(s): 8022-15-9; 93455-97-1
EINECS number(s): 297-385-2

ISO (INTERNATIONAL ORGANIZATION FOR STANDARDIZATION) STANDARD

ISO number: 8902
ISO name: Essential oil of lavandin grosso, French type
Botanical origin: *Lavandula angustifolia* Mill. x *Lavandula latifolia* Medik. 'Grosso'
Parts of plant used: Flowering top
ISO values: ISO values (minimum and maximum concentrations) are shown in Table 5.47.1.

THE PLANT, THE OIL, AND THEIR USES

This section is discussed in Chapter 46 Lavandin abrial oil.

CHEMICAL COMPOSITION

Lavandin grosso oil is a light yellow, clear mobile liquid which has a fresh, herbal and floral, camphoraceous odor. The yield of essential oil from the flowering tops of *Lavandula angustifolia* Mill. x *Lavandula latifolia* Medik. 'Grosso' varies from 7.0 to 9.5 per cent. The main producing country of this oil is France.

Literature data (up to December 1, 2014) on the chemical composition of lavandin grosso oils and unpublished analytical data from one of us (E.S.) are shown in Table 5.47.2 in alphabetical order. In lavandin grosso oils from various origins, over 100 chemicals have been identified. About 44 per cent of these were found in a single reviewed publication only. Probably, lavandin grosso oils may also contain many of the chemicals mentioned in Table 5.48.2 (constituents of other cultivars, or cultivar used to obtain the oil unknown). These are not mentioned here, as, although a number of studies presented in Table 5.48.2 probably have examined lavandin grosso oils, they were not specifically identified as such. In addition, many studies on lavandin grosso oils have investigated the differences between the oils from cultivars 'Abrial', 'Grosso' and 'Super', thereby focusing on the main constituents only.

The major compounds found in lavandin grosso oils from different sources are shown in Table 5.47.3. They include (highest concentrations in any study given) linalool (47.5%), linalyl acetate (37.5%), camphor (12.2%), 1,8-cineole (10.9%), terpinen-4-ol (5.5%), borneol (4.1%), α-terpineol (3.9%) and lavandulyl acetate (3.7%). A well-known ingredient of lavandin grosso oils that was present in high concentrations (>7%) in one study was myrcene (8.5%).

Commercial oils

The ten chemicals that had the highest maximum concentrations in 148 commercial lavandin grosso essential oil samples (concentration ranges provided) are the following: linalool (26.0-38.6%), linalyl acetate (23.9-36.0%), 1,8-cineole (3.0-9.5%), camphor (5.9-9.0%), terpinen-4-ol (2.0-5.5%), borneol (1.3-3.7%), lavandulyl acetate (1.5-3.2%), β-caryophyllene (1.0-2.5%), (*E*)-β-farnesene (0.6-1.7%) and α-terpineol (0.3-1.7%) (Erich Schmidt, unpublished analytical data).

Data on lavandin 'Grosso' (essential) oils in ref. 41 are not presented in Table 5.47.2, as they include those of an oil sample obtained by supercritical fluid extraction.

Differences between commercial 'Grosso', 'Abrialis' and 'Super' lavandin oils

Comparative compositions of commercial lavandin oils can be found in ref. 44 and on page 41 of ref. 31 (which can be accessed on-line). ISO/AFNOR values also offer important comparative material as do the data of one of us (E.S.), which are shown in Tables 5.46.1 and 5.47.1 (no separate data for 'Super'). Generally speaking, the contents of linalool (preferably high), linalyl acetate (preferably highest) and camphor (preferably low) determine the quality of the oil. Again in general terms, 'Super' has the highest linalyl acetate concentrations (30-45%), followed by 'Grosso', while 'Abrial' has the lowest amounts (19-28%). With linalool, concentrations in 'Super' oils may be quite constant around 30%, its concentrations ranging

Table 5.47.1 ISO values (%) for lavandin grosso oil [a]

Compound	CAS	Minimum	Maximum
Linalyl acetate	115-95-7	25.0	38.0
Linalool (β-)	78-70-6	24.0	37.0
Camphor (*dl*-)	76-22-2	6.0	8.5
1,8-Cineole	470-82-6	4.0	8.0
Terpinen-4-ol	562-74-3	1.5	5.0
Borneol	507-70-0	1.5	3.5
Lavandulyl acetate	25905-14-0	1.5	3.5
Limonene	138-86-3	0.5	1.5
(*Z*)-β-Ocimene	3338-55-4	0.5	1.5
α-Terpineol	98-55-5	0.3	1.3
Myrcene	123-35-3	0.3	1.0
Lavandulol	498-16-8	0.2	1.0
(*E*)-β-Ocimene	3779-61-1	0.0	1.0
Hexyl butanoate (butyrate)	2639-63-6	0.3	0.5

[a] ISO 8902 Essential oil of Lavandin grosso, French type ©ISO 2009; Geneva, Switzerland, www.iso.org

Table 5.47.2 Constituents identified in lavandin grosso oils

Constituent	CAS	Percentage and range				
		A	B (8)	C (7)	D (43)	E
Acetone	67-64-1				tr	
α-Amorphene	20085-19-2		0.05-0.4			
trans-α-Bergamotene	13474-59-4	0.1-0.2				
β-Bergamotene	6895-56-3				tr	
Bisabolene	495-62-5					
(E)-α-Bisabolene	25532-79-0					<0.1°; 0.1-0.3[m]
β-Bisabolene	495-61-4	0.09-1.2	0.2-0.3			<0.1°; 0.1-0.2[m]
α-Bisabolol	515-69-5	0.2-0.5	0.5-3.9			<0.1°; tr-0.1[m]; 0.2[b]
β-Bisabolol	15352-77-9				0.4	
Borneol	507-70-0	1.3-3.7	0.6-3.5	1.6-2.9	2.9	2.4-3.5[m]; 2.8[l]; 2.9[p]; 3.1[p]; 3.2[q]; 3.4[a]; 4.1[n]
Bornyl acetate	76-49-3	0.01-0.09	0.2			
Butyl butanoate	109-21-7	tr-0.04			0.04	
Butyl isobutanoate	97-87-0				0.01	
Butyl tiglate	7785-66-2				0.1	
γ-Cadinene	39029-41-9	0.02-0.6			0.3	0.3-0.7[m]
α-Cadinol	481-34-5		0.3-1.7			
γ-Cadinol	50895-55-1	0.08-0.3				
τ-Cadinol	5937-11-1				0.2	tr-1.1[m]
Camphene	79-92-5	0.2-0.6	0.04-0.2		0.3	<0.1°; 0.1[b]; 0.3-0.5[m]; 0.4[p]; 0.5[h]
Camphor	76-22-2	5.9-9.0	4.3-6.9	6.3-8.8	12.2	7.6-9[m]; 7.8[p]; 7.9[a]; 8.1[f]; 8.9[p]; 9.7[l]; 10.8[n]
δ3-Carene	13466-78-9	0.03-0.1			0.05	<0.1°; 0.08[p]; tr-0.3[m]
Carvone	99-49-0		0.4-0.8			
β-Caryophyllene	87-44-5	1.0-2.5		1.4-2.3	2.7	0.5-1.5[m]; 0.6[b]; 1.5°; 1.8[h]; 1.9[f,i]; 2.3[e]
Caryophyllene oxide	1139-30-6	0.04-0.1			0.1	<0.1°; tr-0.4[m]; 0.6[b]
1,8-Cineole	470-82-6	3.0-9.5	3.2-4.7	4.0-8.7	10.2	6-10[m]; 6.7[e]; 6.8 [q]; 7.2[p]; 7.4[h]; 10.7[f]; 10.9[c]
Coumarin	91-64-5	0.04-0.2				<0.1°; tr-0.2[m]
Cryptone	500-02-7					+[z3]; 0.6[c,d]; 1.3[h]
Cubenol	21284-22-0				tr	
Cuminaldehyde	122-03-2				0.02	
o-Cymene	527-84-4	tr-0.02				
p-Cymene	99-87-6	0.01-0.4			0.2	<0.1°; 0.1-0.4[m]; 0.4[p]
p-Cymen-8-ol	1197-01-9				0.02	tr-0.1[m]; 0.6[c,d]
Epoxylinalyl acetate	41610-76-8				0.03	
α-Farnesene	502-61-4					0.6[b]
(E,E)-α-Farnesene	502-61-4					<0.1°
β-Farnesene	502-60-3				1.1	
(E)-β-Farnesene	18794-84-8	0.6-1.7				0.7-0.8[m]
(Z)-β-Farnesene	28973-97-9		0.3-0.6			1.1°; 1.1-1.5[m]; 1.6[i]
α-Fenchene	471-84-1	0.02-0.05				
Fenchone	1195-79-5				0.01	
β-Funebrene	79120-98-2		0.2-0.3			
Geraniol	106-24-1	0.02-0.5			0.2	0.5-1.1[m]; 0.7[h]; 0.8°
(E)-Geraniol (β-)	106-24-1		0.2-1.7			
Geranyl acetate	105-87-3	0.1-0.7	0.03-1.6		1.2	0.3[p]; 0.6°; 0.8-1.3[m]; 1.4[b,p]; 1.5[h]
Germacrene D	23986-74-5	0.3-0.8			1.1	0.2-0.5[m]; 0.6°
Hexanol (1-; n-)	111-27-3	0.03-0.3			0.02	<0.1°
(Z)-3-Hexen-1-ol	928-96-1	0.01-0.1				<0.1°
Hexyl acetate	142-92-7	tr-0.7			0.2	tr-0.1[m]; 0.1[b]; 0.5°
Hexyl butyrate	2639-63-6	0.2-0.6			0.3	tr-0.4[m]; 0.3°; 0.4[b,p]
Hexyl isobutyrate	2349-07-7	0.09-0.3			0.04	<0.1°; tr-0.2[m]; 0.2[b,p]
Hexyl isovalerate	10032-13-0	0.1-0.3				
Hexyl 2-methylbutyrate	10032-15-2	0.05-0.2			0.05	<0.1°; tr-0.3[m]
Hexyl propanoate	2445-76-3				0.02	
Hexyl tiglate	16930-96-4	0.02-0.3	0.1-0.3		0.1	<0.1°
α-Humulene	6753-98-6				0.2	
Isofenchone	6541-58-8				0.08	
Lavandulol	498-16-8	0.2-1.2	4.8-5.5	0.2-0.7	0.8	tr-0.3[m]; 0.4[k,l]; 0.5[h,j,o]; 1.5[p]
Lavandulyl acetate	25905-14-0	1.5-3.2	1.6-2.2	1.6-2.9	2.3	2.4[i,q]; 2.5[k]; 2.6[l]; 2.9[p]; 3.0[b]; 3.1[f]; 3.7[n]
Lavandulyl butyrate	59550-35-5				0.02	
Lavandulyl 2-methylbutyrate		0.06-0.4				
Limonene	138-86-3	0.1-1.4	0.1-0.3	0.4-0.6	0.9	0.8[p]; 0.9-1.2[m]; 1.0[j]; 1.1[h]; 1.9[e]; 2.0[n]

Table 5.47.2 Constituents identified in lavandin grosso oils (*continued*)

Constituent	CAS	Percentage and range				
		A	B (8)	C (7)	D (43)	E
Linalool	78-70-6	26.0-38.6	25.2-47.5	27.1-35.5	22.5	21.5-33.0[m]; 29.6[h]; 29.9[e]; 30.6[n]; 31.4[p]; 31.8[k]; 32.3[o]; 33.4[b]; 34.1[a,p]; 34.3[c]; 42.5[p]
Linalyl hexanoate	7779-23-9				tr	
cis-Linalool oxide	11063-77-7	0.08-0.3	0.4-0.7		0.2	0.1[c]; 1.0[p]
cis-Linalool oxide, furanoid	5989-33-3					<0.1[o]; tr-0.1[m]
trans-Linalool oxide	11063-78-8		0.4-0.6		tr	<0.1[o]; 0.3[p]
trans-Linalool oxide, furanoid	34995-77-2					tr-0.1[m]
Linalyl acetate	115-95-7	23.9-36.0	11.8-25.5	28.3-35.4	26.2	21.7-29.0[m]; 22.4[p]; 30.5[o]; 31.6[e]; 31.9[p]; 33.0[j,p]; 34.6[q]; 35.0[a]; 35.3[k]; 36.2[b]; 37.5[i]
Linalyl isovalerate	1118-27-0		0.2-0.3			
Methoxyhexane	4747-07-3	0.02-0.08			0.05	
5-Methyl-3-heptanone	541-85-5				0.1	
Myrcene	123-35-3	0.3-0.9	0.04-0.2		1.5	0.7-1.0[m]; 0.8[o]; 0.9[h]; 1.1[p]; 1.3[b]; 8.5[n]
Nerol	106-25-2		0.07-0.6		0.05	0.2-0.3[m]; 0.3[o]
Neryl acetate	141-12-8	0.08-1.1	0.3-0.7		0.1	0.3[o]; 0.3-1.3[m]; 0.7[b]; 2.6[h]
(*E*)-β-Ocimene	3779-61-1	0.1-1.3	0.1-0.3	0.2-0.7	0.5	0.2[i]; 0.2-0.8[m]; 0.5[b,o]; 0.6[p]; 4.4[n]
(*Z*)-β-Ocimene	3338-55-4	0.4-1.5	0.03-0.3	0.6-1.4	1.1	0.4-0.9[m]; 0.9[i]; 1.0[j]; 1.1[p]; 1.2[p]; 1.9[c]; 3.4[n]
Octanal	124-13-0				tr	
n-Octanol (1-)	111-87-5	0.03-0.2				
3-Octanone	106-68-3	0.01-0.6			0.04	+[z3]; 0.2-1.0[m]; 0.5[h]; 0.6[p]
1-Octen-3-ol	3391-86-4	0.2-0.5			0.2	<0.1[o]; tr-0.7[m]; 0.5[p]
1-Octen-3-yl acetate	2442-10-6	0.2-0.7	0.03-0.2		0.3	0.1-0.3[m]; 0.3[b]; 0.4[o]; 0.5[p]
Octyl acetate	112-14-1				tr	
α-Phellandrene	99-83-2	0.02-0.07			0.07	<0.1[o]; tr-0.1[m];
α-Pinene	80-56-8	0.2-1.3	0.02-0.1		0.6	0.1[b]; 0.3[o]; 0.6[p]; 1.0[c]; 0.7-1.1[m]
β-Pinene	127-91-3	0.1-0.7	0.04-0.2		0.4	0.2[b]; 0.3[o]; 0.5[p]; 0.9[h]; 0.6-1.1[m]; 1.1[c]
trans-Pinocarveol	1674-08-4	0.01-0.08				
Plinol (isomer II)	72402-00-7				0.2	
Sabinene	3387-41-5	0.09-0.4	0.02-0.1		0.1	<0.1[o]; 0.2[p]; 0.2-0.4[m]
Sabinene hydrate	546-79-2				0.2	
cis-Sabinene hydrate	15537-55-0	0.07-0.3				
trans-Sabinene hydrate	17699-16-0	0.05-0.07				
α-Santalene	512-61-8	0.1-0.3	0.1-1.1		0.2	
β-Sesquiphellandrene	20307-83-9					0.2-0.3[m]
α-Terpinene	99-86-5	0.03-0.08			0.05	<0.1[o]; tr-0.1[m]
γ-Terpinene	99-85-4	0.1-0.3			0.4	<0.1[o]; tr-0.1[m]; 1.6[h]
Terpinen-4-ol	562-74-3	2.0-5.5	1.6-1.8	1.5-3.4	3.3	2.3[c]; 2.8[p]; 3.0[o]; 3.1[p,q]; 3.3[j,n]; 3.6[a]; 3.9[p]
α-Terpineol	98-55-5	0.3-1.7		0.8-1.7	1.2	1.1[l]; 1.1-1.7[m]; 1.2[o]; 1.6[h]; 2.6[k]; 3.3[n]; 3.9[p]
γ-Terpineol	586-81-2				0.4	
Terpinolene (α-)	586-62-9	0.09-0.4			0.3	0.1-0.4[m]; 0.3[o,p]; 0.4[h]
α-Thujene	2867-05-2	0.06-0.2			0.1	<0.1[o]; tr-0.2[m]
Tricyclene	508-32-7	tr-0.02			0.01	
(3*E*,5*Z*)-1,3,5-Undecatriene	19883-27-3				tr	

A 148 lavandin grosso essential oil samples from France, analyzed between 1998 and 2013, plus over 1000 analyses of the 28 major components of lavandin grosso oil investigated for establishing the ISO standard and several revisions; lowest and highest concentrations given (E. Schmidt, unpublished data)

B eight essential oils obtained with eight different hydrodistillation and steam-distillation techniques from lavandin 'Grosso' cultivar flowers from south France; lowest and highest concentrations given (ref. 8)

C eighty-three commercial lavandin grosso oils from various producers and from various parts of France; lowest and highest concentrations given (ref. 7)

D one lavandin grosso oil sample (ref. 43)

E data from other studies (indicated with superscript letters); highest concentrations found in any study reviewed here given; when two or more oils were investigated, only the highest concentrations are mentioned, unless indicated otherwise

[a] three commercial oil samples (ref. 1); [b] one steam-distilled oil sample from France (ref. 10); [c] one sample obtained from a local grower/distiller in Australia (ref. 5); [d] cryptone and *p*-cymen-8-ol combined; [e] one industrially steam-distilled oil from flowers cultivated in Spain (ref. 33); [f] one lab-hydrodistilled oil from plants grown in Norway (ref. 3); [g] one lab-hydrodistilled oil from Italy (ref. 15); [h] one sample of Lavandin grosso oil (ref. 20); [i] one commercial oil from France (ref. 23); [j] one oil sample (ref. 24); [k] unknown number of commercial lavandin oils (ref. 27); [l] one Argentinian lavandin grosso oil (ref. 27); [m] unknown number of oils produced from the 'Grosso' cultivar of lavandin over a season in North Carolina, USA; highest or lowest and highest concentrations given (ref. 37); [n] one sample of lavandin oil 'grosso' produced in British Columbia, Canada (ref. 39); [o] one commercial oil sample of lavandin 'Grosso' (ref. 38); [p] data from various older (<1995) studies cited in ref. 42; [q] forty-one steam distilled oil samples from flowers harvested in New Zealand in 1986 (n=7), 1987 (n=25) and 1988 (n=9); highest mean concentrations for any year given (ref. 1)

tr: trace (in column E[m]: <0.05)

Table 5.47.3 Major constituents of lavandin grosso oils

Constituent	CAS	Percentage and range				
		A	B (8)	C (7)	D (43)	E
Linalool	78-70-6	26.0-38.6	25.2-47.5	27.1-35.5	22.5	31.8[k]; 32.3[o]; 33.4[b]; 34.1[a,p]; 34.3[c]; 42.5[p]
Linalyl acetate	115-95-7	23.9-36.0	11.8-25.5	28.3-35.4	26.2	33.0[j,p]; 34.6[q]; 35.0[a]; 35.3[k]; 36.2[b]; 37.5[i]
Camphor	76-22-2	5.9-9.0	4.3-6.9	6.3-8.8	12.2	7.6-9[m]; 7.8[p]; 7.9[a]; 8.1[f]; 8.9[p]; 9.7[l]; 10.8[n]
1,8-Cineole	470-82-6	3.0-9.5	3.2-4.7	4.0-8.7	10.2	6-10[m]; 6.7[e]; 6.8[q]; 7.2[p]; 7.4[h]; 10.7[f]; 10.9[c]
Terpinen-4-ol	562-74-3	2.0-5.5	1.6-1.8	1.5-3.4	3.3	2.3[c]; 2.8[p]; 3.0[o]; 3.1[p,q]; 3.3[j,n]; 3.6[a]; 3.9[p]
Borneol	507-70-0	1.3-3.7	0.6-3.5	1.6-2.9	2.9	2.4-3.5[m]; 2.8[l]; 2.9[p]; 3.1[p]; 3.2[q]; 3.4[a]; 4.1[n]
α-Terpineol	98-55-5	0.3-1.7		0.8-1.7	1.2	1.1[l]; 1.1-1.7[m]; 1.2[o]; 1.6[h]; 2.6[k]; 3.3[n]; 3.9[p]
Lavandulyl acetate	25905-14-0	1.5-3.2	1.6-2.2	1.6-2.9	2.3	2.4[j,q]; 2.5[k]; 2.6[l]; 2.9[p]; 3.0[b]; 3.1[f]; 3.7[n]

LEGEND: SEE UNDER TABLE 5.47.2

20-40% in 'Abrialis' and 26-34% in 'Grosso' lavandin oils. 'Abrialis' oils have the highest concentrations of camphor (8-12%), and 'Super' the lowest (4-6%) with 'Grosso' taking a middle position (6-8%) (31,44). In the data from one of us (E.S.), however, linalool concentrations in 'Abrialis' were generally higher, ranging from 35 to 56% (Table 5.46.2).

CONTACT ALLERGY/ALLERGIC CONTACT DERMATITIS
No reports on contact allergy to or allergic contact dermatitis from lavandin oil, specified to have been obtained from *Lavandula angustifolia* Mill. x *Lavandula latifolia* Medik. 'Grosso', were found. The literature on 'lavandin oil' (botanical source not specified) is discussed in Chapter 5.48 Lavandin oil.

LITERATURE
For Literature references see Chapter 5.48 Lavandin oil.

Chapter 5.48 LAVANDIN OIL (OTHER CULTIVARS AND CULTIVAR NOT SPECIFIED)

DEFINITION

Lavandin oil is the essential oil obtained from the flowering tops of the lavandin, *Lavandula angustifolia* Mill. x *Lavandula latifolia* Medik. In this section, lavandin oils from cultivars other than 'Grosso' and 'Abrial' and from oils obtained from lavandins of unspecified cultivars are discussed.

INCI NOMENCLATURE

Description/definition: Lavandula hybrida oil is the essential oil obtained from the flowers of the lavandin, La*vandula hybrida*, Labiatae
INCI name EU &USA: Lavandula hybrida oil
CAS registry number(s): 8022-15-9; 91722-69-9
EINECS number(s): 294-470-6

Description/definition: Lavandula hybrida herb oil is an essential oil distilled from the flowering herbs of the lavandin, *Lavandula hybrida*, Labiatae
INCI name EU: Lavandula hybrid herb oil
INCI name USA: Not in the Personal Care Products Council Ingredient Database
Synonym(s): Lavandin oil
CAS registry number(s): 91722-69-9
EINECS number(s): 294-470-6

Description/definition: Lavandula intermedia oil is the volatile oil obtained from the whole plant of the lavender, *Lavandula intermedia*, Labiatae
INCI name EU & USA: Lavandula intermedia oil
CAS registry number(s): 92623-76-2
EINECS number(s): 296-408-3

ISO (INTERNATIONAL ORGANIZATION FOR STANDARDIZATION) STANDARD

Currently there is no ISO standard for 'lavandin oil' (unspecified). AFNOR (Association Française de Normalisation) values (for the 'Super' cultivar) are shown in Table 5.48.1.

THE PLANT, THE OIL, AND THEIR USES

This section is discussed in Chapter 5.46 Lavandin abrial oil.

CHEMICAL COMPOSITION

Lavandin oil is a colorless to slightly yellow, clear mobile liquid which has a floral, fresh lavender-like or herbal fresh camphoraceous odor, depending on the type of lavandin used to prepare the oil. The yield of essential oil from the flowering tops of *Lavandula angustifolia* Mill. x *Lavandula latifolia* Medik. generally varies from 1.5 to 5.5 per cent. The main producing countries of this oil are France, Russia, South Africa, Hungary and India.

Literature data (up to December 1, 2014) on the chemical composition of lavandin oils and unpublished analytical data from one of us (E.S.) are shown in Table 5.48.2 in alphabetical order. In lavandin oils from various origins, over 180 chemicals have been identified. About 61 per cent of these were found in a single reviewed publication only.

The major compounds found in lavandin oils from different sources are shown in Table 5.48.3. They include (highest concentrations in any study given) linalool (60.1%), linalyl acetate (52.2%), camphor (32.7%), 1,8-cineole (26.9%), borneol (16.7%), (*E*)-β-ocimene (12.2%), (*Z*)-β-ocimene (6.5%), terpinen-4-ol (5.6%), β-caryophyllene (4.9%) and lavandulyl acetate (3.4%). Well-known ingredients of lavandin oils that were present in high concentrations (>7%) in one study were sabinene (16.9%), myrcene (16.8%), α-terpineol (9.1%) and α-pinene (7.9%). Uncommon or rare constituents

Table 5.48.1 AFNOR values (%) for lavandin oil super[a]

Compound	CAS	Minimum	Maximum
Linalyl acetate	115-95-7	35.0	47.0
Linalool	78-70-6	25.0	37.0
1,8-Cineole + β-Phellandrene	470-82-6 555-10-2	3.0	7.0
Camphor (*dl*-)	76-22-2	3.5	6.5
Borneol	507-70-0	1.4	3.0
Lavandulyl acetate	25905-14-0	0.6	2.3
Hexyl butanoate (butyrate)	2639-63-6	0.4	1.2
3-Octanone	106-68-3	0.4	1.2
Terpinen-4-ol	562-74-3	tr	1.0
Lavandulol	498-16-8	tr	0.7

[a]AFNOR NF T 75-305 Huile essentielle de lavandin super © AFNOR 1992; La Plaine Saint-Denis Cedex, France, www.afnor.org

of lavandin oils found in high concentrations (>7%) in single studies include isononyl acetate (33.8%) and β-phellandrene (7.9%).

Commercial oils

The ten chemicals that had the highest maximum concentrations in 26 commercial lavandin essential oil samples (concentration ranges provided) are the following: linalool (35.1-56.2%), linalyl acetate (14.9-26.1%), 1,8-cineole (5.4-9.9%), camphor (7.6-9.9%), borneol (2.3-4.2%), (Z)-β-ocimene (1-3.7%), β-caryophyllene (1.4-3.7%), (E)-β-ocimene (1.4-3.3%), β-pinene (0.5-1.9%) and (E)-β-farnesene (0.09-1.2%) (Erich Schmidt, unpublished analytical data).

Data on lavandin 'Super' (essential) oils in ref. 41 are not presented in Table 5.48.2, as it included those of an oil sample obtained by supercritical fluid extraction. The analysis of lavandin oil in ref. 45 is not presented in Table 5.48.2 but can be accessed on-line.

CONTACT ALLERGY/ALLERGIC CONTACT DERMATITIS

General

Contact allergy to lavandin oil has been reported in two publications and there is one case report of possible allergic contact dermatitis (in which a false-positive patch test to lavandin oil due to the excited skin syndrome cannot be excluded).

Testing in groups of patients

Two hundred dermatitis patients from Poland were tested with lavandin oil 2% in petrolatum and one (0.5%) reacted. The same authors also patch tested 51 patients allergic to *Myroxylon pereirae* resin (balsam of Peru) and/or turpentine and/or wood tar and/or colophony and one (2.0%) had a positive patch test; relevance data were not provided (48). A group of 86 patients from Poland previously reacting to the fragrance mix was tested with lavandin oil and four (4.6%) had a positive patch test reaction; relevance data were not provided (49).

Table 5.48.2 Constituents identified in lavandin oils (other cultivars or cultivars not specified)

Constituent	CAS	Percentage and range				
		A	B (15)	C (12)	D (28)	E
Acetaldehyde	75-07-0			tr		
Acetic acid	64-19-7					+[z2]
Acetone	67-64-1			0.01		
Benzaldehyde	100-52-7				0-0.6	0.1[z1]
Benzyl benzoate	120-51-4					1.9[z1]
α-Bergamotene	17699-05-7					0.2[i]; 0.3[l]
α-cis-Bergamotene	18252-46-5					0.1[z5]
α-trans-Bergamotene	13474-59-4	0.07-0.2				0.1[z5]
(Z)-α-Bisabolene	29837-07-8					0.6[z6]
β-Bisabolene	495-61-4	0.08-0.2	0.4-0.7[g]			0.4[i]
α-Bisabolol	515-69-5	0-0.04			0-1.0	0.1[z5]; 0.4[p]; 0.8[q]; 4.2[l]
epi-α-Bisabolol	78148-59-1					2.0[h]
Bisabolone						1.9[v]
Borneol	507-70-0	2.3-4.2	1.8-2.4[b]	2.9	1.7-10.1	2.7[p]; 3.2[z5]; 3.3[z6]; 3.7[i]; 4.2[q]; 4.6-9.2[n]; 5.5[o]; 6.0[j]; 7.1[l]; 16.7[h]
Bornyl acetate	76-49-3	0.01-0.2				1.9[z6]
n-Butyl acetate	123-86-4			0.03		0.1[h]
Butyl methyl ether	628-28-4			<0.01		
Butyric acid	107-92-6					+[z2]
γ-Cadinene	39029-41-9		tr	tr		0.1[h]; 0.2[z5]; 0.6[i]; 1.4[o]
δ-Cadinene	483-76-1	0.1-0.3				0.06[i]; 0.08[h]
Cadinol	11070-72-7				0-0.5	
α-Cadinol	481-34-5					0.1[h]
epi-α-Cadinol	5937-11-1	0.1-0.2				0.3[h]
Camphene	79-92-5	0.3-0.5	0.09-0.3	0.3		0.1[o]; 0.2[i]; 0.3[m,r]; 0.4[s]; 0.6[h,u]; 1.3[l]; 2.2[x]
α-Campholenal	4501-58-0					+[z3]
Camphor	76-22-2	7.6-9.9	3.8-8.1	8.9	5.3-19.8	2.1-13.3[n]; 7.0[h]; 7.1[p]; 7.5[m]; 7.9[z5]; 8.5[i]; 11.7[z6]; 13.3[u]; 13.9[z6]; 14.8[z7]; 20.3[j]; 32.7[l]
Carene	74806-04-5					1.8[l]
δ3-Carene	13466-78-9		tr-0.06	0.02		0.2[i]; 0.3[h]
(E)-Carveol	1197-07-5					0.1[h]
Carvone	99-49-0	0.05-0.1				+[z3]; 0.1[z1]; 0.2[z5]; 0.3[h]
β-Caryophyllene	87-44-5	1.4-3.7		0.7	0-0.9	2.1[n]; 2.2[z5]; 2.3[t]; 3.1[o]; 3.5[z6]; 4.4[z6]; 4.9[l]

Table 5.48.2 Constituents identified in lavandin oils (other cultivars or cultivars not specified) (*continued*)

Constituent	CAS	Percentage and range				
		A	B (15)	C (12)	D (28)	E
Caryophyllene oxide	1139-30-6	0-tr	tr	0.3		0.3[h,q]; 1.1[j]
β-Chamigrene	18431-82-8		0.4-0.7[g]			
1,8-Cineole	470-82-6	5.4-9.9	2.1-8.5	7.6	0-26.1	4.8-17.1[n]; 7.3[r]; 7.6[p]; 8.4[h]; 9.0[z6]; 10.9[z7]; 11.1[z6]; 11.4[z5]; 11.6[j]; 15.3[k]; 21.1[z]; 26.9[l]
Citronellal	106-23-0		tr			
Citronellol	106-22-9		0.1-0.2[d]			6.7[v]
α-Copaene	3856-25-5					0.1[z5]
Coumarin	91-64-5	0.04-0.5		0.03		+[z2]; 0.7[n]; 1.4[o]
Cryptone	500-02-7				0-0.8	+[z2]
α-Cubebene	17699-14-8		tr			
β-Cubebene	13744-15-5		0.6-0.9			0.1[z5]
1-epi-Cubenol	81939-29-9					tr[h]
Cuminaldehyde	122-03-2					+[z2,z3]; 0.5[h]
1,6-Cyclodecadiene, 1-methyl-5-methyl-ene-8-(1-methylethyl)-	37839-63-7					0.5[l]
p-Cymene	99-87-6	0.09-0.3	tr	0.04	1.5-4.8	0.08[i]; 0.1[h,t]; 0.3[z5]; 0.7[z7]
p-Cymen-8-ol	1197-01-9					+[z3]; tr[h]
Daucene	16661-00-0					0.08[h]
(E,E)-2,4-Decadienal	25152-84-5					+[z3]
Decanal	112-31-2					+[z3]
Diacetyl	431-03-8					+[z2]
Diisopropyl ketone	565-80-0					+[z2]
2,6-Dimethyl-6-acetoxyocta-1,7-dien-3-one						+[z6]
2,6-Dimethyl-6-acetoxyoct-7-en-3-one						+[z6]
Dimethyl sulfide	75-18-3			<0.01		
Epoxylinalyl acetate	41610-76-8					+[z2]
Eugenol	97-53-0					+[z2]; 1.9[h]
Eugenyl acetate	93-28-7					3.1[h]
α-Farnesene	502-61-4		0.8-1.4[f]		0.4-1.8	0.5[q]
β-Farnesene	502-60-3					0.6[r]; 1.1[n]; 1.9[p,z3]
(E)-β-Farnesene	18794-84-8	0.09-1.2		0.3		0.7[i]; 2.2[h]
(Z)-β-Farnesene	28973-97-9					0.9[w]
Farnesyl acetone	1117-52-8					<0.01[q]
Furfural	98-01-1					+[z2]
Geranial ((E)-Citral)	141-27-5		tr[e]			0.6[z5]
Geraniol	106-24-1	0.1-0.3	tr		0-1.5	<0.05[z5]; 0.08[t]; 0.3[q]; 0.4[z7]; 1.4[v]
Geranyl acetate	105-87-3	0.2-0.3	0.7-0.9	0.3	0.3-2.2	0.3[z1]; 0.4[h,t]; 0.7[i,r]; 1.0[z6]; 1.9[z7]; 4.3[s]; 4.9[n]
Germacrene D	23986-74-5	0.1-0.5				0.1[h]; 0.4[m]; 0.5[r,z5]; 0.7[i,p]
Germacrene D-4-ol	198991-79-6					0.05[h]
α-Gurjunene	489-40-7					0.05[h]
Heneicosane	629-94-7				0-1.1	
(E,E)-2,4-Heptadienal	4313-03-5					+[z3]
n-Heptane	142-82-5			tr		
Heptenal ((E-)-2-)						+[z3]
2-Heptenal-7-hydroxy-5-isoprenyl-2-methyl acetate						+[z3]
Hexadecanoic acid	57-10-3					1.1[n]
Hexanal	66-25-1					+[z2]
Hexanoic acid	142-62-1					+[z2]
Hexanol (1-)	111-27-3	0.04-0.1	0.05-0.09			+[z2]
(E)-3-Hexanol						0.3[z5]
(E)-2-Hexenal	6728-26-3					+[z3]
3-Hexen-1-ol	544-12-7		0.05-0.06			
(Z)-3-Hexen-1-ol	928-96-1	0.01-0.06				
(Z)-3-Hexenylbutyrate	16491-36-4					1.9[j]

Table 5.48.2 Constituents identified in lavandin oils (other cultivars or cultivars not specified) (*continued*)

Constituent	CAS	Percentage and range				
		A	B (15)	C (12)	D (28)	E
Hexyl acetate	142-92-7	0.08-0.4	0.3-0.7			0.1[z1]; 0.6[r]
Hexyl butyrate	2639-63-6	0.2-0.8	1.3-1.8[c]	0.4	0-0.7	0.4[u]; 0.5[z5]; 0.6[h]; 0.7[z7]
Hexyl isobutyrate	2349-07-7	0.09-0.3	0.05-0.1	0.09		0.07[h]; 0.2[z7]
Hexyl isovalerate	10032-13-0	0-0.03				
Hexyl 2-methylbutyrate	10032-15-2	0.09-0.2				0.1[z5]
n-Hexyl methyl ether	4747-07-3	0.02-0.1				
Hexyl tiglate	16930-96-4	0.2-0.9	0.2-0.3	0.3		0.06[h]
α-Humulene	6753-98-6	0.2-0.4	0.8-1.4[f]			1.0[v]
Isoborneol	124-76-5		tr			
Isobornyl acetate	125-12-2		tr			0.1[z5]
Isocaryophyllene oxide				0.04		
Isononyl acetate	40379-24-6					33.8[z1]
p-Isopropylacetophenone	645-13-6					+[z3]
2-Isopropyl-5-methyl-9-methylene-bicyclo-[4.4.0]dec-1-ene	150320-52-8					0.2[l]
Isovaleraldehyde	590-86-3			tr		
Lavandulol	498-16-8	0.03-0.2	1.8-2.4[b]	0.6		0-1.6[n]; 0.6[t]; 0.8[z6]; 0.9[z7]; 1.0[z7]; 1.9[h]
Lavandulyl acetate	25905-14-0	0.4-0.6	2.0	1.0		0-3.4[n]; 1.9[z6]; 2.0[m]; 2.2[z]; 2.3[i]; 2.4[h]; 2.6[p]
Lavandulyl benzoate	59550-37-7					+[z2]
Lavandulyl butyrate	59550-35-5			0.2		
Lavandulyl caproate	59550-36-6					+[z2]
Lavandulyl isovalerate	51117-21-6					0.8[h]
Lavandulyl α-methylbutyrate	921210-84-6			0.7		+[z2]
Limonene	138-86-3	0.7-1.1	0.6-1.0[a]	0.7	1.0-1.8	1.0[i]; 1.3[z6]; 1.4[n]; 1.7[u]; 1.9[r]; 2.7[n]; 3.1[l]
Limonene dioxide	96-08-2					0.2[q]
Linalool	78-70-6	35.1-56.2	28.1-30.1	35.0	34.8-43.3	36.1[k]; 36.2[x]; 37.1[z6]; 38.5[z]; 42.2[h]; 46.0[n,s]; 47.9[z7]; 49.2[z5]; 54.0[m]; 60.1[z6]
Linalool oxide	1365-19-1					0.4[q]
cis-Linalool oxide	11063-77-7	0.1-0.3	0.04-0.2	0.1		0.08[z7]; 0.2[z5]; 4.6[j]
cis-Linalool oxide, furanoid	5989-33-3					0.1[k]; 2.3[u]
trans-Linalool oxide	11063-78-8	0.01-0.1	0.2-0.3	0.2		0.3[z7]
trans-Linalool oxide, furanoid	34995-77-2					4.2[j]
Linalyl acetate	115-95-7	14.9-26.1	20.5-30.2	27.0	3.8-42.5	30.4[z6]; 32.1[z6]; 37.2[r]; 38.6[w]; 39.5[z7]; 39.9[o]; 40.6[y]; 47.7[q]; 49.6[x]; 52.2[z7]
Linalyl anthranilate	7149-26-0					<0.05[l]
Linalyl butyrate	78-36-4					+[z2]
Linalyl formate	115-99-1					+[z2]
p-Mentha-1,4-dien-7-al	22580-90-1					+[z3]
7-Methoxycoumarin	531-59-9					+[z3]
p-Methylacetophenone	122-00-9					+[z3]
2-Methylbutanal	96-17-3			tr		
2-Methyl-3-buten-2-ol	115-18-4			<0.01		+[z2]
2-Methylfuran	534-22-5			tr		
Methyl heptadienone	73209-52-6					+[z2]
6-Methyl-3,5-heptadien-2-one	1604-28-0					+[z3]
6-Methyl-3-heptanone	624-42-0					+[z3]
Methylheptenone	409-02-9					+[z2]
6-Methyl-5-hepten-2-one	110-93-0					+[z3]
2-Methylpropanal	78-84-2			tr		
Methyl propyl ketone	107-87-9			tr		
γ-Methyl-γ-vinylbutyrolactone						+[z3]
γ-Muurolene	30021-74-0					0.08[h]

Table 5.48.2 Constituents identified in lavandin oils (other cultivars or cultivars not specified) (*continued*)

Constituent	CAS	Percentage and range				
		A	B (15)	C (12)	D (28)	E
Myrcene (β-)	123-35-3	0.5-0.6	0.9-1.0	0.3	0-1.2	1.3[r]; 1.4[l]; 1.5[z6]; 1.7[n]; 2.4[z7]; 4.0-16.8[n]
Myrtenal	564-94-3					+[z3]
1(2H)-Naphthalenone, 3,4-dihydro-4,7-dimethyl-						+[z3]
Nerol	106-25-2		0.1-0.2[d]		0-1.0	0.2[h,z6]; 0.3[t]; 0.6[z7]
(*E*)-Nerolidol	40716-66-3					tr[h]
Neryl acetate	141-12-8	0.09-0.2	0.2-0.4	0.7	0.5-2.7	<0.05[z5]; 0.2[m]; 0.3[r]; 0.4[h,t]; 1.2[o]; 2.4[q]; 2.6[i]
Nonadecane	629-92-5				0-1.7	
Nonanal	124-19-6					+[z3]
Nonanone	30642-09-2		tr			
Nopinone	24903-95-5					+[z2,z3]
nor-α-Santalenone						+[z6]
(*E*)-β-Ocimene	3779-61-1	1.4-3.3	0.4-5.5	3.0		1.4[r]; 1.5[z6]; 1.9[o]; 2.0[z7]; 4.6[z6]; 6.2-12.2[n]
(*Z*)-β-Ocimene	3338-55-4	1.0-3.7	1.1-2.9	2.6		1.3[o]; 1.4[i]; 1.5[w]; 1.8[m]; 3.1-7.8[n]; 3.2[h]; 6.5[k]
(*E*)-allo-Ocimene	3016-19-1			tr		
(*Z*)-allo-Ocimene	17202-20-9			tr		
neo-allo-Ocimene	7216-56-0					0.9[h]
2,7-Octadienal-6-hydroxy-2,6-dimethyl acetate						+[z3]
1,7-Octadiene-3,6-diol-2,6-dimethyl-6-acetate						+[z3]
3,7-Octadiene-2,6-diol-2,6-dimethyl-6-acetate						+[z3]
3,7-Octadien-2-one-6-hydroxy-6-methyl acetate						+[z3]
n-Octanal	124-13-0					+[z2]
n-Octanol (1-)	111-87-5	0-tr				+[z2]
3-Octanol	589-98-0		0.04-0.7	tr		+[z2]; 0.4[z6]
2-Octanone	111-13-7					1.2[z1]
3-Octanone	106-68-3	0.09-0.2		1.0		+[z2]; tr[s]; <0.05[z5]; 0.2[o]; 0.4[z7]
1-Octen-3-ol-6,7-epoxy-3,7-dimethyl acetate						+[z3]
1-Octen-3-ol	3391-86-4	0.1-0.2	tr	0.3	0-0.7	0.2[u]; 0.3[z7]; 0.6[r]
1-Octen-3-yl acetate	2442-10-6	0.08-0.3	0.5-0.7	0.3		0.5[u,z7]
1-Octen-3-yl butyrate	16491-54-6					+[z2]
Octyl acetate	112-14-1					+[z2]
Perillaldehyde	2111-75-3		tr[e]			+[z3]
Perillyl alcohol	536-59-4					+[z2]
Phellandral	21391-98-0					+[z2,z3]
α-Phellandrene	99-83-2	0-0.06	tr			0.2[m,z5]
β-Phellandrene	555-10-2		0.6-1.0[a]			3.9[l]; 4.4[h]; 7.9[k]
α-Photosantalol (A)	98113-14-5					+[z6]
α-Pinene	80-56-8	0.3-0.9	0.09-0.4	0.4	0-0.5	0.4[i]; 0.6[s,z7]; 0.7[k]; 0.8[m]; 1.2[z6]; 2.3[l]; 7.9[z1]
β-Pinene	127-91-3	0.5-1.9	0.1-0.5	0.3	0-0.6	0-0.6[n]; 0.2[o,u]; 0.4[it]; 0.5[h]; 0.7[z1]; 1.3[k]; 1.8[l]
Pinocarvone	30460-92-5					0.1[h]
cis-Piperitol	16721-38-3					0.1[h]
Sabina ketone	513-20-2					+[z2,z3]
Sabinene	3387-41-5	0.06-0.3	0.04-0.2	0.1	0-16.9	0.06[i]; 0.1[s,u]; 0.2[h,r,z5]; 0.3[z1]; 0.8[l]
Sabinene hydrate	546-79-2				0-5.5	
cis-Sabinene hydrate	15537-55-0	0.08-0.1				0.2[z5]
trans-Sabinene hydrate	17699-16-0	tr-0.06				
α-Santalene	512-61-8	0.1-0.2		0.2		+[z6]; 0.3[i]; 0.9[l]
β-Sesquiphellandrene	20307-83-9					0.3[i]

Table 5.48.2 Constituents identified in lavandin oils (other cultivars or cultivars not specified) (*continued*)

Constituent	CAS	Percentage and range				
		A	B (15)	C (12)	D (28)	E
α-Terpinene	99-86-5					tr[h]; 0.2[z5]
γ-Terpinene	99-85-4	0.08-0.3	tr-0.09	tr		0.1[z5]; 0.3[h]; 0.4[l]
Terpinen-4-ol	562-74-3	0.2-0.9	0.2-2.6		1.0-3.4	0-4.1[n]; 3.1[t]; 3.3[p]; 3.5[k]; 4.5[z5]; 5.1[n]; 5.6[h]
1-Terpineol	586-82-3					0.3[z1]
α-Terpineol	98-55-5	0.4-0.8	1.3-1.8[c]	0.5	1.3-3.5	1.5[p]; 1.6[n]; 2.8[v]; 3.6[s]; 4.6[z6]; 6.3[z7]; 9.1[z1]
β-Terpineol	138-87-4					0.5[z1]
Terpinolene (α-)	586-62-9	0.08-0.3	0.2-0.3	0.2		0.1[z7]; 0.3[i]; 0.5[h]; 1.1[l]
α-Thujene	2867-05-2	0-0.05				tr[h]; 0.1[u,z5]; 0.4[l]
α-Thujone	546-80-5		tr			
Thymol	89-83-8					tr[h]
Tricyclene	508-32-7	0.01-0.07		0.03		
Tridecane	629-50-5		0.1			
Undecane	1120-21-4		0.1-0.2			

A Twenty-six lavandin essential oil samples from France, Hungary and Russia analyzed between 1998 and 2012; lowest and highest concentrations given (E. Schmidt, unpublished data)

B three oils from Italy (one abrial, one grosso, one super A); lowest and highest concentrations given (ref. 15); these data are not presented under lavandin abrial and grosso, as the data were not specified to the individual cultivars

C (probably) one lavandin oil from France (ref. 12)

D six oils from three cultivars ('Dutch', 'Giant Hidcote', 'Super A') harvested in two years in Turkey; lowest and highest concentrations given (ref. 28)

E data from other studies (indicated with superscript letters); highest concentrations found in any study reviewed here given; when two or more oils were investigated, only the highest concentrations are mentioned, unless indicated otherwise

[a] β-phellandrene and limonene combined; [b] borneol and lavandulol combined; [c] α-terpineol and hexyl butyrate combined; [d] citronellol and nerol combined; [e] perillaldehyde and geranial combined; [f] α-humulene and α-farnesene combined; [g] β-chamigrene and β-bisabolene combined; [h] four oils from fresh and dried flowers of lavandin cultivated in Macedonia; it should be appreciated these were very atypical oils, as the concentration of linalyl acetate was 3.2% at the maximum (ref. 34); [i] one commercial oil (ref. 32); [j] one oil from Australia, cultivar 'Miss Donnington' (ref. 5); [k] one oil from Australia, cultivar 'Seal' (ref. 5); [l] one oil from Romania; this is an extremely atypical oil, as it contains only traces of linalool and no linalyl acetate (ref. 29); [m] one commercial oil from Italy (ref. 16); [n] two steam-distilled lavandin oil samples from Turkey (ref. 13); [o] twelve steam-distilled oils from Spain produced from lavandin 'Super' under various experimental designs with bed porosity and steam flow rate as parameters (ref. 14); [p] one commercial lavandin 'Super' oil (refs.6,9); [q] one oil produced by steam distillation from flowers of L*avandula x intermedia* 'Super' cultivated in Turkey (ref. 2); [r] one lab-hydrodistilled oil from Greek lavandin flowers (ref. 4); [s] one lab steam-distilled oil from Hungary (ref. 18); [t] one lab-distilled lavandin oil (ref. 19); [u] one sample of lavandin oil (ref. 21); [v] one lab-prepared Turkish lavandin oil (ref. 22); [w] one commercial lavandin oil from lavandin cultivar 'Super' (ref. 23); [x] five oils from lavandin cultivar 'Super' (ref. 12); [y] one industrially steam distilled oil from Spanish lavandin cultivar 'Super' (ref. 33); [z] one oil prepared from lavandin 'Provence' and one from 'Super' grown in Norway (ref. 3); [z1] one lavandin oil prepared in China with a very atypical composition (ref. 25); it has been suggested that this may be due to adulteration of the oil, incorrect identification of constituents, or wrong identification of lavandin flowers used to prepare the oil (ref. 26); [z2] data cited in ref. 12, literature from before 1967; [z3] one lavandin oil from India (ref. 35); [z4] four lavandin oils from British Columbia, Canada, three of 'Super' cultivars and one of the cultivar 'Hidcote Giant'; lowest and highest concentrations given (ref. 39); [z5] one steam-distilled oil from *L. x intermedia* (cultivar unknown) from a farm in India (ref. 40); [z6] data from various older (<1995) studies cited in ref. 42, all cultivars except 'Super'; [z7] data from various older (<1995) studies cited in ref. 42, cultivar 'Super'

tr: trace (in column B: <0.04; in column C: not quantified; in column E[h]: <0.02); +[z2] data cited in ref. 12, literature from before 1967; +[z3] qualitative data from ref. 35

Case reports

An aromatherapist had non-occupational contact dermatitis with allergies to multiple essential oils used at work, including lavandin oil. The patient also reacted to geraniol, linalool, linalyl acetate, α-pinene, the fragrance mix and various other fragrance materials. α-Pinene was demonstrated by GC-MS in the lavandin oil, but curiously enough not linalool (47); α-pinene is not an important component of lavandin oils.

LITERATURE (CHAPTERS 5.46, 5.47 AND 5.48)

1 Lammerink J, Wallace AR, Porter NG. Effects of harvest time and postharvest drying on oil from lavandin (*Lavandula* × *intermedia*). NZ J Crop Hortic Sc 1989;17:315-326

2 Baydar H, Kineci S. Scent composition of essential oil, concrete, absolute and hydrosol from lavandin (*Lavandula* x *intermedia* Emeric ex Loisel.). J Essent Oil Bear Plants 2009;12:131-136

Table 5.48.3 Major constituents of lavandin oils (other cultivars or cultivars not specified)

Constituent	CAS	Percentage and range				
		A	B (15)	C (12)	D (28)	E
Linalool	78-70-6	35.1-56.2	28.1-30.1	35.0	34.8-43.3	42.2[h]; 46[s]; 47.9[z7]; 49.2[z5]; 54.0[m]; 60.1[z6]
Linalyl acetate	115-95-7	14.9-26.1	20.5-30.2	27.0	3.8-42.5	39.5[z7]; 39.9[o]; 40.6[y]; 47.7[q]; 49.6[x]; 52.2[z7]
Camphor	76-22-2	7.6-9.9	3.8-8.1	8.9	5.3-19.8	11.7[z6]; 13.3[u]; 13.9[z6]; 14.8[z7]; 20.3[j]; 32.7[l]
1,8-Cineole	470-82-6	5.4-9.9	2.1-8.5	7.6	0-26.1	11.1[z6]; 11.4[z5]; 11.6[j]; 15.3[k]; 21.1[z]; 26.9[l]
Borneol	507-70-0	2.3-4.2	1.8-2.4[b]	2.9	1.7-10.1	4.2[q]; 4.6-9.2[n]; 5.5[o]; 6.0[j]; 7.1[l]; 16.7[h]
(E)-β-Ocimene	3779-61-1	1.4-3.3	0.4-5.5	3.0		1.4[r]; 1.5[z6]; 1.9[o]; 2.0[z7]; 4.6[z6]; 6.2-12.2[n]
(Z)-β-Ocimene	3338-55-4	1.0-3.7	1.1-2.9	2.6		1.3[o]; 1.4[l]; 1.5[w]; 1.8[m]; 3.1-7.8[n]; 3.2[k]; 6.5[k]
Terpinen-4-ol	562-74-3	0.2-0.9	0.2-2.6		1.0-3.4	0-4.1[n]; 3.1[t]; 3.3[p]; 3.5[k]; 4.5[z5]; 5.1[n]; 5.6[h]
β-Caryophyllene	87-44-5	1.4-3.7		0.7	0-0.9	2.1[n]; 2.2[z5]; 2.3[t]; 3.1[o]; 3.5[z6]; 4.4[z6]; 4.9[l]
Lavandulyl acetate	25905-14-0	0.4-0.6	2.0	1.0		0-3.4[n]; 1.9[z6]; 2.0[m]; 2.2[z]; 2.3[j]; 2.4[h]; 2.6[z3]

LEGEND: SEE UNDER TABLE 5.48.2

3 Renaud ENC, Charles DJ, Simon JE. Essential oil quantity and composition from 10 cultivars of organically grown lavender and lavandin. J Essent Oil Res 2001;13:269-273

4 Gitsopoulos TK, Chatzopoulou P, Georgoulas I. Effects of essential oils of Lavandula x hybrida Rev., Foeniculum vulgare Mill. and Thymus capitatus L. on the germination and radical length of Triticum aestivum L., Hordeum vulgare L., Lolium rigidum L. and Phalaris brachystachys L..J Essent Oil Bear Plants 2013;16:817-825

5 Moon T, Cavanagh HMA, Wilkinson JM. Antifungal activity of Australian grown Lavandula spp. essential oils against Aspergillus nidulans, Trichophyton mentagrophytes, Leptosphaeria maculans and Sclerotinia sclerotiorum. J Essent Oil Res 2007;19:171-175

6 Varona S, Rodríguez Rojo S, Martín A, Cocero MJ, Serra AT, Crespo T, Duarte CMM. Antimicrobial activity of lavandin essential oil formulations against three pathogenic food-borne bacteria. Ind Crops Prod 2013;42:243-250. Data also present in ref. 9

7 Bombarda I, Dupuy N, Van Da JP, Gaydou EM. Comparative chemometric analyses of geographic origins and compositions of lavandin var. Grosso essential oils by mid infrared spectroscopy and gas chromatograpy. Anal Chim Acta 2008;613:31-39

8 Périno-Issartier S, Ginies C, Cravotto G, Chemat F. A comparison of essential oils obtained from lavandin via different extraction processes: Ultrasound, microwave, turbohydrodistillation, steam and hydrodistillation. J Chromatogr A 2013;1305:41-47

9 Varona S, Martin A, Cocero MJ, Gamse T. Supercritical carbon dioxide fractionation of Lavandin essential oil: Experiments and modeling. J Supercrit Fluids 2008;45:181-188. Data also present in ref. 6

10 Barocelli E, Calcina F, Chiavarini M, Impicciatore M, Bruni R, Bianchi A, Ballabeni V. Antinociceptive and gastroprotective effects of inhaled and orally administered Lavandula hybrida Reverchon 'Grosso' essential oil. Life Sci 2004;76:213-223. Data also published in ref. 11

11 Ballabeni V, Tognolini M, Chiavarini M, Impicciatore M, Bruni R, Bianchi A, Barocelli E. Novel antiplatelet and antithrombotic activities of essential oil from Lavandula hybrida Reverchon "grosso". Phytomed 2004;11:596-601. Data also published in ref. 10

12 Steltenkamp RJ, Casazza WT. Composition of the essential oil of lavandin. J Agric Food Chem 1967;15:1063-1069

13 Oszágyan M, Simándi B, Sawinsky J, Kery A, Lemberkovics E. Supercritical fluid extraction of volatile compounds from lavandin and thyme. Flavour Fragr J 1996;11:157-165

14 Cerpa MG, Mato RB, Cocero MJ. Modeling steam distillation of essential oils: Application to lavandin super oil. AIChE Journal 2008;54:909-917

15 Piccaglia R, Marotti M. Characterization of several aromatic plants grown in northern Italy. Flavour Fragr J 1993;8:115-122

16 Giamperi L, Fraternale D, Ricci D. The in vitro action of essential oils on different organisms. J Essent Oil Res 2002;14:312-318

17 Lawrence BM. Progress in essential oils. Perfum Flavor 2008;33(7):44-? (last page unknown)

18 Simandi B, Kery A, Lemberkovics E, Oszagyan M, Hethelyi E. Supercritical fluid extraction of essential oils from Mentha piperita and Lavandula intermedia. Planta Med 1993;59(Suppl):A626.

19 Canaud F, Martineau M-O. Aspic lavande et lavandin. Bull Union Physiciens 1996;90:1941-1950. Data cited in ref. 17

20 Plotto A, Roberts D, Kim H, McDaniel M. Aroma quality of lavender water: a comparative study. Perfum Flavor 2001;24(3):44-46

21 Kubeczka K-H, Formacek V. Essential oils analysis by capillary gas chromatography and carbon-13 NMR spectroscopy, 2nd Ed. New York: J Wiley & Sons, 2002:155-160. Data cited in ref. 17

22 Aridogan BC, Baydar H, Kaya S, Demirci M, Özbasar D, Mumcu E. Antimicrobial activity and chemical composition of some essential oils. Arch Pharm Res 2002;25:860-864. Data cited in ref. 17

23 Milchard MJ, Clery R, DaCosta N, Flowerdew M, Gates L, Moss N, et al. Application of gas-liquid chromatography to the analysis of essential oil. Fingerprints of 12 essential oils. Perfum Flavor 2004;29(5):28-36

24 Ranade GS. Profile: Lavandin oil (Grosso). FAFAI J 2004;6:87. Data cited in ref. 17

25 Zhu L-F, Li Y-H, Li B-L, Lu B-Y, Zhang W-L. Aromatic plants and essential constituents (Suppl 1). South China Institute of Botany, Chinese Academy of Sciences, Hai Feng Publ. Co., distributed by Peace Book Co. Ltd., Hong Kong, 1995. Data cited in ref. 26

26 Lawrence BM. Progress in essential oils. Perfum Flavor 2001;26(1):36-51

27 Mizrahi I, Juárez MA, Elechosa MA, Bandoni AL, Náñez A. Composition and quality of essential oils obtained from crops of *Lavandula* spp. of Argentina. In: Proceedings of the World Congress on Medicinal and Aromatic Plants for Human Welfare. Acta Hort 1999;500:119-125. Data cited in ref. 26

28 Kara N, Baydar H. Determination of lavender and lavandin cultivars (*Lavandula* sp.) containing high quality essential oil in Isparta, Turkey. Turk J Field Crops 2013;18:58-65

29 Jianu C, Pop G, Gruia AT, Horhat FG. Chemical composition and antimicrobial activity of essential oils of lavender (*Lavandula angustifolia*) and lavandin (*Lavandula x intermedia*) grown in western Romania. Int J Agric Biol 2013;15:772-776

30 Directorate Plant Production, Republic of South Africa. Lavender Production. Pretoria, South Africa: Directorate Agricultural Information Services, Department of Agriculture, Forestry and Fisheries, 2009. Available at: http://www.nda.agric.za/docs/brochures/essoilslavender.pdf

31 Bosilcov A. Lavender: A key perfumery material. Dissertation for the Postgraduate Diploma in Aroma Trades Studies of the International Federation of Essential Oils and Aroma Trades. University of Plymouth, 2010. Available at: http://www.ifeat.org/wp-content/uploads/2012/10/Bosilcov+-+Lavender.pdf

32 Morgan T, Morden W, Al-Muhareb E, Herod A, Kandiyoti R. Essential oils investigated by size exclusion chromatography and gas chromatography-mass spectrometry. Energy Fuels 2006;20:734-737

33 Usano-Alemany J, Herraiz Peñalver D, Cuadrado Ortiz J, de Benito López B, Sánchez Ruiz O, Palá-Paúl J. Ecological production of lavenders in Cuenca province (Spain). A study of yield, production and quality of the essential oils. Botanica Complutensis 2011;35:147-152

34 Karapandzova M, Cvetkovikj I, Stefkov G, Stoimenov V, Crvenov M, Kulevanova S. The influence of duration of the distillation of fresh and dried flowers on the essential oil composition of lavandin cultivated in Republic of Macedonia. Maced Pharm Bull 2012;58:31-38

35 Mookherjee BD, Trenkle RW. Isolation, identification, and biogenesis of bifunctional compounds in lavandin oil. J Agric Food Chem 1973;21:298-302

36 Lawrence BM. Progress in essential oils. Perfum Flavor 2014;39(Oct):46-52

37 Lawrence BM. Lavandin oil 'Grosso' Unpublished Data, 1983. Data cited in ref. 36

38 Williams DG. The chemistry of essential oils, 2nd Ed. Port Washington, NY, USA: Micelle Press, 2008:190-191. Data cited in ref. 36

39 Lane WA, Mahmoud SS. Composition of essential oil from *Lavandula angustifolia* and *L. intermedia* varieties grown in British Columbia, Canada. Nat Prod Comm 2008;3:1361-1366

40 Shawl AS, Rouf A, Kumar T, Qurieshi MA. Chemical investigation of essential oil of *Lavandula* x *intermedia*. Indian Perfum 2008;52:33-35. Data cited in ref. 36

41 Flores G, Blanch GP, Ruiz del Castillo ML, Herraiz M. Enantiomeric composition studies in *Lavandula* species using supercritical fluids. J Sep Sci 2005;28:2333-2338

42 Boelens MH. Chemical and sensory evaluation of *Lavandula* oils. Perfum Flavor 1995;20(2):23-51

43 Naef R, Morris AF. Lavender-Lavandin. A comparison. Rivista Ital EPPOS 1992;(Numero Speciale):364-377. Data cited in ref. 42

44 Lawrence BM. Essential oils 2005-2007. Carol Stream, USA: Allured Publishing Corporation, 2008:182

45 Santana O. Cabrera R, Giménez C, González-Coloma A, Sánchez-Vioque R, de los Mozos-Pascual M, et al. Perfil químico y biológico de aceites esenciales de plantas aromáticas de interés agro-industrial en Castilla-La Mancha (España). Grasas y Aceites 2012;63:214-222

46 Bosilcov A. Lavender: A key perfumery material. Dissertation. International Centre for Aroma Trades Studies, Plymouth University, Plymouth, UK, 2010

47 Dharmagunawardena B, Takwale A, Sanders KJ, Cannan S, Roger A, Ilchyshyn A. Gas chromatography: an investigative tool in multiple allergies to essential oils. Contact Dermatitis 2002;47:288-292

48 Rudzki E, Grzywa Z, Bruo WS. Sensitivity to 35 essential oils. Contact Dermatitis 1976;2:196-200

49 Rudzki E, Grzywa Z. Allergy to perfume mixture. Contact Dermatitis 1986;15:115-116

Chapter 5.49   LAVENDER OIL

DEFINITION
Lavender oil (essential oil of lavender) is the essential oil obtained from the flowering tops of the (common, English, garden, true) lavender, *Lavandula angustifolia* Mill.

INCI NOMENCLATURE
Description/definition: Lavandula angustifolia oil is the volatile oil obtained from the flowers of the lavender, *Lavandula angustifolia*, Labiatae (synonym: Lamiaceae)
INCI name EU: Lavandula angustifolia oil
INCI name USA: Lavandula angustifolia (lavender) oil
CAS registry number(s): 8000-28-0; 90063-37-9
EINECS number(s): 289-995-2

Description/definition: Lavandula officinalis flower oil is an essential oil obtained from the fresh flowering tops of the lavender, *Lavandula officinalis* (synonym: *L. vera*), Labiatae (syn. Lamiaceae)
INCI name EU: Lavandula officinalis flower oil (perfuming name, not an INCI name proper)
INCI name USA: Not in the Personal Care Products Council Ingredient Database
CAS registry number(s): 84776-65-8
EINECS number(s): 283-994-0

Although CosIng distinguishes Lavandula officinalis oil from Lavandula angustifolia oil, they are synonyms of which *Lavandula angustifolia* is the currently used name.

ISO (INTERNATIONAL ORGANIZATION FOR STANDARDIZATION) STANDARD
ISO number: 3515
ISO name: Essential oil of lavender
Botanical origin: *Lavandula angustifolia* Mill.
Parts of plant used: Flowering tops
ISO values: ISO values (minimum and maximum concentrations) are shown in Table 5.49.1.

Other species of the genus *Lavandula*, from which lavender oils are produced include *Lavandula hybrida* (synonym: *Lavandula intermedia*, lavandin oil, Chapters 5.46, 5.47 and 5.48), and *Lavandula spica* (spike lavender oil, Chapter 5.80).

THE PLANT, THE OIL, AND THEIR USES
The lavender *Lavandula angustifolia* Mill. (synonyms: *L. officinalis*, *L. vera*), also known as medicinal lavender, true lavender, English lavender or common lavender, is an evergreen dwarf shrub, which can grow up to 75 centimeters. The plant is native to the Mediterranean region (France, Spain, Andorra, Italy), but is cultivated in many other countries as an ornamental and aromatic plant, including in Europe (especially France and Bulgaria, but also England, Moldova, Ukraine and former Yugoslavia), Africa, Australia and China (9,12,41, GRIN Taxonomy for

Plants). Lavender flowers, buds and leaves are edible and are used to flavor broths and jellies. The scent of lavender is said to deter moths and flies, so the dried flowers are placed in closets and drawers. Because of their delightful odor, lavender products are commonly used as fragrances in perfumes, soaps, moisturizers, other cosmetics, cleansers and room air fresheners (47). Infusions and tinctures of lavender flowers are claimed to have sedative and analgesic properties and lavender tincture is used to alleviate depression, headaches, and anxiety (41). The plant is also employed in the therapy of several gastrointestinal and rheumatic disorders (44).

The essential oil of lavender, which is obtained from the flowering tops by steam-distillation, is a very common fragrance ingredient. In food manufacturing, it is employed in flavoring beverages, ice cream, candy, baked goods, and chewing gums (9,47,62). In traditional herbal medicine, lavender oil is considered to have carminative, sedative, anti-depressive, antimicrobial, antifungal, hypnotic (effect of lavender oil aroma inhalation on sleep has been reviewed in ref. 69), analgesic, aphrodisiac, and acaricidal properties (1,41,44,75). In dermatology, lavender oil has been used for wounds (39), eczema, and psoriasis and even to promote hair growth in alopecia areata, albeit with little supporting evidence (2). Recent studies have investigated the effectiveness of lavender oils (sometimes in combination with other essential oils) in many health conditions, including menopausal symptoms (3), dysmenorrhea (4), aphthous ulcers (5), head lice (6), acute otitis externa (42), post-tonsillectomy pain in pediatric patients (53) and agitated behavior in dementia (7).

One of the major applications of lavender oil is in aromatherapy (9,41,47,62). It has been suggested that lavender may have an action similar to the benzodiazepines, linalyl acetate having narcotic effects and linalool acting as sedative (37). Therefore, lavender oil aromatherapy is widely used for relief of anxiety. In fact,

Table 5.49.1 ISO values (%) for lavender oil [a]

Compound	CAS	Minimum	Maximum
Linalyl acetate	115-95-7	25.0	47.0
Linalool	78-70-6 2	0.0	45.0
(Z)-β-Ocimene	3338-55-4	0.0	10.0
Lavandulyl acetate	25905-14-0	0.0	8.0
Terpinen-4-ol	562-74-3	0.0	8.0
(E)-β-Ocimene	3779-61-1	0.0	6.0
3-Octanone	106-68-3	0.0	5.0
1,8-Cineole	470-82-6	0.0	3.0
Lavandulol	498-16-8	0.0	3.0
α-Terpineol	98-55-5	0.0	2.0
Camphor	76-22-2	0.0	1.5
Limonene	138-86-3	0.0	1.0
β-Phellandrene	555-10-2	0.0	1.0

[a] ISO 3515 Essential oil of lavender ©ISO 2001; Geneva, Switzerland, www.iso.org

lavender oil is being put into the front-line as a cure-all for mind, body and spirit and is used by aromatherapists for every possible malady, although the few clinical trials performed have not offered satisfactory evidence of efficacy (37,65). The supposed calming actions of lavender may be the origin of the traditional use of a lavender herb pillow to help induce sleep (37). Claimed biological properties of lavender and lavender oil have been reviewed in refs. 41 and 118.

CHEMICAL COMPOSITION

Lavender oil is a pale yellowish liquid which has a fresh and floral to aromatic and woody odor. The yield of essential oil from the flowering tops of *Lavandula angustifolia* Mill. generally varies from 0.8 to 1.5 per cent. The

main producing countries of this oil are France, Bulgaria, Russia, Australia and the United Kingdom.

Literature data (up to December 30, 2014) on the chemical composition of lavender oils and unpublished analytical data from one of us (E.S.) are shown in Tables 5.49.2 and 5.49.3 in alphabetical order. In lavender oils from various origins, over 450 chemicals have been identified. Nearly 110 of these were found in studies from before 1993 only (Table 5.49.3). Of the more recent chemicals identified in lavender essential oils (Table 5.49.2), about 52 per cent were found in a single reviewed publication only.

The major compounds found in lavender oils from different sources are shown in Table 5.49.4. They include (highest concentrations in any study given) linalool

Table 5.49.2 Constituents identified in lavender oils

Constituent	CAS number	Percentage and range					
		A	B (102)	C (8)	D (111)	E (11)	F
Acetone	67-64-1	0.01-0.03					
8-Acetoxylinalool							0.2[u1]
α-Acoradiene	24048-44-0						<0.05[w9]
β-Acoradiene	28477-64-7						<0.05[w9]
α-Amorphene	20085-19-2						0.4[y9]; 1.0[x]
δ-Amorphene	189165-79-5						0.7[w9]
Aromadendrene	489-39-4						+[u5]; 1.1[v3]
allo-Aromadendrene	25246-27-9						0.1[w9]
Benzaldehyde	100-52-7						0.3[x4]; 0.9[x7]
Benzyl benzoate	120-51-4						0.01[m]
Bergamotene			0.2		0.04		
α-Bergamotene	17699-05-7						tr[e]; 0.1[j]; 0.2[i,x4]
cis-α-Bergamotene	18252-46-5	0.02-0.1		0-0.1		0.05	0.07[v1]; 0.1[f]; 0.2[m]; 0.4[v8]
trans-α-Bergamotene	13474-59-4			0-0.2			0.07[w1]; 0.09[m]; 0.1[e]; 0.3[w2]; 0.5[x]; 1.0[g]
trans-β-Bergamotene	15438-94-5						0.2[f]
Bicyclogermacrene	24703-35-3			0-0.1			0.02[x6]
Bicycloheptan-3-ol							0.1[a,j]
Bicyclosesquiphellandrene	54324-03-7	0.05-0.6					
epi-Bicyclosesquiphellandrene	54274-73-6				0.7		2.0[v2]
β-Bisabolene	495-61-4		0.2				tr[f]; 0.03[m]; 0.1[w2]; 0.3[v1]; 0.4[x6]
Bisabolene epoxide	121467-35-4				0.1		
α-Bisabolol	515-69-5		0.2	0-0.7			1.3[x]; 1.4[o]; 1.5[r]; 2.2[x5]; 3.0[z4]; 8.0[w2]
β-Bisabolol	15352-77-9	0.01-0.1					
α-Bisabolol oxide	22567-36-8		0.02				
Bisabolol oxide B	55399-12-7			0-0.08			
α-Bisabolol oxide B	26184-88-3						0.1[u1];
Borneol	507-70-0	0.3-2.7	1.1	0.3-14.0	3.7	2.6	6.4[w2]; 6.6[v]; 7.5[v1]; 9.3[y9]; 10.2[w4]; 12.1[x5]; 13.1[q]; 14.5[z4]; 15.2[x]; 19.6[w9]
Bornyl acetate	76-49-3	0.02-0.3	0.1	0.03-0.3			0.3[e]; 0.4[j]; 0.5[j]; 1.0[l]; 1.1[v1]; 5.2[v8]; 5.3[z]
Bornyl formate	7492-41-3	0.01-0.02	0.1				0.1[e]
α-Bourbonene	5208-58-2						+[u5]
β-Bourbonene	5208-59-3	0.01-0.04	0.2	0.02-0.09			<0.1[v8]; 0.03[m]; 0.2[w1]
α-Bulnesene	3691-11-0			0-0.1			
n-Butyl acetate	123-86-4	0.01-0.1					
Butyl butanoate	109-21-7						0.09[d]; 0.1[f]; 1.3[v2]
Cadalene	483-78-3						0.4[d]; 0.5[e]
Cadina-1,6-diene							0.1[w9]
β-Cadinene	523-47-7		0.01				0.1[v9]
γ-Cadinene	39029-41-9	0.05-0.2	0.1	0-0.4	0.1		0.3[g]; 0.4[f]; 0.5[w2]; 0.6[e]; 1.2[t]; 4.2[z1]
trans-γ-Cadinene						0.02	0.4[w1]
δ-Cadinene	483-76-1			0.03-0.4		0.03	0.06[w]; 0.1[e,i]; 0.3[v8]

Table 5.49.2 Constituents identified in lavender oils *(continued)*

Constituent	CAS number	A	B (102)	C (8)	D (111)	E (11)	F
α-Cadinol	481-34-5			0-0.4			0.3[w9]; 0.6[w3]; 0.8[y9]; 1.2[t]; 1.6[v2]; 2.0[x]
δ-Cadinol	19435-97-3						0.1[e]; 0.2[r]
τ-Cadinol	58580-31-7	0.01-0.08	0.08				0.3[v]; 0.7[y9]; 1.1[f]; 1.8[e]; 2.5[w2]; 7.7[w9]
α-Calacorene	21391-99-1						0.2[g]
cis-Calamenene	72937-55-4						0.03[m]
trans-Calamenene	73209-42-4						0.1[f]; 0.2[e]; 0.3[w9]
Camphene	79-92-5	0.03-0.5	0.6	0.02-0.5		1.1	0.3[g]; 0.4[f]; 0.5[l]; 0.6[t]; 1.2[h]; 4[y6]; 13.7[z]
α-Campholenal	4501-58-0			0-0.08			0.2[y9]; 0.4[w9,x]
Camphor	76-22-2	0.01-0.7	13.6	0.09-7.1	0.6	10.6	9.7[v]; 10.2[x3]; 10.6[y9]; 11.0[o]; 11.3[v7]; 11.8[w1]; 13.6[u1]; 13.8[z]; 14.3[x7]; 14.6[x]; 0.05[w]
δ2-Carene (= δ4-)	554-61-0						
δ3-Carene	13466-78-9	0.01-0.7		0-0.3			1.5[y9]; 1.6[x]; 1.8[h]; 2.6[s]; 2.9[w9]; 17.1[z2]
3-Caren-2-one							0.1[w9]
Carotol	465-28-1						0.6[t]
Carvacrol	499-75-2			0-0.2			tr[d,e]; 3.2[l]; 26.2[z9]
cis-Carveol	1197-06-4			0.02-0.1		0.04	0.1[w1]
trans-Carveol	1197-07-5				0.5		tr[e]; 0.02[m]; 0.1[f]; 0.2[t]; 0.6[w9]
Carvone	99-49-0			0.02-0.2	0.4		0.06[j]; 0.1[d,f]; 0.3[n,w4]; 0.4[x]; 0.5[w9]
β-Caryophyllene	87-44-5	1.8-5.9	0.5	0.5-2.8	0.2	4.4	4.3[j]; 4.7[e]; 4.9[z5]; 5.0[y1]; 5.7[w6]; 6.1[v3]; 6.2[v2]; 6.3[w5]; 6.9[v8]; 7.6[y8]; 24.1[z1]
γ-Caryophyllene	118-65-0						+[u3]
Caryophyllene oxide	1139-30-6	0.01-0.6	1.6		1.9		1.9[n]; 2[f]; 2.1[e]; 2.2[g]; 4.0[d]; 5.1[v2]; 5.2[t]
α-Cedrene	469-61-4			0.01-0.09			tr[e,x8]; 7.4[x6]
α-Chamigrene	19912-83-5						0.5[z4]
cis-Chrysanthenol	55722-60-6			0-0.06			
1,4-Cineole	470-67-7			0.0-0.09		1.2	0.06[v1]; 0.2[u2]; 0.3[w1]
1,8-Cineole	470-82-6	0.01-2.4	31.3	0.1-20.3	1.0	12.0	13.7[w4]; 15.6[w]; 15.7[z1]; 16[x7]; 18.5[z4]; 18.8[w9]; 20.3[x5]; 29.0[y9]; 38.0[x]; 46.0[z3]
Citrol	624-15-7						1.6[v2]; 10.4[y6]
Citronellal	106-23-0						2.3[w4]
Citronellol	106-22-9						3.7[z2]; 10[k]
α-Citronellol							17.4[x6]
Citronellyl acetate	150-84-5						0.3[w3]
α-Copaene	3856-25-5			0-0.05			tr[e]; 0.1[f]; 0.8[d]
Coumarin	91-64-5	0.01-0.3			0.2		<0.05[w9]; 0.4[u6]
Cryptone	500-02-7	0.01-0.3			3.0		0.8[y9]; 0.9[f]; 1.4[l]; 1.7[x7]; 2.7[t]; 3.7[w9]
α-Cubebene	17699-14-8				0.06		
β-Cubebene	13744-15-5				0.2		
Cubebol	23445-02-5						tr[e]
Cubenene	29837-12-5				0.06		
1-epi-Cubenol	81939-29-9						0.2[e]
1,10-di-epi-Cubenol	73365-77-2						0.01[m]; 0.1[f]; 0.7[w9]
Cumene	98-82-8						1.0[k]
Cuminaldehyde	122-03-2			0.04-0.5	0.8		0.1[j]; 0.2[f]; 0.3[e]; 0.4[d]; 0.8[x]; 1.6[n]
Cuminyl alcohol	536-60-7						tr[e]; 0.04[m]; 0.1[d]; 0.9[w9]; 1.0[t]
Curcumene	644-30-4		0.01				tr[e]; 0.01[m,w4]
γ-Curcumene	28976-68-3						
Cyclobutanecarboxylic acid, hexyl ester							+[u5]; 0.7[w2]
β-Cyclocitral	432-25-7						0.1[f]; 0.3[t]
1,6-Cyclodecadiene, 1-methyl-5-methylene-8-(1-methylethyl)-	37839-63-7						4.7[z1]
Cyclohexane, 1,3-diiodopropenyl-6-methyl-							2.6[v8]
2-Cyclohexen-1-ol	822-67-3				0.4		
2-Cyclohexen-1-one, 2-methyl-5-(1-methylethane)-							0.1[u1]
m-Cymene	535-77-3						0.1[f,t]; 0.2[e]
o-Cymene	527-84-4	0.01-0.2		0.03-0.1	0.1	0.08	<0.05[w9]; 0.09[w1]

Table 5.49.2 Constituents identified in lavender oils *(continued)*

Constituent	CAS number	Percentage and range					
		A	B (102)	C (8)	D (111)	E (11)	F
p-Cymene	99-87-6	0.07-0.5		0.1-0.5		1.2	0.8[m]; 1.0[x]; 1.8[h]; 4.2[z3]; 9.1[z7]; 11.3[x7]
p-Cymenene	1195-32-0	0.03-0.08					
m-Cymen-8-ol	5208-37-7			0.02-0.2		0.06	0.09[w1]
p-Cymen-8-ol	1197-01-9			0-0.3	0.5	0.4	0.06[v9]; 0.1[m,o]; 0.3[x8]; 0.6[w1]; 0.9[x]
Dauca-5,8-diene	142928-08-3						0.4[q]
Daucene	16661-00-0			0-0.1		0.2	0.3[w1]; 0.6[v1]
Daucol	887-08-1						0.3[t]
2,3-Dehydro-1,8-cineole	92760-25-3						0.1[w9]
cis-Dehydroxylinalool oxide	73413-94-2						0.2[d]
Diethyl phthalate [c]	84-66-2						13.8[z2]
Dihydrocarveol	38049-26-2			0-0.5			
Dihydrocarvyl acetate	57287-13-5						1.8[x6]
Dihydrolinalyl acetate	61476-73-1			0.02-0.2			tr[d]; 0.05[w1]
Dihydromyrcenol	53219-21-9						0.6[o]
4,4-Dimethyl-2-buten-4-olide	20019-64-1				0.2		
2,4-Dimethylfuran	3710-43-8						+[u5]
2,6-Dimethyl-1,7-octadiene-3,6-diol	51276-33-6				0.2		
(*E*)-4,8-Dimethyl-3,8-octadiol							1.5[y6]
Dimethyloctatriene	29714-87-2						0.5[a,j]
2,6-Dimethyl-3,5,7-octatrien-2-ol					0.7		5.4[u1]
2,7-Dimethyloxepine	1487-99-6						0.5[u1]
Dodecane (*n*-)	112-40-3						0.1[w]
β-Duprezianene	178443-10-2						0.04[m]
β-Elemene	33880-83-0						<0.05[w9]
Elemol (α-)	639-99-6						0.01[d]; 0.1[v9]
Epicubenol	19912-67-5			0-0.03			
Ethyl hexanoate	123-66-0						0.1[w3]
1-Ethyl-2-methylbenzene	611-14-3						0.1[y6]
Ethyl-2-methyl butyrate	7452-79-1						0.02[j]
Ethyl nerolate	32659-20-4						1.1[x9]
Eucarvone	503-93-5				0.2		
Eugenol	97-53-0						<0.05[w9]
α-Farnesene	502-61-4	0.01-0.02	0.1		0.4		0.4[w2]
(*E,E*)-α-Farnesene	502-61-4						tr[d]
(*Z,E*)-α-Farnesene	26560-14-5				0.09		
β-Farnesene	502-60-3					2.9	1.3[x2]; 1.8[v1]; 2.4[s]; 3.9[v2]; 4.0[w1]; 5.6[x4]
(*E*)-β-Farnesene	18794-84-8	0.4-4.5		0.2-1.7			0.9[w]; 1.0[e]; 2.5[q,w8]; 3.1[j]; 4.5[x5]; 5.4[u]
(*Z*)-β-Farnesene	28973-97-9	0.01-0.02			0.2		0.2[f]; 0.6[w2]; 0.7[j]; 0.8[m]; 3.6[w2]; 5.2[w6]
Farnesyl alcohol	4602-84-0		0.03				12.2[x6]
α-Fenchene	471-84-1						16.8[z2]
Fenchone	1195-79-5						0.6[x1]; 33.9[z]
α-Fenchyl alcohol	512-13-0			0.1-0.5		1.5	0.05[u2]; 0.3[w3]; 0.5[w1]
β-Fenchyl alcohol	22627-95-8						0.4[v6]
Geranial	141-27-5						0.06[j]; 0.2[t]; 0.3[w3]; 1.1[f]
Geraniol	106-24-1						1.4[u6]; 1.7[g]; 2.2[x7]; 3.3[v4]; 5.3[e]; 11.0[y6]
β-Geraniolene							+[u5]
Geranyl acetate	105-87-3	0.3-1.0		0.2-2.4	2.4	0.3	2.6[f]; 3.0[x1]; 3.2[r]; 4.3[w2]; 8.8[x7]; 10.6[v2]
Geranyl butyrate	106-29-6						+[u5]
Geranyl formate	105-86-2						1.5[w3]
Geranyl propionate	105-90-8						0.05[v7]
Germacrene A	28387-44-2		0.08				
Germacrene D	23986-74-5	0.01-0.7		0.2-0.9		0.2	0.3[e,f]; 0.4[v1]; 0.5[i,w]; 1.5[w1]; 1.7[w2]
Germacrene D-4-ol	198991-79-6						0.4[w9]
Globulol	489-41-8			0-0.05			0.1[e]
α-Gurjunene	489-40-7						0.3[w9]

Table 5.49.2 Constituents identified in lavender oils (continued)

Constituent	CAS number	A	B (102)	C (8)	D (111)	E (11)	F
Heneicosane	629-94-7						2.0[x7]
1,2,3,4,4a,7-Hexahydro-1,6-dimethyl-4(1-methylethyl)naphthalene	16728-99-7						0.5[z1]
Hexahydrofarnesyl acetone	502-69-2						0.01[m]
Hexanol (1-; n-)	111-27-3						0.1[i]; 0.2[w4]; 0.5[u1]
(Z)-3-Hexenol	928-96-1	0.01-0.2					0.06[j]
4-Hexen-1-ol	928-92-7						1.2[x4]
Hexenyl butyrate	26912-31-2		0.05				
3-Hexenyl butyrate	2142-93-0						0.7[f]
(Z)-3-Hexenyl butyrate	16491-36-4						0.2[y7]; 0.3[g]; 2.9[d]
(Z)-3-Hexenyl formate	33467-73-1						0.04[j]
(Z)-3-Hexenyl hexanoate	31501-11-8	0.01-0.02					
(Z)-3-Hexenyl isobutyrate	41519-23-7						0.1[w3]
n-Hexyl acetate	142-92-7	0.08-0.6					0.1[m]; 0.2[v]; 0.3[j]; 0.4[e,f]; 0.5[w]; 1.8[v2]
Hexyl butyrate	2639-63-6		0.4	0.1-1.7			0.7[v6]; 0.9[x]; 1.0[r]; 1.1[w4]; 3.0[u1]; 4.0[w2]
Hexyl hexanoate	6378-65-0		0.03				0.2[f]; 0.3[w4]
Hexyl isobutyrate	2349-07-7	0.02-0.2	0.2	0.09-0.2			tr[u]; 0.06[v1]; 0.2[r,w]; 0.3[w2]
Hexyl isovalerate	10032-13-0	0.01-0.05					0.1[m,v1]; 0.2[r,w]; 0.4[w2]
Hexyl-2-methylbutyrate	10032-15-2		0.4	0.03-0.3		0.4	0.06[v1]; 0.1[d,w]; 0.2[q,r]; 0.3[w1]; 0.6[u1]
n-Hexyl methyl ether	4747-07-3	0.01-0.2					0.02[j]
Hexyl tiglate	16930-96-4	0.01-0.04	0.1	0.05-0.2			tr[e]; 0.01[m]; 0.06[d]; 0.2[r]; 0.6[w4]; 0.7[u1]
Hexyl valerate	1117-59-5		0.2				+[u5]
β-Himachalene	1461-03-6						1.2[v3]
Hotrienol	20053-88-7						0.1[e]; 2.1[l]; 4.4[u1]
α-Humulene	6753-98-6	0.03-0.2	0.03				0.1[e]; 0.2[f]; 0.3[d]; 0.4[g]; 1.5[v5]; 6.2[v2]
Humulene epoxide II	19888-34-7						tr[e]; 0.08[d]; 0.1[f]
14-Hydroxy-(Z)-caryophyllene							0.2[d]
(Z)-8-Hydroxylinalool	103619-06-3						+[u5]
Hydroxy-α-terpinyl acetate					0.1		
(E)-β-Ionone	79-77-6						0.1[w9]
Isoborneol	124-76-5			0-0.1			0.4[h,v6]; 2.2[w3]; 3.4[v2]
Isobornyl acetate	125-12-2						0.3[w4]; 0.9[x6]
Isobornyl formate	1200-67-5			0.1-0.5			0.05[v1]; 0.1[j]; 0.2[f,u]; 0.5[q]; 0.6[w9]; 1.0[x]
Isobornyl 2-methylbutanoate	233665-92-4		0.2				
Isobutyl butyrate	539-90-2						0.05[j]
Isoeugenol	97-54-1						tr[y3]
(E)-Isoeugenol	5932-68-3						0.2[v9]
Isopulegol	89-79-2						tr[e]
Isoterpineol							6.8[y6]
Isoterpinolene	586-63-0						0.2[w9]
Isothujyl acetate							tr[e]
Lavandulol	498-16-8	0.08-2.4		0.05-3.3		0.3	1.9[s]; 3.0[n]; 3.6[z8]; 4.6[v2]; 4.9[d]; 5.1[w7]
Lavandulyl acetate	25905-14-0	0.4-6.3		0.7-6.2	3.2	0.3	5.3[d]; 5.7[f]; 6.1[y7]; 6.4[y]; 6.5[v5]; 7.0[x4]; 9.4[w7]; 10.8[y6]; 10.9[n]; 12.0[u6]; 12.3[y5]
Lavandulyl isobutyrate	51117-20-5	0.01-0.02		0-0.08		0.2	0.2[w1]
Lavandulyl isovalerate	51117-21-6			0-0.7			0.1[r,v1]; 0.3[w]; 0.6[q]
Lavandulyl 2-methylbutyrate							0.2[r]
Lavender lactone	1073-11-6						tr[f]
Ledene oxide II							0.1[e]; 2.6[v8]
Ledol	577-27-5				0.07		0.1[e]; 0.2[g]

Table 5.49.2 Constituents identified in lavender oils *(continued)*

Constituent	CAS number	A	B (102)	C (8)	D (111)	E (11)	F
Limonene	138-86-3	0.07-0.7		0.2-3.9	0.2	4.5	1.3[x7]; 1.5[w]; 2.1[z1]; 2.4[w1]; 2.9[r]; 4.1[v3]; 4.4[z]; 5.9[q]; 7.4[w7]; 8.5[x1]; 11.0[k]; 19.6[z9]
Limonene oxide	1195-92-2						0.6[k]
trans-Limonene oxide	4959-35-7						+[u5]
Linalool	78-70-6	26.0-44.8	36.9	23.0-57.5	30.7	45.0	44.9[w2]; 46.9[x3]; 47.8[w7]; 48.4[d]; 48.7[r]; 49.2[y5]; 49.7[o]; 50.2[y4]; 52.6[v1]; 71.9[u6]
α-Linalool	598-07-2						0.1[u2]
Linalool oxide	1365-19-1		0.07		1.2		0.7[w2]; 0.8[v8]; 4.7[v2]
cis-Linalool oxide	11063-77-7	0.01-0.3	0.8	0.3-1.0		0.01	0.4[n]; 0.9[e]; 1.3[q]; 1.6[f]; 2.4[x1]; 3.2[t]
cis-Linalool oxide, furanoid	5989-33-3				5.8		0.5[g]; 0.8[w]; 1.5[w4]; 1.6[d]; 1.9[v]; 2.1[l]
cis-Linalool oxide, pyranoid	14009-71-3						0.1[r]; 0.2[g]; 0.4[w]
trans-Linalool oxide	11063-78-8	0.02-0.7	0.05	0.3-1.0		0.04	0.3[u]; 0.7[e]; 0.9[w4]; 1.2[q]; 1.4[f]; 2.1[t];
trans-Linalool oxide, furanoid	34995-77-2				5.3		0.4[g]; 0.5[j]; 1.1[w]; 1.2[d]; 2.5[r]; 3.0[l]
trans-Linalool oxide, pyranoid	39028-58-5						0.2[w]
Linalyl acetate	115-95-7	26.1-43.3	0.1	4.0-35.4	17.7	31.7	43.8[w6]; 44.9[j]; 45.3[k]; 47.4[y4]; 47.6[v9]; 49.1[y2]; 50.6[g]; 50.8[i]; 51.0[z6]; 59.6[u6]
Linalyl anthranilate	7149-26-0						<0.05[z1]; 12.8[v2]
Linalyl butyrate	78-36-4		0.03				
Linalyl formate	115-99-1	0.1-0.3					0.1[v7]
Linalyl isovalerate	1118-27-0						0.01[d]; 0.1[w9]; 1.2[w2]
Longifolenal	66537-42-6						0.01[m]
Longifolene	475-20-7						0.1[w3]
p-Mentha-1,5-dien-8-ol	1686-20-0						16.2[x6]
trans-p-Mentha-2,8-dien-1-ol	4017-77-0				0.09		
p-Mentha-1,3,8-triene	18368-95-1				0.5		+[u5]
p-Menth-2-en-1-ol	619-62-5						0.2[g]
cis-p-Menth-2-en-1-ol	29803-82-5			0-0.04			tr[f]; 0.1[d]; 0.4[t,w9]
Menthol	89-78-1						0.3[w]
p-Menth-8-yl acetate							0.4[v9]
3-Methyl-2-butenal	107-86-8				0.07		
Methyl chavicol	140-67-0						0.9[x]
5-Methylene-9-decen-2-one							+[u5]
Methyl eugenol	93-15-2						tr[y3]
6-Methyl-5-hepten-2-ol	1569-60-4						+[u3]
4-Methyl-4-vinylbu-tyrolactone					0.8		
Mustakone							0.03[d]
Muurola-4(14),5-diene	157477-72-0						0.2[h]
γ-Muurolene	30021-74-0				0.5		0.2[d,w1]; 0.4[g]
cis-Muurol-5-en-4β-ol							0.1[w9]
cis-14-Muurol-5-en-4-one							tr[e]; 0.8[w9]
α-Muurolol	104245-48-9			0-0.5			0.2[w1]
Myrcene	123-35-3	0.1-1.5	0.2	0.3-1.2	0.4	0.3	1.9[h]; 2.0[w]; 2.7[e]; 3.1[u]; 5.3[y8]; 12.9[w7]
(*E*)-Myroxide	28977-57-3						0.1[f]
Myrtenal	564-94-3				0.7		0.1[e]
Myrtenol	515-00-4						0.1[f,t]; 0.2[w9,y9]; 0.4[x]
Neoisomenthol	491-02-1			0.1-4.3		1.2	1.3[w1]
Neoisopulegol	29141-10-4			0.03-0.09			
Nerol	106-25-2				2.4		1.0[i]; 1.2[g]; 1.8[s]; 3.3[w3]; 4.5[x9]; 8.8[v2]
(*E*)-Nerolidol	40716-66-3						tr[e]
Nerolidyl acetate	56001-43-5	0.01-0.02					
Nerol oxide	1786-08-9				0.3		tr[f]; 0.2[m]; 0.7[u1]

Table 5.49.2 Constituents identified in lavender oils *(continued)*

Constituent	CAS number	A	B (102)	C (8)	D (111)	E (11)	F
Neryl acetate	141-12-8		0.02	0.07-1.2	1.4	0.2	1.4[f]; 1.6[r]; 2.0[x1]; 2.4[h]; 2.6[v2]; 6.5[x7]
Neryl propionate	105-91-9						2.8[v2]
Nonadecane	629-92-5						2.8[x7]
3-Nonyne	20184-89-8						0.1[u1]
Nopinone	24903-95-5						0.1[f]
Nopyl acetate	128-51-8						3.8[z2]
Norbornyl acetate	34640-76-1			0-0.5			
(E)-β-Ocimene	3779-61-1	0.7-4.7		0.3-2.4	1.0	1.7	2.4[u]; 2.6[w8]; 3.0[w5]; 3.6[y]; 4.0[v2]; 4.5[w]; 4.8[x4]; 6.8[v3]; 7.1[x9]; 9.7[w7]; 10.2[u6]
(Z)-β-Ocimene	3338-55-4	0.3-7.5		1.0-6.2	0.5	2.6	3.6[w]; 3.9[e]; 4.1[x4]; 4.3[v5]; 5.8[w8]; 6.0[x2]; 7.6[u1]; 7.8[y]; 8.1[i]; 9[w6]; 14.3[w7]; 18.1[u6]
allo-Ocimene	673-84-7	0.01-0.08					0.1[t]; 0.5[x8]; 0.6[y]; 1.2[u]
(E,E)-allo-Ocimene	3016-19-1						0.2[m]
Ocimenol [(E)-]	28977-58-4	0.01-0.02					
Octadienediol	141581-27-3						0.05[a,j]
2,6-Octadien-1-ol							0.9[v8]
3-Octanol	589-98-0	0.01-0.2		0-0.9		0.2	0.1[j,w1]; 0.2[x4]; 0.3[w4]; 0.4[u,v4]
2-Octanone	111-13-7						0.6[w3]
3-Octanone	106-68-3	0.07-1.2		0.3-3.5	0.2	1.5	1.2[w4]; 1.5[t]; 1.6[u]; 1.7[f]; 2.6[w6]; 6.0[u6]
Octan-3-yl acetate	103-09-3						0.1[r,w]; 0.3[d]; 0.4[f]; 0.5[x9]; 0.6[x2]
1-Octen-3-ol	3391-86-4	0.04-0.6		0.03-1.1	0.7	2.2	0.02[j]; 0.2[t]; 0.3[f]; 0.4[n]; 0.5[v9]; 1.0[e]
Octenyl acetate (1-)	37366-04-4				3.0		0.1[v1]; 0.8[x8]; 0.9[x4]; 1.6[i]; 1.9[x2]; 2.1[v2]
1-Octen-3-yl acetate	2442-10-6	0.02-1.8		0.2-0.4		0.07	1.0[e]; 1.4[g]; 1.6[v8]; 2.7[u6]; 3.4[t]; 3.8[f]
Octyl acetate	112-14-1	0.01-0.01					0.07[w]
3-Octyl acetate	4864-61-3						0.3[u1]
3-Oxo-p-menth-1-en-7-al	160152-34-1						0.09[d]
Palustrol	5986-49-2						0.2[w9]
Perillene	539-52-6			0.0-0.08			0.1[w9]; 0.2[w1]
α-Phellandrene	99-83-2			0-0.1		1.7	0.1[h,v8]; 0.2[g]; 0.9[w1]
β-Phellandrene	555-10-2	0.03-0.6					1.7[x4]; 6.6[y9]; 7[x]; 7.6[v7]; 7.7[w9]; 16.0[z1]
Photosantalol							0.09[d]
α-Pinene	80-56-8	0.02-0.4	1.1	0.08-0.7		0.6	2.1[x]; 2.2[w3]; 3.5[h]; 4.2[z9]; 4.8[z2]; 6.8[z]
β-Pinene	127-91-3		1.5	0.03-1.2		0.3	1.3[h]; 1.4[n]; 1.5[x4]; 2.1[x]; 2.2[w7]; 2.5[w9]
α-Pinene oxide	1686-14-2						0.2[v7]
Pinocamphone	547-60-4		0.04				
trans-Pinocarveol	1674-08-4			0-0.3		0.6	tr[e]; 0.1[d]; 0.2[f,w2]; 0.5[x]; 0.6[t]
Pinocarvone	30460-92-5				0.3		0.08[w4]; 0.1[e,f]; 0.2[w9]
trans-Pinocarvyl acetate	1686-15-3						0.2[x1]
trans-Pinocarvyl formate	186607-19-2	0.01-0.02					
Piperitenone	491-09-8						0.2[f]
trans-Piperitol	16721-39-4						0.5[w9]
Piperitone	89-81-6						0.1[e]; 0.2[w9]; 2.0[n]
Plinol C	4028-60-8						0.4[v7]
Plinol D	4028-58-4						0.06[v7]
Pseudolimonene	499-97-8						4.0[q]; 5.4[v2]
Rosefuran	15186-51-3						0.1[f]; 34.4[m]
trans-Rose oxide	5258-11-7						0.3[e]
Sabina ketone	513-20-2			0-0.1			
Sabinene	3387-41-5	0.01-0.1	0.5	0-0.4		1.5	0.5[q]; 0.8[w1]; 0.9[x]; 1.0[x5]; 1.3[w9]; 13.3[x7]
cis-Sabinene hydrate	15537-55-0	0.01-0.1	0.6			0.3	0.1[m]; 0.3[u]; 0.4[o,w1]; 0.8[w9]; 1.1[x6]
trans-Sabinene hydrate	17699-16-0	0.01-0.02		0-0.2			tr[f]; 0.1[e]; 3.8[v8]
cis-Sabinene hydrate acetate	77318-48-0						0.1[d]
trans-Sabinol	471-16-9	0.01-0.02					
α-Santalene	512-61-8	0.3-0.7			0.6		0.3[n]; 0.4[e]; 0.6[f]; 0.7[i]; 1.4[v2]; 4.5[z1]
7-epi-α-Santalene		0.01-0.04					
epi-β-Santalene	25532-78-9						0.02[m]
α-Selinene	473-13-2				0.1		

Table 5.49.2 Constituents identified in lavender oils *(continued)*

Constituent	CAS number	A	B (102)	C (8)	D (111)	E (11)	F
β-Selinene	17066-67-0						0.4[v6]
β-Sesquiphellandrene	20307-83-9	0.1-0.2					0.4[w3,z1]; 2.5[v2]
Sesquisabinene	58319-04-3	0.03-0.06					
trans-Sesquisabinene hydrate	145512-84-1						0.1[e]
Sesquithujene	58319-06-5	0.01-0.05					
7-epi-Sesquithujene	159407-35-9						0.03[m]; 0.1[w]
Spathulenol	6750-60-3			0-0.06		0.2	0.3[w1]; 4.9[z9]
(*E*)-Tagetenone	33746-72-4						tr[e]
α-Terpinene	99-86-5	0.01-0.07	0.3				<0.1[n]; 0.1[i]; 0.2[u]; 0.3[v1]; 0.5[v8]; 2.8[p]
β-Terpinene	99-84-3						<0.1[v8]
γ-Terpinene	99-85-4	0.01-0.2		0.04-0.2		0.2	tr[f]; 0.06[d]; 0.1[e]; 0.2[i]; 0.4[q]; 0.5[n]; 3.2[u6]
Terpinen-4-ol	562-74-3	0.07-5.9	0.2	0.1-8.1	4.0	2.0	7.9[y]; 9.6[z1]; 9.7[v2]; 12.8[u6]; 13.5[x2]; 14.9[x8]; 16.1[u]; 19.0[w7]; 19.5[x7]; 21.7[x]
Terpinen-4-yl acetate	4821-04-9	0.01-0.02					
1-Terpineol	586-82-3	0.01-0.03					0.3[w3]
α-Terpineol	98-55-5	0.4-1.9	0.8	0.1-6.0	11.5		4.4[s]; 4.6[f]; 4.9[g]; 5.0[i]; 5.6[w7]; 5.7[v5]; 6.0[z1]; 6.2[r]; 6.3[v4]; 7.6[k]; 7.9[v2]; 15.3[p]
γ-Terpineol	586-81-2						1.9[x9]
Terpinolene (α-)	586-62-9	0.01-0.3	0.7	0.03-0.3		0.6	0.2[h]; 0.3[e]; 0.4[w1]; 0.5[o]; 0.7[v8]; 0.8[w]
Terpinyl acetate	8007-35-0					0.2	tr[f]; 0.2[w1]
α-Terpinyl acetate	80-26-2						0.1[g]; 3.0[z2]; 3.3[v8]
n-Tetradecane	629-59-4						tr[w]
2,3,4,6-Tetramethyl-phenol	3238-38-8				0.2		
Thuja-2,4(10)-diene	36262-09-6			0-0.03		0.6	0.3[w1]
α-Thujene	2867-05-2	0.01-0.3	0.02	0.02-0.2		0.3	0.1[e,f]; 0.2[i]; 0.3[q]; 0.4[z1]; 0.6[x1]; 1.6[w2]
β-Thujene	28634-89-1						2.4[v2]
β-Thujone	471-15-8						0.1[w9]
Thujyl acetate	72747-24-1	0.01-0.03					
Thujyl alcohol	513-23-5			0-0.03			0.02[j]
Thymol	89-83-8						0.9[z4]; 3.2[l]
Thymyl acetate	528-79-0						0.1[v9]
Tricyclene	508-32-7	0.01-0.1	0.05	0-0.04			0.02[j]; 0.08[h]; 0.1[f]
Tridecane	629-50-5						0.1[w]
(*Z,E*)-1,3,5-Undeca-triene	19883-27-3	0.01-0.03					0.2[i]
Valencene	4630-07-3						<0.1[v8]
Valeraldehyde	110-62-3						0.1[i]
γ-Valerolactone	108-29-2						0.7[u1]
cis-Verbenol	1845-30-3						0.1[f]; 0.2[t]
trans-Verbenol	1820-09-3				0.9		
Verbenone	80-57-9		0.1				0.07[j]; 0.2[f]; 0.3[d]; 0.4[t]
trans-Verbenyl acetate	1203-21-0						0.6[u1]
p-Vinylguaiacol	7786-61-0						tr[d]
Viridiflorol	552-02-3						0.3[w9]
α-Ylangene	14912-44-8		0.02				
Zingiberene	495-60-3	0.01-0.09					

A 374 samples from France, Bulgaria, Australia, Ukraine, Moldavia, England and China from 2001 to 2013; lowest and highest concentrations given (E. Schmidt, partly unpublished data, partly previously published in ref. 14)

B one steam-distilled oil sample from fully flowering lavender cultivated in Spain (ref. 102); probably the entire flowering plant was used for preparing the oil, which may explain the high concentration (31.3%) of 1,8-cineole (see text under Chemotypes)

C nine samples of commercial lavender oils from different lavender cultivars in Australia; lowest and highest concentrations given (ref. 8); the presence of dihydrolinalyl acetate, which apparently does not occur naturally in any essential oil, may indicate adulteration of the essential oils with synthetic linalyl acetate, in which dihydrolinalyl acetate is an impurity (ref. 12)

D three steam-distilled oils from flowers of lavender cultivated in Poland with 3 distillation times; 40 minutes, 1 hour and 2 hours; highest concentrations given (ref. 111)

E three lab-hydrodistilled oils from the flowers of lavender cultivated at heights of 20, 150 and 500 meters above sea level at one location in northeast Italy; highest concentrations given (ref. 11)

F data from other studies (indicated with superscript letters); highest concentrations found in any study reviewed here given; when two or more oils were investigated, only the highest concentrations are mentioned, unless indicated otherwise

Table 5.49.2 Constituents identified in lavender oils *(continued)*

[a] incorrect identification (ref. 32); [b] the presence of dihydrolinalyl acetate, which apparently does not occur naturally in any essential oil, may indicate adulteration of the essential oils with synthetic linalyl acetate, in which dihydrolinalyl acetate is an impurity (ref. 12); [c] this chemical has not been found in nature up to now; [d] one lab-hydrodistilled flower oil from lavender cultivated in Serbia (ref. 45); [e] one lab-hydrodistilled lavender flower oil from Poland (ref. 9); [f] two lab-hydrodistilled lavender flower samples from plants cultivated in Poland, one from fresh and one from dried material (ref. 75); [g] unknown number of lavender oils produced from plants grown in India (ref. 13); [h] one commercial oil sample, probably from Germany (ref. 40); [i] unknown number of Bulgarian lavender oils from two cultivars (ref. 15); [j] one Indian lavender oil sample (ref. 10); [k] one lavender oil sample from India (ref. 16); [l] one oil produced from lavender plants collected in Crete (Greece) (ref. 17); [m] one lavender oil of Russian origin; with this high concentration, the oil should smell tremendously of caramel (ref. 18); [n] one lab-hydrodistilled oil from the aerial parts of *L. angustifolia* cultivated in Iran (ref. 19); [o] three lab-distilled lavender flower oils produced with and without microwave assistance (ref. 21); [p] one French lavender oil sample (ref. 22); [q] one commercial lavender oil purchased in Bosnia and Herzegovina (ref. 43); [r] seven lab-hydrodistilled oils from lavender cultivar 'Etherio' from Greece harvested at seven dates between June 5 and July 5 (ref. 47); [s] two lab-prepared oils from the aerial parts of Hungarian lavender, one hydrodistilled, the other microwave-assisted hydrodistilled (ref. 49); [t] one lab-hydrodistilled oil from lavender flowers cultivated in Poland (ref. 35); [u] one lab-hydrodistilled oil from the flowering parts of lavender cultivated in Italy (ref. 54); [u1] one lab-hydrodistilled lavender flower oil from Iraq (ref. 109); [u2] one commercial lavender oil of unknown origin (ref. 114); [u3] one commercial lavender oil sample from Korea (ref. 73); [u4] one commercial oil sample from Slovakia (ref. 46); [u5] one commercial oil sample from Italy; the presence of chemicals is indicated in Table 5.49.2 with +[u5], as the amounts of each chemical were expressed in 'arbitrary units' and can therefore not be compared with the other data (ref. 48); only chemicals that have been found in no or one other study are indicated in Table 5.49.2; [u6] data from various studies cited in ref. 12; [v] one lavender oil sample of unknown origin (ref. 58); [v1] one lab-hydrodistilled flower oil from Australia (ref. 59); [v2] five lab-hydrodistilled oils from the flowering spikes of 5 lavender cultivars grown in Poland (ref. 60); [v3] one lab-hydrodistilled oil from the flowering aerial parts of lavender cultivated in Spain (ref. 63); [v4] one lab-hydrodistilled oil from the flowering spikes of lavender growing wild in India (ref. 64); [v5] one steam-distilled oil from the flowering tops of lavender cultivated in the USA (ref. 66); [v6] one lab-hydrodistilled oil from the aerial parts of lavender cultivated in Serbia (ref. 67); [v7] one lab-hydrodistilled oil from flowering lavender cultivated in Egypt (ref. 74); [v8] one commercial lavender oil sample from Australia (ref. 80); [v9] one lab-hydrodistilled lavender oil from the flowering spikes of plants cultivated in India (ref. 81); [w] two industrially steam-distilled oils from lavender flowers of the cultivar 'Etherio' grown at two locations in Greece (ref. 82); [w1] two lab-hydrodistilled oils from flowering wild Italian lavender collected at heights of 20 and 500 meters (ref. 85); [w2] four steam-distilled oils from flowering lavender harvested from a private garden in south France with distillation times varying from 5 to 20 minutes (ref. 86); [w3] one commercial lavender oil from Italy (ref. 30); [w4] one steam-distilled lavender flower oil from the south of France (ref. 29); [w5] one lab-hydrodistilled lavender oil from India (ref. 27); [w6] six Bulgarian commercial lavender oils from different cultivars (ref. 28); [w7] ten lab-distilled oils from ten lavender cultivars grown in Canada with an atypical composition, e.g., low linalyl acetate and high myrcene concentrations (ref. 31); [w8] one commercial Bulgarian lavender oil (ref. 119); [w9] one lab-hydrodistilled oil from the aerial parts of Iranian lavender; atypical composition with borneol and 1,8-cineole as dominant ingredients; see under Chemotypes (ref. 44); [x] six oils from the aerial parts of lavender grown in Iran with methods of drying and distillation times as variables; highly atypical composition dominated by 1,8-cineole, camphor, borneol and terpinen-4-ol; see under Chemotypes (ref. 62); [x1] one lab-distilled oil from Serbia (ref. 93); [x2] one commercial oil purchased from a UK company (ref. 94); [x3] one laboratory steam-distilled oil from the flowers of Italian lavender (ref. 20); [x4] seven steam-distilled oils from the inflorescences of 7 Bulgarian *L. angustifolia* cultivars (ref. 107); [x5] two lavender flower oils from plants cultivated in Jordan, one from fresh, the other prepared from dry flowers (ref. 116); [x6] one lab-hydrodistilled oil from the aerial parts of lavender cultivated in Egypt with an extremely atypical composition: low linalool (4.4%) and linalyl acetate (19.0%), high concentrations of α-citronellol (17.4%) and *p*-mentha-1,5-dien-8-ol (16.2%) (ref. 106); [x7] eight lab-hydrodistilled oils from lavender flowering spikes of 4 Turkish cultivars, harvested in 2 successive years (ref. 108); [x8] one commercial lavender oil of unknown origin (ref. 117); [x9] two lab-hydrodistilled oils from lavender flowers cultivated in Brazil and harvested in two successive years (ref. 55); [y] one lab-hydrodistilled oil from the aerial parts of lavender grown in Greece (ref. 56); [y1] various steam-distilled oils from lavender cultivated in the USA, harvested at full flowering with a second and third harvest 15-20 days later (from other accessions); mean concentrations were reported for different harvests and dry or fresh material (ref. 61); [y2] unknown number of commercial oils from Bulgaria, France and USA; mean concentrations for certain subgroups were reported (ref. 61); [y3] one lab-hydrodistilled oil from the flowering aerial parts of lavender growing wild in Morocco (ref. 79); [y4] one commercial lavender oil sample from France (ref. 89); [y5] one lab-hydrodistilled lavender flower oil from Iran (ref. 26); [y6] one lab-hydrodistilled oil from *Lavandula angustifolia* from China (ref. 90); [y7] two commercial lavender oil samples, one from Australia, the other one purchased from a UK company (ref. 24); [y8] one commercial lavender oil sample purchased in the UK (ref. 92); [y9] one lab-hydrodistilled oil from the flowering aerial parts of *L. angustifolia* cultivated in Iran (ref. 110); [z] one lab-hydrodistilled oil from lavender cultivated on the isle of Elba, Italy; this oil has an extremely atypical composition with no linalool, virtually no linalyl acetate but high concentrations of fenchone (34%), camphor (14%) and camphene (14%) (ref. 71); [z1] one steam-distilled oil from the inflorescences of lavender cultivated in Romania with an extremely atypical composition dominated by β-caryophyllene (24%), β-phellandrene (16%) and 1,8-cineole (16%) (ref. 57); [z2] one lab-prepared oil from lavender flowers cultivated in Egypt with an extremely atypical composition, dominated by δ3-carene (17.1%), α-fenchene (16.8%) and diethyl phthalate (13.8%, which has not been found in nature thus far) (ref. 38); [z3] one commercial *Lavandula officinalis* oil sample from Brazil with an extremely atypical composition, containing 46% 1,8-cineole (ref. 50); [z4] one lab-hydrodistilled flower oil from lavender cultivated in Iran (ref. 95); [z5] one commercial oil sample from an Italian company (ref. 25); [z6] one commercial lavender oil purchased from a UK company (ref. 96); [z7] one commercial oil sample purchased in Italy (ref. 101); [z8] fifty-five commercial lavender oil samples from France; only the mean concentrations of 14 components were shown (ref. 104); [z9] one lab-hydrodistilled oil from the flowering aerial parts of *L. angustifolia* harvested in a botanical garden in Iran with an extremely atypical composition dominated by carvacrol (26%), limonene (20%) and 1,8-cineole (ref. 105)

tr: traces (in column F[d]: <0.01; in column F[e] and F[f]: <0.05)

Table 5.49.3 Constituents of lavender oils identified in earlier (<1993) investigations only (23,103,120-123)

Constituent	CAS	Constituent	CAS
β-Bergamotene	6895-56-3	Formic acid	64-18-6
4-Butanolide	96-48-0	Furfural	98-01-1
Butyl benzoate	136-60-7	cis-2-(Z)-(1-Heptenyl)-3-cyclohexenyl methyl ketone	
Butyl isobutanoate	97-87-0	trans-2-(E)-(1-Heptenyl)-3-cyclohexenyl methyl ketone	
Butyl propanoate	590-01-2		
Butyl tiglate	7785-66-2	trans-2-(Z)-(1-Heptenyl)-3-cyclohexenyl methyl ketone	
Butyric acid	107-92-6		
Cadina-4,10(15)-dien-3-one	39765-72-5	Hexanoic acid	142-62-1
Camphenilone	13211-15-9	2-Hexenal	505-57-7
Caryophylla-2(12),5-dien-13-al		(Z)-3-Hexenyl nonanoate	88191-46-2
Caryophylla-2(12),5-dien-7-one		Hexyl propanoate	2445-76-3
Caryophylla-2(12),6-dien-5-one		13-Hydroxy-α-santalan-12-one	
Caryophylla-2(12),6(13)-dien-5-one		Isobutyric acid	79-31-2
Caryophylla-2(12),6-dien-7-one		4-Isopropyl-4-butanolide	38624-29-2
Cubenol	21284-22-0	4-Isopropyl-6-methyl-1-tetralone	57494-10-7
γ-Decalactone	706-14-9	Kobusone	24173-71-5
Dehydrolinalool [a]	29171-20-8	cis-Linalool oxide-5	
Dehydrolinalyl acetate [a]		trans-Linalool oxide-5	
Dihydrocoumarin	119-84-6	cis-Linalool oxide-6	
Dihydrolinalyl acetate [a]	61476-73-1	trans-Linalool oxide-6	
2,6-Dimethyl-5-acetoxymethylhepta-1,6-dien-3-one		Linalyl hexanoate	7779-23-9
2,6-Dimethyl-5-acetoxymethylhept-6-en-3-one		Linalyl isobutyrate	78-35-3
		Linalyl pentanoate	10471-96-2
2,6-Dimethyl-6-acetoxyocta-1,7-dien-3-one		1-Methoxyhexane	4747-07-3
		2-Methyl-4-butanolide	1679-47-6
4,4-Dimethyl-4-buten-2-olide		2-Methyl-3-buten-1-ol	4516-90-9
Dimethyl sulfide	75-18-3	3-Methyl-2-buten-1-ol	
(E,E)-4,5-Dimethyl-2(2-methyl-1-propenyl)-3-cyclohexenyl methyl ketone		Methylheptenone	409-02-9
(E,Z)-4,5-Dimethyl-2(2-methyl-1-propenyl)-3-cyclohexenyl methyl ketone		Methyl jasmonate	1211-29-6
(Z,E)-4.5-Dimethyl-2(2-methyl-1-propenyl)-3-cyclohexenyl methyl ketone		Methyl-cis-3-methyl-2-(3-methyl-2-butenyl)-3-cyclohexenyl ketone	
(Z,Z)-4,5-Dimethyl-2(2-methyl-1-propenyl)-3-cyclohexenyl methyl ketone		Methyl-trans-3-methyl-2-(3-methyl-2-butenyl)-3-cyclohexenyl ketone	
(E)-3-(4,8-Dimethyl-3,7-nonadienyl)-3-cyclohexenyl methyl ketone		Methyl-cis-4-methyl-5-(3-methyl-2-butenyl)-3-cyclohexenyl ketone	
(E)-4-(4,8-Dimethyl-3,7-nonadienyl)-3-cyclohexenyl methyl ketone		Methyl-trans-4-methyl-5-(3-methyl-2-butenyl)-3-cyclohexenyl ketone	
4,7-Dimethyl-1-tetralone	28449-86-7	Methyl-3-(4-methyl-3-pentenyl)-3-cyclohexenyl ketone	
8,9-Dinorborn-5-en-2-yl (endo) methyl ketone		Methyl-4-(4-methyl-3-pentenyl)-3-cyclohexenyl ketone	
8,9-Dinorborn-5-en-2-yl (exo) methyl ketone		2-Methyl-6-(4'-methylphenyl)heptan-3-one	
8,9-Dinorborn-5-en-3-yl (endo) methyl ketone		6-Methyl-5-(3'-methylphenyl)heptan-2-one	
8,9-Dinorborn-5-en-3-yl (exo) methyl ketone		2-Methyl-6-(4-methylphenyl)hept-1-en-3-one	
8,9-Dinorborn-5-en-2-yl (endo) pentyl ketone		Methyl-5-methyl-8,9,10-trinorborn-5-en-2-yl (endo) ketone	
8,9-Dinorborn-5-en-2-yl (exo) pentyl ketone		Methyl-5-methyl-8,9,10-trinorborn-5-en-2-yl (exo) ketone	
8,9-Dinorborn-5-en-3-yl (endo) pentyl ketone		Methyl-6-methyl-8,9,10-trinorborn-5-en-2-yl (endo) ketone	
8,9-Dinorborn-5-en-3-yl (exo) pentyl ketone		Methyl-6-methyl-8,9,10-trinorborn-5-en-2-yl (exo) ketone	
Epoxylinalyl acetate	41610-76-8	3-(4-Methyl-3-pentenyl)-3-cyclohexenyl pentyl ketone	
3-Ethylbutanal	15877-57-3	4-(4-Methyl-3-pentenyl)-3-cyclohexenyl pentyl ketone	
Ethyl-3-(4-methyl-3-pentenyl)-3-cyclohexenyl propyl ketone		3-(4-Methyl-3-pentenyl)-3-cyclohexenyl propyl ketone	
Ethyl-4-(4-methyl-3-pentenyl)-3-cyclohexenyl propyl ketone		4-(4-Methyl-3-pentenyl)-3-cyclohexenyl propyl ketone	
(E,E)-Farnesal	502-67-0	4-(4-Methylphenyl)pentanal	
(Z,E)-Farnesal	4380-32-9		

Table 5.49.3 Constituents of lavender oils identified in earlier (<1993) investigations only (23,103,120-123) *(continued)*

Constituent	CAS	Constituent	CAS
5-Methyl-8,9,10-trinorborn-5-en-2-yl (*endo*) pentyl ketone		Octanal	124-13-0
5-Methyl-8,9,10-trinorborn-5-en-2-yl (*exo*) pentyl ketone		Octyl propionate	142-60-9
		Pentanal	110-62-3
6-Methyl-8,9,10-trinorborn-5-en-2-yl (*endo*) pentyl ketone		Pentanol	30899-19-5
		5-Pentyl-5-pentanolide	705-86-2
6-Methyl-8,9,10-trinorborn-5-en-2-yl (*exo*) pentyl ketone		α-Photosantalol (A)	98113-14-5
		α-Photosantalol (B)	
4-Methyl-4-vinyl-4-butanolide	1073-11-6	α-Santalan-12-one	
4-Methyl-4-vinyl-4-but-2-enolide		β-Santalene	511-59-1
Nonanal	124-19-6	α-Santalenic acid	
11-Norbourbonan-1-one		nor-α-Santalenone	
Norcadin-5-en-4-one		α-Santal-13-en-12-one	
Norcadin-5-en-4-one isomer		Tricycloekasantalal	16933-18-9
12-Norcaryophyllen-2-one		1,3,5,8-Undecatetraene	50277-31-1
15-Norcedran-8-one		Valeric acid	109-52-4

[a] presence of this chemical indicates that the oil is adulterated with synthetic linalool and/or linalyl acetate (according to ref. 103)

(71.9%), linalyl acetate (59.6%), 1,8-cineole (46.0%), β-caryophyllene (24.1%), terpinen-4-ol (21.7%), borneol (19.6%), (*Z*)-β-ocimene (18.1%), camphor (14.6%), lavandulyl acetate (12.3%), (*E*)-β-ocimene (10.2%) and lavandulol (5.1%). Well-known ingredients of lavender oils that were present in high concentrations (>10%) in one or two studies were limonene (11.0 and 19.6%), δ3-carene (17.1%), β-phellandrene (16.0%), α-terpineol (11.5% and 15.3%), camphene (13.7%), sabinene (13.3%), myrcene (12.9%), *p*-cymene (11.3%), geraniol (11.0%) and geranyl acetate (10.6%). Uncommon or rare constituents of lavender oils found in high concentrations (>10%) in single studies include rosefuran (34.4%), fenchone (33.9%), carvacrol (26.2%), α-citronellol (17.4%), α-fenchene (16.8%), *p*-mentha-1,5-dien-8-ol (16.2%), diethyl phthalate (13.8%), linalyl anthranilate (12.8%), farnesyl alcohol (12.2%), citrol (10.4%) and citronellol (10.0).

Commercial oils

The ten chemicals that had the highest maximum concentrations in 374 commercial lavender essential oil samples (concentration ranges provided) are the following:

linalool (26.0-44.8%), linalyl acetate (26.1-43.3%), (*Z*)-β-ocimene (0.3-7.5%), lavandulyl acetate (0.4-6.3%), terpinen-4-ol (0.07-5.9%), β-caryophyllene (1.8-5.9%), (*E*)-β-ocimene (0.7-4.7%), (*E*)-β-farnesene (0.4-4.5%), borneol (0.3-2.7%) and 1,8-cineole (0.01-2.4%) (Erich Schmidt (mostly), unpublished analytical data).

The quality of essential oil of lavender depends both on the high content of linalool and linalyl acetate and their mutual proportions (higher linalyl acetate concentration is preferred). The price of lavender essential oil is high and because of this it is often falsified by adding cheaper oil of *Lavandula latifolia* (spike lavender oil) or the oil of hybrids of *L. angustifolia* and *L. latifolia* (lavandin oil), or by the addition of synthetic chemicals (41).

Chemotypes

Lavender flower oils are virtually always dominated by linalool and linalyl acetate. However, there have been some reports of lavender oils in which 1,8-cineole (46%, ref. 50; 48.4%, ref. 87; 43.1%, ref. 124; 38.0%, ref. 62) or borneol (19.6%, ref. 44, also high concentration of 1,8-cineole: 18.8%) had the highest concentrations. One oil sample

Table 5.49.4 Major constituents of lavender oils

Constituent	CAS number	Percentage and range					
		A	B (102)	C (8)	D (111)	E (11)	F
Linalool	78-70-6	26.0-44.8	36.9	23.0-57.5	30.7	45.0	49.2[y5]; 49.7[o]; 50.2[y4]; 52.6[v1]; 71.9[u6]
Linalyl acetate	115-95-7	26.1-43.3	0.1	4.0-35.4	17.7	31.7	49.1[y2]; 50.6[g]; 50.8[i]; 51.0[z6]; 59.6[u6]
1,8-Cineole	470-82-6	0.01-2.4	31.3	0.1-20.3	1.0	12.0	18.8[w9]; 20.3[x5]; 29.0[y9]; 38.0[x]; 46.0[z3]
β-Caryophyllene	87-44-5	1.8-5.9	0.5	0.5-2.8	0.2	4.4	6.2[v2]; 6.3[w5]; 6.9[v8]; 7.6[y8]; 24.1[z1]
Terpinen-4-ol	562-74-3	0.07-5.9	0.2	0.1-8.1	4.0	2.0	14.9[x8]; 16.1[u]; 19.0[w7]; 19.5[x7]; 21.7[x]
Borneol	507-70-0	0.3-2.7	1.1	0.3-14.0	3.7	2.6	12.1[x5]; 13.1[q]; 14.5[z4]; 15.2[x]; 19.6[w9]
(*Z*)-β-Ocimene	3338-55-4	0.3-7.5		1.0-6.2	0.5	2.6	7.6[u1]; 7.8[y]; 8.1[i]; 9[w6]; 14.3[w7]; 18.1[u6]
Camphor	76-22-2	0.01-0.7	13.6	0.09-7.1	0.6	10.6	11.8[w1]; 13.6[u1]; 13.8[z]; 14.3[x7]; 14.6[x];
Lavandulyl acetate	25905-14-0	0.4-6.3		0.7-6.2	3.2	0.3	9.4[w7]; 10.8[y6]; 10.9[n]; 12.0[u6]; 12.3[y5]
(*E*)-β-Ocimene	3779-61-1	0.7-4.7		0.3-2.4	1.0	1.7	4.8[x4]; 6.8[v3]; 7.1[x9]; 9.7[w7]; 10.2[u6]
Lavandulol	498-16-8	0.08-2.4		0.05-3.3		0.3	1.9[s]; 3.0[n]; 3.6[z8]; 4.6[v2]; 4.9[d]; 5.1[w7]

LEGEND: SEE UNDER TABLE 5.49.2

was commercial, from Brazil (50), in the other cases the oils were obtained from the aerial parts of *Lavandula angustifolia*, which probably includes the leaves and stems. Indeed, borneol and 1,8-cineole have repeatedly been observed to be the dominant components of lavender *leaf* oils (36,52,78,87,97,110, see below). In lavenders cultivated in Belgrade, Serbia, it was shown that in the shoots with flowers, inflorescences and fruits of clade I, linalool is dominant, whereas in the young leaves before flowering and in old leaves of clade II, 1,8-cineole is dominant. In the young and incompletely developed leaves of clade III, β-phellandrene was found to be the dominant component. Thus, the composition of essential oil of lavender much depends on the plant part used and the stage of development (87). Short distillation times of flowers also result in higher 1,8-cineole concentrations (51).

In single reports, other chemicals were found to be the major component of oils prepared from *Lavandula angustifolia/officinalis*, including fenchone (71), δ3-carene (38), rosefuran (18), carvacrol (105) and β-caryophyllene (57). It has been suggested that reports of lavender flower oils with low linalyl acetate but high 1,8-cineole and camphor content (e.g., ref. 29) may in fact have concerned *spike* lavender oils (33).

Lavender leaf oils

The composition of lavender oils obtained from *leaves* of *Lavandula angustifolia* (which are not the oils of commerce) can be found in refs. 36,45,52,68,70,72,76,78,97,99 and 110 (not presented in Table 5.49.2). In most, linalool or linalyl acetate, which are the main components of the flower oil, were also the dominant ingredients of leaf oils (45,70,72,76,99). In some reports, however, the oils were dominated by 1,8-cineole (65.4%, ref. 97; 48.4%, ref. 87; 41.0%, ref. 110; 37.8%, ref. 78), borneol (22.4%, ref. 36; 8.6%, ref. 52), β-caryophyllene (24.1%) or α-pinene (49.4%, one commercial leaf oil sample from Turkey, ref. 68).

The results of analyses of *Lavandula angustifolia* essential oils in refs. 34,77,84,88,100 and 112 are not presented here, as their qualitative nor quantitative composition qualifies for inclusion of data from these studies in column F of Table 5.49.2 (they provide no new data). Ref. 91 is not discussed, as the authors managed to spell the names of 20 of the 32 chemicals in a commercial oil sample incorrectly (91). A review of various aspects of lavenders (lavender, lavandin, spike lavender), written from the industry's perspective, can be found (on-line) in a dissertation from the International Centre for Aroma Trades Studies, Plymouth, UK (37).

CONTACT ALLERGY / ALLERGIC CONTACT DERMATITIS

General

Contact allergy to / allergic contact dermatitis from lavender oil has been reported in at least 40 publications. In groups of consecutive patients suspected of contact dermatitis, prevalence rates of up to 1.2% positive patch test reactions have been observed, but definite relevance appears to be uncommon. The oil (2% in petrolatum) has been included in the screening series of the North American Contact Dermatitis Group (NACDG) since 2009. There have been many case reports of allergic contact

dermatitis from lavender oil. Most cases were the result of their presence in topical pharmaceutical products. There are at least eleven cases of occupational allergic contact dermatitis from lavender oil, mostly in aromatherapists and other professionals who massage clients with the oil at their work. Cosmetics are an infrequent source of lavender contact allergy. In several cases, there was a co-reaction to linalool, presumably the main allergen in lavender oil. Caryophyllene may also have been an allergen. In the past, lavender oil was a frequent cause of pigmented cosmetic dermatitis in Japanese women.

Fresh lavender oil is probably little allergenic. However, the oil lacks natural protection against autoxidation, and strong contact allergens such as linalool hydroperoxides are formed on air exposure from linalool, which is, together with linalyl acetate, the main component of lavender oil (151).

Testing in groups of patients

The results of patch tests with lavender oil in routine testing (consecutive patients suspected of contact dermatitis) and in groups of selected patients are shown in Table 5.49.5. In routine testing, rates of positive reactions ranged from 0.2% to 1.2%, whereas between 2.8% and 30% of patients in selected groups had positive patch tests. The very high positivity rate of 30% was found in a very small group of 10 patients strongly suspected of fragrance allergy and reacting to the fragrance mix (145).

Case reports

Occupational contact dermatitis

Two aromatherapists had contact dermatitis (one occupational) with allergies to multiple essential oils used at their work, including lavender oil. Both patients also reacted to geraniol, α-pinene, the fragrance mix and various other fragrance materials. In addition, one proved to be allergic to linalool and linalyl acetate, the other to caryophyllene; α-pinene, linalool, geraniol, and caryophyllene were demonstrated by GC-MS in the lavender oils (134). Linalool is a major ingredient of lavender oil and caryophyllene may be present in commercial lavender oils in concentrations up to 6%, but α-pinene and geraniol are not important components (Table 5.49.2, column A). Another aromatherapist had occupational contact dermatitis from multiple essential oils, including lavender oil; the patient also reacted to lavender absolute (136). Yet another aromatherapist developed contact dermatitis from allergy to various essential oils, including lavender oil, used at work (137). Three cases of occupational allergic contact dermatitis in a masseuse, a physiotherapist and a reflexologist were caused by contact allergy to lavender oil and other essential oils (135). An aromatherapist developed occupational contact dermatitis and proved to be allergic to lavender oil; she also reacted to neroli oil (125). A female 'complementary therapist' developed occupational contact dermatitis from a multitude of essential oils used at work, including lavender oil (138).

A hairdresser who had occupational contact dermatitis of the hands developed contact allergy to lavender oil which was present in a shampoo used at work (130). Another hairdresser developed occupational contact

Table 5.49.5 Results of testing groups of patients with lavender oil

Years and Country	Test conc. & vehicle	Number of patients tested positive (%)			Selection of patients (S); Relevance (R); Comments (C)	Ref.
Routine testing						
2011-12 USA, Canada	2% pet.	4,229	16	(0.4%)	R: definite + probable relevance: 69%	153
2009-10 USA, Canada	2% pet.	4,302	9	(0.2%)	R: definite + probable relevance: 30%	143
2002-2003 Korea	2% pet.	422	5	(1.2%)	R: not stated	154
Testing in groups of selected patients						
<2002 Japan, USA and 5 European countries	10% pet.	218	6	(2.8%)	S: patients previously shown to be allergic to fragrances; R: not stated	157
1990-1998 Japan	20% pet.	1,443	53	(3.7%)	S: patients suspected of cosmetic contact dermatitis; virtually all were women; R: 13/25 (52%) in 1997 and 1998; C: annual rates ranged from 0 to 13.9%; sudden increase in 1997 and 1998 from the placing of dried lavender flowers in pillows, drawers, cabinets or rooms and aromatherapy with lavender oil	127
1996-1997 UK	2% pet.	10	3	(30%)	S: patients strongly suspected of fragrance allergy; all also reacted to the fragrance mix; R: not stated	145
<1986 Poland	2% pet.	86	3	(3.5%)	S: patients previously reacting to the fragrance mix; R: not stated	144
<1974 Japan	?	183	6	(3.3%)	S: patients suspected of cosmetic dermatitis; R: unknown; in many, there was co-reactivity with geraniol, which is not an important component of commercial lavender oils (Table 5.49.2, column A)	158

pet.: petrolatum

allergy to lavender oil from its presence in an eau de cologne (140). One case of occupational allergic contact dermatitis in a porcelain painter due to lavender oil, oil of turpentine and oil of anise oil, which were mixed with pigments for painting (165).

Cosmetics

In a group of 70 patients with proven allergic cosmetic dermatitis, lavender oil was the allergen in two (163).

Five positive patch test reactions to lavender oil, which was present in cosmetics that had given a positive patch test or had been positive in a usage test, were seen in a 9-year period in one clinic in Belgium (149). One patient had allergic contact dermatitis from 'Rose Absolute' perfume and a body lotion containing *Rosa centifolia*; she reacted to her own products, lavender oil, *Rosa centifolia*, several indicators of fragrance allergy, geranium oil Bourbon and various individual fragrance chemicals including linalool (150), which is a major component of commercial lavender oils, and limonene, which may be present in lower concentrations in commercial lavender oils (Table 5.49.2, column A).

Pharmaceutical products

Twenty-four cases of contact sensitization to lavender oil present in topical pharmaceutical preparations were observed in Belgium (156). Two positive patch test reactions to lavender oil which was present in a topical NSAID preparation used by patients suspected to be allergic to the NSAID; possible overlap with data from ref. 156 (155). Allergy to lavender oil was the cause of allergic contact dermatitis from Phenergan cream in one

patient (129). Another case of contact allergy to lavender oil in an antihistamine pharmaceutical cream was reported from France (139).

Other products

One patient developed airborne contact dermatitis from contact allergy to lavender oil and two other products (jasmine absolute, rosewood oil) from aromatherapy lamps, whereby the water vapor produced distributes the essential oils as an aerosol in the air; the patient also reacted to linalool, which is an important constituent of all three oils (132). Two patients presented with dermatitis on their cheeks which proved to be caused by lavender drops applied on the pillow for their presumed hypnotic effects; a positive patch test to lavender oil was observed (133). One woman had eczema of the groin, vulva, and perianal area caused by tea tree oil and lavender gel containing lavender oil; the lavender gel reacted upon patch testing, as did lavender *absolute* 10% and 50% in petrolatum (147).

Positive patch tests (relevance unknown, uncertain or not stated)

One positive patch test reaction to lavender oil 2% in petrolatum was found in a group of 31 patients allergic to oil of turpentine (126). Six positive patch test reactions to lavender oil were collected from various European patch test centers (128). A patient with airborne contact dermatitis from tea tree oil also reacted to lavender oil in patch testing (131). One beautician with occupational allergic hand dermatitis from products containing citral and certain essential oils also reacted to lavender oil and geranium oil; citral (neral + geranial) is not an important component of lavender oil (148). One patient with erythema on the face reacted to lavender oil and other essential oils; their relevance was uncertain (160). Of seven patients allergic to the fragrance farnesol, 4 (57%) co-reacted to lavender oil (and various other fragrances) (161). In a group of 21 patients with dermatitis caused by fragrances and tested with a series of essential oils, five (24%) reacted to lavender *absolute* (162). In a group of 91 patients with cosmetic allergy, there was one patch test reaction to lavender oil; the relevance was not ascertained, but felt by the authors to be 'indicative' (164).

Pigmented cosmetic dermatitis

In Japan, in the 1960s and 1970s, many female patients developed pigmentation following dermatitis of the face (142). This so-called pigmented cosmetic dermatitis was shown to be caused by contact allergy to components of cosmetic products, notably essential oils, other fragrance materials, antimicrobials, preservatives and coloring materials (141,142). In a group of 620 Japanese patients with this condition investigated between 1970 and 1980, 4-7% had positive patch test reactions to lavender oil 20% in purified lanolin (141). The number of patients with pigmented cosmetic dermatitis decreased strongly after 1978, when major cosmetic companies began to eliminate strong contact sensitizers from their products (141).

Photocontact allergy to lavender oil has been observed rarely (152). A useful review of various aspects of lavender (oil) is provided in ref. 146.

LITERATURE

1 Thomson KF, Wilkinson SM. Allergic contact dermatitis to plant extracts in patients with cosmetic dermatitis. Br J Dermatol 2000;142:84-88

2 Wu PA, James WD. Lavender. Dermatitis 2011;22:344-347

3 Darsareh F, Taqvoni S, Joolaee S, Haghani H. Effect of aromatherapy massage on menopausal symptoms: a randomized placebo-controlled clinical trial. Menopause 2012;19:995-999

4 Apay SE, Arslan S, Akpinar RB, Celebiioglu D. Effect of aromatherapy massage on dysmenorrhea in Turkish students. Pain Manag Nurs 2012;13:236-240

5 Altaei DT. Topical lavender oil for the treatment of recurrent aphthous ulceration. Am J Dent 2012;25:39-43

6 Barker SC, Altman PM. An *ex vivo*, assessor blind, randomised, parallel group, comparative efficacy trial of the ovicidal activity of three pediculicides after a single application — melaleuca oil and lavender oil, eucalyptus oil and lemon tea tree oil, and a "suffocation" pediculicide. BMC Dermatol 2011;11:14

7 O'Connor DW, Eppingstall B, Taffe J, van der Ploeg ES. A randomized, controlled cross-over trial of dermally-applied lavender (*Lavandula angustifolia*) oil as a treatment of agitated behaviour in dementia. BMC Compl Alter Med 2013;13:315 (7 pages)

8 Shellie R, Mondello L, Marriott P, Dugo G. Characterisation of lavender essential oils by using gas chromatography-mass spectrometry with correlation of linear retention indices and comparison with comprehensive two-dimensional gas chromatography. J Chromatogr A 2002;970:225-234

9 Smigielski K, Ra A, Krosowiak K, Gruska R. Chemical composition of the essential oil of *Lavandula angustifolia* cultivated in Poland. J Essent Oil Bear Plants 2009;12:338-347

10 Shawl AS, Kumar T, Shabir SD, et al. Lavender—a versatile industrial crop in Kashmir. Indian Perfum 2005;49:235-238. Data cited in ref. 32

11 Da Porto C, Decorti D. Analysis of the volatile compounds of flowers and essential oils from *Lavandula angustifolia* cultivated in northeastern Italy by headspace solid-phase microextraction coupled to gas chromatography-mass spectrometry. Planta Med 2008;74:182-187

12 Lawrence BM. Progress in essential oils. Perfum Flavor 2004;29(4):70-? (last page unknown)

13 Mallavarapu GR, Mehta VK, Sastry KP, Krishnan KR, Ramesh S, Kumar S. Composition of lavender oils produced in Kashmir and Kodaikanal. J Med Arom Plant Sci 2000;22:768-770. Data cited in ref. 32

14 Schmidt E. The characteristics of lavender oils from eastern Europe, Ukraine, Moldova and Bulgaria. Perfum Flavor 2003;28(4):48-59

15 Konakchiev A, Tsankova E. Chemical composition of lavender oils from the cultivars Druzhba and Hemus. Plant Sci (Sofia) 2003;40:38-41. Data cited in ref. 32

16 Ranade GS. Profile. Lavender oil. FAFAI 2003;5:79. Data cited in ref. 32

17 Daferera D, Ziogas BN, Polissiou MG. The effectiveness of plant essential oils on the growth of *Botrytis cinerea*, *Fusarium* sp. and *Clavibacter michiganensis* subsp. *michiganensis*. Crop Protect 2003;22:39-44. Data also presented in ref. 98

18 Anonymous. GC-MS analysis of lavender oil (Russian). Indian Perfum 2004:48:244. Data cited in ref. 32

19 Fakhari AR, Salehi P, Heydari R, Ebrahimi SN, Haddad PR. Hydrodistillation-headspace solvent microextraction, a new method for analysis of the essential oil components of *Lavandula angustifolia* Mill. J Chromatogr A 2005;1098:14-18

20 Chemat F, Lucchesi ME, Smadja J, Favretto L, Colnaghi G, Visinoni F. Microwave accelerated steam distillation of essential oil from lavender: a rapid, clean, and environmentally friendly approach. Anal Chim Acta 2006;555:157-160

21 Iriti M, Colnaghi G, Chemat F, Histocytochemistry and scanning electron microscopy of lavender glandular trichomes following conventional and microwave assisted hydro distillation of essential oils: a comparative study. Flav Fragr J 2006;21:704–712

22 Pavela R. Insecticidal activity of essential oils against cabbage aphid *Brevicoryne brassicae*. J Essent Oil Bear Plants 2006;9:99-106

23 Timmer R, ter Heide R, de Valois PJ, Wobben HJ. Analysis of the lactone fraction of lavender oil (*Lavandula vera* D.C.). J Agricult Food Chem 1975;23:53-56

24 Moon T, Cavenagh HMA, Wilkinson JM. Antifungal activity of Australian grown *Lavandula* spp. essential oils against *Aspergillus nidulans*, *Trichophyton mentagrophytes*, *Leptosphaeria maculans* and *Sclerotina sclerotiorum*. J Essent Oil Res 2007;19:171–175

25 Evandri MG, Battinelli L, Daniele C, Mastrangelo S, Bolle P, Mazzanti G. The antimutagenic activity of *Lavandula angustifolia* (lavender) essential oil in the bacterial reverse mutation assay. Food Chem Toxicol 2005;43:1381-1387

26 Hadian J, Ghasemnezhad M, Ranjbar H, Frazane M, Ghorbanpouret M. Antifungal potency of some essential oils in control of postharvest decay of strawberry caused by *Botrytis cinera*, *Rhizopus stolonifer* and *Aspergillus niger*. J Essent Oil Bear Plants 2008;11:553-562

27 Kiran Babu GD, Singh B. Comparative chemical composition of direct, recovered and combined essential oils of *Lavandula angustifolia* Mill. Indian Perfum 2007;51:50–53. Data cited in ref. 33

28 Stoyanova A, Grozeva E. Traditional Bulgarian essential oil-bearing raw materials 2. Lavender (*Lavandula angustifolia* Mill.) Indian Perfum 2008;52:50-55. Data cited in ref. 33

29 Sahraoui N, Abert Vian M, Bornard I, Boutekedjiret C, Chemat F. Improved microwave steam distillation apparatus for isolation of essential oils. Comparison with conventional steam distillation. J Chromatogr A 2008;1210:229-233

30 Romeo FV, DeLuca S, Piscopo A, Poiana M. Antimicrobial effect of some essential oils. J Essent Oil Res 2008;20:373-379

31 Lane WA, Mahmoud SS. Composition of essential oil from *Lavandula angustifolia* and L. *intermedia* varieties grown in British Columbia, Canada. Nat Prod Commun 2008;3:1361-1366

32 Lawrence BM. Essential oils 2008-2011., Carol Stream, IL: Allured Publishing Corporation, 2012:29-33 (reprinted from Perfum Flavor 2008;33(1):38-? (last page unknown))

33 Lawrence BM. Essential oils 2008-2011. Carol Stream, IL: Allured Publishing Corporation, 2012:177-179 (reprinted from Perfum Flavor 2010;35(12):49-? (last page unknown))

34 Renaud ENC, Charles DJ, Simon JE. Essential oil quantity and composition from 10 cultivars of organically grown lavender and lavandin. J Essent Oil Res 2001;13:269-273

35 Śmigielski KB, Prusinowska R, Krosowiak K, Sikora M. Comparison of qualitative and quantitative chemical composition of hydrolate and essential oils of lavender (*Lavandula angustifolia*). J Essent Oil Res 2013;25:291-299

36 Mantovani ALL, Vieira GPG, Cunha WR, Groppo M, Santos RA, Rodrigues V, et al. Chemical composition, antischistosomal and cytotoxic effects of the essential oil of *Lavandula angustifolia* grown in Southeastern Brazil. Revista Brasileira de Farmacognosia 2013;23:877-884

37 Bosilcov A. Lavender: A key perfumery material. Dissertation. International Centre for Aroma Trades Studies, Plymouth University, Plymouth, UK, 2010

38 Tarek N, Hassan HM, AbdelGhani SMM, Radwan IA, Hammouda O, El-Gendy AO. Comparative chemical and antimicrobial study of nine essential oils obtained from medicinal plants growing in Egypt. Beni-Suef Univ J Basic Appl Sci 2014;3:149-156

39 Süntar I, Akkol EK, Tosun A, Keleş H. Comparative pharmacological and phytochemical investigation on the wound-healing effects of the frequently used essential oils. J Essent Oil Res 2014;26:41-49

40 Stupar M, Grbić ML, Džamić A, Unković N, Ristić M, Jelikić A, Vukojević J. Antifungal activity of selected essential oils and biocide benzalkonium chloride against the fungi isolated from cultural heritage objects. South Afr J Bot 2014;93:118-124

41 Prusinowska R, Śmigielski KB. Composition, biological properties and therapeutic effects of lavender (*Lavandula angustifolia* L.). A review. Herba Polonica 2014;60(2):56-66

42 Panahi Y, Akhavan A, Sahebkar A, Hosseini SM, Taghizadeh M, Akbari H, Sharif MR, et al. Investigation of the effectiveness of *Syzygium aromaticum*, *Lavandula angustifolia* and *Geranium robertianum* essential oils in the treatment of acute external otitis: A comparative trial with ciprofloxacin. J Microbiol Immunol Infect 2014;47:211-216

43 Nikolić M, Jovanović KK, Marković T, Marković D, Gligorijević N, Radulović S, Soković M. Chemical composition, antimicrobial, and cytotoxic properties of five Lamiaceae essential oils. Ind Crops Prod 2014;61:225-232

44 Tayarani-Najaran Z, Amiri A, Karimi G, Emami SA, Asili J, Mousavi SH. Comparative studies of cytotoxic and apoptotic properties of different extracts and the essential oil of *Lavandula angustifolia* on malignant and normal cells. Nutrit Cancer 2014;66:424-434

45 Nađalin V, Lepojević Z, Ristić M, Vladić J, Nikolovski B, Adamović D. Investigation of cultivated lavender (*Lavandula officinalis* L.) extraction and its extracts. Chem Ind Chem Eng Q 2014;20:71-86

46 Kačániová M, Vukovič N, Horská E, Šalamon I, Bobková A, Hleba L, et al. Antibacterial activity against *Clostridium* genus and antiradical activity of the essential oils from different origin. J Environ Sci Health, Part B: Pesticides, Food Contaminants, and Agricultural Wastes 2014;49:505-512

47 Hassiotis CN, Ntana F, Lazari DM, Poulios S, Vlachonasios KE. Environmental and developmental factors affect essential oil production and quality of *Lavandula angustifolia* during flowering period. Ind Crops Prod 2014;62:359-366

48 Gismondi A, Canuti L, Grispo M, Canini A. Biochemical composition and antioxidant properties of *Lavandula angustifolia* Miller essential oil are shielded by propolis against UV radiations. Photochem Photobiology 2014;90:702-708

49 Calinescu I, Gavrila AI, Ivopol M, Ivopol GC, Popescu M, Mircioaga N. Microwave assisted extraction of essential oils from enzymatically pretreated lavender (*Lavandula angustifolia* Miller). Cent Eur J Chem 2014;12:829-836

50 Murbach Teles Andrade BF, Nunes Barbosa L, da Silva Probst I, Fernandes A Júnior. Antimicrobial activity of essential oils. J Essent Oil Res 2014;26:34-40

51 Zheljazkov VD, Cantrell CL, Astatkie T, Jeliazkova E. Distillation time effect on lavender essential oil yield and composition. J Oleo Sci 2013;62(4):195-199

52 Yazdani E, Jalali Sendi J, Aliakbar A, Senthil-Nathan S. Effect of *Lavandula angustifolia* essential oil against lesser mulberry pyralid *Glyphodes pyloalis* Walker (Lep: Pyralidae) and identification of its major derivatives. Pest Biochem Physiol 2013;107:250-257

53 Soltani R, Soheilipour S, Hajhashemi V, Asghari G, Bagheri M, Molavi M. Evaluation of the effect of aromatherapy with lavender essential oil on post-tonsillectomy pain in pediatric patients: A randomized controlled trial. Int J Pediat Otorhinolaryngol 2013;77:1579-1581

54 Maietti S, Rossi D, Guerrini A, Useli C, Romagnoli C, Poli F, et al. A multivariate analysis approach to the study of chemical and functional properties of chemodiverse plant derivatives: lavender essential oils. Flavour Fragr J 2013;28:144-154

55 Pereira Machado M, Ciotta N, Deschamps C, Zanette F, Coceo LC, Biasi LA. *In vitro* propagation and chemical characterization of the essential oil of *Lavandula angustifolia* cultivated in Southern Brazil. Ciencia Rural, Santa Maria 2013;43:283-289 (in Portuguese)

56 Karamaouna F, Kimbaris A, Michaelakis A, Papachristos D, Polissiou M, Papatsakona P, Tsora E. Insecticidal activity of plant essential oils against the vine mealybug, *Planococcus ficus*. J Insect Sci 2013;13: Article 142 (12 pages)

57 Jianu C, Pop G, Gruia AT, Horhat FG. Chemical composition and antimicrobial activity of essential oils of lavender (*Lavandula angustifolia*) and lavandin (*Lavandula x intermedia*) grown in Western Romania. Int J Agric Biol 2013;15:772-776

58 Canale A, Benelli G, Conti B, Lenzi G, Flamini G, Francini A, Cioni PL. Ingestion toxicity of three Lamiaceae essential oils incorporated in protein baits against the olive fruit fly, *Bactrocera oleae* (Rossi) (Diptera Tephritidae). Nat Prod Res 2013;27:2091-2099

59 Danh LT, Han LN, Anh Triet ND, Zhao J, Mammucari R, Foster N. Comparison of chemical composition, antioxidant and antimicrobial activity of lavender (*Lavandula angustifolia* L.) essential oils extracted by supercritical CO_2, hexane and hydrodistillation. Food Bioprocess Technol 2013;6:3481-3489

60 Adaszyńska M, Swarcewicz M, Dzięcioł M, Dobrowolska A. Comparison of chemical composition and antibacterial activity of lavender varieties from Poland. Nat Prod Res 2013;27:1497-1501

61 Zheljazkov VD, Astatkie T, Hristov AN. Lavender and hyssop productivity, oil content, and bioactivity as a function of harvest time and drying. Ind Crops Prod 2012;36:222-228

62 Tarakemeh A, Rowshan V, Najafian S. Essential oil content and composition of *Lavandula angustifolia* Mill. as affected by drying method and extraction time. Anal Chem Lett 2012;2:244-249

63 Santana O, Cabrera R, Giménez C, González-Coloma A, Sánchez-Vioque R, de los Mozos-Pascual M, et al. Perfil químico y biológico de aceites esenciales de plantas aromáticas de interés agro-industrial en Castilla-La Mancha (España). Grasas y Aceites, 2012;63:214-222 (in Spanish)

64 Raina AP, Negi KS. Comparative essential oil composition of *Lavendula* species from India. J Herbs Spices Med Plants 2012;18:268-273

65 Perry R, Terry R, Watson LK, Ernst E. Is lavender an anxiolytic drug? A systematic review of randomised clinical trials. Phytomedicine 2012;19:825-835

66 Pavela R. Sublethal effects of some essential oils on the cotton leafworm *Spodoptera littoralis* (Boisduval). J Essent Oil Bear Plants 2012;15:144-156

67 Miladinović D, Ilić BD, Mihajilov-Krstev TM, Nikolić ND, Miladinović LC, Cvetković OG. Investigation of the chemical composition–antibacterial activity relationship of essential oils by chemometric methods. Anal Bioanal Chem 2012;403:1007-1018

68 Kütükoğlu F, Girişgin AO, Aydin L. Varroacidal efficacies of essential oils extracted from *Lavandula officinalis*, *Foeniculum vulgare*, and *Laurus nobilis* in naturally infested honeybee (*Apis mellifera* L.) colonies. Turk J Vet Anim Sci 2012;36:554-559

69 Fismer KL, Pilkington K. Lavender and sleep: A systematic review of the evidence. Eur J Integr Med 2012;e436-e447

70 Djenane D, Aïder M, Yangüela J, Idir L, Gómez D, Roncalés P. Antioxidant and antibacterial effects of *Lavandula* and *Mentha* essential oils in minced beef inoculated with *E. coli* O157:H7 and *S. aureus* during storage at abuse refrigeration temperature. Meat Science 2012;92:667-674

71 Bertoli A, Conti B, Mazzoni V, Meini L, Pistelli L. Volatile chemical composition and bioactivity of six essential oils against the stored food insect *Sitophilus zeamais* Motsch. (Coleoptera Dryophthoridae). Nat Prod Res 2012;26:2063-2071. Data also presented in ref. 83

72 Benelli G, Flamini G, Canale A, Cioni PL, Conti B. Toxicity of some essential oil formulations against the Mediterranean fruit fly *Ceratitis capitata* (Wiedemann) (Diptera Tephritidae). Crop Protection 2012;42:223-229

73 Yang S-O, Kim Y, Kim H-S, Hyun S-H, Kim S-H, Choi HK, Marriott PJ. Rapid sequential separation of essential oil compounds using continuous heart-cut multi-dimensional gas chromatography–mass spectrometry. J Chromatogr A 2011;1218:2626-2634

74 Viuda-Martos M, Mohamady MA, Fernández-López J, Abd ElRazik KA, Omer EA, Pérez-Alvarez JA, Sendra E. *In vitro* antioxidant and antibacterial activities of essentials oils obtained from Egyptian aromatic plants. Food Control 2011;22:1715-1722

75 Smigielski K, Prusinowska R, Raj A, Sikora M, Wolińska K, Gruska R. Effect of drying on the composition of essential oil from *Lavandula angustifolia*. J Essent Oil Bear Plants 2011;14:532-542

76 Hussain AI, Anwar F, Nigam PS, Sarker SD, Moore JE, Rao JR, Mazumdar A. Antibacterial activity of some Lamiaceae essential oils using resazurin as an indicator of cell growth. LWT - Food Sci Technol 2011;44:1199-1206

77 Hassanzadeh MK, Emami SA, Asili J, Najaran ZT. Review of the essential oil composition of Iranian Lamiaceae. J Essent Oil Res 2011;23:35-74

78 Djenane D, Lefsih K, Yangüela J, Roncalés P. Composition chimique et activité anti-*Salmonella enteritidis* CECT 4300 des huiles essentielles d'*Eucalyptus globulus*, de *Lavandula angustifolia* et de *Satureja hortensis*. Tests *in vitro* et efficacité sur les oeufs entiers liquides conservés à 7 ± 1 °C. Phytothérapie 2011;9:343-353 (in French)

79 Alnamer R, Alaoui K, Bouidida EH, Benjouad A, Cherrah Y. Toxicity and psychotropic activity of essential oils of *Rosmarinus officinalis* and *Lavandula officinalis* from Morocco. J Biol Active Prod Nat 2011;1:262-272

80 Yang S-A, Jeon S-K, Lee E-J, Shim C-H, Lee I-S. Comparative study of the chemical composition and antioxidant activity of six essential oils and their components. Nat Prod Res 2010;24:140-151

81 Verma RS, Rahman LU, Chanotiya GS, Verma RK, Chauhan A, Yadav A, et al. Essential oil composition of *Lavandula angustifolia* Mill. cultivated in the mid hills of Uttarakhand, India. J Serb Chem 2010;75:343-348

82 Hassiotis CN, Lazari DM, Vlachonasios KE. The effects of habitat type and diurnal harvest on essential oil yield and composition of *Lavandula angustifolia* Mill. Fresenius Environ Bull 2010;19:1491-1498

83 Conti B, Canale A, Bertoli A, Gozzini F, Pistelli L. Essential oil composition and larvicidal activity of six Mediterranean aromatic plants against the mosquito *Aedes albopictus* (Diptera: Culicidae). Parasitol Res 2010;107:1455-1461. Data also presented in ref. 71

84 Roller S, Ernest N, Buckle J. The antimicrobial activity of high-necrodane and other lavender oils on methicillin-sensitive and -resistant *Staphylococcus aureus* (MSSA and MRSA). J Alter Compl Med 2009;15:275-279

85 Da Porto C, Decorti D, Kikic I. Flavour compounds of *Lavandula angustifolia* L. to use in food manufacturing: Comparison of three different extraction methods. Food Chem 2009;112:1072-1078

86 Farhat A, Ginies C, Romdhane M, Chemat F. Eco-friendly and cleaner process for isolation of essential oil using microwave energy. Experimental and theoretical study. J Chromatogr A 2009;1216:5077-5085

87 Lakusić B, Lakusić D, Ristić M, Marcetić M, Slavkovska V. Seasonal variations in the composition of the essential oils of Lavandula angustifolia (Lamiacae). Nat Prod Commun 2014;9:859-862

88 Mayaud L, Carricajo A, Zhiri A, Aubert G. Comparison of bacteriostatic and bactericidal activity of 13 essential oils against strains with varying sensitivity to antibiotics. Lett Appl Microbiol 2008;47:167-173

89 Kaloustian J, Chevalier J, Mikail C, Martino M, Abou L, Vergnes M-F. Étude de six huiles essentielles: composition chimique et activité antibactérienne. Phytothérapie 2008;6:160-164

90 Cong Y, Abulizi P, Zhi L, Wang X, Mirensha. Chemical composition of the essential oil of *Lavandula angustifolia* from Xinjiang, China. Chem Nat Comp 2008;44:810

91 Xu F, Uebaba K, Ogawa H, Tatsuse T, Wang B-H, Hisajima T, Venkatraman S. Pharmaco-physio-psychologic effect of Ayurvedic oil-dripping treatment using an essential oil from *Lavandula angustifolia*. J Alter Compl Med 2008;14:947-956

92 Umezu T, Nagano K, Ito H, Kosakai K, Sakaniwa M, Morita M. Anticonflict effects of lavender oil and identification of its active constituents. Pharmacol Biochem Behavior 2006;85:713-721

93 Soković M, van Griensven LJLD. Antimicrobial activity of essential oils and their components against the three major pathogens of the cultivated button mushroom, *Agaricus bisporus*. Eur J Plant Pathol 2006;116:211-224. Data also presented in refs. 113 and 115

94 Morgan TJ, Morden WE, Al-muhareb E, Herod A, Kandiyoti R. Essential oils investigated by size exclusion chromatography and gas chromatography-mass spectrometry. Energy & Fuels 2006;20:734-737

95 Afsharypuor S, Azarbayejany N. Chemical constituents of the flower essential oil of *Lavandula officinalis* Chaix. from Isfahan (Iran). Iran J Pharm Sci 2006;2:169-172

96 Prashar A, Locke IC, Evans CS. Cytotoxicity of lavender oil and its major components to human skin cells. Cell Prolif 2004;37:221-229

97 Hajhashemi V, Ghannadi A, Sharif B. Anti-inflammatory and analgesic properties of the leaf extracts and essential oil of *Lavandula angustifolia* Mill. J Ethnopharmacol 2003;89:67-71

98 Daferera DJ, Tarantilis PA, Polissiou MG. Characterization of essential oils from Lamiaceae species by Fourier transform Raman spectroscopy. J Agric Food Chem 2002;50:5503-5507. Data also presented in ref. 17

99 Adam K, Sivropoulou A, Kokkini S, Lanaras T, Arsenakis M. Antifungal activities of *Origanum vulgare* subsp. *hirtum*, *Mentha spicata*, *Lavandula angustifolia*, and *Salvia fruticosa* essential oils against human pathogenic fungi. J Agric Food Chem 1998;46:1739-1745

100 Milhau G, Valentin A, Benoit F, Mallié M, Bastide J-M, Pélissier Y, Bessière J-M. *In vitro* antimalarial activity of eight essential oils. J Essent Oil Res 1997;9:329-333

101 Perrucci S, Macchioni G, Cioni PC, Flamini G, Morelli I, Taccini F. The activity of volatile compounds from *Lavandula angustifolia* against *Psoroptes cuniculi*. Phytother Res 1996;10:5-8

102 Guillen MD, Cabo N. Characterisation of the essential oils of some cultivated aromatic plants of industrial interest. J Sci Food Agric 1996;70:359-363

103 Boelens MH. Chemical and sensory evaluations of *Lavandula* oils. Perfum Flavor 1995;20(May/June):23-51

104 Dupuy N, Gaydou V, Kister J. Quantitative analysis of lavender (Lavandula angustifolia) essential oil using multiblock data from infrared spectroscopy. Am J Anal Chem 2014;5:633-645

105 Bakhsha F, Mazandarani M, Aryaei M, Jafari SY, Bayate H. Phytochemical and anti-oxidant activity of *Lavandula angustifolia* Mill. essential oil on preoperative anxiety in patients undergoing diagnostic curettage. Int J Women's Health Reprod Sci 2014;2(4):268-271

106 Abdel-Hady NM, Abdallah GM, Idris NF. Phytochemical studies and *in vivo* antioxidant activity of two *Lavandula* species (Lamiaceae) against streptozotocin induced oxidative stress in albino rats. J Biomed Pharm Res 2014;3(4):30-40

107 Zagorcheva T, Stanev S, Rusanov K, Atanassov I. Comparative GC/MS analysis of lavender (*Lavandula angustifolia* Mill.) inflorescence and essential oil volatiles. Agricult Sci Technol 2013;5:459-462

108 Kara N, Baydar H. Determination of lavender and lavandin cultivars (*Lavandula* sp.) containing high quality essential oil in Isparta, Turkey. Turk J Field Crops 2013;18:58-65

109 Hamad KJ, Al-Shaheen SJA, Kaskoos RA, Ahamad J, Jameel M, Mir SR. Essential oil composition and antioxidant activity of *Lavandula angustifolia* from Iraq. Int Res J Pharm 2013;4(4):117-120

110 Najafian S, Rowshan V, Tarakemeh A. Comparing essential oil composition and essential oil yield of *Rosemarinus officinalis* and *Lavandula angustifolia* before and full flowering stages. Int J Appl Biol Pharm Technol 2012;3:212-218

111 Wesołowska A, Jadczak D, Grzeszczuk M. Influence of distillation time on the content and composition of essential oil isolated from lavender (*Lavandula angustifolia* Mill.). Herba Polonica 2010;56(3):24-36

112 de Rapper S, Kamatou G, Viljoen A, van Vuuren S. The *in vitro* antimicrobial activity of *Lavandula angustifolia* essential oil in combination with other aroma-therapeutic oils. Evidence-Based Compl Alter Med Volume 2013, Article ID 852049, 10 pages. http://dx.doi.org/10.1155/2013/852049

113 Soković M, Glamočlija J, Marin PD, Brkić D, van Griensven LJLD. Antibacterial effects of the essential oils of commonly consumed medicinal herbs using an *in vitro* model. Molecules 2010;15:7532-7546. Data also presented in refs. 93 and 115

114 Şerban ES, Socaci SA, Tofană M, Maier SC, Bojiţă MT. Chemical composition of some essential oils of Lamiaceae family. Clujul Medical 2010;83(2):286-289

115 Soković M, Marin PD, Brkić D, van Griensven LJLD. Chemical composition and antibacterial activity of essential oils of ten aromatic plants against human pathogenic bacteria. Food 2008;2: (exact page number unknown, article can be accessed on-line). Data also presented in refs. 93 and 113

116 Ihsan SA. Essential oil composition of *Lavandula officinalis* L.. grown in Jordan. J Kerbala Univ Scientific 2007;5:18-21

117 de Rapper S. Synergistic interactions of lavender essential oil. Ph.D. Thesis, University of the Witwatersrand, Johannesburg, South Africa, 2013

118 Cavanagh HMA, Wilkinson JM. Biological activities of lavender essential oil. Phytother Res 2002;16:301-308

119 Williams DG. The chemistry of essential oils, 2nd Ed. Port Washington, NY, USA: Micelle Press, 2008: 186-187. Data cited in ref. 33

120 Kaiser R, Lamparsky D. New carbonyl compounds in the high boiling fraction of lavender oil. I. Helv Chim Acta 1983;66:1835-1842. Data cited in ref. 103

121 Kaiser R, Lamparsky D. New carbonyl compounds in the high boiling fraction of lavender oil. II. Helv Chim Acta 1983;66:1843-1849. Data cited in ref. 103

122 Kaiser R, Lamparsky D. New carbonyl compounds in the high boiling fraction of lavender oil. III. Helv Chim Acta 1984;67:1184-1197. Data cited in ref. 103

123 Naef R, Morris AF. Lavender-Lavandin. A comparison. Rivista Ital EPPOS (Numero Speciale):364-377. Data cited in ref. 103 (not ascertained that the oil was an *essential* oil as per our definition)

124 Nazari F, Shaabani Sh. Phytochemical analysis of essential oil from *Lavandula angustifolia* of Iran. Planta Med 2009;75-PD60. DOI: 10.1055/s-0029-1234539

125 Keane FM, Smith HR, White IR, Rycroft RJG. Occupational allergic contact dermatitis in two aromatherapists. Contact Dermatitis 2000;43:49-51

126 Rudzki E, Grzywa Z, Bruo WS. Sensitivity to 35 essential oils. Contact Dermatitis 1976;2:196-200

127 Sugiura M, Hayakawa R, Kato Y, Sugiura K, Hashimoto R. Results of patch testing with lavender oil in Japan. Contact Dermatitis 2000;43:157-160

128 Calnan CD. Oil of cloves, laurel, lavender, peppermint. Contact Dermatitis Newsletter 1970;7:148

129 Zina G, Bonu G. Phenergan cream (role of base constituents). Contact Dermatitis Newsletter 1969;6:117

130 Brandao FM. Occupational allergy to lavender oil. Contact Dermatitis 1986;15:249-250

131 de Groot AC. Airborne allergic contact dermatitis from tea tree oil. Contact Dermatitis 1996;35:304-305

132 Schaller M, Korting HC. Allergic airborne contact dermatitis from essential oils used in aromatherapy. Clin Exp Dermatol 1995;20:143-145

133 Coulson IH, Ali Khan AS. Facial 'pillow' dermatitis due to lavender oil allergy. Contact Dermatitis 1999;41:111

134 Dharmagunawardena B, Takwale A, Sanders KJ, Cannan S, Roger A, Ilchyshyn A. Gas chromatography: an investigative tool in multiple allergies to essential oils. Contact Dermatitis 2002;47:288-292

135 Trattner A, David M, Lazarov A. Occupational contact dermatitis due to essential oils. Contact Dermatitis 2008;58:282-284

136 Bleasel N, Tate B, Rademaker M. Allergic contact dermatitis following exposure to essential oils. Australas J Dermatol 2002;43:211-213

137 Boonchai W, Lamtharachai P, Sunthonpalin P. Occupational allergic contact dermatitis from essential oils in aromatherapists. Contact Dermatitis 2007; 56:181-182

138 Newsham J, Rai S, Williams JDL. Two cases of allergic contact dermatitis to neroli oil. Br J Dermatol 2011;165(Suppl.1):76

139 Le Coulant P, Texier L, Malleville J, Doussy NN. Sensibilization à une crème antihistaminique. Bull Soc Franç Derm Syph 1964;71:234-237

140 Ménard E. Les dermatoses professionelles. Concours Médicale 1961;83:4308-4311

141 Nakayama H, Matsuo S, Hayakawa K, Takahashi K, Shigematsu T, Ota S. Pigmented cosmetic dermatitis. Int J Dermatol 1984;23:299-305

142 Nakayama H, Harada R, Toda M. Pigmented cosmetic dermatitis. Int J Dermatol 1976;15:673-675

143 Warshaw EM, Belsito DV, Taylor JS, Sasseville D, DeKoven JG, Zirwas MJ, et al. North American Contact Dermatitis Group patch test results: 2009 to 2010. Dermatitis 2013;24:50-59

144 Rudzki E, Grzywa Z. Allergy to perfume mixture. Contact Dermatitis 1986;15:115-116

145 Thomson KF, Wilkinson SM. Allergic contact dermatitis to plant extracts in patients with cosmetic dermatitis. Br J Dermatol 2000;142:84-88

146 Wu PA, James WD. Lavender. Dermatitis 2011;22:344-347

147 Varma S, Blackford S, Statham BN, Blackwell A. Combined contact allergy to tea tree oil and lavender oil complicating chronic vulvovaginitis. Contact Dermatitis 2000;42:309-310

148 De Mozzi P, Johnston GA. An outbreak of allergic contact dermatitis caused by citral in beauticians working in a health spa. Contact Dermatitis 2014;70:377-379

149 Nardelli A, Drieghe J, Claes L, Boey L, Goossens A. Fragrance allergens in 'specific' cosmetic products. Contact Dermatitis 2011;64:212-219

150 Nardelli A, Thijs L, Janssen K, Goossens A. Rosa centifolia in a 'non-scented' moisturizing body lotion as a cause of allergic contact dermatitis. Contact Dermatitis 2009;61:306-309

151 Hagvall L, Sköld M, Brared-Christensson J, Börje A, Karlberg A-T. Lavender oil lacks natural protection against autoxidation, forming strong contact allergens on air exposure. Contact Dermatitis 2008;59:143-150

152 Goiriz R, Delgado-Jiménez Y, Sánchez-Pérez J, García-Diez A. Photoallergic contact dermatitis from lavender oil in topical ketoprofen. Contact Dermatitis 2007;57:381-382

153 Warshaw EM, Maibach HI, Taylor JS, Sasseville D, DeKoven JG, Zirwas MJ, et al. North American Contact Dermatitis Group patch test results: 2011-2012. Dermatitis 2015;26:49-59

154 An S, Lee AY, Lee CH, Kim D-W, Hahm JH, Kim K-J, et al. Fragrance contact dermatitis in Korea: a joint study. Contact Dermatitis 2005;53:320-323

155 Matthieu L, Meuleman L, van Hecke E, Blondeel A, Dezfoulian B, Constandt L, Goossens A. Contact and photocontact allergy to ketoprofen. The Belgian experience. Contact Dermatitis 2004;50:238-241

156 Nardelli A, D'Hooge E, Drieghe J, Dooms M, Goossens A. Allergic contact dermatitis from fragrance components in specific topical pharmaceutical products in Belgium. Contact Dermatitis 2009;60:303-313

157 Larsen W, Nakayama H, Fischer T, Elsner P, Frosch P, Burrows D, et al. Fragrance contact dermatitis – a worldwide multicenter investigation (Part III). Contact Dermatitis 2002;46:141-144

158 Nakayama H, Hanaoka H, Ohshiro A. Allergen Controlled System (ACS). Tokyo, Japan: Kanehara Shuppan, 1974:42. Data cited in ref. 159

159 Mitchell JC. Contact hypersensitivity to some perfume materials. Contact Dermatitis 1975;1:197-199

160 Srivastava PK, Bajaj AK. Ylang-ylang oil not an uncommon sensitizer in India. Indian J Dermatol 2014;59:200-201

161 Goossens A, Merckx L. Allergic contact dermatitis from farnesol in a deodorant. Contact Dermatitis 1997;37:179-180

162 Meynadier JM, Meynadier J, Peyron JL, Peyron L. Formes cliniques des manifestations cutanées d'allergie aux parfums. Ann Dermatol Venereol 1986;113:31-39

163 Schorr WF. Cosmetic allergy: Diagnosis, incidence, and management. Cutis 1974;14:844-850

164 Ngangu Z, Samsoen M, Foussereau J. Einige Aspekte zur Kosmetika-allergie in Strassburg. Dermatosen 1983;31:126-129

165 Vente C, Fuchs T. Contact dermatitis due to oil of turpentine in a porcelain painter. Contact Dermatitis 1997;37:187

Chapter 5.50   LEMON OIL

DEFINITION
Lemon oil (essential oil of lemon) is the essential oil obtained from the pericarp (peel) of the lemon, *Citrus limon* (L.) Burm. f.

INCI NOMENCLATURE
Description/definition: Citrus limon peel oil is the volatile oil obtained from the fresh peel of the lemon, *Citrus limon* (L.), Rutaceae
INCI name EU: Citrus limon peel oil
INCI name USA: Citrus limon (lemon) peel oil
CAS registry number (s): 84929-31-7; 8008-56-8
EINECS number(s): 284-515-8

ISO (INTERNATIONAL ORGANIZATION FOR STANDARDIZATION) STANDARD
ISO number: 855
ISO name: Essential oil of lemon
Botanical origin: *Citrus limon* (L.) Burm. f.
Parts of plant used: Pericarp (peel)
ISO values: ISO values (minimum and maximum concentrations) are shown in Table 5.50.1.

By ISO definition, all citrus essential oils except lime oil are produced by expression; oils obtained from citrus fruits by distillation may not be called essential oils according to ISO criteria (except lime oil). In industry, however, sometimes residues from expression of the citrus peels undergo steam distillation, to obtain the remaining oil; these volatile oils may then be added to the expressed oil. As some of the compounds undergo changes forced by high temperature during distillation, this addition (which is an adulteration) changes the composition of the essential oil. Because of this, and also because it cannot be excluded that oils entirely produced by hydrodistillation reach the market, both the ingredients found in 'genuine' lemon essential oils obtained by expression and those that may be present in hydrodistilled oils are presented here.

THE PLANT, THE OIL, AND THEIR USES
Citrus limon (L.) Burm. f. is a small evergreen tree that produces ellipsoidal yellow fruit, the lemon. The origin of the lemon tree is unknown, likely south China and Myanmar. It is cultivated in many subtropical countries. The main producing countries are China, Mexico and India, followed by Argentina, Brazil, the USA, Turkey, Spain, Iran and Italy, notably the south (Calabria, Sicily) (6,16, Mansfeld's World Database of Agriculture and Horticultural Crops, http://mansfeld.ipk-gatersleben. de; Food and Agriculture Organization of the United Nations, http://faostat.fao.org/). Most of the production of lemons is destined for extracting the juice from the fruits (6). Lemon juice has many applications in food and beverages, serves cosmetic, pharmaceutical and several technical purposes, is exploited as a source of vitamin C, pectin, and enzymes and may be used for cleaning (due to its acidity; the juice may contain up to 6% citric acid). Lemon juice is said to be useful for rheumatic affections and pain, cough, dyspepsia, vomiting, some forms of headache, sore throat and bleeding from the nose (28).

After extraction of the juice many 'waste' products, including peel, membranes and seeds, remain. These are an important source of desiccated pulp, dietary fiber, essential oils, limonene, ethanol, oils obtained from the seeds, and pectin (13). 'Non-nutritional' compounds, such as limonoids and flavonoids (13), vitamins, minerals, and dietary fiber, are used as functional ingredients in the development of healthy foods (functional foods) (29).

From the peel of the fruits the essential oil of lemon is obtained by cold-pressing, and is widely used in the food and beverage industries as flavoring agent, e.g., in baked goods, confectionery, desserts, ice creams, and soft and alcoholic drinks (30). In the pharmaceutical industries it is used as flavoring agent to mask unpleasant tastes of drugs. In perfumery, it forms the base of many fragrance compositions and is employed in other cosmetic products as well (6,13). The essential oil of lemon is also useful as a wood cleaner and polish, where its solvent properties are employed to dissolve old wax, fingerprints, and grime. In addition, lemon oil may serve as a nontoxic insecticide treatment (Mansfeld's World Database of Agriculture and Horticultural Crops, http://mansfeld.ipk-gatersleben.de; Food and Agriculture Organization of the United Nations, http://faostat.fao.org/). In addition, oil of lemon is perceived to be stomachic and carminative, and is said to be used in leprotic ulcers, skin diseases, sunburn and pruritus (28). It is also employed in aromatherapy practices (45,46).

The history, global distribution and nutritional importance of lemon and other *Citrus* fruits have been reviewed in ref. 2.

CHEMICAL COMPOSITION
Lemon oil is a pale yellow to dark green, clear mobile liquid, sometimes cloudy at lower temperature, which has a fresh citrusy green peel note odor. The yield of

Table 5.50.1 ISO values (%) for lemon oil [a]

Compound	CAS	Minimum	Maximum
Limonene	138-86-3	59.0	80.0
β-Pinene	127-91-3	7.0	16.5
γ-Terpinene	99-85-4	6.0	12.0
α-Pinene	80-56-8	1.4	3.0
Sabinene	3387-41-5	1.3	3.0
Geranial	141-27-5	0.5	2.0
Neral	106-26-3	0.2	1.2
β-Bisabolene	495-61-4	0.2	0.9
Geranyl acetate	105-87-3	tr	0.65
Neryl acetate	141-12-8	0.1	0.6
α-Thujene	2867-05-2	0.2	0.5
p-Cymene	99-87-6	tr	0.4
α-Terpineol	98-55-5	0.0	0.4

[a] ISO 855 Essential oil of lemon, obtained by expression ©ISO 2003; Geneva, Switzerland, www.iso.org

essential oil from the peel of *Citrus limon* (L.) Burm. f. generally varies from 0.5 to 1.5 per cent. The main producing countries of this oil are USA, Argentina, Brazil, India, Mexico, Spain, Italy, South Africa and Israel. Cold-pressed lemon and other *Citrus* species essential oils consist of a volatile fraction, which represents 85–99% of the oil, and a non-volatile fraction 1–15%. The latter fraction contains oxygen heterocyclic compounds, especially coumarins, psoralens and polymethoxyflavones (Table 5.50.4). These components play an important role in the characterization of cold-pressed *Citrus* oils, since the composition of this fraction is characteristic of each oil. Moreover, many of the pharmacological and toxicological activities possessed by *Citrus* oils are related to these components (1).

Lemon oil is typically obtained from lemon peel by cold pressing, and it may then be subjected to distillation and other processing steps in order to isolate, purify and refine the oil (30,35). Commercially, lemon oils are available in a number of different forms ('folds'), which differ depending on the degree of distillation used. Lemon oils extracted by cold pressing are usually referred to as single-fold oils, while those that have undergone further processing (such as distillation) are referred to as higher fold oils. Lemon oils may be blended with each other, mixed with distilled lemon oils (31) or solvents, or otherwise be adulterated to create higher fold oils, e.g., threefold (3), fivefold (5), or tenfold (10) oils. Thus, the chemical composition of lemon oils

changes appreciably as a result of distillation and later processing (30)

Literature data (up to June 2, 2014) on the chemical composition of lemon oils and unpublished analytical data from one of us (E.S.) are shown in Table 5.50.2 in alphabetical order. In lemon oils from various origins, over 245 chemicals have been identified. The major compounds found in lemon oils from different sources are shown in Table 5.50.3. They include (highest concentrations in any study given) limonene (73.0%), β-pinene (19.3%), γ-terpinene (12.7%), sabinene (5.8%), geranial (5.4%), α-pinene (4.4%), nerol (4.3%) and myrcene (3.5%). Well-known ingredients of lemon oils that were present in high concentrations (>7%) in one study were linalyl acetate (23.3%), linalool (16.0%) and *p*-cymene (7.8%).

Commercial oils

The ten chemicals that had the highest maximum concentrations in 178 commercial lemon essential oil samples (concentration ranges provided) are the following: limonene (53.5-73.0%), β-pinene (7.4-17.8%), γ-terpinene (3.1-12.4%), sabinene (1.1-2.9%), α-pinene (1.5-2.4%), myrcene (1.2-2.2%), *p*-cymene (0.1-2.2%), geranial (0.6-2.1%), neral (0.3-1.4%) and geranyl acetate (0.1-1.3%) (Erich Schmidt, unpublished analytical data).

It should be realized that values for chemicals in hydrodistilled oils may vary considerably from genuine cold-pressed lemon oils. According to the authors of

Table 5.50.2 Constituents identified in lemon oils

Constituent	CAS	Percentage and range						
		A	B (3)	C (3)	D (3)	E (37)	F	G
Acetic acid	64-19-7		+					
cis-1-Acetoxy-3,7-di-methyl-2,7-octadien-6-ol	112362-25-1		+					
trans-1-Acetoxy-3,7-dimethyl-2,7-octadien-6-ol	33766-43-7		+					
β-Acoradiene	28477-64-7		+					
allo-Aromadendrene	25246-27-9						0-0.05	
α-Bergamotene	17699-05-7		0.4	0-0.4			0.2-0.5	
cis-α-Bergamotene	18252-46-5		0.04			0.01-0.02	0-0.08	0.2[k]
trans-α-Bergamotene	13474-59-4	0.2-0.6	0.8	0-0.4	0.2-1.0	0.2-0.4	0-0.8	0.3[f,g]
trans-β-Bergamotene	15438-94-5					0.01-0.02	0-0.03	
Bergaptene	484-20-8	tr-0.03						
Bicyclogermacrene	24703-35-3	tr-0.1	0.07	0-0.06	0-1.8	tr-0.05	0-0.1	0.04[g]; 0.5[k]
Bicyclo[3.2.1]oct-3-en-6-one, 4,7-dimethyl-, *exo-cis-*	116764-38-6		0.01					
α-Bisabolene	17627-44-0				0.04-0.07			0.5[g]
(*E*)-α-Bisabolene	25532-79-0		0.02	0-0.01	0-0.03	0.01	0-0.01	
(*Z*)-α-Bisabolene	29837-07-8		0.04	0-0.04	0-0.07	0.03-0.05	0.02-0.08	
β-Bisabolene	495-61-4	0.3-0.9	1.0	0-0.6	0-1.6	0.3-0.7	0-0.9	0.5[f]; 3.6[k]
(*E*)-γ-Bisabolene	53585-13-0		+		0-0.01	tr	0-0.01	
(*Z*)-γ-Bisabolene	13062-00-5		tr	0-0.02	0-0.02		0-0.02	
trans-α-Bisabolene epoxide	111536-37-9			0-tr				
α-Bisabolol	515-69-5		0.03	0-0.03	0-0.05	0.01-0.03	0-0.06	0.2[k]
epi-α-Bisabolol	78148-59-1				0-0.02		0-0.02	
β-Bisabolol	15352-77-9		0.01		0-tr			

Table 5.50.2 Constituents identified in lemon oils *(continued)*

Constituent	CAS	Percentage and range						
		A	B (3)	C (3)	D (3)	E (37)	F	G
Borneol	507-70-0		0.02	0-0.01	0-0.03	tr	0-2.5	
Bornyl acetate	76-49-3		0.01	0-0.01	0-0.06	tr-0.01	0-0.07	
Cadina-1,4-diene	29837-12-5					0-1.3		
γ-Cadinene	39029-41-9					0-0.4		
δ-Cadinene	483-76-1			0-0.02		0-0.4		
α-Cadinol	481-34-5					0-0.01		
δ-Cadinol	19435-97-3			tr-0.01				
Camphene	79-92-5	0.04-0.1	0.1	0-0.1	0-0.7	0.04-0.06	0-0.06	tr[f]; 0.05[g]
Campherenol	18530-03-5		0.03	0-0.02	0-0.04	tr-0.02	0-0.03	
Camphor	76-22-2		0.01	0-0.01	0-0.02	tr-0.01	0-0.6	
(+)-2-Carene	4497-92-1						0-0.7	
(+)-4-Carene	13837-63-3						0-1.3	
δ3-Carene	13466-78-9	0.01-0.04	0.04	0-0.01	0-0.07	tr	0-0.09	0.02[g]
Carvacrol	499-75-2		+				0-0.03	0.2[k]
Carveol	99-48-9						0-0.1	
(E)-Carveol	1197-07-5		+	0-tr	0-tr		0-0.04	0.1[k]
(Z)-Carveol	1197-06-4		+	0-tr			0-0.02	
Carvone	99-49-0			0-tr	0-tr	tr-0.01	0-0.1	0.06[k]
β-Caryophyllene	87-44-5	0.1-0.4	0.8	0.1-0.2	0.1-0.7	0.1-0.3	0.05-0.6	0.3[f]; 1.0[k]
Caryophyllene alcohol	56747-96-7				0-0.01		0-0.01	
Caryophyllene oxide	1139-30-6				0-0.1		0-0.2	0.1[k]
1,4-Cineole	470-67-7		+					
1,8-Cineole	470-82-6		+	0-0.1	0-tr		0-tr	0.1[k]
Citronellal	106-23-0	0.02-0.2	0.2	0-0.1	tr-0.3	0.04-0.1	0-0.2	0.1[f]; 0.7[k]
(R)-(+)-Citronellal (D-)	2385-77-5						0.02-0.2	
Citronellol (β-, DL-)	106-22-9		0.2	0-0.02	0-0.04	tr-0.08	0-0.5	0.09[k]
α-Citronellol							0-0.3	
Citronellyl acetate	150-84-5		0.1	0-0.03	0-0.1	0.01-0.05	0-0.1	
Citronellyl propionate	141-14-0		tr					
α-Copaene	3856-25-5						0-0.07	
Curcumene (ar-; α-)	644-30-4						0-0.03	
γ-Curcumene	28976-68-3		+		0-0.03	tr	0-0.03	
1,2-Cyclohexanediol	931-17-9			0-0.2				
p-Cymene	99-87-6	0.1-2.2	1.7	0.1-7.8[b]	0-1.9	0.1-1.5	0-2.9	0.6[g]; 3.3[k]
p-Cymenene	1195-32-0		+				0-tr	
p-Cymen-8-ol	1197-01-9		+	0-0.01			0-0.02	
Decanal	112-31-2	0.05-0.2	0.1	0-0.06	0-0.1	tr-0.05	0-0.2	0.04[g]
Decane	124-18-5					tr		
(E)-2-Decenal	3913-81-3		+		0-tr			
Decyl acetate	112-17-4		0.01	0-0.03	0-0.05		0-0.02	
cis-Dihydrocarvone	3792-53-8						0-tr	
2,3-Dihydro-1,8-cineole							0-tr	
Dill apiole	484-31-1				0-0.1		0-0.1	
5,7-Dimethoxycoumarin	487-06-9	0-0.06						
2,3-Dimethyl-3-(4-methyl-3-pentenyl)-2-norbornanol	98205-40-4					tr-0.02	0-0.02	
2,7-Dimethyl-2,6-octadien-1-ol	22410-74-8				0-tr			
2,6-Dimethyl-1,3,5,7-octatetraene	90973-78-7						+[h];	
Dodecanal	112-54-9		0.06	0-0.01	0-0.03		0-0.05	
Dodecane	112-40-3					tr-0.03		
Dodecanol	112-53-8						0-0.2	
2-Dodecenal	4826-62-4						0-tr	
β-Elemene	33880-83-0						0-0.07	
γ-Elemene	29873-99-2		0.03	0-0.02	0-0.02			
δ-Elemene	20307-84-0		0.03		0-0.05		0-0.06	
2,3-Epoxycarane	62413-92-7							+[h]
Ethyl benzoate	93-89-0		+					
Eudesmol	51317-08-9						0-0.06	

Table 5.50.2 Constituents identified in lemon oils *(continued)*

Constituent	CAS	Percentage and range						
		A	B (3)	C (3)	D (3)	E (37)	F	G
(E,E)-α-Farnesene	502-61-4			0-tr			0-0.01	
β-Farnesene	502-60-3						0-0.03	
(E)-β-Farnesene	18794-84-8		0.05	0-1.0	0-0.2		0-0.07	
(Z)-β-Farnesene	28973-97-9		0.07	0-0.04	0.03-0.04	0.02-0.04		0.1[k]
(E,E)-Farnesol	106-28-5						0-0.05	
(Z,E)-Farnesol	3790-71-4						0-0.03	
(Z,Z)-Farnesol	16106-95-9						0-0.08	
endo-Fenchol (α-)	512-13-0				0-0.02		0-0.02	
Fenchyl alcohol	1632-73-1						0.01-0.07	
Geranial	141-27-5	0.6-2.1	2.7	0.9-1.7	0.7-4.3	0.7-1.5	0.03-5.4	1.2[f]; 2.1[k]
Geranic acid	459-80-3						0-0.04	
Geraniol	106-24-1	0.02-0.2	0.1	0-0.04	0-0.1	tr-0.02	0-1.3	
(E)-Geraniol (β-)	106-24-1						0.3-2.4	
5-Geranoxypsoralen	7380-40-7	tr-0.5						
Geranyl acetate	105-87-3	0.1-1.3	0.9	0.2-3.2[b]	0.2-1.7	0.2-0.5	0.2-1.0	0.2[f]; 0.3[g]
Geranyl propanoate	105-90-8		+	0-tr	0-0.02	tr-0.01	0-0.1	
Germacrene A	28387-44-2						0-0.01	
Germacrene B	15423-57-1		0.1					
Germacrene D	23986-74-5		0.02	0-0.02	0-0.03		0-1.4	0.2[k]
Germacrene D-4-ol	198991-79-6		0.02		0-0.06	0.01-0.06	0-0.06	
Heptanal	111-71-7		0.01		0-0.01		0-0.08	
Heptanol	53535-33-4		+					
Heptyl acetate	112-06-1		tr		0-tr	tr-0.01		
Hexadecanal	629-80-1		+					
Hexadecanoic acid	57-10-3						0-0.4	
Hexadecanol	51260-59-4				0-tr			
Hexanal	66-25-1		+					
Hexenal	1335-39-3						0-0.01	
(E)-2-Hexenal	6728-26-3						0-0.01	
Hexenol	910923-87-4						0-0.01	
(E)-2-Hexen-1-ol	928-95-0						0-0.02	
(Z)-3-Hexen-1-ol	928-96-1						0-0.04	
α-Humulene	6753-98-6		0.05	0-0.05	0-0.04	0.01	0-1.1	
Hydrocarbons C$_{21}$-C$_{33}$, linear chained			+					
Hydrocarbons C$_{21}$-C$_{33}$, linear chained, isomers			+					
1H-Indole	120-72-9						0-tr	
γ-Isogeraniol	13066-51-8		+				0-tr	
cis-Isogeraniol	5944-20-7						0-0.01	
cis-Isopiperitenol	4017-76-9		+					
trans-Isopiperitenol	4017-77-0		+					
Isopiperitone	58615-39-7		+					
Isopulegol	89-79-2		+				0-0.1	
Limonene	138-86-3	53.5-73.0	71.8	52.6-69.9	56.6-76.0	60.3-70.8	46.5-75.7	67.0[f]; 69.8[g]
Limonene dioxide	96-08-2			0-tr	0-tr			
cis-Limonene oxide	13837-75-7	0.0-0.2	0.02	0-tr	0-0.1	tr-0.04	0-0.2	0.1[k]
trans-Limonene oxide	4959-35-7	tr-0.01	0.02	0-0.1	0-0.02	tr-0.04	0-0.2	
Limonen-10-ol	3269-90-7		+					
Limonen-10-yl-acetate	15111-97-4		+					
Linalool	78-70-6	0.09-0.7	0.5	0.07-16.0[c]	0.1-0.8	0.06-0.2[e]	0-0.9	0.2[k]; 0.4[f]
Linalyl acetate	115-95-7		0.05	23.3[c]			0-tr	
Linalyl anthranilate	7149-26-0				0-tr			
Linalyl isobutyrate	78-35-3			0-0.01				
Linoleic acid	60-33-3				0-0.1		0-0.1	
cis-p-Mentha-1(7),8-dien-2-ol	22626-43-3		+					1.3[k]
trans-p-Mentha-1(7),8-dien-2-ol	21391-84-4		+					
p-Mentha-1,8-dien-9-ol	1946-01-6			0-tr	0-0.01			

Table 5.50.2 Constituents identified in lemon oils *(continued)*

Constituent	CAS	Percentage and range						
		A	B (3)	C (3)	D (3)	E (37)	F	G
cis-p-Mentha-2,8-dien-1-ol	3886-78-0		+					0.1[k]
trans-p-Mentha-2,8-dien-1-ol	4017-77-0		+				0-0.07	
trans-p-Mentha-2,8-dien-9-ol				0-tr				
p-Menth-1-en-9-ol	18479-68-0				0-tr			
cis-p-Menth-2-en-1-ol	29803-82-5						0-0.05	
trans-p-Menth-2-en-1-ol	29803-81-4						0-0.04	
Menthone (*p*-)	89-80-5						0-0.03	
p-Menth-8-one				0-tr				
Methyl geranate	2349-14-6		0.01	0-0.01	0-0.02	tr-0.02	0-0.05	
6-Methyl-5-hepten-2-one	110-93-0		0.4	0-0.01	0-tr	tr-0.01	0-0.2	
Methyl jasmonate	1211-29-6		+					
Methyl epi-jasmonate	42536-97-0		+					+[i]
Methylnaphthalene	1321-94-4						0-0.1	
2-Methylnaphthalene	91-57-6						0-0.05	
Methyl thymol	1076-56-8						0-2.3	
α-Muurolene	10208-80-7				0-0.01		0-0.01	
γ-Muurolene	30021-74-0		0.02	0-0.04	0-0.01		0-0.02	
δ-Muurolene	120021-96-7				tr-0.01			
α-Muurolol	104245-48-9				0-0.02		0-0.02	
τ-Muurolol (epi-α-)	19912-62-0						0-0.02	
Myrcene	123-35-3	1.2-2.2	2.2	1.3-1.7	1.0-2.0	1.0-1.3	0.9-3.5	1.5[g]; 1.8[f]
Myrtenal	564-94-3		+				0-0.04	
Myrtenol	515-00-4						0-0.01	
Neral	106-26-3	0.3-1.4	1.3	0-1.0	0.6-2.0	0.4-0.8	0.02-2.7	0.7[f,g]
Neric acid	37349-29-4						0-0.04	
Nerol	106-25-2	0.01-0.5	0.2	0-0.07	0-0.4		0-4.3	
(*E*)-Nerolidol	40716-66-3							0-tr
(*Z*)-Nerolidol	3790-78-1				0-tr			
Neryl acetate	141-12-8	0.2-0.8	0.9	0-3.9[b]	0-1.7	0.3-0.6	0-1.1	0.4[f,g]
Nerylacetone	3879-26-3				0-0.01			
Neryl propionate	105-91-9		+				0-0.01	
Nonanal	124-19-6	0.03-0.2	0.2	0-0.2	0-0.4	0.07-0.1	0-0.1	0.1[k]
Nonanol	28473-21-4		0.2				0-0.1	0.06[g]
2-Nonenal	2463-53-8						0-tr	
(*E*)-2-Nonenal	18829-56-6		+					
Nonyl acetate	143-13-5		0.02	0-0.01	0-0.09	tr-0.01	0-0.02	
Nootkatone	4674-50-4		0.03	0-0.01	0-0.03	tr-0.01	0-0.03	0.04[k]
Norbornanol	86368-39-0		0.04	0-0.02	0-0.03			
Ocimene (β-)	13877-91-3		0.08	0-0.1	0-0.03		0-0.3	
(*E*)-β-Ocimene	3779-61-1	0.08-0.5	0.6	0-0.1	0-0.4	0.05-0.1	0-0.5	0.08[g]; 0.1[k]
(*Z*)-β-Ocimene	3338-55-4	0.02-0.2	0.2	0-0.08	0-0.4	0.01-0.09	0-0.1	0.07[g]
allo-Ocimene	673-84-7		tr					
Octanal	124-13-0	tr-0.1	0.4	0-0.2	0-0.1	0.04-0.07	0-0.2	0.01[f]; 0.06[k]
Octanoic acid	124-07-2		+					
Octanol	111-87-5		0.09	0-tr	0-0.01	0.01-0.06[d]	0-0.1	
2-Octen-1-ol, 3,7-dimethyl	40607-48-5						0-0.2	
Octyl acetate	112-14-1		0.03	0-tr	0-0.02	tr-0.01	0-0.01	
Oxypeucedanin	26091-73-6	tr-0.8						
Pentadecanal	2765-11-9		+					
Pentane	109-66-0							+[h]
Perillaldehyde	2111-75-3	0-0.05	0.05	0-0.02	0-0.09	0.01-0.04	0-0.01	0.05[k]
Perillyl acetate	15111-96-3		+					
Perillyl alcohol	536-59-4		+				0-0.06	
α-Phellandrene	99-83-2	0.02-0.09	0.1	0-0.08	tr-1.6	0.01-0.03	0-0.06	0.02[f]
β-Phellandrene	555-10-2	0-0.5	0.5	0-0.3	0-0.6		0-0.4	0.2[k]

Table 5.50.2 Constituents identified in lemon oils *(continued)*

Constituent	CAS	Percentage and range						
		A	B (3)	C (3)	D (3)	E (37)	F	G
(*E*)-Photocitral							0-0.05	
Phytol	7541-49-3						0-0.01	
α-Pinene	80-56-8	1.5-2.4	4.4	1.8-2.2	0.9-2.6	1.5-2.0	0.6-2.9	1.8[f]; 2.1[g]
β-Pinene	127-91-3	7.4-17.8	17.8	11.2-12.8	8.1-17.2	10.6-19.3	0.6-13.4	12.0[f]; 12.5[g]
Pinene oxide							0-0.3	
trans-Pinocarveol	1674-08-4			0-0.1				
Piperitone	89-81-6		0.01	0-tr	0-tr	tr	0-0.01	
Psoralen	66-97-7	0-tr						
Sabinene	3387-41-5	1.1-2.9	2.8	1.2-2.0	0-3.0	1.5-2.4	0-5.8	1.8[g]; 1.9[f]
cis-Sabinene hydrate	15537-55-0	0.04-0.07	0.07	0-0.06	0-0.1	0.01-0.06[d]	0-0.1	0.06[k]
trans-Sabinene hydrate	17699-16-0	0-0.08	0.07	0-0.04	0-0.2	0.06-0.2[e]	0-0.1	
α-Santalene	512-61-8			0-tr	tr-0.01			
β-Santalene	511-59-1		0.02	0-0.02	0-0.02		0-0.07	0.2[k]
cis-β-Santalene						tr-0.01		
α-Selinene	473-13-2		+					
β-Selinene	17066-67-0				tr-0.01			
Selin-11-en-4α-ol	16641-47-7		0.01					
β-Sesquiphellandrene	20307-83-9		0.04					
cis-Sesquisabinene hydrate	58319-05-4					tr-0.01		
Sesquithujene	58319-06-5			0-0.02				
β-Sinensal	17909-87-4						0-0.09	
Solanone	1937-54-8				0-tr			
Spathulenol	6750-60-3			0-tr	0-0.1	tr-0.04	0-0.09	0.2[k]
α-Terpinene	99-86-5	0.09-0.3	0.3	0.1-0.2	0.01-0.2	0.02-0.2	0-1.1	0.1[k]; 0.2[g]
γ-Terpinene	99-85-4	3.1-12.4	11.4	7.2-10.4	6.3-12.1	7.5-9.7	4.8-12.7	9.0[f]; 9.7[k]
Terpinen-4-ol	562-74-3	tr-0.06	0.1	0-0.07	0-0.05	0.01-0.05	0.07-2.2	0.04[g]; 0.07[k]
1-Terpineol	586-82-3						0-0.08	
α-Terpineol	98-55-5	0.05-0.5	0.8	0.1-0.2	0.1-0.4	0.07-0.2	0.3-2.9	0.3[f]; 0.4[k]
β-Terpineol	138-87-4						0-0.2	
(*E*)-β-Terpineol				0-tr				
(*Z*)-β-Terpineol	7299-40-3			0-tr				
δ-Terpineol	7299-42-5		+					
Terpinolene	586-62-9	0.1-0.5	0.5	0.2-0.4	0.3-0.6	0.1-0.3	0-1.4	0.3[g,k]; 1.4[f]
Terpinyl acetate	8007-35-0		0.5					
α-Terpinyl acetate	80-26-2						0-0.6	
Tetradecanal	124-25-4		0.02	0-0.01	0-0.05			
Tetradecane	629-59-4					tr-0.01		
(*E*)-2-Tetradecenal	51534-36-2		+					
Tetradecene	26952-13-6					tr-0.01		
α-Thujene	2867-05-2	0.2-0.6	0.5	0-0.5	0-0.5	0.3-0.4	0-0.5	0.2[f]; 0.4[g]
α-Thujone	546-80-5						0-tr	
cis-Thujopsene	32435-95-3							2.4[k]
Thymol	89-83-8						0-0.04	0.09[k]
Tricyclene	508-32-7		0.01	0-0.01	0-0.03	0.01	0-0.02	
Tridecanal	10486-19-8		tr					
Tridecane	629-50-5					tr-0.01		
Undecanal	112-44-7	0-0.09	0.08	0-0.04	0-0.1	tr-0.04	0-0.06	
Undecane	1120-21-4			0-tr				
Undecanol	112-42-5						0-tr	
Undecyl acetate	1731-81-3		0.01					
Valencene	4630-07-3		0.2	0-0.05	0-0.1	tr-0.07	0-0.2	0.8[k]
Verbenol	473-67-6				0-0.01			
cis-Verbenol	1845-30-3						0-0.03	
Verbenone	80-57-9						0-0.05	

Table 5.50.2 Constituents identified in lemon oils *(continued)*

A 178 lemon essential oil samples from Italy, Argentina, USA, Spain, Brazil and Israel, analyzed between 1998 and 2013; lowest and highest concentrations given (E. Schmidt, unpublished data)
B review of studies published between 1979 and 1999 on the composition of industrial cold-pressed lemon oils; data cited in ref. 3; highest concentrations given
C review of studies published between 1998 and 2009 on the composition of industrial cold-pressed lemon oils (refs. 7-13); data cited in ref. 3; lowest and highest concentrations given
D review of studies published between 1998 and 2009 on the composition of laboratory cold-pressed lemon oils (refs. 14-21); data cited in ref. 3; lowest and highest concentrations given
E ninety-two samples of industrially produced cold-pressed lemon peel oils extracted by various methods from lemons produced from September 2008 to June 2009 in Sicily, Italy; lowest and highest concentrations given (ref. 37)
F data from studies published in the period 1998-2014 on hydro-/steam distilled lemon peel oils; concentration ranges in these studies as a group given; ref. 20 (one hydrodistilled oil sample from Algeria); ref. 22 (two lemon oils from Cameroon); ref. 23 (one steam-distilled oil from Sicily); ref. 24 (two oils from Japan); ref. 27 (one oil from Venezuela with an unusual amount of methyl thymol); ref. 29 (ten biological oils and ten conventional distilled oils from Italy); ref. 31 (6 industrially distilled oils from Sicily and one from Spain); ref. 34 (twelve oils from Italy); ref. 36 (one oil from Greece); ref. 43 (one oil from Tunisia)
G data from miscellaneous recent studies on cold-pressed lemon oils, indicated with superscript letters; highest concentrations found in any study reviewed here given; when two or more oils were investigated, only the highest concentrations are mentioned, unless indicated otherwise; recent studies on commercial lemon oils of *unknown origin and mode of production* are *not* presented in Table 5.50.2 and include refs. 35,40 and 41

[a] incorrect identification; [b] high value found in ref. 19, indicating 'alteration or peculiar composition' according to the authors of ref. 3; [c] high value found (only) in ref. 19, from *C. limon* cultivar Barum, reminding of bergamot oil (ref. 3); [d] octanol and *cis*-sabinene hydrate combined; [e] linalool and *trans*-sabinene hydrate combined; [f] one commercial cold-pressed oil of unknown origin (ref. 30); [g] four commercial oils, probably cold-pressed, from Argentina, Spain and South Africa (ref. 31); [h] one commercial oil, possibly not cold-pressed, but discussed here as some unique components were identified in this oil, indicated with +[h] (ref. 32); [i] (+)-methyl epijasmonate identified and isolated in a cold-pressed oil from Spain (ref. 33); [k] one commercial cold-pressed oil from Spain (ref. 42)

tr: trace (in column E: <0.01); + present in the oil investigated, but quantity not stated

ref. 3, the composition of these oils is characterized by higher amounts of monoterpene alcohols than those seen in the cold-pressed ones due to hydration phenomena of monoterpene hydrocarbons, and by the absence of aliphatic esters that are present in small amounts in cold-pressed oils, presumably due to hydrolytic phenomena (3). In addition, the hydrodistilled oils lack the non-volatile oxygenated heterocyclic compounds (notably coumarins and psoralens) which may be present in the cold-pressed oils (Table 5.50.4).

Studies on laboratory cold-pressed oils and hydrodistilled oils published between 1979 and 1999 (which are not presented in Table 5.50.2) have been reviewed in ref. 5. The literature on the chemical composition of lemon (peel) oils has also been discussed in ref. 6. This is a review of 37 studies and articles on lemon (peel) oil composition published between 1979 and 2002. Chemicals cited in this article as having been present in 'lemon oils' not mentioned in table 5.50.2 include limonenediol (CAS 1946-00-5), dihydrolinalool (CAS 18479-51-1, not naturally occurring according to ref. 6), *p*-menth-1(7)-en-9-ol (CAS 29548-16-1), thymal (unknown chemical), 1-heptenyl acetate (CAS 35468-97-4), α-terpinyl propionate (CAS 80-27-3), and *cis*- and *trans*-limonene-8,9-epoxide.

Recent studies on commercial lemon oils of unknown origin and mode of production are not presented in Table 5.50.2 and include refs. 35,40 and 41. Some studies on lemon oils not presented in Table 5.50.2 because of a very atypical composition include refs. 28 and 38. Studies on lab-hydrodistilled lemon oils not discussed in Table 5.50.2 include refs. 26 (uncertain whether the lemon peel was used for obtaining the oil) 44 and 25 (not comparable to the other data).

Table 5.50.3 Major constituents of lemon oils

Constituent	CAS	Percentage and range						
		A	B (3)	C (3)	D (3)	E (37)	F	G
Limonene	138-86-3	53.5-73.0	71.8	52.6-69.9	56.6-76.0	60.3-70.8	46.5-75.7	67.0[f]; 69.8[g]
β-Pinene	127-91-3	7.4-17.8	17.8	11.2-12.8	8.1-17.2	10.6-19.3	0.6-13.4	12.0[f]; 12.5[g]
γ-Terpinene	99-85-4	3.1-12.4	11.4	7.2-10.4	6.3-12.1	7.5-9.7	4.8-12.7	9.0[f]; 9.7[k]
Sabinene	3387-41-5	1.1-2.9	2.8	1.2-2.0	0-3.0	1.5-2.4	0-5.8	1.8[g]; 1.9[f]
Geranial	141-27-5	0.6-2.1	2.7	0.9-1.7	0.7-4.3	0.7-1.5	0.03-5.4	1.2[f]; 2.1[k]
α-Pinene	80-56-8	1.5-2.4	4.4	1.8-2.2	0.9-2.6	1.5-2.0	0.6-2.9	1.8[f]; 2.1[g]
Nerol	106-25-2	0.01-0.5	0.2	0-0.07	0-0.4		0-4.3	
Myrcene	123-35-3	1.2-2.2	2.2	1.3-1.7	1.0-2.0	1.0-1.3	0.9-3.5	1.5[g]; 1.8[f]

LEGEND: SEE UNDER TABLE 5.50.2

Table 5.50.4 Heterocyclic oxygenated compounds present in cold-pressed lemon oils (1,4,37)

Name	Synonym(s)	CAS
Coumarins and psoralens		
Byakangelicol		26091-79-2
Cnidicin		14348-21-1
5,7-Dimethoxycoumarin	Citropten; Limetin	487-06-9
6,7-Dimethoxycoumarin	Scoparone	120-08-1
Dimethoxy-6,7-furanocoumarin	Isopimpinellin	482-27-9
8-(6′,7′-Epoxy)-geranyloxypsoralen		
5-(2′,3′-Epoxyisopentenyloxy)-7-methoxycoumarin		
5-(2′,3′-Epoxyisopentenyloxy)psoralen		
7-Geranyloxycoumarin	Aurapten	495-02-3
5-Geranyloxy-7-methoxycoumarin		7380-39-4
5-Geranyloxy-8-methoxypsoralen		69239-53-8
5-Geranoxypsoralen	Bergamottin; Bergaptin	7380-40-7
8-Geranyloxypsoralen		7437-55-0
Gosferenol [a]		
Heraclenin	Prangenin	2880-49-1
Heraclenol	Prangenin hydrate	31575-93-6
7-Hydroxycoumarin	Umbelliferone	93-35-6
5-Hydroxy-6,7-furanocoumarin	Bergaptol	486-60-2
Isoimperatorin		482-45-1
7-(3′-Isopentenyloxy)coumarin		
5-Isopentenyloxy-8-(2′,3′-dihydroxyisopentenyloxy)psoralen		
5-Isopentenyloxy-8-(2′,3′-epoxyisopentenyloxy)psoralen		117030-02-1
5-Isopentenyloxy-7-methoxycoumarin		35590-41-1
8-Isopentenyloxypsoralen	Imperatorin; Marmelide; Ammidin	482-44-0
5-Methoxy-8-(2,3-dihydroxy-3-methylbutoxy)psoralen	Bjakangelicin; Byakangelicin	482-25-7
5-Methoxy-8-geranyloxypsoralen		
6-Methoxy-7-hydroxycoumarin	Scopoletin	92-61-5
5-Methoxy-8-isopentenyloxypsoralen		
5-Methoxypsoralen	Bergapten	484-20-8
8-Methoxypsoralen	Xanthotoxin; Ammodin	298-81-7
Neobyakangenicol		
Oxypeucedanin		26091-73-6
Oxypeucedanin hydrate	Prangol	2643-85-8
Pabulenol [a]		33889-70-2
Phellopterin		2543-94-4

[a] Pabulenol or Gosferenol

CONTACT ALLERGY / ALLERGIC CONTACT DERMATITIS

General

Contact allergy to / allergic contact dermatitis from lemon oil has been reported in 20 publications. In groups of consecutive patients suspected of contact dermatitis, prevalence rates of up to 0.9% positive patch test reactions have been observed, but adequate relevance data are lacking. There have been several case reports of allergic contact dermatitis from lemon oil, mostly from occupational exposure. In one such report, limonene and β-pinene may have been allergens in lemon oil.

Testing in groups of patients

The results of patch tests with lemon oil in routine testing (consecutive patients suspected of contact dermatitis) and in groups of selected patients are shown in Table 5.50.5. In routine testing, rates of positive reactions ranged from 0.5% to 0.9%, whereas between 0.3% and 10% of patients

in selected groups had positive patch tests.

Case reports

Occupational contact dermatitis

An aromatherapist had chronic hand dermatitis and was patch test positive to 17 of 20 oils used at her work (tested 1% and 5% in petrolatum), including lemon oil (50). Two other aromatherapists had contact dermatitis (one occupational) with allergies to multiple essential oils used at their work, including lemon oil. Both patients also reacted to geraniol, α-pinene, the fragrance mix and various other fragrance materials. In addition, one proved to be allergic to linalool and linalyl acetate, the other to caryophyllene; α-pinene and caryophyllene were demonstrated by GC-MS in one of the lemon oils, but limonene was not tested (51). α-Pinene may be present in commercial lemon oils in concentrations up to 2.5% (Table 5.50.2, column A). One physiotherapist developed occupational contact dermatitis, which proved to be caused by contact allergy to lemon oil and other essential oils (6). Two cases of occupational allergic

Table 5.50.5 Results of testing groups of patients with lemon oil

Years and Country	Test conc. & vehicle	Number of patients tested positive (%)			Selection of patients (S); Relevance (R); Comments (C)	Ref.
Routine testing						
2000-2007 USA	2% pet.	345	3	(0.9%)	R: 100%; C: weak study: a. high rate of macular erythema and weak reactions, b. relevance figures include 'questionable' and 'past' relevance	47
<1976 Poland	2% pet.	200	1	(0.5%)	R: not stated	53
Testing in groups of selected patients						
2000-2008 IVDK	2% pet.	6,467	17	(0.3%)	S: patients with dermatitis suspected of causal exposure to fragrances; R: not stated; C: no frequent co-reactivity to limonene, the main component of citrus oils	58
1997-2000 Austria	2% pet.	747	2	(0.3%)	S: patients suspected of fragrance allergy; R: not stated	49
1989-1999 Portugal	2% pet.	67	3	(4.5%)	S: patients who had a positive patch test to the fragrance mix; R: not stated	59
1996-1997 UK	2% pet.	10	1	(10%)	S: patients suspected of cosmetic dermatitis and reacting to the fragrance mix; R: not stated	48
1986 France	1% pet.	21	2	(10%)	S: patients with dermatitis caused by fragrances; R: not stated	63
<1986 Poland	2% pet.	86	2	(2.3%)	S: patients previously reacting to the fragrance mix; R: not stated	55
<1976 Poland	2% pet.	51	1	(2.0%)	S: patients allergic to *Myroxylon pereirae* resin (balsam of Peru) and/or turpentine and/or wood tar and/or colophony	53

IVDK Information Network of Departments of Dermatology, Germany, Switzerland, Austria (www.ivdk.org); pet.: petrolatum

contact dermatitis to 'lemon oil terpenes' (tested 1% in olive oil) in bottle fillers in a perfume factory (56). A positive patch test to lemon oil considered to be relevant was observed in a case of occupational hand dermatitis in a cook/ barman/fruit grower, who also reacted to lemongrass oil, neroli oil and geraniol; the patient was patch tested with 'essences' (no concentration mentioned), but from the text it seems that this would indicate essential oils. Geraniol is not an important constituent of lemon oil (57). One case of occupational contact dermatitis of the fingers of both hands considered by the author to be caused by allergy to geraniol present in lemon oil and lemon peel, with which she came in contact while working in a company for baking ingredients (61); however, geraniol is not an important component of lemon oils (Table 5.50.2, column A). A porter became sensitized to oil of lemon, which he used occupationally; patch tests were positive to oil of lemon, limonene, oil of citronella, Java, oil of citronella, Ceylon (= Sri Lanka), turpentine and its component β-pinene (62). Limonene is the major ingredient in lemon oil and β-pinene can also be present in these oils in high concentrations (maximum concentration found in commercial lemon oils: 17.8%, Table 5.50.2, column A).

Non-occupational allergic contact dermatitis
One patient had developed pigmented contact dermatitis from contact allergy to lemon oil, geraniol and hydroxycitronellal present in a face compact powder (60). Another patient suffering from oral lichenoid

mucositis had relevant positive patch test reactions to lemon oil, clove oil and metals (64).

Positive patch tests (relevance unknown, uncertain or not stated)
One positive patch test reaction to lemon oil in a patient with airborne allergic contact dermatitis from tea tree oil and lavender oil (55). Two positive patch test reactions among 20 workers in a perfume factory without dermatitis (56). One patient had allergic contact dermatitis from a 'natural' deodorant containing lichen acids; reactions were seen to the deodorant, lichen acid mix, usnic acid and some essential oils including lemon oil (65). One positive patch test to lemon oil, possibly from its presence in a topical pharmaceutical preparation (66).

LITERATURE
1 Russo M, Torre G, Carnovale C, Bonaccorsi I, Mondello L, Dugo P. A new HPLC method developed for the analysis of oxygen heterocyclic compounds in Citrus essential oils. J Essent Oil Res 2012;24:119-129
2 Liu YQ, Heying E, Tanumihardjo SA. History, global distribution, and nutritional importance of Citrus fruits. Compreh Rev Food Sci Food Saf 2012;11:530-545
3 Dugo G, Cotroneo A, Bonaccorsi I, Trozzi A. Composition of the volatile fraction of citrus peel oils. In: G Dugo and L Mondello, eds. Citrus oils:

composition, advanced analytical techniques, contaminants, and biological activity. Boca Raton, USA: CRC Press, Taylor & Francis Group, 2011:115-142

4 Dugo P, Russo M. The oxygen heterocyclic components of Citrus essential oils. In: G Dugo and L Mondello, eds. Citrus oils: composition, advanced analytical techniques, contaminants, and biological activity. Boca Raton, USA: CRC Press, Taylor & Francis Group, 2011:405-444

5 Dugo G, Cotroneo A, Verzera A, Bonaccorsi I. Composition of the volatile fraction cold pressed citrus peel oils. In: G Dugo, A Di Giacomo, Eds. Citrus. London, New York: Taylor & Francis, 2002:201-317

6 Kirbaşlar I, Boz I, Kirbaşlar G. Composition of Turkish lemon and grapefruit peel oils. J Essent Oil Res 2006;18:525-543

7 Baratta MT, Dorman HJD, Deans SG, Figueiredo AC, Barroso JG, Ruberto G. Antimicrobial and antioxidant properties of some commercial essential oils. Flavour Fragr J 1998;13:235-244

8 Verzera A, Dugo P, Mondello L, Trozzi A, Cotroneo A. Extraction technology and lemon oil composition. Ital J Food Sci 1999;11:361-370. Data cited in ref. 3

9 Mondello L, Casilli A, Tranchida PQ, Cicero L, Dugo P, Dugo G. Comparison of fast and conventional GC analysis for citrus essential oils. J Agric Food Chem 2003;51:5602-5606

10 Mondello L, Casilli A, Tranchida PQ, Costa R, Dugo P, Dugo G. Fast GC for the analysis of citrus oils. J Chromatogr Sci 2004;42:410-416

11 Verzera A, Trozzi A, Dugo G, Di Bella G, Cotroneo A. Biological lemon and sweet orange essential oil composition. Flavour Fragr J 2004;19:544-548

12 Sawamura M, Son U-S, Choi H-S, Kim MS-L, Phi NTL, Fears M, et al. Compositional changes in commercial lemon essential oil for aromatherapy. Int J Aromather 2004;14:27-36

13 Viuda-Martos M, Ruiz-Navajas Y, Fernández-López J, Pérez-Álvarez JA. Chemical composition of mandarin (C. reticulata L.), grapefruit (C. paradisi L.), lemon (C. limon L.) and orange (C. sinensis L.) essential oils. J Essent Oil Bear Plants 2009;12:236-243

14 Sawamura M. Volatile components of essential oils of the Citrus genus. Recent Res Dev Agric Food Chem 2000;4:131-164

15 Corleone V, Corleone P, La Scala G. New findings on the influence of different lemon varieties in the composition of essential oil obtained from winter fruits and comparison with existing literature. Essenz Deriv Agrum 2000;70:67-79. Data cited in ref. 3

16 Verzera A, Russo C, La Rosa G, Bonaccorsi I, Cotroneo A. Influence of cultivar on lemon oil composition. J Essent Oil Res 2001;13:343-347

17 Gionfriddo F, Mangiola C, Catalfamo M, Manganaro R, Castaldo D. Determinazione delle caratteristiche analitiche e della composizione enantiomerica di oli essenziali agrumari ai fini dell'accertamento della purezza e della qualità. Nota II—Essenza di limone. Essenz Deriv Agrum 2004;74:25-30. Data cited in ref. 3

18 Verzera A, Trozzi A, Zappalà M, Condurso C, Cotroneo A. Essential oil composition of Citrus meyerii Y. Tan. and Citrus medica L. cv. Diamante and their lemon hybrids. J Agric Food Chem 2007;53:4890-4894

19 Lota M-L, de Rocca Serra D, Tomi F, Jacquemond C, Casanova J. Volatile components of peel and leaf oils of lemon and lime species. J Agric Food Chem 2002;50:796-805

20 Ferhat MA, Meklati BY, Chemat F. Comparison of different isolation methods of essential oil from Citrus fruits: cold pressing, hydrodistillation and microwave 'dry' distillation. Flavour Fragr J 2007;22:494-504

21 Choi H-S. Lipolytic effects of citrus peel oils and their components. J Agric Food Chem 2006;54:3254-3258

22 Zollo Amvam PH, Dongmo Jazet PM, Boyom Fekam F, et al. Chemical composition of essential oils from Citrus species grown in Cameroon. Riv Ital EPPOS 1998;(Numero speciale):156-162. Data cited in ref. 3

23 Ruberto G, Starrantino A, Rapisarda P. Essential oils from new citrus fruits. Essenz Deriv Agrum 1999;69:15-26. Data cited in ref. 3

24 Fujita S, Kajiyama K, Takabayashi M, Nonaka A. Constituents of the essential oils of lemon and its hybrid. Data presented at the 52nd TEAC, Itakura, Japan, 2008. Data cited in ref. 3

25 Vekiari SA, Protopapadakis EE, Papadopoulou P, Papanicolau D, Panou C, Vamvakias M. Composition and seasonal variation of the essential oil from leaves and peel of a Cretan lemon variety. J Agric Food Chem 2002;50:147-153

26 Charai M, Faid M, Chaouch A. Essential oils from aromatic plants (Thymus broussonetti Boiss., Origanum compactum Benth., and Citrus limon (L.) N.L. Burm.) as natural antioxidants for olive oil. J Essent Oil Res 1999;11:517-521

27 de Gonzalez CN, Sánchez F, Quintero A. Chemotaxonomic value of essential oil compounds in Citrus species. Acta Hortic 2002;576:49-51. Data cited in ref. 3

28 Mahalwal VS, Ali M. Volatile constituents of the fruit peels of Citrus limon (Linn) Burm. F. J Essent Oil Bear Plants 2003;6:31-35

29 Spadaro F, Circosta C, Costa R, Pizzimenti F, Palumbo DR, Occhiuto F. Volatile fraction composition and biological activity of lemon oil (Citrus limon L. Burm.): Comparative study of oils extracted from conventionally grown and biological fruits. J Essent Oil Res 2012;24:187-193

30 Rao J, McClements DJ. Impact of lemon oil composition on formation and stability of model food and beverage emulsions. Food Chem 2012;134:749-757

31 Schipilliti L, Dugo P, Bonaccorsi I, Mondello L. Authenticity control on lemon essential oils employing gas chromatography—combustion-isotope ratio mass spectrometry (GC–C-IRMS). Food Chem 2012;131:1523–1530

32 Veriotti T, Sacks R. High-speed GC and GC/time-of-flight MS of lemon and lime oil samples. Anal Chem 2001;73: 4395-4402

33 Del Mar Caja M, Blanch GP, Ruiz del Castillo ML. Online RPLC-GC via TOTAD method to isolate (+)-methyl epijasmonate from lemon (*Citrus limon* Burm.). J Agric Food Chem 2008;56:5475–5479

34 Settanni L, Randazzo W, Palazzolo E, Moschetti M, Aleo A, Guarrasi V, et al. Seasonal variations of antimicrobial activity and chemical composition of essential oils extracted from three *Citrus limon* L. Burm. cultivars. Nat Prod Res 2014;28:383-391

35 Misharina TA, Terenina MB, Krikunova NI, Kalinichenko MA. The Influence of the composition of essential lemon oils on their antioxidant properties and the stability of the components. Russ J Bioorg Chem 2011;37:883-887

36 Giatropoulos QA, Papachristos DP, Kimbaris A, Koliopoulos G, Polissiou MG, Emmanouel N, et al. Evaluation of bioefficacy of three Citrus essential oils against the dengue vector *Aedes albopictus* (Diptera: Culicidae) in correlation to their components enantiomeric distribution. Parasitol Res 2012;111:2253-2263

37 Dugo P, Ragonese C, Russo M, Sciarrone D, Santi L, Cotroneo A. Sicilian lemon oil: Composition of the volatile fraction and oxygen heterocyclic fraction and enantiomeric distribution of volatile compounds. J Sep Sci 2010;33:3374-3385

38 Haggag EG, Wahab SMA, El-Zalabany SM, Moustafa EAA, El-Kherasy EM, Mabry TJ. Volatile oils and pectins from *Citrus aurantifolia* (lime) and *Citrus limonia* (lemon). Asian J Chem 1998;10:828-833. Data cited in ref. 39

39 Lawrence BM. Progress in essential oils. Perfum Flavor 2006;31(3):40-? (last page unknown)

40 Kubeczka K-H, Formacek V. Essential oils analysis by capillary gas chromatography and carbon-13 NMR spectroscopy, 2nd edition. New York, USA: John Wiley and Sons, 2002:167-172. Data cited in ref. 39

41 Białon M, Krzysko-Łupicka T, Koszałkowska M, Wieczorek PP. The Influence of chemical composition of commercial lemon essential oils on the growth of *Candida* strains. Mycopathologia 2014;177:29-39

42 Espina L, Somolinos M, Lorán S, Conchello P, García D, Pagán R. Chemical composition of commercial citrus fruit essential oils and evaluation of their antimicrobial activity acting alone or in combined processes. Food Control 2011;22: 896-902

43 Bourgou S, Rahali FZ, Ourghemmi I, Tounsi MS. Changes of peel essential oil composition of four Tunisian Citrus during fruit maturation. The Scientific World Journal 2012, Article ID 528593, 10 pages; doi:10.1100/2012/528593

44 Hamdan D, Ashour M, Mulyaningsih S, El-Shazly A, Wink M. Chemical composition of the essential oils of variegated pink-fleshed lemon (*Citrus x limon* L. Burm. f.) and their anti-Inflammatory and antimicrobial activities. Z Naturforsch C 2013;68:275-284

45 Rhind JP. Essential oils. A handbook for aromatherapy practice, 2nd Edition. London: Singing Dragon, 2012

46 Lawless J. The encyclopedia of essential oils, 2nd Edition. London: Harper Thorsons, 2014

47 Wetter DA, Yiannias JA, Prakash AV, Davis MD, Farmer SA, el-Azhary RA, et al. Results of patch testing to personal care product allergens in a standard series and a supplemental cosmetic series: an analysis of 945 patients from the Mayo Clinic Contact Dermatitis Group, 2000-2007. J Am Acad Dermatol 2010;63:789-798

48 Thomson KF, Wilkinson SM. Allergic contact dermatitis to plant extracts in patients with cosmetic dermatitis. Br J Dermatol 2000;142:84-88

49 Wöhrl S, Hemmer W, Focke M, Götz M, Jarisch R. The significance of fragrance mix, balsam of Peru, colophonium and propolis as screening tools in the detection of fragrance allergy. Br J Dermatol 2001;145:268-273

50 Selvaag E, Holm J, Thune P. Allergic contact dermatitis in an aromatherapist with multiple sensitizations to essential oils. Contact Dermatitis 1995;33:354-355

51 Dharmagunawardena B, Takwale A, Sanders KJ, Cannan S, Roger A, Ilchyshyn A. Gas chromatography: an investigative tool in multiple allergies to essential oils. Contact Dermatitis 2002;47:288-292

52 Trattner A, David M, Lazarov A. Occupational contact dermatitis due to essential oils. Contact Dermatitis 2008;58:282-284

53 Rudzki E, Grzywa Z, Bruo WS. Sensitivity to 35 essential oils. Contact Dermatitis 1976;2:196-200

54 Rudzki E, Grzywa Z. Allergy to perfume mixture. Contact Dermatitis 1986;15:115-116

55 De Groot AC. Airborne allergic contact dermatitis from tea tree oil. Contact Dermatitis 1996;35:304-305

56 Schubert HJ. Skin diseases in workers at a perfume factory. Contact Dermatitis 2006;55:81-83

57 Audicana M, Bernaola G. Occupational contact dermatitis from citrus fruits: Lemon essential oils. Contact Dermatitis 1994;31:183-185

58 Uter W, Schmidt E, Geier J, Lessmann H, Schnuch A, Frosch P. Contact allergy to essential oils: current patch test results (2000–2008) from the Information Network of Departments of Dermatology (IVDK). Contact Dermatitis 2010;63:277-283

59 Manuel Brites M, Goncalo M, Figueiredo A. Contact allergy to fragrance mix — a 10-year study. Contact Dermatitis 2000;43:181-182

60 Serrano G, Pujol C, Cuadra J, Gallo S, Aliaga A. Riehl's melanosis: Pigmented contact dermatitis caused by fragrances. J Am Acad Dermatol 1989;21:1057-1060

61 Hausen BM, Kulenkamp D. Geraniol-Kontaktallergie. Z Hautkr 1990;65:492-494

62 Keil H. Contact dermatitis due to oil of citronell. J Invest Dermatol 1947;8:327-334

63 Meynadier JM, Meynadier J, Peyron JL, Peyron L. Formes cliniques des manifestations cutanées d'allergie aux parfums. Ann Dermatol Venereol 1986;113:31-39

64 Yiannias JA, el-Azhary RA, Hand JH, Pakzad SY, Rogers RS III. Relevant contact sensitivities in patients with the diagnosis of oral lichen planus. J Am Acad Dermatol 2000;42:177-182

65 Sheu M, Simpson EL, Law SV, Storrs FJ. Allergic contact dermatitis from a natural deodorant: A report of 4 cases associated with lichen acid mix allergy. J Am Acad Dermatol 2006;55:332-337

66 Nardelli A, D'Hooge E, Drieghe J, Dooms M, Goossens A. Allergic contact dermatitis from fragrance components in specific topical pharmaceutical products in Belgium. Contact Dermatitis 2009;60:303-313

Chapter 5.51   LEMONGRASS OIL, EAST INDIAN

There are two main lemongrass oils, derived from two species of the Cymbopogon family, lemongrass oil derived from *Cymbopogon flexuosus* (East Indian type) and West Indian type lemongrass oil, obtained from *Cymbopogon citratus*. In this chapter, lemongrass oil *ex Cymbopogon flexuosus* is discussed. The West Indian type essential oil is presented in Chapter 5.52.

DEFINITION

Lemongrass oil (essential oil of lemongrass) East Indian is the essential oil obtained from the aerial part (leaves) of the (East Indian, Malabar) lemongrass, *Cymbopogon flexuosus* (Nees ex Steudel) J.F. Watson.

INCI NOMENCLATURE

Description/definition: Cymbopogon flexuosus oil is the volatile oil obtained from the dried lemon grass, *Cymbopogon flexuosus*, Poaceae
INCI name EU & EU: Cymbopogon flexuosus oil
CAS registry number(s): 8007-02-1; 91844-92-7 (refer both to Cymbopogon citratus leaf oil in the INCI USA nomenclature)
EINECS number(s): 295-161-9

ISO (INTERNATIONAL ORGANIZATION FOR STANDARDIZATION) STANDARD

ISO number: 4718
ISO name: Essential oil of lemongrass
Botanical origin: *Cymbopogon flexuosus* (Nees ex Steudel) J.F. Watson
Parts of plant used: Aerial part (leaves)
ISO values: ISO values (minimum and maximum concentrations) are shown in Table 5.51.1.

Another species of the genus *Cymbopogon* from which lemongrass can be derived is *Cymbopogon schoenanthus* (CAS 89998-16-3; EINECS 289-754-1). For the essential oil of *Cymbopogon martinii* see Chapter 5.66 Palmarosa oil and for the oil of *Cymbopogon nardus* see Chapters 5.23 and 5.24 citronella oil.

Table 5.51.1 ISO values (%) for lemongrass oil [a]

Compound	CAS	Minimum	Maximum
Geranial	141-27-5	35.0	47.0
Neral	106-26-3	25.0	35.0
Geraniol	106-24-1	1.5	8.0
Geranyl acetate	105-87-3	0.5	6.0
Limonene	138-86-3	0.5	3.5
β-Caryophyllene	87-44-5	0.2	3.5
6-Methyl-5-hepten-2-one	110-93-0	0.1	2.0

[a] ISO 4718 Oil of lemongrass ©ISO 2004; Geneva, Switzerland, www.iso.org

THE PLANT, THE OIL, AND THEIR USES

Cymbopogon flexuosus (Nees ex Steudel) J.F. Watson is a perennial grass with stems that may measure 200-300 cm and aromatic leaves that may be 50-100 cm long. This so-called East Indian lemongrass is native to China, India, Nepal, Myanmar, Malaysia and Thailand and is cultivated in India, southeast Asia, Equatorial Africa, in the Caribbean and in Guatemala for the production of (East Indian) lemongrass oil (synonyms: Cochin lemongrass oil, Malabar oil) (1, GRIN Taxonomy for Plants). Fresh bulbous stems and leaves are used in oriental cooking for their distinct lemon flavor. The grass is considered a diuretic, tonic, antiseptic and stimulant and is used to treat diarrhea, stomach-ache, headaches, fevers, and flu. It is also perceived as helpful in treating muscular pain, poor circulation, muscle tone and slack tissue (1).

The volatile essential oil obtained from lemongrass leaves by steam distillation has a variety of uses in the perfumery (16,19), cosmetics and pharmaceutical industries and also for flavoring curries, wines, beverages and herbal teas (1,12,16). It may also be added to waxes, polishes, detergents and insecticides (1). However, lemongrass oil is mainly used for the isolation of citral. Citral is the major active component found in lemongrass oil and the quality of the oil is measured by the citral content (12). Citral is a mixture of geranial (citral A = (*E*)-citral = α-citral) and neral (citral B = (*Z*)-citral = β-citral), of which large quantities are being utilized for production of a number of chemicals such as α- and β-ionones, vitamins A and E and β-carotene (1,2,3,12,16). With the availability of cheaper synthetic citral and competition from Chinese May Chang (*Litsea cubeba* L.) essential oil rich in citral (70-85 %), the production of lemongrass essential oil decreased in India (3).

Lemongrass oil is said to possess biological properties such as anticancer, antimicrobial, antifungal, antibacterial and mosquito repellant activities (12,16,19). It is used as anxiolytic, sedative or anticonvulsive agent (19). The oil is perceived to have efficacy to treat a wide variety of health conditions such as acne, athlete's foot, excessive perspiration, flatulence, muscle aches, oily skin and scabies (12,16,19). In aromatherapy, lemongrass oil is used to improve circulation and skin and muscle tone (1).

CHEMICAL COMPOSITION

Lemongrass oil is a clear, mobile liquid with tender yellow to light yellowish brown color, which has a fresh citrusy, slightly green metallic odor, reminding of citral. The yield of essential oil from lemongrass generally varies from 0.7 to 1.3 per cent. The main producing countries of this oil are China, India, Haiti, Comoros islands, Bhutan and Indonesia.

Literature data (up to December 2, 2014) on the chemical composition of lemongrass oils and unpublished analytical data from one of us (E.S.) are shown in Table 5.51.2 in alphabetical order. In lemongrass oils from various origins, over 175 chemicals have been identified. About 52 per cent of these were found in a single reviewed publication only. The major compounds

Table 5.51.2 Constituents identified in lemongrass oils

Constituent	CAS	A	B (23)	C (4)	D (7)	E
α-Amorphene	20085-19-2					0.9[n]
1,5-di-epi-Aristolochene						tr[c]
1,2-Benzenedicarboxylic acid	88-99-3					1.6[l]
α-Bergamotene	17699-05-7					1.0[l]
cis-α- Bergamotene	18252-46-5					0.2[b]
trans-α-Bergamotene	13474-59-4					0.1[b,c,f]; 0.4[z4]
β-Bisabolene	495-61-4				0.05-0.09	0.1[k]
(E)-γ-Bisabolene	53585-13-0					0.2[z4]
α-Bisabolol	515-69-5					1.9[r]; 39.8[q]
α-Bisabolol oxide B	26184-88-3					8.0[q]
Borneol	507-70-0	0.06-0.6	0-0.8	0.2	0-0.03	<0.1[k]; 0.2[c,n]; 0.4[g]; 1.7[z4]; 1.8[b]; 4.9[d]; 17.1[q]
Bornyl acetate	76-49-3					0.3[g]; 3.7[q]
β-Bourbonene	5208-59-3		0-0.6			tr[b]; 0.2[z4]; 2.0[w]
γ-Cadinene	39029-41-9			0.5	0-0.6	0.3[f]; 0.6[o]; 0.7[z4]; 1.0[g]; 1.2[c]
δ-Cadinene	483-76-1	0.09-1.9	tr-0.2		0-0.04	0.2[b,c,n]; 0.3[g]; 1.5[z4]
α-Cadinol	481-34-5		0-0.3		0-0.4	tr[b]; 1.5[z4]
epi-α-Cadinol	5937-11-1	0-0.3				
δ-Cadinol	19435-97-3					tr[b]
cis-Calamenene	72937-55-4					0.2[z4]
Camphene	79-92-5	0.08-2.6		0.3		0.6[o]; 1.3[g]; 2.0[n]; 2.4[z4]; 4.1[b,f]; 10.3[q]; 13.5[d]
Camphor	76-22-2		0-0.8	0.3	0-0.2	tr[b]; 0.2[k]; 0.3[z4]; 1.2[l]; 2.3[q]
trans-3(10)-Caren-2-ol						0.1[a3,n]
(E)-Carveol	1197-07-5					0.1[z4]
Carvone	99-49-0			0.2		0.2[z4]
β-Caryophyllene	87-44-5	0.5-1.9	tr-0.9	1.1	0.1-0.7	1.3[i,n]; 1.8[o]; 2.0[b]; 2.5[l]; 2.7[q,z4]; 3.2[u]
Caryophyllene oxide	1139-30-6	0.1-0.8	tr-0.9	0.2	0.09-1.6	1.2[k]; 1.6[c]; 1.8[z4]; 2.3[l]; 2.9[q]; 5.0[u]; 5.4[i]; 5.6[j]
α-Chamigrene	19912-83-5					tr[b]
β-Chamigrene	18431-82-8					0.1[b,z4]
Citral (geranial + neral)	5392-40-5					60[s]; 75.9[p]; 82[z1]; 85-90[z1]
Citrol	624-15-7					10.2[j]
Citronellal	106-23-0	0.1-1.5	tr-0.5	1.5	0.06-0.2	0.1[k]; 0.2[f]; 0.3[d,o]; 0.4[m]; 0.7[u]; 2.1[z4]; 4.0[g]
Citronellol (β-, DL-)	106-22-9	0.08-1.0	0-0.3	0.1	0-0.6	0.5[n]; 0.8[e]; 0.9[j]; 1.2[d]; 4.5[q]; 24.1[g]; 32.1[z4]
Citronellyl acetate	150-84-5		tr-0.4	4.2	0-1.0	0.9[k]; 2.2[g]; 2.2[q]; 6.1[z4]
Citronellyl butyrate	141-16-2		0-0.2			
Citronellyl formate	105-85-1					2.0[j]
Citronellyl propionate	141-14-0					0.5[l]
α-Copaene	3856-25-5				0-0.04	0.1[a2,n]
β-Costol	515-20-8					0.08[h]
β-Cubebene	13744-15-5					0.5[g]
Cubebol	23445-02-5					2.0[z4]
Cubenol	21284-22-0					0.5[z4]
1,10-di-epi-Cubenol	73365-77-2					7.6[z4]
1-epi-Cubenol	81939-29-9					0.9[z4]; 3.9[b]
Cuparene	16982-00-6					0.1[c]; 0.4[z4]
Cyclosativene	22469-52-9					0.3[z4]
p-Cymene	99-87-6	0.01-0.3	0-0.2		0-0.05	0.1[c,k]
p-Cymen-8-ol	1197-01-9				0-0.1	
Decanal	112-31-2	0.1-0.5			0-0.2	tr[b]; 0.2[c,f,g]; 0.3[o,z4]
Decanol	36729-58-5					0.1[n]
(Z)-4-Decenal	21662-09-9					0.1[z4]
Dehydro-1,8-cineole	92760-25-3	0.02-0.2				
4,5-Dimethyl-2,6-octadiene	18476-57-8					0.2[a2,n]
Dodecanal	112-54-9					0.1[f]
β-Elemene	33880-83-0		0-0.3	0.1	0-0.3	<0.1[f]; 0.3[b,c,k]; 0.5[h]; 0.7[z4]
Elemicin	487-11-6	0.05-0.1		0.1		17.9[z3]; 53.0[b]; 71.5[z4]
Elemol (α-)	639-99-6	0.04-0.5			0-0.09	0.1[f]; 0.8[z4]; 1.1[r]
β-Elemol	32142-08-8					0.1[c]
(E)-4,5-Epoxycarane						2.1[a3,n]; 2.3[i]; 13.7[j]

Table 5.51.2 Constituents identified in lemongrass oils *(continued)*

Constituent	CAS	A	B (23)	C (4)	D (7)	E
2,3-Epoxygeranial						1.6[h]
α-Eudesmol	473-16-5					0.1[z4]
β-Eudesmol	473-15-4		0-0.2			0.1[z4]
γ-Eudesmol	1209-71-8			0.1		0.2[z4]
10-epi-γ-Eudesmol	15051-81-7		tr-0.6			
Eugenol	97-53-0	0.02-0.2		0.1		0.01[h]; 0.2[g,z4]; 0.4[f]; 0.5[c]
(*E,E*)-Farnesal	502-67-0					0.3[z4]
(*E,E*)-α-Farnesene	502-61-4					tr[b]
(*Z,E*)-α-Farnesene	26560-14-5					1.5[b]
β-Farnesene	502-60-3					0.7[h]
(*E*)-β-Farnesene	18794-84-8		0-0.1			tr[b]; 0.2[z4]
(*Z*)-β-Farnesene	28973-97-9					<0.1[f]
Farnesol	4602-84-0					1.1[m]; 1.9[l]
(*E,E*)-Farnesol	106-28-5					0.1[c]; 0.5[l]
(*Z,Z*)-Farnesol	16106-95-9					0.5[l]
β-Fenchyl alcohol	22627-95-8					0.2[a3,n]
Geranial	141-27-5	35.8-46.3	2.8-52.0	33.1	0-54.5	51.2[m]; 53.6[k]; 60.0[e]; 61.0[h]; 62.2[z2]
Geranic acid	459-80-3					1.3[c]; 2.7[j]
Geraniol	106-24-1	1.5-6.6	0.4-87.9	5.0	0.2-80.2	9.0[w]; 19.6[z4]; 30.5[g]; 74.7[d]; 92.3[z5]; 94.7[v]
Geranyl acetate	105-87-3	1.0-6.0	0.1-0.8	12.0	0.1-4.6	2.0[o]; 2.1[g]; 5.3[l]; 7.7[d]; 10.0[w]; 18.9[z4]; 79.8[z]
Geranyl butyrate	106-29-6					0.1[f,n]; 0.4[z4]; 0.6[b]; 0.8[h]; 5.3[d]
Geranyl formate	105-86-2					2.6[h]
Geranyl propionate	105-90-8					0.5[l]; 2.0[d]
Germacrene D	23986-74-5	0.03-1.3	0-0.1	0.2		0.1[f,n]; 1.5[b]; 4.1[z4]
Germacrene D-4-ol	198991-79-6		0-0.3			0.4[z4]
Globulol	489-41-8					1.0[l]
1-Hexene, 4,5-di-methyl-	16106-59-5					0.02[h]
(*Z*)-3-Hexen-1-ol	928-96-1				0-0.08	0.2[h]
(*E*)-2-Hexenyl butyrate	53398-83-7					0.1[b]; 0.3[z4]
α-Humulene	6753-98-6	0.02-0.2	0-1.2	0.1	0.08-0.2	0.1[k,n]; 0.2[b]; 0.5[z4]
cis-7-Hydroxy-3,7-dimethyl-3,6-oxy-octanal						4.4[c]
trans-7-Hydroxy-3,7-dimethyl-3,6-oxy-octanal						4.5[c]
Intermedeol	6168-59-8					50.6[t]
Isoborneol	124-76-5					tr[b]
(*E*)-Isocitral	72203-98-6	0.5-2.3				1.1[o]
(*Z*)-Isocitral	72203-97-5	0.2-1.5				0.3[o]
exo-Isocitral	55050-40-3					1.9[o]
(*E*)-Isoelemicin	5273-85-8					8.6[z4]
Isoeugenol	97-54-1	0.04-0.5				0.4[n]
(*E*)-Isoeugenol	5932-68-3					0.3[f]
trans-Isolimonene	6876-12-6					0.5[l]
Isopulegol	89-79-2		0-0.3		0-0.5	0.5[k]
Limonene	138-86-3	0.1-5.7	tr-1.8	2.0	0.1-0.8	2.2[e]; 2.4[m]; 3.0[d]; 4.4[v]; 7.6[z4]; 10.2[q]; 11.6[b]
trans-Limonene oxide	4959-35-7	0.01-0.5				
Linalool	78-70-6	0.4-1.8	0-0.8	2.6	0.4-1.4	1.5[n]; 1.6[b]; 3.3[j]; 3.4[h]; 4.4[l]; 6.7[j]; 10.0[d]
Linalool oxide	1365-19-1					0.08[h]; 0.3[l]
cis-Linalool oxide	11063-77-7					0.06[h]
Linalyl acetate	115-95-7		0-0.5			0.04[h]
p-Mentha-1(7),8-dien-2-ol (isocarveol)	35907-10-9					0.9[n]
p-Menth-2-en-1-ol	619-62-5	0.02-1.4				
Methyl eugenol	93-15-2	0.03-0.07		0.2		tr[s]; 0.3[b]; 2.1[q]; 20[t]; 34.2[z4]
Methylheptenone	409-02-9					0.9[r]
6-Methyl-5-hepten-2-one	110-93-0	0.4-2.1	tr-0.7	0.6	0.07-0.5	1.0[f]; 1.5[m]; 2.0[l]; 2.1[n,z4]; 2.2[o]; 2.5[i]; 3.1[j]
(*E*)-Methylisoeugenol	6379-72-2					0.1[f]; 21.1[z4]
(*Z*)-Methylisoeugenol	6380-24-1			0.4		0.1[f]; 0.4[z4]

Table 5.51.2 Constituents identified in lemongrass oils *(continued)*

Constituent	CAS	Percentage and range				
		A	B (23)	C (4)	D (7)	E
Methyl geraniate	2349-14-6					0.1[z4]; 0.7[c]
4-Methyl-2-(3-methyl-2-butenyl)furan						0.1[a3,n]
2-Methyl-6-methylene-3,7-octadien-2-ol	6994-89-4					4.0[h]
cis-Muurola-4(14),5-diene	157477-72-0					0.3[b]
γ-Muurolene	30021-74-0		0-0.1			
cis-Muurol-5-en-4β-ol	157374-46-4					1.0[z4]
Myrcene	123-35-3	0.07-5.0		0.1	0.03-4.0	1.6[b]; 2.2[v]; 4.3[d]; 4.4[n]; 5.7[j]; 8.8[x]
α-Myrcene	1686-30-2					2.2[i]
Myrcenol	543-39-5					0.2[h]
cis-Myrtanol	15358-92-6					0.2[a3,n]
4-Nanonene						0.3[z4]
Neral	106-26-3	27.5-35.5	1.9-32.9	30.0	0-36.1	33.6[o]; 36.1[k]; 36.2[j]; 36.4[f]; 38.0[e]; 39.6[i]
Nerol	106-25-2	0.04-1.0	tr-0.5			0.1[c]; 0.2[o]; 0.3[d]; 0.7[l]; 2.2[m]; 2.8[i]; 6.8[e]
Nerolidol	7212-44-4					0.7[h]
(E)-Nerolidol	40716-66-3					0.6[z4]
Neryl acetate	141-12-8	0.02-0.8			0-0.06	<0.1[f]; 2.5[n]; 4.6[i]; 6.4[j]; 15.7[e]
Neryl propionate	105-91-9					1.2[l]
2-Nonanene						0.2[z4]
4-Nonanone	4485-09-0					0.3[g]; 0.6[b]; 0.9[o]; 1.3[n]
Ocimene	13877-91-3					1.5[i]
(E)-β-Ocimene	3779-61-1	0.03-0.6	0-0.5	0.1	0.03-1.4	<0.1[k]; 0.1[f,m]; 0.2[n]; 0.3[o]; 0.9[z4]; 3.0[b]; 20.1[d]
(Z)-β-Ocimene	3338-55-4	0.04-0.9	0-0.8	0.4	0-1.0	<0.1[k]; 0.1[f,m]; 0.2[d]; 0.4[n]; 0.5[o]; 1.3[z4]; 4.6[b]
allo-Ocimene	673-84-7					0.3[b]
n-Octanal	124-13-0	0.01-0.2				0.1[c,f]; 0.2[z4]
Octanol	29063-28-3					0.1[n]
Octyl butyrate	110-39-4					0.2[z4]
3-Octyne	15232-76-5					0.6[l]
Oplopanone	1911-78-0					0.4[z4]
α-Phellandrene	99-83-2					tr[b]; <0.1[k]; 0.2[d]
β-Phellandrene	555-10-2	0.02-0.9			0-0.07	0.2[g]; 0.4[b,d]
Photocitral A						0.6[l]
(Z,Z)-Photocitral	55253-28-6					0.1[a1,n]
epi-Photocitral						0.4[l]
α-Pinene	80-56-8	0.03-1.5	tr-0.1	0.3		0.02[h]; 0.1[f]; 0.2[g,m]; 0.3[n]; 0.9[z4]; 1.9[b]; 2.7[d]
β-Pinene	127-91-3		0-0.1			tr[b]; 1.2[d]
β-Pinene oxide	6931-54-0	0.03-0.2				
Piperitone	89-81-6			0.1	0-0.07	tr[b]; 0.1[n,z4]; 0.2[c]
cis-Rose oxide	3033-23-6		0-0.3			
Sabinene	3387-41-5	0.01-0.06				0.1[b,z4]
trans-Sabinene hydrate	17699-16-0		tr-0.9	0.1		tr[b]; 0.1[z4]
α-Sinensal	17909-77-2					12.3[h]
Spathulenol	6750-60-3		tr-0.4			
Tagetone	23985-25-3					1.6
α-Terpinene	99-86-5	0.02-2.3				
γ-Terpinene	99-85-4	0.01-0.06			0-0.5	tr[b]; 9.9[d]
Terpinen-4-ol	562-74-3	0.04-0.4	tr-0.9	0.5	0-0.6	<0.1[k]; 0.2[b]; 0.5[z4]
α-Terpineol	98-55-5	0.2-1.9	0-0.9	0.6	0-0.1	0.3[g]; 0.4[m]; 0.5[c]; 0.7[b,z4]; 1.4[d]; 6.1[q]
Terpinolene	586-62-9	0.01-0.08	0-0.5		0-0.05	0.05[m]; 0.1[k]; 0.3[b]; 0.4[d]
α-Thujene	2867-05-2					tr[b]; 0.03[m]
Thymol	89-83-8					tr[b]
Tricyclene	508-32-7	0.01-0.6		0.5		0.1[f]; 0.2[g,n]; 0.5[z4]; 0.8[b]
3,4,5-Trimethoxybenzaldehyde	86-81-7					0.2[z4]
Trimethoxycinnamaldehyde	34346-90-2					0.8[z4]
3,3,6-Trimethyl-1,5-heptadiene	35387-63-4					0.3[l]
2-Undecanone	112-12-9					0.1[f]; 0.6[z4]; 0.8[l]
10-Undecen-1-ol	112-43-6					1.2[i]

Table 5.51.2 Constituents identified in lemongrass oils *(continued)*

Constituent	CAS	Percentage and range				
		A	B (23)	C (4)	D (7)	E
3-Undecyne	60212-30-8					1.2[j]
Vanillin, acetate	881-68-5					0.2[z4]
Veratraldehyde	120-14-9					0.3[z4]
Verbenol	473-67-6					2.0[i]
trans-Verbenol	1820-09-3					0.2[z4]
Vinylcyclooctane	61142-41-4					1.2[l]
Viridifloral						0.5[l]

A fifty-four essential oil samples from China, India, Comoros and Bhutan analyzed in the period 1999-2013; lowest and highest concentrations given (E. Schmidt, unpublished data)
B seven lab-hydrodistilled oils from 7 cultivars raised in India; lowest and highest concentrations given (ref. 23)
C one lab-hydrodistilled oil from leaves of *C. flexuosus* collected in the wild in India (ref. 4)
D five lab-hydrodistilled oils from 5 *C. flexuosus* cultivars from an experimental farm in India; lowest and highest concentrations given (ref. 7)
E data from other studies (indicated with superscript letters); highest concentrations found in any study reviewed here given; when two or more oils were investigated, only the highest concentrations are mentioned, unless indicated otherwise

[a1] artifact formed during distillation according to ref. 25; [a2] incorrect identity based on GC elution order according to ref. 25; [a3] not components of lemongrass oils according to ref. 25; [b] one lab-hydrodistilled oil from an experimental garden in India with elemicin as dominant ingredient (ref. 5); [c] one lab-hydrodistilled oil from *C. flexuosus* cultivated in Vietnam (ref. 18); [d] twelve lab-hydrodistilled oils from five variants from M2 generation derived from mutagenic treatment of selfed seeds of a natural variant of *C. flexuosus* with low citral content plus the original plant; harvested after 6 and 9 months (ref. 9); [e] eighteen lab-hydrodistilled oils from five accessions collected from different parts of northeast India and subsequently cultivated plus a standard cultivar, harvested in three seasons; only the eight most important components constituting collectively >90% of the oil were investigated (ref. 2); [f] one lab-hydrodistilled oil from lemongrass collected at an experimental field in Bangladesh (ref. 6); [g] one lab-hydrodistilled oil from *C. flexuosus* leaves growing wild in India (ref. 8); [h] one lab-hydrodistilled oil from leaves of *C. flexuosus* collected at a university garden in India (ref. 11); [i] ten lab-hydrodistilled oils from lemongrass leaves collected at an agricultural university in India with extraction time, power intensity, size of plant material and solid to water ratio as parameters (ref. 12); [j] one lab-hydrodistilled oil and twenty microwave-assisted hydrodistilled lemongrass oils from material collected at an agricultural university in India with weight of raw material, volume of water, rehydration time, extraction time and power as parameters (ref. 19); [k] one lab-hydrodistilled oil from a research farm in India (ref. 20); [l] one lab-hydrodistilled oil from East Indian lemongrass cultivated in India (ref. 22); [m] one lemongrass oil ex *C. flexuosus* of Indian origin (ref. 26); [n] one lab-hydrodistilled oil from a *C. flexuosus* cultivar grown in India (ref. 27); [o] one commercial lemongrass oil sample from India (ref. 28); [p] seven lab-hydrodistilled oils from an experimental farm in India with times of storage and amounts of water added as parameters; only citral (neral + geranial) and geraniol were investigated (ref. 3); [q] two lab-hydrodistilled oils from germplasm collected in the wild at two places in India and subsequently cultivated (ref. 10); [r] one lab-hydrodistilled oil from *C. flexuosus* leaves collected in the wild in India (ref. 13); [s] one oil of unknown origin (ref. 14); [t] older data cited in ref. 14; [u] eight lemongrass oils produced with a semi-professional steam distillation apparatus from *C. flexuosus* cultivated in the USA with distillation times varying from 2.5 minutes to 4 hours; mean values were given for four components only (ref. 15); [v] three commercial lemongrass oils ex *C. flexuosus* of unknown origin; [w] one lab-hydrodistilled oil from Columbia (ref. 21); [x] one steam-distilled oil from lemongrass collected in the wild in Brazil (ref. 24); [y] data from ref. 29, cited in ref. 4; [z] data from ref. 30, cited in ref. 4; [z1] older literature data cited in ref. 4; [z2] data from ref. 31, cited in ref. 4; [z3] data from ref. 32, cited in ref. 5; [z4] four lab-hydrodistilled oils from four accessions of *C. flexuosus* from India; there were some morphological differences, and the dominant components were different in the four oils: citral, citronellol, methyl eugenol ether and elemicin and therefore, the authors consider that the plants producing atypical oils may warrant recognition as distinct taxa (ref. 33); [z5] oils from three Indian *C. flexuosus* cultivars, of which one had a very high geraniol content (ref. 35)

tr: trace (in columns B and E[b]: <0.1); + present in the oil investigated, but quantity not stated

found in lemongrass oils from different sources are shown in Table 5.51.3. They include (highest concentrations in any study given) geraniol (94.7%), citral (geranial + neral) (90%), geranyl acetate (79.8%; one exceptionally high value, usually <20%), geranial (62.2%), neral (39.6%), limonene (11.6%) and myrcene (8.8%). Well-known ingredients of lemongrass oils that were present in high concentrations (>7%) in one or two studies were methyl eugenol (20% and 34.2%), citronellol (24.1% and 32.1%), (*E*)-β-ocimene (20.1%), borneol (17.1%), neryl acetate (15.7%), camphene (10.3% and 13.5%), linalool (10.0%) and γ-terpinene (9.9%). Uncommon or rare constituents of lemongrass oils found in high concentrations (>7%) in single studies include elemicin (three studies: 17.9%, 53.0% and 71.5%), intermedeol (50.6%), α-bisabolol (39.8%), (*E*)-methylisoeugenol (21.1%), (*E*)-4,5-epoxycarane (13.7%), (*E*)-isoelemicin (8.6%) and 1,10-di-epi-cubenol (7.6%).

Commercial oils

The ten chemicals that had the highest maximum concentrations in 54 commercial lemongrass East Indian essential oil samples (concentration ranges provided) are the following: geranial (35.8-46.3%), neral (27.5-35.5%), geraniol (1.5-6.6%), geranyl acetate (1.0-6.0%), limonene (0.1-5.7%), myrcene (0.07-5.0%), camphene (0.08-2.6%), (*E*)-isocitral (0.5-2.3%), α-terpinene (0.02-2.3%)

Table 5.51.3 Major constituents of lemongrass oils

Constituent	CAS	Percentage and range				
		A	B (23)	C (4)	D (7)	E
Geraniol	106-24-1	1.5-6.6	0.4-87.9	5.0	0.2-80.2	9.0[w]; 19.6[z4]; 30.5[g]; 74.7[d]; 94.7[y]; 92.3[z5];
Citral (geranial + neral)						60[s]; 75.9[p]; 82[z1]; 85-90[z1]
Geranyl acetate	105-87-3	1.0-6.0	0.1-0.8	12.0	0.1-4.6	2.0[o]; 2.1[g]; 5.3[j]; 7.7[d]; 10.0[w]; 18.9[z4]; 79.8[z]
Geranial	141-27-5	35.8-46.3	2.8-52.0	33.1	0-54.5	51.2[m]; 53.6[k]; 60.0[e]; 61.0[h]; 62.2[z2]
Neral	106-26-3	27.5-35.5	1.9-32.9	30.0	0-36.1	33.6[o]; 36.1[k]; 36.2[j]; 36.4[f]; 38.0[e]; 39.6[i]
Limonene	138-86-3	0.1-5.7	tr-1.8	2.0	0.1-0.8	2.2[e]; 2.4[m]; 3.0[d]; 4.4[v]; 7.6[z4]; 10.2[q]; 11.6[b]
Myrcene	123-35-3	0.07-5.0		0.1	0.03-4.0	1.6[b]; 2.2[v]; 4.3[d]; 4.4[n]; 5.7[j]; 8.8[x]

LEGEND: SEE UNDER TABLE 5.51.2

and 6-methyl-5-hepten-2-one (0.4-2.1%) (Erich Schmidt, unpublished analytical data).

Chemotypes

In most oils, citral (neral + geranial) is the dominant ingredient and this is certainly the case in the lemongrass oils of commerce. In fact, the quality of the oil is measured by the citral content (12). In some studies, however, other chemicals have been found to be the major ingredients, including geraniol (7,8,9,23,29), elemicin (5,33), (E)-β-ocimene (9), α-bisabolol (10), intermedeol (14), citronellol (33), methyl eugenol ether (33) and geranyl acetate (30). The following chemotypes have been suggested: citral type, α-bisabolol type, geraniol type, geranyl acetate type, methyl eugenol ether type, geraniol/citronellol/citral type and geraniol/citronellyl acetate/geranyl acetate chemotype (5); however, these authors appear to have received little support for this. The existence of a chemotype with geraniol as the dominant component

seems to have been confirmed sufficiently (7,8,9,23,29), e.g., in some cultivars (35,36). Some authors have suggested that some lemongrass accessions identified as 'Cymbopogon flexuosus', but dominated by chemicals other than citral (neral + geranial), in their study citronellol, methyl eugenol ether and elemicin, warrant recognition as distinct taxa (33).

An additional analysis not discussed in Table 5.51.2 can be found in ref. 34 (with virtually all IUPAC names), which can be accessed on-line.

CONTACT ALLERGY / ALLERGIC CONTACT DERMATITIS

Lemongrass oil (unspecified)

General

Contact allergy to / allergic contact dermatitis from lemongrass oil has been reported in over 20 publications.

Table 5.51.4 Results of testing groups of patients with lemongrass oil

Years and Country	Test conc. & vehicle	Number of patients tested positive (%)			Selection of patients (S); Relevance (R); Comments (C)	Ref.
Routine testing						
2000-2008 IVDK	2% pet.	2,435	14	(0.6%)	R: not stated	48
2000-2007 USA	2% pet.	868	11	(1.3%)	R: 100%; C: weak study: a. high rate of macular erythema and weak reactions, b. relevance figures include 'questionable' and 'past' relevance	37
1999-2000 Denmark	2% pet.	318	4	(1.3%)	R: not specified; C: this study was part of the international study mentioned below (ref. 51); the test preparation was East Indian lemongrass oil	55
1998-2000 six European countries	2% pet.	1,606	25	(1.6%)	R: not specified for individual oils/chemicals; C: the test preparation was East Indian lemongrass oil	51
Testing in groups of selected patients						
2001-2010 Australia	2% pet.	837	45	(5.4%)	S: not specified; R: 56%	54
2000-2008 IVDK	2% pet.	8,445	190	(2.2%)	S: patients with dermatitis suspected of causal exposure to fragrances; R: not stated; C: 51% of the patients allergic to lemongrass oil who were also tested with citral co-reacted to this important component of the oil	48
1997-2000 Austria	2% pet.	747	6	(0.8%)	S: patients suspected of fragrance allergy; R: not stated	39
1996-1997 UK	2% pet.	10	2	(20%)	S: patients suspected of cosmetic dermatitis and reacting to the fragrance mix; R: not stated	38
1994-1995 UK	2% pet.	40	3	(7.5%)	S: patients previously reacting to the fragrance mix; R: not stated; C: there was also one positive reaction in a patient negative to the fragrance mix	52

IVDK Information Network of Departments of Dermatology, Germany, Switzerland, Austria (www.ivdk.org); pet.: petrolatum

With two exceptions (51,55: test substances recorded as East Indian lemongrass oil) all studies have related to 'lemongrass oil', without specifying the botanical origin. In groups of consecutive patients suspected of contact dermatitis, prevalence rates of up to 1.6% positive patch test reactions have been observed, but adequate relevance data are lacking. There have been several case reports of allergic contact dermatitis from lemongrass oil, mostly from occupational exposure. In one case of a cosmetic reaction, citral (the combination of neral and geranial, both the dominant ingredients in lemongrass oils) was the allergenic component; in another report, geraniol may have been an allergen. Undiluted lemongrass oil is a strong irritant and may sensitize (57).

Testing in groups of patients
The results of patch tests with lemongrass oil in routine testing (consecutive patients suspected of contact dermatitis) and in groups of selected patients are shown in Table 5.51.4. In routine testing, rates of positive reactions ranged from 0.6% to 1.6%, whereas between 0.8% and 20% of patients in selected groups had positive patch tests. The high positivity rate of 20% was seen in a very small group of 10 patients with fragrance contact allergy (38).

Case reports

Occupational allergic contact dermatitis
Two aromatherapists had occupational contact dermatitis from multiple essential oils and reacted to lemongrass oil; one had used the oil, in the other patient this was uncertain (40). Another aromatherapist had occupational contact dermatitis from various essential oils, including lemongrass oil (41). Yet another aromatherapist had chronic hand dermatitis and was patch test positive to 17 of 20 oils used at her work (tested 1% and 5% in petrolatum); she also reacted to lemongrass oil, which was probably the primary sensitizer (42). A massage therapist had occupational contact dermatitis from allergies to many essential oils; she reacted to lemongrass oil in the fragrance series, which reaction was considered to be relevant (43). A female 'complementary therapist' developed occupational contact dermatitis from a multitude of essential oils used at work, including lemongrass oil (45).

Eight of thirty men working on a boat developed dermatitis which was ascribed to lemongrass oil which had been spilled; four were patch tested with pine wood contaminated with the lemongrass oil (as there were no samples of the oil itself) and all reacted, with negative reactions to pinewood without the oil (56).

Non-occupational allergic contact dermatitis
One positive patch test reaction to lemongrass oil was seen in a patient who had allergic contact cheilitis from citral in a lip balm (50); citral is the combination of neral and geranial, which are the main components of lemongrass oil, both the West and the East Indian varieties (Table 5.51.2, column A). Two cases of contact allergy to lemongrass oil present in topical pharmaceutical preparations have been reported (58).

Positive patch tests (relevance unknown, uncertain or not stated)
Irritant contact dermatitis of the hands was observed in a male masseur who worked with essential oils regularly; he had positive patch test reactions to lemongrass oil and orange oil, but it was not mentioned whether he had used these oils (44). One positive patch test to lemongrass oil was seen in a case of occupational hand dermatitis in a cook / barman / fruit grower caused by lemon oil, who also reacted to neroli oil and geraniol; the patient was patch tested with 'essences' (no concentration mentioned), but from the text it seems that this would indicate essential oils (46). Geraniol may be present in commercial lemongrass oils in concentrations up to 6% (Table 5.51.2, column A). One positive patch test reaction in a patient allergic to ascaridole, an allergen in tea tree oil (negative to tea tree oil itself) (47). One positive patch test reaction to lemongrass oil in 53 women with chronic anogenital dermatitis (49). One positive patch test reaction to lemongrass oil in a group of 475 patients with contact allergy to ingredients of cosmetics seen from January to April 1996 in 5 European dermatology centers; data on perfumes, toilet waters, aftershave lotions and deodorants were, however, excluded (53).

Lemongrass oil, East Indian
Two studies have used East Indian lemongrass oil for patch testing (51,55). Their results are discussed in Table 5.51.4.

LITERATURE
1 Lemongrass production. Department of Agriculture, Forestry and Fisheries, Republic of South Africa, 2009. Available at: https://yenmarketplace.org/sites/default/files/Lemon%20Grass%20Production.pdf
2 Sarma AQ, Sarma TCh. Studies on the morphological characters and yields of oil and citral of certain lemongrass [*Cymbopogon Flexuosus* (Steud) Wats] accessions grown under agro-climatic conditions of northeast India. J Essent Oil Bear Plants 2005;8:250-257
3 Rajeswara Rao BR, Rajput DK, Patel RP. Storage of essential oils: Influence of presence of water for short periods on the composition of major constituents of the essential oils of four economically Important aromatic crops. J Essent Oil Bear Plants 2011;14:673-678
4 Chowdhury SR, Tandon PK, Chowdhury AR. Chemical composition of the essential oil of *Cymbopogon flexuosus* (Steud) Wats. growing in Kumaon region. J Essent Oil Bear Plants 2010;13:588-593
5 Sarma KK, Nath SC, Leclercq PA. The essential oil of a variant of *Cymbopogon flexuosus* (Nees ex Steud) Wats. from northeast India. J Essent Oil Res 1999;11:381-385
6 Chowdhury JU, Yusuf M, Begum J, Mondello L, Previti P, Dugo G. Studies on the essential oil bearing plants of Bangladesh. Part IV. Composition of the leaf oil of three *Cymbopogon* species: *C. flexuosus* (Nees ex Steud.) Wats., *C. nardus* (L.) Rendle var. *confertiflorus* (Steud.) N. L. Bor and *C. martinii* (Roxb.) Wats. var. *martini*. J Essent Oil Res 1998;10:301-306

7 Bhattacharya AK, Kaul PN, Rajeswara Rao BR, Mallavarapu GR, Ramesh SI. Inter-specific and inter-cultivar variations in the essential oil profiles of lemongrass. J Essent Oil Res 1997;9:361-364

8 Nath SC, Saha BN, Bordoloi DN, Mathur RK, Leclercq PA. The chemical composition of the essential oil of *Cymbopogon flexuosus* (Steud) Wats. growing in northeast India. J Essent Oil Res 1994;6:85-87

9 Kulkarni RN, Mallavarapu GR, Ramesh S. The oil content and composition of new variants of *Cymbopogon flexuosus*. J Essent Oil Res 1992;4:511-514

10 Thappa RK, Agarwal SG. *Cymbopogon flexuosus* oil a rich source of (+)-α-Bisabolol. J Essent Oil Res 1989;1:107-110

11 Kumar A, Shukla R, Singh P, Dubey NK. Biodeterioration of some herbal raw materials by storage fungi and aflatoxin and assessment of *Cymbopogon flexuosus* essential oil and its components as antifungal. Int Biodeter Biodegrad 2009;63:712-716

12 Desai MA, Parikh J, Kumar De A. Modelling and optimization studies on extraction of lemongrass oil from *Cymbopogon flexuosus* (Steud.) Wats. Chem Engineer Res Design 2014;92:793-803

13 Khanuja SPS, Shasany AK, Pawar A, Lal RK, Darokar MP, Naqvi AA, et al. Essential oil constituents and RAPD markers to establish species relationship in *Cymbopogon* Spreng. (Poaceae). Biochem System Ecol 2005;33:171-186

14 Taskinen J, Mathela DK, Mathela CS. Composition of the essential oil of *Cymbopogon flexuosus*. J Chromatogr 1983;262:364-366

15 Cannon JB, Cantrell CL, Astatkie T, Zheljazkov VD. Modification of yield and composition of essential oils by distillation time. Ind Crops Prod 2013;41:214-220

16 Desai MA, Parikh J. Hydrotropic extraction of citral from *Cymbopogon flexuosus* (Steud.) Wats. Ind Eng Chem Res 2012;51:3750-3757

17 Schaneberg BT, Khan IA. Comparison of extraction methods for marker compounds in the essential oil of lemon grass by GC. J Agric Food Chem 2002;50:1345-1349

18 Ottavioli J, Bighelli A, Casanova J, Bang BT, Van Y P. GC (retention indices), GC-MS, and 13C NMR of two citral-rich *Cymbopogon* leaf oils: *C. Flexuosus* and *C. tortilis*. Spectroscopy Letters 2009;42:506-512

19 Desai MA, Parikh J. Microwave assisted extraction of essential oil from *Cymbopogon Flexuosus* (Steud.) Wats.: A parametric and comparative study. Separ Sci Technol 2012;47:1963-1970

20 Rajeswara Rao BR, Kaul PN, Bhattacharya AK. Yield and chemical composition of the essential oils of three *Cymbopogon* species suffering from iron chlorosis. Flavour Fragr J 1996;11:289-293

21 Smith Vera S, Zambrano DF, Méndez-Sanchez SC, Rodríguez-Sanabria F, Stashenko EE, Duque Luna JE. Essential oils with insecticidal activity against larvae of *Aedes aegypti* (Diptera: Culicidae). Parasitol Res 2014;113:2647-2654

22 Pandey AK, Rai MK, Acharya D. Chemical composition and antimycotic activity of the essential oils of corn mint (*Mentha arvensis*) and lemon grass (*Cymbopogon flexuosus*) against human pathogenic fungi. Pharm Biol 2003;41:421-425

23 Padalia RC, Verma RS, Chanotiya CS, Yadav A. Chemical fingerprinting of the fragrant volatiles of nineteen Indian cultivars of *Cymbopogon* Spreng. (Poaceae). Rec Nat Prod 2011;5:290-299

24 Mello da Silveira WS, Cunha Júnior A, Neudí Scheuermann G, Secchi FL, Werneck Vieira CR. Chemical composition and antimicrobial activity of essential oils from selected herbs cultivated in the south of Brazil against food spoilage and foodborne pathogens. Ciência Rural, Santa Maria 2012;42:1300-1306

25 Lawrence BM. Essential oils 2008-2011. Carol Stream, USA: Allured Publishing Corporation, 2012:249-251

26 Ranade GS. Profile: Lemongrass oil (E.I.). FAFAI J 2004;6:89. Data cited in ref. 25

27 Malik et al. 2008. Article plus its data cited in ref. 25, but full bibliography not provided.

28 Milchard M, Clery R, Esdale R, Gates L, Judge F, Moss N, et al. Application of gas-liquid chromatography to the analysis of essential oils. GLC fingerprint chromatograms of six essential oils. Perfum Flavor 2010;35(6):34-40

29 Patra NK, Srivastava RK, Chauhan SP, Ahmed A, Misra LM. Chemical features and productivity of a geraniol rich variety (GRL1) of *Cymbopogon flexuosus*. Planta Medica 1990;56:239-240. Data cited in ref. 4

30 Mathew S, Chittam GJ, Thomas J. A lemon grass chemotype rich in geranyl actate. Indian Perfumer 1996;40: 9-12. Data cited in ref. 4

31 Verma V, Sobti SN, Atal CK. Chemical composition and inheritance pattern of five *Cymbopogon* species. Indian Perfumer 1987;31:293-303. Data cited in ref. 4

32 Mallavarapu GR, Ramesh S, Kulkarni RN, Syamasundar KV. Composition of the essential oil of *Cymbopogon travencorensis*. Planta Medica 1992;58:219-220. Data cited in ref. 5. *C. travancorensis* is considered a synonym of *C. flexuosus* by The Plant List

33 Nath SC, Sarma KK, Vajezikova I, Leclercq PA. Comparison of volatile inflorescence oils and taxonomy of certain Cymbopogon taxa described as Cymbopogon flexuosus (Nees ex Steud.) Wats. Biochem Syst Ecol 2002;30:151-162

34 Vinutha M, Thara Saraswathi KJ. Study on the essential oil of aerial and sub-aerial parts of *Cymbopogon flexuosus* (Nees ex Steud) Wats. Int J Curr Sci 2013;7:E42-E47

35 Ganjewala D. RAPD characterization of three selected cultivars OD-19, GRL-1 and Krishna of East Indian lemongrass (*Cymbopogon flexuosus* Nees ex Steud) Wats. Amer-Euras J Bot 2008;1(2):53-57

36 Ganjewala D, Luthra R. Geranyl acetate esterase controls and regulates the level of geraniol in lemongrass (*Cymbopogon flexuosus* Nees ex Steud.) mutant cv. GRL-1 leaves. Z Naturforsch c 2009;64:251-259

37 Wetter DA, Yiannias JA, Prakash AV, Davis MD, Farmer SA, el-Azhary RA, et al. Results of patch testing to personal care product allergens in a standard series and a supplemental cosmetic series: an analysis of 945 patients from the Mayo Clinic Contact Dermatitis Group, 2000-2007. J Am Acad Dermatol 2010;63:789-798

38 Thomson KF, Wilkinson SM. Allergic contact dermatitis to plant extracts in patients with cosmetic dermatitis. Br J Dermatol 2000;142:84-88

39 Wöhrl S, Hemmer W, Focke M, Götz M, Jarisch R. The significance of fragrance mix, balsam of Peru, colophonium and propolis as screening tools in the detection of fragrance allergy. Br J Dermatol 2001;145:268-273

40 Bleasel N, Tate B, Rademaker M. Allergic contact dermatitis following exposure to essential oils. Australas J Dermatol 2002;43:211-213

41 Boonchai W, Lamtharachai P, Sunthonpalin P. Occupational allergic contact dermatitis from essential oils in aromatherapists. Contact Dermatitis 2007;56:181-182

42 Selvaag E, Holm J, Thune P. Allergic contact dermatitis in an aromatherapist with multiple sensitizations to essential oils. Contact Dermatitis 1995;33:354-355

43 Cockayne SE, Gawkrodger DJ. Occupational contact dermatitis in an aromatherapist. Contact Dermatitis 1997; 37:306-307

44 Jung P, Sesztak-Greinecker G, Wantke F, Götz M, Jarisch R, Hemmer W. Mechanical irritation triggering allergic contact dermatitis from essential oils in a masseur. Contact Dermatitis 2006;54:297-299

45 Newsham J, Rai S, Williams JDL. Two cases of allergic contact dermatitis to neroli oil. Br J Dermatol 2011;165(Suppl.1):76

46 Audicana M, Bernaola G. Occupational contact dermatitis from citrus fruits: Lemon essential oils. Contact Dermatitis 1994;31:183-185

47 Christoffers WA, Blömeke B, Coenraads P-J, Schuttelaar M-LA. The optimal patch test concentration for ascaridole as a sensitizing component of tea tree oil. Contact Dermatitis 2014;71:129-137

48 Uter W, Schmidt E, Geier J, Lessmann H, Schnuch A, Frosch P. Contact allergy to essential oils: current patch test results (2000–2008) from the Information Network of Departments of Dermatology (IVDK). Contact Dermatitis 2010;63:277-283

49 Vermaat H, Smienk F, Rustemeyer Th, Bruynzeel DP, Kirtschig G. Anogenital allergic contact dermatitis, the role of spices and flavour allergy. Contact Dermatitis 2008;59:233-237

50 Hindle E, Ashworth J, Beck MH. Chelitis from contact allergy to citral in lip salve. Contact Dermatitis 2007;57:125-126

51 Frosch PJ, Johansen JD, Menné T, Pirker C, Rastogi SC, Andersen KE, et al. Further important sensitizers in patients sensitive to fragrances. II. Reactivity to essential oils. Contact Dermatitis 2002;47:279-287

52 Katsarma G, Gawkrodger DJ. Suspected fragrance allergy requires extended patch testing to individual fragrance allergens. Contact Dermatitis 1999;41:193-197

53 Goossens A, Beck MH, Haneke E, McFadden JP, Nolting S, Durupt G, Ries G. Adverse cutaneous reactions to cosmetic allergens. Contact Dermatitis 1999;40:112-113

54 Toholka R, Wang Y-S, Tate B, Tam M, Cahill J, Palmer A, Nixon R. The first Australian Baseline Series: Recommendations for patch testing in suspected contact dermatitis. Australas J Dermatol 2014, Sept. 7. doi: 10.1111/ajd.12186

55 Paulsen E, Andersen KE. Colophonium and Compositae mix as markers of fragrance allergy: Cross-reactivity between fragrance terpenes, colophonium and Compositae plant extracts. Contact Dermatitis 2005;53:285-291

56 Mendelsohn HV. Dermatitis from lemon grass oil (*Cymbopogon citratus* or *Andropogon citratus*). Arch Dermatol 1944;50:34-35

57 Mendelsohn HV. Lemon grass oil. Arch Dermatol 1946;53:94-98

58 Nardelli A, D'Hooge E, Drieghe J, Dooms M, Goossens A. Allergic contact dermatitis from fragrance components in specific topical pharmaceutical products in Belgium. Contact Dermatitis 2009;60:303-313

Chapter 5.52 LEMONGRASS OIL, WEST INDIAN

There are two main lemongrass oils, derived from two species of the *Cymbopogon* family, lemongrass oil, East Indian, derived from *Cymbopogon flexuosus,* and West Indian lemongrass oil, obtained from *Cymbopogon citratus*. In this chapter, the West Indian type is discussed. Lemongrass essential oil ex *Cymbopogon flexuosus* is presented in Chapter 5.51.

DEFINITION

Lemongrass oil, West Indian type (essential oil of lemongrass) is the essential oil obtained from the whole aerial part (leaves) of the West Indian lemongrass, *Cymbopogon citratus* (DC) Stapf.

INCI NOMENCLATURE

Description/definition: Cymbopogon citratus leaf oil is an essential oil obtained from the leaves of the lemon grass, *Cymbopogon citratus*, Poaceae
INCI name EU & USA: Cymbopogon citratus leaf oil
CAS registry number(s): 8007-02-1; 89998-14-1
EINECS number(s): 289-752-0

ISO (INTERNATIONAL ORGANIZATION FOR STANDARDIZATION) STANDARD

ISO number: 3217
ISO name: Essential oil of lemongrass
Botanical origin: *Cymbopogon citratus* (DC) Stapf.
Parts of plant used: Whole aerial part (leaves) ISO values:
ISO values: (minimum and maximum concentrations) are shown in Table 5.52.1.

Another species of the genus *Cymbopogon* from which lemongrass can be derived is *Cymbopogon schoenanthus* (CAS 89998-16-3; EINECS 289-754-1). For the essential oil of *Cymbopogon martinii* see Chapter 5.66 Palmarosa oil and for the oil of *Cymbopogon nardus* see Chapters 5.23 and 5.24 Citronella oil.

THE PLANT, THE OIL, AND THEIR USES

Cymbopogon citratus (DC) Stapf. is a tall, aromatic, perennial grass with culms (stems) up to two meters tall and linear leaves, up to one meter long and two centimeters wide. Lemongrass is native to India and Indonesia, and introduced and cultivated in most of the tropics, including Africa (Algeria, Egypt, Morocco), South America (including the Carribean islands and Central America) and Indo-China (7, GRIN Taxonomy for Plants). For the use of lemongrass and lemongrass oils see the Chapter 5.51 East Indian lemongrass oil.

CHEMICAL COMPOSITION

West Indian lemongrass oil is a clear mobile liquid with a soft yellow to yellowish orange color which has a strong lemon note, reminding of citral odor. The yield of essential oil from the grass of *Cymbopogon citratus* (DC) Stapf. generally varies from 0.3 to 0.9 per cent. The main producing countries of this oil are Sri Lanka, China, Indonesia (Java), Madagascar, Guinea and Haiti.

Literature data (up to August 13, 2014) on the chemical composition of West Indian lemongrass oils and unpublished analytical data from one of us (E.S.) are shown in Table 5.52.2 in alphabetical order. In lemongrass oils, West Indian type, from various origins, over 245 chemicals have been identified. About 57 per cent of these were found in a single reviewed publication only. The major compounds found in West Indian lemongrass oils from different sources are shown in Table 5.52.3. They include (highest concentrations in any study given) myrcene (67.0%; one very high concentrations, rest <18.6%), geranial (56.2%), geraniol (40.2%), neral (39.4%), geranyl acetate (9.9%), limonene (8.9%) and linalool (3.5%). Well-known ingredients of West Indian lemongrass oils that were present in high concentrations (>7%) in one or two studies were nerol (12.5%), camphene (10.7%), caryophyllene oxide (9.7%) and 6-methyl-5-hepten-2-one (8.4%). Uncommon or rare constituents of West Indian lemongrass oils found in high concentrations (>7%) in single studies include *cis*-verbenol (37.9%), *trans*-verbenol (32.1%), *cis*-pinocarveol (20.2%), neryl acetate (three studies: 7.5%, 10.8% and 13.6%), α-oxobisabolene (12.1%), β-pinene (three studies: 8.1%, 10.1% and 12.0%) and 1,8-cineole (7.5%).

Commercial oils

The ten chemicals that had the highest maximum concentrations in 32 commercial lemongrass West Indian essential oil samples (concentration ranges provided) are the following: geranial (41.8-46.3%), neral (30.3-33.3%), geraniol (3.0-7.9%), limonene (0.4-5.0%), geranyl acetate (1.0-4.3%), β-caryophyllene (0.4-3.0%), citronellal (0.2-1.9%), elemol (0.02-1.8%), (*E*)-isocitral (1.4-1.7%) and linalool (0.6-1.4%) (Erich Schmidt, unpublished analytical data).

Chemotypes

In virtually all lemongrass oils *ex C. citratus*, citral (geranial + neral) is the dominant ingredient, with geranial in the higher amounts. In a few studies, however, other chemicals have been found to have a concentration higher than both geranial and neral separately: geraniol (21,69) and myrcene (20,28,30,48).

Additional analyses of *C. citratus* leaf oils not presented in Table 5.52.2 can be found in refs. 63-68; most of these can be accessed on-line.

Table 5.52.1 ISO values (%) for lemongrass oil, West Indian type [a]

Compound	CAS	Minimum	Maximum
Geranial	141-27-5	40.0	50.0
Neral	106-26-3	31.0	40.0
Geraniol	106-24-1	1.5	7.0
β-Caryophyllene	87-44-5	1.0	4.0
Nerol	106-25-2	0.01	1.0

[a] ISO 3217 Essential oil of lemongrass ©ISO 2000; Geneva, Switzerland, www.iso.org

Table 5.52.2 Constituents identified in West Indian lemongrass oils

Constituent	CAS	Percentage and range				
		A	B (1)	C (11,58)	D (52)	E
α-Amorphene	20085-19-2					0.1[c]; 0.2[h,x7]
(E)-Anethole	4180-23-8				0-1.5	
Aromadendrene	489-39-4					0.06[t]; 0.2[u]; 0.6[e]
Atrimesol						0.3[x6]
α-Bergamotene	17699-05-7					0.1[d,i]; 0.3[h]; 0.6[c]
cis-α-Bergamotene	18252-46-5			0.1-0.8		
trans-α-Bergamotene	13474-59-4		0.1		0-0.2	<0.05[x8]; 0.1[g,j]; 0.2[w]; 0.4[x2]
Bicyclo[2.2.1]heptane-2,5-diol, 1,7,7-trime-thyl- (2-endo, 5-exo)-						0.09[y2]
Bicyclopentan-2-one	4884-24-6					0.2[c]
(E)-α-Bisabolene	25532-79-0					0.7[x7]
β-Bisabolene	495-61-4					0.2[h]
β-Bisabolenol	147126-90-7					0.5[g]
Borneol	507-70-0					0.7[y5]; 0.8[k]; 2.0[x5]; 2.2[z3]; 3.7[e]; 5.0[x4]
Bornyl acetate	76-49-3			tr-0.3		0.4[x3]; 0.7[x7]
β-Bourbonene	5208-59-3					0.5[i]
α-Cadinene	24406-05-1					0.3[x7]
γ-Cadinene	39029-41-9			0.1-0.6	0-0.7	<0.05[s]; 0.1[h]; 0.3[r]; 0.8[x1]; 1.3[m]
δ-Cadinene	483-76-1		0.2	0-0.1	0-0.7	0.2[h,u]; 1.0[c,x7]; 1.1[k]
σ-Cadinene						0.3[x7]
α-Cadinol	481-34-5		0.4	0-0.4		0.2[u]; 1.3[c]
τ-Cadinol	58580-31-7					0.2[c]
Camphene	79-92-5	0.1-0.8				0.9[e,p]; 1.0[x1]; 1.2[l]; 1.5[m]; 3.5[x5]; 10.7[z7]
Camphene hydrate	465-31-6					1.3[x5]
Camphor	76-22-2					0.02[t]; 0.1[f,y6]; 0.2[x4,x7]; 1.1[y4]
δ3-Carene	13466-78-9					0.1[m,x4]
trans-3(10)-Caren-2-ol	6909-15-5					0.1[i]
Carvacrol	499-75-2					0.09[d]
Carveol	99-48-9					0.7[x6]
(E)-Carveol	1197-07-5					0.3[c]
(Z)-Carveol	1197-06-4					0.2[c,m]; 1.5[x6]
D-Carvone	2244-16-8					0.2[c]
β-Caryophyllene	87-44-5	0.4-3.0	0.2		0-1.0	1.9[m]; 2.2[s]; 2.7[z3]; 2.8[r,x2]; 3.9[z7]
Caryophyllene oxide	1139-30-6	0.05-1.2	0.1		0-0.3	0.8[k,m]; 1.0[s]; 1.5[r]; 2.2[x2]; 2.9[z7]; 9.7[z5]
trans-Chrysanthemal	20104-05-6					0.06[c]; 0.3[y2]
1,4-Cineole	470-67-7					0.5[k]
1,8-Cineole	470-82-6			tr-0.1		0.2[e,f]; 0.4[v]; 1.0[j3]; 1.5[l]; 2.9[b]; 7.5[k]
Citronellal	106-23-0	0.2-1.9	0.2	tr-0.1		0.7[h]; 0.8[l]; 1.1[f]; 1.3[z3]; 1.7[j]; 2.8[v]; 3.0[x6]
Citronellol (β-, DL-)	106-22-9		0.3	0.1-0.2	0-1.0	0.4[f,i]; 0.6[e]; 0.8[h]; 0.9[m]; 2.0[x9]; 2.8[z4]
Citronellyl acetate	150-84-5	0.04-0.2				0.1[j]; 0.2[h]; 1.3[p]
α-Copaene	3856-25-5					1.1[x7]
(E,E)-Cosmene	460-01-5					1.3[x6]
β-Cubebene	13744-15-5					0.3[v]
Cubenol	21284-22-0					0.4[k]
1-epi-Cubenol	81939-29-9					0.3[m]
Cucumber alcohol						0.4[o]
Cuparene	16982-00-6				0-0.5	
α-Cuparenene						<0.1[r]
Curcumene (ar-; α-)	644-30-4					2.9[w]
γ-Curcumene	28976-68-3					0.7[w]
α-Cyclocitral	432-24-6					0.4[k]; 0.5[m]
α-Cyclogeraniol	6627-74-3					0.3[g]
α-Cyclogeraniol ace-tate	68406-89-3					1.8[y]
3-Cyclohexene-1-carboxaldehyde, 1,3,4-trimethyl-	40702-26-9					0.09[y2]

Table 5.52.2 Constituents identified in West Indian lemongrass oils *(continued)*

Constituent	CAS	Percentage and range				
		A	B (1)	C (11,58)	D (52)	E
Cycloisolongifolene	28380-07-6					0.4[c]
(+)-Cycloisosativene						0.6[k]
o-Cymene	527-84-4					0.5[k]
p-Cymene	99-87-6					0.3[x1]; 0.5[i,x7]; 0.9[d]
p-Cymen-8-ol	1197-01-9					0.08[k]; 1.1[x1]
Decanal	112-31-2	0.1-0.4				<0.05[s]; <0.1[r]; 0.3[k]
n-Decane	124-18-5					1.0[a1,x1]
Dendrolasin	23262-34-2					0.3[a1,z8]
Dibutyl phthalate	84-74-2					0.2[a2,u]
Dichloromethane	75-09-2					0.04[y2]
2,2-Dimethylcyclo-hexanone	1193-47-1					0.1[y2]
2,2-Dimethyl-3,4-octadiene	590-71-6					0.3[g]
2,6-Dimethyl-2(3),7-octadienal						0.1[a1,x1]
2,6-Dimethyloctane	2051-30-1					0.1[y6]; 1.4[z1]
(E)-2-Dodecenal	20407-84-5					0.05[y2]
Elemene	11029-06-4	0.01-1.3				
β-Elemene	33880-83-0					0.1[y6]; 0.2[h]; 0.3[c]
Elemol (α-)	639-99-6	0.02-1.8		0-0.7		0.05[c]; <0.1[r]; 0.1[h]; 0.2[x1]
2,3-Epoxygeranial						0.1[x8]
6,7-Epoxymyrcene			0.2			0.2[n]; 0.8[y1]
2,3-Epoxyneral						0.1[n,x8]
(Z)-Epoxyocimene						0.2[x8]
α-Eudesmol	473-16-5			0-0.1		
β-Eudesmol	473-15-4			0-0.2		0.2[c]
Eugenol	97-53-0	0.1-0.4				<0.1[r]; 0.3[e]; 0.4[d]; 0.7[f]
cis-Farnesal	3790-68-9					0.1[c]; 0.2[h]
α-Farnesene	502-61-4					0.3[c,u]
(Z)-β-Farnesene	28973-97-9					0.3[c]; 0.4[w]
Farnesol	4602-84-0					0.3[u]; 2.4[p]
(E,E)-Farnesol	106-28-5					0.03[d]; 0.1[h]; 0.3[u]
(E,E)-Farnesyl acetate	4128-17-0					0.3[g]
(Z,E)-Farnesyl acetate	40266-29-3					0.2[g]
Fenchone	1195-79-5					0.2[y6]; 0.3[x4]
Fenchyl alcohol	1632-73-1					0.5[k]
Geijerene	6902-73-4			0-0.4		
Geranial	141-27-5	41.8-46.3	41.3	30.5-56.2	29.2-51.7	45.9[y9]; 46.9[v]; 48.7[r]; 49.0[y4]; 49.5[g]; 49.8[x3]; 49.9[y8]; 51.3[y7]; 52.3[x1]; 52.9[y2]
Geranic acid	459-80-3					0.1[i,n]; 0.2[y8]; 0.3[d]; 0.4[f]; 0.5[k]; 0.9[c]; 3.0[u]
Geraniol	106-24-1	3.0-7.9	3.3	3.0-7.7	0.4-11.9	2.7[h]; 3.3[x5,q]; 3.4[w]; 3.8[o]; 4.0[u,z]; 4.3[l]; 4.6[z1]; 5.6[k]; 6.7[b]; 7.0[v]; 10.4[y8]; 11.5[y1]; 40.2[x4]
(Z)-Geraniol (β-nerol)	106-25-2					4.3[i]
Geranyl acetate	105-87-3	1.0-4.3	0.6	0.2-3.5	0-0.5	1.8[r]; 1.9[b,h]; 2.2[c]; 2.4[y4]; 4.1[x2]; 4.2[k]; 9.9[p]
Geranyl butyrate	106-29-6					0.1[d]; 0.2[m]
Geranyl formate	105-86-2					0.4[g]
Geranyl hexanoate	10032-02-7					0.02[d]
Germacrene D	23986-74-5		0.1			0.04[t]; 0.3[c,x3]; 0.5[m]; 1.3[k,x7]
Germacrene D-4-ol	198991-79-6			0-0.2		
α-Guaiene	3691-12-1					0.1[c]
1(5)-Guaien-11-ol	13822-35-0					0.2[c]
α-Gurjunene	489-40-7					0.5[c]; 0.7[h]
β-Gurjunene	73464-47-8					0.1[h]
γ-Gurjunene	22567-17-5					0.1[h]; 0.4[u]
τ-Gurjunene						0.1[i]
Heptanal	111-71-7					2.0[l]

Table 5.52.2 Constituents identified in West Indian lemongrass oils *(continued)*

Constituent	CAS	Percentage and range				
		A	B (1)	C (11,58)	D (52)	E
Hexadecanoic acid	57-10-3					0.2[u]; 0.9[x1]
1,4-Hexadiene, 5-methyl-3-(1-methylidene)-						0.1[c]
1,4-Hexadiene, 3,3,5-trimethyl-						0.2[y2]
α-Humulene	6753-98-6	0.1-0.4				0.03[t]; 0.1[x1]; 0.2[w]; 0.3[c,r]; 1.6[x7]
β-Humulene	116-04-1					0.1[h]
α-Humulene epoxide	96638-51-6					0.1[h]
(E)-β-Ionone	79-77-6				0-0.2	
Isoborneol	124-76-5					1.0[f]
Isocaryophyllene	118-65-0					1.2[c]
Isocaryophyllene oxide						0.1[b]
(E)-Isocitral	72203-98-6	1.4-1.7				1.2[s]
(Z)-Isocitral	72203-97-5	0.7-1.1				0.2[s]
exo-Isocitral	55050-40-3					1.9[s]
Isoeugenol	97-54-1	0.2-0.2				0.5[m]
(Z)-Isoeugenol	5912-86-7					0.7[k]
Isolongifolene, 4,5,9,10-dehydro-	156747-45-4					0.02[c]
Isopentyl nerolate						0.1[d]
Isophorone	78-59-1					<0.1[g]
Isophyllocladene	511-85-3					0.4[x]
Isopinocampheol	27779-29-9					0.08[c]
Isopulegol	89-79-2					1.7[v]
Isopulegol II	1370348-44-9					0.1[y2]
Isopulegone	29606-79-9					1.7[v]
Juniper camphor	473-04-1		0.1		0-0.4	0.1[i]; 2.8[c]
Ledene epoxide						0.4[x1]
Ledol	577-27-5					0.1[h]
Limonene	138-86-3	0.4-5.0	0.4	tr-0.2		2.2[k]; 3.1[e]; 4.2[m]; 4.4[z3]; 5.8[x2]; 8.4[s]; 8.9[z4]
Limonene-1,2-epoxide	1195-92-2					3.6[x1]
Limonene oxide	1195-92-2					0.2[k]; 0.3[x3]
Linalool	78-70-6	0.6-1.4	1.1	0.6-1.3	0-1.7	1.5[b]; 1.7[d,j]; 2.4[x6]; 2.8[z4]; 3.2[q]; 3.4[x4]; 3.5[v]
Linalool oxide	1365-19-1					1.0[p]
cis-Linalool oxide	11063-77-7				0-0.2	0.01[t]; 0.02[d]; 0.1[h]
cis-Linalool oxide, furanoid	11063-77-7					<0.1[g]; 0.3[c]
trans-Linalool oxide	11063-78-8					0.1[h,t]; 0.4[v]
trans-Linalool oxide, furanoid	34995-77-2					<0.1[g]
Linalyl acetate	115-95-7					2.3[y6]
β-Maaliene	489-29-2					0.4[c]
p-Mentha-1(7),8(10)-dien-9-ol	29548-13-8					0.1[i]
cis-p-Mentha-2,8-dien-1-ol	3886-78-0					0.1[i]
p-Mentha-1,3,8-triene	18368-95-1					0.1[d]; 0.6[x1]
Menthol	89-78-1					0.5[x4]; 0.6[a1,p]
Menthone	89-80-5					0.2[x4]
Methyl chavicol	140-67-0					0.8[x3]
Methyl geranate	2349-14-6		0.3			0.2[d]
Methylheptenone	409-02-9					0.5[n]; 1.8[x2]
3-Methyl-2-heptanone	2371-19-9					2.0[p]
6-Methyl-5-hepten-2-one	110-93-0	0.2-0.8	1.3	1.2-2.5	0.09-1.6	2.4[v]; 2.5[l]; 2.6[b]; 2.7[j]; 2.9[m]; 3[z2]; 3.2[q]; 8.4[s]

Table 5.52.2 Constituents identified in West Indian lemongrass oils *(continued)*

Constituent	CAS	Percentage and range				
		A	B (1)	C (11,58)	D (52)	E
2-Methyl-4-octanone	7492-38-8	0.7-1.1				
α-Muurolene	10208-80-7		0.2			0.2[c,h]
γ-Muurolene	30021-74-0					0.1[o]; 0.9[c]; 2.1[k]
α-Muurolol	104245-48-9			0-0.1		0.2[b]
epi-α-Muurolol	19912-62-0			0-0.2		0.3[c]; 0.4[h,u]
Myrcene	123-35-3	0.05-0.2	14.2	6.2-22.6	0.2-7.3	12.8[y9]; 15.5[w,x]; 16.2[x6]; 18.0[y3]; 18.6[b]; 25.3[y6]; 27.0[x5]; 27.8[l]; 28.0[z]; 67.0[z2]
Myrcenol	543-39-5					0.09[c]; 0.2[o,y6]; 0.4[i]
Myrtanal	4764-14-1					0.2[m]
Myrtenyl acetate	1079-01-2					0.09[d]
Neointermedeol	5945-72-2		0.1			
Neomenthol	3623-51-6					3.3[a1,y6]
Neral	106-26-3	30.3-33.3	32.1	21.3-32.5	11.4-31.5	33.0[f]; 33.1[n]; 33.4[b]; 34.4[x3]; 34.7[y7]; 34.8[v]; 35.0[x6]; 35.5[i]; 38.2[g]; 39.0[x2]; 39.4[y2]
Neric acid	37349-29-4					0.1[i]; 0.2[f]
Nerol	106-25-2	0.01-0.05		0.1-0.4		1.5[f]; 2.2[y1]; 3.7[c]; 4.2[x9]; 4.5[x4]; 5.1[m]; 12.5[e]
Nerolic acid	4613-38-1					0.8[m]
Neryl acetate	141-12-8					3.0[z3]; 4.0[m]; 6.3[e]; 7.5[p]; 10.8[q]; 13.6[z5]
Nonanal	124-19-6					0.1[f]
4-Nonanone	4485-09-0					0.5[r,y2]; 1.2[s]; 1.3[k,x2]
Nopol	128-50-7					0.4[i]
Ocimene	13877-91-3					0.3[c]
β-Ocimene	13877-91-3					0.3[m]; 0.4[k]; 0.6[c]
(E)-β-Ocimene	3779-61-1	0.1-0.1	0.3	0.1-0.4		0.1[f]; 0.2[d,i]; 0.4[o]; 0.6[b]; 0.7[j,y1]; 1.3[u]
(Z)-β-Ocimene	3338-55-4		0.3	0.2-0.4	0-0.2	0.3[l]; 0.4[i]; 0.5[b]; 0.6[x2]; 0.9[j]; 1.0[x6]; 1.1[y1]
(Z)-Ocimene	27400-71-1					1.2[x1]
allo-Ocimene	673-84-7					tr[j]
(E)-allo-Ocimene	3016-19-1					0.1[y6]
1,6-Octadiene, 2,6-dimethyl-						0.2[c]
n-Octanal	124-13-0	0.05-0.1				0.06[y2]; <0.1[r]; 0.1[m]
2-Octanone	111-13-7					1.1[h]
di-n-Octyl phthalate	117-84-0					2.0[c]
1-Octyn-3-ol	818-72-4					0.2[c]
Oleic acid	112-80-1					0.3[a1,x1]; 2.1[x2]
Oxiranmethanol, 3-methyl-3-(4-methyl-3-pentenyl)-						0.08[c]
α-Oxobisabolene						12.1[x4]
β-Patchoulene	514-51-2					0.2[y6]
1-Pentanol, 5-cyclo-propylidene-	162377-97-1					0.05[c]
2-Pentene						3.6[a3,u]
Perillene	539-52-6					0.1[x8]
α-Phellandrene	99-83-2					0.04[d]; 0.1[k]; 6.9[y5]
α-Phellandren-8-ol	1686-20-0					0.5[i]
Photocitral						0.4[d]
Photocitral A						0.2[x8]; 1.2[y1]
Photocitral B	6040-45-5					0.1[x8]
Pimelyl dihydrazide	13043-98-6					0.06
α-Pinene	80-56-8	0.01-0.2	0.3			0.1[d,g]; 0.2[b]; 0.5[l]; 1.1[e]; 1.3[p]; 3.0[y4]; 4.7[k]
β-Pinene	127-91-3					0.1[y5]; 1.5[p]; 1.6[k]; 8.1[y1]; 10.1[i]; 12.0[o]
α-Pinene oxide	1686-14-2					0.9[c]
β-Pinene oxide	6931-54-0		1.7			0.2[o]; 1.1[y2]; 1.7[a3,t]
Pinocarveol	5947-36-4					0.2[d]
cis-Pinocarveol	6712-79-4					20.2[a4,x1]

Table 5.52.2 Constituents identified in West Indian lemongrass oils *(continued)*

Constituent	CAS	Percentage and range				
		A	B (1)	C (11,58)	D (52)	E
trans-Pinocarveol	1674-08-4					0.4[x5]
trans-Pinocarvone	19890-00-7					0.6[x5]
trans-Pinocarvyl acetate	1686-15-3					0.7[x3]
Piperitenone	491-09-8					<0.1[g]; 0.2[x1]
Piperitone	89-81-6					0.01[c]; <0.1[r]; 0.1[y2]
Piperonal	120-57-0					0.3[f]
Rosefuran epoxide	92356-06-4					0.1[x8]; 0.2[n]; 0.8[h]
Sabinene	3387-41-5			tr-0.3		0.04[y2]; <0.05[b]; 0.3[v]; 0.5[o]; 1.1[l]; 2.2[x3]
Sabinol						0.1[y6]; 0.5[m]
cis-Sabinol	3310-02-9					0.3[d]
α-Selinene	473-13-2					0.2[c]; 1.2[x7]
β-Selinene	17066-67-0					0.1[h]; 0.9[x7]
γ-Selinene	515-17-3					0.7[x7]
δ-Selinene	473-14-3					0.1[h]
Selin-6-en-4-ol	118173-08-3		0.2			0.4[n]; 0.5[k]; 8.9[x7]
Selin-7(11)-en-4α-ol	16641-47-7					0.3[u]
β-Sesquiphellandrene	20307-83-9					0.08[c]
α-Terpinene	99-86-5					0.4[d]
γ-Terpinene	99-85-4					0.7[k]
Terpinen-4-ol	562-74-3					0.3[f]; 0.5[e]; 0.9[x7]; 1.0[l]; 1.4[v]; 1.7[o]
Terpinen-4-yl acetate	4821-04-9					0.5[v]
1-Terpineol	586-82-3					0.4[y6]
α-Terpineol	98-55-5		0.2			<0.1[r]; 0.2[f]; 0.3[y3]; 0.5[x9]; 0.9[k]; 5.2[z4]
γ-Terpineol	586-81-2					0.1[o]
δ-Terpineol	7299-42-5					0.8[v]
α-Terpinolene	586-62-9				0-1.8	0.1[x4]; 0.2[i]; 1.1[x6]; 1.2[x9]
Tetrahydrolinalool	78-69-3					0.3[a2,y6]
α-Thujene	2867-05-2		0.3			0.05[d]; 0.1[h]; 1.9[x5]; 2.3[l]
α-Thujone	546-80-5					0.1[a3,t]; 0.2[g]
β-Thujone (*cis*-)	471-15-8					0.2[g]
Torreyol	19435-97-3				0-0.8	
Tricyclene	508-32-7					0.2[m]
Tricyclo[4.3.1.1(3,8)]-undecan-1-ol	31061-64-0					0.04[y2]
2-Tridecanone	593-08-8					<0.05[x8]; <0.1[y8]; 0.1[d,f]; 0.2[y3]; 0.4[g]
3,6,6-Trimethylcyclo-hexen-2-ol						0.2[c]
3,5,5-Trimethyl-2(5*H*)-furanone	50598-50-0					2.5[a1,x1]
1,3,5-Trimethyl-5-pyrazolone						0.1[a2,d]
2-Undecanone	112-12-9		0.1		0-0.9	0.1[c,d,f]; 0.4[t]; 0.5[g,y3]; 0.7[y1]; 1.6[y]
3-Undecyne	60212-30-8					6.1[c]
Valencene	4630-07-3					0.09[c]; 1.9[u]
β-Vatirenene						0.1[c]
Verbenol	473-67-6					tr[y3]; 0.6[k]; 1.2[o]
cis-Verbenol	1845-30-3					0.2[x6]; 0.5[d]; 0.6[y2]; 1.7[i]; 37.9[z6]
trans-Verbenol	1820-09-3					1.2[b]; 2.7[y]; 32.1[z7]
Verbenone	80-57-9			0.7-1.9		0.2[m]
Viridiflorol	552-02-3					0.05[t]; 0.2[c]
Zingiberene	495-60-3					0.1[m]; 6.9[w]

Table 5.52.2 Constituents identified in West Indian lemongrass oils *(continued)*

A thirty-two essential oil samples from China, Java, Guinea and Haiti analyzed between 1999 and 2013; lowest and highest concentrations given (E. Schmidt, unpublished data)

B three lab-hydrodistilled oils from Brazil, two from cultivated *C. citratus* leaves and one from material purchased at a local market; highest concentrations given; the compositions of the oils were rather similar (ref. 1)

C sixteen laboratory steam-distilled lemongrass leaf oils from plants growing wild in Mali (n=12) and Ivory Coast (n=4); lowest and highest concentrations given (refs. 11,58)

D fifteen lab-hydrodistilled oils from Brazil, one from fresh *C. citratus* leaves, 9 from commercial dried leaves and five from lemongrass tea bags; lowest and highest concentrations given (ref. 52)

E data from other studies (indicated with superscript letters); highest concentrations found in any study reviewed here given; when two or more oils were investigated, only the highest concentrations are mentioned, unless indicated otherwise

[a1] incorrect identity based on GC elution order (refs. 53,59); [a2] does not occur naturally (ref. 53,59); [a3] probably misidentification (ref. 59); [a4] probably misidentification for neral (ref. 53); [b] seven steam-distilled oils from *C. citratus* cultivated in Zimbabwe and harvested in 4 consecutive years (ref. 14); [c] three lab-hydrodistilled oils from the aerial parts (sheaths plus leaves) of lemongrass cultivated in Malaysia and harvested after 5.5, 6.5 and 7.5 months (ref. 5); [d] one lab-hydrodistilled oil from plant material collected in south west Nigeria (ref. 2); [e] one lab-hydrodistilled oil from lemongrass leaves of *C. citratus* growing wild in the Democratic Republic of Congo (ref. 10); [f] one lab-hydrodistilled oil from lemongrass growing wild in Burkina Faso (ref. 12); [g] one lab-hydrodistilled oil from the aerial parts of *C. citratus* cultivated in Cuba (ref. 13); [h] one lab-hydrodistilled oil from plant material growing wild in Malaysia (ref. 18); [i] one steam-distilled oil from fresh leaves collected at a university botanical garden in Benin (ref. 27); [j] one commercial oil purchased from a German company; no neral analyzed, extremely atypical composition (ref. 46); [k] one commercial oil purchased from an Indian company (ref. 47); [l] one lab-hydrodistilled oil from the leaves of *C. citratus* cultivated at a university botanical garden in Benin (ref. 48); [m] one commercial *C. citratus* lemongrass oil from an Indian company (ref. 49); [n] one lab-hydrodistilled oil from West Indian lemongrass growing wild in Benin (ref. 50); [o] one lab-hydrodistilled oil from wild Benin lemongrass (ref. 51); [p] one oil from fresh leaves harvested in Nigeria (ref. 54); [q] one sample of lemongrass oil ex *Cymbopogon citratus* from India (refs. 4,55); [r] one lemongrass oil ex *Cymbopogon citratus* sample from Guatemala (ref. 56); [s] one commercial oil from Guatemala (ref. 57); [t] one oil produced from *C. citratus* plants grown in Argentina (ref. 60); [u] one oil sample from China (ref. 61); [v] one commercial oil sample purchased in Austria (ref. 62); [w] one commercial oil sample of unknown origin (ref. 3); [x] one commercial *C. citratus* oil from Ecuador (ref. 6,32); [x1] one lab-hydrodistilled lemongrass oil from leaves collected at a university campus in India (ref. 8); [x2] one commercial oil purchased from an Iranian company (ref. 16); [x3] one lab-hydrodistilled oil from *C. citratus* collected in the wild in Cameroon (ref. 17); [x4] one steam-distilled oil from the whole plant collected in the wild in Ethiopia (ref. 21); highly atypical composition; [x5] one lab-hydrodistilled oil from the fresh leaves of lemongrass collected in Cameroon (ref. 22); [x6] four oils from Egypt, one from fresh leaves and the other three from leaves dried in different manners (ref. 23); [x7] one lab-hydrodistilled oil from plant material growing wild in Cameroon (ref. 25); [x8] one lab-hydrodistilled oil from the aerial parts of *C. citratus* purchased at a local market in Angola (ref. 26); [x9] one steam-distilled oil from *Cymbopogon citratus* cultivated in the garden of a research center in Pakistan (refs. 34,36); [y] one lab-distilled oil from leaves purchased in Italy (ref. 35); [y1] one microwave-assisted hydrodistilled oil from the leaves of *C. citratus* grown at an experimental agro-industrial station in Colombia (ref. 39); [y2] one lab-hydrodistilled oil from fresh leaves collected in the wild in India (ref. 41); [y3] one lab-hydrodistilled oil from leaves collected at a research center in Zambia (ref. 42); [y4] one lab-hydrodistilled oil from an Indian cultivar of *C. citratus* (ref. 43); [y5] one lab-hydrodistilled oil from the aerial parts of lemongrass cultivated in India (ref. 44); [y6] one lab-hydrodistilled oil from leaves collected at a university campus, Lagos, Nigeria (ref. 45); [y7] one lab-hydrodistilled oil from a cultivar in India; only geranial, neral and myrcene were investigated (ref. 9); [y8] one lab-hydrodistilled oil from the leaves of *Cymbopogon citratus* cultivated in Peru (ref. 15); [y9] one lab-hydrodistilled oil from Cameroon (ref. 19); [z] one lab-hydrodistilled oil from wild growing lemongrass leaves in Benin (ref. 20); [z1] one lab-hydrodistilled oil from leaves of lemongrass ex *C. citratus* grown at an experimental field in Thailand (ref. 24); [z2] one extremely atypical lab-hydrodistilled oil from the Ivory Coast with 67% myrcene and no citral (ref. 28); [z3] one commercial lemongrass oil from a UK supplier (ref. 29); [z4] one commercial oil from a South African supplier (ref. 31); [z5] one lab-hydrodistilled oil from *C. citratus* growing wild in Argentina (ref. 37); [z6] one very atypical commercial oil from China with an extremely high concentration of *cis*-verbenol (ref. 38); [z7] one very atypical commercial oil purchased in Canada with an extremely high concentration of *trans*-verbenol (ref. 40); [z8] one steam-distilled oil from cultivated plants in Egypt; data cited in ref. 53

tr: trace (in column C: <0.05)

Table 5.52.3 Major constituents of West Indian lemongrass oils

Constituent	CAS	Percentage and range				
		A	B (1)	C (11,58)	D (52)	E
Myrcene	123-35-3	0.05-0.2	14.2	6.2-22.6	0.2-7.3	25.3[y6]; 27.0[x5]; 27.8[l]; 28.0[z]; 67.0[z2]
Geranial	141-27-5	41.8-46.3	41.3	30.5-56.2	29.2-51.7	49.8[x3]; 49.9[y8]; 51.3[y7]; 52.3[x1]; 52.9[y2]
Geraniol	106-24-1	3.0-7.9	3.3	3.0-7.7	0.4-11.9	5.6[k]; 6.7[b]; 7.0[y]; 10.4[y8]; 11.5[y1]; 40.2[x4]
Neral	106-26-3	30.3-33.3	32.1	21.3-32.5	11.4-31.5	35.0[x6]; 35.5[i]; 38.2[g]; 39.0[x2]; 39.4[y2]
Geranyl acetate	105-87-3	1.0-4.3	0.6	0.2-3.5	0-0.5	1.8[r]; 1.9[b,h]; 2.2[c]; 2.4[y4]; 4.1[x2]; 4.2[k]; 9.9[p]
Limonene	138-86-3	0.4-5.0	0.4	tr-0.2		2.2[k]; 3.1[e]; 4.2[m]; 4.4[z3]; 5.8[x2]; 8.4[s]; 8.9[z4]
Linalool	78-70-6	0.6-1.4	1.1	0.6-1.3	0-1.7	1.5[b]; 1.7[d,j]; 2.4[x6]; 2.8[z4]; 3.2[q]; 3.4[x4]; 3.5[v]

LEGEND: SEE UNDER TABLE 5.52.2

CONTACT ALLERGY / ALLERGIC CONTACT DERMATITIS

No reports on contact allergy to or allergic contact dermatitis from lemongrass oil, specified to have been obtained from *Cymbopogon citratus* (DC) Stapf. (West Indian lemongrass oil) were found. The literature on 'lemongrass oil' (botanical source not specified) is discussed in Chapter 5.51 East Indian lemongrass oil.

LITERATURE

1 Andrade EHA, das Graças B. Zoghbi M, da Paz Lima M. Chemical composition of the essential oils of *Cymbopogon citratus* (DC.) Stapf cultivated in north of Brazil. J Essent Oil Bear Plants 2009;12:41-45

2 Owolabi MS, Oladimeji MO, Lajide L, Singh G, Marimuthu P, Isidorov VA. *Cymbopogon citratus* (DC) Stapf volatile oil from South West Nigeria. J Essent Oil Bear Plants 2008;11:335-341

3 Tognolini M, Barocelli E, Ballabeni V, Bruni B, Bianchi A, Chiavarini M, Impicciatore M. Comparative screening of plant essential oils: Phenylpropanoid moiety as basic core for antiplatelet activity. Life Sciences 2006;78:1419-1432

4 Pawar VC, Thaker VS. *In vitro* efficacy of 75 essential oils against *Aspergillus niger.* Mycoses 2006;49:316-323. Data also presented in ref. 55

5 Tajidin NE, Ahmad SH, Rosenani AB, Azimah H, Munirah M. Chemical composition and citral content in lemongrass (*Cymbopogon citratus*) essential oil at three maturity stages. Afr J Biotechnol 2012;11:2685-2693

6 Sacchetti G, Maietti S, Muzzoli M, Scaglanti M, Manfredini S, Radice M, Bruni R. Comparative evaluation of 11 essential oils of different origin as functional antioxidants, antiradicals and antimicrobials in foods. Food Chem 2005;91:621-632. Data partly also presented in ref. 32

7 Lemongrass production. Department of Agriculture, Forestry and Fisheries, Republic of South Africa, 2009. Available at: https://yenmarketplace.org/sites/default/files/Lemon%20Grass%20Production.pdf

8 Ali M, Sahrawat I, Singh O. Volatile constituents of *Cymbopogon citratus* (DC.) Stapf. leaves. J Essent Oil Bear Plants 2004;7:56-59

9 Boruah P, Misra BP, Pathak MG, Ghosh AC. Dynamics of essential oil of *Cymbopogon citratus* (DC) Stapf. under rust disease Indices. J Essent Oil Res 1995;7:337-338

10 Cimanga K, Apers S, de Bruyne T, Van Miert S, Hermans N, Totté J, Chemical composition and antifungal activity of essential oils of some aromatic medicinal plants growing in the Democratic Republic of Congo. J Essent Oil Res 2002;14:382-387. Data also presented in ref. 33

11 Sidibé L, Chalchat J-C, Garry R-P, Lacombe L, Harama M. Aromatic plants of Mali (IV): Chemical composition of essential oils of *Cymbopogon citratus* (DC) Stapf and *C. giganteus* (Hochst.) Chiov. J Essent Oil Res 2001;13:110-112. Data also presented in ref. 58

12 Menut C, Bessiére JM, Samaté D, Djibo AK, Buchbauer G, Schopper B. Aromatic plants of tropical West Africa. XI. Chemical composition, antioxidant and antiradical properties of the essential oils of three *Cymbopogon* species from Burkina Faso. J Essent Oil Res 2000;12:207-212

13 Pino JA, Rosado A. Chemical composition of the essential oil of *Cymbopogon citratus* (DC.) Stapf. from Cuba. J Essent Oil Res 2000;12:301-302

14 Chagonda LS, Makanda C, Chalchat J-C. Essential oils of cultivated *Cymbopogon winterianus* (Jowitt) and of *C. citratus* (DC) (Stapf) from Zimbabwe. J Essent Oil Res 2000;12:478-480

15 Leclercq P, Silva Delgado H, Garcia J, Hidalgo JE, Cerruttti T, Mestanza M, et al. Aromatic plant oils of the Peruvian Amazon. Part 2. *Cymbopogon citratus* (DC) Stapf., *Renealmia* sp., *Hyptis recurvata* Poit. and *Tynanthus panurensis* (Bur.) Sandw. J Essent Oil Res 2000;12:14-18

16 Vazirian M, Taheri Kashani S, Shams Ardekani MR, Khanavi M, Jamalifar H, Fazeli MR, et al. Antimicrobial activity of lemongrass (*Cymbopogon citratus* (DC) Stapf.) essential oil against food-borne pathogens added to cream-filled cakes and pastries. J Essent Oil Res 2012;24:579-582

17 Nguefack J, Nguikwie SK, Fotio D, Dongmo B, Amvam Zollo PH, Leth V, et al. Fungicidal potential of essential oils and fractions from *Cymbopogon*

citratus, Ocimum gratissimum and *Thymus vulgaris* to control *Alternaria padwickii* and *Bipolaris oryzae*, two seed-borne fungi of rice (*Oryza Sativa* L.). J Essent Oil Res 2007;19:581-587

18 Piaru SP, Perumal S, Cai LW, Mahmud R, Shah AM, Majid A, et al. Chemical composition, anti-angiogenic and cytotoxicity activities of the essential oils of *Cymbopogan citratus* (lemon grass) against colorectal and breast carcinoma cell lines. J Essent Oil Res 2012;24:453-459

19 Chalchat J-C, Garry RP, Menut C, Lamaty G, Malhuret R, Chopineau J. Correlation between chemical composition and antimicrobial activity. VI. Activity of some African essential oils. J Essent Oil Res 1997;9:67-75

20 Fandohan P, Gbenou JD, Gnonlonfin B, Hell K, Marasas WFO, Wingfield MJ. Effect of essential oils on the growth of *Fusarium verticillioides* and fumonisin contamination in corn. J Agric Food Chem 2004;52:6824-6829

21 Abegaz, Yohannes PG, Dieter RK. Constituents of the essential oil of Ethiopian *Cymbopogon citratus* Stapf. J Nat Prod 1983;46:424-426

22 Tchinda ES, Jazet PMD, Tatsadjieu LN, Ndongson BD, Amvam PHZ, Menut C. Antifungal activity of the essential oil of *Cymbopogon citratus* (Poaceae) against *Phaeoramularia angolensis*. J Essent Oil Bear Plants 2009;12:218-224

23 Hanaa ARM, Sallam YI, El-Leithy AS, Aly SE. Lemongrass (*Cymbopogon citratus*) essential oil as affected by drying methods. Ann Agricult Sci 2012;57:113-116

24 Poonpaiboonpipat T, Pangnakorn U, Suvunnamek U, Teerarak M, Charoenying P, Laosinwattana C. Phytotoxic effects of essential oil from *Cymbopogon citratus* and its physiological mechanisms on barnyard grass (*Echinochloa crus-galli*). Ind Crops Prod 2013;41:403-407

25 Nguefack J, Tamgue O, Lekagne Dongmo JB, Dakole CD, Leth V, Vismer HF, et al. Synergistic action between fractions of essential oils from *Cymbopogon citratus, Ocimum gratissimum* and *Thymus vulgaris* against *Penicillium expansum*. Food Control 2012;23:377-383

26 Machado M, Pires P, Dinis AM, Santos-Rosa M, Alves V, Salgueiro L, et al. Monoterpenic aldehydes as potential anti-*Leishmania* agents: Activity of *Cymbopogon citratus* and citral on *L. infantum, L. tropica* and *L. major*. Exp Parasitol 2012;130:223-231

27 Kpoviessi S, Bero J, Agbani P, Gbaguidi F, Kpadonou-Kpoviessi B, Sinsin B, et al. Chemical composition, cytotoxicity and *in vitro* antitrypanosomal and antiplasmodial activity of the essential oils of four *Cymbopogon species* from Benin. J Ethnopharmacol 2014;151:652-659

28 Kanko C, El-Hadj Sawaliho B, Kone S, Koukoua G, N'Guessan YT. Étude des propriétés physico-chimiques des huiles essentielles de *Lippia multiflora, Cymbopogon citratus, Cymbopogon nardus, Cymbopogon giganteus*. CR Chimie 2004;7:1039-1042

29 Tzortzakis NG, Economakis CD. Antifungal activity of lemongrass (*Cympopogon citratus* L.) essential oil against key postharvest pathogens. Innov Food Sci Emerg Technol 2007;8:253-258

30 Kasumov FY, Babaev RI. Components of the essential oil of *Cymbopogon citratus* Stapf. Khim Prir Soedin 1983;1:108-109. Data cited in ref. 15

31 Regnier T, Combrinck S, Veldman W, Du Plooy W. Application of essential oils as multi-target fungicides for the control of *Geotrichum citri-aurantii* and other postharvest pathogens of citrus. Ind Crops Prod 2014;61:151-159

32 Rolli E, Marieschi M, Maietti S, Sacchetti G, Bruni R. Comparative phytotoxicity of 25 essential oils on pre- and post-emergence development of *Solanum lycopersicum* L.: A multivariate approach. Ind Crops Prod 2014;60:280-290. Data also presented in ref. 6

33 Cimanga K, Kambu K, Tona L, Apers S, De Bruyne T, Hermans N, et al. Correlation between chemical composition and antibacterial activity of essential oils of some aromatic medicinal plants growing in the Democratic Republic of Congo. J Ethnopharmacol 2002;79:213-220. Data also presented in ref. 10

34 Saleem M, Afza N, Anwar MA, Hai SMA, Ali MS, Shujaat S, et al. Chemistry and biological significance of essential oils of *Cymbopogon citratus* from Pakistan. Nat Prod Res 2003;17:159-163. Data also presented in ref. 36

35 Marongiu B, Piras A, Porcedda S, Tuveri E. Comparative analysis of the oil and supercritical CO_2 extract of *Cymbopogon citratus* Stapf. Nat Prod Res 2006;20:455-459

36 Saleem M, Afza N, Anwar MA, Hai SMH, Ali MS. A comparative study of essential oils of *Cymbopogon citratus* and some members of the genus *Citrus*. Nat Prod Res 2003;17:369-373. Data also presented in ref. 34

37 Stefanazzi N, Stadler T, Ferrero A. Composition and toxic, repellent and feeding deterrent activity of essential oils against the stored-grain pests *Tribolium castaneum* (Coleoptera: Tenebrionidae) and *Sitophilus oryzae* (Coleoptera: Curculionidae). Pest Manag Sci 2011;67:639-646

38 Jiang ZL, Akhtar Y, Zhang X, Bradbury R, Isman MB. Insecticidal and feeding deterrent activities of essential oils in the cabbage looper, *Trichoplusia ni* (Lepidoptera: Noctuidae). J Appl Entomol 2012;136:191-202

39 Olivero-Verbel J, Nerio LS, Stashenko EE. Bioactivity against *Tribolium castaneum* Herbst (Coleoptera: Tenebrionidae) of *Cymbopogon citratus* and *Eucalyptus citriodora* essential oils grown in Colombia. Pest Manag Sci 2010;66:664-668

40 Machial CM, Shikano I, Smirle M, Bradbury R, Isman MB. Evaluation of the toxicity of 17 essential oils against *Choristoneura rosaceana* (Lepidoptera: Tortricidae) and *Trichoplusia ni* (Lepidoptera: Noctuidae). Pest Manag Sci 2010;66:1116-1121

41 Sonker N, Pandey AK, Singh P, Tripathi NN. Assessment of *Cymbopogon citratus* (DC.) Stapf. essential oil as herbal preservatives based on

antifungal, antiaflatoxin, and antiochratoxin activities and *in vivo* efficacy during storage. J Food Sci 2014;79:M628-M634

42 Chisowa EH, Hall DR, Farman DI. Volatile constituents of the essential oil of *Cymbopogon citratus* Stapf grown in Zambia. Flavour Fragr J 1998;13:29-30

43 Shahi AK, Kaul MK, Gupta R, Dutt P, Chandra S, Qazi GN. Determination of essential oil quality index by using energy summation indices in an elite strain of *Cymbopogon citratus* (DC) Stapf [RRL(J)CCA12]. Flavour Fragr J 2005;20:118-121

44 Mahanta JJ, Chutia M, Bordoloi M, Pathak MG, Adhikary RK, Sarma TC. *Cymbopogon citratus* L. essential oil as a potential antifungal agent against key weed moulds of *Pleurotus* spp. spawns. Flavour Fragr J 2007;22:525-530

45 Kasali AA, Oyedeji AO, Ashilokun AO. Volatile leaf oil constituents of *Cymbopogon citratus* (DC) Stapf. Flavour Fragr J 2001;16:377-378

46 Baratta MT, Dorman HJD, Deans SG, Figueiredo AC, Barroso JAG, Ruberto G. Antimicrobial and antioxidant properties of some commercial essential oils. Flavour Fragr J 1998;13:235-244

47 Kumar P, Mishra S, Malik A, Satya S. Housefly (*Musca domestica* L.) control potential of *Cymbopogon citratus* Stapf. (Poales: Poaceae) essential oil and monoterpenes (citral and 1,8-cineole). Parasitol Res 2013;112:69-76

48 Gbenou JD, Ahounou JF, Akakpo HB, Laleye A, Yayi E, Gbaguidi F, et al. Phytochemical composition of *Cymbopogon citratus* and *Eucalyptus citriodora* essential oils and their anti-inflammatory and analgesic properties on Wistar rats. Mol Biol Rep 2013;40:1127-1134

49 Tyagi AK, Malik A. Liquid and vapour-phase antifungal activities of selected essential oils against *Candida albicans*: microscopic observations and chemical characterization of *Cymbopogon citratus*. BMC Complem Altern Med 2010;10:65 (11 pages)

50 Bossou AD, Mangelinckx S, Yedomonhan H, Boko PM, Akogbeto MC, De Kimpe N, et al. Chemical composition and insecticidal activity of plant essential oils from Benin against *Anopheles gambiae* (Giles). Parasites & Vectors 2013;6:337 (17 pages)

51 Tchobo FP, Alitonou GA, Soumanou MM, Barea B, Bayrasy C, Laguerre M, et al. Chemical composition and ability of essential oils from six aromatic plants to counteract lipid oxidation in emulsions. J Am Oil Chem Soc 2014;91:471-479

52 Almeida Barbosa LC, Alves Pereira U, Martinazzo AP, Álvares Maltha CR, Ricardo Teixeira R, de Castro Melo E. Evaluation of the chemical composition of Brazilian commercial *Cymbopogon citratus* (D.C.) Stapf samples. Molecules 2008;13:1864-1874

53 Lawrence BM. Essential oils 2008-2011. Carol Stream, USA: Allured Publishing Corporation, 2012:251-254

54 Ekundayo O. Composition of the leaf volatile oil of *Cymbopogon citratus*. Fitoterapia 1985;56:339-341. Data cited in ref. 53

55 Pawar VC, Thaker VS. Evaluation of the anti-*Fusarium oxysporum* f. sp *cicer* and anti-*Alternaria porri* effects of some essential oils. World J Microbial Technol 2007;23:1099-1106. Data also presented in ref. 4

56 Williams DG. The chemistry of essential oils, 2nd Ed. Port Washington, NY, USA: Micelle Press, 2008: 194-195. Data cited in ref. 53

57 Milchard M, Clery R, Esdale R, Gates L, Judge F, Moss N, et al. Application of gas-liquid chromatography to the analysis of essential oils. GLC fingerprint chromatograms of six essential oils. Perfum Flavor 2010;35(6):34-40. Data cited in ref. 53

58 Chalchat JC, Garry RPh, Harama M, Sidibé L. Etude des plantes aromatiques du Mali: Composition chemiques des huiles essentielles de deux *Cymbopogon*: *Cymbopogon citratus* et *Cymbopogon giganteus*. Rivista Ital EPPOS 1998;(Numero Speciale):741-752. Data cited in ref. 59. Data also presented in ref. 11

59 Lawrence BM. Progress in essential oils. Perfum Flavor 2002;27(6):46-? (last page unknown)

60 Viturro CI, Bucu CW. Composicion del aceite esencial de lemongrass de Jujuy. An Asoc Quim Argentina 1998;86:45-48. Data cited in ref. 59

61 Liu J-X, Jiang J-B, Yang Z-X, Gu L. Studies on the chemical constituents of essential oil of Xiangxi *Cymbopogon* by capillary gas chromatography-mass spectrometry. Jishou Daxue Xuebao Ziran Kexueban 1998;19(3):43-45. Data cited in ref. 59

62 Oberhofer B, Nikiforov A, Buchbauer G, Jirovetz L, Bicchi C. Investigation of the alteration of the composition of various essential oils used in aroma lamp applications. Flavour Fragr J 1999;14:293-299

63 Farhang V, Amini J, Javadi T, Nazemi J, Ebadollahi A. Chemical composition and antifungal activity of essential oil of *Cymbopogon citratus* (DC.) Stapf. against three *Phytophthora* species. Greener J Biol Sci 2013;3:292-298

64 Sessou P, Farougou S, Kaneho S, Djenontin S, Alitonou AG, et al. Bioefficacy of *Cymbopogon citratus* essential oil against foodborne pathogens in culture medium and in traditional cheese wagashi produced in Benin. Int Res J Microbiol 2012;3:406-415

65 Koba K, Sanda K, Guyon C, Raynaud C, Chaumont J-P, Nicod L. In vitro cytotoxic activity of *Cymbopogon citratus* L. and *Cymbopogon nardus* L. essential oils from Togo. Bangladesh J Pharmacol 2009;4:29-34

66 Shaaban HA, Ramadan MM, Amer MM, El-Sideek L, Osman F. Chemical composition of *Cymbopogon citratus* essential oil and antifungal activity against some species of mycotoxigenic *Aspergillus* fungi. J Appl Sci Res 2013;9:5770-5779

67 Tadtong S, Watthanachaiyingcharoen R, Kamkaen N. Antimicrobial constituents and synergism effect of the essential oils from *Cymbopogon citratus* and *Alpinia galanga*. Nat Prod Commun 2014;9:277-280

68 LiSha X, QianGZ, Wei OY, Yuan HZ. A comparison of chemical components of essential oils extracted from *Cymbopogon citratus* (DC.) Stapf. by SFE-CO2 and SD. J Med Plant 2012;3(9):56-58

69 Ouamba JM, Menut C, Bessiere JM, Silou T, Lamaty G. Composition chimique des huiles essentielles de trois espèces du genre *Cymbopogon* (Poaceae): *Cymbopogon citratus* (DC) Stapf., C. *densiflorus* (Steud) Stapf. et C. *giganteus* Chiov. Plantes Aromatiques — Huiles Essentielles, Revue Scientifique Panafricaine 1993;1:45-51. Data cited in ref. 12

Chapter 5.53   LITSEA CUBEBA OIL

DEFINITION

Litsea cubeba oil (essential oil of litsea cubeba) is the essential oil obtained from the fruit of the litsea (mountain pepper), *Litsea cubeba* (Lour.) Pers.

INCI NOMENCLATURE

Description/definition: Litsea cubeba fruit oil is the volatile oil obtained from the berries of the *Litsea cubeba*, Lauraceae

INCI name EU &USA: Litsea cubeba fruit oil

CAS registry number(s): 68855-99-2; 90063-59-5

EINECS number(s): 290-018-7

ISO (INTERNATIONAL ORGANIZATION FOR STANDARDIZATION) STANDARD

ISO number: 3214

ISO name: Essential oil of litsea cubeba

Botanical origin: *Litsea cubeba* (Lour.) Pers.

Parts of plant used: Fruit

ISO values: ISO values (minimum and maximum concentrations) are shown in Table 5.53.1.

THE PLANT, THE OIL, AND THEIR USES

Litsea cubeba (Lour.) Pers. is an evergreen, rarely deciduous, shrub or small tree 5-12 meters high. The plant is native to Asia (China, Japan, Taiwan, Bangladesh, Bhutan, India, Nepal, Cambodia, Laos, Myanmar, Thailand, Vietnam, Indonesia, Malaysia) and is cultivated in China, Indo-China, Japan, Indonesia and Taiwan (GRIN Taxonomy for Plants, http://efloras.org, 18). It is sometimes referred to as the May Chang tree or Chinese pepper, and has been used in traditional Chinese medicine for many diseases for thousands of years. Preparations of the root, stem, leaves and fruits of the plant have been used to treat fatigue, fever, chills, headache, dizziness, cholera, diarrhea, constipation, muscle pain, depression, hysteria and paralysis (1,3). The edible fruit is used as spice.

Litsea cubeba essential oil, also known as May Chang oil, is obtained from the small peppery-like fruits of the *Litsea cubeba* tree. It is widely used as a fragrance, especially in bar soaps, perfumes, household sprays and fresheners because of its intense lemon-like, fresh, sweet odor, which results from its high concentrations of citral (neral + geranial). Another major use of the oil is as a raw material source for the isolation of citral. This is used in its own right for flavor and fragrance purposes or converted by the chemical industry into a number of important derivatives including geranyl nitrile and the ionones, such as β-ionone, pseudoionone and methylionone (which are popular for their violet-like fragrance), and vitamins A, E and K (3,4,18,19,25). Litsea cubeba oil is also a popular food-flavoring agent and preservative, particularly for sugar, cakes, drinks and spices (3). Other applications include its use as flavor enhancer in cigarettes, as a botanical antimicrobial and insecticidal agent (3,12,13,19) and in aromatherapy (29,30).

CHEMICAL COMPOSITION

Litsea cubeba oil is a clear and mobile liquid with pale to darker yellowish color, which has a fresh citrusy, slightly green odor. The yield of essential oil from the fruits and leaves of *Litsea cubeba* (Lour.) Pers. generally varies from 2.0 to 3.5 per cent. The main producing countries of this oil are China and Indonesia.

Literature data (up to October 30, 2014) on the chemical composition of litsea cubeba oils and unpublished analytical data from one of us (E.S.) are shown in Table 5.53.2 in alphabetical order. In litsea cubeba oils from various origins, over 170 chemicals have been identified. About 61 per cent of these were found in a single reviewed publication only. The major compounds found in litsea cubeba from different sources are shown in Table 5.53.3. They include (highest concentrations in any study given) citronellal (78.2%; three very high concentrations, all others <6.2%), citral (geranial + neral) (69.8%), neral (63.6%), geranial (50.0%), limonene (8.1%), 6-methyl-5-hepten-2-one (8.9%) and linalool (8.8%). Well-known ingredients of litsea cubeba oils that were present in high concentrations (>7%) in one or two studies were γ-terpinene (43.6%), sabinene (three studies: 11.6%, 11.4% and 14.0%), β-phellandrene (9.2%), 1,8-cineole (9.0%) and α-pinene (8.1% and 8.6%). Uncommon or rare constituents of litsea cubeba oils found in high concentrations (>7%) in single studies include limonol (44.2%) and *p*-menthane-3,8-diol (23.1%).

COMMERCIAL OILS

The ten chemicals that had the highest maximum concentrations in 101 commercial litsea cubeba essential oil samples (concentration ranges provided) are the following: geranial (35.0-42.3%), neral (28.7-34.5%), limonene (6.4-15.6%), sabinene (0.2-5.1%), methylheptenone (0.8-5.0%), citronellal (0.4-3.8%), α-pinene (0.6-3.7%), (Z)-isocitral (0.6-3.0%), linalool (1.0-2.4%) and (E)-isocitral (0.9-2.0%) (Erich Schmidt, unpublished analytical data).

Chemotypes

Most litsea cubeba fruit oils investigated have geranial ((E)-citral, α-citral) and neral ((Z)-citral, β-citral) as main

Table 5.53.1 ISO values (%) for litsea cubeba oil [a]

Compound	CAS	Minimum	Maximum
Geranial ((*E*)-citral, α-citral)	141-27-5	38.0	45.0
Neral ((*Z*)-citral, β-citral)	106-26-3	25.0	33.0
Limonene	138-86-3	9.0	15.0
Linalool	78-70-6	1.5	3.0
Citronellol	106-22-9	0.5	1.5
Geraniol	106-24-1	0.5	1.5
Citronellal	106-23-0	traces	1.5
α-Pinene	80-56-8	traces	1.5
Nerol	106-25-2	0.2	1.2

[a] ISO 3214 Essential oil of litsea cubeba ©ISO 2000; Geneva, Switzerland, www.iso.org

Table 5.53.2 Constituents identified in litsea cubeba oils

Constituent	CAS	Percentage and range				
		A	B (12)	C (21)	D (1)	E
Benzeneacetaldehyde	122-78-1					tr[n]
Bergamal	106-72-9				0.05-0.6	
β-Bisabolene	495-61-4		0-tr			
Borneol	507-70-0	0.03-0.05	0-0.2			<0.1[f]; 0.1[j]
Bornyl acetate	76-49-3					0.1[n]
Cadinene	29350-73-0		0-0.1			0.04[b]
α-Cadinol	481-34-5		0-tr			
Camphene	79-92-5	0.2-0.7	0-0.2	1.1-3.5	0-tr	tr[n]; 0.3[d,e,f]; 0.7[l]; 0.8[b]; 2.2[t]; 3.1[m]
Camphor	76-22-2	0.03-0.1	0-tr	0.1-0.9		tr[m]; 0.2[l]; 0.4[n]; 2.2[u]
(+)-4-Carene	13837-63-3		0-tr			
δ2-Carene (δ4-carene)	554-61-0					<0.6[q]; 2.4[s]
δ3-Carene	13466-78-9		0-tr			tr[m]; <0.6[q]; 4.5[s]
(E)-Carveol	1197-07-5				0.02-0.04	tr[n]
(Z)-Carveol	1197-06-4		0-0.1		0-0.3	0.05[b]; 0.3[v]
D-Carvone ((+)-)	2244-16-8					0.2[v]
Carvone hydrate	7712-46-1					0.4[v]
β-Caryophyllene	87-44-5	0.1-1.7	0.1-0.8		tr-0.1	0.1[s]; 0.4[n]; 0.6[d]; 0.7[e,j]; 0.8[b]; 1.0[f]; 1.4[l]
Caryophyllene oxide	1139-30-6		tr-0.3		0-0.07	tr[n]; 0.2[b]; <0.6[q]; 0.6[v]
cis-Chrysanthenol	55722-60-6				0-1.6	1.0[n]
1,8-Cineole	470-82-6	0.4-1.8	0.1-0.4		0.1-0.6	1.3[f]; 1.6[b]; 2.1[v]; 3.2[u]; 4.2[t]; 5.4[q]; 9.0[s]
Citral (geranial + neral)	5392-40-5					69.8[c]
Citronellal	106-23-0	0.4-3.8	0-6.2	0.7-7.6	2.8-78.2	1.7[n]; 2.3[l]; 2.5[q]; 6.2[j]; 76.6[p]; 77.2[o]
Citronellic acid	502-47-6					0.3[o]
Citronellol (β-, DL-)	106-22-9	0.08-0.3			0.02-11.6	0.06[l]; 0.07[c]; 0.1[f]; 0.2[e]; 11.4[p]; 14.0[o]
(R)-Citronellol (d-)	1117-61-9		0.1-0.8			
(S)-Citronellol (l-)	7540-51-4		0-tr			
Citronellol epoxide	1564-98-3				0-0.4	
Citronellyl formate	105-85-1				0-0.02	tr[o]
α-Copaene	3856-25-5		0-0.1		0.02	0.07[l]; 0.1[b]; 0.7[n]
β-Copaene	18252-44-3					0.1[n]
m-Cymene	535-77-3					<0.6[q]
o-Cymene	527-84-4		0-tr			0.5[v]
p-Cymene	99-87-6	0.01-0.06	0-tr	0.1-0.3	0-0.04	0.08[l]; 0.1[o]; <0.6[q]; 2.1[m]; 4.4[r]
p-Cymen-8-ol	1197-01-9				tr	
Dehydro-1,8-cineole	92760-25-3	0.03-0.1	0-0.1			0.09[b]; 0.1[g]; 0.2[v]
Deoxygeraniol	2609-23-6					2.6[q]
Dihydromyrcenyl acetate	88969-41-9				0-2.2	
2,6-Dimethyl-5-heptenal	106-72-9				0-0.2	
Elemene	11029-06-4					<0.1[s]
β-Elemene	33880-83-0		0-0.1			tr[m]; 0.06[b]; <0.6[q]
γ-Elemene	29873-99-2		0-tr			
Elemicin	487-11-6					3.9[q]
6,7-Epoxylinalool	15249-35-1					<0.6[q]
Eremophilene	10219-75-7		0-tr			
Ethyl geranate	32659-21-5					0.05[b]
α-Farnesene	502-61-4					0.2[m]
β-Farnesene	502-60-3		0-2.3			0.05[b]
(E)-β-Farnesene	18794-84-8	0.02-0.2			0.08	
(Z)-β-Farnesene	28973-97-9				0.01	
Farnesol	4602-84-0					0.06[b]
Fenchone	1195-79-5					1.8[u]
Geranial	141-27-5	35.0-42.3	44.4-50.0	26.6-37.8	0-44.2	39.2[d]; 39.6[e]; 40.3[l]; 41.0[k]; 41.3[u]
Geranic acid	459-80-3	0.02-0.1	0-0.2			0.1[l]; 0.6[v]
Geraniol	106-24-1	0.6-1.5	0.4-2.6	0.1-1.1	0.09	0.8[e]; 1[k]; 1.0[m]; 1.2[n]; 1.4[u]; 1.6[q]; 4.6[o]
(E)-Geraniol (β-)	106-24-1					0.4[v]; 0.6[b]; 1.4[u]
Geranyl acetate	105-87-3	0.04-0.06	0-tr	0-0.9		0.06[l]; 0.3[c]; 0.4[n]; 1.0[m]
Germacrene A	28387-44-2		0-0.1			
Germacrene D	23986-74-5		0-tr			

Table 5.53.2 Constituents identified in litsea cubeba oils (*continued*)

Constituent	CAS	Percentage and range				
		A	B (12)	C (21)	D (1)	E
Globulol	489-41-8				0-0.03	
(E)-Hasmigone	868693-38-3					0.2[n]
Hexadecyl acetate	629-70-9				0-0.03	
α-Humulene	6753-98-6	0.01-0.1	0-0.1	0.4-1.2		tr[n]; <0.1[f]; 0.1[l]; 0.6[c]
Humulene epoxide II	19888-34-7		0-tr			
Hydroxycitronellal	107-75-5				0-0.03	
(E)-Isocitral	72203-98-6	0.9-2.0				0.8[e]; 1.1[l]
(Z)-Isocitral	72203-97-5	0.6-3.0				0.2[e]; 0.7[l]
exo-Isocitral	55050-40-3					1.4[e]
Isogeraniol (cis-)	5944-20-7		0-tr			
Isoisopulegol	18674-65-2					1.2[o]
Isomenthone	491-07-6				0-0.1	
2-Isopropenyl-5-me-thylhex-4-enal	6544-40-7					1.0[v]
Isopulegol	89-79-2		0-1.1			0.1[n]; 0.8[c]; 1.0[p]; 2.1[o]
Isopulegone	29606-79-9		1.8-2.5			
3-Isothujanol	35732-36-6				0.03	
Limonene	138-86-3	6.4-15.6	0.7-5.3	11.6-17.5	0.7-2.1	12.7[c]; 14.6[d]; 25.3[i]; 26.3[b,g]; 48.1[t]
D-Limonene ((R)-)	5989-27-5					18.8[v]; 27.2[r]
cis-Limonene oxide	3837-75-7		0-tr			tr[n]; 0.1[b]
trans-Limonene oxide	4959-35-7		0-0.1		0-0.4	0.2[b]; 0.6[v]
Limonen-6-ol, pivalate						0.03[b]
Limonol	989-61-7					44.2[q]
Limonol acetate						0.7[q]
Limonol formate						<0.6[q]
Linalool	78-70-6	1.0-2.4	1.2-1.6	2.5-4.9	1.8-2.0	1.5[e]; 1.7[n]; 2.0[k]; 2.2[u]; 2.5[j]; 2.9[i]; 8.8[q]
Linalool oxide	1365-19-1					0.3[v]
cis-Linalool oxide	11063-77-7				0-0.07	
cis-Linalool oxide, furanoid	5989-33-3					tr[o]
trans-Linalool oxide	11063-78-8				0-0.1	
trans-Linalool oxide, furanoid	34995-77-2					tr[n,o]
Linalyl acetate	115-95-7				0-0.9	0.6[u]; 2[k]
epi-13-Manool					0-0.02	
13-epi-Manoyl oxide	1227-93-6				0-0.07	
trans-p-Mentha-1(7),8-dien-2-ol	21391-84-4				2.6	
cis-p-Mentha-2,8-dien-1-ol	3886-78-0					0.1[n]
trans-p-Mentha-2,8-dien-1-ol	4017-77-0					tr[n]; 0.3[v]
p-Menthane-3,8-diol	42822-86-6					0.2-23.1[o]
p-Mentha-1,3,8-triene	18368-95-1				0.02	
p-Menth-1-en-7-al	21391-98-0				0-0.02	
trans-p-Menth-2-en-1-ol	29803-81-4					tr[m]
Methyl chavicol	140-67-0					2.8[q]
Methyl citronellate	2270-60-2				0.07-2.0	
3-Methylcyclohexa-none	591-24-2				0-0.01	
3-Methylcyclopenta-nol	18729-48-1				0-0.03	
Methyl eugenol	93-15-2					3.8[q]
Methyl geranate	2349-14-6					0.1[n]
2-Methyl-6-hepten-1-ol	67133-86-2		0-0.1			
Methylheptenone	409-02-9					5.6[u]
6-Methyl-5-hepten-2-ol	1569-60-4				0-0.03	tr[n]
6-Methyl-5-hepten-2-one	110-93-0	0.8-5.0	0.3-0.9	3.1-7.4	0-2.6	1.9[e]; 2.4[v]; 2.6[n]; 4[k]; 4.3[m]; 6.2[t]; 8.9[i]

Table 5.53.2 Constituents identified in litsea cubeba oils (*continued*)

Constituent	CAS	A	B (12)	C (21)	D (1)	E
Methyl isoeugenol	93-16-3					<0.6[q]
1-Methyl-3-(1-methyl-ethyl)-cyclopentane	53771-88-3				0-0.5	
Methyl-2-methyl-3-methylenecyclopen-tanecarboxyate						0.2[v]
Methyl nerolate	1862-61-9					0.2[n]
Methyl salicylate	119-36-8				0.02	
Methyl 10,11-tetra-decadienoate						0.2[v]
Myrcene	123-35-3	0.6-1.9	0.3-0.8		0.02-1.3	1.3[n]; 1.5[u]; 1.8[l]; 2.4[s]; 2.8[t]; 3[k]; 4.7[s]
Myrcene epoxide	29414-55-9				0.02	
Myristicin	607-91-0					1.8[q]
cis-Myrtenol					0.1	
Neoisocarvomen-thyl acetate	51407-21-7				0-0.09	
Neoisoisopulegol	21290-09-5				0.01-0.2	0.2[o]
Neoisoisopulegol acetate	256332-34-0					tr[n]
Neoisopulegol	29141-10-4				0.02-0.3	
Neral	106-26-3	28.7-34.5	34.2-37.0	20.5-24.0	0-39.7	30.8[l]; 31.3[n]; 34[k]; 50.4[j]; 63.6[m]
Nerol	106-25-2	0.3-1.0	0.2-1.3		0-0.4	0.3[c,d]; 0.4[f]; 0.5[l]; 0.7[n]; 1[k]; 1.4[e]
(*E*)-Nerolidol	40716-66-3					0.7[m]
Neryl acetate	141-12-8					0.04[b]; 1.0[n]
α-Ocimene	502-99-8					1.3[q]
β-Ocimene	13877-91-3					0.6[q]
(*E*)-Ocimene	27400-72-2					1.0
(*Z*)-Ocimene	27400-71-1					4.3[s]
(*E*)-β-Ocimene	3779-61-1	0.01-0.1	0-0.1		0.05-0.1	0.05[l]; 0.8[o]
(*Z*)-β-Ocimene	3338-55-4	0.01-0.08		0-0.3	0-0.05	0.09[n]; 0.4[o]
β-Patchoulene	514-51-2					<0.6[q]
Perillene	539-52-6					tr[n]
α-Phellandrene	99-83-2	0.2-0.4	0-0.2			0.05[b]; <0.1[f]; 0.1[g]; 0.3[s]; 0.4[l]
β-Phellandrene	555-10-2	0.2-1.8	0-0.4			0.04[l]; 1.0[m]; 4.5[b,g] **wrong**; 9.2[s]
α-Phellandrene-8-ol	1686-20-0					0.1[n]
Phytol	7541-49-3				0-0.02	
cis-2,3-Pinanediol	18680-27-8				0-0.04	
trans-2,3-Pinanediol					0-0.05	
α-Pinene	80-56-8	0.6-3.7	0.1-0.3	1.6-3.6	0.2-0.6	2.4[r]; 2.8[i]; 2.9[m,f]; 3.8[b]; 8.1[s]; 8.6[t]
β-Pinene	127-91-3	0.6-2.0	0.1-0.3	0-3.2	0.2-0.4	1.3[j,l]; 1.9[i]; 2.9[v]; 3.0[s]; 5.9[t]; 7.0[r]
cis-Piperitol	16721-38-3		0-tr			tr[n]
trans-Piperitol	16721-39-4					0.1[n]
Piperitone	89-81-6		tr-0.1	0-0.1		0.07[v]; <0.1[f]
2-Propenyl-3-me-thyl-4-hydroxy-2-cyclopentene				0-1.4		
Pseudoionone	141-10-6			0-0.1[a]		
Pulegone	89-82-7			0-3.0		
Rosefuran epoxide	92356-06-4				0.02	
Rose oxide	16409-43-1					<0.6[q]
Sabinene	3387-41-5	0.2-5.1		0-2.3	0.4-2.0	0.9[e,f]; 1.5[d]; 2.4[o]; 4.5[b,h]; 9.5[t]; 40.2[s]
cis-Sabinene hydrate	15537-55-0	0.01-0.1			0.02-0.05	0.1[o]
trans-Sabinene hydrate	17699-16-0					tr[o]
Safrole	94-59-7			0-0.9		0.6[q]; 1.1[m]
α-Selinene	473-13-2					0.04[b]
β-Selinene	17066-67-0					<0.6[q]

Table 5.53.2 Constituents identified in litsea cubeba oils (*continued*)

Constituent	CAS	A	B (12)	C (21)	D (1)	E
Selin-6-en-4-ol	118173-08-3		0-tr			
Spathulenol	6750-60-3				0-0.02	0.2[v]
α-Terpinene	99-86-5	0.01-0.1	0-tr		0-0.06	tr[o]; 5.3[r]
γ-Terpinene	99-85-4	0.01-0.09			0-0.1	<0.1[f]; 0.1[b]; 0.2[m]; <0.6[q]; 2.0[s]; 43.6[r]
Terpinen-4-ol	562-74-3	0.06-0.5	0-0.3		tr-0.5	0.4[o]; 0.5[b]; 0.8[m]; 1.4[s]; 1.9[n]; 2.0[j,v]
α-Terpineol	98-55-5	0.08-0.7	0.1-0.4	1.5-2.3	0.04-0.1	0.4[d]; 0.5[b,j]; 0.7[v]; 0.9[o]; 1.2[c]; 1.4[q]
(Z)-β-Terpineol	7299-40-3		0-tr			0.07[b]; 0.4[v]
Terpinolene (α-)	586-62-9	0.08-0.1	0-0.1		0-0.06	tr[n,o]; 0.05[v]; <0.1[f]; 0.1[b,j]; 0.5[s]
γ-Terpinolene	85188-59-6		0-tr			
α-Terpinyl acetate	80-26-2					0.8[n]; 0.09[b]
α-Thujanol	406160-72-3					1.7[m]
α-Thujene	2867-05-2	0.01-0.6			0-0.06	tr[m,o]; 0.2[b,v]; 2.8[s]
β-Thujene	28634-89-1					3.3[v]
(E)-9-Undecenal	324541-83-5				0-0.2	
Verbenol	473-67-6		1.2-1.8			0.4[d]; 1.0[b]
cis-Verbenol	1845-30-3	0.05-1.7			0-0.06	0.5[b]; 1.8[v]
trans-Verbenol	1820-09-3	0.09-1.8			0.02	
Widdrol	6892-80-4					0.04[b]

A 101 essential oil samples from China and Indonesia, analyzed between 1998 and 2013; lowest and highest concentrations given (E. Schmidt, unpublished data)
B eight lab-hydrodistilled oils from fruits of *Litsea cubeba* harvested in eight Chinese regions; lowest and highest concentrations given (ref. 12)
C three oils from China; lowest and highest concentrations given (ref. 21)
D three lab-hydrodistilled oil from fruits collected in India, one of the citral and two of the citronellal chemotype oils; lowest and highest concentrations given (ref. 1)
E data from other studies (indicated with superscript letters); highest concentrations found in any study reviewed here given; when two or more oils were investigated, only the highest concentrations are mentioned, unless indicated otherwise

[a] incorrect identity based on GC elution order/questionable identity (ref. 19); [b] one lab-hydrodistilled oil from Chinese *L. cubeba* fruits (ref. 4); [c] one lab-hydrodistilled fruit oil from Taiwan (ref. 8); [d] one commercial oil from Vietnamese *L. cubeba* fruits, obtained from a USA company (ref. 2); [e] one commercial litsea cubeba oil (ref. 17); [f] one commercial oil sample (ref. 15); [g] must be sabinene (ref. 14); [h] incorrectly indicated as β-phellandrene in ref. 4 (according to ref. 14); [i] one lab-prepared oil from Thailand (ref. 20); [j] one oil from *L. cubeba* fruits growing in Vietnam (ref. 22); [k] one *L. cubeba* oil (ref. 23); [l] one *L. cubeba* oil (ref. 5); [m] one lab-hydrodistilled oil from Chinese *L. cubeba* fruits (ref. 11); [n] one lab-hydrodistilled oil from Taiwanese *L. cubeba* fruits (ref. 10); [o] three lab-hydrodistilled oils of the citronellal chemotype from fruits of *L. cubeba* growing wild in India, from two locations (ref. 6); [p] one lab-hydrodistilled oil from fruits of an Indian *L. cubeba* tree (ref. 7); [q] one lab-hydrodistilled oil from fruits collected in Tibet, China (ref. 13); [r] one commercial oil purchased in China with a very atypical composition (ref. 9); [s] one lab-hydrodistilled oil from fresh *Litsea cubeba* berries harvested in Laos with sabinene as dominant ingredient (ref. 24); [t] one lab-hydrodistilled oil from fresh *Litsea cubeba* berries harvested in Laos with limonene as dominant ingredient (ref. 26); [u] one lab-hydrodistilled *Litsea cubeba* fruit oil from Thailand (ref. 27); [v] one lab-hydrodistilled oil from dried and powdered mature berries from Chinese *Litsea cubeba* (ref. 28)

tr: trace (in columns B, E[n] and E[o]: <0.1)

Table 5.53.3 Major constituents of litsea cubeba oils

Constituent	CAS	A	B (12)	C (21)	D (1)	E
Citronellal	106-23-0	0.4-3.8	0-6.2	0.7-7.6	2.8-78.2	1.7[n]; 2.3[j]; 2.5[q]; 6.2[j]; 76.6[p]; 77.2[o]
Citral (geranial + neral)						69.8[c]
Neral	106-26-3	28.7-34.5	34.2-37.0	20.5-24.0	0-39.7	30.8[l]; 31.3[n]; 34[k]; 50.4[j]; 63.6[m]
Geranial	141-27-5	35.0-42.3	44.4-50.0	26.6-37.8	0-44.2	39.2[d]; 39.6[e]; 40.3[l]; 41.0[k]; 41.3[u]
Limonene	138-86-3	6.4-15.6	0.7-5.3	11.6-17.5	0.7-2.1	12.7[c]; 14.6[d]; 25.3[j]; 26.3[b,g]; 48.1[t]
D-Limonene ((R)-)	5989-27-5					18.8[v]; 27.2[r]
6-Methyl-5-hepten-2-one	110-93-0	0.8-5.0	0.3-0.9	3.1-7.4	0-2.6	1.9[e]; 2.4[v]; 2.6[n]; 4[k]; 4.3[m]; 6.2[t]; 8.9[i]
Linalool	78-70-6	1.0-2.4	1.2-1.6	2.5-4.9	1.8-2.0	1.5[e]; 1.7[n]; 2.0[k]; 2.2[u]; 2.5[j]; 2.9[j]; 8.8[q]

LEGEND: SEE UNDER TABLE 5.53.2

ingredients. In this type, the citral chemotype, concentrations of geranial generally range from 35 to 50% and of neral from 30 to 40% or higher (Table 5.53.2). This is the composition type of commercially traded litsea cubeba oils (18). However, there may also be certain other chemotypes. Citronellal was the main compound in fruit oils in four publications, with concentrations ranging from 45 to 83% (1,6,7,16). In one study, high concentrations were found of linalool (62.0%) and sabinene (61-62%) in a number of oils (16,24). These chemotypes were all reported in oils obtained from the fruits of *L. cubeba* growing wild in India (1,6,7,16) or Laos (24). Other dominant components of litsea cubeba oils in single publications have included limonol (13), limonene (26) and γ-terpinene (9).

CONTACT ALLERGY / ALLERGIC CONTACT DERMATITIS

General
Contact allergy to litsea cubeba oil has been reported in two publications, but no cases of allergic contact dermatitis from the oil have been identified.

Testing in groups of patients
Two hundred dermatitis patients from Poland were tested with litsea cubeba oil 2% in petrolatum and three (1.5%) reacted. The same authors also patch tested 51 patients allergic to *Myroxylon pereirae* resin (balsam of Peru) and/or turpentine and/or wood tar and/or colophony and two (3.9%) had a positive patch test; relevance data were not provided (31). A group of 86 patients from Poland previously reacting to the fragrance mix was tested with litsea cubeba oil and seven (8.1%) had a positive patch test reaction; relevance data were not provided (32).

LITERATURE

1 Saikia AK, Chetia D, D'Arrigo M, Smeriglio A, Strano T, Ruberto G. Screening of fruit and leaf essential oils of *Litsea cubeba* Pers. from north-east India – chemical composition and antimicrobial activity. J Essent Oil Res 2013;25:330-338
2 Seo SM, Kim J, Lee SG, Shin C-H, Shin S-C, Park IK. Fumigant antitermitic activity of plant essential oils and components from ajowan (*Trachyspermum ammi*), allspice (*Pimenta dioica*), caraway (*Carum carvi*), dill (*Anethum graveolens*), geranium (*Pelargonium graveolens*), and litsea (*Litsea cubeba*) oils against Japanese termite (*Reticulitermes speratus* Kolbe). J Agric Food Chem 2009;57:6596–6602
3 Chen Y, Wang Y, Han X, Si L, Wu Q, Lin L. Biology and chemistry of *Litsea cubeba*, a promising industrial tree in China. J Essent Oil Res 2013;25:103-111
4 Wang Y, Jiang Z-T, Li R. Antioxidant activity, free radical scavenging potential and chemical composition of *Litsea cubeba* essential oil. J Essent Oil Bear Plants 2012;15:134-143
5 Kubeczka K-H, Formacek V. Essential oils analysis by capillary gas chromatography and carbon-13 NMR spectroscopy, 2nd edition. New York, USA: John Wiley and Sons, 2002:191-195. Data cited in ref. 19
6 Choudhury S, Ahmed R, Barthel A, Leclercq PA. Composition of the stem, flower and fruit oils of *Litsea cubeba* Pers. from two locations of Assam, India. J Essent Oil Res 1998;10:381-386
7 Nath SC, Hazarika AK, Baruah A, Sarma KK. Essential oils of *Litsea cubeba* Pers. — an additional chemotype of potential industrial value from northeastern India. J Essent Oil Res 1996;5:575-576
8 Liu T-T, Yang T-S. Antimicrobial impact of the components of essential oil of *Litsea cubeba* from Taiwan and antimicrobial activity of the oil in food systems. Int J Food Microbiol 2012;156:68-75
9 Jiang Z, Akhtar Y, Bradbury R, Zhang X, Isman MB. Comparative toxicity of essential oils of *Litsea pungens* and *Litsea cubeba* and blends of their major constituents against the cabbage looper, *Trichoplusia ni*. J Agric Food Chem 2009;57:4833-4837
10 Ho C-L, Jie-Ping O, Liu Y-C, Hung C-P, Tsai M-C, Liao P-C, et al. Compositions and *in vitro* activities of the leaf and fruit oils of *Litsea cubeba* from Taiwan. Nat Prod Comm 2010;5:617-620
11 Wang H, Liu Y. Chemical composition and antibacterial activity of essential oils from different parts of *Litsea cubeba*. Chem Biodivers 2010;7:229-235
12 Si L, Chen Y, Han X, Zhan Z, Tian S, Cui Q, et al. Chemical composition of essential oils of *Litsea cubeba* harvested from its distribution areas in China. Molecules 2012;17:7057-7066
13 Yang Y, Jiang J, Qimei L, Yan X, Zhao J, Yuan H, Qin Z, Wang M. The fungicidal terpenoids and essential oil from *Litsea cubeba* in Tibet. Molecules 2010;15:7075-7082
14 Lawrence BM. Progress in essential oils. Perfum Flavor 2013;38(Feb):42-53
15 Williams DG. The chemistry of essential oils, 2nd Edition. Port Washington: Micelle Press, 2008:196-197. Data cited in ref. 14
16 Choudhury SN, Deka DK, Baruah AKS, Rao PG, Duke MF, Riggins A. Constituents of essential oil of *Litsea cubeba* from north Indian germoplasm. Indian Perfum 2009;53:38-43. Data cited in ref. 14
17 Milchard MJ, Clery R, Esdale R, Gates L, Judge F, Moss N, et al. Application of gas-liquid chromatography to the analysis of essential oils. GLC fingerprint chromatograms of six essential oils. Perfum Flavor 2010;35(6):34-40
18 Food and Agriculture Organization of the United Nations. Flavours and fragrances of plant origin. Chapter 7, Litsea cubeba oil. Rome: FAO, 1995. http://www.fao.org/docrep/018/v5350e/v5350e.pdf
19 Lawrence BM. Progress in essential oils. Perfum Flavor 2006;31(Jan/Feb):54-? (last page unknown)
20 Punyarajun S, Nandhasri P. Volatile oil from *Litsea cubeba* in Thailand. Varasasn Paesachosarthara 1981;8:65-70. Data cited in ref. 19

21 Zhu LF, Ding DS. New resources of essential oils in China. Perfum Flavor 1991;16(4):1-5. Data cited in ref. 19

22 Thappa RK, Agarwal SG, Dhar KL, Hai LH. Some useful aromatic plants of Vietnam. Indian Perfum 1996;40:13-33. Data cited in ref. 19

23 Wright J. Essential oils. In: Ashurst PR, Ed. Food flavorings, 3rd ed. Gaithersburg, MD, USA: Aspen Publ, 1999. Data cited in ref. 19

24 Vongsombath C, Pålsson K, Björk L, Borg-Karlson A-K, Jaenson TGT. Mosquito (Diptera: Culicidae) repellency field tests of essential oils from plants traditionally used in Laos. J Med Entomol 2012;49:1398-1404

25 Hu L, Du M, Zhang J, Wang Y. Chemistry of the main component of essential oil of *Litsea cubeba* and its derivatives. Open Journal of Forestry 2014;4:457-466

26 Andersson K. Mosquito repellency of essential oils derived from Lao plants. Degree project in Biology, Bachelor of Science, Uppsala University, Sweden, 2010. http://www.diva-portal.org/smash/get/diva2:315427/ FULLTEXT01.pdf

27 Ko K, Juntarajumnong W, Chandrapatya A. Repellency, fumigant and contact toxicities of *Litsea cubeba* (Lour.) Persoon against *Sitophilus zeamais* Motschulsky and *Tribolium castaneum* (Herbst). Kasetsart J (Nat Sci) 2009;43:56-63

28 Yang K, Wang CF, You CX, Geng ZF, Sun RQ, Guo SS, et al. Bioactivity of essential oil of *Litsea cubeba* from China and its main compounds against two stored product insects. J Asia-Pacific Entomol 2014;17:459-466

29 Rhind JP. Essential oils. A handbook for aromatherapy practice, 2nd Edition. London: Singing Dragon, 2012

30 Lawless J. The encyclopedia of essential oils, 2nd Edition. London: Harper Thorsons, 2014

31 Rudzki E, Grzywa Z, Bruo WS. Sensitivity to 35 essential oils. Contact Dermatitis 1976;2:196-200

32 Rudzki E, Grzywa Z. Allergy to perfume mixture. Contact Dermatitis 1986;15:115-116

Chapter 5.54 LOVAGE OIL

DEFINITION

Lovage oil (essential oil of roots of lovage) is the essential oil obtained from the roots of the lovage, *Levisticum officinale* W.D.J. Koch.

INCI NOMENCLATURE

Description/definition: Levisticum officinale oil is the volatile oil distilled from the roots of the lovage, *Levisticum officinale*, Apiaceae
INCI name EU & USA: Levisticum officinale oil
CAS registry number (s): 8016-31-7; 84837-06-9
EINECS number(s): 284-292-7

ISO (INTERNATIONAL ORGANIZATION FOR STANDARDIZATION) STANDARD

ISO number: 11019
ISO name: Essential oil of roots of lovage
Botanical origin: *Levisticum officinale* Koch
Parts of plant used: Roots
ISO values: This ISO standard is without a chromatographic profile, no values for individual components are available

THE PLANT, THE OIL, AND THEIR USES

Levisticum officinale Koch is an erect, tall perennial herbaceous plant, growing to 1.5 m or higher. It is native to Afghanistan and Iran. Lovage has escaped from cultivation and is sometimes naturalized in much of Europe, except the extreme north and south, and in eastern North America. The plant, all parts of which are strongly aromatic and have a characteristic celery-like flavor and smell, is cultivated in Europe and China as a spice and medicinal plant, mainly for its leaves and roots (Mansfeld's World Database of Agriculture and Horticultural Crops, http://mansfeld.ipk-gatersleben.de; GRIN Taxonomy for Plants). The fresh or dried leaves, sometimes also the roots, are used for flavoring soups, salads, meats and vegetable dishes. Young leaves and shoots are also eaten blanched. Roots are peeled and cooked as a vegetable. The stems are used for candied products. The seeds are used as a spice, especially in southern European cuisine. Powdered root is sometimes also used as a spice. Lovage is generally recognized as safe for human consumption as a natural seasoning and flavoring agent (16).

Levisticum officinale has been used as a folk remedy and is said to have stimulating, diaphoretic, expectorant, stomachic, carminative, emmenagogue (menstruation promoting) and aphrodisiac properties. The roots of the lovage are officinal for its diuretic effects and have also been known for centuries as a traditional medicine possessing carminative and spasmolytic activity (1,2,3,4,5,16). Recent main producers of the drug are Poland, Central Germany, Belgium, The Netherlands and several southeastern European countries (1,2,4). The essential oil from the roots, leaves and seeds of *L. officinale* are used in the food (condiments), beverage (liqueurs), perfumery, cosmetic (soaps and creams) and

Table 5.54.1 Constituents identified in lovage oils

Constituent	CAS	Percentage and range					
		A	B (2)	C (14)	D (4)	E (5)	F
Aristolene	6831-16-9						0.6[j]
Bornyl acetate	76-49-3	0-0.6	0-0.5				0.3[b]
Bornyl isovalerate	76-50-6					0.3	
6-Butyl-1,4-cyclohep-tadiene	22735-58-6					2.9[a]	
Butylidenephthalide	551-08-6					1.5	32[f]; 47[g]
(E)-Butylidene-phthalide	76681-73-7	0.4-2.8	0.4-3.9		1.8-4.5	0.3	0.5[i]; 1.3[b]
(Z)-Butylidene-phthalide	72917-31-8	1.1-28.6	0.8-28.6		0-6.1	0.8	0.3[c]; 2.2[b]; 2.4[i]; 3.1[h]
3-n-Butylphthalide	6066-49-5						1.5[h]; 85.2[j]
δ-Cadinene	483-76-1					0.2	0.03[j]; 0.9[e]
Camphene	79-92-5	0.07-1.5	0-0.6	1.6-2.5	0.8-2.7	3.0	0.2[j]; 0.4[i]; 0.8[e]; 1.1[f]
Carvacrol	499-75-2			1.1-1.9		1.2	
(E)-Carveol	1197-07-5		0-0.2				
β-Caryophyllene	87-44-5		0-1.2				
α-Copaene	3856-25-5	0-0.3	0-0.2			0.2	0.2[b]; 0.7[e]
β-Cubebene	13744-15-5					0.2	0.9[b]
p-Cymene	99-87-6	0-0.9	0-0.1				0.9[b]; 3.9[e]
2,3-Dehydro-1,8-cineole	92760-25-3		0-0.8				
2,3-Dihydro-2,7-dime-thyl-4H-1-benzopyr-an-4-one							2.4[i]

Table 5.54.1 Constituents identified in lovage oils *(continued)*

Constituent	CAS	Percentage and range					
		A	B (2)	C (14)	D (4)	E (5)	F
β-Elemene	33880-83-0	tr-0.3	0-0.7	2.1-4.8		0.2	0.03[j]; 0.6[i]
Elemol	639-99-6		0-0.8			0.1	
β-Eudesmol	473-15-4						0.03[j]
Eugenol	97-53-0						0.5[e]
Falcarinol	21852-80-2			22.6-32.6			
(Z)-Falcarinol	21852-80-2						0.07[i]
(Z,Z)-α-Farnesene	28973-99-1						0.1[j]
β-Farnesene	502-60-3			5.3-14.6			
(E)-β-Farnesene	18794-84-8						+[d]; 0.8[e]
(Z)-β-Farnesene	28973-97-9		0-0.1				
Geraniol	106-24-1		0-0.5				
Geranyl acetate	105-87-3		0-0.3				
Germacrene B	15423-57-1						0.5[d]; 0.8[e]
Germacrene D	23986-74-5		0-0.6	0.1-0.6		0.3	0.2[i]; 1.6[e]
β-Gurjunene	73464-47-8						1.3[e]
Hexadecanoic acid	57-10-3						0.02[i]
β-Himachalene	1461-03-6						0.5[e]
Isoamyl 2-methyl-butyrate	27625-35-0						1.1[e]
Isobornyl isovalerate	7779-73-9						0.8[i]
Isobornyl 2-methyl-butyrate	94200-10-9		0-0.9	0-0.3		0.3	0.04[j]; 1.0[i]
Isothujyl alcohol			0-0.1				
Ligustilide	4431-01-0	0.1-1.6					2.7[j]
(E)-Ligustilide	81944-09-4	0.4-2.6	0-0.5	0-0.8		1.2	0.3[c]; 0.6[b]; 1.5[i]; 1.9[h]
(Z)-Ligustilide	81944-09-4	4.9-85.2	9.4-70.9	23.4-33.6	37.3-62.5	36.8	44.5[d]; 60.4[i]; 61.8[b]; 79.7[c]
Limonene	138-86-3	0.08-1.9	0-0.4	1.4-2.3	0.5-1.8	1.4	0.8[b]; 1.2[f]
Linalool	78-70-6	tr-4.4	0-0.5				2.8[b]
Longifolene (junipene)	475-20-7						0.08[j]
p-Mentha-1,5-dien-8-ol	1686-20-0		0-0.1				
trans-p-Menth-2-en-1-ol	29803-81-4		0-0.4			0.1	
Methyl chavicol	140-67-0						7.0[c]
2-Methylene-6,6-di-methylbicyclo[3.2.0]heptan-3-ol	1005276-05-0		0-0.7				
Methyl eugenol	93-15-2						7.7[e]
Methyl hexadeca-dienoate	29961-54-4		0-0.5				
Methyl hexadece-noate	29960-49-4		0-6.1				
2-Methyl-6-methyle-ne-1,7-octadiene	1686-30-2		0-1.0				
Methyl pentadeca-noate	7132-64-1		0-1.2				
γ-Muurolene	30021-74-0						0.9[e]
τ-Muurolol	19912-62-0					0.1	0.5[i]; 0.9[j]
Myrcene	123-35-3	tr-1.6	tr-1.6	0.4-1.3	0.3-0.8	1.3	0.05[j]; 0.4[h]; 0.5[b]; 0.9[f]
Nonane	111-84-2		0-3.0			0.2	0.3[e]
(E)-β-Ocimene	3779-61-1	0.04-0.6			0-0.7		0.3[e]; 0.6[b]
(Z)-β-Ocimene	3338-55-4	0.08-0.5			0-0.5	0.4	0.2[b]; 0.4[f]; 8.9[e]
allo-Ocimene	673-84-7						0.01[j]; 1.0[e]
n-Octanal	124-13-0					0.1	
Patchoulane	25491-20-7						0.4[j]
Pentylbenzene	538-68-1	0.05-0.9	0-0.9			0.5	0.1[j]; 0.3[f]; 0.5[b]; 1.2[h]
Pentyl cyclohexadiene	76700-97-5					9.1	7.8[c]; 8.3[i]; 12.7[f]; 19[h]
1-Pentyl-1,3-cyclo-hexadiene	76346-02-6				12.2-29.3		10.8[d]
Pentyl-1,5-cyclo-hexadiene		0.09-6.4	0-12.3				11.3[g]

Table 5.54.1 Constituents identified in lovage oils *(continued)*

Constituent	CAS	Percentage and range					
		A	B (2)	C (14)	D (4)	E (5)	F
Pentylcyclohexane	4292-92-6						0.1[b]
Perillaldehyde	2111-75-3		0-0.4				
Perillyl alcohol	536-59-4		0-0.6				
α-Phellandrene	99-83-2	0-1.6	0-1.3			0.1	0.2[b]; 0.5[f]; 2.2[e]; 2.7[j]
β-Phellandrene	555-10-2	0.07-15.5	0.4-48.9	6.4-16.5	0.7-15.5	4.6	2.7[j]; 7.0[e]; 10.7[f]; 11.9[d]
Phenylacetaldehyde	122-78-1		0-17.2				
Phytol	7541-49-3						0.1[j]
α-Pinene	80-56-8	0.2-3.1	0-1.7	0.3-1.7	2.1-7.9	7.8	0.7[j]; 1.2[i]; 2.2[h]; 3.8[e]; 4.6[f]
β-Pinene	127-91-3	tr-12.3	0-1.6	0.5-4.3	2.5-8.7	9.3	0.8[j]; 1.8[i]; 3.4[h]; 8.0[f]; 8.5[e]
trans-Pinocarveol	1674-08-4						0.01[j]
(Z)-3-Propylidene phthalide	17369-59-4						0.9[h]
Sabinene	3387-41-5	tr-0.6	0-0.6	0.9-1.7	0.4-2.0	1.1	0.1[h]; 5.1[e]
trans-Sabinyl acetate	139757-62-3		0-12.1				
Sedanenolide	63038-10-8						8.1[h]
Sedanolide	6415-59-4						10.0[h]
Selina-3,7(11)-diene	6813-21-4						0.4[e]
α-Selinene	473-13-2						0.7[e]
β-Selinene	17066-67-0	tr-2.2	0-0.6			0.3	0.9[i]; 3.1[e]; 5.9[b,i]
γ-Selinene	515-17-3						1.6[b]
Terpinene	8013-00-1						0.7[b]
α-Terpinene	99-86-5						0.1[f]; 0.5[e]
γ-Terpinene	99-85-4	0-0.7	0-0.1			0.4	0.01[j]; 12.6[e]
Terpinen-4-ol	562-74-3	0-0.3	0-0.2			0.6	0.6[e]
α-Terpineol	98-55-5	tr-1.4	0-0.9			0.3	1.5[e]; 4.1[b]
Terpinolene	586-62-9	tr-1.6	0-1.6	2.6-4.1	0-25.9	4.1	0.4[b]; 0.7[i]; 1.5[f]; 9.2[e]
α-Terpinyl acetate	80-26-2	0.06-1.5	0-26.1				0.2[b,d]; 0.5[j]; 1.2[i]; 6.2[e]
α-Thujene	2867-05-2					0.1	0.7[e]
Tricyclene	508-32-7					0.1	
cis-3-Validene-3,4-dihydrophthalide							0.4[b]
3-Validene-4,5-dihydrophthalide						0.5	

A seventeen lovage essential oil samples from Poland, Hungary, the Netherlands, Germany and China, analyzed between 1998 and 2013; lowest and highest concentrations given (E. Schmidt, unpublished data)

B six lab-prepared oils from commercial lovage root powders obtained from retail pharmacies and health shops in five European countries; lowest and highest concentrations given (ref. 2)

C six oils from roots of lovage plants cultivated in Egypt under different fertilization circumstances; lowest and highest concentrations given (ref. 14)

D seven oils of lovage roots cultivated in south Finland of various ages in 2 years; lowest and highest concentrations given (ref. 4)

E one lab-hydrodistilled lovage root oil from Scotland (ref. 5)

F data from other studies (indicated with superscript letters); highest concentrations found in any study reviewed here given; when two or more oils were investigated, only the highest concentrations are mentioned, unless indicated otherwise

[a] incorrect identification (ref. 11); [b] one lab-hydrodistilled oil from France (ref. 6); the same data are also presented in ref. 18; [c] one lab-hydrodistilled oil from the roots of lovage cultivated in Iran (ref. 1); [d] one lovage root oil lab-hydrodistilled in Germany (ref. 7); [e] one lab-hydrodistilled oil from roots of lovage growing wild in Iran; this is a very atypical oil of lovage, as it does not contain any phthalides (ref. 8); [f] unknown number of lovage root oils from unknown origin (ref. 12); [g] one Polish commercial lovage root oil (ref. 9); [h] two oils from fresh and dried lovage roots from Lithuania (ref. 15); [i] one lab-hydrodistilled oil from lovage roots cultivated in Lithuania (ref. 19); [j] one lab-hydrodistilled oil from the roots of *Levisticum officinale* cultivated in Iran (ref. 3)

tr: trace (in column B: <0.05); + present in the oil investigated, but quantity not stated

tobacco industries (1,3,5). The root oil is stated to have antispasmodic and anti-asthmatic properties (1) and is also used in aromatherapy (20).

CHEMICAL COMPOSITION

Lovage oil is a yellow to dark brown, clear mobile liquid which has an aromatic, herbaceous-spicy odor with celery aspects. The yield of essential oil from the roots of *Levisticum officinale* Koch generally varies from 0.05 to 1.2 per cent, depending on the dryness of the roots. The main producing countries of this oil are Hungary, China, the Netherlands, Germany and Poland.

Literature data (up to October 3, 2014) on the chemical composition of lovage oils and unpublished analytical

Table 5.54.2 Major constituents of lovage oils

Constituent	CAS	Percentage and range					
		A	B (2)	C (14)	D (4)	E (5)	F
(Z)-Ligustilide	81944-09-4	4.9-85.2	9.4-70.9	23.4-33.6	37.3-62.5	36.8	44.5[d]; 60.4[i]; 61.8[b]; 79.7[c]
β-Phellandrene	555-10-2	0.07-15.5	0.4-48.9	6.4-16.5	0.7-15.5	4.6	2.7[i]; 7.0[e]; 10.7[f]; 11.9[d]
Butylidenephthalide	551-08-6					1.5	32[f]; 47[g]
1-Pentyl-1,3-cyclo-hexadiene	76346-02-6				12.2-29.3		10.8[d]
(Z)-Butylidene-phthalide	72917-31-8	1.1-28.6	0.8-28.6		0-6.1	0.8	0.3[c]; 2.2[b]; 2.4[i]; 3.1[h]
Pentyl cyclohexadiene	76700-97-5					9.1	7.8[c]; 8.3[i]; 12.7[f]; 19[h]
Pentyl-1,5-cyclo-hexadiene		0.09-6.4	0-12.3				11.3[g]
β-Pinene	127-91-3	tr-12.3	0-1.6	0.5-4.3	2.5-8.7	9.3	0.8[j]; 1.8[i]; 3.4[h]; 8.0[f]; 8.5[e]
α-Pinene	80-56-8	0.2-3.1	0-1.7	0.3-1.7	2.1-7.9	7.8	0.7[j]; 1.2[i]; 2.2[h]; 3.8[e]; 4.6[f]
(E)-Butylidene-phthalide	76681-73-7	0.4-2.8	0.4-3.9		1.8-4.5	0.3	0.5[i]; 1.3[b]

LEGEND: SEE UNDER TABLE 5.54.1

data from one of us (E.S.) are shown in Table 5.54.1 in alphabetical order. In lovage oils from various origins, over 95 chemicals have been identified. About 52 per cent of these were found in a single reviewed publication only. The major compounds found in lovage oils from different sources are shown in Table 5.54.2. They include (highest concentrations in any study given) (Z)-ligustilide (85.2%), β-phellandrene (48.9%; one very high concentration, all others <16.5%), butylidene phthalides (isomer unspecified, (E)- or (Z)) (47%), pentylcyclohexadienes (pentyl cyclohexadiene, 1-pentyl-1,3-cyclohexadiene and 1-pentyl-1,5-cyclohexadiene: 29.3%), β-pinene (12.3%) and α-pinene (7.9%). Well-known ingredients of lovage oils that were present in high concentrations (>7%) in one study were α-terpinyl acetate (26.1%) and terpinolene (25.9%). Uncommon or rare constituents of lovage oils found in high concentrations (>7%) in single studies include 3-n-butylphthalide (85.2%), falcarinol (32.6%), phenylacetaldehyde (17.2%), β-farnesene (14.6%), trans-sabinyl acetate (12.1%), sedanolide (10.0%), sedanenolide (8.1%), methyl eugenol (7.7%) and methyl chavicol (7.0%).

Commercial oils

The ten chemicals that had the highest maximum concentrations in 17 commercial lovage essential oil samples (concentration ranges provided) are the following: (Z)-ligustilide (4.9-85.2%), (Z)-butylidenephthalide (1.1-28.6%), β-phellandrene (0.07-15.5%), β-pinene (tr-12.3%), pentyl-1,5-cyclohexadiene (0.09-6.4%), linalool (tr-4.4%), α-pinene (0.2-3.1%), (E)-butylidenephthalide (0.4-2.8%), (E)-ligustilide (0.4-2.6%) and β-selinene (tr-2.2%) (Erich Schmidt, unpublished analytical data).

CONTACT ALLERGY / ALLERGIC CONTACT DERMATITIS

General

There have been two case reports of allergic contact dermatitis from lovage oil only.

Case reports

A female patient became allergic to lovage oil from undiluted topical application; she reacted to the oil 5% in ethanol and to another essential oil product (mixture), to cananga oil, jasmine absolute (present in the essential oil mixture), eugenol and isoeugenol (21); lovage oil does not contain appreciable amounts of eugenol or Isoeugenol (Table 5.54.2, column A). Lovage oil sensitized one worker in the fragrance industry (22).

LITERATURE

1 Heidarpour O, Souri MK, Omidbaigi R. Changes in content and constituents of essential oil in different plant parts of lovage (Levisticum officinale Koch. Cv. Budakalaszi) cultivated in Iran. J Essent Oil Bear Plants 2013;16:318-322

2 Raal A, Arak E, Orav A, Kailas T, Müürisepp M. Composition of the essential oil of Levisticum officinale W.D.J. Koch from some European countries. J Essent Oil Res 2008;20:318-322

3 Dayeni M, Omidbaigi R, Bastan MR. Essential oil content and composition of Levisticum officinale large scale cultivated in Iran. J Essent Oil Bear Plants 2006;9:152-155

4 Szebeni-Galambosi Z, Galambosi B, Holm Y. Growth, yield and essential oil of lovage grown in Finland. J Essent Oil Res 1992;4:375-380

5 Santos PAG, Figueiredo AC, Oliveira MM, Barroso JG, Pedro LG, Deans SG, Scheffer JJC. Growth and essential oil composition of hairy root cultures of Levisticum officinale W.D.J. Koch (lovage). Plant Sci 2005;168:1089-1096

6 Perineau F, Ganou L, Vilarem G. Studying production of lovage essential oils in a hydrodistillation pilot unit equipped with a cohobation system. J Chem Tech Biotechnol 1992;53:165-171. Data also presented in ref. 18

7 Stahl-Biskup E, Wichtmann E-M. Composition of the essential oil from roots of some Apiaceae in relation to the development of their oil duct systems. Flavour Fragr J 1991;6:249-255

8   Moradalizadeh M, Akhgar MR, Rajaei P, Faghihi-Zarandi A. Chemical composition of the essential oils of *Levisticum officinale* growing wild in Iran. Chem Nat Comp 2012;47:1007-1008

9   Gora J, Majda T, Lis A, Tichek A, Kurowska A. Chemical composition of some Polish commercial essential oils. Rivista Ital EPPOS (Numero Speciale) 1997:761-766. Data cited in ref. 10

10 Lawrence BM. Progress in essential oils. Perfum Flavor 2004;29(6):80-90

11 Lawrence BM. Progress in essential oils. Perfum Flavor 2009;34(October):56-59

12 Fehr D. Über das ätherishe Öl von *Levisticum officinale*. I. Untersuchung der ätherischen Öle aus Frucht, Blatt, Stengel und Wurzel. Planta Medica 1980;40 (Suppl.):34-40. Data cited in ref. 13

13 European Medicines Agency. Committee on Herbal Medicinal Products (HMPC). Assessment report on *Levisticum officinale* Koch, radix. 27 March 2012, EMA/HMPC/524623/2011. Available at: http://www.ema.europa.eu/docs/en_GB/document_library/Herbal_-_HMPC_assessment_report/2012/05/WC500126833.pdf

14 Ezz El-Din A, Hendawy SF. Comparative efficiency of organic and chemical fertilizers on herb production and essential oil of lovage plants grown in Egypt. Am-Euras J Agric Environ Sci 2010;8:60-66

15 Venskutonis PR. Essential oil composition of some herbs cultivated in Lithuania. In: KHC Başer, Ed. Proceedings of the 13th International Congress of Flavours, Fragrances and Essential Oils, Vol. 2. Istanbul, Turkey: AREP Publishers, 1995:108-123. Data cited in ref. 17

16 Ravindran PN, Pillai GS. Under-utilized herbs and spices. In: Peter KV, Ed. Handbook of herbs and spices, 2nd Ed., Vol. 2. Oxford-Cambridge-Philadelphia-New Delhi: Woodhead Publishing Ltd, 2012:Chapter 5, 96-97

17 Lawrence BM. Progress in essential oils. Perfum Flavor 1999;24(2):35-47

18 Cu J-Q, Perineau F, Delmas M, Gaset A. Extraction of volatile compounds of lovage root, celery seed and carrot seed by different solvents. In: SC Bhattacharrya, N Sen, KL Sethi, Eds. Proceedings of 11th Congress of essential oils, fragrances and flavors, New Delhi, November 1989. New Delhi: Oxford and IBBT Publ Co, 1989, vol 4:89-97. Data cited in ref. 17, and also presented in ref. 6

19 Kemzuraite A, Venskutonis PR, Baranauskiene R, Navikiene D. Optimization of supercritical CO2 extraction of different anatomical parts of lovage (*Levisticum officinale* Koch.) using response surface methodology and evaluation of extracts composition. J Supercr Fluids 2014;87:93-103

20 Lawless J. The encyclopedia of essential oils, 2nd Edition. London: Harper Thorsons, 2014

21 Lapeere H, Boone B, Verhaeghe E, Ongenae K, Lambert J. Contact dermatitis caused by lovage (Levisticum officinalis) essential oil. Contact Dermatitis 2013;69:181-182

22 Calnan CD. Lovage sensitivity. Contact Dermatitis Newsletter 1969;5:99

Chapter 5.55 MANDARIN OIL

DEFINITION
Mandarin oil (essential oil of mandarin, Italian type) is the essential oil obtained from the pericarp (peel) of the mandarin (mandarin orange), *Citrus reticulata* Blanco (synonym: *Citrus nobilis* Andrews).

INCI NOMENCLATURE
Description/definition: Citrus nobilis peel oil is the oil expressed from the peel of the mandarin, *Citrus nobilis*, Rutaceae
INCI name EU: Citrus nobilis peel oil
INCI name USA: Citrus nobilis (mandarin orange) peel oil
CAS registry number(s): 8008-31-9; 84929-38-4
EINECS number(s): 284-521-0

ISO (INTERNATIONAL ORGANIZATION FOR STANDARDIZATION) STANDARD
ISO number: 3528
ISO name: Essential oil of mandarin, Italian type
Botanical origin: *Citrus reticulata* Blanco (synonym: *Citrus nobilis* Andrews)
Parts of plant used: Pericarp (peel)
ISO values: ISO values (minimum and maximum concentrations) are shown in Table 5.55.1.

By ISO definition, all citrus essential oils except lime oil are produced by expression; oils obtained from citrus fruits by distillation may not be called essential oils according to ISO criteria (except lime oil). In industry, however, sometimes residues from expression of the citrus peels undergo steam distillation, to obtain the remaining oil; these volatile oils may then be added to the expressed oil. As some of the compounds undergo changes forced by high temperature during distillation, this addition (which is an adulteration) changes the composition of the essential oil. Because of this, and also because it cannot be excluded that oils entirely produced by hydrodistillation reach the market, both the ingredients found in 'genuine' mandarin essential oils obtained by expression and those that may be present in hydrodistilled oils are presented here.

From a botanical point of view, the mandarins form a complex group. The authors of the standard work '*Citrus* oils: composition, advanced analytical techniques, contaminants, and biological activity' (3) state that the group consists of the Mediterranean mandarin (*Citrus deliciosa* Ten.), the tangerines (*C. tangerina* Hort. ex Tan.), the clementines (*C. clementina* Hort. ex Tan.), *Citrus temple* Hort. ex Tan., *Citrus nobilis* Lour., *Citrus unshiu* (Mak.) Marc and *Citrus reticulata* Blanco. They have experienced that often, when describing the composition of mandarin oil, the authors of scientific publications refer to oils obtained from different mandarin species, that the botanical origin is rarely specified, and when it is, it is sometimes uncertain or questionable. Finally, mandarins are also commonly described using botanical classifications that differ from the one mentioned above (3). That there are taxonomic uncertainties shows from checking the species mentioned above in three plant databases: GRIN Taxonomy for Plants (http://www.ars-grin.gov/cgi-bin/npgs/html/tax_search.pl?), The Plant List (http://www.theplantlist.org/) and the Plant Name Database of the Melbourne University, section 'Sorting Citrus Names' (http://www.plantnames.unimelb.edu.au/Sorting/Citrus.html#ref). As can be seen in Table 5.55.2, there is no consensus on the botanical status of some of the species mentioned in the *Citrus* standard work (3).

Despite these taxonomic uncertainties, all species are discussed together here, with the exception of *C. tangerina* Hort. ex Tan., which is discussed in Chapter 5.82 Tangerine oil. As a consequence, there will most likely be overlap with tangerine oils as presented there. Nevertheless, in mandarin oils as discussed in this chapter, the amounts of limonene appear to be somewhat lower, but the concentration of γ-terpinene is clearly higher than in tangerine oils obtained from *C. tangerina* Hort. ex Tan.

THE PLANT, THE OIL, AND THEIR USES
Citrus reticulata Blanco, the mandarin orange (synonyms: mandarine, mandarin) is an evergreen thorny *Citrus* tree growing to 4.5-6 meters, which gives fruits resembling oranges. The mandarin orange probably originated from Asia and is widely cultivated in the tropics and subtropics. The main producing countries are China (>50% of the world's production), Spain and Brazil, followed by Japan, Turkey, Italy, Egypt, South Korea, USA and Pakistan. In traditional Chinese medicine, the dried peel of the fruit is used in the regulation of ch'I ('natural energy', 'life force', or 'energy flow') and is utilized to treat abdominal distension and to enhance digestion. Mandarins are also part of Ayurveda, the traditional medicine of India. The flowers, fruits, endocarps, pericarps and seeds all are considered to have multiple medical applications (Wikipedia).

The essential oil expressed from the peel is employed commercially in flavoring hard candy, gelatins, ice cream,

Table 5.55.1 ISO values (%) for mandarin oil [a]

Compound	CAS	Minimum	Maximum
Limonene	138-86-3	65.0	75.0
γ-Terpinene	99-85-4	16.0	22.0
α-Pinene	80-56-8	1.6	3.0
Myrcene	123-35-3	1.4	2.0
β-Pinene	127-91-3	1.0	2.0
p-Cymene	99-87-6	n.d.	0.8
Methyl N-methyl anthranilate	85-91-6	0.3	0.7
α-Sinensal	17909-77-2	0.1	0.5
Linalool	78-70-6	0.03	0.2
Decanal	112-31-2	0.04	0.14
n-Octanal	124-13-0	0.03	0.14

[a] ISO 3528 Essential oil of mandarin, Italian type ©ISO 2012; Geneva, Switzerland, www.iso.org
n.d. not detectable

Table 5.55.2 Botanical status of mandarins in various plant databases

Mandarin species	GRIN	Plant List	Melbourne University
(1) *Citrus deliciosa* Ten.	Accepted	Synonym of (7)	Accepted
(2) *C. tangerina* Hort. ex Tan.	Accepted, or *C. reticulata* 'Tangerina' [a]	Synonym of (7) [d]	Synonym of (7)
(3) *C. clementina* Hort. ex Tan.	Accepted, or *C. reticulata* 'Clementine'	Unresolved name [e]	Synonym of (7)
(4) *Citrus temple* Hort. ex Tan.	Accepted [b]	Unresolved name [f]	Accepted [h]
(5) *Citrus nobilis* Lour.	Accepted, or *C. aurantium* Tangor Group	Accepted	Accepted
(6) *Citrus unshiu* (Mak.) Marc	Accepted, or *C. reticulata* 'Unshiu' [c]	Synonym of (7) [g]	Accepted
(7) *Citrus reticulata* Blanco	Accepted	Accepted	Accepted

Accepted: Accepted species name
[a] GRIN name: *Citrus tangerina* Tanaka
[b] GRIN name: *Citrus temple* hort. ex Yu. Tanaka
[c] GRIN name: *Citrus unshiu* Marcow.
[d] Plant List name: *Citrus tangerina* Yu. Tanaka
[e] Plant List name: *Citrus* clementina hort.
[f] Plant List name: *Citrus × temple* Yu. Tanaka
[g] Plant List name: *Citrus unshiu* (Yu. Tanaka ex Swingle) Marcow.
[h] Melbourne database name: *Citrus temple* hort. ex Yu. Tanaka

chewing gum, bakery goods, beverages and liqueurs. Mandarin essential oil is valued in perfume manufacturing, particularly in the formulation of floral compounds and colognes, where the low volatile essential oil components play an important role as head notes. In pharmaceutical industries they are employed as flavoring agents to mask unpleasant tastes of drugs (11,41, GRIN Taxonomy for Plants; New Crop Resource Online Program, http://www.hort.purdue.edu/newcrop/). Mandarin oils are also used in aromatherapy (46,47).

The history, global distribution, and nutritional importance of *Citrus* fruits have been reviewed in ref. 2.

CHEMICAL COMPOSITION

Mandarin oil is a clear mobile liquid with a light green to dark green, light yellow to dark orange or reddish to dark red color with blue influences, depending on the harvest time. The odor will vary from slight citrus to fresh fruity citrus with peel note. The yield of essential oil from the peel of *Citrus reticulata* Blanco generally varies from 0.4 to 0.6 per cent. The main producing countries of this oil are China, Brazil, Italy, Argentina, Spain, Turkey and USA.

Cold-pressed *Citrus* essential oils consist of a volatile fraction, which represents 85-99% of the oil, and a non-volatile fraction 1-15%, which mainly contains oxygen heterocyclic compounds, especially polymethoxyflavones, coumarins, and psoralens (Table 5.55.5). These components play an important role in the characterization of cold-pressed *Citrus* oils, since the composition of this fraction is characteristic for each oil. Moreover, many of the pharmacological and toxicological activities possessed by *Citrus* oils have been demonstrated to be related to these components (1).

Literature data (up to July 4, 2014) on the chemical composition of mandarin oils and unpublished analytical data from one of us (E.S.) are shown in Table 5.55.3 in alphabetical order. In mandarin oils from various origins, over 255 chemicals have been identified. The major compounds found in mandarin oils from different sources are shown in Table 5.55.4. They include (highest concentrations in any study given) limonene (96.0%), γ-terpinene (22.8%), *p*-cymene (16.1%), myrcene (7.6%), α-pinene (5.2%), terpinolene (4.2%) and β-pinene (2.5%). Well-known ingredients of mandarin oils that were present in high concentrations (>7%) in one study were δ3-carene (10.1%) and decanal (7.7%).

Commercial oils

The ten chemicals that had the highest maximum concentrations in 98 commercial mandarin essential oil samples (concentration ranges provided) are the following: limonene (64.0-76.4%), γ-terpinene (10.4-21.1%), terpinolene (0.06-4.2%), α-pinene (1.2-3.1%), myrcene (0.08-2.2%), β-pinene (0.5-2.2%), *p*-cymene (0.3-1.2%), methyl *n*-methyl anthranilate (0.08-1.1%), *trans*-sabinene hydrate (0.01-1.1%) and α-terpineol (0.07-1.1%) (Erich Schmidt, unpublished analytical data).

It should be realized that values in hydrodistilled oils may (but not necessarily do) vary considerably from genuine cold-pressed mandarin oils because of the many hydrolytic reactions that take place during oil isolation; in addition, the hydrodistilled oils lack the non-volatile oxygenated heterocyclic compounds which may be present in the cold-pressed oils (Table 5.55.5).

Ref. 42 is not discussed here because of an extremely atypical composition of the mandarin oil investigated (limonene 46.7%, geranial 19.0%, neral 14.5%, geranyl acetate 3.9%, geraniol 3.5%, β-caryophyllene 2.6%,

Table 5.55.3 Constituents identified in mandarin oils

Constituent	CAS	Percentage and range					
		A	B (3)	C (3)	D (3)	E (11)	F
Acetic acid	64-19-7		tr				
Acetophenone	98-86-2		tr				
4-Acetyl-1-methylcyclohexene	6090-09-1						0.04[o]
Allyl isovalerate	2835-39-4						0.4[l]
(E)-Anethole	4180-23-8			tr			
Benzaldehyde	100-52-7		0.02				
α-Bergamotene	17699-05-7			tr			
trans-α-Bergamotene	13474-59-4		0.01		0.1		tr[f]
Bergaptene	484-20-8	0-tr					
Bicyclogermacrene	24703-35-3		0.02	0.01	tr		tr[c,f]
β-Bisabolene	495-61-4		0.01	0.1	tr		0.09[f]; 0.1[g]
(E)-γ-Bisabolene	53585-13-0		tr				
(Z)-γ-Bisabolene	13062-00-5		tr				
α-Bisabolol	515-69-5						tr[f]
Borneol	507-70-0	0-0.05			0.02		0.01[k]; 0.04[h]
Bornyl acetate	76-49-3						0.07[h]
γ-Cadinene	39029-41-9		0.05				
δ-Cadinene	483-76-1		0.07	0.03	0.2	tr-0.01	tr[e,m]; 0.01[f]; 0.06[c]; 0.08[r]; 0.9[l]
α-Cadinol	481-34-5						1.5[l]
α-Calacorene	21391-99-1						0.1[k]
Camphene	79-92-5	0-0.2	0.05	0.02	0.01	0.01	tr[c,e,k]; 0.01[g,m]; 0.09[h]
Campherenol	18530-03-5						tr[f]
Camphor	76-22-2		tr	0.01	0.03	tr-0.02	0.4[h]
δ2-Carene (= δ4-)	554-61-0		0.03				
δ3-Carene	13466-78-9	0.01-0.1	0.08	tr	0.1	0-tr	0.08[k]; 0.1[f]; 0.2[c]; 0.5[r]; 10.1[o]
Carvacrol	499-75-2		tr		0.05		0.1[h]; 0.3[d]; 1.4[l]
(E)-Carveol	1197-07-5		tr	tr	0.1		0.01[k]; 1.8[d]
(Z)-Carveol	1197-06-4		0.06	0.01	0.1	0-0.08	tr[c]; 0.9[d]; 1.3[l]
Carvone	99-49-0		0.02	0.01	0.3	tr	0.02[m]; 0.08[c]; 0.1[o]; 0.3[r]; 0.6[l]; 1.9[d]
D-Carvone	2244-16-8						0.6[q]
cis-Carvyl acetate	1205-42-1						tr[c]
β-Caryophyllene	87-44-5	0-0.2	0.1	0.08	0.3	0.04-0.08	tr[e]; 0.01[f]; 0.05[c,m]; 0.07[g]; 1.4[l]
Caryophyllene oxide	1139-30-6		+				tr[h]
1-Chlorooctane	111-85-3						0.07[o]
1,8-Cineole	470-82-6	0-0.2	tr	tr	tr		0.2[h]
Citronellal	106-23-0	0-0.1	0.2	0.05	0.3	tr-0.03	0.01[g]; 0.08[c,d]; 0.2[o,r]; 0.5[f]; 0.8[l]
Citronellol	106-22-9	0-0.2	0.1	0.01	0.2	0-0.02	tr[c]; 0.06[k]; 0.07[h]; 0.1[m]; 0.2[o]; 1.2[l]
Citronellyl acetate	150-84-5		0.03	0.01	0.02	0-0.01	tr[c,f]; 0.09[o]
Citronellyl propanoate	141-14-0						
Copaene							0.9[l]
α-Copaene	3856-25-5		0.07	0.1	0.1	tr-0.02	tr[e,m]; 0.03[k]; 0.05[f,r]; 0.1[c]
β-Copaene	18252-44-3						0.01[k]; 0.07[d]
Cresol	1319-77-3		+				
Cryptone	500-02-7			tr			0.09[d]
α-Cubebene	17699-14-8			tr			0.8[l]
β-Cubebene	13744-15-5		0.05	0.03	0.02	0-0.01	<0.06[f]; 0.1[c]; 0.5[l]
Cubebol	23445-02-5			tr			0.06[k]
β-Cubenene							0.1[d]
Cuminaldehyde	122-03-2		tr				
Cuminyl alcohol	536-60-7		tr		0.07		
p-Cymene	99-87-6	0.3-1.2	4.3	16.1	6.9	0.2-0.9	0.05[c]; 0.2[d]; 0.4[m]; 0.7[h]; 1.2[g]
p-Cymenene	1195-32-0		tr			0-tr	
p-Cymen-8-ol	1197-01-9		tr	tr	0.05	tr-0.03	
2,4-Decadienal	2363-88-4				0.01		
(E,E)-2,4-Decadienal	25152-84-5		0.02	0.1	0.01	0-0.01	tr[c]; 0.02[k]
(E,Z)-2,4-Decadienal	25152-83-4		tr				
Decanal	112-31-2	0.02-0.4	0.5	0.2	0.5	0.01-0.2	0.1[e,m]; 0.4[d,f]; 0.6[o,r]; 0.7[c]; 7.7[l]
Decane	124-18-5					0-tr	
Decanoic acid	334-48-5			tr			
Decanol	36729-58-5		0.03				0.02[k]

Table 5.55.3 Constituents identified in mandarin oils *(continued)*

Constituent	CAS	Percentage and range					
		A	B (3)	C (3)	D (3)	E (11)	F
(E)-2-Decenal	3913-81-3	0-0.01	0.03	0.01	0.02	0-0.01	
Decyl acetate	112-17-4		tr		0.01		
Diethyl succinate	123-25-1		+				
cis-Dihydrocarvone	3792-53-8		tr				0.06[h]
trans-Dihydrocarvone	5948-04-9		tr				
3,3-Dimethyl-1-octene					0.02		
(E,Z)-2,6-Dodecadienal	21662-13-5			0.01		tr-0.01	
Dodecanal	112-54-9	0-0.08	0.1	0.05	0.2		tr[e]; 0.02[m]; 0.08[f]; 0.2[c,o]; 0.7[l]; 3.0[r]
Dodecanoic acid	143-07-7						tr[c]
2-Dodecenal	4826-62-4				0.08		0.02[m]
(E)-2-Dodecenal	20407-84-5		0.04	0.02	0.04	0.01-0.03	0.05[c]
3-Dodecen-1-al	68083-57-8				0.1		
β-Elemene	33880-83-0		0.07	0.01	tr		tr[e]; 0.05[c]; <0.06[f]; 0.4[p]
γ-Elemene	29873-99-2			0.1			tr[e,f]
δ-Elemene	20307-84-0						tr[e]
Elemol (α-)	639-99-6			0.01	tr		tr[c]; 0.9[l]
β-Eudesmol	473-15-4			tr			
γ-Eudesmol	1209-71-8						1.1[l]
Eugenol	97-53-0						2.4[q]
Farnesene					tr		
α-Farnesene	502-61-4						tr[e]; 0.5[l]
(E,E)-α-Farnesene	502-61-4	tr-0.2	0.3	0.1	0.8	0.06-0.2	0.03[m]; 0.05[f]; 0.2[c]; 0.5[p]
β-Farnesene	502-60-3		0.03				
(E)-β-Farnesene	18794-84-8		0.1		0.1		0.1[c]
(Z)-β-Farnesene	28973-97-9		0.03	0.02	tr		<0.03[f]; 0.1[n]
Farnesol	4602-84-0						1.1[l]
(Z,E)-Farnesol	3790-71-4		tr	tr	0.03		0.8[h]
β-Fenchyl acetate (exo-)	76109-40-5						0.3[p]
Geranial	141-27-5		0.2	0.04	0.1	tr-0.02	tr[c,e,m]; 0.01[k]; 0.05[f]
Geraniol	106-24-1		0.01	0.01	0.05	0-tr	tr[f]
5-Geranoxypsoralen	7380-40-7	0-0.01					
Geranyl acetate	105-87-3		0.09	0.3	0.2		tr[f,m]; 0.06[c]; 0.2[h]
Geranyl isobutyrate	2345-26-8				0.03		
Germacrene	28028-64-0						1.1[l]
Germacrene A	28387-44-2			0.02			
Germacrene B	15423-57-1			0.3			
Germacrene C	34323-15-4			0.1			
Germacrene D	23986-74-5		0.06	0.3	0.1	0-0.01	tr[c,e]; 0.01[f]; 0.03[k]; 0.2[h]
α-Guaiene	3691-12-1			tr			
β-Gurjunene	73464-47-8		0.04		tr		0.01[f]; 0.08[c]
Heneicosane	629-94-7						2.4[q]
Heptadecanal	629-90-3		tr				
Heptadecane (n-)	629-78-7						0.3[q]
Heptanal	111-71-7			tr			
n-Heptane	142-82-5						0.4[n]
Heptanol	53535-33-4		tr				0.04[o]
2-Heptanol	543-49-7						0.04[o]
Heptyl acetate	112-06-1		0.02	tr			
Hexadecanal	629-80-1		tr				
Hexadecanol	51260-59-4				0.02		
(E)-2-Hexadecenal	22644-96-8		tr				
Hexanal	66-25-1		+				
n-Hexane	110-54-3						5.3[r]
Hexanol	111-27-3		0.02				
(E)-2-Hexenal	6728-26-3		tr				
(E)-2-Hexenyl butyrate	53398-83-7		0.1				
α-Humulene	6753-98-6		0.04	0.01	0.08	tr-0.01	tr[c,e,f,m]; 0.03[h]; 0.08[k]
Humulene oxide II	19888-34-7		+				
Hydrocarbons C_{21}-C_{33}, linear chained			+				

Table 5.55.3 Constituents identified in mandarin oils *(continued)*

Constituent	CAS	Percentage and range					
		A	B (3)	C (3)	D (3)	E (11)	F
Hydrocarbons C$_{21}$-C$_{33}$, linear chained, iso isomers			+				
β-Ionone	79-77-6						0.01[k]
(*E*)-β-Ionone	79-77-6			tr			
γ-Isogeraniol	13066-51-8		tr				
cis-Isopiperitenol	4017-76-9		tr				
trans-Isopiperitenol	4017-77-0		tr				
Isopiperitenone	529-01-1		tr				tr[e]
4-Isopropenyl-1-me-thyl-1,2-cyclohexane-diol				tr			
Isopulegol	89-79-2		tr				
Limonene	138-86-3	64.0-76.4	96.0	78.3	95.5	68.5-77.6[a]	90.1[r]; 90.8[e]; 92.4[n]; 92.6[k]; 92.9[f]
Limonene dioxide	96-08-2		0.1				
Limonene glycol	1946-00-5						6.6[q]
Limonene oxide	1195-92-2		0.1				0.6[o]
cis-Limonene oxide	13837-75-7		0.03	0.01	0.4	tr-0.02	tr[f]; 0.04[g]; 0.1[c]; 2.8[d]
trans-Limonene oxide	4959-35-7		tr	0.01	0.1	tr-0.02	tr[e,f]; 0.03[g]; 0.07[c]
Limonen-4-ol	3419-02-1		tr				
Limonen-10-ol	3269-90-7		tr				
Limonen-10-yl-acetate	15111-97-4		tr				
Linalool	78-70-6	0.01-0.7	1.5	1.0	4.7	0.06-0.2	0.4[f]; 0.6[o]; 0.7[h,n]; 1.0[c]; 2.6[l]; 3.0[p]
Linalool oxide	1365-19-1						0.6[l]
cis-Linalool oxide	11063-77-7		+	tr			0.1[h]
trans-Linalool oxide	11063-78-8		0.01	tr	0.2		
Linalyl acetate	115-95-7		0.5		tr		tr[c]; 0.1[h]; 0.2[o]; 0.4[n]
Linalyl anthranilate	7149-26-0				0.06		
Longipinene	5989-08-2						0.4[n]
Menthadien-1-ol							0.4[l]
cis-p-Mentha-1(7),8-dien-2-ol	22626-43-3		tr				
trans-p-Mentha-1(7),8-dien-2-ol	21391-84-4		tr				
p-Mentha-1,8-dien-9-ol	1946-01-6				0.02		
cis-p-Mentha-2,8-dien-1-ol	3886-78-0		tr	tr			1.8[q]; 2.3[d]
trans-p-Mentha-2,8-dien-1-ol	4017-77-0		tr	tr	0.2		0.05[c]; 0.7[d]; 4.0[q]
(+)-*p*-Mentha-1,8-dien-3-one	16750-82-6						0.2[d]
p-Mentha-1,3,8-triene	18368-95-1		tr	tr	tr	tr-0.03	0.07[r]
p-Menth-1-en-9-al	29548-14-9		+				
p-Menth-1-en-9-ol	18479-68-0		tr		tr		
cis-p-Menth-1-en-9-ol	18479-68-0						tr[c]
p-Menth-2-en-1-ol	619-62-5				0.04		
cis-p-Menth-2-en-1-ol	29803-82-5		tr				
trans-p-Menth-2-en-1-ol	29803-81-4						0.02[k]
p-Menth-1-en-4,5-oxide			tr				
p-Menth-4-en-1,2-oxide			tr				
Methylacetophenone	26444-19-9		tr				
p-Methylacetophe-none	122-00-9			tr			
Methyl anthranilate	134-20-3		tr				
3-Methylbenzalde-hyde	620-23-5		tr				
Methylcarvacrol	6379-73-3						0.01[k]

Table 5.55.3 Constituents identified in mandarin oils *(continued)*

Constituent	CAS	Percentage and range					
		A	B (3)	C (3)	D (3)	E (11)	F
Methyl chavicol	140-67-0						3.7[p]
Methyl geranate	2349-14-6						tr[f]
6-Methyl-5-hepten-2-one	110-93-0		tr	tr			
Methyl *N*-methyl anthranilate	85-91-6	0.08-1.1	0.7	0.5	3.4	0.2-0.5	0.06[m]
Methyl thymol	1076-56-8		tr	0.3	0.03	tr-0.01	0.2[o]
γ-Muurolene	30021-74-0		0.05				tr[c]; 1.2[l]
Myrcene	123-35-3	0.08-2.2	7.6	5.3	2.0	1.3-1.6	1.9[e]; 2.4[p]; 2.5[f]; 3.5[r]; 4.1[l]; 4.6[c]; 5.1[o]
(*E*)-Myroxide	28977-57-3			0.02			
Myrtenol	515-00-4		tr				
Myrtenyl acetate	1079-01-2						tr[k]
Neral	106-26-3		0.06	0.09	0.1	tr-0.05	tr[c,e]; 0.02[f]
Nerol	106-25-2		0.07	0.01	0.05		<0.04[f]; 0.06[h]
(*E*)-Nerolidol	40716-66-3		0.01	tr	0.03		0.01[f]; 0.05[c]
Neryl acetate	141-12-8		0.03	0.01	0.05	tr-0.01	tr[e]; 0.06[c]; 0.08[f]
(*Z*)-5-Nonadecane							3.2[q]
Nonanal	124-19-6	0.01-0.07	0.09	0.03	0.5	tr-0.03	tr[c]; 0.02[f,m]; 0.05[h]; 0.06[o]; 0.3[e]
Nonanoic acid	12-05-0				0.04		
Nonanol	28473-21-4		tr				tr[c]
(*E*)-2-Nonenal	18829-56-6		tr	tr			
Nonyl acetate	143-13-5		0.01				
Nootkatone	4674-50-4		0.03	0.01			tr[f]; 0.05[c]; 4.0[l]
Nopinone	24903-95-5		tr				
Norbornanol	86368-39-0						tr[f]
(*E*)-β-Ocimene	3779-61-1		0.2	0.02	0.3	0.01-0.02	tr[c]; 0.01[g,m]; 0.02[f]; 0.03[k]; 0.6[r]; 1.1[h]
(*Z*)-β-Ocimene	3338-55-4		0.3	0.1	0.2[a]		tr[f]; 0.3[l]
Octanal	124-13-0	0.03-0.3	0.6	0.2	1.6	0.05-0.1	0.1[e]; 0.2[d]; 0.4[c,f,m]; 1.2[l]
Octanoic acid	124-07-2		tr		0.05		
Octanol	29063-28-3	0-0.2	0.1	0.05	0.4	0.01-0.04[b]	0.01[f]; 0.05[c,h]; 1.0[l]
3-Octanol	589-98-0		0.1				
Octyl acetate	112-14-1		0.01	tr	0.05		tr[c]; 0.01[f]
Octyl formate	112-32-3			tr			
(*E*)-Patchoulenol	17806-54-1						0.8[d]
(*Z*)-Patchoulenol	17806-54-1						1.2[d]
Pentadecanal	2765-11-9		tr		0.01		
Perillaldehyde	2111-75-3	tr-0.2	0.1	0.03	0.3	0.02-0.04	0.05[c]; 0.1[m]; 0.2[d]; 0.3[r]; 0.4[p]; 1.7[l]
Perillene	539-52-6		+				
Perillyl alcohol	536-59-4		0.04				tr[c]; 0.9[d]
α-Phellandrene	99-83-2	0-0.1	0.5	0.06	0.1	0.03-0.05	tr[e]; 0.05[c]; 0.06[k]; 0.07[m]; 0.2[f]; 0.4[o]
β-Phellandrene	555-10-2	0.03-0.5	0.5		0.4		1.8[n]
α-Pinene	80-56-8	1.2-3.1	5.2	2.8	1.9	1.6-2.1	1.3[c,h]; 1.4[m]; 1.6[q]; 1.7[l]; 1.8[o]; 2.1[g]
β-Pinene	127-91-3	0.5-2.2	2.5	1.8	1.6	1.1-1.5	0.2[n]; 0.4[e,l]; 0.8[h,p]; 1.2[m]; 1.6[g,k]
cis-Pinene hydrate	17974-51-5		tr		0.06		
cis-Pinocarveol	6712-79-4						0.2[d]
trans-Pinocarveol	1674-08-4		tr				0.01[k]
Piperitenone	491-09-8						0.05[c]
Piperitone	89-81-6		tr	tr	0.02		tr[f]
Sabinene	3387-41-5	0.2-0.6	1.3	0.5	4.0	0.2	0.3[k,m]; 0.4[d]; 0.5[l]; 0.7[o]; 0.8[c]; 1.4[r]
cis-Sabinene hydrate	15537-55-0	0-0.06	0.07	0.04	0.1	0.01-0.04[b]	tr[c,f]; 0.03[g]; 0.04[h]
trans-Sabinene hydrate	17699-16-0	0.01-1.1	0.1	0.04	0.3	tr-0.08	0.02[f]
cis-Sabinene hydrate acetate	77318-48-0				tr		
β-Santalene	511-59-1						<0.03[f]
α-Selinene	473-13-2		0.06	0.03	0.3	tr-0.04	0.02[m]
β-Selinene	17066-67-0		tr		0.03		
Sinensal	3779-62-2						0.2[r]
α-Sinensal	17909-77-2	0.07-0.7	0.7	0.5	1.1	0.2-0.3	0.06[k]; 0.1[e,m]; 0.2[f]; 0.3[c]; 5.0[l]
β-Sinensal (*cis*-)	17909-87-4		0.2	0.04	0.3		tr[f]; 0.1[c]; 0.7[l]
Spathulenol	6750-60-3		tr	tr			2.5[h]

Table 5.55.3 Constituents identified in mandarin oils *(continued)*

Constituent	CAS	Percentage and range					
		A	B (3)	C (3)	D (3)	E (11)	F
α-Terpinene	99-86-5	0.1-0.5	0.5	0.4	0.4	0.2-0.3	tr[f]; 0.02[g]; 0.03[m]; 0.05[c]; 0.7[h]
γ-Terpinene	99-85-4	10.4-21.1	22.8	20.7	19.8	13.1-20.7	3.4[k]; 4.3[e]; 8.2[p]; 14.1[h]; 16.1[m]; 17.8[g]
α-Terpinenol	0.07-1.1						1.5[l]
Terpinen-4-ol	562-74-3	0.01-0.7	0.3	0.05	1.6	tr-0.04	0.1[n]; 0.3[o]; 0.4[m]; 0.7[p]; 1.0[l]; 2.4[r]
Terpinen-4-yl acetate	4821-04-9		0.01				
1-Terpineol	586-82-3		tr				
α-Terpineol	98-55-5		0.5	0.2	4.8	tr-0.2	0.1[o]; 0.2[c,g,n]; 0.4[m]; 1.1[l]; 1.3[h]; 2.2[r]
(E)-β-Terpineol			tr				
(Z)-β-Terpineol	7299-40-3		tr				
γ-Terpineol	586-81-2		0.01				
Terpinolene	586-62-9	0.06-4.2	1.4	0.9	1.1	0.5-0.7	0.1[e,n]; 0.2[g]; 0.3[f]; 0.7[p]; 0.8[m,o]; 1.6[r]
α-Terpinyl acetate	80-26-2		0.06	0.03	0.03		tr[c,f]; 0.01[k]; 0.7[h]
Tetradecanal	124-25-4		0.01	0.01	0.03	tr-0.01	tr[c,f]
Tetradecanol	27196-00-5		0.01				tr[c]
(E)-2-Tetradecenal	51534-36-2		tr	tr		tr-0.02	
α-Thujene	2867-05-2	0.2-1.1	1.1	1.0	0.7	0.5-0.7	tr[c,f]; 0.2[e,p]; 0.4[h]; 0.5[m]; 0.8[g]
Thujol	35732-37-7						1.4[q]
α-Thujone	546-80-5		tr				0.2[h]
Thymol	89-83-8	0-0.07	0.2	0.2	0.7	tr-0.06	tr[e]; 0.09[m]; 0.4[d]
Toluene	108-88-3						0.07[r]
Tricyclene	508-32-7				tr		0.04[m]; 0.09[k]; 0.2[h]
Tridecanal	10486-19-8		0.01			0-0.01	
(E)-2-Tridecenal	7069-41-2		tr				
(Z)-Trimenal	300733-87-3						0.3[d]
Umbellulone	24545-81-1			tr			
Undecanal	112-44-7	0-0.07	0.05	0.03	0.08	tr-0.05	tr[e]; 0.02[f]
(3E,5Z)-1,3,5-Undeca-triene	19883-27-3		+				
(E)-2-Undecenal	53448-07-0		tr				
(Z)-8-Undecenol						0-tr	
Valencene	4630-07-3		0.5	tr	0.09		tr[c]; 0.01[f]; 0.09[d]; 0.2[h]
Verbenone	80-57-9						0.2[k]
p-Vinylguaiacol	7786-61-0						2.3[l]
Viridiflorene (ledene)	21747-46-6					0-0.01	

A ninety-eight mandarin essential oil samples from Italy, Argentina, Spain and Brazil analyzed between 1998 and 2013; lowest and highest concentrations given (E. Schmidt, unpublished data)
B review of studies published between 1979 and 1999 on the composition of industrial cold-pressed, commercial and laboratory-extracted mandarin oils; data cited in ref. 3; highest concentrations given; the laboratory-extracted oils had no chemicals which were not also present in the other oils nor did they have significantly higher concentrations
C review of studies published between 1999 and 2009 on the composition of industrially extracted mandarin oils (refs. 5-10, 12-19,26); data cited in ref. 3; highest concentrations given; most chemicals were not identified in all studies and, consequently, their lower limit concentration is zero; chemicals found in all investigations with their lowest concentrations were β-caryophyllene (tr), p-cymene (0.1), decanal (tr), geranyl acetate (tr), limonene (68.6), linalool (0.02), myrcene (1.4), α-pinene (1.6), β-pinene (1.0), sabinene (tr), α-terpinene (tr), γ-terpinene (7.2), terpinolene (0.5) and α-thujene (0.3)
D review of studies published between 1999 and 2009 on the composition of laboratory cold-extracted and distilled mandarin oils (refs. 20-25,27,28); data cited in ref. 3; highest concentrations given; most chemicals were not identified in all studies and, consequently, their lower limit concentration is zero; chemicals found in all investigations with their lowest concentrations were limonene (65.4), linalool (0.1), myrcene (tr), octanal (tr), α-pinene (0.3), β-pinene (0.2), and α-terpineol (tr)
E one hundred and twenty-four industrially cold-pressed samples of green, yellow and red mandarin oils from Sicily, Italy; lowest and highest concentrations given (ref. 11)
F data from other studies (indicated with superscript letters); highest concentrations found in any study reviewed here given; when two or more oils were investigated, only the highest concentrations are mentioned, unless indicated otherwise

[a] limonene and (Z)-β-ocimene combined; [b] octanol and cis-sabinene hydrate combined; [c] one sample of Turkish cold-pressed clementine oil (ref. 36); [d] one industrially cold-pressed mandarin oil prepared from various mandarin varieties (Clementine, Clemenule, Satsuma) (ref. 29); [e] one hand cold-pressed peel oil from C. reticulata cv. Dalandan from the Philippines (ref. 34); [f] one laboratory cold-pressed oil from C. clementina cv. Nules from Italy (ref. 37); [g] one hand cold-pressed C. reticulata peel oil from Italy (ref. 38); [h] one lab-hydrodistilled oil from C. reticulata cv. Elarbi from Tunisia; results of oils from unripe fruits are not mentioned here (ref. 41); [k] one lab-hydrodistilled oil from mandarin dried peel grown in Tunisia (ref. 30); [l] three lab-hydrodistilled oils from fresh, air-dried and oven-dried peels of Pakistani C. reticulata (ref. 31); [m] one lab-hydrodistilled peel oil from C. reticulata cv. Avana from Italy (ref. 33); [n] one Citrus reticulata peel oil, lab-hydrolyzed, from India (ref. 40); [o] one mandarin peel oil from China; mode of extraction not specified (ref. 39); [p] one lab-hydrodistilled oil from Turkish C. nobilis (Lour.) (ref. 35); [q] one lab-hydrodistilled oil from Pakistan obtained from the peel of C. reticulata var. Kinnow (ref. 32); [r] four lab-hydrodistilled oils from four Egyptian Clementine cultivars (ref. 45)

tr: trace (in columns E and F[m]: <0.01; in column F[c]: <0.05; in column F[e]: <0.1); + present in the oil investigated, but quantity not stated

Table 55.5.4 Major constituents of mandarin oils

Constituent	CAS	Percentage and range					
		A	B (3)	C (3)	D (3)	E (11)	F
Limonene	138-86-3	64.0-76.4	96.0	78.3	95.5	68.5-77.6[a]	90.1[r]; 90.8[e]; 92.4[n]; 92.6[k]; 92.9[f]
γ-Terpinene	99-85-4	10.4-21.1	22.8	20.7	19.8	13.1-20.7	3.4[k]; 4.3[e]; 8.2[p]; 14.1[h]; 16.1[m]; 17.8[g]
p-Cymene	99-87-6	0.3-1.2	4.3	16.1	6.9	0.2-0.9	0.05[c]; 0.2[d]; 0.4[m]; 0.7[h]; 1.2[g]
Myrcene	123-35-3	0.08-2.2	7.6	5.3	2.0	1.3-1.6	1.9[e]; 2.4[p]; 2.5[f]; 3.5[r]; 4.1[l]; 4.6[c]; 5.1[o]
α-Pinene	80-56-8	1.2-3.1	5.2	2.8	1.9	1.6-2.1	1.3[c,h]; 1.4[m]; 1.6[q]; 1.7[l]; 1.8[o]; 2.1[g]
Terpinolene	586-62-9	0.06-4.2	1.4	0.9	1.1	0.5-0.7	0.1[e,n]; 0.2[g]; 0.3[f]; 0.7[p]; 0.8[m,o]; 1.6[r]
β-Pinene	127-91-3	0.5-2.2	2.5	1.8	1.6	1.1-1.5	0.2[n]; 0.4[e,l]; 0.8[h,p]; 1.2[m]; 1.6[g,k]

LEGEND: SEE UNDER TABLE 55.5.3

Table 5.55.5 Heterocyclic oxygenated compounds present in cold-pressed mandarin oils (1,4,10,13)

Name	Synonym(s)	CAS
Hydroxylated polymethoxyflavones		
4',5-Dihydroxy-7,8-dimethoxyflavone		
5-Hydroxy-3',4',6,7,8-pentamethoxyflavone	5-Hydroxyauranetin	50439-46-8
5-Hydroxy-6,7,8,3',4'-pentamethoxyflavone	5-Demethylnobiletin	2174-59-6
4'-Hydroxy-5,6,7,8-tetramethoxyflavone		36950-98-8
5-Hydroxy-3',4',6,7-tetramethoxyflavone	Santaflavone	21763-80-4
5-Hydroxy-4',6,7,8-tetramethoxyflavone	5-Demethyltangeretin	2798-20-1
5-Hydroxy-4',7,8-trimethoxyflavone		
4',5,7-Trihydroxy-3',6,8-trimethoxyflavone		
Polymethoxyflavones		
Heptamethoxyflavone		119279-30-0
3,3',4',5,6,7,8-Heptamethoxyflavone		1178-24-1
3,3',4',5,6,7-Hexamethoxyflavone		
3',4',5,6,7,8-Hexamethoxyflavone	Nobiletin	478-01-3
3',4',5,6,7-Pentamethoxyflavone	Sinensetin	2306-27-6
3',4',5,7,8-Pentamethoxyflavone	Isosinensetin	17290-70-9
4',5,6,7,8-Pentamethoxyflavone	Tangeretin	481-53-8
4',5,6,7-Tetramethoxyflavone	Tetra-O-methylscutellarein	1168-42-9
4',5,7,8-Tetramethoxyflavone	Tetra-O-methylisoscutellarein	6601-66-7
4',5,7-Trimethoxyflavone	Trimethylapigenin	5631-70-9
Coumarins and psoralens		
5,7-Dimethoxycoumarin	Citropten; Limetin	487-06-9
7-Geranyloxycoumarin	Aurapten	495-02-3

nerol 2.3%, citronellal 1.3%, neryl acetate 1.1% (42); this also applies to ref. 43 and 44, the latter of which contains many mistakes in nomenclature.

CONTACT ALLERGY / ALLERGIC CONTACT DERMATITIS

General
Contact allergy to and possible allergic contact dermatitis from mandarin oil has been reported in one publication only. A false-positive patch test reaction due to the excited skin syndrome cannot be excluded.

Case reports
An aromatherapist had occupational contact dermatitis with allergies to multiple essential oils used at work, including mandarin oil; the patient also reacted to geraniol,

linalool, linalyl acetate, α-pinene, the fragrance mix and various other fragrance materials; α-pinene was demonstrated by the mandarin oil; limonene was not tested (48). α-Pinene may be present in commercial mandarin oils in concentrations up to 3% (Table 5.55.3, column A).

LITERATURE
1 Russo M, Torre G, Carnovale C, Bonaccorsi I, Mondello L, Dugo P. A new HPLC method developed for the analysis of oxygen heterocyclic compounds in *Citrus* essential oils. J Essent Oil Res 2012;24:119-129

2 Liu YQ, Heying E, Tanumihardjo SA. History, global distribution, and nutritional importance of *Citrus* fruits. Compreh Rev Food Sci Food Saf 2012;11:530-545

3 Dugo G, Cotroneo A, Bonaccorsi I, Trozzi A. Composition of the volatile fraction of citrus peel oils. Mandarin (*Citrus deliciosa* Ten.), tangerine (*Citrus tangerina* Hort. ex Tan.), and clementine (*Citrus clementina* Hort. ex Tan.) oils. In: G Dugo and L Mondello, eds. Citrus oils: composition, advanced analytical techniques, contaminants, and biological activity. Boca Raton, USA: CRC Press, Taylor & Francis Group, 2011:48-74

4 Dugo P, Russo M. The oxygen heterocyclic components of Citrus essential oils. In: G Dugo and L Mondello, eds. Citrus oils: composition, advanced analytical techniques, contaminants, and biological activity. Boca Raton, USA: CRC Press, Taylor & Francis Group, 2011:405-444

5 Oberhofer B, Nikiforov A, Buchbauer G, Jirovetz L, Bicchi C. Investigation of the alteration of the composition of various essential oils used in aroma lamp applications. Flavour Fragr J 1999;14:293-299

6 Mondello L, Casilli A, Tranchida PQ, Cicero L, Dugo P, Dugo G. Comparison of fast and conventional GC analysis for citrus essential oils. J Agric Food Chem 2003;51:5602-5606

7 Mondello L, Casilli A, Tranchida PQ, Costa R, Dugo P, Dugo G. Fast GC for the analysis of citrus oils. J Chromatogr Sci 2004;42:410-416

8 Pino JA, Muñoz Y, Quijano-Celís CE. Analysis of cold pressed mandarin peel oil from Cuba. J Essent Oil Bear Plants 2006;9:271-276

9 Pino JA, Quijano-Celís CE. Chromatographic deterpenation of mandarin cold pressed oil. J Essent Oil Bear Plants 2007;10:504-509

10 Bonaccorsi I, Dugo P, Trozzi A, Cotroneo A, Dugo G. Characterization of mandarin (*Citrus deliciosa* Ten.) essential oil. Determination of volatiles, non-volatiles, physicochemical indices and enantiomeric ratios. Nat Prod Comm 2009;4:1595-1600

11 Dugo P, Bonaccorsi I, Ragonese C, Russo M, Donato P, Santi L, Mondello L. Analytical characterization of mandarin (*Citrus deliciosa* Ten.) essential oil. Flavour Fragr J 2011;26:34-46

12 Viuda-Martos M, Ruiz-Navajas Y, Fernández-López J, Pérez-Álvarez A. Chemical composition of mandarin (*C. reticulata* L.), grapefruit (*C. paradisi* L.), lemon (*C. limon* L.) and orange (*C. sinensis* L.) essential oils. J Essent Oil Bear Plants 2009;12:236-243

13 Schipilliti L, Tranchida PQ, Sciarrone D, Russo M, Dugo P, Dugo G, Mondello L. Genuineness assessment of mandarin essential oils employing gas chromatography-combustion-isotope ratio mass spectrometry (GC-C-IRMS). J Sept Sci 2010;33:617-625

14 Feger W, Brandauer H, Ziegler H. Germacrenes in citrus peel oils. J Essent Oil Res 2001;13:274-277

15 Steuer B, Schulz H, Läger E. Classification and analysis of citrus oils by NIR spectroscopy. Food Chem 2001;72:113-117

16 Schulz H, Schrader B, Quilitzsch R, Steuer B. Quantitative analysis of various citrus oils by ATR/FT-IR and NIR-FT Raman spectroscopy. Appl Spectrosc 2002;56:117-124

17 Veriotti T, Sacks R. High-speed characterization and analysis of orange oils with tandem-column stop-flow GC and time-of-flight MS. Anal Chem 2002;74:5635-5640

18 Reeve D, Arthur D. Riding the citrus trail: when is a mandarine a tangerine? Perfum Flav 2002;27(4):20-23

19 Sciarrone D, Schipilliti L, Ragonese C, Tranchida PQ, Dugo P, Dugo G, Mondello L. Thorough evaluation of the validity of conventional enantio-gas chromatography in the analysis of volatile chiral compounds in mandarin essential oil: a comparative investigation with multidimensional gas chromatography. J Chromatogr A 2010;1217:1101-1105

20 Sawamura M. Volatile components of essential oils of the *Citrus* genus. Recent Res Dev Agric Food Chem 2000;4:131-164

21 Lota M-L, de Rocca Serra D, Tomi F, Casanova J. Chemical variability of peel and leaf essential oils of 15 species of mandarins. Biochem Syst Ecol 2001;29:77-104

22 Catalfamo M, Gionfriddo F, Mangiola C, Manganaro R, Castaldo D. Determinazione delle caratteristiche analitiche e della composizione enantiomerica di oli essenziali agrumari ai fini dell'accertamento della purezza e della qualità. Nota III – Essenza di mandarino. Essenz Deriv Agrum 2004;74:59-64. Data cited in ref. 3

23 Frizzo CD, Lorenzo D, Dellacassa E. Composition and seasonal variation of the essential oils from two mandarin cultivars of southern Brazil. J Agric Food Chem 2004;52:3036-3041

24 Ruberto G, Starrantino A, Rapisarda P. Essential oils from new citrus fruits. Essenz Deriv Agrum 1999;69:15-26. Data cited in ref. 3

25 Karioti A, Skaltsa H, Gbolade A. Constituents of the distilled essential oils of *Citrus reticulata* and *C. paradisi* from Nigeria. J Essent Oil Res 2007;19:520-522

26 Dugo G, Bartle KD, Bonaccorsi I, et al. Advanced analytical techniques for the analysis of citrus essential oils. Part I. Volatile fraction: HRGC/MS analysis. Essenz Deriv Agrum 1999;69:79-111. Data cited in ref. 3

27 Merle H, Morón M, Blázquez MA, Boira H. Taxonomical contribution of essential oils in mandarins cultivars. Biochem Syst Ecol 2004;32:491-497

28 Ruberto G, Rapisarda P. Essential oils of new pigmented citrus hybrids: *Citrus sinensis* L. Osbeck x *C. clementina* Hort. ex Tanaka. J Food Sci 2002;67:2778-2780

29 Espina L, Somolinos M, Lorán S, Conchello P, García D, Pagán R. Chemical composition of commercial citrus fruit essential oils and evaluation of their antimicrobial activity acting alone or in combined processes. Food Control 2011;22:896-902

30 Hosni K, Zahed N, Chrif R, Abid I, Medfei W, Kallel M, et al. Composition of peel essential oils from four selected Tunisian *Citrus* species: Evidence for the genotypic influence. Food Chem 2010;123:1098-1104

31 Kamal GM, Anwar F, Hussain AI, Sarri N, Ashraf MY. Yield and chemical composition of *Citrus* essential oils as affected by drying pretreatment of peels. Int Food Res J 2011;18:1275-1282

32 Khan MM, Iqbal M, Hanif MA, Mahmood MS, Naqvi SA, Shahid M, Jaskani MJ. Antioxidant and antipathogenic activities of *Citrus* peel oils. J Essent Oil Bear Plants 2012;15:972-979

33 Fabroni S, Ruberto G, Rapisarda P. Essential oil profiles of new *Citrus* hybrids, a tool for genetic citrus improvement. J Essent Oil Res 2012;24:159-169

34 Dharmawan J, Kasapis S, Curran P. Characterization of volatile compounds in selected *Citrus* fruits from Asia—Part II: Peel oil. J Essent Oil Res 2008;20:21-24

35 Gursoy N, Tepe B, Sokmen M. Evaluation of the chemical composition and antioxidant activity of the peel oil of *Citrus nobilis*. Int J Food Prop 2010;13:983-991

36 Kirbaşlar SI, Gök A, Kirbaşlar FG, Tepe S. Volatiles in Turkish clementine (*Citrus clementina* Hort.) peel. J Essent Oil Res 2012;24:153-157

37 Verzera A, Tripodi G, Cotroneo A. Characteristics of a new Citrus hybrid essential oil, *Citrus clementina* cv. Nules x *Citrus limon* cv. Cavone. J Essent Oil Bear Plants 2009;12:293-299

38 Campolo O, Malacrinò A, Zappalà L, Laudani F, Chiera E, Serra D, et al. Fumigant bioactivity of five *Citrus* essential oils against *Tribolium confusum*. Phytoparasitica 2014;42:223-233

39 Tao NG, Liu YJ, Tang YF, Zhang JH, Zhang ML, Zeng HY. Essential oil composition and antimicrobial activity of *Citrus reticulata*. Chem Nat Comp 2009;45:437-438

40 Sultana HS, Ali M, Panda BP. Influence of volatile constituents of fruit peels of *Citrus reticulata* Blanco on clinically isolated pathogenic microorganisms under *In-vitro*. Asian Pac J Trop Biomed 2012;S1299-S1302

41 Bourgou S, Rahali FZ, Ourghemmi I, Tounsi MS. Changes of peel essential oil composition of four Tunisian *Citrus* during fruit maturation. The Scientific World Journal 2012, Article ID 528593, 10 pages

42 Chutia M, Deka Bhuyan P, Pathak MG, Sarma TC, Boruah P. Antifungal activity and chemical composition of *Citrus reticulata* Blanco essential oil against phytopathogens from North East India. LWT Food Sci Technol 2009;42:777-780

43 Johnson OO, Ayoola GA, Adenipekun T. Antimicrobial activity and the chemical composition of the volatile oil blend from *Allium sativum* (garlic clove) and *Citrus reticulata* (tangerine fruit). IJPSDR 2013;5(4):187-193

44 Das DR, Sachan AK, Shuaib M, Imtiyaz M. Chemical characterization of volatile oil components of *Citrus reticulata* by GC-MS analysis. World J Pharm Sci 2014;3:1197-1204

45 El-hawary SS, Taha KF, Abdel-Monem AR, Kirillos FN, Mohamed AA. Chemical composition and biological activities of peels and leaves essential oils of four cultivars of *Citrus deliciosa* var. *tangarina*. Am J Essent Oils Nat Prod 2013;1(2): 1-6

46 Rhind JP. Essential oils. A handbook for aromatherapy practice, 2nd Edition. London: Singing Dragon, 2012

47 Lawless J. The encyclopedia of essential oils, 2nd Edition. London: Harper Thorsons, 2014

48 Dharmagunawardena B, Takwale A, Sanders KJ, Cannan S, Roger A, Ilchyshyn A. Gas chromatography: an investigative tool in multiple allergies to essential oils. Contact Dermatitis 2002;47:288-292

Chapter 5.56   MARJORAM OIL (SWEET)

DEFINITION
Sweet marjoram oil is the essential oil obtained from the flowering tops of the sweet marjoram, *Origanum majorana* L. (synonym: *Majorana hortensis* Moench).

INCI NOMENCLATURE
Description/definition: Origanum majorana flower oil is the volatile oil obtained from the flowers of the sweet marjoram, *Origanum majorana* L., Labiatae
INCI name EU & USA: Origanum majorana flower oil
CAS registry number(s): 8015-01-8; 82082-58-6
EINECS number(s): 282-004-4

ISO (INTERNATIONAL ORGANIZATION FOR STANDARDIZATION) STANDARD
Currently there is no ISO standard for sweet marjoram oil.

Other species of the genus *Origanum,* from which oils can be obtained and which are mentioned in CosIng include *Origanum cretium* (Spanish marjoram, CAS 91722-83-7; EINECS 294-485-8, unknown to GRIN Taxonomy for Plants), *Origanum vulgare* (wild marjoram, CAS 84012-24-8; EINECS 281-670-3) and *Origanum heracleoticum* (CAS 91721-63-0; EINECS 294-363-4). Sweet marjoram oil should not be confused with origanum oil, Spanish type, obtained from *Coridothymus capitatus* (synonym: *Thymus capitatus*).

THE PLANT, THE OIL, AND THEIR USES
Origanum majorana L., popularly known as sweet marjoram, is an herb or undershrub, which can grow to 40-60 cm. The plant is native to Cyprus and Turkey and sometimes naturalized, especially in the Mediterranean region. The sweet marjoram is one of the most important culinary herbs of the western world, widely cultivated in Europe, Africa, southern Asia and the Americas. Annual, biannual and herbaceous perennial forms occur in cultivation. In temperate regions, the plants often do not survive the winter and are usually treated as annuals (4, Plants for a Future, http://www.pfaf.org; Mansfeld's World Database of Agriculture and Horticultural Crops, http://mansfeld.ipk-gatersleben.de).

The leaves and flowering herbs are employed for flavoring meat, sausages, soups, salads, vegetable dishes and oils. The seeds serve as a condiment (1,5,19,32). Sweet marjoram is also used in folk medicine and considered to have antiseptic, antispasmodic, carminative, cholagogue, diaphoretic, diuretic, emmenagogue, expectorant, stimulant, stomachic, aphrodisiac, sudorific and mildly tonic properties. It is taken internally in the treatment of bronchial complaints, tension headaches, insomnia, rheumatic complaints, anxiety, minor digestive upsets and painful menstruation (1,19,32, Plants for a Future, http://www.pfaf.org; Mansfeld's

World Database of Agriculture and Horticultural Crops, http://mansfeld.ipk-gatersleben.de).

The flowering tops of marjoram are steam-distilled to produce the essential oil of sweet marjoram, which is used in commercial food flavoring (e.g., baked goods, processed vegetables, condiments, soups, snack foods and gravies [5,19]), liqueurs, perfumes, cosmetics, pharmaceuticals and industrial products such as fungicides or insecticides (1,4,19). Marjoram oil possesses antimicrobial properties against food borne bacteria and mycotoxigenic fungi and strong antioxidant activity (32), mainly because of its high content of phenolic acids and flavonoids, which is useful in health supplements and food preservation (19). Medicinal uses include external application for sprains and bruises; marjoram oils are also employed in aromatherapy (74,75),

CHEMICAL COMPOSITION
Sweet marjoram oil is a yellowish to yellow-green, clear mobile liquid which has an aromatic, herbaceous and spicy odor. The yield of essential oil from the flowering tops of *Origanum majorana* L. generally varies from 0.5 to 2.0 per cent, depending on the origin of the plants. The main producing countries of this oil are Hungary, Spain, Egypt, southern France, Germany, Morocco, Tunisia, and the USA.

Literature data (up to August 19, 2014) on the chemical composition of sweet marjoram oils and unpublished analytical data from one of us (E.S.) are shown in Table 5.56.1 in alphabetical order. In sweet marjoram oils from various origins, over 240 chemicals have been identified. About 55 per cent of these were found in a single reviewed publication only. The major compounds found in sweet marjoram oils from different sources are shown in Table 5.56.2. They include (highest concentrations in any study given) linalool (88.0%), carvacrol (79.5%), sabinene hydrate (64.2%), terpinen-4-ol (55.6%), 1,8-cineole (51.0%), *trans*-sabinene hydrate (47.7%), α-terpineol (45.4%), *cis*-sabinene hydrate (44.2%), linalyl acetate (26.1%), *p*-cymene (23.4%), sabinene (21.5%), terpinolene (18.5%), γ-terpinene (18.4%), limonene (17.3%) and α-terpinene (13.1%). Well-known ingredients of sweet marjoram oils that were present in high concentrations (>8%) in one or two studies were β-caryophyllene (26.0% and 30%), *trans*-sabinene hydrate acetate (15.5%), thymol (12.2%), caryophyllene oxide (11.9%), *cis*-sabinene hydrate acetate (10.7%), α-pinene (10.2%) and *trans*-*p*-menth-2-en-1-ol (8.0%). Uncommon or rare constituents of sweet marjoram oils found in high concentrations (>7%) in single studies include 3-cyclohexen-1-ol (41.7%), isocaryophyllene (15.7%) and δ2-carene (=δ4-) (8.6%).

Commercial oils
The ten chemicals that had the highest maximum concentrations in 49 commercial marjoram essential oil samples (concentration ranges provided) are the following: β-phellandrene (1.1-50.2%), terpinen-4-ol (0.5-29.7%), *trans*-sabinene hydrate (0.6-24.0%), linalool (0.2-23.5%), γ-terpinene (0.2-16.7%), *trans*-sabinene hydrate acetate

Table 5.56.1 Constituents identified in sweet marjoram oils

Constituent	CAS	Percentage and range				
		A	B (6)	C (60)	D (30)	E
Abietatriene	19407-28-4		0-0.1			0.7[a,y4]
Acorenone	5956-05-8					<0.02[m]
α-Amorphene	20085-19-2					0.1[d]
Anethole	104-46-1					0.5[x1]
(E)-Anethole	4180-23-8					0.05[g]
p-Anisyl acetone	104-20-1					+[z6]
Aromadendrene	489-39-4		0-0.1			0.1[q,r]; 0.3[f]; 0.7[d,y4]; 0.8[x9]
allo-Aromadendrene	25246-27-9		0-0.02	tr		0.4[d]
Ascaridole	512-85-6		0-0.1			
trans-Ascaridole glycol	6790-83-6				0-0.1	
Benzaldehyde	100-52-7			tr		<0.5[n]
trans-α-Bergamotene	13474-59-4					0.1[t]
(E)-α-Bergamotol			0-0.1			
Bicycloelemene	32531-56-9		0-0.03	tr		0.2[x6]; 0.4[x9]
Bicyclogermacrene	24703-35-3	0.04-2.0	1.6-3.9			1.4[y5]; 1.5[v]; 1.7[x6]; 1.9[w]; 2.0[r]; 2.6[q]; 3.3[e]
β-Bisabolene	495-61-4		0-0.04	tr		tr[e]; 0.2[y]; 0.8[x1]; 2.8[y6]
Borneol (= endo-)	507-70-0		0.3-2.5		1.1-5.3	0.1[q]; 0.3[i]; 0.4[c]; 0.5[w1]; 0.6[g]; 1.0[z1]; 1.4[y8]
Bornyl acetate	76-49-3	0.03-0.2			0.1-1.2	0.5[e]; 1.4[x]; 2.2[f]; 2.3[x2]; 2.9[k]; 3.0[z1]; 3.3[d]
Bornyl formate	7492-41-3		0-0.5			
β-Bourbonene	5208-59-3			tr		0.05[y]; 0.1[e]
Butyl acetate	123-86-4					<0.1[c]
α-Cadinene	24406-05-1			0.1		0.1[x9]
γ-Cadinene	39029-41-9		0-0.02	tr		1.7[k]
δ-Cadinene	483-76-1	0-0.7	0-0.03	0.2		tr[e,i]; <0.05[y]; 0.7[d]
α-Cadinol	481-34-5		0-0.03	0.1	0.1	tr[e]; 0.2[t]
epi-α-Cadinol	5937-11-1			0.1		0.2[r]
α-Calacorene	21391-99-1			0.1		
β-Calacorene	50277-34-4			0.1		
Calamenene	483-77-2					0.4[y4]
cis-Calamenene	72937-55-4			tr		
Calarene epoxide	68926-75-0					0.4[a,y4]
Camphene	79-92-5	0-0.06	0-0.1	tr	0-tr	0.1[w]; 0.2[y8]; 0.3[g]; 0.7[w1]; 0.8[e]; 5.8[y9]
α-Campholenal	4501-58-0					+[z6]; 1.5[f]
Camphor	76-22-2		0-0.1		0.1	0.2[d]; 0.4[z2]; 0.6[y6]; 0.7[y8]; 1.8[x7]; 3.4[e]
5-Caranol						<0.02[m]
(+)-trans,trans-5-Caranol	6909-22-4					0.9[n]
δ2-Carene (=δ4-)	554-61-0					8.6[u]
δ3-Carene	13466-78-9					0.08[g]; 0.4[n]; 1.1[x]; 3.9[k]
Carvacrol	499-75-2		0.6-2.1	2.7		0.4[f]; 0.5[e]; 0.9[i]; 3.6[n]; 68.5[x7]; 79.5[g] 45.1[x8]; 65.0[z11]
Carvenone	499-74-1		0-0.03			
Carveol	99-48-9					0.5[x1a]
(E)-Carveol	1197-07-5		0-0.1			1.0[y8]
(Z)-Carveol	1197-06-4					0.4[k]
Carvone	99-49-0					tr[r]; 0.1[q]; 0.2[g]; 0.3[n]; 0.5[b]; 0.6[z7]; 0.8[x1]
Carvyl acetate	97-42-7					0.4[x5]
Caryophylladienol II			0-0.1			
β-Caryophyllene	87-44-5	0.03-3.5		1.8	0-0.2	3.4[x1]; 3.9[y6]; 4.2[q]; 4.5[r]; 4.8[z]; 26.0[d]; 30[z10]
Caryophyllene oxide	1139-30-6	0.04-0.3	0.1-0.4	1.3	0.5-2.1	0.3[i]; 0.7[e]; 0.8[r]; 0.9[w]; 1.3[d]; 1.8[s]; 11.9[y4]
Caryophyllenol II	32214-89-4		0-0.1			
Cedrol	77-53-2					1.1[a,y4]
1,8-Cineole	470-82-6		0-0.4		0.4-1.7	2.2[z10]; 2.3[s]; 2.4[k]; 8.2[i]; 41.5[o]; 48[w1]; 51.0[z1]

Table 5.56.1 Constituents identified in sweet marjoram oils (*continued*)

Constituent	CAS	Percentage and range				
		A	B (6)	C (60)	D (30)	E
Citronellol	106-22-9					0.7[o]; 2.7[y3]
Citronellyl acetate	150-84-5					0.1[i,r]
Citronellyl formate	105-85-1					<0.05[y]
α-Copaene	3856-25-5					<0.05[y]; 1.9[y4]
Cryptone	500-02-7					tr[d]
1-epi-Cubenol	81939-29-9			0.2		
Cuminaldehyde	122-03-2		0-0.04	tr		
Cuminyl alcohol	536-60-7		0-0.01			
Cyclohexanol	108-93-0					0.3[x]
3-Cyclohexen-1-ol	822-66-2					41.7[z9]
o-Cymene	527-84-4					2.0[u,z7]
p-Cymene	99-87-6	0.7-5.2	0.4-2.2	0.2	0.4-3.9	4.7[g]; 5.8[x8]; 6.3[s]; 6.8[w]; 7.0[m]; 7.7[x4]; 8.1[f]; 9.6y2; 9.8y9; 10.4b; 11.6y1; 12.1n; 23.4z
p-Cymenene	1195-32-0		0-0.02	tr		
p-Cymenol	25497-27-2					0.5[y4]
p-Cymen-8-ol	1197-01-9		0-0.2			0.04[w]; 0.07[g]
Dehydro-1,8-cineole	92760-25-3				0-tr	
cis-Dihydrocarvone	3792-53-8		0-0.02			+[z6]; 0.08[g]; 0.3[j]; 0.4[q]
trans-Dihydrocarvone	5948-04-9		0-0.1			+[z6]; 0.1[d,r]; 0.2[y]; 0.3[x6]
Dihydrolinalyl acetate	61476-73-1					0.1[q]
cis-Dihydro-α-terpinyl acetate	80-25-1					<0.02[m]
Dihydroumbellulone	2506-61-8					1.0[a,y4]
1-(1,4-Dimethyl-3-cyclohexen-1-yl)-etha-none	43219-68-7					+[z6]; 0.1[n]
(E,E)-2,6-Dimethyl-3,5,7-octatrien-2-ol			0-0.02			
Dimethylpent-2-enal (4,4-)	22597-46-2					1.1[a,y4]
Eicosane	112-95-8					0.4[k]
β-Elemene	33880-83-0					0.3[x2]; 1.4[x]; 1.6[p]; 2.0[k]
γ-Elemene	29873-99-2					1.8[x4]; 3.7[x5]
δ-Elemene	20307-84-0					0.04[g]; <0.05[e]; 0.1[t]; 0.2[q]; 0.4[y4]
Elemol	639-99-6	0-0.3				0.2[y3]
Elixene	3242-08-8					1.0[z7]
Epiglobulol	88728-58-9			0.1		
Ethylcyclohexanone	4423-94-3					0.3[a,y4]
Ethyl hexadecanoate	628-97-7					<0.5[n]
Ethyl linoleate	544-35-4					<0.5[n]
Ethyl tetradecanoate	124-06-1					<0.5[n]
α-Eudesmol	473-16-5					0.3[d]
β-Eudesmol	473-15-4					0.3[y4]
Eugenol	97-53-0					0.1[s]; 0.3[c]; 1.5[i]
(E,E)-Farnesol	106-28-5					0.02[w]
(Z,Z)-Farnesol	16106-95-9					0.05[w]
exo-Fenchol (β-)	22627-95-8					0.1[i]
Fenchone	1195-79-5					1.2[x1]; 4.9[f]
Fenchyl alcohol	1632-73-1					1.7[f]
Geranial	141-27-5					0.3[x1]
Geraniol	106-24-1	0.04-0.6	0.1-0.5			tr[r]; <0.1[c]; 0.2[m]; 0.3[i]; 0.5[n]; 0.7[k]; 2.5[f]; 3.2[o]
Geranyl acetate	105-87-3	0.05-0.2	0.1-0.6			0.2[p]; 0.3[i]; <0.5[n]; 0.6[c]; 2.3[d]; 4.1[x]; 7.1[k]
(E)-Geranylacetone	3796-70-1		0-0.01			
Germacrene	28028-64-0					1.4[i]
Germacrene D	23986-74-5		0-0.1	0.2		tr[i]; 0.1[y]; 0.2[y6]; 1.2[e]
Globulol	489-41-8		0-0.1	0.1		0.09[e]; 0.1[t]

Table 5.56.1 Constituents identified in sweet marjoram oils (*continued*)

Constituent	CAS	Percentage and range				
		A	B (6)	C (60)	D (30)	E
Guaiol	489-86-1					0.7[a,y4]
α-Gurjunene	489-40-7					tr[d]
2-Heptanone	110-43-0					0.03[x]
Hexadecane	544-76-3					0.2[x]
Hexadecanoic acid	57-10-3					0.2[a,y4]; <0.5[n]
1,2,3,4,4a,7-Hexahy-dro-1,6-dimethyl-4(1-methylethyl) naphthalene	16728-99-7					<0.5[n]
Hexahydrofarnesyl acetone	502-69-2		0-0.03			
Hexanol (1-; *n*-)	111-27-3					0.2[x2]
(*E*)-2-Hexenal	6728-26-3					1.5[k]
(*Z*)-2-Hexenal	16635-54-4		0-0.03			
(Z)-3-Hexenol	928-96-1					<0.05[y]
3-Hexenyl benzoate				tr		
α-Himachalene	3853-83-6					0.4[a,y4]
α-Humulene	6753-98-6	0.06-1.3	0.2	0.2	0-tr	0.2[q]; 0.4[x]; 0.6[f]; 0.8[p]; 2.0[k]; 3.7[y8]; 7.3[x2]
α-Humulene epoxide	96638-51-6					0.4[y4]
Humulene epoxide II	19888-34-7			0.1	0-0.1	
Isoaromadendrene epoxide	499134-59-7					0.4[a,y4]
Isoborneol	124-76-5					0.03[m]; 0.2[i]
Isobornyl acetate	125-12-2					0.5[u]; 2.8[w1]
Isocarvacrol	1740-97-2		0-0.1			
Isocaryophyllene	118-65-0					15.7[y2]
Isomenthone	491-07-6		0-0.1			
cis-Isopulegone			0-0.01			
trans-Isopulegone			0-0.03			
Isospathulenol	88395-46-4					0.2[y]
Isothymol	4427-56-9		0-0.1			
Ledol	577-27-5					0.5[f]
Limonene	138-86-3	0.02-2.9	0.8-1.6	14.3	0-1.1	2.4[w]; 2.6[d]; 2.8[u]; 3.0[v]; 4.2[j]; 5.3[k]; 5.7[z5]; 6.4[z1]; 8.1[w1]; 9.3[y3]; 10.9[x3]; 16.7[y1]; 17.3[i]
Limonen-4-ol	3419-02-1		0-0.1			
Linalool	78-70-6	0.2-23.5	1.6-5.0	2.3	2.3-4.6	12.1[y9]; 15.5[i]; 15.7[y1]; 16.2[h]; 16.4[e]; 22.7[w1]; 24.0[z1]; 28.8[x5]; 32.7[f]; 54.25[z5]; 88.0[z8]
Linalool oxide	1365-19-1					0.1[x1]
cis-Linalool oxide	11063-77-7			tr		+[z6]; 0.2[e,w1]
cis-Linalool oxide, furanoid	11063-77-7				0-0.1	0.02[g]
trans-Linalool oxide	11063-78-8					+[z6]; 0.2[w1]
trans-Linalool oxide, furanoid	34995-77-2				0-0.1	
Linalyl acetate	115-95-7	0.1-3.8	6.8-9.6	2.2	0.1-3.1	3.6[z]; 3.9[c]; 4.1[y7]; 5.1[x]; 5.8[k]; 14.1[i]; 14.2[z5]; 14.5[y1]; 19.9[x3]; 22.4[y3]; 26.1[y6,z12]
Linalyl anthranilate	7149-26-0					0.9[z7]
Longifolenaldehyde	66537-42-6					0.3[a,y4]
Longifolene (junipene)	475-20-7					0.1[i]
trans-Longipinocarveol	889109-69-7					1.1[a,y4]
p-Mentha-1(7),8-diene	499-97-8		0-0.03	tr		
p-Menthane-1,2,3-triol						2.5[a,y4]
p-Menth-2-en-1-ol	619-62-5					3.0[z7]
cis-*p*-Menth-2-en-1-ol	29803-82-5	0.5-2.3	0.2-0.6	0.7	0.3-1.6	1.9[t]; 2.0[c]; 2.3[y5]; 2.4[j]; 2.6[r]; 2.8[y2]; 4.9[x2]
trans-*p*-Menth-2-en-1-ol	29803-81-4	0.8-1.8	0.5-1.2	0.2	0.1-0.9	1.5[z4]; 1.6[j]; 1.8[m]; 1.9[s,w]; 2.1[l]; 8.0[p]
p-Menth-8-yl acetate						0.2[x6]
Menthofuran	494-90-6		0-0.01			

Table 5.56.1 Constituents identified in sweet marjoram oils (*continued*)

Constituent	CAS	Percentage and range				
		A	B (6)	C (60)	D (30)	E
Menthone	89-80-5					2.1^{z10}
Methylcarvacrol	6379-73-3			tr		
Methyl chavicol	140-67-0					0.1^n; 0.2^i; 0.4^{x1}; 1.0^{w1}; 2.6^{y3}
(*E*)-Methyl cinnamate	1754-62-7		0-0.1			
Methyl eugenol	93-15-2					0.4^k
6-Methyl-3-heptanol	18720-66-6					0.09^g
6-Methyl-3-heptanone	624-42-0					$+^{z6}$
Methylheptenone	409-02-9					0.2^w
Methyl 2-hydroxy-3-methyl-pentanoate	41654-19-7					0.03^g
Methyl 2-methylbutyrate	868-57-5					$+^{z6}$; 0.1^g
4-Methyl-3-pentenoic acid	504-85-8					$+^{z6}$
Methyl thymol	1076-56-8			0.5		
α-Muurolene	10208-80-7			tr		
γ-Muurolene	30021-74-0					0.3^{y4}
τ-Muurolol (=epi-α-)	19912-62-0		0-0.01	0.5	0-0.1	
Myrcene	123-35-3	1.4-5.3	0.5-1.6	1.1		1.6^w; 1.8^b; 1.9^f; 2.1^{y5}; 2.3^{y2}; 2.4^{x4}; 4.7^d
Myristicin	607-91-0					tr^{x5}
Myrtenal	564-94-3					0.5^p
Myrtenol	515-00-4					0.06^k
Myrtenyl acetate	1079-01-2					0.6^k
Nerol	106-25-2		0-0.2			$0.2^{k,m,x2}$; 1.2^x
Nerolidol	7212-44-4					0.2^{y3}
Nerol oxide	1786-08-9		0-0.01			
Neryl acetate	141-12-8	0.01-0.3	0.1-0.4			0.1^d; 0.2^t; 0.3^c; 0.6^{x9}; 0.8^p; 3.5^{x4}
Nonadecane	629-92-5					0.07^k
Nonanal	124-19-6		0-0.01			
(*E*)-Ocimene	27400-72-2					0.04^m; 0.1^s; 0.3^o
(*Z*)-Ocimene	27400-71-1					0.3^m; 0.4^s; 0.6^o
(*E*)-β-Ocimene	3779-61-1	0.01-0.3	0-0.2	tr		tr^d; 0.04^g; 0.05^w; 0.2^e; 0.9^n; 4.1^{z2}
(*Z*)-β-Ocimene	3338-55-4	0.02-0.08	0-0.1	0.1		tr^d; 0.04^w; 0.2^g; 1.4^n; 1.8^e; 4.1^{y6}
allo-Ocimene	673-84-7					0.4^d
Octanal	124-13-0					0.5^{x2}
3-Octanol	589-98-0		0-0.03			$+^{z6}$; 0.1^{x9}; $<0.5^n$; 1.1^{y7}
1-Octen-3-ol	3391-86-4	0-0.4				$+^{z6}$; $<0.05^e$; 0.3^g; 2.1^n
1-Octen-3-one	4312-99-6			tr		
1-Octen-3-yl acetate	2442-10-6					$+^{z6}$; 2.1^e
Paraldehyde	123-63-7					$+^{z6}$
Patchoulane	25491-20-7					0.1^g
Patchoulene	1405-16-9					2.1^{y3}
α-Phellandrene	99-83-2	0.3-0.8	0-0.8	tr	0-tr	1.0^t; 1.1^b; 1.5^u; 2.4^{x9}; 2.8^{x1}; 3.6^{x6}; 4.8^{y3}
β-Phellandrene	555-10-2	1.1-50.2	0.8-1.7	1.8		1.5^c; 1.6^y; 1.7^{x5}; 1.8^{x8}; 2.2^{y5}; 2.4^{z3}; 3.6^q
Phenylacetaldehyde	122-78-1					$+^{z6}$; 0.1^x
Phytone	16825-16-4			tr		$1.0^{a,y4}$
α-Pinene	80-56-8	0.6-3.3	0.1-0.3	0.5	0-tr	2.0^{x3}; 2.5^{z1}; 4.4^{w1}; 4.8^{x9}; 5.9^{y9}; 10.2^u
β-Pinene	127-91-3		0-0.2	tr		2.2^{y9}; 2.4^x; 2.8^u; 3.4^{z1}; 4.1^{w1}; 4.6^e; 7.5^{x3}
α-Pinene oxide	1686-14-2					$<0.02^m$
trans-Pinocarveol	1674-08-4		tr-0.1			0.5^{y5}
Pinocarvone	30460-92-5		0-0.1			
Piperitenone	491-09-8		0-0.6			
cis-Piperitol	16721-38-3	0.1-1.0	0-0.5		0-0.8	0.2^e; 0.3^m; $0.5^{c,j}$; 0.6^l; 0.8^y; 1.0^r; 7.5^{y6}

Table 5.56.1 Constituents identified in sweet marjoram oils (*continued*)

Constituent	CAS	Percentage and range				
		A	B (6)	C (60)	D (30)	E
trans-Piperitol	16721-39-4	0.2-3.9	0.1-0.3	tr	0-0.7	0.2[n]; 0.3[v]; 0.4[m]; 0.5[c]; 0.6[j]; 1.2[z7]; 1.3[r]
cis-Piperitol acetate	78774-33-1					1.3[y8]
trans-Piperitol acetate						0.5[y8]
Piperitone	89-81-6					0.1[t]
cis-Piperitone						0.3[x]
Propyl acetate	109-60-4					0.2[a,y4]
Pulegone	89-82-7		0-2.7			0.2[z7]
Sabina ketone	513-20-2		0-0.01			
Sabinene	3387-41-5	1.0-9.4	1.8-3.2	4.9	0-0.8	5.6[z4]; 6.9[w]; 7.5[s]; 7.8[q]; 8.0[x4]; 8.3[z9]; 8.8[j]; 9.3[z3]; 9.5[d]; 10.9[z]; 12.0[y6]; 17.3[z5]; 21.5[p]
Sabinene hydrate	546-79-2				25.5-64.2	
cis-Sabinene hydrate	15537-55-0	0.01-6.8	30.4-44.2		3.3-9.5	9.6[j]; 14.6[x1]; 15.0[m]; 15.8[w]; 20.9[b]; 23.6[y2]; 27.1[s]; 29.3[k]; 30.2[c]; 36.1[p]; 43.7[x2]
trans-Sabinene hydrate	17699-16-0	0.6-24.0	5.5-7.4			5.9[y2]; 7.5[s]; 8.2[x2]; 10.1[h]; 10.8[t]; 11.6[k]; 11.7[j]; 13.2[y8]; 14.8[v]; 15.5[y5]; 39.2[x6]; 47.7[x]
cis-Sabinene hydrate acetate	77318-48-0	0.1-0.4				tr[r]; 0.2[w]; 0.5[y6]; 0.6[j]; 3.4[q]; 5.0[y2]; 10.7[v]
trans-Sabinene hydrate acetate	77318-47-9	0.06-15.5			0-1.1	0.2[q]; 1.1[t]; 1.2[w]; 1.3[x6]
trans-Sabinyl acetate	139757-62-3					0.1[m]
Santalol	11031-45-1					0.3[n]
β-Selinene	17066-67-0					0.3[k]
Spathulenol	6750-60-3	0.04-0.2	0-0.4	1.5	0.1-1.5	0.8[t]; 1.0[b]; 1.1[e]; 1.3[r]; 2.9[z]; 6.0[y4]; 6.2[y2]
α-Terpinene	99-86-5	0.07-10.3	1.6-3.9	0.1	0-tr	5.8[h]; 6.8[t]; 7.3[v,z4]; 7.7[l]; 8.3[x1]; 8.6[j]; 8.9[y5]; 9.2[z3]; 9.5[b]; 10.9[x4]; 11.4[q]; 13.1[z9]
γ-Terpinene	99-85-4	0.2-16.7	3.1-6.3	0.5	0-0.3	11.3[t]; 11.4[z4]; 12.5[h]; 13.1[j]; 13.9[l]; 14.0[y5]; 14.1[u]; 16.0[x4]; 16.4[y9]; 16.8[b]; 18.4[q]
δ-Terpinene	586-62-9					6.7[x6]
Terpinen-3-ol						0.7[a,j]
Terpinen-4-ol	562-74-3	0.5-29.7	8.2-14.3	47.1	7.6-23.1	31.2[w]; 35.5[y2]; 37.1[n]; 37.7[x3]; 38.4[m]; 39.5[s]; 41.4[z2]; 46.0[b]; 49.7[r]; 55.1[y8]; 55.6[z7]
Terpinen-4-yl acetate	4821-04-9	0-0.08	0-0.1		0-tr	0.1[x6]; 1.3[n]; 2.9[b]
1-Terpineol	586-82-3					0.5[t]; 1.8[x9]
α-Terpineol	98-55-5	1.4-3.9	3.9-6.9	4.6	4.8-45.4	5.6[h]; 6.5[y6]; 6.7[y4]; 6.9[c]; 7.2[n]; 8.0[r]; 8.8[b]; 9.1[y8]; 9.5[z7]; 11.1[u]; 13.0[j]; 26.9[y3]
β-Terpineol	138-87-4					1.6[x5]
(*E*)-β-Terpineol						0.1[i]; 1.3[x4]
(*Z*)-β-Terpineol	7299-40-3					0.4[w]; 0.6[i]; 0.9[x4]; 1.2[z]
σ-Terpineol						0.2[p]
Terpinolene (α-)	586-62-9	0.2-6.8	0.6-1.6	tr		3.1[j]; 3.7[b]; 3.8[q]; 4.6[u]; 6.4[x5]; 14.7[d]; 18.5[x9]
Terpinyl acetate	8007-35-0					0.1[x1]; 0.7[y9]; 0.8[z7]; 2.3[y4]
α-Terpinyl acetate	80-26-2					0.02[m]; 0.04[k]; 0.3[b]; 0.4[z1]
Terpinyl propionate	80-27-3					0.6[x9]; 3.3[a,y4]
α-Thujene	2867-05-2	0.1-1.3	0.1-0.5	0.2		1.1[b]; 1.2[l]; 1.3[x4]; 1.4[q]; 2.4[z1]; 3.7[z5]; 4.1[u]
β-Thujene	28634-89-1					0.4[a,y4]
β-Thujone (= *cis*-)	471-15-8					tr[d]
trans-Thujone	33766-30-2			tr		
Iso-3-thujyl acetate						1.5[q]
Thymol	89-83-8		0.1-1.1	1.6		1.5[g]; 2.0[y]; 4.0[z11]; 11.6[e]; 12.0[z8]; 12.2[x7]
Tricyclene	508-32-7			tr		0.07[y8]; 0.1[x2]; 0.2[x]; 1.0[u]
Tridecane	629-50-5					0.03[x]
Verbenone	80-57-9					0.4[n]

Table 5.56.1 Constituents identified in sweet marjoram oils (*continued*)

Constituent	CAS	Percentage and range				
		A	B (6)	C (60)	D (30)	E
Viridiflorene (ledene)	21747-46-6		0-0.1			tr[d]; <0.05[v]; 0.6[x9]
Viridiflorol	552-02-3					0.1[q]; 0.2[d,t]; 0.5[r]; 0.7[w1]

A forty-nine sweet marjoram essential oil samples from Hungary, Egypt, Germany, Morocco, Spain, Tunisia and Serbia analyzed between 1998 and 2013; lowest and highest concentrations given (E. Schmidt, unpublished data)

B four lab-hydrodistilled oils from the aerial parts of *O. majorana* collected in the wild in Turkey from two locations and at different times; lowest and highest concentrations given (ref. 6)

C one lab-hydrodistilled oil from *O. majorana* var. *majorana* plants in full bloom, cultivated from seeds collected in the wild in France (ref. 60)

D six lab-hydrodistilled oils from six populations growing wild in Cyprus; lowest and highest concentrations given (ref. 30)

E data from other studies (indicated with superscript letters); highest concentrations found in any study reviewed here given; when two or more oils were investigated, only the highest concentrations are mentioned, unless indicated otherwise

[a] incorrect identification (ref. 54); [b] eight lab-hydrodistilled oils from air-dried commercial flowering tops and leaves purchased from different companies in Poland (ref. 48); [c] one lab-hydrodistilled marjoram leaf oil from plants collected in the wild in Venezuela at 3300 meters above sea level (ref. 1); [d] one lab-hydrodistilled oil from the 'total plant' of *O. majorana* growing wild in Algeria (ref. 3); [e] one lab-hydrodistilled oil from the flowering parts of sweet marjoram cultivated at an experimental station in Cuba (ref. 8); [f] one steam-distilled oil from whole above-ground plants cultivated in Morocco (ref. 9); [g] two lab-hydrodistilled oils from the aerial parts of plants growing wild in Turkey of the carvacrol chemotype (ref. 11); [h] one lab-hydrodistilled oil from flowering cultivated plants in Egypt (ref. 15); [i] one commercial oil purchased in Germany (ref. 16); [j] one oil purchased from the Essential Oil University in Charlestown, USA and obtained from flowers and leaves of Egyptian *Origanum majorana* (ref. 17); [k] four oils from the aerial parts of plants grown in Tunisia in four growth stages from early vegetative to full flowering (ref. 19); [l] one lab-hydrodistilled oil from commercial Egyptian dried leaves (ref. 22); [m] one lab-hydrodistilled oil from fresh flowering plants growing wild at Reunion Island (ref. 23); [n] one steam-distilled oil from flowering plants cultivated in Greece (ref. 24); [o] one commercial oil purchased from a USA company with an atypical composition and high content of 1,8-cineole (ref. 28); [p] one lab-hydrodistilled oil from sweet marjoram cultivated in Tunisia and harvested in the late vegetative state (ref. 29); [q] one commercial oil sample from an Austrian company (ref. 31); [r] one lab-hydrodistilled oil from marjoram cultivated in Lithuania and in full bloom (ref. 32); [s] eight lab-hydrodistilled oils from plants in the flowering stage cultivated in Egypt under different fertilization regimes and harvested in two seasons (ref. 33); [t] one lab-hydrodistilled oil from the leaves of sweet marjoram growing wild in the Sinai Desert, Egypt (ref. 37); [u] one lab-hydrodistilled oil from leaves collected in the wild in Egypt (ref. 38); [v] one lab-hydrodistilled oil from plants in full bloom cultivated in Iran (ref. 39); [w] one lab-hydrodistilled oil from plants in full bloom growing wild in India (ref. 40); [w1] one commercial marjoram oil sample from a Brazilian supplier (ref. 73); [x] one lab-hydrodistilled oil from fresh shoots of *O. majorana* cultivated in Tunisia (ref. 41); [x1] one commercial oil from Hungary (ref. 43); [x2] one lab-hydrodistilled oil from the shoots of sweet marjoram cultivated in Tunisia (ref. 45); [x3] several oils from Egypt produced from various harvests and collected in various seasons (ref. 69); [x4] unknown number of oils produced from plants cultivated at an Egyptian horticultural research center (ref. 65); [x5] one sweet marjoram oil sample (ref. 66); [x6] one lab-hydrodistilled oil from pre-flowering marjoram cultivated in Iran (ref. 47); [x7] one lab-produced oil from *O. majorana* collected in the wild in Italy (ref. 55); [x8] one carvacrol-rich lab-hydrodistilled oil from plants collected in the wild in Greece (ref. 56); [x9] various oils produced in Egypt from plants under different compost and fertilizer regimens (ref. 57); [y] one commercial oil sample produced in India (ref. 59); [y1] one commercial oil sample from marjoram grown in Poland (ref. 62); [y2] five lab-distilled oils from India, one from fresh material and four from herbs dried in different manners (ref. 63); [y3] one oil from Egyptian sweet marjoram with an atypical composition (ref. 64); [y4] one lab-hydrodistilled oil from dried marjoram leaves purchased from a food company in China (ref. 2); [y5] one commercial oil produced by a German company from Albanian plant material (ref. 4); [y6] one marjoram oil from flowers of plants cultivated in Iran (ref. 7); [y7] one lab-hydrodistilled oil from *O. majorana* in the flowering stage cultivated in Tunisia (ref. 18); [y8] one lab-hydrodistilled leaf oil from marjoram plants cultivated in Argentina (ref. 20); [y9] one lab-hydrodistilled oil from dried, finely ground marjoram from Hungary (ref. 21); [z] one lab-hydrodistilled oil from marjoram whole plant cultivated in Egypt (ref. 25); [z1] one commercial oil from a US company with an extremely atypical composition and very high 1,8-cineole content (ref. 26); [z2] one lab-hydrodistilled oil from marjoram cultivated in Egypt (ref. 27); [z3] one commercial oil obtained from a UK company (ref. 34); [z4] one essential oil of unknown origin; data cited in ref. 35; [z5] six oils from marjoram cultivated in Austria and exposed to different day lengths (ref. 36); [z6] various oils from marjoram cultivated in Finland (ref. 42); [z7] one lab-hydrodistilled oil from dried leaves purchased in Cairo, Egypt (ref. 44); [z8] one lab-hydrodistilled oil from leaves collected in the wild in Turkey with an extremely atypical composition and apparently consisting of virtually only linalool (88%) and thymol (12%) (ref. 46); [z9] one oil from Venezuela (ref. 70); [z10] one oil from Lebanon (ref. 71); [z11] one carvacrol-rich marjoram oil, origin unknown (ref. 67); [z12] one oil from Iran (ref. 72)

tr: trace (in column B: <0.1 [possibly <0.01]; in columns C and D: <0.05; in column E[d]: <0.1); + present in the oil investigated, but quantity not stated

Table 5.56.2 Major constituents of sweet marjoram oils

Constituent	CAS	Percentage and range				
		A	B (6)	C (60)	D (30)	E
Linalool	78-70-6	0.2-23.5	1.6-5.0	2.3	2.3-4.6	24.0[z1]; 28.8[x5]; 32.7[f]; 54.25[z5]; 88.0[z8]
Carvacrol	499-75-2		0.6-2.1	2.7		3.6[n]; 68.5[x7]; 79.5[g]; 45.1[x8]; 65.0[z11]
Sabinene hydrate	546-79-2				25.5-64.2	
Terpinen-4-ol	562-74-3	0.5-29.7	8.2-14.3	47.1	7.6-23.1	39.5[s]; 41.4[z2]; 46.0[b]; 49.7[r]; 55.1[y8]; 55.6[z7]
1,8-Cineole	470-82-6		0-0.4		0.4-1.7	2.2[z10]; 2.3[s]; 2.4[k]; 8.2[l]; 41.5[o]; 48[w1]; 51.0[z1]
trans-Sabinene hydrate	17699-16-0	0.6-24.0	5.5-7.4			11.7[j]; 13.2[y8]; 14.8[v]; 15.5[y5]; 39.2[x6]; 47.7[x]
α-Terpineol	98-55-5	1.4-3.9	3.9-6.9	4.6	4.8-45.4	8.8[b]; 9.1[y8]; 9.5[z7]; 11.1[u]; 13.0[i]; 26.9[y3]
cis-Sabinene hydrate	15537-55-0	0.01-6.8	30.4-44.2		3.3-9.5	23.6[y2]; 27.1[s]; 29.3[k]; 30.2[c]; 36.1[p]; 43.7[x2]
p-Cymene	99-87-6	0.7-5.2	0.4-2.2	0.2	0.4-3.9	9.6[y2]; 9.8[y9]; 10.4[b]; 11.6[y1]; 12.1[n]; 23.4[z]
Linalyl acetate	115-95-7	0.1-3.8	6.8-9.6	2.2	0.1-3.1	14[i]; 14.2[z5]; 14.5[y1]; 19.9[x3]; 22.4[y3]; 26.1[y6]
Sabinene	3387-41-5	1.0-9.4	1.8-3.2	4.9	0-0.8	9.3[z3]; 9.5[d]; 10.9[z]; 12.0[y6]; 17.3[z5]; 21.5[p]
Terpinolene (α-)	586-62-9	0.2-6.8	0.6-1.6	tr		3.1[j]; 3.7[b]; 3.8[q]; 4.6[u]; 6.4[x5]; 14.7[d]; 18.5[x9]
γ-Terpinene	99-85-4	0.2-16.7	3.1-6.3	0.5	0-0.3	14.0[y5]; 14.1[u]; 16.0[x4]; 16.4[y9]; 16.8[b]; 18.4[q]
Limonene	138-86-3	0.02-2.9	0.8-1.6	14.3	0-1.1	6.4[z1]; 8.1[w1]; 9.3[y3]; 10.9[x3]; 16.7[y1]; 17.3[i]
α-Terpinene	99-86-5	0.07-10.3	1.6-3.9	0.1	0-tr	8.9[y5]; 9.2[z3]; 9.5[b]; 10.9[x4]; 11.4[q]; 13.1[z9]

LEGEND: SEE UNDER TABLE 5.56.1

(0.06-15.5%), α-terpinene (0.07-10.3%), sabinene (1.0-9.4%), *cis*-sabinene hydrate (0.01-6.8%) and terpinolene (0.2-6.8%) (Erich Schmidt, unpublished analytical data).

Chemotypes

There are at least two chemotypes of marjoram essential oils (4,8,11). In the first chemotype, *cis*-sabinene hydrate, *trans*-sabinene hydrate and the acetate of *cis*-sabinene hydrate are the main constituents of the volatile fraction. These unstable compounds rearrange during hydrodistillation, forming terpinen-4-ol, α-terpineol, and α- and γ-terpinene (5). For that reason, terpinen-4-ol with *cis*- and *trans*-sabinene hydrate, α-terpineol, α- and γ-terpinene are the main constituents of the essential oil (48). This 'European type' is most frequently encountered in literature and is the oil of commerce. The oil of the second chemotype contains the phenols carvacrol (up to 80%) or thymol (one case only, 68) as major constituents (4). This chemotype is usually obtained from plants growing in the wild, e.g., in Turkey (10,11,67), Italy (55), and Greece (56).

However, in some analyzed marjoram oils other chemicals had the highest concentrations, including linalool (9,36,46,66), limonene (16,62), linalyl acetate (7,72), β-caryophyllene (3,71), β-phellandrene (50.2%, unpublished data from one of us [E.S.]) and 3-cyclohexen-1-ol (70). In two reports (26,28), in which the same Korean institutions (and some authors) were involved, 1,8-cineole was the dominant ingredient (51% and 42%). In both cases, a commercial marjoram oil purchased in the USA (both in New Jersey, but apparently from different companies) had been investigated (26,28).

Thus, a third chemotype of marjoram oil rich in linalool or linalyl acetate has been suggested (48) and a recently proposed classification system of marjoram essential oils is shown in Table 5.56.3. It is not quite clear why only *trans*- but not *cis*-sabinene hydrate is included, but it does take linalool, linalyl acetate and 1,8-cineole as important ingredients into account (48).

Table 5.56.3 Clusters of essential oils of cultivated *Origanum majorana* based on their main components (adapted from ref. 30)

Cluster	Main components
Oils rich in cymyl compounds	Carvacrol
Oils rich in 1,8-cineole followed by linalool	1,8-Cineole; linalool
Oils rich in linalool followed by sabinyl compounds or α-terpineol	Linalool; *trans*-sabinene hydrate; α-terpineol
Oils rich in terpinen-4-ol followed by various amounts of cymyl and/or sabinyl compounds	Terpinen-4-ol; γ-terpinene *p*-cymene; *trans*-Sabinene hydrate
Oils rich in sabinyl compounds followed by terpinen-4-ol	*trans*-Sabinene hydrate; terpinen-4-ol
Oils rich in sabinyl compounds	*trans*-Sabinene hydrate

Additional analyses of sweet marjoram oils not discussed in Table 5.56.1 are provided in refs. 49-53; these can all be accessed on-line.

CONTACT ALLERGY / ALLERGIC CONTACT DERMATITIS

General

Allergic contact dermatitis from marjoram oil has been reported in two publications only, both of which concerned occupational contact dermatitis in aromatherapists. Linalool and possibly α-pinene and caryophyllene may have been allergens in one study.

Case reports

An aromatherapist had chronic hand dermatitis and was patch test positive to 17 of 20 oils used at her work (tested 1% and 5% in petrolatum), including marjoram oil (76). Two other aromatherapists had contact dermatitis (one occupational) with allergies to multiple essential oils used at their work, including marjoram oil. Both patients also reacted to geraniol, α-pinene, the fragrance mix and various other fragrance materials. In addition, one proved to be allergic to linalool and linalyl acetate, the other to caryophyllene. α-Pinene, linalool, geraniol and caryophyllene were demonstrated by GC-MS in both marjoram oils used by the patients (77). Linalool has been found in concentrations of up to 23.5% in commercial marjoram oils, α-pinene and caryophyllene in concentrations over 3% (Table 5.56.1, column A).

LITERATURE

1 Ramos S, Rojas LB, Lucena ME, Meccia G, Usubillaga A. Chemical composition and antibacterial activity of *Origanum majorana* L. essential oil from the Venezuelan Andes. J Essent Oil Res 2011;23:45-49

2 Jiang Z-T, Li R, Wang Y, Chen S-H, Guan W-Q. Volatile oil composition of natural spice, *Origanum majorana* L. grown in China. J Essent Oil Bear Plants 2011;14:458-462

3 Brada M, Saadi A, Wathelet JP, Lognay G. The essential oils of *Origanum majorana* L. and *Origanum floribundum* Munby in Algeria. J Essent Oil Bear Plants 2012;15:497-502

4 Schmidt E, Bail S, Buchbauer G, Stoilova I, Krastanov A, Stoyanova A, Jirovetz L. Chemical composition, olfactory evaluation and antioxidant effects of the essential oil of *Origanum majorana* L. from Albania. Nat Prod Commun 2008;3:1051-1056. Data also presented in ref. 58

5 Novak J, Lukas L, Franz CM. The essential oil composition of wild growing sweet marjoram (*Origanum majorana* L., Lamiaceae) from Cyprus—three chemotypes. J Essent Oil Res 2008;20:339-341

6 Tabanca N, Özek T, Baser KHC, Tümen G. Comparison of the essential oils of *Origanum majorana* L. and *Origanum x majoricum* Cambess. J Essent Oil Res 2004;16:248-252

7 Barazandeh MM. Essential oil composition of *Origanum majorana* L. from Iran. J Essent Oil Res 2001;13:76-77

8 Pino JA, Rosado A, Estarrón M, Fuentes V. Essential oil of majoram (*Origanum majorana* L.) grown in Cuba. J Essent Oil Res 1997;9:479-480

9 Charai M, Mosaddak M, Faid M. Chemical composition and antimicrobial activities of two aromatic plants: *Origanum majorana* L. and *O. compactum* Benth. J Essent Oil Res 1996;8:657-664

10 Baser KHC, Özek T, Tümen G, Sezik E. Composition of the essential oils of Turkish *Origanum* species with commercial importance. J Essent Oil Res 1993;5:619-623. Data also presented in ref. 11

11 Baser KHC, Kirimer N, Tümen G. Composition of the essential oil of *Origanum majorana* L. from Turkey. J Essent Oil Res 1993;5:577-579. Data also partly presented in ref. 10

12 Lawrence BM. Progress in essential oils. Perfum Flavor 2000;25(5):52-71

13 Lawrence BM. Progress in essential oils. Perfum Flavor 1997;22(1):43-56

14 Hassanzadeh MH, Emami SA, Asili J, Tayarani Najaran Z. Review of the essential oil composition of Iranian Lamiaceae. J Essent Oil Res 2011;23:35-74

15 Pavela R. Sublethal effects of some essential oils on the cotton leafworm *Spodoptera littoralis* (Boisduval). J Essent Oil Bear Plants 2012;15:144-156

16 Šipailieneė A, Venskutonis PR, Baranauskienė R, Šarkinas A. Antimicrobial activity of commercial samples of thyme and marjoram oils. J Essent Oil Res 2006;18:698-703

17 Pavela R. Insecticidal activity of essential oils against cabbage aphid *Brevicoryne brassicae*. J Essent Oil Bear Plants 2006;9:99-106

18 Ben Hamida-Ben Ezzeddine N, Abdelkéfi NM, Ben Aissa R, Chaabouni MM. Antibacterial screening of *Origanum majorana* L. oil from Tunisia. J Essent Oil Res 2001;13:295-297

19 Sellami IH, Maamouri E, Chahed T, Wannes WA, Kchouk ME, Marzouk B. Effect of growth stage on the content and composition of the essential oil and phenolic fraction of sweet marjoram (Origanum majorana L.). Ind Crop Prod 2009;30:395-402

20 Banchio E, Bogino PC, Zygadlo J, Giordano W. Plant growth promoting rhizobacteria improve growth and essential oil yield in Origanum majorana L. Biochem. Syst Ecol 2008;36:766-771

21 Vági E, Simándi B, Suhajda A, Héthelyi E. Essential oil composition and antimicrobial activity of *Origanum majorana* L. extracts obtained with ethyl alcohol and supercritical carbon dioxide. Food Res Int 2005;38:51-57

22 Busatta C, Vidal RS, Popiolski AS, Mossi AJ, Dariva C, Rodrigues MRA. Application of Origanum majorana L. essential oil as an antimicrobial agent in sausage. Food Microbiol 2008;25:207-211

23 Vera RR, Chane-Ming J. Chemical composition of the essential oil of marjoram (*Origanum majorana* L.) from Reunion Island. Food Chem 1999;66:143-145

24 Komaitis ME, Ifanti-Papatragianni N, Melissari-Panagiotou E. Composition of the essential oil of marjoram (*Origanum majorana* L.). Food Chem 1992;45:117-118

25 El-Moneim MR Afify, Abd Fatma FR, Turky AF. Control of *Tetranychus urticae* Koch by extracts of three essential oils of chamomile, marjoram and *Eucalyptus*. As Pacif J Trop Biomed 2012;2:24-30

26 Yang Y-C, Lee SI, Clark JM, Ahn Y-J. Ovicidal and adulticidal activities of *Origanum majorana* essential oil constituents against insecticide-susceptible and pyrethroid/malathion-resistant *Pediculus humanus capitis* (Anoplura: Pediculidae). J Agric Food Chem 2009;57:2282-2287

27 Viuda-Martos M, El-Nasser G. S. EL Gendy A, Sendra E, Fernandez-Lopez J, El Razik KAA, et al. Chemical composition and antioxidant and anti-*Listeria* activities of essential oils obtained from some Egyptian plants. J Agric Food Chem 2010;58:9063-9070

28 Jang Y-S, Yang Y-C. Choi D-S, Ahn Y-J. Vapor phase toxicity of marjoram oil compounds and their related monoterpenoids to *Blattella germanica* (Orthoptera: Blattellidae). J Agric Food Chem 2005;53:7892-7898

29 Baatour O, Kaddour R, Tarchoun I, Nasri N, Mahmoudi H, Zaghdoudi M, et al. Modification of fatty acid, essential oil and phenolic contents of salt-treated sweet marjoram (*Origanum majorana* l.) according to developmental stage. J Food Sci 2012;77:C1047-C1054

30 Karousou R, Efstathiou C, Lazari D. Chemical diversity of wild growing *Origanum majorana* in Cyprus. Chem Biodivers 2012;9:2210-2217

31 Novak J, Langbehn J, Pank F, Franz CM. Essential oil compounds in historical sample of marjoram (*Origanum majorana* L. (Lamiaceae). Flavour Fragr J 2002;17:175-180

32 Baranauskiene R, Venskutonis PR, Demyttenaere JCR. Sensory and instrumental evaluation of sweet marjoram (*Origanum majorana* L.) aroma. Flavour Fragr J 2005;20:492-500

33 Edris AE, Shalaby QA, Fadel HM. Effect of organic agriculture practices on the volatile aroma components of some essential oil plants growing in Egypt II: sweet Marjoram (*Origanum marjorana* L.) essential oil. Flavour Fragr J 2003;18:345-351

34 Baratta MT, Dorman HJD, Deans SG, Figueiredo AC, Barroso JG, Ruberto G. Antimicrobial and antioxidant properties of some commercial essential oils. Flavour Fragr J 1998;13:235-244

35 Fischer N, Nitz S, Drawert F. Original flavour compounds and the essential oil composition of marjoram (*Majorana hortensis* Moench). Flavour Fragr J 1987;2:55-61

36 Circella G, Franz Ch, Novak J, Resch H. Influence of day length and leaf insertion on the composition of marjoram essential oil. Flavour Fragr J 1995;10:371-374

37 Abbassy MA, Abdelgaleil SAM, Rabie RYA. Insecticidal and synergistic effects of *Majorana hortensis* essential oil and some of its major constituents. Entomol Exper Appl 2009;131:225-232

38 El-Ghorab AH, Mansour AF, El-Massry KF. Effect of extraction methods on the chemical composition and antioxidant activity of Egyptian marjoram (*Majorana hortensis* Moench). Flavour Fragr J 2004;19:54-61

39 Omidbaigi R, Bastan MR. Essential oil composition of marjoram cultivated in north of Iran. J Essent Oil Bear Plants 2005;8:56-60

40 Raina AP, Singh Negi K. Essential oil composition of *Origanum majorana* and *Origanum vulgare* ssp. *hirtum* growing in India. Chem Nat Comp 2012;47:1015-1017

41 Baatour O, Kaddour R, Aidi Wannes W, Lachaal M, Marzouk B. Salt effects on the growth, mineral nutrition, essential oil yield and composition of marjoram (*Origanum majorana*). Acta Physiol Plant 2010;32:45-51

42 Nykänen I. High resolution gas chromatographic mass spectrometric determination of the flavour composition of marjoram (Origanum majorana L.) cultivated in Finland. Z Lebensm Unters Forsch 1986;183:172-176

43 Misharina TA, Polshkov AN, Ruchkina EL, Medvedeva IB. Changes in the composition of the essential oil of marjoram during storage. Appl Biochem Microbiol 2003;39:311-316

44 El-Seedi HR, Khalil NS, Azeem M, Taher EA, Göransson U, Pålsson K, et al. Chemical composition and repellency of essential oils from four medicinal plants against *Ixodes ricinus* nymphs (Acari: Ixodidae). J Med Entomol 2012;49:1067-1075

45 Baatour O, Tarchoune I, Mahmoudi H, Nassri N, Abidi W, Kaddouri R, et al. Culture conditions and salt effects on essential oil composition of sweet marjoram (*Origanum majorana*) from Tunisia. Acta Pharm 2012;62:251-261

46 Karabörklü S, Ayvaz A, Yilmaz S, Akbulut M. Chemical composition and fumigant toxicity of some essential oils against *Ephestia kuehniella*. J Econ Entomol 2011;104:1212-1219

47 Alizadeh A, Khosh-khui M, Javidnia K, Firuzi O, Jokar SM. Chemical composition of the essential oil, total phenolic content and antioxidant activity in *Origanum majorana* L. (Lamiaceae) cultivated in Iran. Adv Environm Biology 2011;5:2326-2331

48 Lis A, Piter S, Góra J. A comparative study on the content and chemical composition of essential oils in commercial aromatic seasonings. Herba Polonica 2007;53:21-26

49 Mangai SA, Subban R. Seasonal variation in the essential oil composition of *Majorana hortensis* Moench from western ghats region of south India. Asian J Pharm Clin Res 2014;7:173-177

50 Verma RS, Sashidhara KV, Anju Y, Naqvi AA. Essential oil composition of *Majorana hortensis* (Moench) from subtropical India. Acta Pharmaceutica Sciencia 2010;52:19-22

51 Verma RS, Verma RK, Chauhan A, Yadav AK. Changes in the essential oil composition of *Majorana hortensis* Moench. cultivated in India during plant ontogeny. J Serb Chem Soc 2010;75:441-447

52 Alarmalmangai S, Subban R. Composition and antibacterial activity of essential oil of *Majorana hortensis* Moench of western ghats of south India. J Pharm Res 2012;5:471-473

53 Soliman FM, Yousif MF, Zaghloul SS, Okba MM. Seasonal variation in the essential oil composition of *Origanum majorana* L. cultivated in Egypt. Z Naturforsch (C) 2009;64:611-614

54 Lawrence BM. Progress in essential oils. Perfum Flavor 2013;38(June):58-61

55 Gionfriddo F, Mangiola C, Pirrello A, Manganaro R. Studio preliminare sulla composizione degli oli essenziali di alcane piante officinali calabresi. Essenz Deriv Agrum 2001;71:67-70. Data cited in ref. 54

56 Daferera D, Ziogas BN, Polissiou MG. The effectiveness of plant essential oils on the growth of *Botrytis cinera*, *Fusarium* sp. and *Clavibacter michiganensis* subsp. *michiganensis*. Crop Protect 2003;22:39-44

57 Gharib FA, Moussa LA, Massoud NO. Effect of compost and bio-fertilizers on growth, yield and essential oil of sweet marjoram (*Majorana hortensis*) plant. Int J Agric Biol 2008;10:381-387

58 Jirovetz L, Bail S, Buchbauer G, Denkova Z, Slavchev A, Stoyanova A, Schmidt E. Chemical composition, antimicrobial activities and olfactory evaluations of an essential marjoram oil from Albania as well as some target compounds. Ernährung 2008;32:197-201. Data also presented in ref. 4

59 Anonymous. Som extracts, Marjoram oil, Kashmir (India). GC-MS analysis. Indian Perfum 2008;52:76. Data cited in ref. 54

60 Figuérédo G, Cabassu P, Chalchat J-C, Pasquier B. Studies of Mediterranean oregano populations. VIII—Chemical composition of essential oils of oreganos of various origins. Flavour Fragr J 2006;21:134-139

61 Lawrence BM. Progress in essential oils. Perfum Flavor 2004;29(4):70-? (last page unknown)

62 Gora J, Majda T, Lis A, Tichek A, Kurowska A. Chemical composition of some Polish commercial essential oils. Rivista Ital EPPOS 1997;(Numero Speciale):761-766. Data cited in ref. 61

63 Rao LJ. Quality of essential oils and process materials of selected spices and herbs. J Med Arom Plant Sci 2000;22:808-816. Data cited in ref 61.

64 Daw ZY, El-Baroty GE, Ebtesam AM. Inhibition of *Aspergillus parasiticus* growth and aflatoxin production by some essential oils. Chem Mikrobiol Technol Lebensmitt 1994;16(516):129-135. Data cited in ref. 12

65 El-Masry H, Charles DJ, Simon JE. Bentazon and Terbacil as postemergent herbicides for sweet basil and sweet marjoram. J Herbs Spices Med Plants 1995;3:19-26. Data cited in ref. 13

66 Bourrel C, Vilaren G, Michel G, Gaset A. Étude des propriétés bacteriostatiques et fongistatiques en milieu solide de 24 huiles essentielles préambulement analysées. Rivista Ital EPPOS 1995;16:3-12. Data cited in ref. 13

67 Sarer E, Scheffer JJC, Baerheim Svendsen A. Monoterpenes in the essential oil of *Origanum majorana*. Planta Med 1982;46:236-239. Data cited in ref. 11

68 Jolivet J, Rey P, Boussarie MF. Differenciation de quelques huiles essentielles présentant une constitution voisine. II. Essence de marjolaine. Plantes Med Phytother 1971;5:199-208. Data cited in ref. 19

69 Omer EA, Ouda HE, Ahmed SS. Cultivation of sweet marjoram *Majorana hortensis* in newly reclaimed lands of Egypt. J Herbs Spices Med Plants 1994;2:9-16. Data cited in ref. 13

70 Meza M, González N, Usubillaga A. Composición del aceite esencial de *Origanum majorana* L. extraído por diferentes técnicas y su actividad biológica. Rev Fac Agron 2007;24:725-738. Data cited in ref. 1

71 Hilan C, Sfeir R, Jawish D, Aitour S. Huiles essentielles de certaines plantes médicinales Libanaises de la famille des Lamiaceae. Leb Sci J 2006;(7)2:14-22. Data cited in ref. 3

72 Baranzadeh MM. Essential oil composition of *Origanum majorana* L. (in Persian). Iran Med Arom Plant Res 2001;10:65-74. Data cited in ref. 14

73 Murbach Teles Andrade BF, Nunes Barbosa L, da Silva Probst I, Fernandes A Júnior. Antimicrobial activity of essential oils. J Essent Oil Res 2014;26:34-40

74 Rhind JP. Essential oils. A handbook for aromatherapy practice, 2nd Edition. London: Singing Dragon, 2012

75 Lawless J. The encyclopedia of essential oils, 2nd Edition. London: Harper Thorsons, 2014

76 Selvaag E, Holm J, Thune P. Allergic contact dermatitis in an aromatherapist with multiple sensitizations to essential oils. Contact Dermatitis 1995;33:354-355

77 Dharmagunawardena B, Takwale A, Sanders KJ, Cannan S, Roger A, Ilchyshyn A. Gas chromatography: an investigative tool in multiple allergies to essential oils. Contact Dermatitis 2002;47:288-292

Chapter 5.57 MELISSA OIL (LEMON BALM OIL)

DEFINITION

Melissa oil (lemon balm oil) is the essential oil obtained from the aerial parts of the melissa (lemon balm), *Melissa officinalis* L.

INCI NOMENCLATURE

Description/definition: Melissa officinalis leaf oil is a volatile oil obtained from the leaves and tops of the lemon balm (balm mint), *Melissa officinalis* L., Labiatae
INCI name EU & USA: Melissa officinalis leaf oil
CAS registry number(s): 8014-71-9; 84082-61-1
EINECS number(s): 282-007-0

ISO (INTERNATIONAL ORGANIZATION FOR STANDARDIZATION) STANDARD

There is currently no ISO standard for melissa oil.

THE PLANT, THE OIL, AND THEIR USES

Melissa officinalis L. (lemon balm) is a perennial herb which can grow to a height of 70-150 cm and which has leaves with a gentle lemon scent. The lemon balm is native to the Madeira and Canary Islands, Morocco, Tunisia, temperate regions of Asia, Pakistan, and various countries in South and East Europe. The plant is cultivated for the herb as a spice, and as a medicinal, aromatic ornamental and bee plant in Europe (Italy, Hungary), north-western Africa (Morocco, Algeria), Egypt, North, Central and South America (Brazil, Peru), temperate Asia (Turkey, China, Korea, Russia), and Australia (19, Mansfeld's World Database of Agriculture and Horticultural Crops, http://mansfeld.ipk-gatersleben.de; http://efloras.org; GRIN Taxonomy for Plants). It has escaped cultivation and established itself in England, continental Europe, North America and northern Iran (1).

The lemon-scented leaves are used in herbal teas and for flavoring salads, sauces, herb vinegar and fruit cups.

The herb, the essential oil and extracts have many medicinal applications (4,14,16,19,24) and melissa is listed in a number of European Pharmacopoeias (14). Traditionally, the plant has been used to heal wounds, sores (e.g., from genital or oral herpes), bee and wasp stings, to relieve tension, calming nerves, and for headaches (14). It is also used to treat asthma, stomach ailments, indigestion, menstrual cramps, and fevers. Aqueous and alcoholic extracts from the aerial part of *Melissa officinalis* have traditional use in the treatment of fevers and colds, indigestion associated with nervous tension, hyperthyroidism, depression, insomnia, epilepsy, headaches, toothaches, and other ailments (6,19,24). Melissa can also be used as insect repellent (14). The plant is said to possess antibacterial, antimicrobial, antiviral, antifungal, anti-mutagenic, antitumor, antihistaminic, antispasmodic, and antioxidant

properties (1,14,19). Medicinal aspects of *Melissa officinalis* have been reviewed (2).

The essential oil, which is obtained from the (often flowering) aerial parts of *M. officinalis*, is employed in liqueurs (Benedictine, Chartreuse), ice cream, perfumes, cosmetics and furniture polishes (3). It is also used by the food industry to inhibit the spoilage yeast growth (3). The oil is very popular in aromatherapy (46,47) against nervous tension and may be used as a home remedy for headaches and toothaches (19). The yield of essential oil of lemon balm is very low, so the production cost and price of the oil are very high. As a consequence, lemon balm oil is sometimes adulterated with other oils, e.g., from *Cymbopogon* spp. or *Citrus* peel oil (24).

CHEMICAL COMPOSITION

Melissa oil is a slightly yellowish to light brownish, clear mobile liquid which has a fresh, herbal, citrusy odor which can sometimes be recognized as lemon odor. The yield of essential oil from the aerial parts of *Melissa officinalis* L. generally varies from 0.01 to 0.25 per cent. The main producing countries of this oil are Algeria, France, Morocco, Germany, Turkey, USA and Australia.

Literature data (up to October 8, 2014) on the chemical composition of melissa oils (including a number of oils produced from melissa *leaves* only: refs. 8,14,15,24,27,35,43) and unpublished analytical data from one of us (E.S.) are shown in Table 5.57.1 in alphabetical order. In melissa oils from various origins, over 310 chemicals have been identified. About 49 per cent of these were found in a single reviewed publication only. The major compounds found in melissa oils from different sources are shown in Table 5.57.2. They include (highest concentrations in any study given) geranial (74.1%), limonene (57.5%), citronellal (55.8%), neral (43.8%), germacrene D (32.5%), β-caryophyllene (29.4%), caryophyllene oxide (25.3%), geraniol (25.0%), geranyl acetate (25%), citronellol (21.8%), linalool (19.5%) and nerol (16.2%). Well-known ingredients of melissa oils that were present in high concentrations (>7%) in one or two studies were β-pinene (18.2% and 44.5%), α-terpinene (21.9%), methyl citronellate (19.2%), sabinene (12.9% and 17.4%), α-pinene (12.1% and 12.8%), α-humulene (10.2%), (*E*)-β-ocimene (9.3%), α-farnesene (9.1%), isomenthone (8.9%), γ-terpinene (8.2%) and methyl chavicol (7.5%). Uncommon or rare constituents of melissa oils found in high concentrations (>9%) in one or two studies include carvacrol (13.1% and 31.8%), α-humulene epoxide (25.4%), citral (17.4%), terpinen-4-ol (7.7% and 15.8%), cedrane (14.1%), (*E*)-anethole (12.3% and 13.2%), *p*-mentha-1,3,8-triene (12.7%), 2,2,8,8-tetramethyl-5-nonane (12.6%), thymol (7.9% and 10.5%), fenchyl acetate (9.6%) and 1(10),5-germacradien-4-ol (9.6%).

Commercial oils

The ten chemicals that had the highest maximum concentrations in 53 commercial melissa essential oil samples (concentration ranges provided) are the following: geranial (0.2-41.7%), neral (0.5-31.0%), β-caryophyllene (0.7-29.4%), α-citronellal (0.5-29.2%), germacrene D

Table 5.57.1 Constituents identified in melissa oils

Constituent	CAS	A	B (3)	C (4)	D (5)	E (26)	F (6)	G
α-Acoradiene	24048-44-0							5.9[z1]
α-Amorphene	20085-19-2							0.2[i]
(E)-Anethole	4180-23-8							12.3[z8]; 13.2[i]
(E)-Anhydrolinalool oxide	54750-70-8							<0.01[z4]
(Z)-Anhydrolinalool oxide	54750-69-5							<0.01[z4]
Anisaldehyde								0.1[z8]
Aromadendrene	489-39-4					0.3		tr[b]; 0.2[z5]; 1.9[k]
allo-Aromadendrene	25246-27-9						tr	tr[b]; 0.1[z2]; 1.3[z9]
Artemiseole	60485-46-3							1.0[x]
Bakerol	157744-23-5							1.3[k]
Benzaldehyde	100-52-7						tr	tr[j]; <0.01[z4]
Bergamal	106-72-9	0-0.3						
α-Bergamotene	17699-05-7							0.1[z4]; 0.2[h]
cis-α-Bergamotene	18252-46-5							0.1[l]
trans-α-Bergamotene	13474-59-4							tr[j]; 0.1[l,s]
Bicyclogermacrene	24703-35-3							tr[z1]; 0.1[g,z6]
Bicyclo[2.2.1]heptane	279-23-2					0.1		
Bicyclo[2.2.1]heptan-2-ol	1632-68-4					0.4		
Bicyclo[2.2.1]heptan-2-one	497-38-1					0.09		
epi-Bicyclosesquiphellandrene	54274-73-6							0.3[i]
β-Bisabolene	495-61-4					0.2		tr[z1]; 0.1[l,s]; 0.2[i]
Borneol	507-70-0		0.1-1.2	0.8	0.4			tr[j,z1]; 0.2[b]; 4.2[z]
Bornyl acetate	76-49-3							tr[b,z1]; 0.2[c]
β-Bourbonene	5208-59-3	0-0.8						0.2[g]; 0.4[i]; 1.1[z2]; 2.8[z1]
2-Butanone	78-93-3						0.1	
α-Cadinene	24406-05-1				0.8			tr[g,z2,z6]
β-Cadinene	523-47-7							0.7[n]
γ-Cadinene	39029-41-9	0.07-2.4	0.2-1.4	0.3	0.4		tr	0.3[h]; 0.5[z2]; 0.9[z4]; 1.4[n]
δ-Cadinene	483-76-1	0.08-2.6	0.1-1.2	0.3			0.1	0.2[c]; 0.8[z9]; 1.0[l]; 1.8[g]; 2.6[x]
ε-Cadinene	1080-67-7							0.4[h]
α-Cadinol	481-34-5	0.07-0.4					0.2	0.4[z6]; 1.5[g]; 1.7[92]; 2.0[z5]
epi-α-Cadinol	5937-11-1	0.05-0.7						0.1[z6]; 0.2[z5]; 0.3[g]; 0.6[z2]
δ-Cadinol	19435-97-3	0.09-0.9						0.2[g]
Calamenene	483-77-2							0.06[z4]
trans-Calamenene	73209-42-4						tr	
Camphene	79-92-5		0.3-2.1	1.0	1.9		tr	tr[j,z1]; 0.1[e]; 0.2[c]
Camphor	76-22-2		2.0-4.3	0.2	0.8		tr	0.4[z7]; 1.1[b]
δ3-Carene	13466-78-9							0.3[c]; 3.8[m]; 5.0[x]
Carvacrol	499-75-2					1.0	tr	2.8[y]; 4.1[i]; 13.1[b]; 31.8[t]
Carvacryl acetate	6380-28-5							<0.1[i]
Carveol	99-48-9		0.1-1.1	2.2	2.2			
(E)-Carveol	1197-07-5							tr[j]; 0.2[i]
(Z)-Carveol	1197-06-4					0.3		
L-Carvone	6485-40-1							1.5[i]
Carvyl acetate	97-42-7		0.1-2.3	0.9	1.6			
cis-Carvyl acetate	1205-42-1							0.1[c]
Caryophylla-4,8-dien-5-ol	423765-30-4							2.4[j]; 2.9[i]
Caryophylla-4(14),8 (15)-dien-5α-ol								0.5[z9]
β-Caryophyllene	87-44-5	0.7-29.4	0.09-1.8	0.7	6.8	7.3	1.5	11.2[p]; 12.2[n]; 13.2[x]; 15.3[z1]; 16.6[z9]; 21.7[z5]
γ-Caryophyllene	118-65-0							tr[z2]
Caryophyllene oxide	1139-30-6	0.09-2.3	0.1-0.7	0.2	7.3		12.1	10.0[e]; 12.4[i]; 13.2[j]; 13.5[u]; 23.5[z2]; 24.4[z1]; 25.3[n]

Table 5.57.1 Constituents identified in melissa oils (*continued*)

Constituent	CAS	A	B (3)	C (4)	D (5)	E (26)	F (6)	G
Caryophyllene oxide II								0.5[g]
Caryophyllenol	38284-26-3							tr[w]
Caryophyllenol II	32214-89-4							5.0[z1]
Cedrane	13567-54-9							14.1[k]
5-Cedranone								0.2[u]
α-Cedrene	469-61-4							tr[z2]
β-Cedrene	546-28-1							0.3[b]
trans-Chrysanthemal	20104-05-6					0.3		
cis-Chrysanthenol	55722-60-6							1.3[z9]; 1.7[z7]
1,8-Cineole	470-82-6	0-0.1				0.2	tr	tr[z1]; 0.1[w]; 0.2[b]; 0.9[o]; 2.0[j]
Citral	5392-40-5					17.4		
Citronellal	106-23-0	0.5-29.2	27.7-42.6	37.8	37.8	43.8	0.2	19.2[z5]; 20.3[z]; 23[z6]; 36.2[g]; 39.6[b]; 43.8[t]; 46.8[n]; 55.8[z3]
Citronellic acid	502-47-6							tr[z6]; 0.3[g]
Citronellol	106-22-9	0.05-13.1	10.1-18.3	18.4	18.3	0.6	tr	6.2[b,j]; 15.0[z2]; 19.5[s]; 21.8[m]
Citronellyl acetate	150-84-5	0.02-3.9					tr	1.4[s]; 1.6[b]; 1.8[l]; 3.7[x]; 4.0[h]
Citronellyl butyrate	141-16-2				0.1			
Citronellyl formate	105-85-1							0.2[u]
α-Copaene	3856-25-5	0.06-2.8	0.01-2.8	0.8	0.5		tr	0.8[z2]; 1.1[z5]; 1.4[x]; 1.8[z7]
β-Copaene	18252-44-3							tr[j]; 0.1[z4,z9]
cis-Copaen-4α-ol								0.6[i]
Coumarin	91-64-5						tr	
α-Cubebene	17699-14-8		0.1-2.2	0.4	0.7			0.2[s]; 0.3[w]; 0.9[26]; 1.1[m]
β-Cubebene	13744-15-5		0.1-0.6	3.8	0.6			0.1[z7]; 0.3[g]; 0.6[z2]
Cubenene	29837-12-5	0.02-0.9						tr[g,z6]
Cubenol	21284-22-0							0.1[z2]
Cuminaldehyde	122-03-2							tr[j]
Curcumene	644-30-4						tr	
β-Cyclocitral	432-25-7							0.1[z2]
o-Cymene	527-84-4							2.3[b]
p-Cymene	99-87-6	0-0.4				0.2	tr	0.8[e]; 3.6[m]; 4.0[z1]; 6.7[j]
p-Cymenene	1195-32-0						tr	tr[j]
p-Cymen-8-ol	1197-01-9						tr	tr[b,j]
(*E*)-β-Damascenone	23726-93-4							0.06[w]; 0.1[c]; 0.2[i]; 2.8[z1]
(*Z*)-β-Damascenone	59739-63-8		0.1-1.2	0.6	0.8			tr[j]
(*E,E*)-2,4-Decadienal	25152-84-5							
Decahydronaphthalene	91-17-8							0.2[i]
Decanal	112-31-2						tr	
Decane	124-18-5							0.5[k]
Decyl acetate	112-17-4							1.2[k]
Dehydro-1,8-cineole	92760-25-3						tr	
2,5-Diethyltetrahydrofuran	41239-48-9							0.08[w]
Dihydrocarveol	38049-26-2							2.9[j]
Dihydrocitronellyl acetate	20780-49-8							0.3[z7]
1,2-Dihydrotrimethylnaphthalene	133439-48-2							0.06[w]
2,6-Dimethyl-5-heptanal						0.2		
6,10-Dimethyl-3-(1-methylethylidene)-1-cyclodecene	69239-71-0							tr[w]
3,7-Dimethyl-2,6- octadien-1-ol (citrol)	624-15-7							5.7[k]
Diosphenol	490-03-9							0.1[i]
Dodecanal	112-54-9		0.01-1.1	0.2	2.1			

Table 5.57.1 Constituents identified in melissa oils (*continued*)

Constituent	CAS	Percentage and range						
		A	B (3)	C (4)	D (5)	E (26)	F (6)	G
Dodecenal	82107-89-9							0.4[c]
(*E*)-2-Dodecenal	20407-84-5							0.1[z8]
11,14,17-Eicosatrienoic acid, methyl ester	55682-88-7							tr[w]
Eicosane	112-95-8			3.1				0.5[c]; 0.6[z7]; 4.6[k]
1-Eicosene	3452-07-1							0.1[c]
β-Elemene	33880-83-0	0-4.7					tr	0.3[i,m]; 0.4[g]; 0.6[b,h]; <1.0[f]
Elemicin	487-11-6							0.2[z2]
Elemol (α-)	639-99-6							0.8[s]; <1.0[f]; 1.0[h]; 1.1[l]
2*H*-1,4a-Ethanonaphthalen-1-ol								0.3[i]
Ethyl ner(ol)ate	32659-20-4							0.1[z8]
Eugenol	97-53-0						tr	0.5[i]; 0.6[b,s]; 0.8[l]; 1.9[m]
(*E,E*)-Farnesal	502-67-0							0.2[z9]
α-Farnesene	502-61-4	0.2-9.1						2.3[x]
(*E,E*)-α-Farnesene	502-61-4							tr[j]; 0.1[z8]; 0.4[z6]; 0.5[g]
(*E*)-β-Farnesene	18794-84-8	0.06-2.4						0.3[g,i]; 0.4[z6]
(*Z*)-β-Farnesene	28973-97-9		0.2-0.8	0.3	0.4			0.1[c]
(*Z,E,E*)-allo-Farnesene								tr[z6]
(*Z,Z*)-Farnesol	16106-95-9							3.0[z2]
(*E,E*)-Farnesyl acetate	4128-17-0							0.3[d]
(*Z,E*)-Farnesyl acetate	40266-29-3							0.2[d]
Fenchyl acetate	13851-11-1		0.1-3.9	1.0	9.6			
Geranial	141-27-5	0.2-41.7	0.3-4.2	1.8	1.4		39.0	39.9[o]; 41.0[d]; 41.1[z4]; 43.1[v]; 43.2[z8]; 44.2[z7]; 74.1[z]
Geranic acid	459-80-3						tr	tr[z6]; 1.1[w]
(*E*)-Geranic acid	4698-08-2							tr[g]
Geraniol	106-24-1	0.2-22.2	0.9-3.9	1.6	2.2		1.0	3.4[z6]; 4.8[s]; 5.1[l]; 5.3[u]; 5.7[b]; 8.1[e]; 10.3[o]; 23.2[z]; 25.0[y]
Geranyl acetate	105-87-3	0-7.9	7.7-17.4	0.4	15.5		5.3	2.3[u]; 2.9[v]; 3.3[e]; 4.4[d]; 5.4[y]; 5.9[w]; 6.3[s]; 7.7[l]; 8.7[o]; 25[z5]
(*E*)-Geranylacetone	3796-70-1							0.2[w]; 0.3[i]
Geranyl formate	105-86-2						tr	0.2[d,s]; 0.8[i]; 1.5[l]
Geranyllinalool	77368-82-2							0.2[v]
1(10),5-Germacradien-4-ol	74841-87-5	0.1-3.6						2.3[z6]; 9.6[z5]
[1(10)*E*,5*E*]-Germacradien-4α-ol	207221-31-6							2.9[g]
Germacrene B	15423-57-1							0.1[a,z2]
Germacrene D	23986-74-5	0.2-24.0	0.3-2.1	3.1	0.7	0.3		5.1[z9]; 8.3[z]; 10.4[z6]; 13.5[g]; 14.0[z1]; 32.5[z5]
Germacrene D-4-ol	198991-79-6							2.5[z9]
Globulol	489-41-8							6.8[t,u]
β-Guaiene	88-84-6							tr[g]
cis-β-Guaiene	372162-07-7							0.2[b]
trans-β-Guaiene	192053-49-9							tr[z1]
α-Gurjunene	489-40-7							0.4[b]
β-Gurjunene	73464-47-8							tr[z6]; 0.2[z2]
γ-Gurjunene	22567-17-5							0.2[b]
Heneicosane	629-94-7							0.3[c]; 0.4[z7]
Heptacosane	593-49-7							tr[w]
Heptadecane	629-78-7							0.8[v]
(*E,E*)-2,4-Heptadienal	4313-03-5							0.4[z2]
(*E,Z*)-2,4-Heptadienal	4313-02-4							0.6[z2]
5-Hepten-1-ol	89794-36-5				0.2			

Table 5.57.1 Constituents identified in melissa oils (*continued*)

Constituent	CAS	Percentage and range						
		A	B (3)	C (4)	D (5)	E (26)	F (6)	G
Hepten-2-one	30640-40-5							0.3[k]
1-Hepten-3-one	2918-13-0							tr[b]
Hexadecanal	629-80-1							7.4[k]
Hexadecanoic acid	57-10-3							0.6[v]; 1.0[w]
1-Hexadecene	629-73-2							<0.1[z7]
Hexahydrofarnesyl acetate	99624-94-9							0.1[n]
Hexahydrofarnesyl acetone	502-69-2							tr[w]; 0.1[v]; 0.2[f,z9]; 2.2[z1]
1-Hexanol	111-27-3	0-0.09					tr	
Hexenal	1335-39-3							0.05[w]
(*E*)-2-Hexenal	6728-26-3							0.6[v]
3-Hexen-1-ol	544-12-7							1.0[z2]
(*Z*)-3-Hexen-1-ol	928-96-1	0-0.1					tr	tr[g]; <0.01[z4]; 0.2[z6]
(*Z*)-3-Hexenyl acetate	3681-71-8	0-0.1						
Hexyl acetate	142-92-7	0-0.6						
Hexyl tiglate	16930-96-4							0.2[l]
α-Humulene	6753-98-6	0.2-3.3	0.02-1.2	3.1	1.9	0.4	0.2	1.7[z9]; 1.9[x]; 2.2[h]; 10.2[t]
α-Humulene epoxide	96638-51-6							tr[z6]; 0.2[n]; 0.3[u]; 25.4[t]
α-Humulene oxide I	19888-33-6						tr	
α-Humulene oxide II	19888-34-7						0.4	tr[g]
14-Hydroxy-(*E*)-caryophyllene								tr[z7]; 1.2[j]
14-Hydroxy-(*Z*)-caryophyllene								tr[j]
14-Hydroxy-9-epi-(*E*)-caryophyllene	244226-09-3							0.1[d]; 0.3[z9]
4-Hydroxy-4-methyl-2-pentanone	123-42-2							2.8[k]
β-Ionene	84607-57-8					0.09		0.5[i]
β-Ionone	79-77-6							0.1[a,z2]
(*E*)-β-Ionone	79-77-6						tr	tr[z1,z7]; 0.09[n]; 0.1[v]
Isoborneol	124-76-5					0.2		0.6[b]
Isobornyl acetate	125-12-2							tr[b,j]
Isocaryophyllene oxide								tr[z6]; 0.5[z5]
(*Z*)-Isocitral	72203-97-5	0.1-1.4						0.3[z8]
Isogeranial	72203-98-6	0.09-1.1						0.2[g]; 0.3[z6]; 0.5[z8]; 1.2[v]
Isoledene	95910-36-4							tr[b]
trans-Isolimonene	6876-12-6							0.7[c]
Isomenthol	3623-52-7							1.4[v]; 2.4[z7]
Isomenthone	491-07-6	0-0.3	0.2-2.9	0.9	1.2			0.4[w]; 3.0[c]; 8.9[b]
Isopinocamphone	15358-88-0							tr[b]
Isopulegol	89-79-2	tr-1.1				0.5		0.9[g]; 1.9[q]; 2.0[z4]; 2.8[s]; 3.5[l]
Khusinol	24268-34-6							0.4[z9]
Ledol	577-27-5		0.2-0.5	3.2	1.5			0.2[c]
Limonene	138-86-3	0.06-3.9	1.2-6.6	16.8	11.6	0.06	tr	5.8[l]; 23.0[r]; 39.4[m]; 57.5[q]
cis-Limonene epoxide	13837-75-7							0.4[c]
Limonene oxide	1195-92-2		0.3-1.5	0.4	0.7			
trans-Limonene oxide	4959-35-7							<0.1[z7]
Linalool	78-70-6	0.1-2.8	0.7-2.7	19.5	6.9	3.7	1.6	2.7[x]; 3.2[z3]; 4.5[l]; 5.1[z1]; 9.0[z]
Linalool oxide	1365-19-1							0.2[w]
cis-Linalool oxide, furanoid	11063-77-7						0.1	0.3[d,z2]
trans-Linalool oxide, furanoid	34995-77-2						0.1	0.1[l]; 0.2[z2]; 0.4[d]
trans-Linalool oxide, pyranoid	39028-58-5						tr	
Linalyl acetate	115-95-7							2.3[b]
Longifolene (junipene)	475-20-7							0.6[l]; 0.9[b]
Manoyl oxide	596-84-9							3.0[z1]
p-Mentha-1,5-dien-8-ol	1686-20-0							tr[z1]

Table 5.57.1 Constituents identified in melissa oils (*continued*)

Constituent	CAS	Percentage and range						
		A	B (3)	C (4)	D (5)	E (26)	F (6)	G
trans-p-Mentha-1(7), 8-dien-2-ol	21391-84-4						1.2	1.7[z9]
cis-p-Mentha-2,8-dien-1-ol	3886-78-0						0.1	
trans-p-Mentha-2,8-dien-1-ol	4017-77-0						0.1	
p-Mentha-1,3,8-triene	18368-95-1					12.7		
cis-p-Menth-2-en-1-ol	29803-82-5							tr[z1]; 1.1[j]
Menthol	89-78-1	0-0.3	0.7-2.9	0.8	1.3			0.1[w]; 0.3[z7]; 2.9[c]
Menthone (*p*-)	89-80-5	0-2.1						0.3[s]; 0.6[w]
Menthyl acetate	16409-45-3							0.3[w]
Methyl chavicol	140-67-0		0.1-2.3	0.5	1.2			0.1[c,z8]; 7.5[z2]
Methyl citronellate	2270-60-2	0.03-19.2						2.2[z6]; 2.7[c]; 4.6[n]; 4.9[g]
Methyl decanoate	110-42-9							0.9[y]
Methyl eugenol	93-15-2							tr[b]; 0.2[z4]; 0.3[z2]
Methyl geranate	2349-14-6	0.1-1.3					0.2	0.3[v]; 0.4[g,w]; 0.7[n]; 1.3[x]
6-Methyl-3,5-heptadien-2-one	1604-28-0						tr	tr[z2]
5-Methyl-3-heptanone	541-85-5							0.6[n]
6-Methyl-5-hepten-2-ol	1569-60-4						tr	0.09[z4]; 3.8[e]
Methylheptenone	409-02-9	0.09-3.8						+[z3]
6-Methyl-5-hepten-2-one	110-93-0						3.8	2.5[d]; 3.1[h]; 4.5[f]; 5.1[y] 3.7[x]
3-Methyl-2-(2-methyl-2-butenyl)	`					0.1		
1-Methyl-2(3,4)-(1-methylethyl)-benzene								0.3[w]
Methyl salicylate	119-36-8							0.2[z2]
cis-Muurola-4(14), 5-diene	157477-72-0							2.3[b]
α-Muurolene	10208-80-7					0.7	tr	tr[j]; 0.2[s,w]; 0.3[g,l]; 2.3[z4]
γ-Muurolene	30021-74-0							tr[j,z2]; 0.1[l,s]; 0.3[w]
Muurolol	119757-72-1	0.05-1.2						
α-Muurolol	104245-48-9						tr	tr[z6]; 0.1[n]; 0.5[z5]
τ-Muurolol	19912-62-0							0.3[z5]; 0.5[g]; 0.9[z9]; 2.0[i]
Myrcene	123-35-3	tr-0.4	0.2-5.7	2.0	2.3	0.07	tr	0.1[b,g]; 0.2[e,n]; 0.5[m]
Myrcenone	539-70-8							0.3
Myrtenal	564-94-3							2.5[z1]
Myrtenol	515-00-4							2.2[i]; 2.4[j]; 2.8[x]
Neophytadiene	504-96-1					0.3		
Neral	106-26-3	0.5-31.0	0.1-2.2	1.6	4.2		30.4	28.8[z4]; 29.9[d]; 30.2[z7]; 32.1[z]; 33.4[v]; 39.3[h]; 43.8[u]
Neric acid	37349-29-4						tr	
Nerol	106-25-2	0.1-16.2	0.2-2.7	5.6	5.6			1.3[e]; 2.4[z4]; 4.1[h]; 4.3[q]; 4.7[r]
(*E*)-Nerolidol	40716-66-3							tr[j]; <0.1[z7]; 4.5[o]
(*Z*)-Nerolidol	3790-78-1						tr	0.2[a,z2]
Nerol oxide	1786-08-9						tr	1.1[z4]
Neryl acetate	141-12-8	0-2.9					tr	0.6[m]; 0.8[d]; 1.6[c]; 4.0[e]; 4.8[p]
Nonadecane	629-92-5			0.2				0.2[c]
Nonanal	124-19-6	0-0.1						0.1[z7]
Ocimene	13877-91-3							0.1[z2]; 0.3[w]; 1.7[p]
(*E*)-β-Ocimene	3779-61-1	0.05-9.3					tr	0.5[z4]; 3.1[g]; 4.6[z6]; 5.8[z5]
(*Z*)-β-Ocimene	3338-55-4	0.1-1.0	0.1-0.5	0.2	2.2		tr	0.2[c]; 0.3[g]; 0.6[x]; 0.7[z5]; 2.0[o]
(2*Z*),(4*E*),(6*E*)-allo-Ocimene	3016-19-1							tr[g]
Octadecanoic acid	57-11-4							3.3[k]
2,6-Octadienoic acid	83592-56-7					0.6		
3,6-Octadienoic acid	70080-68-1					0.2		
Octanal	124-13-0						tr	0.1[e]
3-Octanal								0.1[w]
2,3-Octanedione	585-25-1							tr[w]

Table 5.57.1 Constituents identified in melissa oils (*continued*)

Constituent	CAS	Percentage and range						
		A	B (3)	C (4)	D (5)	E (26)	F (6)	G
3-Octanol	589-98-0	0-0.1					tr	tr[j]; 0.1[g,z6]; 0.2[x]; 0.4[c]
2-Octanone	111-13-7							0.1[l]
3-Octanone	106-68-3	0-0.3						tr[z6]; 0.1[g,w,z4]
1,3,5-Octatriene	26555-19-1							tr[w]
1,3,6-Octatriene	929-20-4				0.3			
3-Octenal	60671-71-8			2.3				
1-Octen-3-ol	3391-86-4	0.03-1.0				0.4	0.2	0.6[z9]; 0.7[g]; 0.9[n,x]; 1.0[z2]
4-Octen-3-one	14129-48-7							0.2[w]
3-Octyne	15232-76-5							1.0[k]
Pentacosane	629-99-2							0.5[c]
Pentadecanal	2765-11-9							4.7[v]
α-Phellandrene	99-83-2	0-0.2	0.2-0.5	4.6	3.9			tr[b,j,z1]; 1.0[z8]
β-Phellandrene	555-10-2	0-0.2						0.3[b]
Phenylacetaldehyde	122-78-1						0.2	tr[z1]; 0.03[z9]; 0.8[z2]
Phytol	7541-49-3							0.2[i]
α-Pinene	80-56-8	0-0.6	0.3-0.6	3.1	0.9		tr	6.9[z1]; 7.1[i]; 12.1[j]; 12.8[z]
β-Pinene	127-91-3	0-0.4	0.2-0.6	3.1	2.0			1.6[i]; 3.6[z5]; 18.2[z1]; 44.5[z]
trans-Pinocamphone	547-60-4							0.4[b]
trans-Pinocarveol	1674-08-4							1.1[i]; 2.1[z1]; 2.8[j]
Pinocarvone	30460-92-5							tr[b,j]; 1.5[z1]
trans-Piperitol	16721-39-4							tr[j]
Piperitone	89-81-6						0.8	0.1[l,s]; 0.2[i]; 2.6[z4]
Piperitone oxide	148879-33-8							0.3[i]
cis-Piperitone oxide	57130-28-6							1.0[i]
Pulegol	529-02-2	0-0.6						0.2[z4]
Rimuene	1686-67-5							0.2[i]
Rosefuran	15186-51-3							tr[v]
Rosefuran epoxide	92356-06-4						tr	0.2[v]
cis-Rose oxide	3033-23-6	0.02-0.4	0.1-2.7			0.2	tr	0.1[g,z6,z7]; 0.2[z2]; 0.5[x]
trans-Rose oxide	5258-11-7	0.03-0.5		12.9	1.7		tr	0.05[g]; 0.1[z7]; 0.3[c]; 0.4[z2]
Sabinene	3387-41-5	0-0.02	0.4-0.9	0.4	0.9			1.3[i]; 5.8[z5]; 12.9[j]; 17.4[z1]
cis-Sabinene hydrate	15537-55-0							0.5[i]; 0.8[j]; 1.0[z1]
trans-Sabinene hydrate	17699-16-0							tr[j]; 2.1[z1]; 7.8[k]
Safranal	116-26-7							<0.01[z4]
7-epi-α-Selinene	123123-37-5							0.6[b]
β-Selinene	17066-67-0		0.01-1.2	0.3	0.4			0.1[c]; 0.6[i]
Spathulenol	6750-60-3							tr[z2]
α-Terpinene	99-86-5		0.3-4.6	21.9	8.2			tr[j]; 0.2[b]; 0.6[i]; 4.9[z1]
γ-Terpinene	99-85-4	tr-0.3				0.3	tr	0.4[b]; 0.7[m]; 0.8[j]; 1.4[i]; 8.2[z1]
Terpinen-4-ol	562-74-3						tr	0.1[b,l]; 4.0[i]; 7.7[z1]; 15.8[j]
1-Terpineol	586-82-3							1.4[e]
α-Terpineol	98-55-5						0.5	0.1[b,s]; 1.4[e]
δ-Terpineol	7299-42-5							<0.01[z4]
Terpinolene (α-)	586-62-9	0-0.6						tr[j]; 0.1[b,l]; 0.4[i]; 2.1[z1]
α-Terpinyl acetate	80-26-2							0.2[z2]
Tetradecanoic acid	544-63-8							0.8[w]
1,2,3,4-Tetrahydro-1,1,6-trimethyl-naphthalene	475-03-6							0.08[w]
4,8,12,16-Tetramethyl-heptadecan 4-olide	200272-61-3							tr[w]
2,2,8,8-Tetramethyl-5-nonanone	5709-95-5							12.6[k]
α-Thujene	2867-05-2							0.1[b]; 1.7[j]; 5.3[z1]
α-Thujone	546-80-5		0.2-1.2	0.2	1.2			0.1[z8]; 0.2[c]

Table 5.57.1 Constituents identified in melissa oils (*continued*)

Constituent	CAS	Percentage and range						
		A	B (3)	C (4)	D (5)	E (26)	F (6)	G
β-Thujone (*cis-*)	471-15-8		0.2-0.5	3.1	1.0	0.4		tr[b]; 0.2[z9]; 0.8[c]
β-Thujone (*trans-*)	33766-30-2						tr	tr[z1]; 0.6[z8]
Thymol	89-83-8					11.9		0.1[b]; 0.7[i]; 3.7[r]; 7.9[u]; 10.5[t]
ar-Turmerone	532-65-0							0.4[d]
Valencene	4630-07-3							0.1[z7]
Verbenene	4080-46-0							tr[b]
cis-Verbenol	1845-30-3	0-0.8						0.7[x]
trans-Verbenol	1820-09-3	tr-0.9						tr[j,z1]
Verbenone	80-57-9							tr[j]; 0.2[i]
Viridiflorene (ledene)	21747-46-6							0.8[z9]

A fifty-three melissa essential oil samples from Germany, Algeria, France, Austria, USA and Australia analyzed between 1998 and 2013; lowest and highest concentrations given (E. Schmidt, unpublished data)
B twenty oils obtained from the aerial parts of *Melissa officinalis* cultivated in Shanghai, China, harvested at different times of the day; 10 oils were from the first and 10 oils from the second cutting; lowest and highest concentrations given (ref. 3)
C six oils obtained from the aerial parts of *Melissa officinalis* cultivated in Shanghai, China, harvested in two years and obtained by three different methods: hydrodistillation, hydro-steam distillation and steam distillation; highest concentrations given (ref. 4)
D eight oils obtained from the aerial parts of *Melissa officinalis* cultivated in Shanghai, China; two harvests, four methods of drying (fresh, shade-dried, sun-dried and oven-dried); highest concentrations given (ref. 5)
E three oils of lemon balm aerial parts obtained from three lines of *M. officinalis* cultivated in Turkey; highest concentrations given (ref. 26)
F one lab-hydrodistilled oil from *Melissa officinalis* cultivated in Cuba (ref. 6)
G data from other studies (indicated with superscript letters); highest concentrations found in any study reviewed here given; when two or more oils were investigated, only the highest concentrations are mentioned, unless indicated otherwise

[a] incorrect identification (ref. 37); [b] one commercial lemon balm oil from an Italian producer; mean of three analyses (ref. 23); [c] one lab-hydrodistilled lemon balm oil from *M. officinalis* aerial parts from Serbia (ref. 16); [d] one lab-hydrodistilled oil from lemon balm cultivated in Cuba (ref. 7); [e] two oils from fresh and air-dried aerial parts of *Melissa* plants at the beginning of the blooming stage, cultivated in Egypt (ref. 9); [f] one sample of melissa oil (ref. 11); [g] one sample of melissa oil (ref. 12); [h] one sample of melissa oil produced in India (ref. 13); [i] one lab-hydrodistilled oil from *M. officinalis* aerial parts at the flowering stage growing wild in Greece (ref. 17); [j] one lab-hydrodistilled oil from *M. officinalis* aerial parts at the flowering stage growing wild in Greece (ref. 18); [k] one lab-hydrodistilled oil from *M. officinalis* aerial parts growing wild in Iran (ref. 1); [l] one commercial lemon balm oil obtained in Italy (ref. 21); [m] one commercial lemon balm oil purchased in Austria (ref. 20); [n] various oils obtained from *M. officinalis* in different life cycles in Poland (ref. 28); [o] various oils obtained from *M. officinalis* in India under different fertilization regimens (ref. 29); [p] one German commercial sample of melissa oil (ref. 30); [q] an oil of *Melissa officinalis* produced in the laboratory from plants grown in Scotland (ref. 32); [r] several lab-produced oils from Italy (ref. 33); [s] one commercial melissa oil (ref. 34); [t] samples of melissa oils from various parts of Iran (ref. 36); [u] one lab-hydrodistilled leaf oil from plants collected in the wild in Iran (ref. 27); [v] one lab-hydrodistilled oil from the stems and leaves of *M. officinalis* grown in a greenhouse in Iran (ref. 25); [w] one lab-hydrodistilled leaf oil from melissa cultivated in the Slovak Republic (ref. 8); [x] one lab-hydrodistilled oil from the leaves of *M. officinalis* cultivated in India (ref. 14); [y] two lab-distilled oils from the leaves of *M. officinalis* from Iran before and during flowering with a very atypical composition (ref. 15); [z] 22 lab-hydrodistilled oils of *M. officinalis* leaves cultivated in two ecologically different locations in Turkey and harvested in 3 years; the eleven different seeds from which the plants were cultivated came from Europe (n=8) and Turkey (n=3); many were collected from wild populations; 7 out of 8 chemicals investigated were not found in all 22 samples and, consequently, their lowest value was zero; only geranial was present in all samples, lowest concentration was 4.5, highest 74.1 (ref. 24); [z1] three lab-hydrodistilled leaf oils from three wild populations collected at different locations in Greece with a very atypical composition (ref. 35); [z2] one lab-hydrodistilled oil from Lithuania (ref. 38); [z3] one melissa oil from India (ref. 39); [z4] two oils of cultivated Sardinian (Italy) *M. officinalis* ssp. *officinalis* (synonym of *M. officinalis* L. according to The Plant List) (ref. 40); [z5] four oils from German melissa, one harvested before flowering, one during and two after the flowering period (ref. 41); [z6] one oil from the whole *M. officinalis* plant from Germany; the oils of leaves from different stem positions were also analyzed and were quite different, also from each other (ref. 42); [z7] one lab-hydrodistilled oil from the dried leaves of *Algerian* (wild or cultivated?) *Melissa officinalis* (ref. 43); [z8] one lab-hydrodistilled oil from the aerial parts of melissa growing wild in Tajikistan (ref. 44); [z9] one lab-hydrodistilled oil from the flowering aerial parts of *M. officinalis* from Iran (cultivated or wild?) (ref. 45)

tr: trace (in columns F, G[b] and G[i]: <0.1; in columns G[w] and G[z1]: <0.05); + present in the oil investigated, but quantity not stated

Table 5.57.2 Major constituents of melissa oils

Constituent	CAS	Percentage and range						
		A	B (3)	C (4)	D (5)	E (26)	F (6)	G
Geranial	141-27-5	0.2-41.7	0.3-4.2	1.8	1.4		39.0	43.1[v]; 43.2[z8]; 44.2[z7]; 74.1[z]
Limonene	138-86-3	0.06-3.9	1.2-6.6	16.8	11.6	0.06	tr	5.8[l]; 23.0[r]; 39.4[m]; 57.5[q]
Citronellal	106-23-0	0.5-29.2	27.7-42.6	37.8	37.8	43.8	0.2	39.6[b]; 43.8[t]; 46.8[n]; 55.8[z3]
Neral	106-26-3	0.5-31.0	0.1-2.2	1.6	4.2		30.4	32.1[z]; 33.4[v]; 39.3[h]; 43.8[u]
Germacrene D	23986-74-5	0.2-24.0	0.3-2.1	3.1	0.7	0.3		10.4[z6]; 13.5[g]; 14[z1]; 32.5[z5]
β-Caryophyllene	87-44-5	0.7-29.4	0.09-1.8	0.7	6.8	7.3	1.5	15.3[z1]; 16.6[z9]; 21.7[z5]
Caryophyllene oxide	1139-30-6	0.09-2.3	0.1-0.7	0.2	7.3		12.1	13.5[u]; 23.5[z2]; 24.4[z1]; 25.3[n]
Geraniol	106-24-1	0.2-22.2	0.9-3.9	1.6	2.2		1.0	8.1[e]; 10.3[o]; 23.2[z]; 25.0[y]
Geranyl acetate	105-87-3	0-7.9	7.7-17.4	0.4	15.5		5.3	5.9[w]; 6.3[s]; 7.7[l]; 8.7[o]; 25[z5]
Citronellol	106-22-9	0.05-13.1	10.1-18.3	18.4	18.3	0.6	tr	6.2[b,j]; 15.0[z2]; 19.5[s]; 21.8[m]
Linalool	78-70-6	0.1-2.8	0.7-2.7	19.5	6.9	3.7	1.6	2.7[x]; 3.2[z3]; 4.5[l]; 5.1[z1]; 9.0[z]
Nerol	106-25-2	0.1-16.2	0.2-2.7	5.6	5.6			1.3[e]; 2.4[z4]; 4.1[h]; 4.3[q]; 4.7[r]

LEGEND: SEE UNDER TABLE 5.57.1

(0.2-24.0%), geraniol (0.2-22.2%), methyl citronellate (0.03-19.2%), nerol (0.1-16.2%), citronellol (0.05-13.1%) and (E)-β-ocimene (0.05-9.3%) (Erich Schmidt, unpublished analytical data).

Chemotypes

There appear to be 2 chemotypes of melissa oil. Most oils are dominated by geranial and neral; their combined share in the oil is usually in the 50-70% range (8,11,13,25,27,29,30,40,43,44,45). The second chemotype has citronellal as the major ingredient (19-43%) (3,4,5,12,21,23,24,26,34,41), sometimes combined with citronellol in the range of 5-18% (3,4,21,34). Rarely, limonene was found to be the dominant ingredient in a melissa oil sample (32).

CONTACT ALLERGY/ALLERGIC CONTACT DERMATITIS

General

Only two case reports of allergic contact dermatitis from melissa oils have been found. In neither, false-positive patch tests due to the excited skin syndrome can be excluded. In one case geraniol and caryophyllene may have been allergens in melissa oil.

Case reports

An aromatherapist had non-occupational contact dermatitis with allergies to multiple essential oils used at work, including 'melissa blend'. The patient also reacted to geraniol, linalool, linalyl acetate, α-pinene, the fragrance mix and various other fragrance materials. Linalool, geraniol and caryophyllene were demonstrated by GC-MS in the melissa blend (48). High concentrations of geraniol (up to 22.2%) and caryophyllene (up to 29.4%) have been found in commercial melissa oils (Table 5.57.1, column A). A female 'complementary therapist' developed occupational contact dermatitis from a multitude of essential oils used at work, including melissa oil (49).

LITERATURE

1   Esmaeili A, Rohani S. The in vitro antioxidative properties and essential oil composition of Melissa officinalis L. J Essent Oil Bear Plants 2012;15:868-875

2   Moradkhani H, Sargsyan E, Bibak H, Naseri B, Sadat-Hosseini M, Fayazi-Barjin A, Meftahizade H. Melissa officinalis L., a valuable medicine plant: A review. J Med Plants Res 2010;4:2753-2759

3   Khalid KA, Hu W, Cai W, Hussien MS. Influence of cutting and harvest day time on the essential oils of lemon balm (Melissa officinalis L.). J Essent Oil Bear Plants 2009;12:348-357

4   Khalid KA, Cai W, Ahmed AMA. Effect of harvesting treatments and distillation methods on the essential oil of lemon balm and apple geranium plants. J Essent Oil Bear Plants 2009;12:120-130

5   Khalid KA, Hu W, Cai W. The effects of harvesting and different drying methods on the essential oil composition of lemon balm (Melissa officinalis L.). J Essent Oil Bear Plants 2008;11:342-349

6   León-Fernández M, Sánchez-Govín E, Quijano-Celis CE, Pino JA. Effect of planting practice and harvest time in oil content and its composition in Melissa officinalis L. cultivated in Cuba. J Essent Oil Bear Plants 2008;11:62-68

7   Pino JA, Rosado A, Fuentes V. Composition of the essential oil of Melissa officinalis L. from Cuba. J Essent Oil Res 1999;11:363-364

8   Hollá M, Svajdlenka E, Tekel J, Vaverková S, Havránek E. Composition of the essential oil from Melissa officinalis L. cultivated in Slovak Republic. J Essent Oil Res 1997;9:481-484

9   Shalaby AS, El-Gengaihi S, Khattab M. Oil of Melissa officinalis L., as affected by storage and herb drying. J Essent Oil Res 1995;7:667-669

10 Lawrence BM. Progress in essential oils. Perfum Flavor 1989;14(May/June):71-80

11 Tittel G, Wagner H, Bos R. Über die chemische Zusammensetzung von Melissenölen. Planta Med 1982;46:91-98. Data cited in ref. 10

12 Schultze W, Zänglein A, Klose R, Kubeczka KH. Die Melisse. Dünnschichtchromatografische Untersuchung des ätherischen Öles. Deut Apoth Ztg 1989;129:155-163. Data cited in ref. 10

13 Nigam MC, Duhan SPS, Naqui AA. Terpenoid composition of essential oil of Melissa officinalis. PAFAI 1988;10:28-29. Data cited in ref. 10

14 Rehman S-U, Latief R, Bhat KA, Khuroo MA, Shawl AS, Chandra S. Comparative analysis of the aroma chemicals of Melissa officinalis using hydrodistillation and HS-SPME techniques. Arab J Chem (2013). Available at: http://dx.doi.org/10.1016/j.arabjc.2013.09.015

15 Saeb K, Gholamrezae S. Variation of essential oil composition of Melissa officinalis L. leaves during different stages of plant growth. Asian Pac J Trop Biomed 2012:S547-549

16 Mimica-Dukic N, Bozin B, Sokovic M, Simin N. Antimicrobial and antioxidant activities of Melissa officinalis L. (Lamiaceae) essential oil. J Agric Food Chem 2004;52:2485-2489

17 Ntalli NG, Ferrari F, Giannakou I, Menkissoglu-Spiroudi U. Phytochemistry and nematicidal activity of the essential oils from 8 Greek Lamiaceae aromatic plants and 13 terpene components. J Agric Food Chem 2010;58:7856-7863

18 Koliopoulos G, Pitarokili D, Kioulos E, Michaelakis A, Tzakou O. Chemical composition and larvicidal evaluation of Mentha, Salvia, and Melissa essential oils against the West Nile virus mosquito Culex pipiens. Parasitol Res 2010;107:327-335

19 Lemon balm. In: Charles DJ. Antioxidant properties of spices, herbs and other sources. New York, Heidelberg: Springer, 2013:371-376

20 Oberhofer B, Nikiforov A, Buchbauer G, Jirovetz L, Bicchi C. Investigation of the alteration of the composition of various essential oils used in aroma lamp applications. Flavour Fragr J 1999;14:293-299

21 Romeo FV, de Luca S, Piscopo A, Poiana M. Antimicrobial effects of some essential oils. J Essent Oil Res 2008;20:373-379

22 Lawrence BM. Progress in essential oils. Perfum Flavor 2008;33(September):66-? (last page unknown)

23 De Martino L, De Feo V, Nazzaro F. Chemical composition and in vitro antimicrobial and mutagenic activities of seven Lamiaceae essential oils. Molecules 2009;14:4213-4230

24 Sarý AO, Ceylan A. Yield characteristics and essential oil composition of lemon balm (Melissa officinalis L.) grown in the Aegean Region of Turkey. Turk J Agric Forest 2002;22:217-224

25 Sharafzadeh S, Khosh-Khui M, Javidnia K. Aroma profile of leaf and stem of lemon balm (Melissa Officinalis L.) grown under greenhouse conditions. Adv Environm Biol 2011;5:547-550

26 Cosge B, Ipek A, Gurbuz B. GC/MS analysis of herbage essential oil from lemon balms (Melissa officinalis L.) grown in Turkey. J Appl Biol Sci 2009;3:149-152

27 Meftahizade H, Sargsyan E. Investigation of antioxidant capacity of Melissa officinalis L. essential oils. J Med Plant Res 2010;4:1391-1395

28 Klimek B, Majda T, Gora J, Patoa J. Investigation of the essential oil from lemon catnip (Nepeta cataria var. citriodora) in comparison to the oil from lemon balm (Melissa officinalis L.). Herba Polon 2000;46:226-234. Data cited in ref. 22

29 Harshavardhan PG, Vsundhara M, Sirnivasappa KN, Biradar SL, Rao GGE, Gayithri HN. Effect of spacing and integral nutrient management on biomass and oil yield in Melissa officinalis L. Indian Perfum 2005;49:349-354. Data cited in ref. 22

30 Reichling J, Harkenthal M, Saller R. Wirking ausgewählter ätherischer Öle. Erfahrungsheilkunde 1999;6:357-366. Data cited in ref. 31

31 Lawrence BM. Progress in essential oils. Perfum Flavor 2011;36(April):52-59

32 Dorman HJD, Surai P, Deans SG. In vitro antioxidant activity of a number of plant essential oils and phytoconstituents. J Essent Oil Res 2000;12:241-248

33 Gionfriddo F, Mangiola C, Pirrello A, Manganaro R. Studio preliminare sulla composizione degli oli essenziali di alcane piante officinali calabresi. Essenz Deriv Agrum 2001;71:67-70. Data cited in ref. 31

34 Romeo RV, De Luca S, Piscopo A, De Salvo E, Poiana M. Effect of some essential oils as natural food preservatives on commercial grade carrots. J Essent Oil Res 2010;22:283-287

35 Basta A, Tzakou O, Couladis M. Composition of the leaves essential oil of Melissa officinalis s.l. from Greece. Flavour Fragr J 2005;20:642-644

36 Askari F, Sefidkon F. Essential oil composition of Melissa officinalis L. from different regions. Ir J Med Arom Plant Res 2004;20:229-239

37 Lawrence BM. Progress in essential oils. Perfum Flavor 1999;24(3):47-64

38 Venskutonis PR, Dapkevicius A, Baranauskiene M. Flavour composition of some lemon-like aroma herbs from Lithuania. In: G Charalambous, Ed. Food flavors: generation, analysis and process influence. Amsterdam: Elsevier Science BV, 1995:833-847. Data cited in ref. 37

39 Srivastava VK, Singh BM, Negi KS, Pant KC, Suneja P. Gas chromatographic examination of some aromatic plants of Uttar Pradesh hills. Indian Perfum 1997;41(4):129-139

40 Usai M, Atzei AD, Picci V, Furesti A. Essential oil from Sardinian Melissa officinalis L. and Melissa romana Mill. In: Ch Franz A Mathé , G Buchbauer, Eds. Proceedings of 27th International Symposium on Essential Oils, Vienna, 1996. Carol Stream, IL, USA: Allured Publishing, 1997:221-223. Data cited in ref. 37

41 Van den Berg T, Fruendl E, Czygan F-C. Melissa officinalis subsp. altissima: characteristics and possible adulteration of lemon balm. Pharmazie 1997;52:802-808. Data cited in ref. 37

42 Hose S, Zänglein A, van den Berg Th, Schulze W, Kubeczka K-H, Czygan F-C. Ontogenetic variation of the essential leaf oil of *Melissa officinalis* L. Pharmazie 1997;52:247-253. Data cited in ref. 37

43 Abdellatif F, Boudjella H, Zitouni A, Hassani A. Chemical composition and antimicrobial activity of the essential oil from leaves of Algerian *Melissa officinalis* L. EXCLI Journal 2014;13:772-781

44 Sharopov FS, Wink M, Khalifaev DR, Zhang H, Dosoky NS, Setzer WN. Composition and bioactivity of the essential oil of *Melissa officinalis* L. growing wild in Tajikistan. Int J Trad Nat Med 2013;2(2): 86-96

45 Najafian S. Rapid extraction and analysis of volatile oils components of *Melissa officinalis* using headspace and gas chromatography/mass spectrometry. J Herbs Spices Med Plants 2013;19:340-347

46 Rhind JP. Essential oils. A handbook for aromatherapy practice, 2nd Edition. London: Singing Dragon, 2012

47 Lawless J. The encyclopedia of essential oils, 2nd Edition. London: Harper Thorsons, 2014

48 Dharmagunawardena B, Takwale A, Sanders KJ, Cannan S, Roger A, Ilchyshyn A. Gas chromatography: an investigative tool in multiple allergies to essential oils. Contact Dermatitis 2002;47:288-292

49 Newsham J, Rai S, Williams JDL. Two cases of allergic contact dermatitis to neroli oil. Br J Dermatol 2011;165(Suppl.1):76

Chapter 5.58   MYRRH OIL

DEFINITION

Myrrh oil is the essential oil obtained from the wood exudate of the (African) myrrh, *Commiphora myrrha* (Nees) Engl. and related species such as *Commiphora molmol* Engl. ex Tschirch (actually a synonym for *Commiphora myrrha* (Nees) Engl.), *Commiphora gileadensis* L. (Mecca myrrh, balm-of-Gilead), and *Commiphora abyssinica* (O. Berg) Engl. (preferred name: *Commiphora habessinica* (O. Berg) Engl.), also called Abyssinian myrrh or Yemen myrrh.

INCI NOMENCLATURE

Description/definition: Commiphora myrrha gum oil is an essential oil obtained from the gum-resin of the myrrh, *Commiphora myrrha,* Burseraceae

INCI name EU: Commiphora myrrha gum oil (perfuming name, not an INCI name proper)

INCI name USA: Not in the Personal Care Products Council Ingredient Database

CAS registry number(s): 8016-37-3; 84929-26-0; 9000-45-7

EINECS number(s): 284-510-0

Description/definition: Commiphora abyssinica gum oil is an essential oil obtained from the gum-resin of the Yemen myrrh, *Commiphora abyssinica*, Burseraceae

INCI name EU: Commiphora abyssinica gum oil (perfuming name, not an INCI name proper)

INCI name USA: Not in the Personal Care Products Council Ingredient Database

CAS registry number(s): 9000-45-7

EINECS number(s): 232-543-6

ISO (INTERNATIONAL ORGANIZATION FOR STANDARDIZATION) STANDARD

Currently, there is no ISO standard for myrrh oil.

THE PLANT, THE OIL, AND THEIR USES

The *Commiphora* genus contains up to 200 species of often thorny shrubs or small- to medium-sized dioecious trees with a peeling, papery bark growing in sandy and rocky areas distributed across Africa (especially northern Africa: Somalia, Sudan, Ethiopia, Eritrea) and the Arabian peninsula (Yemen, Oman, Saudi Arabia), with four species also found in India (4, GRIN Taxonomy for Plants; www.cropwatch.org). Myrrh is the air-dried gum oleoresin principally obtained from the schizogenous gum-oleoresin cavities in the stem or branches of the small tree *Commiphora myrrha* (Nees) Engl. (which produces true myrrh, also called Somalia myrrh, heerabol myrrh) or other *Commiphora* spp. such as *C. abyssinica* (*habessinica*) (Berg.) Engl., *Commiphora gileadensis* L. (produces Mecca balsam, also called mecca myrrh, opobalsam, balsam of Gilead), *Commiphora schimperi* (O. Berg) Engl. (CAS 89997-88-6; EINECS 289-725-3) and *Commiphora wildii* (CAS 1082996-27-7). The myrrh balsam (oleoresin, gum) is obtained from these trees by incisions in the bark and is a yellowish exudate. On exposure to the air this exudate darkens, hardens and dries to rounded irregular tears, brownish-yellow to reddish-yellow in color. Myrrh comprises 30-60 per cent water-soluble gum, 25-40% alcohol-soluble resin and 3-8% essential oil and is used for perfume, medicinal purposes and as incense (9, www.cropwatch.org).

Myrrh is used in traditional Chinese and in Indian Ayurvedic medicine. In China, the resin and its products (essential oil, ethanol and ether extracts) are utilized for the treatment of skin ulcers, wounds, oral ulcers, pain, toothache, fracture, rheumatism, arthritis, inflammatory diseases, diseases caused by blood stagnation, dysmenorrhea, hemiplegia, and tumors (4). In western medicine, myrrh may be utilized as an antiseptic in mouthwashes, gargles, and toothpastes for prevention and treatment of gum disease. Myrrh can also be applied in liniments for bruises, aches, and sprains. The resin is said to have antitumor, anti-inflammatory, analgesic, antioxidant and antimicrobial properties (4,5). Medical uses of myrrh have been reviewed (5).

Myrrh oil is the essential oil obtained by hydro- or steam-distillation of myrrh wood exudate (resin). Myrrh essential oil has been reported to have anti-inflammatory and antischistosomal activity. Myrrh essential oil is not toxic and is used in aromatherapy, perfumery, to flavor cosmetics such as toothpaste and mouthwash and is a flavor in alcoholic drinks, soft drinks, and food (1,6).

CHEMICAL COMPOSITION

Myrrh oil is a yellowish-brown to brownish clear viscous liquid, which has a warm-balsamic, sweet and spicy aromatic odor. The yield of essential oil from the gum of *Commiphora myrrha* (Nees) Engl. generally varies from 2.5 to 8.0 per cent. The main producing countries of this oil are Ethiopia, Somalia, Djibouti, Kenya and Saudi Arabia.

Literature data (up to December 2, 2014) on the chemical composition of myrrh oils and unpublished analytical data from one of us (E.S.) are shown in Table 5.58.1 in alphabetical order. In myrrh oils from various origins, over 110 chemicals have been identified. About 69 per cent of these were found in a single reviewed publication only.

The major compounds found in myrrh oils from different sources are shown in Table 5.58.2. They include (highest concentrations in any study given) furanoeudesma-1,3-diene (65.6%), curzerene (40.1%), β-elemene (32.1%), lindestrene (29.8%), furanodiene (19.7%), α-copaene (5.5%) and germacrene B (4.3%). Well-known ingredients of myrrh oils that were present in high concentrations (>7%) in one or two studies were α-selinene (18.9%), α-pinene (15.8%), curzerenone (12.0%), α-humulene (8.8%), β-bourbonene (8.3%) and 2-methoxyfuranodiene (7.4%). Uncommon or rare constituents of myrrh oils found in high concentrations (>7%) in single studies include 1(2H)phenanthrenone, 3,4,4a,9,10,10a-hexahydro-4a-methyl- (49.4%), benzenemethanol, 3-methoxy-α-phenyl- (13.7%), α-elemene (12.9%), 7-isopropyl-1,4-dimethyl-2-azulenol (12.2%) and cadina-1,4-diene (7.5%).

Table 5.58.1 Constituents identified in myrrh oils

Constituent	CAS	Percentage and range						
		A	B (3)	C (6)	D (7)	E (1)	F (2)	G
2-Acetoxyfuranodiene		0-tr						
2-O-Acetyl-8,12-epoxy-germacra-1(10),4,7,11-tetraene, isomer I			6.5					
2-O-Acetyl-8,12-epoxy-germacra-1(10),4,7,11-tetraene, isomer II			0.3					
allo-Aromadendrene	25246-27-9	0-0.9	1.7		0.2		0.9	+[d]; 5.4[c]
Atractylone	6989-21-5			2.5				
Benzenemethanol, 3-methoxy-α-phenyl-								13.8[e,h]
Benzyl salicylate	118-58-1			0.2				
trans-α-Bergamotene	13474-59-4					0.3		
Bicyclogermacrene	24703-35-3				0.2			3.0[a,b]
α-Bourbonene	5208-58-2							2.6[f]
β-Bourbonene	5208-59-3	0.3-2.1	0.6	0.8	1.2	0.7		+[d]; 0.5[b]; 8.3[c]
3-tert-Butyl-2-hydroxy-5-vinylbenzaldehyde								0.7[f]
Cadina-1,4-diene	29837-12-5						7.5	
α-Cadinene	24406-05-1	0.3-2.2						
β-Cadinene	523-47-7							1.3[a,b]
γ-Cadinene	39029-41-9	0.4-2.4	0.8		1.2	0.5		0.5[f]
δ-Cadinene	483-76-1	0.4-1.5	0.3	0.5	0.4			+[d]; 0.3[b]; 1.5[f]; 2.6[c]
α-Cadinol	481-34-5						2.2	
τ-Cadinol	5937-11-1	0.5-1.8			1.6	0.7		1.0[f]
Camphene	79-92-5					0.1		
β-Caryophyllene	87-44-5	0.2-1.3	0.7	0.6	1.3	0.3		0.a3[e]; 0.5[f]
9-epi-Caryophyllene			0.4					
Caryophyllene alcohol	56747-96-7		0.4					
Caryophyllene oxide	1139-30-6	0.04-0.3	0.2				2.9	+[d]
α-Copaene	3856-25-5	0.1-5.5		0.3	0.2	0.2	3.5	0.06[b]; 3.9[c]
β-Copaene	18252-44-3	0.4-2.9						
α-Cubebene	17699-14-8							1.8[f]
β-Cubebene	3744-15-5							0.4[f]; 0.5[a,b]
Curzerene	17910-09-7	8.5-34.2	40.1	17.5		24.9		11.6[f]; 12.7[c]; 13.4[b]; 26.6[e]
Curzerenone	20493-56-5	1.2-12.0		1.2				+[g]; 1.9[c]
Cyclohexanemethanol, 4-ethenyl-α,α,4-trimethyl-3-(1-methyl-ethenyl)-, [1R-(1α, 3α,4α)]-								0.6[e]
1H-Cycloprop[e]azulene, decahydro-1,1,7-trimethyl-4-methylene-, [1aR-(1aα,4aα, 7α,7aα,7bα)]-								0.6[e]
p-Cymene	99-87-6	0.2-2.6						1.4[c]
Dehydroaromadendrane	72747-25-2		0.1					
6-(1,3-Dimethylbuta-1,3-dienyl)-1,5,5-tri-methyl-7-oxabicyclo-[4.1.0]hept-2-ene								1.2[f]
5,8a-Dimethyl-3-me- thylene-3a,7,8,8a,9, 9a-hexahydro-3H- naphtho[2,3B]furan-2-one								0.4[f]
α-Elemene	5951-67-7							+[d]; 12.9[f]
β-Elemene	3880-83-0	0.5-8.7	8.4	4.3	8.7	4.0	32.1	+[d]; 3.8[b]
γ-Elemene	29873-99-2	0.1-4.0	2.6	1.1	1.1	0.6		0.5[b]
δ-Elemene	20307-84-0	0.5-2.7	0.5	1.0	2.1	0.8		+[d]; 0.3[e]; 0.9[b]; 2.8[c]; 5.6[f]
cis-β-Elemenone	32663-57-3		0.8					
Elemol (α-)	639-99-6	0.2-1.5	0.2	0.2	0.2			+[g]; 0.5[f]

Table 5.58.1 Constituents identified in myrrh oils (*Continued*)

Constituent	CAS	Percentage and range						
		A	B (3)	C (6)	D (7)	E (1)	F (2)	G
Elemol acetate	60031-93-8					1.4		
Eremophilene	10219-75-7							3.4[f]
3-Ethyl-6-(methoxycar-bonyl)-2-naphthol								0.5[f]
Eudesma-4(14),7(11)-diene								1.2[f]
7-epi-α-Eudesmol	123123-38-6		2.2					
γ-Eudesmol (selinenol)	1209-71-8		2.7					0.5[f]
β-Eudesmol acetate	40882-95-9			0.5				
10-epi-γ-Eudesmol acetate			0.3					
Furanodiene	19912-61-9	4.3-17.0	1.1		19.7			+[g]; 1.4[f]
Furanodienone	24268-41-5							+[g]
Furanoeudesma-1,3-diene	115526-32-4	5.9-36.2	15.0	38.6	34.0	42.7	65.6	+[d]
Furanoeudesma-1,4-diene	631868-96-7				1.2			
Furanoeudesmatriene								+[d]
Furanogermacrene								3.0[c]
Germacra-1(10),7,11-trien-15-oic acid, 8,12-epoxy-6-hydroxy-, γ-lactone-								6.2[f]
Germacrene A	28387-44-2			0.8				
Germacrene B	15423-57-1	2.0-4.3		1.9	4.3		3.6	4.0[f]
Germacrene D	23986-74-5	0.5-3.3		1.3	3.2	1.0		0.3[e]; 0.5[f]
Germacrone	6902-91-6			3.4		0.6		1.4[f]
Guaia-3,9-diene	855270-07-4							1.0[f]
α-Gurjunene	489-40-7	0.9-3.8					0.5	
γ-Himachalene	53111-25-4			0.5				
α-Humulene	6753-98-6	0.3-8.8	0.3	0.3	0.6	0.2	0.7	+[d]; 0.2[b]; 1.2[f]
meso-Hydrobenzoin	579-43-1							0.3[e]
2-Hydroxyfuranodiene			0.2					
6-Hydroxyisobornyl isobutyrate				3.8				
8-Hydroxyisobornyl isobutyrate				2.0				
Isocaryophyllene	118-65-0							0.5[b]
Isoledene	95910-36-4							0.5[f]
7-Isopropyl-1,4-di-methyl-2-azulenol								12.2[f]
Isovelleral	37841-91-1							0.7[f]
Limonene	138-86-3	0.02-0.8						
Lindestrene	2221-88-7	2.7-9.6		14.4	12.0			+[g]; 29.8[c]
α-Longifolene	475-20-7							1.0[e]
Longifolene-(V4)								4.0[e]
(1*RS*,2*RS*,1*SR*)-1-(1-Methoxyethyl)-2-vinylcyclobutane								2.3[f]
2-Methoxyfuranodiene		1.7-7.4			2.1			
4-(2'-Methoxyphenyl)]-4-methylcyclo-hex-3-en-1-one								1.5[f]
6α-(2-Methylcyclo-pent-1-enyl)-3,3-di-methyl-1α-bicyclo-[3.1.0]hexan-2-one								0.7[f]
2-*O*-Methyl-8,12-epoxy-germacra-1(10),4,7,11-tetraene, isomer I			0.6					
2-*O*-Methyl-8,12-epoxy-germacra-1(10),4,7,11-tetraene, isomer II			3.9					
Methyl-7-methoxy-5-methyl-2-hydroxyl-1-naphthoate								0.6[f]
α-Muurolene	10208-80-7						0.4	
γ-Muurolene	30021-74-0		0.3	0.6	0.2			+[d]; 0.1[b]; 1.5[f]
τ-Muurolol	19912-62-0						3.0	
Myrcene	123-35-3	0-0.1						

Table 5.58.1 Constituents identified in myrrh oils (*Continued*)

Constituent	CAS	Percentage and range						
		A	B (3)	C (6)	D (7)	E (1)	F (2)	G
δ-Neoclovene								5.6[f]
Occidentalol	473-17-6							+[d]
Occidol	5986-36-7			0.4				
(*E*)-β-Ocimene	3779-61-1					0.1		
Octyl acetate	112-14-1							2.4[f]
1(2*H*)Phenanthrenone, 3,4, 4a,9,10,10a-hexahydro-4a-methyl-	62318-99-4							49.4[e]
α-Pinene	80-56-8	0.3-15.8				0.1		
β-Pinene	127-91-3	0.4-1.6						
1-(1-Propynyl)-2-cyclohexen-1-ol	79688-55-4							2.0[f]
Sabinene	3387-41-5	0-0.3						
Selina-3,7(11)-diene	6813-21-4		0.2					
α-Selinene	473-13-2	0.4-1.8			0.5	18.9		
β-Selinene	17066-67-0	0.6-2.3			0.6		1.6	+[d]; 0.7[b]
β-Sesquiphellandrene	20307-83-9		0.2					
Terpinen-4-ol	562-74-3	0.09-0.6						
Testosterone	58-22-0							2.1[e]
α-Thujene	2867-05-2	0-2.2						
2,10,10-Trimethyltricyclo-[7.1.1.0(2,7)]-undec-6-en-8-one								3.2[f]
Undeca-4,6-diyne								2.3[f]
Valencene	4630-07-3			1.3				0.4[e]
Verticiol	70000-19-0							1.1[f]
α-Ylangene	14912-44-8							0.7[f]
β-Ylangene	20479-06-5				0.3			

A forty-six myrrh essential oil samples from Ethiopia, Somalia, Djibouti, Kenya and Saudi Arabia, analyzed between 2002 and 2013; lowest and highest concentrations given (E. Schmidt, unpublished data)

B one lab-hydrodistilled oil from myrrh oleoresin obtained from *C. myrrha* growing wild in Iran (ref. 3)

C one lab-hydrodistilled and one steam-distilled oil from Ethiopian *C. myrrha* exudate; highest concentrations given (ref. 6)

D one lab-hydrodistilled oil from gum resin of *C. myrrha* obtained in Ethiopia (ref. 7, data also published in ref. 13)

E one commercial oil of *C. myrrha* resin obtained from a German manufacturer (ref. 1)

F one lab-hydrodistilled oil from *Commiphora habessinica* oleoresin growing wild in Yemen (ref. 2)

G data from other studies (indicated with superscript letters); highest concentrations found in any study reviewed here given; when two or more oils were investigated, only the highest concentrations are mentioned, unless indicated otherwise

[a] incorrect identification (ref. 9); [b] one lab-hydrodistilled oil from gum oleoresin of Somali origin purchased in China (ref. 11); [c] one sample of myrrh oil (ref. 12); [d] one oil obtained from *C. myrrha* gum from Kenya (ref. 10); [e] one steam-distilled oil from *Commiphora myrrha* resin purchased at a local market in Sudan (ref. 14); [f] one lab-hydrodistilled oil from myrrha resin purchased from an Egyptian herb company (ref. 15); [g] (probably) one sample of essential oil from *C. abyssinica* (*habessinica*) (Berg.) Engl. (only Summary data available) (ref. 16); [h] mistaken identity, this chemical has a boiling point of 367°C, so cannot be steam-distilled

tr: trace; + present in the oil investigated, but quantity not stated or unknown (16)

Commercial oils

The ten chemicals that had the highest maximum concentrations in 46 commercial myrrh essential oil samples (concentration ranges provided) are the following: furanoeudesma-1,3-diene (5.9-36.2%), curzerene (8.5-34.2%), furanodiene (4.3-17.0%), α-pinene (0.3-15.8%), curzerenone (1.2-12.0%), lindestrene (2.7-9.6%), α-humulene (0.3-8.8%), β-elemene (1-8.7.0%), 2-methoxyfuranodiene (1.7-7.4%) and α-copaene (0.1-5.5%) (Erich Schmidt, unpublished analytical data).

A review of the chemistry of myrrh is provided in ref. 8. Commercial myrrh oils from *C. myrrha* may be adulterated with oils from other *Commiphora* species, such as *C. sphaerocarpa*, *C. holtziana*, and *C. kataf* (13). The composition of an essential oil from the resin of *Commiphora pyracanthoides* Eng. can be found (on-line) in ref. 17.

Table 5.58.2 Major constituents of myrrh oils

Constituent	CAS	Percentage and range						
		A	B (3)	C (6)	D (7)	E (1)	F (2)	G
Furanoeudesma-1,3-diene	115526-32-4	5.9-36.2	15.0	38.6	34.0	42.7	65.6	+[d]
Curzerene	17910-09-7	8.5-34.2	40.1	17.5		24.9		11.6[f]; 12.7[c]; 13.4[b]; 26.6[e]
β-Elemene	33880-83-0	0.5-8.7	8.4	4.3	8.7	4.0	32.1	+[d]; 3.8[b]
Lindestrene	2221-88-7	2.7-9.6		14.4	12.0			+[g]; 29.8[c]
Furanodiene	19912-61-9	4.3-17.0	1.1		19.7			+[g]; 1.4[f]
α-Copaene	3856-25-5	0.1-5.5		0.3	0.2	0.2	3.5	0.06[b]; 3.9[c]
Germacrene B	15423-57-1	2.0-4.3		1.9	4.3		3.6	4.0[f]

LEGEND: SEE UNDER TABLE 5.58.1

CONTACT ALLERGY / ALLERGIC CONTACT DERMATITIS

General

There are only three publications on contact allergy to myrrh oil. In two, aromatherapists reacted to the oil they used at work, but in neither of them can a false-positive patch test reaction be excluded.

Case reports and positive patch tests

An aromatherapist had chronic hand dermatitis and was patch test positive to 17 of 20 oils used at her work (tested 1% and 5% in petrolatum), including myrrh oil (19). Another aromatherapist had occupational contact dermatitis with allergies to multiple essential oils used at work, including myrrh oil. The patient also reacted to geraniol, α-pinene, caryophyllene, the fragrance mix and various other fragrance materials; these chemicals could not be identified by GC-MS in the myrrh oil (20). One positive patch test to myrrh oil was seen in an aromatherapist with occupational contact dermatitis from (multiple) essential oils; however, this oil had not been used by the patient (18).

LITERATURE

1 Wanner J, Schmidt E, Bail S, Jirovetz L, Buchbauer G, Gochevd V, et al. Chemical composition and antibacterial activity of selected essential oils and some of their main compounds. Nat Prod Comm 2010;5:1359-1364

2 Awadh Ali NA, Wurster W, Lindequist U. Chemical composition of essential oil from the oleogum resin of Commiphora habessinica (Berg.) Engl. from Yemen. J Essent Oil Bear Plants 2009;12:244-249

3 Morteza-Semnani K, Saeedi M. Constituents of the essential oil of Commiphora myrrha (Nees) Engl. var. molmol. J Essent Oil Res 2003;15:50-51

4 Shen T, Li G-H, Wang X-N, Lou H-X. The genus Commiphora: A review of its traditional uses, phytochemistry and pharmacology. J Ethnopharmacol 2012;142:319-330

5 El Ashry ES, Rashed N, Salama OM, Saleh A. Components, therapeutic value and uses of myrrh. Pharmazie 2003;8:163-168

6 Marongiu B, Piras A, Porcedda S, Scorciapino A. Chemical composition of the essential oil and supercritical CO2 extract of Commiphora myrrha (Nees) Engl. and of Acorus calamus L. J Agric Food Chem 2005;53:7939-7943

7 Can Başer KH, Demirci B, Dekebo A, Dagne E. Essential oils of some Boswellia ssp., Myrrh and Opopanax. Flavour Fragr J 2003;18:153-156

8 Hanuš LO, Řezanka T, Dembitsky VM, Moussaieff A. Myrrh — Commiphora Chemistry. Biomedical Papers 2005;149:3-28

9 Lawrence BM. Progress in essential oils. Perfum Flavor 2004;29(October):88-? (last page unknown)

10 Wang W, Zhu Y-Z, Qin X-G, Tian J-G. Analysis of the chemical constituents of essential oil of myrrha from Kenya. Yaowa Fenxi Zahzi 1995;15:33-36. Data cited in ref. 9

11 Tian J-G, Shi S-G. Studies on the constituents of essential oil of imported myrrh and gum opoponax. Zhongguo Zhongyao Zazhi 1996;21:235-235. Data cited in ref. 9

12 Moyler DA, Clery RA. The aromatic resins: their chemistry and uses. In: KAD Swift, Ed. Flavours and fragrances. Cambridge, UK: Royal Soc Chem, 1997:96-115. Data cited in ref. 9

13 Dekebo A, Dagne E, Sterner O. Furanosesquiterpenes from Commiphora sphaerocarpa and related adulterants of true myrrh. Fitoterapia 2002;73:48-55

14 Gadir SA, Ahmed IM. Commiphora myrrha and commiphora Africana essential oils. J Chem Pharm Res 2014;6:151-156

15 Mohamed AA, Ali SI, EL-Baz FK, Hegazy AK, Kord MA. Chemical composition of essential oil and in vitro antioxidant and antimicrobial activities of crude extracts of Commiphora myrrha resin. Ind Crops Prod 2014;57:10-16

16 Brieskorn CH, Noble P. Inhaltsstoffe des etherischen Öls der Myrrhe. II: Sesquiterpene und Furanosesquiterpene. Planta Med 1982;44:87-90

17 Chen Y, Zhou C, Ge Z, Liu Y, Liu Y, Feng W, et al. Composition and potential anticancer activities of essential oils obtained from myrrh and frankincense. Oncol Lett 2013;6:1140-1146

18 Bleasel N, Tate B, Rademaker M. Allergic contact dermatitis following exposure to essential oils. Australas J Dermatol 2002;43:211-213

19 Selvaag E, Holm J, Thune P. Allergic contact dermatitis in an aromatherapist with multiple sensitizations to essential oils. Contact Dermatitis 1995;33:354-355

20 Dharmagunawardena B, Takwale A, Sanders KJ, Cannan S, Roger A, Ilchyshyn A. Gas chromatography: an investigative tool in multiple allergies to essential oils. Contact Dermatitis 2002;47:288-292

Chapter 5.59   NEEM OIL

DEFINITION

Neem oil (margosa oil) is the essential oil obtained from the seed of the neem (neem tree, Indian lilac, margosa tree), *Azadirachta indica* A. Juss. (synonym: *Melia azadirachta* L.).

INCI NOMENCLATURE

Description/definition: Azadirachta indica seed extract is the extract of the seeds of *Azadirachta indica*, Meliaceae
INCI name EU & USA: Azadirachta indica seed extract
Other names: Margosa extract
CAS registry number(s): 84696-25-3
EINECS number(s): 283-644-7

Description/definition: Melia azadirachta seed extract is an extract of the seeds of the neem tree, *Melia azadirachta* L., Meliaceae
INCI name EU & USA: Melia azadirachta seed extract
CAS registry number(s): 90063-92-6
EINECS number(s): 290-052-2

ISO (INTERNATIONAL ORGANIZATION FOR STANDARDIZATION) STANDARD

Currently there is no ISO standard for neem oil.

Neem *essential* oil should not be confused with the INCI entry Melia azadirachta seed oil (CAS 8002-65-1), which is the fixed oil expressed or extracted from the seeds of the neem tree. This oil consists primarily of fatty acids including oleic acid (25-58%), palmitic acid (17-34%), stearic acid (7-24%), linoleic acid (6-24%), behenic acid (0.2-1.7%), arachidic acid (1.3%), and palmitoleic acid (0.1-0.2%) (2). It is uncertain whether the INCI nomenclature mentioned above pertains to essential oils or solvent-extracted neem products.

THE PLANT, THE OIL, AND THEIR USES

Azadirachta indica A. Juss. is a fast-growing evergreen tree from the mahogany family Meliaceae that can reach a height of 15-20 meters. The tree is indigenous to the dry forests of south and south-east Asia and is widely distributed in India, Nepal, Pakistan, Bangladesh, Sri Lanka, Myanmar, Thailand, Malaysia, Indonesia and Iran. Its original natural distribution is obscured by widespread cultivation (also in Africa, Central and Southern America, the Caribbean and the Philippines), but neem is believed to be native to at least Myanmar, Bangladesh and northeast India (6, GRIN Taxonomy for Plants).

The neem tree is the holy tree of the Hindus and is considered a major component in Ayurvedic and Unani medicine. The leaves, bark, seeds and roots of neem contain various pharmacologically active constituents, which may possess antiseptic, anti-inflammatory, anti-diabetic, antibacterial and antifungal effects (1). The bitter and astringent bark (neem bark, margosa bark, *Cortex margosae*) is used to treat fever and worms. The leaves of the neem are antiseptic; teas and infusions of leaves are used to alleviate intestinal complaints, and

to treat skin diseases. The flowers are used as tonic and seed extracts may be employed in treating head lice.

The bitter seed oil (vegetable, fixed oil), which has a strong alliaceous (garlic-like) odor, is used to treat skin diseases such as eczema and furuncles, and to relieve intestinal worm infections (5). The fatty oil is also used as lamp oil and to make soap. The seed oil, leaves and other parts of the tree are effective insecticides, repellent and antifeedant agents in plant protection, for which their ingredient azadirachtin is responsible. Biological, pharmacological and medicinal aspects of the (various parts of the) neem tree and its seed oil have been reviewed (3,4).

Neem essential oil (margosa oil) is obtained by hydrodistillation of the seeds (5,7). There is very little information in literature on neem essential oils; virtually all studies on 'neem oils' pertain to the fixed (vegetable) oil of the seeds, which may be obtained by either cold-pressing or extraction with various solvents, or to essential oils from the leaves or flowers of the neem tree. The essential oil is apparently not used in aromatherapy (10).

CHEMICAL COMPOSITION

Neem oil is a light brownish to brownish clear mobile liquid which has an oily, woody and dusty slightly animalic odor. The yield of essential oil from the seed of *Azadirachta indica* A. Juss. generally varies from 0.03-0.07 per cent. The main producing country of this oil is India. Neem essential oil has also been described as having a repulsive garlic-like odor (7). We doubt whether essential oils from *Azadirachta indica* are commercially available.

Literature data (up to October 24, 2014) on the chemical composition of neem oils are shown in Table 5.59.1 in alphabetical order. In two neem oils obtained by hydrodistillation, over 65 chemicals have been identified. The major compounds found in neem essential oil (one quantitative publication only [5]) are hexadecanoic acid (palmitic acid) (34.0%), oleic acid (15.7%), 5,6-dihydro-2,4,6-tri-ethyl-(4*H*)-1,3,5-dithiazine (11.7%), tricosane (10.5%), tetradecanoic acid (6.8%), pentacosane (4.9%), methyl oleate (3.8%), octadecanoic acid (stearic acid) (2.9%), eudesm-7(11)-en-4-ol (2.7%), and linoleic acid (2.4%). Thus, over 80% of the oil is composed of fatty acids (>60%) and alkanes (5).

The analysis presented in ref. 8 is not discussed, as it probably does not adhere to our strict definition of essential oils. The composition of a *leaf* oil of *Azadirachta indica* can be found in ref. 9.

CONTACT ALLERGY / ALLERGIC CONTACT DERMATITIS

General

There are only three publications on contact allergy to neem oil. Three patients became sensitized from therapeutic use of the oil, two from pure or diluted oils and one from the use of an antipsoriatic herbal preparation containing neem oil.

Case reports

One patient developed allergic contact dermatitis from neem oil in a poultice containing neem oil and tea tree

Table 5.59.1 Constituents identified in neem oils

Constituent	CAS	Percentage A (5)	B (7)	Constituent	CAS	Percentage A (5)	B (7)
Abietatriene	19407-28-4	0.3		Menthol	89-78-1		+
Anthracene	120-12-7	0.4		Menthone	89-80-5		+
Aristolene	6831-16-9	0.2		Methylbenzothiophene	31393-23-4	0.2	
Azulene	275-51-4		+	(*E,E*)-Methyl farnesoate	10485-70-8		+
5,6-Azulenedimethanol			+	Methyl linolenoate		1.1	
1*H*-Benzotriazole	95-14-7	+		Methyl octadecanoate	112-61-8	0.2	
Bergamotene		0.3		Methyl oleate	112-62-9	3.8	
β-Bisabolene	495-61-4	0.3		Methyl palmitate	112-39-0	1.3	
3,5-di-*t*-Butylphenol	1138-52-9	0.2		2-Methyl-2-pentanal		0.2	
Curcumene (*ar-*, α-)	644-30-4	0.2		Nerylacetone	3879-26-3	0.2	
1*H*-Cyclopro[e]azulene			+	Nonacosane	630-03-5	0.5	
2,6-Diethylpyridine	935-28-4	0.2		Nonanoic acid	12-05-0	0.3	
cis-3,5-Diethyl-1,2,4-trithiolane	38348-25-3	0.4		Octadecanoic acid (stearic acid)	57-11-4	2.9	
trans-3,5-Diethyl-1,2,4-trithiolane	38348-26-4	0.7		Octadecanol	26762-44-7	0.8	
Dihydromyrcenol	53219-21-9		+	*n*-Octanol (1-)	111-87-5	0.1	
5,6-Dihydro-2,4,6-triethyl-(4*H*)-1,3,5-dithiazine		11.7		1-Octen-3-ol	3391-86-4	0.1	
Docosane	629-97-0	0.2		Oleic acid	112-80-1	15.7	
Dodecanamide	1120-16-7		+	Pentacosane	629-99-2	4.9	
Dodecanoic acid	143-07-7	1.4		Pentadecane (*n*-)	629-62-9	0.1	
Eicosane (*n*-)	112-95-8		+	Pentadecanoic acid	1002-84-2	1.2	
Eudesm-7(11)-en-4-ol	473-04-1	2.7		2-Pentadecanone	2345-28-0	0.2	
Ethyl hexadecanoate	628-97-7	0.2	+	2-Pentanone, 4-hydroxy-4-methyl-	123-42-2	0.3	
Globulol	489-41-8	0.4		Propyl palmitate	2239-78-3	0.07	
Heneicosane	629-94-7	0.3		1*H*-Pyrazole	288-13-1		+
Heptacosane	593-49-7	1.6		Sabinene	3387-41-5	0.08	
Hexacosane	630-01-3	0.2		β-Terpineol	138-87-4		+
Hexadecane	544-76-3	0.1		Tetracosane	646-31-1	0.7	
Hexadecanoic acid (palmitic acid)	57-10-3	34.0		Tetradecanoic acid	544-63-8	6.8	
9-Hexadecenoic acid	2091-29-4	1.8		Thymol	89-83-8	0.3	
α-Hexylcinnamaldehyde	101-86-0		+	Tricosane	638-67-5	10.5	
Isomenthone	491-07-6		+	Tridecanoic acid	638-53-9	1.4	
Italicene	94535-52-1	0.7		2-Tridecanone	593-08-8	0.2	
Limonene	138-86-3	0.4		1,3,5-Trithiolane		1.1	
Linoleic acid	60-33-3	2.4		Valencene	4630-07-3	0.3	

A one neem oil sample obtained by hydrodistillation ('a kind of steam distillation') from broken neem seeds collected in India (ref. 5)
B one sample of steam-distilled neem oil from Pakistan; in this study, the oils of vetiver (*Vetiveria zizanoides*) and mint (*M. arvensis*) were also examined; the analytical data were presented only qualitatively as 'Major components of three oils identified by GC-MS analysis'; hence, it is uncertain whether these components were present in all three oils or possibly in one or two, and in that case, unknown in which of these three (ref. 7)

+ present in the oil investigated, but quantity not stated

oil for treatment of a furuncle; a patch test with pure neem oil was positive (11,13). Another patient became sensitized to neem oil which was used to treat alopecia areata (12). The antipsoriatic herbal preparation Psorigon® caused allergic contact dermatitis in one patient, who showed positive patch tests to the product and to '*Azadirachta indica*' 1%, 2%, 5% and 10%. It is not certain that the ingredient was an essential oil (14).

LITERATURE

1 Dastan D, Pezhmanmehr M, Askari N, Ebrahimi SN, Hadian J. Essential oil compositions of the leaves of *Azadirachta indica* A. Juss from Iran. J Essent Oil Bear Plants 2010;13:357-361

2 Djenontin TS, Wotto VD, Avlessi F, Lozano P, Sohounhloué DKC, Pioch D. Composition of *Azadirachta indica* and *Carapa procera* (Meliaceae)

seed oils and cakes obtained after oil extraction. Ind Crops Prod 2012;38:39-45

3 Brahmachari G. Neem — an omnipotent plant: A retrospection. Chem Bio Chem 2004;5:408-421

4 Atawodi SE, Atawodi JC. *Azadirachta indica* (neem): a plant of multiple biological and pharmacological activities. Phytochem Rev 2009;8:601-620

5 Kurose K, Yatagai M. Components of the essential oils of *Azadirachta indica* A. Juss, *Azadirachta siamensis* Velton, and *Azadirachta excelsa* (Jack) Jacobs and their comparison. J Wood Sci 2005;51:185-188

6 APFORGEN Priority Species Information Sheet. *Azadirachta indica* A. Juss. Available at: http://webcache.googleusercontent.com/search?q=cache:7177odwLP-cJ:www.apforgen.org/apfCD/Information%2520Sheet/InfoSheet_Azadirachta.pdf+&cd=1&hl=nl&ct=clnk&gl=nl

7 Manzoor F, Naz N, Malik S, Siddiqui BS, Syed A, Perwaiz S. Chemical analysis and comparison of antitermitic activity of essential oils of Neem (*Azadirachta indica*), Vetiver (*Vetiveria zizanioides*) and Mint (*Mentha arvensis*) against *Heterotermes indicola* (Wasmann) from Pakistan. Asian J Chem 2012;24:2069-2072

8 Riar SS, Devakumar C, Ilavazhagan G, Bardhan J, Kain AK, Thomas P, et al. Volatile fraction of neem oil as a spermicide. Contraception 1990;42:479-489

9 Ebrahimi M, Safaralizade MH, Valizadegan O. Contact toxicity of *Azadirachta indica* (Adr. Juss.), *Eucalyptus camaldulensis* (Dehn.) and *Laurus nobilis* (L.) essential oils on mortality of cotton aphids, *Aphis gossypii* Glover (Hem.: Aphididae). Arch Phytopathol Plant Prot 2013;46:2153-2162

10 Rhind JP. Essential oils. A handbook for aromatherapy practice, 2nd Edition. London: Singing Dragon, 2012

11 Greenblatt DT, Banerjee P, White JML. Allergic contact dermatitis to neem oil. Brit J Derm 2011;165(Suppl. 1):74-75

12 Reutemann P, Ehrlich A. Neem oil: an herbal therapy for alopecia areata causes dermatitis. Dermatitis 2008;19:E12-E15

13 Greenblatt DT, Banerjee P, White JML. Allergic contact dermatitis caused by neem oil. Contact Dermatitis 2012;67:242-243

14 Ahmed I, Charles-Holmes R. Contact allergy to Psorigon®. Contact Dermatitis 2000;42:276

Chapter 5.60   NEROLI OIL

DEFINITION

Neroli oil (essential oil of neroli) is the essential oil obtained from the flowers of the bitter orange, *Citrus aurantium* L. (synonyms: *Citrus aurantium* subsp. *amara* (Link) Engl., *Citrus amara* Link, *Citrus bigaradia* Loisel, *Citrus vulgaris* Risso).

INCI NOMENCLATURE

Description/definition: Citrus aurantium amara flower oil is the volatile oil obtained from the flowers of the bitter orange, *Citrus aurantium* L. var. *amara* L., Rutaceae

INCI name EU: Citrus aurantium amara flower oil

INCI name USA: Citrus aurantium amara (bitter orange) flower oil

CAS registry number(s): 8016-38-4; 72968-50-4; 68916-04-1

EINECS number(s): 277-143-2

ISO (INTERNATIONAL ORGANIZATION FOR STANDARDIZATION) STANDARD

ISO number: 3517

ISO name: Essential oil of neroli

Botanical origin: *Citrus aurantium* L. (synonyms: *Citrus amara* Link, *Citrus bigaradia* Loisel, *Citrus vulgaris* Risso)

Parts of plant used: Flower

ISO values: ISO values (minimum and maximum concentrations) are shown in Table 5.60.1.

Citrus aurantium L. is not only used for the production of neroli oil, but is also the source for obtaining bitter orange oil from cold-pressing of the fruit peels (Chapter 5.64) and petitgrain bigarade oil from steam-distillation of the leaves, twigs and unripe fruit (Chapter 5.69) (1).

THE PLANT, THE OIL, AND THEIR USES

Citrus aurantium L. is an evergreen shrub or tree, 2-3 to 7-8 meters tall, with short and sharp spines, which produces ovoid or rounded greenish-yellow fruits with bitter acidic pulp. It probably has a multiple hybrid origin in China and elsewhere, and is widely cultivated in the tropics and subtropics. The countries producing the bitter orange trees and their essential oils include Spain, Italy, France, Egypt, Tunisia, Morocco, Uganda, Ivory Coast, Paraguay, Brazil, Argentina, and China (1, GRIN Taxonomy for Plants).

The fruits serve the production of marmalade and soft drinks; in Mexico, they are consumed fresh with salt and chili. The juice of the fruit is added to meat and fish for flavoring and is used as a vinegar substitute. The flowers are used in teas and have been approved by the FDA in small amounts in some medications (6,16). The oils and extracts obtained from the bitter orange tree are also used in the food and alcoholic beverage industries (1,2).

Bitter orange is employed in herbal medicine as a stimulant and appetite suppressant (to induce weight loss), on account of its active ingredient, synephrine. It is also employed as a nasal decongestant (2). However, bitter orange supplements may have caused some serious side effects.

The neroli essential oil, obtained from the flowers by steam- or hydrodistillation is used in the perfume (mainly fine fragrances) and soap industry. The volatile oil contains a sensual fragrance and forms the heart of one the world's most enduring perfumes, 'Eau de Cologne' (16). It also has limited use in flavoring candy, soft drinks and liqueurs, ice cream, baked goods and chewing gum. In addition, neroli oil has applications in aromatherapy (24,25). It is reported to have sedative, calmative, cytophylactic (stimulating leukocytes to fight infections), aphrodisiac, anti-depressant and antispasmodic properties (1,6,16, Mansfeld's World Database of Agriculture and Horticultural Crops, http://mansfeld.ipk-gatersleben.de).

CHEMICAL COMPOSITION

Neroli oil is a clear mobile liquid with pale yellow to light amber color, sometimes with slight blue influences and which has a flowery, sometimes harsh but fresh odor, reminding of natural orange blossoms. The yield of essential oil from the flowers of *Citrus aurantium* L. generally varies from 0.07 to 0.15 per cent. The main producing countries of this oil are Tunisia, Morocco, Egypt and Italy.

Literature data (up to December 2, 2014) on the chemical composition of neroli oils and unpublished analytical data from one of us (E.S.) are shown in Table 5.60.2 in alphabetical order. In neroli oils from various origins, over 190 chemicals have been identified. About 50 per cent of these were found in a single reviewed publication only.

The major compounds found in neroli oils from different sources are shown in Table 5.60.3. They include (highest concentrations in any study given) linalool (73.7%), limonene (46.6%), linalyl acetate (29.3%), geraniol (26.6%), β-pinene (23.4%), α-terpineol (20.7%), (*E*)-β-ocimene (7.9%) and geranyl acetate (4.9%). Well-known ingredients of neroli oils that were present in high concentrations (>7%) in one or two studies were (*E*)-nerolidol (17.5%), sabinene (15.0%), methyl anthranilate (11.8%), α-terpinyl acetate (11.7%), myrcene (9.3%) and (*E,E*)-farnesol (8.0%). Uncommon or rare constituents of neroli oils found in high concentrations (>7%) in single studies include linalyl anthranilate (16.4%), farnesol (three studies: 7.4%, 7.6% and 9.9%) and *n*-octanol (8.6%).

Commercial oils

The ten chemicals that had the highest maximum concentrations in 79 commercial neroli essential oil samples (concentration ranges provided) are the following: linalool (31.9-57.7%), limonene (10.1-20.3%), β-pinene (3.8-16.9%), linalyl acetate (1.4-15.1%), (*E*)-β-ocimene (3.5-7.9%), α-terpineol (3.2-7.6%), (*E*)-nerolidol (1.7-4.9%), geranyl acetate (1.7-3.7%), geraniol (2.0-3.6%) and (*E,E*)-farnesol (0.3-3.5%) (Erich Schmidt, unpublished analytical data).

A full review of the literature on the chemical composition of neroli oils is provided in ref 18.

Table 5.60.1 ISO values (%) for neroli oil [a]

Compound	CAS	Minimum	Maximum
Linalool	78-70-6	26.0	55.0
Linalyl acetate	115-95-7	1.5	20.0
Limonene	138-86-3	7.0	18.0
β-Pinene	127-91-3	2.0	17.0
(E)-β-Ocimene	3779-61-1	3.0	9.0
α-Terpineol	98-55-5	2.0	8.0
Neryl acetate	141-12-8	0.0	7.0
Geraniol	106-24-1	1.0	5.0
Geranyl acetate	105-87-3	1.0	5.0
(E)-Nerolidol	40716-66-3	0.5	5.0
Myrcene	123-35-3	1.0	4.0
(E,E)-Farnesol	106-28-5	0.5	4.0
Sabinene	3387-41-5	0.0	3.0
Nerol	106-25-2	0.5	2.0
α-Pinene	80-56-8	0.0	2.0
Methyl anthranilate	134-20-3	0.0	1.0
1H-Indole	120-72-9	0.0	0.5

[a] ISO 3517 Essential oil of neroli ©ISO 2013; Geneva, Switzerland, www.iso.org

CONTACT ALLERGY/ALLERGIC CONTACT DERMATITIS

General

Contact allergy to/allergic contact dermatitis from neroli oil has been reported in over 15 publications. In routine testing (one study), a prevalence rate of only 0.3% positive patch test reactions was observed. There have been several case reports of allergic contact dermatitis from neroli oil, mostly from its presence in topical pharmaceutical preparations and some from cosmetics. Also, several cases of occupational contact dermatitis from neroli oil have been documented, both in people who use the oils for massage (including aromatherapists) and patients sensitized from their work as bottle fillers in a perfume factory. In one case, geraniol may have been an allergen in neroli oil.

Testing in groups of patients

The results of patch tests with neroli oil in routine testing (consecutive patients suspected of contact dermatitis) and in groups of selected patients are shown in Table 5.60.4. In routine testing, the rate of positive reactions was 0.3% (only one study performed), whereas between 0.7% and 20% of patients in selected groups had positive patch tests. The high positivity rate of 20% was

Table 5.60.2 Constituents identified in neroli oils

Constituent	CAS	Percentage and range					
		A	B	C (1,17)	D (10)	E (14)	F
Anethole (p-)	104-46-1						0.1[n]
Anhydrolinalool oxide	84616-87-5		0.05				
Aromadendrene	489-39-4		0.3	tr-0.01			
Benzaldehyde	100-52-7		0.9			0.01	
Benzyl alcohol	100-51-6		0.2				
cis-α-Bergamotene	18252-46-5			0-0.01			
trans-α-Bergamotene	13474-59-4					0.01	
Bicycloelemene	32531-56-9			tr-0.01			
Bicyclogermacrene	24703-35-3	0.04-0.2		0.05-0.2			
Bois de rose oxide	7392-19-0						0.04[p]
Bornyl acetate	76-49-3	0.01-0.1		tr-0.01			0.03[l]
γ-Cadinene	39029-41-9			tr-0.01			
δ-Cadinene	483-76-1	0.03-0.1	tr	0.02-0.04	0.01-0.03	0.03	0.03[p]
α-Cadinol	481-34-5			0.01-0.02		0.02	0.1[h]
δ-Cadinol	19435-97-3						0.02[p]
Camphene	79-92-5	0.03-0.1	5.5	0.01	0.05-0.07	0.04	tr[j]; 0.04[e,l]; 0.2[l,n]
Camphene hydrate	465-31-6						0.01[l]
δ3-Carene	13466-78-9		2.5	0.03-0.09		0.5	0.1[n]; 0.2[k]; 2.4[h]; 6.9[a,f]
Carvacrol	499-75-2						4.3[n]
(E)-Carveol	1197-07-5						0.01[l]; 0.2[d,n]
(Z)-Carveol	1197-06-4			0-tr			0.6[n]
Carvone	99-49-0			0.01-0.03			0.04[l]; 0.1[d]; 0.3[n]
β-Caryophyllene	87-44-5	0.4-1.2	0.5	0.6-0.9	0.5-0.7	0.7	0.4[e,k,m]; 0.5[p]; 1.1[l]; 1.2[o]
9-epi-(E)-Caryophyllene	68832-35-9			tr-0.01			
Caryophyllene oxide	1139-30-6			0.02-0.04			0.01[p]; 0.1[h]
1,8-Cineole	470-82-6					0.2	0.1[l]; 15.9[n]

Table 5.60.2 Constituents identified in neroli oils (*continued*)

Constituent	CAS	Percentage and range					
		A	B	C (1,17)	D (10)	E (14)	F
Citronellal	106-23-0		0.6			0.07	0.01[p]
Citronellol	106-22-9		4.1				
Citronellyl acetate	150-84-5			0-0.02		0.03	0.09[p]
Cuminaldehyde	122-03-2						0.1[d]
3-Cyclohexenylcarbinol			0.4				
o-Cymene	527-84-4			tr			
p-Cymene	99-87-6	0.02-0.1	2.5	0.06-0.1	0.01-0.1	0.7	0.1[d]; 0.2[h,l,j,n,p]
p-Cymenene	1195-32-0		tr				
p-Cymen-8-ol	1197-01-9			0.01-0.03			
Decanal	112-31-2		+				0.03[p]
Decane	124-18-5						0.2[n]
cis-Dehydrolinalool oxide				tr-0.02			
cis-Dehydroxylinalool oxide	73413-94-2						0.3[n]
Dihydrocarveol	38049-26-2						0.5[n]
2,3-Dihydrofarnesol	51411-24-6			0.01-0.04			
4,8-Dimethyl-1,3(*E*), 7-nonatriene	19945-61-0			0.03-0.05[b]			
2,5-Dimethyl-2-vinyl-4-hexenal	56134-05-5		+				
Docosane	629-97-0						0.1[h]
Dodecane	112-40-3						0.5[n]
Eicosane	112-95-8						0.03[j]
β-Elemene	3880-83-0		0.1	0.09-0.2		0.3	0.05[p]; 0.3[h]
γ-Elemene	29873-99-2				0.01-0.02		3.4[h]
δ-Elemene	20307-84-0	0.02-0.1		0.03-0.06		0.02	0.1[k]
Ethylbenzene	100-41-4						0.1[h]
Eugenol	97-53-0						4.6[n]
Farnesal	19317-11-4						0.9[j]
(*E,E*)-Farnesal	502-67-0			0.02-0.04			0.9[h]
(2*E*,6*Z*)-Farnesal	3790-67-8			0.01-0.03			
(*Z,E*)-Farnesal	4380-32-9						0.1[d]
Farnesane	3891-98-3						0.3[n]
α-Farnesene	502-61-4	0.03-0.1					0.06[e,m]
(*E,E*)-α-Farnesene	502-61-4			0.01-0.05	0.05-0.1	0.08	
(*E*)-β-Farnesene	18794-84-8	0.04-0.3		0.09-0.2	0.05-0.2		0.1[p]
(*Z*)-β-Farnesene	28973-97-9					0.1	0.08[p]
Farnesol	4602-84-0		5.1				2.3[i]; 5.4[e]; 7.4[m]; 7.6[o]; 9.9[j]
(*E,E*)-Farnesol	106-28-5	0.3-3.5		1.1-2.0		0.02	1.5[p]; 2.3[f]; 5.1[k]; 8.0[h]
(*E,Z*)-Farnesol	3879-60-5				0.7-1.6		
(*Z,E*)-Farnesol	3790-71-4					1.0	0.04[p]; 0.4[d]
Farnesyl acetate	29548-30-9		0.03	0.02-0.04			
(*E,E*)-Farnesyl acetate	4128-17-0	0.03-0.7					
(*E,Z*)-Farnesyl acetate	24163-98-2				0.01-0.09		
Fenchyl alcohol	1632-73-1			0-tr			
Geranial	141-27-5	0.03-0.1	0.04	0.05-0.07	0.05-0.1	0.7	0.08[i]; 0.2[j]; 0.3[h]; 0.5[d]; 1.1[o]
Geraniol	106-24-1	2.0-3.6	4.3	2.9-3.8	0.8-2.3		3.4[m]; 4.2[e]; 4.3[k]; 9.0[f]; 26.6[n]
Geranyl acetate	105-87-3	1.7-3.7	4.2	3.0-3.1		1.4	3.7[f,g]; 3.8[m]; 4.0[e,i]; 4.8[j]; 4.9[l]
Geranylacetone (*E*-)	3796-70-1		0.05	0.03-0.04			0.05[p]; 0.1[h]; 0.2[l]
Geranyl formate	105-86-2			tr-0.04			0.2[l]
(*E,E*)-Geranyllinalool	1113-21-9			0.02-0.03			
Geranyl propanoate	105-90-8			tr-0.01			
Germacrene D	23986-74-5	0.04-0.2	0.1	0.03-0.08	0.05	0.05	0.05[p]; 0.1[k]; 0.7[h]
Globulol	489-41-8			0.01-0.02		0.01	

Table 5.60.2 Constituents identified in neroli oils (*continued*)

Constituent	CAS	Percentage and range					
		A	B	C (1,17)	D (10)	E (14)	F
(Z)-8-Heptadecene	16369-12-3		+				
trans-Herboxide							0.2[n]
Hexadecanoic acid	57-10-3		0.4				
Hexanol	111-27-3	0.01-0.08					
(E)-2-Hexen-1-ol	928-95-0		0.09				
(E)-2-Hexenyl acetate	2497-18-9			0-0.01			
(Z)-3-Hexenyl butyrate	16491-36-4			tr-0.02			
Hexyl acetate	142-92-7				0.01-0.05		
α-Humulene	6753-98-6	0.04-0.1	0.1	0.06-0.1		0.1	0.2[p]
1H-Indole	120-72-9	0.08-0.2	1.0		0.1-0.2	0.06	0.1[i]; 0.2[p]; 0.7[e]; 0.8[m]; 1.9[n]
Isocaryophyllene	118-65-0		0.09				
Isomenthone	491-07-6						0.1[n]
2-Isopropenyl-5-methyl-5-vinyltetrahydrofuran							0.08[l]
Isopulegol	89-79-2						1.0[h,l]
(E)-Jasmone	6261-18-3			0.01-0.02			
(Z)-Jasmone	488-10-8		0.05		0.01-0.05		0.05[p]
Lilac aldehyde	67920-63-2						2.1[n]
Limonene	138-86-3	10.1-20.3	18.1	7.9-11.9	12.2-17.9	24.3	20.9[f]; 24.6[g]; 27.5[h,l]; 46.6[n]
cis-Limonene oxide	13837-75-7						0.07[l]
trans-Limonene oxide	4959-35-7			0.01-0.1			0.2[l]
Linalool	78-70-6	31.9-57.7	73.7	43.7-53.3	31.4-47.1	15.5	44.1[n]; 44.3[d]; 53.2[o]; 55.2[g]
Linalool oxide	1365-19-1						0.05[l]; 0.1[n]
cis-Linalool oxide, furanoid	5989-33-3	0.1-0.4	3.0	0.1-0.3	0.01-0.2	0.02	0.1[m]; 0.2[l]; 0.3[i]; 1.5[d]; 6.1[n]
trans-Linalool oxide, furanoid	34995-77-2	0.02-0.3	3.0		0.06-0.2	0.01	0.2[e,j,m]; 1.3[d]; 3.8[n]
cis-Linalool oxide, pyranoid	14009-71-3		0.09				0.2[d]
trans-Linalool oxide, pyranoid	39028-58-5		0.09				0.2[d]
Linalyl acetate	115-95-7	1.4-15.1	18.4	2.2-14.6	0.6-10.0	9.9	11[i]; 12.1[e]; 14.3[m]; 14.9[j]; 29.3[g]
Linalyl anthranilate	7149-26-0						16.4[a,d]
Linalyl propionate	144-39-8			0-0.02			0.01[l]
trans-p-Mentha-2,8-dien-1-ol	4017-77-0			0-tr			
p-Mentha-1,8(9)-dienyl acetate							0.02[p]
cis-p-Menth-2-en-1-ol	29803-82-5			tr-0.03		0.08	
trans-p-Menth-2-en-1-ol	29803-81-4			0.01-0.03		0.2	
Methyl anthranilate	134-20-3	0.01-0.4	5.0	0.04-0.1	0.09-0.2	0.1	0.1[p]; 0.2[i,k]; 1.2[h]; 3.7[g]; 11.8[n]
Methyl eugenol	93-15-2						0.1[n]
Methyl geranate	2349-14-6			tr-0.02			
6-Methyl-5-hepten-2-one	110-93-0			0.02		0.1	
Methyl jasmonate	1211-29-6		tr				0.01[p]
(Z)-Methyl jasmonate	1211-29-6				0.01		
Methyl N-methyl anthranilate	85-91-6		tr	0.01-0.04		3.2	0.1[p]; 1.3[i]
α-Muurolene	10208-80-7			0.01-0.02			
γ-Muurolene	30021-74-0		tr				
Myrcene	123-35-3	1.0-2.6	9.3	1.3-1.7	1.4-3.1	2.4	1.6[k]; 1.7[i]; 1.8[e,m]; 2.5[p]
trans-Myrtanol	15358-91-5			tr-0.02			
Myrtenal	564-94-3						0.02[l]
Myrtenol	515-00-4						0.01[l]

Table 5.60.2 Constituents identified in neroli oils (*continued*)

Constituent	CAS	Percentage and range					
		A	B	C (1,17)	D (10)	E (14)	F
Neral	106-26-3		0.6	tr-0.03	0.01-0.03	0.5	0.04[j]; 0.2[d]; 3.7[o]
Nerol	106-25-2	0.8-1.5	1.3	1.0-1.3	0.3-0.9	0.7	1.4[e]; 1.5[i]; 1.6[m]; 1.7[m]; 5.6[f]
Nerolidol	7212-44-4		13.0				0.6[d]; 1.5[e]; 2.6[p]; 3.0[i]; 4.9[j]
(*E*)-Nerolidol	40716-66-3	1.7-4.9	0.7	1.2-3.2	2.2-3.4	1.8	1.0[n]; 1.8[k]; 3.8[f]; 5.6[g]; 17.5[h]
(*Z*)-Nerolidol	3790-78-1		1.2			<0.01	0.2[h]
Nerolidyl propionate	74646-28-9						0.1[h]
Neryl acetate	141-12-8	0.8-2.0	3.7	1.4-1.5	0.3-1.6	0.8	1.8[e,m]; 1.9[f,g]; 2.1[i]; 2.4[j]; 2.5[l]
Nonanal	124-19-6		tr				0.01[p]
Nonane	111-84-2						0.4[n]
(*E*)-α-Ocimene	6874-10-8						0.1[n]
(*Z*)-α-Ocimene	6874-44-8						0.1[n]
β-Ocimene	13877-91-3		7.8				1.6[l]
(*E*)-β-Ocimene	3779-61-1	3.5-7.9	6.0	3.3-5.1	5.6-7.0	3.7	4.3[h]; 5.0[i]; 5.6[o]; 6.1[k]; 6.4[e]
(*Z*)-β-Ocimene	3338-55-4	0.4-1.8	1.0	0.5-0.6	0.1-0.7	0.4	0.7[e]; 0.8[h,k,p]; 0.9[i]; 6.5[m]
Octanal	124-13-0						0.02[p]
n-Octanol (1-)	111-87-5		8.6				
Pentacosane	629-99-2						0.02[j]
Pentadecane (*n*-)	629-62-9						0.9[n]
Perillene	539-52-6		tr				0.02[p]
Perillyl acetate	15111-96-3			0-0.01			
α-Phellandrene	99-83-2		0.05	0.01-0.02		0.09	0.04[p]
β-Phellandrene	555-10-2	0.09-0.3	0.4				0.2[e,m]
2-Phenethyl alcohol	60-12-8	0.02-0.1	0.2	0.03-0.05[b]	0-0.3	0.01	0.2[p]
Phenylacetaldehyde	122-78-1		2.0				5.5[n]
Phenylacetic acid	103-82-2						0.1[n]
Phenylacetonitrile	140-29-4	0.1-0.5					
2-Phenylethyl acetate	103-45-7	0.03-0.1			0.03		
Phenyl ethyl ketone	93-55-0		0.9				
Phthalic acid	88-99-3		0.09				
α-Pinene	80-56-8	0.3-1.2	6.7	0.2-0.3	0.8-1.1	1.4	0.6[e,m]; 0.7[i]; 0.9[f]; 1.4[k]; 3.1[l]
β-Pinene	127-91-3	3.8-16.9	23.4	1.9-3.7	10.5-14.6	20.2[c]	11.9[i]; 14.6[g]; 15.2[f]; 17.7[l]; 19.1[k]
trans-Pinene hydrate	3247-40-3						0.05[l]; 0.2[n]
Pinocarveol	5947-36-4						0.06[l]
trans-Pinocarveol	1674-08-4						0.2[n]
trans-Piperitol	16721-39-4			0.01-0.02		0.03	0.01[l]
Piperitone	89-81-6						0.2[n]
Pulegone	89-82-7						0.8[n]
Sabinene	3387-41-5	0.5-1.4	15.0	0.9-1.6	1.4-2.8	20.2[c]	1.6[l]; 1.9[n]; 2.0[k]; 2.3[p]; 4.1[e]
cis-Sabinene hydrate	15537-55-0					0.03	
trans-Sabinene hydrate	17699-16-0					0.1	0.02[l]
trans-Sabinyl acetate	139757-62-3						0.1[h]
Santolinatriene	2153-66-4						0.1[n]
β-Sesquiphellandrene	20307-83-9			0.01			
α-Sinensal	17909-77-2			0.01-0.02			
β-Sinensal	17909-87-4			0.08-0.09			
Spathulenol	6750-60-3			0.03-0.05		0.04	0.6[n]
Terpenediol	80-53-5						0.1[n]
α-Terpinene	99-86-5	0.02-0.2	0.07	0.04-0.1	0.1-0.5	0.5	0.07[m]; 0.1[i]; 0.2[o]; 1.0[n]; 5.6[o]
γ-Terpinene	99-85-4	0.09-0.3	8.3	0.1-0.3	0.01-0.5	3.8	0.1[m]; 0.2[e]; 0.3[h,l,p]; 0.4[k]
Terpinen-4-ol	562-74-3	0.2-0.5	2.4	0.4-0.8	0.3-1.3	1.2	0.4[h]; 0.5[i]; 0.6[e]; 0.7[k]; 0.9[l]; 2.7[n]

Table 5.60.2 Constituents identified in neroli oils (*continued*)

Constituent	CAS	Percentage and range					
		A	B	C (1,17)	D (10)	E (14)	F
1-Terpineol	586-82-3						0.2^n
α-Terpineol	98-55-5	3.2-7.6	4.8	4.9-6.2	1.1-4.5	1.8	7.2^i; 9.3^l; 9.8^o; 14.0^h; 20.7^n
β-Terpineol	138-87-4		0.9				
Terpinolene	586-62-9	0.3-0.5	0.7	0.3-0.4	0.4-0.6	0.6	$0.4^{e,l,m,p}$; 0.7^i
Terpinyl acetate	8007-35-0						0.1^n
α-Terpinyl acetate	80-26-2		0.2	0.05-0.07	0.01-0.3	0.05	0.06^p; 0.2^k; 11.7^h
β-Terpinyl acetate	10198-23-9						1.7^h
α-Terpinyl formate	2153-26-6						0.3^h
Tetracosane	646-31-1						0.09^j
n-Tetradecane	629-59-4			tr-0.01			1.5^n
Tetradecanol	27196-00-5			0.02-0.03			
Tetratetracontane	7098-22-8						0.1^h
α-Thujene	2867-05-2	0.01-0.04	0.1	0.01-0.02	0.01-0.05	0.3	tr^j; 0.03^i; 0.05^p; 0.1^l
Thymol	89-83-8						1.0^n
Tricyclene	508-32-7			0-tr		0.01	
Valencene	4630-07-3		tr				0.05^p

A seventy-nine essential oil samples from Egypt, Tunisia and Morocco, analyzed between 1998 and 2013; lowest and highest concentrations given (E. Schmidt, unpublished data)
B data from studies from before 1980 and summarized in ref. 18 (highest concentrations in any study given)
C one hydrodistilled and four steam-distilled neroli oils from Egypt; lowest and highest concentrations given (ref. 1); the same data are presented in ref. 17
D several neroli oils from Spain and one from Tunisia; lowest and highest concentrations given (ref. 10)
E one neroli oil produced in Italy (ref. 14); data also published in refs. 20 and 21
F data from other studies (indicated with superscript letters); highest concentrations found in any study reviewed here given; when two or more oils were investigated, only the highest concentrations are mentioned, unless indicated otherwise

[a] incorrect identity based on GC elution order (ref. 9); [b] 4,8-dimethyl-1,3(*E*),7-nonatriene and 2-phenethyl alcohol combined; [c] sabinene and β-pinene combined; [d] one sample of neroli oil produced in China (ref. 11); [e] one Greek and one Tunisian sample of neroli oil (ref. 12); the data of the Tunisian sample would be reproduced 11 years later (ref. 3); [f] four commercial neroli essential oils (ref. 13); [g] nine oils obtained in the commercial or retail trade; most of the oils had been adulterated to some extent (ref. 15); [h] one lab-hydrodistilled oil from the flowers of Tunisian *C. aurantium* (ref. 16); [i] two samples of neroli oil of unknown origin (ref. 8); [j] one lab-hydrodistilled oil from flowers collected in Lebanon (ref. 7); [k] one lab-hydrodistilled oil from flowers picked in a Greek citrus orchard (ref. 6); [l] one sample of neroli oil of unknown origin (ref 4); [m] four lab-hydrodistilled oils from flowers collected from *Citrus aurantium* provenances sampled in Greece (n=2), Brazil and Corsica (France) (ref.2); [n] three Iranian neroli oils including one commercial sample (ref. 19); [o] one sample of neroli oil from Italy (ref. 22); [p] one oil sample from Spain (ref. 23)

tr: trace (in column C: <0.005); + present in the oil investigated, but quantity not stated

Table 5.60.3 Major constituents of neroli oils

Constituent	CAS	Percentage and range					
		A	B	C (1,17)	D (10)	E (14)	F
Linalool	78-70-6	31.9-57.7	73.7	43.7-53.3	31.4-47.1	15.5	44.1^n; 44.3^d; 53.2^o; 55.2^g
Limonene	138-86-3	10.1-20.3	18.1	7.9-11.9	12.2-17.9	24.3	20.9^f; 24.6^g; $27.5^{h,l}$; 46.6^n
Linalyl acetate	115-95-7	1.4-15.1	18.4	2.2-14.6	0.6-10.0	9.9	11^i; 12.1^e; 14.3^m; 14.9^j; 29.3^g
Geraniol	106-24-1	2.0-3.6	4.3	2.9-3.8	0.8-2.3		3.4^m; 4.2^e; 4.3^k; 9.0^f; 26.6^n
β-Pinene	127-91-3	3.8-16.9	23.4	1.9-3.7	10.5-14.6	20.2^c	11.9^i; 14.6^g; 15.2^f; 17.7^l; 19.1^k
α-Terpineol	98-55-5	3.2-7.6	4.8	4.9-6.2	1.1-4.5	1.8	7.2^i; 9.3^l; 9.8^o; 14.0^h; 20.7^n
(*E*)-β-Ocimene	3779-61-1	3.5-7.9	6.0	3.3-5.1	5.6-7.0	3.7	4.3^h; 5.0^i; 5.6^o; 6.1^k; 6.4^e
Geranyl acetate	105-87-3	1.7-3.7	4.2	3.0-3.1		1.4	$3.7^{f,g}$; 3.8^m; $4.0^{e,i}$; 4.8^j; 4.9^l

LEGEND: SEE UNDER TABLE 5.60.2

seen in a very small group of 10 patients with fragrance contact allergy (28).

Case reports

Occupational allergic contact dermatitis

Occupational allergic contact dermatitis developed in an aromatherapist who worked with many essential oils diluted to 5%; she reacted to neroli oil and lavender oil (31). A female 'complementary therapist' developed occupational contact dermatitis from a multitude of essential oils used at work, including neroli oil (33). One masseuse / reflexologist was seen with occupational dermatitis from contact allergy to neroli oil and other essential oils (32). Three people working in a perfume factory as bottle fillers developed occupational allergic contact dermatitis from neroli oil they worked with and from various other essential oils; there was also one positive patch test reaction among 20 workers without dermatitis (34).

Non-occupational allergic contact dermatitis

Two positive patch test reactions to neroli oil, which was present in cosmetics that had given a positive patch test or had been positive in a usage test, were seen in a 9-year period in one clinic in Belgium (38). One case of allergic contact dermatitis from a facial moisturizer was due to its ingredient neroli oil (33). Fourteen cases of contact sensitization to neroli oil present in topical pharmaceutical preparations have been observed (26). Five cases of contact allergy to 'oil of orange flowers' in topical pharmaceutical preparations were reported (29). Two positive patch test reactions to neroli oil (of which one was photoaggravated) from its presence in a topical NSAID preparation used by patients suspected to be allergic to the NSAID (40). There may be some overlap in these three reports from Belgium.

Positive patch tests (relevance unknown, uncertain or not stated)

One positive patch test to neroli oil in an aromatherapist with occupational contact dermatitis from (multiple) essential oils; this oil had not been used by the patient (30). A patient suffering from airborne contact dermatitis caused by (other) essential oils had a positive patch test to 'pomerance flower oil' (36). An aromatherapist had chronic hand dermatitis and was patch test positive to 17 of 20 oils used at her work (tested 1% and 5% in petrolatum) and to neroli oil, with which she had no contact (35). One positive patch test to neroli oil in a case of occupational hand dermatitis in a cook/barman / fruit grower from contact allergy to lemon (oil), who also reacted to lemongrass oil and geraniol. The patient was patch tested with 'essences' (no concentration mentioned), but from the text it seems that this would indicate essential oils (37); geraniol may be present in commercial neroli oils in concentrations up to 3.6% (Table 5.60.2, column A).

LITERATURE

1. Dugo G, Bonaccorsi I, Sciarrone D, Costa R, Dugo P, Mondello L, Santi L, Fakhry HA. Characterization of oils from the fruits, leaves and flowers of the bitter orange tree. J Essent Oil Res 2011;23:45-59. Data also presented in ref. 17
2. Boussaada O, Skoula M, Kokkalou E, Chemli R. Chemical variability of flowers, leaves, and peel oils of four sour orange provenances. J Essent Oil Bear Plants 2007;10:453-464
3. Boussaada O, Chemli R. Chemical composition of essential oils from flowers, leaves and peel of *Citrus aurantium* L. var. *amara* from Tunisia. J Essent Oil Bear Plants 2006;9:133-139. Data reproduced from ref. 12

Table 5.60.4 Results of testing groups of patients with neroli oil

Years and Country	Test conc. & vehicle	Number of patients tested | positive (%)			Selection of patients (S); Relevance (R); Comments (C)	Ref.
Routine testing						
2000-2007 USA	2% pet.	324	1	(0.3%)	R: 100%; C: weak study: a. high rate of macular erythema and weak reactions, b. relevance figures include 'questionable' and 'past' relevance	27
Testing in groups of selected patients						
2001-2010 Australia	2% pet.	477	20	(4.2%)	S: not specified; R: 65%	42
2000-2008 IVDK	2% pet.	6,220	44	(0.7%)	S: patients with dermatitis suspected of causal exposure to fragrances; R: not stated	39
1989-1999 Portugal	2% pet.	67	4	(6.6%)	S: patients who had a positive patch test to the fragrance mix; R: not stated	41
1996-1997 UK	2% pet.	10	2	(20%)	S: patients suspected of cosmetic dermatitis and reacting to the fragrance mix; R: not stated	28

IVDK Information Network of Departments of Dermatology, Germany, Switzerland, Austria (www.ivdk.org); pet.: petrolatum

4   Wang K, Zhu R-Z, Qu R-F, Li Z-Y. Comprehensive two-dimensional gas chromatography–time-of-flight mass spectrometry for the analysis of volatile components in neroli essential oil. Mendeleev Commun 2012;22:45-46

5   Ben Hsouna A, Hamdi N, Ben Halima N, Abdelkafi S. Characterization of essential oil from *Citrus aurantium* L. flowers: antimicrobial and antioxidant activities. J Oleo Sci 2013;62:763-772. The analytical data in this article have previously been published in ref. 16 by other authors

6   Sarrou E, Chatzopoulou P, Dimassi-Theriou K, Therios I. Volatile constituents and antioxidant activity of peel, flowers and leaf oils of *Citrus aurantium* L. growing in Greece. Molecules 2013;18:10639-10647

7   Makhoul S, Bakkour Y, El-Nakat H, El Omar F. The Lebanese *Citrus aurantium*: A promising future in medicinal phytochemistry. J Pharmacogn Phytochem 2012;1:63-67

8   Buccellato F. Grapefruit flower. Origins, composition used and more. Perfum Flavor 2013;38(March):24-27

9   Lawrence BM. Progress in essential oils. Perfum Flavor 2003;27(Nov/Dec):58-? (last page unknown)

10 Boelens MH. Differences in chemical and sensory properties of orange flower and rose oils obtained from hydrodistillation and from supercritical CO_2 extraction. Perfum Flav 1997;22:31-35. Data cited in ref. 18

11 Zhu L-F, Li Y-H, Li B-L, Lu BY, Xia NH. Aromatic plants and essential constituents. South China Institute of Botany. Chinese Academy of Sciences, Hai Feng Publishing Co., distributed by Peace Book Co. Ltd., Hong Kong, 1993. Data cited in ref. 9

12 Bussaada O. Variation of essential oil yield and composition of *Citrus aurantium* var. *amara*. MSc Thesis, Mediterranean Agronomic Institute, Chania, Greece, 1995. Data cited in ref. 9. Part of the data were reproduced 11 years later in ref. 3

13 Hethelyi E, Palfine-Ledniczky M, Korany K, Bernath J, Banatfy R. GC, GC-MS determination of neroli bigarade oils from orange blossoms. Olaj Szappan Kosmet 1998;47:222-228. Data cited in ref. 9

14 Dugo G, Bartle KD, Bonaccorsi I, Catalfamo M, Cotreono A, Dugo P, et al. Advanced analytical techniques for the analysis of Citrus essential oils. Part 2. Volatile fraction: LC-HRGC and MDGC. Essenz Deriv Agrum 1999;69:159-217. Data cited in ref. 9. Data also published in refs. 20 and 21

15 Braun M, Franz G. Qualität Ätherischer Öle. Chirale Säulen decken Verfälschungen auf. Pharm Ztg 2001;146:11-17. Data cited in ref. 9

16 Ammar AH. Bouajila J, Lebrihi A, Mathieu F, Romdhane M, Zagrouba F. Chemical composition and in vitro antimicrobial and antioxidant activities of *Citrus aurantium* L. flowers essential oil (Neroli oil). Pak Journal Biol Sci 2012;15:1034-1040. The analytical data in this report later published by other authors in ref. 5

17 Bonaccorsi I, Sciarrone D, Schipilliti L, Trozzi A, Fakhry HA, Dugo G. Composition of Egyptian nerolì oil. Nat Prod Commun 2011;6:1009-1014. Data also presented in ref. 1

18 Dugo G, Peyron L, Bonaccorsi I. Extracts from the bitter orange flowers (*Citrus aurantium* L.): Composition and adulteration. In: G Dugo and L Mondello, Eds. Citrus oils. Composition, advanced analytical techniques, contaminants, and biological activity. Boca Raton, USA: CRC Press, 2010:333-348

19 Monsef-Esfahani HR, Amanzade Y, Alhani Z, Hajimehdipour H, Faramarzi MA. GC/MS analysis of *Citrus aurantium* L. hydrolate and its comparison with the commercial samples. Iran J Pharm Res 2004;3:177-179

20 Mondello L, Dugo P, Bartle KD, Frere B, Dugo G. On-line high performance liquid chromatography coupled with high resolution gas chromatography and mass spectrometry (HPLC–HRGC–MS) for the analysis of complex mixtures containing highly volatile compounds. Chromatographia 1994;39:529-538. Data also published in refs. 14 and 21

21 Mondello L, Dugo G, Dugo P, Bartle KD. On-line HPLC–HRGC in the analytical chemistry of citrus essential oils. Perfum Flav 1996;21:25-49 (issue number unknown). Data also published in refs. 14 and 20

22 Germanà, MA, De Pasquale F, Bazan E, Palazzolo E. Indagine sugli oli essenziali contenuti nei fi ori, nelle foglie e nei germogli di 5 specie di citrus. Essenz Deriv Agrum 1990;60:297-312. Data cited in ref. 18

23 Boelens MH, Oporto A. Natural isolates from Seville bitter orange tree. Perfum Flav 1991;16:2-7

24 Rhind JP. Essential oils. A handbook for aromatherapy practice, 2nd Edition. London: Singing Dragon, 2012

25 Lawless J. The encyclopedia of essential oils, 2nd Edition. London: Harper Thorsons, 2014

26 Nardelli A, D'Hooge E, Drieghe J, Dooms M, Goossens A. Allergic contact dermatitis from fragrance components in specific topical pharmaceutical products in Belgium. Contact Dermatitis 2009;60:303-313

27 Wetter DA, Yiannias JA, Prakash AV, Davis MD, Farmer SA, el-Azhary RA, et al. Results of patch testing to personal care product allergens in a standard series and a supplemental cosmetic series: an analysis of 945 patients from the Mayo Clinic Contact Dermatitis Group, 2000-2007. J Am Acad Dermatol 2010;63:789-798

28 Thomson KF, Wilkinson SM. Allergic contact dermatitis to plant extracts in patients with cosmetic dermatitis. Br J Dermatol 2000;142:84-88

29 Devleeschouwer V, Roelandts R, Garmyn M, Goossens A. Allergic and photoallergic contact dermatitis from ketoprofen: results of (photo) patch testing and follow-up of 42 patients. Contact Dermatitis 2008;58:159-166

30 Bleasel N, Tate B, Rademaker M. Allergic contact dermatitis following exposure to essential oils. Australas J Dermatol 2002;43:211-213

31 Keane FM, Smith HR, White IR, Rycroft RJG. Occupational allergic contact dermatitis in two aromatherapists. Contact Dermatitis 2000;43:49-51

32 Trattner A, David M, Lazarov A. Occupational contact dermatitis due to essential oils. Contact Dermatitis 2008;58:282-284

33 Newsham J, Rai S, Williams JDL. Two cases of allergic contact dermatitis to neroli oil. Br J Dermatol 2011;165(Suppl.1):76

34 Schubert HJ. Skin diseases in workers at a perfume factory. Contact Dermatitis 2006;55:81-83

35 Selvaag E, Holm J, Thune P. Allergic contact dermatitis in an aromatherapist with multiple sensitizations to essential oils. Contact Dermatitis 1995;33:354-355

36 Schaller M, Korting HC. Allergic airborne contact dermatitis from essential oils used in aromatherapy. Clin Exp Dermatol 1995;20:143-145

37 Audicana M, Bernaola G. Occupational contact dermatitis from citrus fruits: Lemon essential oils. Contact Dermatitis 1994;31:183-185

38 Nardelli A, Drieghe J, Claes L, Boey L, Goossens A. Fragrance allergens in 'specific' cosmetic products. Contact Dermatitis 2011;64:212-219

39 Uter W, Schmidt E, Geier J, Lessmann H, Schnuch A, Frosch P. Contact allergy to essential oils: current patch test results (2000–2008) from the Information Network of Departments of Dermatology (IVDK). Contact Dermatitis 2010;63:277-283

40 Matthieu L, Meuleman L, van Hecke E, Blondeel A, Dezfoulian B, Constandt L, Goossens A. Contact and photocontact allergy to ketoprofen. The Belgian experience. Contact Dermatitis 2004;50:238-241

41 Manuel Brites M, Goncalo M, Figueiredo A. Contact allergy to fragrance mix — a 10-year study. Contact Dermatitis 2000;43:181-182

42 Toholka R, Wang Y-S, Tate B, Tam M, Cahill J, Palmer A, Nixon R. The first Australian Baseline Series: Recommendations for patch testing in suspected contact dermatitis. Australas J Dermatol 2014, Sept. 7. doi: 10.1111/ajd.12186

Chapter 5.61   NIAOULI OIL

DEFINITION

Niaouli oil is the essential oil obtained from the leaves and twigs of the broad-leaved paperbark (paper bark tea tree, niaouli), *Melaleuca quinquenervia* (Cav.) S.T. Blake.

INCI NOMENCLATURE

Description/definition: Melaleuca quinquenervia oil is an essential oil hydrodistilled from the leaves of the plant, *Melaleuca quinquenervia*, Myrtaceae

INCI name EU: Melaleuca quinquenervia oil (perfuming name, not an INCI name proper)

INCI name USA: Not in the Personal Care Products Council Ingredient Database

CAS registry number(s): 8022-72-8 (refers to *Melaleuca bracteata*, a different species)

ISO (INTERNATIONAL ORGANIZATION FOR STANDARDIZATION) STANDARD

There is currently no ISO standard for niaouli oil.

It is generally agreed that niaouli oils are produced from the leaves of *Melaleuca quinquenervia* (Cav.) S.T. Blake (1,4,9,10,11,16,17,18) (niaouli is the New Caledonian name for this species). However, some commercial niaouli oils offered on the internet (e.g., essentialoils.co.za; aroma-pure.com; aromatherapybible.com) are claimed to be obtained from a related species, *Melaleuca viridiflora* (official name: *Melaleuca viridiflora* Sol. ex Gaertn.). We have found no indications for commercial production of niaouli oil from *M. viridiflora* and think this is a misinterpretation caused by the fact that *M. viridiflora* and *M. leucadendron* were formerly names used for the species *M. quinquenervia* (2,4,10). Indeed, *M. viridiflora* has frequently been confused with *M. quinquenervia*, and intermediates exist between these two species (10,17). For these reasons, although we cannot exclude that there may be some oils from *M. viridiflora* designated as 'niaouli oil' on the market, we limit the discussion here to *M. quinquenervia* leaf oils.

THE PLANT, THE OIL, AND THEIR USES

Melaleuca quinquenervia (Cav.) S.T. Blake, also called niaouli (4), is a small to medium-sized evergreen tree, which grows as a spreading tree up to 20 m high, with the trunk covered by a white, beige and grey thick papery bark and grey-green ovate (egg-shaped) leaves. It is native to Papua New Guinea, the coastal areas of North and East Australia, Indonesia and New Caledonia (where it is one of the most widespread trees [4]). The niaouli tree is naturalized in southern Africa, India, Malaysia, the Philippines, USA (Florida, Hawaii, where it is an invasive weed [3,10]), the Caribbean (West Indies) and it is cultivated in Madagascar (10,11,18, GRIN taxonomy for plants).

The essential oil of the leaves and twigs of *Melaleuca quinquenervia* is called niaouli oil, which is used in a variety of cosmetic products, especially in Australia. The oil is reported in herbalism and natural medicine to work as an antiseptic and antibacterial agent, to help with bladder infections, respiratory troubles, coughs, colds, neuralgia, rheumatism and catarrh (1, Wikipedia). It has been used in aromatherapy for a long time (1,4). Oils having high concentrations of (*E*)-nerolidol and linalool are used in fragrances (10).

CHEMICAL COMPOSITION

Niaouli oil of the 1,8-cineole type is a clear, colorless mobile liquid which has a fresh, herbal odor reminding of soft eucalyptus oil. The yield of essential oil from the leaves of *Melaleuca quinquenervia* (Cav.) S.T. Blake generally varies from 1.1 to 2.0 per cent. The main producing countries of this oil are Madagascar and New Caledonia.

Literature data (up to December 3, 2014) on the chemical composition of niaouli oils and unpublished analytical data from one of us (E.S.) are shown in Table 5.61.2 in alphabetical order. In niaouli oils from various origins, over 150 chemicals have been identified. About 46 per cent of these were found in a single reviewed publication only.

The major compounds found in niaouli oils from different sources are shown in Table 5.61.3. They include (highest concentrations in any study given) (*E*)-nerolidol (95.6%), 1,8-cineole (76.3%), viridiflorol (71.0%), γ-terpinene (32.4%), α-terpineol (29.5%), β-caryophyllene (28.1%), α-pinene (27.7%), limonene (15.0%), ledol (9.4%) and β-pinene (5.5%). Well-known ingredients of niaouli oils that were present in high concentrations (>7%) in one or two studies were *p*-cymene (40.0% and 27.1%), globulol (39.7% and 25.5%), citronellol (36.0%), linalool (three studies: 33.1%, 14.2% and 15.2%), terpinen-4-ol (10.9% and 24.6%), terpinolene (19.2% and 8.3%), caryophyllene oxide (19.0%), viridiflorene (17.1%) and allo-aromadendrene (9.5%). Uncommon or rare constituents of niaouli oils found in high concentrations (>7%) in single studies include longifolene (33.0%), 10-epi-γ-eudesmol (22.6%), guaiol (21.5%), spathulenol (18.3%), α-eudesmol (16.6%), β-eudesmol (15.2%), calamenene (13.6%) and neral (7.4%).

Commercial oils

The ten chemicals that had the highest maximum concentrations in 39 commercial niaouli essential oil samples (concentration ranges provided) are the following: (*E*)-nerolidol (0.3-80.3%), 1,8-cineole (45.3-61.2%), viridiflorol (1.0-22.6%), α-pinene (4.7-15.0%), limonene (0.8-9.9%), α-terpineol (2.1-9.2%), γ-terpinene (0.02-5.2%), ledol (0.6-4.8%), β-caryophyllene (0.5-4.1%) and β-myrcene (0.3-3.8%) (Erich Schmidt, unpublished analytical data).

Chemotypes

As can be seen in Table 5.61.2, the composition of niaouli oils may show extreme variations. On the basis of dominant ingredients, several chemotypes were proposed by various authors, but no consensus has been reached. Examples are shown in Table 5.61.1. The most important chemotypes are those dominated by either 1,8-cineole (eucalyptol), viridiflorol or (*E*)-nerolidol. However, there is considerable overlap. Thus, certain 1,8-cineole

Table 5.61.1 Chemotypes of *M. quinquenervia* suggested in literature (examples)

Chemotypes	Origin	Percentages (single, maximum, range, or means) and literature references
Main chemotypes		
1,8-Cineole (eucalyptol)	Australia	64.9 (16)
	Benin	60.0 (6)
	Egypt	57.2 (13)
	India	48.3 (7)
	Madagascar	34.9-71.1 (2); 41.8 (9); 51.5 (11)
	New Caledonia	76.3 (4)
Viridiflorol	Benin	62.7 (6)
	Brazil	71.0 (5,12)
	Madagascar	17.1-36.3 (2); 47.7 (9); 66.4 (11)
(*E*)-Nerolidol	Australia	95.0 (16)
	Madagascar	43.8-79.6 (2); 92.4 (11)
Other chemotypes		
1,8-Cineole + viridiflorol	Australia	1,8-cineole: 35, (*E*)-nerolidol: 24 (14)
	Benin	1,8-cineole: 33.2 (mean), viridiflorol: 32.0 (mean) (6)
	Benin	1,8-cineole: 33.1 (mean), viridiflorol: 29.8 (mean) (8)
1,8-Cineole, viridiflorol and terpinolene	Madagascar	1,8-cineole: 23, viridiflorol: 20, terpinolene: 5 (means) (11)
Terpinene chemotype	New Caledonia	*p*-cymene: 0-40, γ-terpinene: 0-32.4 and terpinolene: 0.05-19.2; may also contain up to 47% 1,8-cineole and up to 50% viridiflorol (4)
Viridiflorol and α-pinene	New Caledonia	viridiflorol: 67.4, α-pinene: 27.7 (4)

Table 5.61.2 Constituents identified in niaouli oils

Constituent	CAS	Percentage and range				
		A	B (10)	C (4)	D (11)	E
Allylisothiocyanate	57-06-7					0.05[m]
Aromadendrene	489-39-4	0.08-0.4		0-0.4	0-0.5	<0.1[g,k]; 0.3[r]; 1.7[l]
allo-Aromadendrene	25246-27-9	0.2-0.7	0-4.5	0-1.2	0-0.5	0.2[j,r]; 0.3[d,f,l]; 9.5[s]
Benzaldehyde	100-52-7	0.1-1.8	0.1-4.3	0-0.9	0-1.3	0.1[l]; 0.2[d,g]; 2.5[e]
Bicycloelemene	32531-56-9		0-1.3			
Bicyclogermacrene	24703-35-3		0.1-4.1			0.3[l,r]
Borneol	507-70-0	0.01-0.2				1.3[q]
α-Bulnesene	3691-11-0		0.2-0.6			0.1[l]
Bulnesol	22451-73-6		0-2.1	0-1.2		
Cadina-1,4-diene	29837-12-5		0-0.9		0-0.6	0.05[r]
α-Cadinene	24406-05-1				0-0.3	
β-Cadinene	523-47-7					0.05[r]
γ-Cadinene	39029-41-9	0.1-0.9		0-0.2	0-2.2[c]	<0.1[l]; 0.03[r]; 0.2[c,h]; 0.3[g]; 1.0[d]; 1.6[c,e]
δ-Cadinene	483-76-1	0.1-0.8	0-0.5	0-0.4	0-0.2[c]	<0.1[l]; 0.1[d]; 0.2[c,h,r]; 1.6[c,e]
α-Cadinol	481-34-5		0.1-2.6			0.8[f]; 0.9[p]; 2.1[d]
epi-α-Cadinol	5937-11-1	0.07-0.1	0-5.9			<0.1[g]; 0.05[r]; 0.4[d]; 1.0[f]; 3.5[p]
δ-Cadinol	19435-97-3		0-1.1		0-2.4	1.2[h]
Calacorene	38599-17-6		0-0.4			
trans-Calamene					0-0.3	
Calamenene	483-77-2		0-13.6			<0.1[l]
cis-Calamenene	72937-55-4			0-0.08		
Camphene	79-92-5	0.05-0.2	0-0.3		0-0.5	<0.1[l]; 0.05[r]; 0.1[g,j]; 0.2[d]
Camphor	76-22-2					0.4[d]; 1.7[q]

Table 5.61.2 Constituents identified in niaouli oils (*continued*)

Constituent	CAS	A	B (10)	C (4)	D (11)	E
δ3-Carene	13466-78-9			0-0.4	0-0.2	
(E)-Carveol	1197-07-5					<0.1[g]
Caryophylla-2(12), 6(13)-dien-5-ol						<0.1[g]
β-Caryophyllene	87-44-5	0.5-4.1	0-28.1	0-6.5	0-24.6	0.8[r]; 1.1[f]; 1.3[g,j,l]; 1.7[d]; 5.9[e]; 8.5[h]
Caryophyllene oxide	1139-30-6	0.2-0.6	0-19.0	0-2.4[b]	0.06-6.1	0.7[g]; 0.9[d]; 1.1[p]; 1.3[e,h]; 2.6[f]; 3.5[i]
α-Cedrane	13567-54-9					0.02[r]
epi-Cedrol	19903-73-2					0.4[r]
1,8-Cineole	470-82-6	45.3-61.2	0-75.1	0.1-76.3	0.03-51.5	59.3[p]; 63.2[q]; 64.9[l]; 70.4[k]; 71.1[e]
Citronellal	106-23-0					0.1[g,r]
Citronellic acid	502-47-6		0-2.3			
Citronellol	106-22-9	0.04-0.1	0.1-36.0			0.2[g]
Citronellyl acetate	150-84-5		0-1.4			
Citronellyl butyrate	141-16-2					0.03[r]
α-Copaene	3856-25-5		0-0.2		0-0.3	0.1[g]; 0.2[r]
Cubeban-11-ol	220766-71-2		0.2-3.2			
α-Cubebene	17699-14-8		0-1.6	0-0.3	0-0.2	0.2[r]
γ-Cubebene	147413-90-9					0.5[k]
Cubenol	21284-22-0		0.1-2.3			0.03[r]
ar-Curcumene	644-30-4			0-0.6		
p-Cymene	99-87-6		0.1-7.8	0-40.0	0-3.3	0.7[f]; 0.9[g]; 1.0[d]; 2.1[k]; 5.0[e]; 27.1[p]
p-Cymenene	1195-32-0	0.4-3.1				<0.1[g,l]
p-Cymen-8-ol	1197-01-9					0.1[g]
Dimethyl sulfide	75-18-3					<0.1[g]
β-Elemene	33880-83-0	0.2-0.3	0.1-0.5			
δ-Elemene	20307-84-0					0.1[l]
Epicubenol	19912-67-5		0-1.0			0.04[r]
Epiglobulol	88728-58-9		0-1.4			
Ethyl benzoate	93-89-0					0.1[g]
α-Eudesmol	473-16-5		0-16.6	0-4.7		<0.1[g]; 0.3[d]
β-Eudesmol	473-15-4		0.1-15.2	0-0.3		0.1[g]; 0.4[d]
γ-Eudesmol	1209-71-8		0-6.9	0-0.9		0.1[d]
10-epi-γ-Eudesmol	15051-81-7			0-22.6		0.1[r]
Eugenol	97-53-0					0.2[g]
(E,E)-α-Farnesene	502-61-4		0-0.7			0.1[l]
β-Farnesene	502-60-3					<0.1[l]; 0.1[l]
Farnesol	4602-84-0			0-0.03		
(E,E)-Farnesol	106-28-5		0-4.4			0.2[l]
α-Fenchene	471-84-1		0.01-0.2			0-0.3
α-Fenchol	512-13-0		0-4.0		0-0.3	0.1[g]; 0.7[d]
Fenchone	1195-79-5				0-0.05	
α-Fenchyl acetate	111821-74-0					0.1[r]
Geraniol	106-24-1		0-1.4			
Geranyl acetate	105-87-3		0-0.7			
Germacrene D	23986-74-5				0-0.5	0.03[r];
Globulol	489-41-8	0.06-0.3	0-39.7	0-2.4[b]	0-1.1	0.1[g]; 0.6[r]; 3.4[l]; 25.5[p]
Guaiol	489-86-1		0-1.0	0-21.5		0.2[d]
α-Gurjunene	489-40-7	0.07-2.8	0-0.8	0-0.04	0-1.2	0.1[l,r]; 0.5[h]; 3.1[e]
β-Gurjunene	73464-47-8					0.02[r]
Hexenyl butyrate	26912-31-2		0-3.3			
α-Himachalene	3853-83-6					0.04[r]; 0.3[d]
α-Humulene	6753-98-6	0.2-0.5	0-5.5	0-0.2	0-0.5	0.02[r]; 0.1[h]; 0.2[g,l]; 0.3[f]; 0.4[j]; 0.6[d]

Table 5.61.2 Constituents identified in niaouli oils (*continued*)

Constituent	CAS	Percentage and range				
		A	B (10)	C (4)	D (11)	E
4'-Hydroxy-3'-methylacetophenone	876-02-8					<0.1[g]
Isocaryophyllene	118-65-0			0-2.1		
Isoisopulegol	18674-65-2					0.2[g]
Isoledene	95910-36-4					0.03[r]
13-Isopropyl podocarpa-8,13-dien-15-ol	21414-53-9					0.2[d]
Isopulegol	89-79-2		0-0.2			0.4[g]
Ledol	577-27-5	0.6-4.8	0.1-7.7	0-8.8	0-9.4	0.2[l]; 0.3[p]; 0.4[g]; 1.5[d]; 4.4[h]; 7.8[e]
Limonene	138-86-3	0.8-9.9	0.1-14.3	0.1-15.0	0.9-6.5	6.1[f]; 6.8[l]; 7.4[k]; 7.6[d]; 8.2[p]; 11.7[e]
Linalool	78-70-6	0.03-3.3	0-33.1		0-2.3	0.5[d]; 0.6[g]; 6.3[p]; 6.6[e]; 14.2[r]; 15.2[m]
cis-Linalool oxide, furanoid	5989-33-3		0-0.3			
trans-Linalool oxide, furanoid	34995-77-2		0-0.3			
Linalyl acetate	115-95-7					2.7[p]
Longifolene	475-20-7					33.0[s]
trans-Longipinalol	66141-14-8					0.02[r]
p-Mentha-1(7),8-dien-2-ol (isocarveol)	35907-10-9					<0.1[g]
p-Mentha-1,3,8-triene	18368-95-1					<0.1[g]
cis-*p*-Menth-2-en-1-ol	29803-82-5	0.02-0.08	0-0.6			0.03[r]
trans-*p*-Menth-2-en-1-ol	29803-81-4					0.02[r]
Methyl benzoate	93-58-3		0-0.5		0-0.6	0.1[g]
Methyl citronellate	2270-60-2		0.1-0.8			
Methyl 2-methylbutyrate	868-57-5					<0.1[g]
Methyl thiobenzoate	5873-86-9			0-0.3		0.2[g]
α-Muurolene	10208-80-7					0.02[r]
α-Muurolol	104245-48-9					0.02[r]; 1.6[e]
τ-Muurolol (epi-α-)	19912-62-0		0-1.0			
Myrcene	123-35-3	0.3-3.8	0-2.1	0.03-2.0	0-3.0	0.6[g,k]; 0.7[p]; 0.8[d]; 1.2[j,l]; 2.5[e]; 2.6[q]
trans-Myrtanol	15358-91-5					0.06[r]
Myrtenol	515-00-4					<0.1[g]
Naphthalene	91-20-3		0-2.8			
Neral	106-26-3		0.1-7.4		0-1.0	
Nerol	106-25-2					<0.1[g]
(*E*)-Nerolidol	40716-66-3	0.3-80.3	0-95.6		0-92.4	1.2[h]; 3.2[d]; 9.1[g]; 24.2[m]; 95.0[e,l]
Neryl butyrate	999-40-6					<0.1[g]
(*E*)-α-Ocimene	6874-10-8					0.2[d]
(*E*)-β-Ocimene	3779-61-1		0-0.6	0-0.3		0.07[r]; 0.2[g]; 0.6[k]
(*Z*)-β-Ocimene	3338-55-4		0-0.4			
Palustrol	5986-49-2		0-1.0	0-0.03		<0.1[l]
5-epi-Paradisol	136734-27-5					0.1[d]
β-Patchoulene	514-51-2					0.02[r]
Phellandral	21391-98-0					<0.1[g]
α-Phellandrene	99-83-2	0.05-0.2	0.1-2.3	0-1.9		<0.05[d]; 0.07[r]; 0.1[l]; 1.0[p]
β-Phellandrene	555-10-2		0.1-1.1			
α-Pinene	80-56-8	4.7-15.0	0-17.9	0-27.7	0-21.5	5.6[g]; 6.2[f,j]; 6.9[k]; 9.2[n]; 12.5[d]; 16.9[e]
β-Pinene	127-91-3	0.2-3.3	0.1-5.3	0.1-5.5	0-1.7	1.9[g]; 2.1[r]; 2.3[j]; 2.7[d]; 4.5[e]; 4.8[p]
Pinocarveol	5947-36-4					0.1[g]
cis-Piperitol	16721-38-3		0-0.4			
trans-Piperitol	16721-39-4		0-1.7			0.03[r]
Rose oxide	16409-43-1		0-0.3			
Sabinene	3387-41-5	0.01-0.03	0-1.8		0-4.1	<0.1[k]; 0.1[l]

Table 5.61.2 Constituents identified in niaouli oils (*continued*)

Constituent	CAS	Percentage and range				
		A	B (10)	C (4)	D (11)	E
cis-Sabinene hydrate	15537-55-0					0.3[d]
trans-Sabinene hydrate	17699-16-0					0.04[r]
trans-Sabinol	471-16-9					0.03[r]
α-Selinene	473-13-2		0.1-4.0			0.2[g]
β-Selinene	17066-67-0		0-3.5	0-1.4	0-0.4	0.08[h]; 0.2[g]
Selin-11-en-4-ol	16641-47-7	0.09-0.3			0-0.6	
Spathulenol	6750-60-3		0-18.3			0.04[r]; 0.3[l]; 0.7[f]; 3.6[d]
α-Terpinene	99-86-5	0.01-0.4	0-2.1	0-6.7	0-2.1	0.2[d]; 0.3[g]; 0.4[r]; 0.5[l]; 3.8[p]
γ-Terpinene	99-85-4	0.02-5.2	0-8.1	0-32.4	0-13.5	0.7[l]; 0.9[g]; 1.0[d,f,j]; 7.4[e]; 19.0[p]
Terpinen-4-ol	562-74-3	0.03-3.4	0.1-10.9	0-5.0	0-24.6	1.3[m]; 1.4[f]; 1.5[r]; 1.8[d,f]; 4.4[p]; 4.6[e]
Terpinen-4-yl acetate	4821-04-9					<0.1[g]
α-Terpineol	98-55-5	2.1-9.2	0-14.1	0-24.5	0-17.1	13.3[g]; 15.0[d]; 20.3[p]; 29.1[q]; 29.5[n]
β-Terpineol	138-87-4					0.2[k]
γ-Terpineol	586-81-2					0.3[j]; 0.7[k]
δ-Terpineol	7299-42-5	0.09-0.1	0-0.9			0.1[l]; 0.2[r]; 0.5[g]
Terpinolene	586-62-9	0.01-2.4	0-1.1	0.04-19.2	0-8.3	0.2[g]; 0.3[d,h,j,k]; 0.5[l]; 3.7[e]; 4.9[p]
α-Terpinyl acetate	80-26-2	0.4-1.7	0-4.4	0-11.4		<0.1[l]; 3.7[p]; 4.4[d]
α-Thujene	2867-05-2	0.1-0.3		0-9.7		0.1[r]; 4.5[p]
3-Thujopsanone	25966-79-4					0.06[r]
Thujyl alcohol	513-23-5				0-0.2	
Valencene	4630-07-3					0.05[r]
Verbenone	80-57-9		0.1-0.7			
Viridiflorene (ledene)	21747-46-6	0.4-2.1	0-1.6	0-1.1	0-17.1	0.7[f,p]; 1.1[d]; 1.5[l]; 2.6[e]
Viridiflorol	552-02-3	1.0-22.6	0.1-65.7	0-67.4	0.07-66.4	36.3[e]; 47.7[h]; 47.9[f,o]; 62.7[d]; 71.0[i]
α-Ylangene	14912-44-8				0-0.2	

A thirty-nine niaouli essential oil samples from Madagascar, analyzed between 1998 and 2012; lowest and highest concentrations given (E. Schmidt, unpublished data)

B 136 niaouli oils prepared by steam-distillation with cohobation from trees growing in Australia and Papua New Guinea; lowest and highest concentrations given (ref. 10)

C 133 lab-hydrodistilled leaf oils from seven harvesting locations in New Caledonia; lowest and highest concentrations given (ref. 4)

D 144 oil samples from steam-distilled leaves from 48 trees growing in Madagascar collected during three seasons; lowest and highest concentrations given (ref. 11)

E data from other studies (indicated with superscript letters); highest concentrations found in any study reviewed here given; when two or more oils were investigated, only the highest concentrations are mentioned, unless indicated otherwise

[a] incorrect identification; [b] caryophyllene oxide and globulol combined; [c] γ-cadinene and δ-cadinene combined; [d] three hundred steam- or hydrodistilled oils from trees growing wild in Benin (ref. 6); [e] 159 industrially produced steam-distilled oils from leaves and terminal branchlets from Madagascar (ref. 2); [f] seventy-five lab-hydrodistilled oil samples from 25 trees growing wild in Benin (ref. 8); [g] one lab-hydrodistilled oil from India (ref. 7); [h] sixty-three leaf oil samples from 21 trees from Madagascar (ref. 9); [i] one lab-hydrodistilled leaf oil from Brazil (ref. 12); [j] one lab-hydrodistilled oil from Egypt (ref. 13); [k] one commercial oil purchased from an Italian company (ref. 15); [l] one cineole and one (*E*)-nerolidol chemotype laboratory steam-distilled oil from Australia (ref. 16); [m] one commercial oil from Australia (ref. 14); [n] one commercial oil of unknown origin (ref. 19); [o] nine lab-hydrodistilled oils from New Caledonia and one commercial sample (ref. 20); [p] unknown number of New Caledonian niaouli oils (ref 22); [q] unknown number of commercial oils of unknown origin (ref. 23); [r] one commercial *M. quinquenervia* oil sample from Australia (ref. 24); [s] one lab-hydrodistilled oil sample from Cuba (ref. 25)

tr: trace

Table 5.61.3 Major constituents of niaouli oils

Constituent	CAS	Percentage and range				
		A	B (10)	C (4)	D (11)	E
(*E*)-Nerolidol	40716-66-3	0.3-80.3	0-95.6		0-92.4	1.2[h]; 3.2[d]; 9.1[g]; 24.2[m]; 95.0[e,l]
1,8-Cineole	470-82-6	45.3-61.2	0-75.1	0.1-76.3	0.03-51.5	59.3[p]; 63.2[q]; 64.9[l]; 70.4[k]; 71.1[e]
Viridiflorol	552-02-3	1.0-22.6	0.1-65.7	0-67.4	0.07-66.4	36.3[e]; 47.7[h]; 47.9[f,o]; 62.7[d]; 71.0[i]
γ-Terpinene	99-85-4	0.02-5.2	0-8.1	0-32.4	0-13.5	0.7[l]; 0.9[g]; 1.0[d,f,j]; 7.4[e]; 19.0[p]; 29.5[n]
α-Terpineol	98-55-5	2.1-9.2	0-14.1	0-24.5	0-17.1	13.3[g]; 15.0[d]; 20.3[p]; 29.1[q]; 29.5[n]
β-Caryophyllene	87-44-5	0.5-4.1	0-28.1	0-6.5	0-24.6	0.8[r]; 1.1[f]; 1.3[g,j,l]; 1.7[d]; 5.9[e]; 8.5[h]
α-Pinene	80-56-8	4.7-15.0	0-17.9	0-27.7	0-21.5	5.6[g]; 6.2[f,j]; 6.9[k]; 9.2[n]; 12.5[d]; 16.9[e]
Limonene	138-86-3	0.8-9.9	0.1-14.3	0.1-15.0	0.9-6.5	6.1[f]; 6.8[l]; 7.4[k]; 7.6[d]; 8.2[p]; 11.7[e]
Ledol	577-27-5	0.6-4.8	0.1-7.7	0-8.8	0-9.4	0.2[l]; 0.3[p]; 0.4[g]; 1.5[d]; 4.4[h]; 7.8[e]
β-Pinene	127-91-3	0.2-3.3	0.1-5.3	0.1-5.5	0-1.7	1.9[g]; 2.1[r]; 2.3[j]; 2.7[d]; 4.5[e]; 4.8[p]

LEGEND: SEE UNDER TABLE 5.61.2

chemotype oils may contain fairly high concentrations of (*E*)-nerolidol (up to 20.2%) and viridiflorol (up to 17.5%; possibly higher, ref. 4); viridiflorol chemotype oils may contain up to 31% 1,8-cineole and up to 24.2% (*E*)-nerolidol; and the (*E*)-nerolidol chemotype oils may contain up to 16.2% 1.8-cineole and up to 11.8% viridiflorol (2,4,6,8,9). In fact, in Benin a chemotype was found which contains equal amounts of 1,8-cineole and viridiflorol (both approximately 30% (6,8). Some other chemotypes proposed in literature are shown in Table 5.61.1.

Other authors have suggested that all previously described chemotypes can be catered for under two, all encompassing, chemotypes (10). Their chemotype 1 comprises (*E*)-nerolidol (74-95%) without or with significant quantities of linalool (14-40%). Chemotype 2 contains 1,8-cineole (10-75%), viridiflorol (13-66%), α-terpineol (0.5-14%) and β-caryophyllene (0.5-28%) in varying proportions and order of dominance in the oils (10).

In one report, longifolene (33%) was the dominant ingredient of niaouli oil (25). Earlier reports of chemotypes rich in (*E*)-methylisoeugenol (up to 88%) or methyl eugenol (up to 99%), and a linalool-type (cited in ref. 11) may have been due to botanical confusion. In the case of the (*E*)-methylisoeugenol and methyl eugenol types, for example, *M. leucadendra* was analyzed, not *M. quinquenervia* (12). Most commercial oils are (said to be) rich in 1,8-cineole (1,2,17), but other chemotypes are also produced (2).

An additional analysis of niaouli oil not presented in Table 5.61.2 can be found in ref. 26.

CONTACT ALLERGY/ALLERGIC CONTACT DERMATITIS

General

There have been only four publications on allergic contact dermatitis from niaouli oil. In one case, α-pinene may have been an allergen in niaouli oil.

Case reports

An aromatherapist had non-occupational contact dermatitis with allergies to multiple essential oils used at work, including niaouli oil. The patient also reacted to geraniol, linalool, linalyl acetate, α-pinene, the fragrance mix and various other fragrance materials. α-Pinene was demonstrated by GC-MS in niaouli oil (28). In commercial oils, α-pinene has been found in a maximum concentration of 15.0% (Table 5.61.2, column A). One positive patch test reaction to niaouli oil, which was present in a cosmetic that had given a positive patch test or had been positive in a usage test in patients, was seen in a 9-year period in one clinic in Belgium (29). Two positive patch tests to niaouli oil, probably from its presence in a topical pharmaceutical preparation, have been observed (27).

Additional information on contact allergy to niaouli oil may be found in ref. 30 (article not read).

LITERATURE

1 Barbosa LCA, Silva CJ, Teixeira RR, Meira RMSA, Pinheiro AL. Chemistry and biological activities of essential oils from *Melaleuca* L. species. Agriculturae Conspectus Scientificus 2013;78:11-23

2 Ramanoelina PAR, Bianchini JP, Gaydou EM. Main industrial niaouli (*Melaleuca quinquenervia*) oil chemotype productions from Madagascar. J Essent Oil Res 2008;20:261-266

3 Wheeler GS, Pratt PD, Giblin-Davis RM, Ordung KM. Intraspecific variation of *Melaleuca quinquenervia* leaf oils in its naturalized range in Florida, the Caribbean, and Hawaii. Biochem Syst Ecol 2007;35:489-500

4 Trilles BL, Bombarda I, Bouraïma-Madjebi S, Raharivelomanana P, Bianchini J-P, Gaydou EM. Occurrence of various chemotypes in niaouli [*Melaleuca quinquenervia* (Cav.) S.T. Blake] essential oil from New Caledonia. Flavour Fragr J 2006; 21:677-682

5 Silva CJ. Morfoanatomia foliar e composição química dos óleos essenciais de sete espécies de Melaleuca L. (Myrtaceae) cultivadas no Brasil. Tese de Mestrado. Universidade Federal de Viçosa, Brasil, 2007. Data cited in ref. 1

6 Gbenou JD, Moudachirou M, Chalchat J-C, Figuérédo G. Chemotypes in *Melaleuca quinquenervia* (Cav.) S.T. Blake (niaouli) from Benin using multivariate statistical analysis of their essential oils. J Essent Oil Res 2007;19:101-104

7 Philippe J, Goeb P, Suvarnalatha G, Sankar R, Suresh S. Chemical composition of *Melaleuca quinquenervia* (Cav.) S.T. Blake leaf oil from India. J Essent Oil Res 2002;14:181-182

8 Moudachirou M, Gbenou JD, Garneau F-X, Jean F-I, Gagnon H, Koumaglo KH, et al. Leaf oil of *Melaleuca quinquenervia* from Benin. J Essent Oil Res 1996;8:67-69

9 Ramanoelina PAR, Bianchini J-P, Andriantsiferana M, Viano J, Gaydou EM. Chemical composition of niaouli essential oils from Madagascar. J Essent Oil Res 1992;4:657-658

10 Ireland BF, Hibbert DB, Goldsack RJ, Doran JC, Brophy JJ. Chemical variation in the leaf oil of *Melaleuca quinquenervia* (Cav.) S.T. Blake. Biochem Syst Ecol 2002;30:457-470

11 Ramanoelina PAR, Viano J, Bianchini J-P, Gaydou EM. Occurrence of various chemotypes in niaouli (*Melaleuca quinquenervia*) essential oils from Madagascar using multivariate statistical analysis. J Agric Food Chem 1994;42:1177-1182

12 Silva VJ, Barbosa LCA, Maltha CRA, Pinheiro AL, Ismail FMD. Comparative study of the essential oils of seven *Melaleuca* (Myrtaceae) species grown in Brazil. Flavour Fragr J 2007;22:474-478

13 Aboutabl EA, Tohamy SFE, De Pooter HL, De Buyck LF. A comparative study of the essential oils from three *Melaleuca* species growing in Egypt. Flavour Frag J 1991;6:139-141

14 Lee Y-S, Kim J, Shin S-C, Lee S-G, Park IK. Antifungal activity of Myrtaceae essential oils and their components against three phytopathogenic fungi. Flavour Fragr J 2008;23:23-28

15 Monti D, Tampucci S, Chetoni P, Burgalassi S, Bertoli A, Pistelli L. Niaouli oils from different sources: Analysis and influence on cutaneous permeation of estradiol *in vitro*. Drug Deliv 2009;16:237-242

16 Brophy JJ, Doran JC. Essential oils of tropical *Asteromyrtus*, *Callistemon* and *Melaleuca* species, ACIAR Monograph No. 40. Canberra, Australia: Australian Centre for International Agricultural Research, 1996:95-97

17 Doran JC, Turnbull JW. Australian trees and shrubs: species for land rehabilitation and farm planting in the tropics. ACIAR Monograph No. 24. Canberra, Australia: Australian Centre for International Agricultural Research, 1997:320-323

18 Lawrence BM. Progress in essential oils. Perfum Flavor 2013;38(5):44-48

19 Christoph F, Stahl-Biskup E, Kaulfers P-M. Death kinetics of *Staphylococcus aureus* exposed to commercial tea tree oils s.l. J Essent Oil Res 2001;13:98-102

20 Radoias G, Bosilcov A, Delubriat J-L. Study on the complete extraction technique of the essential oil of niaouli – *Melaleuca quinquinervia* (Cav.) S.T. Blake – from New Caledonia. Poster presented at the 39th ISEO Meeting, Quedlinburg, Germany, 2008. Data cited in ref. 18

21 Lawrence BM. Progress in essential oils. Perfum Flavor 2004;29(8):52-? (last page unknown)

22 Trilles B, Bouraïma-Madjebiand S, Valet G. *Melaleuca quinquenervia* (Cavanilles) S.T. Blake, niaouli. In: I Southwell and R Lowe, Eds. The genus Melaleuca. London: Harwood Academic Publ, 1999:237-245. Data cited in ref. 21

23 Hethelyi E, Takacs G, Ledniczky MP, Domokos J. Gas chromatographic investigation of the biologically active components of *Melaleuca* species and of natural cosmetic components containing tea tree oil. Olaj Szappan Kozmet 2000;49:25-37. Data cited in ref. 21

24 Shellie R, Marriott P, Zappia G, Mondello L, Dugo G. Interactive use of linear retention indices on polar and apolar columns with an MS-library for reliable characterization of Australian tea tree and other *Melaleuca* sp. oils. J Essent Oil Res 2003;15:305-312

25 Pino O, Sánchez Y, Rojas M, Rodríguez H, Abreu Y, Duarte Y, et al. Chemical composition and pesticidal activity of Melaleuca quinquenervia (Cav) S.T. Blake essential oil. Revista de Proteccion Vegetal 2011;26(3):177-186 (in Spanish)

26 Harkenthal M, Reichling J, Geiss HK, Saller R. Comparative study on the *in vitro* antibacterial activity of Australian tea tree oil, cajuput oil, niaouli oil, manuka oil, kanuka oil, and eucalyptus oil. Pharmazie 1999;54(6):460-463

27 Nardelli A, D'Hooge E, Drieghe J, Dooms M, Goossens A. Allergic contact dermatitis from fragrance components in specific topical pharmaceutical products in Belgium. Contact Dermatitis 2009;60:303-313

28 Dharmagunawardena B, Takwale A, Sanders KJ, Cannan S, Roger A, Ilchyshyn A. Gas chromatography: an investigative tool in multiple allergies to essential oils. Contact Dermatitis 2002;47:288-292

29 Nardelli A, Drieghe J, Claes L, Boey L, Goossens A. Fragrance allergens in 'specific' cosmetic products. Contact Dermatitis 2011;64:212-219

30 Escande JP, Foussereau J, Lantz JP, Basset A. Le problème des fausses sensibilisations croisées dans les allergies de groupe aux allergènes vegetaux. Rev Fr Allergol 1973;13:70-75

Chapter 5.62   NUTMEG OIL

DEFINITION
Nutmeg oil (essential oil of nutmeg) is the essential oil obtained from the seed of the nutmeg, *Myristica fragrans* Houtt.

INCI NOMENCLATURE
Description/definition: Myristica fragrans kernel oil is the oil obtained from the kernel of the nutmeg, *Myristica fragrans*, Myristicaceae
INCI name EU: Myristica fragrans kernel oil
INCI name USA: Myristica fragrans (nutmeg) kernel oil
Other names: Myristica oil; mace oil
CAS registry number(s): 8008-45-5; 84082-68-8; 8007-12-3 (wrong CAS number in INCI, refers to the fixed oil, nutmeg butter)
EINECS number(s): 282-013-3

ISO (INTERNATIONAL ORGANIZATION FOR STANDARDIZATION) STANDARD
ISO number: 3215
ISO name: Essential oil of nutmeg, Indonesian type
Botanical origin: *Myristica fragrans* Houtt.
Parts of plant used: Seed
ISO values: ISO values (minimum and maximum concentrations) are shown in Table 5.62.1.

Nutmeg essential oil should not be confused with the fixed oil (vegetable oil) of the nutmeg, called nutmeg butter.

Nutmegs usually contain about 33% of fixed oil and 4.5% of essential oil. The fixed oil is obtained by expressing the crushed nuts or by extracting with solvents. Fixed oil is a semi-solid reddish brown or yellow fat with both the aroma and taste of nutmeg. Its main constituent (>80%) is trimyristin (myristic acid triglycerides); other ingredients are unsaponifiable constituents (10%) , oleic acid (as glyceride), resinous material, linolenic acid (as glyceride) and very small amounts of formic, acetic and cerotic acids (22,27). Nugmeg butter is used to make candles and is added to certain salves and in medicines for external application to treat sprains and rheumatism (27).

THE PLANT, THE OIL, AND THEIR USES
Myristica fragrans Houtt. is a spreading, medium- to large-sized, aromatic evergreen tree usually growing to around 5-13 meters high, occasionally 20 meters. The origin of the nutmeg is uncertain but may include the Indonesian Molucca Islands (the 'spice islands'). The tree is cultivated in India, Sri Lanka, Indonesia, Malaysia, the Caribbean (Grenada) and to a lesser extent elsewhere, mainly for its nutmegs, the hard kernels of its seed (21, 27, GRIN Taxonomy for Plants, http://www.worldagroforestry.org). The export of nutmeg is dominated by Indonesia and Grenada, and broadly speaking both nutmeg and its derivatives are classified as East Indian (Indonesia, Malaysia, Sri Lanka) or West Indian (Carribean, Grenada) (17,21). The East Indian nutmeg is superior in flavor to the West Indian (21,27).

Nutmeg is the hard kernel of the seed of *M. fragrans*; dried crimson aril (bright red cover of the seed) is the spice mace. Nutmeg is a mild baking spice with a sweet smell commonly added to sausages, meats, fish, soups, et cetera. Nutmeg is widely employed as a traditional medicine in the Middle East and Asia. In Western folk medicine, nutmeg may be used as a stomachic, stimulant, carminative as well as for intestinal catarrh and colic, headaches, diarrhea, vomiting (also during pregnancy), nausea, fever, bad breath, rheumatism, to stimulate appetites and to control flatulence. In Ayurvedic medicine it is used to treat mild fever, asthma, and to reduce the catarrh of the respiratory tract (21,27). Nutmeg is believed to have aphrodisiac, stimulant, narcotic, carminative, astringent, hypolipidemic, antithrombotic, anti-platelet aggregation, antifungal, anti-dysenteric, and anti-inflammatory properties (15,20,29). In high doses, nutmeg has (mild) hallucinogenic effects, which is ascribed to the phenylpropanoids myristicin and elemicin contained in them (1,11,21, 29). Cases of serious nutmeg poisoning, which can occur after ingestion of 5 grams of nutmeg (20), with gastrointestinal symptoms (nausea, abdominal pain, vomiting), dry mouth, variations in the pupil width (both myosis and mydriasis) and central nervous system and cardiovascular symptoms, have repeatedly been described (13,14).

The essential oil of nutmeg is obtained by steam distillation of ground nutmeg. Nutmeg oil is used in the flavoring of meat products, pastry, liqueurs, in ketchups and soft drinks, and in the pharmaceutical, perfumery and cosmetics (mainly for men) industry and to flavor tobacco (11,17). The oil is used externally for rheumatism and internally as a carminative; it is also used in soups as postpartum medications (1). The use of essential oils through aromatherapy is gaining in importance (27,41). Nutmeg essential oil has been reported to exhibit anti-inflammatory, antimicrobial, anti-obesity, larvicidal, and molluscicidal activities (21).

Mace oil (the essential oil obtained from the aril [i.e., the bright red cover of the seed]) has a composition which is similar to nutmeg oil and therefore is a direct substitute for nutmeg oil (27). However, its composition (40) is not discussed here.

CHEMICAL COMPOSITION
Nutmeg oil is a nearly colorless to pale yellow clear mobile liquid which has a spicy, aromatic, slight peppery sweet odor. The yield of essential oil from the nut of *Myristica fragrans* Houtt. generally varies from 6.0 to 16.0 per cent. The main producing countries of this oil are Indonesia, Sri Lanka and Malaysia.

Literature data (up to December 7, 2014) on the chemical composition of nutmeg oils and unpublished analytical data from one of us (E.S.) are shown in Table 5.62.2 in alphabetical order. In nutmeg oils from various origins, over 120 chemicals have been identified. The major compounds found in nutmeg oils from different sources are shown in Table 5.62.3. They include (highest concentrations in any study given) sabinene (57.0%), myristicin (45.6%), terpinen-4-ol (31.3%), α-pinene (27.6%),

Table 5.62.1 ISO values (%) for nutmeg oil [a]

Compound	CAS	Minimum	Maximum
Sabinene	3387-41-5	14.0	29.0
α-Pinene	80-56-8	15.0	28.0
β-Pinene	127-91-3	13.0	18.0
Myristicin	607-91-0	5.0	12.0
Limonene	138-86-3	2.0	7.0
γ-Terpinene	99-85-4	2.0	6.0
Terpinen-4-ol	562-74-3	2.0	6.0
Safrole	94-59-7	1.0	2.5
δ3-Carene	13466-78-9	0.5	2.0

[a] ISO 3215 Essential oil of nutmeg, Indonesian type ©ISO 1998; Geneva, Switzerland, www.iso.org

β-pinene (23.9%), γ-terpinene (11.2%), α-terpinene (9.8%), limonene (8.8%) and p-cymene (7.4%). Well-known ingredients of nutmeg oils that were present in high concentrations (>7%) in one or two studies were elemicin (29.7%), safrole (22.1%), eugenol (19.9%), methyl eugenol (16.7% [wrong identification] and 17.7%), myrcene (10.5%) and linalool (7.4%). An uncommon constituent of nutmeg oils found in a high concentration (>7%) in a single study was methyl Isoeugenol (16.8%).

Commercial oils

The ten chemicals that had the highest maximum concentrations in 51 commercial nutmeg essential oil samples (concentration ranges provided) are the following: sabinene (16.5-36.7%), α-pinene (12.6-25.3%), β-pinene (2.6-16.5%), myrcene (2.3-10.5%), myristicin

Table 5.62.2 Constituents identified in nutmeg oils

Constituent	CAS	A	B (2)	C (2)	D (2)	E (16)	F
Acetic acid	64-19-7						+[r]
β-Asarone	5273-86-9						0.03[c]; 1.1[b]
α-Bergamotene	17699-05-7						+[e]
α-cis Bergamotene	18252-46-5		tr				
α-trans-Bergamotene	13474-59-4	0.05-0.1				0.1	<0.05[f]; 0.1[x]; 0.2[b]
β-trans-Bergamotene	15438-94-5		tr	tr			
Bicyclogermacrene	24703-35-3						0.1[b]
β-Bisabolene	495-61-4	0.01-0.06	tr	tr	tr	0.1	+[e]
Borneol	507-70-0						+[e]; 0.1[i]; 0.3[q,w]; 2.1[l]
Bornyl acetate	76-49-3	0.05-0.2	tr	tr	0.1		+[e]; 0.1[q,w]; 0.2[c]; 0.4[f]; 1.1[b]
Butyric acid	107-92-6						+[r]
δ-Cadinene	483-76-1	0.04-0.1	tr	tr	tr-0.3	0.1	+[e]
α-Cadinol	481-34-5						0.7[l]
Camphene	79-92-5	0.2-0.4	0.2	tr-0.2	0.1-0.6		0.3[h]; 0.4[e,q]; 0.5[m]; 1.0[d]; 4.4[l]
Camphor	76-22-2						+[r]
δ2-Carene (δ4-carene)	554-61-0						1.3[a,o]; 2.9[d]
δ3-Carene	13466-78-9	1.0-1.5	tr	0.3	0.5-1.5		0.8[x]; 0.9[d]; 1.0[o]; 1.4[e]; 1.6[s]; 1.7[f]
β-Caryophyllene	87-44-5	0.05-0.6	tr	tr	0.4-0.6	0.3	+[e]; 0.2[t,x]; 0.3[s]; 0.4[u]; 0.8[b]; 0.9[f]
β-Cedrene	546-28-1		tr				
Chavicol	501-92-8						5.4[l]
1,8-Cineole	470-82-6						0.1[i]; 0.8[p]; 1.5[l]; 2.2[y]; 3.2[n]; 3.5[h]
Citronellol	106-22-9		tr	tr	0.2	0.2	+[e]; <0.05[f]; 0.8[c]; 1.2[l]
Citronellyl acetate	150-84-5	0.01-0.1	tr	tr	tr		+[e]; 0.2[g]; 0.3[m]; 0.5[f]
α-Copaene	3856-25-5	0.1-0.5	0.1	tr-0.8	tr-0.5		+[e]; <0.1[b]; 0.1[i]; 0.3[n]; 0.6[f]; 0.7[k]
α-Cubebene	17699-14-8	0.04-0.2	tr	tr	tr-0.2		+[e]; <0.05[f]; 0.1[m]; 0.2[b,g,t]
β-Cubebene	3744-15-5						<0.1[b]; 0.1[x]; 0.3[g]
Cumene	98-82-8						+[r]
Cyclamen aldehyde	103-95-7						+[r]
o-Cymene	527-84-4						0.8[t]; 2.0[v]
p-Cymene	99-87-6	0.5-2.9	0.2	0.7-3.2	0.3-2.7		1.9[q]; 3.2[n,y]; 3.3[b]; 5.6[d]; 6.5[i]; 7.4[l]
p-Cymenene	1195-32-0						+[e]; 0.1[i]
p-Cymen-8-ol	1197-01-9						+[e]; <0.05[f]; 0.3[i]
β-Damascenone (E)-	23726-93-4						tr[v]

Table 5.62.2 Constituents identified in nutmeg oils (*continued*)

Constituent	CAS	Percentage and range					
		A	B (2)	C (2)	D (2)	E (16)	F
2,6-Dimethyl-2,6-octadiene	2792-39-4					0.4	
Docosane	629-97-0						<0.05f
Eicosane	112-95-8					0.1	
Elemicin	487-11-6	0.2-2.2				0.6	2.9[d]; 4.6[n]; 4.8[i]; 5.6[f]; 29.7[j]
cis-1,2-Epoxyterpinen-4-ol	1753-41-9						1.1[i]
Ethyl hexadecanoate	628-97-7						0.07[c]
Ethyl oleate	111-62-6						0.01[c]
Ethyl tetradecanoate	124-06-1						0.04[c]
Eugenol	97-53-0	0.02-0.3	tr	tr-0.2	<0.3-0.7	0.2	0.7[n]; 0.8[b]; 1.1[s]; 2.6[j]; 19.9[t]
α-Farnesene	502-61-4		tr				+[e]
β-Farnesene	502-60-3		tr	tr			
(*E*)-β-Farnesene	18794-84-8	0-0.1				0.1	
Fenchyl alcohol	1632-73-1					8.1	+[e]; 0.05[c]
Formic acid	64-18-6						+[r]
Geraniol	106-24-1	0-0.08	tr	tr	tr		+[e]; 0.1[q,w]
Geranyl acetate	105-87-3	0.07-0.4	tr	0.2	tr-0.3	0.4	+[e]; 0.1[m]; 0.2[i]; 0.3[b]; 0.6[g]; 0.9[h]
Germacrene D	23986-74-5	0.01-0.1	tr		tr	0.1	+[e]; 0.1[x]; 0.3[b]
Guaiol	489-86-1					0.1	
Heptacosane	593-49-7					0.1	
Hexacosane	630-01-3					0.1	
Hexadecanoic acid	57-10-3						0.03[c]
α-Humulene	6753-98-6					0.1	+[e]; tr[b]; <0.05[f]
Isobornyl acetate	125-12-2					0.2	
(*E*)-Isoelemicin	5273-85-8						+[e]; 0.03[h]
(*Z*)-Isoelemicin	5273-84-7						+[e]
Isoeugenol	97-54-1	0.03-0.7					1.7[c]
(*E*)-Isoeugenol	5932-68-3						+[e]; 0.4[o]; 1.3[d]
(*Z*)-Isoeugenol	5912-86-7						+[e]
Isosylvestrene	61557-13-9						0.5[v]
Isoterpinolene	586-63-0						1.4[t]
Limonene	138-86-3	3.8-4.5	3.6	2.9-4.4	2.0-7.0	8.8	4.5[s]; 5.6[c]; 6.3[g]; 7.5[k]; 7.9[d]
Linalool	78-70-6	0.1-0.3	0.4	0.2-0.9	0.2-7.4	0.7	0.6[g,h]; 0.8[c]; 1.0[e]; 2.2[s]; 3.1[l]
Linalyl acetate	115-95-7	0.01-0.03					+[e]; 0.06[c]
trans-p-Menth-2-ene-1,4-diol	21473-37-0						+[e]; 0.2[i]
cis-p-Menth-2-en-1-ol	29803-82-5	0.04-0.4	tr	0.1-0.4	0.1-1.2		0.1[i]; 0.2[t]; 0.4[b]; 0.5[g]; 0.7[f]; 1.2[h]
trans-p-Menth-2-en-1-ol	29803-81-4	0.06-0.6	tr		0.2-0.3		+[e]; 0.2[i]; 0.3[b,x]; 0.4[f,t]
Menthone	89-80-5						+[r]
Menthyl isovalerate	16409-46-4						+[r]
5-Methoxyeugenol	90377-06-3						+[e]
6-Methoxyeugenol	6627-88-9						0.1[c]
3-Methyl-4-decen-1-ol (*E*-)	24404-71-5						+[e]
3-Methyl-4-decenyl acetate							+[e]
Methyl eugenol	93-15-2	0.1-0.7	tr	0.1-0.2	0.5-1.2	0.1	1.8[k]; 2.4[u]; 2.6[g]; 16.7[t,z]; 17.7[d]
Methyl isoeugenol	93-16-3	0.02-0.4					16.8[t]
(*E*)-Methylisoeugenol	6379-72-2					0.1	+[e]; 0.1[i]; 0.2[x]; 0.5[u]
Methyl tetradecanoate	124-10-7						0.3[q,w]
α-Muurolene	10208-80-7					0.4	
Myrcene	123-35-3	2.3-10.5	2.6	2.2-3.4	0.3-4.0	2.5	2.2[k]; 2.4[p]; 2.5[g,m]; 2.9[n]; 3.1[b]
α-Myrcene	1686-30-2						0.8[t]; 2.4[c]
Myristicin	607-91-0	1.7-9.9	tr	0.5-0.8	3.3-13.5	6.8	14.0[e]; 16.2[k]; 17.7[y]; 31.3[v]; 45.6[j]
Nerol	106-25-2						+[e]

Table 5.62.2 Constituents identified in nutmeg oils (*continued*)

Constituent	CAS	Percentage and range					
		A	B (2)	C (2)	D (2)	E (16)	F
Neryl acetate	141-12-8						+[e]; 0.2[q]
(*Z*)-α-Ocimene	6874-44-8		tr				
β-Ocimene	13877-91-3						0.03[c]
(*E*)-β-Ocimene	3779-61-1	0-0.03	tr			0.7	
(*Z*)-β-Ocimene	3338-55-4	0.03-0.2	tr				
Octacosane	630-02-4					0.1	
Octadecane	593-45-3					1.3	
Octadecanoic acid	57-11-4						0.01[c]
Octanoic acid	124-07-2						+[r]
α-Phellandrene	99-83-2	0.8-1.2	0.3	0.3-0.7	0.4-5.8		0.7[p,s]; 1.0[n,u]; 1.1[d]; 1.6[e]; 4.3[h]
β-Phellandrene	555-10-2	1.9-3.2	tr		tr-2.4		4.0[v]; 4.5[o]; 5.5[l]; 6.7[b]; 7.5[u]
α-Pinene	80-56-8	12.6-25.3	22.2	1.6-12.6	18.0-21.2	8.5	22.0[q]; 22.8[d]; 25.1[s]; 27.6[o]
β-Pinene	127-91-3	2.6-16.5	16.4	7.8-12.1	8.7-17.7	3.5	17.7[n]; 18.8[s]; 21.5[q,w]; 23.9[k]
cis-Piperitol	16721-38-3	0.03-0.2	tr	0.4-1.2	0.1-0.6		+[e]; <0.05[f]; 0.1[b,l,t]; 0.5[h]; 0.6[n]
trans-Piperitol	16721-39-4		tr		tr		+[e]; <0.05[f]; 0.1[b,l]; 2.1[i]
Sabinene	3387-41-5	16.5-36.7	37.0	42.0-57.0	14.0-44.1		49.1[b]; 50.1[j]; 50.7[n]; 56.5[p]
cis-Sabinene hydrate	15537-55-0	0.2-1.0	0.3	0.2-0.8	tr-0.6		0.2[t]; 0.6[v]; 0.7[n]; 0.8[p]; 1.6[b]; 2.3[f]
trans-Sabinene hydrate	17699-16-0	0.2-0.7	0.4	0.3-2.4	tr-0.6		+[e]; 0.03[c]; 0.5[u]; 0.8[h,n]; 2.4[p]; 3.1[f]
cis-Sabinene hydrate acetate	77318-48-0		tr				
trans-Sabinene hydrate acetate	77318-47-9		tr				
Safrole	94-59-7	0.7-1.8	tr	0.1-0.5	0.3-3.3	0.4	3.3[e]; 3.4[d]; 3.9[k]; 4.3[c,v]; 22.1[j]
Seychellene	20085-93-2					0.1	
α-Terpinene	99-86-5	2.4-4.6	1.4	0.8-4.2	tr-5.2	9.8	2.7[c]; 3.5[i]; 3.8[k]; 4.0[n]; 4.8[d]; 6.4[u]
γ-Terpinene	99-85-4	3.6-7.4	2.4	1.7-4.7	1.3-7.7	9.9	4.2[v]; 6.8[k,n]; 7.3[d]; 7.8[i]; 11.2[u]
Terpinen-4-ol	562-74-3	0.2-8.4	4.0	3.0-6.4	1.0-10.9	21.3	10.9[n]; 12.2[d]; 13.9[c]; 15[u]; 31.3[i]
Terpinen-4-yl acetate	4821-04-9						+[e]; 0.6[h]
α-Terpineol	98-55-5	0.5-0.9		0.3	0.1-1.4		1.3[d]; 1.4[e]; 1.5[v]; 3.1[c]; 4.7[l]; 5.2[i]
cis-α-Terpineol							0.2[t]
Terpinolene (α-)	586-62-9	1.5-2.2	0.7	0.4-1.7	0.6-2.6	5.1	1.3[g]; 1.6[c]; 1.8[k]; 2.4[l]; 2.6[n]; 2.9[v]
β-Terpinolene	1400450-37-4						0.3[g]; 0.5[m]
α-Terpinyl acetate	80-26-2	0.04-0.3					+[e]; 0.1[i,m,q]; 0.2[b]; 0.3[f]; 0.4[v]
β-Terpinyl acetate	10198-23-9						3.1[t]
Tetradecanoic acid	544-63-8						0.1[c]; 0.8[l]; 2.9[i]
α-Thujene	2867-05-2	1.3-3.1	tr	1.2	0.9-2.7		1.3[i]; 1.5[g]; 2.2[v]; 2.8[m]; 3.1[d]; 6.4[u]
α-Thujone (*cis*-)	546-80-5						1.3[s]
Toluene	108-88-3						+[r]
Vanillin	121-33-5						+[e]
Vanillin, acetate	881-68-5						0.1[m]; 0.2[g]

A fifty-one nutmeg essential oil samples from Indonesia, Sri Lanka and Malaysia, analyzed between 1998 and 2013; lowest and highest concentrations given (E. Schmidt, unpublished data)

B one lab-hydrodistilled oil from Jamaican (West Indian) nutmegs (ref. 2)

C literature from before 2002 of studies on chemical composition of *West Indian* nutmeg oils performed in refs. 3-10 and summarized in ref. 2; lowest and highest concentrations given

D literature from before 2002 of studies on chemical composition of *East Indian* nutmeg oils performed in refs. 3-10 and summarized in ref. 2; lowest and highest concentrations given

E one lab-hydrodistilled oil from fresh nutmegs purchased at a local market in Malaysia (ref. 16)

F data from other studies (indicated with superscript letters); highest concentrations found in any study reviewed here given; when two or more oils were investigated, only the highest concentrations are mentioned, unless indicated otherwise

Table 5.62.2 Constituents identified in nutmeg oils *(continued)*

[a] incorrect identity based on GC elution order (ref. 23); [b] one lab-hydrodistilled oil from nutmegs collected in three parts of Nigeria (ref. 1); [c] one lab-hydrodistilled oil from West-Java (ref. 20); [d] fourteen oil samples from various parts of Indonesia (Sulawesi and Molucca) (ref. 11); [e] one commercial nutmeg oil purchased from a French company (ref. 12); [f] one lab-hydrodistilled oil from nutmegs purchased at a local market in India; possibly the aril (lacy covering of the kernel) was still present (ref. 15); [g] one lab-hydrodistilled oil from nutmegs cultivated in Southern Bahia, Brazil (ref. 18); [h] one 'typical' laboratory steam-distilled oil and one 'typical' commercial sample from Sri Lanka (ref. 19); [i] one lab-hydrodistilled oil from nutmegs purchased at a local market in Pakistan (ref. 25); [j] 65 essential nutmeg oils from *M. fragrans* germplasm accessions from an Indian institute; only myristicin, elemicin, safrole and sabinene were investigated (ref. 28); [k] one lab-hydrodistilled oil from nutmegs purchased from an Austrian company (ref. 29); [l] one lab-hydrodistilled oil from Malaysia (ref. 30); curiously, α-pinene, β-pinene and sabinene were missing from the list, and only 55% of the content was accounted for! [m] one lab-hydrodistilled oil from India (ref. 31); [n] data cited in ref. 27 on both West Indian and East Indian oils previously investigated (ref. 27); [o] data cited in ref. 23 on a sample of nutmeg oil obtained from plants cultivated in Guangdong, China (ref. 23); [p] one steam-distilled nutmeg oil (ref. 24); [q] one lab-distilled oil from nutmegs purchased in the United Kingdom (ref. 26); [r] data cited in ref. 22; [s] one commercial nutmeg oil sample from a Swiss chemical company (ref. 39); [t] one lab-hydrodistilled Chinese nutmeg oil sample (ref. 38); [u] one lab-hydrodistilled oil sample from nutmegs purchased in a Brazilian supermarket (ref. 37); [v] one lab-hydrodistilled oil from nutmegs bought in Italy (ref. 36); [w] one nutmeg oil sample, lab-hydrodistilled (ref. 35); [x] one lab-hydrodistilled Indian nutmeg oil (ref. 34); [y] one commercial nutmeg oil from a Brazilian company (ref. 32); [z] wrong identification

tr: trace (in columns B,C,D: <0.1); + present in the oil investigated, but quantity not stated

(1.7-9.9%), terpinen-4-ol (0.2-8.4%), γ-terpinene (3.6-7.4%), α-terpinene (2.4-4.6%), limonene (3.8-4.5%) and β-phellandrene (1.9-3.2%) (Erich Schmidt, unpublished analytical data).

Differences between East Indian and West Indian nutmeg oils

East Indian nutmeg oil is considered to be superior in flavor to the West Indian variety (21,27). Whereas large variations are observed among individual samples, the review presented in ref. 2 (columns C and D in Table 5.62.2) suggests that myristicin concentrations are generally higher in oils from the East Indies than the West Indies (<1%) and the same goes for α-pinene. On the other hand, sabinene tends to be higher in oils from the West Indies (42-57%) compared to the East Indies (14-44%) (2).

CONTACT ALLERGY/ALLERGIC CONTACT DERMATITIS

General
There has been only one publication on contact allergy to/allergic contact dermatitis from nutmeg oil.

Positive patch tests (relevance unknown, uncertain or not stated)
Four positive patch test reactions to nutmeg oil were observed in a group of 14 patients (29%) with oral lichen planus and contact allergy to spearmint oil (42). The high percentage of co-reactivity may possibly be explained by their common ingredients limonene (highest concentrations in commercial spearmint and nutmeg oils 23.7% resp. 4.5%) and myrcene (highest concentration in commercial spearmint and nutmeg oils 2.6% resp. 10.5%) (Table 5.62.2, column A).

Table 5.62.3 Major constituents of nutmeg oils

Constituent	CAS	Percentage and range					
		A	B (2)	C (2)	D (2)	E (16)	F
Sabinene	3387-41-5	16.5-36.7	37.0	42.0-57.0	14.0-44.1		49.1[b]; 50.1[j]; 50.7[n]; 56.5[p]
Myristicin	607-91-0	1.7-9.9	tr	0.5-0.8	3.3-13.5	6.8	14.0[e]; 16.2[k]; 17.7[v]; 31.3[v]; 45.6[j]
Terpinen-4-ol	562-74-3	0.2-8.4	4.0	3.0-6.4	1.0-10.9	21.3	10.9[n]; 12.2[d]; 13.9[c]; 15[u]; 31.3[i]
α-Pinene	80-56-8	12.6-25.3	22.2	1.6-12.6	18.0-21.2	8.5	22.0[q]; 22.8[d]; 25.1[s]; 27.6[o]
β-Pinene	127-91-3	2.6-16.5	16.4	7.8-12.1	8.7-17.7	3.5	17.7[n]; 18.8[s]; 21.5[q,w]; 23.9[k]
γ-Terpinene	99-85-4	3.6-7.4	2.4	1.7-4.7	1.3-7.7	9.9	4.2[v]; 6.8[k,n]; 7.3[d]; 7.8[i]; 11.2[u]
α-Terpinene	99-86-5	2.4-4.6	1.4	0.8-4.2	tr-5.2	9.8	2.7[c]; 3.5[i]; 3.8[k]; 4.0[n]; 4.8[d]; 6.4[u]
Limonene	138-86-3	3.8-4.5	3.6	2.9-4.4	2.0-7.0	8.8	4.5[s]; 5.6[c]; 6.3[g]; 7.5[k]; 7.9[d]
p-Cymene	99-87-6	0.5-2.9	0.2	0.7-3.2	0.3-2.7		1.9[q]; 3.2[n,v]; 3.3[b]; 5.6[d]; 6.5[i]; 7.4[l]

LEGEND: SEE UNDER TABLE 5.62.2

LITERATURE

1 Ogunwande IA, Olawore NO, Adeleke KA, Ekundayo O. Chemical composition of essential oil of *Myristica fragrans* Houtt (nutmeg) from Nigeria. J Essent Oil Bear Plants 2003;6:21-26

2 Simpson GIC, Jackson YA. Comparison of the chemical composition of East Indian, Jamaican and other West Indian essential oils of *Myristica fragrans* Houtt. J Essent Oil Res 2002;14:6-9

3 Lawrence BM. Progress in essential oils. Perfum Flavor 1976;1(4):34. Data cited in ref. 2

4 Lawrence BM. Essential oils 1976-1978. Carol Stream, USA: Allured Publishing Corporation, 1979: 50-103. Data cited in ref. 2

5 Lawrence BM. Progress in essential oils. Perfum Flavor 1979;4(1):49. Data cited in ref. 2

6 Lawrence BM. Progress in essential oils. Perfum Flavor 1981;6(3):48-49. Data cited in ref. 2

7 Lawrence BM. Progress in essential oils. Perfum Flavor 1985;10(4):47-48. Data cited in ref. 2

8 Lawrence BM. Progress in essential oils. Perfum Flavor 1990;15(6):62-66. Data cited in ref. 2

9 Lawrence BM. Progress in essential oils. Perfum Flavor 1992;17(5):46. Data cited in ref. 2

10 Lawrence BM. Progress in essential oils. Perfum Flavor 1997;22(4):57-74. Data cited in ref. 2

11 Dupuy N, Molinet J, Mehl F, Nanlohy F, Le Dréau Y, Kister J. Chemometric analysis of mid infrared and gas chromatography data of Indonesian nutmeg essential oils. Ind Crops Prod 2013;43:596-601

12 Schenk HP, Lamparsky D. Analysis of nutmeg oil using chromatographic methods. J Chromatogr 1981;204:391-395

13 Ehrenpreis JE, DesLauriers C, Lank P, Armstrong PK, Leikin JB. Nutmeg poisonings: A retrospective review of 10 years experience from the Illinois Poison Center, 2001–2011. J Med Toxicol 2014;10:148-151

14 Barceloux DG. Nutmeg (*Myristica fragrans* Houtt.). Disease-a-Month 2009;55(6):373-379

15 Kapoor IPS, Singh B, Singh G, De Heluani CS, De Lampasona MP, Catalan CAN. Chemical composition and antioxidant activity of essential oil and oleoresins of nutmeg (*Myristica fragrans* Houtt.) fruits. Int J Food Prop 2013;16:1059-1070

16 Piaru SP, Mahmud R, Abdul Majid AMS, Ismail S, Man CN. Chemical composition, antioxidant and cytotoxicity activities of the essential oils of *Myristica fragrans* and *Morinda citrifolia*. J Sci Food Agric 2012;92:593-597

17 Ehlers D, Kirchhoff J, Gerard D, Quirin K-W. High-performance liquid chromatography analysis of nutmeg and mace oils produced by supercritical CO_2 extraction – comparison with steam-distilled oils – comparison of East Indian, West Indian and Papuan oils. Int J Food Sci Technol 1998;33:215-223

18 Moreira Valente VM, Jham GN, Dhingra OD, Ghiviriga I. Composition and antifungal activity of the Brazilian *Myristica fragrans* Houtt essential oil. J Food Safety 2011;31:197-202

19 Sarath-Kumara SJ, Jansz ER, Dharmadasa HM. Some physical and chemical characteristics of Sri Lankan nutmeg oil. J Sci Food Agric 1985;36:93-100

20 Muchtaridi, Subarnas A, Apriyantono A, Mustarichie R. Identification of compounds in the essential oil of nutmeg seeds (*Myristica fragrans* Houtt.) that inhibit locomotor activity in mice. Int J Mol Sci 2010;11:4771-4781

21 Charles DJ. Nutmeg. In: Antioxidant properties of spices, herbs and other sources. New York: Springer Science + Business Media, 2013:427-433

22 Daniel D. Nutmeg and derivatives. Food and Agriculture Organization of the United Nations, Rome, Working Paper MISC/94/7. Available at: http://www.fao.org/docrep/v4084e/v4084e00.htm

23 Lawrence BM. Progress in essential oil. Perfum Flavor 2005;30(5):52-? (last page unknown)

24 Borges P, Pino J. Nota. Obtención de oleorresina de nuez moscado (*Myristica fragrans* H.). Revista Espan Cienc Tecnol Aliment 1993;33:209-215. Data cited in ref. 23

25 Atta-ur-Rahman, Choudhary ML, Farooq A, Ahmed A, Zafar MI, Demirci B, et al. Antifungal activities and essential oil constituents of some spices from Pakistan. Pak J Chem Soc 2000;22:60-65

26 Dorman HJD, Surai P, Deans SG. In vitro antioxidant activity of a number of plant essential oils and phytoconstituents. J Essent Oil Res 2000;12:241-248. Data also presented in ref. 35

27 Rema J, Rishnamoorthy B. Nutmeg and Mace. In: Peter KV, Ed. Handbook of herbs and spices, 2nd Ed, Vol. 1. Oxford-Cambridge-Philadelhia-New Delhi: Woodhead Publishing Ltd, 2012:399-416

28 Maya KM, Zachariah TJ, Krishnamoorthy B. Chemical composition of essential oil of nutmeg (*Myristica fragrans* Houtt.) accessions. J Spic Arom Crops 2004;13:135-139

29 Jukić M, Politeo O, Miloš M. Chemical composition and antioxidant effect of free volatile aglycones from nutmeg (*Myristica fragrans* Houtt.) compared to its essential oil. Croat Chem Acta 2006;79:209-214

30 Sim SF, Sinang FM, Kertini D, Collick F, Dankan ME, Renting Nixion Luncha V, et al. Combining essential oils of *Piper betle* and *Myristica fragrans* for enhanced antimicrobial properties. Borneo J Res Sci Technol 2013;3:35-42

31 Pal M, Srivastava M, Soni DK, Kumar A, Tewari SK. Composition and anti-microbial activity of essential oil of *Myristica fragrans* from Andaman Nicobar Island. Int J Pharm Life Sci (IJPLS) 2011;2:1115-1117

32 Murbach Teles Andrade BF, Nunes Barbosa L, da Silva Probst I, Fernandes A Júnior. Antimicrobial activity of essential oils. J Essent Oil Res 2014;26:34-40

33 Lawrence BM. Progress in essential oils. Perfum Flavor 2000;25(Sept/October):52-71

34 Mallavarapu GR, Ramesh S. Composition of essential oils of nutmeg and mace. J Med Arom Plant Sci 1998;20:746-748. Data cited in ref. 33

35 Dorman HJD, Deans SG. Chemical composition, antimicrobial and *in vitro* antioxidant properties of *Monarda citriodora* var. citriodora, *Myristica fragrans*, *Origanum vulgare* ssp. *hirtum*, *Pelargonium* sp. and *Thymus zygis* oils. J Essent Oil Res 2004;16:145-150. Nutmeg data previously presented in ref. 26

36 Piras A, Rosa A, Marongiu B, Atzeri A, Dessi MA, Falconieri D, et al. Extraction and separation of volatile and fixed oils from seeds of *Myristica fragrans* by supercritical CO_2: Chemical composition and cytotoxic activity on Caco-2 cancer cells. J Food Sci 2012;77:C448-C453

37 Lima RK, das Gracas Cardoso M, Andrade MA, Guimaraes PL, Batista LR, Nelson DL. Bactericidal and antioxidant activity of essential oils from *Myristica fragrans* Houtt and *Salvia microphylla* H.B.K. J Am Oil Chem Soc 2012;89:523-528

38 Du S-S, Yang K, Wang C-F, You C-X, Geng Z-F, Guo S-S, Deng Z-W, Liu Z-L. Chemical constituents and activities of the essential oil from *Myristica fragrans* against cigarette beetle *Lasioderma serricorne*. Chem Biodivers 2014;11:1449-1456

39 Kostić I, Petrović O, Milanović S, Popović Z, Stanković S, Todorović G, et al. Biological activity of essential oils of *Athamanta haynaldii* and *Myristica fragrans* to gypsy moth larvae. Ind Crops Prod 2013;41:17-20

40 Singh G, Marimuthu P, de Heluani CS, Catalan C. Antimicrobial and antioxidant potentials of essential oil and acetone extract of *Myristica fragrans* Houtt. (aril part). J Food Sci 2005;70:M141-M148

41 Lawless J. The encyclopedia of essential oils, 2nd Edition. London: Harper Thorsons, 2014

42 Gunatheesan S, Tam MM, Tate B, Tversky J, Nixon R. Retrospective study of oral lichen planus and allergy to spearmint oil. Australas J Dermatol 2012;53:224-228

Chapter 5.63 OLIBANUM OIL (FRANKINCENSE OIL)

DEFINITION

Olibanum oil (frankincense oil) is the essential oil obtained from the wood exudate of the olibanum tree (frankincense), *Boswellia sacra* Flueck. (synonym: *Boswellia carteri* Birdw.) and other *Boswellia* species.

INCI NOMENCLATURE

Description/definition: Boswellia carterii gum oil is an essential oil obtained from the dried, ground gum of the olibanum, *Boswellia carterii*, Burseraceae
INCI name EU: Boswellia carterii gum oil (perfuming name, not an INCI name proper)
INCI name USA: Not in the Personal Care Products Council Ingredient Database
Other names: Frankincense oil
CAS registry number(s): 8016-36-2; 89957-98-2
EINECS number(s): 289-620-2

Description/definition: Boswellia carterii oil is the volatile oil obtained from the *Boswellia carterii*, Burseraceae.
INCI name EU & USA: Boswellia carterii oil
Other names: Frankincense oil
CAS registry number(s): 8016-36-2; 89957-98-2
EINECS number(s): 289-620-2

The name *Boswellia carterii*, found not only in INCI but also in (most) other publications, is incorrect and should read *Boswellia carteri* (GRIN Taxonomy for Plants, The Plant List). Apart from *Boswellia sacra* Flueck., the gum oleoresin olibanum can also be obtained from other members of the genus *Boswellia* including *Boswellia frereana* Birdw. (a Somalian species), *Boswellia papyrifera* (Delile ex Caill.) Hochst. (Somalia, Ethiopia [especially Eritrea], Sudan and the other east African countries), *Boswellia neglecta* S. Moore (Kena, Ethiopia), *Boswellia rivae* Engl. (Ethiopia), *Boswellia odorata* Hutch., *Boswellia dalzielii* Hutch. (both in tropical regions of Africa), and *Boswellia serrata* Roxb. ('Indian olibanum', found in the middle and northern parts of East India) (2). The main producing olibanum species are *Boswellia sacra* Flueck., *Boswellia serrata* Roxb. and *Boswellia frereana* Birdw. ('Elemi olibanum', the most expensive olibanum in the trade) (3,8). These oils are all sold under the same name as 'frankincense oil' and to a lesser extent 'olibanum oil' (3).

As most commercial olibanum essential oils appear to be obtained from *Boswellia sacra/Boswellia carteri,* the discussion of the chemical composition of olibanum oils here is limited to studies in which this species is specifically mentioned as the source of the oils. A short discussion of the composition of olibanum oils obtained from other species is provided below.

ISO (INTERNATIONAL ORGANIZATION FOR STANDARDIZATION) STANDARD

Currently, there is no ISO standard for olibanum oil.

THE PLANT, THE OIL, AND THEIR USES

Boswellia sacra Flueck. is a small deciduous tree, which reaches a height of 2 to 8 meters, with one or more trunks. Its bark has the texture of paper and can be removed easily. It is native to Somalia, Ethiopia, Oman and Yemen and is cultivated in Somalia and Oman (9) for the production of a gum oleoresin. The resin, which is called olibanum or frankincense, is harvested by making a small, shallow incision on the trunk or branches of the tree or by removing a portion of the crust of it. This allows a milky substance to seep from the wounds. This material is left on the tree to dry in the sun for a few days, after which the so-called 'resin tears' can be scraped off and the olibanum is collected by hand. The color of the material varies from light yellow to dark brown (3, 8, GRIN Taxonomy for Plants; www.kew.org). Olibanum consists of 60% resin (of which 50% are boswellic acids), 20% rubber, 6-8% bassorine (an α-D-galacturonic acid polysaccharide), 5-15% essential oils and 0.5% bitter and mucilage compounds (8).

Olibanum (either from *Boswellia sacra* Flueck. or other species of the genus *Boswellia*, see below) has a long history of medicinal, religious (e.g., as incense in the Catholic Church and other religious ceremonies) and social uses (9). In ancient medicine and traditional medicine of China and India it has been used both internally and externally for almost any medical condition including inflammation, wounds, skin diseases, urinary tract infections, gynecological disorders, and respiratory infections. The oil is said to exhibit antiseptic, antibacterial, antifungal, astringent, immune-stimulant, anticancer and sedative properties and to alleviate the pain caused by rheumatism (1,3,8). Pharmacological and medical aspects of olibanum have been reviewed (10). Today, frankincense is used in several industries including pharmacy, food, flavor, liqueur, beverage, cosmetic, and perfumery industries (9).

The essential oil of olibanum is obtained by steam-distillation of the olibanum oleoresin and is one of the most commonly used oils in aromatherapy practice (1,11) where it is employed against respiratory disorders such as asthma, colds, cough, bronchitis, and laryngitis (9). In perfumery, frankincense oil is used as a fixative and for its fresh balsamic, dry, resinous, somewhat green note fragrance (9). It is also added as a fixative and fragrance to soaps, creams, lotions and detergents (11). The chemistry and biology of essential oils of the genus *Boswellia* are described in ref. 20.

CHEMICAL COMPOSITION

Olibanum oil is a mobile, colorless to yellowish, sometimes slightly green liquid which has a balsamic, slightly terpenic, spicy warm and tender sweet odor. The yield of essential oil from the wood exudate of several *Boswellia* species generally varies from 4 to 10 per cent. The main producing countries of this oil are Saudi Arabia, Somalia, Ethiopia and India (*Boswellia serrata*).

Literature data (up to December 3, 2014) on the chemical composition of *Boswellia sacra* olibanum oils and unpublished analytical data from one of us (E.S.) are

Table 5.63.1 Constituents identified in olibanum oils obtained from *Boswellia sacra*

Constituent	CAS	Percentage and range						
		A	B (1)	C (4)	D (9)	E (15)	F (11)	G
Allopregnane-7α, 11α-diol-3,20-dione	1204662-29-2		0.04					
4-Allyloxyimino-2-carene								0.7[e]
α-Amorphene	20085-19-2		0.2					0.1[d]
δ-Amorphene	189165-79-5				0.1-0.9			
Aromadendrene	489-39-4							2.6[e]
allo-Aromadendrene	25246-27-9			0.1				0.01[d]; 0.06[h]
allo-Aromadendrene oxide-(2)			0.3					
Benzene, 1-ethyl-3,5-dimethyl-	934-74-7							11.8[e]
Benzyl benzoate	120-51-4					0.2		
Benzyl butyl ether	588-67-0		0.02					
α-*trans*-Bergamotene	13474-59-4	0.02-0.3						
(Z)-*trans*-α-Bergamotol	88034-74-6		0.05					
Beyerene	3564-54-3					1.0		
Bicyclosesquiphellandrene	54324-03-7			0.1				
β-Bisabolene	495-61-4		0.06			0.2		
Borneol	507-70-0	0.05-0.08		0.04				0.06[h]
Bornyl acetate	76-49-3	0.09-1.6		0.4	0.4-0.7	0.09	0.3	+[k]; 0.1[d]; 0.5[h]; 2.2[i]
Bornylene	464-17-5		0.03					
α-Bourbonene	5208-58-2			0.03				
β-Bourbonene	5208-59-3	0.05-0.2		1.0b	0.2-0.3		0.4	+[i]; 0.1[d]
α-Bulnesene	3691-11-0		0.1					
(E)-Cadina-1(6),4-diene	931410-54-7				0-0.2			
γ-Cadinene	39029-41-9	0.2-0.3	0.1	0.3	0-0.2		1.7	0.07[h]; 0.1[d]; 0.5[e]
δ-Cadinene	483-76-1	0.9-3.7		0.5			2.6	0.1[h]; 2.5[e]; 3.2[f]
epi-α-Cadinol	5937-11-1	0.3-1.7		0.2	0-0.1		1.6	
δ-Cadinol	19435-97-3		0.09					
cis-Calamenene	72937-55-4					0.01		
Camphene	79-92-5	0.2-0.6		2.1	1.4-2.4		2.1	+[j]; <0.1[e]; 0.6[d]; 1.0[i]; 2.0[l]
4-Camphenylbutan- 2-one			0.1					
α-Campholenal	4501-58-0	0.01-0.7		0.6	0.3-0.4		0.1	+[j]; 0.2[d]; 0.7[h]; 1.5[l]
Camphor	76-22-2			0.04				0.2[e]
δ2-Carene (δ4-carene)	554-61-0							0.8[d]
δ3-Carene	13466-78-9	0.1-1.8		3.9			0.8	0.9[h]
(Z)-4-Carene				0.2				
Carvacrol	499-75-2	0.03-0.1						
(E)-Carveol	1197-07-5	0.4-0.8			0-0.3		0.2	0.1[d]
(Z)-Carveol	1197-06-4	0.09-0.2		0.1		0.04		0.1[h]
Carvomenthene	5502-88-5			0.1				
Carvone	99-49-0	0.3-0.6		0.04	0-0.2	0.3	0.1	0.1[d]
β-Caryophyllene	87-44-5	1.1-4.7		2.8	0.5-1.5		0.8	0.4[e]; 0.6[h]; 0.9[d]; 10.5[f]
Caryophyllene oxide	1139-30-6	1.1-1.3	0.1	0.3			1.8	0.01[d]; 0.05[h]; 5.4[f]
α-Cedrene	469-61-4						6.1	
Cedrol	77-53-2						0.9	
Cembra-1,3,7,11-tetraene								+[k]
Cembra-3,7,11,15-tetraene								+[k]
Cembrene	1898-13-1	0.08-0.5	0.2	0.08		0.3		+[i]; 0.8[g]
Cembrene A	31570-39-5				0-0.2		0.1	2.1[i]; 5.5[g]
Cembrene C	64363-64-0							0.1[i]
Cembrenol	67921-02-2			0.04				

Table 5.63.1 Constituents identified in olibanum oils obtained from *Boswellia sacra* (*continued*)

Constituent	CAS	Percentage and range						
		A	B (1)	C (4)	D (9)	E (15)	F (11)	G
Chrysanthenone	473-06-3							+[i]
1,8-Cineole	470-82-6	0.2-0.8	0.09	<0.01	0.3-0.5			+[i]; 0.5[g]; 1.0[l]; 1.2[i]; 2.5[e]
Citronellol	106-22-9					0.2		
Citronellyl acetate	150-84-5		0.4			0.7		+[i]
α-Copaene	3856-25-5	0.2-1.2		0.8	0-0.4	0.4	1.2	0.1[h]; 0.3[d,e]; 1.6[f]
Cubebene	11012-64-9		0.08					
α-Cubebene	17699-14-8	0.2-0.8		0.03			0.6	0.1[d]
β-Cubebene	13744-15-5	0.6-0.7					0.5	
Cubebol	23445-02-5			0.3				
Cubenene	29837-12-5							
Cycloartanyl acetate	4575-74-0		0.05					
Cyclobuta[1,2:3,4]dicyclopentene, decahydro-3a-methyl-6-methylene-1-(1-methylethyl)-, [1S-(1α, 3aα,3bβ, 6aβ,6αβ,6bα)]-			0.2					
Cycloheptane,4-methylene-1-methyl-2-(2-methyl-1-propen-1-yl)-1-vinyl-	826337-63-7		0.8					
1,4-Cyclohexadiene	628-41-1							0.1[d]
3-Cyclohexene-1-methanol	1679-51-2		0.09					
Cyclopentane, 1 acetoxymethyl-3-isopro-penyl-2-methyl-			0.07					
m-Cymene	535-77-3							5.0[n]
o-Cymene	527-84-4		0.03					+[i]; 0.5[l]
p-Cymene	99-87-6	2.7-4.9		4.0	1.6-2.7		6.2	0.2[g]; 1.2[h]; 1.4[i]; 5.7[f]; 7.5[l]
p-Cymenene	1195-32-0							0.08[h]
p-Cymen-8-ol	1197-01-9	0.1-0.4			0-0.1		0.3	0.1[h]
Decamethylene glycol	112-47-0		0.1					
n-Decanol	112-30-1		0.09			0.2		
Decyl acetate	112-17-4		0.7			1.2		+[i]
2,6-Dimethoxytoluene	5673-07-4					0.6		
3,5-Dimethoxytoluene	4179-19-5			0.05			0.2	+[k]; 0.1[h]
1,3-Dimethylcyclohexene	2808-76-6		0.6					
6,6-Dimethyl-2-me-thylenebicyclo[2.2.1] heptane				0.3				
Dodecanoic acid, 4-penten-1-yl ester	607361-53-5		0.04					
2-Dodecanone	6175-49-1		0.02					
1-Dodecene (α-)	112-41-4		0.02					
Duva-3,9,13-triene-1,5α-diol						0.06		+[i]
4,8,13-Duvatriene-1,3-diol	7220-78-2		0.04					
α-Duva-4,8,13-triene-1,3-diol	57605-80-8					0.2		
Duva-3,9,13-triene-1,5α-diol-1-acetate						21.4		
Duva-3,9,13-trien-1α-ol-5,8-oxide-1-acetate						0.5		
β-Elemene	33880-83-0	0.6-1.2	5.6	1.0[b]	0.9-2.6		1.3	0.3[d]; 0.9[h]
γ-Elemene	29873-99-2		0.2					
δ-Elemene	20307-84-0	0.02-0.1	0.7					
Elemol	639-99-6			0.01				
Elixene	3242-08-8		2.3					
Epiglobulol	88728-58-9		0.2					

Table 5.63.1 Constituents identified in olibanum oils obtained from *Boswellia sacra* (*continued*)

Constituent	CAS	Percentage and range						
		A	B (1)	C (4)	D (9)	E (15)	F (11)	G
3-Ethyl-3-hydroxyandrostan-17-one			0.2					
2-(4-Ethyl-4-methyl-3-(isopropenyl) cyclohexyl)propan-2-ol			0.08					
β-Eudesmol	473-15-4			0.07	0-0.3			
Farnesyl acetate	29548-30-9		0.06			0.01		
Formic acid, 3,7,11-trimethyl-1,6,10-dodecatrien-3-yl ester			9.6					
Geraniol	106-24-1	0.2						
(Z)-Geraniol (β-Nerol)	106-25-2		0.03					
Geranyl acetate	105-87-3	0.06-1.2				0.6	0.9	+[k]
Germacrene	28028-64-0		0.8					
Germacrene D	23986-74-5	0.4-0.9		0.5				0.09[h]; 0.1[d]; 0.4[i]
Globulol	489-41-8		0.1					
α-Guaiene	3691-12-1		0.5					
β-Guaiene	88-84-6							2.2[e]
α-Gurjunene	489-40-7			0.1				0.1[d]
γ-Gurjunene	22567-17-5	0.2-0.9						
Hedycaryol	21657-90-9						0.4	
Hexahydrobenzylacetone	2316-85-0		0.03					
Hexyl acetate	142-92-7		0.1					+[i]
Hexyl hexanoate	6378-65-0					0.09		0.3[i]
Hexyl octanoate	1117-55-1		0.6					
α-Himachalene (α-*cis*-)	3853-83-6		0.04					
Hinesol	23811-08-7			0.08				
α-Humulene	6753-98-6	0.3-2.2	0.07	0.7	0.2-0.5		0.8	0.2[d,h]; 4.4[f]
Humulene oxide	96638-51-6			0.1				
(2E,6E,10E)-12-Hydroxy-3,7,11-trimethyl-2,6,10-dodecatrienyl acetate			0.1					
Incensole	22419-74-5			0.06			1.3[c]	+[k]; 6.1[g]
Incensole acetate	34701-53-6						1.3[c]	+[k]; 1.0[i]; 13.0[g]
Isoamyl caprylate	2035-99-6		0.03					
Isoamyl 2-methylbutyrate	27625-35-0							0.1[d]
Isoamyl valerate	2050-09-1							0.1[d]
Isoborneol	124-76-5		0.03					<0.1[e]
Isocaryophyllene	118-65-0			0.03				
Isocembrene	25269-16-3					0.3		
Isokaurene	5947-50-2							1.1[g]; 2.3[i]
Isophyllocladene	511-85-3		0.7			0.6		
Isophytol	505-32-8		0.03					
Isopinocampheol	27779-29-9					0.1		
10-Isopropenyl-3,7-cyclodecadien-1-one		0.06						
6-Isopropenyl-4,8a-dimethyl-1,2,3,5,6,7,8,8a-octahydronaphthalen-2-ol		0.1						
Isoterpinolene	586-63-0					0.04		
Kaurene	34424-57-2							0.9[g]
(Z)-Lanceol	859202-95-2		0.07					
Ledol	577-27-5						0.5	
Levopimaradiene	122712-77-0							0.7[g]
Limonene	138-86-3	5.5-18.5	0.3	14.4[a]	1.7-15.9	7.6	18.2	5.5[n]; 9.0[h]; 20.4[f]; 33.5[d]
cis-Limonene oxide	13837-75-7							0.1[e]
Linalool	78-70-6	0.06-0.3	0.4	0.2	0-0.1		0.3	2.1[i]
Linalool oxide	1365-19-1							<0.1[e]

Table 5.63.1 Constituents identified in olibanum oils obtained from *Boswellia sacra* (*continued*)

Constituent	CAS	Percentage and range						
		A	B (1)	C (4)	D (9)	E (15)	F (11)	G
Longicyclene	1137-12-8		0.07					
α-Longifolene	475-20-7		0.4					
Maaliane	527-91-3					0.02		
p-Mentha-1,3,8-triene	18368-95-1			0.1				
p-Menth-2-en-1-ol	619-62-5							34.5[e]
o-Methylanisole	578-58-5			0.1		0.2		0.4[h]
Methyl (4Z,7Z,10Z, 13Z,16Z,19Z)-4,7,10, 13,16,19-docosahexaenoate			0.2					
Methyl dodecanoate	111-82-0		0.03					
Methyl eugenol	93-15-2	0.05-0.1						
α-Muurolene	10208-80-7		0.09	0.1	0-0.1		0.5	0.1[d]
γ-Muurolene	30021-74-0		0.05	0.3	0-0.3		0.5	0.1[d,h]
α-Muurolol	104245-48-9					0.03		
Myrcene	123-35-3	1.8-10.4	0.03	7.3	1.0-8.9	0.2	8.2	0.5[i]; 6.9[d]; 7.5[h]; 22.4[f]
Myrcenol	543-39-5							0.1[h]
Myrtenal	564-94-3	0.07-0.3		0.2				+[i]; 0.1[d]
Myrtenol	515-00-4	0.07-1.3		0.1				0.1[h]
(-)-Myrtenyl acetate	36203-31-3		0.04					
Naphthalene	91-20-3		0.09					
Naphthalene, decahydro-1,1, 4a-trimethyl-6-methylene-5-(3-methyl-2-pentenyl)	78548-63-7					5.7		
Naphthalene, 1,2,3, 4,4a,7-hexahydro-1, 6-dimethyl-4-(1-methylethyl)-	16728-99-7						0.1	
(E)-Nerolidol	40716-66-3							+[k]; 0.2[i]
(Z)-Nerolidol	3790-78-1					0.07		
Nerolidyl isobutyrate	1263759-11-0		18.3					
Neryl acetate	141-12-8	0.04-0.1	0.8			0.5		+[i,k]
Nonanal	124-19-6		0.02					
Nonyl acetate	143-13-5		0.03					
(E)-β-Ocimene	3779-61-1	0.05-0.3	0.04		0.1-0.5		0.1	1.7[i]; 32.3[d]
(Z)-β-Ocimene	3338-55-4	0.01-4.5	0.1	0.3	0.2-0.5	0.4	0.1	0.2[d]; 0.3[h]; 0.4[i]
allo-Ocimene	673-84-7							0.1[d]
Octanal	124-13-0		0.03					
Octane	111-65-9		0.03					
Octanoic acid, octyl ester	2306-88-9		0.3					
Octanoic acid, phenylmethyl ester	10276-85-4		0.05					
n-Octanol	111-87-5		3.3			1.1		+[k]; 3.1[g]; 11.9[i]
1-Octan-3-ol								2.4[n]
Octanol acetate	79517-25-2							45.2[g]
Octyl acetate	112-14-1	0.04-7.8	34.7			13.4	0.3	+[j,k]; 1.5[l]; 39.3[i]
Octyl formate	112-32-3					1.4		
Octyl heptanoate	5132-75-2		0.04					
1-Pentadecanol	629-76-5		0.05					
Perillene	539-52-6				0-0.2			0.1[d]
Phellandral	21391-98-0	0.2-0.7						
Phellandrene	1329-99-3							22.8[n]
α-Phellandrene	99-83-2	0.8-2.8		3.3	0.6-3.2	0.03	1.9	0.8[e]; 1.3[h]
β-Phellandrene	555-10-2	0.2-1.0		14.4[a]		0.2		
α-Phellandrene epoxide	288393-04-4							0.1[d]
α-Phellandren-8-ol	1686-20-0			0.3	1.1-3.4			0.2[d]; 0.8[h]

Table 5.63.1 Constituents identified in olibanum oils obtained from *Boswellia sacra* (*continued*)

Constituent	CAS	Percentage and range						
		A	B (1)	C (4)	D (9)	E (15)	F (11)	G
β-Phellandren-8-ol	65293-09-6			0.6				0.8[m]
Phenanthrene, 7-ethenyl-1,2,3,4,4a, 5,6, 7,8,9,10,10a-dodecahydro-1,1,4a,7-tetramethyl-	55255-56-6					4.1		
Phyllocladene	469-86-3							13.2[g]
α-Pinene	80-56-8	14.8-46.5	0.07	68.2	46.8-76.0	3.1	15.1	42.0[n]; 43.0[i]; 65.5[m]; 78.5[h]
β-Pinene	127-91-3	0.8-2.4	0.02	2.0	1.3-2.0	0.3	2.1	0.6[e]; 0.7[i]; 1.5[l]; 2.4[h]; 4.2[f]
2-β-Pinene								0.1[d]
(S)-β-Pinene (laevo-)	18172-67-3							1.8[d]
α-Pinene oxide	1686-14-2			0.07				0.5[i]
Pinocamphone	547-60-4			0.1	0-0.1			0.1[h]
Pinocarveol	5947-36-4	1.6-3.2						
L-Pinocarveol	547-61-5		0.02					
trans-Pinocarveol	1674-08-4			0.7	0.5-1.4		0.6	+[i]; 0.1[d]; 0.8[h]
Pinocarvone	30460-92-5			0.07	0-0.1			
Piperitone	89-81-6					0.03		
Piramidene								0.3[g]
9-cis-Retinal	514-85-2					2.8		
Sabinene	3387-41-5	0.5-7.2	0.02	4.9	0-0.8	0.2	4.2	+[i]; 0.7[i]; 2.9[n]; 3.6[h]; 5.6[f]
cis-Sabinene hydrate	15537-55-0	0.02-0.1						
trans-Sabinene hydrate	17699-16-0	0.01-0.09						
cis-Sabinol	3310-02-9	0.1-0.2					0.7	
Sabinyl acetate	53833-85-5							0.1[d]
Sandaracopimaradiene	1686-56-2							0.5[g]
α-Santalol	115-71-9		0.3					
Sativene	6813-05-4			0.01				
(+)-Sativene	3650-28-0		0.05					
Sclarene	511-02-4					2.9		0.8[g]
α-Selinene	473-13-2			0.3	0.5-1.1		0.6	0.1[d]; 0.2[h]
β-Selinene	17066-67-0			0.4	1.2-1.8			0.1[d]; 0.5[h]
δ-Selinene	473-14-3					0.2		
Spathulenol	6750-60-3			0.05				
(-)-Spathulenol (β-)	77171-55-2		0.2			0.03		
α-Terpinene	99-86-5	0.2-0.5						7.9[e]
β-Terpinene	99-84-3							0.7[e]
γ-Terpinene	99-85-4	0.1-0.5		0.4	0-0.3		0.5	1.0[d]; 0.4[h]; 16.9[e]
trans-Terpinene						0.5		
Terpinen-4-ol	562-74-3	0.2-5.8	0.07	1.8	0-0.5		1.4	0.2[d]; 0.4[i]; 0.5[h]
α-Terpineol	98-55-5	0.1-0.4		0.4			0.9	0.1[d]
Terpinolene	586-62-9	0.09-0.3		0.2	0-0.1		0.2	0.3[h]; 0.4[d]; 3.4[e]
α-Terpinyl acetate	80-26-2	0.01-0.2		0.1	0-0.3			0.2[h]; 2.7[e]
(Z)-11-Tetradecen-1-yl acetate	20711-10-8		0.3					
Thuja-2,4(10)-diene	36262-09-6	0.01-0.05			0.2-0.6			0.2[d]
α-Thujene	2867-05-2	1.6-25.8		7.9	0.1-1.0		7.3	0.3[e]; 1.0[h]; 6.6[d]; 52.4[f]
α-Thujone	546-80-5	0.01-0.3						
β-Thujone (cis-)	471-15-8	0.1-0.6		0.2			0.5	+[i]; 0.1[d]
Thujopsene	470-40-6						1.0	

Table 5.63.1 Constituents identified in olibanum oils obtained from *Boswellia sacra* (*continued*)

Constituent	CAS	Percentage and range						
		A	B (1)	C (4)	D (9)	E (15)	F (11)	G
Thunbergol	25269-17-4		0.5			4.1		
Toluene	108-88-3							0.1[m]; 0.2[h]
Tricyclene	508-32-7	0.05-0.07	0.01	0.06	0-0.1			
3,6,6-Trimethylbicyclo[3.1.1] hept-2-ene	4889-83-2			0.09				
1,7,7-Trimethylbicyclo[2.2.1]hept-2-yl acetate	92618-89-8		1.1					
(2,2,6-Trimethylbicyclo[4.1.0] hept-1-yl)-methanol	78996-11-9		0.02					
3,7,7-Trimethyl-1, 3,5-cycloheptatriene	3479-89-8			0.1				
1,3,6-Trimethylenecyclohexane								0.1[d]
Umbellulone	24545-81-1	0.01-0.03						
Verbenene	4080-46-0	0.04-0.05		0.6				
cis-Verbenol	1845-30-3	0.05-0.08		0.6	0-0.2			+[i]; 0.5[h]
trans-Verbenol	1820-09-3				0-0.2		0.5	0.4[i]
Verbenone	80-57-9	0.5-2.0		0.4	0.4-1.3		0.3	+[j]; 0.1[d]; 0.4[h]; 6.5[l]
Verticilla-4(20),7,11-triene								+[k]; 6.0[i]
Verticiol	70000-19-0		1.3			1.2		
Viridiflorene (ledene)	21747-46-6							1.9[e]
Viridiflorol	552-02-3			0.2		0.06		
α-Ylangene	14912-44-8			0.04				
β-Ylangene	20479-06-5			0.03				

A twenty-eight essential oil samples from *Boswellia sacra* from Ethiopia and Somalia analyzed between 1998 and 2013; lowest and highest concentrations given (E. Schmidt, unpublished data)

B one lab-hydrodistilled oil from Ethiopian olibanum resin (ref. 1)

C three commercial oils from Oman (obtained from *B. sacra*) and three commercial oils from Somalia (obtained from *B. carteri*); although it is widely accepted (The Plant List, GRIN Taxonomy for Plants) that these are identical, the authors dispute this; mean values were provided for 3 samples from Oman and 3 from Somalia; in the table, the highest mean concentration for each chemical is given (ref. 4)

D four lab-hydrodistilled oils from four different commercial grades *B. sacra* olibanum obtained in Oman; lowest and highest concentrations given (ref. 9)

E one lab-hydrodistilled oil obtained from oleoresin from *B. sacra* obtained at a spice market in Egypt (ref. 15)

F one commercial steam-distilled oil obtained from Somalian *B. sacra* gum oleoresin (ref. 11)

G data from other studies (indicated with superscript letters); highest concentrations found in any study reviewed here given; when two or more oils were investigated, only the highest concentrations are mentioned, unless indicated otherwise

[a] limonene and β-phellandrene combined; [b] β-Bourbonene and β-elemene combined; [c] isoincensole and isoincensole acetate combined; [d] one lab-hydrodistilled oil from Omani *B. sacra* gum oleoresin (ref. 6); [e] one commercial oil produced in Korea from *B. sacra* oleoresin gum of Somalian origin (ref. 5); [f] eleven commercial oils from *B. sacra* or *B. carteri* purchased at various herbal shops or pharmacies in South Africa; only 12 major components were investigated (ref. 3); [g] one lab-hydrodistilled olibanum oil (ref. 12); [h] four fractions of a lab-hydrodistilled olibanum oil with processing times of 0-2h, 8-10h, 11-12h (78°C) and 11-12h (100°C); highest concentrations in any fraction given (ref. 14); [i] one lab-hydrodistilled oil from Ethiopian olibanum (ref. 2); [j] probably one commercial oil (ref. 16, data cited in ref. 8); [k] one lab-hydrodistilled oil obtained from Ethiopian *B. carteri* oleoresin gum (ref. 7); [l] older data cited in ref. 13; only analytical data from olibanum oils that were very likely to have been obtained from B. *sacra* are mentioned here; [m] two lab-hydrodistilled oils from *B. sacra* gum resin from Oman, one distilled at a temperature of 78°C and the other at 100°C (ref. 19); the data (i.e., highest concentrations) are virtually identical to those presented in ref. 14, only those that differed have been implemented in the table; [n] one commercial *Boswellia carterii* gum oil from India (ref. 18)

+ present in the oil investigated, but quantity not stated

Table 5.63.2 Major constituents of olibanum oils

Constituent	CAS	Percentage and range						
		A	B (1)	C (4)	D (9)	E (15)	F (11)	G
α-Pinene	80-56-8	14.8-46.5	0.07	68.2	46.8-76.0	3.1	15.1	42.0[n]; 43.0[l]; 65.5[m]; 78.5[h]
α-Thujene	2867-05-2	1.6-25.8		7.9	0.1-1.0		7.3	0.3[e]; 1.0[h]; 6.6[d]; 52.4[f]
Octyl acetate	112-14-1	0.04-7.8	34.7			13.4	0.3	+[j,k]; 1.5[l]; 39.3[i]
Limonene	138-86-3	5.5-18.5	0.3	14.4[a]	1.7-15.9	7.6	18.2	5.5[n]; 9.0[h]; 20.4[f]; 33.5[d]
Myrcene	123-35-3	1.8-10.4	0.03	7.3	1.0-8.9	0.2	8.2	0.5[i]; 6.9[d]; 7.5[h]; 22.4[f]
p-Cymene	99-87-6	2.7-4.9		4.0	1.6-2.7		6.2	0.2[g]; 1.2[h]; 1.4[i]; 5.7[f]; 7.5[l]

LEGEND: SEE UNDER TABLE 5.63.1

shown in Table 5.63.1 in alphabetical order. In *B. sacra* olibanum oils from various origins, over 245 chemicals have been identified. About 58 per cent of these were found in a single reviewed publication only. The major compounds found in *Boswellia sacra* olibanum oils from different sources are shown in Table 5.63.2. They include (highest concentrations in any study given) α-pinene (78.5%), α-thujene (52.4%), octyl acetate (39.3%), limonene (33.5%), myrcene (22.4%) and p-cymene (7.5%). Well-known ingredients of *Boswellia sacra* olibanum oils that were present in high concentrations (>7%) in one or two studies were γ-terpinene (16.9%) and β-caryophyllene (10.5). Uncommon or rare constituents of *Boswellia sacra* olibanum oils found in high concentrations (>7%) in single studies include octanol acetate (45.2%), p-menth-2-en-1-ol (34.5%), (*E*)-β-ocimene (32.3%), phellandrene (22.8%), duva-3,9,13-triene-1,5α-diol-1-acetate (21.4%), nerolidyl isobutyrate (18.3%), phyllocladene (13.2%), incensole acetate (13.0%), *n*-octanol (11.9%), benzene, 1-ethyl-3,5-dimethyl- (11.8%), formic acid, 3,7,11-trimethyl-1,6,10-dodecatrien-3-yl ester (9.6%) and α-terpinene (7.9%).

Commercial oils

The ten chemicals that had the highest maximum concentrations in 28 commercial olibanum (frankincense) essential oil samples (concentration ranges provided) are the following: α-pinene (14.8-46.5%), α-thujene (1.6-25.8%), limonene (5.5-18.5%), myrcene (1.8-10.4%), octyl acetate (0.04-7.8%), sabinene (0.5-7.2%), terpinen-4-ol (0.2-5.8%), p-cymene (2.7-4.9%), β-caryophyllene (1.1-4.7%) and (*Z*)-β-ocimene (0.01-4.5%) (Erich Schmidt, unpublished analytical data).

Other *Boswellia* species

The volatile constituents of other *Boswellia* species, of which *B. serrata* and *B. frereana* are the most important, have been summarized in reference 8. The chemical composition of oils from *Boswellia serrata* from India may differ considerably from that of the *Boswellia sacra* oils. The main ingredients of Indian olibanum oil are α-thujene (5.8-65.3%), α-pinene (2.2-41.2%), myrcene (0-38%), limonene (0.7-16.7%), p-cymene (1.0-12.5%) and sabinene (0.4-8.9%) (3,8,11,17; Erich Schmidt, unpublished analytical data of 21 essential oils samples from *Boswellia serrata* from India analyzed between 1998

and 2013). The main constituents of *B. frereana* have included α-pinene (2.0-64.7%), α-thujene (0-33.1%), p-cymene (5.4-16.9%) and sabinene (2.6-7.0%) (2,3).

CONTACT ALLERGY/ALLERGIC CONTACT DERMATITIS

General

There are only three publications on contact allergy to / allergic contact dermatitis from olibanum (frankincense) oil, all concerning occupational contact dermatitis in aromatherapists. In two and possibly all, a false-positive patch test reaction due to the excited skin syndrome cannot be excluded. In one report, α-pinene and possibly caryophyllene may have been allergens in the olibanum oil.

Case reports and positive patch test reactions

An aromatherapist had chronic hand dermatitis and was patch test positive to 17 of 20 oils used at her work (tested 1% and 5% in petrolatum), including olibanum oil (22). Another aromatherapist had occupational contact dermatitis with allergies to multiple essential oils used at work, including frankincense oil. The patient also reacted to geraniol, α-pinene, caryophyllene, the fragrance mix and various other fragrance materials. α-Pinene and caryophyllene were demonstrated by GC-MS in the olibanum oil (23). α-Pinene is the major ingredient in olibanum oil: concentrations in commercial olibanum oils were found to be 14.8-46.5%, whereas caryophyllene may be present in such oils in concentrations of up to 4.7% (Table 5.63.1, column A). One positive patch test reaction to frankincense oil in an aromatherapist with occupational contact dermatitis from multiple essential oils (21); apparently she had no previous contact with olibanum oil.

Contact dermatitis from contact allergy to olibanum *extract* (ex *Boswellia serrata*) present in a 'naturopathic cream' is described in ref. 24.

LITERATURE

1 Chen Y, Zhou C, Ge Z, Liu Z, Liu Y, Feng W, et al. Composition and potential anticancer activities of essential oils obtained from myrrh and frankincense. Oncol Lett 2013;6:1140-1146

2   Basar S. Phytochemical investigations on *Boswellia* species. Comparative studies on the essential oils, pyrolysates and boswellic acids of *Boswellia carterii* Birdw., *Boswellia serrata* Roxb., *Boswellia frereana* Birdw., *Boswellia neglecta* S. Moore *and Boswellia rivae* Engl. PhD thesis, University of Hamburg, Germany, 2005

3   Van Vuuren SF, Kamatou GPP, Viljoen AM. Volatile composition and antimicrobial activity of twenty commercial frankincense essential oil samples. S Afr J Bot 2010;76:686–691

4   Woolley CL, Suhail MM, Smith BL, Boren KE, Taylor LC, Schreuder MF, et al. Chemical differentiation of *Boswellia sacra* and Boswellia *carterii* essential oils by gas chromatography and chiral gas chromatography–mass spectrometry. J Chromatogr A 2012;1261:158-163

5   Yang S-A, Jeon S-K, Lee E-J, Shim C-H, Lee I-S. Comparative study of the chemical composition and antioxidant activity of six essential oils and their components. Nat Prod Res 2010;24:140-151

6   Al-Harrasi A, Al-Saidi S. Phytochemical analysis of the essential oil from botanically certified oleogum resin of *Boswellia sacra* (Omani Luban). Molecules 2008;16:2181-2189

7   Basar S, Koch A, König WA. A verticillane-type diterpene from *Boswellia carterii* essential oil. Flavour Fragr J 2001;16:315-318

8   Mertens M, Buettner A, Kirchhoff E. The volatile constituents of frankincense – a review. Flavour Fragr J 2009;24:279-300

9   Al-Saidi S, Rameshkumar KB, Hisham A, Sivakumar N, Al-Kindy S. Composition and antibacterial activity of the essential oils of four commercial grades of Omani Luban, the oleo-gum resin of *Boswellia sacra* Flueck. Chem Biodivers 2012;9:615-624

10 Moussaieff A, Mechoulam R. Boswellia resin: from religious ceremonies to medical uses; a review of *in-vitro, in-vivo* and clinical trials. J Pharm Pharmacol 2009;61:1281-1293

11 Camarda L, Dayton T, Di Stefano V, Pitonzo R, Schillaci D. Chemical composition and antimicrobial activity of some oleogum resin essential oils from *Boswellia* spp. Ann Chim 2007;97:837-844

12 Marongiu B, Piras A, Porcedda S, Tuveri E. Extraction of *Santalum album* and *Boswellia carterii* Birdw. volatile oil by supercritical carbon dioxide: influence of some process parameters. Flavour Fragr J 2006;21:718-724

13 Tucker AO. Frankincense and myrrh. Econ Bot 1986;40:425-433

14 Ni X, Suhail MM, Yang Q, Cao A, Fung K-M, Postier RG, et al. Frankincense essential oil prepared from hydrodistillation of *Boswellia sacra* gum resins induces human pancreatic cancer cell death in cultures and in a xenograft murine model. BMC Compl Alt Med 2012;12:253 (14 pages). Almost identical data presented in ref. 19

15 Mikhaeil BR, Maatoog GT, Badria FA, Amer MM. Chemistry and immunomodulatory activity of frankincense oil. Z Naturforsch C 2003;58:230-238

16 Obermann H. Die chemischen und geruchlichen Unterschiede von Weihrauchharzen. Dragoco Rep 1977;24:260-265. Data cited in ref. 8

17 Baratta MT, Dorman HJD, Deans SG, Figueiredo AC, Barroso JG, Ruberto G. Antimicrobial and antioxidant properties of some commercial essential oils. Flavour Fragr J 1998;13:235-244

18 Chudasama KS, Thaker VS. Screening of potential antimicrobial compounds against *Xanthomonas campestris* from 100 essential oils of aromatic plants used in India: an ecofriendly approach. Arch Phytopathol Plant Prot 2012;45:783-795

19 Suhail MM, Wu W, Cao A, Mondalek FG, Fung K-M, Shih P-T, et al. *Boswellia sacra* essential oil induces tumor cell-specific apoptosis and suppresses tumor aggressiveness in cultured human breast cancer cells. BMC Compl Altern Med 2011;11:129. Almost identical data presented in ref. 14

20 Hussain H, Al-Harrasi A, Al-Rawahi Ahmed, Hussain Javid. Chemistry and biology of essential oils of genus *Boswellia*. Evidence-based Complementary & Alternative Medicine (eCAM) 2013:1-12

21 Bleasel N, Tate B, Rademaker M. Allergic contact dermatitis following exposure to essential oils. Australas J Dermatol 2002;43:211-213

22 Selvaag E, Holm J, Thune P. Allergic contact dermatitis in an aromatherapist with multiple sensitizations to essential oils. Contact Dermatitis 1995;33:354-355

23 Dharmagunawardena B, Takwale A, Sanders KJ, Cannan S, Roger A, Ilchyshyn A. Gas chromatography: an investigative tool in multiple allergies to essential oils. Contact Dermatitis 2002;47:288-292

24 Acebo E, Raton JA, Sautua S, Eizaguirre X, Trébol I, Díaz Pérez JL. Allergic contact dermatitis from *Boswellia serrata* extract in a naturopathic cream. Contact Dermatitis 2004;51:91-92

Chapter 5.64   ORANGE OIL, BITTER

DEFINITION

Bitter orange oil (essential oil of bitter orange) is the essential oil obtained from the pericarp (peel) of the bitter orange, *Citrus aurantium* L. (synonyms: *Citrus amara* Link, *Citrus bigaradia* Loisel, *Citrus vulgaris* Risso).

INCI NOMENCLATURE

Description/definition: Citrus aurantium amara (bitter orange) peel oil is the volatile oil obtained from the peel of *Citrus aurantium amara*, Rutaceae
INCI name EU & USA: Citrus aurantium amara (bitter orange) peel oil
Other names: Bitter orange oil
CAS registry number(s): 68916-04-1; 72968-50-4
EINECS number(s): 277-143-2

ISO (INTERNATIONAL ORGANIZATION FOR STANDARDIZATION) STANDARD

ISO number: 9844
ISO name: Essential oil of bitter orange
Botanical origin: *Citrus aurantium* L. (synonyms: *Citrus amara* Link, *Citrus bigaradia* Loisel, *Citrus vulgaris* Risso)
Parts of plant used: Pericarp (peel)
ISO values: ISO values (minimum and maximum concentrations) are shown in Table 5.64.1.

Citrus aurantium L. is not only used for the production of bitter orange (peel) oil, but also for obtaining neroli oil from the flowers (Chapter 5.60) and petitgrain bigarade oil from the leaves, twigs and unripe fruits (Chapter 5.69) (9).

By ISO definition, all citrus essential oils except lime oil are produced by expression; oils obtained from citrus fruits by distillation may not be called essential oils according to ISO criteria (except lime oil). In industry, however, sometimes residues from expression of the citrus peels undergo steam distillation, to obtain the remaining oil; these volatile oils may then be added to the expressed oil. As some of the compounds undergo changes forced by high temperature during distillation, this addition (which is an adulteration) changes the composition of the essential oil. Because of this, and also because it cannot be excluded that oils entirely produced by hydrodistillation reach the market, both the ingredients found in 'genuine' bitter orange essential oils obtained by expression and those that may be present in hydrodistilled oils are presented here.

THE PLANT, THE OIL, AND THEIR USES

Citrus aurantium L., commonly named bigarade or bitter (sour) orange, is an evergreen shrub or tree, 2-3 to 7-8 meters tall, with short and sharp spines, which produces ovoid or rounded greenish-yellow fruit, 25-70 mm in diameter. It probably has a multiple hybrid origin in China and elsewhere, and is widely cultivated in the tropics and subtropics, important areas of cultivation being Paraguay, Morocco, and Spain (GRIN Taxonomy for Plants, Mansfeld's

World Database of Agriculture and Horticultural Crops, http://mansfeld.ipk-gatersleben.de, 16).

The bitter orange is used for fruit production and as a common rootstock for other *Citrus* species in many Mediterranean countries because of its high tolerance to low temperature, salt, humidity and various soil conditions, resistance to several viral diseases and improvement in the fruit quality of the grafted plant. *Citrus aurantium* may also be cultivated as an ornamental tree (14,16,19,32).

The fruits of the bitter orange serve the production of marmalade (32) and soft drinks (orangeade). Due to the acidity of its pulp and the bitterness of its rind it is not normally consumed as fresh fruit (26), although in Mexico it is eaten fresh with salt and chili. The juice of the fruit is added to meat and fish for flavoring and is used as a vinegar substitute. In addition, bitter orange products are extensively used as a fragrance component in soaps, detergents, cosmetics and perfumes and may be employed to aromatize certain drugs (16,19,26).

The peel, flower, leaves, and fruit are used in both traditional herbal and modern medicine on account of the various bioactive compounds that they contain, such as phenolics, flavonoids, essential oils, the alkaloid *p*-synephrine and vitamins. It is employed as a stimulant, weight loss remedy (as it suppresses appetite) and as a nasal decongestant. It is also suggested to be able to diminish the side effects of many medications (16,19). Bitter orange supplements have, however, been linked to a number of serious side effects and deaths, but a causal relationship has not been ascertained (20).

The bitter orange oil, obtained from the peel, is used for flavoring baked goods, chewing gum, ice cream, soft drinks, liqueurs (Cointreau, Curaçao, Licor Beirão) and as a fragrance in perfumery and cosmetics (9,32). The role of essential oils and antioxidants derived from *Citrus* by-products in food protection and medicine has been reviewed (ref. 25). Besides these uses, undiluted essential oils are sold on the international market of aromatherapy (32,40,41).

The history, global distribution and nutritional importance of *Citrus* fruits have been reviewed in ref. 37.

CHEMICAL COMPOSITION

Bitter orange oil is a clear mobile liquid, which has a light, pale yellow to brownish green color and a fresh-citrusy note with characteristic odor of the scratched peel of the bitter orange. The yield of essential oil from the peel of *Citrus aurantium* L. generally varies from 0.7 to 1.2 per cent. The main producing countries of this oil are Brazil, Spain, Italy, Ivory Coast and Dominican Republic. Other such countries include France, Egypt, Tunisia, Morocco, Uganda, Paraguay, Argentina and China (9).

Cold-pressed *Citrus* essential oils consist of a volatile fraction, which represents 85-99% of the oil, and a non-volatile fraction 1-15%, which mainly contains oxygen heterocyclic compounds, especially coumarins, psoralens and polymethoxyflavones (Table 5.64.4). These components play an important role in the characterization of cold-pressed *Citrus* oils, since the composition of this fraction is characteristic of each oil. Moreover,

Table 5.64.1 ISO values (%) for bitter orange oil [a]

Compound	CAS	Minimum	Maximum
Limonene	138-86-3	93.0	95.0
Myrcene	123-35-3	1.5	3.0
β-Pinene	127-91-3	0.2	1.2
Linalyl acetate	115-95-7	traces	1.0
α-Pinene	80-56-8	0.2	0.7
Linalool	78-70-6	0.1	0.4
Geranyl acetate	105-87-3	0.1	0.3
Decanal	112-31-2	traces	0.3
(E)-Nerolidol	40716-66-3	traces	0.3
n-Octanal	124-13-0	traces	0.3
β-Caryophyllene	87-44-5	traces	0.2
Germacrene D	23986-74-5	traces	0.2

[a] ISO 9844 Essential oil of bitter orange ©ISO 2006; Geneva, Switzerland, www.iso.org

many of the pharmacological and toxicological activities possessed by *Citrus* oils have been demonstrated to be related to these components (23).

Literature data on the chemical composition of bitter orange oils and unpublished analytical data from one of us (E.S.) are shown in Table 5.64.2 in alphabetical order. In bitter orange oils from various origins, over 215 chemicals have been identified. The major compounds found in bitter orange oils (which always contain >90% limonene)

from different sources are shown in Table 5.64.3. They include (highest concentrations in any study given) limonene (98.3%), myrcene (8.9%), linalool (5.2%), linalyl acetate (5.0%) and β-pinene (4.0%).

Commercial oils

The ten chemicals that had the highest maximum concentrations in 72 commercial bitter orange essential oil samples (concentration ranges provided) are the following: limonene (92.2-95.6%), myrcene (1.5-2.5%), β-pinene (0.2-0.8%), α-pinene (0.4-0.7%), (E)-β-ocimene (0.02-0.6%), β-phellandrene (0.2-0.6%), linalyl acetate (0.05-0.4%), decanal (0.02-0.3%), linalool (0.1-0.3%) and n-octanal (0.01-0.3%) (Erich Schmidt, unpublished analytical data).

It should be realized that values for chemicals in hydrodistilled oils may vary considerably from genuine cold-pressed bitter orange oils because of the many hydrolytic reactions that take place during oil isolation. In addition, the hydrodistilled oils lack the non-volatile oxygenated heterocyclic compounds (notably coumarins and psoralens) which may be present in the cold-pressed oils (Table 5.64.4).

CONTACT ALLERGY/ALLERGIC CONTACT DERMATITIS

Orange oil (unspecified)

General

Contact allergy to/allergic contact dermatitis from orange oil (unspecified whether bitter or sweet orange oil) has

Table 5.64.2 Constituents identified in bitter orange oils

Constituent	CAS	Percentage and range						
		A	B (1)	C (1)	D (1)	E (6-9)	F (10-15)	G
Acetophenone	98-86-2			<0.01[b]				
Alcohols, linear chained, C$_{12}$, C$_{14}$-C-$_{17}$, C$_{19}$-C$_{24}$					2.9			
Aromadendrene	489-39-4						0.01	tr[j]
Benzaldehyde	100-52-7			0.08[b]				
Benzyl benzoate	120-51-4			0.7[b]				
trans-α-Bergamotene	13474-59-4	0.0-0.01	0.02			0.3	0.02	
Bergaptene	484-20-8	0.002-0.035						
Bicyclogermacrene	24703-35-3		tr		0.01	0.01	0.3	0.01[i]
Bicyclopentan-2-one	4884-24-6						tr	
(Z)-α-Bisabolene	29837-07-8						0.1	
β-Bisabolene	495-61-4		0.02			0.5	0.3	
α-Bisabolol	515-69-5			tr				
Borneol	507-70-0		tr		tr			0.08[o]
Bornyl acetate	76-49-3				tr			tr[o]
γ-Cadinene	39029-41-9				tr		0.01	
δ-Cadinene	483-76-1		0.01		0.1		0.03	tr[j]
δ-Cadinol	19435-97-3						tr	
τ-Cadinol	5937-11-1						0.01	
α-Calacorene	21391-99-1							0.1[g]
Camphene	79-92-5		0.01	0.2[b]	0.01	0.01	0.01	tr[m]; 0.02[o]; <0.05[e,f]
Camphor	76-22-2		0.01		tr		tr	0.2[o]
δ2-Carene (= δ4-)	554-61-0						tr	

Table 5.64.2 Constituents identified in bitter orange oils (*continued*)

Constituent	CAS	Percentage and range						
		A	B (1)	C (1)	D (1)	E (6-9)	F (10-15)	G
δ3-Carene	13466-78-9		0.01	tr	0.01	tr	0.02	0.01°
Carvacrol	499-75-2				0.02			0.7°
Carveol	99-48-9				0.05			
(*E*)-Carveol	1197-07-5			0.6[b]		0.03	0.2	
(*Z*)-Carveol	1197-06-4			0.3[b]	tr	0.02	0.05	
Carvone	99-49-0		0.02	1.2[b]	0.03	0.03	tr	tr[l]
cis-Carvyl acetate	1205-42-1						tr	
β-Caryophyllene	87-44-5	0.04-0.2	0.1	0.1	0.3	0.06	0.2	0.02[m]; <0.05[e]; 0.05[h,q]; 0.1[s]
Caryophyllene oxide	1139-30-6					0.01		tr[j]; 0.08°
Cedrol	77-53-2				tr		tr	
1,8-Cineole	470-82-6							0.3°
Citronellal	106-23-0		0.02		0.07	tr	0.1	0.02[r]
Citronellol	106-22-9		0.01		0.07		0.01	0.03°
Citronellyl acetate	150-84-5		0.05			0.01	0.03	
α-Copaene	3856-25-5		+		0.05		0.01	
β-Copaene	18252-44-3							0.2[r]
α-Cubebene	17699-14-8				tr			
β-Cubebene	13744-15-5				0.05		tr	
o-Cymene	527-84-4							0.06[q]
p-Cymene	99-87-6		0.01	2.1[b]	0.5	0.01	0.5	0.04[q]; 0.08°; 1.7[h]
p-Cymenene	1195-32-0		tr	1.4[b]				
α-Cyperone	473-08-5							0.1[g]
(*Z*)-β-Damascenone	59739-63-8					tr		
2,4-Decadienal	2363-88-4						tr	tr[j]
(*E,E*)-2,4-Decadienal	25152-84-5		0.01		0.01			
(*E,Z*)-2,4-Decadienal	25152-83-4		0.01		0.01			
Decanal	112-31-2	0.02-0.3	0.2	0.2	0.2	0.1	0.2	<0.06[k]; 0.08[q]; 0.2[n]; 0.4[j,s]; 0.5[r]
Decanoic acid	334-48-5		0.05		0.03			
Decanol	36729-58-5		0.05		0.05	tr		
(*E*)-2-Decenal	3913-81-3		0.02		0.02	0.01	0.01	
Decyl acetate	112-17-4		0.07		0.02	0.03	0.1	tr[j]
cis-Dehydrolinalool oxide			tr					
trans-Dehydrolinalool oxide			tr					
cis-Dihydrocarvone	3792-53-8							0.03°
trans-Dihydrocarvone	5948-04-9		0.1[b]					
2,3-Dimethyl-3-(4-methyl-3-pentenyl)-2-norbornanol	98205-40-4					tr		
4,8-Dimethyl-1,3(*E*),7-nonatriene	19945-61-0					0.01		tr[j]
2,7-Dimethyl-2,6-octadien-1-ol	22410-74-8						tr	
3,3-Dimethyl-1-octene							0.01	
(*E,Z*)-2,6-Dodecadienal	21662-13-5					0.01		
Dodecanal	112-54-9	0.01-0.05	0.04	tr	0.03	0.1	0.2	0.05[j]; 0.04[s]; 0.4[r]
Dodecanol	112-53-8		0.01			0.2		
2-Dodecenal	4826-62-4						0.01	
(*E*)-2-Dodecenal	20407-84-5		0.01			tr		0.02[j]
Dodecyl acetate	112-66-3		0.01		tr	tr	tr	
β-Elemene	33880-83-0		0.03		0.01		0.09	tr[j]; 0.1[g]
γ-Elemene	29873-99-2				0.04		0.05	
δ-Elemene	20307-84-0		0.05		0.05	0.03	0.03	
Elemol (α-)	639-99-6				tr		tr	
(*E*)-Epoxyocimene	255832-06-5							tr[j]

Table 5.64.2 Constituents identified in bitter orange oils (*continued*)

Constituent	CAS	Percentage and range						
		A	B (1)	C (1)	D (1)	E (6-9)	F (10-15)	G
Ethyl acetate	141-78-6						0.3	
Farnesal	19317-11-4						tr	
Farnesene	502-61-4				0.2			0.1[e,f,m]
α-Farnesene	502-61-4						0.01	
(*E,E*)-α-Farnesene	502-61-4		tr		tr		0.03	
(*E*)-β-Farnesene	18794-84-8		0.01			0.01	0.02	
(*Z*)-β-Farnesene	28973-97-9		0.03			0.02		
Farnesol	4602-84-0				0.6			tr[e,f]; 0.01[m]
(*E,E*)-Farnesol	106-28-5		0.01					
(*Z,E*)-Farnesol	3790-71-4		0.01					tr[o]
Geranial	141-27-5	0.02-0.08	0.1	0.06	0.1	0.1	0.2	0.02[h]; 0.04[g]; 0.1[j,q,s]; 0.3[r]
Geraniol	106-24-1	0.01-0.05	0.2	tr	0.2	0.01	0.3	tr[j]; 0.1[e]; 0.2[f,m]
(*E*)-Geraniol (β-)	106-24-1							0.1[q]
Geranyl acetate	105-87-3	0.04-0.2	0.2	0.1	0.4	0.5	0.6	0.08[q]; 0.1[h,s]; 0.2[e]; 0.3[f,m]; 0.4[j]
Geranyl isobutyrate	2345-26-8						0.2	
Geranyl propanoate	105-90-8		0.02		tr		tr	
Germacrene A	28387-44-2							0.01[i]
Germacrene B	15423-57-1		0.01			tr	0.01	0.02[i]
Germacrene C	34323-15-4							0.02[i]
Germacrene D	23986-74-5	0.03-0.1	0.1		0.1	0.2	2.1	0.04[j,o]; 0.06[q]; 0.1[i,s]
Guaiacol	90-05-1							2.2[c]
Heptanal	111-71-7		0.01	0.08[b]				
(*E*)-2-Heptenal	18829-55-5		tr					
Heptyl acetate	112-06-1				tr	tr	tr	
Hexanal	66-25-1		0.03					
Hexadecanol	51260-59-4							0.01
Hexanol	111-27-3		tr					tr
(*E*)-2-Hexenal	6728-26-3		tr					
(*Z*)-3-Hexen-1-ol	928-96-1							0.02
(*E*)-2-Hexyl acetate							tr	
α-Humulene	6753-98-6		0.04		0.04	tr	0.09	0.01[j]; 0.1[o]; 0.2[g]
α-Ionone	127-41-3				0.04			
β-Ionone	79-77-6				0.4			
Isodihydrocarveol	18675-35-9			0.06[b]				
Isomenthol	3623-52-7						tr	
Isopiperitone	58615-39-7		0.01		0.02			
Isopulegol	89-79-2		0.01				tr	
Limonene	138-86-3	92.2-95.6	94.3	93.6	95.5	96.5	96.3	95.5[l]; 95.9[s]; 96.7[k]; 96.9[g]; 98.3[p]
Limonene dioxide	96-08-2			0.09				
Limonene oxide	1195-92-2			0.2				
cis-Limonene oxide	13837-75-7		0.01	0.5[b]	0.05	0.08	0.04	0.03[j]
trans-Limonene oxide	4959-35-7		0.01	0.3[b]	0.05	0.08	0.1	0.03[j]
Limonen-10-ol	3269-90-7					tr		
Linalool	78-70-6	0.1-0.3	0.4	0.5[b]	3.2	0.3	5.2	1.5[e]; 1.8[q]; 2.0[f,m]; 2.5[l]; 2.9[d]
Linalool oxide	1365-19-1							0.2[k]; 0.4[q]
cis-Linalool oxide	11063-77-7		0.01	0.02[b]	0.2		0.04	0.09[o]; 0.2[e,f,m]
cis-Linalool oxide, furanoid	5989-33-3	0.0-0.07						
trans-Linalool oxide	11063-78-8		0.01	tr	0.3			0.2[e,f,m]
trans-Linalool oxide, furanoid	34995-77-2	0.0-0.02						
Linalyl acetate	115-95-7	0.05-0.4	1.2	0.7[b]	2.7	1.0	5.0	0.2[e]; 0.4[s]; 0.5[f,m]; 0.9[j]; 2.7[l]; 4.0[r]
Linalyl anthranilate	7149-26-0						2.4	
p-Mentha-1,8-dien-9-ol	1946-01-6						0.03	

Table 5.64.2 Constituents identified in bitter orange oils (*continued*)

Constituent	CAS	Percentage and range						
		A	B (1)	C (1)	D (1)	E (6-9)	F (10-15)	G
p-Mentha-1,8(9)-dien-10-ol			0.01					
trans-*p*-Mentha-2,8-dien-1-ol	4017-77-0					0.03		tr[j]
trans-*p*-Mentha-2,8-dien-9-ol							0.02	
p-Mentha-1,8(9)-dien-10-yl acetate			0.02		0.01			
cis-Mentha-1,3,8-triene				0.3[b]				
p-Mentha-1,3,8-triene	18368-95-1			0.1[b]				
p-Menth-1-en-9-ol	18479-68-0						tr	
p-Menth-1-en-9-yl acetate	28839-13-6				tr		tr	
Methyl benzoate	93-58-3			0.6[b]				
Methyl eugenol	93-15-2			0.8[b]				
Methyl salicylate	119-36-8			0.1[b]				
α-Muurolene	10208-80-7		tr				tr	
γ-Muurolene	30021-74-0				0.03		0.03	
Myrcene	123-35-3	1.5-2.5	3.1	2.4	2.3	4.9	1.9	2.0[l]; 2.1[h]; 2.6[q]; 4.4[j]; 8.9[s]
Myrcene oxide	29414-55-9		0.01					
(*E*)-Myroxide	28977-57-3		tr					
Myrtanol	514-99-8				0.2			
Myrtenal	564-94-3		0.01					
cis-Myrtenol					0.6			
Naphthalene	91-20-3				tr			
Neral	106-26-3		0.05	tr	0.1	0.03	0.1	0.04[h,j]; 0.05[r]; 0.06[s]
Nerol	106-25-2		0.04		0.2		0.04	tr[j]; 0.1[e,f,m,o]; 0.2[r,q]
Nerolidol	7212-44-4						0.1	0.04[j]; 0.1[f,m]; 0.9[r]
(*E*)-Nerolidol	40716-66-3	0.01-0.1	0.2		0.2	0.09	3.2	0.05[q]; 0.1[e]
(*Z*)-Nerolidol	3790-78-1					0.2	0.06	
Neryl acetate	141-12-8	0.01-0.04	0.05	0.09	0.09	0.5	0.2	0.04[q]; 0.09[m]; 0.1[e,f,k,n]; 0.2[r]
Neryl propionate	105-91-9		0.01					
Nonanal	124-19-6	0.01-0.04	0.1	0.07[b]	0.08	0.03	0.03	0.05[j]; 0.2[o]
Nonanol	28473-21-4		0.03			tr		
(*E*)-2-Nonenal	18829-56-6		tr					
Nonyl acetate	143-13-5		0.01		0.05	0.01	0.01	
Nootkatone	4674-50-4	0.01-0.05	0.4		0.1	0.03	tr	<0.06[k]; 0.4[r]
β-Ocimene	13877-91-3							0.2[k]
(*E*)-Ocimene	27400-72-2			0.06[b]				<0.05[e]; 1.2[r]
(*Z*)-Ocimene	27400-71-1			0.04[b]				
(*E*)-β-Ocimene	3779-61-1	0.02-0.6	0.4	0.09	0.8	0.6	0.5	0.02[o]; 0.1[s]; 0.07[j]; 0.3[g,m]; 0.9[d]
(*Z*)-β-Ocimene	3338-55-4	0.0-0.01	0.05		0.05	0.01	0.01	0.02[m]; 0.1[q]; 0.2[e]; 0.3[f,n]
n-Octanal	124-13-0	0.01-0.3	0.2`	0.1	0.2	0.07	0.6	0.03[h]; 0.1[j,q]; 0.2[n]; 0.3[s]; 0.5[p]
Octanol	111-87-5		0.2	0.1	0.3	0.02	0.1	0.02[o]; 0.03[j]
(*E*)-2-Octenal	2548-87-0		0.01					
Octyl acetate	112-14-1	0.02-0.05	0.05		0.1	0.03	0.07	tr[j]; 0.1[q]; 0.3[r]
Octyl propyl ether	29379-41-7						tr	
Pentadecanal	2765-11-9						0.02	
Perillaldehyde	2111-75-3	0.01-0.03	0.02	0.2	0.5	0.02	0.1	0.02[j,s]
Perillene	539-52-6		tr					
Perillyl acetate	15111-96-3				0.02	0.01		
Perillyl alcohol	536-59-4				0.01	0.01	0.02	
α-Phellandrene	99-83-2	0.02-0.06	0.08	0.3[b]	0.02		0.1	0.03[j]; 0.1[s]
β-Phellandrene	555-10-2	0.2-0.6	tr	1.5	0.3		0.5	tr[j]; 0.01[q]; 0.3[e,f,m]
Phenol	108-95-2							1.6[c]
α-Pinene	80-56-8	0.4-0.7	0.9	0.8[b]	1.2	0.6	0.9	0.4[e]; 0.5[n,r]; 0.6[d]; 0.7[s]; 0.9[h]; 1.2[j]

Table 5.64.2 Constituents identified in bitter orange oils (*continued*)

Constituent	CAS	Percentage and range						
		A	B (1)	C (1)	D (1)	E (6-9)	F (10-15)	G
β-Pinene	127-91-3	0.2-0.8	1.3	2.5[b]	1.6	1.0	4.0	0.6[n,s]; 0.9[h]; 1.2[l]; 1.4[g]; 3.3[d]; 3.4[r]
trans-Pinocarveol	1674-08-4				tr			
Pinocarvone	30460-92-5			0.04[b]				
cis-Piperitol	16721-38-3					0.3		
Piperitone	89-81-6		tr					
Psoralen	66-97-7	0.001-0.007						
Sabinene	3387-41-5	0.1-0.3	0.5	0.3	0.2	0.3	0.7	0.1[e,f,m]; 0.2[j,n,o,q]; 0.5[d]; 0.6[r]
cis-Sabinene hydrate	15537-55-0		0.01	0.7[b]		tr	tr	0.09[o]
trans-Sabinene hydrate	17699-16-0		tr				0.06	
β-Santalene	511-59-1		0.01					
7-epi-α-Selinen-2-one			0.02		0.02			
β-Sesquiphellandrene	20307-83-9		0.01					
α-Sinensal	17909-77-2		tr				tr	
β-Sinensal (*cis*-)	17909-87-4		0.01				0.01	
Solanone	1937-54-8						tr	
Spathulenol	6750-60-3					tr	0.1	tr[o]
α-Terpinene	99-86-5		0.02	0.1[b]	4.3	tr	0.01	0.02[m]; <0.05[e,f]; 1.7[o]
γ-Terpinene	99-85-4	0.0-0.2	0.1	0.3	0.7	0.08	4.4	<0.05[e,f]; 0.05[r]; 0.3[o]; 0.7[l]
Terpinen-4-ol	562-74-3		0.03	0.2[b]	0.2	0.1	0.4	0.01[o]; 0.09[q]; 0.1[e,r]; 0.2[f,m]
Terpinen-4-yl acetate	4821-04-9					0.7		
α-Terpineol	98-55-5	0.06-0.3	2.9	0.2	0.7	0.4	0.2	0.2[k]; 0.3[l]; 0.4[j,o]; 0.6[e,f,m]; 0.8[q]
β-Terpineol	138-87-4				0.02		tr	
(*E*)-β-Terpineol				0.2[b]				
Terpinolene	586-62-9	0.0-0.2	0.1	tr[b]	0.7	0.1	0.2	0.02[o]; 0.03[s]; 0.6[e,m,f]
α-Terpinyl acetate	80-26-2		0.03	<0.01[b]	0.03	0.01	0.1	tr[o]; 0.07[g]
Tetradecanal	124-25-4		0.03		0.5	tr	tr	
Tetradecane	629-59-4				0.2			0.2[l]
(*E*)-2-Tetradecenal	51534-36-2		tr			tr		tr[j]
(*E*)-2-Tridecenal	7069-41-2		tr					
α-Thujene	2867-05-2		0.02	<0.01[b]	0.2	0.02	0.1	0.02[i]; 0.4[o]
α-Thujone	546-80-5							0.09[o]
Thujyl alcohol	513-23-5				tr		tr	
Thymol	89-83-8				0.1		tr	1.4[c]
Tridecanal	10486-19-8		0.01		0.01			
2,5,6-Trimethyl-1,3, 6-heptatriene	42123-66-0						tr	
Undecanal	112-44-7		0.01	tr	0.01	0.1	0.03	
Undecane	1120-21-4				0.06			
Undecyl acetate	1731-81-3		tr					
Valencene	4630-07-3		0.3	tr	0.01		0.01	0.04[o]; 0.5[r]
Verbenone	80-57-9							0.1[g]

A seventy-two bitter orange essential oil samples from Brazil, Italy, Spain and Egypt analyzed between 1998 and 2013; lowest and highest concentrations given (E. Schmidt, unpublished data)

B review of studies published between 1979 and 1999 on the composition of cold-pressed bitter orange oils; data cited in ref. 1; highest concentrations given

C review of studies published between 1979 and 1999 on the composition of commercial bitter orange oils; data cited in ref. 1; highest concentrations given. This column also contains data from ref. 2, indicated by the superscript letter [b]

D review of studies published between 1979 and 1999 on the composition of laboratory-extracted bitter orange oils; data cited in ref. 1; highest concentrations given. The data in refs. 3,4 and 5 were not provided in ref. 1, as the compositions of the analyzed oils were (very) atypical for bitter orange peel oils

E review of studies published between 1998 and 2009 on the composition of industrial bitter orange oils (refs. 6-9); data cited in ref. 1; highest concentrations given

Table 5.64.2 Constituents identified in bitter orange oils (*continued*)

F review of studies published between 1998 and 2009 on the composition of laboratory cold-extracted bitter orange oils (refs. 10-15); data cited in ref. 1; highest concentrations given

G data from other studies (indicated with superscript letters); highest concentrations found in any study reviewed here given; when two or more oils were investigated, only the highest concentrations are mentioned, unless indicated otherwise

[a] incorrect identification; [b] one commercial cold-pressed oil (ref. 2); [c] one lab-hydrodistilled oil from spontaneously growing fruit in Sicily with very atypical composition (ref. 17); [d] one bitter orange oil from Venezuela (ref. 18); [e] one lab-hydrodistilled oil from Tunisia (ref. 16); [f] four lab-hydrodistilled peel oils from four bitter orange tree provenances (Chanion, Brazil, Roxani, Corsica) (ref. 19); [g] one lab-hydrodistilled peel oil from Tunisia (ref. 21); [h] one commercial bitter orange sample purchased from a USA company (ref. 22); [i] two commercial oils from Italy and Brazil; only the germacrenes were investigated (ref. 24); [j] one cold-pressed oil from the West Indies (ref. 26); [k] one lab-hydrodistilled oil from Greece (ref. 27); [l] unknown number of bitter orange peel oils from Japan (ref. 29); [m] four cold-pressed oils from peels of bitter oranges grown in Greece, Brazil and Corsica (ref. 30); [n] one lab-hydrodistilled bitter orange oil from northern Greece (ref. 32); [o] three lab-hydrodistilled peel oils from Tunisian bitter oranges at various ripening stages; data from the oil obtained from ripe fruits given (ref. 33); [p] four lab-hydrodistilled oil samples from Brazil (ref. 34); [q] one oil of unknown origin and mode of production (ref. 35); [r] one cold-pressed oil from Venezuela (ref. 36); [s] 10 commercial oil samples from different geographic origins (ref. 39)

tr: trace (in column G[j]: <0.005)

Table 5.64.3 Major constituents of bitter orange oils

Constituent	CAS	Percentage and range						
		A	B (1)	C (1)	D (1)	E (6-9)	F (10-15)	G
Limonene	138-86-3	92.2-95.6	94.3	93.6	95.5	96.5	96.3	95.5[l]; 95.9[s]; 96.7[k]; 96.9[g]; 98.3[p]
Myrcene	123-35-3	1.5-2.5	3.1	2.4	2.3	4.9	1.9	2.0[l]; 2.1[h]; 2.6[q]; 4.4[j]; 8.9[s]
Linalool	78-70-6	0.1-0.3	0.4	0.5[b]	3.2	0.3	5.2	1.5[e]; 1.8[q]; 2.0[f,m]; 2.5[l]; 2.9[d]
Linalyl acetate	115-95-7	0.05-0.4	1.2	0.7[b]	2.7	1.0	5.0	0.2[e]; 0.4[s]; 0.5[f,m]; 0.9[j]; 2.7[l]; 4.0[r]
β-Pinene	127-91-3	0.2-0.8	1.3	2.5[b]	1.6	1.0	4.0	0.6[n,s]; 0.9[h]; 1.2[l]; 1.4[g]; 3.3[d]; 3.4[r]

LEGEND: SEE UNDER TABLE 5.64.2

Table 5.64.4 Heterocyclic oxygenated compounds present in cold-pressed bitter orange oils (9,23,31,38)

Name	Synonym(s)	CAS
Hydroxylated polymethoxyflavones		
5-Hydroxy-6,7,8,3',4'-pentamethoxyflavone	5-Demethylnobiletin	2174-59-6
Polymethoxyflavones		
Heptamethoxyflavone		119279-30-0
3',4',5,6,7,8-Hexamethoxyflavone	Nobiletin	478-01-3
3',4',5,6,7-Pentamethoxyflavone	Sinensetin	2306-27-6
3,4',6,7,8-Pentamethoxyflavone	Auranetin	522-16-7
4',5,6,7,8-Pentamethoxyflavone	Tangeretin	481-53-8
4',5,6,7-Tetramethoxyflavone	Tetra-*O*-methylscutellarein	1168-42-9
Coumarins and psoralens		
Auraptenol		108354-46-7
Coumarin, 8-(2,3-dihydroxy-3-methylbutyl)-7-methoxy-	Meranzin hydrate	5875-49-0
Coumarin, 8-(2,3-epoxy-3-methylbutyl)-7-methoxy-, (-)-	Meranzin	23971-42-8
Coumarin, 7-methoxy-8-(3-methyl-2-oxobutyl)-	Isomeranzin	1088-17-1
5,7-Dimethoxycoumarin	Citropten; Limetin	487-06-9
5-(6',7'-Epoxy)geranyloxypsoralen	Epoxybergamottin	206978-14-5
5-(6',7'-Epoxy)geranyloxypsoralen hydrate	Epoxybergamottin hydrate	
7-Geranyloxycoumarin	Aurapten	495-02-3
5-Geranoxypsoralen	Bergamottin; Bergaptin	7380-40-7
8-Geranyloxypsoralen		7437-55-0

Table 5.64.4 Heterocyclic oxygenated compounds present in cold-pressed bitter orange oils (9,23,31,38) (*continued*)

Name	Synonym(s)	CAS
7-Hydroxycoumarin	Umbelliferone	93-35-6
5-Hydroxy-6,7-furanocoumarin	Bergaptol	486-60-2
Isoimperatorin		482-45-1
7-Methoxy-8-isopentenylcoumarin	Osthole	484-12-8
5-Methoxypsoralen	Bergapten	484-20-8
Oxypeucedanin[a]		26091-73-6
Phellopterin[a]		2543-94-4

[a] false-positive (38)

been reported in over 10 publications. In groups of consecutive patients suspected of contact dermatitis, prevalence rates of up to 3.2% positive patch test reactions have been observed, but adequate relevance data are lacking. Most case reports concern occupational contact dermatitis in aromatherapists and others who massage clients with oils and in patients working in a perfume factory.

Testing in groups of patients

The results of patch tests with orange oil in routine testing (consecutive patients suspected of contact dermatitis) and in groups of selected patients are shown in Table 5.64.5. In routine testing, rates of positive reactions ranged from 0.6% to 3.2%, whereas between 0.1% and 4.5% of patients in selected groups had positive patch tests. However, relevance data are lacking.

Case reports

An aromatherapist had chronic hand dermatitis and was patch test positive to 17 of 20 oils used at her work (tested 1% and 5% in petrolatum), including orange oil (44). A physiotherapist had occupational contact dermatitis from orange oil and other essential oils (46). Two cases of occupational allergy to 'orange oil terpenenes'

(tested 1% in olive oil) in bottle fillers in a perfume factory were seen; another two positive patch test reactions were observed among 20 workers in the same factory who did not have dermatitis (49). Irritant contact dermatitis of the hands was diagnosed in a male masseur who worked with essential oils regularly; he had positive patch test reactions to orange oil and lemongrass oil but it was not mentioned whether he had used these oils (48). One positive patch test reaction to orange oil, which was present in a cosmetic that had given a positive patch test or had been positive in a usage test in patients, was seen in a 9-year period in one clinic in Belgium (50).

Additional information on allergy to orange oil can be found in ref. 45 (article not read).

Orange oil, bitter

General

Contact allergy to bitter orange oil has been reported in two publications only. In a group of consecutive patients suspected of contact dermatitis, the prevalence rate of positive patch test reactions was 1.5%, but relevance data are lacking.

Table 5.64.5 Results of testing groups of patients with orange oil

Years and Country	Test conc. & vehicle	Number of patients tested	Number of patients positive (%)	Selection of patients (S); Relevance (R); Comments (C)	Ref.
Routine testing					
2000-2007 USA	2% pet.	678	4 (0.6%)	R: 100%; C: weak study: a. high rate of macular erythema and weak reactions, b. relevance figures include 'questionable' and 'past' relevance	42
1967-1970 Italy	20% pet.	590	19 (3.2%)	R: not stated, but in the population tested (which included patients with dermatitis other than contact dermatitis) many had regular contact with oranges	47
Testing in groups of selected patients					
2000-2008 IVDK	2% pet.	6,246	10 (0.2%)	S: patients with dermatitis suspected of causal exposure to fragrances; R: not stated; C: no frequent co-reactivity to limonene, the main component of all citrus oils	51
1997-2000 Austria	2% pet.	747	1 (0.1%)	S: patients suspected of fragrance allergy: R: not stated	43
1989-1999 Portugal	2% pet.	67	3 (4.5%)	S: patients who had a positive patch test to the fragrance mix; R: not stated	52

IVDK Information Network of Departments of Dermatology, Germany, Switzerland, Austria (www.ivdk.org); pet.: petrolatum

Testing in groups of patients

Two hundred dermatitis patients from Poland were tested with bitter orange oil 2% in petrolatum and three (1.5%) reacted. The same authors also patch tested 51 patients allergic to *Myroxylon pereirae* resin (balsam of Peru) and/or turpentine and/or wood tar and/or colophony and three (5.9%) had a positive patch test; relevance data were not provided (54). A group of 86 patients from Poland previously reacting to the fragrance mix was tested with bitter orange oil and two (2.3%) had a positive patch test reaction; relevance data were not provided (53).

LITERATURE

1 Dugo G, Cotroneo A, Bonaccorsi I, Trozzi A. Composition of the volatile fraction of citrus peel oils. In: G Dugo and L Mondello, Eds. Citrus oils: composition, advanced analytical techniques, contaminants, and biological activity. Boca Raton, USA: CRC Press, Taylor & Francis Group, 2011:10-20

2 Chouchi D, Barth D, Reverchon E, Della Porta G. Bigarade peel oil fractionation by supercritical carbon dioxide desorption. J Agric Food Chem 1996;44:1100-1104

3 El-Samahy SK, Askar A, Abd El-Fadeel MG. Quantitative analysis of some citrus peel oils. Riechstoffe, Aromen, Kosmetica 1982;3:68-70. Data cited in ref. 1

4 Kusunose H, Sawamura M. Aroma constituents of some sour citrus oil. Nippon Shokuhin Kogyo Gakkaishi 1980;27:517-521. Data cited in ref. 1

5 Zhu L-F, Li H-Y, Li B-L, Lu B-Y, Zhang W-L. Aromatic plants and essential constituents (supplement 1). Hong Kong: Peace Book Co Ltd, 1995. Data cited in ref 1

6 Pino JA, Rosado A. Composition of cold pressed bitter orange oil from Cuba. J Essent Oil Res 2000;12:675-676

7 Mondello L, Casilli A, Tranchida PQ, Cicero L, Dugo P, Dugo G. Comparison of fast and conventional GC analysis for citrus essential oils. J Agric Food Chem 2003;51:5602-5606

8 Mondello L, Casilli A, Tranchida PQ, Costa R, Dugo P, Dugo G. Fast GC for the analysis of citrus oils. J Chromatogr Sci 2004;42:410-416

9 Dugo G, Bonaccorsi I, Sciarrone D, Costa R, Dugo P, Mondello L, et al. Characterization of oils from the fruits, leaves and flowers of the bitter orange tree. J Essent Oil Res 2011;23:45-59

10 Huang Y, Wu Y. Chemical components of essential oils from peels of 25 citrus species and cultivars. Tianran Chanwu Yanjiu Yu Kaifa 1998;10:48-54. Data cited in ref. 1

11 Song HS, Sawamura M, Ito T, Ido A, Ukeda H. Quantitative determination and characteristic flavour of Daidai (*Citrus aurantium* L. var. *cyathifera* Y. Tanaka) peel oil. Flavour Fragr J 2000;15:323-328

12 Sawamura M. Volatile components of essential oils of the *Citrus* genus. Recent Res Dev Agric Food Chem 2000;4:131-164

13 Lota M-L, de Rocca Serra D, Jacquemond C, Tomi F, Casanova J. Chemical variability of peel and leaf essential oils of sour orange. Flavour Fragr J 2001;16:89-96

14 Kirbaslar FG, Kirbaslar SI. Composition of cold pressed bitter orange peel oil from Turkey. J Essent Oil Res 2003;15:6-9

15 Gionfriddo F, Catalfamo M, Siano F, Mangiola C, Cautela D, Castaldo D. Determinazione delle caratteristiche analitiche e della composizione enantiomerica di oli essenziali agrumari ai fini dell'accertamento della purezza e della qualità. Nota I – Essenze di arancia amara, arancia dolce e bergamotto. Essenz Deriv Agrum 2003;73:29-39. Data cited in ref. 1

16 Boussaada O, Chemli R. Chemical composition of essential oils from flowers, leaves and peel of *Citrus aurantium* L. var. *amara* from Tunisia. J Essent Oil Bear Plants 2006;9:133-139

17 Giamperi L, Fraternali D, Ricci D. Essential oil composition and antioxidant activity of peels of *Citrus sinensis* blood and blond and *Citrus aurantium*. Riv Ital EPPOS 2002;34:21-28. Data cited in ref. 1.

18 de Gonzalez CN, Sánchez F, Quintero A. Chemotaxonomic value of essential oil compounds in *Citrus* species. Acta Hortic 2002;576:49-51. Data cited in ref. 1

19 Boussaada O, Skoula M, Kokkalou E, Chemli R. Chemical variability of flowers, leaves, and peels oils of four sour orange provenances. J Essent Oil Bear Plants 2007;10:453-464. Virtually the same data as in ref. 30

20 Stohs SJ. Assessment of the adverse event reports associated with *Citrus aurantium* (bitter orange) from April 2004 to October 2009. J Funct Foods 2010;2:235-238

21 Hosni K, Zahed N, Chrif R, Abid I, Medfei W, Kallel M, et al. Composition of peel essential oils from four selected Tunisian *Citrus* species: Evidence for the genotypic influence. Food Chem 2010;123:1098-1104

22 Veriotti T, Sacks R. High-speed characterization and analysis of orange oils with tandem-column stop-flow GC and time-of-flight MS. Anal Chem 2002;74:5635-5640

23 Russo R, Torre G, Carnovale C, Bonaccorsi I, Mondello L, Dugo P. A new HPLC method developed for the analysis of oxygen heterocyclic compounds in Citrus essential oils. J Essent Oil Res 2012;24:119-129

24 Feger W, Brandauer H, Ziegler H. Germacrenes in Citrus peel oils. J Essent Oil Res 2001;13:274-277

25 Hardin A, Crandall PG, Stankus T. Essential oils and antioxidants derived from Citrus by-products in food protection and medicine: An Introduction and review of recent literature. J Agric Food Inform 2010;11:99-122

26 Deterre S, Rega B, Delarue J, Decloux M, Lebrun M, Giampaoli P. Identification of key aroma compounds from bitter orange (*Citrus aurantium* L.) products: essential oil and macerate–distillate extract. Flavour Fragr J 2012;27:77-88

27 Papachristos DP, Kimbaris AC, Papadopoulos NT, Polissiou MG. Toxicity of citrus essential oils against *Ceratitis capitata* (Diptera: Tephritidae) larvae. Ann Appl Biol 2009;155:381-389

28 Lawrence BM. Progress in essential oils. Perfum Flavor 2000;25(March/April):46-49

29 Namba T, Araki I, Mikage M, Hattori M. Fundamental studies on the evaluation of crude drugs. VIII. Monthly variations in anatomical characteristics and chemical components of the dried fruit peels of *Citrus unshiu*, *C. aurantium* and *C. natsudaidai*. Shoyakugaku Zasshi 1985;39:52-62. Data cited in ref. 28

30 Bussada O. Variation of essential oil yield and composition of *Citrus aurantium* var. *amara*. M.Sc. Thesis, Mediterranean Agronomic Institute, Chania, Greece, 1995.Data cited in ref. 28. Virtually the same data as in ref. 19

31 Buiarelli F, Cartoni GP, Coccioli F, Leone T. Analysis of bitter essential oils from orange and grapefruit by high performance liquid chromatography with microbore columns. J Chromatogr 1996;730:9-16

32 Sarrou E, Chatzopoulou P, Dimassi-Theriou K, Therios I. Volatile constituents and antioxidant activity of peel, flowers and leaf oils of *Citrus aurantium* L. growing in Greece. Molecules 2013;18:10639-10647

33 Bourgou S, Zohra Rahali F, Ourghemmi I, Saidani Tounsi M. Changes of peel essential oil composition of four Tunisian Citrus during fruit maturation. Sci World J 2012, Article ID 528593, 10 pages: doi:10.1100/2012/528593

34 Moraes TM, Kushima H, Moleiro FC, Santos RC, Rocha LRM, Marques MO, et al. Effects of limonene and essential oil from Citrus aurantium on gastric mucosa: Role of prostaglandins and gastric mucus secretion. Chemico-Biological Interactions 2009;180:499-505

35 Ali J, Khan A, Ullah N, Amin M, Zia-ur-Rahman, Umar K, Sahibzada M, et al. Essential oil chemical composition and antimicrobial activity of sour oranges (*Citrus aurantium*) peels. J Pharm Res 2012;5: 1690-1695

36 Quintero A, de Gónzalez CN, Sánchez F, Usubillaga A, Rojas L. Constituents and biological activity of *Citrus aurantium amara* L. essential oil. In: J Bernáth, et al, Eds. Proc Int Conf on MAP. Acta Hort 597, ISHS 2003, 115-117. Available at: http://wwwlib.teiep.gr/images/stories/acta/Acta%20597/597_14.pdf

37 Liu YQ, Heying E, Tanumihardjo SA. History, global distribution, and nutritional importance of Citrus fruits. Compreh Rev Food Sci Food Saf 2012;11:530–545

38 Dugo P, Russo M. The oxygen heterocyclic components of Citrus essential oils. In: G Dugo and L Mondello, eds. Citrus oils: composition, advanced analytical techniques, contaminants, and biological activity. Boca Raton, USA: CRC Press, Taylor & Francis Group, 2011:405-444

39 Deterre SC, Rega B, Delarue J, Teillet E, Giampaoli P. Classification of commercial bitter orange essential oils (*Citrus aurantium* L.), based on a combination of chemical and sensory analyses of specific odor markers. J Essent Oil Res 2014;26:254-262

40 Rhind JP. Essential oils. A handbook for aromatherapy practice, 2nd Edition. London: Singing Dragon, 2012

41 Lawless J. The encyclopedia of essential oils, 2nd Edition. London: Harper Thorsons, 2014

42 Wetter DA, Yiannias JA, Prakash AV, Davis MD, Farmer SA, el-Azhary RA, et al. Results of patch testing to personal care product allergens in a standard series and a supplemental cosmetic series: an analysis of 945 patients from the Mayo Clinic Contact Dermatitis Group, 2000-2007. J Am Acad Dermatol 2010;63:789-798

43 Wöhrl S, Hemmer W, Focke M, Götz M, Jarisch R. The significance of fragrance mix, balsam of Peru, colophonium and propolis as screening tools in the detection of fragrance allergy. Br J Dermatol 2001;145:268-273

44 Selvaag E, Holm J, Thune P. Allergic contact dermatitis in an aromatherapist with multiple sensitizations to essential oils. Contact Dermatitis 1995;33:354-355

45 Rivasseau J. A propos des sensibilisations de groupe. Bull Soc Franç Derm Syph 1956;63:83-84

46 Trattner A, David M, Lazarov A. Occupational contact dermatitis due to essential oils. Contact Dermatitis 2008;58:282-284

47 Meneghini CL, Rantuccio F, Lomuto M. Additives, vehicles and active drugs of topical medicaments as causes of delayed-type allergic dermatitis. Dermatologica 1971;143:137-147

48 Jung P, Sesztak-Greinecker G, Wantke F, Götz M, Jarisch R, Hemmer W. Mechanical irritation triggering allergic contact dermatitis from essential oils in a masseur. Contact Dermatitis 2006;54:297-299

49 Schubert HJ. Skin diseases in workers at a perfume factory. Contact Dermatitis 2006;55:81-83

50 Nardelli A, Drieghe J, Claes L, Boey L, Goossens A. Fragrance allergens in 'specific' cosmetic products. Contact Dermatitis 2011;64:212-219

51 Uter W, Schmidt E, Geier J, Lessmann H, Schnuch A, Frosch P. Contact allergy to essential oils: current patch test results (2000–2008) from the Information Network of Departments of Dermatology (IVDK). Contact Dermatitis 2010;63:277-283

52 Manuel Brites M, Goncalo M, Figueiredo A. Contact allergy to fragrance mix — a 10-year study. Contact Dermatitis 2000;43:181-182

53 Rudzki E, Grzywa Z. Allergy to perfume mixture. Contact Dermatitis 1986;15:115-116

54 Rudzki E, Grzywa Z, Bruo WS. Sensitivity to 35 essential oils. Contact Dermatitis 1976;2:196-200

Chapter 5.65 ORANGE OIL, SWEET

DEFINITION

Sweet orange oil (essential oil of sweet orange) is the essential oil obtained by expression from the pericarp (peel) of the (sweet) orange, *Citrus sinensis* (L.) Osbeck.

INCI NOMENCLATURE

Description/definition: Citrus aurantium dulcis (orange) peel oil is the volatile oil obtained by expression from the peel of *Citrus sinensis*, Rutaceae
INCI name EU & USA: Citrus aurantium dulcis (orange) peel oil
Other names: Sweet orange oil
CAS registry number(s): 8008-57-9; 8028-48-6
EINECS number(s): 232-433-8

Description/definition: Citrus sinensis peel oil expressed is an essential oil expressed from fresh epicarps of the sweet orange Valencia, *Citrus sinensis* (syn: *Citrus aurantium dulcis*), Rutaceae
INCI name EU: Citrus sinensis peel oil expressed (not officially an INCI name but perfuming name)
INCI name USA: Not in the Personal Care Products Council Ingredient Database
Other names: Sweet orange oil; orange oil
CAS registry number(s): 97766-30-8; 8028-48-6
EINECS number(s): 307-891-8; 232-433-8

It should be realized that – contrary to what is suggested in the description/definition of *Citrus sinensis* peel oil above — *Citrus aurantium dulcis* is *not* a synonym for *Citrus sinensis* (L.) Osbeck.

Description/definition: Citrus sinensis Valencia peel oil expressed is an essential oil expressed from the epicarps of the sweet orange Valencia, *Citrus sinensis*, Rutaceae
INCI name EU: Citrus sinensis Valencia peel oil expressed (not officially an INCI name but perfuming name)
INCI name USA: Not in the Personal Care Products Council Ingredient Database
CAS registry number(s): 97766-30-8
EINECS number(s): 307-891-8

ISO (INTERNATIONAL ORGANIZATION FOR STANDARDIZATION) STANDARD

ISO number: 3140
ISO name: Essential oil of sweet orange
Botanical origin: *Citrus sinensis* (L.) Osbeck
Parts of plant used: Pericarp (peel)
ISO values: ISO values (minimum and maximum concentrations) are shown in Table 5.65.1.

By ISO definition, all citrus essential oils except lime oil are produced by expression; oils obtained from citrus fruits by distillation may not be called essential oils according to ISO criteria (except lime oil). In industry,

Table 5.65.1 ISO values (%) for sweet orange oil [a]

Compound	CAS	Minimum	Maximum
Limonene	138-86-3	93.0	96.0
Myrcene	123-35-3	1.5	3.5
α-Pinene	80-56-8	0.4	0.8
Sabinene	3387-41-5	0.2	0.8
Linalool	78-70-6	0.15	0.7
n-Decanal	112-31-2	0.1	0.7
n-Octanal	124-13-0	0.1	0.4
Geranial	141-27-5	0.05	0.2
β-Pinene	127-91-3	0.02	0.15
Neral	106-26-3	0.03	0.1
n-Nonanal	124-19-6	0.01	0.06
β-Sinensal	17909-87-4	0.01	0.06
Valencene	4630-07-3	0.01	0.06

[a] ISO 3140 Essential oil of sweet orange ©ISO 2010: Geneva, Switzerland, www.iso.org

however, sometimes residues from expression of the citrus peels undergo steam distillation, to obtain the remaining oil; these volatile oils may then be added to the expressed oil. As some of the compounds undergo changes forced by high temperature during distillation, this addition (which is an adulteration) changes the composition of the essential oil. Because of this, and also because it cannot be excluded that oils entirely produced by hydrodistillation reach the market, both the ingredients found in 'genuine' sweet orange essential oils obtained by expression and those that may be present in hydrodistilled oils are presented here.

THE PLANT, THE OIL, AND THEIR USES

Citrus species may have originated in India 30 to 40 million years ago. *Citrus* fruits subsequently spread to East Asia and the Mediterranean region, where a great number of varieties were born by numerous mutations. The most popular citrus fruit is the sweet orange, *Citrus sinensis* (L.) Osbeck. This is a tree, up to 10 m tall, with a few slender and flexible spines and aromatic leaves of 7.5-11 cm. China is the origin of sweet oranges: *sinensis* in its botanical name *Citrus sinensis* (L.) Osbeck refers to China. Sweet oranges are categorized into the common oranges, acidless oranges, pigmented oranges (blood oranges and navel oranges) (21). Among these categories, the common sweet oranges comprise the most popular varieties cultivated in many countries, such as Valencia, Barao, Salustiana, Belladonna and Berna. Oranges are grown widely in the tropics and subtropics. Brazil produces the greatest volume of this species, followed by the United States, China, Mexico, Spain, India, Egypt and Morocco (21,22).

Orange peel oil is a by-product of orange juice production. During commercial juice extraction the peel is macerated and oil glands in the flavedo are ruptured. Depending on the type of juice extractor, most peel oil is collected using an exterior water spray during juice extraction. The resulting oil–water emulsion is recovered separately from the expressed juice using a series of centrifuges. It is a non-heated oil product, often called 'cold-pressed peel oil' (14). These essential oils have wide commercial applications in food processing, pharmaceutical preparations, perfumery and cosmetics because of their flavor and fragrance (21). Sweet orange oils are also employed in aromatherapy practices (62,63).

CHEMICAL COMPOSITION

Sweet orange oil is a clear mobile, yellow to orange-brown liquid which has a fresh citrusy odor reminding of fresh scratched orange peel. The yield of essential oil from cold pressing the peel of *Citrus sinensis* (L.) Osbeck generally ranges from 0.3 to 0.6 per cent. The main producing countries of this oil are USA (California, Florida), Brazil, Argentina, South Africa, Spain, Israel, Italy and Guinea. The amount of essential oil obtained by hydro-distillation (non-commercial oils) ranges from 0.1% to 2.5% (30,35).

Cold-pressed *Citrus* essential oils consist of a volatile fraction, which represents 85-99% of the oil, and a non-volatile fraction 1-15%, which mainly contains oxygen heterocyclic compounds, especially coumarins, psoralens and polymethoxyflavones (Table 5.65.4). These components play an important role in the characterization of cold-pressed *Citrus* oils, since the composition of this fraction is characteristic of each oil. Moreover, many of the pharmacological and toxicological activities possessed by *Citrus* oils have been demonstrated to be related to these components (52).

Literature data (up to December 10, 2014) on the chemical composition of sweet orange oils and unpublished analytical data from one of us (E.S.) are shown in Table 5.65.2 in alphabetical order. In sweet orange oils from various origins, over 335 chemicals have been identified. The major compounds found in sweet orange oils from different sources are shown in Table 5.65.3. In all studies, limonene is by far the most important compound, with concentrations ranging in 'genuine' cold-pressed oils up to 97.3 per cent. Other important constituents include (highest concentration in any study given) myrcene (22.4%; high concentration in one study only, all others have a concentration <7.2%), linalool (15.8%; high concentration in one study only, all others have a concentration <5.9%) and α-pinene (14.8%; high

Table 5.65.2 Constituents identified in sweet orange oils

Constituent	CAS	Percentage and range						
		A	B (1)	C (1)	D (1)	E (2-15)	F (16-28)	G
Acetaldehyde	75-07-0							0.6[z10]
Acetic acid	64-19-7				+			6.0[x2]
5-Acetyl-2-methyl-pyridine	36357-38-7							0.7[x3]
δ-Amorphene	189165-79-5							0.05[z4]
Amyl alcohol	71-41-0							0.07[z10]
Aromadendrene	489-39-4						tr	0.02[y]
allo-Aromadendrene	25246-27-9				0.01	0.02[h]		
Artemiseole	60485-46-3							0.2[x]
Benzaldehyde	100-52-7				+			
cis-α-Bergamotene	18252-46-5			tr	tr			tr[z9]
trans-α-Bergamotene	13474-59-4			tr		0.03		tr[y,z9]
Bicyclogermacrene	24703-35-3	tr			0.01	0.01[l]	tr	0.04[y]
β-Bisabolene	495-61-4		0.01	1.5	0.02	tr[e]	tr	tr[z9]; 0.3[z11]
α-Bisabolol	515-69-5							tr[y]
β-Bisabolol	15352-77-9							tr[y]
Bisabol-1-one	61432-71-1							0.02[y]
Borneol	507-70-0					tr[j]		tr2[z11]; 0.7[z3]
Bornyl acetate	76-49-3		0.01		0.04	0.01	0.03	tr2[z11]; 4.2[z3]
Bornyl isobutyrate	24717-86-0							0.03[y]
2,3-Butanediol	513-85-9							tr2[z11]
Butanol	35296-72-1				+			
Butyl butanoate	109-21-7				+			
2,6-di-tert-Butyl-p-cresol	128-37-0				+			
Butyl hexanoate	626-82-4				+			
Cadinene	29350-73-0						tr[p]	
α-Cadinene	24406-05-1		0.01					
γ-Cadinene	39029-41-9		0.1	0.1	0.01	0.02[j]	0.03	0.02[z5]
δ-Cadinene	483-76-1		0.05	0.07	0.03	0.03	0.2	tr[z9]; 0.1[z7,x5]
δ-Cadinol	19435-97-3			tr				
Camphene	79-92-5	0.0-0.01	0.01	tr	3.9	0.01	0.7	0.07[z9]; 1.2[z3]
α-Campholenal	4501-58-0							tr[y]
Camphor	76-22-2				+	tr[j]		tr[z9]; 4.8[z3]
δ2-Carene (δ=4-)	554-61-0							0.9[z12]

Table 5.65.2 Constituents identified in sweet orange oils *(continued)*

Constituent	CAS	Percentage and range						
		A	B (1)	C (1)	D (1)	E (2-15)	F (16-28)	G
δ3-Carene	13466-78-9	0.03-0.3	0.2	0.2	0.2	0.3	0.3	tr2[z11]; 0.5[z9]; 0.7[w]; 1.0[z12]
(+)-4-Carene	13837-63-3							0.8[x4]
Carvacrol	499-75-2						tr[u]	0.3[z3]
Carvacryl methyl oxide	6379-73-3							0.01[y]
Carveol	99-48-9						tr[t]	0.2[x4]
(E)-Carveol	1197-07-5		0.1	0.06	tr	0.3	0.3	tr2[z11]; 0.6[z2]
(Z)-Carveol	1197-06-4		tr	tr	0.01	0.3	0.5	0.3[z2]; 4.5[z12]
Carvone	99-49-0	0.01-0.08	0.09	0.07	2.5[c]	0.08	0.5	tr[z8]; 0.4[z12]; 0.6[z9]; 0.7[z2]; 4.5[x3] +[y1,x2]
Carvone camphor								
Carvone oxide	33204-74-9						tr[q]	
trans-Carvyl acetate	1134-95-8						tr[u]	
β-Caryophyllene	87-44-5	0.01-0.05	0.04	0.1	0.04	0.05	0.2	tr2[z11]; 0.07[z9]; 0.1[z8]; 0.2[w]; 0.3[x5]
Caryophyllene oxide	1139-30-6			tr	tr		0.2	0.07[z9]; 0.09[z2]
α-Cedrene	469-61-4						0.7	
Cedrol	77-53-2						tr	
1,4-Cineole	470-67-7				+			
1,8-Cineole	470-82-6				+		tr	tr[z3]
Cinnamaldehyde	104-55-2							tr[z11]
Citral	5392-40-5						+[n]	0.2[x6]
Citronellal	106-23-0	0.03-0.1	0.1	0.2	0.06	0.05	0.4	tr2[z11]; 0.07[z9]; 0.08[z2]; 0.2[z13]
Citronellol	106-22-9	0.01-0.03	0.02	0.2	0.02	0.02	0.1	tr2[z11]; 0.1[z8]; 0.2[w,x4]; 1.6[z12]
Citronellyl acetate	150-84-5		0.07	0.09	0.01	0.01	0.05	0.07[z9]; 0.1[w]; 0.2[z11]
α-Copaene	3856-25-5	0.01-0.08	0.04	0.08	0.02	0.02	0.1	tr[z11]; 0.05[y]; 0.06[z9]
β-Copaene	18252-44-3	0.00-0.02	0.03		4.5[c]	0.02[i]		0.1[z2]
Cryptone	500-02-7							0.06[z2]
α-Cubebene	17699-14-8		0.02	0.1			0-tr	0.1[x4]; 0.7[z12]
β-Cubebene	13744-15-5	0.01-0.05	0.02	0.3	0.03	0.02	0.1	0.1[z2]
Cubebol	23445-02-5							0.03[y]
epi-Cubebol	38230-60-3							tr[y]
epi-Cubenol	19912-67-5							0.04[y]
Cuminaldehyde	122-03-2						0.1	
Curcumene	644-30-4		tr					0.5[z10]
α-Cyclocitral	432-24-6							+[y1,x2]
β-Cyclocitral	432-25-7							0.01[y]
3-Cyclohexene-1-methanol	1679-51-2							0.02[x]
p-Cymene	99-87-6	0.01-0.04	0.06	0.2	tr	0.2	0.1	0.09[z9]; 0.3[z3]; 0.7[x3]; 2.3[x4]
p-Cymen-8-ol	1197-01-9							tr[z9]
Cyperene	2387-78-2							tr[y]
α-Cyperone	473-08-5							0.03[y]
2,4-Decadienal	2363-88-4						tr[o]	
(E,E)-2,4-Decadienal	25152-84-5		0.01	tr		+[m]	tr[u]	0.01[y]
(E,Z)-2,4-Decadienal	25152-83-4		0.02			+[m]	tr[u]	
Decanal	112-31-2	0.09-0.6	0.4	0.3	0.4	0.4	1.9	0.2[z11]; 0.3[z9,x6]; 0.4[z2,z8]; 1.0[z12]
n-Decanoic acid	334-48-5			tr				tr[z11]; 0.1[z6]
Decanol	36729-58-5		tr		tr		0.3	tr2[z11]; 0.1[z6]; 0.2[z8]
(E)-2-Decenal	3913-81-3		tr		tr	+[m]	tr[u]	tr2[z11]
(Z)-2-Decenal	2497-25-8					+[m]		
Decyl acetate	112-17-4		0.03	tr	0.03	0.01	tr[o]	tr[z5]; 0.1[z11]
Dehydrocarveol	28982-60-7						tr[r]	
Dehydrolinalool	29171-20-8							0.4[w]
Dendrolasin ((E)-)	23262-34-2							tr[z8]
Dihydrocarveol	38049-26-2						tr[q]	
Dihydrocarvone	5948-04-9				+			
cis-Dihydrocarvone	3792-53-8							0.1[z3]

Table 5.65.2 Constituents identified in sweet orange oils *(continued)*

Constituent	CAS	Percentage and range						
		A	B (1)	C (1)	D (1)	E (2-15)	F (16-28)	G
N,N-Dimethylaniline	121-69-7		tr					
endo-trans-4,7-Dimethylbicyclo[3.2.1]oct-en-3-en-6-one								+[x1]
exo-cis-4,7-Dimethyl-bicyclo[3.2.1]octen-3-en-6-one								+[x1]
2,3-Dimethyl-3-(4-methyl-3-pentenyl)-2-norbornanol	98205-40-4					0.01[i]		
2,6-Dimethyl-2,6-octadiene	2792-39-4							0.1[x4]
2,7-Dimethyl-2,6-octadien-1-ol	22410-74-8							0.01[z5]
3,7-Dimethyl-2,6-octadien-1-ol (citrol)	624-15-7				+			
2,6-Dimethyl-1,3,5,7-octatetraene	90973-78-7							0.03[x]
Dodecanal	112-54-9	0.01-0.1	0.1	0.1	0.06	+[m]	0.1	0.03[z4]; 0.1[z8]
Dodecanoic acid	143-07-7			tr			tr[q]	tr[z11]; 0.1[z6]
Dodecanol	112-53-8				+	0.1		
(*E*)-2-Dodecenal	20407-84-5						tr	0.1[z8]
β-Elemene	33880-83-0	0.00-0.01	0.02	0.1	0.03	0.03	0-0.1	tr2[z11]; 0.1[z13]; 0.2[w]; 0.3[x4];
γ-Elemene	29873-99-2			tr				
δ-Elemene	20307-84-0		tr	0.06			tr[q]	
Elemol (α-)	639-99-6		tr	tr	tr		tr	0.01[z5]
trans-4,5-Epoxide-(*E*)-2-decenal						+[m]		
Ethanol	64-17-5						tr[q]	
Ethyl acetate	141-78-6						tr[s]	0.1[z11]
Ethyl heptanoate	106-30-9					+[m]		
Ethyl octanoate	106-32-1				+			
Eucarvone	503-93-5			0.1				
Eugenol	97-53-0				+	+[m]	tr	0.5[z12]
α-Farnesene	502-61-4							0.7[x4]
(*E,E*)-α-Farnesene	502-61-4		0.3	0.7	0.03	0.1	0.1	3.6[w]
β-Farnesene	502-60-3		0.02	0.2		0.02[g]	tr	
(*E*)-β-Farnesene	18794-84-8		0.01	0.02	0.03	0.02	0.1	0.03[z5]; 0.05[z9]
(*Z*)-β-Farnesene	28973-97-9	0.00-0.01						0.01[z5]; 0.06[z9]
Farnesol	4602-84-0		tr	tr			tr[r]	
(*E,E*)-Farnesol	106-28-5						tr[u]	
(*Z,E*)-Farnesol	3790-71-4						tr	tr[z3]
Fenchol	1632-73-1				+			
Fenchyl acetate	13851-11-1							0.01[y]
Formaldehyde	50-00-00				+			7.1[z10]
Formic acid	64-18-6				+			3.6[z10]
3-Furanacetic acid	123617-80-1							3.0[z12]
Geranial	141-27-5	0.03-0.3	0.2	0.3	3.5[c]	0.1	0.5	tr2[z11]; 0.2[z5,z9]; 0.3[x8]
Geraniol	106-24-1		0.02	0.08	0.09	0.02	0.3	tr2[z11]; 0.06[z9]; 0.3[w]
Geranyl acetate	105-87-3	0.00-0.01	0.03	0.2	0.02	0.03	0.04	tr2[z11]; 0.09[z9]; 0.1[z3]
Geranyl acetone	3796-70-1				+			
Geranyl formate	105-86-2			tr				0.7[x]
Geranyl propanoate	105-90-8							tr[z11]
Geranyl-α-terpinene								0.01[y]
Germacrene A	28387-44-2					0.02[l]	tr[u]	
Germacrene B	15423-57-1		tr					
Germacrene D	23986-74-5	0.01-0.07	0.02	tr	0.02	0.02	0.1	0.06[z9]; 0.1[z8]; 0.3[w]
Germacrene D-4-ol	198991-79-6		tr					
Globulol	489-41-8						tr[q]	0.4[x3]

Table 5.65.2 Constituents identified in sweet orange oils *(continued)*

Constituent	CAS	Percentage and range						
		A	B (1)	C (1)	D (1)	E (2-15)	F (16-28)	G
β-Guaiene	88-84-6			tr				
β-Gurjunene	73464-47-8			tr	0.01		0.1	0.01[z4]
Heneicosane (*n*-)	629-94-7							3.8[x3]
Heptadecanal	629-90-3		+	tr	tr			
Heptanal	111-71-7		+		tr			
Heptanol	53535-33-4				+		tr[q]	
2-Heptanone	110-43-0				+			
Heptyl acetate	5921-82-4				0.01			
Heptyl formate	112-23-2							tr[x2,y2]
Hexadecanal	629-80-1		tr	tr				
n-Hexadecane	544-76-3							0.2[z13]; 0.5[x5]
Hexadecanoic acid	57-10-3						tr[h]	
Hexadecanol	51260-59-4							tr[z5]
Hexanal	66-25-1		+	0.02	tr			
Hexanol	111-27-3				+		0.2 (ref.28)	
(*E*)-2-Hexenal	6728-26-3				+			
(*E*)-2-Hexen-1-ol	928-95-0				+			tr[x2,y2]
(*Z*)-3-Hexen-1-ol	928-96-1				+			
(*Z*)-3-Hexenyl butyrate	16491-36-4				+			
(*Z*)-3-Hexenyl hexanoate	31501-11-8				+			
Hexyl acetate	142-92-7				+			
Hexyl butyrate	2639-63-6				+			1.2[x5]
α-Himachalene	3853-83-6			tr				
α-Humulene	6753-98-6		0.04	0.07	0.01	0.02	0.01	0.07[z9]; 0.2[x4]; 0.3[z3]
β-Ionone	79-77-6				+			0.4[y]
Isoamyl acetate	123-92-2					0.1[d]		tr[z11]
Isoamyl isovalerate	659-70-1							tr[z11]
Isoborneol	124-76-5						tr	
Isodihydrocarveol	18675-35-9					0.01[g]		
Isoisopulegol	18674-65-2							+[y1,x2]
Isomenthone	491-07-6							+[y1,x2]
2-Isononenal	53966-58-8							tr2[z11]
Isophorone	78-59-1							2.9[z12]
Isopiperitenol	491-05-4							+[y1,x2]
Isopiperitone	58615-39-7			tr			0-tr[r]	
Isopulegol	89-79-2				+			0.3[x]
Isosafrole	120-58-1							tr[z11]
Isosylvestrene	61557-13-9							0.1[z2]
Limonene	138-86-3	94.7-95.7	96.1	94.5	96.8	96.1	96.6	91.0[x6]; 92.1[z9]; 96.6[x8]; 97.3[y]
Limonene dioxide	96-08-2		0.2	0.05			0-tr	
Limonene oxide	1195-92-2		0.4	0.09		0.5[k]	0.4	0.8[z12]
Limonene glycol	1946-00-5							7.7[x3]
cis-Limonene oxide	13837-75-7	0.01-0.05	0.01	tr	0.02	0.5	0.03	tr[z9]; 1.0[z3]
trans-Limonene oxide	4959-35-7	0.01-0.04	0.02		0.02	0.5		tr[z9]; tr2[z11]; 0.2[w]
Limonen-4-ol	3419-02-1							+[y1,x2]
Limonen-10-yl acetate	15111-97-4				+		tr[u]	
Linalool	78-70-6	0.04 -0.07	0.8	1.2	15.8[c]	0.4	5.9	1.0[z8]; 1.2[z11]; 2.1[z12]; 3.0[x9]; 3.5[x3]
Linalool oxide	1365-19-1			0.08		+[m]	0.2[s]	1.1[z12]
cis-Linalool oxide	11063-77-7				tr		0.2	tr2[z11]; 0.8[z2]
cis-Linalool oxide, furanoid	5989-33-3							0.5[z8]
trans-Linalool oxide	11063-78-8			tr	tr			tr2[z11]
trans-Linalool oxide, furanoid	34995-77-2							0.9[z8]
Linalyl acetate	115-95-7		0.06	0.2	tr	tr[d,k]	0.1	2.8[x]

Table 5.65.2 Constituents identified in sweet orange oils *(continued)*

Constituent	CAS	Percentage and range						
		A	B (1)	C (1)	D (1)	E (2-15)	F (16-28)	G
Linalyl formate	115-99-1						0.2^v	
cis-p-Mentha-1(7),8-dien-2-ol	22626-43-3							0.04^{z2}
trans-p-Mentha-1(7),8-dien-2-ol	21391-84-4							$+^{y1,x2}$
p-Mentha-1(7),8(10)-dien-9-ol	29548-13-8				+		0.1^u	0.01^{z5}
trans-p-Mentha-2,8-dienol	4017-77-0							0.3^{z2}; $1.1^{z12}1.6^{x3}$
p-Mentha-2,8-dien-1-ol	22771-44-4					0.02^h		
cis-p-Mentha-2,8-dien-1-ol	3886-78-0							0.8^{z2}; 1.4^{x3}
cis-p-Mentha-2,8-dien-9-ol							2.0^v	
trans-p-Mentha-2,8-dien-9-ol							0.4^v	
(+)-p-Mentha-1,8-dien-3-one	16750-82-6							0.09^{z2}
p-Mentha-1,3,8-triene	18368-95-1							1.9^{z12}
p-Menth-1-en-9-al	29548-14-9						tr^q	
p-Menth-1-en-9-ol	18479-68-0						tr^q	
cis-p-Menth-2-en-1-ol	29803-82-5						tr^u	
trans-p-Menth-2-en-1-ol	29803-81-4							0.01^y
p-Menth-4(8)-en-9-ol	15714-11-1				+			0.01^{x2}
Menthone	89-80-5						tr^p	tr^{z11}
Menthyl acetate	16409-45-3							$tr2^{z11}$
2-(Methylamino)benzyl alcohol	29055-08-1		tr					
Methyl anthranilate	134-20-3		tr					$tr2^{z11}$
2-Methylbutanol	137-32-6				+			
3-Methylbutanol					+			
2-Methyl-3-buten-2-ol	115-18-4				+			
3-Methyl-2-buten-1-ol	556-82-1				+			
2-Methylbutyrate							tr^p	
Methyl decanoate	110-42-9							tr^{z11}
Methyl eugenol	93-15-2				+			
Methylheptenone	409-02-9							0.04^{z10}
6-Methyl-5-hepten-2-one	110-93-0		0.04	0.05	tr	tr^j		tr^{z9}; 0.01^x
Methyl isobutyrate	547-63-7							tr^{z11}
Methyl isoeugenol	93-16-3				+			
3-Methyl-6-isopropenyl-2-cyclohexen-1-one								0.3^x
Methyl N-methyl anthranilate	85-91-6			0.01		0.01^h		
Methyl nopinone				0.7				
Methyl octanoate	111-11-5				+			
4-Methylproline	3005-85-4							2.2^{x3}
α-Muurolene	10208-80-7		+		tr	0.02^h	tr^u	0.02^{z2}
γ-Muurolene	30021-74-0		0.02		0.02	0.01^j		
Myrcene	123-35-3	1.4-3.5	2.0	2.3	22.4^c	2.5	5.3	2.5^{z13}; 4.4^{z12}; 4.7^{z8}; 6.3^w; 7.2^{x3}
Myrcene epoxide	29414-55-9						tr^q	
Myrcenol	543-39-5					tr^k		tr^{z11}
Myrtenal	564-94-3		0.03					
Myrtenyl acetate	1079-01-2							0.06^x
Naphthalene	91-20-3							0.04^{x6}

Table 5.65.2 Constituents identified in sweet orange oils *(continued)*

Constituent	CAS	Percentage and range						
		A	B (1)	C (1)	D (1)	E (2-15)	F (16-28)	G
Neral	106-26-3	0.01-0.2	0.2	0.3	0.2	0.2	0.5	0.07[z9]; 0.1[z7,x]; 0.3[x8]
Nerol	106-25-2	0.01-0.1	0.05	0.1	0.02	0.02	0.6	0.06[z9]; 0.1[z7,z11]; 0.2[z8]
Nerolidol			0.01				tr[o]	
(E)-Nerolidol	40716-66-63		tr	tr	0.01		tr[r]	0.03[y]; 0.1[zB]
(Z)-Nerolidol	3790-78-1					0.01[j]	tr	tr2[z11]; 0.07[z9]
Nerolidyl acetate	56001-43-5						0.02[q]	
Neryl acetate	141-12-8	0.01-0.02	0.06	0.09	0.01	0.04	0.01	0.03[y]; 0.06[z9]; 0.3[z8]; 0.5[x4]
Neryl propionate	105-91-9			0.1				tr[z9]
Nikkol								5.7[z12]
(Z)-5-Nonadecane								2.9[x3]
(E,Z)-2,6-Nonadienol	28069-72-9						tr[q]	
Nonanal	124-19-6	0.01-0.1	0.1	0.1	0.1	0.1	0.3	tr[z8]; 0.06[z9,x6]; 0.1[z11]; 0.2[z3]
Nonane	111-84-2							0.2[z13]; 0.4[x5]
Nonanoic acid	12-05-0			tr			0.01[q]	
Nonanol	28473-21-4		0.3	0.8	tr			0.04[z7]
(E)-2-Nonenal	18829-56-6					+[m]		tr2[z11]
Nonyl acetate	143-13-5		tr		0.01	0.01	0.01	tr2[z11]
Nootkatone	4674-50-4		0.03		0.04	0.02	0.1	0.05[s2]
β-Ocimene	13877-91-3					0.05[e]		0.3[w]
(E)-β-Ocimene	3779-61-1	0.01-0.07	0.1	0.2	0.06	0.06	0.1	0.07[z9]; 0.1[z13]; 0.2[x]
(Z)-β-Ocimene	3338-55-4		0.03	tr	0.01	0.01	0.07	tr[z9,z11]; 0.3[z4,z8,z12]
Octadecanal	638-66-4		tr				tr[q]	
Octadecane	593-45-3						tr	0.1[x5]
(Z)-9-Octadecenal	2423-10-1			tr				
Octanal	124-13-0	0.07-0.5	0.3	0.4	0.7	0.4	1.4	0.3[z8]; 0.4[x]; 0.8[z13]; 1.3[z11]; 1.4[z9]
Octanoic acid	124-07-2			tr			0.02	
Octanol	111-87-5	0.01-0.1	0.2	0.3	0.2	0.1	1.1	0.08[z9]; 0.1[z6,z11]; 0.3[z8]; 0.9[x3]
1-Octen-3-ol	3391-86-4					+[m]		
Octyl acetate	112-14-1	0.01-0.02	0.03	+	0.01	0.03	0.1	0.1[z7]
Octyl butyrate	110-39-4				+			
Octyl hexanoate	4887-30-3				+			
Octyl propyl ether	29379-41-7							0.01[z5]
(E)-Patchoulenol	17806-54-1							0.4[z2]
Patchouli alcohol	5986-55-0				+			
Pentadecane (n-)	629-62-9							0.2[z13]
Pentanol	30899-19-5				+			
1,4,7,10,13-Pentaoxa-cyclopentadecane	33100-27-5						tr[q]	
Perilla aldehyde	2111-75-3	0.01-0.07	0.04	0.05	0.01	0.06	0.2	0.08[z2]; 0.1[z8,z11]; 0.4[x3]
Perillene	539-52-6		0.01					
Perillyl acetate	15111-96-3					0.02[h]	tr[q]	
Perillyl alcohol	536-59-4	0.00-0.01	0.01	0.02	tr	0.01[h]	0.1	tr[z11]; 0.08[z2]
α-Phellandrene	99-83-2	0.02-0.3	0.07	0.05	0.09	0.2	2.9	tr2[z11]; 0.07[z9]; 0.1[z8]; 0.3[y]; 0.5[z12]
β-Phellandrene	555-10-2	0.02-0.7	0.2	1.5	0.2	0.2[f]	0.6	0.07[z2]
Phenylacetaldehyde	122-78-1				+			
Photocitral A								+[y1,x2]
α-Pinene	80-56-8	0.2-1.0	1.4	0.9	14.8[c]	0.7	1.6	1.0[z8]; 1.4[z13]; 1.6[x5]; 2.4[z11]; 2.5[w]
β-Pinene	127-91-3	0.01-0.5	0.1	1.3	1.1	0.5	2.7	0.1[z11]; 0.5[x8]; 1.3[x4]; 1.8[y]; 2.2[x7]
(S)-β-Pinene (laevo-)	18172-67-3							0.5[x]
α-Pinene epoxide	1686-14-2						tr[q]	
cis-Pinocarveol	6712-79-4							0.06[z2]
trans-Pinocarveol	1674-08-4							0.05[y]
trans-Pinocarvyl acetate	1686-15-3							0.01[y]

Table 5.65.2 Constituents identified in sweet orange oils *(continued)*

Constituent	CAS	A	B (1)	C (1)	D (1)	E (2-15)	F (16-28)	G
Piperitenone	491-09-8							0.2[x2]; 0.4[z12]
cis-Piperitol	16721-38-3						tr[q]	
Piperitone	89-81-6					tr[j]		
Pulegone	89-82-7				+			0.03[z10]
trans-Rose oxide	5258-11-7							tr[y]
Sabina ketone	513-20-2						0.2[q]	
Sabinene	3387-41-5	0.02-0.7	0.8	2.5	0.9	0.5	1.2	0.5[x6]; 0.9[x5]; 1.0[z9]; 1.6[z11]; 1.8[z5]
cis-Sabinene hydrate	15537-55-0		0.01	tr	0.05	0.04		tr[z9]; 0.4[z3]
trans-Sabinene hydrate	17699-16-0			tr	tr	tr	0.6	tr[z11]; 0.4[x]
7-epi-α-Santalene			tr					
Sativene	6813-05-4							0.02[z2]
β-Selinene	17066-67-0						tr[u]	0.02[z2]
7-epi-α-Selinene	123123-37-5						tr[u]	0.02[z2]
β-Sesquiphellandrene	20307-83-9						tr	
α-Sinensal	17909-77-2	0.01-0.07	0.04	0.2	0.06	0.04	0.7	0.04[y]; 0.1[z11]; 0.5[z12]
β-Sinensal	17909-87-4	0.01-0.1	0.1	0.2	0.05	0.04	0.2	tr[z9]; tr2[z11]; 0.1[z7]; 0.8[z12]
Spathulenol	6750-60-3			tr				0.01[y]; 0.07[z9]
α-Terpinene	99-86-5		0.02	0.09	0.3	0.02	1.7	0.1[z9]; 0.9[z3]
β-Terpinene	99-84-3							0.3[a,x1]
γ-Terpinene	99-85-4	0.01-0.6	0.3	4.7	0.2	0.3	0.07	0.2[z13]; 0.4[x5]; 3.3[w]; 10.3[x4]
Terpinen-4-ol	562-74-3	0.00-0.01	0.01	0.01	0.3	0.01	0.9	tr[z11]; 0.06[z9]; 0.1[z8]; 0.4[w]; 0.6[x4]
α-Terpineol	98-55-5	0.01-0.3	0.1	0.3	1.4[c]	0.2	1.3	0.7[x4]; 0.8[z8]; 3.1[z12]; 21.4[z5]
β-Terpineol	138-87-4				+			
(Z)-β-Terpineol	7299-40-3							3.2[z6]
trans-β-Terpineol	7299-41-4							1.3[z6]
γ-Terpineol	586-81-2							0.3[x7]
Terpinolene	586-62-9	0.01-0.06	0.07	0.2	2.6[c]	0.06	0.3	tr2; 0.07[z9]; 1.1[z3]
α-Terpinyl acetate	80-26-2		tr	tr	0.01	tr[j]	0.01	tr2[z11]; 0.02[y]; 0.06[z9]
Tetradecanal	124-25-4		0.01		tr	0.01	0.1	
Tetradecane	629-59-4					0.01		0.2[z13]; 0.6[x5]
1,4,7,10-Tetraoxacyclo-decane						tr[q]		
α-Thujene	2867-05-2	0.01-0.03	0.01	1.7	0.01	0.03	0.01	tr[z9]; 0.4[z3]
β-Thujene	28634-89-1							5.4[x4]
Thujol	35732-37-7							tr2[z11]; 1.8[x3]
α-Thujone	546-80-5							0.06[z3]
Thymol	89-83-8							0.1[z2]; 0.2[x4]
Thymyl isobutyrate	5451-67-2							0.02[y]
Tricyclene	508-32-7					tr[j]		0.09[z3]
Tridecanal	10486-19-8		0.01		tr	tr[j]	0.01[o]	
Tridecane	629-50-5						0.2[s]	
Undecanal	112-44-7	0.01-0.4	0.06	0.07	0.02	0.01	0.1	tr[z9]; 0.1[z7]
Undecane (*n*-)	1120-21-4							0.5[x5]
Undecanoic acid	112-37-8			tr			tr	
(E)-2-Undecenal	53448-07-0						tr	tr2[z11]
Valencene	4630-07-3	0.01-0.1	0.4	0.05	0.3	0.2	0.4	0.3[z2]; 0.4[x5]; 1.2[z12]
Valerianol	20489-45-6							0.01[y]
Verbenol	473-67-6					0.2[k]		
Verbenone	80-57-9							0.4[y]
p-Vinylguaiacol	7786-61-0							1.3[z12]
Viridiflorene (ledene)	21747-46-6							0.01[y]
Wine lactone	182699-77-0					+[m]		
α-Ylangene	14912-44-8					0.02[a,h]		

Table 5.65.2 Constituents identified in sweet orange oils *(continued)*

A 422 commercial sweet orange oils coming from California, Brazil, Florida and Italy analyzed between 1990 and 2013; lowest and highest concentrations given (E. Schmidt, unpublished data)

B review of studies published between 1979 and 1999 on the composition of industrial cold-pressed 'genuine' sweet orange oils; data cited in ref. 1; highest concentrations given

C review of studies published between 1979 and 1999 on the composition of commercial and 'unusual' sweet orange oils; data cited in ref. 1; highest concentrations given

D review of studies published between 1979 and 1999 on the composition of laboratory cold-pressed and extracted oils; data cited in ref. 1; highest concentrations given

E review of studies published between 1999 and 2009 on the composition of industrial cold-pressed sweet orange oils (refs. 2-15); data cited in ref. 1; highest concentrations given

F review of studies published between 1999 and 2009 on the composition of laboratory cold-pressed and extracted oils (refs. 16-28); data cited in ref. 1; highest concentrations given

G data from other studies (indicated with superscript letters); highest concentrations found in any study reviewed here given; when two or more oils were investigated, only the highest concentrations are mentioned, unless indicated otherwise

[a] probably a misidentification / incorrect identity based on GC elution order/does not occur nature (55); [b] deleted ; [c] uncharacteristically high concentration in one study (ref. 1); [d] data from ref. 2; [e] data from ref. 3; [f] data from ref. 4; [g] data from ref. 5; [h] data from ref. 6; [i] data from refs. 7 and 8; [j] data from ref. 9; [k] data from ref. 10; [l] data from ref. 11; [m] data from ref. 14; [n] data from ref. 15; [o] data from ref. 18; [p] data from ref. 19; [q] data from ref. 21; [r] data from ref. 22; [s] data from ref. 23; [t] data from ref. 25; [u] data from ref. 26; [v] data from ref. 27; [w] one lab-hydrodistilled sweet orange oil from Bintang sweet orange (ref. 30); [x] one lab-distilled sweet orange oil from India (ref. 31); [x1] data from various studies cited in ref. 55; as there is considerable overlap with the data presented in ref. 1 (columns C and D in Table 5.65.2), only the chemicals that have *not* been found in other studies are mentioned here (ref. 55); [x2] data from various studies cited in ref. 56; because of the abundance of literature, only the chemicals that have *not* been found in other studies are mentioned here (ref. 56); [x3] three lab-hydrodistilled oils from the peels of 3 cultivars of *C. sinensis* grown in Pakistan (ref. 54); [x4] two lab-hydrodistilled peel oils from the same batch of oranges from China, one produced with a conventional and the other with an improved Clevenger apparatus (ref. 58); [x5] one laboratory cold-pressed orange peel oil from Iran (ref. 60); [x6] one commercial sweet orange oil sample from China (ref. 59); [x7] one lab-hydrodistilled peel oil from Argentina with – for hydrodistillation – a very high limonene concentration (ref. 53); [x8] one lab-hydrodistilled *C. sinensis* peel oil from Mexico with – for hydrodistillation – a very high limonene concentration (ref. 57); [x9] one lab-hydrodistilled peel oil from Egypt (ref. 61); [y] four lab distilled oils from four *C. sinensis* cultivars from Tunisia; many of the chemicals found in this study were *tentatively* identified (ref. 32); [y1] formed under artificial aging conditions of sweet orange oil (ref. 56); [y2] found in a commercially deterpinated orange oil sample (ref. 56); [z] one sample of "genuine cold-pressed sweet orange oil' (ref. 33); [z2] one commercial Spanish sample of cold-pressed *Citrus sinensis* oil (ref. 34); [z3] three samples of Tunisian lab-hydrodistilled oil from oranges in various stages of maturation (ref. 35); [z4] one sample of lab prepared cold-distilled peel oil from Jinchen sweet orange fruit (China) (ref. 36); [z5] cold-pressed orange oils from six cultivars in Japan (ref. 41, cited in ref. 37); [z6] one lab-prepared hydrodistilled oil (ref. 42, cited in ref. 37); [z7] three lab-prepared cold-pressed, hydrodistilled and microwave-assisted hydrodistilled orange oils (ref. 43, cited in ref. 37); [z8] one lab-hydrodistilled oil from cultivar Cajel orange cultivated in the Yucatan Peninsula, Mexico (ref. 40); [z9] five cold-pressed peel oils from Washington navel cultivars of *C. sinensis* cultivated in various locations in Turkey (ref. 39); [z10] Data cited in ref. 39; [z11] one cold-pressed oil from Rwanda and one from Uganda (ref. 38); [z12] three lab-hydrodistilled oils from fresh, air-dried, and oven-dried peels from Pakistani *C. sinensis* (ref. 49); [z13] one oil obtained from cold pressing *C. sinensis* cv. *Valencia* peels from Iran (ref. 50)

tr: trace (in columns B,C,D,E,F: not quantified; columns G[y], G[z2] <0.01; column G[z9] <0.05; column G[z11]<0.005); tr2: <0.05 but higher than 0.005; + present in the oil investigated, but quantity not stated

concentration in one study only, all others have a concentration <2.5%). Well-known ingredients of sweet orange oils that were present in high concentrations (>7%) in one or two studies were α-terpineol (21.4%) and γ-terpinene (10.3%). A rare constituent of sweet orange oils found in a high concentration (>7%) in a single study includes formaldehyde (7.1%).

Commercial oils

The ten chemicals that had the highest maximum concentrations in 422 commercial sweet orange essential oil samples (concentration ranges provided) are the following: limonene (94.7-95.7%), myrcene (1.4-3.5%), α-pinene (0.2-1.0%), sabinene (0.02-0.7%), β-phellandrene (0.02-0.7%), decanal (0.09-0.6%), γ-terpinene (0.01-0.6%),

Table 5.65.3 Major constituents of sweet orange oils

Constituent	CAS	Percentage and range						
		A	**B (1)**	**C (1)**	**D (1)**	**E (2-15)**	**F (16-28)**	**G**
Limonene	138-86-3	94.7-95.7	96.1	94.5	96.8	96.1	96.6	91.0[x6]; 92.1[z9]; 96.6[x8]; 97.3[y]
Myrcene	123-35-3	1.4-3.5	2.0	2.3	22.4[c]	2.5	5.3	2.5[z13]; 4.4[z12]; 4.7[z8]; 6.3[w]; 7.2[x3]
Linalool	78-70-6	0.04-0.07	0.8	1.2	15.8[c]	0.4	5.9	1.0[z8]; 1.2[z11]; 2.1[z12]; 3.0[x9]; 3.5[x3]
α-Pinene	80-56-8	0.2-1.0	1.4	0.9	14.8[c]	0.7	1.6	1.0[z8]; 1.4[z13]; 1.6[x5]; 2.4[z11]; 2.5[w]

LEGEND: SEE UNDER TABLE 5.65.2

Table 5.65.4 Heterocyclic oxygenated compounds present in cold-pressed sweet orange oils (33,37,44-48,52,55)

Name	Synonym(s)	CAS
Hydroxylated polymethoxyflavones		
5,8-Dihydroxy-3,3',4',7-tetramethoxyflavone	Gossypetin 3,3',4',7-*O*-tetramethyl ether	7380-44-1
5-Hydroxy-3,3',4',6,7,8-hexamethoxyflavone		1176-88-1
3-Hydroxy-4',5,6,7,8-pentamethoxyflavone	3-Hydroxytangeretin	
5-Hydroxy-3, 3',4',7,8-pentamethoxyflavone		
5-Hydroxy-3',4',6,7,8-pentamethoxyflavone	5-Demethylnobiletin	2174-59-6
3-Hydroxy-4',5,6,7-tetramethoxyflavone		
5-Hydroxy-3,3',4',7- tetramethoxyflavone	Retusin; Tetra-*O*-methylquercetin	1245-15-4
5-Hydroxy-4',6,7,8-tetramethoxyflavone	5-De(s)methyltangeretin; Gardenin B	2798-20-1
5-Hydroxy-4',6,7-trimethoxyflavone	Salvigenin	19103-54-9
Polymethoxyflavones		
3,3',4,4',5,6,7-Heptamethoxyflavone		
3, 3',4',5,6,7,8-Heptamethoxyflavone	3-Methoxynobiletin	1178-24-1
3,3',4',5,6,7-Hexamethoxyflavone	3-Methoxysinensetin; Hexa-*O*-methylquercetagetin	1251-84-9
3,3',4',5,7,8-Hexamethoxyflavone	Hexa-*O*-methylgossypetin	7741-47-1
3',4',5,6,7,8-Hexamethoxyflavone	Nobiletin	478-01-3
3',3',5,7,8-Pentamethoxyflavone		99801-93-1
3',4',5,6,7-Pentamethoxyflavone	Sinensetin	2306-27-6
3,4',6,7,8-Pentamethoxyflavone	Auranetin	522-16-7
4',5,6,7,8-Pentamethoxyflavone	Tangeretin	481-53-8
4',5,6,7-Tetramethoxyflavone	Tetra-*O*-methylscutellarein	1168-42-9
4',5,7,8-Tetramethoxyflavone	6-Demethoxytangeritin; Tetra-*O*-methylisoscutellarein	6601-66-7
(Hydroxylated) polymethoxyflavanones and polymethoxychalcones		
2'-Hydroxy-3,3',4,4',5',6'-hexamethoxychalcone		
2'-Hydroxy-3,4,4',5',6'-pentamethoxychalcone		
5-Hydroxy-3',4',6,7,8-pentamethoxyflavanone	Desmethyl-5-citromitine	15512-52-4
4',5,6,7-Tetramethoxyflavanone		
Coumarins and psoralens		
8-(2,3-Dihydroxy-3-methylbutyl)-7-methoxycoumarin	Meranzin hydrate	5875-49-0
5,7-Dimethoxycoumarin	Citropten; Limetin	487-06-9
5,8-Dimethoxy-6,7-furanocoumarin	Isopimpinellin	482-27-9
Epoxybergamottin hydrate		
7-Geranyloxycoumarin	Aurapten	495-02-3
5-Geranyloxypsoralen	Bergamottin; Bergaptin	7380-40-7
5-Methoxypsoralen	Bergapten	484-20-8

n-octanal (0.07-0.5%), β-pinene (0.01-0.5%) and undecanal (0.01-0.4%) (Erich Schmidt, unpublished analytical data).

It should be realized that values in hydrodistilled oils may (but not necessarily do) vary considerably from genuine cold-pressed orange oils because of the many hydrolytic reactions that take place during oil isolation (42). The content of limonene, by far the most important compound in sweet orange oil, is usually (but not always 53,57) lower in hydrodistilled oils (sometimes as low as 65%), and that of some other components consequently higher. In addition, the hydrodistilled oils lack the non-volatile oxygenated heterocyclic compounds which are present in the cold-pressed oils (table 5.65.4).

Reviews of studies on the chemical composition of *Citrus sinensis* (L.) Osbeck peel oils have been provided in refs. 1,37,39 (cold-pressed oils), 55 and 56. The history, global distribution and nutritional importance of *Citrus* fruits have been reviewed in ref. 51.

CONTACT ALLERGY/ALLERGIC CONTACT DERMATITIS

General
Contact allergy to/allergic contact dermatitis from sweet orange oil has been reported in a few publications only. In a group of consecutive patients suspected of contact dermatitis (one study only), the prevalence rate of positive patch test reactions was 0.5%, but relevance data are lacking. In one patient, limonene may have been an allergen in (a positive patch test to) sweet orange oil.

Patch testing in groups of patients
Two hundred dermatitis patients from Poland were tested with sweet orange oil 2% in petrolatum and one (0.5%) reacted. The same authors also patch tested 51 patients allergic to *Myroxylon pereirae* resin (balsam of Peru) and/or turpentine and/or wood tar and/or colophony and two (3.9%) had a positive patch test;

relevance data were not provided (64). A group of 86 patients from Poland previously reacting to the fragrance mix was tested with sweet orange oil and four (4.7%) had a positive patch test reaction; relevance data were not provided (66).

Case reports and positive patch tests

An aromatherapist had non-occupational allergic contact dermatitis with allergies to multiple essential oils used at work, including sweet orange oil. The patient also reacted to geraniol, linalool, linalyl acetate, α-pinene, the fragrance mix and various other fragrance materials; α-pinene was demonstrated by GC-MS in the orange oil, limonene was not tested (65). α-Pinene is not an important component of sweet orange oil. A female patient developed allergic contact dermatitis from the perfume in an eye cream; she was patch tested with all 94 components of the perfume and reacted to 'orange sweet, terpeneless' (test concentration unknown) and eleven of the other chemicals in the perfume (67). One patient with allergic contact dermatitis from tea tree oil co-reacted to sweet orange oil and its main ingredient limonene (69); limonene may also be an allergen in tea tree oil. Additional information on allergy to sweet orange oil can be found in ref. 68 (article not read).

LITERATURE

1 Dugo G, Cotroneo A, Bonaccorsi I, Trozzi A. Composition of the volatile fraction of citrus peel oils. In: G Dugo and L Mondello, eds. Citrus oils: composition, advanced analytical techniques, contaminants, and biological activity. Boca Raton, USA: CRC Press, Taylor & Francis Group, 2011:75-89

2 Oberhofer B, Nikiforov A, Buchbauer G, Jirovetz L, Bicchi C. Investigation of the alteration of the composition of various essential oils used in aroma lamp applications. Flavour Fragr J 1999;14:293-299

3 Di Giacomo G, Gionfriddo F, Kunkar C. Caratteristiche dell'olio essenziale concentrato di arancia dolce pigmentata. Essenz Deriv Agrum 1999;69:5-13. Data cited in ref. 1

4 Kubeczka KH, Formaček V. Essential oils analysis by capillary gas chromatography and carbon-13 NMR spectroscopy, 2nd Ed. New York: John Wiley & Sons, 2002. Data cited in ref. 1

5 Shen Z, Mishra V, Imison B, Palmer M, Fairclough R. Use of adsorbent and supercritical carbon dioxide to concentrate flavor compounds from orange oil. J Agric Food Chem 2002;50:154-160

6 Lopes D, Raga AC, Stuart GR, de Oliveira JV. Influence of vacuum distillation parameters on the chemical composition of five–fold sweet orange oil (Citrus sinensis Osbeck). J Essent Oil Res 2003;15:408-411

7 Mondello L, Casilli A, Tranchida PQ, Cicero L, Dugo P, Dugo G. Comparison of fast and conventional GC analysis for citrus essential oils. J Agric Food Chem 2003;51:5602-5606

8 Mondello L, Casilli A, Tranchida PQ, Costa R, Dugo P, Dugo G. Fast GC for the analysis of citrus oils. J Chromatogr Sci 2004;42:410-416

9 Verzera A, Trozzi A, Dugo G, Di Bella G, Cotroneo A. Biological lemon and sweet orange essential oil composition. Flavour Fragr J 2004;19:544-548

10 Viuda-Martos M, Ruiz-Navajas Y, Fernandez-Lopez J, Perez-Alvarez JA. Chemical composition of mandarin (C. reticulata L.), grapefruit (C. paradisi L.), lemon (C. limon L.) and orange (C. sinensis L.) essential oils. J Essent Oil Bear Plants 2009;12:236-243

11 Feger W, Brandauer H, Ziegler H. Germacrenes in citrus peel oils. J Essent Oil Res 2001;13:274–27

12 Veriotti T, Sacks R. High-speed characterization and analysis of orange oils with tandem-column stop-flow GC and time-of-flight MS. Anal Chem 2002;4:5635-5640

13 Schulz H, Schrader B, Quilitzsch R, Steuer B. Quantitative analysis of various citrus oils by ATR/FT-IR and NIR-FT Raman spectroscopy. Appl Spectrosc 2002;56:117-124

14 Elston A, Lin J, Rousseff R. Determination of the role of valencene in orange oil as a direct contributor to aroma quality. Flavour Fragr J 2005;20:381-386

15 Ogawa CA, Diagone CA, Lancas FM. Separation of monoterpenes in orange essential oil by capillary liquid chromatography and micellar electrokinetic chromatography. J Liq Chrom Rel Technol 2002;25:1651-1659

16 Marongiu B, Piras A, Porcedda S. Extraction of volatile fractions and carotenoids from orange and kumquat peel by supercritical carbon dioxide. J Essent Oil Bear Plants 2003;6:86-96

17 Thao NT, Kashiwagi T, Sawamura M. Characterization by GC-MS of Vietnamese citrus species and hybrids based on the isotope ratio of monoterpene hydrocarbons. Biosci Biotechnol Biochem 2007;71:2155-2161

18 Mitiku SB, Sawamura M, Itoh T, Ukeda H. Volatile components of peel cold pressed oils of two cultivars of sweet orange (Citrus sinensis (L.) Osbeck) from Ethiopia. Flavour Fragr J 2000;15:240-244

19 Minh Tu NT, Thanh LX, Une A, Ukeda H, Sawamura M. Volatile constituents of Vietnamese pummelo, orange, tangerine and lime peel oils. Flavour Fragr J 2002;17:169-174

20 Gionfriddo F, Catalfamo M, Siano F, Mangiola C, Cautela D, Castaldo D. Determinazione delle caratteristiche analitiche e della composizione enantiomerica di oli essenziali agrumari ai fini dell'accertamento della purezza e della qualita. Nota I – Essenze di arancia amara, arancia dolce e bergamotto. Essenz Deriv Agrum 2003;73:29-39. Data cited in ref. 1

21 Njoroge SM, Koaze H, Karanja PN, Sawamura M. Essential oil constituents of three varieties of Kenyan sweet oranges (Citrus sinensis). Flavour Fragr J 2005;20:80-85

22 Sawamura M, Minh Tu NT, Yu X, Xu B. Volatile constituents of the peel oils of several sweet oranges in China. J Essent Oil Res 2005;17:2-6

23 Choi H-S. Lipolytic effects of citrus peel oils and their components. J Agric Food Chem 2006;54:3254-3258

24 Dharmawan J, Kasapis S, Curran P. Characterization of volatile compounds in selected citrus fruits from Asia – Part II: Peel oil. J Essent Oil Res 2008;20:21-24

25 Ruberto G, Starrantino A, Rapisarda P. Essential oils from new citrus fruits. Essenz Deriv Agrum 1999;69:15-26

26 Ruberto G, Rapisarda P. Essential oils of new pigmented citrus hybrids: *Citrus sinensis* L. Osbeck x *C. clementina* Hort. ex Tanaka. J Food Sci 2002;67:2778-2780

27 Giamperi L, Fraternali D, Ricci D. Essential oil composition and antioxidant activity of peels of *Citrus sinensis* blood and blond and *Citrus aurantium*. Riv Ital EPPOS 2002;34:21-28. Data cited in ref. 1

28 Ferhat MA, Meklati BY, Smadja J, Chemat F. An improved microwave Clevenger apparatus for distillation of essential oils from orange peel. J Chromatogr A 2006;1112:121-126

29 Liu K, Chen Q, Liu Y, Zhou X, Wang X. Isolation and biological activities of decanal, linalool, valencene, and octanal from sweet orange oil. J Food Sci 2012;77: C1156-1161

30 Tao N-G, Liu Y-J, Zhang M-L. Chemical composition and antimicrobial activities of essential oil from the peel of bingtang sweet orange (*Citrus sinensis* Osbeck). Int J Food Sci Technol 2009;44:1281-1285

31 Singh P, Shukla R, Prakash B, Kumar A, Singh S, Mishra PK, Dubey NK. Chemical profile, antifungal, antiaflatoxigenic and antioxidant activity of *Citrus maxima* Burm. and *Citrus sinensis* (L.) Osbeck essential oils and their cyclic monoterpene, dl-limonene. Food Chem Toxicol 2010;48:1734-1740

32 Hosni K, Zahed N, Chrif R, Abid I, Medfei W, Kallel M, et al. Composition of peel essential oils from four selected Tunisian *Citrus* species: evidence for the genotypic influence. Food Chem 2010;123:1098-1104

33 Dugo P, Mondello L, Dugo L, Stancanelli R, Dugo G. LC-MS for the identification of oxygen heterocyclic compounds in citrus essential oils. J Pharm Biomed Anal 2000;24:147-154

34 Espina L, Somolinos M, Lorán S, Conchello P, García D, Pagán R. Chemical composition of commercial citrus fruit essential oils and evaluation of their antimicrobial activity acting alone or in combined processes. Food Control 2011;22:896-902

35 Bourgou S, Rahali FS, Ourghemmi I, Saïdani Tounsi M. Changes of peel essential oil composition of four Tunisian *Citrus* during fruit maturation. Scient World J 2012, Article ID 528593, doi:10.1100/2012/528593

36 Qiao Y, Xie BJ, Zhang Y, Zhang Y, Fan G, Yao XL, et al. Characterization of aroma active compounds in fruit juice and peel oil of Jinchen sweet orange fruit (*Citrus sinensis* (L.) Osbeck) by GC-MS and GC-O. Molecules 2008;13:1333-1344

37 Lawrence BM. Essential oils 2008-2011. Carol Stream, USA: Allured Publishing Corporation, 2012:79-93. (Reprinted from: Lawrence BM. Progress in essential oils. Perfum Flavor 2009;34(January):48-57 and Perfum Flav 2009;34(March): 52-56)

38 Njoroge SM, Thi Lan Phi N, Sawamura M. Chemical composition of peel essential oils of sweet oranges (*Citrus sinensis*) from Uganda and Rwanda. J Essent Oil Bear Plants 2009;12:26-33

39 Kirbaslar FG, Kirbaslar SI, Pozan G, Boz I. Volatile constituents of Turkish orange (*Citrus sinensis* (L.) Osbeck) peel oils. J Essent Oil Bear Plants 2009;12:586-604

40 Pino JA, Cuevas-Glory L, Sauri-Duch E. Volatile constituents of peel and leaf oils of Cajel orange (*Citrus sinensis* [L.] Osbeck). J Essent Oil Bear Plants 2010;13:742-746

41 Sawamura M. Volatile components of essential oils of the *Citrus* genus. Recent Res Dev Agric Food Chem 2000;4:131-164. Data cited in ref. 37

42 Singh G, Kapoor IPS, Kaur J, Singh OP, Vernwal SK, Yadav RSS, et al. Biotransformation of *Citrus sinensis* peel oil and commercial limonene by various fungi and *Musa paradisiacal* stem enzyme extract. J Med Arom Plant Sci 2002;24:23-27. Data cited in ref. 37

43 Ferhat MA, Meklati BY, Colnagui G, Visinoni F, Chemat F. Green chemical processing in the teaching laboratory: a convenient solvent free microwave extraction of natural products. Recent Progr Genie Procédés 2007;94:1-8. Data cited in ref. 37

44 Manthey JA, Grohmann K. Concentrations of hesperidin and other orange peel flavonoids in citrus processing byproducts. J Agric Food Chem 1996;44:811-814

45 Dugo P, Mondello L, Lamonica G, Dugo G. OPLC analysis of heterocyclic oxygen compounds from citrus fruit essential oils. J Planar Chromatogr 1996;9:120-125

46 Dugo P, McHale D. The oxygen heterocyclic compounds of citrus essential oils. In: Citrus. The Genus Citrus. G Dugo and A DiGiacomo, Eds. London: Taylor and Francis, 2002:355-390

47 Li S-M, Lo Ch-Y, Ho C-T. Hydroxylated polymethoxyflavones and methylated flavonoids of sweet orange (*Citrus sinensis*) peel. J Agric Food Chem 2006;54:4176-4185

48 Bonaccorsi IL, McNair HM, Brunner LA, Dugo P, Dugo G. Fast HPLC for the analysis of oxygen heterocyclic compounds of citrus essential oils. J Agric Food Chem 1999;47:4237-4239

49 Kamal GM, Anwar F, Hussain AI, Sarri N, Ashraf MY. Yield and chemical composition of Citrus essential oils as affected by drying pretreatment of peels. Int Food Res J 2011;18:1275-1282

50 Azar PA, Nekoei M, Larijani K, Bahraminasab S. Chemical composition of the essential oils of *Citrus sinensis* cv. Valencia and a quantitative structure–retention relationship study for the prediction of retention indices by multiple linear regression. J Serb Chem Soc 2011;76:1627-1637

51 Liu YQ, Heying E, Tanumihardjo SA. History, global distribution, and nutritional importance of *Citrus* fruits. Compreh Rev Food Sci Food Saf 2012;11:530–545

52 Russo M, Torre G, Carnovale C, Bonaccorsi I, Mondello L, Dugo P. A new HPLC method developed for the analysis of oxygen heterocyclic compounds in *Citrus* essential oils. J Essent Oil Res 2012;24:119-129

53 Rossi YE, Palacios SM. Fumigant toxicity of *Citrus sinensis* essential oil on *Musca domestica* L. adults in the absence and presence of a P450 inhibitor. Acta Tropica 2013;127:33-37

54 Khan MM, Iqbal M, Hanif MA, Mahmood MS, Naqvi SA, Shahid S, et al. Antioxidant and antipathogenic activities of citrus peel oils. J Essent Oil Bear Plants 2012;15:972-979

55 Lawrence BM. Progress in essential oils. Perfum Flavor 2000;25(Sept/Oct):52-72

56 Lawrence BM. Progress in essential oils. Perfum Flavor 1992;17(5):131-146

57 Velázquez-Nuñez MJ, Avila-Sosa R, Palou E, López-Malo A. Antifungal activity of orange (*Citrus sinensis* var. Valencia) peel essential oil applied by direct addition or vapor contact. Food Control 2013;31:1-4

58 Chen Y, Wu J, Xu Y, Fu M, Xiao G. Effect of second cooling on the chemical components of essential oils from orange peel (*Citrus sinensis*). J Agric Food Chem 2014;62:8786-8790

59 Xiong Y, Zhao Z, Zhu L, Chen Y, Ji H, Yang D. Removal of three kinds of phthalates from sweet orange oil by molecular distillation. LWT - Food Sci Technol 2013;53:487-491

60 Nekoei M, Mohammadhosseini M. Application of HS-SPME, SDME and cold-press coupled to GC/MS to analysis of the essential oils of *Citrus sinensis* cv. Thomson navel and QSRR study for prediction of retention indices by stepwise and genetic algorithm-multiple linear regression approaches. Anal Chem Letters 2014;4:93-103

61 Mohareb ASO, Badawy MEI, Abdelgaleil SAM. Antifungal activity of essential oils isolated from Egyptian plants against wood decay fungi. J Wood Sci 2013;59:499-505

62 Rhind JP. Essential oils. A handbook for aromatherapy practice, 2nd Edition. London: Singing Dragon, 2012

63 Lawless J. The encyclopedia of essential oils, 2nd Edition. London: Harper Thorsons, 2014

64 Rudzki E, Grzywa Z, Bruo W-S. Sensitivity to 35 essential oils. Contact Dermatitis 1976;2:196-200

65 Dharmagunawardena B, Takwale A, Sanders KJ, Cannan S, Roger A, Ilchyshyn A. Gas chromatography: an investigative tool in multiple allergies to essential oils. Contact Dermatitis 2002;47:288-292

66 Rudzki E, Grzywa Z. Allergy to perfume mixture. Contact Dermatitis 1986;15:115-116

67 Larsen WG. Cosmetic dermatitis due to a perfume. Contact Dermatitis 1975;1:142-145

68 Michel PJ. Dermite des mains récidivante par contacts avec des oranges (intolerance à l'essence d'orange douce). Bull Soc Franç Derm Syph 1953;60:320

69 Kränke B. Allergisierende Potenz von Teebaumöl. Hautarzt 1997;48:203-204

Chapter 5.66   PALMAROSA OIL

DEFINITION
Palmarosa oil (essential oil of palmarosa) is the essential oil obtained from the aerial parts (leaves) of the motia grass (palmarosa), *Cymbopogon martini* (Roxb.) Will. Watson.

INCI NOMENCLATURE
Description/definition: Cymbopogon martini motia herb oil is an essential oil obtained from the herbs of the lemon grass (wrong name!), *Cymbopogon martini* var. *motia*, Gramineae
INCI name EU: Cymbopogon martini motia herb oil (perfuming name, not an INCI name proper)
INCI name USA: Not in the Personal Care Products Council Ingredient Database
CAS registry number(s): 91722-54-2; 8014-19-5, 84649-81-0
EINECS number(s): 294-453-3; 283-461-2

ISO (INTERNATIONAL ORGANIZATION FOR STANDARDIZATION) STANDARD
ISO number: 4727
ISO name: Essential oil of palmarosa
Botanical origin: *Cymbopogon martinii* (Roxburgh) W. Watson var. *motia*
Parts of plant used: Aerial part (leaves)
ISO values: For this oil no chromatographic profile is currently available

The AFNOR standard also has no chromatographic profile.

According to ISO, palmarosa oil is obtained from the variety *motia* of palmarosa. In several investigations, however, palmarosa oil was obtained from *Cymbopogon martini* (Roxb.) W. Watson var. *martini*. Whereas this species is acknowledged by the multilingual multiscript plant name database of the University of Melbourne, the variety *motia* Burk is considered an 'unresolved taxon', but which is not synonymous with the basic species. Both the terms *martini* and *martinii* are used in the botanical literature, but as the surname of the botanist was 'Martin', the epithet should be spelled 'martini' (http://www.plantnames.unimelb.edu.au/Sorting/Cymbopogon.html).

THE PLANT, THE OIL, AND THEIR USES
Cymbopogon martini (Roxb.) Will. Watson, commonly known as palmarosa, is a perennial, tufted, multi-harvest grass with numerous erect culms, 150-300 cm long, that arise from a short, stout and woody rhizome, and which has linear or lanceolate glaucous aromatic leaves, 25-50 cm long. It is native to India and Pakistan. The plant is naturalized in China, India, Indo-China, Malaysia, the western Indian Ocean, and Australia. The grass is cultivated in India, Indonesia, the Seychelles, Brazil and Madagascar (2, GRIN Taxonomy for Plants; http://www.globinmed.com; www.kew.org).

The essential oil of palmarosa is obtained by steam distillation of the leaves (ISO criterion) or the leaves plus inflorescences (flowering herbs) (11,22) of *Cymbopogon martini*, notably the variety *motia*. This oil has a high content of geraniol, up to 90%. Palmarosa oil has a sweet floral, rose-like odor with an herbal note (22). It is widely used as a perfumery raw material in soaps, floral, rose-like perfumes and cosmetics preparations. It is also used for flavoring tobacco products, foods, non-alcoholic beverages, for masking the odor of botanical pesticides (10) and in the preparation of mosquito repellent products. In medicine, the oil is used as a remedy for lumbago, stiff joints, muscular pains, for bilious complaints, and in aromatherapy (1,5,10,11,21). Palmarosa oil is considered to have antiseptic and wound-healing properties and is therefore used in curing boils and other skin diseases (5). In aromatherapy, it is used as an aphrodisiac for elevating the senses (10). The essential oil of palmarosa is highly valued in the perfumery industry as a source of high grade geraniol *ex* palmarosa, which is separated through fractional distillation (2,10); the olfactory note of this geraniol is considered superior to geraniol prepared from other sources (22).

CHEMICAL COMPOSITION
Palmarosa oil is a pale yellow to yellow, clear mobile liquid which has a sweet floral odor with aspects of geranium and rose. The yield of essential oil from the leaves of *Cymbopogon martini* (Roxb.) Will. Watson var. *motia* generally varies from 0.8 to 1.3 per cent. The main producing countries of this oil are India and Nepal.

Literature data (up to August 23, 2014) on the chemical composition of palmarosa oils and unpublished analytical data from one of us (E.S.) are shown in Table 5.66.1 in alphabetical order. In palmarosa oils from various origins, over 155 chemicals have been identified. About 49 per cent of these were found in a single reviewed publication only. The major compounds found in palmarosa oils from different sources are shown in Table 5.66.2. They include (highest concentrations in any study given) geraniol (96.7%), geranyl acetate (22.6%), linalool (4.9%) and β-caryophyllene (3.4%). Well-known ingredients of palmarosa oils that were present in high concentrations (>7%) in one or two studies were nerol (15.0% and 24.8%), citronellol (19.1%), β-elemene (12.3%), geranial (8.8%) and α-pinene (8.0%). Uncommon or rare constituents of palmarosa oils found in high concentrations (>7%) in single studies include camphene (11.4%), isomenthyl acetate (10.1%), isoasarone (9.2%) and elemicin (7.9%).

Commercial oils
The ten chemicals that had the highest maximum concentrations in 34 commercial palmarosa essential oil samples (concentration ranges provided) are the following: geraniol (74.2-86.9%), geranyl acetate (3.4-12.5%), linalool (2.3-4.5%), β-caryophyllene (1.7-2.5%), (*E*)-β-ocimene (0.6-2.1%), (*Z,E*)-farnesol (0.5-1.5%), geranyl hexanoate (0.4-1.5%), limonene (0.1-1.4%), (*Z*)-β-ocimene (0.3-0.7%) and geranial (0.2-0.7%) (Erich Schmidt, unpublished analytical data).

The composition of palmarosa *seed* oils is discussed in refs. 21 and 22 and that of inflorescences in refs. 2 and 21.

Table 5.66.1 Constituents identified in palmarosa oils

Constituent	CAS	A	B (23)	C (22)	D (16)	E
α-Amorphene	20085-19-2					+[z5]
Amyl hexanoate	540-07-8					tr[z9]
Asarone	2883-98-9					0.3[i]
trans-Bergamotene						1.1[i]
trans-α-Bergamotene	13474-59-4			0.05		
Bicyclogermacrene	24703-35-3				0-0.02	+[z5]
β-Bisabolene	495-61-4			tr		+[z5]; 0.1[h]
γ-Bisabolene	495-62-5				0-0.04	+[z5]
Borneol	507-70-0		0-0.1	tr		
α-Cadinene	24406-05-1		0-0.3		0.01-0.02	+[z5]
γ-Cadinene	39029-41-9				0-0.1	+[z5]; 0.4[o]
δ-Cadinene	483-76-1		0-tr	0.09		+[z5]
trans-Cadinol						0.2[i]
Calacorene	38599-17-6				0.09-0.3	+[z5]; 0.3[z7]
cis-Calamenene	72937-55-4					+[z5]
trans-Calamenene	73209-42-4					+[z5]
Camphene	79-92-5			tr		tr[f]; 11.4[i]
Camphor	76-22-2			tr		0.3[i]
δ3-Carene	13466-78-9					0.6[o]; 1.5[y]
β-Caryophyllene	87-44-5	1.7-2.5	0.1-1.1	2.0	1.0-1.8	1.6[f]; 1.8[t]; 2.0[h]; 2.5[y]; 2.6[i]; 3.3[j]; 3.4[v]
Caryophyllene oxide	1139-30-6	0.05-0.3	0.1-1.1	0.2	0-0.2	0.1[h,o,z9]; 0.2[b,l]; 0.3[i,z4]; 0.6[j]
1,8-Cineole	470-82-6			0.2[a]		0.09[d]; 0.3[i]; 0.4[z4]; 0.7[o]
Citronellal	106-23-0			tr		tr[e,j,z9]; 0.07[m]
Citronellol	106-22-9		0-2.1	<0.06		tr[e]; 0.1[h,k]; 0.2[g]; 0.4[z]; 0.5[n]; 19.1[z1]
Citronellyl acetate	150-84-5		0-1.1	tr		tr[z9]; 0.1[k]; 0.2[m]; 2.3[z3]
Citronellyl formate	105-85-1			tr		
α-Cubebene	17699-14-8		0-0.2			
β-Cubebene	13744-15-5			tr	0-0.02	+[z5]
Cubenene	29837-12-5					+[z5]
Cuparene	16982-00-6					0.4[i]
Curcumene (= ar-; α-)	644-30-4					+[z5]
β-Curcumene	28976-67-2		0-0.1			+[z5]
Cyclohexanone	108-94-1		0-tr			
Cymbodiacetal						+[u]
m-Cymene	535-77-3		0-tr			+[z5]
o-Cymene	527-84-4					+[z5]
p-Cymene	99-87-6		0-0.2	tr	0-0.01	+[z5]; tr[b,l]; 0.06[i,m]; 0.1[h,z4]
p-Cymen-8-ol	1197-01-9					1.1[i]
Decan-2-ol (2-Decanol)	1120-06-5		tr-0.2			
β-Elemene	33880-83-0	0-0.08	0-0.4	0.1	0-0.01	+[z5]; 0.06[m]; 0.1[b,h,l]; 12.3[z]
γ-Elemene	29873-99-2					+[z5]
Elemicin	487-11-6					7.9[i]
Elemol (α-)	639-99-6			0.2		1.7[d]
2,3-Epoxygeranial						1.6[z]
Farnesal	19317-11-4					tr[j]
α-Farnesene	502-61-4					+[z5]; 0.3[z7]
β-Farnesene	502-60-3		0-0.1		0.01	+[z5]; 0.7[z]
(E)-β-Farnesene	18794-84-8			0.1		0.1[j]
Farnesol	4602-84-0					0.2[o]; 1.2[v]; 1.6[j,z7]; 1.9[l]
(E,E)-Farnesol	106-28-5		0.5-0.7		0-0.2	tr[z9]; 1.9[b]
(E,Z)-Farnesol	3879-60-5			0.5	0-0.1	0.5[z4]; 1.9[g]; 3.4[f]
(Z,E)-Farnesol	3790-71-4	0.5-1.5			0.4-1.2	
(Z,Z)-Farnesol	16106-95-9			0.2		0.3[f]
Farnesyl acetate	29548-30-9			0.05		0.1[l,z9]; 0.6[z1]
(E,E)-Farnesyl acetate	4128-17-0					0.1[b]
(E,Z)-Farnesyl acetate	24163-98-2				0.05-0.3	
(Z,E)-Farnesyl acetate	40266-29-3	0.04-0.1				
Geranial	141-27-5	0.2-0.7	1.0-8.8	1.9		0.2[b,l]; 0.3[z4]; 0.5[e]; 2.1[t]; 4.0[z]
Geraniol	106-24-1	74.2-86.9	67.6-83.6	75.9	76.3-82.8	81.4[t,z4]; 82.2[b,l]; 82.6[q]; 82.8[z7]; 85.4[z3]; 85.7[g]; 86.3[d]; 87.6[y]; 91.6[c]; 96.7[p]

Table 5.66.1 Constituents identified in palmarosa oils *(continued)*

Constituent	CAS	Percentage and range				
		A	B (23)	C (22)	D (16)	E
Geranyl acetate	105-87-3	3.4-12.5	2.2-15.8	10.7	5.1-11.8	9.2[z3]; 10.7[f]; 11.3[q]; 12.0[g,z2]; 12.5[s]; 12.9[j]; 13.6[x,y1]; 15.8[h]; 15.9[c]; 20.0[r]; 20.1[m]; 22.6[v]
Geranyl butyrate	106-29-6	0.08-0.2	0.1-0.3	0.2	0.1-0.3	0.1[b,f]; 0.2[g,h,z9]; 0.3[e,m]; 0.4[j,z7,z8]; 1.0[o]
Geranyl formate	105-86-2	0.03-0.2	0.1-0.9	0.1		tr[b]; 0.08[m]; 0.1[z9]; 0.2[e,j]; 0.5[z3]
Geranyl heptanoate	73019-15-5		0-0.1	0.06		0.1[h,k]; 0.2[m]
Geranyl hexanoate	10032-02-7	0.4-1.5	0.2-0.6	0.6		0.4[o]; 0.6[z9]; 0.7[k]; 0.8[f,g,m]; 1.9[v]; 3.0[z1]
Geranyl isobutyrate	2345-26-8		0-0.1	0.06		0.2[m]; 0.8[f]
Geranyl isovalerate	109-20-6			tr	0.01-0.1	0.1[g,j]; 0.4[z7,z8]
Geranyllinalool	77368-82-2					0.1[j]
Geranyl octanoate	51532-26-4		0.1-0.2			1.3[o]
Geranyl propionate	105-90-8			tr		0.1[k,m]; 0.2[h]
Geranyl tiglate	7785-33-3		0-0.1			
Geranyl valerate	10402-47-8		0-0.1	tr		tr[j,z9]; 0.07[m]; 0.1[h]
Germacrene A	28387-44-2					tr[b]
Germacrene B	15423-57-1				0-0.02	+[z5]
Germacrene D	23986-74-5		0-0.1			+[z5]
Gymnomitrene						0.3[i]
β-Helmiscapene	66141-12-6				0.05-0.1	+[z5]
2-Heptadecanone	2922-51-2		0-0.1			
2-Heptanol	543-49-7		0-0.1			
2-Heptanone	110-43-0					tr[z9]
1-Hepten-3-ol	4938-52-7		0-0.1			
(Z)-3-Hexen-1-ol	928-96-1		0-0.1			0.06[m]
Hotrienol	20053-88-7					tr[z9]
α-Humulene	6753-98-6	0.07-0.2		0.2	0.1-2.1	0.1[b,l,z4]; 0.2[j]; 0.6[z7]
α-Humulene epoxide	96638-51-6		0-0.1			
Isoamyl acetate	123-92-2					tr[z9]
Isoamyl formate	110-45-2					tr[z9]
Isoasarone						9.2[i]
Isoborneol	124-76-5		0-0.1			
Isomenthyl acetate	20777-45-1					10.1[t]
Isopropyl *n*-butyrate	638-11-9		0-tr			
Isopropyl propionate	637-78-5		0-tr			
Isovaleraldehyde	590-86-3					tr[z9]
Lavandulyl acetate	25905-14-0					tr[z9]
Limonene	138-86-3	0.1-1.4	0-0.1	0.2[a]	0.2-2.2	0.4[f]; 0.5[g]; 0.6[n]; 0.7[z]; 1.0[m]; 1.1[i]; 2.3[c]
Linalool	78-70-6	2.3-4.5	1.6-3.4	2.6	2.3-3.9	2.9[z9]; 3.2[m]; 3.4[z]; 3.6[c,o]; 3.8[y]; 4.6[j]; 4.9[v]
Linalool oxide	1365-19-1					0.1[z]
cis-Linalool oxide, furanoid	11063-77-7			tr		
trans-Linalool oxide, furanoid	34995-77-2			tr		
Linalyl acetate	115-95-7					0.04[m]
Linalyl formate	115-99-1		0-0.2			
Linalyl propionate	144-39-8		0-0.3			
trans-p-Menth-2-en-1-ol	29803-81-4		0-0.4			
Methyl chavicol	140-67-0			tr		
Methyl eugenol	93-15-2					0.1[i]
Methylheptenone	409-02-9	0.03-0.1				tr[h]; 0.09[z]; 2.2[d]
6-Methyl-5-hepten-2-one	110-93-0			tr		tr[b,l,z9]; 0.1[f,j]
Methyl isoeugenol	93-16-3				0.01-0.08	0.2[i]
γ-Muurolene	30021-74-0					+[z5]
Myrcene	123-35-3	0.1-0.6	0.1-0.3	0.6	0.1-0.3	0.3[f]; 0.5[c]; 0.7[i]; 1.0[h]; 2.6[z]; 3.3[k,m]

Table 5.66.1 Constituents identified in palmarosa oils *(continued)*

Constituent	CAS	\multicolumn{5}{c}{Percentage and range}
		A	B (23)	C (22)	D (16)	E
Neointermedeol	5945-72-2					1.0[i]
Neophytadiene	504-96-1					tr[z9]
Neral	106-26-3	0.06-0.3		0.2	0.3-0.6	tr[b]; 0.1[l]; 0.2[e,t]; 0.4[z4]; 0.6[j,z7]; 2.2[z]
Nerol	106-25-2	0.1-0.5		<0.06	0-0.01	0.1[d,l,t]; 0.2[e,z3]; 0.6[z4]; 2.3[z]; 15.0[z6]; 24.8[w]
Nerolidol	7212-44-4					0.3[j]; 2.4[z1]
(*E*)-Nerolidol	40716-66-3	0.1-0.4	0.1-0.3			0.1[b,l]
Neryl acetate	141-12-8		0-0.3	tr	0.01-0.09	tr[e]; 0.1[j]; 0.4[f]; 0.5[m]; 0.8[z3]; 1.3[z4]
Neryl formate	2142-94-1		0-0.1	tr		0.1[j]
Neryl heptanoate			0-0.3			
Nonadecane	629-92-5		0-0.2			
Nonadecanol	1454-84-8		0-0.1			
Nonanal	124-19-6					tr[b,l]
α-Ocimene	502-99-8					0.4[z9]
β-Ocimene	13877-91-3					0.3[x,y1]; 1.8[z9]
(*E*)-Ocimene	27400-72-2					0.1[z3]; 1.7[t]
(*Z*)-Ocimene	27400-71-1		0-0.2			0.4[t]
(*E*)-β-Ocimene	3779-61-1	0.6-2.1	0.1-0.7	1.3		1.4[z4]; 1.6[g]; 1.7[j]; 2.0[k]; 2.1[f]; 2.5[m,v]
(*Z*)-β-Ocimene	3338-55-4	0.3-0.7		tr		0.3[b,l]; 0.4[g]; 0.5[j]; 0.8[n]; 0.9[i,k]; 1.0[m]
allo-Ocimene	673-84-7					0.1[j]
(*E,E*)-allo-Ocimene	3016-19-1					tr[z9]
Octanol	29063-28-3					tr[b,l,z9]
α-Panasinsene	56633-28-4					tr[z9]
α-Phellandrene	99-83-2		0-tr	tr		+[z5]; tr[h]; 0.1[k]; 0.2[m]
α-Pinene	80-56-8			tr	0-0.02	0.05[m]; 0.4[f]; 0.7[o]; 4.3[w]; 8.0[i]
β-Pinene	127-91-3				0-0.02	+[z5]; tr[b,l]; 0.04[m]; 0.2[i,o]
Piperitone	89-81-6			tr		
Sabinene	3387-41-5		0-0.1			0.1[g,k,m]; 0.3[i]
7-epi-α-Santalene						tr[b,l]
Selina-4,7-diene					0.1-0.3	+[z5]
α-Selinene	473-13-2			tr	0.1-0.4	tr[h,j]; 0.1[b,l]; 0.3[z7]
β-Selinene	17066-67-0			0.2	0-0.05	+[z5]; 0.1[h,j,z9]; 5.4[i]
γ-Selinene	515-17-3				0.01-0.3	
δ-Selinene	473-14-3					+[z5]
α-Sinensal	17909-77-2					0.2[z]
α-Terpinene	99-86-5				0.04-0.2	+[z5]; tr[h]; 0.06[m]; 0.1[k]
γ-Terpinene	99-85-4		0-0.1			tr[b,l]; 0.1[k]; 0.2[m]; 0.3[d]
Terpinen-4-ol	562-74-3		0-2.0	tr		0.1[m]
α-Terpineol	98-55-5	0.02-0.2	0-0.4	0.09		tr[h,z9]; 0.09[m]; 0.1[f]
β-Terpineol	138-87-4					0.3[h]
δ-Terpineol	7299-42-5					0.9[i]
Terpinolene (α-)	586-62-9	0.0-0.09	0-0.1	tr	0.01-0.2	tr[b,l,z9]; 0.1[k,m]; 0.2[i]
α-Terpinyl acetate	80-26-2					tr[h]
Tricyclene	508-32-7			tr		
3,4,5-Trimethoxy-benzaldehyde	86-81-7					0.3[i]
1-Undecanol	112-42-5		0-0.1			
Valencene	4630-07-3					tr[z9]; 0.1[j]
cis-Verbenol	1845-30-3		0.1-0.5			
Zingiberene	495-60-3					1.4[i]

A thirty-four palmarosa essential oil samples from India and Nepal, analyzed between 1998 and 2013; lowest and highest concentrations given (E. Schmidt, unpublished data)

B three lab-hydrodistilled oils from the flowering herbs of *C. martini* var. *motia* cultivated at three stations in India; lowest and highest concentrations given (ref. 23)

C one lab-hydrodistilled oil from the flowering herbs of palmarosa var. *motia* cultivated in India (ref. 22)

D twelve industrially steam-distilled oils from freshly cut herb of *C. martini* var. *martini* collected in Madagascar during four years; lowest and highest concentrations given (ref. 16);

E data from other studies (indicated with superscript letters); highest concentrations found in any study reviewed here given; when two or more oils were investigated, only the highest concentrations are mentioned, unless indicated otherwise

Table 5.66.1 Constituents identified in palmarosa oils *(continued)*

[a] limonene and 1,8-cineole combined; [b] one lab-hydrodistilled oil from the leaves of the *martini* variety cultivated in Bangladesh (ref. 7); [c] eight lab-hydrodistilled oils from fresh herbage of eight selections of *C. martini* var. *martini* cultivated in India; only the results of the major components were reported (ref. 6); [d] four lab-hydrodistilled oils from 4 cultivars of palmarosa var. *martini* (ref. 4); [e] one commercial oil from *C. martini* var. *motia* obtained from a German company (ref. 3); one lab-hydrodistilled oil from the whole palmarosa herb (var. *motia*) cultivated in India (ref. 2); [g] four oils from fresh flowering palmarosa (var. *motia*), one obtained by lab-hydrodistillation and three in a field-distillation unit (ref. 10); [h] one lab-hydrodistilled oil of palmarosa from India (ref. 11); [i] one completely atypical oil obtained by hydrodistillation of palmarosa leaves collected in the wild in India containing no geraniol (ref. 19); [j] one commercial palmarosa oil from Grasse, France (ref. 24); [k] one lab-hydrodistilled oil from *C. martini* var. *motia* cultivated in India (ref. 25); [l] one commercial palmarosa oil from India (ref. 26); [m] four oils of palmarosa from the *motia* variety from India, biomass harvested in two seasons and distilled with either hydrodistillation or steam-distillation (ref. 35); [n] one oil of palmarosa of south Indian origin (ref. 34); [o] one lab-hydrodistilled oil of leaves + stems of *C. martini* var. *martini* cultivated in India (ref. 8); [p] four lab-hydrodistilled leaf oils from four Indian palmarosa var. *martini* varieties; only geraniol and geranyl acetate content were examined (ref. 5); [q] six oils from *C. martini* var. *motia* from India obtained in a field-distillation unit; various amounts of water were added and oils were stored for 1-15 days before analysis; only geraniol, geranyl acetate and linalool were examined (ref. 1); [r] one commercial oil obtained from a UK company; only the content of geraniol and geranyl acetate was presented (ref. 9); [s] nine oils of palmarosa cultivated in a greenhouse in the USA and obtained by distillation, with distillation times varying from 1.25 minute to 4 hours; only the content of geraniol and geranyl acetate was presented (ref. 12); [t] one commercial palmarosa oil purchased in Brazil (ref. 14); [u] one steam-distilled oil from wild Indian palmarosa in the flowering stage (ref. 15); [v] one commercial oil from *C. martini* var. *martini* from Nepal (ref. 17); [w] one lab-hydrodistilled oil from fresh palmarosa leaves growing wild in India (ref. 18); [x] one commercial oil of palmarosa from Brazil (ref. 20); [y] one commercial palmarosa oil purchased from a Spanish company (ref. 27); [y1] one commercial palmarosa oil *ex Cympogon martini* purchased from a Brazilian company (ref. 40); [z] one lab-hydrodistilled oil from palmarosa leaves harvested in an Indian university botanical garden (ref. 28); [z1] one lab-hydrodistilled oil from Indian *C. martini* var. *martini* (ref. 29); [z2] one lab-hydrodistilled oil from palmarosa leaves purchased at a local Taiwanese market (ref. 30); [z3] one commercial palmarosa oil of unknown origin (ref. 31); [z4] one commercial oil of palmarosa of unknown origin (ref. 32); [z5] one steam-distilled oil from Madagascar; the components of the hydrocarbon fraction were examined; as the percentages cannot be compared with the other data, their presence in the hydrocarbon fraction is indicated with +[z5]; only chemicals which were found in only three or less other studies discussed in Table 5.66.2 are presented (ref. 13); [z6] one oil from Guatemala (ref. 36); [z7] one typical palmarosa oil produced in India (ref. 38); [z8] geranyl butyrate and geranyl isovalerate combined; [z9] one oil of palmarosa of unknown origin (ref. 39)

tr: trace (in columns C, E[b], E[f] and E[h]: <0.1; in column E[e]: <0.01; + present in the oil investigated, but quantity not stated or comparable to the other data

Table 5.66.2 Major constituents of palmarosa oils

Constituent	CAS	Percentage and range				
		A	B (23)	C (22)	D (16)	E
Geraniol	106-24-1	74.2-86.9	67.6-83.6	75.9	76.3-82.8	85.4[z3]; 85.7[g]; 86.3[d]; 87.6[y]; 91.6[c]; 96.7[p]
Geranyl acetate	105-87-3	3.4-12.5	2.2-15.8	10.7	5.1-11.8	13.6[x,y1]; 15.8[h]; 15.9[c]; 20.0[r]; 20.1[m]; 22.6[v]
Linalool	78-70-6	2.3-4.5	1.6-3.4	2.6	2.3-3.9	2.9[z9]; 3.2[m]; 3.4[z]; 3.6[c,o]; 3.8[y]; 4.6[j]; 4.9[v]
β-Caryophyllene	87-44-5	1.7-2.5	0.1-1.1	2.0	1.0-1.8	1.6[f]; 1.8[t]; 2.0[h]; 2.5[y]; 2.6[j]; 3.3[j]; 3.4[v]

LEGEND: SEE UNDER TABLE 5.66.1

CONTACT ALLERGY/ALLERGIC CONTACT DERMATITIS

General

Contact allergy to/allergic contact dermatitis from palmarosa oil has been reported in two publications only.

Case reports and positive patch tests

Contact allergy to palmarosa oil with cross-reactivity to citronella oil (42, article not read). One positive patch test to palmarosa oil in an aromatherapist with occupational contact dermatitis from multiple (other) essential oils; palmarosa oil itself had not been used by the patient. Geraniol, the major component of palmarosa oil, was negative (41).

LITERATURE

1 Rajeswara Rao BR, Rajput DK, Patel RP. Storage of essential oils: influence of presence of water for short periods on the composition of major constituents of the essential oils of four economically important aromatic crops. J Essent Oil Bear Plants 2011;14:673-678

2 Rajeswara Rao BR, Rajput DK, Patel RP. Essential oil profiles of different parts of palmarosa (*Cymbopogon martinii* (Roxb.) Wats. var. *motia* Burk.). J Essent Oil Res 2009;21:519-521

3 Jirovetz L, Eller G, Buchbauer G, Schmidt E, Denkova Z, Stoyanova AS, et al. Chemical composition, antimicrobial activities and odor descriptions of some essential oils with characteristic floral-rosy scent and of their principal aroma compounds. Recent Res Devel Agron Horticult 2006;2:1-12

4 Srivastava HK, Satpute GK, Naqvi AA. Induced mutants in M2 generation and selection for enhanced essential oil yield and quality in palmarosa (*Cymbopogon martinii*, Roxb.) Wats., var. *martini*. J Essent Oil Res 2000;12:501-506

5 Fatima S, Abad Farooqi AH, Ansari SR, Sharma S. Effect of water stress on growth and essential oil metabolism in *Cymbopogon martinii* (palmarosa) cultivars. J Essent Oil Res 1999;11:491-496

6 Sarma PC, Baruah P, Pathak MG, Kanjilal PB. Comparison of the major components of the oils of eight selections of *Cymbopogon martinii* (Roxb.) Wats. var. *martini*. J Essent Oil Res 1998;10:673-674

7 Chowdhury JU, Yusuf M, Begum J, Mondello L, Previti P, Dugo G. Studies on the essential oil bearing plants of Bangladesh. Part IV. Composition of the leaf oil of three *Cymbopogon* species: *C. flexuosus* (Nees ex Steud.) Wats., *C. nardus* (L.) Rendle var. *confertiflorus* (Steud.) N. L. Bor and *C. martinii* (Roxb.) Wats. var. *martini*. J Essent Oil Res 1998;10:301-306

8 Siddiqui N, Garg SC. Chemical composition of *Cymbopogon martinii* (Roxb.) Wats. var. *martini*. J Essent Oil Res 1990;2:93-94

9 Prashar A, Hili P, Veness RG, Evans CS. Antimicrobial action of palmarosa oil (*Cymbopogon martinii*) on *Saccharomyces cerevisiae*. Phytochem 2003;63:569-575

10 Rajeswara Rao BR, Kaul PN, Syamasundar KV, Ramesh S. Chemical profiles of primary and secondary essential oils of palmarosa (*Cymbopogon martinii* (Roxb.) Wats. var. *motia* Burk.). Ind Crops Prod 2005;21:121-127

11 Rajeswara Rao BR. Biomass and essential oil yields of rainfed palmarosa (*Cymbopogon martinii* (Roxb.) Wats. var. *motia* Burk.) supplied with different levels of organic manure and fertilizer nitrogen in semi-arid tropical climate. Ind Crops Prod 2001;14:171-178

12 Cannon JB, Cantrell CL, Astatkie T, Zheljazkov VD. Modification of yield and composition of essential oils by distillation time. Ind Crops Prod 2013;4:214-220

13 Gaydou EM, Randriamiharisoa RP. Hydrocarbons from the essential oil of *Cymbopogon martini*. Phytochem 1986;26:183-185

14 Katiki LM, Chagas ACS, Bizzo HR, Ferreira JFS, Amarante AFT. Anthelmintic activity of *Cymbopogon martinii*, *Cymbopogon schoenanthus* and *Mentha piperita* essential oils evaluated in four different *in vitro* tests. Vet Parasitol 2011;183:103-108

15 Bottini AT, Dev V, Garfagnoli DJ, Hope H, Joshi P, Lohani H, et al. Isolation and crystal structure of a novel dihemiacetal bis-monoterpenoid from *Cymbopogon martini*. Phytochem 1987;26:2301-2302

16 Randriamiharisoa RP, Gaydou EM. Composition of palmarosa (*Cymbopogon martinii*) essential oil from Madagascar. J Agric Food Chem 1987;35:62-66

17 Yonzon M, Lee DJ, Yokochi T, Kawano Y, Nakahara T. Antimicrobial activities of essential oils of Nepal. J Essent Oil Res 2005;17:107-111

18 Nirmal SA, Girme AS, Bhalke RD. Major constituents and anthelmintic activity of volatile oils from leaves and flowers of *Cymbopogon martini* Roxb. Nat Prod Res 2007;21:1217-1220

19 Tyagi BK, Shahi AK, Kaul BL. Evaluation of repellent activities of *Cymbopogon* essential oils against mosquito vectors of malaria, filariasis and dengue fever in India. Phytomed 1998;5:324-329

20 Murbach Teles Andrade BF, Conti BJ, Basso Santiago K, Fernandes Júnior A, Sforcin JM. *Cymbopogon martinii* essential oil and geraniol at noncytotoxic concentrations exerted immunomodulatory/anti-inflammatory effects in human monocytes. J Pharm Pharmacol 2014;66:1491-1496

21 Dubey VS, Mallavarapu GR, Luthra R. Changes in the essential oil content and its composition during palmarosa (*Cymbopogon martinii* (Roxb.) Wats. var. *motia*) inflorescence development. Flavour Fragr J 2000;15:309-314

22 Mallavarapu GR, Rajeswara Rao BR, Kaul PN, Ramesh S, Bhattacharya AK. Volatile constituents of the essential oils of the seeds and the herb of palmarosa (*Cymbopogon martinii* (Roxb.) Wats. var. *motia* Burk.). Flavour Fragr J 1998;13:167-169

23 Raina VK, Srivastava SK, Aggarwal KK, Syamasundar KV, Khanuja SPS. Essential oil composition of *Cymbopogon martinii* from different places in India. Flavour Fragr J 2003;18:312-315

24 Ramilijaona J, Raynaud E, Bouhlel C, Sarrazin E, Fernandez X, Antoniotti A. Enzymatic modification of palmarosa essential oil: chemical analysis and olfactory evaluation of acylated products. Chem Biodivers 2013;10:2291-2301

25 Rajeswara Rao BR, Kaul PN, Bhattacharya AK, Mallavarapu GR, Ramesh S. Yield and chemical composition of the essential oils of three *Cymbopogon* species suffering from iron chlorosis. Flavour Fragr J 1996;11:289-293

26 Daniel DK, Malik S, Albert K. Lipase-catalyzed esterification of palmarosa oil. Eng Life Sci 2011;11:195-200

27 Velluti A, Sanchis V, Ramos AJ, Marin S. Effect of essential oils of cinnamon, clove, lemon grass, oregano and palmarosa on growth of and fumonisin B1 production by *Fusarium verticillioides* in maize. J Sci Food Agric 2004;84:1141-1146

28 Prasad CS, Shukla R, Kumar A, Dubey NK. In vitro and in vivo antifungal activity of essential oils of *Cymbopogon martini* and *Chenopodium ambrosioides* and their synergism against dermatophytes. Mycoses 2009;53:123-129

29 Rajendrudu G, Rama Das VS. Interspecific differences in the constituents of essential oils of *Cymbopogon*. Proceedings: Plant Sciences 1983;92:331-334

30 Tsai M-L, Lin C-C, Lin W-C, Yang C-H. Antimicrobial, antioxidant, and anti-inflammatory activities of essential oils from five selected herbs. Biosci Biotechnol Biochem 2011;75:1977-1983

31 Delespaul Q, de Billerbeck VG, Roques CG, Michel G, Marquier-Viñuales C, Bessière J-M. The antifungal activity of essential oils as determined by different screening methods. J Essent Oil Res 2000;12:256-266

32 Lawrence BM. Progress in essential oils. Perfum Flavor 2002;27(1):42-? (last page unknown)

33 Lawrence BM. Progress in essential oils. Perfum Flavor 2004;29(8):52-? (last page unknown)

34 Narayanan P. Comparison of quality of essential oils from south India. FAFAI 2003;5(1):47-56. Data cited in ref 33

35 Kaul PN, Bhattacharya AK, Rajeswara Rao BR. Influence of method of distillation on quality of palmarosa (Cymbopogon martinii (Roxb.) Wats, var. motia Burk.). J Essent Oil Bear Plants 1998;1:39-44

36 Srinivas SH. Atlas of essential oils. New York, USA: Published by the author, 1986. Data cited in ref. 7

37 Lawrence BM. Progress in essential oils. Perfum Flavor 2012;37(February):54-61

38 Ranade GS. Profile essential oil of palmarosa. FAFAI 2007;11(22):77. Data cited in ref. 37

39 Anonymous. Palmarisa oil, GC-MS analysis. Indian Perfumer 2009;53(1):59. Data cited in ref. 37

40 Murbach Teles Andrade BF, Nunes Barbosa L, da Silva Probst I, Fernandes Júnior A. Antimicrobial activity of essential oils. J Essent Oil Res 2014;26:34-40

41 Bleasel N, Tate B, Rademaker M. Allergic contact dermatitis following exposure to essential oils. Australas J Dermatol 2002;43:211-213

42 Paschoud JM. Quelques cas d'eczema de contact avec sensibilisation de groupe. Dermatologica 1963;127:349

Chapter 5.67 PATCHOULI OIL

DEFINITION
Patchouli oil (essential oil of patchouli) is the essential oil obtained from the leaves of the patchouli plant, *Pogostemon cablin* (Blanco) Benth. (synonyms: *Mentha cablin* Blanco; *Pogostemon patchouli* Pellet).

INCI NOMENCLATURE
Description/definition: Pogostemon cablin leaf oil is the volatile oil obtained from the leaves of *Pogostemon cablin*, Lamiaceae
INCI name EU & USA: Pogostemon cablin leaf oil
CAS registry number(s): 8014-09-3; 84238-39-1
EINECS number(s): 282-493-4

ISO (INTERNATIONAL ORGANIZATION FOR STANDARDIZATION) STANDARD
ISO number: 3757
ISO name: Essential oil of patchouli
Botanical origin: *Pogostemon cablin* (Blanco) Benth. (synonym: *Mentha cablin* Blanco)
Parts of plant used: Leaf
ISO values: ISO values (minimum and maximum concentrations) are shown in Table 5.67.1.

THE PLANT, THE OIL, AND THEIR USES
Pogostemon cablin (Blanco) Benth. is a perennial aromatic herb or subshrub with erect stems, which can grow to a height of 1 meter. It is (possibly) native to the Philippines and is cultivated in the Western Indian Ocean Islands (Madagascar, Seychelles), China, Taiwan, India, Indonesia, Malaysia, Brazil and on a smaller scale in many other countries (1,6,18,29). The patchouli has occasionally escaped from cultivation and is now widely naturalized in southeast Asia, southern China, India, Sri Lanka, Mauritius, and Florida (3, GRIN Taxonomy for Plants; Mansfeld's World Database of Agriculture and Horticultural Crops, http://mansfeld.ipk-gatersleben.de).

The leaves of the patchouli are widely used in Asia for incense, body and garment perfumes and insect repellents.

Patchouli oil, the essential oil obtained by steam distillation of the dried leaves of *Pogostemon cablin* (19), is one of the best fixatives for heavy perfumes (1,3,6). In fact, patchouli oil is a perfume by itself and is highly valued in the perfume, soap and cosmetics industries (1,4,6,19). The main component of the essential oil, patchoulol (patchouli alcohol), is important for the duration of the odor and is frequently used as an indicator of essential oil quality (24). Other important aroma compounds in the oil are α-guaiene, α-patchoulene, seychellene, α-bulnesene, norpatchoulenol and pogostol (35).

Patchouli essential oil is also used in incense, and for flavoring tobacco, chewing gum, beverages, frozen dry desserts, candy, baked goods, gelatin, and meat and meat products. It is equally employed in scented industrial products such as paper towels, laundry detergents and air fresheners (1,19). Patchouli oil Is perceived to have antidepressant, antiphlogistic, antiseptic, aphrodisiac, astringent, deodorant, diuretic, febrifuge, fungicide, insecticide, sedative, tonic, and euphoric properties (3,4,18,24). The oil is used medicinally, notably in India, China and Japan, for the treatment of skin infections, dandruff, eczema, wounds, scars, constipation and as temporary antidote to insect bites (1,3,14,18,24). Patchouli oil is also widely used in aromatherapy, e.g., to calm nerves, control appetite and relieve depression and stress (3,4,19).

CHEMICAL COMPOSITION
Patchouli oil is a mobile to highly viscose liquid, which has a woody, earthy, slightly camphoraceous and balsamic odor. The yield of essential oil from the leaves of *Pogostemon cablin* (Blanco) Benth. generally varies from 0.2 to 3.7 per cent. The main producing countries of this oil are Indonesia, Madagascar, India, China and Brazil. The aroma of freshly distilled patchouli is slightly harsh. As the oil ages, the aroma mellows considerably, becoming sweeter and more balsamic. Thus, patchouli is one of very few oils that improve with age (19).

Literature data (up to December 12, 2014) on the chemical composition of patchouli oils and unpublished analytical data from one of us (E.S.) are shown in Table 5.67.2 in alphabetical order. In patchouli oils from various origins, over 210 chemicals have been identified. About 66 per cent of these were found in a single reviewed publication only. The major compounds found in patchouli oils from different sources are shown in Table 5.67.3. They include (highest concentrations in any study given) patchouli alcohol (patchoulol) (71.8%), α-bulnesene (25.0%), guaiene (23.6%), α-guaiene (23.3%), β-patchoulene (16.2%), α-patchoulene (12.3%), γ-patchoulene (11.7%), seychellene (9.6%), δ-patchoulene (8.0%) and pogostol (5.1%). A well-known ingredient of patchouli oils that was present in a high concentration (>7%) in two studies was β-caryophyllene (9.2% and 20.0%). Uncommon

Table 5.67.1 ISO values (%) for patchouli oil [a]

Compound	CAS	Minimum	Maximum
Patchouli alcohol (patchoulol)	5986-55-0	27.0	35.0
α-Bulnesene	3691-11-0	13.0	21.0
α-Guaiene	3691-12-1	11.0	16.0
β-Caryophyllene	87-44-5	2.0	5.0
β-Patchoulene	514-51-2	1.8	3.5
Pogostol	21698-41-9	1.0	2.5
Norpatchoulenol	41429-52-1	0.35	1.0
α-Copaene	3856-25-5	0.01	1.0

[a] ISO 3757 Essential oil of patchouli ©ISO 2002; Geneva, Switzerland, www.iso.org

Table 5.67.2 Constituents identified in patchouli oils

Constituent	CAS	A	B (25)	C (24)	D (1)	E
Acetophenone	98-86-2					+[z4]
Aciphyllene	87745-31-1	2.6-3.6	2.5			1.2[x1]; 2.4[o]; 2.6[i]; 2.9[x5]; 3.4[z2]
Anthracene	120-12-7					0.3-0.6[y4]
4,5-di-epi-Aristolochene	54868-40-5					+[y5]
Aristolone	160568-09-2					0.6[y3]; 1.2[m]
Aromadendrene	489-39-4					2.8[z8]
α-Aromadendrene	146389-60-8					0.4[z6]
allo-Aromadendrene	25246-27-9					+[z7]; 0.6[x6]; 1.8[o]; 2.1[y6]; 5.0[k,y1]
Aromadendrene oxide	85710-39-0					1.6[n]
β-Atlantol	420109-31-5			0-1.5		
(E)-γ-Atlantone	108549-47-9					0.4[o]
Azulene	275-51-4					15.0[p]
Benzaldehyde	100-52-7					+[z4,z7]; <0.1[x8]
Benzyl alcohol	100-51-6					6.7[z8]
Benzyl benzoate	120-51-4					1.7[x1]
α-Bergamotene	17699-05-7					5.8[z3]
β-Bergamotene	6895-56-3					4.8[z3]
trans-β-Bergamotene	15438-94-5					0.6[i]
Bicyclogermacrene	24703-35-3		0.05			0.7[y6]
β-Bisabolene	495-61-4					0.2[n]
cis-α-Bisabolene epoxide	121467-35-4					0.6-0.9[y4]
Borneol	507-70-0				tr	
β-Bourbonene	5208-59-3					0.1[x8]
α-Bulnesene	3691-11-0	14.8-18.9	19.4	5.0-13.0	8.7	11.6[y2]; 15.8[y6]; 16.8[x5]; 17.0[x4]; 19.5[n]; 20.7[k]; 21.4[x3]; 22.0[m]; 22.4[f]; 23.1[x6]; 23.3[r]; 25.0[l]
Bulnesol	22451-73-6				0.1	0.7[y3]
Bulnesoxide	33784-90-6	0.1-0.3				0.3[i]; 4.0[l]
But-3-enal, 2-methyl-4-(2,6,6-trimethyl-1-cyclohexenyl)-			0.04			
Cadinene	29350-73-0					1.0[m]
α-Cadinene	24406-05-1				tr	
γ-Cadinene	39029-41-9					1.0[z3]
δ-Cadinene	483-76-1				0.1	<0.1[x5]; 1.2[e]
Camphene	79-92-5					+[z4,z7]
α-Campholenal	4501-58-0					+[z4]
β-Caryophyllene	87-44-5	3.0-4.7	3.4	1.2-3.2	2.4	3.8[k]; 3.9[s]; 4.0[x4]; 4.5[b]; 4.8[x3]; 5.1[j]; 9.2[d]; 20.0[l]
9-epi-(E)-Caryophyllene	68832-35-9					+[z4]; 0.7[r]; 1.6[o,y6]
Caryophyllene oxide	1139-30-6	0.2-0.4	0.9		0.3	0.5[d]; 0.6[x8]; 0.7[g]; 0.8[y3]; 0.9[o]; 1.6[y4]; 2.0[j]; 2.6[z6]
Cedranediol	88588-48-1					0.8[q]
α-Cedrol	77-53-2					2.6[z8]
β-Chamigrene	18431-82-8					0.3[z6]
Chiloscyphone	23538-45-6		0.06			
1,8-Cineole	470-82-6					+[z4]
Citronellol	106-22-9					+[z4]
α-Copaene	3856-25-5	0.2-1.6		0.1		0.2[z2]; 0.3[i]; 0.5[x,x5]; 0.6[z5]; 0.7[y6]
β-Copaen-4α-ol	124753-76-0					0.8[z']
α-Costol	65018-15-7					+[z7]
α-Cubebene	17699-14-8					+[z4]
β-Cubebene	13744-15-5					0.6[x3]; 1.3[t]
Cuminaldehyde	122-03-2					0.5[z6]
γ-Curcumene	28976-68-3					+[y5]
Cyclohexane, trisubstituted						3.5[g]
Cyclohexanone, 2,3,3-trimethyl-2-(3-methyl-1,3-butadienyl)-, (E)-	69296-91-9		0.4			
Cyclohexene, trisubstituted						3.1[g]
Cyclolongifolene oxide, dehydro-			<0.01			

Table 5.67.2 Constituents identified in patchouli oils *(continued)*

Constituent	CAS	Percentage and range				
		A	B (25)	C (24)	D (1)	E
Cyclopentanone, 3,3-dimethyl-2-(3-methyl-1,3-butadie-nyl)-, (*E*)-	88725-86-4		0.08			
Cycloseychellene	52617-34-2	0.5-0.7			0.4	0.6^{x5}; $0.7^{i,o,z2}$; $0.8^{z5,y6}$
p-Cymene	99-87-6				tr	$+^{z4}$
p-Cymen-8-ol	1197-01-9					$+^{z4}$
Dibutyl phthalate	84-74-2					$0-1.0^{y4}$
Dihydroaromaden-drene						1.1^{z}
Dihydrocaranone	112529-25-6					1.1^{z6}
2,4-Diisopropenyl-1-methylvinylcyclo-hexone						1.0^{z3}
4,4-Dimethyl-3-(3-methylbut-3-enyli-dene)-2-methylene-bicyclo[4.1.0]heptane	79718-83-5		0.02			
6,8a-Dimethyl-9-me-thylidene-2,5-metha-no-1,2,3,3a,4,5,8,8a-octahydroazulene						1.2^{z3}
4,8-Dimethyl-1,3(*E*),7-nonatriene	19945-61-0					$+^{z4,z7}$
Dispiro[2.0.2.5]unde-cane, 8-methylene-	51567-09-0		<0.01			
Dodecanoic acid	143-07-7					0.3^{y4}
Elaidic acid	112-79-8					$0.7-1.1^{y4}$
α-Elemene	5951-67-7					1.4^{m}
β-Elemene	33880-83-0		3.9	0.4-1.1	0.8	0.6^{z3}; 0.7^{e}; 0.9^{x8}; 1.0^{i}; 1.1^{z2}; $1.2^{g,t}$; 1.3^{k}; 2.6^{r}
γ-Elemene	29873-99-2		0.02			4.8^{r}
δ-Elemene	20307-84-0	0.1-1.2	0.07			$<0.01^{u}$; 0.07^{i}; 0.1^{x8}; $0.2^{z2,x7}$; 0.3^{z5}; 1.3^{n}; 1.7^{x}
trans-β-Elemenone	20303-60-0					2.7^{n}
Elemol	639-99-6					$0-0.5^{y4}$; $+^{z7}$; 3.0^{y3}
1,10-Epoxy-11-bulne-sene						0.2^{z2}
Eremophila-1(10),8,11-triene						$+^{z7}$
Eremophilene	10219-75-7		1.3			0.6^{g}; 1.4^{n}; 4.3^{z3}
2-Ethylhexanol	104-76-7					$+^{z7}$
cis-Eudesma-4(15),11-dien-5-ol						$+^{z7}$
Eugenol	97-53-0					$+^{z4,z7}$; 2.0^{y6}
Farnesal	19317-11-4					$0.8-1.8^{y4}$
(*E,E*)-Farnesal	502-67-0					$+^{z7}$
α-Farnesene	502-61-4					$0-1.9^{y4}$
(*E,E*)-α-Farnesene	502-61-4					$+^{y5}$
(*Z,E*)-α-Farnesene	26560-14-5		0.01			
(*Z,Z*)-α-Farnesene	28973-99-1		<0.01			
(*E*)-β-Farnesene	18794-84-8					$+^{y5}$
Farnesol	4602-84-0					$0-2.5^{y4}$; 1.6^{n}
(*E,E*)-Farnesol	106-28-5					$+^{y5}$
(6*Z*,10*E*)-Farnesol						$0-2.3^{y4}$
(*E,E*)-Farnesyl acetate	4128-17-0					$0-1.9^{y4}$
Farnesyl pyrophos-phate	13058-04-3					$+^{y5}$
Geranial	141-27-5					$+^{z7}$
Geranic acid	459-80-3					$0-0.7^{y4}$
Germacrene A	28387-44-2					11.7^{u}
Germacrene B	15423-57-1					19.0^{v}
Germacrene C	34323-15-4					$+^{y5}$
Germacrene D	23986-74-5				0.1	$0.2^{u,z2,x8}$
Globulol	489-41-8		0.3			0.8^{y3}; 4.6^{u}; 6.3^{s}

Table 5.67.2 Constituents identified in patchouli oils *(continued)*

Constituent	CAS	Percentage and range				
		A	B (25)	C (24)	D (1)	E
Guaia-3,9-diene	855270-07-4					0.5[z6]
Guaiene	88-84-6					18.8[x]; 23.6[m]
α-Guaiene	3691-12-1	12.4-15.6	12.8	3.7-9.5	10.1	7.7[x2]; 7.8[x1]; 14.5[x4]; 15.2[x5]; 15.3[k]; 15.4[i,n]; 15.9[x6]; 18.2[c]; 19.8[g]; 20.0[r]; 20.6[t]; 23.3[x3]
β-Guaiene	88-84-6		0.2			0.1-2.7[y2]; 3.5[z3]; 3.9[s]; 14.2[y3];
cis-β-Guaiene	372162-07-7					0.7[y3] 14.4[y6]
trans-β-Guaiene	192053-49-9					0.9[z]
γ-Guaiene	145267-53-4					20.5[c]
α-Guaiene oxide						0.2[x7]; 0.3[d]; 1.0[l]
Guaiol	489-86-1				tr	
Guai-11-ol						6.3[m]
α-Gurjunene	489-40-7					0.2[g]; 1.5[x6]; 3.9[v]; 11.3[z8]
β-Gurjunene	73464-47-8					3.3[y3]
γ-Gurjunene	22567-17-5				0.2	0.3[x4]; 0.6[y6]; 1.4[x5]; 2.2[q,z2]; 3.7[y3]
2-Heptanone	110-43-0					+[z4]
Hexadecanoic acid	57-10-3					<0.1[x8]; 1.5-6.9[y4]
cis-11-Hexadecenoic acid						0-0.6[y4]
3-Hexadecyne	61886-62-2					0.9[z3]
Hexahydropseudo-ionone	1604-34-8		0.01			
α-Himachalene	3853-83-6					0.9[r]
δ-Himachalene	135447-48-2					0.6[r]
α-Humulene	6753-98-6	0.5-1.0		0.3-0.7		0.5[o,u,z]; 0.6[x]; 0.7[b,z3]; 0.8[z2,y6]; 0.9[d]; 1.0[i,x5]
β-Humulene	116-04-1					13.2[p]
Humulene (ep)oxide II	19888-34-7		0.08			
Intermedeol	6168-59-8					+[z7]
Isoaromadendrene epoxide	499134-59-7		0.4			0.3-1.6[y4]
Isocaryophyllene	118-65-0		0.4			0.5[x5]; 6.3[y3]
Isoledene	95910-36-4		<0.01			
Isopatchoulenone	3466-15-7					0.1[z2]
Isovaleric acid	503-74-2					+[z4]
Isovanillic acid	645-08-9					0-12.1[y4]
Ledol	577-27-5					0-0.8[y4]
Limonene	138-86-3	0.02-0.5				+[z7]; <0.1[x8]; 0.1[z2]
Linalool	78-70-6					+[z4,z7]
Linoleic acid	60-33-3					0.5-3.2[y4]
Longicamphenylone	38647-26-6					0.8[a,x3]
Longifolene (junipene)	475-20-7					18.4[x]
α-Longifolene	475-20-7					1.4-6.9[y4]
Longipinanol	66141-14-8				tr	<0.01[u]
β-Longipinene	41432-70-6				tr	
trans-Longipinocar-veol	889109-69-7					0.6[y3]
β-Maaliene	489-29-2		0.06			
7-Menthanoazulene						6.9[p]
o-Methylanisole	578-58-5					0-0.6[y4]
Methyl benzoate	93-58-3					14.3[c]
2-Methylbutyric acid	116-53-0					+[z4]
5-Methyl-2-furalde-hyde	620-02-0					+[z4]
Methyl linoleate	112-63-0					0-2.7[y4]
Methyl salicylate	119-36-8					+[z4,z7]
γ-Muurolene	30021-74-0				0.2	
Myrcene	123-35-3				tr	+[z4]
Myrtenol	515-00-4					+[z4]
2(1H)-Naphthalenone, 3,5,6,7,8,8a-hexahy-dro-4,8a-dimethyl-6-(1-methylethenyl)-	725240-70-0		0.1			
Nerolidol	7212-44-4					+[y5]

Table 5.67.2 Constituents identified in patchouli oils *(continued)*

Constituent	CAS	A	B (25)	C (24)	D (1)	E
(Z)-Nerolidol	3790-78-1					0-1.9[y4]
Nonadecane	629-92-5					1.5[n]
Nootkatene	5090-61-9					0.5[o]
Norpatchoulenol	41429-52-1	0.2-0.9	0.8		1.6	0.3[x7]; 0.4[x4]; 0.5[l]; 0.6[k,z2]; 0.9[i]; 1.0[z5]; 1.1[x]; 1.2[d]
Nortetrapatchoulol						0.3[z2]
3-Octanol	589-98-0					+[z4]
3-Octanone	106-68-3					+[z4,z7]
2-Octenoic acid, 4-isopropylidene-7-methyl-6-methyl-ene-, methyl ester			0.03			
1-Octen-3-ol	3391-86-4				0.4	+[z4,z7]; 0.2[u]; 2.0[z]
1-Octen-3-one	4312-99-6					+[z4]
4-Oxo-14-norvitrane	77284-02-7					0.7[z3]
9-Oxopatchoulol						0.1[z2]
α-Panasinsene	56633-28-4		0.1			9.6[p]
Patchoulene	1405-16-9					1.2[g]; 5.0[m]
α-Patchoulene	560-32-7	4.7-8.0	6.8	1.7-3.9	1.6	3.8[x8]; 4.5[x4]; 5.5[x6]; 5.6[y6]; 5.7[i]; 5.9[f]; 6.2[x5]; 7.5[g]; 7.9[j,k]; 8.0[e]; 8.6[y3]; 9.2[q]; 10.5[t]; 12.3[s]
β-Patchoulene	514-51-2	2.0-4.1		1.7-3.8	1.0	1.1[x8]; 1.8[x1]; 2.0[x5]; 2.3[x3]; 3.2[y6]; 3.7[j]; 4.1[c]; 4.2[b]; 6.5[s]; 7.0[y3]; 7.5[x2]; 9.3[k,y1]; 12.1[t]; 12.9[n]; 16.2[m]
γ-Patchoulene	508-55-4	5.9-8.0			1.3	0.3[g]; 0.6[x8]; 0.9[x5]; 1.1[i]; 2.8[x4]; 3.2[z]; 3.6[y6]; 3.7[x1]; 3.9[u]; 5.5[q]; 6.3[x3]; 6.7[k]; 8.5[x4]; 9.4[x2]; 11.7[n]
δ-Patchoulene	53823-16-8	1.7-7.5				0.7[y3]; 8.0[r]
Patchoulenone	5986-54-9					0.2[z2]
Patchouli alcohol (patchoulol)	5986-55-0	22.3-33.9	33.4	47.3-69.0	61.6	32.6[x6]; 42.7[s]; 42.9-64.3[y2]; 44.5[z3]; 47.8[d]; 54.3[v]; 54.9[x8]; 60.3[u]; 62.4[x2]; 70.7[z]; 71.8[x1]
Pentadecanoic acid	1002-84-2					0-0.4[y4]
2-Pentylfuran	3777-69-3					+[z4]
α-Phellandrene	99-83-2					+[z4]
2-Phenethyl alcohol	60-12-8					+[z4]
Phytol	7541-49-3					+[z7]
α-Pinene	80-56-8	0.05-1.2	0.03	tr		0.07[i]; 0.08[g]; 0.1[z2]; 0.2[b,z5]; 0.5[n]; 1.1[s]; 1.2[x1]
β-Pinene	127-91-3	0.1-3.3	0.1			<0.1[x8]; 0.1[g]; 0.2[i,x3]; 0.3[z2]; 0.4[b,z5]; 1.0[t]
trans-Pinocarveol	1674-08-4					+[z4]
Pogostol	21698-41-9	1.6-4.5	2.6	3.3-5.1	0.5	0.4[d]; 1.0[l]; 2.0[x4]; 2.4[e,y1]; 3.1[o]; 3.3[z5]; 5.1[x1]
Pogostone	23800-56-8				1.0	0.5[g,h]; 1.1[k,y1]; 10.3-27.7[y2]; 21.1-38.3[y4];
Sabinene	3387-41-5	0.03-0.08				+[z4,z7]
Selina-3,7(11)-diene	6813-21-4					2.9[m]; 5.1[x1]
Selinene	27104-12-7					3.9[b]
α-Selinene	473-13-2				tr	0.2[u]; 1.8[x8]; 3.5[s]; 3.9[r]; 4.2[t]
7-epi-α-Selinene	123123-37-5				0.1	+[z7]; 0.2[u,z2,x8]
β-Selinene	17066-67-0					+[z4,z7]; <0.01[u]; 0.4[y3]; 0.8[z3]
δ-Selinene	473-14-3					1.8[m]
Seychellene	20085-93-2	3.2-9.4	8.2	3.0-6.5	3.1	7.4[j]; 7.5[e]; 7.9[o]; 8.1[y6]; 8.5[v]; 8.9[i]; 9.5[d]; 9.6[z8]
Spathulenol	6750-60-3					0.5[z6]; 0.5-4.1[y4]
Squalene	111-02-4					0-2.8[y4]
α-Terpinene	99-86-5					+[z4]
γ-Terpinene	99-85-4				tr	+[z4]
Terpinen-4-ol	562-74-3					+[z4]
α-Terpineol	98-55-5				tr	+[z4,z7]
Terpinolene (α-)	586-62-9					+[z4]
Tetradecanoic acid	544-63-8					0-0.6[y4]
α-Thujene	2867-05-2					+[z4]
Thujopsadiene	24048-40-6				0.3	
Thujopsene	470-40-6					0-1.1[y4]; 0.4[x8]; 3.1[z8]
3-Thujopsene						1.6[m]
cis-Thujopsene	32435-95-3				tr	0.3[u] 0.9[y3];
Toluene	108-88-3					+[z4]

Table 5.67.2 Constituents identified in patchouli oils *(continued)*

Constituent	CAS	Percentage and range				
		A	B (25)	C (24)	D (1)	E
4,6,6-Trimethyl-2-(3-methylbuta-1,3-dienyl)-3-oxatricyclo-[5.1.0.0(2,4)]octane			0.04			
Valencene	4630-07-3				0.3	0.9[u]
Verbenone	80-57-9					+[z4]
p-Vinylguaiacol	7786-61-0					+[z4]
Viridiflorene (ledene)	21747-46-6					0.4-1.6[y4]; 1.9[u]; 3.9[k]
Viridiflorol	552-02-3					1.4[x1]
β-Ylangene	20479-06-5					+[y5]

A fifty-two essential oil samples from Indonesia, Madagascar, India, China and Brazil analyzed between 1998 and 2013; lowest and highest concentrations given (E. Schmidt, unpublished data)
B one lab-hydrodistilled oil from China (ref. 25)
C twenty-eight lab-hydrodistilled oils from seven accessions of patchouli cultivated in Brazil and harvested in four periods over two years; lowest and highest concentrations given (ref. 24)
D one lab-hydrodistilled oil from patchouli leaves collected at an experimental field of an Indian aromatic plant research center (ref. 1)
E data from other studies (indicated with superscript letters); highest concentrations found in any study reviewed here given; when two or more oils were investigated, only the highest concentrations are mentioned, unless indicated otherwise

[a] incorrect identification (ref. 41); [b] one lab-hydrodistilled oil from patchouli leaves cultivated in India (ref 3); [c] one commercial patchouli essential oil sample of unknown origin (ref. 5); from the presence of methyl benzoate and γ-guaiene it was concluded that this oil sample was adulterated (ref. 41); [d] one lab-hydrodistilled oil from a herbal garden in India (ref. 4); [e] one steam-distilled oil from patchouli leaves cultivated in Vietnam (ref. 8); [f] one commercial oil from China; the data presented do not correspond with the original publication as they have been recalculated (ref. 7); [g] one Chinese patchouli oil (ref. 10); [h] probably patchoulenone (ref. 9); [i] one commercial oil sample of unknown origin (ref. 11); [j] two commercial oils from India (of which one was adulterated with diethyl phthalate [ref.12]) and one from Indonesia (ref. 12); [k] one Chinese and two Indonesian commercial oils of different quality (ref. 13); [l] one oil of unknown origin (ref. 15); [m] three commercial patchouli oils from China, Indonesia and India (ref. 16); [n] one lab-hydrodistilled oil from India (ref. 17); [o] one lab-hydrodistilled oil from the leaves of patchouli plants cultivated in Brazil (ref. 22); [p] a very atypical oil from Indonesia (ref. 23); [q] one commercial oil of unknown origin (ref. 26); [r] one steam-distilled patchouli oil from Brazil (ref. 27); [s] one steam-distilled oil from Indonesian patchouli dried leaves (ref. 28); [t] one commercial oil purchased in Taiwan (ref. 29); [u] one lab-hydrodistilled oil from plants cultivated in a university garden in Thailand (ref. 30); [v] fourteen lab-hydrodistilled oils from India from two harvests of patchouli leaves with irrigation level, amount of organic mulch and nitrogen level as parameters (ref. 32); [w] one steam-distilled oil from the Philippines (ref. 33); [x] three steam-distilled oils from three Philippine patchouli cultivars (ref. 34); [x1] three oils from patchouli leaves collected in different parts of Papua New Guinea (ref. 42); [x2] hydrodistilled oils of Indian patchouli leaves in various states of dryness (ref. 43); [x3] one lab-hydrodistilled oil sample from an Indonesian patchouli population grown in India (ref. 44); [x4] 'typical composition' of patchouli oil according to ref. 45; [x5] one patchouli oil sample of unknown origin (ref. 46); [x6] one patchouli oil sample produced in Indonesia (ref. 47); [x7] one patchouli oil sample from India, as control in a plant irradiation experiment (ref. 48); [x8] one lab-hydrodistilled oil from patchouli leaves harvested in India and dried for 15 days (ref. 49); [x9] three lab-hydrodistilled oils from India, one produced from dry leaves, one from leaves soaked in water for 12 hours and the third from 24-hours soaked patchouli leaves (ref. 50); [y] one commercial oil from Australia (ref. 38); [y1] three commercial patchouli oils, one from China and two from Indonesia (ref. 51); [y2] 16 oils from leaves of P. cablin collected in various provinces of China; highest, or lowest and highest concentrations given (ref. 52); [y3] one commercial oil purchased in China (ref. 53); [y4] oils from the leaves and stems of three populations of P. cablin from China, of which two had pogostone as dominant ingredient; highest, or lowest and highest concentration given (ref. 54); [y5] one lab-hydrodistilled patchouli oil from China; only qualitative data provided (ref. 56); [y6] one commercial patchouli oil sample from Indonesia (ref. 55); [z] eight lab-hydrodistilled oils from leaves of patchouli plants cultivated in Brazil and harvested every three hours for 24 hours (ref. 31); [z2] one sample of patchouli oil from Indonesia (ref. 6); [z3] one commercial oil from China (ref. 20); [z4] three commercial patchouli oils from China, India and Indonesia; the compounds indicated with +[z4] were present in one, two or all three oils, but their quantities were indicated in such a manner that they cannot be compared to the other data (ref. 2); [z5] three commercial patchouli oils from China, India and Indonesia (ref. 2); [z6] five oils obtained from leaves of the same samples of Indonesian patchouli plants with steam distillation, but with distillation times varying from 2 to 10 hours (ref. 21); [z7] two lab-hydrodistilled oils from Malaysia and two commercial oils from Indonesia and Malaysia; the quantity of the compounds found in the oils cannot be compared with the other data; only the chemicals not or rarely found in other studies are mentioned here and indicated with +[z7] (ref. 39); [z8] one commercial patchouli oil sample obtained from a Brazilian company (ref. 40)

tr: trace; + present in the oil investigated, but quantity not stated; +[z4] see under [z4]; +[z7] see under [z7]

Table 5.67.3 Major constituents of patchouli oils

Constituent	CAS	Percentage and range				
		A	B (25)	C (24)	D (1)	E
Patchouli alcohol (patchoulol)	5986-55-0	22.3-33.9	33.4	47.3-69.0	61.6	32.6[x6]; 42.7[s]; 42.9-64.3[y2]; 44.5[z3]; 47.8[d]; 54.3[v]; 54.9[x8]; 60.3[u]; 62.4[x2]; 70.7[z]; 71.8[x1]
α-Bulnesene Guaiene	3691-11-0 88-84-6	14.8-18.9	19.4	5.0-13.0	8.7	21.4[x3]; 22.0[m]; 22.4[f]; 23.1[x6]; 23.3[r]; 25.0[l] 18.8[x]; 23.6[m]
α-Guaiene	3691-12-1	12.4-15.6	12.8	3.7-9.5	10.1	15.9[x6]; 18.2[c]; 19.8[g]; 20.0[r]; 20.6[t]; 23.3[x3]
β-Patchoulene	514-51-2	2.0-4.1		1.7-3.8	1.0	6.5[s]; 7.0[y3]; 7.5[x2]; 9.3[k,y1]; 12.1[t]; 12.9[n]; 16.2[m]
α-Patchoulene	560-32-7	4.7-8.0	6.8	1.7-3.9	1.6	7.5[g]; 7.9[j,k]; 8.0[e]; 8.6[y3]; 9.2[q]; 10.5[t]; 12.3[s]
γ-Patchoulene	508-55-4	5.9-8.0			1.3	3.9[q]; 5.5[q]; 6.3[x3]; 6.7[k]; 8.5[x4]; 9.4[x2]; 11.7[n]
Seychellene	20085-93-2	3.2-9.4	8.2	3.0-6.5	3.1	7.4[j]; 7.5[e]; 7.9[o]; 8.1[y6]; 8.5[v]; 8.9[j]; 9.5[d]; 9.6[z8]
δ-Patchoulene	53823-16-8	1.7-7.5				0.7[y3]; 8.0[r]
Pogostol	21698-41-9	1.6-4.5	2.6	3.3-5.1	0.5	0.4[d]; 1.0[l]; 2.0[x4]; 2.4[e,y1]; 3.1[o]; 3.3[z5]; 5.1[x1]

LEGEND: SEE UNDER TABLE 5.67.2

or rare constituents of patchouli oils found in high concentrations (>7%) in single studies include pogostone (two studies: 27.7% and 38.3%), γ-guaiene (20.5%), germacrene B (19.0%), longifolene (18.4%), azulene (15.0%), *cis*-β-guaiene (14.4%), methyl benzoate (14.3%), β-guaiene (14.2%), β-humulene (13.2%), isovanillic acid (12.1%), germacrene A (11.7%), α-gurjunene (11.3%) and α-panasinsene (9.6%).

Commercial oils

The ten chemicals that had the highest maximum concentrations in 52 commercial patchouli essential oil samples (concentration ranges provided) are the following: patchouli alcohol (patchoulol) (22.3-33.9%), α-bulnesene (14.8-18.9%), α-guaiene (12.4-15.6%), seychellene (3.2-9.4%), α-patchoulene (4.7-8.0%), γ-patchoulene (5.9-8.0%), δ-patchoulene (1.7-7.5%), β-caryophyllene (3.0-4.7%), pogostol (1.6-4.5%) and β-patchoulene (2.0-4.1%) (Erich Schmidt, unpublished analytical data).

Chemotypes

In virtually all analyses of commercial and lab-prepared oils, patchouli alcohol (patchoulol) was the major component. In certain parts of China, however, a 'pogostone type' patchouli oil is found, in which pogostone is the dominant ingredient, with concentrations of up to 38% (54,57). The existence of this pogostone chemotype patchouli oil seems to be corroborated by DNA profiling of *P. cablin* taxa from China (58).

For older literature not reviewed here see refs. 36 and 37.

Table 5.67.4 Results of testing groups of patients with patchouli oil

Years and Country	Test conc. & vehicle	Number of patients tested	positive (%)	Selection of patients (S); Relevance (R); Comments (C)	Ref
Routine testing					
2000-2008 IVDK	10% pet.	2,446 14	(0.6%)	R: not stated	64
1999-2000 Denmark	10% pet.	318 3	(0.9%)	R: not specified; C: this study was part of the international study mentioned below (ref. 65)	70
1998-2000 six European countries	10% pet.	1,606 13	(0.8%)	R: not specified for individual oils/chemicals	65
Testing in groups of selected patients					
2000-2008 IVDK	10% pet.	828 12	(1.4%)	S: patients with dermatitis suspected of causal exposure to fragrances; R: not stated	64
<1996 Japan, Ireland, USA, UK, Switzerland, Sweden	10% pet.	167 5	(3.0%)	S: patients known or suspected to be allergic to fragrances; R: not stated	59
<1986 France	3% pet.	21 3	(14%)	S: patients with dermatitis caused by fragrances; R: not stated	69
1971-1980 Japan	20% pet.	477 2	(0.4%)	S: patients with dermatoses other than pigmented cosmetic dermatitis and volunteers; R: not stated	62
<1974 Japan	?	183 11	(6.0%)	S: patients suspected of cosmetic dermatitis; R: unknown	66

IVDK Information Network of Departments of Dermatology, Germany, Switzerland, Austria (www.ivdk.org); pet.: petrolatum

CONTACT ALLERGY/ALLERGIC CONTACT DERMATITIS

General

Contact allergy to/allergic contact dermatitis from patchouli oil has been reported in more than 10 publications. Pigmented cosmetic dermatitis from essential oils including patchouli oil used to be frequent in Japanese women.

In groups of consecutive patients suspected of contact dermatitis, prevalence rates of up to 0.9% positive patch test reactions have been observed, but relevance data are lacking. Of four patients with allergic contact dermatitis from patchouli oil, three had occupational allergic contact dermatitis.

Testing in groups of patients

The results of patch tests with patchouli oil in routine testing (consecutive patients suspected of contact dermatitis) and in groups of selected patients are shown in Table 5.67.4. In routine testing, rates of positive reactions ranged from 0.6% to 0.9%, whereas between 0.4% and 14% of patients in selected groups had positive patch tests.

Case reports

A massage therapist had occupational contact dermatitis from allergies to various essential oils, including patchouli oil (60). Two patients, bottle fillers in a perfume factory, developed occupational contact allergy to patchouli oil and various other essential oils they worked with; in this study, there was also one positive patch test reaction among 20 workers in the same factory who did not have dermatitis (1). One patient developed allergic contact dermatitis from the application of pure patchouli oil to the skin behind the ears and in the axillae (68).

Pigmented cosmetic dermatitis

In Japan, in the 1960s and 1970s, many female patients developed facial pigmentation following dermatitis of the face (63). This so-called pigmented cosmetic dermatitis was shown to be caused by contact allergy to components of cosmetic products, notably essential oils, other fragrance materials, antimicrobials, preservatives and coloring materials (62,63). In a group of 620 Japanese patients with this condition investigated between 1970 and 1980, 1-3% had positive patch test reactions to patchouli oil 20% in petrolatum (62). The number of patients with pigmented cosmetic dermatitis decreased strongly after 1978, when major cosmetic companies began to eliminate strong contact sensitizers from their products (63).

LITERATURE

1 Verma RS, Padalia RC, Chauhan A. Assessment of similarities and dissimilarities in the essential oils of patchouli and Indian Valerian. J Essent Oil Res 2012;24:487-491

2 Cornwell CP. Notes on the composition of patchouli oil (*Pogostemon cablin* (Blanco) Benth.). J Essent Oil Res 2010;22:360-364

3 Sundaresan V, Singh SP, Mishra AN, Shasany AK, Darokar MP, Kalra A, et al. Composition and comparison of essential oils of *Pogostemon cablin* (Blanco) Benth. (patchouli) and *Pogostemon travancoricus* Bedd. var. *travancoricus*. J Essent Oil Res 2009;21:220-222

4 Rekha K, Bhan MK, Dhar AK. Development of erect plant mutant with improved patchouli alcohol in patchouli [*Pogostemon cablin* (Blanco) Benth.]. J Essent Oil Res 2009;21:135-137

5 Wei A, Shibamoto T. Antioxidant activities and volatile constituents of various essential oils. J Agric Food Chem 2007;55:1737-1742

6 Buré CM, Sellier NM. Analysis of the essential oil of Indonesian patchouli (*Pogostemon cablin* Benth.) using GC/MS (EI/CI). J Essent Oil Res 2004;16:17-19

7 Rakotonirainy O, Gaydou EM, Faure R, Bombarda I. Sesquiterpenes from patchouli (*Pogostemon cablin*) essential oil. Assignment of the proton and carbon-13 NMR spectra. J Essent Oil Res 1997;9:321-327

8 Dũng NX, Leclercq PA, Thai TH, Moi LD. Chemical composition of patchouli oil from Vietnam. J Essent Oil Res 1989;1:99-100

9 Lawrence BM. Progress in essential oils. Perfum Flavor 2007;32(1):48-56 (patchouli oil 52-55)

10 Zhang J-W, Lan W-J, Su J-G, Zeng L-M, Yang D-P, Wang F-S. Chemical constituents and their antifungal and antibacterial activities of essential oil of *Pogostemon cablin* II. Zhongcaoyao 2002;33:210-212. Data cited in ref. 9

11 Kubeczka K-H, Formacek V. Essential oils analysis by capillary gas chromatography and carbon-13 NMR spectroscopy, 2nd Ed. New York, USA: John Wiley and Sons, 2002:233-238. Data cited in ref. 9

12 Narayanan P. Comparison of quality of essential oils from South India. FAFAI J 2003;5:47-56. Data cited in ref. 9

13 Milchard MJ, Clery R, DaCosta N, Flowerdew M, Gates L, Moss N, et al. Application of gas-liquid chromatography to the analysis of essential oils. Fingerprints of 12 essential oils. Perfum Flavor 2004;29(5):28-36. Data cited in ref. 9

14 Lawrence BM. Progress in essential oils. Perfum Flavor 2002;27(3):48-69

15 Mookherjee BD, Light KK, Hill ID. A study on the odor-structure relationship of patchouli compounds. In: BD Mookherjee and CJ Mussinan, Eds. Essential oils. Wheaton, USA: Allured Publishing, 1980:247-272. Data cited in ref. 14

16 Yang D, Michel D, Mandin D, Andriamboavonjy H, Poitry P, Chaumont J-P, et al. Antifungal and antibacterial properties *in vitro* of three patchouli oils from different origins. Acta Bot Gallica 1996;143:29-35

17 Gokulakrishnan J, Kuppusamy E, Shanmugam D, Appavu A, Kaliyamoorthi K. Pupicidal and repellent activities of *Pogostemon cablin* essential oil chemical compounds against medically important human vector mosquitoes. Asian Pac J Trop Dis 2013;3:26-31

18 Chakrapani P, Venkatesh K, Sekhar C, Singh B, Jyothi A, Kumar BP, et al. Phytochemical, pharmacological importance of patchouli (*Pogostemon cablin* (Blanco) Benth) an aromatic medicinal plant. Int J Pharm Sci Rev Res 2013;21:7-15

19 Ramya HG, Palanimuthu V, Rachna S. An introduction to patchouli (*Pogostemon cablin* Benth.) – A medicinal and aromatic plant: It's importance to mankind. Agric Eng Int: CIGR Journal 2013;15:243-250

20 Kocevski D, Du M, Kan J, Jing C, Lacanin I, Pavlović H. Antifungal effect of *Allium tuberosum*, *Cinnamomum cassia*, and *Pogostemon cablin* essential oils and their components against population of *Aspergillus* species. J Food Sci 2013;78:M731-M737

21 Yahya A, Yunus RM. Influence of sample preparation and extraction time on chemical composition of steam distillation derived patchouli oil. Procedia Engineering 2013;53:1-6

22 Albuquerque ELD, Lima JKA, Souza FHO, Silva IMA, Santos AA, Araújo APA, et al. Insecticidal and repellence activity of the essential oil of *Pogostemon cablin* against urban ants species. Acta Tropica 2013;127:181-186

23 Pujiarti R, Ohtani Y, Widowati TB, Kasmudjo W, Herath NK, Wang CN. Effect of *Melaleuca leucodendron*, *Cananga odorata* and *Pogostemon cablin* oil odors on human physiological responses. Wood Res J 2012;3:100-105

24 Blank AF, Pergentino Sant'ana PC, Santana Santos P, Arrigoni-Blank MF, do Nascimento Prata AP, Ramos Jesus HC, Barreto Alves P. Chemical characterization of the essential oil from patchouli accessions harvested over four seasons. Ind Crops Prod 2011;34:831-837

25 Zhang L-G, Zhang C, Ni L-J, Yang Y-J, Wang C-M. Rectification extraction of Chinese herbs' volatile oils and comparison with conventional steam distillation. Separ Purific Technol 2011;77:261-268

26 Machial CM, Shikano I, Smirle M, Bradbury R, Isman MB. Evaluation of the toxicity of 17 essential oils against *Choristoneura rosaceana* (Lepidoptera: Tortricidae) and *Trichoplusia ni* (Lepidoptera: Noctuidae). Pest Manag Sci 2010;66:1116-1121

27 Donelian A, Carlson LHC, Lopes TJ, Machado RAF. Comparison of extraction of patchouli (*Pogostemon cablin*) essential oil with supercritical CO2 and by steam distillation. J Supercrit Fluids 2009;48:15-20

28 Pavela R. Insecticidal properties of several essential oils on the house fly (*Musca domestica* L.). Phytother Res 2008;22:274-278

29 Tsai Y-C, Hsu H-C, Yang W-C, Tsai W-J, Chen C-C, Watanabe T. α-Bulnesene, a PAF inhibitor isolated from the essential oil of *Pogostemon cablin*. Fitoterapia 2007;78:7-11

30 Bunrathep S, Lockwood GB, Songsak T, Ruangrungsi N. Chemical constituents from leaves and cell cultures of *Pogostemon cablin* and use of precursor feeding to improve patchouli alcohol level. ScienceAsia 2006;32:293-296

31 Silva MAS, Ehlert PAD, Ming LC, Marques MOM. Composition and chemical variation during daytime of constituents of the essential oil of *Pogostemon patchouli* pellet leaves. In: LE Craker et al., Eds. Proc. XXVI IHC – Future for Medicinal and Aromatic Plants. Acta Hort 629, ISHS 2004, Publication supported by Can Int Dev Agency (CIDA).

32 Singh M, Sharma S, Ramesh S. Herbage Oil yield and oil quality of patchouli [*Pogostemon cablin* (Blanco) Benth.] influenced by irrigation, organic mulch and nitrogen application in semi-arid tropical climate. Ind Crops Prod 2002;16:101-107

33 Hasegawa Y, Tajima K, Toi N. An additional constituent occurring in the oil from a patchouli cultivar. Flavour Fragr J 1992;7:333-335

34 Sugimura Y, Ichikawa Y, Otsuji K, Fujita M, Toi N, Kamata N, et al. Cultivarietal comparison of patchouli plants in relation to essential oil production and quality. Flavour Fragr J 1990;5:109-114

35 Nikiforov A, Jirovetz L, Buchbauer G, Raverdino V. GC-FFIR and GC-MS in odour analysis of essential oils. Mikrochim Acta (Vienna) 1988;II:193-198

36 Lawrence BM. Progress in essential oils. Perfum Flavor 1990;15(2):75-79 (possibly 76-77)

37 Lawrence BM. Progress in essential oils. Perfum Flavor 1981;6(4):73-76 (possibly 1981;6(5):73-76)

38 Betts TJ. Evaluation of a "Chirasil-Val" capillary for the gas chromatography of volatile oil constituents, including sesquiterpenes in patchouli oil. J Chromatogr A 1994;664:295-300

39 Hussin NI, Mondello L, Costa R, Dugo P, Yusoff NI, Yarmo MA, et al. Quantitative and physical evaluation of patchouli essential oils obtained from different sources of *Pogostemon cablin*. Nat Prod Commun 2012;7:927-930

40 Murbach Teles Andrade BF, Nunes Barbosa L, da Silva Probst I, Fernandes Júnior A. Antimicrobial activity of essential oils. J Essent Oil Res 2014;26:34-40

41 Lawrence BM. Progress in essential oils. Perfum Flavor 2014;39(July):46-52

42 Wossa SW, Rali T, Leach DN. Volatile chemical constituents of patchouli (*Pogostemon cablin* [Blanco]) Benth.: Labiatae from three localities in Papua New Guinea. Papua New Guinea J Agric Forest Fish 2006;49:49-54. Data cited in ref. 41

43 Sharma TC. Patchouli—an aromatic oil crop for development of perfumery industry in Northeast India. In: L Jirovetz, NX Dung and VK Varshey, Eds. Aromatic plants from Asia, their chemistry and application in food and therapy. Dehradun, India: Har Krishan Bhalla & Sons, 2007:187-201. Data cited in ref. 41

44 Prakash O, Joshi S, Shukla AK, Pant AK. Sesquiterpenoid rich essential oil from the leaves of *Pogostemon patchouli* Pellet grown organically under Tarai conditions. J Essent Oil Bear Plants 2007;10:157-161

45 Ranade GS. Profile patchouli oil. FAFAI 2007;9(3):99. Data cited in ref. 41

46 Williams DG. The chemistry of essential oils, 2nd Ed. Port Washington, NY, USA: Micelle Press, 2008:198-199. Data cited in ref. 41

47 Aisyah Y, Hastuti P, Sastrohamidjojo H, Hidayat C. Chemical composition and antibacterial properties of the essential oil of *Pogostemon cablin*. Majalah Farm Indonesia 2008;19(3):151-156. Data cited in ref. 41

48 Ramachandra KM, Farooqi AA, Srinivasappa KN. Effect of gamma rays on growth, yield and oil composition in patchouli (*Pogostemon patchouli* Pellet.). Indian Perfum 2008;52(2):58-60. Data cited in ref. 41

49 Rana VS, Blazquez MA. Volatile leaf essential oil commercial *Pogostemon cablin* Benth. cultivated in Manipur. Indian Perfum 2009;53(1):30-32. Data cited in ref. 41

50 Kumar A. Effect of soaking time of patchouli leaves on oil yield and chemical composition. Indian Perfum 2009;53(4):27-28. Data cited in ref. 41

51 Milchard MJ, Clery R, Esdale R, Gates L, Judge F, Moss N, et al. Application of gas-liquid chromatography to the analysis of essential oils. Perfum Flavor 2010;35(5):34-42

52 Wu L-H, Wu Y-G, Guo Q-S, Li S-P, Zhou K-B, Zhang J-F. Comparison of genetic diversity in *Pogostemon cablin* from China revealed by RAPD, morphological and chemical analysis. J Med Plants Res 2011;5:4549-4559. Data cited in ref. 41

53 He J-J, Chen H-M, Li C-W, Wu D-W, Wu X-L, Shi S-J, et al. Experimental study on antinociceptive and anti-allergy effects of patchouli oil. J Essent Oil Res 2013;25:488-496

54 Wu Y, Li C, Li X, Yuan M, Hu X. Comparison of the essential oil compositions between *Pogostemon cablin* and *Agatache rugosa* used as herbs. J Essent Oil Bear Plants 2013;16:705-713

55 Rolli E, Marieschi M, Maietti S, Sacchetti M, Bruni R. Comparative phytotoxicity of 25 essential oils on pre- and post-emergence development of Solanum lycopersicum L.: A multivariate approach. Ind Crops Prod 2014;60:280-290

56 Yang X, Zhang X, Yangand S-P, Liu W-Q. Evaluation of the antibacterial activity of patchouli oil. Iran J Pharm Res 2013;12:307-316

57 Luo JP, Liu YP, Feng YF, Guo XL, Cao H. Two chemotypes of *Pogostemon cablin* and influence of region of cultivation and harvesting time on volatile oil composition. Acta Pharmaceutica Sinica [Yao Xue Xue Bao] 2003;38(4):307-310 (in Chinese, data obtained from English Abstract)

58 Liu YP, Luo JP, Feng YF, Guo XL, Cao H. DNA profiling of *Pogostemon cablin* chemotypes differing in essential oil composition. Acta Pharmaceutica Sinica [Yao Xue Xue Bao] 2002;37(4):304-308 (in Chinese, data obtained from English Abstract)

59 Larsen W, Nakayama H, Lindberg M, Fischer T, Elsner P, Burrows D, et al. Fragrance contact dermatitis: A worldwide multicenter investigation (Part 1). Am J Cont Derm 1996;7:77-83

60 Cockayne SE, Gawkrodger DJ. Occupational contact dermatitis in an aromatherapist. Contact Dermatitis 1997;37:306-307

61 Schubert HJ. Skin diseases in workers at a perfume factory. Contact Dermatitis 2006;55:81-83

62 Nakayama H, Matsuo S, Hayakawa K, Takhashi K, Shigematsu T, Ota S. Pigmented cosmetic dermatitis. Int J Dermatol 1984;23:299-305

63 Nakayama H, Harada R, Toda M. Pigmented cosmetic dermatitis. Int J Dermatol 1976;15:673-675

64 Uter W, Schmidt E, Geier J, Lessmann H, Schnuch A, Frosch P. Contact allergy to essential oils: current patch test results (2000–2008) from the Information Network of Departments of Dermatology (IVDK). Contact Dermatitis 2010;63:277-283

65 Frosch PJ, Johansen JD, Menné T, Pirker C, Rastogi SC, Andersen KE, et al. Further important sensitizers in patients sensitive to fragrances. II. Reactivity to essential oils. Contact Dermatitis 2002;47:279-287

66 Nakayama H, Hanaoka H, Ohshiro A. Allergen Controlled System (ACS). Tokyo, Japan: Kanehara Shuppan, 1974:42. Data cited in ref. 9

67 Mitchell JC. Contact hypersensitivity to some perfume materials. Contact Dermatitis 1975;1:197-199

68 Hausen BM, Kunze B. Kontaktallergie auf Patchouli-Öl. Akt Dermatol 1991;17:199-202

69 Meynadier JM, Meynadier J, Peyron JL, Peyron L. Formes cliniques des manifestations cutanées d'allergie aux parfums. Ann Dermatol Venereol 1986;113:31-39

70 Paulsen E, Andersen KE. Colophonium and Compositae mix as markers of fragrance allergy: Cross-reactivity between fragrance terpenes, colophonium and Compositae plant extracts. Contact Dermatitis 2005;53:285-291

Chapter 5.68   PEPPERMINT OIL

DEFINITION
Peppermint oil (essential oil of peppermint) is the essential oil obtained from the flowering aerial parts and leaves of the peppermint, *Mentha x piperita* L.

INCI NOMENCLATURE
Description/definition: Mentha piperita oil is the volatile oil obtained from the whole plant of the peppermint, *Mentha piperita* (L.), *Labiatae*
INCI name EU: Mentha piperita oil
INCI name USA: Mentha piperita (peppermint) oil
CAS registry number(s): 8006-90-4; 84082-70-2
EINECS number(s): 282-015-4

ISO (INTERNATIONAL ORGANIZATION FOR STANDARDIZATION) STANDARD
ISO number: 856
ISO name: Essential oil of peppermint
Botanical origin: *Mentha x piperita* L.
Parts of plant used: Aerial parts, leaf
ISO values: ISO values (minimum and maximum concentrations) are shown in Table 5.68.1.

Other *Mentha* species from which mint oils are obtained and which are mentioned in CosIng include *Mentha spicata* L. (spearmint oil, Chapter 5.79), *Mentha arvensis* (corn mint), *Mentha aquatica* (water mint), *Mentha viridis* (synonym of *Mentha spicata*), *Mentha pulegium* (pennyroyal) and *Mentha citrata* (bergamot mint).

THE PLANT, THE OIL, AND THEIR USES
Mentha x piperita L. is an herbaceous rhizomatous perennial plant growing to 30-90 cm tall. It is a sterile hybrid cross between *Mentha aquatica* and *Mentha spicata* (1,15,24). The peppermint is naturalized in the Azores, Siberia, Australia, New Zealand, all across Europe except the Scandinavian countries, Canada and the USA. The plant is cultivated in large parts of Europe, USA, Canada, Chile, Argentina, Australia, some African countries, Brazil and Japan (GRIN Taxonomy for Plants; Mansfeld's World Database of Agriculture and Horticultural Crops, http://mansfeld.ipk-gatersleben.de).

Mentha piperita is widely used as medicinal plant. The herb has a strong, pepper-like, pungent odor, hence the specific name 'piperita' (16). Peppermint is commonly used to soothe or treat symptoms of the gastrointestinal tract such as nausea, vomiting, abdominal pain, indigestion, irritable bowel, and bloating (1). The plant is described to have a wide range of properties including anti-oxidant, antitumor, anti-allergic, adstringent, antiseptic, antipruritic, anti-catarrhal, carminative, anti-inflammatory, anti-spasmodic, diaphoretic, anti-emetic, nervine, antimicrobial, analgesic, stimulant, stomachic and rubefacient (1,2,16). Peppermint has a high menthol content, and is often used as a flavoring agent in tea, ice cream, confectionery, chewing gum and toothpaste (1).

The essential oil of peppermint, obtained by steam-distillation of the leaves, has many pharmaceutical applications (2,24,49). It has vasoconstrictive and cooling properties and is one of the main oils used as an external application for relieving muscle spasms, pain, neuralgia, headache and tooth-ache (15,49). Furthermore, it is often used orally for relieving stomach upset, nausea and other intestinal disorders (2,15). Sometimes peppermint oil may be used in steam inhalation or by other routes to relieve the symptoms of a cold (2,24). The European Medicines Agency recently reviewed the pharmacological and clinical literature of peppermint oil and considered two indications as proven and well established: minor spasms of the gastrointestinal tract, flatulence and abdominal pain, especially in patients with irritable bowel syndrome (oral use) and mild tension type headache (cutaneous use) (23). Peppermint oil is also widely employed for flavoring chewing gum, cough drops, sweets, alcoholic drinks, toothpaste, mouth freshener and is also used for perfumes, other cosmetic products and in the tobacco industry (5,15,16,49). It is also a popular oil in aromatherapy (2,81).

Peppermint has been thoroughly reviewed in references 18 and 19. The biological properties of menthol, the main component of peppermint oil, are discussed in ref. 33.

CHEMICAL COMPOSITION
Peppermint oil is a colorless to pale greenish-yellow, clear mobile liquid which has a fresh, minty, cooling, green and sweetish odor, with variations depending on its origin. The yield of essential oil from the flowering aerial tops and leaves of *Mentha x piperita* L. generally varies from 0.3 to 0.9 per cent. The main producing countries of this oil are USA, India, Russia, China, France, Italy, Bulgaria, Japan, Spain, Morocco, England, Poland, Canada, Hungary and Australia.

Literature data (up to September 21, 2014) on the chemical composition of peppermint oils and unpublished

Table 5.68.1 ISO values (%) for peppermint oil[a]

Compound	CAS	Minimum	Maximum
Menthol	89-78-1	32.0	49.0
Menthone (*p-*)	89-80-5	13.0	28.0
1,8-Cineole	470-82-6	3.0	8.0
Isomenthone	491-07-6	2.0	8.0
Menthyl acetate	16409-45-3	2.0	8.0
Menthofuran	494-90-6	1.0	8.0
Neomenthol	3623-51-6	2.0	6.0
β-Caryophyllene	87-44-5	1.0	3.5
Limonene	138-86-3	1.0	3.0
Pulegone	89-82-7	0.5	3.0
trans-Sabinene hydrate	17699-16-0	0.5	2.3
3-Octanol	589-98-0	0.1	0.5

[a] ISO 856 Essential oil of peppermint ©ISO 2006; Geneva, Switzerland, www.iso.org

Table 5.68.2 Constituents identified in peppermint oils

Constituent	CAS	Percentage and range				
		A	B (3)	C (16)	D (49)	E
Acetaldehyde	75-07-0					+[b]
Acetic acid	64-19-7					+[b]
Acetophenone	98-86-2					+[b]
p-Acetylanisole	100-06-1					0.2[y4]
Amorphene						0.5[l]
Amyl alcohol	71-41-0					+[b]
Amyl isovalerate	25415-62-7		0.1-0.4			0.2[w4]
(E)-Anethole	4180-23-8	0-0.2				0.1[h]
p-Anisaldehyde	123-11-5					+[b]
Aromadendrene	489-39-4				0.1-0.2	0.2[k]; 0.7[l]; 1.8[w]; 10.2[o]
allo-Aromadendrene	25246-27-9					+[b,p]
Benzaldehyde	100-52-7					+[b]
Benzyl alcohol	100-51-6					+[b]
α-Bergamotene	17699-05-7					tr[w4]
Bicycloelemene	32531-56-9					0.6[w7]
Bicyclogermacrene	24703-35-3	0-0.7			0-0.1	0.08[p]; 0.1[w1,w3]; 0.2[c,m,y3]; 1.3[r]
epi-Bicyclosesqui-phellandrene	54274-73-6					1.7[w4]
β-Bisabolene	495-61-4				tr-0.7	
Borneol	507-70-0					0.01[y9]; 0.05[w2]
Bornyl acetate	76-49-3					tr[w5]; 0.02[j]; 1.1[x5]
endo-1-Bourbonalol						+[b]
α-Bourbonene	5208-58-2		0.1-0.4			0.5[e]
1,5-di-epi-α-Bourbonene						0.2[c]
β-Bourbonene	5208-59-3	0.02-0.5		2.5		0.4[r]; 0.5[z6]; 0.6[f,k]; 0.7[w7]; 1.2[y7]; 1.8[l]; 2.6[o]
β-Bourbonene isomer						+[b]
2-Bromocyclohexanol	2425-33-4					0.08[d]
Butanal	123-72-8					+[b]
2-Butanol	78-92-2					+[b]
Butyl isovalerate	109-19-3					+[b]
Butyl 2-methylbutyrate	15706-73-7		0-0.1			
α-Cadinene	24406-05-1				0-tr	0.1[z6]; 0.3[l]
γ-Cadinene	39029-41-9		tr-0.2			+[b]; tr[w4]; 0.8[z6]; 3.7[w5]
δ-Cadinene	483-76-1	0.01-0.2	0.1		0-0.1	tr[w5]; 0.1[p]; 0.2[d,k]; 0.5[o,y4]; 0.8[r]; 1.2[l]
α-Cadinol	481-34-5		tr-0.1		tr-0.1	0.1[f,w3]; 0.2[p]; 0.3[z6]
epi-α-Cadinol	5937-11-1				0-0.2	0.2[p]; 0.5[o,w5]
δ-Cadinol	19435-97-3		tr-0.1			
α-Calacorene	21391-99-1					0.08[l]
trans-Calamenene	73209-42-4					0.2[z7]
Camphene	79-92-5	tr-0.6	0-tr		0-tr	tr[h,k]; 0.02[w1]; 0.3[e]; 1.1[y9]; 1.2[y7]
Camphor	76-22-2		0-tr		0-tr	0.4[w2]
4-Carene	29050-33-7					0.05[l]
δ3-Carene	13466-78-9	0-0.06			0-0.3	tr[w4]; 0.4[e,y6]
Carvacrol	499-75-2				0.1-0.2	
Carveol	99-48-9					0.2[e]
(E)-Carveol	1197-07-5		tr-0.1		tr-0.2	14.5[y7]
(Z)-Carveol	1197-06-4		0-tr		tr-0.1	+[b]; 0.2[l]
Carvone	99-49-0	0.03-0.9	0-0.2	0.5		0.8[w3]; 0.9[l]; 1.7[t]; 3.1[z6]; 5.0[y4]; 23.4[v4]
D-Carvone	2244-16-8					0.1[c]
Carvone oxide	33204-74-9				0-0.2	
Carvotanacetone	499-71-8					2.3[z7]
cis-Carvyl acetate	1205-42-1		0-0.1		0-0.1	
(E)-Carvyl formate	29239-07-4					+[b]
(Z)-Carvyl formate						+[b]
β-Caryophyllene	87-44-5	0.1-5.2	1.2-2.4	4.9	0.1-1.3	4.2[x3]; 4.3[z6]; 4.9[v1]; 6.6[w7]; 7.6[y7]; 19.2[x]
Caryophyllene oxide	1139-30-6	0-0.2	0.1-0.2	0.3	tr-0.3	0.3[d,w3]; 0.5[z6]; 0.8[f]; 1.0[l]; 1.2[x1]; 1.7[x5]
α-Cedrol	77-53-2					0.2[l]
Chromene	254-04-6					0.6[o]
1,8-Cineole	470-82-6	0.3-9.9	0-4.8	8.9	0.5-6.5	0-8.9[v]; 1.0-13.5[q]; 6.0[n]; 6.2[x6]; 6.4[x1]; 6.6[c]; 7.0[u]; 7.5[s]; 8.8[z3]; 9.6[y5]; 11.6[m]; 13.9[v1]

Table 5.68.2 Constituents identified in peppermint oils *(continued)*

Constituent	CAS	Percentage and range				
		A	B (3)	C (16)	D (49)	E
Citronellol	106-22-9					+[b]
Citronellyl acetate	150-84-5				tr-1.7	+[b]
α-Copaene	3856-25-5				0-0.1	+[b]; tr[w4]; 0.1[z6]; 0.4[l]
β-Copaene	18252-44-3					0.1[f]
α-Cubebene	17699-14-8			tr	0-tr	+[b]; tr[w5]
β-Cubebene	13744-15-5					+[b]; tr[w4]; 0.5[w7]; 0.6[l]; 1.1[o]
Cubenol	21284-22-0					0.2[e]
Cuparene	16982-00-6					0.2[l]
m-Cymene	535-77-3					0.6[y6]
o-Cymene	527-84-4					0.6[e]
p-Cymene	99-87-6	0-07-0.6	0.1	0.3		0.08-1.0[v]; 0.5[l]; 0.6[k]; 0.7[d]; 0.8[v1]; 0.9[f]
p-Cymen-8-ol	1197-01-9					0.1[e]
(*E*)-β-Damascenone	23726-93-4					+[b]
3,4-Didehydro-7,8-dihydro-γ-ionone						+[b]
2,5-Diethyltetrahydrofuran	41239-48-9	0-0.03				+[b]; tr[w4]
Dihydrocarvone	5948-04-9					0.2[z7]
cis-Dihydrocarvone	3792-53-8		0.5-0.8			
Dihydrocarvyl acetate	57287-13-5		0-0.1			
Dihydroedulan						0.06[p]; 0.1[c]; 0.4[w3]
Dihydroedulan I	74006-61-4		0-0.1			+[y1]
trans-2,6-Dimethyl-3,5,7-octatrien-2-ol			0.1			
Dimethyl sulfide	75-18-3					+[b]
α-Elemene	5951-67-7					1.1[w]
β-Elemene	33880-83-0	0.02-0.3	0.1-0.2		0.1-0.2	+[p]; tr[w5]; 0.1[h,y8]; 0.2[w6]; 0.3[w]; 0.5[z6]; 1.0[l]
γ-Elemene	29873-99-2		0.3-0.6			0.3[w5]
Elemol	639-99-6					0.06[l]
Epiglobulol	88728-58-9		0.6-1.1			0.3[d]; 0.8[e]
1,5-Epoxysalvial-4(14)-ene	88395-47-5					+[b]
Ethanol	64-17-5					+[b]
2-Ethylfuran	3208-16-0					+[b]; tr[w4]
Ethyl isovalerate	108-64-5					+[b]
Ethyl-2-methyl butyrate	7452-79-1					+[b]
β-Eudesmol	473-15-4			0.3		0.2[l]
γ-Eudesmol (selinenol)	1209-71-8				0-0.1	
Eugenol	97-53-0					+[b]; 0.1[w4]; 0.4[y4]
β-Farnesene	502-60-3					0.4[w4]; 0.9[d]
(*E*)-β-Farnesene	18794-84-8	0.06-0.6	0.2-0.4	0.5		0.1[p]; 0.2[h,w1]; 0.3[y3]; 0.5[y4]; 1.2[w7]
(*Z*)-β-Farnesene	28973-97-9					0.2[c]; 0.4[w6]; 0.7[r]
Furfural(dehyde)	98-01-1					+[b]
Geranial	141-27-5				0-0.1	+[b]
Geraniol	106-24-1				0-0.4	0.05[w2]; 3.2[v5];
Geranyl acetate	105-87-3		tr-0.1			0.1[y7]
Geranyl formate	105-86-2					0.6[e]
Germacrene A	28387-44-2					0.4[w6]; 0.5[r]
Germacrene B	15423-57-1					0.3[z2]; 0.4[n]
Germacrene D	23986-74-5	tr-2.7	1.2-3.3	1.7	0-0.6	2.4[z2]; 2.5[v1]; 2.6[n]; 3.0[u]; 3.1[z6]; 6.2[w7]; 9.7[o]
Germacrene D-4-ol	198991-79-6			6.3		
Globulol	489-41-8	0-0.1				0.2[c]; 0.3[f]; 0.4[z5]
α-Gurjunene	489-40-7					+[p]; 0.1[y7]; 8.9[x7]
β-Gurjunene	73464-47-8					+[b]; 0.1[w5]
γ-Gurjunene	22567-17-5					1.4[x7]
Heneicosane (*n*-)	629-94-7				0-0.5	
n-Heptadecane	629-78-7				0-0.1	
Heptadecen-2-one					0-tr	
Heptanal	111-71-7					+[b]
1-Heptanol	111-70-6	0-0.1			0-tr	
3-Heptanol	589-82-2					+[b]; tr[w4]

Table 5.68.2 Constituents identified in peppermint oils (continued)

Constituent	CAS	Percentage and range				
		A	B (3)	C (16)	D (49)	E
Heptan-3-yl acetate						+[b]
Hexadecanal	629-80-1				0-0.1	
Hexadecane	544-76-3					+[y1]
Hexadecanoic acid	57-10-3				0-0.1	
cis-9-Hexadecenoic acid	373-49-9				0-0.1	
Hexadecen-1-ol	37822-83-6				0-0.4	
Hexadecyl acetate	629-70-9				0-0.1	
Hexanal	66-25-1					+[b]
Hexanol (1-; n-)	111-27-3	0-0.04				+[b]; tr[z5]
3-Hexanol	623-37-0					+[b]
(Z)-3-Hexanol					0-0.1	
(E)-2-Hexenal	6728-26-3					+[b]; tr[w4]
(Z)-2-Hexenal	16635-54-4					+[b]
(E)-2-Hexen-1-ol	928-95-0					+[b]
(Z)-3-Hexen-1-ol	928-96-1	tr-0.2				+[b]; tr[w4]; 0.1[z5]; 0.2[v1]; 0.5[t]
(Z)-3-Hexenyl acetate	3681-71-8					+[b]
(Z)-3-Hexenyl iso-valerate	35154-45-1	tr-0.6				+[b]; 0.06[w2]; 0.1[e]
(Z)-3-Hexenyl methyl-butyrate	53398-85-9					0.1[w4]
Hexyl isovalerate	10032-13-0					+[b]
α-Humulene	6753-98-6		tr-0.1	0.1	0.1-1.4	0.08[c]; 0.1[w3]; 0.2[w2]; 0.4[n]; 0.7[l]; 1.5[x7]
Humulene epoxide II	19888-34-7				0-0.1	
Isoamyl acetate	123-92-2					+[b]
Isoamyl alcohol	123-51-3					+[b]
Isoamyl isobutyrate	2050-01-3					+[b]
Isoamyl isovalerate	659-70-1					+[b]; tr[w4]; 0.1[w3]
Isoamyl 2-methylbutyrate	27625-35-0					0.1[w3]
Isoamyl methyl ketone	110-12-3				0-0.1	
Isobornyl acetate	125-12-2					tr[k]
Isobutanal	78-84-2					+[b]
Isobutyl alcohol	78-83-1					+[b]
Isobutyl isovalerate	589-59-3					+[b]
Isobutyl-2-methyl-butyrate						+[b]
Isocaryophyllene oxide						0.1[p]
Isodihydrocarvyl acetate			0-tr			
Isoeugenol	97-54-1					0.3[x5]
Isoisopulegol	18674-65-2					0.8[x7]
Isomenthol	3623-52-7	0.1-3.0	0.2-0.7			1.0[d]; 1.7[w5]; 2.5[w]; 3.0[g]; 4.9[v1]; 6.4[e]; 6.5[x6]
Isomenthone	491-07-6	2.2-10.6	2.4-7.0	3.9	4.4-8.5	4.1[w5]; 4.8[i]; 5.3[k]; 5.4[w2]; 5.5[z4]; 5.8[n]; 6.5[p]; 7.4[f]; 7.6[l]; 9.0[w3]; 11.8[y5]; 14.8[e]; 15.5[z6]
Isomenthyl acetate	20777-45-1	0.01-0.3	0.1-0.2			0.8[z2]; 2.5[w3]; 3.1[w8]; 7.3[y5]; 10.3[y7]; 30.5[j]
Isoneomenthol						0.02[j]
Isopiperitenone	529-01-1					+[b]
Isopiperitone	58615-39-7				tr-0.2	
Isopropyl 2-methyl-butyrate	66576-71-4					0.1[w4]
Isopulegol	89-79-2	0.09-0.8		0.6	0-5.1	0.1[w3]; 0.3[d]; 0.5[y6]; 0.6[l]; 1.3[y6]; 1.5[v1]; 3.0[e]
Isopulegol acetate	57576-09-7				0-0.1	+[b]; 0.4[l]
Isopulegone	29606-79-9					tr[h]; 0.2[y6]; 0.3[w4]
Isothujone	59573-80-7					+[b]
Isovaleraldehyde	590-86-3					+[b]
Isovaleric acid	503-74-2					0.2[d]
Italicene	94535-52-1					0.1[w3]
(E)-Jasmone	6261-18-3					+[b]
(Z)-Jasmone	488-10-8		0.1		0-0.1	0.04[w2]; 1.0[l]
Ledene oxide II						0.2[k]

Table 5.68.2 Constituents identified in peppermint oils (continued)

Constituent	CAS	Percentage and range				
		A	B (3)	C (16)	D (49)	E
Limonene	138-86-3	0.3-18.5	2.8-9.8	2.9	0.1-0.7	0-4.5[v]; 0.2-6.8[q]; 4.2[u]; 5.2[x5]; 5.8[y6]; 6.0[p]; 6.2[v1]; 6.9[r]; 7.7[l]; 9.6[s]; 10.0[m]; 10.6[e]; 10.8[x]
Limonene-1,2-epoxide isomer						+[b]
Limonene oxide	1195-92-2					0.1[e]
Linalool	78-70-6	0.03-0.7		0.6	tr-1.1	0.4[d]; 0.5[f]; 0.7[z]; 0.9[n]; 1.5[l]; 4.8[x3]; 7.3[y5] 12.3[v3]; 51.0[v4]; 60.7[v5]
Linalool oxide	1365-19-1					tr[k]; 0.03[j]
cis-Linalool oxide	11063-77-7				0-tr	
trans-Linalool oxide	11063-78-8					0.02[j]
Linalyl acetate	115-95-7	0.08-0.6				0.03[y9]; 0.2[w5]; 20.7[v5]; 62.7[o]; 72.0[v3]
Linalyl butyrate	78-36-4					+[y1]
Linalyl isobutyrate	78-35-3					0.1[j]
Linalyl propionate	144-39-8				0-0.2	
Longifolene (junipene)	475-20-7					0.5[l]
p-Mentha-1(7),8-diene	499-97-8					+[b]
Menthalactone	13341-72-5	0-0.2				
cis-p-Mentha-2,8-dienol	3886-78-0		0-0.1			+[b]
trans-p-Mentha-2,8-dienol	4017-77-0					+[b]
p-Menthane-3,8-diol	42822-86-6					0.2[y7]
3-Menthene	500-00-5					+[y1]; 0.3[k]
p-Menth-1-en-9-ol	18479-68-0					0.2[y7]
p-Menth-2-en-1-ol	619-62-5					0.1[w5]; 0.6[k]; 1.6[n,z2]
cis-p-Menth-2-en-1-ol	29803-82-5			tr	0.1	+[b]; 0.05[p]; 0.1[w3]
trans-p-Menth-2-en-1-ol	29803-81-4	tr-0.5			tr-0.1	0.04[c]
p-Menth-1-en-9-yl acetate	28839-13-6					+[b]
Menthofuran	494-90-6	0.07-7.0	3.3-4.5	14.6	0-0.8	0.3-13.7[q]; 4-7.3[v]; 11.2[y3]; 11.8[x4]; 14.9[x2]; 17.9[v1]; 19.5[u]; 21.3[z2]; 27.2[z3]; 46.8[z4]
Menthofurolactone						+[y]
Menthofurolactone isomer						+[b]
Menthol	89-78-1	23.0-47.9	26.4-47.7	47.8	29.8-37.3	27.1-60.6[q]; 41.5[v2]; 51.8[w]; 51.9[n]; 52.4[w]; 54.2[y4]; 55.3[y3]; 56.9[t]; 59.2[x5]; 76.7[x]
Mentholactone	68330-67-6					0.2[z5]
Menthone	89-80-5	10.6-38.5	13.6-31.9	48.6	24.4-33.0	2-31.9[v]; 2-44.5[q]; 28.2[n]; 29.8[i]; 31.8[y5]; 32.5[o]; 32.8[p]; 33.874.6[s]; 55.0[t]; 74.6[s]
p-Menthone-1,2,3-triol						0.05[j]
Menthyl acetate	16409-45-3	0.5-7.7	0.6-0.9	9.5	1.6-4.3	0.5-10.3[v]; 0.8-23.6[q]; 12.4[t]; 15.1[y3]; 16.8[z3]; 17.2[x]; 17.4[r]; 29.5[o]; 32.8[v1]
cis-Menthyl acetate						0.6[w5]
trans-Menthyl acetate						13.3[w5]
Menthyl formate	2230-90-2					+[b]
2-Methylbutanal	96-17-3					+[b]; tr[w4]
2-Methyl-3-buten-2-ol	115-18-4					+[b]
3-Methyl-3-butenol	763-32-6					+[b]
2-Methylbutyl acetate	624-41-9					+[b]
2-Methylbutyl iso-butyrate	2445-69-4					+[b]
2-Methylbutyl 2-methylbutyrate	2445-78-5					0.06[p,w2]
2-Methylbutyl 3-methylbutyrate	2445-77-4					+[b,p]; 0.08[w2]
Methyl chavicol	140-67-0				0-0.1	+[b]; 0.7[y4]
Methyl cinnamate	103-26-4					0.3[x5]
3-Methylcyclohexanol	591-23-1					+[b]
3-Methylcyclohexanone	591-24-2					+[b]; tr[w4]

Table 5.68.2 Constituents identified in peppermint oils *(continued)*

Constituent	CAS	Percentage and range				
		A	B (3)	C (16)	D (49)	E
(*E*)-4-Methylcyclohexyl acetate						+[b]
(*E*)-4-Methylcyclohex-yl isovalerate						+[b]
Methylcyclopentane	96-37-7					4.9[y6]
3-Methylcyclopen-tanone	1757-42-2					+[b]
2-Methylfuran	534-22-5					+[b]
Methyl thymol	1076-56-8					+[b]
Mint furanone						0.06[j]; 0.3[d]
Mintsulfide	72445-42-2					+[b]
cis-Muurola-3,5-diene	157374-44-2		0-0.1			
cis-Muurola-4(14),5-diene	157477-72-0		0.1			
trans-Muurola-4(14), 5-diene	262352-87-4		0.1			
Muurolene	69671-15-4					0.1[e]
α-Muurolene	10208-80-7					0.8[l]; 0.9[z6]; 3.7[o]
γ-Muurolene	30021-74-0		0-1.6	tr		+[b]; 0.2[w5]; 0.3[n]; 1.0[l]
τ-Muurolene	152287-43-9					1.3[o]
τ-Muurolol (epi-α-)	19912-62-0					+[p]; 0.1[w5]
Myrcene	123-35-3	0.05-3.1	0.3-0.4	3.0	0-0.9	1.0[c]; 1.1[y9]; 1.3[e]; 4.1[m]; 4.4[z6]; 4.9[z3]; 15.5[v1]
Myrtenal	564-94-3				0.3-0.5	
Myrtenol	515-00-4			0.1		+[b]; 0.08[p]; 0.1[w4]; 0.2[e]
Myrtenyl acetate	1079-01-2					+[b]; tr[w4]
Neoisoisopulegol	21290-09-5	tr-0.4				
Neoisoisopulegol acetate	256332-34-0					0.1[w3]
Neoisomenthol	491-02-1	0.08-3.0				0.5[g]; 0.6[f]; 1.1[x7]; 1.5[e]; 2.9[w9]; 6.5[l]
Neoisomenthyl acetate		0-0.3				0.2[z5]; 0.4[y2]
Neoisopulegol	29141-10-4	0.2-1.0				0.3[c]
Neoisopulegyl acetate	57576-10-0					0.05[p]
Neomenthol	3623-51-6	0.2-7.4	2.0-2.7	4.9	1.9-4.4	5.2[z4]; 6.0[w1]; 6.3[y4]; 6.5[x7]; 6.9[x9]; 7.1[t]
Neomenthyl acetate	2230-87-7		0.1-0.2			0.6[c]; 0.7[w4]; 0.9[f]; 1.0[y5]; 1.1[w3]; 1.5[z6]
Nerol	106-25-2					+[b]
Nerolidol	7212-44-4					4.3[d]
(*E*)-Nerolidol	40716-66-3				tr-0.1	+[b]
(*Z*)-Nerolidol	3790-78-1					0.06[w2]
Neryl acetate	141-12-8					0.1[e]; 0.4[e]
Nonadecane	629-92-5				0-0.1	
Nonanal	124-19-6					+[b]
1-Nonen-3-ol	21964-44-3					+[b]
(*E*)-Ocimene	27400-72-2					0.2[r]
(*Z*)-Ocimene	27400-71-1					0.04[y9]; 0.1[r]
β-Ocimene	13877-91-3					0.3[x5]
(*E*)-β-Ocimene	3779-61-1	tr-0.2	0.1-0.2			tr[h,n]; 0.06[i,p]; 0.1[w1]; 0.7[u]; 1.7[v1];
(*Z*)-β-Ocimene	3338-55-4	tr-0.3	0.4-0.7	0.5	0-0.1	0.1[w3]; 0.2[i,p]; 0.3[n]; 0.9[u]; 1.8[z6]
Octadecanal	638-66-4				0-0.1	
Octadecane	593-45-3				0-0.1	tr[w5]
(*Z*)-1,5-Octadien-3-ol	50306-18-8					+[b]
Octanoic acid	124-07-2				0-0.3	
3-Octanol	589-98-0	0.05-2.4	0.2-0.4	0.3		0.2[c]; 0.3[m]; 0.4[d]; 0.8[y6]; 1.2[v1]; 3.5[e]; 10.1[v4]
n-Octanol (1-)	111-87-5	0.02-0.4				
3-Octanone	106-68-3					0.2[e]
Octan-3-yl acetate	103-09-3					+[b]; tr[k]; 0.04[j]
1-Octen-3-ol	3391-86-4	0.03-0.4	0-0.1	0.1		+[b]; 0.04[y9]; 0.2[w4]; 0.3[w2]; 0.7[y4]; 0.8[v1]
trans-2-Octen-1-ol	18409-17-1		0.1			
1-Octen-3-one	4312-99-6					+[b]
Octenyl acetate (1-)	37366-04-4		0-0.1			
1-Octen-3-yl acetate	2442-10-6					0.05[c]
7-Octen-1-yl acetate						1.1[y4]

Table 5.68.2 Constituents identified in peppermint oils (continued)

Constituent	CAS	A	B (3)	C (16)	D (49)	E
3-Octyl acetate	4864-61-3		tr-0.1			0.07^{w2}; 0.09^{p}; $0.1^{w3,w4}$
Oleic acid	112-80-1				0-0.2	
Pentadecanal	2765-11-9				0-0.1	
Pentadecanoic acid	1002-84-2				0-0.1	
1-Penten-3-ol	616-25-1					$+^{b}$
(E)-2-Penten-1-ol	1576-96-1					$+^{b}$
trans-2-(2-Pentenyl) furan	70424-14-5					$+^{b}$
2-Pentylfuran	3777-69-3					$+^{b}$
Perillene	539-52-6					0.1^{e}
Perillyl alcohol	536-59-4		0-0.2			tr^{w4}
α-Phellandrene	99-83-2				0-0.4	$+^{b}$; $tr^{h,k}$; 0.2^{z2}
β-Phellandrene	555-10-2					0.03^{c}; 0.2^{n}; 0.4^{k}; $0.8^{d,y6}$; 2.8^{e}; 5.6^{w6}
Phenylacetaldehyde	122-78-1					$+^{b}$
2-Phenyl-(E)-2-butenal	4411-89-6					$+^{b}$
2-Phenylethyl acetate	103-45-7					$+^{b}$
2-Phenylethyl isovalerate	140-26-1					$+^{b}$
2-Phenylethyl 2-methylbutyrate	24817-51-4					$+^{b}$
α-Pinene	80-56-8	0.06-9.7	0.7-0.8	0.9	0.3-0.6	$0.2-1.1^{v}$; 1.7^{v1}; 2.2^{w8}; 2.4^{t}; 3.5^{l}; 4.8^{e}
β-Pinene	127-91-3	0.2-6.5	1.1-1.3	1.3	0.5-0.8	$0.8-3.7^{v}$; 1.9^{m}; 2.0^{x6}; 2.3^{w8}; 2.5^{x8}; 5.7^{e}
Pinocarveol	5947-36-4					0.2^{e}
trans-Pinocarveol	1674-08-4					$+^{p}$; 0.2^{w3}
Piperitenone	491-09-8	0-0.4			0-0.1	$+^{b}$; 0.5^{l}
Piperitenone oxide	90582-88-0				0-tr	0.2^{j}; 19.3^{y7}; 55.6^{m}
cis-Piperitol	16721-38-3				tr-0.1	$+^{b}$; 0.5^{x7}
trans-Piperitol	16721-39-4				0-0.1	$+^{b}$; $<0.1^{w3}$
Piperitone	89-81-6	0.2-5.4	0.7-0.8	1.8	0.4-0.6	$0.5-2.3^{v}$; 1.6^{o}; 2.1^{e}; 2.2^{z4}; 2.3^{z3}; 5.9^{z6}
Piperitone oxide	148879-33-8		0-0.1			0.07^{y9}; 0.5^{y5}; 0.7^{m}; 1.9^{y7}
Propyl isovalerate	557-006					$+^{b}$
Pulegol	529-02-2	0.04-1.1				0.1^{z5}; 0.2^{h}
Pulegone	89-82-7	0.2-5.4	0.2-2.2	1.2	0.2-7.5	$1.2-15.4^{v}$; 3.9^{z2}; 4.1^{d}; 4.2^{n}; 4.4^{w6}; 5.0^{z4}; 5.4^{w7}; 6.4^{l}; 9.7^{t}; 13.0^{z6}; 14.4^{x3}; 15.4^{z3}
Pyridine	110-86-1					$+^{b}$
Rosefuran oxide	92356-06-4					0.05^{w2}
Sabinene	3387-41-5	0.1-2.1	0.4-0.7	0.3	0.2-0.9	0.6^{t}; 0.7^{g}; 1.8^{y7}; 1.0^{m}; 1.1^{y9}; 1.2^{y5}; 2.5^{r}
cis-Sabinene hydrate	15537-55-0	0.05-1.7	2.6-3.2	2.6	0-tr	0.3^{h}; $0.4-4.8^{v}$; 0.5^{h}; 0.8^{g}; 1.4^{f}; 1.5^{w3}
trans-Sabinene hydrate	17699-16-0	0-0.2	0.3-0.7		0.2-1.8	0.4^{w3}; 1.1^{k}; 1.2^{w4}; 1.4^{w7}; 1.5^{w6}; 4.4^{z6}
Sabinol						$+^{b}$; 0.2^{d}
cis-Sabinol	3310-02-9					tr^{w4}
trans-Sabinol	471-16-9					0.03^{c}
Sabinyl acetate	53833-85-5				0-tr	$+^{b}$; 0.2^{e}
Santolinatriene	2153-66-4					0.5^{k}
Spathulenol	6750-60-3		0-0.1	0.8	0-0.1	$0.1^{p,w3}$; 0.2^{e}; 0.5^{f}; 0.8^{l}; 1.0^{o}
α-Terpinene	99-86-5	0.03-0.5	0.2	0.3		0.2^{j}; 0.3^{k}; 0.4^{n}; 0.5^{w3}; 0.7^{v1}; 1.0^{d}; 19.7^{y7}
β-Terpinene	99-84-3					0.7^{k}
γ-Terpinene	99-85-4	0.02-0.7	0.3-0.4	1.7	0-0.2	$0-1.1^{v}$; 0.3^{c}; 0.4^{h}; 0.5^{j}; 0.6^{u}; 0.9^{w3}; 2.3^{v1}
δ-Terpinene	586-62-9					0.3^{w1}
τ-Terpinene						0.9^{o}
Terpinen-4-ol	562-74-3	0.1-2.7	tr-0.1	2.3	0.2-0.6	1.3^{y8}; 1.5^{f}; 1.6^{w3}; 1.7^{y4}; 2.7^{l}; 3.0^{v1}; 3.8^{x3}
α-Terpineol	98-55-5	0.05-0.6	tr-0.1	0.5	0.3-0.5	0.6^{w2}; 0.8^{e}; 0.9^{x7}; 1.3^{y4}; 1.4^{d}; 1.9^{w}; 6.1^{l}
β-Terpineol	138-87-4					0.6^{d}
γ-Terpineol	586-81-2					2.6^{l}; 2.7^{y7}
δ-Terpineol	7299-42-5					0.1^{p}
Terpinolene (α-)	586-62-9	tr-0.2	0.1	tr	0-0.7	0.09^{c}; 0.1^{r}; 0.2^{d}; 0.3^{e}; 0.4^{j}; 1.1^{o}; 3.7^{y5}
β-Terpinolene	1400450-37-4					0.02^{y9}
Tetradecanoic acid	544-63-8				0-0.1	
Tetradecanol	27196-00-5				0-tr	

Table 5.68.2 Constituents identified in peppermint oils *(continued)*

Constituent	CAS	Percentage and range				
		A	B (3)	C (16)	D (49)	E
α-Thujene	2867-05-2	tr-0.2	0-0.1		tr-0.1	tr[k,w4]; 0.04[w1]; 0.05[w]; 0.1[z5]; 0.2[z6]; 0.4[u]
α-Thujone	546-80-5				0-0.1	tr[h]
Thujol	35732-37-7					0.4[e]
Thujopsene	470-40-6					0.4[l]
Thymol	89-83-8				0.1-0.3	+[b]; 0.1[w4]
Tricyclene	508-32-7					1.2[y7]
Valeraldehyde	110-62-3					+[b]
Verbenene	4080-46-0					+[b]
Verbenone	80-57-9					0.1[e]
Viridifloral					0.2-0.6	0.05[j]
Viridiflorene (ledene)	21747-46-6				0-0.1	0.7[l]
Viridiflorol	552-02-3	0-0.5		1.9		0.2[c,p]; 0.3[w1,w4]; 0.4[w6]; 1.0[z6]; 2.0[w7]
Widdrol	6892-80-4					1.9[x1]
α-Ylangene	14912-44-8					0.3[l]

A 157 peppermint essential oil samples from USA, India, Russia, China, France, Italy, Bulgaria, Spain, Morocco and Canada, analyzed between 1998 and 2013; lowest and highest concentrations given (E. Schmidt, unpublished data)
B three lab-hydrodistilled oils from leaves of *Mentha* x *piperita* cultivated in home gardens (n=2) and commercially (n=1) in Estonia; lowest and highest concentrations given (ref. 3)
C eleven lab-hydrodistilled oils from the aerial parts of five Indian cultivars and three wild accessions; highest concentrations given; for most components, the lower concentration was zero (i.e., absent in one or more oils); the following chemicals were found in all oils (with their lowest concentrations in brackets): menthol (30.3%), menthone (4.5%), 1,8-cineole (4.1%), neomenthol (1.5%), isomenthone (1.0%), menthyl acetate (1.0%), β-caryophyllene (<0.1%), (*E*)-β-farnesene (<0.1%), menthofuran (<0.1%), pulegone (<0.1%) and α-terpineol (<0.1%) (ref. 16)
D five lab-hydrodistilled oils from 5 accessions of *M. piperita* cultivars from India; lowest and highest concentrations given (ref. 49)
E data from other studies (indicated with superscript letters); highest concentrations found in any study reviewed here given; when two or more oils were investigated, only the highest concentrations are mentioned, unless indicated otherwise

[a] incorrect identification; [b] four lab-hydrodistilled oils from fresh and dried leaves of two harvests from *M. piperita* cultivated in Oregon, USA; as the amounts were expressed in a manner different from the other data ('ppt', not explained), they are indicated here with +[b]; only the less common ingredients are indicated in the table (ref. 59); [c] one lab-hydrodistilled leaf oil from plants harvested in a medicinal garden in Poland (ref. 24); [d] one commercial oil purchased from an Indian company (ref. 25); [e] one commercial oil from an Indian company (ref. 26); this analysis contains several mistakes, as indicated in ref. 65; [f] one commercial oil purchased from a USA company (ref. 32); [g] one commercial oil from Provence, France (ref. 35); [h] one commercial oil from a German company (refs. 17,38); [i] one lab-hydro-distilled oil from *M. piperita* cultivated in India (ref. 39); [j] one lab-hydrodistilled oil from the stems and leaves of peppermint cultivated in Egypt (ref. 40); [k] one commercial oil from Korea (ref. 42); [l] two lab-hydrodistilled oils from the leaves of plants cultivated in Pakistan and harvested in winter and summer (ref. 43); [m] one lab-hydrodistilled peppermint oil from Italy with an extremely atypical composition: high piperitenone oxide (55.6%) content and no menthol or menthone (ref. 48); [n] two lab-hydrodistilled oils from two successive harvests of the same crop cultivated in Italy (ref. 50); [o] four lab-hydrodistilled oils from 4 cultivars from a botanical garden in Russia; one had an extremely atypical composition with 62.7% linalyl acetate and only 0.5% menthol (ref. 53); [p] one lab-hydrodistilled oil from peppermint cultivated in West Siberia, Russia (ref. 54); [q] 82 oil samples produced in Tasmania (n=29), King Island (n=14), USA (n=3) and various other countries; the data from the oils other than from Tasmania and King Island were retrieved from older literature; lowest and highest concentrations given (ref. 55); [r] one lab-hydrodistilled oil from the aerial parts of *M. piperita* collected in Serbia (ref. 60); [s] five lab-hydrodistilled oils from 5 accessions of various origins cultivated in Finland and harvested in various stages; only the main components were investigated (ref. 21); [t] seven industrially steam-distilled oils from plants cultivated in France and harvested at 7 times between June and October (ref. 8); [u] one commercial oil from an Italian company (ref. 20); [v] twenty lab-hydrodistilled oils from 20 germplasm accessions of *M. piperita* from a gene bank in India; lowest and highest concentrations given (ref. 14); [v1] seventeen lab-hydrodistilled oils from three *Mentha piperita* cultivars grown in India with age (30-180 days after transplanting) as variables (ref. 63); [v2] one lab-hydrodistilled peppermint oil from Argentina (ref. 64); [v3] one steam-distilled oil sample from a 'pleasant smelling' cultivar of *Mentha x piperita* grown in Portugal with an extremely atypical composition and dominated by linalyl acetate (72%) and linalool (12.3%) (ref. 78); [v4] one lab-hydrodistilled oil sample prepared from *Mentha piperita* grown at a University Research Center in Brazil with an extremely atypical composition and dominated by linalool (51%), carvone (23.4%) and 3-octanol, without any menthone, isomenthone, menthyl acetate, neomenthol or menthol detected (ref. 79); [v5] one oil sample from Morocco with an extremely atypical composition dominated by linalool (60.7%) and linalyl acetate (20.7%) (ref. 72); [w] one lab-hydrodistilled oil from fresh and one from dried aerial parts of peppermint from a botanical garden in Belarus (ref. 13); [w1] one lab-hydrodistilled oil from the above ground parts of *M. piperita* cultivated in Iran and harvested at 50% bloom (ref. 12); [w2] one steam-distilled oil from peppermint cultivated in Mongolia (ref. 10); [w3] two lab-hydrodistilled leaf oils from two Bulgarian peppermint cultivars (ref. 9); [w4] one steam-distilled oil from the fresh parts of flowering plants from Cuba (ref. 7);

Table 5.68.2 Constituents identified in peppermint oils *(continued)*

[w5] one lab-hydrodistilled oil from the aerial parts of *M. piperita* growing in South Italy (ref. 6); [w6] one lab-hydrodistilled leaf oil from flowering peppermints cultivated in Iran (ref. 5); [w7] one lab-hydrodistilled oil from the leaves of *M. piperita* from Egypt (ref. 4); [w8] one commercial peppermint oil purchased in Poland (ref. 56); [w9] one lab-hydrodistilled oil from the aerial parts of peppermint from Portugal (ref. 47); [x] eighteen lab-hydrodistilled oils from peppermint biomass harvested in Slovakia in 3 years and from April to September; only the main components were investigated (ref. 46); [x1] one commercial oil from a German company (ref. 45); [x2] twelve lab-hydrodistilled oils from dried and fresh *M. piperita* from the USA with 6 harvest times; only menthol, menthofuran and 1,8-cineole were investigated (ref. 37); [x3] four commercial oils from various countries (ref. 36); [x4] one laboratory steam-distilled oil from Egypt (ref. 34); [x5] one commercial oil from Lithuania (ref. 30); [x6] one lab-hydrodistilled oil from the aerial parts of peppermint growing wild in Algeria (ref. 29); [x7] one lab-hydrodistilled oil from dried leaves of Persian peppermint (ref. 15); [x8] three steam-distilled oils from the above ground parts of 3 cultivars of peppermint in the flowering phase grown in Bulgaria; only the main components were investigated (ref. 2); [x9] fourteen lab-hydrodistilled oils from peppermint plants cultivated in India under different levels of farmyard manure and spacing and harvested in two years; only the main components were investigated (ref. 1); [y] menthofurolactone demonstrated in peppermint oil and synthesized (ref. 51); [y1] two lab-hydrodistilled oils from fresh and dried peppermint leaves cultivated in Italy; as the amounts of the components cannot be compared to the other data, they are indicated with +[y1]; only the less common components found are presented (ref. 58); [y2] one lab-hydrodistilled oil from stems and leaves of peppermint harvested in a botanical garden in Burkina Fasso (ref. 52); [y3] one lab-hydrodistilled oil from the aerial flowering parts of *M. piperita* growing wild in Iran (ref. 57); [y4] one lab-hydrodistilled leaf oil from Poland (ref. 44); [y5] one lab-hydrodistilled oil from *M. piperita* cultivated in the USA (ref. 41); [y6] one commercial mint oil from Italy (ref. 31); [y7] one lab-hydrodistilled oil from *M. piperita* growing wild in Iran with a very atypical composition (ref. 28); [y8] one commercial oil from Brazil (ref. 27); [y9] one lab-hydrodistilled leaf oil from peppermint growing wild in Morocco (ref. 62); [z] twelve steam-distilled oils from the above ground parts of *M. piperita* cultivated in Canada under different nitrogen fertilization regimes and harvested in two years (ref. 11); [z2] data from before 2000,cited in ref. 65; [z3] five lab-hydrodistilled peppermint oils from five cultivars grown in India (refs. 66,67); [z4] nine oils from the top, the lower part and the whole plant of three Indian peppermint cultivars (ref. 68); [z5] one 50-year-old Bulgarian and one fresh USA commercial oil sample (ref. 69); [z6] eight lab-hydrodistilled oils from commercial peppermint samples from 7 European countries purchased in 2000, 2001 and 2002 (ref. 74); [z7] data from various studies published before 1997, cited in ref. 77; only chemicals not already present in other studies discussed are included (ref. 77)

tr: trace (in columns B,C,D,E[k] and E[w4]: <0.1; in column E[l] and E[w5]: <0.05) + present in the oil investigated, but quantity not stated or presented in a manner which cannot be compared to the other data

analytical data from one of us (E.S.) are shown in Table 5.68.2 in alphabetical order. In peppermint oils from various origins, over 335 chemicals have been identified. About 59 per cent of these were found in a single reviewed publication only. The major compounds found in peppermint oils from different sources are shown in Table 5.68.3. They include (highest concentrations in any study given) menthol (76.7%), menthone (74.6%), menthofuran (46.8%), menthyl acetate (32.8%), limonene (18.5%), isomenthone (15.5%), pulegone (15.4%), 1,8-cineole (13.9%), *trans*-menthyl acetate (13.3%) and neomenthol (7.4%). Well-known ingredients of peppermint oils that were present in high concentrations (>7%) in one or two studies were linalyl acetate (three studies: 20.7%, 62.7% and 72.0%, see under Chemotypes), linalool (four studies: 7.3%, 12.3%, 51.0% and 60.7%, see under Chemotypes), isomenthyl acetate (three studies: 7.3%, 10.3% and 30.5%), carvone (23.4%), α-terpinene (19.7%), β-caryophyllene (19.2%), myrcene (15.5%), 3-octanol (10.1%), germacrene D (9.7%) and α-pinene

Table 5.68.3 Major constituents of peppermint oils

Constituent	CAS	Percentage and range				
		A	B (3)	C (16)	D (49)	E
Menthol	89-78-1	23.0-47.9	26.4-47.7	47.8	29.8-37.3	54.2[y4]; 55.3[y3]; 56.9[t]; 59.2[x5]; 76.7[x]
Menthone	89-80-5	10.6-38.5	13.6-31.9	48.6	24.4-33.0	32.5[o]; 32.8[p]; 33.874.6[s]; 55.0[t]; 74.6[s]
Menthofuran	494-90-6	0.07-7.0	3.3-4.5	14.6	0-0.8	17.9[v1]; 19.5[u]; 21.3[z2]; 27.2[z3]; 46.8[z4]
Menthyl acetate	16409-45-3	0.5-7.7	0.6-0.9	9.5	1.6-4.3	16.8[z3]; 17.2[x]; 17.4[r]; 29.5[o]; 32.8[v1]
Limonene	138-86-3	0.3-18.5	2.8-9.8	2.9	0.1-0.7	6.2[v1]; 6.9[r]; 7.7[l]; 9.6[s]; 10.0[m]; 10.6[e]; 10.8[x]
Isomenthone	491-07-6	2.2-10.6	2.4-7.0	3.9	4.4-8.5	7.4[f]; 7.6[l]; 9.0[w3]; 11.8[y5]; 14.8[e]; 15.5[z6]
Pulegone	89-82-7	0.2-5.4	0.2-2.2	1.2	0.2-7.5	5.4[w7]; 6.4[l]; 9.7[t]; 13.0[z6]; 14.4[x3]; 15.4[z3]
1,8-Cineole	470-82-6	0.3-9.9	0-4.8	8.9	0.5-6.5	7.0[u]; 7.5[s]; 8.8[z3]; 9.6[y5]; 11.6[m]; 13.9[v1]
trans-Menthyl acetate						13.3[w5]
Neomenthol	3623-51-6	0.2-7.4	2.0-2.7	4.9	1.9-4.4	5.2[z4]; 6.0[w1]; 6.3[y4]; 6.5[x7]; 6.9[x9]; 7.1[t]

LEGEND: SEE UNDER TABLE 5.68.2

(9.7%). Uncommon or rare constituents of peppermint oils found in high concentrations (>7%) in single studies include piperitenone oxide (55.6%), (*E*)-carveol (14.5%), aromadendrene (10.2%) and α-gurjunene (8.9%).

Commercial oils

The ten chemicals that had the highest maximum concentrations in 157 commercial peppermint essential oil samples (concentration ranges provided) are the following: menthol (23.0-47.9%), menthone (10.6-38.5%), limonene (0.3-18.5%), isomenthone (2.2-10.6%), 1,8-cineole (0.3-9.9%), α-pinene (0.06-9.7%), menthyl acetate (0.5-7.7%), neomenthol (0.2-7.4%), menthofuran (0.07-7.0%) and β-pinene (0.2-6.5%) (Erich Schmidt, unpublished analytical data).

Chemotypes

Virtually all peppermint oils are dominated by menthol and menthone and/or related chemicals, including isomenthone. There have, however, been several reports where high concentrations of linalool (up to 60%) and/or linalyl acetate (up to 72%) were found with menthol/menthone in low concentrations or being absent (53,72,78,79). Whether this represents a specific chemotype or a mistake (botanical misidentification) was made, e.g., the plant investigated was not *Mentha piperita* but *Mentha citrata* (bergamot mint, known to be rich in linalool and linalyl acetate [25]) is at the moment unknown, though their smell should be quite different. We tend to believe this to be a botanical misidentification.

Additional analyses of peppermint oil not discussed in Table 5.68.2 can be found in refs. 22 (summary of published compositions of *M. piperita* grown in Iran), 70,71,73 and 75; these can all be accessed on-line. Older literature is reviewed in refs. 76 and 77.

CONTACT ALLERGY/ALLERGIC CONTACT DERMATITIS

General

Contact allergy to/allergic contact dermatitis from peppermint oil has been reported in over 45 publications. The oil (2% in petrolatum) has been included in the screening series of the North American Contact Dermatitis Group (NACDG) since 2009. In groups of consecutive patients suspected of contact dermatitis, prevalence rates of up to 1.8% positive patch test reactions have been observed. In most studies, no relevance data were provided, but in the NACDG investigations, 'definite' + 'probable' relevance was only 36-39%. There have been many case reports of allergic reactions to products with peppermint oil. Often, toothpastes, other oral products and peppermint in foods are causative products, which may induce oral discomfort, cheilitis, stomatitis, oral ulcers, burning mouth syndrome, lip swelling and lichenoid reactions of the oral mucosa. Some causative products were cosmetics; occupational contact dermatitis does occur, but its share appears to be smaller than in some other essential oils. The main allergen is probably menthol (which is also the main ingredient with concentrations of 23-48% in commercial oils, see Table

5.68.2 column A), co-reacting with peppermint oil in 13 patients. Other possible allergens, which have shown co-reactions with peppermint oil and may be present in concentrations >5% are α-pinene, limonene, caryophyllene, piperitone and pulegone.

Testing in groups of patients

The results of patch tests with peppermint oil in routine testing (consecutive patients suspected of contact dermatitis) and in groups of selected patients are shown in Table 5.68.4. In routine testing, rates of positive reactions ranged from 0.3% to 1.8%, whereas between 0.1% and 19% of patients in selected groups had positive patch tests. The high positivity rate of 19% was seen in a small group of 16 patients known to be allergic to propolis and *Myroxylon pereirae* (126).

Case reports

Occupational allergic contact dermatitis

An aromatherapist had chronic hand dermatitis and was patch test positive to 17 of 20 oils used at her work (tested 1% and 5% in petrolatum), including peppermint oil (88). Another aromatherapist had occupational contact dermatitis with allergies to multiple essential oils used at work, including peppermint oil. The patient also reacted to geraniol, α-pinene, caryophyllene, the fragrance mix and various other fragrance materials; α-pinene and caryophyllene were demonstrated by GC-MS in the peppermint oil. Menthol, the main ingredient in peppermint oil, was not tested (89). α-Pinene has been found in commercial peppermint oils in a maximum concentration of 9.7% and caryophyllene in a maximum concentration of 5.2% (Table 5.68.2, column A). Another aromatherapist had occupational contact dermatitis from contact allergy to various essential oils, including 'mint oil' (87). Occupational allergic contact dermatitis from peppermint oil occurred in an unknown number of food handlers (127, article not read).

Cosmetics

Four patients developed allergic contact dermatitis of the lips (allergic contact cheilitis) and the perioral skin from peppermint oil in a lip balm (93). One positive patch test reaction to peppermint oil, which was present in a cosmetic that had given a positive patch test or had been positive in a usage test, was seen in a 9-year period in one clinic in Belgium (82). One patient developed allergic cosmetic dermatitis which proved to be caused by peppermint oil in a skin care product (130). In a group of 39 patients with cosmetic allergy, where the causative allergen was identified with certainty or high probability, peppermint oil was the allergen in one case (124). One case of allergic contact dermatitis caused by a depilatory product from its ingredient peppermint oil has been reported (125).

Toothpastes and other oral products

Swelling of the tongue, lips and gingival mucosa from contact allergy to peppermint oil in an antiseptic spray

Table 5.68.4 Results of testing groups of patients with peppermint oil

Years and Country	Test conc. & vehicle	Number of patients		Selection of patients (S); Relevance (R); Comments (C)	Ref.
		tested |	positive (%)		
Routine testing					
2011-12 USA, Canada	2% pet.	4,230	18 (0.4%)	R: definite + probable relevance: 39%	117
2009-10 USA, Canada	2% pet.	4,303	26 (0.6%)	R: definite + probable relevance: 36%	99
2000-2007 USA	2% pet.	500	5 (1.0%)	R: 100%; C: weak study: a. high rate of macular erythema and weak reactions, b. relevance figures include 'questionable' and 'past' relevance	83
2002-2003 Korea	2% pet.	422	5 (1.2%)	R: not stated	96
1999-2000 Denmark	2% pet.	318	2 (0.6%)	R: not specified; C: this study was part of the international study mentioned below (ref. 118)	128
1998-2000 six European countries	2% pet.	1,606	9 (0.6%)	R: not specified for individual oils/chemicals	118
1983-1984 Italy	2% pet.	1,200	3 (0.3%)	R: not stated	129
<1977 Poland	2% pet.	400	7 (1.8%)	R: not stated	104
<1976 Poland	2% pet.	200	1 (0.5%)	R: not stated	90
<1970 UK	?	1,147	4 (0.3%)	data cited in ref. 95, no details known	96
Testing in groups of selected patients					
2011-2012 Italy	1% pet.	122	3 (2.5%)	S: patients who reported adverse cutaneous reactions to products (notably cosmetics) containing botanical ingredients in a questionnaire; they were tested with a 'botanical series'; R: all three reactions were relevant	114
2000-2008 IVDK	2% pet.	6,546	39 (0.6%)	S: patients with dermatitis suspected of causal exposure to fragrances; R: not stated	115
<2005 USA	2% pet.	160	7 (4.4%)	S: patients using consumer products containing peppermint oil; R: 'at least possibly relevant'	110
1997-2000 Austria	2% pet.	747	1 (0.1%)	S: patients suspected of fragrance allergy; R: not stated	85
1997-1998 Italy	2% pet.	54	2 (3.7%)	S: patients with cheilitis suspected of toothpaste allergy; R: both reactions were relevant	119
1996-1997 UK	2% pet.	10	1 (10%)	S: patients suspected of cosmetic dermatitis and reacting to the fragrance mix; R: not stated	84
1994-1995 UK	2% pet.	40	2 (5%)	S: patients previously reacting to the fragrance mix; R: not stated; C: there was also one positive reaction in a patient negative to the fragrance mix	120
<1986 Poland	2% pet.	86	6 (7.0%)	S: patients previously reacting to the fragrance mix; R: not stated	103
<1983 Poland	2% pet.	16	3 (19%)	S: patients known to be allergic to propolis and *Myroxylon pereirae*; R: not stated	126
<1978 Denmark	5% pet.	38	2 (5.3%)	S: see text under Case reports	97
<1976 Poland	2% pet.	51	1 (2.0%)	S: patients allergic to *Myroxylon pereirae* resin (balsam of Peru) and/or turpentine and/or wood tar and/or colophony	90

IVDK Information Network of Departments of Dermatology, Germany, Switzerland, Austria (www.ivdk.org); pet.: petrolatum

used in dentistry, a mouthwash and candies; the patient, who had been sensitized primarily by turpentine oil, also reacted to limonene and α-pinene (94). Limonene has been found in commercial peppermint oils in concentrations up to 18.5% and α-pinene in a maximum concentration of 9.7% (Table 5.68.2, column A). Six cases of allergic reactions to peppermint oil in patients with burning mouth syndrome (n=2), recurrent oral ulcers (n=2) or lichenoid mucosal reactions (n=2); five also reacted to menthol (main ingredient), two to eugenol and isoeugenol and one to anethole, but these are not important components of peppermint oils. The reactions to peppermint oil were considered to be relevant, but the causative products were not specified (95). One patient with oral burning and discomfort and a lichenoid reaction of the oral mucosa had positive patch test reactions to peppermint oil and menthol; the symptoms improved after avoiding mint-flavored mouthwashes and food (102). One patient had allergic contact cheilitis from peppermint oil in toothpaste (111). Another patient developed stomatitis and lip dermatitis from contact allergy to peppermint oil and menthol present in toothpastes (121). Six cases of contact stomatitis and contact dermatitis from contact allergy to oil of peppermint, oil of spearmint, carvone and anethole in toothpastes have been described; unknown how many cases were caused by peppermint oil (123) (article not read).

Other products

A patient developed allergic contact dermatitis from peppermint oil and its main component menthol in a transdermal therapeutic system (92). Three patients with allergic contact dermatitis from peppermint oil were seen in Japan (article not read); the allergens were menthol (the main component of peppermint oil), piperitone and pulegone (106). Both piperitone and pulegone have been found in a maximum concentration of 5.4% in commercial peppermint oils (Table 5.68.2, column A). One patient had orofacial granulomatosis, mainly of the lower lip; he was allergic to peppermint oil and menthol. An exclusion diet resulted in reduction of the swelling; upon re-exposure to menthol, further episodes of lip swelling occurred (108). One positive patch test reaction to peppermint oil was observed in a group of 53 women with chronic anogenital dermatitis and was considered to be relevant (116).

Positive patch tests (relevance unknown, uncertain or not stated)

Two positive patch test reactions to peppermint oil were seen in a group of 32 patients with sore mouth, stomatitis and/or dermatitis around the mouth or dentist personnel. One also reacted to spearmint oil and its ingredient carvone. The causative products were supposed to be toothpastes (97). A positive patch test to peppermint oil was observed in an aromatherapist with occupational contact dermatitis from multiple essential oils; it was uncertain whether this oil had been used by the patient (86). A positive patch test reaction to peppermint oil was seen in a patient who had airborne allergic contact dermatitis from tea tree oil and lavender oil (91). One positive patch test to peppermint oil was observed in a women who drank large amounts of peppermint tea, which appeared to induce vulvar dermatitis (100). Four positive patch test reactions to peppermint oil occurred in 7 patients with allergic contact dermatitis from compound tincture of benzoin (104). Two patients with cheilitis, who both reacted to peppermint oil and to menthol have been presented; the causative products were not specified, but presumably toothpastes were implicated (107). One patient had prolonged lip swelling, apparently from contact allergy to peppermint oil and menthol (112). One positive patch test reaction to peppermint oil in a patient allergic to ascaridole, an allergen in tea tree oil (negative to tea tree oil itself) (113).

A useful review of peppermint oil allergy is provided in ref. 98. Menthol, the main ingredient of peppermint oil, is said to have a low sensitizing potential (101). Older literature on contact allergy to peppermint oil can be found in ref. 109 (article not read).

LITERATURE

1 Meena GRL, Nag M, Pathania VL, Kaul VK, Singh B, Singh RD, Ahuja PS. Effect of organic manure and plant spacing on biomass and quality of *Mentha piperita* L. in Himalaya in India. J Essent Oil Res 2013;25:354-357

2 Jirovetz L, Wlcek K, Buchbauer G, Gochev V, Girova T, Dobreva A, et al. Chemical composition and antifungal activity of essential oils from various Bulgarian *Mentha x piperita* L. cultivars against clinical isolates of *Candida albicans*. J Essent Oil Bear Plants 2007;10:412-420

3 Orav A, Kapp K, Raal A. Chemosystematic markers for the essential oils in leaves of *Mentha* species cultivated or growing naturally in Estonia. Proc Eston Acad Sci 2013;62(3):175-186. doi: 10.3176/proc.2013.3.03. Available online at www.eap.ee/proceedings

4 Elansary HO, Ashmawy NA. Essential oils of mint between benefits and hazards. J Essent Oil Bear Plants 2013;16:429-438

5 Moghaddam M, Pourbaige M, Tabar HK, Farhadi N, Mohammad S, Hosseini A. Composition and antifungal activity of peppermint (*Mentha piperita*) essential oil from Iran. J Essent Oil Bear Plants 2013;16:506-512

6 Senatore F, De Fusco R, Grassia A, Moro CO, Rigano D, Napolitano F. Chemical composition and antibacterial activity of essential oils from five culinary herbs of the Lamiaceae family growing in Campania, Southern Italy. J Essent Oil Bear Plants 2003;6:166-173

7 Pino JA, Borges P, Martinez MA, Vargas M, Flores H, Martín del Campo ST, Fuentes V. Essential oil of *Mentha piperita* L. grown in Jalisco. J Essent Oil Res 2002;14:189-190

8 Chalchat J-C, Garry R-P, Michet A. Variation of the chemical composition of essential oil of *Mentha piperita* L. during the growing time. J Essent Oil Res 1997;9:463-465

9 Stojanova A, Paraskevova P, Anastassov Ch. A comparative investigation on the essential oil composition of two Bulgarian cultivars of *Mentha piperita* L. J Essent Oil Res 2000;12:438-440

10 Karasawa D, Shatar S, Erdenechimeg A, Okamoto Y, Tateba H, Shimizu S. A study on Mongolian mints. A new chemotype from *Mentha asiatica* Borriss and constituents of *M. arvensis* L. and *M. piperita* L. J Essent Oil Res 1995;7:255-260

11 Court WA, Roy RC, Pocs R, More AF, White PH. Optimum nitrogen fertilizer rate for peppermint (*Mentha piperita* L.) in Ontario, Canada. J Essent Oil Res 1993;5:663-666

12 Saharkhiz MJ, Goudarzi T. Foliar application of salicylic acid changes essential oil content and chemical compositions of peppermint (*Mentha piperita* L.). J Essent Oil Bear Plants 2014;17:435-440

13 Shutava HG, Kavalenka NH, Supichenka HN, Leontiev VN, Shutava TG. Essential oils of Lamiaceae with high content of α-, β-pinene and limonene enantiomers. J Essent Oil Bear Plants 2014;17:18-25

14 Shasany AK, Gupta S, Gupta MK, Singh AK, Naqvi AA, Khanuja SPS. Chemotypic comparison of AFLP analyzed Indian peppermint germplasm to selected peppermint oils of other countries. J Essent Oil Res 2007;19:138-145

15 Fatemi F, Dini S, Rezaei MB, Dadkhah A, Dabbagh R, Naij S. The effect of γ-irradiation on the chemical composition and antioxidant activities of peppermint essential oil and extract. J Essent Oil Res 2014;26:97-104

16 Padalia RC, Verma RS, Chanotiya CS. Variability in volatile terpenoid compositions of peppermint cultivars and some wild accession from northern India. J Essent Oil Res 2011;23:29-33

17 Jirovetz L, Buchbauer G, Bail S, Denkova Z, Slavchev A, Stoyanova A, Schmidt E, Geissler M. Antimicrobial activities of essential oils of mint and peppermint as well as some of their main compounds. J Essent Oil Res 2009;21:363-366. Data also presented in ref. 38

18 Lawrence BM. The genus Mentha. Boca Raton, FL, USA: CRC Press, 2006

19 Lawrence BM. Peppermint oil. Carol Stream, IL, USA: Allured Publishing Corp, 2007

20 Giamperi L, Fraternale D, Ricci D. The in vitro action of essential oils on different organisms. J Essent Oil Res 2002;14:312-318

21 Aflatuni A, Heikkinen K, Tomperi P, Jalonen J, Laine K. Variation in the extract composition of mints of different origin cultivated in Finland. J Essent Oil Res 2000;12:462-466

22 Hassanzadeh MK, Emami SA, Asili J, Najaran ZT. Review of the essential oil composition of Iranian Lamiaceae. J Essent Oil Res 2011;23:35-74

23 European Medicines Agency. Assessment report on Mentha x piperita L., aetheroleum. London, UK, Doc Ref EMEA/HMPC/349465/2006: 2008. Available at: http://www.ema.europa.eu

24 Skalicka-Wozniak K, Walasek M. Preparative separation of menthol and pulegone from peppermint oil (Mentha piperita L.) by high-performance counter-current chromatography. Phytochem Lett 2014;10:94-98

25 Kumar P, Mishra S, Malik A, Satya S. Efficacy of Mentha × piperita and Mentha citrata essential oils against housefly, Musca domestica L. Ind Crops Prod 2012;39:106-112

26 Tyagi AK, Malik A. Antimicrobial potential and chemical composition of Mentha piperita oil in liquid and vapour phase against food spoiling microorganisms. Food Control 2011;22:1707-1714

27 Katiki LM, Chagas ACS, Bizzo HR, Ferreira JFS, Amarante AFT. Anthelmintic activity of Cymbopogon martinii, Cymbopogon schoenanthus and Mentha piperita essential oils evaluated in four different in vitro tests. Vet Parasitol 2011;183:103-108

28 Yadegarinia D, Gachkar L, Rezaei MB, Taghizadeh M, Astaneh SA, Rasooli I. Biochemical activities of Iranian Mentha piperita L. and Myrtus communis L. essential oils. Phytochem 2006;67:1249-1255. Data also presented in ref. 80

29 Djenane D, Aïder M, Yangüela J, Idir L, Gómez D, Roncalés P. Antioxidant and antibacterial effects of Lavandula and Mentha essential oils in minced beef inoculated with E. coli O157:H7 and S. aureus during storage at abuse refrigeration temperature. Meat Science 2012;92:667-674

30 Lazutka JR, Mierauskiene J, Slapsyte G, Dedonyte V. Genotoxocity of dill (Anethum graveolens L.), peppermint (Mentha x piperita L.) and pine (Pinus sylvestris L.) essential oils on human lymphocytes and Drosophila melanogaster. Food Chem Toxicol 2001;39:485-492

31 Tyagi AK, Gottardi D, Malik A, Guerzoni ME. Anti-yeast activity of Mentha oil and vapours through in vitro and in vivo (real fruit juices) assays. Food Chem 2013;137:108-114

32 Nikolić M, Jovanović KK, Marković T, Marković D, Gligorijević N, Radulović S, Soković, M. Chemical composition, antimicrobial, and cytotoxic properties of five Lamiaceae essential oils. Ind Crops Prod 2014;61:225-232

33 Kamatou GPP, Vermaak I, Viljoen AM, Lawrence BM. Menthol: A simple monoterpene with remarkable biological properties. Phytochem 2013;96:15-25

34 Oraby MM, El-Borollosy AM. Essential oils from some Egyptian aromatic plants as an antimicrobial agent and for prevention of potato virus Y transmission by aphids. Ann Agricult Sci 2013;58:97-103

35 Rolli E, Marieschi M, Maietti S, Sacchetti M, Bruni R. Comparative phytotoxicity of 25 essential oils on pre- and post-emergence development of Solanum lycopersicum L.: A multivariate approach. Ind Crops Prod 2014;60:280-290

36 Işcan G, Kirimer N, Kürkcüoğlu M, Başer KHC, Demirci F. Antimicrobial screening of Mentha piperita essential oils. J Agric Food Chem 2002;50:3943-3946

37 Zheljazkov VD, Cantrell CL, Astatkie T, Hristov A. Yield, content, and composition of peppermint and spearmint as a function of harvesting time and drying. J Agr Food Chem 2010;58:11400-11407

38 Schmidt E, Bail S, Buchbauer G, Stoilova I, Atanasova T, Stoyanova A, Krastanov A, Jirovetz L. Chemical composition, olfactory evaluation and antioxidant effects of essential oil from Mentha × piperita. Nat Prod Commun 2009;8:1107-1112. Data also presented in ref. 17

39 Verma RS, Pandey V, Padalia RC, Saikia D, Krishna B. Chemical composition and antimicrobial potential of aqueous distillate volatiles of Indian peppermint (Mentha piperita) and spearmint (Mentha spicata). J Herbs Spic Med Plants 2011;17:258-267

40 Aziz EE, Craker LE. Essential oil constituents of peppermint, pennyroyal, and apple mint grown in a desert agrosystem. J Herbs Spic Med Plants 2010;15:361-367

41 Aziz EE, Al-Amier H, Craker LE. Influence of salt stress on growth and essential oil production in peppermint, pennyroyal, and apple mint. J Herbs Spic Med Plants 2008;14:177-87

42 Yang S-A, Jeon S-K, Lee E-J, Shim C-H, Lee I-S. Comparative study of the chemical composition and antioxidant activity of six essential oils and their components. Nat Prod Res 2010;24:140-151

43 Hussain AI, Anwar F, Nigam PS, Ashraf M, Gilani AH. Seasonal variation in content, chemical composition and antimicrobial and cytotoxic activities of essential oils from four *Mentha* species. J Sci Food Agric 2010;90:1827-1836

44 Freire MM, Jham GN, Dhingra OK, Jardim CM, Coura Barcelos R, Moreira Valente VM. Composition, antifungal activity and main fungitoxic components of the essential oil of *Mentha piperita* L. J Food Saf 2012;32:29-36

45 Akhtary Y, Pages E, Stevens A, Bradbury R, da Camara CAG, Ismany MB. Effect of chemical complexity of essential oils on feeding deterrence in larvae of the cabbage looper. Physiol Entomol 2012;37:81-91

46 Grulova D, De Martino L, Mancini E, Salamon I, De Feo V. Seasonal variability of the main components in essential oil of *Mentha × piperita* L. J Sci Food Agric 2015;95:621-627

47 Machado M, Santoro G, Sousa MC, Salgueiro L, Cavaleiro C. Activity of essential oils on the growth of *Leishmania infantum* promastigotes. Flavour Fragr J 2010;25:156-160

48 Fiocco D, Fiorentino D, Frabboni L, Benvenuti S, Orlandini G, Pellati F, Gallone A. Lavender and peppermint essential oils as effective mushroom tyrosinase inhibitors: a basic study. Flavour Fragr J 2011;26:441-446

49 Dwivedi S, Khan M, Srivastava SK, Syamasunnder KV, Srivastava A. Essential oil composition of different accessions of *Mentha piperita* L. grown on the northern plains of India. Flavour Fragr J 2004;19:437-440

50 Piccaglia R, Marotti M. Characterization of several aromatic plants grown in Northern Italy. Flavour Fragr J 1993;8:115-122

51 Frérot E, Bagnoud A, Vuilleumier C. Menthofurolactone: a new *p*-menthane lactone in *Mentha piperita* L.: analysis, synthesis and olfactory properties. Flavour Fragr J 2002;17:218-226

52 Bassolé IH, Lamien-Meda A, Bayala B, Tirogo S, Franz C, Novak J, et al. Composition and antimicrobial activities of *Lippia multiflora* Moldenke, *Mentha × piperita* L. and *Ocimum basilicum* L. essential oils and their major monoterpene alcohols alone and in combination. Molecules 2010;15:7825-7839

53 Kurilov DV, Kirichenko EB, Bidyukova GF, Olekhnovich LS, Ku LD. Composition of the essential oil of introduced mint forms of *Mentha piperita* and *Mentha arvensis* species. Doklady Biological Sciences 2009;429:538-540

54 Myadelets MA, Domrachev DV, Cheremushkina VA. A study of the chemical composition of essential oils of some species from the Lamiaceae L. family cultivated in the western Siberian region. Russian J Bioorgan Chem 2013;39:733-738

55 Clark RJ, Menary RC. Variations in composition of peppermint oil in relation to production areas. Economic Botany 1981;35:59-69

56 Golebiowski M, Ostrowski B, Paszkiewicz M, Czerwicka M, Kumirska J, Halinski L, et al. Chemical composition of commercially available essential oils from blackcurrant, ginger, and peppermint. Chem Nat Comp 2008;44:794-796

57 Saharkhiz MJ, Motamedi M, Zomorodian K, Pakshir K, Miri R, Hemyari K. Chemical composition, antifungal and antibiofilm activities of the essential oil of *Mentha piperita* L. ISRN Pharmaceutics 2012, Article ID 718645, 6 pages. doi:10.5402/2012/718645

58 Orio L, Cravotto G, Binello A, Pignata G, Nicola S, Chemat F. Hydrodistillation and in situ microwave generated hydrodistillation of fresh and dried mint leaves: a comparison study. J Sci Food Agric 2012;92: 3085-3090

59 Chen MZ, Trinnaman L, Bardsley K, St Hilaire CJ, Da Costa NC. Volatile compounds and sensory analysis of both harvests of double-cut Yakima peppermint (*Mentha piperita* L.). J Food Sci 2011;76:C1032-C1038

60 Soković M, Glamočlija J, Marin PD, Brkić D, van Griensven LJLD. Antibacterial effects of the essential oils of commonly consumed medicinal herbs using an in vitro model. Molecules 2010;15:7532-7546. Data also presented in ref. 61

61 Sokovic MD, Vukojevic J, Marin PD, Brkic DD, Vajs V, van Griensven LJLD. Chemical composition of essential oils of *Thymus* and *Mentha* species and their antifungal activities. Molecules 2009;14:238-249. Data also presented in ref. 60

62 Derwich E, Chabir R, Taouil R, Senhaji O. *In-vitro* antioxidant activity and GC/MS studies on the leaves of *Mentha piperita* (Lamiaceae) from Morocco. Int J Pharm Sci Drug Res 2011;3:130-136

63 Verma RS, Rahman L, Verma RK, Chauhan A, Yadav AK, Singh A. Essential oil composition of menthol mint (*Mentha arvensis*) and peppermint (*Mentha piperita*) cultivars at different stages of plant growth from Kumaun region of western Himalaya. Open Access J Med Arom Plants 2010;1:13-18

64 Palacios SM, Bertoni A, Rossi Y, Santander R, Urzúa A. Efficacy of essential oils from edible plants as insecticides against the house fly, *Musca domestica* L. Molecules 2009;14:1938-1947

65 Lawrence BM. Progress in essential oils. Perfum Flavor 2013;38(April):47-54

66 Khanuja SPS, Shasany AK, Kalsa A, Patra NK, Darokar MP, Padmapriya T, et al. A sweet smelling peppermint plant, 'CIM-Madhuras.' J Med Arom Plant Sci 2004;26:790-794. Data cited in ref. 65

67 Khanuja SPS, Patra NK, Shasany AK, Kumar B, Gupta S, Gupta MK, et al. High menthofuran chemotype 'CIM-Indus' of *Mentha x piperita*. J Med Arom Plant Sci 2005;27:721-726. Data cited in ref. 65

68 Singh AK, Yadav A, Singh Chanotiya C, Gupta AK, Bahl JR, Khanuja SPS. Quality evaluation of *Mentha piperita* leaf oils and their relation to position on stem, pre-storage temperature shock treatment and storage. J Med Arom Plant Sci 2008;30:298-303. Data cited in ref. 65

69 Schmidt E, Wanner J, Bail S, Jirovetz L, Buchbauer G, Gocher V, et al. Comparative analysis of historical peppermint oil from Bulgaria and a commercial oil of North American origin. Perfum Flavor 2009;34(11):46-50

70 Golparvar AR, Hadipanah A. Chemical compositions of the essential oil from peppermint (*Mentha piperita* L.) cultivated in Isfahan conditions. J Herb Drugs 2013;4:75-80

71 Moghtader M. *In vitro* antifungal effects of the essential oil of *Mentha piperita* L. and its comparison with synthetic menthol on *Aspergillus niger*. Afr J Plant Sci 2013;7:521-527

72 Debbab A, Mosaddak B, Aly AH, Hakiki A, Mosaddak M. Chemical characterization and toxicological evaluation of the essential oil of *Mentha piperita* L. growing in Morocco. Scientific Study & Research 2007;8:281-288. Data cited in ref. 44

73 Tsai M-L, Wu C-T, Lin T-F, Lin W-C, Huang Y-C, Yang C-H. Chemical composition and biological properties of essential oils of two mint species. Trop J Pharm Res 2013;12:577-582

74 Orav A, Raal A, Arak E. Comparative chemical composition of the essential oil of *Mentha × piperita* L. from various geographical sources. Proc Estonian Acad Sci Chem 2004;53:174-181

75 Mkolo NM, Olowoyo JO, Sako KB, Mdakane STR, Mitonga MMA, Magano SR. Repellency and toxicity of essential oils of *Mentha piperita* and *Mentha spicata* on larvae and adult of *Amblyomma hebraeum* (Acari: Ixodidae). Sci J Microbiol 2011, available at: www.sjpub.org/sjmb/N.M.%20Mkoloa%20et%20al.pdf

76 Lawrence BM. Progress in essential oils. Perfum Flavor 1993;18(4):59-72

77 Lawrence BM. Progress in essential oils. Perfum Flavor 1997;22(July/Aug):57-74

78 Martins MM, Costa SB, Neves C, Cavaleiro C, Salgueiro I, Costa MLB. Olive oil flavoured by the essential oils of Mentha piperita and Thymus mastichina L. Food Qual Prefer 2004;15:447-452

79 Duarte MCT, Figueira GM, Sartoratto A, Rehder VLG, Delarmelina C. Anti-*Candida* activity of Brazilian medicinal plants. J Ethnopharmacol 2005;97:305-311

80 Rasooli I, Gachkar L, Yadegarinia D, Rezaei MB, Alipoorastaneh SD. Antibacterial and antioxidative characterization of essential oils from *Mentha piperita* and *Mentha spicata* grown in Iran. Acta Aliment 2008;37: 41-52. Data cited in ref. 44. Data also presented in ref. 28

81 Rhind JP. Essential oils. A handbook for aromatherapy practice, 2nd Edition. London: Singing Dragon, 2012

82 Nardelli A, Drieghe J, Claes L, Boey L, Goossens A. Fragrance allergens in 'specific' cosmetic products. Contact Dermatitis 2011;64:212-219

83 Wetter DA, Yiannias JA, Prakash AV, Davis MD, Farmer SA, el-Azhary RA, et al. Results of patch testing to personal care product allergens in a standard series and a supplemental cosmetic series: an analysis of 945 patients from the Mayo Clinic Contact Dermatitis Group, 2000-2007. J Am Acad Dermatol 2010;63:789-798

84 Thomson KF, Wilkinson SM. Allergic contact dermatitis to plant extracts in patients with cosmetic dermatitis. Br J Dermatol 2000;142:84-88

85 Wöhrl S, Hemmer W, Focke M, Götz M, Jarisch R. The significance of fragrance mix, balsam of Peru, colophonium and propolis as screening tools in the detection of fragrance allergy. Br J Dermatol 2001;145:268-273

86 Bleasel N, Tate B, Rademaker M. Allergic contact dermatitis following exposure to essential oils. Australas J Dermatol 2002;43:211-213

87 Boonchai W, Lamtharachai P, Sunthonpalin P. Occupational allergic contact dermatitis from essential oils in aromatherapists. Contact Dermatitis 2007;56:181-182

88 Selvaag E, Holm J, Thune P. Allergic contact dermatitis in an aromatherapist with multiple sensitizations to essential oils. Contact Dermatitis 1995;33:354-355

89 Dharmagunawardena B, Takwale A, Sanders KJ, Cannan S, Roger A, Ilchyshyn A. Gas chromatography: an investigative tool in multiple allergies to essential oils. Contact Dermatitis 2002;47:288-292

90 Rudzki E, Grzywa Z, Bruo WS. Sensitivity to 35 essential oils. Contact Dermatitis 1976;2:196-200

91 De Groot AC. Airborne allergic contact dermatitis from tea tree oil. Contact Dermatitis 1996;35:304-305

92 Foti C, Conserva A, Antelmi A, Lospalluti L, Angelini G. Contact dermatitis from peppermint and menthol in a local action transcutaneous patch. Contact Dermatitis 2003;49:312-313

93 Tran A, Pratt M, DeKoven J. Acute allergic contact dermatitis of the lips from peppermint oil in a lip balm. Dermatitis 2010;21:111-115

94 Dooms-Goossens A, Degreef H, Holvoet C, Maertens M. Turpentine-induced hypersensitivity to peppermint oil. Contact Dermatitis 1977;3:304-308

95 Morton CA, Garioch J, Todd P, Lamey PJ, Forsyth A. Contact sensitivity to menthol and peppermint in patients with intra-oral symptoms. Contact Dermatitis 1995;32:281-284

96 Calnan CD. Oil of cloves, laurel, lavender, peppermint. Contact Dermatitis Newsletter 1970;7:148

97 Andersen KE. Contact allergy to toothpaste flavors. Contact Dermatitis 1978;4:195-198

98 Herro E, Jacob SE. *Mentha piperita* (peppermint). Dermatitis 2010;21:327-329

99 Warshaw EM, Belsito DV, Taylor JS, Sasseville D, DeKoven JG, Zirwas MJ, et al. North American Contact Dermatitis Group patch test results: 2009 to 2010. Dermatitis 2013;24:50-59

100 Vermaat H, van Meurs T, Rustemeyer T, Bruynzeel DP, Kirtschig G. Vulval allergic contact dermatitis due to peppermint oil in herbal tea. Contact Dermatitis 2008;58:364-365

101 Ale SI, Hostynek JJ, Maibach HI. Menthol: a review of its sensitisation potential. Exog Dermatol 2002:1:74-80

102 Fleming CJ, Forsyth A. D5 patch test reactions to menthol and peppermint. Contact Dermatitis 1998;38:337

103 Rudzki E, Grzywa Z. Allergy to perfume mixture. Contact Dermatitis 1986;15:115-116

104 Fettig J, Taylor J, Sood A. Post-surgical allergic contact dermatitis to compound tincture of benzoin and association with reactions to fragrances and essential oils. Dermatitis 2014;25:211-212

105 Rudzki E, Grzywa Z. Balsam of Peru as screening agent for essential oils sensitivity. Dermatologica 1977;155:115-121

106 Saito F, Oka K. Allergic contact dermatitis due to peppermint oil. Skin Res 1990;32:161-167

107 Wilkinson SM, Beck MH. Allergic contact dermatitis from menthol in peppermint. Contact Dermatitis 1994;30:42-43

108 Lewis FM, Shah M, Gawkrodger DJ. Contact sensitivity to food additives can cause oral and perioral symptoms. Contact Dermatitis 1995;33:429-430

109 Smith IL. Acute allergic reaction following the use of toothpaste—a case report. Br Dent J 1968;125:304-305

110 Guin JD. Use of consumer product ingredients for patch testing. Dermatitis 2005;16:71-77

111 Freeman S, Stephens R. Cheilitis: Analysis of 75 cases referred to a contact dermatitis clinic. Am J Cont Derm 1999;10:198-200

112 Shah M, Lewis M, Gawkrodger DJ. Contact allergy in patients with oral symptoms: a study of 47 patients. Am J Cont Derm 1996;7:146-151

113 Christoffers WA, Blömeke B, Coenraads P-J, Schuttelaar M-LA. The optimal patch test concentration for ascaridole as a sensitizing component of tea tree oil. Contact Dermatitis 2014;71:129-137

114 Corazza M, Borghi A, Gallo R, Schena D, Pigatto P, Lauriola MM, et al. Topical botanically derived products: use, skin reactions, and usefulness of patch tests. A multicentre Italian study. Contact Dermatitis 2014;70:90-97

115 Uter W, Schmidt E, Geier J, Lessmann H, Schnuch A, Frosch P. Contact allergy to essential oils: current patch test results (2000–2008) from the Information Network of Departments of Dermatology (IVDK). Contact Dermatitis 2010;63:277-283

116 Vermaat H, Smienk F, Rustemeyer Th, Bruynzeel DP, Kirtschig G. Anogenital allergic contact dermatitis, the role of spices and flavour allergy. Contact Dermatitis 2008;59:233-237

117 Warshaw EM, Maibach HI, Taylor JS, Sasseville D, DeKoven JG, Zirwas MJ, et al. North American Contact Dermatitis Group patch test results: 2011-2012. Dermatitis 2015;26:49-59

118 Frosch PJ, Johansen JD, Menné T, Pirker C, Rastogi SC, Andersen KE, et al. Further important sensitizers in patients sensitive to fragrances. II. Reactivity to essential oils. Contact Dermatitis 2002;47:279-287

119 Francalanci S, Sertoli A, Giorgini S, Pigatto P, Santucci B, Valsecchi R. Multicentre study of allergic contact cheilitis from toothpastes. Contact Dermatitis 2000;43:216-222

120 Katsarma G, Gawkrodger DJ. Suspected fragrance allergy requires extended patch testing to individual fragrance allergens. Contact Dermatitis 1999;41:193-197

121 Downs AMR, Lear JT, Sansom JE. Contact sensitivity in patients with oral symptoms. Contact Dermatitis 1998; 39:258-259

122 Magnusson B, Wilkinson DS. Cinnamic aldehyde in toothpaste. 1. Clinical aspects and patch tests. Contact Dermatitis 1975;1:70-76

123 Hjorth N, Jervoe P. Allergisk Kontaktstomatitis og Kontaktdermatitis fremkaldt of smagsstoffer i tandpasta. Tandlaegebladet 1967;71:937-942. Data cited in ref. 122

124 de Groot AC. Contact allergy to cosmetics: causative ingredients. Contact Dermatitis 1987;17:26-34

125 Travassos AR, Claes L, Boey L, Drieghe J, Goossens A. Non-fragrance allergens in specific cosmetic products. Contact Dermatitis 2011;65:276-285

126 Rudzki E, Grzywa Z. Dermatitis from propolis. Contact Dermatitis 1983;9:40-45

127 Peltonen L, Wickstrom G, Vaahtoranta M. Occupational dermatoses in the food industry. Dermatosen 1985;33:166-169

128 Paulsen E, Andersen KE. Colophonium and Compositae mix as markers of fragrance allergy: Cross-reactivity between fragrance terpenes, colophonium and Compositae plant extracts. Contact Dermatitis 2005;53:285-291

129 Santucci B, Cristaudo A, Cannistraci C, Picardo M. Contact dermatitis to fragrances. Contact Dermatitis 1987;16:93-95

130 de Groot AC. Contact allergy to cosmetics: causative ingredients. Contact Dermatitis 1987;17:26-34

Chapter 5.69 PETITGRAIN BIGARADE OIL

DEFINITION

Petitgrain bigarade oil (essential oil of bitter orange petitgrain, cultivated) is the essential oil obtained from the leaves and twigs with little green (unripe) fruits of the bitter orange, *Citrus aurantium* L. (synonyms: *Citrus amara* link, *Citrus bigaradia* Loisel, *Citrus vulgaris* Risso).

INCI NOMENCLATURE

Description/definition: Citrus aurantium amara leaf/twig oil is the volatile oil obtained from the leaves and twigs of the bitter orange, *Citrus aurantium* L. var. *amara* L., Rutaceae
INCI name EU: Citrus aurantium amara leaf/twig oil
INCI name USA: Citrus aurantium amara (bitter orange) leaf/twig oil
CAS registry number(s): 8014-17-3; 72968-50-4; 68916-04-1
EINECS number(s): 277-143-2

Description/definition: Citrus aurantium leaf oil is an essential oil obtained from the leaves of the sour orange, *Citrus aurantium*, Rutaceae
INCI name EU: Citrus aurantium leaf oil
INCI name USA: not in the Personal Care Products Council Ingredient Database
CAS registry number(s): 72968-50-4
EINECS number(s): 277-143-2

ISO (INTERNATIONAL ORGANIZATION FOR STANDARDIZATION) STANDARD

ISO number: 8901
ISO name: Essential oil of bitter orange petitgrain, cultivated
Botanical origin: *Citrus aurantium* L. (synonyms: *Citrus amara* link, *Citrus bigaradia* Loisel, *Citrus vulgaris* Risso)
Parts of plant used: Leaf, twig with little green fruit
ISO values: ISO values (minimum and maximum concentrations) are shown in Table 5.69.1.

There is also an ISO standard entitled 'Essential oil of petitgrain, Paraguayan type' (ISO 3064, 2000, revision 2014, Table 5.69.2). The botanical source for this oil is stated to be *Citrus aurantium* L. var. *Paraguay* (syn. *Citrus aurantium* var. *bigaradia* Hook f.). However, the species name *Citrus aurantium* L. var. *Paraguay* cannot be traced in botanical databases and a Google search yields only ISO(-related) websites. Also, its synonym *Citrus aurantium* var. *bigaradia* Hook f. is generally indicated to be synonymous with *Citrus aurantium* L., which is the botanical source for petitgrain bigarade oil, cultivated, the oil discussed in ISO 8901 and presented above. Therefore, ISO 3064 may in fact give standards for the same oils as ISO 8901, but obtained from trees in Paraguay, unconvincingly being presented as a species different from *Citrus aurantium* L. In literature, the

Table 5.69.1 ISO values (%) for petitgrain bigarade oil [a]

Compound	CAS	Minimum	Maximum
Linalyl acetate	115-95-7	40.0	72.0
Linalool	78-70-6	10.0	32.0
α-Terpineol	98-55-5	1.0	7.0
Limonene	138-86-3	1.0	6.0
Geranyl acetate	105-87-3	1.5	5.5
Geraniol	106-24-1	1.0	4.0
(E)-β-Ocimene	3779-61-1	1.0	4.0

[a] ISO 8901 Essential oil of bitter orange petitgrain, cultivated ©ISO 2003; Geneva, Switzerland, www.iso.org

botanical source for petitgrain (bigarade) oils analyzed and presented here, has been either *Citrus aurantium* L., *Citrus aurantium* ssp. *aurantium* or *Citrus aurantium* var./ssp. *amara* (both names are generally considered to be a synonym for *C. aurantium* L.). The botanical names as mentioned in ISO 3064, however, were never found in these articles, although a number of the oils analyzed originated from Paraguay.

Citrus aurantium L. is not only used for the production of petitgrain bigarade oil, but is also the source for obtaining bitter orange oil from cold-pressing of the fruit peels (Chapter 5.64) and neroli oil from the flowers (Chapter 5.60).

THE PLANT, THE OIL, AND THEIR USES

For a general introduction to *Citrus aurantium* L., its fruits and bitter orange oil, which is the essential oil obtained by cold-pressing the peels of ripe bitter orange fruits, see Chapter 5.64 Orange oil, bitter. For the essential oil produced by steam- or hydrodistillation of the *flowers* of the bitter orange tree, see Chapter 5.60 Neroli oil.

Petitgrain bigarade oil is obtained by steam-distillation of the leaves, or of the leaves, young branches

Table 5.69.2 ISO values (%) for petitgrain oil, Paraguayan type[a]

Compound	CAS	Minimum	Maximum
Linalyl acetate	115-95-7	40.0	60.0
Linalool	78-70-6	15.0	30.0
α-Terpineol	98-55-5	3.2	6.8
Geranyl acetate	105-87-3	2.0	5.0
Geraniol	106-24-1	2.0	4.5
Neryl acetate	141-12-8	1.0	3.0
(E)-β-Ocimene	3779-61-1	1.0	3.0
Myrcene	123-35-3	1.3	2.7
Nerol	106-25-2	0.5	2.0
β-Pinene	127-91-3	0.5	2.0
β-Caryophyllene	87-44-5	0.3	1.5
Sabinene	3387-41-5	0.1	0.5

[a] ISO/DIS 3064 Essential oil of petitgrain, Paraguayan type ©ISO 2000; Geneva, Switzerland, www.iso.org

and immature fruits of the bitter orange tree (5,7). It is widely used in perfumery for the sweet and fresh note it gives to colognes and lotions. The oils produced in the Mediterranean area are said to have better odor properties than those produced in Paraguay (6,7). Petitgrain bigarade oil is also employed in the fabrication of soaps because of its good resistance to an alkaline medium (5). In addition, bitter orange leaf oil may be added to foods and drinks as flavoring. The oil may also be employed in aromatherapy practices (44,45).

CHEMICAL COMPOSITION

Petitgrain bigarade oil is a pale yellow to amber yellow clear mobile liquid with a slight blue fluorescence, which has a fresh ethereal, slightly orange and green odor. The yield of essential oil from the leaves and twigs with little green fruit of *Citrus aurantium* L. generally varies from 0.2 to 0.6 per cent. The main producing countries of this oil are Paraguay (assuming the trees there are indeed *C. aurantium* L., see above), Egypt, Morocco and Spain.

Literature data (up to September 3, 2014) on the chemical composition of petitgrain bigarade oils and unpublished analytical data from one of us (E.S.) (oils both from Paraguay and from other countries) are shown in Table 5.69.3 in alphabetical order. In petitgrain bigarade oils from various origins, over 157 chemicals have been identified. About 45 per cent of these were found in a single reviewed publication only. The major compounds found in petitgrain bigarade oils from different sources are shown in Table 5.69. They include (highest concentrations in any study given) linalool (94.1%), linalyl acetate (73.1%), sabinene (55.1%), β-pinene (49.8%), limonene (26.8%), (*E*)-β-ocimene (18.6%), α-terpineol (16.8%), geranyl acetate (12.5%) and geraniol (8.4%). Well-known ingredients of petitgrain bigarade oils that were present in high concentrations (>7%) in one or two studies were myrcene (three studies: 8.3%, 9.8% and 42.6%), γ-terpinene (39.9%), terpinen-4-ol (19.2% and 20.9%), neryl acetate (12.5%), *p*-cymene (8.6%) and citronellal (7.0%). A rare constituent of petitgrain bigarade oils found in a high concentration (>7%) in a single study includes linalool oxide (12.1%).

Commercial oils

The ten chemicals that had the highest maximum concentrations in 47 commercial petitgrain bigarade essential oil samples (concentration ranges provided) are the following: linalyl acetate (41.3-54.0%), linalool (19.8-34.0%), α-terpineol (4.6-7.5%), geranyl acetate (2.5-4.8%), geraniol (0.9-4.4%), (*E*)-β-ocimene (0.9-4.1%), myrcene (0.4-3.6%), neryl acetate (1.6-2.8%), limonene (0.3-2.6%), β-pinene (0.09-1.9%) (Erich Schmidt, unpublished analytical data).

Chemotypes

Virtually all petitgrain bigarade oils have linalool and/or linalyl acetate as main components, with a combined content of >60% of oil. This applies to both commercial samples and oils extracted in the laboratory (9). Several compositions with other dominant constituents have

been observed infrequently and include (summarized in ref. 9): (a) β-pinene associated with nerol and/or linalool; (b) sabinene, followed by β-ocimene and linalool; (c) limonene alone or associated with citronellol + geranial, nerol, β-ocimene and neral; (d) α-phellandrene, followed by decanal and limonene; (e) β-ocimene; (f) myrcene; (g) linalool/terpinen-4-ol. In the investigation of the leaf oils from 26 *C. aurantium* cultivars from Corsica by the authors of that review, 24 were of the linalool/linalyl acetate chemotype, one had the atypical sabinene/(*E*)-β-ocimene (type b above) and one the β-pinene/linalool composition (type a) (9). Chemotypes with high sabinene content and with high β-pinene concentrations have also been observed in some cultivars in China (ref. 8). One may wonder whether high limonene contents (>6-7%) in petitgrain bigarade oils are indicative of adulteration with peel oils.

A review of the literature on petitgrain bigarade oils from before 1996 is provided in ref. 6, a review up to 2009 in ref. 7. Additional analyses of petitgrain bigarade oils not presented in Table 5.69.3 include refs. 40-43. These can all be accessed on-line.

CONTACT ALLERGY/ALLERGIC CONTACT DERMATITIS

General

Contact allergy to/allergic contact dermatitis from petitgrain bigarade oil has been reported in four publications only. In a group of consecutive patients suspected of contact dermatitis, a prevalence rate of 0.5% positive patch test reactions has been observed; relevance data are lacking. Two published case reports both describe occupational contact dermatitis from petitgrain bigarade oil. In one case, linalool and possibly geraniol may have been allergens in the oil.

Testing in groups of patients

Two hundred dermatitis patients from Poland were tested with petitgrain bigarade oil 2% in petrolatum and one (0.5%) reacted. The same authors also patch tested 51 patients allergic to *Myroxylon pereirae* resin (balsam of Peru) and/or turpentine and/or wood tar and/or colophony and three (5.9%) had a positive patch test; relevance data were not provided (47). A group of 86 patients from Poland previously reacting to the fragrance mix was tested with petitgrain bigarade oil and seven (8.1%) had a positive patch test reaction. Relevance data were not provided (49); there were also four reactions to 'petitgrain Paraguay oil' (49).

Case reports

An aromatherapist had non-occupational contact dermatitis with allergies to multiple essential oils used at work, including petitgrain oil. The patient also reacted to geraniol, linalool, linalyl acetate, α-pinene, the fragrance mix and various other fragrance materials; α-pinene, linalool and geraniol were demonstrated by GC-MS in petitgrain oil (46). Linalool is an important chemical in petitgrain

Table 5.69.3 Constituents identified in petitgrain bigarade oils

Constituent	CAS	Percentage and range				
		A	B (9)	C (6)	D (8)	E
(E)-Anhydrolinalool oxide	54750-70-8					0.1[z6]
(Z)-Anhydrolinalool oxide	54750-69-5					0.1[z6]
Apiole	523-80-8					4.1[k]
Aromadendrene	489-39-4					+[x]
Benzaldehyde	100-52-7					0.7[z]
trans-α-Bergamotene	13474-59-4			tr-0.01		tr[f]; 0.01[z2]
Bicycloelemene	32531-56-9					0.2[x]
Bicyclogermacrene	24703-35-3	0.04-0.4		0.04-0.3		0.2[d,i]; 0.3[z2]
(Z)-α-Bisabolene	29837-07-8		0-0.3			
β-Bisabolene	495-61-4			0-0.01		tr[f]; 1.0[z4]
2,6-di-tert-Butyl-p-cresol	128-37-0					0.6[h]
δ-Cadinene	483-76-1	0-0.08		0.02-0.03	0.02-0.5	0.02[d]; 0.04[z2]; 0.07[r]; 0.1[f,h]
δ-Cadinol	19435-97-3					0.01[r]
Camphene	79-92-5		0-0.2	tr-0.01	tr-0.2	tr[b,c]; 0.01[d]; 0.1[a,o,z5]; 1.2[v]
Camphor	76-22-2					tr[y]
δ3-Carene	13466-78-9	tr-0.8	0-1.8	0.2-0.7	0-3.1	0.3[f]; 0.4[o]; 0.7[d]; 1.2[z2]; 2.6[k]; 3.4[a]
Carvacrol	499-75-2					0.9[k]
(Z)-Carveol	1197-06-4					1.0
β-Caryophyllene	87-44-5	0.5-1.1	0-7.0	0.5-0.6	0.1-2.8	1.1[i]; 1.6[p]; 1.8[r]; 2.2[z4]; 3.7[h]; 4.4[k]
Caryophyllene oxide	1139-30-6	0-0.05		0.02-0.07	0.03-0.8	tr[f]; 0.02[d,z2]; 0.04[r]; 2.8[h]
8S,13-Cedraniol						0.9[h]
1,8-Cineole	470-82-6		0-2.4	0.02-0.05	0-1.1	tr[f]; 0.01[d]; 0.06[z2]; 1.9[h]
Citronellal	106-23-0	0-0.1	0-1.0	0.01-0.04	0.03-4.2	tr[f]; 0.04[d]; 0.05[r]; 0.9[v]; 3.5[k]; 7.0[a]
Citronellol	106-22-9				tr-0.9	0.02[m]; 0.5[u]; 1.8[v]; 2.3[g]; 4.7[k]
Citronellyl acetate	150-84-5			tr	0.03-0.7	0.07[r]; 0.1[q,z2]; 1.0[a]
α-Copaene	3856-25-5			tr-0.01		0.01[z2]
β-Copaene	18252-44-3					0.2[g]
α-Cubebene	17699-14-8			tr-0.01		0.02[z2]
Cyclofenchene	488-97-1					0.1[q]; 0.6[x]
α-Cyclogeraniol	6627-74-3					2.0[h]
m-Cymene	535-77-3					0.5[q]; 4.6[k]
o-Cymene	527-84-4			0-0.01		0.9[h]
p-Cymene	99-87-6	tr-0.1	0-0.1	0.03-0.08	0-8.6	0.1[a]; 0.2[x]; 0.4[j]; 0.8[u]; 1.5[k]; 2.7[m]
p-Cymenene	1195-32-0			0.01-0.08	0-1.5	tr[p]
p-Cymen-8-ol	1197-01-9					0.01[d]
Decanal	112-31-2	tr-0.1			0-0.06	tr[d,m]; 0.02[j,r]; 0.1[u]
2,3-Dehydro-1,8-cineole	92760-25-3					0.2[h]
2,6-Dimethoxy-1-methylbenzene	5673-07-4					0.6[h]
(E)-10-Dodecenyl acetate	35153-09-4					0.7[h]
β-Elemene	33880-83-0			tr-0.02	tr-0.8	tr[f]; 0.02[d]; 0.03[r]; 0.04[z2]
δ-Elemene	20307-84-0			0.01	0-0.9	0.01[d]; 0.02[z2]
Elemol	639-99-6				0-1.9	
trans-4,5-Epoxide-(E)-2-decenal	134454-31-2					4.0[h]
trans-4,5-Epoxy-(E)-2-nonenal						0.3[h]
Eugenol	97-53-0					1.2[k]
Farnesene						tr[b,c]; 0.1[e,z1]
(E,E)-α-Farnesene	502-61-4			0.01-0.06	0-1.2	tr[f]; 0.05[z2]
(E)-β-Farnesene	18794-84-8					tr[q]; 0.04[d]; 0.1[i]; 0.5[r]
(Z)-β-Farnesene	28973-97-9			0.04-0.07	0-0.7	0.07[z2]; 0.08[h,r]; 0.1[f]
Farnesol	4602-84-0					0.2[y,z]
Farnesyl acetate	29548-30-9					0.09[x]
α-Fenchene	471-84-1			tr		
Fenchyl acetate	13851-11-1					2.8[s]
Geranial	141-27-5	tr-0.1	0-1.0	0.4-0.6	tr-3.5	0.3[d]; 0.6[a,f]; 3.9[l]

Table 5.69.3 Constituents identified in petitgrain bigarade oils *(continued)*

Constituent	CAS	Percentage and range				
		A	B (9)	C (6)	D (8)	E
Geraniol	106-24-1	0.9-4.4	0-6.7	0.7-1.0	tr-6.1	6.1[b]; 6.4[c]; 6.6[g]; 7.1[e,z1]; 8.4[z3]
Geranyl acetate	105-87-3	2.5-4.8	0.1-5.5	1.9-3.2	0-5.6	5.5[z3]; 6.0[e]; 6.2[b]; 6.4[c]; 8.7[i]; 12.5[k]
Geranyl formate	105-86-2					0.2[g]; 2.0[x]
Germacrene B	15423-57-1				0-1.2	0.2[h]
Germacrene D	23986-74-5					0.04[r]; 0.1[p]
Heptanal	111-71-7					0.02[u]
Hexanal	66-25-1			tr		
(Z)-3-Hexanol						tr[g]
(E)-2-Hexenal	6728-26-3					0.3[z6]
(E)-2-Hexen-1-ol	928-95-0					tr[z6]
(Z)-3-Hexen-1-ol	928-96-1					tr[m]; 0.1[z5]; 2.3[j]
(E)-2-Hexenyl acetate	2497-18-9					0.01[d]
α-Humulene	6753-98-6	0.01-0.1	0-tr	0.04-0.06	0.02-0.7	0.05[d]; 0.1[f,i]; 0.2[p]; 1.4[y]; 2.5[h]; 4.7[k]
14-Hydroxy-1-epi-caryophyllene						0.2[h]
8-Hydroxymenthol						3[s]
Indole	120-72-9					
Isoborneol	124-76-5					0.2[h,s]
Isodihydronepeta-lactone						1.7[h]
Isopulegol	89-79-2				0-0.4	tr[z6]
Isoterpinolene	586-63-0			0.02-0.06		tr[f]
Lavandulyl acetate	25905-14-0					0.9[h]
Lilac alcohol A	33081-34-4					0.7[h]
Limonene	138-86-3	0.3-2.6	0.6-2.3	0.4-2.2	0.5-6.9	1.7[h]; 1.9[d]; 2.3[k]; 2.5[f]; 4.0[p]; 4.4[q]; 5.4[r]; 6.3[a]; 6.7[i]; 8.0[n]; 11.1[z3]; 26.8[v]
Linalool	78-70-6	19.8-34.0	1.0-37.7	21.7-32.6	0.4-39.8	36.1[x]; 36.8[e]; 37.3[z4]; 39.4[t]; 58.2[i]; 64.1[s]; 66.0[h]; 66.1[g]; 71.0[j]; 94.1[z]
Linalool oxide	1365-19-1					0.2[p]; 12.1[l]
cis-Linalool oxide	11063-77-7	tr-0.08			tr-0.1	tr[b,c,q]; 0.04[d]; 0.06[r]; 1.0[h]; 0.2[t]
cis-Linalool oxide, furanoid	11063-77-7		0-0.2	0.03-0.09		0.05[z2]; 0.08[z5]; 0.1[f]; 3.2[x]; 5.4[y]
cis-Linalool oxide, pyranoid	14009-71-3					0.2[y]; 1.1[h]
trans-Linalool oxide	11063-78-8	0-0.1				tr[b,c,q]; 0.04[r,z1]; 1.6[h]
trans-Linalool oxide, furanoid	34995-77-2		0-0.1	0.01-0.03		tr[f]; 0.03[z2]; 0.07[x]; 0.1[z5]; 0.3[y]
trans-Linalool oxide, pyranoid	39028-58-5					0.6[y]
Linalyl acetate	115-95-7	41.3-54.0	0-36.8	50.7-62.6	0-34.7	44.3[w]; 45.5[p]; 48.9[r]; 50.1[f]; 50.8[o]; 54.6[d]; 55.0[m]; 56.8[q]; 71.0[n]; 73.1[a]
Linalyl propionate	144-39-8			0.02-0.04		tr[f]; 0.04[d]
p-Mentha-1,4-dien-7-ol	22539-72-6					0.07[y]
p-Mentha-1,8(9)-di-enyl acetate						0.01[r]
cis-p-Menth-2-en-1-ol	29803-82-5		0-0.4	tr-0.01		0.01[d]
trans-p-Menth-2-en-1-ol	29803-81-4			tr-0.02		tr[f]
p-Menth-6-en-2-one	43205-82-9					tr[q]
2-Methoxy-3-isobu-tylpyrazine	24683-00-9					tr[z6]
Methyl anthranilate	134-20-3					0.1[p,r,z]; 2.4[v]; 3.5[y]
Methylcarvacrol	6379-73-3					8.1[s]
Methyl geranate	2349-14-6			tr-0.03	0-0.04	tr[f]; 0.03[d]; 0.3[a]
6-Methyl-5-hepten-2-one	110-93-0			0.01-0.1		tr[f]; 0.05[z5]; 0.06[d]; 0.08[z2]
Methyl N-methyl anthranilate	85-91-6			tr-0.1		tr[f,p]; 0.05[r]; 0.2[z2]; 0.3[d]; 2.5[u]
γ-Muurolene	30021-74-0					0.5[y]

Table 5.69.3 Constituents identified in petitgrain bigarade oils *(continued)*

Constituent	CAS	Percentage and range				
		A	B (9)	C (6)	D (8)	E
Myrcene	123-35-3	0.4-3.6	2.0-3.5	0.6-1.2	0.5-3.1	1.8[f]; 2.0[h]; 2.3[e]; 2.4[n]; 2.5[b,c]; 2.6[r]; 2.7[i]; 5.4[w]; 5.5[m]; 8.3[k]; 9.8[a]; 42.6[l]
Neral	106-26-3	0.01-0.2	0-0.7	0.2-0.4	tr-2.4	0.3[f]; 0.4[a]; 1.8[v]; 2.2[j]; 3.9[j];5.1[k]
Nerol	106-25-2	0.6-1.5	tr-2.3	0.8-1.0	0.06-1.9	1.9[z3]; 2.2[b,c]; 2.4[e]; 2.9[j]; 3.3[k]; 4.3[j]
Nerolidol	7212-44-4					0.1[q]; 0.2[b,c,p]; 0.3[z1]
(E)-Nerolidol	40716-66-3	0-0.06	0-0.4	0.05-0.08	0.02-4.0	tr[e]; 0.06[z2]; 0.07[d]; 0.1[f,i]
(Z)-Nerolidol	3790-78-1					tr[y]
Neryl acetate	141-12-8	1.6-2.8	0.1-2.9	1.0-1.7	0.2-3.1	2.6[m]; 3.0[n]; 3.3[b]; 3.4[c]; 4.5[i]; 12.5[k]
Neryl formate	2142-94-1					+[x]
Nonanal	124-19-6		0-tr	0.02-0.05	0-0.2	tr[f]; 0.03[u]; 0.5[y]
Nonane	111-84-2					6.9[a]
Nootkatone	4674-50-4					0.03[r]
β-Ocimene	13877-91-3					1.8[v]; 1.9[x]; 2.6[j]
(E)-β-Ocimene	3779-61-1	0.9-4.1	2.5-15.1	0.6-1.8	1.6-18.6	2.5[o]; 2.9[b,c]; 3.3[m]; 3.6[z2]; 4.1[i]; 5.9[a]
(Z)-β-Ocimene	3338-55-4	0.3-1.8	0.2-1.1	0.2-0.4	0.08-1.0	1.2[i]; 1.7[w]; 2.2[h]; 2.9[a]; 2.6[z3]; 3.0[z1]
allo-Ocimene	673-84-7		0-tr			0.4[a]
n-Octanal	124-13-0		0-0.1			0.01[r]; 0.02[u]
Octanol	29063-28-3				0-0.04	0.3[g]
γ-Patchoulene	508-55-4					+[x]
(Z)-2-Penten-1-ol	1576-95-0					0.1[z6]
Perillaldehyde	2111-75-3					0.01[r]
Perillene	539-52-6					tr[p]; 0.01[r]
α-Phellandrene	99-83-2	0-0.1	0-0.2	tr-0.03	0-0.1	0.04[d]; 0.05[z2]; 0.1[g]; 0.2[m]; 0.5[a]; 2.7[j]
β-Phellandrene	555-10-2	0-0.01	0-0.8	0.03-0.04	0.03-0.9	tr[f,p]; 0.05[b,c]; 0.1[h,o]
2-Phenethyl alcohol	60-12-8					0.2[p,r]
Phytol	7541-49-3					0.2[x]
α-Pinene	80-56-8	0.01-0.2	0.1-1.9	0.03-0.3	0.1-4.0	0.1[f]; 0.2[c,e]; 0.4[h]; 1.2[z4]; 1.4[a]; 1.7[k]
β-Pinene	127-91-3	0.09-1.9	2.1-36.7	0.7-1.2	0.8-49.8	2.5[j]; 2.7[n]; 3.0[k]; 3.6[i]; 3.8[t]; 23.4[a]
trans-Pinocarveol	1674-08-4					tr[y]
Pinocarvone	30460-92-5					0.8[h]
Pseudolimonene	499-97-8					0.01[h]; 1.0[a]
Sabinene	3387-41-5	0.02-0.4	0.3-52.6	0.1-0.2	0.3-55.1	0.6[j]; 1.0[v]; 1.4[n]; 1.6[x]; 3.6[h]; 8.9[a]
cis-Sabinene hydrate	15537-55-0			tr-0.01	tr-0.4	tr[f]
trans-Sabinene hydrate	17699-16-0		0-0.6			
(Z)-α-Santalol	115-71-9					0.09[h]
β-Sesquiphellandrene	20307-83-9					0.01[d]
α-Sinensal	17909-77-2					0.1[x]
Spathulenol	6750-60-3		0-0.1	0.03-0.1	0.02-0.8	0.02[d]; 0.03[z2]; 0.1[f,r]
α-Terpinene	99-86-5		0-2.0	tr-0.02	tr-0.7	tr[b,c]; 0.01[d]; 0.3[p]; 1.5[h]; 1.9[a]
γ-Terpinene	99-85-4	0.02-0.09	0-3.2	0.01-0.09	0.03-39.9	1.1[m]; 1.6[u]; 2.5[v]; 4.1[k]; 6.5[l]; 7.7[a]
Terpinen-4-ol	562-74-3	0.05-0.2		0.05-0.08	0.08-2.3	0.4[a]; 0.6[j]; 0.7[u]; 0.8[m]; 19.2[s]; 20.9[g]
Terpinen-4-yl acetate	4821-04-9					1.5[h]
α-Terpineol	98-55-5	4.6-7.5	0.3-11.8	3.1-5.6	0.08-10.4	5.9[j]; 6.8[n]; 7.2[t]; 7.6[m]; 7.8[w]; 7.9[z3]; 9[b,c]; 11.7[e,z1]; 12.1[h]; 12.9[i]; 16.8[k]
Terpinolene	586-62-9	0.2-0.7	0.4-0.8	0.08-0.2	0.08-1.8	0.3[d]; 0.4[j]; 0.5[e]; 0.6[b,c]; 0.7[i]; 1.1[a]
Terpinyl acetate	8007-35-0	0-0.2				0.1[a]
α-Terpinyl acetate	80-26-2		0-0.1	0.08-0.2	0-0.2	0.06[z2]; 0.1[p,r]; 0.2[f]; 1.1[v]; 2.2[m]; 2.3[w]
α-Thujene	2867-05-2	0.03-0.09	0-0.5	tr-0.01	tr-1.8	tr[p,y]; 0.01[d]; 0.02[r,z2]; 0.3[a]; 0.7[z4]
Thymol	89-83-8				0-8.1	tr[j,m]; 0.5[k]; 3.2[a]
Tricyclene	508-32-7			tr		tr[f,z2]; 0.3[a]
1,3,5-Trimethylbenzene	108-67-8					0.04[h]
1-(2,3,4,5-Trimethyl-phenyl)-butan-1-one						0.05[h]

Table 5.69.3 Constituents identified in petitgrain bigarade oils *(continued)*

Constituent	CAS	Percentage and range				
		A	B (9)	C (6)	D (8)	E
2,2,6-Trimethyl-6-vi-nyltetrahydropyran	7392-19-0					0.01[r]
Undecanal	112-44-7				0-0.08	
4-Undecanone	14476-37-0					0.3[h]
2-Undecenal	2463-77-6			0-tr		0.7[h]
Valencene	4630-07-3				0-0.5	0.03[r]; 0.5[a]
α-Zingiberene	495-60-3					5.3[h]

A forty-seven bitter orange petitgrain essential oil samples (both cultivated and Paraguayan type) from Paraguay, Morocco, Spain and Egypt, analyzed between 1998 and 2013; lowest and highest concentrations given (E. Schmidt, unpublished data)
B twenty-six lab-hydrodistilled leaf oils from 26 cultivars grown at an agricultural research center in Corsica, France; lowest and highest concentrations given (ref. 9)
C five industrially prepared petitgrain bigarade oils from Sicily, Italy; lowest and highest concentrations given (ref. 6)
D sixteen steam-distilled oils from 16 cultivars grown in China; lowest and highest concentrations given (ref. 8)
E data from other studies (indicated with superscript letters); highest concentrations found in any study reviewed here given; when two or more oils were investigated, only the highest concentrations are mentioned, unless indicated otherwise

[a] five leaf oils from 5 clones of C. aurantium cultivated in Italy (ref. 12); [b] four lab-hydrodistilled oils from the leaves of 4 provenances of bitter orange trees (ref. 2); [c] seven lab-hydrodistilled oils of leaves harvested between November and May in Tunisia (ref. 3); [d] one petitgrain bigarade oil from Egypt (ref. 1); [e] one lab-hydrodistilled leaf oil from Tunisia (ref. 4); [f] one commercial steam-distilled leaf oil from a citrus plantation in Turkey (ref. 5); [g] one lab-hydrodistilled leaf oil from Mauritius (ref. 11); [h] five lab-hydrodistilled oils from the leaves of the bitter orange tree *C. aurantium* ssp. aurantium from five harvests between April and January (ref. 14); [i] two lab-hydrodistilled oils, one from old leaves and one from young leaves of the bitter orange tree collected in Greece (ref. 16); [j] eleven laboratory steam-distilled oils from 11 bitter orange cultivars produced in the USA (ref. 30); [k] one laboratory steam-distilled sample of petitgrain bigarade oil from Egypt (ref. 31); [l] one lab-produced oil from China with extremely high concentration of myrcene and absence of linalyl acetate (ref. 38); [m] several commercial oils (ref. 32); [n] ten commercial oil samples (5 from France, 4 from Paraguay, 1 from Egypt) (ref. 33); [o] one commercial oil sample (ref. 34); [p] one oil sample from Spain (ref. 35); [q] one commercial oil sample (ref. 36); [r] one commercial oil sample (ref. 37); [s] one lab-hydrodistilled leaf oil from Tunisia (ref. 13); [t] one lab-hydrodistilled oil from C. aurantium leaves from Iran (ref. 39); [u] one Italian bitter orange petitgrain oil sample (ref. 17); [v] one bitter orange petitgrain oil produced in Egypt (ref. 19); some results appear to be unreliable (ref. 18); [w] one redistilled Paraguayan petitgrain oil (ref. 20); [x] one steam-distilled oil produced in Algeria (ref. 21); [y] one bitter orange leaf oil produced in China (ref. 22); [z] one leaf oil from Sicily, Italy, containing no linalyl acetate (ref. 23); [z1] two oils, one from Greece and one from Tunisia (ref.24); [z2] one petitgrain bigarade oil from Sicily (ref. 26); [z3] four lab-hydrodistilled leaf oils from 4 bitter orange cultivars (ref. 27); [z4] one essential oil from south Italy (ref. 28); [z5] one oil of petitgrain bigarade (ex *C. aurantium*) of European origin (ref. 29); [z6] data from a literature review cited in ref. 6

tr: trace (in columns B, E[b], E[c], E[e], E[f]: <0.05; in column C: < 0.01); + present in the oil investigated, but quantity not stated

Table 5.69.4 Major constituents of petitgrain bigarade oils

Constituent	CAS	Percentage and range				
		A	B (9)	C (6)	D (8)	E
Linalool	78-70-6	19.8-34.0	1.0-37.7	21.7-32.6	0.4-39.8	64.1[s]; 66.0[h]; 66.1[g]; 71.0[j]; 94.1[z]
Linalyl acetate	115-95-7	41.3-54.0	0-36.8	50.7-62.6	0-34.7	54.6[d]; 55.0[m]; 56.8[q]; 71.0[n]; 73.1[a]
Sabinene	3387-41-5	0.02-0.4	0.3-52.6	0.1-0.2	0.3-55.1	0.6[j]; 1.0[v]; 1.4[n]; 1.6[x]; 3.6[h]; 8.9[a]
β-Pinene	127-91-3	0.09-1.9	2.1-36.7	0.7-1.2	0.8-49.8	2.5[j]; 2.7[n]; 3.0[k]; 3.6[j]; 3.8[t]; 23.4[a]
Limonene	138-86-3	0.3-2.6	0.6-2.3	0.4-2.2	0.5-6.9	5.4[r]; 6.3[a]; 6.7[j]; 8.0[n]; 11.1[z3]; 26.8[v]
(E)-β-Ocimene	3779-61-1	0.9-4.1	2.5-15.1	0.6-1.8	1.6-18.6	2.5[o]; 2.9[b,c]; 3.3[m]; 3.6[z2]; 4.1[j]; 5.9[a]
α-Terpineol	98-55-5	4.6-7.5	0.3-11.8	3.1-5.6	0.08-10.4	9[b,c]; 11.7[e,z1]; 12.1[h]; 12.9[j]; 16.8[k]
Geranyl acetate	105-87-3	2.5-4.8	0.1-5.5	1.9-3.2	0-5.6	5.5[z3]; 6.0[e]; 6.2[b]; 6.4[c]; 8.7[j]; 12.5[k]
Geraniol	106-24-1	0.9-4.4	0-6.7	0.7-1.0	tr-6.1	6.1[b]; 6.4[c]; 6.6[g]; 7.1[e,z1]; 8.4[z3]

LEGEND: SEE UNDER TABLE 5.69.3

bigarade oils and has been found in commercial oils in concentrations of 20-34%; geraniol had a highest concentration of 4.4% (Table 5.69.3, column A). Two bottle fillers in a perfume factory developed occupational allergic contact dermatitis from petitgrain bigarade oil and other essential oils they worked with; there was also one positive patch test reaction to petitgrain bigarade oil among twenty workers who did not have dermatitis (48).

LITERATURE

1 Dugo G, Bonaccorsi I, Sciarrone D, Costa R, Dugo P, Mondello L, et al. Characterization of oils from the fruits, leaves and flowers of the bitter orange tree. J Essent Oil Res 2011;23(2):45-59

2 Boussaada O, Skoula M, Kokkalou E, Chemli R. Chemical variability of flowers, leaves, and peels oils of four sour orange provenances. J Essent Oil Bear Plants 2007;10:453-464

3 Boussaada O, Chemli R. Seasonal variation of essential oil composition of Citrus aurantium L. var. amara. J Essent Oil Bear Plants 2007;10:109-120

4 Boussaada O, Chemli R. Chemical composition of essential oils from flowers, leaves and peel of Citrus aurantium L. var. amara from Tunisia. J Essent Oil Bear Plants 2006;9:133-139

5 Kirbaslar G, Kirbaslar SI. Composition of Turkish bitter orange and lemon leaf oils. J Essent Oil Res 2004;16:105-108

6 Mondello L, Dugo G, Dugo P, Bartle KD. Italian citrus petitgrain oils. Part I. Composition of bitter orange petitgrain oil. J Essent Oil Res 1996;8:597-609. Data also cited in ref. 25

7 Dugo G, Cotroneo A, Bonaccorsi I. Composition of petitgrain oils. In: G Dugo and L Mondello, Eds. Citrus oils: composition, advanced analytical techniques, contaminants, and biological activity. Boca Raton, USA: CRC Press, Taylor & Francis Group, 2011:253-330

8 Huang Y, Pu Z, Chen Q. The chemical composition of the leaf essential oils from 110 Citrus species, cultivars, hybrids and varieties of Chinese origin. Perfum Flav 2000;25(1):53-66

9 Lota M-L, de Rocca Serra D, Jacquemond C, Tomi F, Casanova J. Chemical variability of peel and leaf essential oils of sour orange. Flavour Fragr J 2001;16:89-96. Data also partly presented in ref. 15 (cited in ref. 7)

10 Juchelka D, Steil A, Witt K, Mosandl A. Chiral compounds of essential oils. XX. Chirality evaluation and authenticity profiles of neroli and petitgrain oils. J Essent Oil Res 1996;8:487-497

11 Gurib-Fakim A, Demarne F. Aromatic plants of Mauritius: Volatile constituents of the leaf oils of Citrus aurantium L., Citrus paradisi Macfad and Citrus sinensis (L.) Osbeck. J Essent Oil Res 1995;7:105-109

12 De Pasquale F, Siragusa M, Abbate L, Tusa N, De Pasquale C, Alonzo G. Characterization of five sour orange clones through molecular markers and leaf essential oils analysis. Scientia Horticulturae 2006;109:54-59

13 Hosni K, Hassen I, M'Rabet Y, Sebei H, Casabianca H. Genetic relationships between some Tunisian Citrus species based on their leaf volatile oil constituents. Biochem Syst Ecol 2013;50:65-71

14 Ellouze I, Abderrabba M, Sabaou N, Mathieu F, Lebrihi A, Bouajila J. Season's variation impact on Citrus aurantium leaves essential oil: Chemical composition and biological activities. J Food Sci 2012;77(9):T173-T180

15 de Rocca Serra D, Lota M-L, Tomi F, Casanova J. Essential oils and taxonomy among Citrus. Example of bergamot. Riv Ital EPPOS 1998;(Numero speciale):38-43. Data also presented in ref. 9 (cited in ref. 7)

16 Sarrou E, Chatzopoulou P, Dimassi-Theriou K, Therios I. Volatile constituents and antioxidant activity of peel, flowers and leaf oils of Citrus aurantium L. growing in Greece. Molecules 2013;18:10639-10647

17 Calvarano I. Le essenze Italiane di petitgrain. Nota II. Petitgrain bigarade e bergamotto. Essenz Deriv Agrum 1968;38:31-48. Data cited in ref. 7

18 Lawrence BM. Progress in essential oils. Perfum Flavor 2003;28(3):54-? (last page unknown)

19 Karawya MS, Balbaa SI, Hifnawy MS. Study of the leaf essential oils of bitter orange and bergamot growing in Egypt. Amer Perfum 1970;85(11):29-32. Data cited in ref. 18

20 Urbeita-Rehnfeld JC, Jennings WG. Paraguayan petitgrain oil. Chem Mikrobiol Technol Lebensm 1974;3:36-38. Data cited in ref. 18

21 Baaliouamer A, Meklati BY. Analysis of bitter orange petitgrain essential oil by combined gas chromatography-mass spectrometry. Agric Biol Chem 1986;50:2111-2114. Data cited in ref. 7

22 Lin ZK, Hua F, Gu YH. The chemical constituents of the essential oil from the flowers, leaves and peels of Citrus aurantium. Act Bot Sinica 1986;28:641-645. Data cited in ref. 18

23 Germanà MA, De Pasquale F, Bazan E, Palazzolo E. Indagine sugli oli essenziali contenuti nei fiori, nelle foglie e nei germogli di cinque specie di citrus. Essenz Deriv Agrum 1990;60:297-312. Data cited in ref. 7

24 Boussaada O. Variation of essential oil yield and composition of Citrus aurantium var. amara. MSc Thesis Mediterranean Agronomic Institute, Chania, Greece, 1995. Data cited in refs. 7 and 18

25 Dugo G, Mondello L, Cotroneo A, Stagno d'Alcontres I, Basile A, Previti P, et al. Characterization of Italian Citrus petitgrain oils. Perfum Flavor 1996;21(3):17-28. Data also cited in ref. 6

26 Mondello L, Dugo P, Dugo G, Bartle KD. On-line HPLC-HRGC-MS for the analysis of natural complex mixtures. J Chromat Sci 1996;34:174-180

27 Protopapadakis E, Papanikolau X. Characterization of Citrus aurantium and C. taiwanica rootstocks by isoenzyme and essential oil analysis. J Hort Sci Biotechnol 1998;73:81-85

28 Adami M, Arcuri L, Di Giacomo G, Ferri D, Gazzola G, Gionfriddo F, et al. Estrazione di petit-grain con CO_2 supercritica. Essenz Deriv Agrum 2000;70:193-200. Data cited in ref. 18

29 Kubeczka K-H, Formacek V. Essential oils analysis by capillary gas chromatography and carbon-13 NMR spectroscopy, 2nd edition. New York, USA: John Wiley and Sons, 2002. Data cited in ref. 18

30 Ortiz JM, Kumamoto J, Scora RW. Possible relationships among sour oranges by analysis of their essential oils. Int Flav Food Addit 1978;5:224-226

31 Haggag EG, Mahmoud II, Abou-Moustafa EA, Mabry TJ. Coumarins, fatty acids, volatile and non-volatile terpenoids from the leaves of *Citrus aurantium* L. (sour orange) and *Citrus sinensis* (L.) Osbek (sweet orange). Asian J Chem 1999;11:784-789

32 Lawrence BM. Progress in essential oils. Perfum Flav 1980;5(6):27-32. Data cited in ref. 7

33 Prager MJ, Miskiewicz MA. Gas chromatographic-mass spectrometric analysis, identification and detection of adulteration of perfumery products from bitter orange trees. J Assoc Off Anal Chem 1981;64:131-138. Data cited in ref. 7

34 Formàček V, Kubeczka KH. Essential oils analysis by capillary chromatography and C-13 NMR spectroscopy. New York: Wiley & Sons, 1982. Data cited in ref. 7

35 Boelens MH, Sindreu RJ. Essential oils from Seville bitter orange (*Citrus aurantium* L. ssp. *amara* L.). In: BM Lawrence, BD Mookherjee and BJ Willis, Eds. Flavors and fragrances: A world perspective. Amsterdam: Elsevier Scientific Publishers BV, 1988:551-565. Data cited in ref. 7

36 Haubruge E, Lognay G, Marlier M, Danhier P, Gilson, J-C, Gaspar Ch. Étude de la toxicité de cinq huiles essentielles extraites de *Citrus* sp. á l'égard de *Sitophilus zeamais* Motsch (Col., Curculionidae), *Prostephanus truncatus* (Horn) (Col., Bostrychidae) et *Tribolium castaneum* Herbst (Col., Tenebrionidae). Med Fac Landbouww Rijksuniv Gent 1989;54:1083-1093. Data cited in ref. 7

37 Boelens MH, Oporto A. Natural isolates from Seville bitter orange tree. Perfum Flav 1991;16:2-7 (issue number missing). Data cited in ref. 7

38 Lin Z, Hua, Y. Systematic evolutional relation of chemical components of the essential oils from 11 taxa of citrus leaves. Acta Botan Sin 1992;34:133-139. Data cited in ref. 7

39 Azadi B, Nickavar B, Amin G. Volatile constituents of the peel and leaf of *Citrus aurantium* L. cultivated in the north of Iran. J Pharm Health Sci 2012;1(3):37-41

40 Periyanayagam K, Dhanalakshmi S, Karthikeyan V, Jagadeesan M. Chemical composition, cytotoxicity and antioxidant activities of the essential oil from the leaves of *Citrus aurantium* L. J Nat Prod Plant Resour 2013;3:19-23

41 Majnooni M-B, Mansouri K, Gholivand MB, Mostafaie A, Mohammadi-Motlagh HR, Afnanzade N-S, Abolghasemi MM, et al. Chemical composition, cytotoxicity and antioxidant activities of the essential oil from the leaves of *Citrus aurantium* L. Afr J Biotechnol 2012;11:498-503

42 Najafian S, Rowshan V. Comparative of HS SPME and HD techniques in *Citrus aurantium* L. Int J Med Arom Plants 2012;2:488-494

43 Trabelsi D, Ammar AH, Bouabdallah F, Zagrouba F. Antioxidant and antimicrobial activities of essential oils and methanolic extracts of Tunisian *Citrus aurantium* L. IOSR J Environ Sci Toxicol Food Technol 2014;8(5):18-27

44 Rhind JP. Essential oils. A handbook for aromatherapy practice, 2nd Edition. London: Singing Dragon, 2012

45 Lawless J. The encyclopedia of essential oils, 2nd Edition. London: Harper Thorsons, 2014

46 Dharmagunawardena B, Takwale A, Sanders KJ, Cannan S, Roger A, Ilchyshyn A. Gas chromatography: an investigative tool in multiple allergies to essential oils. Contact Dermatitis 2002;47:288-292

47 Rudzki E, Grzywa Z, Bruo WS. Sensitivity to 35 essential oils. Contact Dermatitis 1976;2:196-200

48 Schubert HJ. Skin diseases in workers at a perfume factory. Contact Dermatitis 2006;55:81-83

49 Rudzki E, Grzywa Z. Allergy to perfume mixture. Contact Dermatitis 1986;15:115-116

Chapter 5.70   PINE NEEDLE OIL (SCOTS PINE OIL)

DEFINITION
Pine needle oil is the essential oil obtained from the needles and twigs of the Scotch pine, *Pinus sylvestris* L.

INCI NOMENCLATURE
Description/definition: Pinus sylvestris leaf oil is the volatile oil obtained from the needles of the Scotch pine, *Pinus sylvestris* L., Pinaceae
INCI name EU & USA: Pinus sylvestris leaf oil
Other names: Scots pine oil; pine scotch oil
CAS registry number(s): 8023-99-2; 84012-35-1
EINECS number(s): 281-679-2

ISO (INTERNATIONAL ORGANIZATION FOR STANDARDIZATION) STANDARD
Currently, there is no ISO standard for pine needle oil.

Other species of the genus *Pinus* from which pine needle oils are made and that are mentioned in CosIng include *Pinus nigra* (CAS 90082-74-9; EINECS 290-165-7) and *Pinus mugo* (see Chapter 5.32 Dwarf pine oil). It is highly likely that some oils commercially available under the name 'pine needle oil' are in fact obtained from other *Pinus* species, e.g., from *Pinus yunnanensis* (the Yunnan pine from China) and from *Pinus pinaster*, without the botanical origin being specifically mentioned.

THE PLANT, THE OIL, AND THEIR USES
Pinus sylvestris L. (sometimes termed *Pinus silvestris*– botanically incorrect, but linguistically correct, sylvestris being derived from the Latin word silva, forest) is an evergreen coniferous tree that can grow up to 25-40 meters tall and 0.5-1.2 meters in diameter. It is native to the temperate regions of Asia and Europe, grows naturally from Scotland to the Pacific Ocean and from above the Arctic Circle to the Mediterranean. *P. sylvestris* has been widely planted in New Zealand and in the colder regions of North America and is listed as an invasive species in some areas there. It is the world's most widespread pine (www.conifers.org; GRIN Taxonomy for Plants; 1,6).

Pine needle essential oils are widely used as fragrances in perfumes and other cosmetics, in pharmaceuticals (1), as flavoring additives for food and beverages, as scenting agents in a variety of household products, and as intermediates in the synthesis of perfume chemicals (17). Scots pine oil is believed to possess a wide range of pharmacological properties including antibacterial, antifungal, antiviral, antiseptic, anti-neuralgic, cholagogue, choleretic, diuretic, and expectorant. It is used in illnesses of the respiratory system such as cough or catarrh and applied in medical baths, compresses and massages for its warmth effect. Its insecticidal and deodorant properties are also well known (6). In aromatherapy, the oils are used for medicinal purposes as carminative, rubefacient, emmenagogue, and abortifacient agents (1,17).

CHEMICAL COMPOSITION
Pine needle oil is a colorless clear mobile liquid, which has a harsh, fresh and coniferous terpeny and woody odor. The yield of essential oil from the needles of *Pinus sylvestris* L. generally varies from 0.3 to 0.6 per cent. The main producing countries of this oil are China, USA, Canada, Austria, Hungary, Poland and Russia.

Literature data (up to October 25, 2014) on the chemical composition of pine needle oils and unpublished analytical data from one of us (E.S.) are shown in Table 5.70.1 in alphabetical order. In pine needle oils from various origins, over 255 chemicals have been identified. About 56 per cent of these were found in a single reviewed publication only. The major compounds found in pine needle oils from different sources are shown in Table 5.70.2. They include (highest concentrations in any study given) α-pinene (89.5%), δ3-carene (77.1%), myrcene (34.7%), δ-cadinene (34.0%; one very high concentration, all others <11.5%), β-pinene (33.4%), bornyl acetate (32.7%), limonene (30.1%; one very high concentration, all others <9.4%), camphene (27.4%) and γ-cadinene (5.5%). Well-known ingredients of pine needle oils that were present in high concentrations (>7%) in one or two studies were α-phellandrene (31.4%), β-phellandrene (three studies: 10.9%, 13.9% and 29.1%), terpinolene (18.9%), germacrene D (10.3% and 16.1%), α-terpinene (15.5%), α-muurolene (11.6% and 13.9%), caryophyllene oxide (12.7%), β-caryophyllene (10.8%), sabinene (10.8%), borneol (10%), α-cadinol (8.9%) and p-cymene (7.6%).

Uncommon or rare constituents of pine needle oils found in high concentrations (>7%) in single studies include aromadendrene (44.7%), manoyl oxide (30.2%), 1-terpineol (21.8%), β-terpineol (17.9%), geranyl acetate (12.3%), *trans*-verbenol (9.6%), 6-camphenone (9.5%), α-cubebene (9.3%), nerolidol (8.0% and 8.1%) and germacrene A (7.5%).

Commercial oils
The ten chemicals that had the highest maximum concentrations in 112 commercial pine needle essential oil samples (concentration ranges provided) are the following: limonene (3.1-30.1%), bornyl acetate (8.4-29.9%), camphene (4.2-27.4%), α-pinene (13.3-20.4%), myrcene (0.4-13.5%), δ3-carene (1.4-13.0%), β-pinene (0.9-7.1%), β-phellandrene (0.1-6.0%), α-phellandrene (0.04-5.3%) and γ-cadinene (0-5.3%) (Erich Schmidt, unpublished analytical data).

Chemotypes
P. sylvestris trees can be divided in two groups: a low δ3-carene group and a high δ3-carene type, according to low and high concentrations of this compound in the trees' needles; this is mirrored in the amount of δ3-carene found in pine needle essential oils (6). Oils

Table 5.70.1 Constituents identified in pine needle oils

Constituent	CAS	Percentage and range				
		A	B (17)	C (3)	D (18)	E
Abietadiene	36312-33-1					$+^{z4}$
Abieta-8(14),13(15)-diene						$+^{z4}$
Acetophenone	98-86-2					0.7^r
α-Acoradiene	24048-44-0					0.2^k
3-Adamantanecarboxylic acid, phenyl ester						0.2^u
δ-Amorphene	189165-79-5				tr	
p-Anethole	104-46-1					0.2^u
Aromadendrene	489-39-4			0.1-0.4	0.5	0.2^l; 0.4^w; 44.7^r
allo-Aromadendrene	25246-27-9			0.1-0.4		0.5^k
Benzaldehyde	100-52-7	$0-0.5^c$				
Benzene, 1,3-bis(phe-noxyphenoxy)-	34012-02-7					0.2^u
Benzenemethanol, α-ethyl-α-2,5,7-octa-trienyl-	74685-43-1					0.5^u
Benzyl benzoate	120-51-4			tr-0.2		
Benzyl butyrate	103-37-7					3.4^r
Benzyl salicylate	118-58-1			0-0.1		
trans-α-Bergamotene	13474-59-4			0-0.1		3.6^k; 4.4^r
Bicycloelemene	32531-56-9				tr	
Bicyclogermacrene	24703-35-3			2.1-6.2^f	6.0	$+^{z4}$
β-Bisabolene	495-61-4	0.08-0.5				0.2^v; 1.2^y
(Z)-β-Bisabolene				0.1-0.2		
homo-γ-Bisabolene						4.1^r
α-Bisabolol	515-69-5	0.03-0.3		0.1-0.2		0.2^x
Borneol	507-70-0	0.3-3.3				0.3^v; 0.7^y; 3.1^{z9}; 5.1^j; 6.7^u; 10^{z5}
Bornyl acetate	76-49-3	8.4-29.9	0.08-3.9	0.5-3.0		$0.2-5.6^o$; 1.9^w; 3.2^j; $4.2^{v,z7}$; 8.8^x; 32.7^{z9}
endo-1-Bourbonalol						1.1^w
β-Bourbonene	5208-59-3	$0-0.5^c$	0.1-0.2		tr	$+^{z4}$; 0.2^l; 1.4^w; 2.4^t
α-Bulnesene	3691-11-0					0.5^k
trans-Cadina-1,4-diene	20085-13-6				tr	
trans-Cadina-1(2),4-diene	1395047-77-4					$+^{z4}$
trans-Cadina-1(6),4- diene	931410-54-7				tr	
Cadina-1(10),4-dien-8α-ol	151513-79-0					0.3^k
α-Cadinene	24406-05-1			0.2-0.4	tr	$+^{z4}$; 0.3^j; 0.7^w
γ-Cadinene	39029-41-9	0-5.3		3.0-5.5	3.8	$+^{z4}$; 0.8^j; 2.9^k; 5.4^v
γ2-Cadinene	5957-56-2					$+^{z7}$; 2.1^j
δ-Cadinene	483-76-1	0-3.8	0.6-7.3	4.7-11.6	7.1	$2.4^{j,w}$; $2.7-8.2^o$; 4.0^j; 7.7^l; $9.5^{v,z7}$; 34.0^k
Cadinol	11070-72-7					3.7^l
α-Cadinol	481-34-5	0-1.2		1.9-7.7	3.5	1.3^s; 1.6^j; 3.1^t; $3.9-9.8^{o,q}$; 8.9^k
epi-α-Cadinol (τ-)	5937-11-1				1.8	$+^{z4}$; 0.6^s; 0.9^j; $3.9-9.8^{o,q}$
α-Calacorene	21391-99-1					$+^{z4,z7}$
β-Calacorene	50277-34-4					$+^{z4}$
Calamenene	483-77-2					$+^{z7}$; 0.4^j
Camphene	79-92-5	4.2-27.4	0.4-16.8	1.5-5.3	2.1	$2.6-7.6^m$; 7.0^y; 7.8^r; 9.4^v; 14.9^{z5}; 21.7^{29}
Camphene hydrate	465-31-6	0.01-0.1				
Camphenilone	13211-15-9		0.01-0.4			
Camphenone						0.3^k
6-Camphenone	53803-33-1					1.2^k; 9.5^r
Campholenal	23727-15-3	tr-1.3				$+^{z6}$
Camphor	76-22-2	0.03-0.7	0.09-0.7^b			0.2^{z9}; 0.3^w; 2.9^y
(+)-4-Carene	13837-63-3					0.2^u
β-Carene	554-60-9					$+^{z7}$
δ2-Carene (δ4-carene)	554-61-0			0-0.1		0.02^{z7}; 0.3^w
δ3-Carene	13466-78-9	1.4-13.0	0-1.4	9.1-24.6		$4.9-22.9^o$; 43.4^j; 58.3^{z2}; 77.1^{z5}
(E)-Carveol	1197-07-5		0.07-1.1			$+^{z6}$; 0.7^i
γ-Carveol						0.2^n
Carvone	99-49-0					3.1^w
Carvyl acetate	97-42-7					0.4^w
Caryophylla-3,8-dien-5-ol						$+^{z6}$

Table 5.70.1 Constituents identified in pine needle oils (*continued*)

Constituent	CAS	Percentage and range				
		A	B (17)	C (3)	D (18)	E
β-Caryophyllene	87-44-5	0.1-3.0	0.6-6.0[d]	2.6-4.9	4.7	1.3[n]; 2.9-7.9[o]; 3.3[l]; 3.8[j,v]; 10.8[i]
Caryophyllene oxide	1139-30-6	0-0.05	0.4-2.9	0.1-0.3[h]		+[z4]; 1.1[i]; 1.9[w]; 12.7[t]
Chamazulene	529-05-5					1.7[z7]
β-Chamigrene	18431-82-8					0.8[k]; 1.7[r]
Cholesta-3,5-diene	747-90-0					0.2[u]
Chroman-2-one	119-84-6					0.6[k]
1,4-Cineole	470-67-7					8.1[k]
1,8-Cineole	470-82-6			0.6-1.2[e]		3.2[z9]
Citronellal	106-23-0					0.2[z7]
Citronellol	106-22-9					0.1[z1]; 0.2[x]
Copaborneol	21966-93-8					+[z7]
α-Copaene	3856-25-5	0.02-1.7	0.2-0.6		0.3	+[z4]; 0.3[n]; 0.5[w]; 0.7[l]; 1.3[j]
β-Copaene	18252-44-3				tr	+[z4]; 0.5[w]
α-Cubebene	17699-14-8				tr	+[z3,z4]; 0.4[w]; 0.7[k]; 9.3[r]
β-Cubebene	13744-15-5				tr	+[z3,z4]; 0.2[n]
Cubebol	23445-02-5					0.3[i]
epi-Cubebol	38230-60-3			2.1-6.2[f]		0.4[t]
Cubenene	29837-12-5			0.1-0.3		+[z6]; 0.07[z7]
Cubenol	21284-22-0			2.0-5.1		
1-epi-Cubenol	81939-29-9			0.2-0.5	tr	+[z4]
1,10-di-epi-Cubenol	73365-77-2			0.1-0.3	tr	+[z4]
Cuminaldehyde	122-03-2					+[z6]
Cuparene	16982-00-6					1.9[k]
Curcumene (ar-, α-)	644-30-4					0.3[z7]
(+)-Cycloisosativene						0.2[u]
p-Cymene	99-87-6	0.06-5.2	0.05-0.9	0-0.2		0.1-0.6[m]; 0.1[s,z2]; 0.3[j]; 0.4[x]; 2.0[w]; 7.6[z7]
p-Cymenene	1195-32-0					0.8[w]
p-Cymen-7-ol	536-60-7					+[z6]
m-Cymen-8-ol	5208-37-7					+[z4]
p-Cymen-8-ol	1197-01-9		tr-0.8	0-tr		+[z4,z6]; 0.4[w]
Cyperene	2387-78-2					3.6[k]
(E,E)-2,4-Decadienal	25152-84-5					+[z4]; 0.4[i]
n-Decanoic acid	334-48-5					+[z6]
n-Decanol (1-)	112-30-1					+[z4]
2-Decanone	693-54-9					+[z4]
(E)-2-Decenal	3913-81-3					0.4[i]
Dehydroabietane	19407-28-4					+[z6]
Dichloroacetic acid, 1-adamantylmethyl ester						0.2[u,z10]
Dihydroactinidiolide	17092-92-1					+[z6]
Dihydrocarveol acetate	20777-49-5					0.5[w]
4,10-Dimethyl-7-iso-propyl[4.4.0]bicyclo-1,4-decadiene						0.2[j]
Dodecanal	112-54-9	tr-0.2				
n-Dodecane	112-40-3					+[z4]
Dodecanoic acid	143-07-7					+[z6]; 2.0[w]
Elemazulene	529-08-8					[z7]
β-Elemene	33880-83-0			0.8-2.7	1.6	+[z4]; 0.5[w]; 0.7[t]; 1.8[l]
δ-Elemene	20307-84-0					+[z4]; 0.3[l]; 0.4[j]
Ethanone, 1-(2,5-di-hydroxydiphenyl)-						0.09[u]
Eudesma-4(15),7-dien-1β-ol	119120-23-9					+[z4]
β-Farnesene	502-60-3					0.06[z7]; 0.4[j]
(E)-β-Farnesene	18794-84-8					0.7[w]
(Z)-β-Farnesene	28973-97-9					0.2[u]
Fenchene	471-84-1					0.1-0.2[m]
α-Fenchol	512-13-0					0.2[y]; 0.3[z7]
Fenchone	1195-79-5					1.9[y]

Table 5.70.1 Constituents identified in pine needle oils (*continued*)

Constituent	CAS	Percentage and range				
		A	B (17)	C (3)	D (18)	E
Fenchyl acetate	13851-11-1	0.02-1.5				
Fenchyl alcohol	1632-73-1					5.3[u]
α-Funebrene	50894-66-1					0.8[t]
Geranial	141-27-5					0.4[z7]
Geraniol	106-24-1					0.3[z7]
Geranyl acetate	105-87-3	0.02-0.4				12.3[z9]
[1(10)*E*,5*E*]-Germa-cradien-4α-ol	207221-31-6					1.9[s]
Germacrene	28028-64-0					3.0[l]
Germacrene A	28387-44-2				0.7	7.5[r]
Germacrene D	23986-74-5	0-1.2	0.2-10.3	1.4-6.5	3.7	+[z4]; 0.4[z1]; 1.3[w]; 1.5[s]; 16.1[i]
Germacrene D-4-ol	198991-79-6			2.6-13.2[g]	tr	+[z4]; 2.4[k]
Gleenol	72203-99-7					+[z4]
Globulol	489-41-8			0.1-0.3[h]		
β-Guaiene	88-84-6					0.7[z7]; 1.0[j]
trans-β-Guaiene	192053-49-9					+[z4]
α-Gurjunene	489-40-7	0.1-0.3				
β-Gurjunene	73464-47-8			0.1-1.0		+[z4]; 0.2[l]
γ-Gurjunene	22567-17-5			0-0.3		
Heptacosane	593-49-7					1.1[t]
Heptafluorobutanoic acid, 2-(1-adamantyl)-ethyl ester						0.1[u,z10]
Hexadecanoic acid	57-10-3					1.3[u]
Hexadecanoic acid, 2-hydroxy-, methyl ester	16742-51-1					1.2[u]
Hexanal	66-25-1					0.4[i]
Hexenal	1335-39-3	0.02-0.3				
(*E*)-Hexenal	85761-70-2			tr-1.0		
(*E*)-2-Hexenal	6728-26-3					0.3[l]
(*Z*)-3-Hexen-1-ol	928-96-1		0-0.1			
(*Z*)-3-Hexenyl benzoate	25152-85-6		0-0.4			+[z4]
α-Humulene	6753-98-6	0.2-0.6	0.6-1.5	0.9		0.4[y]; 0.6[n,w]; 0.8[l]; 1.1[j]; 1.7[i]; 1.8[v,z7]
Humulene oxide II	19888-34-7					2.5[t]
(*E*)-β-Ionone	79-77-6					0.07[k]
Isoborneol	124-76-5					2.2[y]; 2.6[u]
Isobornyl acetate	125-12-2	0.01-0.2				
Isofenchone	6541-58-8					0.5[y]
Isolongifolene	1135-66-6					6.5[r]
Isomenthol	3623-52-7					1.3[w]
Liguloxide	21764-22-7		0-tr			
Ligustilide	4431-01-0					+[z6]
Limonene	138-86-3	3.1-30.1	0.4-2.3	0.6-1.2[e]	0.7	2.0-12.0[m]; 4.4[z9]; 5.2[s]; 8.0[z5]; 8.7[y]; 9.4[w]
Linalool	78-70-6	0.07-1.2				2.3[k]
Linalyl acetate	115-95-7	0.05-0.4				0.1[n]; 0.2[z7]; 0.6[w]
Longifolene (α-)	475-20-7	0-0.6				<0.1[j]; 0.1[z1]; 0.3[n]; 0.9[y]; 3.5[v,z7]; 5.5[u]
Longipinene	5989-08-2					0.5[u]; 0.9[z1]
Manoyl oxide	596-84-9					+[z4]; 1.2[i]; 30.2[t]
epi-Manoyl oxide	1227-93-6					1.2[i]
p-Mentha-2,4(8)-diene	586-63-0			0-tr		
p-Mentha-1,5-dien-8-ol	1686-20-0	0.3-1.6		0-0.2		+[z4,z6]; 0.9[i]
cis-β-Mentha-6,8-dien-2-ol		0.04-0.2[a1]				
Menthol	89-78-1					+[z6]; 0.9[w]
Menthone	89-80-5					+[z6]; 0.3[w]
Menthyl acetate	16409-45-3					+[z3]

Table 5.70.1 Constituents identified in pine needle oils (*continued*)

Constituent	CAS	Percentage and range				
		A	B (17)	C (3)	D (18)	E
5-Methoxy-2,8,8-trimethyl-dipyran-4-one	35930-31-5					+[z6]
Methyl dehydroabietate	1235-74-1					+[z6]
Methyl eugenol	93-15-2			0-tr		+[z6]
5-Methyl-2-furaldehyde	620-02-0					+[z6]
5-Methylpentadecane	25117-33-3					0.7[r]
Methyl thymol	1076-56-8					+[z4]
trans-Muurola-3,5-diene	262352-88-5					+[z4]
cis-Muurola-4(14),5-diene	157477-72-0			0-0.1		+[z4]
trans-Muurola-4(14),5-diene	262352-87-4				tr	+[z4]
α-Muurolene	10208-80-7			0-1.3	1.1	0.3[j]; 0.4[i,n]; 0.9[t]; 1.4[y]; 2.6[z7]; 11.6[v]; 13.9[r]
γ-Muurolene	30021-74-0	0-1.1		0.5-1.5	0.9	+[z4]; 0.6[k]; 0.8[l]; 0.9[j,v]; 1.8[i]; 2.5[t]
ε-Muurolene	30021-46-6					+[z7]; 0.7[y]
cis-Muurol-5-en-4α-ol	157374-45-3			0-0.1		
cis-Muurol-5-en-4β-ol	157374-46-4			0-0.1		
Muurolol	119757-72-1					+[z6]; 4.4[l]
α-Muurolol	104245-48-9	0-0.4		0-0.7		+[z4]; 0.7[t]; 4.0-9.1[o,p]
τ-Muurolol (epi-α-)	19912-62-0	0-0.5	0.4-1.4		1.8	+[z4]; 0.6[s]; 0.9[j]; 1.9[t]; 4.0-9.1[o,p]
Myrcene	123-35-3	0.4-13.5	0.7-3.4	0.8-2.3	0.8	2.3-5.6[m]; 3.8[v]; 4.8[z1]; 5.9[y]; 14.0[n]; 34.7[r]
Myrtenal	564-94-3		0.2-1.8			+[z6]; 0.6[i,w]
Myrtenol	515-00-4		0.2-1.0			+[z6]; 0.3[n]; 2.5[i]
Neral	106-26-3					0.2[z7]
Nerol	106-25-2					0.4[z7]
Nerolidol	7212-44-4					8.0[z7]; 8.1[z5]
(*E*)-Nerolidol	40716-66-3	0.01-0.09				
Nonacosane	630-03-5					6.8[t]
2-Norpinene-2-carboxaldehyde						+[z3,a3]
β-Ocimene	13877-91-3					1.1[l]
(*E*)-β-Ocimene	3779-61-1		0.2-2.9	0.3-1.8	1.3	+[z4]; 0.3[w]; 1.3[j]; 3.3[z2]; 3.9[r]
(*Z*)-β-Ocimene	3338-55-4			0-tr		+[z4]; 0.3[w]
trans-13-Octadecenoic acid	693-71-0					4.7[u]
Oleic acid	112-80-1					0.6[u]
Oplopanone	1911-78-0			0-0.1		+[z4]; 2.8[t]
β-Oplopenone	28305-60-4					
Oxabicyclopentadecan-3-one				0-0.1		
Oxidohimachalene	64825-84-9			2.6-13.2[g]		
α-Patchoulene	560-32-7					0.8[k]
γ-Patchoulene	508-55-4					0.2[z7]; 0.9[j]
2-Pentadecanone	2345-28-0					0.6[a2,w]
2-Pentylfuran	3777-69-3		0.02-0.1			
Perilla alcohol	536-59-4					2.3[r]
α-Phellandrene	99-83-2	0.04-5.3				0.1-0.4[m]; 0.1[n]; 1.9[r]; 31.4[x]
β-Phellandrene	555-10-2	0.1-6.0	tr-0.7	0.6-1.2[e]		0.9[j,l]; 1.6[z2]; 4.3[y]; 10.9[v]; 13.9[z8]; 29.1[n]
2-Phenethyl alcohol	60-12-8					+[z6]
Phenethyl isovalerate	140-26-1					0.4[r]; 2.0[r]
trans-Pinane	10281-53-5			0-tr		
α-Pinene	80-56-8	13.3-20.4	19.4-56.9	18.5-33.0	34.4	28.8-68.6[m]; 57.5[z2]; 58.6[r]; 69.1[x]; 89.5[z5]
β-Pinene	127-91-3	0.9-7.1	2.9-17.1	0.6-2.1	6.8	1.9-10.5[m]; 18.4[j]; 20.6[z7]; 28.0[z5]; 33.4[z7]
β-Pinene oxide	6931-54-0		0.04-0.3			
Pinocamphor			0.09-0.7[b]			
Pinocarveol	5947-36-4					0.8[z7]
trans-Pinocarveol	1674-08-4		0.3-2.2			1.3[w]; 1.5[i]
Pinocarvone	30460-92-5		0.1-1.0			+[z6]; 0.7[i]

Table 5.70.1 Constituents identified in pine needle oils (*continued*)

Constituent	CAS	Percentage and range				
		A	B (17)	C (3)	D (18)	E
trans-Pinocarvyl acetate	1686-15-3					0.3[i]
Piperitone	89-81-6					0.3[z7]
Sabinene	3387-41-5	0.6-3.3	0.06-1.0	0.3-0.7	tr	0.2-1.3[m]; 0.2[z1]; 0.5[j,l]; 0.7[s]; 2.1[z2]; 10.8[r]
Sandaracopimaradiene	1686-56-2					0.7[i]
Sandaracopimarinal	3855-14-9					6.7[i]
epi-β-Santalene	25532-78-9					2.4[r]
Santene	529-16-8					<0.1[j]; 0.2[z7]; 3.4[y]
Santolina alcohol	35671-15-9					4.2[r]
β-Selinene	17066-67-0			0.5-1.3	1.6	+[z4]; 0.9[t]
Sesquisabinene	58319-04-3				tr	
Spathulenol	6750-60-3		0.08-3.5	2.6-13.2[g]		+[z4,z6]; 0.9[w]; 4.3[t]
Sylvestrene	1461-27-4				0.6	
α-Terpinene	99-86-5	0.05-0.2		tr-0.1		0.1-0.3[m]; 0.4[z2]; 3.0[z5]; 3.2[v,z7]; 15.5[r]
β-Terpinene	99-84-3					+[z3,a3]
γ-Terpinene	99-85-4	0.09-1.1	0.01-0.2	0.1-0.3		0.2-4.3[m]; 0.4[v]; 0.5[w]; 1.1[j]; 1.2[y]; 1.6[z2]
Terpin(ene) hydrate	2451-01-6					0.5[u]
Terpinen-4-ol	562-74-3	0.04-0.2	0.6-6.0[d]	tr-0.1		1.1[j]; 1.6[w,y,z9]
Terpinen-4-yl acetate	4821-04-9			0.2-0.4		
1-Terpineol	586-82-3					21.8[u]
α-Terpineol	98-55-5	0.1-0.7		0-0.1		+[z3]; 0.4[z1]; 0.7[j]; 1.8[i]; 1.9[v,z7]; 2.3[z7]; 6.7[x]
β-Terpineol	138-87-4					17.9[u]
Terpinolene	586-62-9	0.2-1.2	0.09-0.2	1.2-2.4	0.7	0.4-4.3[m]; 1.7[s]; 2.5[z5]; 4.1[j]; 4.6[z2]; 18.9[z8]
α-Terpinyl acetate	80-26-2	0.2-1.0		tr-1.2		+[z4]; 0.08[l]; 0.5[y]; 0.9[n]
n-Tetradecane	629-59-4		0-0.2			+[z4]
Tetramethylbicyclo-[2.2.2] oct-2-ene						+[z6]
Thuja-2,4(10)-diene	36262-09-6					0.6[w]
α-Thujene	2867-05-2			0.1-0.7	0.6	+[z4]; 0.4[k]; 0.5[i]; 1.1[z2]
β-Thujene	28634-89-1					+[z3,a3]
Thunbergol	25269-17-4					+[z6]
Torreyol	19435-97-3		0.2-0.6			
Tricyclene	508-32-7	0.9-2.7	0.1-4.3	0.2-1.2	0.5	0.8-2.2[m]; 0.8[j]; 1.0[s,w]; 1.1[z2]; 3.1[y]
Tridecanoic acid	638-53-9					+[z6]
3-Tridecanone	1534-26-5					0.6[w]
Undecane	1120-21-4					+[z4]
Undecanone	53452-70-3			tr-0.1		0.05[l]
2-Undecanone	112-12-9			0.1-1.1		+[z4]; 0.3[l]; 0.9[w]
cis-Verbenol	1845-30-3		0.05-1.8			
trans-Verbenol	1820-09-3		0.3-9.6			+[z6]
Verbenone	80-57-9					+[z6]
4-Vinylphenol	2628-17-3					+[z6]
Viridiflorol	552-02-3			0-0.1		
α-Ylangene	14912-44-8					+[z7]; 0.1[j]

A 112 pine needle essential oil samples from China, Russia, Austria, USA, Canada and Hungary, analyzed between 1998 and 2013; lowest and highest concentrations given (E. Schmidt, unpublished data); these data also contain analyses from oils of *Pinus yunnanensis* (pine oil China, pine oil east Asia) and of *Pinus pinaster*, as these oils are also sold under the name 'pine needle oil'
B sixteen lab-hydrodistilled oils from needles of *Pinus sylvestris* collected in four regions of Turkey in four seasons; lowest and highest concentrations given (ref. 17)
C fifteen lab-hydrodistilled pine needle oils from *P. sylvestris* needles harvested in various parts of Lithuania (ref. 3)
D one lab-hydrodistilled oil produced from needles of *P. sylvestris* collected at the Botanical Garden of the University of Cambridge, UK (ref. 18)
E data from other studies (indicated with superscript letters); highest concentrations found in any study reviewed here given; when two or more oils were investigated, only the highest concentrations are mentioned, unless indicated otherwise

[a1] incorrect identity based on GC elution order (ref. 23); [a2] incorrect identity based on GC elution order (ref. 21); [a3] compound does not occur naturally; identification likely to be in error (ref. 27); [b] camphor and pinocamphor combined; [c] benzaldehyde and β-bourbonene combined; [d] β-caryophyllene and terpinen-4-ol combined; [e] β-phellandrene, limonene and 1,8-cineole combined; [f] epi-cubebol and bicyclogermacrene combined; [g] oxidohimachalene, germacrene-D-4-ol and spathulenol combined; [h] caryophyllene oxide and globulol combined; [i] two lab-hydrodistilled oils from Turkish needles (ref. 12); [j] two lab-hydrodistilled oils from France (ref. 10); [k] one (atypical) laboratory steam-distilled pine needle oil from Poland (ref. 13); [l] one commercial oil from Lithuania; ingredients as supplied by the manufacturer (ref. 11); [m] fifteen lab-hydrodistilled oils from the needles of *Pinus sylvestris* specimens growing in various parts of Belarus under different ecological circumstances; lowest and highest concentrations given (ref. 19); [n] one lab-hydrodistilled

Table 5.70.1 Constituents identified in pine needle oils (*continued*)

oil from twigs (presumably with needles attached) harvested in Greece (ref. 20); °thirty-two lab-hydrodistilled oils from eight trees in Lithuania; lowest and highest concentrations given (ref. 14); [p] α-muurolol and τ-muurolol (= epi-α-muurolol) combined; [q] α-cadinol and epi-α-cadinol (= τ-cadinol) combined; [r] six lab-distilled oils from *P. sylvestris* needles collected in six cities in Europe (ref. 9); [s] one lab-hydrodistilled oil from needles of a German pine tree (ref. 15); [t] one lab-hydrodistilled oil from needles of *Pinus sylvestris* harvested in Scotland with a very atypical composition (ref. 2); [u] one laboratory steam-distilled oil from needles harvested in Nigeria; very atypical oil (ref. 30); [v] various pine essential oils from Russia (Caucasus and Siberia) (ref. 7); [w] one commercial oil made of French *P. sylvestris* needles (ref. 22); [x] unknown number of oils from Slovakia (ref. 8); [y] unknown number of oils from Siberian needles harvested in July and September (ref. 25); [y1] six lab-distilled pine oils from *P. sylvestris* needles collected in six European cities (ref. 34); [z1] one oil from the former Yugoslavia (ref. 26); [z2] five pine needle oils from Estonia (ref. 4); [z3] one oil sample from Romania (ref. 29); [z4] see +[z4] (below); [z5] data from various studies published before 1988, cited in ref. 16; [z6] see +[z6] (below); [z7] data from various studies published before 1990, cited in ref. 31; [z8] one commercial oil from Poland (ref. 33); [z9] one commercial oil from Brazil (ref. 32); [z10] wrong identification, does not occur in nature

tr: trace (in column B: <0.01; in column C: <0.1; in column D: <0.050; + present in the oil investigated, but quantity not stated; +[z4] oils from Lithuania; quantitative analysis was provided, but amounts were expressed in μg/g dry matter and are therefore not comparable to other data (ref. 5); +[z6] an oil prepared by steam distillation of needles purchased in Korea; quantitative analysis was provided, but amounts were expressed in μg/g dry matter and are therefore not comparable to other data (ref. 28); it was suggested that the botanical origin of the needles in this study is questionable (ref. 23)

Table 5.70.2 Major constituents of pine needle oils

Constituent	CAS	Percentage and range				
		A	B (17)	C (3)	D (18)	E
α-Pinene	80-56-8	13.3-20.4	19.4-56.9	18.5-33.0	34.4	28.8-68.6[m]; 57.5[z2]; 58.6[r]; 69.1[x]; 89.5[z5]
δ3-Carene	13466-78-9	1.4-13.0	0-1.4	9.1-24.6		4.9-22.9[o]; 43.4[j]; 58.3[z2]; 77.1[z5]
Myrcene	123-35-3	0.4-13.5	0.7-3.4	0.8-2.3	0.8	2.3-5.6[m]; 3.8[v]; 4.8[z1]; 5.9[v]; 14.0[n]; 34.7[r]
δ-Cadinene	483-76-1	0-3.8	0.6-7.3	4.7-11.6	7.1	2.4[i,w]; 2.7-8.2[o]; 4.0[j]; 7.7[l]; 9.5[v,z7]; 34.0[k]
β-Pinene	127-91-3	0.9-7.1	2.9-17.1	0.6-2.1	6.8	1.9-10.5[m]; 18.4[j]; 20.6[z7]; 28.0[z5]; 33.4[z7]
Bornyl acetate	76-49-3	8.4-29.9	0.08-3.9	0.5-3.0		0.2-5.6[o]; 1.9[w]; 3.2[y]; 4.2[v,z7]; 8.8[x]; 32.7[z9]
Limonene	138-86-3	3.1-30.1	0.4-2.3	0.6-1.2[e]	0.7	2.0-12.0[m]; 4.4[z9]; 5.2[s]; 8.0[z5]; 8.7[y]; 9.4[w]
Camphene	79-92-5	4.2-27.4	0.4-16.8	1.5-5.3	2.1	2.6-7.6[m]; 7.0[y]; 7.8[r]; 9.4[v]; 14.9[z5]; 21.7[z9]
γ-Cadinene	39029-41-9	0-5.3		3.0-5.5	3.8	+[z4]; 0.8[j]; 2.9[k]; 5.4[v]

LEGEND: SEE UNDER TABLE 5.70.1

with low δ3-carene usually have high concentrations of α-pinene; thus two chemotypes of pine needle oils may be distinguished, α-pinene and δ3-carene (10). In the great majority of the oils, α-pinene is the dominant chemical, with concentrations generally ranging from 25 to 60%. In such oils, concentrations of δ3-carene range from zero (8,12,13,16,17,18,20,22) to a maximum of 25% (3,4,7,11,14,15,16,17,24,26). In only a few oils, δ3-carene was the dominant ingredient, with concentrations ranging from 41 to 77% (4,10,16,19).

CONTACT ALLERGY/ALLERGIC CONTACT DERMATITIS

General

Contact allergy to/allergic contact dermatitis from pine needle oil has been reported in 10 publications. In groups of consecutive patients suspected of contact dermatitis, prevalence rates of up to 2.0% positive patch test reactions have been observed, but relevance data are lacking. In one case report, α-pinene may have been an allergen in the oil.

Testing in groups of patients

Routine testing

Between 1973 and 1977, 3500 patients were tested in Spain with 'pine oil' and 15 (0.4%) had a positive patch test (40). Four positive patch test reactions to pine needle oil were seen in 200 (2.0%) consecutive patients with dermatitis (38). Relevance data were not provided in either study.

Testing in groups of selected patients

In a group of 51 patients allergic to *Myroxylon pereirae* resin (balsam of Peru) and/or turpentine and/or wood tar and/or colophony, there were 15 (29.4%) positive reactions to pine needle oil (38). The high percentage is likely related to the fact that all these products are obtained from trees, three of them (pine needle oil, turpentine and colophony) from *Pinus* species. In another study from these authors, a group of 86 patients from Poland previously reacting to the fragrance mix was tested with pine needle oil and three (3.4%) had a positive patch test reaction (39). Relevance data were not provided in either study (38,39). In a group of 21 patients with dermatitis caused by fragrances and tested with a series of essential oils, three (14%) reacted to pine (*Pinus sylvestris*) oil; relevance data were not provided (41).

Case reports

An aromatherapist had non-occupational contact dermatitis with allergies to multiple essential oils used at work, including pine oil. The patient also reacted to geraniol, linalool, linalyl acetate, α-pinene, the fragrance mix and various other fragrance materials; α-pinene was demonstrated by GC-MS in the pine oil (37). α-Pinene is an important component of pine needle oil and was found in concentrations of 13.3-20.4% in commercial needle oils (Table 5.70.1, column A). Four cases (34) and two (35) of contact sensitization to pine needle oil in topical pharmaceutical preparations have been reported; there may be overlap between these patients. One patient working as painter and car mechanic developed occupational contact dermatitis from a wax polish. He reacted to the wax polish and its ingredients pine oil (5% and 10% in olive oil) and dipentene (*d,l*-limonene) and co-reacted to turpentine oil, which also contains limonene and is obtained from *Pinus* species, just as pine oil (44).

Positive patch tests (relevance unknown, uncertain or not stated)

One positive patch test was obtained to pine needle oil and dwarf pine needle oil in a patient allergic to *Melaleuca alternifolia* (tea tree) oil (36). One case of allergic contact dermatitis from a 'natural' deodorant containing lichen acids has been reported; the patient reacted to the deodorant, lichen acid mix, usnic acid and some essential oils including pine oil (42). In Poland, 30% of patients allergic to turpentine oil also reacted to pine needle oil; α-pinene is the major ingredient of turpentine oil and also an important component of pine needle oil (43).

LITERATURE

1 Judzentiene A, Kupcinskiene E. Chemical composition on essential oils from needles of *Pinus sylvestris* L. grown in northern Lithuania. J Essent Oil Res 2008;20:26-29

2 Tsitsimpikou C, Petrakis PV, Ortiz A, Harvala C, Roussis V. Volatile needle terpenoids of six *Pinus* species. J Essent Oil Res 2001;13:174-178

3 Venskutonis PR, Vyskupaityte C, Plausinaitis R. Composition of essential oils of *Pinus sylvestris* L. from different locations of Lithuania. J Essent Oil Res 2000;12:559-565

4 Orav A, Kailas T, Liiv M, Aav R. Capillary gas chromatographic analysis of the monoterpenoic fraction of Estonian conifer needle oil. Proc Estonian Acad Sci Chem 1995;44:149-155. Data cited in ref. 27

5 Kupcinskiene E, Stikliene E, Judzentiene A. The essential oil qualitative and quantitative composition in the needles of *Pinus sylvestris* L. growing along industrial transects. Environ Poll 2008;155:481-491

6 Maciąg A, Milaković D, Christensen HH, Antolović V, Kalemba D. Essential oil composition and

7 Góra J, Lis A. Olejki sosnowe. Aromaterapia 2000;2:5-12. Data cited in refs. 6 and 21

8 Berta F, Spuka J, Chladna A. The composition of terpenes in needles of *Pinus sylvestris* in a relatively clear and in a city environment. Biologia Bratislawa 1997;52:71-78. Data cited in refs. 6 and 27

9 Dawidowicz AL, Czapczynska NB, Wianowska D. Relevance of the sea sand disruption method (ssdm) for the biometrical differentiation of the essential-oil composition from conifers. Chem Biodivers 2013;10:241-250

10 Chalchat J-C, Garry R-P, Michet A, Remery A. The essential oils of two chemotypes of *Pinus sylvestris*. Phytochem 1985;24:2443-2444

11 Lazutka JR, Mierauskienė J, Slapšytė G, Dedonytė V. Genotoxicity of dill (*Anethum graveolens* L.), peppermint (*Mentha* x *piperita* L.) and pine (*Pinus sylvestris* L.) essential oils in human lymphocytes and *Drosophila melanogaster*. Food Chem Toxicol 2001;39:485-492

12 Ustun O, Senol FS, Kurkcuoglu M, Orhan IE, Kartal M, Baser KHC. Investigation on chemical composition, anticholinesterase and antioxidant activities of extracts and essential oils of Turkish *Pinus* species and pycnogenol. Ind Crops Prod 2012;38:115-123

13 Dawidowicz AL, Czapczyńska NB. Sea sand disruption method (SSDM) as a valuable tool for isolating essential oil components from conifers. Chem Biodivers 2011;8:2045-2056

14 Judzentiene A, Stikline A, Kupcinskiene E. Changes in the essential oil composition in the needles of Scots pine (*Pinus sylvestris* L.) under anthropogenic stress. Sci World J 2007;7(S1):141-150

15 Kubeczka K-H, Schultze W. Biology and chemistry of conifer oils. Flavour Fragr J 1987;2:137-148

16 Ekundayo O. Volatile constituents of *Pinus* needle oils. Flavour Fragr J 1988;3:1-11

17 Ustun O, Sezik E, Kurkcuoglu M, Baser KHC. Study of the essential oil composition of *Pinus sylvestris* from Turkey. Chem Nat Comp 2006;42:26-31

18 Ioannou E, Koutsaviti A, Tzakou O, Roussis V. The genus *Pinus*: a comparative study on the needle essential oil composition of 46 pine species. Phytochem Rev 2014;13:741-768

19 Shpak SI, Lamotkin SA, Lamotkin AI. Chemical composition of *Pinus silvestris* essential oil from contaminated areas. Chem Nat Comp 2007;43:55-58

20 Koukos PK, Papadopoulou KI, Papagiannopoulos AD, Patiaka DTh. Essential oil of the twigs of some conifers grown in Greece. Holtz Roh Werkst 2001;58:437-438

21 Lawrence BM. Progress in essential oils. Perfum Flav 2013;38(July):48-54

22 Tognolini M, Barocelli E, Ballabeni V, Bruni R, Bianchi A, Chiavarini M, et al. Comparative screening of plant essential oils: phenylpropanoid moiety

plant-insect relations in Scots pine (*Pinus sylvestris* L.). Scientific bulletin of the Technical University of Lodz. Food Chem Biotechnol 2007;71:71-95. Available at: http://www.bfs.p.lodz.pl/get_file. php?fileId=32

as basic core for antiplatelet activity. Life Sciences 2006;78:1419-1432

23 Lawrence BM. Progress in essential oils. Perfum Flavor 2007;32 (August):32-? (last page unknown)

24 Stepen RA. The composition of essential oil and volatile terpenoids of *Pinus sylvestris* L. shoots in middle Siberia. Rast Resur 1995;31:63-70. Data cited in ref. 25

25 Lawrence BM. Progress in essential oils. Perfum Flavor 2001;26(3):66- ? (last page unknown)

26 Ruzie NLJ, Petrovic SS, Stevanovic-Janezic T, Petrovic SD. Optimization of the preparation of the essential oil from technical foliage of white pine (*Pinus sylvestris* L.) by steam distillation. Hem Ind 1999;53:231-234. Data cited in ref. 25

27 Lawrence BM. Progress in essential oils. Perfum Flavor 2003;28(5):70- ? (last page unknown)

28 Ka M-H, Choi EH, Chun H-S, Lee K-G. Antioxidative activity of volatile extracts isolated from *Angelica tenuissimae*, peppermint leaves, pine needles and sweet flag leaves. J Agric Food Chem 2005;53:4124-4129

29 Marculescu A, Gleizes M. Composition of volatile substances in conifer species from the montane zone. 3. Pine (*Pinus sylvestris* L.) leaves. Revista Chim (Bucharest) 2002;53:185-189. Data cited in ref. 27

30 Fayemiwo KA, Adeleke MA, Okoro OP, Awojide SH, Awoniyi IO. Larvicidal efficacies and chemical composition of essential oils of *Pinus sylvestris* and *Syzygium aromaticum* against mosquitoes. Asian Pac J Trop Biomed 2014;4:30-34

31 Lawrence BM. Progress in essential oils. Perfum Flavor 1991;16(2):59-67

32 Murbach Teles Andrade BF, Nunes Barbosa L, da Silva Probst I, Fernandes A Júnior. Antimicrobial activity of essential oils. J Essent Oil Res 2014;26:34-40

33 Golebiowski M, Paszkiewicz M, Halinski L, Malinski E, Stepnowski P. Chemical composition of commercially available essential oils from eucalyptus, pine, ylang, and juniper. Chem Nat Comp 2009;45:234-235

34 Nardelli A, D'Hooge E, Drieghe J, Dooms M, Goossens A. Allergic contact dermatitis from fragrance components in specific topical pharmaceutical products in Belgium. Contact Dermatitis 2009;60:303-313

35 Devleeschouwer V, Roelandts R, Garmyn M, Goossens A. Allergic and photoallergic contact dermatitis from ketoprofen: results of (photo) patch testing and follow-up of 42 patients. Contact Dermatitis 2008;58:159-166

36 Hausen BM, Reichling J, Harkenthal M. Degradation products of monoterpenes are the sensitizing agents in tea tree oil. Am J Contact Dermat 1999;10:68-77

37 Dharmagunawardena B, Takwale A, Sanders KJ, Cannan S, Roger A, Ilchyshyn A. Gas chromatography: an investigative tool in multiple allergies to essential oils. Contact Dermatitis 2002;47:288-292

38 Rudzki E, Grzywa Z, Bruo WS. Sensitivity to 35 essential oils. Contact Dermatitis 1976;2:196-200

39 Rudzki E, Grzywa Z. Allergy to perfume mixture. Contact Dermatitis 1986;15:115-116

40 Romaguera C, Grimalt F. Statistical and comparative study of 4600 patients tested in Barcelona (1973–1977). Contact Dermatitis 1980;6:309-315

41 Meynadier JM, Meynadier J, Peyron JL, Peyron L. Formes cliniques des manifestations cutanées d'allergie aux parfums. Ann Dermatol Venereol 1986;113:31-39

42 Sheu M, Simpson EL, Law SV, Storrs FJ. Allergic contact dermatitis from a natural deodorant: A report of 4 cases associated with lichen acid mix allergy. J Am Acad Dermatol 2006;55:332-337

43 Rudzki E, Berova N, Czernielewski A, Grzywa Z, Hegyi E, Jirásek J, et al. Contact allergy to oil of turpentine: a 10-year retrospective view. Contact Dermatitis 1991;24:317-318

44 Martins C, Gonçalo M, Gonçalo S. Allergic contact dermatitis from dipentene in wax polish. Contact Dermatitis 1995;33:126-127

Chapter 5.71   RAVENSARA OIL

DEFINITION
Ravensara oil is the essential oil obtained from the twigs with leaves of the ravensara, *Ravensara aromatica* Sonn. (synonym: *Cryptocarya agathophylla* van der Werff).

INCI NOMENCLATURE
Description/definition: Ravensara aromatica leaf oil is the volatile oil distilled from the leaves of *Ravensara aromatica*, Lauraceae
INCI name EU & USA: Ravensara aromatica leaf oil
CAS registry number(s): 91770-56-8
EINECS number(s): 294-842-8

Description/definition: Ravensara aromatica twig oil is an essential oil obtained from the twigs with fruits of the plant, *Ravensara aromatica*, Lauraceae
INCI name EU: Ravensara aromatica twig oil (perfuming name, not an INCI name proper)
INCI name USA: Not in the Personal Care Products Council Ingredient Database
CAS registry number(s): 91770-56-8
EINECS number(s): 294-842-8

ISO (INTERNATIONAL ORGANIZATION FOR STANDARDIZATION) STANDARD
There is currently no ISO standard for ravensara oil.

Ravensara oil should not be confused – but often is – with ravintsara oil, the essential oil obtained from the species *Cinnamomum camphora*. As will be shown below, many commercial oils sold as 'ravensara' are in fact ravintsara oils. These oils usually contain high concentrations of 1,8-cineole (eucalyptol) or camphor, chemicals mostly not present or in low concentrations in oils obtained from *R. aromatica* (6,12). Adding to the confusion are reports on ravensara (anisata) oils obtained from *Ravensara anisata* Danguy. Several authors consider this not to be a species of its own, but a synonym for *R. aromatica* (6,7,11,12,16). However, the status of *R. anisata* Danguy as a plant species is unresolved according to The Plants List (www.theplantlist.org). In the GRIN Taxonomy for Plants Database (http://www.ars-grin.gov/) the name *R. anisata* cannot be found and taxonomists appear not to recognize *R. anisata* as a species (2). Some authors have suggested that commercial *Ravensara anisata* oils are (usually) in fact essential oils obtained from the bark of *R. aromatica* (1,2,16). Because of these uncertainties and because the name *Ravensara anisata* is hardly ever mentioned on labels of commercial ravensara *leaf* oils (6), studies on *R. anisata* oils (3,9,13,14,16) are not discussed here.

THE PLANT, THE OIL, AND THEIR USES
Ravensara aromatica Sonnerat is an evergreen aromatic tree which can grow up to 20 meters and has a deep rich reddish brown bark. It is native to the wet forests of east Madagascar and is cultivated in Mauritius (GRIN Taxonomy for Plants, 1,2,8). The whole tree is strongly aromatic. The fruits and the bark smell of anise; the bark is used by the Madagascan inhabitants for the preparation of rum. In addition, Ravensara is used by the locals as a universal remedy for physical and mental disorders. The leaves are used for the preparation of ointments and cough mixtures and extracts of the bark and leaves are used to treat indigestion (8). Both the leaves and bark of *Ravensara aromatica* are rich in essential oils, to which antimicrobial, antifungal and antiviral properties have been attributed. These oils are commercialized in Madagascar and in northern hemisphere countries under the name of 'Ravensara' or 'Ravensare' (2), mainly (in Europe and the USA) for aromatherapy (1,19,20).

CHEMICAL COMPOSITION
Ravensara oil is a yellowish to greenish yellow easily mobile liquid, which has an aromatic, phenolic to spicy anise odor. The yield of essential oil from the twigs with leaves of *Ravensara aromatica* Sonn. generally varies from 0.6 to 2.8 per cent. The main producing country of this oil is Madagascar. Literature data (up to October 14, 2014) on the chemical composition of ravensara oils (only seven suitable publications found) and unpublished analytical data from one of us (E.S.) are shown in Table 5.71.1 in alphabetical order. In ravensara oils from various origins, over 95 chemicals have been identified. About 39 per cent of these were found in a single reviewed publication only. The major compounds found in ravensara oils from different sources are shown in Table 5.71.2. They include (highest concentrations in any study given) methyl chavicol (94.8%), methyl eugenol (86.6%), sabinene (34.4%), limonene (27.8%), α-terpinene (27.7%), linalool (21.4%), β-pinene (15.7%), terpinen-4-ol (12.0%), α-pinene (8.3%), myrcene (7.5%) and β-caryophyllene (7.4%). Well-known ingredients of ravensara oils that were present in high concentrations (>7%) in one study were α-terpineol (14.7%) and α-phellandrene (11.8%).

Commercial oils
The ten chemicals that had the highest maximum concentrations in 41 commercial ravensara essential oil samples (concentration ranges provided) are the following: 1,8-cineole (0.1-68.0%), sabinene (0.1-25.5%), methyl eugenol (tr-21.4%), methyl chavicol (0.04-19.9%), limonene (0.08-19.4%), β-pinene (0.1-15.7%), α-terpineol (0.2-14.7%), linalool (0.05-12.4%), α-phellandrene (0.04-11.8%) and α-pinene (0.3-8.3%) (Erich Schmidt, unpublished analytical data).

Chemotypes
Five chemotypes of *R. aromatica* leaf oils have been distinguished (1,2,6). Chemotype 1 is characterized by a large amount of methyl chavicol (86-98%); chemotype 2 has a high content of methyl eugenol (72-95%), and chemotype 3 is dominated by α-terpinene (18-58%) and limonene (4-22%). In chemotype 4 the major ingredient is sabinene (20-63%); limonene dominates in chemotype 5 (20-63%) (1,2,6). The composition of the various chemotypes is presented in Table 5.71.3.

In a number of studies, 1,8-cineole (eucalyptol) has been found as the dominant ingredient, with percentages ranging from 31 to 66% (3,8,9,10,15,18). However, it is highly likely that these oils were from *Cinnamomum camphora* (ravintsara oil) which are often characterized by a high 1,8-cineole content, rather than from *R. aromaticum* (ravensara oil) (2,5,11). Therefore, 'ravensara' oils with a high 1,8-cineole content are not discussed here. Indeed, it has been well demonstrated that the essential oils of ravensara and ravintsara are frequently misidentified in the marketplace. Whereas commercial oils sold in Madagascar as ravensara or ravintsara oil are usually well labeled, only a minority (36%) of oils sold as ravensara oil in Europe and the USA are actually obtained from *R. aromaticum* (6); most are either obtained from *Cinnamomum camphora* (ravintsara oil) or adulterated with this oil rich in 1,8-cineole (6,12,16). Consequently, the possibility that commercial oils labeled as 'ravensara oil' may contain high concentrations of 1,8-cineole (or camphor) should always be considered (6).

Table 5.71.1 Constituents identified in ravensara oils

Constituent	CAS	Percentage and range					
		A	B (1)	C (2)	D (12)	E (17)	F
(E)-Anethole	4180-23-8		0-0.1	0-0.1		0.2	
Aromadendrene	489-39-4	tr-0.1				tr	
allo-Aromadendrene	25246-27-9	0-0.7	0-0.1	0-0.1	0.1-1.1		
cis-α-Bergamotene	18252-46-5					tr	
trans-α-Bergamotene	13474-59-4					tr	
Bicyclogermacrene	24703-35-3	0.02-0.4				0.1	
β-Bisabolene	495-61-4					0.1	
(E)-γ-Bisabolene	53585-13-0					tr	
Borneol	507-70-0	0-0.2	0-0.5	0-0.5		0.1	
Bornyl acetate	76-49-3	0-0.4	0-0.7	0-0.7		tr	
β-Bourbonene	5208-59-3	0-0.04				tr	
α-Bulnesene	3691-11-0					tr	
Cadina-1(6),4-diene	16729-00-3					tr	
γ-Cadinene	39029-41-9					tr	
δ-Cadinene	483-76-1	0.02-0.6			0.4-1.1	0.4	0.1[b]
trans-Calamenene	73209-42-4					0.1[a]	
Calarene	17334-55-3					tr	
Camphene	79-92-5	0.04-1.3	0-0.8	0-0.8	0.4-0.9	0.6	0.1-0.2[c]; 0.4[b]
Camphor	76-22-2	0.08-2.8					
δ3-Carene	13466-78-9	1.1-6.4			2.5-4.9	2.6	4.1[b]
β-Caryophyllene	87-44-5	0.1-3.6	0-2.2	0-1.5	2.1-7.4	3.8	1.1-1.8[c]; 1.5[d]; 1.7[b]
9-epi-(E)-Caryophyllene	68832-35-9					0.1	
Caryophyllene oxide	1139-30-6	0-0.1	0-0.1	0-0.6		0.1	
1,8-Cineole	470-82-6						0.1[d]; 1.2[b]
Citronellal	106-23-0						<0.1[d]
Citronellol	106-22-9						0.3[d]
Citronellyl acetate	150-84-5	0-0.4					<0.1[d]
α-Copaene	3856-25-5	0.01-0.7	0-1.9	0-1.9	0.5-1.9	3.0	0.4[b]
Cubebene	11012-64-9	0-0.3					
α-Cubebene	17699-14-8	0.02-0.5				0.1	
β-Cubebene	13744-15-5	0.09-4.2				0.2	0.2[b]
Cubebol	23445-02-5					tr	
Cyclosativene	22469-52-9					tr	
p-Cymene	99-87-6	0.08-2.8	0-3.0	0-3.0	1.2-3.4	1.5	0.4[d]; 1.7[b]
β-Elemene	33880-83-0	0.1-0.7			0.4-0.9	0.1	
γ-Elemene	29873-99-2	0.05-0.6					
δ-Elemene	20307-84-0		0-0.1	0-tr	0.1-0.9	0.1	
Elemicin	487-11-6	tr-2.7	0-1.6	tr-1.7		tr	0.1-0.4[c]; 2.7[d]
Elemol	639-99-6	0-0.4	0-0.5	0-0.4		0.1	
Epicubenol	19912-67-5					tr	
γ-Eudesmol (selinenol)	1209-71-8					tr	
Eugenol	97-53-0	0.04-0.6	0-tr	0-tr	0.2-0.7	tr	0.3[d]
α-Fenchene	471-84-1					tr	
Fenchyl alcohol	1632-73-1					tr	
Germacrene B	15423-57-1					tr	
Germacrene D	23986-74-5	0.1-4.8	0-2.2	tr-2.2	3.0-6.9	0.8	1.4-1.6[c]; 2.3[b]
Guaiol	489-86-1	0-0.1				tr	
α-Gurjunene	489-40-7					tr	
Heptyl acetate	112-06-1					0.1	
α-Humulene	6753-98-6	0.5-1.4	0-0.5	0-1.5	0.5-1.7	0.4	0.1-0.4[c]; 0.2[b]
Humulene epoxide	96638-51-6					tr	

Table 5.71.1 Constituents identified in ravensara oils (*continued*)

Constituent	CAS	Percentage and range					
		A	B (1)	C (2)	D (12)	E (17)	F
Limonene	138-86-3	0.08-19.4	0.6-21.8	tr-27.8	13.9-20.9	27.5	2.9-3.5[c]; 21.5[b]
Linalool	78-70-6	0.05-12.4	0.7-21.4	0.4-21.4	3.9-8.9		1.1-1.5[c]; 1.2[d]; 3.5[b]
cis-Linalool oxide, furanoid	5989-33-3		0-1.1	0-1.1			
trans-Linalool oxide, furanoid	34995-77-2		0-1.5	0-1.5			
cis-p-Menth-2-en-1-ol	29803-82-5	0-0.3	0-0.6	0-0.6		tr	
trans-p-Menth-2-en-1-ol	29803-81-4	0-0.09	0-1.1	0-0.4		tr	
Methyl chavicol	140-67-0	0.04-19.9	1.4-94.8	1.4-94.5	2.8-16.8	7.2	3.2[d]; 6.5[b]; 76.6-83.4[c]
Methyl eugenol	93-15-2	tr-19.4	0-80.3	tr-81.6	7.8-16.9	2.7	7.0-9.9[c]; 18.7[b]; 86.6[d]
(*E*)-Methylisoeugenol	6379-72-2		0-0.1	0-0.1			
α-Muurolene	10208-80-7					0.1	
γ-Muurolene	30021-74-0	0-0.3				0.1	
α-Muurolol	104245-48-9					tr	
τ-Muurolol	19912-62-0					tr	
Myrcene	123-35-3	0.1-4.2	0.1-7.5	0-3.4	1.9-3.4	2.3	0.1-0.4[c]; 0.2[d]; 2.5[b]
Naphthalene, 1,2,3, 4,4a, 7-hexahydro-1,6-dimethyl-4-(1-methyl-ethyl)-, (1α,4β,4αβ)-						tr	
Nerol	106-25-2	tr-1.7					
Nerolidol	7212-44-4						<0.1[d]
(*E*)-Nerolidol	40716-66-3		0-0.7	0-0.7			
(*E*)-β-Ocimene	3779-61-1	0-0.2	0-0.5	tr-2.5	0.2-0.3	0.3	0.1[b]
(*Z*)-β-Ocimene	3338-55-4	0.01-2.5			0.7-1.2	1.0	0.6[d]; 1.1[b]
allo-Ocimene	673-84-7	tr-0.2				tr	
α-Phellandrene	99-83-2	0.04-11.8	0-2.8	tr-2.8	0.2-1.9	1.2	0.1-0.2[c]; 1.3[b]
α-Pinene	80-56-8	0.3-8.3	0.1-8.1	0-8.1	2.9-5.2	3.0	0.1-0.6[c]; 0.3[d]; 4.9[b]
β-Pinene	127-91-3	0.1-15.7	0-5.5	0-7.7	2.0-2.9	2.1	0.1-0.4[c]; 0.3[d]; 2.5[b]
Piperitenone	491-09-8		0-0.2	0-0.1			
cis-Piperitol	16721-38-3		0-1.0	0-<0.1			
trans-Piperitol	16721-39-4	tr-1.0	0-<0.1	0-tr		tr	
Piperitone	89-81-6		0-0.1	0-0.2			
Sabinene	3387-41-5	0.1-25.5	0-34.4	tr-34.4	8.4-16.2	4.5	0.7-1.6[c]; 14.7[b]
cis-Sabinene hydrate	15537-55-0	0.08-4.2				tr	
trans-Sabinene hydrate	17699-16-0	0.08-4.0					
Safrole	94-59-7	0-2.0					
α-Selinene	473-13-2	0.07-3.9					
β-Selinene	17066-67-0	0.07-0.8				tr	
Spathulenol	6750-60-3	0-0.4	0-0.1	0-0.3			
α-Terpinene	99-86-5	0.3-8.2	0-27.7	tr-27.7	0.1-7.0	9.4	0.1-0.5[c]; 0.4[d]; 5.0[b]
γ-Terpinene	99-85-4	0.3-2.5	0-5.8	0-5.9	0.2-2.6	2.9	0.1-0.2[c]; 1.7[b]
Terpinen-4-ol	562-74-3	0.03-6.3	0-12.0	0-12.0	0.1-0.3	2.0	0.1-1.0[c]; 0.5[d]; 2.0[b]
α-Terpineol	98-55-5	0.2-14.7			2.4-4.4	0.2	0.2[b]
δ-Terpineol	7299-42-5	0-0.8					
Terpinolene	586-62-9	0.2-0.9			0.1-0.5	0.3	0.3[b]
α-Terpinyl acetate	80-26-2	tr-0.3	0-1.3	0-1.3		tr	
α-Thujene	2867-05-2	0.2-1.7	0-1.6	0-1.6	0.4-1.1	0.6	0.9[b]
Tricyclene	508-32-7					tr	
Zonarene	41929-05-9					0.1[a]	

A forty-one ravensara essential oil samples from Madagascar, analyzed between 2002 and 2013; lowest and highest concentrations given (E. Schmidt, unpublished data)

B sixteen lab-hydrodistilled oils from leaves of 16 ravensara trees growing wild in Madagascar; lowest and highest concentrations given (ref. 1); same authors as in C (ref. 2)

C twenty-eight lab-hydrodistilled ravensara leaf oils from wild growing Madagascan *R. aromatica* trees (ref. 2); lowest and highest concentrations given; same authors as in B (ref. 1)

D five commercial oils purchased from a Madagascan essential oil company; lowest and highest concentrations given (ref. 12)

E one commercial leaf oil prepared by growers of *R. aromatica* in Madagascar (ref. 17)

F data from other studies (indicated with superscript letters); highest concentrations found in any study reviewed here given; when two or more oils were investigated, only the highest concentrations are mentioned, unless indicated otherwise

Table 5.71.1 Constituents identified in ravensara oils (*continued*)

[a] *trans*-calamenene and zonarene combined; [b] one authenticated commercial ravensara leaf oil from Madagascar (ref. 16); [c] five lab-hydrodistilled oils from leaves of ravensara trees growing wild in Madagascar; lowest and highest concentrations given (ref. 4); [d] one commercial leaf oil from Madagascar (ref. 5)

tr: trace (in columns B and C: <0.1; in column E: <0.05%)

Table 5.71.2 Major constituents of ravensara oils

Constituent	CAS	Percentage and range					
		A	B (1)	C (2)	D (12)	E (17)	F
Methyl chavicol	140-67-0	0.04-19.9	1.4-94.8	1.4-94.5	2.8-16.8	7.2	3.2[d]; 6.5[b]; 76.6-83.4[c]
Methyl eugenol	93-15-2	tr-19.4	0-80.3	tr-81.6	7.8-16.9	2.7	7.0-9.9[c]; 18.7[b]; 86.6[d]
Sabinene	3387-41-5	0.1-25.5	0-34.4	tr-34.4	8.4-16.2	4.5	0.7-1.6[c]; 14.7[b]
Limonene	138-86-3	0.08-19.4	0.6-21.8	tr-27.8	13.9-20.9	27.5	2.9-3.5[c]; 21.5[b]
α-Terpinene	99-86-5	0.3-8.2	0-27.7	tr-27.7	0.1-7.0	9.4	0.1-0.5[c]; 0.4[d]; 5.0[b]
Linalool	78-70-6	0.05-12.4	0.7-21.4	0.4-21.4	3.9-8.9		1.1-1.5[c]; 1.2[d]; 3.5[b]
β-Pinene	127-91-3	0.1-15.7	0-5.5	0-7.7	2.0-2.9	2.1	0.1-0.4[c]; 0.3[d]; 2.5[b]
Terpinen-4-ol	562-74-3	0.03-6.3	0-12.0	0-12.0	0.1-0.3	2.0	0.1-1.0[c]; 0.5[d]; 2.0[b]
α-Pinene	80-56-8	0.3-8.3	0.1-8.1	0-8.1	2.9-5.2	3.0	0.1-0.6[c]; 0.3[d]; 4.9[b]
Myrcene	123-35-3	0.1-4.2	0.1-7.5	0-3.4	1.9-3.4	2.3	0.1-0.4[c]; 0.2[d]; 2.5[b]
β-Caryophyllene	87-44-5	0.1-3.6	0-2.2	0-1.5	2.1-7.4	3.8	1.1-1.8[c]; 1.5[d]; 1.7[b]

LEGEND: SEE UNDER TABLE 5.71.1

Table 5.71.3 Five chemotypes of ravensara oils (6) [a,b]

Constituent	CAS	Percentage and range				
		Type 1[c]	Type 2[d]	Type 3[e]	Type 4[f]	Type 5[g]
β-Caryophyllene	87-44-5	0.1-0.6	0-0.5	0-0.8	0-tr	tr-5.8
Elemicin	487-11-6	0.1-0.2	0.5-2.6			tr-4.8
Germacrene D	23986-74-5	0-1.4	0-0.5	0.5-1.1	0-0.3	tr-7.4
Limonene	138-86-3	0.6-4.2	0.7-5.0	4.4-21.8	7.2-8.5	19.8-62.8
Linalool	78-70-6	0.7-1.6	0.8-1.9	3.3-4.5	6.5-21.4	2.3-11.7
trans-p-Menth-2-en-1-ol	29803-82-5	0-0.1	0-1.1	0-0.2	6.2-12.0	0.0
Methyl chavicol	140-67-0	85.6-98.3	2.6-5.0	0.1-3.4	0.1-10.3	0.1-8.1
Methyl eugenol	93-15-2	0-0.9	72.2-95.4	0-1.1	0.3-1.3	0.6-0.9
Myrcene	123-35-3	0.1-0.5	0.2-7.5	1.3-12.8	1.8-9.3	1.1-13.9
α-Phellandrene	99-83-2	0-0.3	0.1-0.2	0-9.2	0.2-8.9	0.1-17.5
α-Pinene	80-56-8	0.1-0.5	0.2-1.4	tr-3.0	2.9-10.9	0.6-7.9
β-Pinene	127-91-3	0-0.5	0.2-0.6	2.7-2.9	2.7-6.7	0.1-6.0
Sabinene	3387-41-5	0-3.4	0.4-5.1	0.4-16.0	17.0-45.7	0.7-45.7
α-Terpinene	99-86-5	0-0.8	3.1-7.8	18.2-58.1	0-13.3	1.5-3.1
γ-Terpinene	99-85-4	0-0.1	tr-1.6	5.7-5.8	1.9-4.4	tr-0.2
Terpinen-4-ol	562-74-3	0-2.2	0.3-2.2	0.1-9.8	2.4-11.7	0.5-1.0
δ-Terpineol	7299-42-5				1.8-5.0	

[a] only ingredients are shown that were present in concentrations >3% in any chemotype oil; [b] the sources of the data are partly refs. 1 and 2 and were partly unspecified; however, in a PowerPoint presentation prepared by the senior author, it was stated that over 500 ravensara oil samples have been analyzed (http://www.fofifa.mg/f_Ravaromatica.ppt.); [c] chemotype 1, methyl chavicol; [d] chemotype 2, methyl eugenol; [e] chemotype 3, α-terpinene and limonene; [f] chemotype 4, sabinene; [g] chemotype 5, limonene (ref. 6)

tr: trace (amount not specified)

CONTACT ALLERGY / ALLERGIC CONTACT DERMATITIS

General

There have been two case reports of allergic contact dermatitis from ravensara oils only, both in an occupational setting, in an aromatherapist and a complementary therapist. In one, α-pinene and possibly linalool may have been allergens in the ravensara oils.

Case reports

An aromatherapist had non-occupational contact dermatitis with allergies to multiple essential oils used at work, including ravensara oil. The patient also reacted to geraniol, linalool, linalyl acetate, α-pinene, the fragrance mix and various other fragrance materials; α-pinene was demonstrated by GC-MS in the ravensara oil (21). This chemical has been found in commercial ravensara oils in concentrations of up to 8.3% and linalool in a maximum concentration of 12.4% (Table 5.71.1 column A). A female 'complementary therapist' developed occupational contact dermatitis from a multitude of essential oils used at work, including ravensara oil (22).

LITERATURE

1 Andrianoelisoa HS, Menut C, Ramanoelina P, Raobelison F, Collas de Chatelperron P, Danthu P. Chemical composition of essential oils from bark and leaves of individual trees of *Ravensara aromatica* Sonnerat. J Essent Oil Res 2010;22:66-70

2 Andrianoelisoa HS, Menut C, Collas de Chatelperron P, Saracco J, Ramanoelina P, Danthu P. Intraspecific chemical variability and highlighting of chemotypes of leaf essential oils from *Ravensara aromatica* Sonnerat, a tree endemic to Madagascar. Flavour Fragr J 2006;21:833-838

3 Tucker AO, Maciarello MJ. Two commercial oils of Ravensara from Madagascar: *R. anisata* Danguy and *R. aromatica* Sonn. (Lauraceae). J Essent Oil Res 1995;7:327-329

4 Ramanoelina PAR, Rasoarahona JRE, Gaydou EM. Chemical composition of *Ravensara aromatica* Sonn. leaf essential oils from Madagascar. J Essent Oil Res 2006;18:215-217

5 Möllenbeck S, König T, Schreier P, Schwab W, Rajaonarivony J, Ranarivelo L. Chemical composition and analyses of enantiomers of essential oils from Madagascar. Flavour Fragr J 1997;12:63-69

6 Andrianoelisoa H, Menut C, Danthu P. *Ravensara aromatica* ou *Ravintsara*: une confusion qui perdure parmi les distributeurs d'huiles essentielles en Europe et en Amérique du Nord. Phytothérapie 2012;10: 161-169

7 Rasoanaivo P, De La Gorce P. Essential oils of economical value in Madagascar: present state of knowledge. Herbal Gram 1998;43:31-39, 58-59

8 Holm Y, Hiltunen R. Chemical composition of a commercial oil of *Ravensara aromatica* Sonn. used in aromatherapy. J Essent Oil Res 1999;11:677-678

9 Théron E, Holeman M, Potin-Gautier M, Pinel R. Authentication of *Ravensara aromatica* and *R. anisata*. Planta Med 1994;60:489-491

10 Théron E, Holeman M, Potin-Gautier M, Pinel R. Studies of the aging of Malagasy essential oils rich in 1,8-cineole. Part I. *Helicrysum gymnocephalum*. Part II. *Ravensara aromatica*. Riv Ital EPPOS 1994;131:33-38. Data cited in ref. 8

11 Behra O, Rakotoarison C, Rhiannon H. Ravintsara vs Ravensara. A taxonomic clarification. Int J Aromather 2001;11:4-7

12 Juliani HR, Kapteyn J, Jones D, Koroch AR, Wang M, Charles D, Simon JE. Application of near-infrared spectroscopy in quality control and determination of adulteration of African essential oils. Phytochem Anal 2006;17:121-128

13 De Medici D, Pierretti S, Salvatore G, Nicoletti M, Rasoanaivo P. Chemical analysis of essential oils of Malagasy medicinal plants by gas chromatography and NMR spectroscopy. Flavour Fragr J 1992;7:275-281

14 Raharivelomanana PJ. Contribution à l'étude des huiles essentielles de *Laurus nobilis*, *Cinnamomum zeylanicum*, *Ravensara anisata* (Lauracées). Composition chimique, inhibition microbienne. Postgraduate Certificate, University of Antananarivo, Faculty of Sciences, Antananarivo, Madagascar: 1988

15 Lawrence BM. Progress in essential oils. Perfum Flavor 2000;25(5):68-71

16 Juliani H, Behra O, Moharram H, Ranarivelo L, Ralijerson B, Andriatsiferana M, et al. Searching for the real Ravensara (*Ravensara aromatica* Sonn.) essential oil. Perfum Flavor 2005;30(1):60-65

17 Costa R, Pizzimenti F, Marotta F, Dugo P, Santi L, Mondello L. Volatiles from steam-distilled leaves of some plant species from Madagascar and New Zealand and evaluation of their biological activity. Nat Prod Commun 2010;5:1803-1808

18 Lis-Balchin M, Deans SG, Eaglesham E. Relationship between bioactivity and chemical composition of commercial essential oils. Flavour Fragr J 1998;13:98-104

19 Rhind JP. Essential oils. A handbook for aromatherapy practice, 2nd Edition. London: Singing Dragon, 2012

20 Lawless J. The encyclopedia of essential oils, 2nd Edition. London: Harper Thorsons, 2014

21 Dharmagunawardena B, Takwale A, Sanders KJ, Cannan S, Roger A, Ilchyshyn A. Gas chromatography: an investigative tool in multiple allergies to essential oils. Contact Dermatitis 2002;47:288-292

22 Newsham J, Rai S, Williams JDL. Two cases of allergic contact dermatitis to neroli oil. Br J Dermatol 2011;165(Suppl.1):76

Chapter 5.72 ROSEMARY OIL

DEFINITION

Rosemary oil (essential oil of rosemary) is the essential oil obtained from the leaves and flowering tops of the rosemary, *Rosmarinus officinalis* L.

INCI NOMENCLATURE

Description/definition: Rosmarinus officinalis leaf oil is the essential oil obtained from the flowering tops and leaves of the rosemary, *Rosmarinus officinalis* L., Lamiaceae

INCI name EU: Rosmarinus officinalis leaf oil

INCI name USA: Rosmarinus officinalis (rosemary) leaf oil

CAS registry number(s): 8000-25-7; 84604-14-8

EINECS number(s): 283-291-9

ISO (INTERNATIONAL ORGANIZATION FOR STANDARDIZATION) STANDARD

ISO number: 1342

ISO name: Essential oil of rosemary

Botanical origin: *Rosmarinus officinalis* L.

Parts of plant used: Flowering top, leaf

ISO values: ISO values (minimum and maximum concentrations) are shown in Table 5.72.1.

Table 5.72.1 ISO values (%) for rosemary oil [a]

Compound	CAS	Minimum	Maximum
1,8-Cineole	470-82-6	16.0	55.0
α-Pinene	80-56-8	9.0	26.0
Camphor	76-22-2	5.0	22.0
Camphene	79-92-5	2.5	13.0
Borneol	507-70-0	1.0	10.0
β-Pinene	127-91-3	2.0	9.0
Limonene	138-86-3	1.5	5.5
Myrcene	123-35-3	1.0	4.5
α-Terpineol	98-55-5	1.0	4.0
p-Cymene	99-87-6	0.5	2.5
Bornyl acetate	76-49-3	0.1	2.5
Linalool	78-70-6	0.0	2.5
Verbenone	80-57-9	0.0	2.5

[a] ISO 1342 Essential oil of rosemary ©ISO 2012; Geneva, Switzerland, www.iso.org

THE PLANT, THE OIL, AND THEIR USES

Rosmarinus officinalis L. is a woody evergreen perennial herb, up to 1.5 meters tall, which has strongly aromatic, needle-like evergreen leaves. Rosemary is native to the Madeira Islands, the Canary Islands, northern Africa, the Mediterranean European countries, Cyprus, Turkey, and the Caucasus (GRIN Taxonomy for Plants). The plant is widely cultivated for medicinal, culinary, cosmetic and ornamental purposes. Rosemary is one of the most prized culinary herbs, especially in Mediterranean cuisine (53). Dried rosemary leaves are used as seasoning for soups, meat, fish, and poultry and also applied in alcoholic beverages (vermouth) and herbal soft drinks

(62). The plant is broadly reputed for its antioxidant activities (1) and is a source of commercial antioxidant derivatives, marketed as oil and/or water miscible formulations used to retard lipid oxidation in foods (53,86).

R. officinalis has been extensively used in traditional folk medicine. The infusion of leaves is perceived to have tonic (after bleeding), antitussive, anti-asthmatic, cholagogue, choleretic, carminative, stomachic, antispasmodic (heart and stomach pains, colic), febrifuge, emmenagogue, antimicrobial, anti-rheumatic and anti-neuralgic properties (1,22). Infusions of the leaves are used externally to cure abscesses, wounds, as a gargle for mouth ulcers and as a hair tonic to improve hair growth (22). The plant is claimed to be useful in the treatment of anxiety, bloat, migraine, hypertension, headache, anorexia and many other medical problems (53,56).

The essential oil of rosemary, which is obtained by steam-distillation from the leaves and flowering tops of the plant, is widely used as an ingredient in rubefacients, liniments, inhalants, perfumes, soaps, deodorants, bath essences, hair lotions, shampoos and other cosmetics, room sprays, detergents, softeners, disinfectants and insecticides (1,2,12,22,53,62). Rosemary oil is also utilized as a seasoning for foodstuffs such as meat dishes, salami and sauces and for flavoring liquors (5,12). The oil has many applications in traditional medicine for its perceived pulmonary antiseptic, cholagogue, choleretic, stomachic, antidiarrheal and anti-rheumatic activities (2,5). In aromatherapy, it is claimed to be effective in treating anxiety-related conditions and to increase alertness (53,94).

CHEMICAL COMPOSITION

Rosemary oil is a colorless to pale yellow clear mobile liquid, sometimes with a greenish touch, which has an herbal, fresh green, terpeny and somewhat aromatic-spicy odor. The yield of essential oil from the leaves with flowering tops of *Rosmarinus officinalis* L. generally varies from 0.3 to 1.2 per cent, depending on the freshness of the source material. The main producing countries of this oil are Tunisia, Morocco, Spain, Portugal and France.

Literature data (up to December 28, 2014) on the chemical composition of rosemary and unpublished analytical data from one of us (E.S.) are shown in Table 5.72.2 in alphabetical order. In rosemary oils from various origins, over 505 chemicals have been identified. About 53 per cent of these were found in a single reviewed publication only. The major compounds found in rosemary oils from different sources are shown in Table 5.72.3. They include (highest concentrations in any study given) 1,8-cineole (71.3%), α-pinene (57.5%), myrcene (46.0%), camphor (44.5%), verbenone (43.5%), bornyl acetate (23.0%), limonene (21.7%), borneol (21.4%), β-caryophyllene (14.5%), camphene (14.3%), α-terpineol (12.8%) and β-pinene (12.0%). Well-known ingredients of rosemary oils that were present in high concentrations (>7%) in one or two studies were p-cymene (10.5% and 44.0%), piperitone (17.0% and 23.7%), linalool (9% and 20.5%), δ3-carene (19.8%), γ-terpinene (16.6%), terpinen-4-ol (10.8%), geraniol (9.2% and 9.3%), α-humulene (8.9%

and α-phellandrene (7.9%). Uncommon or rare constituents of rosemary oils found in high concentrations (>7%) in single studies include viridiflorol (19.5%), α-thujone (19.2%), methyl chavicol (13.3%), 2-ethyl-4,5-dimethylphenol (12.0%), cis-linalool oxide (10.8%), bicyclo[3.1.1]heptan-3-one (8.2%) and β-terpinene (7.1%).

Commercial oils

The ten chemicals that had the highest maximum concentrations in 108 commercial rosemary essential oil samples (concentration ranges provided) are the following: 1,8-cineole (7.6-59.8%), myrcene (0.2-46.0%), α-pinene (1.9-46.0%), camphor (2.9-24.2%), bornyl acetate (0.2-13.1%), camphene (0.06-12.0%), β-pinene (0.3-10.3%), limonene (0.7-6.7%), borneol (0.4-6.3%) and β-caryophyllene (0.7-6.2%) (Erich Schmidt, unpublished analytical data).

Chemotypes

The characteristic compounds in R. officinalis essential oil are 1,8-cineole, α-pinene, camphor and borneol (117). Various classifications of rosemary oil chemotypes have been proposed. Frequently cited is the distinction of two major oil types: oils with over 40% of 1,8-cineole and oils with approximately equal ratios (20-30%) of 1,8-cineole, α-pinene and camphor (2,84,98,114). Other authors distinguish three chemotypes, namely, cineoliferum (high content in 1,8-cineole), camphoriferum (camphor >20%) and verbenoniferum (verbenone >15%) (14,52,53,118). However, several other chemotypes have been claimed according to the relative abundance of other relevant compounds such as α-pinene (many reports of α-pinene being the major component,

e.g., 9,10,16,18,19,25,52,76), linalool (cited in ref. 81), and myrcene (11,16,53,55). American rosemary cultivars have even been grouped into six chemotypes (76). In some reports, other chemicals were found to be the dominant ingredients (i.e., with the highest concentration), including p-cymene (44.0%, ref. 98), limonene (21.7%, ref. 59), viridiflorol (19.5%, ref. 63), piperitone (23.7%, ref.40) and α-thujone (16.4-19.2%, ref. 63).

Analyses of the oils obtained from different parts of R. officinalis (apical, intermediate and lower parts of leaves, flowers and stems) can be found in ref. 62. The analyses in refs. 45 and 91 are not presented in the table. Ref. 115 (with many data) is not discussed, as the 'essential' rosemary oils were obtained by simultaneous hydrodistillation and solvent extraction and consequently are, as per ISO and the definition used by us, not essential oils (115).

The results of analyses of R. officinalis essential oils in refs. 3, 36, 57, 66, 69, 87, 95, 96 and 104 are not presented here, as their qualitative nor quantitative composition qualifies for inclusion of data from these studies in column E of Table 5.72.2. Because of the abundance of available literature on the composition of rosemary oils, recent reviews published in Perfumer and Flavorist and reprinted in refs. 111-113 (which often contain data from sources that are very hard or impossible to access otherwise), were, with one exception (ref. 116) screened only for chemicals which have not been found in the analyses presented in refs. 1-110,114 (of course, most of the articles discussed in the reviews are already included in this group of 111 items). For the same reason (abundance of material), no separate internet search for articles presenting analyses of the composition of rosemary oils was

Table 5.72.2 Constituents identified in rosemary oils

Constituent	CAS	Percentage and range				
		A	B (52)	C (67)	D (14)	E
Abietal	6704-50-3					tr[d]
Abietatriene	19407-28-4			0.4	tr-0.1	tr[d]; 0.02[u2]; 0.4[v8]
Acetic acid	64-19-7					+[b]
1-epi-Acetoxy-2-(1-methylethenyl)-5-methylcyclohexane						0.01[u2]
p-Acetylanisole	100-06-1			0.2		
α-Acoradiene	24048-44-0					0.02[v6]
Adamantine						0.8[v8]
Allopteoxylin methyl ether						0.05[u2]
α-Amorphene	20085-19-2			tr		0.1[t3]; 0.4[h]; 0.6[s5]; 1.3[x6]
γ-Amorphene	6980-46-7		0-0.3			
δ-Amorphene	189165-79-5					0.1[u]
p-Anethole	104-46-1					1.9[v8]
(E)-Anethole	4180-23-8		0-0.2			
Anisole	100-66-3				tr	0.09[d]
Apiole	523-80-8		0-0.2			
Aristolene	6831-16-9					0.1[s1]
Aromadendrene	489-39-4			0.1	0.1-0.2	0.01[d]; 0.02[v7]; 0.03[t1]; 0.07[s7]; 0.3[u5]
allo-Aromadendrene	25246-27-9		0-0.2		tr-0.2	tr[u]; <0.05[e]; 0.05[d]; 0.2[h]; 0.4[u8]
Azulene	275-51-4					0.7[v8]
Benzaldehyde	100-52-7					<0.01[h]; 0.08[v4]
Benzoic acid	65-85-0					+[b]

Table 5.72.2 Constituents identified in rosemary oils (*continued*)

Constituent	CAS	Percentage and range				
		A	B (52)	C (67)	D (14)	E
Benzyl benzoate	120-51-4					0.01[u2]
Berbonone						7.6[u2]
Bergamotene						0.01[t1]
Bicyclogermacrene	24703-35-3					1.2[y4]
Bicyclo[4.1.0]heptane	286-08-8					1.4[v8]
Bicyclo[3.1.1]heptan-3-one						8.2[x4]
α-Bisabolene	17627-44-0					0.05[t9]
(*E*)-α-Bisabolene	25532-79-0					0.2[t4]
(*Z*)-α-Bisabolene	29837-07-8			tr		tr[x6]; 2.1[t5]
β-Bisabolene	495-61-4	0.02-0.4	0-0.1		0.1-0.2	tr[k]; 0.04[d]; 0.1[j]; 0.2[g]; 0.3[v7]; 0.4[t8]; 0.7[v8]
γ-Bisabolene	495-62-5				tr	0.04[d]
α-Bisabolol	515-69-5				0.3-0.5	0.2[t8]; 0.3[k]; 0.4[s1]; 0.7[d]; 1.0[z7]; 1.3[p]; 1.9[v7]
epi-α-Bisabolol	78148-59-1					0.2[u8]; 0.3[u]
Bisabolol-4-ol			0-0.6			
α-Bisabolol oxide	22567-36-8					1.6[v4]
2,5-Bornanediol						0.01[u2]
2-Bornene (α-)	464-17-5					0.08[u2]; 0.8[v8]
Borneol	507-70-0	0.4-6.3	1.4-18.3	10.0	1.1-7.1	12.1[r]; 12.9[w8]; 13.7[v]; 14.2[v8]; 14.7[u9]; 15.9[s]; 16.9[v2]; 17.5[o]; 18.1[u4]; 21.4[i]
Bornyl acetate	76-49-3	0.2-13.1	0-9.3	4.2	0.1-0.8	4.4[n]; 4.7[e]; 5.5[v2]; 5.8[j]; 6.1[t9]; 7.6[u8]; 9.0[t5]; 12.3[t2]; 14.3[k]; 14.9[u4]; 17.0[w3]; 21[u9]; 23.0[m]
Bornyl formate	7492-41-3					0.1[t1]; 0.3[o]; 1.2[w1]
Bornyl valerate	7549-41-9					+[c]
β-Bourbonene	5208-59-3					<0.1[u5]; 0.04[h]; 0.1[x6]
Butyl acetate	123-86-4					1.3[y1]
Butylbenzene	104-51-8					1.1[w7]
Butyric acid	107-92-6					+[b]; 1.0[v8]
Cadalene	483-78-3					+[c]
Cadina-1(6),4-diene	16729-00-3					+[c]
α-Cadinene	24406-05-1			0.1		tr[r]; 0.08[k]; 0.1[a,d]; 0.2[x6]; 0.4[s7]; 1.0[j]
γ-Cadinene	39029-41-9		0-0.5	1.0	0.1-0.3	0.2[g]; 0.3[h]; 0.4[r]; 0.7[s5]; 1.1[x6]; 1.5[v8]; 1.6[v8]
trans-γ-Cadinene						tr[t]
δ-Cadinene	483-76-1	0.03-0.7	0-0.4	1.8	0.1-0.2	0.6[e]; 0.8[j]; 1.2[y1]; 1.6[s5]; 2.0[x6]; 2.3[y7]; 3.2[v8]
cis-Cadinen-4-en-7-ol			0-0.2			
α-Cadinol	481-34-5			0.1		0.07[s1]; 0.7[v7]
epi-α-Cadinol	5937-11-1		0-0.3	1.2	tr	0.1[d,j]; 2.3[v4]
δ-Cadinol	19435-97-3			0.1		0.03[s1]
Calacorene	38599-17-6					0.2[r,s3]
α-Calacorene	21391-99-1			0.3		0.07[s1]; 0.1[u]; 0.2[t7]; 0.3[s5]; 0.6[g]
β-Calacorene	50277-34-4					tr[t4]
Calamene	1406-50-4					+[c]
Calamenene	483-77-2				0.2-0.3	tr[r]; 0.02[d]; 0.09[h]; 0.1[v]
cis-Calamenene	72937-55-4					1.7[t7]
(1*S*)-*cis*-Calamenene	483-77-2					0.1[u8]
trans-Calamenene	73209-42-4					<0.1[q]
Calarene	17334-55-3			0.6		0.03[v7]
Camphene	79-92-5	0.06-12.0	1.5-9.6	11.1	1.3-3.6	8.6[g]; 8.9[s5]; 9.2[v3]; 9.4[s8]; 10.2[i]; 10.7[t9]; 11.3[k]; 11.8[z8]; 12.7[s6]; 13.5[u9]; 14.3[u7]
Camphene hydrate	465-31-6		0-0.2			0.1[u1,u7]
6-Camphenol	3570-04-5					0.1[y]; 0.2[t9]
6-Camphenone	53803-33-1					0.4[v4]
α-Campholenal	4501-58-0		0-0.1	tr		tr[s5]; 0.04[h]; 0.1[t7]; 0.2[f]; 0.3[t3]; 1.1[v6]; 1.2[t8]
α-Campholenic acid	28973-89-9					+[b]
γ-Campholenic acid	67246-55-3					+[b]
α-Campholenol	1901-38-8					0.01[v3]; 0.9[w3]; 1.4[t7]
γ-Campholenol						1.0[v];
α-Campholytic acid	6709-22-4					+[b]
Camphor	76-22-2	2.9-24.2	0.3-26.8	27.5	7.6-18.9	24.1[t7]; 24.9[s6]; 26.0[q]; 26.3[w9]; 26.6[u9]; 28.1[i]; 32.3[v2]; 34.7[g]; 34.8[s3]; 35.3[p]; 44.5[z]
(+)-2-Carene	4497-92-1					+[c]; 0.05[u1]
4-Carene	29050-33-7					0.06[t9]

Table 5.72.2 Constituents identified in rosemary oils (*continued*)

Constituent	CAS	Percentage and range				
		A	B (52)	C (67)	D (14)	E
γ-3-Carene						0.08[t8]
δ2-Carene (= δ4-)	554-61-0					0.2[u3]
δ3-Carene	13466-78-9	0-0.8		tr	0.1-1.1	1.2[k]; 1.3[m]; 1.4[g]; 1.5[u4]; 2.0[x2]; 3.2[t9]; 19.8[x7]
3(10)-Caren-2-ol	93905-77-2					1.2[v4]
Carvacrol	499-75-2		0-4.9	0.3	tr	0.1[h]; 0.2[s]; 0.3[f]; 0.4[u6]; 0.7[t9]; 0.9[w2]; 1.7[o]
Carveol	99-48-9					0.04[h]
(E)-Carveol	1197-07-5		0-0.1			0.5[v6]; 0.7[w2]; 1.0[t7]; 2.4[v4]
(Z)-Carveol	1197-06-4		0-1.1			0.09[t9]
Carvone	99-49-0		0-0.4		tr-0.2	0.3[t7]; 0.4[u7]; 0.5[t2]; 0.7[p]; 1.1[t5]; 1.3[v4]; 1.7[u8]
Carvotanacetone	499-71-8		0-0.1			0.1[v6]
Caryolan-8-ol	178737-45-6		0-0.1			
Caryophylladienol II						0.1[t7]
Caryophylla-3,8(13)-dien-5α-ol				1.0		
Caryophylla-3,8(13)-dien-5β-ol						0.08[v6];
Caryophylla-4(12),8(13)-dien-5β-ol	19431-80-2			0.6		0.1[x6]
β-Caryophyllene	87-44-5	0.7-6.2	0.1-2.9	7.7	0.2-4.2	5.3[u9]; 5.5[u6]; 6.3[s1]; 6.4[v8]; 6.8[y1]; 7.4[y9]; 8.4[s9]; 9.2[h]; 10.9[x6]; 12.9[s7]; 13.6[x1]; 14.5[y7]
γ-Caryophyllene	118-65-0			+[c];		
9-epi-(E)-Caryophyllene	68832-35-9			tr		
Caryophyllene oxide	1139-30-6	0-0.6	0-0.3	1.7	0.5-1.0	1.7[i]; 1.9[v4]; 2.0[p]; 2.3[y1]; 2.4[t6]; 3.1[x6]; 6.0[t7]
Caryophyllenol						0.2[u1]
Caryophyllenol I	32214-88-3					0.5[t7]
Caryophyllenol II	32214-89-4					0.2[t7]
Cedranediol	88588-48-1					tr[d]; 0.7[v]
Cembrene	1898-13-1			0.4		
Chavicol	501-92-8					+[c]
cis-Chrysanthenol	55722-60-6					1.9[z4]
Chrysanthenone	473-06-3		0-0.9			0.2[t7]; 0.3[t]; 0.6[i]; 0.7[g,t5]; 0.8[w]; 1.0[t9,y7]
1,4-Cineole	470-67-7					0.5[v5]
1,8-Cineole	470-82-6	7.6-59.8	5.7-67.0	52.6	41.2-63.3	52.4[r]; 52.8[y3]; 55.9[t4]; 57.5[j]; 58.6[x]; 58.7[u3]; 59.5[s6]; 60.9[t7]; 63.3[d]; 64.5[y8]; 71.3[w6]
Citronellal	106-23-0				tr	0.08[d]
Citronellic acid	502-47-6					+[b]
Citronellol	106-22-9		0-0.4			tr[q]; 0.01[v6]; 0.1[s3,x6]; 0.3[f]; 0.4[w3]
Clovenol						0.2[t7]
α-Copaene	3856-25-5	0.03-0.4	0-0.9	0.6	tr	0.1[g]; 0.2[s4]; 0.6[r]; 0.7[s5]; 1.3[x6]; 1.6[y7]
β-Copaene	18252-44-3		0-0.2			
β-Copaen-4α-ol	124753-76-0					tr[s5]
α-Corocalene	20129-39-9					+[c]
Coumarin	91-64-5					1.6[v4]
m-Cresol	108-39-4					+[b]
o-Cresol	95-48-7					+[b]; 0.04[v4]
p-Cresol	106-44-5					+[b];
Cryptone	500-02-7					0.1[x5]
α-Cubebene	17699-14-8	0.01-0.4	0-0.1	0.4	0.1-0.2	0.08[h]; 0.1[d]; 0.2[v]; 0.3[x6]; 0.5[f]
β-Cubebene	13744-15-5					0.03[h]
Cubebol	23445-02-5					<0.05[t4]
Cubenene	29837-12-5			0.1		
Cubenol	21284-22-0					tr[s1]; 0.5[v4]
p-Cumenol	99-89-8					0.5[v4]
Cuminaldehyde	122-03-2				tr	tr[w4]; 0.04[d]; 0.1[t3]; 3.4[v4]
Cuminyl alcohol	536-60-7					2.5[v4]
Curcumene (ar-; α-)	644-30-4					+[c]; tr[w3]; 0.1[v6]; 0.2[t7]; 0.4[v];
γ-Curcumene	28976-68-3					0.04[v7]
Cyclohexane	110-82-7					1.1[v8]
3-Cyclohexane-1-carboxylic acid, 3,7-dimethylethyl ester						0.08[u2]

Table 5.72.2 Constituents identified in rosemary oils (*continued*)

Constituent	CAS	Percentage and range				
		A	B (52)	C (67)	D (14)	E
Cyclohexane, 1,3-di-iodopropenyl-6-methyl-						5.2[u5]
Cyclohexanepropanol, 2,2-dimethyl-6-methylene-						0.01[u2]
1-Cyclohexene-1-propanol, 2,6,6-trimethyl-						0.03[u2]
o-Cymene	527-84-4			0.2		0.2[t7]; 0.6[x8]; 0.7[z7]; 1.8[x3]; 2.0[u1]; 2.6[v7]
p-Cymene	99-87-6	0.02-3.8	0.4-3.8	2.7	0.2-1.5	2.5[j]; 2.6[p]; 2.7[m]; 2.9[t3]; 3.1[u3]; 3.4[g]; 3.8[s3]; 4.3[u4]; 4.5[t1]; 4.8[y6]; 6.3[u8]; 10.5[o]; 44.0[w4]
m-Cymenene	1124-20-5					0.3[w]
p-Cymenene	1195-32-0				0.1-0.5	0.03[s7]; 0.07[t9]; 0.1[u7]; 0.2[t7]; 0.3[s3]; 0.5[d]
Cymen-8-ol			0-0.2			0.1[h]
m-Cymen-8-ol	5208-37-7					0.5[t7]
p-Cymen-8-ol	1197-01-9		0-0.2			0.3[w1]; 0.4[t6]; 0.7[g]; 1.8[w2]; 2.0[t7]; 4.1[v7]
Decanoic acid	334-48-5					+[b]
n-Decanol	112-30-1					0.2[v7]
3-Decanone	928-80-3					<0.05[w]
2,3-Dehydro-1,8-cineole	92760-25-3					0.3[v4]; 0.8[t7]
Diazene, acetylphenyl	13443-97-5					0.01[u2]
Dihydrocarveol	38049-26-2					0.1[t9]; 0.5[v]
Dihydrocarvone	5948-04-9					1.6[a,y5]
cis-Dihydrocarvone	3792-53-8					0.2[u1]
Dihydrocarvyl acetate	57287-13-5					2.0[t9]
Dihydroeudesmol	6770-16-7					0.6[t5]
Dihydroeugenol	2785-87-7					0.3[t5]
Dihydropinocarvone						0.9[t4]
trans-Dihydro-α-terpineol	5114-00-1					
6,6-Dimethylbicyclo-[3.1.1]hept-3-ene-2-butylene						tr[j]; 0.8[k]
4,4-Dimethyl-2-buten-4-olide	20019-64-1					0.09[v4]
(3E)-2,6-Dimethyl-3,5-heptadien-2-ol						0.3[s1]
2,6-Dimethyl-1,6-heptadien-3-yl acetate	74902-74-2					0.02[u1]
2,7-Dimethyl-2,6-octadien-1-ol	22410-74-8					4.0[x5]
1,7-Dimethyloctanol						0.4[a,y5]
2,6-Dimethylphenol	576-26-1					1.5[u7]
2,5-Dimethylstyrene	2039-89-6					0.5[u8]
Dioctyl phthalate	117-81-7					tr[a2,d]
1,5-Diphenyl-2H-1,2,4-triazoline						0.1[z6]
Docosanoic acid	112-85-6					0.3[z6]
Dodecane	112-40-3				tr	0.02[d]
Eicosane	112-95-8				tr	tr[d]; 0.1[u1]
Eicosene	27400-78-8				0.1-0.2	tr[d]
β-Elemene	33880-83-0				0.1-0.2	0.03[d]; 0.2[s8]; 0.6[u5]
Epoxy allo-aromadendrene	85760-81-2					1.4[x5]
Eremophilene	10219-75-7					0.5[v8]
7-Ethenyl-1,2,3,4,4a,5,6,7,8,9,10,10a-dodecahydro-1,1,4a,7-tetramethylphenanthrene	55255-56-6			0.2		
Ethyl alcohol	64-17-5					

Table 5.72.2 Constituents identified in rosemary oils (*continued*)

Constituent	CAS	Percentage and range				
		A	B (52)	C (67)	D (14)	E
2-Ethyl-1,1-dimethyl-3-methylenecyclo-hexane						0.02[u1]
2-Ethyl-4,5-dimethyl-phenol	2219-78-5					12.0[x5]
Eucarvone	503-93-5					0.2[v8]; 1.1[t9]; 1.8[v4]
α-Eudesmol	473-16-5			3.0	0.1-0.3	0.08[d]; 0.4[s5]
7-epi-α-Eudesmol	123123-38-6					tr[s5]; 0.2[y3]
β-Eudesmol	473-15-4			1.3	0.1-0.3	1.1[d,s5]
γ-Eudesmol	1209-71-8			0.9	tr	0.09[d]
Eugenol	97-53-0		0-0.1	0.2	0.1-0.2	tr[r]; 0.01[h]; 0.03[s7]; 0.1[d,g]; 0.2[s3]; 0.6[t2]
α-Farnesene	502-61-4					<0.05[s]; 0.01[v6]; 0.1[r]
β-Farnesene	502-60-3				0.1-0.2	0.2[k]; 0.5[j]
(E)-β-Farnesene	18794-84-8	0-0.1				0.01[v6]; 0.02[s1]; 1.1[w]
(Z)-β-Farnesene	28973-97-9					0.1[d]; 0.3[x8]; 1.3[z7]
Farnesyl acetate	29548-30-9				0.1-0.3	0.02[d]
Farnesyl acetone	1117-52-8					1.3[v8]
Fenchene	514-14-7					<0.01[h]; 2.3[t4]
α-Fenchene	471-84-1	0.01-0.6				0.09[w2]; 0.1[w3]; 0.2[t7]
β-Fenchene	497-32-5					+[c]
Fenchone	1195-79-5				0.1-0.7	tr[w3,w4]; 0.3[z9]; 0.7[d]
α-Fenchone						+[c]
α-Fenchyl acetate	111821-74-0				tr	0.06[d]; 1.7[u2]; 2.3[s3]
Fenchyl alcohol	1632-73-1	0-0.2		tr		tr[w4]; 0.1[t7,u3]; 0.2[x5]
α-Fenchyl alcohol	512-13-0		0-1.2		tr-0.4	0.2[u1]; 0.3[w1]
β-Fenchyl alcohol	22627-95-8		0-0.1			0.02[u1]; 0.1[u2]
Ferruginol	514-62-5					0.1[v7]
trans-Ferruginol				0.4		0.06[s1]
Filifolone						0.07[u1]; 0.5[w1]; 0.9[t7]
Geijerene	6902-73-4					0.3[u4]
Geranial	141-27-5		0-0.2			<0.05[t4]; 0.2[w3]; 0.3[v9]; 0.4[n]
Geranic acid	459-80-3					+[b]
Geraniol	106-24-1	0-0.2	0-0.8		tr	0.2[f]; 0.4[t]; 0.7[t8]; 0.8[x7]; 2.0[u2]; 2.5[m]; 2.7[w5]; 4.3[n]; 4.7[k]; 5.1[v9]; 5.8[t2]; 6.2[w3]; 9.2[u9]; 9.3[e]
Geranyl acetate	105-87-3		0-0.1			0.2[z4]; 0.3[v4]; 0.4[n]; 0.6[t4]; 0.7[k]; 0.8[p]; 0.9[e]
Geranyl acetone	3796-70-1		0-0.1		0.1-0.2	0.02[u2]; 0.2[v6]
Germacrene B	15423-57-1					0.01[a,d]
Germacrene D	23986-74-5	0-0.4	0-0.3			tr[k]; 0.01[v3]; 0.04[s1]; 0.1[g,j]; 0.3[r]; 0.5[u5] 1.7[y7];
Germacrene D-4-ol	198991-79-6					0.09[s1]
Globulol	489-41-8				tr	1.1[d]
Guaiacol	90-05-1					+[b]
cis-Guaia-3,9-dien-11-ol						0.1[y]
α-Guaiene	3691-12-1					tr[s1]; 0.1[g]
β-Guaiene	88-84-6			tr		0.05[t9]
β-Gurjunene	73464-47-8					0.05[u8]; 0.1[v]; 0.2[f,s]
γ-Gurjunene	22567-17-5			tr		0.2[s]
β-Gurjunene epoxide						0.04[t9]
Heneicosane	629-94-7				tr-0.1	tr[d]
1,6-Heptadien-4-ol	2883-45-6					0.6[v8]
Heptanal	111-71-7					tr[w4]
Heptanoic acid	111-14-8					+[b]
n-Hexadecane	544-76-3					tr[s5]
Hexadecanoic acid	57-10-3					+[b]; 0.1[r]
Hexadecanol	51260-59-4				0.1-0.4	tr[d]; 0.05[v7]
Hexanal	66-25-1					tr[w4]
Hexanoic acid	142-62-1					+[b]
3-Hexanone	589-38-8					+[c]
(E)-2-Hexenal	6728-26-3					tr[s5,u]
(Z)-5-Hexenal oxime						0.07[u1]
(E)-2-Hexenoic acid	13419-69-7					+[b]
(E)-3-Hexenoic acid	1577-18-0					+[b]

Table 5.72.2 Constituents identified in rosemary oils (*continued*)

Constituent	CAS	A	B (52)	C (67)	D (14)	E
4-Hexenoic acid	35194-36-6					+[b]
(Z)-3-Hexen-1-ol	928-96-1	0-0.08	0-0.1			0.1[t7]
Hexyl acetate	142-92-7					0.01[v3]; 0.2[t4]
α-Himachalene	3853-83-6					0.1[u6]
β-Himachalene	1461-03-6					0.3[u6]
Hinesol	23811-08-7					tr[s5]
Humuladienol						tr[r]
α-Humulene	6753-98-6	0.1-1.6	0-0.6	1.1	0.1-0.2	1.9[s7]; 2.0[y1]; 2.1[g]; 2.9[s9]; 3.0[x6]; 5.4[j]; 8.9[h]
α-Humulene epoxide	96638-51-6					tr[r,w4]; 0.02[v6]; 0.1[s4]
Humulene epoxide II	19888-34-7		0-0.2			0.1[s1]; 0.2[g]; 0.3[t4]; 0.9[i]
14-Hydroxy-9-epi-(E)-caryophyllene	244226-09-3		0-0.3			
exo-2-Hydroxycineole acetate	72257-53-5					0.1[u3]
4-Hydroxy-4-methyl-2-pentanone	123-42-2					0.7[u5]
Isoaromadendrene epoxide	499134-59-7					0.02[u2]; 0.07[t9]; 0.7[x5]
Isoborneol	124-76-5					1[v8]; 3.1[x2]; 3.3[x7]; 3.5[u5]; 4.3[z3] 4.7[y1]; 5.3[w2]
Isobornyl acetate	125-12-2					0.3[y]; 0.4[u9]; 0.5[u6]; 1.4[u5]; 1.7[s]; 2.0[v]; 5.3[t]
Isobornyl formate	1200-67-5		0-1.2			
Isocaryophyllene oxide						0.1[r]; 0.3[y4]
Isodihydrocarvyl acetate	220329-20-4					0.2[w]
Isoeugenol	97-54-1			tr		0.02[u2]; 0.5[v4]
Isofenchone	6541-58-8					0.3[z9]
Isohexanoic acid	646-07-1					+[b]
Isoisopulegol	18674-65-2					tr[q]; 0.02[t9]; 0.1[g]; 0.3[w]
Isoledene	95910-36-4					0.6[t5]
trans-Isolimonene	6876-12-6					0.3[u1]
Isolongifolol	1139-17-9					0.02[u2]
Isomenthol	3623-52-7					0.1[a,y5]
Isopinocampheol	27779-29-9					0.3[w3,y4]
Isopinocamphone	15358-88-0	0.01-1.4	0-2.0			0.1[h]; 0.2[w]; 0.3[t1]; 0.6[k]; 1.1[t9]; 2.8[u8]; 3.5[t7]
Isopiperitenone	529-01-1					0.03[u2]
4-Isopropylbenzoic acid	536-66-3					+[b]
Isopulegol	89-79-2		0-0.1			0.08[v7]; 0.2[f]; 1.0[z4]
Isopulegol II	1370348-44-9					1.2[x9]
Isopulegol acetate	57576-09-7					1.8[t9]
Isoterpinolene	586-63-0					0.7[x7]
Isovaleric acid	503-74-2					+[b]
Jasmone	488-10-8					0.1[w3]
(Z)-Jasmone	488-10-8		0-0.2			0.07[z6]; 0.2[v7]
Ketone, 1-[4,4-(4-methylpentyl)-3-cyclohexane-1-yl)]						0.01[u2]
Lavandulol	498-16-8					0.06[k]; 0.2[j]; 0.3[w3]
Lavandulyl acetate	25905-14-0					0.04[u2]; 0.3[j]; 0.8[k]
Ledol	577-27-5				0.2-0.5	1.7[d]
Levulinic acid	123-76-2					+[b]
Limonene	138-86-3	0.7-6.7	0-5.8		2.0-6.7	5.8[v6]; 5.9[f]; 6.3[t5]; 6.5[w7]; 6.6[n]; 6.7[d]; 7.1[u8]; 7.8[s8]; 9.3[v1]; 11.0[p]; 14.9[z7]; 21.7[u1]
Limonen-4-ol	3419-02-1					0.1[t7]
Limonene oxide	1195-92-2					1.1[d]
cis-Limonene oxide	13837-75-7				0.1-1.1	
Linalool	78-70-6	0.2-2.1	0-4.7	0.7	0.8-2.7	3.1[s3]; 3.2[i]; 3.8[u9]; 3.9[r]; 4.0[v2]; 4.9[e]; 5.0[s6]; 5.3[w7]; 5.4[u8]; 5.9[x4]; 6.5[t7]; 9.4[k]; 20.5[w4]
Linalool oxide	1365-19-1					<0.1[u5]
cis-Linalool oxide	11063-77-7					0.04[v4]; 0.1[w2]; 10.8[u1]
cis-Linalool oxide, furanoid	11063-77-7					tr[t]; 0.8[t7]

Table 5.72.2 Constituents identified in rosemary oils (*continued*)

Constituent	CAS	Percentage and range				
		A	B (52)	C (67)	D (14)	E
trans-Linalool oxide, furanoid	34995-77-2					0.1[w2]; 0.6[t7]
Linalyl acetate	115-95-7					0.5[p]; 0.6[t4]; 0.7[u9]; 0.8[v]; 0.9[s]; 1.3[t1]; 5.5[k]
Linalyl isobutyrate	78-35-3					tr[u]
Linalyl propionate	144-39-8				tr	0.01[d]; 3.4[v8]
Longifolene	475-20-7					0.02[u2]
Lyral	130066-44-3					0.1[a2,y5]
Lyratal						0.07[t4]
β-Maaliene	489-29-2					0.3[v8]
Manool	596-85-0					13.1[h]
Manoyl oxide	596-84-9			0.3		
Megastigmatrienone	13215-88-8					0.6[v4]
p-Mentha-1,3-dien-7-al	1197-15-5					<0.05[t4]; 1.4[v5]
p-Mentha-7,8-diene			0-2.4			
p-Mentha-1,5-dien-8-ol	1686-20-0					0.2[t7]
cis-p-Mentha-1(7),8-dien-2-ol	22626-43-3					6.3[t5]
p-Mentha-1,8-dien-3-one	16750-82-6					<0.1[q]
3-Menthene	500-00-5					<0.1[u5]
p-Menth-1(7)-en-4-ol						+[c]
p-Menth-1-en-9-ol	18479-68-0					<0.05[t4]
cis-p-Menth-2-en-1-ol	29803-82-5					tr[w4]; 0.1[t]; 3.2[d]
trans-p-Menth-2-en-1-ol	29803-81-4					tr[w4]; 0.04[s1]
Menthone	89-80-5					0.4[u1]
o-Methoxybenzoic acid	529-75-9					+[b]
2-Methoxy-3,8-di-oxocephalotax-1-ene	114942-83-5					1.1[z6]
Methyl bornyl ether	10395-54-7					0.06[u2]
3-Methyl-3-butenyl benzoate	5205-12-9					0.01[u2]
2-Methylbutyric acid	116-53-0					+[b]
Methylcamphenoate	52557-97-8					2.0[t9]
Methylcarvacrol	6379-73-3		0-0.4			1.4[t5]
Methyl chavicol	140-67-0		0-13.3			tr[w4]; 0.1[t]; 0.3[w3]
Methyl eicosanoate	1120-28-1					tr[d]
Methyl eugenol	93-15-2	0-0.2	0-1.3	1.1	tr	0.1[t8]; 0.2[g]; 0.3[u8]; 0.4[f]; 0.8[t2]; 1.0[x4]
5-Methylheptanoic acid	1070-68-4					+[b]
5-Methyl-3-heptanone	541-85-5					0.3[t7]
2-Methylisoborneol	2371-42-8					0.3[t5]
Methyl isoeugenol	93-16-3					0.3[s3]
Methyl jasmonate	1211-29-6				0.1-0.3	tr[w4]; 0.09[d]; 0.2[v6]; 0.3[v7]
(*Z*)-Methyl jasmonate	1211-29-6					0.1[t8]
8-Methylnonanoic acid	5963-14-4					+[b]
Methyl octadecanoate	112-61-8					tr[d]
7-Methyloctanoic acid	26896-18-4					+[b]
2-Methyl-2-pentenoic acid	3142-72-1					+[b]
4-Methyl-3-penten-2-one	141-79-7					0.01[u1]
5-Methyl-2-pyrithione						1.4[v8]
2-Methyl-1-thioindan	6383-15-9					0.2[u2]
Methyl thymol	1076-56-8		0-0.1			
3-Methylvaleric acid	105-43-1					+[b]
α-Muurolene	10208-80-7		0-0.1	tr		tr[v]; <0.1[q]; 0.1[s]; 0.2[h]; 0.4[x6]; 0.7[e]; 0.8[v8]
γ-Muurolene	30021-74-0		0-0.2	0.6	tr	0.07[u6]; 0.1[g]; 0.2[s1]; 0.3[v7]; 0.9[v4]; 1.2[y1]

Table 5.72.2 Constituents identified in rosemary oils (*continued*)

Constituent	CAS	Percentage and range				
		A	B (52)	C (67)	D (14)	E
δ-Muurolene	120021-96-7					tr[k]; 0.1[j]
ε-Muurolene	30021-46-6					<0.1[a,y5];4.8[a,y4]
α-Muurolol	104245-48-9		0-0.1			
τ-Muurolol (epi-α-)	19912-62-0		0-0.7	0.3		0.08[u8]; 0.4[f]
Myrcene	123-35-3	0.2-46.0	0.7-3.3	0.7	1.0-3.1	6.5[y2]; 7.0[w2]; 8.6[v7]; 9.5[t1]; 10.0[y6]; 11.5[i]; 12.4[q]; 21.5[s9]; 22.1[z2]; 22.7[n]; 29.5[f]; 31.1[u]
Myrcenol	543-39-5					0.8[z7]
cis-Myrtanol	15358-92-6					0.1[u2]; 0.4[w]; 1.2[u8]; 1.9[w1]
trans-Myrtanol	15358-91-5					0.4[w]; 1.6[w1]; 2.3[u8]
cis-Myrtanyl acetate	29021-36-1					0.2[w1]
trans-Myrtanyl acetate	90934-53-5					0.3[w1]
Myrtenal	564-94-3		0-0.7		tr-0.2	0.05[h]; 0.1[t7]; 0.2[d]; 0.6[p]; 1.9[v4]; 3.1[x7]
Myrtenic acid	19250-17-0					+[b]
Myrtenol	515-00-4		0-1.9		tr	0.7[x4]; 0.9[t7]; 1.1[i]; 1.3[z5]; 1.4[z6]; 1.6[t5]; 1.7[w3]
cis-Myrtenol			0-0.7			
Myrtenyl acetate	1079-01-2		0-0.2			0.03[u2]
Naphthalene	91-20-3					0.06[d]
Neodihydrocarvyl acetate	56422-50-5					0.2[w]
Neothujol	35732-36-6					0.1[u8]
Neral	106-26-3					0.05[u2]; 0.2[w3]; 0.4[n]
Nerol	106-25-2				tr	tr[s1]; 0.05[d]; 0.1[j]; 0.3[t9]; 1.3[u9]
(E)-Nerolidol	40716-66-3					tr[u]; 0.1[s1]
Neryl acetate	141-12-8		0-0.1			0.1[t1]; 0.2[u2]
Nerylacetone	3879-26-3		0-0.2			0.8[v8]
Neryl propionate	105-91-9					0.04[t1]
Nonadecane	629-92-5				0.1-0.2	tr[d]; 0.3[u1]
Nonadecanol	1454-84-8				tr	tr[d]
Nonanal	124-19-6					0.03[v7]; 0.2[t7]
Nonanoic acid	12-05-0	12-05-0				+[b]
Nonanol	28473-21-4				0.1-0.4	0.4[d,s]; 0.6[v]
2-Nonanone	821-55-6					0.3[t4]
Nopol	128-50-7					0.4[f]; 0.5[u2]; 1.3[x4]
(E)-β-Ocimene	3779-61-1	0-0.3	0-0.4	tr	tr	0.05[h]; 0.07[d]; 0.1[p]; 0.2[n]; 1.1[v5]; 1.2[f]; 1.5[k]
(Z)-β-Ocimene	3338-55-4	0-0.2	0-1.0	tr		0.1[v1]; 0.2[t]; 0.3[h]; 0.4[g]; 2.3[x7]; 2.6[n]; 3.3[f]
allo-Ocimene	673-84-7					0.1[u1]; 0.4[y5]
(E)-allo-Ocimene	3016-19-1					tr[w3]; 1.8[w7]
cis-Ocimene, 8-oxo-						0.6[v4]
n-Octadecane	593-45-3				tr-0.2	0.01[d]; 0.2[u1]
Octadecanol	26762-44-7				tr	tr[d]
1-Octadecene	112-88-9				tr-0.1	0.08[d]
9-Octadecenoic acid	2027-47-6					0.3[z6]
3,6-Octadienoic acid, 3,7-dimethyl, methyl ester						0.07[v6]; 0.08[u2]
Octanoic acid	124-07-2					+[b]
Octanol	29063-28-3					+[c]
n-Octanol (1-)	111-87-5					<0.05[t4]; 0.1[v3]
3-Octanol	589-98-0		0-0.1			0.02[h]; 0.1[t3]; 0.2[g]; 0.3[t5]; 0.5[w1]; 0.9[i]
3-Octanone	106-68-3		0-3.1			1.5[x8]; 3.2[g]; 3.4[i]; 4.7[s2]; 4.8[z1]; 5.1[w1]; 6.2[y2]
4-Octanone	589-63-9					0.9[t3]; 2.9[t6]
(E)-3-Octenoic acid	5163-67-7					+[b]
(Z)-3-Octenoic acid	5169-51-7					+[b]
1-Octen-3-ol	3391-86-4	0.02-0.2	0-1.5			0.2[v3]; 0.3[f,g]; 0.4[s3]; 0.5[s]; 0.6[x6]; 0.7[t7]; 2.0[t4]
3-Octen-1-ol	18185-81-4					tr[u]; 1.3[u1]
7-Octen-4-ol	53907-72-5					0.4[u3]
1-Octen-3-one	4312-99-6					0.1[w3]
1-Octen-3-yl acetate	2442-10-6					<0.1[u5]; 0.03[v3]
β-Oplopenone	28305-60-4					tr[s5]
4-Oxo-β-isodamascol						0.2[u2]
Patchoulane	25491-20-7					3.5[v8]

Table 5.72.2 Constituents identified in rosemary oils (*continued*)

Constituent	CAS	Percentage and range				
		A	B (52)	C (67)	D (14)	E
β-Patchoulene	514-51-2					0.4[v8]
Pentacosane	629-99-2					<0.05[s]
Pentadecanol	629-76-5				0.1-0.2	0.03[d]
1-Pentene	109-67-1					0.8[v8]
α-Phellandrene	99-83-2	0.05-2.1	0.1-0.7	0.2	tr-1.3	0.9[u4]; 1.0[v3]; 1.1[g]; 2.0[i]; 2.2[u8]; 5.8[m]; 7.9[x3]
β-Phellandrene	555-10-2					0.9[u1]; 1.0[z4]; 1.2[x9]; 3.0[f]; 3.9[u4]; 5.2[v3]; 5.5[i]
L-Phellandrene	6153-17-9					0.08
α-Phellandren-8-ol	1686-20-0					+[c]
β-Phellandren-8-ol	65293-09-6					+[c]
Phenethyl iodide	17376-04-4					0.01[u2]
Phenol	108-95-2					+[b]; 2.1[v8]
Phenoxyacetic acid	122-59-8					+[b]
Phenylacetaldehyde	122-78-1		0-0.2			tr[s5]; 0.04[h]; 0.1[t]; 0.3[g,t7]
Phenylacetic acid	103-82-2					+[b]
Phthalic acid	88-99-3					0.4[a2,z6]
Phyllocladene	469-86-3				0.1-0.2	0.01[d]
Phytol	7541-49-3				tr	0.04[d]; 0.5[v8]
Phytone	16825-16-4					0.2[v8]
Pimaradiene	1686-61-9					0.09[s1]
α-Pinene	80-56-8	1.9-46.0	9.8-46.1	10.4	2.2-9.2	39.0[e]; 39.4[k]; 42.8[z1]; 43.5[v6]; 44.1[t8]; 45.1[w5]; 46.1[w]; 46.2[n]; 47.1[m]; 57.5[u9]
β-Pinene	127-91-3	0.3-10.3	0-7.0	3.2	2.1-7.8	5.9[v3]; 6.2[t]; 6.5[e]; 6.9[s2]; 7.0[f]; 7.5[g]; 7.8[d]; 8.3[v1]; 8.5[x6]; 8.7[k]; 9.2[y1]; 10.8[w6]; 12.0[x3]
2-β-Pinene						1.4[u1]
(S)-β-Pinene (laevo-)	18172-67-3					0.08[u1]
trans-Pinene hydrate	3247-40-3		0-0.1			1.0[f]
α-Pinene oxide	1686-14-2				tr	0.08[d]; 0.2[t9]
α-Pinen-10-ol						0.7[v8]
Pinocamphone	547-60-4				0-tr	0.1[p]; 0.2[s1]; 0.3[h,n]; 1.9[v8]; 4.1[w7]
cis-Pinocamphone						0.2[w]; 0.3[g,q]; 0.5[t]; 1.0[u4]; 1.1[t2]; 1.4[v8]
trans-Pinocamphone	547-60-4		0-0.7			0.1[g]; 0.2[w]; 0.3[q]; 0.5[u7]; 1.0[w1]
Pinocarveol	5947-36-4	0-0.1	0-0.1			
cis-Pinocarveol	6712-79-4					0.02[t9]
trans-Pinocarveol	1674-08-4					0.07[t9]; 0.3[t]; 0.4[t7]; 1.5[v4]; 2.3[u8];
Pinocarvone	30460-92-5		0-0.4		tr	0.3[t]; 0.5[f]; 0.9[q]; 1.0[t9]; 1.1[w2]; 1.3[t8]; 1.5[v6]
Pinocarvyl acetate	1078-95-1					0.4[w3]
cis-Piperitol	16721-38-3					0.1[x6]; 0.5[o]
trans-Piperitol	16721-39-4					tr[w4]; 0.4[o]
Piperitenone	491-09-8		0-0.3			0.04[u7]; 0.07[u2]; 0.1[u]; 0.2[w1]; 0.3[u8]; 0.4[t7]
Piperitone	89-81-6				tr	0.3[v6]; 0.4[u]; 0.7[o]; 6.7[x8]; 17.0[s2]; 23.7[z7]
Pseudolimonene	499-97-8					tr[w4]
(Z)-Pulegol			0-1.1			
Pulegone	89-82-7					<0.1[q]; 0.07[v7]; 0.1[u1]; 1.0[s3]
Quinoline	91-22-5					0.6[v8]
Rimuene	1686-67-5					0.4[v8]
Sabinene	3387-41-5	0.04-2.3	0-0.1		0.1-1.7	1.4[k]; 1.7[d]; 2.0[u1]; 2.9[t1]; 3.7[u4]; 4.6[w]; 5.1[s3]
Sabinene hydrate	546-79-2				tr	
cis-Sabinene hydrate	15537-55-0	0.01-0.4	0-0.2	0.2		0.1[d]; 0.2[t]; 0.3[h]; 0.4[g]; 0.5[s1]; 0.6[i]; 1.9[o]
trans-Sabinene hydrate	17699-16-0	tr-0.1	0-0.1	tr		0.1[k]; 0.2[j]; 0.3[h]; 0.4[r]; 0.6[g]; 0.9[u5]; 1.4[o]
Sabinol						+[c]; 0.05[u2]
trans-Sabinol	471-16-9					0.8[i]
Safrole	94-59-7					+[c]
Salicylaldehyde	90-02-8					+[b]
Salicylic acid	69-72-7					+[b]
α-Santalene	512-61-8					tr[k]; 0.2[v]; 0.8[j]
Santene	529-16-8					+[c]
α-Selinene	473-13-2					+[c]; 0.05[t9]
β-Selinene	17066-67-0			0.1		tr[t]; 0.2[x6]
γ-Selinene	515-17-3					0.8[v8]
δ-Selinene	473-14-3			0.2		
β-Sesquicyclogeraniol						0.08[v6]

Table 5.72.2 Constituents identified in rosemary oils (*continued*)

Constituent	CAS	Percentage and range				
		A	B (52)	C (67)	D (14)	E
Sesquiphellandrene	73744-93-1					+[c]
β-Sesquiphellandrene	20307-83-9					0.07[t8]; 0.1[s1,v7]
Spathulenol	6750-60-3			0.3	0.1-0.4	0.02[u7]; 0.1[d]; 0.2[w3,y3]
Spiro[2.4]heptane						0.02[u1]; 0.7[v8]
(E)-Tagetenone	33746-72-4					tr[w4]; <0.1[q]
α-Terpinene	99-86-5	0.03-1.2	0.3-1.2	0.6	0.3-1.1	0.7[f]; 0.8[j]; 0.9[n]; 1.0[e]; 1.1[d]; 1.8[t1]; 3.3[o]
β-Terpinene	99-84-3					0.1[y3]; 7.1[u5]
γ-Terpinene	99-85-4	0.2-1.5	0-1.5	0.8	0.1-1.0	1.9[u]; 2.1[i]; 2.2[e]; 2.3[f]; 2.8[n]; 3.1[t8]; 16.6[w4]
τ-Terpinene						0.6[t9]
Terpinen-4-ol	562-74-3	0.05-1.6	0.6-1.6	1.2	1.0-3.8	2.9[u8]; 3.6[o]; 3.8[d]; 4.0[r]; 4.6[t4]; 5.4[w2]; 10.8[e]
α-Terpineol	98-55-5	0.3-5.0	0-5.0	3.5	3.1-8.1	4.4[i]; 4.9[u8]; 5.7[v4]; 5.9[d]; 6[s8]; 6.8[t7]; 7.8[p]; 8.0[w6]; 8.1[d]; 8.8[s5]; 9.2[x5]; 11.2[e]; 12.8[r]
cis-α-Terpineol						1.5[x3]
β-Terpineol	138-87-4				0.1-0.9	0.9[d]
(E)-β-Terpineol	7299-41-4					5.7[v5]
(Z)-β-Terpineol	7299-40-3					0.07[s1]; 0.3[s7]
γ-Terpineol	586-81-2	tr-0.2				
δ-Terpineol	7299-42-5	0.03-0.9				0.2[w1]; 0.4[u]; 0.7[e]; 0.8[t7]; 0.9[u3]
Terpinolene	586-62-9	0.05-1.2	0-1.7	1.0	tr	0.9[f]; 1.1[i]; 1.2[u4]; 1.3[g]; 1.5[e]; 1.7[t8]; 1.8[n]
α-Terpinyl acetate	80-26-2					tr[u]; 0.1[t4]; 1.6[u5]; 2.8[u8]
δ-Terpinyl acetate						tr[w4]
Tetracosane	646-31-1					<0.05[s]
Tetracyclo[6.3.2.0(2,5).0(1,8)]tridecan-9-ol, 4,4-dimethyl-						0.02[u2]
Tetradecanoic acid	544-63-8					tr[r]
2,2,6,8-Tetramethyl-7-oxatricyclo[6.1.0.0(1,6)]nonane						0.01[u2]
Thuja-2,4(10)-diene	36262-09-6		0-1.3			0.09[s1]; 0.1[t]; 0.2[q]; 0.4[g]; 0.5[i]; 1.1[t7]
α-Thujene	2867-05-2	0.02-5.3	0-0.8	tr	0.1-0.4	0.2[q]; 0.3[f]; 0.4[h]; 0.5[u4]; 0.6[n]; 0.7[s8]; 2.2[y1]
Thujen-2-one	546-78-1					0.2[y3]
α-Thujone	546-80-5					0.06[t4]; 6.4[v7]; 19.2[h]
β-Thujone	471-15-8					0.2[t4]; 0.9[u8]; 2.6[v7]; 6.0[h]
Thymol	89-83-8		0-11.5	0.3	tr-0.2	0.4[t2]; 0.5[t9]; 0.9[o]; 1.0[s]; 1.1[v]; 1.8[w4,w8]
Thymoquinone	490-91-5					1.4[v4]
Totarol	511-15-9					tr[d]
cis-Totarol	511-15-9				tr-0.2	
trans-Totarol	511-15-9					0.5[v8]
Tricosane	638-67-5				tr	tr[d,s]
Tricyclene	508-32-7	0.04-0.4	0-0.3	0.5	tr-0.2	0.05[h]; 0.1[d]; 0.2[j]; 0.3[g]; 0.4[i,t3]; 0.5[w]; 1.0[y1]
2-Tridecane						0.2[x5]
Trifluoroacetyl-α-terpineol	28664-18-8					2.0[x3]
3,3,6-Trimethylbicyclo-[2.2.1]hept-1-en-7-ol						0.04[k]; 0.08[j]
Trimethylenenor-bornane	2825-82-3					0.2[v8]
Undecane	1120-21-4					0.07[t4]
1-Undecanol	112-42-5					0.2[v7]
2-Undecanone	112-12-9			0.1		
Valencene	4630-07-3		0-0.1	0.3		<0.1[q]; 0.01[u2]
Valeric acid	109-52-4					+[b]
Vanillin	121-33-5					0.03[d]
Verbanone						+[c]
Verbenene	4080-46-0	0-0.6	0-0.2			0.1[u4]; 0.5[t5]; 0.8[t8]; 1.2[w3]; 1.3[v6]; 1.8[x4]
Verbenol	473-67-6	0.02-0.4				0.7[e]; 2.2[w]
cis-Verbenol	1845-30-3					0.1[t]; 0.2[s1]; 0.5[w2]; 0.6[w3]; 5.7[t4]
trans-Verbenol	1820-09-3					0.2[s1]; 0.3[w2]; 0.4[w1]; 0.6[w3]; 0.7[t7]
Verbenone	80-57-9	0.03-2.8	0-22.9		0.1-1.1	11.3[v7]; 12[u4]; 12.8[k]; 13.0[s6]; 15.3[v9]; 15.6[m]; 17.4[z6]; 21.8[t2]; 24.1[z5]; 24.9[w3]; 43.5[t7]
4-Vinylphenol	2628-17-3					+[c]

Table 5.72.2 Constituents identified in rosemary oils (*continued*)

Constituent	CAS	Percentage and range				
		A	B (52)	C (67)	D (14)	E
Viridiflorene	21747-46-6					0.1[j]; 0.2[k]
Viridiflorol	552-02-3					19.5[h]
Widdrol	6892-80-4					0.01[u2]
Ylangene			0-0.1			0.1[t3]; 0.2[t6]; 0.4[v8]
α-Ylangene	14912-44-8			0.1	tr	0.09[d]; 0.1[q,u]; 0.2[g,v7]
β-Ylangene	20479-06-5			tr		
α-Zingiberene	495-60-3					0.1[v7]; 0.5[x6]

A one hundred and eight rosemary essential oil samples from Tunisia, Morocco, Spain, Portugal and New Zealand, analyzed between 1998 and 2014; lowest and highest concentrations given (E. Schmidt, unpublished data)

B forty-one lab-hydrodistilled oils from flowering wild Sicilian rosemary collected at various sites; lowest and highest concentrations given (ref. 52)

C eight lab-distilled oils from 8 Tunisian *R. officinalis* populations growing in different bioclimatic areas; highest concentrations given; for most, the lower value was zero; chemicals with a lowest concentration >1% were 1,8-cineole (23.2%), camphor (7.2%), α-pinene (7.1%), borneol (3.3%), camphene (2.1%), α-terpineol (1.6%) and β-caryophyllene (1.1%) (ref. 67)

D unknown number (120?) of lab-hydrodistilled oils from Moroccan rosemary of the 1,8-cineole chemotype; lowest and highest concentrations given (ref. 14)

E data from other studies (indicated with superscript letters); highest concentrations found in any study reviewed here given; when two or more oils were investigated, only the highest concentrations are mentioned, unless indicated otherwise

[a] incorrect identity based on GC elution order / (probably) incorrect identification (refs. 111,112,116); [a2] this chemical has not been found in nature up to now; [b] investigation of the phenolic and acidic fractions of rosemary oil (ref. 33); [c] various studies cited in ref. 117; [d] unknown number of lab-hydrodistilled oils from Morocco (ref. 116); [e] eleven laboratory steam-distilled oils from rosemary collected in Lebanon at three locations, in two successive years and both during and after full flowering (ref. 10); [f] five laboratory steam-distilled oils from wild rosemary collected at five locations in Portugal; the oils of plants grown in nurseries from wild vegetation were also investigated (ref. 11); [g] twelve lab-hydrodistilled oils from twigs of rosemary collected in the wild in Spain at 4 different locations and during three stages (flowering, fruiting, hibernation) (ref. 9); [h] four oils from the flowering aerial parts of rosemary cultivated in four different regions in Tunisia (ref. 63); [i] thirteen lab-hydrodistilled oils from rosemary collected at the Adriatic coast, the isle of Zakynthos (Greece) and in Belgrade (Serbia) (ref. 81); [j] eight oils obtained from distillers from various parts of Morocco (ref. 25); [k] five commercial rosemary oils, three from France and two from Spain (ref. 25); [l] sixteen oils from *R. officinalis* from wild plants from Sardinia and from plants cultivated from their cuttings and grown on two soil types and harvested each month between May and November (ref. 19); [m] eleven semi-industrially distilled oils from branches and leaves of rosemary collected from various parts of Sardinia, Italy (ref. 18); [n] four lab-hydrodistilled leaf oils from rosemary harvested in Uruguay and Brazil, of which 3 populations were cultivated and one wild (ref. 16); [o] one lab-hydrodistilled oil from the aerial parts of rosemary growing wild in Turkey (ref. 24); [p] one rosemary oil from Spain and one from Italy (ref. 22); [q] one lab-hydrodistilled rosemary leaf oil from Brazil (ref. 13); [r] one lab-hydrodistilled and one laboratory steam-distilled oil from rosemary growing wild in Algeria (ref. 8); [s] one lab-hydrodistilled oil from the aerial parts of *R. officinalis* growing wild in southern Italy (ref. 6); [s1] one lab-hydrodistilled rosemary leaf oil from plants collected in a garden in Ethiopia (ref. 5); [s2] one lab-prepared and one commercial oil from Hungary (ref. 21); [s3] one lab-hydrodistilled oil from the aerial parts of rosemary cultivated in Cuba (ref. 20); [s4] one lab-hydrodistilled oil from flowering rosemary harvested in Tunisia (ref. 17); [s5] one lab-hydrodistilled oil from the flowering aerial parts of rosemary growing wild on the isle of Zakynthos, Greece (ref. 7); [s6] nine lab-hydrodistilled oils from leaves of rosemary cultivated in gardens on the isle of Djerba, Tunisia (ref. 2); [s7] one lab-hydrodistilled oil from flowering rosemary growing wild in Lebanon (ref. 1); [s8] three commercial rosemary oils from France, Spain and north Africa (ref. 26); [s9] four lab-prepared oils from fully flowering *R. officinalis* cultivated at two locations in Argentina (ref. 26); [t] one rosemary leaf oil from plants cultivated in Italy (ref. 27); [t1] one industrially steam-distilled oil from *R. officinalis* in full bloom, cultivated in Spain (ref. 28); [t2] one commercial oil sample from Sardinia (ref. 30); [t3] 150 lab-hydrodistilled oils from the aerial parts of 150 rosemary specimens from 31 wild Spanish populations; the results were presented as averages for 5 bioclimatic zones (ref. 37); [t4] one lab-hydrodistilled rosemary oil from plants growing wild in Tunisia (ref. 38); [t5] two lab-hydrodistilled oils from cultivated Iranian pre-flowering rosemary (ref. 39); [t6] four oils from wild rosemary collected in Spain in 2 bioclimatic zones and two phenological stages (ref. 46); [t7] twelve oils from *R. officinalis* specimens collected from three locations in Turkey and harvested in December, March, June and September (ref. 43); [t8] one lab-hydrodistilled rosemary leaf oil from Algeria (ref. 54); [t9] three industrially steam-distilled oils from rosemary cultivated in Italy and harvested in 3 phenological stages: flowering aerial parts, seeds + leaves, and leaves only (ref. 41); [u] two lab-hydrodistilled oils from flowering *R. officinalis* cultivated in Argentina, of which one was of the myrcene chemotype (ref. 55); [u1] one lab-hydrodistilled rosemary leaf oil from Serbia with an atypical composition and limonene as the dominant ingredient (ref. 59); [u2] one lab-hydrodistilled rosemary oil from Iran (ref. 56); [u3] six leaf oils from rosemary growing wild on the isle of Zakynthos, Greece, collected 3 times per year in two successive years (ref. 60); [u4] seventeen semi-industrially steam-distilled oils from flowering parts of rosemary collected at 17 sites in Sardinia; the results were presented as means of 4 harvesting regions (ref. 61); [u5] one commercial rosemary oil sample from Tunisia (ref. 70); [u6] one lab-hydrodistilled rosemary leaf oil from Morocco (ref. 72); [u7] one lab-hydrodistilled oil from the aerial parts of rosemary purchased at a market in Turkey (ref. 73); [u8] two lab-hydrodistilled oil samples from Egypt, one obtained from wild, the other from cultivated rosemary (ref. 75); [u9] eleven lab-prepared oil samples from 11 rosemary cultivars from the USA (ref. 76); [v] one steam-distilled rosemary leaf oil from southern Italy (ref. 80); [v1] one commercial rosemary oil sample purchased from a German company (ref. 82); [v2] three lab-hydrodistilled oils from 3 different wild Moroccan rosemary populations (ref. 83); [v3] one lab-hydrodistilled oil sample from the aerial parts of flowering rosemary growing wild in India (ref. 86); [v4] one lab-hydrodistilled oil from rosemary leaves of plants growing wild in Crete, Greece (ref. 88); [v5] three

Table 5.72.2 Constituents identified in rosemary oils (*continued*)

lab-hydrodistilled oils from flowers, leaves and young twigs of rosemary collected in various parts of Morocco (ref. 90); [v6] one lab-hydro-distilled oil from wild flowering Algerian rosemary; the effect of different distillation times on the major components was also studied (ref. 94); [v7] one lab-hydrodistilled rosemary oil sample from Brazil (ref. 100); [v8] one lab-hydrodistilled oil sample from Turkey (ref. 102); [v9] one steam-distilled rosemary leaf oil from Mexico (ref. 103); [w] two lab-hydrodistilled oils from flowering rosemary growing wild in Iran (ref. 106); [w1] one lab-hydrodistilled leaf oil from Iran (ref. 107); [w2] one lab-hydrodistilled oil from the leaves of wild growing rosemary in Tunisia (ref. 108); [w3] two commercial oil samples from Corsica (France) and two lab-hydrodistilled oils from rosemary collected in the wild in Sardinia (Italy) at two different sites (ref. 114); [w4] one oil from flowering *R. officinalis* from Turkey with an atypical composition, dominated by *p*-cymene (ref. 98); [w5] nineteen industrially steam-distilled oils from the aerial parts of rosemary cultivated in Brazil and harvested in 3 successive years; the data were shown as averages for each year (ref. 110); [w6] ten commercial rosemary oils from various USA suppliers (ref. 97); [w7] three steam-distilled oils from leaves cultivated at experimental farms and collected at three different dates (ref. 77); [w8] two lab-hydrodistilled rosemary leaf oils from Italy (ref. 64); [w9] eleven lab-hydrodistilled oils from plants cultivated in India with harvesting stage (n=4) and fertilizing regimens (n=7) as variables; only 5 chemicals were investigated (ref. 49); [x] two lab-hydrodistilled oils from the flowering tops of wild rosemary growing in south Italy (ref. 12); [x1] four lab-hydrodistilled oils from rosemary harvested in Algeria at different stages of maturity (full budding, beginning of flowering, full flowering, after flowering) (ref. 15); [x2] one lab-hydrodistilled oil from the aerial parts of rosemary cultivated in Tunisia (ref. 109); [x3] one lab-hydrodistilled oil from the aerial parts of Mexican rosemary (ref. 99); [x4] one lab-hydrodistilled leaf oil from rosemary cultivated in Turkey (ref. 92); [x5] one lab-hydrodistilled oil from rosemary cultivated in Algeria (ref. 85); [x6] one lab-hydrodistilled oil from the aerial parts of rosemary growing wild in Algeria (ref. 84); [x7] one lab-hydrodistilled leaf oil sample from Argentina (ref. 71); [x8] one commercial rosemary oil sample from Serbia (ref. 51); [x9] one commercial oil sample from China (ref. 48); [y] one lab-hydrodistilled oil from commercial dried rosemary leaves from Brazil (ref. 47); [y1] one commercial rosemary oil purchased in Brazil (ref. 34); [y2] one lab-hydrodistilled oil sample from flowering *R. officinalis* collected in the wild in Spain (ref.23); [y3] one lab-hydrodistilled oil from rosemary leaves purchased at an Egyptian market (ref. 4); [y4] data cited in ref. 112; [y5] data cited in ref. 111; [y6] one lab-hydrodistilled leaf oil sample from rosemary cultivated in Brazil (ref. 105); [y7] one lab-hydrodistilled oil from the aerial parts of rosemary grown in Tunisia (ref. 101); [y8] one commercial oil purchased from a Brazilian company (ref. 93); [y9] one lab-hydrodistilled leaf oil from Turkey (ref. 89); [z] six lab-hydrodistilled oils from wild rosemary growing in France of the camphor chemotype (ref. 79); [z1] one oil obtained from an institute of medicinal plants in Iran (ref. 78); [z2] two lab-prepared rosemary leaf oils from air-dried and oven-dried plant material (ref. 68); [z3] one lab-hydrodistilled oil from flowering rosemary cultivated in Egypt (ref. 65); [z4] one steam-distilled oil from commercial rosemary leaves from China (ref. 58); [z5] one lab-hydrodistilled oil from commercial rosemary aerial parts; the influence of storage conditions was also investigated (ref. 50); [z6] one lab-hydrodistilled leaf oil sample from rosemary plants harvested in a botanical garden in South Africa (ref. 42); [z7] one lab-hydrodistilled oil sample from Iran with an atypical composition, dominated by piperitone (ref. 40); [z8] one steam-distilled oil from flowering wild Spanish rosemary (ref. 35); [z9] data cited in ref. 31

tr: trace (in columns C, E[r] and E[u]: <0.05; in columns D and E[s5]:< 0.1; in column E[d]: <0.01); + present in the oil investigated, but quantity not stated

Table 5.72.3 Major constituents of rosemary oils

Constituent	CAS	Percentage and range				
		A	B (52)	C (67)	D (14)	E
1,8-Cineole	470-82-6	7.6-59.8	5.7-67.0	52.6	41.2-63.3	59.5[s6]; 60.9[t7]; 63.3[d]; 64.5[y8]; 71.3[w6]
α-Pinene	80-56-8	1.9-46.0	9.8-46.1	10.4	2.2-9.2	45.1[w5]; 46.1[w]; 46.2[n]; 47.1[m]; 57.5[u9]
Myrcene	123-35-3	0.2-46.0	0.7-3.3	0.7	1.0-3.1	12.4[q]; 21.5[s9]; 22.1[z2]; 22.7[n]; 29.5[f]; 31.1[u]
Camphor	76-22-2	2.9-24.2	0.3-26.8	27.5	7.6-18.9	28.1[i]; 32.3[y2]; 34.7[g]; 34.8[s3]; 35.3[p]; 44.5[z]
Verbenone	80-57-9	0.03-2.8	0-22.9		0.1-1.1	17.4[z6]; 21.8[t2]; 24.1[z5]; 24.9[w3]; 43.5[t7]
Bornyl acetate	76-49-3	0.2-13.1	0-9.3	4.2	0.1-0.8	12.3[t2]; 14.3[k]; 14.9[u4]; 17.0[w3]; 21[u9]; 23.0[m]
Limonene	138-86-3	0.7-6.7	0-5.8		2.0-6.7	7.1[u8]; 7.8[s8]; 9.3[v1]; 11.0[p]; 14.9[z7]; 21.7[u1]
Borneol	507-70-0	0.4-6.3	1.4-18.3	10.0	1.1-7.1	15.9[s]; 16.9[v2]; 17.5[o]; 18.1[u4]; 21.4[i]
β-Caryophyllene	87-44-5	0.7-6.2	0.1-2.9	7.7	0.2-4.2	8.4[s9]; 9.2[h]; 10.9[x6]; 12.9[s7]; 13.6[x1]; 14.5[y7]
Camphene	79-92-5	0.06-12.0	1.5-9.6	11.1	1.3-3.6	11.3[k]; 11.8[z8]; 12.7[s6]; 13.5[u9]; 14.3[u7]
α-Terpineol	98-55-5	0.3-5.0	0-5.0	3.5	3.1-8.1	8.0[w6]; 8.1[d]; 8.8[s5]; 9.2[x5]; 11.2[e]; 12.8[r]
β-Pinene	127-91-3	0.3-10.3	0-7.0	3.2	2.1-7.8	8.3[v1]; 8.5[x6]; 8.7[k]; 9.2[y1]; 10.8[w6]; 12.0[x3]

LEGEND: SEE UNDER TABLE 5.72.2

performed. A review of early literature (before 1985) on the composition of rosemary oils is provided in ref. 117 (from this review, only chemicals which have *not* been found in any analysis from refs. 1-110,114 are presented in Table 5.72.2).

CONTACT ALLERGY / ALLERGIC CONTACT DERMATITIS

General
Contact allergy to / allergic contact dermatitis from rosemary oil has been reported in a few publications only. Routine testing has not been performed. Three case reports of allergic contact dermatitis from rosemary oil were from occupational exposure in an aromatherapist, a masseuse and a physiotherapist. In one patient, α-pinene and possibly linalool may have been allergens in rosemary oil.

Testing in groups of patients
A group of 86 patients from Poland previously reacting to the fragrance mix was tested with rosemary oil and three (3.4%) had a positive patch test reaction; relevance data were not provided (121).

Case reports and positive patch test reactions
An aromatherapist had occupational contact dermatitis with allergies to multiple essential oils used at work, including rosemary oil; the patient also reacted to geraniol, linalool, linalyl acetate, α-pinene, the fragrance mix and various other fragrance materials; α-pinene and linalool were detected in the rosemary oil by GC-MS (119). α-Pinene has been found in commercial rosemary oils in a maximum concentration of 5.0%, linalool in lower concentrations up to 2.1% (Table 5.72.2, column A). Two masseuses and one physiotherapist had occupational allergic contact dermatitis from rosemary oil and other essential oils (120). One positive patch test reaction to rosemary oil 0.5% petrolatum, relevance not stated (cited in ref. 122).

LITERATURE

1. Apostolides NA, El Beyrouthy M, Dhifi W, Najm S, Cazier F, Najem W, Labaki M, Aboukaïs A. Chemical composition of aerial parts of *Rosmarinus officinalis* L. essential oil growing wild in Lebanon. J Essent Oil Bear Plants 2013;16:274-282

2. Akrout A, Hajlaoui H, Mighri H, Najjaa H, El Jani H, Zaidi S, Neffati M. Chemical and biological characteristics of essential oil of Rosmarinus officinalis cultivated in Djerba. J Essent Oil Bear Plants 2010;13:398-411

3. Mahboubi M, Kazempour N. *In vitro* antimicrobial activity of some essential oils from Labiatae family. J Essent Oil Bear Plants 2009;12:494-508

4. El-Massry KF, Farouk A, Abou-Zeid M. Free radical scavenging activity and lipoxygenase inhibition of rosemary (*Rosmarinus officinalis* L) volatile oil. J Essent Oil Bear Plants 2008;11:536-543

5. Mikre W, Rohloff J, Hymete A. Volatile constituents and antioxidant activity of essential oils obtained from important aromatic plants of Ethiopia. J Essent Oil Bear Plants 2007;10:465-474

6. Senatore F, De Fusco R, Grassia A, Moro CO, Rigano D, Napolitano F. Chemical composition and antibacterial activity of essential oils from five culinary herbs of the Lamiaceae family growing in Campania, Southern Italy. J Essent Oil Bear Plants 2003;6:166-173

7. Pitarokili D, Tzakou O, Loukis A. Composition of the essential oil of spontaneous *Rosmarinus officinalis* from Greece and antifungal activity against phytopathogenic fungi. J Essent Oil Res 2008;20:457-459

8. Boutekedjiret C, Belabbes R, Bentahar F, Bessière J-M, Rezzoug SA. Isolation of rosemary oils by different processes. J Essent Oil Res 2004;16:195-199. Data also presented in ref. 74

9. Salido S, Altarejos J, Nogueras M, Saánchez A, Luque P. Chemical composition and seasonal variations of rosemary oil from southern Spain. J Essent Oil Res 2003;15:10-14

10. Diab Y, Auezova L, Chebib H, Chalchat J-C, Figueredo G. Chemical composition of Lebanese rosemary (*Rosmarinus officinalis* L.) essential oil as a function of the geographical region and the harvest time. J Essent Oil Res 2002;14:449-452

11. Serrano E, Palma J, Tinoco T, Venâncio F, Martins A. Evaluation of the essential oils of rosemary (*Rosmarinus officinalis* L.) from different zones of "Alentejo" (Portugal). J Essent Oil Res 2002;14:87-92

12. Guazzi E, Maccioni S, Monti G, Flamini G, Cioni PL, Morelli I. *Rosmarinus officinalis* L. in the gravine of Palagianello (Taranto, South Italy). J Essent Oil Res 2001;13:231-233

13. Porte A, Luiz de O Godoy R, Lopes D, Koketsu M, Limp Gonçalves SL, Torquilho HS. Essential oil of *Rosmarinus officinalis* L. (rosemary) from Rio de Janeiro, Brazil. J Essent Oil Res 2000;12:577-580

14. Elamrani A, Zrira S, Benjilali B, Berrada M. A study of Moroccan rosemary oils. J Essent Oil Res 2000;12:487-495. The data are virtually identical to those presented in ref. 116

15. Boutekedjiret C, Belabbes R, Bentahar F, Bessiere J-M. Study of *Rosmarinus officinalis* L. essential oil yield and composition as a function of the plant life cycle. J Essent Oil Res 1999;11:238-240

16. Dellacassa E, Lorenzo D, Moyna P, Frizzo CD, Serafini LA, Dugo P. *Rosmarinus officinalis* L. (Labiatae) essential oils from the south of Brazil and Uruguay. J Essent Oil Res 1999;11:27-30

17. Boutekedjiret C, Bentahar F, Belabbes R, Bessiere J-M. The essential oil from *Rosmarinus officinalis* L. in Algeria. J Essent Oil Res 1998;10:680-682

18. Tuberose CIG, Satta M, Cabras P, Garau VL. Chemical composition of *Rosmarinus officinalis* oils of Sardinia. J Essent Oil Res 1998;10:660-664

19. Moretti MDL, Peana AT, Sanna Passino G, Solinas V. Effects of soil properties on yield and composition of *Rosmarinus officinalis* essential oil. J Essent Oil Res 1998;10:261-267

20. Pino JA, Estarrón M, Fuentes V. Essential oil of rosemary (*Rosmarinus officinalis* L.) from Cuba. J Essent Oil Res 1998;10:111-112

21 Domokos J, Héthelyi E, Pálinkás J, Szirmai S, Tulok MH. Essential oil of rosemary (*Rosmarinus officinalis* L.) of Hungarian origin. J Essent Oil Res 1997;9:41-45

22 Arnold N, Valentini G, Bellomaria B, Hocine L. Comparative study of the essential oils from *Rosmarinus eriocalyx* Jordan & Fourr. from Algeria and *R. officinalis* L. from other countries. J Essent Oil Res 1997;9:167-175

23 Tomei PE, Cioni PL, Flamini G, Stefani A. Evaluation of the chemical composition of the essential oils of some Lamiaceae from Serrania de Ronda (Andaluçia, Spain). J Essent Oil Res 1995;7:279-282

24 Pérez-Alonso MJ, Velasco-Negueruela A, Emin Duru M, Harmandar M, Esteban JL. Composition of the essential oils of *Ocimum basilicum* var. *glabratum* and *Rosmarinus officinalis* from Turkey. J Essent Oil Res 1995;7:73-75

25 Chalchat J-C, Garry R-P, Michet A, Benjilali B, Chabart JL. Essential oils of rosemary (*Rosmarinus officinalis* L.). The chemical composition of oils of various origins (Morocco, Spain, France). J Essent Oil Res 1993;5:613-618

26 Mizrahi I, Juarez MA, Bandoni AL. The essential oil of *Rosmarinus officinalis* growing in Argentina. J Essent Oil Res 1991;3:11-15

27 Benelli G, Flamini G, Canale A, Cioni PL, Conti B. Toxicity of some essential oil formulations against the Mediterranean fruit fly *Ceratitis capitata* (Wiedemann) (Diptera Tephritidae). Crop Protection 2012;42:223-229

28 Guillén MD, Cabo N, Burillo J. Characterisation of the essential oils of some cultivated aromatic plants of industrial interest. J Sci Agric 1996;70:359-363

29 Tognolini M, Barocelli E, Ballabeni V, Bruni B, Bianchi A, Chiavarini M, Impicciatore M. Comparative screening of plant essential oils: Phenylpropanoid moiety as basic core for antiplatelet activity. Life Sciences 2006;78:1419-1432. Data also presented in ref. 30

30 Sacchetti G, Maietti S, Muzzoli M, Scaglianti M, Manfredini S, Radice M, Bruni R. Comparative evaluation of 11 essential oils of different origin as functional antioxidants, antiradicals and antimicrobials in foods. Food Chem 2005;91:621-632. Data also presented in ref. 29

31 Lawrence BM. Progress in essential oils. Perfum Flavor 1995;20(1):47-53

32 Lawrence BM. Progress in essential oils. Perfum Flavor 1991;16(March/April):59-67

33 ter Heide R, de Valois PJ, de Rijke D, Bednarczyk AA. Acids and phenols in seven spice essential oils. Paper presented at American Chemical Society Meeting, New York, USA, April 13-18, 1986. Data cited in ref. 32

34 Barreto HM, Silva Filho EC, de O. Lima E, Coutinho HDM, Morais-Braga MFB, Tavares CCA, et al. Chemical composition and possible use as adjuvant of the antibiotic therapy of the essential oil of *Rosmarinus officinalis* L. Ind Crops Prod 2014;59:290-294

35 Laborda R, Manzano I, Gamón M, Gavidia I, Pérez-Bermúdez P, Boluda R. Effects of *Rosmarinus officinalis* and *Salvia officinalis* essential oils *on Tetranychus urticae* Koch (Acari: Tetranychidae). Ind Crops Prod 2013;48:106-110

36 Machado DG, Cunha MP, Neis VB, Balen GO, Colla A, Bettio LEB, et al. Antidepressant-like effects of fractions, essential oil, carnosol and betulinic acid isolated from *Rosmarinus officinalis* L. Food Chem 2013;136:999-1005

37 Jordán MJ, Lax V, Rota MC, Lorán S, Sotomayor JA. Effect of bioclimatic area on the essential oil composition and antibacterial activity of *Rosmarinus officinalis* L. Food Control 2013;30:463-468

38 Hosni K, Hassen I, Chaâbane H, Jemli M, Dallali S, Sebei H, Casabianca H. Enzyme-assisted extraction of essential oils from thyme (*Thymus capitatus* L.) and rosemary (*Rosmarinus officinalis* L.): Impact on yield, chemical composition and antimicrobial activity. Ind Crops Prod 2013;47:291-299

39 Rowshan V, Farhadi F, Najafian S. The essential oil of *Dodonaea viscosa* leaves is allelopathic to rosemary (*Rosmarinus officinalis* L.). Ind Crops Products 2014;56:241-245

40 Rasooli I, Fakoor MH, Yadegarinia D, Gachkar L, Allameh A, Rezaei MB. Antimycotoxigenic characteristics of *Rosmarinus officinalis* and *Trachyspermum copticum* L. essential oils. Int J Food Microbiol 2008;122:135-139. Data also presented in ref. 44

41 Beretta G, Artali R, Maffei Facino R, Gelmini F. An analytical and theoretical approach for the profiling of the antioxidant activity of essential oils: The case of *Rosmarinus officinalis* L. J Pharm Biomed Anal 2011;55:1255-1264

42 Okoh OO, Sadimenko AP, Afolayan AJ. Comparative evaluation of the antibacterial activities of the essential oils of *Rosmarinus officinalis* L. obtained by hydrodistillation and solvent free microwave extraction methods. Food Chem 2010;120:308-312

43 Celiktas OY, Kocabas EEH, Bedir E, Sukan FV, Ozek T, Baser KHC. Antimicrobial activities of methanol extracts and essential oils of *Rosmarinus officinalis*, depending on location and seasonal variations. Food Chem 2007;100: 553-559

44 Gachkar L, Yadegari D, Rezaei MB, Taghizadeh M, Astaneh SA, Rasooli I. Chemical and biological characteristics of *Cuminum cyminum* and *Rosmarinus officinalis* essential oils. Food Chem 2007;102:898-904. Data also presented in ref. 40

45 Szumny A, Figiel A, Gutiérrez-Ortíz A, Carbonell-Barrachina AA. Composition of rosemary essential oil (*Rosmarinus officinalis*) as affected by drying method. J Food Engin 2010;97:253-260

46 Jordán MJ, Lax V, Rota MC, Lorán S, Sotomayor JA. Effect of the phenological stage on the chemical composition, and antimicrobial and antioxidant properties of *Rosmarinus officinalis* L. essential oil and its polyphenolic extract. Ind Crops Products 2013;48:144-152

47 da Silva Bomfim N, Polis Nakassugi L, Faggion Pinheiro Oliveira J, Kohiyama CY, Galerani Mossini SA, Grespan R, et al. Antifungal activity and inhibition of fumonisin production by *Rosmarinus officinalis* L. essential oil in *Fusarium verticillioides* (Sacc.) Nirenberg. Food Chem 2015;166:330-336

48 Wang W, Wu N, Zu YG, Fu YJ. Antioxidative activity of *Rosmarinus officinalis* L. essential oil compared to its main components. Food Chem 2008;108:1019-1022

49 Singh M, Guleria N. Influence of harvesting stage and inorganic and organic fertilizers on yield and oil composition of rosemary (*Rosmarinus officinalis* L.) in a semi-arid tropical climate. Ind Crops Prod 2013;42:37-40

50 Usai M, Marchetti M, Foddai M, Del Caro A, Desogus R, Sanna I, Piga A. Influence of different stabilizing operations and storage time on the composition of essential oil of thyme (*Thymus officinalis* L.) and rosemary (*Rosmarinus officinalis* L.). LWT - Food Sci Technol 2011;44:244-249

51 Dimitrijević SI, Mihajlovski KR, Antonović DG, Milanović-Stevanović MR, Mijin DZ. A study of the synergistic antilisterial effects of a sub-lethal dose of lactic acid and essential oils from *Thymus vulgaris* L., *Rosmarinus officinalis* L. and *Origanum vulgare* L. Food Chem 2007;104:774-782

52 Napoli EM, Curcuruto G, Ruberto G. Screening of the essential oil composition of wild Sicilian rosemary. Biochem System Ecol 2010;38:659-670

53 Kabouche Z, Boutaghane N, Laggoune S, Kabouche A, Ait-Kaki Z, Benlabed Z. Comparative antibacterial activity of five Lamiaceae essential oils from Algeria. Int J Aromather 2005;15:129-133. Data also presented in ref. 85

54 Bousbia N, Abert Vian M, Ferhat MA, Petitcolas E, Meklati BY, Chemat F. Comparison of two isolation methods for essential oil from rosemary leaves: Hydrodistillation and microwave hydrodiffusion and gravity. Food Chem 2009;114:355-362

55 Ojeda-Sana AM, van Baren CM, Elechosa MA, Juárez MA, Moreno S. New insights into antibacterial and antioxidant activities of rosemary essential oils and their main components. Food Control 2013;31:189-195

56 Jalali-Heravi M, Sadat Moazeni R, Sereshti H. Analysis of Iranian rosemary essential oil: Application of gas chromatography–mass spectrometry combined with chemometrics. J Chromatogr A 2011;1218:2569-2576

57 Rodríguez-Rojo S, Varona S, Núnez M, Cocero MJ. Characterization of rosemary essential oil for biodegradable emulsions. Ind Crops Prod 2012;37:137-140

58 Jiang Y, Wu N, Fu Y-J, Wang W, Luo W, Zhao C-J, et al. Chemical composition and antimicrobial activity of the essential oil of rosemary. Environ Toxicol Pharmacol 2011;32:63-68

59 Bozin B, Mmimica-Dukic N, Samojlik I, Jovin E. Antimicrobial and antioxidant properties of rosemary and sage (*Rosmarinus officinalis* L. and *Salvia officinalis* L., Lamiaceae) essential oils. J Agric Food Chem 2007;55:7879-7885

60 Papageorgiou V, Gardeli C, Mallouchos A, Papaioannou M, Komaitis M. Variation of the chemical profile and antioxidant behavior of *Rosmarinus officinalis* L. and *Salvia fruticosa* Miller grown in Greece. J Agric Food Chem 2008;56:7254-7264

61 Angioni A, Barra A, Cereti E, Barile D, Coisson JD, Arlorio M, et al. Chemical composition, plant genetic differences, antimicrobial and antifungal activity investigation of the essential oil of *Rosmarinus officinalis* L. J Agric Food Chem 2004;52:3530-3535

62 Flamini G, Cioni PL, Morelli I, Macchia M, Ceccarini L. Main agronomic-productive characteristics of two ecotypes of *Rosmarinus officinalis* L. and chemical composition of their essential oils. J Agric Food Chem 2002;50:3512-3517

63 Ben Farhat M, Jordan MAJ, Chaouech-Hamada R, Landoulsi A, Sotomayor JA. Variations in essential oil, phenolic compounds, and antioxidant activity of Tunisian cultivated *Salvia officinalis* L. J Agric Food Chem 2009;57:10349-10356

64 Angelini LG, Carpanese G, Cioni PL, Morelli I, Macchia M, Flamini G. Essential oils from Mediterranean Lamiaceae as weed germination inhibitors. J Agric Food Chem 2003;51:6158-6164

65 Viuda-Martos M, El-Nasser GS, El Gendy A, Sendra E, Fernandez-Lopez J, El Razik KAA, Omer EA, et al. Chemical composition and antioxidant and anti-*Listeria* activities of essential oils obtained from some Egyptian plants. J Agric Food Chem 2010;58:9063-9070

66 Alnamer R, Alaoui K, Bouidida EH, Benjouad A, Cherrah Y. Toxicity and psychotropic activity of essential oils of *Rosmarinus officinalis* and *Lavandula officinalis* from Morocco. J Biol Active Prod Nature 2011;1:262-272

67 Ben Jemia M, Tundis R, Pugliese A, Menichini F, Senatore F, Bruno M, et al. Effect of bioclimatic area on the composition and bioactivity of Tunisian *Rosmarinus officinalis* essential oils. Nat Prod Res 2015;29:213-222

68 Maggi M, Gende L, Russo K, Fritz R, Eguaras M. Bioactivity of *Rosmarinus officinalis* essential oils against *Apis mellifera*, *Varroa destructor* and *Paenibacillus* larvae related to the drying treatment of the plant material. Nat Prod Res 2011;25:397-406

69 Bertoli A, Conti B, Mazzoni V, Meini L, Pistelli L. Volatile chemical composition and bioactivity of six essential oils against the stored food insect *Sitophilus zeamais* Motsch. (Coleoptera Dryophthoridae). Nat Product Res 2012;26:2063-2071

70 Yang S-A, Jeon S-K, Lee E-J, Shim C-H, Lee I-S. Comparative study of the chemical composition and antioxidant activity of six essential oils and their components. Nat Prod Res 2010;24:140-151

71 Sanchez Chopa C, Descamps LR. Composition and biological activity of essential oils against *Metopolophium dirhodum* (Hemiptera: Aphididae) cereal crop pest. Pest Manag Sci 2012;68:1492-1500

72 Ait-Ouazzou A, Loran S, Bakkali M, Laglaoui A, Rota C, Herrera A, Pagan R, Conchello P. Chemical composition and antimicrobial activity of essential oils of *Thymus algeriensis*, *Eucalyptus globulus* and *Rosmarinus officinalis* from Morocco. J Sci Food Agric 2011;91:2643-2651

73 Figueredo G, Ünver A, Chalchat J-C, Arslan D, Özcan MM. A research on the composition of essential oil isolated from some aromatic plants by microwave and hydrodistillation. J Food Biochem 2012;36:334-343

74 Boutekedjiret C, Bentahar F, Belabbes R, Bessiere JM. Extraction of rosemary essential oil by steam distillation and hydrodistillation. Flavour Fragr J 2003;18:481-484. Data also presented in ref. 8

75 Soliman FM, El-Kashoury EA, Fathy MM, Gonaid MH. Analysis and biological activity of the essential oil of *Rosmarinus officinalis* L. from Egypt. Flavour Fragr J 1994;9:29-33

76 Tucker AO, Maciarello MJ. The essential oils of some rosemary cultivars. Flavour Fragr J 1986;1:137-142

77 Papachristos DP, Karamanoli KI, Stamopoulos DC, Menkissoglu-Spiroudi U. The relationship between the chemical composition of three essential oils and their insecticidal activity against *Acanthoscelides obtectus* (Say). Pest Manag Sci 2004;60:514–520

78 Azizkhani M, Akhondzadeh Basti A, Misaghi A, Tooryan F. Effects of *Zataria multiflora* Boiss., *Rosmarinus officinalis* L. and *Mentha longifolia* L. essential oils on growth and gene expression of enterotoxins c and e in *Staphylococcus aureus* ATCC 29213. J Food Saf 2012;32:508-516

79 Kaloustian J, Portugal H, Pauli AM, Pastor J. Chemical, chromatographic, and thermal analysis of rosemary (*Rosmarinus officinalis*). J Appl Polymer Sci 2002;83:747-756

80 Reverchon E, Senatore F. Isolation of rosemary oil: comparison between hydrodistillation and supercritical CO_2 extraction. Flavour Fragr J 1992;7:227-230

81 Lakušić DV, Ristić MS, Slavkovska VN, Šinžar-Sekulić JB, Lakušić BS. Environment-related variations of the composition of the essential oils of rosemary (*Rosmarinus officinalis* L.) in the Balkan Peninsula. Chem Biodivers 2012;9:1286-1302

82 Baratta MT, Dorman HJD, Deans SG, Figueiredo AC, Barroso JA, Ruberto G. Antimicrobial and antioxidant properties of some commercial essential oils. Flavour Fragr J 1998;13:235-244

83 Lahlou M, Berrada R. Composition and niticidal activity of essential oils of three chemotypes of *Rosmarinus officinalis* L. acclimatized in Morocco. Flavour Fragr J 2003;18:124-127

84 Djeddi S, Bouchenah N, Settar I, Skaltsa HD. Composition and antimicrobial activity of the essential oil of *Rosmarinus officinalis* from Algeria. Chem Nat Comp 2007;43:487-490

85 Touafek O, Nacer A, Kabouche A, Kabouche Z, Bruneau C. Chemical composition of the essential oil of *Rosmarinus officinalis* cultivated in the Algerian Sahara. Chem Nat Comp 2004;40:28-29. Data also presented in ref. 53

86 Tantry MA, Shabir S, Khan R, Habib A, Akbar S. Determination of essential oil composition of *Rosmarinus officinalis* growing as exotic species in Kashmir Valley. Chem Nat Comp 2012;47:1012-1014

87 Tahri M, Imelouane B, Aouinti F, Amhamdi H, Elbachiri A. The organic and mineral compounds of the medicinal aromatics, *Rosmarinus tournefortii* and *Rosmarinus officinalis*, growing in eastern Morocco. Res Chem Intermed 2014;40:2651-2658

88 Katerinopoulos HE, Pagona G, Afratis A, Stratigakis N, Roditakis N. Composition and insect attracting activity of the essential oil of *Rosmarinus officinalis*. J Chem Ecol 2005;31:111-122

89 Isikber AA, Alma MH, Kanat M, Karci A. Fumigant toxicity of essential oils from *Laurus nobilis* and *Rosmarinus officinalis* against all life stages of *Tribolium confusum*. Phytoparasitica 2006;34:167-177

90 Khia A, Ghanmi M, Satrani B, Aafi A, Aberchane M, Quaboul B, et al. Effet de la provenance sur la qualité chimique et microbiologique des huiles essentielles de *Rosmarinus officinalis* L. du Maroc. Phytothérapie 2014;12:341-347

91 Karakaya S, Nehir El S, Karagozlu N, Sahin S, Sumnu G, Bayramoglu B. Microwave-assisted hydrodistillation of essential oil from rosemary. J Food Sci Technol 2014;51:1056-1065

92 Arslan M, Dervis S. Antifungal activity of essential oils against three vegetative compatibility groups of *Verticillium dahlia*. World J Microbiol Biotechnol 2010;26:1813-1821

93 Fonseca AOS, Pereira DIB, Jacob RG, Maia Filho FS, Oliveira DH, Maroneze BP. *In vitro* susceptibility of Brazilian *Pythium insidiosum* isolates to essential oils of some Lamiaceae family species. Mycopathologia 2014, DOI 10.1007/s11046-014-9841-6

94 Tigrine-Kordjani N, Meklati BY, Chemat F, Guezil FZ. Kinetic investigation of rosemary essential oil by two methods: solvent-free microwave extraction and hydrodistillation. Food Anal Methods 2012;5:596-603

95 Conti B, Canale A, Bertoli A, Gozzini F, Pistelli L. Essential oil composition and larvicidal activity of six Mediterranean aromatic plants against the mosquito *Aedes albopictus* (Diptera: Culicidae). Parasitol Res 2010;107:1455-1461

96 Martac I-M, Podea P. Determination of antimicrobial and antioxidant activities of essential oil isolated from *Rosmarinus officinalis* L. Carpath J Food Sci Technol 2012;4:40-45

97 Isman MB, Wilson JA, Bradbury R. Insecticidal activities of commercial rosemary oils (*Rosmarinus officinalis*) against larvae of *Pseudaletia unipuncta* and *Trichoplusia ni* in relation to their chemical compositions. Pharm Biol 2008;46:82-87

98 Özcan MM, Chalchat J-C. Chemical composition and antifungal activity of rosemary (*Rosmarinus officinalis* L.) oil from Turkey. Int J Food Sci Nutrit 2008;59:691-698

99 Martínez AL, González-Trujano ME, Pellicer F, López-Muñoz FJ, Navarrete A. Antinociceptive effect and GC/MS analysis of *Rosmarinus officinalis* L. essential oil from its aerial parts. Planta Med 2009;75:508-511

100 Bernardes WA, Lucarini R, Tozatti MG, Bocalon Flauzino LG, Souza MGM, Turatti ICC. Antibacterial activity of the essential oil from *Rosmarinus officinalis* and its major components against oral pathogens. Z Naturforsch C 2010;65:588-593

101 Sebai H, Selmi S, Rtibi K, Gharbi N, Sakly M. Protective effect of *Lavandula stoechas* and *Rosmarinus officinalis* essential oils against reproductive damage and oxidative stress in alloxan-induced diabetic rats. J Med Food 2014, DOI: 10.1089/jmf.2014.0040, 9 pages

102 Topal U, Sasaki M, Goto M, Otles S. Chemical compositions and antioxidant properties of essential oils from nine species of Turkish plants obtained by supercritical carbon dioxide extraction and steam distillation. Int Food Sci Nutrit 2008;59:619-634

103 Martinez-Velazquez M, Rosario-Cruz R, Castillo-Herrera G, Flores-Fernandez JM, Alvarez AH, Lugo-Cervantes E. Acaricidal effect of essential oils from *Lippia graveolens* (Lamiales: Verbenaceae), *Rosmarinus officinalis* (Lamiales:Lamiaceae), and *Allium sativum* (Liliales: Liliaceae) against *Rhipicephalus (Boophilus) microplus* (Acari: Ixodidae). J Med Entomol 2011;48:822-827

104 Panizzi L, Flamini G, Cioni PL, Morelli I. Composition and antimicrobial properties of essential oils of four Mediterranean Lamiaceae. J Ethnopharmacol 1993;39:167-170

105 Nogueira de Melo GA, Grespan R, Pitelli Fonseca J, Oliveira Farinha T, Leite Silva E, Lopes Romero A, et al. *Rosmarinus officinalis* L. essential oil inhibits *in vivo* and *in vitro* leukocyte migration. J Med Food 2011;14:944-949

106 Jamshidi R, Afzali Z, Afzali D. Chemical composition of hydrodistillation essential oil of rosemary in different origins in Iran and comparison with other countries. Amer-Euras J Agr Envriron Sci 2009;5:78-81

107 Moghtader M, Afzali D. Study of the antimicrobial properties of essential oil of rosemary. Amer-Euras J Agr Environ Sci 2009;5:393-397

108 Marzouk Z, Neffati A, Marzouk B, Chraief I, Fathia K, Chekir Ghedira L, Boukef K. Chemical composition and antibacterial and antimutagenic activity of Tunisian *Rosmarinus officinalis* L. oil from Kasrine. J Food Agricult Environ 2006;4(3&4):61-65

109 Zoubiri S, Baaliouamer A. Chemical composition and insecticidal properties of some aromatic herbs essential oils from Algeria. Food Chem 2011;29:179-182.

110 Atti-Santos AC, Marcelo R, Pauletti GF, Rora LD, Rech JC, Pansera MR. Physico-chemical evaluation of *Rosmarinus officinalis* L. essential oils. Braz Arch Biol Technol 2005;48:1035-1039.

111 Lawrence BM. Essential oils 2001-2004. Carol Stream, USA: Allured Publishing Corporation, 2006:46-54

112 Lawrence BM. Essential oils 2005-2007. Carol Stream, USA: Allured Publishing Corporation, 2008:170-181

113 Lawrence BM. Essential oils 2008-2011. Carol Stream, USA: Allured Publishing Corporation, 2012 (no data on rosemary oils)

114 Pintore G, Usai M, Bradesi P, Juliano C, Boatto G, Tomi F, et al. Chemical composition and antimicrobial activity of *Rosmarinus officinalis* L. oils from Sardinia and Corsica. Flavour Fragr J 2002;17:15-19

115 Boutekedjiret C, Buatois B, Bessiere JM. Characterisation of rosemary essential oil of different areas of Algeria. J Essent Oil Bear Plants 2005;8:65-70

116 El-Amrani A, Zrira S, Ismaili-Alaoui M, Berrada M, Benjilali B. Chemical characterization of Moroccan rosemary essential oil. Rivista Ital EPPOS 1998; (Numero Speciale):646-654. Data cited in ref. 111. The data are virtually identical to those presented in ref. 14

117 Boelens MH. The essential oil from *Rosmarinus officinalis* L. Perfum Flavorist 1985;10:(Oct/Nov):21-37

118 Granger R, Passet J, Arbousset G. L'Essence de *Rosmarinus officinalis* L. II. Influence des facteurs écologiques et individuels. Parf Cosm Sav 1973;3(6):307-312. Data cited in ref. 25

119 Dharmagunawardena B, Takwale A, Sanders KJ, Cannan S, Roger A, Ilchyshyn A. Gas chromatography: an investigative tool in multiple allergies to essential oils. Contact Dermatitis 2002;47:288-292

120 Trattner A, David M, Lazarov A. Occupational contact dermatitis due to essential oils. Contact Dermatitis 2008;58:282-284

121 Rudzki E, Grzywa Z. Allergy to perfume mixture. Contact Dermatitis 1986;15:115-116

122 Hjorther AB, Christophersen C, Hausen BM, Menné T. Occupational allergic contact dermatitis from carnosol, a naturally-occurring compound present in rosemary. Contact Dermatitis 1997;37:99-100

Chapter 5.73   ROSE OIL

DEFINITION
Rose oil (essential oil of rose) is the essential oil obtained from the flowers of the damask rose, *Rosa* x *damascena* Mill.

INCI NOMENCLATURE
Description/definition: Rosa damascena flower oil is the volatile oil obtained from the flowers of the damask rose, *Rosa damascena*, Rosaceae
INCI name EU & USA: Rosa damascena flower oil
CAS registry number(s): 8007-01-0; 90106-38-0
EINECS number(s): 290-260-3

ISO (INTERNATIONAL ORGANIZATION FOR STANDARDIZATION) STANDARD
ISO number: 9842
ISO name: Essential oil of rose
Botanical origin: *Rosa* x *damascena* Mill.
Parts of plant used: Flower
ISO values: ISO values (minimum and maximum concentrations) are shown in Table 5.73.1.

Table 5.73.1 ISO values (%) for rose oil [a]

Compound	CAS	Minimum	Maximum
Citronellol	106-22-9	20.0	49.0
Geraniol	106-24-1	6.0	29.0
Nonadecane	629-92-5	6.0	16.0
Nerol	106-25-2	3.0	12.0
Ethanol	64-17-5	0.0	7.0
Heneicosane	629-94-7	1.5	5.5
n-Heptadecane	629-78-7	0.6	4.0
2-Phenethyl alcohol	60-12-8	0.0	3.5

[a] ISO 9842 Essential oil of rose ©ISO 2003; Geneva, Switzerland, www.iso.org

CosIng mentions various rose flower essential oils from other *Rosa* species: *Rosa canina* (CAS 84696-47-9; EINECS 283-652-0); *Rosa centifolia* (CAS 84604-12-6; EINECS 283-289-8) (which is most often used for the production of rose concretes [4]); *Rosa gallica* (CAS 84604-13-7; EINECS 283-290-3); *Rosa moschata* (no CAS or EINECS number); *Rosa rugosa* (CAS 92347-25-6; EINECS 296-213-3); and a general oil for all *Rosa* species: Rose flower oil (CAS 8007-01-1). As by far most commercial rose oils (including all oils from Bulgaria and Turkey, where 80-90% of the world's production of rose oils is realized [5]) are produced from *Rosa damascena*, the discussion here is limited to this species.

THE PLANT, THE OIL, AND THEIR USES
Rosa x *damascena* Mill. is a deciduous shrub growing to 2.2 meters tall, the stems densely armed with stout, curved prickles and stiff bristles. It is commonly called the damask rose, as the plant was originally brought to Europe from Damascus, Syria (5). This rose species is a hybrid from *R. gallica* L. and *R. phoenicea* Boiss., probably originating in Syria or south Turkey (5). The damask rose is cultivated in Bulgaria, Turkey, India, Iran, southern France, southern Italy, Morocco, Ukraine, Caucasus, Syria, and China (5,28, Mansfeld's World Database of Agriculture and Horticultural Crops, http://mansfeld.ipk-gatersleben.de). The cultivated variety is called 'trigintipetala', meaning 'having 30 petals' (13,48). The whole rose oil industry in Bulgaria and Turkey, which are the main producers of rose essential oil (5), is based on a single genotype which has been vegetatively propagated for centuries (18,26).

The damask rose is one of the most important *Rosa* species for flowers and garden cultivation (5). Roses have been used since the earliest times in rituals, cosmetics, perfumes, medicines and aromatherapy (4,77,78). *Rosa damascena* has many culinary uses as a flavor additive. Rose water (a by-product of the oil production process) is often sprinkled on meat dishes. The most popular use is in the flavoring of foods such as ice cream, jam, rice pudding, cake, yoghurt and for flavoring alcoholic liquors and soft drinks (5,20).

Rose essential oil is obtained from the fresh flowers of *R. damascena* by hydro- or steam-distillation (4,26,28). Because of the labor-intensive production process and the low content of oil in rose blooms (about 3,500 to 4,000 kilograms of rose flowers are necessary to produce one kilogram of rose oil [26]), rose oil is very expensive and is often called 'liquid gold' (4). Rose oil is primarily used as a fragrance ingredient in the perfumery and cosmetics industry and is included in a large number of fine fragrances, creams, soaps, lotions and other cosmetic products (9,15,20). The main constituents of roses and their oils responsible for the typical rose scent are 2-phenethyl alcohol, citronellol, the rose oxides, geraniol, nerol, β-ionone, β-damascone and β-damascenone (7,22,29).

However, rose oil also has pharmacological applications. Claimed biological properties of *R. damascena* essential oil include antioxidant, antimicrobial, anti-inflammatory, antitussive, bronchodilatory, anti-convulsive, anti-HIV, analgesic, hypnotic, antispasmodic and psychosomatic activities (1,5,16,24). Traditionally rose oil is used as a remedy for anxiety, depression and for the treatment of stress-related conditions (8,15). The oil has also been used in poor circulation, asthma and coughs, irregular menstruation, gall stones and uterine disorders. On the skin, it is said to be useful for dry skin, eczema and sensitive skin (1,8,10). Vapor therapy of rose oil is considered to be helpful for some allergies, headaches and migraine (1). In aromatherapy, it is widely applied because of its claimed psychosomatic properties and for the treatment of cardiac diseases (5).

CHEMICAL COMPOSITION
Rose oil is a light yellow, crystallized or mobile liquid (depending on the temperature), which has a strong floral odor with subtle fruity or green, sweet honey-like background notes. The yield of essential oil from the

flowers of *Rosa* x *damascena* Mill. generally varies from 0.015 to 0.035 per cent. The main producing countries of this oil are Bulgaria, Turkey, Russia, Morocco and Iran.

Literature data (up to October 10, 2014) on the chemical composition of rose oils and unpublished analytical data from one of us (E.S.) are shown in Table 5.73.2 in alphabetical order. In rose oils from various origins, over 440 chemicals have been identified. About 56 per cent of these were found in a single reviewed publication only. The major compounds found in rose oils from different sources are shown in Table 5.73.3. They include (highest concentrations in any study given) citronellol (67.9%), geraniol (44.7%), nonadecane (39.7%), heneicosane (32.4%), nerol (17.4%), nonadecene (9.1%) and linalool (8.4%). Well-known ingredients of rose oils that were present in high concentrations (>7%) in one or two studies were 2-phenethyl alcohol (five studies: 9.6%, 10.4%, 15.7%, 37.9% and 44.8%), eicosane (29.9%),

phenethyl acetate (14.8%), docosane (7.3% and 14.1%), ethanol (13.4%), *trans*-rose oxide (10.1%), tricosane (8.5%), geranyl acetate (8.2%), methyl eugenol (8.1%), β-caryophyllene (7.8%), α-terpineol (7.5%) and heptadecane (7.2%). Uncommon or rare constituents of rose oils found in high concentrations (>7%) in single studies include heptacosane (12.7%) and 9-nonadecene (7.3%)

Commercial oils
The ten chemicals that had the highest maximum concentrations in 51 commercial rose essential oil samples (concentration ranges provided) are the following: citronellol (0.5-44.8%), geraniol (4.9-23.8%), heneicosane (0.8-21.0%), nonadecane (6.3-15.9%), nerol (0.6-11.0%), α-terpineol (0.06-7.5%), linalool (0.02-6.4%), ethanol (0.03-4.9%), methyl eugenol (0.06-4.3%) and eicosane (*n*-) (0.1-4.0%) (Erich Schmidt, unpublished analytical data).

Table 5.73.2 Constituents identified in rose oils

Constituent	CAS	Percentage and range					
		A	B (28)	C (14)	D (27)	E (2)	F
Acetaldehyde	75-07-0			0.04			$+^{y4}$; 0.02^p
Acetaldehyde diethyl acetal	105-57-7			0.005			$+^{y4}$
Acetic acid	64-19-7						0.1^{y4}; 0.3^{x3}
Acetone	67-64-1			0.003	0-tr		
Acetylenedicarbonic acid, DL-(-)-methyl-							$0.03^{a2,x5}$
Aciphyllene	87745-31-1					0.1	0.2^{y5}
Amyl alcohol	71-41-0	0.01-0.9					
(*E*)-Anethole	4180-23-8			0.008			
Aromadendrene	489-39-4						0.2^p
Benzaldehyde	100-52-7	0.01-0.4	0-0.1	0.02	0-0.2	tr	tr^{x3}; 0.1^o; 0.3^w; 0.5^p; 2.5^r
Benzene	71-43-2						$0.3^{a3,x5}$
Benzophenone	119-61-9						0.1^w
Benzyl acetate	140-11-4						$+^{y4}$; 0.1^t; 0.3^{x2}
Benzyl alcohol	100-51-6	tr-0.1		0.01			$<0.1^t$; 0.5^{y4}; 2.7^w
Benzyl benzoate	120-51-4			0.05		tr	0.06^{x5}; 0.1^t; $0.2^{f,t}$; 0.3^e
Benzyl isovalerate	103-38-8						$+^{y4}$
Benzyl 2-methylbutyrate	56423-40-6						$+^{y4}$
Benzyl propionate	122-63-4						$+^{y4}$
Benzyl tiglate	37526-88-8			0.2			$+^{y4}$; $0.1^{t,x2}$; 0.3^t
Benzyl valerate	10361-39-4						$+^{y4}$
β-Bergamotene	6895-56-3						$+^{y4}$
β-Bourbonene	5208-59-3	0.04-1.1				0.1	tr^{x3}; $0.1^{e,x4}$; $0.2^{c,d}$; 0.3^{y9}; 1.2^y
α-Bulnesene	3691-11-0	0.08-0.4	0-1.0		0.2-1.3	0.5	0.1^{x5}; $0.5^{d,y5}$; 0.6^p
Butanal	123-72-8						$+^{y4}$
Butanol	35296-72-1			0.003	0-tr		0.01^{x3}
2-Butanol	78-92-2			0.0002	0-0.06		
Butyl methyl sulfide	628-29-5						$+^{y4}$
Butyl tiglate	7785-66-2						tr^{x3}
α-Cadinene	24406-05-1						tr^{x3}; 0.4^{x9}; 0.6^y
γ-Cadinene	39029-41-9	tr-0.05		0.1			0.07^{y5}; 0.1^p
δ-Cadinene	483-76-1	0.02-0.3					tr^d; 0.02^{x5}; 0.09^{y5}; 0.1^e
α-Cadinol	481-34-5					0.1	0.1^p
δ-Cadinol	19435-97-3						0.5^{x1}
Camphene	79-92-5						0.01^{y4}; 0.4^{y4}; 0.6^r
Camphor	76-22-2						$+^n$
δ3-Carene	13466-78-9						0.02^{x3}
Carveol	99-48-9						0.1^p
Carvomenthyl acetate	5256-66-6						$+^{y4}$

Table 5.73.2 Constituents identified in rose oils (*continued*)

Constituent	CAS	Percentage and range					
		A	B (28)	C (14)	D (27)	E (2)	F
β-Caryophyllene	87-44-5	0.1-2.1	0.2-1.1	0.5		0.7	1.1[t]; 1.2[x7]; 1.3[y7]; 1.6[x2]; 7.8[x4]
Caryophyllene oxide	1139-30-6					0.1	tr[d]; 0.07[y5]; 0.1[e]; 0.2[x4]
Cembrane							0.05[a2,x5]
1,4-Cineole	470-67-7						+[y4]
1,8-Cineole	470-82-6			0.03	0-0.09	0.1	tr[x3]; 0.2[t,u]
Cinnamal(dehyde)	104-55-2						+[y4]
(E)-Cinnamaldehyde	14371-10-9						1.3[r]
Cinnamyl formate	104-65-4						+[y4]
Citrol	624-15-7						0.1[a2,x5]
Citronellal	106-23-0				0-tr	tr	tr[d]; 0.03[x3]; 0.06[p]; 0.4[j]
α-Citronellal	141-26-4						0.01[p]
Citronellal diethyl acetal							+[y4]
Citronellic acid	502-47-6			0.01			
Citronellol	106-22-9	0.5-44.8	25.5-55.3	32.0	24.5-43.0	27.5	42.6[y6]; 43.9[s]; 45.0[o,p]; 46.7[h]; 48.0[f]; 48.2[x1]; 53.6[x4]; 67.9[y3]
α-Citronellol							0.1[p]
Citronellyl acetate	150-84-5	0.09-1.6	0.4-1.3	0.5	tr-0.8	0.3	0.7[o]; 0.8[u]; 0.9[f]; 1.4[s]; 1.6[g]; 2.8[r]
Citronellyl benzoate	10482-77-6						+[y4]
Citronellyl butyrate	141-16-2						+[y4]
Citronellyl decanoate	72934-06-6						+[y4]
Citronellyl formate	105-85-1				0-0.1	0.2	tr[x3]; 0.1[f,j]
Citronellyl heptanoate							+[y4]
Citronellyl hexanoate	10580-25-3			0.005			+[y4]
Citronellyl nerate	72934-19-1						+[y4]
Citronellyl nonanoate							+[y4]
Citronellyl octanoate						tr	+[y4]
Citronellyl phenyl-acetate	139-70-8						0.1[x5]
Citronellyl propionate	141-14-0						+[y4]
α-Copaene	3856-25-5						0.05[p]
α-Costol	65018-15-7						0.1[y5]
o-Cresol	95-48-7						0.1[q]
α-Cubebene	17699-14-8						tr[x3]
Cuminaldehyde	122-03-2						+[y4]
Cycloheptanone	502-42-1						+[y4]
Cycloheptatriene	544-25-2						0.05[a3,x5]
Cyclotetracosane	297-03-0						0.09[a2,x5]
o-Cymene	527-84-4						0.1[q]
p-Cymene	99-87-6	0.02-0.2		0.01	0-tr	0.2	tr[x3]; 0.1[p]
p-Cymenene	1195-32-0			0.02		tr	+[y4]
α-Damascenone							0.7[i]
β-Damascenone (E)-	23726-93-4						0.03[s]; 0.05[x3]; 0.1[t,y5]; 0.5[x2]
(Z)-β-Damascenone	59739-63-8						0.04[x3]
β-Damascone (E-)	23726-91-2	0.0-tr		0.1			<0.01[p,x3]; 0.1[w]; 0.4[i]; 1.6[y4]
2,4-Decadienal	2363-88-4						0.09[y5]
Decanal	112-31-2			0.005			tr[x2]; 0.02[x]; 0.03[p]; 0.1[t]; 0.3[r]
Decanal diethyl acetal	34764-02-8						+[y4]
Decane	124-18-5						+[y4]
Decanoic acid	334-48-5			0.002			
Decanol	36729-58-5			tr			
2-Decen-1-ol	22104-80-9						0.2[w]
Dibenzothiophene	132-65-0						+[y4]
Dibutyl phthalate	84-74-2			0.2[z]			
1,1-Diethoxyethane	105-57-7	0.0-0.2					
Dihydro-α-terpinyl acetate	80-25-1						+[y4]
Dimethyl disulfide	624-92-0						+[y4]

Table 5.73.2 Constituents identified in rose oils (*continued*)

Constituent	CAS	Percentage and range					
		A	B (28)	C (14)	D (27)	E (2)	F
1,4-Dimethyl-3-(2-methyl-1)-1-cyclo-heptene							$0.04^{a2,x5}$
3-(4,8-Dimethyl-3,7-nonadienyl) thiophene							$+^{y4}$
(Z)-2,6-Dimethyl-2,6-octadiene	2492-22-0						0.3^{x1}
(2E,5E)-3,7-Dimethyl-2,5-octadiene-1,7-diol	93079-92-6					0.1	0.02^{s}
(2Z,5E)-3,7-Dimethyl-2,5-octadiene-1,7-diol							0.1^{s}
(E)-3,7-Dimethyl-5-octene-1,7-diol						0.1	0.4^{s}
Dimethyl sulfide	75-18-3						$+^{y4}$
Dimethyl trisulfide	3658-80-8						$+^{y4}$
2,4-Dinitroanisole	119-27-7						0.8^{x1}
Dipropyl sulfide	111-47-7						$+^{y4}$
Docosane	629-97-0			0.07	0.05-0.3	0.2	1.0^{t}; 1.4^{x}; 2.5^{x4}; 7.3^{y2}; 14.1^{x5}
1,22-Docosanediol	22513-81-1						$0.03^{a2,x5}$
1-Docosanol	661-19-8						4.3^{x4}
1-Docosene	1599-67-3						0.1^{f}; 0.2^{t}
Dodecanal	112-54-9						$+^{y4}$
Dodecanal diethyl acetal	53405-98-4						$+^{y4}$
Dodecane	112-40-3			0.005			0.2^{x4}; 0.3^{w}; 0.4^{r}
Dodecanoic acid	143-07-7						0.1^{y5}
Eicosadiene							$+^{y4}$
Eicosane	112-95-8	0.1-4.0	0.2-1.9	0.7	0.6-2.2	2.0	2.5^{y5}; 2.7^{d}; 3.2^{x}; 4.5^{y2}; 29.9^{x5}
1-Eicosanol	629-96-9						0.3^{x5}
1-Eicosene	3452-07-1	0.0-1.0					$0.1^{t,x2}$; 0.4^{g}; 0.7^{t}
(E)-5-Eicosene	74685-30-6				0.07-0.3		
9-Eicosene	42448-90-8						2.7^{x9}; 4.7^{x3}; 5.0^{y}
(E)-9-Eicosene					0.1-0.4	tr	1.0^{d}
β-Elemene	33880-83-0						$0.2^{f,y5}$; 1.5^{x3}; 1.7^{v}
δ-Elemene	20307-84-0						tr^{x3}
Elemol (α-)	639-99-6						0.1^{d}; 0.2^{p}
4,5-Epithiocaryophyl-lene	65563-95-3						$+^{y4}$
1,2-Epithiohumulene							$+^{y4}$
4,5-Epithiohumulene							$+^{y4}$
Ethanol	64-17-5	0.03-4.9		1.6	tr-13.4		1.0^{x7}; 1.4^{r}; 1.9^{v}; 3.0^{p}; 5.1^{o}
Ethyl acetate	141-78-6				0-0.5		$+^{y4}$
Ethyl benzoate	93-89-0			0.02			$+^{y4}$
Ethyl decanoate	110-38-3					tr	$+^{y4}$
Ethyl dodecanoate	106-33-2						$+^{y4}$
Ethyl geranate	32659-21-5						$+^{y4}$
Ethyl geranyl ether	40267-72-9						$+^{y4}$
Ethyl hexadecanoate	628-97-7						0.05^{p}
Ethyl laurate	106-33-2			0.03			$+^{y4}$
1-Ethyl-2-methylcyclo-decane							$0.03^{a2,x5}$
Ethyl nerate	32659-20-4						$+^{y4}$
Ethyl neryl ether	22882-89-9						$+^{y4}$
Ethyl nonenoate							$+^{y4}$
Ethyl octadecanoate	111-61-5						$+^{y4}$
Ethyl octanoate	106-32-1						$+^{y4}$
Ethyl pentadecanoate	41114-00-5						$+^{y4}$
Ethyl salicylate	118-61-6						$+^{y4}$
p-Ethylstyrene	3454-07-7						$+^{y4}$
Ethyl tridecanoate	28267-29-0						$+^{y}$
α-Eudesmol	473-16-5					tr	0.4^{y5}

Table 5.73.2 Constituents identified in rose oils (*continued*)

Constituent	CAS	Percentage and range					
		A	B (28)	C (14)	D (27)	E (2)	F
β-Eudesmol	473-15-4						tr[d]; 0.1[e]; 0.2[y5]; 0.4[p]
γ-Eudesmol (selinenol)	1209-71-8						0.1[d]
10-epi-γ-Eudesmol	15051-81-7						0.1[x4]
Eugenol	97-53-0	0.04-3.2	0.4-1.7	0.6	0.3-0.9	1.1	2.3[r]; 2.7[y7]; 3.3[y]; 3.5[q]; 5.1[x]
Eugenyl acetate	93-28-7						0.01[x3]
(E,E)-Farnesal	502-67-0						0.03[a2,x5]
α-Farnesane							0.3[w]
α-Farnesene	502-61-4	0.04-0.8					0.2[y5]; 0.8[y9]
(E,E)-α-Farnesene	502-61-4	0.01-2.1			0.2-1.3	0.1	0.3[x4]; 0.4[g]; 0.6[y8]; 0.7[c]; 1.1[y7]
(Z,E)-α-Farnesene	26560-14-5	0.02-1.6					
β-Farnesene	502-60-3						0.7[e]
(E)-β-Farnesene	18794-84-8						0.7[g]; 2.0[x3]
Farnesol	4602-84-0		0.4-2.2				1.0[x9]; 2.0[f]; 2.1[m]; 5.4[y]; 6.3[y4]
(E,E)-Farnesol	106-28-5	0.2-1.4		1.0		1.3	1.6[o]; 1.7[c]; 1.9[x6]; 2.1[t]; 3.2[x2]
(E,Z)-Farnesol	3879-60-5					0.2	1.5[p]
(Z,E)-Farnesol	3790-71-4			0.2			0.2[x4]; 0.5[y5]
(Z,Z)-Farnesol	16106-95-9						0.6[y5]; 1.8[d]
(Z,E)-Farnesyl acetate	40266-29-3						0.2[g]
Fenchyl alcohol	1632-73-1						0.2[x4]
β-Fenchyl alcohol	22627-95-8						1.1[x1]
Geranial ((E)-Citral)	141-27-5	0.2-1.8	0.2-1.5	0.03	0.4-1.1	1.0	0.7[m]; 0.8[j]; 0.9[f]; 1.1[u]; 1.3[p]
Geranic acid	459-80-3					0.1	0.2[x1]; 0.5[w]; 0.6[y5]
(E)-Geranic acid	4698-08-2			0.2			
(Z)-Geranic acid				0.004			
Geraniol	106-24-1	4.9-23.8	8.3-28.5	15.7	2.1-18.0	19.9	29.3[c]; 30.2[x2]; 31.7[m]; 35.6[y6]; 35.8[r]; 36.2[w]; 39.3[u]; 44.7[y3]
Geraniol oxide							+[y4]
Geranyl acetate	105-87-3	0.2-2.2	0.5-3.3	0.7	0.4-2.3	1.0	2.1[y5]; 2.2[s]; 2.3[y9]; 2.5[c]; 2.8[x2]; 3.0[y7]; 4.1[y]; 4.2[m]; 4.7[r]; 8.2[y6]
(E)-Geranylacetone	3796-70-1			0.1			
Geranyl butyrate	106-29-6						0.3[j]
Geranyl formate	105-86-2					0.3	0.02[p]; 0.05[y5]; 0.1[j]; 1.0[x2]
Geranyllinalool	77368-82-2						+[y4]
Geranyl octanoate	51532-26-4						+[y4]
Geranyl propionate	105-90-8						+[y4]
Geranyl salicylate							+[y4]
Germacrene D	23986-74-5	0.1-1.3	0.4-4.2			1.3	1.4[s]; 1.8[c,y2]; 2.0[y5]; 2.1[d]; 2.6[y7]
α-Guaiene	3691-12-1	0.2-2.0	0.2-1.4	0.2	tr-2.9	0.7	0.6[c]; 0.7[y9]; 0.8[y5]; 1.0[m]; 2.0[x2]
Guaiol	489-86-1						0.02[p]
β-Gurjunene	73464-47-8					0.1	
Heneicosane	629-94-7	0.8-21.0	0.6-9.3	3.0	2.3-8.9	7.2	10.6[y6]; 11.4[y4]; 11.5[d]; 13.3[y9]; 15.1[x]; 16.0[y7]; 16.1[y5]; 32.4[y2]
1-Heneicosanol	15594-90-8						0.3[x5]
1-Heneicosene	1599-68-4					0.1	0.1[t]; 0.2[s,v]; 0.4[t]; 0.5[q]; 5.0[t]
9-Heneicosene	629-95-8						0.5[x4]
Heptacosane	593-49-7			0.02		0.4	0.3[d]; 0.8[q]; 2.5[x9]; 2.6[t]; 12.7[x]
8-Heptadanene							0.2[f]
Heptadecadiene							+[y4]
Heptadecanal	629-90-3						+[y4]
Heptadecane	629-78-7	0.9-2.5	0.6-4.2	1.6	0.9-2.6	2.2	2.7[x7]; 3.3[y7]; 3.5[q]; 3.7[l]; 3.9[g]; 4.0[e]; 4.5[x6]; 4.6[x8]; 6.0[y]; 7.2[y5]
1-Heptadecene	6765-39-5				0.1-0.5	1.6	+[y4]; 0.8[e]
7-Heptadecene	54290-12-9			0.02			
8-Heptadecene	16369-12-3						0.2[x4]; 0.3[d]; 0.5[y5]
(Z)-8-Heptadecene	16369-12-3	0.1-0.4		0.1			
Heptanal	111-71-7	0.03-0.4	0-0.3	0.03	tr-0.2	tr	tr[d,x2]; 0.04[x3]; 0.1[p]; 0.2[v]
Heptanal diethyl acetal	688-82-4						+[y4]
Heptane	142-82-5						0.05[a2,x5]
Heptanoic acid	111-14-8			0.002			

Table 5.73.2 Constituents identified in rose oils (*continued*)

Constituent	CAS	Percentage and range					
		A	B (28)	C (14)	D (27)	E (2)	F
Heptanol	53535-33-4	0.01-0.4		0.02		tr	0.02[o]; 0.05[p]; 0.1[q]
1-Hepten-3-one	2918-13-0						+[y4]
8-Heptylpentadecane	71005-15-7						0.1[a2,x]
Hexacosane	630-01-3						0.3[x]; 0.5[q,x5]
Hexadecanal	629-80-1						+[y4]
Hexadecane	544-76-3	0.0-1.6		0.05	0-0.2		0.08[x5]; 0.1[d,x3]; 0.2[c]; 0.3[e]; 0.8[y9]
Hexadecanol	51260-59-4						3.3[q]
Hexanal	66-25-1	tr-0.2		0.006			tr[x3]; 0.1[w]
Hexanal diethyl acetal	3658-93-3						+[y4]
Hexanoic acid	142-62-1			0.0006			
Hexanol (1-)	111-27-3	0.02-0.5	0-0.4	0.2		0.1	0.2[p,x3]; 0.3[x9]
3-Hexanol	623-37-0				0-tr		0.3[r]
2-Hexenal	505-57-7						
3-Hexenal	4440-65-7						0.3[o]
1-Hexene	592-41-6						0.2[x1]
(E)-3-Hexenol	928-97-2			0.05	0-tr		0.2[x3]
(Z)-3-Hexenol	928-96-1	0.01-0.3			0-0.08	tr	0.1[x3]; 0.2[t]; 0.3[k,r]
Hexyl acetate	142-92-7						+[y4]; 0.01[o]
2-Hexyl-1-decanol	2425-77-6						0.9[w]
Hexyl isovalerate	10032-13-0						+[y4]
Hexyl 2-methylbu-tyrate	10032-15-2						+[y4]
Hexyl methyl sulfide	20291-60-5						+[y4]
8-Hexylpentadecene							0.04[a2,x5]
β-Himachalene	1461-03-6						+[y4]
α-Humulene	6753-98-6	0.04-1.1	0.1-1.0	0.2	tr-1.5	0.5	0.8[y5]; 0.9[y8]; 1.1[m,y7]; 1.7[x4]
Hydroxycitronellal	107-75-5						tr[a1,j]
3-Hydroxy-β-damas-cenone							+[y4]
(E)-α-Ionone	127-41-3						+[n]
β-Ionone	79-77-6						tr[x3,y]; 0.08[y4]; 0.1[t]; 0.6[t]; 1.0[y4]
Isoamyl alcohol	123-51-3	0.0-0.4		0.05		tr	tr[x3]; 0.1[p]; 0.2[v]
Isoborneol	124-76-5						0.2[w]
Isobutyl alcohol	78-83-1			0.008	0-0.2		0.01[x3]
Isogeraniol	5944-20-7					0.1	0.1[x1]; 0.2[x3]
Isomenthol	3623-52-7						1.1[q]
Isomenthone	491-07-6	0.004					0.1[q]; 0.2[m]
Isonerol oxide							0.02[s]
Isopentanal	590-86-3						+[y4]
Isopropyl hexade-canoate	142-91-6						+[y4]
Isopropyl myristate	110-27-0						+[y4]
Lavandulyl acetate	25905-14-0						+[y4]; 0.2[x4]
Limonene	138-86-3	0.02-0.5	0-0.3	0.04		0.2	0.2[p]; 0.3[x]; 0.6[y]; 0.8[x2]; 1.0[e]
Linalool	78-70-6	0.02-6.4	0.5-2.9	2.7	0.2-1.7	7.5	2.5[v]; 2.7[l]; 2.8[e]; 3.8[r]; 8.4[i]
cis-Linalool oxide	11063-77-7	0.01-0.3					0.02[x3]; 0.1[t]
cis-Linalool oxide, furanoid	11063-77-7					0.1	
trans-Linalool oxide	11063-78-8			0.04			tr[x3]; 0.1[t]
trans-Linalool oxide, furanoid	34995-77-2					tr	
Linalyl acetate	115-95-7						0.03[x3]
Linoleic acid	60-33-3						+[y4]
Linolenic acid	463-40-1						+[y4]
Longifolene (junipene)	475-20-7						0.3[w]
Marrubine							2.4[x4]
p-Mentha-1,4-dien-7-ol	22539-72-6						0.03[a2,x5]
p-Menth-1-en-9-al	29548-14-9			0.008			
Menthol	89-78-1			0.008			0.1[q]

Table 5.73.2 Constituents identified in rose oils (*continued*)

Constituent	CAS	Percentage and range					
		A	B (28)	C (14)	D (27)	E (2)	F
Menthone	89-80-5			0.009			0.5^q
Menthyl acetate	16409-45-3			0.01			+^{y4}; 0.06^{x5}
Methanol	67-56-1			0.08			
Methyl benzoate	93-58-3						0.06^{x5}
Methyl benzyl ether	538-86-3			0.01	0-tr		+^{y4}
2-Methylbutanal	96-17-3			0.004			
3-Methylbutanal	590-86-3			0.004			
1-Methyl-1-butanol				tr-0.8			
2-Methyl-1-butanol	137-32-6			0.05			0.1^p; 0.2^v
3-Methyl-2-buten-1-ol	556-82-1			0.006			
Methyl citronellate	2270-60-2						+^{y4}
3-Methyldecane	13151-34-3						0.02^{a2,x5}
1-Methyldibenzo-thiophene	31317-07-4						+^{y4}
2-Methyldibenzo-thiophene	20928-02-3						+^{y4}
3-Methyldibenzo-thiophene	16587-52-3						+^{y4}
4-Methyldibenzo-thiophene	7372-88-5						+^{y4}
Methyl dihydro-chaulmoograte							0.01^{a2,x5}
Methyl eugenol	93-15-2	0.06-4.3	0.4-2.7	2.3	0.6-3.3	1.4	2.3^d; 2.4^k; 2.5^v; 2.9^p; 3.4^{x2}; 3.6^{x1}; 3.7^{y5}; 4.0^{l,s}; 4.3^{y6}; 8.1^{x7}
Methyl geranate	2349-14-6	0.04-0.2	0-0.3				0.1^p; 0.5^{x4}
(*E*)-Methylgeranate	2349-14-6			0.06			
(*Z*)-Methylgeranate	1862-61-9			0.07			
Methylheptenol				0.02			
6-Methyl-5-hepten-2-ol	1569-60-4				0-tr	tr	
Methylheptenone	409-02-9			0.03			
6-Methyl-5-hepten-2-one	110-93-0				0-0.06	tr	0.04^o
Methyl hexadeca-noate	112-39-0						+^{y4}
3-Methyl-2-(3-methyl-2-butenyl)thiophene							+^{y4}
Methyl nonanoate	1731-84-6						+^{y4}; 0.02^p
2-Methyloctadecyne							0.03^{a2,x5}
Methyl octanoate	111-11-5						+^{y4}
Methyl pentadeca-noate	7132-64-1						+^{y4}
4-Methylpentanol	626-89-1						+^{y4}
4-(4-Methyl-3-pente-nyl)-1,2-dithiacyclo-hex-4-ene	73188-23-5						+^{y4}
3-(4-Methyl-3-pen-tenyl)thiophene	62429-57-6						+^{y4}
Methyl phenylacetate	101-41-7						0.2^q
Methyl salicylate	119-36-8			0.01			+^{y4}
Methyltetracosane							1.4^q
Methyl tetradeca-noate	124-10-7						+^{y4}
Methyl tridecanoate	1731-88-0						+^{y4}
Mintsulfide	72445-42-2						+^{y4}
β-Mintsulfide							+^{y4}
γ-Muurolene	30021-74-0				tr-3.0		
α-Muurolol	104245-48-9						0.01^p
Myrcene	123-35-3		0-2.3	0.09	0.06-0.9	0.5	0.8^q; 1.3^c; 1.4^t; 2.2^{y7}; 2.4^{x2}
Myrtanal	4764-14-1						0.01^p

Table 5.73.2 Constituents identified in rose oils (*continued*)

Constituent	CAS	Percentage and range					
		A	B (28)	C (14)	D (27)	E (2)	F
Myrtenol	515-00-4						0.04^p
Myrtenyl acetate	1079-01-2						0.5^q
Neral	106-26-3	0.1-0.9	0.1-1.0	0.01	0.2-0.7	0.4	0.4^j; 0.6^c; 0.7^{y9}; 0.8^p; 2.2^t
Nerol	106-25-2	0.6-11.0	0.9-3.3	8.7	0.8-7.6	10.5	10.3^{y1}; 11.2^w; 11.8^t; 11.9^h; 13.1^m; 14.5^r; 16.3^{y3}; 17.4^{y6}
Nerolidol	7212-44-4						0.07^{x3}; 0.1^d; 1.3^{y4}
(E)-Nerolidol	40716-66-3			0.2		0.2	0.3^t; 0.5^{y4}; 1.0^{x7}
(Z)-Nerolidol	3790-78-1				tr-0.4		0.2^e
Nerol oxide	1786-08-9	0.01-0.1		0.09		tr	tr^d; 0.1^p; 0.2^s
Neryl acetate	141-12-8	0.06-1.1		0.07		0.3	0.3^c; 0.5^{z7}; 0.7^t; 1.0^{y9}; 1.3^{x4}
Neryl butyrate	999-40-6						0.1^j
Neryl formate	2142-94-1					0.1	0.01^p
Neryl octanoate							$+^{y4}$
Neryl phenylacetate	10522-32-4						$0.04^{a2,x5}$
Neryl propionate	105-91-9						0.1^j
Nonacosane	630-03-5						$0.02^{a2,x5}$
Nonadecadiene							$+^{y4}$
Nonadecane	629-92-5	6.3-15.9	2.6-23.2		6.4-19.0	15.7	17.9^{y6}; 22.9^j; 24.3^{y9}; 24.7^y; 24.8^{y5}; 25.8^d; 26.5^{y7}; 39.7^{y2}
Nonadecan-9-ene							3.1^f
Nonadecene	27400-77-7	0.1-3.6	0.8-4.8	9.1	1.8-5.4	4.2	0.6^{x4}; 2.6^v; 3.0^j; 3.6^f; 4.9^s; 5.5^{y6}; 6.5^{x5}; 6.9^g; 7.0^e
(Z)-5-Nonadecene							1.5^{x1}; 1.7^{y5}; 6.7^d
9-Nonadecene	31035-07-1						0.4^d; 0.5^{x4}; 2.8^{y1}; $5.7^{t,y2}$; 7.3^{x6}
(Z)-9-Nonadecene	51865-02-2			3.1			0.8^{x2}; 1.2^t
2,6-Nonadienal	557-48-2						0.1^{y5}
Nonanal	124-19-6	0.01-0.2		0.03	tr-0.1	0.1	tr^d; 0.05^p; 0.09^o; 0.1^t; 0.2^{x2}
Nonanal diethyl acetal	54815-13-3						$+^{y4}$
Nonane	111-84-2						$+^{y4}$
Nonanoic acid	12-05-0			0.006			
Nonanol	28473-21-4					tr	0.03^p; 0.09^r
2-Nonanol	628-99-9						0.1^q
Nonenyl phenyl ether							$+^{y4}$
Nonyl phenyl ether							$+^{y4}$
β-Ocimene	13877-91-3						0.3^q
(E)-Ocimene	27400-72-2			0.04			
(Z)-Ocimene	27400-71-1			0.01			
(E)-β-Ocimene	3779-61-1	0.02-0.2			0-0.07	tr	tr^{x3}; 0.1^p; 0.4^t; 0.8^{x2}; 1.4^e
(Z)-β-Ocimene	3338-55-4				0-tr	tr	0.1^p; 0.2^t; $0.9^{e,x2}$
allo-Ocimene	673-84-7						0.2^e
Octadecadiene							$+^{y4}$
Octadecanal	638-66-4						$+^{y4}$; 0.03^{x5}; 0.08^{x5}
Octadecane	593-45-3	0.04-0.2	0-0.4	0.2	0-1.4	0.4	0.4^{x4}; 0.5^{x9}; 0.6^d; 0.7^g; 0.8^e
1-Octadecanol	112-92-5						0.3^{x5}
16-Octadecenal	56554-87-1						$0.03^{a2,x5}$
1-Octadecene	112-88-9	0.02-0.2					$+^{y4}$; 0.2^g; 0.3^e
(E)-3-Octadecene	7206-19-1						0.1^{y5}
Octanal	124-13-0						tr^{x2}; 0.01^p; 0.05^o; $<0.1^t$; 0.2^t
Octanal diethyl acetal	54889-48-4						$+^{y4}$
Octanoic acid	124-07-2			0.004			
Octanol	29063-28-3			0.01			0.07^o; 0.1^q; 0.4^r
3-Octanol	589-98-0						0.7^q
Octyl benzoate	94-50-8						$+^{y4}$
Octyl phenyl ether	1818-07-1						$+^{y4}$
di-n-Octyl phthalate	117-84-0						$0.9^{x1,z}$
Oxaiceane	55092-18-7						$0.02^{a2,x5}$
Pentacosane	629-99-2			0.1		0.5	0.8^g; 0.9^t; 1.5^q; 2.1^{x2}; 3.6^u
1-Pentacosanol	26040-98-2						0.05^{x5}

Table 5.73.2 Constituents identified in rose oils (*continued*)

Constituent	CAS	Percentage and range					
		A	B (28)	C (14)	D (27)	E (2)	F
Pentadecanal	2765-11-9						+[y4]
Pentadecane	629-62-9	0.05-0.8	tr-0.7	0.3	tr-0.9	0.4	0.5[c]; 0.6[f]; 0.7[y5]; 0.9[g]; 1.0[q]
2-Pentadecanone	2345-28-0			0.03			
Pentane	109-66-0						0.08[x5]
Pentanol	30899-19-5			0.02	tr-0.2[b]		
2-Pentylfuran	3777-69-3				0-tr		
Perillene	539-52-6					tr	
α-Phellandrene	99-83-2						0.1[e]
Phenethyl acetate	103-45-7	0.2-1.6		0.05	0.3-0.8	0.2	0.9[w]; 1.1[t]; 1.6[y6]; 3.0[x4]; 14.8[q]
2-Phenethyl alcohol	60-12-8	0.02-2.7	0.3-3.4	1.2	0.3-2.2	2.9	3.9[r]; 4.1[x9]; 5.1[x1]; 9.6[y,z1]; 10.4[q,z1] 15.7[i,z1]; 37.9[y4,z1]; 44.8[x,z1]
2-Phenethyl dodecanoate	6309-54-2						0.05[p]
Phenethyl isovalerate	140-26-1						+[y4]
2-Phenethyl propionate	122-70-3						+[y4]
Phenethyl valerate	7460-74-4						+[y4]
Phenylacetaldehyde	122-78-1						+[y4]; 0.4[w]
Phenylacetic acid	103-82-2						1.4[y4]
2-Phenylethyl benzoate	94-47-3						+[y4]; 0.1[f]; 0.4[x4]; 5.4[x1]
2-Phenylethyl formate	104-62-1						+[y4]
(*E*)-2-Phenylethyl geranate				0.01			
2-Phenylethyl hexanoate	6290-37-5			0.003			+[y4]
2-Phenylethyl isobutyrate	103-48-0			0.1			
2-Phenylethyl 2-methylbutyrate	24817-51-4			0.04			+[y4]
Phenylethyl tiglate	55719-85-2	0.04-0.1					+[y4]; 0.1[x4]
Phytol	7541-49-3						+[y4]
α-Pinene	80-56-8	0.06-1.7	0-5.9	0.2	0.08-2.2	1.3	1.5[x3]; 1.6[y7]; 1.7[y5]; 1.9[y9]; 2.8[x2]
β-Pinene	127-91-3	0.02-1.1	0-0.1	0.04	tr-0.5	0.3	0.4[c]; 0.5[y9]; 0.6[x2,y7]; 2.0[e]
Pinene epoxide	1686-14-2						+[y4]
Propanal	123-38-6						+[y4]
2-Propen-1-ol	107-18-6				0-tr		
Pulegol	529-02-2						0.1[q]
Rosefuran	15186-51-3			0.01		tr	tr[d]; 0.02[p]; 0.04[s]; 0.1[s]
cis-Rose oxide	3033-23-6	0.01-0.6	0-0.5	0.4	0.2-0.9	0.3	0.7[j,y5]; 0.8[y7]; 1.0[s]; 1.1[y8]; 2.6[c]
cis-trans-Rose oxide							0.06[x5]
trans-Rose oxide	5258-11-7	0.02-0.1	0-0.3	0.2	0.07-0.3	0.1	0.3[i]; 0.4[d]; 0.5[j]; 0.7[x2]; 10.1[u]
Sabinene	3387-41-5	tr-2.7	0-0.5		tr-0.2	0.1	tr[d]; 0.05[v]; 0.1[p]; 0.2[t]; 0.4[m]
α-Santalol							0.02[a2,x5]
α-Selinene	473-13-2						0.08[y5]; 0.2[d]
β-Selinene	17066-67-0						0.1[x4]; 0.4[q]
δ-Selinene	473-14-3						0.03[a2,x5]
trans-trans-Sesquicitronellene				0.01			
trans-Sesquimyrcene				0.02			
Sesquirosefuran	39007-93-7						0.03[a2,x5]
α-Sinensal	17909-77-2						0.02[a2,x5]
α-Terpinene	99-86-5	0.02-0.09		0.004	tr-0.06	tr	tr[d]; 0.05[p]; 0.1[g]
γ-Terpinene	99-85-4	0.03-0.4	0-0.3	0.02	tr-0.2[b]	0.1	tr[d]; 0.05[p]; 0.09[y5]; 0.1[e]; 1.9[m]
Terpinen-4-ol	562-74-3	0.06-2.6	0.1-0.6	0.3	0.06-0.4	0.4	0.5[f]; 0.6[t]; 0.7[y8]; 1.0[x2]; 3.7[y5]
α-Terpineol	98-55-5	0.06-7.5	0.1-0.7	0.7	tr-0.4	2.6	0.5[t]; 0.8[p,y7]; 1.1[e]; 2.7[y]; 3.4[x2]
γ-Terpineol	586-81-2						0.1[p]; 0.4[c]; 0.5[y8]; 0.7[y7]
Terpinolene (α-)	586-62-9	0.01-0.05	0-0.1	0.02	tr	tr	tr[d]; 0.05[p]; 0.2[w]; 0.3[g]

Table 5.73.2 Constituents identified in rose oils (*continued*)

Constituent	CAS	Percentage and range					
		A	B (28)	C (14)	D (27)	E (2)	F
Terpinolene epoxide							+[y4]
α-Terpinyl acetate	80-26-2						0.1[p,q]
Tetracosane	646-31-1			0.03		0.1	0.3[x3]; 0.4[x2]; 0.5[t]; 0.6[x5]; 1.5[q]
9-Tetracosene							0.2[x4]
Tetradecanal	124-25-4						+[y4]; 0.1[d]
Tetradecane	629-59-4			0.006		tr	0.1[p,x4]; 0.4[w]
Tetradecanol	27196-00-5						0.3[x9]; 0.4[q]
Tetradecenyl acetate							+[y4]
Tetradecyl acetate	638-59-5						+[y4]
Tetrahydrolinalool	78-69-3						tr[x3]
Theaspirane A	36431-72-8						+[n]
Theaspirane B	36431-72-8						+[n]
Thujyl alcohol	513-23-5						+[y4]
Tricosane	638-67-5			0.5	0.3-1.9	0.4	2.2[x6]; 2.8[y6]; 4.0[y5]; 5.6[t]; 8.5[x]
Tricosene	56924-46-0	0.08-1.3					0.3[g]
(Z)-9-Tricosene	27519-02-4					0.2	0.2[d]
Tridecanal	10486-19-8						+[y4]
Tridecanal diethyl acetal	72934-16-8						+[y4]
Tridecane	629-50-5			0.006		tr	0.2[x9]; 0.4[w]
2-Tridecanone	593-08-8			0.02			
1-Tridecene	2437-56-1						+[y4]
Trieicosane							+[y4]
Trieicosene							+[y4]
1,1,3-Trimethylbenzyl alcohol				0.1			
Trimethyl vinyl tetra-hydropyran	7392-19-0						+[y4]
Undecanal	112-44-7			0.01			+[y4]
Undecanal diethyl acetal	53405-97-3						+[y4]
Undecane	1120-21-4			0.001			+[y4]
2-Undecanone	112-12-9			0.03			
Undecatriene	16356-11-9						+[y4]
Valencene	4630-07-3						1.0[q]
Valeraldehyde	110-62-3	tr-0.2		0.02	0-0.2		+[y4]; 0.07[o]; 0.05[x3]; 0.1[p,v]; 0.3[w]
Valeraldehyde diethyl acetal	3658-79-5						+[y4]
Valerianol (kusunol)	20489-45-6						0.1[p]
Valeric acid	109-52-4				0.0007		
Verbenone	80-57-9						5.6[q]
Viridiflorene (ledene)	21747-46-6						0.3[d]
α-Ylangene	14912-44-8			0.01			
β-Ylangene	20479-06-5					tr	+[y4]
Zingiberene	495-60-3			0.008			

Table 5.73.2 Constituents identified in rose oils (*continued*)

A fifty-one rose essential oil samples from Bulgaria, Turkey, Russia and Iran, analyzed between 1998 and 2013; lowest and highest concentrations given (E. Schmidt, unpublished data)
B nineteen commercial rose oil samples, 8 from Bulgaria, 8 from Turkey and from Iran (lowest and highest concentrations given (ref. 28)
C one rose oil from Bulgaria (ref. 14)
D fifteen oils from Turkey, produced in two years, of which 5 were lab-prepared decanted oils from fresh petals and 10 were commercial blended oils from paled (fermented) petals; lowest and highest concentrations given (ref. 27)
E two commercial rose oils produced from *R. damascena* Mill. var. *trigintipetala* Dieck cultivated in Saudi Arabia (ref. 2)
F data from other studies (indicated with superscript letters); highest concentrations found in any study reviewed here given; when two or more oils were investigated, only the highest concentrations are mentioned, unless indicated otherwise

[a1] not a naturally occurring compound (ref. 35); [a2] incorrect identification (ref. 35); [a3] artifact or contaminant (ref. 35); [b] γ-terpinene and pentanol combined; [c] five lab-prepared oils from five varieties cultivated in India (ref. 1); [d] one lab-hydrodistilled rose oil and one microwave-assisted hydrodistilled oil from Iran (ref. 4); [e] one lab-hydrodistilled rose oil from India (ref. 9); [f] two oils from Turkey, of which one was produced by a private-type distillation unit and the other in a modern factory (ref. 8); [g] one historical (>50 year old) rose oil sample from Iran (ref. 7); [h] one commercial Turkish oil sample (ref. 36); [i] two lab-hydrodistilled oils from India (ref. 37); [j] one commercial rose oil sample (ref. 38); [k] one sample of Indian rose oil (ref. 39); [l] five commercial oils from Bulgaria, China, Iran, Morocco and Turkey (ref. 40); [m] two rose oils from India (ref. 41); [n] one rose oil from Bulgaria (ref. 43); [o] unknown number of rose oils from Bulgaria and Turkey (ref. 44); [p] several hydrodistilled Bulgarian rose oil-plus one Turkish rose oil (ref. 45); [q] one steam-distilled oil of rose (ref. 46); [r] five oils from India and one from Bulgaria (ref. 47); [s] unknown number of rose oils from Turkey from one supplier over a period of 16 years (ref. 48); [t] one lab-hydrodistilled oil from fresh flowers of *R. damascena* cultivated in India (ref. 49); [u] two lab-hydrodistilled oils from two *R. damascena* cultivars from India (ref. 50); [v] one Turkish and one Bulgarian commercial oil sample (ref. 51); [w] one lab-hydrodistilled oil from Turkey (ref. 21); [x] seven lab-hydrodistilled *Rosa damascena* oils from various parts of Lebanon (ref. 23); [x1] one lab-hydrodistilled oil from Iran (ref. 24); [x2] three lab-hydrodistilled oils from roses cultivated at two locations in India and harvested in two years (ref. 25); [x3] one rose oil from China obtained by steam-distillation in an industrial setting (ref. 29); [x4] two rose oils from Iran (ref. 31); [x5] one lab-hydrodistilled rose flower oil from Iran (ref. 34); this report contains many mistakes in identification (ref. 35); chemicals wrongly identified are only added to the table when no other study refers to them, but with a footnote that they were mistakenly identified; [x6] five lab-hydrodistilled oils from 5 Bulgarian *R. damascena* cultivars (ref. 26); [x7] one commercial oil of rose (ref. 3); [x8] one historical (>60 years old) rose oil from a distillery in Bulgaria and 4 oils produced from Bulgarian *R. damascena* cultivars (ref. 10); [x9] one lab-hydrodistilled oil from India (ref. 12); [y] five oils obtained from roses cultivated in India in a field distillation unit under different temperature and pressure regimens (ref. 30); [y1] one commercial rose oil sample from Turkey (ref. 32); [y2] one very atypical lab-hydrodistilled oil from Iran with nonadecane and heneicosane as dominant ingredients (ref. 33); [y3] eight lab-hydrodistilled oils from Turkish *R. damascena* with harvest dates and duration of fermentation as parameters (ref. 17); only 5 components were investigated; [y4] data from several studies from 1988 and older, cited in ref. 13; [y5] twenty-eight lab-hydrodistilled essential oils from roses cultivated in Iran which had been stored for varying times (0-4 days) under various parameters (dry or immersed in distilled water, temperatures ranging from -20°C to 10°C) (ref. 6); [y6] thirteen lab-hydrodistilled oils from *R. damascena* cultivated in Turkey with different fermentation times (0-36 hours) and distillation times (30-180 minutes); analytical data of the distillation *fractions* were also shown but are not presented here (ref. 11); [y7] twenty-seven lab-hydrodistilled oils from *R. damascena* cultivated in India with times of harvesting (8x during a period of 24 hours) and storage conditions (time, temperature) as parameters (ref. 15); [y8] fourteen lab-hydrodistilled oils from India with levels of pruning and time of pruning as parameters (ref. 19); [y9] fifteen lab-hydrodistilled oils from *R. damascena* produced in India with harvesting date, flower harvesting conditions and storage conditions as parameters (ref. 16); [z] contaminant from packing; [z1] impossible to obtain this high concentration with steam-distillation

tr: trace (in columns D and F[d]: <0.05; in column F[x2]: <0.1; in column F[x3]: <0.01); + present in the oil investigated, but quantity not stated

Table 5.73.3 Major constituents of rose oils

Constituent	CAS	Percentage and range					
		A	B (28)	C (14)	D (27)	E (2)	F
Citronellol	106-22-9	0.5-44.8	25.5-55.3	32.0	24.5-43.0	27.5	48.0[f]; 48.2[x1]; 53.6[x4]; 67.9[y3]
Geraniol	106-24-1	4.9-23.8	8.3-28.5	15.7	2.1-18.0	19.9	35.8[r]; 36.2[w]; 39.3[u]; 44.7[y3]
Nonadecane	629-92-5	6.3-15.9	2.6-23.2		6.4-19.0	15.7	24.8[y5]; 25.8[d]; 26.5[y7]; 39.7[y2]
Heneicosane	629-94-7	0.8-21.0	0.6-9.3	3.0	2.3-8.9	7.2	15.1[x]; 16.0[y7]; 16.1[y5]; 32.4[y2]
Nerol	106-25-2	0.6-11.0	0.9-3.3	8.7	0.8-7.6	10.5	13.1[m]; 14.5[r]; 16.3[y3]; 17.4[y6]
Nonadecene	27400-77-7	0.1-3.6	0.8-4.8	9.1	1.8-5.4	4.2	4.9[s]; 5.5[y6]; 6.5[x5]; 6.9[g]; 7.0[e]
Linalool	78-70-6	0.02-6.4	0.5-2.9	2.7	0.2-1.7	7.5	2.5[v]; 2.7[l]; 2.8[e]; 3.8[r]; 8.4[i]

LEGEND: SEE UNDER TABLE 5.73.2

CONTACT ALLERGY/ALLERGIC CONTACT DERMATITIS

General

Contact allergy to/allergic contact dermatitis from rose oils has been reported in over 25 publications. In a group of consecutive patients suspected of contact dermatitis, a prevalence rate of 1.6% positive patch test reactions has been observed, but reliable relevance data are lacking. There have been several case reports on allergic contact dermatitis from rose oil, mostly from its presence in perfumes or other cosmetics. Geraniol may have been an allergen in rose oil in several cases, citronellol (the main ingredient) and linalool each in one patient. In addition, there are many descriptions of positive patch test reactions to rose oil (which is often tested in a fragrance series) where the relevance was uncertain. Most of these occurred in aromatherapists and others working occupationally with (other) essential oils.

Testing in groups of patients

The results of patch tests with rose oil in routine testing (consecutive patients suspected of contact dermatitis) and in groups of selected patients are shown in Table 5.73.4. In routine testing, the rate of positive reactions was 1.6% (one study only), whereas between 0.4% and 30% of patients in selected groups had positive patch tests. The high positivity rate of 30% was found in a very small group of 10 patients strongly suspected of fragrance allergy and reacting to the fragrance mix (55).

Case reports

Two aromatherapists developed occupational contact dermatitis from contact allergy to multiple essential oils; they both reacted to rose oil Bulgarian in the fragrance series, which reactions were considered to be relevant (58). A massage therapist had occupational contact dermatitis from allergies to many essential oils; she reacted to rose oil, Bulgarian, in the fragrance series, judged to be clinically relevant (60). A patient with hand dermatitis reacted to rose oil, geraniol and several other fragrances and essential oils; she used a 'fragrance-free' hand soap containing rose oil (62); commercial rose oils contained up to 23.8% geraniol (Table 5.73.2, column A). One patient developed allergic

cosmetic dermatitis from rose oil in a face cream (70). Two positive patch test reactions to rose oil, which was present in cosmetics that had given a positive patch test or had been positive in a usage test in patients, were seen in a 9-year period in one clinic in Belgium (63). One patient had allergic contact dermatitis from 'Rose Absolute' perfume and a body lotion containing *Rosa centifolia*; she reacted to her own products, *Rosa centifolia*, rose oil Bulgarian, several indicators of fragrance allergy, two other essential oils and various individual fragrance chemicals including linalool and citral (the combination of geranial and neral) (65); linalool was found in commercial rose oils in concentrations up to 6.4%; geranial has been found in a lower concentration (max. 1.8%), but neral is not an important constituent of rose oils (Table 5.73.2, column A). One patient developed contact allergy to a perfume; she reacted to the fragrance mix, Bulgarian rose oil and geraniol; chromatographic analysis of the perfume showed it to contain 33% citronellol and 20% geraniol (63). Five cases of contact sensitization to rose oil present in topical pharmaceutical preparations have been reported (52).

Positive patch tests (relevance unknown, uncertain or not stated)

In a group of 819 patients suspected of contact dermatitis, there were four positive patch test reactions to Bulgarian rose oil (54). Four positive patch tests to rose oil Bulgarian were observed in four massage therapists/aromatherapists with occupational contact dermatitis from (multiple) essential oils; one had not used this oil, in the other three this was uncertain (57). Two aromatherapists with occupational contact dermatitis from other essential oils co-reacted to rose oil Bulgarian in the fragrance series (58). A naturopathic therapist and a masseuse with occupational contact dermatitis to various essential oils also reacted to Bulgarian rose oil in the fragrance series (59). Two positive patch tests to Bulgarian rose oil were seen in two aromatherapists allergic to multiple essential oils; they did not use rose oils themselves, but were allergic to geraniol, which is, after citronellol, the most important component of rose oils (61). A patient with allergic contact cheilitis from geraniol in ice cream, candy and gum had a positive patch test to rose oil and the fragrance mix I, both

Table 5.73.4 Results of testing groups of patients with rose oil

Years and Country	Test conc. & vehicle	Number of patients tested	positive (%)	Selection of patients (S); Relevance (R); Comments (C)	Ref.
Routine testing					
2000-2007 USA	2% pet.	679	11 (1.6%)	R: 100%; C: weak study: a. high rate of macular erythema and weak reactions, b. relevance figures include 'questionable' and 'past' relevance	53
Testing in groups of selected patients					
2004-2008 Spain	2% pet.	86	6 (7.0%)	S: patients previously reacting to the fragrance mix I or *Myroxylon pereirae* (n=54) or suspected of fragrance contact allergy (n=32); R: not stated; C: almost all patients also reacted to geraniol, one of its major components	64
<2004 Israel	2% pet.	91	2 (2.2%)	S: patients who had shown a doubtful or positive reaction to the fragrance mix I and/or *Myroxylon pereirae* resin and/or one or two commercial fine fragrances; R: not stated	69
1989-1999 Portugal	2% pet.	67	3 (4.5%)	S: patients who had a positive patch test to the fragrance mix; R: not stated	71
1990-1998 Japan	5% pet.	1,483	6 (0.4%)	S: patients suspected of cosmetic contact dermatitis, virtually all were women; range of annual frequency of sensitization: 0-0.9%; R: not stated; C: tested was 'rose oil Bulgarian'	56
1996-1997 UK	2% pet.	10	3 (30%)	S: patients strongly suspected of fragrance allergy and reacting to the fragrance mix; R: not stated; C: tested was 'rose oil, Bulgarian'	55
<1994 Japan	?	?	? (3.9%)	S and R: unknown. Possibly routine testing	72
<1974 Japan	?	137	4 (2.9%)	S: patients suspected of cosmetic dermatitis; R: unknown	75

of which contain geraniol (66). One patient with allergic contact cheilitis from lime reacted to lime, rose oil, geranium oil Bulgarian and their important component geraniol (67). One patient developed disseminated allergic contact dermatitis from geraniol and lavender essence in a cream; the patient also reacted to Bulgarian rose oil and geranium oil Bourbon, both of which contain high concentrations of geraniol (76).

LITERATURE

1 Kumar R, Sharma S, Sood S, Agnihotri VK, Singh V, Singh V. Evaluation of several *Rosa damascena* varieties and *Rosa bourboniana* accession for essential oil content and composition in western Himalayas. J Essent Oil Res 2014;26:147-152

2 Kürkçüoglu M, Abdel-Megeed A, Başer KHC. The composition of Taif rose oil. J Essent Oil Res 2013;25:364-367

3 Wei A, Shibamoto T. Antioxidant activities and volatile constituents of various essential oils. J Agric Food Chem 2007;55:1737-1742

4 Mohamadi M, Shamspur T, Mostafavi A. Comparison of microwave-assisted distillation and conventional hydrodistillation in the essential oil extraction of flowers *Rosa damascena* Mill. J Essent Oil Res 2013;25:55-61

5 Pal PK. Evaluation, genetic diversity, recent development of distillation method, challenges and opportunities of *Rosa damascena*: A review. J Essential Oil Bear Plants 2013;16:1-10

6 Mohamadi M, Mostafavi A, Shamspur T. Effect of storage on essential oil content and composition of *Rosa damascena* Mill. petals under different conditions. J Essent Oil Bear Plants 2011;14:430-441

7 Almasirad A, Amanzadeh Y, Taheri A, Iranshahi M. Composition of a historical rose oil sample (*Rosa damascena* Mill., Rosaceae). J Essent Oil Res 2007;19:110-112

8 Chalchat J-C, Özcan MM. A comparative investigation on the composition of rose (*Rosa damascena* Mill.) oil produced by using two different methods. J Essent Oil Bear Plants 2009;12:447-452

9 Chowdhury SR, Tandon PK, Chowdhury AR. Rose oil from *Rosa damascena* Mill., raised on alkaline soils. J Essent Oil Bear Plants 2009;12:213-217

10 Gochev V, Jirovetz L, Wlcek K, Buchbauer G, Schmidt E, Stoyanova A, Dobreva A. Chemical composition and antimicrobial activity of historical rose oil from Bulgaria. J Essent Oil Bear Plants 2009;12:1-6

11 Baydar H, Schulz H, Krüger H, Erbas S, Kineci S. Influences of fermentation time, hydro-distillation time and fractions on essential oil composition of damask rose (*Rosa damascena* Mill.). J Essent Oil Bear Plants 2008;11:3:224-232

12 Sood RP, Singh B, Singh V. Constituents of rose oil from Kangra valley, H. P. (India). J Essent Oil Res 1992;4:425-426

13 Lawrence BM. Progress in essential oils. Perfum Flavor 1991;16(May/June):43-77

14 Kovats E. Composition of essential oils. Part 7. Bulgarian oil of rose (*Rosa damascena* Mill.) J. Chromatogr 1987;406:185-222

15 Kumar R, Sharma S, Sood S, Agnihotri VK, Singh B. Effect of diurnal variability and storage conditions on essential oil content and quality of damask rose (*Rosa damascena* Mill.) flowers in north western Himalayas. Scientia Hort 2013;154:102-108

16 Kumar R, Sharma S, Sood S, Agnihotri VK. Agronomic interventions for the improvement of essential oil content and composition of damask rose (*Rosa damascena* Mill.) under western Himalayas. Ind Crops Prod 2013;48:171-177

17 Baydar H, Baydar NG. The effects of harvest date, fermentation duration and Tween 20 treatment on essential oil content and composition of industrial oil rose (*Rosa damascena* Mill.). Ind Crops Prod 2005;21:251-255

18 Rusanov K, Kovacheva N, Rusanova M, Atanassov I. Low variability of flower volatiles of *Rosa damascena* Mill. plants from rose plantations along the Rose Valley, Bulgaria. Ind Crops Prod 2012;37:6-10

19 Pal PK, Agnihotri VK, Gopichand, Singh RD. Impact of level and timing of pruning on flower yield and secondary metabolites profile of *Rosa damascena* under western Himalayan region. Ind Crops Prod 2014;52:219-227

20 Rusanov K, Kovacheva N, Rusanova M, Atanassov I. Traditional *Rosa damascena* flower harvesting practices evaluated through GC/MS metabolite profiling of flower volatiles. Food Chem 2011;129:1851-1859

21 Ozel MZ, Gogus F, Lewis AC. Comparisons of direct thermal desorption with water distillation and superheated water extraction for the analysis of volatile components of *Rosa damascena* Mill. using GCxGC-TOF/MS. Anal Chim Acta 2006;566:172-177

22 Mannschreck A, von Angerer E. The scent of roses and beyond: molecular structures, analysis, and practical applications of odorants. J Chem Educ 2011;88:1501-1506

23 Najem W, El Beyrouthy M, Wakim LH, Neema C, Ouaini N. Essential oil composition of *Rosa damascena* Mill. from different localities in Lebanon. Acta Bot Gall: Botany Letters 2011;158:365-373

24 Mahboubi M, Kazempour N, Khamechian T, Fallah MH, Kermani MM. Chemical composition and antimicrobial activity of *Rosa damascena* Mill essential oil. J Biol Active Prod Nat 2011;1:19-26

25 Verma RS, Padalia RC, Chauhan A, Singh A, Yadav AK. Volatile constituents of essential oil and rose water of damask rose (*Rosa damascena* Mill.) cultivars from North Indian hills. Nat Prod Res 2011;25:1577-1584

26 Rusanov KE, Kovacheva NM, Atanassov II. Comparative GC/MS analysis of rose flower and distilled oil volatiles of the oil bearing rose *Rosa damascena*. Biotechnol Biotechnol Equipm 2011;25:2210-2216

27 Bayrak A, Akgul A. Volatile oil composition of Turkish rose (*Rosa damascena*). J Sci Food Agric 1994;64:441-448

28 Pellati F, Orlandini G, van Leeuwen KA, Anesin G, Bertelli D, Paolini M, et al. Gas chromatography combined with mass spectrometry, flame ionization detection and elemental analyzer/isotope ratio mass spectrometry for characterizing and detecting the authenticity of commercial essential oils of *Rosa damascena* Mill. Rapid Commun Mass Spectrom 2013;27:591-602

29 Jirovetz L, Buchbauer G, Stoyanova A, Balinova A, Guangjiun Z, Xihan M. Solid phase microextraction/ gas chromatographic and olfactory analysis of the scent and fixative properties of the essential oil of *Rosa damascena* L. from China. Flavour Fragr J 2005;20:7-12

30 Babu KGD, Singh B, Joshi VP, Singh V. Essential oil composition of Damask rose (*Rosa damascena* Mill.) distilled under different pressures and temperatures. Flavour Fragr J 2002;17:136-140

31 Safaei-Ghomi J, Akhoondi S, Batooli H, Dackhili M. Chemical variability of essential oil components of two *Rosa x damascena* genotypes growing in Iran. Chem Nat Comp 2009;45:262-264

32 Ulusoy S, Boşgelmez-Tinaz G, Seçilmiş-Canbay H. Tocopherol, carotene, phenolic contents and antibacterial properties of rose essential oil, hydrosol and absolute. Curr Microbiol 2009;59:554-558

33 Moein M, Karami F, Tavallali H, Ghasemi Y. Composition of the essential oil of *Rosa damascena* Mill. from south of Iran. Iran J Pharmaceut Sci 2010;6:59-62.

34 Jalali-Heravi M, Parastar H, Sereshti H. Development of a method for analysis of Iranian damask rose oil: combination of gas chromatography—mass spectrometry with chemometric techniques. Anal Chim Acta 2008;623:11-21

35 Lawrence BM. Essential oils 2008-2011. Carol Stream, USA: Allured Publishing Corporation, 2012:257-264

36 Aridogan BC, Baydar H, Kaya S, Demirci M, Özbasar D, Mumcu E. Antimicrobial activity and chemical composition of some essential oils. Arch Pharm Res 2002;25:860-864

37 Agarwal SG, Gupta A, Kapahi BK, Baleshwar, Thappa RK, Suri OP. Chemical composition of rose water volatiles. J Essent Oil Res 2005;17:265-267

38 Jirovetz L, Eller G, Buchbauer G, Schmidt E, Denkova Z, Stoyanova AS, et al. Chemical composition, antimicrobial activities and some odor descriptions of some essential oils with characteristic floral-rosy scent and of their principal aroma compounds. Recent Res Devel Agron Hort 2006;2:1-12. Data cited in ref. 35

39 Ranade GS. Profile essential oil rose oil (Indian). FAFAI J 2008;10:97. Data cited in ref. 35

40 Gochev V, Wlcek K, Buchbauer G, Stoyanova A, Dobreva A, Schmidt E, Jirovetz L. Comparative evaluation of antimicrobial activity and composition

of rose oils from various geographical origins, in particular Bulgaria rose oil. Nat Prod Comm 2008;3:1063-1068

41 Kumar A, Aggarwal S, Singh GR, Chauhan N. Chemical composition of Rose absolute and oil from flowers of *Rosa damascena*. Indian Perfum 2010;54:60-62. Data cited in ref. 35

42 Lawrence BM. Progress in essential oils. Perfum Flavor 2005;30(4):60-? (last page unknown)

43 Lindström M, Jaquier J. A gas chromatographic study of the enantiomeric composition of some volatile compounds in Bulgarian rose oil (*Rosa damascena* Mill.). Rivista Ital EPPOS 1993;(Numero Speciale):129-139. Data cited in ref. 42

44 Brud W, Konpacka-Brud I. Rose oils. Variations on a theme. Int J Aromather 1994;6(2):12-15. Data cited in ref. 42

45 Boelens MH, Boelens H. Differences in chemical and sensory properties of orange flower and rose oils obtained from hydrodistillation and from supercritical CO_2 extraction. Perfum Flavor 1997;22(3):31-35

46 Reverchon E, Della Porta G, Gorgoglione D. Supercritical CO_2 extraction of volatile oil from rose concrete. Flavour Fragr J 1997;12:37-41

47 Naqvi AA, Mandal S. Investigation of rose oils from different places in India by capillary gas chromatography. J Med Arom Plant Sci 1997;19:1000-1002. Data cited in ref. 42

48 Başer KHC, Kürkçüoglu M, Özek T. Turkish rose oil research: recent results. Perfum Flavor 2003;28:34-41. Data cited in ref. 42

49 Gupta R, Mallavarapu GR, Ramesh S, Kumar S. Composition of flower essential oils of *Rosa damascena* and *Rosa indica* grown in Lucknow. J Med Arom Plant Sci 2000;22(4a):9-12. Data cited in ref. 42

50 Patra NK, Kalra A, Singh HB, Singh HP, Mengi N, Reddy V, et al. A genotype with impressive oil productivity developed via half-sib progeny selection in Kannauj-Damask rose race. J Med Arom Plant Sci 2001;22:259-266. Data cited in ref. 42.

51 Kubeczka K-H, Formacek V. Essential oils analysis by capillary gas chromatography and carbon-13 NMR spectroscopy, 2nd edition. New York, USA: John Wiley and Sons, 2002:273-284. Data cited in ref. 42

52 Nardelli A, D'Hooge E, Drieghe J, Dooms M, Goossens A. Allergic contact dermatitis from fragrance components in specific topical pharmaceutical products in Belgium. Contact Dermatitis 2009;60:303-313

53 Wetter DA, Yiannias JA, Prakash AV, Davis MD, Farmer SA, el-Azhary RA, et al. Results of patch testing to personal care product allergens in a standard series and a supplemental cosmetic series: an analysis of 945 patients from the Mayo Clinic Contact Dermatitis Group, 2000-2007. J Am Acad Dermatol 2010;63:789-798

54 Kohl L, Blondeel A, Song M. Allergic contact dermatitis from cosmetics: retrospective analysis of 819 patch-tested patients. Dermatology 2002;204:334-337

55 Thomson KF, Wilkinson SM. Allergic contact dermatitis to plant extracts in patients with cosmetic dermatitis. Br J Dermatol 2000;142:84-88

56 Sugiura M, Hayakawa R, Kato Y, Sugiura K, Hashimoto R. Results of patch testing with lavender oil in Japan. Contact Dermatitis 2000;43:157-160

57 Bleasel N, Tate B, Rademaker M. Allergic contact dermatitis following exposure to essential oils. Australas J Dermatol 2002;43:211-213

58 Boonchai W, Lamtharachai P, Sunthonpalin P. Occupational allergic contact dermatitis from essential oils in aromatherapists. Contact Dermatitis 2007;56:181-182

59 Trattner A, David M, Lazarov A. Occupational contact dermatitis due to essential oils. Contact Dermatitis 2008;58:282-284

60 Cockayne SE, Gawkrodger DJ. Occupational contact dermatitis in an aromatherapist. Contact Dermatitis 1997;37:306-307

61 Dharmagunawardena B, Takwale A, Sanders KJ, Cannan S, Roger A, Ilchyshyn A. Gas chromatography: an investigative tool in multiple allergies to essential oils. Contact Dermatitis 2002;47:288-292

62 Scheinman PL. Is it really fragrance free? Am J Cont Derm 1997;8:239-242

63 Nardelli A, Drieghe J, Claes L, Boey L, Goossens A. Fragrance allergens in 'specific' cosmetic products. Contact Dermatitis 2011;64:212-219

64 Cuesta L, Silvestre JF, Toledo F, Lucas A, Pérez-Crespo M, Ballester I. Fragrance contact allergy: a 4-year retrospective study. Contact Dermatitis 2010;63:77-84

65 Nardelli A, Thijs L, Janssen K, Goossens A. *Rosa centifolia* in a 'non-scented' moisturizing body lotion as a cause of allergic contact dermatitis. Contact Dermatitis 2009;61:306-309

66 Tamagawa-Mineoka R, Katoh N, Kishimoto S. Allergic contact cheilitis due to geraniol in food. Contact Dermatitis 2007;56:242-243

67 Thomson MA, Preston PW, Prais L, Foulds IS. Lime dermatitis from gin and tonic with a twist of lime. Contact Dermatitis 2007;56:114-115

68 An S, Lee AY, Lee CH, Kim D-W, Hahm JH, Kim K-J, et al. Fragrance contact dermatitis in Korea: a joint study. Contact Dermatitis 2005;53:320-323

69 Trattner A, David M. Patch testing with fine fragrances: comparison with fragrance mix, balsam of Peru and a fragrance series. Contact Dermatitis 2004;49:287-289

70 Penchalaiah K, Handa S, Lakshmi SB, Sharma VK, Kumar B. Sensitizers commonly causing allergic contact dermatitis from cosmetics. Contact Dermatitis 2000;43:311-313

71 Manuel Brites M, Goncalo M, Figueiredo A. Contact allergy to fragrance mix—a 10-year study. Contact Dermatitis 2000;43:181-182

72 Sugai T. Group study IV—farnesol and lily aldehyde. Environ Dermatol 1994;1:213-214

73 Vilaplana J, Romaguera C, Grimalt F. Contact dermatitis from geraniol in Bulgarian rose oil. Contact Dermatitis 1991;24:301

74 Nakayama H, Hanaoka H, Ohshiro A. Allergen Controlled System (ACS). Tokyo, Japan: Kanehara Shuppan, 1974:42. Data cited in ref. 75

75 Mitchell JC. Contact hypersensitivity to some perfume materials. Contact Dermatitis 1975;1:197-199

76 Juarez A, Goiriz R, Sanchez-Perez J, Garcia-Diez A. Disseminated allergic contact dermatitis after exposure to a topical medication containing geraniol. Dermatitis 2008;19:163

77 Rhind JP. Essential oils. A handbook for aromatherapy practice, 2nd Edition. London: Singing Dragon, 2012

78 Lawless J. The encyclopedia of essential oils, 2nd Edition. London: Harper Thorsons, 2014

Chapter 5.74   ROSEWOOD OIL

DEFINITION

Rosewood oil (essential oil of rosewood, Brazilian type) is the essential oil obtained from the wood of the rosewood, *Aniba rosaeodora* Ducke (and possibly from other *Aniba* species such as *Aniba parviflora* (Meisn.) Mez.).

INCI NOMENCLATURE

Description/definition: Aniba rosaeodora wood oil is the volatile oil obtained from the wood of the rosewood tree, *Aniba rosaeodora, Lauraceae*
INCI name EU: Aniba rosaeodora wood oil
INCI name USA: Aniba rosaeodora (rosewood) wood oil
Other names: Bois de rose oil
CAS registry number(s): 8015-77-8; 83863-32-5
EINECS number(s): 281-093-7

ISO (INTERNATIONAL ORGANIZATION FOR STANDARDIZATION) STANDARD

ISO number: 3761
ISO name: Essential oil of rosewood, Brazilian type
Botanical origin: *Aniba rosaeodora* Ducke or *Aniba parviflora* (Meisn.) Mez.
Parts of plant used: Wood
ISO values: ISO values (minimum and maximum concentrations) are shown in Table 5.74.1.

The Brazilian rosewood *Aniba rosaeodora* Ducke should not be confused with *Dalbergia nigra* (Vell.) Allemão ex Benth., which is also called Brazilian rosewood, but which is a hardwood used (amongst other things) for musical instrument manufacture.

THE PLANT, THE OIL, AND THEIR USES

Aniba rosaeodora Ducke (synonyms: *Aniba rosaeodora* var. *amazonica* Ducke, *Aniba duckei* Kostermans) is a massive, evergreen fragrant tree, up to 30 meters in height and 2 meters in diameter. It is native to French Guyana, Columbia, Ecuador, Suriname, Venezuela, Brazil and Peru (1,12). Because of overuse of the trees in the past (notably for production of essential oil), *A. rosaeodora* is now an endangered species (International Union for Conservation of Nature and Natural Resources, http://www.iucnredlist.org; Convention on International Trade in Endangered Species of Wild Fauna and Flora (CITES, www.cites.org), 12).

Rosewood oil is obtained by felling wild *Aniba rosaeodora* Ducke trees ('bois de rose' or 'bois de rose fenelle' in French; 'pau-rosa' in Brazilian) and possibly other species of *Aniba* such as *Aniba parviflora* (5) and steam distilling the comminuted trunk wood. The oil possesses a characteristic aroma and is a long-established ingredient in the more expensive perfumes and a perfume fixative (1,2,12). It is cited to be used in food production also (1). Rosewood oil is rich in linalool, a chemical which can be transformed into a number of derivatives of value to the flavor and fragrance industries (12), and up until the 1960s rosewood oil was an important source of natural linalool. With the advent of synthetic linalool this use largely disappeared. For those applications where natural linalool is preferred, rosewood oil has been displaced by cheaper alternatives (Chinese Ho oils from *Cinnamomum camphora*). There does remain, however, a small market for the preparation of linalool derivatives possessing an 'ex rosewood' character and rosewood essential oils are still used as the main scent in a few top perfumes (1,6). The oil is used in aromatherapy formulations, but has become less attractive as environmental concerns have grown over the destructive nature of rosewood oil production in Brazil (1).

Rosewood oils were produced in French Guyana between 1900 and 1970, and had a very good reputation among perfumers. However, they have disappeared from the market and were substituted by lesser quality Brazilian oil. Brazil is currently the sole producer of rosewood oil (1). Since the 1960s, when production reached its peak, the export of rosewood oil has experienced a long decline. Many factors have contributed to this, of which plant sources exhaustion, logistics and costs of production, government regulations and synthetic linalool trade are the main factors (9,12). Rosewood *leaf* oils are suggested as environmentally friendly alternatives to the wood oil, having equivalent properties (6).

CHEMICAL COMPOSITION

Rosewood oil is a colorless to pale yellow clear mobile liquid which has a fresh, floral and tender woody odor. The yield of essential oil from the wood of either *Aniba rosaeodora* Ducke or *Aniba parviflora* (Meisn.) Mez. generally varies from 0.6 to 1.4 per cent. The main (and probably sole) producing country of this oil is Brazil.

Literature data (up to December 5, 2014) on the chemical composition of rosewood oils and unpublished analytical data from one of us (E.S.) are shown in Table 5.74.2 in alphabetical order. In rosewood oils from various origins, over 60 chemicals have been identified. About 42 per cent of these were found in a single reviewed

Table 5.74.1 ISO values (%) for rosewood oil [a]

Compound	CAS	Minimum	Maximum
Linalool	78-70-6	70.0	90.0
α-Terpineol	98-55-5	2.0	7.0
1,8-Cineole	470-82-6	t	3.0
α-Copaene	3856-25-5	t	3.0
Geraniol	106-24-1	0.5	2.5
cis-Linalool oxide, furanoid	5989-33-3	0.5	2.0
trans-Linalool oxide, furanoid	34995-77-2	0.5	2.0
Benzyl benzoate	120-51-4	0.2	1.6
α-Pinene	80-56-8	t	1.0

[a] ISO 3761 Essential oil of rosewood ©ISO 2005; Geneva, Switzerland, www.iso.org

publication only. The major compounds found in rosewood oils from different sources are shown in Table 5.74.3. They include (highest concentrations in any study given) linalool (99.0%), α-terpineol (18.8%), linalool oxide (14.1%), geraniol (7.8%), cis-linalool oxide (5.8%), trans-linalool oxide (5.2%), benzyl benzoate (3.7%), α-pinene (3.2%) and α-copaene (2.1%). A well-known ingredient of rosewood oils that was present in a high concentration (>7%) in one study was limonene (19.2%).

Commercial oils

The ten chemicals that had the highest maximum concentrations in 36 commercial rosewood essential oil samples (concentration ranges provided) are the following: linalool (73.0-88.4%), α-terpineol (1.9-8.8%), trans-sabinene hydrate (0.01-3.3%), geraniol (0.7-2.3%), α-pinene (0.2-2.2%), cis-linalool oxide (0.9-2.1%), α-copaene (0.2-2.1%), cis-linalool oxide, furanoid (0.03-2%), trans-linalool oxide (0.9-2.0%) and trans-linalool oxide, furanoid (0.06-1.7%) (Erich Schmidt, unpublished analytical data).

CONTACT ALLERGY/ALLERGIC CONTACT DERMATITIS

General
Contact allergy to/allergic contact dermatitis from rosewood oil has been reported in two case reports only. In one, linalool may have been an allergen in rosewood oil.

Case reports
One patient had airborne contact dermatitis from contact allergy to rosewood oil and two other essential oils in aromatherapy lamps, whereby the water vapor produced distributes the essential oils as an aerosol in the air; the patient also reacted to linalool, which is an important constituent of all three oils (14). In commercial rosewood oils, linalool was the dominant ingredient with concentrations of 74-88% (Table 5.74.2, column A). An aromatherapist had occupational contact dermatitis from multiple essential oil sensitizations, including rosewood (15).

Table 5.74.2 Constituents identified in rosewood oils

Constituent	CAS	Percentage and range				
		A	B (9)	C (3)	D (5)	E
allo-Aromadendrene	25246-27-9		tr-0.2			
Benzaldehyde	100-52-7	tr-0.2	0-0.4			
Benzyl benzoate	120-51-4	0.3-1.3	0.6-3.7	0.2	0.3-0.6	0.3[a]
δ-Cadinene	483-76-1	0.08-0.5		0.1		0.6[f]
epi-α-Cadinol	5937-11-1		tr-0.3			
δ-Cadinol	19435-97-3	0.01-0.07				
Camphene	79-92-5	tr-0.08				
Camphor	76-22-2				0-0.09	tr[a]
β-Caryophyllene	87-44-5	0.02-0.3	tr			
Caryophyllene oxide	1139-30-6		tr-0.1		0.1-0.3	
1,8-Cineole	470-82-6	0.4-1.2	0-0.8	0.6	0-0.8	0.8[a]
α-Copaene	3856-25-5	0.2-2.1	tr-0.4	0.8		0.3[a]; 1.9[f]
1-epi-Cubenol	81939-29-9		0-0.5			
Cyclosativene	22469-52-9				0-0.9	
p-Cymene	99-87-6	0.02-0.2		0.1	0-0.09	4.1[c]
(E,Z)-2,6-Dimethyl-5,7-octadien-2-ol						0.02[a]
β-Elemene	33880-83-0	0.05-0.2	tr-0.6	0.2		
Eremophilene	10219-75-7	0.6-1.2			0.3-1.7	0.2[a]; 1.2[d]
Geraniol	106-24-1	0.7-2.3	0.1-1.8	1.3		0.2[a]; 6.4[e]; 7.8[c]
Geranyl acetate	105-87-3	0.02-0.3				
α-Guaiene	3691-12-1	0.01-0.2				
α-Gurjunene	489-40-7	0.03-0.2				
Hotrienol	20053-88-7	0.1-0.5			0.1-0.2	0.1[a]
α-Humulene	6753-98-6		tr			
Limonene	138-86-3	0.1-1.2	tr-0.7	0.9	0.1-0.6	0.2[a]; 2.4[b]; 19.2[c]
Linalool	78-70-6	73.0-88.4	75.3-84.8	81.3	72.0-92.0	76.0[f]; 85.0[b]; 87.7[d]; 89.9[a]; 99.0[e]
Linalool oxide	1365-19-1			0.2	0.1-0.6	0.02[a]; 14.1[b]
cis-Linalool oxide	11063-77-7	0.9-2.1		2.1	1.0-5.8	0.7[d]; 1.0[f]; 1.6[a]
cis-Linalool oxide, furanoid	5989-33-3	0.03-2.0	0.5-2.7			
cis-Linalool oxide, pyranoid	14009-71-3				0.03-0.4	
trans-Linalool oxide	11063-78-8	0.9-2.0		2.1	0.9-5.2	0.7[d]; 1.1[f]; 1.4[a]

Table 5.74.2 Constituents identified in rosewood oils (*continued*)

Constituent	CAS	Percentage and range				
		A	B (9)	C (3)	D (5)	E
trans-Linalool oxide, furanoid	34995-77-2	0.06-1.7	0.7-2.6			
trans-Linalool oxide, pyranoid	39028-58-5		tr-0.3			
Methylheptenone	409-02-9					tr[a]
epi-α-Muurolol	19912-62-0		0-0.2			
Myrcene	123-35-3	0.03-0.3			0-0.5	
Nerol	106-25-2	0.2-0.7	tr-0.4	0.4		0.7[a]; 1.3[f]; 1.7[e]
Nerolidol	7212-44-4			0.2		0.08[a]
(*E*)-Nerolidol	40716-66-3	0.02-0.3	tr-0.2			
(*E*)-β-Ocimene	3779-61-1	0.05-0.2	tr-0.2		0-0.07	tr[a]
(*Z*)-β-Ocimene	3338-55-4	0.06-0.1	0-0.2		0.03-0.2	tr[a]
(*E*)-Ocimenol	28977-58-4			0.1		
3-Octanol	589-98-0					tr[a]
α-Pinene	80-56-8	0.2-2.2	0.1-0.5			3.2[c]
β-Pinene	127-91-3	0.07-0.9	0-0.3	0.1	0-0.4	0.3[a]
Pseudowiddrene	32540-28-6					1.9[f]
Sabinene	3387-41-5			0.5		
trans-Sabinene hydrate	17699-16-0	0.01-3.3				
α-Selinene	473-13-2	0.09-1.1	0.4-0.9	0.7		0.4[a]; 2.0[b]
β-Selinene	17066-67-0	0.1-1.5	0.5-1.0	0.8		4.4[b]
γ-Selinene	515-17-3			0.2	0.2-0.4	
Spathulenol	6750-60-3	0.05-0.6	tr		0-1.0	0.2[a]
γ-Terpinene	99-85-4	0.02-0.09				
Terpinen-4-ol	562-74-3	0.08-0.1	tr	0.2	0.1-0.3	0.2[a]
α-Terpineol	98-55-5	1.9-8.8	0.7-5.6	4.8		2.6[a]; 3.1[d]; 5.6[f]; 18.8[e]
Terpinolene	586-62-9	0.01-0.3				
α-Thujene	2867-05-2			0.3		
Viridiflorene	21747-46-6					1.3[f]

A thirty-six rosewood essential oil samples from Brazil analyzed between 1998 and 2013; lowest and highest concentrations given (E. Schmidt, unpublished data)
B nine lab-hydrodistilled rosewood oils; lowest and highest concentrations given (ref. 9)
C one commercial rosewood oil sample purchased in Serbia (ref. 3)
D unknown number of lab-distilled and commercial rosewood oils; lowest and highest concentrations given (ref. 5)
E data from other studies (indicated with superscript letters); highest concentrations found in any study reviewed here given; when two or more oils were investigated, only the highest concentrations are mentioned, unless indicated otherwise

[a] one rosewood oil sample of unknown origin (ref. 7); [b] one commercial rosewood oil (ref. 8); [c] one commercial oil sample purchased in Italy (ref. 10); [d] one rosewood oil sample (uncertain whether it was lab-prepared or commercial) (ref. 4); [e] eighty-two lab-hydrodistilled rosewood oils from *A. rosaeodora* from French Guiana (ref. 2); [f] one commercial rosewood oil sample from Ecuador (ref. 13)

tr: trace (in column B: not quantified)

Table 5.74.3 Major constituents of rosewood oils

Constituent	CAS	Percentage and range				
		A	B (9)	C (3)	D (5)	E
Linalool	78-70-6	73.0-88.4	75.3-84.8	81.3	72.0-92.0	76.0[f]; 85.0[b]; 87.7[d]; 89.9[a]; 99.0[e]
α-Terpineol	98-55-5	1.9-8.8	0.7-5.6	4.8		2.6[a]; 3.1[d]; 5.6[f]; 18.8[e]
Linalool oxide	1365-19-1			0.2	0.1-0.6	0.02[a]; 14.1[b]
Geraniol	106-24-1	0.7-2.3	0.1-1.8	1.3		0.2[a]; 6.4[e]; 7.8[c]
cis-Linalool oxide	11063-77-7	0.9-2.1		2.1	1.0-5.8	0.7[d]; 1.6[a]
trans-Linalool oxide	11063-78-8	0.9-2.0		2.1	0.9-5.2	0.7[d]; 1.4[a]
Benzyl benzoate	120-51-4	0.3-1.3	0.6-3.7	0.2	0.3-0.6	0.3[a]
α-Pinene	80-56-8	0.2-2.2	0.1-0.5			3.2[c]
α-Copaene	3856-25-5	0.2-2.1	tr-0.4	0.8		0.3[a]; 1.9[f]

LEGEND: SEE UNDER TABLE 5.74.2

LITERATURE

1 Coppen JJW. Rosewood oil. In: Non-wood forest products. Flavours and fragrances of plant origin. Rome: Food and Agriculture Organization of the United Nations, 1995:29-36

2 Chantraine J-M, Dhénin J-M, Moretti C. Chemical variability of rosewood (*Aniba rosaeodora* Ducke) essential oil in French Guiana. J Essent Oil Res 2009;21:486-495

3 Simić A, Soković MD, Ristić M, Grujić-Jovanović S, Vukojević J, Marin PD. The chemical composition of some Lauraceae essential oils and their antifungal activities. Phytother Res 2004;18:713-717

4 Nóbrega de Almeida R, Machado Araújo DE, Ramos Goncalvesa JC, Costa Montenegro F, Pergentino de Sousa D, Leite JR, et al. Rosewood oil induces sedation and inhibits compound action potential in rodents. J Ethnopharmacol 2009;124:440-443

5 Santana A, Ohashi S, de Rosa L, Green CL. Brazilian rosewood oil: The prospect for sustainable production and oil quality management. Int J Aromather 1997;8:16-20

6 Souza RCZ, Marques Eiras M, Cabral EC, Barata LES, Eberlin MN, Catharino RR. The famous Amazonian rosewood essential oil: Characterization and adulteration monitoring by electrospray ionization mass spectrometry fingerprinting. Anal Lett 2011;44:2417-2422

7 Burfield T, Sheppard-Hanger S. Substituting for rosewood oil *Aniba rosaeodora* var. *amazonica* Ducke – a look at other high linalool containing oils, June 2003. Adapted and modified from Aromatherapy Today 2003;26(June):30-37. Available at: http://www.users.globalnet.co.uk/~nodice/new/magazine/subrosewood/subrose.htm

8 Lupe F, Souza R, Barata O. Seeking a sustainable alternative to Brazilian rosewood. Perfum Flavor 2008;33(July):40-43

9 Maia JGS, Andrade EHA, Couto HAR, da Silva AC, Marx F, Henke C. Plant sources of Amazon rosewood oil. Quim Nova 2007;30:906-910

10 Rosato A, Piarulli M, Corbo F, Muraglia M, Carone A, Vitali ME, et al. *In vitro* synergistic action of certain combinations of gentamicin and essential oils. Curr Med Chem 2010;17:3289-3295

11 Lawrence BM. A preliminary report on the world production of some selected essential oils and countries. Perfum Flavor 2009;34(1):38-44

12 Convention on International Trade in Endangered Species of Wild Fauna and Flora. Fifteenth meeting of the Conference of the Parties, Doha (Qatar), 13-25 March 2010. Available at: http://www.cites.org/eng/cop/15/prop/E-15-Prop-29.pdf

13 Rolli E, Marieschi M, Maietti S, Sacchetti G, Bruni R. Comparative phytotoxicity of 25 essential oils on pre- and post-emergence development of *Solanum lycopersicum* L.: A multivariate approach. Ind Crops Prod 2014;60:280-290

14 Schaller M, Korting HC. Allergic airborne contact dermatitis from essential oils used in aromatherapy. Clin Exp Dermatol 1995;20:143-145

15 Selvaag E, Holm J, Thune P. Allergic contact dermatitis in an aromatherapist with multiple sensitizations to essential oils. Contact Dermatitis 1995;33:354-355

Chapter 5.75 SAGE OIL, DALMATIAN

There are two major commercial sage oils: sage oil, Dalmatian, obtained from *Salvia officinalis* L. and sage oil, Spanish, which is obtained from *Salvia lavandulifolia* Vahl. In this chapter, Dalmatian sage oil is discussed. The Spanish type essential oil is presented in Chapter 5.76.

DEFINITION

Dalmatian sage oil (essential oil of Dalmatian sage) is the essential oil obtained from the flowering tops of the sage, *Salvia officinalis* L.

INCI NOMENCLATURE

Description/definition: Salvia officinalis oil is the essential oil derived from the herbal plant the sage, *Salvia officinalis* L., Lamiaceae
INCI name EU: Salvia officinalis oil
INCI name USA: Salvia officinalis (sage) oil
CAS registry number(s): 8022-56-8; 84082-79-1; 84776 73-8 (deleted CAS registry number)
EINECS number(s): 282-025-9

ISO (INTERNATIONAL ORGANIZATION FOR STANDARDIZATION) STANDARD

ISO number: 9909
ISO name: Essential oil of Dalmatian sage
Botanical origin: *Salvia officinalis* L.
Parts of plant used: Flowering top
ISO values: ISO values (minimum and maximum concentrations) are shown in Table 5.75.1.

Other species of the genus *Salvia* from which sage oils are obtained and which are mentioned in CosIng include *Salvia hispanica* (CAS 93384-40-8; EINECS 297-250-8; *not*

Table 5.75.1 ISO values (%) for Dalmatian sage oil [a]

Compound	CAS	Minimum	Maximum
α-Thujone (*cis-*)	546-80-5	18.0	43.0
Camphor	76-22-2	4.5	24.5
1,8-Cineole	470-82-6	5.5	13.0
α-Humulene	6753-98-6	0.0	12.0
β-Thujone (*trans-*)	471-15-8	3.0	8.5
Camphene	79-92-5	1.5	7.0
α-Pinene	80-56-8	1.0	6.5
Limonene	138-86-3	0.5	3.0
Bornyl acetate	76-49-3	0.0	2.5
Linalool + Linalyl acetate	78-70-6 115-95-7	0.0	1.0

[a] ISO 9909 Essential oil of Dalmatian sage ©ISO 1997; Geneva, Switzerland, www.iso.org

identical with sage, Spanish type [=*Salvia lavandulifolia* Vahl]) and *Salvia sclarea* (see Chapter 5.25 Clary sage oil). According to ISO, sage oils should be produced from the flowering tops of *S. officinalis*. However, in most publications, oils had been obtained from the *leaves* only.

THE PLANT, THE OIL, AND THEIR USES

Salvia officinalis L. is a perennial, evergreen subshrub, up to 60 cm tall, with a woody base, soft gray-green oval leaves and a mass of blue or violet inflorescences (4). The name Salvia comes from the Latin salvere, which means 'to be well, to be in good health', indicating the (perceived) medical value of the plant (5). The Dalmatian sage is native to southeastern Europe (Albania, former Yugoslavia, Greece, Italy) and is now widely cultivated in many (warm)-temperate regions of the world, mainly to obtain dried leaves to be used as a raw material in medicine, perfumery, and the food industry (3, GRIN Taxonomy for Plants; Mansfeld's World Database of Agriculture and Horticultural Crops, http://mansfeld.ipk-gatersleben.de).

The leaves and extracts of *Salvia officinalis* L. are used as spices and for healing of different diseases. The plant is said to be appetizing and to promote digestion of food (4,5). In folk medicine, Dalmatian sage is used to treat inflamed throat and gingivitis (by gargling), cure laryngitis and hoarse voice, reduce sweat gland activity, treat fevers, reduce digestive tract disturbances, stimulate micturition, purify the colon and liver, and to strengthen the nervous system (60). A plethora of pharmacological activities has been ascribed to sage (5,21,46,48,66). Some authors claim that extracts of *S. officinalis* are efficacious in the management of mild to moderate Alzheimer's disease and age-associated memory loss (46,60); as yet, this is unproven (74).

The essential oil of sage, which is obtained by steam-distillation of the flowering tops or the leaves, is used in traditional medicine and by the pharmaceutical, perfumery, liqueur and food industries (5,48). Essential oils of sage and their preparations are externally used for inflammations and infections of the mucous membranes of the throat and mouth (stomatitis, gingivitis and pharyngitis). Internally, they are used for dyspeptic symptoms and excessive perspiration (46,48). They are also employed in incense and for aromatherapy, though often considered too toxic for that practice because of the high thujone content (97,98,99). A variety of pharmacological activities has been described to the essential oils of sage, including antioxidant, anti-inflammatory, antispasmodic, antimicrobial, stimulant, insecticidal, antiviral and antifungal properties, depending on their composition (see under Chemotypes) (4,33,59). However, the European Medicines Agency in 2009 assessed the safety and efficacy of their use in minor indications that do not require supervision of a medical practitioner, and concluded that the benefit-risk analysis of sage essential oil is negative, warning for the toxicity of thujone, usually the main ingredient of commercial sage oils (ref. 96).

CHEMICAL COMPOSITION

Dalmatian sage oil is a colorless to yellow clear mobile liquid, which has an herbaceous, aromatic-spicy green camphoraceous odor. The yield of essential oil from the flowering tops of *Salvia officinalis* L. generally varies from 0.05 to 0.2 per cent. The main producing countries of this oil are USA, Morocco, Tunisia, Albania, Hungary and Turkey.

Literature data (up to October 26, 2014) on the chemical composition of Dalmatian sage oils and unpublished analytical data from one of us (E.S.) are shown in Table 5.75.2 in alphabetical order. In Dalmatian sage oils from various origins, over 310 chemicals have been identified. About 47 per cent of these were found in a single reviewed publication only. The major compounds found in Dalmatian sage oils from different sources are shown in Table 5.75.3. They include (highest concentrations in any study given) α-thujone (*cis-*) (67.9%), 1,8-cineole (67.1%), camphor (50.2%), β-thujone (*trans-*) (49.7%), α-humulene (33.2%), viridiflorol (30.3%), α-pinene (24.8%), borneol (23.9%), manool (20.9%), β-pinene (19.0%), β-caryophyllene (15.8%), camphene (10.6%), bornyl acetate (10.3%) and limonene (7.6%). Well-known ingredients of Dalmatian sage oils that were present in high concentrations (>7%) in one or two studies were γ-muurolene (9.3% and 13.8%), *p*-cymene (11.8%), 1-octen-3-ol (8.5%) and sabinene (8.2%). Uncommon or rare constituents of Dalmatian sage oils found in high concentrations (>7%) in single studies include sclareol (23.1%, probably botanical misinterpretation, sclareol is part of *Salvia sclarea*, clary sage), hexadecanoic acid (12.5%), globulol (10.4%) and pimaradiene (10.2%).

Commercial oils

The ten chemicals that had the highest maximum concentrations in 55 commercial Dalmatian sage essential oil samples (concentration ranges provided) are the following: α-thujone (14.0-39.9%), camphor (8.5-22.6%), α-humulene (2.5-12.4%), 1,8-cineole (2.1-12.1%), β-thujone (*cis-*) (2.5-9.6%), β-pinene (1-9.2%), camphene (1.2-7.9%), β-caryophyllene (1.9-7.7%), α-pinene (1.1-6.5%) and limonene (0.4-6.2%) (Erich Schmidt, unpublished analytical data).

Chemotypes

Chemotypes of *S. officinalis* with either α-thujone, β-thujone, camphor, or 1,8-cineole as the main constituent or based on their α /β-thujone ratios have been proposed by some authors (51), but the concept was discarded by others (50). In 2014, three chemotypes were presented on the basis of the investigation of 12 Dalmatian sage populations in Montenegro: Chemotype A: high total thujone content, low content of borneol,

Table 5.75.2 Constituents identified in Dalmatian sage oils

Constituent	CAS	Percentage and range				
		A	B (65)	C (48)	D (66)	E
m-Acetanisole	5451-83-2					0.4[r]
2-Acetylpyrrole	1072-83-9					tr[r]
Adamantane	281-23-2					0.6[v3]
α-Amorphene	20085-19-2	0.03-0.3				0.2[w2,y4]; 0.4[w3]; 0.5[w4]; 2.2[w9]
Aromadendrene	489-39-4					1.2[r]; 1.3[c]; 1.8[h]; 1.9[w4]; 3.5[t]; 5.2[w9]
allo-Aromadendrene	25246-27-9					0.3[n]; 0.4[y8]; 0.5[c]; 0.6[e]; 0.7[q]; 1.5[r]
Benzaldehyde	100-52-7					tr[r]; <0.01[w3]
Benzene	71-43-2					0.4[d]
Benzyl alcohol	100-51-6					tr[r]
trans-α-Bergamotene	13474-59-4					0.6[x2]
(*E*)-α-Bergamotol						0.2[c]; 0.4[e]
(*Z*)-α-*trans*-Bergamotol	88034-74-6					0.1[i]
Bergamotol acetate						0.9[u2]
trans-α-Bergamotol acetate						0.2[w]; 0.3[c]; 0.8[e]
Bicyclogermacrene	24703-35-3					0.2[y9]; 0.6[w4]; 4.2[w9]
Bicyclo[3.1.1]heptan-3-one						0.4[z7]
(*Z*)-α-Bisabolene	29837-07-8					0.9[v6]
β-Bisabolene	495-61-4					0.4[l]; 0.6[q]; 0.8[v6]
(*E*)-γ-Bisabolene	53585-13-0					0.2[i]
α-Bisabolol	515-69-5					0.1[y3]; 0.2[c]; 0.3[e]
epi-α-Bisabolol	78148-59-1					1.6[i]
β-Bisabolol	15352-77-9					0.5[c]; 0.8[e]
α-Bisabolol acetate					0-0.2	0.2[f]
Bornanedione						0.6[v3]
Borneol	507-70-0	0.8-4.6	0.5-7.9	1.6-11.8	0.9-6.3	8.8[f]; 9.7[x4]; 11.1[z8]; 13.4[w1]; 14.6[x2]; 15.6[g]; 16.4[w7]; 16.5[v9]; 16.9[p]; 23.9[v8]

Table 5.75.2 Constituents identified in Dalmatian sage oils (*continued*)

Constituent	CAS	Percentage and range				
		A	B (65)	C (48)	D (66)	E
Bornyl acetate	76-49-3	0.3-2.9	tr-2.0	0.1-7.8	0.3-6.0	2.6[z6]; 2.7[z9]; 2.9[w9]; 3.1[h]; 3.2[o]; 3.6[f]; 4.9[d]; 5.3[w7]; 5.7[k]; 5.9[v8]; 6.8[v9]; 10.3[p]
β-Bourbonene	5208-59-3					tr[r]; 0.04[w3]; 0.1[q]; 0.2[c]; 0.3[e]; 0.4[i]
n-Butyl acetate	123-86-4					0.04[w]; 0.2[w1]; 0.7[x5]
Cadalene	483-78-3					0.1[c]
trans-Cadina-1(6),4-diene	931410-54-7					0.2[r]
α-Cadinene	24406-05-1					tr[x3]; 0.1[c]; 0.2[e]; 0.3[q]
β-Cadinene	523-47-7					0.1[o]
γ-Cadinene	39029-41-9				0-tr	0.2[q]; 0.3[w3]; 0.4[c]; 0.6[e]; 1.0[i]; 4.9[l]
δ-Cadinene	483-76-1	0.02-0.4	tr-1.5	0-0.2		1.9[k]; 2.0[e]; 2.3[w9]; 3.3[y2]; 3.5[i]; 3.8[v7]
α-Cadinol	481-34-5			0-0.8		tr[q]; <0.1[j]; 0.2[t]; 0.3[c]; 0.4[e]
epi-α-Cadinol	5937-11-1			0-0.9		tr[q]
α-Calacorene	21391-99-1					tr[q]; 0.1[x2]; 0.2[c]; 0.3[e]; 0.4[i]
β-Calacorene	50277-34-4					tr[q]; 0.1[c]
Calamenene	483-77-2					0.09[w3]
cis-Calamenene	72937-55-4					0.1[c]; 0.2[q]; 0.3[v3]
trans-Calamenene	73209-42-4					tr[v]; 0.2[i,x3]
Calarene ((+)-)	17334-55-3					0.1[w2]
Camphene	79-92-5	1.2-7.9	0.5-5.9	0.1-7.1	1.0-9.9	5.9[g]; 6.6[m]; 6.9[w1]; 7.3[t]; 7.6[z9]; 8.5[w7]; 8.6[v8]; 8.8[x]; 9.1[z5]; 9.3[f]; 10.3[k]; 10.6[x1]
α-Campholenal	4501-58-0				tr-0.6	0.08[w1]; 0.1[v]; 0.5[x7]; 0.9[x5]; 1.1[x2]
Camphor	76-22-2	8.5-22.6	3.2-12.3	11.3-29.8	5.2-36.5	27.0[g]; 29.0[x1]; 29.1[h]; 29.2[x6]; 32.6[r]; 33.2[y7]; 35.4[f]; 45.7[z3]; 46.3[x5]; 50.2[v8]
δ3-Carene	13466-78-9					tr[j]; 0.02[x5]; 0.05[q]; 0.5[h]
Carvacrol	499-75-2			0-3.4		tr[j]; 0.06[z7]; 0.1[e]; 0.2[c]; 0.3[v]; 0.8[s]
Carveol	99-48-9					tr[q]; 0.04[w3]; 0.2[h]
(E)-Carveol	1197-07-5					tr[d,x3]; 0.1[c]; 0.2[j]
(Z)-Carveol	1197-06-4				0-0.1	tr[c,x3]
Carvone	99-49-0					<0.1 (ref.?); 0.2[y5]; 1.4[u2]
cis-Carvone oxide						tr[r]
Carvotanacetone	499-71-8					tr[r]; 0.3[u]
Carvyl acetate	97-42-7					0.1[c]
cis-Carvyl acetate	1205-42-1					<0.01[w]; 0.05[w1]
trans-Carvyl acetate	1134-95-8					tr[x3]; 0.03[w1]; 0.2[w]
Caryophylla-4,8-dien-5-ol	423765-30-4					0.3[c]; 0.4[e]
Caryophylla-4(12),8(13)-dien-5β-ol	19431-80-2					0.4[d]
β-Caryophyllene	87-44-5	1.9-7.7	1.5-15.8	0.9-7.5	tr-1.5	9.4[d]; 9.7[v8]; 10.6[x2]; 10.7[x4]; 10.8[e]; 11.2[z3]; 11.3[y4]; 11.5[w4]; 11.8[r]; 12.2[w9]
γ-Caryophyllene (cis-)	118-65-0					tr[d]; 0.1[c]
9-epi-(E)-Caryophyllene	68832-35-9				0-tr	
Caryophyllene oxide	1139-30-6	0.04-0.5	tr-1.7	0.1-4.3	tr-0.4	1.3[r]; 1.6[t]; 2.1[p]; 2.2[v8]; 2.4[z3]; 2.9[w4]
Caryophyllenol	38284-26-3					tr[d]; 0.4[v6]
Caryophyllenyl alcohol	913176-41-7					tr[r]
8,14-Cedranoxide	18319-31-8					0.3[s]
α-Cedrene	469-61-4					0.4[q]
β-Cedrene	546-28-1					0.2[r,x2]
1,8-Cineole	470-82-6	2.1-12.1	6.7-20.5	2.7-45.3	5.0-15.7	19.4[g]; 20.9[k]; 22.5[w7]; 33.3[q]; 36.4[x5]; 43.6[w8]; 54.6[x1]; 56.3[z3]; 58.4[x]; 67.1[v]
Cinnamyl alcohol	104-54-1					2.2[x5]
Citronellal	106-23-0			0-0.2		
Citronellyl propionate	141-14-0					0.6[x2]; 1.2[o]
α-Copaene	3856-25-5	0-0.2			0-0.1	0.7[e]; 0.8[n]; 0.9[y2]; 1.1[w9]; 1.2[g]; 1.8[i]
β-Copaene	18252-44-3			tr-1.6	0-0.3	
Cresol	1319-77-3					0.3[v3]

Table 5.75.2 Constituents identified in Dalmatian sage oils (*continued*)

Constituent	CAS	Percentage and range				
		A	B (65)	C (48)	D (66)	E
p-Cresol	106-44-5					0.1[r]
α-Cubebene	17699-14-8		tr			tr[q]; 0.08[w3]; 0.1[c]; 0.2[e]; 0.7[y2]
β-Cubebene	13744-15-5					tr[c]; 0.03[w3]; 0.1[q]; 0.2[k]; 0.4[w4]; 1.2[w9]
Cubenene	29837-12-5					0.2[c]; 0.4[q]
1,10-di-epi-Cubenol	73365-77-2					0.3[c]; 0.5[e]
Cuminyl alcohol	536-60-7					0.1[c]
o-Cymene	527-84-4					0.03[z7]; 0.04[r]; <0.05[x6]; 0.3[v6]; 0.9[u2]
p-Cymene	99-87-6	0.08-2.5	tr-0.6	0.1-1.0	0.5-1.3	1.5[z5]; 1.7[v]; 1.8[w7]; 1.9[d]; 2.8[f]; 11.8[z2]
p-Cymenene	1195-32-0					0.1[l]; 0.2[m]
p-Cymen-8-ol	1197-01-9				tr-0.2	tr[q,w3]; 0.1[c]; 0.2[x5]; 0.3[e]; 0.4[v3]; 1.0[r]
Damascone (β-)	23726-91-2					0.09[x5]
Decahydronaph-thalene	91-17-8					0.8[y9]
n-Decane	124-18-5					<0.01[w]; 0.04[w1]
Dehydrosabina ketone	147043-52-5					0.3[s]
cis-Dihydrocarvone	3792-53-8					0.4[x5]
Dihydrolinalyl acetate	61476-73-1				tr-0.1	0.1[e]
2,5-Dimethylstyrene	2039-89-6					0.2[q]
Eicosane	112-95-8					1.5[v3]; 2.0[v6]
β-Elemene	33880-83-0					0.06[x5]; 0.09[o]; 2.0[x7]; 3.7[y1]; 4.8[q]
γ-Elemene	29873-99-2					0.5[x2]; 0.8[l]; 6.5[p]
δ-Elemene	20307-84-0					0.07[w]; 0.1[w1]
Elemol	639-99-6					0.6[q]
Epilaurene	18452-45-4					0.4[c]; 0.6[e]
Ethanol	64-17-5					0.2[x5]
Ethyl (*E*)-cinnamate	4192-77-2					0.5[u2]
2-Ethylhexanoic acid	149-57-5					0.1[i]
2-Ethylpyrazine	13925-00-3					tr[r]
α-Eudesmol	473-16-5					tr[d]
β-Eudesmol	473-15-4					tr[d,x2]; 0.3[v3]
Eugenol	97-53-0					tr[d]; 0.01[w3]; 0.1[c]; 0.2[e]; 0.3[x5]
Eugenyl acetate	93-28-7					0.8[x5]
α-Farnesene	502-61-4					1.2[o]
(*E,E*)-α-Farnesene	502-61-4					0.3[y1]; 2.5[v6]
(*Z*)-β-Farnesene	28973-97-9					0.3[n]
Farnesol	4602-84-0					0.9[x5]
(*Z,E*)-Farnesol	3790-71-4					1.2[i]
(*Z,Z*)-Farnesol	16106-95-9					0.3[i]; 0.7[n]
Farnesyl acetate	29548-30-9					0.2[v3]
(*Z,E*)-Farnesyl acetate	40266-29-3					0.2[i]
Fenchene	471-84-1					<0.01[w3]
α-Fenchene	471-84-1	tr-0.3				tr[q]
endo-Fenchol (α-)	512-13-0					0.2[y]
Fenchone	1195-79-5					0.2[i,y1]
Fenchyl acetate	13851-11-1					1.6[q]
α-Fenchyl acetate	111821-74-0					0.2[i]; 2.2[y1,y5]
Fenchyl alcohol	1632-73-1					0.1[x5]
Ferruginol	514-62-5					0.5[c]
cis-Ferruginol						0.9[e]
trans-Ferrugional						0.4[d]
β-Funebrene	79120-98-2					0.1[w2]
Furfural	98-01-1					tr[r]
Geraniol	106-24-1					<0.1[j]
Geranyl acetate	105-87-3					tr[d]; 0.06[x5]; 0.4[j]; 0.6[h]
Geranyllinalool	77368-82-2					0.5[v3]
Geranyl 2-methylbu-tyrate	68705-63-5					0.08[x5]
Germacrene	28028-64-0					0.2[z7]
Germacrene B	15423-57-1					tr[q]
Germacrene D	23986-74-5					0.1[c]; 0.4[w1]; 1.0[w9]; 1.2[w4]; 2.5[x7]
Germacrene D-4-ol	198991-79-6					0.1[c]
Globulol	489-41-8					0.1[w2]; 3.0[x9]; 3.3[u2]; 10.4[x2]

Table 5.75.2 Constituents identified in Dalmatian sage oils (*continued*)

Constituent	CAS	Percentage and range				
		A	B (65)	C (48)	D (66)	E
o-Guaiacol	90-05-1					0.1[r]
α-Guaiene	3691-12-1					0.02[z7]; 0.1[c]
cis-β-Guaiene	372162-07-7					0.4[c]
trans-β-Guaiene	192053-49-9					0.1[c]
Guaiol	489-86-1					3.3[m]
α-Gurjunene	489-40-7					tr[r]; 0.1[c]; 0.3[w4]; 0.6[t]; 0.8[q]; 1.1[w9]
β-Gurjunene	73464-47-8					0.2[c]; 0.3[e,v3]; 0.5[q]
γ-Gurjunene	22567-17-5					0.1[c,x2]; 1.2[w8]
δ-Gurjunene	1786-19-2					8.2[z8]
Heptadecane	629-78-7					0.1[v6]
2-*n*-Heptylfurane	3777-71-7					1.7[r]
Hexadecane	544-76-3					1.1[l]
Hexadecanoic acid	57-10-3					12.5[v3]
(*E*)-2-Hexenal	6728-26-3					0.2[x5]
(*E*)-2-Hexen-1-ol	928-95-0					tr[x3]
(*Z*)-3-Hexen-1-ol	928-96-1					tr[x3]; 0.02[x5]
(*E*)-2-Hexenyl isovalerate					0-0.1	
Humuladienone	24405-90-1					3.0[h]
α-Humulene	6753-98-6	2.5-12.4	2.0-16.0	0.4-8.5	0.6-2.2	11.6[w1,w4]; 12.5[d]; 12.7[i]; 13.6[h]; 14.0[z4]; 14.9[x4]; 17.3[w9]; 19.0[k]; 33.2[y2]
α-Humulene epoxide	96638-51-6	0.03-0.5		tr-3.5		0.4[m]; 1.0[t]; 1.2[w9]; 1.4[y]; 2.2[y5]; 2.3[w4]
Humulene (ep)oxide I	19888-33-6				tr-0.3	
Humulene (ep)oxide II	19888-34-7		0.7-3.3		0.4-1.2	0.7[x3]; 0.9[n]; 1.0[f]; 1.6[d]; 2.6[c]; 3.7[e]
Humulene (ep)oxide III	21624-36-2					<0.05[x6]
Hydroxycaryophyllene	78683-81-5					0.3[o]
14-Hydroxy-9-epi-β-caryophyllene					0-0.2	0.5[c]; 0.6[m]; 0.9[d]
14-Hydroxy-α-humulene	108043-85-2					0.1[c]
14-Hydroxy-α-muurolene	105661-29-8					0.2[e]
Isoborneol	124-76-5					0.05[r]; 0.2[t]; 2.4[w7]; 4.5[v7]
Isobornyl acetate	125-12-2					0.6[x2,x3]; 2.2[u2]
Isobornyl formate	1200-67-5				0-tr	
α-Isocomene	65372-78-3					tr[r]
(*E*)-Isoeugenol	5932-68-3					tr[r]
Isoledene	95910-36-4					tr[r]
Isomenthone	491-07-6					0.7[u2]
Isopimara-9(11),15-diene	39702-28-8				0-tr	
Isopinocamphone	15358-88-0					0.04[w]; 0.1[o]; 0.2[z7]; 0.3[g]; 0.4[w1]
Isopulegol	89-79-2					0.2[m]
Isoterpinolene	586-63-0					0.3[y3]
Iso-3-thujanol	7712-79-0				tr-0.5	<0.1[j]; 3.1[f]
Iso-3-thujanol acetate					0-tr	
Jasmone (*Z*-)	488-10-8					1.0[u2]
(*E*)-Jasmone	6261-18-3					tr[r]
(*Z*)-Lanceol	859202-95-2					0.2[i,s]
Ledol	577-27-5					0.3[y6]; 0.6[l]; 0.8[t]; 4.2[y2]
Limonene	138-86-3	0.4-6.2			1.1-4.2	2.4[m]; 3.6[w7]; 4.2[y]; 6.5[f]; 6.6[y4]; 7.6[k]
Linalool	78-70-6	0.08-0.7	tr-0.8			0.4[t]; 0.5[w6]; 1.1[x8]; 1.8[57]; 2.0[y4]; 4.7[w7]
cis-Linalool oxide	11063-77-7					0.1[t]; 0.2[w]; 0.3[w1]
trans-Linalool oxide	11063-78-8					tr[r]; 0.5[x5]
Linalyl acetate	115-95-7	0-0.1				0.3[x7]; 0.4[q]; 0.7[x8]; 0.8[w6]; 1.0[z6]; 3.5[w7]
Longifolene (junipene)	475-20-7					0.2[q]; 0.8[v6]; 1.0[v3]
β-Longipinene	41432-70-6					0.2[r]
Lyral [b]	130066-44-3					0.1[b,m]
Maltol	118-71-8					tr[r]
Manool	596-85-0	tr-4.9	3.5-10.4		0.3-1.2	8.9[z]; 9.2[o]; 10.4[c]; 13.1[w3]; 13.3[h]; 14.4[v3]; 14.6[x4]; 14.7[y5]; 15.9[d]; 20.9[e]

Table 5.75.2 Constituents identified in Dalmatian sage oils (*continued*)

Constituent	CAS	Percentage and range				
		A	B (65)	C (48)	D (66)	E
13-epi-Manool						0.1[c]; 0.4[e]; 1.5[x9]
Manoyl oxide	596-84-9					0.5[c]; 0.8[e]
cis-p-Menth-2-en-1-ol	29803-82-5				0-0.1	
Menthol	89-78-1					0.3[y3]; 0.7[x6]
Methanol	67-56-1					0.07[x5]
Methyl chavicol	140-67-0					0.3[u2]
Methyl eugenol	93-15-2					0.7[x5]
5-Methylfurfural	620-02-0					tr[r]
Methyl hexadeca-noate	112-39-0					0.3[y5]; 7.1[v3]
2-Methyl-4-nitroso-resorcinol	65882-00-0					4.2[v3]
Methyl oleate	112-62-9					0.1[y5]
Methyl palmitoleate	1120-25-8					0.06[y5]
Methylpyrazine	109-08-0					tr[r]
Muurola-4,10(14)-dien-1β-ol					0.2-0.8	0.4[f]
Muurolene	69671-15-4					1.5[k]
α-Muurolene	10208-80-7					0.2[w3]; 0.3[c]; 0.4[e]; 1.7[i]; 2.0[n]; 4.3[y4]
γ-Muurolene	30021-74-0				0-0.3	1.2[e]; 1.3[n]; 2.0[y2]; 2.9[i]; 9.3[d]; 13.8[p]
α-Muurolol	104245-48-9					0.1[c]; 1.4[q]
γ-Muurolol	138068-73-2		tr-1.2			
τ-Muurolol (epi-α-)	19912-62-0					tr[q]; 0.2[c]; 0.5[e]
Myrcene	123-35-3	0.5-3.2	tr-0.6	0-4.2	0.4-2.0	1.5[v4]; 1.6[z5]; 2.1[v]; 2.5[z7]; 3.1[k]; 3.8[w2]
Myrtanol	514-99-8					1.0[m]
Myrtenal	564-94-3					tr[q]; 0.05[w3]; 0.1[u]
Myrtenol	515-00-4				tr-0.5	0.2[g]; 0.3[c]; 0.4[x3]; 0.5[x8]; 0.7[s]; 1.0[h]
Myrtenyl acetate	1079-01-2				0-0.2	<0.05[x6]; 0.1[r]; 3.7[y2]
Naphthalene	91-20-3					0.9[z7]
Neomenthol	3623-51-6					0.7[v6]
Neothujan-3-ol acetate					0-tr	
Neral	106-26-3			0-3.5	0-tr	tr[x3]; <0.1[j]
Nerol	106-25-2					0.2[x7]; 0.3[x5]
(Z)-Nerolidyl acetate	91050-14-5				0-0.1	
Neryl acetate	141-12-8					tr[d]; 0.04[w]; 0.1[w1,x5]
Nonadecane	629-92-5					0.2[v3]; 0.9[x5]
(E)-α-Ocimene	6874-10-8					1.7[o]
(E)-β-Ocimene	3779-61-1	0-1.0	tr-1.7	0-0.2		0.2[x5]; 0.3[g]; 0.4[i]; 0.7[d]; 0.9[x7]; 1.3[x2]
(Z)-β-Ocimene	3338-55-4	0.02-0.6	tr-5.0	0-0.4		0.09[r]; 0.1[i,y1]; 0.4[t]; 0.5[q]; 1.1[z4]; 1.8[d]
allo-Ocimene	673-84-7					0.2[y3]
n-Octadecane	593-45-3					0.5[v3]
Octadecanoic acid	57-11-4					4.3[v3]
Octanal	124-13-0					0.08[x5]
3-Octanal						tr[x3]
Octanol	29063-28-3					0.06[x5]
3-Octanol	589-98-0					tr[r]; 0.02[w3]; 0.1[i]
3-Octanone	106-68-3					0.02[w3]; 0.2[r]
1-Octen-3-ol	3391-86-4	0.02-0.2				0.08[w3]; 0.2[u]; 0.3[t]; 0.5[y9]; 0.7[k]; 8.5[o]
β-Oplopenone	28305-60-4					0.2[c]; 0.3[e]
6-Oxobornyl acetate						0.1[o]
Palustrol	5986-49-2					<0.3[w4,w9]; 0.9[y2]; 1.1[t]
Pentadecane (n-)	629-62-9					0.4[v6]
α-Phellandrene	99-83-2	0.05-0.8		0-0.1	0.9-1.3	0.1[q]; 0.2[f]; 0.3[x3]; 1.1[j]; 1.2[y5]; 4.5[x9]
β-Phellandrene	555-10-2					0.1[y4,z7]
L-Phellandrene	6153-17-9					0.2[o]
Phenylacetaldehyde	122-78-1					tr[r]; 0.04[w3]
Pimaradiene	1686-61-9					0.3[c]; 0.5[e]; 10.2[r]
α-Pinene	80-56-8	1.1-6.5	1.2-5.9	0.1-6.4	tr-0.1	5.7[z3]; 6.6[w7]; 6.7[v2]; 7.1[v9]; 7.2[k]; 7.3[g]; 7.5[h]; 8.0[v8]; 8.2[v]; 8.7[x1]; 22.3[x]; 24.8[x1]
β-Pinene	127-91-3	1.0-9.2	1.2-5.3	0-4.9	1.2-11.9	9.9[z8]; 10.1[v4]; 11.3[y2]; 11.6[v9]; 13.7[k]; 14.0[t]; 15.0[w9]; 16.4[x2]; 17.8[u2]; 19.0[w4]
cis-Pinene hydrate	17974-51-5					0.05[u2]

Table 5.75.2 Constituents identified in Dalmatian sage oils (*continued*)

Constituent	CAS	Percentage and range				
		A	B (65)	C (48)	D (66)	E
Pinocamphone	547-60-4					tr[x6]; 0.2[i]; 0.3[w3]; 1.1[m]; 2.0[y2]
cis-Pinocamphone						0.1[c]; 0.3[d]; 0.5[z]
trans-Pinocamphone	547-60-4				0-0.5	0.1[e]; 0.2[c]; 0.3[d]; 0.4[f]; 1.8[t]
Pinocarveol	5947-36-4					6.7[z2]
trans-Pinocarvyl acetate	1686-15-3				0-0.7	tr[x3]
cis-Piperitol	16721-38-3					0.2[j]
trans-Piperitol	16721-39-4					tr[x3]; <0.1[j]
Piperitone	89-81-6					1.1[u2]
Pulegone	89-82-7					0.9[y5]
Pyrazine, 2-ethyl-3-methyl-	15707-23-0					tr[r]
Quinolinone	59-31-4					0.3[v3]
Rimuene	1686-67-5					0.8[c]; 1.1[e]
Sabinene	3387-41-5	0-0.7		0-0.2		0.4[w1]; 0.7[k]; 0.8[x5]; 3.9[u2]; 7.0[w8]; 8.2[x9]
Sabinene hydrate	546-79-2					0.5[v3]; 2.3[u1]
cis-Sabinene hydrate	15537-55-0	tr-0.4				0.07[z7]; 0.1[z]; 0.2[f]; 0.3[w3]; 0.4[r]
trans-Sabinene hydrate	17699-16-0	0-0.3	tr		0.4-1.1	0.04[o]; 0.2[g,x5]; 0.3[r]; 0.4[x2]
trans-Sabinene hydrate acetate	77318-47-9					tr[x3]
cis-Sabinol	3310-02-9					2.9[y2]
trans-Sabinol	471-16-9					0.3[u2]
Sabinyl acetate	53833-85-5					0.4[y3]
cis-Sabinyl acetate	53833-85-5			0-0.4	0-0.1	0.2[w]; 0.4[w1]
trans-Sabinyl acetate	139757-62-3					0.1[c]; 0.2[e]; 0.4[f]; 0.6[d]; 0.8[f]
Salvene	33746-69-9			0-0.6		0.3[m,y9]; 0.9[v4]; 1.1[y3]
cis-Salvene		tr-0.8			0.2-0.7	0.3[d]; 0.4[w]; 0.6[y8]; 0.8[k]; 1.2[f]; 2.7[w1]
trans-Salvene	33746-69-9	tr-0.1			tr-0.1	tr[d]; 0.04[w]; 0.06[o]; 0.1[y8]; 0.2[f]; 0.3[w1]
α-Santalol						0.3[m,o]
(Z)-α-Santalol acetate						0.1[x2]
Sclarene	511-02-4					0.9[u2]; 1.9[u2]
Sclareol	515-03-7					0.1[v6]; 0.2[z9]; 1.2[l]; 2.0[i]; 5.2[y2]; 23.1[p]
Selina-3,11-dien-6α-ol	75521-07-2					1.6[c]; 2.6[e]
α-Selinene	473-13-2					0.06[x5]; 0.09[w]; 0.3[t]
β-Selinene	17066-67-0					tr[d]; 0.05[q]; 0.08[z7]; 0.1[c]; 1.2[w2]
Selin-11-en-4-ol (α-)	16641-47-7				0.2	0.2[f]; 5.4[y6]
(E)-Sesquilavandulol	104121-84-8					5.0[y4]
β-Sesquiphellandrene	20307-83-9					0.1[y4]
cis-Sesquisabinene hydrate	58319-05-4					0.1[c]
Spathulenol	6750-60-3			0-0.2		tr[r]; <0.1[j]; 0.1[x2]; 0.2[w2]; 0.6[c]; 0.9[l]
Syringol	91-10-1					0.5[r]
α-Terpinene	99-86-5	0.04-0.4	tr	0-0.5	0.1-0.3	0.2[j]; 0.3[d]; 0.4[f]; 0.5[w3]; 0.8[w2]; 1.1[x5]
β-Terpinene	99-84-3					0.2[y2]
γ-Terpinene	99-85-4	0.1-3.4	tr-0.5	0-0.7	0.2-0.4	0.7[f]; 0.9[w3]; 1.0[k]; 1.1[q]; 1.2[z7]; 1.4[w2]
δ-Terpinene	586-62-9					0.1[o]
Terpinen-4-ol	562-74-3	0.2-0.9	tr-0.6	0.1-2.1	0.4-0.7	0.7[m]; 1.0[r]; 1.1[z7]; 1.3[w2]; 2.3[x7]; 2.4[x5]
α-Terpineol	98-55-5	0.02-0.7	tr	0-2.3	0.1-0.5	0.6[j]; 0.7[z5]; 0.8[r]; 1.2[v]; 3.2[w8]; 4.5[w2]
δ-Terpineol	7299-42-5					0.8[v]
Terpinolene	586-62-9	0.06-0.9	tr	0-0.5	0.2-0.5	0.2[g]; 0.3[j]; 0.4[d]; 0.5[w]; 0.6[f]; 1.0[k]
α-Terpinyl acetate	80-26-2					0.1[j]; 0.4[v]; 0.5[z2]; 0.9[z6]; 1.0[x5]; 1.5[w2]
β-Thujaplicinol	4356-35-8					tr[r]
α-Thujene	2867-05-2	0.05-0.5		0-0.2	0-0.4	0.8[z7]; 1.4[w7]; 1.6[i]; 1.8[w2]; 2.8[y2]; 4.1[j]
α-Thujone (*cis*-)	546-80-5	14.0-39.9	1.3-36.9	3.0-26.6	11.0-49.7	48.6[v8]; 48.9[x4]; 50[w5]; 53.2[w4]; 55.2[w1]; 57.0[s]; 64.4[v5]; 65.5[z1]; 67.9[v9]
β-Thujone (*trans*-)	471-15-8	2.5-6.9	tr-28.7	1.5-12.9	11.0-49.7	16.4[p]; 16.5[l]; 17.8[z7]; 18.4[q]; 19.1[v8]; 31.9[w7]; 32[w5]; 36.8[k]; 40.1[w6]; 46.9[v9]
Thujyl acetate	72747-24-1					0.2[y8]; 0.3[u2]
Thujyl alcohol	513-23-5					0.2[x6]; 0.4[y3]
Thymol	89-83-8			0-2.0		tr[d,x3]; 0.1[c]; 0.2[m]; 0.4[x5]

Table 5.75.2 Constituents identified in Dalmatian sage oils (*continued*)

Constituent	CAS	Percentage and range				
		A	B (65)	C (48)	D (66)	E
Tricyclene	508-32-7	0.08-0.8	tr	0-0.2	tr	tr[c]; 0.1[j]; 0.2[d]; 0.3[f]; 0.4[t]; 0.8[k]
5-(Trimethylsilyl)furfural						3.7[v3]
Undecanal	112-44-7					1.0[x5]
Undecane (*n-*)	1120-21-4					0.2[w]
2-Undecanone	112-12-9					2.4[x5]
Valencene	4630-07-3					0.06[x5]; 0.1[u2]; 0.2[w2]; 0.3[o]
Verbenene	4080-46-0					0.3[t]; 1.6[m]
cis-Verbenol	1845-30-3					0.2[x5]; 0.3[t]
trans-Verbenol	1820-09-3					0.2[v3]; 0.4[t]
Verbenone	80-57-9					0.1[t,v3,x8]
Viridiflorene (ledene)	21747-46-6	0.03-0.2			tr-0.2	0.7[s]; 0.8[r]; 1.3[c]; 1.5[e]; 2.5[v2]; 2.7[h]
Viridiflorol	552-02-3	0.07-2.7	3.0-20.4	1.1-15.7	1.5-3.4	12.3[x5]; 13.5[y5]; 14.2[s]; 16.5[e]; 16.9[x4]; 19.0[v4]; 19.5[w3]; 24.0[h]; 26.1[d]; 30.3[v3]
Widdrol	6892-80-4					0.4[w]; 1.0[w1]
α-Ylangene	14912-44-8					<0.1[x2]; 0.1[q]; 0.2[c]; 0.3[e]; 0.6[i]
Zonarene	41929-05-9				0-0.5	

A fifty-five Dalmatian sage essential oil samples from Albania, Bulgaria, Egypt, Morocco, Tunisia, Bosnia-Herzegovina, Germany, Hungary and France, analyzed between 1998 and 2013; lowest and highest concentrations given (E. Schmidt, unpublished data)

B 38 lab-hydrodistilled oils from the aerial parts at the beginning of the flowering period harvested in two years of nineteen *S. officinalis* accessions cultivated in Germany of material from a gene bank, originating from ten countries; lowest and highest concentrations given (ref. 65)

C twelve *Salvia officinalis* oils lab-hydrodistilled from commercial sage leaves from nine European countries obtained in retail pharmacies; lowest and highest concentrations given (ref. 48)

D 25 lab-hydrodistilled Dalmatian sage leaf oils from plants cultivated from the branches of 25 indigenous populations in Croatia (n=23) and Bosnia and Herzegovina (n=2); the biomass was harvested in the dormant vegetative state; lowest and highest concentrations given (ref. 66)

E data from other studies (indicated with superscript letters); highest concentrations found in any study reviewed here given; when two or more oils were investigated, only the highest concentrations are mentioned, unless indicated otherwise

[a] incorrect identification; [b] synthetic fragrance material; [c] five lab-hydrodistilled oils from the aerial parts of *S. officinalis* harvested in five gardens in the Vilnius district of Lithuania (ref. 73); [d] eleven lab-hydrodistilled leaf oils from wild populations of *S. officinalis* growing in Serbia and Montenegro; flowers were also analyzed (data not shown here) (ref. 64); [e] eight lab-hydrodistilled oils from the aerial parts of wild sage populations growing in eastern Lithuania (ref. 59); [f] twelve lab-hydrodistilled oils from the leaves of 12 populations of wild sage collected in Montenegro (ref. 60); [g] six leaf oils from Polish *S. officinalis* cv. Bona from first and second crop in three successive years (ref. 9); [h] eight lab-hydrodistilled oils from wild sage collected at two locations in Bosnia and Herzegovina and analyzed in four stages (vegetative, pre-flowering, flowering, after flowering) (ref. 10); [i] one lab-hydro- and one laboratory steam-distilled oil from the leaves of sage collected in the wild in Serbia during the flowering phase; analyses were also made of the flower and the stem oils (data not presented here) (ref. 11); [j] one lab-hydrodistilled oil from flowers collected in the garden of an agricultural high school at Reunion Island in the flowering phase and analyzed by GC on two columns (ref. 13); [k] eight lab-hydrodistilled oils from 8 cultivars grown in the USA and harvested as the top 13 cm of non-flowering plants (ref. 17); [l] one lab-hydrodistilled oil sample from the aerial parts of *S. officinalis* collected in the wild in Italy (ref. 21); [m] one commercial sage oil obtained from an Italian company (refs. 22,27); [n] one lab-hydrodistilled oil from wild sage from Serbia (ref. 31); [o] one lab-hydrodistilled oil from commercial plant material from a local Romanian supplier (ref. 32); [p] eighteen lab-hydrodistilled *S. officinalis* oils from plants cultivated at 13 experimental sites in Italy (ref. 33); [q] one lab-hydrodistilled oil from flowering aerial parts of sage cultivated in a home garden in Tunisia (ref. 35); [r] two lab-hydrodistilled oils, one from sage grown in Italy and one from plants cultivated by micropropagation from the mother plant from the former cultivars (ref. 36); [s] two lab-hydrodistilled oils from wild Croatian populations, collected after flowering (ref. 38); [t] three lab-hydrodistilled oils from flowering wild growing sage from Portugal (ref. 42); [u] one commercial oil purchased in Portugal (ref. 42); [u1] 580 lab-hydrodistilled oils from sage collected in 8 zones of Albania and in eight different months of one year; the results were given as average per month per zone (ref. 12); [u2] data from various studies, cited in ref. 95; [v] four lab-hydrodistilled oils from commercial dried aerial parts of sage obtained from a local herbal shop in Portugal; four different distillation times were used (30 minutes to 3 hours) (ref. 46); [v1] one lab-hydrodistilled oil from Argentinian *S. officinalis* (ref. 41); [v2] one laboratory steam-distilled oil from flowering sage cultivated in Spain (ref. 43); [v3] one laboratory steam-distilled oil of Dalmatian sage cultivated in Tunisia under experimental conditions, with an atypical composition (ref. 44); [v4] four steam-distilled oils from sage cultivated in Italy and harvested at various developmental stages (ref. 45); [v5] eighteen oils from sage leaves and whole plants cultivated in Israel with nine harvest dates and two cutting heights; data were presented as the means of the nine harvests; data from oils of leaf primordia, inflorescences and stems were also shown but are not presented here (ref. 49); [v6] one lab-hydrodistilled leaf oil from wild Serbian sage (ref. 53); [v7] one commercial oil of sage obtained from a USA company (ref. 55); [v8] 25 lab-hydrodistilled oils from commercial sage leaves from Germany, Turkey and Austria (ref. 61); [v9] literature data cited in ref. 61; [w] one lab-hydrodistilled oil from leaves of *S. officinalis* cultivated in Portugal (ref. 50);

Table 5.75.2 Constituents identified in Dalmatian sage oils (*continued*)

[w1] twelve oils from the aerial parts of sage harvested in the wild in Portugal in six different months and two successive years (ref. 50); [w2] one lab-hydrodistilled oil from the aerial parts of wild Greek flowering sage (ref. 52); [w3] four lab-hydrodistilled oils from the flowering aerial parts of *Salvia officinalis* cultivated in four Tunisian regions (ref. 54); [w4] thirty lab-hydrodistilled oils from the 'total herb' of sage cultivated at two sites in New Zealand and harvested each month for 15 months (ref. 51); [w5] 67 laboratory steam-distilled oils from the stems and leaves of sage accessions cultivated in New Zealand and originating from 8 different countries; only minimum and maximum levels for the thujones were presented (ref. 67); [w6] seven lab-hydrodistilled oils from sage growing wild in the former Yugoslavia and harvested 7 times between June and December (ref. 57); [w7] thirteen steam-distilled oils from wild Yugoslavian sage leaves collected at three different locations and with various harvesting dates (ref. 58); [w8] one lab-hydrodistilled sage leaf oil from plants growing wild in Lebanon (ref. 62); [w9] oils from thirteen *S. officinalis* accessions cultivated in New Zealand, 9 from flowering and 4 from non-flowering cultivars and harvested after full flowering; results were expressed as means for the 9 flowering and for the 4 non-flowering accessions (ref. 63); [x] eighteen lab-hydrodistilled oils from dried commercial leaves obtained in the UK and originating from Italy, Turkey, UK, Dalmatia, Greece and unknown origin; only six components were analyzed (ref. 67); [x1] eleven commercial sage oils from the UK, Dalmatia, Spain and unknown origin obtained in the UK; only six components were analyzed (ref. 67); [x2] five lab-hydrodistilled oils from *S. officinalis* cultivated in Iran and analyzed 5 times during the phenological cycle (ref. 68); [x3] one lab-hydrodistilled oil from the over-ground parts of *S. officinalis* collected at the flowering stage in Algiers, Algeria (ref. 69); [x4] seven lab-hydrodistilled oils from plants cultivated in Turkey and dried in seven different manners (ref. 70); [x5] seven lab-hydrodistilled oils from the aerial parts of sage grown in a house garden in Tunisia and dried in seven different manners (ref. 71); [x6] one lab-hydrodistilled oil from commercial dried sage leaves from Poland (ref. 1); [x7] one lab-hydrodistilled oil from sage in full bloom cultivated in Egypt (ref. 3); [x8] one steam-distilled oil from *Salvia officinalis* cultivated in Turkey (ref. 2); [x9] one lab-hydrodistilled oil from sage collected in the wild in Libya at the full flowering stage (ref. 4); [v] one lab-hydrodistilled sage leaf oil from wild plants collected in Montenegro (ref. 5); [v1] one lab-hydrodistilled leaf oil from sage cultivated in Egypt and harvested in the flowering phase (ref. 7); [v2] three leaf oils from *S. officinalis* cultivars grown in Hungary (ref. 75); [v3] one commercial oil from the Provence, France (ref. 6); [v4] two lab-hydrodistilled oils from the aerial parts of *S. officinalis* collected at two sites in Tunisia (ref. 8); [v5] one lab-hydrodistilled oil from Cuban cultivated sage (ref. 14); [v6] one lab-hydrodistilled sage leaf oil from plants cultivated in Bulgaria and harvested at the flowering stage (ref. 15); [v7] twelve lab-hydrodistilled oils from plants cultivated in Israel from the seeds of wild Dalmatian populations and harvested four times; only the thujones and camphor were analyzed (ref. 16); [v8] unknown number of lab-distilled oils from sage plants cultivated in Italy under different fertilization regimens; results were expressed in w/v % (ref. 18); [v9] one lab-hydrodistilled oil from the aerial parts of sage from south Tunisia (ref. 19); [z] one lab-hydrodistilled oil from leaves and stems of sage grown in a botanical garden in Tunisia (ref. 20); [z1] one lab-hydrodistilled oil from the aerial parts of *S. officinalis* growing wild in south France (ref. 23); [z2] one commercial oil of unknown origin (ref. 24); [z3] eight lab-hydrodistilled sage leaf oils from plants cultivated in Greece from seeds of wild populations (ref. 29); [z4] five lab-hydrodistilled oils from sage cultivated in Italy, planted with different densities and harvested in different months (two in the flowering phase, three in the vegetative state) (ref. 25); [z5] one lab-hydrodistilled sage oil from plants of unknown origin (probably France; wild? cultivated?) (ref. 28); [z6] one steam-distilled oil from wild sage growing in Tunisia (ref. 34); [z7] four lab-hydrodistilled oils from *S. officinalis* harvested in four home gardens in Tunisia; results were expressed as means of the 4 determinations (ref. 37); [z8] one lab-hydrodistilled oil from sage cultivated in Brazil (ref. 29); [z9] one lab-hydrodistilled oil from Albanian sage leaves (ref. 40)

tr: trace (in column B: <0.5%; in columns D, E[c], E[q] and E[x3]: <0.1; in columns E[d] and E[v]: <0.05; in column E[j]: <0.02)

Table 5.75.3 Major constituents of Dalmatian sage oils

Constituent	CAS	Percentage and range				
		A	B (65)	C (48)	D (66)	E
α-Thujone (*cis*-)	546-80-5	14.0-39.9	1.3-36.9	3.0-26.6	11.0-49.7	55.2[w1]; 57.0[s]; 64.4[v5]; 65.5[z1]; 67.9[v9]
1,8-Cineole	470-82-6	2.1-12.1	6.7-20.5	2.7-45.3	5.0-15.7	43.6[w8]; 54.6[x1]; 56.3[z3]; 58.4[x]; 67.1[v]
Camphor	76-22-2	8.5-22.6	3.2-12.3	11.3-29.8	5.2-36.5	33.2[v7]; 35.4[f]; 45.7[z3]; 46.3[x5]; 50.2[v8]
β-Thujone (*trans*-)	471-15-8	2.5-6.9	tr-28.7	1.5-12.9	11.0-49.7	31.9[w7]; 32[w5]; 36.8[k]; 40.1[w6]; 46.9[v9]
α-Humulene	6753-98-6	2.5-12.4	2.0-16.0	0.4-8.5	0.6-2.2	14.0[z4]; 14.9[x4]; 17.3[w9]; 19.0[k]; 33.2[v2]
Viridiflorol	552-02-3	0.07-2.7	3.0-20.4	1.1-15.7	1.5-3.4	19.0[v4]; 19.5[w3]; 24.0[h]; 26.1[d]; 30.3[v3]
α-Pinene	80-56-8	1.1-6.5	1.2-5.9	0.1-6.4	tr-0.1	7.5[h]; 8.0[v8]; 8.2[v]; 8.7[v1]; 22.3[x]; 24.8[x1]
Borneol	507-70-0	0.8-4.6	0.5-7.9	1.6-11.8	0.9-6.3	15.6[g]; 16.4[w7]; 16.5[v9]; 16.9[p]; 23.9[v8]
Manool	596-85-0	tr-4.9	3.5-10.4		0.3-1.2	14.4[v3]; 14.6[x4]; 14.7[v5]; 15.9[d]; 20.9[e]
β-Pinene	127-91-3	1.0-9.2	1.2-5.3	0-4.9	1.2-11.9	14.0[t]; 15.0[w9]; 16.4[x2]; 17.8[u2]; 19.0[w4]
β-Caryophyllene	87-44-5	1.9-7.7	1.5-15.8	0.9-7.5	tr-1.5	11.2[z3]; 11.3[y4]; 11.5[w4]; 11.8[r]; 12.2[w9]
Camphene	79-92-5	1.2-7.9	0.5-5.9	0.1-7.1	1.0-9.9	8.6[v8]; 8.8[x]; 9.1[z5]; 9.3[f]; 10.3[k]; 10.6[x1]
Bornyl acetate	76-49-3	0.3-2.9	tr-2.0	0.1-7.8	0.3-6.0	4.9[d]; 5.3[w7]; 5.7[k]; 5.9[v8]; 6.8[v9]; 10.3[p]
Limonene	138-86-3	0.4-6.2			1.1-4.2	2.4[m]; 3.6[w7]; 4.2[v]; 6.5[f]; 6.6[y4]; 7.6[k]

LEGEND: SEE TABLE 5.75.2

camphor, camphene, and α-pinene; Chemotype B: contents of α- and β-thujone, α-pinene, camphene, and camphor intermediate between those of the oils of chemotypes A and C, with a higher borneol content; Chemotype C: dominated by camphor, camphene and α-pinene; lowest contents of limonene, total thujones, α-humulene and viridiflorol. However, these results may well be explained by the large natural variations found in essential oils rather than distinct chemotypes. An enormous inhomogeneity (variability) has, for example, been found in the oils from a single batch of commercial *Salvia officinalis* (61). Other chemotype classifications have also been suggested (17,66).

In some publications, other chemicals were shown to be the dominant component in *S. officinalis* essential oils, including viridiflorol (8,10,44,64,65), sclareol (33), manool (14,59), α-humulene (63,75,85), linalool, germacrene D, α-pinene, limonene (all cited in ref. 65), and borneol (33, 86 and cited in ref. 65).

Literature on sage oils from before 1997 has been reviewed in ref. 56. Additional analyses of sage essential oils not presented in Table 5.75.2 can be found in refs. 77-93; these studies can all be accessed on-line. Refs. 30,72 and 76 are not discussed, as they provide no qualitative or quantitative data not already shown in Table 5.75.2. In ref. 95 such additional studies are cited; information from this review that *is* relevant, however, is presented in column E[u2].

CONTACT ALLERGY/ALLERGIC CONTACT DERMATITIS

Sage oil, unspecified

General
Contact allergy to and possible allergic contact dermatitis from sage oil has been reported in one publication only. A false-positive patch test reaction due to the excited skin syndrome cannot be excluded.

Case reports
An aromatherapist had chronic hand dermatitis and was patch test positive to 17 of 20 oils used at her work (tested 1% and 5% in petrolatum), including sage oil (100).

Sage oil, Dalmatian
No reports on contact allergy to sage oil, specifically mentioned to be obtained from *Salvia officinalis*, have been found. Allergy to sage oils (botanical source unspecified) is discussed above.

LITERATURE

1 Occhipinti A, Capuzzo A, Arceusz A, Maffei ME. Comparative analysis of α- and β-thujone in the essential oil and supercritical CO2 extract of sage (*Salvia officinalis* L.). J Essent Oil Res 2014;26:85-90

2 Baydar H, Sangun MK, Erbas S, Kara N. Comparison of aroma compounds in distilled and extracted products of sage (*Salvia officinalis* L.). J Essent Oil Bear Plants 2013;16:39-44

3 Khalid KA. Evaluation of *Salvia officinalis* L. essential oil under selenium treatments. J Essent Oil Res 2011;23: 57-60

4 Awen BZ, Unnithan CR, Ravi S, Kermagy A, Prabhu V, Hemlal H. Chemical composition of *Salvia officinalis* essential oil of Libya. J Essent Oil Bear Plants 2011;14:89-94

5 Damjanovic-Vratnica B, Đakov T, Šukovic D, Damjanovic J. Chemical composition and antimicrobial activity of essential oil of wild-growing *Salvia officinalis* L. from Montenegro. J Essent Oil Bear Plants 2008;11:79-89

6 Tognolini M, Barocelli E, Ballabeni V, Bruni R, Bianchi A, Chiavarini M, Impicciatore M. Comparative screening of plant essential oils: Phenylpropanoid moiety as basic core for antiplatelet activity. Life Sciences 2006;78:1419-1432

7 Edris AE, Jirovetz L, Buchbauer G, Denkova Z, Stoyanova A, Slavchev A. Chemical composition, antimicrobial activities and olfactive evaluation of a *Salvia officinalis* L. (sage) essential oil from Egypt. J Essent Oil Res 2007;19:186-189

8 Fellah S, Diouf PN, Petrissans M, Perrin D, Romdhane M, Abderrabba M. Chemical composition and antioxidant properties of *Salvia officinalis* L. oil from two culture sites in Tunisia. J Essent Oil Res 2006;18:553-556

9 Zawiślak G, Dyduch J. The analysis of the content and chemical composition of essential oil in the leaves of sage (*Salvia officinalis* L.) cv. 'Bona' in the second year of cultivation. J Essent Oil Res 2006;18:402-404

10 Maric S, Maksimovic M, Milos M. The impact of the locality altitudes and stages of development on the volatile constituents of *Salvia officinalis* L. from Bosnia and Herzegovina. J Essent Oil Res 2006;18:178-180

11 Velickovic DT, Ristic MS, Randjelovic NV, Smelcerovic AA. Chemical composition and antimicrobial characteristic of the essential oils obtained from the flower, leaf and stem of *Salvia officinalis* L. originating from southeast Serbia. J Essent Oil Res 2002;14:453-458

12 Asllani U. Chemical composition of Albanian sage oil (*Salvia officinalis* L.). J Essent Oil Res 2000;12:79-84

13 Vera RR, Chane-Ming J, Fraisse DJ. Chemical composition of the essential oil of sage (*Salvia officinalis* L.) from Reunion Island. J Essent Oil Res 1999;11:399-402

14 Pino JA, Estarrón M, Fuentes V. Essential oil of sage (*Salvia officinalis* L.) grown in Cuba. J Essent Oil Res 1997;9:221-222

15 Tsankova ET, Konaktchiev AN, Genova EM. Constituents of essential oils from three *Salvia* species. J Essent Oil Res 1994;6:375-378

16 Putievsky E, Ravid U, Sanderovich D. Morphological observations and essential oils of sage (*Salvia officinalis* L.) under cultivation. J Essent Oil Res 1992;4:291-293

17 Tucker AO, Maciarello MJ. Essential oils of cultivars of Dalmatian sage (*Salvia officinalis* L.). J Essent Oil Res 1990;2:139-144

18 Piccaglia R, Marotti M, Galletti GC. Effect of mineral fertilizers on the composition of *Salvia officinalis* oil. J Essent Oil Res 1989;1:73-83

19 Lamari A, Teyeb H, Ben Cheikh H, Douki W, Neffati M. Chemical composition and insecticidal activity of essential oil of *Salvia officinalis* L. cultivated in Tunisia. J Essent Oil Bear Plants 2014;17:506-512

20 Charchari S, Chafaa I, Kassoussi K, Zekri MM, Benhalla A, Boudina N. Glandular trichomes, secretory cavities and essential oil of sage (*Salvia officinalis* L.). J Essent Oil Bear Plants 2010;13:267-274

21 Senatore F, De Fusco R, Grassia A, Moro CO, Rigano D, Napolitano F. Chemical composition and antibacterial activity of essential oils from five culinary herbs of the Lamiaceae family growing in Campania, Southern Italy. J Essent Oil Bear Plants 2003;6:166-173

22 Giamperi L, Fraternale D, Ricci D. The *in vitro* action of essential oils on different organisms. J Essent Oil Res 2002;14:312-318. The same data had been previously reported in ref. 27 by entirely different authors!

23 Foray L, Bertrand C, Pinguet F, Soulier M, Astre C, Marion C, et al. In vitro cytotoxic activity of three essential oils from *Salvia* species. J Essent Oil Res 1999;11:522-526

24 Moretti MDL, Peana AT, Franceschini A, Carta C. In vivo activity of *Salvia officinalis* oil against *Botrytis cinerea*. J Essent Oil Res 1998;10:157-160. Data also shown in ref. 26

25 Piccaglia R, Marotti M, Dellacecca V. Effect of planting density and harvest date on yield and chemical composition of sage oil. J Essent Oil Res 1997;9:187-191

26 Carta C, Moretti NDL, Peana AT. Activity of the oil of *Salvia officinalis* L. against *Botrytis cinerea*. J Essent Oil Res 1996;8:399-404. Data also presented in ref. 24

27 Baratta MT, Dorman HJD, Deans SG, Biondi DM, Ruberto G. Chemical composition, antimicrobial and antioxidative activity of laurel, sage, rosemary, oregano and coriander essential oils. J Essent Oil Res 1998;10:618-627. The data were later also reported in ref. 22 by entirely different authors!

28 Milhau G, Valentin A, Benoit F, Mallié M, Bastide J-M, Pélissier Y, Bessière JM. In vitro antimalarial activity of eight essential oils. J Essent Oil Res 1997;9:329-333

29 Kanias GD, Souleles C, Loukis A, Philotheoupanou E. Statistical study of essential oil composition in three cultivated sage species. J Essent Oil Res 1998;10:395-403

30 Altindal N, Altindal D. Allelopathic effect of sage and Turkish oregano volatile oils on *in vitro* Sainfoin (*Onobrychis viciifolia*). J Essent Oil Bear Plants 2013;16:328-333

31 Veličković DT, Milenović DM, Ristić MS, Veljković VB. Ultrasonic extraction of waste solid residues from the Salvia sp. essential oil hydrodistillation. Biochem Engineering J 2008;42:97-104

32 Radulescu V, Chiliment S, Oprea E. Capillary gas chromatography–mass spectrometry of volatile and semi-volatile compounds of Salvia officinalis. J Chromatogr A 2004;1027:121-126

33 Russo A, Formisano C, Rigano D, Senatore F, Delfine S, Cardile V, Rosselli S, et al. Chemical composition and anticancer activity of essential oils of Mediterranean sage (*Salvia officinalis* L.) grown in different environmental conditions. Food Chem Toxicol 2013;55(May):42-47

34 Bayrak A, Akgül A. Composition of essential oils from Turkish *Salvia* species. Phytochem 1987;26:846-847

35 Hayouni EA, Chraief I, Abedrabba M, Bouix M, Leveau J-Y, Mohammed H, Hamdi M. Tunisian *Salvia officinalis* L. and *Schinus molle* L. essential oils: Their chemical compositions and their preservative effects against *Salmonella* inoculated in minced beef meat. Int Journal Food Microbiol 2008;125:242-251. Data partly also presented in ref. 47

36 Avato P, Fortunato IM, Ruta C, D'Elia R. Glandular hairs and essential oils in micropropagated plants of *Salvia officinalis* L. Plant Science 2009;169:29-36

37 Bouaziz M, Yangui T, Sayadi S, Dhouib A. Disinfectant properties of essential oils from *Salvia officinalis* L. cultivated in Tunisia. Food Chem Toxicol 2009;47:2755-2760

38 Maksimović M, Vidic D, Miloš M, Šolić ME, Abadžić S, Siljak-Yakovlev S. Effect of the environmental conditions on essential oil profile in two Dinaric *Salvia* species: *S. brachyodon* Vandas and *S. officinalis* L. Biochem System Ecol 2007;35:473-478

39 Longaray Delamare AP, Moschen-Pistorello IT, Artico L, Atti-Serafini L, Echeverrigaray S. Antibacterial activity of the essential oils of *Salvia officinalis* L. and *Salvia triloba* L. cultivated in South Brazil. Food Chem 2007;100:603-608

40 Aleksovski SA, Sovová H. Supercritical CO2 extraction of *Salvia officinalis* L. J Supercrit Fluids 2007;40:239-245

41 Gañán N, Brignole EA. Supercritical carbon dioxide fractionation of *T. minuta* and *S. officinalis* essential oils: Experiments and process analysis. J Supercrit Fluids 2013;78:12-20

42 Pinto E, Ribeiro Salgueiro L, Cavaleiro C, Palmeira A, Gonçalves MJ. In vitro susceptibility of some species of yeasts and filamentous fungi to essential oils *of Salvia officinalis*. Ind Crops Prod 2007;26:135-141

43 Laborda R, Manzano I, Gamón M, Gavidia I, Pérez-Bermúdez P, Boluda R. Effects of *Rosmarinus officinalis* and *Salvia officinalis* essential oils on *Tetranychus urticae* Koch (Acari: Tetranychidae). Ind Crops Prod 2013;48:106-110

44 Tounekti T, Munné-Bosch S, Vadel AM, Chtara C, Khemira H. Influence of ionic interactions on essential oil and phenolic diterpene composition of Dalmatian sage (*Salvia officinalis* L.). Plant Physiol Biochem 2010;48:813-821

45 Marino M, Bersani C, Comi G. Impedance measurements to study the antimicrobial activity of essential oils from Lamiaceae and Compositae. Int J Food Microbiol 2001;67:187-195

46 Miguel G, Cruz C, Faleiro ML, Simões MTF, Figueiredo AC, Barroso JG, Pedro LG. *Salvia officinalis* L. essential oils: effect of hydrodistillation time on the chemical composition, antioxidant and antimicrobial activities. Nat Prod Res 2011;25:526-541

47 Bouajaj S, Benyamna A, Bouamama H, Romane A, Falconieri D, Piras A, Marongiu B. Antibacterial, allelopathic and antioxidant activities of essential oil of *Salvia officinalis* L. growing wild in the Atlas Mountains of Morocco. Nat Prod Res 2013;27:1673-1676. Data partly also presented in ref. 35

48 Raal A, Orav A, Arak E. Composition of the essential oil of *Salvia officinalis* L. from various European countries. Nat Prod Res 2007;21:406-411

49 Zutic I, Putievsky E, Dudai N. Influence of harvest dynamics and cut height on yield components of sage (*Salvia officinalis* L.). J Herbs Spices Med Plants 2004;10:49-61

50 Santos-Gomes PC, Fernandes-Ferreira M. Organ- and season-dependent variation in the essential oil composition of *Salvia officinalis* L. cultivated at two different sites. J Agric Food Chem 2001;49:2908-2916

51 Perry NB, Anderson RE, Brennan NJ, Douglas MH, Heaney AJ, McGimpsey JA, Smallfield BM. Essential oils from Dalmatian sage (Salvia officinalis L.): Variations among Individuals, plant parts, seasons, and sites. J Agric Food Chem 1999;47:2048-2054

52 Ntalli NG, Ferrari F, Giannakou I, Menkissoglu-Spiroudi U. Phytochemistry and nematicidal activity of the essential oils from 8 Greek Lamiaceae aromatic plants and 13 terpene components. J Agric Food Chem 2010;58:7856-7863

53 Bozin B, Mimica-Dukic N, Samojlik I, Jovin E. Antimicrobial and antioxidant properties of rosemary and sage (*Rosmarinus officinalis* L. and *Salvia officinalis* L., Lamiaceae) essential oils. J Agric Food Chem 2007;55:7879-7885

54 Ben Farhat M, Jordán MJ, Chaouech-Hamada R, Landoulsi A, Sotomayor JA. Variations in essential oil, phenolic compounds, and antioxidant activity of Tunisian cultivated *Salvia officinalis* L. J Agric Food Chem 2009;57:10349-10356

55 Wei A, Shibamoto T. Antioxidant/lipoxygenase inhibitory activities and chemical compositions of selected essential oils. J Agric Food Chem 2010;58:7218-7225

56 Boelens MH, Boelens H. Chemical and sensory evaluation of sage oils. Perfum Flavor 1997;22(2):19-40

57 Pitarević I, Kuftinec J, Blažević N, Kuštrak D. Seasonal variation of essential oil yield and composition of Dalmatian sage, *Salvia officinalis*. J Nat Prod 1984;47:409-412

58 Kuštrak D, Kuftinec J, Blažević N. Yields and composition of sage oils from different regions of the Yugoslavian Adriatic Coast. J Nat Prod 1984;47:520-524

59 Bernotiene G, Nivinskiene O, Butkiene R, Mockute D. Essential oil composition variability in sage (Salvia officinalis L.). Chemija 2007;18:38-43

60 Stešević D, Ristić M, Nikolić V, Nedović M, Caković D, Šatović Z. Chemotype diversity of indigenous Dalmatian sage (*Salvia officinalis* L.) populations in Montenegro. Chem Biodivers 2014;11:101-114

61 Länger R, Mechtler Ch, Jurenitsch J. Composition of the essential oils of commercial samples of *Salvia officinalis* L. and *S. fruticosa* Miller: A comparison of oils obtained by extraction and steam distillation. Phytochem Anal 1996;7:289-293

62 Loizzo MR, Saab AM, Tundis R, Statti GA, Menichini F, Lampronti I, et al. Phytochemical analysis and *in vitro* antiviral activities of the essential oils of seven Lebanon species. Chem Biodiversity 2008;5:461-470

63 Perry NB, Baxter AJ, Brennan NJ, van Klink JW, McGimpsey JA, Douglas MH, Joulain D. Dalmatian sage. Part 1. Differing oil yields and compositions from flowering and non-flowering accessions. Flavour Fragr J 1996;11:231-238

64 Couladis M, Tzakou O, Mimica-Dukić N, Jančić R, Stojanović D. Essential oil of *Salvia officinalis* L. from Serbia and Montenegro. Flavour Fragr J 2002;17:119-126

65 Lamien-Meda A, Schmiderer C, Lohwasser U, Börner A, Franz C, Novak J. Variability of the essential oil composition in the sage collection of the Genebank Gatersleben: a new viridiflorol chemotype. Flavour Fragr J 2010;25:75-82

66 Jug-Dujaković M, Ristić M, Pljevljakušić D, Dajić-Stevanović Z, Liber Z, Hančević K, et al. High diversity of indigenous populations of Dalmatian sage (*Salvia officinalis* L.) in essential-oil composition. Chem Biodivers 2012;9:2309-2323

67 Svoboda KP, Deans SG. A study of the variability of rosemary and sage and their volatile oils on the British market: their antioxidative properties. Flavour Fragr J 1992;7:81-87

68 Mirjalili MH, Salehi P, Sonboli A, Vala MM. Essential oil variation of *Salvia officinalis* aerial parts during its phenological cycle. Chem Nat Comp 2006;42:19-23

69 Dob T, Berramdane T, Dahmane D, Benabdelkader T, Chelghoum C. Chemical composition of the essential oil of *Salvia officinalis* from Algeria. Chem Nat Comp 2007;43:491-494

70 Esturk O. Intermittent and continuous microwave-convective air-drying characteristics of sage (*Salvia officinalis*) leaves. Food Bioprocess Technol 2012;5:1664-1673

71 Sellami IH, Bettaieb Rebey I, Sriti J, Zohra Rahali F, Limam F, Marzouk B. Drying sage (*Salvia officinalis* L.) plants and its effects on content,

chemical composition, and radical scavenging activity of the essential oil. Food Bioprocess Technol 2012;5:2978-2989

72 Sagareishvili TG, Grigolava BL, Gelashvili NE, Kemertelidze EP. Composition of essential oil from *Salvia officinalis* cultivated in Georgia. Chem Nat Compounds 2000;36:360-361

73 Mockutë D, Nivinskienë O, Bernotienë G. Butkienë R. The cisthujone chemotype of *S. officinalis* L. essential oils. Chemija 2003;14:216-220

74 Miroddi M, Navarra M, Quattropani MC, Calapai F, Gangemi S, Calapai S. Systematic review of clinical trials assessing pharmacological properties of *Salvia* species on memory, cognitive impairment and Alzheimer's disease. CNS Neurosci Ther 2014;20:485-495

75 Böszörményi A, Héthelyi E, Farkas A, Horváth G, Papp N, Lemberkovics E, Szőke E. Chemical and genetic relationships among sage (*Salvia officinalis* L.) cultivars and Judean sage (*Salvia judaica* Boiss.) J Agric Food Chem 2009;57:4663-4667

76 Soković M, Glamočlija J, Marin PD, Brkić D, van Griensven LJLD. Antibacterial effects of the essential oils of commonly consumed medicinal herbs using an *in vitro* model. Molecules 2010;15:7532-7546

77 Viuda-Martos M, Ruíz-Navajas Y, Fernández-López J, Pérez-Álvarez JA. Chemical composition of the essential oils obtained from some spices widely used in Mediterranean region. Acta Chim Slov 2007;54:921-926

78 Aziz EE, Sabry RM, Ahmed SS. Plant growth and essential oil production of sage (*Salvia officinalis* L.) and curly-leafed parsley (*Petroselinum crispum* ssp. *crispum* L.) cultivated under salt stress conditions. World Appl Sci J 2013;28:785-796

79 Grigore A, Colceru-Mihul S, Paraschiv I, Nita S, Christof R, Iuksel R, Ichim M. Chemical analysis and antimicrobial activity of indigenous medicinal species volatile oils. Romanian Biotechnol Lett 2012;17(5):7620-7627

80 Porte A, Godoy RLO, Maia-Porte LH. Chemical composition of sage (*Salvia officinalis* L.) essential oil from the Rio de Janeiro State (Brazil). Rev Bras Pl Med Campinas 2013;15(3):438-441

81 Baj T, Ludwiczuk A, Sieniawska E, Skalickawoèniak K, Widelski J, Zieba J, et al. GC-MS analysis of essential oils from *Salvia officinalis* L.: comparison of extraction methods of the volatile components. Acta Poloniae Pharmaceutica - Drug Research 2013;70:35-40

82 Stefkov G, Cvetkovikj I, Karapandzova M, Kulevanova S. Essential oil composition of wild growing sage from R. Macedonia. Macedonian Pharmaceutical Bulletin 2011;57(1,2):71-76

83 Rowshan V, Najafian S. Evaluation of the essential oil composition of *Salvia officinalis* L. by different drying methods in full flowering stages. Intl J Farm & Alli Sci 2013;2(12):350-354

84 Anghel I, Grumezescu V, Andronescu E, Anghel GA, Grumezescu AM, Mihaiescu DE, Chifiriuc MC. Protective effect of magnetite nanoparticle/*Salvia officinalis* essential oil hybrid nanobiosystem against fungal colonization on the Provox® voice section prosthesis. Digest J Nanomat Biostruct 2012;7:1205-1212

85 Lakušić BS, Ristić MS, Slavkovska VN, Stojanović DL, Lakušić DV. Variations in essential oil yields and compositions of *Salvia officinalis* (Lamiaceae) at different developmental stages. Botanica Serbica 2013;37:127-139

86 Badiee P, Nasirzadeh AR, Motaffaf M. Comparison of *Salvia officinalis* L. essential oil and antifungal agents against *Candida* species. Pharm Technol & Drug Res 2012. doi: 10.7243/2050-120X-1-7

87 Tosun A, Khan S, Kim YS, Calín-Sánchez A, Hysenaj X, Carbonell-Barrachina AA. Essential oil composition and anti-inflammatory activity of *Salvia officinalis* L (Lamiaceae) in murin macrophages. Trop J Pharm Res 2014;13:937-942

88 Abu-Darwish MS, Cabral C, Ferreira IV, Gonçalves JM, Cavaleiro C, Cruz MT, et al. Essential oil of common sage (*Salvia officinalis* L.) from Jordan: assessment of safety in mammalian cells and its antifungal and anti-inflammatory potential. BioMed Res Int, Volume 2013, Article ID 538940, 9 pages; http://dx.doi.org/10.1155/2013/538940

89 Raina AP, Negi KS, Dutta M. Variability in essential oil composition of sage (*Salvia officinalis* L.) grown under North Western Himalayan Region of India. J Med Plants Res 2013;7:683-688

90 Politeo O, Jukić M, Miloć M. Chemical composition and antioxidant activity of essential oils of twelve spice plants. Croatica Chemica Acta 2006;79:545-552

91 Lakhal H, Ghorab H, Chibani S, Kabouche A, Semra Z, Smati F, et al. Chemical composition and biological activities of the essential oil of *Salvia officinalis* from Batna (Algeria). Der Pharmacia Lettre 2013;5:310-314

92 Arraiza MP, Arrabal C, López JV. Seasonal variation of essential oil yield and composition of sage (*Salvia officinalis* L.) grown in Castilla - La Mancha (Central Spain). Not Bot Horti Agrobo 2012;40(2):3 pages

93 Jirovetz L, Buchbauer G, Denkova Z, Slavchev A, Stoyanova A, Schmidt E. Chemical composition, antimicrobial activities and odor descriptions of various *Salvia* sp. and *Thuja* sp. essential oils. Ernährung/Nutrition 2006;30:152-159

95 Lawrence BM. Progress in essential oils. Perfum Flavor 2001;26(3):66-? (last page unknown)

96 Committee on Herbal Medicinal Products. Public statement on Salvia officinalis L., Aetheroleum. London: European Medicines Agency, 2009 http://www.ema.europa.eu/docs/en_GB/document_library/Public_statement/2010/02/WC500070841.pdf

97 Rhind JP. Essential oils. A handbook for aromatherapy practice, 2nd Edition. London: Singing Dragon, 2012

98 Lawless J. The encyclopedia of essential oils, 2nd Edition. London: Harper Thorsons, 2014

99 Davis P. Aromatherapy. An A-Z, 3rd Edition. London: Vermilion, 2005

100 Selvaag E, Holm J, Thune P. Allergic contact dermatitis in an aromatherapist with multiple sensitizations to essential oils. Contact Dermatitis 1995;33:354-355

Chapter 5.76 SAGE OIL, SPANISH

There are two major commercial sage oils: sage oil, Dalmatian, obtained from *Salvia officinalis* L. and sage oil, Spanish, which is obtained from *Salvia lavandulifolia* Vahl. In this chapter, the Spanish type sage oil is discussed. The Dalmatian sage essential oil is presented in Chapter 5.75.

DEFINITION

Spanish sage oil (essential oil of sage, Spanish) is the essential oil obtained from the flowering tops of the Spanish sage, *Salvia lavandulifolia* Vahl.

INCI NOMENCLATURE

Description/definition: Salvia lavandulaefolia leaf oil is the volatile oil obtained from the leaves of the *Salvia lavandulaefolia*, Lamiaceae
INCI name EU & USA: Salvia lavandulaefolia leaf oil
CAS registry number(s): 90106-49-3; 95371-15-6
EINECS number(s): 290-272-9

Description/definition: Salvia lavandulifolia herb oil is an essential oil obtained from the herbs of the sage, *Salvia lavandulifolia*, Lamiaceae
INCI name EU: Salvia lavandulifolia herb oil (perfuming name, not an INCI name proper)
INCI name USA: Not in the Personal Care Products Council Ingredient Database
CAS registry number(s): 90106-49-3; 95371-15-6
EINECS number(s): 290-272-9

ISO (INTERNATIONAL ORGANIZATION FOR STANDARDIZATION) STANDARD

ISO number: 3526
ISO name: Essential oil of sage, Spanish
Botanical origin: *Salvia lavandulifolia* Vahl
Parts of plant used: Flowering top
ISO values: ISO values (minimum and maximum concentrations) are shown in Table 5.76.1.

The Spanish sage is sometimes termed *Salvia lavandulaefolia*, but correct is *Salvia lavandulifolia*. Other species of the genus *Salvia* from which sage oils may be obtained and which are mentioned in CosIng include *Salvia hispanica* (CAS 93384-40-8; EINECS 297-250-8; *not* identical with Spanish sage) and *Salvia sclarea* (see Chapter 5.25 Clary sage oil).

THE PLANT, THE OIL, AND THEIR USES

Salvia lavandulifolia Vahl is a small woody and densely branched herbaceous perennial with wiry stems, growing 30-50 cm tall and up to 60 cm wide (2). The plant is native to northern Africa (Algeria, Morocco) and southwestern Europe (France, Spain) and is endemic in Turkey (5). The Spanish sage grows in the wild and inhabits stony slopes, rock crevices, and oak and pine woodlands (1). It is cultivated on a small scale in Spain (GRIN Taxonomy for Plants). In western cuisine, the Spanish sage is traditionally used as a condiment, for preserving food and for imparting a slightly spicy flavor to foods including meat, soups, cheese and liquors (3). Leaves of *S. lavandulifolia* are used medicinally as an analgesic, spasmolytic, choleretic, sedative, antiseptic, astringent and as a hypoglycemic drug (1,3,4). They are believed to be effective against mouth ulcers and sore throat, colds, menstrual pains and for the treatment of some menopause symptoms such as flushing or sweating (2). Experimental studies have shown some potential value of *S. lavandulifolia* in dementia therapy attributed to its (perceived) sedative, antioxidant, anti-inflammatory, estrogenic and anticholinesterase activities, all of which are currently believed to be relevant to the treatment of Alzheimer's disease (3,8). However, the beneficial effects of enhancing cognitive performance both in healthy subjects and in patients with dementia or cognitive impairment need to be investigated further in clinical trials with higher methodological standards (8). The plant and its products are also used in the cosmetics industry (1,10).

The essential oil of Spanish sage, which is obtained mostly from wild plants in Spain (province of Granada) is used for cosmetics (soap, toothpaste), desserts, soft drinks, alcoholic beverages and as medicine, and in Spain also for cooking (Mansfeld's World Database of Agriculture and Horticultural Crops, http://mansfeld.ipk-gatersleben.de). It is also employed in aromatherapy practices (26,27).

CHEMICAL COMPOSITION

Spanish type sage oil is a colorless to pale yellow liquid, which has a fresh herbaceous, camphoraceous odor, reminding of eucalyptus. The yield of essential oil from the flowering tops of *Salvia lavandulifolia* Vahl. generally varies from 0.6 to 0.9 per cent. The main producing country of this oil is Spain.

Literature data (up to October 29, 2014) on the chemical composition of Spanish type sage oils and unpublished analytical data from one of us (E.S.) are shown in Table 5.76.2 in alphabetical order. In Spanish sage oils from various origins, over 120 chemicals have been identified. About 41 per cent of these were found in a single reviewed publication only. The major compounds found in Spanish sage oils from different sources are shown in Table 5.76.3. They include (highest concentrations in any study given) 1,8-cineole (52.5%), camphor (46.1%), limonene (25.1%), α-pinene (24.5%), β-pinene (19.2%), borneol (14.5%),

Table 5.76.1 ISO values (%) for sage oil, Spanish [a]

Compound	CAS	Minimum	Maximum
Camphor	76-22-2 1	1.0	36.0
1,8-Cineole	470-82-6	10.0	30.0
α-Pinene	80-56-8	4.0	11.0
Sabinyl acetate	53833-85-5	0.5	9.0
Terpinyl acetate	8007-35-0	0.5	9.0
Borneol	507-70-0	1.0	7.0
Limonene	138-86-3	2.0	6.0
Linalyl acetate	115-95-7	0.1	5.0
Linalool	78-70-6	0.3	4.0
Sabinene	3387-41-5	0.1	3.5
Terpinen-4-ol	562-74-3	0.0	2.0

[a] ISO 3526 Oil of sage, Spanish ©ISO 2005; Geneva, Switzerland, www.iso.org

viridiflorol (11.8%), camphene (10.9%), β-caryophyllene (8.9%) and myrcene (7.6%). Well-known ingredients of Spanish sage oils that were present in high concentrations (>7%) in one or two studies were α-thujone (11.1% and 19.0%, possibly Dalmatian sage), sabinyl acetate (12.8%), linalool (11.0%), bornyl acetate (10.2%) and linalyl acetate (10.2%). Uncommon or rare constituents of Spanish sage oils found in high concentrations (>7%) in single studies include β-thujone (cis-) (20.0%, possibly Dalmatian sage), δ-terpineol (12.0%), α-terpinyl acetate (11.2%), ledol (10.8%) and β-phellandrene (9.3%).

Commercial oils

The ten chemicals that had the highest maximum concentrations in 42 commercial Spanish sage essential oil samples (concentration ranges provided) are the following: camphor (11.0-36.0%), 1,8-cineole (10.0-30.0%), α-pinene (4.0-11.0%), sabinyl acetate (0.5-9.0%), terpinyl acetate (0.5-9.0%), borneol (1.0-7.0%), limonene (2.0-6.0%), linalyl acetate (0.1-5.0%), linalool (0.3-4.0%) and sabinene (0.1-3.5%) (Erich Schmidt, unpublished analytical data).

Table 5.76.2 Constituents identified in sage oils, Spanish type

Constituent	CAS	Percentage and range				
		A	B (5)	C (1)	D (22)	E
Aromadendrene	489-39-4		0-0.8			0.02[a]; 0.5[a]
allo-Aromadendrene	25246-27-9		0-0.8	0.2-0.4		0.2[r]
cis-α-Bergamotene	18252-46-5			0.1-0.2		0.2[r]
trans-α-Bergamotene	13474-59-4			0.2-0.3		0.3[r]
Bicyclogermacrene	24703-35-3		0.2-1.4			
Bisabolene	495-62-5					3.0[f]
α-Bisabolol	515-69-5					0.3[f]
β-Bisabolol	15352-77-9			0.1-0.3		0.2[r]
Bornane-2,5-dione						0.1[a]
Borneol (endo-)	507-70-0	1.4-3.0	1.4-8.7	4.7-5.8	0.1-4.5	3.4[a]; 3.5[u]; 3.7[e]; 4.9[r]; 5.0[b]; 14.5[c]
Bornyl acetate	76-49-3	0.2-3.2	0.2-2.0	0.5-2.4		0.7[q]; 0.9[j]; 1.0[a]; 1.1[o]; 1.4[d]; 10.2[c]
endo-Bornyl propionate						0.1[a]
Cadina-1,4-diene	29837-12-5					0.06[a]
Cadinene	29350-73-0					0.5[f]
β-Cadinene	523-47-7					0.2[a]
γ-Cadinene	39029-41-9			0-0.1		0.1[r]; 0.2[a]; 0.9[a]
δ-Cadinene	483-76-1		0.1-1.7	0.1-0.3		0.1[a]; 0.2[j]; 0.3[r]; 0.4[g]; 1.8[a]
epi-α-Cadinol	5937-11-1			0.2-0.4		0.3[f]; 0.5[r]
δ-Cadinol	19435-97-3					0.3[f]
Camphene	79-92-5	1.9-9.2	1.4-4.6	5.4-10.9	0.2-4.9	7.0[g]; 7.1[f]; 7.5[j]; 7.7[l]; 7.9[u]; 10.7[r]
Camphene hydrate	465-31-6			0-0.2		0.1[r]
α-Campholenal	4501-58-0		0-0.1			
Camphor	76-22-2	5.7-30.8	0-15.4	16.3-30.4	0.6-19.9	30.1[l]; 32.6[j]; 35.9[s]; 39.0[j]; 46.1[b]
δ3-Carene	20296-50-8					+[h]; 1.2[m]
Carvacrol	499-75-2		0-0.2			
Carvone	99-49-0					0.1[e]
Carvotanacetone	499-71-8					+[h]
β-Caryophyllene	87-44-5	0.5-1.5	1.5-8.1	1.6-6.1	1.4-8.9	1.4[s,u]; 2.4[f]; 3.6[m]; 4.9[n]; 5.6[a]; 5.8[o]
Caryophyllene oxide	1139-30-6		0.4-5.6	1.5-2.2	0.3-4.1	0.3[n]; 0.4[a]; 0.5[s]; 1.2[q]; 1.6[r]
1,8-Cineole	470-82-6	11.8-47.7	6.4-34.5	13.3-18.7	3.0-52.5	30.6[m]; 31.3[e]; 31.9[f]; 41.2[k]; 47.0[n]
Clovene	469-92-1					0.06[a]; 0.07[a]
α-Copaene	3856-25-5					tr[g]; 0.06[a]; 0.7[a]
α-Cubebene	17699-14-8		0-0.5			tr[g]; 0.2[a]
β-Cubebene	13744-15-5					0.1[a]
Cuminaldehyde	122-03-2					0.1[f]
Cuminyl alcohol	536-60-7		0.1-0.4			
Curcumene (ar-; α-)	644-30-4			0.2-0.4		0.01[a]; 0.1[r]
p-Cymene	99-87-6	0.3-3.2	0.8-5.2	0.8-0.9		0.8[j,l]; 1.0[r]; 1.1[u]; 1.4[o]; 1.9[g]; 3.6[a]
p-Cymenene	1195-32-0					0.3[g]
p-Cymen-8-ol	20834059-7		0-0.2			
Dihydrocarvone	5948-04-9		0.1-0.8			
3-Ethenyl-3-methyl-2-(1-methylethenyl)-6-(1-methylethyl)-cyclohexanol	174955-50-1					0.1[a]
β-Eudesmol	473-15-4			0.4-0.5		0.4[r]
Eugenol	97-53-0		0-0.1			0.4[f]
α-Farnesene	502-61-4					0.2[a]; 0.5[f]

Table 5.76.2 Constituents identified in sage oils, Spanish type (*continued*)

Constituent	CAS	Percentage and range				
		A	B (5)	C (1)	D (22)	E
α-Fenchene	471-84-1	0.07-0.1				0.1[d]
endo-Fenchol (α-)	512-13-0		0-0.1			
Fonenol	1275572-21-8					0.4[a]
Geraniol	106-24-1	0.08-0.4	0-0.1			0.4[c,d,g]
Geranyl acetate	105-87-3		0-0.1			4.9[f]
Geranyl propanoate	105-90-8	tr-0.5				0.5[d]
Germacrene D	23986-74-5					0.07[a]
α-Gurjunene	489-40-7		0-0.3			0.1[a]; 0.7[a]
γ-Gurjunene	22567-17-5					6.2[p]
α-Humulene	6753-98-6	0.07-0.4	0.1-3.9	0.7-3.7		0.4[d,f,i]; 0.5[r]; 1.0[n]; 3.3[p]; 4.2[a]; 6.1[m]
Humulene oxide II	19888-34-7			1.3-3.1		0.3[r]
β-Ionone	79-77-6		0-0.2			
Isoborneol	124-76-5	0.02-0.7				0.2[e]; 0.8[u]
Isobornyl acetate	125-12-2	tr-0.4				
Isopinocamphone	15358-88-0					0.05[a]; 0.6[a]
2-Isopropyl-5-methyl-9-methylene-bicyclo[4.4.0]dec-1-ene	150320-52-8					0.3[a]
Lavandulol	498-16-8					0.2[e]
Ledol	577-27-5					10.8[f]
Limonene	138-86-3	2.0-10.5	0.8-16.6	2.8-3.5	2.5-25.1	3.4[e,r]; 3.5[m]; 4.0[o]; 4.5[g]; 5.0[d]; 5.1[u]
trans-Limonene oxide	4959-35-7		0-0.3			
Linalool	78-70-6	1.8-5.2	0.1-0.3			0.8[q]; 1.5[j]; 2.4[d]; 3.1[l]; 4.2[u]; 11.0[o]
Linalool oxide	1365-19-1					0.2[e]
Linalyl acetate	115-95-7	0.3-4.2				0.5[e]; 0.9[j]; 2.1[o]; 3.4[u]; 4.2[d]; 10.2[i]
Linalyl isobutyrate	78-35-3					+[h]
Manool	596-85-0					1.7[l]
Methyl dodecanoate	111-82-0					0.4[a]
3-Methyl-4-isopropyl-phenol	3228-02-2		0.1-0.3			
α-Muurolene	10208-80-7		0-0.2			0.3[a]
τ-Muurolene	152287-43-9		0.1-0.3			
Myrcene	123-35-3	1.3-6.2	0.9-2.4	1.3-2.4	8.1-25.0[t]	1.6[r]; 2.5[d,j]; 3.2[g]; 5.5[b]; 5.8[a,n]; 7.6[e]
Myrtenal	564-94-3			0.3-0.3		0.2[r]
Myrtenol	19894-97-4					0.1[f]
Myrtenyl acetate	1079-01-2		0-0.1			
Neoiso-3-thujanol acetate						0.1[u]
Nerol	106-25-2		0-0.1			0.1[g]
Nerolidol	7212-44-4					1.3[e]
Neryl acetate	141-12-8					0.06[e]
(*E*)-β-Ocimene	3779-61-1	0-0.1	0-0.2			0.1[d,l]; 0.2[f]; 0.4[g]
(*Z*)-β-Ocimene	3338-55-4	tr-0.2	0.1-1.2			0.1[d]; 0.2[l]; 0.3[u]
(*Z*)-allo-Ocimene	17202-20-9					0.1[g]
n-Octanol (1-)	111-87-5					0.2[e]
2-Octanol	123-96-6		0-0.2			
3-Octanol	589-98-0		0-0.1			0.2[e]
α-Phellandrene	99-83-2	0.04-0.2	0-0.3			0.1[d,l,u]
β-Phellandrene	555-10-2					9.3[f]
Phenyl acetate	122-79-2					0.1[e]
α-Pinene	80-56-8	4.9-24.5	6.7-23.2	4.8-5.0	1.5-34.3	7.0[a]; 8.2[u]; 8.6[g]; 9.5[o]; 10.9[e]; 11.7[f]
β-Pinene	127-91-3	2.1-7.2	3.8-19.2	7.9-12.5	8.1-25.0[t]	5.5[q]; 6.8[n]; 6.9[m]; 9.4[r]; 11.8[e]; 12.0[c]
α-Pinene oxide	1686-14-2					+[h]
Pinocamphone	547-60-4					0.08[a]
Sabinene	3387-41-5	0.3-1.3	0.1-2.7	0.5-0.8	0.1-4.4	0.5[i]; 0.6[r]; 1.3[d,g]; 1.5[j]; 2.8[p]; 3.8[a]
cis-Sabinene hydrate	15537-55-0	tr-0.1	0.1-0.1			0.1[d]; 0.2[b]; 0.4[a]
trans-Sabinene hydrate	17699-16-0		0.1-0.5			0.5[g]; 0.6[f]
trans-Sabinol	471-16-9	tr-0.9				0.8[d]
Sabinyl acetate	53833-85-5	0.5-3.4	0-0.5			2.5[e]; 3.4[d]; 3.6[g]; 12.8[j,s]
trans-Sabinyl acetate	139757-62-3					1.6[u]
Sativene	6813-05-4					0.1[a]

Table 5.76.2 Constituents identified in sage oils, Spanish type (continued)

Constituent	CAS	Percentage and range				
		A	B (5)	C (1)	D (22)	E
α-Selinene	473-13-2			0.1-0.1		0.06[a]; 0.1[r]
γ-Selinene	515-17-3		0.1-0.2			0.3[a]
Spathulenol	6750-60-3		0.1-2.4	0.7-0.7	0-2.8	0.04[a]; 1.3[r]
α-Terpinene	99-86-5	0.02-1.0	0.1-1.2			0.1[d,f,l]; 0.2[b]; 0.4[n]; 1.8[a]
γ-Terpinene	99-85-4	0.2-0.8	1.1-6.9	0.1-0.3		0.3[d]; 0.4[u]; 0.5[b]; 1.1[a,f,n]; 1.4[e]; 1.6[o]
Terpinen-4-ol	562-74-3	0.2-0.6	0.5-1.0	0.5-0.9		0.2[u]; 0.3[a]; 0.4[j]; 0.5[d]; 0.6[r]; 0.7[a,n]
Terpineol	586-82-3					0.4[f]; 0.8[e]
α-Terpineol	98-55-5	0.4-3.4	0.3-1.0			tr[g]; 0.3[n]; 0.5[a]; 0.7[d]; 1.2[a]; 1.4[m]; 1.9[l]
β-Terpineol	138-87-4					+[h]; 0.1[d]
(Z)-β-Terpineol	7299-40-3	0.06-0.2				
γ-Terpineol	586-81-2	0.3-0.6				0.2[u]
δ-Terpineol	7299-42-5					0.3[d]; 12.0[f]
Terpinolene	586-62-9	0.1-0.4	0.2-0.7			tr[g]; 0.2[l,u]; 0.3[a,d]; 0.5[n]
Terpinyl acetate	8007-35-0	0.2-1.1	0-0.7			1.0[d]; 3.0[e]
α-Terpinyl acetate	80-26-2					0.8[i]; 2.0[u]; 11.2[g]
α-Thujene	2867-05-2		0.1-0.9	0.3-0.6		0.1[b,f,u]; 0.2[g,j]; 0.3[a]; 0.5[r]
α-Thujone	546-80-5	0-tr				0.4[n]; 1.1[o]; 3.2[m]; 11.1[a]; 19.0[p]
β-Thujone (cis-)	471-15-8		0-0.2			0.3[c]; 0.6[m]; 2.9[a]; 20.0[p]
Thymol	89-83-8					1.1[e]
Tricyclene	508-32-7	0.1-0.7	0.1-0.1			0.2[a]; 0.4[d]; 0.5[l,u]
Verbenone	80-57-9		0-0.4	0.3-0.3		0.2[e,r]; 0.5[a]
Viridiflorene (ledene)	21747-46-6			0.2-0.2		0.3[r]
Viridiflorol	552-02-3	0.1-4.6	0.1-9.7	0.4-11.8	0-5.4	0.1[f]; 0.3[a,r]; 1.1[a]; 1.6[o]; 4.5[n]
α-Ylangene	14912-44-8					0.04[a]

A forty-two Spanish sage essential oil samples from Spain, analyzed between 1998 and 2013; lowest and highest concentrations given (E. Schmidt, unpublished data)
B twenty S. lavandulifolia essential oils obtained by steam distillation with cohobation from the aerial parts of 20 wild Spanish plant populations in the flowering stage; lowest and highest concentrations given (ref. 5)
C ten lab-hydrodistilled oils obtained from the aerial parts of wild growing populations from two regions in Morocco (each five samples) and harvested between March and June; mean values for each of the two regions given (only means were provided in the article) (ref. 1)
D twenty-two lab-hydrodistilled oils from S. lavandulifolia plants in full bloom cultivated from four wild populations in Central Spain, harvested in two successive years (each eleven oils); lowest and highest concentrations given (ref. 22)
E data from other studies (indicated with superscript letters); highest concentrations found in any study reviewed here given; when two or more oils were investigated, only the highest concentrations are mentioned, unless indicated otherwise

[a] one steam-distilled oil from flowering plants cultivated in Spain (ref. 11); [b] one lab-hydrodistilled oil from leaves and flowers from Spain (ref. 6); [c] one commercial Spanish sage oil (ref. 20); [d] one sample of S. lavandulifolia oil (ref. 19); [e] one lab-hydrodistilled oil from the aerial parts of fully flowering plants from an experimental field in Spain and one oil obtained from plants in the vegetative state (ref. 4); [f] two lab-hydrodistilled oils from the aerial parts of Salvia lavandulifolia plants cultivated in Spain (ref. 3); [g] one commercial oil of unknown origin (ref. 7); [h] data from before 1970, cited in ref. 7; [i] one lab-hydrodistilled oil from the flowering aerial parts of Spanish sage growing wild in the south of France (ref. 9); [j] one commercial oil sample from France (ref. 10); [k] unknown number of essential oils from Salvia lavandulaefolia Vahl; only 1,8-cineole was analyzed (ref. 15); [l] one commercial oil sample (ref. 14); [m] one sample of Spanish sage oil (ref. 17); [n] one oil of S. lavandulaefolia (ref. 18); [o] one lab-hydrodistilled oil from Spain (ref. 24); [p] one lab-hydrodistilled oil from the flowering aerial parts of plants cultivated in Brazil (ref. 23); because of the extremely high concentrations of both α- and β-thujone, there may have been a botanical misidentification in this report; high thujone values may be found in S. officinalis (Dalmatian sage) oils; [q] one commercial steam-distilled leaf oil (ref. 12); [r] one lab-hydrodistilled leaf oil from Morocco (ref. 1); [s] unknown number of commercial oil samples of unknown geographical origin (ref. 21); [t] β-pinene and myrcene combined; [u] one commercial oil of Spanish sage obtained from a Serbian company (ref. 25)

tr: trace; + present in the oil investigated, but quantity not stated

Table 5.76.3 Major constituents of Spanish type sage oils

Constituent	CAS	Percentage and range				
		A	B (5)	C (1)	D (22)	E
1,8-Cineole	470-82-6	11.8-47.7	6.4-34.5	13.3-18.7	3.0-52.5	30.6[m]; 31.3[e]; 31.9[f]; 41.2[k]; 47.0[n]
Camphor	76-22-2	5.7-30.8	0-15.4	16.3-30.4	0.6-19.9	30.1[l]; 32.6[j]; 35.9[s]; 39.0[i]; 46.1[b]
Limonene	138-86-3	2.0-10.5	0.8-16.6	2.8-3.5	2.5-25.1	3.4[e,r]; 3.5[m]; 4.0[o]; 4.5[g]; 5.0[d]; 5.1[u]
α-Pinene	80-56-8	4.9-24.5	6.7-23.2	4.8-5.0	1.5-34.3	7.0[a]; 8.2[u]; 8.6[g]; 9.5[o]; 10.9[e]; 11.7[f]
β-Pinene	127-91-3	2.1-7.2	3.8-19.2	7.9-12.5	8.1-25.0[t]	5.5[q]; 6.8[n]; 6.9[m]; 9.4[r]; 11.8[e]; 12.0[c]
Borneol (endo-)	507-70-0	1.4-3.0	1.4-8.7	4.7-5.8	0.1-4.5	3.4[a]; 3.5[u]; 3.7[e]; 4.9[r]; 5.0[b]; 14.5[c]
Viridiflorol	552-02-3	0.1-4.6	0.1-9.7	0.4-11.8	0-5.4	0.1[f]; 0.3[a,r]; 1.1[a]; 1.6[o]; 4.5[n]
Camphene	79-92-5	1.9-9.2	1.4-4.6	5.4-10.9	0.2-4.9	7.0[g]; 7.1[f]; 7.5[j]; 7.7[l]; 7.9[u]; 10.7[r]
β-Caryophyllene	87-44-5	0.5-1.5	1.5-8.1	1.6-6.1	1.4-8.9	1.4[s,u]; 2.4[f]; 3.6[m]; 4.9[n]; 5.6[a]; 5.8[o]
Myrcene	123-35-3	1.3-6.2	0.9-2.4	1.3-2.4	8.1-25.0[t]	1.6[r]; 2.5[d,j]; 3.2[g]; 5.5[b]; 5.8[a,n]; 7.6[e]

LEGEND: SEE UNDER TABLE 5.76.2

CONTACT ALLERGY/ALLERGIC CONTACT DERMATITIS

No reports on contact allergy to sage oil, specifically mentioned to be obtained from *Salvia lavandulifolia*, have been found. Literature on contact allergy to/allergic contact dermatitis from 'sage oil' (botanical source not specified) is discussed in Chapter 5.75 Sage oil, Dalmatian.

LITERATURE

1 Zrira S, Menut C, Bessière CM, Elamrani A, Benjilali B. A study of the essential oil of *Salvia lavandulifolia* Vahl from Morocco. J Essent Oil Bear Plants 2004;7:232-238

2 Usano-Alemany J, Palá-Paúl J, Herráiz-Peñalver D. Temperature stress causes different profiles of volatile compounds in two chemotypes of *Salvia lavandulifolia* Vahl. Biochem System Ecol 2014;54:166-171

3 Porres-Martínez M, González-Burgos E, Carretero Accame ME, Gómez-Serranillos MP. Phytochemical composition, antioxidant and cytoprotective activities of essential oil of *Salvia lavandulifolia* Vahl. Food Res Int 2013;54:523-531

4 Porres-Martínez M, González-Burgos E, Carretero ME, Gómez-Serranillos MP. Influence of phenological stage on chemical composition and antioxidant activity of *Salvia lavandulifolia* Vahl. essential oils. Ind Crops Prod 2014;53:71-77

5 Herraiz-Peñalver D, Usano-Alemany J, Cuadrado J, Jordán MJ, Lax V, Sotomayor JA, Palá-Paúl J. Essential oil composition of wild populations of *Salvia lavandulifolia* Vahl. from Castilla-La Mancha (Spain). Biochem Syst Ecol 2010;38:1224-1230

6 Langa E, Della Porta G, Palavra AMF, Urieta JS, Mainar AM. Supercritical fluid extraction of Spanish sage essential oil: Optimization of the process parameters and modelling. J Supercrit Fluids 2009;49:174-181

7 Lawrence BM, Hogg JW, Terhune SJ. Essential oils and their constituents. III. Some new trace constituents in the essential oil of *Salvia lavandulaefolia*, Vahl. J Chromatogr 1970;50:59-65

8 Miroddi M, Navarra M, Quattropani MC, Calapai F, Gangemi S, Calapai G. Systematic review of clinical trials assessing pharmacological properties of *Salvia* species on memory, cognitive Impairment and Alzheimer's disease. CNS Neurosci Therap 2014;20:485-495

9 Foray L, Bertrand C, Pinguet F, Soulier M, Astre C, Marion C, et al. In vitro cytotoxic activity of three essential oils from *Salvia* Species. J Essent Oil Res 1999;11:522-526

10 Pages N, Fournier G, Velut V, Imbert C. Potential teratogenicity in mice of the essential oil of *Salvia lavandulifolia* Vahl. Study of a fraction rich in sabinyl acetate. Phytother Res 1992;6:80-83

11 Guillén MD, Cabo N, Burillo J. Characterisation of the essential oils of some cultivated aromatic plants of industrial interest. J Sci Food Agric 1996;70:359-363

12 Savelev S, Okello E, Perry NSL, Wilkins RM, Perry EK. Synergistic and antagonistic interactions of anticholinesterase terpenoids in *Salvia lavandulaefolia* essential oil. Pharmacol Biochem Behav 2003;75:661-668

13 Lawrence BM. Progress in essential oils. Perfum Flavor 2007;32(1):48-? (last page unknown)

14 Kubeczka K-H, Formacek V. Essential oils analysis by capillary gas chromatography and carbon-13 NMR spectroscopy, 2nd edition. New York, USA: John Wiley and Sons, 2002:303-313. Data cited in ref. 13

15 De Vincenzi M, Silano M, De Vincenzi A, Maialetti F, Scazzocchio B. Constituents of aromatic plants: eucalyptol. Fitoterapia 2002;73:269-275

16 Lawrence BM. Progress in essential oils. Perfum Flavor 2001;26(2):22-? (last page unknown)

17 Zani F, Massimo G, Venvenuti S, Biandri A, Albasini A, Melegari M, et al. Studies on the genotoxic properties of essential oils with *Bacillus subtilis* rec-assay and *Salmonella*/microsome reversion assay. Planta Med 1991;57:237-241

18 Mathé I, Nagy G, Dobos A, Miklossy VV, Janicsak G. Comparative studies of the essential oils of some species of sect. Salvia. In: Frantz Ch, Mathé A,

Buchbauer G, Eds. Proceedings of 27th International Symposium on Essential Oils, Vienna, 1996. Carol Stream, IL, USA: Allured Publ., 1997:244-247. Data cited in ref. 16

19 Jirovetz L, Wlcek K, Buchbauer G, Gochev V, Girova T, Stoyanova A, et al. Antifungal activities of essential oils of *Salvia lavandulifolia*, *Salvia officinalis* and *Salvia sclarea* against various pathogenic *Candida* species. J Essent Oil Bear Plants 2007;10:430-439

20 Perry NSL, Houghton PJ, Jenner P, Keith A, Perry EK. *Salvia lavandulaefolia* essential oil inhibits cholinesterase in vivo. Phytomed 2002;9:48-51

21 Fournier G, Pages N, Cosperec I. Contribution to the study of *Salvia lavandulifolia* essential oil: potential toxicity attributable to sabinyl acetate. Planta Med 1993;59:96-97

22 Usano-Alemany J, Palá-Paúl J, Santa-Cruz Rodríguez M, Herraiz-Peñalver D. Chemical description and essential oil yield variability of different accessions of *Salvia lavandulifolia*. Nat Prod Comm 2014;9:273-276

23 Pierozan MK, Fernandes Pauletti G, Rorota L, Atti dos Santos AC, Lerin LA, Di Luccio M, et al. Chemical characterization and antimicrobial activity of essential oils of *Salvia* L. species. Ciênc Tecnol Aliment Campinas 2009;29:764-770

24 Santana O, Cabrera R, Giménez C, González-Coloma A, Sánchez-Vioque R, de los Mozos-Pascual M, et al. Chemical and biological profiles of the essential oils from aromatic plants of agro industrial interest in Castilla-La Mancha (Spain). Grasas y Aceites, 2012;63:214-222 (in Spanish)

25 Nikolić M, Jovanović KK, Marković T, Marković D, Gligorijević N, Radulović S, Soković M. Chemical composition, antimicrobial, and cytotoxic properties of five Lamiaceae essential oils. Ind Crops Prod 2014;61:225-232

26 Rhind JP. Essential oils. A handbook for aromatherapy practice, 2nd Edition. London: Singing Dragon, 2012

27 Lawless J. The encyclopedia of essential oils, 2nd Edition. London: Harper Thorsons, 2014

Chapter 5.77   SANDALWOOD OIL

There are three varieties of sandalwood oil: East Indian sandalwood oil obtained from the wood of *Santalum album* L., Australian sandalwood oil obtained from *Santalum spicatum* (R. Br.) A. DC., and New Caledonian sandalwood oil, prepared from the wood of *Santalum austrocaledonicum* Vieill. The East Indian variety has not commercially been available since 2008, as *Santalum album* is on the endangered species list of the International Union for the Conservation of Nature (IUCN) and Indian law prohibits the export of sandalwood oil.

SANDALWOOD OIL (EAST INDIA)

DEFINITION
Sandalwood oil (East India) (essential oil of sandalwood) is the essential oil obtained from the wood of the East Indian sandalwood, *Santalum album* L.

INCI NOMENCLATURE
Description/definition: Santalum album oil is the volatile oil obtained from the heartwood of the sandalwood, *Santalum album* L., Santalaceae
INCI name EU: Santalum album oil
INCI name USA: Santalum album (sandalwood) oil
Other names: Oil of santal
CAS registry number(s): 8006-87-9; 84787-70-2
EINECS number(s): 284-211-1

ISO (INTERNATIONAL ORGANIZATION FOR STANDARDIZATION) STANDARD
ISO number: 3518
ISO name: Essential oil of sandalwood
Botanical origin: *Santalum* album L.
Parts of plant used: Wood
ISO values: ISO values (minimum and maximum concentrations) are shown in Table 5.77.1.

THE PLANT, THE OIL, AND THEIR USES
Santalum album L. is a small evergreen tree, which can grow to a height of 20 meters with a girth of up to 2.4 meters. The plant is found distributed in India, Indonesia, New Caledonia, Philippine Islands, Malaysia and Sri Lanka. India used to account for virtually all of the world's production of sandalwood and its oil (3). *Santalum album* is a partial parasite that attaches to the roots of other trees; it needs 'nurse' species in the area of planting out. About 500 potential host plants are known; for sandalwood cultivation *Cassia siamea* has been favored in general (3, GRIN Taxonomy for Plants; http://www.worldagroforestry.org).

Powder from the heartwood and roots is used to make incense sticks, burnt as perfumes in houses and temples, or is ground into a paste and used as a cosmetic. Distillation of the powdered heartwood yields the East Indian sandalwood oil (Oleum santali) (the sapwood contains little, if any, detectable oils) (1). Its main use is in the creation of perfumes. *S. album* produces excellent fragrant material, and is one of the oldest and most expensive perfumery raw materials. The oil is highly rated for its fixative properties and for its persistent, heavy, sweet, woody odor (3,4). The oil is also used extensively in the cosmetics industry in the manufacture of soaps, face creams, toilet powders and air fresheners (3). It also has applications in many food products, alcoholic and non-alcoholic beverages, for flavoring chewing tobacco and is a very popular oil in aromatherapy (3,6,7).

Sandalwood essential oil is considered to have antiviral, bactericidal, anti-inflammatory, antimitotic, anti-hypertensive, diuretic, expectorant, antipyretic, stimulant and sedative properties (6,12). The oil is used in Ayurveda and other medicinal systems for common colds, fever, bronchitis, inflammation of the mouth and pharynx, infections of urinary tract, liver and gall bladder complaints and other maladies (6,12).

Since 2008 *Santalum album* is on the International Union for the Conservation of Nature (IUCN) list of endangered species. As a consequence, the governments of the two sandalwood production areas of India, Madras and Mysore, decided to stop the production of the oil and any export. However, small quantities are illegally being sold and are subsequently adulterated.

CHEMICAL COMPOSITION
Sandalwood oil is a clear, slightly viscous, and colorless to pale yellow liquid, which has a soft, sweet and woody note with long lasting powdery odor.

The yield of essential oil from the wood of *Santalum album* L. generally varies from 1.9 to 6.2 per cent but may be as high as 8% (1). The main producing country of this oil used to be India. *S. album* essential oil from legitimate sources is hardly available anymore.

Literature data (up to December 7, 2014) on the chemical composition of sandalwood oils East India and unpublished analytical data from one of us (E.S.) are shown in Table 5.77.2 in alphabetical order. In sandalwood oils from various origins, over 125 chemicals have been identified. About 69 per cent of these were found in a single reviewed publication only. The major compounds found in sandalwood oils East India from different sources are shown in Table 5.77.3. They include (highest concentrations in any study given) (Z)-α- + (Z)-β-santalol (87.6%), (Z)-α-santalol (56.7%), (Z)-β-santalol (29.6%), (Z)-*trans*-α-bergamotol (8.6%), epi-β-santalol (6.6%), β-santalene (3.6%) and (Z)-lanceol

Table 5.77.1 ISO values (%) for East Indian sandalwood oil[a]

Compound	CAS	Minimum	Maximum
(Z)-α-Santalol	115-71-9	41.0	55.0
(Z)-β-Santalol	42495-69-2	16.0	24.0

[a] ISO 3518 Essential oil of sandalwood ©ISO 2002; Geneva, Switzerland, www.iso.org

(3.3%). A well-known ingredient of sandalwood oils East India that was present in a high concentration (>7%) in one study was(E)-α-santalal (12.8%). Uncommon or rare constituents of sandalwood oils East India found in high concentrations (>7%) in single studies include (E)-β-santalal (8.2%) and β-curcumen-12-ol (7.2%).

Commercial oils

The ten chemicals that had the highest maximum concentrations in 39 commercial East Indian sandalwood essential oil samples (concentration ranges provided) are the following: (Z)-α-santalol (43.4-53.3%), (Z)-β-santalol (15.6-23.6%), (Z)-trans-α-bergamotol (4.5-8.6%), epi-β-santalol (3.2-6.6%), β-santalene (0.5-3.6%), (Z)-lanceol (0.7-3.3%), (E)-nuciferol (0.3-2.9%), (E)-α-santalal (1.0-2.8%), epi-β-santalene (0.3-2.3%), (E)-β-santalol (0.5-2.0%) (Erich Schmidt, unpublished analytical data).

A comprehensive review of the constituents of the heartwood from fragrant sandalwood species including S. album obtained by hydrodistillation and solvent extraction is presented in ref. 13 (essential oil composition not separated from solvent extracts).

Table 5.77.2 Constituents identified in East Indian sandalwood oils

Constituent	CAS	Percentage and range				
		A	B (2)	C (4)	D (5)	E
Acetyldihydroalbene				<0.1		<0.01[e]
α-Acoradiene	24048-44-0			<0.1	+	
β-Acoradiene	28477-64-7		0.7		+	
10-epi-β-Acoradiene	847374-86-1				+	
γ-Acoradiene	28400-12-6			tr		
β-Alaskene	28908-21-6			0.1		
Albene	38451-64-8					<0.01[e]
(E)-Amylcinnamyl alcohol			1.0			
cis-Arteannuic alcohol	147648-62-2		5.1			
α-Bergamotal	148031-27-0			0.2		
(Z)-trans-α-Bergamotal						+[i]
cis-α-Bergamotene	18252-46-5			tr		
trans-α-Bergamotene	13474-59-4	0.05-0.9		0.2	+	0.1[i]
trans-β-Bergamotene	15438-94-5			0.1		
α-Bergamotenic acid	124439-27-6					+[i]
(Z)-α-Bergamotol	88034-74-6			0.2		
(Z)-trans-α-Bergamotol	88034-74-6	4.5-8.6		6.4	2.3-9.3[j]	0.2-1.8[c]; 3.9[i]; 4.7[i]
α-Bisabolene	17627-44-0	0.08-0.9				
(E)-α-Bisabolene	25532-79-0			tr	+	
(Z)-α-Bisabolene	29837-07-8		0.7			
β-Bisabolene	495-61-4	0.8-1.6		0.1	+	0.07[i]
(E)-γ-Bisabolene	53585-13-0			tr		
(Z)-γ-Bisabolene	13062-00-5			tr		
Bisabolenol A	120913-67-9					0.7[i,k]
Bisabolenol B	120913-68-0					0.7[i,k]
Bisabolenol C	120913-69-1					0.5[i,k]
Bisabolenol D	120913-70-4					1.8[i,k]
Bisabolenol E	120913-71-5					0.5[i,k]
α-Bisabolol	515-69-5			0.2	+	0.3[i]
epi-α-Bisabolol	78148-59-1				0.2-0.8[j]	0.5-1.8[c]
β-Bisabolol	15352-77-9			0.7[a]		0.6[i]
epi-β-Bisabolol	235421-59-7			0.7[a]		
Bulnesol	22451-73-6				+	
Caryophyllene oxide	1139-30-6		<0.1			
α-Cedrene	469-61-4			0.1	+	
Citronellol	106-22-9					+[g]
Curcumene (ar-; α-)	644-30-4	0.1-0.4	1.0	0.3	+	0.3[i]; 0.5[i]
β-Curcumene	28976-67-2		0.7	0.2	+	
γ-Curcumene	28976-68-3	0.04-0.2		0.1	+	0.04[i]
β-Curcumen-12-ol	698365-10-5					0.1[i]; 2.5-7.2[c]
γ-Curcumen-12-ol						1.2-2.2[c]
Cyclosantalal	168099-27-2			0.4		0.2-2.3[e]
epi-Cyclosantalal	168252-33-3			0.3		0.1-1.5[e]
Cyclosantalic acid						<0.01[e]
epi-Cyclosantalic acid						<0.01[e]
(E)-Dendrolasin	23262-34-2	0.01-0.08		<0.1		

Table 5.77.2 Constituents identified in East Indian sandalwood oils (*continued*)

Constituent	CAS	Percentage and range				
		A	B (2)	C (4)	D (5)	E
Dihydroalbene						<0.01[e]
Dihydro-ar-norcurcumenic acid						+[i]
Dihydro-α-santalic acid	25342-86-3					+[i]
Dihydro-α-santalol	126209-93-6			0.6		+[i]; 0.4[i]
Dihydro-β-santalol	34289-89-9					+[i]
endo-2, *endo*-3-Dimethylnorbornan-*exo*-2-ol	59432-92-7					+[g]
5,6-Dimethyl-5-norbornen-*exo*-2-ol	59300-41-3					+[g]
(*E*)-5-(2,3-Dimethyl-3-nortricyclyl)-pent-3-en-2-one						+[g]
α-Ekasantalal	887498-75-1					0.07[i]
β-Ekasantalal	13827-98-0					0.01[i]
α-Ekasantalic acid	1837-15-6					+[i]
α-nor-Ekasantalic acid						+[i]
(*E*)-β-Farnesene	18794-84-8			tr		
(*E,E*)-Farnesol	106-28-5				tr-3.9[j]	
Fokienol	33440-00-5			0.5		
α-Funebrene	50894-66-1				+	
N-Furfurylpyrrole (1-)	1438-94-4					+[g]
Geraniol	106-24-1					+[g]
Germacrene C	34323-15-4		<0.1			
Globulol	489-41-8		0.6			
Heliofolen-12-al					+	
Himachalol	1891-45-8		0.3			
β-Himachalol					+	
α-Humulene	6753-98-6		1.0			
(*E*)-α-Ionone	127-41-3		4.2			
12-Isoitalicenol					+	
trans-Isolongifolanone	14727-47-0		1.3			
3-Isothujopsanone	25966-81-8		0.6			
11-Keto-dihydro-α-santalic acid						+[i]
(*Z*)-Lanceol	859202-95-2	0.7-3.3		1.4	0.5-2.1[j]	0.5-0.8[c]; 1.5[i]; 1.7[i]
Limonenal	6784-13-0		0.3			
Longifolene (junipene)	475-20-7		0.3			
β-Longipinene	41432-70-6		0.6			
p-Methylacetophenone	122-00-9					+[g]
4,4-Methyl-cyclohexa-1,3-dien-1-yl methyl ketone						+[g]
4-Methylcyclohex-3-en-1-yl methyl ketone						+[g]
endo-2-Methyl-3-methylidenenorbornan-*exo*-2-ol	59300-40-2					+[g]
Muurol-5-en-4α-ol	157374-45-3		0.7			
(*E*)-Nerolidol	40716-66-3			0.1		0.06[i]
(*E*)-Nerolidyl acetate	85611-33-2		4.6			
(*Z*)-Nerolidyl acetate	91050-14-5		1.4			
Nortricycloekasantalic acid	59300-52-6					+[g]
Nortricycloekasantalol						+[g]
(*E*)-Nuciferal	25532-74-5					0.7-1.6[f]
(*E*)-Nuciferol	1786-15-8	0.3-2.9		0.1	+	
(*Z*)-Nuciferol	78339-53-4			3.4	0.8-4.9[j]	0.7-1.3[c]; 1.1[i]; 1.2[i]
β-Oplopenone	28305-60-4		0.4			
6,11-Oxido-acor-4-ene			0.6			

Table 5.77.2 Constituents identified in East Indian sandalwood oils (*continued*)

Constituent	CAS	Percentage and range				
		A	B (2)	C (4)	D (5)	E
trans-α-Photosantalol						4.7-8.4[f]
trans-β-Photosantalol						2.8-6.6[f]
trans-epi-β-Photo-santalol						1.4.1-1[f]
α-Santalal	13827-97-9		1.9			0.5[i]; 2.9[i]
(*E*)-α-Santalal	19903-70-9	1.0-2.8				+[g]; 12.8-7.7[f]
β-Santalal	13827-98-0			0.6		0.6[i]; 1.9[i]
(*E*)-β-Santalal	59331-82-7					+[g]; 8.2-4.0[f]
α-Santalene	512-61-8	0.2-1.9	0.6	0.7	+	0.8[i]; 1.1[i]; 2.2[l]
β-Santalene	511-59-1	0.5-3.6	1.0	1.2	+	1.0[i]; 1.4[i]; 1.8[l]
epi-β-Santalene	25532-78-9	0.3-2.3	1.4	0.8	+	1.0[i]
Santalene oxide	61754-00-5					3.1-1.6[f]
β-Santalic acid	73590-17-7					+[i]
(*E*)-α-Santalol	14490-17-6			0.4		0.6[i]
(*Z*)-α-Santalol	115-71-9	43.4-53.3	35.0	41.1	4.1-46.7[j]	17.8-33.0[c]; 38.8-52.5[b]; 40.6-48.0[d]; 50[i]; 56.7[l]
(*E*)-β-Santalol	37172-32-0	0.5-2.0	1.4	1.5		1.5[i]; 1.6[i]
(*Z*)-β-Santalol	42495-69-2	15.6-23.6	14.0	19.8	1.3-20.9[j]	7.5-16.4[c]; 16.5-32.9[b]; 18.3-26.7[c]; 20.9[i]; 29.6[l]
(*Z*)-α- + (*Z*)-β -Santalol						87.6[m]
epi-β-Santalol		3.2-6.6	1.7	3.5	0.7-3.7[j]	1.5-2.7[c]; 4.1[i]
α-Santalol acetate	41414-75-9				+	
Santalone	59300-51-5					+[g]
Santene	529-16-8	0.02-0.5	0.4	0.2		0.01[i]
Sesquisabinene	58319-04-3			tr	+	
epi-Sesquithujene (7-)	159407-35-9			tr	+	
α-Sinensal			0.5[h]			
β-Sinensal			5.7[h]			
Spirosantalol	117020-19-6	0.2-1.0		0.9		1.0[i]; 1.2[i]
Teresantalal	59300-39-9					+[g]
α-Terpineol	98-55-5					+[g]
α-Tetrasantalic acid						+[i]
4-(*p*-Tolyl)valeral-dehyde	4895-19-6					+[g]
Tricycloekasantalal	16933-18-9					0.3[i]
Tricycloekasantalic acid	59300-52-6					+[g]
Tricycloekasantalol	16933-12-3					+[g]
α-Zingiberene	495-60-3		0.5			

A thirty-nine sandalwood essential oil samples from India and Indonesia, analyzed between 1998 and 2004; lowest and highest concentrations given (E. Schmidt, unpublished data)
B one lab-hydrodistilled oil from *Santalum album* heartwood (ref. 2)
C one commercial oil sample from Tamil Nadu, India, purchased from a German company (ref. 4)
D one commercial oil, presence indicated by +, quantity not stated; four commercial oils, qualitatively and quantitatively investigated for the presence of eight constituents; lowest and highest concentrations given; the concentrations are expressed as percentages (v/v) of the sesquiterpenes alcoholic fraction (ref. 5)
E data from other studies (indicated with superscript letters); highest concentrations found in any study reviewed here given; when two or more oils were investigated, only the highest concentrations are mentioned, unless indicated otherwise

[a] β-bisabolol and epi-β-bisabolol combined; [b] lowest and highest concentrations for (*Z*)-α-santalol and (*Z*)-β-santalol in three commercial *S. album* oils (ref. 16); [c] nine lab-hydrodistilled oils from two *S. album* trees from Sri Lanka obtained from heartwood samples from 50 cm to three m above ground; lowest and highest concentrations given (ref. 19); [d] twenty-two commercial East Indian sandalwood oils; lowest and highest concentrations given (ref. 15); [e] unspecified number of essential oils from India (ref. 14); [f] two lab-hydrodistilled oils of *Santalum album* wood harvested in India and Indonesia (Java); both concentrations given (first East Indian, second Javanese); only some uncommon constituents were presented (ref. 21); [g] one commercial East Indian sandalwood oil sample (ref. 17); [h] wrongful determination (E. Schmidt); [i] data from various studies published before 1991, cited in ref. 23; [j] concentrations expressed as percentages (v/v) of the sesquiterpenes alcoholic fraction (ref. 5); [k] the presence of bisabolenols A,B, C and D is considered to be unlikely (ref. 23); [l] four lab-hydrodistilled or steam-distilled oils from heartwood material of Indian *S. album* prepared or pretreated in different manners (ref. 28); [m] one commercial sandalwood oil from *S. album* purchased in Taiwan (ref. 26)

tr: trace (in column C: <0.01); + present in the oil investigated, but quantity not stated

Table 5.77.3 Major constituents of East Indian sandalwood oils

Constituent	CAS	Percentage and range				
		A	B (2)	C (4)	D (5)	E
(Z)-α- + (Z)-β-Santalol						87.6[m]
(Z)-α-Santalol	115-71-9	43.4-53.3	35.0	41.1	4.1-46.7[j]	17.8-33.0[c]; 38.8-52.5[b]; 40.6-48.0[d]; 50[i]; 56.7[l]
(Z)-β-Santalol	42495-69-2	15.6-23.6	14.0	19.8	1.3-20.9[j]	7.5-16.4[c]; 16.5-32.9[b]; 18.3-26.7[c]; 20.9[i]; 29.6[l]
(Z)-trans-α-Berga-motol	88034-74-6	4.5-8.6		6.4	2.3-9.3[j]	0.2-1.8[c]; 3.9[i]; 4.7[i]
epi-β-Santalol		3.2-6.6	1.7	3.5	0.7-3.7[j]	1.5-2.7[c]; 4.1[i]
β-Santalene	511-59-1	0.5-3.6	1.0	1.2	+	1.0[i]; 1.4[i]; 1.8[l]
(Z)-Lanceol	859202-95-2	0.7-3.3		1.4	0.5-2.1[j]	0.5-0.8[c]; 1.5[i]; 1.7[i]

LEGEND: SEE UNDER TABLE 5.77.2

SANDALWOOD OIL (AUSTRALIA)

DEFINITION

Sandalwood oil (Australia) (essential oil of Australian sandalwood) is the essential oil obtained from the wood of the Australian sandalwood, *Santalum spicatum* (R. Br.) A. DC.

INCI NOMENCLATURE

Description/definition: Santalum spicata wood oil is the volatile oil obtained from the wood of the *Santalum spicata* (R. Br.) A. DC., Santalaceae

INCI name EU: Santalum spicata wood oil (perfuming name, not an INCI name proper)

INCI name USA: not in the Personal Care Products Council Ingredient Database

CAS registry number(s): 8024-35-9; 92875-02-0; 1175539-50-0

EINECS number(s): 296-618-5

ISO (INTERNATIONAL ORGANIZATION FOR STANDARDIZATION) STANDARD

ISO number: 22769

ISO name: Essential oil of Australian sandalwood

Botanical origin: *Santalum spicatum* (R. Br.) A. DC.

Parts of plant used: Wood

ISO values: ISO values (minimum and maximum concentrations) are shown in Table 5.77.4.

THE PLANT, THE OIL, AND THEIR USES

The West Australian sandalwood *Santalum spicatum* (R. Br.) A. DC. (synonyms: *Eucarya spicata* (R. Br.) Sprague & Summerh.; *Fusanus spicatus* R. Br.) is an evergreen tree that grows from 3 up to 8 meters tall. It is native to West Australia. *Santalum spicatum* is used for the production of furniture but mainly for producing an essential oil. Distillation of the stem wood, butt wood and roots yields the Australian sandalwood oil. This oil is employed for the production of cosmetics such as soaps, powder and creams (3) as well as for medicinal purposes (11). Sandalwood oils are widely used in aromatherapy; their presumed properties and medicinal applications have been reviewed (7).

Table 5.77.4 ISO values (%) for Australian sandalwood oil [a]

Compound	CAS	Minimum	Maximum
(Z)-α-Santalol	115-71-9	15.0	25.0
(Z)-β-Santalol	42495-69-2	5.0	20.0
(E,E)-Farnesol	106-28-5	2.5	15.0
(Z)-Nuciferol	1786-15-8	2.0	15.0
epi-α-Bisabolol	515-69-5	2.0	12.5
(Z)-trans-α-Bergamotol	88034-74-6	2.0	10.0
(Z)-Lanceol	859202-95-2	1.0	10.0
epi-β-Santalol		0.5	3.5

[a] ISO 22769 Essential oil of Australian sandalwood ©ISO 2009; Geneva, Switzerland, www.iso.org

CHEMICAL COMPOSITION

Australian sandalwood oil is a clear, slightly viscous, colorless to brownish liquid which has a soft, woody and sweet note with long lasting dry powdery odor. The yield of essential oil from the wood of *Santalum spicatum* (R. Br.) A. DC. generally varies from 1.7 to 3.5 per cent. The main producing country of this oil is Australia.

Literature data (up to December 7, 2014) on the chemical composition of Australian sandalwood oils (only 8 publications found) and unpublished analytical data from one of us (E.S.) are shown in Table 5.77.5 in alphabetical order. In Australian sandalwood oils from various origins, over 90 chemicals have been identified. About 70 per cent of these were found in a single reviewed publication only. The major compounds found in Australian sandalwood oils from different sources are shown in Table 5.77.6. They include (highest concentrations in any study given) (Z)-α-santalol (42.9%), (E,E)-farnesol (31.6%), (Z)-β-santalol (17.7%), epi-α-bisabolol (12.8%), (Z)-lanceol (10.8%), (Z)-nuciferol (10.7%), (Z)-trans-α-bergamotol (8.9%) and cis-β-curcumen-12-ol (7.8%). A rare constituent of Australian sandalwood oils found in a high concentration (>7%) in a single study includes farnesol (14.1%).

Table 5.77.5 Constituents identified in Australian and New Caledonian sandalwood oils

Constituent	CAS	AUSTRALIAN SANDALWOOD OILS (*Santalum spicatum*) Percentage and range				NEW CALEDIONIAN OILS (*Santalum austrocaledonicum*) Percentage and range		
		A	B (10)	C (11)	D	E	F (9)	G
Acetyldihydroalbene							<0.1	
α-Acoradiene	24048-44-0	0.06-0.3	0.1			0.04-0.3	<0.1	
β-Acoradiene	28477-64-7	0.07-0.1	<0.1			0.07-0.2	<0.1	
γ-Acoradiene	28400-12-6		0.1				tr	
α-Acorenol	28296-85-7				+f		tr	
epi-α-Acorenol					+f		tr	
β-Acorenol	28400-11-5				+f		tr	
epi-β-Acorenol					+f		tr	
β-Alaskene	28908-21-6		tr				<0.1	
Amorpha-4,11-diene	92692-39-2		tr					
(Z)-*trans*-α-Berga-motal							0.2	
cis-α-Bergamotene	18252-46-5		0.1				<0.1	
trans-α-Bergamotene	13474-59-4	0.08-0.9	0.4			0.06-0.9	0.2	
trans-β-Bergamotene	15438-94-5		0.1					
Bergamotol								7.2j
(Z)-*trans*-α-Berga-motol	88034-74-6	1.0-8.9	6.0	0.4	4.3e; 5.2c	5.0-8.6	9.9	9.2i
Bisabola-2,10-diene-6,13-diol			0.3				0.1	
Bisabola-2,10-diene-7,13-diol			tr				<0.1	
(E)-α-Bisabolene	25532-79-0		0.1				tr	
(Z)-α-Bisabolene	29837-07-8		<0.1					
β-Bisabolene	495-61-4	0.1-1.4	0.1	0.6		0.2-1.3	0.3	
γ-Bisabolene	495-62-5			tr				
(E)-γ-Bisabolene	53585-13-0		tr				<0.1	
(Z)-γ-Bisabolene	13062-00-5		<0.1				<0.1	
(Z)-γ-Bisabolen-12-ol							0.6	
α-Bisabolol	515-69-5		4.8	0.5		0.6-2.9	0.7	
epi-α-Bisabolol	515-69-5	0.4-12.8		10.7	5.9e; 6.6b			0.4i
β-Bisabolol	15352-77-9	0.4-1.2	2.1a			0.2-1.1	0.8a	
epi-β-Bisabolol	235421-59-7		2.1a				0.8a	
α-Bulnesene	3691-11-0		tr					
Campherenol	18530-03-5		0.6				0.5	
α-Cedrene	469-61-4	0.05-0.9	0.1			0.06-0.2	tr	
di-epi-α-Cedrene			tr					
epi-α-Cedrene	35944-22-0						tr	
β-Cedrene	546-28-1		tr					
Citronellol	106-22-9						tr	
Curcumene (ar-; α-)	644-30-4	0.1-1.2	0.3	0.5		0.1-0.5	0.3	
β-Curcumene	28976-67-2	0.08-0.6	0.7	0.8		0.07-0.2	0.2	
γ-Curcumene	28976-68-3	0.04-0.2	0.4			0.04-0.3	0.1	
cis-β-Curcumen-12-ol	698365-10-5	1.0-3.6	7.8			0.4-1.0	1.1	
cis-γ-Curcumen-12-ol			4.7				0.6	
Cyclosantalal	168099-27-2		0.1				0.7	
epi-Cyclosantalal	168252-33-3		0.1				0.4	
p-Cymene	99-87-6						tr	
p-Cymenene	1195-32-0		tr					
Dendrolasin	23262-34-2		1.0	2.0			0.1	
Dihydroalbene							tr	
Dihydro-α-santalol	126209-93-6						0.4	
1,3-Dimethyl-3-cyclo-hexenecarboxalde-hyde			tr					
endo-2, *endo*-3-Di-methylnorbornan-*exo*-2-ol	59432-92-7		<0.1					
α-Ekasantalal	887498-75-1						0.1	
β-Ekasantalal	13827-98-0						tr	

Table 5.77.5 Constituents identified in Australian and New Caledonian sandalwood oils (*continued*)

Constituent	CAS	AUSTRALIAN SANDALWOOD OILS (*Santalum spicatum*) Percentage and range				NEW CALEDIONIAN OILS (*Santalum austrocaledonicum*) Percentage and range		
		A	B (10)	C (11)	D	E	F (9)	G
11-epi-6,10-Epoxy-bisabol-2-en-12-ol	235421-61-1		1.2				tr	
(*Z,E*)-α-Farnesene	26560-14-5		tr				<0.1	
(*E*)-β-Farnesene	18794-84-8		0.1					
Farnesol	4602-84-0				14.1[d]			
(*E,E*)-Farnesol	106-28-5	1.4-18.4	19.8	31.6	11.0[b]; 11.5[e]	0.9-1.7	1.0	0.7[i]
(*E,E*)-Farnesyl acetate	4128-17-0		0.2					
Furfural	98-01-1		tr				tr	
N-Furfurylpyrrole (=1-)	1438-94-4						<0.1	
Geraniol	106-24-1		tr					
Geranylacetone (*E*-)	3796-70-1		tr	1.1			<0.1	
Guaiol	489-86-1		0.1					
nor-Helifolen-12-al I			tr		+[f]		<0.1	
nor-Helifolen-12-al II					+[f]		<0.1	
Humuladienone	24405-90-1						<0.1	
(*Z*)-12-Hydroxysesqui-cineole	699007-30-2		1.2				tr	
(*Z*)-Lanceol	859202-95-2	0.8-10.8	1.7	3.9	2.0[e]; 5.2[c]	5.0-15.2	9.1	3.9[i]; 9.1[j]
Limonene	138-86-3		tr				<0.1	
Linalool	78-70-6						tr	
p-Methylacetophe-none	122-00-9		tr					
1-(4-Methyl-3-cyclo-hexenyl)ethanone			tr				<0.1	
6-Methyl-3-hepten-2-one	2009-74-7		tr					
2-*exo*-Methyl-3-me-thylene-*endo*-2-norbornanol			tr					
α-Methyl teresantalic acid			tr					
Muurola-4,11-diene	235091-39-1		tr					
(*E*)-Nerolidol	40716-66-3	0.1-0.7	0.6	1.3		0.1-0.3	0.2	
(*E*)-Nuciferol	1786-15-8	3.9-9.1				0.6-2.6		
(*Z*)-Nuciferol	1786-15-8	0.6-10.7	4.4	6.5	6.9[b]; 8.0[e]		2.1	1.7[i]
1,11-Oxidocalamenene	143785-42-6		tr					
α-Photosantalol (A)	98113-14-5						0.1	
β-Photosantalol (A)	63569-02-8						0.1	
epi-β-Photosantalol							0.1	
α-Pinene	80-56-8		tr				tr	
α-Santalal	13827-97-9		0.4				0.9	
(*E*)-α-Santalal						0.1-0.7		
(*E*)-β-Santalal	59331-82-7					0.02-0.08	0.2	
(*Z*)-β-Santalal	887491-74-9						0.1	
α-Santalene	512-61-8	0.4-4.9	1.0	0.3		0.2-6.5	1.1	
β-Santalene	511-59-1	0.4-4.3	0.9	0.1		0.3-1.6	0.9	
epi-β-Santalene	25532-78-9	0.3-4.8	0.6	0.2		0.5-2.8	0.9	
Santalenone							tr	
nor-α-Santalenone							0.1	
nor-β-Santalenone	176777-64-3						0.1	
(*E*)-α-Santalol	14490-17-6		0.5				0.4	
(*Z*)-α-Santalol	115-71-9	17.0-42.9	14.5	9.1	18.7[g]; 22[c]; 26.5[d]	38.6-46.6	38.2	44.9[i]; 45.8[k]; 47.5[j]
(*E*)-β-Santalol	37172-32-0	0.2-1.3	0.7	0.6		0.2-0.9	0.4	
(*Z*)-β-Santalol	42495-69-2	6.4-17.7	8.4	5.4	7.3[g]; 8.1[b]; 13.5[d]	13.2-19.2	18.2	18.7[j] 19.4[i];20.6[k]
epi-β-Santalol		2.0-3.7	1.6	2.9	1.7[c]; 1.8[e]	2.2-4.4	3.8	3.1[i]; 3.3[k]; 3.6[j]

Table 5.77.5 Constituents identified in Australian and New Caledonian sandalwood oils (*continued*)

Constituent	CAS	AUSTRALIAN SANDALWOOD OILS (*Santalum spicatum*) Percentage and range				NEW CALEDIONIAN OILS (*Santalum austrocaledonicum*) Percentage and range		
		A	B (10)	C (11)	D	E	F (9)	G
(Z)-Santalol acetate		0.1-0.6				0.09-0.7		
Santene	529-16-8	0.2-1.9	0.1			0.1-2.3	0.1	
α-Selinene	473-13-2		<0.1					
β-Selinene	17066-67-0		<0.1					
Sesquicineole	90131-02-5		0.2				<0.1	
Sesquiphellandrene	73744-93-1						tr	
β-Sesquiphellandrene	20307-83-9		0.2					
Sesquisabinene A			tr					
Sesquisabinene B	1367879-38-6		0.3				tr	
Sesquisabinene hydrate	139341-65-4		0.2					
Sesquithujene	58319-06-5		0.1				tr	
epi-Sesquithujene (7-)	159407-35-9		tr				tr	
β-Sinensal (*cis*-)	17909-87-4							9.4[k]
Spirosantalol	117020-19-6		0.7				0.8	
Teresantalal	59300-39-9		tr				tr	
α-Teresantalic acid	562-66-3		0.4				tr	
Terpinen-4-ol	562-74-3						tr	
Tricycloekasantalol	16933-12-3						<0.1	
Zingiberene (α-)	495-60-3		tr					

AUSTRALIAN SANDALWOOD OILS (*Santalum spicatum*)
A twenty-three sandalwood essential oil samples from Australia analyzed between 2002 and 2013; lowest and highest concentrations given (E. Schmidt, unpublished data)
B one lab-hydrodistilled oil from the wood of butts and roots of *S. spicatum* (ref. 10)
C one laboratory steam-distilled oil from *S. spicatum* wood (ref. 11)
D data from other studies (indicated with superscript letters); highest concentrations found in any study reviewed here given; when two or more oils were investigated, only the highest concentrations are mentioned, unless indicated otherwise

[a] β-bisabolol and epi-β-bisabolol combined; [b] one laboratory steam-distilled *S. spicatum* essential oil (ref. 18); [c] one commercial Australian sandalwood oil purchased from an Australian company (ref. 20); [d] four commercial *S. spicatum* oils (ref. 16); [e] one commercial Australian sandalwood oil; the concentrations are expressed as percentages (v/v) of the sesquiterpenes alcoholic fraction (ref. 5); [f] one commercial Australian sandalwood oil; only qualitative data provided (ref. 24); [g] three commercial sandalwood oils *ex S. spicatum* (ref. 25)

NEW CALEDONIAN SANDALWOOD OILS (*Santalum austrocaledonicum*)
E thirty-nine New Caledonian sandalwood essential oil samples from Vanuatu and Nouméa analyzed between 2005 and 2013; lowest and highest concentrations given (E. Schmidt, unpublished data)
F one commercial oil sample obtained from New Caledonian *S. austrocaledonicum* var. *austrocaledonicum* (ref. 9)
G data from other studies on New Caledonian sandalwood oils (indicated with superscript letters); highest concentrations found in any study reviewed here given

[i] one commercial oil from New Caledonian *S. austrocaledonicum*; the concentrations are expressed as percentages (v/v) of the sesquiterpenes alcoholic fraction (ref. 5); [j] one commercial sandalwood oil *ex S. austrocaledonicum* purchased from a French company (ref. 27); [k] one commercial New Caledonian sandalwood oil purchased in Korea (ref. 29)

tr: trace (in columns B and F: <0.01); + present in the oil investigated, but quantity not stated

Commercial oils

The ten chemicals that had the highest maximum concentrations in 23 commercial Australian sandalwood essential oil samples (concentration ranges provided) are the following: (Z)-α-santalol (17.0-42.9%) (E,E)-farnesol (1.4-18.4%), (Z)-β-santalol (6.4-17.7%), epi-α-bisabolol (0.4-12.8%), (Z)-lanceol (0.8-10.8%), (Z)-nuciferol (0.6-10.7%), (E)-nuciferol (3.9-9.1%), (Z)-*trans*-α-bergamotol (1.0-8.9%), α-santalene (0.4-4.9%) and epi-β-santalene (0.3-4.8%) (Erich Schmidt, unpublished analytical data).

A comprehensive review of the constituents of the heartwood from fragrant sandalwood species including *S. spicatum* obtained by hydrodistillation and solvent extraction is presented in ref. 13 (essential oil composition not separated from solvent extracts).

SANDALWOOD OIL (NEW CALEDONIA)

DEFINITION

Sandalwood oil (New Caledonia) is the essential oil obtained from the wood of the New Caledonian sandalwood, *Santalum austrocaledonicum* Vieill.

Table 5.77.6 Major constituents of Australian sandalwood oils

Constituent	CAS	Australian Sandalwood Oils (*Santalum spicatum*) Percentage and range			
		A	B (10)	C (11)	D
(Z)-α-Santalol	115-71-9	17.0-42.9	14.5	9.1	18.7[g]; 22[c]; 26.5[d]
(E,E)-Farnesol	106-28-5	1.4-18.4	19.8	31.6	11.0[b]; 11.5[e]
(Z)-β-Santalol	42495-69-2	6.4-17.7	8.4	5.4	7.3[g]; 8.1[b]; 13.5[d]
epi-α-Bisabolol	515-69-5	0.4-12.8		10.7	5.9[e]; 6.6[b]
(Z)-Lanceol	859202-95-2	0.8-10.8	1.7	3.9	2.0[e]; 5.2[c]
(Z)-Nuciferol	1786-15-8	0.6-10.7	4.4	6.5	6.9[b]; 8.0[e]
(Z)-trans-α-Bergamotol	88034-74-6	1.0-8.9	6.0	0.4	4.3[e]; 5.2[c]
cis-β-Curcumen-12-ol	698365-10-5	1.0-3.6	7.8		

LEGEND: SEE UNDER TABLE 5.77.5

INCI NOMENCLATURE

Description/definition: Santalum austrocaledonicum oil is the volatile oil obtained from the heartwood of the *Santalum austrocaledonicum* Vieill., Santalaceae

INCI name EU and USA: Santalum austrocaledonicum wood oil

Other names: Oil of santal

CAS registry number(s): 91845-48-6

EINECS number(s): 295-223-5

ISO (INTERNATIONAL ORGANIZATION FOR STANDARDIZATION) STANDARD

There is currently no ISO standard for sandalwood oil from New Caledonia.

THE PLANT, THE OIL, AND THEIR USES

Santalum austrocaledonicum Vieill. is an evergreen tree that grows from 3 up to 8 meters tall. It is native to the islands of Vanuatu and Nouméa (New Caledonia). It is a partial parasite that attaches to the roots of other trees; it needs 'nurse' species in the area of planting out. In the forest, 'false guaiac' (*Acacia spirobis)* is observed as symbiosis plant, but also other plants can be linked with it. The New Caledonian sandalwood is used for the production of furniture, but mainly for producing essential oil. Distillation of the stem wood, butt wood and roots yields the New Caledonian sandalwood oil. This oil is employed in the production of cosmetics such as soaps, powder and creams (3). Sandalwood oils are widely used in aromatherapy; their presumed properties and medicinal applications have been reviewed (7).

CHEMICAL COMPOSITION

Sandalwood oil is a clear, slightly viscous, colorless to pale yellow liquid which has a soft, woody and sweet note with long lasting powdery odor. The yield of essential oil from the wood of *Santalum austrocaledonicum* Vieill. varies from 1.0 to 2.0 per cent. The main producing countries of this oil are New Caledonia (Islands of Nouméa, Lifou and Maré) and the Republic of Vanuatu.

Literature data (up to December 7, 2014) on the chemical composition of New Caledonian sandalwood oils (only 4 publications found) and unpublished analytical data from one of us (E.S.) are shown in Table 5.77.5 in alphabetical order. In sandalwood oils from various origins, over 90 chemicals have been identified. The major compounds found in New Caledonian sandalwood oils from different sources are shown in Table 5.77.7. They include (highest concentrations in any study given) (Z)-α-santalol (47.5%), (Z)-β-santalol (20.6%), (Z)-lanceol (15.2%), (Z)-trans-α-bergamotol (9.9%), α-santalene (6.5%) and epi-β-santalol (4.4%). A rare constituent of New Caledonian sandalwood oils found in a high concentration (>7%) in a single study is β-sinensal (cis-)(9.4%).

Commercial oils

The ten chemicals that had the highest maximum concentrations in 39 commercial New Caledonian sandalwood essential oil samples (concentration ranges

Table 5.77.7 Major constituents of New Caledonian sandalwood oils

Constituent	CAS	New Caledonian Oils (*Santalum austrocaledonicum*) Percentage and range		
		E	F (9)	G
(Z)-α-Santalol	115-71-9	38.6-46.6	38.2	44.9[i]; 45.8[k]; 47.5[j]
(Z)-β-Santalol	42495-69-2	13.2-19.2	18.2	18.7[j]; 19.4[i]; 20.6[k]
(Z)-Lanceol	859202-95-2	5.0-15.2	9.1	3.9[i]; 9.1[j]
(Z)-trans-α-Bergamotol	88034-74-6	5.0-8.6	9.9	9.2[i]
α-Santalene	512-61-8	0.2-6.5	1.1	
epi-β-Santalol		2.2-4.4	3.8	3.1[i]; 3.3[k]; 3.6[j]

LEGEND: SEEE UNDER TABLE 5.77.5

Table 5.77.8 Differences between New Caledonian, Australian and East Indian sandalwood oils[a]

Component	New Caledonian oils	Australian oils	East Indian oils
(Z)-trans-α-Bergamotol	5.0-8.6	1.0-8.9	4.5-8.6
α-Bisabolol	0.6-2.9	–	–
β-Bisabolol	0.2-1.1	0.4-1.2	–
(E,E)-Farnesol	0.9-1.7	1.4-18.4	–
(Z)-Lanceol	5.0-15.2	0.8-10.8	0.7-3.3
(E)-Nuciferol	0.6-2.6	3.9-9.1	0.3-2.9
(Z)-Nuciferol	-	0.6-10.7	–
(Z)-α-Santalol	38.6-46.6	17.0-42.9	43.4-53.3
(E)-β-Santalol	0.2-0.9	0.2-1.3	0.5-2.0
(Z)-β-Santalol	13.2-19.2	6.4-17.7	15.6-23.6
epi-β-Santalol	2.2-4.4	2.0-3.7	3.2-6.6

[a] Erich Schmidt, unpublished data 1998-2013

provided) are the following: (Z)-α-santalol (38.6-46.6%), (Z)-β-santalol (13.2-19.2%), (Z)-lanceol (5.0-15.2%), (Z)-trans-α-bergamotol (5.0-8.6%), α-santalene (0.2-6.5%), epi-β-santalool (2.2-4.4%), α-Bisabolol (0.6-2.9%), epi-β-santalene (0.5-2.8%), (E)-nuciferol (0.6-2.6%) and santene (0.1-2.3%) (Erich Schmidt, unpublished analytical data).

A comprehensive review on the constituents of the heartwood from fragrant sandalwood species including *S. austrocaledonicum* obtained by hydrodistillation and solvent extraction is presented in ref. 13 (essential oil composition not separated from solvent extracts).

Differences between the three sandalwood oils

The major differences among the three sandalwood oils can be seen in Table 5.77.8. The comparison of the compositions of New Caledonian, Australian and East Indian sandalwood oils is based on the analyses of 39 New Caledonian, 23 Australian and 39 East Indian commercial oil samples (Erich Schmidt, unpublished data).

The percentage of the santalols is the highest in the East Indian oil and the lowest in the Australian variety. The Australian sandalwood oils may have high concentrations of (E,E)-farnesol and (Z)-nuciferol, the former being absent in the East Indian oils and the latter chemical being absent in both other oils. α-Bisabolol was present in commercial oils from *S. austrocaledonicum* only and the New Caledonian oils had the highest concentrations of (Z)-lanceol.

CONTACT ALLERGY/ALLERGIC CONTACT DERMATITIS

General

Contact allergy to/allergic contact dermatitis from sandalwood oil (botanical source rarely stated) has been reported in over 30 publications. Pigmented cosmetic dermatitis from sandalwood oil used to be frequent in Japanese women. In groups of consecutive patients suspected of contact dermatitis, prevalence rates of up to

2.4% positive patch test reactions have been observed, but reliable relevance data are lacking. There are only a few case reports of allergic contact dermatitis from sandalwood oil. In one of these, santalol may have been an allergen in the oil.

Testing in groups of patients

The results of patch tests with sandalwood oil in routine testing (consecutive patients suspected of contact dermatitis) and in groups of selected patients are shown in Table 5.77.9. In routine testing, rates of positive reactions ranged from 0.1% to 2.4%, whereas between 0.6% and 20% of patients in selected groups had positive patch tests. The high positivity rate of 20% was seen in a very small group of 10 patients strongly suspected of fragrance allergy and reacting to the fragrance mix (37).

Case reports

Sandalwood oil was responsible for 3 out of 399 cases of cosmetic (photo)allergy where the causal allergen was identified in a study of the NACDG, USA, 1977-1983 (33). One patient suspected to be allergic to incense had positive patch tests to two brands of incense, sandalwood oil, musk ambrette and santalol; gas chromatography of pentane:ether extracts of the incense showed 9% and 34% musk ambrette and 8% santalol in both incenses. The sandalwood oil extract contained 73% santalol, which is the dominant component of sandalwood oil (65). One bottle filler in a perfume factory developed occupational allergy to East Indian sandalwood oil and some other essential oils he had contact with at work (44). Four positive patch tests to sandalwood and three photopatch tests were seen in an investigation in New York, 1986-1993; some (number not specified) of these were considered to be relevant (32).

Positive patch tests (relevance unknown, uncertain or not stated)

Three positive patch tests to sandalwood oil occurred in massage therapists/aromatherapists

Table 5.77.9 Results of testing groups of patients with sandalwood oil

Years and Country	Test conc. & vehicle	Number of patients		Selection of patients (S); Relevance (R); Comments (C)	Ref.
		tested	positive (%)		
Routine testing					
2000-2008 IVDK	10% pet.	3,671	46 (1.3%)	R: not stated	50
2000-2007 USA	2% pet.	870	17 (2.0%)	R: 82%; C: weak study: a. high rate of macular erythema and weak reactions, b. relevance figures include 'questionable' and 'past' relevance	36
2002-2003 Korea	2% pet.	422	10 (2.4%)	R: not stated	51
1999-2000 Denmark	2% pet.	318	4 (1.3%)	R: not specified; C: this study was part of the international study mentioned below (ref. 52)	63
	10% pet.	318	5 (1.6%)		
1998-2000 six	10% pet.	1,606	15 (0.9%)	R: not specified for individual oils/chemicals; the test substance was prepared from East Indian sandalwood oil	52
European countries	2% pet.	1,606	7 (0.4%)		
1979-1990 Japan	2% pet.	3,152	44 (1.4%)	R: not stated	55
1983-1984 Italy	2% pet.	1,200	1 (0.1%)	R: not stated	56
Testing in selected groups of patients					
2006-2010 USA	2% pet.	100	4 (4%)	S: patients with eyelid dermatitis; R: not stated	39
2001-2010 Australia	2% pet.	986	51 (5.2%)	S: not specified; R: 28%	60
2001-2010 Canada	not stated	160	20 (12.5%)	S: patients with suspected photosensitivity and patients who developed pruritus or a rash after sunscreen application; R: not stated; C: weak study: inadequate reading of test results, erythema only was considered to represent a positive patch test reaction	63
2000-2008 IVDK	10% pet.	1,002	18 (1.8%)	S: patients with dermatitis suspected of causal exposure to fragrances; R: not stated	50
2004-2008 Spain	2% pet.	86	2 (2.3%)	S: patients previously reacting to the fragrance mix I or *Myroxylon pereirae* (n=54) or suspected of fragrance contact allergy (n=32); R: not stated	49
1993-2006 USA	2% pet.	76	1 (1.3%)	S: not stated; R: not specified	35
2000-2005 USA	2% pet.	182	2 (1.1%)	S: patients with suspected photodermatoses and/or with suspected allergic reactions to sunscreen products; R: one reaction was relevant; C: there were also three patients with *photo*allergic reactions to sandalwood oil	48
1989-1999 Portugal	2% pet.	67	4 (6.0%)	S: patients who had a positive patch test to the fragrance mix; R: not stated	53
1996-1997 UK	2% pet.	10	2 (20%)	S: patients suspected of cosmetic dermatitis and reacting to the fragrance mix; R: not stated	37
1990-1998 Japan	2% pet.	1,483	12 (0.8%)	S: patients suspected of cosmetic contact dermatitis, virtually all were women; range of annual frequency of sensitization: 0-1.9%; R: not stated	40
<1996 Japan, Ireland, USA, UK, Switzerland, Sweden	10% pet.	167	11 (6.6%)	S: patients known or suspected to be allergic to fragrances; R: not stated	38
<1994 Japan	?	?	? (0.9%)	S and R: unknown. Possibly routine testing	54
1985-1990 USA	undiluted	176	1 (0.6%)	S: patients with history of photosensitivity; R: not stated	31
<1986 Poland	2% pet.	86	2 (2.3%)	S: patients previously reacting to the fragrance mix	47
<1986 France	2% pet.	21	4 (19%)	S: patients with dermatitis caused by fragrances; R: not stated	62
<1976 France	2% pet.	51	2 (3.9%)	S: patients allergic to *Myroxylon pereirae* resin (balsam of Peru) and/or turpentine and/or wood tar and/or colophony; R: not stated	43
<1974 Japan	?	137	14 (10.2%)	S: patients suspected of cosmetic dermatitis; R: unknown	58

IVDK Information Network of Departments of Dermatology, Germany, Switzerland, Austria (www.ivdk.org); pet.: petrolatum

with occupational contact dermatitis from multiple essential oils; it is uncertain whether sandalwood oil had been used by the patients (42). Of seven patients allergic to the fragrance farnesol, 4 (57%) co-reacted to sandalwood oil (and various other fragrances) (61). One positive patch test reaction to 'sandal oil' was seen in a group of 460 patients with positive patch tests related to cosmetics (57).

Pigmented cosmetic dermatitis

In Japan, in the 1960s and 1970s, many female patients developed pigmentation following dermatitis of the face (46). This so-called pigmented cosmetic dermatitis was shown to be caused by contact allergy to components of cosmetic products, notably essential oils, other fragrance materials, antimicrobials, preservatives and coloring materials (45,46). In a group of 620 Japanese patients with this condition investigated between 1970 and 1980, 3-5% had positive patch test reactions to sandalwood oil 10% in petrolatum (45). The number of patients with pigmented cosmetic dermatitis decreased strongly after 1978, when major cosmetic companies began to eliminate strong contact sensitizers from their products (45).

Sandalwood oil has also been responsible for a number of cases of *photo*contact allergy (30,31,32,33,34,35,41). Additional information on contact allergy to sandalwood oil can be found in ref. 64 (article not read).

LITERATURE

1   Jones CG, Plummer JA, Barbour EL. Non-destructive sampling of Indian sandalwood (*Santalum album* L.) for oil content and composition. J Essent Oil Res 2007;19:157-164

2   Marongiu B, Piras A, Porcedda S, Tuveri E. Extraction of *Santalum album* and *Boswellia carterii* Birdw. volatile oil by supercritical carbon dioxide: influence of some process parameters. Flavour Fragr J 2006;21:718-724

3   Venkatesha Gowda VS, Patil KB, Ashwath DS. Manufacturing of sandalwood oil, market potential demand and use. J Essent Oil Bear Plants 2004;7:293-297

4   Braun NA, Meier M, Pickenhagen W. Isolation and chiral GC analysis of β-bisabolols—trace constituents from the essential oil of *Santalum album* L. (Santalaceae). J Essent Oil Res 2003;15:63-65

5   Sciarrone D, Costa R, Ragonese C, Tranchida PQ, Tedone L, Santi L, et al. Application of a multidimensional gas chromatography system with simultaneous mass spectrometric and flame ionization detection to the analysis of sandalwood oil. J Chromatogr A 2011;1218:137-142

6   Kuriakose S, Joe IH. Feasibility of using near infrared spectroscopy to detect and quantify an adulterant in high quality sandalwood oil. Spectrochim Acta Part A: Mol Biomol Spectrosc 2013;115:568-573

7   Erligmann A. Sandalwood oils. Int J Aromather 2001;11:186-192

8   Burdock GA, Carabin IG. Safety assessment of sandalwood oil (*Santalum album* L.). Food Chem Toxicol 2008;46:421-432

9   Braun NA, Meier M, Hammerschmidt F-J. New Caledonian sandalwood oil—a substitute for East Indian sandalwood oil? J Essent Oil Res 2005;17:477-480

10 Valder C, Neugebauer M, Meier M, Kohlenberg B, Hammerschmidt F-J, Braun NA. Western Australian sandalwood oil — new constituents of *Santalum spicatum* (R. Br.) A. DC. (Santalaceae). J Essent Oil Res 2003;15:178-186

11 Brophy JJ, Fookes CJR, Lassak EV. Constituents of *Santalum spicatum* (R. Br.) A. DC. wood oil. J Essent Oil Res 1991;3:381-385

12 Misra BB, Dey S. Comparative phytochemical analysis and antibacterial efficacy of *in vitro* and *in vivo* extracts from East Indian sandalwood tree (*Santalum album* L.). Lett Appl Microbiol 2012;55:476-486

13 Baldovini N, Delasalle C, Joulain D. Phytochemistry of the heartwood from fragrant *Santalum* species: a review. Flavour Fragr J 2011;26:7-26

14 Brunke E-J, Vollhardt J, Schmaus G. Cyclosantalal and epicyclosantalal—new sesquiterpenes aldehydes from East Indian sandalwood oil. Flavour Fragr J 1995;10:211-219

15 Verghese J, Sunny TP, Balakrishnan KV. (Z)-(+)-α-santalol and (Z)-(-)-β-santalol concentration, a new quality determinant of East Indian sandalwood oil. Flavour Fragr J 1990;5:223-226

16 Jirovetz L, Buchbauer G, Denkova Z, Stoyanova A, Murgov I, Gearon V, et al. Comparative study on the antimicrobial activities of different sandalwood essential oils of various origin. Flavour Fragr J 2006;21:465-468

17 Demole E, Demole C, Enggist P. A chemical Investigation of the volatile constituents of East Indian sandalwood oil (*Santalum album* L.). Acta Helv Chim 1976;59:737-747

18 Piggot MJ, Ghisalberti EL, Trengove RD, Western Australian sandalwood oil: extraction by different techniques and variations of the major components in different sections of a single tree. Flavour Fragr J 1997;12:43-46

19 Subasinghe U, Gamage M, Hettiarachchi DS. Essential oil content and composition of Indian sandalwood (*Santalum album*) in Sri Lanka. J Forest Res 2013;24:127-130

20 Shellie R, Marriott P, Morrison P. Comprehensive two-dimensional gas chromatography with flame ionization and time-of-flight mass spectrometry detection: qualitative and quantitative analysis of West Australian sandalwood oil. J Chromatogr Sc 2004;42:417-422

21 Mookherjee BD, Trenkle RW, Wilson RA. New insights in the three most important natural fragrance products: wood, amber and musk. In: H Woidich and G. Buchbauer, Eds. Proceedings of the 12th International Congress of Flavours, Fragrances and Essential Oils, Oct 4-8, 1992. Vienna, Austria: Austrian Assoc Flav Frag Industry, 1992:234-262. Data cited in ref. 22

22 Lawrence BM. Progress in essential oils. Perfum Flavor 2009;34 (May):52-56

23 Lawrence BM. Progress in essential oils. Perfum Flavor 1991;16(6):50-52

24 Braun NA, Meier M, Kohlenberg B, Valder C, Neugebauer M. *Santalum spicatum* (R. Br.) A. DC. (Santalaceae)—nor-helifolenal and acorenol isomers: Isolation and biogenetic considerations. J Essent Oil Res 2003;15:381-386

25 Howes M-JR, Simmonds MSJ, Kite GC. Evaluation of the quality of sandalwood essential oils by gas chromatography—mass spectrometry. J Chromatogr 2004;1028:307-312

26 Yen H-F, Wang S-Y, Wu C-C, Lin W-Y, Wu T-Y, Chang F-R, et al. Cytotoxicity, anti-platelet aggregation assay and chemical components analysis of thirty-eight kinds of essential oils. J Food Drug Anal 2012;20:478-483

27 Roh HS, Kim JH, Shin E-S, Lee DW, Choo HY, Park CG. Bioactivity of sandalwood oil (*Santalum austro-caledonicum*) and its main components against the cotton aphid, *Aphis gossypii*. J Pest Sci 2014. DOI 10.1007/s10340-014-0631-1

28 Nautiyal OH. Analytical and Fourier transform infrared spectroscopy evaluation of sandalwood oil extracted with various process techniques. J Nat Prod 2011;4:150-157

29 Roh HS, Lim EG, Kim J, Park CG. Acaricidal and oviposition deterring effects of santalol identified in sandalwood oil against two-spotted spider mite, *Tetranychus urticae* Koch (Acari:Tetranychidae). J Pest Sci 2011;84:495-501

30 Pigatto PD, Legori A, Bigardi AS, Guarrera M, Tosti A, Santucci B, et al. Gruppo Italiano recerca dermatiti da contatto ed ambientali Italian multicenter study of allergic contact photodermatitis: epidemiological aspects. Am J Cont Derm 1996;7:158-163

31 DeLeo VA, Suarez SM, Maso MJ. Photoallergic contact dermatitis. Results of photopatch testing in New York, 1985 to 1990. Arch Dermatol 1992;128:1513-1518

32 Fotiades J, Soter NA, Lim HW. Results of evaluation of 203 patients for photosensitivity in a 7.3 year period. J Am Acad Dermatol 1995;33:597-602

33 Adams RM, Maibach HI. A five-year study of cosmetic reactions. J Am Acad Dermatol 1985;13:1062-1069

34 Starke JC. Photoallergy to sandalwood oil. Arch Dermatol 1967;96:62-63

35 Victor FC, Cohen DE, Soter NA. A 20-year analysis of previous and emerging allergens that elicit photoallergic contact dermatitis. J Am Acad Dermatol 2010;62:605-610

36 Wetter DA, Yiannias JA, Prakash AV, Davis MD, Farmer SA, el-Azhary RA, et al. Results of patch testing to personal care product allergens in a standard series and a supplemental cosmetic series: an analysis of 945 patients from the Mayo Clinic Contact Dermatitis Group, 2000-2007. J Am Acad Dermatol 2010;63:789-798

37 Thomson KF, Wilkinson SM. Allergic contact dermatitis to plant extracts in patients with cosmetic dermatitis. Br J Dermatol 2000;142:84-88

38 Larsen W, Nakayama H, Lindberg M, Fischer T, Elsner P, Burrows D, et al. Fragrance contact dermatitis: A worldwide multicenter investigation (Part 1). Am J Cont Derm 1996;7:77-83

39 Wenk KS, Ehrlich AE. Fragrance series testing in eyelid dermatitis. Dermatitis 2012;23:22-26

40 Sugiura M, Hayakawa R, Kato Y, Sugiura K, Hashimoto R. Results of patch testing with lavender oil in Japan. Contact Dermatitis 2000;43:157-160

41 Greenspoon J, Ahluwalia R, Juma N, Rosen CF. Allergic and photoallergic contact dermatitis: A 10-year experience. Dermatitis 2013;24:29-32

42 Bleasel N, Tate B, Rademaker M. Allergic contact dermatitis following exposure to essential oils. Australas J Dermatol 2002;43:211-213

43 Rudzki E, Grzywa Z, Bruo WS. Sensitivity to 35 essential oils. Contact Dermatitis 1976;2:196-200

44 Schubert HJ. Skin diseases in workers at a perfume factory. Contact Dermatitis 2006;55:81-83

45 Nakayama H, Matsuo S, Hayakawa K, Takhashi K, Shigematsu T, Ota S. Pigmented cosmetic dermatitis. Int J Dermatol 1984;23:299-305

46 Nakayama H, Harada R, Toda M. Pigmented cosmetic dermatitis. Int J Dermatol 1976;15:673-675

47 Rudzki E, Grzywa Z. Allergy to perfume mixture. Contact Dermatitis 1986;15:115-116

48 Scalf LA, Davis MDP, Rohlinger AL, Connolly SM. Photopatch testing of 182 patients: a 6-year experience at the Mayo Clinic. Dermatitis 2009;20:44-52

49 Cuesta L, Silvestre JF, Toledo F, Lucas A, Pérez-Crespo M, Ballester I. Fragrance contact allergy: a 4-year retrospective study. Contact Dermatitis 2010;63:77-84

50 Uter W, Schmidt E, Geier J, Lessmann H, Schnuch A, Frosch P. Contact allergy to essential oils: current patch test results (2000–2008) from the Information Network of Departments of Dermatology (IVDK). Contact Dermatitis 2010;63:277-283

51 An S, Lee AY, Lee CH, Kim D-W, Hahm JH, Kim K-J, et al. Fragrance contact dermatitis in Korea: a joint study. Contact Dermatitis 2005;53:320-323

52 Frosch PJ, Johansen JD, Menné T, Pirker C, Rastogi SC, Andersen KE, et al. Further important sensitizers in patients sensitive to fragrances. II. Reactivity to essential oils. Contact Dermatitis 2002;47:279-287

53 Manuel Brites M, Goncalo M, Figueiredo A. Contact allergy to fragrance mix—a 10-year study. Contact Dermatitis 2000;43:181-182

54 Sugai T. Group study IV – farnesol and lily aldehyde. Environ Dermatol 1994;1:213-214

55 Utsumi M, Sugai T, Shoji A, Watanabe K, Asoh S, Hashimoto Y. Incidence of positive reactions to sandalwood oil and its related fragrance materials in patch tests and a case of contact allergy to natural and synthetic sandalwood oil in a museum worker. Skin Res 1992;34(Suppl. 14):209-213

56 Santucci B, Cristaudo A, Cannistraci C, Picardo M. Contact dermatitis to fragrances. Contact Dermatitis 1987;16:93-95

57 Romaguera C, Camarasa JMG, Alomar A, Grimalt F. Patch tests with allergens related to cosmetics. Contact Dermatitis 1983;9:167-168

58 Nakayama H, Hanaoka H, Ohshiro A. Allergen Controlled System (ACS). Tokyo, Japan: Kanehara Shuppan, 1974:42. Data cited in ref. 59

59 Mitchell JC. Contact hypersensitivity to some perfume materials. Contact Dermatitis 1975;1:197-199

60 Toholka R, Wang Y-S, Tate B, Tam M, Cahill J, Palmer A, Nixon R. The first Australian Baseline Series: Recommendations for patch testing in suspected contact dermatitis. Australas J Dermatol 2014, Sept. 7. doi: 10.1111/ajd.12186

61 Goossens A, Merckx L. Allergic contact dermatitis from farnesol in a deodorant. Contact Dermatitis 1997;37:179-180

62 Meynadier JM, Meynadier J, Peyron JL, Peyron L. Formes cliniques des manifestations cutanées d'allergie aux parfums. Ann Dermatol Venereol 1986;113:31-39

63 Paulsen E, Andersen KE. Colophonium and Compositae mix as markers of fragrance allergy: Cross-reactivity between fragrance terpenes, colophonium and Compositae plant extracts. Contact Dermatitis 2005;53:285-291

64 Sugai T. Historical data of the JSCD Group study III – Fragrance materials. Environ Dermatol 1994;1:209-212

65 Hayakawa R, Matsunaga K, Arima Y. Depigmented contact dermatitis due to incense. Contact Dermatitis 1987;16:272-274

Chapter 5.78   SILVER FIR OIL

DEFINITION

Silver fir oil is the essential oil obtained from the needles of the (European) silver fir, *Abies alba* Mill.

INCI NOMENCLATURE

Description/definition: Abies alba leaf oil is the volatile oil derived from the needles of the fir, *Abies alba*, Pinaceae

INCI name EU & USA: Abies alba leaf oil

Other names: Fir needle oil

CAS registry number(s): 8021-27-0; 90028-76-5

EINECS number(s): 289-870-2

ISO (INTERNATIONAL ORGANIZATION FOR STANDARDIZATION) STANDARD

There is currently no ISO standard available for silver fir oil.

THE PLANT, THE OIL, AND THEIR USES

The common silver fir *Abies alba* Mill., also known as the Christmas tree, is an evergreen tree up to 45-55 m tall with a diameter of 200-260 cm. It is native to middle and south European countries, Ukraine and Belarus. The tree is found at altitudes of 300-1,700 m on mountains with a rainfall of over 1,000 mm. The silver fir is widely cultivated (GRIN Taxonomy for Plants; www.conifers.org; 2). In Europe, the essential oil obtained from the needles is said to be used in perfumes, room sprays, deodorants and bath preparations (we doubt whether this oil is used in fragrances and cosmetics). Silver fir oil is also employed in inhalants for the treatment of colds, and in medicinal preparations against rheumatism and similar ailments and is said to have soothing qualities (1). It is also employed in aromatherapy practices (13,14).

CHEMICAL COMPOSITION

Silver fir oil is a colorless mobile liquid which has a fresh coniferous, slightly camphoraceous odor. The yield of essential oil from the needles of *Abies alba* Mill. generally varies from 0.1 to 0.2 per cent. The main producing countries of this oil are France (incl. Corsica), Montenegro, Poland, and Austria.

Literature data (up to October 17, 2014) on the chemical composition of silver fir oils and unpublished analytical data from one of us (E.S.) are shown in Table 5.78.1 in alphabetical order. In silver fir oils from various origins, over 110 chemicals have been identified. About 42 per cent of these were found in a single reviewed publication only. The major compounds found in silver fir oils from different sources are shown in Table 5.78.2. They include (highest concentrations in any study given) limonene (68.3%), α-pinene (38.9%), β-pinene (32.8%), bornyl acetate (30.3%) and camphene (25.2%). Well-known ingredients of silver fir oils that were present in high concentrations (>7%) in one or two studies were β-phellandrene (15.1% and 14.4%) and tricyclene (12.9%). Uncommon or rare constituents of silver fir oils found in high concentrations (>7%) in single studies include α-fenchyl acetate (14.2%) and δ3-carene (13.9%).

Commercial oils

The ten chemicals that had the highest maximum concentrations in 16 commercial silver fir essential oil samples (concentration ranges provided) are the following: limonene (6.1-54.7%), α-pinene (0.5-2.8%), β-pinene (7.4-31.7%), camphene (5.8-17.3%), bornyl acetate (0.4-14.2%), β-phellandrene (0.01-4.9%), β-caryophyllene (0.1-4.2%), tricyclene (0.5-2.6%), myrcene (0.7-2.5%) and α-terpineol (0.07-2.3%) (Erich Schmidt, unpublished analytical data).

Chemotypes

In an investigation of the oils of over 50 trees from Corsica (France), two clusters were found: I (64%) and II (36%). These two types of essential oils were distinguished on the basis of their contents of limonene, camphene, α-pinene, β-phellandrene and β-pinene. The chemical composition of samples of cluster I was characterized by a very high content of limonene (mean 46.1%) and appreciable contents of camphene (mean 16.9) and α-pinene (mean 12.2). Similar oils dominated by limonene have been found elsewhere, including in Austria (55%: ref. 5), Greece (46%: ref. 9) and France (34%: ref. 6). The essential oils of cluster II from Corsica were characterized by comparable proportions of camphene (mean 23.7), α-pinene (mean 18.5) and limonene (15.6), combined with appreciable concentrations of β-phellandrene (up to 23%) and β-pinene (up to 12%). In Albania, *Abies alba* oils seem to be dominated by β-pinene (10), which was also the component with the highest concentration in some other oils (1,7). Occasionally, bornyl acetate was found to be the dominant ingredient in silver fir oils (8,9).

Table 5.78.1 Constituents identified in silver fir oils

Constituent	CAS	Percentage and range						
		A	B (4)	C (4)	D (2)	E (1)	F (7)	G
Androstan-17-one, 1,3-ethyl-3-hydroxy-, (5α)-	57344-99-7							0.1[f]
Aristolene	6831-16-9				0.7			
Aromadendrene	489-39-4							0.05[f]
β-Bisabolene	495-61-4			0.1		tr		0.1[f]
Borneol	507-70-0	0.05-1.9		0.1		2.1[a]	0.2	0.4-5.5[k]; 1.0[m]; 1.5[l]; 1.7[f]
Bornyl acetate	76-49-3	0.4-14.2	2.7	2.3		9.0	7.1	1.3[d]; 1.6-17.6[k]; 4.5[g]; 9.9[m]; 18.8[g]; 24.4[i]; 30[l]; 30.3[f]
Cadina-1,4-diene	29837-12-5							0.1-0.3[k]
α-Cadinene	24406-05-1							tr-0.3[k]
γ-Cadinene	39029-41-9				0.8	1.1		tr-1.2[k]; 0.4[e]; 0.7[e]
δ-Cadinene	483-76-1	0.1-1.1	0.1	0.3	0.9	0.4		0.1[m]; 0.3-3.5[k]; 0.4[e]; 0.7[e]
α-Cadinol	481-34-5	tr-0.3		0.1				
epi-α-Cadinol	5937-11-1	0-0.4		0.1				0.3[m]
α-Calacorene	21391-99-1							tr-0.3[k]
trans-Calamenene	73209-42-4	tr						
Camphene	79-92-5	5.8-17.3	9.9	20.6	15.3	16.7	15.2	3.0-17.2[k]; 14.8[c]; 15.7[m]; 18[l]; 19.6[i]; 19.8[f]; 21.3[g]; 25.2[g]
Camphene hydrate	465-31-6	tr-0.4						tr-0.4[k]
2-exo-Camphene hydrate			0.2					
Campholenal	23727-15-3							
α-Campholenal	4501-58-0		0.1	0.1		0.1	0.09	tr-0.5[k]
Camphor	76-22-2					0.2	0.05	tr-0.2[k]
δ2-Carene (= δ4-)	554-61-0						0.02	
δ3-Carene	13466-78-9	0.05-0.3					0.08	0.3[d]; 2.0[g]; 13.9[f]
(E)-Carveol	1197-07-5		0.1					
Carvone	99-49-0		0.2					
β-Caryophyllene	87-44-5	0.1-4.2	0.4	0.7	4.0	1.3	0.2	0.6-12.8[k]; 2.2[f]; 2.3[c]; 4.2[d]
9-epi-(E)-Caryophyllene (β-)	68832-35-9			0.1				tr-0.4[k]
Caryophyllene oxide	1139-30-6	0-0.2	tr			0.2		
β-Cedrene	546-28-1			0.1				
Cembrene	1898-13-1		tr	0.2				
1,8-Cineole	470-82-6		tr	tr			0.06	
Citronellal	106-23-0			0.2				0.2[m]
Citronellol	106-22-9		0.2	0.5				0.3[m]
Citronellyl acetate	150-84-5	0-0.1	0.5	1.5				1.2[m]
α-Copaene	3856-25-5	tr-0.08				tr		
Cryptone	500-02-7					0.1		
Cubebol	23445-02-5		0.1					
m-Cymene	535-77-3					0.3[b]		
o-Cymene	527-84-4					0.3[b]		
p-Cymene	99-87-6	0.02-1.4	0.1	tr		0.1		0.1[m]; 0.6[f]
p-Cymenene	1195-32-0		tr					
Decanal	112-31-2	0-0.2	0.1	0.1				0.1[m]
Dodecanal	112-54-9		0.2	0.4				0.4[m]
β-Elemene	33880-83-0		tr					0.7[f]
γ-Elemene	29873-99-2				0.5			
β-Eudesmol	473-15-4				0.3			
(Z,E)-α-Farnesene	26560-14-5		0.1					
(E)-β-Farnesene	18794-84-8		tr	0.4		0.2		
(Z)-β-Farnesene	28973-97-9					tr		
α-Fenchol (endo-)	512-13-0			0.1	tr			tr-0.4[k]
α-Fenchyl acetate	111821-74-0				14.2			
Geraniol	106-24-1			0.1				0.2[m]
Geranyl acetate	105-87-3	0-0.08	0.5		0.2			2.0[m]
Germacrene A	28387-44-2							tr-0.4[k]
Germacrene D	23986-74-5	tr-0.7			0.8			0.1-0.3[k]; 0.2[m]
Globulol	489-41-8				1.5			

Table 5.78.1 Constituents identified in silver fir oils (*continued*)

Constituent	CAS	Percentage and range						
		A	B (4)	C (4)	D (2)	E (1)	F (7)	G
cis-β-Guaiene	372162-07-7							tr-1.2[k]
α-Gurjunene	489-40-7				0.3	0.4		
γ-Gurjunene	22567-17-5							tr-0.4[k]
Hexanol (1-; *n*-)	111-27-3	0-0.04						
(Z)-3-Hexenol	928-96-1	0-0.03						
(Z)-3-Hexenyl acetate	3681-71-8	0-0.09						
Himachala-2,4-diene						0.1		
α-Himachalene (α-*cis*-)	3853-83-6					tr		tr-0.2[k]
β-Himachalene	1461-03-6		0.2	0.2				0.1[m]; 0.2-1.2[k]; 2.6[d]
Himachalenol				0.1				
α-Humulene	6753-98-6	0.02-1.2	0.1	0.2	1.9	0.6	0.2	0.2[f]; 0.3-7.2[k]; 0.7[m]; 0.8[c]; 2.0[d]
γ-Humulene	26259-79-0			0.2				0.1[m]
α-Humulene epoxide	96638-51-6	0-0.1						
4-Hydroxy-4-methyl-2-pentanone	123-42-2							0.06[f]
α-Ionone	127-41-3				0.4			
Isocembrene	25269-16-3		0.1					
Isophyllocladene	511-85-3				tr			
Isopimaradiene	54605-21-9		0.1					
Isopinocamphone	15358-88-0							tr-0.3[k]
Limonene	138-86-3	6.1-54.7	68.3	9.3	11.0	6.1	12.4	6.5-17.5[k]; 27.8[g]; 34.1[d]; 34.5[g]; 35.7[j]; 43.5[m]; 54.7[c]
Linalool	78-70-6	0-0.3	0.3	0.2	0.1			tr-0.9[k]; 0.4[m]
Linalyl acetate	115-95-7		0.1		0.1		0.05	tr-1.0[k]; 0.1[m]
Longiborneol	465-24-7		0.1	0.3				0.7[m]
Longicyclene	1137-12-8			tr				
Longifolene (junipene)	475-20-7	0-0.6	0.5	0.2		0.6		0.4[m]
α-Longipinene	5989-08-2	0-0.4	0.3	tr		0.2		0.5[m]
Manoyl oxide	596-84-9		tr	tr				tr[m]
p-Menth-1-en-9-yl acetate	28839-13-6							tr-0.7[k]
Methyl geranate	2349-14-6			0.1				
Methyl thymol	1076-56-8				0.1			
α-Muurolene	10208-80-7	0-0.04				tr		
γ-Muurolene	30021-74-0					0.3		tr-0.9[k]
τ-Muurolol (epi-α-)	19912-62-0			0.1				
Myrcene	123-35-3	0.7-2.5	2.2	1.0	1.1	1.0	0.9	0.6-1.2[k]; 1.0[d]; 1.1[g]; 1.2[m]; 1.9[c]
Myrtanyl acetate	29021-36-1			0.1				
Myrtenol	515-00-4					tr		
Neryl acetate	141-12-8				0.1			0-0.6[k]
Ocimene (β-)	13877-91-3				0.2			
(E)-β-Ocimene	3779-61-1			tr				
(Z)-β-Ocimene	3338-55-4							tr-0.2[k]
α-Phellandrene	99-83-2			0.2	0.9		0.2	0.1-0.2[k]; 0.2[m]
β-Phellandrene	555-10-2	0.01-4.9	0.4	15.1		4.9		0.2[c]; 0.5[g]; 1.8[g]; 2.1[f]; 14.4[m]
α-Pinene	80-56-8	0.5-32.8	6.4	19.0	10.9	17.3	30.2	3[j]; 8.4-15.2[k]; 15.8[g]; 16.1[g]; 18.0[m]; 21.2[j]; 31.7[d]; 38.9[j]
β-Pinene	127-91-3	7.4-31.7	0.8	11.6	19.8	32.8	31.1	2.7[g]; 3.0[d]; 3.4[g,j]; 8.1[m]; 14.5-31.6[k]; 28.5[g]
trans-Pinocarveol	1674-08-4					tr		tr-1.0[k]
Pinocarvone	30460-92-5					0.1		
Sabinene	3387-41-5	tr-0.5	tr	tr		0.1		
Santene	529-16-8	0.01-2.0	1.3	4.3		1.5		0.3-3.2[k]; 1.6[f]; 3.6[m]; 5.0[c]
α-Selinene	473-13-2					0.1		0-1.0[k]
β-Selinene	17066-67-0					tr		tr-1.0[k]
β-Sesquiphellandrene	20307-83-9							0.8[m]
α-Terpinene	99-86-5	0.04-0.2		0.1	0.2			1.2[f,l]
γ-Terpinene	99-85-4	tr-0.1	tr	0.1	0.2	tr	0.04	0.1[m]
Terpinen-4-ol	562-74-3		0.1	0.1	0.1			tr-0.4[k]
α-Terpineol	98-55-5	0.07-2.3	0.3	0.8	0.2	2.1[a]	0.3	1.2[m]; 1.2-6.5[k]
Terpinolene	586-62-9	0.1-0.4		0.8	0.9	0.3	0.4	0.3[d]; 0.4[c]; 0.5[m]; 0.6-1.7[k]

Table 5.78.1 Constituents identified in silver fir oils (*continued*)

Constituent	CAS	Percentage and range						
		A	B (4)	C (4)	D (2)	E (1)	F (7)	G
α-Terpinyl acetate	80-26-2	0.06-0.8	0.1		0.4			0.2-1.4[k]; 0.5[d]
α-Thujene	2867-05-2				2.8			
Tricyclene	508-32-7	0.5-2.6	1.1	3.0	2.1	2.6		0.3-1.8[k]; 0.5[d]; 2.1[c,m]; 12.9[f]
Valencene	4630-07-3							0.1[f]

A sixteen silver fir essential oil samples from France, Montenegro and Poland analyzed between 1998 and 2008; lowest and highest concentrations given (E. Schmidt, unpublished data)
B one lab-hydrodistilled oil from Corsica, selected from 53 oil samples on the basis of a very different chromatographic profile than the oil presented from the same study in column C (ref. 4)
C one lab-hydrodistilled oil from Corsica, selected from 53 oil samples on the basis of a very different chromatographic profile than the oil presented from the same study in column B (ref. 4)
D one lab-hydrodistilled oil from West Serbia (ref. 2)
E one lab-hydrodistilled oil from Montenegrin *A. alba* twigs and leaves (ref. 1)
F one commercial steam-distilled oil from Romania, analyzed with GC-MS, using two different techniques of injection; highest concentrations given (ref. 7)
G data from other studies (indicated with superscript letters); highest concentrations found in any study reviewed here given; when two or more oils were investigated, only the highest concentrations are mentioned, unless indicated otherwise

[a] α-terpineol and borneol combined; [b] *m*-cymene and *o*-cymene combined; [c] one lab-hydrodistilled oil from needles and twigs of *A. alba* harvested in a German university botanic garden (ref. 5); [d] one commercial oil from Germany (ref. 5); [e] γ-cadinene and δ-cadinene combined; [f] one commercial silver fir essential oil obtained by steam distillation from leaf and twig of *Abies alba* from Korea (ref. 9); [g] one lab-hydrodistilled oil from Germany (ref. 3); [h] one commercial silver fir oil from Germany (ref. 3); [i] four lab-hydrodistilled oils from Romania obtained from material harvested in the four seasons (ref. 8); [j] one French silver fir oil (ref. 6); [k] eighty lab-steam distilled oils from 4 locations in Albania, 40 harvested in winter and 40 in summer; lowest and highest concentrations given (ref. 10); [l] one commercial *Abies alba* oil from Slovakia (ref. 11); [m] two lab-hydrodistilled oil samples from the twigs with leaves from *Abies alba* trees harvested in two clusters in Corsica, France (ref. 12)

tr: trace (in column G[k]: <0.1)

Table 5.78.2 Major constituents of silver fir oils

Constituent	CAS	Percentage and range						
		A	B (4)	C (4)	D (2)	E (1)	F (7)	G
Limonene	138-86-3	6.1-54.7	68.3	9.3	11.0	6.1	12.4	34.5[g]; 35.7[j]; 43.5[m]; 54.7[c]
α-Pinene	80-56-8	0.5-32.8	6.4	19.0	10.9	17.3	30.2	18.0[m]; 21.2[i]; 31.7[d]; 38.9[j]
β-Pinene	127-91-3	7.4-31.7	0.8	11.6	19.8	32.8	31.1	8.1[m]; 14.5-31.6[k]; 28.5[g]
Bornyl acetate	76-49-3	0.4-14.2	2.7	2.3		9.0	7.1	9.9[m]; 18.8[g]; 24.4[i]; 30[l]; 30.3[f]
Camphene	79-92-5	5.8-17.3	9.9	20.6	15.3	16.7	15.2	18[l]; 19.6[i]; 19.8[g]; 21.3[g]; 25.2[g]

LEGEND: SEE UNDER TABLE 5.78.1

CONTACT ALLERGY/ALLERGIC CONTACT DERMATITIS

General
Contact allergy to silver fir oil has been reported in two publications, but no cases of allergic contact dermatitis from the oil have been identified.

Testing in groups of patients
Two hundred dermatitis patients from Poland were tested with silver fir oil 2% in petrolatum and two (1%) reacted. The same authors also patch tested 51 patients allergic to *Myroxylon pereirae* resin (balsam of Peru) and/or turpentine and/or wood tar and/or colophony and nine (17.6%) had a positive patch test; relevance data were not provided (15). This high percentage can likely be explained by the fact that all materials tested originate from trees. A group of 86 patients from Poland previously reacting to the fragrance mix was tested with silver fir oil and two (2.3%) had a positive patch test reaction; relevance data were not provided (16).

LITERATURE

1 Chalchat J-C, Sidibé L, Maksimovic ZA, Petrovic SD, Gorunovic MS. Essential oil of *Abies alba* Mill., Pinaceae, from the pilot production in Montenegro. J Essent Oil Res 2001;13:288-289
2 Roussis V, Couladis M, Tzakou O, Loukis A, Petrakis PV, Dukic NM, et al. A comparative study on the needle volatile constituents of three *Abies* species grown in south Balkans. J Essent Oil Res 2000;12:41-46

3 Schales C, Gerlach H, Kösters J. Investigations on the antibacterial effect of conifer needle oils on bacteria isolated from the feces of captive Capercaillies (*Tetrao urogallus* L., 1758). J Vet Med B 1993;40:381-390

4 Duquesnoy E, Castola V, Casanova J. Composition and chemical variability of the twig oil of *Abies alba* Miller from Corsica. Flavour Fragr J 2007;22:293-299

5 Kubeczka KH, Schultze W. Biology and chemistry of conifer oils. Flavour Fragr J 1987;2:137-148

6 Chalchat J-C, Garry RP, Michet A. Huiles essentielles de résineux d'Auvergne: pin sylvestre, épicéa, sapins pectine et de Vancouver, Douglas. Parf Cosm Arômes 1986;69:55-58. Data cited in ref. 1

7 Şerban ES, Socaci SA, Tofană M, Maier SC, Bojiţă MT. Advantages of "headspace" technique for GC/MS analysis of essential oils. Farmacia 2012;60:249-256

8 Baath MH, Burzo I. Quantitative and qualitative seasonal variation of volatile oil from 16 conifer species. Analele ştiinţifice ale Universităţii "Al. I. Cuza" Iaşi, Tomul LV, fasc. 2, s.II a. Biologie vegetală, 2009:103-110. Available at: http://www.bio.uaic.ro/publicatii/anale_vegetala/issue/2009F2/13-2009F2.pdf

9 Yang S-A, Jeon S-K, Lee E-J, Im N-K, Jhee KH, Lee S-P, Lee I-S. Radical scavenging activity of the essential oil of silver fir (*Abies alba*). J Clin Biochem Nutr 2009;44:253-259

10 Zeneli G, Tsitsimpikou C, Petrakis PV, Naxakis G, Habili D, Roussis V. Foliar and cortex oleoresin variability of silver fir (*Abies alba Mill.*) in Albania. Z Naturforsch c 2001;56:531-539

11 Kačániová M, Vukovič N, Horská E, šalamon I, Bobková A, Hleba L, et al. Antibacterial activity against *Clostridium genus* and antiradical activity of the essential oils from different origin. J Environ Sci Health, Part B: Pesticides, Food Contaminants, and Agricultural Wastes 2014;49:505-512

12 Duquesnoy E, Marongiu B, Castola V, Piras A, Porcedda S, Casanova J. Combined analysis by GC (RI), GC-MS and 13C NMR of the supercritical fluid extract of *Abies alba* twigs. Nat Prod Commun 2010;5:1995-1998

13 Rhind JP. Essential oils. A handbook for aromatherapy practice, 2nd Edition. London: Singing Dragon, 2012

14 Lawless J. The encyclopedia of essential oils, 2nd Edition. London: Harper Thorsons, 2014

15 Rudzki E, Grzywa Z, Bruo WS. Sensitivity to 35 essential oils. Contact Dermatitis 1976;2:196-200

16 Rudzki E, Grzywa Z. Allergy to perfume mixture. Contact Dermatitis 1986;15:115-116

Chapter 5.79 SPEARMINT OIL

DEFINITION
Spearmint oil (essential oil of spearmint) is the essential oil obtained from the flowering aerial parts and leaves of the spearmint, *Mentha spicata* L.

INCI NOMENCLATURE
Description/definition: Mentha spicata herb oil is an essential oil obtained from the herbs of the spearmint, *Mentha spicata* L., Labiatae
INCI name EU: Mentha spicata herb oil (perfuming name, not officially an INCI name)
INCI name USA: Not in the Personal Care Products Council Ingredient Database
CAS registry number(s): 84696-51-5
EINECS number(s): 283-656-2

Description/definition: Mentha viridis leaf oil is the volatile oil obtained from the dried tops and leaves of the garden mint (spearmint), *Mentha viridis* L., Labiatae
INCI name EU: Mentha viridis leaf oil
INCI name USA: Mentha viridis (spearmint) leaf oil
CAS registry number(s): 8008-79-5; 84696-51-5
EINECS number(s): 283-656-2

ISO (INTERNATIONAL ORGANIZATION FOR STANDARDIZATION) STANDARD
ISO number: 3033
ISO name: Essential oil of spearmint
Botanical origin: *Mentha spicata* L./*Mentha viridis* L. var. *crispa* Benth. / *Mentha x gracilis* Sole
Parts of plant used: Flowering aerial part, leaf

ISO has four standards for spearmint oil:
ISO/DIS 3033-1 Oil of spearmint–Part 1: Native type (*Mentha spicata* L.)
ISO/DIS 3033-2 Oil of spearmint–Part 2: Chinese type (*Mentha viridis* L. var. *crispa* Benth.), redistilled oil
ISO/DIS 3033-3 Oil of spearmint–Part 3: Indian type (*Mentha spicata* L.), redistilled oil
ISO/DIS 3033-4 Oil of spearmint–Part 4: Scotch variety (*Mentha x gracilis* Sole)

The first three oils are from *Mentha spicata* L. (*Mentha viridis* L. var. *crispa* Benth. Is considered by The Plant List to be a synonym for *Mentha spicata* L.), two of which are redistilled oils. Part four concerns the Scotch variety of spearmint oil, which is obtained from *Mentha x gracilis* Sole, a plant infrequently mentioned in literature as a source of spearmint oil. ISO values (minimum and maximum concentrations) for these four oils *as a group* are shown in Table 5.79.1.

Synonyms of *Mentha spicata* include: *Mentha viridis, Mentha spicata* ssp. *spicata* and *Mentha crispa* L. Other mint oils include peppermint oil (*Mentha x piperita* L., Chapter 5.68) and corn mint oil (*Mentha canadensis* L., synonym: *Mentha arvensis*, not discussed here).

Table 5.79.1 ISO values (%) for spearmint oils [a]

Compound	CAS	Minimum	Maximum
Carvone	99-49-0	57.0	84.0
Limonene	138-86-3	0.0	22.0
cis-Dihydrocarvone	3792-53-8	1.0	4.0
trans-Dihydrocarvyl acetate	20777-49-5	0.1	4.0
β-Bourbonene	5208-59-3	0.5	2.0
Menthone	89-80-5	0.0	2.0
trans-Sabinene hydrate	17699-16-0	0.1	1.0
(Z)-Jasmone	488-10-8	0.0	0.7
cis-Carvyl acetate	1205-42-1	0.1	0.6
Viridiflorol	552-02-3	0.0	0.5
3-Octanol	589-98-0	tr	0.4

[a] ISO 3033 Parts 1-4 Essential oil of spearmint ©ISO 2005; Geneva, Switzerland, www.iso.org

THE PLANT, THE OIL, AND THEIR USES
Mentha spicata L. is an herbaceous creeping rhizomatous perennial plant growing 30-100 cm tall with a strong aromatic odor (7). Its leaves have serrated margins and pointed tips, which explains the 'spear' in the name spearmint. The plant is native to western Asia (Cyprus, Lebanon, Syria, Turkey) and southeastern Europe (Albania, Bulgaria, former Yugoslavia, Greece, Italy). It is naturalized widely and found in many places around the world as garden escapes (11). The spearmint is cultivated in Africa, Asia (Cyprus, Turkey, China, Japan, India, Pakistan), Australia, New Zealand, Europe, Canada, USA and the Caribbean (GRIN Taxonomy for Plants; www.efloras.org).

Dried tops and leaves of the spearmint are used medicinally as a stimulant, carminative, and for relieving disorders of the nerves. Leaves are eaten in the form of chutney and are popular as tea flavoring (7,27). *Mentha spicata* and its essential oil are said to have antifungal, antiviral, antimicrobial, insecticide, and antioxidant properties (4,7,23, 27).

Spearmint essential oil, obtained by distillation of aerial flowering parts and the leaves of *Mentha spicata* L., is rich in carvone and presents a characteristic spearmint odor (7). This oil (as well as other mint oils) and carvone are used in perfumery, cosmetics, pharmaceuticals and in the food industries. As flavour or fragrance, spearmint oils are widely added to products such as toothpastes, mouth washes, cigarettes, chewing gum and alcoholic drinks (1,7,10,23,27). They are also employed in aromatherapy practices (76,77).

CHEMICAL COMPOSITION
Spearmint oil is a colorless to pale yellow clear mobile liquid, which has a fresh herbal, minty soft aromatic odor. The yield of essential oil from the leaves of *Mentha spicata* L. generally varies from 0.3 to 1.8 per cent, depending on the dryness of the spearmint leaves used. The main producing countries of this oil are USA, Canada, India and China.

Literature data (up to August 31, 2014) on the chemical composition of spearmint oils and unpublished analytical data from one of us (E.S.) are shown in Table 5.79.2 in alphabetical order. In spearmint oils from various origins, over 250 chemicals have been identified. About 45 per cent of these were found in a single reviewed publication only. The major compounds found in spearmint oils from different sources are shown in Table 5.79.3. They include (highest concentrations in any study given) linalool (93.9%), carvone (83.0%), piperitenone oxide (80%), pulegone (72.1%), menthone (62%), piperitone (49.3%), limonene (48%), 1,8-cineole (40.5%) and isomenthone (39.1%). Well-known ingredients of spearmint oils that were present in high concentrations (>8%) in one or two studies were p-cymene (48.9%), trans-carvyl acetate (32.2%), menthyl acetate (28.1%), trans-sabinene hydrate (22%), dihydrocarvone (three studies: 9.7%, 19.1% and 21.5%), β-pinene (10.6%) and α-pinene (10.2%). Uncommon or rare constituents of spearmint oils found in high concentrations (>8%) in single studies include piperitone oxide (72.3%), terpinen-4-yl acetate (55.3%), carvone oxide (52.2%), carvacrol (49.6%), trans-piperitone oxide (33.5% and 45.0%) and cis-piperitone oxide (30%).

Commercial oils

The ten chemicals that had the highest maximum concentrations in 71 commercial spearmint essential oil samples (concentration ranges provided) are the following: carvone (60.6-82.3%), limonene (0.4-23.7%), dihydrocarveol (0.05-5.1%), 1,8-cineole (0.01-4.4%), trans-dihydrocarvyl acetate (0.1-3.7%), cis-dihydrocarvone (0.3-3.3%), β-caryophyllene (0.09-3.0%), myrcene (0.01-2.6%), 3-octanol (0.01-2.2%) and menthol (0.2-2.2%) (Erich Schmidt, unpublished analytical data).

Chemotypes

The spearmint oils of commerce are obtained from M. spicata cultivars with high carvone content. However, wild growing plants of this species are very polymorphic, both in their morphology and essential oils. In Greece, for example, four chemotypes were found, each with their distinctive smell (41). The first chemotype is dominated by linalool (65-90%), the second has a high content of carvone (35-68%) and dihydrocarvone (5-21%), the third chemotype is characterized by high percentages of piperitone oxide (up to 89%) and/or piperitenone oxide (up to 70%) (this is the most common type, found all over the country), whereas the fourth chemotype is rich in menthone, isomenthone and pulegone (41,61). In northern Turkey, five chemotypes were found in naturally growing M. spicata populations: piperitone oxide, piperitenone oxide, carvone, linalool, and pulegone/menthone/isomenthone (17); this is virtually identical to the Greek classification. Previously, three chemotypes with several subgroups of wild M. spicata had been proposed: carvone-type, dihydrocarvone type and dihydrocarveol-type (chemotype I); piperitenone/piperitone, pulegone/menthone and/or isomenthone, isomenthone/menthone (chemotype II); and piperitone oxide and piperitenone oxide (chemotype III) (63).

Classifications have also been suggested by the authors of references 10,17,39 and 54. Other chemicals which have been found in high concentrations in spearmint oils, sometimes considered by the authors as indicative of chemotype, include β-caryophyllene (19,54), carvacrol (10), (Z)-carveol (5), carvone oxide (10), trans-carvyl acetate (10), 1,8-cineole (8,17; additionally cited in ref. 18), p-cymene (10), limonene (6), menthyl acetate (39), trans-sabinene hydrate (17; additionally cited in ref. 18), terpinen-4-ol (17), terpinen-4-yl acetate (10) and α-terpinyl acetate (54).

A review of the composition of commercial mint oil is provided in ref. 2. Additional recent analytical studies not presented in Table 5.79.2 include refs. 49-53 and 55-58. These can all be found on the internet and accessed on-line. Several articles cited in ref. 59 are not discussed, as they do not contain information not already present in Table 5.79.2.

CONTACT ALLERGY/ALLERGIC CONTACT DERMATITIS

General

Contact allergy to/allergic contact dermatitis from spearmint oil has been reported in over 20 publications. In groups of consecutive patients suspected of contact dermatitis, prevalence rates of up to 1.6% positive patch test reactions have been observed, but relevance data are lacking. At least ten case reports of contact allergic reactions to spearmint oil have been reported. Nearly all were from its presence in toothpastes, causing stomatitis, cheilitis, perioral dermatitis, sore mouth and possibly oral lichenoid reactions (Table 5.79.4). In about half the cases where carvone, the dominant (60-80%) ingredient of spearmint oils, was also tested, co-reactivity occurred, and this is likely to be the main allergen.

Testing in groups of patients

The results of patch tests with spearmint oil in routine testing (consecutive patients suspected of contact dermatitis) and in groups of selected patients are shown in Table 5.79.4. In routine testing, rates of positive reactions ranged from 0.8% to 1.6%, whereas between 0.1% and 10% of patients in selected groups had positive patch tests.

Case reports

An aromatherapist had occupational contact dermatitis with allergies to multiple essential oils used at work, including spearmint oil; carvone was not tested (79). One case of occupational contact allergy to spearmint oil in a chewing gum finisher has been reported (85). Four positive patch test reactions to spearmint oil were seen in a group of 40 patients with sore mouth, stomatitis and/or dermatitis around the mouth or who were dentist personnel. Two also reacted to anethole (not an important component), two to carvone (the main ingredient of spearmint oil, 60-80% of the total oil) and one to peppermint oil. The causative products were supposed to be toothpastes (80). One patient with oral lichen

Table 5.79.2 Constituents identified in spearmint oils

Constituent	CAS	Percentage and range				
		A	B (54)	C (37)	D (41)	E
α-Amorphene	20085-19-2					0.3[x3]; 0.5[e]
Amyl isovalerate	25415-62-7		0-0.4			
Amyl valerate	2173-56-0				0-0.1	
p-Anethole	104-46-1		0-0.1			
(E)-Anethole	4180-23-8					0.5[k]
Aromadendrene	489-39-4					0.3[w]
allo-Aromadendrene	25246-27-9		0-tr			
Benzaldehyde	100-52-7					0.1[e]
Benzyl alcohol	100-51-6		0-tr			1.4[q]
α-Bergamotene	17699-05-7					1.5[e]
trans-β-Bergamotene	15438-94-5					0.3[x3]
Bicycloelemene	32531-56-9					0.06[c]; 0.1[y3,z15]; 0.3[q,u,x6]; 0.9[h]
Bicyclogermacrene	24703-35-3					
Bicyclo[3.1.1]hept-2-en-4-ol, 2,6,6-trime-thyl-, hexanoate			0-0.5			
epi-Bicyclosesqui-phellandrene	54274-73-6					0.3[w]; 2.0[h]
α-Bisabolol	515-69-5					0.2[f]
Borneol	507-70-0			0.7		0.3[c]; 0.5[x6]; 1.2[e]; 1.4[g]; 1.9[z16]; 5.9[q]; 7.2[z]
Bornyl acetate	76-49-3		0-tr	0.6	0-0.7	<0.05[x4]; 0.1[w]; 0.3[q]; 0.6[z1]
α-Bourbonene	5208-58-2		tr-2.6			
β-Bourbonene	5208-59-3	0.2-1.7			0-3.5	1.5[x6]; 1.6[i]; 1.8[w]; 2.0[y8]; 2.2[y7]; 2.4[f]; 2.9[h]
2-Butanol	78-92-2					0.01[c]
Butyl 2-methylbutyrate	15706-73-7		0-0.2			
α-Cadinene	24406-05-1			0.2		0.1[u]; 0.7[m]
β-Cadinene	523-47-7		0-0.1			
γ-Cadinene	39029-41-9		tr-0.2	1.1		0.8[q]; 0.9[w]; 1.1[m]
δ-Cadinene	483-76-1		0.1-0.3	0.3	0-1.6	0.2[g]; 0.3[w,x]; 0.4[x3]; 0.5[v]; 0.7[e]; 0.9[z11]
α-Cadinol	481-34-5		tr-0.5			0.3[w]; 0.4[e]; 0.5[z14]; 1.0[v]; 1.5[z]; 2.1[m]
epi-α-Cadinol	5937-11-1					tr[m]; 0.05[u]; 0.7[z14]
α-Calacorene	21391-99-1					tr[m]
trans-Calamene						6.4[m]
Calamenene	483-77-2			0.2		+[y9]; 0.1[z11]; 0.2[z7]; 0.7[q]
cis-Calamenene	72937-55-4		0-0.1			0.1[e]; 0.3[w,z15]; 0.6[x3]; 0.7[u]
Camphene	79-92-5	0-0.7	0-0.1		0-0.1	tr[m]; 0.04[s]; 0.1[e]; 0.4[c]; 0.6[z]; 0.9[v]; 1.7[x1]
Camphor	76-22-2			3.7		0.2[x]; 0.3[f]; 1.0[n]
δ3-Carene	13466-78-9			0.3		0.02[c]; 0.1[h]
Carvacrol	499-75-2					0.1[x]; 0.4[e]; 49.6[x1]
Carvenone	499-74-1					0.1[e]; 0.2[j]
Carveol	99-48-9			0.8	0-tr	0.1[x]; 2.3[i]
(E)-Carveol	1197-07-5	0.08-1.0	0-0.1			0.3[q]; 0.4[l]; 0.5[z11]; 0.7[c]; 0.8[r]; 1.0[y6]; 1.3[x6]
(Z)-Carveol	1197-06-4	0.1-2.0	0-0.1			2.3[q]; 3.9[s]; 10.6[z1]; 11.6[y2]; 21.3[z1]; 24.3[x4]
Carvomenthol	499-69-4					1.9[x3]
Carvone	99-49-0	60.6-82.3	2.3-67.4	50.5	0-49.7	75.9[z2]; 76.7[y6]; 77.0[x2]; 78.9[y3]; 88.4[x7] 68.4[y4]; 69.3[c]; 72.9[y]; 73.2[o]; 79.9[x8]; 83.0[n]
Carvone oxide	33204-74-9					0.1[h]; 0.3[q]; 0.4[w]; 52.2[x1]
trans-Carvone oxide	18383-49-8					0.03[u]; 0.3[f]
Carvyl acetate	97-42-7	0.08-1.2				0.1[t]; 0.2[u]; 1.4[i]; 1.6[f]; 2.1[x4]
cis-Carvyl acetate	1205-42-1		0-0.7			0.4[w]; 0.9[o]; 1.2[x6]; 1.6[z9]; 2.1[z1]; 6.4[z11]
trans-Carvyl acetate	1134-95-8		0-0.2			1.2[c]; 2.6[z11]; 2.7[z14]; 4.0[h]; 5.9[p]; 32.2[x1]
β-Caryophyllene	87-44-5	0.09-3.0	1.9-11.4	3.0	0-6.3	1.6[o]; 1.7[g]; 1.8[i]; 2.5[f]; 2.8[q]; 3.1[y7]; 3.3[w]; 5.4[z12]; 6.1[x3]; 7.9[x5]; 8.0[v]; 9.5[h]; 20.3[y2]
Caryophyllene oxide	1139-30-6	0-0.1	tr-0.4	0.9		0.2[g]; 0.3[e]; 0.6[x4]; 0.7[q]; 1.0[w]; 1.5[h]; 3.0[v]

Table 5.79.2 Constituents identified in spearmint oils (*continued*)

Constituent	CAS	Percentage and range				
		A	B (54)	C (37)	D (41)	E
Cedrenol	28231-03-0					+[z3]
1,8-Cineole	470-82-6	0.01-4.4	0-4.1	9.1	0.4-6.8	6.4[q]; 7[x2]; 7.4[y7]; 7.8[v]; 9.0[h]; 14.5[m]; 14.8[x1]; 19[y1]; 21.3[w]; 22.7[z13]; 33.8[n]; 40.5[g]
α-Copaene	3856-25-5		0-0.3		0-1.7	0.09[u]; 0.1[e]; 0.3[f]; 0.4[v]; 0.8[c]; 1.0[p]
β-Copaene	18252-44-3					tr[m]; 0.5[f]
α-Cubebene	17699-14-8		0-0.1	0.1		tr[e]
β-Cubebene	13744-15-5		0-0.4			0.2[v]; 0.5[x3]; 0.7[y7]; 3.2[i]
Cubenene	29837-12-5		0-0.1			+[y9]
β-Cubenene						+[y9]
Cubenol	21284-22-0		0-0.2			0.1[z11]
1,10-di-epi-Cubenol	73365-77-2					0.3[u]; 0.6[q]; 3.3[m]
Cuminaldehyde	122-03-2					0.4[e]
m-Cymene	535-77-3					2.8[y4]
o-Cymene	527-84-4					0.1[j]
p-Cymene	99-87-6	0-0.5	0-tr	0.3	tr-0.3	tr[h]; 0.1[f]; 0.2[e]; 0.3[o,v]; 0.5[c,k]; 48.9[x1]
p-Cymenene	1195-32-0				0-0.1	tr[h]; 0.2[e]; 1.5[m]
p-Cymen-8-ol	1197-01-9		0-0.2			tr[m]
2,5-Diethyltetrahydrofuran	41239-48-9					0.06[z6]; 0.3[e]
Dihydrocarveol	38049-26-2	0.05-5.1	0-1.5	0.5	0-2.0	3.2[x6]; 5.9[p]; 6.3[s]; 7.1[o]; 15.7[i]; 16.5[z106]
cis-Dihydrocarveol						0.3[c]
trans-Dihydrocarveol						0.2[c]
Dihydrocarvone	5948-04-9				0-21.5	0.8[t]; 1.0[u]; 9.7[z10]; 19.1[z12]
cis-Dihydrocarvone	3792-53-8	0.3-3.3	0-4.8			3.2[x6]; 4.6[y2]; 4.8[q]; 4.9[s]; 6.8[x3]; 7.1[z14]; 7.6[l]
trans-Dihydrocarvone	5948-04-9	0.07-0.7	0-0.1			2.8[f]; 2.9[w]; 4[x2]; 4.0[z14]; 6.1[z11]; 7.2[y4]
Dihydrocarvyl acetate	20777-49-5				0-12.3[b]	1.7[y7]; 2.3[f]; 3.0[p]; 3.8[z10]; 6.1[z11]; 7.0[i]
trans-Dihydrocarvyl acetate	20777-49-5	0.1-3.7				0.01[r]; 0.8[y4]
Dihydroedulan I	74006-61-4		0-0.9			0.1[w]; 0.2[h]
Dihydroedulan II	41678-32-4		0-0.5			0.2[w]; 0.3[h]; 4.4[m]
Dihydroeugenol	2785-87-7			0.2		
Dihydrolinalyl acetate	61476-73-1					0.3[j]
Dimethylfuran lactone				0.1		
cis-2,6-Dimethyl-3,5,7-octatrien-2-ol						
trans-2,6-Dimethyl-3,5,7-octatrien-2-ol			0-tr			
β-Elemene	33880-83-0		0.3-0.9	1.0		0.4[u]; 0.5[c]; 0.7[h]; 0.8[y8]; 0.9[w]; 1.2[x]; 1.7[z15]
δ-Elemene	20307-84-0		0-0.1			0.05[u]
Elemicin	487-11-6					+[z3]
Elixene	3242-08-8		0.2-0.9			0.4[i]; 0.9[v]
Epiglobulol	88728-58-9		0-0.2			
3,9-Epoxy-1-*p*-menthene	13955-48-1		0-0.1			
Eucarvone	503-93-5					0.8[e]
Eugenol	97-53-0		0-0.4	0.3		0.2[h]; 1.2[z15]
α-Farnesene	502-61-4					0.3[w]
β-Farnesene	502-60-3					0.7[i]
(*E*)-β-Farnesene	18794-84-8	0.08-0.7	0.1-1.0			0.07[c]; 0.2[r,t]; 0.4[h]; 0.7[z6]
(*Z*)-β-Farnesene	28973-97-9					1.4[f]
Farnesyl acetate	29548-30-9				0-tr	
Furyl propyl ketone						1.2[e]
Geraniol	106-24-1				0-0.2	0.9[c]
Geranyl acetate	105-87-3					<0.05[x]; 0.2[z7]
Germacrene A	28387-44-2					0.2[c,z11]; 0.5[k]; 1.2[h]
Germacrene D	23986-74-5	0.09-1.1	0-3.3		0-7.5	1.4[q]; 1.7[y7]; 1.8[c]; 3.4[x3]; 4.1[p]; 4.2[x]; 5.3[v]
Germacrene D-4-ol	198991-79-6		0-0.1			0.1[u,z11]
Globulol	489-41-8					0.4[z14]; 0.7[f]

Table 5.79.2 Constituents identified in spearmint oils (*continued*)

Constituent	CAS	Percentage and range				
		A	B (54)	C (37)	D (41)	E
Grandisol	26532-22-9			0.3		
α-Guaiene	3691-12-1			1.0		
β-Guaiene	88-84-6					+[z3]
α-Gurjunene	489-40-7					0.1[e,z11]; 0.2[u]
β-Gurjunene	73464-47-8			0.5		0.2[u]
3-Hepten-2-one	1119-44-4					0.1[e]
Hexadecanoic acid	57-10-3		0-0.1			<0.01[x9]
Hexanol (1-; *n*-)	111-27-3					0.1[c]
(*E*)-1-Hexen-3-ol						0.7[c,d]
(*Z*)-3-Hexen-1-ol	928-96-1					0.04[c]; 0.05[z6]; 0.1[e,h]
(*Z*)-3-Hexenyl isovalerate	35154-45-1		0-0.4			
cis-3-Hexenyl valerate	35852-46-1		0-0.6		0-tr	
Hexyl isovalerate	10032-13-0		0-0.3		0-0.6	
Hexyl valerate	1117-59-5				0-0.2	
α-Himachalene	3853-83-6			0.2		
α-Humulene	6753-98-6		0-0.3	0.6		tr[m]; 0.1[y8]; 0.2[c,e]; 0.3[f,v]; 0.4[y7]; 0.6[h]; 0.8[q]
Humulene epoxide II	19888-34-7					0.9[m]
8-Hydroxy-*p*-menth-4-en-3-one						0.7[z11]
β-Ionone	79-77-6					0.1[e]
Isoamyl isovalerate	659-70-1		tr-0.3			
Isobutyl isovalerate	589-59-3		0-0.1			
Isocaryophyllene	118-65-0					0.4[h]
Isodihydrocarveol	18675-35-9					1.4[z11]
Isodihydrocarvyl acetate	57287-13-5		0-1.4			0.1[y6]; 0.8[x6]
Isomenthol	3623-52-7		0-0.1			1.1[g]
Isomenthone	491-07-6	0.01-0.8	0.1-0.3	1.1	0-15.2	0.3[g]; 1.5[i]; 1.6[y3]; 2.6[j]; 17[y1]; 32[x2]; 39.1[x1]
Isopiperitenone	529-01-1					0.6[e]; 0.9[g]; 2.8[l]
Isopiperitone	58615-39-7					0.5[v]
Isopulegol	89-79-2		0.3			0.2[i]; 0.4[z8]
Isopulegol acetate	57576-09-7					8.4[z14]
Isopulegone	29606-79-9					1.6[i]
cis-Isopulegone						1.7[z]
Isoterpinolene	586-63-0					0.2[j]
Isovaleraldehyde	590-86-3					0.03[t]; 0.07[z6]; 0.3[c]
Jasmone	488-10-8					0.3[p]; 0.7[f]
(*Z*)-Jasmone	488-10-8	tr-0.9	0.1-1.1	1.2		0.2[h,s]; 0.3[q,z6]; 0.8[c]
Lavandulol	498-16-8					0.4[z8]; 1.1[z7]
Limonene	138-86-3	0.4-23.7	0.6-13.9	4.9	0.2-4.8	14.5[h]; 20.7[y2]; 21.0[x3]; 22.1[n]; 22.3[r]; 23[x2]; 25.6[y5]; 26.9[y8]; 30.0[t]; 34.8[o]; 48[x5]
Linalool	78-70-6		tr-0.1	0.2	0-75.3	0.5[c]; 0.8[f,p]; 1.0[e]; 1.1[m]; 3.1[q]; 6.9[i]; 11.3[z14]; 11.5[y2]; 17[y1]; 65[z4]; 82.8[n]; 93.9[x]
Linalool oxide	1365-19-1				0-1.0	
cis-Linalool oxide, furanoid	11063-77-7					tr[m]; 0.1[x]
trans-Linalool oxide, furanoid	34995-77-2					0.2[x]
Linalyl acetate	115-95-7					0.1[c,x]; 1.0[f]; 2.1[i]
Linoleic acid	60-33-3					<0.01[x9]
p-Mentha-1,3-dien-7-al	1197-15-5					0.3[z14]
p-Mentha-1(7),8-diene	499-97-8					0.1[e]
p-Mentha-3,8-diene	586-67-4					0.1[j]
p-Mentha-1(7),8(10)-dien-9-ol	29548-13-8		0-0.3			

Table 5.79.2 Constituents identified in spearmint oils (*continued*)

Constituent	CAS	Percentage and range				
		A	B (54)	C (37)	D (41)	E
p-Mentha-2,8-dien-1-ol	22771-44-4					0.2[w]
cis-*p*-Mentha-2,8-dienol	3886-78-0		0-0.3			
trans-*p*-Mentha-2,8-dienol	4017-77-0		0-0.4			0.2[q]
p-Mentha-2,4(8),6-triene-2,3-diol						7.2[e]
p-Menth-2-en-1-ol	619-62-5			1.2		0.2[z11]
cis-*p*-Menth-2-en-1-ol	29803-82-5					0.1[j]
p-Menth-3-en-8-ol	18479-65-7					0.1[j]
Menthofuran	494-90-6		0-0.1			0.2[g]; 0.5[f]; 0.7[z14]; 1.6[n]; 2.7[x1]; 4.7[z13]
Menthol	89-78-1	0.2-2.2	0-0.2		0-8.2	0.2[c]; 0.5[k]; 1.2[o]; 1.5[g]; 4.7[x5]; 5.5[f]; 6.5[l]; 7.9[y2]; 8.7[l]; 11.4[z]; 13.4[n]; 15.7[x1]
Menthone	89-80-5	0.08-1.8	0-0.1		0-44.5	0.2[e,h]; 0.3[l]; 0.7[g]; 1.6[f]; 2.4[n]; 6.9[l]; 14[x2]; 15.6[j]; 21.9[k]; 32.7[z]; 62[y1]
Menthyl acetate	16409-45-3	0.1-0.6	0-0.1			0.4[z14]; 0.6[g]; 0.8[f]; 1.4[z8]; 5.9[z13]; 28.1[n]
2-Methylbenzaldehyde	529-20-4		0-0.7			
2-Methylbutanal	96-17-3					0.03[z6]; 0.1[c]
Methyl eugenol	93-15-2			0.4		
4-Methylisopulegone						1.3[i]
cis-Muurola-3,5-diene	157374-44-2		0.1-0.4			
cis-Muurola-4(14),5-diene	157477-72-0		0.1-0.5			<0.01[y6]; 0.7[u]; 0.8[m]; 0.9[x6]
trans-Muurola-4(14),5-diene	262352-87-4		0.2-0.8			
α-Muurolene	10208-80-7					0.1[e]
γ-Muurolene	30021-74-0		0-3.2			tr[e,m]; 0.4[z6]; 0.6[q]; 2.3[u]; 2.4[z15]
cis-Muurol-5-en-4α-ol	157374-45-3					0.03[u]
cis-Muurol-5-en-4β-ol	157374-46-4					0.02[u]
α-Muurolol	104245-48-9					0.2[e,w]; 1.6[q]
τ-Muurolol (epi-α-)	19912-62-0			0.5		
Myrcene (β-)	123-35-3	0.01-2.6	2.3-5.8	0.2	tr-5.9	1.5[p]; 1.9[f]; 2.1[u]; 2.3[k]; 2.5[x6]; 5.1[r]; 7.4[g]
Myrcenyl acetate	1118-39-4		0-0.7			
Myrtenal	564-94-3		0-0.4			
Naphthalene	91-20-3				0-tr	
Neodihydrocarveol	18675-33-7	0.08-0.6		0.7		<0.1[z8]; 0.1[l]; 0.2[s]; 0.6[z6]; 1.2[x]
Neoisodihydrocarveol						0.3[l]; 2.1[s]; 3.9[p]
Neoisodihydrocarvyl acetate						0.2[l]; 0.6[y4]; 0.9[s]
Neoisomenthol	491-02-1					1.1[g]
Neomenthol	3623-51-6		0-0.1			0.2[l]; 1.3[g]; 2.4[z8]
Neomenthyl acetate	2230-87-7		0-0.1			
Nerol	106-25-2				0-tr	
Nerolidyl acetate	56001-43-5					0.8[i]
Neryl acetate	141-12-8					<0.05[x]
Nonanal	124-19-6				0-0.1	<0.1[z16]
(*E*)-Ocimene	27400-72-2				0-tr	0.1[f]
(*Z*)-Ocimene	27400-71-1		0.5-1.6		0-1.0	0.02[t]; 0.1[f,y3]
β-Ocimene	13877-91-3				0-0.2	tr[y4]; 0.3[y8]
(*E*)-β-Ocimene	3779-61-1	tr-1.0	0.2-0.5			tr[j]; 0.03[s]; 0.1[h]; 0.2[e]; 0.3[v]; 0.5[o]; 0.6[c]
(*Z*)-β-Ocimene	3338-55-4	0-0.2				0.1[s]; 0.2[c,o]; 0.3[h]; 0.4[w]; 0.6[v]; 0.9[x6]; 1.1[g]
allo-Ocimene	673-84-7					0.03[u]
neo-allo-Ocimene	7216-56-0					0.1[u]
3-Octanol	589-98-0	0.01-2.2	0-0.9	0.5	0-0.6	1.0[v]; 1.2[y8]; 1.3[p]; 1.6[z16]; 1.7[e]; 2.0[t]; 2.6[m]
3-Octanol, acetate	4864-61-3		0.1-1.3			
3-Octanone	106-68-3					0.1[e]
1-Octen-3-ol	3391-86-4		0-0.1			tr[m]; 0.04[t]; 0.08[s]; 0.1[h]; 0.6[x]; 0.7[e]; 0.9[p]
trans-2-Octen-1-ol	18409-17-1		0-0.4			
Octenyl acetate (1-)	37366-04-4		0-2.7			0.2[h]

Table 5.79.2 Constituents identified in spearmint oils (*continued*)

Constituent	CAS	Percentage and range				
		A	B (54)	C (37)	D (41)	E
1-Octen-3-yl acetate	2442-10-6					<0.1[z16]
Octyl acetate	112-14-1		0-0.4		0-0.1	
3-Octyl acetate	4864-61-3	tr-0.4			tr-0.4	<0.1[z16]; 0.2[t]; 0.3[h]; 0.4[z6,z9]
Perillyl alcohol	536-59-4		0-0.4			
α-Phellandrene	99-83-2		0-0.2			0.4[c]
β-Phellandrene	555-10-2					0.07[s]; 3.0[v]
Phenylacetaldehyde	122-78-1					tr[m]; 0.3[a,z16]
α-Pinene	80-56-8	0-1.0	0.3-0.7	0.5	0.1-0.9	0.9[h]; 1.5[y7]; 1.7[r]; 1.8[m]; 2.2[z13]; 2.5[g]; 10.2[n]
β-Pinene	127-91-3	0-1.0	0.5-1.2		0.2-2.0	1.8[r]; 2.6[m]; 3.1[z9]; 3.5[w]; 5.0[g]; 7.9[z13]; 10.6[n]
trans-Pinocarveol	1674-08-4				0-0.1	
Piperitenone	491-09-8				0-24.2	0.1[h]; 0.2[g]; 0.3[l]; 0.7[w]; 0.8[j]
Piperitenone oxide	90582-88-0		0-61.9	0.5	0-70.3	15.7[l]; 26.2[g]; 35.7[m]; 52.3[e]; 80[z5]
Piperitone	89-81-6	0.02-0.6	0-0.7		0-1.8	1.0[m]; 1.3[z]; 20.3[x5]; 28.2[v]; 48.0[y2]; 49.3[x1]
Piperitone oxide	148879-33-8		0-0.2		0-72.3	0.3[i]; 23.3[x1]
cis-Piperitone oxide	57130-28-6					5.6[l]; 5.9[g]; 30[x2]
trans-Piperitone oxide						33.5[g]; 45.0[l]
Pulegone	89-82-7			1.1	0-30.8	0.03[l]; 0.2[c]; 0.3[g,y8]; 2.2[m]; 3.1[o]; 4.0[h]; 7.9[i]; 11.3[z13]; 26.7[z]; 29.6[v]; 49.2[n]; 72.1[j]
Sabinene	3387-41-5	0-0.5	0.5-1.7	0.2	0-1.0	0.5[e]; 0.6[z]; 0.7[k]; 1.2[h,s,y8]; 5.0[g,x1]
Sabinene hydrate	546-79-2					0.2[y4]; 0.7[y8]; 0.8[p]; 2.8[v]
cis-Sabinene hydrate	15537-55-0	tr-1.0	tr-1.3			0.4[s,y3]; 1.0[x4]; 1.2[z1]; 1.8[f]; 2.5[z15]; 2.7[z2]
trans-Sabinene hydrate	17699-16-0	tr-1.0				0.9[y7]; 1.0[x3]; 1.3[r]; 2.2[z6]; 2.8[z16]; 4[y1]; 22[y1]
trans-Sabinol	471-16-9					0.1[j]
(Z)-α-Santalol	115-71-9			0.4		
Santolinyl acetate	79507-88-3					0.3[u]
Selin-6-en-8-ol			0-0.3			
α-Selinene	473-13-2			0.5		
β-Selinene	17066-67-0			1.9		tr[m]
δ-Selinene	473-14-3					tr[m]
β-Sesquiphellandrene	20307-83-9					0.7[z11]
Spathulenol	6750-60-3		0.1-0.1			0.1[x]; 0.3[h]; 0.4[q]; 5.2[m]
α-Terpinene	99-86-5	0-0.2	0-0.1			0.03[t]; 0.1[h,j]; 0.2[c]; 0.3[f]; 0.4[y7]; 0.5[s]; 0.6[y8]
γ-Terpinene	99-85-4	0-0.4	0-0.2	0.8	tr-0.4	0.2[g]; 0.4[f]; 0.7[r]; 0.8[z15]; 0.9[y8]; 1.3[y7]; 1.4[k,o]
Terpinen-4-ol	562-74-3	0.07-1.5	0-0.3	3.0	0-0.3	1.7[v]; 1.8[z2]; 2[y1]; 2.6[z15]; 2.7[y7]; 5[y1]; 6.1[p]
Terpinen-4-yl acetate	4821-04-9					55.3[x1]
α-Terpineol	98-55-5	0.1-0.4	0-0.2	2.0	0-3.0	0.3[e]; 0.5[c]; 0.6[z1]; 1.0[v]; 1.3[q]; 2.1[z15]; 2.7[p]
β-Terpineol	138-87-4			0.3	0-3.0	
δ-Terpineol	7299-42-5					+[y9]
α-Terpinolene	586-62-9	0-0.1	0.1-0.3	0.05	0-tr	tr[h]; 0.02[r]; 0.05[c]; 0.1[e]; 0.2[p,f]; 0.3[k,y8]
Terpinyl acetate	8007-35-0				0-12.3[b]	
α-Terpinyl acetate	80-26-2		0-32.3	0.4		0.1[x]
β-Terpinyl acetate	10198-23-9		0-0.5			
Thuja-2,4(10)-diene	36262-09-6					0.3[z16]
3-Thujanol	513-23-5					1.8[j]
2-Thujene						+[y9]
α-Thujene	2867-05-2		tr-0.1			tr[e,g]; 0.08[z6]; 0.1[k,w]; 0.2[j]; 0.7[p]
Thujone (α-)	546-80-5					+[y9]; 0.7[z14]
Thymol	89-83-8					0.1[e,y3]; 0.6[g]; 5.2[z15]
Toluene	108-88-3				0-tr	

Table 5.79.2 Constituents identified in spearmint oils (*continued*)

Constituent	CAS	Percentage and range				
		A	B (54)	C (37)	D (41)	E
Tricyclene	508-32-7					0.3^k; 1.3^{x1}
Viridiflorol	552-02-3	0-0.2				0.1^x; 0.2^{z6}; 0.3^r; 0.4^w; 4.3^m
α-Ylangene	14912-44-8					tr^e

A seventy-one spearmint essential oil samples from USA, Canada, India and China, analyzed between 1998 and 2013; lowest and highest concentrations given (E. Schmidt, unpublished data)
B five lab-hydrodistilled oils from leaves of spearmint cultivated in Estonia, four in home gardens and one in an herb farm; lowest and highest concentrations given (ref. 54)
C one lab-hydrodistilled oil from fresh *M. spicata* leaves cultivated in Tunisia (ref. 37)
D twenty-four lab-hydrodistilled oils from the aerial parts of flowering *M. spicata* growing wild in north Greece and collected at 17 different sites; lowest and highest concentrations given (ref. 41)
E data from other studies (indicated with superscript letters); highest concentrations found in any study reviewed here given; when two or more oils were investigated, only the highest concentrations are mentioned, unless indicated otherwise

[a] incorrect identity based on GC elution order (ref. 59); [b] dihydrocarvyl acetate and terpinyl acetate combined; [c] one oil produced from *M. spicata* collected in Cameroon (ref. 65); [d] cannot exist in (*E*)- or (*Z*)-form (ref. 59); [e] one lab-hydrodistilled oil from leaves and stalks of spearmint plants cultivated at an experimental station in Cuba (refs. 13,15); [f] one commercial steam-distilled oil from plants cultivated in the Provence, France (ref. 3); [g] three lab-hydrodistilled leaf oils from three wild populations growing at Zakynthos Island, Greece (ref. 8); [h] one steam-distilled oil from leaves and stalks of a wild growing spearmint population in Mexico (ref. 12); [i] eight lab-hydrodistilled oils from the aerial parts of spearmint in the flowering period, one from cultivated *M. spicata* and 7 from wild populations collected in various parts of China (ref. 28); [j] one lab-hydrodistilled oil from leaves collected in a medicinal garden in India (ref. 42); [k] one lab-hydrodistilled oil from Serbia (refs. 44,46); [l] twenty lab-hydrodistilled oils from plants cultivated from wild populations in Greece; ten were of the piperitone oxide chemotype and ten of the carvone chemotype; mean concentrations per chemotype were provided (ref. 48); [m] one lab-hydrodistilled oil from fully flowering spearmint growing wild in Greece (ref. 43); [n] 138 oils produced from 14 natural populations and 62 cultivated *M. spicata* landraces from various parts of Turkey; there were three chemotypes: carvone (31-83%), pulegone (2 samples: 45-49%) and linalool (one sample: >80%) (ref. 39); [o] six lab-hydrodistilled oils from *Mentha spicata* cultivated at 6 locations in Egypt (ref. 38); [p] two steam-distilled oils from spearmint plants in bloom cultivated in Italy and the USA from the same cultivar (ref. 36); [q] two lab-hydrodistilled leaf oils from spearmint cultivated in Pakistan and harvested in two seasons (ref. 35); [r] one lab-hydrodistilled oil from fresh leaves of *M. spicata* cultivated in India (ref. 31); [s] one lab-hydrodistilled oil from the leaves of spearmint collected in the wild in Greece in the flowering phase (ref. 30); [t] two steam-distilled oils from stems + leaves of spearmint from two USA locations (ref. 25); [u] one lab-hydrodistilled oil from fresh aerial parts of *M. spicata* cultivated in Brazil (ref. 23); [v] four lab-hydrodistilled oils rich in pulegone and piperitone from four accessions cultivated in Turkey at four locations and harvested at the earliest flowering stage (ref. 22); [w] one lab-hydrodistilled oil from the aerial parts of flowering spearmint collected in the wild in eastern Turkey (ref. 18); [x] four lab-hydrodistilled oils from the leaves of wild-growing *M. spicata* of the linalool chemotype and harvested from May to October (ref. 11); [x1] nine oils from nine accessions of *M. spicata* in full bloom cultivated from landraces (ref. 10); [x2] seven lab-hydrodistilled oils from *M. spicata* ssp. *spicata* growing wild in Turkey (ref. 1); [x3] one lab-hydrodistilled leaf oil from spearmint cultivated in Egypt (ref. 4); [x4] one lab-hydrodistilled oil from the aerial flowering parts of spearmint collected in the wild in Pakistan (ref. 5); [x5] one atypical commercial oil from Iran with an extremely high content of limonene (48%) (ref. 6); [x6] one lab-hydrodistilled oil from the aerial parts of wild Indian spearmint harvested in the pre-flowering phase (ref. 7); [x7] four lab-hydrodistilled oils from plants cultivated in northern Finland and harvested in different seasons; only carvone was investigated (ref. 9); [x8] one water-distilled oil from south Algeria; only carvone and 1,8-cineole were investigated (ref. 14); [x9] one lab-hydrodistilled oil from Cuba; only the data which were not present in ref. 13 (column E[e]) are shown here (ref. 15); [y] sixteen lab-hydrodistilled spearmint oils from *M. spicata* planted at different times and harvested twice; only carvone was investigated (ref. 16); [y1] three lab-hydrodistilled oils from spearmint collected in the wild in north Turkey (ref. 17); [y2] data from Iranian investigations, cited in ref. 19; [y3] one lab-hydrodistilled oil from *Mentha spicata* plants at full flowering stage cultivated in Sudan from the rhizomes of wild spearmint plants (ref. 20); [y4] one lab-hydrodistilled oil from the fully flowering aerial parts of *M. spicata* growing wild at Crete, Greece (ref. 21); [y5] one lab-hydrodistilled spearmint oil from the pre-flowering aerial parts of plants cultivated in India (ref. 24); [y6] one lab-hydrodistilled oil sample from the flowering aerial parts of a wild spearmint population in the western Himalayas, India (ref. 27); [y7] three lab-hydrodistilled oils from leaves and stems of wild Turkish spearmint populations; average concentrations given (ref. 29); [y8] one steam-distilled oil from leaves plus stems of *M. spicata* var. *viridis* cultivated in Italy (ref. 32); [y9] four oils from fresh and dried leaves of spearmint cultivated in Italy and obtained by hydrodistillation or microwave-generated distillation; indicated with +[y9], as the amounts cannot be compared with the other data (ref. 33); [z] one lab-hydrodistilled oil from the leaves of *M. spicata* harvested in the flowering phase of plants cultivated in Tunisia (ref. 34); [z1] one lab-hydrodistilled leaf oil from India (ref. 40); [z2] one lab-hydrodistilled oil from spearmint purchased at a traditional market in Portugal (ref. 45); [z3] one lab-hydrodistilled oil from fresh leaves of spearmint growing wild in northern Italy; indicated with +[z3], cannot be compared with the other data (ref. 26); [z4] data from ref. 62; [z5] data from ref. 64; [z6] one commercial spearmint oil produced in the USA (ref. 60); [z7] one commercial *M. spicata* oil from China (ref. 66); [z8] one commercial spearmint oil sample (ref. 67); [z9] one oil from India (ref. 68); [z10] one spearmint oil produced in Moldova (ref. 69); [z11] one spearmint oil sample from *M. spicata* cultivated in Italy (ref. 70); [z12] several spearmint oils from plants grown in Hungary from Hungarian and German clones (ref. 71); [z13] spearmint oil samples from plants grown in two locations in Turkey over two seasons and with two harvesting times (ref. 72); [z14] one oil from *M. spicata* grown in an experimental garden in Iran (ref. 73); [z15] two oils from *M. crispa* (= *M. spicata*) from Reunion produced by microwave distillation and hydrodistillation (ref. 74); [z16] two oils from plants collected in the north and the south of Algeria (ref. 75)

tr: trace (in columns B, E[e], E[g], E[m]: <0.1), + present in the oil investigated, but quantity not stated or expressed in values which cannot be compared with the other data

Table 5.79.3 Major constituents of spearmint oils

Constituent	CAS	Percentage and range				
		A	B (54)	C (37)	D (41)	E
Linalool	78-70-6		tr-0.1	0.2	0-75.3	11.5[y2]; 17[y1]; 65[z4]; 82.8[n]; 93.9[x]
Carvone	99-49-0	60.6-82.3	2.3-67.4	50.5	0-49.7	68.4[y4]; 69.3[c]; 72.9[y]; 73.2[o]; 79.9[x8]; 83.0[n]
Piperitenone oxide	90582-88-0		0-61.9	0.5	0-70.3	15.7[l]; 26.2[g]; 35.7[m]; 52.3[e]; 80[z5]
Pulegone	89-82-7			1.1	0-30.8	11.3[z13]; 26.7[z]; 29.6[v]; 49.2[n]; 72.1[j]
Menthone	89-80-5	0.08-1.8	0-0.1		0-44.5	6.9[i]; 14[x2]; 15.6[j]; 21.9[k]; 32.7[z]; 62[y1]
Piperitone	89-81-6	0.02-0.6	0-0.7		0-1.8	1.0[m]; 1.3[z]; 20.3[x5]; 28.2[v]; 48.0[y2]; 49.3[x1]
Limonene	138-86-3	0.4-23.7	0.6-13.9	4.9	0.2-4.8	23[x2]; 25.6[y5]; 26.9[y8]; 30.0[t]; 34.8[o]; 48[x5]
1,8-Cineole	470-82-6	0.01-4.4	0-4.1	9.1	0.4-6.8	14.8[x1]; 19[y1]; 21.3[w]; 22.7[z13]; 33.8[n]; 40.5[g]
Isomenthone	491-07-6	0.01-0.8	0.1-0.3	1.1	0-15.2	0.3[g]; 1.5[i]; 1.6[y3]; 2.6[j]; 17[y1]; 32[x2]; 39.1[x1]
(Z)-Carveol	1197-06-4	0.1-2.0	0-0.1			2.3[q]; 3.9[s]; 10.6[z1]; 11.6[y2]; 21.3[z1]; 24.3[x4]
β-Caryophyllene	87-44-5	0.09-3.0	1.9-11.4	3.0	0-6.3	5.4[z12]; 6.1[x3]; 7.9[x5]; 8.0[v]; 9.5[h]; 20.3[y2]
Dihydrocarveol	38049-26-2	0.05-5.1	0-1.5	0.5	0-2.0	3.2[x6]; 5.9[p]; 6.3[s]; 7.1[o]; 15.7[i]; 16.5[z106]
Menthol	89-78-1	0.2-2.2	0-0.2		0-8.2	5.5[f]; 6.5[l]; 7.9[y2]; 8.7[i]; 11.4[z]; 13.4[n]; 15.7[x1]

LEGEND: SEE UNDER TABLE 5.79.2

Table 5.79.4 Results of testing groups of patients with spearmint oil

Years and Country	Test conc. & vehicle	Number of patients tested	positive	(%)	Selection of patients (S); Relevance (R); Comments (C)	Ref.
Routine testing						
2000-2007 USA	2% pet.	500	5	(1.0%)	R: 100%; C: weak study: a. high rate of macular erythema and weak reactions, b. relevance figures include 'questionable' and 'past' relevance	78
1999-2000 Denmark	2% pet.	318	5	(1.6%)	R: not specified; C: this study was part of the international study mentioned below (ref. 86)	100
1998-2000 six European countries	2% pet.	1,606	13	(0.8%)	R: not specified for individual oils/chemicals	86
Testing in groups of selected patients						
1999-2011 Australia	5% pet.	1,467	73	(5.0%)	S: patients tested with the 'toothpaste, essential oils and fragrance (rare) series'; R: 19/73 (26%) relevant; 14/19 had biopsy-proven oral lichen planus (OLP); in ten of these, the OLP improved >80% after avoidance of spearmint; C: 50% had a positive patch test to carvone and 8/14 (57%) to 'sassafras'; the latter reactions were considered cross-reactions to spearmint oil	98, 99
2001-2010 Australia	5% pet.	1,383	68	(4.9%)	S: not specified; R: 31%	97
<2005 USA	2% pet.	111	4	(3.6%)	S: patients using consumer products containing spearmint oil; R: 'at least possibly relevant'	81
2000 USA, Japan and 4 European countries	5% pet.	178	9	(5.1%)	S: patients previously shown to be allergic to fragrances; R: not stated	88
1997-2000 Austria	2% pet.	747	1	(0.1%)	S: patients suspected of fragrance allergy; R: not stated	82
1997-1998 Italy	2% pet.	54	4	(7.4%)	S: patients with cheilitis suspected of toothpaste allergy; R: all reactions were relevant	89
1996-1997 UK	2% pet.	10	1	(10%)	S: patients suspected of cosmetic dermatitis and reacting to the fragrance mix; R: not stated	81
<1978 Denmark	5% pet.	40	4	(10%)	S: see text under case reports	80

pet.: petrolatum

planus developed allergic contact stomatitis from spearmint oil in mouth rinse and chewing gum (83). In one patient, contact allergy to spearmint oil in toothpaste caused sore mouth, fissuring of the lips and dermatitis of the surrounding skin; a patch test with anethole 5% in petrolatum was negative (90). One patient had erosive cheilitis from contact allergy to spearmint oil and its main component carvone in toothpaste (91). Four patients developed stomatitis and dermatitis from contact allergy to spearmint oil in toothpastes (96). Some additional cases of contact allergy to spearmint oil in toothpastes have been reported (93,94, data cited in ref. 92). One positive patch test reaction to spearmint oil (and peppermint oil) was seen in a patient with contact dermatitis from compresses with an infusion of fresh leaves of *Mentha spicata* (87). One patient had a positive patch test reaction to spearmint oil 0.1% in petrolatum, considered to be clinically relevant, in a group of 146 patients referred for cheilitis to one UK hospital between 1982 and 2001 (82).

LITERATURE

1 Başer KHC, Kürkçüoğlu M, Demirci B, Özek T, Tarımcılar G. Essential oils of *Mentha* species from Marmara region of Turkey. J Essent Oil Res 2012;24:265-272
2 Lawrence BM. The composition of commercially important mints. In: BM Lawrence, Editor. Mint. The Genus *Mentha* – Medicinal and Aromatic Plants – Industrial Profiles. Boca Raton, Florida, USA: CRC Press, 2007:217-232
3 Tognolini M, Barocelli E, Ballabeni V, Bruni R, Bianchi A, Chiavarini M, Impicciatore M. Comparative screening of plant essential oils: Phenylpropanoid moiety as basic core for antiplatelet activity. Life Sciences 2006;78:1419-1432
4 Elansary HO, Ashmawy NA. Essential oils of mint between benefits and hazards. J Essent Oil Bear Plants 2013;16:429-438
5 Hussain AI, Anwar F, Shahid M, Ashraf M, Przybylski R. Chemical composition, and antioxidant and antimicrobial activities of essential oil of spearmint (*Mentha spicata* L.) from Pakistan. J Essent Oil Res 2010;22:78-84
6 Sharafi SE, Rasooli I, Owlia P, Nadoushan MJ, Ghazanfari T, Taghizadeh M. Phytochemical bioactivities from *Mentha spicata* essential oil for health promotion. J Essent Oil Bear Plants 2010;13:237-249
7 Chauhan RS, Nautiyal MC, Tava A. Essential oil composition from aerial parts of *Mentha spicata* L. J Essent Oil Bear Plants 2010;13:353-356
8 Cook CM, Kokkini S, Lanaras T. *Mentha spicata* essential oils rich in 1,8-cineole and 1,2-epoxy-*p*-menthane derivatives from Zakynthos (Ionian Island, W Greece). J Essent Oil Res 2007;19:225-230
9 Aflatuni A, Uusitalo Sari EK J, Hohtola A. Optimum harvesting time of four *Mentha* species in Northern Finland. J Essent Oil Res 2006;18:134-138
10 Zeinali H, Arzani A, Razmjoo K, Rezaee MB. Evaluation of oil compositions of Iranian mints (*Mentha* ssp.). J Essent Oil Res 2005;17:156-159
11 Kofidis G, Bosabalidis A, Kokkini S. Seasonal variation of essential oils in a linalool-rich chemotype of *Mentha spicata* grown wild in Greece. J Essent Oil Res 2004;16:469-472
12 Pino J, Borges P, Martínez M, Vargas M, Flores H, Estarrón M, Fuentes V. Essential oil of *Mentha spicata* L. from Jalisco. J Essent Oil Res 2001;13:409-410
13 Pino JA, Rosado A, Sánchez E. Essential oil of *Mentha spicata* L. from Cuba. J Essent Oil Res 1998;10:657-659. Data partly also presented in ref. 15
14 Khalfi O, Benyoussef E-H, Yahiaoui N. Extraction, analysis and insecticidal activity of spearmint essential oil from Algeria against *Rhyzopertha dominica* (F.). J Essent Oil Bear Plants 2006;9:17-21
15 Pino JA, Garcia J, Martinez MA. Comparison of solvent extract and supercritical carbon dioxide extract of spearmint leaf. J Essent Oil Res 1999;11:191-193. Data also presented in ref. 13
16 Singh M, Singh PV, Singh DV. Effect of planting time on growth, yield and quality of spearmint (*Mentha spicata* L.) under subtropical climate of central Uttar Pradesh. J Essent Oil Res 1995;7:621-626
17 Baser KHC, Kürkçüoglu M, Tarimcilar G, Kaynak G. Essential oils of *Mentha* species from Northern Turkey. J Essent Oil Res 1999;11:579-588
18 Şarer E, Toprak SY, Otlu B, Durmaz R. Composition and antimicrobial activity of the essential oil from *Mentha spicata* L. subsp. *spicata*. J Essent Oil Res 2011;23:105-108
19 Hassanzadeh MK, Emami SA, Asili J, Najaran ZT. Review of the essential oil composition of Iranian Lamiaceae. J Essent Oil Res 2011;23:35-74
20 Younis YMH, Beshir SM. Carvone-rich essential oils from *Mentha longifolia* (L.) Huds. ssp. *schimperi* Briq. and *Mentha spicata* L. grown in Sudan. J Essent Oil Res 2004;16:539-541
21 Kokkini S, Karousou R, Lanaras T. Essential oils of spearmint (Carvone-rich) plants from the island of Crete (Greece). Biochem Syst Ecol 1995;23:425-430
22 Telci I, Demirtas I, Bayram E, Arabaci O, Kacare O. Environmental variation on aroma components of pulegone/piperitone rich spearmint (*Mentha spicata* L.). Ind Crops Prod 2010;32:588-592
23 Scherer R, Fumiere Lemos M, Fumiere Lemos M, Coimbra Martinelli G, Damasceno Lopes Martins J, Gomes da Silva Universidade A. Antioxidant and antibacterial activities and composition of Brazilian spearmint (*Mentha spicata* L.). Ind Crops Prod 2013;50:408-413
24 Kedia A, Prakash B, Mishra PK, Chanotiya CS, Dubey NK. Antifungal, antiaflatoxigenic, and insecticidal efficacy of spearmint (*Mentha spicata* L.) essential oil. Int Biodeter Biodegrad 2014;89:29-36
25 Barton P, Hughes RE Jr, Hussein MM. Supercritical carbon dioxide extraction of peppermint and spearmint. J Supercrit Fluids 1992;5:157-162

26 Da Porto C, Decorti D. Ultrasound-assisted extraction coupled with under vacuum distillation of flavour compounds from spearmint (carvone-rich) plants: Comparison with conventional hydrodistillation. Ultrasonics Sonochemistry 2009;16:795-799

27 Chauhan RS, Kaul MK, Shahi AK, Kumar A, Ram G, Tawa A. Chemical composition of essential oils in Mentha spicata L. accession [IIIM(J)26] from North-West Himalayan region, India. Ind Crops Prod 2009;29:654-656

28 Zhao D, Xu YW, Yang GL, Husaini AM, Wu W. Variation of essential oil of Mentha haplocalyx Briq. and Mentha spicata L. from China. Ind Crops Prod 2013;42:251-260

29 Sertkaya E, Kaya K, Soylu S. Acaricidal activities of the essential oils from several medicinal plants against the carmine spider mite (Tetranychus cinnabarinus Boisd.) (Acarina: Tetranychidae). Ind Crops Prod 2010;31:107-112. Data also presented in ref. 47

30 Adam K, Sivropoulou A, Kokkini S, Lanaras T, Arsenakis M. Antifungal activities of Origanum vulgare subsp. hirtum, Mentha spicata, Lavandula angustifolia, and Salvia fruticosa essential oils against human pathogenic fungi. J Agric Food Chem 1998;46:1739-1745

31 Verma RS, Pandey V, Padalia RC, Saikia D, Krishna B. Chemical composition and antimicrobial potential of aqueous distillate volatiles of Indian peppermint (Mentha piperita) and spearmint (Mentha spicata). J Herbs Spices Med Plants 2011;17:258-267

32 Tibaldi G, Fontana E, Nicola S. Postharvest management affects spearmint and calamint essential oils. J Sci Food Agric 2013;93:580-586

33 Orio L, Cravotto G, Binello A, Pignata G, Nicola S and Chemat F. Hydrodistillation and in situ microwave-generated hydrodistillation of fresh and dried mint leaves: a comparison study. J Sci Food Agric 2012;92:3085-3090

34 Dhifi W, Jelali N, Mnif W, Litaiem M, Hamdi N. Chemical composition of the essential oil of Mentha spicata L. from Tunisia and its biological activities. J Food Biochem 2013;37:362-368

35 Hussain AI, Anwar F, Nigam PS, Ashraf M, Gilani AH. Seasonal variation in content, chemical composition and antimicrobial and cytotoxic activities of essential oils from four Mentha species. J Sci Food Agric 2010;90:1827-1836

36 Maffei M, Codignola A, Fieschi M. Essential oil from Mentha spicata L.(spearmint) cultivated in Italy. Flavour Fragr J 1986;1:105-109

37 Mkaddem M, Bouajila J, Ennajar M, Lebrihi A, Mathieu F, Romdhane M. Chemical composition and antimicrobial and antioxidant activities of Mentha (longifolia L. and viridis) essential oils. J Food Sci 2009;74: M358-M363

38 Edris AE, Shalaby AS, Fadel HM, Abdel-Wahab MA. Evaluation of a chemotype of spearmint (Mentha spicata L.) grown in Siwa Oasis, Egypt. Eur Food Res Technol 2003;18:74-78

39 Telci I, Sahbaz N(I), Yilmaz G, Tugay ME. Agronomical and chemical characterization of spearmint (Mentha spicata L.) originating in Turkey. Econ Bot 2004;58:721-728

40 Govindarajan M, Sivakumar R, Rajeswari M, Yogalakshmi K. Chemical composition and larvicidal activity of essential oil from Mentha spicata (Linn.) against three mosquito species. Parasitol Res 2012;110:2023-2032

41 Kokkini S, Vokou D. Mentha spicata (Lamiaceae) chemotypes growing wild in Greece. Econ Bot 1989;43:192-202

42 Joshi RK. Pulegone and menthone chemotypes of Mentha spicata Linn. from western Ghats region of north west Karnataka, India. Natl Acad Sci Lett 2013;36:349-352

43 Koliopoulos G, Pitarokili D, Kioulos E, Michaelakis A, Tzakou O. Chemical composition and larvicidal evaluation of Mentha, Salvia, and Melissa essential oils against the West Nile virus mosquito Culex pipiens. Parasitol Res 2010;107:327-335

44 Soković MD, Vukojević J, Marin PD, Brkić DD, Vajs V, van Griensven LJLD. Chemical composition of essential oils of Thymus and Mentha species and their antifungal activities. Molecules 2009;14:238-249. Data also published in ref. 46

45 Mata AT, Proenca C, Ferreira AR, Serralheiro MLM, Nogueira JMF, Araujo MEM. Antioxidant and antiacetylcholinesterase activities of five plants used as Portuguese food spices. Food Chem 2007;103:778-786

46 Sokovic M, Van Griensven LJLD. Antimicrobial activity of essential oils and their components against the three major pathogens of the cultivated button mushroom Agaricus bisporus. Eur J Plant Pathol 2006;116:211-224. Data also published in ref. 44

47 Sertkaya E, Kaya K, Soylu S. Acaricidal activities of the essential oils from several medicinal plants against the carmine spider mite (Tetranychus cinnabarinus Boisd.) (Acarina: Tetranychidae). Indust Crops Prod 2010;31:107-112. Data also presented in ref. 29

48 Karousou R, Grammatikopoulos G, Lanars T, Manetas Y, Kokkini S. Effects of enhanced UV-B radiation on Mentha spicata essentials oils. Phytochem 1998;49:2273-2277

49 Chowdhury JU, Nandi NC, Uddina M, Rahman M. Chemical constituents of essential oils from two types of spearmint (Mentha spicata L. and M. cardiaca L.) introduced in Bangladesh. Banglad J Sci Ind Res 2007;42:79-82

50 Boukhebti H, Chaker AN, Belhadj H, Sahli F, Ramdhani M, Laouer H, Harzallah D. Chemical composition and antibacterial activity of Mentha pulegium L. and Mentha spicata L. essential oils. Der Pharmacia Lettre 2011;3:267-275

51 Znini M, Bouklah M, Majidi L, Kharchouf S, Aouniti A, Bouyanzer A, et al. Chemical composition and inhibitory effect of Mentha spicata essential oil on

the corrosion of steel in molar hydrochloric acid. Int J Electrochem Sci 2011;6:691-704

52 Padalia RC, Verma RS, Chauhan A, Sundaresan V, Chanotiya CS. Essential oil composition of sixteen elite cultivars of *Mentha* from western Himalayan region, India. Maejo Int J Sci Technol 2013;7:83-93

53 Teixeira ML, das G. Cardoso M, Figueiredo ACS, Moraes JC, Assis FA, de Andrade J, et al. Essential oils from *Lippia origanoides* Kunth. and *Mentha spicata* L.: Chemical composition, insecticidal and antioxidant activities. Amer J Plant Sci 2014;5:1181-1190

54 Orav A, Kapp K, Raal A. Chemosystematic markers for the essential oils in leaves of *Mentha* species cultivated or growing naturally in Estonia. Proceed Eston Acad Sci 2013;62:175-186. Available online at www.eap.ee/proceedings

55 Adelpoor MJ, Golparvar AR. Chemical composition of essential oils of three ecotypes of *Mentha spicata* L. from Kohgiluyeh va Boyer-Ahmad Province, Iran. J Herb Drugs 2013;4:143-146

56 Martins MR, Tinoco MT, Almeida AS, Cruz-Morais J. Chemical composition, antioxidant and antimicrobial properties of three essential oils from Portuguese flora. J Pharmacogn 2012;3:39-44

57 Pavela R, Kaffková K, Kumšta M. Chemical composition and larvicidal activity of essential oils from different *Mentha* L. and *Pulegium* species against *Culex quinquefasciatus* Say (Diptera: Culicidae). Plant Protect Sci 2014;50:36-42

58 Moosavy M-H, Shavisi N. Determination of antimicrobial effects of nisin and *Mentha spicata* essential oil against *Escherichia coli* O157:H7 under various conditions (pH, temperature and NaCl concentration). Pharm Sciences 2013;19:61-67

59 Lawrence BM. Progress in essential oils. Perfum Flavor 2008;33(1):36-? (last page unknown)

60 Kubeczka KH, Formacek V. Essential oils analysis by capillary gas chromatography and carbon-13 NMR spectroscopy. New York, USA: John Wiley and Sons, 2002:321-326. Data cited in ref. 59.

61 Kokkini S. Chemical races within the genus *Mentha* L. In: HF Linskens and JF Jackson, Eds. Modern methods of plant analysis, New Series 12. Heidelberg, Germany: Springer-Verlag, 1991:63-78

62 Gora J, Kalemba D. Chemical composition of essential oil from Mentha spicata L. Herba Polonica. Poznan 1979;25:269-275. Data cited in ref. 39

63 Lawrence BM. A study of the monoterpene Interrelationship in the genus *Mentha* with special reference to the origin of pulegone and menthofuran. PhD Dissertation, Groningen State University, Groningen, Netherlands, 1978

64 Misra LN, Tyag BR, Thakur RS. Chemotypic variation in Indian spearmint. Planta Medica 1989;55:575-576. Data cited in ref 38

65 Jirovetz L, Buchbauer G, Shahabi M, Ngassoum MB. Comparative investigations of the essential oil and volatiles of spearmint. Perfum Flavor 2002;27(6):16-21. Data cited in ref. 59

66 Zhu L-F, Li Y-H, Li B-L, Lu B-Y, Zhang W-L. Aromatic Plants and Essential Constituents. Supplement I. South China Institute of Botany, Chinese Academy of Sciences. Hong Kong: Hoi Feng Publ., Peace Book Co, 1995. Data cited in ref. 59

67 Dimandja J-MD, Stanfill SB, Grainger J, Patterson DG. Application of comprehensive two-dimensional gas chromatography (GC x GC) to the qualitative analysis of essential oils. J High Resol Chromatogr 2000;23:208-214

68 Ranade GS. Spearmint oil (*Mentha spicata*). FAFAI 2005;7(4):73. Data cited in ref. 59

69 Shikimaka AP, Vorobiova EA, Kubrak MN, Buga TV, Terpenoid composition of the essential oil of the chemotype mint carvone. Bul Acad Stiinte Repub Stiinte Biol Chim 1993;5:75-76. Data cited in ref. 59

70 Avato P, Sgarra G, Casadoro G. Chemical composition of the essential oils of *Mentha* species cultivated in Italy. Sci Pharm 1993;63:223-230. Data cited in ref. 59

71 Hethelyi EB, Stoeva T, Bernath J. Investigation of the characteristics of Bulgarian and Hungarian *Mentha spicata* L. Huds spearmint oil by using GC, GC/MS techniques. Olaj Szappan Kozmet 2002;51:26-32. Data cited in ref. 59

72 Ozguven M, Kirici S. Research on yield essential oil content and components of mint (*Mentha*) species in different ecologies. Turk J Agric For 1999;23:465-472. Data cited in ref. 59

73 Hadjïakhoondi A, Aghel N, Zamanizadeh-Nadgar N, Vatandoost H. Chemical and biological study of *Mentha spicata* essential oil from Iran. Daru 2000;8(1/2):19-21. Data cited in ref. 59

74 Lucchesi ME, Chemat F, Smadja J. Solvent-free microwave extraction of essential oil from aromatic herbs: comparison with conventional hydrodistillation. J Chromatogr A 2004;1043:323-327

75 Benyoussef E-H, Yahiaoui N, Nacer-Bey N, Khelfaoui A, Belhadj M. Essential oil of *Mentha spicata* L. from Algeria. Rivista Ital EPPOS 2004;C37:31-35. Data cited in ref. 59

76 Rhind JP. Essential oils. A handbook for aromatherapy practice, 2nd Edition. London: Singing Dragon, 2012

77 Lawless J. The encyclopedia of essential oils, 2nd Edition. London: Harper Thorsons, 2014

78 Wetter DA, Yiannias JA, Prakash AV, Davis MD, Farmer SA, el-Azhary RA, et al. Results of patch testing to personal care product allergens in a standard series and a supplemental cosmetic series: an analysis of 945 patients from the Mayo Clinic Contact Dermatitis Group, 2000-2007. J Am Acad Dermatol 2010;63:789-798

79 Dharmagunawardena B, Takwale A, Sanders KJ, Cannan S, Roger A, Ilchyshyn A. Gas chromatography: an investigative tool in multiple allergies to essential oils. Contact Dermatitis 2002;47:288-292

80 Andersen KE. Contact allergy to toothpaste flavors. Contact Dermatitis 1978;4:195-198

81 Guin JD. Use of consumer product ingredients for patch testing. Dermatitis 2005;16:71-77

82 Strauss RM, Orton DI. Allergic contact cheilitis in the United Kingdom: a retrospective study. Dermatitis 2003;14:75-77

83 Clayton R, Orton D. Contact allergy to spearmint oil in a patient with oral lichen planus. Contact Dermatitis 2004;51:314-315

84 Tomson N, Murdoch S, Finch TM. The dangers of making mint sauce. Contact Dermatitis 2004;51:92-93

85 Morris GE. Dermatoses among food handlers. Ind Med Surg 1954;23:343.

86 Frosch PJ, Johansen JD, Menné T, Pirker C, Rastogi SC, Andersen KE, et al. Further important sensitizers in patients sensitive to fragrances. II. Reactivity to essential oils. Contact Dermatitis 2002;47:279-287

87 Bonamonte D, Mundo L, Daddabbo M, Foti C. Allergic contact dermatitis from *Mentha spicata* (spearmint). Contact Dermatitis 2001;45:298

88 Larsen W, Nakayama H, Fischer T, Elsner P, Frosch P, Burrows D, et al. Fragrance contact dermatitis: a worldwide multicenter investigation (Part II). Contact Dermatitis 2001;44:344-346

89 Francalanci S, Sertoli A, Giorgini S, Pigatto P, Santucci B, Valsecchi R. Multicentre study of allergic contact cheilitis from toothpastes. Contact Dermatitis 2000;43:216-222

90 Skrebova N, Brocks K, Karlsmark T. Allergic contact cheilitis from spearmint oil. Contact Dermatitis 1998;39:35-36

91 Worm M, Jeep S, Sterry W, Zuberbier T. Perioral contact dermatitis caused by L-carvone in toothpaste. Contact Dermatitis 1998;38:338

92 Sainio E-L, Kanerva L. Contact allergens in toothpastes and a review of their hypersensitivity. Contact Dermatitis 1995;33:100-105

93 Grattan CEH, Peachy RD. Contact sensitization to toothpaste flavouring. J Royal Coll Gen Pract 1985;35:498. Data cited in ref. 92

94 Baer ON. Toothpaste allergies. J Clin Pediatr Dent 1992;16:230-231

95 Magnusson B, Wilkinson DS. Cinnamic aldehyde in toothpaste. 1. Clinical aspects and patch tests. Contact Dermatitis 1975;1:70-76

96 Hjorth N, Jervoe P. Allergisk Kontaktstomatitis og Kontaktdermatitis fremkaldt of smagsstoffer i tandpasta. Tandlaegebladet 1967;71:937-942. Data cited in ref. 95

97 Toholka R, Wang Y-S, Tate B, Tam M, Cahill J, Palmer A, Nixon R. The first Australian Baseline Series: Recommendations for patch testing in suspected contact dermatitis. Australas J Dermatol 2014, Sept. 7. doi: 10.1111/ajd.12186

98 Gunatheesan S, Tam MM, Tate B, Tversky J, Nixon R. Retrospective study of oral lichen planus and allergy to spearmint oil. Australas J Dermatol 2012;53:224-228

99 Cahill J, Gunatheesan S, Tam M, Tate B, Nixon R. Oral lichen planus and allergy to spearmint oil. Contact Dermatitis 2012;66(Suppl. 2):38 (FC1.03)

100 Paulsen E, Andersen KE. Colophonium and Compositae mix as markers of fragrance allergy: Cross-reactivity between fragrance terpenes, colophonium and Compositae plant extracts. Contact Dermatitis 2005;53:285-291

Chapter 5.80 SPIKE LAVENDER OIL

DEFINITION

Spike lavender oil (essential oil of spike lavender, Spanish type) is the essential oil obtained from the flowering top of the spike lavender, *Lavandula latifolia* Medik. (synonym: *Lavandula spica* L.).

INCI NOMENCLATURE

Description/definition: Lavandula latifolia herb oil is an essential oil distilled from the flowering herbs of the lavender, *Lavandula latifolia* (syn: *Lavandula spica*), Labiatae
INCI name EU: Lavandula latifolia herb oil (perfuming name, not an INCI name proper)
INCI name USA: Not in the Personal Care Products Council Ingredient Database
CAS registry number(s): 84837-04-7
EINECS number(s): 284-290-6

Description/definition: Lavandula spica flower oil is the volatile oil obtained from the flowers of the spikenard, *Lavandula spica*, Labiatae
INCI name EU: Lavandula spica flower oil
INCI name USA: Lavandula spica (lavender) flower oil
CAS registry number(s): 8016-78-2; 84837-04-7; 97722-12-8
EINECS number(s): 307-762-6

ISO (INTERNATIONAL ORGANIZATION FOR STANDARDIZATION) STANDARD

ISO number: 4719
ISO name: Essential oil of spike lavender, Spanish type
Botanical origin: *Lavandula latifolia* Medik.
Parts of plant used: Flowering top
ISO values: ISO values (minimum and maximum concentrations) are shown in Table 5.80.1.

Other oils obtained from *Lavandula* species include lavender oil (Chapter 5.49) and lavandin oil (Chapters 5.46, 5.47 and 5.48).

Table 5.80.1 ISO values (%) for spike lavender oil [a]

Compound	CAS	Minimum	Maximum
Linalool	78-70-6	34.0	50.0
1,8-Cineole	470-82-6	16.0	39.0
Camphor (dl-)	76-22-2	8.0	16.0
Limonene	138-86-3	0.5	3.0
(*E*)-α-Bisabolene	25532-79-0	0.4	2.5
α-Terpineol	98-55-5	0.2	2.0
Linalyl acetate	115-95-7	0.0	1.6

[a] ISO 4719 Essential oil of spike lavender, Spanish type ©ISO 2012; Geneva, Switzerland, www.iso.org

THE PLANT, THE OIL, AND THEIR USES

Lavandula latifolia Medik., commonly known as spike lavender, is a strongly aromatic evergreen shrub growing to 30-80 cm tall. It is native to Italy, France and Spain and is naturalized in other Mediterranean countries (1). The plant is cultivated as an essential oil plant, for scent, ornament, and as a bee plant (GRIN Taxonomy for Plants; 5). In Spain, the culture of spike lavender is said to have been largely replaced in the last years by the more productive species lavandin (*Lavandula x intermedia* Emeric ex Loisel; *Lavandula hybrida*) (1). Spike lavender is mainly used for essential oil production. In addition, a great range of medical uses of this plant has been reported on account of its perceived antispasmodic, sedative, antihypertensive, antiseptic, healing and anti-inflammatory properties, which render it highly appreciated in phytotherapy and aromatherapy (1).

The essential oil of spike lavender is obtained by steam distillation of the flowering tops of both cultivated and wild *L. latifolia* populations, but often, stems and leaves are also distilled together with the inflorescences (21). Spike lavender oils are used in perfumery, cosmetics, as flavor in food products, in technical preparations such as room sprays and disinfectants, and in veterinary medicine (liniments) (1,4). Spike lavender oil is believed to be antibacterial, antifungal, carminative, sedative, anti-depressive and effective for burns and insect bites and is widely used in aromatherapy (1,4,5). The most important ingredients of the oil are linalool, 1,8-cineole and camphor, together accounting for more than 80% of the oil (5). The relative concentration of camphor and linalool determines the quality and price of the product. The most appreciated oils for the perfume and cosmetic industries are those with high content in linalool and low content in camphor, while those richer in camphor are mainly used in aromatherapy and phytotherapy (1).

CHEMICAL COMPOSITION

Spike lavender oil is a light yellow to orange-yellow liquid which has a fresh and floral but also camphoraceous minty odor. The yield of essential oil from the flowering tops of *Lavandula latifolia* Medik. generally varies from 1.5 to 2.2 per cent. The main producing country of this oil is Spain.

Literature data (up to October 15, 2014) on the chemical composition of spike lavender oils and unpublished analytical data from one of us (E.S.) are shown in Tables 5.80.2 and 5.80.3 in alphabetical order. Table 5.80.3 shows chemicals which have been identified in earlier studies (notably in ref. 29: 15 years qualitative investigation of Spanish spike lavender oils, cited in refs. 21 and 22), but which have not – possibly with a few exceptions – been demonstrated in more recent investigations. In spike lavender oils from various origins, over 395 chemicals have been identified; this includes 225 compounds found before 1986 (Table 5.80.3). Of the >170 chemicals identified more recently (Table 5.80.2), about 38 per cent were found in a single reviewed publication only. The major compounds found in spike lavender oils

Table 5.80.2 Constituents identified in spike lavender oils

Constituent	CAS	Percentage and range					
		A	B (1)	C (5)[b]	D (10)	E (8)	F
Acetone	67-64-1						0.1[v]
(E)-Anethole	4180-23-8						0.1[f]
Aromadendrene	489-39-4						0.3[e]
Bergamotene					0.5		0.1[f]
cis-α-Bergamotene	18252-46-5						
trans-α-Bergamotene	13474-59-4					0.1-0.2	0.1[f]
β-Bergamotene	6895-56-3						tr[v]
Bicyclo[2.2.1]heptan-2-one,5,5,6-trimethyl-	3292-05-5						0.5[v]
α-Bisabolene	17627-44-0						1.6[x]; 2.1[u]
(E)-α-Bisabolene	25532-79-0	0.1-1.5				1.3-2.3	1.9[l]
β-Bisabolene	495-61-4	0.2-0.4			0.2	0.4-0.5	0.1[f]; 0.2[x]; 0.3[e]; 0.4[l]; 1.6[o]
α-Bisabolol	515-69-5		0.06-1.5		0.2	0.1-0.8	2.6[f]; 2.7[g]
β-Bisabolol	15352-77-9						0.02[v]
α-Bisabolol oxide	22567-36-8				0.02		
Borneol (endo-)	507-70-0	0.07-1.3	0.2-5.9	0.9-1.8[b]	1.1	1.3-3.6	4.9[l]; 6.9[x]; 7.0[i]; 10[f]; 14.2[g]
Bornyl acetate	76-49-3	tr-0.3			0.1	tr	tr[t]; 0.08[v]; 0.3[p,u]; 0.7[x]; 4.2[e]
Bornyl formate	7492-41-3				0.1	tr-0.2	
β-Bourbonene	5208-59-3				0.2		<0.1[e]
Butyl acetate	123-86-4						0.1[u]
β-Cadinene	523-47-7				0.01		
γ-Cadinene	39029-41-9				0.1	0.2	0.1[w]; 0.2[v]; 0.3[u]; 0.8[l]
δ-Cadinene	483-76-1	0.09-0.3	0.09-0.7			tr	tr[t]; 0.5[u]; 1.0[l]
α-Cadinol	481-34-5		0.06-1.2		0.08	tr	
epi-α-Cadinol	5937-11-1					0.4-0.5	0.06[v]; 0.1[f]; 0.9[g]
cis-Calamenene	72937-55-4						0.3[l,m]
trans-Calamenene	73209-42-4						0.3[l,m]
Camphene	79-92-5	0.09-0.8	0.2-5.3	0.3-0.5[b]	0.6	0.5-0.6	0.7[e]; 0.8[i]; 1.0[o,v]; 1.8[x]; 2.1[x]
α-Campholenal	4501-58-0					0.1	+[q]; 0.3[e]
α-Campholenic acid	28973-89-9						+[q]
Camphor	76-22-2	8.7-35.1	1.1-46.7	11.4-18.6[b]	13.6	11.2-14.9	17.3[l]; 18.5[i]; 21.5[s]; 23.5[j]
δ3-Carene	13466-78-9	0-0.02					0.05[v]; 0.1[g,w]; 0.4[f]; 0.5[e]
Carvacrol	499-75-2						0.2[e]
Carveol	99-48-9						0.4[l]
(E)-Carveol	1197-07-5					0.1	
(Z)-Carveol	1197-06-4						0.05[x]; 0.1[t]; 0.5[e]
Carvone	99-49-0					tr-0.1	0.1[t]; 0.5[x]
β-Caryophyllene	87-44-5	0.03-1.9	0.05-2.0	0.5-1.1[b]	0.5	1.0-1.9	1.6[i]; 1.7[x]; 1.9[p]; 2.2[v]; 2.3[o]
γ-Caryophyllene [(Z)-]	118-65-0		0.05-1.7				
9-epi-(E)-Caryophyllene	68832-35-9						0.2[g]
Caryophyllene oxide	1139-30-6	tr-1.7		0.2-0.7[b]	1.6	0.3-0.4	0.1[f,w]; 0.2[t,u,x]; 0.3[h,v]; 1.0[l]
1,4-Cineole	470-67-7						0.1[e]
1,8-Cineole	470-82-6	14.2-31.2	6.6-57.1	20.8-47.9[b]	31.3	29.1-34.9	36.3[t]; 37.4[x]; 40.5[n]; 42.4[p]
Citronellol	106-22-9						0.4[x]
α-Copaene	3856-25-5						0.7[e]
Coumarin	91-64-5						0.1[u]; 0.2[w,x]; 1.1[l]; 2.4[t]
Cryptone	500-02-7						+[x]; 0.05[w]; 0.07[v]; 2.5[p]
α-Cubebene	17699-14-8						0.1[e]
β-Cubebene	13744-15-5				0.02		0.7[e]
Cuminaldehyde	122-03-2						+[x]; 1.0[g]
Cuminyl alcohol	536-60-7					tr-0.1	0.2[l,t]
Curcumene (ar-; α-)	644-30-4				0.01		
p-Cymene	99-87-6	0.05-1.1		0.1-0.2[b]		tr-0.2	0.4[h]; 0.6[g,s]; 0.9[l]; 1.4[v]; 1.5[e]
Cymenol							0.4[l]
p-Cymen-8-ol	1197-01-9					0.2-0.4	0.09[v]; 0.1[w]; 1.0[t]
Dihydrocoumarin	119-84-6						0.05[w]; 0.2[x]
Dihydroeugenol	2785-87-7						0.1[e]
β-Elemene	33880-83-0						2.1[e]
Elemicin	487-11-6		0.04-1.8				

Table 5.80.2 Constituents identified in spike lavender oils (*continued*)

Constituent	CAS	Percentage and range					
		A	B (1)	C (5)[b]	D (10)	E (8)	F
Eugenol	97-53-0						0.08[x]
Farnesene					0.2		
α-Farnesene	502-61-4		0.05-0.7				
(*E,E*)-α-Farnesene	502-61-4						<0.1[g]
β-Farnesene	502-60-3						tr[s]; 0.03[l]; 0.2[w]; 0.3[v]
(*E*)-β-Farnesene	18794-84-8	0.2-4.3				0.4-0.7	0.2[t]; 4.8[f]
(*Z*)-β-Farnesene	28973-97-9						0.1[f]
Farnesol	4602-84-0		0.04-5.0		0.03		
α-Fenchene	471-84-1	0.02-1.0					
Fenchone	1195-79-5						0.01[v]; 0.05[w]; 0.7[e]
Fenchyl acetate	13851-11-1		0.1-0.6				0.2[e]
Geraniol	106-24-1	tr-1.0	0.06-0.8	0.1-0.4[b]			0.2[h,l]; 0.9[t]; 1.2[x]; 1.8[s]; 2.2[s]
Geranyl acetate	105-87-3	0.06-0.2		0.1-0.2[b]			tr[t]; 0.07[x]; 0.1[w]; 0.7[e]; 1.2[v]
Germacrene A	28387-44-2				0.08		
Germacrene D	23986-74-5	0.05-0.4				0.5-1.0	0.05[h]; 0.1[w]; 0.2[v]; 0.4[f]
α-Gurjunene	489-40-7						0.5[l]
Hexanol	111-27-3	tr-0.1				tr-0.1	0.01[v]; 0.07[x]; 0.2[u]
2-Hexenal	505-57-7						tr[v]
(*E*)-2-Hexenal	6728-26-3					tr	
(*Z*)-3-Hexen-1-ol	928-96-1					tr	
Hexenyl butyrate	26912-31-2				0.05		
Hexyl acetate	142-92-7	0.01-0.1					0.1[u]; 0.2[f]
Hexyl butyrate	2639-63-6	0.07-0.1			0.4		0.02[v]; 0.1[f]
Hexyl hexanoate	6378-65-0				0.03		0.2[f]
Hexyl isobutyrate	2349-07-7	0.05-0.3			0.2		tr[v]; 0.1[w]
Hexyl isovalerate	10032-13-0					0.2	0.06[v]
Hexyl-2-methyl butyrate	10032-15-2			0-0.1[b]	0.4		0.2[f]
Hexyl tiglate	16930-96-4	0.03-0.4			0.1		tr[v]; 0.1[f]; 0.8[x]
Hexyl valerate	1117-59-5				0.2		
β-Himachalene	1461-03-6						1.3[o]
α-Humulene	6753-98-6	0.1-0.2			0.3	tr	0.1[w]; 0.2[h,v]; 0.4[u]; 0.5[x]
Isoborneol	124-76-5						0.09[n]; 0.1[e]; 0.3[r,t,u]; 0.4[x]
Isobornyl formate	1200-67-5						0.3[g]
Isobornyl isovalerate	7779-73-9				0.2		
Lavandulol	498-16-8	tr-0.9		0.2-0.3[b]			0.6[x]; 0.7[u]; 0.9[i]; 1.5[x]; 8.7[e]
Lavandulyl acetate	25905-14-0	0.05-0.7					0.1[w]; 0.3[i]; 0.6[g]; 0.8[f]; 1.3[i]
Ledol	577-27-5			0.4-1.3			
Limonene	138-86-3	0.2-3.0	0.03-3.4	0.2-0.9[b]		0.2-0.9	1.2[p]; 2.2[x]; 2.6[o]; 2.7[x]; 3.2[v]
Linalool	78-70-6	24.9-42.3	3.7-61.1	15.1-54.7[b]	36.9	27.2-32.4	43.8[x]; 46.4[x]; 47.9[h]; 53.9[x]
Linalool oxide	1365-19-1				0.07		tr[s]; 0.5[l,t,x]; 0.7[u]
cis-Linalool oxide	11063-77-7				0.8		0.1[v]; 8.5[x]
cis-Linalool oxide, furanoid	11063-77-7	0-0.3				tr-0.1	0.4[h]
trans-Linalool oxide	11063-78-8				0.05		0.2[v]; 6.8[x]
trans-Linalool oxide, furanoid	34995-77-2	0.01-0.4				0.3-0.4[c]	6.4[h]
Linalyl acetate	115-95-7	0.6-0.9	0.05-0.8		0.1		1.6[i]; 1.8[e,g]; 2.9[i]; 3.4[f]; 9.3[o]
Linalyl butyrate	78-36-4				0.03		
Longifolene (junipene)	475-20-7						+[x]; 0.3[e]
7-Methoxycoumarin	531-59-9						1.1[l]
p-Methylacetophenone	122-00-9						+[x]
Methylcarvacrol	6379-73-3						1.9[e]
Methylheptenone	409-02-9						+[x]
6-Methyl-5-hepten-2-one	110-93-0						0.07[x]
α-Muurolene	10208-80-7		0.1-1.6				
γ-Muurolene	30021-74-0		0.1-1.1				

Table 5.80.2 Constituents identified in spike lavender oils (*continued*)

Constituent	CAS	Percentage and range					
		A	B (1)	C (5)[b]	D (10)	E (8)	F
Myrcene	123-35-3	0.05-0.9			0.2	0.8	1.0[g]; 1.1[e]; 1.3[s]; 1.4[f,s]; 1.5[n]
Myrtenal	564-94-3			0.2-0.3[b]		0.2-0.3[d]	
Myrtenol	515-00-4			0.1-0.2[b]		0.2-0.3[d]	0.07[v]; 0.1[w]; 0.2[l]; 0.8[t]
Nerol	106-25-2	tr-0.3	0.09-0.8				0.04[u]; 0.08[x]; 0.3[w]; 0.7[t,v]
Neryl acetate	141-12-8	1.1-1.3			0.02		tr[s]; 0.05[w]; 0.4[x]; 0.9[v]
Nopinone	24903-95-5						+[x]
(*E*)-Ocimene	27400-72-2						0.3[x]; 0.4[x]; 0.6[u]
(*Z*)-Ocimene	27400-71-1						0.01[x]; 0.2[u]
(*E*)-β-Ocimene	3779-61-1	0.08-0.2	0.06-1.3			tr-0.1	0.1[w]; 0.6[g]; 0.8[e]; 1.3[f]
(*Z*)-β-Ocimene	3338-55-4	0.09-1.0	0.2			0.3	0.08[v]; <0.1[g]; 0.5[v]; 4.2[f]
Octanal	124-13-0						0.2[w]
Octanol	111-87-5						0.1[n]
3-Octanol	589-98-0						0.01[x]; 0.1[u]
3-Octanone	106-68-3	tr-0.4				tr	0.05[v]; 0.07[h,x]; 0.08[w]; 0.2[u]
1-Octen-3-ol	3391-86-4	0.04-2.6		0.2-0.3[b]		tr	0.05[w]; 0.09[v]; 0.3[f]
1-Octen-3-yl acetate	2442-10-6						tr[v]; 0.2[u]
Oxydihydrocampho-lenic acid							+[q]
α-Phellandrene	99-83-2					tr-0.1	tr[r,v]; 0.1[f,p,u,x]; 0.3[e]
β-Phellandrene	555-10-2		0.2-3.7	0.5-1.8[b]			0.3[g]; 0.8[n]
α-Pinene	80-56-8	0.6-3.6	0.2-3.2		1.1	1.2-1.9	2.8[p]; 3.7[x]; 4.2[l,x]; 4.8[v]; 6.8[x]
β-Pinene	127-91-3	0.4-2.6	0.04-2.1	0.8-2.9[b]	1.5	1.7-2.4	2.6[n]; 3.2[p]; 3.6[x]; 4.1[v]; 4.5[x]
Pinic acid	473-73-4						+[q]
Pinocamphone	547-60-4				0.04		
trans-Pinocarveol	1674-08-4	0.05-0.5		0.1-0.3[b]			0.4[l]
Pinocarvone	30460-92-5					0.1	
Pinonaldehyde	2704-78-1						+[q]
Plinol A (isomer 1)	4028-59-5						0.4[v]
Plinol B (isomer 2)	4099-07-4						0.4[v]
Plinol C (isomer 3)	4028-60-8						tr[v]
Sabinene	3387-41-5	0.02-0.8	0.02-1.2	0.3-1.0[b]	0.5	0.6-0.8	0.5[w]; 0.6[l]; 0.7[u]; 0.9[f,l]; 1.6[p]
Sabinene hydrate	546-79-2			0.3-1.1[b]			0.5[n]
cis-Sabinene hydrate	15537-55-0	tr-0.3			0.6	0.4-0.5	
trans-Sabinene hydrate	17699-16-0						0.8[g]; 0.9[f]
α-Terpinene	99-86-5	tr-0.2		0-0.1[b]	0.3	0.1	0.05[p,v]; 0.2[n]; 0.3[u];
γ-Terpinene	99-85-4	0.2-0.3		0.1-0.2[b]		0.2	0.2[v]; 0.3[e,n]; 0.5[u]; 0.7[p]
Terpinen-4-ol	562-74-3	0.4-0.8	0.3-7.1		0.2	0.4	0.9[g]; 1.6[s]; 2.3[f]; 3.3[k]; 7.7[e]
Terpinen-4-yl acetate	4821-04-9						0.2[e]
Terpineol	8000-41-7						1.5[n]
α-Terpineol	98-55-5	0.08-1.4	0.1-3.0	0.6-2.2[b]	0.8	1.4-1.6	1.9[x]; 2.4[p]; 2.5[f]; 2.6[g,t]; 2.7[v]
γ-Terpineol	586-81-2	tr-0.2					0.5[v]
δ-Terpineol	7299-42-5	0.06-0.6				0.4-1.2	1.0[l]
Terpinolene (α-)	586-62-9	0.09-0.4	0.3-0.7	0.1-0.3[b]	0.7	0.3-0.4[c]	0.07[v]; 0.2[n,u]; 0.3[p,x]; 0.5[f]
α-Terpinyl acetate	80-26-2						+[x]
Thuja-2,4(10)-diene	36262-09-6					tr	
α-Thujene	2867-05-2		0.9-1.8		0.02	tr	0.05[u]; 0.1[e,f]; 0.2[p,x]; 0.6[g]
α-Thujone	546-80-5						tr[t]; 0.08[x]
β-Thujone (*cis*-)	471-15-8						tr[t]; 0.05[x]
Thymol	89-83-8		0.1-0.8				0.01[n]
Tricyclene	508-32-7	0.02-0.08			0.05	tr	tr[v]; 0.05[w]
1,1,3-Trimethyl-2-oxa-bicyclo[2.2.2]octan-5-one					0.2		
cis-Verbenol	1845-30-3		0.06-0.6	0-0.1[b]			

Table 5.80.2 Constituents identified in spike lavender oils (*continued*)

Constituent	CAS	A	B (1)	C (5)[b]	D (10)	E (8)	F
Verbenone	80-57-9		0.08-0.7		0.1	0.1	
Viridiflorol	552-02-3		0.04-1.9				
α-Ylangene	14912-44-8				0.02		0.1[e]

A twenty-four spike lavender essential oil samples from Spain, analyzed between 1998 and 2013; lowest and highest concentrations given (E. Schmidt, unpublished data)
B one hundred and ninety-four lab-hydrodistilled oils from 194 samples of at least 25 individual plants in full bloom growing wild in six different biogeographic provinces in Spain; lowest and highest concentrations given (ref. 1)
C seventy-two lab-hydrodistilled oils from *L. latifolia* flowers collected from seven wild populations from different bioclimatic belts in Spain; lowest and highest *average* concentrations given as shown for the seven populations (ref. 5)
D one lab-hydrodistilled oil from plants cultivated in northeastern Spain (ref. 10)
E three steam-distilled oils of *L. latifolia* populations growing wild in Spain in full bloom; lowest and highest concentrations given (ref. 8)
F data from other studies (indicated with superscript letters); highest concentrations found in any study reviewed here given; when two or more oils were investigated, only the highest concentrations are mentioned, unless indicated otherwise

[a] incorrect identification; [b] in seven groups, the lowest and highest concentrations in each of these groups were presented; here, the lowest and highest values of these *average* concentrations are given (ref. 5); [c] *trans*-linalool oxide, furanoid and terpinolene combined; [d] myrtenal and myrtenol combined; [e] one steam-distilled oil sample from the flowering tops of spike lavenders growing wild in north Tunisia (ref. 6); [f] two distilled oils from fresh and dried spike lavender flowers cultivated in the national botanical garden, Tehran, Iran (ref. 9); [g] one lab-hydrodistilled oil from plants cultivated in Iran (ref. 4); [h] one spike lavender oil of Spanish origin (ref. 27); [i] one spike lavender oil produced in France (ref. 25); [j] one spike lavender oil sample produced in Argentina (ref. 26); [k] one steam-distilled oil from flowering plants cultivated in Israel (ref. 23); [l] one oil of Spanish origin obtained by steam distillation, possibly from *L. latifolia* plants in the post-flowering stage (ref. 11); [m] *cis*- and *trans*-calamenene combined; [n] one lab-hydrodistilled oil from Spain (ref. 3); [o] one lab-hydrodistilled spike lavender oil from plants cultivated in Spain (ref. 2); [p] six oils prepared from Spanish cultivated plants 1-6 years old and one oil of wild growing spike lavender (ref. 30); [q] one sample of Spanish spike lavender oil (ref. 31); [r] one spike lavender oil of unknown origin (ref. 28); [s] two lab-distilled oils prepared from spike lavender plants collected in Coimbra and Alto do Vieura, Portugal (ref. 33); [t] one commercial oil of unknown origin (ref. 7); [u] three Spanish spike lavender oils from three seasons (ref. 21); [v] one spike lavender oil of unknown origin and mode of production (ref. 12); [w] one spike lavender oil of unknown origin and mode of production (ref. 22); [x] data from various studies from before 1980, cited in ref. 22; highest concentrations given (ref. 22)

tr: trace (in column E: <0.1); + present in the oil investigated, but quantity not stated

from different sources are shown in Table 5.80.4. They include (highest concentrations in any study given) linalool (61.1%),1,8-cineole (57.1%), camphor (46.7%), borneol (14.2%), α-pinene (6.8%) and β-pinene (4.5%). Well-known ingredients of spike lavender oils that were present in high concentrations (>7%) in one or two studies were linalyl acetate (9.3%), lavandulol (8.7%) and terpinen-4-ol (7.7%). A rare constituent of spike lavender oils found in a high concentration (>7%) in a single study was *cis*-linalool oxide (8.5%).

Commercial oils

The ten chemicals that had the highest maximum concentrations in 24 commercial spike lavender essential oil samples (concentration ranges provided) are the following: linalool (0.6-42.3%), camphor (8.0-35.1%), 1,8-cineole (3.2-31.2%), (*E*)-β-farnesene (0.2-4.3%), α-pinene (0.6-3.6%), limonene (0.2-3.0%), 1-octen-3-ol (0.04-2.6%), β-pinene (0.4-2.6%), β-caryophyllene (0.03-1.9%) and caryophyllene oxide (tr-1.7%) (Erich Schmidt, unpublished analytical data).

Analysis of *L. latifolia leaves* (without flowers) has been presented in ref. 5. Analysis of oils obtained from plants in the *fruiting* period is described in ref. 8. Older literature on spike lavender oils has been summarized in

refs. 13-22. A review of various aspects of lavenders (lavender, lavandin, spike lavender), written from the industry's perspective, can be found (on-line) in a dissertation from the International Centre for Aroma Trades Studies, Plymouth, UK (34).

CONTACT ALLERGY/ALLERGIC CONTACT DERMATITIS

General
Contact allergy to spike lavender oil has been reported in two publications, but no cases of allergic contact dermatitis from the oil have been identified.

Testing in groups of patients
Two hundred dermatitis patients from Poland were tested with spike lavender oil 2% in petrolatum and one (0.5%) reacted. The same authors also patch tested 51 patients allergic to *Myroxylon pereirae* resin (balsam of Peru) and/or turpentine and/or wood tar and/or colophony and one (2.0%) had a positive patch test; relevance data were not provided (35). A group of 86 patients from Poland previously reacting to the fragrance mix was tested with spike lavender oil and eight (9.3%) had a positive patch test reaction; relevance data were not provided (36).

Table 5.80.3 Constituents of spike lavender oils identified in earlier (<1986) investigations only (21,22,29)

Constituent	CAS	Constituent	CAS
Acetic acid	64-19-7	2-Ethylfuran	3208-16-0
2-Acetyl-5-isopropylpyridine		2-Ethyl-3-methylbutanoic acid	32444-32-9
3-Acetyl-6-methylpyridine	36357-38-7	o-Ethylphenol	90-00-6
2-Acetylpyridine	1122-62-9	2-Ethylpyridine	100-71-0
Allyl acetate	591-87-7	3-Ethylpyridine	536-78-7
Benzaldehyde	100-52-7	4-Ethylpyridine	536-75-4
Benzoic acid	65-85-0	2-Furancarboxylic acid	88-14-2
Bornyl isobutyrate	24717-86-0	Furfural (dehyde)	98-01-1
Bornyl isovalerate	76-50-6	Geranic acid	459-80-3
Bornyl 2-methylbutyrate	94200-10-9	Geranyl acetone	3796-70-1
Bornyl propanoate	78548-53-5	Geranyl butyrate	106-29-6
Butanol	35296-72-1	Geranyl formate	105-86-2
2-Butanol	78-92-2	Geranyl 3-methylbutanoate	109-20-6
4-Butanolide	96-48-0	Geranyl propanoate	105-90-8
Butanone (2-)	78-93-3	Guaiacol	90-05-1
3-Buten-2-one	78-94-4	Heptanal	111-71-7
2-Butyl acetate	105-46-4	Heptanoic acid	111-14-8
3-sec-Butylpyridine	25224-14-0	Heptanol	53535-33-4
Butyric acid	107-92-6	2-Heptanone	110-43-0
δ-Cadinol	19435-97-3	Hexanal	66-25-1
10-epi-α-Cadinol (τ-Cadinol)	5937-11-1	Hexanoic acid	142-62-1
Camphene hydrate	465-31-6	2-Hexanone	591-78-6
Camphenilone	13211-15-9	3-Hexanone	589-38-8
γ-Campholenic acid	67246-55-3	5-Hexen-2-one	109-49-9
α-Campholytic acid	6709-22-4	Hexyl benzoate	6789-88-4
3-Carboxy-4,4-dimethyl-cyclobutane-1-acetic acid		4-Hexyl-4-butanolide	706-14-9
		(E)-Hotrienol	53834-70-1
Carvotanacetone	499-71-8	2-Hydroxyacetophenone (α-) (ω-)	582-24-1
trans-Carvotan alcohol		2-Hydroxy-4-isopropenylbenzaldehyde	
α-Cedrene	469-61-4	2-Hydroxy-3-isopropyl-2-cyclohexen-1-one	
Cinnamic acid	621-82-9		
Citronellal	106-23-0	6-Hydroxy-3-isopropyl-6-methyl-2-cyclohexen-1-one	
m-Cresol	108-39-4		
o-Cresol	95-48-7	2'-Hydroxy-4'-methylacetophenone	6921-64-8
p-Cresol	106-44-5	2'-Hydroxy-5'-methylacetophenone	1450-72-2
p-Cymenene	1195-32-0	2'-Hydroxy-6'-methylacetophenone	
Decanoic acid	334-48-5	α-Ionone	127-41-3
Decanol	36729-58-5	β-Ionone	79-77-6
Diacetone alcohol	123-42-2	Isoamyl alcohol	123-51-3
cis-Dihydrolinalool oxide		Isobutanoic acid	79-31-2
trans-Dihydrolinalool oxide		Isobutyl alcohol	78-83-1
2,5-Dimethylbenzaldehyde	5779-94-2	Isofenchone	6541-58-8
2,3-Dimethylbutanoic acid	14287-61-7	Isoprene	78-79-5
3,3-Dimethylbutanoic acid	1070-83-3	4-Isopropenylbenzaldehyde	10133-50-3
4,4-Dimethyl-4-but-2-enolide	20019-64-1	4-Isopropenyl-3-methyl-2-cyclohexen-1-one	
2,2-Dimethyl-3-cyclopenten-1-ethanal			
5,5-Dimethyl-2-cyclopenten-1-ethanal		5-Isopropenyl-2-methyltetrahydrofuran-2-ethanal	
3,3-Dimethyl-1,6-hexanedioic acid			
1,6-Dimethyl-4-isopropylstyrene		4-Isopropylbenzoic acid	536-66-3
2,6-Dimethyl-6-methoxy-7-octen-2-ol	91243-33-3	4-Isopropyl-4-butanolide	38624-29-2
(E)-2,7-Dimethyl-1,4,6-octatrien-3-ol		3-Isopropylcyclopentanone	10264-56-9
2,4(5)-Dimethylphenol		3-Isopropylfuran	15012-74-5
2,3-Dimethylpropanoic acid		2-Isopropylglutaric acid	32806-63-6
3,3-Dimethylpropenoic acid		3-Isopropyl-1,6-hexanedioic acid	
2,3-Dimethylpyrazine	5910-89-4	5-Isopropyl-2-methyl-1-cyclopentene-1-carboxaldehyde	
2,5-Dimethylpyrazine	123-32-0		
2,6-Dimethylpyridine	108-48-5	5-Isopropyl-2-methyl-2-vinyltetrahydrofuran	71635-17-1
1-Dodecene (α-)	112-41-4		
endo-1,8-Epoxy-p-menthan-2-ol	18679-48-6	4-Isopropylpyridine	696-30-0
exo-1,8-Epoxy-p-menthan-2-ol	18679-48-6	2-Isopropylsuccinic acid	2338-45-6
1,8-Epoxy-p-menth-2-en-4-ol		Lavandulyl isobutyrate	51117-20-5
Ethanol	64-17-5	Lavandulyl isovalerate	51117-21-6
Ethyl acetate	141-78-6	Lavandulyl 2-methylbutyrate	

Table 5.80.3 Constituents of spike lavender oils identified in earlier (<1986) investigations only (21,22,29) (*continued*)

Constituent	CAS	Constituent	CAS
Lavandulyl propionate		(*E*)-Nerolidol	40716-66-3
cis-Linalool oxide, pyranoid	14009-71-3	Neryl hexanoate	68310-59-8
trans-Linalool oxide, pyranoid	39028-58-5	Nerylic acid	
Maleic acid	110-16-7	Neryl 2-methylbutanoate	51117-19-2
Malonic acid	141-82-2	Neryl 3-methylbutanoate	3915-83-1
p-Mentha-1,3-dien-7-al	1197-15-5	Neryl propionate	105-91-9
p-Menth-1-en-9-al	29548-14-9	(*2E,6Z*)-Nonadienal	557-48-2
3-(2-Methoxyphenyl)propanoic acid	6342-77-4	Nonanal	124-19-6
2-Methoxy-3,5,5-trimethyl-2-cyclo-hexene-1,4-dione	41654-27-7	Nonanoic acid	12-05-0
		Octanoic acid	124-07-2
2-Methylbutanal	96-17-3	3-Octanone	106-68-3
3-Methylbutanal	590-86-3	(*E*)-2-Octenal	2548-87-0
Methyl butanoate	623-42-7	Octyl acetate	112-14-1
2-Methylbutanoic acid	116-53-0	Octyl 2-methylbutanoate	29811-50-5
3-Methylbutanoic acid	503-74-2	Octyl 3-methylbutanoate	7786-58-5
2-Methyl-4-butanolide	1679-47-6	Octyl propionate	142-60-9
3-Methyl-2-butanone	563-80-4	1-Oxo-4,4,5-trimethylcyclopentane-3-acetic acid	
(*E*)-2-Methyl-2-butenal	497-03-0		
(*E*)-2-Methyl-2-butenoic acid	80-59-1	Pentanal	110-62-3
2-Methyl-3-buten-2-ol	115-18-4	Pentanoic acid	109-52-4
3-Methyl-2-buten-1-ol	556-82-1	Pentanol	30899-19-5
3-Methylbutyl 2-methylbutanoate	27625-35-0	2-Pentanone	107-87-9
3-Methyl-2-cyclohexen-1-one	1193-18-6	3-Pentanone	96-22-0
2-Methyl-1-cyclopentene-1-carbox-aldehyde		1-Penten-2-ol	61923-56-6
		1-Penten-3-ol	616-25-1
2-Methylfuran	534-22-5	5-Pentyl-5-pentanolide	705-86-2
3-Methylfurfural	33342-48-2	Perillaldehyde (perilla aldehyde)	2111-75-3
5-Methylfurfural	620-02-0	Phellandral	21391-98-0
2-Methyl-3,6-heptadien-2-one		Phenol	108-95-2
6-Methyl-3,5-heptadien-2-one	1604-28-0	Phenylacetaldehyde	122-78-1
4-Methylhexanoic acid	1561-11-1	Phenylacetic acid	103-82-2
5-Methylhexanoic acid	628-46-6	3-Phenylpropanoic acid	501-52-0
α-Methylionone	93302-56-8	Pinonaldehyde	2704-78-1
5-Methyl-2-isopropenylpyridine	56057-93-3	*cis*-Piperitol	16721-38-3
2-Methyl-5-isopropylpyridine	20194-71-2	*trans*-Piperitol	16721-39-4
5-Methyl-2-isopropylpyridine		Propanal	123-38-6
2-Methylpentanoic acid	97-61-0	Propanoic acid	79-09-4
3-Methylpentanoic acid	105-43-1	2-Propanol	67-63-0
4-Methylpentanoic acid	646-07-1	4-Propylbenzoic acid	2438-05-3
2-Methyl-3-pentanone	565-69-5	Pyridine	110-86-1
4-Methyl-2-pentanone	108-10-1	Quinoline	91-22-5
2-Methyl-2-pentenal	623-36-9	Sabina ketone	513-20-2
4-Methyl-3-pentenoic acid	504-85-8	Safrole	94-59-7
4-Methyl-3-penten-2-one	141-79-7	Salicylaldehyde	90-02-8
4-(4-Methylphenyl)pentanal		Salicylic acid	69-72-7
2-(4-Methylphenyl)propanal	99-72-9	β-Selinene	17066-67-0
2-Methylpyridine	109-06-8	Spathulenol	6750-60-3
3-Methylpyridine	108-99-6	Succinic acid	110-15-6
Methyl thymol (thymol methyl ether)	1076-56-8	*cis*-1,8-Terpin hydrate	2451-01-6
4-Methyl-4-vinyl-4-butanolide		Tetrahydropyranyl-2-ethanal	
4-Methyl-4-vinyl-4-but-2-enolide		Toluene	108-88-3
5-Methyl-5-vinyltetrahydrofuran-2-(2-methylethanal)		3,4,4-Trimethyl-2-cyclopenten-1-one	30434-65-2
		2,4,5-Trimethyloxazoline	
Myrcenol	543-39-5	2,2,6-Trimethyl-6-vinyltetrahydropyran	7392-19-0
Myrtenic acid	19250-17-0	2-Undecanone	112-12-9
Neral	106-26-3		

Table 5.80.4 Major constituents of spike lavender oils

Constituent	CAS	Percentage and range					
		A	B (1)	C (5) [b]	D (10)	E (8)	F
Linalool	78-70-6	24.9-42.3	3.7-61.1	15.1-54.7[b]	36.9	27.2-32.4	43.8[x]; 46.4[x]; 47.9[h]; 53.9[x]
1,8-Cineole	470-82-6	14.2-31.2	6.6-57.1	20.8-47.9[b]	31.3	29.1-34.9	36.3[t]; 37.4[x]; 40.5[n]; 42.4[p]
Camphor	76-22-2	8.7-35.1	1.1-46.7	11.4-18.6[b]	13.6	11.2-14.9	17.3[l]; 18.5[l]; 21.5[s]; 23.5[j]
Borneol (endo-)	507-70-0	0.07-1.3	0.2-5.9	0.9-1.8[b]	1.1	1.3-3.6	4.9[l]; 6.9[x]; 7.0[l]; 10[f]; 14.2[g]
α-Pinene	80-56-8	0.6-3.6	0.2-3.2		1.1	1.2-1.9	2.8[p]; 3.7[x]; 4.2[l,x]; 4.8[v]; 6.8[x]
β-Pinene	127-91-3	0.4-2.6	0.04-2.1	0.8-2.9[b]	1.5	1.7-2.4	2.6[n]; 3.2[p]; 3.6[x]; 4.1[v]; 4.5[x]

LEGEND: SEE UNDER TABLE 5.80.2

LITERATURE

1 Herraiz-Peñalver D, Ángeles Cases M, Varela F, Navarrete P, Sánchez-Vioque R, Usano-Alemany J. Chemical characterization of *Lavandula latifolia* Medik. essential oil from Spanish wild populations. Biochem System Ecol 2013;46:59-68

2 Santana O, Cabrera R, Giménez C, González-Coloma A, Sánchez-Vioque R, de los Mozos-Pascual M, et al. Chemical and biological profiles of the essential oils from aromatic plants of agro industrial interest in Castilla-La Mancha (Spain). Grasas y Aceites, 2012;63:214-222 (in Spanish)

3 Rodrigues N, Malheiro R, Casal S, Asensio-S.-Manzanera MC, Bento A, Pereira JA. Influence of spike lavender (*Lavandula latifolia* Med.) essential oil in the quality, stability and composition of soybean oil during microwave heating. Food Chem Toxicol 2012;50:2894-2901

4 Eikani MH, Golmohammad F, Shokrollahzadeh S, Mirza M, Rowshanzamir S. Superheated water extraction of *Lavandula latifolia* Medik volatiles: Comparison with conventional techniques. J Essent Oil Res 2008;20:482-487

5 Munoz-Bertomeu J, Arrillaga I, Segura J. Essential oil variation within and among natural populations of *Lavandula latifolia* and its relation to their ecological areas. Biochem System Ecol 2007;35:479-488

6 Alatrache A, Jamoussi B, Tarhouni R, Abdrabba M. Analysis of the essential oil of *Lavandula latifolia* from Tunisia. J Essent Oil Bear Plants 2007;10:446-452

7 De Pascual Teresa J, Ovejero J, Anaya J, Caballero E, Hernandez JM. Chemical composition of the Spanish spike oil. Planta Medica 1989;55:398-399. Data cited in ref. 22

8 Salido S, Altarejos J, Nogueras M, Sánchez A, Luque P. Chemical composition and seasonal variations of spike lavender oil from southern Spain. J Essent Oil Res 2004;16:206-210

9 Barazandeh MM. Essential oil composition of *Lavandula latifolia* Medik from Iran. J Essent Oil Res 2002;14:103-104

10 Guillén MD, Cabo N, Burillo J. Characterisation of the essential oils of some cultivated aromatic plants of industrial interest. J Sci Food Agric 1996;70:359-363

11 de Pascual Teresa J, Caballero E, Caballero C, Machin G. Constituents of the essential oil of *Lavandula latifolia*. Phytochem 1983;22:1033-1034

12 Naef R, Morris AF. Lavender-Lavandin. A Comparison. Rivista Ital EPPOS 1992;(Numero Speciale):364-377. Data cited in ref. 22.

13 Lawrence BM. Progress in essential oils. Perfum Flavor 1976;1(2):17-18

14 Lawrence BM. Progress in essential oils. Perfum Flavor 1978;3(3):48-49

15 Lawrence BM. Progress in essential oils. Perfum Flavor 1980;5(2):38

16 Lawrence BM. Progress in essential oils. Perfum Flavor 1983;8(5):20-22

17 Lawrence BM. Progress in essential oils. Perfum Flavor 1984;9(1):56

18 Lawrence BM. Progress in essential oils. Perfum Flavor 1987;12(5):62

19 Lawrence BM. Progress in essential oils. Perfum Flavor 1990;15(5):58-59

20 Lawrence BM. Progress in essential oils. Perfum Flavor 1993;18(6):53

21 Boelens MH. The essential oil of spike lavender *Lavandula latifolia* Vill. (*L. spica* D.C.). Perfum Flavor 1986;11(5):43-63

22 Boelens MH. Chemical and sensory evaluation of Lavandula oils. Perfum Flavor 1995;20(2):23-51

23 Ravid U, Putievski E, Katzir I, Ikan R. Determination of the enantiomeric composition of terpinen-4-ol in essential oils using a permethylated β-cyclodextrin coated capillary column. Flavour Fragr J 1992;7:49-52

24 Lawrence BM. Progress in essential oils. Perfum Flavor 2001;26(3):66-? (last page unknown)

25 Canaud F, Martineau M-O. Aspic lavande et lavandin. Bull Union Physiciens 1996;90:1941-1950. Data cited in ref. 24

26 Mizrahi I, Juárez MA, Elechosa MA, Bandoni AL, Núñez M. Composition and quality of essential oils obtained from crops of *Lavandula* ssp. of Argentina. In: Proceedings of the World Congress of Medicinal and Aromatic Plants for Human Welfare. Acta Hort 1999;500:119-125. Data cited in ref. 24

27 Kubeczka K-H, Formacek V. Essential oils analysis by capillary gas chromatography and carbon-13 NMR spectroscopy, 2nd edition. New York, USA: John Wiley and Sons, 2002:333-338. Data cited in ref. 32

28 Formacek V, Kubeczka K-H. Essential oils analysis by capillary gas chromatography and carbon-13 NMR spectroscopy. New York, USA: John Wiley and Sons, 1982:137-142. Data cited in ref. 22

29 Ter heide R, et al. Qualitative analysis of spike lavender oil Spanish. Paper no. 645. IX International Congress on Essential Oils, Singapore, 1983. Data cited in ref. 22

30 Carrasco J, et al. Investigation analytique sur l'huile essentielle d'Aspic cultive. Paper no. 112. VIII International Congress on Essential Oils, Cannes, France, 1980. Data cited in ref. 22

31 de Rijke D, et al. New compounds with small rings in essential oils. Paper no. 179. VIII International Congress on Essential Oils, Cannes, France, 1980. Data cited in ref. 22

32 Lawrence BM. Progress in essential oils. Perfum Flavor 2006;31(6):40-? (last page unknown)

33 Proenca da Cunha A. Estudio cariologico do Oleo Essencial de *Lavandula latifolia* Medicus da Regio de Coimbra. Bol Fac Pharm Coimbra 1985;27-34. Data cited in ref 32

34 Bosilcov A. Lavender: A key perfumery material. Dissertation. International Centre for Aroma Trades Studies, Plymouth University, Plymouth, UK, 2010

35 Rudzki E, Grzywa Z, Bruo WS. Sensitivity to 35 essential oils. Contact Dermatitis 1976;2:196-200

36 Rudzki E, Grzywa Z. Allergy to perfume mixture. Contact Dermatitis 1986;15:115-116

Chapter 5.81   STAR ANISE OIL

DEFINITION

Star anise oil (essential oil of star anise, Chinese type) is the essential oil obtained from the fruit of the (Chinese) star anise, *Illicium verum* Hook. f.

INCI NOMENCLATURE

Description/definition: Illicium verum fruit/seed oil is the volatile oil obtained from the dried, ripe fruits and seeds of star anise, *Illicium verum,* Illiciaceae
INCI name EU: Illicium verum fruit/seed oil
INCI name USA: Illicium verum (anise) fruit/seed oil
CAS registry number(s): 8007-70-3; 84650-59-9; 68952-43-2
EINECS number(s): 283-518-1

ISO (INTERNATIONAL ORGANIZATION FOR STANDARDIZATION) STANDARD

ISO number: 11016
ISO name: Essential oil of star anise, Chinese type
Botanical origin: *Illicium verum* Hook. f.
Parts of plant used: Fruit
ISO values: ISO values (minimum and maximum concentrations) are shown in Table 5.81.1.

Star anise oil should not be confused with aniseed oil, obtained from *Pimpinella anisum* L. (see Chapter 5.3 Aniseed oil).

THE PLANT, THE OIL, AND THEIR USES

Illicium verum Hook. f. is an aromatic evergreen tree that grows up to 15 m tall. It is native to southern China and Vietnam and is cultivated mainly in these countries. The tree can also be found in Jamaica, Laos, Philippines, Korea, Japan, and Taiwan (25). Star anise, the star-shaped dried composite fruit of *Illicium verum*, is widely used in Chinese (as an ingredient of the traditional five-spice powder of Chinese cooking), Indian, Malaysian and Indonesian cuisines and also in the production of alcoholic beverages such as sambuca, pastis and some types of absinthe (1,4,8,10,11,15,20,25). Star anise is very important in Chinese traditional medicine as a stimulant and expectorant, to relieve flatulence, to treat vomiting, stomach aches, insomnia, skin inflammation and rheumatic pain, and to increase libido (1,4,10,14). Star anise is believed to possess anti-carcinogenic, fumigant, anti-HIV, anti-inflammatory, antibacterial and antifungal activities (1). It is commonly used to flavor cough mixtures and pastilles (11). Furthermore, star anise is the industrial source of shikimic acid, a primary ingredient used to create the antiviral drug Tamiflu (oseltamivir phosphate), which is regarded as a remedy for the bird flu H5N1 strain of virus (1,4).

The essential oil of star anise, obtained by steam distillation of the dried fruits, is located in the meso- and pericarp and has a hot, sweet and aniseed-like taste (12). The main ingredient of the oil is (*E*)-anethole, generally present in concentrations ranging from 79 to 90% (7) and sometimes >90% (5,6,10,11,18,29). Chinese star anise essential oils have a wide range of commercial applications in the production of perfumes, cosmetics (including toothpastes), soaps, food and beverage flavorings and pharmaceutical preparations such as cough lozenges (12,14,25). The oil of star anise is considered to be stimulant, stomachic, carminative, mildly expectorant and diuretic and has antifungal and antioxidant properties (10,11,25). It may be employed in gastrointestinal problems and insomnia and is applied topically to treat rheumatism and as an antiseptic (25). It is employed in aromatherapy practices, but aromatherapy experts caution about the risk of toxicity due to the high anethole content (38,39).

The seeds account for some 20% of the total fruit and are a by-product of the essential oil production processes. Usually they are treated as waste. The seeds mainly contain C18:2, C18:1, C18:3, C20:4, C16, C18 and C20 fatty acids (12,14).

CHEMICAL COMPOSITION

Star anise oil is a clear, mobile liquid, which will become a solid crystalline mass when cold (below 15°C), and which has an intense typical anise and fennel seed odor. The yield of essential oil from the fruit of *Illicium verum* Hook. f. generally varies from 2.5 to 3.5 per cent from fresh material and up to 9 per cent from dried fruits. The main producing countries of this oil are China, Vietnam and North Korea.

Literature data (up to December 6, 2014) on the chemical composition of star anise oils and unpublished analytical data from one of us (E.S.) are shown in Table 5.81.2 in alphabetical order. In star anise oils from various origins, over 160 chemicals have been identified. About 54 per cent of these were found in a single reviewed publication only. The major compounds found in star anise oils from different sources are shown in Table 5.81.3. They include (highest concentrations in any study given) (*E*)-anethole (94.4%), foeniculin (*E*-) (14.6%), limonene (11.6%), methyl chavicol (8.9%) and linalool (2.8%).

Table 5.81.1 ISO values (%) for star anise oil [a]

Compound	CAS	Minimum	Maximum
(*E*)-Anethole	4180-23-8	86.0	93.0
Methyl chavicol	140-67-0	0.6	6.0
Limonene	138-86-3	0.2	6.0
Foeniculin (*E*-)	78259-41-3	0.1	3.0
Linalool	78-70-6	0.2	2.5
α-Pinene	80-56-8	0.1	1.5
(*Z*)-Anethole	25679-28-1	0.1	1.0
β-Caryophyllene	87-44-5	0.0	0.8
α-Phellandrene	99-83-2	0.0	0.7
α-*trans*-Bergamotene	13474-59-4	0.06	0.6
p-Anisaldehyde	123-11-5	0.1	0.5
α-Terpineol	98-55-5	0.0	0.3
α-*cis*-Bergamotene	18252-46-5	0.04	0.09

[a] ISO 11016 Essential oil of star anise ©ISO 1999; Geneva, Switzerland, www.iso.org

795

Table 5.81.2 Constituents identified in star anise oils

Constituent	CAS	Percentage and range					
		A	B (16)	C (20)	D (8)	E (17)	F
Acetaldehyde	75-07-0						0.01[d]
Acethydrazide	1068-57-1					0.08	
Acetone	67-64-1						1.0[a]
2-Acetonylcyclo-hexanone	6126-53-0					0.4	
Allyl benzyl ether	14593-43-2			1.8			
N-Amino-1,2,3,4-tetrahydroquinoline							0.2[a]
(E)-Anethole	4180-23-8	84.3-90.1	85.7	75.0	80.8	74.1	90.8[n]; 91.8[v]; 92.3[p]; 93.7[w]; 93.9[e]; 94.4[d]
(Z)-Anethole	25679-28-1	0.1-0.4	0.07	0.6		2.1	0.1[e,k]; 0.3[p]; 0.4[n,o,q]; 0.7[b]; 1.1[a]; 1.6[d]
m-Anisaldehyde	591-31-1			0.06		0.06	0.2[m]
p-Anisaldehyde	123-11-5	0.2-0.9	1.8	1.8	3.3	1.7	0.6[k]; 0.8[j]; 1.6[c]; 1.8[h,s]; 1.9[r]; 2.0[g]; 2.7[l]
Anisic acid	1335-08-6	tr-0.1					
p-Anisoin	119-52-8					0.1	
Anisyl acetone (p-)	104-20-1	tr-0.2	0.2			0.3	0.2[n]; 1.2[i]
Anisyl alcohol	1331-81-3						<0.2[o]
Anisyl isobutyrate	66989-82-0						+[f]
Aromadendrene	489-39-4				0.1		
allo-Aromadendrene	25246-27-9		tr				0.06[a]
Benzenemethanol, 2-(2-aminopropoxy)-3-methyl-	53566-98-6					0.07	
α-Bergamotene	17699-05-7			0.1		0.1	0.01[e]; 0.04[m]
cis-α-Bergamotene	18252-46-5	0.03-0.1			tr		0.1[o]
trans-α-Bergamotene	13474-59-4	0.2-0.5	0.3		0.3		0.2[a,n]; 0.3[o]; 0.5[d,h]; 0.7[i]; 1.0[c]
Bicyclo[2.2.1]heptane-2,3-dione, 6-(acetyl-oxy)-1,5,5-trime-thyl-, endo-	55044-71-8					0.07	
Bicyclo[3.3.1]hept-2-ene, 2,6-diene-							0.4[a]
Bicyclo[3.1.1]hept-2-ene-2-ethanol, 6,6-dimethyl-	128-50-7					0.06	
Bisabolene	495-62-5	0.08-0.2					
β-Bisabolene	495-61-4		0.1	0.06	0.1		+[f]; tr[o]; 0.2[a,o]; 0.5[c]
(±)-β-Bisabolene	4891-79-6					0.2	
Borneol	507-70-0						0.02[d]
γ-Cadinene	39029-41-9				tr		
δ-Cadinene	483-76-1						tr[o]; 0.03[d]; 0.04[e]
α-Cadinol	481-34-5		0.2		0.1		0.02[e]
epi-α-Cadinol	5937-11-1		0.06				
α-Calacorene	21391-99-1		tr				
Camphene	79-92-5						0.03[p]; 0.04[f]; 0.2[b,c]
Camphor	76-22-2	0.01-0.2		0.05			
δ2-Carene (δ4-)	554-61-0						0.06[a]
δ3-Carene	13466-78-9	0.03-0.7	0.2	0.1			0.03[a]; 0.1[r]; 0.2[e]; 0.3[b]; 0.4[c]; 0.8[g]; 0.9[o,q]
Carvacrol	499-75-2		tr				
Carveol	99-48-9		tr				
Carvone	99-49-0						tr[o]
Caryophyllene				1.5			0.1[a]
β-Caryophyllene	87-44-5	0.2-0.8		0.05	0.3	0.5	0.1[b,e]; 0.2[n]; 0.5[d,k,o,s]; 0.6[f]; 0.8[c]
Caryophyllene oxide	1139-30-6		0.08		tr		4.8[t]
β-Cedrene	546-28-1			0.3			
Chavicol	501-92-8		2.7			3.8	
1,4-Cineole	470-67-7						0.01[d]
1,8-Cineole	470-82-6	0.06-0.3	0.4	0.6	0.2		0.05[d]; 0.2[m,r]; 1.5[g]
Cinnamyl acetate	103-54-8		0.5				0.3[f]
(E)-Cinnamyl acetate	21040-45-9				tr		
Cinnamyl alcohol	104-54-1						0.1[o]
α-Copaene	3856-25-5		0.06	0.09	tr	0.3	tr[f,o]; 0.08[o]; 0.2[c,d]; 0.4[a]
β-Copaene	18252-44-3						0.5[h]

Table 5.81.2 Constituents identified in star anise oils *(continued)*

Constituent	CAS	Percentage and range					
		A	B (16)	C (20)	D (8)	E (17)	F
Cryptone	500-02-7		tr				
Cubebene	11012-64-9						0.1[b]
α-Cubebene	17699-14-8						0.1[e]; 0.3[t]
Cuminaldehyde	122-03-2			0.7		1.7	
1,3-Cyclohexadiene	592-57-4						0.1[a]
Cyclohexanolthyl-4-	108-93-0						0.02[a]
2-Cyclohexen-1-one	930-68-7						0.05[a]
2-(1-Cyclopentenyl) furan							0.9[b]
Cymene	25155-15-1		0.2		0.2		
p-Cymene	99-87-6	0.05-0.4		0.2			0.05[e,m]; 0.06[n]; 0.1[b,f,h,k]; 0.2[o,r]; 0.4[c]
p-Cymen-8-ol	1197-01-9						<0.8[o]
Cyperene	2387-78-2						0.2[b]
5-Decyne-4,7-diol, 4,7-dimethyl-	126-87-4					0.1	
Dehydro-p-cymene	1195-32-0		tr				
Elemene	11029-06-4			0.05			
β-Elemene	33880-83-0		tr				0.01[e]; 0.1[b]
γ-Elemene	29873-99-2					0.07	
Elemol	639-99-6		tr		tr		
(+)-9-Epiledene							0.1[b]
4-(1-Ethoxyethenyl) benzaldehyde	839721-49-2						0.1[a]
Eudesmene	34766-40-0						0.05[p]
α-Farnesene	502-61-4					0.2	0.1[a]
(E,E)-α-Farnesene	502-61-4			0.03	tr		0.7[f]
(Z,E)-α-Farnesene	26560-14-5			0.2			
β-Farnesene	502-60-3			0.06		0.3	
(E)-β-Farnesene	18794-84-8	0.02-0.3					tr[o]; 0.3[a]
(Z)-β-Farnesene	28973-97-9		0.1				tr[o]; 0.08[o]
Farnesol	4602-84-0		0.2				
Fenchone	1195-79-5						tr[o]; 2.1[c]
Foeniculin (E-)	78259-41-3	0.09-1.7		1.1	1.5	0.2	1.2[v]; 1.3[m]; 2.3[h]; 4.6[c]; 5.1[o]; 6.7[l]; 14.6[f]
Geraniol	106-24-1	tr-0.1					
Geranyl acetate	105-87-3		tr			0.1	
Geranyl isobutyrate	2345-26-8					0.1	
Germacrene B	15423-57-1		0.02				
Germacrene D	23986-74-5			0.06		0.06	
Gurjunene	80599-13-9			0.01			
α-Gurjunene	489-40-7						+[f]
Herboxide	13679-86-2		tr				
Hexadecanoic acid	57-10-3			0.01			
Hexatriacontane	630-06-8						1.2[h]
Hexyl oleate	20290-84-0					0.1	
α-Humulene	6753-98-6		tr				tr[o]; 0.1[b]; 1.4[t]
Hydrazinecarboxylic acid, ethyl ester	4114-31-2					0.1	
Isobornyl thiocyano-acetate	115-31-1						0.4[b]
Isocaryophyllene	118-65-0			0.8		0.3	
Isolongifolene	1135-66-6						+[f]
Limonene	138-86-3	0.2-3.3	2.9	1.9	0.4	0.8	3.1[w]; 3.3[c]; 4.7[l]; 6.0[i]; 7.8[j]; 10.4[o,q]; 11.6[g]
Linalool	78-70-6	0.2-1.3	0.4	1.7	1.4	0.5	0.9[c]; 1.0[k,v]; 1.5[r]; 1.1[g]; 1.4[j]; 2.3[h]; 2.8[o]
cis-Linalool oxide	11063-77-7						0.02[a]
trans-Linalool oxide, furanoid	34995-77-2						0.02[d]
Longifolene	475-20-7						+[f]; tr[o]
trans-p-Mentha-1(7), 8-dien-2-ol	21391-84-4		tr				

Table 5.81.2 Constituents identified in star anise oils *(continued)*

Constituent	CAS	Percentage and range					
		A	B (16)	C (20)	D (8)	E (17)	F
cis-*p*-Menth-1-ene-3,6-diol							+[o]
p-Methoxyaceto-phenone (4-)	100-06-1		0.02				<0.2[o]
p-Methoxycinnam-aldehyde	1963-36-6			tr			
4-Methoxy-5-[3-(4-methoxyphenyl)oxaziridin-2-yl]pentan-1-ol							4.9[a]
p-Methoxyphenyl-acetone	122-84-9						0.1[l]; 0.2[o]; 0.4[h,r]
p-Methoxyvalero-phenone	1671-76-7		tr				
Methyl *p*-anisate	121-98-2		0.07			0.07	tr[f]; 0.6[o]
1-(3-Methylbutyl)-2,3,5,6-tetramethyl-benzene							0.07[a]
Methyl chavicol	140-67-0	0.2-5.9	0.5	5.2	4.0	4.8	4.0[r]; 4.1[p]; 5.0[f]; 5.5[o]; 6.3[c]; 6.6s; 6.7[m]; 8.9[l]
Methyl isoeugenol	93-16-3		0.09	0.2		0.3	0.2[a]
(*E*)-Methylisoeugenol	6379-72-2				tr		0.1[o]
2-Methyl-3-phenyl-propanal	5445-77-2			1.0		3.0	
α-Muurolene	10208-80-7		0.1				tr[o]
γ-Muurolene	30021-74-0						0.06[a]
α-Muurolol	104245-48-9						0.07[n]
α-Myrcene	1686-30-2			0.08			
Myrcene (β-)	123-35-3	0.05-0.3	0.05		tr		0.02[e]; 0.1[b,h]; 0.2[c]; 0.4[f,o,q,t]
Naphthalene	91-20-3						0.2[a]
Nerolidol	7212-44-4						0.8[f]
(*E*)-Nerolidol	40716-66-3			0.05	0.1	0.2	
(*Z*)-Nerolidol	3790-78-1						0.2[h]
(*E*)-Ocimene	27400-72-2						0.09[e,f]; 0.1[b]
(*Z*)-Ocimene	27400-71-1						0.4[f]
β-Ocimene	13877-91-3			0.03			
(*Z*)-β-Ocimene	3338-55-4						0.01[e]; 0.1[c]
Octadecanoic acid	57-11-4			0.02			
1,6-Octadien-3-ol	51361-43-4						0.3[a]
α-Phellandrene	99-83-2	0.2-0.6	0.1	0.3	0.1		0.04[e]; 0.1[b,c,m]; 0.2[f,t]; 0.4[h,k,p]; 0.5[q]; 0.9[o]
β-Phellandrene	555-10-2	0.08-0.5					0.06[r]; 0.4[k]; 0.5[o,p]; 0.8[h,s]; 1.7[o]
Phenylethanolamine	7568-93-6			0.02		0.06	
α-Pinene	80-56-8	0.3-1.8	0.4	0.9	0.5		0.4[h]; 0.5[r]; 0.6[k]; 0.7[w]; 0.8[j]; 2.1[g,q]; 2.6[o]
β-Pinene	127-91-3	0.03-0.2	0.06	0.09	0.2		tr[f]; 0.03[e]; 0.05[k]; 0.2[c,o,q,t]
1-Propanone							0.3[a]
Pyridine	110-86-1						0.06[a]
Sabinene	3387-41-5	tr-0.2	tr	0.2			tr[k]; 0.07[m]; 0.1[f]; 0.2[c,o,q]
cis-Sabinol	3310-02-9		tr				
Salicylaldehyde	90-02-8		0.02				
α-Santalene	512-61-8						tr[o]
β-Selinene	17066-67-0						tr[o]
Spathulenol	6750-60-3		0.1	<0.01	tr		
Spiro[4.5]dec-1-ene	697-27-8					0.1	
α-Terpinene	99-86-5	0.03-0.1	tr				tr[d]; 0.02[e]; 0.09[f]; 0.1[c,o]; 0.2[q]
γ-Terpinene	99-85-4	0.02-0.2	0.06	0.1	tr	1.2	tr[d]; 0.02[a]; 0.04[e]; 0.07[m]; 0.1[b]; 0.2[o]; 0.5[c]
Terpinen-4-ol	562-74-3	0.07-0.3	0.3	0.2	0.2	0.5	0.09[e]; 0.1[d,n]; 0.2[a,b]; 0.3[c,h,u]; 0.5[m]; 0.7[j]
1-Terpineol	586-82-3		tr				0.02[d]
α-Terpineol	98-55-5	0.04-0.4	0.2	0.1			0.08[e]; 0.1[r]; 0.2[f]; 0.3[h,t]; 0.4[d,o]; 0.5[a]; 1.0[g]
β-Terpineol	138-87-4						+[f]; 0.02[d]
γ-Terpineol	586-81-2						0.04[d]; 0.1[e]; 0.4[b]; 0.5[t]
Terpinolene	586-62-9	0.03-0.2	tr	0.04			0.03[e]; 0.04[d]; 0.08[l,m]; 0.1[b]; 0.2[o]; 0.9[o]
β-Thujaplicine	499-44-5		0.06				
Thujene	58037-87-9			0.1			

Table 5.81.2 Constituents identified in star anise oils *(continued)*

Constituent	CAS	Percentage and range					
		A	B (16)	C (20)	D (8)	E (17)	F
α-Thujene	2867-05-2		tr		tr		
Thujopsene	470-40-6		tr				
2,4,6-Trimethylben-zenesulfonamide	4543-58-2						0.1[a]
4,8,8-Trimethyl-4-vinylcaryophyllene							0.3[a]
3-Undecyne	60212-30-8			0.2		0.2	
p-Vinylanisole	637-69-4		tr				

A forty-one star anise essential oil samples from China, analyzed between 1998 and 2013; lowest and highest concentrations given (E. Schmidt, unpublished data)
B one lab-hydrodistilled oil from star anise purchased at a local Chinese market (ref. 16)
C one lab-hydrodistilled oil from a Good Agricultural Practice base in China (ref. 20)
D one commercial star anise oil (ref. 8)
E one lab-hydrodistilled oil from China (ref. 17)
F data from other studies (indicated with superscript letters); highest concentrations found in any study reviewed here given; when two or more oils were investigated, only the highest concentrations are mentioned, unless indicated otherwise

[a] one lab-hydrodistilled star anise oil from fruits purchased at a local market in India (ref. 24); [b] one oil of unknown origin and type (ref. 22); [c] one hydrodistilled oil and one microwave-assisted hydrodistilled oil, lab-prepared from Chinese star anise (ref. 19); [d] one lab-hydrodistilled oil from star anise purchased at a local Indian market (ref. 11); [e] one steam-distilled oil from star anise purchased at a local Indian market (ref. 6); [f] one lab-hydrodistilled fruit oil from Chinese star anise (ref. 2); [g] one lab-hydrodistilled oil (ref. 26); [h] one commercial star anise oil purchased in Italy (ref. 27); [i] one steam-distilled fruit oil (ref. 23); [j] one lab-hydrodistilled oil (ref. 21); [k] one lab-hydrodistilled oil sample (ref. 13); [l] one lab-prepared, microwave-assisted hydrodistilled oil (ref. 9); [m] one lab-hydrodistilled fruit oil from star anise purchased at a local Shanghai market (ref. 1); [n] one oil of unknown origin and mode of preparation (ref. 18); [o] literature data from 1966 to 1982, reviewed and cited in ref. 3; [p] one commercial oil from China (ref. 29); [q] one Chinese and one Vietnamese star anise oil (ref. 30); [r] one commercial star anise oil sample obtained from a USA company (ref. 31); [s] one commercial star anise oil sample purchased from a French company (ref. 33); [t] one lab-hydrodistilled oil sample from star anise fruits purchased at a local market in Egypt (ref. 32); [u] one oil sample produced by steam-distillation of star anise purchased from a Brazilian company (ref. 35); [v] one commercial oil from a German company (ref. 34); [w] one commercial oil sample from an Italian company (ref. 37)

tr: trace (in column B: <0.05; in column D: < 0.1; in columns F[d] and F[f]: <0.01); + present in the oil investigated, but quantity not stated

COMMERCIAL OILS

The ten chemicals that had the highest maximum concentrations in 41 commercial star anise essential oil samples (concentration ranges provided) are the following: (E)-anethole (84.3-90.1%), methyl chavicol (0.2-5.9%), limonene (0.2-3.3%), α-pinene (0.3-1.8%), foeniculin (E-) (0.09-1.7%), linalool (0.2-1.3%), p-anisaldehyde (0.2-0.9%), β-caryophyllene (0.2-0.8%), δ3-carene (0.03-0.7%) and α-phellandrene (0.2-0.6%) (Erich Schmidt, unpublished analytical data).

An additional analysis of star anise oil not presented in Table 5.81.2 can be found on-line in ref. 36.

CONTACT ALLERGY / ALLERGIC CONTACT DERMATITIS

General

Contact allergy to star anise oil has been reported in one publication only. No clinical records of allergic contact dermatitis from star anise oils have been found. The main compound, anethole (84-90% of the entire oil) is also the principal allergen; possible other allergens are methyl chavicol and limonene.

Table 5.81.3 Major constituents of star anise oils

Constituent	CAS	Percentage and range					
		A	B (16)	C (20)	D (8)	E (17)	F
(E)-Anethole	4180-23-8	84.3-90.1	85.7	75.0	80.8	74.1	90.8[n]; 91.8[v]; 92.3[p]; 93.7[w]; 93.9[e]; 94.4[d]
Foeniculin (E-)	78259-41-3	0.09-1.7		1.1	1.5	0.2	1.2[v]; 1.3[m]; 2.3[h]; 4.6[c]; 5.1[o]; 6.7[l]; 14.6[f]
Limonene	138-86-3	0.2-3.3	2.9	1.9	0.4	0.8	3.1[w]; 3.3[c]; 4.7[l]; 6.0[j]; 7.8[j]; 10.4[o,q]; 11.6[g]
Methyl chavicol	140-67-0	0.2-5.9	0.5	5.2	4.0	4.8	4.0[r]; 4.1[p]; 5.0[f]; 5.5[o]; 6.3[c]; 6.6s; 6.7[m]; 8.9[l]
Linalool	78-70-6	0.2-1.3	0.4	1.7	1.4	0.5	0.9[c]; 1.0[k,v]; 1.5[r]; 1.1[g]; 1.4[j]; 2.3[h]; 2.8[o]

LEGEND: SEE UNDER TABLE 5.81.2

Testing in groups of patients

One hundred consecutive patients with dermatitis were tested with star anise oil 0.5%, 1% and 2% in petrolatum; over 1/3 had positive reactions to the 1% and the 2% concentration, indicating that these concentrations are irritant. In five, patch test sensitization occurred, with flare-up of the 1% and 2% concentrations. Three were tested with anethole 1% in petrolatum (the main ingredient of star anise oil, present in concentrations of 84-90% in commercial oils; Table 5.81.2, column A), and all reacted. Another of these actively sensitized patients was tested with 9 components of star anise oil and reacted to anethole, α-pinene (maximum concentration in commercial essential oils 1.8%), safrole and methylchavicol (maximum concentration in commercial star anise essential oils 5.9%, Table 5.81.2, column A), all tested 1% in petrolatum. Fifteen patients positive to 1% star anise oil, negative to 0.5% but positive to one or more 'balsams' (*Myroxylon pereirae*, turpentine, wood tars, colophony) were also tested with these 9 components with the following results: 5 reactions to anethole, 8 to α-pinene (who all co-reacted to turpentine; α-pinene is probably the main allergen in turpentine oil), 3 to limonene and one to safrole (limonene was found in commercial star anise oils in a maximum concentration of 3.3%, Table 5.81.2, column A). A concentration of 0.5% star anise oil may not be can irritant, but detected only one sensitization out of the five actively sensitized patients. There was no cross- or pseudo-cross-reactivity to other essential oils. No mention was made of any clinical relevance of the reactions, which presumably there was not (40).

LITERATURE

1   Huang B, Liang J, Wang G, Qin L. Comparison of the volatile components of *Illicium verum* and *I. lanceolatum* from East China. J Essent Oil Bear Plants 2012;15:467-475

2   Cu J-Q, Perineau F, Goepfert G. GC/MS analysis of star anise oil. J Essent Oil Res 1990;2:91-92

3   Lawrence BM. Progress in essential oils. Perfum Flavor 1984;9(April/May):23-31

4   Wang G-W, Hu W-T, Huang BK, Qin L-P. *Illicium verum*: A review on its botany, traditional use, chemistry and pharmacology. J Ethnopharmacol 2011;136:10-20

5   Tuan, DQ, Ilangantileke SG. Liquid CO2 extraction of essential oil from star anise fruits (*Illicium verum*). J Food Engineer 1997;31:47-57

6   Padmashree A, Roopa N, Semwal AD, Sharma GK, Agathian G, Bawa AS. Star-anise (*Illicium verum*) and black caraway (*Carum nigrum*) as natural antioxidants. Food Chem 2007;104:59-66

7   Lederer I, Schulzki G, Gross J, Steffen J-P. Combination of TLC and HPLC-MS/MS methods. Approach to a rational quality control of Chinese star anise. J Agric Food Chem 2006;54:1970-1974

8   Howes M-JR, Kite GC, Simmonds MSJ. Distinguishing Chinese star anise from Japanese star anise using thermal desorption-gas chromatography-mass spectrometry. J Agric Food Chem 2009;57:5783-5789

9   Zhai Y, Sun S, Wang Z, Cheng J, Sun Y, Wang L, et al. Microwave extraction of essential oils from dried fruits of *Illicium verum* Hook. f. and *Cuminum cyminum* L. using ionic liquid as the microwave absorption medium. J Sep Sci 2009;32:3544-3549

10 Cai M, Guo X, Liang H, Sun P. Microwave-assisted extraction and antioxidant activity of star anise oil from *Illicium verum* Hook. f. Int J Food Sci Technol 2013;48:2324-2330

11 Singh G, Maurya S, de Lampasona MP, Catalan C. Chemical constituents, antimicrobial investigations and antioxidative potential of volatile oil and acetone extract of star anise fruits. J Sci Food Agric 2006;86:111-121

12 Bernard T, Perineau F, Delmas M, Gaset A. Extraction of essential oils by refining plant materials. II. Processing of products in the dry state: *Illicium verum* Hooker (fruit) and *Cinnamomum zeylanicum* Nees (bark). Flavour Fragr J 1989;4:85-90

13 Formacek V, Kubeczka KH. Essential oil analysis by capillary gas chromatography and carbon 13 NMR spectroscopy. New York: John Wiley, 1982:291-299. Data cited in ref. 3

14 Li G, Sun Z, Xia L, Shi J, Liu Y, Suoa Y, et al. Supercritical CO_2 oil extraction from Chinese star anise seed and simultaneous compositional analysis using HPLC by fluorescence detection and online atmospheric CI-MS identification. J Sci Food Agric 2010;90:1905-1913

15 Tonutti I, Liddle P. Aromatic plants in alcoholic beverages. A review. Flavour Fragr J 2010;25:341-350

16 Gholivand MB, Rahimi-Nasrabadi M, Chalabi H. Determination of essential oil components of star anise (*Illicium verum*) using simultaneous hydro-distillation-static headspace liquid-phase microextraction-gas chromatography mass spectrometry. Analytical Letters 2009;42:1382-1397

17 Yan JH, Xiao XX, Huang KL. Component analysis of volatile oil from *Illicium verum* Hook. f. J Cent South Univ of Technol 2002;9:173-176

18 Dzamic A, Sokovic M, Ristic MS, Grijic-Jovanovic S, Vukojevic J, Marin PD. Chemical composition and antifungal activity of *Illicium verum* and *Eugenia caryophyllata* essential oils. Chem Nat Comp 2009;45:259-261

19 Wang Z, Wang L, Li T, Zhou X, Ding L, Yu Y, et al. Rapid analysis of the essential oils from dried *Illicium verum* Hook. f. and *Zingiber officinale* Rosc. by improved solvent-free microwave extraction with three types of microwave-absorption medium. Anal Bioanal Chem 2006;386:1863-1868

20 Wang Q, Jiang L, Wen Q. Effect of three extraction methods on the volatile component of *Illicium verum* Hook. f. analyzed by GC–MS. Wuhan Univ J Nat Sci 2007;12:529-534

21 Kimbaris AC, Koliopoulos G, Michaelakis A, Konstantopoulou MA. Bioactivity of *Dianthus caryophyllus*, *Lepidium sativum*, *Pimpinella anisum*, and *Illicium verum* essential oils and their major components against the West Nile vector *Culex pipiens*. Parasitol Res 2012;111:2403-2410

22 Huang Y, Zhao J, Zhou L, Wang J, Gong Y, Chen X, et al. Antifungal activity of the essential oil of *Illicium verum* fruit and its main component *trans-anethole*. Molecules 2010;15:7558-7569

23 Lee SO, Park IK, Choi GJ, Lim HK, Jang KS, Cho KY, et al. Fumigant activity of essential oils and components of *Illicium verum* and *Schizonepeta tenuifolia* against *Botrytis cinerea* and *Colletotrichum gloeosporioides*. J Microbiol Biotechnol 2007;17:1568-1572

24 Ariamuthu S, Balakrishnan V, Srinivasan ML. Chemical composition and antibacterial activity of essential oil from fruits of *Illicium verum* Hook. f. Int J Res Phytochem Pharmacol 2013;3:85-89

25 Orwa C, Mutua A, Kindt R, Jamnadass R, Simons A. Agroforestree Database: a tree reference and selection guide version 4.0 (2009). Available at http://www.worldagroforestry.org/treedb2/AFTPDFS/Illicium_verum.pdf

26 Lucchesi ME, Chemat F, Smadja J. An original solvent free microwave extraction of essential oils from spices. Flavour Fragr J 2004;19:134-138

27 Guerrini A, Sacchetti G, Muzolli M, Moreno Rueda G, Medici A, Besco E, et al. Composition of the volatile fraction of *Ocotea bofo* Kunth (Lauraceae) calyces by GC-MS and NMR fingerprinting and its antimicrobial and antioxidant activity. J Agric Food Chem 2006;54:7778-7788

28 Lawrence BM. Progress in essential oils. Perfum Flavor 2003;28(6):58-75

29 Xie L, Xu S-Y. Analysis of star anise oil using GC/MS. Zhongguo Youzhi 1997;22:43-45. Data cited in ref. 28

30 Kubeczka K-H, Formacek V. Essential oils analysis by capillary gas chromatography and carbon-13 NMR spectroscopy, 2nd Edition. New York: John Wiley and Sons, 2002:339-347. Data cited in ref. 28

31 Wei A, Shibamoto T. Antioxidant/lipoxygenase inhibitory activities and chemical compositions of selected essential oils. J Agric Food Chem 2010;58:7218-7225

32 Aly SE, Sabry BA, Shaheen MS, Hathout AS. Assessment of antimycotoxigenic and antioxidant activity of star anise (*Illicium verum*) in vitro. J Saudi Soc Agricult Sci 2014. http://dx.doi.org/10.1016/j.jssas.2014.05.003

33 Rolli E, Marieschi M, Maietti S, Sacchetti G, Bruni R. Comparative phytotoxicity of 25 essential oils on pre- and post-emergence development of *Solanum lycopersicum* L.: A multivariate approach. Ind Crops Prod 2014;60:280-290

34 Stević T, Berić T, Šavikin K, Soković M, Godevac D, Dimkić I, Stanković S. Antifungal activity of selected essential oils against fungi isolated from medicinal plant. Ind Crops Prod 2014;55:116-122

35 Freire, JM, Cardoso MG, Batista LR, Andrade MA. Essential oil of *Origanum majorana* L., *Illicium verum* Hook. f. and *Cinnamomum zeylanicum* Blume: chemical and antimicrobial characterization. Rev Bras Pl Med Botucatu v.13, 2011;13:209-214

36 Yan J-H, Xiao X-X, Huang K-L. Component analysis of volatile oil from *Illicium verum* Hook. f. J Cent South Univ Technol 2002;9(3):173-177. Available at: http://www.zndxzk.com.cn/down/upfile/soft/200991/y2002-03-07.pdf

37 Pistelli L, Mancianti F, Bertoli A, Cioni PL, Leonardi M, Pisser F, et al. Antimycotic activity of some aromatic plants essential oils against canine isolates of *Malassezia pachydermatis*: An in vitro assay. The Open Mycology Journal 2012;6:17-21

38 Rhind JP. Essential oils. A handbook for aromatherapy practice, 2nd Edition. London: Singing Dragon, 2012

39 Lawless J. The encyclopedia of essential oils, 2nd Edition. London: Harper Thorsons, 2014

40 Rudzki E, Grzywa Z. Sensitizing and irritating properties of star anise oil. Contact Dermatitis 1976;2:305-306

Chapter 5.82   TANGERINE OIL

DEFINITION
Tangerine oil is the essential oil obtained from the peri-carp (peel) of the tangerine, *Citrus tangerina* Hort. ex Tan. (synonym: *Citrus tangerina* Tanaka).

INCI NOMENCLATURE
Description/definition: Citrus tangerina peel oil is the volatile oil expressed from the peel of the ripe fruit of the tangerine, *Citrus tangerina*, Rutaceae
INCI name EU: Citrus tangerina peel oil
INCI name USA: Citrus tangerina (tangerine) peel oil
CAS registry number(s): 223748-44-5

ISO (INTERNATIONAL ORGANIZATION FOR STANDARDIZATION) STANDARD
There is currently no ISO standard for tangerine oil available.

There is some taxonomical uncertainty about *Citrus tangerina,* which is called both *Citrus tangerina* Hort. ex Tanaka and *C. tangerina* Tanaka. Recognition as a species has not been accepted universally (4). Mansfeld's World Database of Agriculture and Horticultural Crops (http://mansfeld.ipk-gatersleben.de) states that the former is an accepted name. GRIN Taxonomy for Plants lists it under the latter name, but gives as comment: 'or *C. reticulata* Tangerina'. The Plant List (www.theplantlist.org) considers *Citrus tangerina* as a synonym for *Citrus reticulata* Blanco (the mandarin, but also called tangerine). The University of Melbourne, Australia botanical database (www.plantnames.unimelb.edu.au) mentions *Citrus reticulata* Blanco cv. Tangerine as a synonym of *Citrus tangerina* Hort. ex Tanaka. Some use the name *C. tangerina* exclusively for the Dancy cultivar of tangerine (4,5). In this chapter, we have included only those (few) studies that mention or presumably have used *Citrus tangerina* as the source of the oils analyzed. As a consequence, there will most likely be overlap with mandarin oils, where results of *Citrus reticulata* Blanco 'Tangerina'/cv. Tangerine are presented (Chapter 5.55). Nevertheless, in tangerine oils ex *C. tangerina* as presented in this chapter, the amounts of limonene appear to be somewhat higher, but the concentration of γ-terpinene clearly lower than in (other) mandarin oils.

THE PLANT, THE OIL, AND THEIR USES
Citrus tangerina Hort. Ex Tan. (tangerine) is an orange-colored citrus fruit. It originates presumably in China, where it can also be found in the wild state. The tangerine is cultivated in China, Japan and the USA (Mansfeld's World Database of Agriculture and Horticultural Crops, http://mansfeld.ipk-gatersleben.de). Varieties include the cultivar Dancy (formerly the most popular form) and the hybrids Sunburst, Robinson and Murcott. The name "Tangerine" comes from Tangiers, a city in Morocco, where the first shipment of tangerines was allegedly sent to mainland Europe around 1845 (6). Tangerines are most commonly peeled and eaten out of hand. The fresh fruit is also used in salads, desserts and main dishes. The peel is dried and used in Sichuan cuisine. The fruit is popular throughout the world and highly revered; the tangerine is a symbol of luck in China and regarded as a remedy for indigestion in France (6). In botanical terms, the tangerine is a type of mandarin orange, closely related to *Citrus reticulata* Blanco. However, in the flavor and fragrance industry, differences are said to exist between the varieties both in terms of juice and essential oil (6). Tangerine oil is employed in aromatherapy practices (13,14).

The history, global distribution, and nutritional importance of *Citrus* fruits have been reviewed in ref. 11.

CHEMICAL COMPOSITION
Tangerine oil is a yellowish orange to deep orange, clear mobile liquid which has a fresh juicy, citrusy peel note odor. The yield of essential oil from the peel of *Citrus tangerina* Hort. ex Tan. generally varies from 0.4 to 0.8 per cent, depending on the dryness of the fruit. The main producing countries of this oil are China, USA, Brazil and Japan.

Cold-pressed *Citrus* essential oils consist of a volatile fraction, which represents 85-99% of the oil, and a non-volatile fraction 1-15%, which mainly contains oxygen heterocyclic compounds, especially coumarins, psoralens and polymethoxyflavones. These components play an important role in the characterization of cold-pressed *Citrus* oils, since the composition of this fraction is characteristic of each oil. Moreover, many of the pharmacological and toxicological activities possessed by *Citrus* oils have been demonstrated to be related to these components (5,12).

Literature data (up to October 14, 2014) on the chemical composition of tangerine oils and unpublished analytical data from one of us (E.S.) are shown in Table 5.82.1 in alphabetical order. In tangerine oils from various origins, over 125 chemicals have been identified. About 46 per cent of these were found in a single reviewed publication only. The major compounds found in tangerine oils from different sources are shown in Table 5.82.2. They include (highest concentrations in any study given) limonene (97.8%), γ-terpinene (5.6%) and myrcene (3.2%).

Commercial oils
The ten chemicals that had the highest maximum concentrations in 28 commercial tangerine essential oil samples (concentration ranges provided) are the following: limonene (81.8-97.8%), γ-terpinene (0.1-5.6%), terpinolene (0.2-2.9%), myrcene (0.1-2.2%), α-pinene (0.08-1.9%), *p*-cymene (0.4-1.1%), α-thujene (0.08-1.2%), β-pinene (tr-0.9%), sabinene (0.1-0.6%) and decanal (0.1-0.5%) (Erich Schmidt, unpublished analytical data).

It should be realized that values for chemicals in hydrodistilled oils may vary considerably from genuine cold-pressed tangerine oils because of the many hydrolytic reactions that take place during oil isolation. In addition, the hydrodistilled oils lack the non-volatile oxygenated heterocyclic compounds (notably polymethoxylated flavonoids), which may be present in the cold-pressed oils (Table 5.82.3).

Table 5.82.1 Constituents identified in tangerine oils

Constituent	CAS	Percentage and range					
		A	B (7)	C (1)	D (5)	E (2)	F
Bicyclogermacrene	24703-35-3			0.01			
α-Bisabolene	17627-44-0						0.01[a,f]
β-Bisabolene	495-61-4					0-0.1	
α-Bulnesene	3691-11-0			tr			
β-Cadinene	523-47-7						<0.01[a,f]
δ-Cadinene	483-76-1		tr-0.02	0.04	0.01-0.02		<0.01[g]
Camphene	79-92-5	0-0.02	tr	tr	tr-0.01		tr[h]; <0.01[g]
δ3-Carene	13466-78-9	tr-0.09	0.04-1.3	0.01			
(E)-Carveol	1197-07-5		tr-0.02	0.01			tr[h]
(Z)-Carveol	1197-06-4			tr	tr-0.02		
Carvone	99-49-0			0.01	0.01-0.03		0.02[h]
β-Caryophyllene	87-44-5	0-0.08	0.02-0.1	0.01	tr-0.01		tr[h]; 0.01[f]; 0.06[g]
Caryophyllene oxide	1139-30-6			tr			
Citronellal	106-23-0	0-0.1	0.02-0.06	0.1	0.03-0.06	0-tr	0.05[f]; 0.07[g,h]
Citronellol	106-22-9	0.01-0.2	0-0.2	0.06	0.03-0.05	0-tr	0.04[h]; 0.05[f]; 0.08[g]
Citronellyl acetate	150-84-5	tr-0.02	0-0.03	0.02	tr		0.01[f,h]; 0.04[g]
α-Copaene	3856-25-5	tr-0.02	tr-0.03	0.03	0.01		<0.01[f]; 0.02[h]; 0.03[g]
β-Copaene	18252-44-3			<0.01			
α-Cubebene	17699-14-8			tr			0.03[a,f]
β-Cubebene	13744-15-5			0.04[c]	0.02-0.04[c]		0.01[h]
Cubebol	23445-02-5			tr			
p-Cymene	99-87-6	0.4-1.1		0.01	tr-1.1	0.-0.4	3.3[g]
(E,E)-2,4-Decadienal	25152-84-5		0.04	0.01	0.02		
Decanal	112-31-2	0.1-0.5	0.1-0.2	0.4	0.1		0.1[g,h]
Decanoic acid	334-48-5			tr			
Decanol	36729-58-5			0.04[b]			
(E)-2-Decenal	3913-81-3			0.01	0.01-0.02		
Decyl acetate	112-17-4						0.01[h]
Dodecanal	112-54-9	0-0.05	0.03-0.05	0.07	0.02-0.03		0.01; 0.02[h]; 0.03[g]
Dodecanol	112-53-8			tr			
(E)-2-Dodecenal	20407-84-5				tr		
Elemene	11029-06-4						<0.01[f]
β-Elemene	33880-83-0	0-0.02		0.04[c]	0.02-0.04	0-0.1	0.01[h]; 0.04[g]
γ-Elemene	29873-99-2				tr-0.01		<0.01[g]; 0.01[h]
δ-Elemene	20307-84-0						0.07[g]
Elemol (α-)	639-99-6			0.01	tr-0.01		tr[h]
Farnesene	502-61-4						0.01[f]
α-Farnesene	502-61-4						0.04[g]
(E,E)-α-Farnesene	502-61-4	tr-0.1	tr-0.03	0.07	0.04-0.05		0.06[h]
β-Farnesene	502-60-3						0.02[g]
(E)-β-Farnesene	18794-84-8	0-0.03		0.05			0.03[h]
Geranial	141-27-5	0-0.1	0-0.05	0.03	0.01-0.02		0.02[g]
Geraniol	106-24-1		0-0.02	tr	tr		
γ-Geraniol	13066-51-8			0.01			
Geranyl acetate	105-87-3	0-tr	tr-0.02	0.01	tr-0.01	0-tr	tr[h]; <0.01[g]
Geranyl isobutyrate	2345-26-8						0.02[h]
5-Geranyloxypsoralen	7380-40-7	0-tr					
Germacrene A	28387-44-2			0.01			
Germacrene B	15423-57-1				0.03-0.09		
Germacrene D	23986-74-5			0.02			0.01[h]
Germacrene D-4-ol	198991-79-6				tr-0.01		
α-Guaiene	3691-12-1			tr			
Heptanal	111-71-7				tr-0.01[e]		
Hexadecanal	629-80-1			tr			
Hexanol (1-)	111-27-3						<0.01[g]
(Z)-3-Hexenol	928-96-1						0.03[g]
2-Hexylcyclopropane acetic acid				<0.01			
α-Humulene	6753-98-6			0.01	0.01		0.02[g]
Isopiperitenone	529-01-1			tr			
Limonene	138-86-3	81.8-97.8	87.4-91.7	94.6	89.6-90.9	87.1-90.9	89.2[g]; 91.7[f]; 95.7[h]

Table 5.82.1 Constituents identified in tangerine oils *(continued)*

Constituent	CAS	Percentage and range					
		A	B (7)	C (1)	D (5)	E (2)	F
Limonene dioxide	96-08-2		tr-0.07				
Limonene oxide	1195-92-2		tr-0.07				
cis-Limonene oxide	13837-75-7			<0.01	0.03-0.2		
trans-Limonene oxide	4959-35-7			0.01	0.03-0.1	0-tr	tr[h]
Linalool	78-70-6	0.03-0.3	0.3-1.2	0.4	0.5-0.7	0.4-2.2	0.03[g]; 0.5[h]; 1.1[f]
Linalyl acetate	115-95-7		tr-0.07			0-tr	
p-Mentha-1(2),8-dien-10-ol	3269-90-7			0.02			
cis-*p*-Mentha-1(7),8-dien-2-ol	22626-43-3			tr			
p-Mentha-1,8-dien-9-ol	1946-01-6						0.02[h]
cis-*p*-Mentha-2,8-dien-1-ol	3886-78-0			<0.01			
trans-*p*-Mentha-2,8-dien-1-ol	4017-77-0			0.01			
p-Mentha-1(2),8-dien-10-yl acetate				0.01			
Methyl *N*-methyl anthranilate	85-91-6		0-0.07				
Methyl thymol	1076-56-8		0.1		0.03-0.05		0.05[f]; 0.1[g]
α-Muurolene	10208-80-7			tr			
γ-Muurolene	30021-74-0			tr	0.02-0.08		0.07[g]
Myrcene	123-35-3	0.1-2.2	1.9-3.2	1.9	1.8-2.3	1.5-1.7	1.7[h]; 1.9[g]
Myroxide	28977-57-3				0.01-0.02		
Neral	106-26-3	tr-0.09	tr-0.07	0.08	tr-0.01		tr[h]; <0.01[g]; 0.01[f]
Nerol	106-25-2		tr	0.01			<0.01[g]; 0.01[h]
(*E*)-Nerolidol	40716-66-3			tr			
Neryl acetate	141-12-8	0-0.05	0-0.02	0.04	0.01-0.03	0-tr	0.02[g]
Neryl propionate	105-91-9						0.01[a,f]
Nonanal	124-19-6	0.01-0.4	tr-0.06	0.1	0.03-0.06		0.03[h]
Nonane	111-84-2				tr-0.01[e]		
Nonanol	28473-21-4		tr-0.2	tr			
Nootkatone	4674-50-4			tr			
β-Ocimene	13877-91-3						0.07[f]
(*E*)-β-Ocimene	3779-61-1	0.03-0.1	0.1-0.2	0.04	0.02-0.06	tr-0.2	0.07[g]; 0.08[h]
(*Z*)-β-Ocimene	3338-55-4						tr[h]; 0.3[g]
Octanal	124-13-0	0.02-0.3	0.2	0.4	0.08-0.1	0.1-0.2	0.3[h]
Octanoic acid	124-07-2			tr			
Octanol	111-87-5	0-0.09	tr-0.02	0.04	0.02-0.06[d]		<0.01[g]; 0.01[h]
Octyl acetate	112-14-1			0.01			
Octyl formate	112-32-3				tr		
Perillaldehyde	2111-75-3	0-0.02	0.04	0.04[b]	0.03-0.05		0.02[f,h]; 0.03[g]
α-Phellandrene	99-83-2	0-0.09	tr-0.05	0.03	0.03-0.04		tr[h]; 0.4[g]
β-Phellandrene	555-10-2	0.05-0.4	tr	0.3		0-0.3	tr[h]; 0.1[g]
α-Pinene	80-56-8	0.08-1.9	0.8-2.0	0.5	0.8-1.2	0.5-0.9	0.5[h]; 0.6[f]; 0.9[g]
β-Pinene	127-91-3	tr-0.9	0.06-0.6	0.03	0.3	0.3-0.4	0.03[h]; 0.3[g]; 1.2[f]
Sabinene	3387-41-5	0.1-0.6	0.2-1.4	0.3	0.1-0.2	0.1-0.2	0.2[f,g]; 0.4[h]
cis-Sabinene hydrate	15537-55-0	0-0.09			0.02-0.06[d]		0.2[g]
trans-Sabinene hydrate	17699-16-0	0-0.03		tr			0.01[h]; 0.02[g]
β-Selinene	17066-67-0						0.04[a,f]
7-epi-α-Selinene	123123-37-5			tr			
β-Sesquiphellandrene	20307-83-9			tr			
α-Sinensal	17909-77-2	0.07-0.3	0.09-0.3	tr	0.1-0.2	tr-0.2	0.1[g]; 0.3[f]
β-Sinensal (*cis*-)	17909-87-4			tr			
Spathulenol	6750-60-3				0.03-0.04		
α-Terpinene	99-86-5	0.02-0.1	tr-0.02		0.02-0.06	0.1	0.06[f]
γ-Terpinene	99-85-4	0.1-5.6	0.02-4.5		2.2-3.5	3.9-5.2	tr[h]; 0.03[g]
Terpinen-4-ol	562-74-3	tr-0.05	tr-0.05	tr	0.01-0.02		<0.01[g]; 0.03[f]; 0.04[h]
α-Terpineol	98-55-5	0.06-0.3	0.04-0.09	0.04	0.07-0.2	tr-0.1	0.08[f]; 0.1[g]
Terpinolene	586-62-9	0.2-2.9	0.07-0.2	0.01	0.1-0.2	0.2	0.01[h]; 1.3[g]; 2.8[f]

Table 5.82.1 Constituents identified in tangerine oils *(continued)*

Constituent	CAS	A	B (7)	C (1)	D (5)	E (2)	F
α-Terpinyl acetate	80-26-2			tr	0.01-0.04		
Tetradecanal	124-25-4			0.01	tr		
Tetradecane	629-59-4			tr			
Tetradecanol	27196-00-5				tr-0.01		
Tetradecene	26952-13-6				tr-0.01		
α-Thujene	2867-05-2	0.08-1.2	0.1-0.2	tr	0.1-0.2	0.1-0.2	
Thymol	89-83-8	0-0.06	0-0.07		0.05-0.07		0.09[g]
Tridecanal	10486-19-8			tr			
Tridecane	629-50-5			tr			
n-Tridecanol	112-70-9				tr		
Undecanal	112-44-7	0-0.03	0.02-0.05	0.03	0.01-0.02		<0.01[g]; 0.01[h]
Undecane	1120-21-4			tr			
Valencene	4630-07-3		0.02-0.05	0.01			tr[f]

A twenty-eight tangerine essential oil samples from Brazil, China and USA, analyzed between 2006 and 2013; lowest and highest concentrations given (E. Schmidt, unpublished data)
B review of six studies published between 1979 and 1999 on the composition of industrial cold-pressed and commercial tangerine oils (not for all certain that the source was indicated as *C. tangerina*); lowest and highest concentrations given; data cited in ref. 7
C one industrially cold-pressed Murcott (honey) tangerine peel oil from Brazil (ref. 1)
D six samples of cold-pressed Dancy tangerine oil from different geographic origins in Mexico; lowest and highest concentrations given (ref. 5)
E six samples of hand-pressed *C. tangerina* oils from six French tangerine cultivars; lowest and highest concentrations given (ref. 2)
F data from other studies (indicated with superscript letters); highest concentrations found in any study reviewed here given; when two or more oils were investigated, only the highest concentrations are mentioned, unless indicated otherwise

[a] incorrect identity based on GC elution order (ref. 4); [b] decanol and perillaldehyde combined; [c] β-cubebene and β-elemene combined; [d] octanol and *cis*-sabinene hydrate combined; [e] nonane and heptanal combined; [f] one tangerine peel oil produced in China (ref. 8); [g] one sample of cold-pressed tangerine oil produced in China (ref. 9); [h] one sample of hand-pressed Murcott peel tangerine essential oil (ref. 10)

tr: trace (in columns C, D and F[h]: <0.01; in column E: <0.1)

Table 5.82.2 Major constituents of tangerine oils

Constituent	CAS	A	B (7)	C (1)	D (5)	E (2)	F
Limonene	138-86-3	81.8-97.8	87.4-91.7	94.6	89.6-90.9	87.1-90.9	89.2[g]; 91.7[f]; 95.7[h]
γ-Terpinene	99-85-4	0.1-5.6	0.02-4.5		2.2-3.5	3.9-5.2	tr[h]; 0.03[g]
Myrcene	123-35-3	0.1-2.2	1.9-3.2	1.9	1.8-2.3	1.5-1.7	1.7[h]; 1.9[g]

LEGEND: SEE UNDER TABLE 5.82.1

Table 5.82.3 Heterocyclic oxygenated compounds present in the non-volatile fraction of cold-pressed tangerine oils (3,5,13)

Name	Synonym	CAS
Heptamethoxyflavone (ref. 5, not specified)		119279-30-0
3,3',4',5,6,7,8-Heptamethoxyflavone	3-Methoxynobiletin	1178-24-1
3,3',4',5,6,7-Hexamethoxyflavone		1251-84-9
3',4',5,6,7,8-Hexamethoxyflavone	Nobiletin	478-01-3
7-Hydroxy-3,3',4',5,6,8-hexamethoxyflavone		185678-89-1
5-Hydroxy-3',4',6,7,8-pentamethoxyflavone		50439-46-8
7-Hydroxy-3,3',4',5,6-pentamethoxyflavone		57393-68-7
3',4',5,6,7-Pentamethoxyflavone	Sinensetin	2306-27-6
3',4',5,7,8-Pentamethoxyflavone	Isosinensetin	17290-70-9
4',5,6,7,8 -Pentamethoxyflavone	Tangeretin	481-53-8
4',5,6,7-Tetramethoxyflavone	Tetra-*O*-methylscutellarein	1168-42-9
4',5,7,8-Tetramethoxyflavone	Tetra-*O*-methylisoscutellarein	6601-66-7

CONTACT ALLERGY/ALLERGIC CONTACT DERMATITIS

General

Allergic contact dermatitis from tangerine oil has been reported in one publication only.

Case reports

One patient developed allergic contact dermatitis from a perfume; all (coded) ingredients were tested, and there were strongly positive patch test reactions to tangerine oil 2% and 10% in petrolatum only (15).

LITERATURE

1 Feger W, Brandauer H, Ziegler H. Analytical investigation of Murcott (honey) tangerine peel oil. J Essent Oil Res 2003;15:143-147
2 Lota M-L, de Rocca Serra D, Tomi F, Casanova J. Chemical variability of peel and leaf essential oils of 15 species of mandarins. Biochem Syst Ecol 2001;29:77-104
3 Chen J, Montanari AM, Widmer WW. Two new polymethoxylated flavones, a class of compounds with potential anticancer activity, isolated from cold pressed Dancy tangerine peel oil solids. J Agric Food Chem 1997;45:364-368
4 Lawrence BM. Progress in essential oils. Perfum Flav 2006;31(10):49-? (last page unknown).
5 Dugo P, Mondello L, Favoino O, Cicero L, Rodriguez Zenteno NA, Dugo G. Characterization of cold-pressed Mexican Dancy tangerine oils. Flavour Fragr J 2005;20: 60-66
6 Reeve D, Arthur D. Riding the citrus trail: When is a mandarin a tangerine? Perfum Flavor 2002;27(July/August): 20-22
7 Dugo G, Mondello L. Citrus Oils. Composition, advanced analytical techniques, contaminants, and biological activity. Boca Raton, FL, USA: CRC Press, Taylor & Francis Group, 2011:65-69
8 Li C, Qi C. Studies on the chemical constituents of the essential oil from fruit peel of Citrus tangerina Hort. ex Tanaka. Fujian Fenxi Ceshi 1997;6:716-718. Data cited in ref. 4
9 Huang Y, Wu Y. Chemical components of essential oils from peels of 25 citrus species and cultivars. Tianran Chanwu Yanjiu Yu Kaifa 1998;10(4):48-54. Data cited in ref. 7
10 Sawamura M. Volatile components of essential oils of the Citrus genus. Recent Res Dev Agric Food Chem 2000;4:131-164. Data cited in ref. 7
11 Liu YQ, Heying E, Tanumihardjo SA. History, global distribution, and nutritional importance of Citrus fruits. Compreh Rev Food Sci Food Saf 2012;11:530-545
12 Russo M, Torre G, Carnovale C, Bonaccorsi I, Mondello L, Dugo P. A new HPLC method developed for the analysis of oxygen heterocyclic compounds in Citrus essential oils. J Essent Oil Res 2012;24:119-129
13 Rhind JP. Essential oils. A handbook for aromatherapy practice, 2nd Edition. London: Singing Dragon, 2012
14 Lawless J. The encyclopedia of essential oils, 2nd Edition. London: Harper Thorsons, 2014
15 Vilaplana J, Romaguera C. Contact dermatitis from the essential oil of tangerine in fragrance. Contact Dermatitis 2002;46:108

Chapter 5.83   TEA TREE OIL

DEFINITION

Tea tree oil (essential oil of Melaleuca, terpinen-4-ol type) is the volatile oil obtained from the leaves and terminal branchlets of either the narrow-leaf tea tree *Melaleuca alternifolia* (Maiden et Betche) Cheel, the flax-leaf (narrow-leaf) tea tree *Melaleuca linariifolia* Smith, or the creek tea tree *Melaleuca dissitiflora* F. Muell.

INCI NOMENCLATURE

Description/definition: Melaleuca alternifolia leaf oil is the oil distilled from the leaves of the tea tree, *Melaleuca alternifolia*, Myrtaceae
INCI name EU: Melaleuca alternifolia leaf oil
INCI name USA: Melaleuca alternifolia (tea tree) leaf oil
CAS registry number(s): 68647-73-4; 85085-48-9; 8022-72-8 (mentioned by CosIng, but refers to *Melaleuca bracteata*)
EINECS number(s): 285-377-1

ISO (INTERNATIONAL ORGANIZATION FOR STANDARDIZATION) STANDARD

ISO number: 4730
ISO name: Essential oil of melaleuca, terpinen-4-ol type (tea tree oil)
Botanical origin: *Melaleuca alternifolia* (Maiden et Betche) Cheel; *Melaleuca linariifolia* Smith; *Melaleuca dissitiflora* F. Muell.
Parts of plant used: Leaf, terminal branchlets
ISO values: ISO values (minimum and maximum concentrations) are shown in Table 5.83.1.

According to ISO, essential oil of Melaleuca, terpinen-4-ol type (tea tree oil) is obtained by steam distillation of the foliage and terminal branchlets of *Melaleuca alternifolia* (Maiden et Betche) Cheel, *Melaleuca linariifolia* Smith, and *Melaleuca dissitiflora* F. Mueller. However, in practice, commercial tea tree oil is produced from *M. alternifolia* (Maiden and Betche) Cheel (8,37), which is an extremely fast growing tree and a constantly renewable source of oil (41). Therefore, only literature pertaining to this species is discussed in this chapter. However, the other two *Melaleuca* species provide similar oils. In fact, *Melaleuca alternifolia* was raised to species rank from *M. linariifolia* var. *alternifolia* (Maiden & Betche) in 1924 by Cheel (37,56). The composition of the oils of *Melaleuca linariifolia* can be found in refs. 41,48 and 56 and that of *Melaleuca dissitiflora* in refs. 48 and 52. Many oils discussed in literature and also a number that can be purchased on-line, are stated to have been produced from leaves only (whereas it is produced from leaves and terminal branchlets when conforming to ISO).

THE PLANT, THE OIL, AND THEIR USES

Melaleuca alternifolia is a tall shrub or small tree up to 15 meters high with a bushy crown and papery bark. This tree is native to Australia; the natural occurrence area is the northern coastal region of New South Wales, bordering Queensland. Tea tree oil, which is obtained from the leaves (and terminal branchlets) by steam-distillation, has been reported to have multiple biological activities such as anti-inflammatory, antitumoral, analgesic and biocidal. Tea tree oil is considered to be an effective bactericide, antiviral, insecticidal and acaricidal agent (19,26,27,37). It is seen by many as a universal remedy for acne, eczema, skin infections including herpes simplex and warts, wounds, burns, insect bites, dandruff (1) and nail mycoses (2). It is marketed as a 'natural' topical antimicrobial (its antimicrobial effects are well-documented) and anti-inflammatory agent (6,46). The product is present in many different formulations including pure oil (also for aromatherapy [40,74,75]), ointments, wart-paint (3), acne treatments (4) and household products such as fabric softeners, detergents and cleansers (5,40,46). In a monograph by the European Medicines Agency (25), tea tree oil was considered to be suitable for the treatment of small superficial wounds and insect bites, small boils (furuncles and mild acne), itching and irritation in cases of mild athlete's foot, and minor inflammation of oral mucosa (25). The oil is also used in many types of cosmetic products (6,40,46).

Tea tree oil is sold diluted and highly concentrated up to undiluted to the public (6). However, as products with high concentrations and especially aged (oxidized) oils may induce allergic reactions, the European Cosmetics Association COLIPA in 2002 recommended that tea tree oil should not be used in cosmetic products in a way that results in a concentration greater than 1% oil being applied to the body. Also, manufacturers were advised to consider the use of antioxidants and/or specific packaging to minimize exposure to light (7).

Useful reviews on various aspects of tea tree oil are provided in refs. 5,6,9,10,11,12,17,25,27,37 and 40.

Table 5.83.1 ISO values (%) for tea tree oil [a]

Compound	CAS	Minimum	Maximum
Terpinen-4-ol	562-74-3	30.0	48.0
γ-Terpinene	99-85-4	10.0	28.0
1,8-Cineole	470-82-6	tr	15.0
α-Terpinene	99-86-5	5.0	13.0
α-Terpineol	98-55-5	1.5	8.0
p-Cymene	99-87-6	0.5	8.0
α-Pinene	80-56-8	1.0	6.0
Sabinene	3387-41-5	tr	3.5
Aromadendrene	489-39-4	tr	3.0
δ-Cadinene	483-76-1	tr	3.0
Viridiflorene (ledene)	21747-46-6	tr	3.0
Limonene	138-86-3	0.5	1.5
Globulol	489-41-8	tr	1.0
Viridiflorol	552-02-3	tr	1.0

[a] ISO 4730 Essential oil of melaleuca, terpinen-4-ol type ©ISO 2004; Geneva, Switzerland, www.iso.org

809

CHEMICAL COMPOSITION

Tea tree oil is a colorless to pale yellow, clear mobile liquid which has a terpeny, coniferous and minty-camphoraceus odor. The yield of essential oil from the leaves with terminal branches of either *Melaleuca alternifolia* (Maiden et Betche) Cheel, *Melaleuca linariifolia* Smith, or *Melaleuca dissitiflora* F. Muell. generally varies from 1.0 to 1.8 per cent. The main producing country of this oil is Australia; minor quantities come from China, South Africa and Vietnam.

Literature data (up to September 11, 2014) on the chemical composition of tea tree oils and unpublished analytical data from one of us (E.S.) are shown in Table 5.83.2 in alphabetical order. In tea tree oils from various origins, over 220 chemicals have been identified. About 55 per cent of these were found in a single reviewed publication only. The major compounds found in tea tree oils from different sources are shown in Table 5.83.3.

They include (highest concentrations in any study given) 1,8-cineole (64.1%), terpinen-4-ol (57.9%), terpinolene (45.7%), *p*-cymene (35.3%), γ-terpinene (28.3%), α-terpinene (12.9%) and α-terpineol (11.8%). Well-known ingredients of tea tree oils that were present in high concentrations (>7%) in one or two studies were *cis*-sabinene hydrate (19.4%), α-phellandrene (12.2%), α-pinene (9.2%) and limonene (7.9%). A rare constituent of tea tree oil found in a high concentration (>7%) in a single study is 2-phenethyl alcohol (15.3%).

Commercial oils

The ten chemicals that had the highest maximum concentrations in 97 commercial tea tree essential oil samples (concentration ranges provided) are the following: terpinolene (0.04-45.7%), terpinen-4-ol (6.2-44.9%), γ-terpinene (3.1-23.0%), *cis*-sabinene hydrate

Table 5.83.2 Constituents identified in tea tree oils

Constituent	CAS	Percentage and range					
		A	B (41)	C (57)	D (60)	E (59)	F
8-α-Acetoxyelemol	41370-57-4					+3	
4-Acetyloxycycloheptanone							0.3[n,o]
α-Amorphene	20085-19-2			tr			0.3[e]
δ-Amorphene	189165-79-5						2.0[h]
Apo vertenex						+1	
Aromadendrene	489-39-4	0.1-2.0	0.1-2.0	1.5	1.5-1.6	+3	0.1-6.6[c]; 1.5[k]; 1.6[e]; 1.7[v]; 1.8[i]
allo-Aromadendrene	25246-27-9		0.4-0.8	0.3	0.6-0.7	+3	0.5[h,s]; 0.6[f]; 0.7[e]; 0.8[g]
Ascaridole	512-85-6						0.2[u]
Ascaridole epoxide						+3	
Ascaridole glycol				0-0.7			
cis-Ascaridole glycol	6790-83-6					+2	
trans-Ascaridole glycol	6790-83-6					+3	0.2[w]
Bicycloelemene	32531-56-9					+1	
Bicyclogermacrene	24703-35-3	0-1.2	0.3-0.6	0.1	0-1.2	+3	0.5[y6]; 0.6[k]; 0.7[g]; 1.1[i]; 6.2[x4]
Bicyclosesquiphellandrene	54324-03-7						tr[k]
epi-Bicyclosesquiphellandrene	54274-73-6						0.5[e]
Bicyclo[7.2.0]undecan-3-ol <11,11-dimethyl-, 4,8-*bis*(methylene)>	79580-01-1					+2	
Borneol	507-70-0					+1	tr[j]
α-Bulnesene	3691-11-0			tr			0.2[e]
β-Bulnesene							0.2[y6]
Cadina-1,4-diene	29837-12-5		0.1-0.3	0.1	0.2-0.3		tr[k]; 0.2[m,t]; 0.3[e]
trans-Cadina-1(2),4-diene	1395047-77-4					+3	0.3[h]
7,10-Cadina-1,(6),4-diene					0.3-0.6		
trans-Cadina-1(6),4-diene	931410-54-7						0.3[k]; 0.4[h]; 0.6[i]
Cadina-3,5-diene	267665-20-3				0-0.2		
α-Cadinene	24406-05-1						0.8[h]
β-Cadinene	523-47-7		0.2-0.6				

Table 5.83.2 Constituents identified in tea tree oils *(continued)*

Constituent	CAS	Percentage and range					
		A	B (41)	C (57)	D (60)	E (59)	F
γ-Cadinene	39029-41-9		0.02-0.04			+3	tr[k]
trans-γ-Cadinene							tr[h]
δ-Cadinene	483-76-1	0.2-1.9	0.8-2.0	1.3	1.9-2.2		0.1-7.5[c]; 1.9[e]; 2.1[y6]; 2.4[x2]
τ-Cadinol	5937-11-1		0.03-0.06				0.1[g]
α-Calacorene	21391-99-1					+3	
Calamenene	483-77-2	tr-0.2		0.1			0.1[x1]; 0.2[m]; 0.3[e]; 0.5[y6]; 0.6[i]
α-Calamenene							tr[k]
Camphene	79-92-5	tr-0.07	0-0.2	tr		+3	tr[h]; 0.1[j]
Camphene hydrate	465-31-6					+3	
Camphor	76-22-2						tr[h,j]
Carvacrol	499-75-2					+3	
Carvenone	499-74-1					+2	
Carvone hydrate	7712-46-1					+3	
Carvone oxide	33204-74-9					+2	
β-Caryophyllene	87-44-5	0.2-1.5	0.3-0.7	0.1	0.4-0.6	+3	0.4[e]; 0.5[f]; 0.6[g]; 0.8[i]; 3.1[n,o]
9-epi-(E)-Caryophyllene	68832-35-9					+3	
Caryophyllene oxide	1139-30-6					+2	tr[j]; 0.1[w]
Cedrane	13567-54-9		0.06-0.1				
β-Cedrene	546-28-1		0.02-0.05				
Cedrol	77-53-2					+2	
1,4-Cineole	470-67-7			tr		+3	tr[e,j]
1,8-Cineole	470-82-6	0.5-18.3	2.5-3.2	5.1	5.1-5.4	+3	0.5-17.7[c]; 5.5[h]; 6.0[y]; 6.4[u]; 6.6[y6]; 6.7[x5]; 7.3[v]; 24.3[q]; 64.1[y2]
Citronellal	106-23-0		0-0.03				
Citronellyl butyrate	141-16-2		0.2-0.4				
α-Copaene	3856-25-5		0.08-0.2	tr	0.2-0.3	+3	tr[k]; 0.1[m]; 0.2[e,w]; 0.3[g]
Cryptone	500-02-7					+2	
Cubeban-11-ol	220766-71-2		0.09-0.3				
α-Cubebene	17699-14-8		0.04-0.06	tr		+1	tr[h,k]; 0.05[m]; 0.09[e]; 0.1[g]; 0.2[f]
β-Cubebene	13744-15-5		0.02			+2	
Cubebol	23445-02-5						0.07[y6]
Cubenol	21284-22-0		0.05-0.1	0.1	0.2	+1	0.2[m]; 0.3[h]; 0.5[e]
1-epi-Cubenol	81939-29-9		0.09-0.3		0.3		0.1[w]; 0.2[m]; 0.6[h]
Cuminyl alcohol	536-60-7						tr[k]
3,5-Cycloheptadine-1-one							0.7[n,o]
1,3-Cyclohexadiene	592-57-4					+1	
o-Cymene	527-84-4		0.03-0.06				4.3[x4]
p-Cymene	99-87-6	0.3-19.4	2.5-3.2	2.9	2.9-19.9	+3	0.4-12.4[c]; 7.3[n]; 8.9[y9]; 9.6[w]; 11.5[y8]; 13.9[y2]; 14.4[u]; 35.3[y2,y3]
p-Cymenene	1195-32-0	0.04-3.1	0.05-0.1	tr			tr[k]; 0.1[j]; 0.2[e]
p-Cymen-8-ol	1197-01-9		0.04-0.07	tr		+2	0.05[x1]; 0.08[m]; 0.4[v]; 0.6[y6]
Daucene	16661-00-0		0.02-0.04				
Dehydrolinalool	29171-20-8					+2	
Dehydrosesquicineole	211237-38-6					+2	
1,2-Diacetylethane	110-13-4					+2	
Dihydrolinalyl acetate	61476-73-1					+2	
2,2-Dimethyl-4,5-di-1-propenyl-1,3-dioxolane	36334-88-0					+3	
3,6-Dimethyl-1,5-heptadiene	34891-10-6					+2	
2,5-Dimethyl-2,4-hexadiene	764-13-6					+1	
4,6-Dimethyl-2-octanone						+3	

Table 5.83.2 Constituents identified in tea tree oils *(continued)*

Constituent	CAS	Percentage and range					
		A	B (41)	C (57)	D (60)	E (59)	F
trans,trans-2,4-Dodecadienal	21662-16-8					+3	
α-Elemene	5951-67-7						0.3^m
β-Elemene	33880-83-0			0.1		+3	
γ-Elemene	29873-99-2		0.01-0.03				0.7^n
Epicedrol	19903-73-2		0.07-0.2				
Epicubenol	19912-67-5					+3	
Epiglobulol	88728-58-9				0.1		tr^k; 0.2^e
6,7-Epoxide citral	79083-70-8					+2	
Epoxy allo-aromadendrene	85760-81-2						0.1^h
5-Ethyl-3,4-dihydro-1(2*H*)-naphthalenone						+2	
α-Eudesmol	473-16-5	0.03-0.5					0.3^j
10-epi-γ-Eudesmol	15051-81-7		0.07-0.2				
5-epi-7-epi-α-Eudesmol	446050-56-2						0.3^h
α-Fenchyl acetate	111821-74-0		0-0.03				
Fenchyl alcohol	1632-73-1					+3	0.2^{x6}
β-Fenchyl alcohol	22627-95-8						3.3^{x4}
Flourensadiol	55812-89-0					+2	
Geranial	141-27-5						2.4^v
trans-Geraniol	106-24-1					+1	
Germacrene B	15423-57-1						0.2^f
Germacrene D	23986-74-5		0.04-0.09			+1	0.4^n; 0.5^{x4}
Globulol	489-41-8	0.02-0.6	0.2-0.6	0.2	0.5-0.7	+3	0.1-3.0^c; 0.8^e; 1.2^h; 2.0^{x7}; 3.1^n
Guai-5-en-11-ol					0.2		0.09^m
α-Guaiene	3691-12-1						1.3^h; 1.9^g
cis-β-Guaiene	372162-07-7				0.4-0.5		0.2^h; 1.4^g
trans-β-Guaiene	192053-49-9						0.1^h
Guaiol	489-86-1						0.2^g
α-Gurjunene	489-40-7	0.2-1.0	0.03-0.07	0.2	0.5	+3	0.3^k; 0.4^{h,m}; 0.5^e; 0.6^g
β-Gurjunene	73464-47-8		0.06-0.1	0.1			tr^e; 0.1^g; 0.4^f
γ-Gurjunene	22567-17-5					+3	0.1^w; 0.2^e
Hexadecanol	51260-59-4						0.6^{n,o}
Hexanediol							4.9^{n,o}
cis-3-Hexen-1-ol	928-96-1	0.01-0.07				+3	tr^k; 0.1^j
cis-3-Hexenyl acetate	3681-71-8	0-0.02				+3	
α-Himachalene	3853-83-6		0.1-0.2				
α-Humulene	6753-98-6	tr-0.2	0.07-0.1	tr	0.1-0.2	+3	tr^{j,k}; 0.09^m; 0.1^h; 0.2^w; 0.7^{y6}
Humulene epoxide	96638-51-6						0.5^{n,o}
Hydroxycaryophyllene	78683-81-5					+2	
Hydroxycitronellal	107-75-5					+2	
5-Hydroxyisobornyl isobutanoate						+2	
cis-8-Hydroxylinalool						+1	
Isoascaridole	1619-26-7					+1	
Isocaucalol	5172-21-4					+2	
Isoledene	95910-36-4		0.06-0.1			+3	tr^h
Isospathulenol	88395-46-4					+1	
Isothujyl acetate						+2	
Laciniatafuranone E	199115-10-1					+2	
Ledene oxide	882187-44-2						1.2^{n,o}
Ledol	577-27-5	0.02-0.3		tr		+2	tr^k; 0.1^j; 0.2^{g,h}
Limonene	138-86-3	0.5-3.0	1.6-2.1	1.0	1.4-1.9^b	+3	0.4-2.7^c; 2.3^x; 2.6^s; 3.4^q; 7.9^{y2}
Limonene epoxide	1195-92-2					+3	

Table 5.83.2 Constituents identified in tea tree oils *(continued)*

Constituent	CAS	Percentage and range					
		A	B (41)	C (57)	D (60)	E (59)	F
cis-Limonene oxide	13837-75-7					+3	
Linalool	78-70-6	0.06-0.8	0.05-0.07	tr		+3	0.07[m]; 0.2[g]; 0.3[h]; 0.9[q]; 3.2[y6]
cis-Linalool oxide	11063-77-7					+3	
Longicyclene	1137-12-8					+2	
Longifolene (junipene)	475-20-7						1.3[n,o]
trans-Longipinalol			0.06-0.1				
β-Maaliene	489-29-2						tr[e]
γ-Maaliene	20071-49-2					+2	
cis-*p*-Mentha-1(7),8-dien-2-ol	22626-43-3						tr[h]
cis-*p*-Mentha-2,8-dien-1-ol	3886-78-0						tr[h]
1-*p*-Menthen-9-al	29548-14-9					+1	
p-Menth-6-ene-2,3-diol						+2	
p-Menth-2-en-1-ol	619-62-5	0.04-0.7					
cis-*p*-Menth-2-en-1-ol	29803-82-5		0.09-0.4	0.1	0.2-0.3	+3	0.1[j,w]; 0.2[d]; 0.8[h]
trans-*p*-Menth-2-en-1-ol	29803-81-4		0.06-0.2	0.2	0.1-0.3	+3	0.1[j,w]; 0.3[m]; 0.4[y6]; 0.5[h]; 0.6[d]
Menthone	89-80-5					+2	
3-Methyl-3-cyclohex-en-1-ol	53783-91-8					+2	
2-Methyl-5-decanone	54410-89-8						3.3[n,o]
4-Methyl-3-decen-5-ol	81782-77-6					+3	
α-Methylenedodeca-nal						+2	
Methyl eugenol	93-15-2	0.01-0.4		tr		+3	tr[e,k]; 0.1[j]
6-Methyl-3,5-hepta-dien-2-one	1604-28-0					+1	
6-Methyl-5-hepten-2-one	110-93-0					+2	
2-Methylisoborneol	2371-42-8					+3	
3-Methyloctane	2216-33-3					+1	
α-Muurolene	10208-80-7		0.1-0.2	0.1			0.1[f]; 0.2[g]; 0.3[e]; 0.6[y6]; 0.7[y6]
γ-Muurolene	30021-74-0	0-0.3		tr			tr[h]; 0.6[g]
α-Muurolol	104245-48-9		0.02-0.06				tr[g]; 0.2[h]
Myrcene	123-35-3	0.2-4.1	0.8-1.0	0.5	0.2-0.7	+3	0.1-1.8[c]; 1.0[g]; 1.2[x1]; 2.0[s]; 2.5[y2]
trans-Myrtanol	15358-91-5		0-0.03				
Neodihydrocarveol	18675-33-7						6.3[n,o]
Nerol	106-25-2			tr		+1	0.6[y6]
Nonanal	124-19-6						tr[h]
Nonane	111-84-2					+3	tr[h]
Nootkatone	4674-50-4						0.3[n,o]
trans-β-Ocimene	3779-61-1		0-0.02	tr	0-0.02	+3	0.2[h,n]
allo-Ocimene diepoxide						+1	
1-Octadecanamine	124-30-1						0.3[n,o]
Octanal	124-13-0						0.4[x5]
Oleic acid	112-80-1						1.7[n,o]
β-Oplopenone	28305-60-4						tr[h]
Palustrol	5986-49-2			tr		+3	0.06[e]; 0.5[y6]; 2.2[y6]
β-Patchoulene	514-51-2		0.05-0.1				tr[h]
α-Phellandrene	99-83-2	0.2-0.6	0.3-0.5	0.3	0.1-0.5	+3	0.1-1.9[c]; 0.4[f]; 0.5[g]; 3.1[q]; 12.2[y2]
β-Phellandrene	555-10-2	tr-5.2		0.9	1.4-1.9[b]		0.4-1.6[c]; 0.7[l]; 0.8[i]; 1.0[q]; 1.5[y6]

Table 5.83.2 Constituents identified in tea tree oils *(continued)*

Constituent	CAS	Percentage and range					
		A	B (41)	C (57)	D (60)	E (59)	F
2-Phenethyl alcohol	60-12-8						15.3[y6,y7]
α-Pinene	80-56-8	1.8-9.2	2.5-2.8	2.6	0.5-2.5	+3	0.8-3.6[c]; 2.5[f]; 2.8[m]; 3.0[y1]; 3.4[x]
β-Pinene	127-91-3	0.3-1.7	0.7-0.8	0.3	0.7	+3	0.1-1.6[c]; 0.8[g]; 1.0[x1]; 1.1[j]; 1.8[y6]
cis-Pinene hydrate	17974-51-5					+3	
trans-Pinocarveol	1674-08-4						tr[j]
Piperitol	491-04-3	0.05-0.3					
cis-Piperitol	16721-38-3			tr	0.2-0.3	+3	0.1[h,y6]
trans-Piperitol	16721-39-4		0.03-0.08	tr	0.2-0.3	+3	0.1[j]; 0.3[h]
Piperitone	89-81-6					+1	
cis-Piperityl acetate	78774-33-1		0.02-0.07				
Pogostol	21698-41-9					+3	
Rosifoliol	63891-61-2			tr	0.1	+3	0.05[x1]
Sabinene	3387-41-5	0.03-1.3	0.03-0.6	0.2	0-0.3	+3	0-3.2[c]; 0.9[v]; 1.0[p]; 1.3[x1]; 1.6[x9]
Sabinene hydrate	546-79-2						0.4[r]
cis-Sabinene hydrate	15537-55-0	tr-19.4	0.01-0.2	tr	0-0.2	+3	tr[k]; 0.1[h]; 1.7[d]; 2.3[v]
trans-Sabinene hydrate	17699-16-0	0.01-0.3	0-0.09	tr	0-0.1	+3	tr[k]; 0.1[j,m]
cis-Sabinene hydrate acetate	77318-48-0		0.03-0.1				
trans-Sabinol	471-16-9		0.02-0.03				
β-Santalol	77-42-9					+3	
Santene	529-16-8					+1	
Selina-5,11-diene	52026-55-8				0.2	+2	
α-Selinene	473-13-2						1.7[h]
β-Selinene	17066-67-0						0.1[h]; 0.3[g]; 0.5[x4]
Selin-11-en-4α-ol	16641-47-7						tr[h]
Spathulenol	6750-60-3	tr-1.1	0.06-0.2	tr	0-0.2	+3	0.1[j]; 0.2[e]; 0.3[h]; 1.3[n,o]
α-Terpinene	99-86-5	2.3-11.7	8.5-10.6	10.4	1.2-10.2	+3	10.3[y1]; 10.7[h]; 10.9[x5]; 11.3[y5]; 11.9[x1]; 12.7[v]; 12.9[y6]
γ-Terpinene	99-85-4	3.1-23.0	21.1-22.5	23.0	6.9-19.2	+3	9.5-28.3[c]; 21.2[e]; 21.5[x2]; 22.4[x5]; 22.5[y1]; 22.8[y4]; 23.0[j]; 23.2[y5]
Terpinen-4-ol	562-74-3	6.2-44.9	34.9-44.0	40.1	30.1-33.1	+3	28.6-57.9[c]; 47.7[n]; 48.7[v]; 49.3[x9] 50.0[y6]; 52.9[y2]; 53.4[y9]; 53.7[x7]
1-Terpineol	586-82-3					+3	
α-Terpineol	98-55-5	1.9-4.2	2.4-2.8	2.4	3.4-4.8	+3	1.5-7.6[c]; 3.3[d]; 3.4[j]; 3.8[y1]; 3.9[y4]; 4.1[h]; 4.5[q]; 4.6[p]; 9.6[y3]; 11.8[y2]
(Z)-β-Terpineol	7299-40-3						0.2[f]
γ-Terpineol	586-81-2					+3	
δ-Terpineol	7299-42-5		0-0.02				tr[h,j]
Terpinolene	586-62-9	0.04-45.7[z]	3.1-3.6	3.1	1.6-3.8	+3	1.6-5.4[c]; 3.2[e]; 3.3[j]; 3.4[j]; 3.5[h]; 3.6[d]; 3.8[p]; 4.0[g]; 4.2[x1,y]; 45.6[q]
Terpinolene-4,8-diol					0-0.3		
α-Terpinyl acetate	80-26-2						6.0[q]
α-Thujene	2867-05-2	0.05-1.4	0.9-1.1	0.9	0.9-1.0	+3	0.1-2.1[c]; 1.1[h]; 1.2[n]; 1.4[q]; 1.7[y6]
3-Thujopsanone	25966-79-4		0.02-0.03				
Torreyol	19435-97-3					+2	
4-Tridecanol	26215-92-9						0.3[n,o]
1,2,4-Trihydroxymenthane	66767-24-6			tr	0-1.6	+3	+[y2]; tr-4.6[y3]; 0.01[u]

Table 5.83.2 Constituents identified in tea tree oils *(continued)*

Constituent	CAS	Percentage and range					
		A	B (41)	C (57)	D (60)	E (59)	F
2,3,6-Trimethyl-1,5-heptadiene	33501-88-1					+2	
2-Undecene	2244-02-2					+1	
Undecone							0.9[n,o]
Valencene	4630-07-3		0.08-0.1				
Valerena-4,7(11)-diene	351222-66-7					+2	
Verbenene	4080-46-0					+1	
cis-Verbenyl acetate	29135-27-1		0-0.08				
trans-Verbenyl acetate	1203-21-0		0-0.04				
β-Vetispirene	28908-27-2					+2	
Viridiflorene (ledene)	21747-46-6	0.3-2.1	0.7-1.6	1.0	1.4-2.2	+3	0.3-6.1[c]; 1.3[f]; 1.5[d]; 1.8[e]; 2.0[y6]
Viridiflorol	552-02-3	0.08-0.8	0.06-0.2	0.1	0.4-0.5	+3	0.1-1.4[c]; 0.4[h]; 0.6[e]; 1.1[y6]; 1.6[x7]
α-Ylangene	14912-44-8			tr			0.06[m]
Zonarene	41929-05-9				0-0.4	+1	tr[k]

A ninety-seven tea tree essential oil samples from Australia and Vietnam, analyzed between 1998 and 2013; lowest and highest concentrations given (E. Schmidt, unpublished data)
B six commercial *M. alternifolia* oils from Australia; lowest and highest concentrations given (ref. 41)
C 'typical' composition of tea tree oil from Australia, terpinen-4-ol type, based on the analysis of 'numerous' oil samples (probably commercial) (ref. 57)
D one unoxidized, one partially oxidized and one substantially oxidized commercial tea tree oil from Australia selected for testing for a tea tree oil safety dossier; lowest and highest concentrations given (ref. 60)
E one fresh and one aged (oxidized) commercial oil from Australia; +1 = only present in the fresh oil; +2 = only present in the aged oil; +3 = present in both the fresh and the aged oil; only qualitative determinations (ref. 59)
F data from other studies (indicated with superscript letters); highest concentrations found in any study reviewed here given; when two or more oils were investigated, only the highest concentrations are mentioned, unless indicated otherwise

[a] incorrect identification; [b] β-phellandrene and limonene combined; [c] 'hundreds' of commercial Australian tea tree oil samples, terpinen-4-ol type; lowest and highest concentrations given (ref. 57); [d] one steam-distilled oil from Australia (ref. 57); [e] one oil sample purchased at a local retail outlet in Australia (ref. 53); [f] one commercial oil purchased from an Italian oil company (ref. 33); [g] one tea tree oil sample of unknown origin and mode of production (ref. 32); [h] one lab-hydrodistilled oil from the aerial parts of 5-month-old cultivated *M. alternifolia* plants (ref. 26); [i] one commercial tea tree oil sample purchased in Italy (ref. 18); [j] one commercial oil sample (ref. 62); [k] one sample of tea tree oil of unknown origin (ref. 64); [l] one commercial tea tree oil sample from Australia (ref. 65); [m] one sample of commercial tea tree oil (ref. 67); [n] one lab-hydrodistilled oil from the leaves and branches of *M. alternifolia* grown in India (ref. 38), of which many were misidentified or tentatively incorrectly identified (ref. 61); [o] misidentified or tentatively incorrectly identified (ref. 61); [p] three Australian oils produced from resp. whole branches, young leaves and old leaves by steam-distillation (ref. 51); [q] two lab-hydrodistilled oils, one from cultivated and one from wild growing *M. alternifolia* from Australia (ref. 54); [r] one commercial tea tree oil from Australia (ref. 46); [s] one commercial oil from Australia (ref. 45); [t] one commercial tea tree oil obtained from a UK company (ref. 35); [u] one oil produced at a research facility in Australia (ref. 31); [v] one commercial oil sample purchased in India (ref. 28); [w] one lab-hydrodistilled oil from leaves collected at a botanical garden in Brazil (ref. 23); [x] one commercial oil from Australia and one purchased in the UK (ref. 22); [x1] one steam-distilled oil from the leaves and the terminal branchlets of *M. alternifolia* cultivated in India (ref. 49); [x2] one commercial tea tree oil sample from Australia (ref. 47); [x3] one commercial Australian tea tree oil (ref. 44); [x4] one laboratory steam-distilled oil from leaves purchased at a local Taiwan market (ref. 21); [x5] six oils from foliage collected at 2 sites in Australia, of which two were steam-distilled at day 1, two at day 7 and two obtained by cohobation (ref. 55); [x6] one commercial tea tree oil purchased from an Italian company (ref. 63); [x7] one lab-hydrodistilled oil from leaves collected at a *Melaleuca* plantation in Brazil (ref. 34); [x8] one commercial oil purchased from a French company (ref. 30); [x9] one lab-hydrodistilled leaf oil from an experimental station in Taiwan (ref. 24); [y] one commercial oil purchased in Brazil (ref. 20); [y1] one commercial oil sample purchased in Spain (ref. 19); [y2] data from 1986-1988, cited in ref. 58; [y3] five tea tree oil samples, aged 1-10 years, either slow, moderate or rapid (ref. 57); [y4] various oil samples of tea tree leaf grown in South Africa and harvested during one vegetative cycle (ref. 70); [y5] data from various studies cited in ref. 66; [y6] data from various studies cited in ref. 69; [y7] the oil was probably adulterated (ref. 69); [y8] one aged tea tree oil sample (ref. 16); [y9] one commercial tea tree oil sample from a Brazilian supplier (ref. 73); [z] the very high concentration of 45.7% for terpinolene was found in one sample from China only; the median value for all oils was 3.1%

tr: trace (in columns F[h], F[j]: <0.1); + present in the oil investigated, but quantity not stated; +1 = only present in the fresh oil; +2 = only present in the aged oil; +3 = present in both the fresh and the aged oil (ref. 59)

Table 5.83.3 Major constituents of tea tree oils

Constituent	CAS	Percentage and range					
		A	B (41)	C (57)	D (60)	E (59)	F
1,8-Cineole	470-82-6	0.5-18.3	2.5-3.2	5.1	5.1-5.4	+3	6.6[y6]; 6.7[x5]; 7.3[v]; 24.3[q]; 64.1[y2]
Terpinen-4-ol	562-74-3	6.2-44.9	34.9-44.0	40.1	30.1-33.1	+3	28.6-57.9[c]; 53[y2]; 53.4[y9]; 53.7[x7]
Terpinolene	586-62-9	0.04-45.7	3.1-3.6	3.1	1.6-3.8	+3	3.6[d]; 3.8[p]; 4.0[g]; 4.2[x1,y]; 45.6[q]
p-Cymene	99-87-6	0.3-19.4	2.5-3.2	2.9	2.9-19.9	+3	11.5[y8]; 13.9[y2]; 14.4[u]; 35.3[y2,y3]
γ-Terpinene	99-85-4	3.1-23.0	21.1-22.5	23.0	6.9-19.2	+3	9.5-28.3[c]; 22.8[y4]; 23.0[j]; 23.2[y5]
α-Terpinene	99-86-5	2.3-11.7	8.5-10.6	10.4	1.2-10.2	+3	11.3[y5]; 11.9[x1]; 12.7[v]; 12.9[y6]
α-Terpineol	98-55-5	1.9-4.2	2.4-2.8	2.4	3.4-4.8	+3	4.1[h]; 4.5[q]; 4.6[p]; 9.6[y3]; 11.8[y2]

LEGEND: SEE UNDER TABLE 5.83.2

(tr-19.4%), p-cymene (0.3-19.4%), 1,8-cineole (0.5-18.3%), α-terpinene (2.3-11.7%), α-pinene (1.8-9.2%), β-phellandrene (tr-5.2%) and α-terpineol (1.9-4.2%) (Erich Schmidt, unpublished analytical data).

Chemotypes

Currently, six chemotypes of M. alternifolia leaf oil are commonly distinguished as follows (adapted from ref. 13) (numbers are concentrations in %):

	Terpinen-4-ol	1,8-Cineole	Terpinolene
Type 1	22-40	0-17	2-6
Type 2	<3	22-44	41-60
Type 3	10-14	34-46	16-24
Type 4	6-14	41-63	0-3
Type 5	<1	72-86	<1
Type 6	<1	65-80	6-14

There is an obvious terpinen-4-ol chemotype (type 1), an obvious terpinolene chemotype (type 2), and an obvious 1,8-cineole chemotype (type 5). The three remaining chemotypes (3, 4 and 6) are dominated by the oil component 1,8-cineole and are considered to be 1,8-cineole chemotypes that differ in the levels of either terpinen-4-ol or terpinolene present (13). In wild populations, chemotype 1 is the most frequent and this high terpinen-4-ol oil is the only oil produced commercially (13,37,43).

The composition of tea tree oil changes particularly in the presence of atmospheric oxygen but also when the oil is exposed to light, humidity and higher temperatures. Under these conditions, the antioxidants α-terpinene, γ-terpinene and terpinolene oxidize to p-cymene. Consequently, the levels of α-terpinene, γ-terpinene and terpinolene decrease whereas the level of p-cymene increases up to tenfold (16,57). Hence, the concentration of p-cymene is a good measure of the oxidative degradation of tea tree oil (60). Oxidation processes further lead to the formation of peroxides, endoperoxides and epoxides such as ascaridole (14) and 1,2,4-trihydroxymenthane (15,16,57), which are known sensitizers (31).

Additional analytical studies not discussed here include refs. 29,36,42,50,68,71 and 72; these studies provide only information which is already shown in Table 5.83.2.

CONTACT ALLERGY/ALLERGIC CONTACT DERMATITIS

General

Of all essential oils, tea tree oil has caused most allergic reactions since the first case reports were published in 1991 from Australia, where tea tree oil is produced. The oil has been extensively investigated. Neat tea tree oil is a moderate sensitizer in humans (87,89,94,115,116,120). Undiluted oils and formulations containing 5% tea tree oil can also induce irritation of the skin/irritant contact dermatitis (89,114,115,120). Contact allergy to/allergic contact dermatitis from tea tree oil has been reported frequently. There are many reports of routine testing; tea tree oil 5% was added to the screening series of the North American Contact Dermatitis Group (NACDG) in 2003. In groups of consecutive patients suspected of contact dermatitis, prevalence rates of up to 2.5% positive patch test reactions have been observed. In two well-documented studies, current relevance was found in 41% and 56% of the positive patch tests (84,91). In the NACDG studies, 'definite' + 'probable' relevance ranged from 20% to 56%. Many case reports of allergic contact dermatitis have been documented. Nearly 2/3 were caused by the application of pure tea tree oil on damaged skin for therapeutic purposes. Cosmetics are the cause in a minority of cases (25%). Exposure to light and air leads to the formation of allergenic chemicals and increases the allergenicity of tea tree oils. The most important sensitizers in tea tree oil appear to be terpinolene, ascaridole, α-terpinene (and its oxidation products), 1,2,4-trihydroxymenthane, α-phellandrene, d-limonene and myrcene. A review of toxicity from systemic administration of tea tree oil is provided in refs. 121 and 115 (also a review of allergic contact dermatitis in the latter).

Testing in groups of patients

The results of patch tests with tea tree oil in routine testing (consecutive patients suspected of contact dermatitis) and in groups of selected patients are shown in Table 5.83.4. In routine testing, rates of positive reactions ranged from 0.1% to 2.5%, whereas between 1.6% and 41% of patients in selected groups had positive patch tests. The very high positivity rate of 41% was seen in a small group of 17 patients suspected of cosmetic dermatitis and tested with the undiluted oil,

Table 5.83.4 Results of testing groups of patients with tea tree oil

Years and Country	Test conc. & vehicle	Number of patients		Selection of patients (S); Relevance (R); Comments (C)	Ref.
		tested	positive (%)		
Routine testing					
2011-2013 The Netherlands	5% pet.	221	2 (0.9%)	R: not relevant; C: both patients also reacted to ascaridole	130
2011-12 USA, Canada	5% pet., oxidized	4,231	36 (0.9%)	R: definite + probable relevance: 56%	135
2009-10 USA, Canada	5% pet.	4,299	43 (1.0%)	R: definite + probable relevance: 50%; the test material was oxidized	82
2001-2010 Australia	10% pet.	5,087	129 (2.5%)	R: 33%	138
2007-8 USA, Canada	5% pet.	5,078	71 (1.4%)	R: definite + probable relevance: 37%	99
2000-2007 USA	5% pet.	869	18 (2.1%)	R: 100%; C: weak study: a. high rate of macular erythema and weak reactions, b. relevance figures include 'questionable' and 'past' relevance	78
<2006 USA, Canada	5% pet.	1,603	5 (0.3%)	R: definite + probable relevance: 20%	129
2005-6 USA, Canada	5% pet.	4,435	62 (1.4%)	R: definite + probable relevance: 36%	98
2003-4 USA, Canada	5% pet.	5,137	45 (0.9%)	R: not stated	100
2000-2004 Australia	10% and 5% pet.	2,320	41 (1.8%)	R: 17/41 (41%); only 4 patients had used cosmetic products containing tea tree oil (soap, hand cream, face cream, deodorant and hand lotion, one product each); 66% of the 41 patients recalled prior use of tea tree oil and 20% specified application of neat (100%) tea tree oil	84
<2004 USA	5% pet.	1,603	5 (0.3%)	C: no details known	101
2002-2003 Denmark	10% pet.	377	1 (0.3%)	R: probably relevant	114
1999-2003 Germany	5% DEP, oxidized	2,284	21 (0.9%)	R: percentage not specified; some patients had used (self-made) cosmetics containing tea tree oil, others had used the neat oil for eczema, acne, flea bites, muscle pain, and for evaporation in the sauna or indoors to banish wasps	118
2001 United Kingdom	pure, oxidized	550	13 (2.4%)	R: 4 relevant, 5 possibly relevant, 4 relevance unknown; C: 2 cases of occupational allergy in a beauty therapist and a complementary therapist; other exposures included the use of a shaving gel and children's shampoo; 38% irritant patch test reactions to pure oxidized tea tree oil	122
< 2000 Italy	5%, 1% and 0.1%, pet., undiluted	725	1 (0.1%)	C: details not known; irritant reactions to undiluted tea tree oil	90
1999-2000 Germany, Austria	5% DEP, oxidized	3,375	36 (1.1%)	R: current relevance 56%; range of positive patch tests per center: 0%-2.3%; co-reactivity to oil of turpentine: 39%; 10 allergic patients were tested with allergenic constituents: see Table 5.83.6 for results	91
1999 Australia	?	477	12 (2.5%)	R: not stated; C: in a group of 45 patients reacting to compound tincture of benzoin, there were 15 (33%) reactions to tea tree oil	134
1997 France	5-10-50% in arachis oil and pure	1,216	7 (0.6%)	R: the patients used pure oils, creams and hair products containing tea tree oil	96
Testing in groups of selected patients					
2014 The Netherlands	5% pet., oxidized	29	4 (13.8%)	S: patients with dermatitis who had previously been tested with ascaridole and had a (doubtful) positive or irritant reaction to ascaridole at that time; R: no relevance found; C: all four were also allergic to ascaridole	130
2011-2012 Italy	5% pet.	122	2 (1.6%)	S: patients who reported adverse cutaneous reactions to products (notably cosmetics) containing botanical ingredients in a questionnaire; they were tested with a 'botanical series'; R: both reactions were relevant	132
2001-2010 Australia	5% pet.	794	28 (3.5%)	S: not specified; R: 43%	138

Table 5.83.4 Results of testing groups of patients with tea tree oil *(continued)*

Years and Country	Test conc. & vehicle	Number of patients		Selection of patients (S); Relevance (R); Comments (C)	
		tested	positive (%)		Ref.
2001-2002 Sweden	5% alc.	1,075	29 (2.7%)	S: patients referred for routine testing willing to participate in a study on cosmetic use and adverse reactions; R: not stated	79
1998-1999 Australia	pure and 10% pet.	216	6 (2.8%)	S: healthy adult volunteers; R: not stated; C: the patients were patch tested with ten different samples. When 'indistinguishable' reactions were counted, the percentage of positive reactions rose to 4.8%; in the subgroup of patients (63%) who had previously come into contact with tea tree oil, the percentages were 4.6% (without 'indistinguishable' reactions) and 7.6% (with such reactions); probably an overestimation	88
1996-1997 UK	pure	17	7 (41%)	S: patients suspected of cosmetic dermatitis; R: 6/7 relevant	80

DEP: diethyl phthalate; pet.: petrolatum

which may give rise to irritant reactions (although 6/7 reactions were considered to be relevant) (80). In two well-documented studies, current relevance was found in 41% and 56% of the positive patch tests (84,91). In the NACDG studies, 'definite' + 'probable' relevance ranged from 20% to 56%.

Case reports

Details of published case reports of allergic contact dermatitis to tea tree oil are summarized in Table 5.83.5. At least 85 patients have been reported. Of the cases where the products responsible for the allergic reactions were specified, most (63%) related to pure tea tree oil applied for therapeutic purposes on a variety of skin conditions including acne, eczema, sunburn, wounds (of any cause), warts, herpes and fungal infections. There were also some cases caused by topical pharmaceutical preparations containing tea tree oil. Six patients (two aromatherapists, a complementary therapist, two pedicurists and a beautician) had occupational allergic contact dermatitis from tea tree oil. Thus, 75% of all cases are caused by the use of undiluted oil or products with high concentrations usually applied on damaged skin. Cosmetics are the cause of tea tree allergic contact dermatitis in only some 25% of all cases (see also refs. 84 and 118, data provided in Table 5.83.5). Products with low concentrations (<2%) of tea tree oil will infrequently induce contact allergy or elicit allergic reactions. Thus, of 27 cases of contact dermatitis to products with tea tree oil that were reported to the Swedish MPA (Medicinal Products Agency), all had a tea tree oil concentration of 2% or higher (119).

Positive patch test reactions

In some articles, patients with allergic contact dermatitis from other sources co-reacted to tea tree oil, which either had no relevance or the relevance of which was not mentioned or uncertain. Because of the abundance of (other) literature on tea tree oil, these articles are not specifically mentioned here.

The allergens in tea tree oil

Melaleuca oil is a moderate sensitizer in animal and human experiments (89,94,115,116,120); skin sensitization may be enhanced by irritancy (115). Oil stored in open bottles or in a bottle opened several times suffers an aging process resulting in photo-oxidation of the oil leading to degradation products (peroxides, epoxides and endoperoxides), which are strong sensitizers (85,102). Air exposure leads to a 3-fold increase in the sensitization potency for tea tree oil (85). Auto-oxidation of α-terpinene to allylic epoxides and other oxidation products may be contributory (127).

In 1994 and 1999, the main sensitizers were identified (81,85). Since then, especially German investigators have tested a considerable number of patients allergic to tea tree oil with one ingredient or a battery of its constituents to identify the main sensitizers. The results are shown in Table 5.83.6.

The most important sensitizers in tea tree oil appear to be terpinolene, ascaridole, α-terpinene (and its oxidation products, [127]), 1,2,4-trihydroxymenthane, α-phellandrene, *d*-limonene and myrcene. Other chemicals which may be responsible for tea tree oil allergy, albeit less frequently, include aromadendrene, *d*-carvone, *l*-carvone, terpinen-4-ol, viridiflorene, sabinene, *p*-cymene and possibly 1,8-cineole (83,85,91,92,93,118,130,131,133,139). Most of these have been found in low concentrations or not at all in commercial tea tree oils, which can be explained by the fact that these were fresh oil samples (Table 5.83.2, column A).

Conversely, of 14 patients with occupational contact dermatitis from *d*-limonene and patch tested with tea tree oil 5% in petrolatum, 5 (36%) had a positive (n=4) or doubtful positive (n=1) reaction to tea tree oil. This indicates that previous contact allergy to limonene may result in a positive patch test to tea tree oil (139).

Table 5.83.5 Case reports of allergic contact dermatitis from tea tree oil

Years and country	No. of patients allergic	Causative products, clinical data and comments	Ref.
2013 Netherlands	2	Soap and cream containing tea tree oil in one patient, shaving oil in the second patient who had the clinical picture of folliculitis barbae; both patients also reacted to ascaridole	133
2011 UK	1	Essential oil used by a 'complementary therapist' with contact allergy to many other oils	128
2000-2010 Belgium	5	Skin care products; this represented 0.5% of 959 cases of cosmetic allergy where the causal allergen was found	77
2000-2009 Belgium	1	Skin care product	137
1978-2008 Belgium	2	Topical pharmaceutical preparations	76
2007 USA	1	Pure oil used for aromatherapy	126
2007 Australia	1	The patient was sensitized by pure oil used for acne, and later developed allergic contact dermatitis of the eyelids from using a tea tree oil-containing shampoo	104
2004 Canada	1	Pure oil for aromatherapy	125
2004 Germany	1	Pure oil on the face of a 12-year-old boy for a 'minimal skin affection'	117
2003 United Kingdom	1	Pure oil on a piercing wound; contact allergy may have precipitated linear IgA disease	95
2002 United Kingdom	1	Pure tea tree oil; the patient was a professional aromatherapist, who also reacted to many other essential oils	124
2000 United Kingdom	1	'Tea tree oil products' used for vulvovaginitis	111
2000 Germany	1	No details known	123
2000 USA	1	Erythema multiforme-like contact dermatitis ('id-') from application of pure oil to a wound	112
1999 Germany	8	Pure oil in seven patients for treatment of eczema, plantar warts and sunburn	92
<1999 Germany	16	Ten patients had used pure oil for skin disorders such as eczema, warts, sunburn and herpes (n=9) and for 'hygiene and cosmetic purposes' (n=1); one patient developed dermatitis from shampoo to which pure oil had been added; no data for the other 5 cases	85
1998 Germany	1	Pure oil on psoriasis	108
1997 United Kingdom	1	Wart paint with tea tree oil	103
1997 France	7	Pure oils and cosmetics containing tea tree oil	96
1997 Sweden	1	Pure oil on skin irritation	109
1997 Germany	2	Pure oil, in one patient used on basal cell carcinoma; one also reacted to limonene and sweet orange oil	110
1996 USA	12	Details not known	105
1996 Netherlands	1	Airborne allergic contact dermatitis from inhalation of aqueous solution of tea tree oil; source of primary sensitization not mentioned	113
1995 Norway	1	Hand dermatitis in an aromatherapist, primarily sensitized to lemongrass oil; positive patch test reaction to tea tree oil used at her work. Cajeput was mentioned as a synonym, so possibly it was not the oil from *Melaleuca alternifolia*	106
1994 Norway	1	Pure oil for acne	?
1994 Germany	7	Pure oil on skin disorders such as fungal infection, dog scratches, insect bites, and hand rashes	93
1994 Netherlands	3	Pure oil; occupational contact dermatitis in two pedicurists and a beautician	97
1994 Norway	1	Pure oil for treatment of acne	136
1992 Netherlands	1	Pure oil for treatment of dermatitis; systemic contact dermatitis after oral administration; the patient co-reacted to 1,8-cineole, an ingredient of the oil	83
1992 Australia	2	Undiluted oil; first two cases of contact allergy reported	86

Table 5.83.6 Testing with ingredients in patients with positive patch test reactions to tea tree oil

Years and country	No. tested allergic to tea tree oil (test conc./veh.)	Ingredients tested, test concentration and vehicle, numbers positive, percentage positive (in brackets) and comments	Ref.
2011-2013 Netherlands	6 (5% pet.)	All reacted to ascaridole 1% and/or 2% and/or 5% in petrolatum	130
2009-2013 Spain	4 (5% pet. and pure)	All reacted to oxidized d-limonene (concentration/vehicle unknown)	131
1999-2003 Germany	20 (5% DEP)	Terpinolene 5% DEP: n=17 (85%); Ascaridole 5% DEP: n=15 (75%); α-Terpinene 5% DEP: n=16 (80%); 1,2,4-Trihydroxymenthane 5% pet.: n=13 (65%); α-Phellandrene 5% DEP: n=7 (35%); d-Limonene 5% DEP: n=11 (55%); Myrcene 5% DEP: n=7 (35%); Viridiflorene 5% DEP: n=1 (5%); d-Carvone 5% DEP: n=4 (20%); l-Carvone 5% DEP: n=4 (20%); Aromadendrene 5% DEP: n=1 (5%); Sabinene 5% DEP: n=2 (10%); Terpinen-4-ol 5% DEP: n=1 (5%)	118
2000 Germany	8 (20% olive oil)	Terpinolene 10% aqua: n=7 (88%); Ascaridole (5% aqua): n=7 (88%); α-Terpinene 5% aqua: n=6 (75%); α-Phellandrene 5% aqua: n= 5 (63%); 1,2,4-Trihydroxymenthane 5% pet.: n=2 (25%); d-Carvone (5% aqua): n=1 (13%); Terpinen-4-ol 10% aqua: n=1 (13%)	92
2000 Germany	15 (test conc./veh. not specified)	All were tested with 1,2,4-dihydroxymenthane and 11 (73%) reacted positively	102
1999-2000 Germany, Austria	10 (5% DEP)	Terpinolene 10% DEP: n= 10 (100%); Ascaridole 5% DEP: n=10 (100%); α-Terpinene 5% DEP: n= 10 (100%); 1,2,4-Trihydroxymenthane 5% DEP: n=9 (90%); α-Phellandrene 5% DEP: n=6 (60%); d-Limonene 5% DEP: n=4 (40%); Myrcene 5% DEP: n=1 (10%); Viridiflorene 5% DEP: n= 1 (10%)	91
1999 Germany	16 (test vehicle not mentioned)*	Terpinolene 10%: n=16 (100%); Ascaridole 5%: n=12 (75%); α-Terpinene 5%: n=11 (69%); 1,2,4-Trihydroxymenthane 5%: n=8 (50%); α-Phellandrene 5%: n= 5 (31%); Myrcene 5%: n=2 (13%); d-Limonene 5%: n=1 (6%); Viridiflorene 5%: n=1 (6%)	85
1998 Germany	1 (conc./veh. ?)	1 reaction to ascaridole; article not read	107
1997 Australia	3 (varying test concentrations)	α-Terpinene: n=1; 3 patients reacted to a sesquiterpenoid hydrocarbon fraction and sesquiterpenoid mixed with paraffin to obtain a concentration as in 25% tea tree oil	89, 94
1994 Germany	7 (1% solution)	Limonene 1% alc.: n= 6 (86%); α-Terpinene 1% alc.: n= 5 (71%); Aromadendrene 1% alc.: n=5 (71%); Terpinen-4-ol 1% and 5% alc.: n=2 (29%); p-Cymene 1% alc.: n=1 (14%); α-Phellandrene 1% alc.: n=1 (14%)	93
1992 Netherlands	1 (pure)	1,8 Cineole (eucalyptol)	83

conc.: concentration; DEP: diethyl phthalate; No.: number; veh.: vehicle

* Test concentrations were probably 5% DEP for all allergens except 1,2,4-trihydroxymenthane, which was tested in petrolatum (118)

LITERATURE

1 Satchell AC, Saurajen A, Bell C, Barnetson RS. Treatment of dandruff with 5% tea tree oil shampoo. J Am Acad Dermatol 2002;47:852-855

2 Pazyar N, Yaghoobi R, Bagherani N, Kazerouni A. A review of applications of tea tree oil in dermatology. Int J Dermatol 2013;52:784-790

3 Bhushan M, Beck MH. Allergic contact dermatitis from tea tree oil in a wart paint. Contact Dermatitis 1997;36:117-118

4 Enshaieh S, Jooya A, Siadat AH, Iraji F. The efficacy of 5% topical tea tree oil gel in mild to moderate acne vulgaris: a randomized, double-blind placebo-controlled study. Indian J Dermatol Venereol Leprol 2007;73:22-25

5 Crawford GH, Sciacca JR, James WD. Tea tree oil: cutaneous effects of the extracted oil of Melaleuca alternifolia. Dermatitis 2004;15:59-66

6 Scientific Committee on Consumer products (SCCP). Opinion on Tea Tree Oil. Adopted by the SCCP during the 18th plenary meeting of 16 December 2008. SCCP Report 1155/08. Available at: http://ec.europa.eu/health/ph_risk/committees/04_sccp/docs/sccp_o_160.pdf

7 The European Cosmetics Association Colipa. Recommendations on Tea-tree oil. Available at: https://www.cosmeticseurope.eu/publications-cosmetics-europe-association/recommendations.html?view=item&id=45%3And-12-tea-tree-oil&catid=47%3Arecommendations

8 Carson CF, Hammer KA, Riley TV. Compilation and review of published and unpublished tea tree oil literature. A report for the Rural Industries Research and Development Corporation. Barton, Australia: RIRDC Publication No 05/151, 2005, ISBN 1 74151

214 X. Available at https://rirdc.infoservices.com.au/downloads/05-151

9 Carson CF, Riley TV. Safety, efficacy and provenance of tea tree (*Melaleuca alternifolia*) oil. Contact Dermatitis 2001;45:65-67

10 Southwell IA. Tea tree oil stability and evaporation rate. An addendum to *p*-cymene and peroxides, indicators of oxidation in tea tree oil. A report for the Rural Industries Research and Development Corporation by Ian Southwell, September 2006, RIRDC Publication No 06/112, RIRDC Project No ISO-2A., Annex 9 of the dossier

11 Hausen BM. "Wundermittel" mit Tücken: Teebaumöl. Ärzt Prax Dermatol 1999:9-10

12 Beckmann B, Ippen H. Teebaum-Öl. Dermatosen 1998;46:120-124

13 Homer LE, Leach DN, Lea D, Slade Lee L, Henry RJ, Baverstock PR. Natural variation in the essential oil content of *Melaleuca alternifolia* Cheel (Myrtaceae). Bioch System Ecol 2000;28:367-282

14 Hausen BM. Kontaktallergie auf Teebaumöl und Ascaridol. Akt Derm 1998;24:60-62

15 Harkenthal M, Reichling J, Geiss HK, Saller R. Oxidationsprodukte als mögliche Ursache von Kontaktdermati-tiden. Pharmazeut Z 1998;47:4092

16 Harkenthal M, Hausen BM, Reichling J. 1,2,4-Trihydroxy menthane, a contact allergen from oxidized Australian tea tree oil. Pharmazie 2000;55:153-154

17 Brophy JJ, Craven LA, Doran JC. Melaleucas: their botany, essential oils and uses. ACIAR Monograph No. 156. Canberra: Australian Centre for International Agricultural Research, 2013. Available at: http://aciar.gov.au/publication/mn156

18 Rolli E, Marieschi M, Maietti S, Sacchetti G, Bruni R. Comparative phytotoxicity of 25 essential oils on pre- and post-emergence development of *Solanum lycopersicum* L.: A multivariate approach. Ind Crops Prod 2014;60:280-290

19 Gómez-Rincón C, Langa E, Murillo P, Valero MS, Berzosa C, López V. Activity of tea tree (*Melaleuca alternifolia*) essential oil against L3 larvae of *Anisakis simplex*. BioMed Res Int Volume 2014, Article ID 549510, 6 pages. http://dx.doi.org/10.1155/2014/549510

20 Baldissera MD, Da Silva AS, Oliveira CB, Santos RCV, Vaucher RA, Raffin RP, et al. Trypanocidal action of tea tree oil (*Melaleuca alternifolia*) against *Trypanosoma evansi in vitro* and *in vivo* used mice as experimental model. Exp Parasitol 2014;141:21-27

21 Yang J-Y, Cho K-S, Chung N-H, Kim C-H, Suh J-W, Lee H-S. Constituents of volatile compounds derived from *Melaleuca alternifolia* leaf oil and acaricidal toxicities against house dust mites. J Korean Soc Appl Biol Chem 2013;56:91-94

22 Thomsen NA, Hammer KA, Riley TV, Van Belkum A, Carson CF. Effect of habituation to tea tree (*Melaleuca alternifolia*) oil on the subsequent susceptibility of *Staphylococcus* spp. to antimicrobials, triclosan, tea tree oil, terpinen-4-ol and carvacrol. Int J Antimicr Agents 2013;4:343-351

23 Pereira TS, Rochade Sant'Anna J, Leite Silva E, Lelis Pinheiro A, Alves de Castro-Prado MA. *In vitro* genotoxicity of *Melaleuca alternifolia* essential oil in human lymphocytes. J Ethnopharmacol 2014;151:852-857

24 Lee C-J, Chen L-W, Chen L-G, Chang T-L, Huang C-W, Huang M-C, Wang C-C. Correlations of the components of tea tree oil with its antibacterial effects and skin irritation. J Food Drug Anal 2013;21:169-176

25 European Medicines Agency. Assessment report on *Melaleuca alternifolia* (Maiden and Betch) Cheel, *M. linariifolia* Smith, *M. dissitiflora* F. Mueller and/or other species of *Melaleuca*, aetheroleum. Committee on Herbal Medicinal Products (HMPC) 2013, EMA/HMPC/320932/2012

26 Benelli G, Canale A, Flamini G, Cioni PL, Demia F, Ceccarini L, et al. Biotoxicity of *Melaleuca alternifolia* (Myrtaceae) essential oil against the Mediterranean fruit fly, *Ceratitis capitata* (Diptera: Tephritidae), and its parasitoid *Psyttalia concolor* (Hymenoptera: Braconidae). Ind Crops Prod 2013;50:596-603

27 Almeida Barbosa LC, Silva CJ, Teixeira RR, Strozi Alves Meira RM, Lelis Pinheiro A. Chemistry and biological activities of essential oils from *Melaleuca* L. species. Agriculturae Conspectus Scientificus 2013;78:11-23

28 Kulkarni A, Jan N, Nimbarte S. Monitoring of antimicrobial effect of GC-MS standardized *Melaleuca alternifolia* oil (tea tree oil) on multidrug resistant uropathogens. IOSR J Pharm Biol Sci (IOSRJPBS) 2012;2(2):6-14

29 Callander JT, James PJ. Insecticidal and repellent effects of tea tree (*Melaleuca alternifolia*) oil against *Lucilia cuprina*. Vet Parasitol 2012;184:271-278

30 Noumi E, Snoussi M, Hajlaoui H, Trabelsi N, Ksouri R, Valentin E, et al. Chemical composition, antioxidant and antifungal potential of *Melaleuca alternifolia* (tea tree) and *Eucalyptus globulus* essential oils against oral *Candida* species. J Med Plants Res 2011;5:4147-4156

31 Sciarrone D, Ragonese C, Carnovale C, Piperno A, Dugo P, Dugo G, Mondello L. Evaluation of tea tree oil quality and ascaridole: A deep study by means of chiral and multi heart-cuts multidimensional gas chromatography system coupled to mass spectrometry detection. J Chromatogr A 2010;1217:6422-6427

32 Nardoni S, Bertoli A, Pinto L, Mancianti F, Pisseri F, Pistelli L. *In vitro* effectiveness of tea tree oil against *Trichophyton equinum*. Journal de Mycologie Médicale 2010;20:75-79

33 Angelini P, Pagiotti R, Granetti B. Effect of antimicrobial activity of *Melaleuca alternifolia* essential oil on antagonistic potential of *Pleurotus* species against *Trichoderma harzianum* in dual culture. World J Microbiol Biotechnol 2008;24:197-202

34 Silva CJ, Barbosa LCA, Maltha CRA, Pinheiro AL, Ismail FMD. Comparative study of the essential oils of seven *Melaleuca* (Myrtaceae) species grown in Brazil. Flavour Fragr J 2007;22:474-478

35 Morgan TJ, Morden WE, Al-muhareb E, Herod AA, Kandiyoti R. Essential oils investigated by size exclusion chromatography and gas chromatography-mass spectrometry. Energy & Fuels 2006;20:734-737

36 Caldefie-Chézet F, Fusillier C, Jarde T, Laroye H, Damez H, Vasson M-P, Guillot J. Potential anti-inflammatory

effects of *Melaleuca alternifolia* essential oil on human peripheral blood leukocytes. Phytother Res 2006;20:364-370

37 Carson CF, Hammer KA, Riley TV. *Melaleuca alternifolia* (tea tree) oil, a review of antimicrobial and other medicinal properties. Clin Microbiol Rev 2006;19:50-62

38 Ansari SH, Mukhtar HM, Mir SR, Abdin MZ, Singh P. Analysis of Indian tea tree (*Melaleuca alternifolia* cheel.) essential oil, a vital aromatherapy oil. J Essent Oil Bear Plants 2006;9:70-74

39 Iori A, Grazioli D, Gentile E, Marano G, Salvatore G. Acaricidal properties of the essential oil of *Melaleuca alternifolia* Cheel (tea tree oil) against nymphs of *Ixodes ricinus*. Vet Parasitol 2005;129:173-176

40 Hartford O, Zug KA. Tea tree oil. Cutis 2005;76:178-180

41 Shellie R, Marriott P, Zappia G, Mondello L, Dugo G. Interactive use of linear retention indices on polar and apolar columns with an MS-library for reliable characterization of Australian tea tree and other *Melaleuca* sp. oils. J Essent Oil Res 2003;15:305-312

42 Hammer KA, Carson CF, Riley TV. Antifungal activity of the components of *Melaleuca alternifolia* (tea tree) oil. J Appl Microbiol 2003;95:853-860

43 Lee LS, Brooks LO, Homer LE, Rossetto M, Henry RJ, Baverstock BR. Geographic variation in the essential oils and morphology of natural populations of *Melaleuca alternifolia* (Myrtaceae). Biochem System Ecol 2002;30:343-360

44 Caboi F, Murgia S, Monduzzi M, Lazzari P. NMR investigation on *Melaleuca alternifolia* essential oil dispersed in the monoolein aqueous system: phase behavior and dynamics. Langmuir 2002;18:7916-7922

45 Christoph F, Stahl-Biskup E, Kaulfers P-M. Death kinetics of *Staphylococcus aureus* exposed to commercial tea tree oils s.l. J Essent Oil Res 2001;13:98-102

46 Cox SD, Mann CM, Markham JL. Interactions between components of the essential oil of *Melaleuca alternifolia*. J Appl Microbiol 2001;91:492-497

47 Hart PH, Brand C, Carson CF, Riley TV, Prager RH, Finlay-Jones JJ. Terpinen-4-ol, the main component of the essential oil of *Melaleuca alternifolia* (tea tree oil), suppresses inflammatory mediator production by activated human monocytes. Inflamm Res 2000;49:619-626

48 Cornwell CP, Leach DN, Wyllie SG. The origin of terpinen-4-ol in the steam distillates of *Melaleuca argentea*, *M. dissitiflora* and *M. linariifolia*. J Essent Oil Res 1999;11:49-53

49 Verghese J, Jacob CV, Kunjunni Kartha CV, McCarron M, Mills AJ, Whittaker D. Indian tea tree (*Melaleuca alternifolia* Cheel) essential oil. Flavour Fragr J 1996;11:219-221

50 Bishop CD, Thornton IB. Evaluation of the antifungal activity of the essential oils of *Monarda citriodora* var. *citriodora* and *Melaleuca alternifolia* on post-harvest pathogens. J Essent Oil Res 1997;9:77-82

51 Cornwell CP, Leach DN, Wyllie SG. Incorporation of oxygen-18 into terpinen-4-ol from the H_2 ^{18}O steam distillates of *Melaleuca alternifolia* (tea tree). J Essent Oil Res 1995;7:613-620

52 Williamson LR, Lusunzi I. Essential oil from *Melaleuca dissitiflora*: a potential source of high quality tea tree oil. Ind Crops Prod 1994;2:211-217

53 Leach DN, Wyllie SG, Hall JG, Kyratzis I. Enantiomeric composition of the principal components of the oil of *Melaleuca alternifolia*. J Agric Food Chem 1993;41:1627-1632

54 Southwell IA, Stiff IA, Brophy JJ. Terpinolene varieties of *Melaleuca*. J Essent Oil Res 1992;4:363-367

55 Murtagh GJ, Curtis A. Post-harvest retention of oil in tea tree foliage. J Essent Oil Res 1991;3:179-184

56 Southwell IA, Stiff IA. Differentiation between *Melaleuca alternifolia* and *M. linariifolia* by monoterpenoid comparison. Phytochem 1990;29:3529-3533

57 Brophy JJ, Davies NW, Southwell IA. Gas chromatographic quality control for oil of *Melaleuca* terpinen-4-ol type (Australian tea tree). J Agric Food Chem 1989;37:1330-1335

58 Lawrence BM. Progress in essential oils. Perfum Flavor 1989;14(May/June):71-80

59 Tranchida PQ, Shellie RA, Purcaro G, Conte LS, Dugo P, Dugo G, et al. Analysis of fresh and aged tea tree essential oils by using GCxGCC-qMS. J Chromatogr Sci 2010;48:262-266

60 Southwell I. *p*-Cymene and organic peroxides as indicators of oxidation in tea tree oil. A report for the Rural Industries Research and Development Corporation. Australian Government, Rural Industries Research and Development Corporation, 2006. RIRDC Publication No 06/112, RIRDC Project No ISO-2A. Available at: https://rirdc.infoservices.com.au/downloads/06-112

61 Lawrence BM. Progress in essential oils. Perfum Flavor 2012;37(April):56-62

62 Jirovetz L, Buchbauer G, Denkova Z, Stoyanova A, Murgov I, Schmidt E, Geissler M. Antimicrobial testing and gas chromatographic analysis of pure oxygenated monoterpenes 1,8-cineole, α-terpineol, terpinen-4-ol and camphor as well as target compounds in essential oils of pine (*Pinus pinaster*), rosemary (*Rosmarinus officinalis*), tea tree (*Melaleuca alternifolia*). Sci Pharm 2005;73:27-39. Data cited in ref. 61

63 Evandri MG, Battinelli L, Daniele C, Mastrangelo S, Bolle P, Mazzanti G. The antimutagenic activity of *Lavandula angustifolia* (lavender) essential oil in the bacterial reverse mutation assay. Food Chem Toxicol 2005;43:1381-1387

64 Williams DG. The chemistry of essential oils, 2nd Ed. Port Washington, NY, USA: Micelle Press, 2008. Data cited in ref. 61

65 Milchard MJ, Clery R, Esdale R, Gates L, Judge F, Moss N, et al. Application of gas-liquid chromatography to the analysis of essential oils. Perfum Flavor 2010;35(5):34-42

66 Lawrence BM. Progress in essential oils. Perfum Flavor 2006;31(4):52-? (last page unknown)

67 Kubeczka K-H, Formacek V. Essential oils analysis by capillary gas chromatography and carbon-13 NMR spectroscopy, 2nd edition. New York, USA: John Wiley and Sons, 2002:349-354. Data cited in ref. 66

68 Hayes AJ, Leach DN, Markham JL, Markovic B. In vitro cytotoxicity of Australian tea tree oil using human cell lines. J Essent Oil Res 1997;9:575-582

69 Lawrence BM. Progress in essential oils. Perfum Flavor 2001;26(6):44-? (last page unknown)

70 de Figueiredo M. Chemical composition and oil concentration of tea tree leaf oil grown in South Africa during one vegetative cycle. J Essent Oil Res (Special Edition) 2006;18:52-53. Data cited in ref. 61

71 Miyazawa M, Yamafuji C. Inhibition of acetylcholinesterase activity by tea tree oil and constituent terpenoids. Flavour Fragr J 2006;21:198-201

72 Hayes AJ, Markovic B. Toxicity of Australian essential oil *Backhousia citriodora* (lemon myrtle). Part 1. Antimicrobial activity and in vitro cytotoxicity. Food Chem Toxicol 2002;40:535-543

73 Murbach Teles Andrade BF, Nunes Barbosa L, da Silva Probst I, Fernandes A Júnior. Antimicrobial activity of essential oils. J Essent Oil Res 2014;26:34-40

74 Rhind JP. Essential oils. A handbook for aromatherapy practice, 2nd Edition. London: Singing Dragon, 2012

75 Lawless J. The encyclopedia of essential oils, 2nd Edition. London: Harper Thorsons, 2014

76 Nardelli A, D'Hooge E, Drieghe J, Dooms M, Goossens A. Allergic contact dermatitis from fragrance components in specific topical pharmaceutical products in Belgium. Contact Dermatitis 2009;60:303-313

77 Travassos AR, Claes L, Boey L, Drieghe J, Goossens A. Non-fragrance allergens in specific cosmetic products. Contact Dermatitis 2011;65:276-285

78 Wetter DA, Yiannias JA, Prakash AV, Davis MD, Farmer SA, el-Azhary RA, et al. Results of patch testing to personal care product allergens in a standard series and a supplemental cosmetic series: an analysis of 945 patients from the Mayo Clinic Contact Dermatitis Group, 2000-2007. J Am Acad Dermatol 2010;63:789-798

79 Lindberg M, Tammela M, Bostrom A, Fischer T, Inerot A, Sundberg K, Berne B. Are adverse skin reactions to cosmetics underestimated in the clinical assessment of contact dermatitis? A prospective study among 1075 patients attending Swedish patch test clinics. Acta Derm Venereol 2004;84:291-295

80 Thomson KF, Wilkinson SM. Allergic contact dermatitis to plant extracts in patients with cosmetic dermatitis. Br J Dermatol 2000;142:84-88

81 Knight TE, Hausen BM. Melaleuca oil (tea-tree oil) dermatitis. J Am Acad Dermatol 1994;30:423-427

82 Warshaw EM, Belsito DV, Taylor JS, Sasseville D, DeKoven JG, Zirwas MJ, et al. North American Contact Dermatitis Group patch test results: 2009 to 2010. Dermatitis 2013;24:50-59

83 de Groot AC, Weijland JW. Systemic contact dermatitis from tea tree oil. Contact Dermatitis 1992;27:279-280

84 Rutherford T, Nixon R, Tam M, Tate B. Allergy to tea tree oil: retrospective review of 41 cases with positive patch tests over 4.5 years. Australas J Dermatol 2007;48:83-87

85 Hausen BM, Reichling J, Harkenthal M. Degradation products of monoterpenes are the sensitizing agents in tea tree oil. Am J Cont Derm 1999;10:68-77

86 Apted JH. Contact dermatitis associated with the use of tea-tree oil. Australas J Dermatol 1991;32:177

87 Satchell AC, Saurajen A, Bell C, Barnetson RS. Treatment of interdigital tinea pedis with 25% and 50% tea tree oil solution: a randomized, placebo-controlled, blinded study. Australas J Dermatol 2002;43:175-178

88 Greig JE, Carson CF, Stuckey MS, Riley TV. Skin sensitivity testing for tea tree oil – A Report for the Rural Industries Research and Development Corporation. Rural Industries Research and Development Corporation, Barton Act Australia Report no. 99. 1999. Available at: https://rirdc.infoservices.com.au/items/99-102

89 Southwell I, Freeman S, Rubel DM. Skin irritancy of tea tree oil. J Essent Oil Res 1997;9:47-52

90 Lisi P, Meligeni L, Pigatto P, et al. The prevalence of sensitivity to melaleuca essential oil. It Ann Clin Exp Allergol Dermat 2000;54:141-144

91 Pirker C, Hausen BM, Uter W, Hillen U, Brasch J, Bayerl C, et al. Sensitization to tea tree oil in Germany and Austria. A multicenter study of the German Contact Dermatitis group. J Dtsch Dermatol Ges 2003;1:629-634

92 Lippert U, Walter A, Hausen BM, Fuchs Th. Increasing incidence of contact dermatitis to tea tree oil. J Allergy Clin Immunol 2000;105;S43 (abstract 127)

93 Knight TE, Hausen BM. Melaleuca oil (tea tree oil) dermatitis. J Am Acad Dermatol 1994;30:423-427

94 Rubel DM, Freeman S, Southwell I. Tea tree oil allergy: what is the offending agent? Report of three cases of tea tree oil allergy and review of the literature. Australas J Dermatol 1998;39:244-247

95 Perrett CM, Evans AV, Russell-Jones R. Tea tree oil dermatitis associated with linear IgA disease. Clin Exp Dermatol 2003;28:167-170

96 Fritz TM, Burg G, Krasovec M. Allergic contact dermatitis to cosmetics containing *Melaleuca alternifolia* (tea tree) oil. Ann Dermatol Venereol 2001;128:123-126

97 Van der Valk PG, de Groot AC, Bruynzeel DP, Coenraads PJ, Weijland JW. Allergic contact eczema due to tea tree oil. Ned Tijdschr Geneeskd 1994;138:823-825 (in Dutch)

98 Zug KA, Warshaw EM, Fowler JF Jr, Maibach HI, Belsito DL, Pratt MD, et al. Patch-test results of the North American Contact Dermatitis Group 2005-2006. Dermatitis 2009;20:149-160

99 Fransway AF, Zug KA, Belsito DV, DeLeo VA, Fowler JF Jr, Maibach HI, et al. North American Contact Dermatitis Group patch test results for 2007-2008. Dermatitis 2013;24:10-21

100 Warshaw EM, Belsito DV, DeLeo VA, Fowler JF Jr, Maibach HI, Marks JG, et al. North American Contact Dermatitis Group patch-test results, 2003-2004 study period. Dermatitis 2008;19:129-136

101 Crawford GH, Sciacca JR, James WD. Tea tree oil: cutaneous effects of the extracted oil of *Melaleuca alternifolia*. Dermatitis 2004;15:59-66

102 Harkenthal M, Hausen BM, Reichling J. 1,2,4-Trihydroxymenthane, a contact allergen from oxidized Australian tea tree oil. Pharmazie 2000;55:153-154

103 Bhushan M, Beck MH. Allergic contact dermatitis from tea tree oil in a wart paint. Contact Dermatitis 1997;36:117-118

104 Williams JD, Nixon RL, Lee A. Recurrent allergic contact dermatitis due to allergen transfer by sunglasses. Contact Dermatitis 2007;57:120-121

105 Fransway A. Allergy to oil of Melaleuca: Report of 12 cases. American Contact Dermatitis Society Meeting. 9.02.1996, Washington DC, USA (abstract)

106 Selvaag E, Holm J, Thune P. Allergic contact dermatitis in an aromatherapist with multiple sensitizations to essential oils. Contact Dermatitis 1995;33:354-355

107 Hausen BM. Kontaktallergie auf Teebaumöl und Ascaridol. Akt Derm 1998;24:60-62

108 Fritz TM, Elmer P. Allergisches Kontaktekzem auf Teebaumöl bei einer Patientin mit Psoriasis. Akt Derm 1998;24:7-10

109 Hackzell-Bradley M, Bradley T, Fischer T. Kontaktallergi av 'tea tree-oil'. Lakartidningen 1997;94:4359-4361

110 Kränke B. Allergisierende Potenz von Teebaumöl. Hautarzt 1997;48:203-204

111 Varma S, Blackford S, Statham BN, Blackwell A. Combined contact allergy to tea tree oil and lavender oil complicating chronic vulvovaginitis. Contact Dermatitis 2000;42:309-310

112 Khanna M, Qasem K, Sasseville D. Allergic contact dermatitis to tea tree oil with erythema multiforme-like id reaction. Dermatitis 2000;11:238-242

113 de Groot AC. Airborne allergic contact dermatitis from tea tree oil. Contact Dermatitis 1996;35:304-305

114 Veien NK, Rosner K, Skovgaard GL. Is tea tree oil an important contact allergen? Contact Dermatitis 2004;50:378-379

115 Scientific Committee on Consumer products (SCCP). Opinion on Tea Tree Oil. Adopted by the SCCP during the 18th plenary meeting of 16 December 2008. SCCP Report 1155/08. Available at: http://ec.europa.eu/health/ph_risk/committees/04_sccp/docs/sccp_o_160.pdf

116 Anonymous. Human studies Draize method, study no. DT-029. Skin & Cancer Foundation, Australia, 1997

117 Kütting B, Brehler R, Traupe H. Allergic contact dermatitis in children – strategies of prevention and risk management. Eur J Dermatol 2004;14:80-85

118 Hausen BM. Evaluation of the main contact allergens in oxidized tea tree oil. Dermatitis 2004;15:213-214

119 Anonymous. Tea Tree Oil (TTO) Monograph on active ingredient being used in cosmetic products, prepared by the Norwegian delegation to the Council of Europe Committee of experts on cosmetic products, 2001, RD 4-3/35.

120 Aspres N, Freeman S. Predictive testing for irritancy and allergenicity of tea tree oil in normal human subjects. Exogen Dermatol 2003;2:258-261

121 Hammer KA, Carson CF, Riley TV, Nielsen JB. A review of the toxicity of Melaleuca alternifolia (tea tree) oil. Food Chem Toxicol 2006;44:616-625

122 Coutts I, Shaw S, Orton D. Patch testing with pure tea tree oil—12 months experience. Br J Dermatol 2002;147(Suppl. 62):70

123 Reindl H, Gall H, Hausen BM, Peter U. Akutes Kontaktekzem nach Anwendung von Teebaumöl. Allergo J 2000;9:100-103

124 Dharmagunawardena B, Takwale A, Sanders KJ, Cannan S, Roger A, Ilchyshyn A. Gas chromatography:

125 Monthrope Y, Shaw J. A 'natural' dermatitis: Contact allergy to tea tree oil. Univ Toronto Med J 2004;82:59-60

126 Stonehouse A, Studdiford J. Allergic contact dermatitis from tea tree oil. Department of Family & Community Medicine Faculty Papers. Paper 12:2007 (not further specified, cited by Posadzki P, Alotaibi A, Ernst E. Adverse effects of aromatherapy: A systematic review of case reports and case series. Int J Risk Saf Med 2012;24:147-161)

127 Rudbäck J, Andresen Bergström M, Börje A, Nilsson U, Karlberg AT. α-Terpinene, an antioxidant in tea tree oil, autoxidizes rapidly to skin allergens on air exposure. Chem Res Toxicol 2012;25:713-721

128 Newsham J, Rai S, Williams JDL. Two cases of allergic contact dermatitis to neroli oil. Br J Dermatol 2011;165(Suppl.1):76

129 Belsito DV, Fowler JF Jr, Sasseville D, Marks JG Jr, De Leo VA, Storrs FJ. Delayed-type hypersensitivity to fragrance materials in a select North American population. Dermatitis 2006;17:23-28

130 Christoffers WA, Blömeke B, Coenraads P-J, Schuttelaar M-LA. The optimal patch test concentration for ascaridole as a sensitizing component of tea tree oil. Contact Dermatitis 2014;71:129-137

131 Santesteban R, Loidi L, Agulló A, Hervella M, Larrea M, Yanguas I. Allergic contact dermatitis to tea tree oil. Contact Dermatitis 2014;70 (Suppl.1):102

132 Corazza M, Borghi A, Gallo R, Schena D, Pigatto P, Lauriola MM, et al. Topical botanically derived products: use, skin reactions, and usefulness of patch tests. A multicentre Italian study. Contact Dermatitis 2014;70:90-97

133 Christoffers WA, Blömeke B, Coenraads P-J, Schuttelaar M-LA. Co-sensitization to ascaridole and tea tree oil. Contact Dermatitis 2013;69:187-189

134 Scardamaglia L, Nixon R, Fewings J. Compound tincture of benzoin: a common contact allergen? Australas J Dermatol 2005;44:180-184

135 Warshaw EM, Maibach HI, Taylor JS, Sasseville D, DeKoven JG, Zirwas MJ, et al. North American Contact Dermatitis Group patch test results: 2011-2012. Dermatitis 2015;26:49-59

136 Selvaag E, Eriksen B, Thune P. Contact allergy due to tea tree oil and cross-sensitization to colophony. Contact Dermatitis 1994;31:124-125

137 Nardelli A, Drieghe J, Claes L, Boey L, Goossens A. Fragrance allergens in 'specific' cosmetic products. Contact Dermatitis 2011;64:212-219

138 Toholka R, Wang Y-S, Tate B, Tam M, Cahill J, Palmer A, Nixon R. The first Australian Baseline Series: Recommendations for patch testing in suspected contact dermatitis. Australas J Dermatol 2014, Sept. 7. doi: 10.1111/ajd.12186

139 Pesonen M, Suomela S, Kuuliala O, Henriks-Eckerman M-L, Aalto-Korte K. Occupational contact dermatitis caused by D-limonene. Contact Dermatitis 2014;71:273-279

Chapter 5.84   THUJA OIL

DEFINITION
Thuja oil is the essential oil obtained from the twigs with leaves of the (northern) white cedar, *Thuja occidentalis* L. and from other *Thuja* species.

INCI NOMENCLATURE
Description/definition: Thuja occidentalis leaf oil is the volatile oil expressed from the leaves and twigs of the thuja, *Thuja occidentalis* L., Cupressaceae
INCI name EU & USA: Thuja occidentalis leaf oil
Other names: Cedar leaf oil
CAS registry number(s): 90131-58-1; 8007-20-3
EINECS number(s): 290-370-1

ISO (INTERNATIONAL ORGANIZATION FOR STANDARDIZATION) STANDARD
There is currently no ISO standard for thuja oil.

Thuja leaf oil may also be obtained from other species of the genus *Thuja*, including *Thuja plicata* (CAS 94334-32-4; EINECS 305-105-8).

THE PLANT, THE OIL, AND THEIR USES
Thuja occidentalis L. is an evergreen coniferous tree which grows up to 25 m tall and 1 m in diameter. It is native to Canada and the USA and is cultivated in China, Korea, Russia (European part), and Europe, often for ornamental purposes (GRIN Taxonomy for Plants; www.conifers.org, 7). One of its common names is 'Arborvitae', which is particularly used in the horticultural trade in the United States and is Latin for 'tree of life'—due to the supposed medicinal properties of the sap, bark and twigs. In folk medicine, *Thuja occidentalis* has been used to treat bronchial catarrh, enuresis, cystitis, psoriasis, uterine carcinomas, amenorrhea and rheumatism (7). It has a myriad of other applications in Western herbal medicine, traditional Chinese medicine, homeopathy and aromatherapy. The pharmaceutical, pharmacological and clinical properties of *Thuja occidentalis* (products) have been reviewed (8,9).

Cedar leaf essential oil, made from the twigs with leaves of *Thuja occidentalis L.* by steam distillation or hydrodistillation, has been used for perfumes, cleansers, disinfectants, hair preparations, room sprays, deodorants and soft soaps (7). It is also approved for use in food in the USA and the EU (10). Some aromatherapists recommend the oil for treating acne and dandruff. It is still used in some mainstream over-the-counter preparations to relieve congestion in the upper respiratory tract,

the best-known of which is Vicks VapoRub™. Thuja leaf oil is also added to pest repellent sprays and paints to protect against mites, moths, and rodents (9). The major constituent of the oil of *T. occidentalis* foliage, thujone, is used pharmacologically as an active ingredient in the production of nasal decongestants and cough suppressants, perfumes, shoe polishes and soaps (3,7). It is, however, considered too toxic (due to the high thujone content) for aromatherapy practices (19,20,21).

CHEMICAL COMPOSITION
Thuja oil is a clear, colorless to slightly greenish mobile liquid, which has a strong, fresh camphoraceous and greenish odor. The yield of essential oil from the leaves of *Thuja occidentalis* L. generally varies from 0.9 to 1.5 per cent. The main producing countries of this oil are the USA and Canada.

Literature data (up to December 5, 2014) on the chemical composition of thuja oils and unpublished analytical data from one of us (E.S.) are shown in Table 5.84.1 in alphabetical order. In thuja oils from various origins, over 120 chemicals have been identified. About 42 per cent of these were found in a single reviewed publication only.

The major compounds found in thuja oils from different sources are shown in Table 5.84.2. They include (highest concentrations in any study given) α-thujone (69.8%), α-pinene (35.5%), δ3-carene (33.2%), fenchone (15.7%), sabinene (13.9%), β-thujone (12.2%), bornyl acetate (6.3%) and terpinen-4-ol (5.4%). Well-known ingredients of thuja oils that were present in high concentrations (>7%) in one or two studies were limonene (27.6%), beyerene (15.0%), rimuene (7.9%) and α-terpinyl acetate (7.0%). Uncommon or rare constituents of thuja oils found in high concentrations (>7%) in single studies include sabinyl acetate (16.7%), cedrol (10.3%) and α-cedrol (9.1%).

Commercial oils
The ten chemicals that had the highest maximum concentrations in 44 commercial thuja essential oil samples (concentration ranges provided) are the following: α-thujone (47.8-66.3%), fenchone (12.4-15.7%), β-thujone (*cis*-thujone) (6.8-10.8%), δ3-carene (0.7-8.2%), sabinene (1.7-4.4%), α-pinene (1.4-3.6%), terpinen-4-ol (1.5-3.4%), (Z)-3-hexenol (0.07-3.0%), limonene (1.1-2.7%), and bornyl acetate (0.9-2.5%) (Erich Schmidt, unpublished analytical data).

An additional analysis of steam-distilled *T. occidentalis* oil not discussed in Table 5.84.1 can be found in ref. 17.

Table 5.84.1 Constituents identified in thuja oils

Constituent	CAS	Percentage and range					
		A	B (3)	C (1)	D (2)	E (7)	F
Abietadiene	36312-33-1		tr-0.6		tr-0.09		
Abietal	6704-50-3		0-0.09		tr-0.08		
Abietatriene	19407-28-4		tr-0.09		tr-1.0		<0.05[g]
trans-α-Bergamotene	13474-59-4						tr[c]
Beyerene	3564-54-3		1.3-6.0		4.3-15.0	11.2	1.7[d]
(-)-15-Beyerene	2359-73-1			8.3-13.7			+[j]
(+)-Beyeren-19-ol						1.5	
Bicyclogermacrene	24703-35-3						0.2[c]
β-Bisabolene	495-61-4						0.1[c]
2-Bornene (= α-)	464-17-5						0.1[c]
Borneol	507-70-0	0.1-0.2				0.3	0.2[b]; 0.3[f]; 0.4[g]
Bornyl acetate	76-49-3	0.9-2.5	1.2-4.4	2.8-6.3	1.2-2.2	2.5	0.8[d]; 2.3[f]; 3.0[g]; 3.2[e]; 5.9[l]
α-Bulnesene	3691-11-0						tr[c]
γ-Cadinene	39029-41-9						0.2[b]
trans-γ-Cadinene							0.2[c]
δ-Cadinene	483-76-1		tr-0.07		0.05-0.2[a]	0.1	<0.02[d]; 0.3[c]; 1.3[b]
α-Cadinol	481-34-5						0.02[d]; 0.2[c,e]
epi-α-Cadinol	5937-11-1						tr[c]
δ-Cadinol	19435-97-3		tr-0.6		0.05-0.2[a]		
Camphene	79-92-5	0.07-2.5	0.4-1.8	0.6-1.2	0.4-0.8	2.5	0.6[g]; 2.0[l]; 2.5[c]; 2.6[b]; 3.6[f]
Camphene hydrate	465-31-6	0.2-0.3					0.2[b]
α-Campholenal	4501-58-0						0.1[g]
Camphor	76-22-2	0.02-2.1	1.5-3.3	0.3-2.4	1.0-3.0	4.5	1.2[b]; 1.5[d]; 2.2[e]; 2.5[l]; 2.6[f]
δ2-Carene (= δ4-)	554-61-0						1.3[l]
δ3-Carene	13466-78-9	0.7-8.2					0.1[h]; 1.0[b]; 25.4[k]; 33.2[c]
Carvacrol	499-75-2					0.2	
Carveol	99-48-9						0.1[e]
(*E*)-Carveol	1197-07-5		tr-0.3				
Carvone	99-49-0					0.3	
β-Caryophyllene	87-44-5	0.1-0.2	0.1-0.3				1.2[b]; 1.8[c]
Caryophyllene oxide	1139-30-6		0.2-0.6	0.8-2.4	0.05-0.2	1.4	0.1[b]; 0.2[c]; 0.3[e]; 0.9[g]
Cedrol	77-53-2						10.3[c]
α-Cedrol	77-53-2						9.1[k]
1,8-Cineole	470-82-6	0.04-0.1					tr[c]; 0.08[b]
(*E*)-Cinnamyl acetate	21040-45-9					0.08	0.1[g]
Cuminaldehyde	122-03-2					0.07	
Cuminyl alcohol	536-60-7						0.08[e]
Cyclofenchone	488-97-1					0.4	
p-Cymene	99-87-6	0.6-1.5	0.07-1.0	0.2-0.8	0.3-0.6	0.2	0.3[g]; 0.8[d]; 0.9[e]; 1.3[f]; 2.4[b]
p-Cymen-8-ol	1197-01-9	0.1-0.2				0.3	0.1[b,g]; 0.2[e]; 2.7[d]
β-Elemene	33880-83-0	0.05-0.3					0.1[b]
Elemol (α-)	639-99-6	0.2-1.3					0.8[b]
Ethyl cinnamate	103-36-6			0-0.1			<0.02[d]
Ethyl hexanoate	123-66-0						<0.05[g]
Ethyl isovalerate	108-64-5	0.03-0.08					0.01[b]
Ethyl-2-methyl butyrate	7452-79-1	0.03-0.2					0.05[b]; 1.3[e]
Ethyl pentanoate	539-82-2						
Ethyl senecionate							0.05[e]
Fenchene	471-84-1		0.3-1.9		0.4-0.8		1.1[e]
α-Fenchene	471-84-1	0.09-1.4					<0.02[d]; 0.5[g]; 1.5[f]; 2.0[b]
β-Fenchene	497-32-5						<0.01[b]
Fenchone	1195-79-5	12.4-15.7	9.0-15.0	6.7-11.1	6.2-13.0	4.2	7.8[d]; 12.8[e]; 12.9[b]; 14.6[f]
Fenchyl acetate	13851-11-1	0.2-1.3	0.4-0.5				0.3[b]; 0.4[e]; 1.1[d]
α-Fenchyl acetate (endo-)	111821-74-0			0.2-0.5	0.1-0.3	1.0	0.2[g]
Fenchyl alcohol	1632-73-1						0.2[b]
β-Funebrene	79120-98-2						0.4[c]
Geranial	141-27-5						0.3[c]
Geranyl acetate	105-87-3		0.4-1.0	0-0.2	0.6-1.1	0.5	0.05[e]
Germacrene D	23986-74-5						0.7[c]
Globulol	489-41-8						0.9[c]
Hexanol (1-; n-)	111-27-3	0.01-1.3					

Table 5.84.1 Constituents identified in thuja oils (*continued*)

Constituent	CAS	Percentage and range					
		A	B (3)	C (1)	D (2)	E (7)	F
(Z)-3-Hexen-1-ol	928-96-1	0.07-3.0					0.1[g]
α-Humulene	6753-98-6						0.2[b]; 1.3[c]
Humulene epoxide II	19888-34-7						0.2[g]
Humulene oxide	96638-51-6						<0.01[b]
Hydroxyisopimarene			tr		tr-0.09		
Isoborneol	124-76-5						<0.01[b]
Isobornyl acetate	125-12-2						0.5[c]; 0.8[d]
Isokaurene	5947-50-2					0.1	
Limonene	138-86-3	1.2-2.7	1.0-3.2	0.8-1.4	0.6-1.2	3.2	1.7[f]; 2.4[b]; 3.1[c]; 3.6[l]; 27.6[h]
Linalool	78-70-6	0.05-0.9					0.2[c]; 0.3[f]; 1.9[b]
Linalyl acetate	115-95-7						0.3[e]; 1.2[b]
p-Mentha-2,4(8)-diene	586-63-0						0.5[c]
cis-p-Menth-2-en-1-ol	29803-82-5		0.4-0.6		0.2-0.6		tr[c]; 0.1[d,e]; 0.3[g]
Menthol	89-78-1			0-0.8			
Methylcarvacrol	6379-73-3	0.03-0.09		tr-0.2			0.1[e]
Methyl chavicol	140-67-0	0.2-0.5					
τ-Muurolol (epi-α-)	19912-62-0					0.08	
Myrcene	123-35-3	1.3-2.2	tr-2.0	0.7-2.2	0.6-1.1	1.2	1.4[f]; 1.8[e]; 2.1[c]; 4.1[b]; 4.7[h]
Neryl acetate	141-12-8						tr[c]
(E)-β-Ocimene	3779-61-1						<0.01[b]; 0.1[c]
(Z)-β-Ocimene	3338-55-4						tr[c]; <0.01[b]; 0.4[h]
3-Octenyl acetate	7380-48-5						tr[c]
α-Phellandrene	99-83-2	0.04-0.3					tr[c]; <0.05[g]; 0.1[b,d]
β-Phellandrene	555-10-2	0.2-0.3	tr-0.5				0.1[e]; 0.4[h]; 1.7[b]
α-Pinene	80-56-8	1.4-3.6	1.0-5.0	1.9-2.9	0.5-1.3	1.5	3.3[b,f]; 10.0[h]; 27.7[c]; 35.5[k]
β-Pinene	127-91-3	0.1-0.3	tr-0.3	0.1	tr-0.09		0.1[g]; 0.2[d,f]; 1.1[b]; 1.5[c]
trans-Pinocarveol	1674-08-4						tr[c]
Pinocarvone	30460-92-5						tr[c]
trans-Piperitol	16721-39-4					0.2	
Piperitone	89-81-6					0.1	
Pulegone	89-82-7						0.1[d]
Rimuene	1686-67-5		0.6-1.9	1.5-6.6	1.6-7.9	5.6	0.7[d]; 0.8[g]
Sabina ketone	513-20-2					0.2	0.2[g]
Sabinene	3387-41-5	1.7-4.4	2.1-5.9	3.2-7.8	2.7-5.0	4.6	4.1[g]; 5.0[i]; 12.1[b]; 13.9[h]
cis-Sabinene hydrate	15537-55-0						tr[c]; 0.1[b]; 0.4[g]
trans-Sabinene hydrate	17699-16-0						0.03[d]; 1.1[b]
cis-Sabinol	3310-02-9						0.3[b]
trans-Sabinol	471-16-9						0.03[d]; 0.8[b]
Sabinyl acetate (cis-)	53833-85-5			0.2		0.3	16.7[b]
Santene	529-16-8						
β-Selinene	17066-67-0						tr[c]
Sclarene	511-02-4		tr		tr		
Spathulenol	6750-60-3						1.7[c]
α-Terpinene	99-86-5	0.3-0.9	0.5-1.2	0.2-0.6	0.2-0.5	0.6	<0.3[g]; 0.5[f]; 1.2[h]; 1.8[b]
γ-Terpinene	99-85-4	0.6-1.5	tr-0.4	0.5-1.1	0.3-0.5	1.0	0.4[c]; 0.5[g]; 0.8[f]; 2.3[b]; 3.5[h]
Terpinen-4-ol	562-74-3	1.5-3.4	1.3-2.5	1.8-3.3	0.4-0.8	3.3	1.8[f]; 2.5[e]; 2.7[d]; 3.3[b]; 5.4[h]
Terpinen-4-yl acetate	4821-04-9		tr-0.4				0.1[d]; 0.2[e]
1-Terpineol	586-82-3						0.2[d]
α-Terpineol	98-55-5	0.1-0.4	tr-0.3			1.4	tr[c]; 0.1[d]; 0.3[b]; 0.4[f]; 0.6[e]
β-Terpineol	138-87-4						0.2[d]
Terpinolene (α-)	586-62-9	0.2-0.7	tr-1.5		0.1-0.2		0.2[e,f]; 1.5[h]; 2.3[b]; 5.7[c]
α-Terpinyl acetate	80-26-2	0.2-1.2	1.0-1.5	1.6-2.9	0.2-0.4		0.9[g]; 1.2[b,f]; 1.8[e]; 7.0[h]
Thuja-2,4(10)-diene	36262-09-6						tr[c]
α-Thujene	2867-05-2	0.3-1.4	0.3-0.7		0.1-0.3	0.3	<0.02[d]; 0.2[c,g]; 0.8[l]; 1.5[b]
α-Thujone	546-80-5	47.8-66.3	45.0-56.0	30.4-40.5	43.9-56.7	51.6	49.6[f]; 52.5[b]; 65.0[i]; 69.8[d]
β-Thujone (cis-)	471-15-8	6.8-10.8	7.2-9.0	6.5-9.0	8.5-12.2	5.6	9.2[l]; 9.0[f]; 9.5[b,d]; 10.4[g]
Thymol	89-83-8	0.2-0.4		0-1.2			0.3[f]
Totarol (trans-)	511-15-9		tr-0.08		tr-0.1	1.4	
Tricyclene	508-32-7	0.03-0.06					<0.05[g]; 0.05[e]; 0.1[b]; 0.2[c]
cis-Verbenol	1845-30-3						tr[c]
Verbenone	80-57-9						tr[c]

Table 5.84.1 Constituents identified in thuja oils (*continued*)

A forty-four thuja essential oil samples from the USA and Canada analyzed between 1998 and 2013; lowest and highest concentrations given (E. Schmidt, unpublished data)
B two USA commercial oils and three lab-prepared oils from the leaves, buds and twigs of *T. occidentalis*; lowest and highest concentrations given (ref. 3)
C seven lab-steam distilled oils prepared from the needles of two *T. occidentalis* cultivars ('malonyana' and 'malonyana' [skeleton misák]) harvested in an arboretum in Slovakia; lowest and highest concentrations given (ref. 1)
D ten lab-steam distilled oils from needles of *T. occidentalis* trees located at a tree research center in Michigan, USA; lowest and highest concentrations given (ref. 2)
E two lab-hydrodistilled oils from two cultivars ('globosa' and 'aurea') grown in Poland; highest concentration given (ref. 7)
F data from other studies (indicated with superscript letters); highest concentrations found in any study reviewed here given; when two or more oils were investigated, only the highest concentrations are mentioned, unless indicated otherwise

[a] δ-cadinol and δ-cadinene combined; [b] one commercial oil produced in Germany obtained from the needles of *T. occidentalis* grown in the Balkans (ref. 10); [c] one lab-hydrodistilled oil from Italian *T. occidentalis* needles (ref. 5); [d] one lab-hydrodistilled oil from Poland (ref. 6); [e] one lab, steam distilled oil sample from Japan (ref. 12); [f] one commercial thuja needle oil from Canada (ref. 11); [g] one lab-hydrodistilled oil from western Canada (ref. 15); [h] one leaf oil produced from *T. occidentalis* grown in Norway with a very atypical composition (ref. 16); [i] data cited in ref. 8; [j] data taken from ref. 13; [k] one oil from Egypt (ref. 4); [l] one steam-distilled oil from the foliage of *T. occidentalis* harvested in Canada (ref. 18)

tr: trace; + present in the oil investigated, but quantity not stated

Table 5.84.2 Major constituents of thuja oils

Constituent	CAS	Percentage and range					
		A	B (3)	C (1)	D (2)	E (7)	F
α-Thujone	546-80-5	47.8-66.3	45.0-56.0	30.4-40.5	43.9-56.7	51.6	49.6[f]; 52.5[g]; 65.0[i]; 69.8[d]
α-Pinene	80-56-8	1.4-3.6	1.0-5.0	1.9-2.9	0.5-1.3	1.5	3.3[b,f]; 10.0[h]; 27.7[c]; 35.5[k]
δ3-Carene	13466-78-9	0.7-8.2					0.1[h]; 1.0[b]; 25.4[k]; 33.2[c]
Fenchone	1195-79-5	12.4-15.7	9.0-15.0	6.7-11.1	6.2-13.0	4.2	7.8[d]; 12.8[e]; 12.9[b]; 14.6[f]
Sabinene	3387-41-5	1.7-4.4	2.1-5.9	3.2-7.8	2.7-5.0	4.6	4.1[i]; 5.0[i]; 12.1[b]; 13.9[h]
β-Thujone (*cis-*)	471-15-8	6.8-10.8	7.2-9.0	6.5-9.0	8.5-12.2	5.6	9.2[i]; 9.0[f]; 9.5[b,d]; 10.4[g]
Bornyl acetate	76-49-3	0.9-2.5	1.2-4.4	2.8-6.3	1.2-2.2	2.5	0.8[d]; 2.3[f]; 3.0[g]; 3.2[e]; 5.9[l]
Terpinen-4-ol	562-74-3	1.5-3.4	1.3-2.5	1.8-3.3	0.4-0.8	3.3	1.8[f]; 2.5[e]; 2.7[d]; 3.3[b]; 5.4[h]

LEGEND: SEE UNDER TABLE 5.84.1

CONTACT ALLERGY/ALLERGIC CONTACT DERMATITIS

General

Allergic contact dermatitis from thuja oil has been reported in one publication only.

Case report

One patient developed erythema multiforme-like contact dermatitis from the application of thuja essential oil on hemorrhoids. He reacted to thuja oil (pure and 1% in petrolatum), the fragrance mix, colophony, juniper tar and pine tar, which are, with the exception of the fragrance mix, all tree products (22).

LITERATURE

1 Svajdlenka E, Mártonfi P, Tomasko I, Grancai D, Nagy M. Essential oil composition of *Thuja occidentalis* L. samples from Slovakia. J Essent Oil Res 1999;11:532-536

2 Kamdem PD, Hanover JW. Inter-tree variation of essential oil composition of *Thuja occidentalis* L. J Essent Oil Res 1993;5:279-282

3 Kamdem PD, Hanover JW, Gage DA. Contribution to the study of the essential oil of *Thuja occidentalis* L. J Essent Oil Res 1993;5:117-122

4 Badawy MEI, Abdelgaleil SAM. Composition and antimicrobial activity of essential oils isolated from Egyptian plants against plant pathogenic bacteria and fungi. Ind Crops Prod 2014;52:776-782

5 Benelli G, Flamini G, Canale A, Cioni PL, Conti B. Toxicity of some essential oil formulations against the Mediterranean fruit fly *Ceratitis capitata* (Wiedemann) (Diptera: Tephritidae). Crop Prot 2012;42:223-229

6 Szołyga B, Gniłka R, Szczepanik M, Szumny A. Chemical composition and insecticidal activity of *Thuja occidentalis* and *Tanacetum vulgare* essential oils against larvae of the lesser mealworm, *Alphitobius diaperinus*. Entomologia Experimentalis et Applicata 2014:151:1-10

7   Tsiri D, Graikou K, Poblocka-Olech L, Krauze-Baranowska M, Spyropoulos C, Chinou I. Chemosystematic value of the essential oil composition of *Thuja* species cultivated in Poland—antimicrobial activity. Molecules 2009;14:4707-4715

8   Naser B, Bodinet C, Tegtmeier M, Lindequist U. *Thuja occidentalis* (Arbor vitae): A review of its pharmaceutical, pharmacological and clinical properties. eCAM 2005;2:69-78

9   Brijesh K, Ruchi R, Sanjita D, Saumya D. Phytoconstituents and therapeutic potential of *Thuja occidentalis*. Res J Pharm Biol Chem Sci 2012;3:354-362

10 Jirovetz L, Buchbauer G, Denkova Z, Slavchev A, Stoyanova A, Schmidt E. Chemical composition, antimicrobial activities and odor descriptions of various *Salvia* sp. and *Thuja* sp. essential oils. Ernährung/Nutrition 2006;30:152-159

11 Kéïta SM, Vincent C, Schmidt JP, Arnason JT. Insecticidal effects of *Thuja occidentalis* (Cupressaceae) essential oil on *Callosobruchus maculatus* (Coleoptera: Bruchidae). Can J Plant Sci 2001;81:173-177

12 Yatagai M, Sato T, Takahashi T. Terpenes of leaf oils from Cupressaceae. Biochem Syst Ecol 1985;13:377-385

13 Pietsch M, König WA. Enantiomeric composition of the chiral constituents of essential oils—Part 3: Diterpene hydrocarbons. J High Resol Chromatogr 1997;20: 257-260

14 Lawrence BM. Progress in essential oils. Perfum Flavor 2007;32(5):40-? (last page unknown)

15 Lopes D, Kolodziejczyk P. Essential oils from western Canada. Poster presentation P-47. 34th International Symposium on Essential Oils, September 7-10, 2003, Würzburg, Germany. Data cited in ref. 14

16 Rohloff J, Langleite BO. Monoterpene patterns of industrially produced needle tree oils. In: L Jirovetz and G Buchbauer, Eds. Processing, analysis and application of essential oils. Dehradun, India: Har Krishnan Bhalla and Sons, 2005. Data cited in ref. 14

17 Simon DZ, Beliveau J, Aube C. Cedarleaf oil (*Thuja occidentalis* L.) extracted by hydrodiffusion and steam distillation a comparison of oils produced by both processes. Pharmaceutical Biology 1987;25:4-6

18 Shaw AC. The essential oil of *Thuja occidentalis* L. Can J Chem 1953;31:277-283 10.1139/v53-039

19 Rhind JP. Essential oils. A handbook for aromatherapy practice, 2nd Edition. London: Singing Dragon, 2012

20 Lawless J. The encyclopedia of essential oils, 2nd Edition. London: Harper Thorsons, 2014

21 Davis P. Aromatherapy. An A-Z, 3rd Edition. London: Vermilion, 2005

22 Puig L, Alomar A, Randazzo L, Cuatrecasas M. Erythema multiformlike reaction caused by topical application of thuja essential oil. Am J Cont Derm 1994;5:94-97

Chapter 5.85   THYME OIL

There are two major thyme oils: thyme oil obtained from *Thymus vulgaris* L. and oil of thyme containing thymol, Spanish type, obtained from *Thymus zygis* L. In this chapter, thyme oil from *Thymus vulgaris* is discussed.

The Spanish type essential oil is presented in Chapter 5.86.

DEFINITION
Thyme oil is the essential oil obtained from the flowering tops of the (English, common, garden) thyme, *Thymus vulgaris* L.

INCI NOMENCLATURE
Description/definition: Thymus vulgaris flower/leaf oil is the volatile oil obtained from the flowers and leaves of the thyme, *Thymus vulgaris* L., Lamiaceae
INCI name EU: Thymus vulgaris flower/leaf oil
INCI name USA: Thymus vulgaris (thyme) flower/leaf oil
CAS registry number(s): 8007-46-3; 84929-51-5
EINECS number(s): 284-535-7

ISO (INTERNATIONAL ORGANIZATION FOR STANDARDIZATION) STANDARD
There is currently no ISO standard for thyme oil obtained from *Thymus vulgaris*.

THE PLANT, THE OIL, AND THEIR USES
Thymus vulgaris L. (common thyme, garden thyme or just 'thyme') is a bushy, woody-based evergreen subshrub with small, highly aromatic, grey-green leaves growing to 15-30 cm tall by 40 cm wide. It is native to Italy, France, Spain and Morocco. The thyme is widely cultivated as a spice and medicinal plant in the Mediterranean area, Europe and many other countries (3,4,46). In Europe and North America the plant sometimes escaped from cultivation and became naturalized (GRIN Taxonomy for Plants; http://mansfeld.ipk-gatersleben.de).

The flowering herb is one of the most popular herbs throughout the world and is used to flavor meat dishes, sauces, vegetables, liqueurs and cheese and in scenting herbal teas (3). Leaves, flowers and oil are employed medicinally. Thyme is stated to possess carminative, antispasmodic, antitussive, expectorant, bactericidal, anthelmintic, antifungal and astringent properties (19,45). In traditional medicine, it has been used for dyspepsia, chronic gastritis, headache, dysmenorrhea, asthma, diarrhea in children, enuresis in children, laryngitis, tonsillitis (as a gargle), and specifically for pertussis and bronchitis (19,20,33, http://obtrandon.files.wordpress.com/2010/ 05/thymus-vulgaris-thyme.pdf).

Thyme oil is obtained by steam distillation of fresh or dried leaves and flowering tops of the plant (4). Most commercial oils are of the thymol chemotype and generally contain 35-55% thymol. This oil is considered to have fungicidal, bactericidal, anthelminthic, antispasmodic, antioxidant, insecticidal, rubefacient, anti-inflammatory and anesthetic properties (5,8,20,33,46). It is used in foods (not only to flavor, but also to preserve meat and fats) (5), cosmetics, perfumery (to create spicy, leathery notes) and pharmaceuticals such as oral hygiene products (8,5,45,46). In addition, It is extensively used in phytotherapy and aromatherapy, though usually in low concentrations in the case of high carvacrol content (5,85,86).

CHEMICAL COMPOSITION
Thyme oil is a colorless to dark red, clear mobile liquid which has a characteristic odor, ranging from aromatic, spicy and medicinal-phenolic, over citrusy fresh and aromatic, to natural floral, woody and herbaceous, depending on the chemotype. The yield of essential oil from the flowering tops of *Thymus vulgaris* L. generally varies from 0.2 to 1.3 per cent, depending on whether fresh or dried biomass is used. The main producing countries of this oil are France, Hungary, Morocco, Spain, Portugal, Germany and USA.

Literature data (up to October 18, 2014) on the chemical composition of thyme oils and unpublished analytical data from one of us (E.S.) are shown in Table 5.85.2 in alphabetical order. In thyme oils from various origins, over 325 chemicals have been identified. About 50 per cent of these were found in a single reviewed publication only.

The major compounds found in thyme oils from different sources are shown in Table 5.85.3. They include (highest concentrations in any study given) geraniol (97.1%), linalool (93.8%), α-terpineol (90.4%), carvacrol (89.7%), *p*-cymene (80%), thymol (78.2%), geranyl acetate (68.6%), terpinyl acetate (α-) (62%), *trans*-sabinene hydrate (54.8%), sabinene hydrate (isomer unspecified) (52.2%), 1,8-cineole (36.5%), *cis*-sabinene hydrate (33.4%), terpinen-4-ol (32%), γ-terpinene (27.6%) and camphene (17.2%). Well-known ingredients of thyme oils that were present in high concentrations (>8%) in one or two studies were camphor (11.2% and 38.5%), borneol (9.8% and 33.9%), α-thujene (25.9%), α-terpinene (24.0% and 25%), linalyl acetate (8.1% and 18%), β-caryophyllene (three studies: 8.3%, 12.3% and 17.6%), α-pinene (9.4% and 12.4%), germacrene D (11.7%), β-myrcene (8.0%) and menthol (8.1%). Uncommon or rare constituents of thyme oils found in high concentrations (>10%) in single studies include geranyl formate (48.4%), β-cyclocitral (38.3%), *cis*-verbenol (33.5%), myrcen-8-ol (three studies: 12.3%, 15% and 21.2%), *o*-cymene (three studies: 13.2%, 15.5% and 18.3%), citronellyl acetate (17.5%), tetradecane (16.1%), δ-terpinene (14.9%), diethyl phthalate (14.8%), myrcenyl acetate (12.8%) and germacrone (*E,E*-) (10.6%).

Commercial oils
The ten chemicals that had the highest maximum concentrations in 25 commercial thyme essential oil samples (concentration ranges provided) are the following: carvacrol (tr-77.8%), linalool (0.03-68.5%), thymol (0.2-47.8%), 1,8-cineole (0.2-36.5%), *cis*-sabinene hydrate (0.07-32.7%), geraniol (0-26.0%), α-thujene (0.2-25.9%), *p*-cymene (0.5-25.7%), geranyl acetate (0-21.8%) and γ-terpinene (0.2-21.2%) (Erich Schmidt, unpublished analytical data).

Chemotypes

In the beginning of the 1970s, the investigation of the essential oils of over 150 wild thyme populations in the south of France revealed the existence of six genetically distinct chemotypes that can be distinguished on the basis of the dominant monoterpenes: geraniol (G), α-terpineol (A), 4-thujanol (sabinene hydrate) (U), linalool (L), carvacrol (C), and thymol (T) (30,60). These results were confirmed in a more recent investigation of >300 oils of wild thyme at the end of the flowering stage, collected in the south of France from 15 sites in two distinct areas. The minimum and maximum concentrations found for each chemotype determinant are shown in Table 5.85.1.

The thymol chemotype is most often accompanied by high concentrations of p-cymene (up to 30%) and γ-terpinene (up to 24%) (59). Thymol is biosynthesized by the aromatization of y-terpinene to p-cymene, followed

by hydroxylation of p-cymene to thymol (21,36). The thujanol chemotype is further characterized by (fairly) high concentrations of three other monoterpenes plus their acetates: terpinen-4-ol (2.2-29.6%), linalool (2.4-32.5%) and myrcenol-8 (0-21.5%) (10,59).

In Spain, a 1,8-cineole chemotype was found, which appears to be exclusive to this country (21,31,52). Some authors state that there are in addition a terpinen-4-ol, a borneol and a p-cymene chemotype (ref. 53). Indeed, these chemicals have been found as dominant components in Thymus vulgaris essential oils in several studies: p-cymene (45% [11]; 32.2% [25]; 38.2% [39]; 32.4% [53]; 55.9% [55]; 25.3% [56]; 39.9% [65]; 31.5% [66]); terpinen-4-ol (21.8% [26]); and borneol (33.9% [53]). Many thyme oils, both from wild plants and commercial samples, do not fit in any of the chemotypes and may be considered mixed chemotypes (34).

Currently, most analytical studies reported in literature concern the thymol chemotype. Other chemotypes (or at least the relevant chemicals being dominant) are reported less frequently, including carvacrol (13,33,34, 47,51,53,62,78), linalool (20,34,35,52,53,62,68), 1,8-cineole (21,52 [both from Spain]), geraniol (34,35,53, 62,68), α-terpineol (34,62) and thujanol (sabinene hydrate) (10,52, 53,68). Sometimes, other chemicals were found to be the dominant component of Thymus vulgaris essential oils, including camphor (38.5% [32]), β-cyclocitral (38.3% [61]) and geranyl formate (48.4% [53]). Most commercial oils are said to be of the thymol chemotype (76); other chemotype oils are, however, also commercially available (68).

The results of analyses of T. vulgaris essential oils from refs. 41 and 48 are not presented here, as their qualitative nor quantitative composition qualifies for inclusion of data from these studies in column F of Table 5.85.2. The literature on the composition of Thymus vulgaris and other Thymus species from between 1960 and 1989 has been reviewed in ref. 14.

Table 5.85.1 Chemotypes of *Thymus vulgaris* found wild in the south of France with their minimum and maximum concentrations in the essential oils (59)

Chemotype	Number of oils Investigated	Minimum concentration	Maximum concentration
Carvacrol[c]	100	21.5	84.1
Geraniol[a]	26	23.5	84.1
Linalool[a]	75	32.2	93.8
α-Terpineol[a]	15	40.9	90.4
Thujanol[a,b]	29	1.6 [d]	52.2
Thymol[c]	88	21.4	72.9

[a] percentages are the sum of this monoterpene and its acetate; [b] the sum of *cis*- and *trans*-thujanol; [c] oils that had equal proportions of carvacrol and thymol were classified as carvacrol chemotype; [d] this must be a mistake; an oil with a thujanol concentration of 1.6% can hardly be classified as thujanol chemotype

Table 5.85.2 Constituents identified in thyme oils

Constituent	CAS	Percentage and range					
		A	B (68)	C (7)	D (34)	E (21)	F
Abietatriene	19407-28-4						tr[x6]
African-1-en							0.1[y4]
α-Amorphene	20085-19-2						0.3[i]
γ-Amorphene	6980-46-7						0.3[u]
δ-Amorphene	189165-79-5						0.3[u]
(*E*)-Anethole	4180-23-8						0.1[h,w5]
Apiole	523-80-8						0.4[y8]
Aromadendrene	489-39-4	tr-0.8		tr		0.03	0.03[x9]; 0.1[w1,y9]; 1.0[w5]
allo-Aromadendrene	25246-27-9	0-0.4				0.2	tr[e,w4]; 0.07[f]; 0.2[u]
Aromadendrene oxide							0.08[v]; 0.1[i]
Benzaldehyde	100-52-7			tr			
1,4-Benzenediamine, *N,N*-dimethyl-	99-98-9						0.4[z8]
Benzyl acetate	140-11-4						tr[w4]
Benzyl alcohol	100-51-6						0.5[r]
Bergamal	106-72-9			tr			
α-Bergamotene	17699-05-7						0.1[i]; 0.5[v]
trans-α-Bergamotene	13474-59-4			tr			
Bicyclogermacrene	24703-35-3	0-0.3	0-0.2				1.7[j]
α-Bisabolene	17627-44-0	0.08-0.3					2.9[u2]

Table 5.85.2 Constituents identified in thyme oils (*continued*)

Constituent	CAS	Percentage and range					
		A	B (68)	C (7)	D (34)	E (21)	F
(Z)-α-Bisabolene	29837-07-8					0.01	
β-Bisabolene	495-61-4			0.1			0.1[e]; 0.3[n]; 0.4[w5]; 1.0[i]; 1.4[u]
(E)-γ-Bisabolene	53585-13-0						0.2[w]
Bisabolol	515-69-5						0.4[v]
Borneol	507-70-0	0.2-3.8	0.2-1.3	4.1	0.3-3.6	4.8	2.6[w1]; 2.7[u]; 3.3[p]; 3.7[t]; 4.7[x8]; 5.7[u2]; 6.6[i]; 6.7[v]; 9.8[y2]; 33.9[d]
Bornyl acetate	76-49-3	0-0.9	0-0.2		0-0.4	1.1	0.8[u]; 1.0[t]; 1.9[g]; 2.3[z6]; 5.0[z]
Bornyl propanoate	78548-53-5						tr[y7]
β-Bourbonene	5208-59-3			0.1			0.05[k]; 0.1[e]; 0.2[f]; 0.3[c]; 0.4[d]
Butanol	35296-72-1					tr	
Butanone (2-)	78-93-3					0.07	
Butyl caprylate	589-75-3					0.03	
Cadalene	483-78-3			0.2			0.1[y4]; 1.8[y8]
α-Cadinene	24406-05-1			0.1			2.2[v6]
γ-Cadinene	39029-41-9	0-0.7	0-0.1	0.4		0.1	0.2[e]; 0.3[f]; 0.4[v]; 0.5[y8]; 0.6[d]
cis-γ-Cadinene							0.5[w6]
δ-Cadinene	483-76-1	0-0.4		0.5		0.1	0.4[i]; 0.7[f]; 0.9[v]; 1.8[w5]; 2.3[w2]
Cadinol	11070-72-7						0.5[v]
α-Cadinol	481-34-5			0.1			0.05[x9]; 0.1[y7]; 0.2[e]; 0.9[u2]
δ-Cadinol	19435-97-3						0.09[x9]; 0.4[x7]
τ-Cadinol	5937-11-1						0.2[h]; 0.3[v]; 0.4[e]; 0.5[x2]; 0.7[w6]
α-Calacorene	21391-99-1			tr			
Calamenene	483-77-2						0.1[v7]; 0.5[w5]
Calamenene B							0.2[y4]
cis-Calamenene	72937-55-4						0.7[y8]
trans-Calamenene	73209-42-4						0.3[y5,y7]
Calarene ((+)-)	17334-55-3					0.01	
Camphene	79-92-5	0.08-8.0	0.3-0.8	0.5	0.2-1.2	3.2	1.9[c]; 2.1[f]; 5.2[r]; 9.3[d]; 17.2[v9]
Camphene hydrate	465-31-6			0.1			
α-Campholenal	4501-58-0						0.1[v8]
Camphor	76-22-2	0-11.2	0-0.9	0.5	0-0.9	2.2	2.1[i]; 2.9[v]; 3.1[u]; 3.4[c]; 38.5[v9]
δ3-Carene	13466-78-9	0.07-3.9				0.02	0.2[f]; 1.0[c]; 1.1[w5]; 1.7[y8]; 4.1[v6]
(+)-4-Carene	13837-63-3						1.1[z8]
Carotol	465-28-1						0.2[w5]
Carvacrol	499-75-2	tr-77.8	tr-1.7	4.4	0.2-37.6	0.02	42.0[d]; 46.6[w9]; 70.3[x6]; 71.6[u2]; 72[w8]; 84.1[y1]; 86.1[y9]; 89.7[x4]
Carvacryl acetate	6380-28-5						0.08[w1]; 0.09[b]; 0.3[u2,w2]
Carveol	99-48-9					0.05	
(E)-Carveol	1197-07-5						0.2[x1]
(Z)-Carveol	1197-06-4						1.4[d]
Carvone	99-49-0					0.01	0.1[y2]; 0.2[h]
trans-Carvone oxide	18383-49-8						0.5[y9]
Caryophylla-4(12),8(13)-dien-5α-ol							0.4[w1]
Caryophylla-4(14),8(15)-dien-5β-ol				0.1			
β-Caryophyllene	87-44-5	0.05-0.6	2.2-7.0	1.3	0.9-2.0	2.1	3.2[u]; 3.5[x7]; 4.3[x9]; 5.3[f]; 5.6[u2]; 6.6[c]; 8.3[d,x]; 12.3[y1]; 17.6[q]
(Z)-β-Caryophyllene							0.4[r]
γ-Caryophyllene (cis-)	118-65-0						0.1[u]; 1.6[w1]
Caryophyllene oxide	1139-30-6	0.4-7.0	0.1-0.4	1.8		0.5	2.4[j]; 2.6[u2]; 3.4[c]; 3.5[w1]; 5.5[z7]
Caryophyllenol	38284-26-3						1.7[o]
bis(2-Chloroisopropyl)ether	39638-32-9						0.4[w5]
1,4-Cineole	470-67-7						8.9[x4]
1,8-Cineole	470-82-6	0.2-36.5	0.2-0.9	1.5	0-4.8	36.4	4.1[x4]; 9.4[v6]; 22.1[c]; 24.1[u2]
Citral	5392-40-5						0.3[x]
Citronellal	106-23-0	0-0.4				0.01	
Citronellol	106-22-9					<0.06	0.1[v9]; 0.6[x]
Citronellyl acetate	150-84-5	0-17.5	0-0.1				
α-Copaene	3856-25-5			0.1		0.5	0.06[k]; 0.08[f]; 0.1[x7]; 0.3[e]; 0.8[d]
β-Copaene	18252-44-3			tr			0.2[y5,y7]

Table 5.85.2 Constituents identified in thyme oils (*continued*)

Constituent	CAS	Percentage and range					
		A	B (68)	C (7)	D (34)	E (21)	F
cis-Copaen-8-ol	58569-25-8						0.1[w5]
α-Cubebene	17699-14-8						<0.1[v6]
β-Cubebene	13744-15-5						0.1[e]; 2.4[y8]
Cubenol	21284-22-0						tr[h]
1,10-di-epi-Cubenol	73365-77-2			0.1			0.1[e]
Cuminaldehyde	122-03-2			tr			
Cuminyl alcohol	536-60-7						0.08[x9]; 0.1[w5]
Cuparene	16982-00-6						0.1[y4]
Curcumene (ar-; α-)	644-30-4			0.1			0.1[x7]
β-Cyclocitral	432-25-7						38.3[y2]
o-Cymene	527-84-4						tr[w1]; 13.2[i]; 15.5[v]; 18.3[v4]
p-Cymene	99-87-6	0.5-25.7	0.5-24.0	26.1	1.8-27.7	1.0	38.9[v5]; 39.4[z4]; 39.9[v1]; 43.8[u3]; 44.8[r]; 46.4[u2]; 55.9[x8]; 80[w8]
p-Cymenene	1195-32-0			0.2			0.1[y4]; 0.2[w5]; 0.4[z8]
p-Cymen-8-ol	1197-01-9	0.05-0.2				0.03	tr[x6]; <0.03[x9]; 0.1[e,u,w4]
Decanal	112-31-2					0.02	tr[w4]
Diethyl phthalate	84-66-2						14.8[a,v2]
Dihydrocarveol	38049-26-2						0.1[y9]
Dihydrocarvone	5948-04-9					0.03	0.1[h]; 0.3[i]; 0.6[v]
cis-Dihydrocarvone	3792-53-8		0-0.1				0.3[y5]
trans-Dihydrocarvone	5948-04-9	0-0.1					0.5[c]
2,5-Dimethylstyrene	2039-89-6						tr[x6]
Docosane	629-97-0						1.3[w9]
n-Dodecane	112-40-3						3.0[z9]
Elemene	11029-06-4					0.3	
β-Elemene	33880-83-0						0.1[e]
γ-Elemene	29873-99-2						0.3[i]; 0.4[v7]; 0.7[v]; 1.4[c]
δ-Elemene	20307-84-0						3.8[c]
(*E*)-β-Elemenene							1.5[u2]
β-Elemol	32142-08-8						1.2[c]
Ethyl acetate	141-78-6					tr	
Ethyl butyrate	105-54-4					tr	
2-Ethyl-4,5-dimethyl-phenol	2219-78-5						0.2[w5]
4-Ethyl-2-methoxy-6-methylphenol	120550-70-1						1.2[y5,y7]
Ethyl octanoate	106-32-1					0.01	
Eudesm-3-en-7-ol							0.1[y4]
β-Eudesmol	473-15-4			tr			0.8[w5]
γ-Eudesmol	1209-71-8			0.2			0.08[y]; 0.1[l,y4]; 0.2[e,u2]
10-epi-γ-Eudesmol	15051-81-7						0.1[y7]
Eugenol	97-53-0			0.5		0.04	0.1[b]; 0.2[h]; 0.3[z5]; 0.4[u3]; 0.5[i,v]
(*E,E*)-α-Farnesene	502-61-4						0.1[x7]
β-Farnesene	502-60-3						0.2[h]
(*E*)-β-Farnesene	18794-84-8						0.1[x7]
(*Z*)-β-Farnesene	28973-97-9						0.04[x9]
Farnesol	4602-84-0						0.1[e]
endo-Fenchol (α-)	512-13-0			tr			
Fenchyl alcohol	1632-73-1					0.02	
β-Fenchyl alcohol	22627-95-8						<0.1[v6]
Furfural	98-01-1			tr		0.04	
Geranial	141-27-5	0-0.8	0-0.6	tr	0-3.3		0.1[h]; 0.2[e]; 0.5[d]; 0.6[u]; 7.3[u2]
Geraniol	106-24-1	0-26.0	0-26.0	tr	0-32.9	0.8	3.5[t]; 8.3[u]; 26.0[d]; 37.8[x]; 41[w8]; 57.1[y3]; 84.1[y1]; 97.1[y2]
Geranyl acetate	105-87-3	0-21.8	0-21.8		0-11.7	0.2	13[w8]; 20.9[u2]; 54.4[y3]; 68.6[d]
Geranyl acetone	3796-70-1						0.2[i]
Geranyl butyrate	106-29-6	0-0.6	0-0.6	0.1			tr[y7]
Geranyl formate	105-86-2						48.4[d]
Geranyl isobutyrate	2345-26-8			tr			
Geranyl isovalerate	109-20-6			tr			
Geranyl propionate	105-90-8	0-0.09	0-0.9	0.2			0.08[y]; 0.2[z5]; 0.3[x2]
Germacrene B	15423-57-1						1.9[u2]

Table 5.85.2 Constituents identified in thyme oils (*continued*)

Constituent	CAS	Percentage and range					
		A	B (68)	C (7)	D (34)	E (21)	F
Germacrene D	23986-74-5	0.07-0.6	0.1-0.6		0-0.4		0.6[x5]; 1.1[f]; 1.2[x8]; 11.7[j,u1]
Germacrone (*E,E*-)	6902-91-6						10.6[u2]
Globulol	489-41-8						0.6[w5]
cis-β-Guaiene	372162-07-7						0.1[y4]
Guaiol	489-86-1						0.7[w5]
α-Gurjunene	489-40-7						0.1[u]; 0.5[w5]; 1.4[z]
β-Gurjunene	73464-47-8						tr[e]; 0.04[x9]; 0.1[x7]
γ-Gurjunene	22567-17-5						0.1[e]
Heptadecane (*n*-)	629-78-7						3.0[z9]
Heptanal	111-71-7					tr	
2-Heptanone	110-43-0			tr			
n-Hexadecane	544-76-3						3.2[z9]
(*E,E*)-2,4-Hexadienal	142-83-6			tr			
Hexanal	66-25-1			tr			0.1[h]
Hexanol (1-; *n*-)	111-27-3			tr	0.02		
(*E*)-2-Hexenal	6728-26-3			tr	tr		
(*E*)-2-Hexen-1-ol	928-95-0				0.04		
(*Z*)-3-Hexen-1-ol	928-96-1			tr	0.03		
(*Z*)-3-Hexenyl buty-rate	16491-36-4						0.1[w5]
β-Himachalene	1461-03-6						0.1[y9]
Hotrienol	20053-88-7			tr			
α-Humulene	6753-98-6	0.06-0.6	0.1-0.2	0.1		0.1	0.2[e]; 0.3[b]; 0.4[w2]; 0.6[x5]; 1.3[d]
β-Humulene	116-04-1						1.5[u]
Humulene (ep)oxide I	19888-33-6			tr			
Humulene (ep)oxide II	19888-34-7			0.1			
14-Hydroxy-9-epi-(*E*)-caryophyllene	244226-09-3						0.6[w1]
α-Ionone	127-41-3					0.01	
β-Ionone	79-77-6					0.02	
Isoamyl alcohol	123-51-3	0-0.04					
Isoascaridole	1619-26-7						0.02[y]
Isoborneol	124-76-5			0.1		0.07	tr[w4]; 0.3[v6]; 0.4[e]; 0.7[w6]; 4.2[i,v]
Isobornyl acetate	125-12-2						<1.1[c]; 0.2[k,y9]; 0.7[f]; 3.8[x1]
Isobornyl propionate	2756-56-1			0.2			
Isoeugenol	97-54-1					0.02	
Isopinocamphone	15358-88-0						2.3 (ref.?)
Isoprenyl isovalerate				0.2			
Isopulegol	89-79-2						0.1[x]; 0.3[i]; 0.4[v]
Isothymol	4427-56-9						0.1[v3]
Isovaleraldehyde	590-86-3	0-0.05					
(*Z*)-Jasmone	488-10-8					0.02	
Lavandulol	498-16-8	0-0.1	0-0.1				2.6[t]
Ledol	577-27-5						0.2[w5]; 1.4[y8]
Limonene	138-86-3	0.3-7.5	0.5-3.0	tr	0-2.9		1.2[s]; 1.4[f]; 2.0[u]; 2.3[z2]; 2.4[n]; 3.4[d]; 4.0[t]; 4.3[q]; 7.6[c]
cis-Limonene oxide	13837-75-7					0.02	
trans-Limonene oxide	4959-35-7					0.02	tr[w4]
Limonen-10-ol	3269-90-7	0-9.7					
Linalool	78-70-6	0.03-68.5	2.0-68.5	5.8	0.2-67.2	3.7	44.0[w7]; 58.0[x]; 67[w8]; 69.0[y2]; 74.6[c]; 75.4[y3]; 77.7[d]; 93.8[y1]
cis-Linalool oxide	11063-77-7						tr[w4]; 0.1[w1,y5]; 0.7[d]
cis-Linalool oxide, furanoid	11063-77-7		0-0.1	0.1	0-0.4	0.02	0.9[o]
cis-Linalool oxide, pyranoid	14009-71-3	0-0.2	0-0.1				
trans-Linalool oxide	11063-78-8						0.1[w1,y5]; 0.4[d]
trans-Linalool oxide, furanoid	34995-77-2			0.2	0-0.4		
trans-Linalool oxide, pyranoid	39028-58-5	0-0.1					
Linalyl acetate	115-95-7	0-5.2	0-5.2		0-3.4		4.0[c]; 4.1[t]; 6.0[d]; 8.1[y3]; 18[w8]

Table 5.85.2 Constituents identified in thyme oils (*continued*)

Constituent	CAS	Percentage and range					
		A	B (68)	C (7)	D (34)	E (21)	F
Linalyl propionate	144-39-8						2.7[v8]
Longifolene (junipene)	475-20-7						<0.1[v6]
Mayurone	4677-90-1						0.1[w5]
p-Mentha-1(7),8(10)-dien-9-ol	29548-13-8		0-9.7				
1-*p*-Menthene	5502-88-5						0.3[r]
cis-p-Menth-2-en-1-ol	29803-82-5	0-0.6	0-0.5	0.1			0.2[w5]; 1.0[d]; 1.1[q]
trans-p-Menth-2-en-1-ol	29803-81-4	0-0.5	0-0.2				0.2[w5]; 0.4[d]
Menthol	89-78-1						0.2[x,u2]; 1.3[y8]; 8.1[z9]
Menthone (*p*-; *trans*-)	89-80-5						2.2[y8]
Methyl benzoate	93-58-3			tr			
3-Methyl-2-buten-1-ol	556-82-1					tr	
3-Methyl-3-buten-2-ol	10473-14-0					tr	
Methylcarvacrol	6379-73-3	0-1.2	0-0.1	1.0	0-0.7		1.2[l]; 1.5[u]; 1.7[k,y6]; 4.0[x8]; 5.2[f]
Methyl chavicol	140-67-0						0.04[y2]; 0.1[h]
3-Methyl-3-cyclo-hexen-1-ol	53783-91-8						0.3[w5]
2-Methyldecane	6975-98-0						0.6[z9]
Methyl eugenol	93-15-2			0.2			0.06[b]; 0.1[h]; 0.2[u3]
6-Methyl-5-hepten-2-one	110-93-0	0-0.1	0-0.1				0.3[v8]
Methyl 2-methyl-butyrate	868-57-5	tr-0.5	0-0.4				
Methyl salicylate	119-36-8		tr				
Methyl thymol	1076-56-8	0.4-0.9	0-0.1	2.1	0-0.7		1.5[p,x9]; 1.6[i]; 2.8[x8]; 3.8[f]; 5.5[d]
α-Muurolene	10208-80-7			0.1			0.08[x9]; 0.1[e]; 0.2[w5]; 0.5[f]
γ-Muurolene	30021-74-0			0.1			0.07[x9]; 0.1[e,x2]; 0.2[f]; 0.3[y5]
Muurolol	119757-72-1						0.3[v]
α-Muurolol	104245-48-9			tr			
epi-α-Muurolol	19912-62-0			0.3			
Myrcene (β-)	123-35-3	0.3-6.4	0.4-6.4		0.2-9.4	3.2	2.5[s]; 3.3[x5]; 4.0[v7]; 6.9[w7]; 8.0[q]
α-Myrcene	1686-30-2						7.3[c]
Myrcen-8-ol							12.3[q]; 15[w8]; 21.2[y1]
Myrcenyl acetate	1118-39-4						10[w8]; 12.8[q]
Myristicin	607-91-0						0.7[y8]
trans-Myrtanyl acetate						tr	
Myrtenal	564-94-3						tr[w4]
Myrtenol	515-00-4						0.1
Neomenthol	3623-51-6						2.8[y8]
Neral	106-26-3	tr-0.6	0-0.4		0-2.5	0.02	0.5[d]; 4.1[u2]
Nerol	106-25-2	0-0.2	0-0.2		0-4.4	<0.06	2.1[u2]
(*E*)-Nerolidol	40716-66-3	0-0.1	0-0.1	tr			0.2[x1]; 0.3[x4]
(*Z*)-Nerolidol	3790-78-1						0.1[x1]
Nerol oxide	1786-08-9			tr			
Neryl acetate	141-12-8	0-1.2	0-0.2		0-0.6	0.02	1.1[u2]
Neryl isobutyrate	2345-24-6			tr			
Neryl isovalerate	3915-83-1		0-0.1				
Nonanal	124-19-6					0.05	
Nonanol	28473-21-4						tr[y7]
3-Nonanol	624-51-1						0.2[v8]
2-Nonanone	821-55-6						0.8[v8]
β-Ocimene	13877-91-3						0.1[c]
(*E*)-Ocimene	27400-72-2						0.2[u]; 1.3[y6]
(*Z*)-Ocimene	27400-71-1						0.1[u]; 1.2[k]
(*E*)-β-Ocimene	3779-61-1	tr-1.0	0-0.1	tr		4.1	tr[w,x6]; 0.1[e]; 0.2[q]; 1.4[w2]; 3.6[z]
(*Z*)-β-Ocimene	3338-55-4					0.4	tr[e,w]
allo-Ocimene	673-84-7						0.1[u]; 0.2[y8]
n-Octadecane	593-45-3						2.0[z9]
9-Octadecenoic acid	2027-47-6						1.5[w9]
3-Octanol	589-98-0	0.06-0.3		0.1		tr	0.08[y]; 0.2[h,w3]; 0.3[u]; 0.4[y7]; 1.5[k]
3-Octanone	106-68-3			0.1		0.01	tr[w4]; 0.2[y]; 0.6[u]

Table 5.85.2 Constituents identified in thyme oils (*continued*)

Constituent	CAS	Percentage and range					
		A	B (68)	C (7)	D (34)	E (21)	F
1-Octen-3-ol	3391-86-4	0.1-1.2	0.2-0.7	1.7	0-0.7	0.06	1.1[q]; 1.3[z3]; 1.6[p]; 2.5[z5]; 2.7[w3]
Octenyl acetate (1-)	37366-04-4				0-0.7		
Oxacyclotetradecan-2-one							0.9[w5]
Pentacosane	629-99-2						6.3[w9]
Pentadecane (*n*-)	629-62-9						tr[w1]; 2.6[z9]
Pentanal	110-62-3			tr			
1-Penten-3-ol	616-25-1					tr	
3-Penten-2-ol (*trans*-)	1569-50-2					tr	
Perillaldehyde	2111-75-3	0-3.0	0-3.0				
Perillyl acetate	15111-96-3						5.7[c]
α-Phellandrene	99-83-2	tr-0.4	0-0.2			0.06	0.2[e]; 0.3[k]; 0.4[f]; 0.5[w9]; 2.1[v]
β-Phellandrene	555-10-2	0.1-1.2	0.1-0.6				0.1[h]; 0.3[x6]; 0.6[w9]; 0.7[j]; 1.3[w5]
Phencone							2.7[x]
2-Phenethyl alcohol	60-12-8			tr			
2-Phenylethyl isovalerate	140-26-1			tr			
Phytol	7541-49-3			tr			2.9[z9]
α-Pinene	80-56-8	0.4-5.1	0.4-2.0	0.7	0.5-1.4	3.0	2.5[f]; 2.7[w]; 3.1[q,x1]; 3.3[v5]; 5.4[v6]; 5.7[d]; 9.4[v9]; 12.4[u2]
β-Pinene	127-91-3	0-7.5	0-0.1	0.2	0.2-1.8	3.7	0.7[b]; 1.2[d]; 1.3[w]; 1.8[n]; 2.2[t]; 3.3[c]
trans-Pinocamphone	547-60-4						4.2[w5]
trans-Pinocarveol	1674-08-4	0-0.04		0.1		0.2	
Pinocarvone	30460-92-5			tr			
Pinocarvyl acetate	1078-95-1	0-4.3					
trans-Pinocarvyl acetate			0-4.3				
Piperitenone oxide	90582-88-0						0.1[u]; 2.0[z]
cis-Piperitol	16721-38-3	0-0.1	0-0.1				
trans-Piperitol	16721-39-4	0-0.1	0-0.1				
Piperitone	89-81-6						1.4[y8]
Pulegone	89-82-7						0.1[x]; 1.1[y8]
trans-Rose oxide	5258-11-7			tr			
Sabinene	3387-41-5	0.06-2.8	0.1-2.3		0-2.6	3.3	1.8[s]; 2.4[d]; 2.5[u2]; 2.7[c]; 6.1[q]
Sabinene hydrate (*cis-* + *trans-*)	546-79-2						52.2[y1]
cis-Sabinene hydrate	15537-55-0	0.07-32.7	0.5-32.7		0.3-33.4	0.3	0.6[n]; 1.3[j]; 9.2[d]; 11.8[q]; 21.4[c]
trans-Sabinene hydrate	17699-16-0	0.1-6.8	0.2-5.5		0-3.3	0.7	1.1[h]; 1.2[e]; 29.8[d]; 42[w8]; 54.8[q]
Salicylaldehyde	90-02-8			tr			
α-Selinene	473-13-2						0.1[w5]; 0.3[y8]
β-Selinene	17066-67-0			0.1		0.4	0.2[w5]
γ-Selinene	515-17-3						0.09[w9]
(*E*)-Sesquilavandulol	104121-84-8			0.1			
Spathulenol	6750-60-3	0.06-0.6				1.0	0.1[e]; 0.7[i]; 1.0[y8]; 1.3[w5]; 1.9[c]
α-Terpinene	99-86-5	0.1-4.9	0.1-2.1	tr	0.3-1.8	0.2	2.1[s]; 2.2[w5]; 2.3[c]; 2.6[t]; 3.0[d]; 3.2[w]; 3.5[f]; 6.7[z]; 24.0[x]; 25[w8]
γ-Terpinene	99-85-4	0.2-21.2	0.4-9.5	0.9	2.6-19.6	1.1	17.8[e]; 19.7[n]; 21[w8]; 22.3[x3]; 22.7[x7]; 22.8[z5]; 23.3[f]; 27.6[u3]
δ-Terpinene	586-62-9						0.2[x2]; 14.9[s]
Terpinen-4-ol	562-74-3	0.2-13.5	0.2-6.5	3.1	0-4.7	0.9	5.2[v]; 7.3[q]; 11.9[w7]; 16.1[c]; 21.1[d]; 21.8[w5]; 29.6[y1]; 32[w8]
Terpinen-4-yl acetate	4821-04-9						8[w8]
1-Terpineol	586-82-3	0.02-0.1		tr			
α-Terpineol	98-55-5	0.1-2.8	0.2-2.3	0.6	0-11.5	2.8	8.6[d]; 12.8[v2]; 30.3[y1]; 90.4[y1]
β-Terpineol	138-87-4						0.1[x1]; 0.3[g]
(*E*)-β-Terpineol							0.6[v6]
(*Z*)-β-Terpineol	7299-40-3	0.08-0.4					0.1[v6]; 1.3[i]; 1.4[v]
γ-Terpineol	586-81-2		0-0.3			0.03	0.1[u]
Terpinolene (α-)	586-62-9	0.1-1.9	0.2-1.0	0.1	0-2.9	0.4	0.5[g]; 1.4[d]; 2.2[w]; 3.2[j]; 6.5[x1]
γ-Terpinolene	85188-59-6						0.9[c]
Terpinyl acetate (α-)	80-26-2	0.4-0.7			0-33.4	23.2	1.1[u]; 26.0[c]; 40.7[y3]; 62[w8]
α-Terpinyl isobutyrate	7774-65-4			tr			
Tetracosane	646-31-1						3.6[w9]

Table 5.85.2 Constituents identified in thyme oils (*continued*)

Constituent	CAS	Percentage and range					
		A	B (68)	C (7)	D (34)	E (21)	F
Tetradecane	629-59-4						tr[w1]; 16.1[z9]
α-Thujene	2867-05-2	0.2-25.9	0.2-1.6	0.2	0.5-1.5	0.3	2.0[x3]; 2.2[j]; 2.8[z]; 2.9[f]; 3.0[v7]
α-Thujone	546-80-5						0.06[x4]; 1.0[y8]; 1.2[y9]; 0.5[x9]
β-Thujone (*cis*-)	471-15-8			tr		0.02	0.2[x9,y8]; 0.7[y9]; 7.6[u]
(-)-β-Thujone (*trans*-)	33766-30-2						0.7[u]
Thymodihydroquinone							0.1[y9]; 0.5[f]
Thymohydroquinone	2217-60-9						0.1[y4]
Thymol	89-83-8	0.2-47.8	0-38.8	37.4	0.8-36.6	0.1	66.9[t]; 71.0[z]; 71.2[w6]; 72.9[y1]; 75.7[z2]; 77.7[y2]; 77.8[z]; 78.2[u2]
Thymoquinone	490-91-5					0.01	tr[w1,y5]; 0.03[y]; 0.7[w1]; 0.8[f]
Thymyl acetate	528-79-0	0-0.04					0.1[x2]; 0.2[e,w3]; 0.3[f]; 0.4[w2]
Thymyl formate							0.3[y5,y7]
Tolyldimethylcarbinol							0.4[w5]
Tricosane	638-67-5						0.8[w9]
Tricyclene	508-32-7	0.3-0.4		tr		0.1	tr[e]; 0.1[u]; 0.4[g]; 0.9[r]; 1.2[s]
α-Trimethyldodecane							2.1[w5]
2-Undecanone	112-12-9						<0.1[v7]
Valencene	4630-07-3			0.1		2.9	0.3[f]
Verbenene	4080-46-0					0.03	
Verbenol	473-67-6	0-0.2					0.5[w5]
cis-Verbenol	1845-30-3					0.3	tr[w4]; 33.5[y2]
trans-Verbenol	1820-09-3					0.02	tr[w4]
Verbenone	80-57-9	0-0.4				0.05	tr[w4]; 0.3[i,v]
Viridiflorene (ledene)	21747-46-6						0.1[x9]; 0.3[u]
Viridiflorol	552-02-3						0.4[l]; 2.1[j]
α-Ylangene	14912-44-8			0.1			

A twenty-five thyme essential oil samples from France, Germany, Hungary, Morocco and Portugal analyzed between 2005 and 2013; lowest and highest concentrations given (E. Schmidt, unpublished data)
B four commercial oils from France of 4 chemotypes: geraniol, 4-thujanol/terpinen-4-ol, thymol and linalool; lowest and highest concentrations given (ref. 68)
C one lab-hydrodistilled oil from fully flowering *T. vulgaris* cultivated in Cuba (ref. 7)
D thirteen lab-hydrodistilled oils from 13 *Thymus vulgaris* accessions cultivated in Austria from seeds of various origins, notably France; lowest and highest concentrations given (ref. 34)
E two lab-hydrodistilled oils from *Thymus vulgaris* 1,8-cineole type cultivated in Spain, one harvested in the vegetative state and the other during full bloom; post-bloom analytical data were also provided, but are not presented here (ref. 21)
F data from other studies (indicated with superscript letters); highest concentrations found in any study reviewed here given; when two or more oils were investigated, only the highest concentrations are mentioned, unless indicated otherwise

[a] incorrect identification or adulteration; [b] 130 lab-hydrodistilled oils from thyme collected in the wild in six regions of Albania in a five year period in each month of the year; the results were expressed as average results for each region (ref. 9); [c] eight lab-hydrodistilled oils from wild growing thyme collected at 4 different sampling sites in different climatic zones in Spain, some in the vegetative state, others in (full) flowering (ref. 52); [d] seven commercial oils from France from 7 chemotypes; only β-caryophyllene was present in all samples (ref. 53); [e] sixteen lab-hydrodistilled oils from *Thymus vulgaris* cultivated in Lithuania, 8 from fresh and 8 from dried herbs (ref. 45); [f] six lab-hydrodistilled oils from the aerial parts of thyme cultivated in Italy and harvested at different growth stages, including full flowering, from two and five year old plants (ref. 25); [g] one lab-hydrodistilled oil from the UK (ref. 81); [h] one steam-distilled oil from *Thymus vulgaris* in full bloom cultivated in Mongolia (ref. 82); [i] one lab-hydrodistilled oil from thyme growing wild in Iran (ref. 1); [j] one lab-hydrodistilled oil from thyme growing wild in Iran (ref. 2); [k] four lab-hydrodistilled oils from thyme leaves of plants cultivated in Egypt under different fertilization regimens (ref. 3); [l] one lab-hydrodistilled oil from the flowering aerial parts of thyme cultivated in Spain (ref. 4); [m] four lab-hydrodistilled oils from Spanish thyme (*T. vulgaris*) during vegetative stand, initial flowering, full flowering and after flowering (ref. 4); only 6 components were analyzed; [n] three lab-hydrodistilled oils from *T. vulgaris* cultivated in Iran and harvested before flowering, at the beginning of flowering and at full flowering; analytical data from fruit set and seed ripening stages were also presented but not discussed here (ref. 5); [o] one lab-hydrodistilled oil from the aerial parts (probably flowering) of thyme cultivated in Hungary (ref. 6); [p] nine lab-hydrodistilled oils from thyme cultivated in Brazil and harvested each month between November and July (ref. 8); [q] (probably) 47 lab-hydrodistilled oils from sabinene hydrate (thujanol)-rich clones cultivated at an experimental station in France and harvested in spring and autumn of two successive years (ref. 10); [r] one commercial oil obtained from a US company (ref. 11); [s] one commercial oil from an Iranian company (ref. 12); [t] five lab-hydrodistilled oils from 5 thyme cultivars in the vegetative state from Canada and Germany, cultivated in Canada (ref. 13); [u] one commercial thyme oil produced in the Provence, France (ref. 15); [u1] data taken from (the summary of) ref. 32; [u2] data from various studies, cited in ref. 83; [u3] >180 oil samples from Dutch and German thyme cultivars grown in Albania and collected in each month of the year in a 4-year period; the results were expressed as the average values for each month, the number of oils investigated per month varying from >5 to >30 (ref. 9); [v] four lab-hydrodistilled oils from *T. vulgaris* collected in the wild in Iran and harvested in a period of 40 days in various stages of development

Table 5.85.2 Constituents identified in thyme oils (*continued*)

(ref. 17); [v1] one lab-hydrodistilled oil from the flowering aerial parts of *T. vulgaris* cultivated in Turkey (ref. 65); [v2] two commercial oils from Hungary (ref. 66); [v3] one commercial oil from France (ref. 67); [v4] one lab-hydrodistilled oil from thyme cultivated in Serbia (ref. 70); [v5] one commercial oil from Brazil (ref. 71); [v6] one commercial oil sample purchased in Germany (ref. 77); [v7] one lab-hydrodistilled thyme oil from plants cultivated in Cuba (ref. 79); [v8] one commercial thyme oil sample from Italy (ref. 80); [v9] data from ref. 84, cited in ref. 29; [w] three lab-hydrodistilled oils from the flowering aerial parts of *Thymus vulgaris* cultivated at three locations in Iran from the same batch of seeds (ref. 18); [w1] one lab-hydrodistilled oil from the flowering parts of thyme collected in the wild in Iran (ref. 19); [w2] one lab-hydrodistilled oil sample from thyme cultivated in Serbia (ref. 20); [w3] one lab-hydrodistilled and one microwave-assisted lab-hydrodistilled oil from flowering cultivated Iranian *T. vulgaris* (ref. 22); [w4] one lab-hydrodistilled oil from thyme cultivated in Spain (ref. 26); [w5] two lab-hydrodistilled oils from the aerial parts of two clonal-stock thyme lines cultivated in Canada (ref. 27); [w6] six lab-hydrodistilled oils from the flowering tops of *T. vulgaris* cultivated in Hungary, one fresh and five dried in several ways (natural, oven with different temperatures, lyophilized) (ref. 28); [w7] one steam-distilled commercial oil from leaves, stem and flowers from Spain (ref. 29); [w8] unknown number of oils produced in a period of 4-5 years from thyme collected in the wild at 150 sites in the south of France; this publication was used to establish 6 *T. vulgaris* chemotypes (ref. 30); [w9] one lab-hydrodistilled oil from flowering wild *T. vulgaris* from Iran (ref. 33); [x] four lab-hydrodistilled oils from 4 accessions growing wild in Italy (ref. 35); [x1] one lab-hydrodistilled oil from flowering thyme cultivated in Egypt (ref. 40); [x2] two lab-hydrodistilled oils from plants cultivated in Greece, one from leaves and the other from leaves and flowers (ref. 43); [x3] six lab-hydrodistilled oils from flowering thyme cultivated in Brazil grown from 6 commercial seeds from USA, Spain, France, Italy, EU and Brazil (ref. 46); [x4] nine lab-hydrodistilled oils from plants cultivated in Jordan, 3 in the vegetative state, 3 at the beginning of blooming and 3 in full bloom, planted with different (15-45 cm) inter-row spacing; fruit maturation stage data were also shown but are not presented here (ref. 47); [x5] one lab-hydrodistilled oil from the aerial fully flowering parts of thyme cultivated in Iran (ref. 50); [x6] one oil from Portugal, described as both commercial and lab-hydrodistilled (?) (ref. 51); [x7] one lab-hydrodistilled oil from plants growing wild in Cameroon (ref. 54); [x8] fifty-two steam-distilled oils from thyme growing wild in New Zealand and harvested each month for 13 months (ref. 55); [x9] two lab-hydrodistilled oils from plants cultivated in Italy from wild samples and harvested in two successive years (ref. 56); [y1] >300 lab-hydrodistilled oils from *T. vulgaris* collected in the wild at various sites in the south of France, comprising the 6 major chemotypes (ref. 59); see also Table 5.85.1; the concentrations for geraniol, α-terpineol, linalool and sabinene hydrate (4-thujanol) were the sum of the percentage for these monoterpenes *plus* their acetates; [y2] 4 lab-hydrodistilled oils obtained from plants 'growing in distinct areas of commercial plantations in southern France' (were the plants commercial or wild?) with three known and one unknown chemotype: *cis*-verbenol/β-cyclocitral (ref. 61); [y3] seven lab-hydrodistilled oils from 7 thyme accessions from French gene banks, cultivated in Germany (ref. 62); [y4] one commercial oil of thyme obtained from a Polish company (ref. 64); [y5] one lab-hydrodistilled thyme oil of unknown origin (ref. 69); [y6] one lab-hydrodistilled oil from commercial thyme purchased in Serbia (ref. 73); [y7] three lab-hydrodistilled oils from the aerial parts of flowering thyme cultivated in Spain, with 3 different particle sizes (ref. 72); [y8] one lab-hydrodistilled thyme leaf oil from plants cultivated in Serbia and Montenegro (ref. 75); [y9] four lab-hydrodistilled thyme oils, three from plants growing wild and one from *T. vulgaris* cultivated in Jordan (ref. 78); [z] one lab-hydrodistilled leaf oil from Iran (ref. 23); [z1] one laboratory steam-distilled leaf of from *T. vulgaris* cultivated in Poland (ref. 24); [z2] one lab-hydrodistilled thyme oil from Egypt (ref. 37); [z3] one lab-hydrodistilled oil from the aerial parts of thyme cultivated in Sardinia, Italy; the effects of different stabilizing operations and of storage on the composition of the oil were also investigated, but are not reported here (ref. 38); [z4] one commercial oil from India (ref. 39); [z5] one lab-hydrodistilled oil from thyme obtained at a local market at Reunion Island (ref. 42); [z6] one lab-hydrodistilled oil from the aerial parts of thyme growing wild in Italy (ref. 44); [z7] one lab-hydrodistilled thyme leaf oil from Argentina (ref. 49); [z8] one commercial oil sample from South Africa (ref. 58); [z9] one steam-distilled oil from the former Yugoslavia (ref. 63)

tr: trace (in column C: <0.1; in columns E, F[e], F[w4] and F[y7]: <0.05)

Table 5.85.3 Major constituents of thyme oils

Constituent	CAS	Percentage and range					
		A	B (68)	C (7)	D (34)	E (21)	F
Geraniol	106-24-1	0-26.0	0-26.0	tr	0-32.9	0.8	41[w8]; 57.1[y3]; 84.1[y1]; 97.1[y2]
Linalool	78-70-6	0.03-68.5	2.0-68.5	5.8	0.2-67.2	3.7	74.6[c]; 75.4[y3]; 77.7[d]; 93.8[y1]
α-Terpineol	98-55-5	0.1-2.8	0.2-2.3	0.6	0-11.5	2.8	8.6[d]; 12.8[v2]; 30.3[y1]; 90.4[y1]
Carvacrol	499-75-2	tr-77.8	tr-1.7	4.4	0.2-37.6	0.02	72[w8]; 84.1[y1]; 86.1[y9]; 89.7[x4]
p-Cymene	99-87-6	0.5-25.7	0.5-24.0	26.1	1.8-27.7	1.0	44.8[r]; 46.4[u2]; 55.9[x8]; 80[w8]
Thymol	89-83-8	0.2-47.8	0-38.8	37.4	0.8-36.6	0.1	75.7[z2]; 77.7[y2]; 77.8[z]; 78.2[u2]
Geranyl acetate	105-87-3	0-21.8	0-21.8		0-11.7	0.2	13[w8]; 20.9[u2]; 54.4[y3]; 68.6[d]
Terpinyl acetate (α-)	80-26-2	0.4-0.7			0-33.4	23.2	1.1[u]; 26.0[c]; 40.7[y3]; 62[w8]
trans-Sabinene hydrate	17699-16-0	0.1-6.8	0.2-5.5		0-3.3	0.7	1.1[h]; 1.2[e]; 29.8[d]; 42[w8]; 54.8[q]
Sabinene hydrate	546-79-2						52.2[y1]
1,8-Cineole	470-82-6	0.2-36.5	0.2-0.9	1.5	0-4.8	36.4	4.1[x4]; 9.4[v6]; 22.1[c]; 24.1[u2]
cis-Sabinene hydrate	15537-55-0	0.07-32.7	0.5-32.7		0.3-33.4	0.3	0.6[n]; 1.3[j]; 9.2[d]; 11.8[q]; 21.4[c]
Terpinen-4-ol	562-74-3	0.2-13.5	0.2-6.5	3.1	0-4.7	0.9	16[c]; 21.1[d]; 21.8[w5]; 29.6[y1]; 32[w8]
γ-Terpinene	99-85-4	0.2-21.2	0.4-9.5	0.9	2.6-19.6	1.1	22.7[x7]; 22.8[z5]; 23.3[f]; 27.6[u3]
Camphene	79-92-5	0.08-8.0	0.3-0.8	0.5	0.2-1.2	3.2	1.9[c]; 2.1[f]; 5.2[r]; 9.3[d]; 17.2[v9]

LEGEND: SEE UNDER TABLE 5.85.2

CONTACT ALLERGY/ALLERGIC CONTACT DERMATITIS

Thyme oil (unspecified)

General
Contact allergy to thyme oil (botanical origin not specified) has been reported in a few publications only. Routine testing has not been performed; there are some studies in which groups of selected patients were tested, but without data on relevance.

Patch testing in groups of patients
One positive patch test reaction to thyme oil was observed in a group of 51 patients (2%) allergic to *Myroxylon pereirae* resin (balsam of Peru) and/or turpentine and/or wood tar and/or colophony, tested with thyme oil 2% in petrolatum (88). Another group of 86 patients from Poland previously reacting to the fragrance mix was tested with thyme oil and four (4.6%) had a positive patch test reaction (89). In neither of these studies were data on relevance provided. In a group of 100 patients with ulcus cruris and tested with a special series including thyme oil, five (5%) had a positive patch test reaction; relevance is unknown (article not read) (91). Additional information on contact allergy to thyme oil may be found in ref. 90 (article not read).

Thyme oil ex *Thymus vulgaris*

General
Allergic contact dermatitis from thyme oil ex *Thymus vulgaris* has been reported in one publication only. The patient reacted to 'sweet thyme oil', which probably is the oil from *Thymus vulgaris* of the linalool chemotype. Linalool and possibly α-pinene may have been allergens in the oil in this case report.

Case reports and positive patch tests
An aromatherapist had non-occupational contact dermatitis with allergies to multiple essential oils used at work, including sweet thyme oil. The patient also reacted to geraniol, linalool, linalyl acetate, α-pinene, the fragrance mix and various other fragrance materials; α-pinene and linalool were demonstrated by GC-MS in the thyme oil (87). Sweet thyme oil is presumably the essential oil obtained from the linalool chemotype of *Thymus vulgaris*. In commercial thyme oils, linalool concentrations of up to 68.5% have been found and α-pinene had a maximum concentration of 5.1% (Table 5.85.2, column A).

LITERATURE

1 Nezhadali A, Nabavi M, Rajabian M. Chemical composition of the essential oil of *Thymus vulgaris* L. from Iran. J Essent Oil Bear Plants 2012;15:368-372

2 Shafaghat A, Shafaghatlonba M. Comparison of biological activity and chemical constituents of the essential oils from leaves of *Thymus caucasicus*, *T kotschyanus* and *T vulgaris*. J Essent Oil Bear Plants 2011;14:786-791

3 Edris AE, Shalaby AS, Fadel HM. Effect of organic agriculture practices on the volatile flavor components of some essential oil plants growing in Egypt: III. *Thymus vulgaris* L. essential oil. J Essent Oil Bear Plants 2009;12:319-326

4 Arraiza MP, Andrés MP, Arrabal C, López JV. Seasonal variation of essential oil yield and composition of thyme (*Thymus vulgaris* L.) grown in Castilla—La Mancha (central Spain). J Essent Oil Res 2009;21:360-362

5 Omidbaigi R, Kazemi Sh, Daneshfar E. Harvest time affecting the essential oil content and compositions of *Thymus vulgaris*. J Essent Oil Bear Plants 2008;11: 162-167

6 Horváth G, Szabó LG, Héthelyi E, Lemberkovics E. Essential oil composition of three cultivated *Thymus* chemotypes from Hungary. J Essent Oil Res 2006;18:315-317

7 Martínez-Pérez Y, Quijano-Celís CE, Pino JA. Volatile constituents of Cuban thyme oil (*Thymus vulgaris* L.). J Essent Oil Bear Plants 2007;10:179-183

8 Atti-Santos AC, Pansera MR, Paroul N, Atti-Serafini L, Moyna P. Seasonal variation of essential oil yield and composition of *Thymus vulgaris* L. (Lamiaceae) from south Brazil. J Essent Oil Res 2004;16:294-295

9 Asllani U, Toska V. Chemical composition of Albanian thyme oil (*Thymus vulgaris* L.). J Essent Oil Res 2003;15:165-167

10 Delpit B, Lamy J, Holland F, Chalchat J-C, Garry RP. Clonal selection of sabinene hydrate-rich thyme (*Thymus vulgaris*). Yield and chemical composition of essential oils. J Essent Oil Res 2000;12:387-391

11 Wei A, Shibamoto T. Antioxidant/lipoxygenase inhibitory activities and chemical compositions of selected essential oils. J Agric Food Chem 2010;58:7218-7225

12 Yahyazadeh M, Omidbaigi R, Zare R, Taheri H. Effect of some essential oils on mycelial growth of *Penicillium digitatum* Sacc. World J Microbiol Biotechnol 2008;24: 1445-1450

13 Letchamo W, Gosselin A, Hoelzl J, Marquard R. The selection of *Thymus vulgaris* cultivars to grow in Canada. J Essent Oil Res 1999;11:337-342

14 Stahl-Biskup E. The chemical composition of *Thymus* oils: A review of the literature 1960–1989. J Essential Oil Res 1991;3:61-82

15 Tognolini M, Barocelli E, Ballabeni V, Bruni R, Bianchi A, Chiavarini M, Impicciatore M. Comparative screening of plant essential oils: Phenylpropanoid moiety as basic core for antiplatelet activity. Life Sciences 2006;78:1419-1432. Data also presented in ref. 16

16 Sacchetti G, Maietti S, Muzzoli M, Scaglanti M, Manfredini S, Radice M, Bruni R. Comparative evaluation of 11 essential oils of different origin as functional antioxidants, antiradicals and antimicrobials in foods. Food Chem 2005;91:621-632. Data also presented in ref. 15

17 Nezhadali A, Nabavi M, Rajabian M, Akbarpour M, Pourali P, Amini F. Chemical variation of leaf essential oil at different stages of plant growth and in vitro antibacterial activity of *Thymus vulgaris* Lamiaceae, from Iran. Beni-Suef Univ J Basic Appl Sci 2014;3:87-92

18 Pirbalouti AG, Hashemi M, Ghahfarokhi FT. Essential oil and chemical compositions of wild and cultivated *Thymus daenensis* Celak and *Thymus vulgaris* L. Ind Crops Prod 2013;48:43-48

19 Gavahian M, Farahnaky A, Javidnia K, Majzoobi M. Comparison of ohmic-assisted hydrodistillation with traditional hydrodistillation for the extraction of essential oils from *Thymus vulgaris* L. Innov Food Sci Emerg Technol 2012;14:85-91

20 Nikolić M, Glamoclija J, Ferreira ICFR, Calhelha RC, Fernandes A, Marković T, et al. Chemical composition, antimicrobial, antioxidant and antitumor activity of *Thymus serpyllum* L., *Thymus algeriensis* Boiss. and Reut and *Thymus vulgaris* L. essential oils. Ind Crops Prod 2014;52:183-190

21 Jordan MJ, Martinez RM, Goodner KL, Baldwin EA, Sotomayor JA. Seasonal variation of *Thymus hyemalis* Lange and Spanish *Thymus vulgaris* L. essential oils composition. Ind Crops Prod 2006;24:253-263

22 Golmakani M-T, Rezaei K. Comparison of microwave-assisted hydrodistillation with the traditional hydrodistillation method in the extraction of essential oils from *Thymus vulgaris* L. Food Chem 2008;109:925-930

23 Razzaghi-Abyaneh M, Shams-Ghahfarokhi M, Rezaee M-B, Jaimand K, Alinezhad S, Saberi R, et al. Chemical composition and antiaflatoxigenic activity of *Carum carvi* L., *Thymus vulgaris* and *Citrus aurantifolia* essential oils. Food Control 2009;20:1018-1024

24 Dawidowicz AL, Rado E, Wianowska D, Mardarowicz M, Gawdzik J. Application of PLE for the determination of essential oil components from *Thymus vulgaris* L. Talanta 2008;76:878-884

25 Hudaib M, Speroni E, Di Pietra AM, Cavrini V. GC/MS evaluation of thyme (*Thymus vulgaris* L.) oil composition and variations during the vegetative cycle. J Pharm Biomed Anal 2002;29:691-700

26 Rota MC, Herrera A, Martinez RM, Sotomayor JA, Jordan MJ. Antimicrobial activity and chemical composition of *Thymus vulgaris*, *Thymus zygis* and *Thymus hyemalis* essential oils. Food Control 2008;19:681-687

27 Bhaskara Reddy MV, Angers P, Gosselin A, Arul J. Characterization and use of essential oil from *Thymus vulgaris* against *Botrytis cinerea* and *Rhizopus stolonifer* in strawberry fruits. Phytochem 1998;47:1515-1520

28 Sárosi Sz, Sipos L, Kókai Z, Pluhár Zs, Szilvássy B, Novák I. Effect of different drying techniques on the aroma profile of *Thymus vulgaris* analyzed by GC–MS and sensory profile methods. Ind Crops Prod 2013;46:210-216

29 Ballester-Costa C, Sendra E, Fernández-López J, Pérez-Álvarez JA, Viuda-Martos M. Chemical composition and *in vitro* antibacterial properties of essential oils of four *Thymus* species from organic growth. Ind Crops Prod 2013;50:304-311

30 Granger R, Passet J. Thymus vulgaris spontane de France: races chimiques et chemotaxonomie. Phytochemistry 1973;12:1683-1691

31 Adzet T, Granger R, Passet J, San Martin R. Le polymorphisme chimique dans le genre *Thymus*: sa signification taxonomique. Biochem. Syst 1977;5:269-272

32 Asbaghian S, Shafaghat A, Zarea K, Kasimov F, Salimi F. Comparison of volatile constituents, and antioxidant and antibacterial activities of the essential oils of *Thymus caucasicus*, *T. kotschyanus* and *T. vulgaris*. Nat Prod Commun 2011;6:137-140

33 Behbahani MH, Ghasemi Y, Khoshnoud MJ, Faridi P, Moradli G, Montazeri Najafabady N. Volatile oil composition and antimicrobial activity of two *Thymus* species. Pharmacogn J 2013;5:77-79

34 Chizzola R, Michitsch H, Franz C. Antioxidative properties of *Thymus vulgaris* leaves: Comparison of different extracts and essential oil chemotypes. J Agric Food Chem 2008;56:6897-6904

35 De Lisi A, Tedone L, Montesano V, Sarli G, Negro D. Chemical characterisation of *Thymus* populations belonging from Southern Italy. Food Chemistry 2011;125:1284-1286

36 Poulose AJ, Croteau R. Biosynthesis of aromatic monoterpenes. Conversion of γ-terpinene to *p*-cymene and thymol in *Thymus vulgaris* L. Arch Biochem Biophys 1978;187:307-314

37 Oraby MM, El-Borollosy AM. Essential oils from some Egyptian aromatic plants as an antimicrobial agent and for prevention of potato virus Y transmission by aphids. Ann Agricult Sci 2013;58:97-103

38 Usai M, Marchetti M, Foddai M, Del Caro A, Desogus R, Sanna I, Piga A. Influence of different stabilizing operations and storage time on the composition of essential oil of thyme (*Thymus officinalis* L.) and rosemary (*Rosmarinus officinalis* L.). LWT - Food Sci Technol 2011;44:244-249

39 Pandey A, Chattopadhyay P, Banerjee S, Pakshirajan K, Singh L. Antitermitic activity of plant essential oils and their major constituents against termite *Odontotermes assamensis* Holmgren (Isoptera: Termitidae) of North East India. Int Biodeter Biodegrad 2012;75:63-67

40 Viuda-Martos M, Mohamady MA, Fernández-López J, Abd ElRazik KA, Omer EA, Pérez-Alvarez JA, Sendra E. *In vitro* antioxidant and antibacterial activities of essentials oils obtained from Egyptian aromatic plants. Food Control 2011;22:1715-1722

41 Sajfrtova M, Sovova H, Karban J, Rochova J, Pavela R, Barnet M. Effect of separation method on chemical composition and insecticidal activity of Lamiaceae isolates. Ind Crops Prod 2013;47:69-77

42 Lucchesi ME, Chemat F, Smadja J. Solvent-free microwave extraction of essential oil from aromatic herbs: comparison with conventional hydro-distillation. J Chromatogr A 2004;1043:323-327

43 Rodríguez-Solana R, Daferera DJ, Mitsi M, Trigas P, Polissiou M, Tarantilis PA. Comparative chemotype determination of Lamiaceae plants by means of GC–MS, FT-IR, and dispersive-Raman spectroscopic techniques and GC-FID quantification. Ind Crops Prod 2014;62:22-33

44 Panizzi L, Flamini G, Cioni PL, Morelli L. Composition and antimicrobial properties of essential oil of four Mediterranean Lamiaceae. J Ethnopharmacol 1993;39:167-170

45 Baranauskiene R, Venskutonis PR, Viskelis P, Dambrauskiene E. Influence of nitrogen fertilizers on the yield and composition of thyme (*Thymus vulgaris*). J Agric Food Chem 2003;51:7751-7758

46 Echeverrigaray G, Agostini G, Tai-Serfeni L, Paroul N, Pauletti GF, Atti dos Santos AC. Correlation between the chemical and genetic relationships among commercial thyme cultivars. J Agric Food Chem 2001;49: 4220-4223

47 Abu-Darwish MS, Aludatt MH, Al-Tawaha AR, Ereifej K, Almajwal A, Odat N, et al. Seasonal variation in essential oil yield and composition from *Thymus vulgaris* L. during different growth stages in the south of Jordan. Nat Prod Res 2012;26:1310-1317

48 Machial CM, Shikano I, Smirle M, Bradbury R, Isman MB. Evaluation of the toxicity of 17 essential oils against *Choristoneura rosaceana* (Lepidoptera: Tortricidae) and *Trichoplusia ni* (Lepidoptera: Noctuidae). Pest Manag Sci 2010;66:1116-1121

49 Werdin Gonzalez JO, Gutierrez MM, Murray AP, Ferrero AA. Composition and biological activity of essential oils from Labiatae against *Nezara viridula* (Hemiptera: Pentatomidae) soybean pest. Pest Manag Sci 2011;67:948-955

50 Rustaiee AR, Yavari A, Nazeri V, Shokrpour M, Sefidkon F, Rasouli M. Genetic diversity and chemical polymorphism of some *Thymus* species. Chem Biodivers 2013;10:1088-1098

51 Pina-Vaz C, Rodrigues AG, Pinto E, Costa-de-Oliveira S, Tavares C, Satgueiro L, et al. Antifungal activity of *Thymus* oils and their major compounds. JEADV 2004;18:73-78

52 Torras J, Grau MD, Lopez JF, de las Heras FXC. Analysis of essential oils from chemotypes of *Thymus vulgaris* in Catalonia. J Sci Food Agric 2007;87:2327-2333

53 Kaloustian J, Abou L, Mikail C, Arniot MJ, Portugal H. Southern French thyme oils: chromatographic study of chemotypes. J Sci Food Agric 2005;85:2437-2444

54 Amvam Zollo PH, Biyiti L, Tchoumbougnang F, Menut C, Lamaty G, Bouchet Ph. Aromatic plants of tropical Central Africa. Part XXXII. Chemical composition and antifungal activity of thirteen essential oils from aromatic plants of Cameroon. Flavour Fragr J 1998;13:107-114

55 McGimpsey JA, Douglas MH, van Klink JW, Beauregard DA, Perry NB. Seasonal variation in essential oil yield and composition from naturalized *Thymus vulgaris* L. in New Zealand. Flavour Fragr J 1994;9:347-352

56 Piccaglia R, Marotti M. Composition of the essential oil of an Italian *Thymus vulgaris* L. ecotype. Flavour Fragr J 1991;6:241-244

57 Dawidowicz AL, Rado E, Wianowska D. Static and dynamic superheated water extraction of essential oil components from *Thymus vulgaris* L. J Sep Sci 2009;32:3034-3042

58 Sellamuthu PS, Sivakumar D, Soundy P. Antifungal activity and chemical composition of thyme, peppermint and citronella oils in vapor phase against avocado and peach postharvest pathogens. J Food Saf 2013;33:86-93

59 Thompson JD, Chalchat J-C, Michet A, Linhart YB, Ehlers B. Qualitative and quantitative variation in monoterpenes co-occurrence and composition in the essential oil of *Thymus vulgaris* chemotypes. J Chem Ecol 2003;29:859-880

60 Passet J. *Thymus vulgaris* L.: Chémotaxonomie et biogénèse monoterpénique. PhD Thesis, University of Montpellier, France, 1971

61 Pavela R, Vrchotová N, Tříska J. Mosquitocidal activities of thyme oils (*Thymus vulgaris* L.) against *Culex quinquefasciatus* (Diptera: Culicidae). Parasitol Res 2009;105:1365-1370

62 Mewes S, Krüger H, Pank F. Physiological, morphological, chemical and genomic diversities of different origins of thyme (*Thymus vulgaris* L.). Genet Resour Crop Evol 2008;55:1303-1311

63 Zeković Z, Lepojeviíc Z, Vujić Dj. Supercritical extraction of thyme (*Thymus vulgaris* L.). Chromatographia 2000;51(3-4):175-179

64 Lysakowska M, Denys A, Sienkiewicz M. The activity of thyme essential oil against *Acinetobacter* spp. Cent Eur J Biol 2011;6:405-413

65 Arslan M, Dervis S. Antifungal activity of essential oils against three vegetative compatibility groups of *Verticillium dahliae*. World J Microbiol Biotechnol 2010;26:1813-1821

66 Móricz AM, Ott PG, Böszörményi A, Lemberkovics E, Mincsovics E, Tyihák E. Bioassay-guided isolation and identification of antimicrobial compounds from thyme essential oil by means of overpressured layer chromatography, bioautography and GC–MS. Chromatographia 2012;75:991-999

67 Ahmad A, van Vuuren S, Viljoen A. Unravelling the complex antimicrobial interactions of essential oils — the case of *Thymus vulgaris* (thyme). Molecules 2014; 19:2896-2910

68 Schmidt E, Wanner J, Hiiferl M, Jirovetz L, Buchbauer G, Gochev V, et al. Chemical composition, olfactory analysis and antibacterial activity of *Thymus vulgaris* chemotypes geraniol, 4-thujanol/terpinen-4-ol, thymol and linalool cultivated in southern France. Nat Prod Commun 2012;7:1095-1098

69 Coelho JP, Cristino AF, Matos PG, Rauter AP, Nobre BP, Mendes RL, et al. Extraction of volatile oil from aromatic plants with supercritical carbon dioxide: experiments and modeling. Molecules 2012;17:10550-10573

70 Miladinović DL, Ilić BS, Mihajilov-Krstev TM, Nikolić ND, Miladinović LC, Cvetković OG. Investigation of the chemical composition–antibacterial activity relationship of essential oils by chemometric methods. Anal Bioanal Chem 2012;403:1007-1018

71 Samara de Lira Mota K, de Oliveira Pereira F, Araújo de Oliveira W, Oliveira Lima I, de Oliveira Lima E.

Antifungal activity of *Thymus vulgaris* L. essential oil and its constituent phytochemicals against *Rhizopus oryzae*: Interaction with ergosterol. Molecules 2012;17: 14418-14433

72 Grosso C, Figueiredo AC, Burillo J, Mainar AM, Urieta JS, Barroso JG, et al. Composition and antioxidant activity of *Thymus vulgaris* volatiles: Comparison between supercritical fluid extraction and hydrodistillation. J Sep Sci 2010;33:2211-2218

73 Soković MD, Vukojević J, Marin PD, Brkić DD, Vajs V, Van Griensven LJLD. Chemical composition of essential oils of *Thymus* and *Mentha* species and their antifungal activities. Molecules 2009;14:238-249. Virtually the same data are presented in ref. 74

74 Soković M, Glamočlija J, Ćirić A, Kataranovski D, Marin PD, Vukojević J, Brkić D. Antifungal activity of the essential oil of *Thymus vulgaris* L. and thymol on experimentally induced dermatomycoses. Drug Devel Indus Pharm 2008;34:1388-1393. Contains virtually the same data as ref. 73

75 Bozin B, Mimica-Dukic N, Simin N, Anackov G. Characterization of the volatile composition of essential oils of some Lamiaceae spices and the antimicrobial and antioxidant activities of the entire oils. J Agric Food Chem 2006;54:1822-1828

76 Lawrence BM. Progress in essential oils. Perfum Flavor 2008;33(1):58-60

77 Sipailiene A, Venskutonis PR, Baranauskiene R, Sarkinas A. Antimicrobial activity of commercial samples of thyme and marjoram oils. J Essent Oil Res 2006;18:698-703

78 Hudaib M, Aburjai T. Volatile components of *Thymus vulgaris* L. from wild-growing and cultivated plants in Jordan. Flavour Fragr J 2007;22:322-327

79 Pino JA, Estarrón M, Fuentes V. Essential oil of thyme (*Thymus vulgaris* L.) grown in Cuba. J Essent Oil Res 1997;9:609-610

80 Juliano C, Mattana A, Usai M. Composition and *in vitro* antimicrobial activity of the essential oil of *Thymus herba-barona* Loisel growing wild in Sardinia. J Essent Oil Res 2000;12:516-522

81 Dorman HJD, Surai P, Deans SG. *In vitro* antioxidant activity of a number of plant essential oils and phytoconstituents. J Essent Oil Res 2000;12:241-248

82 Shatar S, Altantsetseg S. Essential oil composition of some plants cultivated in Mongolian climate. J Essent Oil Res 2000;12:745-750

83 Lawrence BM. Progress in essential oils. Perfum Flavor 2004;29(3):44-? (last page unknown)

84 Imelouane B, Amhamdi H, Wathelet JP, Ankit M, Khedid, K, El-Bachiri A. Chemical composition and antimicrobial activity of essential oil of thyme (*Thymus vulgaris*) from eastern Morocco. Int J Agric Biol 2009;11:205-208. Data cited in ref. 29

85 Rhind JP. Essential oils. A handbook for aromatherapy practice, 2nd Edition. London: Singing Dragon, 2012

86 Lawless J. The encyclopedia of essential oils, 2nd Edition. London: Harper Thorsons, 2014

87 Dharmagunawardena B, Takwale A, Sanders KJ, Cannan S, Roger A, Ilchyshyn A. Gas chromatography: an investigative tool in multiple allergies to essential oils. Contact Dermatitis 2002;47:288-292

88 Rudzki E, Grzywa Z, Bruo WS. Sensitivity to 35 essential oils. Contact Dermatitis 1976;2:196-200

89 Rudzki E, Grzywa Z. Allergy to perfume mixture. Contact Dermatitis 1986;15:115-116

90 Escande JP, Foussereau J, Lantz JP, Basset A. Le problème des fausses sensibilisations croisées dans les allergies de groupe aux allergènes vegetaux. Rev Fr Allergol 1973;13:70-75

91 Le Roy R, Grosshans E, Foussereau J. Recherche d'allergie de contact dans 100 cas d'ulcère de jambe. Dermatosen in Beruf und Umwelt 1981;29(6):168-170

Chapter 5.86   THYME OIL, SPANISH

There are two major thyme oils: thyme oil obtained from *Thymus vulgaris* L. and oil of thyme containing thymol, Spanish type, obtained from *Thymus zygis* L. In this chapter, the Spanish type is discussed. The thyme oil from *Thymus vulgaris* is presented in Chapter 5.85.

DEFINITION

Thyme oil containing thymol, Spanish type (essential oil of thyme containing thymol, Spanish type) is the essential oil obtained from the flowering tops of the Spanish thyme, *Thymus zygis* L.

INCI NOMENCLATURE

Description/definition: Thymus zygis flower oil is the volatile oil obtained from the flower tips of the thyme, *Thymus zygis* L., Lamiaceae
INCI name EU & USA: Thymus zygis flower oil
CAS registry number(s): 85085-75-2
EINECS number(s): 285-397-0

Description/definition: Thymus zygis herb oil is an essential oil obtained from the herbs of the thyme, *Thymus zygis* L., Lamiaceae
INCI name EU: Thymus zygis herb oil (perfuming name, not officially an INCI name)
INCI name USA: Not in the Personal Care Products Council Ingredient Database
Other names: red thyme oil
CAS registry number(s): 85085-75-2
EINECS number(s): 285-397-0

ISO (INTERNATIONAL ORGANIZATION FOR STANDARDIZATION) STANDARD

ISO number: 14715
ISO name: Essential oil of thyme containing thymol, Spanish type
Botanical origin: *Thymus zygis* (Loefl.) L.
Parts of plant used: Flowering top
ISO values: ISO values (minimum and maximum concentrations) are shown in Table 5.86.1.

THE PLANT, THE OIL, AND THEIR USES

Thymus zygis L. (Spanish thyme, sauce thyme, red thyme) is an evergreen shrub growing to 0.3 m high. It is native to Morocco, Portugal and Spain (GRIN Taxonomy for Plants). Three subspecies are often mentioned in literature: *Thymus zygis* subsp. *gracilis* (Boiss.) R. Morales, *Thymus zygis* subsp. *sylvestris* (Hoffmanns & Link) Cout., and *Thymus zygis* subsp. *zygis* (4,5), but the latter is considered a synonym for *Thymus zygis* L. by The Plant List (www.theplantlist.org). The Spanish thyme, notably of the *gracilis* subspecies, is sometimes cultivated but is also collected on a large scale in the wild, for the production of leaves and essential oil (12,26). Spain is responsible for some 85% of the world's production of *Thymus zygis* products (5).

The dried leaves of the Spanish thyme are used as a condiment in southwestern Europe and are also sold by herb-shops and used in the food (mainly the meat) industry (7,10). Thyme is considered a medicinal plant, which is inserted in many Pharmacopoeias. Thyme (both *Thymus vulgaris* and *Thymus zygis*) is stated to possess carminative, antispasmodic, antitussive, expectorant, bactericidal, anthelmintic and astringent properties. It has been used for dyspepsia, chronic gastritis, asthma, diarrhea, enuresis, laryngitis, tonsillitis (as a gargle), and for pertussis and bronchitis (http://obtrandon.files.wordpress.com/2010/05/thymus-vulgaris-thyme.pdf).

The flowering herb is employed in the production of Spanish (red) thyme essential oil, which is used as aroma in the food industry, utilized in the cosmetics and fragrances industries and also employed for pharmaceutical purposes (20). The oils with high thymol content are highly valued. Thyme oils are also employed in aromatherapy practices, but usually in low concentrations in the case of high thymol content (34,35).

CHEMICAL COMPOSITION

Thyme oil containing thymol, Spanish type, is a yellowish to red clear mobile liquid, which has an intense aromatic–phenolic, spicy and slightly woody odor. The yield of essential oil from the flowering tops of *Thymus zygis* (Loeffl.) L. generally varies from 12.0 to 15.0 per cent. Only dried herb is used for distillation. The main producing country of this oil is Spain.

Literature data (up to December 5, 2014) on the chemical composition of thyme oils, Spanish and unpublished analytical data from one of us (E.S.) are shown

Table 5.86.1 ISO values (%) for thyme oil containing thymol, Spanish type[a]

Compound	CAS	Minimum	Maximum
Thymol	89-83-8 3	7.0	55.0
p-Cymene	99-87-6 1	4.0	28.0
γ-Terpinene	99-85-4	4.0	11.0
Linalool	78-70-6	3.0	6.5
Carvacrol	499-75-2	0.5	5.5
Myrcene	123-35-3	1.0	2.8
α-Terpinene	99-86-5	0.9	2.6
α-Pinene	80-56-8	0.5	2.5
Terpinen-4-ol	562-74-3	0.1	2.5
β-Caryophyllene	87-44-5	0.5	2.0
α-Thujene	2867-05-2	0.2	1.5
Methylcarvacrol	6379-73-3	0.1	1.5
trans-Sabinene hydrate	17699-16-0	tr	0.5

[a] ISO 14715 Essential oil of thyme containing thymol, Spanish type ©ISO 2008; Geneva, Switzerland, www.iso.org

in Table 5.86.2 in alphabetical order. In Spanish thyme oils from various origins, over 170 chemicals have been identified. About 47 per cent of these were found in a single reviewed publication only. The major compounds found in Spanish thyme oils from different sources are shown in Table 5.86.3. They include (highest concentrations in any study given) thymol (84.9%), linalool (82.3%), geranyl acetate (68.6%), carvacrol (68%), p-cymene (50.6%), γ-terpinene (40%), 1,8-cineole (34.3%), geraniol (23.5%), terpinen-4-ol (17.0%) and camphene (8.5%). Well-known ingredients of Spanish thyme oils that were present in high concentrations (>7%) in one or two studies were α-terpinyl acetate (70.3% and 79%), β-caryophyllene (30%), α-terpineol (22.2% and 28%), borneol (9.7% and 20%), camphor (20%), trans-sabinene hydrate (18.2% and 18.4%), caryophyllene oxide (13%), myrcene (10.0% and 12%), (E)-β-ocimene (11.9%) and linalyl acetate (8.6%). Uncommon or rare constituents of Spanish thyme oils found in high concentrations (>7%) in single studies include o-cymene (32.0%) and ledol (9.6%).

Commercial oils

The ten chemicals that had the highest maximum concentrations in 38 commercial Spanish thyme essential oil samples (concentration ranges provided) are the following: thymol (39.2-56.2), p-cymene (12.8-25.4), γ-terpinene (3.4-9.8), carvacrol (0.2-5.3), linalool (3.2-5.2), camphene (0.3-4.2), α-terpinene (0.7-3.3), limonene (1.8-2.3), β-caryophyllene (0.6-2.2) and borneol (0.3-2.1) (Erich Schmidt, unpublished analytical data).

Chemotypes

Based on investigation of 33 lab-hydrodistilled oils (13 from T. zygis subsp. sylvestris, 20 from subsp. gracilis) collected from 21 wild populations in south Spain, the following chemotype characterization has been proposed (5):

Chemotype	Main constituents and concentrations
Thymol	Thymol ≥ 43%
Thymol/p-cymene/γ-terpinene	Thymol 28-45%; p-cymene ≥ 13%; γ-terpinene ≥ 8%; linalool < 14%; Carvacrol < 25%
Carvacrol	Carvacrol ≥ 25%
Linalool	Linalool ≥ 14%; carvacrol < 38%
1,8-Cineole/myrcene/spathulenol	1,8-Cineole ≥ 5%; myrcene ≥ 10%; spathulenol ≥ 2%
1,8-Cineole /α-terpineol	1,8-Cineole ≥ 30%; α-terpineol ≥ 28%
α-Terpinyl acetate	α-Terpinyl acetate ≥ 79%

In the investigated populations, the thymol type was the most frequent (n=15, 45%), followed by the thymol/p-cymene/γ-terpinene chemotype (n=5, 15%), carvacrol (n=4, 12%) and linalool (n=3, 9%). In the subspecies gracilis, thymol is far more often the dominant ingredient (17/20 [85%], concentrations ranging from 28 to 80%) than in the sylvestris subspecies (3/13 [23%], concentrations ranging from 43 to 59%) (5). In a more recent study by the same authors, the dominance of thymol in T. zygis subsp. gracilis populations (37/45 = 82%) and to a lesser extent the subspecies sylvestris (7/19, 37%) was confirmed, albeit in the same populations of wild Spanish thyme (4). Earlier authors had described chemotypes based on one dominant ingredient (thymol, linalool, carvacrol) or combinations of chemicals (e.g., thymol/carvacrol, geraniol/geranyl acetate, 1,8-cineole/linalool, 1,8-cineole/thymol, linalool/thymol, α-terpineol/terpinyl acetate and 1,8-cineole/linalool/thymol/) (8,14,17).

A 1960-1989 literature review on Thymus zygis L. and other Thymus species is provided in ref. 3.

Table 5.86.2 Constituents identified in thyme oils, Spanish type

Constituent	CAS	Percentage and range				
		A	B (5)	C (12)	D (6)	E
(E)-Anethole	4180-23-8			0.2		
Aromadendrene	489-39-4	0-0.04		0.4		tr[w,x]; 0.2[c,q]; 0.3°; 0.5[z]
allo-Aromadendrene	25246-27-9	0-tr		0.04	tr-0.1	tr[c]
Benzyl acetate	140-11-4			0.05		0.2[c]
α-Bergamotene	17699-05-7					<0.01
Bicyclogermacrene	24703-35-3				tr-0.1	
α-Bisabolene	17627-44-0					0.04[j]
(Z)-α-Bisabolene	29837-07-8			0.01		
β-Bisabolene	495-61-4				0.1-0.8	0.2[f]; 0.5[j]
Borneol	507-70-0	0.3-2.1	20 (63)	0.7	2.9-3.5	0-4.1[b]; 2.5[j,p]; 3.1[i]; 4.2[o,x]; 5.9[l]; 9.7[v,y]
Bornyl acetate	76-49-3	tr-0.3	1.5 (?)	0.06	0.1-0.2	tr[c]; 0.1[d,e,l,q,x]; 0.2[l]; 0.4[s]; 1.9[k]
Bornyl formate	7492-41-3					+[z]
β-Bourbonene	5208-59-3				tr-0.1	0.1[g]
2-Butanone	78-93-3			<0.01		
Butyl caprylate	589-75-3			0.03		
Cadinane-1,4-diol						0.2[w]
β-Cadinene	523-47-7		4.3 (24)			

Table 5.86.2 Constituents identified in thyme oils, Spanish type (*continued*)

Constituent	CAS	Percentage and range				
		A	B (5)	C (12)	D (6)	E
γ-Cadinene	39029-41-9			0.08	tr-0.1	tr[e,w]; 0.1[g]; 0.2
δ-Cadinene	483-76-1	0-0.01		0.2	tr-0.1	<0.05[y]; tr[f,w]; 0.1[c,d,h]; 0.2[g,q]
Cadinenol	17910-08-6					0.1[w]
α-Cadinol	481-34-5				tr	0.02-0.1[b]; 0.1[w]
τ-Cadinol (10-epi-α-)	58580-31-7					0.3[w]
Calamenene	483-77-2					0.1[x]
cis-Calamenene	72937-55-4					tr[g]
Camphene	79-92-5	0.3-4.2	8.5 (59)	0.4	3.3-4.5	0.03-3.6[b]; 1.4[i]; 1.6[j]; 3.7[z]; 5.2[l]; 5.3[v]; 6.2[y]
Campholenal	23727-15-3				tr	tr[e]
Camphor	76-22-2	0.4-1.1	20 (50)	0.02	1.4-4.1	tr[c]; 0.07-1.4[b]; 0.4[f]; 1.2[z]; 2.0[z]; 3.1[y]; 3.9[o]
δ2-Carene (=δ4-)	554-61-0					0.08[h]
δ3-Carene	13466-78-9			0.09		tr[c]; 0.05[j]; 0.08[q]; 0.09[h]; 0.2[f]
Carvacrol	499-75-2	0.2-5.3	68 (62)	3.6	1.3-25.0	0.03-22.8[b]; 3.5[c]; 8.1[l]; 26.0[o]; 43.6[v]; 43.9[z]
Carvacryl acetate	6380-28-5					tr[i,w]
(E)-Carveol	1197-07-5				tr-0.1	0.03[j]
Carvone	99-49-0				tr-0.1	
Carvotanacetone	499-71-8					0.09[j]
β-Caryophyllene	87-44-5	0.6-2.2	30 (62)	1.2	1.0-3.3	0-1.9[b]; 1.5[r]; 1.8[p]; 3.0[n]; 3.6[g]; 4.3[i]; 4.6[o]
Caryophyllene oxide	1139-30-6	tr-0.4	13 (61)	0.3	0.3-0.8	0.07-1.3[b]; 0.3[i]; 0.4[c,d,e]; 0.6[h]; 0.9[v]; 1.0[g]
1,4-Cineole	470-67-7					0.2[i]; 1.5[h]
1,8-Cineole	470-82-6	tr-1.1	31 (50)	0.03	0.7-12.0	0.08-34.5[b]; 0.6[p]; 1.3[k]; 1.8[s]; 3.0[j]; 20.0[o]
Citral	5392-40-5		2.1 (25)			0.1[n]
Citronellol (β-, DL-)	106-22-9			<0.05		0.06-0.2[b]; 0.1[j]; 0.8[n]
Citronellyl acetate	150-84-5			0.01		
α-Copaene	3856-25-5				tr-0.1	tr[x]; 0.1[g]
β-Cubebene	13744-15-5					0.2[j]
Cuminaldehyde	122-03-2					1.2[z]
Cuminyl alcohol	536-60-7				tr-0.1	
o-Cymene	527-84-4					32.0[h]
p-Cymene	99-87-6	12.8-25.4	25 (62)	22.4	11.0-17.0	0.2-28.2[b]; 36.6[f]; 39.4[v,y]; 42.2[o]; 50.6[l]
Cymenene	26444-18-8					0.1[f]
m-Cymenene	1124-20-5					0.2[h]
p-Cymenene	1195-32-0				tr-0.1	0.04[j]
p-Cymen-8-ol	1197-01-9			0.08	tr-0.6	tr[c]; 0.1[d,e,z]; 0.2[g,h]; 0.3[p]; 1.1[z]
Decanal	112-31-2			0.02		0.2[c]
1-Decanol	112-30-1					0.7[i]
Dihydrocarveol	38049-26-2					0.04[n]
trans-Dihydrocarvone	5948-04-9			0.06	tr-0.4	tr[e,l]; 0.2[c]
2,5-Dimethylstyrene	2039-89-6					0.2[g]
trans-β-Elemenone	20303-60-0					0.3[h]
Elemol (α-)	639-99-6					0-0.3[b]
Ethylcarvacrol	4732-13-2					0.3[h]
α-Eudesmol	473-16-5					<0.05[y]
β-Eudesmol	473-15-4					<0.05[y]; tr[w]
γ-Eudesmol	1209-71-8					<0.05[y]
Eugenol	97-53-0			0.1		0.1[d]
Fenchone	1195-79-5					tr[w]; 0.07[h]; 1.4[n]
Geranial	141-27-5			0.4	tr-1.0	0.07[j]; 0.1[e]; <0.3[d]; 0.3[c]
Geraniol	106-24-1			0.01	0.1-19.8	0.03-0.7[b]; 0.9[j]; 5.8[z]; 16.1[z]; 18.2[i]; 23.5[o]
Geranyl acetate	105-87-3				0.5-20.8	0.02-5.2[b]; 12.9[z]; 17.3[i]; 20.5[o]; 68.6[z]
Geranyl propanoate	105-90-8					0.4[j]
Germacrene D	23986-74-5				tr-0.2	<0.05[y]; 0.1[f,w]; 0.8[h]
Germacrene D-4-ol	198991-79-6					0.2[h]
Globulol	489-41-8					1.7[j]
1-Heptanol	111-70-6					tr[i]
3-Heptanone	106-35-4			0.01		
Hexanol (1-)	111-27-3			0.04		
3-Hexanone	589-38-8			0.01		
(Z)-3-Hexen-1-ol	928-96-1			0.05		
α-Humulene	6753-98-6	0-0.07		0.2	tr-0.1	tr[f]; 0.05[j]; 0.1[e,g]; 0.2[d,z]; 0.4[i]
β-Humulene	116-04-1					0.2[h]

Table 5.86.2 Constituents identified in thyme oils, Spanish type (*continued*)

Constituent	CAS	Percentage and range				
		A	B (5)	C (12)	D (6)	E
Humulene epoxide	96638-51-6					0.2[w]
α-Ionone	127-41-3			0.01		
β-Ionone	79-77-6			0.02		
Isoamyl alcohol	123-51-3			0.01		
Isoborneol	124-76-5			0.01		tr[w]; 0.1[c]
Isobornyl acetate	125-12-2			0.02		0.08-1.2[b]
Isoeugenol	97-54-1			0.04		
Isogeranial	72203-98-6					0.05[j]
Isopulegol	89-79-2					0.06[n]
(Z)-Jasmone	488-10-8			0.03		
Lavandulol	498-16-8		1.0 (?)			
Ledol	577-27-5		9.6 (19)			
Limonene	138-86-3	1.8-2.3	5.3 (54)	0.5	0.6-2.2	0-19.0[b]; 1.6[j]; 1.7[g]; 1.9[o]; 2.5[c]; 2.8[y]; 3.0[f,p,z]
cis-Limonene oxide	13837-75-7			0.01		
trans-Limonene oxide	4959-35-7					0.5[c]
Linalool	78-70-6	3.2-5.2	49 (62)	3.4	3.5-30.0	1.1-91.4[b]; 49.2[o]; 73.6[w]; 79.0[z]; 82.3[c]
cis-Linalool oxide	11063-77-7			0.03		<0.05[y]; 0.03[j]; 0.09[h]; 0.8[z]; 1.1[c]
cis-Linalool oxide, furanoid	11063-77-7					0.5[p]
cis-Linalool oxide, pyranoid	14009-71-3				0.1-0.3	
trans-Linalool oxide	11063-78-8			0.01		<0.05[y]; 0.1[g]; <0.2[d]; 0.4[z]
trans-Linalool oxide, furanoid	34995-77-2				tr-0.1	0.4[p]; 0.5[l]
Linalyl acetate	115-95-7	tr-0.1			tr-0.5	0.04-1.6[b]; 0.7[l]; 1.0[o,z]; 1.3[j]; 3.4[w]; 8.6[z]
cis-p-Menth-2-en-1-ol	29803-82-5					0.1[f]; 0.2[g]; 0.9[p]
trans-p-Menth-2-en-1-ol	29803-81-4					0.1[g]; 0.2[j]; 0.5[p]
trans-p-Menth-5-en-2-ol						0.2[p]
Menthol	89-78-1					0.2[n]
Methyl acetate	79-20-9			0.03		
3-Methyl-2-buten-1-ol	556-82-1			0.02		
Methylcarvacrol	6379-73-3	0.06-0.9		0.5		tr[c]; 0.1[l]; 0.5[q]; 0.8[e]; 1.1[d]
Methyl eugenol	93-15-2			0.01		
Methyl thymol	1076-56-8	0.09-0.4	6.2 (20)	0.02		tr[c]; 0.5[j]; 1.5[z]
α-Muurolene	10208-80-7					tr[w]; 0.1[g]; 0.3[z]
Myrcene (β-)	123-35-3	0.1-1.9	12 (62)	1.3	1.0-2.3	0.08-3.1[b]; 2.0[o]; 2.1[r]; 3.0[f,g]; 5.2[c]; 10.0[p]
Myrcenol	543-39-5					0.5[o]; 0-28.6[z];
Myrtenal	564-94-3					0.2[c]
Myrtenol	515-00-4				tr-0.1	
Neral	106-26-3			0.01	tr-0.5	0.07[j]; 0.4[i]
Nerol	106-25-2			<0.05	tr-0.8	0.08[j]; 0.3[i]
Neryl acetate	141-12-8			0.01	tr-0.3	
Nonanal	124-19-6			0.02		
β-Ocimene	13877-91-3		1.9 (?)			0.07[j]
(E)-β-Ocimene	3779-61-1			0.02	tr	tr[c]; 0.1[f,z]; 0.2[g]; 11.9[h]
(Z)-β-Ocimene	3338-55-4			0.01	tr-0.1	0-1.1[b]; tr[f]; <0.05[y]; 0.05[h]
Octanol (1-)	111-87-5	0.04-0.1			0.1	tr[i]
3-Octanol	589-98-0		3.1(22)	0.2		0.04[j]; 0.2[d]; 1.0[y]; 1.7[l]
3-Octanone	106-68-3	tr-0.4	2.1 (55)	0.4	0.1-0.2	0.1[c,e]; 0.2[d,z]; 0.4[f]; 0.5[g]
1-Octen-3-ol	3391-86-4			0.5	0.1	0.1[z]; 0.2[j]; 0.3[c,d,e]; 0.6[f,g]
Octyl acetate	112-14-1			0.01		
Perillaldehyde	2111-75-3	0-0.1		0.02		<0.3[d]
α-Phellandrene	99-83-2		0.2 (?)	0.1	0.1-0.3	0.09[q]; 0.1[e,x]; 0.2[c,d,g,h,i]; 0.3[f,z]
β-Phellandrene	555-10-2					0.2[q]; 0.3[f]; 1.6[y]
trans-Pinane	10281-53-5					0.2[h]
cis-Pinan-2β-ol	4948-29-2					0.1[h]
α-Pinene	80-56-8	0.04-0.4	3.7 (60)	0.7	0.5-1.3	0-4.8[b]; 2.1[k]; 3.0[y]; 3.6[c]; 4.0[p]; 4.9[o]; 5.2[l]
β-Pinene	127-91-3	tr-0.4	7 (59)	0.2	0.5-1.3	0.04-2.3[b]; 0.3[j]; 0.4[f]; 0.7[l]; 0.9[z]; 2.0[o]; 2.1[h]
trans-Pinocarveol	1674-08-4			0.01	tr-0.2	tr[e]
Pinocarvone	30460-92-5			<0.01	tr	
Pulegone	89-82-7					2.4[n]
Sabinene	3387-41-5			0.01	0.1-0.7	0-1.2[b]; 0.2[f]; 0.3[j]; 0.4[h,o]; 1.3[p]; 1.9[z]; 2.0[c]

Table 5.86.2 Constituents identified in thyme oils, Spanish type (*continued*)

Constituent	CAS	Percentage and range				
		A	B (5)	C (12)	D (6)	E
cis-Sabinene hydrate	15537-55-0	0.08-0.7		0.1	0.1-0.8	tr[c,e]; 0.1[d]; 0.2[l,q]; 0.5[j]; 0.9[y]; 1.2[p]; 1.8[z]
trans-Sabinene hydrate	17699-16-0	0-0.08		0.7	0.1-1.5	0.2-1.7[b]; 0.6[d]; 1.1[y]; 6.9[z]; 18.2[c]; 18.4[p]
β-Selinene	17066-67-0					tr[w]
α-Sesquiphellandrene						0.04[j]
β-Sesquiphellandrene	20307-83-9				tr-0.1	
(*E*)-Sesquisabinene hydrate	145512-84-1					0.04[j]
Shyobunol	35727-45-8					0.06[j]
Spathulenol	6750-60-3		2.8 (50)	0.1	tr-0.2	0-0.4[b]; tr[e]; 0.1[d]; 0.2[c]; 0.3[j]
α-Terpinen-7-al	1197-15-5					0.3[h]
α-Terpinene	99-86-5	0.7-3.3	4.3 (60)	1.2	0.8-2.7	0-1.9[b]; 2.1[o]; 2.7[f]; 3.1[c]; 3.5[p]; 3.9[j]; 6.9[n]
γ-Terpinene	99-85-4	3.4-9.8	40 (62)	3.3	3.8-11.5	0.2-11.0[b]; 12[z]; 13.6[j]; 15.8[v]; 16.9[o]; 21.0[f]
Terpinen-4-ol	562-74-3	0.2-2.1	2.5 (54)	0.6	0.9-3.9	0.03-17.0[b]; 2.5[j]; 3.1[v]; 10.1[z]; 10.9[p]; 11.0[c]
1-Terpineol	586-82-3					0.7[i]
α-Terpineol	98-55-5	tr-0.6	28 (28)	0.09	0.2-0.8	0-1.1[b]; 0.3[z]; 0.4[g]; 1.4[y]; 0.9[k]; 2.3[c]; 22.2[p]
β-Terpineol	138-87-4					+[z]; 0.3[k]
γ-Terpineol	586-81-2			0.04		
δ-Terpineol	7299-42-5				tr-0.3	0-0.9[b]
Terpinolene (α-)	586-62-9	0.09-0.3	5 (35)	0.2	0.1-0.6	0-0.8[b]; 0.2[e]; 0.5[k]; 0.9[z]; 1.5[c]; 3.2[v]
Terpinyl acetate	8007-35-0			0.01		
α-Terpinyl acetate	80-26-2	tr-0.2	79 (10)		tr-0.1	+[z]; 1.3[z]; 70.3[z]
2,4-Thujadiene	119205-49-1					0.03[j]
Thuja-2,4(10)-diene	36262-09-6					0.1[v]
α-Thujene	2867-05-2	tr-0.4	1.1 (?)	1.1	0.9-1.1	0-3.2[b]; 0.7[c]; 0.9[j]; 1.0[d,l]; 1.1[m]; 1.6[y]; 2.6[f]
α-Thujone	546-80-5					0.1[w]
β-Thujone (*cis*-)	471-15-8			0.01		
Thymol	89-83-8	39.2-56.2	80 (62)	64.2	5.2-23.8	0.02-71.8[b]; 62.1[e]; 62.9[z]; 68.1[c]; 84.9[t]
Thymoquinone	490-91-5			0.02		
Thymyl acetate	528-79-0					tr[w]
Tricyclene	508-32-7			0.05	0.1-0.2	0.2[c,y]; 1.3[h]
Valencene	4630-07-3					tr[e]; 0.4[c]
Verbenene	4080-46-0			0.01		tr[z]; 0.03[h]
cis-Verbenol	1845-30-3			0.02		tr[c,e,w]
trans-Verbenol	1820-09-3			0.01	0.1-0.3	tr[c]; 1.4[y]
Verbenone	80-57-9	tr-0.1		0.02		tr[e]; 0.2[z]
Viridiflorene (ledene)	21747-46-6		1.4 (?)			
Viridiflorol	552-02-3					0-0.8[b]; 0.1[y]

A thirty-eight thyme, Spanish type, essential oil samples from Spain analyzed between 1998 and 2013; lowest and highest concentrations given (E. Schmidt, unpublished data)

B sixty-three lab-hydrodistilled oils from the aerial parts (after removal of the woody stems) of two subspecies of *Thymus zygis* (*Thymus zygis* subsp. *sylvestris* Hoffmanns & Link Brot. ex Coutinho and *Thymus zygis* subsp. *gracilis* Boiss. R. Morales) from 21 populations growing wild in South Spain, collected in the flowering phase; the data of 34 samples were presented; except for borneol, which was present in all 63 samples and had a lowest concentration of 1%, all chemicals were absent in at least one oil, and therefore the lowest concentration was zero; in brackets the number of samples (out of 63) in which the chemical was present (ref. 5); the data presented by the same authors in a more recent study (4) bear a great resemblance to these results and are therefore not presented separately

C three groups of four oils obtained from *Thymus zygis* ssp. *gracilis* plants of the thymol chemotype cultivated in Spain under three different watering levels and harvested at the end of the blooming stage; for each group, the means were provided; highest means for any group given (ref. 12)

D four lab-hydrodistilled oils from wild growing *T. zygis* subsp. *sylvestris* in the flowering stage collected at four different sites in central Portugal; lowest and highest concentrations given (ref. 6)

E data from other studies (indicated with superscript letters); highest concentrations found in any study reviewed here given; when two or more oils were investigated, only the highest concentrations are mentioned, unless indicated otherwise

[a] incorrect identity; [b] three lab-hydrodistilled oils from *T. zygis* subsp. *gracilis* and five from *T. zygis* subsp. *sylvestris*, growing wild in southeast Spain and harvested in the flowering stage; lowest and highest concentrations given; the oils came from a group of 20 investigated samples and were chosen as representative for the various chemotypes; data from (apparent) hybrids provided by the author are not shown here (ref. 8); [c] three lab-hydrodistilled oils from the aerial parts (not specified whether blooming or not) of Spanish thyme cultivated in Spain; one was of the thymol chemotype, two of the linalool type (ref. 9); [d] eighty lab-hydrodistilled oils obtained from eighty *T. zygis* subsp. *gracilis* plants cultivated in Spain and harvested between blooming stage and beginning of fruit maturation; there were four groups of twenty plants each grown under different watering levels; the data were presented as the means of each

Table 5.86.2 Constituents identified in thyme oils, Spanish type (*continued*)

group; highest mean concentrations given (ref. 7); [e] one oil obtained from *T. zygis* subsp. *gracilis* (ref. 29); [f] one lab-hydrodistilled oil from the aerial parts of the *sylvestris* subspecies, thymol chemotype (ref. 19); [g] one lab-prepared oil from *Thymus zygis* subsp. *zygis* cultivated in Portugal (ref. 22); [h] one lab-hydrodistilled oil from the aerial parts of *Thymus zygis* in the flowering stage growing wild in Portugal (ref. 20); [i] one lab steam distilled and one lab-hydrodistilled oil from flowering *T. zygis* subsp. *sylvestris* growing wild in the north of Portugal (ref. 17); [j] one lab-hydrodistilled oil from pre-flowering *T. zygis* subsp. *gracilis* collected from the wild in Portugal (ref. 13); [k] one lab-hydrodistilled oil from the aerial parts of *Thymus zygis*, origin unknown (ref. 1); [l] one steam-distilled oil from the aerial parts in full bloom of Spanish thyme growing wild in Morocco; very poor quality study (ref. 2); [m] one industrially steam-distilled oil from leaves, stem and flowers of *T. zygis* from Spain (ref. 10); [n] one lab-hydrodistilled oil from plants cultivated in Italy (ref. 11); [o] unspecified number of lab-hydrodistilled oils from *T. zygis* subsp. *sylvestris* growing wild in Portugal (ref. 14); [p] three steam-distilled oils, two from *T. zygis* subsp. *gracilis* cultivated in Spain and one from wild *T. zygis*, 'linalool-type' ('red-verbena') (ref. 15); [q] one commercial red thyme oil ex *Thymus zygis* L. (ref. 27); [r] one Spanish *Thymus zygis* thyme oil (ref. 28); [s] one commercial oil from Spain (ref. 16); [t] one oil from *Thymus zygis* chemotype thymol obtained from a local producer in South France (ref. 21); [u] one commercial oil of unknown origin (ref. 23); [v] five lab-hydrodistilled oils from the aerial parts of *Thymus zygis* in the flowering phase collected in the wild in Portugal, four of subspecies *sylvestris* and one of subspecies *zygis* (ref. 24); [w] one linalool-rich oil obtained from *T. zygis* subspecies *zygis* harvested from their natural habitat in Spain (ref. 31); [x] one thymol-rich oil obtained from *T. zygis* subspecies *zygis* harvested from their natural habitat in Spain (ref. 31); [y] five lab-hydrodistilled Portuguese *T. zygis* oils, 4 from ssp. *sylvestris*, one from ssp. *zygis* (ref. 32); [z] data from various studies from before 1984 cited in ref. 33

tr: trace (in columns D,E[c] and E[g]: <0.05)

Table 5.86.3 Major constituents of thyme oils, Spanish types

Constituent	CAS	Percentage and range				
		A	B (5)	C (12)	D (6)	E
Thymol	89-83-8	39.2-56.2	80 (62)	64.2	5.2-23.8	0.02-71.8[b]; 62.1[e]; 62.9[z]; 68.1[c]; 84.9[t]
Linalool	78-70-6	3.2-5.2	49 (62)	3.4	3.5-30.0	1.1-91.4[b]; 49.2[o]; 73.6[w]; 79.0[z]; 82.3[c]
Geranyl acetate	105-87-3				0.5-20.8	0.02-5.2[b]; 12.9[z]; 17.3[j]; 20.5[o]; 68.6[z]
Carvacrol	499-75-2	0.2-5.3	68 (62)	3.6	1.3-25.0	0.03-22.8[b]; 3.5[c]; 8.1[l]; 26.0[o]; 43.6[v]; 43.9[z]
p-Cymene	99-87-6	12.8-25.4	25 (62)	22.4	11.0-17.0	0.2-28.2[b]; 36.6[f]; 39.4[v,y]; 42.2[o]; 50.6[l]
γ-Terpinene	99-85-4	3.4-9.8	40 (62)	3.3	3.8-11.5	0.2-11.0[b]; 12[z]; 13.6[i]; 15.8[v]; 16.9[o]; 21.0[f]
1,8-Cineole	470-82-6	tr-1.1	31 (50)	0.03	0.7-12.0	0.08-34.5[b]; 0.6[p]; 1.3[k]; 1.8[s]; 3.0[j]; 20.0[o]
Geraniol	106-24-1			0.01	0.1-19.8	0.03-0.7[b]; 0.9[j]; 5.8[z]; 16.1[i]; 18.2[i]; 23.5[o]
Terpinen-4-ol	562-74-3	0.2-2.1	2.5 (54)	0.6	0.9-3.9	0.03-17.0[b]; 2.5[l]; 3.1[y]; 10.1[z]; 10.9[p]; 11.0[c]
Camphene	79-92-5	0.3-4.2	8.5 (59)	0.4	3.3-4.5	0.03-3.6[b]; 1.4[i]; 1.6[j]; 3.7[z]; 5.2[i]; 5.3[v]; 6.2[y]

LEGEND: SEE UNDER TABLE 5.86.2

CONTACT ALLERGY/ALLERGIC CONTACT DERMATITIS

No reports on contact allergy to thyme oil, specifically mentioned to be obtained from *Thymus zygis*, have been found. Literature on contact allergy to/allergic contact dermatitis from 'thyme oil' (botanical source not specified) is discussed in Chapter 5.85 Thyme oil (*Thymus vulgaris*).

LITERATURE

1 Dorman HJD, Deans SG. Chemical composition, antimicrobial and in vitro antioxidant properties of *Monarda citriodora* var. *citriodora*, *Myristica fragrans*, *Origanum vulgare* ssp. *hirtum*, *Pelargonium* sp. and *Thymus zygis* oils. J Essent Oil Res 2004;16:145-150

2 Tantaoui-Elaraki A, Lattaoui N, Errifi A, Benjilali B. Composition and antimicrobial activity of the essential oils of *Thymus broussonettii*, *T. zygis* and *T. satureioides*. J Essent Oil Res 1993;5:45-53

3 Stahl-Biskup E. The chemical composition of thymus oils: A review of the literature 1960–1989. J Essent Oil Res 1991;3:61-82

4 Pérez-Sánchez R, Gálvez C, Ubera JL. Bioclimatic influence on essential oil composition in South Iberian Peninsular populations of *Thymus zygis*. J Essent Oil Res 2012;24:71-81

5 Pérez-Sánchez R, Ubera JL, Lafont F, Gálvez C. Composition and variability of the essential oil in *Thymus zygis* from Southern Spain. J Essent Oil Res 2008;20:192-200

6 Gonçalves MJ, Cruz MT, Cavaleiro C, Lopes MC, Salgueiro L. Chemical, antifungal and cytotoxic evaluation of the essential oil of *Thymus zygis* subsp. *sylvestris*. Ind Crops Prod 2010;32:70-75

7 Jordán MJ, Martínez RM, Martínez C, Moñino I, Sotomayor JA. Polyphenolic extract and essential oil quality of *Thymus zygis* ssp. *gracilis* shrubs cultivated under different watering levels. Ind Crops Prod 2009;29:145-153

8 Sáez F. Essential oil variability of *Thymus zygis* growing wild in southeastern Spain. Phytochem 1995;40:819-825

9 Rota MC, Herrera A, Martínez RM, Sotomayor JA, Jordán MJ. Antimicrobial activity and chemical composition of *Thymus vulgaris, Thymus zygis* and *Thymus hyemalis* essential oils. Food Control 2008;19:681-687

10 Ballester-Costa C, Sendra E, Fernández-López J, Pérez-Álvarez JA, Viuda-Martos M. Chemical composition and in vitro antibacterial properties of essential oils of four *Thymus* species from organic growth. Ind Crops Prod 2013;50:304-311

11 De Lisi A, Tedone L, Montesano V, Sarli G, Negro D. Chemical characterisation of *Thymus* populations belonging from Southern Italy. Food Chem 2011;125:1284-1286

12 Sotomayor JA, Martínez RM, García, AJ, Jordán MJ. Thymus *zygis* subsp. *gracilis*: watering level effect on phytomass production and essential oil quality. J Agric Food Chem 2004;52:5418–5424

13 Teixeira MA, Rodrigues AE. Coupled extraction and dynamic headspace techniques for the characterization of essential oil and aroma fingerprint of *Thymus* species. Ind Eng Chem Res 2014;53:9875-9882

14 Proença da Cunha A, Salgueiro LR. The chemical polymorphism of *Thymus zygis* ssp. *sylvestris* from Central Portugal. J Essent Oil Res 1991;3:409-412

15 Sánchez Gómez P, Sotomayor Sánchez JA, Soriano Cano MC, Correal Castellanos E, García Vallejo MC. Chemical composition of the essential oil of *Thymus zygis* ssp. *gracilis* c.v. "linalool type," and its performance under cultivation. J Essent Oil Res 1995;7:399-402

16 Youdim KA, Deans SG, Finlayson HJ. The antioxidant properties of thyme (*Thymus zygis* L.) essential oil: an inhibitor of lipid peroxidation and a free radical scavenger. J Essent Oil Res 2002;14:210-215

17 Moldao-Martins M, Bernardo-Gil MG, Beirao da Costa ML, Rouzet M. Seasonal variation in yield and composition of *Thymus zygis* L. subsp. *sylvestris* essential oil. Flavour Fragr J 1999;14:177-182. Data partly also published in ref. 18

18 Moldao-Martins M, Bernardo-Gil MG, Beirao-da-Costa ML. Sensory and chemical evaluation of *Thymus zygis* L. essential oil and compressed CO2 extracts. Eur Food Res Technol 2002;214:207-211

19 Machado M, Dinis AM, Salgueiro L, Cavaleiro C, Custódio JBA, do Céu Sousa M. Anti-*Giardia* activity of phenolic-rich essential oils: effects of *Thymbra capitata, Origanum virens, Thymus zygis* subsp. *sylvestris*, and *Lippia graveolens* on trophozoites growth, viability, adherence, and ultrastructure. Parasitol Res 2010;106:1205-1215. Data partly also presented in ref. 25

20 Amarti F, El Ajjouri M, Ghanmi M, Satrani B, Aafi A, Farah A, et al. Composition chimique, activité antimicrobiennne et antioxydante de l'huile essentielle de *Thymus zygis* du Maroc. Phytothérapie 2011;9:149-157

21 Kaloustian J, Chevalier J, Mikail C, Martino M, Abou L, Vergnes M-F. Étude de six huiles essentielles: composition chimique et activité antibactérienne. Phytothérapie 2008;6:160-164

22 Pina-Vaz C, Gonçalves Rodrigues A, Pinto E, Costa-de-Oliveira S, Tavares C, et al. Antifungal activity of *Thymus* oils and their major compounds. JEADV 2004; 18:73-78

23 Penalver P, Huerta B, Borge C, Astorga R, Romero R, Perea A. Antimicrobial activity of five essential oils against origin strains of the Enterobacteriaceae family. APMIS 2005;113:1-6

24 Dandlen SA, Lima AS, Mendes MD, Miguel MG, Faleiro L, Sousa MJ, et al. Antioxidant activity of six Portuguese thyme species essential oils. Flavour Fragr J 2010;25:150-155. Data also presented in ref. 32

25 Machado M, Santoro G, Sousa MC, Salgueiroa L, Cavaleiro C. Activity of essential oils on the growth of *Leishmania infantum* promastigotes. Flavour Fragr J 2010;25:156-160

26 Lawrence BM. Progress in essential oils. Perfum Flavor 2008;(May):58-? (last page unknown)

27 Kubeczka K-H, Formacek V. Essential oils analysis by capillary gas chromatography and carbon-13 NMR spectroscopy, 2nd edition. New York, USA: John Wiley and Sons, 2002:355-360. Data cited in ref. 26

28 Milchard MJ, Clery R, Da Costa N, Flowerdew M, Gates L, Moss N, et al. Application of gas-liquid chromatography to the analysis of essential oils. Fingerprints of 12 essential oils. Perfum Flavor 2004;29(5):28-36. Data cited in ref. 26

29 Martinéz S, Madrid J, Hernandez F, Megías MF, Sotomayor JA, Jordan MJ. Effect of thyme essential oils (*Thymus hyemalis* and *Thymus zygis*) and monensin on in vitro ruminal degradation and volatile fatty acid production. J Agric Food Chem 2006;54:6598-6602

30 Lawrence BM. Progress in essential oils. Perfum Flavor 2004;29(3):44-? (last page unknown)

31 Velasco-Negueruela A, Perez-Alonso MJ. Nuevos datos sobre la composicion quimica de aceites esenciales procedentes de tomillos ibericos. Bot Complutensis 1990;16:91-97. Data cited in ref. 30

32 Dandlen SA, Miguel MG, Duarte J, Faleiro ML, Sousa MJ, Lima AS, et al. Acetylcholinesterase inhibition activity of Portuguese *Thymus* species essential oils. J Essent Oil Bear Plants 2011;14:140-150. Data partly also presented in ref. 24

33 Lawrence BM. Progress in essential oils. Perfum Flavor 1984;9(April/May):23-31

34 Rhind JP. Essential oils. A handbook for aromatherapy practice, 2nd Edition. London: Singing Dragon, 2012

35 Lawless J. The encyclopedia of essential oils, 2nd Edition. London: Harper Thorsons, 2014

Chapter 5.87 TURPENTINE OIL

DEFINITION

Turpentine oil (essential oil of turpentine) is the essential oil obtained by steam distillation of the gum resin of *Pinus massoniana* Lamb. (Chinese turpentine oil), *Pinus pinaster* Aiton (Iberian turpentine oil), and other *Pinus* species.

INCI NOMENCLATURE

Description/definition: No description given in CosIng
INCI name EU: Turpentine, steam distilled (*Pinus* spp.)
INCI name USA: Turpentine (defined as: Turpentine is a mixture of terpene hydrocarbons obtained from various species of *Pinus*)
CAS registry number(s): 8006-64-2
EINECS number(s): 232-350-7

ISO (INTERNATIONAL ORGANIZATION FOR STANDARDIZATION) STANDARD

ISO number: 21389
ISO name: Essential oil of gum turpentine, Chinese type
Botanical origin: Mainly from *Pinus massoniana* Lamb.
Parts of plant used: Oleoresin
ISO values: ISO values (minimum and maximum concentrations) are shown in Table 5.87.1

ISO number: 11020
ISO name: Essential oil of turpentine, Iberian type
Botanical origin: *Pinus pinaster* Aiton
Parts of plant used: Oleoresin
ISO values: ISO values (minimum and maximum concentrations) are shown in Table 5.87.2

THE PLANT, THE OIL, AND THEIR USES

Turpentine oil (often simply called 'turpentine') is obtained by steam distillation of the oleoresin of *Pinus pinaster* Aiton (Iberian turpentine oil), *Pinus massoniana* Lamb. (Chinese turpentine oil) and other *Pinus* species. What is left after distillation (the non-volatile residue) is called rosin (colophony, colophonium). Oil of turpentine formerly was widely used as a paint-thinner and for cleaning paint brushes, but this application has largely been abandoned and has been replaced with other solvents. Turpentine oil is an ingredient in many liniments, cold remedies, and veterinary medications. It may also be used in topical NSAID pharmaceutical preparations. Other commonly used products traditionally containing turpentine include varnishes, sealing wax, dry cleaning materials, shoe and floor polishes, printers' ink, and various adhesives, including adhesive tape (2,9,10,45). The oil is not applied per se in perfumes and cosmetics, but is used for the production of α- and β-pinene and limonene. Despite its strong terpeny odor and known risk of sensitization, turpentine oils may also be employed in aromatherapy practices (47). In earlier days it was utilized for cutting pine needle oils and other oils with high concentrations of pinenes (adulteration).

CHEMICAL COMPOSITION

Turpentine oil, *Chinese* type, is a colorless, clear mobile liquid, which has a strong terpeny and resinous odor.

The yield of essential oil from this turpentine oil varies from 17.0 to 22.0 per cent, depending on the processing method. The main producing country of this oil is China.

Turpentine oil, *Iberian* type, is a colorless, clear mobile liquid, which has a strong terpeny, mildly coniferous, resinous odor. The yield of essential oil from turpentine oil varies from 20.0 to 23.0 per cent, depending on the processing method. The main producing countries of this oil are Spain and Portugal.

Turpentine oils consist mainly of α-pinene, with variable amounts of β-pinene, limonene, β-caryophyllene, camphene, *p*-cymene, longifolene and myrcene (Tables 5.87.1 and 5.87.2). The relative proportions of these chemicals vary with the country of origin and the *Pinus* species from which the gum is obtained (5). An important variable is the amount of δ3-carene (should be low, as oxidized δ3-carene causes allergic reactions), which is high in turpentines from Scandinavia, the East-European countries, Russia and Indonesia (46) and low in turpentine oils from the south of France, Spain, Portugal and China (21). In Table 5.87.3 a comparison is given between oils from various East-European countries and China (21). Analyses of 52 commercial *Chinese* turpentine oils are given in Table 5.87.4. The results of analyses of 47 commercial *Iberian* type turpentine oils are shown in Table 5.87.5.

CONTACT ALLERGY/ALLERGIC CONTACT DERMATITIS

General

Turpentine oil is often loosely referred to as 'turpentine'. However, 'turpentine' does not always indicate the *essential oil of* turpentine, which is obtained by steam-distillation of the gum resin of *Pinus* species and which is also known as gum turpentine or balsam oil. Turpentine can also be a byproduct of the sulfate extraction process in which pine wood is converted into paper pulp. This oil (*not* essential) of turpentine is known as sulfate oil or sulfate turpentine (27). In literature on contact allergy to / allergic contact dermatitis from turpentine oil presented here (which is largely restricted to the period after 1975), a distinction between these two turpentine products can usually not be made.

Turpentine oil is both an irritant and a sensitizer. Old, oxidized turpentine is more irritating and sensitizing than is the freshly made product. When turpentine oil is allowed to stand, especially with exposure to light, oxidation results in the formation of formic acid and aldehydes, which may be irritating to the skin. Oxidation products of turpentine may also cause allergic contact sensitization (2,10). Indeed, up to the mid-1970s, occupational dermatitis from turpentine oil was well known in painters and home decorators, workers in the pottery industries (8), mechanics using impregnated soaps and shoe repairers (5,6,10). The main allergens were considered to be oxidation products of δ3-carene, δ3-carene hydroperoxides (also sometimes called turpentine peroxides), notably in oils from Sweden and Finland, which contained high concentrations of this component (7,8,10,28). However,

Table 5.87.1 ISO values (%) for turpentine oil, Chinese type [a]

Compound	CAS	Minimum	Maximum
α-Pinene	80-56-8	65.0	90.0
β-Pinene	127-91-3	3.0	18.0
Limonene	138-86-3	traces	5.0
β-Caryophyllene	87-44-5	traces	3.0
Camphene	79-92-5	traces	2.5
p-Cymene	99-87-6	traces	2.5
Longifolene	475-20-7	traces	2.5
Myrcene	123-35-3	traces	1.5
Caryophyllene oxide	1139-30-6	traces	0.4
δ3-Carene	13466-78-9	traces	0.3

[a] ISO/DIS 21389 Essential oil of gum turpentine, Chinese type ©ISO 2003; Geneva, Switzerland, www.iso.org

Table 5.87.2 ISO values (%) for turpentine oil, Iberian type [a]

Compound	CAS	Minimum	Maximum
α-Pinene	80-56-8	71.0	85.0
β-Pinene	127-91-3	3.0	18.0
Limonene	138-86-3	traces	5.0
β-Caryophyllene	87-44-5	traces	3.0
Camphene	79-92-5	traces	2.5
p-Cymene	99-87-6	traces	2.5
Longifolene	475-20-7	traces	2.5
Myrcene	123-35-3	traces	1.5
Caryophyllene oxide	1139-30-6	traces	0.4
δ3-Carene	13466-78-9	traces	0.3

[a] ISO 11020 Essential oil of turpentine, Iberian type ©ISO 1998; Geneva, Switzerland, www.iso.org

Table 5.87.3 Composition of turpentine oil samples from various East-European countries and China (21)

Components	China	Hungary	Russia	Poland	East Germany (former)
α-Pinene	81%	60%	66%	47%	50%
Carenes	–	14%	15%	37%	17%
β-Pinene	6.5%	15%	7.5%	5%	6%
Camphene	2%	1.5%	1.4%	1%	–
Dipentene	2.5%	5%	4%	3%	0.5%
Unidentified	8%	4.5%	6.1%	7%	26.5%

Table 5.87.4 Constituents identified in commercial turpentine oils from *Pinus massoniana* Lamb. [a]

Constituent	CAS	Range (%)	Constituent	CAS	Range (%)
α-Amorphene	20085-19-2	0-0.05	α-Longipinene	5989-08-2	0.1-0.2
Borneol	507-70-0	0.03-0.06	Methyl chavicol	140-67-0	0-0.04
Bornyl acetate	76-49-3	0.09-0.2	α-Muurolene	10208-80-7	tr-0.1
δ-Cadinene	483-76-1	0.1-0.3	Myrcene (β-)	123-35-3	0.6-1.2
Camphene	79-92-5	0.7-1.1	Myrtenol	515-00-4	tr-0.05
δ3-Carene	13466-78-9	tr-0.02	α-Phellandrene	99-83-2	0.02-0.06
β-Caryophyllene	87-44-5	0.9-4.1	β-Phellandrene	555-10-2	0.4-1.1
Caryophyllene oxide	1139-30-6	0.08-0.2	α-Pinene	80-56-8	57.1-74.4
1,4-Cineole	470-67-7	0.01-0.04	β-Pinene	127-91-3	11.4-20.8
1,8-Cineole	470-82-6	0.05-0.09	α-Pinene oxide	1686-14-2	tr-0.04
α-Copaene	3856-25-5	0.1-0.4	Pinocarveol	5947-36-4	0.08-0.4
α-Cubebene	17699-14-8	0.07-0.3	Sabinene	3387-41-5	0.01-0.07
p-Cymene	99-87-6	0.04-0.4	α-Terpinene	99-86-5	0.02-0.1
α-Fenchene	471-84-1	0.01-0.2	γ-Terpinene	99-85-4	0.02-0.1
Fenchyl alcohol	1632-73-1	0.02-0.08	α-Terpineol	98-55-5	0.2-1.0
α-Humulene	6753-98-6	0.1-0.5	Terpinolene (α-)	586-62-9	0.05-0.9
α-Humulene epoxide	96638-51-6	0.01-0.05	α-Thujene	2867-05-2	0.01-0.1
Isolongifolene	1135-66-6	0.1-0.7	Tricyclene	508-32-7	0.02-0.09
Limonene	138-86-3	1.3-13.6	*trans*-Verbenol	1820-09-3	0.02-0.08
Longifolene	475-20-7	0.1-2.1			

[a] Fifty-two *Pinus massoniana* Lamb. essential oil samples (Chinese turpentine) from China, analyzed between 1998 and 2014; lowest and highest concentrations given (E. Schmidt, unpublished data)

Table 5.87.5 Constituents identified in commercial turpentine oils from *Pinus pinaster* Aiton [a]

Constituent	CAS	Range (%)	Constituent	CAS	Range (%)
Bornyl acetate	76-49-3	0.1-0.3	α-Longipinene	5989-08-2	0.08-0.3
δ-Cadinene	483-76-1	0.08-0.2	Myrcene (β-)	123-35-3	0.01-1.1
Camphene	79-92-5	0.04-1.9	Myrtenol	515-00-4	0-0.03
δ3-Carene	13466-78-9	0-0.06	α-Phellandrene	99-83-2	0.02-0.06
β-Caryophyllene	87-44-5	0.3-4.4	β-Phellandrene	555-10-2	0.07-0.9
Caryophyllene oxide	1139-30-6	tr-0.9	α-Pinene	80-56-8	68.2-81.9
1,4-Cineole	470-67-7	0-0.02	β-Pinene	127-91-3	4.8-17.3
1,8-Cineole	470-82-6	tr-0.03	Pinocarveol	5947-36-4	0.05-1.0
α-Copaene	3856-25-5	0.08-0.3	Sabinene	3387-41-5	tr-0.08
α-Cubebene	17699-14-8	0.05-0.4	α-Terpinene	99-86-5	0.01-0.08
p-Cymene	99-87-6	tr-1.7	γ-Terpinene	99-85-4	0.02-0.08
α-Fenchene	471-84-1	tr-0.1	α-Terpineol	98-55-5	tr-1.2
α-Humulene	6753-98-6	0.2-0.7	Terpinolene (α-)	586-62-9	tr-0.04
α-Humulene epoxide	96638-51-6	0.05-0.1	α-Thujene	2867-05-2	tr-0.08
Limonene	138-86-3	0.5-3.9	Tricyclene	508-32-7	tr-0.08
Longifolene	475-20-7	0.1-2.4			

[a] Forty-seven *Pinus pinaster* Aiton essential oil samples (Iberian turpentine) from Spain and Portugal, analyzed between 1998 and 2014; lowest and highest concentrations given (E. Schmidt, unpublished data)

other components were also allergens in turpentine oil, including α-pinene, dipentene (*dl*-limonene) and β-pinene (see below: The allergens in turpentine oil).

When turpentine oils rich in δ3-carene (from Sweden, Finland, Indonesia, Hungary, Poland, Russia) (21) were gradually replaced with oils with low concentrations of δ3-carene (Spain, Portugal, China) (4,21) and turpentine oil was also widely replaced with cheaper substitutes such as petroleum-based white spirit and other organic solvents, the incidence of allergic contact dermatitis to turpentine oil subsequently dropped (27,29,38). In 1979, turpentine peroxides (tested since 1972 instead of turpentine oil to avoid irritant reactions) were removed from the European standard series by the ICDRG (International Contact Dermatitis Research group) because of low rates of positive reactions (in 1975 and 1976 0.7% in 11,798

patients patch tested in 5 European countries and the USA) (27). However, turpentine oil continued to be used in the pottery industry and Indonesian turpentine oil caused a cluster of cases of occupational contact dermatitis in 1996 in the UK (5). From 1996 to 1998, a sudden increase of rates of sensitization to turpentine oil was observed in Germany and Austria to 4.4% in 1998 (10). It was assumed that the oil of turpentine patch test was a surrogate marker for terpenes present in alternative cosmetic and medical products recovered from plants, e.g., tea tree oil, which showed frequent concomitant reactions with oil of turpentine (12). However, thereafter, the prevalence rates steadily decreased again from 4.4% in 1998 to 1.6% in 2002 (11).

In groups of consecutive patients suspected of contact dermatitis, prevalence rates of up to 4.2% positive patch

Table 5.87.6 Results of testing groups of patients with turpentine oil

Years and Country	Test conc. & vehicle	Number of patients tested | positive			Selection of patients (S); Relevance (R); Comments (C)	Ref.
Routine testing						
2007-2008 five European countries	10% pet.	6647	102	(1.5%)	R: not stated	3
2005-2008 IVDK	10% pet.	37,163	669	(1.8%)	R: not stated	15
2004-2008 Iran	10% pet.	469	7	(1.5%)	R: not stated	44
2002-2003 some European countries	10% pet.	3767	60	(1.6%)	R: not stated; rates ranged from 0.4% to 4.3% in various centers (both in Germany)	14
1996-2002 IVDK	10% pet.	60,737	1595	(2.6%)	R: not stated; annual prevalence was highest in 1998 (4.4%) and thereafter steadily declined to 1.6% in 2002	11
1992-1997 IVDK	10% pet.	45,005	560	(1.2%)	R: not specified; C: the prevalence rates were low from 1992-1995 (0.3-0.6%), but rose to 1.8% in 1996 and 3.2% in 1997; the cause of this increase was unknown	10

Table 5.87.6 Results of testing groups of patients with turpentine oil (*continued*)

Years and Country	Test conc. & vehicle	Number of patients tested	positive		Selection of patients (S); Relevance (R); Comments (C)	Ref.
1984-1988 five East-European countries	10% o.o.	48,020	887	(1.8%)	R: not specified; 30% of the patients from Poland co-reacted to pine needle oil, which also contains high concentrations of α-pinene	21
<1986, Spain	?	1610	67	(4.2%)	R: not specified, but occupations were mentioned, including 7 painters; Scandinavian turpentine was used for patch testing; there were 36 co-reactions to α-pinene, 25 to β-pinene, 16 to δ3-carene, 7 to *l*-limonene and 5 to *d*-limonene in patients allergic to turpentine	22
1979-1983 five East-European countries	10% o.o.	36,431	596	(1.6%)	R: not specified	21
1979-1983 Portugal	10% pet.	4316	99	(2.3%)	R: not stated; C: the prevalence of sensitization declined steadily from 3.6% in 1979 to 1.3% in 1983; turpentine oil from Portugal is made from *P. pinaster* and *P. pinea*, both of which do not contain δ3-carene; 22 patients allergic to turpentine oils were tested with a number of its ingredients and δ3-carene; there were 17 reactions to α-pinene, 15 to dipentene (*dl*-limonene), 4 to δ3-carene, 3 to α-terpineol and 2 to β-pinene	4
1973-1977 Denmark	?	3225	45	(1.4%)	R: not stated	26
1973-1977 Spain	10% lanette wax	4600	35	(0.8%)	R: not stated	25
1972-1976 Canada	1% o.o.	1075	28	(2.6%)	R: not stated	24
Testing in groups of selected patients						
2003-2012 IVDK	10% pet.	2046	36	(1.8%)	S: nurses with occupational contact dermatitis; R: not stated; C: the prevalence did not differ significantly from the group of nurses without occupational contact dermatitis	39
1996-2009 IVDK	10% pet.	744	17	(2.3%)	S: the group consisted of female cleaners; R: not stated; the rate of positive reactions was not higher than in control groups of females (without cleaners) with occupational contact dermatitis and females without occupational contact dermatitis	17
1993-2003 IVDK	10% pet.	1224	15	(1.2%)	S: patients with scalp dermatitis; R: not specified	42
1998-2002 IVDK	10% pet.	304	18	(5.9%)	S: patients with a positive patch test to skin care creams; R: not stated; C: this percentage was significantly higher than in the group of patients who did *not* react to skin creams (2.9%)	18
1998-2002 IVDK	10% pet.	70	6	(8.6%)	S: patients with a positive patch test to bath and shower products; R: not stated; C: the percentage was higher than in patients without positive reactions to bath and shower products (3.5%), but the difference was not significant	18
1995-1999 IVDK	10% pet.	969	21	(2.2%)	S: patients with periorbital allergic contact dermatitis; R: not stated; C: the percentage positive was about the same as in patients with dermatitis other than periorbital	1
<1996 UK	10% pet.	24	14	(58%)	S: patients with occupational hand dermatitis working in the pottery industry and exposed to Indonesian turpentine oil; 7 co-reacted to α-pinene, 4 to δ3-carene	5
1975-1980 Denmark	?	?	36	(?)	S: not specified for this particular allergen; R: 39%, of which 36% was occupational (occupations not specified)	16

IVDK Information Network of Departments of Dermatology, Germany, Switzerland, Austria (www.ivdk.org); o.o.: olive oil; pet.: petrolatum

test reactions have been observed, but reliable relevance data are lacking (Table 5.87.6). In the general German adult population, 2.5% of 1141 test subjects reacted to oil of turpentine, 4.3% of the women and 0.7% of the men (13). The *estimated* 10-year prevalence rate of sensitization to turpentine oil in the general German population in the period 1992-2002 varied from 0.3% (medium case scenario) to 0.8% (worst case scenario) (41).

Testing in groups of patients

The results of patch tests with turpentine oil in routine testing (consecutive patients suspected of contact dermatitis) and groups of selected patients are shown in Table 5.87.6. In routine testing, rates of positive reactions ranged from 1.2 to 4.2%, whereas between 0.8% and 58% (aimed testing in pottery workers with dermatitis) of patients in selected groups had positive patch tests.

Case reports

Occupational allergic contact dermatitis

An artist painter developed occupational hand dermatitis from contact allergy to turpentine oil he used to dilute paints and clean brushes (2). One case of occupational allergic contact dermatitis in a porcelain painter due to oil of turpentine, oil of aniseed and lavender oil, that were mixed with pigments for painting (19). In a group of 37 painters, varnishers and lacquerers seen between 1970 and 1993 in a clinic in Portugal and suspected of occupational contact dermatitis, 8 (22%) reacted to turpentine, which reactions were probably (but not implicitly) being considered relevant (20). Occupational contact dermatitis of the hands developed in a patient working in a laboratory and handling the oil at work; later, he developed stomatitis from pharmaceutical products containing peppermint oil; he reacted to turpentine peroxides, peppermint oil, limonene (*d*- and *l*-) and α-pinene, chemicals which are present in both peppermint and turpentine oils (32). An oil painter had contact allergy to turpentine and developed dermatitis from its use at work; he also reacted to epoxy resin in a spray varnish (36).

Non-occupational allergic contact dermatitis

A woman developed hand dermatitis from washing the clothes of her husband, which were impregnated with an NSAID solution that he used daily as a sports trainer; the patient reacted to the topical pharmaceutical product and its ingredient oil of turpentine (9). Two patients with allergic contact dermatitis from turpentine in topical pharmaceutical products (30). Hand dermatitis from contact allergy to Portuguese turpentine oil in a paint brush cleaner in a patient who painted his own house (31). A patient developed allergic contact dermatitis on the scalp, trunk and hands from turpentine present in a hair piece adhesive (34). Two out of three patients known to be sensitive to turpentine developed perianal dermatitis when suppositories containing turpentine were inserted (35). One case of airborne allergic contact dermatitis from turpentine in a patient who had previously worked as a painter (43). One case of allergic contact dermatitis from turpentine oil used to dilute paints and clean brushes in a hobby painter (45).

Positive patch tests (relevance unknown, uncertain or not stated)

In a group of 460 patients who were considered to have positive patch tests related to cosmetics, 37 (8%) had positive tests to turpentine (test concentration and relevance not mentioned) (23). One positive patch test to turpentine oil in a car mechanic and painter, who had allergic contact dermatitis from wax polish; the patient reacted to its ingredients pine oil and dipentene (*d,l*-limonene); turpentine oil also contains limonene and is obtained from *Pinus* species, just as pine oil (40).

The allergens in turpentine oil

Traditionally, δ3-carene hydroperoxides are considered to be the main allergens in turpentine oil, at least in Scandinavian turpentine oil (7). There can be no doubt, however, that there are also other sensitizers, some of which are important. Already in 1957 it was shown that oxidized limonene could elicit positive reactions in patients sensitized to turpentine (37). In Spain, where mainly Scandinavian turpentine was used, 67 patients allergic to turpentine oil were tested with a battery of its ingredients. There were 36 reactions to α-pinene (15% in olive oil), 25 to β-pinene (15% in olive oil), 16 to δ3-carene (15% in olive oil), 7 to *l*-limonene (1% in alcohol) and 5 to *d*-limonene (1% in alcohol) (22). In Portugal, of 22 turpentine-allergic patients, 17 reacted to α-pinene, 15 to dipentene (*dl*-limonene), 3 to α-terpineol, 2 to β-pinene and 4 to δ3-carene. Portugal was an important producer of turpentine oil with low or no δ3-carene content, obtained from *P. pinaster* and *P. pinea* (4). In 14 pottery workers with occupational contact dermatitis after contact with Indonesian (high δ3-carene concentration) turpentine oil, 7 co-reacted to α-pinene and 4 to δ3-carene (5). Of 40 Polish patients allergic to turpentine oil and tested with both Polish turpentine (high carene content) and Chinese turpentine oil (*no* carenes), 19 reacted to both (probably pinenes and limonene as allergens) and 17 only to Polish turpentine (presumably δ3-carene was the allergen) (21). Thus, it appears that α-pinene may be equally important if not more important than δ3-carene as a contact allergen in turpentine oils. In Bulgarian turpentine oil, α-phellandrene was shown to be a sensitizer (33).

LITERATURE

1 Herbst RA, Uter W, Pirker C, Geier J, Frosch PJ. Allergic and non-allergic periorbital dermatitis: patch test results of the Information Network of the Departments of Dermatology during a 5-year period. Contact Dermatitis 2004;51:13-19

2 Laube S, Tan BB. Contact dermatitis from turpentine in a painter. Contact Dermatitis 2004;51:41-42

3 Uter W, Aberer W, Armario-Hita JC, Fernandez-Vozmediano JM, Ayala F, Balato A, et al. Current patch test results with the European baseline series and extensions to it from the 'European Surveillance System on Contact Allergy' network, 2007–2008. Contact Dermatitis 2012;67:9-19

4 Cachao P, Menezes Brandao F, Carmo M, Frazao S, Silva M. Allergy to oil of turpentine in Portugal. Contact Dermatitis 1986;14:205-208

5 Lear JT, Heagerty AH, Tan BB, Smith AG, English JS. Transient re-emergence of oil of turpentine allergy in the pottery industry. Contact Dermatitis 1996;35:169-172

6 Pirilä V, Pirilä L. Terpentinallergie. Berufsdermatosen 1964;12:163-167

7 Pirilä V, Kilpio O, Olkknen A, Pirilä L, Siltanen E. On the chemical nature of the eczematogens in oil of turpentine. V. Dermatologica 1969;139:183-194

8 Benezra C, Fousserea J, Maleville J. L'identification chimique des allergènes vegetaux et son interêt dans la prévention de nombreux eczémas allergiques professionels. Maladies Professionelles de Médecine du Travail et de Sécurite Sociale 1970;31:539-543 (data cited in ref. 5)

9 Borrego L, Hernandez N, Martel R, Almeida P. Turpentine sensitization in a nonsteroidal anti-inflammatory solution user. Dermatitis 2012;23:182-183

10 Treudler R, Richter G, Geier J, Schnuch A, Orfanos CE, Tebbe B. Increase in sensitization to oil of turpentine: recent data from a multicenter study on 45,005 patients from the German-Austrian information network of departments of dermatology (IVDK). Contact Dermatitis 2000;42:68-73

11 Schnuch A, Lessmann H, Geier J, Frosch PJ, Uter W. Contact allergy to fragrances: frequencies of sensitization from 1996 to 2002. Results of the IVDK. Contact Dermatitis 2004;50:65-76

12 Pirker C, Hausen BM, Uter W, Hillen U, Brasch J, Bayerl C, et al. Sensitization to tea tree oil in Germany and Austria. A multicenter study of the German Contact Dermatitis Group. J Dtsch Dermatol Ges 2003;1:629-634

13 Schäfer T, Böhler E, Ruhdorfer S, Weigl L, Wessner D, Filipiak B, et al. Epidemiology of contact allergy in adults. Allergy 2001;56:1192-1196

14 Uter W, Hegewald J, Aberer W, Ayala F, Bircher AJ, Brasch J, et al. The European standard series in 9 European countries, 2002/2003 – first results of the European Surveillance System on Contact Allergies. Contact Dermatitis 2005;53:136-145

15 Uter W, Geier J, Frosch PJ, Schnuch A. Contact allergy to fragrances: current patch test results (2005 to 2008) from the IVDK network. Contact Dermatitis 2010; 63:254-261

16 Veien NK, Hattel T, Justesen O, Norholm A. Patch testing with substances not included in the standard series. Contact Dermatitis 1983;9:304-308

17 Liskowsky J, Geier J, Bauer A. Contact allergy in the cleaning industry: analysis of contact allergy surveillance data of the Information Network of Departments of Dermatology. Contact Dermatitis 2011;65:159-166

18 Uter W, Balzer C, Geier J, Frosch PJ, Schnuch A. Patch testing with patients' own cosmetics and toiletries – results of the IVDK, 1998–2002. Contact Dermatitis 2005;53:226-233

19 Vente C, Fuchs T. Contact dermatitis due to oil of turpentine in a porcelain painter. Contact Dermatitis 1997;37:187

20 Moura C, Dias M, Vale T. Contact dermatitis in painters, polishers and varnishers. Contact Dermatitis 1994;31:51-53

21 Rudzki E, Berova N, Czernielewski A, Grzywa Z, Hegyi E, Jirásek J, et al. Contact allergy to oil of turpentine: a 10-year retrospective view. Contact Dermatitis 1991;24:317-318

22 Romaguera C, Alomar A, Conde-Salazar L, Camarasa JMG, Grimalt F, Martin Pascual A, et al. Turpentine sensitization. Contact Dermatitis 1986;14:197

23 Romaguera C, Camarasa JMG, Alomar A, Grimalt F. Patch tests with allergens related to cosmetics. Contact Dermatitis 1983;9:167-168

24 Lynde CW, Warshawski L, Mitchell JC. Screening patch tests in 4190 eczema patients 1972-81. Contact Dermatitis 1982;8:417-421

25 Romaguera C, Grimalt F. Statistical and comparative study of 4600 patients tested in Barcelona (1973-1977). Contact Dermatitis 1980;6:309-315

26 Hammershøy O. Standard patch test results in 3225 consecutive Danish patients from 1973 to 1977. Contact Dermatitis 1980;6:263-268

27 Cronin E. Oil of turpentine – a disappearing allergen. Contact Dermatitis 1979;5:308-311

28 Pirilä V, Siltanen E. On the chemical nature of the eczematogenic agent in oil of turpentine. III. Dermatologica 1958;117:1-8

29 Foussereau J. Allergy to turpentine, lanolin and nickel in Strasbourg. Contact Dermatitis 1978;4:300

30 Nardelli A, D'Hooge E, Drieghe J, Dooms M, Goossens A. Allergic contact dermatitis from fragrance components in specific topical pharmaceutical products in Belgium. Contact Dermatitis 2009;60:303-313

31 Calnan CD. Turpentine in paint brush cleaner. Contact Dermatitis 1978;4:57-58

32 Dooms-Goossens A, Degreef H, Holvoet C, Maertens M. Turpentine-induced hypersensitivity to peppermint oil. Contact Dermatitis 1977;3:304-308

33 Michailov P, Berowa N, Zuzulowa A. Klinische und biochemische Untersuchungen über die berufsbedingten allergischen und toxischen Erscheinungen durch Terpentin. Allergie und Asthma 1970;16:201-205

34 Kanof NB. Eczematous contact dermatitis to turpentine signaled by a reaction to a hair piece adhesive. Contact Dermatitis 1977;3:108

35 Klaschka F. Allergy to turpentine: Examination of systemic trigger action. Contact Dermatitis 1975;1:319-320

36 Conde-Salazar L, Romero L, Guimaraens D, Harto A. Contact dermatitis in an oil painter. Contact Dermatitis 1982;8:209-210

37 Hellerström S, Thyresson N, Widmark G. Chemical aspects on turpentine eczema. Dermatologica 1957;115:277-286

38 Gollhausen R, Enders F, Przybilla B, Burg G, Ring J. Trends in allergic contact sensitization. Contact Dermatitis 1988;18:147-154

39 Molin S, Bauer A, Schnuch A, Geier J. Occupational contact allergy in nurses: results from the Information Network of Departments of Dermatology 2003–2012. Contact Dermatitis 2015;72:164-171

40 Martins C, Gonçalo M, Gonçalo S. Allergic contact dermatitis from dipentene in wax polish. Contact Dermatitis 1995;33:126-127

41 Thyssen JP, Uter W, Schnuch A, Linneberg A, Johansen JD. 10-year prevalence of contact allergy in the general population in Denmark estimated through the CE-DUR method. Contact Dermatitis 2007;57:265-272

42 Hillen U, Grabbe S, Uter W. Patch test results in patients with scalp dermatitis: analysis of data of the Information Network of Departments of Dermatology. Contact Dermatitis 2007;56:87-93

43 Dooms-Goossens AE, Debusschere KM, Gevers DM, Dupré KM, Degreef HJ, et al. Contact dermatitis caused by airborne agents: A review and case reports. J Am Acad Dermatol 1986;15:1-10

44 Firooz A, Nassiri-Kashani M, Khatami A, Gorouhi F, Babakoohi S, Montaser-Kouhsari L, et al. Fragrance contact allergy in Iran. J Eur Acad Dermatol Venereol 2010;24:1437-1441

45 Barchino-Ortiz L, Cabeza-Martínez R, Leis-Dosil VM, Suárez-Fernández RM, Lázaro-Ochaita P. Allergic contact hobby dermatitis from turpentine. Allergol Immunopathol (Madr) 2008;36:117-119

46 Sukarno A, Hardiyanto EB, Marsoem SN, Na'iem M. Oleoresin production, turpentine yield and components of *Pinus merkusii* from various Indonesian provenances. J Trop Forest Sci 2015;27:136-141

47 Lawless J. The encyclopedia of essential oils, 2nd Edition. London: Harper Thorsons, 2014

Chapter 5.88   VALERIAN OIL

DEFINITION
Valerian oil is the essential oil obtained from the rhizomes and roots of the valerian, *Valeriana officinalis* L.

INCI NOMENCLATURE
Description/definition: Valeriana officinalis root oil is the volatile oil obtained from the root of the valerian, *Valeriana officinalis*, Valerianaceae (Caprifoliaceae)
INCI name EU & USA: Valeriana officinalis root oil
CAS registry number(s): 8008-88-6; 97927-02-1 (CAS number for *Valeriana officinalis* ssp. c*ollina*)
EINECS number(s): 308-322-6

ISO (INTERNATIONAL ORGANIZATION FOR STANDARDIZATION) STANDARD
There is currently no ISO standard for valerian oil.

THE PLANT, THE OIL, AND THEIR USES
Valeriana officinalis L. (the common valerian) is a hardy perennial flowering plant, which reaches a height of 0.6-1.2 m. It is native to western Asia (Iran, Turkey), the Caucasus, Siberia, China, Japan, Korea, Taiwan, and all across Europe; the valerian is naturalized elsewhere. Cultivars selected for high contents of pharmacologically active substances are cultivated in temperate Europe, Korea, Japan, as well as Africa and USA (GRIN Taxonomy for Plants; Mansfeld's World Database of Agriculture and Horticultural Crops, http://mansfeld.ipk-gatersleben.de).

The valerians are highly respected medicinal plants listed in many Pharmacopoeias (14). Their underground parts, the rhizomes and roots, contain pharmacologically active substances, notably the ester iridoids (also named valepotriates) valerenic acid, its derivatives and an essential oil (i.e., substances which can be obtained as an essential oil by hydrodistillation). It is generally adopted, though unproven, that the whole mixture of these components is responsible for the pharmacological activity of the valerian (5,8). The highest contents of valerenic acids and essential oil may be located in the adventitious roots, followed by the lateral roots and the rhizomes (29).

Valerian roots (*Valerianae radix*) and their isolates (including essential oils) are employed in folk medicine, modern phytotherapy and aromatherapy, but also as official drugs (2,8,10,14). They are widely used in western Europe and many other countries in the world, mainly for their sedative, calming and hypnotic effects (2,3,5,8,19). The *Valerianae radix* is often perceived as a milder alternative or a possible substitute for the stronger synthetic sedatives in the treatment of states of nervous excitation and anxiety-induced sleep disturbances (3). In addition, the valerian is believed to have antispasmodic, carminative, analgesic, antihypertensive, cardiovascular, and antitussive properties (1,3). Traditional indications for its use are hysterical states, excitability, insomnia, hypochondriasis, migraine, cramp, intestinal colic, and rheumatic pains (3,19). Valerian roots and their products may also be used for food flavoring, e.g., in ice cream, baked goods and condiments (3,14,19), occasionally for perfumery and in cosmetic products (1,2,8,10), as poison antidotes and deodorants (5), as natural repellents of insects, pests, and some rodents (8,10) and in veterinary practice (2,8,10,14). In our opinion, valerian root products are rarely used in perfumery, and always in very low concentrations (traces).

CHEMICAL COMPOSITION
Valerian oil is a yellowish-green to brownish yellow, clear mobile liquid when fresh, which turns into a dark-brown and viscous liquid when aging; it has a slightly aromatic, tobacco-amber and woody odor when fresh, but acidic sharp, cheesy and disgusting odor when older. The yield of essential oil from the roots and rhizomes of *Valeriana officinalis* L. generally varies from 0.5 to 1.3 per cent, depending on the dryness of the roots. The main producing countries of this oil are Serbia, Bosnia, Holland, Germany, Russia and USA.

The discussion of the composition of *Valeriana officinalis* roots here is limited to studies on *V. officinalis* L., two subspecies, *V. officinalis* ssp. *officinalis* and *V. officinalis* ssp. *latifolia*, which are synonymous with *V. officinalis* L. (The Plant List, www.theplantlist.org) and *V. alternifolia* Bunge, which is also *V. officinalis* according to The Plant List. The composition of subspecies *Valeriana officinalis* subsp. *collina* (Wallr.) Nyman is discussed in refs. 6, 26 and 28 and that of *Valeriana officinalis* var. *sambucifolia* (correct name: *Valeriana sambucifolia* [The Plant List]) in refs. 6 and 27.

Literature data (up to December 5, 2014) on the chemical composition of valerian oils and unpublished analytical data from one of us (E.S.) are shown in Table 5.88.1 in alphabetical order. In valerian oils from various origins, over 330 chemicals have been identified. About 50 per cent of these were found in a single reviewed publication only.

The major compounds found in valerian oils from different sources are shown in Table 5.88.2. They include (highest concentrations in any study given) valerianol (kusunol) (57.3%), bornyl acetate (50.6%), camphene (32.1%), α-fenchene (28.3%), valerenal (15.6%), α-pinene (14.8%), valeranone (12.4%) and myrtenyl acetate (9.1%). Well-known ingredients of valerian oils that were present in high concentrations (>7%) in one or two studies were kessanyl acetate (29.5%), borneol (14.6%), spathulenol (7.3% and 13.3%), isovaleric acid (13.1% and 13.0%), α-kessyl acetate (12.6% and 11.1), *cis*-valerenyl acetate (12.2%), myrtenyl isovalerate (10.5%), δ-elemene (9.3%), α-humulene (7.3% and 8.5%), allo-aromadendrene (7.6%) and β-caryophyllene (7.3%). Uncommon or rare constituents of valerian oils found in high concentrations (>7%) in single studies include 4-methylene-1-methyl-2-(2-methyl-1-propen-1-yl)-1-vinylcycloheptane (36.8%), patchouli alcohol (16.8%), longipinocarvone (14.5%), α-campholenal (11.5%), 6,7-dimethoxy-*m*-cymene (11.0%), 15-acetoxyvaleranone (8.7%), vulgarone B (8.4%), nerolidol epoxyacetate (7.8%), farnesal (7.7%) and acetoxyvaleranone (7.6%).

Table 5.88.1 Constituents identified in valerian oils

Constituent	CAS	Percentage and range				
		A	B (6)	C (3)	D (19)	E
Acetoxyvaleranone						7.6[g]
15-Acetoxyvaleranone						8.7[s]
Alismol	87827-55-2					+[r]
δ-Amorphene	189165-79-5					0.4[f]
(Z)-Anethole	25679-28-1					<0.04[c]
Apollanol						0.1[b]
Aristola-1,9-diene						0.4[a2,t]
Aristolene	6831-16-9					0.8[b]
Aromadendrene	489-39-4	0-0.4		0-4.0	0-0.2	0.2[p]; 0.6[m]; 1.1[a,s]; 1.5[j]
allo-Aromadendrene	25246-27-9	0-0.4		0.3-6.9	0.3-7.6	1.3[j]; 1.4[s]; 1.7[f]; 2.5[e]; 4.1[g]; 6.6[k]
Aromadendrene oxide	85710-39-0					0.2[o]; 0.4[h]
allo-Aromadendrene oxide	85760-81-2					1.9[f]
Benzodiazepine	12794-10-4					+[u]; 0.4[m]; 1.1[a,s]
Benzyl benzoate	120-51-4					0.05[b]; 0.1[c]
trans-α-Bergamotene	13474-59-4					tr[d]; 0.3[k]
Bicyclo[4.4.0]dec-5-ene,1,5-dimethyl-3-hydroxy-8-(1-methylene-2-hydroxy)-						0.1[o]
Bicyclo[10.8.0] eicosa-1,14,18-triene						0.9[b]
Bicycloelemene	32531-56-9					0.4[f]
Bicyclogermacrene	24703-35-3			0-1.2	0.1-1.4	0.1[g]; 0.8[m]; 2.2[s]; 2.6[f]; 4.3[j]
epi-Bicyclosesquiphellandrene	54274-73-6			0-4.2	0.1-0.9	
β-Bisabolene	495-61-4		0.2-0.8			0.1[c]; 0.3[d]; 0.4[b,g]; 0.9[e]; 1.6[k]
(Z)-γ-Bisabolene	13062-00-5					
α-Bisabolol	515-69-5			tr-1.3	0.2-0.7	0.7[c]; 0.9[s]; 1.4[m]; 1.9[i]
epi-α-Bisabolol	78148-59-1	0-0.6	0.5-1.5			tr[d]; 0.2[k]; 0.5[e]
Borneol	507-70-0	0.06-4.3	0.1-0.6	0-2.5	0-0.6	0.5[c]; 1.0[e,i]; 4.5[h]; 4.9[o]; 6.6[m]; 14.6[k]
Bornyl acetate	76-49-3	9.8-21.4	2.3-33.5	2.9-33.5	8.8-33.5	23.9[p]; 32.1[j]; 39.1[k]; 44.2[t]; 50.6[n,q]
Bornyl isovalerate	76-50-6	0.5-1.1	0.5-1.2	0-3.3	0.2-2.0	0.1[c]; 0.4[s]; 0.5[e]; 0.8[o]; 1.1[m]; 1.3[l]
Boronia butenal	3155-71-3					0.1[h]
α-Bulnesene	3691-11-0					0.2[b]; 0.3[c]; 1.1[m]; 7.1[l]
p-tert-Butylbenzyl alcohol	877-65-6					0.4[a2,t]
2-tert-Butyl-1,4-dimethoxybenzene	21112-37-8					0.4[h]
Cadinene	29350-73-0					0.07[b]
α-Cadinene	24406-05-1					+[u]; 0.2[c]; 0.9[m]
γ-Cadinene	39029-41-9			0-1.7	0-1.1	0.5[g]; 0.6[s]
δ-Cadinene	483-76-1		0.1-0.3	0-1.2	tr-0.6	0.3[f]; <0.5[e,i]; 0.7[k]; 0.8[g]; 1.2[c]; 1.9[s]
α-Cadinol	481-34-5					<0.1[k]; 1.8[i]; 3.5[s]
epi-α-Cadinol	5937-11-1					2.8[g]
γ-Cadinol	50895-55-1					0.5[b]
δ-Cadinol	19435-97-3					0.05[b]; 0.2[b]
α-Calacorene	21391-99-1					+[u]; 1.2[m]
β-Calacorene	50277-34-4					+[u]; 0.9[m]
Camphene	79-92-5	0.6-24.1	0.1-6.4	0-11.0	0.6-5.9	8.6[c]; 11.0[m]; 12.2[h]; 16.2[n,q]; 32.1[t]
Camphene hydrate	465-31-6					tr[e,f]; 2.3[i]
4-Camphenylbutan-2-one						0.4[o]
α-Campholenal	4501-58-0					2.3[s]; 11.5[i]
Camphor	76-22-2	0.06-0.3	tr-0.1	0-0.4	0-0.1	tr[d]; <0.04[c]; 0.05[n,q]; 0.1[f]; 0.5[k]; 0.8[m]
d-Camphor	464-49-3					0.1[h]
δ2-Carene (δ4-carene)	554-61-0					0.04[n]; 1.4[s]
δ3-Carene	13466-78-9					0.04[a,q]

Table 5.88.1 Constituents identified in valerian oils (*continued*)

Constituent	CAS	Percentage and range				
		A	B (6)	C (3)	D (19)	E
3(10)-Caren-4-ol aceto acetic acid ester						3.4[o]
Carvacryl acetate	6380-28-5					0.2[f]
(*E*)-Carveol	1197-07-5		tr-0.1	0-2.8	0.1-1.7	0.3[m]; 0.5[b]
(*Z*)-Carveol	1197-06-4		0.1-0.2			0.1[b,c]
Carvyl acetate	97-42-7					5.5[n,q]
cis-Carvyl acetate	1205-42-1				0-0.4	+[u]; 0.6[m]
β-Caryophyllene	87-44-5	0.02-1.5		0.2-3.4	1.2-3.8	0.5[m]; 1.2[l]; 1.3[h]; 2.5[e]; 4.2[f]; 5.1[g]; 7.3[j]
γ-Caryophyllene	118-65-0					+[u]
2-epi-(*E*)-Caryophyllene (= 9-epi-)	68832-35-9					+[r]
α-Caryophyllene alcohol	4586-22-5					0.7[h]
Caryophyllene oxide	1139-30-6			0-2.0	0.2-0.7	0.5[e]; 0.6[f,h]; 1.0[s]; 1.1[i]
α-Cedrene	469-61-4					+[u]; 0.8[m]; 3.7[s]
β-Cedrene	546-28-1					+[u]; 1.1[m]
Cedrenol	28231-03-0					2.4[p]
Cedr-8-en-13-ol	18319-35-2					0.3[b]; 1.1[h]; 2.3[o]
Cembrene	1898-13-1					0.7[b]
Chamecynone	10208-54-5					1.3[o]
β-Chamigrene	18431-82-8					0.4[g]
1,8-Cineole	470-82-6	0-0.7	tr-0.2			tr[e]; 0.01[p]; <0.04[c]; 0.3[g]; 0.5[b]
Citronellol	106-22-9					tr[d]; 0.8[g]
Citronellyl acetate	150-84-5		tr-0.1			tr[d]; 0.8[g]
Citronellyl isovalerate	68922-10-1	0.02-0.09	0.2-2.1			
α-Copaene	3856-25-5		tr-0.2	0-0.3	0-0.4	0.1[c]; 0.2[e,k]; 0.4[m]
α-Copaen-11-ol	41370-56-3					0.8[b]
β-Copaen-4α-ol	124753-76-0					1.2[j]
Cryptofauronol	2212-90-0		0.4-3.9			
Cryptofauronyl acetate			tr-2.9			
α-Cubebene	17699-14-8					0.3[b]
β-Cubebene	13744-15-5					+[u]; 0.6[m]; 1.2[s]; 2.4[j]
Cubenol	21284-22-0					1.0[b]
Cuminaldehyde	122-03-2		tr-0.1			
Curcumene (ar-, α-)	644-30-4	0.09-1.8	0.7-2.1	0-1.0	0.4-0.7	0.2[c]; 0.5[k]; 0.6[i,p]; 1.1[d]; 1.3[b]
β-Curcumene	28976-67-2					tr[d]; <0.1[k]; 0.2[e]
γ-Curcumene	28976-68-3					0.9[d]
Cycloisosativene	406485-43-6					0.5[m]; 1.8[j]
1*H*-Cyclopropanol azulen-4-ol						0.8[a,s]
Cyclosativene	22469-52-9					+[u]
p-Cymene	99-87-6	0.02-0.8	0.1-1.5	0-1.0	tr-0.2	0.1[c,g]; 0.2[e]; 0.3[f,h,k,l,s]; 0.5[m]; 0.8[t]
p-Cymenene	1195-32-0					tr[e]
Cyperene	2387-78-2					0.08[a,p]; 0.3[f]
2-Decen-1-ol	22104-80-9					0.05[b]
Dehydroaromadendrene	698388-95-3					0.1[o]; 2.6[h]; 2.7[g]
8,9-Dehydrocycloisolongifolene						0.5[o]
Dehydroisolongifolene				0-2.0		
2,3-Dehydro-4-oxo-α-ionol						0.2[b]
Deoxysericealactone	19892-19-4					5.1[a,p]
Dihydrocarvyl acetate	57287-13-5					1.6[n,q]
Dihydroisolongifolene					0-1.0	+[u]; 0.3[m]
6,7-Dimethoxy-*m*-cymene						11.0[a,s]
2,5-Dimethoxy-*p*-cymene	14753-08-3	0.06-0.6				tr[d]; 0.3[o]; 0.5[c]; 2.9[j]
2,6-Dimethoxy-*p*-cymene	291774-65-7		0.2-1.4	0-0.5	0-0.2	+[u]; 0.6[m]

Table 5.88.1 Constituents identified in valerian oils (*continued*)

Constituent	CAS	A	B (6)	C (3)	D (19)	E
		\multicolumn{5}{c}{Percentage and range}
1,2-Dimethyl-4-formyl-1-cyclohexene	18022-66-7					1.3[l]
Docosane	629-97-0					0.1[b]
Drimenol	468-68-8		0.1-0.4			0.02[b]; 0.1[c]
Elemene	11029-06-4			0-0.4		
α-Elemene	5951-67-7					0.8[s]; 1.2[l]
β-Elemene	33880-83-0	0.1-0.4	0.1-0.4		0-0.3	0.2[e,g]; 0.3[d]; 0.4[c]; 0.5[s]; 0.8[l]; 0.9[f]
γ-Elemene	29873-99-2	0.03-0.1	0.1-0.4			0.4[h]; 0.6[b]; 0.9[m]; 1.1[c]; 1.8[f]; 5.4[j]
δ-Elemene	20307-84-0	0.2-0.9	0.2-2.9	0-0.6	0-1.8	0.4[g]; 0.6[d,s]; 1.1[e]; 3.4[c]; 5.2[j]; 9.3[f]
Elemol (α-)	639-99-6					<0.1[k]; 0.4[g]; 0.5[f]; 0.7[c]; 1.2[d]
Elixene	3242-08-8					0.4[o]
Epiglobulol	88728-58-9			0-1.6	0.2-0.6	+[u]; 1.3[h]
Epizonarene	41702-63-0					2.9[s]
Eremophilene	10219-75-7					0.3[p]; 1.1[a,s]
Ethyl hexadecanoate	628-97-7		tr-0.6			0.8[b]
Eudesma-2,6,8-triene			0.4-7.6			0.2[c]; 0.5[b]
α-Eudesmol	473-16-5					+[u]; 1.0[c]
5-epi-7-epi-α-Eudesmol	446050-56-2					0.8[f]
β-Eudesmol	473-15-4	1.3-4.3	0.6-6.9	0-1.7	0-1.1	0.08[b]; 0.2[e]; 0.5[m]; 0.8[c]; 1.8[g]
γ-Eudesmol	1209-71-8					0.07[b]; 0.4[o]; 0.5[h,m]
10-epi-γ-Eudesmol	15051-81-7	0.09-1.5				0.6[g]; 1.8[d]
Eugenol	97-53-0		tr-0.1			tr[d,e]; 0.1[b,c]; 0.6[k]
Eugenyl isovalerate	61114-24-7		0.1-0.4			0.07[b]; 0.1[c]
Farnesal	19317-11-4					7.7[a,p]
α-Farnesene	502-61-4			0-1.9	0.3-2.3	0.2[c]; 0.8[m]
(*E,E*)-α-Farnesene	502-61-4					0.3[l]; 0.9[m]; 1.5[s]
(*E*)-β-Farnesene	18794-84-8					tr[d]; 5.4[e]
(*Z*)-β-Farnesene	28973-97-9					0.1[c]; 2.7[k]
(*E,E*)-Farnesol	106-28-5					+[u]; 2.6[s]
(*Z,E*)-Farnesol	3790-71-4			0-0.7	0-0.2	
Faurinone	21682-87-1		0.4-0.6			0.3[c]
α-Fenchene	471-84-1	0.8-4.0		0-28.3	0.6-5.8	4.5[d]; 6.1[m]; 7.2[j]; 8.2[c]
Fonenol	1275572-21-8					0.3[b]
Geraniol	106-24-1					0.02[b]
Geranyl isovalerate	109-20-6			0-1.9	0-0.3	+[u]; 0.4[m]
Geranyl valerate	10402-47-8			0-3.0	0-0.5	
Germacrene	28028-64-0					0.6[b]
Germacrene A	28387-44-2					6.5[j]
Germacrene B	15423-57-1					0.1[f]; 0.3[j]; 0.6[d]; 1.0[g]; 2.4[m]; 2.8[s]
Germacrene D	23986-74-5	0.2-0.6	0.2-1.4	0-1.0	0-0.7	0.7[g]; 1.1[c,m]; 1.4[f]; 1.7[s]; 6.5[j]
Globulol	489-41-8					tr[h]; 0.4[b]; 0.6[f]; 2.1[g]
Guaia-3,9-diene	855270-07-4					0.06[b]
α-Guaiene	3691-12-1			0-3.2	0-0.3	0.2[f]; 0.6[o]; 0.7[m]; 1.0[j]; 1.1[c]; 3.6[l]
cis-β-Guaiene	372162-07-7					<0.1[g]; 0.5[c]
trans-β-Guaiene	192053-49-9					0.2[c]
Guaiol	489-86-1					<0.1[k]; 0.7[m]; 1.0[b]; 3.0[j]
α-Gurjunene	489-40-7			0-1.1	0-1.5	0.4[c]; 0.7[f]; 1.2[g]; 2.9[j]; 3.0[j]; 4.1[s]
β-Gurjunene	73464-47-8	0.2-0.9	tr-1.0	0-0.8	0-0.4	0.1[c]; 0.2[b]; 0.4[g]; 1.1[m]; 5.2[s]
γ-Gurjunene	22567-17-5					3.1[c]
β-Helmiscapene	66141-12-6					1.1[l]
Hexadecanoic acid	57-10-3		0.1-1.9	0-5.0	1-1.3	0.3[h]; 1.7[b]; 3.9[c]
Hexanal	66-25-1					tr[e]; 0.1[h]

Table 5.88.1 Constituents identified in valerian oils (*continued*)

Constituent	CAS	A	B (6)	C (3)	D (19)	E
Hexanoic acid	142-62-1					0.3[o]
α-Hexylcinnamaldehyde	101-86-0					0.3[b]
Hexyl isovalerate	10032-13-0			0-2.0	0.1-0.3	tr[e]; 0.4[m]
Hinesol	23811-08-7					0.2[b]; 2.0[j]; 3.1[c]
α-Humulene	6753-98-6	0.2-1.1	0.1-0.9	0-2.5	0.3-2.2	0.7[g]; 0.8[o]; 1.2[c]; 1.9[j]; 2.9[k]; 7.3[j]; 8.5[s]
β-Humulene	116-04-1					8.2[l]
Humulene (ep)oxide II	19888-34-7					+[u]
8-Hydroxycycloisolongifolene						3.1[b]
Hydroxyneoisolongifolane						0.6[b]
3-Hydroxypregn-5-en-20-one	145-13-1					1.2[o,x]
α-Ionene	475-03-6					0.07[b]
β-Ionol	22029-76-1					0.2[b]
(*E*)-α-Ionone	127-41-3					0.1[o]
β-Ionone	79-77-6		0.1-0.7	0-2.1	0-3.7	0.08[b]; 0.7[h]; 1.0[f,j,s]; 5.3[k]
(*E*)-β-Ionone	79-77-6					0.4[g]
Isoamyl isovalerate	659-70-1	0.01-0.1		0-0.5	tr-0.1	0.06[e]; 0.1[f]
Isoborneol	124-76-5		tr-0.1			tr[e]; 0.1[c]; 0.2[p]
Isobornyl acetate	125-12-2					6.6[p]
Isobornyl isovalerate	7779-73-9					0.6[h]
Isobornyl 2-methylbutanoate	233665-92-4					0.5[s]
Isocaryophyllene	118-65-0					2.0[s]
Isocurcumenol	24063-71-6					1.4[b]
Isocyclosativene						0.8[a,s]
Isoeugenyl isovalerate	61114-23-6		0.1-1.1			0.5[c]
Isolongifolan-8-ol	1139-08-8					0.3[o]; 0.4[b]
Isomenthyl acetate	20777-45-1			0-2.0		
2-Isopropyltricyclo-[4.3.1.1(2,5)] undec-3-en-10-ol						0.6[o]
Isospathulenol	88395-46-4					0.8[a,s]
Isothymol methyl ether	31574-44-4					tr[f]; 1.7[j]
Isovaleric acid	503-74-2	0.09-2.4		0-13.1	0-2.1	0.2[n]; 0.9[g]; 1.1[h]; 3.2[k]; 6.0[f]; 13.0[e]
Isovelleral	37841-91-1					1.8[b]
Italicene	94535-52-1					+[u]; 0.5[m]; 0.8[s]
trans-(*E*)-Jasmonol						0.4[f]
Juniper camphor	473-04-1					tr[h]; 1.2[b,o]
Kessane	3321-66-2	2.3-6.7	0.2-3.8	0-2.6	0-1.5	0.7[e]; 1.3[c]; 3.3[g]
Kessanyl acetate	17806-59-6	1.3-29.5[v]	0.1-3.5		0-2.0	1.1[c]
Kessoglycyl monoacetate			0.1-0.7			
α-Kessyl acetate	3925-77-7		0.1-12.6	0-3.8	0.4-2.3	2.1[c]; 11.1[b]
α-Kessyl alcohol	3321-65-1			0-4.7	0-1.2	0.04[b]; 0.1[c]; 0.4[m]; 2.1[s]
(*E*)-Lanceol	198828-12-5					0.7[b]
(*Z*)-Lanceol	859202-95-2					0.9[b]; 1.8[j]; 2.0[h]
Ledene oxide	882187-44-2					1.1[o]
Ledol	577-27-5	0.6-2.0	0.5-2.1	0-12.0	0.2-1.7	0.03[b]; 0.4[o]; 0.6[c]; 1.7[a,p]
Limonene	138-86-3	0.2-2.4	0.1-1.2	0-4.0	0.2-2.3	1.0[g]; 1.2[h]; 1.5[c]; 1.7[t]; 1.8[n,q]; 2.4[j]; 6.6[j]
Limonen-6-ol, pivalate						0.5[o]
Linalool	78-70-6		tr			tr[h]; 0.1[b,c]; 0.3[k]
Linalyl acetate	115-95-7					0.2[b]
Linalyl isovalerate	1118-27-0			0-1.7	0.7-3.0	
Linoleic acid	60-33-3					0.3[b]

Table 5.88.1 Constituents identified in valerian oils (*continued*)

Constituent	CAS	Percentage and range				
		A	B (6)	C (3)	D (19)	E
Longiborneol acetate	36204-27-0					2.2[d]
Longicyclene	1137-12-8					tr[d]
Longipinanol	66141-14-8					+[u]; 0.4[s]
α-Longipinene	5989-08-2					3.4[i]
trans-Longipinocarveol	889109-69-7					0.08[b]
Longipinocarvone	65556-52-7					3.4o; 14.5[h]
Maaliol	527-90-2		0.1-1.6			0.7[c]
Mayurone	4677-90-1					0.06[a,p]
p-Mentha-1(7),8(10)-dien-9-ol	29548-13-8					0.3[p]
trans-p-Mentha-2,8-dien-1-ol	4017-77-0					tr[h]
Menthone	89-80-5			0-0.7		
Methylcarvacrol	6379-73-3	0.06-0.2	0.1-0.2			tr[d]; 0.04[b]; 0.1[c,e]; 0.2[f]; 0.7[m]
Methyl eicosa-5,8, 11,14, 17-pentaenoate	2734-47-6					1.8o
4-Methylene-1-methyl-2-(2-methyl-1-propen-1-yl)-1-vinylcycloheptane	826337-63-7					36.8o
Methyl isovalerate	556-24-1					0.1[e]
Methyl palmitate	112-39-0		0.1-0.9			0.05[b]
2-Methyl-6-(1-phenylethyl) phenol	17959-01-2					0.3[h]
Methyl thymol	1076-56-8	0.04-0.4	tr-0.4			tr[d,e]; 0.1[c,n]; 0.2[b]; 0.3[h]; 0.6[m]; 1.3[j]
3-Methylvaleric acid	105-43-1		0.1-0.2	0-4.0	0-0.2	0.2q; 0.7[m]
3-Methyl-4-(2,5-xylyl)-butyric acid	30275-76-4					0.2[b]
cis-Muurola-3,5-diene	157374-44-2					0.1[f]
α-Muurolene	10208-80-7					0.1[e]
γ-Muurolene	30021-74-0					0.2[c,f]
α-Muurolol	104245-48-9					2.6[f]
(*E*)-Muurolol						0.4[b]
τ-Muurolol (epi-α-)	19912-62-0			0-0.7	0.2-1.6	+[u]; 0.3[b]; 0.9[m]
Myrcene (β-)	123-35-3		tr-0.1			0.1[c]; 0.2[h]; 1.0[n,q]
Myrcenol	543-39-5					0.1[n,q]
Myrtenol	515-00-4	0.2-0.8	0.1-4.0	0-1.9	0-0.5	0.03[b]; 0.1[c]; 0.2[e]; 0.7[d]
Myrtenyl acetate	1079-01-2	0.9-6.1	tr-9.1	0.2-7.2	2.0-7.2	0.7[m]; 1.1[e]; 1.3[g]; 2.6[h]; 4.1[c]; 6.2[j]
Myrtenyl hexanoate		0.1-0.2				
Myrtenyl isovalerate	33900-84-4	0.6-1.9	1.8-10.5	0.3-5.0	1.1-2.5	0.7[c,m]
Naphthalene	91-20-3					+[u]; 0.8[a,s]
β-Neoclovene	56684-96-9					0.01[b]
Nerolidol	7212-44-4					1.0[h]
Nerolidol epoxyacetate						7.8[j]
Neryl isovalerate	3915-83-1			0-2.7	0-0.4	+[u]
Nojigiku acetate	55627-02-6		tr-0.8			
Nootkatone	4674-50-4					1.0[p]
(*E*)-Nuciferol	1786-15-8					0.07[b]
(*Z*)-Nuciferol	78339-53-4					0.1[b]
(*Z*)-Ocimene	27400-71-1					6.7[j]
(*E*)-β-Ocimene	3779-61-1					tr[d]
(*Z*)-β-Ocimene	3338-55-4					tr[d,e]
allo-Ocimene	673-84-7					0.01q
neo-allo-Ocimene	7216-56-0					1.5[j]
Oleic acid	112-80-1					0.5[b]
Pacifigorgiadiene A	351222-63-4			0-0.4		

Table 5.88.1 Constituents identified in valerian oils (*continued*)

Constituent	CAS	Percentage and range				
		A	B (6)	C (3)	D (19)	E
Pacifigorgiadiene B	351222-64-5			0-1.1		
Pacifigorgia-1(6),10-diene						+[r]
Pacifigorgia-1(9),10-diene						+[r]
Pacifigorgiadiene isomers					0.1-0.8	+[u]; 1.0[m]
Pacifigorgiol	84014-68-6	0.2-2.1	0.5-3.2			+[r]
(-)-Pacifigorgiol	90988-77-5					+[r]; 0.4[c]
Panasinsene						0.09[b]
α-Panasinsene	56633-28-4					3.2[l]
α-Patchoulene	560-32-7					tr[d]; 0.1[c]; 3.8[f]
β-Patchoulene	514-51-2					<0.1[k]; 0.2[c]; 2.3[l]
γ-Patchoulene	508-55-4					tr[d]; <0.04[c]; 0.05[a,p]; 0.1[b]
Patchouli alcohol[w]	5986-55-0					0.4[b]; 1.2[h]; 16.8[l,w]
7,10-Pentadecadiynoic acid	22117-06-2					0.7[b]
7,11-Pentadecadiynoic acid						1.0[b]
Pentadecane (*n*-)	629-62-9					tr[h]
7,10-Pentadiynoic acid						4.4[a,p]
Perillaldehyde	2111-75-3					tr[h]
α-Phellandrene	99-83-2					tr[e]; 0.1[c]
β-Phellandrene	555-10-2		tr-2.4	0-0.1	tr-0.7	0.5[k]; 1.2[s]; 3.5[c]
α-Pinene	80-56-8	0.2-2.4	0.1-0.2	0-14.0	0.4-3.6	4.7[h]; 5.2[c]; 6.8[n,q]; 7.4[t]; 14.8[l]
β-Pinene	127-91-3	0.3-4.9	tr-0.9	0-3.5	0.2-1.2	1.4[e]; 1.5[c]; 1.6[j]; 3.4[h]; 4.4[t]; 6.5[n,q]
2-Pinen-10-ol	515-00-4					0.6[o]; 1.8[h]
cis-Pinocarveol	6712-79-4					+[u]; 0.4[m]; 0.7[a,s]
trans-Pinocarveol	1674-08-4					tr[d]
trans-Pinocarvyl acetate	1686-15-3			0-6.2	0.1-0.5	+[u]; tr[d]; 0.6[m]
2-Propenal, 3-(2,6,6-trimethyl-1-cyclo-hexen-1-yl)-	4951-40-0					0.2[o]
Pulegone	89-82-7					+[u]; 0.3[a,s]; 1.5[m]
Retinyl acetate	127-47-9					0.2[o]
Sabinene	3387-41-5	0.01-0.2	tr-0.1	0-0.2	tr-0.2	tr[d,f]; 0.1[b,e]; 0.3[c]; 0.6[n,q]; 0.8[m]
cis-Sabinene hydrate	15537-55-0					tr[e]
trans-Sabinene hydrate	17699-16-0					tr[e]
Sabinol						0.6[p]
trans-Sabinyl acetate	139757-62-3					0.1[c]
α-Santalene	512-61-8					0.5[k]
α-Santalol						1.5[a,p]
(Z)-α-Santalol	115-71-9					0.1[b]; 0.2[f]
(Z)-β-Santalol	42495-69-2					+[u]; 0.9[m]
Santolinatriene	2153-66-4					tr[f]
Selinadiene alcohol	35688-11-0			0-4.3	0-0.3	+[u]
Selina-4,11-diene	17627-30-4					0.4[l]
α-Selinene	473-13-2					+[u]; tr[d]; <0.1[k]; 0.6[m]; 1.0[s]
7-epi-α-Selinene	123123-37-5					1.2[d]
β-Selinene	17066-67-0					0.2[g]; 1.8[s]
γ-Selinene	515-17-3					1.7[s]
δ-Selinene	473-14-3					0.2[f]; 2.3[a,s]; 2.7[l]
Seychellene	20085-93-2					0.04[b]
Sibirene	14029-18-6					0.2[f]
Spathulenol	6750-60-3			0.3-7.3	0.7-4.1	1.8[e]; 2.9[o]; 3.1[h]; 4.7[g]; 4.8[s]; 13.3[j]
Tamariscene	351413-96-2					+[r]

Table 5.88.1 Constituents identified in valerian oils (*continued*)

Constituent	CAS	Percentage and range				
		A	B (6)	C (3)	D (19)	E
α-Terpinene	99-86-5					tr[d,e,f]; <0.1[k]; 0.1[c]; 1.0[s]
γ-Terpinene	99-85-4	0.02-0.2	tr-0.2	0-0.8	0-0.3	tr[d]; 0.02[b]; 0.1[e,h]; 0.2[f]; 0.4[i]; 0.6[c]
Terpinen-4-ol	562-74-3	0.04-0.5	0.1-0.6	0-1.2	0.1-0.4	0.1[f]; 0.2[e,q]; 0.3[h]; 0.4[m]; 0.5[c]; 0.6[k]
1-Terpineol	586-82-3					<0.04[c]
α-Terpineol	98-55-5	0.04-0.3	tr-0.1	0-0.5	0-0.5	tr[e]; 0.1[c]; 0.2[o]; 0.5[h]; 0.6[m]; 1.2[k]
Terpinolene (α-)	586-62-9		tr-0.1			tr[d]; <0.1[h,k]; 0.1[c,f]
α-Terpinyl acetate	80-26-2	0.1-0.9		0.2-1.7	0.4-1.1	0.2[g]; 0.4[c]; 0.5[f]; 0.7[b,d,o]; 1.7[h]; 5.5[k]
α-Thujene	2867-05-2	0-0.1		0-0.7	0-0.1	0.08[e]; 0.1[c,f,n,q]
β-Thujol						1.7[a2,t]
Thujopsene	470-40-6	0.2-1.9				
Thymol	89-83-8		0.1			0.1[b]; 0.3[c]
Tricyclene	508-32-7	0.2-0.4	tr-0.1	0-0.7	tr-0.1	0.1[c,e,f]; 0.5[h]
2,8,8-Trimethyl-4-methylene-2-vinylbicyclo[5.2.0]nonane						0.3[o]
Valencene	4630-07-3			0-0.8	0-0.8	0.9[b]; 1.1[m]; 2.0[j]; 2.1[d]
Valencene ketone	137764-27-3			0-2.9	0.4-3.0	+[u]; 0.8[m]
Valeranone	55528-90-0	1.3-4.0	0.5-8.2	0.5-10.9	0.5-9.4	4.1[b]; 7.4[c]; 0.9[s]; 1.3[j]; 5.8[m]; 12.4[p];
Valerena-4,7(11)-diene	351222-66-7					+[r]
Valerenal	4176-16-3	0.6-8.6	0.4-12.4	tr-15.6	0-14.7	8.3[j]; 11.3[b]; 12.9[m]; 14[s]; 14.7[f]; 15.5[j]
Valerenic acid	3569-10-6		0.3-2.8		0-0.9	tr[e]; 0.6[m]; 1.6[c]; 4.0[b]; 5.9[s]
Valerenol	101628-22-2	0.2-0.4	0.3-1.2	0-0.7	0-0.8	0.3[e]; 0.9[c,i]; 1.1[j]; 2.2[s]; 4.3[g]
cis-Valerenol						0.6[b]
trans-Valerenol						3.8[b]
cis-Valerenyl acetate	101527-78-0		0.2-0.8	0-1.6	0-1.6	0.8[e]; 1.0[j]; 2.7[s]; 12.2[b]
trans-Valerenyl acetate	101527-74-6		0.1-1.5	0-1.1	0-0.8	0.5[e]; 0.6[b]; 1.0[c]; 1.4[c]; 3.6[j]; 3.7[h]
cis-Valerenyl isovalerate			tr-0.7			<0.04[c]; 0.06[e]; 0.2[b]; 3.6[i]
trans-Valerenyl isovalerate	101527-75-7		tr-1.6		0-1.1	0.2[b]; 0.5[c,e]; 4.1[s]
Valerianol (kusunol)	20489-45-6	6.3-19.8	1.5-14.2	0.2-18.2	0.3-16.7	0.5[e]; 1.1[m]; 2.6[c]; 12.6[b]; 57.3[d]
Valeric acid	109-52-4	0.2-1.6				
Verbenol	473-67-6					0.06[b]
Verbenyl acetate	33522-69-9					1.5[o]
Verrucarol	2198-92-7					0.8[h]
Viridiflorol	552-02-3			0-1.7	0.1-0.6	+[u]; 0.5[m]
Vulgarone B	64180-68-3					8.4[i]
Widdrol	6892-80-4					0.5[h]
α-Ylangene	14912-44-8		0.1-0.4			+[r]; 0.1[c]
α-Zingiberene	495-60-3			0-1.6	0.1-1.5	1.0[e]; 1.1[c,m]

A eleven valerian essential oil samples from Serbia, Bosnia, Russia and USA, analyzed between 1998 and 2014; lowest and highest concentrations given (E. Schmidt, unpublished data)

B sixteen lab-hydrodistilled oils from sixteen *V. officinalis* root and rhizome samples collected in botanical gardens in Germany, The Netherlands, Austria and Poland; lowest and highest concentrations given (ref. 6)

C fifteen lab-hydrodistilled valerian root oils from fifteen root samples obtained from retail pharmacies in different European countries; lowest and highest concentrations given (ref. 3)

D five lab-hydrodistilled oils from root samples obtained in Estonian pharmacies or from plants cultivated in Estonia; lowest and highest concentrations given (ref. 19)

E data from other studies (indicated with superscript letters); highest concentrations found in any study reviewed here given; when two or more oils were investigated, only the highest concentrations are mentioned, unless indicated otherwise

Table 5.88.1 Constituents identified in valerian oils (*continued*)

[a] / [a2] incorrect identification (refs. 7,20); [b] one lab-hydrodistilled oil from roots collected in Iran (ref. 14); [c] two lab-hydrodistilled oils from fresh and dried root and rhizome of *V. officinalis* collected in Lithuania (ref. 11); [d] one lab-hydrodistilled oil from valerian roots growing wild in western Serbia (ref. 5); [e] one oil from valerian roots from western Canada steam-distilled in a commercial unit (ref. 4); [f] one lab-hydrodistilled oil from Austrian valerian root (ref. 1); [g] one lab-hydrodistilled oil from the underground parts of *V. officinalis* cultivated in Iran (ref. 8); [h] one lab-hydrodistilled oil from roots and rhizomes of *V. officinalis* var. *latifolia* plants growing wild in China (ref. 13); [i] one lab-hydrodistilled oil from plants cultivated in Iran (ref. 15); [j] four lab-hydrodistilled oils from valerian roots of plants cultivated in Iran with various production methods (aeroponic, floating, growing media, or soil) (ref. 16); [k] one lab-hydrodistilled oil from Chinese roots of *V. alternifolia* Bunge (= *V. officinalis*) (ref. 17); [l] one lab-hydrodistilled oil from valerian growing wild in China (ref. 18); [m] one lab-hydrodistilled oil from roots of *V. officinalis* cultivated in Iran (ref. 23); [n] one steam-distilled oil from roots of *V. officinalis* var. *latifolia* plants growing in China (ref. 24); data also published in ref. 22; [o] one very atypical oil of *V. officinalis* var. *latifolia* from China (ref. 12); [p] one oil of valerian root of Chinese origin (ref. 21); [q] one oil from the root of Chinese *V. officinalis* var. *latifolia* (ref. 22); data also published in ref. 24; [r] one commercial valerian root oil obtained from a German company (ref. 9); [s] four lab-hydrodistilled oils from two *V. officinalis* cultivars, 'Anthose' and 'Select' of two different ages (8 and 14 months); (ref. 10); this report contains many mistakes (ref. 20); [t] one valerian root oil of Chinese origin (ref. 25); [u] unknown number of valerian oil samples of the roots of *V. officinalis* cultivated in Iran; only qualitative data provided (ref. 30); [v] or α-kessyl acetate; [w] patchouli alcohol is a component of *Valeriana javanica*, not of *V. officinalis*; [x] incorrect identification, compound has a far too high a boiling point

tr: trace (in columns B, C, D, E[e], E[f]: <0.05; in column E[h]: <0.1); + present in the oil investigated, but quantity not stated

Commercial oils

The ten chemicals that had the highest maximum concentrations in 11 commercial valerian essential oil samples (concentration ranges provided) are the following: kessanyl acetate (1.3-29.5%), camphene (0.6-24.1%), bornyl acetate (9.8-21.4%), valerianol (kusunol) (6.3-19.8%), valerenal (0.6-8.6%), kessane (2.3-6.7%), myrtenyl acetate (0.9-6.1%), β-pinene (0.3-4.9%), borneol (0.06-4.3%) and β-eudesmol (1.3-4.3%) (Erich Schmidt, unpublished analytical data).

CONTACT ALLERGY/ALLERGIC CONTACT DERMATITIS

General

Contact allergy to and possible allergic contact dermatitis from valerian oil has been reported in one publication only. A false-positive patch test reaction due to the excited skin syndrome cannot be excluded.

Case reports

An aromatherapist had occupational contact dermatitis with allergies to multiple essential oils used at work, including valerian oil. The patient also reacted to geraniol, α-pinene, caryophyllene, the fragrance mix and various other fragrance materials. None of these were identified with GC-MS in the valerian oil sample (31).

LITERATURE

1 Huynh L, Pacher T, Tran H, Novak J. Comparative analysis of the essential oils of *Valeriana hardwickii* Wall. from Vietnam and *Valeriana officinalis* L. from Austria. J Essent Oil Res 2013;25:409-414

2 Seidler-Lozykowska K, Mielcarek S, Baraniak M. Content of essential oil and valerenic acids in valerian (*Valeriana offcinalis* L.) roots at the selected developmental phases. J Essent Oil Res 2009;21:413-416

Table 5.88.2 Major constituents of valerian oils

Constituent	CAS	Percentage and range				
		A	B (6)	C (3)	D (19)	E
Valerianol (kusunol)	20489-45-6	6.3-19.8	1.5-14.2	0.2-18.2	0.3-16.7	0.5[e]; 1.1[m]; 2.6[c]; 12.6[b]; 57.3[d]
Bornyl acetate	76-49-3	9.8-21.4	2.3-33.5	2.9-33.5	8.8-33.5	23.9[p]; 32.1[j]; 39.1[k]; 44.2[t]; 50.6[n,q]
Camphene	79-92-5	0.6-24.1	0.1-6.4	0-11.0	0.6-5.9	8.6[c]; 11.0[m]; 12.2[h]; 16.2[n,q]; 32.1[t]
α-Fenchene	471-84-1	0.8-4.0		0-28.3	0.6-5.8	4.5[d]; 6.1[m]; 7.2[j]; 8.2[c]
Valerenal	4176-16-3	0.6-8.6	0.4-12.4	tr-15.6	0-14.7	8.3[i]; 11.3[b]; 12.9[m]; 14[s]; 14.7[f]; 15.5[j]
α-Pinene	80-56-8	0.2-2.4	0.1-0.2	0-14.0	0.4-3.6	4.7[h]; 5.2[c]; 6.8[n,q]; 7.4[t]; 14.8[l]
Valeranone	55528-90-0	1.3-4.0	0.5-8.2	0.5-10.9	0.5-9.4	4.1[b]; 7.4[c]; 0.9[s]; 1.3[j]; 5.8[m]; 12.4[p];
Myrtenyl acetate	1079-01-2	0.9-6.1	tr-9.1	0.2-7.2	2.0-7.2	0.7[m]; 1.1[e]; 1.3[g]; 2.6[h]; 4.1[c]; 6.2[j]

LEGEND: SEE UNDER TABLE 5.88.1

3 Raal A, Arak E, Orav A, Kailas T, Müürisepp M. Variation in the composition of the essential oil of commercial *Valeriana officinalis* L. roots from different countries. J Essent Oil Res 2008;20:524-529

4 Lopes D, Strobl H, Kolodziejczyk P. influence of drying and distilling procedures on the chemical composition of valerian oil (*Valeriana officinalis* L.). J Essent Oil Bear Plants 2005;8:134-139

5 Pavlovic M, Kovacevic N, Tzakou O, Couladis M. The essential oil of *Valeriana officinalis* L. *s.l.* growing wild in western Serbia. J Essent Oil Res 2004;16:397-399

6 Bos R, Woerdenbag HJ, Hendriks H, Scheffer JJC. Composition of the essential oils from underground parts of *Valeriana officinalis* L. *s.l.* and several closely related taxa. Flavour Fragr J 1997;12:359-370

7 Lawrence BM. Progress in essential oils. Perfum Flavor 1999;24(3):47-64

8 Safaralie A, Fatemi S, Sefidkon F. Essential oil composition of *Valeriana officinalis* L. roots cultivated in Iran. Comparative analysis between supercritical CO_2 extraction and hydrodistillation. J Chromat A 2008;1180:159-164

9 Paul C, König AW, Muhle H. Paciforgianes and tamariscene as constituents of *Frullania tamarisci* and *Valeriana officinalis*. Phytochem 2001;57:307-313

10 Letchamo W, Ward W, Heard B, Heard D. Essential oil of *Valeriana officinalis* L. cultivars and their antimicrobial activity as influenced by harvesting time under commercial organic cultivation. J Agric Food Chem 2004;52:3915-3919

11 Baranauskiene R. Essential oil composition of *Valeriana officinalis* ssp. *officinalis* grown in Lithuania. Chem Nat Comp 2007;43:331-333

12 Huang B, Qin L, Liu Y, Zhang Q, Rahman K, Zheng H. Chemical composition and hypnotic activities of the essential oil from roots of *Valeriana officinalis* var. *latifolia* in China. Chem Nat Comp 2009;45:560-561

13 Huang B, Qin L, Chu Q, Zhang Q, Gao L, Zheng H. Comparison of headspace SPME with hydrodistillation and SFE for analysis of the volatile components of the roots of *Valeriana officinalis* var. *latifolia*. Chromatographia 2009;69:489-496

14 Asadollahi-Baboli M. Comprehensive analysis of *Valeriana officinalis* L. essential oil using GC-MS coupled with integrated chemometric resolution techniques. Int J Food Prop 2015;18:597-607

15 Samaneh RT, Tayebeh R, Hassan E, Vahid N. Composition of essential oils in subterranean organs of three species of *Valeriana* L. Nat Prod Res 2010;24:1834-1842

16 Tabatabaei SJ. Effects of cultivation systems on the growth, and essential oil content and composition of valerian. J Herbs Spic Med Plants 2008;14:54-67

17 Cui L, Wang Z-Y, Zhou XH. Volatile constituents of essential oils from roots and rhizomes of *Valeriana fauriei* Briq. and *V. alternifolia* Bunge from Changbai Mountain. J Essent Oil Bear Plants 2011;14:329-333

18 Wang J, Zhao J, Liu H, Zhou L, Liu Z, Wang J, et al. Chemical analysis and biological activity of the essential oils of two valerianaceous species from China: *Nardostachys chinensis* and *Valeriana officinalis*. Molecules 2010;15: 6411-6422

19 Raal A, Orav A, Arak E, Kailas T, Müürisepp M. Variation in the composition of the essential oil of *Valeriana officinalis* L. roots from Estonia. Proc Eston Acad Sci (Chem) 2007;56:67-74. Available at: http://www.kirj.ee/public/Chem/2007/issue_2/chem-2007-2-2.pdf.

20 Lawrence BM. Progress in essential oils. Perfum Flavor 2009;34(7):54-59

21 Zhang H-L, Zhang Z-J, Wang Y-M, Zhao X-Y. Study on the chemical constituents of essential oil from *Valeriana officinalis* Linn. Tiaran Chanwa Yanjiu Yu Kaifa 1993;5:21-27. Data cited in ref. 20

22 Zhu L-F, Li Y-H, Li B-L, Lu B-Y, Zhang W-L. Aromatic plants and essential constituents (suppl. I). Hong Kong: Peace Book Co Ltd, 1995:46. Data cited in ref. 20. Data also published in ref. 24

23 Morteza E, Joorabloo A. Evaluation of medicinal plant valerian (*Valeriana officinalis* L.) essential oil compositions cultivated at Garmsar zone in Iran. J Pharm Scientif Innov (JPSI) 2012;1:87-88

24 Long C, Xiao H, Peng J. The chemical constituents of the essential oil from the roots of *Valeriana officinalis* var. *latifolia*. Acta Botanica Yunannica 1987;9:109-112. Data also published in ref. 22

25 Zhu L-F, Li Y-H, Li B-L, Lu B-Y, Xia N-H. Aromatic plants and essential constituents. South China Institute of Botany, Chinese Academy of Sciences, Hang Feng Publishing Co., 1993, distributed by Peace Book Co. Data cited in ref. 7

26 Georgiev EV, Stojanova AS, Tchapkanov VA. On the Bulgarian valerian essential oil. J Essent Oil Res 1999;11:352-354

27 Gränicher F, Christen P, Kapetanidis I. Essential oils from normal and hairy roots of *Valeriana officinalis* var. *sambucifolia*. Phytochemistry 1995;40:1421-1424

28 Bos R, Hendriks H, Pras N, Stojanova AS, Georgiev EV. Essential oil composition of *Valeriana officinalis* ssp. *collina* cultivated in Bulgaria. J Essent Oil Res 2000;12:313-316

29 Penzkofer M, Ziegler E, Heuberge H. Contents of essential oil, valerenic acids and extractives in different parts of the rootstock of medicinal valerian (*Valeriana officinalis* L. s.l.). J Appl Res Med Arom Plants 2014;1:98-106

30 Morteza E, Akbari GA, Modares Sanavi SAM, Foghi B, Abdoli M, Farahani HA. Planting density influence on variation of the essential oil content and compositions in valerian (*Valeriana officinalis* L.) under different sowing dates. Afr J Microbiol Res 2009;3:897-902

31 Dharmagunawardena B, Takwale A, Sanders KJ, Cannan S, Roger A, Ilchyshyn A. Gas chromatography: an investigative tool in multiple allergies to essential oils. Contact Dermatitis 2002;47:288-292

Chapter 5.89   VETIVER OIL

DEFINITION
Vetiver oil (essential oil of vetiver) is the essential oil obtained from the roots of the vetiver, *Chrysopogon zizanioides* (L.) Roberty (synonym: *Vetiveria zizanioides* (L.) Nash).

INCI NOMENCLATURE
Description/definition: Vetiveria zizanoides root oil is an essential oil distilled from the dried roots of the vetiver, *Vetiveria zizanoides,* Poaceae
Other names: khas khas oil; khus oil; cus cus oil
CAS registry number(s): 8016-96-4; 84238-29-9
EINECS number(s): 282-490-8

ISO (INTERNATIONAL ORGANIZATION FOR STANDARDIZATION) STANDARD
ISO number: 4716
ISO name: Essential oil of vetiver
Botanical origin: *Chrysopogon zizanioides* (L.) Roberty (synonym: *Vetiveria zizanioides* (L.) Nash)
Parts of plant used: Root
ISO values: ISO values (minimum and maximum concentrations) are shown in Table 5.89.1.

The genus *Vetiver* was reclassified some 15 years ago and is now combined with *Chrysopogon*, based on their overlapping genetic and morphological data. Although this has led to the recognition of *Chrysopogon zizanioides* (L.) Roberty as the proper classification for *Vetiveria zizanioides* (L.) Nash (13,32), most publications continue to use the latter name (and often incorrectly spelled zizanoides instead of zizanioides, e.g., in INCI nomenclature).

THE PLANT, THE OIL, AND THEIR USES
Chrysopogon zizanioides (L.) (synonym: *Vetiveria zizanioides* (L)), commonly known as vetiver, is a perennial grass of the Poaceae family, native to India. In western and northern India, it is popularly known as khus. Vetiver can grow up to 1.5 meters high and form clumps as wide. Unlike most grasses, which form horizontally spreading, mat-like root systems, vetiver's roots grow downward, two to four meters in depth. It can be found, either wild or cultivated, in a wide range of areas from highlands to lowlands widely spread in subtropical and tropical regions of Asia, Africa, Oceania, and Central and South America. Its primary uses in agricultural and non-agricultural fields include applications in water and soil conservation, slope stabilization, erosion control, environmental protection, absorption of heavy metals, wastewater treatment and other related uses. It is also used for livestock grazing and as an ornamental plant (1-5,24).

Dried roots from *Chrysopogon zizanioides* are used in India for making mats which, when sprinkled with water and hung like curtains, cool the air and give off a pleasant aroma. The roots are perceived to be stimulant, tonic, cooling, stomachic, diuretic, antispasmodic, and emmenagogue, and they are used in fevers, inflammations, and irritability of the stomach (23). Various tribal people in India use different parts of the grass for many of their ailments, such as boils, burns, epilepsy, fever, scorpion sting, snakebite, and sores in the mouth. The root paste is used for headache and toothache, the leaf paste is used for lumbago, sprain, and rheumatism, the stem decoction for urinary tract infection, the leaf juice as an anthelmintic, the vapors for malarial fever, and the root ash is given for acidity relief (23).

Steam-distillation of the roots produces vetiver essential oil. The oil is used as a perfumery source, for aromatherapy, as an aroma in food and as a flavor agent in some beverages (1,2,3,4,5,21,23,24). Essential oil of vetiver is one of the most viscous oils, with an extremely slow rate of volatility, which makes it very persistent and one of the finest fixatives known. The oil and its constituents are used extensively for blending oriental type perfumes and floral compounds and are a main ingredient in about one-third of all western quality perfumes and 20% of all men's fragrances. The concentration of vetiver oils used ranges from 0.5% to nearly 10% in such products (6). Vetiver essential oil is also very popular in aromatherapy. Claimed biological properties of vetiver oils include anti-inflammatory, antiseptic, aphrodisiac, cicatrisant, nervine, sedative, tonic, healing and calming. It may be used in the treatment of rheumatism, gout, arthritis, muscle aches, cramps, dry skin, depression, nervous tension, insomnia and many stress-related diseases (20,21).

CHEMICAL COMPOSITION
Vetiver oil is a yellowish to reddish brown viscous liquid, which has a woody, earthy, slightly dusty and balsamic odor. The yield of essential oil from the roots of *Chrysopogon zizanioides* (L.) Roberty generally varies from 1.5 to 2.5 per cent. The main producing countries of this oil are China, Madagascar, Java, India, Haiti, Brazil, Mexico, Nepal, Vietnam and Reunion ('Bourbon type vetiver oil').

Literature data (up to December 13, 2014) on the chemical composition of vetiver oils and unpublished analytical data from one of us (E.S.) are shown in Table 5.89.2 in alphabetical order. Nitrogen-containing compounds found in commercial Haitian vetiver oil (on the order of <0.1-2.7 ppm) are shown in Table 5.89.3 (8). In vetiver oils from various origins, over 445 chemicals have been identified. About 55 per cent of these were found in a single reviewed publication only. The major compounds found in vetiver oils from different sources are shown in Table 5.89.4. They include (highest concentrations in any study given) khusimol (36%), (*E*)-isovalencenol (26.2%), isovalencenol (15.3%), β-vetivenene (10.9%), α-vetivone (10.6%) and β-vetivone (8.3%). Well-known ingredients of vetiver oils that were present in high concentrations (>7%) in one or two studies were khusinol (4 studies: 13.0%, 19.2%, 31.4% and 56.2%), khusenic acid (24.0% and 32.3%), vetiselinenol (three studies: 7.8%, 11.0% and 19.5%), khusimone (11.1%), nootkatone (8.3% and

Table 5.89.1 ISO values (%) for vetiver oil [a]

Compound	CAS	Minimum	Maximum
Khusimol	16223-63-5	5.0	18.0
(E)-Isovalencenol	22387-74-2	1.0	16.0
β-Vetivene	27840-40-0	0.7	9.0
α-Vetivone (Isonootkatone)	15764-04-2	1.0	6.0
β-Vetivone	18444-79-6	2.0	5.0

[a] ISO 4716 Essential oil of vetiver ©ISO 2013; Geneva, Switzerland, www.iso.org

10.2%), β-eudesmol (8.4%), α-amorphene (7.8%) and vetiselinol (7.8%). Uncommon or rare constituents of vetiver oils found in high concentrations (>7%) in single studies include khusol (49.3%), khusinal (33.6%), bicyclovetivenol (two studies: 11.5% and 21.3%), 8-cedren-13-ol (12.4%), cycloisolongifolene (10.9%), valerenol (10.5%), dehydroaromadendrene (9.7%), isokhusimol (9.2%), hexadecanoic acid (8.4%), viridiflorol (8.2%), cyclocopacamphenol (7.5%) and vetiverenol (7.5%).

Commercial oils

The ten chemicals that had the highest maximum concentrations in 51 commercial vetiver essential oil

Table 5.89.2 Constituents identified in vetiver oils

Constituent	CAS	Percentage and range				
		A	B (12)	C (1)	D (9-11)	E
10-epi-Acora-3,11-dien-15-al					+	+^{t}
α-Acoradiene	24048-44-0		0.1-0.4			+^{y7}; 0.2^{x6}
β-Acoradiene	28477-64-7		0-0.4			
Acora-2,4-diene (epimer A)	67304-11-4			0.2-0.9	0.3	
Acora-2,4-diene (epimer B)	67304-11-4			0.2-0.4	0.2	
Acora-3,5-diene	89955-08-8					+^{t}
Acora-3,7(14)-diene	255062-41-0					+^{t}
Acora-3,9-diene			0-0.2			+^{t}
Acora-4,7-diene						+^{z1}
Acora-4,9-diene	38229-83-3					+^{z1}
10-epi-Acora-3,11-dien-15-ol					+	
β-Acorenol	28400-11-5	0-0.9				0.6^y
Acorenone	5956-05-8					+^{t}
10-epi-Acor-3-en-5-one	139758-04-6				+	
α-Agarofuran	5956-12-7				+	
Amorpha-4,7(11)-diene	203057-25-4		0-0.3			
α-Amorphene	20085-19-2	0.4-3.4	1.9-4.1		+	2.5^x; 3.0^{x2}; 3.5^{y3}; 4.2^w; 7.8^{y6}
γ-Amorphene	6980-46-7	0-1.5	0-3.3			+^{t}; 0.1^{x3}; 0.3^{x6}; 0.9^{y3}
δ-Amorphene	189165-79-5	0-1.4	0-2.5	0.4-1.8	+	0.2^{x6}; 0.4^{x7}; 2.0^x; 3.5^w; 4.3^{y3}
Amorph-4-en-10-ol (epimer A)	300350-22-5			0.2-0.5^f 5.0-6.4^h	+	
Amorph-4-en-10-ol (epimer B)	300350-22-5				+	
Aromadendrene	489-39-4					+^{z1}; 0.2^u; 1.2^{z3}
allo-Aromadendrene	25246-27-9					+^{z1}
allo-Aromadendrene oxide						1.0^{x5}
β-Atlantol	420109-31-5					0.5^{x3}
Azulene	275-51-4					0.9^{z3}; 4.4^{x2}
trans-α-Bergamotene	13474-59-4					+^{t}
Bicyclovetivenol	76250-14-1					1.1^w; 3.6^y; 11.5^{x1}; 21.3^{z4}
Bisabola-3(15),10-dien-7-ol	194553-32-7				+	+^{t}
1,3,5-Bisabolatrien-7-ol				0.2-0.3		
β-Bisabolene	495-61-4				+	0.7^u; 1.2^{x4}; 1.5^{z3}
α-Bisabolol	515-69-5				+	0.2^u; 1.1^{x4}
6-epi-α-Bisabolol					+	
β-Bisabolol	15352-77-9		0-0.7		+	+^{t}; 0.5^u; 3.1^{y1}; 4.7^{x4}
Borneol	507-70-0					1.5^{x4}
Bornyl acetate	76-49-3					0.2^{x4}
β-Bourbonene	5208-59-3					0.2^{x4}
α-Bulnesene	3691-11-0					1.0^{z3}

Table 5.89.2 Constituents identified in vetiver oils (*continued*)

Constituent	CAS	Percentage and range				
		A	B (12)	C (1)	D (9-11)	E
Bulnesol	22451-73-6					+[t]
Butyl decanoate	30673-36-0					0.6[x1]
Cadina-4α,10β-diol						+[z1]
Cadinenal						0.1[z1]
α-Cadinene	24406-05-1					0.8[x6]
γ-Cadinene	39029-41-9	0-2.8	0-3.4	tr-.6	0.6	0.3[u]; 0.5[x5]; 0.7[w]; 3.6[x4]
δ-Cadinene	483-76-1			1.8-2.5[d]	+	0.3[u]; 1.1[x5]; 1.9[y3]; 2.6[y6]; 3.1[x2]
α-Cadinol	481-34-5	0-3.5	0-6.5			2.7[x1]; 4.0[x8]; 4.6[x]; 4.8[y1]; 5.0[x3]
δ-Cadinol	19435-97-3					2.6[x5]
τ-Cadinol	58580-31-7					6.0[x1]
α-Calacorene	21391-99-1	0.2-1.6	0.4-1.7	0.3-0.8		+[t,y7,z1]; 0.3[x6]; 0.7[y3]; 0.8[w]
β-Calacorene	50277-34-4		0-3.5	1.6-3.7	+	+[z1]; 0.3[x1]
γ-Calacorene	24048-45-1					+[t,z1]
cis-Calamenene	72937-55-4			tr-0.1[e]		+[z1]
trans-Calamenene	73209-42-4					+[t,z1]
Calarene	17334-55-3					0.4[x6]
Camphene	79-92-5					0.7[x4]
β-Caryophyllene	87-44-5		0-0.3			+[t,z1]; 0.1[x4]; 0.8[z3]; 1.4[y3]; 3.7[x5]
γ-Caryophyllene	118-65-0					1.6[z3]
trans-Caryophyllene oxide	1139-30-6					1.7[x5]
Cascarilladiene	59742-39-1		0-0.3		0.3	+[t]
Cascarilladienol					+	
1,7-di-epi-α-Cedrenal						0.4[x3]
Cedren-15-al			0-0.9		+	
α-Cedrene	469-61-4			0.2-0.3	0.2	+[y7,z1]
β-Cedrene	546-28-1		0-0.6			+[t]
β-Cedrene epoxide					+	
γ-Cedrene						3.2[z3]
α-Cedrenol						+[z1]
8-Cedren-13-ol	18319-35-2					2.7[x5]; 12.4[y6]
Chamigrene	18431-82-8					+[t]
1,8-Cineole	470-82-6					<0.1[y2]
Clovene	469-92-1					2.2[z3]
α-Copaene	3856-25-5		0-0.1			0.2[x4]; 1.4[x2]
β-Copaene	18252-44-3		0-0.8			+[t]; 0.3[x6]
α-Costol	65018-15-7					0.3[x7]
m-Cresol	108-39-4					+[z]
o-Cresol	95-48-7					+[z]
p-Cresol	106-44-5					+[z]
α-Cubebene	17699-14-8		0-0.1	tr-0.1	tr	+[t]; 0.3[u]
β-Cubebene	13744-15-5		0-0.1	0.2-0.3		+[t]
Cubebol	23445-02-5					+[t]
10-epi-Cubebol	176589-53-0					+[t]
Cubenol	21284-22-0					0.8[x3]; 0.9[x5]; 2.1[y6]
1-epi-Cubenol	81939-29-9		0-2.4	1.0-1.4	+	+[t]; 0.5[x1]
1,10-di-epi-Cubenol	73365-77-2					3.1[x1]
10-epi-Cubenol				1.4-3.4[g]		
Cuparene	16982-00-6					+[t]; 0.3[x6]
Curcumene (ar-; α-)	644-30-4			1.4-3.7	+	+[t,y5]; 0.7[x2]; 2.0[z3]; 2.4[y6]; 2.7[x5]
β-Curcumene	28976-67-2	0-0.3				0.3[y]
cis-β-Curcumen-12-ol	698365-10-5					1.5[x3]
Cyclocopacamphan-12-al					+	

Table 5.89.2 Constituents identified in vetiver oils (*continued*)

Constituent	CAS	Percentage and range				
		A	B (12)	C (1)	D (9-11)	E
Cyclocopacamphanol A	30810-34-5		0-6.7	5.0-6.4[h]	2.4	+[t]
Cyclocopacamphanol B	28052-00-8		1.1-2.7	5.0-6.4[h]	3.2	+[t]
Cyclocopacamphan-12-ol					+	1.7[w]; 4.0[x8]
Cyclocopacamphenol						7.5[z1,z2]
1,7-Cyclogermacra-1(10),4-dien-15-al			0-0.4		+	+[t]
1,7-Cyclogermacra-1(10),4-dien-15-ol					+	+[t]
Cyclohexadecanolide	109-29-5					0.02[x7]
Cycloisolongifolene	28380-07-6					10.9[x2]
(+)-Cycloisosativene						+[y5]
1-Cyclopentenone	28982-58-3					0.8[z3]
Cyclosativene	22469-52-9	0-0.3	0-0.2			+[t]; 0.3[y]
p-Cymene	99-87-6					<0.1[y2]; 0.6[z3]
Cyperene	2387-78-2		0-0.1			
α-Cyperene	17627-30-4					0.3[u]
α-Cyperone	473-08-5					0.7[z3]
β-Cyperone	23665-63-6				+	
Decanal	112-31-2					0.2[u]
Decane	124-18-5					+[z1]
Dehydroaromadendrene	698388-95-3					0.5[z3]; 5.5[y6]; 9.7[x5]
Dehydrocadinenal						0.2[z1]
8,9-Dehydrocycloisolongifolene						1.8[x5]
4,5-Dehydroisolongifolene						0.5[z3]
Dehydronigritene						+[t]
Dehydro-β-vetivone isomers						+[z1]
cis-Dihydro-nor-zizaene						1.4[y3]
trans-Dihydro-nor-zizaene						0.9[y3]
Dimethyl-6,7-bicyclo[4.4.0]-dec-10-en-4-one			0-2.0			
1,10-Dimethylbicyclo[4.4.0]-dec-6-en-3-one	39850-88-9					+[z1]
[6S,10S]-Dimethylbicyclo-[4.4.0]-1-decen-3-one						+[z1]
3,8-Dimethyl-4-(1-methylethylidene)-2,4,6,7,8,8a-hexahydro-5(1H)-azulenone						4.9[y6]
Dimethyloctalone-1						0.3[z1]
Dimethyloctalone-11						1.3[z1]
Docosane	629-97-0					+[z1]; 0.5[u]
Dodecane	112-40-3					+[z1]
trans-Dracunculifoliol	162657-70-7				+	
7-epi-cis-Dracunculifoliol					+	
α-Duprezianene	79801-29-9					0.1[x]
Eicosane	112-95-8					+[z1]; 0.3[u]
Elema-1,11-dien-15-al (epimers A and B)					+	
β-Elemene	33880-83-0	0-0.2				+[t]; 0.1[x]; 0.2[u]; 0.3[x4]
Elemol	639-99-6	0.3-1.2	0.3-1.2	0.2-0.5		0.1[u]; 0.2[x1]; 0.5[x]; 1.6[x4]; 2.3[z1]
β-Elemol	32142-08-8				+	
Epiglobulol	88728-58-9					2.2[y6]; 4.0[z3]
Epilaurene	18452-45-4					0.2[x1]
Epizizanal	82463-21-6		0-1.0		+	+[t]; 0.8[z1]
Epizizanol	28624-26-2	0.2-0.9		0.7-1.2	+	+[t,z1]
Epizizanone	28624-27-3			2.3-3.2	+	+[t,y7,z1]; 0.1[z1]; 2.3[x1]; 2.5[x6]; 6.0[x7]
1,10-Epoxyamorph-4-ene					+	

Table 5.89.2 Constituents identified in vetiver oils (*continued*)

Constituent	CAS	Percentage and range				
		A	B (12)	C (1)	D (9-11)	E
6,12-Epoxy-elema-1,3-diene					+	
7,10-Epoxy-eremophila-1,11-diene			0-0.1		+	
7,10-Epoxy-eremophila-1,9-dien-8α-ol				tr-0.3		
7,11-Epoxy-eremophila-1,9-dien-8α-ol					+^{s}	
10,11-Epoxy-eremophil-1-ene					+	
7,11;8,12-di-Epoxy-eremophil-9-ene				0.2-0.4^n	+	
7,11;8,12-di-Epoxy-eremophil-9-ene (epimer A)					+	
7,11;8,12-di-Epoxy-eremophil-9-ene (epimer B)				1.0-2.3^q	0.2	
13-nor-7,8-Epoxy-eremophil-1(10)-en-11-one					+^{s}	
5,11-Epoxy-eudesmane					0.5	
6,12,7(11)-di-Epoxy-eudesm-4-(epimer A)ene					+	
6,12,7(11)-di-Epoxy-eudesm-4-ene (epimer B)					+	
6,12,11-di-Epoxy-eudesm-4-ene				0.3-0.5 0.2-0.4^n	+	
13-nor-4,5-Epoxy-eudesm-6-en-11-one					+^{s}	
13-nor-7,8-*trans*-Epoxy-eudesm-4(15)-en-11-one					+	+^{t}
7,10-Epoxy-10-hydroxysalvialane	300349-36-4				+	
7,10-Epoxy-10-methoxysalvialane					+	
4α,7-Epoxy-11-methoxy-10βH-spirovetivene					0.3	
3,10-Epoxy-muurol-4-ene					+	
7,15-Epoxy-prezizaane	87059-19-6				+	+^{t}
6,12-Epoxy-spiroax-4-ene			0-0.3		+	+^{t}
4,7-Epoxy-spirovetiva-2,11-diene					+	+^{t}
4,7-Epoxy-spiroventiv-2-en-11-ol					+	
12-nor-2,3-Epoxyziza-6(13)-ene	300349-30-8				+	
Eremoligenol						2.3^y
Eremophila-1(10),6-dien-12-al			0-1.2	0.9-1.3	+	+^{t}
Eremophila-1(10),7(11)-diene		0.4-1.6	0.9-2.1			+^{t}; 1.6^y
Eremophila-1(10),4(15)-dien-2α-ol			0-1.1	0.1-0.9^l	+	
Eremophila-1(10),6-dien-12-ol					+	
(E)-Eremophila-1(10),7(11)-dien-12-yl acetate	352461-71-3		<0.1	0.2-0.4	+	
(E)-Eremophila-1(10),7(11)-dien-12-yl formate				0.1-0.4	+	
Eremophil-7(11)-en-10β-ol					+	
12,13-di-nor-6(7→8)-*abeo*-Eremophil-1(10)-en-7-one					+	
13-nor-Eremophil-1(10)-en-11-one			0-0.2	0.2-0.5^f	+	+^{t}
11,12,13-tri-nor-Eremo-phil-1(10)-en-7-one					0.2	+^{t,z1}
Ethyl 4-(4-methylphenyl)-4-pentenoate						2.1^{y6}
Eudesma-4(15),7(11)-diene					+	
cis-Eudesma-6,11-diene	194607-93-7	0.3-2.7	0-2.9	1.5-2.1^b	+	+^{t}; 1.3^y; 2.4^w
Eudesma-3,5-dien-1α-ol					1.2	
Eudesma-4(15),7-dien-3β-ol	300349-35-3				+	+^{t}

Table 5.89.2 Constituents identified in vetiver oils (*continued*)

Constituent	CAS	Percentage and range				
		A	B (12)	C (1)	D (9-11)	E
cis-Eudesma-6,11-dien-3β-ol			0-1.4		+	
nor-Eudesmadienone	137695-19-3					+[y7]; 0.7[z1]
13-nor-Eudesma-4,6-dien-11-one			1.0-1.5	0.6-0.7	+	+[t,z1]
13-nor-Eudesma-4(15),7-dien-11-one						+[z1]
13-nor-*trans*-Eudesma-4(15),7-dien-11-one				5.0-6.4[h]	+	
11,12,13-tri-nor-*cis*-Eudesma-5,8-dien-7-one					+	
trans-Eudesma-4(15),7-dien-12-yl formate				0.3-0.4	+	+[t]
13,4-di-nor-Eudesma-5,7,9-trien-11-one						+[z1]
cis-Eudesm-6-en-12-al					+	
15-nor-Eudesm-11-en-4-al						+[z1]
Eudesm-4(10)-en-4-ol						2.3[y]
Eudesm-4(15)-en-5β-ol				1.4-3.4[g]	+	+[t]
Eudesm-4(15)-en-6-ol		0-0.6				0.6[y]
Eudesm-6-en-4β-ol	116538-31-9			1.7-2.4	3.1	+[t]
cis-Eudesm-6-en-11-ol	194607-96-0	0.2-2.3	1.1-3.3		+	1.7[y]; 2.4[w]
cis-Eudesm-6-en-12-ol epimer A		0-1.2				
cis-Eudesm-6-en-12-ol epimer B		0-0.9				
13-nor-Eudesm-5-en-11-one			0-0.3		+	+[t]; 0.3[x8]
13-nor-*cis*-Eudesm-6-en-8-one					+	
13-nor-*cis*-Eudesm-6-en-11-one					+	+[t]
cis-11,12,13-trinor-Eudesm-5-en-7-one					+	+[t,z1]
trans-11,12,13-tri-nor-Eudesm-5-en-7-one				0.5-1.0	+	+[t,z1]
α-Eudesmol	473-16-5	0.08-0.8			+	+[t,y7]; 0.4[u]
7-epi-α-Eudesmol	123123-38-6					+[t]; 1.1[x6]; 1.6[x1]
β-Eudesmol	473-15-4		0-5.2		+	+[t,y7]; 0.1[x4]; 0.9[z3]; 1.1[x5]; 8.4[z1]
γ-Eudesmol	1209-71-8					+[t]
10-epi-γ-Eudesmol	15051-81-7	0-0.9	0-1.8		+	+[t,z1]; 0.5[x1,x3]; 0.7[x4]; 2.2[z1]
Eugenol	97-53-0					+[z]; 1.1[z3]
Farnesene	502-61-4					1.3[z3]
Funebran-15-al			0-1.0	0.1-0.4	+	+[t]
15-nor-Funebran-3-one			0-0.2		+	+[t]
α-Funebrenal						+[z1]
α-Funebren-15-al	57766-59-3				+	+[t,z1]
α-Funebrene	50894-66-1	0.01-0.5	0.1-0.7	0.1-1.1	0.2	+[t,z1]
β-Funebrene	79120-98-2	0.05-0.3	0.2-0.4	0.2-0.5		+[t]; 0.2[u]; 0.3[y]
β-Funebrene epoxide	22472-83-9				+	
α-Funebrenic acid						+[z1]
α-Funebrenol						+[z1]
β-Funebren-14-ol					+	
Furfural	98-01-1	0-0.1				
5,6-*seco*-6,7-Furoeudesman-5-one					+	+[t]
Furopelargone A	1143-45-9				+	
Geraniol	106-24-1					0.2[x4]
Geranyl acetate	105-87-3					0.1[x4]
Guaia-6,9-diene	37839-64-8		0-0.2			+[t]; 0.1[x]
Guaia-1(5),11-dien-3α-ol					+	
α-Guaiene	3691-12-1		0-0.6			
β-Guaiene	88-84-6					1.0[z3]; 1.3[x5]; 4.3[y6]
cis-β-Guaiene	372162-07-7					0.8[w]

Table 5.89.2 Constituents identified in vetiver oils (*continued*)

Constituent	CAS	Percentage and range				
		A	B (12)	C (1)	D (9-11)	E
γ-Guaiene	145267-53-4					0.1^u
Guaienol	300349-29-5		0-0.4		+	+^t
cis-Guai-6-en-10-ol					+	
α-Gurjunene	489-40-7		0-0.5			+^z1; 5.9^y6
β-Gurjunene	73464-47-8			0.4		0.7^u; 5.2^x5
γ-Gurjunene	22567-17-5					0.1^u; 0.7^x3; 2.8^x5
Helifol-1-en-14-ol					+	
Heneicosane	629-94-7					1.2^u
Heptadecane	629-78-7					1.0^u
Hexadecane	544-76-3					+^z1; 0.6^u
Hexadecanoic acid	57-10-3					0.3^x6; 0.6^x7; 8.4^u
2,3,5,5,8,8-Hexamethyl-1,3,6-cyclooctatriene						2.3^y6
β-Himachalene	1461-03-6					1.5^z3
α-Humulene	6753-98-6					0.9^z3
14-Hydroxy-γ-cadinene						0.3^x1
14-Hydroxy-δ-cadinene	153408-92-5					0.4^x6; 2.4^x
10-Hydroxycalamenene	153408-93-6			tr-0.3	+	+^t
Hydroxyjunipene	469-27-2					+^y5
Hydroxyvalencene						4.6^x3
13-Hydroxyvalencene						0.3^x7; 1.2^x1
Intermedeol	6168-59-8		0-1.3		+	+^t
Isocedranol	13567-45-8					+^t; 4.0^y1
Isoeugenol	97-54-1		0-1.3			+^t; 0.3^u; 0.5^z3
(E)-Isoeugenol	5932-68-3					+^z; 0.3^x1; 2.1^x
(Z)-Isoeugenol	5912-86-7					+^z
Isokhusenol	16202-80-5		0-3.3			
Isokhusimol	26128-00-7					+^t; 2.0^z3; 9.2^y1
trans-Isolimonene	6876-12-6					0.3^x6
Isolongifolene	1135-66-6					0.8^z3
Isonootkatol	57422-86-3		0-1.1		+	+^t
α-Isonootkatol	1380573-94-3			5.6-6.9^p		+^t
β-Isonootkatol				5.6-6.9^p		+^t
6-Isopropenyl-4,8a-dimethyl-1,2,3,5,6,7,8,8a-octahydronaphthalen-2-ol						2.0^y6
2-Isopropenyl-1,3,5-trimethylbenzene	14679-13-1					2.4^y6
3-Isopropyl-2(3-methylcyclopent-2-enyl)-tetrahydrofurfuran					tr	
3-Isopropyl-6-methyl-2-(3-methylcyclopent-2-enyl)-3,4-dihydro-2H-pyran					+	+^t
3-(2-Isopropyl-5-methylphenyl)-2-methylpropionic acid						3.2^y6
Isovalencenal					+ (E,Z)-	2.5^w
(E)-Isovalencenal	137695-18-2		0.7-1.4	1.0-2.3^q	+	+^t; 0.4^z1; 1.4^x8
(Z)-Isovalencenal	137695-20-6		0-0.1	0.2-0.6	+	+^t,y7; 0.2^z1; 0.9^x8
Isovalencenol					+	4.6^z1; 7.3^x6; 7.4^y; 8.0^z1; 15.3^w
(E)-Isovalencenol	22387-74-2	1.0-17.7	0-1.5	0.1-0.4	+	+^t; 2.1-16.1^x9; 5.6-6.9^p; 8.5^x7; 11.8^x; 13.2^x3; 16.5^v; 26.2^u
(Z)-Isovalencenol	22387-74-2			5.6-6.9^p		+^t
Isovalencenyl methyl ether	300349-26-2		0-1.1			
Isovellerdiol	37841-93-3					2.4^y6
Isovetiselinenol	67690-36-2					+^z1

Table 5.89.2 Constituents identified in vetiver oils (continued)

Constituent	CAS	Percentage and range				
		A	B (12)	C (1)	D (9-11)	E
Isovetiselinol				1.9-2.4^j	+	
Isozizanoic acid	16202-79-2		0-1.3			+^{t,y7,z1}
Junenol	472-07-1				+	+^{t}; 3.6^{x3}; 3.9^u
Junicedranol	168180-13-0					1.9^{x3}
Juniper camphor	473-04-1		0-1.2	1.9-2.4^j	0.3	+^{t}; 0.8^x; 1.0^{x1}; 1.1^u; 3.1^{x2}
Kessane	3321-66-2				+	+^{t}
Khusenediol						+^{z1}
Khusenic acid	16203-25-1	0-4.3	0-4.8	1.9-4.6^r		0-25.2^{x9}; 1.6^{x3}; 4.3^y; 4.4^{z1}; 5.7^u; 0-6.7^v; 24.0^w; 32.3^{y2}
2-epi-Khusian-2-ol	300349-32-0					+^{t}
Khusimene	18444-94-5	0.2-2.5				1.2^u; 1.3^x; 1.7^{x7}; 3.0^w; 3.6^{x2}
Khusimol	16223-63-5	4.8-18.8	3.4-13.7	16.1-19.2^m	+	4.5-36.0^{x9}; 16.7^{z1}; 19.6^{x3}; 20.4^{x7}; 21.5^{x4}; 24.6^x; 25.6^{x1}; 27.6^{z1}; 27.9^{x1}; 29.5^{y1}; 31.4^v
Khusimone	30557-76-7	0.09-3.3	0.5-1.1	1.5-2.1	+	1.4^{x6}; 1.8^{x1}; 3.3^y; 3.9^w; 11.1^{z3}
Khusinal						33.6^{z1}
Khusinol	24268-34-6	0-0.2	0-2.8			3.4^w; 4.3^{z3}; 4.5-31.4^v; 13.0^{x1}; 19.2^{x5}; 56.2^{z1}
epi-Khusinol						+^{z1}
Khusinol acetate	78405-34-2					3.8^{x1}; 6.1^{z3}
nor-Khusinol oxide						+^{y9,z1}
Khusiol	66512-56-9	0.1-2.3	1.5-2.6		+	+^{t}; 0.5^{x1}; 2.5^y; 3.4^w
allo-Khusiol	312296-11-0		0-0.1		+	+^{t,z1}
Khusione	66397-72-6		0-0.8		+	
Khusitoneol	102818-81-5					+^{z1}
Khusol	18045-73-3					49.3^{z1}
Ledene oxide	882187-44-2					1.1^{x5}
Levojunenol	30951-17-8					+^{z1}
Limonene	138-86-3					<0.1^{y2}; 0.6^{x4}
Linalool	78-70-6					<0.1^{y2}; 0.5^{x4}
Linalyl acetate	115-95-7					0.08^{x4}
Longicyclene	1137-12-8					0.1^x
Longifolenaldehyde	66537-42-6					1.2^{x2}
Longifolene	475-20-7					1.9^{z3}; 5.7^{x5}
α-Longipinene	5989-08-2					4.9^{z3}
Maaliol	527-90-2					1.6^{x3}
epi-Marsupellol	486452-15-7					0.8^y
4-epi-Marsupellol						0.7^y
10-Methoxy-7,10-epoxysalvialane	300349-03-5					+^{t}
Methoxyphenol (4-; p-)	150-76-5					+^{z}
Methyl cyclocopacamphanoate					+	
Methyl epizizanoate	22387-78-6				+	
Methyl (E)-eremophila-1(10), 7(11)-dien-12-oate					+	
8α-Methyl-11,12,13-tri-noreremophil-1(10)-en-7-one					+	
8-Methylheptadecane	13287-23-5					0.7^{z3}
Methyl isozizanoate						+^{z1}
Methyl zizanoate	18444-89-8				+	+^{z1}
Mustakone						2.8^{x3}
α-Muurolene	10208-80-7					0.6^u; 1.9^{y1}
γ-Muurolene	30021-74-0					0.3^{x2}; 2.9^{x5}

Table 5.89.2 Constituents identified in vetiver oils (*continued*)

Constituent	CAS	Percentage and range				
		A	B (12)	C (1)	D (9-11)	E
α-Muurolol	104245-48-9					1.8^{z3}
τ-Muurolol (epi-α-)	19912-62-0					2.4^{x5}
Myrcene	123-35-3					0.1^{x4}
Myrtenol	515-00-4					0.8^{u}
Nerol	106-25-2					0.2^{x4}
Neryl acetate	141-12-8					0.05^{x4}
Nigritene						$+^{y4}$
6-epi-Nigritene						$+^{y4}$
Nigritene A						$+^{t}$
Nigritene B						$+^{t}$
Nonadecane	629-92-5					$+^{z1}$; 1.1^{u}
Nootkatene	5090-61-9	0-1.0	0-0.9	$1.8\text{-}2.5^{d}$		$+^{t}$; 0.7^{x4}; 1.8^{y3}
Nootkatol	50763-67-2			$16.1\text{-}19.2^{m}$	1.7	$+^{t}$; 0.5^{x3}; 1.8^{x}; 2.1^{x7}; 3.5^{x1}
epi-Nootkatol						1.9^{x6}
Nootkatone	4674-50-4	0-0.3	0-0.9	0.7-1.6	+	1.1^{w}; 1.4^{x7}; 1.6^{x3}; 2.2^{x2}; 4.1^{x4}; 5.3^{x1}; 5.7^{x6}; 8.3^{u}; 10.2^{y1}
(Z)-β-Ocimene	3338-55-4					0.08^{x4}
Octadecane	593-45-3					$+^{z1}$; 0.5^{u}
Oplopanone	1911-78-0					$+^{t}$
(E)-Opposita-4(15),7(11)-dien-12-al	224794-18-7		0-1.7	$1.9\text{-}2.4^{j}$	+	$+^{t}$
(E)-Opposita-4(15),7(11)-dien-12-ol			0-1.5	$3.7\text{-}5.9^{k}$	+	
13-nor-Opposit-4(15)-en-11-one					+	
Pentacosane	629-99-2					1.5^{u}
5,10-Pentadecadiyn-1-ol	64275-50-9					$+^{y5}$
Pentadecane	629-62-9					$+^{z1}$; 0.2^{u}
α-Phellandrene	99-83-2					0.1^{x1}
Phenylacetaldehyde	122-78-1					0.3^{u}
α-Pinene	80-56-8					$<0.1^{y2}$; 0.2^{x4}
β-Pinene	127-91-3					$<0.1^{y2}$; 0.2^{u}
trans-Pinocarvone	19890-00-7					0.8^{u}
Pogostol	21698-41-9				+	0.9^{x6}
Porosa-4,6-dien-7-one					+	
Prekhusenic acid			0-1.6			
Prezizaan-15-al	87059-20-9		0-0.5		+	$+^{t}$
nor-Prezizaan-7α-ol					+	
Prezizaan-15-ol	312284-45-0				+	
15-nor-Prezizaan-7-one					+	$+^{t}$
Prezizaene	31145-21-8	0.1-1.5			0.7	0.8^{6}; 1.0^{w}; 1.4^{y}; 1.7^{u}; 2.7^{y3}
Preziza-7(15)-en-3α-ol					+	
Preziza-7(15)-en-12-ol			0.4-0.8		1.5	0.7^{x8}
12-nor-Preziza-7(15)-en-2-one			0-0.7	0.2-0.3	+	$+^{t}$
Preziza-7(15)-en-12-yl acetate					+	
Salvial-4(14)-en-1-one	73809-82-2		0-2.9			
(E)-β-Santalol	37172-32-0					0.6^{x1}
(Z)-β-Santalol	42495-69-2					1.3^{x1}
(Z)-β-Santalol acetate	77-43-0					2.7^{x1}
(+)-Sativene	3650-28-0					2.3^{x2}; 2.8^{y6}
Selina-3,7(11)-diene	6813-21-4		0-0.2		+	$+^{t,y5}$
Selina-4,7(11)-diene	41071-31-2		0.3-0.8			
Selina-4(14),17-diene						0.2^{x4}
Selina-4(15),7(11)-diene			0-0.4			$+^{t}$
α-Selinene	473-13-2			0.5-0.8		

Table 5.89.2 Constituents identified in vetiver oils (*continued*)

Constituent	CAS	Percentage and range				
		A	B (12)	C (1)	D (9-11)	E
β-Selinene	17066-67-0		0-3.1			
γ-Selinene	515-17-3				+	
δ-Selinene	28624-28-4		1.1-2.3	3.9-6.1^c	+	+^{t}; 0.4^{x2}; 0.8^{x6}
Selin-6-en-4α-ol	118173-08-3		0-2.7			
Sesquicineole	90131-02-5			1.2-2.0	+	+^{t}; 0.2^y
(E)-Sesquisabinene hydrate	145512-84-1					0.2^{x1}
6-epi-Shyobunol					+	
Solavetivone	54878-25-0					4.2^{y6}
Spathulenol	6750-60-3					0.2^{x1}; 2.5^{y6}; 6.0^{z3}
Spirovetiva-3,7(11)-dien-12-al (epimer A)	300349-33-1					+^{t}
Spirovetiva-3,7(11)-dien-12-al (epimer B)	300349-33-1					+^{t}
Spirovetiva-1(10),7(11)-diene	137941-80-1	0-0.8	0-1.5			+^{t}
Spirovetiva-3,7(11)-dien-2α-ol						+^{t}
Spirovetiva-3,7(11)-dien-12-ol		0-0.1			+	+^{t}; 2.1^{x3}
Spirovetiva-1(10),11-dien-2-one						+^{z1}
α-Terpinene	99-86-5					0.06^{x4}
Terpinen-4-ol	562-74-3					<0.1^{y2}; 0.1^{x4}
α-Terpineol	98-55-5					0.1^{y2}; 0.4^{x4}
(E)-β-Terpineol	7299-41-4					0.09^{x4}
(Z)-β-Terpineol	7299-40-3					0.2^{x4}
Tetracosane	646-31-1					+^{z1}; 0.5^u
Tetradecane	629-59-4					+^{z1}
1,5,9,9-Tetramethyl-2-methyl-enespiro[3.5]non-5-ene						4.0^{y6}
3,3,8,8-Tetramethyltricyclo-[5.1.0.0(2,4)]oct-5-ene-5-propanoic acid						4.8^{y6}
α-Thujene	2867-05-2					0.2^{x4}
Thujopsene	470-40-6					+^{t}
13-Thujopsene						6.0^{x5}
Tricosane	638-67-5					1.2^u
4,8,8-Trimethyl-2-methylene-4-vinylbicyclo[5.2.0]nonane						4.5^{y6}
2,6,10-Triphenyldodecane						0.6^{z3}
Undecane	1120-21-4					+^{z1}
Valencene	4630-07-3	0.4-1.9	0-0.5			0.3^{u,x1}; 0.4^y; 0.7^{x4}; 0.8^{x2}; 2.8^{y1}
Valencene ketone	137764-27-3		0-0.2	0.6-1.2	+	+^{t,y7}; 0.1^{z1}
Valencen-12-ol	438536-20-0					+^{t}
Valerenal	4176-16-3					1.5^{x2}
Valerenol	101628-22-2					3.8^{x2}; 10.5^{x2}
cis-Valerenyl acetate	101527-78-0					1.3^{x2}; 1.7^{x2}
Valerianol	20489-45-6				+	1.8^{x1}
Vanillin	121-33-5					+^{z}; 1.0^{z3}
β-Vatirenene						5.9^{y6}
Verbenone	80-57-9					0.5^u
Vetidiol						+^{y8}
Vetiselinene	105497-53-8		0.6-1.4			
Vetiselinenol	28102-68-3		1.7-3.7	3.7-5.9^k	+	1.8-5.2^{x9}; 3.0^x; 3.7^{x6}; 5.1^{x3}; 5.8^{y1}; 7.8^{x8}; 11.0^v; 19.5^{z1}
Vetiselinenol epimer						+^{t}
Vetiselinol	32433-12-8	1.5-4.9	1.9-7.8		0.9	4.0^y; 5.6^{x4}
α-Vetispirene	28908-28-3	0-3.5	0-2.2	1.5-2.1^b		+^{t}; 3.1^{y3}

Table 5.89.2 Constituents identified in vetiver oils (*continued*)

Constituent	CAS	Percentage and range				
		A	B (12)	C (1)	D (9-11)	E
β-Vetispirene	28908-27-2	0.8-5.2	1.5-4.5	3.9-6.1^c	+	+^{t}; 0.3^{x3}; 0.6^{x6}; 2.2^y; 2.3^x; 2.7^w
β-Vetivenene	27840-40-0	0.3-10.4	0-5.7	0.8-9.4	+	3.7^x; 5.2^w; 6.0^{y3}; 9.8^{x5}; 10.9^{y3}
γ-Vetivenene	28908-26-1	0.2-4.8	0.2-4.3	tr-0.1^e	+	+^{t}; 0.5^x; 0.7^{x1}; 4.3^{y3}; 5.1^w
Vetiverenol						7.5^{z1,z2}
Vetiverol	68129-81-7					2.3^{x4}
β-Vetivol						+^{t}
α-Vetivone	15764-04-2	1.8-6.1	2.5-6.4	1.9-4.6^r	3.1	2.3-5.7^v; 6.4^{x8}; 6.6^{z4}; 7.2^{z1}; 7.3^{x6}; 7.7^{x1}; 9.8^{y1}; 10.6^u
β-Vetivone	18444-79-6	1.9-5.7	2.0-4.9	0.8-6.6	2.9	2.6^{x5}; 2.7^x; 3.5^{z1}; 4.1^y; 4.2^{x8}; 4.9^{x3}; 5.2^{z4}; 5.6^{z1}; 8.0^w; 8.3^{x4}
p-Vinylguaiacol	7786-61-0					+^{z}
4-Vinylphenol	2628-17-3					+^{z}
Viridiflorene	21747-46-6					4.8^{y6}
Viridiflorol	552-02-3					0.3^{x3}; 0.6^{x6}; 2.5^{z3}; 8.2^{x2}
Widdrol	6892-80-4					2.1^{y6}
α-Ylangene	14912-44-8	0.06-0.5	0-0.6	0.2-0.3	0.2	+^{t}; 0.1^w
Zierone						4.7^{x2}
Zizaane				1.3-2.0	1.3	
2-epi-Ziza-6(13)-en-12-al			0-1.5		+	+^{t}; 1.0^{x8}
2-nor-Zizaene						2.1^{y3}
trans-2-nor-Zizaene			0-1.1			
12-nor-Zizaene						+^{y4}
2-epi-Ziza-6(13)-en-3-ol						2.8^{x3}
2-epi-Ziza-6(13)-en-3α-ol	87068-39-1		1.0-2.7			1.9^w
2-epi-Ziza-6(13)-en-3β-ol	300350-23-6		0-0.4			+^{t}
12-nor-Ziza-6(13)-en-2α-ol	300539-81-5		0.3-0.9		0.8	+^{t}
12-nor-Ziza-6(13)-en-2β-ol			0-2.2		1.1	+^{t}
Ziza-6(13)-en-4-one						+^{z1}
Ziza-6(13)-en-12-yl acetate	61474-33-7		0-0.9	0.1-0.4	0.2	+^{t}
Ziza-6(13)-en-12-yl formate				0.1-0.4^o	0.3	+^{t}
Ziza-6(13)-en-12-yl methyl ether	300349-20-6		0-0.4			
Zizanal	82509-29-3			0.6-4.6^i		+^{t}; 0.3^{x7}; 0.9^{z1}; 1.7^{x6}; 2.7^{x1}
Zizanol	28102-79-6	0.8-3.2		0.5-2.4	+	+^{t,y7}; 2.4^{z1}; 2.5^y; 5.3^{z1}
Zizanone	28051-97-0		0-2.3	0.6-4.6^i	+	+^{t,y7}; 0.2^{z1}; 2.8^w
Zizanyl acetate						+^{z1}

A Fifty-one vetiver essential oil samples from India, Haiti, Nepal, China, Madagascar, Java and Vietnam analyzed between 2004 and 2014; lowest and highest concentrations given (E. Schmidt, unpublished data)

B twenty-seven commercial samples of vetiver oil from nine different countries: Brazil (n=4), China (n=2), Haiti (n=4), India (n=2), Indonesia (n=4), Madagascar (n=3), Mexico (n=2), Reunion Islands (n=4) and El Salvador (n=2); lowest and highest concentrations given (ref. 12)

C two lab-hydrodistilled oils and two commercial vetiver oils from India; lowest and highest concentrations given (ref. 1)

D in-depth investigation of commercial Haitian vetiver oil; where no concentration is mentioned, but a + is shown, the amount of the particular chemical was only stated *in combination with* one or more other ingredients, no quantification was provided, or the amounts cannot be compared with the other data, as quantification was in a particular fraction of the oil (refs. 9,10,11,39,40)

E data from other studies (indicated with superscript letters); highest concentrations found in any study reviewed here given; when two or more oils were investigated, only the highest concentrations are mentioned, unless indicated otherwise

^{b} α-vetispirene and *cis*-eudesma-6,11-diene combined; ^{c} δ-selinene and β-vetispirene combined; ^{d} nootkatene and δ-cadinene combined; ^{e} *cis*-calamene and γ-vetivenene combined; ^{f} amorph-4-en-10-ol (epimer A) and 13-nor-eremophil-1(10)-en-11-one combined; ^{g} eudesma-4(15)-en-5β-ol and 10-epi-cubenol combined; ^{h} cyclocopacamphanol A + B and 13-nor-*trans*-eudesma-4(15),7-dien-11-one and amorph-4-en-10-ol combined; ^{i} ziza-6(13)-en-3-one and zizanal combined; ^{j} eudesm-7(11)-en-4α-ol, (*E*)-opposita-4(15),7(11)-dien-12-al and eudesma-4(15),7-dien-2β-ol combined; ^{k} *trans*-eudesma-4(15),7-dien-12-ol and (*E*)-opposita-4(15),7(11)-dien-12-ol combined; ^{l} eremophila-1(10),4(15)-dien-2α-ol and 4β*H*,5α-eremophila-1(10),7-dien-2α-ol

Table 5.89.2 Constituents identified in vetiver oils (*continued*)

combined; [m] khusimol and nootkatone combined; [n] 7,11,8,12-di-epoxy-eremophil-9-ene and 6,12,7,11-di-epoxy-eudesma-4-ene combined; [o] ziza-6(13)-en-12-yl formate and (*Z*)-isovalencenol combined; [p] isonootkatol and isovalencenol combined; [q] 7,11;8,12-di-epoxy-eremophil-9-ene (epimer B) and (*E*)-isovalencenal combined; [r] α-vetivone and khusenic acid combined; [s] presumably an artifact; [t] fifteen vetiver oils from Haiti (n=5), Indonesia (Java) (n=6), Brazil (n=3), and Reunion Islands (n=1); the chemicals are indicated with +[t], they cannot be compared to the other data (ref. 4); [u] twelve lab-hydrodistilled oils from cultivar 'Sunshine' harvested every two months for 2 years after transplantation in south Italy, plus a reference oil of unknown origin (ref. 13); [v] 34 laboratory steam-distilled root oils from thirteen vetiver accessions grown in test plots in Florida (USA), Nepal and Portugal (ref. 32); [w] four commercial oils from Brazil, Haiti, Java, and Reunion (Bourbon) and one lab-hydrodistilled vetiver root oil from Brazil (ref. 3); [x] three lab-hydrodistilled oils from the roots of 5-month-old vetiver pot plants grown in the USA (ref. 31); [x1] one lab-hydrodistilled oil from vetiver roots from Comoros (ref. 17); [x2] one commercial vetiver oil from Indonesia; this oil has a very atypical composition (e.g., no khusimol, no isovalencenol, no vetivenes or vetivones), but the analysis is also unreliable (e.g., valerenol is mentioned twice as constituent) (ref. 18); [x3] one lab-hydrodistilled oil from Brazil (ref. 22); [x4] one lab-hydrodistilled oil from India; in this study, oils from in vitro generated vetiver morphotypes were also investigated (ref. 25); [x5] two lab-hydrodistilled oils from vetiver cultivated in Turkey and harvested in two successive years (ref. 26); [x6] one lab-hydrodistilled vetiver root oil from Australia (ref. 27); [x7] one lab-hydrodistilled root oil from vetiver grown in the USA; oils of tissue-cultured fungi- and bacteria-free plantlets in sterile and non-sterile soil were also investigated (ref. 29); [x8] one commercial vetiver oil from Reunion (Bourbon vetiver) (ref. 30); [x9] thirteen commercial oils from Indonesia (n=6), Haiti (n=3), France, El Salvador, China and origin unknown (1 each); lowest and highest concentrations given (ref. 32); [y] one commercial oil from El Salvador (ref. 6); [y1] twelve lab-hydrodistilled vetiver root oils with harvest time and fertilizing as variables (ref. 34); [y2] one commercial vetiver root oil from Madagascar; incomplete analysis; less than 35% of the oil components was identified (ref. 33); [y3] one commercial vetiver oil sample from Indonesia (ref. 14); [y4] one commercial oil from Indonesia; three norsesquiterpe-nes were isolated: 12-norzizaene, nigritene and 6-epi-nigritene; these compounds were generated upon thermic degradation of the vetiver acids ziza-noic acid (khusenic acid) and isozizanoic acid (ref. 19); [y5] one steam-distilled oil sample produced from vetiver roots cultivated at an experimental station in India (ref. 23); [y6] one commercial oil purchased in Taiwan which has a highly atypical composition (15,24); [y7] one commercial Chinese vetiver oil sample (ref. 35); [y8] one vetiver oil from North India (probably commercial) (ref. 36); [y9] one oil from north India, probably commercial (ref. 37); [z] one lab-hydrodistilled vetiver root oil from Japan; the acidic fraction was investigated (ref. 38); [z1] data from various studies published between 1970 and 1996 and cited in ref. 5; [z2] cyclocopacamphenol and vetiverenol combined; [z3] one lab-distilled oil from India (ref. 38); [z4] one commercial vetiver oil sample (ref. 41)

tr: traces; + present in vetiver oil, but concentration mentioned *in combination with* one or more other ingredients, no quantification was provided, or data were quantitated in a manner which cannot be compared to the other data

Table 5.89.3 Nitrogen-containing compounds found in Haitian vetiver oils (adapted from 8)

Compound	CAS	Compound	CAS
2-Acetylpyridine	1122-62-9	2-Ethyl-6-methylpyrazine	13925-03-6
2-Aminoacetophenone	551-93-9	Ethyl pyrazine	13925-00-3
1,3-Benzothiazole	128366-28-9	Isocyanato-methylbenzene	25550-57-6
2-Butylpyridine	5058-19-5	2-Isopropyl-3-methoxypyrazine	25773-40-4
Dimethyl anthranilate	85-91-6	Methyl anthranilate	134-20-3
2,5-Dimethyl-3-(3-methylbutyl)pyrazine	18433-98-2	Methylindoline	824-21-5
2,5-Dimethyl-3-pentylpyrazine	56617-69-7	Methylpyrazine	109-08-0
2,5-Dimethyl-3-(1-propenyl)pyrazine		4-Methylpyridine	108-89-4
3,5-Dimethyl-2-(1-propenyl)pyrazine		Methylquinoline	
2,5-Dimethylpyrazine	123-32-0	2-Pentylpyridine	2294-76-0
2,6-Dimethylpyrazine	108-50-9	2-Phenylpyridine	1008-89-5
2,3-Dimethylpyridine	583-61-9	3-Phenylpyridine	1008-88-4
Dimethylquinoline		1-(Pyridin-2-yl)propan-1-one	3238-55-9
2-Ethyl-3,5-dimethylpyrazine	27043-05-6	Quinoline	91-22-5
3-Ethyl-2,5-dimethylpyrazine	13360-65-1	Trimethylindoline	
2-Ethyl-3-methylpyrazine	15707-23-0	2,3,5-Trimethylpyrazine	14667-55-1
2-Ethyl-5-methylpyrazine	36731-41-6		

Table 5.89.4 Major constituents of vetiver oils

Constituent	CAS	Percentage and range				
		A	B (12)	C (1)	D (9-11)	E
Khusimol	16223-63-5	4.8-18.8	3.4-13.7	16.1-19.2^m	+	27.6^{z1}; 27.9^{z1}; 29.5^{y1}; 31.4^v
(E)-Isovalencenol	22387-74-2	1.0-17.7	0-1.5	0.1-0.4	+	2.1-16.1^{x9}; 13.2^{x3}; 16.5^v; 26.2^u
Isovalencenol					+	4.6^{z1}; 7.3^{x6}; 7.4^v; 8.0^{z1}; 15.3^w
β-Vetivenene	27840-40-0	0.3-10.4	0-5.7	0.8-9.4	+	3.7^x; 5.2^w; 6.0^{y3}; 9.8^{x5}; 10.9^{y3}
α-Vetivone	15764-04-2	1.8-6.1	2.5-6.4	1.9-4.6^r	3.1	7.2^{z1}; 7.3^{x6}; 7.7^{x1}; 9.8^{y1}; 10.6^u
β-Vetivone	18444-79-6	1.9-5.7	2.0-4.9	0.8-6.6	2.9	4.9^{x3}; 5.2^{z4}; 5.6^{z1}; 8.0^w; 8.3^{x4}

LEGEND: SEE UNDER TABLE 5.89.2

samples (concentration ranges provided) are the following: khusimol (4.8-18.8%), (E)-isovalencenol (1.0-17.7%), β-vetivene (0.3-10.4%), α-vetivone (isonootkatone) (1.8-6.1%), β-vetivone (1.9-5.7%), β-vetispirene (0.8-5.2%), vetiselinol (1.5-4.9%), γ-vetivenene (0.2-4.8%), zizanoic acid (khusenic acid) (0-4.3%) and α-cadinol (0-3.5%) (Erich Schmidt, unpublished analytical data).

Additional analyses of vetiver oil not presented in Table 5.89.2 can be found on-line in refs. 16 (very atypical composition) and 28 (mostly IUPAC names).

CONTACT ALLERGY/ALLERGIC CONTACT DERMATITIS

General
Contact allergy to vetiver oil has been reported in two publications, and two possible cases of allergic contact dermatitis from the oil have been identified (in both of which false-positive patch test reactions due to the excited skin syndrome cannot be excluded).

Testing in groups of patients
Two hundred dermatitis patients from Poland were tested with vetiver oil 2% in petrolatum and one (0.5%) reacted. The same authors also patch tested 51 patients allergic to *Myroxylon pereirae* resin (balsam of Peru) and/or turpentine and/or wood tar and/or colophony and two (3.9%) had a positive patch test; relevance data were not provided (44). A group of 86 patients from Poland previously reacting to the fragrance mix was tested with vetiver oil and nine (10.4%) had a positive patch test reaction; relevance data were not provided (45).

Case reports
An aromatherapist had chronic hand dermatitis and was patch test positive to 17 of 20 oils used at her work (tested 1% and 5% in petrolatum), including vetiver oil (42). Another aromatherapist had occupational contact dermatitis with allergies to multiple essential oils used at work, including vetiver oil. The patient also reacted to geraniol, α-pinene, caryophyllene, the fragrance mix and various other fragrance materials, but none of these chemicals was identified by GC-MS in the vetiver oil sample (43).

LITERATURE

1 Mallavarapu GR, Syamasundar KV, Ramesh S, Rao BR. Constituents of south Indian vetiver oils. Nat Prod Commun 2012;7:223-225

2 Lavinia UC. Other uses, and utilization of vetiver: Vetiver Oil. Proceedings of the Third International Conference on Vetiver and Exhibition, Guangzhou, China, 6-9 October 2003. Beijing: China Agricultural Press, 2003:486-491. Available at: http://www.vetiver.com/ICV3-Proceedings/IND_vetoil.pdf

3 Martinez J, Rosa PTV, Menut C. Valorization of Brazilian vetiver (*Vetiveria zizanoides* (L.) Nash ex Small) oil. J Agric Food Chem 2004;52:6578-6584

4 Filippi J-J, Belhassena E, Baldovinia N, Brevard H, Meierhenrich UJ. Qualitative and quantitative analysis of vetiver essential oils by comprehensive two-dimensional gas chromatography and comprehensive two-dimensional gas chromatography/mass spectrometry. J Chromatogr A 2013;1288:127-148

5 Lawrence BM. Essential oils 2008-2011. Carol Stream, IL: Allured Books, 2012:57-72

6 Schmidt E. Vetiveröl – Duft und Analytik. Forum 2008;32:41-43

7 Shibamoto T, Nishimura O. Isolation and identification of phenols in oil of vetiver. Phytochem 1982;21:793

8 Clery RA, Hammond CJ, Wright AC. Nitrogen compounds from Haitian vetiver oil. J Essent Oil Res 2005;17:591-592

9 Weyerstahl P, Marschall H, Splittgerber U, Wolf D, Surburg H. Constituents of Haitian vetiver oil. Flavour Fragr J 2000;15:395-412

10 Weyerstahl P, Marschall H, Splittgerber U, Wolf D. 1,7-Cyclogermacra-1(10),4-dien-15-al, a sesquiterpene with a novel skeleton, and other sesquiterpenes from Haitian vetiver oil. Flavour Fragr J 2000;15:61-83

11 Weyerstahl P, Marschall H, Splittgerber U, Wolf D. Analysis of the polar fraction of Haitian vetiver oil. Flavour Fragr J 2000;15:153-173

12 Champagnat P, Figueredo G, Chalchat J-C, Carnat A-P, Bessière J-M. A study on the composition of commercial *Vetiveria zizanoides* oils from

different geographical origins. J Essent Oil Res 2006;18:416-422

13 Massardo DR, Senatore F, Alifano P, Del Giudice L, Pontieri P. Vetiver oil production correlates with early root growth. Biochem System Ecol 2006;34:376-382

14 Rolli E, Marieschi M, Maietti S, Sacchetti G, Bruni R. Comparative phytotoxicity of 25 essential oils on pre- and post-emergence development of *Solanum lycopersicum* L.: A multivariate approach. Ind Crops Prod 2014;60:280-290

15 Peng H-Y, Lai C-C, Lin CC, Chou S-T. Effect of *Vetiveria zizanioides* essential oil on melanogenesis in melanoma cells: downregulation of tyrosinase expression and suppression of oxidative stress. Sci World J 2014, Article ID 213013, 9 pages. Data also presented in ref. 24

16 Kadarohman A, Eko SR, Dwiyanti G, Lailatul KL, Kadarusman E, Nur FA. Quality and chemical composition of organic and non-organic vetiver oil. Indonesian J Chem 2014;14:43-50

17 Soidrou SH, Farah A, Satrani B, Ghanmi M, Jennan S, Hassane SOS, et al. Fungicidal activity of four essential oils from *Piper capense*, *Piper borbonense* and *Vetiveria zizanoides* growing in Comoros against fungi decay wood. J Essent Oil Res 2013;25:216-223

18 Kadarohman A, Sardjono RE, Aisyah S, Khumaisah LL. Biolarvicidal of vetiver oil and ethanol extract of vetiver root distillation waste (*Vetiveria zizanoides*) effectiveness toward *Aedes aegypti*, *Culex* sp., and *Anopheles sundaicus*. J Essent Oil Bear Plants 2013;16:749-762

19 Filippi J-J. Norsesquiterpenes as markers of overheating in Indonesian vetiver oil. Flavour Fragr J 2014;29:137-142

20 Bharat B, Sharma SK, Singh T, Singh L, Arya H. *Vetiveria zizanoides* (Linn.) Nash: a pharmacological overview. Int Res J Pharm 2013;4:18-20

21 Balasankar D, Vanilarasu K, Preetha PS, Umadevi SRM, Bhowmik D. Traditional and medicinal uses of vetiver. J Med Plants Studies 2013;1:191-200

22 Lima GM, Quintans-Júnior LJ, Thomazzi SM, Almeida EMSA, Melo MS, Searfini MR, et al. Phytochemical screening, antinociceptive and anti-inflammatory activities of *Chrysopogon zizanioides* essential oil. Rev Bras Farmacogn Braz J Pharmacogn 2012;22:443-450

23 Gupta S, Dwivedi GR, Darokar MP, Srivastava SK. Antimycobacterial activity of fractions and isolated compounds from *Vetiveria zizanioides*. Med Chem Res 2012;21:1283-1289

24 Chou S-T, Lai C-P, Lin C-C, Shih Y. Study of the chemical composition, antioxidant activity and anti-inflammatory activity of essential oil from *Vetiveria zizanioides*. Food Chem 2012;134:262-268. Data also presented in ref. 15

25 Saraswathi KJT, Jayalakshmi NR, Vyshali P, Kameshwari MNS. Comparitive study on essential oil in natural and *in vitro* regenerated plants of *Vetiveria zizanioides* (Linn.) Nash. American-Eurasian J Agric Environ Sci 2011;10:458-463

26 Kirici S, Inan M, Turk M, Giray ES. To study of essential oil and agricultural properties of vetiver (*Vetiveria zizanioides*) in the southeastern of Mediterranean. Adv Environ Biol 2011;5:447-451

27 Danh LT, Truong P, Mammucari R, Foster N. Extraction of vetiver essential oil by ethanol-modified supercritical carbon dioxide. Chem Engineer J 2010;165:26-34

28 Bhuiyan NI, Chowdhury JU, Begum J. Essential oil in roots of *Vetiveria zizanioides* (l.) Nash ex. Small from Bangladesh. Bangladesh J Bot 2008;37:213-215

29 Adams RP, Nguyen S, Johnston DA, Park S, Provin TL, Habte M. Comparison of vetiver root essential oils from cleansed (bacteria- and fungus-free) vs. non-cleansed (normal) vetiver plants. Biochem System Ecol 2008;36:177-182

30 Champagnat P, Sidibé L, Forestier C, Carnat A, Chalchat J-C, Lamaison J-L. Antimicrobial activity of essential oils from *Vetiveria nigritana* and *Vetiveria zizanioides* roots. J Essent Oil Bear Plants 2007;10:519-524

31 Adams RP, Habte M, Park S, Dafforn MR. Preliminary comparison of vetiver root essential oils from cleansed (bacteria- and fungus-free) versus non-cleansed (normal) vetiver plants. Biochem System Ecol 2004;32:1137-1144

32 Adams RP, Pandey RN, Dafforn MR, James SA. Vetiver DNA fingerprinted cultivars: effects of environment on growth, oil yields and composition. J Essent Oil Res 2003;15:363-371

33 Möllenbeck S, König T, Schreier P, Schwab W, Rajaonarivony J, Ranarivelo L. Chemical composition and analyses of enantiomers of essential oils from Madagascar. Flavour Fragr J 1997;12:63-69

34 Dethier M, Sakubu S, Ciza A, Cordier Y, Menut C, Lamaty G. Aromatic plants of tropical central Africa. XXVIII. Influence of cultural treatment and harvest time on vetiver oil quality in Burundi. J Essent Oil Res 1997;9:447-451

35 Sellier N, Cazaussus A, Budzinski H, Lebon M. Structure determination of sesquiterpenes in Chinese vetiver oil by gas chromatography-tandem mass spectrometry. J Chromatogr 1991;557:451-458

36 Kalsi PS, Talwar KK. Stereostructure of vetidiol, a new antipodal sesquiterpene diol from vetiver oil; a novel role of biological activity to predict the position and stereochemistry of one of the hydroxyl group. Tetrahedron 1987;43:2985-2988

37 Kalsi PS, Kaur B, Talwar KK. Stereostructure of nor-khusinoloxide, a new antipodal C_{14} terpenoid from vetiver oil. Confirmation of stereostructural features by biological evaluation, a new tool for prediction of stereostructure in cadinanes. Tetrahedron 1985;41:3387-3390

38 Chowdhury AR, Kumar D, Lohani H. GC-MS analysis of essential oils of *Vetiveria zizanoides* (Linn.) Nash roots. FAFAI 2002;April/June:33-35. Data cited in ref. 5

39 Weyerstahl P, Marschall H, Splittgerber U, Wolf D. New cis-Eudesm-6-ene derivatives from vetiver oil. Liebigs Ann 1997:1783-1787

40 Weyerstahl P, Marschall H, Splittgerber U. New sesquiterpene ethers from vetiver oil. Liebigs Ann 1996;1195-1199

41 Kim H-J, Chen F, Wang X, Chung H-Y, Jin Z-Y. Evaluation of antioxidant activity of vetiver (*Vetiveria zizanoides* L.) oil and identification of its antioxidant constituents. J Agric Food Chem 2005;53:7691-7695

42 Selvaag E, Holm J, Thune P. Allergic contact dermatitis in an aromatherapist with multiple sensitizations to essential oils. Contact Dermatitis 1995;33:354-355

43 Dharmagunawardena B, Takwale A, Sanders KJ, Cannan S, Roger A, Ilchyshyn A. Gas chromatography: an investigative tool in multiple allergies to essential oils. Contact Dermatitis 2002;47:288-292

44 Rudzki E, Grzywa Z, Bruo WS. Sensitivity to 35 essential oils. Contact Dermatitis 1976;2:196-200

45 Rudzki E, Grzywa Z. Allergy to perfume mixture. Contact Dermatitis 1986;15:115-116

Chapter 5.90   YLANG-YLANG OIL

DEFINITION
Ylang-ylang oil (essential oil of ylang-ylang) is the essential oil obtained from the flowers of the ylang-ylang tree, *Cananga odorata* (Lam.) Hook. f. et Thomson, forma *genuina*.

INCI NOMENCLATURE
Description/definition: Cananga odorata flower oil is the oil obtained from the flower, *Cananga odorata*, Anonaceae
INCI name EU & USA: Cananga odorata flower oil
CAS registry number(s): 8006-81-3; 83863-30-3
EINECS number(s): 281-092-1

ISO (INTERNATIONAL ORGANIZATION FOR STANDARDIZATION) STANDARD
ISO number: 3063
ISO name: Essential oil of ylang-ylang
Botanical origin: *Cananga odorata* (Lam.) Hook f. et Thomson, forma *genuina*
Parts of plant used: Flower
ISO values: ISO values (minimum and maximum concentrations) for the various grades of ylang-ylang oil are shown in Tables 5.90.1 ('extra'), 5.90.4 ('first'), 5.90.7 ('second') and 5.90.10 ('third').

Ylang-ylang oil should not be confused with cananga oil, obtained from *Cananga odorata* (Lam.) Hook f. et Thomson, forma **macrophylla** (3). These oils were originally thought to be identical and are still often used – incorrectly – as synonyms. See also Chapter 5.11 Cananga oil.

THE PLANT, THE OIL AND THEIR USES
Cananga odorata Hook. fil. et Thomson, commonly known as kenanga (Indonesia), ylang-ylang (Philippines), canang odorant (French) and cananga (English), is an evergreen medium-sized tree growing 10-30 meters in height. Its flowers blossom throughout the year and are very fragrant. The cananga is native to Malaysia and Indonesia, but is now naturalized to most of the larger Pacific Island groups, northern Australia, Thailand, and Vietnam. The tree has long been cultivated on a large scale in southeast Asia and some islands of the Indian Ocean, mainly Madagascar and the Comoro Islands, for the production of essential oil from the flowers (1,2,3,28,31). There are two forms of *Cananga odorata*: the forma *genuina* and the forma *macrophylla*. From both species essential oils are produced, mostly by steam distillation of fresh mature flowers (27): cananga oil from the *macrophylla* form and the more precious ylang-ylang essential oil from *C. odorata* forma *genuina* (27,29,30). The applications of both oils, which were formerly considered to be identical and are still often used – incorrectly – as synonyms, are broadly the same. However, the ylang-ylang oil is generally preferred, notably in the fragrance industry.

Ylang-ylang oil is mainly used for fine (floral) fragrances, but also for other cosmetics including soaps and skin lotions and detergents. The oil may also be utilized as a flavoring agent for beverages, ice cream, candies, chewing gums and baked goods (15,29). Aromatherapy with ylang-ylang oil is claimed to be useful for depression (4), breathing problems, hypertension (4), and anxiety and is applied as an aphrodisiac. In folk medicine, the oil is used for a variety of infectious and skin diseases, insect bites, high blood pressure, and as a sedative and antidepressant (1,2,3).

Ylang-ylang essential oil production has the particularity of relying on a fractionation based on distillation times, resulting in four to five grades (fractions) of oil that have different commercial applications: 'extra super', 'extra' (30 minutes distillation time), 'first' (2-5 hours), 'second' (6-10 hours), and 'third' (up to 12 hours). The commercial grades differ considerably in their chemical composition; the first fractions are the most valuable for the production of fine fragrances (2,5,12). There are also commercial 'complete' ylang-ylang oils, which result from 24 hours distillation of the flowers without removal of earlier fractions. The 'extra' fraction constitutes from 17 to 30% of the oil yield; relative quantities for the other fractions (oils from the Comoro Islands) are: 'first' 10-14%, 'second' 7-8% and 'third' 58-60% (32). Most of the oils on the market are mixtures of various fractions, but are sold as one particular fraction. As the various grades (fractions, qualities) of ylang-ylang oils have different compositions and there are different ISO norms for each grade, they are discussed here separately. It should be realized that there are no absolute boundaries between the various qualities and that overlap of a quality in some studies with different grades in other investigations may be considerable.

Older reviews of ylang-ylang oils include refs. 7-11. It should be appreciated that there is a real possibility that, especially in older scientific literature, ylang-ylang and cananga oils have been mixed up.

YLANG-YLANG OIL 'EXTRA'
Ylang-ylang oil 'extra' is the first (and most valuable) fraction of the essential oil obtained with 30 minutes distillation time. ISO values are shown in Table 5.90.1.

CHEMICAL COMPOSITION
Ylang-ylang oil 'extra' is a clear mobile liquid with pale yellowish to darker yellow color, which has an intense exotic, somewhat jasmine-like, floral odor with an aromatic woody note in the background. The yield of essential oil from the flowers of *Cananga odorata* (Lam.) Hook f. et Thomson, forma *genuina*, fraction 'extra', generally varies from 1.9 to 2.3 per cent. The main producing countries of this oil are the Comoro Islands, Madagascar and Vietnam.

Literature data (up to December 8, 2014) on the chemical composition of ylang-ylang oils 'extra' and unpublished analytical data from one of us (E.S.) are shown in Table 5.90.2 in alphabetical order. In ylang-ylang oils 'extra' from various origins, over 190 chemicals have been

Table 5.90.1 ISO values (%) for ylang-ylang oils 'extra' [a]

Compound	CAS	Minimum	Maximum
Linalool	78-70-6	7.0	24.0
Germacrene D	23986-74-5	5.0	20.0
Benzyl acetate	140-11-4	5.5	17.5
p-Cresyl methyl ether	104-93-8	5.0	16.0
(E,E)-α-Farnesene	502-61-4	1.0	15.0
Geranyl acetate	105-87-3	2.5	14.0
Methyl benzoate	93-58-3	4.0	9.0
β-Caryophyllene	87-44-5	2.5	8.5
Benzyl benzoate	120-51-4	3.0	8.0
(E)-Cinnamyl acetate	21040-45-9	0.5	6.5
Benzyl salicylate	118-58-1	1.2	4.0
(E,E)-Farnesol	106-28-5	0.8	3.0
(E,E)-Farnesyl acetate	29548-30-9	0.5	3.0
Geraniol	106-24-1	0.1	3.0
3-Methyl-2-butenyl acetate (prenyl acetate)	1191-16-8	0.6	2.3

a ISO 3063 Essential oil of ylang-ylang 'extra' ©ISO 2004; Geneva, Switzerland, www.iso.org

identified. About 56 per cent of these were found in a single reviewed publication only. The major compounds found in ylang-ylang oils 'extra' from different sources are shown in Table 5.90.3. They include (highest concentrations in any study given) linalool (52.8%), benzyl acetate (44.4%), geranyl acetate (26.0%), p-cresyl methyl ether (23.0%), germacrene D (21.5%), benzyl benzoate (16.0%), methyl benzoate (11.6%) and (E,E)-α-farnesene (10.3%). Well-known ingredients of ylang-ylang oils 'extra' that were present in high concentrations (>7%) in one or two studies were (E)-cinnamyl acetate (11.9%) and β-caryophyllene (10.1%).

Commercial oils

The ten chemicals that had the highest maximum concentrations in 51 commercial ylang-ylang 'extra' essential oil samples (concentration ranges provided) are the following: germacrene D (11.7-21.3%), benzyl acetate (3.8-16.9%), linalool (5.8-13.5%), p-cresyl methyl ether (3.5-11.4%), (E,E)-α-farnesene (5.6-10.3%), β-caryophyllene (3.6-10.1%), geranyl acetate (2.9-9.4%), benzyl benzoate (2.7-9.3%), methyl benzoate (2.1-6.0%) and (E)-cinnamyl acetate (1.3-5.6%) (Erich Schmidt, unpublished analytical data).

Table 5.90.2 Constituents identified in ylang-ylang oils 'extra'

Constituent	CAS	Percentage and range					
		A	B (12)	C (1)	D (5)	E (34)	F
4-Allylphenyl acetate	61499-22-7		0-tr				
α-Amorphene	20085-19-2					0.2-0.2	
τ-Amorphol	53947-91-4						0.8[a]
Anethole	104-46-1					0.5-0.6	
(E)-Anethole	4180-23-8	0.05-0.7	0.1-0.3	0.4			0.4[a]
p-Anisyl acetate	1200-06-2		0-0.06	0.04			
Aromadendrene	489-39-4			0.06			
Benzaldehyde	100-52-7		0.03-0.08	0.2			
Benzyl acetate	140-11-4	3.8-16.9	5.6-38.6	27.5	3.7-44.4	4.6-1.3	12.4[a]
Benzyl alcohol	100-51-6	0.03-0.3	tr-0.08	0.5		0.8-0.1	0.1[b]
Benzyl benzoate	120-51-4	2.7-9.3	2.7-4.5	1.0	1.1-16.0	4.9-6.1	4.8[a]; 5.2[b]
Benzyl butyrate	103-37-7	0.03-0.6	0.03-0.07	0.06			
Benzyl cinnamate	103-41-3		0-tr				
Benzyl 4-methyl-pentanoate	77509-00-3			tr			
Benzyl salicylate	118-58-1	0.9-4.7	0.7-1.9	0.2	0.5-5.8	0.2-0.3	1.9[b]; 2.4[a]
Bicycloelemene	32531-56-9		tr-0.04	0.03			
Bicyclogermacrene	24703-35-3	0.09-0.7	tr-0.05				
α-Bisabolol	515-69-5					0.2-0.07	
β-Bourbonene	5208-59-3	0.1-0.4	0-0.04	tr			
Bulnesol	22451-73-6			tr			
Butyl benzoate	136-60-7		tr-0.07	0.04			
trans-Cadina-1(6),4-diene	931410-54-7		0-tr				
Cadina-3,5-diene	267665-20-3		0-0.02				
γ-Cadinene	39029-41-9	0.5-1.1	0.06-0.2	tr			0.6[a]
δ-Cadinene	483-76-1	1.2-2.5	0.2-0.3	0.3		3.4-0.4	1.1[a]
ε-Cadinene	1080-67-1					0.3-0.3	
τ-Cadinene	152287-05-3					0.6-0.9	
α-Cadinol	481-34-5	0.6-2.2	0.1-0.3	0.07		0.4-0.5	
δ-Cadinol	19435-97-3	0.1-1.0				0.3-1.1	

Table 5.90.2 Constituents identified in ylang-ylang oils 'extra' (*continued*)

Constituent	CAS	Percentage and range					
		A	B (12)	C (1)	D (5)	E (34)	F
τ-Cadinol (10-epi-α-)	58580-31-7	0.04-4.0				0.2-0.2	
Calamene	1406-50-4					0.2-0.3	
Camphene	79-92-5					0.04-0.04	
β-Caryophyllene	87-44-5	3.6-10.1	0.4-1.1	0.4	0.3-1.7	6.3-6.6	5.2[a]
Caryophyllene oxide	1139-30-6	0.04-0.2	tr-0.05	0.06			
α-Cedrene	469-61-4					0.2-0.2	
β-Cedrene	546-28-1						0.1[a]
α-Cedrol	77-53-2					0.3-0.5	
1,8-Cineole	470-82-6	0.1-0.4	0.2-0.9	1.1		0.7-0.2	
(Z)-Cinnamaldehyde	57194-69-1						+[b]
Cinnamyl acetate	103-54-8			0.9		1.5-2.0	4.0[a]
(E)-Cinnamyl acetate	21040-45-9	1.3-5.6	1.0-5.9		0.9-11.9		
Cinnamyl alcohol	104-54-1			tr			
(E)-Cinnamyl alcohol	4407-36-7		0-0.02				
(Z)-Cinnamyl alcohol	4510-34-3						+[b]
Copaborneol	21966-93-8		tr-0.03	tr			
α-Copaene	3856-25-5	0.6-1.2	0-0.05	0.1		0.9-0.8	0.6[a]
β-Copaene	18252-44-3		0.06-0.2	0.7			
m-Cresol	108-39-4						+[b]
p-Cresol	106-44-5	0.01-0.05	0.1-0.2	0.2			+[b]
m-Cresyl acetate	122-46-3						+[b]
p-Cresyl acetate	140-39-6		0-0.3				+[b]
p-Cresyl methyl ether	104-93-8	3.5-11.4	11.8-15.5	9.7	2.7-23.0	9.0-3.6	8.4[b]; 8.5[a]
α-Cubebene	17699-14-8			tr			
β-Cubebene	13744-15-5		0.05-0.1	0.1			
1,10-di-epi-Cubenol	73365-77-2		0-tr				
1-epi-Cubenol	81939-29-9		0-0.05				
β-Curcumene	28976-67-2		tr-0.2	0.4			
Decane	124-18-5			tr			
Diethyl (2E)-3-methyl-2-pentanedioate				tr			
Diethyl 1,5-pentanedioate	818-38-2			tr			
2,5-Dimethyl-3-methyl-ene-1,5-heptadiene	74663-83-5			tr			
Docosane	629-97-0			tr			
Dodecane	112-40-3			tr			
Eicosane	112-95-8		0-tr	tr			
β-Elemene	33880-83-0	0.3-0.5					
Elemol (α-)	639-99-6	0.1-0.2	tr-0.05	tr			
Ethyl benzoate	93-89-0		0.02-0.1	0.4		1.0-1.5	
Eugenol	97-53-0	0.02-0.3	tr-1.1	tr			0.08[a]
(E,E)-Farnesal	502-67-0			tr			
(E,E)-α-Farnesene	502-61-4	5.6-10.3	0.3-1.0	1.6	0.3-2.4	1.7-7.4	8.2[a]
Farnesol	4602-84-0						2.0[b]
(E,E)-Farnesol	106-28-5	0.7-2.3	0.6-1.7		0.2-2.2	1.0-0.9	
(Z,E)-Farnesol	3790-71-4		0-tr				
(Z,Z)-Farnesol	16106-95-9			0.09			
Farnesyl acetate	29548-30-9						1.6[b]
(E,E)-Farnesyl acetate	4128-17-0	1.1-4.3	0.08-0.2	0.05	0.05-0.5	0.8-1.0	1.7[a]
(Z,Z)-α-Farnesyl acetate	24163-97-1		·				1.3[a]
Geranial	141-27-5		tr-0.3	0.03			
Geraniol	106-24-1	0.09-1.6	0.2-4.4	0.4	0.02-5.3	0.3-0.2	0.3[a]; 0.7[b]
Geranyl acetate	105-87-3	2.9-9.4	4.7-26.0	2.0	0.4-12.5	3.9-4.5	4.5[a]
Geranyl benzoate	94-48-4		0.03-0.2	0-0.02			
1(10),5-Germacradien-4-ol	74841-87-5		0.2-0.3				
Germacrene D	23986-74-5	11.7-21.3	1.3-2.5	tr	0.6-5.5	7.7-21.5	17.1[a]
Germacrene D-4-ol	198991-79-6			0.06			
Guaiacylacetone	2503-46-0	tr					

Table 5.90.2 Constituents identified in ylang-ylang oils 'extra' (*continued*)

Constituent	CAS	Percentage and range					
		A	B (12)	C (1)	D (5)	E (34)	F
Guaiol	489-86-1		0-0.04	tr			
Heneicosane	629-94-7			tr			
Heptacosane	593-49-7			tr			
Heptanal	111-71-7		0-0.2	0.05			
Hexacosane	630-01-3			tr			
n-Hexadecane	544-76-3			tr			
1-Hexadecene	629-73-2			tr			
Hexanal	66-25-1		0-0.02				
Hexanol	111-27-3		0-tr				
(*E*)-2-Hexenol	928-95-0					0.1-0	
3-Hexen-1-ol	544-12-7		0-tr	tr		0.06-0	
2-Hexenyl acetate	2497-18-9					0.2-0.1	
3-Hexenyl acetate	1708-82-3					0.4-0.2	
(*Z*)-3-Hexenyl acetate	3681-71-8	0.06-0.2	0.1-0.2	0.5			0.1[a]
(*Z*)-3-Hexenyl benzoate	25152-85-6			tr			
Hexyl acetate	142-92-7	0.05-0.3	0.1-0.3	1.0			0.2[a]
α-Humulene	6753-98-6	1.2-3.3	0.3-0.6			2.0-2.2	0.9[b]; 1.8[a]
Hydrocinnamyl acetate	122-72-5			0.03			
5-Indanol	1470-94-6		0-tr	tr			
1*H*-Indole	120-72-9			tr			
α-Ionene	475-03-6			tr			
Isoamyl acetate	123-92-2		0-tr				
Isocaryophyllene	118-65-0						0.2[b]
Isodauca-6,9-diene	317819-81-1		0-tr				
Isoeugenol	97-54-1	0.02-0.7		tr			0.5[a]
(*E*)-Isoeugenol	5932-68-3		0.4-1.5	0.6			
Isoeugenol acetate	93-29-8		tr-0.05	tr			0.1[a]
Isogermacrene D	317819-80-0		tr-0.05	0.03			
Isoprenyl acetate	17616-47-6	0.06-1.1					0.8[a]
Junenol	472-07-1		0-tr				
Ledane	28580-43-0			tr			
Levoglucosenose	37112-31-5			tr			
Limonene	138-86-3	0.02-0.1				0.1-0.09	
β-Limonene	5989-54-8			0.1			
Linalool	78-70-6	5.8-13.5	12.9-46.2	9.0	8.5-52.8	29.2-20.8	8.6[a]; 11.0[b]
trans-Linalool oxide acetate	56469-40-0		0-tr				
cis-Linalool oxide, furanoid	5989-33-3		tr-0.03	0.02			
trans-Linalool oxide, furanoid	34995-77-2		0-tr	tr			
trans-Linalool oxide, pyranoid	39028-58-5		0-tr				
Linalyl acetate	115-95-7		0-tr	0.03			
4-Methoxybenzaldehyde	123-11-5			tr			
2-Methoxy-4-methyl-phenol	93-51-6		0-0.02	0.03			
2-Methoxyphenol	90-05-1		0-tr	0.7			
1-Methoxy-4-propyl-benzene	104-45-0			tr			
Methyl *p*-anisate	121-98-2		tr-0.09	0.08			
Methyl anthranilate	134-20-3			tr			0.4[b]
Methyl benzoate	93-58-3	2.1-6.0	6.7-10.4	6.1	4.0-11.6	4.2-1.5	5.0[a]
2-Methyl-3-buten-2-ol	115-18-4					0.4-0.03	
3-Methyl-2-buten-1-ol	556-82-1		0-0.04			0.07-0.06	
3-Methyl-3-buten-1-ol	763-32-6		0-tr				
3-Methyl-2-butenyl acetate (prenyl acetate)	1191-16-8	0.2-2.3	0.07-1.3	4.2	0-2.9		1.6[a]
3-Methyl-3-butenyl acetate	5205-07-2		0.05-0.8	2.2			

Table 5.90.2 Constituents identified in ylang-ylang oils 'extra' (*continued*)

Constituent	CAS	Percentage and range					
		A	B (12)	C (1)	D (5)	E (34)	F
3-Methyl-2-butenyl benzoate	5205-11-8	0.9-2.2	0.2-0.8	0.4			0.8[a]
3-Methyl-3-butenyl benzoate	5205-12-9			0.03			
2-Methylbutyl acetate	624-41-9		0-tr				
Methyl chavicol	140-67-0	0.08-0.2	0.02-0.3	0.06			
Methyl eugenol	93-15-2		0-0.07			0.2-0.2	0.09[b]
6-Methyl-5-hepten-2-one	110-93-0		tr-0.02	0.07			
Methyl 2-methoxyben-zoate	606-45-1		0-tr	tr			
Methyl 3-methylbutanoate	556-24-1			tr			
cis-2-Methyl-7-octadecene	35354-39-3			tr			
Methyl octanoate	111-11-5			0.4			
Methyl salicylate	119-36-8	0.02-0.2	0.2-0.3	0.3			0.1-tr
α-Muurolene	10208-80-7		tr-0.05				
γ-Muurolene	30021-74-0	0.9-2.2	tr-0.04				
τ-Muurolene	152287-43-9						0.2-0.6
τ-Muurolol	19912-62-0	0.3-1.1	0.09-0.2	0.06			0.1-0.07
Myrcene	123-35-3	0.04-0.5	tr-0.08	0.3			0.1-0.2
Neral	106-26-3		0-0.1				
Nerol	106-25-2		0-tr	tr			1.3[b]
Nerolidol	7212-44-4	0.02-0.05					0.06[b]
(E)-Nerolidol	40716-66-3					0.3-0.3	
Neryl acetate	141-12-8		tr-0.02	2.7			
Nonacosane	630-03-5			tr			
Nonadecane	629-92-5			tr			
β-Ocimene	13877-91-3			0.07			
(E)-β-Ocimene	3779-61-1					0.7-tr	
Octacosane	630-02-4			tr			
Octadecanal	638-66-4			tr			
Octadecane	593-45-3			tr			
Pentacosane	629-99-2			tr			
Pentadecane	629-62-9			tr			
2-Phenethyl alcohol	60-12-8			0.2			
Phenylacetaldehyde	122-78-1		0.04-0.2	0.02			
Phenylacetonitrile	140-29-4		0-tr	0.02			
1-Phenylallyl acetate	7217-71-2			tr			
1-Phenylethyl acetate	93-92-5	0.04-0.3					
2-Phenylethyl acetate	103-45-7	0.08-0.3	0.6			0.2-0.1	0.2[a]
2-Phenylnitroethane	6125-24-2		0.1-0.2	0.3			
1-Phenyl-2-propen-1-ol	4393-06-0			tr			
α-Pinene	80-56-8	0.2-0.4	0.03-0.09	0.1		2.3-2.0	0.2[a,b]
β-Pinene	127-91-3	0.05-0.1	tr-0.05			0.4-0.5	1.8[b]
Plinol A	4028-59-5			0.1			
Plinol D	4028-58-4			0.05			
4-(2-Propenyl)phenol	501-92-8		0-tr				
α-Pyronene	514-94-3			0.04			
Sabinene	3387-41-5			tr			
Terpinen-4-ol	562-74-3		0-tr				
α-Terpineol	98-55-5	0.03-0.2	0.08-0.3	0.3		0.2-0.2	
Terpinolene (α-)	586-62-9			0.07			
Tetracosane	646-31-1			tr			
Tetradecane	629-59-4			tr			
Tricosane	638-67-5			tr			
Vanillin	121-33-5		0.02-0.1	tr			
Veratrole	91-16-7		tr-0.03	0.07			
Vinyl butyrate	123-20-6			tr			
p-Vinylguaiacol	7786-61-0			tr			
α-Ylangene	14912-44-8					0.8-0.9	

Table 5.90.2 Constituents identified in ylang-ylang oils 'extra' (*continued*)

Constituent	CAS	Percentage and range					
		A	B (12)	C (1)	D (5)	E (34)	F
β-Ylangene	20479-06-5			1.7			
Zonarene	41929-05-9		0-tr	tr			

A fifty-one ylang-ylang essential oil samples 'extra' from Comoros, Mayotte and Madagascar analyzed between 1998 and 2013; lowest and highest concentrations given (E. Schmidt, unpublished data)
B twenty-eight essential oils obtained by lab-hydrodistillation of mature flowers from Grande Comore, Mayotte, Nossi Bé (Madagascar) and Ambanja (Madagascar); quality 'extra', hydrodistillation time 25 minutes; mean percentages for ingredients of the oils from each individual island were provided; lowest and highest mean percentages given (ref. 12)
C one lab-hydrodistilled ylang-ylang essential oil from mature flowers collected on the isle of Mayotte; distillation time 25 minutes (ref. 1)
D twenty-three lab-hydrodistilled essential oil samples (25 minutes distillation time) from mature flowers collected at 11 plantations on three islands: Grande Comore, Madagascar, and Mayotte; only the data of the 15 compounds mentioned in ISO 3063:2004 were presented; lowest and highest concentrations given (ref. 5)
E two ylang-ylang oils obtained by laboratory steam-distillation for 1 hour of Columbian flowers, one batch being fresh, the other 24 hours old; both values are mentioned, left from fresh flowers, right from flowers plucked 24 hours earlier (ref. 34)
F data from other studies (indicated with superscript letters) in which the 'extra' fraction has been analyzed; highest concentrations found in in any study reviewed here given

[a] one commercial oil from Comoro Islands (ref. 17); [b] one commercial ylang-ylang oil sample form Comoro Islands (ref. 18)

tr: trace (in columns B and C: <0.02%); + present in the oil investigated, but quantity not stated

YLANG-YLANG OIL 'FIRST'

Ylang-ylang oil 'first' is the second fraction of the essential oil obtained with 2-5 hours distillation time. ISO values are shown in Table 5.90.4.

CHEMICAL COMPOSITION

Ylang-ylang oil 'first' is a clear mobile liquid with pale yellowish to darker yellow color, which has a strong exotic, somewhat jasmine-like, floral odor with an aromatic woody note in the background. The yield of essential oil from the flowers of *Cananga odorata* (Lam.) Hook f. et Thomson, forma *genuina*, fraction 'first', generally varies from 1.9 to 2.3 per cent. The main producing countries of this oil are the Comoro Islands, Madagascar and Vietnam.

Literature data (up to December 8, 2014) on the chemical composition of ylang-ylang oils 'first' and unpublished analytical data from one of us (E.S.) are shown in Table 5.90.5 in alphabetical order. In ylang-ylang oils 'first' from various origins, over 145 chemicals have been identified. About 55 per cent of these were found in a single reviewed publication only. The major compounds found in ylang-ylang oils 'first' from different sources are shown in Table 5.90.6. They include (highest concentrations in any study given) linalool (30.0%), benzyl acetate (26.3%), geranyl acetate (26.0%), germacrene D (25.7%), benzyl benzoate (14.9%), β-caryophyllene (11.2%), p-cresyl methyl ether (10.7%) and (E,E)-α-farnesene (10.7%). Well-known ingredients of ylang-ylang oils 'first' that were present in high concentrations (>7%) in one or two studies were methyl salicylate (10.4%), (E)-cinnamyl acetate (10.3), and benzyl salicylate (7.1%).

Commercial oils

The ten chemicals that had the highest maximum concentrations in 12 commercial ylang-ylang 'first' essential oil samples (concentration ranges provided) are the following: germacrene D (13.6-15.8%), benzyl acetate

Table 5.90.3 Major constituents identified in ylang-ylang oils 'extra'

Constituent	CAS	Percentage and range					
		A	B (12)	C (1)	D (5)	E (34)	F
Linalool	78-70-6	5.8-13.5	12.9-46.2	9.0	8.5-52.8	29.2-20.8	8.6[a]; 11.0[b]
Benzyl acetate	140-11-4	3.8-16.9	5.6-38.6	27.5	3.7-44.4	4.6-1.3	12.4[a]
Geranyl acetate	105-87-3	2.9-9.4	4.7-26.0	2.0	0.4-12.5	3.9-4.5	4.5[a]
p-Cresyl methyl ether	104-93-8	3.5-11.4	11.8-15.5	9.7	2.7-23.0	9.0-3.6	8.4[b]; 8.5[a]
Germacrene D	23986-74-5	11.7-21.3	1.3-2.5	tr	0.6-5.5	7.7-21.5	17.1[a]
Benzyl benzoate	120-51-4	2.7-9.3	2.7-4.5	1.0	1.1-16.0	4.9-6.1	4.8[a]; 5.2[b]
Methyl benzoate	93-58-3	2.1-6.0	6.7-10.4	6.1	4.0-11.6	4.2-1.5	5.0[a]
(E,E)-α-Farnesene	502-61-4	5.6-10.3	0.3-1.0	1.6	0.3-2.4	1.7-7.4	8.2[a]

LEGEND: SEE UNDER TABLE 5.90.2

Table 5.90.4 ISO values (%) for ylang-ylang oils 'first' [a]

Compound	CAS	Minimum	Maximum
Germacrene D	23986-74-5	9.5	24.0
Linalool	78-70-6	3.0	19.0
(E,E)-α-Farnesene	502-61-4	3.0	18.0
Geranyl acetate	105-87-3	2.0	15.0
Benzyl acetate	140-11-4	2.8	14.0
β-Caryophyllene	87-44-5	4.0	12.0
p-Cresyl methyl ether	104-93-8	3.0	10.0
Benzyl benzoate	120-51-4	4.2	8.0
Methyl benzoate	93-58-3	1.5	5.5
(E)-Cinnamyl acetate	21040-45-9	0.5	5.0
Benzyl salicylate	118-58-1	1.6	4.0
(E,E)-Farnesyl acetate	29548-30-9	1.0	4.0
Geraniol	106-24-1	0.1	2.6
(E,E)-Farnesol	106-28-5	0.1	2.5
3-Methyl-2-butenyl acetate (prenyl acetate)	1191-16-8	0.2	1.8

[a] ISO 3063 Essential oil of ylang-ylang 'first' ©ISO 2004; Geneva, Switzerland, www.iso.org

(12.0-15.1%), linalool (9.8-11.3%), p-cresyl methyl ether (8.0-10.7%), (E,E)-α-farnesene (8.1-10.2%), methyl benzoate (4.5-5.6%), geranyl acetate (4.4-5.5%), (E)-cinnamyl acetate (3.9-4.6%), benzyl benzoate (4.0-4.5%) and β-caryophyllene (3.7-4.2%) (Erich Schmidt, unpublished analytical data).

YLANG-YLANG OIL 'SECOND'
Ylang-ylang oil 'second' is the third fraction of the essential oil obtained with 6-10 hours distillation time. ISO values are shown in Table 5.90.7.

CHEMICAL COMPOSITION
Ylang-ylang oil 'second' is a clear mobile liquid with yellowish to darker yellow color, which has a typical exotic floral, somewhat jasmine-like, and woody odor. The yield of essential oil from the flowers of *Cananga odorata* (Lam.) Hook f. et Thomson, forma *genuina*, fraction 'second', generally varies from 1.9 to 2.3 per cent. The main producing countries of this oil are the Comoro Islands, Madagascar and Vietnam.

Literature data (up to December 8, 2014) on the chemical composition of ylang-ylang oils 'second' and unpublished analytical data from one of us (E.S.) are shown in

Table 5.90.5 Constituents identified in ylang-ylang oils 'first'

Constituent	CAS	Percentage and range				
		A	B (12)	C (14)	D (34)	E
4-Allylphenyl acetate	61499-22-7		0-tr			
α-Amorphene	20085-19-2				0.4-0.3	
τ-Amorphol	53947-91-4					1.1[e]
Amyl acetate	628-63-7	0.04-0.05				
Anethole	104-46-1				0.4-0.5	
(E)-Anethole	4180-23-8	0.3-0.4	0.1-0.4			0.2[e]
p-Anisyl acetate	1200-06-2		0.05-0.1			
Benzaldehyde	100-52-7		0.03-0.06			<0.1[g]
Benzyl acetate	140-11-4	12.0-15.1	3.4-26.3	3.3-8.0	4.6-0.4	4.6[g]; 7.7[e]
Benzyl alcohol	100-51-6	0.07-0.09	tr-0.07		0.6-tr	0.1[g]; 0.4[f]
Benzyl benzoate	120-51-4	4.0-4.5	7.1-14.5	4.3-14.9	12.3-14.5	5.8[e]; 7.6[g]
Benzyl butyrate	103-37-7	0.07-0.09	0.03-0.09			
Benzyl cinnamate	103-41-3		0-0.03			
Benzyl salicylate	118-58-1	1.7-2.2	2.5-7.1	0.3-3.4	0.3-1.8	1.2[f]; 2.6[e]
Bicycloelemene	32531-56-9		0.1-0.2			
γ-Bisabolene	495-62-5					<0.1[g]
α-Bisabolol	515-69-5				0.08-tr	
β-Bourbonene	5208-59-3	0.1-0.3	0-0.06			
Butyl benzoate	136-60-7		0.03-0.1			
trans-Cadina-1,4-diene	20085-13-6		0-tr			
trans-Cadina-1(6),4-diene	931410-54-7		0.04-0.07			
Cadina-3,5-diene	267665-20-3		tr-0.05			
α-Cadinene	24406-05-1		tr-0.03			
γ-Cadinene	39029-41-9	0.7-1.0	0.2-0.3			0.8[e]
δ-Cadinene	483-76-1	0.6-1.2	0.5-0.7	0.2-2.8	0.4-0.7	1.7[e]; 2.3[g]
ε-Cadinene	1080-67-7			1.4-3.7[b]	0.3-0.6	
τ-Cadinene	152287-05-3				1.3-1.2	
α-Cadinol	481-34-5	0.7-1.0	0.5-0.9	0.2-2.0	0.2-1.0	0.6[g]
γ-Cadinol	50895-55-1			0.09-2.4[d]		0.4[g]

Table 5.90.5 Constituents identified in ylang-ylang oils 'first' (*continued*)

Constituent	CAS	Percentage and range				
		A	B (12)	C (14)	D (34)	E
δ-Cadinol	19435-97-3	0.2-1.2		0.1-3.0	0.2-0.06	0.8[g]
τ-Cadinol (10-epi-α-)	58580-31-7	0.02-1.9			0.5-1.0	
Calamene	1406-50-4				0.1-0.2	
Calamenene	483-77-2					0.3[g]
Camphene	79-92-5				tr-tr	
β-Caryophyllene	87-44-5	3.7-4.2	1.0-2.4	1.1-11.2	7.9-8.1	8.0[e]; 10.7[g]
Caryophyllene oxide	1139-30-6	0.01-0.09	0.07-0.2			
α-Cedrene	469-61-4				0.3-0.9	<0.1[g]
β-Cedrene	546-28-1					0.2[e]
Cedrol	77-53-2				tr-0.2	
1,8-Cineole	470-82-6	0.2-0.3	0.1-0.4	0.4-0.8	0.4-tr	0.2[f]; 0.7[g]
(Z)-Cinnamaldehyde	57194-69-1					+[f]
Cinnamyl acetate	103-54-8				1.5-2.5	1.1[g]; 2.8[e]
(E)-Cinnamyl acetate	21040-45-9	3.9-4.6	1.6-10.3	0.6-1.9		
(E)-Cinnamyl alcohol	4407-36-7		0-0.03			
(Z)-Cinnamyl alcohol	4510-34-3					+[f]
Copaborneol	21966-93-8		0.05-0.07			
α-Copaene	3856-25-5	0.5-0.7	0.04-0.08	0.4-1.6	1.2-1.1	0.7[g]; 0.8[e]
β-Copaene	18252-44-3		0.3-0.4			
m-Cresol	108-39-4		0-tr			+[f]
p-Cresol	106-44-5	0.01-0.02	0.09-0.3			+[f]
m-Cresyl acetate	122-46-3					+[f]; 0.4[g]
p-Cresyl acetate	140-39-6		0-0.1			+[f]
p-Cresyl methyl ether	104-93-8	8.0-10.7	2.2-7.1	1.1-10.4	2.1-0.4	6.5[f]; 8.4[g]
β-Cubebene	13744-15-5		0.2-0.3			<0.1[g]
1,10-di-epi-Cubenol	73365-77-2		tr-0.03			
1-epi-Cubenol	81939-29-9		0.04-0.1			
β-Curcumene	28976-67-2		0.02-0.4			
p-Cymene	99-87-6					0.09[f]
Eicosane	112-95-8		0-tr			
β-Elemene	33880-83-0	0.3-0.6				
Elemol	639-99-6	0.03-0.08	0.04-0.2			
Ethyl benzoate	93-89-0		tr-0.03		0.7-0.4	
Eugenol	97-53-0	0.01-0.2	0.04-1.7			0.2[e]; 0.4[g]
(2E,6Z)-Farnesal	3790-67-8		tr-0.03			
(E,E)-α-Farnesene	502-61-4	8.1-10.2	1.9-4.0	0.3-4.9[c]	5.5-8.0	10.7[e]
Farnesol	4602-84-0					1.5[f]
(E,E)-Farnesol	106-28-5	0.9-1.1	1.8-5.9		4.0-3.4	1.3[g]
(Z,E)-Farnesol	3790-71-4		0-0.06			
Farnesyl acetate	29548-30-9					2.0[f]
(E,E)-Farnesyl acetate	4128-17-0	1.7-2.6	0.3-0.8	0.5-7.8	1.8-4.5	1.8[g]; 1.9[e]
(Z,E)-Farnesyl acetate	40266-29-3		0-tr			
(Z,Z)-α-Farnesyl acetate	24163-97-1					1.6[e]
Furfuryl alcohol	98-00-0					0.4[g]
Geranial	141-27-5		0.03-0.3			
(E)-Geranic acid	4698-08-2		0.2-4.4			
Geraniol	106-24-1	0.2-0.3		0.9-3.0	0.1-0.2	1.1[f]; 1.6[g]
Geranyl acetate	105-87-3	4.4-5.5	4.7-26.0	6.2-11.0	3.2-3.0	4.3[e]; 7.6[g]
Geranyl benzoate	94-48-4	0.01-0.09	0-0.1			
1(10),5-Germacradien-4-ol	74841-87-5	0.1-0.2				
Germacrene D	23986-74-5	13.6-15.8	7.3-8.8	0.1-13.5	23.8-25.7	10.3[g]; 20.1[e]
Guaiol(E)-2-	489-86-1		0.04-0.1			
Hexen-1-ol	928-95-0		0-tr		tr-0	
2-Hexenyl acetate	2497-18-9				0.1-0.08	
3-Hexenyl acetate	1708-82-3				0.08-0.09	
(Z)-3-Hexenyl acetate	3681-71-8	0.1-0.2	tr-0.09			
(Z)-3-Hexenyl benzoate	25152-85-6		0-tr			
Hexyl acetate	142-92-7	0.2-0.3	0.05-0.2			
α-Humulene	6753-98-6	1.3-2.0	0.9-1.7	1.4-3.7[b]	2.7-2.6	1.7[f]; 2.7[e]
5-Indanol	1470-94-6		tr-0.04			
Isodauca-6,9-diene	317819-81-1		0-0.02			

Table 5.90.5 Constituents identified in ylang-ylang oils 'first' (*continued*)

Constituent	CAS	Percentage and range				
		A	B (12)	C (14)	D (34)	E
Isoeugenol	97-54-1	0.4-0.8				0.5[e]
(*E*)-Isoeugenol	5932-68-3		1.1-3.7			
Isoeugenol acetate	93-29-8		0.07-0.2			
Isogermacrene D	317819-80-0		0.1			
Isoprenyl acetate	17616-47-6	0.7-2.1				0.4[e]
Junenol	472-07-1		tr-0.04			
Limonene	138-86-3	0.0-0.02			0.06-tr	
Linalool	78-70-6	9.8-11.3	6.7-17.3	11.7-30.0	15.0-4.6	12.2[f]; 19.0[g]
cis-Linalool oxide, furanoid	5989-33-3		tr			
Linalyl acetate	115-95-7		tr-0.04			
Menthone (*p*-)	89-80-5					0.2[g]
2-Methoxy-4-methylphenol phenol	93-51-6		0-0.03			
Methyl *p*-anisate	121-98-2		0.03-0.1			
Methyl anthranilate	134-20-3					0.8[f]
Methyl benzoate	93-58-3	4.5-5.6	2.3-5.2	1.7-5.6	2.2-0.2	3.4[e]; 3.6[g]
2-Methyl-3-buten-2-ol	115-18-4			0.3-1.3	0.2-tr	0.5[g]
3-Methyl-2-buten-1-ol	556-82-1		0-0.02	0.05-0.2	tr-0.07	0.2[g]
3-Methyl-3-buten-1-ol	763-32-6					0.2[g]
bis(3-Methyl-3-buten-1-ol)			0.01-0.2			
3-Methyl-2-butenyl acetate (prenyl acetate)	1191-16-8	1.6-2.1	0.04-1.3	0.06-0.4		0.3[g]; 0.9[e]
3-Methyl-3-butenyl acetate	5205-07-2		0.02-0.6			
3-Methyl-2-butenyl benzoate	5205-11-8	0.4-1.3	0.3-1.5			0.1[e]
Methyl chavicol	140-67-0	0.05-0.1	tr-0.3			
Methyl eugenol	93-15-2		tr-0.1		0.2-0.5	<0.1[g]; 0.1[f]
6-Methyl-5-hepten-2-one	110-93-0		0-tr			
Methyl salicylate	119-36-8	0.08-0.1	0.09-0.2	1.7-10.4	tr-0	0.2[g]
α-Muurolene	10208-80-7		0.06-0.1	0.2-1.1		0.6[g]
γ-Muurolene	30021-74-0	0.5-1.4	0.06-0.1	0.5-1.9		
τ-Muurolene	152287-43-9				0.7-0.6	
τ-Muurolol	19912-62-0	0.5-0.7	0.4-0.8	0.09-2.4[d]	0.4-0.6	1.7[g]
Myrcene	123-35-3	0.05-0.07	tr-0.08	0.05-0.2	0.09-0.09	0.1[g]
Neral	106-26-3		0-0.09			
Nerol	106-25-2		0-tr			0.1[g]; 2.0[f]
Nerolidol	7212-44-4	tr-0.02				0.08[f]
(*E*)-Nerolidol	40716-66-3				0.2-0.2	0.1[g]
Neryl acetate	141-12-8		0.03-0.04			
(*E*)-β-Ocimene	3779-61-1				tr-0	
2-Phenethyl alcohol	60-12-8					<0.1[g]
Phenylacetaldehyde	122-78-1		0.03-0.2			
Phenylacetonitrile	140-29-4		0-tr			
1-Phenylethyl acetate	93-92-5	0.09-0.2			0.1-tr	
2-Phenylethyl acetate	103-45-7		0.07-0.2			0.1[e]
2-Phenylnitroethane	6125-24-2		0.2			
α-Pinene	80-56-8	0.2-0.3	0.04-0.08	0.09-0.4	1.2-1.2	0.2[e]; 0.4[g]
β-Pinene	127-91-3	0.08-0.1	tr-0.08	0.06-0.2	0.4-0.4	0.2[f,g]
4-(2-Propenyl)phenol	501-92-8		0-tr			
Safrole	94-59-7					<0.1[g]
γ-Terpinene	99-85-4					0.9[f]
Terpinen-4-ol	562-74-3		0-tr			
α-Terpineol	98-55-5	0.09-0.1	0.08-0.2		0.1-0.03	0.6[g]
Terpinolene (α-)	586-62-9		0-tr			0.4[f]
Tetradecane	629-59-4		0-tr			
Vanillin	121-33-5		0.07-0.2			
Veratrole	91-16-7		tr-0.02			
p-Vinylguaiacol	7786-61-0		0-tr			
α-Ylangene	14912-44-8				0.9-1.0	0.1[g]
Zonarene	41929-05-9		0-0.02			

Table 5.90.5 Constituents identified in ylang-ylang oils 'first' (*continued*)

A twelve ylang-ylang essential oil samples 'first' from Comoros, Mayotte and Madagascar analyzed between 1998 and 2013; lowest and highest concentrations given (E. Schmidt, unpublished data)
B twenty-eight essential oils obtained by lab-hydrodistillation of mature flowers from Grande Comore, Mayotte, Nossi Bé (Madagascar) and Ambanja (Madagascar); hydrodistillation time 1 hour; mean percentages for ingredients of the oils from each individual island were provided; lowest and highest mean percentages given (ref. 12)
C twenty commercial ylang-ylang oil samples of first grade from Madagascar; lowest and highest concentrations given (ref. 14)
D two ylang-ylang oils obtained by laboratory steam-distillation for 2 hours of Columbian flowers, one batch being fresh, the other 24 hours old; both values are mentioned, left from fresh flowers, right from flowers plucked 24 hours earlier (ref. 34)
E data from other studies (indicated with superscript letters) in which fraction 'first' has been analyzed; highest concentrations found in any study reviewed given

[b] α-humulene and ε-cadinene combined; [c] γ-cadinene and α-farnesene combined; [d] τ-muurolol and γ-cadinol combined; [e] one commercial oil from Comoro Islands (ref. 17); [f] one commercial ylang-ylang oil sample form Comoro Islands (ref. 18); [g] one commercial first grade oil from Madagascar (ref. 13)

tr: trace (in column B: <0.02)

Table 5.90.6 Major constituents identified in ylang-ylang oils 'first'

Constituent	CAS	Percentage and range				
		A	B (12)	C (14)	D (34)	E
Linalool	78-70-6	9.8-11.3	6.7-17.3	11.7-30.0	15.0-4.6	12.2[f]; 19.0[g]
Benzyl acetate	140-11-4	12.0-15.1	3.4-26.3	3.3-8.0	4.6-0.4	4.6[g]; 7.7[e]
Geranyl acetate	105-87-3	4.4-5.5	4.7-26.0	6.2-11.0	3.2-3.0	4.3[e]; 7.6[g]
Germacrene D	23986-74-5	13.6-15.8	7.3-8.8	0.1-13.5	23.8-25.7	10.3[g]; 20.1[e]
Benzyl benzoate	120-51-4	4.0-4.5	7.1-14.5	4.3-14.9	12.3-14.5	5.8[e]; 7.6[g]
β-Caryophyllene	87-44-5	3.7-4.2	1.0-2.4	1.1-11.2	7.9-8.1	8.0[e]; 10.7[g]
p-Cresyl methyl ether	104-93-8	8.0-10.7	2.2-7.1	1.1-10.4	2.1-0.4	6.5[f]; 8.4[g]
(*E,E*)-α-Farnesene	502-61-4	8.1-10.2	1.9-4.0	0.3-4.9[c]	5.5-8.0	10.7[e]

LEGEND: SEE UNDER TABLE 5.90.5

Table 5.90.7 ISO values (%) for ylang-ylang oils 'second' [a]

Compound	CAS	Minimum	Maximum
Germacrene D	23986-74-5	13.0	28.0
(*E,E*)-α-Farnesene	502-61-4	5.0	21.0
β-Caryophyllene	87-44-5	4.8	17.0
Geranyl acetate	105-87-3	1.7	12.0
Benzyl benzoate	120-51-4	4.5	10.0
Linalool	78-70-6	2.0	9.5
Benzyl acetate	140-11-4	0.5	8.8
p-Cresyl methyl ether	104-93-8	1.0	5.0
(*E*)-Cinnamyl acetate	21040-45-9	0.4	4.8
Benzyl salicylate	118-58-1	1.8	4.0
(*E,E*)-Farnesyl acetate	29548-30-9	1.0	3.5
Methyl benzoate	93-58-3	1.0	3.5
(*E,E*)-Farnesol	106-28-5	0.8	3.5
Geraniol	106-24-1	0.1	2.4
3-Methyl-2-butenyl acetate (prenyl acetate)	1191-16-8	0.1	0.9

[a] ISO 3063 Essential oil of ylang-ylang 'second' ©ISO 2004; Geneva, Switzerland, www.iso.org

Table 5.90.8 in alphabetical order. In ylang-ylang oils 'second' from various origins, over 120 chemicals have been identified. About 52 per cent of these were found in a single reviewed publication only. The major compounds found in ylang-ylang oils 'second' from different sources are shown in Table 5.90.9. They include (highest concentrations in any study given) germacrene D (26.4%), (*E,E*)-α-farnesene (19.7%), β-caryophyllene (19.6%), geranyl acetate (19.0%), benzyl benzoate (18.2%), linalool (13.2%) and benzyl acetate (8.4%). Well-known ingredients of ylang-ylang oils 'second' that were present in high concentrations (>7%) in one or two studies were *p*-cresyl methyl ether (11.4%), benzyl salicylate (9.5%), (*E*)-cinnamyl acetate (7.8%) and (*E,E*)-farnesol (7.7%). A rare constituent of ylang-ylang oils 'second' found in a high concentration (>7%) in a single study is (*E*)-isoeugenol (8.0%).

Commercial oils

The ten chemicals that had the highest maximum concentrations in 42 commercial ylang-ylang 'second' essential oil samples (concentration ranges provided) are the following: germacrene D (12.2-26.4%), (*E,E*)-α-farnesene (5.2-19.7%), β-caryophyllene (4.9-18.8%), linalool (0.1-13.2%), geranyl acetate (3.0-11.8%), *p*-cresyl

Table 5.90.8 Constituents identified in ylang-ylang oils 'second' and in complete oils and oils of unknown quality

Constituent	CAS	YLANG-YLANG OILS 'SECOND'				COMPLETE OILS AND OILS OF UNKNOWN QUALITY
		Percentage and range				
		A	B (12)	C (14)	D	E
Acetaldehyde	75-07-0					+[e]
Acetic acid	64-19-7					+[e]
Acetone	67-64-1					+[e]
α-Amorphene	20085-19-2					0.2[c,e]; 0.3[b]
τ-Amorphol	53947-91-4				2.0[o]	
Amyl acetate	628-63-7	tr-0.04				+[e]; 0.02[q]
(E)-Anethole	4180-23-8	0.04-0.2	0.03-0.2		0.2[o]	0.1[q]; 0.3[c]; 0.4[f]; 0.6[b]
p-Anisyl acetate	1200-06-2		0.2-0.3			
Benzaldehyde	100-52-7		tr-0.03			+[e]
Benzoic acid	65-85-0					+[e]
Benzyl acetate	140-11-4	0.3-8.4	0.8-8.2	0.6-3.1	1.7[o]	9.6[c]; 9.8[j]; 10.3[r]; 13.5[t]; 14.7[f]; 25.1[e]
Benzyl alcohol	100-51-6	0.02-0.3	tr-0.06		0.2[p]	0.02[q]; 0.1[b]; 0.2[c]; 0.5[e]; 1.9[j]
Benzyl benzoate	120-51-4	4.3-11.0	12.2-18.2	5.3-12.3	6.3[o]; 6.5[p]	9.8[q]; 13.4[k]; 14.1[c]; 14.5[b]; 16.4[e]
Benzyl butyrate	103-37-7		0-0.04			0.01[q]; 0.1[c]
Benzyl cinnamate	103-41-3		0-0.05			
Benzyl salicylate	118-58-1	1.4-3.1	4.0-9.5	1.0-3.9	1.2[p]; 2.6[o]	0.8[k]; 1.8[b]; 2.1[f]; 2.3[c]; 4.7[q]; 5.2[e]
α-Bergamotene	17699-05-7					5.4[s]
Bicycloelemene	32531-56-9		0.2-0.3			
Bicyclogermacrene	24703-35-3	0.3-0.4	0.2			0.5[f]; 0.7[q]
γ-Bisabolene	495-62-5					tr[e]
α-Bisabolol	515-69-5					0.02[c]; 0.07[b]
β-Bourbonene	5208-59-3	0.04-0.1	0-0.07			0.2[q]
Butyl benzoate	136-60-7		tr-0.06			0.03[c]
Butyraldehyde	123-72-8					+[e]
Cadina-1,4-diene	29837-12-5					1.0[e]
trans-Cadina-1,4-diene	20085-13-6		0.04-0.05			
trans-Cadina-1(6),4-diene	931410-54-7		0.1			
Cadina-3,5-diene	267665-20-3	0.03-0.1				
α-Cadinene	24406-05-1	0.05-0.06				0.1[e]
β-Cadinene	523-47-7					1.0[d]; 1.7[s]
γ-Cadinene	39029-41-9	0.6-1.3	0.3-0.4	1.7-12.7[m]	1.1[o]	0.08[c]; 0.2[d]; 1.1[q]
δ-Cadinene	483-76-1	1.3-4.6	0.9-1.1	2.1-5.2	3.1[o]	0.3[k]; 0.4[c,f]; 0.7[b,q]; 2.3[e]; 5.6[j]; 27.8[k,l]
ε-Cadinene	1080-67-7			3.4-5.8[a]		
τ-Cadinene	152287-05-3					1.2[b]
α-Cadinol	481-34-5	0.6-2.2	1.3-1.7	0.5-1.9		0.2[c]; 0.6[f]; 0.8[e]; 1.0[b,g]; 1.4[q]; 1.6[k]
γ-Cadinol	50895-55-1			0.4-1.5[n]		0.4[e]
δ-Cadinol	19435-97-3	0.4-2.0		0.2-1.0		0.02[q]; 0.2[c]; 0.8[d]; 0.9[e]; 1.1[b]
τ-Cadinol (10-epi-α-)	58580-31-7	0.1-0.7				0.05[q]
Calamene	1406-50-4					0.3[b]
Calamenene	483-77-2					0.03[c]; 1.0[e]
Camphene	79-92-5					0.03[c]; 0.04[b]
δ2-Carene (δ4-carene)	554-61-0					0.1[d]
δ3-Carene	13466-78-9					0.06[d]
β-Caryophyllene	87-44-5	4.9-18.8	1.5-2.8	1.8-19.6	12.9[o]	10.6[e]; 11[g]; 12.6[j]; 12.9[r]; 25.7[d]; 33.0[s]
Caryophyllene oxide	1139-30-6	0.08-0.3	0.1-0.2			0.02[q]
α-Cedrene	469-61-4					0.03[e]; 0.3[c]; 0.9[b]
β-Cedrene	546-28-1				0.4[o]	
Cedrol	77-53-2					0.03[c]; 0.5[b]
1,8-Cineole	470-82-6	0.04-0.4	0.06-0.2	0.1-0.3		0.01[d]; 0.08[q]; 0.1[j]; 0.2[b,f]; 0.4[e]
(Z)-Cinnamaldehyde	57194-69-1				+[p]	
Cinnamyl acetate	103-54-8				1.1[o]	1.0[e]; 2.5[b]; 4.1[c]; 4.6[f,t]; 4.8[g]
(E)-Cinnamyl acetate	21040-45-9	0.4-4.2	0.8-7.8	0.3-1.9		1.3[q]
Cinnamyl alcohol	104-54-1					0.07[c]
(E)-Cinnamyl alcohol	4407-36-7		0-tr			
(Z)-Cinnamyl alcohol	4510-34-3				+[p]	
Citronellol	106-22-9					0.02[d]

Table 5.90.8 Constituents identified in ylang-ylang oils 'second' and in complete oils and oils of unknown quality (*continued*)

Constituent	CAS	YLANG-YLANG OILS 'SECOND'				COMPLETE OILS AND OILS OF UNKNOWN QUALITY
		Percentage and range				
		A	B (12)	C (14)	D	E
Citronellyl acetate	150-84-5					0.02[d]; 0.04[c]
Copaborneol	21966-93-8		0.05-0.08			
α-Copaene	3856-25-5	0.1-1.3	0.1-0.2	1.9-3.6	1.2[o]	0.4[d]; 0.5[s]; 0.6[f]; 0.8[q]; 1.0[e]; 1.1[b]
β-Copaene	18252-44-3		0.8-0.9			0.1[e]
m-Cresol	108-39-4		0-tr		+[p]	
p-Cresol	106-44-5	0.01-0.05	0.04-0.1		+[p]	0.02[q]; 0.1[c,e]
m-Cresyl acetate	122-46-3				+[p]	
p-Cresyl acetate	140-39-6		0-0.3		+[p]	+[e]
p-Cresyl methyl ether	104-93-8	3.5-11.4	0.8-3.2	0.6-5.3	1.2[o]; 2.7[p]	2.4[s]; 3.6[b]; 4.9[q]; 6.8[c]; 9.7[f]; 18.5[e]
α-Cubebene	17699-14-8	0.1-0.2				0.1[e]
β-Cubebene	13744-15-5	0.2-1.2	0.5			0.1[e]; 0.2[f]; 27.8[k,l]
1,10-di-epi-Cubenol	73365-77-2		0.07-0.1			
1-epi-Cubenol	81939-29-9		0.2-0.3			
Cuparene	16982-00-6					
β-Curcumene	28976-67-2		0.06-0.9			
Cyclocopacamphene	24112-86-5					0.1[e]
p-Cymene	99-87-6					0.06[j]; 0.6[d]
Dihydrolinalool	18479-51-1					0.5[j]
β-Elemene	33880-83-0	0.5-0.6				tr[e]; 0.1[d]; 0.2[f]; 0.4[q]
δ-Elemene	20307-84-0	0.06-0.08				0.1[e]
Elemol (α-)	639-99-6		0.07-0.2			tr[q]; 0.05[d]
Ethyl benzoate	93-89-0	0.04-0.3	0-tr			0.2[c]; 1.5[b]
Eugenol	97-53-0	0.04-1.0	0.04-1.0		0.3[o]	0.02[c]; 0.1[q]; 0.8[e]
(2*E*,6*Z*)-Farnesal	3790-67-8		tr-0.04			
(*E,E*)-α-Farnesene	502-61-4	5.2-19.7	4.2-10.6	1.7-12.7[m]	15.7[o]	7.4[f]; 8.0[b]; 10.0[q]; 12.6[g]; 34.0[j]; 38.7[d]
(*Z,E*)-α-Farnesene	26560-14-5					12.4[g,h]
β-Farnesene	502-60-3					3.3[s]
(*E*)-β-Farnesene	18794-84-8					0.5[j]
Farnesol	4602-84-0				1.6[p]	2.3[k]; 4.8[s]; 8.4[d]
(*E,E*)-Farnesol	106-28-5	0.7-3.3	3.1-7.7			1.0[g]; 1.3[f]; 1.8[c,e]; 3.4[b]; 3.9[q]
(*Z,E*)-Farnesol	3790-71-4		0-0.2			2.3[e]
Farnesyl acetate	29548-30-9				2.0[p]	0.6[c]; 0.9[s]; 1.1[k]; 1.2[d]
(*E,E*)-Farnesyl acetate	4128-17-0	1.2-4.9	0.9-1.8	0.67-6.2	2.7[o]	1.5[f]; 1.7[g]; 1.9[e]; 4.5[b]; 4.9[j]; 5.5[q]
(*Z,E*)-Farnesyl acetate	40266-29-3		0-0.08			
(*Z,Z*)-α-Farnesyl acetate	24163-97-1				2.0[o]	
Formaldehyde	50-00-00					+[e]
Formic acid	64-18-6					+[e]
Furfural	98-01-1					+[e]
Geranial	141-27-5		0-0.05			0.04[s]
(*E*)-Geranic acid	4698-08-2					+[e]
Geraniol	106-24-1	0.1-2.0	0.1-1.3	0.1-1.2	0.4[o]; 0.9[p]	0.3[f]; 0.7[k]; 0.9[q]; 2.5[s]; 2.9[e]; 4[t]; 5.1[d]
Geranyl acetate	105-87-3	3.0-11.8	3.7-19.0	2.6-7.2	3.0[o]	5.9[f]; 6.0[q]; 6.2[k]; 6.7[e]; 9.9[r]; 13.7[d]
Geranyl benzoate	94-48-4	0.04-0.1	tr-0.2			+[e]; 0.01[q]
1(10),5-Germacradien-4-ol	74841-87-5	0.2-0.3				0.05[q]
Germacrene D	23986-74-5	12.2-26.4	16.6-19.4	1.5-19.3	21.8[o]	19.1[g]; 20.1[q]; 25.7[b]; 27.8[k,l]; 35.3[j]
Guaiol	489-86-1		0.08-0.2			
α-Gurjunene	489-40-7					4.6[j]
Heptanal	111-71-7		0-tr			
Heptanoic acid	111-14-8					+[e]
Hexanoic acid	142-62-1					+[e]
Hexanol	111-27-3					+[e]
3-Hexen-1-ol	544-12-7					0.2[c]
(*Z*)-3-Hexen-1-ol	928-96-1					+[e]
2-Hexenyl acetate	2497-18-9					0.1[b]
3-Hexenyl acetate	1708-82-3					0.08[c]; 0.2[b]
(*Z*)-3-Hexenyl acetate	3681-71-8	0.03-0.09	0-0.04		0.02[o]	0.08[q]; 0.2[f]

Table 5.90.8 Constituents identified in ylang-ylang oils 'second' and in complete oils and oils of unknown quality (*continued*)

Constituent	CAS	YLANG-YLANG OILS 'SECOND'			COMPLETE OILS AND OILS OF UNKNOWN QUALITY	
		Percentage and range				
		A	B (12)	C (14)	D	E
(*Z*)-3-Hexenyl-benzoate	25152-85-6		tr-0.05			
Hexyl acetate	142-92-7	0.02-0.1	0-0.08		0.01[o]	0.08[q]; 0.3[f]
α-Humulene	6753-98-6	1.6-4.8	1.8-3.6	3.4-5.8[a]	2.6[p]; 3.4[o]	2.4[c]; 2.6[b]; 2.8[q]; 3.6[g]; 3.7[i]; 6.5[d]; 7.7[s]
5-Indanol	1470-94-6		tr-0.03			
Isocaryophyllene	118-65-0					3.3[e]
Isodauca-6,9-diene	317819-81-1		0-tr			
Isoeugenol	97-54-1	0.08-0.8			0.3[o]	0.06[q]; 0.2[e]
(*E*)-Isoeugenol	5932-68-3		2.9-8.0			0.5[f]
Isoeugenol acetate	93-29-8		0.1-0.2			
Isogermacrene D	317819-80-0		0.3			
Isoprenyl acetate	17616-47-6	0.01-0.4			0.04[o]	0.2[q]
Isosafrole	120-58-1					0.07[e]
Isovaleraldehyde	590-86-3					+[e]
Junenol	472-07-1		0.08-0.2			
Limonene	138-86-3	0.04-0.7				0.1[q]; 0.2[j]
β-Limonene	5989-54-8					0.07[c]; 0.09[b]
Linalool	78-70-6	0.1-13.2	2.0-4.8	3.9-12.2	2.6[o]; 6.1[p]	14.5[t]; 19.0[e]; 20.8[b,c]; 21.3[k]; 24.5[j]
cis-Linalool oxide, furanoid	5989-33-3		0-tr			
Linalyl acetate	115-95-7		0-0.03			0.2[e]; 8.3[g]
Linalyl benzoate	126-64-7					+[e]
2-Methoxy-4-methyl-phenol	93-51-6		0-tr			0.05[c]
Methyl *p*-anisate	121-98-2		tr-0.07			
Methyl anthranilate	134-20-3				0.9[p]	+[e]; tr[c]
Methyl benzoate	93-58-3	0.8-5.4	0.5-1.7	0.6-2.3	0.7[o]	1.4[k]; 1.5[b,j]; 2.9[q]; 4.1[c]; 5.7[f]; 8.7[e]
2-Methylbutenol	60766-00-9					+[e]
2-Methyl-3-buten-2-ol	115-18-4			0.03-0.2		0.03[b]; 0.2[c]; 0.4[e]
3-Methyl-2-buten-1-ol	556-82-1			0.02-0.1		0.08[b]; 0.2[e]
3-Methyl-3-buten-1-ol	763-32-6					0.1[e]
bis(3-Methyl-3-buten-1-ol)				0-0.01		
2-Methyl-2-butenyl acetate	33425-30-8					+[e]
2-Methylbut-3-en-2-yl acetate	24509-88-4					1.6[e]
3-Methyl-2-butenyl acetate (prenyl acetate)	1191-16-8	0.01-1.8	0.02-0.7	0.01-0.2	0.08[o]	0.8[q]; 3.2[e]
3-Methyl-3-butenyl acetate	5205-07-2		tr-0.3			1.0[f]; 1.6[e]; 2.1[f]
3-Methyl-2-butenyl benzoate	5205-11-8	0.1-0.9	0.3-1.3		0.3[o]	0.9[f]; 1.3[q]
2-Methylbutyric acid	116-53-0					+[e]
3-Methylbutyric acid	503-74-2					+[e]
Methyl chavicol	140-67-0	0.01-0.06	0-0.07			0.04[s]; 0.1[d,q]
Methyl eugenol	93-15-2		0-0.04		0.2[p]	0.1[e]
Methyl salicylate	119-36-8	0.07-0.09	tr-0.05	0.6-5.3		tr[b]; 0.03[c]; 0.2[f]; 1.3[e]; 2.8[j]
Mintsulfide	72445-42-2					0.001[s]
α-Muurolene	10208-80-7		0.1-0.2	0.7-1.9		1.0[e]
γ-Muurolene	30021-74-0	0.8-1.0	0.2	1.5-3.8		0.2[c]; 0.4[f]; 0.9[q]; 1.1[e]; 17.1[d]; 19.8[s]
τ-Muurolene	152287-43-9					0.6[b]
τ-Muurolol	19912-62-0	0.4-2.0	1.0-1.4	0.4-1.5[n]		0.1[q]; 1.2[k,l]; 1.5[e,i]
Myrcene	123-35-3	0.03-0.1	0.02-0.1	0.02-0.09		0.05[q]; 0.1[j]; 0.2[b,c]; 0.6[d]; 1.0[e]

Table 5.90.8 Constituents identified in ylang-ylang oils 'second' and in complete oils and oils of unknown quality (*continued*)

Constituent	CAS	YLANG-YLANG OILS 'SECOND'				COMPLETE OILS AND OILS OF UNKNOWN QUALITY
		Percentage and range				
		A	B (12)	C (14)	D	E
Myristicin	607-91-0					0.2[j]
Neral	106-26-3		0-tr			
Nerol	106-25-2		0-tr		2.9[p]	0.1[e]; 1.5[g]
Nerolidol	7212-44-4				0.07[p]	tr[q]; 0.4[c]
(E)-Nerolidol	40716-66-3					0.3[b]; 0.7[e]
Neryl acetate	141-12-8		tr			
n-Nonanoic acid	12-05-0					+[e]
Nonanyl acetate	143-13-5		0-tr			
allo-Ocimene	673-84-7					0.07
(E)-β-Ocimene	3779-61-1					tr[b]; 0.06[c]; 1.4[d]
(Z)-β-Ocimene	3338-55-4					1.5[d]
Octanoic acid	124-07-2					+[e]
α-Phellandrene	99-83-2					0.01[d]
2-Phenethyl alcohol	60-12-8					+[e]
Phenylacetaldehyde	122-78-1	tr-0.07				+[e]
Phenylacetic acid	103-82-2					+[e]
Phenylacetonitrile	140-29-4		0-tr			+[e]; tr[c]
1-Phenylethyl acetate	93-92-5	0.04-0.1			0.04[o]	0.04[q]
2-Phenylethyl acetate	103-45-7		tr-0.08			0.1[b]; 0.2[f]
2-Phenylnitroethane	6125-24-2		0.09-0.1			+[e]; tr[c]
Phthalic acid	88-99-3					49.7[t];
α-Pinene	80-56-8	0.07-0.3	0.08-0.1	0.1-1.2	0.1[o,p]	0.07[q]; 0.2[d]; 0.4[j]; 1.3[e]; 1.9[c]; 2.0[b]
β-Pinene	127-91-3	0.01-0.06	tr-0.04	0.03-0.2		0.02[s]; 0.04[q]; 0.1[d]; 0.2[e]; 0.5[b,c]
4-(2-Propenyl)phenol	501-92-8		0-tr			
Propionaldehyde	123-38-6					+[e]
Sabinene	3387-41-5					0.3[d]; 0.6[j]
Safrole	94-59-7					0.4[e]
Salicylic acid	69-72-7					+[e]
γ-Terpinene	99-85-4					0.2[j]
Terpinen-4-ol	562-74-3					0.1[d]; 0.2[j]
α-Terpineol	98-55-5	0.04-0.2	0.04-0.08			0.02[c]; 0.07[d]; 0.1[q]; 0.2[b]; 0.8[e]
γ-Terpineol	586-81-2				0.2[p]	0.4[j]
Tetradecane	629-59-4					8.1[c]
α-Thujene	2867-05-2					0.1[d]
Thujopsene	470-40-6					0.9[j]
Valeric acid	109-52-4					+[e]
Vanillin	121-33-5		0.08-0.2			
Veratrole	91-16-7		0-tr			
p-Vinylguaiacol	7786-61-0		tr-0.04			
Viridiflorene (ledene)	21747-46-6					0.1[e]
α-Ylangene	14912-44-8		0-tr			0.1[e]; 0.3[c]; 1.0[b]
β-Ylangene	20479-06-5					0.1[e,f]
Zonarene	41929-05-9		0.03-0.04			

YLANG-YLANG OILS QUALITY 'SECOND'

A forty-two ylang-ylang essential oil samples 'second' from Comoros, Mayotte and Madagascar analyzed between 1998 and 2013; lowest and highest concentrations given (E. Schmidt, unpublished data)

B twenty-eight essential oils obtained by lab-hydrodistillation of mature flowers from Grande Comore, Mayotte, Nossi Bé (Madagascar) and Ambanja (Madagascar); fraction 'second', hydrodistillation time 3 hours; mean percentages for ingredients of the oils from each individual island were provided; lowest and highest mean percentages given (ref. 12)

C thirteen 'second' quality commercial oils from Madagascar; lowest and highest concentrations given (ref. 14)

D data from other studies (indicated with superscript letters) with ylang-ylang essential oils, quality 'second'; highest concentrations found in any study reviewed here given

UNFRACTIONATED YLANG-YLANG ESSENTIAL OILS AND OILS OF UNKNOWN QUALITY

E data from other studies (indicated with superscript letters) with unfractionated ylang-ylang essential oils and oils of unknown quality; highest concentrations found in any study reviewed here given

Table 5.90.8 Constituents identified in ylang-ylang oils 'second' and in complete oils and oils of unknown quality (*continued*)

[a] α-humulene and ε-cadinene combined; [b] two lab-hydrodistilled oils (1 hour and 2 hours distillation time) from Columbia (ref. 16); [c] one laboratory steam-distilled (2 hours) oil from Colombia (ref. 6); [d] three oils from China, Laos and Thailand (ref. 19); [e] various publications from before 1986 (ref. 8, data cited in ref. 11); [f] one commercial oil from unknown origin (ref. 20); [g] one commercial ylang-ylang oil purchased in the USA (ref. 22); [h] either (Z,Z)-α-farnesene or (Z,E)-α-farnesene; [i] one commercial oil from unknown origin purchased in Poland (ref. 23); [j] one commercial oil purchased in Italy (ref. 24); the oil was termed 'cananga oil', but because of its high content of linalool (24.5%), it was much more likely ylang-ylang oil; [k] one oil from *Cananga odorata* flowers collected in Vietnam (ref. 25); [l] δ-cadinene, β-cubebene and germacrene D combined; [m] γ-cadinene and α-farnesene combined; [n] τ-muurolol and γ-cadinol combined; [o] one commercial oil from Comoro Islands (ref. 17); [p] one commercial ylang-ylang oil sample from Comoro Islands (ref. 18); [q] two commercial oils from Madagascar analyzed by one of us (E.S.) in 2001 and 2002; [r] one commercial oil purchased from a Brazilian company (ref. 33); [s] data from various older studies cited in ref. 10; [t] one commercial ylang-ylang oil from India with an extremely atypical composition and dominated by nearly 50% phthalic acid, which should be erroneous, as phthalic acid is not known to occur in nature (ref. 35)

tr: trace (in column B: <0.02); + present in the oil investigated, but quantity not stated

Table 5.90.9 Major constituents identified in ylang-ylang oils 'second'

Constituent	CAS	Percentage and range			
		A	B (12)	C (14)	D
Germacrene D	23986-74-5	12.2-26.4	16.6-19.4	1.5-19.3	21.8[o]
(E,E)-α-Farnesene	502-61-4	5.2-19.7	4.2-10.6	1.7-12.7[m]	15.7[o]
β-Caryophyllene	87-44-5	4.9-18.8	1.5-2.8	1.8-19.6	12.9[o]
Geranyl acetate	105-87-3	3.0-11.8	3.7-19.0	2.6-7.2	3.0[o]
Benzyl benzoate	120-51-4	4.3-11.0	12.2-18.2	5.3-12.3	6.3[o]; 6.5[p]
Linalool	78-70-6	0.1-13.2	2.0-4.8	3.9-12.2	2.6[o]; 6.1[p]
Benzyl acetate	140-11-4	0.3-8.4	0.8-8.2	0.6-3.1	1.7[o]

LEGEND: SEE UNDER TABLE 5.90.8

methyl ether (3.5-11.4%), benzyl benzoate (4.3-11.0%), benzyl acetate (0.3-8.4%), methyl benzoate (0.8-5.4%) and (E,E)-farnesyl acetate (1.2-4.9%) (Erich Schmidt, unpublished analytical data).

YLANG-YLANG OIL 'THIRD'

Ylang-ylang oil 'third' is the fourth fraction of the essential oil obtained with distillation times up to 12 hours. ISO values are shown in Table 5.90.10.

CHEMICAL COMPOSITION

Ylang-ylang oil 'third' is a clear mobile liquid with yellow to yellow brownish color, which has a typical exotic floral, somewhat jasmine-like, and woody odor. The yield of essential oil from the flowers of *Cananga odorata* (Lam.) Hook f. et Thomson, *forma genuina*, fraction 'third', generally varies from 1.9 to 2.3 per cent. The main producing countries of this oil are the Comoro Islands, Madagascar and Vietnam.

Literature data (up to December 8, 2014) on the chemical composition of ylang-ylang oils 'third' and unpublished analytical data from one of us (E.S.) are shown in Table 5.90.11 in alphabetical order. In

Table 5.90.10 ISO values (%) for ylang-ylang oils 'third' [a]

Compound	CAS	Minimum	Maximum
Germacrene D	23986-74-5	15.0	35.0
(E,E)-α-Farnesene	502-61-4	9.0	29.0
β-Caryophyllene	87-44-5	5.0	19.0
Benzyl benzoate	120-51-4	4.0	8.5
Geranyl acetate	105-87-3	0.4	6.6
Benzyl salicylate	118-58-1	2.0	5.0
(E,E)-Farnesyl acetate	29548-30-9	1.5	5.0
(E,E)-Farnesol	106-28-5	0.8	4.0
Linalool	78-70-6	0.1	4.0
Benzyl acetate	140-11-4	0.1	3.0
(E)-Cinnamyl acetate	21040-45-9	0.1	2.5
p-Cresyl methyl ether	104-93-8	0.1	1.4
Methyl benzoate	93-58-3	0.1	0.9
Geraniol	106-24-1	tr	0.8
3-Methyl-2-butenyl acetate (prenyl acetate)	1191-16-8	tr	0.2

[a] ISO 3063 Essential oil of ylang-ylang 'third' ©ISO 2004; Geneva, Switzerland, www.iso.org

Table 5.90.11 Constituents identified in ylang-ylang oils 'third'

Constituent	CAS	Percentage and range					
		A	B (12)	C (1)	D (5)	E (14)	F
τ-Amorphol	53947-91-4						1.9[e]
(E)-Anethole	4180-23-8	0.02-0.1	0-0.04	tr			0.09[e]
p-Anisyl acetate	1200-06-2		0.4-0.6	tr			
Aromadendrene	489-39-4			1.5			
Benzaldehyde	100-52-7			0.05			
Benzyl acetate	140-11-4	0.3-6.3	0.1-2.6	0.07	0.03-4.6	0.4-1.2	0.7[e]
Benzyl alcohol	100-51-6	0.01-0.1	tr-0.04	tr			0.1[f]
Benzyl benzoate	120-51-4	3.5-14.0	7.5-17.8	1.2	6.2-23.4	5.9-12.8	5.4[e]; 5.5[f]
Benzyl butyrate	103-37-7			tr			
Benzyl cinnamate	103-41-3		0-0.06				
Benzyl 4-methyl pentanoate	77509-00-3			tr			
Benzyl salicylate	118-58-1	1.5-3.6	3.2-8.2	4.2	2.8-12.0	0.7-2.7	1.9[f]; 2.8[e]
Bicycloelemene	32531-56-9		0.3	0.3			
Bicyclogermacrene	24703-35-3	0.3-0.4	0.3-0.4				
cis-α-Bisabolene epoxide	121467-35-4			tr			
β-Bourbonene	5208-59-3		0-0.3	tr			
Bulnesol	22451-73-6			0.05			
Butyl benzoate	136-60-7			tr			
trans-Cadina-1,4-diene	20085-13-6		0.1				
trans-Cadina-1(6), 4-diene	931410-54-7		0.3				
Cadina-3,5-diene	267665-20-3		0.06-0.2				
α-Cadinene	24406-05-1		0.1-0.2				
γ-Cadinene	39029-41-9	0.6-1.0	0.5-0.7	2.1			1.3[e]
δ-Cadinene	483-76-1	2.2-5.6	1.7-2.2	0.6		3.1-4.8	3.7[e]
ε-Cadinene	1080-67-7					3.9-5.8[b]	
α-Cadinol	481-34-5	1.0-2.6	2.6-2.9	1.5		0.8-1.5	
γ-Cadinol	50895-55-1					0.4-0.9[d]	
δ-Cadinol	19435-97-3	0.5-1.8				0.2-0.8	
τ-Cadinol (10-epi-α-)	58580-31-7	0.1-0.8					
trans-Calamenene	73209-42-4		0-0.07				
β-Caryophyllene	87-44-5	5.2-16.3	2.0-6.5	0.3	0.8-8.1	14.8-21.5	12.4[e]
Caryophyllene oxide	1139-30-6	0.1-0.5	0.1-0.2	tr			
β-Cedrene	546-28-1						0.5[e]
1,8-Cineole	470-82-6	0.01-0.2	tr-0.08	0.2		0-0.1	
(Z)-Cinnamaldehyde	57194-69-1						+[f]
Cinnamyl acetate	103-54-8			1.6			0.9[e]
(E)-Cinnamyl acetate	21040-45-9	0.6-3.1	0.2-2.4		0.1-4.0	0.01-0.5	
Cinnamyl alcohol	104-54-1			tr			
(E)-Cinnamyl alcohol	4407-36-7		0-tr				
(Z)-Cinnamyl alcohol	4510-34-3						+[f]
Copaborneol	21966-93-8		0.04-0.08	0.03			
α-Copaene	3856-25-5	0.1-2.2	0.3-0.4	0.8		2.1-3.5	1.3[f]; 1.6[e]
β-Copaene	18252-44-3		1.0-1.5	0.1			
m-Cresol	108-39-4		0-tr				+[f]
p-Cresol	106-44-5		tr-0.07	0.05			+[f]
m-Cresyl acetate	122-46-3						+[f]
p-Cresyl acetate	140-39-6		0-tr				+[f]
p-Cresyl methyl ether	104-93-8	0.2-3.9	0.2-1.7	1.6	0.02-2.6	0.4-1.9	0.4[e]; 0.7[f]
α-Cubebene	17699-14-8	0.1-2.2		0.09			
β-Cubebene	13744-15-5	0.2-0.8	0.6-0.7	0.6			
1-epi-Cubenol	81939-29-9		0.4-0.5				
1,10-di-epi-Cubenol	73365-77-2		0.1-0.2				
β-Curcumene	28976-67-2		0.1-1.9	2.7			
Cyperene	2387-78-2		0-tr				
Decane	124-18-5			0.03			
Diethyl (2E)-3-methyl- 2-pentanedioate				0.02			

Table 5.90.11 Constituents identified in ylang-ylang oils 'third' (*continued*)

Constituent	CAS	Percentage and range					
		A	B (12)	C (1)	D (5)	E (14)	F
Diethyl 1,5-pentane dioate	818-38-2			tr			
2,5-Dimethyl-3-methylene-1,5-heptadiene	74663-83-5			0.04			
Docosane	629-97-0			tr			
Dodecane	112-40-3			tr			
Eicosane	112-95-8			tr			
β-Elemene	33880-83-0	0.4-1.2					
δ-Elemene	20307-84-0	0.02-0.04					
Elemol (α-)	639-99-6		0.05-0.07	0.02			
Ethyl benzoate	93-89-0	0.01-0.1					
Eugenol	97-53-0	0.03-0.4	0.04-0.5				0.06[e]
(*E,E*)-Farnesal	502-67-0			tr			
(2*E*,6*Z*)-Farnesal	3790-67-8		tr-0.02				
(*E,E*)-α-Farnesene	502-61-4	6.3-28.9	7.8-21.8	10.1	6.4-29.6	6.5-17.4[c]	23.8[e]
(*Z,E*)-α-Farnesene	26560-14-5			0.2			
Farnesol	4602-84-0						1.5[f]
(*E,E*)-Farnesol	106-28-5	0.9-2.1	1.9-4.8	0.03	0.9-6.0		
(*Z,E*)-Farnesol	3790-71-4		0-0.06	0.02			
(*Z,Z*)-Farnesol	16106-95-9			1.4			
Farnesyl acetate	29548-30-9						2.9[f]
(*E,E*)-Farnesyl acetate	4128-17-0	1.3-5.5	1.7-3.3	2.0	1.2-6.3	1.6-3.1	3.5[e]
(*Z,E*)-Farnesyl acetate	40266-29-3		0-tr				
(*Z,Z*)-α-Farnesyl acetate	24163-97-1						1.6[e]
Geranial	141-27-5			tr			
Geraniol	106-24-1	0.05-1.9	0.03-0.2	tr	0-0.3	0.1-1.7	0.2[e]; 0.8[f]
Geranyl acetate	105-87-3	0.8-6.2	1.2-4.1		0.08-4.7	2.0-4.4	1.2[e]
Geranyl benzoate	94-48-4	0.03-0.05	0.02-0.4	tr			
1(10),5-Germacradien-4-ol	74841-87-5	0.2-0.3					
Germacrene D	23986-74-5	11.7-25.8	21.0-25.9	2.8	12.5-27.7	15.1-25.1	21.7[e]
Germacrene D-4-ol	198991-79-6			tr			
Globulol	489-41-8			tr			
Guaiacylacetone	2503-46-0			tr			
Guaiol	489-86-1		0.1-0.3	0.5			
Heneicosane	629-94-7			tr			
Heptacosane	593-49-7			tr			
Heptanal	111-71-7		0-tr				
Hexacosane	630-01-3			tr			
n-Hexadecane	544-76-3			tr			
Hexadecanoic acid	57-10-3			0.3			
1-Hexadecene	629-73-2			0.04			
(*Z*)-3-Hexenyl acetate	3681-71-8	0.01-0.03	0-tr	0.03			
(*Z*)-3-Hexenyl benzoate	25152-85-6		0.02-0.06	tr			
Hexyl acetate	142-92-7	0.01-0.08	0-0.04	0.07			
α-Humulene	6753-98-6	1.6-4.6	2.4-5.4	6.2		3.9-5.8[b]	1.8[f]; 3.6[e]
Hydrocinnamyl acetate	122-72-5			tr			
5-Indanol	1470-94-6		0-tr	tr			
1*H*-Indole	120-72-9			tr			
α-Ionene	475-03-6			tr			
Isodauca-6,9-diene	317819-81-1		0-tr				
Isoeugenol	97-54-1	0.2-0.8		tr			0.4[e]
(*E*)-Isoeugenol	5932-68-3		5.2-12.0	0.4			
Isoeugenol acetate	93-29-8		0.1-0.2	tr			
Isogermacrene D	317819-80-0		0.3-0.5	1.8			
Isoprenyl acetate	17616-47-6	0.01-0.3					0.01[e]
Junenol	472-07-1		0.4				
Ledane	28580-43-0			0.2			
Levoglucosenose	37112-31-5			0.07			
Limonene	138-86-3	0.01-0.2					
β-Limonene	5989-54-8			0.2			

Table 5.90.11 Constituents identified in ylang-ylang oils 'third' (*continued*)

Constituent	CAS	Percentage and range					
		A	B (12)	C (1)	D (5)	E (14)	F
Linalool	78-70-6	0.5-6.7	0.4-1.1	0.3	0.04-1.8	1.3-4.8	0.8[e]; 2.1[f]
cis-Linalool oxide, furanoid	5989-33-3		0-tr				
2-Methoxy-4-methyl-phenol	93-51-6			tr			
2-Methoxyphenol	90-05-1			tr			
1-Methoxy-4-propyl-benzene	104-45-0			tr			
Methyl *p*-anisate	121-98-2			tr			
Methyl anthranilate	134-20-3			tr			1.0[f]
Methyl benzoate	93-58-3	0.1-2.1	0.08-0.7	0.7	0-0.1	0.2-0.6	0.2[e]
2-Methyl-3-buten-2-ol	115-18-4					0-0.07	
3-Methyl-2-buten-1-ol	556-82-1					0-0.02	
bis(3-Methyl-3-buten-1-ol)						0-0.01	
3-Methyl-2-butenyl acetate (prenyl acetate)	1191-16-8	0.01-0.7	0-0.2	0.3	0-0.3	0-0.02	0.02[e]
3-Methyl-3-butenyl acetate	5205-07-2		0-0.1	0.2			
3-Methyl-2-butenyl benzoate	5205-11-8	0.03-0.05	0.3-0.5	0.2			0.3[e]
3-Methyl-3-butenyl benzoate	5205-12-9			tr			
Methyl chavicol	140-67-0	0.03-0.04					
Methyl eugenol	93-15-2		0-tr				0.09[f]
6-Methyl-5-hepten-2-one	110-93-0		0-tr	0.03			
Methyl 2-methoxy-benzoate	606-45-1			tr			
cis-2-Methyl-7-octa-decene	35354-39-3			0.05			
Methyl octanoate	111-11-5			7.2			
Methyl salicylate	119-36-8	0.05-0.07	0-tr	0.2		0.4-1.9	
α-Muurolene	10208-80-7		0.2-0.3	0.3		0.8-1.6	
γ-Muurolene	30021-74-0	0.9-3.3	0.4-0.5			2.0-3.2	
τ-Muurolol	19912-62-0	0.3-0.5		2.3-2.6	4.4	0.4-0.9[d]	
Myrcene	123-35-3	0.01-0.1	tr-0.1	tr		0.01-0.1	
Nerol	106-25-2						3.0[f]
Nerolidol	7212-44-4						0.06[f]
Neryl acetate	141-12-8		0-tr	0.2			
Nonadecane	629-92-5			tr			
β-Ocimene	13877-91-3			tr			
Octadecanal	638-66-4			0.02			
Octadecane	593-45-3			tr			
Pentacosane	629-99-2			tr			
Pentadecane	629-62-9			tr			
α-Phellandrene	99-83-2			tr			
Phenylacetaldehyde	122-78-1		0-tr	tr			
Phenylacetonitrile	140-29-4			tr			
1-Phenylallyl acetate	7217-71-2			tr			
1-Phenylethyl acetate	93-92-5	0.01-0.08					
2-Phenylethyl acetate	103-45-7		0-tr	tr			0.02[e]
2-Phenylnitroethane	6125-24-2		0.04-0.05	0.03			
1-Phenyl-2-propen-1-ol	4393-06-0			tr			
α-Pinene	80-56-8	0.05-0.2	0.07-0.1	0.04		0.03-0.2	0.08[e]; 0.1[f]
β-Pinene	127-91-3	0.01-0.1	tr-0.06			0.01-0.06	
4-(2-Propenyl)phenol	501-92-8		0-tr				
Selina-4(15),5-diene	1107026-89-0		0-tr				
α-Terpineol	98-55-5	0.02-0.3	tr-0.05	tr			

Table 5.90.11 Constituents identified in ylang-ylang oils 'third' (*continued*)

Constituent	CAS	A	B (12)	C (1)	D (5)	E (14)	F
Tetracosane	646-31-1		0-tr	tr			
Tetradecane	629-59-4		0-tr	tr			
Tricosane	638-67-5		0-tr	tr			
Undecane	1120-21-4			tr			
Vanillin	121-33-5		0.1-0.2	0.05			
Veratrole	91-16-7			tr			
Vinyl butyrate	123-20-6			tr			
p-Vinylguaiacol	7786-61-0		0.03-0.07	tr			
α-Ylangene	14912-44-8		0.03-0.06	0.06			
β-Ylangene	20479-06-5			0.7			
Zonarene	41929-05-9		0.05-0.09	tr			

A twenty-two essential oil samples 'third' from Comoros and Madagascar analyzed between 1998 and 2013; lowest and highest concentrations given (E. Schmidt, unpublished data)

B 28 essential oils obtained by lab-hydrodistillation of mature flowers from Grande Comore, Mayotte, Nossi Bé (Madagascar) and Ambanja (Madagascar); quality 'third', hydrodistillation time 8 hours; mean percentages for ingredients of the oils from each individual island were provided; lowest and highest mean percentages given (ref. 12)

C one lab-hydrodistilled ylang-ylang essential oil (quality 'third', distillation time 8 hours) from mature flowers collected on the isle of Mayotte (ref. 1)

D twenty-four lab-hydrodistilled essential oil samples (quality 'third', distillation time 8 hours) from mature flowers collected in 11 plantations on three islands: Grande Comore, Madagascar, and Mayotte; only the data of the 15 compounds mentioned in ISO 3063:2004 were presented; lowest and highest concentrations given (ref. 5)

E eleven commercial oils of quality 'third' from Madagascar; lowest and highest concentrations given (ref. 14)

F data from other studies (indicated with superscript letters) with ylang-ylang essential oils, quality 'third'; highest concentrations found in any study reviewed here given

[b] α-humulene and ε-cadinene combined; [c] γ-cadinene and α-farnesene combined; [d] τ-muurolol and γ-cadinol combined; [e] one commercial oil from Comoro Islands (ref. 17); [f] one commercial ylang-ylang oil sample form Comoro Islands (ref. 18); [g] one commercial third grade oil from Madagascar (ref. 13)

tr: trace (in column B: <0.02); + present in the oil investigated, but quantity not stated

ylang-ylang oils 'third' from various origins, over 170 chemicals have been identified. About 58 per cent of these were found in a single reviewed publication only. The major compounds found in ylang-ylang oils 'third' from different sources are shown in Table 5.90.12. They include (highest concentrations in any study given) (*E,E*)-α-farnesene (29.6%), germacrene D (27.7%), benzyl benzoate (23.4%), β-caryophyllene (21.5%), geranyl acetate (6.2%) and α-humulene (6.2%). A well-known ingredient of ylang-ylang oils 'third' that was present in a high concentration (>7%) in two studies is benzyl salicylate (8.2% and 12.0%). Uncommon or rare constituents of ylang-ylang oils 'third' found in high concentrations (>7%) in single studies include (*E*)-isoeugenol (12.0%) and methyl octanoate (7.2%).

Commercial oils

The ten chemicals that had the highest maximum concentrations in 22 commercial ylang-ylang 'third' essential oil samples (concentration ranges provided) are the following: (*E,E*)-α-farnesene (6.3-28.9%), germacrene D (11.7-25.8%), β-caryophyllene (5.2-16.3%), benzyl benzoate (3.5-14.0%), linalool (0.5-6.7%), benzyl acetate (0.3-6.3%), geranyl acetate (0.8-6.2%), δ-cadinene (2.2-5.6%), (*E,E*)-farnesyl acetate (1.3-5.5%) and α-humulene (1.6-4.6%) (Erich Schmidt, unpublished analytical data).

Table 5.90.12 Major constituents identified in ylang-ylang oils 'third'

Constituent	CAS	A	B (12)	C (1)	D (5)	E (14)	F
(*E,E*)-α-Farnesene	502-61-4	6.3-28.9	7.8-21.8	10.1	6.4-29.6	6.5-17.4[c]	23.8[e]
Germacrene D	23986-74-5	11.7-25.8	21.0-25.9	2.8	12.5-27.7	15.1-25.1	21.7[e]
Benzyl benzoate	120-51-4	3.5-14.0	7.5-17.8	1.2	6.2-23.4	5.9-12.8	5.4[e]; 5.5[f]
β-Caryophyllene	87-44-5	5.2-16.3	2.0-6.5	0.3	0.8-8.1	14.8-21.5	12.4[e]
Geranyl acetate	105-87-3	0.8-6.2	1.2-4.1		0.08-4.7	2.0-4.4	1.2[e]
α-Humulene	6753-98-6	1.6-4.6	2.4-5.4	6.2		3.9-5.8[b]	1.8[f]; 3.6[e]

LEGEND: SEE UNDER TABLE 5.90.11

YLANG-YLANG OIL, UNFRACTIONATED (COMPLETE) AND OILS OF UNKNOWN QUALITY

Unfractionated (complete) ylang-ylang oils are obtained by distillation up to 24 hours without removal of earlier fractions. There are currently no ISO values for this ylang-ylang essential oil quality. Ylang-ylang oil complete is a clear mobile liquid with pale yellowish to darker yellow color, which has an intense exotic floral, somewhat jasmine-like with aromatic soft woody odor. The main producing countries of this oil are the Comoro islands, Mayotte, Madagascar and Vietnam. This oil is mainly used in aromatherapy; in perfumery usually the single fractions from extra to three are employed.

Literature data (up to December 8, 2014) on the chemical composition of unfractionated ylang-ylang oils or oils of unknown qualities are shown in Table 5.90.8 in alphabetical order. Due to the heterogeneity of this group and the often unknown quality, no summary of the major compounds is provided.

DIFFERENCES IN COMPOSITION OF THE VARIOUS QUALITIES (FRACTIONS) OF YLANG-YLANG OILS

The four fractions (qualities) of ylang-ylang essential oils differ mainly in the quantities of their main ingredients.

This is readily shown in the ISO values for the four qualities: 'extra', 'first', 'second' and 'third' (Table 5.90.13). The main constituents of the 'extra' quality ylang-ylang oils (upper ISO limits given) are linalool (24%), germacrene D (20%), benzyl acetate (17.5%), p-cresyl methyl ether (16.0%), (E,E)-α-farnesene (15.0%), and geranyl acetate (14.0%). For some of these chemicals, there is a steady decrease of their concentrations towards the quality 'third': linalool (from 24.0 to 4.0%), benzyl acetate (from 17.5% to 3.0%),

p-cresyl methyl ether (from 16.0% to 1.4%), geranyl acetate (from 14.0 to 6.6%) and methyl benzoate (from 9.0% to 0.9%). Conversely, other chemicals increase in concentration from the 'extra' to the 'third' quality: germacrene D (from 20.0 to 35.0%), (E,E)-α-farnesene (from 15.0 to 29.0%), and β-caryophyllene (from 8.5% to 19.0%). Thus, the third quality is dominated by these three compounds: germacrene D, (E,E)-α-farnesene and β-caryophyllene. The other ingredients mentioned in ISO remain more or less stable in all fractions and/or are quantitatively less important.

CONTACT ALLERGY/ALLERGIC CONTACT DERMATITIS

General

Contact allergy to/allergic contact dermatitis from ylang-ylang oil has been reported in over 50 publications. Ylang-ylang oil used to be a frequent sensitizer in Japan, causing many cases of pigmented cosmetic dermatitis. In the 1990s, the frequency of allergy to this oil decreased, presumably from the elimination of its main sensitizer, dihydro-isoeugenol (50,51). The oil (2% in petrolatum) has been included in the screening series of the North American Contact Dermatitis Group (NACDG) since 2001. In groups of consecutive patients suspected of contact dermatitis, prevalence rates of up to 2.5% positive patch test reactions have been observed, but percentages of 'definite' relevance and even 'probable' relevance are low. There have been 13 descriptions in case reports of allergic contact dermatitis from ylang-ylang oil; in 11, the cause was occupational exposure to the oil (8 in aromatherapists/massagists, 2 in workers in the cosmetic industry and one in a beautician). In one patient each, linalool and caryophyllene may have been an allergen in ylang-ylang oil. There are also many reports of positive patch test reactions to ylang-ylang oil with no or unknown relevance, usually in patients reacting to other essential oils and fragrances.

Testing in groups of patients

The results of patch tests with ylang-ylang oil in routine testing (consecutive patients suspected of contact dermatitis) and in groups of selected patients are shown in Table 5.90.14. In routine testing, rates of positive reactions ranged from 0.7 to 2.6%, whereas between 0.8% and 38% of patients in selected groups had positive patch tests. The very high positivity rate of 38% was seen in a group of 21 patients with dermatitis caused by fragrances (84).

Case reports

Occupational allergic contact dermatitis

Three massage therapists/aromatherapists with occupational contact dermatitis from (multiple) essential oils had positive patch tests to ylang-ylang oil; one patient had used the oil, in the other two it was uncertain (46).

Table 5.90.13 ISO values for the four qualities (fractions) of ylang-ylang essential oils

Compound	Extra	First	Second	Third
Benzyl acetate	5.5-17.5	4.0-12.0	0.5-8.8	0.1-3.0
Benzyl benzoate	3.0-6.0	4.2-8.0	4.5-10.0	4.0-8.5
Benzyl salicylate	1.2-4.0	1.6-4.0	1.8-4.0	2.0-5.0
β-Caryophyllene	2.5-8.5	4.0-12.0	4.8-17.0	5.0-19.0
(E)-Cinnamyl acetate	0.5-6.5	0.5-5.0	0.4-0.8	0.1-2.5
p-Cresyl methyl ether	5.0-16.0	3.0-10.0	1.0-5.0	0.1-1.4
(E,E)-α-Farnesene	1.0-15.0	3.0-18.0	5.0-21.0	9.0-29.0
(E,E)-Farnesol	0.8-3.0	0.1-2.5	0.8-3.5	0.8-4.0
(E,E)-Farnesyl acetate	0.5-3.0	1.0-4.0	1.0-3.5	1.5-5.0
Geraniol	0.1-3.0	0.1-2.6	0.1-2.4	tr-0.8
Geranyl acetate	2.5-14.0	2.0-15.0	1.7-12.0	0.4-6.6
Germacrene D	5.0-20.0	9.5-24.0	13.0-28.0	15.0-35.0
Linalool	7.0-24.0	3.0-19.0	2.0-9.5	0.1-4.0
Methyl benzoate	4.0-9.0	1.5-5.5	1.0-3.5	0.1-0.9
3-Methyl-2-butenyl acetate	0.6-2.3	0.2-1.8	0.1-0.9	tr-0.2

Table 5.90.14 Results of testing groups of patients with ylang-ylang oil

Years and Country	Test conc. & vehicle	Number of patients		Selection of patients (S); Relevance (R); Comments (C)	Ref.
		tested	positive (%)		
Routine testing					
2011-12 USA, Canada	2% pet.	4,230	30 (0.7%)	R: definite + probable relevance: 27%	72
2009-10 USA, Canada	2% pet.	4,303	56 (1.3%)	R: definite + probable relevance: 27%	49
2000-2008 IVDK	10% pet.	3,175	80 (2.5%)	R: not stated	71
2007-8 USA, Canada	2% pet.	5,080	71 (1.4%)	R: definite + probable relevance: 17%	42
2000-2007 USA	2% pet.	870	8 (0.9%)	R: 100%; C: weak study: a. high rate of macular erythema and weak reactions, b. relevance figures include 'questionable' and 'past' relevance	36
2005-6 USA, Canada	2% pet.	4,434	67 (1.5%)	R: definite + probable relevance: 15%	41
<2006 USA, Canada	2% pet.	1,603	11 (0.7%)	R: definite + probable relevance: 0%	66
2003-4 USA, Canada	2% pet.	5,137	64 (1.2%)	R: not stated	43
2001-2 USA, Canada	2% pet.	4,893	49 (1.0%)	R: definite + probable relevance: 22%	67
1999-2000 Denmark	10% pet. (I)	318	5 (1.6%)	R: not specified; C: this study was part of the international study mentioned below (ref. 74)	85
	10% pet. (II)	318	6 (1.9%)		
1998-2000 six European countries	10% pet. (I)	1,606	42 (2.6%)	R: not specified for individual oils/chemicals	74
	10% pet. (II)	1,606	41 (2.6%)		
1998-9 Netherlands	4% pet.	1,825	18 (1.0%)	R: not stated; C: 15/18 also reacted to the fragrance mix	76
<1976 Poland	2% pet.	200	4 (2.0%)	R: not stated	57
Testing in groups of selected patients					
2001-2010 Australia	2% pet.	1,020	64 (6.3%)	S: not specified; R: 30%	82
2006-2010 USA	2% pet.	100	6 (6.0%)	S: patients with eyelid dermatitis; R: not stated	44
2000-2008 IVDK	10% pet.	2,155	85 (3.9%)	S: patients with dermatitis suspected of causal exposure to fragrances; R: not stated	71
2004-2008 Spain	2% pet.	86	12 (13.9%)	S: patients previously reacting to the fragrance mix I or *Myroxylon pereirae* (n=54) or suspected of fragrance contact allergy (n=32); R: not stated	70
<2004 Israel	2% pet.	91	2 (2.2%)	S: patients who had shown a doubtful or positive reaction to the fragrance mix I and/or *Myroxylon pereirae* resin and/or one or two commercial fine fragrances; R: not stated	73
1989-1999 Portugal	2% pet.	67	9 (13.4%)	S: patients who had a positive patch test to the fragrance mix; R: not stated	75
1990-1998 Japan	5% pet.	1,483	33 (2.2%)	S: patients suspected of cosmetic contact dermatitis, virtually all were women; range of annual frequency of sensitization: 0-4.3%; R: not stated	45
1989-1998 India	2% pet.	10	2 (20%)	S: patients previously reacting to the fragrance mix and/or *Myroxylon pereirae* resin; R: not stated	77
1996-1997 UK	2% pet.	10	3 (30%)	S: patients strongly suspected of fragrance allergy; all also reacted to the fragrance mix; R: not stated	39
<1996 Japan, Ireland, USA, UK, Switzerland, Sweden	10% pet.	167	29 (17.4%)	S: patients known or suspected to be allergic to fragrances; R: not stated	40
<1986 Poland	2% pet.	86	8 (9.3%)	S: patients previously reacting to the fragrance mix; R: not stated	62
<1986 France	2.5% pet.	21	8 (38%)	S: patients with dermatitis caused by fragrances; R: not stated	84
1971-1980 Japan	5% pet.	477	4 (0.8%)	S: patients with dermatoses other than pigmented cosmetic dermatitis and volunteers; R: not stated	60
<1976 Poland	2% pet.	51	3 (5.9%)	S: patients allergic to *Myroxylon pereirae* resin (balsam of Peru) and/or turpentine and/or wood tar and/or colophony	57
<1974 Japan	?	183	25 (13.7%)	S: patients suspected of cosmetic dermatitis; R: unknown; in many, there was co-reactivity with benzyl salicylate, which may be present in commercial ylang-ylang oils in concentrations of up to 4.7% (Table 5.90.2) and to geraniol, which may reach concentrations of up to 2.0% in such oils (Table 5.90.8)	78

(I) fraction 1 of ylang-ylang oil
(II) fraction 2 of ylang-ylang oil
IVDK Information Network of Departments of Dermatology, Germany, Switzerland, Austria (www.ivdk.org); pet.: petrolatum

An aromatherapist developed occupational contact dermatitis from contact allergy to multiple essential oils; she reacted to both ylang-ylang and cananga oil in the fragrance series, which reactions were considered to be relevant (47). Another aromatherapist developed occupational contact dermatitis from ylang-ylang oil; she also reacted to angelica oil and geraniol (54), which is not an important component of ylang-ylang oil. Yet another two aromatherapists had contact dermatitis (one occupational) with allergies to multiple essential oils used at work, including ylang-ylang oil; both patients also reacted to geraniol, α-pinene, the fragrance mix and various other fragrance materials. In addition, one proved to be allergic to linalool and linalyl acetate, the other to caryophyllene; α-pinene, linalool, and caryophyllene were demonstrated by GC-MS in many essential oils (55).

Caryophyllene is an important component of ylang-ylang oil, which has been present in certain grades of commercial ylang-ylang oils in a concentration of 18.8% (Table 5.90.8, column A); linalool may reach concentrations >13% (Table 5.90.8, column A). A massage therapist had occupational contact dermatitis from allergies to many essential oils including ylang-ylang oil, while cananga oil reacted in the fragrance series (58).

A patient working in a cosmetic factory had occupational dermatitis from contact allergy to ylang-ylang oil in a fragrance mixture he was handling daily; he also reacted to aniseed oil, with which he had no contact (63). A woman packing cosmetics developed occupational allergic contact dermatitis oil from ylang-ylang oil; she also reacted to other fragrance materials (52). A beautician developed occupational contact dermatitis from contact allergy to ylang-ylang oil in a massage lotion (59).

Non-occupational allergic contact dermatitis

A patient developed allergic contact dermatitis from the perfume in an eye cream; she was patch tested with all 94 components of the perfume and reacted to ylang-ylang oil (test concentration unknown) and eleven of the other chemicals in the perfume (80). Two positive patch tests to ylang-ylang oil, possibly from its presence in topical pharmaceutical preparations, have been observed (86).

Positive patch tests (relevance unknown, uncertain or not stated)

In a group of 611 men with (presumed) cosmetic allergy, 17 (2.8%) had positive patch test reactions to ylang-ylang oil in a study of the NACDG, 2001-2004; specific relevance data were not provided. The figure in women is unknown, as only the top 20 allergens were shown and ylang-ylang oil was not in that list (37). In a group of 819 patients suspected of contact dermatitis, four had positive patch test reactions to ylang-ylang oil (38). A naturopathic therapist and a masseuse with occupational contact dermatitis from various (other) essential oils reacted to ylang-ylang oil in the fragrance series (56). Positive patch test reactions to ylang-ylang oil were seen in three patients with allergic contact dermatitis of the lips (cheilitis) and the perioral skin from

peppermint oil in lip balm (61). A patient had hand dermatitis from contact allergy to geraniol and rose oil in a 'fragrance-free' hand soap; she also reacted to ylang-ylang oil and several other fragrances and essential oils (64). Four positive patch test reactions to ylang-ylang oil were observed in 7 patients with allergic contact dermatitis from compound tincture of benzoin (65). A beautician with occupational allergic hand dermatitis from products containing citral and certain essential oils co-reacted to ylang-ylang oil (68); citral (neral + geranial) is not an important component of ylang-ylang oil. Three positive reactions were seen to ylang-ylang oil, which was tested in seven patients out of a group of 63 who were patch test positive to their own shaving product/eau de toilette/perfume (69). Two positive reactions were observed to ylang-ylang oil, which was tested in 43 patients out of a group of 819 who were patch test *negative* to their own shaving product/eau de toilette/perfume (69). Five female patients from India, two with facial hyperpigmentation, three with erythema of the face, reacted to ylang-ylang oil and other fragrances; two used aromatic oils on the face for headaches, in one a Repeated Open Application Test (ROAT) was positive. It was uncertain whether these products contained ylang-ylang oil (81). Of seven patients allergic to the fragrance farnesol, 3 (43%) co-reacted to ylang-ylang oil (and various other fragrances) (83).

Co-reactivity to cananga oil

Co-reactivity between ylang-ylang oil and cananga oil and vice versa is discussed in Chapter 5.11 Cananga oil.

Pigmented cosmetic dermatitis

In Japan, in the 1960s and 1970s, many female patients developed facial pigmentation following dermatitis of the face (48). This so-called pigmented cosmetic dermatitis was shown to be caused by contact allergy to components of cosmetic products, notably essential oils, other fragrance materials, antimicrobials, preservatives and coloring materials (48,60). In a group of 620 Japanese patients with this condition investigated between 1970 and 1980, 6-14% had positive patch test reactions to ylang-ylang oil 5% in petrolatum (60). In a group of 222 patients with pigmented cosmetic dermatitis investigated in the period 1975-1977, the rate of positive reactions was even 20% (50). The number of patients decreased strongly after 1978, when major cosmetic companies began to eliminate strong contact sensitizers from their products, including dihydroisoeugenol from ylang-ylang oil (50,51,60). Yet, even today cases of (presumed) pigmented cosmetic dermatitis from ylang-ylang oil are being reported (81).

LITERATURE

1　Benini C, Danflous J-P, Wathelet J-P, du Jardin P, Fauconnier M-L. Le point sur: L'ylang-ylang [*Cananga odorata* (Lam.) Hook.f. & Thomson]: une plante à huile essentielle méconnue dans une filière en danger. Biotechnol Agron Soc Environ 2010;4:693-705

2 Brokl M, Fauconnier M-L, Benini C, Lognay G, du Jardin P, Focant J-F. Improvement of ylang-ylang essential oil characterization by GC×GC-TOFMS. Molecules 2013;18:1783-1797

3 Burdock GA, Carabin IG. Safety assessment of Ylang–Ylang (*Cananga* spp.) as a food ingredient. Available at: http://www.aseanfood.info/Articles/11022798.pdf

4 Hongratanaworakit T, Buchbauer G. Relaxing effect of ylang ylang oil on humans after transdermal absorption. Phytother Res 2006;20:758-763

5 Benini C, Mahy G, Bizoux J-P, Wathelet J-P, du Jardin P, Brostaux Y, et al. Comparative chemical and molecular variability of *Cananga odorata* (Lam.) Hook. f. & Thomson *forma genuina* (Ylang-Ylang) in the western Indian Ocean Islands: Implication for valorization. Chem Biodivers 2012;9:1389-1402

6 Stashenko EE, Quiroz Prada N, Martinez JR. RGC/FID/NPD and HRGCMSD study of Colombian ylang-ylang (*Cananga odorata*) oils obtained by different extraction techniques. J High Resol Chromatogr 1996;19:353-358

7 Fournier G, Leboeuf M, Cavé A. Annonaceae essential oils: A review. J Essent Oil Res 1999;11:131-142

8 Lawrence BM. Progress in essential oils. Perfum Flavor 1986;11:111-125

9 Lawrence BM. Progress in essential oils. Perfum Flavor 1989;14:71-80

10 Lawrence BM. Progress in essential oils. Perfum Flavor 1995;20:49-58

11 Ekundayo O. A review of the volatiles of the Annonaceae. J Essent Oil Res 1989;1:223-245

12 Benini C, Ringue M, Wathelet JP, Lognay G, du Jardin P, Fauconnier ML. Variations in the essential oils from ylang-ylang (*Cananga odorata* [Lam.] Hook f. & Thomson *forma genuina*) in the Western Indian Ocean islands. Flavour Fragr J 2012;27:356-366

13 Gaydou EM, Randriamiharisoa R, Bianchini JP. Composition of the essential oil of ylang-ylang (*Cananga odorata* Hook Fil. & Thomson *forma genuina*) from Madagascar. J Agric Food Chem 1986;34:481-487

14 Gaydou EM, Randriamiharisoa RP, Bianchini J-P, Llinas J-R. Multidimensional data analysis of essential oils. Application to ylang-ylang *(Cananga odorata* Hook Fil. et Thomson, *forma genuina)* grades classification. J Agric Food Chem 1986;36:574-579

15 Burdock GA, Carabin LG. Safety assessment of ylang-ylang oil as a food ingredient. Food Chem Toxicol 2008;46:433-445

16 Stashenko E, Torres W, Morales JRM. A study of the compositional variation of the essential oil of ylang-ylang (*Cananga odorata* Hook Fil. & Thomson, *forma genuina*) during flower development. J High Resolut Chromatogr 1995;18:101-104

17 Benveniste B, Azzo N. Ylang Ylang. Kato Technical Bulletin and Newsletter 1992, Vol. VI, June 30, Kato Worldwide Ltd., Mount Vernon, NY, USA. Data cited in ref. 10

18 Srinivas SR. Atlas of essential oils. New York, USA, published by the author, 1986. Data cited in ref. 9

19 Ding J-K, Yi Y-F, Ding Z-H, Sun H-D, Liu Z-G, Dao S-H. Studies on the constituents of the essential oils from *Cananga odorata* in the different varieties and the flowered periods. Acta Bot Yunnanica 1988;10:331-334. Data cited in ref. 10

20 Kubeczka K-H, Formacek V. Essential oils analysis by capillary gas chromatography and carbon-13 NMR spectroscopy, 2nd edition. New York, USA: John Wiley and Sons, 2002:361-368. Data cited in ref. 21

21 Lawrence BM. Progress in essential oils. Perfum Flavor 2006;31(6):40-? (last page unknown)

22 Wei A, Shibamoto T. Antioxidant activities and volatile constituents of various essential oils. J Agric Food Chem 2007;55:1737-1742

23 Golebiowski M, Paszkiewicz M, Halinski L, Malinski E, Stepnowski P. Chemical composition of commercially available essential oils from eucalyptus, pine, ylang, and juniper. Chem Nat Comp 2009;45:278-279

24 Sacchetti G, Maietti S, Muzzoli M, Scaglianti M, Manfredini S, Radice M, Bruni R. Comparative evaluation of 11 essential oils of different origin as functional antioxidants, antiradicals and antimicrobials in foods. Food Chem 2005;91:621-632

25 Phan TS, Phan MG, Nguyen DH. Study of the chemical components of the essential oil from the flowers of *Cananga odorata* (Lamb.) Hook. f. et Thomas (Annonaceae) in Vietnam. Tap Chi Duoc Hoc 2001;7:9-11. Data cited in ref. 26

26 Lawrence BM. Progress in essential oils. Perfum Flavor 2004;29(6):80-90

27 Kristiawan M, Sobolik V, Allaf K. Isolation of Indonesian cananga oil by instantaneous controlled pressure drop. J Essent Oil Res 2008;20:135-146

28 Kristiawan M, Sobolik V, Al-Haddad M, Allaf K. Effect of pressure-drop rate on the isolation of cananga oil using instantaneous controlled pressure-drop process. Chemical Engineering and Processing 2008;47:66-75

29 Kristiawan M, Sobolik V, Allaf K. Isolation of Indonesian cananga oil using multi-cycle pressure drop process. J Chromatogr A 2008;1192:306-318

30 Kristiawan M, Sobolik V, Allaf K. Yield and composition of Indonesian cananga oil obtained by steam distillation and organic solvent extraction. Int J Food Engin 2012;8(3):article 28 (19 pages). DOI: 10.1515/1556-3758.1412

31 Saputra MD. A combination of water-steam distillation and solvent extraction of *Cananga odorata* essential oil. IOSR J Engin 2012;2:5-12

32 Lawrence BM. A preliminary report on the world production of some selected essential oils and countries. Perfum Flavor 2009;34(Jan):38-44

33 Murbach Teles Andrade BF, Nunes Barbosa L, da Silva Probst I, Fernandes Júnior A. Antimicrobial activity of essential oils. J Essent Oil Res 2014;26:34-40

34 Stashenko E, Martinez JR, Macku C, Shibamoto T. HRGC and GC-MSs analysis of essential oil from Colombian ylang-ylang (*Cananga odorata* Hook Fil. et Thomson, forma *genuina*). J High Resol Chromatogr 1993;16:441-444

35 Chudasama KS, Thaker VS. Biological control of phytopathogenic bacteria *Pantoea agglomerans* and *Erwinia chrysanthemi* using 100 essential oils. Arch Phytopathol Plant Prot 2014;47:2221-2232

36 Wetter DA, Yiannias JA, Prakash AV, Davis MD, Farmer SA, el-Azhary RA, et al. Results of patch testing to personal care product allergens in a standard series and a supplemental cosmetic series: an analysis of 945 patients from the Mayo Clinic Contact Dermatitis Group, 2000-2007. J Am Acad Dermatol 2010;63:789-798

37 Warshaw EM, Buchholz HJ, Belsito DV, Maibach HI, Folwer JF Jr, Rietschel RL, et al. Allergic patch test reactions associated with cosmetics: Retrospective analysis of cross-sectional data from the North American Contact Dermatitis Group, 2001-2004. J Am Acad Dermatol 2009;60:23-38

38 Kohl L, Blondeel A, Song M. Allergic contact dermatitis from cosmetics: retrospective analysis of 819 patch-tested patients. Dermatology 2002;204:334-337

39 Thomson KF, Wilkinson SM. Allergic contact dermatitis to plant extracts in patients with cosmetic dermatitis. Br J Dermatol 2000;142:84-88

40 Larsen W, Nakayama H, Lindberg M, Fischer T, Elsner P, Burrows D, et al. Fragrance contact dermatitis: A worldwide multicenter investigation (Part 1). Am J Cont Derm 1996;7:77-83

41 Zug KA, Warshaw EM, Fowler JF Jr, Maibach HI, Belsito DL, Pratt MD, et al. Patch-test results of the North American Contact Dermatitis Group 2005-2006. Dermatitis 2009;20:149-160

42 Fransway AF, Zug KA, Belsito DV, DeLeo VA, Fowler JF Jr, Maibach HI, et al. North American Contact Dermatitis Group patch test results for 2007-2008. Dermatitis 2013;24:10-21

43 Warshaw EM, Belsito DV, DeLeo VA, Fowler JF Jr, Maibach HI, Marks JG, et al. North American Contact Dermatitis Group patch-test results, 2003-2004 study period. Dermatitis 2008;19:129-136

44 Wenk KS, Ehrlich AE. Fragrance series testing in eyelid dermatitis. Dermatitis 2012;23:22-26

45 Sugiura M, Hayakawa R, Kato Y, Sigiura K, Hashimoto R. Results of patch testing with lavender oil in Japan. Contact Dermatitis 2000;43:157-160

46 Bleasel N, Tate B, Rademaker M. Allergic contact dermatitis following exposure to essential oils. Australas J Dermatol 2002;43:211-213

47 Boonchai W, Lamtharachai P, Sunthonpalin P. Occupational allergic contact dermatitis from essential oils in aromatherapists. Contact Dermatitis 2007;56:181-182

48 Nakayama H, Harada R, Toda M. Pigmented cosmetic dermatitis. Int J Dermatol 1976;15:673-675

49 Warshaw EM, Belsito DV, Taylor JS, Sasseville D, DeKoven JG, Zirwas MJ, et al. North American Contact Dermatitis Group patch test results: 2009 to 2010. Dermatitis 2013;24:50-59

50 Sugawara M, Nakayama H, Watanabe S. Contact hypersensitivity to ylang-ylang oil. Contact Dermatitis 1990;23:248-249

51 Toyoda T, Watanabe S, Kawasaki M, et al. Dihydroisoeugenol found in ylang-ylang oil. Skin Res 1989;31 (Suppl. 7):35-43 (in Japanese)

52 Kanerva L, Estlander T, Jolanki R. Occupational allergic contact dermatitis caused by ylang-ylang oil. Contact Dermatitis 1995;33:198-199

53 Selvaag E, Holm J, Thune P. Allergic contact dermatitis in an aromatherapist with multiple sensitizations to essential oils. Contact Dermatitis 1995;33:354-355

54 Keane FM, Smith HR, White IR, Rycroft RJG. Occupational allergic contact dermatitis in two aromatherapists. Contact Dermatitis 2000;43:49-51

55 Dharmagunawardena B, Takwale A, Sanders KJ, Cannan S, Roger A, Ilchyshyn A. Gas chromatography: an investigative tool in multiple allergies to essential oils. Contact Dermatitis 2002;47:288-292

56 Trattner A, David M, Lazarov A. Occupational contact dermatitis due to essential oils. Contact Dermatitis 2008;58:282-284

57 Rudzki E, Grzywa Z, Bruo WS. Sensitivity to 35 essential oils. Contact Dermatitis 1976;2:196-200

58 Cockayne SE, Gawkrodger DJ. Occupational contact dermatitis in an aromatherapist. Contact Dermatitis 1997;37:306-307

59 Romaguera C, Vilaplana J. Occupational contact dermatitis from ylang-ylang oil. Contact Dermatitis 2000;43:251

60 Nakayama H, Matsuo S, Hayakawa K, Takhashi K, Shigematsu T, Ota S. Pigmented cosmetic dermatitis. Int J Dermatol 1984;23:299-305

61 Tran A, Pratt M, DeKoven J. Acute allergic contact dermatitis of the lips from peppermint oil in a lip balm. Dermatitis 2010;21:111-115

62 Rudzki E, Grzywa Z. Allergy to perfume mixture. Contact Dermatitis 1986;15:115-116

63 Rudzki E, Rebandel P, Grzywa Z. Occupational dermatitis from cosmetic creams. Contact Dermatitis 1993;29:210

64 Scheinman PL. Is it really fragrance free? Am J Contact Dermatitis 1997;8:239-242

65 Fettig J, Taylor J, Sood A. Post-surgical allergic contact dermatitis to compound tincture of benzoin and association with reactions to fragrances and essential oils. Dermatitis 2014;25:211-212

66 Belsito DV, Fowler JF Jr, Sasseville D, Marks JG Jr, De Leo VA, Storrs FJ. Delayed-type hypersensitivity to fragrance materials in a select North American population. Dermatitis 2006;17:23-28

67 Pratt MD, Belsito DV, DeLeo VA, Fowler JF Jr, Fransway AF, Maibach HI, et al. North American Contact Dermatitis Group patch-test results, 2001-2002 study period. Dermatitis 2004;15:176-183

68 De Mozzi P, Johnston GA. An outbreak of allergic contact dermatitis caused by citral in beauticians working in a health spa. Contact Dermatitis 2014;70:377-379

69 Uter W, Geier J, Schnuch A, Frosch PJ. Patch test results with patients' own perfumes, deodorants and shaving lotions: results of the IVDK 1998–2002. J Eur Acad Dermatol Venereol 2007;21:374-379

70 Cuesta L, Silvestre JF, Toledo F, Lucas A, Pérez-Crespo M, Ballester I. Fragrance contact allergy: a 4-year retrospective study. Contact Dermatitis 2010;63:77-84

71 Uter W, Schmidt E, Geier J, Lessmann H, Schnuch A, Frosch P. Contact allergy to essential oils: current patch test results (2000–2008) from the Information Network of Departments of Dermatology (IVDK). Contact Dermatitis 2010;63:277-283

72 Warshaw EM, Maibach HI, Taylor JS, Sasseville D, DeKoven JG, Zirwas MJ, et al. North American Contact Dermatitis Group patch test results: 2011-2012. Dermatitis 2015;26:49-59

73 Trattner A, David M. Patch testing with fine fragrances: comparison with fragrance mix, balsam of Peru and a fragrance series. Contact Dermatitis 2004;49:287-289

74 Frosch PJ, Johansen JD, Menné T, Pirker C, Rastogi SC, Andersen KE, et al. Further important sensitizers in patients sensitive to fragrances. II. Reactivity to essential oils. Contact Dermatitis 2002;47:279-287

75 Manuel Brites M, Goncalo M, Figueiredo A. Contact allergy to fragrance mix—a 10-year study. Contact Dermatitis 2000;43:181-182

76 de Groot AC, Coenraads PJ, Bruynzeel DP, Jagtman BA, van Ginkel CJW, Noz K, van der Valk PGM, et al. Routine patch testing with fragrance chemicals in The Netherlands. Contact Dermatitis 2000;42:184-185

77 Gupta N, Shenoi SD, Balachandran C. Fragrance sensitivity in allergic contact dermatitis. Contact Dermatitis 1999;40:53-54

78 Nakayama H, Hanaoka H, Ohshiro A. Allergen Controlled System (ACS). Tokyo, Japan: Kanehara Shuppan, 1974:42. Data cited in ref. 79

79 Mitchell JC. Contact hypersensitivity to some perfume materials. Contact Dermatitis 1975;1:197-199

80 Larsen WG. Cosmetic dermatitis due to a perfume. Contact Dermatitis 1975;1:142-145

81 Srivastava PK, Bajaj AK. Ylang-ylang oil not an uncommon sensitizer in India. Indian J Dermatol 2014;59:200-201

82 Toholka R, Wang Y-S, Tate B, Tam M, Cahill J, Palmer A, Nixon R. The first Australian Baseline Series: Recommendations for patch testing in suspected contact dermatitis. Australas J Dermatol 2014, Sept. 7. doi: 10.1111/ajd.12186

83 Goossens A, Merckx L. Allergic contact dermatitis from farnesol in a deodorant. Contact Dermatitis 1997;37:179-180

84 Meynadier JM, Meynadier J, Peyron JL, Peyron L. Formes cliniques des manifestations cutanées d'allergie aux parfums. Ann Dermatol Venereol 1986;113:31-39

85 Paulsen E, Andersen KE. Colophonium and Compositae mix as markers of fragrance allergy: Cross-reactivity between fragrance terpenes, colophonium and Compositae plant extracts. Contact Dermatitis 2005;53:285-291

86 Nardelli A, D'Hooge E, Drieghe J, Dooms M, Goossens A. Allergic contact dermatitis from fragrance components in specific topical pharmaceutical products in Belgium. Contact Dermatitis 2009;60:303-313

Chapter 5.91   ZDRAVETZ OIL

DEFINITION
Zdravetz oil is the essential oil obtained from the aerial parts (above ground plant) of the big root (Bulgarian) geranium (zdravetz) *Geranium macrorrhizum* L.

INCI NOMENCLATURE
Description/definition: Geranium macrorrhizum herb oil is an essential oil obtained from the herbs of the plant, *Geranium macrorrhizum* (L.), Geraniaceae
INCI name EU: Geranium macrorrhizum herb oil (perfuming name, not officially an INCI name)
INCI name USA: Not in the Personal Care Products Council Ingredient Database
CAS registry number(s): 92347-05-2; 68991-32-2
EINECS number(s): 296-192-0

ISO (INTERNATIONAL ORGANIZATION FOR STANDARDIZATION) STANDARD
There is currently no ISO standard for zdravetz oil.

THE PLANT, THE OIL AND THEIR USES
Geranium macrorrhizum L. (sometimes erroneously spelled *Geranium macrorhizum*) (zdravetz) is a hardy, flowering, perennial plant that grows to a height of 30-50 cm and has a thick, succulent rhizome (underground stem, hence the name 'macrorrhizum', big root) and soft aromatic leaves. It is native to middle Europe (Austria), southeastern Europe (Albania, Bulgaria, Croatia, Greece, Italy, Romania, Serbia, Slovenia) and southwestern Europe (France) and naturalized elsewhere. In Bulgaria, *G. macrorrhizum* is widely grown as a cultural symbol associated with health and good luck and for its aromatic properties (www.kew.org; http://mansfeld.ipk-gatersleben.de; GRIN Taxonomy for Plants, 6). 'Zdrave' is the Bulgarian word for 'health' (7); the common name of the plant in Serbian is 'zdravats', which could be translated as 'health' or 'to be healthy' (1).

Like a number of other species belonging to the genus *Geranium*, *G. macrorrhizum* is highly appreciated in the traditional medicine of Serbs and other Slavic people from the Balkan Peninsula (9). The plant, fresh leaves, roots, and infusions may be used for the treatment of stomach disorders, as an aphrodisiac, for the treatment of diarrhea, skin inflammation and for tampons in nose bleeding. The extracts of *G. macrorrhizum* have been reported to possess a broad spectrum of antimicrobial, hypotensive, spasmolytic, astringent, cardiotonic, antioxidant, and sedative activities (1,5,6,9).

Steam distillation of the aerial parts of *Geranium macrorrhizum* yields the essential oil which is called zdravetz oil (sometimes spelled zdravets). The oil is used in traditional medicine as both a stimulant and as a carminative, and is also used in limited quantities in cigarette flavorings and is (rarely) employed in aromatherapy practices (10). The commercial oil, produced solely in Bulgaria, is mainly obtained from wild-growing plants and the total annual amount exported from Bulgaria may be very limited (possibly only several hundred kilograms) (6,7).

CHEMICAL COMPOSITION
Zdravetz oil at a temperature of over 20°C is a light to dark green clear liquid, which has a floral herbaceous, soft earthy odor. At lower temperatures, over half of the oil volume is taken up by large colorless prismatic crystals with no odor, formed by the ingredient germacrone. The liquid portion called 'eleoptene' contains all odorous components (6,7). The yield of essential oil from the aerial parts of *Geranium macrorrhizum* L. generally varies from 0.3 to 0.45 per cent. The main producing country of this oil is Bulgaria.

Very little research on the composition of zdravetz oil has been performed. In fact, only one analytical study investigating the essential oil (one lab-hydrodistilled oil sample) of *G. macrorrhizum* has been published in the last 10 years (6).

Literature data (up to November 9, 2014) on the chemical composition of zdravetz oils (very few publications were found) and unpublished analytical data from one of us (E.S.) are shown in Table 5.91.1 in alphabetical order. In zdravetz oils from various origins, over 275 chemicals have been

Table 5.91.1 Constituents identified in zdravetz oils

Constituent	CAS	Percentage and range			
		A	B (6)	C (1)	D
2-Acetylfuran	1192-62-7		tr		
α-Acoradiene	24048-44-0		tr		
β-Acoradiene	28477-64-7		tr		
Acora-3,5-dien-11-ol	94480-77-0		tr		
Amorpha-4,7-dien-11-ol		07-1.3	1.1		
Amyl isovalerate	25415-62-7		tr		
4,5-di-epi-Aristolochene	54868-40-5		tr		
β-Barbatene	39863-73-5		tr		
Benzaldehyde	100-52-7		tr		
Benzothiazole	95-16-9		tr		
Benzyl alcohol	100-51-6		tr		
Benzyl benzoate	120-51-4		tr		
Benzyl butyrate	103-37-7		tr		
Benzyl hexanoate	6938-45-0		tr		

Table 5.91.1 Constituents identified in zdravetz oils (*continued*)

Constituent	CAS	Percentage and range			
		A	B (6)	C (1)	D
Benzyl isobutyrate	103-28-6		tr		
Benzyl isovalerate	103-38-8		tr		
Benzyl 2-methylbutanoate	56423-40-6		tr		
Benzyl salicylate	118-58-1		tr		
β-Bisabolene	495-61-4		tr		
(*E*)-γ-Bisabolene	53585-13-0		0.2		
(*Z*)-γ-Bisabolene	13062-00-5		tr		
α-Bisabolol	515-69-5	0.06-2.8			
epi-α-Bisabolol	78148-59-1		tr		
Borneol	507-70-0	0.06-0.1	0.1	0.2[a]	+[b,e,f]
Bornyl acetate	76-49-3	0.08-0.4			
di-epi-α-Bourbonene	317831-19-9		tr		
β-Bourbonene	5208-59-3		0.1		
α-Bulnesene	3691-11-0	0.2-1.2	1.3		
Butyl hexanoate	626-82-4		tr		
Cabreuva oxide B	107602-53-9		tr		
Cadalene	483-78-3		tr		
trans-Cadina-1,4-diene	20085-13-6		tr		
(*E*)-Cadina-1(6),4-diene	931410-54-7		tr		
Cadina-1(10),6,8-triene	1460-96-4		tr		
α-Cadinene	24406-05-1		0.4		
γ-Cadinene	39029-41-9		0.2		
δ-Cadinene	483-76-1	0.1-0.4	0.3		+[b,e,f]
α-Calacorene	21391-99-1		tr		
Calamenene	483-77-2				+[b,e]
Camphene	79-92-5	0.05-0.1		0.1	
Camphene hydrate	465-31-6		tr		
α-Campholenal	4501-58-0	0.01-0.3	tr		
α-Campholenol	1901-38-8		tr		
Camphor	76-22-2		tr		
δ2-Carene (δ4-carene)	554-61-0		tr		
δ3-Carene	13466-78-9		0.1		+[b,e]
Carvacrol	499-75-2		tr		
(*E*)-Carveol	197-07-5		tr		
(*Z*)-Carveol	1197-06-4		tr		
β-Caryophyllene	87-44-5	0.2-0.5	0.4		+[b,e,f]
Caryophyllene oxide	1139-30-6		0.6		
α-Cedrene	469-61-4		tr		
1,8-Cineole	470-82-6		tr		
Citronellal	106-23-0		0.1		
Citronellic acid	502-47-6		tr		
Citronellol	106-22-9		0.3		
Citronellyl acetate	150-84-5		tr		
Citronellyl butyrate	141-16-2		tr		
Citronellyl formate	105-85-1		tr		
Citronellyl hexanoate	10580-25-3		tr		
β-Copaene	18252-44-3		tr		
p-Cresyl methyl ether	104-93-8		tr		
α-Cubebene	17699-14-8		tr		
ar-Curcumene	644-30-4	0.3-1.5	0.6	0.3	+[b,e,f]
β-Curcumene	28976-67-2		0.6		
γ-Curcumene	28976-68-3	1.0-2.4	4.1		
α-Cyclocitral	432-24-6		tr		
Cyclosativene	22469-52-9		tr		
p-Cymene	99-87-6	0.08-0.7	0.4		+[b,e,f]
p-Cymenene	1195-32-0		tr		
p-Cymen-8-ol	1197-01-9	0.1	0.1		
(*E*)-β-Damascenone	23726-93-4		tr		
Decanal	112-31-2		tr		
2-Decanone	693-54-9		tr		
2,5-Diethylfuran	10504-06-0		0.1		

Table 5.91.1 Constituents identified in zdravetz oils (*continued*)

Constituent	CAS	Percentage and range			
		A	B (6)	C (1)	D
Dihydroedulan II	41678-32-4		tr		
Dihydrogermazol					+[b]
2,6-Dimethyl-3,7-octadiene-2,6-diol	13741-21-4		tr		
Docosane	629-97-0		tr		
Dodecanal	112-54-9		tr		
Dodecane	112-40-3		tr		
1-Dodecanol	112-53-8		0.2		
2-Dodecanone	6175-49-1		tr		
1-Dodecene (α-)	112-41-4		tr		
Eicosane	112-95-8		tr		
α-Elemene	5951-67-7				+[b,e,f]
β-Elemene	33880-83-0	0.4-0.7	0.6	0.2	+[b,e,f]
γ-Elemene	29873-99-2	0.9-1.3	1.3		
δ-Elemene	20307-84-0		tr		
cis-β-Elemenone	32663-57-3			0.4	+[b]
trans-β-Elemenone	20303-60-0	1.4-5.9	1.6	3.3	+[b]
Elemol (α-)	639-99-6				+[b,e]
1,10-Epoxygermacrone					+[g]
4,5-Epoxygermacrone					+[g]
4,8-Epoxyterpinolene	4584-23-0		tr		
Eremophila-1(10),11-dien-9β-ol		0.8-1.2		1.1	
cis-Eudesma-6,11-diene	194607-93-7		tr		
Eudesma-3,7(11)-dien-8-one	54707-46-9	0-0.02	tr		
(4βH,5αH)-*cis*-Eudesm-6-en-11-ol				tr	
Eudesm-11-en-4α-ol			tr		
α-Eudesmol	473-16-5	0-5.3			
β-Eudesmol	473-15-4	0.4-2.1			+[b,e]
γ-Eudesmol	1209-71-8	0.4-2.5	0.3		
10-epi-γ-Eudesmol	15051-81-7	0-0.3			
Eugenol	97-53-0		tr		
(2E,6Z)-Farnesal	3790-67-8		tr		
Farnesane	3891-98-3		tr		
(E,E)-α-Farnesene	502-61-4		tr		
(Z,E)-α-Farnesene	26560-14-5		tr		
(E)-β-Farnesene	18794-84-8		tr		
Furfural	98-01-1		tr		
Geranial	141-27-5		tr		
Geraniol	106-24-1		0.1		+[b]
Geranylacetone (E-)	3796-70-1		tr	0.5	
Germacrene B	15423-57-1	4.5-11.2	11.3		
Germacrene D	23986-74-5		0.6		
Germacrone (E,E-)	6902-91-6	45.5-66.6	49.7	37.4	+[d,e]
Germazane					+[b,e]
Germazene					+[b]
Germazol					+[b]
Germazone	62332-96-1				+[b,d]
Globulol	489-41-8		tr		
Guaiazulene	489-84-9				+[b]
α-Guaiene	3691-12-1		0.1		
Heneicosane	629-94-7		tr		
Heptadecane	629-78-7		tr		
1-Heptadecene	6765-39-5		tr		
Heptanal	111-71-7		tr		
Heptanoic acid	111-14-8				+[b]
Hexadecanal	629-80-1		tr		
Hexadecane	544-76-3		tr		
Hexadecanoic acid	57-10-3		tr		
Hexahydrofarnesyl acetone	502-69-2		tr		
Hexanal	66-25-1		tr		
Hexanoic acid	142-62-1		tr		+[b]
Hexanol (1-)	111-27-3		tr		

Table 5.91.1 Constituents identified in zdravetz oils (*continued*)

Constituent	CAS	Percentage and range			
		A	B (6)	C (1)	D
(*E*)-3-Hexenol	928-97-2		tr		
(*Z*)-3-Hexenyl benzoate	25152-85-6		tr		
(*Z*)-3-Hexenyl (*Z*)-3-hexenoate	61444-38-0		tr		
n-Hexyl acetate	142-92-7		0.1		
Hexyl benzoate	6789-88-4		tr		
Hexyl isovalerate	10032-13-0		tr		
Homofarnesane	155635-42-0		tr		
Hotrienol	20053-88-7		tr		
α-Humulene	6753-98-6		0.1		+[b,e,f]
Humulene epoxide II	19888-34-7		tr		
Isoamyl benzoate	94-46-2		tr		
Isoamyl butyrate	106-27-4		tr		
Isoamyl isovalerate	659-70-1		tr		
Isoamyl 2-methylbutyrate	27625-35-0		tr		
Isoamyl phenylacetate	102-19-2		tr		
Isobornyl acetate	125-12-2		tr		
Isobutyl benzoate	120-50-3		tr		
Isogermacrene D	317819-80-0		tr		
Isogermacrone	5975-50-8				+[c,d]
Isopulegol	89-79-2		tr		
Italicene	94535-52-1	0.08-0.3	0.3		
Italicene ether	104188-25-2		tr		
10-epi-Italicene ether	104265-25-0		tr		
Junenol	472-07-1				+[b,e]
Juniper camphor	473-04-1		tr		+[b,e]
Limonene	138-86-3	0.07-0.8	0.4		+[b,e,f]
Linalool	78-70-6	0.2-0.7	0.9	0.8	
cis-Linalool oxide, furanoid	11063-77-7		0.1		
trans-Linalool oxide, furanoid	34995-77-2		tr		
13-epi-Manoyl oxide	1227-93-6		tr		
p-Mentha-1,8-dien-4-ol	3419-02-1		0.3		
p-Mentha-1,4,8-triene	28233-65-0		0.1		
p-Menth-1-en-9-al	29548-14-9		tr		
trans-p-Menth-2-en-7-ol	19898-87-4		tr		
p-Methylacetophenone	122-00-9		tr		
Methyl *p*-anisate	121-98-2		tr		
2-Methylbenzaldehyde	529-20-4		tr		
3-Methylbutanoic acid	503-74-2		tr		
3-Methyl-2-butenal	107-86-8		tr		
3-Methyl-3-butenylbenzoate	5205-12-9		tr		
3-Methyl-2-butenyl 3-methylbutanoate			tr		
3-Methyl-3-butenyl 3-methylbutanoate	54410-94-5		tr		
3-Methylbutyl hexanoate	2198-61-0		tr		
Methyl chavicol	140-67-0		tr		
(*E*)-6-Methyl-3,5-heptadien-2-one	16647-04-4		tr		
6-Methyl-5-hepten-2-one	110-93-0		tr		
4-Methylpentanoic acid	646-07-1		tr		
2-Methyl-2-pentenal	623-36-9		tr		
Methyl 2-phenylacetate	101-41-7		tr		
Methyl salicylate	119-36-8		tr		
Methyl thymol	1076-56-8		tr		
cis-Muurola-4(14), 5-diene	157477-72-0		tr		
α-Muurolene	10208-80-7		tr		+[f]
γ-Muurolene	30021-74-0				+[b,e,f]
Myrcene	123-35-3	0.03-0.2	0.2		
Myrtenol	515-00-4		0.1		
Neral	106-26-3		tr		
(*E*)-Nerolidol	40716-66-3		tr		
Nonadecane	629-92-5		tr		
Nonanal	124-19-6		0.1		
2-Nonanone	821-55-6		tr		
(*E*)-β-Ocimene	3779-61-1	0.09-0.2	0.2		

Table 5.91.1 Constituents identified in zdravetz oils (*continued*)

Constituent	CAS	Percentage and range			
		A	B (6)	C (1)	D
(Z)-β-Ocimene	3338-55-4	0.8-1.6	1.5		
(E)-Ocimenol	28977-58-4		0.1		
Octacosane	630-02-4		tr		
Octadecane	593-45-3		tr		
Octanal	124-13-0		tr		
Octane	111-65-9		tr		
Octanoic acid	124-07-2				+[b]
n-Octanol (1-)	111-87-5		tr		
Octyl acetate	112-14-1		tr		
Pentadecanal	2765-11-9		tr		
Pentadecane	629-62-9		tr		
Pentanoic acid	109-52-4		tr		
Pentyl benzoate	2049-96-9		tr		
2-Pentylfuran	3777-69-3		tr		
Pentyl 2-methylbutanoate	68039-26-9		tr		
Perilla alcohol	536-59-4		tr		
Perillaldehyde	2111-75-3		tr		
α-Phellandrene	99-83-2		tr		+[b,e]
β-Phellandrene	555-10-2		tr		
Phenanthrene	85-01-8		tr		
2-Phenethyl alcohol	60-12-8		0.1		
2-Phenethyl benzoate	94-47-3		tr		
Phenethyl isovalerate	140-26-1		tr	1.6	
2-Phenylethyl butanoate	103-52-6		tr		
2-Phenylethyl hexanoate	6290-37-5		tr		
2-Phenylethyl 2-methylbutyrate	24817-51-4		tr		
2-Phenylethyl 2-phenylacetate	102-20-5		tr		
(E)-Phytol	150-86-7		tr		
Pimaradiene	1686-61-9		tr		
α-Pinene	80-56-8	0.05-2.1	0.2		+[b,e]
β-Pinene	127-91-3	0.01-0.03	0.1		
Pinenol	31619-93-9				+[b]
trans-Pinocarveol	1674-08-4		tr		
Piperitone	89-81-6			3.0	
Rosefuran	15186-51-3		tr		
cis-Rose oxide	3033-23-6		tr		
Sabinene	3387-41-5	0-0.02	0.1		
Safranal	116-26-7		tr		
Salvial-4(14)-en-1-one	73809-82-2		tr		
α-Santalene	512-61-8				+[b,e,f]
Selina-3,7(11)-diene	6813-21-4	0.2-0.5	0.5		+[b,e,f]
Selina-4(15),7-diene	105497-53-8	0.06-0.7	tr		
Selina-4(15),7(11)-diene			tr		+[b,e,f]
Selina-4,11-diene	17627-30-4		tr		
α-Selinene	473-13-2	0.1-0.3	0.3		
β-Selinene	17066-67-0	0.2-0.6	0.4	0.5	
γ-Selinene	515-17-3				+[b]
δ-Selinene	473-14-3		0.2		
Selin-6-en-4-ol	118173-08-3		tr		
Spirovetiva-1(10),7(11)-diene	137941-80-1		tr		
α-Terpinene	99-86-5	0.06-0.3	0.3		
γ-Terpinene	99-85-4	0.2-1.4	1.4		+[b,e,f]
Terpinen-4-ol	562-74-3	0.06-0.3	0.1		
α-Terpineol	98-55-5	0.3-0.8	0.8	0.2[a]	
Terpinolene (α-)	586-62-9	0.05-1.5	1.6		+[b,e,f]
Tetradecanal	124-25-4		tr		
1-Tetradecanol	112-72-1		0.4		
Tetradecyl acetate	638-59-5		0.2		
α-Thujene	2867-05-2	0.02-0.1	0.1		
Thymol	89-83-8		tr		
n-Triacontane	638-68-6				+[b]
Tricosane	638-67-5		tr		

Table 5.91.1 Constituents identified in zdravetz oils (*continued*)

Constituent	CAS	Percentage and range			
		A	B (6)	C (1)	D
Tridecanal	10486-19-8		tr		
Tridecane	629-50-5		tr		
1-Tridecanol	112-70-9		tr		
2-Tridecanone	593-08-8		tr		
1-Tridecene	2437-56-1		tr		
Undecanal	112-44-7		tr		
Undecane	1120-21-4		tr		
2-Undecanone	112-12-9		tr		
1-Undecene	821-95-4		tr		
trans-Verbenol	1820-09-3		tr		
Verbenone	80-57-9		tr		
β-Vetivenene	27840-40-0		tr		
p-Vinylguaiacol	7786-61-0		tr		
α-Ylangene	14912-44-8		tr		
β-Ylangene	20479-06-5		tr		
α-Zingiberene	495-60-3		tr		

A eight zdravetz essential oil samples from Bulgaria analyzed between 1998 and 2014; lowest and highest concentrations given (E. Schmidt, unpublished data)
B one lab-hydrodistilled oil from *G. macrorrhizum* growing wild in southwest Serbia (ref. 6)
C one lab-hydrodistilled oil from the aerial parts of *G. macrorrhizum* in full bloom, growing wild in west Serbia (ref. 1)
D data from other studies (indicated with superscript letters); highest concentrations found in any study reviewed here given; when two or more oils were investigated, only the highest concentrations are mentioned, unless indicated otherwise

[a] α-terpineol and borneol combined; [b] data cited in ref. 5; there is overlap with data cited in refs. 7 and 8; [c] data cited in ref. 2; [d] data from a Bulgarian zdravetz oil and data cited in ref. 3; [e] data cited in ref. 7; there is overlap with data cited in refs. 5 and 8; [f] data cited in ref. 8; there is overlap with data cited in refs. 5 and 7; [g] one lab-hydrodistilled oil from Serbia; only new epoxygermacrones were analyzed (ref. 9)

tr: trace (in column B: <0.05); + present in the oil investigated, but quantity not stated

identified. About 79 per cent of these (notably chemicals found in traces in ref. 6, one lab-hydrodistilled oil from plants growing wild in Serbia) were found in a single reviewed publication only. The high percentage is strongly influenced by the paucity of recently reported studies on *G. macrorrhizum* essential oil. The major compounds found in zdravetz oils from different sources are shown in Table 5.91.2. They include (highest concentrations in any study given) germacrone (*E,E*-) (66.6%), germacrene B (11.3%), *trans*-β-elemenone (5.9%), α-eudesmol (5.3%), γ-curcumene (4.1%), α-bisabolol (2.8%) and γ-eudesmol (2.5%).

Commercial oils

The ten chemicals that had the highest maximum concentrations in eight commercial zdrawetz essential oil samples (concentration ranges provided) are the following: germacrone (*E,E*-) (45.5-66.6%), germacrene B (4.5-11.2%), *trans*-β-elemenone (1.4-5.9%), α-eudesmol (0-5.3%), α-bisabolol (0.06-2.8%), γ-eudesmol (0.4-2.5%), γ-curcumene (1.0-2.4%), α-pinene (0.05-2.1%), β-eudesmol (0.4-2.1%) and (*Z*)-β-ocimene (0.8-1.6%) (Erich Schmidt, unpublished analytical data).

Table 5.91.2 Major constituents of zdravetz oils

Constituent	CAS	Percentage and range			
		A	B (6)	C (1)	D
Germacrone (*E,E*-)	6902-91-6	45.5-66.6	49.7	37.4	+[d,e]
Germacrene B	15423-57-1	4.5-11.2	11.3		
trans-β-Elemenone	20303-60-0	1.4-5.9	1.6	3.3	+[b]
α-Eudesmol	473-16-5	0-5.3			
γ-Curcumene	28976-68-3	1.0-2.4	4.1		
α-Bisabolol	515-69-5	0.06-2.8			
γ-Eudesmol	1209-71-8	0.4-2.5	0.3		

LEGEND: SEE UNDER TABLE 5.91.1

CONTACT ALLERGY/ALLERGIC CONTACT DERMATITIS

General

Contact allergy to zdravetz oil has been reported in two publications, but no cases of allergic contact dermatitis from the oil have been identified.

Testing in groups of patients

Two hundred dermatitis patients from Poland were tested with zdravetz oil 2% in petrolatum and one (0.5%) reacted. The same authors also patch tested 51 patients allergic to *Myroxylon pereirae* resin (balsam of Peru) and/ or turpentine and/or wood tar and/or colophony and one (2.0%) had a positive patch test; relevance data were not provided (11). A group of 86 patients from Poland previously reacting to the fragrance mix was tested with zdravetz oil and four (4.6%) had a positive patch test reaction; relevance data were not provided (12).

LITERATURE

1 Chalchat J-C, Petrovic SD, Maksimovic ZA, Gorunovic MS. A comparative study on essential oils of *Geranium macrorrhizum* L. and *Geranium phaeum* L., Geraniaceae from Serbia. J Essent Oil Res 2002;14:333-335

2 Orahovats AS, Bozhkova NV, Hipert H. Photolysis and pyrolysis products of isogermacrone. Tetrahedron Letters 1983;24:947-950

3 Tsankova E, Ognyanov I. Germazone, a novel tricyclic sesquiterpene ketone in the essential oil from *Geranium macrorrhizum* L. Tetrahedron Letters 1976;17:3833-3836

4 Ivancheva S, Stantcheva B. Ethnobotanical inventory of medicinal plants in Bulgaria. J Ethnopharmacol 2000;69:165-172

5 Brud WS, Ognyanov I. Zdravetz. Int J Aromather 1995;7:10-11

6 Radulović NS, Dekić MS, Stojanović-Radić ZZ, Zoranić SK. *Geranium macrorrhizum* L. (Geraniaceae) essential oil: A potent agent against *Bacillus subtilis*. Chem Biodivers 2010;7:2783-2800

7 Ognyanov I. Bulgarian zdravets oil. Perfum Flavor 1985;10(Oct/Nov):39-44

8 Lawrence BM. Progress in essential oils. Perfum Flavor 1976;3(2):49-50

9 Radulovic NS, Zlatkovic D, Dekic M, Stojanovic-Radic Z. Further antibacterial *Geranium macrorrhizum* L. metabolites and synthesis of epoxygermacrones. Chem Biodivers 2014;11:542-550

10 Rhind JP. Essential oils. A handbook for aromatherapy practice, 2nd Edition. London: Singing Dragon, 2012

11 Rudzki E, Grzywa Z, Bruo WS. Sensitivity to 35 essential oils. Contact Dermatitis 1976;2:196-200

12 Rudzki E, Grzywa Z. Allergy to perfume mixture. Contact Dermatitis 1986;15:115-116

Chapter 5.92   CARDAMOM OIL

DEFINITION

Cardamom oil (essential oil of cardamom, cardamom seed oil) is the essential oil obtained from the fruits of the cardamom, *Elettaria cardamomum*, Zingiberaceae.

INCI NOMENCLATURE

Description/definition: Elettaria cardamomum seed oil is the volatile oil obtained from the dried ripe seeds of *Elettaria cardamomum*, Zingiberaceae
INCI name EU & USA: Elettaria cardamomum seed oil
CAS registry number(s): 8000-66-6; 85940-32-5 (seed extract)
EINECS number(s): 288-922-1

ISO (INTERNATIONAL ORGANIZATION FOR STANDARDIZATION) STANDARD

ISO number: 4733
ISO name: Essential oil of cardamom
Botanical origin: *Elettaria cardamomum* (L.) Maton
Parts of plant used: Fruit
ISO values: ISO values (minimum and maximum concentrations) are shown in Table 5.92.1.

Table 5.92.1 ISO values (%) for cardamom oil [a]

Compound	CAS	Minimum	Maximum
Terpinyl acetate	8007-35-0	32.0	45.0
1,8-Cineole	470-82-6	23.0	35.0
Linalyl acetate	115-95-7	4.0	9.0
Limonene	138-86-3	2.0	7.0
Linalool	78-70-6	3.0	7.0
α-Terpineol	98-55-5	tr	7.0
Sabinene	3387-41-5	2.0	5.0
Terpinen-4-ol	562-74-3	0.8	3.0
Myrcene	123-35-3	tr	2.5
(*E*)-Nerolidol	40716-66-3	0.5	2.0
α-Pinene	80-56-8	1.0	2.0

[a] ISO 4733 Essential oil of cardamom ©ISO 2004; Geneva, Switzerland, www.iso.org

THE PLANT, THE OIL, AND THEIR USES

Cardamom (*Elettaria cardamomum* L. (Maton)) is a tall, perennial herbaceous plant belonging to the Zingiberaceae family. Its dried fruit is one of the most highly priced spices in the world. It takes 3-4 years before the plant starts bearing the yellow-grey capsules containing many small black seeds. Fruits are gathered just before they are ripe in order to conserve the seeds inside the capsule. The dried fruit is used either whole or in ground form as a flavoring agent, in medicinal preparations and for the production of essential oil of cardamom. Cardamom fruits are widely used for flavoring purposes in foods such as cakes, pastries, sausages, sauces, gingerbread, candies and curry, especially in the Nordic countries Sweden and Norway, the United Kingdom, and in Asia. In Arab countries and India, it is a common flavoring ingredient for coffee. Medically, they are used for flatulent indigestion and to stimulate the appetite. The seeds are also prescribed in Ayurvedic medicine for coughs, colds, bronchitis, and asthma.

Cardamom essential oil, extracted from the fruits by distillation, is utilized in perfumery and in tobacco products as well as for flavoring liqueur and food. According to some researchers, cardamom oil has antibacterial, antiseptic, carminative and antispasmodic properties (2,3, Mansfeld's World Database of Agricultural and Horticultural crops: http://mansfeld.ipk-gatersleben.de). It also has applications in aromatherapy (4,5).

CHEMICAL COMPOSITION

As we found the literature (one article only) on contact allergy to cardamom oil at the very last stage of writing this book, there was no more opportunity to document a full literature review on the composition of cardamom oils. However, we do present the compositional data of 101 commercial oils analyzed by one of us (Erich Schmidt).

Cardamom oil is a colorless to pale, mobile liquid which has a strong aromatic, spicy and eucalyptus-like, discreet woody odor. The yield of essential oil from *Elettaria cardamomum* L. varies from 2.5 to 6.5 per cent, depending on the origin of the source plant. The main producing countries of this oil are India, Sri Lanka, Guatemala, Costa Rica and El Salvador. The components found in 101 commercial cardamom essential oils are shown in Table 5.92.2. The ten major constituents are terpinyl acetate (31.3-46.1%), 1,8-cineole (20.9-38.2%), linalool (0.2-8.7%), linalyl acetate (0.4-8.5%), α-terpineol (1.7-6.7%), limonene (1.4-6.0%), sabinene (1.8-5.5%), α-terpinolene (0.09-5.4%), α-pinene (1.0-4.0%) and myrcene (0.04-3.2%).

CONTACT ALLERGY/ALLERGIC CONTACT DERMATITIS

General

Contact allergy to cardamom oil has been reported in one case report only. Dipentene (*dl-* limonene) may have been an allergen in that case.

Case report

A male confectioner developed occupational contact dermatitis which was ascribed to working with cardamom powder. He reacted to cardamom powder, cardamom oil 2% in petrolatum, δ-carene (δ2, δ3?), dipentene (*dl*-limonene), bergamot oil and turpentine peroxides (1). Limonene is the main component of bergamot oil, can be present in cardamom oil in concentrations up to 6% (Table 5.92.2) and has been found in commercial turpentine oils in concentrations up to 13.6%. The reaction to δ-carene may also explain the co-reactivity to turpentine peroxides.

Table 5.92.2 Constituents identified in cardamom oils[a]

Constituent	CAS	Percentage	Constituent	CAS	Percentage
β-Caryophyllene	87-44-5	0.07-0.9	β-Pinene	127-91-3	0.3-0.6
1,8-Cineole	470-82-6	20.9-38.2	Sabinene	3387-41-5	1.8-5.5
p-Cymene	99-87-6	0.04-1.3	cis-Sabinene hydrate	15537-55-0	0.1-0.8
Geranial	141-27-5	0.2-1.4	trans-Sabinene hydrate	17699-16-0	0.1-0.5
Geraniol	106-24-1	0.2-1.2	α-Terpinene	99-86-5	0.08-0.6
Geranyl acetate	105-87-3	0.1-0.6	γ-Terpinene	99-85-4	0.3-1.0
Limonene	138-86-3	1.4-6.0	Terpinen-4-ol	562-74-3	0.8-2.5
Linalool	78-70-6	0.2-8.7	Terpinen-4-yl acetate	4821-04-9	0-0.2
Linalyl acetate	115-95-7	0.4-8.5	α-Terpineol	98-55-5	1.7-6.7
Myrcene	123-35-3	0.04-3.2	δ-Terpineol	7299-42-5	0.05-1.0
Neral	106-26-3	0.1-0.9	α-Terpinolene	586-62-9	0.09-5.4
(E)-Nerolidol	40716-66-3	0.5-1.7	Terpinyl acetate	8007-35-0	31.3-46.1
(E)-β-Ocimene	3779-61-1	0.06-0.4	α-Thujene	2867-05-2	0.04-0.5
n-Octanal	124-13-0	0.05-0.3			
α-Pinene	80-56-8	1.0-4.0			

[a] one hundred and one cardamom essential oil samples from India, Sri Lanka, Guatemala, Costa Rica and El Salvador analyzed between 1998 and 2014; lowest and highest concentrations given (E. Schmidt, unpublished data)

LITERATURE

1 Mobacken H, Fregert S. Allergic contact dermatitis from cardamom. Contact Dermatitis 1975;1:175-176

2 Morsy NFS. A short extraction time of high quality hydrodistilled cardamom (*Elettaria cardamomum* L. Maton) essential oil using ultrasound as a pretreatment. Ind Crops Prod 2015;65:287-292

3 Lucchesi ME, Smadja J, Bradshaw S, Louw W, Chemat F. Solvent free microwave extraction of *Elletaria cardamomum* L.: A multivariate study of a new technique for the extraction of essential oil. J Food Eng 2007;79:1079-1086

4 Rhind JP. Essential oils. A handbook for aromatherapy practice, 2nd Edition. London: Singing Dragon, 2012

5 Lawless J. The encyclopedia of essential oils, 2nd Edition. London: Harper Thorsons, 2014

Chapter 5.93   HIMALAYAN CEDARWOOD OIL

There are five major cedarwood essential oils: cedarwood oil Atlas, cedarwood oil Himalaya, cedarwood oil Texas, cedarwood oil Virginia and cedarwood oil China. These are obtained from different botanical species, and as a consequence, their chemical compositions differ both qualitatively and quantitatively. Unfortunately, in non-botanical literature, usually the term 'cedarwood oil' is used, lacking information on the botanical origin.

DEFINITION

Himalayan cedarwood oil (essential oil of cedarwood Himalaya) is the essential oil obtained from the wood of the Himalayan cedar (deodar, deodar cedar) *Cedrus deodara* (Roxb. ex D. Don) G. Don.

INCI NOMENCLATURE

Description/definition: Cedrus deodara wood oil is the volatile oil obtained by steam distillation of the stumps of the deodar cedar, *Cedrus deodara*, Pinaceae
INCI name EU & USA: Cedrus deodara wood oil
CAS registry number(s): 91771-47-0
EINECS number(s): 294-939-5

ISO (INTERNATIONAL ORGANIZATION FOR STANDARDIZATION) STANDARD

There is currently no ISO standard for Himalayan cedarwood oil available.

THE PLANT, THE OIL, AND THEIR USES

Cedrus deodara (Himalayan cedar, deodar, deodar cedar) is a graceful, ornamental evergreen tree which can grow up to 60 meters and develop a diameter of 3 meters. It is native to the temperate (Afghanistan, China) and tropical (India, Nepal, Pakistan) regions of Asia and grows extensively on the slopes of the western Himalayas at altitudes of 1200-3000 meters. The tree is cultivated in China as an ornamental plant and for its timber, which is utilized in shipbuilding, furniture, bridges, and construction. Various parts of the deodar are used in traditional medicine for the treatment of a wide range of ailments such as fever, inflammation, pain, ulcers, hyperglycemia, infections, insomnia, mental disorders and diseases of the skin and blood. Anti-inflammatory, analgesic, anti-hyperglycemic, antispasmodic, insecticidal, anti-apoptotic, anti-cancer, immune-modulatory, molluscicidal, anxiolytic and anticonvulsant properties have been attributed to the plant (3). The essential oil of the Himalayan cedar finds use in Ayurvedic medicine, as an anthelmintic and in aromatherapy, but not in perfumes or other cosmetics. The oil sold to aromatherapists is usually rectified (5). The traditional and folklore

medicinal uses, phytochemistry and biological activities of *Cedrus deodara* have been reviewed (3).

CHEMICAL COMPOSITION

Cedarwood oil Himalaya is a yellowish to brownish slightly viscous clear liquid, which has a pleasant, typical, mild and woody cedar connotation odor. The yield of essential oil from cedarwood oil Himalaya varies from 2.3 to 4.7 per cent, depending on the dryness of the wood and its pre-distillation grinding. The main producing country of this oil is Nepal.

Literature data on the chemical composition of Himalayan cedarwood oils and unpublished analytical data from one of us (E.S.) are shown in Table 5.93.1 in alphabetical order. In Himalayan cedarwood oils from various origins, over 65 chemicals have been identified. About 60 per cent of these were found in a single reviewed publication only, but it should be realized that very few analytical studies have been performed on this essential oil. The major compounds found in Himalayan cedarwood oils from different sources include (highest concentrations in any study given) β-himachalene (46.7%), α-himachalene (30.8%), γ-himachalene (12.6%), (*E*)-α-atlantone (9.7%), himachalol (8.5%) and deodarone (5.4%). Well-known ingredients of *Cedrus deodara* essential oils that were present in concentrations of 4% or higher in one study were 'cis-atlantone and α-atlantone' (19%) and α-cedrene (15.8%).

Commercial oils

The ten chemicals that had the highest maximum concentrations in 8 commercial Himalayan cedarwood essential oil samples (concentration ranges provided) are the following: β-himachalene (32.8-46.7%), α-himachalene (16.6-20.3%), γ-himachalene (9.8-11.6%), (*E*)-α-atlantone (4.4-9.7%), himachalol (0.5-8.5%), (*Z*)-γ-atlantone (0.6-2.7%), (*Z*)-α-atlantone (0.8-2.6%), (*E*)-γ-atlantone (05-2.1%), deodarone (0.7-2.2%), and δ-cadinene (0.3-1.9% (Erich Schmidt, unpublished observations).

CONTACT ALLERGY/ALLERGIC CONTACT DERMATITIS

General

Contact allergy to Himalayan cedarwood oil has been reported in two publications only.

Testing in groups of patients

Two hundred dermatitis patients from Poland were tested with 'cedarwood oil' 2% in petrolatum and three (1.5%) reacted (1). Although not explicitly stated, from the information given in this study and from a later study by the same authors (2) it is very likely that the cedarwood oil tested was Himalayan cedarwood oil. A group

Table 5.93.1 Constituents identified in Himalayan cedarwood oils

Constituent	CAS	Percentage and range			
		A	B (7)	C (8)	D
8α-Acetoxyelemol	41370-57-4				+[a]
4-Acetyl-1-methylcyclohexene	6090-09-1				+[a]
Albicanol	54632-04-1		0.2		
Aromadendrene	489-39-4		0.6		+[a]
allo-Aromadendrene	25246-27-9		0.3		
Aromadendrene oxide					+[a]
(E)-α-Atlantone	26294-59-7	4.4-9.7	8.6	6.5	+[a]
(Z)-α-Atlantone	56192-70-2	0.8-2.6	1.4	2.4	+[a]; 19[b]
β-Atlantone	38331-79-2		0.4		+[a]
(E)-γ-Atlantone	108549-47-9	0.5-2.1	2.4		+[a]
(Z)-γ-Atlantone	108549-48-0	0.6-2.7	2.3		+[a]
(E)-α-Bisabolene	25532-79-0	0.6-1.4			
(Z)-α-Bisabolene	29837-07-8		2.2		+[a]
β-Bisabolol	15352-77-9				+[a]
2-Butyl-1-methyl-1,2,3,4-tetrahydro-naphthalen-1-ol					+[a]
γ-Cadinene	39029-41-9		0.7		
δ-Cadinene	483-76-1	0.3-1.9	0.4	0.7	
α-Calacorene	21391-99-1	0.06-0.7			
Calarene epoxide	68926-75-0				+[a]
β-Caryophyllene [(E)-, trans-]	87-44-5			0.3	
Caryophyllene oxide	1139-30-6		0.4		+[a]
α-Cedrene	469-61-4	0.5-1.0		15.8	
β-Cedrene	546-28-1			1.4	
Cedrenol	28231-03-0			2.4	
β-Cedren-9α-ol	13567-41-4				+[a]
8-Cedren-13-ol acetate			0.4		+[a]
α-Cedrol	77-53-2			1.4	
Centdarol	57308-24-4				+[c]
α-Copaene	3856-25-5	0-0.04			
Cubinene [c]			2.3		
Curcumene (ar-; α-)	644-30-4		0.9		
β-Curcumene	28976-67-2				+[a]
3-Cyclohexene-1-methanol	1679-51-2				+[a]
α-Dehydro-ar-himachalene	78204-62-3	0.3-1.2	0.7		+[a]
γ-Dehydro-ar-himachalene	51766-65-5	0.6-1.8	0.7		+[a]
4,5-Dehydroisolongifolene			0.2		+[a]
9,10-Dehydroisolongifolene	67530-11-4		0.3		
Deodarone	41943-81-1	0.7-2.2	0.3	5.4	+[a]
Deodarone isomer			0.3		
7β,3α-Dihydroxy-1α,2,6-cyclohima-chalane					+[a]
(Z)-β-Farnesene	28973-97-9		0.4		
(E,E)-Farnesol	106-28-5		0.2		
α-Gurjunene	489-40-7	0.04-0.3			
Himachalene					59[b]
α-Himachalene	3853-83-6	16.6-20.3	17.1	30.8	+[a]
β-Himachalene	1461-03-6	32.8-46.7	38.8	12.3	+[a]
γ-Himachalene	53111-25-4	9.8-11.6	12.6		+[a]
α-Himachalene epoxide	64825-84-9	0.1-0.5	0.3		
β-Himachalene oxide	31560-66-4	0.5-1.3	0.3		+[a]
Himachalol	1891-45-8	0.5-8.5	1.0	1.3	+[a]
Humulane-1,6-dien-3-ol	915392-38-0				+[a]
α-Humulene	6753-98-6		0.5		+[a]
14-Hydroxy-9-epi-(E)-caryophyllene	244226-09-3				+[a]
Isocentdarol	57308-23-3				+[c]
Limonene	138-86-3	0-0.02		0.2	
Longiborneol	465-24-7		0.2		+[a]
Longifolenaldehyde	66537-42-6				+[a]
Longifolene	475-20-7	0.04-0.6	0.6		+[a]
α-Longipinene	5989-08-2	0-0.08			
m-Methylacetophenone	585-74-0				+[a]

Table 5.93.1 Constituents identified in Himalayan cedarwood oils (*continued*)

Constituent	CAS	Percentage and range			
		A	B (7)	C (8)	D
p-Methylacetophenone	122-00-9		0.3		
4-Methyl-δ3-tetraacetophenone					+c
p-Methyl-3,4,5,6-tetrahydroaceto-phenone			0.6		
Nerolidol	7212-44-4				+a
α-Pinene	80-56-8	0.04-0.2		0.08	
α-Terpineol	98-55-5	0-0.08			
Thujopsene	470-40-6	0-0.05			
m-Tolyldimethylacetaldehyde					+a
Vestitenone			0.4		+a
β-Vetivenene	27840-40-0		0.2		+a

A eight Himalayan cedarwood essential oil samples from Nepal, analyzed between 2001 and 2013; lowest and highest concentrations given (E. Schmidt, unpublished data)
B one lab-hydrodistilled Himalayan cedarwood oil from India (ref. 7)
C one lab-hydrodistilled oil from sawdust of Indian *Cedrus deodara* (ref. 8)
D data from other studies (indicated with superscript letters); highest concentrations found in any study reviewed here given

a present in a pentane fraction, himachalene enriched fraction, atlantone enriched fraction and/or acetonitrile fraction of a steam-distilled Himalayan cedarwood oil; as the concentrations cannot be compared with the other data, the presence of these chemicals is indicated with +a (ref. 4); b one lab-hydrodistilled oil sample from India; atlantone was defined as 'cis-atlantone and α-atlantone' (ref. 6); c possibly cubenol
+ present in the oil investigated, but quantity not stated

of 86 patients reacting to the fragrance mix was tested with Himalayan cedarwood oil 2% in petrolatum and 3 (3.5%) reacted (2). In neither of these studies were relevance data provided.

LITERATURE
1 Rudzki E, Grzywa Z, Bruo WS. Sensitivity to 35 essential oils. Contact Dermatitis 1976;2:196-200
2 Rudzki E, Grzywa Z. Allergy to perfume mixture. Contact Dermatitis 1986;15:115-116
3 Chaudhary AK, Ahmad S, Mazumder A. *Cedrus deodara* (Roxb.) Loud.: A review on its ethnobotany, phytochemical and pharmacological profile. Pharmacogn J 2011;3:12-17
4 Chaudhary A, Sharma P, Nadda G, Tewary DK, Singh B. Chemical composition and larvicidal activities of the Himalayan cedar, *Cedrus deodara* essential oil and its fractions against the diamondback moth, *Plutella xylostella*. Journal of Insect Science 2011;11:157
5 Burfield T. Cedarwood oils. Part 1. www.cropwatch.org
6 Makhaik M, Naik SN, Tewary DK. Evaluation of antimosquito properties of essential oils. J of Scientific and Industrial Research 2005;64:129-133
7 Chaudhary A, Kaur P, Singh B, Pathania V. Chemical composition of hydrodistilled and solvent volatiles extracted from woodchips of Himalayan cedrus: *Cedrus deodara* (Roxb.) Loud. Nat Prod Commun 2009;4:1257-1260
8 Nigam MC, Ahmad A, Misra LN. Composition of the essential oil of *Cedrus deodara*. Indian Perfumer 1990;34:278-281

Chapter 6 CHEMICALS IN ESSENTIAL OILS: ALPHABETICAL LIST AND OILS IN WHICH THEY HAVE BEEN IDENTIFIED

This chapter presents an alphabetical list of all chemicals which have been identified in the 91 essential oils and two jasmine absolutes discussed in this book (Chapters 5.1-5.93); if available, CAS numbers are given. In addition, it is indicated in which oils the chemicals have been demonstrated (Table 6.1). In literature, many chemicals are described under different names. For each compound, we have chosen a 'preferred name' (or 'main entry') the other names are synonyms. Approximately 4,350 main entry chemicals are shown in Table 6.1. An ˢ following a compound name means that the chemical has one or more synonyms, which have appeared in the analytical literature reviewed. These synonyms, of which there are approximately 930, are all listed in Table 6.1 in their respective alphabetical sequence, but refer to the preferred name ('see...'). The synonym(s) of any *main entry* can be found in Chapter 7 'List of synonyms'.

An important feature of this chapter is that for each chemical it shows in which essential oils the compound has been identified. Some substances are present in one or a few oils only; others can be found more often, and some chemicals, e.g., limonene and β-caryophyllene,

are components of nearly all oils. The essential oils in which the chemicals have been identified are indicated in the right column with numbers. These numbers correspond to the individual oil files in Chapter 5. Thus, the number 20 signifies the oil in Chapter 5.20, 55 means Chapter 5.55 etc. A conversion list of (oil) numbers mentioned in Table 6.1 to oil names is shown at the end of this chapter in Table 6.2.

We are well aware that the list contains (quite a few) wrong names and that it likely also contains chemicals with 'preferred names' that are actually synonyms which we have not recognized as such. In some studies, names for chemicals have been used that we could not find in any chemical database or with the aid of any search engine, and some were probably or certainly wrong. We have chosen *not* to exclude these chemicals from presentation in the table. The same goes for compounds which are synthetic and therefore should not be present in essential oils. Obviously, their 'presence' may have been the result of a misidentification by the investigator. However, they can also have entered the oil as the result of adulteration, contamination (e.g., phthalates from plastic containers) or be impurities from the production method, and have been correctly identified. These chemicals are marked with an ᵃ, meaning that 'this chemical has not been found in nature up to now'. A list of such chemical compounds (we do not pretend it to be exhaustive) is presented in Table 6.3 at the end of this chapter.

Table 6.1 Alphabetical list of all chemicals and the essential oils in which they have been identified

Chemical	CAS number	Essential oils in which the chemical has been identified. Legend of numbers: see Table 6.2 at the end of this chapter
Abienol	1616-86-0	31,32
Abieta-8,13(15)-dien-18-al		32
Abietadiene	36312-33-1	15,16,31,32,44,70,84
Abieta-8,12-diene	see Levopimaradiene	
Abieta-8(14),13(15)-diene		32,70
8,13-Abietadien-18-ol		32
Abietal	6704-50-3	32,72,84
4-epi-Abietal	34223-60-4	44
Abietatriene	19407-28-4	15,16,31,32,44,56,59,70,72,84,85
Acarone		9
Acetaldehydeˢ	75-07-0	8,14,34,37,48,65,68,73,81,90
Acetaldehyde diethyl acetalˢ	105-57-7	14,73
m-Acetanisole	5451-83-2	3,41,75
o-Acetanisoleˢ	579-74-8	37,41
Acethydrazide	1068-57-1	81
Acetic acid	64-19-7	4,6,10,13,14,25,30,37,48,50,55,62,65,68,72,73,80,90
Aceteugenol	see Eugenyl acetate	
Acetic acid, phenylmethyl ester	see Benzyl acetate	
Acetoeugenol	see Eugenyl acetate	
Acetoneˢ	67-64-1	4,15,37,42,43,46-49,73,80,81,90
p-Acetonylanisole	see *p*-Anisyl acetone	
2-Acetonylcyclohexanone	6126-53-0	81
Acetophenone	98-86-2	13,14,21,22,26,31,55,64,67,68,70
2-Acetoxy-1,8-cineole	438619-71-7	45
(*E*)-1-Acetoxy-2,6-dimethyl-2,7-octa-dien-6-ol		43

Table 6.1 Alphabetical list of all chemicals and the essential oils in which they have been identified (*continued*)

Chemical	CAS number	Essential oils in which the chemical has been identified. Legend of numbers: see Table 6.2 at the end of this chapter
cis-1-Acetoxy-3,7-dimethyl-2,7-octa-dien-6-ol	112362-25-1	50
trans-1-Acetoxy-3,7-dimethyl-2,7-octadien-6-ol	33766-43-7	50
2-Acetoxydodecane		26
8α-Acetoxyelemol	41370-57-4	83,93
2-Acetoxyfuranodiene		58
Acetoxylinalool	1301265-88-2	25
8-Acetoxylinalool		49
1-epi-Acetoxy-2-(1-methylethenyl)-5-methylcyclohexane		72
1-Acetoxy-2-propanol	627-69-0	26
1-Acetoxy-2-propanone[s]	592-20-1	26
9β-Acetoxy-3,5α,8-trimethyltricyclo-[6.3.1.0(1,5)]dodec-3-ene		26
Acetoxyvaleranone		88
15-Acetoxyvaleranone		88
p-Acetylanisole[s]	100-06-1	5,45,68,72,81
Acetylcedrene	32388-55-9	12
Acetyl chavicol[s]	61499-22-7	5,90
Acetyldihydroalbene		77
Acetylenedicarbonic acid, DL-(-)-methyl-		73
2-*O*-Acetyl-8,12-epoxygermacra-1(10),4,7,11-tetraene		58
Acetyl eugenol	see Eugenyl acetate	
2-Acetylfuran	1192-62-7	27,91
1-Acetyl-3-isopropylcyclopent-5-ene		35
2-Acetyl-5-isopropylpyridine		80
N-Acetyl methylanthranilate	see Methyl *N*-acetylanthranilate	
4-Acetyl-1-methylcyclohexane		7,15
4-Acetyl-1-methylcyclohexene	6090-09-1	55,93
3-Acetyl-6-methylpyridine	see 5-Acetyl-2-methylpyridine	
5-Acetyl-2-methylpyridine	36357-38-7	65,80
4-Acetyloxycycloheptanone		83
17-(Acetyloxy)kauran-18-al[s]	1421058-94-7	27
2-Acetylpyridine	1122-62-9	80,89
2-Acetylpyrrole	1072-83-9	75
Acetylthymol	see Thymyl acetate	
Aciphyllene[s]	87745-31-1	67,73
10-epi-Acora-3,11-dien-15-al		89
Acoradiene	24048-44-0	13
α-Acoradiene	24048-44-0	4,10,12,16-18,49,57,70,72,77,89,91
β-Acoradiene	28477-64-7	4,10,12,16-18,49,50,77,89,91
10-epi-β-Acoradiene	847374-86-1	4,16,18,77
γ-Acoradiene	28400-12-6	6,10,16-18,77
Acora-2,4-diene (epimer A)	67304-11-4	89
Acora-2,4-diene (epimer B)	67304-11-4	89
Acora-3,5-diene	89955-08-8	89
Acora-3,7(14)-diene	see Acora-4,10-diene	
Acora-3,9-diene		89
Acora-3(10),14-diene	see β-Alaskene	
Acora-4,7-diene		89
Acora-4,9-diene	38229-83-3	12,89
Acora-4,10-diene[s]	255062-41-0	12,89
β-Acoradienol	149496-35-5	7
Acora-3,5-dien-11-ol	94480-77-0	91
10-epi-Acora-3,11-dien-15-ol		89
Acora-7(11),9-dien-2-one		10
Acorafuran		10
Acoragermacrone	50281-45-3	10
α-Acorenol	28296-85-7	16-18,31,36,77

Table 6.1 Alphabetical list of all chemicals and the essential oils in which they have been identified (*continued*)

Chemical	CAS number	Essential oils in which the chemical has been identified. Legend of numbers: see Table 6.2 at the end of this chapter
epi-α-Acorenol		77
β-Acorenol	28400-11-5	16-18,31,77,89
epi-β-Acorenol		77
Acorenone	5956-05-8	10,15,56,89
1,4-*trans*-1,7-*trans*-Acorenone		10
Acorenone B	21653-33-8	38
4-epi-Acorenone (B)	56363-00-9	10
10-epi-Acor-3-en-5-one	139758-04-6	89
Acorone	10121-28-5	10
epi-Acorone	185303-18-8	10
Adamantane	281-23-2	8,13,75
3-Adamantanecarboxylic acid, phenyl ester		70
Adamantane tricyclo[3.3.1.1]		45
Adamantine		72
3(1-Adamantyl)sydnone		25
African-1-en		85
α-Agarofuran	5956-12-7	37,89
Agarospirol[s]	1460-73-7	35,37,38
α-Alaskene	see γ-Acoradiene	
β-Alaskene[s]	28908-21-6	10,16-18,77
Albene	38451-64-8	77
Albicanol	54632-04-1	93
Alcanfor	see *d*-Camphor	
Alcohols, linear chained, C_{12}, C_{14}-C_{17}, C_{19}-C_{24}		64
Alismol	87827-55-2	3,88
Allohimachalol	see allo-Himachalol	
Allopregnane-7α,11α-diol-3,20-dione	1204662-29-2	63
Allopteoxylin methyl ether		72
4-Alloxyimino-2-carene		39
Allyl acetate	591-87-7	80
p-Allyl anisole	see Methyl chavicol	
Allyl benzyl ether	14593-43-2	81
4-Allyl-2,6-dimethoxyphenol	see Methoxyeugenol	
4-Allylguaiacol	see Eugenol	
Allylisothiocyanate	57-06-7	61
Allyl isovalerate	2835-39-4	55
3-Allyl-6-methoxyphenol	see *m*-Eugenol	
4-Allyloxyimino-2-carene		63
p-Allylphenol	see Chavicol	
4-Allylphenyl acetate	see Acetyl chavicol	
4-Allylsyringol	see Methoxyeugenol	
Ambrettolide	7779-50-2	19
2-Aminoacetophenone	551-93-9	89
3-Amino-1-phenylbutane	22374-89-6	42
N-Amino-1,2,3,4-tetrahydroquinoline		81
7-epi-Amiteol	147383-87-7	4
Amorpha-4,7(11)-diene	203057-25-4	89
Amorpha-4,11-diene	92692-39-2	77
Amorpha-4,7-dien-11-ol		91
Amorpha-4,9-dien-2-ol	394251-66-2	44
α-Amorphane		3
Amorphene		68
α-Amorphene	20085-19-2	4,5,7,8,12,13,19,21,22,24-26,36-38,44-46,49,51,52,56,57,62,66,72,75,79,83,85,87,89,90
γ-Amorphene	6980-46-7	13,23,34,41,72,85,89
δ-Amorphene	189165-79-5	4,25,32,37,45,49,62,65,70,72,83,85,88,89
α-Amorphenic acid	69793-64-2	30
Amorph-4-en-10-ol (epimer A)	300350-22-5	89
Amorph-4-en-10-ol (epimer B)	300350-22-5	89

Table 6.1 Alphabetical list of all chemicals and the essential oils in which they have been identified (*continued*)

Chemical	CAS number	Essential oils in which the chemical has been identified. Legend of numbers: see Table 6.2 at the end of this chapter
τ-Amorphol	53947-91-4	90
Amyl acetate[s]	628-63-7	38,43,90
Amyl alcohol	71-41-0	65,68,73
n-Amyl angelate	7785-63-9	20
(*E*)-Amylcinnamyl alcohol		77
2-Amylfuran	see 2-Pentylfuran	
Amyl hexanoate	540-07-8	66
Amyl isobutyrate	2445-72-9	25
Amyl isovalerate[s]	25415-62-7	10,68,79,91
n-Amyl methacrylate	2849-98-1	20
S-sec-Amyl 3-methyl-2-butenethioate		36
Amyl propionate	624-54-5	20
S-sec-Amyl thiotiglate		36
Amyl valerate	2173-56-0	79
α-Amyrin	638-95-9	42
β-Amyrin	559-70-6	43
Androstan-17-one,1,3-ethyl-3-hydroxy-,(5α)-	57344-99-7	63,78
Anethemol		7
p-Anethole[s]	104-46-1	5,19,29,30,38,56,60,70,72,79,90
(*E*)-Anethole	4180-23-8	1-4,7,8,10,12,13,19,21,23,25,26,29,32,35,36,38,41,42,45,52, 55-57,68,71-73,79-81,85,86,90
(*Z*)-Anethole	25679-28-1	3,4,29,81,88
Angelic acid	565-63-9	1
Angelicin	523-50-2	2
Angelyl acetate	41414-68-0	20
Angelyl angelate		20
Anhydrolinalool oxide	84616-87-5	35,37,60
(*E*)-Anhydrolinalool oxide	54750-70-8	57,69
(*Z*)-Anhydrolinalool oxide	54750-69-5	57,69
Anhydrooplopanone	108654-35-9	45
Anisaldehyde		57
m-Anisaldehyde[s]	591-31-1	81
p-Anisaldehyde[s]	123-11-5	3,4,7,13,14,35,68,81,90
p-Anisaldehyde dimethyl acetal	2186-92-7	3
Anisic acid	1335-08-6	3,81
p-Anisic acid[s]	100-09-4	3,4
p-Anisoin	119-52-8	81
Anisole	100-66-3	11,13,72
p-Anisyl acetate	1200-06-2	90
p-Anisyl acetone[s]	104-20-1	3,56,81
Anisyl alcohol	1331-81-3	3,81
Anisyl isobutyrate	66989-82-0	81
Anisyl methyl ketone	see *p*-Methoxyphenylacetone	
Anthracene	120-12-7	59,67
Anymol		7
Apiole	523-80-8	7,12,23,69,72,85
Apollanol		88
Apo vertenex		83
Arachidic acid	506-30-9	24
ar-Curcumene	see Curcumene	
Aristola-1,9-diene		88
Aristolene	6831-16-9	10,12,35,37,41,54,59,72,78,88
Aristolen-1α-ol		9
1,5-di-epi-Aristolochene		51
4,5-di-epi-Aristolochene	54868-40-5	10,67,91
5-epi-Aristolochene	115888-31-8	10
Aristolone	160568-09-2	10,12,67
Aromadendrene	489-39-4	2-6,8-10,12,13,15,19,21-23,25,26,31,32,34,35,37-39,41,43-45, 49,52,56,57,60,61,63-73,75,76,78-81,83,85,86,88-91
Aromadendrene VI		37
α-Aromadendrene	146389-60-8	67
β-Aromadendrene[s]	72747-25-2	58

Table 6.1 Alphabetical list of all chemicals and the essential oils in which they have been identified (*continued*)

Chemical	CAS number	Essential oils in which the chemical has been identified. Legend of numbers: see Table 6.2 at the end of this chapter
Aromadendrene oxide		12,67,85,88,93
Aromadendrene oxide 1		11
Aromadendrene oxide 2	85710-39-0	11
allo-Aromadendrene	25246-27-9	4,5,9-11,15,16,19,22,25-27,30,31,34-38,41,44,45,49,50,56-58, 61,63,65,67,68,70-72,74-76,79,81,83,85,86,88,89,93
allo-Aromadendrene epoxide	see Epoxy allo-aromadendrene	
iso-Aromadendrene epoxide	see Isoaromadendrene epoxide	
allo-Aromadendrene oxide	see Epoxy allo-aromadendrene	
allo-Aromadendrene oxide-(2)		63
cis-Arteannuic alcohol	147648-62-2	76
Artemiseole	60485-46-3	57,65
Artemisia alcohol	29887-38-5	12,19
Artemisia ketone	546-49-6	1,7,19,36
Artemisyl acetate	3465-88-1	19
Asaronaldehyde	4460-86-0	10
Asarone	see α-Asarone	
α-Asarone ((*E*)-; *trans*-)	2883-98-9	10,12,26-28,44,66
β-Asarone ((*Z*)-; *cis*-)	5273-86-9	10,34,62
γ-Asarone[s]	5353-15-1	10
Ascaridole	512-85-6	6,45,56,83
Ascaridole epoxide		83
Ascaridole glycol		83
cis-Ascaridole glycol	6790-83-6	83
trans-Ascaridole glycol	6790-83-6	44,56,83
β-Atlantol	420109-31-5	19,34,44,67,89
cis-Atlantone		93
(*E*)-α-Atlantone	26294-59-7	15,44,93
(*Z*)-α-Atlantone	56192-70-2	15,93
β-Atlantone	38331-79-2	15,93
γ-Atlantone	532-66-1	2,15
(*E*)-γ-Atlantone	108549-47-9	15,67,93
(*Z*)-γ-Atlantone	108549-48-0	15,93
Atractylone	6989-21-5	58
Atrimesol		52
Auranetin[s]	522-16-7	64,65
Aurantiumal[s]		39
Aurapten[s]	495-02-3	39,50,55,64,65
Auraptenol	108354-46-7	39,64
Azulene	275-51-4	4,8,13,19,37,38,59,67,72,89
5,6-Azulenedimethanol		59
Azulene, 1,4-dimethyl-7-(1-methyl- ethyl)-	see Guaiazulene	
Bakerol	157744-23-5	57
α-Barbatene	53060-59-6	16
β-Barbatene	39863-73-5	16,91
Bargamol	see Linalyl acetate	
BBT	see 5-(3-Buten-1-ynyl)-2, 2'-bithiophene	
BBTOAc	see 5-(4-Acetoxy-1-butynyl)-2, 2'-bithiophene	
Behenic acid	see Docosanoic acid	
Benzaldehyde	100-52-7	4,7-14,19-22,25-28,32,37,38,41-45,48,49,55-57,60,61,64,65, 67-70,72,74,75,79,80,85,90,91
Benzenaminium	17032-11-0	10
Benzene	71-43-2	35,45,73,75
Benzeneacetaldehyde	see Phenylacetaldehdye	
1,4-Benzenediamine, *N,N*-dimethyl-	99-98-9	85
1,2-Benzenedicarboxylic acid	see Phthalic acid	
Benzene, 1,2-dimethoxy-4-(1-propenyl)-	see Methyl isoeugenol	
Benzene, 1-(1,5-dimethyl-4-hexenyl)- 4-methyl-	see Curcumene	
Benzeneethanol	see Phenethyl alcohol	
Benzene, (3-ethoxy-1,5-hexadien-1-yl)-[s]	67323-95-9	22
Benzene, 1-ethyl-3,5-dimethyl-	934-74-7	63
Benzenemethanol	see Benzyl alcohol	

Table 6.1 Alphabetical list of all chemicals and the essential oils in which they have been identified (*continued*)

Chemical	CAS number	Essential oils in which the chemical has been identified. Legend of numbers: see Table 6.2 at the end of this chapter
Benzenemethanol, 2-(2-aminopropoxy)-3-methyl-	53566-98-6	80
Benzenemethanol, α-ethyl-α-2,5,7-octatrienyl-	74685-43-1	70
Benzenemethanol, 3-methoxy-α-phenyl-		58
Benzene, 1-methoxy-4-(1-propenyl)-	see *p*-Anethole	
Benzene, 1-methyl-2-(1-methylethenyl)-	see *o*-Methyl-α-methylstyrene	
Benzene, 1-methyl-4-(1-methylethenyl)-	see *p*-Cymenene	
Benzene, 1,3-bis(phenoxyphenoxy)-	34012-02-7	70
Benzenepropanal	see Hydrocinnamaldehyde	
Benzenepropanol	see Hydrocinnamyl alcohol	
Benzene propanol, acetate	see Hydrocinnamyl acetate	
1*H*-Benzocycloheptene		3
Benzodiazepine[a]	12794-10-4	88
5-Benzodiazepine[a]		8
Benzofuran	271-89-6	13
Benzofuran, 6-ethenyl-4,5,6,7-tetrahydro-3,6-dimethyl-5-isopropenyl-, *trans*-	see Curzerene	
Benzoic acid	65-85-0	4,8,13,14,21,22,26,27,37,42,43,72,80,90
Benzophenone	119-61-9	12,73
Benzothiazole	95-16-9	91
1,3-Benzothiazole	128366-28-9	89
1*H*-Benzotriazole	95-14-7	59
Benzyl acetate[s]	140-11-4	11,22,26,28,42,43,73,85,86,90
Benzyl alcohol[s]	100-51-6	4,8,11,13,14,19,21,22,26-28,42,43,45,60,67,68,73,75,79,85,90,91
Benzyl angelate	37526-87-7	20
Benzyl benzoate	120-51-4	4,8-11,13-15,21,22,25-27,29,32,37,42,43,48,49,63,64,67,70,72-74,88,90,91
Benzyl butanoate	see Benzyl butyrate	
Benzyl butyl ether	588-67-0	63
Benzyl butyrate[s]	103-37-7	42,43,70,90,91
Benzyl cinnamate	103-41-3	21,90
Benzyl crotonate	65416-24-2	42
Benzyl cyanide	see Phenylacetonitrile	
Benzyl docosanoate	85263-74-7	42
Benzyl eicosanoate	77509-04-7	42,43
Benzyl formate[s]	104-57-4	13,21,42,43
Benzyl hexanoate	6938-45-0	91
Benzylidenemalonaldehyde	82700-43-4	13,14
Benzyl isobutyrate[s]	103-28-6	20,34,91
Benzyl isovalerate[s]	103-38-8	1,43,73,91
Benzyl linoleate	47557-83-5	42,43
Benzyl linolenate	77509-02-5	42,43
Benzyl 2-methylbutanoate	see Benzyl 2-methylbutyrate	
Benzyl 3-methylbutanoate	see Benzyl isovalerate	
Benzyl 2-methylbutyrate	56423-40-6	73,91
Benzyl methyl ether	see Methyl benzyl ether	
Benzyl 4-methylpentanoate	77509-00-3	90
Benzyl 2-methylpropanoate	see Benzyl isobutyrate	
Benzyl nitrile	see Phenylacetonitrile	
Benzyl *n*-octanoate[s]	10276-85-4	27,63
Benzyl palmitate	41755-60-6	42,43
Benzyl phenylacetate	102-16-9	43
Benzyl propionate	122-63-4	42,73
Benzyl salicylate	118-58-1	9,11,26,42,43,58,70,90,91
Benzyl stearate	5531-65-7	42,43
Benzyl tiglate	37526-88-8	25,26,37,42,43,73
Benzyl valerate	10361-39-4	73
Berbonone		72
Bergamal[s]	106-72-9	12,19,23,24,34,38,43,53,57,85

Table 6.1 Alphabetical list of all chemicals and the essential oils in which they have been identified (*continued*)

Chemical	CAS number	Essential oils in which the chemical has been identified. Legend of numbers: see Table 6.2 at the end of this chapter
2,12-Bergamotadien-14-al		30
α-Bergamotal	148031-27-0	76
(Z)-*trans*-α-Bergamotal		77
Bergamotene		5,6,38,42,43,49,59,72,80
α-Bergamotene[s]	17699-05-7	4,6,10-13,38,40,45,48-52,55,57,62,67,68,79,80,85,86,90
cis-α-Bergamotene	18252-46-5	2-4,6-8,12,19,24,38,41,44,48-52,57,60,62,65-77,80,81
trans-Bergamotene		66,71
trans-α-Bergamotene	13474-59-4	3,4,6-8,10,12,13,16,19,21,23-27,30,31,35,37-39,41,42, 46-52,55-58,60,62-66,69-71,75-77,80,81,84,85,88,89
β-Bergamotene[s]	6895-56-3	21,45-47,49,67,73,80
trans-β-Bergamotene	15438-94-5	6,49,50,62,67,77,79
α-*trans*-β-Bergamotene		37
α-Bergamotenic acid	124439-27-6	77
Bergamotol		77
(E)-α-Bergamotol		10,37,56,75
(Z)-α-Bergamotol	88034-74-6	19,38,77
(Z)-α-*trans*- Bergamotol	88034-74-6	15,30,38,63,75,77
Bergamotol acetate		75
trans-α-Bergamotol acetate[s]		75
α-Bergamotyl acetate		38
α-*trans*-Bergamotyl acetate	see *trans*-α-Bergamotol acetate	
Bergapten[s]	484-20-8	1,2,6,39,50,55,64,65
Bergaptol[s]	486-60-2	6,39,50,64
Berkheyaradulene[s]	see α-Isocomene	
Beyerene	3564-54-3	63,84
(-)-15-Beyerene	2359-73-1	84
(+)-Beyeren-19-ol		84
Bicyclo[5.3.0]decane, 2-methylene-5- (1-methylvinyl)-8-methyl-		8
Bicyclo[4.3.1]dec-1(9)-ene		45
Bicyclo[4.4.0]dec-5-ene, 1,5-dimethyl- 3-hydroxy-8-(1-methylene-2-hydroxy)-		88
Bicyclo[10.8.0]eicosa-1,14,18-triene		88
Bicycloelemene	32531-56-9	1,4,25,32,36,37,56,60,61,68-70,79,83,88,90
Bicyclogermacrene	24703-35-3	1,2,4,6,8,10-12,16,19,22-25,29,31,32,34,36,37,39,41,43-45,49, 50,55-58,60-62,64,66-72,75,76,79,82-86,88,90
Bicyclo[2.2.1]heptane	279-23-2	57
Bicyclo[4.1.0]heptane	286-08-8	72
Bicyclo[3.1.1]heptane, 6,6-dimethyl- 2-methylene-	see β-Pinene	
Bicyclo[2.2.1]heptane-2,5-diol, 1,7,7- trimethyl-, (2-*endo*, 5-*exo*)-		52
Bicyclo[2.2.1]heptane-2,3-dione, 6-(acetyloxy)-1,5,5-trimethyl-, *endo*-	55044-71-8	80
Bicyclo[3.3.1]heptane, 6-methyl-2-me- thylidene-6-(4-methyl-3-penten-1-yl)-	see β-Bergamotene	
Bicycloheptanol		41
Bicyclo[2.2.1]heptan-2-ol	1632-68-4	57
Bicycloheptan-3-ol		49
Bicyclo[2.2.1]heptan-2-ol, 1,3,3- trimethyl-	see Fenchyl alcohol	
Bicyclo[2.2.1]heptan-2-one	see 2-Norbornanone	
Bicyclo[3.1.1]heptan-3-one		72,75
Bicyclo[2.2.1]heptan-2-one, 5,5-di- methyl-3-methylene-	499126-76-0	35
Bicyclo[2.2.1]heptan-3-one, 6,6-dime- thyl-2-methylene-	see Camphenilone	
Bicyclo[2.2.1]heptan-2-one, 5,5,6- trimethyl-	see 5,5,6-Trimethylbi-cyclo[2.2.1]heptan-2-one	
Bicycloheptene carboxaldehyde		39
Bicyclo[3.3.1]hept-2-ene-2,6-diene		81

Table 6.1 Alphabetical list of all chemicals and the essential oils in which they have been identified (*continued*)

Chemical	CAS number	Essential oils in which the chemical has been identified. Legend of numbers: see Table 6.2 at the end of this chapter
Bicyclo[3.1.1]hept-2-ene-2-carboxa		45
Bicyclo[3.1.1]hept-2-ene, 2,6-dimethyl-6-(4-methyl-3-pentenyl)-	see α-Bergamotene	
Bicyclo[3.1.1]hept-2-ene-2-ethanol, 6,6-dimethyl-	see Nopol	
Bicyclo[3.1.1]hept-2-ene-2-methanol, 6,6-dimethyl-	see Myrtenol	
Bicyclo[3.1.1]hept-2-ene,2,6,6-tri-methyl-	see α-Pinene	
Bicyclo[3.1.1]hept-2-en-4-ol, 2,6,6-trimethyl-, acetate	see Verbenyl acetate	
Bicyclo[3.1.1]hept-2-en-4-ol, 2,6,6-trimethyl-, hexanoate		79
Bicyclo[3.1.1]hept-3-en-2-ol, 4,6,6-trimethyl-, [1S-(1α,2β,5α)]-	see *cis*-Verbenol	
Bicyclo[3.1.0]hex-2-ene, 2 methyl-5-isopropyl-	see α-Thujene	
Bicyclo[3.2.1]oct-3-en-6-one, 4,7-di-methyl-, *exo-cis-*	116764-38-6	50
Bicyclopentan-2-one[s]	4884-24-6	39,52,64
Bicyclopentylone	see Bicyclopentan-2-one	
[Bicyclopentyl]-2-one	see Bicyclopentan-2-one	
epi-Bicyclophellandrene		4
Bicyclosesquiphellandrene	54324-03-7	11,31,44,49,63,83
epi-Bicyclosesquiphellandrene	54274-73-6	4,10,25,26,31,36,38,41,44,49,57,68,79,83,88
Bicyclo[7.2.0]undecan-3-ol <11,11-di-methyl-, 4,8-bis(methylene)>	79580-01-1	83
Bicyclo[7.2.0]undec-4-ene	6671-82-5	34
Bicyclovetivenol	76250-14-1	89
Biisocrotyl	see 2,5-Dimethyl-2,4-hexadiene	
Binesol	see Agarospirol	
β-Biotol	19902-26-2	16
β-Biotone	19902-29-5	16
1,1'-Biphenyl, 3,3'-dimethyl-	612-75-9	19
Bisabola-2,10-diene-6,13-diol		77
Bisabola-2,10-diene-7,13-diol		77
Bisabola-3(15),10-dien-7-ol	194553-32-7	89
1,3,5-Bisabolatrien-7-ol		89
Bisabolene	495-62-5	12,38,46,76,77,81
α-Bisabolene	17627-44-0	4,6-8,10,12,16,36,41,45,50,72,77,80,82,85,86
(*E*)-α-Bisabolene	25532-79-0	4,6,12,13,19,38,45,47,50,52,72,77,80,93
(*Z*)-α-Bisabolene	29837-07-8	3,4,6,7,10,12,13,19,30,45,48,50,64,72,75,77,85,86,93
β-Bisabolene	495-61-4	1-4,6-8,10-14,16,19-21,23,25,27,29,31,34,36-39,41,43-53, 55-57,59,62-72,75,77,78,80-82,84-86,88,89,91
(±)-β-Bisabolene[s]	4891-79-6	81
(*E*)-β-Bisabolene		38
(*Z*)-β-Bisabolene		40,70
γ-Bisabolene	495-62-5	4,16,36,66,72,90
(*E*)-γ-Bisabolene	53585-13-0	4,6,16,19,38,50,51,55,71,75,76,85,91
(*Z*)-γ-Bisabolene	13062-00-5	4,6,7,12,19,38,39,50,55,76,88,91
homo-γ-Bisabolene		70
(*Z*)-δ-Bisabolene		4
Bisabolene epoxide	121467-35-4	49
cis-α-Bisabolene epoxide	121467-35-4	26,67,90
trans-α-Bisabolene epoxide	111536-37-9	38,50
Bisabolenol A	120913-67-9	77
Bisabolenol B	120913-68-0	77
Bisabolenol C	120913-69-1	77
Bisabolenol D	120913-70-4	77
Bisabolenol E	120913-71-5	77
β-Bisabolenol	147126-90-7	52
(*Z*)-γ-Bisabolen-12-ol		77

Table 6.1 Alphabetical list of all chemicals and the essential oils in which they have been identified (*continued*)

Chemical	CAS number	Essential oils in which the chemical has been identified. Legend of numbers: see Table 6.2 at the end of this chapter
Bisabolol	515-69-5	12,85
α-Bisabolol	515-69-5	1,4,6-10,12-14,16-19,21,24,25,32,35,36,38,41,44,46-51,55,64,65,70,72,75-77,79,80,88-91
α-(-)-Bisabolol	see Levomenol	
epi-α-Bisabolol[s]	78148-59-1	1,4,6,12,16-18,19,44,45,48,50,72,75,77,88,91
6-epi-α-Bisabolol		89
(-)-(4S,8R)-8-epi-α-Bisabolol	see epi-α-Bisabolol	
β-Bisabolol	15352-77-9	4,6-8,11-14,19,23,29,38,45-47,49,50,65,75-77,80,88,93
epi-β-Bisabolol	235421-59-7	6,16,77
(E)-Bisabol-11-ol		12
α-Bisabolol acetate		75
Bisabolol oxide A	see α-Bisabolol oxide	
α-Bisabolol oxide[s]	22567-36-8	19,36,45,49,72,80
Bisabolol oxide B	55399-12-7	19,49
α-Bisabolol oxide B	26184-88-3	36,49,51
Bisabolol-4-ol		72
Bisabolone		2,48
1-Bisabolone	see Bisabol-1-one	
(6R,7R)-Bisabolone	72441-71-5	1
Bisabol-1-one	61432-71-1	65
α-Bisabolone oxide	22567-38-0	36
α-Bisabolone oxide A	58985-73-2	19
Bis(2-ethylhexyl) phthalate[a,s]	117-81-7	19,24
Bjakangelicin	see 5-Methoxy-8-(2,3-dihydroxy-3-methylbutoxy)psoralen	
Bois de rose oxide[s]	7392-19-0	37,60,69,73,80
2,5-Bornanediol		72
Bornanedione		75
Bornane-2,5-dione	4230-32-4	76
2-Bornanol, 2-methyl-	91278-70-5	13
2-Bornanone	see Camphor	
2-Bornene (α-)[s]	464-17-5	8,24,31,63,72,84
Born-5-en-2-ol		12
Borneol[s]	507-70-0	1,2,4-8,10,12-14,16,19-25,29-32,34-38,41,44-53,55-57,61,62-68,70-72,75,76,78-81,83-89,91
endo-Borneol	see Borneol	
l-Borneol	464-45-9	19
Bornyl acetate[s]	76-49-3	1,2,4,6-8,10,12,13,16,19,21-26,29,31,32,35,36,38,41,44-57,60,62-65,68,70-72,75,76,78-80,84-89,91
endo-Bornyl acetate	see Bornyl acetate	
exo-Bornyl acetate	see Isobornyl acetate	
6-Oxobornyl acetate		75
Bornylene	see 2-Bornene	
Bornyl formate	7492-41-3	49,56,72,80,86
Bornyl isobutyrate	24717-86-0	65,80
Bornyl isovalerate[s]	76-50-6	32,54,80,88
Bornyl 2-methylbutyrate	94200-10-9	54,80
Bornyl 3-methylbutyrate	see Bornyl isovalerate	
Bornyl propanoate[s]	see Bornyl propionate	
Bornyl propionate	78548-53-5	76,80,85
endo-Bornyl propionate	see Bornyl propionate	
Bornyl valerate	7549-41-9	72
Boronia butenal	3155-71-3	88
endo-1-Bourbonalol		23,37,44,68,70
α-Bourbonene	5208-58-2	37,49,58,63,68,79
di-epi-α-Bourbonene	317831-19-9	91
1,5-di-epi-α-Bourbonene		68
1,5-di-epi-Bourbonene		37
β-Bourbonene	5208-59-3	1-4,7,19,20,23-25,27,31,32,35-39,41,44,45,49,51,52,56-58,63,67,68,70-73,75,79,80,85,86,89-91
β-Bourbonene isomer		68

Table 6.1 Alphabetical list of all chemicals and the essential oils in which they have been identified (*continued*)

Chemical	CAS number	Essential oils in which the chemical has been identified. Legend of numbers: see Table 6.2 at the end of this chapter
nor-Bourbonone	see Norbourbonone	
1-Bromoadamantane	768-90-1	25
2-Bromocyclohexanol	2425-33-4	68
Bufa-20,22-dienolide, 14-hydroxy-3-oxo-, (5α-)-	39845-12-0	8
γ-Bulgarene	68000-46-4	38
Bulnesene	164108-17-2	4,35
α-Bulnesene[s]	3691-11-0	4,8-10,20,35,37,40,45,49,61,63,67,70,71,73,77, 82-84,88,89,91
β-Bulnesene		83
α-Bulnesene epoxide[s]	33784-90-6	67
Bulnesol[s]	22451-73-6	9,12,23,33,36,38,40,41,45,61,67,77,89,90
Bulnesol isomer		40
Bulnesoxide	see α-Bulnesene epoxide	
Butanal[s]	123-72-8	4,68,73,90
Butane	106-97-8	37
2,3-Butanediol	513-85-9	42,44,65
Butanoic acid	see Butyric acid	
Butanoic acid, 3,7-dimethyl-2,6-octa-dienyl ester	see Geranyl butyrate	
Butanoic acid, 3-hydroxy-2-methyl-ene-, 2-methylpropyl ester	80758-68-5	20
Butanoic acid, 2-methyl-, butyl ester	see Butyl 2-methylbutyrate	
Butanoic acid, 2-methyl-, ethyl ester	see Ethyl-2-methyl butyrate	
Butanoic acid, 2-methyl-, hexyl ester	see Hexyl 2-methylbutyrate	
Butanoic acid, 2 methyl-, 4-methoxy-2-(3-methyloxiranyl)phenyl ester	97180-28-4	3
Butanoic acid, 2 methyl-, 1,7,7-trime-thylbicyclo[2.2.1]hept-2-yl ester	see Bornyl isovalerate	
Butanol	35296-72-1	38,65,73,80,85
2-Butanol	78-92-2	68,73,79,80
t-Butanol	75-65-0	37
4-Butanolide	96-48-0	49,80
1-Butanol, 3-methyl, acetate	see Isoamyl acetate	
1-Butanol, 3 methyl, propanoate	see Isoamyl propionate	
Butanone (2-)	78-93-3	37,57,80,85
2-Butenal	4170-30-3	26
But-3-enal, 2-methyl-4-(2,6,6-trime-thyl-1-cyclohexenyl)-		67
3-Buten-1-ol, 3-methyl-, benzoate	see 3-Methyl-3-butenyl benzoate	
3-Buten-2-one	78-94-4	80
2-Buten-4-one		38
2-Butenyl angelate		20
(Z)-3-(1-Butenyl)-pyridine		37
5-(3-Buten-1-ynyl)-2,2′-bithienyl	see 5-(3-Buten-1-ynyl)-2, 2′-bithiophene	
1-Butyl acetate	see n-Butyl acetate	
n-Butyl acetate (1-)	123-86-4	6,48,49,56,72,75,80
2-Butyl acetate	105-46-4	80
n-Butyl angelate	7785-64-0	1,20
Butylbenzene	104-51-8	45,72
t-Butylbenzene	98-06-6	31
Butyl benzoate	136-60-7	26,42,43,49,90
p-tert-Butylbenzoic acid	98-73-7	9
S-sec-Butyl benzothioate		36
4-Butylbenzyl alcohol	60834-63-1	14
p-tert-Butylbenzyl alcohol	877-65-6	88
Butyl butanoate[s]	109-21-7	20,46,47,49,65
Butyl 2-butenoate	7299-91-4	20
Butyl butyrate	see Butyl butanoate	
Butyl caprylate	589-75-3	85,86
2,6-di-tert-Butyl-p-cresol s	128-37-0	65,69
6-Butyl-1,4-cycloheptadiene	22735-58-6	54

Table 6.1 Alphabetical list of all chemicals and the essential oils in which they have been identified (*continued*)

Chemical	CAS number	Essential oils in which the chemical has been identified. Legend of numbers: see Table 6.2 at the end of this chapter
Butyl decanoate	30673-36-0	89
2-*tert*-Butyl-1,4-dimethoxybenzene	21112-37-8	88
Butyl formate	592-84-7	37
2-Butylfuran	4466-24-4	37
Butyl hexanoate	626-82-4	65,91
3-*tert*-Butyl-2-hydroxy-5-vinylbenz-aldehyde		58
Butylidene dihydrophthalide	see Ligustilide	
Butylidenephthalide (3-)	551-08-6	30,54
(*E*)-Butylidenephthalide	76681-73-7	54
(*Z*)-Butylidenephthalide	72917-31-8	54
Butyl isobutanoate	97-87-0	46,47,49
Butyl isovalerate[s]	109-19-3	19,36,68
Butyl methacrylate	97-88-1	20
S-*sec*-Butyl 3-methylbutanethioate	2432-91-9	36
S-*sec*-Butyl 3-methyl-2-but-2-enethioate	34322-09-3	36
Butyl 2-methylbutyrate[s]	15706-73-7	13,19,68,79
Butyl 3-methylbutyrate	see Butyl isovalerate	
Butyl 2-methylcrotonate	see Butyl angelate	
Butyl methyl ether	628-28-4	42,48
S-*sec*-Butyl 2-methyl-2-hexenethioate		36
S-*sec*-Butyl 2-methyl-2-pentenethioate		36
2,4-di-*t*-Butyl-6-methylphenol	616-55-7	20
2,6-di-*t*-Butyl-4-methylphenol	see 2,6-di-*tert*-Butyl-*p*-cresol	
2-Butyl-1-methylpyrrolidone		26
2-Butyl-1-methyl-1,2,3,4-tetrahydro-naphthalen-1-ol		93
Butyl methyl sulfide	628-29-5	73
Butyl octadecanoate	123-95-5	7
3,5-di-*t*-Butylphenol	1138-52-9	59
Butyl phthalate	see Dibutyl phthalate	
Butyl propanoate	590-01-2	49
3-*n*-Butylphthalide	6066-49-5	12,54
2-Butylpyridine	5058-19-5	89
3-*sec*-Butylpyridine	25224-14-0	80
S-*sec*-Butyl thiotiglate		36
n-Butyl tiglate	7785-66-2	20,46,47,49,73
Butyraldehyde	see Butanal	
n-Butyric acid[s]	107-92-6	4,8,37,48,49,62,72,80
Byakangelicin	see 5-Methoxy-8-(2,3-dihydroxy-3-methylbutoxy)psoralen	
Byakangelicol	26091-79-2	6,50
Cabreuva oxide B	107602-53-9	91
Cadala-1,4,9-triene	71609-04-6	10
Cadalene	483-78-3	7,8,10,13,15,16,21,25,27,31,37,44,45,49,72,75,85,91
Cadina-1,4-diene	see Cubenene	
cis-Cadina-1,4-diene	see Cubenene	
trans-Cadina-1,4-diene	20085-13-6	23,32,34,37,44,45,70,90,91
Cadina-1(2),4-diene	see Cubenene	
trans-Cadina-1(2),4-diene	1395047-77-4	70,83
Cadina-1(6),4-diene	16729-00-3	71,72
7,10-Cadina-1(6),4-diene		83
7a*H*,10b*H*-Cadina-1(6),4-diene		37
10β-Cadina-1(6),4-diene		41
cis-Cadina-1(6),4-diene	1187195-00-1	4,23,32
trans-Cadina-1(6),4-diene	931410-54-7	12,19,23,37,38,44,63,70,75,83,90,91
Cadina-1(10),4-diene	see δ-Cadinene	
Cadina-1,6-diene		49
Cadina-3,5-diene	267665-20-3	4,11,83,90
Cadina-3,9-diene	see β-Cadinene	
Cadina-4(15),6-diene		8
Cadina-1(10),4-dien-8α-ol	151513-79-0	70

Table 6.1 Alphabetical list of all chemicals and the essential oils in which they have been identified (*continued*)

Chemical	CAS number	Essential oils in which the chemical has been identified. Legend of numbers: see Table 6.2 at the end of this chapter
Cadina-4,10(15)-dien-12-ol	see Khusol	
Cadina-5,10(15)-dien-4-ol		8
Cadina-4,10(15)-dien-3-one	39765-72-5	49
Cadina-4α,10β-diol		89
Cadinane-1,4-diol		86
Cadina-1(10),6,8-triene	1460-96-4	91
Cadinenal		89
Cadinene	29350-73-0	8,24,26,32,39,53,65,67,76,88
α-Cadinene	24406-05-1	1,2,4,5,7,10-12,16,19,23-26,31,32,34,35,38,41,44,45,52,56-58, 61,65-68,70,72,73,75,78,79,83,85,88-91
β-Cadinene[s]	523-47-7	4,5,8,10,11,13,23,25,26,35-37,41,43,44,49,57,58,61,65,75, 76,79,80,82,83,86,90
γ-Cadinene[s]	39029-41-9	1-5,7-13,16,19-29,31,32,34-39,41-52,55-58,60,61,63-68,70, 71-73,75,76,78-81,83-86,88-91,93
γ2-Cadinene	5957-56-2	70
cis-γ-Cadinene		85
trans-γ-Cadinene		41,49,72,83,84
δ-Cadinene[s]	483-76-1	1-16,19-39,42-46,48-52,54-58,60-76,78-91,93
d-Cadinene	880143-55-5	4
ε-Cadinene	1080-67-7	57,90
σ-Cadinene		52
τ-Cadinene	152287-05-3	37,90
χ-Cadinene	855779-65-6	12
ω-Cadinene	17627-21-3	3
Cadinene-5,8-diene		13
cis-Cadinene ether		12
trans-Cadinene ether		12
Cadinenol	17910-08-6	31,86
Cadin-4-en-1-ol	see Cubenol	
cis-Cadinen-4-en-7-ol	see cis-4-Cadinen-7-ol	
cis-4-Cadinen-7-ol[s]	217650-27-6	72
Cadin-4-en-10-ol	see α-Cadinol	
14-nor-Cadin-5-en 4-one isomer A		36
Cadinol	11070-72-7	4,13,38,48,70,85
trans-Cadinol		38,66
α-Cadinol[s]	481-34-5	1-14,16,21-27,29,31,32,34-38,41-45,47-53,55-58,60-62, 68,70,72
10-α-Cadinol		10
1,10-di-epi-Cadinol		44
epi-α-Cadinol	see τ-Cadinol	
10-epi-α-Cadinol	see τ-Cadinol	
β-Cadinol		
γ-Cadinol	50895-55-1	4,5,19,47,88,90
δ-Cadinol[s]	19435-97-3	2,4,5,8,10,11,19,23,25,27,31,32, 34,35,37,39,41,43-45,49, 50-52,60,61,63-65,68-70,72-74,76,80,83-85,88-90
τ-Cadinol[s]	5937-11-1	4-13,16,19,21-27,29,31,32,34,36-38,41,43-49,51,52,56-58, 61,63,64,68,70,72,74-76,78-81,83-86,88-90
Cajeputol	see 1,8-Cineole	
Cajolone		9
Calacorene	38599-17-6	15,25,26,61,66,72
α-Calacorene	21391-99-1	3,4,7-10,13-16,19,21,23,25,27,31,34-37,44,45,49,55,56,64,68, 70,72,75,78,79,81,83,88,89,91,93
β-Calacorene	50277-34-4	4,10,15,16,23,25,31,37,38,44,45,56,70,72,75,88,89
γ-Calacorene	24048-45-1	89
Calacorene hydrate		10
Calamendiol	30167-28-3	10
Calamene	1406-50-4	12,72,90
cis-Calamene	see cis-Calamenene	
trans-Calamene		4,8,10,61,79
Calamenene (l-; (1S)-cis-)	483-77-2	4,5,8-11,13,15,16,21,23,25-27,31,34,36-38,45,56,57,61,70,72, 75,79,83,84,86,90,91

Table 6.1 Alphabetical list of all chemicals and the essential oils in which they have been identified (*continued*)

Chemical	CAS number	Essential oils in which the chemical has been identified. Legend of numbers: see Table 6.2 at the end of this chapter
(-)-Calamenene		22
α-Calamenene		83
cis-Calamenene[s]	72937-55-4	4,5,8,10-13,15,21,23,31,34-38,41,44,49,51,56,61,63, 66,72,75,79,80,85,86,89
trans-Calamenene	73209-42-4	4,8,10,11,16,19,25,37,38,44,45,49,57,66,68,71,72,75,78,80, 85,88,90,
Calamenene B		85
Calamenene hydrate		41
Calamenenol	52658-10-3	27
Calamenen-1-ol		15
cis-Calamenen-10-ol		44
trans-Calamenen-10-ol	828923-23-5	19,31,34,44
Calamenoic acid		10
cis-Calemen-10-ol		37
Calamenone		10
Calamol	66219-01-0	10
Calamusenone	71305-96-9	10
Calarene ((+)-)	17334-55-3	4,10,12,19,23,25,35,38,71,72,75,85,89
Calarene epoxide	68926-75-0	56,93
Camphene	79-92-5	1-15,19-22,24-27,29-39,41,44-58,60-69,71-76,78-91
Camphene hydrate[s]	465-31-6	10,32,35,36,38,44,52,60,70,72,76,78,80,83-85,88,91
2-*exo*-Camphene hydrate		78
Camphenilone[s]	13211-15-9	24,45,49,70,80
Camphenol (6-)	3570-04-5	4,12,72
Camphenone		60
6-Camphenone	53803-33-1	12,31,36,44,60,72
E-Camphenone		4
4-Camphenylbutan-2-one		63,88
Campherenol[s]	18530-03-5	6,50,55,77
Campholenal[s]	23727-15-3	70,78,86
α-Campholenal[s]	4501-58-0	12,15,20,24,29,31,34-36,41,44,45,48,49,56,63,65,67,72,75, 76,78,80,84,85,88,91
6-Campholenal		10,12
Campholene aldehyde	see Campholenal	
α-Campholene aldehyde	see α-Campholenal	
α-Campholenic acid	28973-89-9	37,72,80
γ-Campholenic acid	67246-55-3	29,72,80
α-Campholenol	1901-38-8	72,91
γ-Campholenol		72
α-Campholytic acid	6709-22-4	72,80
Camphor (*dl*-)	76-22-2	1,3-10,13,14,16-23,25,26,29-33,35-38,41,44-53,55-57,61-66, 68,69,70-76,78-81,83-86,88,91
d-Camphor[s]	464-49-3	27,88
l-Camphor	464-48-2	26
m-Camphorene	20016-73-3	44
p-Camphorene	532-87-6	44
Canferenol	see Campherenol	
Capraldehyde	see Decanal	
Capric acid	see Decanoic acid	
Capric aldehyde	see Decanal	
Caprinaldehyde	see Decanal	
Caproaldehyde	see Hexanal	
Capronic acid	see Hexanoic acid	
Caproic acid	see Hexanoic acid	
Caprylic acid	see Octanoic acid	
Caprylyl acetate	see Octyl acetate	
trans-Carane	554-59-6	37
5-Caranol		56
(+)-*trans,trans*-5-Caranol	6909-22-4	56
1-Carboxaldehyde-3-cyclohexene	see 3-Cyclohexene-1-carboxaldehyde	

Table 6.1 Alphabetical list of all chemicals and the essential oils in which they have been identified (*continued*)

Chemical	CAS number	Essential oils in which the chemical has been identified. Legend of numbers: see Table 6.2 at the end of this chapter
3-Carboxy-4,4-dimethyl-cyclobutane-1-acetic acid		80
2-Caren-10-al	124752-20-1	7,36
3-Caren-10-al	14595-13-2	36
Car-3-ene	see δ3-Carene	
Carene	74806-04-5	8,45,48
2-Carene	see δ2-Carene	
(+)-2-Carene	4497-92-1	31,50,72
Car-2-ene	see δ2-Carene	
3-Carene	see δ3-Carene	
3(10)-Carene	see β-Carene	
β-Carene[s]	554-60-9	70
γ-Carene		34
γ-3-Carene		44,72
δ2-Carene (= δ4-)[s]	554-61-0	2,4,6,8-10,12,24,29,31,37,39,44,45,49,53,55,56,62-65,70,72,78,81,84,86,88,90,91
δ3-Carene[s]	13466-78-9	1-10,12-14,16,19,21,22,24,25,29,31,32,34-39,41,44-50,52,53,55-57,60-66,68-76,78-82,84-88,90,91
4-Carene	29050-33-7	35,44,68,72
(+)-4-Carene	13837-63-3	50,53,65,70,85
(Z)-4-Carene		63
trans-2-Caren-2-ol		5
trans-Caren-2-ol		35
3-Caren-2-ol	93905-79-4	36
3(10)-Caren-2-ol	93905-77-2	12,72
(E)-3-Caren-2-ol	139563-36-3	12
trans-3(10)-Caren-2-ol	6909-15-5	38,51,52
2-Caren-4-ol	6617-35-2	36
trans-3(10)-Caren-4-ol	22626-38-6	8
3(10)-Caren-4-ol acetoacetic acid ester		88
2-Caren-3-one		8
3-Caren-2-one		8,32,49
Car-3-en-2-one	see 3-Caren-2-one	
cis-Caren-3-one		31
Carhydranol	see Dihydrocarveol	
Carotol	465-28-1	12,15,36,49,85
Carvacrol[s]	499-75-2	4-8,10,12,13,16,17,19,21,25,26,29-32,34,35,37,38,41,44,45,49,50,52,54-57,60,63-65,68,69,72,75,76,79-81,83-86,91
Carvacryl acetate	6380-28-5	7,57,85,86,88
Carvacryl methyl ether	see Methylcarvacrol	
Carvacryl methyl oxide	see Methylcarvacrol	
L-Carvenol		7
Carvenone[s]	499-74-1	7,12,34,56,79,83
Carveol[s]	99-48-9	5,6,8,12,29,31,35,36,38,39,41,44,50,52,56,57,64,65,68,72,73,75,79-81,84,85
γ-Carveol		70
(E)-Carveol[s]	1197-07-5	2,4,6-8,12,15,19,29,32,33,35,39,41,44,45,48-57,60,61,63-65,68,70,72,75,78-80,82,84-86,88,91
(Z)-Carveol[s]	1197-06-4	6-8,12,32,33,35,38,39,41,44,45,49,50,52,53,56,57,60,63-65,68,69,72,75,79,80,82,85,88,91
Carvomenthene[s]	5502-88-5	4,34,63,85
4-Carvomenthenol	see Terpinen-4-ol	
3-Carvomenthenone	see Piperitone	
Carvomenthol	499-69-4	41,79
Carvomenthone	499-70-7	12
Carvomenthyl acetate	5256-66-6	73
Carvone	99-49-0	2-4,6-8,10,12-15,19,25,26,29,31-33,35-37,39,41,44,45,47-51,53,56,60,63-65,68,70,72,75,76,78-82,84-86
D-Carvone (+)[s]	2244-16-8	39,52,53,55,68
(E)-Carvone		39

Table 6.1 Alphabetical list of all chemicals and the essential oils in which they have been identified (*continued*)

Chemical	CAS number	Essential oils in which the chemical has been identified. Legend of numbers: see Table 6.2 at the end of this chapter
L-Carvone	6485-40-1	8,57
(Z)-Carvone		39
Carvone camphor		65
Carvone hydrate[s]	7712-46-1	8,53,83
Carvone oxide	33204-74-9	8,36,65,68,79,83
cis-Carvone oxide		12,75
trans-Carvone oxide	18383-49-8	12,38,79,85
Carvotanacetone	499-71-8	3,7,8,13,30,35,36,39,41,68,72,75,76,80,86
trans-Carvotan alcohol		80
Carvyl acetate	97-42-7	2,6,39,45,56,57,70,75,79,88
cis-Carvyl acetate	1205-42-1	2,7,35,39,45,55,57,64,68,75,79,88
trans-Carvyl acetate	1134-95-8	2,39,44,45,65,75,79
(E)-Carvyl formate	29239-07-4	68
(Z)-Carvyl formate		68
Caryolan-8-ol	178737-45-6	16,72
Caryophylla-2(12),5-dien-13-al		49
Caryophylladienol II[s]		26,45,56,61,72
Caryophylla-2(12),6-dien-5α-ol	see Caryophyllenol I	
Caryophylla-2(12),6-dien-5β-ol	see Caryophyllenol II	
Caryophylla-2(12),6(13)-dien-5-ol	see Caryophylladienol II	
Caryophylla-2(12),6(13)-dien-5α-ol	see Caryophylladienol II	
Caryophylla-3(15),7(14)-dien-6-ol	257293-89-3	45
Caryophylla-3,8-dien-5-ol		70
Caryophylla-3,8(13)-dien-5α-ol		72
Caryophylla-3,8(13)-dien-5β-ol		72
Caryophylla-4,8-dien-5-ol	423765-30-4	57,75
Caryophylla-4(12),8(13)-dien-5-ol		32
Caryophylla-4(12),8(13)-dien-5α-ol		85
Caryophylla-4(12),8(13)-dien-5β-ol	19431-80-2	26,37,45,72,75
Caryophylla-4(14),8(15)-dien-5-ol	644981-74-8	12,45
Caryophylla-4(14),8(15)-dien-5α-ol		57
Caryophylla-4(14),8(15)-dien-5β-ol		85
Caryophylla-2(12),5-dien-7-one		49
Caryophylla-2(12),6-dien-5-one		49
Caryophylla-2(12),6(13)-dien-5-one		49
Caryophylla-2(12),6-dien-7-one		49
(2R,5E)-Caryophyll-5-en-12-al		25
(2S,5E)-Caryophyll-5-en-12-al		25
Caryophyllene	see β-Caryophyllene	
α-Caryophyllene	see α-Humulene	
β-Caryophyllene[s]	87-44-5	1-58,60-76,78-93
(E)-Caryophyllene	see β-Caryophyllene	
(Z)-Caryophyllene	see γ-Caryophyllene	
(Z)-β-Caryophyllene		22,85
cis-Caryophyllene	see γ-Caryophyllene	
trans-Caryophyllene	see β-Caryophyllene	
2-epi-(E)-β-Caryophyllene	see 9-epi-(E)-Caryophyllene	
γ-Caryophyllene[s]	118-65-0	2,7-9,11-13,21,25,26,31,37,38,41,43,49,52,56-58,60,61,63, 67,72,75,79-82,85,88-90
9-epi-Caryophyllene		58
9-epi-(E)-Caryophyllene[s]	68832-35-9	7,12,37,40,41,60,67,71,72,75,78,80,83,88
Caryophyllene acetate	32214-91-8	7
Caryophyllene alcohol	56747-96-7	6,27,28,50,58
α-Caryophyllene alcohol	4586-22-5	15,88
Caryophyllene epoxide	see trans-Caryophyllene oxide	
Caryophyllene epoxide II	see Caryophyllene oxide II	
Caryophyllene oxide	see trans-Caryophyllene oxide	
trans-Caryophyllene oxide[s]	1139-30-6	1,2,4-16,19-32,34-39,41,44-53,55-58,60,61,63-80,83-91,93
Caryophyllene oxide II[s]		57
Caryophyllenol[s]	38284-26-3	4,8,57,72,75,85
Caryophyllenol I[s]	32214-88-3	6,45,72
Caryophyllenol II[s]	32214-89-4	6,45,56,57,72

Table 6.1 Alphabetical list of all chemicals and the essential oils in which they have been identified (*continued*)

Chemical	CAS number	Essential oils in which the chemical has been identified. Legend of numbers: see Table 6.2 at the end of this chapter
Caryophyllen-5-ol II		8
Caryophyllenyl alcohol	913176-41-7	7,8,12,13,21,26,44,75
Cascarilladiene	59742-39-1	89
Cascarilladienol[s]		89
3-Casen-2-ol		39
Cedrane	see α-Cedrane	
α-Cedrane	13567-54-9	57,61,83
Cedranediol	88588-48-1	67,72
8S,13-Cedraniol		69
Cedran-8β-ol		26
Cedranone	13567-40-3	15
5-Cedranone		57
8,14-Cedranoxide	18319-31-8	75
α-Cedrenal	28387-62-4	16
Cedren-15-al	69993-59-5	89
1,7-di-epi-α-Cedrenal		16,89
Cedrene	11028-42-5	4,8,13,14,18,38
α-Cedrene[s]	469-61-4	2,4,8,10,15-18,21,22,25,31,32,36-39,41,44,49,57,63,65,75, 77,80,89,90,91,93
di-epi-α-Cedrene		77,88
1,7-di-epi-α-Cedrene	see α-Funebrene	
epi-α-Cedrene	35944-22-0	77
α-Cedrene epoxide	29597-36-2	12,19,38
di-epi-α-Cedrene epoxide		13
β-Cedrene	546-28-1	2,4,10,12,13,16-18,31,34,36-38,41,44,57,62,75,77,78,81,83, 88-90,93
1,7-di-epi-β-Cedrene	see β-Funebrene	
γ-Cedrene		89
Cedr-8-ene	see α-Cedrene	
Cedr-9-ene		27
β-Cedrene epoxide		89
Cedrenol	28231-03-0	8,18,30,31,36,79,88,89,93
α-Cedrenol		89,
epi-α-Cedrenol		38
(+)-8(15)-Cedren-9-ol	see β-Cedren-9α-ol	
Cedr-8(15)-en-9-α-ol	see β-Cedren-9α-ol	
Cedr-8(15)-en-10-ol	138117-22-3	38
8-Cedren-13-ol[s]	18319-35-2	11,17,30,38,88,89
Cedr-8-en -13-ol	see 8-Cedren-13-ol	
8-Cedren-13-ol acetate		93
β-Cedren-9α-ol[s]	13567-41-4	16,38,93
β-Cedren-9-one		12
2-epi-α-Cedren-3-one	288249-25-2	16
Cedrol	see α-Cedrol	
α-Cedrol[s]	77-53-2	6,8,10-12,15-19,22,23,25,31,36-39,41,44,45,56,63-65,67,68, 83,84,90,93
allo-Cedrol	50657-30-2	16-18
epi-Cedrol[s]	19903-73-2	15-18,36,61,83
12-epi-Cedrol		19,25
Cedroxyde	13786-79-3	15
Cedryl acetate	77-54-3	16,19
Cembrane		73
Cembra-1,3,7,11-tetraene		63
Cembra-3,7,11,15-tetraene		63
Cembrene	1898-13-1	16,63,72,78,88
Cembrene A	31570-39-5	25,63
Cembrene B	67737-66-0	63
Cembrene C	64363-64-0	63
Cembrenol	67921-02-2	63
Centdarol	57308-24-4	93
Cephrol	see Citronellol	

Table 6.1 Alphabetical list of all chemicals and the essential oils in which they have been identified (*continued*)

Chemical	CAS number	Essential oils in which the chemical has been identified. Legend of numbers: see Table 6.2 at the end of this chapter
Cetene	see 1-Hexadecene	
n-Cetyl alcohol	see 1-Hexadecanol	
Cetylic acid	see Hexadecanoic acid	
Chamazulene	529-05-5	19,32,70
Chamecynone[s]	10208-54-5	88
Chamigrene	see β-Chamigrene	
α-Chamigrene	19912-83-5	8,12,16-18,31,41,49,51
β-Chamigrene	18431-82-8	5,16-18,23,31,36,40,48,51,67,70,88,89
β-Chamigrene, isomer		16
α-Chamipinene	847374-85-0	16,18
Chavibetol	see *m*-Eugenol	
Chavicol[s]	501-92-8	4,5,12-14,22,26-28,42,62,72,81,90
Chiloscyphone	23538-45-6	67
2-Chlorocyclohexanol	1561-86-0	13
1-Chloroeicosane	42217-02-7	8
bis(2-Chloroisopropyl)ether	39638-32-9	85
1-Chlorooctane	111-85-3	55
Cholesta-3,5-diene	747-90-0	70
Cholest-22-ene-21-ol, 3,5-dehydro-6-methoxy-, pivalate		8
Chroman-2-one	see Dihydrocoumarin	
Chromene	254-04-6	68
trans-Chrysanthemal	20104-05-6	4,52,57
(*E*)-Chrysanthemic acid	4638-92-0	37
(*Z*)-Chrysanthemic acid		37
(*E*)-Chrysanthemol	5617-92-5	19
cis-Chrysanthenol	55722-60-6	12,19,32,49,53,57,72
trans-Chrysanthenol	38043-83-3	4
Chrysanthenone	473-06-3	5,30,35,63,72
Chrysanthenyl acetate		4,7
cis-Chrysanthenyl acetate	67999-48-8	2,7,12,31,41,45
trans-Chrysanthenyl acetate	50764-55-1	2,12,29,44
1,4-Cineole[s]	470-67-7	4,6,9,15,34-36,44,45,49,50,52,65,70,72,73,80,81,83,85-87
1,8-Cineole[s]	470-82-6	3-16,19-39,41,44-50,52,53,55-57,60-76,78-81,83-92
1,8-Cineole-2-yl-acetate		45
Cinnamal	see Cinnamaldehyde	
Cinnamaldehyde[s]	104-55-2	5,22,26,39,45,65,73
(*E*)-Cinnamaldehyde	14371-10-9	7,12-14,21,22,30,73
(*Z*)-Cinnamaldehyde	57194-69-1	13,14,21,22,90
Cinnamaldehyde propyleneglycol acetal	4353-01-9	21
Cinnamic acid	621-82-9	8,13,22,80
(*E*)-Cinnamic acid	140-10-3	13,14,21,43
(*Z*)-Cinnamic acid	102-94-3	13
Cinnamic aldehyde	see Cinnamal(dehyde)	
Cinnamyl acetate[s]	103-54-8	4,13,14,21,22,43,45,81,90
(*E*)-Cinnamyl acetate	21040-45-9	4,13,14,21,22,45,81,84,90
(*Z*)-Cinnamyl acetate		21,22
Cinnamyl alcohol	104-54-1	14,21,22,42,43,45,75,81,90
(*E*)-Cinnamyl alcohol	4407-36-7	13,14,21,22,43,90
(*Z*)-Cinnamyl alcohol	4510-34-3	43,45,90
Cinnamyl alcohol, acetate	see Cinnamyl acetate	
(*E*)-Cinnamyl benzoate	50555-04-9	42,43
(*Z*)-Cinnamyl benzoate		43
Cinnamyl cinnamate	122-69-0	21,38
Cinnamyl formate	104-65-4	73
Citosterol	see Sitosterol	
Citral	5392-40-5	4,9,24,25,29,34,35,39,45,51,53,57,65,85,86
Citral A	see Geranial	
Citral B	see Neral	
α-Citral	see Geranial	
β-Citral	see Neral	
(*E*)-Citral	see Geranial	

Table 6.1 Alphabetical list of all chemicals and the essential oils in which they have been identified (*continued*)

Chemical	CAS number	Essential oils in which the chemical has been identified. Legend of numbers: see Table 6.2 at the end of this chapter
(Z)-Citral	see Neral	
Citrol[s]	624-15-7	34,35,49,51,57,65,73
Citronellal[s]	106-23-0	4-6,8,21-24,29,30,34-39,44,45,48-53,55,57,60,61,64-66,69-73, 75,78,80,82,83,85,91
α-Citronellal[s]	141-26-4	34,73
β-Citronellal	see Citronellal	
(R)-(+)-Citronellal (D-; d-)	2385-77-5	50
Citronellal diethyl acetal		73
β-Citronellene	2436-90-0	22,26,34,45
Citronellic acid[s]	502-47-6	24,34,37,44,53,57,61,72,73,91
Citronellol (β-, DL-)[s]	106-22-9	4,6,8,9,23-26,29,31,32,34-39,43-45,48-53,55-57,60-73,77, 78,80,82,85,86,88,90,91
α-Citronellol		34,49,50,73
(R)-Citronellol (d-)	1117-61-9	53
(S)-Citronellol (l-)	7540-51-4	53
Citronellol epoxide	see Citronellol oxide	
Citronellol oxide	1564-98-3	53
Citronellyl acetate[s]	150-84-5	4-6,23,24,29,32,34,37-39,44,45,49-52,55-57,60-66,68,69,71, 73,78,82,85,86,88,90,91
Citronellyl benzoate	10482-77-6	73
Citronellyl butyrate	141-16-2	7,23,37,38,44,51,57,61,73,83,91
Citronellyl caprylate	see Citronellyl octanoate	
Citronellyl decanoate	72934-06-6	73
Citronellyl diethylamine		37
Citronellyl formate[s]	105-85-1	6,23,24,34,51,53,56,57,66,73,91
Citronellyl heptanoate		37,73
Citronellyl hexanoate	10580-25-3	37,73,91
Citronellyl isobutyrate	97-89-2	32,37,44
Citronellyl isoheptanoate		37
Citronellyl isohexanoate[s]	71662-18-5	37
Citronellyl isooctanoate		37
Citronellyl isovalerate	68922-10-1	32,37,88
Citronellyl 4-methylvalerate	see Citronellyl isohexanoate	
Citronellyl nerate	72934-19-1	73
Citronellyl nonanoate		37,73
Citronellyl octanoate[s]	72934-05-5	37,73
Citronellyl phenylacetate	139-70-8	73
Citronellyl propanoate	see Citronellyl propionate	
Citronellyl propionate[s]	141-14-0	23,24,30,37,50,51,55,73,75
Citronellyl tiglate	24717-85-9	37
(E)-Citronellyl tiglate	24717-85-9	37
(Z)-Citronellyl tiglate	84254-89-7	37
Citronellyl valerate	7540-53-6	32,37
Citropten	see 5,7-Dimethoxycoumarin	
Citrusal	43145-56-8	39
Clovene	469-92-1	8,36,76,89
α-Clovene		27
Clovenol		72
Cnidicin	14348-21-1	50
Cnidilin	see 5-Isopentenyloxy-8-methoxypsoralen	
Coahuilensol methyl ether		32
Coniferyl alcohol	458-35-5	42,43
Coniferyl aldehyde	458-36-6	42
Coniferyl benzoate	4159-29-9	42
Copaborneol	21966-93-8	70,90
α-Copaenal		44
Copaene		12,39
α-Copaene	3856-25-5	1-5,7-16,19-23,25-28,30-33,35-39,41,44,45,48-58,61-65, 67-76,78-83,85-90,93

Table 6.1 Alphabetical list of all chemicals and the essential oils in which they have been identified (*continued*)

Chemical	CAS number	Essential oils in which the chemical has been identified. Legend of numbers: see Table 6.2 at the end of this chapter
β-Copaene	18252-44-3	1,2,4,8,10,20,23,25,31,32,37-39,41,44,45,53,55,57,58,64,65, 68-70,72,75,79,81,82,85,89-91
15-Copaenol	115728-41-1	12
α-Copaen-8-ol (*cis-*)	58569-25-8	2,25,38,45,85
α-Copaen-11-ol	41370-56-3	2,31,88
β-Copaen-4α-ol	124753-76-0	2,25,26,67,72,88
cis-Copaen-4α-ol		57
α-Corocalene	20129-39-9	44,72
(*E,E*)-Cosmene	460-01-5	52
α-Costal	4586-01-0	30
β-Costal		30
γ-Costal		30
α-Costol	65018-15-7	30,37,67,73,89
β-Costol	515-20-8	19,30,51
(+)-β-Costol		25
γ-Costol		30
Costunolide[s]	553-21-9	30
Coumarin[s]	91-64-5	4,7,13,14,21,22,46-49,57,72,80
Coumarin, 7-[(6,7-dihydroxy-3,7-di-methyl-2-octenyl)oxy]-	see Marmin	
Coumarin, 7-[(6,7-epoxy-3,7-dimethyl-2-octenyl)oxy]-, (*E*)-(+)-	see Epoxyaurapten	
Coumarin, 8-(2,3-epoxy-3-methylbu-tyl)-7-methoxy-, (-)-	see Meranzin	
Coumarin, 8-(2-formyl-2-methylpro-[s] pyl)-7-methoxy-	5980-07-4	39
Coumarin, 7-methoxy-8-(3-methyl-2-oxobutyl)-	see Isomeranzin	
Coumarin, 7-methoxy-8-[(2,2,5,5-tetramethyl-1,3-dioxolan-4-yl)methyl]-	see Pranferin	
Cresol	1319-77-3	55,75
m-Cresol	108-39-4	8,72,80,89,90
o-Cresol	95-48-7	8,13,72,73,80,89
p-Cresol	106-44-5	8,26,42,43,72,75,80,89,90
m-Cresyl acetate	122-46-3	26,90
o-Cresyl acetate	533-18-6	26
p-Cresyl acetate	140-39-6	42,90
p-Cresyl methyl ether[s]	104-93-8	11,37,90,91
Croweacin	484-34-4	41
Cryptoacorone	5989-62-8	10
Cryptofauronol	2212-90-0	88
Cryptofauronyl acetate		88
Cryptone[s]	500-02-7	1,2,8,12,30,32,35,37,38,41,45-49,55,56,65,72,78, 80,81,83
Cubeban-11-ol	220766-71-2	61,83
Cubebene	11012-64-9	8,20,38,63,71,81
α-Cubebene	17699-14-8	1,2,4,5,7,8,10-13,19,22-28,31,32,34-39,41,44,45,55, 57,58,61-73,75,76,79-83,85,87-91
β-Cubebene[s]	13744-15-5	1,2,4,7,8,10-13,22-25,31,32,34-39,41,44,45,48,49,51,52, 54,55,57,58,62,63-68,70,72,75,76,79,80,82,83,85,86,88-90
γ-Cubebene	147413-90-9	61
Cubebol	23445-02-5	8,11,19,25,31,32,36,37,44,45,49,51,55,63,65,70-72,78, 82,83,89
epi-Cubebol (4-)	38230-60-3	23,25,37,44,45,65,70
10-epi-Cubebol	176589-53-0	1,4,44,45,89
Cubenene[s]	29837-12-5	2,4,8,9,11,12,21,23,24,26,31,34,36-38,43-45,49,50,57, 58,61,63,66,70,72,75-79,83,90
α-Cubenene	205537-26-4	8
β-Cubenene		55,79
Cubenol[s]	21284-22-0	4,8,11-13,15-19,23-27,31,34-38,41,45-47,49,51,52,57,61, 68,70,72,79,83,85,88,89

Table 6.1 Alphabetical list of all chemicals and the essential oils in which they have been identified (*continued*)

Chemical	CAS number	Essential oils in which the chemical has been identified. Legend of numbers: see Table 6.2 at the end of this chapter
1,10-di-epi-Cubenol[s]	73365-77-2	4,7,20,23,32,37,44,49,51,70,75,79,85,89,90
epi-Cubenol[s]	19912-67-5	12,31,32,34,49,61,65,71,83
1-epi-Cubenol	81939-29-9	4,11,15,16,21,23,24,32,36,37,45,48,49,51,52,56,70,74,83,89,90
10-epi-Cubenol		89
Cubinene		93
Cucumber alcohol		52
Cumarin	see Coumarin	
Cumene[s]	98-82-8	13,38,41,49,62
8-Cumenol		25
o-Cumenol	88-69-7	7
p-Cumenol	99-89-8	72
p-Cumic aldehyde	see Cuminaldehyde	
Cuminal	see Cuminaldehyde	
Cuminaldehyde[s]	122-03-2	1,4,7,8,12-14,21,22,29,32,34-37,39,41,45-49,55-57,60, 65,67,70,72,73,76,79-81,84-86,88
Cuminol	see Cuminyl alcohol	
Cuminyl acetate[s]	59230-57-8	7,45
Cuminyl alcohol[s]	536-60-7	1,2,6-8,12,13,32,35,41,44,45,49,55,56,70,72,75,76,80,83-86
Cumyl methyl ether	see Methyl cumyl ether	
Cuparenal	16982-01-7	17,18
Cuparene	16982-00-6	2,8,10,13,15-18,25,31,36,44,51,52,66,68,70,85,89,90
α-Cuparene		2,3
δ-Cuparene		44
α-Cuparenene		52
α-Cuprenene	29621-78-1	16,25
β-Cuprenene	119683-81-7	16
γ-Cuprenene	4895-23-2	16,18
δ-Cuprenene	66389-22-8	16,18
4-Cuprenen-1-ol		16
Curcumene (*ar-*; *α-*)[s]	644-30-4	1,2,3,6-8,10,12-21,29,30,37,38,41,45,49,50,52,57,59,61,65, 66, 70,72,76,77,80,85,88,89,91,93
β-Curcumene	28976-67-2	4,8,10,66,77,88-91,93
γ-Curcumene	28976-68-3	1,6,10,12,15-18,30,49,50,52,67,72,77,88,91
cis-β-Curcumen-12-ol	698365-10-5	77,89
γ-Curcumen-12-ol		77
cis-γ-Curcumen-12-ol		77
Curcumenyl acetate	19431-85-7	38
Curzerene[s]	17910-09-7	58
Curzerenone	20493-56-5	58
Cyclamen aldehyde	103-95-7	62
Cycloartanyl acetate	4575-74-0	63
Cyclobazzanene	88661-61-4	4
Cyclobuta[1,2:3,4]dicyclopentene, decahydro-3a-methyl-6-methylene-1-(1-methylethyl)-, [1S-(1α,3aα,3bβ, 6aβ,6αβ,6bα)]-		63
Cyclobutane, (1,3-butadienyl)-	80344-48-5	45
Cyclobutanecarboxylic acid, hexyl ester		49
Cyclocitral	52844-21-0	5
α-Cyclocitral	432-24-6	5,52,65,91
β-Cyclocitral	432-25-7	25,31,49,57,65,85
Cyclocolorenone	489-45-2	44
Cyclocopacamphan-12-al		89
Cyclocopacamphanol A	30810-34-5	89
Cyclocopacamphanol B	28052-00-8	89
Cyclocopacamphan-12-ol		89
Cyclocopacamphene	24112-86-5	90
Cyclocopacamphenol		89
1,6-Cyclodecadiene	1124-79-4	4,8
3,7-Cyclodecadiene-1-methanol, α,α,4,8-tetramethyl-,	see Hedycaryol	

Table 6.1 Alphabetical list of all chemicals and the essential oils in which they have been identified (*continued*)

Chemical	CAS number	Essential oils in which the chemical has been identified. Legend of numbers: see Table 6.2 at the end of this chapter
1,6-Cyclodecadiene, 1-methyl-5-methylene-8-(1-methylethyl)-	37839-63-7	24,29,48,49
Cyclodecane	293-96-9	20
Cyclodecene	3618-12-0	45
Cycloenane		34
Cyclofenchene[s]	488-97-1	31,44,69,84
Cyclofenchone	see Cyclofenchene	
α-Cyclogeraniol	6627-74-3	37,52,69
α-Cyclogeraniol acetate[s]	68406-89-3	29,52
Cyclogeranyl acetate	see α-Cyclogeraniol acetate	
1,7-Cyclogermacra-1(10),4-dien-15-al		89
1,7-Cyclogermacra-1(10),4-dien-15-ol		89
3,5-Cycloheptadien-1-one[s]	1121-65-9	83
3,5-Cycloheptadine-1-one	see 3,5-Cycloheptadien-1-one	
Cycloheptane, 4-methylene-1-methyl-2-(2-methyl-1-propen-1-yl)-1-vinyl-[s]	826337-63-7	26,63,88
Cycloheptanone	502-42-1	73
Cycloheptatriene	544-25-2	73
1,3-Cyclohexadiene	592-57-4	81
1,4-Cyclohexadiene	628-41-1	35,63
1,4-Cyclohexadiene-1-methanol	32937-33-0	45
1,6-Cyclodecadiene, 1-methyl-5-methylene-8-(1-methylethyl)-, [s-(E,E)]-	see Germacrene D	
1,4-Cyclohexadiene, 1-methyl-4-(1-methylethyl)-	see γ-Terpinene	
1,3-Cyclohexadiene, 1,3,3,5-tetramethyl-	4724-89-4	26
Cyclohexane	110-82-7	4,72
Cyclohexanecarboxaldehyde	2043-61-0	9
3-Cyclohexane-1-carboxylic acid, 3, 7-dimethylethyl ester		72
Cyclohexane, 1,3-diiodopropenyl-6-methyl-		49,72
1,2-Cyclohexanediol	931-17-9	50
Cyclohexane, 2-ethenyl-1,1-dimethyl-3-methylene-[s]	95452-08-7	38
Cyclohexane, 1-ethenyl-1-methyl-2, 4-bis(1-methylethenyl)-,	110823-68-2	24
Cyclohexane, 1-ethenyl-1-methyl-2, 4-bis(1-methylethenyl)-,[1S-(1α,2α,4α)]-		24
Cyclohexane, 1-ethenyl-1-methyl-2, 4-bis(1-methylethenyl)-,[1S-(1α,2β,4β)]-	see β-elemene	
Cyclohexanemethanol, 4-ethenyl-α,α,4-trimethyl-3-(1-methylethenyl)-, [1R-(1α,3α,4α)]-		24,58
Cyclohexane, 1-methylene-4-(1-methylethenyl)-	see Pseudolimonene	
Cyclohexane, 1-methyl-4-(1-methylethenyl)-, cis-	6252-33-1	34
Cyclohexane, 1-methyl-4-(1-methylethenyl)-, trans-	6252-33-1	34
Cyclohexanepropanol, 2,2-dimethyl-6-methylene-		72
Cyclohexane, trisubstituted		67
Cyclohexanol	108-93-0	4,10,34,56,81
Cyclohexanol, 5-methyl-2-(2-hydroxy-2-propyl)-	see p-Menthane-3,8-diol	
Cyclohexanol, 1-methyl-4-(1-methylethenyl)-	see β-Terpineol	
Cyclohexanone	108-94-1	10,66

Table 6.1 Alphabetical list of all chemicals and the essential oils in which they have been identified (*continued*)

Chemical	CAS number	Essential oils in which the chemical has been identified. Legend of numbers: see Table 6.2 at the end of this chapter
Cyclohexanone, 2-methyl-5-(1-methylethenyl)-	see Dihydrocarvone	
Cyclohexanone, 2,3,3-trimethyl-2-(3-methyl-1,3-butadienyl)-, (*E*)-	69296-91-9	67
Cyclohexene	110-83-8	4,8
3-Cyclohexene-1- carboxaldehyde[s]	100-50-5	9
3-Cyclohexene-1-carboxaldehyde, 1,3,4-trimethyl-	40702-26-9	52
Cyclohexene, 4-ethenyl-4-methyl-3-(1-methylethenyl)-1-(1-methylethyl)-, (3*R-trans*)-	see δ-Elemene	
3-Cyclohexene-1-methanol	1679-51-2	45,63,65,93
3-Cyclohexene-1-methanol, α,α, 4-trimethyl -	see α-Terpineol	
Cyclohexene, 3-methyl-6-(1-isopropyl)-	5256-65-5	34
Cyclohexene, 1-methyl-4-(5-methyl-1-methylene-4-hexenyl)-	see (±)-β-Bisabolene	
1-Cyclohexene-1-propanol, 2,6,6-trimethyl-		72
Cyclohexene, trisubstituted		67
2-Cyclohexen-1-ol	822-67-3	45,49
3-Cyclohexen-1-ol	822-66-2	56
2-Cyclohexen-1-ol, 2-methyl-5-(1-methylethenyl), *cis*-	see (*Z*)-Carveol	
3-Cyclohexen-1-ol, 1-methyl-4-(1-methylethyl)-	see 1-Terpineol	
3-Cyclohexen-1-ol, 4-methyl-3-(1-methylethyl)-	see 4-Methyl-3-(methylethyl)-3-cyclohexen-1-ol	
2-Cyclohexen-1-one	930-68-7	81
2-Cyclohexen-1-one, 2-methyl-5-(1-methylethane)-		45,49
3-Cyclohexenyl carbinol		60
Cyclohexyl benzoate	2412-73-9	43
Cycloisolongifolene	28380-07-6	31,52,89
Cycloisosativene[s]	406485-43-6	38,52,70,88,89
(+)-Cycloisosativene	see Cycloisosativene	
Cyclolongifolene	164108-26-3	35
Cyclolongifolene oxide, dehydro-		67
1,3-Cyclooctadiene	1700-10-3	30
1*H*-Cyclopenta[1,3]cyclopropa-[1,2]-benzene,octahydro-7-methyl-3-methylene-4-(-1-methylethyl), [3a*S*(3aα, 3bβ,4β,7α,7a*S*)]-	see β-Cubebene	
Cyclopentadecane	295-48-7	37
Cyclopentadecanolide[s]	106-02-5	1,2,44
Cyclopentane	287-92-3	45
Cyclopentane, 1 acetoxymethyl-3-isopropenyl-2-methyl-		63
Cyclopentanone, 3,3-dimethyl-2-(3-methyl-1,3-butadienyl)-, (*E*)-	88725-86-4	67
1-Cyclopentenone	28982-58-3	89
2-(1-Cyclopentenyl)furan		81
1-Cyclopentyl-1-hexadecanone	55255-86-2	43
1*H*-Cyclopro[e]azulene		59
1*H*-Cycloprop[e]azulene, decahydro-1,1,7-trimethyl-4-methylene-, [1a*R*-(1aα,4aα,7α,7aα,7bα]-		58
1*H*-Cyclopropanol azulen-4-ol		88
4-Cyclopropyl-2-methoxyphenol	83356-69-8	42

Table 6.1 Alphabetical list of all chemicals and the essential oils in which they have been identified (*continued*)

Chemical	CAS number	Essential oils in which the chemical has been identified. Legend of numbers: see Table 6.2 at the end of this chapter
Cyclosantalal	168099-27-2	77
epi-Cyclosantalal	168252-33-3	77
Cyclosantalic acid		77
epi-Cyclosantalic acid		77
Cyclosativene[s]	22469-52-9	3,7,8,12,13,16,26,34,36,38,41,44,45,51,71,74,88,89,91
(+)-Cyclosativene	see Cyclosativene	
Cycloseychellene	52617-34-2	67
Cyclotetracosane	297-03-0	73
Cyclotridecanolide	see Tridecanolide	
Cyercene 1	136669-16-4	25
Cyercene 4	136669-19-7	30
Cymbodiacetal		66
Cymene[s]	25155-15-1	81
β-Cymene	see *m*-Cymene	
m-Cymene[s]	535-77-3	2,6,8,26,27,35-37,49,53,63,66,68,69,78,79
o-Cymene[s]	527-84-4	2,7-10,12,15,21-23,26,31,35,36,44,45,47,49,52,53, 56-58,60,62-64,66,68,69,72,73,75,78,79,83,85,86
p-Cymene[s]	99-87-6	1-10,12-16,19-25,27-39,41,44-57,60-92
allo-Cymene	see (*E*)-allo-Ocimene	
p-Cymene 8-methyl ether		8
Cymenene	26444-18-8	86
m-Cymenene[s]	1124-20-5	8,35,72,86
p-Cymenene[s]	1195-32-0	2,4-13,20,22,27,31-35,38,39,44,45,49, 50,55-57,60-64,69,70,72,73,75-81,83,85,86,88,91
Cymenol		· 80
p-Cymenol	25497-27-2	29,56
p-Cymen-3-ol	see Thymol	
p-Cymen-4-ol		4
p-Cymen-7-ol	see Cuminyl alcohol	
Cymen-8-ol		72
m-Cymen-8-ol	5208-37-7	2,8,32,44,49,70,72
p-Cymen-8-ol	1197-01-9	1,2,4-10,12,13,20,29,31-37,41,44-53,55-57,60-63,65-70,72,75, 76,79-81,83-86,91
p-Cymen-9-ol	4371-50-0	7,12,41
p-Cymen-7-ol acetate	see Cuminyl acetate	
Cymol	see Cymene	
p-Cymol	see *p*-Cymene	
Cyperene	2387-78-2	11,30,31,35,41,44,65,70,81,88-90
α-Cyperene[s]	17627-30-4	16,30,37,88,89,91
β-Cyperene	see δ-Selinene	
α-Cyperone	473-08-5	11,12,64,65,89
β-Cyperone	see Eudesma-4,6-dien-3-one	
Cypertundone		41
α-Damascenone		73
β-Damascenone (*E*)-	23726-93-4	4,19,20,25,41,57,62,68,73,91
(*Z*)-β-Damascenone	59739-63-8	57,64,73
Damascone (β-)	23726-91-2	4,73,75
Dauca-5,8-diene	142928-08-3	12,23,49
(+)-Dauca-8,11-diene	see Isodaucene	
Daucene	16661-00-0	12,16,48,49,83
trans-Dauc-8-en-4β-ol	255062-40-9	12
Daucol	887-08-1	12,49
2,4-Decadienal	2363-88-4	4,55,64,65,73
(*E,E*)-2,4-Decadienal	25152-84-5	7,19,31,39,48,55,57,64,65,70,82
(*E,Z*)-2,4-Decadienal	25152-83-4	39,55,64,65
(*Z,Z*)-3,6-Decadienol		43
Decahydronaphthalene[s]	91-17-8	57,75
γ-Decalactone[s]	706-14-9	12,49,80
n-Decaldehyde	see Decanal	
Decalin	see Decahydronaphthalene	
Decamethylene glycol[s]	112-47-0	63

Table 6.1 Alphabetical list of all chemicals and the essential oils in which they have been identified (*continued*)

Chemical	CAS number	Essential oils in which the chemical has been identified. Legend of numbers: see Table 6.2 at the end of this chapter
Decanal[s]	112-31-2	4-7,10,13,19,21,23,24,29,37-39,44,48,50-52,55,57,60,64,65, 69,73,78,82,85,86,89,91
(*E*)-2-Decanal		29
n-Decane	124-18-5	4,7,15,19,34,41-43,50,52,55,57,60,73,75,89,90
1,10-Decanediol	see Decamethylene glycol	
Decane, 3-ethyl-3-methyl	see 3-Ethyl-3-methyldecane	
n-Decanoic acid[s]	334-48-5	4,7,10,13,14,19,20,29,37,55,64,65,70,72,73,80,82
Decanol	36729-58-5	4,6,8,10,39,51,55,64,65,73,80,82
2-Decanol[s]	1120-06-5	66
n-Decanol (1-)[s]	112-30-1	9,19,29,63,70,72,86
Decan-2-ol	see 2-Decanol	
3-Decanol	1565-81-7	29
Decanol acetate	see Decyl acetate	
Decanone	693-54-9	32,70,91
3-Decanone	928-80-3	72
Decenal	25447-70-5	23,29
2-Decenal	3913-71-1	29,32
(*E*)-2-Decenal	3913-81-3	6,29,38,39,41,50,55,64,65,70,83
(*Z*)-2-Decenal	2497-25-8	39,65
(*E*)-4-Decenal	65405-70-1	10
(*Z*)-4-Decenal	21662-09-9	5,10,51
Decene	25339-53-1	23
1-Decene	872-05-9	19
2-Decenoic acid	3913-85-7	37
2-Decen-1-ol	22104-80-9	73,88
(*E*)-2-Decenol	18409-18-2	29
(*Z*)-3-Decen-1-ol	10340-22-4	43
cis-7-Decen-5-olide	see δ-Jasmolactone	
trans-7-Decen-5-olide		42
3-Decen-5-one	32064-73-6	20
(*E*)-3-Decenyl acetate	83446-51-9	45
Decyl acetate[s]	112-17-4	6,10,39,50,55,57,63-65,83
Decyl alcohol	see *n*-Decanol	
5-Decyne-4,7-diol, 4,7-dimethyl-[s]	126-87-4	81
Dehydroabietal	13601-88-2	15,32,44
Dehydroabietane	see Abietatriene	
Dehydroabietol		32
Dehydroaromadendrane	see β-Aromadendrene	
Dehydroaromadendrene	698388-95-3	10,15,35,41,45,88,89
Dehydro-β-atlantone		15
Dehydrocadinenal		89
Dehydrocarveol	28982-60-7	65
Dehydro-1,8-cineole	92760-25-3	7,41,51,53,56,57,72
2,3-Dehydro-1,8-cineole	92760-25-3	4,5,45,49,54,69
Dehydrocostunolide	see Dehydrocostus lactone	
Dehydrocostus lactone[s]	477-43-0	30,45
8,9-Dehydrocycloisolongifolene		88,89
Dehydro-*p*-cymene	see *p*-Cymenene	
Dehydrohimachalene		15
α-Dehydro-*ar*-himachalene	78204-62-3	15,93
γ-Dehydro-*ar*-himachalene	51766-65-5	7,15,17,93
Dehydro-β-ionone		30
Dehydroisolongifolene		88
4,5-Dehydroisolongifolene		89,93
8,9-Dehydroisolongifolene	26839-55-4	15
9,10-Dehydroisolongifolene	67530-11-4	93
Dehydrolinalool	29171-20-8	12,29,49,65,83
cis-Dehydrolinalool oxide		64
trans-Dehydrolinalool oxide		64
Dehydrolinalyl acetate[a]		49
8,9-Dehydroneoisolongifolene		15
Dehydronerolidol	2387-68-0	19

Table 6.1 Alphabetical list of all chemicals and the essential oils in which they have been identified (*continued*)

Chemical	CAS number	Essential oils in which the chemical has been identified. Legend of numbers: see Table 6.2 at the end of this chapter
Dehydronigritene		89
2,3-Dehydro-4-oxo-α-ionol		88
Dehydrosabinaketone	147043-52-5	7,12,19,38,41,75
Dehydrosaussurea lactone	28290-35-9	30
Dehydrosesquicineole	211237-38-6	19,83
Dehydro-β-vetivone		89
Dehydroxyisocalamenediol		10
cis-Dehydroxylinalool oxide[s]	73413-94-2	25,49,60
trans-Dehydroxylinalool oxide		25
5-Demethylnobiletin	see 5-Hydroxy-6,7,8,3′,4′-pentamethoxyflavone	
5-Demethyltangeretin[s]	see 5-Hydroxy-4′,6,7,8-tetramethoxyflavone	
Dendralasine	see Dendrolasin	
Dendrolasin[s]	23262-34-2	19,36,52,65,77
(E)-Dendrolasin	23262-34-2	19,77
Deodarone	41943-81-1	15,91
Deodarone isomer		91
Deoxygeraniol	2609-23-6	53
Deoxysericealactone	19892-19-4	88
Desmethyl-5-citromitine	see 5-Hydroxy-3′,4′,6,7, 8-penta-methoxyflavanone	
Deutenyl curcumene		25
Diacetone alcohol	see 4-Hydroxy-4-methyl-2-pentanone	
3,6-Diacetoxy-2,6-dimethyl-1,7-octa-diene		25
Diacetyl	431-03-8	48
1,2-Diacetylethane	110-13-4	83
2,4-Diacetylphloroglucinol	2161-86-6	26
Diazene, acetylphenyl	13443-97-5	72
Dibenzothiophene	132-65-0	73
Dibutyl octanedioate	16090-77-0	4
Dibutyl phthalate[a,s]	84-74-2	19,45,52,67,73
Dichloroacetic acid, 1-adamantyl-methyl ester		70
Dichloromethane	75-09-2	52
1,2-Dicyclopropylcyclobutane	61141-62-6	10
3,4-Didehydro-7,8-dihydro-γ-ionone		68
(2R,5R,6R)-2,12:5,6-Diepoxycaryo-phyllane	60444-80-6	25
1,10-Diepicubenol	see 1,10-di-epi-Cubenol	
Diepoxide allocimene	see allo-Ocimene diepoxide	
5,6-Diethenyl-1-methylcyclohexene		3
1,1-Diethoxyethane	see Acetaldehyde diethyl acetal	
2,5-Diethylfuran	10504-06-0	91
Di(2-ethylhexyl) phthalate	see Bis(2-Ethylhexyl) phthalate	
Diethyl (2E)-3-methyl-2-pentanedioate		90
Diethyl 1,5-pentanedioate	818-38-2	90
Diethyl phthalate[a]	84-66-2	22,25,37,49,85
4,4-Diethyl-2,5-octadiyne	61227-87-0	10
2,6-Diethylpyridine	935-28-4	59
Diethyl succinate	123-25-1	55
2,5-Diethyltetrahydrofuran	41239-48-9	57,68,79
cis-3,5-Diethyl-1,2,4-trithiolane	38348-25-3	59
trans-3,5-Diethyl-1,2,4-trithiolane	38348-26-4	59
Di-α-furylmethane	1197-40-6	3
Dihydroactinidiolide	17092-92-1	41,70
Dihydroalbene		77
Dihydroagarofuran		37
Dihydro-α-agarofuran	20053-66-1	25
Dihydro-β-agarofuran (trans-)	5956-09-2	25
cis-Dihydroagarofuran	150652-94-1	25
Dihydroanethole[s]	104-45-0	4,36,90

Table 6.1 Alphabetical list of all chemicals and the essential oils in which they have been identified (*continued*)

Chemical	CAS number	Essential oils in which the chemical has been identified. Legend of numbers: see Table 6.2 at the end of this chapter
Dihydroaromadendrene		67
10,11-Dihydroatlantone		15
Dihydrocalamenenol		15
Dihydrocampholenic acid		37
Dihydrocaranone	112529-25-6	67
Dihydrocarveol[s]	38049-26-2	7,8,22,25,29,39,45,49,57,60,65,72,85,86
cis-Dihydrocarveol		79
trans-Dihydrocarveol		79
Dihydrocarveol acetate	see Dihydrocarvyl acetate	
Dihydrocarvone[s]	5948-04-9	3,36,68,72,76,79,85
cis-Dihydrocarvone	3792-53-8	3,23,25,29,31,50,55,56,64,65,68,72,75,79,85
trans-Dihydrocarvone	5948-04-9	7,29,31,44,55,56,64,65,79,85,86
Dihydrocarvyl acetate ((*R*-))[s]	57287-13-5	12,34,49,68,72,79,88
iso-Dihydrocarvyl acetate	see Isodihydrocarvyl acetate	
trans-Dihydrocarvyl acetate	20777-49-5	9,45,70,79
2,3-Dihydro-1,8-cineole		50
Dihydrocinnamaldehyde	see Hydrocinnamaldehyde	
Dihydrocinnamic acid	see Hydrocinnamic acid	
Dihydrocitronellol	see Dimethyl octanol	
Dihydrocitronellyl acetate	20780-49-8	57
Dihydro-*cis*-α-copaen-8-ol		38
Dihydrocoumarin[s]	119-84-6	49,70,80
Dihydro-8-cumenol		25
Dihydrocurcumene	1461-02-5	36
Dihydro-*ar*-norcurcumenic acid		77
Dihydrodehydrocostus lactone	4955-03-7	30
(*E*)-β-10,11-Dihydro-10,11-epoxy-farnesene	255062-42-1	12
2,3-Dihydro-2,7-dimethyl-4*H*-1-benzo-pyran-4-one		54
2,5-Dihydro-2,5-dimethylfuran		19
Dihydroedulan		68
Dihydroedulan I	74006-61-4	68,79
Dihydroedulan II	41678-32-4	4,79,91
2,3-Dihydro epoxygeranyl acetate		6
2,3-Dihydro epoxyneryl acetate		6
Dihydroeudesmol	6770-16-7	72
Dihydroeugenol	2785-87-7	72,79,80
Dihydroeugenol acetate	33943-26-9	11
Dihydrofarnesal	32480-08-3	34
Dihydrofarnesol	1335-48-4	43
2,3-Dihydrofarnesol	51411-24-6	7,60
1,7-Dihydrofuropelargone		37
7,8-Dihydrofuropelargone		37
Dihydrogermazol		91
Dihydro-α-ionone	31499-72-6	30
Dihydroisolongifolene		88
Dihydrolinalool	18479-51-1	6,21,90
cis-Dihydrolinalool oxide		80
trans-Dihydrolinalool oxide		80
Dihydrolinalyl acetate[a]	61476-73-1	19,49,56,75,79,83
cis-Dihydromayurone	7129-16-0	16
Dihydromyrcenol	53219-21-9	49,59
Dihydromyrcenyl acetate[a]	88969-41-9	53
Dihydronerolidol	20685-70-5	19
cis-Dihydro-nor-zizaene		89
trans-Dihydro-nor-zizaene		89
Dihydropinocarvone		72
1,2-Dihydropyridine, 1-(1-oxobutyl)-	849947-72-4	8
Dihydrosabinene	471-12-5	44
Dihydro-α-santalic acid	25342-86-3	76
Dihydro-α-santalol	126209-93-6	77

Table 6.1 Alphabetical list of all chemicals and the essential oils in which they have been identified (*continued*)

Chemical	CAS number	Essential oils in which the chemical has been identified. Legend of numbers: see Table 6.2 at the end of this chapter
Dihydro-β-santalol	34289-89-9	77
Dihydrotagetone	1879-00-1	7
Dihydro-α-terpineol	498-81-7	45
trans-Dihydro-α-terpineol	5114-00-1	45,72
Dihydro-α-terpinyl acetate (*cis*-)	80-25-1	45,56,73
5,6-Dihydro-2,4,6-triethyl-(4*H*)-1,3,5-dithiazine		59
1,2-Dihydrotrimethylnaphthalene	133439-48-2	57
Dihydroumbellulone	2506-61-8	56
Dihydroverbenone	18358-52-6	41
Dihydroxybergamottin	145414-76-2	39
6′,7′,-Dihydroxybergamottin	145414-76-2	39
6′,7′,-Dihydroxybergamottin decanal acetal	1181223-80-2	39
6′,7′,-Dihydroxybergamottin octanal acetal	1181223-79-9	39
7β,3α-Dihydroxy-1α,2,6-cyclohimachalane		93
4′,5-Dihydroxy-7,8-dimethoxyflavone		55
5,8-Dihydroxy-3,3′,4′,7-tetramethoxyflavone[s]	7380-44-1	65
2,4-Diisopropenyl-1-methylvinylcyclohexone		67
Diisopropyl ketone	565-80-0	48
Dill apiole	484-31-1	3,7,21,29,50
Dill ether	74410-10-9	4,7
3,4-Dimethoxyallylbenzene	see Methyl eugenol	
Dimethoxy allylphenol		13
1,3-Dimethoxybutane	10143-66-5	9
Dimethoxycinnamaldehyde		13
Dimethoxycitral		29
cis-Dimethoxycitral		38
trans-Dimethoxycitral		38
Dimethoxycoumarin		6
5,7-Dimethoxycoumarin[s]	487-06-9	6,39,50,55,64,65
6,7-Dimethoxycoumarin[s]	120-08-1	50
6,7-Dimethoxy-*m*-cymene		88
1,7-Dimethoxy-*p*-cymene		41
2,5-Dimethoxy-*p*-cymene[s]	14753-08-3	19,45,88
2,6-Dimethoxy-*p*-cymene	291774-65-7	88
5,8-Dimethoxy-6,7-furanocoumarin[s]	482-27-9	6,39,50,65
1,2-Dimethoxy-4-(3-methoxy-1-propenyl) benzene	58045-87-7	13
1,2-Dimethoxy-4-methylbenzene	494-99-5	45
2,6-Dimethoxy-1-methylbenzene	see 2,6-Dimethoxytoluene	
1,4-Dimethoxy-2-methyl-5-isopropylbenzene	see 2,5-Dimethoxy-*p*-cymene	
3,4-Dimethoxyphenethyl alcohol	7417-21-2	13
1,2-Dimethoxy-4-(1-propenyl)benzene	see Methyl Isoeugenol	
3,4-Dimethoxystyrene	6380-23-0	3
2,3-Dimethoxytoluene	4463-33-6	10,45
2,6-Dimethoxytoluene[s]	5673-07-4	63,69
3,5-Dimethoxytoluene[s]	4179-19-5	63
2,6-Dimethyl-5-acetoxymethylhepta-1,6-dien-3-one		49
2,6-Dimethyl-5-acetoxymethyl-hept-6-en-3-one		49
3,7-Dimethyl-3-acetoxy-octa-1,5-dien-7-ol		6
3,7-Dimethyl-3-acetoxy-octa-1,7-dien-6-ol	41610-78-0	6
2,6-Dimethyl-6-acetoxy-octa-1,7-dien-3-one		6,48,49
2,6-Dimethyl-6-acetoxy-oct-1-en-7-one		6

Table 6.1 Alphabetical list of all chemicals and the essential oils in which they have been identified (*continued*)

Chemical	CAS number	Essential oils in which the chemical has been identified. Legend of numbers: see Table 6.2 at the end of this chapter
2,6-Dimethyl-6-acetoxy-oct-7-en-3-one		6,48
cis-1,4-Dimethyladamantane		13
N,N-Dimethylaniline	121-69-7	65
2,5-Dimethylanisole	1706-11-2	26
Dimethyl anthranilate	see Methyl *N*-methylanthranilate	
2,5-Dimethylbenzaldehyde	5779-94-2	80
1,3-Dimethylbenzene	108-38-3	13,21
τ4-Dimethylbenzene butanal		25
α,4-Dimethylbenzenemethanol		13
2,5-Dimethylbenzoic acid	610-72-0	38
(6*S*,10*S*)-Dimethylbicyclo[4.4.0]-1-decen-3-one		89
1,10-Dimethylbicyclo[4.4.0]-dec-6-en-3-one	39850-88-9	89
Dimethyl-6,7-bicyclo[4.4.0]-dec-10-en-4-one		89
2,3-Dimethylbicyclo[2.2.1]hept-2-ene	see Santene	
6,6-Dimethylbicyclo[3.1.1]hept-3-ene-2-butylene		72
endo-trans-4,7-Dimethylbicyclo[3.2.1]-octen-3-en-6-one		65
exo-cis-4,7-Dimethylbicyclo[3.2.1]-octen-3-en-6-one		65
6-(1,3-Dimethylbuta-1,3-dienyl)-1,5,5-trimethyl-7-oxabicyclo-[4.1.0]hept-2-ene		58
2,3-Dimethylbutanoic acid	14287-61-7	80
3,3-Dimethylbutanoic acid	see 3,3-Dimethylbutyric acid	
4,4-Dimethyl-2-buten-4-olide[s]	20019-64-1	49,72,80
4,4-Dimethyl-4-buten-2-olide		49
3,3-Dimethylbutyric acid[s]	1070-83-3	36,80
2-(3,3-Dimethylcyclohexadiene)ethanol		7
2,2-Dimethylcyclohexanone	1193-47-1	52
1,3-Dimethylcyclohexene	2808-76-6	63
1,3-Dimethyl-3-cyclohexenecarbox-aldehyde		77
3,4-Dimethyl-3-cyclohexene-1-carbox-aldehyde	see 1,2-Dimethyl-4-formyl-1-cyclohexene	
1-(1,4-Dimethyl-3-cyclohexen-1-yl)-ethanone	see 1,4-Dimethyl-3-cyclohexenyl methyl ketone	
1,4-Dimethyl-3-cyclohexenyl methyl ketone[s]	43219-68-7	7,56
1,5-Dimethylcyclopentene	16491-15-9	34
2,2-Dimethyl-3-cyclopenten-1-ethanal		80
5,5-Dimethyl-2-cyclopenten-1-ethanal		80
3,9-Dimethyldecanoic acid		37
10,10-Dimethyl-2,6-dimethylene-bicyclo[7.2.0]undecan-5-ol	see Caryophyllenol	
2,2-Dimethyl-4,5-di-1-propenyl-1,3-dioxolane	36334-88-0	83
Dimethyl disulfide	624-92-0	73
Dimethyl dodecene		40
5,5-Dimethylene-3-methylenebicyclo-[2.2.1]heptan-2-one	see Bicyclo[2.2.1]heptan-2-one, 5,5-dimethyl-3-methylene-	
2-(1,1-Dimethylethyl)-2,5-cyclohexa-diene-1,4-dione		26
4,5-Dimethyl-2-ethylphenol	see 2-Ethyl-4,5-dimethylphenol	
1,2-Dimethyl-4-formyl-1-cyclohexene	18022-66-7	9,88
1,5-[(3,6-Dimethyl-6-formyl-2-heptenyl)-oxy]psoralen	see Aurantiumal	
Dimethylfuran lactone		79

Table 6.1 Alphabetical list of all chemicals and the essential oils in which they have been identified (*continued*)

Chemical	CAS number	Essential oils in which the chemical has been identified. Legend of numbers: see Table 6.2 at the end of this chapter
2,4-Dimethylfuran	3710-43-8	49
5,5-Dimethyl-2(5*H*)-furanone	see 4,4-Dimethyl-2-buten-4-olide	
3,6-Dimethyl-1,5-heptadiene	34891-10-6	83
(3*E*)-2,6-Dimethyl-3,5-heptadien-2-ol		72
2,6-Dimethyl-1,6-heptadien-3-yl acetate	74902-74-2	72
2,6-Dimethyl-5-heptanal		57
2,6-Dimethyl-1-heptanol	2768-12-9	4
2,6-Dimethyl-5-heptenal	see Bergamal	
2,6-Dimethyl-5-hepten-1-ol	36806-46-9	37
3,6-Dimethyl-5-hepten-1-ol	51673-46-2	34
2,5-Dimethyl-2,4-hexadiene^s	764-13-6	3,83
3,3-Dimethyl-1,6-hexanedioic acid		80
3,5-Dimethylhexanoic acid	60308-87-4	37
2,5-Dimethyl-2-hexenoic acid		37
Dimethyl-3-hexenoic acid		37
3,7-Dimethyl-3-hydroxy-1,6-octadienyl formate		6
1,2-Dimethylindan	17057-82-8	3
Dimethyl ionene		7
Dimethyl ionone	68555-94-2	12
4,10-Dimethyl-7-isopropyl[4.4.0]bi-cyclo-1,4-decadiene		70
1,6-Dimethyl-4-isopropylstyrene		80
2,6-Dimethyl-6-methoxy-7-octen-2-ol	91243-33-3	80
1,1-Dimethyl-2-(3-methyl-1,3-buta-dienyl)cyclopropane	68998-21-0	8
4,4-Dimethyl-3-(3-methyl-2-buten-1-yl-idene)-2-methylidenebicyclo[4.1.0]-heptane		15
4,4-Dimethyl-3-(3-methylbut-3-enyli-dene)-2-methylenebicyclo[4.1.0]-heptane	79718-83-5	67
2,5-Dimethyl-3-(3-methylbutyl)pyrazine	18433-98-2	89
1,4-Dimethyl-3-(2-methyl-1)-1-cyclo-heptene		73
6,6-Dimethyl-2-methylenebicyclo		45
6,6-Dimethyl-2-methylenebicyclo[2.2.1]-heptane		63
2,5-Dimethyl-3-methylene-1,5-hepta-diene	74663-83-5	90
(5*E*)-2,5-Dimethyl-3-methylene-1,5-heptadiene	1316759-92-8	41
5,8a-Dimethyl-3-methylene-3a,7,8,8a,9,9a-hexahydro-3*H*-naphtho[2,3B]-furan-2-one		58
1,5-Dimethyl-6-methylenespiro[2.4]-heptane	62238-24-8	8
1,1-Dimethyl-3-methylene-2-vinyl-cyclohexane	see Cyclohexane, 2-ethenyl-1,1-dimethyl-3-methylene-	
1,2-Dimethyl-3,5-bis(1-methylethenyl)-cyclohexane	62337-99-9	26
7-Dimethyl-1-(1-methylethyl)		8
6,10-Dimethyl-3-(1-methylethylidene)-1-cyclodecene	69239-71-0	57
3,8-Dimethyl-4-(1-methylethylidene)-2,4,6,7,8,8a-hexahydro-5(1*H*)-azulenone		89
6,8a-Dimethyl-9-methylidene-2,5-methano-1,2,3,3a,4,5,8,8a-octahydro-azulene		67

Table 6.1 Alphabetical list of all chemicals and the essential oils in which they have been identified (*continued*)

Chemical	CAS number	Essential oils in which the chemical has been identified. Legend of numbers: see Table 6.2 at the end of this chapter
2,3-Dimethyl-3-(4-methyl-3-pentenyl)-2-norbornanol	98205-40-4	6,50,64,65
3,5-Dimethyl-4,6-di-*O*-methylphloro-acetophenone	21722-31-6	9
(*E,E*)-4,5-Dimethyl-2(2-methyl-1- pro-penyl)-3-cyclohexenyl methyl ketone		49
(*E,Z*)-4,5-Dimethyl-2(2-methyl-1-prop-enyl)-3-cyclohexenyl methyl ketone		49
(*Z,E*)-4.5-Dimethyl-2(2-methyl-1-pro-penyl)-3-cyclohexenyl methyl ketone		49
(*Z,Z*)-4,5-Dimethyl-2(2-methyl-1-pro-penyl)-3-cyclohexenyl methyl ketone		49
(*E*)-4,8-Dimethyl-3,7-nonadien-2-one	27539-94-2	19
(*Z*)-4,8-Dimethyl-3,7-nonadien-2-one	27576-61-7	19
4,8-Dimethyl-3,8-nonadien-2-one		19
(*E*)-3-(4,8-Dimethyl-3,7-nonadienyl)-3-cyclohexenyl methyl ketone		49
(*E*)-4-(4,8-Dimethyl-3,7-nonadienyl)-3-cyclohexenyl methyl ketone		49
3-(4,8-Dimethyl-3,7-nonadienyl) thiophene		73
4,8-Dimethyl-1,3(*E*),7-nonatriene	19945-61-0	6,60,64,67
4,8-Dimethyl-1,3(*Z*),7-nonatriene	21214-62-0	6
4,8-Dimethyl-7-nonen-2-one	3664-64-0	19
endo-2, *endo*-3-Dimethylnorbornan-*exo*-2-ol	59432-92-7	77
5,6-Dimethyl-5-norbornen-*exo*-2-ol	59300-41-3	77
(*E*)-5-(2,3-Dimethyl-3-nortricyclyl)-pent-3-en-2-one		77
2,2-Dimethyl-3,4-octadienal	590-71-6	52
3,7-Dimethyl-2,6-octadienal	see Citral	
(*E*)-2,6-Dimethyl-3,7-octadien-2,6-diol	see 2,6-Dimethyl-3,7-octadiene-2,6-diol	
2,6-Dimethyl-2,6-octadiene	2792-39-4	24,37,62,65
(*Z*)-2,6-Dimethyl-2,6-octadiene	2492-22-0	24,34,37,44,73
2,6-Dimethyl-2(3),7-octadiene		52
4,5-Dimethyl-2,6-octadiene	18476-57-8	51
2,6-Dimethyl-1,7-octadiene-3,6-diol	51276-33-6	6,45,49
(*E*)-2,6-Dimethyl-2,7-octadiene-1,6-diol	see (*E*)-8-Hydroxylinalool	
(*Z*)-2,6-Dimethyl-2,7-octadiene-1,6-diol	see (*Z*)-8-Hydroxylinalool	
2,6-Dimethyl-3,7-octadiene-2,6-diol[s]	13741-21-4	4,6,42,43,45,91
3,7-Dimethyl-1,5-octadiene-3,7-diol	see 2,6-Dimethyl-3,7-octadiene-2,6-diol	
3,7-Dimethyl-1,6-octadiene-3,5-diol	75654-19-2	43
3,7-Dimethyl-1,7-octadiene-3,6-diol		4
(2*E*,5*E*)-3,7-Dimethyl-2,5-octadiene-1,7-diol	see (*E,E*)-3,7-Dimethyl-2,5-octadiene-1,7-diol	
(*E,E*)-3,7-Dimethyl-2,5-octadiene-1,7-diol	93079-92-6	73
(*Z,E*)-3,7-Dimethyl-2,5-octadiene-1,7-diol		73
2,6-Dimethyl-1,7-octadien-3,6-ol	see 2,6-Dimethyl-1,7-octadiene-3,6-diol	
2,6-Dimethyl-3,7-octadien-2,6-ol	see 2,6-Dimethyl-3,7-octadiene-2,6-diol	
(*E,Z*)-2,6-Dimethyl-5,7-octadien-2-ol		74
2,7-Dimethyl-2,6-octadien-1-ol	22410-74-8	6,39,50,64,65,72
3,7-Dimethyl-1,6-octadien-3-ol	see Linalool	
3,7-Dimethyl-2,6-octadien-1-ol	see Citrol	
3,7-Dimethyl-2,6-octadien-1-ol acetate	16409-44-2	45
(*E*)-4,8-Dimethyl-3,8-octadiol		49
2,7-Dimethyl-3,5-octadione		35
Dimethyloctalene-1		89
Dimethyloctalone-11		89
2,6-Dimethyloctane	2051-30-1	52
3,7-Dimethyloctanoic acid	5698-27-1	37

Table 6.1 Alphabetical list of all chemicals and the essential oils in which they have been identified (*continued*)

Chemical	CAS number	Essential oils in which the chemical has been identified. Legend of numbers: see Table 6.2 at the end of this chapter
Dimethyl octanol[s]	106-21-8	6,34,37
1,7-Dimethyloctanol		72
4,7-Dimethyl-4-octanol	19781-13-6	41
4,6-Dimethyl-2-octanone		83
2,6-Dimethyl-1,3,5,7-octatetraene	90973-78-7	50,65
Dimethyloctatriene	29714-87-2	49
3,7-Dimethyl-1,3,6-octatriene	see β-Ocimene	
3,7-Dimethyl-1,3,7-octatriene	see α-Ocimene	
2,6-Dimethyl-1,5,7-octatrien-3-ol	29414-56-0	35
cis-2,6-Dimethyl-1,5,7-octatrien-3-ol		6
trans-2,6-Dimethyl-1,5,7-octatrien-3-ol		6
2,6-Dimethyl-3,5,7-octatrien-2-ol		49
(*E,E*)-2,6-Dimethyl-3,5,7-octatrien-2-ol	see *trans*-2,6-Dimethyl-3,5,7-octatrien-2-ol	
cis-2,6-Dimethyl-3,5,7-octatrien-2-ol		79
trans-2,6-Dimethyl-3,5,7-octatrien-2-ol[s]		8,56,86,79
(*E*)-2,7-Dimethyl-1,4,6-octatrien-3-ol		80
3,7-Dimethyl-1,5,7-octatrien-3-ol	see Hotrienol	
3,3-Dimethyl-1-octene		55,64
3,7-Dimethyl-1-octene	4984-01-4	38
(*E*)-3,7-Dimethyl-5-octene-1,7-diol		37,73
3,7-Dimethyl-7-octene-1,6-diol	22460-95-3	37
Dimethyloctenoic acid		37
3,7-Dimethyl-6-octenoic acid	see Citronellic acid	
3,7-Dimethyl-6-octen-1-ol acetate	see Citronellyl acetate	
cis-2,7-Dimethyl oct-5-yn-3-ene		31
2,7-Dimethyloxepine	1487-99-6	49
3,7-Dimethyl-6-oxo-octanoic acid	38975-38-1	37
2,3-Dimethyl-1,3-pentadiene	1113-56-0	41
Dimethylpent-2-enal (4,4-)	22597-46-2	56
2,5-Dimethyl-3-pentylpyrazine	56617-69-7	89
2,4(5)-Dimethylphenol		80
2,6-Dimethylphenol	576-26-1	72
(*E*)-1-(2,6-Dimethylphenyl)-2-buten-1-one	80445-59-6	20
2,2-Dimethyl-1-phenyl-1-propanone	938-16-9	43
Dimethyl *o*-phthalate[a]	131-11-3	25
2,3-Dimethylpropanoic acid		80
2,2-Dimethylpropanol	75-84-3	37
3,3-Dimethylpropenoic acid		80
2,5-Dimethyl-3-(1-propenyl)pyrazine		89
3,5-Dimethyl-2-(1-propenyl)pyrazine		89
2,3-Dimethylpyrazine	5910-89-4	80
2,5-Dimethylpyrazine	123-32-0	80,89
2,6-Dimethylpyrazine	108-50-9	80,89
2,3-Dimethylpyridine	583-61-9	89
2,6-Dimethylpyridine	108-48-5	80
Dimethylquinoline		89
Dimethylstyrene	27576-03-0	45
2,5-Dimethylstyrene	2039-89-6	31,72,75,85,86
3,4-Dimethylstyrene	27831-13-6	38
α,*m*-Dimethylstyrene	see *m*-Cymenene	
α,*p*-Dimethylstyrene	see *p*-Cymenene	
Dimethyl sulfide	75-18-3	37,48,49,61,68,73
Dimethyl sulfone[s]	67-71-0	37
Dimethyl sulfoxide	67-68-5	37
Dimethyltetradecanoic acid		37
4,7-Dimethyl-1-tetralone	28449-86-7	49
(*E*)-2,6-Dimethyl-10-(*p*-tolyl)-undeca-2,6-diene	55968-43-9	25
Dimethyl trisulfide	3658-80-8	73
(*E*)-6,10-Dimethyl-5,9-undecadien-2-one	see (*E*)-Geranylacetone	
2,5-Dimethylundecane	17301-22-3	13

Table 6.1 Alphabetical list of all chemicals and the essential oils in which they have been identified (*continued*)

Chemical	CAS number	Essential oils in which the chemical has been identified. Legend of numbers: see Table 6.2 at the end of this chapter
2,5-Dimethyl-2-vinylfuran		37
2,5-Dimethyl-2-vinyl-4-hexenal	56134-05-5	60
Dimyrcene	532-87-6	5
2,4-Dinitroanisole	119-27-7	73
8,9-Dinorborn-5-en-2-yl (*endo*) methyl ketone		49
8,9-Dinorborn-5-en-2-yl (*exo*) methyl ketone		49
8,9-Dinorborn-5-en-3-yl (*endo*) methyl ketone		49
8,9-Dinorborn-5-en-3-yl (*exo*) methyl ketone		49
8,9-Dinorborn-5-en-2-yl (*endo*) pentyl ketone		49
8,9-Dinorborn-5-en-2-yl (*exo*) pentyl ketone		49
8,9-Dinorborn-5-en-3-yl (*endo*) pentyl ketone		49
8,9-Dinorborn-5-en-3-yl (*exo*) pentyl ketone		49
Dioctyl phthalate[a]	117-81-7	72
Diosphenol	490-03-9	57
Dipentene	see *dl*-Limonene	
Diphenylethane	38888-98-1	42
1,2-Diphenylethanol	614-29-9	42
Diphenyl ether	101-84-8	37
1,5-Diphenyl-2*H*-1,2,4-triazoline		72
Dipropyl sulfide	111-47-7	73
Dipropyl sulfone	598-03-8	22
Dispiro[2.0.2.5]undecane, 8-methylene-	51567-09-0	67
1,3-Divinylbenzene	108-57-6	6
Docosane	629-97-0	4,8,25,29,37,43,59,60,62,73,85,88-91
1,22-Docosanediol	22513-81-1	73
Docosanoic acid[s]	112-85-6	24,72
1-Docosanol	661-19-8	73
6,9,12,15-Docosatetranoic acid		37
1-Docosene	1599-67-3	4,25,43,73
(*E,E*)-2,4-Dodecadienal	21662-16-8	83
(*E,Z*)-2,6-Dodecadienal	21662-13-5	55,64
(*Z,Z*)-3,6-Dodecadienol		43
Dodecahydro-3α,6,6,9α-tetramethyl-(2,1β)-furan		25
δ-Dodecalactone	713-95-1	4,25
δ-Dodecanolide	see δ-Dodecalactone	
Dodecanal	112-54-9	6,7,8,29,38,39,50,51,55,57,64,65,70,73,78,82,91
(*E*)-2-Dodecanal		29
(*Z*)-2-Dodecanal		29
Dodecanal diethyl acetal	53405-98-4	73
Dodecanamide	1120-16-7	59
Dodecane (*n*-)	112-40-3	6,13-15,19,26,29,31,35,37,41,43,49,50,60,70,72,73,85,89-91
Dodecane, 2,6,10-trimethyl-	see Farnesane	
Dodecanoic acid[s]	143-07-7	4,7,13,14,29,37,55,59,65,67,70,73
Dodecanoic acid, 4-penten-1-yl ester	607361-53-5	63
1-Dodecanol	112-53-8	4,6,8,9,19,50,64,65,82,91
2-Dodecanone	6175-49-1	45,63,91
3-Dodecanone	1534-27-6	4,91
1,3,6,10-Dodecatetraene, 3,7,11-tri-methyl-, (*Z,E*)-	see (*Z,E*)-α-Farnesene	
1,6,10-Dodecatriene, 7,11-dimethyl-3-methylene-, (*Z*)-	see (*Z*)-β-Farnesene	

Table 6.1 Alphabetical list of all chemicals and the essential oils in which they have been identified (*continued*)

Chemical	CAS number	Essential oils in which the chemical has been identified. Legend of numbers: see Table 6.2 at the end of this chapter
(Z,Z,Z)-3,6,9-Dodecatrienol	81345-02-0	43
1,6,10-Dodecatrien-3-ol, 3,7,11-tri-methyl-, S-(Z)-	142-50-7	34
Dodecenal	82107-89-9	6,29,57
2-Dodecenal	4826-62-4	29,39,50,55,64
(E)-2-Dodecenal	20407-84-5	25,29,52,55,57,64,65,82
3-Dodecenal[s]	68083-57-8	29,55
3-Dodecen-1-al	see 3-Dodecenal	
1-Dodecene (α-)	112-41-4	9,63,80
(E)-10-Dodecenyl acetate	35153-09-4	69
Dodecyl acetate	112-66-3	39,64
trans-Dracunculifoliol	162657-70-7	89
7-epi-cis-Dracunculifoliol		89
Drimenol	468-68-8	88
α-Duprezianene	79801-29-9	7,16-18,89
β-Duprezianene	178443-10-2	17,18,49
Durene	95-93-2	35
Duva-3,9,13-triene-1,5α-diol		63
4,8,13-Duvatriene-1,3-diol	7220-78-2	63
α-Duva-4,8,13-triene-1,3-diol	57605-80-8	63
Duva-3,9,13-triene-1,5α-diol acetate		63
Duva-3,9,13-trien-1α-ol-5,8-oxide-1-acetate		63
Eicosadiene		73
Eicosane (n-)	112-95-8	4,8,13,19,29,34,35,37,56-58,60,62,72,73,75,89-91
1-Eicosanol	629-96-9	73
n-Eicosan-5-ol		34
n-Eicosan-6-ol		34
11,14,17-Eicosatrienoic acid, methyl ester	55682-88-7	57
Eicosene	27400-78-8	72
1-Eicosene	3452-07-1	57,73
(E)-1-Eicosene	3452-07-1	4
(E)-3-Eicosene	74685-33-9	8
(E)-5-Eicosene	74685-30-6	73
9-Eicosene	42448-90-8	73
(E)-9-Eicosene		73
n-Eicos-14-en-2-ol		34
α-Ekasantalal	887498-75-1	77
β-Ekasantalal	see β-Santalal	
α-Ekasantalic acid	1837-15-6	77
α-nor-Ekasantalic acid		77
Elaidic acid[s]	112-79-8	7,67
Elema-1,11-dien-15-al		89
Elema-1,3,11(13)-trien-12-al		30
Elema-1,3,11(13)-trien-12-ol	see β-Elemenol	
Elemazulene	529-08-8	70
Elemene	11029-06-4	3,4,8,23,24,52,53,81,82,85,88
α-Elemene	5951-67-7	8,11,13,25,30,35,43,45,58,67,68,83,88,91
β-Elemene	33880-83-0	1-13,16,18,19,21-26,29-37,39-41,43-45,49-58,60, 61,63-74,78-85,88-91
(E)-β-Elemene		8
(Z)-β-Elemene		8
iso-β-Elemene	783322-21-4	45
γ-Elemene	29873-99-2	1,2,4,6-14,22,23,26,30,31,34,36-39,41,43,44,50,53,55,56,58, 60,63-68,71,75,78,81-83,85,88,91
cis-γ-Elemene		2
δ-Elemene[s]	20307-84-0	1-4,6,8-10,14,19,23,25,26,29,30,35-39,41,44,45,50,55,56,58,60, 61,63-65,67,69-71,73,75,79,82,85,88,90,91
(E)-β-Elemenene		85
β-Elemenol[s]	65018-04-4	10,30
cis-β-Elemenone	32663-57-3	1,58,91

Table 6.1 Alphabetical list of all chemicals and the essential oils in which they have been identified (*continued*)

Chemical	CAS number	Essential oils in which the chemical has been identified. Legend of numbers: see Table 6.2 at the end of this chapter
trans-β-Elemenone	20303-60-0	37,67,86,91
Elemicin	487-11-6	7,8,10,12,15,24,30,33,45,51,53,57,62,66,71,79,80
Elemol (α-)	639-99-6	1,2,4,8-12,21-25,29-34,36-41,44,45,49,51,52,54-58,63-69, 71,73,75,81,82,84,86,88-91
β-Elemol	32142-08-8	30,38,85,89
Elemol acetate	60031-93-8	12,58
Elemophilene	see Eremophilene	
Elixene[s]	3242-08-8	4,26,37,56,63,79,88
Endobornyl acetate	see *endo*-Bornyl acetate	
Endrin	72-20-8	45
(*E*)- En-yn-dicycloether	see *trans*-Spiroether	
(*Z*)-En-yn-dicycloether	see *cis*-Spiroether	
Epicedrol	see epi-Cedrol	
Epicubenol	see epi-Cubenol	
Epiglobulol	88728-58-9	13,26,35,38,41,45,56,61,63,68,79,83,88,89
Epilaurene	18452-45-4	75,89
(+)-9-Epiledene		81
4,5-Epithiocaryophyllene	65563-95-3	73
1,2-Epithiohumulene		73
4,5-Epithiohumulene		73
Epizizanal	82463-21-6	89
Epizizanol[s]	28624-26-2	89
Epizizanone[s]	28624-27-3	89
Epizonalene		37
Epizonarene	41702-63-0	4,7,31,37,38,44,45,88
10-Epizonarene	41702-63-0	8
6,7-Epoxide citral	79083-70-8	83
trans-4,5-Epoxide-(*E*)-2-decenal	134454-31-2	31,39,65,69
Epoxy allo-aromadendrene[s]	85760-81-2	12,19,25,26,27,37,72,83,88,89
1,10-Epoxyamorph-4-ene		89
Epoxyaurapten[s]	21499-17-2	39
Epoxybergamottin	206978-14-5	6,39
6′,7′-Epoxybergamottin	206978-14-5	39,64
Epoxybergamottin hydrate		39,64,65
11-epi-6,10-Epoxybisabol-2-en-12-ol	235421-61-1	77
1,10-Epoxy-11-bulnesene		67
2,3-Epoxycarane	62413-92-7	36,50
4,5-Epoxycarane	27867-36-3	38
(*E*)-4,5-Epoxycarane		51
4,5-Epoxycarene		8
(2*R*,5*E*)-2,12-Epoxycaryophyll-5-ene		25
Epoxydihydrocaryophyllene	see Caryophyllene oxide	
8,13-Epoxy-15,16-dinorlabd-12-ene	14752-13-7	25
7,10-Epoxy-eremophila-1,11-diene		89
7,10-Epoxy-eremophila-1,9-dien-8α-ol		89
7,11-Epoxy-eremophila-1,9-dien-8α-ol		89
7,11;8,12-di-Epoxy-eremophil-9-ene		89
7,11;8,12-di-Epoxy-eremophil-9-ene (epimer A)		89
7,11;8,12-di-Epoxy-eremophil-9-ene (epimer B)		89
10,11-Epoxy-eremophil-1-ene		89
13-nor-7,8-Epoxy-eremophil-1(10)-en-11-one		89
5,11-Epoxy-eudesmane		89
6,12,7(11)-di-Epoxy-eudesm-4-ene (epimer A)		89
6,12,7(11)-di-Epoxy-eudesm-4-ene (epimer B)		89
6,12,11-di-Epoxy-eudesm-4-ene		89
13-nor-4,5-Epoxy-eudesm-6-en-11-one		89

Table 6.1 Alphabetical list of all chemicals and the essential oils in which they have been identified (*continued*)

Chemical	CAS number	Essential oils in which the chemical has been identified. Legend of numbers: see Table 6.2 at the end of this chapter
13-nor-7,8-*trans*-Epoxy-eudesm-4(15)-en-11-one		89
2,3-Epoxygeranial		36,38,51,52,66
5-(6′,7′-Epoxy)geranyloxypsoralen	see Epoxybergamottin	
8-(6′,7′-Epoxy)geranyloxypsoralen		50
1,10-Epoxygermacrone		91
4,5-Epoxygermacrone		91
3,4-Epoxyhexanol	67663-02-9	43
3,4-Epoxy-(*Z*)-5-hexenyl benzoate		43
3,4-Epoxyhexyl acetate	113816-35-6	43
3,4-Epoxyhexyl benzoate	189155-40-6	43
3,4-Epoxyhexyl butyrate	189155-39-3	43
Epoxy-β-himachalene	see β-Himachalene oxide	
6,7-Epoxy-β-himachalene		15
epi-Epoxy-β-himachalene		15
1,2-Epoxyhumulene	see Humulene epoxide II	
7,10-Epoxy-10-hydroxysalvialane	300349-36-4	89
5-(2′,3′-Epoxyisopentenyloxy)-7 methoxycoumarin		50
5-(2′,3′-Epoxyisopentenyloxy)psoralen		50
4,5-Epoxy-1-isopropyl-4-methyl-1-cyclohexene		7
6,7-Epoxylinalool	15249-35-1	53
Epoxylinalyl acetate	41610-76-8	46-49
1,8-Epoxy-*p*-menthane	see 1,8-Cineole	
endo-1,8-Epoxy-*p*-menthan-2-ol	18679-48-6	80
exo-1,8-Epoxy-*p*-menthan-2-ol	18679-48-6	80
1,8-Epoxy-*p*-menth-2-ene		45
3,9-Epoxy-1-*p*-menthene	13955-48-1	79
1,8-Epoxy-*p*-menth-2-en-4-ol		80
Epoxymenthyl acetate (1,2-)	29815-69-8	35
7,10-Epoxy-10-methoxysalvialane		89
4α,7-Epoxy-11-methoxy-10β*H*-spiro-vetivene		89
3,10-Epoxy-muurol-4-ene		89
Epoxymyrcene	see Myrcene epoxide	
6,7-Epoxymyrcene		52
2,3-Epoxyneral		52
trans-4,5-Epoxy-(*E*)-2-nonenal		39,69
(*E*)-Epoxyocimene	255832-06-5	4,64
(*Z*)-Epoxyocimene		4,52
7,15-Epoxy-prezizaane	87059-19-6	89
trans-Epoxypseudoisoeugenyl 2-methyl-butyrate	125028-84-4	3
Epoxyrosefuran	see Rosefuran epoxide	
1,5-Epoxysalvial-4(14)-ene	88395-47-5	25,68
6,12-Epoxy-spiroax-4-ene		89
4,7-Epoxyspirovetiva-2,11-diene		89
4,7-Epoxy-spirovetiv-2-en-11-ol		89
2,3-Epoxysqualene	see Squalene 2,3-oxide	
cis-1,2-Epoxyterpinen-4-ol	1753-41-9	62
4,8-Epoxyterpinolene	4584-23-0	2,91
12-nor-2,3-Epoxyziza-6(13)-ene	300349-30-8	89
Eremanthin	37936-58-6	45
Eremoligenol		89
Eremophila-1(10),6-dien-12-al		89
Eremophila-1(10),7(11)-diene		89
Eremophila-1(10),4(15)-dien-2α-ol		89
Eremophila-1(10),6-dien-12-ol		89
4β*H*,5α-Eremophila-1(10),7-dien-2α-ol	see Vetiselinol	
Eremophila-1(10),11-dien-9β-ol		91

Table 6.1 Alphabetical list of all chemicals and the essential oils in which they have been identified (*continued*)

Chemical	CAS number	Essential oils in which the chemical has been identified. Legend of numbers: see Table 6.2 at the end of this chapter
(*E*)-Eremophila-1(10),7(11)-dien-12-yl acetate	352461-71-3	89
(*E*)-Eremophila-1(10),7(11)-dien-12-yl formate		89
Eremophila ketone		16
Eremophila-1(10),8,11-triene		67
Eremophilene[s]	10219-75-7	4,8,9,19,26,35-38,45,53,58,67,72,74,88
Eremophil-7(11)-en-10β-ol		89
12,13-di-nor-6(7→8)-*abeo*-Eremophil-1(10)-en-7-one		89
13-nor-Eremophil-1(10)-en-11-one		89
11,12,13-tri-nor-Eremophil-1(10)-en-7-one		89
Erucic acid	1072-39-5	7
Esdragole	see Methyl chavicol	
Esteragenol	see Methyl chavicol	
Estragole	see Methyl chavicol	
Estrone	53-16-7	25
Ethanal	see Acetaldehyde	
Ethane	74-84-0	38
1,3a-Ethano(1*H*)inden-4-ol, octahydro-2,2,4,7a-tetramethyl-	117591-80-7	37
Ethanol	see Ethyl alcohol	
2*H*-1,4a-Ethanonaphthalen-1-ol		57
Ethanone, 1-(2,5-dihydroxydiphenyl)-		70
2-Ethenyl-1,1-dimethyl-3-methylcyclohexane		26
7-Ethenyl-1,2,3,4,4a,5,6,7,8,9,10,10a-dodecahydro-1,1,4a,7-tetramethyl-phenanthrene	55255-56-6	63,72
3-Ethenyl-3-methyl-2-(1-methylethenyl)-6-(1-methylethyl)cyclohexanol	174955-50-1	76
1-Ethenyloxyhexadecane	822-28-6	8
4-(1-Ethoxyethenyl)benzaldehyde	839721-49-2	81
(3-Ethoxyhexa-1,5-dienyl)benzene	see Benzene, (3-ethoxy-1,5-hexadien-1-yl)-	
Ethyl acetate	141-78-6	14,42,64,65,73,80,85
trans-2-Ethyl-3-acetoxytetrahydrofuran		43
Ethyl alcohol[s]	64-17-5	14,23,34,37,42,65,68,72,73,75,80
Ethyl angelate		20
α-Ethyl-*p*-anisyl alcohol		3
Ethyl anthranilate	87-25-2	43
Ethylbenzene	100-41-4	14,60
Ethyl benzoate	93-89-0	21,22,25,26,42,43,50,61,73,90
trans-2-Ethyl-3-benzoyloxytetrahydrofuran		43
Ethyl benzyl ether	539-30-0	42
3-Ethylbutanal	15877-57-3	49
Ethyl butyrate	105-54-4	39,44,85
2-Ethylbutyric acid	88-09-5	37
Ethyl caprylate	see Ethyl octanoate	
Ethylcarvacrol	4732-13-2	86
Ethyl cinnamate	103-36-6	13,21,22,26,27,32,45,84,86
Ethyl (*E*)-cinnamate	4192-77-2	13,14,21,22,75
Ethylcyclohexanone (2-)	4423-94-3	56
2-(2-Ethylcyclopropyl)-3-methyl-2-cyclopenten-1-one	85135-71-3	42
Ethyl decanoate	110-38-3	7,19,44,73
5-Ethyl-3,4-dihydro-1(2*H*)-naphthalenone		83
2-Ethyl-3,5-dimethylpyrazine	27043-05-6	89
3-Ethyl-2,5-dimethylpyrazine	13360-65-1	89

Table 6.1 Alphabetical list of all chemicals and the essential oils in which they have been identified (*continued*)

Chemical	CAS number	Essential oils in which the chemical has been identified. Legend of numbers: see Table 6.2 at the end of this chapter
Ethyl dodecanoate[s]	106-33-2	7,44,73
1-Ethyl-2,3-dimethylbenzene	933-98-2	7
2-Ethyl-1,1-dimethyl-3-methylene-cyclohexane		72
2-Ethyl-4,5-dimethylphenol[s]	2219-78-5	72,85
Ethyl-5-ethylnicotinate	68686-59-9	43
Ethyl formate	109-94-4	14,37
2-Ethylfuran	3208-16-0	4,68,80
Ethyl geranate	32659-21-5	37,53,73
Ethyl geranyl ether	40267-72-9	73
4-Ethylguaiacol	2785-89-9	13,14
Ethyl heptanoate	106-30-9	7,44,65
Ethyl hexadecanoate[s]	628-97-7	3,7,19,37,42-44,56,58,62,73,88
Ethyl hexanoate	123-66-0	7,19,26,44,45,49,84
2-Ethylhexanoic acid	149-57-5	75
2-Ethylhexanol	104-76-7	67
bis-(2-Ethylhexyl)-1,2-benzene dicarboxylic acid ester	132969-07-4	24
2(or5)-Ethyl-4-hydroxy-5(or2)-methyl-3(2H)-furanone		39
Ethyl isobutyrate	97-62-1	20
Ethyl isovalerate	108-64-5	4,19,45,68,84
Ethyl jasmonate	54562-26-4	42
Ethyl laurate	see Ethyl dodecanoate	
Ethyl levulinate	539-88-8	44
Ethyl linolate	see Ethyl linoleate	
Ethyl linoleate[s]	544-35-4	7,30,43,56
Ethyl linolenate	1141-91-9	7,42,43
3-Ethyl-6-(methoxycarbonyl)-2-naphthol		58
Ethyl-p-methoxycinnamate[s]	1929-30-2	13,21
4-Ethyl-2-methoxy-6-methylphenol	120550-70-1	85
1-Ethyl-2-methylbenzene[s]	611-14-3	13,49
1-Ethyl-4-methylbenzene	622-96-8	13
2-Ethyl-3-methylbutanoic acid	32444-32-9	80
Ethyl-2-methylbutyrate[s]	7452-79-1	4,19,20,45,49,68,84
1-Ethyl-2-methylcyclodecane		73
3-Ethyl-3-methyldecane[s]	17312-66-2	13,14,21
Ethyl 4-methyl-5-ethylnicotinate		43
2-Ethyl-5-methylfuran	1703-52-2	38
2-(4-Ethyl-4-methyl-3-(isopropenyl)-cyclohexyl)propan-2-ol		63
Ethyl-3-(4-methyl-3-pentenyl)-3-cyclohexenyl propyl ketone		49
Ethyl-4-(4-methyl-3-pentenyl)-3-cyclohexenyl propyl ketone		49
Ethyl 4-(4-methylphenyl)-4-pentenoate		89
2-Ethyl-3-methylpyrazine[s]	15707-23-0	75,89
2-Ethyl-5-methylpyrazine	36731-41-6	89
2-Ethyl-6-methylpyrazine	13925-03-6	89
Ethyl myristate	see Ethyl tetradecanoate	
Ethyl nerate	see Ethyl nerolate	
Ethyl nerolate[s]	32659-20-4	49,57,73
Ethyl neryl ether	22882-89-9	73
Ethyl nicotinate	614-18-6	43
1-Ethyl-3-nitrobenzene	7369-50-8	26
Ethyl nonanoate	123-29-5	7
Ethyl nonenoate		73
Ethyl octadecanoate[s]	111-61-5	7,73
Ethyl octanoate[s]	106-32-1	7,26,44,65,73,85
4-Ethyloctanoic acid	16493-80-4	30
Ethyl oleate	111-62-6	3,7,42,62

Table 6.1 Alphabetical list of all chemicals and the essential oils in which they have been identified (*continued*)

Chemical	CAS number	Essential oils in which the chemical has been identified. Legend of numbers: see Table 6.2 at the end of this chapter
Ethyl palmitate	see Ethyl hexadecanoate	
Ethyl pentadecanoate	41114-00-5	7,73
Ethyl pentanoate	539-82-2	84
o-Ethylphenol	90-00-6	80
Ethyl phenylacetate	101-97-3	43
Ethyl 3-phenylpropionate	2021-28-5	45
2-Ethyl-5-propylphenol	72386-20-0	13
Ethyl pyrazine	13925-00-3	89
2-Ethylpyrazine	13925-00-3	75
2-Ethylpyridine	100-71-0	80
3-Ethylpyridine	536-78-7	43,80
4-Ethylpyridine	536-75-4	80
Ethyl salicylate	118-61-6	43,73
Ethyl senecionate		84
Ethyl stearate	see Ethyl octadecanoate	
p-Ethylstyrene	3454-07-7	73
Ethyl tetradecanoate[s]	124-06-1	7,29,44,56,62
Ethyl tridecanoate	28267-29-0	73
Ethyl-5-vinylnicotinate		43
7-Ethynyl-4a,5,6,7,8,8a-hexahydro-1,4a-dimethyl-2(1*H*)-naphthalenone	see Chamecynone	
Euasarone	see γ-Asarone	
Eucalyptol	see 1,8-Cineole	
Eucarvone	503-93-5	8,36,49,65,72,79
Eucumene		31
4(15),6-Eudesmadiene		31
Eudesma-3,7(11)-diene	see Selina-3,7(11)-diene	
Eudesma-3,11-diene	see α-Selinene	
Eudesma-4,6-diene	see δ-Selinene	
Eudesma-4(14),7(11)-diene		58
Eudesma-4(15),7(11)-diene		89
Eudesma-4(14),11-diene	see β-Selinene	
cis-Eudesma-6,11-diene	194607-93-7	89,91
Eudesma-3,5-dien-1α-ol		89
Eudesma-4(15),7-dien-β-ol	see Eudesma-4(15),7-dien-1β-ol	
Eudesma-4(15),7-dien-1β-ol	119120-23-9	4,25,44,70
Eudesma-4(15),7-dien-3β-ol	300349-35-3	89
cis-Eudesma-4(15),11-dien-5-ol		67
Eudesma-5,7-dien-4-ol	see Cascarilladienol	
cis-Eudesma-6,11-dien-3β-ol		89
nor-Eudesmadienone	137695-19-3	89
Eudesma-3,7(11)-dien-8-one	54707-46-9	91
Eudesma-3,11-dien-2-one	86917-79-5	10
Eudesma-4,6-dien-3-one[s]	23665-63-6	89
13-nor-Eudesma-4,6-dien-11-one		89
13-nor-Eudesma-4(15),7-dien-11- one		89
13-nor-*trans*-Eudesma-4(15),7-dien-11-one		89
11,12,13-tri-nor-*cis*-Eudesma-5,8-dien-7-one		89
trans-Eudesma-4(15),7-dien-12-yl formate		89
Eudesmane	473-11-0	38
Eudesma-2,6,8-triene		88
13,4-di-nor-Eudesma-5,7,9-trien-11-one		89
cis-Eudesm-6-en-12-al		89
15-nor-Eudesm-11-en-4-al		89
Eudesmene	34766-40-0	81
β-Eudesmene	see β-Selinene	
Eudesm-3-en-7-ol		85
Eudesm-4(10)-en-4-ol		89
Eudesm-4(15)-en-5β-ol		89
Eudesm-4(15)-en-6-ol		89

Table 6.1 Alphabetical list of all chemicals and the essential oils in which they have been identified (*continued*)

Chemical	CAS number	Essential oils in which the chemical has been identified. Legend of numbers: see Table 6.2 at the end of this chapter
Eudesm-5-en-11-ol	337981-29-0	40
Eudesm-6-en-4β-ol	116538-31-9	89
cis-Eudesm-6-en-11-ol	194607-96-0	89
(4β*H*,5α*H*)-*cis*-Eudesm-6-en-11-ol		91
cis-Eudesm-6-en-12-ol epimer A		89
cis-Eudesm-6-en-12-ol epimer B		89
Eudesm-7(11)-en-4-ol	see Juniper camphor	
Eudesm-11-en-4-ol	see Intermedeol	
Eudesm-11-en-4α-ol		91
cis-11,12,13-trinor-Eudesm-5-en-7-one		89
trans-11,12,13-trinor-Eudesm-5-en-7-one		89
13-nor-Eudesm-5-en-11-one		89
13-nor-*cis*-Eudesm-6-en-8-one		89
13-nor-*cis*-Eudesm-6-en-11-one		89
Eudesmol	51317-08-9	8,23,25,30,33,38,50
α-Eudesmol[s]	473-16-5	2,4,7-12,16,19,23,25,30,31,33-38,40,41,44,45,51,52,56,61, 72,73,75,83,86,88,89,91
7-epi-α-Eudesmol	123123-38-6	11,25,40,45,58,72,89
5-epi-7-epi-α-Eudesmol	446050-56-2	23,37,83,88
β-Eudesmol[s]	473-15-4	2,4,6-9,11,12,16,19,23-25,29-31,33-38,40,41,44,45,51,52, 54-56,61,63,68,72,73,75,76,78,85,86,88,89,91
γ-Eudesmol[s]	1209-71-8	2-4,8-10,12,16,19,23-25,30,31,34-38,40,41,44,45,51,55,58, 61,68,71-73,85,86,88,89,91
7-epi-γ-Eudesmol	117066-77-0	2,9
8-epi-γ-Eudesmol		31
10-epi-γ-Eudesmol	15051-81-7	4,7,23-25,33,36-38,40,41,45,51,61,73,83,85,88,89,91
τ-Eudesmol		24,38
α-Eudesmol acetate		45
β-Eudesmol acetate	40882-95-9	58
γ-Eudesmol acetate	67996-61-6	19
10-epi-γ-Eudesmol acetate		58
Eudesmyl acetate	51317-10-3	10,45
Eugenol[s]	97-53-0	4,5,7-14,19,21-28,29,31,34,35,39,41-44,48,49,51,52,54-57, 60-62,65,67-69,71-73,75,76,79,80,85,86,88-91
m-Eugenol[s]	501-19-9	26
Eugenol methyl ether	see Methyl eugenol	
Eugenyl acetate[s]	93-28-7	3,4,5,7,9,13,14,21,22,26-29,45,48,73,75
Eugenyl isovalerate	61114-24-7	88
Falcarinol	21852-80-2	54
(*Z*)-Falcarinol	21852-80-2	54
Farnesal	19317-11-4	34,38,60,64,66,67,88
(*E,E*)-Farnesal	502-67-0	36,38,49,51,57,60,67,73,88,90
(2*E,6Z*)-Farnesal	3790-67-8	60,90,91
(*Z,E*)-Farnesal	4380-32-9	49,60
(*Z,Z*)-Farnesal	3790-68-9	38,52
Farnesal D (*cis,trans*-)		7
Farnesane[s]	3891-98-3	13,42,60,91
α-Farnesane		73
Farnesene		24,55,64,69,80,82
α-Farnesene	502-61-4	3-5,9,12,13,19,21,24-26,37-39,44,47-50,52,53,55,57,60,62, 64-67,72,73,75,76,79-82,88,89
(*E,E*)-α-Farnesene	502-61-4	2-6,8,11,12,19,20,22,25-28,31,37-39,41-45,47,49,51,55,57,60, 61,64,65,67,69,73,75,80-82,85,88,90,91
(*E,Z*)-α-Farnesene	28973-98-0	19,30
(*Z,E*)-α-Farnesene	26560-14-5	4,19,22,24,42,43,45,49,51,67,73,77,78,81,90,91
(*Z,Z*)-α-Farnesene	28973-99-1	11,12,25,42,43,54,67
β-Farnesene	502-60-3	2-4,7,8,12,13,19,20,22,25,36,38-40,42,45-49,51,53-55,61,62, 65,66,68,70,72,73,79-82,85,90
(*E*)-β-Farnesene	18794-84-8	1-4,6,8,9-13,15,16,18,19,22,23,25,27,29,32,37-39,41-44,46-51, 53-55,57,60,62,64-70,72,73,77-82,85,88,90,91

Table 6.1 Alphabetical list of all chemicals and the essential oils in which they have been identified (*continued*)

Chemical	CAS number	Essential oils in which the chemical has been identified. Legend of numbers: see Table 6.2 at the end of this chapter
(*Z*)-β-Farnesene[s]	28973-97-9	2,4,6,7,11,12,15,16,18,19,24,25,37,38,39,41,44,46-55,57, 60,64,65,68-70,72,75,78-81,85,88,93
(*Z,E,E*)-allo-Farnesene		57
Farnesene epoxide		38
Farnesol	see Farnesyl alcohol	
α-Farnesol	58181-75-2	10,19
β-Farnesol	58181-76-3	38
(*E,E*)-Farnesol[s]	106-28-5	3,4,11,19,22-28,34,36-39,42,43,50-52,56,60,61,64-66,73, 77,88,90,93
(*E,Z*)-Farnesol	3879-60-5	1,7,19,23,38,39,42,43,60,66,73
(*Z,E*)-Farnesol ((2*Z*,6*E*)-)	3790-71-4	11-13,19,23,24,26-28,37,50,55,60,64-66,73,75,88,90
(6*Z*,10*E*)-Farnesol		67
(*Z,Z*)-Farnesol[s]	16106-95-9	7,25,36,38,42,44,45,50,51,56,57,66,73,75,90
(*E,E*)-α-Farnesol		6,8,38
(*Z,E*)-α-Farnesol		8
(*E*)-β-Farnesol	see (*E,E*)-Farnesol	
(*Z*)-β-Farnesol		41
cis-Farnesol	see (*Z,Z*)-Farnesol	
trans-Farnesol	see (*E,E*)-Farnesol	
Farnesyl acetate	29548-30-9	4,11,12,25,35,43,60,63,66,69,72,75,79,90
(*E,E*)-Farnesyl acetate	4128-17-0	1,7,11,27,37,42,43,52,57,60,66,67,77,90
(*E,Z*)-Farnesyl acetate	24163-98-2	60,66
(*Z,E*)-Farnesyl acetate	40266-29-3	7,37,52,57,66,73,75,90
(*Z,Z*)-α-Farnesyl acetate	24163-97-1	90
Farnesyl acetone	1117-52-8	9,45,48,72
(5*E*,9*E*)-Farnesyl acetone (*E,E*)-	1117-52-8	12,25
(5*E*,9*Z*)-Farnesyl acetone		25
Farnesyl alcohol[s]	4602-84-0	1,4,6,9,11-14,21-24,34,37-39,42,43,45,49,51-53,55,60,61, 64-68,73,75,77,80,81,85,90
(*E,E*)-Farnesyl formate	85633-25-6	43
Farnesyl pyrophosphate	13058-04-3	67
Faurinone	21682-87-1	88
Fenchene[s]	514-14-7	7,72,75,84
α-Fenchene	471-84-1	2,7,8,23,30,31,32,34,35,44,45,47,49,61,69-72,75,76,80,84,87,88
β-Fenchene	497-32-5	32,72,84
Fenchol	see Fenchyl alcohol	
α-Fenchol	see α-Fenchyl alcohol	
β-Fenchol	see β-Fenchyl alcohol	
endo-Fenchol	see α-Fenchyl alcohol	
exo-Fenchol	see β-Fenchyl alcohol	
Fenchone	1195-79-5	3,4,7,8,10,13,15,21,22,25,31,32,35,36,41,44,46,47,49,52,53, 56,61,70,72,75,80,81,84,86
α-Fenchone		72
Fenchyl acetate	13851-11-1	4,57,65,69,70,75,80,84
α-Fenchyl acetate[s]	111821-74-0	3,4,31,36,38,44,45,61,72,75,78,83,84
β-Fenchyl acetate[s]	76109-40-5	4,55
endo-Fenchyl acetate	see α-Fenchyl acetate	
exo-Fenchyl acetate	see β-Fenchyl acetate	
Fenchyl alcohol[s]	1632-73-1	4,6,9,13,16,31,32,35,36,45,50,52,56,60,62,65,70-73,75,83-85
α-Fenchyl alcohol[s]	512-13-0	4,6,7,12,25,32,38,44,45,49,50,61,70,72,75,76,78,85,87
β-Fenchyl alcohol	22627-95-8	4,7,29,41,49,51,56,72,73,83,85
Ferruginol	514-62-5	31,72,75
cis-Ferruginol		75
trans-Ferruginol		72,75
Filifolone		72
9*H*-Fluoren-9-ol	1689-64-1	13
Flourensadiol	55812-89-0	83
Foeniculin (*E*-)[s]	78259-41-3	3,81
Fokienol	33440-00-5	77
Fonenol	1275572-21-8	76,88
Formaldehyde	50-00-00	65,90
Formic acid	64-18-6	25,26,37,41,49,62,65,90

Table 6.1 Alphabetical list of all chemicals and the essential oils in which they have been identified (*continued*)

Chemical	CAS number	Essential oils in which the chemical has been identified. Legend of numbers: see Table 6.2 at the end of this chapter
Formic acid, phenylmethyl ester	see Benzyl formate	
Formic acid, 3,7,11-trimethyl-1,6,10-dodecatrien-3-yl ester		63
N-Formyl methylanthranilate	see Methyl N-formylanthranilate	
Ftalete[a]	see Phthalic acid	
Funebran-15-al		89
15-nor-Funebran-3-one		89
α-Funebrenal		89
α-Funebren-15-al	57766-59-3	89
α-Funebrene[s]	50894-66-1	10,16,18,31,70,77,89
2-epi-α-Funebrene	854154-70-4	10,16-18
β-Funebrene	79120-98-2	10,12,16-18,31,47,75,84,89
2-epi-β-Funebrene		37
β-Funebrene epoxide	22472-83-9	89
α-Funebrenic acid		89
α-Funebrenol		89
β-Funebren-14-ol		89
Furaldehyde	see Furfural	
Furan	110-00-9	37
3-Furanacetic acid	123617-80-1	65
2-Furancarboxylic acid	88-14-2	80
2-Furanmethanol, 5-ethenyltetra-hydro-α,α,5-trimethyl-, cis-	see cis-Linalool oxide	
Furanodiene	19912-61-9	8,58
Furanodienone	24268-41-5	58
Furanoeudesma-1,3-diene	115526-32-4	58
Furanoeudesma-1,4-diene	631868-96-7	58
5,6-seco-6,7-Furoeudesman-5-one		89
Furanoeudesmatriene		58
Furanogermacrene		58
Furfural[s]	98-01-1	4,10,13,19,21,22,26-28,30,37,44,48,49,68,75,77,80,85,89-91
Furfuraldehyde	see Furfural	
2-Furfuryl acid		37
Furfuryl alcohol	98-00-0	44,90
Furfuryl methyl sulfide	40228-18-0	19
Furfuryl octanoate	39252-03-4	15
N-Furfurylpyrrole (1-)	1438-94-4	77
Furopelargone A	1143-45-9	32,37,89
Furopelargone B		37
Furopelargonic acetate		37
Furyl propyl ketone		79
Galbanolene	see Undecatriene	
Gaultheriaoel[d]	see Methyl salicylate	
Geijerene	6902-73-4	3,52,72
Geranial[s]	141-27-5	4-7,11-14,21-25,29,34,35,37-39,41,45,48-53,55,56,60,64-70,72,73,82-86,90-92
cis-Geranial	see Neral	
Geranic acid[s]	459-80-3	4,5,37,38,50,51-53,57,67,72,73,80
(E)-Geranic acid	4698-08-2	37,57,73,90
(Z)-Geranic acid		37,73
Geraniol[s]	106-24-1	3-14,19,21-26,29,34-39,41-57,60-66,68-70,72-83,85,86,88-92
β-Geraniol	see Geraniol	
γ-Geraniol	13066-51-8	82
(E)-Geraniol	see Geraniol	
(Z)-Geraniol	see Nerol	
β-Geraniolene		49
Geraniol oxide		73
Geranoic acid	see Geranic acid	
Geranyl acetate[s]	105-87-3	4-14,19,21,22-25,29,31-35,37-39,41-57,60-66,68-70,72-76,78-82,84-86,89,90,92
cis-Geranyl acetate	see Neryl acetate	
trans-Geranyl acetate	see Geranyl acetate	

Table 6.1 Alphabetical list of all chemicals and the essential oils in which they have been identified (*continued*)

Chemical	CAS number	Essential oils in which the chemical has been identified. Legend of numbers: see Table 6.2 at the end of this chapter
2,3-Geranyl acetate oxide	76638-49-8	6
Geranyl acetone	3796-70-1	4,9,19,25,30,42,45,57,60,65,72,80,85,91
(*E*)-Geranylacetone^s	3796-70-1	7,37,56,73,77
(*Z*)-Geranylacetone	see Nerylacetone	
Geranyl benzoate	94-48-4	11,22,43,90
Geranyl butyrate^s	106-29-6	23,24,37,39,45,49,51,52,66,73,80,85
Geranyl caproate	see Geranyl hexanoate	
Geranyl formate^s	105-86-2	4-6,23,24,25,29,37,43,49,51,52,57,60,65,66,68,69,73,80,85
Geranylgeraniol	24034-73-9	34
Geranyl heptanoate	73019-15-5	37,66
Geranyl hexanoate^s	10032-02-7	24,37,52,66
Geranyl isobutyrate^s	2345-26-8	20,24,37,38,55,64,66,81,82,85
Geranyl isoheptanoate		37
Geranyl isohexanoate		37
Geranyl isooctanoate		37
Geranyl isovalerate^s	109-20-6	19,37,66,80,85,88
Geranyllinalool^s	77368-82-2	42,43,57,60,66,73,75
(*E,E*)-Geranyllinalool	1113-21-9	25
(*E,Z*)-Geranyllinalool		25
(*Z,E*)-Geranyllinalool		25
Geranyl 3-methylbutanoate^s	see Geranyl isovalerate	
Geranyl 2-methylbutyrate	68705-63-5	37,75
Geranyl 3-methylpentanoate		37
Geranyl 3-methylvalerate		37
Geranyl 4-methylvalerate		37
Geranyl nonanoate	68039-29-2	37
Geranyl octanoate	51532-26-4	37,66,73
7-Geranyloxycoumarin	see Aurapten	
5-Geranyloxy-7-methoxycoumarin	7380-39-4	6,39,50
5-Geranyloxy-8-methoxypsoralen	69239-53-8	6,50
5-Geranyloxypsoralen	7380-40-7	6,39,50,55,64,65,82
8-Geranyloxypsoralen	7437-55-0	50,64
Geranyl pentanoate	see Geranyl valerate	
Geranyl propanoate	see Geranyl propionate	
Geranyl propionate^s	105-90-8	6,24,25,35,37,39,49-51,60,64-66,73,76,80,85,86
Geranyl salicylate		73
Geranyl-α-terpinene		65
Geranyl tiglate	7785-33-3	19,37,66
Geranyl valerate^s	10402-47-8	37,66,88
Germacradienol		24
Germacradien-4-ol		43
1(10),5-Germacradien-4-ol	74841-87-5	43,57,90
[1(10)*E*,5*E*]-Germacradien-4α-ol	207221-31-6	43,44,57,70
Germacradien-5-ol		12
Germacra-1(10),7,11-trien-15-oic acid, 8,12-epoxy-6-hydroxy-, γ-lactone-		58
Germacra-4(15),5,10(14)-trien-α-ol		41
(*E,E*)-Germacra-3,7(11),9-trien-6-one	see Germacrone	
Germacrene	28028-64-0	4,7,8,12,13,23,25,38,41,45,55,56,63,68,70,75,88
Germacrene A	28387-44-2	4,6,8,10,19,23,25,27,36-39,45,49,50,53,55,58,64-67, 70,78-80,82,88
Germacrene B	15423-57-1	1-4,6-10,12,19,23,29,31,34,36-41,44,45,50,54,55,57,58, 64-69,71,72,75,81-83,85,88,91
Germacrene C	34323-15-4	6,8,39,55,64,67,77
Germacrene D	23986-74-5	1-13,19-27,29,31-34,36-39,41-58,60-73,75,76,78-86,88,90,91
Germacrene D-1,10-epoxide	65882-77-1	41
Germacrene D-4-ol	198991-79-6	10,12,23-25,32,34,39,44,45,48-52,57,65,68,70,72,75,79,82,86,90
Germacrene D-11-ol		41
β-Germacrenol		23
Germacrone^s	6902-91-6	37,58,85,91
Germazane		91
Germazene		91

Table 6.1 Alphabetical list of all chemicals and the essential oils in which they have been identified (*continued*)

Chemical	CAS number	Essential oils in which the chemical has been identified. Legend of numbers: see Table 6.2 at the end of this chapter
Germazol		91
Germazone	62332-96-1	91
Gleenol	72203-99-7	16,37,44,70
Globulol	489-41-8	4,5,9-11,13,19,27,30,31,34,35,37-39,45,49,51,53,56-58,60,61, 63,65,67,68,70,72,75,77-79,83,84-86,88,90,91
(-)-Globulol	see Globulol	
Glycerin		14
Gosferenol		50
Gossypetin 3,3',4',7-*O*-tetramethyl ether	see 5,8-Dihydroxy-3,3',4',7-tetramethoxyflavone	
Grandisol	26532-22-9	79
Guaiacol[s]	90-05-1	13,14,21,27,37,64,72,75,80,90
o-Guaiacol	see Guaiacol	
β-Guaiacol		38
Guaiacylacetone	2503-46-0	90
Guaiacyl cinnamate		13,14
4a*H*,10a*H*-Guaia-1(5),6-diene		37
4b*H*,10a*H*-Guaia-1(5),6-diene		37
Guaia-1(5),11-diene	see α-Guaiene	
Guaia-3,7-diene	6754-04-7	36,37
Guaia-3,9-diene	855270-07-4	58,67,88
Guaia-4,11-diene	see Aciphyllene	
Guaia-6,9-diene	37839-64-8	32,37,45,89
Guaia-1(5),11-dien-3α-ol		89
cis-Guaia-3,9-dien-11-ol		72
Guaiazulene[s]	489-84-9	10,19,26,40,91
Guaiene	88-84-6	24,37,67
α-Guaiene[s]	3691-12-1	4,8,9,13,14,21,23,25-27,34-38,40,41,45,52,55,63,67,72-74,75, 79,82,83,88,89,91
β-Guaiene	88-84-6	4,8,9,11,13,14,19,23,27,30,35,37,40,57,63,65,67,70,72,79,89
cis-β-Guaiene	372162-07-7	10,11,25,37,40,44,45,57,67,75,78,83,85,88,89
trans-β-Guaiene	192053-49-9	4,8,10,11,19,57,67,70,75,83,88
γ-Guaiene	145267-53-4	4,8,67,89
δ-Guaiene	see α-Bulnesene	
α-Guaiene oxide		67
Guaienol	300349-29-5	89
1(5)-Guaien-11-ol	13822-35-0	52
Guai-1(10)-en-11-ol	see Bulnesol	
Guai-5-en-11-ol	220437-52-5	83
cis-Guai-6-en-10-ol		89
trans-Guai-11-en-10-ol	see Pogostol	
Guaiol	489-86-1	4,9,33-38,40,41,45,56,61,62,67,71,73,75,77,83,85,88,90
Guai-11-ol		67
Guaiol isomer		40
Guaioxide	20149-50-2	40
Guaiyl acetate	134-28-1	4
Gurjunene	80599-13-9	81
α-Gurjunene	489-40-7	3-5,8-10,12,15,16,19,23,25,31,33-35,37,40,41,44,45,48,49,52, 56-58,61,63,67,68,70-71,74-76,78-81,83,85,88-90,93
β-Gurjunene	73464-47-8	4,8,10,12,23,25,29,31,35-39,41,44,45,52,54,55,57,61,65,67, 68,70,72,73,75,79,83,85,88,89
γ-Gurjunene	22567-17-5	4,8-10,16,24,25,30,34,37,39-41,45,52,57,63,67,68,70,72,75-78, 83,85,88,89
δ-Gurjunene	1786-19-2	45,75
τ-Gurjunene		52
β-Gurjunene epoxide		72
Gymnomitrene		66
(-)-Hanamyol	94388-63-3	40
Hanamyol isomer 1		40
Hanamyol isomer 2		40
(*E*)-Hasmigone	868693-38-3	53
Hedycaryol[s]	21657-90-9	24,25,34,36,41,63

Table 6.1 Alphabetical list of all chemicals and the essential oils in which they have been identified (*continued*)

Chemical	CAS number	Essential oils in which the chemical has been identified. Legend of numbers: see Table 6.2 at the end of this chapter
Helifolan-2-ol	see Khusiol	
Helifolan-2-one	see Khusione	
Heliofolen-12-al		77
nor-Helifolen-12-al I		77
nor-Helifolen-12-al II		77
Helifol-1-en-14-ol		89
Heliotropine	see Piperonal	
β-Helmiscapene	66141-12-6	66,88
Heneicosane (*n*-)	629-94-7	4,8,26,29,30,34,37,43,48,49,55,57,58,65,68,72,73,89-91
1-Heneicosanol	15594-90-8	73
1-Heneicosene	1599-68-4	43,73
9-Heneicosene	629-95-8	73
n-Heneicos-3-ene		34
n-Heneicos-4-ene		34
n-Heneicos-8-ene		34
n-Heneicos-10-ene		34
Hentriacontane	630-04-6	4
Heptacosane	593-49-7	7,19,41,57,58,62,70,73,90
8-Heptadanene		73
Heptadecadiene		73
Heptadecanal	629-90-3	42,43,55,65,73
n-Heptadecane	629-78-7	4,7,8,10,26,27,29,32,35,43,55,57,68,73,75,85,91
Heptadecanoic acid	506-12-7	43
Heptadecanolide	5637-97-8	1,2
2-Heptadecanone	2922-51-2	66
n-Heptadeca-1,8-11-14-tetraene	71046-96-3	30
1-Heptadecene	6765-39-5	73,91
7-Heptadecene	54290-12-9	73
8-Heptadecene	16369-12-3	73
(*Z*)-8-Heptadecene	16369-12-3	60,73
Heptadecen-2-one		68
2,4-Heptadienal	5910-85-0	45
(*E,E*)-2,4-Heptadienal	4313-03-5	4,26,48,57
(*E,Z*)-2,4-Heptadienal	4313-02-4	57
3,5-Heptadienal, 2-ethylidene-6-methyl-	99172-18-6	8
(*E,E*)-2,4-Heptadiene	2384-94-3	34
1,6-Heptadien-4-ol	2883-45-6	72
Heptafluorobutanoic acid, 2-(1-ada-mantyl)ethyl ester[a]		70
γ-Heptalactone[s]	105-21-5	6,42
Heptamethoxyflavone	119279-30-0	39,55,64,82
3,3',4,4',5,6,7-Heptamethoxyflavone		65
3,3',4',5,6,7,8-Heptamethoxyflavone[s]	1178-24-1	39,55,65,82
Heptanal	111-71-7	1,2,4,7,12,29,41,42,50,52,55,64,65,68,69,72,73,80,82,85,90,91
Heptanal diethyl acetal	688-82-4	73
n-Heptane	142-82-5	48,73
Heptanene		45
Heptanoic acid	111-14-8	4,13,30,37,72,73,80,90,91
Heptanol	53535-33-4	20,29,55,65,73,80
n-Heptanol	see 1-Heptanol	
1-Heptanol[s]	111-70-6	35,50,68,86
2-Heptanol	543-49-7	26,28,37,38,55,66
3-Heptanol	589-82-2	8,68
2-Heptanol acetate	see 2-Heptyl acetate	
4-Heptanolide	see γ-Heptalactone	
2-Heptanone	110-43-0	12,19,22,26,37,38,56,65-67,80,85
3-Heptanone	106-35-4	86
Heptan-3-yl acetate		68
2-Heptenal	2463-63-0	7
(*E*)-2-Heptenal	18829-55-5	48,64
(*Z*)-2-Heptenal		19

Table 6.1 Alphabetical list of all chemicals and the essential oils in which they have been identified (*continued*)

Chemical	CAS number	Essential oils in which the chemical has been identified. Legend of numbers: see Table 6.2 at the end of this chapter
2-Heptenal-7-hydroxy-5-isoprenyl-2-methyl acetate		48
n-Heptene	25339-56-4	42
Heptenoic acid		37
1-Hepten-3-ol[s]	4938-52-7	66
Hept-1-en-3-ol	see 1-Hepten-3-ol	
5-Hepten-1-ol	89794-36-5	57
Heptenone	see 1-Hepten-3-one	
Hepten-2-one	30640-40-5	7,57
1-Hepten-3-one	2918-13-0	57,73
3-Hepten-2-one	1119-44-4	79
(E)-2-Heptenyl acetate	16939-73-4	7
cis-2-(Z)-(1-Heptenyl)-3-cyclohexenyl methyl ketone		49
trans-2-(E)-(1-Heptenyl)-3-cyclohexenyl methyl ketone		49
trans-2-(Z)-(1-Heptenyl)-3-cyclohexenyl methyl ketone		49
Heptyl acetate	112-06-1	6,26,39,50,55,64,71
2-Heptyl acetate[s]	5921-82-4	26,38,65
Heptyl benzoate	7155-12-6	43
Heptyl formate	112-23-2	65
2-n-Heptylfurane	3777-71-7	75
8-Heptylpentadecane	71005-15-7	73
Heraclenin	2880-49-1	50
Heraclenol	31575-93-6	50
Herboxide	13679-86-2	81
cis-Herboxide	see cis-Dehydroxylinalool oxide	
trans-Herboxide		60
Herniarin	see 7-Methoxycoumarin	
Hexacosane	630-01-3	7,19,58,62,73,90
Hexadecanal	629-80-1	13,29,39,41,50,55,57,65,68,73,82,91
n-Hexadecane	544-76-3	4,7,8,29,32,35,42,43,56,58,65,68,72,73,75,85,89-91
Hexadecanoic acid[s]	57-10-3	3,4,7,9,13,19,21,25,26,29,37,38,41-43,45,48,50,52,54,56-58, 60,62,65,67,68,70,72,75,79,81,88-91
Hexadecanoic acid, 2-hydroxy-, methyl ester	16742-51-1	70
Hexadecanoic acid, methyl ester	see Methyl hexadecanoate	
Hexadecanol	51260-59-4	8,13,24,39,50,55,64,65,72,73,83
Hexadecanolide	109-29-5	2,89
Hexadecatrienal		30
Z,Z,Z-Hexadeca-7,10,13-trienal	56797-43-4	30
Hexadecenal	27104-14-9	13
(E)-2-Hexadecenal	22644-96-8	55
(Z)-7-Hexadecenal	56797-40-1	13
1-Hexadecene[s]	629-73-2	4,57,90
9-Hexadecenoic acid	2091-29-4	59
cis-9-Hexadecenoic acid[s]	373-49-9	68
cis-11-Hexadecenoic acid		67
Hexadecen-1-ol	37822-83-6	68
Hexadecyl acetate[s]	629-70-9	27,42,53,68
3-Hexadecyne	61886-62-2	67
(E,E)-2,4-Hexadienal	142-83-6	85
1,4-Hexadiene,5-methyl-3-(1-methyl-idene)-		52
1,4-Hexadiene, 3,3,5-trimethyl-		52
(Z)-3,5-Hexadienol		43
(Z)-3,5-Hexadienyl acetate		43
(Z)-3,5-Hexadienyl benzoate		43
(Z)-3,5-Hexadienyl butyrate	69925-34-4	43
Hexahydrobenzylacetone	2316-85-0	63

Table 6.1 Alphabetical list of all chemicals and the essential oils in which they have been identified (*continued*)

Chemical	CAS number	Essential oils in which the chemical has been identified. Legend of numbers: see Table 6.2 at the end of this chapter
1,2,3,4,4a,7-Hexahydro-1,6-dimethyl-4-isopropylnaphthalene	see 1,2,3,4,4a,7-Hexahydro-1,6-dimethyl-4(1-methylethyl)naphthalene	
1,2,3,4,4a,7-Hexahydro-1,6-dimethyl-4-(1-methylethyl)naphthalene[s]	16728-99-7	13,26,49,56,63
Hexahydrofarnesyl acetate	99624-94-9	29,57
Hexahydrofarnesyl acetone[s]	502-69-2	4,19,24,29,37,41,42,45,49,56,57,91
1,2,3,4,5,6-Hexahydro-1-methyl-2,2'-bipyridine		
Hexahydronaphthalene	41375-99-9	8,41
Hexahydropseudoionone	1604-34-8	67
3,3',4',5,6,7-Hexamethoxyflavone[s]	1251-84-9	55,65,82
3,3',4',5,7,8-Hexamethoxyflavone[s]	7741-47-1	65
3',4',5,6,7,8-Hexamethoxyflavone	478-01-3	6,39,55,64,65,81
Hexamethylbenzene	87-85-4	45
2,3,5,5,8,8-Hexamethyl-1,3,6-cyclo-octatriene		89
Hexa-*O*-methylgossypetin	see 3,3',4',5,7,8-Hexamethoxyflavone	
Hexanal	66-25-1	2,4,6,7,10,12,13,19,22,29,30,32,38,39,41,42,44,45,48,50,55, 64,65,68-70,72,73,80,85,88,90,91
2*E*-Hexanal	see (*E*)-2-Hexanal	
(*E*)-2-Hexanal[s]		42
Hexanal diethyl acetal	3658-93-3	73
Hexane	92112-69-1	37,38
n-Hexane	110-54-3	44,55
Hexanedioic acid, dioctyl ester	123-79-5	19
Hexanediol		83
Hexanoic acid[s]	142-62-1	4,8,10,13,30,31,37,43,48,49,72,73,80,88,90,91
Hexanol (1-; n-)	111-27-3	4-6,10,12-14,21,22,25-27,29,32,37-39,41-43,45-49,55-57,60, 64,65,68,73,78-80,82,84-86,90,91
2-Hexanol	626-93-7	22,42,43
3-Hexanol	623-37-0	42,43,68,73
(*E*)-3-Hexanol		45,48
(*Z*)-3-Hexanol		19,45,68,69
4-Hexanolide	695-06-7	42
2-Hexanone	591-78-6	26,27,28,37,38,80
3-Hexanone	589-38-8	72,80,86
Hexatriacontane	630-06-8	19,25,81
Hexenal	1335-39-3	10,50,57,70
2-Hexenal	505-57-7	25,44,45,46,49,73,80
(*E*)-Hexenal	85761-70-2	70
(*E*)-2-Hexenal	6728-26-3	5,6,19,25-27,29,34,35,37,39,41,42,45,48,50,55-57,64,65,69, 70,72,75,80,85
(*Z*)-2-Hexenal	16635-54-4	56,68
3-Hexenal	4440-65-7	73
(*Z*)-3-Hexenal	6789-80-6	22,37,42,43
(*Z*)-5-Hexenal oxime		72
1-Hexene	592-41-6	73
1-Hexene, 4,5-dimethyl-	16106-59-5	51
(*E*)-2-Hexenoic acid	13419-69-7	37,72
(*E*)-3-Hexenoic acid	1577-18-0	37,43,72
4-Hexenoic acid	35194-36-6	72
3-Hexenoic acid, butyl ester, (*Z*)-	69668-84-4	34
Hexenol	910923-87-4	50
1-Hexen-3-ol	4798-44-1	22,35,39
(*E*)-1-Hexen-3-ol		79
2-Hexenol	see (*E*)-2-Hexenol	
(*E*)-2-Hexenol[s]	928-95-0	6,8,19,22,25,37,42,43,45,50,60,65,68,69,75,85,90
(*E*)-2-Hexen-1-ol	see (*E*)-2-Hexenol	
(*E*)-3-Hexenol[s]	928-97-2	4,37,42,43,45,73,91
(*Z*)-3-Hexenol[s]	928-96-1	4-6,19,22,25-27,31,32,34,35,37,39,41-43,45-51,56,57,64-66, 68-70,72,73,75,78-80,82-86,90

Table 6.1 Alphabetical list of all chemicals and the essential oils in which they have been identified (*continued*)

Chemical	CAS number	Essential oils in which the chemical has been identified. Legend of numbers: see Table 6.2 at the end of this chapter
3-Hexen-1-ol	544-12-7	13,22,25,34,37,43,45,48,57,90
(*E*)-3-Hexen-1-ol	see (*E*)-3-Hexenol	
(*Z*)-3-Hexen-1-ol	see (*Z*)-3-Hexenol	
cis-4-Hexenol	928-91-6	42
4-Hexen-1-ol	928-92-7	49
5-Hexen-2-one	109-49-9	80
Hexenyl acetate	28933-77-9	25,35
2-Hexenyl acetate	2497-18-9	90
(*E*)-2-Hexenyl acetate	2497-18-9	22,42,43,60,69
3-Hexenyl acetate	1708-82-3	42,90
(*E*)-3-Hexenyl acetate	3681-82-1	4,12,42,43
(*Z*)-3-Hexenyl acetate	3681-71-8	4,6,7,11,19,22,25,31,37,42,43,57,68,78,83,90
(*Z*)-3-Hexenyl angelate	84060-80-0	4,20
(*E*)-2-Hexenyl benzoate	76841-70-8	42,43
3-Hexenyl benzoate		56
(*E*)-3-Hexenyl benzoate	75019-52-2	42,43
(*Z*)-3-Hexenyl benzoate	25152-85-6	4,11,22,25,42,43,70,90,91
3-Hexenyl butanoate	see 3-Hexenyl butyrate	
(*Z*)-3-Hexenyl butanoate	see (*Z*)-3-Hexenyl butyrate	
Hexenyl butyrate	26912-31-2	49,61,80
(*E*)-Hexenyl butyrate		39
(*E*)-2-Hexenyl butyrate	53398-83-7	51,55
(*Z*)-2-Hexenyl butyrate		4
3-Hexenyl butyrate[s]	2142-93-0	45,49
(*E*)-3-Hexenyl butyrate	53398-84-8	29,43
(*Z*)-3-Hexenyl butyrate[s]	16491-36-4	25,29,42,43,48,49,60,65,85
(*Z*)-3-Hexenyl formate	33467-73-1	49
(*Z*)-3-Hexenyl hexanoate	31501-11-8	37,43,49,65
3-Hexenyl hexenoate		43
(*Z*)-3-Hexenyl (*Z*)-3-hexenoate	61444-38-0	4,91
(*Z*)-3-Hexenyl hydroxybutyrate		43
3-Hexenyl isobutyrate	57859-47-9	42
(*Z*)-3-Hexenyl isobutyrate	41519-23-7	49
Hexenyl isovalerate	see (*Z*)-3-Hexenyl isovalerate	
(*E*)-2-Hexenyl isovalerate		75
(*Z*)-2-Hexenyl isovalerate		1,19
(*E*)-3-Hexenyl isovalerate		43
(*Z*)-3-Hexenyl isovalerate	35154-45-1	19,37,43,68,79
(*Z*)-3-Hexenyl methylbutyrate	53398-85-9	43,68
(*Z*)-3-Hexenyl nonanoate	88191-46-2	49
(*Z*)-3-Hexenyl oxyacetaldehyde	68133-72-2	4
cis-3-Hexenyl pentanoate	see *cis*-3-Hexenyl valerate	
3-Hexenyl propionate		42
(*E*)-2-Hexenyl salicylate	68133-77-7	43
(*Z*)-3-Hexenyl salicylate	65405-77-8	43
(*E*)-3-Hexenyl tiglate		43
(*Z*)-3-Hexenyl tiglate	67883-79-8	20,37,43
cis-3-Hexenyl valerate[s]	35852-46-1	79
n-Hexyl acetate	142-92-7	6,11,20,39,43,46-49,57,60,63,65,72,73,80,90,91
(*E*)-2-Hexyl acetate		64
Hexyl angelate	65652-33-7	20
Hexyl benzoate	6789-88-4	42,43,80,91
Hexyl butanoate	see Hexyl butyrate	
4-Hexyl-4-butanolide	see γ-Decalactone	
Hexyl butyrate[s]	2639-63-6	19,20,46,47-49,65,80
α-Hexylcinnamaldehyde[a,s]	101-86-0	22,59,88
2-Hexylcyclopropane acetic acid		82
2-Hexyl-1-decanol	2425-77-6	73
Hexyl formate	629-33-4	37
Hexyl hexanoate	6378-65-0	49,63,80
Hexyl isobutanoate	see Hexyl isobutyrate	

Table 6.1 Alphabetical list of all chemicals and the essential oils in which they have been identified (*continued*)

Chemical	CAS number	Essential oils in which the chemical has been identified. Legend of numbers: see Table 6.2 at the end of this chapter
Hexyl isobutyrate[s]	2349-07-7	46,47,48,49,80
Hexyl isovalerate[s]	10032-13-0	1,36,44,46-49,68,73,79,88,91
Hexyl 3-isovalerate	see Hexyl isovalerate	
Hexyl methacrylate	142-09-6	20
Hexyl 3-methylbutanoate	see Hexyl isovalerate	
Hexyl 2-methylbutyrate[s]	10032-15-2	19,20,46-49,73,80
Hexyl 3-methylbutyrate	see Hexyl isovalerate	
n-Hexyl methyl ether	see 1-Methoxyhexane	
Hexyl 2-methylpropanoate	see Hexyl isobutyrate	
Hexa-*O*-methylquercetagetin	see 3,3',4',5,6,7-Hexamethoxyflavone	
Hexyl methyl sulfide	20291-60-5	73
Hexyl octanoate	1117-55-1	63
Hexyl oleate	20290-84-0	81
8-Hexylpentadecene		73
Hexyl pentanoate	see Hexyl valerate	
α-*n*-Hexyl- α-phenylacrolein	see α-Hexylcinnamaldehyde	
Hexyl propanoate	2445-76-3	20,46,47,49
Hexyl tiglate	16930-96-4	20,46-49,57,80
Hexyl valerate[s]	1117-59-5	49,79,80
Himachala-2,4-diene		15,78
Himachala-1,3,5-trien-5-ol[s]	66656-01-7	15
Himachalene		15,39,93
α-Himachalene (α-*cis*-)[s]	3853-83-6	3,4,7,8,10,12,13,15-18,31,34,38,45,56,61,63,65,67,72,78, 79,83,93
α-*ar*-Himachalene	see α-Himachalene	
β-Himachalene	1461-03-6	2,3,12,15-18,30,31,36,38,49,54,72,73,78,80,85,89,93
γ-Himachalene	53111-25-4	3,7,15,17,58,93
γ-*ar*-Himachalene	see γ-Himachalene	
δ-Himachalene	135447-48-2	67
α-Himachalene epoxide[s]	64825-84-9	7,10,15,70,93
β-Himachalene oxide[s]	31560-66-4	15,93
Himachalenol		78
11α*H*-Himachal-4-en-1β-ol		12
Himachalol	1891-45-8	4,7,15,77,93
β-Himachalol		77
allo-Himachalol[s]	19435-77-9	15
Hinesene	123484-18-4	10
Hinesol	23811-08-7	37,63,72,88
Hinesol acetate	88494-77-3	15
Homofarnesane	155635-42-0	91
γ-Homogeraniol		5
Homomyrtenol	see Nopol	
Homovanillyl alcohol	2380-78-1	45
Hotrienol	20053-88-7	6,10,36,45,46,49,66,74,85,91
(*E*)-Hotrienol	53834-70-1	4,80
Hotrienyl acetate	150447-00-0	6
Humuladienol		72
Humuladienone	24405-90-1	27,75,77
α-Humulane		18
β-Humulane		18
Humulane-1,6-dien-3-ol	915392-38-0	13,25,27,93
α-Humulene	6753-98-6	1-16,18-39,41,43,45-53,55-58,60-72,74-86,87-91,93
β-Humulene	116-04-1	2,4,25,29,52,67,85,86,88
γ-Humulene	26259-79-0	31,78
Humulene (ep)oxide I	19888-33-6	4,6,8,10,12,21,22,26,45,57,75,85,86
α-Humulene epoxide[s]	96638-51-6	2,4,8,9,11,22,26-28,32,52,56,57,63,66,71,72,75,78,83,84,87
Humulene epoxide II[s]	19888-34-7	1,2,4,6,8,10-12,19,21,22,25-27,29,31,32,37,41,44,45,49,53, 55-57,67,68,70,72,75,76,79,84,85,88,91
Humulene (ep)oxide III	21624-36-2	25,75
Humulene 6,7-epoxide	see Humulene (ep)oxide II	
Humulene oxide	see α-Humulene epoxide	
Humulene-1,2-oxide	see Humulene (ep)oxide II	

Table 6.1 Alphabetical list of all chemicals and the essential oils in which they have been identified (*continued*)

Chemical	CAS number	Essential oils in which the chemical has been identified. Legend of numbers: see Table 6.2 at the end of this chapter
Humulenol (II)	19888-00-7	2,10,26
Hydrazinecarboxylic acid, ethyl ester	4114-31-2	81
meso-Hydrobenzoin	579-43-1	58
Hydrocarbons C$_{21}$-C$_{33}$, linear chained		50,55
Hydrocarbons C$_{21}$-C$_{33}$, linear chained, isomers		50,55
Hydrocinnamaldehyde[s]	104-53-0	13,14,21,22
Hydrocinnamic acid	501-52-0	13,14,80
Hydrocinnamyl acetate[s]	122-72-5	13,14,21,22,45,90
Hydrocinnamyl alcohol[s]	122-97-4	13,14,21,22,43
2-Hydroxyacetophenone (α; ω-)	582-24-1	13,80
2α-Hydroxy-amorpha-4,7(11)-diene		44
5-Hydroxyauranetin	see 5-Hydroxy-3′,4′,6,7,8-pentamethoxyflavone	
2-Hydroxybenzaldehyde	see Salicylaldehyde	
14-Hydroxy-γ-cadinene		89
Hydroxy-δ-cadinene		89
14-Hydroxy-δ-cadinene	153408-92-5	15
17-Hydroxy-δ-cadinene		44
10-Hydroxycalamenene	153408-93-6	89
3-Hydroxycarbofuran	16655-82-6	3
8-Hydroxycarvone	see Carvone hydrate	
8-Hydroxycarvotanacetone	see Carvone hydrate	
Hydroxycaryophyllene	78683-81-5	75,83
4-Hydroxy-*cis*-caryophyllene		26
14-Hydroxy-(*E*)-caryophyllene		57
14-Hydroxy-(*Z*)-caryophyllene		49,57
14-Hydroxy-1-epi-caryophyllene		69
14-Hydroxy-epi-(*E*)-caryophyllene		32
14-Hydroxy-9-epi-β-caryophyllene	see 14-Hydroxy-9-epi-(*E*)-caryophyllene	
14-Hydroxy-9-epi-(*E*)-caryophyllene[s]	244226-09-3	7,12,19,25,37,44,45,57,72,75,85,93
exo-2-Hydroxycineole	66965-45-5	35
2-Hydroxy-1,8-cineole	103665-39-0	45
3-Hydroxy-1,8-cineole	118013-29-9	45
exo-2-Hydroxycineole acetate	72257-53-5	4,72
2-Hydroxycinnamaldehyde	3541-42-2	13
4-Hydroxycinnamaldehyde	2538-87-6	21
Hydroxycitronellal[a]	107-75-5	23,25,37,53,73,83
Hydroxycitronellol[a]	107-74-4	34
7-Hydroxycoumarin[s]	93-35-6	50,64
8-Hydroxycycloisolongifolene		88
endo-8-Hydroxycycloisolongifolene		25
3-Hydroxy-β-Damascenone		73
4α-Hydroxydihydroagarofuran		37
7-Hydroxydihydrocitronellol		37
7-Hydroxy-6,7-dihydrogeraniol		37
4-Hydroxy-α,*p*-dimethylcyclohexane-methanol		34
4-Hydroxy-2,5-dimethyl-3(2*H*)furanone	3658-77-3	39
cis-7-Hydroxy-3,7-dimethyl-3,6-oxy-octanal		51
trans-7-Hydroxy-3,7-dimethyl-3,6-oxy-octanal		51
3-Hydroxydodecanoic acid methyl ester	see Methyl 3-hydroxydodecanoate	
5-Hydroxy-6,7-furanocoumarin	see Bergaptol	
2-Hydroxyfuranodiene		58
6α-Hydroxygermacra-1(10),4-diene	20674-02-6	10
6-Hydroxyheptenylcyclopentane		42
2′-Hydroxy-3,3′,4,4′,5′,6′-hexame-thoxychalcone		65
Hydroxyhexamethoxyflavone		39

Table 6.1 Alphabetical list of all chemicals and the essential oils in which they have been identified (*continued*)

Chemical	CAS number	Essential oils in which the chemical has been identified. Legend of numbers: see Table 6.2 at the end of this chapter
5-Hydroxy-3,3',4',6,7,8-hexamethoxyflavone	1176-88-1	65
7-Hydroxy-3,3',4',5,6,8-hexamethoxyflavone	185678-89-1	82
14-Hydroxy-α-humulene	108043-85-2	44,75
3-Hydroxy-β-ionone	116296-75-4	43
5-Hydroxyisobornyl isobutyrate		83
6-Hydroxyisobornyl isobutyrate	107783-32-4	58
8-Hydroxyisobornyl isobutyrate		58
Hydroxyisopimarene		84
2-Hydroxyisopinocamphone		41
2-Hydroxy-4-isopropenylbenzaldehyde		80
2-Hydroxy-3-isopropyl-2-cyclohexen-1-one		80
(*Z*)-(2-Hydroxy-2-isopropyl-5-methyl)-cyclohexanol		34
6-Hydroxy-3-isopropyl-6-methyl-2-cyclohexen-1-one		80
Hydroxyjunipene	469-27-2	89
Hydroxylinalool	256418-61-8	6,25
8-Hydroxylinalool	64142-78-5	29,43,83
(*E*)-8-Hydroxylinalool[s]	75991-61-6	42,43
(*Z*)-8-Hydroxylinalool[s]	103619-06-3	43,49
3α-Hydroxymanool		7
8-Hydroxy-*p*-menth-4-en-3-one		79
8-Hydroxymenthol		69
4-Hydroxy-3-methoxybenzene	see Eugenol	
(*E*)-3-(4-Hydroxy-3-methoxyphenyl)-allyl acetate		42
2'-Hydroxy-4'-methylacetophenone	6921-64-8	80
2'-Hydroxy-5'-methylacetophenone	1450-72-2	80
2'-Hydroxy-6'-methylacetophenone		80
4'-Hydroxy-3'-methylacetophenone	876-02-8	61
4-Hydroxy-4-methylacetophenone		7
2-Hydroxy-2-methyl-3-butenyl angelate		20
4-Hydroxy-4-methyl-2-cyclohexenone	60565-80-2	7
s-(+)-5-(1-Hydroxy-1-methylethyl)-2-cyclohexan-1-one		41
4-Hydroxy-4-methyl-2-pentanone[s]	123-42-2	35,38,39,57,59,72,78,80
3-Hydroxy-2-methyl-4-pyrone[s]	118-71-8	42,43,75
2-Hydroxy-3-methylvalerianic acid methyl ester		10
14-Hydroxy-α-muurolene	105661-29-8	15,44,75
Hydroxyneoisolongifolane		88
8-Hydroxyneomenthol	3564-95-2	23
2'-Hydroxy-3,4,4',5',6'-pentamethoxychalcone		65
Hydroxypentamethoxyflavone		39
3-Hydroxy-4',5,6,7,8-pentamethoxyflavone[s]		65
5-Hydroxy-3',4',6,7,8-pentamethoxyflavone[s]	50439-46-8	55,82
5-Hydroxy-3,3',4',7,8-pentamethoxyflavone		65
5-Hydroxy-6,7,8,3',4'-pentamethoxyflavone[s]	2174-59-6	55,64,65
7-Hydroxy-3,3',4',5,6-pentamethoxyflavone	57393-68-7	82
4-Hydroxy-2-phenethyl alcohol	501-94-0	13
2-Hydroxypinocamphone	10136-65-9	41
trans-2-Hydroxypinocamphone	20536-50-9	41

Table 6.1 Alphabetical list of all chemicals and the essential oils in which they have been identified (*continued*)

Chemical	CAS number	Essential oils in which the chemical has been identified. Legend of numbers: see Table 6.2 at the end of this chapter
6-Hydroxypiperitol		8
3-Hydroxypregn-5-en-20-one	145-13-1	88
13-Hydroxy-α-santalan-12-one		49
(Z)-12-Hydroxysesquicineole	699007-30-2	77
3-Hydroxytangeretin	see 3-Hydroxy-4′,5,6,7,8-pentamethoxyflavone	
Hydroxy-α-terpinyl acetate		49
3-Hydroxy-4′,5,6,7-tetramethoxy-flavone		65
4′-Hydroxy-5,6,7,8-tetramethoxyflavone	36950-98-8	55
5-Hydroxy-3,3′,4′,7- tetramethoxy-flavone[s]	1245-15-4	65
5-Hydroxy-3′,4′,6,7-tetramethoxy-flavone[s]	21763-80-4	55
5-Hydroxy-3′,4′,7,8-tetramethoxyflavone		6
5-Hydroxy-4′,6,7,8-tetramethoxyflavone[s]	2798-20-1	55,65
5-Hydroxy-4′,6,7-trimethoxyflavone[s]	19103-54-9	65
5-Hydroxy-4′,7,8-trimethoxyflavone		55
(2E,6E,10E)-12-Hydroxy-3,7,11-trime-thyl-2,6,10-dodecatrienyl acetate		63
(+)-(1R,2R)-2-Hydroxy-2,6,6-trimethyl-norpinan-3-one		41
Hydroxyvalencene		89
13-Hydroxyvalencene		89
Ibuprofen[a]	15687-27-1	5
Imperatorin	see 8-Isopentenyloxypsoralen	
Incensole	22419-74-5	63
Incensole acetate	34701-53-6	63
5-Indanol	1470-94-6	90
1H-Indole	120-72-9	6,42,43,50,60,69,90
2-Indolinone	59-48-3	42,43
Intermedeol[s]	6168-59-8	17,45,51,67,89
α-Ionene[s]	475-03-6	4,30,57,88,90
β-Ionene	84607-57-8	57
(E)-α-Ionol	25312-34-9	30
β-Ionol	22029-76-1	88
α-Ionone	127-41-3	7,15,30,64,78,80,85,86
(E)-α-Ionone	27-41-3	44,73,77,88
β-Ionone	79-77-6	25,30,39,55,57,64,65,73,76,79,80,85,86,88
(E)-β-Ionone	79-77-6	4,15,19,30,44,49,52,55,57,70,88
β-Ionone epoxide	23267-57-4	35
Ipsdienol	35628-00-3	36,45
Isoabienol	10207-79-1	16,25
Isoacarone		10
Isoacorone	6168-64-5	10,38
Isoamyl acetate[s]	123-92-2	4,19,20,44,65,66,68,90
Isoamyl alcohol[s]	123-51-3	4,10,25,35,37,38,41,65,68,73,80,85,86
Isoamyl angelate	10482-55-0	20
Isoamyl benzoate[s]	94-46-2	2,13,14,91
Isoamyl benzyl ether	122-73-6	1,2
Isoamyl butyrate	106-27-4	7,20,35,91
Isoamyl caprylate	2035-99-6	63
Isoamyl formate[s]	110-45-2	37,66
Isoamyl 3-hydroxy-2-methylenebutyrate		20
Isoamyl isobutyrate[s]	2050-01-3	20,34,68
Isoamyl isovalerate[s]	659-70-1	1,4,13,35,41,65,68,79,88,91
Isoamyl methacrylate[s]	7336-27-8	20
Isoamyl 2-methylcrotonate	see Isoamyl tiglate	
Isoamyl 2-methylbutyrate[s]	27625-35-0	1,20,54,63,68,80,91
Isoamyl methyl ketone[s]	110-12-3	38,68
Isoamyl phenylacetate	102-19-2	91
Isoamyl propionate	105-68-0	20,35
Isoamyl tiglate[s]	41519-18-0	20

Table 6.1 Alphabetical list of all chemicals and the essential oils in which they have been identified (*continued*)

Chemical	CAS number	Essential oils in which the chemical has been identified. Legend of numbers: see Table 6.2 at the end of this chapter
Isoamyl valerate	2050-09-1	63
Isoanethole	see Methyl chavicol	
(Z)-Isoapiole		29
Isoaromadendrene epoxide[s]	499134-59-7	8,13,25,36,56,67,72
Isoasarone		66
Isoascaridole	1619-26-7	83,85
Isobazzanene	88661-59-0	16,18
Isobergaptene	482-48-4	1
Isobicyclogermacral		35
Isoborneol	124-76-5	4,8,13,14,19,24,29,31,32,36-38,44,45,48,49,51,52,56,57,63,65, 66,69,70,72,73,75,76,80,84-86,88
Isobornyl acetate[s]	125-12-2	2,4,6,7,12,25,29,31,32,38,41,45,48,49,56,57,62,68,70,72,75, 76,84-86,88,91
Isobornyl butyrate	58479-55-3	44
Isobornyl formate	1200-67-5	19,35,38,49,72,75,80
Isobornyl isovalerate	7779-73-9	32,54,80,88
Isobornyl 2-methylbutanoate	233665-92-4	49,88
Isobornyl 3-methylbutyrate[s]	see Isobornyl isovalerate	
Isobornyl propionate	2756-56-1	35,85
Isobornyl thiocyanoacetate	115-31-1	81
Isobutanal[s]	78-84-2	37,48,68
Isobutanoic acid	see Isobutyric acid	
Isobutanol	see Isobutyl alcohol	
Isobutyl acetate	110-19-0	20
Isobutyl alcohol[s]	78-83-1	37,42,68,73,80
Isobutyl angelate[s]	7779-81-9	20
Isobutylbenzene	538-93-2	2
Isobutyl benzoate	120-50-3	91
Isobutyl butyrate	539-90-2	20,49
Isobutyl crotonate	589-66-2	20
Isobutyl 3-hydroxy-2-methylenebutyrate		20
Isobutyl isobutyrate	97-85-8	20,34,45
Isobutyl isovalerate	589-59-3	1,20,68,79
Isobutyl methacrylate	97-86-9	20
2-Isobutyl-3-methoxypyrazine	see 2-Methoxy-3-isobutylpyrazine	
Isobutyl-2-methylbutyrate[s]	2445-67-2	20,45,68
Isobutyl phenylacetate	102-13-6	37
Isobutyl propionate[s]	540-42-1	20
Isobutyl tiglate	61692-84-0	20
Isobutyric acid[s]	79-31-2	20,37,49,80
8-Isobutyryloxy isobornyl isobutyrate		19
Isocalamenediol	25330-21-6	10
Isocarvacrol	1740-97-2	56
Isocarveol	see p-Mentha-1(7),8-dien-2-ol	
Isocaryophyllene	see γ-Caryophyllene	
Isocaryophyllene oxide		6,8,23,25,34,48,52,57,68,72
Isocaucalol	5172-21-4	83
Isocaurene	see Isokaurene	
Isocedranol	13567-45-8	15,45,89
5-Isocedrol		15
6-Isocedrol	see epi-Cedrol	
Isocentdarol	57308-23-3	93
Isocembrene	25269-16-3	63,78
Isochavicol		26
Isocineole	see 1,4-Cineole	
Isocitral #1	see (Z)-Isocitral	
Isocitral #2	see (E)-Isocitral	
Isocitral #3	see exo-Isocitral	
(E)-Isocitral	see Isogeranial	
(Z)-Isocitral[s]	72203-97-5	4,6,38,51-53,57
exo-Isocitral[s]	55050-40-3	51-53

Table 6.1 Alphabetical list of all chemicals and the essential oils in which they have been identified (*continued*)

Chemical	CAS number	Essential oils in which the chemical has been identified. Legend of numbers: see Table 6.2 at the end of this chapter
Isocomene[s]	65372-78-3	19,75
α-Isocomene[s]	see Isocomene	
β-Isocomene		31
Isocomenene		10
Isocritonilide	62458-57-5	30
Isocumene	74296-31-4	7
Isocurcumenol	24063-71-6	88
Isocyanatomethylbenzene	25550-57-6	89
Isocyclosativene		88
Isodauca-6,9-diene	317819-81-1	90
Isodaucene[s]	142878-08-8	12
Isodecanoic acid	26403-17-8	37
Isodihydrocarveol	18675-35-9	7,64,65,79
Isodihydrocarvyl acetate[s]	220329-20-4	32,68,72
Isodihydronepetalactone		69
Isoelemicin	487-12-7	24
(E)-Isoelemicin	5273-85-8	10,45,51,62
(Z)-Isoelemicin	5273-84-7	10,62
Isoeucalyptol	see 1,4-Cineole	
Isoeugenol[s]	97-54-1	3-5,10,14,21,22,26-28,42,45,49,51,52,62,68,72,85,86,89,90
(E)-Isoeugenol	5932-68-3	7,10,21,26,27,42,44,45,49,51,62,75,89,90
(Z)-Isoeugenol	5912-86-7	10,13,14,22,27,52,62,89
o-Isoeugenol	1076-55-7	3
Isoeugenyl acetate	93-29-8	22,90
(E)-Isoeugenyl acetate	5912-87-8	10,27,45
(Z)-Isoeugenyl acetate		7
Isoeugenyl isovalerate	61114-23-6	88
(E)-Isoeugenol methyl ether	see (E)-Methylisoeugenol	
(Z)-Isoeugenol methyl ether	see (Z)-Methylisoeugenol	
Isofaurinone	87038-80-0	19
Isofenchone	6541-58-8	35,46,47,70,72,80
Isofuranogermacrene	see Curzerene	
Isogeijerene	5975-49-5	3
Isogeranial[s]	72203-98-6	23,38,51-53,57,86
Isogeranic acid		37
Isogeraniol	5944-20-7	37
γ-Isogeraniol	13066-51-8	50,55,73
cis-Isogeraniol	5944-20-7	50,53
Isogermacrene A	783322-20-3	10
Isogermacrene D	317819-80-0	25,41,90,91
Isogermacrone	5975-50-8	91
Isogermacrone epoxide		37
Isoheptanoic acid	1330-19-4	37
Isohexadecanoic acid	32844-67-0	37
Isohexane	73513-42-5	45
Isohexanoic acid[s]	646-07-1	37,72,80,91
Isohomogenol	see Methyl isoeugenol	
Isoimperatorin	482-45-1	50,64
Isoisopulegol	18674-65-2	23,34,37,53,61,65,68,72
Isoitalicene	94482-89-0	12,17
12-Isoitalicenol		77
Isokaurene[s]	5947-50-2	63,84
Isokhusenic acid	see Isozozanoic acid	
Isokhusenol	see Ziza-5-en-12-ol	
Isokhusimol	26128-00-7	89
Isoledene	95910-36-4	3,8,12-16,21,26,35-37,45,57,58,61,67,72,75,83
Isolimonene[s]	499-99-0	2,12,34,44
trans-Isolimonene	6876-12-6	4,51,57,72,89
Isolongifolane		35
Isolongifolan-8-ol	1139-08-8	88
trans-Isolongifolanone	14727-47-0	36,77

Table 6.1 Alphabetical list of all chemicals and the essential oils in which they have been identified (*continued*)

Chemical	CAS number	Essential oils in which the chemical has been identified. Legend of numbers: see Table 6.2 at the end of this chapter
Isolongifolene	1135-66-6	4,7,12,15,16,18,36,37,70,81,87,89
allo-Isolongifolene	87064-18-4	7
Isolongifolene, 4,5,9,10-dehydro-	156747-45-4	52
Isolongifolen-8-ol		35
Isolongifolol	1139-17-9	7,72
Isomenthol[s]	3623-52-7	37,57,64,68,70,72,73,79
Isomenthone[s]	491-07-6	4,6,19,23,37,53,56-58,60,65,68,73,75,79
Isomenthyl acetate	20777-45-1	6,10,66,68,88
Isomeranzin[s]	1088-17-1	39,64
Isomethyleugenol	see (*E*)-Methylisoeugenol	
Isoneomenthol		4,68
Isoneral	see (*Z*)-Isocitral	
Isonerol oxide		73
2-Isononenal	53966-58-8`	65
Isononyl acetate	40379-24-6	48
Isonootkatol	57422-86-3	89
α-Isonootkatol	1380573-94-3	89
β-Isonootkatol		89
Isonootkatone	15764-04-2	89
Isooctane	592-27-8	37
Isooctanoic acid	25103-52-0	37
Isopatchoulenone	3466-15-7	67
Isopentanal	see Isovaleraldehyde	
7-(3'-Isopentenyloxy)coumarin		50
5-Isopentenyloxy-8-(2',3'-dihydroxy-isopentenyloxy)-psoralen		50
5-Isopentenyloxy-8-(2',3'-epoxyiso-pentenyloxy)-psoralen	117030-02-1	50
5-Isopentenyloxy-7-methoxycoumarin	35590-41-1	6,50
5-Isopentenyloxy-8-methoxypsoralen[s]	14348-22-2	6
8-Isopentenyloxypsoralen[s]	482-44-0	2,50
Isopentenyl isovalerate	231623-80-6	35
Isopentyl	See (also) Isoamyl	
Isopentyl acetate	see Isoamyl acetate	
Isopentyl butyrate	see Isoamyl butyrate	
Isopentyl isovalerate	see Isoamyl isovalerate	
Isopentyl 2-methylbutanoate	see Isoamyl 2-methylbutyrate	
Isopentyl nerolate		52
Isophellopterin	see 5-Isopentenyloxy-8-methoxypsoralen	
Isophorone	78-59-1	39,52,65
Isophthal(al)dehyde	626-19-7	26
Isophyllocladene[s]	511-85-3	52,63,78
Isophytol	505-32-8	19,42,43,63
Isophytyl acetate	58425-36-8	42
Isopimaradiene	54605-21-9	78
Isopimara-9(11),15-diene	39702-28-8	31,32,41,75
Isopimpinellin	see 5,8-Dimethoxy-6,7-furanocoumarin	
Isopinocampheol	27779-29-9	41,52,63,72
Isopinocamphone[s]	15358-88-0	12,31,35,36,41,57,72,75,76,78,85
Isopinocarveol	see *cis*-Pinocarveol	
Isopinol	110268-86-5	45
Isopiperitenol	491-05-4	65
cis-Isopiperitenol	4017-76-9	55
trans-Isopiperitenol	see *trans-p*-Mentha-2,8-dien-1-ol	
Isopiperitenone[s]	529-01-1	55,68,72,79,82
Isopiperitone	58615-39-7	50,64,65,68,79
Isoprene	78-79-5	37,80
Isoprenyl acetate[s]	17616-47-6	90
Isoprenyl isovalerate		85
4-Isopropenylbenzaldehyde	10133-50-3	20,80

Table 6.1 Alphabetical list of all chemicals and the essential oils in which they have been identified (*continued*)

Chemical	CAS number	Essential oils in which the chemical has been identified. Legend of numbers: see Table 6.2 at the end of this chapter
10-Isopropenyl-3,7-cyclodecadien-1-one		63
8-Isopropenyl-1,5-dimethyl-1,5-cyclo-decadiene		9
6-Isopropenyl-4,8a-dimethyl-1,2,3,5, 6,7,8,8a-octahydronaphthalen-2-ol		63,89
1-Isopropenyl-3-methylbenzene	see *m*-Cymenene	
4-Isopropenyl-1-methyl-1,2-cyclohexa-nediol		55
4-Isopropenyl-3-methyl-2-cyclohex-en-1-one		80
2-Isopropenyl-5-methylhex-4-enal	6544-40-7	53
(*E*)-7-Isopropenyl-4-methyl-10-me-thylene-4-cyclodecen-1-one		37
5-Isopropenyl-2-methyltetrahydro-furan-2-ethanal		80
2-Isopropenyl-5-methyl-5-vinyltetra-hydrofuran		60
2-Isopropenyl-1,3,5-trimethylbenzene	14679-13-1	89
Isopropyl acetate	108-21-4	14
p-Isopropylacetophenone	645-13-6	48
Isopropyl angelate	61692-76-0	1,20
p-Isopropyl anisole	4132-48-3	31,38
4-Isopropylbenzoic acid	536-66-3	72,80
S-Isopropyl benzothioate		36
Isopropylbenzyl acetate	see Cuminyl acetate	
4-Isopropylbenzyl alcohol	see Cuminyl alcohol	
5-Isopropylbicyclo[3.1.0]hexan-2-one	see Sabina ketone	
4-Isopropyl-4-butanolide	38624-29-2	49,80
Isopropyl *n*-butyrate	638-11-9	66
Isopropyl cresol	see Carvacrol	
4-Isopropyl-2-cyclohexenone	see Cryptone	
3-Isopropylcyclopentanone	10264-56-9	80
7-Isopropyl-1,4-dimethyl-2-azulenol		58
Isopropyl formate	625-55-8	37
3-Isopropylfuran	15012-74-5	80
2-Isopropylglutaric acid	32806-63-6	80
Isopropyl hexadecanoate[s]	142-91-6	73
3-Isopropyl-1,6-hexanedioic acid		80
4-Isopropylidene-1-vinyl-*o*-menth-8-ene	see Elixene	
Isopropyl isovalerate	32665-23-9	1
2-Isopropyl-3-methoxypyrazine	25773-40-4	36,89
1-Isopropyl-2-methoxy-4-methyl-benzene	see Methyl thymol	
2-Isopropyl-5-methylanisole	see Methyl thymol	
S-Isopropyl 3-methyl-2-butenthioate		36
Isopropyl 2-methylbutyrate	66576-71-4	1,19,20,45,68
4-Isopropyl-1-methyl-2-cyclohexen-1-ol	see *trans*-*p*-Menth-2-en-1-ol	
5-Isopropyl-2-methyl-1-cyclopen-tene-1-carboxaldehyde		80
3-Isopropyl-6-methyl-2-(3-methylcyclo-pent-2-enyl)-3,4-dihydro-2*H*-pyran		89
3-Isopropyl-2(3-methylcyclopent-2-enyl)-tetrahydrofurfuran		89
2-Isopropyl-5-methyl-(2*Z*)-hexanal	66656-67-5	7
2-Isopropyl-5-methyl-9-methylene-bicyclo[4.4.0]dec-1-ene	150320-52-8	35,37,41,48,76
8-Isopropyl-5-methyl-2-methylene-1,2,3,4,4a,5,6,7-octahydronaphthalene		41

Table 6.1 Alphabetical list of all chemicals and the essential oils in which they have been identified (*continued*)

Chemical	CAS number	Essential oils in which the chemical has been identified. Legend of numbers: see Table 6.2 at the end of this chapter
3-(2-Isopropyl-5-methylphenyl)-2-methylpropionic acid		89
2-Isopropyl-4-methylpyridine	4855-56-5	37
4-Isopropyl-6-methyl-1,2,3,4-tetrahydronaphthalen-1-one	see 4-Isopropyl-6-methyl-1-tetralone	
4-Isopropyl-6-methyl-1-tetralone[s]	57494-10-7	10,49
5-Isopropyl-2-methyl-2-vinyltetrahydrofuran	71635-17-1	80
Isopropyl myristate	110-27-0	7,27,41,73
Isopropyl palmitate	see Isopropyl hexadecanoate	
3-Isopropylpentanoic acid	60308-89-6	30
13-Isopropyl podocarpa-8,13-dien-15-ol	21414-53-9	61
Isopropyl propionate	637-78-5	66
4-Isopropylpyridine	696-30-0	80
2-Isopropylsuccinic acid	2338-45-6	80
S-Isopropyl thiotiglate		36
Isopropyl tiglate	1733-25-1	1
2-Isopropyltricyclo[4.3.1.1(2,5)]undec-3-en-10-ol		88
Isopulegol	89-79-2	6,23,24,34-37,39,49-53,55,57,60,61,64,65,68,69,72,75, 79,85,86,91
cis-Isopulegol		34
trans-Isopulegol		34
Isopulegol II[s]	1370348-44-9	24,52,72
Isopulegol acetate	57576-09-7	4,34,35,68,72,79
Isopulegone	29606-79-9	52,53,68,79
cis-Isopulegone		56,79
trans-Isopulegone		56
Isoquinoline	119-65-3	21
Isosafrole	120-58-1	65,90
cis-Isosafrole	17627-76-8	45
Isosativene	24959-83-9	13
Isoshyobunone	21698-46-4	10
Isosinensetin	see 3′,4′,5,7,8-Pentamethoxyflavone	
Isospathulenol	88395-46-4	4,8,10,25,34,35,41,45,56,83,88
Isosqualenol		42
Isosylvestrene	61557-13-9	62,65
Isoterpinene	see α-Terpinolene	
Isoterpineol		49
Isoterpinolene[s]	586-63-0	1,2,4,8-10,12,25,32,35-37,41,45,49,62,63,69,70,72,75,79,84
(+)-3-Isothujanol	see Neothujol	
Isothujol	see Thujyl alcohol	
Isothujone	59573-80-7	68
δ-Isothujone	see β-Thujone	
3-Isothujopsanone	25966-81-8	15,16,77
Iso-3-thujanol acetate		75
Isothujyl acetate		49,83
Isothujyl alcohol		54
Isothymol	4427-56-9	56,85
Isothymol methyl ether	31574-44-4	88
Isoundecanoic acid	2724-56-3	37
Isovalencenal		89
(*E*)-Isovalencenal	137695-18-2	89
(*Z*)-Isovalencenal	137695-20-6	89
Isovalencenol		89
(*E*)-Isovalencenol	22387-74-2	89
(*Z*)-Isovalencenol	22387-74-2	89
Isovalencenyl methyl ether	300349-26-2	89
Isovaleraldehyde[s]	590-86-3	1,4,29,35,37,45,48,66,68,73,79,80,85,90
Isovaleric acid[s]	503-74-2	4,13,30,37,67,68,72,80,88,90,91
Isovanillic acid	645-08-9	67

Table 6.1 Alphabetical list of all chemicals and the essential oils in which they have been identified (*continued*)

Chemical	CAS number	Essential oils in which the chemical has been identified. Legend of numbers: see Table 6.2 at the end of this chapter
Isovelleral	37841-91-1	10,58,88
Isovellerdiol	37841-93-3	89
Isovetiselinenol	67690-36-2	89
Isozizanoic acid[s]	16202-79-2	89
Italicene	94535-52-1	12,16,38,58,68,88,91
Italicene epoxide	104188-24-1	12
Italicene ether	104188-25-2	30,91
10-epi-Italicene ether	104265-25-0	91
Jacksone	1228757-70-7	35
Jaeschkeanadiol	41690-67-9	36
Jasmine ketolactone	70981-24-7	42
Jasmine lactone	see δ-Jasmolactone	
δ-Jasmolactone[s]	25524-95-2	42,43
Jasmone	488-10-8	35,72,79
(*E*)-Jasmone[s]	6261-18-3	4,41,42,60,68,75
(*Z*)-Jasmone[s]	488-10-8	4,34,41-43,45,60,68,72,75,79,85,86
trans-(*E*)-Jasmonol		88
Jasmonyl acetate	149982-46-7	42
Junenol	472-07-1	16,37,44,89-91
laevo-Junenol	see Levojunenol	
Junicedranol	168180-13-0	16,89
Junicedranone		16,18
Junipene	see Longifolene	
Juniper camphor[s]	473-04-1	12,23,27,34,37,38,44,52,58,88,89,91
Junipercedrol	175448-28-9	16
laevo-Jujenol	see Levojunenol	
Karahanaenone	19822-67-4	7,31
Kauran-18-al		8,9
Kauran-18-al, 17-(acetyloxy)-[s]	see 17-(Acetyloxy)kauran-18-al	
Kaurane	1573-40-6	31
Kaurene	34424-57-2	63
Kaur-15-ene	see Isokaurene	
(5α,9α,10β)-Kaur-15-ene	see Isophyllocladene	
Kessane	3321-66-2	10,88,89
Kessanyl acetate	17806-59-6	88
Kessoglycyl monoacetate		88
α-Kessyl acetate	3925-77-7	88
α-Kessyl alcohol	3321-65-1	88
11-Keto-dihydro-α-santalic acid		77
Ketone,1-[4,4-(4-methylpentyl)-3-cyclo-hexane-1-yl)]		72
Khusenediol		89
Khusenic acid[s]	16203-25-1	89
Khusian-2-ol	see Khusiol	
2-epi-Khusian-2-ol	300349-32-0	89
Khusimene[s]	18444-94-5	89
Khusimol	16223-63-5	15,89
Khusimone	30557-76-7	89
Khusinal		89
Khusinol	24268-34-6	10,57,89
epi-Khusinol		89
Khusinol acetate	78405-34-2	89
nor-Khusinol oxide		89
Khusiol[s]	66512-56-9	10,16,89
allo-Khusiol	312296-11-0	89
Khusione[s]	66397-72-6	89
Khusitoneol	102818-81-5	89
Khusol[s]	18045-73-3	89
Kobusone	24173-71-5	49
Kusunol	see Valerianol	
Labda-7,14,-dien-13-ol	40185-30-6	25

Table 6.1 Alphabetical list of all chemicals and the essential oils in which they have been identified (*continued*)

Chemical	CAS number	Essential oils in which the chemical has been identified. Legend of numbers: see Table 6.2 at the end of this chapter
(*E,Z*)-11,13-Labdadien-8-ol		31
13(16),14-Labdadien-8-ol		31
Laciniatafuranone E	199115-10-1	83
Lanceol	10067-29-5	12
(*E*)-Lanceol	198828-12-5	88
(*Z*)-Lanceol	859202-95-2	38,63,75,77,88
Lanceol acetate	199273-99-9	25
Lauric acid	see Dodecanoic acid	
Lavandulol[s]	498-16-8	6,19,24,25,29,37,45-49,72,76,79,80,85,86
Lavandulyl acetate	25905-14-0	4,19,37,46-49,66,69,72,73,80
Lavandulyl benzoate	59550-37-7	48
Lavandulyl butyrate	59550-35-5	46-48
Lavandulyl caproate	59550-36-6	48
Lavandulyl isobutyrate	51117-20-5	46,49,80
Lavandulyl isovalerate[s]	51117-21-6	48,49,80
Lavandulyl 2-methylbutyrate		47,49,80
Lavandulyl 3-methylbutyrate	see Lavandulyl isovalerate	
Lavandulyl α-methylbutyrate	921210-84-6	48
Lavandulyl propionate		80
Lavender lactone	1073-11-6	49
Ledane	28580-43-0	90
Ledene	see Viridiflorene	
(+)-Ledene	see Viridiflorene	
Ledene alcohol	1197210-11-9	23
Ledene epoxide		52
Ledene oxide	882187-44-2	12,23,36,83,88,89
Ledene oxide II		8,49,68
Ledol	577-27-5	4,7,9,13,19,24,26,27,34-37,41,45,49,52,56,57,61,63,67,72, 75,76,80,83,85,86,88
Lemonol	see Geraniol	
Leoidosene		37
Lepalone[s]	80445-58-5	20
Lepidozenal		30
Lepidozene	133005-43-3	19,45
Levoglucosenose	37112-31-5	90
Levojunenol[s]	30951-17-8	89
Levomenol[s]	23089-26-1	8,12
Levopimaradiene[s]	122712-77-0	63
Levulinic acid	123-76-2	72
Liguloxide	21764-22-7	36,70
Ligustilide[s]	4431-01-0	54,70
(*E*)-Ligustilide	81944-09-4	54
(*Z*)-Ligustilide	81944-09-4	54
Lilac alcohol A	33081-34-4	69
Lilac aldehyde	67920-63-2	60
Lilac aldehyde A	53447-46-4	4
Lilial®[a]	80-54-6	6
Limetol		37
Limonenal[s]	6784-13-0	4,77
Limonen-10-al	57074-31-4	35
Limonene[s]	138-86-3	1-39,41-67,69-82,84-93
α-Limonene	see *dl*-Limonene	
β-Limonene	5989-54-8	31,90
D-Limonene ((*R*)-)	5989-27-5	31,53
dl-Limonene[s]	138-86-3	9
Psi-Limonene (ψ-)	see Pseudolimonene	
Limonene aldehyde	see Limonenal	
Limonene di(ep)oxide	96-08-2	9,36,39,48,50,55,64,65,82
Limonene-1,2-diol	see Limonene glycol	
Limonene (ep)oxide	1195-92-2	2,8,33,36,38,39,49,52,55,57,64,65,68,72,82,83
Limonene-1,2-epoxide	see Limonene (ep)oxide	
Limonene-1,2-epoxide isomer		68

Table 6.1 Alphabetical list of all chemicals and the essential oils in which they have been identified (*continued*)

Chemical	CAS number	Essential oils in which the chemical has been identified. Legend of numbers: see Table 6.2 at the end of this chapter
cis-Limonene (ep)oxide[s]	13837-75-7	2,4,6,7,12,23,26,27,35,38,39,45,50,53,55,57,60,63-65,72,82, 83,85,86
cis-Limonene-1,2-epoxide	see *cis*-Limonene (ep) oxide	
trans-Limonene (ep)oxide[s]	4959-35-7	6,7,27,39,45,49-51,53,55,57,60,64,65,76,82,85,86
trans-Limonene-1,2-epoxide	see *trans*-Limonene (ep) oxide	
Limonene glycol[s]	1946-00-5	55,65
Limonen-4-ol[s]	3419-02-1	6,7,35,41,55,56,65,72,91
Limonen-10-ol[s]	3269-90-7	6,39,50,55,64,82,85
Limonen-6-ol, pivalate		8,53,88
Limonen-10-yl-acetate[s]	15111-97-4	6,39,50,55,65
Limonol	989-61-7	53
Limonol acetate		53
Limonol formate		53
α-Linalool	598-07-2	4,35,49
Linalool (β-)[s]	78-70-6	1-14,19,21-39,41-57,60-74,76-86,88-92
Linalool oxide[s]	1365-19-1	4,22,29,34-36,39,43,48,49,51,53,55-57,60,63-66,68, 69,72,73-76,79,80,86
Linalool oxide I	see *cis*-Linalool oxide, furanoid	
Linalool oxide II	see *trans*-Linalool oxide, furanoid	
Linalool oxide A	see *trans*-Linalool oxide	
Linalool oxide A, furanoid	see *trans*-Linalool oxide, furanoid	
Linalool oxide B	see *cis*-Linalool oxide	
Linalool oxide B, furanoid	see *cis*-Linalool oxide, furanoid	
Linalool oxide D	see *cis*-Linalool oxide, pyranoid	
cis-Linalool oxide	11063-77-7	4-6,10-12,15,19,24,25,27,29,34,37,39,41-43,45-49,51-53,55, 56,64,65,68,69,72,74,75,80,81,83,85,86,91
cis-Linalool oxide-5		49
cis-Linalool oxide-6		49
cis-Linalool oxide, furanoid[s]	5989-33-3	5,6,8,13,21-23,25,29,34,35,37-39,42,43,47-49,52,53, 57,60,61,64-66,69,71-74,79,80,85,86,90
cis-Linalool oxide, pyranoid[s]	14009-71-3	4,6,29,43,49,60,69,74,80,85,86
trans-Linalool oxide	11063-78-8	4-6,8,10,12,19,25-27,29,34,37-39,41,44-49,52,53,55,56,64, 65,68,69,73-75,80,85,86
trans-Linalool oxide-5		49
trans-Linalool oxide-6		49
trans-Linalool oxide, furanoid[s]	34995-77-2	4-6,8,11,13,21,22,25,29,34,35,37,39,42,43,47-49,52,53, 56,57,60,61,64-66,69,71-74,79-81,85,86,90,91
trans-Linalool oxide acetate	56469-40-0	38,90
trans-Linalool oxide, pyranoid	39028-58-5	6,29,49,57,60,69,74,80,85,90
Linalyl acetate[s]	115-95-7	1,3,4,6,7,9,10,12,13,21-23,25-27,29,31,32,34,35,37-39,41-43, 45-53,55-57,60-62,64-66,68-70,72,73,75,76,78-80,82,84-86, 88,89-91
Linalyl acetate oxide	477705-86-5	6,51
Linalyl *o*-aminobenzoate	see Linalyl anthranilate	
Linalyl anthranilate[s]	7149-26-0	48,49,50,55,56,60,64
Linalyl benzoate	126-64-7	90
Linalyl butyrate	78-36-4	48,49,68,80
Linalyl formate	115-99-1	4,25,48,49,65,66
Linalyl geranyl ether	see Geranyllinalool	
Linalyl hexanoate	7779-23-9	47,49
Linalyl isobutyrate	78-35-3	21,49,50,68,72,76
Linalyl isovalerate	1118-27-0	47,49,88
Linalyl pentanoate	10471-96-2	49
Linalyl propanoate	see Linalyl propionate	
Linalyl propionate[s]	144-39-8	4,6,9,10,22,29,41,60,66,68,69,72,85
Lindestrene	2221-88-7	58
Linoleic acid[s]	60-33-3	3,7,19,30,37,43,50,59,67,73,79,88
cis-Linoleic acid		19
Linolenic acid	463-40-1	37,42,43,73
Linolic acid	see Linoleic acid	
3-Longibornene	see Longifolene	

Table 6.1 Alphabetical list of all chemicals and the essential oils in which they have been identified (*continued*)

Chemical	CAS number	Essential oils in which the chemical has been identified. Legend of numbers: see Table 6.2 at the end of this chapter
Longiborneol	465-24-7	15,25,44,78,93
Longiborneol acetate	36204-27-0	40,88
Longicamphenylone	38647-26-6	12,16,67
Longicyclene	1137-12-8	1,2,7,34,37,38,45,63,78,83,88,89
Longifolenal	see Longifolenaldehyde	
Longifolenaldehyde[s]	66537-42-6	12,38,49,56,89,93
Longifolene[s]	475-20-7	1,7,9,10,12,15,16,23,25,31,32,34-37,39,44,45,49,54,56-58,61, 63,67,68,70,72,73,75,77,78,80,81,83,85,87,89,93
Longifolene-(V4)		13,58
(+)-Longifolene	see α-Longifolene	
α-Longifolene[s]	475-20-7	30,67,89
Longipinanol	66141-14-8	1,2,4,67,88
trans-Longipinalol	66141-14-8	61,83
epi-Longipinanol		19
Longipinene	see α-Longipinene	
α-Longipinene[s]	5989-08-2	1,3,7,11,12,15,19,34,36,55,70,78,87-89,93
β-Longipinene	41432-70-6	7,31,45,67,75,77
trans-Longipinocarveol	889109-69-7	25,38,56,67,88
Longipinocarvone	65556-52-7	25,88
Lyral[®a]	130066-44-3	45,72,75
Lyratal		72
Maaliane	527-91-3	63
α-Maaliene	489-28-1	9
β-Maaliene	489-29-2	19,37,41,45,52,67,72,83
γ-Maaliene	20071-49-2	4,83
Maaliol	527-90-2	4,88,89
Machilol	see γ-Eudesmol	
Maleic acid	110-16-7	80
Malonic acid	141-82-2	80
Maltol	see 3-Hydroxy-2-methyl-4-pyrone	
Manool	596-85-0	15,25,31,72,75,76
epi-Manool[s]		25,53,75
13-epi-Manool	see epi-Manool	
Manoyl oxide	596-84-9	15,16,25,31,32,44,57,70,72,75,78
epi-Manoyl oxide[s]	1227-93-6	25,31,32,44,53,70,91
13-epi-Manoyl oxide	see epi-Manoyl oxide	
2-keto-Manoyl oxide		32
Marmin[s]	14957-38-1	39
6′,7′-Marmin decanal acetal	1181223-78-8	39
Marrubine		73
epi-Marsupellol	486452-15-7	89
4-epi-Marsupellol		89
Matricin	29041-35-8	19
Mayurone	4677-90-1	16-18,85,88
Megastigma-4,6(*E*),8(*Z*)-triene	71186-24-8	26
Megastigmatrienone	13215-88-8	72
Melonal	see Bergamal	
p-Mentha-1,3-dien-7-al[s]	1197-15-5	7,12,29,41,44,72,79,80,86
p-Mentha-1,4-dien-7-al[s]	22580-90-1	7,29,48
α-*p*-Menthadiene		8
p-Mentha-1,3-diene	see α-Terpinene	
p-Mentha-1,4-diene	see γ-Terpinene	
p-Mentha-1,4(8)-diene	see Terpinolene	
p-Mentha-1,8-diene	see Limonene	
p-Mentha-1(7),8-diene	see Pseudolimonene	
p-Mentha-2,4(8)-diene	see Isoterpinolene	
p-Mentha-2,8-diene	see Isolimonene	
trans-*m*-Mentha-2,8-diene		2,45
m-Mentha-3(8),6-diene	25946-29-6	8
p-Mentha-3,8-diene	586-67-4	23,34,38,79
p-Mentha-7,8-diene		72
Menthadienol		5

Table 6.1 Alphabetical list of all chemicals and the essential oils in which they have been identified (*continued*)

Chemical	CAS number	Essential oils in which the chemical has been identified. Legend of numbers: see Table 6.2 at the end of this chapter
α,*p*-Menthadienol		8
Menthadien-1-ol		55
Menthadien-8-ol	see α-Phellandren-8-ol	
p-Mentha-1(7),2-dien-8-ol	see β-Phellandren-8-ol	
m-Mentha-1,3-dien-8-ol		8
p-Mentha-1,4-dien-7-ol	22539-72-6	12,29,45,69,73
p-Mentha-1,5-dien-7-ol	19876-45-0	35,36,45
p-Mentha-1,5-dien-8-ol	see α-Phellandren-8-ol	
p-Mentha-1(7),5-dien-2-ol	30681-15-3	2
cis-*p*-Mentha-1(7),5-dien-2-ol	30681-15-3	35
trans-*p*-Mentha-1(7),5-dien-2-ol	30681-15-3	35
p-Mentha-1,8-dien-4-ol	see Limonen-4-ol	
p-Mentha-1,8-dien-6-ol	see Carveol	
cis-*p*-Mentha-1,8-dien-6-ol	see (*Z*)-Carveol	
trans-p-Mentha-1,8-dien-6-ol	see (*E*)-Carveol	
Mentha-1,8-dien-7-ol	see Perillyl alcohol	
p-Mentha-1,8-dien-7-ol[s]	see Perillyl alcohol	
p-Mentha-1(2),8-dien-10-ol	see Limonen-10-ol	
cis-*p*-Mentha-1(7),8-dien-1-ol		12
trans-*p*-Mentha-1(7),8-dien-1-ol		12
p-Mentha-1(7),8-dien-2-ol[s]	35907-10-9	35,36,41,51,61
cis-*p*-Mentha-1(7),8-dien-2-ol	22626-43-3	34,35,45,50,55,65,72,82,83
trans-*p*-Mentha-1(7),8-dien-2-ol	21391-84-4	35,45,50,53,55,57,65,81
p-Mentha-1,8-dien-9-ol	1946-01-6	6,50,55,64,82
p-Mentha-1(7),8(10)-dien-9-ol	29548-13-8	12,52,65,79,85,88
p-Mentha-1,8(9)-dien-10-ol		64
p-Mentha-1,8(10)-dien-9-ol	see Limonen-10-ol	
trans-*p*-Mentha-2,8-dienol	see *trans*-*p*-Mentha-2,8-dien-1-ol	
Mentha-2,8-dien-1-ol	58940-40-2	36
p-Mentha-2,8-dien-1-ol	22771-44-4`	65,79
cis-*p*-Mentha-2,8-dien-1-ol	3886-78-0	6-8,10,12,38,39,44,45,50,52,53,55,57,65,68,79,82,83
trans-*p*-Mentha-2,8-dien-1-ol[s]	4017-77-0	6-8,12,35,37,39,41,45,49,50,53,55,57,60,64,65,68,79,82,88
cis-*p*-Mentha-2,8-dien-9-ol		64,65
trans-*p*-Mentha-2,8-dien-9-ol		50,65
cis-β-Mentha-6,8-dien-2-ol		70
p-Mentha-1,3-dien-7-ol, acetate	81893-40-5	6
p-Mentha-1,8-dien-3-one	see Isopiperitenone	
(+)-*p*-Mentha-1,8-dien-3-one	16750-82-6	55,65,72
p-Mentha-1,7-dien-4-yl acetate		6
p-Mentha-1,7(10)-dien-2-yl acetate		6
p-Mentha-1,8-dien-10-yl acetate		65
p-Mentha-1,8(1)-dien-9-yl acetate	see Limonen-10-yl acetate	
p-Mentha-1,8(9)-dienyl acetate		60,69
p-Mentha-1,8(9)-dien-10-yl acetate		39,64
p-Mentha-1,8(10)-dien-9-yl acetate	see Limonen-10-yl-acetate	
p-Mentha-1(2),8-dien-10-yl acetate		82
Menthalactone	13341-72-5	68
p-Menthane	99-82-1	44
p-Menthane-3,8-diol[s]	42822-86-6	34,53,68
p-Menthane-1,2,3-triol		56
p-Menthane-1,2,4-triol	see 1,2,4-Trihydro-xymenthane	
7-Menthanoazulene		67
Menthanol		41
p-Menthanol		41
3-*p*-Menthanol	see Menthol	
trans-*p*-Menthan-8-ol		8
p-Menthan-8-yl acetate	20777-41-7	31
p-Menthatriene	116868-92-9	35
α,*p*-Menthatriene		8
cis-Mentha-1,3,8-triene		64
p-Mentha-1,3,8-triene	18368-95-1	6-9,35,39,44,45,49,52,53,55,57,61,63,64,65
p-Mentha-1,4,8-triene	28233-65-0	91

Table 6.1 Alphabetical list of all chemicals and the essential oils in which they have been identified (*continued*)

Chemical	CAS number	Essential oils in which the chemical has been identified. Legend of numbers: see Table 6.2 at the end of this chapter
p-Mentha-1,5,8-triene	21195-59-5	7
p-Mentha-2,4(8),6-triene-2,3-diol		79
p-Menth-1-en-7-al	see Phellandral	
p-Menth-1-en-9-al[s]	29548-14-9	25,36,37,55,65,73,80,83,91
1-*p*-Menthen-9-al	see *p*-Menth-1-en-9-al	
Menthene	29350-67-2	13,36
1-*p*-Menthene	see 1-Carvomenthene	
3-Menthene	500-00-5	68,72
cis-*p*-Menth-1-ene-3,6-diol		81
trans-*p*-Menth-2-ene-1,4-diol	21473-37-0	62
p-Menth-2-ene-1,8-diol	57030-53-2	36
cis-*p*-Menth-2-ene-1,8-diol	54164-91-9	45
trans-*p*-Menth-2-ene-1,8-diol	54164-90-8	45
cis-*p*-Menth-4-ene-1,2-diol		2
p-Menth-6-ene-2,3-diol		83
o-Menth-8-ene	15193-25-6	25
1-*p*-Menthene-8-ethyl acetate		45
(*E*)-2-Menthenol	see *trans*-*p*-Menth-2-en-1-ol	
(*Z*)-2-Menthenol	see *cis*-*p*-Menth-2-en-1-ol	
cis-*p*-Menthen-1-ol	35376-39-7	41
trans-*p*-Menthen-1-ol	586-23-2	41
p-Menthen-4-ol	see Terpinen-4-ol	
p-Menthen-8-ol	see α-Terpineol	
p-Menth-1-en-4-ol	see Terpinen-4-ol	
p-Menth-1(7)-en-4-ol		72
p-Menth-1-en-8-ol	see α-Terpineol	
p-Menth-1-en-9-ol	18479-68-0	34,39,50,55,64,65,72
cis-*p*-Menth-1-en-9-ol	18479-68-0	55,68
p-Menth-2-en-1-ol[s]	619-62-5	1,4,36,38,45,49,51,55,56,63,68,79,83
cis-*p*-Menth-2-en-1-ol[s]	29803-82-5	1,2,4,6,8,10,12,19,24,25,30,31,32,35,37,38,41,44,45,49,50,55-57,60-62,65,68,69,71,72,75,79,83-86
trans-*p*-Menth-2-en-1-ol[s]	29803-81-4	2,4,6-8,10,24,31,32,35,41,44,45,50,53-56,60-62,65,66,68,69,71,72,83,85,86
p-Menth-2-en-7-ol		7
trans-*p*-Menth-2-en-7-ol	19898-87-4	91
cis-*p*-Menth-2-en-8-ol		34
trans-*p*-Menth-2-en-8-ol		34
p-Menth-3-en-1-ol	see 1-Terpineol	
p-Menth-3-en-8-ol	18479-65-7	23,79
p-Menth-4(8)-en-9-ol	15714-11-1	65
p-Menth-8-en-1-ol	see β-Terpineol	
p-Menth-8-en-2-ol	see Dihydrocarveol	
cis-*p*-Menth-8-en-2-ol		8
trans-p-Menth-8-en-2-ol		8
p-Menth-1-en-3-ol, acetate[s]	1204-30-4	31
p-Menth-1-en-8-ol, acetate	see α-Terpinyl acetate	
p-Menth-8-en-2-ol, acetate	see Dihydrocarvyl acetate	
p-Menth-1-en-3-one	see Piperitone	
p-Menth-3-en-2-one	see Carvenone	
p-Menth-6-en-2-one	43205-82-9	69
p-Menth-1-en-4,5-oxide		6,55
p-Menth-4-en-1,2-oxide		6,55
p-Menth-1-en-3-yl acetate	see *p*-Menth-1-en-3-ol, acetate	
p-Menth-1-en-8-yl acetate	see α-Terpinyl acetate	
p-Menth-1-en-9-yl acetate	28839-13-6	6,64,68,79
Menthofuran	494-90-6	4,45,56,68,79
Menthofurolactone		68
Menthofurolactone isomer		68
Menthol	89-78-1	1-4,6,10,13,25,29,34,35,37,38,41,49,52,57,58,68,70,73,75,79,84-86
cis-Menthol	see Isomenthol	
Mentholactone	68330-67-6	68

Table 6.1 Alphabetical list of all chemicals and the essential oils in which they have been identified (*continued*)

Chemical	CAS number	Essential oils in which the chemical has been identified. Legend of numbers: see Table 6.2 at the end of this chapter
Menthomenthol	98167-53-4	37
Menthone (*p*-; *trans*-)	89-80-5	2,4,6,10,19,25,30,35,37,38,41,45,50,52,56-58,62,65,68, 70,72,73,79,83,85,88,90
cis-Menthone	see Isomenthone	
trans-Menthone	see Menthone	
L-Menthone	14073-97-3	19
p-Menth-8-one		50
p-Menthone-1,2,3-triol		68
Menthyl acetate	16409-45-3	4,10,19,29,57,65,68,70,73,79
cis-Menthyl acetate		68
trans-Menthyl acetate		68
p-Menth-8-yl acetate		49,56
Menthyl formate	2230-90-2	68
Menthyl isovalerate	16409-46-4	62
Meranzin[s]	23971-42-8	39,64
Meranzin hydrate	5875-49-0	39,64,65
4-Mercapto-4-methyl-2-pentanol	31539-84-1	39
Mesityl oxide	see 4-Methyl-3-penten-2-one	
Methacrylic acid	79-71-4	37
Methallyl angelate[s]	61692-78-2	20
Methallyl methacrylate[s]	816-74-0	20
1,4-Methanoazulene	249-73-0	34
1,4-Methanoazulene, decahydro-4,8,8-trimethyl-9-methylene-, [1S-(1α,3aα,4α,8aα)]-	see α-Longifolene	
1,4-Methanoazulene, decahydro-4,8,8-trimethyl-9-methylene-,(1S,3aR,4S,8aS)-	see α-Longifolene	
1,4-Methanoazulen-7-one		25
Methanol	67-56-1	37,73,75
Methional	3268-49-3	39
2-Methoxyacetophenone	see *o*-Acetanisole	
3'-Methoxyacetophenone	586-37-8	45
p-Methoxyacetophenone (4'-)	see *p*-Acetylanisole	
2-Methoxybenzaldehyde	135-02-4	13,14,21
3-Methoxybenzaldehyde	see *m*-Anisaldehyde	
4-Methoxybenzaldehyde	see *p*-Anisaldehyde	
7-Methoxybenzofuran	7168-85-6	31
o-Methoxybenzoic acid	529-75-9	72
p-Methoxybenzoic acid	see *p*-Anisic acid	
9-Methoxycalamenene		8
(*Z*)-2-Methoxycinnamaldehyde	76760-43-5	13
(*E*)-2-Methoxycinnamaldehyde (*o*-)	60125-24-8	13,14,21
3-Methoxycinnamaldehyde	56578-36-0	4
o-Methoxycinnamaldehyde (2-)[s]	1504-74-1	4,13,14,21
p-Methoxycinnamaldehyde	1963-36-6	21,81
(*E*)-*p*-Methoxycinnamaldehyde	24680-50-0	4,21
(*E*)-2-Methoxycinnamic acid	1011-54-7	13
(*Z*)-2-Methoxycinnamic acid	14737-91-8	13
2-Methoxycinnamic alcohol		4,13
(*E*)-2-methoxycinnamyl acetate	38822-47-8	13
o-Methoxycinnamyl acetate (2-)	110823-66-0	13
7-Methoxycoumarin[s]	531-59-9	6,48,80
2-Methoxydihydrocinnamic acid[s]	6342-77-4	13,80
5-Methoxy-8-(2,3-dihydroxy-3-methyl-butoxy)psoralen[s]	482-25-7	6,39,50
2-Methoxy-3,8-dioxocephalotax-1-ene	114942-83-5	72
10-Methoxy-7,10-epoxysalvialane	300349-03-5	89
(1RS,2RS,1SR)-1-(1 Methoxyethyl)-2-vinylcyclobutane		58
Methoxyeugenol (6-)[s]	6627-88-9	5,42,62
5-Methoxyeugenol	90377-06-3	62

Table 6.1 Alphabetical list of all chemicals and the essential oils in which they have been identified (*continued*)

Chemical	CAS number	Essential oils in which the chemical has been identified. Legend of numbers: see Table 6.2 at the end of this chapter
6-Methoxyeugenol	see Methoxyeugenol	
7-Methoxy-8-(2-formyl-2-methylpro-pyl)-coumarin	see Coumarin, 8-(2-formyl-2-methylpropyl)-7-methoxy-	
2-Methoxyfuran	25414-22-6	26
2-Methoxyfuranodiene		58
4-Methoxy-7*H*-furo[3,2-g][1] benzo-pyran-7-one	see Bergapten	
5-Methoxy-8-geranyloxypsoralen		50
1-Methoxyhexane[s]	4747-07-3	46-49
1-Methoxyhexane-3-thiol		25
5-Methoxy-7-hydroxycoumarin	3067-10-5	6
6-Methoxy-7-hydroxycoumarin[s]	92-61-5	50
2-Methoxy-3-isobutylpyrazine[s]	24683-00-9	36,69
7-Methoxy-8-isopentenylcoumarin	see Osthole	
5-Methoxy-8-isopentenyloxypsoralen		50
4-Methoxy-5-[3-(4-methoxyphenyl)-oxaziridin-2-yl]pentan-1-ol		81
7-Methoxy-8-(3-methyl-2-butenyl)-2*H*-1-benzopyran-2-one	see Osthole	
7-Methoxy-6-(3-methyl-2-butenyl)-coumarin[s]	581-31-7	6
1-Methoxy-4-methyl-2-(1-methyl-ethyl)benzene	see Methyl thymol	
2-Methoxy-4-methyl-1-(1-methyl-ethyl)benzene	see Methyl thymol	
2-Methoxy-4-methylphenol	see *p*-Methylguaiacol	
2-Methoxy-3-methylpyrazine	2847-30-5	4
3-Methoxynobiletin	see 3,3′,4′,5,6,7,8-Heptamethoxyflavone	
Methoxyphenol (4-; *p*-)	150-76-5	89
2-Methoxyphenol	see Guaiacol	
p-Methoxyphenylacetone[s]	122-84-9	3,81
4-(2′-Methoxyphenyl)-4-methylcyclo-hex-3-en-1-one		58
3-(2-Methoxyphenyl)propanoic acid	see 2-Methoxydihydro-cinnamic acid	
1-(4-Methoxyphenyl)-2-propanone	see *p*-Methoxyphenyl-acetone	
3-Methoxy-1,2-propanediol	623-39-2	14
1-Methoxy-2-propanol	107-98-2	14
1-Methoxy-4-(1-propenyl)benzene	see *p*-Anethole	
1-Methoxy-4-(2-propenyl)benzene	see Methyl chavicol	
2-Methoxy-3-(2-propenyl)phenol	1941-12-4	24,26
2-Methoxy-4-(1-propenyl)phenol	see Isoeugenol	
1-Methoxy-4-propylbenzene	see Dihydroanethole	
8-Methoxypsoralen[s]	298-81-7	50
3-Methoxysinensetin	see 3,3′,4′,5,6,7-Hexamethoxyflavone	
cis-4-Methoxythujane	1100111-04-3	7
trans-4-Methoxythujane	1100111-06-5	7
2-Methoxy-1,7,7-trimethylbicyclo-[2.2.1]heptane	5331-32-8	38
2-Methoxy-3,5,5-trimethyl-2-cyclo-hexene-1,4-dione	41654-27-7	80
5-Methoxy-2,8,8-trimethyl-dipyran-4-one	35930-31-5	70
p-Methoxyvalerophenone	1671-76-7	81
2-Methoxy-4-vinylphenol	see *p*-Vinylguaiacol	
Methyl acetate	79-20-9	42,86
Methylacetophenone	26444-19-9	6,55
m-Methylacetophenone	585-74-0	26,93
p-Methylacetophenone	122-00-9	7,8,35,38,48,55,77,80,91,93
Methyl *N*-acetylanthranilate[s]	2719-08-6	42,43
1-Methyl-4-acetylcyclohex-1-ene		15
Methyl alaninate	10065-72-2	14

Table 6.1 Alphabetical list of all chemicals and the essential oils in which they have been identified (*continued*)

Chemical	CAS number	Essential oils in which the chemical has been identified. Legend of numbers: see Table 6.2 at the end of this chapter
Methyl 2-aminobenzoate	see Methyl anthranilate	
2-(Methylamino)benzyl alcohol	29055-08-1	65
2-Methyl-3-amino-1-pentene		37
3-Methylamyl angelate[s]	53082-58-9	20
3-Methylamyl isobutyrate[s]	84254-84-2	20
3-Methylamyl methacrylate		20
3-Methylamyl valerate	113615-01-3	20
Methyl *p*-anisate[s]	121-98-2	41,81,90,91
2-Methylanisole	see *o*-Methylanisole	
o-Methylanisole[s]	578-58-5	63,67
p-Methylanisole	see *p*-Cresyl methyl ether	
Methyl anthranilate[s]	134-20-3	43,55,60,65,69,89,90
Methyl arachidate	see Methyl eicosanoate	
Methyl arachidonate	2566-89-4	41
Methyl arachinoate		42,43
2-Methylbenzaldehyde[s]	529-20-4	7,79,91
3-Methylbenzaldehyde	620-23-5	55
4-Methylbenzaldehyde (*p*-)	104-87-0	30
Methylbenzene[s]	108-88-3	2,37,38,44,45,55,62,63,67,79,80
Methyl benzoate	93-58-3	11,13,14,26,42,43,61,64,67,73,85,90
2-Methylbenzofuran	4265-25-2	13,14
Methylbenzothiophene	31393-23-4	59
3-Methylbenzothiophene	1455-18-1	22
Methyl benzyl ether[s]	538-86-3	73
Methyl bornyl ether	10395-54-7	72
3-Methyl-1,3-butadienyl-1-acetate	see Isoprenyl acetate	
2-Methylbutanal	96-17-3	37,48,68,73,79,80
3-Methylbutanal	see Isovaleraldehyde	
3-Methyl-2-butanal		37
Methyl butanoate	see Methyl butyrate	
2-Methylbutanoic acid	see 2-Methylbutyric acid	
3-Methylbutanoic acid	see Isovaleric acid	
1-Methyl-1-butanol		73
2-Methyl-1-butanol	137-32-6	37,73
2-Methyl-2-butanol	75-85-4	43
2-Methylbutanol	137-32-6	65
3-Methylbutanol	see Isoamyl alcohol	
3-Methyl-1-butanol	see Isoamyl alcohol	
2-Methyl-4-butanolide	1679-47-6	49,80
3-Methyl-2-butanone	563-80-4	37,80
(*E*)-2-Methyl-2-butenal	497-03-0	37,80
3-Methyl-2-butenal	107-86-8	37,49,91
(*E*)-2-Methyl-2-butenoic acid	see Tiglic acid	
2-Methylbutenol	60766-00-9	90
2-Methyl-2-buten-1-ol	4675-87-0	20
2-Methyl-3-buten-1-ol	4516-90-9	49
2-Methyl-3-buten-2-ol	115-18-4	5,37,38,48,65,68,80,90
3-Methyl-2-buten-1-ol[s]	556-82-1	37,49,65,73,80,85,86,90
3-Methyl-3-butenol (-1-ol)	763-32-6	26,68,90
3-Methyl-3-buten-2-ol	10473-14-0	85
bis(3-Methyl-3-buten-1-ol)		90
3-Methyl-3-buten-1-ol acetate	5205-07-2	90
1-(3-Methyl-2-butenoxy)-4-(1-propenyl)benzene	see Foeniculin	
2-Methyl-2-butenyl acetate	33425-30-8	20,90
2-Methylbut-3-en-2-yl acetate	24509-88-4	90
3-Methyl-2-butenyl acetate[s]	1191-16-8	20,90
3-Methyl-4-butenyl acetate		26
2-Methyl-2-butenyl angelate		20
(*E*)-2-Methyl-2-butenyl angelate		20
3-Methyl-2-butenyl angelate	83783-82-8	20
3-Methyl-2-butenyl benzoate	5205-11-8	90

Table 6.1 Alphabetical list of all chemicals and the essential oils in which they have been identified (*continued*)

Chemical	CAS number	Essential oils in which the chemical has been identified. Legend of numbers: see Table 6.2 at the end of this chapter
3-Methyl-3-butenyl benzoate[s]	5205-12-9	72,90,91
(*E*)-2-Methyl-2-butenyl isobutyrate	95654-17-4	20
3-Methyl-3-butenyl isovalerate[s]	54410-94-5	1,91
(*E*)-2-Methyl-2-butenyl methacrylate	88142-95-4	20
2-Methyl-2-butenyl 2-methylbutanoate	95654-18-5	20
3-Methyl-2-butenyl 2-methylbutanoate		37
3-Methyl-2-butenyl 3-methylbutanoate		91
3-Methyl-3-butenyl 3-methylbutanoate	see 3-Methyl-3-butenyl isovalerate	
2-(3'-Methyl-2'-butenyl)-3-methylfuran	see Rosefuran	
3-Methyl-3-butenyl valeriate		44
2-Methylbutyl acetate	624-41-9	10,20,45,68,90
3-Methylbutyl acetate[s]	see Isoamyl acetate	
2-Methylbutyl angelate	61692-77-1	20
3-Methylbutyl benzoate	see Isoamyl benzoate	
3-Methylbutyl butanoate	see Isoamyl butyrate	
Methyl 9-(2-[(2-butylcyclopropyl)methyl]cyclopropyl)nonanoate		41
2-Methylbutyl formate	35073-27-9	37
3-Methylbutyl formate	see Isoamyl formate	
3-Methylbutyl hexanoate	2198-61-0	91
2-Methylbutyl isobutyrate[s]	2445-69-4	20,68
3-Methylbutyl isobutyrate	see Isoamyl isobutyrate	
2-Methylbutyl isopentanoate	see 2-Methylbutyl 3-methylbutyrate	
2-Methylbutyl isovalerate	see 2-Methylbutyl 3-methylbutyrate	
2-Methylbutyl methacrylate	60608-94-8	20
3-Methylbutyl methacrylate	see Isoamyl methacrylate	
3-Methylbutyl 2-methylbutanoate	see Isoamyl 2-methylbutyrate	
3-Methylbutyl 3-methylbutanoate	see Isoamyl isovalerate	
2-Methylbutyl 2-methylbutyrate	see 2-Methylbutyl tiglate	
2-Methylbutyl 3-methylbutyrate[s]	2445-77-4	68
2-Methylbutyl 2-methylpropanoate	see 2-Methylbutyl isobutyrate	
6-Methyl-2,4-di-*tert*-butylphenol		34
1-(3-Methylbutyl)-2,3,5,6-tetramethylbenzene		81
2-Methylbutyl tiglate[s]	2445-78-5	1,19,20,68
2-Methylbutyl valerate	55590-83-5	1
Methylbutyrate[s]	623-42-7	37,80
2-Methyl butyrate		26,65
2-Methylbutyric acid[s]	116-53-0	4,13,37,67,72,80,90
3-Methylbutyric acid	see Isovaleric acid	
exo-Methylcamphenilol		24
Methylcamphenoate	52557-97-8	72
Methyl caprate	see Methyl decanoate	
Methyl caprylate	see Methyl octanoate	
Methylcarvacrol[s]	6379-73-3	1,4,16,25,31,36,44,45,55,56,65,69,72,80,84-86,88
3-Methylcatechol	1189946-33-5	7
Methyl chavicol[s]	140-67-0	3,5,7,10,11,20-23,26,27,29,30,34-37,41,42,45,49,52-57,66, 68,71,72,75,81,84,85,87,90,91
Methyl cinnamate	103-26-4	4,5,13,21,22,68
(*E*)-Methyl cinnamate	1754-62-7	4,21,22,43,56
(*Z*)-Methyl cinnamate	19713-73-6	4,21,22,42,43
Methyl citronellate	2270-60-2	8,29,34,37,44,53,57,61,73
3-Methylcoumarin	2445-82-1	21
Methyl cumyl ether	935-67-1	7
Methyl cyclocopacamphanoate		89
4,4-Methyl-cyclohexa-1,3-dien-1-yl methyl ketone		77
3-Methylcyclohexanol	591-23-1	68
3-Methylcyclohexanone	591-24-2	37,53,68
3-Methyl-3-cyclohexen-1-ol	53783-91-8	83,85
3-Methyl-2-cyclohexen-1-one	1193-18-6	80
1-(4-Methyl-3-cyclohexenyl)ethanone		77

Table 6.1 Alphabetical list of all chemicals and the essential oils in which they have been identified (*continued*)

Chemical	CAS number	Essential oils in which the chemical has been identified. Legend of numbers: see Table 6.2 at the end of this chapter
4-Methylcyclohex-3-en-1-yl methyl ketone		77
4-(4-Methylcyclohex-3-enyl) pent-3-en-2-one	94390-70-2	15
(*E*)-4-Methylcyclohexyl acetate		68
(*E*)-4-Methylcyclohexyl isovalerate		68
Methylcyclopentane	96-37-7	26,34,44,68
3-Methylcyclopentadecanone[s]	541-91-3	2
3-Methylcyclopentanol	18729-48-1	53
2-Methylcyclopentanone	1120-72-5	37
3-Methylcyclopentanone	1757-42-2	37,68
3-Methylcyclopentene	1120-62-3	34
2-Methyl-1-cyclopentene-1-carbox-aldehyde		80
1-(3-Methylcyclopent-2-enyl)cyclo-hexene		45
6α-(2-Methylcyclopent-1-enyl)-3,3-dimethyl-1α-bicyclo[3.1.0]hexan-2-one		58
2-Methyldecane	6975-98-0	4,85
3-Methyldecane	13151-34-3	73
Methyl decanoate[s]	110-42-9	7,19,57,65
2-Methyl-5-decanone	54410-89-8	83
3-Methyl-4-decen-1-ol (*E*-)	24404-71-5	62
4-Methyl-3-decen-5-ol	81782-77-6	83
3-Methyl-4-decenyl acetate		62
Methyl dehydroabietate	1235-74-1	15,32,70
Methyl dehydrojasmonate		42
1-Methyldibenzothiophene	31317-07-4	73
2-Methyldibenzothiophene	20928-02-3	73
3-Methyldibenzothiophene	16587-52-3	73
4-Methyldibenzothiophene	7372-88-5	73
Methyl dihydrochaulmoograte		73
Methyl dihydrojasmonate	24851-98-7	35
Methyl 3,7-dimethylocta-2,6-dienoate	see (*Z*)-Methylgeranate	
Methyl (4*Z*,7*Z*,10*Z*,13*Z*,16*Z*,19*Z*)-4,7,10,13,16,19-docosahexaenoate		63
6-Methyldocosane	55124-81-7	43
Methyl docosanoate	929-77-1	43
Methyl dodecanoate	111-82-0	22,63,76
Methyl eicosadienoate		43
Methyl eicosanoate[s]	1120-28-1	43,72
Methyl eicosa-5,8,11,14,17-pentaenoate	2734-47-6	88
Methyl eicosatrienoate	82729-72-4	43
(11*Z*,14*Z*,17*Z*)-Methyl eicosatrienoate	82729-72-4	42
Methyl eicosenoate	76899-35-9	43
(11*Z*)-Methyl eicosenoate	see Methyl-*Z*-11-eicosenoate	
Methyl-*Z*-11-eicosenoate[s]	2390-09-2	42
Methyl elaidinate		42,43
2,3-bis(Methylene)bicyclo[3.2.1]octane	49826-54-2	38
1,3,5-tris(Methylene)cycloheptane		45
5-Methylene-9-decen-2-one		49
2-Methylene-6,6-dimethylbicyclo-[3.2.0]heptan-3-ol	1005276-05-0	54
3,4-Methylenedioxybenzaldehyde	see Piperonal	
α-Methylenedodecanal		83
2-Methylene-5-isopropenylcyclohexanol	see *p*-Mentha-1(7),8-dien-2-ol	
2-Methylene-5-(1-methylethenyl)-cyclohexanol	see *p*-Mentha-1(7),8-dien-2-ol	
4-Methylene-1-methyl-2-(2-methyl-1-propen-1-yl)-1-vinylcycloheptane	see Cycloheptane, 4-methylene-1-methyl-2-(2-methyl-1-propen-1-yl)-1-vinyl-	
2-Methylenepropane-1,3-diyl-1-an-gelate-3-isobutyrate		20

Table 6.1 Alphabetical list of all chemicals and the essential oils in which they have been identified (*continued*)

Chemical	CAS number	Essential oils in which the chemical has been identified. Legend of numbers: see Table 6.2 at the end of this chapter
4-Methylene-2,8,8-trimethyl-2-vinyl-bicyclo[5.2.0]nonane	see 2,8,8-Trimethyl-4-methylene-2-vinylbicyclo[5.2.0]nonane	
Methyl epizizanoate[s]	22387-78-6	89
2-O-Methyl-8,12-epoxygermacra-1(10),4,7,11-tetraene		58
Methyl (E)-eremophila-1(10),7(11)-dien-12-oate		89
8α-Methyl-11,12,13-tri-nor-eremophil-1(10)-en-7-one		89
(1-Methylethyl) benzene	see Cumene	
1-Methyl-2-ethylbenzene	see 1-Ethyl-2-methylbenzene	
3-(1-Methylethyl)benzoic acid	5651-47-8	26
7-(1-Methylethylidene)bicyclo[4.1.0] heptane	53282-47-6	41
(1-Methylethylidene)-cyclohexane	5749-72-4	7
Methyl 5-ethylnicotinate	68686-58-8	43
Methyl ethyl phthalate[a]	34006-77-4	25
4-Methyl-3-ethylpyridine	529-21-5	43
Methyl eugenol[s]	93-15-2	3-5,7-13,21-29,31,33-35,37,41,45,49,51,53,54,56,57,60,62, 63-66,70-73,75,79,83,85,86,90
(E,E)-Methyl farnesoate	10485-70-8	59
Methyl formate	107-31-3	37
Methyl N-formylanthranilate[s]	41270-80-8	43
5-Methyl-2-furaldehyde	620-02-0	10,26,28,67,70,75,80
Methylfuran	27137-41-3	4
2-Methylfuran	534-22-5	48,68,80
3-Methylfurfural	33342-48-2	10,80
5-Methylfurfural	see 5-Methyl-2-furaldehyde	
Methyl geranate[s]	2349-14-6	6,7,37,44,45,50-53,55,57,60,69,73,78
(E)-Methylgeranate	2349-14-6	73
(Z)-Methylgeranate[s]	1862-61-9	38,53,73
Methyl geraniate	see Methyl geranate	
Methylguaiacol	see Veratrole	
p-Methylguaiacol[s]	93-51-6	26,42,90
1-Methylguanine	938-85-2	3
Methyl heneicosadienoate	122768-03-0	43
Methyl heneicosanoate	6064-90-0	43
Methyl heneicosenoate	146407-38-7	43
8-Methylheptadecane	13287-23-5	89
Methyl heptadecanoate	1731-92-6	42,43
Methyl heptadienone	73209-52-6	48
2-Methyl-3,6-heptadien-2-one		80
6-Methyl-3,5-heptadien-2-one	1604-28-0	37,48,57,80,83
(E)-6-Methyl-3,5-heptadien-2-one	16647-04-4	91
5-Methylheptanoic acid	1070-68-4	72
6-Methylheptanoic acid	929-10-2	37
6-Methyl-3-heptanol	18720-66-6	6,56
3-Methyl-2-heptanone	2371-19-9	52
5-Methyl-2-heptanone	18217-12-4	37
5-Methyl-3-heptanone	541-85-5	19,47,57,72
6-Methyl-2-heptanone	928-68-7	42
6-Methyl-3-heptanone	624-42-0	48,56
6-Methyl-5-heptanone	13019-20-0	42
Methylheptenol		73
2-Methyl-6-hepten-1-ol	67133-86-2	53
6-Methyl-5-hepten-2-ol	1569-60-4	12,19,29,37,38,49,53,57,73
Methylheptenone[s]	409-02-9	4,24,25,29,37,38,48,49,51-53,56,57,65,66,73,74,80
2-Methyl-2-hepten-6-one	see 6-Methyl-5-hepten-2-one	
6-Methyl-3-heptenone		4
6-Methyl-3-hepten-2-one	2009-74-7	77

Table 6.1 Alphabetical list of all chemicals and the essential oils in which they have been identified (*continued*)

Chemical	CAS number	Essential oils in which the chemical has been identified. Legend of numbers: see Table 6.2 at the end of this chapter
6-Methyl-5-hepten-2-one[s]	110-93-0	4-6,11,12,19,21,23,24,26,35,37-39,42-45,48,50-53,55,57, 60,65,66,69,73,80,83,85,90,91
Methyl hexadecadienoate	29961-54-4	54
Methyl hexadecanoate[s]	112-39-0	1,7,39,42,43,49,73,75,88
2-Methylhexadecan-1-ol	68526-87-4	35
Methyl hexadecenoate	29960-49-4	54
Methyl-(Z)-9-hexadecenoate	see Methyl palmitoleate	
2-Methyl-2,3-hexadiene	29212-09-7	45
5-Methyl-2-hexanal		19
Methyl hexanoate	106-70-7	26
4-Methylhexanoic acid	1561-11-1	80
5-Methylhexanoic acid	628-46-6	80
4-Methylhexanol	818-49-5	20
4-Methyl-3-hexanol	615-29-2	42
3-Methyl-1,3,5-hexatriene	2196-24-9	45
Methyl (Z)-3-hexenoate	13894-62-7	42
2-Methylhexenoic acid		37
2-Methyl-3-hexenoic acid	62243-57-6	37
5-Methyl-2-hexenol		42
Methyl hexyl ketone	see 2-Octanone	
Methyl 3-hydroxydodecanoate[s]	72864-23-4	25
Methyl 2-hydroxy-3-methylpentanoate	41654-19-7	56
Methylindoline	824-21-5	89
Methylionone	1335-46-2	34
α-Methylionone[a]	93302-56-8	80
6-Methyl-α-ionone	79-69-6	7
6-Methyl-α-(E)-ionone		7
6-Methyl-γ-ionone[a]	79-68-5	15
6-Methyl-γ-(E)-ionone[a]	79-68-5	7
Methyl isoamyl ketone	see Isoamyl methyl ketone	
2-Methylisoborneol	2371-42-8	72,83
Methyl isobutyrate	547-63-7	20,65
Methyl isoeugenol[s]	93-16-3	4,10,21-24,45,53,62,65,66
(E)-Methylisoeugenol[s]	6379-72-2	3,4,10,12,21,22,24,27,45,51,62,71,72,81
(Z)-Methylisoeugenol[s]	6380-24-1	10,22,24,27,45,51
4-Methylisopropenylbenzene	see p-Cymenene	
2-Methyl-5-isopropenylcyclohexanone	see Dihydrocarvone	
3-Methyl-6-isopropenyl-2-cyclohexen-1-one		65
5-Methyl-2-isopropenylpyridine	56057-93-3	80
1-Methyl-2-isopropylbenzene	see o-Cymene	
cis-1-Methyl-3-isopropylcyclopentane		34
trans-1-Methyl-3-isopropylcyclopentane		34
1-Methyl-4-isopropylidene cyclohexane	1124-27-2	34
3-Methyl-4-isopropylphenol	3228-02-2	76
2-Methyl-5-isopropylpyridine	20194-71-2	80
5-Methyl-2-isopropylpyridine		80
4-Methylisopulegone		79
2-Methylisovaleraldehyde		7
Methyl isovalerate[s]	556-24-1	4,88,90
Methyl isozizanoate		89
Methyl jasmine		4
Methyl jasmonate[s]	1211-29-6	4,42,43,49,50,60,72
(E)-Methyl jasmonate		42,43
(Z)-Methyl jasmonate	1211-29-6	42,43,60,72
Methyl epi-jasmonate	42536-97-0	4,50
(Z)-methyl epi-jasmonate		42,43
Methyl levopimarate		32
Methyl linoleate	112-63-0	7,19,30,42,43,67
Methyl linolenate	301-00-8	42,43
Methyl linolenoate		59
Methyl mercaptan	74-93-1	37

Table 6.1 Alphabetical list of all chemicals and the essential oils in which they have been identified (*continued*)

Chemical	CAS number	Essential oils in which the chemical has been identified. Legend of numbers: see Table 6.2 at the end of this chapter
Methyl 2-methoxybenzoate	606-45-1	90
Methyl 4-methoxybenzoate	see Methyl *p*-anisate	
Methyl-7-methoxy-5-methyl-2-hydroxyl-1-naphthoate		58
Methyl *N*-methylanthranilate[s]	85-91-6	6,39,42,43,55,60,65,69,82,89
Methyl 3-methylbutanoate	see Methyl isovalerate	
3-Methyl-2-(2-methyl-2-butenyl)		57
4-Methyl-2-(3-methyl-2-butenyl)furan		51
3-Methyl-2-(3-methyl-2-butenyl)-thiophene		73
Methyl 2-methylbutyrate	868-57-5	4,10,20,41,56,61,85
Methyl-3-methylcyclopentenylketone		37
cis-2-Methyl-3-methylenehept-5-ene	see *cis*-Salvene	
trans-2-Methyl-3-methylenehept-5-ene	see *trans*-Salvene	
2-*exo*-Methyl-3-methylene-*endo*-2-norbornanol		77
2-Methyl-6-methylene-1,7-octadiene	see α-Myrcene	
2-Methyl-6-methylene-3,7-octadiene		45
2-Methyl-6-methylene-3,7-octadien-2-ol	6994-89-4	45,51
1-Methyl-2-(1-methylethenyl)benzene	see *o*-Methyl-α-methylstyrene	
1-Methyl-4-(1-methylethenyl)benzene	see *p*-Cymenene	
5-Methyl-2-(1-methylethenyl)cyclo-hexanol	see Isopulegol II	
(*S*)-2-Methyl-5-(1-methylethenyl)-2-cyclohexen-1-one	see D-Carvone	
2-Methyl-5-(1-methylethenyl)-cyclo-hexyl acetate		26
5-Methyl-2-(1-methylethenyl)-4-hexen-1-ol	see Lavandulol	
1-Methyl-2(3,4)-(1-methylethyl)benzene		57
1-Methyl-3-(1-methylethyl)benzene	see *m*-Cymene	
1-Methyl-4-(1-methylethyl)benzene	see *p*-Cymene	
2-Methyl-5-(1-methylethyl)bicyclo-[3.1.0]hex-2-ene	see α-Thujene	
4-Methyl-1-(1-methylethyl)bicyclo-[3.1.0]hex-2-ene	see β-thujene	
1-Methyl-4-(1-methylethyl)-2-cyclo-hexen-1-ol	see *p*-Menth-2-en-1-ol	
(*Z*)-1-Methyl-4-(1-methylethyl)-2-cy-clohexen-1-ol	see *cis*-p-Menth-2-en-1-ol	
trans-1-Methyl-4-(1-methylethyl)-2-cyclohexen-1-ol	see *trans*-p-Menth-2-en-1-ol	
4-Methyl-3-(methylethyl)-3-cyclo-hexen-1-ol[s]	654053-64-2	9
1-Methyl-3-(1-methylethyl)-cyclo-pentane	53771-88-3	53
1-Methyl-4-(1-methylethylidene)-cyclohexene	see Terpinolene	
Methyl 4-methyl-5-ethylnicotinate		43
Methyl 16-methylheptadecanoate	5129-61-3	43
Methyl-*cis*-3-methyl-2-(3-methyl-2-butenyl)-3-cyclohexenyl ketone		49
Methyl-*trans*-3-methyl-2-(3-methyl-2-butenyl)-3-cyclohexenyl ketone		49
Methyl-*cis*-4-methyl-5-(3-methyl-2-butenyl)-3-cyclohexenyl ketone		49
Methyl-*trans*-4-methyl-5-(3-methyl-2-butenyl)-3-cyclohexenyl ketone		49
Methyl-2-methyl-3-methylenecyclo-pentanecarboxyate		53

Table 6.1 Alphabetical list of all chemicals and the essential oils in which they have been identified (*continued*)

Chemical	CAS number	Essential oils in which the chemical has been identified. Legend of numbers: see Table 6.2 at the end of this chapter
endo-2-Methyl-3-methylidenenorbor-nan-*exo*-2-ol	59300-40-2	77
Methyl 4-methylnicotinate	33402-75-4	43
Methyl 14-methyl pentadecanoate	5129-60-2	43
Methyl-3-(4-methyl-3-pentenyl)-3-cyclohexenyl ketone		49
Methyl-4-(4-methyl-3-pentenyl)-3-cyclohexenyl ketone		49
α-Methyl-α-(4-methyl-3-pentenyl)-oxiranemethanol	see Linalool oxide	
2-Methyl-6-(4'-methylphenyl) heptan-3-one		49
6-Methyl-5-(3'-methylphenyl) heptan-2-one		49
2-Methyl-6-(4-methylphenyl) hept-1-en-3-one		49
o-Methyl-α-methylstyrene[s]	7399-49-7	8,35
Methyl-5-methyl-8,9,10-trinorborn-5-en-2-yl (*endo*) ketone		49
Methyl-5-methyl-8,9,10-trinorborn-5-en-2-yl (*exo*) ketone		49
Methyl-6-methyl-8,9,10-trinorborn-5-en-2-yl (*endo*) ketone		49
Methyl-6-methyl-8,9,10-trinorborn-5-en-2-yl (*exo*) ketone		49
Methyl 4-methyl-5-vinylnicotinate		43
Methyl myristate	see Methyl tetradecanoate	
Methyl myrtenate	30649-97-9	41
Methylnaphthalene	1321-94-4	50
2-Methylnaphthalene	91-57-6	50
Methyl nerate	see (*Z*)-Methylgeranate	
Methyl nerolate	see (*Z*)-Methylgeranate	
Methyl nicotinate	93-60-7	42,43
2-Methyl-4-nitrosoresorcinol	65882-00-0	75
Methyl nonadecanoate	1731-94-8	43
Methyl nonadecenoate	19788-74-0	43
2-Methylnonane	871-83-0	42,43
3-Methylnonane	5911-04-6	7,12
Methyl nonanoate	1731-84-6	73
8-Methylnonanoic acid	5963-14-4	37,72
Methyl *n*-nonyl ketone	see 2-Undecanone	
Methyl nopinone		65
Methyl octadecanoate[s]	112-61-8	1,7,42,43,59
Methyl 6,9,12-octadecantrienoate		43
cis-2-Methyl-7-octadecene	35354-39-3	90
Methyl (*Z*)-9-octadecenoate	see Methyl oleate	
2-Methyloctadecyne		73
7-Methyl-3,4-octadiene	37050-05-8	30
3-Methyloctanoic acid	6061-10-5	37
2-Methyloctane	3221-61-2	19
3-Methyloctane	2216-33-3	19,83
Methyl octanoate[s]	111-11-5	26,43,65,73,90
7-Methyloctanoic acid	26896-18-4	72
2-Methyl-4-octanone	7492-38-8	52
4-Methyl-1-octene	13151-12-7	39
Methyloctenoic acid		37
2-Methyl-3-octenoic acid		37
Methyl oleate[s]	112-62-9	7,39,42,43,59,75
Methyl 15-oxoeicosanoate		35
Methyl palmitate	see Methyl hexadecanoate	

Table 6.1 Alphabetical list of all chemicals and the essential oils in which they have been identified (*continued*)

Chemical	CAS number	Essential oils in which the chemical has been identified. Legend of numbers: see Table 6.2 at the end of this chapter
Methyl palmitoleate[s]	1120-25-8	75
5-Methylpentadecane	25117-33-3	70
Methyl pentadecanoate	7132-64-1	43,54,73
2-Methylpentanal	123-15-9	26
2-Methyl-2-pentanal		59
3-Methylpentane	96-14-0	42
2-Methylpentanoic acid	97-61-0	37,80
3-Methylpentanoic acid	105-43-1	37,72,80,88
4-Methylpentanoic acid	see Isohexanoic acid	
2-Methyl-2-pentanol	590-36-3	42
2-Methyl-3-pentanol	565-67-3	42
3-Methylpentanol	see 3-Methyl-1-pentanol	
3-Methyl-1-pentanol	589-35-5	20,37
3-Methyl-3-pentanol	77-74-7	37,42
4-Methylpentanol	626-89-1	73
4-Methyl-2-pentanol	108-11-2	42
2-Methyl-3-pentanone	565-69-5	37,80
4-Methyl-2-pentanone	108-10-1	37,80
2-Methyl-2-pentenal	623-36-9	80,91
4-Methyl-3-pentenal	5362-50-5	41
2-Methyl-2-pentenoic acid	3142-72-1	72
4-Methyl-3-pentenoic acid	504-85-8	56,80
4-Methyl-3-penten-2-one[s]	141-79-7	37,41,72,80
3-(4-Methyl-3-pentenyl)-3-cyclohexenyl pentyl ketone		49
4-(4-Methyl-3-pentenyl)-3-cyclohexenyl pentyl ketone		49
3-(4-Methyl-3-pentenyl)-3-cyclohexenyl propyl ketone		49
4-(4-Methyl-3-pentenyl)-3-cyclohexenyl propyl ketone		49
3-Methyl-2-(*cis*-2-pentenyl)-2-cyclo-penten-1-one	see (*Z*)-Jasmone	
3-Methyl-2-(*trans*-2-pentenyl)-2-cyclo-penten-1-one	see (*E*)-Jasmone	
4-(4-Methyl-3-pentenyl)-1,2-dithiacy-clohex-4-ene	73188-23-5	73
3-(4-Methyl-3-pentenyl)thiophene	62429-57-6	73
3-Methylpentyl acetate	35897-13-3	20
2-Methylpentyl angelate		20
3-Methylpentyl angelate	see 3-Methylamyl angelate	
3-Methylpentyl formate		37
3-Methylpentyl 3-hydroxy-2-methyl-enebutyrate		20
3-Methylpentyl isobutyrate	see 3-Methylamyl isobutyrate	
3-Methylpentyl isovalerate	35852-41-6	20
3-Methylpentyl methacrylate		20
2-Methylpentyl propionate		20
3-Methylpentyl valerate		20
2-Methyl-5-phenol		29
Methyl phenylacetate	101-41-7	43,73,91
Methyl 1-phenylallyl ether	22665-13-0	3
2-Methyl-6-(1-phenylethyl)phenol	17959-01-2	88
4-(4-Methylphenyl)pentanal		49,80
2-Methyl-3-phenylpropanal[s]	5445-77-2	7,29,45,81
2-(4-Methylphenyl)propanal	99-72-9	80
Methyl phenyl-3-propionate		43
Methyl piperinate		8
4-Methylproline	3005-85-4	65
2-Methylpropanal	see Isobutanal	

Table 6.1 Alphabetical list of all chemicals and the essential oils in which they have been identified (*continued*)

Chemical	CAS number	Essential oils in which the chemical has been identified. Legend of numbers: see Table 6.2 at the end of this chapter
2-Methylpropane-1,3-diyl-1-angelate-3-isobutyrate		20
2-Methylpropanoic acid	see Isobutyric acid	
2-Methylpropanol	see Isobutyl alcohol	
2-Methyl-1-propanol	see Isobutyl alcohol	
2-Methyl-2-propenyl angelate	see Methallyl angelate	
2-Methyl-2-propenyl isobutyrate		20
2-Methyl-2-propenyl methacrylate	see Methallyl methacrylate	
2-Methyl-2-propenyl tiglate	7493-71-2	20
2-Methyl-2-propyl angelate	see Isobutyl angelate	
1-Methyl-3-propyl benzene	1074-43-7	7
2-Methylpropyl formate	542-55-2	37
Methyl propyl ketone	see 2-Pentanone	
2-Methylpropyl-2-methylbutanoate	see Isobutyl 2-methylbutyrate	
2-Methylpropyl-2-propanoate	see Isobutyl propionate	
Methylpyrazine	109-08-0	75,89
2-Methylpyridine	109-06-8	80
3-Methylpyridine	108-99-6	80
4-Methylpyridine	108-89-4	89
5-Methyl-2-pyrithione		72
Methylquinoline		89
2-Methylquinoline	91-63-4	42
Methyl salicylate[c,s]	119-36-8	4,13,19,26-28,30,42,43,45,53,57,64,67,73,85,90,91
Methyl stearate	see Methyl octadecanoate	
Methylsulphonylmethane	see Dimethyl sulfone	
α-Methyl teresantalic acid		77
4-Methyl-δ3-tetraacetophenone		93
Methyltetracosane		73
Methyl 10,11-tetradecadienoate		53
Methyl tetradecanoate[s]	124-10-7	7,43,62,73
p-Methyl-3,4,5,6-tetrahydroaceto-phenone		93
Methyl thiobenzoate	5873-86-9	61
2-Methyl-1-thioindan	6383-15-9	72
Methyl thymol[s]	1076-56-8	1,4,7,31,32,45,50,55,56,68,70,72,78,80,82,85,86,88,91
Methyl tiglate	6622-76-0	20
5-Methyl-8,9,10-trinorborn-5-en-2-yl (*endo*) pentyl ketone		49
5-Methyl-8,9,10-trinorborn-5-en-2-yl (*exo*) pentyl ketone		49
6-Methyl-8,9,10-trinorborn-5-en-2-yl (*endo*) pentyl ketone		49
6-Methyl-8,9,10-trinorborn-5-en-2-yl (*exo*) pentyl ketone		49
5-Methyltricosane	22331-09-5	42,43
2-Methyltridecane	1560-96-9	9
Methyl tridecanoate	1731-88-0	73
12-Methyltridecano-13-lactone	see 12-Methyl-13-tridecanolide	
Methyltridecanolide	see 12-Methyl-13-tridecanolide	
12-Methyl-13-tridecanolide[s]	57092-32-7	2
2-Methylvaleric acid	see 2-Methylpentanoic acid	
3-Methylvaleric acid	105-43-1	88
4-Methylvaleric acid	see Isohexanoic acid	
Methyl vanillin ether	see Veratraldehyde	
4-Methyl-4-vinyl-4-butanolide	1073-11-6	49,80
4-Methyl-4-vinyl-4-but-2-enolide		49,80
4-Methyl-4-vinylbutyrolactone		49
γ-Methyl-γ-vinylbutyrolactone		48
2-Methyl-2-vinyl-4-isopropenyltetra-hydrofuran		25
Methyl-5-vinylnicotinate	38940-67-9	43
4-Methyl-3-vinylpyridine		43

Table 6.1 Alphabetical list of all chemicals and the essential oils in which they have been identified (*continued*)

Chemical	CAS number	Essential oils in which the chemical has been identified. Legend of numbers: see Table 6.2 at the end of this chapter
5-Methyl-5-vinyltetrahydrofuran-2-(2-methylethanal)		80
3-Methyl-4-(2,5-xylyl)-butyric acid	30275-76-4	88
Methyl ziza-6(13)-en-12-oate	18444-89-8	89
Methyl 2-epi-ziza-6(13)-en-12-oate	see Methyl epizizanoate	
Mint furanone		68
Mintoxide		25
Mintsulfide	72445-42-2	4,8,22,25,37,41,68,73,90
β-Mintsulfide		73
β-Monopalmitin	23470-00-0	3
α-Multijugenol	34298-31-2	32
Muscone	see 3-Methylcyclopenta-decanone	
Musk ambrette[a,b]	123-69-3	19
Muskone	see 3-Methylcyclopenta-decanone	
Mustakone		49,89
Muurola-3,5-diene	157374-44-2	31,44
cis-Muurola-3,5-diene	157374-44-2	4,7,44,68,79,88
trans-Muurola-3,5-diene	262352-88-5	1,23,41,44,70
Muurola-4(14),5-diene	157477-72-0	49
cis-Muurola-4(14),5-diene	157477-72-0	4,7,12,23,25,37,44,45,51,57,70,79,91
trans-Muurola-4(14),5-diene	262352-87-4	4,23,32,38,44,68,70,79
Muurola-4,11-diene	235091-39-1	77
Muurola-4,10(14)-dien-1β-ol		12,75
Muurola-4,10(14)-dien-8β-ol		31
Muurolene	69671-15-4	24,25,68,75
α-Muurolene	10208-80-7	1-6,8,10,11,13-16,19,21-27,31,32,34-39,43,44,45,50,52,56-58,60-65,68,70-72,75,76,78-83,85-91
γ-Muurolene	30021-74-0	1-8,10-14,16,19,21,23-27,31,32,34,37-39,41,43-45,48-52,54-58,60,63-73,75,78-83,85,88-91
δ-Muurolene	120021-96-7	6,19,50,72
ε-Muurolene	30021-46-6	4,68,70,72
τ-Muurolene	152287-43-9	4,23,24,76,90
cis-Muurol-5-en-4α-ol	157374-45-3	10,70,77,79
cis-Muurol-5-en-4β-ol	157374-46-4	49,51,70,79
Muurol-5-en-4-one		31
cis-14-Muurol-5-en-4-one		49
cis-14-nor-Muurol-5-en-4-one		4
Muurolol	119757-72-1	57,70,85
α-Muurolol	104245-48-9	1,2,4,8,11,13,14,19,21-24,31,32,34-39,41,44,45,49,50,52,57,61,63,70-73,75,79,81,83,85,88,89
epi-α-Muurolol[s]	19912-62-0	2,4,8,10,11,13,16,19,21,23-26,31,32,34-38,42-45,50,52,54,56-58,61,68,70-72,74,75,78,79,84,85,88-90
γ-Muurolol	138068-73-2	5,75
τ-Muurolol	see epi-α-Muurolol	
(*E*)-Muurolol		88
Myltayl-4(12)-ene	79562-97-3	16
Myrcene[s]	123-35-3	1-13,15,19-26,29-39,41,44-58,60-76,78-92
α-Myrcene[s]	1686-30-2	51,54,62,81,85
β-Myrcene	see Myrcene	
Myrcene epoxide[s]	29414-55-9	12,29,39,53,64,65
Myrcenol	543-39-5	13,36,44,45,51,52,63,65,72,80,86,88
Myrcen-8-ol		85
Myrcenone	539-70-8	57
Myrcenyl acetate	1118-39-4	79,85
Myristic acid	see *n*-Tetradecanoic acid	
Myristicin	607-91-0	7,8,12,29,41,53,56,62,85,90
Myroxide	28977-57-3	81
(*E*)-Myroxide	28977-57-3	4,49,55,64
(*Z*)-Myroxide	33281-83-3	4,7
Myrtanal	4764-14-1	36,52
Myrtanol	514-99-8	36,37,64,73,75
cis-Myrtanol	15358-92-6	34,37,44,51,72

Table 6.1 Alphabetical list of all chemicals and the essential oils in which they have been identified (*continued*)

Chemical	CAS number	Essential oils in which the chemical has been identified. Legend of numbers: see Table 6.2 at the end of this chapter
trans-Myrtanol	15358-91-5	12,37,44,60,61,72,83
Myrtanyl acetate	29021-36-1	78
trans-Myrtanyl acetate	90934-53-5	44,72,85
Myrtenal	564-94-3	2,4,6-8,12,15,20,22,29,31,32,35,36,38,41,44,45,48-50,56, 57,60,63-65,68,70,72,75,76,79,80,85,86
Myrtenic acid	19250-17-0	72,80
Myrtenol[s]	515-00-4	2,4,6-8,10,12,15,20,24,25,29,31,32,34-38,41,44,45,49,50, 55-57,60,61,63,67,68,70,72,73,75,76,78,80,85-89,91
cis-Myrtenol		53,64,72
Myrtenone		26
Myrtenyl acetate	1079-01-2	2,4,7,12,20,24,29,31,36,41,44,45,52,55,56,65,68,72,73,75,76, 88
(-)-Myrtenyl acetate	36203-31-3	63
Myrtenyl angelate		20
Myrtenyl formate		41
Myrtenyl hexanoate		88
Myrtenyl isobutyrate	29021-37-2	20
Myrtenyl isovalerate[s]	33900-84-4	20,88
Myrtenyl 2-methylbutyrate	138530-44-6	20
Myrtenyl 3-methylbutyrate	see Myrtenyl isovalerate	
Myrtenyl methyl ether	10300-03-5	41
Myrtenyl propionate		35
4-Nanonene		51
Naphthalene	91-20-3	4,7-11,16,41,45,61,63-65,72,75,79,81,88
Naphthalene, decahydro-1,1,4a-tri-methyl-6-methylene-5-(3-methyl-2-pentenyl)-	78548-63-7	63
Naphthalene, 1,2,3,4,4a,7-hexahydro-1,6-dimethyl-4-(1-methylethyl)-	see 1,2,3,4,4a,7-Hexahydro-1,6-dimethyl-4-(1-methylethyl)naphthalene	
Naphthalene, 1,2,3,4,4a,7-hexahydro-1,6-dimethyl-4-(1-methylethyl)-(1α,4β,4αβ)-		71
Naphthalene, 1,2,3,5,6,8a-hexahydro-4,7-dimethyl-1-(1-methylethyl)-,(1S-cis)-	see δ-Cadinene	
Naphthalene, 1,2,4a,5,8,8a-hexahydro-4,7-dimethyl-1-(1-methylethyl)-	16509-53-8	26
2-Naphthalenemethanol		45
Naphthalene, 1,2,3,5,6,7,8,8a-octa-hydro-1,8a-dimethyl-7-(1-methyl-ethenyl)-, [1S-(1α,7α,8aα)]-	see Valencene	
Naphthalene, 1,2,3,4,4a,5,6,8a-octa-hydro-7-methyl-4-methylene-1-(1-methylethyl)-,	see γ-Cadinene	
Naphthalene, 1,2,3,4,4a,5,6,8a-octa-hydro-7-methyl-4-methylene-1-(1-methylethyl)-, (1α,4aα,8aα)-	see γ-Muurolene	
1-Naphthalenol, 2-methyl-		21
2(1H)-Naphthalenone	136156-72-4	7
1(2H)-Naphthalenone, 3,4-dihydro-4,7-dimethyl-		48
2(1H)-Naphthalenone, 3,5,6,7,8,8a-hexahydro-4,8a-dimethyl-6-(1-methylethenyl)-	725240-70-0	67
2(3H)-Naphthalenone, 4,4a,5,6,7,8-hexahydro-4a-methyl-	826-56-2	34
1H-Naphtho[2.1.6]pyran		25
Neobyakangenicol		50
Neocallitropsene	729602-94-2	12
α-Neocallitropsene	729602-94-2	10,16

Table 6.1 Alphabetical list of all chemicals and the essential oils in which they have been identified (*continued*)

Chemical	CAS number	Essential oils in which the chemical has been identified. Legend of numbers: see Table 6.2 at the end of this chapter
5-Neocedranol	13567-44-7	7
Neocembrene	see Cembrene A	
Neocembrene A	see Cembrene A	
α-Neoclovene	4545-68-0	7,36
β-Neoclovene	56684-96-9	88
δ-Neoclovene		58
Neodihydrocarveol	18675-33-7	39,79,83
Neodihydrocarvyl acetate	56422-50-5	72
Neohexane	75-83-2	22
Neointermedeol	5945-72-2	4,52,66
5-epi-Neointermedeol[s]	136734-27-5	30,35,61
Neoisocarvomenthyl acetate	51407-21-7	53
Neoisodihydrocarveol		79
Neoisodihydrocarvyl acetate		79
Neoisoisopulegol	21290-09-5	23,34,37,53,68
Neoisoisopulegol acetate	256332-34-0	45,53,68
Neoisomenthol	491-02-1	37,49,68,79
Neoisomenthyl acetate		68
Neoisopulegol	29141-10-4	23,24,34,37,49,53,68
Neoisopulegyl acetate	57576-10-0	19,34,68
Neoiso-3-thujanol acetate		76
Neoisothujyl acetate	62181-91-3	44
Neomenthol	3623-51-6	4,6,52,68,75,79,85
Neomenthyl acetate	2230-87-7	6,68,79
Neopentylidenecyclohexane	39546-80-0	25
Neophytadiene	504-96-1	4,24,42,45,57,66
Neothujan-3-ol acetate		75
Neothujol[s]	35732-36-6	7,53,72
Neral[s]	106-26-3	4-8,11,12,19,21,23-25,29,34-39,41,42,45,50-53,55, 57,60,61,64-66,69,70,72,73,75,80,82,85,86,90-92
Neranoic acid		4
Neric acid	37349-29-4	4,5,37,38,50,57
Nerol[s]	106-25-2	4-7,12,13,19,21-27,29,34-39,41-49,51-53,54,55-56,57,60-66, 68,69-76,80,82,83,85,86,89,90
Nerolic acid	4613-38-1	37,52
Nerolidol	7212-44-4	4,6,8,11,13,14,22,25,26,29,30,36,38,39,42,43,51,52,56,60, 64-71,73,74,76,81,88,90,93
(*E*)-Nerolidol	40716-66-3	1,3-8,10-13,15-17,19,21,22,24,26,27,29-32,34-39,41-43,45, 48-51,53,55,57,60,61,63-66,68-74,77,80-82,85,90-92
(*Z*)-Nerolidol	3790-78-1	4,6-8,11,13,25,35.38-40,42,43,45,50,57,60,63-65,67-69, 73,81,85
Nerolidol epoxyacetate		88
Nerolidol oxide-I		37
Nerolidol oxide-II		37
Nerolidyl acetate	56001-43-5	8,19,27,39,49,51,65
(*E*)-Nerolidyl acetate	85611-33-2	37,77
(*Z*)-Nerolidyl acetate	91050-14-5	37,75,77
Nerolidyl isobutyrate ((*E*)-)	1263759-11-0	63
Nerolidyl propionate	74646-28-9	60
Nerol oxide	1786-08-9	25,45,49,56,57,73,85
Neryl acetate[s]	141-12-8	4-8,12,22-25,29,34,35,37-39,41,44-50,52,53,55-57, 60,62-66,68,69,72,73,75,76,78,80,82,84-86,89,90
2,3-Neryl acetate oxide		6
Nerylacetone[s]	3879-26-3	4,41,50,59,72
Neryl butyrate	999-40-6	37,61,73
Neryl formate	2142-94-1	25,37,39,66,69,73
Neryl heptanoate		66
Neryl hexanoate	68310-59-8	80
Nerylic acid		80
Neryl isobutyrate	2345-24-6	37,85

Table 6.1 Alphabetical list of all chemicals and the essential oils in which they have been identified (*continued*)

Chemical	CAS number	Essential oils in which the chemical has been identified. Legend of numbers: see Table 6.2 at the end of this chapter
Neryl isovalerate[s]	3915-83-1	37,80,85,88
Neryl 2-methylbutanoate	51117-19-2	80
Neryl 3-methylbutanoate	see Neryl isovalerate	
Neryl octanoate		73
Neryl phenylacetate	10522-32-4	73
Neryl propanoate	see Neryl propionate	
Neryl propionate[s]	105-91-9	6,23,37,49,50,51,64-66,72,73,80,82
Neryl tiglate		37
Neryl valerate	10522-33-5	37
Nezukol	14699-32-2	31
Nigritene		89
Nigritene A		89
Nigritene B		89
6-epi-Nigritene		89
Nikkol		65
Nitrobicyclononane		41
2-Nitroethanol	625-48-9	14
1-Nitro-2-phenylethane	see 2-Phenylnitroethane	
Nobiletin	see 3′,4′,5,6,7,8-Hexamethoxyflavone	
Nojigiku acetate	55627-02-6	88
Nonacosane	630-03-5	4,7,9,26,59,70,73,90
Nonadecadiene		73
Nonadecane	629-92-5	4,7,8,12,19,29,32,35,37,42-45,48,49,56,57,66-68,72,73, 75,89-91
(*Z*)-5-Nonadecane		55,65
Nonadecan-9-ene		73
Nonadecanol	1454-84-8	66,72
Nonadecene	27400-77-7	4,37,73
(*Z*)-5-Nonadecene		73
9-Nonadecene	31035-07-1	73
(*Z*)-9-Nonadecene	51865-02-2	73
(2*E*,4*E*)-Nonadienal	5910-87-2	39
(2*E*,6*Z*)-Nonadienal	557-48-2	25,39,73,80
1,7-Nonadiene	13150-91-9	34
(*E*,*Z*)-2,6-Nonadienol	28069-72-9	65
γ-Nonalactone[s]	104-61-0	7,42
Nonanal	124-19-6	3,6-8,10,12-14,19,22,23,25,29,37-39,41,48-50,52, 55-57,60,63-66,68,69,72,73,79,80,82,83,85,86,91
Nonanal diethyl acetal	54815-13-3	73
Nonane (*n-*)	111-84-2	7,8,10,19,29,39,54,60,65,69,73,82,83
4,6-Nonanedione	14090-88-1	35
2-Nonanene		51
n-Nonanoic acid	12-05-0	8,13,14,20,29,37,39,55,59,65,72,73,80,90
Nonanol	28473-21-4	6,39,42,50,55,64,65,72,73,82,85
2-Nonanol	628-99-9	26,27,28,38,73
3-Nonanol	624-51-1	85
4-Nonanolide	see γ-Nonalactone	
Nonanone	30642-09-2	48
2-Nonanone	821-55-6	26,27,31,38,45,72,85,91
4-Nonanone	4485-09-0	51,52
Nonanyl acetate	see Nonyl acetate	
2-Nonenal	2463-53-8	50
(*E*)-2-Nonenal	18829-56-6	5,8,12,50,55,64,65
(*Z*)-2-Nonenal	60784-31-8	7,39
(*Z*)-4-Nonenal	2277-15-8	39
1-Nonen-3-ol	21964-44-3	68
3-Nonen-2-one	14309-57-0	19
Nonenyl phenyl ether		73
Nonyl acetate[s]	143-13-5	6,10,39,50,55,63-65,90
Nonyl phenyl ether		73
3-Nonyne	20184-89-8	49

Table 6.1 Alphabetical list of all chemicals and the essential oils in which they have been identified (*continued*)

Chemical	CAS number	Essential oils in which the chemical has been identified. Legend of numbers: see Table 6.2 at the end of this chapter
Nootkatene	5090-61-9	67,89
Nootkatol	50763-67-2	44,89
epi-Nootkatol		89
Nootkatone	4674-50-4	6,39,44,50,55,64,65,69,82,83,88,89
Nopinene	see β-Pinene	
Nopinone[s]	24903-95-5	8,12,36,41,48,49,55,80
Nopol[s]	128-50-7	52,72,81
Nopyl acetate	128-51-8	49
Norbornanol	86368-39-0	6,50,55
2-Norbornanone[s]	497-38-1	45,57
endo-2-Norborneol acetate		45
Norbornyl acetate	34640-76-1	49
11-Norbourbonan-1-one		49
Norbourbonone[s]	13844-03-6	37
Norcadin-5-en-4-one		49
Norcadin-5-en-4-one isomer		49
12-Norcaryophyllen-2-one		49
12-Norcaryophyll-5-en-2-one		25
(-)-(E)-12-Nor-caryophyll-5-en-2-one		25
15-Norcedran-8-one		49
12-Norcyercene-B		25
(1R)-(+)-Norinone		41
Norpatchoulenol	41429-52-1	67
2-Norpinene-2-carboxaldehyde		70
Nortetrapatchoulol		67
Nortricycloekasantalic acid	see Tricycloekasantalic acid	
Nortricycloekasantalol		77
Nuciferal	25532-74-5	38
(E)-Nuciferal	25532-74-5	77
(E)-Nuciferol	1786-15-8	38,77,88
(Z)-Nuciferol	78339-53-4	77,88
Occidentalol	473-17-6	58
Occidol	5986-36-7	58
Ocimene	see β-Ocimene	
α-Ocimene[s]	502-99-8	38,53,66
β-Ocimene[s]	13877-91-3	4-6,8,10,24-26,29,32,34-39,44,45,50-53,57,60,62,64-66, 69,70,73,78,81,82,85,86,90
(E)-Ocimene	27400-72-2	2,4,5,10,19,21,25,34,41,44,53,56,64,66,68,73,79-81,85
(Z)-Ocimene	27400-71-1	2,4-6,8,10,20,21,25,35,38,41,52,53,56,64,66,68,73,79-81, 85,88
(E)-α-Ocimene	6874-10-8	12,60,61,75
(Z)-α-Ocimene	6874-44-8	25,60,62
(E)-β-Ocimene[s]	3779-61-1	1-10,12,13,19,21-27,29-39,41,43-50,51-58,60-66,68-76, 78-80,82-86,88,90-92
(Z)-β-Ocimene[s]	3338-55-4	1-10,12,13,19,21-26,29,31-39,41,44-57,60-66,68-76, 78-82,84-86,88-91
o-Ocimene		35
p-Ocimene	see β-Ocimene	
cis-p-Ocimene	see (Z)-β-Ocimene	
trans-Ocimene	see (E)-β-Ocimene	
trans-p-Ocimene	see (E)-β-Ocimene	
allo-Ocimene	673-84-7	4,6-8,10,16,23,25,31,35,36,41,44,49-52,54,56,63,66,69, 71-73,75,79,85,88,90
(E)-allo-Ocimene	3016-19-1	10,24,44,48,49,52,57
(Z)-allo-Ocimene	17202-20-9	48,76
neo-allo-Ocimene	7216-56-0	4,10,37,48,79,88
allo-Ocimene diepoxide		83
(E,E)-allo-Ocimene	673-84-7	66,72,90
cis-2,3-Ocimene oxide		6
trans-2,3-Ocimene oxide		6
cis-Ocimene, 8-oxo-		72

Table 6.1 Alphabetical list of all chemicals and the essential oils in which they have been identified (*continued*)

Chemical	CAS number	Essential oils in which the chemical has been identified. Legend of numbers: see Table 6.2 at the end of this chapter
Ocimenol	see (*E*)-Ocimenol	
(*E*)-Ocimenol[s]	28977-58-4	49,74,91
(*Z*)-Ocimenol	39900-51-1	8,22
(*E*)-Ocimenone	see (*E*)-Tagetenone	
(*Z*)-Ocimenone	see (*Z*)-Tagetenone	
Octacosane	630-02-4	7,12,19,62,90,91
9,12-Octadecadienal	26537-70-2	38
Octadecadiene		73
9,12-Octadecadienoic acid	2197-37-7	13
9,12-Octadecadienol	506-43-4	43
Octadecanal	638-66-4	39,65,68,73,90
1-Octadecanamine	124-30-1	83
n-Octadecane	593-45-3	4,7,8,10,12,19,29,32,35,37,43-45,62,65,68,72,73,75,85,89-91
Octadecane-3,12-diol		34
Octadecanoic acid[s]	57-11-4	7,8,26,29,42,57,59,62,75,81
Octadecanol	26762-44-7	44,59,72
1-Octadecanol	112-92-5	73
9,12,15-Octadecatrienal	26537-71-3	38
(*Z*)-9-Octadecenal	2423-10-1	65
16-Octadecenal	56554-87-1	8,73
9-Octadecenamide	3322-62-1	41
1-Octadecene	112-88-9	4,72,73
(*E*)-3-Octadecene	7206-19-1	73
9-Octadecenoic acid	2027-47-6	37,72,85
(*E*)-9-Octadecenoic acid	see Elaidic acid	
trans-13-Octadecenoic acid	693-71-0	70
(*E*)-9-Octadecenoic acid, trimethylsilyl ester[a]	96851-47-7	8
2,4-Octadienal	30361-28-5	4
2,6-Octadienal, 3,7-dimethyl-, (*E*)-	see Geranial	
2,7-Octadienal-6-hydroxy-2,6-dime-thyl acetate		48
2,6-Octadiene	4974-27-0	34
1,6-Octadiene, 2,6-dimethyl-	6874-34-6	52
Octadienediol	141581-27-3	49
1,7-Octadiene-3,6-diol-2,6-dimethyl-6-acetate		48
3,7-Octadiene-2,6-diol-2,6-dimethyl-6-acetate		48
2,4-Octadienoic acid	83615-26-3	37
2,6-Octadienoic acid	83592-56-7	57
3,6-Octadienoic acid	70080-68-1	57
3,6-Octadienoic acid, 3,7-dimethyl, methyl ester		72
(*Z*)-1,5-Octadien-3-ol	50306-18-8	68
1,6-Octadien-3-ol	51361-43-4	34,81
2,6-Octadien-1-ol		49
2,6-Octadien-1-ol, 3,7-dimethyl, (*E*)-	see Geraniol	
1,6-Octadien-3-ol, 3,7-dimethyl-, acetate	see Linalyl acetate	
2,6-Octadien-1-ol, 3,7-dimethyl-, acetate, (*E*)-	see Geranyl acetate	
1,6-Octadien-3-ol, 3,7-dimethyl-, 2-aminobenzoate	see Linalyl anthranilate	
2,6-Octadien-1-ol, 3,7-dimethyl-, formate	see Geranyl formate	
(E,E)-3,5-Octadien-2-one	30086-02-3	19
3,7-Octadien-2-one-6-hydroxy-6-methyl acetate		48
Octadienyl formate		6
Octahydronaphthalene	31244-58-3	8
1,2,3,4,4a,5,6,8a-Octahydronaphthalene	4276-46-4	9

Table 6.1 Alphabetical list of all chemicals and the essential oils in which they have been identified (*continued*)

Chemical	CAS number	Essential oils in which the chemical has been identified. Legend of numbers: see Table 6.2 at the end of this chapter
1a,2,3,5,6,7,7a,7b-Octahydro-1,1,7,7a-tetramethyl-1*H*-cyclopropa[a]naphthalene	see β-Gurjunene	
2,4a,5,6,7,8,9,9a-Octahydro-3,5,5-tri-methyl-9-methylene-1*H*-benzocyclo-heptene	see α-Himachalene	
γ-Octalactone[s]	104-50-7	42
1-Octanal	see *n*-Octanal	
3-Octanal		57
n-Octanal[s]	124-13-0	2,4,6,7,10,12,19,22,29,38,39,41,46-52,54-57,60,63-65,69, 73,75,80,82,83,91,92
Octanal diethyl acetal	54889-48-4	73
n-Octane	111-65-9	4,10,19,26,37,63,91
2,3-Octanedione	585-25-1	57
Octanoic acid[s]	124-07-2	4,6,7,10,13,14,29,30,37,50,55,62,65,68,72,73,80,82,90,91
Octanoic acid, octyl ester	2306-88-9	63
Octanoic acid, phenylmethyl ester	see Benzyl *n*-octanoate	
Octanol	29063-28-3	24,37,39,51,55,65,66,69,72,73,75,80
1-Octan-3-ol		63
2-Octanol	123-96-6	7,19,37,76
3-Octanol	589-98-0	4,5,10,19,41,44,46,48,49,55-57,67,68,72-76,79,80,85,86
n-Octanol (1-)	111-87-5	4-7,27,29,32,38,46-48,50,59,60,63,64,68,72,76,82,86,91
Octanol acetate	79517-25-2	63
3-Octanol, acetate	see 3-Octyl acetate	
4-Octanolide	see γ-Octalactone	
2-Octanone[s]	111-13-7	26,48,49,52,57
3-Octanone	106-68-3	1,4,5,12,23,25,37,41,46-49,57,67,68,72,75,79,80,85,86
4-Octanone	589-63-9	72
Octan-3-yl acetate	see 3-Octanyl acetate	
3-Octanyl acetate[s]	103-09-3	4,49,68
1,3,5-Octatriene	26555-19-1	57
1,3,6-Octatriene	929-20-4	7,57
Octenal	25447-69-2	38
(*E*)-2-Octenal[s]	2548-87-0	4,64,80
trans-Oct-2-en-1-al	see (*E*)-2-Octenal	
3-Octenal	60671-71-8	57
6-Octenal	63826-25-5	34
1-Octene	111-66-0	37
(*E*)-3-Octenoic acid	5163-67-7	72
(*Z*)-3-Octenoic acid	5169-51-7	72
7-Octenoic acid	18719-24-9	30
2-Octenoic acid, 4-isopropylidene-7-methyl-6-methylene-, methyl ester		67
1-Octen-3-ol	3391-86-4	4-6,10,19,20,25,29,31,34,37,38,41,45-49,56,57,59,65, 67,68,72,75,79,80,85,86
trans-2-Octen-1-ol	18409-17-1	68,79
2-Octen-4-ol	4798-61-2	39
3-octenol	see 3-Octen-1-ol	
3-Octen-1-ol[s]	18185-81-4	4,72
6-Octen-1-ol	63768-12-7	34
7-Octen-4-ol	53907-72-5	4,8,10,72
2-Octen-1-ol, 3,7-dimethyl-	40607-48-5	50
6-Octen-1-ol, 3,7-dimethyl-, formate	see Citronellyl formate	
1-Octen-3-ol-6,7-epoxy-3,7-dimethyl acetate		48
1-Octen-3-one	4312-99-6	56,67,68,72
3-Octen-2-one	1669-44-9	19
4-Octen-3-one	14129-48-7	57
Octenyl acetate (1-)	37366-04-4	4,45,49,68,79,85
1-Octen-3-yl acetate	2442-10-6	4,10,25,46-49,56,68,72,79,80
3-Octenyl acetate	7380-48-5	84

Table 6.1 Alphabetical list of all chemicals and the essential oils in which they have been identified (*continued*)

Chemical	CAS number	Essential oils in which the chemical has been identified. Legend of numbers: see Table 6.2 at the end of this chapter
7-Octen-1-yl acetate		68
1-Octen-3-yl butyrate	16491-54-6	48
3-Octen-5-yne, 2,7-dimethyl-, (Z)-	28935-76-4	12
Octyl acetate	112-14-1	4,6,10,39,47-50,55,58,63-65,79,80,82,86,91
2-Octyl acetate	2051-50-5	27
3-Octyl acetate[s]	4864-61-3	46,49,86,79
Octyl aldehyde	see *n*-Octanal	
Octyl benzoate	94-50-8	43,73
Octyl butyrate	110-39-4	51,65
Octyl formate	112-32-3	39,55,63,82
Octyl heptanoate	5132-75-2	63
Octyl hexanoate	4887-30-3	65
Octyl 2-methylbutanoate	29811-50-5	80
Octyl 3-methylbutanoate	7786-58-5	80
Octyl phenyl ether	1818-07-1	73
di-*n*-Octyl phthalate[a]	117-84-0	52,73
Octyl propionate	142-60-9	49,80
Octyl propyl ether	29379-41-7	64,65
3-Octyne	15232-76-5	51,57
1-Octyn-3-ol	818-72-4	52
Oleic acid	112-80-1	3,7,26,27,42,43,52,59,68,70,83,88
Oplopanolol		43
Oplopanone	1911-78-0	10,32,44,51,70,89
β-Oplopenone	28305-60-4	25,32,35,44,45,70,72,75,76,83
(*E*)-Opposita-4(15),7(11)-dien-12-al	224794-18-7	89
(*E*)-Opposita-4(15),7(11)-dien-12-ol		89
13-nor-Opposit-4(15)-en-11-one		89
Orcinol dimethyl ether	see 3,5-Dimethoxytoluene	
Origanene	see α-Thujene	
Osmorhizole	3698-23-5	3
Osthole[s]	484-12-8	1,2,29,39,64
12-Oxabicyclo[9.1.0]dodeca		8
2-Oxabicyclo[9.1.0]dodeca-3,7-diene, 1,5,5,8-tetramethyl-,[1*R*-(1*R**,3*E*,7*E*,11*R**)]-	see Humulene epoxide II	
Oxabicyclopentadecan-3-one		70
Oxacyclotetradecan-2-one		85
Oxaiceane	55092-18-7	73
6,11-Oxido-acor-4-ene		10,77
1,11-Oxidocalamenene	143785-42-6	77
Oxidohimachalene	see α-Himachalene epoxide	
Oxiranmethanol, 3-methyl-3-(4-methyl-3-pentenyl)-		52
α-Oxobisabolene		52
6-Oxo-6,7-dihydrocitronellic acid		37
4-Oxo-β-isodamascol		72
Oxomanoyl oxide		32
3-Oxo-*p*-menth-1-en-7-al	160152-34-1	41,49
8-Oxo-neoisolongifolene		13
4-Oxo-14-norvitrane	77284-02-7	67
9-Oxopatchoulol		67
1-Oxo-4,4,5-trimethylcyclopentane-3-acetic acid		80
Oxydihydrocampholenic acid		80
14-Oxy-α-muurolene	69394-04-3	44
Oxypeucedanin	26091-73-6	6,39,50,64
Oxypeucedanin hydrate[s]	2643-85-8	50
Pabulenol	33889-70-2	50
Pacifigorgiadiene A	351222-63-4	88
Pacifigorgiadiene B	351222-64-5	88
Pacifigorgia-1(6),10-diene		88
Pacifigorgia-1(9),10-diene		88
Pacifigorgiol	84014-68-6	88

Table 6.1 Alphabetical list of all chemicals and the essential oils in which they have been identified (*continued*)

Chemical	CAS number	Essential oils in which the chemical has been identified. Legend of numbers: see Table 6.2 at the end of this chapter
(-)-Pacifigorgiol	90988-77-5	88
Paeonol	552-41-0	30
Palmitic acid	see Hexadecanoic acid	
γ-Palmitolactone	730-46-1	19
Palmitoleic acid	see *cis*-9-Hexadecenoic acid	
Palustrol	5986-49-2	4,13,16,27,34,35,45,49,61,75,83
di-epi-Palustrol		35
Panasinsene		88
α-Panasinsene	56633-28-4	39,66,67,88
β-Panasinsene	56684-97-0	7,35
Paradisiol	148810-80-4	39
5-epi-Paradisol	see 5-epi-Neointermedeol	
Paraffins (high)		42
Paraldehyde	123-63-7	56
Patchenol	see Patchoulenol	
Patchoulane	25491-20-7	9,41,54,56,72
Patchoulene	1405-16-9	8,10,13,14,45,56,67
α-Patchoulene	560-32-7	7,8,12,13,19,37,38,43,67,70,88
β-Patchoulene	514-51-2	7,12,34,40,52,53,61,67,72,83,88
γ-Patchoulene	508-55-4	7,12,40,41,67,69,70,88
δ-Patchoulene	53823-16-8	67
Patchoulenol[s]	17806-54-1	7
(*E*)-Patchoulenol	17806-54-1	55,65
(*Z*)-Patchoulenol	17806-54-1	55
Patchoulenone	5986-54-9	67
Patchouli alcohol[s]	5986-55-0	13,65,67,88
Patchoulol	see Patchouli alcohol	
PBT	see 5-(3-Penten-1-ynyl)-2,2'-bithiophene	
Pentacosadienoic acid		41
Pentacosane	629-99-2	7,19,29,35,37,57,59,60,72,73,85,89,90
1-Pentacosanol	26040-98-2	73
7,10-Pentadecadiynoic acid	22117-06-2	88
7,11-Pentadecadiynoic acid		88
5,10-Pentadecadiyn-1-ol	64275-50-9	89
Pentadecanal	2765-11-9	20,50,55,57,64,68,73,91
Pentadecane (*n*-)	629-62-9	4,6,12,15,16,19,29,35,37,59,60,65,73,75,88-91
Pentadecanoic acid	1002-84-2	7,13,29,37,38,59,67,68
1-Pentadecanol	629-76-5	21,37,63,72
Pentadecano-15-lactone	see Cyclopentadecanolide	
14-Pentadecanolide		2
15-Pentadecanolide	see Cyclopentadecanolide	
2-Pentadecanone	2345-28-0	7,59,70,73
2-Pentadecanone, 6,10,14-trimethyl-	see Hexahydrofarnesyl acetone	
1-Pentadecene	13360-61-7	30
1,3-Pentadiene	504-60-9	37,45
7,10-Pentadiynoic acid		88
3',3,5,7,8-Pentamethoxyflavone	99801-93-1	65
3',4',5,6,7-Pentamethoxyflavone	see Sinensetin	
3',4',5,7,8-Pentamethoxyflavone[s]	17290-70-9	55,82
3,4',6,7,8-Pentamethoxyflavone[s]	see Auranetin	
4',5,6,7,8-Pentamethoxyflavone[s]	481-53-8	6,39,55,64,65,82
(2*E*,6*E*,10*E*,14*E*,18*E*)-2,6,10,15,18-Penta-methyl-2,6,10,14,18-docosapenten-22-al		42
2,2,4,6,6-Pentamethylheptane	13475-82-6	13
Pentanal	see Valeraldehyde	
Pentane	109-66-0	45,50,73
1,3-Pentanediol, 2,2,4-trimethyl-[s]	144-19-4	13,14,21
Pentanoic acid	see Valeric acid	
Pentanol	30899-19-5	6,37,49,65,73,80
2-Pentanol	6032-29-7	6
3-Pentanol	584-02-1	42
1-Pentanol, 5-cyclopropylidene-	162377-97-1	52

Table 6.1 Alphabetical list of all chemicals and the essential oils in which they have been identified (*continued*)

Chemical	CAS number	Essential oils in which the chemical has been identified. Legend of numbers: see Table 6.2 at the end of this chapter
2-Pentanone[s]	107-87-9	37,48,80
3-Pentanone	96-22-0	80
2-Pentanone, 4-hydroxy-4-methyl-	see 4-Hydroxy-4-methyl-2-pentanone	
1,4,7,10,13-Pentaoxacyclopentadecane	33100-27-5	65
1-Pentene	109-67-1	72
2-Pentene		52
Penten-3-ol	77035-93-9	42
1-Penten-2-ol	61923-56-6	80
1-Penten-3-ol	616-25-1	37,68,85
(*E*)-2-Penten-1-ol	1576-96-1	68
(*Z*)-2-Penten-1-ol	1576-95-0	69
3-Penten-1-ol	39161-19-8	42
3-Penten-2-ol (*trans*-)	1569-50-2	85
1-Penten-3-one, 5-(3-furanyl)-2-methyl-	see Lepalone	
(*Z*)-2-Pentenyl benzoate	65466-10-6	42
Pentenyl curcumene		38
2-Pentenylcyclopentanone-3-acetic acid methyl ester	see Methyl jasmonate	
trans-2-(2-Pentenyl)furan	70424-14-5	68
5-(*cis*-2-Pentenyl)-5-pentanolide	see δ-Jasmolactone	
5-(3-Penten-1-ynyl)-2,2′-bithienyl	see 5-(3-Penten-1-ynyl)-2,2′-bithiophene	
Pentyl acetate	see Amyl acetate	
Pentyl alcohol	see Amyl alcohol	
Pentylbenzene	538-68-1	7,12,54
Pentyl benzoate	2049-96-9	43,91
Pentyl cyclohexadiene	76700-97-5	54
Pentyl-1,5-cyclohexadiene		54
1-Pentyl-1,3-cyclohexadiene	76346-02-6	54
Pentylcyclohexane	4292-92-6	54,
2-Pentyl-2-cyclopenten-1-one	25564-22-1	31
2-Pentylfuran[s]	3777-69-3	2,4,10,19,37,67,68,70,73,91
Pentyl 2-methylbutanoate	68039-26-9	20,91
Pentyl 3-methylbutanoate	see Amyl isovalerate	
Pentyl 2-methylcrotonate	7785-65-1	20
5-Pentyl-5-pentanolide	705-86-2	49,80
2-Pentylpyridine	2294-76-0	89
Perilla alcohol	see Perillyl alcohol	
Perillaldehyde (perilla aldehyde)	2111-75-3	4,6,22,23,36,38,39,41,48,50,54,55,64,65,69,80,82,85,86,88,91
Perillene	539-52-6	6,10,25-27,31,36,37,39,44,49,52,53,55,60,63-65,68,69,73
Perillyl acetate	15111-96-3	6,39,50,60,64,65
Perillyl alcohol[s]	536-59-4	2,6,7,35,36,48,54,55,64,65,68,70,79,85,91
Phellandral[s]	21391-98-0	1,2,7,32,36,37,39,41,45,48,53,61,63,80
Phellandrene	1329-99-3	4,29,30,45,63
1-Phellandrene	4221-98-1	45
α-Phellandrene	99-83-2	1-14,19,21-26,27,29-39,41,44-57,60-73,75,76,78-91
β-Phellandrene	555-10-2	1-10,12,13,15,21-27,29-39,42,44,45,48-51,53-57,60-65, 68-70,72,75,76,78-88,91
L-Phellandrene	6153-17-9	35,45,72,75
α-Phellandrene-δ-dimer		33
α-Phellandrene epoxide	288393-04-4	8,35,63
Phellandrenol		8
α-Phellandren-8-ol[s]	1686-20-0	1,2,4,7,8,12,31,32,36,44,45,49,52-54,57,63,70,72
β-Phellandren-6-ol		8
β-Phellandren-8-ol[s]	65293-09-6	8,44,45,63,72
Phellopterin	2543-94-4	50,64
Phenandrene		41
Phenanthrene	85-01-8	31,41,91
Phenanthrene,1-methyl-7-(1-methyl-ethyl)-	483-65-8	9
Phenanthrenol	30774-95-9	31
1(2*H*)Phenanthrenone, 3,4,4a,9,10, 10a-hexahydro-4a-methyl-	62318-99-4	58

Table 6.1 Alphabetical list of all chemicals and the essential oils in which they have been identified (*continued*)

Chemical	CAS number	Essential oils in which the chemical has been identified. Legend of numbers: see Table 6.2 at the end of this chapter
Phencone		85
2-Phenethanol	see 2-Phenethyl alcohol	
Phenethyl	see also Phenylethyl	
Phenethyl acetate	see 2-Phenylethyl acetate	
2-Phenethyl alcohol[s]	60-12-8	4,6,13,14,19,21,22,35,37,42,43,60,67,69,70,73,83,85,90,91
2-Phenethyl anthranilate	133-18-6	22
1-Phenethyl benzoate		43,91
2-Phenethyl benzoate (β-)	see 2-Phenylethyl benzoate	
β-Phenethyl cinnamate[s]	103-53-7	13
2-Phenethyl decanoate	61810-55-7	22
2-Phenethyl dodecanoate	6309-54-2	73
Phenethyl formate	see 2-Phenylethyl formate	
Phenethyl iodide	17376-04-4	72
Phenethyl isovalerate	see 2-Phenylethyl isovalerate	
2-Phenethyl propionate	122-70-3	20,22,35,37,43,73
Phenethyl valerate	7460-74-4	73
Phenol	108-95-2	13,21,22,31,37,43,45,64,72,80
Phenoxyacetic acid	122-59-8	72
Phenylacetaldehyde[s]	122-78-1	4,7,12,13,21,22,25-27,30,37,41-43,53,54-57,60,65,68, 72,73,75,79,80,89,90
(*E*)-Phenylacetaldehyde oxime		43
(*Z*)-Phenylacetaldehyde oxime		43
syn-Phenylacetaldoxime		43
anti-Phenylacetaldoxime		43
Phenyl acetate	122-79-2	76
Phenylacetic acid	103-82-2	37,43,60,72,73,80,90
Phenylacetonitrile[s]	140-29-4	42,43,60,90
2-Phenylalcohol	see 2-Phenethyl alcohol	
1-Phenylallyl acetate	7217-71-2	90
1-Phenyl-1-butanol	614-14-2	7
2-Phenyl-(*E*)-2-butenal	4411-89-6	68
2-Phenylethanol	see 2-Phenethyl alcohol	
β-Phenylethanol	see 2-Phenethyl alcohol	
Phenylethanolamine	7568-93-6	81
1-Phenylethyl acetate (α-)	93-92-5	4,13,14,21,90
2-Phenylethyl acetate (β-)[s]	103-45-7	22,26,29,34,37,42,43,60,68,73,90
Phenylethyl alcohol	see 2-Phenethyl alcohol	
2-Phenylethyl benzoate[s]	94-47-3	8,13,21,22,42,43,73
2-Phenylethyl butanoate	see 2-Phenylethyl butyrate	
2-Phenylethyl butyrate[s]	103-52-6	37,91
2-Phenylethyl formate[s]	104-62-1	13,37,43,73
(*E*)-2-Phenylethyl geranate		73
2-Phenylethyl hexanoate	6290-37-5	73,91
2-Phenylethyl isobutyrate	103-48-0	20,34,37,73
2-Phenylethyl isovalerate[s]	140-26-1	35,37,68,70,73,85,91
Phenyl ethyl ketone	93-55-0	60
2-Phenylethyl 2-methylbutanoate	see 2-Phenylethyl 2-methylbutyrate	
2-Phenylethyl 2-methylbutyrate[s]	24817-51-4	1,37,68,73,91
2-Phenylethyl 3-methylbutanoate	see Phenethyl isovalerate	
2-Phenylethyl 2-phenylacetate	102-20-5	91
2-Phenylethyl salicylate	87-22-9	43
2-Phenylethyl tiglate	55719-85-2	37,73
Phenylmethyl acetate	see Benzyl acetate	
2-Phenylnitroethane[s]	6125-24-2	42,43,90
1-Phenyl-2,4-pentadiyne		25
2-Phenylpropanal	93-53-8	7,13
3-Phenylpropanal	see Hydrocinnamaldehyde	
β-Phenylpropanoate		35
3-Phenylpropanoic acid	see Hydrocinnamic acid	
1-Phenyl-2-propanol	698-87-3	42
3-Phenylpropanol	see Hydrocinnamyl alcohol	
3-Phenyl-2-propenal	see Cinnamaldehyde	

Table 6.1 Alphabetical list of all chemicals and the essential oils in which they have been identified (*continued*)

Chemical	CAS number	Essential oils in which the chemical has been identified. Legend of numbers: see Table 6.2 at the end of this chapter
1-Phenyl-2-propen-1-ol	4393-06-0	90
3-Phenylpropyl acetate	see Hydrocinnamyl acetate	
γ-Phenylpropyl acetate	see Hydrocinnamyl acetate	
3-Phenylpropyl alcohol	see Hydrocinnamyl alcohol	
3-Phenylpropyl isobutyrate	103-58-2	20
2-Phenylpyridine	1008-89-5	89
3-Phenylpyridine	1008-88-4	89
Photocitral		52
epi-Photocitral		51
Photocitral A		37,51,52,65
Photocitral B	6040-45-5	37,52
(*E*)-Photocitral		50
(*E,E*)-Photocitral		37
(*Z,Z*)-Photocitral	55253-28-6	37,51
Photonerol		37
epi-Photonerol A		37
Photosantalol		49
α-Photosantalol (A)	98113-14-5	48,49,77
α-Photosantalol (B)		49,77
trans-α-Photosantalol		77
β-Photosantalol (A)	63569-02-8	77
trans-β-Photosantalol		77
epi-β-Photosantalol		77
trans-epi-β-Photosantalol		77
Phthalate[a]	see Phtalic acid	
Phthalic acid[a,s]	88-99-3	13,24,45,51,60,72,90
Phyllocladene	469-86-3	63,72
Phytadiene	30917-33-0	42,45
Phytol	7541-49-3	4,13,23,25,27,42,43,45,50,53,54,57,67,69,72,73,85
(*E*)-Phytol[s]	150-86-7	12,19,31,42,91
(*Z*)-Phytol	5492-30-8	42
Phytol acetate	see Phytyl acetate	
Phytone	16825-16-4	19,23,41,56,72
Phytyl acetate[s]	10236-16-5	41-43
cis-Phytyl acetate	5016-85-3	19,42
trans-Phytyl acetate	10236-16-5	42
Phytyl benzoate	827598-68-5	42
Pimaradiene[s]	1686-61-9	7,15,31,32,72,75,91
Pimara-8(14),15-diene	see Pimaradiene	
Pimelyl dihydrazide	13043-98-6	52
Pinadiene		8
trans-Pinane	10281-53-5	4,7,70,86
Pinanediol		41
cis-2,3-Pinanediol	18680-27-8	41,53
trans-2,3-Pinanediol		53
2-Pinanol	473-54-1	36
cis-Pinan-2β-ol	4948-29-2	86
Pinan-3-one	see Pinocamphone	
3-Pinanone	see Pinocamphone	
cis-3-Pinanone	see Isopinocamphone	
trans-3-Pinanone	see *trans*-Pinocamphone	
2(10)-Pinene	see β-Pinene	
α-Pinene[s]	80-56-8	1,3-17,19-39,41,44-58,60-93
β-Pinene[s]	127-91-3	1-15,19-33,35-39,41,44-58,60-76,83-92
1-β-Pinene	see (*S*)-β-Pinene	
2-β-Pinene		63,72
(*S*)-β-Pinene (laevo-)[s]	18172-67-3	63,65,72
α-Pinene epoxide[s]	1686-14-2	2,7,8,11,12,32,34-36,44,49,52,56,63,65,72,73,76,87
cis-Pinene hydrate	17974-51-5	55,75,83
trans-Pinene hydrate	3247-40-3	60,72
Pinene oxide		50
α-Pinene oxide	see α-Pinene epoxide	

Table 6.1 Alphabetical list of all chemicals and the essential oils in which they have been identified (*continued*)

Chemical	CAS number	Essential oils in which the chemical has been identified. Legend of numbers: see Table 6.2 at the end of this chapter
β-Pinene oxide	6931-54-0	7,12,51,52,70
Pinenol	31619-93-9	91
Pinen-4-ol	see Verbenol	
2-Pinen-10-ol	see Myrtenol	
α-Pinen-10-ol		72
2(10)-Pinen-3-one	see Pinocarvone	
Pinic acid	473-73-4	80
Pinocampheol		41
Pinocamphone	547-60-4	8,20,25,32,36,41,44,49,63,72,75,76,80
cis-Pinocamphone		12,72,75
trans-Pinocamphone[s]	547-60-4	4,12,15,20,32,57,72,75,85
Pinocamphor		70
Pinocarveol	5947-36-4	8,12,20,31,32,35,45,52,60,61,63,68,70,72,75,86
cis-Pinocarveol[s]	6712-79-4	7,29,36,45,52,55,65,72,88
trans-Pinocarveol	1674-08-4	6-8,12,15,19,20,31,32,34-36,41,44,45,47,49,50,54-57, 60,63-65,67-70,72,78-80,83-86,88,91
L-Pinocarveol	547-61-5	29,63
Pinocarvone[s]	30460-92-5	4,8,12,19,20,29,31,35,36,41,44,45,48,49,52,56,57,63,64,69, 70,72,78,80,84-86
trans-Pinocarvone	19890-00-7	52,89
Pinocarvyl acetate	1078-95-1	29,36,72,85
cis-Pinocarvyl acetate	73366-18-4	29,44,45
trans-Pinocarvyl acetate	1686-15-3	1,6,10,12,41,44,45,49,52,65,70,75,85,88
trans-Pinocarvyl formate	186607-19-2	49
Pinol	2437-97-0	45
Pinonaldehyde	2704-78-1	80
β-Pinone	see Nopinone	
cis-Pinonic acid	61826-55-9	41
Piperidine	110-89-4	8
Piperitenone	491-09-8	12,32,41,49,52,55,56,65,68,71,72,79
Piperitenone oxide	90582-88-0	68,79,85
Piperitol	491-04-3	36,45,83
cis-Piperitol	16721-38-3	2,4,7,8,24,31,32,35,37-39,41,44,45,48,53,56,61,62,64,65, 68,71,72,75,80,83,85
trans-Piperitol	16721-39-4	2,4,7,12,32,35,41,45,49,53,56,57,60-62,68,71,72,75,80,83-85
cis-Piperitol acetate	78774-33-1	56,83
trans-Piperitol acetate		56
Piperitone[s]	89-81-6	2,4,6,8,10,12,19,21,22,25,32,33,35-37,41,45,49-53,55-57,60, 63-66,68,70-72,75,79,83-85,91
cis-Piperitone		56
Piperitone 1-epoxide	5286-38-4	8
Piperitone oxide	148879-33-8	57,68
cis-Piperitone oxide	57130-28-6	57,79
trans-Piperitone oxide		79
Piperonal[s]	120-57-0	8,31,52
Piperonic acid	136-72-1	8
Piramidene		63
Plinol	72402-00-7	4,47
Plinol A	4028-59-5	80,90
Plinol B	4099-07-4	80
Plinol C	4028-60-8	4,49,80
Plinol D	4028-58-4	49,90
Podocephalol	see Himachala-1,3,5-trien-5-ol	
Pogostol[s]	21698-41-9	30,67,83,89
Pogostone	23800-56-8	67
Porosa-4,6-dien-7-one		89
Pranferin[s]	33573-60-3	39
Prangol	see Oxypeucedanin hydrate	
Pregeijerene	20082-17-1	3
Pregeijerene B		32
Pregnane		30
14β-Pregnane		30

Table 6.1 Alphabetical list of all chemicals and the essential oils in which they have been identified (*continued*)

Chemical	CAS number	Essential oils in which the chemical has been identified. Legend of numbers: see Table 6.2 at the end of this chapter
Pregn-5-en-20-one, 3,7-dihydroxy-, 3-acetate	1863-39-4	34
Preisocalamenediol		10
Prekhusenic acid		89
Prenol	see 3-Methyl-2-buten-1-ol	
Prenyl acetate	see 3-Methyl-2-butenyl acetate	
Prenyl isobutyrate	76649-23-5	20
Prezizaan-15-al	87059-20-9	16,89
nor-Prezizaan-7α-ol		89
Prezizaan-7β-ol	see Allokhusiol	
Prezizaan-15-ol	312284-45-0	89
15-nor-Prezizaan-7-one		89
Prezizaene	31145-21-8	10,16,18,89
Prezizaene 2		10
Prezizaene isomer		10
Preziza-7(15)-en-3α-ol		89
Preziza-7(15)-en-12-ol		89
12-nor-Preziza-7(15)-en-2-one		89
Preziza-7(15)-en-12-yl acetate		89
Prochamazulene	489-87-2	19
Propanal[s]	123-38-6	73,80,90
Propanal, 2-methyl-3-phenyl-	see 2-Methyl-3-phenylpropanal	
.... propanoate	see also .... propionate	
Propanoic acid	see Propionic acid	
Propanoic acid, 2-methyl-3,7-dimethyl-2,6-octadienyl ester, (*E*)-	see Geranyl isobutyrate	
2-Propanol	67-63-0	37,80
1-Propanone		81
2-Propanone	see Acetone	
2-Propanone, 1-(acetyloxy)-	see 1-Acetoxy-2-propanone	
2-Propanone, 2-methylhydrazone	5771-02-8	26
2-Propenal, 3-(2-methoxyphenyl)-	see *o*-Methoxycinnam-aldehyde	
2-Propenal, 3-(2,6,6-trimethyl-1-cyclo-hexen-1-yl)-[s]	4951-40-0	88
2-Propenoic acid, 3(4-methoxyphenyl), ethyl ester	see Ethyl-*p*-methoxycinnamate	
2-Propenoic acid, 3-phenyl, 2-phenyl-ethyl ester	see β-Phenethyl cinnamate	
2-Propen-1-ol	107-18-6	73
2-Propen-1-ol, 3-phenyl-, acetate	see Cinnamyl acetate	
5-(2-Propenyl)-1,3-benzodioxole	see Safrole	
2-Propenyl-3-methyl-4-hydroxy-2-cyclopentene		53
4-(2-Propenyl)phenol	see Chavicol	
Propionaldehyde	see Propanal	
.... propionate	see also ....propanoate	
Propionic acid[s]	79-09-4	4,37,80
Propyl acetate	109-60-4	56
Propyl angelate	53082-57-8	20
p-Propylanisole	see Dihydroanethole	
Propylbenzene	103-65-1	13
Propyl benzoate	2315-68-6	43
4-Propylbenzoic acid	2438-05-3	80
Propylene glycol	57-55-6	26
Propyl formate	110-74-7	37
2-Propyl formate		37
(*Z*)-3-Propylidene phthalide	17369-59-4	54
Propyl isobutyrate	644-49-5	20
Propyl isovalerate	557-006	68
Propyl methacrylate	2210-28-8	20

Table 6.1 Alphabetical list of all chemicals and the essential oils in which they have been identified (*continued*)

Chemical	CAS number	Essential oils in which the chemical has been identified. Legend of numbers: see Table 6.2 at the end of this chapter
Propyl 2-methylcrotonate	see Propyl tiglate	
Propyl palmitate	2239-78-3	59
Propyl tiglate[s]	61692-83-9	20
1-(1-Propynyl)-2-cyclohexen-1-ol	79688-55-4	58
Pseudoionone	141-10-6	53
cis-Pseudoisoeugenyl 2-methylbutyrate		3
trans-Pseudoisoeugenyl 2-methyl-butyrate	58989-20-1	3
Pseudolimonene[s]	499-97-8	1,22,41,44,45,49,56,68,69,72,79
Pseudowiddrene	32540-28-6	16-18,74
psi-Limonene	see Pseudolimonene	
Psoralen	66-97-7	2,50,64
PTB	see 5-(3-Penten-1-ynyl)-2,2'-bithiophene	
Pulegol	529-02-2	23,24,34,57,68,73
(*E*)-Pulegol	22472-79-3	34
(*Z*)-Pulegol		72
Pulegone	89-82-7	4,6,7,12,19,23,25,26,31,35,39,45,53,56,60,65,68,72,75,79,84-86,88
2*H*-Pyran-2-one,tetrahydro-6,6-di-methyl-		26
Pyrazine, 2-ethyl-3-methyl-	see 2-Ethyl-3-methylpyrazine	
1*H*-Pyrazole	288-13-1	59
Pyrethrin	88108-26-3	19
3(2*H*)-Pyridazinone	504-30-3	26
Pyridine	110-86-1	42,68,80,81
1-(Pyridin-2-yl)propan-1-one	3238-55-9	89
α-Pyronene	514-94-3	25,90
3-Pyrrolidinol	40499-83-0	26
Quinoline	91-22-5	4,42,72,80,89
Quinolinone	59-31-4	75
Retinal[a,s]	116-31-4	12
9-*cis*-Retinal[a]	514-85-2	63
Retinyl acetate[a]	127-47-9	88
Retusin	see 5-Hydroxy-3,3',4',7-tetramethoxyflavone	
Rhodinal	see α-Citronellal	
Rimuene	1686-67-5	7,57,72,75,84
Rose acetate	90-17-5	37
Rosefuran[s]	15186-51-3	2,37,38,41,49,57,73,91
Rosefuran epoxide	92356-06-4	4,37,52,53,57,68
Rosenoxide	see Rose oxide	
Rose oxide[c]	16409-43-1	36,53,61
cis-Rose oxide	3033-23-6	4,9,23,24,34,37,44,51,57,73,91
cis-trans-Rose oxide		73
trans-Rose oxide	5258-11-7	15,23,34,37,44,49,57,65,73,85
Rosifoliol	63891-61-2	44,83
Sabina acetone		45
Sabina ketone[s]	513-20-2	2,12,29,35,37,39,45,48,49,56,65,80,84
Sabinene[s]	3387-41-5	1-13,15,19-25,29,30-39,41-44-76,78-88,90-92
Sabinene hydrate	546-79-2	4,7,38,46,56,72,75,79,80,85
cis-Sabinene hydrate[s]	15537-55-0	1,4,6-8,10,12,24,29,33,34,36,38,39,41,44-50,53,55-57,60-65, 68,69,71,72,75,76,79,80,82-86,88,92
trans-Sabinene hydrate[s]	17699-16-0	1,2,4,6-8,10,12,19,23,25,26,29,33,36,38,39,41,44,45-51,53, 55-57,60-65,68,69,71,72,74-76,79,80,82-86,88,92
Sabinene hydrate acetate	87553-42-2	25,45
cis-Sabinene hydrate acetate	77318-48-0	6,7,29,55,56,62,83
trans-Sabinene hydrate acetate	77318-47-9	6,7,45,56,62,75
cis-Sabinene hydrate methyl ether		7
trans-Sabinene hydrate methyl ether		7
Sabinene ketone		12
Sabinol		2,22,36,38,52,68,72,88
cis-Sabinol	3310-02-9	12,33,35,41,45,52,68,75,81,84
trans-Sabinol	471-16-9	8,9,12,44,45,49,61,63,68,72,75,76,79,83,84
Sabinyl acetate	53833-85-5	2,41,63,68,75,76
cis-Sabinyl acetate	53833-85-5	2,45,75,84

Table 6.1 Alphabetical list of all chemicals and the essential oils in which they have been identified (*continued*)

Chemical	CAS number	Essential oils in which the chemical has been identified. Legend of numbers: see Table 6.2 at the end of this chapter
trans-Sabinyl acetate	139757-62-3	1,12,29,45,54,56,75,76,88
Safranal	116-26-7	57,91
Safrole[s]	94-59-7	3,8,13,19,21,22,30,53,62,71,72,80,90
Safrone		22
Salicylaldehyde[s]	90-02-8	13,14,21,72,80,81,85
Salicylic acid	69-72-7	4,13,14,19,26,37,72,80,90
Salvene	33746-69-9	75
cis-Salvene[s]		75
trans-Salvene[s]	33746-69-9	75,
Salviadienol	1064085-05-7	25
Salvial-4(14)-en-1-one	73809-82-2	4,19,25,41,44,89,91
Salvigenin	see 5-Hydroxy-4′,6,7-trimethoxyflavone	
Sandaracopimaradiene[s]	1686-56-2	31,32,63,70
Sandaracopimara-8(14),15-diene	see Sandaracopimaradiene	
Sandaracopimarinal	3855-14-9	32,44,70
Santaflavone	see 5-Hydroxy-3′,4′,6,7-tetramethoxyflavone	
α-Santalal	13827-97-9	77
(*E*)-α-Santalal	19903-70-9	77
β-Santalal[s]	13827-98-0	77
(*E*)-β-Santalal	59331-82-7	77
(*Z*)-β-Santalal	887491-74-9	77
α-Santalan-12-one		49
Santalene	see α-Santalene	
α-Santalene	512-61-8	4,6,8,12,34,46-50,72,77,81,88,91
epi-α-Santalene		44
7-epi-α-Santalene		49,65,66
β-Santalene	511-59-1	4,6,12,31,46,49,50,55,64,77
cis-β-Santalene		4,50
epi-β-Santalene	25532-78-9	6,12,49,70,77
Santalene oxide	61754-00-5	77
α-Santalenic acid		49
Santalenone		77
α-Santal-13-en-12-one		49
nor-α-Santalenone		48,49,77
nor-β-Santalenone	176777-64-3	77
β-Santalic acid	73590-17-7	77
Santalol	11031-45-1	56
α-Santalol		36,73,75,88
(*E*)-α-Santalol	14490-17-6	77
(*Z*)-α-Santalol	115-71-9	11,12,19,25,30,63,69,79,88
β-Santalol	77-42-9	83
(*E*)-β-Santalol	37172-32-0	77,89
(*Z*)-β-Santalol	42495-69-2	6,19,77,88,89
epi-β-Santalol		77
α-Santalol acetate	41414-75-9	77
(*Z*)-α-Santalol acetate		75,77
(*Z*)-β-Santalol acetate	77-43-0	89
Santalone	59300-51-5	45,77
Santene[s]	529-16-8	31,32,70,72,77,78,83,84
Santolina alcohol	35671-15-9	25,36,70
Santolina epoxide	60485-45-2	39
Santolinatriene	2153-66-4	39,60,68,88
Santolinyl acetate	79507-88-3	79
Sarisan	18607-93-7	8
Sativene	6813-05-4	12,13,19,34,36,63,65,76
(+)-Sativene	3650-28-0	63,89
Sclarene	511-02-4	44,63,75,84
Sclareol	515-03-7	23,25,75
Sclareol oxide	5153-92-4	25
Scoparone	see 6,7-Dimethoxycoumarin	
Scopoletin	see 6-Methoxy-7-hydroxycoumarin	
Sedanenolide	63038-10-8	12,54

Table 6.1 Alphabetical list of all chemicals and the essential oils in which they have been identified (*continued*)

Chemical	CAS number	Essential oils in which the chemical has been identified. Legend of numbers: see Table 6.2 at the end of this chapter
Sedanolide	6415-59-4	12,54
Selina-3,7(11)-diene[s]	6813-21-4	1,37,38,44,54,58,67,89,91
Selina-4(15),5-diene	1107026-89-0	45,90
Selina-4,7-diene		66
Selina-4,7(11)-diene	41071-31-2	89
Selina-4(14),7(11)-diene	see γ-Selinene	
Selina-4(15),7-diene	see Vetiselinene	
Selina-4(15),7(11)-diene		37,43,89,91
Selina-4,11-diene	see α-Cyperene	
Selina-4(14),17-diene		89
Selina-5,11-diene	52026-55-8	83
Selinadiene alcohol	35688-11-0	88
Selina-3,11-dien-6α-ol	75521-07-2	45,75
11-Selina-4α-ol	see Selin-11-en-4α-ol	
Selinene	27104-12-7	23,67
α-Selinene[s]	473-13-2	4-10,12,16-18,21-24,26,30,32,35-38,40,41,44,45,49,50,52-55, 58,61,63,66,67,71-79,83,85,88,89,91
7-epi-α-Selinene	123123-37-5	7,38,45,57,65,67,82,88
β-Selinene[s]	17066-67-0	2,4-13,16,18-25,26,30-32,35-38,40,41,44,45,49,50,52-58, 61,63,65-67,70-75,77-86,88,89,91
γ-Selinene[s]	515-17-3	8,30,37,45,52,54,66,72,74,76,88,89,91
δ-Selinene[s]	473-14-3	10,12,35-38,41,52,63,66,67,72,73,79,85,88,89,91
Selinenol	see γ-Eudesmol	
α-Selinenol	see α-Eudesmol	
β-Selinenol	see β-Eudesmol	
Selin-6-en-4-ol	see Selin-6-en-4α-ol	
Selin-6-en-4α-ol[s]	118173-08-3	9,52,53,89,91
Selin-6-en-8-ol		79
Selin-7(11)-en-4-ol	see Juniper camphor	
Selin-7(11)-en-4α-ol	see Selin-11-en-4α-ol	
Selin-11-en-4-ol	see Selin-11-en-4α-ol	
Selin-11-en-4α-ol[s]	16641-47-7	4,8,10,12,36,37,38,44,45,50,52,61,75,83
α-Selin-11-en-4-ol	see Selin-11-en-4α-ol	
Selin-11-en-5α-ol		45
7-epi-α-Selinen-2-one		64
Selin-7(11)-en-4-yl acetate		7
Sempervirol	1857-11-0	31
Sesquicineole	90131-02-5	77,89
Trans,trans-Sesquicitronellene		73
β-Sesquicyclogeraniol		72
γ-*cis*-Sesquicyclogeraniol		25
(*E*)-Sesquilavandulol	104121-84-8	1,26,37,75,85
(*Z*)-Sesquilavandulol	121521-16-2	10
trans-Sesquimyrcene		73
Sesquiphellandrene	73744-93-1	72,77
α-Sesquiphellandrene		86
β-Sesquiphellandrene	20307-83-9	2-4,6,8-10,12,15,16,19,24,29,38,40,41,44-46,48-50,52-58,60, 64,65,69,72,75,77-79,82,86
cis-β-Sesquiphellandrol	56144-26-4	38
trans-β-Sesquiphellandrol	56144-27-5	10,38
Sesquirosefuran	39007-93-7	19,73
Sesquisabinene	58319-04-3	32,49,70,77
Sesquisabinene A		77
Sesquisabinene B	1367879-38-6	4,77
Sesquisabinene hydrate	139341-65-4	38,77
(*E*)-Sesquisabinene hydrate	145512-84-1	6,19,38,49,86,89
cis-Sesquisabinene hydrate	58319-05-4	6,10,12,40,50,75
Sesquithujene	58319-06-5	16,38,49,50,77
7-epi-Sesquithujene	159407-35-9	4,12,17,18,38,49,77
Sesquithuriferol	117468-55-0	10,12,16
Seychellene	20085-93-2	13,24,38,40,41,45,62,67,88

Table 6.1 Alphabetical list of all chemicals and the essential oils in which they have been identified (*continued*)

Chemical	CAS number	Essential oils in which the chemical has been identified. Legend of numbers: see Table 6.2 at the end of this chapter
Shikomol	see Safrole	
Shyobunol	35727-45-8	86
6-epi-Shyobunol		89
Shyobunone	21698-44-2	10,12
Shyobunone epimer		10
2,6-di-epi-Shyobunone		10
epi-Shyobunone	39020-72-9	10
6-epi-Shyobunone	65794-23-2	10,12
Sibirene	14029-18-6	44,88
δ-Silenene	see δ-Selinene	
Silphiperfola-4,7(14)-diene	210637-49-3	7
Silphiperfol-5-en-3-ol A		12
Sinensal	3779-62-2	55
α-Sinensal[s]	17909-77-2	6,12,39,51,55,60,64-66,69,73,77,82
β-Sinensal (*cis-*)	17909-87-4	4,36,38,39,50,55,60,64,65,77,82
Sinensetin[s]	2306-27-6	55,64,65,82
Sinularene		7
Sitosterol[s]	12002-39-0	6
β-Sitosterol	83-46-5	42
Solanone	1937-54-8	6,50,64
(*E*)-Solanone	54868-48-3	6
Solavetivone	54878-25-0	89
Soledene		37
Spathulene	116845-09-1	35
β-Spathulene	53526-64-0	15
Spathulenol	6750-60-3	1-8,10,12,13,15,19,21-25,29-32,34-38,41,44,45,49-51,53, 55-57,60,61,63-65,67-72,74-76,79-86,88,89
(-)-Spathulenol (β-)	77171-55-2	3,9,13,63
allo-Spathulenol	99147-40-7	45
Spirafoliolide		45
Spiro[4.5]dec-1-ene	697-27-8	81
cis-Spiroether[s]	4575-53-5	19
trans-Spiroether[s]	50257-98-2	19
Spiro[2.4]heptane		72
Spirosantalol	117020-19-6	77
Spirovetiva-3,7(11)-dien-12-al epimer A	300349-33-1	89
Spirovetiva-3,7(11)-dien-12-al epimer B	300349-33-1	89
Spirovetiva-1(10),7(11)-diene	137941-80-1	89,91
Spirovetiva-3,7(11)-dien-2α-ol	see β-Vetivol	
Spirovetiva-3,7(11)-dien-12-ol		89
Spirovetiva-1(10),11-dien-2-one		89
Squalene	111-02-4	7,26,42,43,45,67
Squalene 2,3-oxide[s]	7200-26-2	42,43
[3]Staffane-3,3-dicarboxylic acid		10
Stearic acid	see Octadecanoic acid	
Styrallyl alcohol	13323-81-4	7
Styrene	100-42-5	13,14,21,22,44
Suberosin	see 7-Methoxy-6-(3-methyl-2-butenyl) coumarin	
Succinic acid	110-15-6	41,80
Sulcatone	see Methylheptenone	
Surfynol 102	see 5-Decyne-4,7-diol, 4,7-dimethyl-	
Sylvestrene	1461-27-4	2,10,32,70
Syringol	91-10-1	75
(*E*)-Tagetenone[s]	33746-72-4	49,72
(*Z*)-Tagetenone[s]	33746-71-3	12,35
Tagetone	23985-25-3	51
Tamariscene	351413-96-2	88
Tangeretin	see 4′,5,6,7,8-Pentamethoxyflavone	
Teresantalal	59300-39-9	77
α-Teresantalic acid	562-66-3	77
Terpenediol[s]	80-53-5	34,60
1,8-Terpin	see Terpenediol	

Table 6.1 Alphabetical list of all chemicals and the essential oils in which they have been identified (*continued*)

Chemical	CAS number	Essential oils in which the chemical has been identified. Legend of numbers: see Table 6.2 at the end of this chapter
α-Terpinen-7-al	see *p*-Mentha-1,3-dien-7-al	
γ-Terpinen-7-al	see *p*-Mentha-1,4-dien-7-al	
Terpinene	8013-00-1	31,45,54
α-Terpinene[s]	99-86-5	1-13,19,21,22,24-39,41,44-57,60-73,75,76,78-89,91,92
β-Terpinene[a]	99-84-3	12,45,49,63,65,68,70,72,75
γ-Terpinene[s]	99-85-4	1-13,19-27,29-39,41,44-57,60-76,78-88,90-92
δ-Terpinene		68,75,85
τ-Terpinene		68,72
trans-Terpinene		63
γ-Terpinene 1,2-epoxide	17023-74-4	7
Terpinene hydrate[s]	2451-01-6	70,80
α-Terpinenol		39,55
Terpinen-1-ol	see 1-Terpineol	
Terpinen-2-ol	see α-Terpineol	
Terpinen-3-ol		56
Terpinen-4-ol[s]	562-74-3	1-10,12-14,16,17,19-27,29-39,41,44,45,47-58,60-86,88-92
γ-Terpinen-7-ol		12
1-Terpinenol	see 1-Terpineol	
1-Terpinen-5-ol	55708-42-4	8
Terpinen-4-yl acetate	4821-04-9	6,23,25,31,44,45,49,52,55,56,61,62,64,69,70,79,80,84,85,92
Terpineol	8000-41-7	80
1-Terpineol[s]	586-82-3	4,7-9,14,19,35,38,45,48-50,52,55-57,60,70,76,81,83-86,88
4-Terpineol	see Terpinen-4-ol	
α-Terpineol[s]	98-55-5	1-16,19-39,41-57,60-82,84-93
cis-α-Terpineol		13,62,72
β-Terpineol[s]	138-87-4	8,9,25-27,29,32,35,36,45,48,50,56,59-61,64-66,68,70,72,76, 79,81,84-86
(*E*)-β-Terpineol	7299-41-4	14,25,28,44,45,50,55,56,64,65,72,85,89
(*Z*)-β-Terpineol	7299-40-3	7,10,14,15,25,38,39,41,44,45,50,53,55,56,65,72,76,83,85,89
γ-Terpineol	586-81-2	4,9,10,14,19,24,25,41,44,45,47,49,52,55,61,65,68,72,73,76, 80,81,83,85,86,90
δ-Terpineol	7299-42-5	4,8,9,12,25,30,31,34,35,41,45,50,52,57,61,66,68,71,72,75,76, 79,80,83,86,92
σ-Terpineol		56
Terpin hydrate	see Terpinene hydrate	
cis-1,8-Terpin hydrate	2451-01-6	80
Terpinolene (α-)[s]	586-62-9	1-10,12-14,16,19,21-26,29-39,41-57,60-74,78-88,90-92
β-Terpinolene	1400450-37-4	62,68
γ-Terpinolene	85188-59-6	53,85
Terpinolene-4,8-diol		83
Terpinolene epoxide		73
Terpinolene 1,2-epoxide	6784-10-7	7
Terpinyl acetate	8007-35-0	9,12,25,32,34,36,39,44,49,50,56,60,69,76,79,86,92
α-Terpinyl acetate[s]	80-26-2	4,6,8,12,16,21,24-27,29,31,32,34-36,38,41,44,45,49,50,53-57, 60,61-66,69-73,75,76,78-80,82-86,88
β-Terpinyl acetate	10198-23-9	35,62,79
γ-Terpinyl acetate	10235-63-9	12
δ-Terpinyl acetate		32,45,72
α-Terpinyl formate	2153-26-6	31,37,60
α-Terpinyl isobutyrate	7774-65-4	6,32,85
α-Terpinyl propionate	80-27-3	45,56
2,2′:5′,2′-Terthiophene	see α-Terthienyl	
Testosterone[a]	58-22-0	58
Tetracosane	646-31-1	7,12,19,25,29,37,42,43,59,60,72,73,85,89,90
9-Tetracosene		73
Tetracyclo[6.3.2.0(2,5).0(1,8)]tridecan-9-ol, 4,4-dimethyl-		26,44,72
Tetradecanal	124-25-4	1,6,7,10,13,21,29,39,50,55,64,65,73,82,91
n-Tetradecane	629-59-4	4,7,8,15,19,29,35,37,39,41-43,45,49,50,60,64,65,70,73,82, 85,89,90
Tetradecanoic acid[s]	544-63-8	4,7,13,27,29,37,45,57,59,62,67,68,72

Table 6.1 Alphabetical list of all chemicals and the essential oils in which they have been identified (*continued*)

Chemical	CAS number	Essential oils in which the chemical has been identified. Legend of numbers: see Table 6.2 at the end of this chapter
Tetradecanol	27196-00-5	29,44,55,60,68,73,82
1-Tetradecanol	112-72-1	91
14-Tetradecanolide[s]	3537-83-5	2
Tetradecenal	54264-02-7	13,39
2-Tetradecenal[s]	64461-99-0	29
(E)-2-Tetradecenal	51534-36-2	50,55,64
(E)-α-Tetradecenal		7
Tetradec-2-enal	see 2-Tetradecenal	
Tetradecene	26952-13-6	4,29,45,50,82
(E)-7-Tetradecenol	37011-95-3	7
Tetradecenyl acetate		73
(Z)-11-Tetradecen-1-yl acetate	20711-10-8	63
(E)-5-Tetradecen-3-yne	74744-48-2	45
Tetradecyl acetate	638-59-5	73,91
Tetrahydrocitronellene		45
Tetrahydroionol	4361-23-3	29
Tetrahydrolinalool	78-69-3	52,73
Tetrahydrolinalyl acetate[a]	20780-48-7	21
Tetrahydro-3-methylfuran		26
Tetrahydropyran	142-68-7	41
Tetrahydropyranyl-2-ethanal		80
1,2,3,6-Tetrahydropyridine	694-05-3	10
1,2,3,4-Tetrahydro-1,1,6-trimethyl-naphthalene	see α-Ionene	
4',5,6,7-Tetramethoxyflavanone		65
4',5,6,7-Tetramethoxyflavone[s]	1168-42-9	6,39,55,64,65,82
4',5,7,8-Tetramethoxyflavone[s]	6601-66-7	55,65,82
Tetramethylbicyclo[2.2.2]oct-2-ene		70
2,3,4,5-Tetramethyl-2-cyclopenten-1-ol		7
1,2,3,3-Tetramethylcyclopenten-4-one		35
4,8,12,16-Tetramethyl-γ-heptade-calactone		42
2,6,10,14-Tetramethylheptadecane	18344-37-1	9
4,8,12,16-Tetramethylheptadecan-4-olide	200272-61-3	57
Tetramethylhexadecane		43
(2E,6E,10E)-3,7,11,15-Tetramethyl-2,6,10,14-hexadecatetraen-1-yl acetate		43
[R-[R*, R*-(E)]]-3,7,11,15-Tetramethyl-2-hexadecen-1-ol	see (E)-Phytol	
Tetra-O-methylisoscutellarein	see 4',5,7,8-Tetramethoxy flavone	
1,5,9,9-Tetramethyl-2-methylene-spiro[3.5]non-5-ene		89
2,2,8,8-Tetramethyl-5-nonanone	5709-95-5	57
1,5,5,8-Tetramethyl-12-oxabicyclo[9.1.0]dodeca-3,7-diene	90820-79-4	26,34
2,2,6,8-Tetramethyl-7-oxatricyclo-[6.1.0.0(1,6)]nonane		72
Tetramethylpentadecane	80297-59-2	43
2,3,4,6-Tetramethylphenol	3238-38-8	49
Tetramethylpyrazine	1124-11-4	4
Tetra-O-methylquercetin	see 5-Hydroxy-3,3',4',7-tetramethoxyflavone	
Tetramethylscutellarein	see 4',5,6,7-Tetramethoxyflavone	
5,5,9,10-Tetramethyltricyclo[7.3.0.0-(1,6)]dodecan-11-one		37
3,3,8,8-Tetramethyltricyclo[5.1.0.0(2,4)]-oct-5-ene-5-propanoic acid		89
6,10,11,11-Tetramethyltricyclo[5.3.0.1(2,3)]undec-1(7)-ene		35
6,10,11,11-Tetramethyltricyclo[5.3.0.1(2,3)]undec-7-ene		35

Table 6.1 Alphabetical list of all chemicals and the essential oils in which they have been identified (*continued*)

Chemical	CAS number	Essential oils in which the chemical has been identified. Legend of numbers: see Table 6.2 at the end of this chapter
1,4,7,10-Tetraoxacyclodecane		65
α-Tetrasantalic acid		77
Tetratetracontane	7098-22-8	60
Theaspirane A	36431-72-8	37,73
Theaspirane B	36431-72-8	37,73
Thiazole	288-47-1	22
2,4-Thujadiene	119205-49-1	86
Thuja-2,4(10)-diene	36262-09-6	2,7,12,35,41,44,45,49,63,70,72,79,80,84,86
α-Thujanol	406160-72-3	53
3-Thujanol	see Thujyl alcohol	
4-Thujanol	see Sabinene hydrate	
cis-Thujanol	see cis-Sabinene hydrate	
cis-4-Thujanol	see cis-Sabinene hydrate	
trans-Thujanol	see trans-Sabinene hydrate	
trans-4-Thujanol	see trans-Sabinene hydrate	
iso-3-Thujanol	7712-79-0	75
4-Thujanyl acetate		45
β-Thujaplicine	499-44-5	4,7,81
γ-Thujaplicine	672-76-4	7
β-Thujaplicinol	4356-35-8	75
α-Thujenal[s]	57129-54-1	7,12,15,20,36,45
Thuj-3-en-10-al	see α-Thujenal	
trans-Thuj-3-en-10-al	see α-Thujenal	
Thujene	58037-87-9	34,81
2-Thujene		79
3-Thujene	see α-Thujene	
4(10)-Thujene	see Sabinene	
α-Thujene[s]	2867-05-2	1,2,4-10,12,14,15,19-23,25,26,29-36,38,39,41,44-58,60-65, 67-72,74-76,78-92
β-Thujene[s]	28634-89-1	8,41,44,49,53,56,65,70
Thujenol		36
Thujen-2-ol		45
Thuj-3-en-10-ol		45
Thujen-2-one (3-)	546-78-1	36,72
4-Thujen-2α-yl acetate		45
Thujilic alcohol	see Thujyl alcohol	
Thujol	35732-37-7	12,38,55,65,68
β-Thujol		88
Thujone	see α-Thujone	
α-Thujone[s]	546-80-5	4,5,7,12,19,22,23,25,32,35,36,41,44,45,48,50,52,55,57,62-65, 68,72,75,76,79,80,84-86
β-Thujone[s]	471-15-8	7,12,19,23,25,35,36,38,41,44,45,49,52,56,57,63,72,75,76, 80,84,85,86
(-)-β-Thujone	33766-30-2	19,41,56,85
cis-Thujone	see α-Thujone	
trans-Thujone	see β-Thujone	
Thujopsadiene	24048-40-6	15-18,67
3-Thujopsanone	25966-79-4	16,61,83
trans-3-Thujopsanone	25966-79-4	17
cis-Thujopsenal	470-41-7	16,18
Thujopsene[s]	470-40-6	7,9,15-17,31,32,35,36,38,44,63,67,68,81,88-90,93
3-Thujopsene		67
13-Thujopsene		89
cis-Thujopsene	32435-95-3	7,17,18,44,50,67
cis-Thujopsenic acid		17
Thujyl acetate	72747-24-1	49,75
iso-3-Thujyl acetate		45,56
neo-3-Thujyl acetate		45
Thujyl alcohol[s]	513-23-5	7-9,31,39,45,49,61,64,73,75,75,79
Thunbergol	25269-17-4	63,70
Thymodihydroquinone		85
Thymohydroquinone	2217-60-9	7,85

Table 6.1 Alphabetical list of all chemicals and the essential oils in which they have been identified (*continued*)

Chemical	CAS number	Essential oils in which the chemical has been identified. Legend of numbers: see Table 6.2 at the end of this chapter
Thymohydroquinone dimethyl ether	see 2,5-Dimethoxy-*p*-cymene	
Thymol[s]	89-83-8	4-7,12,16,21,23,25,26,29-31,35,37,39,41,44,45,48-51,55, 56,57,59,60,64,65,68,69,72,75,76,79,80,82,84-86,88,91
Thymol acetate	see Thymyl acetate	
Thymol methyl ether	see Methyl thymol	
Thymoquinone	490-91-5	7,72,85,86
Thymyl acetate[s]	528-79-0	4,44,49,85,86
Thymyl formate		85
Thymyl isobutyrate	5451-67-2	65
Tiglic acid[s]	80-59-1	37,80
Tiglyl acetate	19248-94-3	20
α-Tocopherol	59-02-9	42,43
o-Tolualdehyde	see 2-Methylbenzaldehyde	
Toluene	see Methylbenzene	
p-Tolylacetaldehyde	104-09-6	13
m-Tolyldimethylacetaldehyde		93
Tolyldimethylcarbinol		85
p-Tolylmethylcarbinol	536-50-5	35
4-(*p*-Tolyl)valeraldehyde	4895-19-6	77
Torilenol	84071-85-2	19,25
Torquatone	3567-96-2	35
Torreyol	see δ-Cadinol	
Totarene		41
Totarol	511-15-9	31,72
cis-Totarol	511-15-9	72
trans-Totarol	511-15-9	72,84
n-Triacontane[s]	638-68-6	4,7,91
Tricontane	see *n*-Triacontane	
Tricosadienoic acid		41
Tricosane	638-67-5	4,7,12,19,25,37,43,59,72,73,85,89-91
Tricosene	56924-46-0	13,43,73
(*Z*)-9-Tricosene	27519-02-4	73
11-Tricosene	52078-56-5	43
(*Z*)-11-Tricosene	52078-37-2	43
Tricyclene	508-32-7	2,4,6-8,10,12,19,23,24,29,31,32,34,36-38,41,44-52,54-56,60, 63,65,66,68-72,75,76,79,80,84-88
Tricycloekasantalal	16933-18-9	49,77
Tricycloekasantalic acid	59300-52-6	77
Tricycloekasantalol	16933-12-3	77
Tricyclo[3.2.2.0]nonane-2-carboxylic acid		26
Tricyclo[4.3.1.1(3,8)]undecan-1-ol	31061-64-0	52
Tricyclo[5.4.0.0(2,8)]undec-9-ene, 2,6,6,9-tetramethyl-	see α-Longipinene	
Tridecanal	10486-19-8	6,7,19,20,29,50,55,64,65,73,82,91
Tridecanal diethyl acetal	72934-16-8	73
Tridecane	629-50-5	6,25,31,35,39,44,48-50,56,65,73,82,91
2-Tridecane		72
Tridecanoic acid	638-53-9	7,29,37,59,70
Tridecanol	26248-42-0	31
1-Tridecanol	112-70-9	82,91
4-Tridecanol	26215-92-9	83
Tridecano-13-lactone	see Tridecanolide	
Tridecanolide[s]	1725-04-8	1,2
13-Tridecanolide	see Tridecanolide	
15-Tridecanolide		2
2-Tridecanone	593-08-8	7,38,45,52,59,73,91
3-Tridecanone	1534-26-5	70
(*E*,*Z*,*Z*)-2,4,7-Tridecatrienal	13552-96-0	26
(*E*)-2-Tridecenal	7069-41-2	55,64
1-Tridecene	2437-56-1	73,91
Trieicosane		73
Trieicosene		73

Table 6.1 Alphabetical list of all chemicals and the essential oils in which they have been identified (*continued*)

Chemical	CAS number	Essential oils in which the chemical has been identified. Legend of numbers: see Table 6.2 at the end of this chapter
Trifluoroacetyl-α-terpineol	28664-18-8	72
1,2,4-Trihydroxymenthane[s]	66767-24-6	83
4′,5,7-Trihydroxy-3′,6,8-trimethoxy-flavone		55
(Z)-Trimenal	300733-87-3	55
2′,3′,4′-Trimethoxyacetophenone	13909-73-4	26
2′,4′,6′-Trimethoxyacetophenone	832-58-6	26
3,4,5-Trimethoxybenzaldehyde	86-81-7	51,66
Trimethoxycinnamaldehyde	34346-90-2	51
4′,5,7-Trimethoxyflavone[s]	5631-70-9	55
2,4,5-Trimethoxyphenylacetone	16603-18-2	10
1-(2,4,5-Trimethoxyphenyl)-1-methoxypropan-2-ol	98205-47	10
3-(3,4,5-trimethoxyphenyl)-propenyl acetate		6
2,4,5-Trimethoxypropiophenone		10
endo-Trimethylamine		41
Trimethylapigenin[s]	5631-70-9	55
2,4,6-Trimethylbenzaldehyde	487-68-3	30
1,3,5-Trimethylbenzene	108-67-8	69
2,4,6-Trimethylbenzenesulfonamide	4543-58-2	81
1,1,3-Trimethylbenzyl alcohol		73
2,6,6-Trimethylbicyclo[3.1.0] heptan-3-one		41
5,5,6-Trimethylbicyclo[2.2.1] heptan-2-one[s]	3292-05-5	46,80
2,6,6-Trimethylbicyclo[3.1.1]hept-2-ene	see α-Pinene	
2,7,7-Trimethylbicyclo[2.2.1]hept-2-ene	see Fenchene	
3,6,6-Trimethylbicyclo[3.1.1]hept-2-ene	4889-83-2	63
1,7,7-Trimethylbicyclo[2.2.1] hept-5-en-2-ol	91055-72-0	35
3,3,6-Trimethylbicyclo[2.2.1] hept-1-en-7-ol		72
1,7,7-Trimethylbicyclo[2.2.1]hept-2-yl acetate	92618-89-8	63
(2,2,6-Trimethylbicyclo[4.1.0] hept-1-yl)-methanol	78996-11-9	63
1,4,4-Trimethylcycloheptadienone		8
3,7,7-Trimethyl-1,3,5-cycloheptatriene	3479-89-8	63
(E)-1-(2,6,6-Trimethyl-1,3-cyclohexa-dien-1-yl)-2-buten-1-one	see (E)-Damascenone	
Trimethylcyclohexanol	1321-60-4	43
2,2,6-Trimethylcyclohexanone	2408-37-9	19
3,3,5-Trimethylcyclohexene	503-45-7	44
3,6,6-Trimethylcyclohexen-2-ol		52
3-(2,6,6-Trimethyl-1-cyclohexen-1-yl)-2-propenal	see 2-Propenal, 3-(2,6,6-trimethyl-1-cyclohexen-1-yl)-	
3,4,4-Trimethyl-2-cyclopenten-1-one	30434-65-2	80
1-(1,2,3-Trimethylcyclopent-2-enyl)-methanone		7
2,5,9-Trimethyldecane	62108-22-9	13
α-Trimethyldodecane		85
2,6,10-Trimethyl-2,6,9,11-dodecate-traenal	see α-Sinensal	
3,7,11-Trimethyl-2,6,10-dodecatrien-1-ol	see Farnesyl alcohol	
1,2,4-Trimethylenecyclohexane	14296-81-2	3
Trimethylenenorbornane	2825-82-3	72
2,2,6-Trimethyl-6-ethenyl-tetrahydro-2H-pyran-3-ol	see Linalool oxide	
3,5,5-Trimethyl-2(5H)-furanone	50598-50-0	52
2,3,6-Trimethyl-1,5-heptadiene	33501-88-1	83

Table 6.1 Alphabetical list of all chemicals and the essential oils in which they have been identified (*continued*)

Chemical	CAS number	Essential oils in which the chemical has been identified. Legend of numbers: see Table 6.2 at the end of this chapter
3,3,5-Trimethyl-1,5-heptadiene	74630-29-8	24
3,3,6-Trimethyl-1,5-heptadiene	35387-63-4	51
2,5,5-Trimethyl-2,6-heptadienoic acid		37
2,5,6-Trimethyl-1,3,6-heptatriene	42123-66-0	9,64
Trimethylindoline		89
4,6,6-Trimethyl-2-(3-methylbuta-1,3-dienyl)-3-oxatricyclo[5.1.0.0(2,4)]octane		67
4,11,11-Trimethyl-8-methylenebicyclo-[7.2.0]undec-4-ene		41
2,8,8-Trimethyl-4-methylene-2-vinyl-bicyclo[5.2.0]nonane^s		88
4,8,8-Trimethyl-2-methylene-4-vinyl-bicyclo[5.2.0]nonane		89
1,4,6-Trimethyl naphthalene	2131-42-2	16
2,3,3-Trimethyl-2-norbornanol	see Camphene hydrate	
Trimethyl norpinen-3-one		41
1,1,3-Trimethyl-2-oxabicyclo[2.2.2]-octan-5-one		80
1,3,3-Trimethyl-2-oxabicyclo[2.2.2]-octan-5-one	see 1,8-Cineole	
2,4,5-Trimethyloxazoline		80
Trimethyl-6,10,14-pentadecanone	see Hexahydrofarnesyl acetone	
2,2,4-Trimethyl-1,3-pentanediol	see 1,3-Pentanediol, 2,2,4-trimethyl-	
1-(2,3,4,5-Trimethylphenyl)-butan-1-one		69
2,3,5-Trimethylpyrazine	14667-55-1	89
1,3,5-Trimethyl-5-pyrazolone		52
5-(Trimethylsilyl)furfural		75
4,8,8-Trimethylspiro[2.6]nona-4,6-diene	81532-24-3	8
2,10,10-Trimethyltricyclo[7.1.1.0(2,7)]-undec-6-en-8-one		58
4,8,8-Trimethyl-4-vinylcaryophyllene		81
Trimethyl vinyl tetrahydropyran	see Bois de rose oxide	
2,2,6-Trimethyl-6-vinyltetrahydropyran	see Bois de rose oxide	
2,6,10-Triphenyldodecane		89
1,3,5-Trithiolane		59
Tritriacontane	630-05-7	4
ar-Turmerol	38142-57-3	38
β-Turmerol		38
α-Turmerone	82508-15-4	29
β-Turmerone	82508-14-3	29
ar-Turmerone	532-65-0	8,21,29,41,57
Umbelliferone	see 7-Hydroxycoumarin	
Umbellulone	24545-81-1	7,12,31,35,45,55,63
Undeca-4,6-diyne		58
Undecanal	112-44-7	6,12,29,38,39,50,55,64,65,69,73,75,82,91
Undecanal diethyl acetal	53405-97-3	73
Undecane (*n*-)	1120-21-4	4,7,15,29,31,35,37,38,43,48,50,64,65,70,72,73,75,82,89-91
Undecane, 2,6-dimethyl-	17301-23-4	13,14
Undecanol	see 1-Undecanol	
1-Undecanol^s	112-42-5	31,50,66,72
2-Undecanol	1653-30-1	31,38
Undecanoic acid	112-37-8	13,14,19,29,37,39,65
Undecanone	53452-70-3	70
2-Undecanone^s	112-12-9	7,8,26,32,38,41,44,45,51,52,70,72,73,75,80,85,91
4-Undecanone	14476-37-0	69
1,3,5,8-Undecatetraene	50277-31-1	49
Undecatriene	16356-11-9	73
1,3,5-Undecatriene	see Undecatriene	
(*E,E*)-1,3,5-Undecatriene	19883-29-5	36
(3*E*,5*Z*)-1,3,5-Undecatriene	19883-27-3	2,36,46,47,49,55
(3*Z*,5*E*)-1,3,5-Undecatriene	51447-08-6	44

Table 6.1 Alphabetical list of all chemicals and the essential oils in which they have been identified (*continued*)

Chemical	CAS number	Essential oils in which the chemical has been identified. Legend of numbers: see Table 6.2 at the end of this chapter
(6Z,8E)-Undeca-6,8,10-trien-3-one	1123751-39-2	36
(6Z,8E)-Undeca-6,8,10-trien-4-one		36
2-Undecenal	2463-77-6	39,69
(E)-2-Undecenal	53448-07-0	55,65
10-Undecenal	112-45-8	10
(E)-9-Undecenal	324541-83-5	53
Undec-(9Z)-en-1-al	see (Z)-Undec-9-en-1-al	
(9Z)-Undecen-1-al	see (Z)-Undec-9-en-1-al	
1-Undecene	821-95-4	91
2-Undecene	2244-02-2	83
(Z)-8-Undecenol		55
10-Undecen-1-ol	112-43-6	51
1-Undecen-10-one		32
5-Undecen-3-yne	74744-31-3	2
Undecone		83
Undecyl acetate	1731-81-3	6,50,64
Undecyl alcohol	see 1-Undecanol	
3-Undecyne	60212-30-8	51,52,81
Ursolic acid	77-52-1	42
Uvidine		7
Valencene[s]	4630-07-3	3,4,6,8,10,17-19,22-25,37-39,41,44,45,49,50,52,55,57-61,64-67, 69,72,73,75,78,82,83,85,86,88,89
Valencene ketone	137764-27-3	88,89
Valencen-12-ol	438536-20-0	89
Valeraldehyde[s]	110-62-3	49,68,73,80,85
Valeraldehyde diethyl acetal	3658-79-5	73
Valeranone	55528-90-0	20,25,88
Valerena-4,7(11)-diene	351222-66-7	83,88
Valerenal	4176-16-3	30,88,89
Valerenic acid	3569-10-6	88
Valerenol	101628-22-2	30,45,88,89
cis-Valerenol		88
trans-Valerenol		88
cis-Valerenyl acetate	101527-78-0	88,89
trans-Valerenyl acetate	101527-74-6	88
cis-Valerenyl isovalerate		88
trans-Valerenyl isovalerate	101527-75-7	88
Valerianol[s]	20489-45-6	25,37,41,45,65,73,88,89
Valeric acid[s]	109-52-4	4,37,49,72,73,80,88,90,91
γ-Valerolactone	108-29-2	49
cis-3-Validene-3,4-dihydrophthalide		54
3-Validene-4,5-dihydrophthalide		54
Vanillin	121-33-5	4,13,14,21,22,26,28,42,45,62,72,89,90
Vanillin, acetate	881-68-5	51,62
β-Vatirenene		52,89
Veratraldehyde[s]	120-14-9	51
Veratrole[s]	91-16-7	15,19,90
Verbenene	4080-46-0	12,15,29,31,35,36,39,41,44,57,63,68,75,83,85,86
Verbenol[s]	473-67-6	4-6,8,12,35,44,50-53,65,72,85,88
cis-Verbenol[s]	1845-30-3	1,2,8,12,32,35,38,41,44,45,49,50,52,53,57,63,66,70,72,75, 80,84-86
trans-Verbenol	1820-09-3	1,2,4,7,12,36,45,49,51-53,57,63,70,72,75,85-87,91
Verbanone		72
Verbenone	80-57-9	8,12,15,29,34-36,41,44,45,49,50,52,55-57,61,63-65,67,68, 70,72,73,75,76,80,84-86,89,91
Verbenyl acetate[s]	33522-69-9	2,88
cis-Verbenyl acetate	29135-27-1	7,12,83
trans-Verbenyl acetate	1203-21-0	34,49,83
Verbenyl ethyl ether		44
Verrucarol	2198-92-7	88
Verticilla-4(20),7,11-triene		63
Verticiol	70000-19-0	58,63

Table 6.1 Alphabetical list of all chemicals and the essential oils in which they have been identified (*continued*)

Chemical	CAS number	Essential oils in which the chemical has been identified. Legend of numbers: see Table 6.2 at the end of this chapter
Vestitenone		93
Vetidiol		89
Vetiselinene[s]	105497-53-8	89,91
Vetiselinenol	28102-68-3	89
Vetiselinenol epimer		89
Vetiselinol[s]	32433-12-8	89
α-Vetispirene	28908-28-3	89
β-Vetispirene	28908-27-2	83,89
β-Vetivene	see β-Vetivenene	
β-Vetivenene[s]	27840-40-0	12,15,25,35,89,91,93
γ-Vetivenene	28908-26-1	89
Vetiverenol		89
Vetiverol	68129-81-7	45,89
β-Vetivol		89
α-Vetivone	see Isonootkatone	
β-Vetivone	18444-79-6	89
p-Vinylanisole	637-69-4	42,81
2-Vinylbenzaldehyde	28272-96-0	13
4-Vinylbenzaldehyde	1791-26-0	13
p-Vinylbenzohydrazide		22
Vinyl butyrate	123-20-6	90
Vinylcyclooctane	61142-41-4	51
p-Vinylguaiacol[s]	7786-61-0	29,39,49,55,65,67,89-91
4-Vinyl-2-methoxyphenol	see p-Vinylguaiacol	
4-Vinyl-4-pentanolide		42
2-Vinylphenol	695-84-1	13,14,21,22
4-Vinylphenol	2628-17-3	70,72,89
2-Vinylpyridine	100-69-6	42
3-Vinylpyridine	1121-55-7	43
Viridifloral		51,68
Viridiflorene[s]	21747-46-6	3-5,9,10,13,23,25,34,35,37,38,41,44,45,56,57,61,63,65,67,68, 72,73-76,83,85,89,90
Viridiflorol	552-02-3	4,9,10,13,19,22,23,26,29-31,34,35-38,41,44,45,49,52,56,61, 63,67,68,70,72,75,76,79,80,83,85,88,89
Vitamin A aldehyde	see Retinal	
Vitispirane	65416-59-3	37
Vulgarol B		25,30
Vulgarone A	62065-10-5	10
Vulgarone B	64180-68-3	88
Widdrene	see Thujopsene	
Widdrol	6892-80-4	7,15-18,36,44,53,68,72,75,88,89
Wine lactone	182699-77-0	39,65
Xanthotoxin	see 8-Methoxypsoralen	
Xylene	1330-20-7	44
o-Xylene	95-47-6	38,44
p-Xylene	106-42-3	19,44
Ylangene		25,72
α-Ylangene	14912-44-8	1,3,4,7-9,11-13,15,19,21-23,25-27,29,31,34,35,37,38,44,45, 49,58,61,63,65,68,70,72,73,75,76,79,80,83,85,88-91
β-Ylangene	20479-06-5	1,7,11,16,23,25,36,37,38,41,44,58,63,67,72,73,90,91
Yomogi alcohol	26127-98-0	19
Yomogi alcohol acetate		19
Zierone		89
Zingerone	see Zingiberone	
Zingiberene (α-)	495-60-3	1-4,7,8,10,12,25,27,29,31,38,45,49,52,66,69,72,73,77,88,91
Zingiberenol	58334-55-7	38
epi-Zingiberenol	72346-47-5	38
Zingiberone	122-48-5	38
Zizaane		89
2-epi-Ziza-6(13)-en-12-al		89
Zizaene	see Khusimene	
Ziza-6(13)-ene	see Zizaane	

Table 6.1 Alphabetical list of all chemicals and the essential oils in which they have been identified (*continued*)

Chemical	CAS number	Essential oils in which the chemical has been identified. Legend of numbers: see Table 6.2 at the end of this chapter
2-nor-Zizaene		89
trans-2-nor-Zizaene		89
12-nor-Zizaene		89
Ziza-5-en-12-ol	16202-80-5	89
Ziza-6(13)-en-3α-ol	see Zizanol	
2-epi-Ziza-6(13)-en-3-ol		89
2-epi-Ziza-6(13)-en-3α-ol	87068-39-1	89
Ziza-6(13)-en-3β-ol	see Epizizanol	
2-epi-Ziza-6(13)-en-3β-ol	300350-23-6	89
12-nor-Ziza-6(13)-en-2α-ol	300539-81-5	89
12-nor-Ziza-6(13)-en-2β-ol		89
Ziza-6(13)-en-3-one	see Zizanone	
2-epi-Ziza-6(13)-en-3-one	see Epizizanone	
Ziza-6(13)-en-4-one		89
Ziza-6(13)-en-12-yl acetate	61474-33-7	89
Ziza-6(13)-en-12-yl formate		89
Ziza-6(13)-en-12-yl methyl ether	300349-20-6	89
Zizanal	82509-29-3	89
Zizanoic acid	see Khusenic acid	
Zizanol	28102-79-6	89
Zizanone	28051-97-0	89
Zizanyl acetate		89
Zonarene	41929-05-9	16,23,37,41,71,75,83,90

[a] this chemical has not been found in nature up to now (see Table 6.3 for a full list)

[b] musk ambrette is a synthetic nitromusk, forbidden in Europe since 1986

[c] can also indicate oil of wintergreen

[s] this chemical has one or more synonyms which were named as such in one or more analytical publications reviewed; the presence of the chemicals described under this synonym name is shown under this entry; the corresponding synonym names can be found in Chapter 7 'List of synonyms'

Table 6.2 Conversion of numbers in table 6.1 to the corresponding essential oils and their chapter numbers

Number in table 6.1	Essential oil name	Chapter	Number in table 6.1	Essential oil name	Chapter
1	Angelica fruit oil	5.1	48	Lavandin oil	5.48
2	Angelica root oil	5.2	49	Lavender oil	5.49
3	Aniseed oil	5.3	50	Lemon oil	5.50
4	Basil oil, sweet	5.4	51	Lemongrass oil, East Indian	5.51
5	Bay oil	5.5	52	Lemongrass oil, West Indian	5.52
6	Bergamot oil	5.6	53	Litsea cubeba oil	5.53
7	Black cumin oil	5.7	54	Lovage oil	5.54
8	Black pepper oil	5.8	55	Mandarin oil	5.55
9	Cajeput oil	5.9	56	Marjoram oil (sweet)	5.56
10	Calamus oil	5.10	57	Melissa oil (lemon balm oil)	5.57
11	Cananga oil	5.11	58	Myrrh oil	5.58
12	Carrot seed oil	5.12	59	Neem oil	5.59
13	Cassia bark oil	5.13	60	Neroli oil	5.60
14	Cassia leaf oil	5.14	61	Niaouli oil	5.61
15	Cedarwood oil, Atlas	5.15	62	Nutmeg oil	5.62
16	Cedarwood oil, China	5.16	63	Olibanum (frankincense) oil	5.63
17	Cedarwood oil, Texas	5.17	64	Orange oil, bitter	5.64
18	Cedarwood oil, Virginia	5.18	65	Orange oil, sweet	5.65
19	Chamomile oil, German	5.19	66	Palmarosa oil	5.66
20	Chamomile oil, Roman	5.20	67	Patchouli oil	5.67
21	Cinnamon bark oil, Sri Lanka	5.21	68	Peppermint oil	5.68
22	Cinnamon leaf oil, Sri Lanka	5.22	69	Petitgrain bigarade oil	5.69
23	Citronella oil, Java	5.23	70	Pine needle oil	5.70
24	Citronella oil, Sri Lanka	5.24	71	Ravensara oil	5.71
25	Clary sage oil	5.25	72	Rosemary oil	5.72
26	Clove bud oil	5.26	73	Rose oil	5.73
27	Clove leaf oil	5.27	74	Rosewood oil	5.74
28	Clove stem oil	5.28	75	Sage oil, Dalmatian	5.75
29	Coriander fruit oil	5.29	76	Sage oil, Spanish	5.76
30	Costus root oil	5.30	77	Sandalwood oil	5.77
31	Cypress oil	5.31	78	Silver fir oil	5.78
32	Dwarf pine oil	5.32	79	Spearmint oil	5.79
33	Elemi oil	5.33	80	Spike lavender oil	5.80
34	Eucalyptus citriodora oil	5.34	81	Star anise oil	5.81
35	Eucalyptus globulus oil	5.35	82	Tangerine oil	5.82
36	Galbanum resin oil	5.36	83	Tea tree oil	5.83
37	Geranium oil	5.37	84	Thuja oil	5.84
38	Ginger oil	5.38	85	Thyme oil	5.85
39	Grapefruit oil	5.39	86	Thyme oil, Spanish	5.86
40	Guaiacwood oil	5.40	87	Turpentine oil	5.87
41	Hyssop oil	5.41	88	Valerian oil	5.88
42	Jasminum grandiflorum absolute	5.42	89	Vetiver oil	5.89
43	Jasminum sambac absolute	5.43	90	Ylang-ylang oil	5.90
44	Juniper berry oil	5.44	91	Zdravetz oil	5.91
45	Laurel leaf oil	5.45	92	Cardamom oil	5.92
46	Lavandin abrial oil	5.46	93	Cedarwood oil Himalaya	5.93
47	Lavandin grosso oil	5.47			

Table 6.3 Chemicals identified in essential oils which have not been found in nature up to now [a]

Benzodiazepine	Lilial ®
5-Benzodiazepine	Lyral ®
Bis(2-ethylhexyl)phthalate	Methyl ethyl phthalate
Butylcyclohexyl phthalate	α-Methylionone
Cyclamen aldehyde	6-Methyl-γ-ionone
Dehydrolinalyl acetate	6-Methyl-γ-(E)-ionone
Dibutyl phthalate	Musk ambrette
Dichloroacetic acid, 1-adamantylmethyl ester	(E)-9-Octadecenoic acid, trimethylsilyl ester
Diethyl phthalate	7,di-n-Octyl phthalate
Dihydrolinalyl acetate	Phthalate
Dihydromyrcenyl acetate	Phthalic acid
Dimethyl-o-phthalate	Retinal (= Vitamin A aldehyde)
Ftalate	9-cis-Retinal
Heptafluorobutanoic acid, 2-(1-adamantyl) ethyl ester	Retinyl acetate
α-Hexylcinnamaldehyde	β-Terpinene
Hydroxycitronellal	Testosteron (not in flora)
Hydroxycitronellol	Tetrahydrolinalyl acetate
Ibuprofen	Vitamin A aldehyde (= Retinal)

[a] we do not pretend this list to be exhaustive

Chapter 7 LIST OF SYNONYMS

In this chapter, an alphabetical list is provided of all chemicals ('main entries') shown in Chapter 6, Table 6.1 for which synonyms have been used in various publications (indicated by an ^{s} following their name). These synonyms can be found in Table 7.1. All synonyms of the main entries are also shown in Table 6.1 in alphabetical order between the main entries; they refer for their preferred names ('see…') in that table.

Table 7.1 List of main entry names and their synonyms as used in various publications

Main entry name	CAS number	Synonyms mentioned for main entries
Abietatriene	19407-28-4	Dehydroabietane
Acetaldehyde	75-07-0	Ethanal
Acetaldehyde diethyl acetal	105-57-7	1,1-Diethoxyethane
o-Acetanisole	579-74-8	2-Methoxyacetophenone
Acetone	67-64-1	2-Propanone
1-Acetoxy-2-propanone	592-20-1	2-Propanone, 1-(acetyloxy)-
p-Acetylanisole	100-06-1	p-Methoxyacetophenone (4'-)
Acetyl chavicol	61499-22-7	4-Allylphenyl acetate
17-(Acetyloxy)kauran-18-al	1421058-94-7	Kauran-18-al, 17-(acetyloxy)-
Aciphyllene	87745-31-1	Guaia-4,11-diene
Acora-4,10-diene	255062-41-0	Acora-3,7(14)-diene
Agarospirol	1460-73-7	Binesol
β-Alaskene	28908-21-6	Acora-3(10),14-diene
Allokhusiol	66512-57-0	Prezizaan-7β-ol
Amyl acetate	628-63-7	Pentyl acetate
Amyl isovalerate	25415-62-7	Pentyl 3-methylbutanoate
p-Anethole	104-46-1	Benzene, 1-methoxy-4-(1-propenyl)- ; 1-Methoxy-4-(1-propenyl)benzene
m-Anisaldehyde	591-31-1	3-Methoxybenzaldehyde
p-Anisaldehyde	123-11-5	4-Methoxybenzaldehyde
p-Anisic acid	100-09-4	p-Methoxybenzoic acid
p-Anisyl acetone	104-20-1	p-Acetonylanisole
β-Aromadendrene	72747-25-2	Dehydroaromadendrane
α-Asarone ((E)-; trans-)	2883-98-9	Asarone
γ-Asarone		Euasarone
Auranetin	522-16-7	3,4',6,7,8-Pentamethoxyflavone
Aurantiumal		1,5-[(3,6-Dimethyl-6-formyl-2-heptenyl)oxy]psoralen
Aurapten	495-02-3	7-Geranyloxycoumarin
Benzene, (3-ethoxy-1,5-hexadien-1-yl)-	67323-95-9	(3-Ethoxyhexa-1,5-dienyl)benzene
Benzyl acetate	140-11-4	Phenylmethyl acetate; Acetic acid, phenylmethyl ester
Benzyl alcohol	100-51-6	Benzenemethanol
Benzyl butyrate	103-37-7	Benzyl butanoate
Benzyl formate	104-57-4	Formic acid, phenylmethyl ester
Benzyl isobutyrate	103-28-6	Benzyl 2-methylpropanoate
Benzyl isovalerate	103-38-8	Benzyl 3-methylbutanoate
Benzyl n-octanoate	10276-85-4	Octanoic acid, phenylmethyl ester
Bergamal	106-72-9	2,6-Dimethyl-5-heptenal; Melonal
α-Bergamotene	17699-05-7	Bicyclo[3.1.1]hept-2-ene, 2,6-dimethyl-6-(4-methyl-3-pentenyl)-
β-Bergamotene	6895-56-3	Bicyclo[3.3.1]heptane,6-methyl-2-methylidene-6-(4-methyl-3-penten-1-yl)-
trans-α-Bergamotol acetate		trans-α-Bergamotyl acetate
Bergapten	484-20-8	4-Methoxy-7H-furo[3,2-g][1]benzopyran-7-one
Bergaptol	486-60-2	5-Hydroxy-6,7-furanocoumarin
Bicyclopentan-2-one	4884-24-6	[Bicyclopentyl]-2-one; Bicyclopentylone
(±)-β-Bisabolene	4891-79-6	Cyclohexene, 1-methyl-4-(5-methyl-1-methylene-4-hexenyl)-
epi-α-Bisabolol	78148-59-1	(-)-(4S,8R)-8-epi-α-Bisabolol
α-Bisabolol oxide	22567-36-8	Bisabolol oxide A
Bisabol-1-one	61432-71-1	1-Bisabolone
Bis(2-ethylhexyl) phthalate	117-81-7	Di(2-ethylhexyl) phthalate
Bois de rose oxide	7392-19-0	Trimethyl vinyl tetrahydropyran; 2,2,6-Trimethyl-6-vinyltetra-hydropyran
2-Bornene (α-)	464-17-5	Bornylene
Borneol	507-70-0	endo-Borneol
Bornyl acetate	76-49-3	endo-Bornyl acetate
Bornyl isovalerate	76-50-6	Bornyl 3-methylbutyrate; Butanoic acid, 2 methyl-, 1,7,7-trimethyl-bicyclo[2.2.1]hept-2-yl ester

Table 7.1 List of main entry names and their synonyms as used in various publications (*continued*)

Main entry name	CAS number	Synonyms mentioned for main entries
Bornyl propionate	78548-53-5	Bornyl propanoate; *endo*-Bornyl propionate
α-Bulnesene	3691-11-0	δ-Guaiene
α-Bulnesene epoxide	33784-90-6	Bulnesoxide
Bulnesol	22451-73-6	Guai-1(10)-en-11-ol
Butanal	123-72-8	Butyraldehyde
n-Butyl angelate	7785-64-0	Butyl 2-methylcrotonate
Butyl butanoate	109-21-7	Butyl butyrate
2,6-di-*tert*-Butyl-*p*-cresol	128-37-0	2,6-di-*t*-Butyl-4-methylphenol
Butyl dodecanoate	106-18-3	Butyl laurate
Butyl isovalerate	109-19-3	Butyl 3-methylbutyrate
Butyl 2-methylbutyrate	15706-73-7	Butanoic acid, 2-methyl-, butyl ester
n-Butyric acid	107-92-6	Butanoic acid
β-Cadinene	523-47-7	Cadina-3,9-diene
γ-Cadinene	39029-41-9	Naphthalene,1,2,3,4,4a,5,6,8a-octahydro-7-methyl-4-methylene-1-(1-methylethyl)-
δ-Cadinene	483-76-1	Naphthalene, 1,2,3,5,6,8a-hexahydro-4,7-dimethyl-1-(1-methylethyl)-, (1*S-cis*)-; Cadina-1(10),4-diene
cis-4-Cadinen-7-ol	217650-27-6	*cis*-Cadinen-4-en-7-ol
α-Cadinol	481-34-5	Cadin-4-en-10-ol
δ-Cadinol	19435-97-3	Torreyol
τ-Cadinol	5937-11-1	10-epi-α-Cadinol; epi-α-Cadinol
cis-Calamenene	72937-55-4	*cis*-Calamene
Camphene hydrate	465-31-6	2,3,3-Trimethyl-2-norbornanol
Camphenilone	13211-15-9	Bicyclo[2.2.1]heptan-3-one, 6,6-dimethyl-2-methylene-
Campherenol	18530-03-5	Canferenol
Campholenal	23727-15-3	Campholene aldehyde
α-Campholenal	4501-58-0	α-Campholene aldehyde
d-Camphor	464-49-3	Alcanfor
β-Carene	554-60-9	3(10)-Carene
δ2-Carene (= δ4-)	554-61-0	Car-2-ene; 2-Carene
δ3-Carene	13466-78-9	Car-3-ene; 3-Carene
3-Caren-2-one		Car-3-en-2-one
Carvacrol	499-75-2	Isopropyl cresol
Carvenone	499-74-1	*p*-Menth-3-en-2-one
Carveol	99-48-9	*p*-Mentha-1,8-dien-6-ol
(*E*)-Carveol	1197-07-5	*trans-p*-Mentha-1,8-dien-6-ol
(*Z*)-Carveol	1197-06-4	2-Cyclohexen-1-ol, 2-methyl-5-(1-methylethenyl), *cis*-; *cis-p*-Mentha-1, 8-dien-6-ol
Carvomenthene	5502-88-5	1-*p*-Menthene
D-Carvone (+−)	2244-16-8	(*S*)-2-Methyl-5-(1-methylethenyl)-2-cyclohexen-1-one
Carvone hydrate	7712-46-1	8-Hydroxycarvone; 8-Hydroxycarvotanacetone
Caryophylladienol II		Caryophylla-2(12),6(13)-dien-5-ol; Caryophylla-2(12),6(13)-dien-5α-ol
β-Caryophyllene	87-44-5	Caryophyllene; *trans*-Caryophyllene; (*E*)-Caryophyllene
γ-Caryophyllene	118-65-0	Isocaryophyllene; (*Z*)-Caryophyllene; *cis*-Caryophyllene
9-epi-(*E*)-Caryophyllene	68832-35-9	2-epi-(*E*)-β-Caryophyllene
trans-Caryophyllene oxide	1139-30-6	Caryophyllene (ep)oxide; Epoxydihydrocaryophyllene
Caryophyllene oxide II		Caryophyllene epoxide II
Caryophyllenol	38284-26-3	10,10-Dimethyl-2,6-dimethylenebicyclo[7.2.0]undecan-5-ol
Caryophyllenol I	32214-88-3	Caryophylla-2(12),6-dien-5α-ol
Caryophyllenol II	32214-89-4	Caryophylla-2(12),6-dien-5β-ol
Cascarilladienol		Eudesma-5,7-dien-4-ol
α-Cedrene	469-61-4	Cedr-8-ene
8-Cedren-13-ol	18319-35-2	Cedr-8-en -13-ol
β-Cedren-9α-ol	13567-41-4	Cedr-8(15)-en-9-α-ol
α-Cedrol	77-53-2	Cedrol
epi-Cedrol	19903-73-2	Epicedrol; 6-Isocedrol
Cembrene A	31570-39-5	Neocembrene; Neocembrene A
Chamecynone	10208-54-5	7-Ethynyl-4a,5,6,7,8,8a-hexahydro-1,4a-dimethyl-2(1*H*)-naphthalenone
β-Chamigrene	18431-82-8	Chamigrene
Chavicol	501-92-8	*p*-Allylphenol; 4-(2-Propenyl)phenol
1,4-Cineole	470-67-7	Isocineole; Isoeucalyptol
1,8-Cineole	470-82-6	Cajeputol; Eucalyptol; 1,8-Epoxy-*p*-menthane
Cinnamaldehyde	104-55-2	Cinnamic aldehyde; Cinnamal; 3-Phenyl-2-propenal

Table 7.1 List of main entry names and their synonyms as used in various publications (*continued*)

Main entry name	CAS number	Synonyms mentioned for main entries
Cinnamyl acetate	103-54-8	Cinnamyl alcohol, acetate; 2-Propen-1-ol, 3-phenyl- , acetate
Citrol	624-15-7	3,7-Dimethyl-2,6-octadien-1-ol
Citronellal	106-23-0	β-Citronellal
α-Citronellal	141-26-4	Rhodinal
Citronellic acid	502-47-6	3,7-Dimethyl-6-octenoic acid
Citronellol (β-, DL-)	106-22-9	Cephrol
Citronellyl acetate	150-84-5	3,7-Dimethyl-6-octen-1-ol acetate
Citronellyl formate	105-85-1	6-Octen-1-ol, 3,7-dimethyl-, formate
Citronellyl isohexanoate	71662-18-5	Citronellyl 4-methylvalerate
Citronellyl octanoate	72934-05-5	Citronellyl caprylate
Citronellyl propionate	141-14-0	Citronellyl propanoate
Costunolide	553-21-9	Costus lactone
Coumarin	91-64-5	Cumarin
Coumarin, 8-(2-formyl-2-methylpropyl)-7-methoxy-	5980-07-4	7-Methoxy-8-(2-formyl-2-methylpropyl)-coumarin
p-Cresyl methyl ether	104-93-8	*p*-Methylanisole
Cryptone	500-02-7	4-Isopropyl-2-cyclohexenone
β-Cubebene	13744-15-5	1*H*-Cyclopenta[1,3]cyclopropa-[1,2]benzene, octahydro-7-methyl-3-methylene-4-(-1-methylethyl), [3a*S*(3aα,3bβ,4β,7α,7a*S*)]-
Cubenene	29837-12-5	*cis*-Cadina-1,4-diene; Cadina-1,4-diene; Cadina-1(2),4-diene
Cubenol	21284-22-0	Cadin-4-en-1-ol
1,10-di-epi-Cubenol	73365-77-2	1,10-Diepicubenol
epi-Cubenol	19912-67-5	Epicubenol
Cumene	98-82-8	(1-Methylethyl) benzene
Cuminaldehyde	122-03-2	*p*-Cumic aldehyde; Cuminal
Cuminyl acetate	59230-57-8	*p*-Cymen-7-ol acetate; Isopropylbenzyl acetate
Cuminyl alcohol	536-60-7	Cuminol; *p*-Cymen-7-ol; 4-Isopropylbenzyl alcohol
Curcumene (*ar*-; α-)	644-30-4	Benzene, 1-(1,5-dimethyl-4-hexenyl)-4-methyl-; *ar*-Curcumene
Curzerene	17910-09-7	Benzofuran, 6-ethenyl-4,5,6,7-tetrahydro-3,6-dimethyl-5-isopropenyl-, *trans*-; Isofuranogermacrene
Cyclofenchene	488-97-1	Cyclofenchone
α-Cyclogeraniol acetate	68406-89-3	Cyclogeranyl acetate
3,5-Cycloheptadien-1-one	1121-65-9	3,5-Cycloheptadine-1-one
Cycloheptane,4-methylene-1-methyl-2-(2-methyl-1-propen-1-yl)-1-vinyl-	826337-63-7	4-Methylene-1-methyl-2-(2-methyl-1-propen-1-yl)-1-vinyl-cycloheptane
Cyclohexane, 2-ethenyl-1,1-dimethyl-3-methylene-	95452-08-7	1,1-Dimethyl-3-methylene-2-vinylcyclohexane
3-Cyclohexene-1-carboxaldehyde	100-50-5	1-Carboxaldehyde-3-cyclohexene
Cycloisosativene	406485-43-6	(+)-Cycloisosativene
Cyclopentadecanolide	106-02-5	15-Pentadecanolide; Pentadecano-15-lactone
Cyclosativene	22469-52-9	(+)-Cyclosativene
Cymene	25155-15-1	Cymol
m-Cymene	535-77-3	1-Methyl-3-(1-methylethyl)benzene
o-Cymene	527-84-4	1-Methyl-2-isopropylbenzene
p-Cymene	99-87-6	*p*-Cymol; 1-Methyl-4-(1-methylethyl)benzene
m-Cymenene	1124-20-5	α,*m*-Dimethylstyrene; 1-Isopropenyl-3-methylbenzene
p-Cymenene	1195-32-0	4-Methylisopropenylbenzene; Dehydro-*p*-cymene; α,*p*-Dimethylstyrene; 1-Methyl-4-(1-methylethenyl)benzene;Benzene,1-methyl-4-(1-methylethenyl)-
α-Cyperene	17627-30-4	Selina-4,11-diene
β-Damascenone (*E*)-	23726-93-4	(*E*)-1-(2,6,6-Trimethyl-1,3-cyclohexadien-1-yl)-2-buten-1-on;
Decahydronaphthalene	91-17-8	Decalin
γ-Decalactone	706-14-9	4-Hexyl-4-butanolide
Decamethylene glycol	112-47-0	1,10-Decanediol
Decanal	112-31-2	Capraldehyde; Capric aldehyde; Caprinaldehyde; *n*-Decaldehyde
n-Decanoic acid	334-48-5	Capric acid
2-Decanol	1120-06-5	Decan-2-ol
n-Decanol (1-)	112-30-1	Decyl alcohol
Decyl acetate	112-17-4	Decanol acetate
5-Decyne-4,7-diol, 4,7-dimethyl-	126-87-4	Surfynol 102
Dehydrocostus lactone	477-43-0	Dehydrocostunolide
cis-Dehydroxylinalool oxide	73413-94-2	*cis*-Herboxide
Dendrolasin	23262-34-2	Dendralasine

Table 7.1 List of main entry names and their synonyms as used in various publications (*continued*)

Main entry name	CAS number	Synonyms mentioned for main entries
Dibutyl phthalate	84-74-2	Butyl phthalate
Dihydroanethole	104-45-0	1-Methoxy-4-propylbenzene; *p*-Propylanisole
Dihydrocarveol	38049-26-2	Carhydranol; *p*-Menth-8-en-2-ol
Dihydrocarvone	5948-04-9	Cyclohexanone, 2-methyl-5-(1-methylethenyl)
Dihydrocarvyl acetate ((*R*-))	57287-13-5	*p*-Menth-8-en-2-ol, acetate
Dihydrocoumarin	119-84-6	Chroman-2-one
3-Hydroxy-4′,5,6,7,8-pentamethoxy-flavone		3-Hydroxytangeretin
5,8-Dihydroxy-3,3′,4′,7-tetramethoxy-flavone	7380-44-1	Gossypetin 3,3′,4′,7-O-tetramethyl ether
5,7-Dimethoxycoumarin	487-06-9	Citropten
6,7-Dimethoxycoumarin	120-08-1	Scoparone
2,5-Dimethoxy-*p*-cymene	14753-08-3	1,4-Dimethoxy-2-methyl-5-isopropylbenzene; Thymohydroquinone dimethyl ether
5,8-Dimethoxy-6,7-furanocoumarin	482-27-9	Isopimpinellin
2,6-Dimethoxytoluene	5673-07-4	2,6-Dimethoxy-1-methylbenzene
3,5-Dimethoxytoluene	4179-19-5	Orcinol dimethyl ether
4,4-Dimethyl-2-buten-4-olide	20019-64-1	5,5-Dimethyl-2(5*H*)-furanone
3,3-Dimethylbutyric acid	1070-83-3	3,3-Dimethylbutanoic acid
1,4-Dimethyl-3-cyclohexenyl methyl ketone	43219-68-7	1-(1,4-Dimethyl-3-cyclohexen-1-yl)-ethanone
2,5-Dimethyl-2,4-hexadiene	764-13-6	Biisocrotyl
2,6-Dimethyl-3,7-octadiene-2,6-diol	13741-21-4	(*E*)-2,6-Dimethyl-3,7-octadien-2,6-diol; 3,7-Dimethyl-1,5-octadiene-3,7-diol; 2,6-Dimethyl-3,7-octadien-2,6-ol
Dimethyl octanol	106-21-8	Dihydrocitronellol
trans-2,6-Dimethyl-3,5,7-octatrien-2-ol		(*E,E*)-2,6-Dimethyl-3,5,7-octatrien-2-ol
Dimethyl sulfone	67-71-0	Methylsulphonylmethane
Docosanoic acid	112-85-6	Behenic acid
Dodecanoic acid	143-07-7	Lauric acid
3-Dodecenal	68083-57-8	3-Dodecen-1-al
Elaidic acid	112-79-8	(*E*)-9-Octadecenoic acid
δ-Elemene	20307-84-0	Cyclohexene,4-ethenyl-4-methyl-3-(1-methylethenyl)-1-(1-methylethyl)-,(3*R-trans*)-
β-Elemenol	65018-04-4	Elema-1,3,11(13)-trien-12-ol
Elixene	3242-08-8	4-Isopropylidene-1-vinyl-*o*-menth-8-ene
Epizizanol	28624-26-2	Ziza-6(13)-en-3β-ol
Epizizanone	28624-27-3	2-epi-Ziza-6(13)-en-3-one
Epoxy allo-aromadendrene	85760-81-2	allo-Aromadendrene (ep)oxide
Epoxyaurapten	21499-17-2	Coumarin, 7-[(6,7-epoxy-3,7-dimethyl-2-octenyl)oxy]-, (*E*)-(+)-
Eremophilene	10219-75-7	Elemophilane
Ethyl alcohol	64-17-5	Ethanol
2-Ethyl-4,5-dimethylphenol	2219-78-5	4,5-Dimethyl-2-ethylphenol
Ethyl dodecanoate	106-33-2	Ethyl laurate
Ethyl hexadecanoate	628-97-7	Ethyl palmitate
Ethyl linoleate	544-35-4	Ethyl linolate
Ethyl-*p*-methoxycinnamate	1929-30-2	2-Propenoic acid, 3(4-methoxyphenyl), ethyl ester
1-Ethyl-2-methylbenzene	611-14-3	1-Methyl-2-ethylbenzene
Ethyl-2-methylbutyrate	7452-79-1	Butanoic acid, 2-methyl-, ethyl ester
3-Ethyl-3-methyldecane	17312-66-2	Decane, 3-ethyl-3-methyl-
2-Ethyl-3-methylpyrazine	15707-23-0	Pyrazine, 2-ethyl-3-methyl-
Ethyl nerolate	32659-20-4	Ethyl nerate
Ethyl octadecanoate	111-61-5	Ethyl stearate
Ethyl octanoate	106-32-1	Ethyl caprylate
Ethyl tetradecanoate	124-06-1	Ethyl myristate
Eudesma-4,6-dien-3-one	23665-63-6	β-cyperone
α-Eudesmol	473-16-5	α-Selinenol
β-Eudesmol	473-15-4	β-selinenol
γ-Eudesmol	1209-71-8	Selinenol; Machilol
Eugenol	97-53-0	4-Hydroxy-3-methoxybenzene; 4-Allylguaiacol
m-Eugenol	501-19-9	3-Allyl-6-methoxyphenol; Chavibetol
Eugenyl acetate	93-28-7	Aceteugenol; Acetyl eugenol; Acetoeugenol
Farnesane	3891-98-3	Dodecane, 2,6,10-trimethyl-
(*Z,E*)-α-Farnesene	26560-14-5	1,3,6,10-Dodecatetraene, 3,7,11-trimethyl-, (*Z,E*)-
(*Z*)-β-Farnesene	28973-97-9	1,6,10-Dodecatriene, 7,11-dimethyl-3-methylene-, (*Z*)-

Table 7.1 List of main entry names and their synonyms as used in various publications (*continued*)

Main entry name	CAS number	Synonyms mentioned for main entries
(*E,E*)-Farnesol	106-28-5	(*E*)-β-Farnesol; *trans*-Farnesol
(*Z,Z*)-Farnesol	16106-95-9	*cis*-Farnesol
Farnesyl alcohol	4602-84-0	3,7,11-Trimethyl-2,6,10-dodecatrien-1-ol; Farnesol
α-Fenchyl acetate	111821-74-0	*endo*-Fenchyl acetate
β-Fenchyl acetate	76109-40-5	*exo*-Fenchyl acetate
Fenchyl alcohol	1632-73-1	Bicyclo[2.2.1]heptan-2-ol, 1,3,3-trimethyl-; Fenchol
α-Fenchyl alcohol	512-13-0	*endo*-Fenchol ; α-Fenchol
β-Fenchyl alcohol	22627-95-8	*exo*-Fenchol; β-Fenchol
Foeniculin ((*E*-))	78259-41-3	1-(3-Methyl-2-butenoxy)-4-(1-propenyl)benzene
α-Funebrene	50894-66-1	1,7-di-epi-α-Cedrene
β-Funebrene	79120-98-2	1,7-di-epi-β-Cedrene
Furfural	98-01-1	Furaldehyde; Furfuraldehyde
Geranial	141-27-5	(*E*)-Citral; Citral A; α-Citral; 2,6-Octadienal, 3,7-dimethyl-, (*E*)-
Geraniol	106-24-1	(*E*)-Geraniol; β-Geraniol; Lemonol; 2,6-Octadien-1-ol, 3,7-dimethyl, (*E*)-
Geranyl acetate	105-87-3	2,6-Octadien-1-ol, 3,7-dimethyl-, acetate, (*E*)-; *trans*-Geranyl acetate
(*E*)-Geranylacetone	3796-70-1	(*E*)-6,10-Dimethyl-5,9-undecadien-2-one
Geranyl butyrate	106-29-6	Butanoic acid, 3,7-dimethyl-2,6-octadienyl ester
Geranyl formate	105-86-2	2,6-Octadien-1-ol, 3,7-dimethyl-, formate
Geranyl hexanoate	10032-02-7	Geranyl caproate
Geranyl isobutyrate	2345-26-8	Propanoic acid, 2-methyl-3,7-dimethyl-2,6-octadienyl ester, (*E*)-
Geranyl isovalerate	109-20-6	Geranyl 3-methylbutanoate
Geranyllinalool	77368-82-2	Linalyl geranyl ether
Geranyl propionate	105-90-8	Geranyl propanoate
Geranyl valerate	10402-47-8	Geranyl pentanoate
Germacrone	6902-91-6	(*E,E*)-Germacra-3,7(11),9-trien-6-one
Guaiacol	90-05-1	2-Methoxyphenol
α-Guaiene	3691-12-1	Guaia-1(5),11-diene
β-Gurjunene	73464-47-8	1a,2,3,5,6,7,7a,7b-Octahydro-1,1,7,7a-tetramethyl-1*H*-cyclopropa[a] naphthalene
Hedycaryol	21657-90-9	3,7-Cyclodecadiene-1-methanol, α,α,4,8-tetramethyl-
γ-Heptalactone	105-21-5	4-Heptanolide
3,3′,4′,5,6,7,8-Heptamethoxyflavone	1178-24-1	3-Methoxynobiletin
1-Heptanol	111-70-6	*n*-Heptanol
1-Hepten-3-ol	4938-52-7	Hept-1-en-3-ol
1-Hepten-3-one	2918-13-0	Heptenone
2-Heptyl acetate	5921-82-4	2-Heptanol acetate
Hexadecanoic acid	57-10-3	Palmitic acid; Cetylic acid
1-Hexadecanol	36653-82-4	*n*-Cetyl alcohol
1-Hexadecene	629-73-2	Cetene
cis-9-Hexadecenoic acid	373-49-9	Palmitoleic acid
Hexadecyl acetate	629-70-9	Palmityl acetate
1,2,3,4,4a,7-Hexahydro-1,6-dimethyl-4-(1-methylethyl)naphthalene	16728-99-7	Naphthalene, 1,2,3,4,4a,7-hexahydro-1,6-dimethyl-4-(1-methylethyl)-
Hexahydrofarnesyl acetone	502-69-2	Trimethyl-6,10,14-pentadecanone; 2-Pentadecanone, 6,10,14- trimethyl-
3,3′,4′,5,6,7-Hexamethoxyflavone	1251-84-9	3-Methoxysinensetin; Hexa-*O*-methylquercetagetin
3,3′,4′,5,7,8-Hexamethoxyflavone	7741-47-1	Hexa-*O*-methylgossypetin
3′,4′,5,6,7,8-Hexamethoxyflavone	478-01-3	Nobiletin 2*E*-Hexanal
(*E*)-2-Hexanal		2*E*-Hexanal
Hexanoic acid	142-62-1	Capronic acid; Caproic acid
(*E*)-2-Hexenol	928-95-0	(*E*)-2-Hexen-1-ol
(*E*)-3-Hexenol	928-97-2	(*E*)-3-Hexen-1-ol
(*Z*)-3-Hexenol	928-96-1	(*Z*)-3-Hexen-1-ol
3-Hexenyl butyrate	2142-93-0	3-Hexenyl butanoate
(*Z*)-3-Hexenyl butyrate	16491-36-4	(*Z*)-3-Hexenyl butanoate
cis-3-Hexenyl valerate	35852-46-1	*cis*-3-Hexenyl pentanoate
Hexyl butyrate	2639-63-6	Hexyl butanoate
α-Hexylcinnamaldehyde	101-86-0	α-*n*-Hexyl-α-phenylacrolein
Hexyl isobutyrate	2349-07-7	Hexyl isobutanoate; Hexyl 2-methylpropanoate
Hexyl isovalerate	10032-13-0	Hexyl 3-methylbutanoate; Hexyl 3-isovalerate; Hexyl 3-methylbutyrate
Hexyl 2-methylbutyrate	10032-15-2	Butanoic acid, 2-methyl-, hexyl ester
Hexyl valerate	1117-59-5	Hexyl pentanoate
Himachala-1,3,5-trien-5-ol	66656-01-7	Podocephalol

Table 7.1 List of main entry names and their synonyms as used in various publications (*continued*)

Main entry name	CAS number	Synonyms mentioned for main entries
α-Himachalene (α-*cis*-)	3853-83-6	2,4a,5,6,7,8,9,9a-Octahydro-3,5,5-trimethyl-9-methylene-1*H*-benzocycloheptene; α-*ar*-Himachalene
γ-Himachalene	53111-25-4	γ-*ar*-Himachalene
α-Himachalene epoxide	64825-84-9	Oxidohimachalene
β-Himachalene oxide	31560-66-4	Epoxy-β-himachalene
allo-Himachalol	19435-77-9	Allohimachalol
α-Humulene epoxide	96638-51-6	Humulene oxide
Humulene epoxide II	19888-34-7	1,2-Epoxyhumulene; Humulene oxide II; Humulene-1,2-oxide; Humulene 6,7-epoxide; 2-Oxabicyclo[9.1.0]dodeca-3,7-diene, 1,5,5,8-tetramethyl-, [1*R*-(1*R**,3*E*,7*E*,11*R**)]-
Hydrocinnamaldehyde	104-53-0	Benzenepropanal; Dihydrocinnamaldehyde; 3-Phenylpropanal
Hydrocinnamic acid	501-52-0	3-Phenylpropanoic acid; Dihydrocinnamic acid
Hydrocinnamyl acetate	122-72-5	Benzenepropanol, acetate; γ-Phenylpropyl acetate; 3-Phenylpropyl acetate
Hydrocinnamyl alcohol	122-97-4	Benzenepropanol; 3-Phenylpropanol; 3-Phenylpropyl alcohol
14-Hydroxy-9-epi-(*E*)-caryophyllene	244226-09-3	14-Hydroxy-9-epi-β-caryophyllene
7-Hydroxycoumarin	93-35-6	Umbelliferone
(*E*)-8-Hydroxylinalool	75991-61-6	(*E*)-2,6-Dimethyl-2,7-octadiene-1,6-diol
(*Z*)-8-Hydroxylinalool	103619-06-3	(*Z*)-2,6-Dimethyl-2,7-octadiene-1,6-diol
4-Hydroxy-4-methyl-2-pentanone	123-42-2	Diacetone alcohol; 2-Pentanone, 4-hydroxy-4-methyl-
3-Hydroxy-2-methyl-4-pyrone	118-71-8	Maltol
5-Hydroxy-3′,4′,6,7,8-pentamethoxy-flavanone	15512-52-4	Desmethyl-5-citromitine
5-Hydroxy-3′,4′,6,7,8-pentamethoxy-flavone	50439-46-8	5-Hydroxyauranetin
5-Hydroxy-6,7,8,3′,4′-pentamethoxy-flavone	2174-59-6	5-Demethylnobiletin
5-Hydroxy-3,3′,4′,7-tetramethoxy-flavone	1245-15-4	Retusin; Tetra-*O*-methylquercetin
5-Hydroxy-3′,4′,6,7-tetramethoxy-flavone	21763-80-4	Santaflavone
5-Hydroxy-4′,6,7,8-tetramethoxyflavone	2798-20-1	5-Demethyltangeretin
5-Hydroxy-4′,6,7-trimethoxyflavone	19103-54-9	Salvigenin
Intermedeol	6168-59-8	Eudesm-11-en-4-ol
α-Ionene	475-03-6	1,2,3,4-Tetrahydro-1,1,6-trimethylnaphthalene
Isoamyl acetate	123-92-2	Isopentyl acetate; 1-Butanol, 3-methyl, acetate; 3-Methylbutyl acetate
Isoamyl alcohol	123-51-3	3-Methylbutanol; 3-Methyl-1-butanol
Isoamyl benzoate	94-46-2	3-Methylbutyl benzoate
Isoamyl butyrate	106-27-4	Isopentyl butyrate; 3-Methylbutyl butanoate
Isoamyl formate	110-45-2	3-Methylbutyl formate
Isoamyl isobutyrate	2050-01-3	3-Methylbutyl isobutyrate
Isoamyl isovalerate	659-70-1	Isopentyl isovalerate; 3-Methylbutyl 3-methylbutanoate
Isoamyl methacrylate	7336-27-8	3-Methylbutyl methacrylate
Isoamyl 2-methylbutyrate	27625-35-0	Isopentyl 2-methylbutanoate; 3-Methylbutyl 2-methylbutanoate
Isoamyl methyl ketone	110-12-3	Methyl isoamyl ketone
Isoamyl tiglate	41519-18-0	Isoamyl 2-methylcrotonate
Isoaromadendrene epoxide	499134-59-7	iso-Aromadendrene epoxide
Isobornyl acetate	125-12-2	*exo*-Bornyl acetate
Isobutanal	78-84-2	2-Methylpropanal
Isobutyl alcohol	78-83-1	Isobutanol; 2-Methylpropanol; 2-Methyl-1-propanol
Isobutyl angelate	7779-81-9	2-Methyl-2-propyl angelate
Isobutyl-2-methylbutyrate	2445-67-2	2-Methylpropyl-2-methylbutanoate
Isobutyl propionate	540-42-1	2-Methylpropyl-2-propanoate
Isobutyric acid	79-31-2	2-Methylpropanoic acid; Isobutanoic acid
(*Z*)-Isocitral	72203-97-5	Isocitral #1; Isoneral
exo-Isocitral	55050-40-3	Isocitral #3
Isocomene	65372-78-3	α-Isocomene; Berkheyaradulene
Isodaucene	142878-08-8	(+)-Dauca-8,11-diene
Isodihydrocarvyl acetate	220329-20-4	iso-Dihydrocarvyl acetate
Isoeugenol	97-54-1	2-Methoxy-4-(1-propenyl)phenol
Isogeranial	72203-98-6	(*E*)-Isocitral; Isocitral #2

Table 7.1 List of main entry names and their synonyms as used in various publications (*continued*)

Main entry name	CAS number	Synonyms mentioned for main entries
Isohexanoic acid	646-07-1	4-Methylpentanoic acid
Isokaurene	5947-50-2	Isocaurene; Kaur-15-ene
Isolimonene	499-99-0	*p*-Mentha-2,8-diene
Isomenthol	3623-52-7	*cis*-Menthol
Isomenthone	491-07-6	*cis*-Menthone
Isomeranzin	1088-17-1	Coumarin, 7-methoxy-8-(3-methyl-2-oxobutyl)-
Isonootkatone	15764-04-2	α-vetivone
5-Isopentenyloxy-8-methoxypsoralen	14348-22-2	Cnidilin; Isophellopterin
8-Isopentenyloxypsoralen	482-44-0	Imperatorin
Isophyllocladene	511-85-3	(5α,9α,10β)-Kaur-15-ene
Isopinocamphone	15358-88-0	*cis*-3-Pinanone
Isopiperitenone	529-01-1	*p*-Mentha-1,8-dien-3-one
Isoprenyl acetate	17616-47-6	3-Methyl-1,3-butadienyl-1-acetate
Isopropyl hexadecanoate	142-91-6	Isopropyl palmitate
4-Isopropyl-6-methyl-1-tetralone	57494-10-7	4-Isopropyl-6-methyl-1,2,3,4-tetrahydronaphthalen-1-one
Isopulegol II	1370348-44-9	5-Methyl-2-(1-methylethenyl)cyclohexanol
Isoterpinolene	586-63-0	*p*-Mentha-2,4(8)-diene
Isovaleraldehyde	590-86-3	Isopentanal; 3-Methylbutanal
Isovaleric acid	503-74-2	3-Methylbutanoic acid; 3-Methylbutyric acid
Isozizanoic acid	16202-79-2	Isokhusenic acid
δ-Jasmolactone	25524-95-2	*cis*-7-Decen-5-olide; Jasmine lactone; 5-(*cis*-2-Pentenyl)-5-pentanolide
(*E*)-Jasmone	6261-18-3	3-Methyl-2-(*trans*-2-pentenyl)-2-cyclopenten-1-one
(*Z*)-Jasmone	488-10-8	3-Methyl-2-(*cis*-2-pentenyl)-2-cyclopenten-1-one
Juniper camphor	473-04-1	Eudesm-7(11)-en-4-ol; Eudesm-7(11)-en-4α-ol; Selin-7(11)-en-4-ol
Khusenic acid	16203-25-1	Zizanoic acid
Khusimene	18444-94-5	Zizaene
Khusiol	66512-56-9	Helifolan-2-ol; Khusian-2-ol
Khusione	66397-72-6	Helifolan-2-one
Khusol	18045-73-3	Cadina-4,10(15)-dien-12-ol
Lavandulol	498-16-8	5-Methyl-2-(1-methylethenyl)-4-hexen-1-ol
Lavandulyl isovalerate	51117-21-6	Lavandulyl 3-methylbutyrate
Lavender lactone	1073-11-6	4-Methyl-4-vinyl-4-butanolide
Lepalone	80445-58-5	1-Penten-3-one, 5-(3-furanyl)-2-methyl-
Levojunenol	30951-17-8	laevo-Junenol
Levomenol	23089-26-1	α-(-)-Bisabolol
Levopimaradiene	122712-77-0	Abieta-8,12-diene
Ligustilide	4431-01-0	Butylidene dihydrophthalide
Limonenal	6784-13-0	Limonene aldehyde
Limonene	138-86-3	*p*-Mentha-1,8-diene
dl-Limonene	138-86-3	Dipentene; α-Limonene
Limonene (ep)oxide	1195-92-2	Limonene-1,2-epoxide
cis-Limonene (ep)oxide	13837-75-7	*cis*-Limonene-1,2-epoxide
trans-Limonene (ep)oxide	4959-35-7	*trans*-Limonene-1,2-epoxide
Limonene glycol	1946-00-5	Limonene-1,2-diol
Limonen-4-ol	3419-02-1	*p*-Mentha-1,8-dien-4-ol
Limonen-10-ol	3269-90-7	*p*-Mentha-1(2),8-dien-10-ol; *p*-Mentha-1,8(10)-dien-9-ol
Limonen-10-yl-acetate	15111-97-4	*p*-Mentha-1,8(1)-dien-9-yl acetate; *p*-Mentha-1,8(10)-dien-9-yl acetate
Linalool (β-)	78-70-6	3,7-Dimethyl-1,6-octadien-3-ol
Linalool oxide	1365-19-1	Epoxylinalool; α-Methyl-α-(4-methyl-3-pentenyl)oxiranemethanol; 2,2,6-Trimethyl-6-ethenyl-tetrahydro-2*H*-pyran-3-ol
cis-Linalool oxide	11063-77-7	2-Furanmethanol, 5-ethenyltetrahydro-α,α,5-trimethyl-, *cis*-Linalool oxide B
cis-Linalool oxide, furanoid	5989-33-3	Linalool oxide I; Linalool oxide B, furanoid
cis-Linalool oxide, pyranoid	14009-71-3	Linalool oxide D
trans-Linalool oxide	11063-78-8	Linalool oxide A
trans-Linalool oxide, furanoid	34995-77-2	Linalool oxide II; Linalool oxide A, furanoid
Linalyl acetate	115-95-7	Bargamol
Linalyl anthranilate	7149-26-0	1,6-Octadien-3-ol, 3,7-dimethyl-, 2-aminobenzoate; Linalyl *o*-aminobenzoate
Linalyl propionate	144-39-8	Linalool propanoate
Linoleic acid	60-33-3	Linolic acid
Longifolenaldehyde	66537-42-6	Longifolenal

Table 7.1 List of main entry names and their synonyms as used in various publications (*continued*)

Main entry name	CAS number	Synonyms mentioned for main entries
Longifolene	475-20-7	3-Longibornene
α-Longifolene	475-20-7	1,4-Methanoazulene, decahydro-4,8,8-trimethyl-9-methylene-, [1*S*-(1α,3aα,4α,8aα)]-; (+)-Longifolene; Junipene; 1,4-Methanoazulene, decahydro-4,8,8-trimethyl-9-methylene-, (1*S*,3a*R*,4*S*,8a*S*)-
α-Longipinene	5989-08-2	Longipinene; Tricyclo[5.4.0.0(2,8)]undec-9-ene, 2,6,6,9-tetramethyl-
epi-Manool		13-epi-Manool
epi-Manoyl oxide	1227-93-6	13-epi-Manoyl oxide
Marmin	14957-38-1	Coumarin, 7-[(6,7-dihydroxy-3,7-dimethyl-2-octenyl)oxy]-
p-Mentha-1,3-dien-7-al	1197-15-5	α-Terpinen-7-al
p-Mentha-1,4-dien-7-al	22580-90-1	γ-Terpinen-7-al
p-Mentha-1(7),8-dien-2-ol	35907-10-9	Isocarveol; 2-Methylene-5-isopropenylcyclohexanol; 2-Methylene-5-(1-methylethenyl)cyclohexanol
trans-p-Mentha-2,8-dien-1-ol	4017-77-0	*trans-p*-Mentha-2,8-dienol; *trans*-Isopiperitenol
p-Menthane-3,8-diol	42822-86-6	Cyclohexanol, 5-methyl-2-(2-hydroxy-2-propyl)-
p-Menth-1-en-9-al	29548-14-9	1-*p*-Menthen-9-al
p-Menth-2-en-1-ol	619-62-5	1-Methyl-4-(1-methylethyl)-2-cyclohexen-1-ol
cis-p-Menth-2-en-1-ol	29803-82-5	(*Z*)-1-Methyl-4-(1-methylethyl)-2-cyclohexen-1-ol
trans-p-Menth-2-en-1-ol	29803-81-4	4-Isopropyl-1-methyl-2-cyclohexen-1-ol; *trans*-1-Methyl-4-(1-methylethyl)-2-cyclohexen-1-ol
p-Menth-1-en-3-ol, acetate	1204-30-4	*p*-Menth-1-en-3-yl acetate
Meranzin	23971-42-8	Coumarin, 8-(2,3-epoxy-3-methylbutyl)-7-methoxy-, (-)-
Methallyl angelate	61692-78-2	2-Methyl-2-propenyl angelate
Methallyl methacrylate	816-74-0	2-Methyl-2-propenyl methacrylate
o-Methoxycinnamaldehyde (2-)	1504-74-1	2-Propenal, 3-(2-methoxyphenyl)-
7-Methoxycoumarin	531-59-9	Herniarin
2-Methoxydihydrocinnamic acid	6342-77-4	3-(2-Methoxyphenyl)propanoic acid
5-Methoxy-8-(2,3-dihydroxy-3-methyl-butoxy)psoralen	482-25-7	Bjakangelicin; Byakangelicin
Methoxyeugenol (6-)	6627-88-9	4-Allylsyringol; 4-Allyl-2,6-dimethoxyphenol; 6-Methoxyeugenol
1-Methoxyhexane	4747-07-3	*n*-Hexyl methyl ether
6-Methoxy-7-hydroxycoumarin	92-61-5	Scopoletin
2-Methoxy-3-isobutylpyrazine	24683-00-9	2-Isobutyl-3-methoxypyrazine
7-Methoxy-6-(3-methyl-2-butenyl)-coumarin	581-31-7	Suberosin
p-Methoxyphenylacetone	122-84-9	1-(4-Methoxyphenyl)-2-propanone; Anisyl methyl ketone
8-Methoxypsoralen	298-81-7	Xanthotoxin
Methyl *N*-acetylanthranilate	2719-08-6	*N*-Acetyl methylanthranilate
3-Methylamyl angelate	53082-58-9	3-Methylpentyl angelate
3-Methylamyl isobutyrate	84254-84-2	3-Methylpentyl isobutyrate
Methyl *p*-anisate	121-98-2	Methyl 4-methoxybenzoate
o-Methylanisole	578-58-5	2-Methylanisole
Methyl anthranilate	134-20-3	Methyl 2-aminobenzoate
2-Methylbenzaldehyde	529-20-4	*o*-Tolualdehyde
Methylbenzene	108-88-3	Toluene
Methyl benzyl ether	538-86-3	Benzyl methyl ether
3-Methyl-2-buten-1-ol	556-82-1	Prenol
3-Methyl-3-butenyl benzoate	5205-12-9	3-Buten-1-ol, 3-methyl-, benzoate
3-Methyl-3-butenyl isovalerate	54410-94-5	3-Methyl-3-butenyl 3-methylbutanoate
2-Methylbutyl isobutyrate	2445-69-4	2-Methylbutyl 2-methylpropanoate
2-Methylbutyl 3-methylbutyrate	2445-77-4	2-Methylbutyl isovalerate; 2-Methylbutyl isopentanoate
2-Methylbutyl tiglate	2445-78-5	2-Methylbutyl 2-methylbutyrate
Methylbutyrate	623-42-7	Methyl butanoate
2-Methylbutyric acid	116-53-0	2-Methylbutanoic acid
Methylcarvacrol	6379-73-3	Carvacryl methyl ether; Carvacryl methyl oxide
Methyl chavicol	140-67-0	Estragole; Esdragole; Esteragenol; Isoanethole; *p*-Allyl anisole; 1-Methoxy-4-(2-propenyl)benzene
3-Methylcyclopentadecanone	541-91-3	Muscone; Muskone
Methyl decanoate	110-42-9	Methyl caprate
Methyl eicosanoate	1120-28-1	Methyl arachidate
Methyl-*Z*-11-eicosenoate	2390-09-2	(11*Z*)-Methyl eicosenoate
Methyl epizizanoate	22387-78-6	Methyl 2-epi-ziza-6(13)-en-12-oate
Methyl eugenol	93-15-2	3,4-Dimethoxyallylbenzene; Eugenol methyl ether

Table 7.1 List of main entry names and their synonyms as used in various publications (*continued*)

Main entry name	CAS number	Synonyms mentioned for main entries
Methyl *N*-formylanthranilate	41270-80-8	*N*-Formyl methylanthranilate
Methyl geranate	2349-14-6	Methyl geraniate
(*Z*)-Methylgeranate	1862-61-9	Methyl 3,7-dimethylocta-2,6-dienoate; Methyl nerolate; Methyl nerate
p-Methylguaiacol	93-51-6	2-Methoxy-4-methylphenol
Methylheptenone	409-02-9	Sulcatone
6-Methyl-5-hepten-2-one	110-93-0	2-Methyl-2-hepten-6-one
Methyl hexadecanoate	112-30-0	Methyl palmitate; Hexadecanoic acid, methyl ester
Methyl 3-hydroxydodecanoate	72864-23-4	3-Hydroxydodecanoic acid methyl ester
Methyl isoeugenol	93-16-3	Benzene, 1,2-dimethoxy-4-(1-propenyl)-; Isohomogenol
(*E*)-Methylisoeugenol	6379-72-2	(*E*)-Isoeugenol methyl ether; Isomethyleugenol
(*Z*)-Methylisoeugenol	6380-24-1	(*Z*)-Isoeugenol methyl ether
Methyl isovalerate	556-24-1	Methyl 3-methylbutanoate
Methyl jasmonate	1211-29-6	2-Pentenylcyclopentanone-3-acetic acid methyl ester
Methyl *N*-methylanthranilate	85-91-6	Dimethyl anthranilate
4-Methyl-3-(methylethyl)-3-cyclo-hexen-1-ol	654053-64-2	3-Cyclohexen-1-ol, 4-methyl-3-(1-methylethyl)-
o-Methyl-α-methylstyrene	7399-49-7	Benzene, 1-methyl-2-(1-methylethenyl)-
Methyl octadecanoate	112-61-8	Methyl stearate
Methyl octanoate	111-11-5	Methyl caprylate
Methyl oleate	112-62-9	Methyl (*Z*)-9-octadecenoate
Methyl palmitoleate	1120-25-8	Methyl-(*Z*)-9-hexadecenoate
4-Methyl-3-penten-2-one	141-79-7	Mesityl oxide
2-Methyl-3-phenylpropanal	5445-77-2	Propanal, 2-methyl-3-phenyl-
Methyl salicylate	119-36-8	Gaultheriaoel
Methyl tetradecanoate	124-10-7	Methyl myristate
Methyl thymol	1076-56-8	2-Methoxy-4-methyl-1-(1-methylethyl)benzene; 1-Isopropyl-2-methoxy-4-methylbenzene; 1-Methoxy-4-methyl-2-(1-methyl-ethyl)benzene; 2-Isopropyl-5-methylanisole
12-Methyl-13-tridecanolide	57092-32-7	12-Methyltridecano-13-lactone; Methyltridecanolide
4-Methyl-4-vinyl-4-butanolide	1073-11-6	Lavender lactone
Musk ambrette	123-69-3	Ambrettolide
γ-Muurolene	30021-74-0	Naphthalene, 1,2,3,4,4a,5,6,8a-octahydro-7-methyl-4-methylene-1-(1-methylethyl)-, (1α,4aα,8aα)-
epi-α-Muurolol	19912-62-0	τ-Muurolol
Myrcene	123-35-3	β-Myrcene
α-Myrcene	1686-30-2	2-Methyl-6-methylene-1,7-octadiene
Myrcene epoxide	29414-55-9	Epoxymyrcene
Myrtenol	515-00-4	2-Pinen-10-ol; Bicyclo[3.1.1]hept-2-ene-2-methanol, 6,6-dimethyl-
Myrtenyl isovalerate	33900-84-4	Myrtenyl 3-methylbutyrate
5-epi-Neointermedeol	136734-27-5	5-epi-Paradisol
Neothujol	35732-36-6	3-Isothujanol
Neral	106-26-3	Citral B; β-Citral; (*Z*)-Citral; *cis*-Geranial
Nerol	106-25-2	(*Z*)-Geraniol
Neryl acetate	141-12-8	*cis*-Geranyl acetate
Nerylacetone	3879-26-3	(*Z*)-Geranylacetone
Neryl isovalerate	3915-83-1	Neryl 3-methylbutanoate
Neryl propionate	105-91-9	Neryl propanoate
γ-Nonalactone	104-61-0	4-Nonanolide
Nonyl acetate	143-13-5	Nonanyl acetate
Nopinone	24903-95-5	β-Pinone
Nopol	128-50-7	Bicyclo[3.1.1]hept-2-ene-2-ethanol, 6,6-dimethyl-
2-Norbornanone	497-38-1	Bicyclo[2.2.1]heptan-2-one
Norbourbonone	13844-03-6	nor-Bourbonone
α-Ocimene	502-99-8	3,7-Dimethyl-1,3,7-octatriene
β-Ocimene	13877-91-3	3,7-Dimethyl-1,3,6-octatriene; Ocimene; *p*-Ocimene
(*E*)-β-Ocimene	3779-61-1	*trans*-Ocimene; *trans-p*-Ocimene
(*Z*)-β-Ocimene	3338-55-4	*cis*-Ocimene; *cis-p*-Ocimene
allo-Ocimene diepoxide		Diepoxide allocimene
(*E*)-Ocimenol	28977-58-4	Ocimenol
Octadecanoic acid	57-11-4	Stearic acid
γ-Octalactone	104-50-7	4-Octanolide
n-Octanal	124-13-0	Octyl aldehyde; 1-Octanal

Table 7.1 List of main entry names and their synonyms as used in various publications (*continued*)

Main entry name	CAS number	Synonyms mentioned for main entries
Octanoic acid	124-07-2	Caprylic acid
2-Octanone	111-13-7	Methyl hexyl ketone
3-Octanyl acetate	103-09-3	Octan-3-yl acetate
(*E*)-2-Octenal	2548-87-0	*trans*-Oct-2-en-1-al
3-Octen-1-ol	18185-81-4	3-Octenol
3-Octyl acetate	4864-61-3	3-Octanol, acetate
Osthole	484-12-8	7-Methoxy-8-(3-methyl-2-butenyl)-2*H*-1-benzopyran-2-one; 7-Methoxy-8-isopentenylcoumarin
Oxypeucedanin hydrate	2643-85-8	Prangol
Patchoulenol	17806-54-1	Patchenol
Patchouli alcohol	5986-55-0	Patchoulol
3′,4′,5,7,8-Pentamethoxyflavone	17290-70-9	Isosinensetin
4′,5,6,7,8-Pentamethoxyflavone	481-53-8	Tangeretin
1,3-Pentanediol, 2,2,4-trimethyl-	144-19-4	2,2,4-Trimethyl-1,3-pentanediol
2-Pentanone	107-87-9	Methyl propyl ketone
5-(3-Penten-1-ynyl)-2,2′-bithiophene	129050-92-6	5-(3-Penten-1-ynyl)-2,2′-bithienyl; PBT; PTB
2-Pentylfuran	3777-69-3	2-Amylfuran
Perillyl alcohol	536-59-4	*p*-Mentha-1,8-dien-7-ol ; Perilla alcohol
Phellandral	21391-98-0	*p*-Menth-1-en-7-al
α-Phellandren-8-ol	1686-20-0	Menthadien-8-ol
β-Phellandren-8-ol	65293-09-6	*p*-Mentha-1(7),2-dien-8-ol
2-Phenethyl alcohol	60-12-8	Benzeneethanol; 2-Phenethanol ; 2-Phenyl alcohol; β-Phenyl-ethanol; 2-Phenylethanol; Phenylethyl alcohol
β-Phenethyl cinnamate	103-53-7	2-Propenoic acid, 3-phenyl, 2-phenylethyl ester
Phenylacetaldehyde	122-78-1	Benzeneacetaldehyde
Phenylacetonitrile	140-29-4	Benzyl cyanide; Benzyl nitrile
2-Phenylethyl acetate (β-)	103-45-7	Phenethyl acetate
2-Phenylethyl benzoate	94-47-3	2-Phenethyl benzoate (β-)
2-Phenylethyl butyrate	103-52-6	2-Phenylethyl butanoate
2-Phenylethyl formate	104-62-1	Phenethyl formate
2-Phenylethyl isovalerate	140-26-1	Phenethyl isovalerate; 2-Phenylethyl 3-methylbutanoate
2-Phenylethyl 2-methylbutyrate	24817-51-4	2-Phenylethyl 2-methylbutanoate
2-Phenylnitroethane	6125-24-2	1-Nitro-2-phenylethane
Phthalic acid	88-99-3	1,2-Benzenedicarboxylic acid
(*E*)-Phytol	150-86-7	[*R*-(*R**, *R**-(*E*))]-3,7,11,15-Tetramethyl-2-hexadecen-1-ol
Phytyl acetate	10236-16-5	Phytol acetate
Pimaradiene	1686-61-9	Pimara-8(14),15-diene
α-Pinene	80-56-8	Bicyclo[3.1.1]hept-2-ene, 2,6,6-trimethyl-; 2,6,6-Trimethylbicyclo[3.1.1]hept-2-ene
β-Pinene	127-91-3	Bicyclo[3.1.1]heptane, 6,6-dimethyl-2-methylene-; Nopinene; 2(10)-Pinene
(*S*)-β-Pinene (laevo-)	18172-67-3	1-β-Pinene
α-Pinene epoxide	1686-14-2	α-Pinene oxide
Pinocamphone	547-60-4	Pinan-3-one; 3-Pinanone
trans-Pinocamphone	547-60-4	*trans*-3-Pinanone
cis-Pinocarveol	6712-79-4	Isopinocarveol
Pinocarvone	30460-92-5	2(10)-Pinen-3-one
Piperitone	89-81-6	3-Carvomenthenone
Piperonal	120-57-0	3,4-Methylenedioxybenzaldehyde; Heliotropine
Pogostol	21698-41-9	*trans*-Guai-11-en-10-ol
Pranferin	33573-60-3	Coumarin, 7-methoxy-8-[(2,2,5,5-tetramethyl-1,3-dioxolan-4-yl)- methyl]-
2-Propenal, 3-(2,6,6-trimethyl-1-cyclo-hexen-1-yl)-	4951-40-0	3-(2,6,6-Trimethyl-1-cyclohexen-1-yl)-2-propenal
Propanal	123-38-6	Propionaldehyde
Propionic acid	79-09-4	Propanoic acid
Propyl tiglate	61692-83-9	Propyl 2-methylcrotonate
Pseudolimonene	499-97-8	Cyclohexane, 1-methylene-4-(1-methylethenyl)-; psi-Limonene; *p*-Mentha-1(7),8-diene; ψ-Limonene
Retinal	116-31-4	Vitamin A aldehyde
Rosefuran	15186-51-3	2-(3′-Methyl-2′-butenyl)-3-methylfuran
Rose oxide	16409-43-1	Rosenoxide
Sabina ketone	513-20-2	5-Isopropylbicyclo[3.1.0]hexan-2-one

Table 7.1 List of main entry names and their synonyms as used in various publications (*continued*)

Main entry name	CAS number	Synonyms mentioned for main entries
Sabinene	3387-41-5	4(10)-Thujene
cis-Sabinene hydrate	15537-55-0	*cis*-Thujanol; *cis*-4-Thujanol
trans-Sabinene hydrate	17699-16-0	*trans*-Thujanol; *trans*-4-Thujanol
Safrole	94-59-7	5-(2-Propenyl)-1,3-benzodioxole ; shikimol
Salicylaldehyde	90-02-8	2-Hydroxybenzaldehyde
cis-Salvene		*cis*-2-Methyl-3-methylenehept-5-ene
trans-Salvene	33746-69-9	*trans*-2-Methyl-3-methylenehept-5-ene
Sandaracopimaradiene	1686-56-2	Sandaracopimara-8(14),15-diene
β-Santalal	13827-98-0	β-Ekasantalal
Santene	529-16-8	2,3-Dimethylbicyclo[2.2.1]hept-2-ene
Selina-3,7(11)-diene	6813-21-4	Eudesma-3,7(11)-diene
α-Selinene	473-13-2	Eudesma-3,11-diene
β-Selinene	17066-67-0	Eudesma-4(14),11-diene; β-Eudesmene
γ-Selinene	515-17-3	Selina-4(14),7(11)-diene
δ-Selinene	473-14-3	β-Cyperene; Eudesma-4,6-diene
Selin-6-en-4α-ol	118173-08-3	Selin-6-en-4-ol
Selin-11-en-4α-ol	16641-47-7	11-Selina-4α-ol; Selin-11-en-4-ol; α-Selin-11-en-4-ol; Selin-7(11)-en-4α-ol
α-Sinensal	17909-77-2	2,6,10-Trimethyl-2,6,9,11-dodecatetraenal
Sinensetin	2306-27-6	3′,4′,5,6,7-Pentamethoxyflavone
Sitosterol	12002-39-0	Citosterol
cis-Spiroether	4575-53-5	(*Z*)-En-yn-dicycloether
trans-Spiroether	50257-98-2	(*E*)-En-yn-dicycloether
Squalene 2,3-oxide	7200-26-2	2,3-Epoxysqualene
(*E*)-Tagetenone	33746-72-4	(*E*)-Ocimenone
(*Z*)-Tagetenone	33746-71-3	(*Z*)-Ocimenone
Terpenediol	80-53-5	1,8-Terpin
α-Terpinene	99-86-5	*p*-Mentha-1,3-diene
γ-Terpinene	99-85-4	1,4-Cyclohexadiene, 1-methyl-4-(1-methylethyl)-; *p*-Mentha-1,4-diene
Terpinene hydrate	2451-01-6	Terpin hydrate
Terpinen-4-ol	562-74-3	4-Carvomenthenol; 4-Terpineol; *p*-Menthen-4-ol; *p*-Menth-1-en-4-ol
1-Terpineol	586-82-3	3-Cyclohexen-1-ol, 1-methyl-4-(1-methylethyl)-; 1-Terpinenol; Terpinen-1-ol; *p*-Menth-3-en-1-ol
α-Terpineol	98-55-5	3-Cyclohexene-1-methanol, α,α,4-trimethyl -; Terpinen-2-ol *p*-Menth-1-en-8-ol
β-Terpineol	138-87-4	Cyclohexanol, 1-methyl-4-(1-methylethenyl-; *p*-Menth-8-en-1-ol
Terpinolene (α-)	586-62-9	*p*-Mentha-1,4(8)-diene; 1-Methyl-4-(1-methylethylidene)-
α-Terpinyl acetate	80-26-2	*p*-Menth-1-en-8-ol acetate
α-Terthienyl	1081-34-1	2,2′:5′,2′-Terthiophene
Tetradecanoic acid	544-63-8	*n*-Myristic acid
14-Tetradecanolide	3537-83-5	Tetradecanolactone
2-Tetradecenal	64461-99-0	Tetradec-2-enal
4′,5,6,7-Tetramethoxyflavone	1168-42-9	Tetramethylscutellarein
4′,5,7,8-Tetramethoxyflavone	6601-66-7	Tetra-*O*-methylisoscutellarein
α-Thujenal	57129-54-1	*trans*-Thuj-3-en-10-al; Thuj-3-en-10-al
α-Thujene	2867-05-2	Bicyclo[3.1.0]hex-2-ene, 2 methyl-5-isopropyl-; Origanene; 2-Methyl-5-(1-methylethyl)bicyclo[3.1.0]hex-2-ene ; 3-Thujene
β-Thujene	28634-89-1	4-Methyl-1-(1-methylethyl)bicyclo[3.1.0]hex-2-ene
α-Thujone	546-80-5	Thujone; *cis*-Thujone
β-Thujone	471-15-8	δ-Isothujone; *trans*-Thujone
Thujopsene	470-40-6	Widdrene
Thujyl alcohol	513-23-5	Isothujol; 3-Thujanol; Thujilic alcohol
Thymol	89-83-8	*p*-Cymen-3-ol
Thymyl acetate	528-79-0	Acetylthymol; Thymol acetate
Tiglic acid	80-59-1	(*E*)-2-Methyl-2-butenoic acid
n-Triacontane	638-68-6	Tricontane
Tricycloekasantalic acid	59300-52-6	Notricycloekasantalic acid
Tridecanolide	1725-04-8	13-Tridecanolide; Cyclotridecanolide; Tridecano-13-lactone
1,2,4-Trihydroxymenthane	66767-24-6	*p*-Menthane-1,2,4-triol
4′,5,7-Trimethoxyflavone	5631-70-9	Trimethylapigenin
Trimethylapigenin	5631-70-9	4′,5,7-Trimethoxyflavone
5,5,6-Trimethylbicyclo[2.2.1]heptan-2-one	3292-05-5	Bicyclo[2.2.1]heptan-2-one, 5,5,6-trimethyl-

Table 7.1 List of main entry names and their synonyms as used in various publications (*continued*)

Main entry name	CAS number	Synonyms mentioned for main entries
2,8,8-Trimethyl-4-methylene-2-vinyl-bicyclo[5.2.0]nonane		4-Methylene-2,8,8-trimethyl-2-vinylbicyclo[5.2.0]nonane
1-Undecanol	112-42-5	Undecanol; Undecyl alcohol
2-Undecanone	112-12-9	Methyl *n*-nonyl ketone
Undecatriene	16356-11-9	Galbanolene; 1,3,5-Undecatriene
(Z)-Undec-9-en-1-al		Undec-(9Z)-en-1-al; (9Z)-Undecen-1-al
Valencene	4630-07-3	Naphthalene, 1,2,3,5,6,7,8,8a-octahydro-1,8a-dimethyl-7-(1-methylethenyl)-, [1S-(1α,7α,8aα)]-
Valeraldehyde	110-62-3	Pentanal
Valerianol	20489-45-6	Kusunol
Valeric acid	109-52-4	Pentanoic acid
Veratraldehyde	120-14-9	Methyl vanillin ether
Veratrole	91-16-7	Methylguaiacol
Verbenol	473-67-6	Pinen-4-ol
cis-Verbenol	1845-30-3	Bicyclo[3.1.1]hept-3-en-2-ol, 4,6,6-trimethyl-, [1S-(1α,2β,5α)]-
Verbenyl acetate	33522-69-9	Bicyclo[3.1.1]hept-2-en-4-ol, 2,6,6-trimethyl-, acetate
Vetiselinene	105497-53-8	Selina-4(15),7-diene
Vetiselinol	32433-12-8	4βH,5α-Eremophila-1(10),7-dien-2α-ol
β-Vetivenene	27840-40-0	β-Vetivene
β-Vetivol		Spirovetiva-3,7(11)-dien-2α-ol
p-Vinylguaiacol	7786-61-0	2-Methoxy-4-vinylphenol; 4-Vinyl-2-methoxyphenol
Viridiflorene	21747-46-6	Ledene; (+)-Ledene
Zingiberone	122-48-5	Zingerone
Ziza-5-en-12-ol	16202-80-5	Isokhusenol
Zizanol	28102-79-6	Ziza-6(13)-en-3α-ol
Zizanone	28051-97-0	Ziza-6(13)-en-3-one

Index

**Individual chemicals can be found in Chapter 6, Table 6.1.
All synonyms can be found in Chapter 7, Table 7.1.**

A

Abies alba Mill., 765
Absolutes, 7–8
Acorus calamus L., 129
Adulteration, 5, 18
Aging, 19, 30
Airborne allergic contact dermatitis, 27
Allergenic components, 2
Allergens in essential oils, 30
Allergic reactions/allergic contact dermatitis, *See* Contact allergies/allergic contact dermatitis
Analytical data, 5–6, 48–49
Analytical investigations of essential oil components, 29–30
Analytical methods, 6, 16–17
Angelica archangelica L., 53, 57
Angelica fruit oil, 53
Angelica root oil, 57
Aniba rosaeodora Ducke, 727
Animal feed, 10
Aniseed oil, 63
Anise oil, star, 795
Aromatherapy, 1, 10, 47
Aromatic waters, 8
Atlas cedarwood oil, 173
Azadirachta indica A. Juss., 571

B

Basil oil, sweet, 69
Bay oil, 85
Bay oil, sweet, 431
Beauticians, 28
Bergamot oil, 85
Bitter orange oil, 611
Bitter orange oil (neroli oil), 575
Bitter orange petitgrain oil (petitgrain bigarade oil), 669
Black cumin oil, 103
Black pepper oil, 113
Blue chamomile oil, 193
Boswellia sacra Flueck., 601
Brown method, 11
Bulnesia sarmientoi Lorentz ex Griseb., 383

C

Cajeput oil, 123
Calamus oil, 129
Cananga odorata (Lam.) Hook. f. et Thomson, forma *genuina*, 887
Cananga odorata (Lam.) Hook. f. et Thomson, forma *macrophylla*, 139
Cananga oil, 29, 139
Canarium luzonicum (Blume) A. Gray, 309
Carbon-13 NMR, 17
Cardamom oil, 921
Carrot seed oil, 147
CAS numbers, 47
Cassia bark oil, 157

Cassia leaf oil, 157, 167
Cassia oils, 157, 167
Cedarwood oil, Atlas, 173
Cedarwood oil, China, 179
Cedarwood oil, Himalayan, 923
Cedarwood oil, Texas, 185
Cedarwood oil, Virginia, 189
Cedarwood oils, 173, 179, 185, 189, 923
Cedrus atlantica (Endl.) G. Manetti ex Carrière, 173
Cedrus deodora (Roxb. ex D. Don) G. Don, 923
Chamaemulum nobile (L.) All., 205
Chamomile oil, German, 193
Chamomile oil, Roman, 205
Chamomile oils, 193, 205
Chamomilla recutita (L.) Rauschert, 193
Cheilitis, 26
Chemical composition
    allergens, 30
    atypical compositions, 51
    chapter format, 47–51
    data provided, 5–6, 48–49
    data selection, 47–48
    dermatologist lack of knowledge, 2
    factors influencing, 11, 15–16
    nomenclature, 49
    overview of some results, 49–51
    products which are not essential oils, 7–8
    types of compounds, 11, 12–14t
Chemistry of essential oils, 11, 12–14t
Chemotypes, 15–16, 19
China cedarwood oil, 179
Chinese cinnamon, 157, 167
Chrysopogon zizanioides (L.) Roberty, 871
Cilantro oil, 277
Cinnamon bark oil, Sri Lanka, 213
Cinnamon leaf oil, Sri Lanka, 223
Cinnamomum cassia (Nees & T. Nees) J. Presl, 157, 167
Cinnamomum zeylanicum Blume, 213, 223
Citronella oil, Java, 231
Citronella oil, Sri Lanka, 239
Citronella oils, 231, 239
Citrus aurantium L., 575, 611, 669
Citrus bergamia (Risso et Poit.), 85
Citrus limon (L.) Burm. f., 483
Citrus paradisi Macfad., 373
Citrus reticulata Blanco, 531
Citrus sinensis (L.) Osbeck., 621
Citrus tangerina Hort. ex Tan, 803
Clary sage oil, 247
Clinical relevance, 2, 25–26
Clove bud oil, 259
Clove essential oils, 259, 269, 273
Clove leaf oil, 269
Clove stem oil, 273
Cold enfleurage, 8
Cold-pressed oils, 7
Colophony, 29
Commercial test preparations, 1

**Individual chemicals can be found in Chapter 6, Table 6.1.
All synonyms can be found in Chapter 7, Table 7.1.**

Individual chemicals can be found in Chapter 6, Table 6.1.
All synonyms can be found in Chapter 7, Table 7.1.

**Individual chemicals can be found in Chapter 6, Table 6.1.
All synonyms can be found in Chapter 7, Table 7.1.**